GEORGIA TEACHER'S EDITION
Algebra 1

HOLT, RINEHART AND WINSTON

A Harcourt Classroom Education Company

Austin • New York • Orlando • Atlanta • San Francisco • Boston • Dallas • Toronto • London

AUTHORS

James E. Schultz, Senior Series Author
Dr. Schultz has over 30 years of experience teaching at the high school and college levels and is the Robert L. Morton Professor of Mathematics Education at Ohio University. He helped to establish standards for mathematics instruction as a coauthor of the NCTM *Curriculum and Evaluation Standards for School Mathematics* and *A Core Curriculum: Making Mathematics Count for Everyone.*

Paul A. Kennedy, Senior Author
Dr. Kennedy is a professor of Mathematics Education, director of the Institute for Mathematics, Science, and Technology Education, and is the holder of the prestigious Adams Chair at Texas Christian University. A leader in mathematics education reform, Dr. Kennedy is often invited to speak and conduct workshops on the teaching of secondary mathematics.

Wade Ellis, Jr.
Professor Ellis has coauthored numerous books and articles on how to integrate technology realistically and meaningfully into the mathematics curriculum. He was a key contributor to the landmark study *Everybody Counts: A Report to the Nation on the Future of Mathematics Education.*

Kathleen A. Hollowell
Dr. Hollowell is an experienced high school mathematics and computer science teacher who currently serves as the Director of the Mathematics & Science Education Resource Center at the University of Delaware. Dr. Hollowell is particularly well versed in the special challenge of motivating students and making the classroom a more dynamic place to learn.

Copyright © 2001 by Holt, Rinehart and Winston

All rights reserved. No part of this publication may be reproduced or transmitted in any form or by any means, electronic or mechanical, including photocopy, recording, or any information storage and retrieval system, without permission in writing from the publisher.

Requests for permission to make copies of any part of the work should be mailed to the following address: Permissions Department, Holt, Rinehart and Winston, 1120 South Capital of Texas Highway, Austin, Texas 78746-6487

(Acknowledgments appear on pages 879–880, which are extensions of the copyright page)

HRW is a registered trademark licensed to Holt, Rinehart and Winston, Inc.

Printed in the United States of America

ISBN: 0-03-066038-6

1 2 3 4 5 6 7 8 9   032   03 02 01 00 99

Algebra 1

Georgia Teacher's Edition

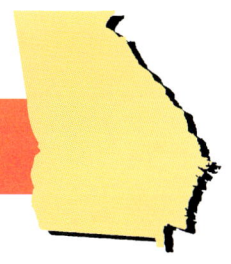

Contents

Using the *Georgia Teacher's Edition* GA4

Georgia Quality Core Curriculum Standards for Algebra 1 .... GA6

Correlation to the Algebra 1 QCC Standards by standard GA8

Correlation to the Algebra 1 QCC Standards by lesson GA12

Correlation to the Georgia High School Graduation Qualifying Test GA18

Overview of *HRW Algebra 1* program A1

***Teacher's Edition* Table of Contents** T6

Using the Algebra 1 Georgia Teacher's Edition

The following pages contain correlations to the Georgia Quality Core Curriculum Standards for Algebra 1. Additionally, a chart provides a correlation to the Georgia High School Graduation Qualifying Test.

Quality Core Curriculum Correlated by Standard

These charts indicate page numbers where each QCC Standard is covered in the *Pupil's Edition* and *Teacher's Edition*.

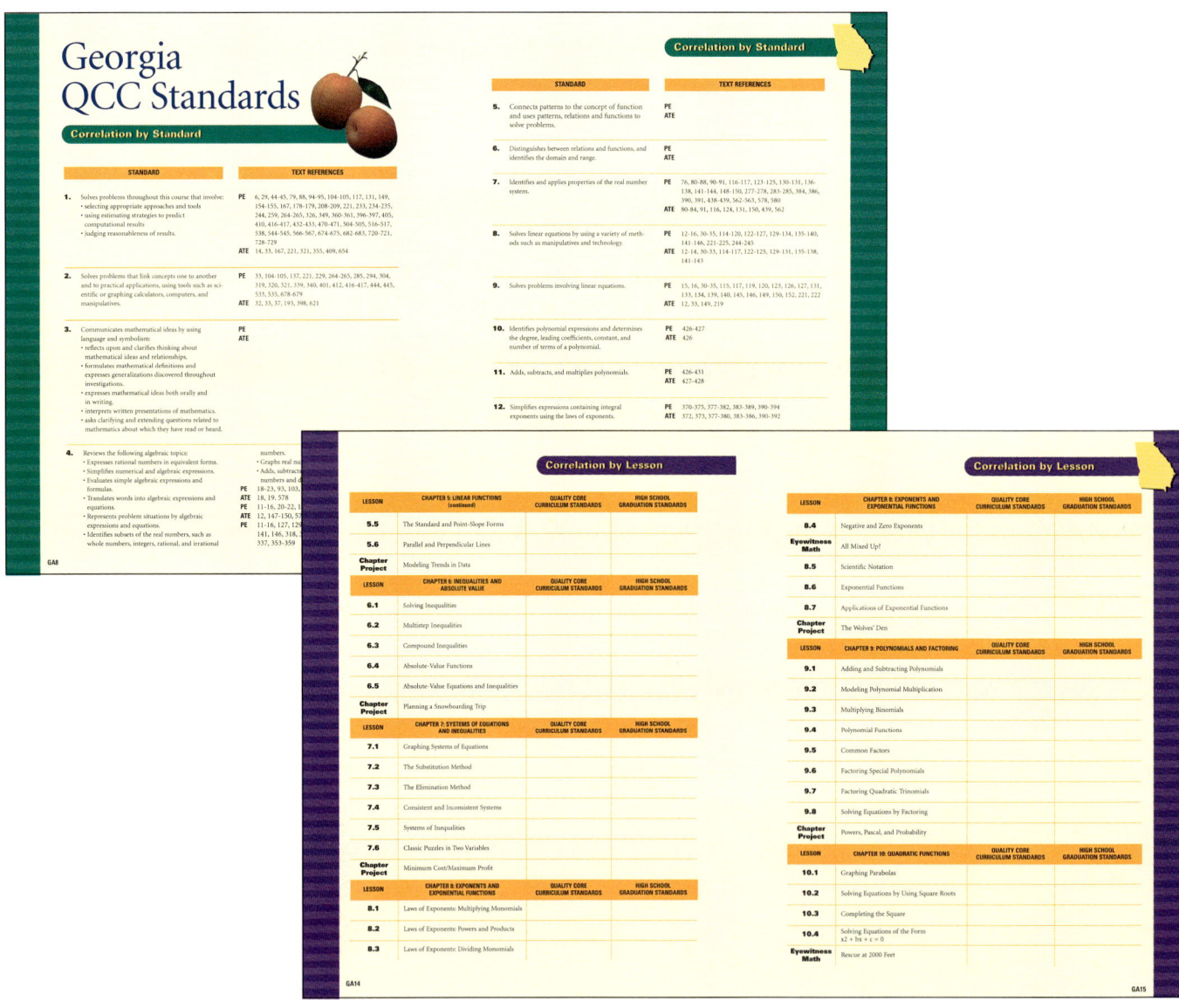

Quality Core Curriculum Correlated by Lesson

These charts indicate the specific QCC Standards covered by each lesson in the *Pupil's Edition* and *Teacher's Edition*.

Using the Algebra 1 Georgia Teacher's Edition

Georgia High School Graduation Qualify Test

These charts indicate page numbers where each objective for the Graduation Qualifying Test is covered in all three Holt, Rinehart and Winston high school mathematics books—*Algebra 1, Geometry,* and *Algebra 2.*

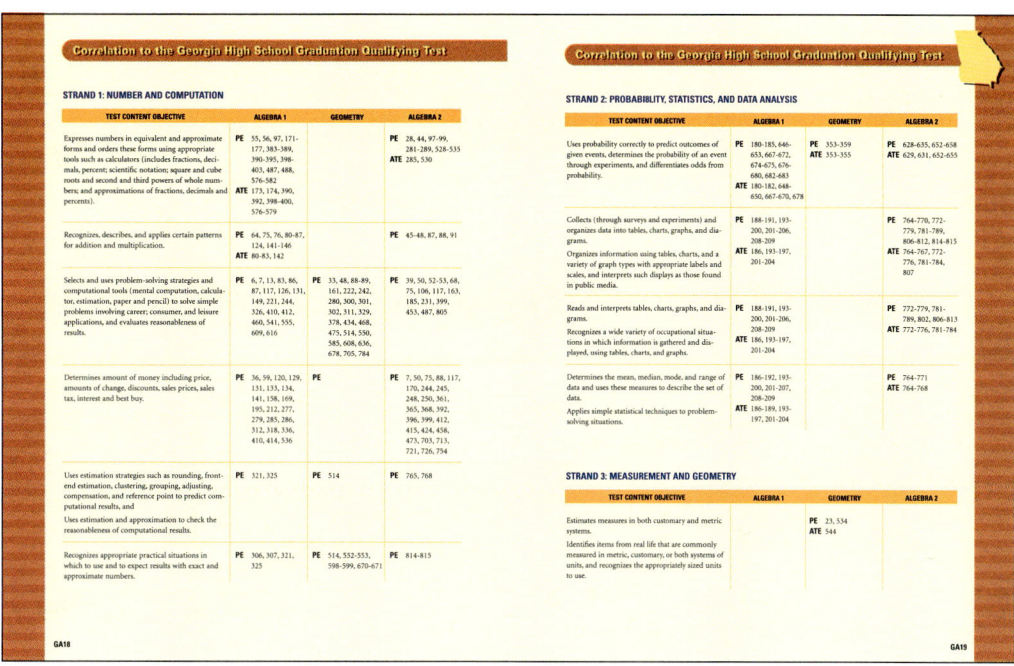

Internet Resources

For additional Georgia-specific teacher and student resources, visit us on the Internet at www.hrw.com/ga.

One-Stop Planner

The *Georgia One-Stop Planner CD-ROM* contains all print-based teaching resources for *Algebra 1,* plus customizable lesson plans and a Test Generator.

Georgia Quality Core Curriculum Standards for Algebra 1

Problem Solving

1. Solves problems throughout this course that involve:
   - selecting appropriate approaches and tools.
   - using estimating strategies to predict computational results.
   - judging reasonableness of results.

2. Solves problems that link concepts one to another and to practical applications, using tools such as scientific or graphing calculators, computers, and manipulatives.

Communication

3. Communicates mathematical ideas by using language and symbolism:
   - reflects upon and clarifies thinking about mathematical ideas and relationships.
   - formulates mathematical definitions and expresses generalizations discovered throughout investigations.
   - expresses mathematical ideas both orally and in writing.
   - interprets written presentations of mathematics.
   - asks clarifying and extending questions related to mathematics about which they have read or heard.

Language of Algebra: Numbers and Variables

4. Reviews the following algebraic topics:
   - expresses rational numbers in equivalent forms.
   - simplifies numerical and algebraic expressions.
   - evaluates simple algebraic expressions and formulas.
   - translates words into algebraic expressions and equations.
   - represents problem situations by algebraic expressions and equations.
   - identifies subsets of the real numbers, such as whole numbers, integers, rational, and irrational numbers.
   - graphs real numbers on a number line.
   - adds, subtracts, multiplies, and divides rational numbers and determines their absolute value.

Connections, Patterns and Functions

5. Connects patterns to the concept of function and uses patterns, relations and functions to solve problems.

6. Distinguishes between relations and functions, and identifies the domain and range.

Real Number System

7. Identifies and applies properties of the real number system.

Solving Equations

8. Solves linear equations by using a variety of methods such as manipulatives and technology.

9. Solves problems involving linear equations.

Polynomials

10. Identifies polynomial expressions and determines the degree, leading coefficients, constant, and number of terms of a polynomial.

11. Adds, subtracts, and multiplies polynomials.

12. Simplifies expressions containing integral exponents using the laws of exponents.

Factorization

13. Determines monomial factors of polynomials and divides a polynomial by a monomial.

14. Factors simple quadratic expressions such as trinomials, perfect-square trinomials, difference of two squares, and polynomials with common factors.

15. Solves quadratic equations using factoring.

Georgia Quality Core Curriculum Standards for Algebra 1

Linear Equations in Two Variables

16. Graphs points (ordered pairs of numbers) in the coordinate plane and identifies the coordinates of given points in the plane.

17. Defines slope as rate of change and calculates the slope given the change of the two variables.

18. Interprets and makes predictions from data displayed in line graphs or scatter plots, by examining patterns and recognizing algebraic concepts such as rate of change (slope), intercepts, range, and domain.

19. Identifies the slope and intercepts of a linear equation.

20. Sketches the graph of a linear equation in two variables given appropriate information, such as slope, x-intercept, y-intercept, two points and the linear equation.

21. Graphs linear equations in two variables and identifies graphs of lines, including special cases such as vertical, horizontal, parallel and perpendicular lines.

22. Writes the equation of a line given information such as slope, x-intercept, y-intercept, two points and the graph.

23. Solves a system of linear equations in two variables by a variety of methods including graphing.

24. Solves problems using systems of linear equations.

Inequalities

25. Solves simple and compound inequalities in one variable and graphs the solution set on the number line.

Linear Inequalities

26. Solves problems involving linear inequalities and graphs the solution on a coordinate plane.

Rational Expressions: Algebraic Fractions

27. Simplifies simple algebraic fractions.

28. Solves simple rational equations.

Variation

29. Solves problems involving direct and inverse variation.

Rational Expressions: Algebraic Fractions

30. Solves problems involving ratios, proportions and percents.

Rational and Irrational Numbers and Radical Expressions

31. Finds the square roots of rational numbers and decimal approximations of irrational numbers, by simplifying radicals and by using a calculator.

32. Multiplies, divides, adds, and subtracts radicals with index of two.

33. Applies the Pythagorean Theorem and its converse in problem solving situations.

34. Solves simple radical equations and solves problems involving simple radical equations.

Statistics

35. Summarizes data in various ways, including mean, median, mode, and range.

Probability

36. Identifies possible outcomes of simple experiments and predicts or describes the probability of a given event expressed as a rational number from 0 through 1.

37. Conducts and interprets a compound probability experiment.

Georgia QCC Standards

Correlation by Standard

STANDARD	TEXT REFERENCES	
1. Solves problems throughout this course that involve: • selecting appropriate approaches and tools. • using estimating strategies to predict computational results. • judging reasonableness of results.	PE	6, 29, 44-45, 79, 88, 94-95, 104-105, 117, 131, 149, 154-155, 167, 178-179, 208-209, 221, 233, 234-235, 244, 259, 264-265, 326, 349, 360-361, 396-397, 405, 410, 416-417, 432-433, 470-471, 504-505, 516-517, 538, 544-545, 566-567, 674-675, 682-683, 720-721, 728-729
	TE	14, 33, 167, 221, 321, 355, 409, 654
2. Solves problems that link concepts one to another and to practical applications, using tools such as scientific or graphing calculators, computers, and manipulatives.	PE	33, 104-105, 137, 221, 229, 264-265, 285, 294, 304, 319, 320, 321, 339, 340, 401, 412, 416-417, 444, 445, 533, 535, 678-679
	TE	32, 33, 37, 193, 621
3. Communicates mathematical ideas by using language and symbolism: • reflects upon and clarifies thinking about mathematical ideas and relationships. • formulates mathematical definitions and expresses generalizations discovered throughout investigations. • expresses mathematical ideas both orally and in writing. • interprets written presentations of mathematics. • asks clarifying and extending questions related to mathematics about which they have read or heard.		*In every lesson's exercises, students begin with communication exercises which address this standard.*
4. Reviews the following algebraic topics: • expresses rational numbers in equivalent forms. • simplifies numerical and algebraic expressions. • evaluates simple algebraic expressions and formulas. • translates words into algebraic expressions and equations. • represents problem situations by algebraic expressions and equations. • identifies subsets of the real numbers, such as whole numbers, integers, rational, and irrational numbers.	PE	11-16, 18-23, 36, 54-56, 57-59, 66, 72, 79, 88, 89-93, 97-102, 103, 117, 119, 120, 122-127, 128, 129-134, 135, 139-140, 141, 145-146, 147-152, 153, 154-155, 167-169, 170, 171-172, 173-177, 185, 192, 207, 225, 233, 244-245, 251, 253, 264-265, 278-280, 285-287, 289-293, 300, 302, 318-325, 328, 330, 331, 336, 337, 345-351, 353-359, 360-361, 372-375, 378-382, 383-388, 390-393, 398-403, 409-414, 416-417, 426-431, 577-582
	TE	11-14, 18, 19, 54, 56, 122, 123, 130, 147-150, 167, 171, 173, 289, 290, 319, 321, 346, 354, 355, 433, 439, 443, 577, 578

Correlation by Standard

STANDARD		TEXT REFERENCES
(4. continued) • graphs real numbers on a number line. • adds, subtracts, multiplies, and divides rational numbers and determines their absolute value.		
5. Connects patterns to the concept of function and uses patterns, relations and functions to solve problems.	PE TE	5-10, 33, 220-225, 241-242, 251, 253, 257, 264-165, 272, 405, 409-415, 443-447, 464, 483, 485, 486, 488, 489, 490, 491, 497, 500, 503, 504-505, 510, 515, 516-517, 551, 554-555, 556, 557, 566-567, 627 30, 219, 245, 253, 405, 410, 411, 499, 507, 512, 532
6. Distinguishes between relations and functions, and identifies the domain and range.	PE	33, 219, 220, 221, 222, 246, 253, 298, 349, 532, 534, 535, 536, 588, 589, 692-698, 693, 694, 695, 707, 728-729, 733
7. Identifies and applies properties of the real number system.	PE TE	76, 80-88, 90-91, 116-117, 123-125, 130-131, 136-138, 141-144, 148-150, 277-278, 283-285, 384, 386, 390, 391, 438-439, 562-563, 578, 580 80-84, 91, 116, 124, 131, 150, 439, 562
8. Solves linear equations by using a variety of methods such as manipulatives and technology.	PE TE	12-16, 30-35, 114-120, 122-127, 129-134, 135-140, 141-146, 221-225, 244-245 12-14, 30-33, 114-117, 122-125, 129-131, 135-138, 141-143
9. Solves problems involving linear equations.	PE TE	15, 16, 30-35, 115, 117, 119, 120, 123, 126, 127, 131, 133, 134, 139, 140, 145, 146, 149, 150, 152, 221, 222 12, 33, 149, 219
10. Identifies polynomial expressions and determines the degree, leading coefficients, constant, and number of terms of a polynomial.	PE TE	426-427 426
11. Adds, subtracts, and multiplies polynomials.	PE TE	426-431, 432-436, 438-442 427-428
12. Simplifies expressions containing integral exponents using the laws of exponents.	PE TE	370-375, 377-382, 383-389, 390-394 372, 373, 377-380, 383-386, 390-392
13. Determines monomial factors of polynomials and divides a polynomial by a monomial.	PE TE	448-451 448, 449

Correlation by Standard

STANDARD	TEXT REFERENCES
14. Factors simple quadratic expressions such as trinomials, perfect trinomials, difference of two squares, and polynomials with common factors.	PE 452-456, 458-463 TE 452-454, 458-461
15. Solves quadratic equations using factoring.	PE 466-467, 499, 502, 503
16. Graphs points (ordered pairs of numbers) in the coordinate plane and identifies the coordinates of given points in the plane.	PE 24-28 TE 24-26
17. Defines slope as rate of change and calculates the slope given the change of the two variables.	PE 31, 32, 33, 34, 35, 36, 226-233, 234-235, 236-243, 237, 257, 261, 265, 612 TE 32, 97, 227, 245, 247, 248
18. Interprets and makes predictions from data displayed in line graphs or scatter plots, by examining patterns and recognizing algebraic concepts such as rate of change (slope), intercepts, range, and domain.	PE 26, 31, 37-43, 45, 193, 199, 217, 225, 237, 242, 257, 263, 264-265, 340, 395, 405, 414, 417, 4941, 515, 605, 720-721, 727 TE 37, 38, 195
19. Identifies the slope and intercepts of a linear equation.	PE 244-251 TE 244-248
20. Sketches the graph of a linear equation in two variables given appropriate information, such as slope, x-intercept, y-intercept, two points and the linear equation.	PE 30-36, 221, 230, 244-251, 252-257, 264-265 TE 245, 246, 252, 255
21. Graphs linear equations in two variables and identifies graphs of lines, including special cases such as vertical, horizontal, parallel and perpendicular lines.	PE 31, 32-33, 34-36, 221, 229, 230, 248, 250, 251, 258-262 TE 260
22. Writes the equation of a line given information such as slope, x-intercept, y-intercept, two points and the graph.	PE 244-251, 252-257, 258-262, 264-265 TE 244, 245, 248, 253
23. Solves a system of linear equations in two variables by a variety of methods including graphing.	PE 318-325, 326-330, 331-337, 338-344 TE 318-322, 326-328, 331-333, 338-341
24. Solves problems using systems of linear equations.	PE 322, 325, 328, 330, 331, 336, 353-359 TE 353-356

Correlation by Standard

STANDARD	TEXT REFERENCES
25. Solves simple and compound inequalities in one variable and graphs the solution set on the number line.	PE 278-281, 282-287, 289-293, 302-307 TE 278, 282-285, 289-290, 302
26. Solves problems involving linear inequalities and graphs the solution on a coordinate plane.	PE 345-351 TE 345-349
27. Simplifies simple algebraic fractions.	PE 532-536, 538-542 TE 532-534, 538-540
28. Solves simple rational equations.	PE 552-557 TE 552-554
29. Solves problems involving direct and inverse variation.	PE 238-242, 528, 530, 531, 543, 571 TE 238, 553
30. Solves problems involving ratios, proportions and percents.	PE 164-170, 171-177, 552-557, 566-567 TE 164-167, 171-174, 553
31. Finds the square roots of rational numbers and decimal approximations of irrational numbers, by simplifying radicals and by using a calculator.	PE 576-577, 580, 581 TE 576
32. Multiplies, divides, adds, and subtracts radicals with index of two.	PE 577-582 TE 576-579
33. Applies the Pythagorean Theorem and its converse in problem solving situations.	PE 591-597, 600 TE 591-594
34. Solves simple radical equations and solves problems involving simple radical equations.	PE 584-589, 598-604 TE 584-587
35. Summarizes data in various ways, including mean, median, mode, and range.	PE 186-192, 193-200, 201-207, 208-209 TE 186-189, 193-197, 201-204
36. Identifies possible outcomes of simple experiments and predicts or describes the probability of a given event expressed as a rational number from 0 through 1.	PE 180-185, 265, 470-471, 544-545, 648-652, 667-672, 673, 674-675, 676-681, 682-683, 687 TE 181, 182, 675
37. Conducts and interprets a compound probability experiment.	PE 661-665, 667-672, 687 TE 662, 663

GA11

Georgia QCC Standards

Correlation by Lesson

LESSON	CHAPTER 1: FROM PATTERNS TO ALGEBRA	QUALITY CORE CURRICULUM STANDARDS	HIGH SCHOOL GRADUATION STANDARDS
1.1	Using Differences to Identify Patterns	1, 3, 5	4, 34
1.2	Variables, Expressions, and Equations	3, 4, 8, 9	35, 37, 38
1.3	The Algebraic Order of Operations	3, 4	26, 27, 35
1.4	Graphing with Coordinates	1, 3, 16, 18	25
1.5	Representing Linear Patterns	2, 3, 4, 5, 6, 8, 9, 17, 18, 20, 21	5, 34, 38, 42
1.6	Scatter Plots and Lines of Best Fit	3, 18	
Chapter Project	Working with Patterns in Data	1, 18	

LESSON	CHAPTER 2: OPERATIONS IN ALGEBRA	QUALITY CORE CURRICULUM STANDARDS	HIGH SCHOOL GRADUATION STANDARDS
2.1	The Real Numbers and Absolute Value	3, 4	2, 5
2.2	Adding Real Numbers	3, 4	3
2.3	Subtracting Real Numbers	3, 4	
2.4	Multiplying and Dividing Real Numbers	1, 3, 4, 7	3, 33
2.5	Properties and Mental Computation	1, 3, 4, 7	3, 4, 26, 27, 33
2.6	Adding and Subtracting Expressions	3, 4, 7	26, 27, 33, 38
Eyewitness Math	Abandon Ship!	1	19, 20
2.7	Multiplying and Dividing Expressions	3, 4	2, 20, 26, 27, 38
Chapter Project	Using a Spreadsheet	1, 2	

Correlation by Lesson

LESSON	CHAPTER 3: EQUATIONS	QUALITY CORE CURRICULUM STANDARDS	HIGH SCHOOL GRADUATION STANDARDS
3.1	Solving Equations by Adding and Subtracting	1, 3, 4, 7, 8, 9	4, 5, 26, 27, 31, 33, 34, 38
3.2	Solving Equations by Multiplying and Dividing	3, 4, 7, 8, 9	3, 4, 33, 34, 37, 38
3.3	Solving Two-Step Equations	1, 3, 4, 7, 8, 9	4, 5, 33, 34, 37, 38
3.4	Solving Multistep Equations	2, 3, 4, 7, 8, 9	26, 27, 33, 34, 37, 38
3.5	Using the Distributive Property	3, 4, 7, 8, 9	3, 5, 26, 27, 33, 34, 37, 38
3.6	Using Formulas and Literal Equations	1, 3, 4, 7, 9	4, 20, 26, 27, 33, 35, 38
Chapter Project	Track Team Schemes	1, 4	26, 27, 38

LESSON	CHAPTER 4: PROPORTIONAL REASONING AND STATISTICS	QUALITY CORE CURRICULUM STANDARDS	HIGH SCHOOL GRADUATION STANDARDS
4.1	Using Proportional Reasoning	1, 3, 4, 30	5, 23, 24, 38, 39, 40
4.2	Percent Problems	3, 4, 30	2, 38, 39
Eyewitness Math	Barely Enough Grizzlies?	1	
4.3	Introduction to Probability	3, 4, 36	10
4.4	Measures of Central Tendency	3, 4, 35	11, 12, 13, 14, 15, 16
4.5	Graphing Data	3, 18, 35	3, 18, 35 5, 11, 12, 13, 14, 15, 16, 30 ,31
4.6	Other Data Displays	3, 4, 35	11, 12, 13, 14, 15, 16
Chapter Project	Designing a Research Study	1, 35	11, 12, 13, 14, 15, 16

LESSON	CHAPTER 5: LINEAR FUNCTIONS	QUALITY CORE CURRICULUM STANDARDS	HIGH SCHOOL GRADUATION STANDARDS
5.1	Linear Functions and Graphs	1, 2, 3, 4, 5, 6, 8, 9, 18, 20, 21	4, 34, 38
5.2	Defining Slope	1, 2, 3, 4, 17, 20, 21	43
Eyewitness Math	Watch Your Step!	1, 17	20
5.3	Rate of Change and Direct Variation	3, 5, 17, 18, 29	20, 26, 27, 36, 39
5.4	The Slope-Intercept Form	1, 3, 4, 5, 6, 8, 19, 20, 21, 22	4, 34, 38, 44

Correlation by Lesson

LESSON	CHAPTER 5: LINEAR FUNCTIONS (continued)	QUALITY CORE CURRICULUM STANDARDS	HIGH SCHOOL GRADUATION STANDARDS
5.5	The Standard and Point-Slope Forms	3, 4, 5, 6, 17, 18, 20, 22	38
5.6	Parallel and Perpendicular Lines	1, 3, 17, 18, 21, 22	19
Chapter Project	Modeling Trends in Data	1, 2, 4, 5, 17, 18, 20, 22, 36	38

LESSON	CHAPTER 6: INEQUALITIES AND ABSOLUTE VALUE	QUALITY CORE CURRICULUM STANDARDS	HIGH SCHOOL GRADUATION STANDARDS
6.1	Solving Inequalities	3, 4, 7, 25	5, 33, 41
6.2	Multistep Inequalities	2, 3, 4, 7, 25	5, 33
6.3	Compound Inequalities	3, 4, 25	
6.4	Absolute-Value Functions	2, 3, 6	
6.5	Absolute-Value Equations and Inequalities	2, 3, 4, 25	9
Chapter Project	Planning a Snowboarding Trip		

LESSON	CHAPTER 7: SYSTEMS OF EQUATIONS AND INEQUALITIES	QUALITY CORE CURRICULUM STANDARDS	HIGH SCHOOL GRADUATION STANDARDS
7.1	Graphing Systems of Equations	2, 3, 4, 23, 24	5, 6, 7, 8, 9, 37
7.2	The Substitution Method	1, 3, 4, 23, 24	4, 26, 27, 37
7.3	The Elimination Method	3, 4, 23, 24	5, 37
7.4	Consistent and Inconsistent Systems	2, 3, 18, 23	
7.5	Systems of Inequalities	1, 3, 4, 6, 26	
7.6	Classic Puzzles in Two Variables	3, 4, 24	20, 37
Chapter Project	Minimum Cost/Maximum Profit	1, 4	

LESSON	CHAPTER 8: EXPONENTS AND EXPONENTIAL FUNCTIONS	QUALITY CORE CURRICULUM STANDARDS	HIGH SCHOOL GRADUATION STANDARDS
8.1	Laws of Exponents: Multiplying Monomials	3, 4, 12	26, 27
8.2	Laws of Exponents: Powers and Products	3, 4, 12	26, 27
8.3	Laws of Exponents: Dividing Monomials	3, 4, 7, 12	2, 26, 27, 33

Correlation by Lesson

LESSON	CHAPTER 8: EXPONENTS AND EXPONENTIAL FUNCTIONS (continued)	QUALITY CORE CURRICULUM STANDARDS	HIGH SCHOOL GRADUATION STANDARDS
8.4	Negative and Zero Exponents	3, 4, 7, 12, 18	2, 33
Eyewitness Math	All Mixed Up?	1	
8.5	Scientific Notation	2, 3, 4	2
8.6	Exponential Functions	1, 3, 5, 18	
8.7	Applications of Exponential Functions	1, 2, 3, 4, 5, 18	4, 5
Chapter Project	The Wolves' Den		

LESSON	CHAPTER 9: POLYNOMIALS AND FACTORING	QUALITY CORE CURRICULUM STANDARDS	HIGH SCHOOL GRADUATION STANDARDS
9.1	Adding and Subtracting Polynomials	3, 4, 10, 11	
9.2	Modeling Polynomial Multiplication	1, 3, 11	20, 26, 27
9.3	Multiplying Binomials	3, 7, 11	20, 26, 27, 33
9.4	Polynomial Functions	2, 3, 5	
9.5	Common Factors	3, 13	
9.6	Factoring Special Polynomials	3, 14	
9.7	Factoring Quadratic Trinomials	3, 14	4
9.8	Solving Equations by Factoring	3, 5, 15	26, 27
Chapter Project	Powers, Pascal, and Probability	1, 36	

LESSON	CHAPTER 10: QUADRATIC FUNCTIONS	QUALITY CORE CURRICULUM STANDARDS	HIGH SCHOOL GRADUATION STANDARDS
10.1	Graphing Parabolas	3, 5	
10.2	Solving Equations by Using Square Roots	3, 5, 18	2, 20, 26, 27
10.3	Completing the Square	3, 5	
10.4	Solving Equations of the Form $x^2 + bx + c = 0$	3, 5, 15	26, 27
Eyewitness Math	Rescue at 2000 Feet	1, 5	

Correlation by Lesson

LESSON	CHAPTER 10: QUADRATIC FUNCTIONS (continued)	QUALITY CORE CURRICULUM STANDARDS	HIGH SCHOOL GRADUATION STANDARDS
10.5	The Quadratic Formula	3, 5	
10.6	Graphing Quadratic Inequalities	3, 5, 18	
Chapter Project	What's the Difference?	1, 5	
LESSON	CHAPTER 11: RATIONAL FUNCTIONS	QUALITY CORE CURRICULUM STANDARDS	HIGH SCHOOL GRADUATION STANDARDS
11.1	Inverse Variation	3, 29	20, 26, 27, 36
11.2	Rational Expressions and Functions	2, 3, 6, 27	5
11.3	Simplifying Rational Expressions	1, 3, 27, 29	4, 26, 27
Eyewitness Math	How Worried Should You Be?	1, 36	
11.4	Operations with Rational Expressions	3, 5	
11.5	Solving Rational Equations	3, 5, 28, 30	4, 20
11.6	Proof in Algebra	3, 7	33
Chapter Project	Designing a Park	1, 5, 30	20
LESSON	CHAPTER 12: RADICAL FUNCTIONS AND COORDINATE GEOMETRY	QUALITY CORE CURRICULUM STANDARDS	HIGH SCHOOL GRADUATION STANDARDS
12.1	Operations with Radicals	3, 4, 7, 31, 32	2, 33
12.2	Square-Root Functions and Radical Equations	3, 6, 34	
12.3	The Pythagorean Theorem	3, 33	32
12.4	The Distance Formula	3, 33, 34	
12.5	Geometric Properties	3, 17, 18	4
12.6	The Tangent Function	3	4
12.7	The Sine and Cosine Functions	3, 5	
12.8	Introduction to Matrices		
Chapter Project	Working the Angles		

Correlation by Lesson

LESSON	CHAPTER 13: PROBABILITY	QUALITY CORE CURRICULUM STANDARDS	HIGH SCHOOL GRADUATION STANDARDS
13.1	Theoretical Probability	3, 36	
13.2	Counting the Elements of Sets		
13.3	The Fundamental Counting Principle	3, 37	
13.4	Independent Events	3, 36, 37	10
Eyewitness Math	Hot Hands or Hoopla?	1, 36	10
13.5	Simulations	2, 3, 36	10
Chapter Project	Happy Birthday!	1, 36	10

LESSON	CHAPTER 14: FUNCTIONS AND TRANSFORMATIONS	QUALITY CORE CURRICULUM STANDARDS	HIGH SCHOOL GRADUATION STANDARDS
14.1	Graphing Functions and Relations	3, 6	19
14.2	Translations	3	21, 22
14.3	Stretches and Compressions	3, 6	
14.4	Reflections	3	21, 22
Eyewitness Math	Is There Order in Chaos?	1, 18	
14.5	Combining Transformations	3, 18	21, 22
Chapter Project	Pick a Number	1, 6	

GA17

Correlation to the Georgia High School Graduation Qualifying Test

The chart below indicates page numbers where each objective for the Georgia High School Graduation Qualifying Test is covered in all three Holt, Rinehart and Winston high school mathematics books—*Algebra 1, Geometry,* and *Algebra 2.*

STRAND 1: NUMBER AND COMPUTATION

TEST CONTENT OBJECTIVE		ALGEBRA 1		GEOMETRY		ALGEBRA 2
Expresses numbers in equivalent and approximate forms and orders these forms using appropriate tools such as calculators (includes fractions, decimals, percent; scientific notation; square and cube roots and second and third powers of whole numbers; and approximations of fractions, decimals and percents).	PE	55, 56, 97, 171-177, 383-389, 390-395, 398-403, 487, 488, 576-582			PE	28, 44, 97-99, 281-289, 528-535
	TE	173, 174, 390, 392, 398-400, 576-579			TE	285, 530
Recognizes, describes, and applies certain patterns for addition and multiplication.	PE	64, 75, 76, 80-87, 124, 141-146			PE	45-48, 87, 88, 91
	TE	80-83, 142				
Selects and uses problem-solving strategies and computational tools (mental computation, calculator, estimation, paper and pencil) to solve simple problems involving career; consumer; and leisure applications, and evaluates reasonableness of results.	PE	6, 7, 13, 83, 86, 87, 117, 126, 131, 149, 221, 244, 326, 410, 412, 460, 541, 555, 609, 616	PE	33, 48, 88-89, 161, 222, 242, 280, 300, 301, 302, 311, 329, 378, 434, 468, 475, 514, 550, 585, 608, 636, 678, 705, 784	PE	39, 50, 52-53, 68, 75, 106, 117, 163, 185, 231, 399, 453, 487, 805
Determines amount of money including price, amounts of change, discounts, sales prices, sales tax, interest and best buy.	PE	36, 59, 120, 129, 131, 133, 134, 141, 158, 169, 195, 212, 277, 279, 285, 286, 312, 318, 336, 410, 414, 536			PE	7, 50, 75, 88, 117, 170, 244, 245, 248, 250, 361, 365, 368, 392, 396, 399, 412, 415, 424, 458, 473, 703, 713, 721, 726, 754
Uses estimation strategies such as rounding, front-end estimation, clustering, grouping, adjusting, compensation, and reference point to predict computational results, and Uses estimation and approximation to check the reasonableness of computational results.	PE	321, 325	PE	514	PE	765, 768
Recognizes appropriate practical situations in which to use and to expect results with exact and approximate numbers.	PE	306, 307, 321, 325	PE	514, 552-553, 598-599, 670-671	PE	814-815
					TE	285

GA18

Correlation to the Georgia High School Graduation Qualifying Test

STRAND 2: PROBABILITY, STATISTICS, AND DATA ANALYSIS

TEST CONTENT OBJECTIVE	ALGEBRA 1	GEOMETRY	ALGEBRA 2
Uses probability correctly to predict outcomes of given events, determines the probability of an event through experiments, and differentiates odds from probability.	PE 180-185, 646-653, 667-672, 674-675, 676-680, 682-683 TE 180-182, 648-650, 667-670, 678	PE 353-359 TE 353-355	PE 628-635, 652-658 TE 629, 631, 652-655
Collects (through surveys and experiments) and organizes data into tables, charts, graphs, and diagrams. Organizes information using tables, charts, and a variety of graph types with appropriate labels and scales, and interprets such displays as those found in public media.	PE 188-191, 193-200, 201-206, 208-209 TE 186, 193-197, 201-204		PE 764-770, 772-779, 781-789, 806-812, 814-815 TE 764-767, 772-776, 781-784, 807
Reads and interprets tables, charts, graphs, and diagrams. Recognizes a wide variety of occupational situations in which information is gathered and displayed, using tables, charts, and graphs.	PE 188-191, 193-200, 201-206, 208-209 TE 186, 193-197, 201-204		PE 772-779, 781-789, 802, 806-813 TE 772-776, 781-784
Determines the mean, median, mode, and range of data and uses these measures to describe the set of data. Applies simple statistical techniques to problem-solving situations.	PE 186-192, 193-200, 201-207, 208-209 TE 186-189, 193-197, 201-204		PE 764-771 TE 764-768

STRAND 3: MEASUREMENT AND GEOMETRY

TEST CONTENT OBJECTIVE	ALGEBRA 1	GEOMETRY	ALGEBRA 2
Estimates measures in both customary and metric systems. Identifies items from real life that are commonly measured in metric, customary, or both systems of units, and recognizes the appropriately sized units to use.		PE 23, 534 TE 544	

GA19

Correlation to the Georgia High School Graduation Qualifying Test

STRAND 3: MEASUREMENT AND GEOMETRY (continued)

TEST CONTENT OBJECTIVE	ALGEBRA 1	GEOMETRY	ALGEBRA 2
Estimates and solves problems involving measurement, including selecting appropriate tools such as calculator or mental calculation.	**PE** 94-95, 258, 262, 699 **TE** 165	**PE** 18-19, 23, 25, 33, 36-38, 52-53, 57, 60-62, 119, 179, 183, 184, 217, 254, 262, 315, 316, 418, 517, 518, 519, 552-553, 574, 599, 601-602, 618-619, 630, 632, 657, 750, 751 **TE** 26, 254, 438, 454, 535, 545, 632	**PE** 544-545
Applies customary or metric units of measure to determine length, area, volume/capacity, weight/mass, time, and temperature (includes evaluating reasonableness and precision of results, and reading different scales).	**PE** 94-95, 102, 149, 152, 234-235, 240, 354-355, 357, 358, 359, 365, 436, 440, 490, 530, 552-553, 556, 567	**PE** 25-33, 171-176, 177-182, 294-302, 303-311, 312-313, 314-319, 321-329, 331-337, 339-345, 430-435, 437-444, 445-451, 453-458, 460-467, 469-475, 580-587, 588-591 **TE** 27, 28, 170, 177, 179, 294-297, 303-307, 314-316, 321-325, 331-334, 339, 340, 342, 430-433, 437-440, 445-448, 453-455, 460-463, 469-472, 584, 588	**PE** 430, 431, 447, 451, 453, 454, 455, 467, 482, 485, 526, 852-856
Identifies and differentiates between similar and congruent figures and identifies figures that have been transformed by rotation, reflection, and translation.	**PE** 700-706, 714-718, 726 **TE** 700-703, 714-716	**PE** 50-51, 53, 55, 56, 57, 58, 62, 64, 66, 154, 197, 252, 279, 281, 503, 504, 507-509, 682, 684, 741, 742, 743 **TE** 2C, 501, 507, 682	**PE** 133-141, 580-581, 861, 909, 912-915 **TE** 912

GA20

Correlation to the Georgia High School Graduation Qualifying Test

TEST CONTENT OBJECTIVE	ALGEBRA 1	GEOMETRY	ALGEBRA 2
Uses proportions to find missing lengths of sides of similar figures and to enlarge or reduce figures Solves problems involving similar figures and scale drawings Applies ratios to similar geometric figures, as in scale drawings, as well as with mixtures and compound applications.	**PE** 167-169	**PE** 468, 533-537, 540-542, 544-551, 552-553 **TE** 545	**PE** 31, 35
Graph points in the coordinate plane, identifies the coordinates, and uses the concept of coordinates in problem situations, such as reading maps.	**PE** 24-28 **TE** 24-26	**PE** 59-67, 190-196, 339-345, 610-611, 680-684 **TE** 61, 62, 192, 347-349, 611-613	**PE** 4-6, 37-43, 81-82 **TE** 38, 39
Finds the perimeter and area of plane figures (such as polygons, circles, composite figures) and surface area and volume of simple solids (such as rectangular prisms, pyramids, cylinders, cones, spheres) Calculates perimeter and area of plane figures; finds appropriate measures of objects and their models prior to such calculations for basic polygons and circles.	**PE** 23, 87, 93, 102, 121, 139, 145, 149, 150, 152, 155, 241, 329, 373, 374, 375, 380, 382, 385, 436, 440, 441, 468, 475, 490, 500-501, 530, 542	**PE** 294-296, 297, 298, 299, 301, 310, 315-317, 319, 430, 431, 432-433, 434-435, 438, 440, 441, 442, 446, 447-448, 449, 450-451, 454, 456-458, 461, 462, 464-467, 472, 474-475, 569, 570, 571, 657-658 **TE** 294, 295, 315, 428C, 437, 438, 446, 453, 454, 455, 462, 473	**PE** 49, 288, 430, 447, 448, 485, 526, 527
Identifies lines, angles, circles, polygons, cylinders, cones, rectangular solids, and spheres in everyday objects.		**PE** 14, 145, 146, 153, 167, 169, 431-435, 439, 443, 444, 446, 448, 451, 453-458, 459, 463, 467, 472, 473, 474, 475 **TE** 432, 438, 454	**PE** 49, 249, 447, 448, 526, 527

GA21

Correlation to the Georgia High School Graduation Qualifying Test

STRAND 3: MEASUREMENT AND GEOMETRY (continued)

TEST CONTENT OBJECTIVE	ALGEBRA 1	GEOMETRY	ALGEBRA 2
Applies geometric properties, such as the sum of the angles of a polygon property, percent of area of a circle, determined by the central angle measure in a pie chart, or parallel sides and angle relations for parallelograms, to practical drawings.	PE 197, 198	PE 148-153, 155-161, 162-167, 170-175, 177-182, 183-188, 190-196 TE 148-150, 155-158, 162, 163, 170-172, 177-179, 183-185, 190-192	PE 49, 776-777
Draws and measures angles; determines the number of degrees in the interior angles of geometric figures, such as right and straight angles, circles, triangles, and quadrilaterals; and classifies angles (right, acute, obtuse, complementary, supplementary) and triangles (right, acute, obtuse, scalene, isosceles, and equilateral).	PE 120, 197, 198	PE 10, 13, 14-16, 17, 20-23, 25-26, 28-32 TE 2C, 2D, 11, 17, 18, 19, 25, 26, 28, 43	PE 49, 776-777, 843-846, 852
Uses the Pythagorean Theorem to solve problems (includes selecting appropriate tools such as the calculator).	PE 591-597 TE 591-594	PE 323-329, 331-335, 339-343, 351, 362, 364, 378, 390-393, 398, 606, 646 TE 292, 323, 332, 333, 340, 398	PE 284-289, 318, 341, 563, 895 TE 285

STRAND 4: ALGEBRA

TEST CONTENT OBJECTIVE	ALGEBRA 1	GEOMETRY	ALGEBRA 2
Simplifies expressions with and without grouping symbols.	PE 76, 80-88, 90-91, 116-117, 123-125, 130-131, 136-138, 141-144, 148-150, 277-278, 283-285, 384, 386, 390, 391, 438-439, 562-563, 578, 580 TE 80-84, 91, 116, 124, 131, 150, 439, 562		PE 88-91, 94-100, 101, 110, 144, 440, 498-504, 505-510, 528-534 TE 88, 96, 506, 508, 528

GA22

Correlation to the Georgia High School Graduation Qualifying Test

TEST CONTENT OBJECTIVE	ALGEBRA 1	GEOMETRY	ALGEBRA 2
Evaluates simple algebraic expressions.	**PE** 12-16, 30-35, 114-120, 122-127, 129-134, 135-140, 141-146, 221-225, 244-245 **TE** 12-14, 30-33, 114-117, 122-125, 129-131, 135-138, 141-143	**PE** 20, 297, 431, 450, 454, 455, 473	**PE** 5, 6, 28, 29 **TE** 5
Substitutes known values in formulas and solves problems with formulas	**PE** 11-16, 20-22, 147-152 **TE** 12, 147-150, 577	**PE** 294-302, 303-311, 312-313, 314-319, 321-329, 331-337, 339-345, 430-435, 437-444, 445-451, 453-458, 460-467, 469-475 **TE** 294-297, 303-307, 314-316, 321-325, 331-334, 339, 340, 342, 430-433, 437-440, 445-448, 453-455, 460-463, 469-472	**PE** 28, 49, 284-289, 307-313, 430, 447, 448, 485, 526, 527, 590 **TE** 285, 308
Identifies and applies mathematics to practical problems requiring direct and inverse proportions.	**PE** 238-242, 526-531 **TE** 238-239, 526-528	**PE** 510, 511, 514, 515, 533-537, 552-553	**PE** 30-36, 480-487 **TE** 30, 31, 481
Translates words into simple algebraic expressions and equations.	**PE** 11-16, 127, 129, 131, 133, 134, 135, 139, 140, 141, 146, 318, 322, 324, 325, 328, 330, 331, 336, 337, 353-359 **TE** 11-14, 321, 354, 355		**PE** 5, 9-10, 18, 22, 25, 27, 28, 50, 51, 56, 59, 68-69, 116-117, 123, 131, 142-143, 158-163, 165, 170, 177-178, 184-185, 187-189, 191-194, 202-203 **TE** 5, 157

Correlation to the Georgia High School Graduation Qualifying Test

STRAND 4: ALGEBRA (continued)

TEST CONTENT OBJECTIVE	ALGEBRA 1		GEOMETRY		ALGEBRA 2	
Solves simple equations, including addition, subtraction, multiplication, division, proportions, and two-step equations.	PE	12-16, 30-35, 89, 99, 102, 114-120, 122-127, 129-134, 135-140, 141-146, 147-152, 154-155, 167-169, 173-177, 221-225, 244-245, 251, 253, 264-265	PE	23, 152, 174, 234, 233, 513, 551, 599, 604, 606, 607	PE	29-35, 45-51, 54-60, 62-69, 93
	TE	12-14, 30-33, 114-117, 122-125, 129-131, 135-138, 141-143, 167, 173			TE	30, 31, 45-47, 54-56, 62-65
Identifies ratio and proportion as they appear in applied situations and solves proportions for missing numbers in applied problems	PE	164-169, 172, 173, 176, 238	PE	510, 511, 514, 515, 533-537, 552-553	PE	29-35
	TE	164-167			TE	30, 31
Solves linear inequalities in one variable and graphs the solution set on the number line.	PE	278-281, 282-287, 289-293, 302-307			PE	54-59, 61-69
	TE	278, 282-285, 289-290, 302			TE	54-56, 61-65
Graphs a linear equation in two variables.	PE	31-36, 220-225, 244-251			PE	4-11, 14-19, 21-28
	TE	32, 33, 244-248			TE	4, 5, 16, 21-23
Finds the slope and intercepts of a graphed line.	PE	229, 230, 248, 250, 251, 258-262			PE	21-27
	TE	260			TE	21-24
Solves problems that involve systems of two linear equations in two variables.	PE	244-251, 252-257, 258-262, 264-265			PE	12-19, 21-27
	TE	244, 245, 248, 253			TE	16, 21-24

GA24

Holt, Rinehart and Winston Mathematics

OUR MISSION:

Holt, Rinehart and Winston has listened to the needs of you and your students and has responded by creating a superior mathematics program, one which helps today's students successfully understand and use mathematics in their everyday lives.

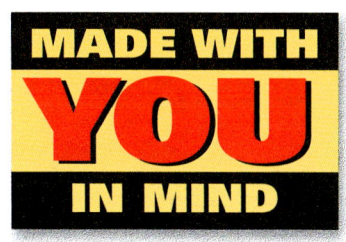
MADE WITH YOU IN MIND

Algebra 1

A1

Program Philosophy

Holt, Rinehart and Winston is committed to excellence in mathematics education and dedicated to fulfilling the needs of our users.

Holt, Rinehart and Winston wants all students to succeed in math. **Holt, Rinehart and Winston Mathematics** is special because it was designed with the needs of you and your students in mind. In conversations with teachers and students across the country, Holt, Rinehart and Winston identified challenges in math education today. The development of the core content and instructional methodology used in this new *Algebra 1*, *Geometry*, and *Algebra 2* series was based on the valuable input received. In order to best serve you and your students, our textbooks were created with the following goals in mind:

- **motivate students through real-world applications**
- **engage students in meaningful activities**
- **foster mathematical reasoning through multiple representations**
- **provide substantial, skill-building practice**
- **accommodate a variety of learning styles and ability levels**
- **segment lessons into manageable bites for easy comprehension**
- **integrate technology**

Algebra 1 Contents

PROGRAM OVERVIEW A4–A16

CHAPTER 1—FROM PATTERNS TO ALGEBRA 2–51
Patterns, variables, expressions, and equations

CHAPTER 2—OPERATIONS IN ALGEBRA 52–111
Add, subtract, multiply, and divide real numbers and expressions

CHAPTER 3—EQUATIONS 112–161
Solve equations, multistep equations, formulas, and literal equations

CHAPTER 4—PROPORTIONAL REASONING AND STATISTICS 162–215
Proportions, percent, and introduction to probability and statistics

CHAPTER 5—LINEAR FUNCTIONS 216–273
Slope, graphs, rate of change, direct variation, and parallel and perpendicular lines

CHAPTER 6—INEQUALITIES AND ABSOLUTE VALUE 274–315
Inequalities, multistep inequalities, compound inequalities, and absolute value

CHAPTER 7—SYSTEMS OF EQUATIONS AND INEQUALITIES 316–367
Solve systems by graphing, substitution, and elimination; and identify consistent and inconsistent systems

CHAPTER 8—EXPONENTS AND EXPONENTIAL FUNCTIONS 368–423
Laws of exponents, zero and negative exponents, and scientific notation

CHAPTER 9—POLYNOMIALS AND FACTORING 424–477
Add, subtract, and multiply polynomials; and factor polynomials and quadratic trinomials

CHAPTER 10—QUADRATIC FUNCTIONS 478–523
Parabolas, completing the square, quadratic formula, and graphing quadratic inequalities

CHAPTER 11—RATIONAL FUNCTIONS 524–573
Inverse variation, rational expressions, and simplifying and solving rational equations

CHAPTER 12—RADICAL FUNCTIONS AND COORDINATE GEOMETRY 574–645
Pythagorean Theorem, distance formula, and sine, cosine, and tangent functions

CHAPTER 13—PROBABILITY 646–689
Theoretical probability, counting principle, simulations, and independent events

CHAPTER 14—FUNCTIONS AND TRANSFORMATIONS 690–735
Parent functions, reflections, stretches, and translations

INFO BANK 736–802
Extra Practice, Calculator Keystroke Guide, and Tables

Planning & Applications

With today's hectic schedules, time and resource management tools are essential. The **A** **CHAPTER PLANNING GUIDE** and the **B** **LESSON PACING GUIDE** manage and organize instruction to save teachers valuable time. Referencing information is quick and easy because both guides are located in one place—no more fumbling through extraneous material.

Organized and presented in a table format, the **CHAPTER PLANNING GUIDE** allows teachers to immediately identify what is available for lesson support. The guide details each chapter's supplements lesson by lesson, taking the worry and work out of planning.

The **LESSON PACING GUIDE** provides instructional flexibility by mapping out lessons for a one-year, two-year, or block scheduled course.

The **C** **CONNECTIONS AND APPLICATIONS** table shows how *Algebra 1* integrates math topics, interdisciplinary connections, cultural connections, and real-world applications into the text. These motivating exercises and activities invite students to apply their math skills in a variety of situations, giving math education additional depth and meaning.

The **D** **BLOCK SCHEDULING** table contains suggested activities and time frames for teaching in a block scheduled environment. The table also provides options to accommodate the needs of a heterogeneous classroom.

Assessment & Technology

E **ALTERNATIVE ASSESSMENT** gives teachers options by providing types of assessment that differ from chapter quizzes and tests. For example, **PERFORMANCE ASSESSMENT** items provide unique opportunities for students to demonstrate their understanding of concepts. Teachers can also rely on **PORTFOLIO PROJECTS,** located on the interleaves of each chapter, for alternative assessment. These mini-projects introduce students to real-world situations that involve algebraic relationships in a fun and exciting way. Students will enjoy investigating solutions to familiar problems, while discovering the benefits of algebra in obtaining these solutions.

Teachers can refer to **F** **INTERNETCONNECT** on the interleaves of each chapter for an overview of using the Internet and Technical Support. Lesson Links identify pages where specific Internet resources for students can be found in the Teacher's Edition. Resource Links direct parents and teachers to www.hrw.com for support and teaching resources. *Holt, Rinehart and Winston monitors the content of this point-of-use system, ensuring that students' use of the Internet is focused on the topics being studied and that further enrichment or research is done with approved Web resources.*

Algebra 1 makes using **G** **TECHNOLOGY** a breeze. For example, graphics calculators are fully supported throughout the program with visual illustrations and specific, detailed keystroke instructions.

Spice up instruction with a variety of **Teaching Resources**, which give you many avenues to help students understand concepts. You can use **Teaching Resources** to reteach, intervene, and diversify instruction. You will find reproductions of key blackline masters at point-of-use in the side columns of the *Teacher's Edition*. In addition, facsimiles of other Teaching Resources can be found on interleaf pages.

 This extensive collection of support materials also comes on CD-ROM.

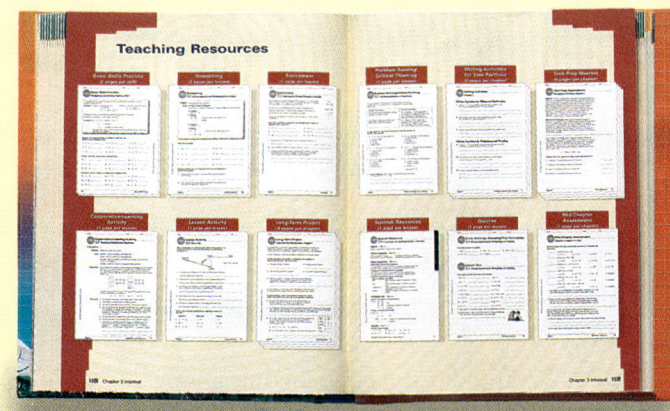

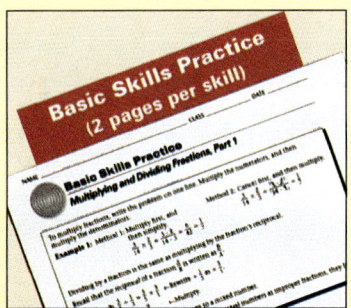

BASIC SKILLS PRACTICE provides instruction and practice in pre-algebra computational skills.

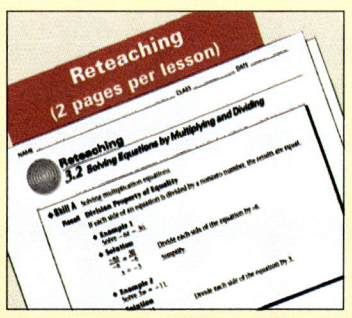

RETEACHING provides alternative teaching strategies for presenting lesson concepts in different ways, making instruction accessible to all students.

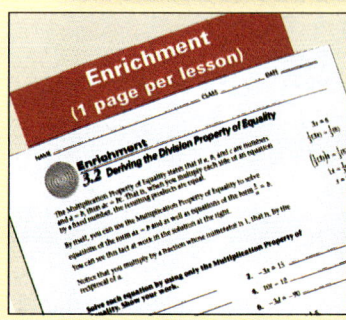

ENRICHMENT is for students who are ready, willing, and able to go beyond the level and depth of the current lesson.

COOPERATIVE-LEARNING ACTIVITIES contain group exercises and activities that are lesson-focused and structured with well-defined roles, facilitating peer education and teamwork.

LESSON ACTIVITIES reinforce lesson content through the use of manipulatives and multiple representations and by promoting student involvement in everyday life.

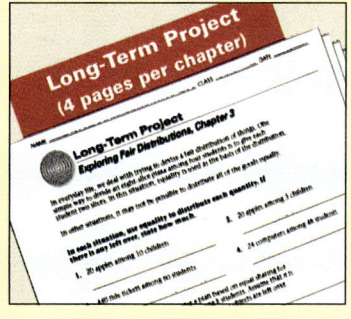

LONG-TERM PROJECTS include hands-on activities and projects that extend learning beyond a single class period.

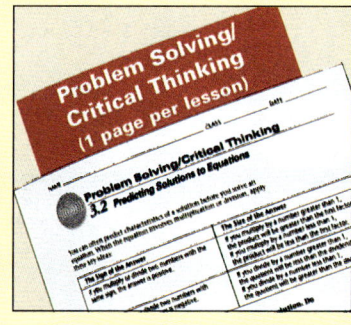

PROBLEM SOLVING/CRITICAL THINKING stimulates higher-order thinking by developing the skill of choosing and implementing a strategy for solving a problem.

WRITING ACTIVITIES provide students with unique opportunities to communicate personal experiences and mathematical knowledge, as well as to think critically about the topics in each chapter.

TECH PREP APPLICATIONS show how chapter concepts are used in technical career applications and contain group projects from various technical fields.

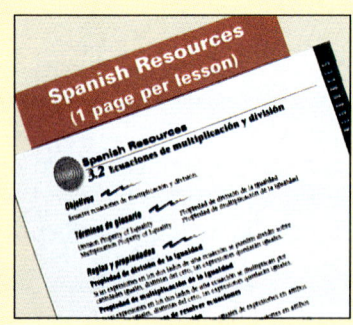

SPANISH RESOURCES contain translations of lesson objectives, glossary terms, and rules and properties, as well as solved examples and additional practice.

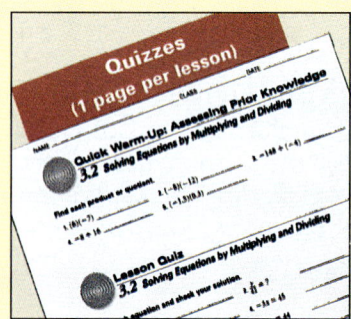

QUICK WARM-UPS and **LESSON QUIZZES** help teachers assess students' basic mastery of prerequisite and current lesson skills and are offered as both transparencies and blackline masters.

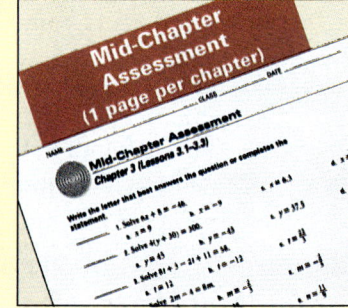

MID-CHAPTER ASSESSMENT is a set of test items that span the first half of a given chapter.

Other **Teaching Resources** not pictured here include **End-of-Course Exam Practice, Student Study Guide,** and **Teaching Transparencies.**

A6

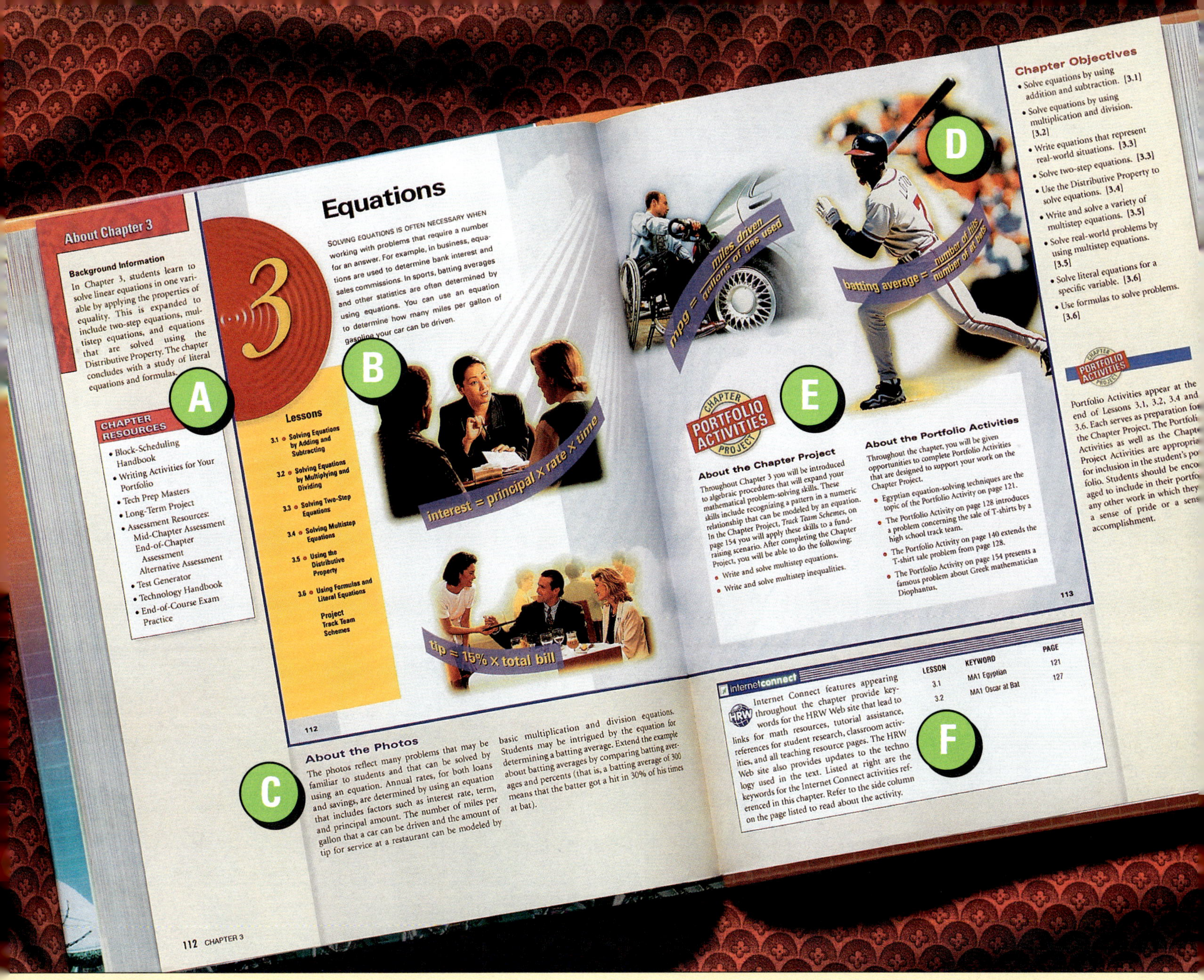

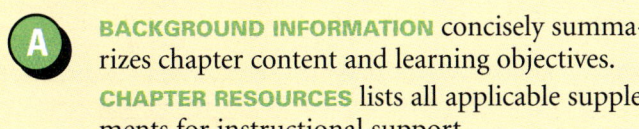
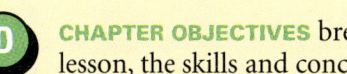
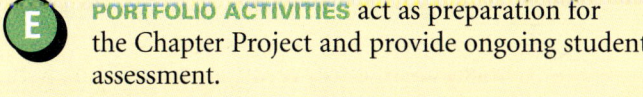
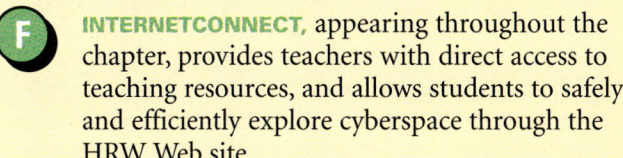

Algebra 1 takes the worry and work out of lesson planning by clearly organizing instructional materials for quick and easy reference. Important information is coherently summarized at the beginning of each chapter.

A — **BACKGROUND INFORMATION** concisely summarizes chapter content and learning objectives.
CHAPTER RESOURCES lists all applicable supplements for instructional support.

B — The **INTRODUCTION** shows the value of chapter content by relating it to scenarios that are meaningful to students.
The **LESSON** list provides a quick overview of chapter content.

C — **ABOUT THE PHOTOS** identifies the mathematical applications behind the images seen in chapter photographs.

Students benefit from a program that brings algebra to life, one which makes learning meaningful to them. This program motivates students by linking instruction to real-world examples and applications. *Algebra 1* helps teachers engage students in the following ways:

D — **CHAPTER OBJECTIVES** break down, lesson-by-lesson, the skills and concepts to be learned.

E — **PORTFOLIO ACTIVITIES** act as preparation for the Chapter Project and provide ongoing student assessment.

F — **INTERNETCONNECT**, appearing throughout the chapter, provides teachers with direct access to teaching resources, and allows students to safely and efficiently explore cyberspace through the HRW Web site.

Prepare

NCTM Principles & Standards
1–3, 5–10

LESSON RESOURCES

- Practice 3.2
- Reteaching Master 3.2
- Enrichment Master 3.2
- Cooperative-Learning Activity 3.2
- Lesson Activity 3.2
- Problem Solving/Critical Thinking Master 3.2
- Teaching Transparency 13

QUICK WARM-UP

Find each product or quotient.

1. $(6)(-7)$ -42
2. $(-8)(-12)$ 96
3. $-148 \div (-4)$ 37
4. $-8 \div 16$ -0.5
5. $(-1.5)(0.3)$ -0.45

Also on Quiz Transparency 3.2

Teach

Why Equations are important tools in real-world problem solving. Give students the equations $12x = 102$ and $\frac{n}{12} = 8.5$. Have them suggest real-world situations that these equations might represent.
Samples: What is your wage per hour if you earn $102 for working 12 hours? What total pay for 12 hours of work gives you a wage of $8.50 per hour?

122 LESSON 3.2

3.2 Solving Equations by Multiplying and Dividing

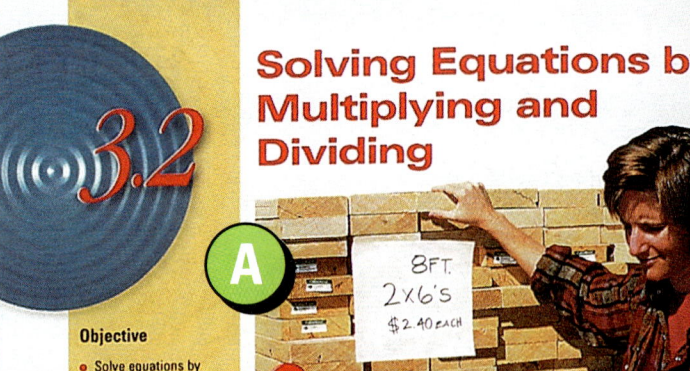

Objective
- Solve equations by using multiplication and division.

Why Carpentry problems are one type of many real-world problems that can be modeled by equations involving multiplication and division.

APPLICATION
CARPENTRY

Alicia needs to purchase twenty 8-foot boards to complete the framing of a house. In order to compare costs, she needs to know the cost per foot. At one store she finds 8-foot boards for $2.40 each. What is the cost per foot?

Let x be the cost in dollars per foot. Then the equation $8x = 2.40$ models this situation.

How can you find the solution to this equation?

Activity
Finding Solutions to Multiplication Equations

1. Explain why the equation $8x = 2.40$ models the situation presented above.

2. Draw a picture of a board divided into eight 1-foot pieces. How does drawing a picture help you to solve the problem?

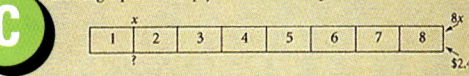

3. Solve $8x = 2.40$ by using guess-and-check.

4. Examine the algebraic method for solving $8x = 2.40$ shown below. Why does this method work?

$$8x = 2.40$$
$$\frac{8x}{8} = \frac{2.40}{8} \quad \text{Divide both sides by 8.}$$
$$x = 0.30 \quad \text{Simplify.}$$

CHECKPOINT ✓   5. Use the method from Step 4 to solve the following equations:
$$8x = 3.20 \qquad 10x = 25$$

122 CHAPTER 3

Alternative Teaching Strategy

USING MODELS Model the process of solving an equation such as $3x = 12$ by using algebra tiles. First place 3 x-tiles on the left side of the equal sign and 12 unit tiles on the right.

Now divide the tiles on each side of the equal sign into groups of equal size.

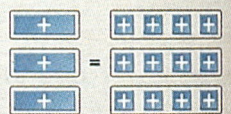

Each group of 1 x-tile can be matched with a group of 4 unit tiles, so the solution is $x = 4$.

Have students create a similar model for the process of solving $4x = -12$. Use 4 x-tiles and 12 negative unit tiles. Each x-tile can be matched with a group of 3 negative unit tiles, so $x = -3$.

A Students get motivated to learn algebra when they see its relevance. **WHY** sections make abstract concepts real to students with interesting, easily identifiable examples. Algebraic concepts represented in lesson photographs provide another source of motivation.

B The **QUICK WARM-UP** is a great way to review previously taught and prerequisite skills.
These optional, meaningful **ACTIVITIES** get students practicing, thinking, and making connections.

C Lesson instruction fosters mathematical reasoning through **MULTIPLE REPRESENTATIONS**—verbal, symbolic, and tabular.

D **CHECKPOINT** exercises, appearing throughout the chapter, assess concept comprehension.
ALTERNATIVE TEACHING STRATEGIES help you reach all your students. These practice problems bring algebra down to earth with connections to real-world situations, hands-on activities, and technology.

A8

Solving Equations by Using Division

You can use the Division Property of Equality to solve equations such as the one presented at the beginning of the lesson.

The Division Property of Equality

For all real numbers a, b, and c, if $a = b$ and $c \neq 0$, then $\frac{a}{c} = \frac{b}{c}$.

Dividing each side of an equation by equal amounts results in an equivalent equation.

EXAMPLE 1 Solve each equation. Check the solution.
  a. $3x = -9$ b. $2 = 7a$ c. $-6m = -3.6$

● SOLUTION

a. Divide each side of the equation by 3.
$$3x = -9$$
$$\frac{3x}{3} = \frac{-9}{3}$$
$$x = -3$$

To check the solution, substitute -3 for x in the original equation.
$$3(-3) \stackrel{?}{=} -9$$
$$-9 = -9 \text{ True}$$

b. Divide each side of the equation by 7.
$$2 = 7a$$
$$\frac{2}{7} = \frac{7a}{7}$$
$$\frac{2}{7} = a$$
$$a = \frac{2}{7} \quad \text{Sym. Prop.}$$

To check the solution, substitute $\frac{2}{7}$ for a in the original equation.
$$2 \stackrel{?}{=} 7\left(\frac{2}{7}\right)$$
$$2 = 2 \text{ True}$$

c. Divide each side of the equation by -6.
$$-6m = -3.6$$
$$\frac{-6m}{-6} = \frac{-3.6}{-6}$$
$$m = 0.6$$

To check the solution, substitute 0.6 for m in the original equation.
$$-6(0.6) \stackrel{?}{=} -3.6$$
$$-3.6 = -3.6 \text{ True}$$

TRY THIS Solve each equation. Check the solution.
  a. $-5x = 4.5$ b. $-30 = -6m$ c. $40 = 5p$

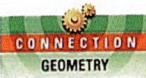

EXAMPLE 2 In a regular pentagon, all of the interior angles are equal in measure. The sum of the angle measures is 540°. Write and solve an equation to find the measure of each angle.

CONNECTION
GEOMETRY

● SOLUTION

The total number of angles is 5. The sum of the measures of the angles is 540°. Let x be the measure of each angle. Then $5x$ is the sum of the measures of all 5 angles and $5x = 540$.

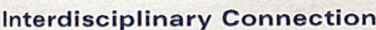

Solve for x.
$$5x = 540$$
$$\frac{5x}{5} = \frac{540}{5} \quad \text{Div. Prop. of Equality}$$
$$x = 108 \quad \text{Simplify.}$$

CHECK
$$5(108) \stackrel{?}{=} 540$$
$$540 = 540 \text{ True}$$

Each angle has a measure of 108°.

LESSON 3.2 SOLVING EQUATIONS BY MULTIPLYING AND DIVIDING **123**

Interdisciplinary Connection

PHYSICS According to Newton's second law of motion, the ratio of the force acting upon a body to the acceleration that the force produces is a constant that is called the *mass* of the body. In symbols, this relationship is given as $\frac{F}{a} = m$, where F represents force, a represents acceleration, and m represents mass. Have students solve this equation for F and then solve for a. $F = ma; a = \frac{F}{m}$

Near the surface of Earth, every object experiences a constant acceleration due to gravity. The force that accelerates the object downward is its weight. Using Newton's second law of motion, written in the form $F = ma$, this relationship becomes $w = mg$, where w represents weight, m represents mass, and g represents acceleration due to gravity. Have students solve this equation for m, then solve for g. $m = \frac{w}{g}; g = \frac{w}{m}$

Activity Notes

In this Activity, students are given an equation that represents a real-world situation and are asked to solve it by using the guess-and-check method. Then they see how to obtain the same result by an algebraic method. Students should arrive at the conclusion that they can solve an equation of the form $ax = b$, where $a \neq 0$, by dividing each side of the equation by a.

CHECKPOINT ✓

5. $8x = 3.20$ $10x = 25$
   $\frac{8x}{8} = \frac{3.20}{8}$ $\frac{10x}{10} = \frac{25}{10}$
   $x = 0.40$ $x = 2.5$

ADDITIONAL EXAMPLE 1

Solve each equation. Check the solution.

a. $-5t = -15$ $t = 3$

b. $4y = 9$ $y = \frac{9}{4}$

c. $8p = -4.8$ $p = -0.6$

TRY THIS

a. $x = -0.9$

b. $m = 5$

c. $p = 8$

Use Teaching Transparency 13.

☞ For Additional Example 2 and Math Connection, see page 124.

LESSON 3.2 **123**

ADDITIONAL EXAMPLE 2

In a regular octagon, all of the interior angles are equal in measure. The sum of the interior angle measures is 1080°. Write and solve an equation to find the measure of each interior angle.
$8x = 1080$; $x = 135$
The measure of each interior angle is 135°.

Math CONNECTION

GEOMETRY A *regular polygon* is a polygon whose angles are all congruent *and* whose sides are all congruent. The sum of the measures of the interior angles of any polygon of n sides is $(n-2)180°$. Thus, the measure of each interior angle is $\frac{(n-2)180°}{n}$.

Teaching Tip

You may need to remind students that the *coefficient* of a term is its numerical factor.

ADDITIONAL EXAMPLE 3

Solve each equation. Check the solution.
a. $\frac{n}{7} = -1.2$ $n = -8.4$
b. $5 = \frac{y}{-6}$ $y = -30$
c. $\frac{r}{-3} = -30$ $r = 90$

TRY THIS
a. $t = -16$
b. $d = -16.2$
c. $m = 4.5$

Solving Equations by Using Multiplication

APPLICATION: SPORTS

To compute a batting average, the number of hits is divided by the number of times at bat.
$$\frac{\text{Hits}}{\text{Times at bat}} = \text{Average}$$

Suppose that Alex batted 150 times during his varsity baseball career. If his final batting average is 0.300, how many hits did he get?

Let H represent the number of hits. The problem can be modeled by the following division equation:
$$\frac{H}{150} = 0.300$$

In algebraic notation, $\frac{H}{150}$ is the same as $\frac{1}{150}(H)$. To solve the equation, multiply both sides of the equation by 150 because 150 is the reciprocal of $\frac{1}{150}$, the coefficient of H.
$$150\left(\frac{H}{150}\right) = 150(0.300)$$
$$H = 45$$

Alex had 45 hits in his varsity baseball career.

The Multiplication Property of Equality

For all real numbers a, b, and c, if $a = b$, then $ac = bc$.

Multiplying each side of an equation by equal amounts results in an equivalent equation.

EXAMPLE 3 Solve each equation. Check the solution.
a. $\frac{m}{5} = 1.3$ b. $-3 = \frac{x}{-4}$ c. $\frac{w}{5} = -10$

SOLUTION

a. Multiply each side of the equation by 5.
$\frac{m}{5} = 1.3$
$(5)\frac{m}{5} = (5)1.3$
$m = 6.5$

b. Multiply each side of the equation by -4.
$-3 = \frac{x}{-4}$
$(-4)(-3) = (-4)\frac{x}{-4}$
$12 = x$
$x = 12$

c. Multiply each side of the equation by 5.
$\frac{w}{5} = -10$
$(5)\frac{w}{5} = (5)(-10)$
$w = -50$

To check the solution, substitute -6.5 for m in the original equation.
$\frac{6.5}{5} \stackrel{?}{=} 1.3$
$1.3 = 1.3$ True

To check the solution, substitute 12 for x in the original equation.
$-3 \stackrel{?}{=} \frac{12}{-4}$
$-3 = -3$ True

To check the solution, substitute -50 for w in the original equation.
$\frac{-50}{5} \stackrel{?}{=} -10$
$-10 = -10$ True

TRY THIS Solve each equation. Check the solution.
a. $\frac{t}{-3.2} = 5$ b. $\frac{d}{3} = -5.4$ c. $3 = \frac{m}{1.5}$

Inclusion Strategies

USING SYMBOLS Some students may become confused when solving the equations in this lesson because familiar symbols for multiplication and division are not evident. You may want to give students a form that utilizes these symbols and show how the method of solving equations in this form parallels the methods used to solve the equations in Lesson 3.1.

$x + 3 = 21 \rightarrow x + 3 - 3 = 21 - 3$
$x - 3 = 21 \rightarrow x - 3 + 3 = 21 + 3$
$3x = 21 \rightarrow 3 \cdot x = 21 \rightarrow 3 \cdot x + 3 = 21 + 3$
$\frac{x}{3} = 21 \rightarrow x \div 3 = 21 \rightarrow x + 3 \cdot 3 = 21 \cdot 3$

Enrichment

Tell students that equations with exactly the same solution are called *equivalent equations*. For example, $-5y = 20$ and $d + 3 = -1$ are equivalent because the solution of each is -4. Give students the equation $4r = -32$ and instruct them to find three equivalent equations: one of the form $x + a = b$, one of the form $x - a = b$, and one of the form $\frac{x}{a} = b$. samples: $n + 2 = -6$; $y - 5 = -13$; $\frac{a}{2} = -4$

 A **APPLICATIONS** demonstrate how algebra concepts and principles are used in the real world.

 B The clear lesson format makes it easy for students to follow along. Moreover, **EXAMPLES** segment each lesson into manageable bites.

 C **INCLUSION STRATEGIES** accommodate all students—those who have trouble mastering concepts, have different learning styles and ability levels, or need more practice or hands-on experience.

ENRICHMENT provides additional activities for students who are ready to go beyond the level and depth of the current lesson.

Using Reciprocals

Recall that in order to divide by a nonzero number, you multiply by the reciprocal of that number. For example, $10 \div 5$ can be rewritten as $10 \times \frac{1}{5}$. The value of both expressions is the same, 2.

Note the following:

$$\frac{a}{b} \cdot \frac{b}{a} = \frac{ab}{ba}$$
$$= \frac{ab}{ab}$$
$$= 1$$

This tells you that the reciprocal of a rational number, $\frac{a}{b}$, is $\frac{b}{a}$. Reciprocals can be used to solve multiplication and division equations.

EXAMPLE 4

Solve each equation by using reciprocals. Check each solution.

a. $\frac{3}{5}y = 10$ **b.** $-\frac{1}{3} = 5x$ **c.** $\frac{2b}{3} = -10$

● SOLUTION

a. The reciprocal of the coefficient of y is $\frac{5}{3}$. Multiply each side of the equation by $\frac{5}{3}$.

$\left(\frac{5}{3}\right)\frac{3}{5}y = \left(\frac{5}{3}\right)10$

$y = \frac{50}{3}$, or $16\frac{2}{3}$

b. The reciprocal of the coefficient of x is $\frac{1}{5}$. Multiply each side of the equation by $\frac{1}{5}$.

$\left(\frac{1}{5}\right)\left(-\frac{1}{3}\right) = \left(\frac{1}{5}\right)5x$

$-\frac{1}{15} = x$

$x = -\frac{1}{15}$ Sym. Prop.

c. The reciprocal of the coefficient of b is $\frac{3}{2}$. Multiply each side of the equation by $\frac{3}{2}$.

$\left(\frac{3}{2}\right)\frac{2b}{3} = \left(\frac{3}{2}\right)(-10)$

$b = -\frac{30}{2}$, or -15

To check the solution, substitute $\frac{50}{3}$ for y in the original equation.

$\frac{3}{5}\left(\frac{50}{3}\right) \stackrel{?}{=} 10$

$\frac{50}{5} \stackrel{?}{=} 10$

$10 = 10$ True

To check the solution, substitute $-\frac{1}{15}$ for x in the original equation.

$-\frac{1}{3} \stackrel{?}{=} 5\left(-\frac{1}{15}\right)$

$-\frac{1}{3} \stackrel{?}{=} -\frac{5}{15}$

$-\frac{1}{3} = -\frac{1}{3}$ True

To check the solution, substitute -15 for b in the original equation.

$\frac{2(-15)}{3} \stackrel{?}{=} -10$

$-\frac{30}{3} \stackrel{?}{=} -10$

$-10 = -10$ True

 **TRY THIS** Solve by using reciprocals. Check the solution.

a. $4m = -\frac{3}{8}$ **b.** $\frac{2d}{5} = \frac{3}{2}$ **c.** $-\frac{4}{5} = 3x$

CRITICAL THINKING Explain how multiplying by a reciprocal instead of dividing is similar to adding the opposite instead of subtracting.

LESSON 3.2 SOLVING EQUATIONS BY MULTIPLYING AND DIVIDING   125

Teaching Tip

In the statement about the reciprocal of a number, be sure students note that you must divide by a nonzero number. Point out that if a real number a were the reciprocal of zero, then the product $0 \cdot a$ would have to equal 1. However, this contradicts the fact that the product of zero and any real number is zero. Therefore, zero has no reciprocal.

ADDITIONAL EXAMPLE 4

Solve each equation by using reciprocals. Check each solution.

a. $-\frac{9}{4}z = 10$

$z = -\frac{40}{9}$, or $z = -4\frac{4}{9}$

b. $-6s = \frac{1}{4}$ $s = -\frac{1}{24}$

c. $-\frac{3c}{4} = -15$ $c = 20$

TRY THIS

a. $m = -\frac{3}{32}$

b. $d = \frac{15}{4}$

c. $x = -\frac{4}{15}$

Teaching Tip

In part **c** of Example 4, some students may wonder why the reciprocal of the coefficient of b is $\frac{3}{2}$. You may want to show them the following justification:

$$\frac{2b}{3} = \frac{2 \cdot b}{3 \cdot 1} = \frac{2}{3} \cdot \frac{b}{1} = \frac{2}{3} \cdot b$$

The expression $\frac{2}{3} \cdot b$ is the same as $\frac{2}{3}b$. This means that the coefficient of b is $\frac{2}{3}$ and the reciprocal of the coefficient is $\frac{3}{2}$.

CRITICAL THINKING

Adding the opposite makes use of the additive inverse of a number. Multiplying by the reciprocal uses the multiplicative inverse of a number. These numbers have unique properties relative to the identity elements for multiplication and addition. Essentially, the result is that there are really only two operations, multiplication and addition. Division and subtraction

Reteaching the Lesson

USING LANGUAGE SKILLS Work with students to translate equations into verbal questions that incorporate the word *multiply* or *divide*. For example, translate the equation $8x = 40$ as "What number do you multiply by eight to get forty?"

Now discuss the fact that, just as addition and subtraction are inverse operations, so are multiplication and division. Remind students that this means they can think of using one operation to "undo" the other. Refer back to the equation $8x = 40$. Since the variable, x, is multiplied by 8, you *divide* each side by 8 to undo the operation. Emphasize that the goal is to isolate the variable on one side of the equal sign.

Repeat the discussion, this time focusing on the equation $\frac{x}{8} = 5$. In this case the translation is "What number do you divide by eight to get five?" Since the variable, x, is divided by 8, you *multiply* each side by 8 to undo the operation.

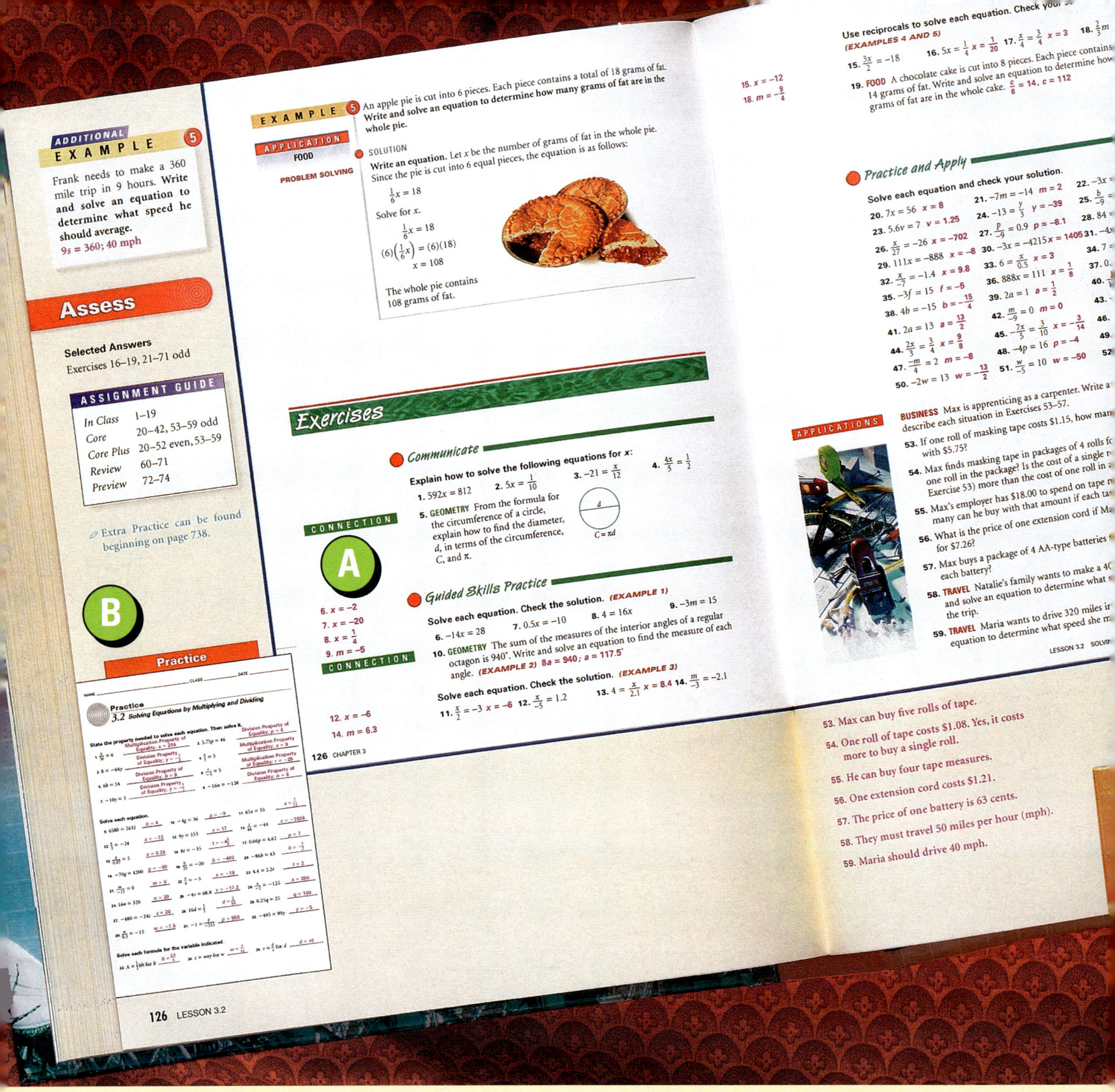

ADDITIONAL EXAMPLE 5

Frank needs to make a 360 mile trip in 9 hours. Write and solve an equation to determine what speed he should average.

$9s = 360$; 40 mph

Assess

Selected Answers
Exercises 16–19, 21–71 odd

ASSIGNMENT GUIDE

In Class	1–19
Core	20–42, 53–59 odd
Core Plus	20–52 even, 53–59
Review	60–71
Preview	72–74

Extra Practice can be found beginning on page 738.

EXAMPLE 5
APPLICATION
FOOD
PROBLEM SOLVING

An apple pie is cut into 6 pieces. Each piece contains a total of 18 grams of fat. Write and solve an equation to determine how many grams of fat are in the whole pie.

SOLUTION
Write an equation. Let x be the number of grams of fat in the whole pie. Since the pie is cut into 6 equal pieces, the equation is as follows:

$$\frac{1}{6}x = 18$$

Solve for x.

$$\frac{1}{6}x = 18$$
$$(6)\left(\frac{1}{6}x\right) = (6)(18)$$
$$x = 108$$

The whole pie contains 108 grams of fat.

Exercises

Communicate

Explain how to solve the following equations for x:

1. $592x = 812$
2. $5x = \frac{1}{10}$
3. $-21 = \frac{x}{12}$
4. $\frac{4x}{5} = \frac{1}{2}$

CONNECTION

5. **GEOMETRY** From the formula for the circumference of a circle, explain how to find the diameter, d, in terms of the circumference, C, and π.

$C = \pi d$

Guided Skills Practice

Solve each equation. Check the solution. (EXAMPLE 1)

6. $-14x = 28$
7. $0.5x = -10$
8. $4 = 16x$
9. $-3m = 15$

CONNECTION

10. **GEOMETRY** The sum of the measures of the interior angles of a regular octagon is 940°. Write and solve an equation to find the measure of each angle. (EXAMPLE 2) $8a = 940$; $a = 117.5°$

Solve each equation. Check the solution. (EXAMPLE 3)

11. $\frac{x}{2} = -3$ $x = -6$
12. $\frac{x}{-5} = 1.2$
13. $4 = \frac{x}{2.1}$ $x = 8.4$
14. $\frac{m}{-3} = -2.1$

6. $x = -2$
7. $x = -20$
8. $x = \frac{1}{4}$
9. $m = -5$
12. $x = -6$
14. $m = 6.3$

126 CHAPTER 3

Use reciprocals to solve each equation. Check your solution. (EXAMPLES 4 AND 5)

15. $\frac{3x}{2} = -18$
16. $5x = \frac{1}{20}$
17. $\frac{x}{4} = \frac{3}{4}x = 3$
18. $\frac{2}{3}x$

19. **FOOD** A chocolate cake is cut into 8 pieces. Each piece contains 14 grams of fat. Write and solve an equation to determine how many grams of fat are in the whole cake. $\frac{c}{8} = 14$, $c = 112$

15. $x = -12$
18. $m = -\frac{9}{4}$

Practice and Apply

Solve each equation and check your solution.

20. $7x = 56$ $x = 8$
21. $-7m = -14$ $m = 2$
22. $-3x$
23. $5.6v = 7$ $v = 1.25$
24. $-13 = \frac{y}{3}$ $y = -39$
25. $\frac{b}{-9}$
26. $\frac{x}{27} = -26$ $x = -702$
27. $\frac{p}{-9} = 0.9$ $p = -8.1$
28. 84
29. $111x = -888$ $x = -8$
30. $-3x = -4215$ $x = 1405$
31. -4
32. $\frac{x}{-7} = -1.4$ $x = 9.8$
33. $6 = \frac{x}{0.5}$ $x = 3$
34. $7 =$
35. $-3f = 15$ $f = -5$
36. $888x = 111$ $x = \frac{1}{8}$
37. 0.
38. $4b = -15$ $b = -\frac{15}{4}$
39. $2a = 1$ $a = \frac{1}{2}$
40.
41. $2a = 13$ $a = \frac{13}{2}$
42. $\frac{m}{-9} = 0$ $m = 0$
43.
44. $\frac{2x}{3} = \frac{3}{4}$ $x = \frac{9}{8}$
45. $-\frac{7x}{5} = \frac{3}{10}$ $x = \frac{3}{14}$
46.
47. $-\frac{m}{4} = 2$ $m = -8$
48. $-4p = 16$ $p = -4$
49.
50. $-2w = 13$ $w = -\frac{13}{2}$
51. $\frac{w}{-5} = 10$ $w = -50$
52.

APPLICATIONS

BUSINESS Max is apprenticing as a carpenter. Write and describe each situation in Exercises 53–57.

53. If one roll of masking tape costs $1.15, how many with $5.75?

54. Max finds masking tape in packages of 4 rolls for one roll in the package? Is the cost of one roll in a Exercise 53) more than the cost of one roll in ta

55. Max's employer has $18.00 to spend on tape m many can he buy with that amount if each ta

56. What is the price of one extension cord if Ma for $7.26?

57. Max buys a package of 4 AA-type batteries each battery?

58. **TRAVEL** Natalie's family wants to make a 40 and solve an equation to determine what s the trip.

59. **TRAVEL** Maria wants to drive 320 miles in equation to determine what speed she m

LESSON 3.2 SOLVIN

53. Max can buy five rolls of tape.
54. One roll of tape costs $1.08. Yes, it costs more to buy a single roll.
55. He can buy four tape measures.
56. One extension cord costs $1.21.
57. The price of one battery is 63 cents.
58. They must travel 50 miles per hour (mph).
59. Maria should drive 40 mph.

A
Can your students articulate what they've learned? In **COMMUNICATE EXERCISES**, students use their own own words to express their understanding of key concepts.

Algebra 1 contains substantial, skill-building practice, beginning in class with **GUIDED SKILLS PRACTICE**.

B
See them without searching: Many reduced teaching resources (Practice Workbook, Student Technology Guide, and Assessment) are at point-of-use with accompanying answers. Samples of other Teaching Resources are conveniently located on the interleaf pages preceding each chapter.

C
Skill-building practice continues–practice exercises for homework can be found in **PRACTICE AND APPLY**, in the **EXTRA PRACTICE** section in the appendix of the student book, and in the accompanying **PRACTICE WORKBOOK**.

Stop mistakes before they happen with **ERROR ANALYSIS**, a guide to common pitfalls associated with certain concepts and skills.

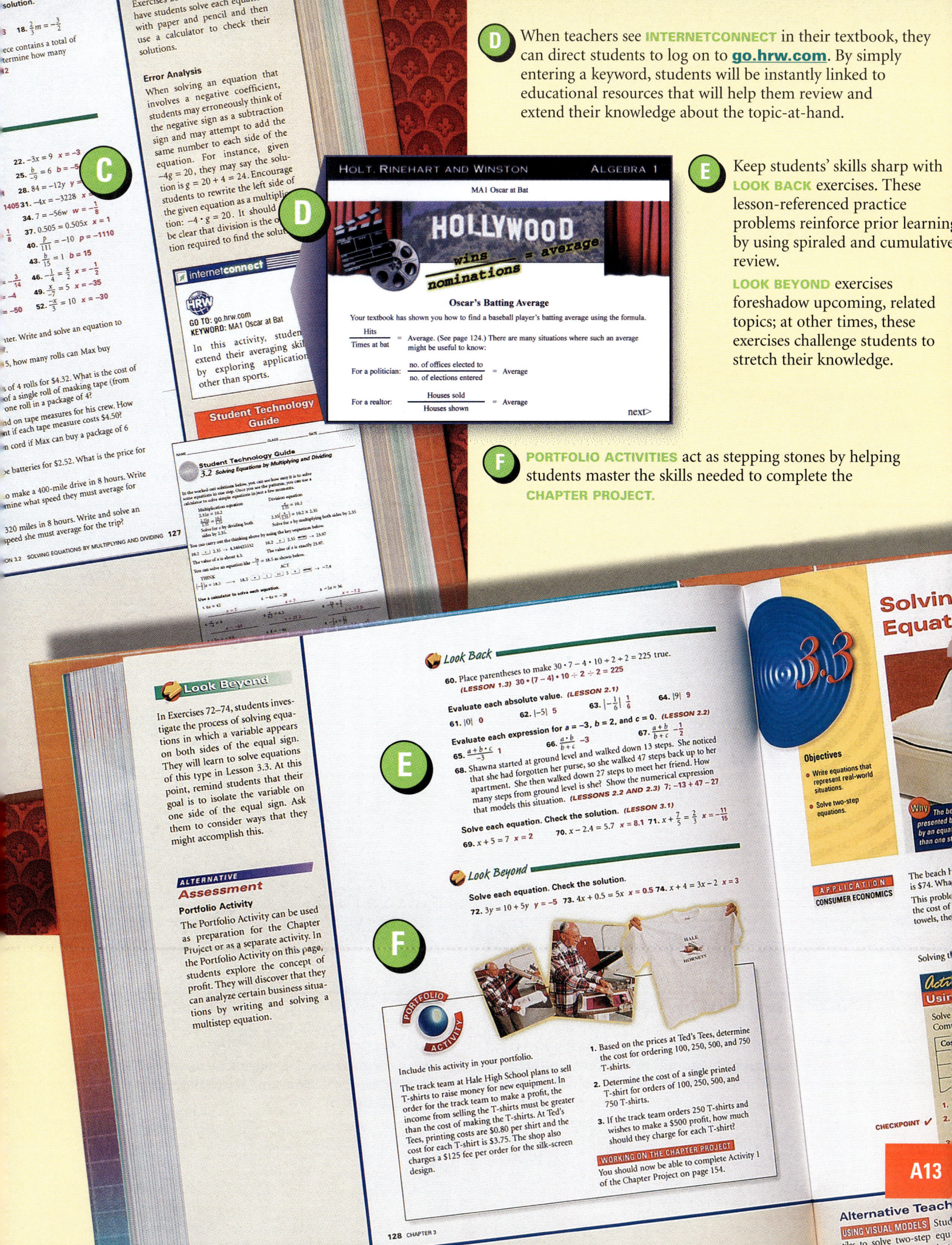

 Federal officials have stated that the grizzly bear population has rebounded after years of near extinction. Environmentalists concede that the population has increased; however, the numbers are still too low to strip the animal of federal protection. Compelling stories and newsworthy events, such as this one in **EYEWITNESS MATH,** engage students and make learning fun. In **COOPERATIVE LEARNING,** students break into groups to solve the question of counting bears in the wild with discussion and a hands-on activity.

 Because students learn in different ways, *Algebra 1* gives teachers numerous assessment options. For example, the **CHAPTER PROJECT** is a long-term, alternative assessment that ties concepts and skills together and that can be assigned at various stages throughout the chapter.

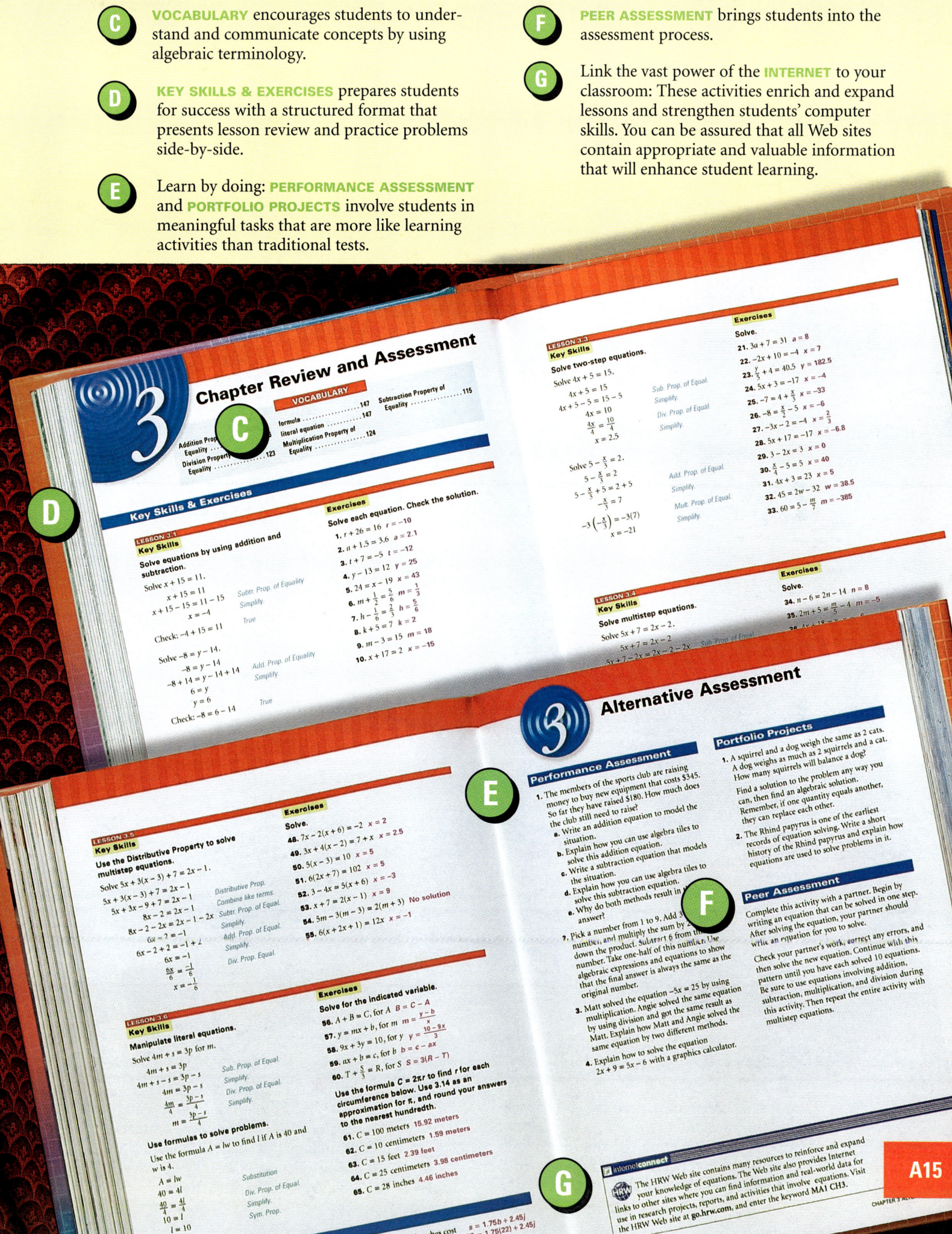

C **VOCABULARY** encourages students to understand and communicate concepts by using algebraic terminology.

D **KEY SKILLS & EXERCISES** prepares students for success with a structured format that presents lesson review and practice problems side-by-side.

E Learn by doing: **PERFORMANCE ASSESSMENT** and **PORTFOLIO PROJECTS** involve students in meaningful tasks that are more like learning activities than traditional tests.

F **PEER ASSESSMENT** brings students into the assessment process.

G Link the vast power of the **INTERNET** to your classroom: These activities enrich and expand lessons and strengthen students' computer skills. You can be assured that all Web sites contain appropriate and valuable information that will enhance student learning.

A15

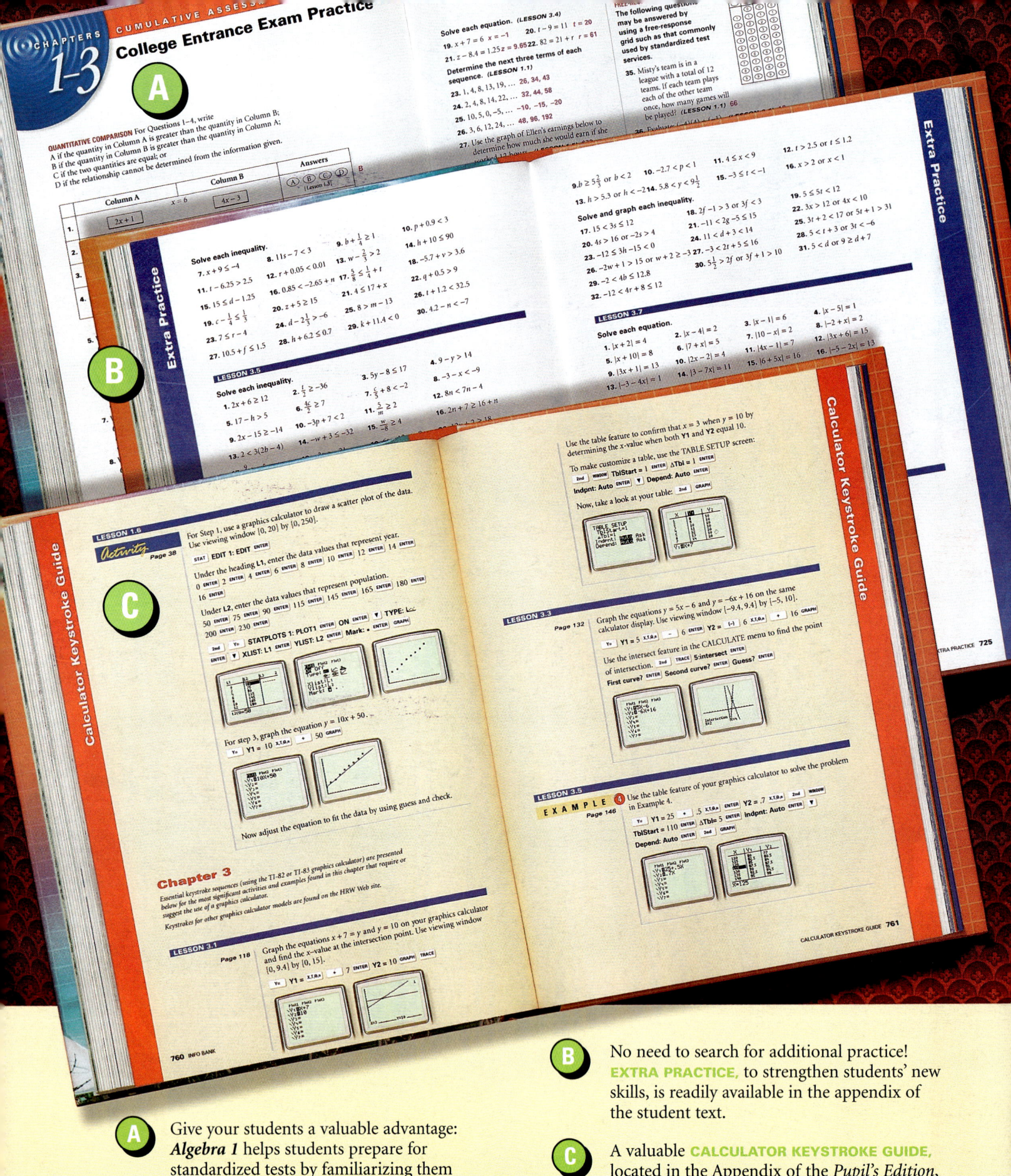

A Give your students a valuable advantage: *Algebra 1* helps students prepare for standardized tests by familiarizing them with assessment formats. At the end of each chapter, **COLLEGE ENTRANCE EXAM PRACTICE** poses lesson-referenced questions in quantitative comparison, free-response grid, and multiple choice formats.

B No need to search for additional practice! **EXTRA PRACTICE,** to strengthen students' new skills, is readily available in the appendix of the student text.

C A valuable **CALCULATOR KEYSTROKE GUIDE,** located in the Appendix of the *Pupil's Edition*, provides full, visual support for using graphics calculators. Without teacher assistance or bulky manuals, students can learn how to use their calculators by following specific, detailed keystroke instructions.

Annotated Teacher's Edition

Algebra 1

HOLT, RINEHART AND WINSTON

A Harcourt Classroom Education Company

Austin • New York • Orlando • Atlanta • San Francisco • Boston • Dallas • Toronto • London

AUTHORS

James E. Schultz, Senior Series Author
Dr. Schultz has over 30 years of experience teaching at the high school and college levels and is the Robert L. Morton Professor of Mathematics Education at Ohio University. He helped to establish standards for mathematics instruction as a coauthor of the NCTM *Curriculum and Evaluation Standards for School Mathematics* and *A Core Curriculum: Making Mathematics Count for Everyone.*

Paul A. Kennedy, Senior Author
Dr. Kennedy is a professor of Mathematics Education, director of the Institute for Mathematics, Science, and Technology Education, and is the holder of the prestigious Adams Chair at Texas Christian University. A leader in mathematics education reform, Dr. Kennedy is often invited to speak and conduct workshops on the teaching of secondary mathematics.

Wade Ellis, Jr.
Professor Ellis has coauthored numerous books and articles on how to integrate technology realistically and meaningfully into the mathematics curriculum. He was a key contributor to the landmark study *Everybody Counts: A Report to the Nation on the Future of Mathematics Education.*

Kathleen A. Hollowell
Dr. Hollowell is an experienced high school mathematics and computer science teacher who currently serves as the Director of the Mathematics & Science Education Resource Center at the University of Delaware. Dr. Hollowell is particularly well versed in the special challenge of motivating students and making the classroom a more dynamic place to learn.

Copyright © 2001 by Holt, Rinehart and Winston

All rights reserved. No part of this publication may be reproduced or transmitted in any form or by any means, electronic or mechanical, including photocopy, recording, or any information storage and retrieval system, without permission in writing from the publisher.

Requests for permission to make copies of any part of the work should be mailed to the following address: Permissions Department, Holt, Rinehart and Winston, 1120 South Capital of Texas Highway, Austin, Texas 78746-6487

(Acknowledgments appear on pages 879-880, which are extensions of the copyright page)

HRW is a registered trademark licensed to Holt, Rinehart and Winston, Inc.

Printed in the United States of America

ISBN: 0-03-052218-8

3 4 5 6 7 8 9   032   03 02 01 00

CONTRIBUTING AUTHOR

Irene "Sam" Jovell
An award winning teacher at Niskayuna High School, Niskayuna, New York, Ms. Jovell served on the writing team for the *New York State Mathematics, Science, and Technology Framework*. A popular presenter at state and national conferences, her workshops focus on technology-based innovative education.

STAFF CREDITS

Project Editors
Joel Riemer, *Senior Editor*
Charles McClelland, *Senior Editor*
Darren Peterson, *Associate Editor*
June Turner, *Associate Editor*

Editorial Staff
Gary Standafer, *Associate Director*
Marty Sopher, *Managing Editor*
Eileen Shihadeh, *Associate Editor*
Laurie Baker, *Senior Copyeditor*
Cindy Foreman, *Copyeditor*

Book Design
Diane Motz, *Senior Art Director*
Lisa Woods, *Designer*
Ed Diaz, *Design Associate*
Lori Male, *Design Associate*

Image Services
Elaine Tate, *Art Buyer Supervisor*
Michelle Rumpf, *Art Buyer Supervisor*

Photo Research
Cindy Verheyden, *Photo Researcher*
Jerry Cheek, *Assistant Photo Researcher*

Photo Studio
Sam Dudgeon, *Senior Staff Photographer*
Victoria Smith, *Photography Specialist*

Editorial Permissions
Carrie Jones

Manufacturing
Jevara Jackson, *Manufacturing Coordinator*

Production
Gene Rumann, *Production Manager*
Rose Degollado, *Senior Production Coordinator*
Susan Mussey, *Senior Production Coordinator*

Cover Design
Pronk&Associates

Design Implementation
Pronk&Associates

Research and Curriculum
Mike Tracy
Joyce Herbert
Kathy McKee
Guadalupe Solis
Jennifer Swift

CONTENT CONSULTANT

Rosemary A. Garmann
Rosemary Garmann is a graduate of Marian College (B.A.) and the University of Notre Dame (M.S. in Mathematics) and has completed post-graduate studies in mathematics and education at the University of Cincinnati, Xavier University, University of Akron, and Butler University. Ms. Garmann has taught elementary, middle school, and high school classes and is currently employed at Miami University in Oxford, Ohio, as a mathematics instructor. Formerly, she was a mathematics supervisor, building administrator, and mathematics teacher in the Cincinnati City Schools, as well as a mathematics consultant for the Hamilton County, Ohio, school districts.

REVIEWERS

Kenneth Adiekweh
LAUSD-Fremont
Los Angeles, California

Anthony Antonelli
St. Ignatius High School
Cleveland, Ohio

Catherine Barker
LAUSD-Carson
Carson, California

Judy B. Basara
St. Hubert High School
Philadelphia, Pennsylvania

Jeffery M. Blount
San Diego Unified School District
San Diego, California

Charene Borden
LAUSD-Fran Polytech-North Hollywood
Sun Valley, California

Kathi Bowers
Palo Alto Unified School District
Palo Alto, California

Paula Brown
San Diego Unified School District
La Jolla, California

Blanche Brownley
District of Columbia Public Schools
Washington, D.C.

Jeanette Burds
LAUSD-Chatsworth-Granada Hills
Northridge, California

Lynn Carlucci
Scottsdale Unified School District
Scottsdale, Arizona

David Casoli
Aliquippa School District
Aliquippa, Pennsylvania

Felisha Cheatem
Akron Public Schools
Akron, Ohio

Portia Clinton
Elk Grove Unified School District
Elk Grove, California

Elden Cozort
Scottsdale Unified School District
Scottsdale, Arizona

Nancy Crouse
San Diego Unified School District
San Diego, California

David Depner
Montour School District
Coraopolis, Pennsylvania

Guy Gnash DeRosa-Reed, Jr.
LAUSD-Crenshaw-Dorsey
Los Angeles, California

Rekha Desai
LAUSD-Belmont
Los Angeles, California

Mak Family
LAUSD-Belmont
Los Angeles, California

Beverly Fleming
Fairfax Public Schools
Fairfax, Virginia

Mark Freathy
Elk Grove Unified School District
Elk Grove, California

Paul Fulkerson
Round Rock Independent School District
Round Rock, Texas

Linda Fulmore
Phoenix Union High School District
Phoenix, Arizona

Rick Hartley
San Diego Unified School District
San Diego, California

Stan Heeb
San Ramon Valley Unified School District
San Ramon, California

John Holmes
West Contra Costa Unified School District
Richmond, California

David Honda
San Diego Unified School District
San Diego, California

Terry Mahaffey
San Diego Unified School District
San Diego, California

Sally Maloney
Worchester Public Schools
Worchester, Massachusetts

Pam Mason
LAUSD-Chatsworth-Granada Hills
Granada Hills, California

Amy Mikolajczyk
South Western City School District
Grove City, Ohio

Cheryl Mockel
Mead School District
Spokane, Washington

Derek Moriuchi
LAUSD-Garfield
Los Angeles, California

Judy Muhlethaler
San Dieguito Union High School District
Encinitas, California

Adrian Ortega
LAUSD-Venice-Westchester
Los Angeles, California

Martin Paco
Albuquerque Public Schools
Albuquerque, New Mexico

Curt Perry
Plymouth-Canton Community Schools
Plymouth, Michigan

Mai Pham
LAUSD-Southgate
South Gate, California

Mary Prater
LAUSD-Fairfax-LA-Hollywood
Los Angeles, California

John Remensky
South Park School District
Library, Pennsylvania

Elaine Riding
McGuffey School District
Claysville, Pennsylvania

Dolores Salvo
Mt. Lebanon School District
Pittsburgh, Pennsylvania

Mike Shakelford
San Ramon Valley Unified School District
San Ramon, California

Karen Smith
Sumner County School District
Gallatin, Tennessee

Kathleen Yamashita
San Ramon Valley Unified School District
San Ramon, California

Diane Young
Austin Independent School District
Austin, Texas

Table of Contents

PROFESSIONAL ARTICLES

Multiple Representations and Active Learning ... T14

Bringing Math to Life ... T16

Technology ... T17

Assessment ... T19

The New Standards ... T22

1 FROM PATTERNS TO ALGEBRA … 2

Chapter 1 Interleaf … 2A-F
- 1.1 Using Differences to Identify Patterns … 4
- 1.2 Variables, Expressions, and Equations … 11
- 1.3 The Algebraic Order of Operations … 18
- 1.4 Graphing With Coordinates … 24
- 1.5 Representing Linear Patterns … 30
- 1.6 Scatter Plots and Lines of Best Fit … 37

Chapter Project: *Working With Patterns in Data* … 44
Chapter Review and Assessment … 46
Alternative Assessment … 49
Cumulative Assessment: *College Entrance Exam Practice* … 50

Extra Practice for Chapter 1 … 738

2 OPERATIONS IN ALGEBRA … 52

Chapter 2 Interleaf … 52A-F
- 2.1 The Real Numbers and Absolute Value … 54
- 2.2 Adding Real Numbers … 60
- 2.3 Subtracting Real Numbers … 67
- 2.4 Multiplying and Dividing Real Numbers … 73
- 2.5 Properties and Mental Computation … 80
- 2.6 Adding and Subtracting Expressions … 89

Eyewitness Math: *Abandon Ship!* … 94
- 2.7 Multiplying and Dividing Expressions … 96

Chapter Project: *Using a Spreadsheet* … 104
Chapter Review and Assessment … 106
Alternative Assessment … 109
Cumulative Assessment: *College Entrance Exam Practice* … 110

Extra Practice for Chapter 2 … 741

MATH CONNECTIONS

Coordinate Geometry 25, 26, 28, 35, 59
Geometry 7, 9, 23, 87, 93, 102, 108
Patterns in Data 71
Statistics 22, 35, 65, 71, 78

APPLICATIONS

Science and Technology
Biology 54; Chemistry 65; Ecology 14, 38; Meteorology 4, 23, 37, 39, 70; Navigation 70; Physics 6, 9; Technology 8, 9, 14, 15, 16, 20, 33, 38, 77; Telecommunications 10, 34; Temperature 72

Social Studies
Demographics 42; Geography 42

Language Arts
Communicate 8, 15, 21, 27, 34, 40, 58, 64, 70, 77, 85, 92, 101; Eyewitness Math 94

Business and Economics
Accounting 87; Banking 65, 72, 73, 78; Business 28, 60, 78, 102; Consumer Economics 16, 31, 36, 59; Economics 41; Income 86, 87, 102; Inventory 72, 80, 82, 87, 93; Manufacturing 93

Life Skills
Health 33, 36; Home Improvement 16, 18, 22; Personal Finance 67

Sports and Leisure
Contests 34; Entertainment 16, 36, 48, 88; Music 42; Recreation 10, 11, 13, 15, 26, 28, 36, 83, 99; Sports 42; Travel 31, 32, 35

Other
Fund-raising 36; Gifts 36; Transportation 34

3 EQUATIONS 112

	Chapter 3 Interleaf	112A-F
3.1	Solving Equations by Adding and Subtracting	114
3.2	Solving Equations by Multiplying and Dividing	122
3.3	Solving Two-Step Equations	129
3.4	Solving Multistep Equations	135
3.5	Using the Distributive Property	141
3.6	Using Formulas and Literal Equations	147
	Chapter Project: *Track Team Schemes*	154
	Chapter Review and Assessment	156
	Alternative Assessment	159
	Cumulative Assessment: *College Entrance Exam Practice*	160
	Extra Practice for Chapter 3	744

4 PROPORTIONAL REASONING AND STATISTICS 162

	Chapter 4 Interleaf	162A-F
4.1	Using Proportional Reasoning	164
4.2	Percent Problems	171
	Eyewitness Math: *Barely Enough Grizzlies?*	178
4.3	Introduction to Probability	180
4.4	Measures of Central Tendency	186
4.5	Graphing Data	193
4.6	Other Data Displays	201
	Chapter Project: *Designing a Research Study*	208
	Chapter Review and Assessment	210
	Alternative Assessment	213
	Cumulative Assessment: *College Entrance Exam Practice*	214
	Extra Practice for Chapter 4	747

MATH CONNECTIONS

Geometry 119, 120, 123, 126, 139, 145, 149, 150, 152, 167, 168, 169, 170

Probability 184
Patterns in Data 212

Statistics 117, 119, 138, 139, 177, 190, 191, 196, 198, 199, 206

APPLICATIONS

Science and Technology
Biology 168, 191, 203, 206; Communication 133; Ecology 133; Health 175, 200; Meteorology 171; Physics 148; Technology 118, 132, 137, 204; Temperature 140, 147

Social Studies
Geography 199; Political Science 177; Student Government 146

Language Arts
Communicate 118, 126, 132, 138, 144, 151, 168, 175, 183, 190, 198, 205; Eyewitness Math 178

Business and Economics
Business 127, 135, 136; Consumer Economics 120, 129, 131, 133, 134, 141, 158, 169, 212; Economics 195; Inventory 167; Investments 148, 195; Small Business 146

Life Skills
Carpentry 122; Cooking 169; Discounts 173, 177; Education 212; Fire Prevention 199; Home Improvement 139; Landscaping 152; Personal Finance 114, 120; Taxes 174, 176, 177

Sports and Leisure
Cats 197; Games 184; Dogs 199; Recreation 192; Sports 120, 124, 164, 187, 189; Theater 152; Travel 120, 127, 152, 169, 207

Other
Calendar 120; Food 126, 127; Gardening 140

LINEAR FUNCTIONS — 216

	Chapter 5 Interleaf	216A-F
5.1	Linear Functions and Graphs	218
5.2	Defining Slope	226
	Eyewitness Math: *Watch Your Step!*	234
5.3	Rate of Change and Direct Variation	236
5.4	The Slope-Intercept Form	244
5.5	The Standard and Point-Slope Forms	252
5.6	Parallel and Perpendicular Lines	258
	Chapter Project: *Modeling Trends in Data*	264
	Chapter Review and Assessment	266
	Alternative Assessment	271
	Cumulative Assessment: *College Entrance Exam Practice*	272
	Extra Practice for Chapter 5	750

INEQUALITIES AND ABSOLUTE VALUE — 274

	Chapter 6 Interleaf	274A-F
6.1	Solving Inequalities	276
6.2	Multistep Inequalities	282
6.3	Compound Inequalities	289
6.4	Absolute-Value Functions	294
6.5	Absolute-Value Equations and Inequalities	300
	Chapter Project: *Planning a Snowboarding Trip*	308
	Chapter Review and Assessment	310
	Alternative Assessment	313
	Cumulative Assessment: *College Entrance Exam Practice*	314
	Extra Practice for Chapter 6	753

MATH CONNECTIONS

Coordinate Geometry 224, 251, 257, 296, 304
Geometry 232, 241, 259, 262, 287
Maximum/Minimum 301
Statistics 294
Transformations 298

APPLICATIONS

Science and Technology
Aviation 236, 242, 290; Chemistry 299; Ecology 251; Meteorology 270, 293; Physics 228, 232, 239, 240, 242, 258, 262; Technology 221, 225, 229, 237, 246, 254, 294, 296, 304; Temperature 281

Social Studies
Geography 232, 270; Government 307; Demographics 225

Language Arts
Communicate 223, 230, 240, 249, 256, 261, 279, 286, 292, 297, 305; Eyewitness Math 234

Business and Economics
Banking 312; Consumer Economics 277, 279, 285, 286, 312; Income 240, 241, 242, 270; Investments 280; Manufacturing 300, 307; Sales 293; Savings 312; Small Business 287

Life Skills
Budget 251; Health 218, 242, 306; Rentals 241

Sports and Leisure
Crafts 302; Entertainment 225; Recreation 221, 227, 242; Sports 244, 245, 246; Theater 287; Travel 270, 287, 299

Other
Auditorium Seating 280; Fund-raising 240, 251, 253, 257; Hobbies 312; Pool Maintenance 289; Transportation 222, 237

SYSTEMS OF EQUATIONS AND INEQUALITIES — 316

Chapter 7 Interleaf 316A-F
7.1 Graphing Systems of Equations 318
7.2 The Substitution Method 326
7.3 The Elimination Method 331
7.4 Consistent and Inconsistent Systems 338
7.5 Systems of Inequalities 345
7.6 Classic Puzzles in Two Variables 353
Chapter Project: *Minimum Cost/Maximum Profit* 360
Chapter Review and Assessment 362
Alternative Assessment 365
Cumulative Assessment: *College Entrance Exam Practice* 366

Extra Practice for Chapter 7 756

EXPONENTS AND EXPONENTIAL FUNCTIONS — 368

Chapter 8 Interleaf 368A-F
8.1 Laws of Exponents: Multiplying Monomials 370
8.2 Laws of Exponents: Powers and Products 377
8.3 Laws of Exponents: Dividing Monomials 383
8.4 Negative and Zero Exponents 390
Eyewitness Math: *All Mixed Up?* 396
8.5 Scientific Notation 398
8.6 Exponential Functions 404
8.7 Applications of Exponential Functions 409
Chapter Project: *The Wolves' Den* 416
Chapter Review and Assessment 418
Alternative Assessment 421
Cumulative Assessment: *College Entrance Exam Practice* 422

Extra Practice for Chapter 8 759

MATH CONNECTIONS

Coordinate Geometry 343, 382, 403

Geometry 329, 330, 335, 343, 351, 364, 373, 374, 375, 380, 382, 385, 388
Number Theory 330, 343, 359, 395

Patterns in Data 340
Probability 394, 414
Statistics 324

APPLICATIONS

Science and Technology
Archaeology 409, 413, 41; Astronomy 398, 399, 400, 402, 403, 420; Aviation 325, 354, 359; Biology 372, 374, 375, 383; Chemistry 356, 358, 359, 364; Physics 400, 408; Technology 319, 320, 321, 324, 337, 338, 339, 341, 346, 372, 375, 388, 394, 401, 403, 412, 414, 415

Social Studies
Demographics 340, 404, 405, 407, 408, 412, 413, 414; Social Studies 402

Language Arts
Communicate 323, 329, 334, 342, 350, 357, 374, 381, 387, 393, 401, 407, 413; Eyewitness Math 396

Business and Economics
Business 336, 359; Consumer Economics 318, 336; Economics 389, 402; Finance 336, 345, 420; Income 336, 344, 349; Investments 336, 411, 413, 414; Packaging 389; Sales 336, 343; Small Business 328

Life Skills
Academics 351; Personal Finance 410; Rental Fees 331, 333

Sports and Leisure
Auto Racing 326; Gardening 350; Hobbies 322; Recreation 330, 336, 358, 382; Sports 351

Other
Fund-raising 330, 352

T9

9 POLYNOMIALS AND FACTORING — 424

	Chapter 9 Interleaf	424A-F
9.1	Adding and Subtracting Polynomials	426
9.2	Modeling Polynomial Multiplication	432
9.3	Multiplying Binomials	438
9.4	Polynomial Functions	443
9.5	Common Factors	448
9.6	Factoring Special Polynomials	452
9.7	Factoring Quadratic Trinomials	458
9.8	Solving Equations by Factoring	464
	Chapter Project: *Powers, Pascal, and Probability*	470
	Chapter Review and Assessment	472
	Alternative Assessment	475
	Cumulative Assessment: *College Entrance Exam Practice*	476

Extra Practice for Chapter 9 762

10 QUADRATIC FUNCTIONS — 478

	Chapter 10 Interleaf	478A-F
10.1	Graphing Parabolas	480
10.2	Solving Equations by Using Square Roots	486
10.3	Completing the Square	492
10.4	Solving Equations of the Form $x^2 + bx + c = 0$	498
	Eyewitness Math: *Rescue at 2000 Feet*	504
10.5	The Quadratic Formula	506
10.6	Graphing Quadratic Inequalities	511
	Chapter Project: *What's the Difference?*	516
	Chapter Review and Assessment	518
	Alternative Assessment	521
	Cumulative Assessment: *College Entrance Exam Practice*	522

Extra Practice for Chapter 10 766

MATH CONNECTIONS

Coordinate Geometry 489

Geometry 428, 436, 441, 442, 446, 449, 451, 452, 456, 463, 474, 490, 500, 503, 510, 520

Maximum/Minimum 511
Number Theory 485
Statistics 457, 469

APPLICATIONS

Science and Technology
Ecology 485; Optics 513; Physics 451, 464, 480, 483, 485, 486, 488, 489, 490, 497, 515; Technology; 437, 444, 445, 446, 447, 465, 467, 468, 482, 487, 501, 515

Language Arts
Communicate 429, 435, 441, 446, 450, 455, 462, 467, 484, 489, 495, 502, 509, 514; Eyewitness Math 504

Business and Economics
Accounting 510; Business 511; Construction 442, 456, 468; Framing 440; Landscaping 474; Packaging 426, 443, 444, 445, 446

Life Skills
Discounts 436; Sales Tax 463

Sports and Leisure
Fine Arts 458; Photography 503; Sports 447, 520

Other
Interior Decorating 431; Marching Band 503

11 RATIONAL FUNCTIONS 524

	Chapter 11 Interleaf	524A-F
11.1	Inverse Variation	526
11.2	Rational Expressions and Functions	532
11.3	Simplifying Rational Expressions	538
	Eyewitness Math: *How Worried Should You Be?*	544
11.4	Operations With Rational Expressions	546
11.5	Solving Rational Equations	552
11.6	Proof in Algebra	558
	Chapter Project: *Designing a Park*	566
	Chapter Review and Assessment	568
	Alternative Assessment	571
	Cumulative Assessment: *College Entrance Exam Practice*	572

Extra Practice for Chapter 11 769

12 RADICAL FUNCTIONS & COORDINATE GEOMETRY 574

	Chapter 12 Interleaf	574A-F
12.1	Operations With Radicals	576
12.2	Square-Root Functions and Radical Equations	583
12.3	The Pythagorean Theorem	591
12.4	The Distance Formula	598
12.5	Geometric Properties	606
12.6	The Tangent Function	613
12.7	The Sine and Cosine Functions	621
12.8	Introduction to Matrices	628
	Chapter Project: *Working the Angles*	636
	Chapter Review and Assessment	638
	Alternative Assessment	643
	Cumulative Assessment: *College Entrance Exam Practice*	644

Extra Practice for Chapter 12 772

MATH CONNECTIONS

Coordinate Geometry 536, 550, 556, 604
Geometry 530, 541, 542, 570, 582, 592, 594, 596, 597, 599, 602, 606, 619, 625
Probability 635
Statistics 565, 634
Transformations 537
Trigonometry 616, 618, 620, 624, 627

APPLICATIONS

Science and Technology
Aviation 604, 613, 616, 626; Astronomy 619; Biology 542; Civil Engineering 620, 626; Ecology 551, 619; Engineering 609; Mechanics 531; Physics 528, 531, 584, 585, 589; Technology 533, 535, 554, 586, 605, 615, 616, 617, 622, 623

Language Arts
Communicate 529, 535, 541, 549, 555, 563, 580, 588, 595, 602, 610, 618, 624, 633; Eyewitness Math 544

Business and Economics
Construction 619, 626, 642; Consumer Economics 536; Income 597; Investments 531; Small Business 554, 557

Life Skills
Discounts 543; Landscaping 582, 596

Sports and Leisure
Contests 565; Fitness 570; Hobbies 620; Recreation 546, 556, 594, 617, 621, 623, 624; Sports 596, 635, 642; Travel 531, 570

Other
Emergency Services 598; Music 531; Transportation 628, 642

13 PROBABILITY   646

	Chapter 13 Interleaf	646A-F
13.1	Theoretical Probability	648
13.2	Counting the Elements of Sets	654
13.3	The Fundamental Counting Principle	661
13.4	Independent Events	667
	Eyewitness Math: *Hot Hands or Hoopla?*	674
13.5	Simulations	676
	Chapter Project: *Happy Birthday!*	682
	Chapter Review and Assessment	684
	Alternative Assessment	687
	Cumulative Assessment: *College Entrance Exam Practice*	688

Extra Practice for Chapter 13 776

14 FUNCTIONS AND TRANSFORMATIONS   690

	Chapter 14 Interleaf	690A-F
14.1	Graphing Functions and Relations	692
14.2	Translations	700
14.3	Stretches and Compressions	707
14.4	Reflections	714
	Eyewitness Math: *Is There Order in Chaos?*	720
14.5	Combining Transformations	722
	Chapter Project: *Pick a Number*	728
	Chapter Review and Assessment	730
	Alternative Assessment	733
	Cumulative Assessment: *College Entrance Exam Practice*	734

Extra Practice for Chapter 14 779

MATH CONNECTIONS

Coordinate Geometry 666, 681, 699   **Geometry** 653, 660, 665, 719, 726   **Statistics** 658, 659, 660, 706, 712, 719

APPLICATIONS

Science and Technology
Astronomy 652; Chemistry 666; Meteorology 678, 680, 699; Physics 718, 732; Technology 676, 700, 707, 708, 709; Telecommunications 665

Social Studies
Geography 652

Language Arts
Communicate 651, 658, 664, 671, 679, 697, 705, 711, 717, 725; Eyewitness Math 674, 720; Language Arts 652, 665

Business and Economics
Architecture 706; Business 732; Consumer Economics 665; Finance 686; Income 713; Inventory 663; Investments 660; Marketing 686; Packaging 726

Life Skills
Academics 666, 673; Clothing 662; Interpersonal Skills 654; Scheduling 661, 667, 669

Sports and Leisure
Art 713; Contests 673; Games 657, 660, 665, 672; Recreation 672, 692, 722, 726; Sports 648, 676, 680; Travel 673, 677, 699

Other
Book Club 704

INFO BANK TABLE OF CONTENTS 736

Extra Practice 738
Calculator Keystroke Guide 782
Table of Squares, Cubes, Square Roots, and Cube Roots 799
Table of Random Digits 800
Glossary 801
Additional Answers 810
Index 871
Credits 879

Multiple Representations and Active Learning

Recent data from the Third International Mathematics and Science Study (TIMMS) have resulted in calls to improve mathematics curriculum and instruction. The first course in algebra is considered to be an anchor point for a worthy mathematics curriculum. Student success in the first course in algebra is often an indication of an elementary curriculum that is infused with concrete connections to algebraic thinking and instruction that is sensitive to the cognitive development of algebra students. Typically, eighth- and ninth-grade students are making a transition in their thinking capabilities from concrete to more abstract reasoning. An algebra curriculum that can accommodate the developmental needs of students has the best chance of success. How can this be accomplished?

School algebra should help students experience the power of formalized algebra. However, the transition from the more concrete domain of arithmetic to the more abstract domain of algebra needs to be mediated through an entry point that the student finds familiar. The algebra curriculum should enable students to grasp the formal structures of algebra by grounding students in familiar contexts and real-world applications of algebraic thinking.

Teaching a student a new concept through an entry point of familiar knowledge builds on the student's previous life experiences, mathematical development, emotional disposition, and confidence. A teaching methodology that considers where a student starts in the learning process attempts to find an anchor point in the familiar in order to build concepts. Perhaps the notion that "learning should be fun" should be recast as "learning should be familiar."

WHAT IS ALGEBRA?

Algebraic thinking has been widespread through the ages. The Rhind papyrus from 1650 B.C.E. presents a series of basic linear equations in one variable for a student to solve—a drill-and-practice worksheet. Egyptians used a technique called the rule of false position to solve such problems (for a description of ancient Egyptian equation-solving techniques, see the Portfolio Activity on page 121). The early Babylonians had tables consisting of Pythagorean triples (many years before Pythagoras). The work of al-Kwarizmi, an Arab mathematician of the ninth century, provided the first practical description of algebraic methods. Finding the value of unknown quantities by applying basic properties of arithmetic was the approach—building on the notion that "equals added to equals yields equals." Algebraic thinking is ancient and multicultural in its heritage.

Only in the last few centuries has algebra been fully formalized. Modern algebra is based on the field properties of the real numbers. The Commutative, Associative, and Distributive Properties, along with the Properties of Equality are the basis for all algebraic manipulation. Most school algebra curricula of the last 40 years have reflected an awareness of these properties as the foundation of modern algebra.

Formalized algebra is powerful. Algebraic procedures can be applied to simplify any problem that utilizes real numbers. Algebra can be extended to complex numbers also, thus giving science a powerful tool for understanding the nature of the universe. However, a completely formal approach to algebra can be a barrier to learning because it assumes an abstract thinking ability that usually is only just beginning to develop in the learner. An instructional approach that uses multiple representations of algebraic concepts and that encourages active learning can help students overcome barriers to understanding.

MULTIPLE REPRESENTATIONS AND *ALGEBRA 1*

Algebraic structure and techniques can be made more familiar to students through the use of multiple representations. Concrete representations such as those provided by algebra tiles can help students who need tactile-kinesthetic reinforcement grasp concepts such as factoring. Tabular and graphical representations can help students to "see" a problem that may be represented by only words or algebraic symbols. Finally, once a conceptual foundation is laid, students can move on to the symbolic manipulation that most of us consider to be the essence of school algebra.

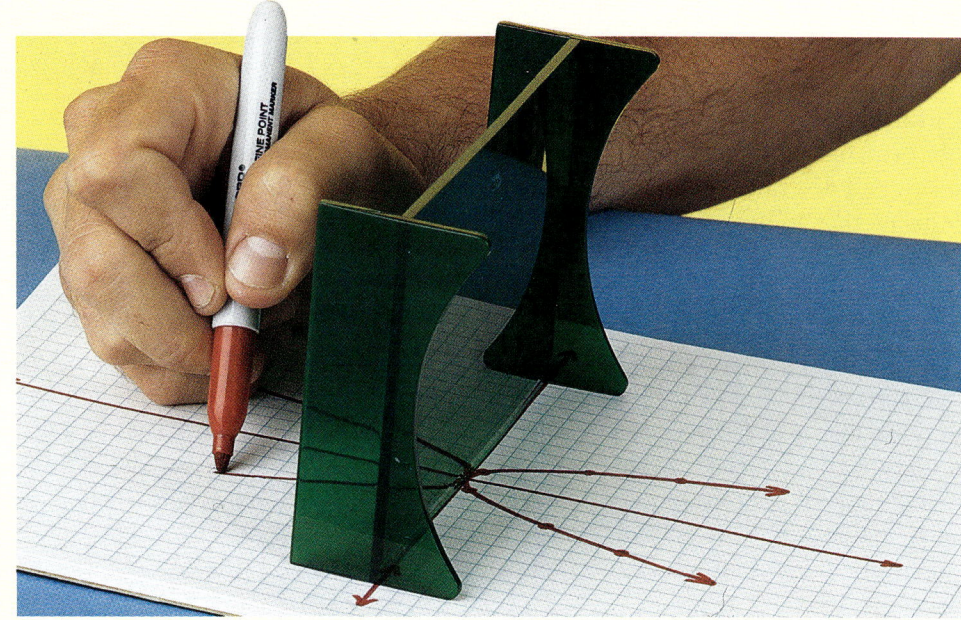

Holt, Rinehart and Winston's *Algebra 1* program provides enough variety of approaches to enable teachers to be successful in the classroom. The formal nature of algebra is stressed through the early introduction of basic equation-solving techniques and the properties of real numbers. Often, examples and exposition are rooted in real-world applications that anchor the student in a familiar experience.

Students are encouraged to use multiple representations to solve problems, such as building a table and a graph for solving linear equations in one and two variables or exploring the number pattern that occurs in an exponential function. Throughout the book, students are asked to look for patterns by organizing observational data into tables; then they make conjectures from the data. As students generalize or synthesize their observations as conjectures, they progress from the concrete to the abstract. Students who do not possess acute memorization skills will be able to grasp algebraic thinking, will find the instructional more relevant, and will gain a deeper understanding of the concepts.

ACTIVE LEARNING AND *ALGEBRA 1*

Those teachers wishing to take advantage of the full capabilities of Holt, Rinehart and Winston's *Algebra 1* can have their students complete the activities that accompany most of the lessons. These activities are exploratory in nature and encourage students to build their own understanding of an algebra concept. An activity-oriented approach gives teachers several advantages. Perhaps the most important of these is the ability to provide for learning styles other than verbal and visual. Students who need hands-on, tactile experiences become conceptually engaged by instructional approaches that encourage active learning.

In many classrooms, a key decision will be how and when to use active learning. Some topics and situations will continue to require direct instruction. Holt, Rinehart and Winston's *Algebra 1* can easily support direct instruction. Clear, concise, and appropriately sequenced examples help guide students through concept development. Most examples are followed by a "Try This" problem embedded in the instructional sequence to give students the opportunity to try their mastery of a concept and to provide the teacher with an ongoing assessment of students' understanding.

Teachers who implement active-learning approaches can expect to see from their students gradual improvement in long-term recall of information, clearer understanding of connections among concepts, more creative approaches to problem solving, and more sophisticated ways of using and communicating mathematical ideas. Teachers can use Holt, Rinehart and Winston's *Algebra 1* to support active learning.

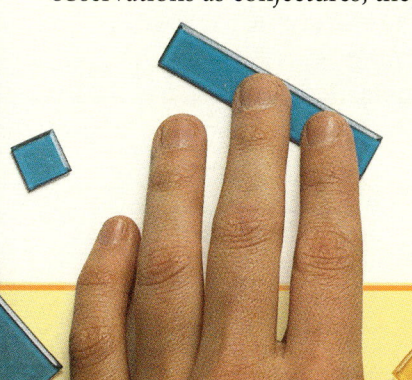

Bringing Math *to Life*

Our lives are touched daily by the application of mathematics. Mathematics is the main tool of science, providing ways to quantify, verify, and describe results and theories. Computers, with circuitry and software that applies mathematical logic, control many of the dimensions of the infrastructure around us. Traffic lights, power grids, and automated teller machines are among the many things that depend on computer-based mathematical logic structures to work properly.

In ancient Egypt, problem sets were written specifically to drill a particular mathematical skill. The drill-and-practice approach may be valid at certain points in the instructional sequence, but when used as the primary means of instruction, it decontextualizes the mathematics and reinforces the impression that mathematics is an abstraction that has no practical application. Most teachers agree that a combination of drill and practice and an instructional context consisting of applications, connections, and problem solving is the best way to teach mathematics.

APPLICATIONS

Throughout Holt, Rinehart and Winston's *Algebra 1*, students are introduced to concepts through real-world applications. These applications provide connections to domains of human activity such as business, consumer economics, science, life skills, vocational skills, and leisure activities. These applications anchor the student in a real-world context that is familiar and provides a springboard to the discovery of the underlying mathematical concepts.

Applications provide evidence that mathematics is important in our world and that the algebra learned in the classroom today does have practical uses. Students may also recognize the power of mathematics to accelerate their opportunities in life.

Appropriate use of technology to support the exploration of applications helps students to prepare for real-world problem-solving situations. With the use of technology, students can explore realistic, rather than contrived, data. Students who learn to tackle algebraic problems by actively exploring various approaches, collaborating with others, and comparing results to look for alternative strategies will be equipped to transfer these approaches to their personal and professional challenges later in life.

CONNECTIONS

Mathematics provides the conceptual basis for the structure of many things around us. For example, the architecture of a building is influenced by formulas that help determine stress and weight factors. Mathematics is the main tool of science, providing statistical modeling to support experimental research. Even so-called "pure" mathematics sometimes finds practical and direct use in helping to describe the nature of things (consider fractal geometry). Our insurance rates, tax rates, and interest rates are all determined by mathematical formulas. Mathematics is itself a connection between domains of knowledge.

Mathematics is also connected within itself. A simple geometric construction of a 45-45-90 right triangle can be used to discover irrational numbers such as $\sqrt{2}$. The distance formula is an algebraic representation of the length of a line segment.

Holt, Rinehart and Winston's *Algebra 1* emphasizes connections within mathematical disciplines and to other domains of knowledge. Instruction that emphasizes curricular connections is more likely to motivate student interest in mathematically oriented subjects such as chemistry, physics, logic, economics, statistics, and computer science.

Technology

TECHNOLOGY GOALS

To meet changing curriculum requirements, algebra teachers are continually searching for the best way to integrate technology into their classroom. They know that technology allows their students to become active participants in the learning process. Through the use of technology, today's mathematics classrooms are being transformed into laboratories where students explore and experiment with mathematical concepts rather than just memorize isolated facts. Students make generalizations and reach conclusions about mathematical concepts and relationships and apply them to real-world situations. Holt, Rinehart and Winston's *Algebra 1* supports instruction that utilizes technology with activities and examples that encourage students to make and test conjectures and to confirm mathematical ideas for themselves.

Teachers also know that technology encourages cooperative learning. Cooperative-learning groups allow students to compare results, brainstorm, and reach conclusions based on group results. As in real life, where scientists and financial analysts often consult each other, students in the technology-oriented classroom learn to communicate and consult with each other. Students learn that such consultations are not "cheating," but are rather a method of sharing information that will be used to solve a problem.

Furthermore, the use of technology provides students with an avenue to explore real-world applications that would not be practical to venture into without the use of such technology. Students are able to study realistic data instead of data that simply works out nicely. For instance, when calculating by hand, most students study only simple interest. The use of technology allows students to explore compound interest, which is how interest is calculated by today's financial institutions.

Real data from student's local town or neighborhood can be used to find the average price or rental cost of a home. Students can study data that are more closely related to their own lives, such as comparing the cost of phone plans and determining when one plan costs less than another.

Many lessons in Holt, Rinehart and Winston's *Algebra 1* utilize the power of technology as an exploration tool. For example, in Lesson 5.4, students explore the concept of a constant. A graphics calculator is used to help students make a conjecture about how the value of *b* affects the graph of $y = 2x + b$.

TYPES OF TECHNOLOGY

The most evident use of technology in Holt, Rinehart and Winston's *Algebra 1* appears in the form of a graphics calculator. Computer spreadsheet software and Internet resources are also utilized.

Most educators agree that technology should be used in the mathematics classroom. However, the reality is that not every student enrolled in a first-year algebra course has access to a graphics calculator or a computer. That is why the use of graphics calculators in *Algebra 1* is

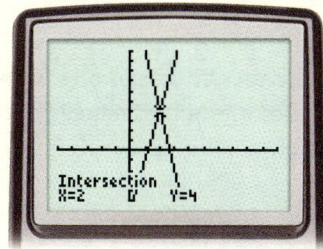

secondary in nature. In most instances when a graphics calculator is used in an example, it is shown as a technological alternative to the traditional paper-and-pencil method.

Graphics calculators perform many specialized and complex mathematical actions, including graphing. These calculators use certain keys to access full-screen menus that show additional operations. The scope of capabilities is wide, including traditional arithmetic calculations, matrix operations, statistics mode, function graphing, parametric graphing, polar graphing, sequence graphing, and more. Graphics calculators are also capable of running user-defined programs. When using the graphing capabilities of the calculators, the use of a range-setting window adds greatly to the usefulness of the technology.

The power of technology is most apparent when students analyze functions that would be very tedious and time consuming to study without the use of a graphics calculator or a computer spreadsheet. In Lesson 1.1, students use a spreadsheet to discover number patterns in data taken during the flight of a small rocket. In Lesson 7.1, students can use the trace or intersection feature of a graphics calculator to find the solution to a system of linear equations.

TEACHING AIDS

Holt, Rinehart and Winston's *Algebra 1* can be used to teach algebra with calculator technology. The text provides a Calculator Keystroke Guide in an appendix. This support feature provides essential keystroke sequences for TI-82 and TI-83 graphics calculators. This support is presented for all activities and examples found in *Algebra 1* that require or suggest the use of a graphics calculator.

Another teaching aid that is gaining wide acceptance is the Internet. Throughout both the Teacher's Edition

and Pupil's Edition of *Algebra 1* you will find references to the Holt, Rinehart and Winston Web site. These "Internet Connects" provide math resources or educational links for tutorial assistance, references for student research, classroom activities, and teaching resources. The HRW Web site also provides updates to the technology used in the text.

An effective, powerful, yet easy to use tool in the classroom is Mathepedia computer software. This tool allows students to explore mathematics by utilizing a dictionary design and over 140 interactive mathematics tools. This comprehensive reference tool contains over 900 terms and definitions with extensive hypertext links to show the connections between math topics. Topics can be arranged alphabetically and according to the strands of the NCTM Principles and Standards. This powerful tool has outstanding graphing capabilities with nine types of available graphs and the option of displaying five different graphs at one time on the same axes. Mathepedia

may be used by students in the classroom, the computer lab, or at home. Students can use it for reviewing concepts, exploring topics they are currently studying, and engaging in exciting enrichment activities.

Assessment

ASSESSMENT GOALS

An essential aspect of any learning environment, such as an algebra classroom, is the process of assessing or evaluating what students have learned. Informally, this has been done with paper-and-pencil tests given by the teacher on a regular basis to measure students' knowledge of the material. Formal evaluations with standardized tests are generally conducted over a period of years to establish performance records for both individuals and groups of students within a school or school district. Both types of tests are very good at measuring the ability of a student to use a particular mathematical skill or to recall a specific fact. They fall short, however, in evaluating other key goals of learning mathematics, such as being able to solve problems, to reason critically, to understand concepts, and to communicate mathematically, both verbally and in writing. Other techniques, usually referred to as alternative assessment, are needed to evaluate students' performance on these goals of instruction.

The goals of an alternative-assessment program are to provide a means of evaluating students' progress in nonskill areas of mathematical learning. Thus, the design and structure of alternative assessment techniques must be quite different from those of the skill-oriented, paper-and-pencil tests of the past.

TYPES OF ALTERNATIVE ASSESSMENT

In the world outside school, a person's work is evaluated by what that person can do (that is, by the results the person achieves) and not by taking a test. For example, a musician may demonstrate skill by making music, a pilot by flying an airplane, a writer by writing a book, and a surgeon by performing an operation. Students learn to think mathematically and to solve problems on a continuous basis over a long period of time as they study mathematics at many grade levels. Students, too, can demonstrate what they have learned by collecting a representative sample of their best work in a portfolio. A portfolio should illustrate achievements in problem solving, critical thinking, writing about mathematics, mathematical applications or connections, and any other activity that demonstrates an understanding of both concepts and skills.

Specific examples of the kinds of work that students can include in their portfolio are solutions to nonroutine problems, graphs, tables or charts, computer printouts, group reports or reports of individual research, simulations, and artwork or models. Each entry should be dated and should be chosen to show the student's growth in mathematical competence and maturity.

A portfolio is just one way for students to demonstrate their performance on a mathematical task. Performance assessment can also be achieved in other ways, such as by asking students questions and evaluating their answers, by observing their work in cooperative-learning groups, by having students give verbal presentations, by working on extended projects and investigations, and by keeping journals.

Peer assessment and self-evaluation are also valuable methods of assessing students' performance. Students should be able to critique their classmates' work and their own work against standards set by the teacher. In order to evaluate their work, students need to know the teacher's goals of instruction and the criteria (scoring rubrics) that have been established for evaluating performance against the goals. Students can help to design their own self-assessment forms that they then fill out on a regular basis and give to the teacher. They can also help to construct test items that are incorporated into tests given to their classmates. This work is ideally done in small groups of four students. The teacher can then choose items from each group to construct the test for the entire class. Another alternative testing technique is to have students work on take-home tests that pose more open-ended and nonroutine questions and problems. Students can devote more time to such tests and, in doing so, demonstrate their understanding of concepts and skills and their ability to do mathematics independently.

T19

SCORING

The use of alternative assessment techniques implies the need to have a set of standards against which students' work is judged. Numerical grades are no longer sufficient because growth in understanding and problem solving cannot be measured by a single number or letter grade. Instead, scoring rubrics or criteria can allow the teacher more flexibility to recognize and comment on all aspects of a student's work, pointing out both strengths and weaknesses that need to be corrected.

A scoring rubric can be created for each major instructional goal, such as being able to solve problems or communicate mathematically. A rubric generally consists of four or five short descriptive paragraphs that can be used to evaluate a piece of work. For example, if a five-point paragraph scale is used, a rating of 5 may denote that the student has completed all aspects of the assignment and has a comprehensive understanding of problems. The content of the fifth paragraph specifies the details of what constitutes the rating of highly satisfactory. On the other hand, a rating of 1 designates an essentially unsatisfactory performance, and the first paragraph would specify the details of what constitutes an unsatisfactory rating. The other three paragraphs provide an opportunity for the teacher to recognize significant accomplishments by the student as well as aspects of the work that need improvement. Thus, scoring rubrics are a far more realistic and educationally substantive way to evaluate a student's performance than a single grade, which is usually determined by an answer being either right or wrong.

The Holistic Scoring Rubric illustrated on the next page is an excellent and effective guide for use by math teachers who are practicing performance assessment in their

classrooms. The scorer gathers evidence about a student's mathematical ability. The rubric is then used to assign a single performance rating based on an overall view of the full contents of the student's portfolio. The Holistic Scoring Rubric lists the types of work and tools for entries that are appropriate for a student to place in his or her portfolio.

Holt, Rinehart and Winston's *Algebra 1* provides numerous opportunities in each chapter of the Pupil's Edition for portfolio assessment. Investigation and discovery are encouraged in all of the activities in the student lessons of the program. Interdisciplinary topics from astronomy to zoology and applications from arts to transportation are designated throughout the lessons. Projects are found at the end of every chapter, and Portfolio Activities are found within the chapter. Communicate exercises offer excellent writing opportunities. Tools such as graphics calculators and algebra tiles can be used to support the activities that students include in their portfolios.

ASSESSMENT AND *ALGEBRA 1*

Throughout the textbook, students are asked to explain their work; describe what they are doing; compare and contrast different approaches; analyze a problem; make sketches, graphs, tables, and other models; hypothesize, conjecture, and look for counterexamples; and make and prove generalizations.

All of these activities, including the more traditional responses to routine problems, provide the teacher with a wealth of assessment opportunities to see how well students are progressing in their understanding and knowledge of algebra. The assessment task can be aligned with the major process goals of instruction, with scoring rubrics established for each goal. For example, a teacher may decide to organize his or her assessment tasks based on the following general skills of mathematics: problem solving, reasoning, communicating, and connecting.

Within each of these areas, specific goals can be written and shared with students. In this way, the assessment process becomes an integral part not only of evaluating students' progress, but also of the instructional process itself. The results of assessment can be used to modify the instructional approach in order to enhance learning for all students.

In addition to the many opportunities for performance assessment found in the lessons of *Algebra 1*, a variety of assessment types are integrated into the chapter-end material. The Chapter Review and Assessment and the Cumulative Assessment include both traditional and alternative assessment. Also, Portfolio Projects and Performance Assessment items are provided on the Alternative Assessment page. The Cumulative Assessments are formatted in the style of college entrance exams.

In addition to multiple-choice and free-response items, each Cumulative Assessment contains quantitative-comparison questions that emphasize the concepts of equality, inequality, and estimation. Other types of college entrance exam questions found in the Cumulative Assessments are student-produced response questions with solutions recorded in answer grids.

Holistic Scoring Rubric

	PERFORMANCE GOALS AND CRITERIA				
	Mathematical Reasoning Selecting and using appropriate types of reasoning and methods of proof through inductive and deductive reasoning	**Problem Solving** Solving of problems through the use of exploration, appropriate strategies, and a systematic approach	**Communication** Communicating ideas, thoughts, and approaches through the use of everyday language, mathematical language and symbols, graphs, tables, charts, and diagrams	**Mathematical Connections** Recognizing connections between different mathematical ideas or between mathematics and other disciplines	**Use of Tools** Using technology (calculators, computers, etc.) and/or manipulatives
LEVEL FOUR: SUPERIOR	• Uses sophisticated mathematical reasoning • Provides strong supporting arguments • Includes examples and counterexamples	• Shows thorough understanding of the problems' mathematical ideas and processes • Uses and synthesizes multiple strategies that lead to correct solutions	• Contains a complete response with clear, precise, and appropriate language • Uses effective diagrams such as graphs, tables, or charts	• Demonstrates a comprehensive knowledge of connections to other mathematical topics or other disciplines	• Makes appropriate use of technology and manipulatives to demonstrate mathematical concepts
LEVEL THREE: COMPETENT	• Uses sound mathematical reasoning • Includes some supporting arguments	• Shows basic understanding of the problems' mathematical ideas and processes • Uses appropriate strategies that lead to correct solutions	• Contains a solid response but is expressed less elegantly and less completely • Uses accurate diagrams	• Demonstrates some knowledge of connections to other mathematical topics or other disciplines	• Uses some technology and manipulatives to demonstrate solutions to problems
LEVEL TWO: MARGINAL	• Uses somewhat appropriate mathematical reasoning • Includes incomplete or faulty arguments	• Indicates a partial understanding of the problem • Selects some appropriate strategies that lead to partially correct solutions	• Contains a fairly complete response but uses unclear language • Uses inappropriate and/or unclear diagrams	• Demonstrates few connections	• Occasionally uses technology and manipulatives appropriately
LEVEL ONE: LIMITED	• Uses limited mathematical reasoning • Includes no arguments	• Shows little understanding of the problems • Uses poor or inappropriate strategies that lead to incorrect solutions	• Uses some appropriate mathematical language • Uses few, if any, diagrams	• Does not demonstrate or demonstrates inappropriate connections to other mathematical topics or other disciplines	• Rarely uses technology and manipulatives appropriately

NCTM Principles and Standards

In 1989, the National Council of Teachers of Mathematics published the Curriculum and Evaluation Standards for School Mathematics, generally referred to as the Standards, with the overall objective of improving students' mathematical education. Just as important, the Standards acknowledged the need for school mathematics to respond to and anticipate the changing needs of an increasingly technological and quantitative world. This document called for a focus across all grade levels in the areas of problem solving, communicating, reasoning, and making connections. Content changes for grades 5 through 8 called for increased attention in the areas of developing number and operation sense, identifying and using functional relationships, developing and using tables, graphs, and rules to describe situations, and developing an understanding of variables, expressions, and equations. Other changes called for increased attention in using statistical methods to describe, analyze, evaluate, and make decisions; creating experimental and theoretical models of situations involving probabilities; and developing an understanding of geometric objects and relationships. Changes in mathematical content for grades 9 through 12 called for increased attention in the use of real-world problems in algebra to motivate students and apply theory, integration of geometry topics at all grade levels, realistic trigonometry applications and modeling, and the integration of functions across topics at all grade levels.

An underlying precept of the original Standards was to provide an ongoing means for improving student performance in mathematical education. Revision of the original NCTM Standards was planned in the early development of the 1989 version. It was assumed that additional research and emerging technologies would necessitate the need for modification of the Standards' themes and messages. It is also important to articulate any language in the original Standards that caused confusion or lead to a misunderstanding of the intended messages. For example, "mathematical topics to receive decreased attention" was sometimes interpreted as areas for elimination. Since the creation of the original document, it has also become clear that certain mandates for change— such as basic skills and conceptual learning—needed adjustment and further clarification. Thus, the major purpose of the new set of standards is based on improvements in the field, research on learning and teaching, and technological advances that have occurred over the last 10 years.

Even though the new document, which will be referred to as Principles and Standards, includes a number of substantive and structural features that are different from the original Standards, a goal of the new Principles and Standards is for students to be able to achieve the goals established in the original Standards. Such goals as students learning to value mathematics, becoming confident in their own ability, becoming mathematical problem solvers, learning to communicate mathematically, and learning to reason mathematically remain important foundations of the new Principles and Standards.

The new document is built around ten standards organized across all grade levels. Five of these standards are made up of mathematical content, and five are made up of process standards. Each standard is elaborated at four groups of grade levels — pre-k–2, grades 3–5, grades 6–8, and grades 9–12. Within each standard, a number of focus areas are identified, and main points are elaborated with detailed descriptions of the content at each group of grade levels.

Five standards describe the mathematical content that students should learn and the goals for students' understanding of concepts and procedures:

- Number and Operation
- Patterns, Functions, and Algebra
- Geometry and Spatial Sense
- Measurement
- Data Analysis, Statistics, and Probability

Five standards describe the mathematical processes through which students should acquire and use their mathematical knowledge. These are described in terms of students' mathematical learning outcomes.

- Problem Solving
- Reasoning and Proof
- Communications
- Connections
- Representation

The first four process standards listed above are present across all grade levels in the original Curriculum and Evaluation Standards. The fifth process standard, Representation, has been added to the new Principles and Standards to highlight the importance of such processes as organizing, recording, and communicating mathematical ideas and using representation in modeling. Representations are used to create, organize, record, and communicate mathematical ideas; develop a repertoire of mathematical representations that can be used purposefully, flexibly, and appropriately; and model and interpret physical, social, and mathematical phenomena. The following table correlates the NCTM Principles and Standards to the lessons contained in *Algebra 1*. Standard 10, Representation, is expanded into six descriptive categories of representations.

HOLT, RINEHART AND WINSTON'S *Algebra 1* and NCTM Principles and Standards Correlation

REPRESENTATIONS

LESSON		NCTM STANDARDS	CONCRETE	DRAWING	GRAPH	SYMBOL	TABLE	VERBAL	VISUAL
1.1	Using Differences to Identify Patterns	1–3, 6–10	●	●	●	●	●	●	●
1.2	Variables, Expressions, and Equations	1, 2, 5, 6, 8–10	●	●		●	●	●	●
1.3	The Algebraic Order of Operations	1–3, 5–10		●		●		●	●
1.4	Graphing With Coordinates	1–3, 6–10			●	●			
1.5	Representing Linear Patterns	1–3, 5, 6, 8–10		●	●	●	●	●	●
1.6	Scatter Plots and Lines of Best Fit	1–3, 5–10		●	●	●	●	●	●
	Chapter Project: *Working with Patterns in Data*	1–6, 8–10		●		●	●	●	●
2.1	The Real Numbers and Absolute Value	1–3, 6–10		●	●	●		●	●
2.2	Adding Real Numbers	1,2, 5–10	●			●	●	●	●
2.3	Subtracting Real Numbers	1,2, 5–10	●	●				●	●
2.4	Multiplying and Dividing Real Numbers	1, 2, 5, 6, 8–10		●		●	●	●	●
2.5	Properties and Mental Computation	1, 2, 5–10		●		●		●	●
2.6	Adding and Subtracting Expressions	1–3, 6–10	●	●		●		●	●
	Eyewitness Math: *Abandon Ship!*	1, 2, 5, 6, 9, 10		●				●	●
2.7	Multiplying and Dividing Expressions	1–3, 6, 8–10	●	●		●	●	●	●
	Chapter Project: *Using a Spreadsheet*	1–3, 5, 6, 8–10		●		●	●	●	●
3.1	Solving Equations by Adding and Subtracting	1–3, 6–10	●	●	●	●	●	●	●
3.2	Solving Equations by Multiplying and Dividing	1–3, 5–10		●		●	●	●	●
3.3	Solving Two-Step Equations	1, 2, 6–10			●	●	●	●	●
3.4	Solving Multistep Equations	1–3, 5–10	●	●	●			●	●
3.5	Using the Distributive Property	1–3, 5–10			●	●		●	●
3.6	Using Formulas and Literal Equations	1–4, 6–10			●	●	●	●	●
	Chapter Project: *Track Team Schemes*	1–4, 6, 8–10		●		●		●	●
4.1	Using Proportional Reasoning	1–3, 5–10			●	●	●	●	●
4.2	Percent Problems	1, 2, 5, 6, 8–10		●		●	●	●	●
	Eyewitness Math: *Barely Enough Grizzlies?*	1, 2, 5, 6, 8–10	●			●	●	●	●
4.3	Introduction to Probability	1, 2, 5, 6, 8–10	●	●		●	●	●	●
4.4	Measures of Central Tendency	1, 2, 5, 6, 8–10				●	●	●	●
4.5	Graphing Data	1–3, 5–10		●	●		●	●	●
4.6	Other Data Displays	1–3, 5, 6, 8–10		●	●		●	●	●
	Chapter Project: *Designing a Research Study*	1–3, 5–10				●	●	●	●
5.1	Linear Functions and Graphs	1–3, 5, 6, 8–10			●	●	●	●	●
5.2	Defining Slope	1–10		●	●	●	●	●	●
	Eyewitness Math: *Watch Your Step!*	1–4, 6, 8, 9		●	●	●	●	●	●
5.3	Rate of Change and Direct Variation	1–6, 8–10			●	●	●	●	●
5.4	The Slope-Intercept Form	1–6, 8–10			●	●	●	●	●
5.5	The Standard and Point-Slope Forms	1–10			●	●	●	●	●
5.6	Parallel and Perpendicular Lines	1–6, 8–10			●	●	●	●	●
	Chapter Project: *Modelling Trends in Data*	1–10		●	●	●	●	●	●

(Continued)

LESSON		NCTM STANDARDS	CONCRETE	DRAWING	GRAPH	SYMBOL	TABLE	VERBAL	VISUAL
6.1	Solving Inequalities	1–4, 6–10		●	●	●	●	●	●
6.2	Multistep Inequalities	1–10		●	●	●	●	●	●
6.3	Compound Inequalities	1–4, 6–10			●	●		●	●
6.4	Absolute-Value Functions	1–10			●	●	●	●	●
6.5	Absolute-Value Equations and Inequalities	1–10		●	●	●	●	●	●
	Chapter Project: *Planning a Snowboarding Trip*	1–3, 5, 6, 8–10				●		●	●
7.1	Graphing Systems of Equations	1–3, 5, 6, 8–10			●	●	●	●	●
7.2	The Substitution Method	1–3, 5, 6, 8–10		●		●		●	●
7.3	The Elimination Method	1–3, 5, 6, 8–10		●		●	●	●	●
7.4	Consistent and Inconsistent Systems	1–3, 5–10		●	●	●	●	●	●
7.5	Systems of Inequalities	1–3, 5–10		●	●	●		●	●
7.6	Classic Puzzles in Two Variables	1, 2, 5, 6, 8–10		●		●	●	●	●
	Chapter Project: *Minimum Cost/Maximum Profit*	1–3, 5, 6, 8–10			●	●	●	●	●
8.1	Laws of Exponents: Multiplying Monomials	1–6, 8–10				●	●	●	●
8.2	Laws of Exponents: Powers and Products	1–4, 6–10		●		●		●	●
8.3	Laws of Exponents: Dividing Monomials	1–4, 6–10		●		●		●	●
8.4	Negative and Zero Exponents	1, 2, 5–10		●		●	●	●	●
	Eyewitness Math: *All Mixed Up?*	1, 2, 5, 7, 9		●	●	●		●	●
8.5	Scientific Notation	1–4, 6, 8–10		●		●	●	●	●
8.6	Exponential Functions	1–6, 8–10		●	●	●	●	●	●
8.7	Applications of Exponential Functions	1–6, 8–10			●	●	●	●	●
	Chapter Project: *The Wolves' Den*	1–3, 5, 6, 8–10	●			●	●	●	●
9.1	Adding and Subtracting Polynomials	1–3, 6, 8–10	●	●		●	●	●	●
9.2	Modeling Polynomial Multiplication	1, 2, 6, 8–10	●	●		●		●	●
9.3	Multiplying Binomials	1–4, 6, 8–10		●		●		●	●
9.4	Polynomial Functions	1–6, 8–10	●	●	●	●		●	●
9.5	Common Factors	1–4, 6, 8–10	●	●		●		●	●
9.6	Factoring Special Polynomials	1–10	●	●		●		●	●
9.7	Factoring Quadratic Trinomials	1–4, 6–10				●	●	●	●
9.8	Solving Equations by Factoring	1–6, 8–10		●	●	●	●	●	●
	Chapter Project: *Powers, Pascal, and Probability*	1–3, 5–10		●		●	●	●	●
10.1	Graphing Parabolas	1–3, 6, 8–10		●	●	●		●	●
10.2	Solving Equations by Using Square Roots	1–3, 5–10	●	●	●	●	●	●	●
10.3	Completing the Square	1–3, 5, 6, 8–10	●	●		●		●	●
10.4	Solving Equations of the Form $x^2 + bx + c = 0$	1–3, 6–10		●	●	●		●	●
	Eyewitness Math: *Rescue at 2000 Feet*	1, 2, 5–10				●		●	●
10.5	The Quadratic Formula	1–3, 6–10				●		●	●
10.6	Graphing Quadratic Inequalities	1–3, 5–10		●	●	●		●	●
	Chapter Project: *What's the Difference?*	1, 2, 5, 6, 8–10		●		●	●	●	●

(Continued)

LESSON		NCTM STANDARDS	CONCRETE	DRAWING	GRAPH	SYMBOL	TABLE	VERBAL	VISUAL
11.1	Inverse Variation	1–6, 8–10		●	●	●	●	●	●
11.2	Rational Expressions and Functions	1–3, 5, 6, 8–10		●	●	●	●	●	●
11.3	Simplifying Rational Expressions	1–10		●		●	●	●	●
	Eyewitness Math: *How Worried Should You Be?*	1, 2, 5–10				●		●	●
11.4	Operations With Rational Expressions	1–6, 8–10				●	●	●	●
11.5	Solving Rational Equations	1–6, 8–10			●	●	●	●	●
11.6	Proof in Algebra	1, 2, 5–10		●		●	●	●	●
	Chapter Project: *Designing a Park*	1–6, 9, 10		●				●	●
12.1	Operations With Radicals	1–10		●		●		●	●
12.2	Square-Root Functions and Radical Equations	1–10		●	●	●	●	●	●
12.3	The Pythagorean Theorem	1–4, 6–10		●		●	●	●	●
12.4	The Distance Formula	1–10		●	●	●		●	●
12.5	Geometric Properties	1–4, 6–10		●	●	●	●	●	●
12.6	The Tangent Function	1–4, 6–10		●	●	●	●	●	●
12.7	The Sine and Cosine Functions	1–4, 6–10		●		●	●	●	●
12.8	Introduction to Matrices	1, 2, 5–10		●		●	●	●	●
	Chapter Project: *Working the Angles*	1, 3–6, 9, 10		●			●	●	●
13.1	Theoretical Probability	1–3, 5, 6, 8–10		●		●	●	●	●
13.2	Counting the Elements of Sets	1–3, 5–10		●		●	●	●	●
13.3	The Fundamental Counting Principle	1–3, 5–10		●		●		●	●
13.4	Independent Events	1–3, 5–10		●		●		●	●
	Eyewitness Math: *Hot Hands or Hoopla?*	1, 2, 5–10		●		●		●	●
13.5	Simulations	1–3, 5–10				●	●	●	●
	Chapter Project: *Happy Birthday!*	1, 2, 5–10			●	●	●	●	●
14.1	Graphing Functions and Relations	1–3, 6, 8–10		●	●	●	●	●	●
14.2	Translations	1–3, 5, 6, 8–10		●	●	●	●	●	●
14.3	Stretches and Compressions	1–3, 5, 6, 8–10		●	●	●	●	●	●
14.4	Reflections	1–3, 5–10		●	●	●	●	●	●
	Eyewitness Math: *Is There Order in Chaos?*	1, 2, 5, 6, 8–10				●		●	●
14.5	Combining Transformations	1–3, 5–10		●	●	●	●	●	●
	Chapter Project: *Pick a Number*	1–3, 5, 6, 8–10		●	●	●	●	●	●

1 From Patterns to Algebra

CHAPTER PLANNING GUIDE

Lesson	1.1	1.2	1.3	1.4	1.5	1.6	Projects and Review
Pupil Edition Pages	4–10	11–17	18–23	24–29	30–36	37–43	44–51
Extra Practice (Pupil Edition)	738	738	739	739	740	740	
Practice Workbook	1	2	3	4	5	6	
Reteaching Masters	1–2	3–4	5–6	7–8	9–10	11–12	
Enrichment Masters	1	2	3	4	5	6	
Cooperative-Learning Activities	1	2	3	4	5	6	
Lesson Activities	1	2	3	4	5	6	
Problem Solving/ Critical Thinking	1	2	3	4	5	6	
Student Study Guide	1	2	3	4	5	6	
Spanish Resources	1	2	3	4	5	6	
Student Technology Guide	1	2	3	4	5	6	
Assessment Resources	5	6	7	9	10	11	8, 12–17
Teaching Transparencies	1			2, 3		4, 5	
Quiz Transparencies	1.1	1.2	1.3	1.4	1.5	1.6	
Writing Activities for Your Portfolio							1–3
Tech Prep Masters							1–4
Long-Term Projects							1–4

LESSON PACING GUIDE

	1.1	1.2	1.3	1.4	1.5	1.6	Projects and Review
Traditional	2 days	2 days	1 day	1 day	2 days	2 days	2 days
Block	1 day	1 day	$\frac{1}{2}$ day	$\frac{1}{2}$ day	1 day	1 day	1 day
Two-Year	4 days	4 days	2 days	2 days	4 days	4 days	4 days

CONNECTIONS AND APPLICATIONS

Lesson	1.1	1.2	1.3	1.4	1.5	1.6	Review
Algebra	4–10	11–17	18–23	24–29	30–36	37–43	44–51
Geometry	7, 9		23	25, 26, 28	35		
Statistics			22		35	38, 41, 42	
Business and Economics		16		28	31, 36	41	
Life Skills		16	18, 22		33, 36		
Science and Technology	4, 6, 8, 9, 10	14, 15, 16	20, 23		33, 34	37, 38, 39	
Sports and Leisure	10	11, 12, 13, 15, 16		26, 28	31, 32, 34, 35, 36	42	48
Cultural Connection: Africa			23				
Cultural Connection: Europe				24, 29			
Other					34, 36	42	

BLOCK-SCHEDULING GUIDE

Day	Lesson	Teacher Directed: Lesson Examples, Teaching Transparencies	Student Guided: Activity, Try This	Cooperative-Learning Activity, Lesson Activity, Student Technology Guide	Practice: Practice & Apply, Extra Practice, Practice Workbook	Assessment: Quiz, Mid-Chapter Assessment	Problem Solving, Reteaching
1	1.1	10 min	15 min	15 min	65 min	15 min	15 min
2	1.2	10 min	15 min	15 min	65 min	15 min	15 min
3	1.3	8 min	8 min	8 min	35 min	8 min	8 min
3	1.4	7 min	7 min	7 min	30 min	7 min	7 min
4	1.5	10 min	15 min	15 min	65 min	15 min	15 min
5	1.6	10 min	15 min	15 min	65 min	15 min	15 min
6	Assess.	50 min PE: Chapter Review	90 min PE: Chapter Project, Writing Activities	90 min Tech Prep Masters	65 min PE: Chapter Assessment, Test Generator	30 min Chap. Assess. (A or B), Alt. Assess. (A or B), Test Generator	

PE: Pupil's Edition

Alternative Assessment

The following are suggestions for an alternative assessment for students who may benefit from a different type of assessment than the regular chapter quizzes and the mid-chapter and end-of-chapter assessment materials. Many of the questions are open-ended, and students' answers will vary.

Performance Assessment

1. Consider the number patterns below.
   a. What do the first and second differences tell you about these number patterns?

   Pattern 1: 3, 7, 11, 15, 19, ...
   Pattern 2: 2, 6, 12, 20, 30, ...

   b. Suppose that you are told that the expressions below model the number patterns above.

   Expression 1: $y = 4n - 3$
   Expression 2: $y = n(n + 1)$

   What does n represent in these expressions?

2. a. Explain how to plot the location of a point whose coordinates are given relative to a pair of horizontal and vertical axes.
   b. Assuming that the city hall is located at the origin, what are the school's coordinates if it is 5 miles east and 4 miles north of the city hall?

3. a. Represent each set of ordered pairs on a coordinate grid.

   Set 1: (1, 1), (2, 2), (3, 3), (4, 4), (5, 5), (6, 6)
   Set 2: (1, 1), (2, 4), (3, 9), (4, 16), (5, 25), (6, 36)
   Set 3: (0, 5), (5, 0), (10, 5), (15, 0), (20, 5)

   b. Make a scatter plot for each set of ordered pairs.
   c. For each set of points, what correlation, if any, does the line of best fit imply?

Portfolio Project

Suggest that students choose one of the following projects for inclusion in their portfolios.

1. A number pattern is often called a *sequence*.
   a. Research different meanings of the word *sequence*.
   b. Find both simple and complex examples of sequences.

2. The distance, d, traveled in miles at a rate of r miles per hour for t hours is modeled by the equation $d = rt$.
   a. Show how you know this relationship is linear. Use graphs, tables, and scatter plots.
   b. Explain the importance of this equation for motorists.

---

internetconnect

The table below identifies the pages in this chapter that contain technology information and support in the side columns.

Content Links	
Lesson Links	pages 9, 15, 41
Portfolio Links	pages 10, 29
Graphics Calculator Support	page 782

Resource Links

For information about teacher and parent resources as well as professional development help, visit **www.hrw.com/math**.

Technical Support HRW has assembled a team of dedicated technical and teaching professionals and a comprehensive service program to provide you with the support you need.

- The HRW Technical Support Line operates from 7 A.M. to 6 P.M. central time, Monday through Friday, at (800) 323-9239.
- The HRW Technical Support Center on the World Wide Web is available 24 hours a day, seven days a week, at **www.hrwtechsupport.com**.
- You can e-mail our Technical Support Center at **tsc@hrwtechsupport.com**.
- The Technical Support Center's fax-on-demand service at (800) 352-1680 offers solutions to common problems and answers to frequently asked questions.

Technology

Lesson Suggestions and Calculator Examples
(Keystrokes are based on a TI-83 calculator.)

Lesson 1.1 Using Differences to Identify Patterns
This lesson shows students how to repeat subtraction in order to find a pattern. The calculator display at right represents the number pattern below.

80, 83, 86, 89, 92, . . .

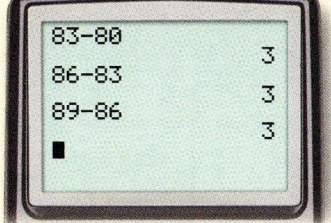

The display shows that each successive difference is the same. From such an example, students can learn how to recognize a linear number pattern.

Lesson 1.2 Variables, Expressions, and Equations
The calculator is useful for evaluating an expression for many values of its variable. To accomplish this, students will need to learn how to do the following:

1. enter an expression into the function list
2. set up the characteristics of a table of values
3. display the table of ordered pairs that results

This procedure can be seen on page 15 of the *Pupil's Edition*.

Lesson 1.3 The Algebraic Order of Operations
To evaluate a numerical expression on a calculator, have students

- become familiar with the operational symbols on the calculator keypad and
- apply the correct order of operations.

In particular, stress the need to use (and) to group quantities that must be kept together. Point out that $\frac{1+3}{5}$ is not equal to $1 + 3 \div 5$.

In this lesson, one of the students' main objectives should be to practice the evaluation of numerical expressions.

Lesson 1.4 Graphing With Coordinates
Although this lesson deals with the fundamental notions of the coordinate plane, students will need to use some statistical features of the graphics calculator to display the graphs of a set of ordered pairs. To graph {(3, 2), (5, 6), (7, 10), and (9, 14)}, students will need to follow the steps below.

Press Y= and clear any existing equations. Press STAT and choose 1:Edit.... Enter the *x*-values under **L1**, and the *y*-values under **L2**. Press WINDOW and choose window settings that will show all of the points. Then press 2nd Y= , turn **Plot 1** ON, and select the scatter plot icon. Now display the graph by pressing GRAPH.

The calculator displays below show the table and graph for the ordered pairs.

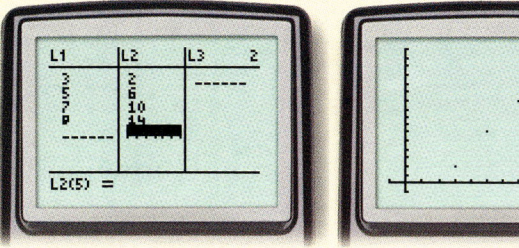

Lesson 1.5 Representing Linear Patterns
This lesson is the students' introduction to graphing an equation on a graphics calculator. Students will need to learn how to do the following:

1. enter an expression into the function list
2. select window settings that will show the graph
3. display the graph

This procedure can be seen in the Keystroke Guide on page 783.

The displays below show the function entry and graph. The window settings are entered by pressing WINDOW and are as follows:

Xmin=0   Ymin=0
Xmax=12   Ymax=16
Xscl=1   Yscl=1

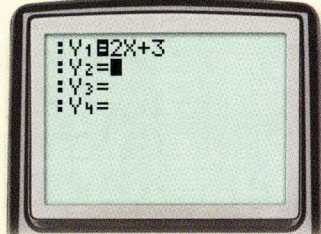 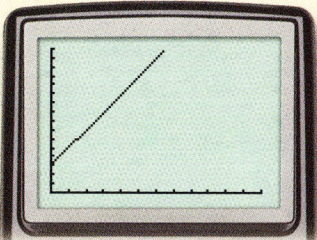

Have students graph different functions by changing the expression for **Y1** and using the same window settings. This will be the students' introduction to editing an expression in the function list.

Lesson 1.6 Scatter Plots and Lines of Best Fit
In this lesson, students will extend and reinforce the skills that they learned from Lesson 1.4, which explained how to plot a set of points in the coordinate plane.

(For an example of how to create a scatter plot, see page 784.)

For further information, refer to the

- technology discussions in the lessons.
- lesson-related teacher's commentary in the side columns of this *Teacher's Edition*.
- lesson-related *Student Technology Guide* masters.
- *HRW Technology Handbook*.

Teaching Resources

Basic Skills Practice
(2 pages per skill)

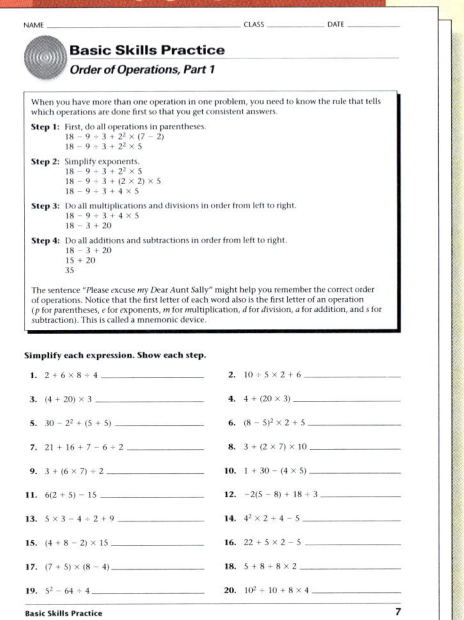

Reteaching
(2 pages per lesson)

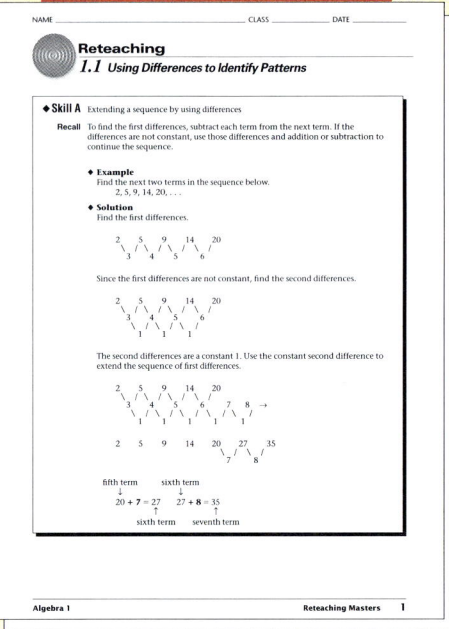

Enrichment
(1 page per lesson)

Cooperative-Learning Activity
(1 page per lesson)

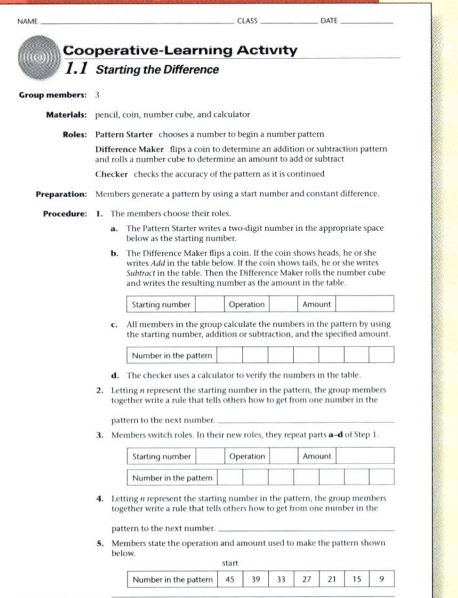

Lesson Activity
(1 page per lesson)

Long-Term Project
(4 pages per chapter)

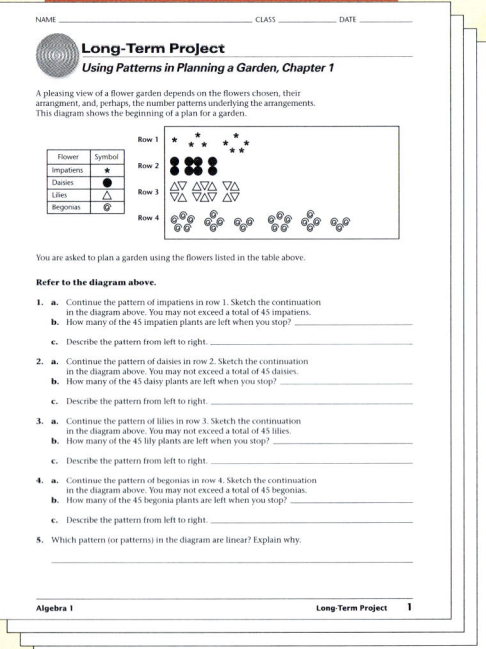

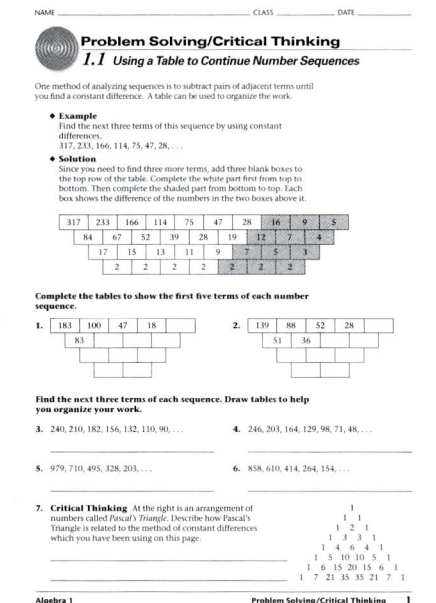
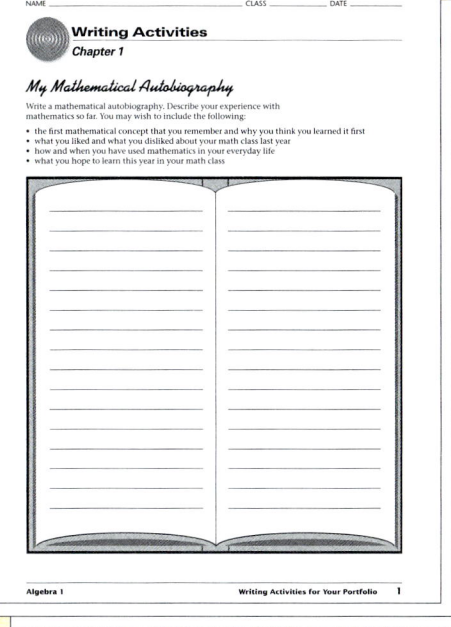
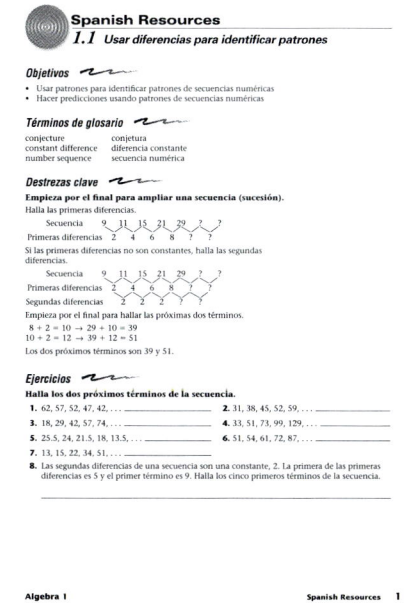

About Chapter 1

Background Information

Patterns are everywhere in the world around us. The study of mathematics is a quest to describe those patterns and to use them as the basis for predictions. In this chapter, students identify and extend the patterns found in number sequences. They use algebraic expressions to describe these patterns, and they use equations to solve problems arising from them. Students then study the order of operations used in evaluating expressions. A review of linear equations and the coordinate plane leads to a final lesson in which scatter plots and lines of best fit are used to investigate patterns in two-variable data.

CHAPTER RESOURCES

- Block Scheduling Handbook
- Writing Activities for Your Portfolio
- Tech Prep Masters
- Long-Term Project
- Assessment Resources:
  Mid-Chapter Assessment
  Chapter Assessments
  Alternative Assessments
- Test and Practice Generator
- Technology Handbook
- End-of-Course Exam Practice

From Patterns to Algebra

THROUGHOUT HISTORY, PEOPLE FROM ALL PARTS of the world have been fascinated with patterns. The early native tribes of the American Southwest were aware of many patterns that influenced their lives. Many of their artifacts show a deep appreciation for the beauty of patterns.

In the technological world of modern society, patterns provide a basis for discoveries in science and mathematics. Scientists use patterns to study, understand, and predict nature. Mathematicians look for regular patterns when investigating systems of numbers. Patterns can provide a powerful tool for solving problems.

Lessons

- 1.1 Using Differences to Identify Patterns
- 1.2 Variables, Expressions, and Equations
- 1.3 The Algebraic Order of Operations
- 1.4 Graphing With Coordinates
- 1.5 Representing Linear Patterns
- 1.6 Scatter Plots and Lines of Best Fit

Project Working With Patterns in Data

Rows of corn kernels

Kiwi Fruit

Zebras

Peacock feathers

The rug, wedding basket, and pot show the rich use of patterns in Navaho artifacts.

About the Photos

The photos show patterns that appear in nature as well as patterns in human artifacts. Have students look for patterns in the classroom. Ask them how many varieties of the same pattern they can find.

Geologists use seismographs to record patterns in earthquake tremors.

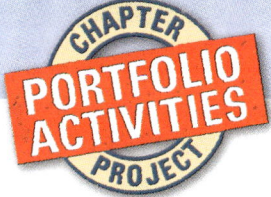

About the Chapter Project

Recognizing patterns in data is useful for solving problems. Algebra provides tools that help you to generalize and make predictions based on patterns. In the Chapter Project, *Working With Patterns in Data* on page 44, you will learn how to recognize mathematical patterns in golden rectangles, in a set of circular objects in your classroom, and in a simulation of a wave cheer. After completing the Chapter Project, you will be able to do the following:

- Analyze patterns in data by using scatter plots and tables.

About the Portfolio Activities

Throughout the chapter, you will be given opportunities to complete Portfolio Activities that are designed to support your work on the Chapter Project.

- Building staircases with cubes to represent triangular numbers is the topic of the Portfolio Activity on page 10.
- Representing a number sequence by using triangles constructed with toothpicks is the topic of the Portfolio Activity on page 17.
- A famous sequence of numbers named for the mathematician Fibonacci is the topic of the Portfolio Activity on page 29.
- Correlations in real-world sets of data are the topic of the Portfolio Activity on page 43.

Chapter Objectives

- Use differences to identify patterns in number sequences. [1.1]
- Make predictions by using patterns in number sequences. [1.1]
- Use variables to represent unknown quantities. [1.2]
- Represent real-world situations with equations and solve by guess-and-check. [1.2]
- Apply the algebraic order of operations. [1.3]
- Plot points and lines on a coordinate plane. [1.4]
- Represent linear patterns with equations. [1.5]
- Represent linear equations with graphs. [1.5]
- Interpret data in a scatter plot. [1.6]
- Find a line of best fit on a scatter plot by inspection. [1.6]

Portfolio Activities appear at the end of Lessons 1.1, 1.2, 1.4, and 1.6. Each serves as preparation for the Chapter Project. The Portfolio Activities and the Chapter Project are appropriate for inclusion in the student's portfolio. Students should be encouraged to include in their portfolio any other work in which they feel a sense of pride or a sense of accomplishment.

internet connect

Internet connect features appearing throughout the chapter provide keywords for the HRW Web site that lead to links for math resources, tutorial assistance, references for student research, classroom activities, and all teaching resource pages. The HRW Web site also provides updates to the technology used in the text. Listed at right are the keywords for the Internet Connect activities referenced in this chapter. Refer to the side column on the page listed to read about the activity.

LESSON	KEYWORD	PAGE
1.1	MA1 College Costs	9
1.1	MA1 Triangular	10
1.2	MA1 Mars Weight	15
1.4	MA1 Fibonacci	29
1.6	MA1 Nose Count	41

CHAPTER 1 **3**

Prepare

NCTM PRINCIPLES & STANDARDS 1–3, 6–10

LESSON RESOURCES

- Practice 1.1
- Reteaching Master 1.1
- Enrichment Master 1.1
- Cooperative-Learning Activity 1.1
- Lesson Activity 1.1
- Problem Solving/Critical Thinking Master 1.1
- Teaching Transparency 1

QUICK WARM-UP

Find each difference.

1. $11 - 5$ **6**
2. $9 - 14$ **−5**

Evaluate each expression.

3. 7^2 **49**
4. 5^3 **125**

Also on Quiz Transparency 1.1

Teach

Why Patterns in sets of data are not always immediately evident. Demonstrate this fact by providing students with a pattern such as 2, 5, 9, 14, 20. Ask them to first study the pattern and then try to devise a rule that describes it. Discuss any rules they suggest. If time permits, wrap up the discussion by showing students that while the first differences are not the same, all of the second differences are 1.

Use Teaching Transparency 1.

1.1 Using Differences to Identify Patterns

Objectives

- Use differences to identify patterns in number sequences.
- Make predictions by using patterns in number sequences.

Why Organizing data and using differences to help identify patterns provide a way to better understand relationships between things like thunder and lightning.

Lightning heats the air, which expands rapidly and creates thunder.

There are many patterns in nature. Analyzing these patterns helps us to understand nature better. For example, past weather patterns are used to predict future weather. Mathematics provides tools for studying the patterns around us.

APPLICATION
METEOROLOGY

Activity
Exploring Differences

1. The speed of light is approximately 186,000 miles per second. The speed of sound is approximately 1116 feet per second. Explain why you see lightning before you hear the resulting thunder.

2. The table below shows the approximate distance from a lightning strike to an observer based on the time it takes the observer to hear the thunder.

Seconds	2	4	6	8	10
Miles	0.4	0.8	1.2	?	?

CHECKPOINT ✓

a. Copy and complete the table.
b. For each 2-second increase in time, what is the increase in miles?
c. Use this rate of change to calculate the distance in miles for 12, 14, and 16 seconds.
d. The sound of thunder travels 4.4 miles in 22 seconds. Find the distance that the sound of thunder travels in 23 seconds.

Alternative Teaching Strategy

USING VISUAL MODELS Have students use graph paper to create visual models of number patterns. For a given pattern, instruct students to shade the number of squares that correspond to each number, starting with the first number and working in order. As they shade the squares for each successive number, have them record the number of additional squares they shaded in comparison to the previous number. Point out that these numbers represent the differences between terms. If the differences are not constant, have them repeat the process, this time shading the number of squares that correspond to each difference.

4 LESSON 1.1

Patterns in Number Sequences

When mathematicians study numbers for patterns, they often make conjectures. A **conjecture** is a statement about observations that is believed to be true. When mathematicians make a conjecture, they try to either prove that the conjecture is true or find a counterexample to show that the conjecture is not true.

You may make conjectures when studying a pattern such as a number sequence. A **number sequence** is a string of numbers, or terms, in a certain order. Often there will be three dots after the last given number. The three dots (…), indicate that there are more terms that are not listed.

If the difference from one term to the next in a number sequence is always the same, the difference is called a **constant difference**.

EXAMPLE 1 Find the next three terms of each sequence by using constant differences.
   a. 80, 73, 66, 59, 52, …
   b. 1, 4, 9, 16, 25, …

● **SOLUTION**

a. Find the first differences.

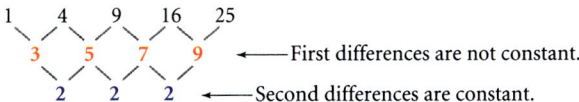
First differences are constant.

Notice that the terms decrease by 7. The first difference is −7. Subtract 7 from the previous term to find each new term.

$52 - 7 = 45 \qquad 45 - 7 = 38 \qquad 38 - 7 = 31$

The next three terms are 45, 38, and 31.

b. Find the first differences. Since the first differences are not constant, find the second differences.

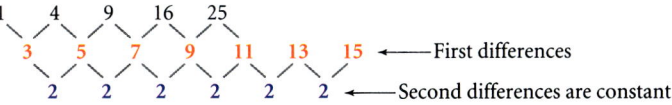
← First differences are not constant.
← Second differences are constant.

Work backward to find the next three first differences. Add 2 to the previous first differences.

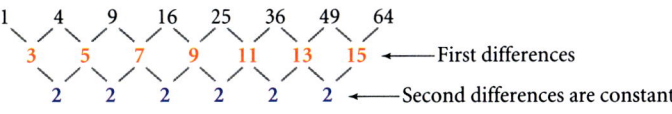
← First differences
← Second differences are constant.

Find the next three terms by adding the first differences to the previous terms.

$25 + 11 = 36 \qquad 36 + 13 = 49 \qquad 49 + 15 = 64$

1 4 9 16 25 36 49 64
 3 5 7 9 11 13 15 ← First differences
  2 2 2 2 2 2   ← Second differences are constant.

The next three terms are 36, 49, and 64.

Activity Notes

Answers to Activity steps can be found in Additional Answers of the Teacher's Edition.

In the Activity on page 4, students correlate number sequences and rates of change. By applying successive differences to observed data, they determine how an increase in the distance between the lightning and an observer changes the amount of time before the observer hears the thunder. Then they use this rate of change to make predictions.

CHECKPOINT ✔
c. 2.4 miles in 12 seconds
   2.8 miles in 14 seconds
   3.2 miles in 16 seconds

Teaching Tip

Note that negative numbers appear in this lesson, although the concept is not presented formally until Lesson 2.1. In this lesson, however, negative numbers are used only to indicate differences between terms of a decreasing sequence.

ADDITIONAL EXAMPLE 1

Find the next three terms of each sequence by using the constant differences.

a. 12.4, 10.7, 9, 7.3, 5.6, …
   3.9, 2.2, 0.5

b. 49, 64, 81, 100, 121, …
   144, 169, 196

Interdisciplinary Connection

SCIENCE Have students research some of the many ways that scientists use mathematical patterns to make predictions. Post a list of the students' findings in the classroom. Have each student choose one item from the list, research it further, and write a report about it. Ask students to share their reports with the class.

Enrichment

Give students this expression for a sequence.

$$a_1, a_2, a_3, a_4, a_5, \ldots, a_n$$

Tell them that a_n denotes the nth term of the sequence. Explain that *formulas* for sequences can be written by using this notation. For example, a formula for the sequence in part **b** of Example 1 is $a_n = n^2$. Have students write a similar formula for the sequence 1, 8, 27, 64, … $a_n = n^3$

Challenge students to write a formula for at least one other sequence presented in the lesson.

LESSON 1.1 **5**

TRY THIS Find the next three terms of each sequence by using constant differences.
a. 1, 4, 7, 10, 13, …
b. 2, 6, 12, 20, 30, …

EXAMPLE 2

APPLICATION
PHYSICS

The table below shows the data from the flight of a small rocket during the first 4 seconds of its flight. The flight of the rocket ends when it hits the ground 14 seconds after takeoff. **Use the method of constant differences to find the maximum height during the rocket's flight.**

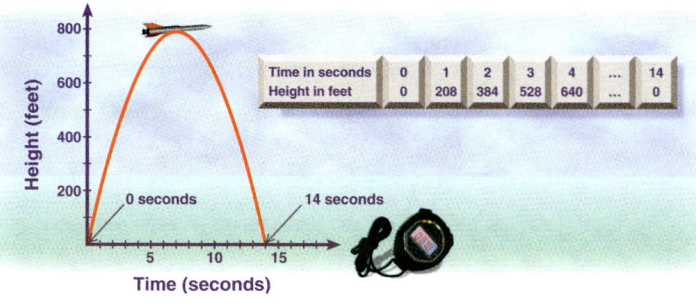

Time in seconds	0	1	2	3	4	…	14
Height in feet	0	208	384	528	640	…	0

SOLUTION
Look at the differences to discover the pattern.

Time in seconds	0	1	2	3	4
Height in feet	0	208	384	528	640

208　176　144　112　← First differences
　−32　−32　−32　← Second differences are constant.

PROBLEM SOLVING

The second differences are a constant −32. This means that the first differences *decrease* by 32 each time. Use the strategy of **working backward** from the second differences to extend the table.

Time in seconds	0	1	2	3	4	5	6	7	8	9	10
Height in feet	0	208	384	528	640	720	768	784	768	720	640

208　176　144　112　80　48　16　−16　−48　−80
　−32　−32　−32　−32　−32　−32　−32　−32　−32

After 7 seconds, the numbers for the height begin to repeat in reverse order. By continuing the table up to 14 seconds, the pattern gives points along the complete path of the rocket.

Time in seconds	11	12	13	14
Height in feet	528	384	208	0

The table shows that the highest point reached is 784 feet. This occurs 7 seconds after takeoff.

TRY THIS Suppose that a small rocket flight takes 15 seconds from the rocket's launch to its return to the ground. After how many seconds will it reach its maximum height?

Try This questions provide an opportunity for ongoing assessment. The answers are provided in the Teacher's Edition side copy as well as in the Pupil's Edition Selected Answers.

TRY THIS
a. 16, 19, 22
b. 42, 56, 72

ADDITIONAL EXAMPLE 2

A projectile is launched from ground level. The data in the table below give its height above ground during the first 4 seconds immediately after the launch. After 10 seconds, the projectile hits the ground.

Time in seconds	Height in feet
0	0
1	144
2	256
3	336
4	384
…	…
10	0

Use the method of constant differences to find the maximum height of the projectile. 400 feet

TRY THIS
7.5 seconds

Inclusion Strategies

INVITING PARTICIPATION Have students work in pairs. Each student should create a sequence in which the first differences are constant and then write the first five terms of the sequence on a sheet of paper. Students should exchange papers with their partner, and each should find the next three terms of the other's sequence. Repeat the activity, with each partner creating a sequence in which the second differences are constant. With the entire class, discuss any problems the students encountered in creating or extending the sequences.

Reteaching the Lesson

USING TABLES Some students may benefit from organizing the differences into a vertical table, such as this table for part **b** of Example 1.

Terms	First differences	Second differences
1		
4	4 − 1 = 3	
9	9 − 4 = 5	5 − 3 = 2
16	16 − 9 = 7	7 − 5 = 2
25	25 − 16 = 9	9 − 7 = 2

6 LESSON 1.1

Using Problem-Solving Strategies

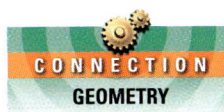

Some sequences can also be studied with diagrams. For example, the sequence 2, 6, 12, 20, 30, … is found by counting the number of dots in the pattern below.

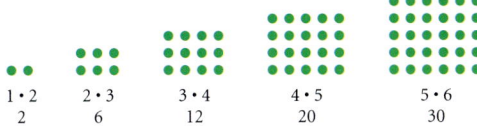

The next three terms are 6 · 7 = 42, 7 · 8 = 56, and 8 · 9 = 72. Why do you think the numbers in the sequence 2, 6, 12, 20, 30, … are called rectangular numbers?

Drawing diagrams is just one problem solving strategy that can be helpful in extending sequences. The next example uses several strategies.

EXAMPLE 3 Suppose that 10 friends have just returned to school. Each friend has exactly one conversation with each of the other friends to talk about what they did during summer break. Use problem-solving strategies to determine how many conversations there will be.

• **SOLUTION**

PROBLEM SOLVING

Problem solving strategies include **solving a simpler problem**, **making a table or chart**, and **looking for a pattern**.

Begin with a smaller number of people. With just one person, no conversations are possible. With two people, one conversation is possible. The dot patterns below represent the number of conversations for up to four people.

no conversations   one conversation   three conversations   six conversations

Arrange the information from the simpler problems in a table.

People	1	2	3	4	5	6
Conversations	0	1	3	6	?	?

Use differences to determine how the number of conversations is increasing. Then extend the pattern to 10 people.

People	1	2	3	4	5	6	7	8	9	10
Conversations	0	1	3	6	10	15	21	28	36	45

1 2 3 4 5 6 7 8 9

With 10 friends, it takes 45 conversations for each person to have exactly one conversation with each other person.

Math CONNECTION

GEOMETRY Many dot patterns are related to areas of geometric figures. For example, finding the total number of dots in the rectangular arrays at the top of this page is related to the formula for the area, A, of a rectangle with length l and width w, $A = lw$.

ADDITIONAL EXAMPLE 3

Find the sum $1 + 3 + 5 + 7$ by using a geometric dot pattern.

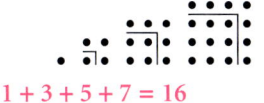

$1 + 3 + 5 + 7 = 16$

Teaching Tip

TECHNOLOGY In the spreadsheet on page 8, be sure that students understand how to identify the cells. Ask them to state the entries in different cells as you name them. Point out that the numbers in columns A and B were entered directly, but the entries in cells C2 through C5 and cells D2 through D4 were computed from formulas.

Teaching Tips about technology refer to keystrokes for the TI-82, TI-83, or TI-83 Plus.

A keystroke guide is provided at the end of each chapter for examples given in the Pupil's Edition. If you are using a different graphics calculator, such as a Casio or Sharp, go to the HRW Internet site at go.hrw.com and enter the keyword MA1 CALC to find keystroke guides.

LESSON 1.1

Communicate exercises provide an opportunity for students to discuss their discoveries, conjectures, and the lesson concepts. Throughout the text, the answers to all Communicate exercises can be found in the Teacher's Edition Additional Answers.

Assess

Selected Answers

Exercises 4–9, 11–43 odd

ASSIGNMENT GUIDE

In Class	1–9
Core	10–20, 22–33, 35
Core Plus	10–20 even, 21–35
Review	36–43
Preview	44

✏ Extra Practice can be found beginning on page 738.

Answers to odd-numbered Extra Practice exercises can be found immediately after Selected Answers in the Pupil's Edition.

Practice

TECHNOLOGY SPREADSHEET

A computer spreadsheet program can extend your ability to use differences to discover number patterns, especially when the differences and calculations become more complicated. Data from Example 2 on page 6 are shown in the spreadsheet below.

Compute the first difference (208, in cell C2) by subtracting the value in cell B2 from the value in cell B3. What instructions do you think are needed to compute the values in cells C3 and D2 of the spreadsheet? **C3 = B4 − B3**
D2 = C3 − C2

	A	B	C	D
1	seconds	height	1st differences	2nd differences
2	0	0	208	−32
3	1	208	176	−32
4	2	384	144	−32
5	3	528	112	
6	4	640		
7	5	720		

Exercises

Communicate

1. Explain how to find the first and second differences for the following sequence: 11, 14, 19, 26, 35, 46, …

2. Describe the method for predicting the next two terms in the following sequence: 88, 76, 64, 52, 40, …

3. Explain how to work backward from a constant second difference to extend the following sequence: 1, 6, 16, 31, 51, 76, …

Guided Skills Practice

Find the next three terms of each sequence by using constant differences. (EXAMPLES 1, 2, AND 3)

4. 1, 3, 5, 7, 9, … **11, 13, 15**
5. 10, 12, 14, 16, 18, … **20, 22, 24**
6. 87, 78, 69, 60, 51, … **42, 33, 24**
7. 1, 4, 9, 16, 25, … **36, 49, 64**
8. 10, 17, 26, 37, 50, … **65, 82, 101**
9. 3, 13, 27, 45, 67, … **93, 123, 157**

Practice and Apply

Find the next three terms of each sequence.

10. 18, 32, 46, 60, 74, … **88, 102, 116**
11. 33, 49, 65, 81, 97, … **113, 129, 145**
12. 100, 94, 88, 82, 76, … **70, 64, 58**
13. 44, 41, 38, 35, 32, … **29, 26, 23**
14. 20, 21, 26, 35, 48, … **65, 86, 111**
15. 30, 31, 35, 42, 52, … **65, 81, 100**
16. 1, 8, 18, 31, 47, … **66, 88, 113**
17. 4, 7, 15, 28, 46, … **69, 97, 130**
18. 1, 2.5, 4, 5.5, 7.0, … **8.5, 10, 11.5**
19. 0.5, 2, 4.5, 8, 12.5, … **18, 24.5, 32**

8 LESSON 1.1

CHALLENGE

21. a. First dif.; 1
    b. Second dif.; 2
    c. Third dif.; 6
    d. The fourth differences are constant

20. If the second differences of a sequence are a constant 3, the first of the first differences is 7, and the first term is 2, find the first five terms of the sequence. **2, 9, 19, 32, 48**

21. For each sequence, determine which differences are constant and what the constant difference is.
    a. 1, 2, 3, 4, 5, …
    b. $1^2, 2^2, 3^2, 4^2, 5^2, …$
    c. $1^3, 2^3, 3^3, 4^3, 5^3, …$
    d. Based on **a**, **b**, and **c**, determine how many differences you have to compute until you find a constant difference for the sequence $1^4, 2^4, 3^4, 4^4, 5^4, …$

TECHNOLOGY Refer to the spreadsheet on page 8. What spreadsheet instructions should you place in the following cells?

22. C3 **B4 − B3**   23. D3 **C4 − C3**   24. D5 **C6 − C5**   25. C6 **B7 − B6**

CONNECTIONS

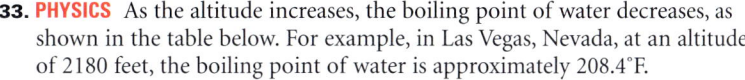

GEOMETRY The sum of the counting numbers from 1 to 5 can be represented by drawing a triangle of dots as shown at left. A rectangle can be formed by copying the triangle and rotating it. The sum of the counting numbers from 1 to 5 is half of the number of dots in the rectangle, or $\frac{5 \cdot 6}{2}$. For Exercises 26–28, sketch similar geometric dot patterns and find each sum.

26. 1 + 2 + 3 + 4 + 5 + 6 + 7 **28**
27. 1 + 2 + 3 + 4 + 5 + 6 + 7 + 8 + 9 + 10 **55**
28. 1 + 2 + 3 + 4 + 5 + 6 + 7 + 8 + 9 + 10 + 11 + 12 **78**

GEOMETRY Find each sum below. Think of a geometric dot pattern, but do not draw a sketch.

29. 1 + 2 + ⋯ + 40 **820**   30. 1 + 2 + ⋯ + 60 **1830**   31. 1 + 2 + ⋯ + 80 **3240**

32. **GEOMETRY** The table below shows the number of diagonals for various types of polygons. Use constant differences to fill in the missing values in the table.

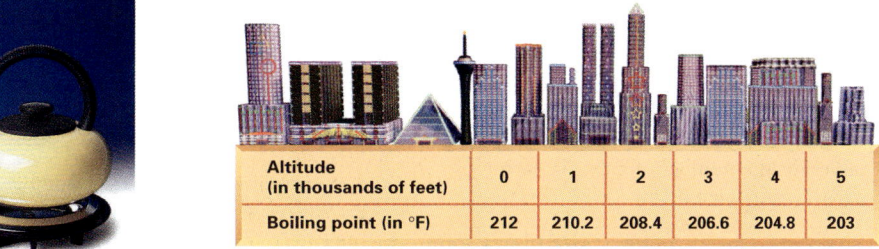

	Triangle	Rectangle	Pentagon	Hexagon	Heptagon	Octagon	Nonagon	Decagon
Sides	3	4	5	6	7	8	9	10
Diagonals	0	2	5	9	14	?	?	?

APPLICATION

33. **PHYSICS** As the altitude increases, the boiling point of water decreases, as shown in the table below. For example, in Las Vegas, Nevada, at an altitude of 2180 feet, the boiling point of water is approximately 208.4°F.

Altitude (in thousands of feet)	0	1	2	3	4	5
Boiling point (in °F)	212	210.2	208.4	206.6	204.8	203

a. Estimate the boiling point of water in Colorado Springs, Colorado, which is at an altitude of 6170 feet. **201.2°**
b. According to the pattern in the table, what is the boiling point of water at an altitude of 18,000 feet? **179.6°**

Internet connect

GO TO: go.hrw.com
KEYWORD: MA1 College Costs

In this activity, students access information on college costs, such as tuition and fees. Then they use this information to make a table and predict the costs of these expenses for upcoming years.

Guided Skills Practice provides an opportunity for students to practice the skills and concepts discovered from each example before working independently on the Practice and Apply exercises. The answers to all Guided Skills Practice can be found in the Pupil's Edition Selected Answers as well as in the Teacher's Edition side copy or Additional Answers.

Technology

A calculator may be helpful for Exercises 36–43. You may wish to have students complete the exercises with paper and pencil first and then use a calculator to verify their answers.

Error Analysis

When a pattern involves second differences, students may give incorrect responses because they are extending the sequence of first differences rather than the original sequence. Remind them that if two steps "backward" are needed to get to a constant difference, then two steps "forward" are needed to get back to the original sequence.

Student Technology Guide

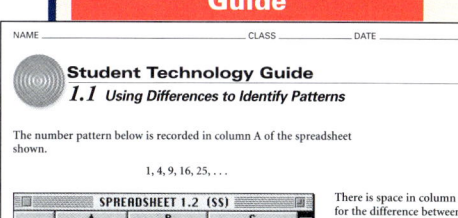

LESSON 1.1 **9**

Look Back exercises regularly review important concepts and skills. It is recommended that they be assigned to all students. Answers to the odd-numbered Look Back exercises can be found in the Selected Answers of the Pupil's Edition. Answers to all of the Look Back exercises can be found in the side copy or Additional Answers of the Teacher's Edition.

Look Beyond exercises are of three types: those that foreshadow upcoming concepts, those that extend concepts introduced in the lesson, and those that involve nonroutine problem solving. Answers to Look Beyond questions can be found in the side copy or Additional Answers of the Teacher's Edition.

Each sequence in Exercise 44 is a *geometric sequence*. Lead students to see that the terms of these sequences are related by a common *ratio* rather than a common difference.

ALTERNATIVE
Assessment
Portfolio Activity

The Portfolio Activity can be used as preparation for the Chapter Project or as a separate activity. In this Portfolio Activity, students construct "stairways" of cubes and use them to investigate a special set of numbers called the *triangular numbers*.

Answers to Portfolio Activites can be found in Additional Answers of the Teacher's Edition.

APPLICATIONS

11 teams, 55 games

12 teams, 66 games

34. RECREATION If each of 11 football teams in a college athletic conference played each of the other teams just once in a season, how many games would there be? As part of your answer, make a table similar to the one in Example 3 on page 7. How many games would there be for 12 teams?

35. TELECOMMUNICATIONS Suppose that 8 cities are to be linked by fiber optic cables, with 2 cables between each pair of cities. How many cables will there be? **56 cables**

 Look Back

Multiply.

36. $2.7 \cdot 3.1$ **8.37** **37.** $4.31 \cdot 2$ **8.62** **38.** $\frac{1}{2} \cdot \frac{2}{3}$ **$\frac{1}{3}$** **39.** $2\frac{2}{3} \cdot 1\frac{1}{4}$ **$3\frac{1}{3}$**

Divide.

40. $3.64 \div 2$ **1.82** **41.** $0.52 \div 1.3$ **0.4** **42.** $\frac{1}{2} \div \frac{1}{2}$ **1** **43.** $2\frac{3}{8} \div 2$ **$1\frac{3}{16}$**

 Look Beyond

44. The method of finding differences works only for certain kinds of number sequences. Try it on the following sequences and tell what happens. Predict the next three terms of each sequence without using differences. Explain your method.

a. 1, 2, 4, 8, 16, 32, …
64, 128, 256

b. 1, 10, 100, 1000, 10,000, …
100,000, 1,000,000, 10,000,000

Include this activity in your portfolio.

You can use cubes to build the sequence of stairways shown above.

1. Sketch a picture of the fourth stairway on graph paper. How many cubes does it take to build the fourth stairway? Build a table of values for the number of cubes in each stairway, as shown below.

Number of steps	1	2	3	4	5	6	7	8	9	10
Cubes in the stairway	1	3	6	?	?	?	?	?	?	?

2. Sketch the stairways up to the 10th stairway.

3. What is the pattern in the sequence of stairways? How does the 10th stairway compare with your answer in the table you made in Step 1?

The sequence of numbers that you have been working with, 1, 3, 6, 10, …, is called the sequence of triangular numbers.

4. Show visually that the square number 16 is the sum of two triangular numbers and state the two triangular numbers.

5. Show visually that 100 is the sum of two triangular numbers and state the two triangular numbers.

 internetconnect

GO TO: go.hrw.com
KEYWORD: MA1 Triangular

To support the Portfolio Activity, students may visit the HRW Web site, where they can explore other figurate numbers such as pentagonal and hexagonal numbers.

1.2 Variables, Expressions, and Equations

Objectives
- Use variables to represent unknown quantities.
- Represent real-world situations with equations and solve by guess-and-check.

Why Using variables and expressions to form equations provides a systematic way of representing and solving many types of problems, such as finding the average speed in a bicycle race.

The annual Leadville Trail 100 Mountain Bike Race is also known as the "Race Across the Sky."

Using Variables, Expressions, and Equations

APPLICATION
RECREATION

Suppose that while you are on vacation you want to rent a mountain bike and the cost is $3 per hour plus a $20 fee. You can write an **algebraic expression** that represents the cost of renting the bike. To write an algebraic expression, replace words with numbers, letters, and symbols such as $+$, $-$, $\cdot$, and $\div$.

$$\underset{3}{\$3} \quad \underset{\cdot}{\text{per}} \quad \underset{h}{\text{hour}} \quad \underset{+}{\text{plus}} \quad \underset{20}{\$20}$$

The expression can be written as $3 \cdot h + 20$, or $3h + 20$.

Variables are letters that are used to represent quantities in algebraic expressions. The letter h is used in the algebraic expression $3h + 20$ to represent the number of hours the bike will be rented.

You can find the value of an expression for a specific value of a variable. For example, to find the value of $3h + 20$ when h is 5, substitute 5 for h and simplify.

$$3h + 20$$
$$3(5) + 20$$
$$15 + 20$$
$$35$$

Prepare

NCTM PRINCIPLES & STANDARDS
1, 2, 5, 6, 8–10

LESSON RESOURCES
- Practice 1.2
- Reteaching Master 1.2
- Enrichment Master 1.2
- Cooperative-Learning Activity 1.2
- Lesson Activity 1.2
- Problem Solving/Critical Thinking Master 1.2

QUICK WARM-UP

Evaluate each expression for $x = 2$ and $x = 3$.

1. $3x - 1$ 5; 8
2. $4(x + 3)$ 20; 24

Also on Quiz Transparency 1.2

Teach

Why Variables can be used in equations and expressions to represent unknown information. The term *variable* means that the value may vary depending on the situation. Remind students that they have already used variables in formulas to find perimeter and area of geometric figures. Ask them if they can think of other ways that they have used variables in the past.

Alternative Teaching Strategy

INVITING PARTICIPATION Students can work in groups of three to act out the situation that begins the lesson by calculating the cost of the rental for 1 hour, 2 hours, 3 hours, and so on. Ask one student in each group to be the person renting the mountain bike, another student to collect and count the money, and a third student to record the results in a table. After each group has gathered enough information, have them write an expression to describe the pattern in the table. Then lead them to see how the expression becomes part of an equation that can be used to calculate the cost of a rental for any whole number of hours.

ADDITIONAL EXAMPLE 1

Make a table to show the value of the expression $4m - 10$ when m is 8, 9, 10, and 11.

m	$4m - 10$	Value
8	$4(8) - 10$	22
9	$4(9) - 10$	26
10	$4(10) - 10$	30
11	$4(11) - 10$	34

Activity Notes

The rental problem that begins this lesson demonstrates the process of writing an expression to describe a pattern and then using the expression to write an equation. In this Activity, students are presented with a different situation, and a series of questions leads them to write their own equation. They should arrive at the conclusion that for a speed of 20 miles per hour, the equation that relates distance, d, and time, t, is $d = 20t$.

Cooperative Learning

Have students work in pairs to complete the Activity. One student can fill in the table, while the other student records the answers to the questions.

CHECKPOINT ✔
2. The distance cycled is equal to the time multiplied by 20. If d represents distance and t represents time, then $d = 20t$.

EXAMPLE 1
Make a table to show the value of the expression $2h + 5$ when h is 1, 2, 3, and 4.

• SOLUTION
Make a table as shown and substitute to find the value of the expression for each value of h.

h	1	2	3	4
$2h + 5$	$2(1) + 5$	$2(2) + 5$	$2(3) + 5$	$2(4) + 5$
Value	7	9	11	13

Two algebraic expressions separated by an equal sign form an **equation**. You can use the algebraic expression $3h + 20$ to write an equation for the following sentence:

The cost is $3 per hour plus $20.

$$c = 3 \cdot h + 20$$

When you substitute the number 4 for the variable h and perform the operations, you can find the cost, c, of renting the bike for 4 hours.

$$c = 3h + 20$$
$$c = 3(4) + 20$$
$$c = 12 + 20$$
$$c = 32$$

The cost is $32.

Activity
Modeling Data With a Table and an Equation

Suppose that the average speed of a mountain biker in a race is 20 miles per hour. Build a table for time and distance.

Table for 20 miles per hour

Time (hours)	0	0.5	1.0	1.5	2.0	2.5	3.0	3.5	4.0
Distance (miles)	0	?	20	?	?	?	?	?	?

1. Copy and complete the table. Describe your method for filling in the distances. How could you find the distance cycled in 7 hours? in 8 hours?

CHECKPOINT ✔ 2. How do you find the distance for a specific time? Rewrite your answer with variables instead of words. Use d for distance and t for time. Use symbols for "equals" and "multiplied by." Your result will be an *equation* with variables.

The biking activity above began with a time of zero. However, in a case like Example 2 below, it does not make sense to rent equipment for zero time. As you work more problems, you will have to judge in which situations it makes sense to start with zero and in which ones it does not.

Interdisciplinary Connection

PHYSICS The descent of an airplane can be calculated with a simple equation. For example, suppose that an airplane descends at an average rate of 500 feet per minute. Let y represent the total number of feet descended, and let x represent the number of minutes of descent. Then $y = 500x$ gives the descent in feet for any given number of minutes. Ask students how many feet the airplane will descend in 11 minutes and how many minutes it will take for the airplane to descend 10,000 feet.
5500 feet; 20 minutes

Now suppose that this airplane begins its descent at an altitude of 20,000 feet. Let z represent the airplane's altitude in feet, and let x represent the number of minutes of descent. Have students write an equation that can be used to calculate the altitude of the airplane at any time during its descent. Ask them to use the equation to find the altitude after 5 minutes of descent.
$z = 20{,}000 - 500x$; **17,500 feet**

EXAMPLE 2

**APPLICATION
RECREATION**

At City Park you can rent in-line skates and protective gear for $2 per hour plus a $3 fee. **Make a table to show the charges for 1 hour through 3 hours. Write an equation for the situation.**

● SOLUTION

Number of hours	Expression for the cost	Cost ($)
1	2(1) + 3	5
2	2(2) + 3	7
3	2(3) + 3	9
h	2(h) + 3	c

The cost equals 2 times the number of hours plus 3.
$$c = 2h + 3$$

TRY THIS At City Park you can rent a skateboard and protective gear for $2.50 per hour plus a $2 fee. Make a table to show the charges by the hour, and write an equation for the situation.

Solutions to Equations

The **solution** to an equation with one variable consists of the values for the variable that make a true statement when substituted into the equation. The solution for $2h = 15$ is $h = 7.5$ because substituting 7.5 for h in the original equation results in a true statement.

$2h = 15$ *Original equation*
$2(7.5) \stackrel{?}{=} 15$ *Substitute h = 7.5.*
$15 = 15$ *True statement*

EXAMPLE 3 Jessica places 3 photos on the cover of her picture album. She places 4 photos on each of the other pages. She has 155 photos in all. **How many pages will she need if she uses all of her photos?**

● SOLUTION

PROBLEM SOLVING Write an equation and use a **guess-and-check** strategy.

Let p equal the number of pages in Jessica's picture album. To answer the question, solve for p in the equation $4p + 3 = 155$. Guess the number of pages, substitute the number you guessed for p, and check your guess.

Guess 1: Try $p = 20$.
$4p + 3 \stackrel{?}{=} 155$
$4(20) + 3 = 83$

83 is too small. Try a larger number for p.

Guess 2: Try $p = 40$.
$4p + 3 \stackrel{?}{=} 155$
$4(40) + 3 = 163$

163 is too large. Try a smaller number for p.

Guess 3: Try $p = 38$.
$4p + 3 \stackrel{?}{=} 155$
$4(38) + 3 = 155$

Correct.

The solution to the problem is 38 pages.

Enrichment

Have students write problems of their own like the one in Example 3. Suggest that they start with the answer and work backward to write the problem. Have students exchange and solve each other's problems.

Inclusion Stragtegies

KINESTHETIC STRATEGIES Have students use counters or index cards to act out the situation described in Example 3.

ADDITIONAL EXAMPLE 2

The cost to rent a set of jet skis is a $20 fee plus $50 per hour. **Make a table to show the total cost to rent the skis for 1 hour through 5 hours. Write an equation for the situation.**

No. of hours	Expression for the cost	Cost (in dollars)
1	20 + 50(1)	70
2	20 + 50(2)	120
3	20 + 50(3)	170
4	20 + 50(4)	220
5	20 + 50(5)	270

equation: $c = 20 + 50h$, where c is the cost in dollars and h is the number of hours

TRY THIS

No. of hours	Expression for the cost	Cost (in dollars)
1	2.5(1) + 2	4.50
2	2.5(2) + 2	7.00
3	2.5(3) + 2	9.50
h	2.5(h) + 2	$2.5h + 2$

The equation is $c = 2.5h + 2$.

ADDITIONAL EXAMPLE 3

Megan places 2 photos on the cover of her picture album. She places 5 photos on each of the other pages. She has 67 photos in all. **How many pages will she need if she uses all her photos?** 13 pages

LESSON 1.2

Critical Thinking questions provide students an opportunity to analyze, assimilate, and expand their understanding of lesson concepts. The answers are provided in the Teacher's Edition side copy.

CRITICAL THINKING
40 pages; there will be only 1 photo on the last page.

TRY THIS
25 students

ADDITIONAL EXAMPLE 4

The Gamez family lives in a farmhouse that has a 1000-gallon water tank. On average, the family uses 80 gallons of water per day. Suppose that the tank is full. **For how many days will the family have an adequate supply of water?**

a. Solve the problem by using the constant feature of a calculator.
   After subtracting 80 gallons from 1000 gallons 12 times, only 40 gallons of water remain. The family has enough water for 12 days.

b. Write an equation and solve it by using the table feature of a graphics calculator. $g = 1000 - 80d$, where g is the number of gallons of water in the tank, and d is the number of days. In the table, when the number of days, d, is 12, the number of remaining gallons, g, is only 40, so there is enough water for 12 days.

See Keystroke Guide, page 782.

CRITICAL THINKING   In Example 3, if Jessica had 160 photos, there would be no whole-number solution to the equation—notice that $4(39) + 3 = 159$ and $4(40) + 3 = 163$. How many pages would she need for 160 photos? Explain.

TRY THIS   Marge's class is going on a field trip to an aquarium. The class rate is $12 per student plus $25. How many students can go if the class has $335 to spend?

EXAMPLE 4

APPLICATION
ECOLOGY

In 1990, the United Nations reported a deforestation figure of around 114,364 acres of trees per day. More recent reports claim that the deforestation figure may be even higher. Regardless of the actual figure, it is very difficult to imagine the size of the area affected because it is spread around the globe. In order to get a better understanding, it helps to think of the changes in terms of a single known area, such as Olympic National Park in Washington. The area of Olympic National Park is about 922,654 acres. **How long would it take to lose an area of forest land equal to the area of Olympic National Park if 114,364 acres are destroyed each day worldwide?**

a. Solve the problem by using the constant feature of a calculator.

b. Write an equation and solve it by using the table feature of a graphics calculator.

SOLUTION

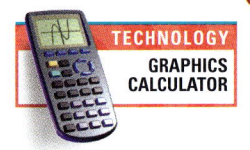

TECHNOLOGY
GRAPHICS CALCULATOR

a. On a calculator with a constant feature that repeats an operation, enter the area of Olympic National Park. Subtract 114,364 repeatedly, counting the number of times you subtract.

  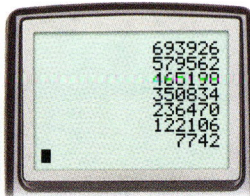

As you continue to subtract, you can see how quickly the forest area is diminished. Continue repeating the subtraction until the entire forest area is depleted.

After subtracting 114,364 eight times, only 7742 acres remain. Therefore, it would take 9 days to lose 922,654 acres of forest land.

Reteaching the Lesson

GUIDED ANALYSIS   Give students the following problem:

The members of a softball team are purchasing shirts that cost $21 each at a local store. Make a table to show the total cost of purchasing from 1 to 12 shirts. Write an equation for the situation.

In making their tables, students should use the headings *Number of shirts*, *Expression for the cost*, and *Cost*. If necessary, provide tables with some entries completed. Discuss students' completed tables and equations.

Now ask students to consider how the solution of the problem changes when the shirts are purchased by mail order. Tell them that a $10 shipping fee is charged in addition to the $21 cost per shirt. Have students amend their tables and equations to reflect this change.

Let s be the number of shirts, and let c be the total cost of all the shirts. The equation for the first situation is $c = 21s$, and the equation for the second situation is $c = 10 + 21s$.

TECHNOLOGY
GRAPHICS CALCULATOR

b. To express this situation as an equation, you need to interpret and relate the information from the problem. The acreage left is 922,654 minus 114,364 acres for each day that passes. If A represents the number of acres left and D the number of days, the situation can be expressed by the following equation:

$$A = 922{,}654 - 114{,}364D$$

Use the [Y] button on some caculators to enter the equation. (Replace A with **Y1** and D with **X**.)
Y1 = 922654 – 114,364X

Use the [TblSet] key on some calculators to set up the table. Then use the [TABLE] key to see the table.

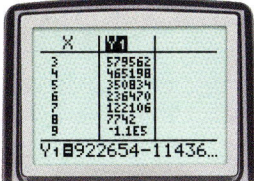

The forest acreage is depleted by day 9.

Exercises

Communicate

1. Give an example of each: *variable*, *expression*, and *equation*.

2. Describe how you would set up the equation for the following problem: If 12 pencils cost $1.92, find the cost of 1 pencil.

3. Describe how you would set up the equation for the following problem: If tickets for a concert cost $10 each, how many can you buy with $35?

4. Explain how you would use the guess-and-check strategy to solve the equation $10x + 3 = 513$.

Guided Skills Practice

5. Make a table to show the value of the expression $5c + 7$ when c is 1, 2, 3, and 4. **(EXAMPLE 1)** 12, 17, 22, 27

APPLICATIONS

6. **RECREATION** At the roller rink, skating costs $2.50 per hour plus $7. Complete the table of values for this situation. Write an equation to represent the situation. **(EXAMPLE 2)** $c = 2.5h + 7$

Hours	1	2	3	4
Cost ($)	?	?	?	?

9.50 12 14.50 17

7. **RECREATION** Write and solve an equation for the following problem: Bob has $32. How many hours can he skate if skating costs $2.50 per hour plus $7? **(EXAMPLE 3)** $2.5h + 7 = 32$, $h = 10$

5.
c	1	2	3	4
$5c + 7$	$5(1) + 7$	$5(2) + 7$	$5(3) + 7$	$5(4) + 7$
Value	12	17	22	27

Teaching Tip

TECHNOLOGY If you are using a TI-82 or TI-83 graphics calculator to create the table in part **b** of Example 4, press [Y=] and set **Y1= 92654 – 114364X** and make the following choices in the **TABLE SETUP** screen:

TblStart=1 (or **TblMin=1**)
ΔTbl=1
Indpnt: Auto
Depend: Auto

To display the table, press [2nd] [GRAPH]. After you display the table, press to scroll down to the part of the table that is pictured in the example.

internetconnect

GO TO: go.hrw.com
KEYWORD: MA1 Mars Weight

In this activity, students explore the operations of division and multiplication by accessing an interactive astronomy site that enables them to calculate their body weight on other planets.

Assess

Selected Answers
Exercises 5–9, 11–47 odd

ASSIGNMENT GUIDE

In Class	1–9
Core	11–35 odd, 37–39, 41, 43
Core Plus	10–36 even, 37–43
Review	44–48
Preview	49

✏ Extra Practice can be found beginning on page 738.

Answers to odd-numbered Extra Practice exercises can be found immediately after Selected Answers in the Pupil's Edition.

LESSON 1.2 **15**

Technology

A calculator may be helpful for Exercises 40–43. You may wish to have students first complete the exercises and then use a calculator to verify their answers.

Error Analysis

When solving the real-world problems presented in this lesson, some students may apply appropriate procedures but interpret their results inaccurately. In Example 4, for instance, they may be confused about whether the area will be destroyed after 8 days or after 9 days. Tell all students that a good check on the reasonableness of their results is to write the solution to such a problem as a complete sentence or even a brief paragraph that places the numerical answer in its real-world context.

8. Write an equation and use the table feature on a graphics calculator to determine the number of the term that has a value of 0 in the sequence below. *(EXAMPLE 4)* **The 28th term of the sequence is zero.**
   13,500, 13,000, 12,500, 12,000, …

9. Write an equation that describes the linear pattern in the data below. Use t for time and d for distance. *(EXAMPLE 4)* **$d = 24t$**

Time (hours)	0	1	2	3	4	5
Distance (miles)	0	24	48	72	96	120

Practice and Apply

10. 4, 8, 12, 16, 20
11. 5, 10, 15, 20, 25
12. 11, 18, 25, 32, 39
13. 3, 6, 9, 12, 15
14. 2, 7, 12, 17, 22
15. 1, 3, 5, 7, 9
16. 15, 21, 27, 33, 39
17. 5, 14, 23, 32, 41
18. 8, 16, 24, 32, 40
19. 5, 8, 11, 14, 17
20. 12, 21, 30, 39, 48
21. 0, 3, 6, 9, 12

APPLICATIONS

Make a table showing the value of each expression when the value of the variable is 1, 2, 3, 4, and 5.

10. $4x$
11. $5y$
12. $7s + 4$
13. $3n$
14. $5p - 3$
15. $2f - 1$
16. $6q + 9$
17. $9g - 4$
18. $8x$
19. $3r + 2$
20. $9y + 3$
21. $3z - 3$

Use guess-and-check to solve the following equations.

22. $4x + 3 = 47$ **$x = 11$**
23. $19x + 13 = 51$ **$x = 2$**
24. $3x + 17 = 56$ **$x = 13$**
25. $7x - 12 = 44$ **$x = 8$**
26. $13x - 3 = 62$ **$x = 5$**
27. $21x - 17 = 193$ **$x = 10$**
28. $5p - 10 = 50$ **$x = 12$**
29. $3s + 2 = 53$ **$s = 17$**
30. $11a - 2 = 64$ **$a = 6$**

CONSUMER ECONOMICS If pencils cost 20 cents each and pens cost 30 cents each, find the cost of the following items.

31. 0 pencils **$0**
32. 4 pencils **$0.80**
33. 10 pencils **$2.00**
34. p pencils **0.20p**
35. r pens **0.30r**
36. r pens and p pencils **$0.30r + $0.20p**

37. **ENTERTAINMENT** If tickets for a concert cost $9 each, how many tickets can you buy if you have $135? Write an equation and solve by guess-and-check. **15 tickets**

38. **ENTERTAINMENT** If tickets for a concert cost $11 each, how many tickets can you buy if you have $132? Write an equation and solve by guess-and-check. **12 tickets**

39. **HOME IMPROVEMENT** Write an equation that describes the linear pattern in the data below. Use g for the number of gallons of paint and c for the total cost.

$c = 16g$

Gallons of paint	0	1	2	3	4	5
Total cost	0	16	32	48	64	80

TECHNOLOGY Tell how you would generate each sequence below on a calculator. Find the 20th term of each sequence.

40. 6, 8, 10, 12, 14, …
41. 15, 25, 35, 45, 55, …
42. 100, 90, 80, 70, 60, …
43. 52, 48, 44, 40, 36, …

40. **Enter 6 and repeatedly add 2. The 20th term is 44.**

41. **Enter 15 and repeatedly add 10. The 20th term is 205.**

42. **Enter 100 and repeatedly subtract 10. The 20th term is –90.**

43. **Enter 52 and repeatedly subtract 4. The 20th term is –24.**

16 LESSON 1.2

 Look Back

The sum will be $\frac{n(n+1)}{2}$.

44. The sum $1 + 2 + 3 + 4 + 5$ is $\frac{5 \cdot 6}{2}$. Examine the next two cases—the sum of the first six integers and the sum of the first seven integers. Make a conjecture about the sum of the first n integers. **(LESSON 1.1)**

45. Find the first and second differences for the sequence 2, 8, 24, 51, 90, ...
(LESSON 1.1) First dif. 6, 16, 27, 39   Second dif. 10, 11, 12

46. Find the next two terms in the sequence 11, 22, 33, 44, 55, 66, ...
(LESSON 1.1) 77, 88

5, 12, 21, 32, 45

47. The first differences of a sequence are 7, 9, 11, and 13. The first term of the sequence is 5. What are the first five terms of the sequence? **(LESSON 1.1)**

48. The first term of a sequence is 11. The first difference is 3, and the second differences are a constant 2. What are the first five terms of the sequence?
(LESSON 1.1) 11, 14, 19, 26, 35

 Look Beyond

49. The expression x^2 means "x times x." Use guess-and-check to solve for x in the equation $x^2 = 256$. $x = 16$ or $x = -16$

Include this activity in your portfolio.

$3 = 2(1) + 1$ $5 = 2(2) + 1$ $7 = 2(3) + 1$

1. Study the sequence of toothpick triangles. Sketch the next two sets in the sequence. How many toothpicks does it take to build each set? Build a table for the sets of triangles and compute the first differences.

Set number, x	0	1	2	3	4	5
Segments, y	?	3	5	?	?	?
		2	2	?	?	

2. How many toothpicks does it take to draw the 100th set? How could you use differences to work back to the number of toothpicks in set 0?

The differences can be used to write an equation, $y = 2x + 1$. If you separate the toothpicks, it is easy to visualize the equation.

WORKING ON THE CHAPTER PROJECT
You should now be able to complete Activities 1 and 2 of the Chapter Project on page 44.

Answers to Portfolio Activities can be found in Additional Answers of the Teacher's Edition.

 Look Beyond

Exercise 49 involves finding the solution to a quadratic equation. At this point, students are expected to solve the problem by using the guess-and-check strategy. The more formal algebraic methods for solving quadratic equations will be studied in Chapter 10.

Understanding why −16 is a solution requires an understanding of the rules for multiplying signed numbers. In this text, these rules are not presented formally until Lesson 2.3. However, students should be familiar with them from their work in previous courses.

ALTERNATIVE
Assessment

Portfolio Activity
The Portfolio Activity can be used as preparation for the Chapter Project or as a separate activity. In this Portfolio Activity, students may benefit by using real toothpicks to build the sequence of triangles. Students can then physically pull apart the sequence to see the pattern that leads to the equation.

Student Technology Guide

 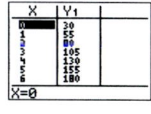

Prepare

NCTM PRINCIPLES & STANDARDS
1–3, 5–10

LESSON RESOURCES

- Practice 1.3
- Reteaching Master 1.3
- Enrichment Master 1.3
- Cooperative-Learning Activity 1.3
- Lesson Activity 1.3
- Problem Solving/Critical Thinking Master 1.3

QUICK WARM-UP

Evaluate each expression.

1. $187 - 18.7$ 168.3
2. $30 + 3.2 + 0.3$ 33.5
3. 3^3 27
4. $40 \cdot 12$ 480

Also on Quiz Transparency 1.3

Teach

Why An established order of operations is necessary to avoid ambiguous results when evaluating an expression that involves multiple operations. To illustrate this point, give students the following expression:

$3 \cdot 6 + 9 \div 3 - 1$

Tell them to imagine that the operations could be performed in any order. Ask them to list possible values for the expression.

See Keystroke Guide, page 782.

1.3 The Algebraic Order of Operations

Objective
- Apply the algebraic order of operations.

Why Calculating how much floor covering is needed for a particular room can involve a series of operations such as addition, subtraction, multiplication, and division. Knowing the correct algebraic order of operations is important.

APPLICATION
HOME IMPROVEMENT

Mary plans to buy floor covering for two rooms. One room is 12 feet by 15 feet, and the other room is 20 feet by 14 feet. To find the number of square feet, she uses a simple four-function calculator and enters the following:

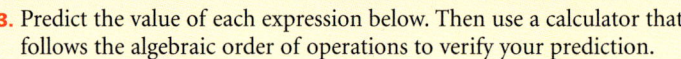

Much to Mary's surprise, her calculator's answer is 2800. Mary knows that 2800 square feet is too much because her entire house is less than 2000 square feet. Did her calculator give a wrong answer?

Activity
Exploring the Order of Operations

You will need: a graphics calculator or a scientific calculator

1. What steps did Mary's simple calculator use to get 2800?
2. Unlike Mary's calculator, graphics and scientific calculators follow a logical system called the **algebraic order of operations**. The solution to Mary's problem is shown at left on a graphics calculator. What order did this calculator follow?
3. Predict the value of each expression below. Then use a calculator that follows the algebraic order of operations to verify your prediction.
   a. $2 + 3 \cdot 5$
   b. $2^3 + 3 \cdot 5$
   c. $15 \cdot 2 - 3^2$
   d. $2 + 3 \cdot 5 - 4$
   e. $6 \div 2 \cdot 3$
   f. $(2 + 3) \cdot 5$
   g. $(2^3 + 3) \cdot 5$
   h. $(15 \cdot 2 - 3)^2$

CHECKPOINT ✓
4. Write a paragraph to describe the algebraic order of operations that the graphics calculator follows.

Alternative Teaching Strategy

HANDS-ON STRATEGIES Write the following expression on the board or overhead:

$12 - 4 \cdot 2 + 70 \div 2$

Beneath the expression, write Steps 3 and 4 of the order of operations, as given on page 19.

Have students write each number and operation in the expression on an index card. Have them place the cards in order on their desks. Ask students to place the first set of cards to be evaluated (4, •, and 2) in a pile, turn the pile over, and write the product on the back of the last card. You may want to have students clip the pile of cards together. Repeat this process for the other operations in the proper order. Be sure that all students arrive at the correct value for the expression, **39**.

After students are comfortable with Steps 3 and 4 of the order of operations, have them work with Step 2 and then Step 1, using the following expressions:

$10^2 - 4 \div 2 \div 2$ 99
$[(12 + 3) - 3 \cdot 4] \div 3$ 1

18 LESSON 1.3

To avoid misunderstandings and errors, mathematicians have agreed on certain rules for computation called the **algebraic order of operations**.

The Algebraic Order of Operations

1. Perform all operations within inclusion symbols from innermost outward.
2. Perform all operations with exponents.
3. Perform all multiplications and divisions in order from left to right.
4. Perform all additions and subtractions in order from left to right.

Symbols like parentheses, ( ), brackets, [ ], braces, { }, and the fraction bar, —, are called **inclusion symbols**. Treat any grouped numbers and variables as a single quantity. Operations should always be done within the innermost inclusion symbols first. Then work outward.

EXAMPLE 1 Use the algebraic order of operations to evaluate each expression.

a. $3 - 12 \div 4$ b. $6^2 \div 4 \cdot 3$ c. $(8 \cdot 7 + 2) + (3 \cdot 2 - 2 \cdot 2)$

SOLUTION

a. Divide first and then subtract: $3 - 12 \div 4 = 3 - 3 = 0$

b. Evaluate the exponent and then divide and multiply in order:
$6^2 \div 4 \cdot 3 = 36 \div 4 \cdot 3 = 9 \cdot 3 = 27$

c. Perform all operations within parentheses first:
$(8 \cdot 7 + 2) + (3 \cdot 2 - 2 \cdot 2) = (56 + 2) + (6 - 4) = 58 + 2 = 60$

TRY THIS Use the algebraic order of operations to evaluate each expression.

a. $13 + 2 \cdot 5$ b. $6^2 - 4 \cdot 3 + 7^2$ c. $(3 \cdot 4 + 7) + (2 \cdot 4 + 3 \cdot 4)$

EXAMPLE 2 Insert inclusion symbols to make $30 + 4 \div 2 - 1 = 16$ true.

SOLUTION

Use parentheses to group $30 + 4$ before dividing by 2.

$[(30 + 4) \div 2] - 1 \overset{?}{=} 16$
$(34 \div 2) - 1 \overset{?}{=} 16$
$17 - 1 \overset{?}{=} 16$
$16 = 16$ *True*

TRY THIS Insert inclusion symbols to make $80 \div 2 + 6 \cdot 5 = 50$ true.

Activity Notes

In this Activity, students use either a graphics calculator or scientific calculator to investigate the order in which operations are performed. After working with all of the given expressions, they should be able to summarize, in their own words, the order of operations formally presented on page 19.

CHECKPOINT ✔

4. Most graphics calculators follow the standard algebraic order of operations. All calculations included by parentheses are performed first, followed by any exponents. Then multiplication and division are performed from left to right, followed by addition and subtraction from left to right.

Teaching Tip

To remember the order of operations, many students find it helpful to use the mnemonic device PEMDAS (Parentheses, Exponents, Multiplication, Division, Addition, Subtraction) or the sentence Please Excuse My Dear Aunt Sally.

ADDITIONAL EXAMPLE 1

Use the algebraic order of operations to evaluate each expression.

a. $6 + 5 \cdot 4$ 26

b. $4^2 \cdot 3 \div 8$ 6

c. $(7 + 2 \cdot 10) + (5 \cdot 3 - 4 \cdot 2)$ 34

☞ For answers to Try This and Additional Example 2, see page 20.

Interdisciplinary Connection

ACCOUNTING Ask students to consider the importance of the order of operations in accounting. Have them write a paragraph describing the effect on an important financial calculation if the order of operations were not used properly. Students may want to talk to an accountant or a banker to get ideas for their work.

Enrichment

Have students write expressions of their own that contain several pairs of inclusion symbols. Encourage them to use nested symbols, as illustrated in the following expression:

$$8\{7[6(5 + 4) + 3] - 2\} + 1$$

Ask students to evaluate each of their own expressions, make any necessary changes, and then exchange expressions with a partner. Students should then evaluate their partner's expressions.

LESSON 1.3

TRY THIS
a. 23
b. 73
c. 39

ADDITIONAL EXAMPLE 2

Insert inclusion symbols to make $18 + 30 \div 2 = 24$ true. $(18 + 30) \div 2 = 24$

TRY THIS
$80 \div (2 + 6) \cdot 5 = 50$

ADDITIONAL EXAMPLE 3

Use the algebraic order of operations to evaluate $2a^2 + b$ for $a = 7$ and $b = 2$.
100

TRY THIS
600 feet

ADDITIONAL EXAMPLE 4

Use a graphics or scientific calculator to evaluate $\frac{80 + 32}{16} - \frac{200 - 50}{30} \cdot 2$.
Describe the keystrokes.

Keystrokes:

TRY THIS

Keystrokes:

54

EXAMPLE 3 Use the algebraic order of operations to evaluate $5x^2 + 7y$ when x is 3 and y is 2.

● **SOLUTION**

Replace x with 3 and y with 2. First square the 3. Perform all of the multiplications. Finally, add the results.

$5x^2 + 7y$
$5 \cdot 3^2 + 7 \cdot 2$
$5 \cdot 9 + 7 \cdot 2$
$45 + 14$
59

TRY THIS The height, in feet, of a falling object dropped from a height of 1000 feet after s seconds is given by $1000 - 16s^2$. What is its height after 5 seconds?

EXAMPLE 4 Use a graphics or scientific calculator to evaluate $\frac{57 + 95}{16} + \frac{220}{88 + 104}$. Describe the keystrokes.

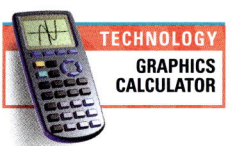

TECHNOLOGY
GRAPHICS CALCULATOR

● **SOLUTION**

Since the entire quantity $57 + 95$ is divided by 16, place parentheses around $57 + 95$. Do the same for $88 + 104$.

$(57 + 95) \div 16 + 220 \div (88 + 104)$

The keystrokes are shown below.

The answer is 10.646 to three decimal places. Is this reasonable? The first fraction is about $\frac{160}{16}$, or 10. The second fraction is about $\frac{200}{200}$, or 1. Thus, the estimated sum is about $10 + 1$, or 11. The answer 10.646 is reasonable.

TRY THIS Use a graphics or scientific calculator to evaluate $\frac{17 + 97}{2} - \frac{42 - 6}{12}$. Describe the keystrokes.

EXAMPLE 5 Evaluate $2\{2[2(3) + 1] + 1\} + 1$.

● **SOLUTION**

When you evaluate an expression that contains several pairs of inclusion symbols, begin with the innermost pair and work outward.

$2\{2[2(3) + 1] + 1\} + 1$
$2[2(7) + 1] + 1$
$2(15) + 1$
31

If you use a calculator or computer, the expression will usually contain only parentheses. The expression would be entered as $2(2(2(3) + 1) + 1) + 1$. This does not affect the way you simplify the expression.

TRY THIS Evaluate $3\{4[5(6) - 3(2)] - 2\} + 7$.

Inclusion Strategies

ENGLISH LANGUAGE DEVELOPMENT Be sure that all students understand the terms used in the statement of the order of operations on page 19. Ask if there are any terms they do not understand, and carefully explain the meanings. If a word is unfamiliar, encourage students to make an association with a word that they recognize from everyday usage. For example, in discussing *inclusion* symbols, ask students what it means to be *included* in a game or other activity.

Reteaching the Lesson

VISUAL STRATEGIES Color-code the steps of the order of operations. For example, write the first step in blue, the second step in red, the third step in green, and the fourth step in purple. Then have students write expressions with the same color-coding. For example, all inclusion symbols should be written in blue, all exponents in red, all multiplication and division signs in green, and all addition and subtraction signs in purple.

Exercises

Communicate

1. Explain how to evaluate $3 + 2 \cdot 4$.
2. What two possible answers might you get for $20 - 2 \cdot 5$? Which answer is correct and why?
3. Describe the steps for evaluating $\{[3(8-4)]^2 - 6\} \div (4-2)$.
4. Explain why the rules called the *algebraic order of operations* are necessary for computation.
5. Explain why 4 is a reasonable estimate for $\frac{173 + 223}{151 - 47}$.

Guided Skills Practice

Use the algebraic order of operations to evaluate each expression. (EXAMPLE 1)

6. $(7 + 3^2) - 2 \cdot 4$ **8**
7. $4 \div 2 + 70 \div 2$ **37**
8. $12 \div 2 \cdot 3 - 4^2$ **2**

Insert inclusion symbols to make each equation true. (EXAMPLE 2)

9. $14 \div 5 + 2 + 8 = 10$
   $14 \div (5 + 2) + 8 = 10$
10. $4 + 7 \cdot 3 + 1 = 34$
    $(4 + 7) \cdot 3 + 1 = 34$
11. $8 + 25 \cdot 2 \div 6 = 11$
    $(8 + 25) \cdot 2 \div 6 = 11$

Evaluate each expression for $a = 4$ and $b = 6$. (EXAMPLE 3)

12. $a^2 + b^2$ **52**
13. $\frac{a+b}{2} - \frac{b-a}{2}$ **4**
14. $a^2 - (b^2 - a^2 - a)$ **0**

15. Evaluate $\frac{3+7}{5^2} + 2$. Describe the scientific or graphics calculator keystrokes that are used to evaluate the expression. (EXAMPLE 4) **2.4**

Evaluate. (EXAMPLE 5)

16. $2\{2[3(2) + 4]\} + 6$ **46**
17. $4\{5[13 - 2(6)]\} - 19$ **1**

Practice and Apply

Evaluate each expression.

18. $57 \cdot 29 + 89$ **1742**
19. $72(98) + 12$ **7068**
20. $89 + 57 \cdot 29$ **1742**
21. $3(15) + 9$ **54**
22. $43 \cdot 32 + 91 \cdot 67$ **7473**
23. $45(75) + 9(24)$ **3591**
24. $157 - 29 + 23 \cdot 9$ **335**
25. $91 \div 7 + 6$ **19**
26. $187 - 34 \div 17$ **185**
27. $2(5 + 4) \div 9$ **2**
28. $12 - 7 \cdot 3 + 9^2$ **72**
29. $3 - 1 + 24 \div 6$ **6**
30. $7 + 6 \div 2 \cdot 10$ **37**
31. $3 \cdot 2 + 7 \cdot 6$ **48**
32. $100 - (3)(6)(4)$ **28**
33. $3^2 + 7 \cdot 2 - 8 \cdot 2$ **7**
34. $80 \div 4 \cdot 2 - 2 \cdot 2$ **36**
35. $2 \cdot 6^2 - 100 \div 50$ **70**
36. $3(2 + 7 - 8) + 16$ **19**
37. $4^2 \cdot 2 + [7 - (3^2 - 5)]$ **35**
38. $[15(10) - 12(10)] \div 10$ **3**
39. $4[(3 + 2 \cdot 3) - 5] + 7$ **23**
40. $(8 - 4) \cdot (12 - 3) \div (2 + 1 \cdot 2)$ **9**
41. $8^2 - 7^2 + 6^2$ **51**

Teaching Tip

TECHNOLOGY If you omit an open parenthesis `(` when using the TI-82 or TI-83 graphics calculator, you will receive an error message. However, you are allowed to omit the close parenthesis `)`. Alert students to the potential for error that arises as a result. For instance, if a student tries to evaluate $(3 + 9) \div 6$ by entering `(` `3` `+` `9` `÷` `6` `ENTER`, the calculator will proceed to give a result of 4.5 rather than the intended result of 2.

See Keystroke Guide, page 782.

ADDITIONAL EXAMPLE 5

Evaluate:

$2\{2[2(5) + 1] + 1\} + 1$ **47**

TRY THIS
289

Assess

Selected Answers
Exercises 6–17, 19–75 odd

ASSIGNMENT GUIDE

In Class	1–17
Core	19–61 odd, 62–64, 67
Core Plus	18–60 even, 62–68
Review	69–75
Preview	76–79

✏ Extra Practice can be found beginning on page 738.

Mid-Chapter Assessment for Lessons 1.1 through 1.3 can be found on page 8 of the *Assessment Resources*.

LESSON 1.3 **21**

Technology

A calculator may be helpful with the exercises in this lesson. You may wish to have students complete the exercises with paper and pencil first and then use a calculator to verify their answers.

Error Analysis

Students who persist in performing all computations in order from left to right should be encouraged to focus first on determining the correct order. Post the order of operations at the front of the classroom with the steps numbered from 1 to 4 as shown on page 19. Have students write a given exercise on a piece of paper and then write the number of the appropriate step next to each inclusion symbol, exponent, and operation sign, as shown below.

④ ② ③   ①   ④   ③
$8 + 4^2 \div (2 + 6) - 9 \times 2$

---

42. $(28 - 2) \cdot 0 = 0$
43. $59 - 4 \cdot (6 - 4) = 51$
44. $(25 - 15) \cdot 6 + 4 = 64$
45. $(22 - 4) \cdot 3 = 54$
46. $81 \div (9 + 18) = 3$
47. $108 - 17 \cdot (2 + 3) = 23$

CHALLENGE

CONNECTION

APPLICATION

423 ft²
531 ft²

Place inclusion symbols to make each equation true.

42. $28 - 2 \cdot 0 = 0$ 43. $59 - 4 \cdot 6 - 4 = 51$ 44. $25 - 15 \cdot 6 + 4 = 64$
45. $22 - 4 \cdot 3 = 54$ 46. $81 \div 9 + 18 = 3$ 47. $108 - 17 \cdot 2 + 3 = 23$

Given $a = 5$, $b = 3$, and $c = 4$, evaluate each expression.

48. $a + b - c$ **4**   49. $a^2 + b^2$ **34**   50. $a^2 - b^2$ **16**   51. $(a + b) \cdot c$ **32**
52. $a^2 - b - c$ **18**   53. $a^2 - (b + c)$ **18**   54. $a^2 + b^2 + c^2$ **50**   55. $b^2 + (c^2 - a)$ **20**

Evaluate each expression. Round to the nearest thousandth if necessary. Describe the graphics or scientific calculator keystrokes used to evaluate each expression.

56. $\frac{28 + 59}{97 - 68}$ **3**   57. $\frac{97 - 17}{72 + 8}$ **1**   58. $\frac{40 \cdot 90}{8 \cdot 25}$ **18**

59. $\frac{28 + 59}{97 - 17}$ **1.088**   60. $\frac{97 - 17}{72 + 7}$ **1.013**   61. $\frac{43 \cdot 91}{8 \cdot 25}$ **19.565**

Evaluate each expression.

62. $3\{4[5(4 + 7)] - 2\}$ **654**   63. $10 + 2\{44 - [3(2 + 7)]\}$ **44**   64. $2\{2[2(2 + 2)]\} - 2$ **30**

65. Place inclusion symbols to make the equation below true.
$100 - 99 \cdot 3 + 7 \cdot 2 + 1 \div 24 = 1$
$(100 - 99) \cdot [3 + 7 \cdot (2 + 1)] \div 24 = 1$

66. **STATISTICS** The table shows the results of an algebra quiz from a teacher's grade book. Find the average for the class and describe your method.

Score	100	90	80	70	60	
Number of students	4	12	7	0	1	**87.5**

67. **HOME IMPROVEMENT** Mr. Hilbert owns two rental properties, house A and house B. He wants to carpet two rooms in house A and three rooms in house B. He finds a store selling carpet for $12.99 per square yard. Use the dimensions indicated in the picture. (Hint: 1 yd² = 9 ft²)

a. Find the total number of square feet of carpet needed for house A.
b. Find the total number of square feet of carpet needed for house B.
c. Find the cost of carpeting the rooms in house A. **$610.53**
d. Find the cost of carpeting the rooms in house B. **$766.41**

Practice

1.3 The Algebraic Order of Operations

Evaluate each expression.

1. $16 + 4 \cdot 8 + 2$ **50**   2. $16 + 4 \cdot (8 + 2)$ **56**
3. $(16 + 4) \cdot (8 + 2)$ **200**   4. $(16 + 4) \cdot 8 + 2$ **162**
5. $0.2(2.5) + 8$ **8.5**   6. $16 \cdot 37 + 88 \cdot 49$ **4904**
7. $16 + 8 \div 2$ **20**   8. $4 \cdot 6 \div 12 + 10$ **12**
9. $\frac{12 + 6}{4 + 2}$ **3**   10. $\frac{5 + 3 \cdot 5}{5}$ **4**
11. $12 + 6 \div 4 + 2$ **15.5**   12. $90 \div 3 + 5$ **35**
13. $36 - 6 \cdot 3 \div 18 \cdot 3$ **33**   14. $9 - 3 \div 4 + 2 \cdot 12 + 6 \div 2 \cdot 3$ **41.25**
15. $4 + 1 \cdot 4^2 - 3$ **17**   16. $6 + 3^3 - 18 \div 6$ **30**

Place inclusion symbols to make each equation true.

17. $27 + 5 \cdot 8 - 6 = 37$ $27 + [5(8 - 6)] = 37$
18. $12 \cdot 1 + 5 \div 12 = 6$ $12[(1 + 5) \div 12] = 6$
19. $44 \cdot 5 - 3 + 2 = 10$ $4(5 - 3) + 2 = 10$
20. $3 \cdot 4 + 2 \div 6 = 3$ $3(4 + 2) \div 6 = 3$

Given $a = 3$, $b = 6$, and $c = 5$, evaluate each expression.

21. $b - a + c$ **8**   22. $a + b - c$ **4**
23. $a \cdot b + a \cdot c$ **33**   24. $b \div a + a \cdot c$ **17**
25. $a^2 + c^2$ **34**   26. $b^2 - a^2$ **27**
27. $(a + b) \div c$ **1.8**   28. $a + b^2 - c$ **34**

22   LESSON 1.3

APPLICATION

68. **METEOROLOGY** The temperature-humidity index, or THI, is a number that measures the degree of discomfort that a person may feel because of the amount of humidity in the air. The higher the index, the greater the discomfort. You can find the THI by using the formula $THI = 0.4(t + s) + 15$, where t is the air temperature and s is the temperature from a thermometer with a wet cloth on its bulb. Find the THI for each pair of Fahrenheit temperatures.
   a. $t = 80, s = 65$ **73**   b. $t = 86, s = 74$ **79**
   c. $t = 100, s = 81$ **87.4**

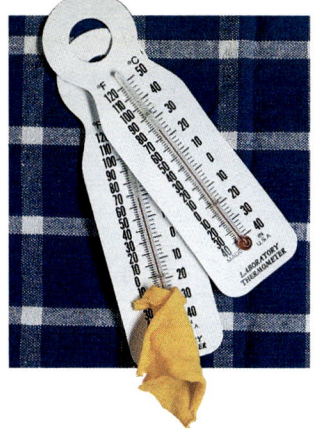

Look Beyond

Exercises 76–79 provide students with a brief look at the process of squaring a binomial. This process will be studied in depth in Chapter 9.

Look Back

69. **CULTURAL CONNECTION: AFRICA** Thousands of years ago Hypatia, the earliest known female mathematician, investigated the relationship between square numbers and triangular numbers.

   a. Find at least three more examples of how a square number can be represented as the sum of two triangular numbers. *(LESSON 1.1)*
   b. Are there any square numbers that cannot be represented as the sum of two triangular numbers? *(LESSON 1.1)* **No**

Find the next two terms of each sequence. *(LESSON 1.1)*

70. 2, 5, 8, 11, 14, … **17, 20**
71. 59, 54, 49, 44, 39, … **34, 29**
72. 3, 6, 9, 12, 15, … **18, 21**

Write an equation for each situation and solve it by using guess-and-check. *(LESSON 1.2)*

73. Jean has $45. How many hours can Jean ice skate if skating costs $5 plus $3 per hour? **Jean can skate for 13 hours and have $1 left.**

74. How many oranges can you buy with $3 if oranges cost 13¢ each? **23 oranges with 1 cent left.**

CONNECTION

75. **GEOMETRY** The perimeter of a rectangle is the number of units in the distance around the border, or the sum of twice the length and twice the width. The area is the product of the length and width. For what length and width will the perimeter and the area both be 16? **length 4, width 4**

Look Beyond

An exponent indicates repeated multiplication. Tell whether each statement is true or false.

76. $(3 + 4)^2 \stackrel{?}{=} 3 + 4^2$ **false**
77. $(3 + 4)^2 \stackrel{?}{=} (3 + 4)(3 + 4)$ **true**
78. $(3 + 4)^2 \stackrel{?}{=} 3^2 + 2(3)(4) + 4^2$ **true**
79. $(3 + 4)^2 \stackrel{?}{=} 3^2 + 4^2$ **false**

69. a.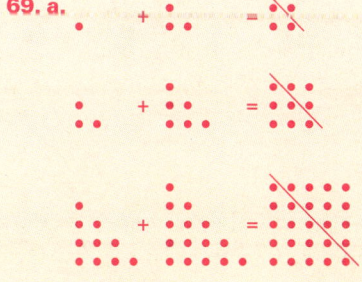

Student Technology Guide

1.3 The Algebraic Order of Operations

When evaluating expressions, graphics calculators and scientific calculators follow a logical system called the algebraic order of operations. For example, study the expressions simplified in the graphics calculator screen shown at right. The answers are different because in the first expression, the multiplication is done before the subtraction. In the second expression, the parentheses force the calculator to perform the subtraction before the multiplication.

Place inclusion symbols to make each equation true. Check with a graphics calculator.

1. $8 - 2 \cdot 3 + 10 = 28$
   $(8 - 2) \cdot 3 + 10 = 28$
2. $2 \cdot 3 + 4 + 9 + 5 \cdot 2 = 42$
   $2 \cdot (3 + 4) + (9 + 5) \cdot 2 = 42$
3. $2 \cdot 6^2 \; 12 \; 1 \; 6 \; 1 \; 8 \; 4 = 104$
   $2 \cdot (6^2 - 12) + (6 + 8) \cdot 4 = 104$
4. $8^2 \; 6 \cdot 9 \; 2 \; 2 = 4$
   $8^2 - 6(9 - 2) \div 2 = 4$

Using a calculator to evaluate an expression with fractions necessitates the use of parentheses as grouping symbols for the numerator and the denominator.

For example, the difference $\frac{64 + 18}{20 + 12} - \frac{45 - 9}{24}$ can be simplified with a calculator if it is rewritten as $(64 + 18) \div (20 + 12) - (45 - 9) \div 24$. The keystrokes are shown below.

(64 + 18) ÷ (20 + 12) - (45 - 9) ÷ 24 ENTER

The answer is 1.0625.

Rewrite each expression so that it can be simplified with a calculator.

5. $\frac{25 + 15}{5} + \frac{18 + 12}{24}$
   $(25 + 15) \div 5 + (18 + 12) \div 24$
6. $\frac{(90 - 15)}{25} - \frac{48 + 24}{8 + 16}$
   $(90 - 15) \div 25 - (48 + 24) \div (8 + 16)$
7. $\frac{3(16 + 8)}{2} + \frac{6(12 - 8)^2}{3}$
   $3(16 + 8) \div 2 + 6(12 - 8)\wedge 2 \div 3$
8. $\frac{4(18 - 12)}{8} + (4 + 2)^3$
   $4(18 - 12) \div 8 + (4 + 2)\wedge 3$

Evaluate each expression.

9. See Exercise 5. **9.25**
10. See Exercise 6. **0**
11. See Exercise 7. **68**
12. See Exercise 8. **219**

LESSON 1.3   23

Prepare

NCTM PRINCIPLES & STANDARDS 1–3, 6–10

LESSON RESOURCES

- Practice 1.4
- Reteaching Master 1.4
- Enrichment Master 1.4
- Cooperative-Learning Activity 1.4
- Lesson Activity 1.4
- Problem Solving/Critical Thinking Master 1.4
- Teaching Transparency 2
- Teaching Transparency 3

Quick Warm-Up

Graph each point on the same number line.

1. $A: -3$
2. $B: 2.5$
3. $C: -\frac{1}{3}$

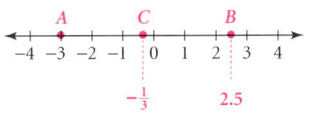

Also on Quiz Transparency 1.4

Teach

Why In graphs, information can be presented in a visual, easy-to-see format. Graphs are used in newspapers, magazines, and other types of media to display data. Have students collect graphs from such sources and bring them to class. Display students' graphs and discuss them as appropriate throughout the lesson.

1.4 Graphing With Coordinates

Objective
- Plot points and lines on a coordinate plane.

Why Patterns in data can sometimes be difficult to recognize. Just as locations can be related to each other with a map, graphing provides a picture of how two variables relate to one another.

Road maps often have a coordinate grid to help you to locate cities. For example, Houston, Texas is located at (M, 11) on the map.

The letter and number grid on many road maps is an example of the use of a coordinate system to help locate a specific point. Mathematicians use coordinates to represent points on a graph.

One basic form of a mathematical graph is a number line.

$$-7 \; -6 \; -5 \; -4 \; -3 \; -2 \; -1 \; 0 \; 1 \; 2 \; 3 \; 4 \; 5 \; 6 \; 7$$

The arrows indicate that the line and the numbers extend infinitely in both directions.

CULTURAL CONNECTION: EUROPE In the seventeenth century, René Descartes, a French mathematician and philosopher, used a horizontal and a vertical number line to divide a plane into four regions, called **quadrants**.

A plane set up according to Descartes' system is known as a **coordinate plane**.

The horizontal line is called the x-**axis**, and numbers on this axis are called x-**coordinates**.

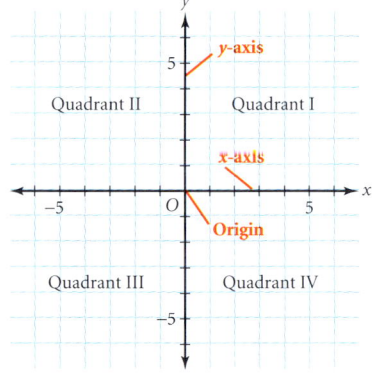

The vertical line is called the y-**axis**, and numbers on this axis are called y-**coordinates**.

Alternative Teaching Strategy

INVITING PARTICIPATION Create a large coordinate plane on the floor. Have one student start at the origin and walk 5 units to the right and then 7 units up. Explain that the point at which the student stopped has the coordinates (5, 7). Have a second student start at the origin and walk 1 unit to the right and 3 units up. Ask students to give the coordinates of the point represented. **(1, 3)**

Now give the students representing the points a piece of string to represent the line connecting the points (5, 7) and (1, 3). Ask students to give the coordinates of two other points that lie on this line. **Samples: (2, 4) and (3, 5)**

Ask students to make a conjecture about the equation of the line represented by the string. $y = x + 2$

Letters other than *x* and *y* can represent the axes. For example, you might see the *x*-axis shown as the *t*-axis when *t* refers to time in a problem.

To locate a point, *A*, on the coordinate plane shown at right, you start at the origin and move 10 units to the right, and then 5 units up. The numbers 10 and 5 are called the coordinates of point *A* because they correspond to the coordinates directly below and to the left on the axes. The number 10 is called the **x-coordinate**, and 5 is called the **y-coordinate**. To name the exact location of point *A*, you write *A*(10, 5).

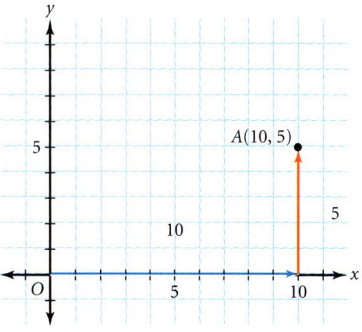

The order in which you name the coordinates is important. This means that a point identified by (10, 5) is different from a point identified by (5, 10). For this reason, the coordinates of a point on a coordinate plane are called an **ordered pair**. The ordered pair (0, 0) is called the **origin**.

EXAMPLE 1

CONNECTION
COORDINATE GEOMETRY

a. Determine the quadrant and coordinates for points *A*, *B*, and *C* on the graph.
b. Graph each of the following points: *D*(2, 6), *E*(10, −3), *F*(−7, −4).

SOLUTION

a. Point *A* is 4 units to the right and 4 units up. It is in quadrant I and the coordinates are (4, 4).

Point *B* is 1 unit to the left and 3 units up. It is in quadrant II and the coordinates are (−1, 3).

Point *C* is 6 units to the right and 2 units down. It is in quadrant IV and the coordinates are (6, −2).

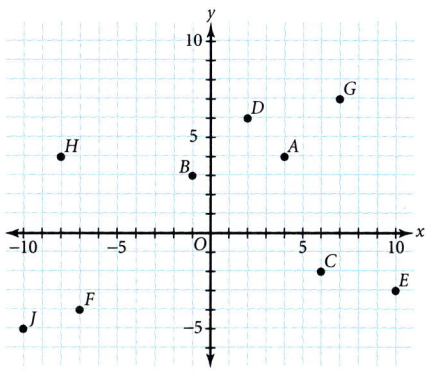

b. Begin at the origin, (0, 0). Move 2 units to the right and 6 units up. Label the point *D*.

Begin at the origin, (0, 0). Move 10 units to the right and 3 units down. Label the point *E*.

Begin at the origin, (0, 0). Move 7 units to the left and 4 units down. Label the point *F*.

TRY THIS
a. Determine the coordinates for points *G*, *H*, and *J* on the graph in Example 1.
b. Graph each of the following points: *K*(−3, 8), *L*(−5, −7), and *M*(8, −2).

Enrichment
The point (0, *y*) describes any point on the *y*-axis. Have students write ordered pairs that describe the following points:
a. 3 units to the right of the *y*-axis   (3, *y*)
b. 5 units to the left of the *y*-axis   (−5, *y*)
c. 7 units above the *x*-axis   (*x*, 7)
d. 2 units below the *x*-axis   (*x*, −2)

Use Teaching Transparency 2.

ADDITIONAL EXAMPLE 1

a. Determine the coordinates of points *G*, *H*, and *J* on the graph.

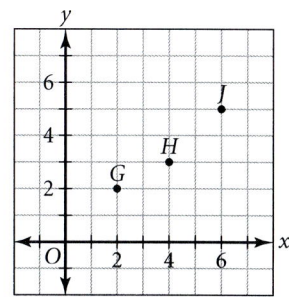

G(2, 2), *H*(4, 3), *J*(6, 5)

b. Graph each of the following points: *K*(1, 5), *L*(3, 2), *M*(7, 4).

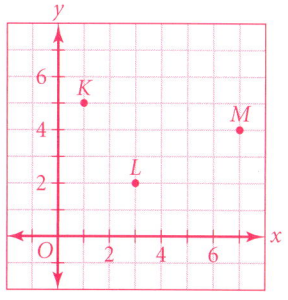

Use Teaching Transparency 3.

Math CONNECTION
COORDINATE GEOMETRY Students use a coordinate plane to determine the location of points and to plot points.

TRY THIS
a. *G*(7, 7), *H*(−8, 4), *J*(−10, −5)
b.

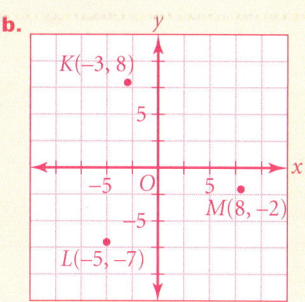

LESSON 1.4   **25**

Activity Notes

In this Activity, students explore how a graph can be used to model data from a real-world situation. Then they use the model to make predictions.

CHECKPOINT ✔

5. $d = 23t$

ADDITIONAL EXAMPLE 2

Make a table for the equation $y = x - 1$, and find values for y by substituting 1, 2, 3, 4, and 5 for x. Graph the ordered pairs on a coordinate plane, and connect the points to make a line.

x	1	2	3	4	5
y	0	1	2	3	4

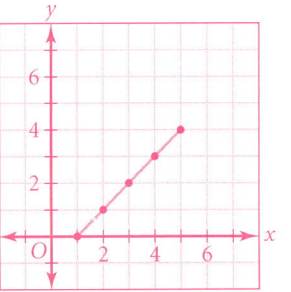

TRY THIS

x	1	2	3	4	5
y	3	6	9	12	15

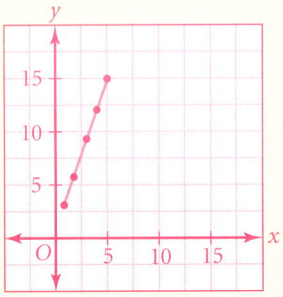

Activity: Graphing Data

APPLICATION: RECREATION

You will need: graph paper and a pencil

You can use the data in the table below to build a graph that shows the distance a bicyclist travels in a given period of time, not counting rest periods.

Time in hours	0.5	1.0	1.5	2.0	2.5
Distance in miles	11.5	23.0	?	?	?

1. Copy and complete the table. Assume that the bicyclist maintains a steady speed.
2. List the five ordered pairs in your table.
3. Graph the ordered pairs on a coordinate plane and connect the points with a line. The resulting graph models the data for the bicyclist.
4. If the bicyclist continues at the same speed, what distance is covered in 10 hours? How far does the bicyclist travel in 5 hours? in 9 hours?

CHECKPOINT ✔

5. Write an equation for the distance, d, the bicyclist travels in terms of time, t.

EXAMPLE 2

CONNECTION: COORDINATE GEOMETRY

Make a table for the equation $y = 2x$, and find the values for y by substituting 1, 2, 3, 4, and 5 for x. Graph the ordered pairs on a coordinate plane, and connect the points to make a line.

SOLUTION

Make a table to show the values of x and y.

x	1	2	3	4	5
y	2	4	6	8	10

Graph the ordered pairs on a coordinate plane, and connect the points to form a line.

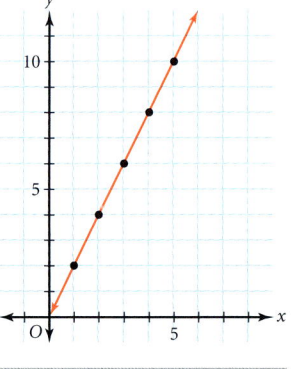

TRY THIS

Make a table for the equation $y = 3x$, and find the values for y by substituting 1, 2, 3, 4, and 5 for x. Graph the ordered pairs on a coordinate plane and connect the points to make a line.

Inclusion Strategies

VISUAL LEARNERS Have students use colored pencils to create their tables and graphs. They can use a different color to record each ordered pair in the table, then use the same color to graph the point on the coordinate plane. A regular pencil can then be used to write the equation and connect the points to form the line.

Reteaching the Lesson

USING DISCUSSION Ask students to think about situations that can be represented on a coordinate plane. You may need to get the discussion started by making a suggestion, such as the number of hours worked and the amount of money earned or the number of days worked and the total number of hours worked. Ask students how they would show the data for each situation in a table and then in a graph.

26 LESSON 1.4

Exercises

Communicate

1. Explain how you identify the coordinates of a given point on a graph.
2. Is (6, 7) the same point as (7, 6)? Explain.
3. Describe the steps for plotting the point with the coordinates (7, 3).
4. Discuss the advantages and disadvantages of displaying data in a table and in a graph.
5. Describe how to make a table of values for x and y for the equation $y = 2x + 5$.

Guided Skills Practice

Determine the quadrant and coordinates of the given points. *(EXAMPLE 1)*

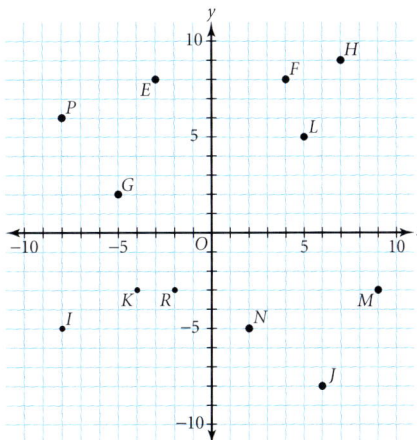

II, (−3, 8)   6. E   7. F   I, (4, 8)
II, (−5, 2)   8. G   9. H   I, (7, 9)
III, (−8, −5)   10. I   11. J   IV, (6, −8)
III, (−4, −3)   12. K   13. L   I, (5, 5)
IV, (9, −3)   14. M   15. N   IV, (2, −5)
III, (−2, −3)   16. R   17. P   II, (−8, 6)

Graph each point on a coordinate plane. *(EXAMPLE 1)*

18. L(4, 3)   19. M(1, 3)   20. N(5, 1)

Make a table for each equation and find values for y by substituting 1, 2, 3, 4, and 5 for x. Graph the ordered pairs, and connect the points to make a line. *(EXAMPLE 2)*

21. $y = 3x$ 22. $y = 2x + 1$ 23. $y = 3x - 2$
    3, 6, 9, 12, 15   3, 5, 7, 9, 11   1, 4, 7, 10, 13

Practice and Apply

Graph each list of ordered pairs. State whether they lie on a straight line.

24. (1, −3), (2, −6), (3, −9)   25. (1, 5), (2, −4), (3, 1)
26. (1, 10), (2, 7), (3, 2)   27. (0, 3), (2, 5), (4, 7)
28. (5, 2), (7, 2), (9, 2)   29. (−4, 1), (−4, 5), (−4, 9)

22.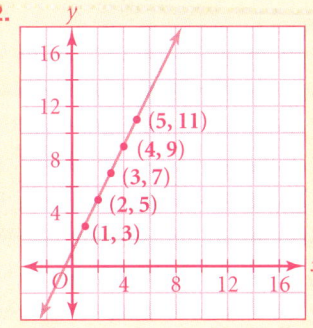

23.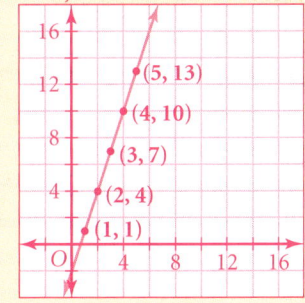

21.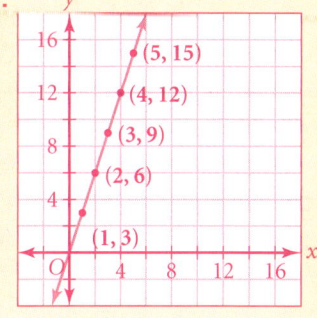

The answers to Exercises 24–29 can be found in Additional Answers beginning on page 810.

Math CONNECTION

COORDINATE GEOMETRY Students graph the equation of a line by creating a table and substituting values for x to find values for y.

Assess

Selected Answers
Exercises 6–23, 25–65 odd

ASSIGNMENT GUIDE

In Class	1–23
Core	24–43, 45–51 odd, 53–59
Core Plus	24–52 even, 53–60
Review	61–65
Preview	66, 67

✎ Extra Practice can be found beginning on page 738.

Answers to odd-numbered Extra Practice exercises can be found immediately after Selected Answers in the Pupil's Edition.

18–20.
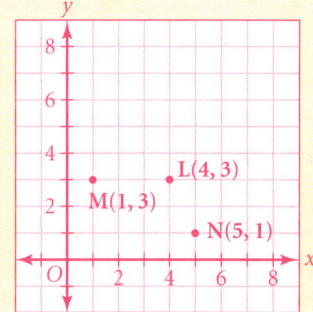

LESSON 1.4   **27**

Technology

A graphics calculator may be helpful for Exercises 44–52. You may wish to have students complete the exercises with paper and pencil first and then use the calculator to verify their answers.

Error Analysis

Students may incorrectly name or locate points because they confuse the order of the *x*- and *y*-coordinates. Encourage them to focus their attention on the labels for the axes, pointing out that the coordinates are named in alphabetical order. Students might also enjoy brainstorming to create a mnemonic device that they can use to remember the order. One possibility is HTV, which stands for Horizontal Then Vertical.

The answers to Exercises 55–60 can be found in Additional Answers beginning on page 810.

Determine the quadrant and coordinates of the given points.

30. *A* III, (−8, −4)	31. *B* I, (10, 7)
32. *C* II, (−9, 4)	33. *D* IV, (7, −8)
34. *E* III, (−6, −9)	35. *F* I, (4, 10)
36. *G* I, (5, 3)	37. *H* II, (−2, 5)
38. *R* IV, (10, −3)	39. *S* III, (−3, −5)
40. *T* II, (−4, 1)	41. *U* II, (−6, 8)
42. *V* I, (7, 9)	43. *W* IV, (4, −3)

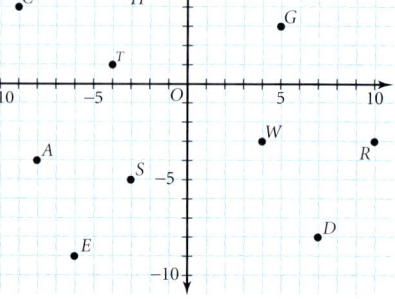

44. 4, 5, 6, 7, 8
45. −3, −2, −1, 0, 1
46. 2, 4, 6, 8, 10
47. 7, 9, 11, 13, 15
48. 1, 5, 9, 13, 17
49. 7, 12, 17, 22, 27
50. 2, 3, 4, 5, 6
51. 2, 9, 16, 23, 30
52. $1\frac{1}{4}, 1\frac{3}{4}, 2\frac{1}{4}, 2\frac{3}{4}, 3\frac{1}{4}$

Make a table for each equation, and find the values for *y* by substituting 1, 2, 3, 4, and 5 for *x*.

44. $y = x + 3$
45. $y = x - 4$
46. $y = 2x$
47. $y = 2x + 5$
48. $y = 4x - 3$
49. $y = 5x + 2$
50. $y = x + 1$
51. $y = 7x - 5$
52. $y = \frac{1}{2}x + \frac{3}{4}$

CONNECTIONS

COORDINATE GEOMETRY Graph and compare the following two equations by making a table and plotting points. Connect the points for each equation to make a line. What effect does the operation before the *x* have on the graph?

53. $y = 3 + x$
54. $y = 3 - x$

COORDINATE GEOMETRY Graph each of the following equations on the same set of coordinate axes by making tables, plotting points, and drawing lines. Then compare the graphs. Explain your conclusions.

55. $y = x + 7$
56. $y = x - 7$
57. $y = 7 - x$
58. $y = -7 - x$

APPLICATIONS

59. **BUSINESS** Suppose that a mail-order T-shirt company charges $8 per T-shirt plus a $3 handling charge per order.
   a. How much does an order of 2 shirts cost? **$19**
   b. How much does an order of 5 shirts cost? **$43**
   c. Make a set of ordered pairs from the given information, and plot them as points on a graph. Do the points lie on a straight line? **yes**
   d. Use your graph to find how many shirts you could order with $75. **9**

60. **RECREATION** Suppose that a bicyclist is riding at a steady speed. The table below represents the distance she travels in a given period of time. Copy and complete the table. List the ordered pairs from the table, and graph them on a coordinate plane. Use the graph to determine how far the bicyclist would travel in 8 hours. **196 miles**

3, 73.5
4, 98
5, 122.5

Time (hours)	1	2	3	4	5
Distance (miles)	24.5	49	?	?	?

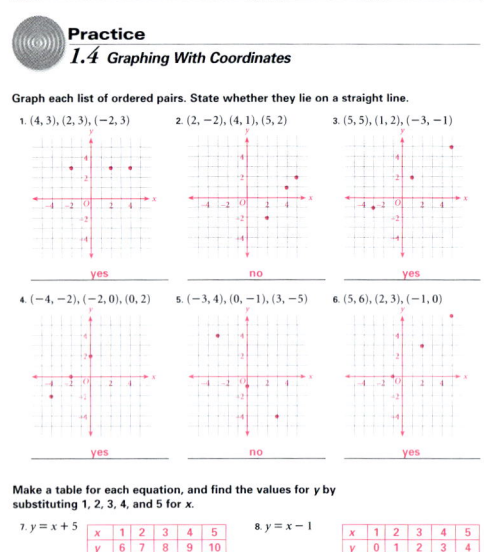

53.
x	1	2	3	4	5
y	4	5	6	7	8

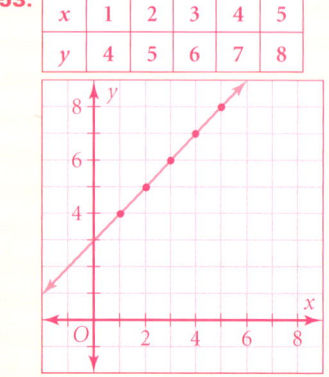

54.
x	1	2	3	4	5
y	2	1	0	−1	−2

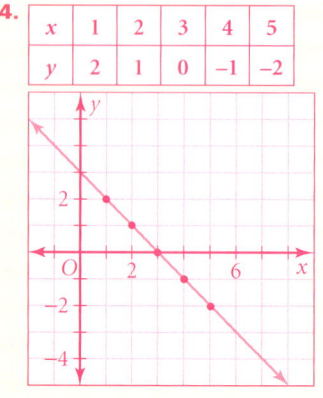

28 LESSON 1.4

Karl Friedrich Gauss (1777–1855) was one of the greatest of all mathematicians.

Look Back

61. Find the sum $1 + 2 + 3 + 4 + \cdots + 18$. **(LESSON 1.1)** **171**

62. Find the next three numbers in the following pattern: 43, 49, 55, 61, 67, … **(LESSON 1.1)** **73, 79, 85**

CULTURAL CONNECTION: EUROPE When Gauss was very young, a teacher told him to find the sum of all of the numbers from 1 to 100, thinking this would keep him very busy for a long time. But Gauss very cleverly paired the numbers and observed that $1 + 99 = 100$, $2 + 98 = 100$, $3 + 97 = 100$, and so on. **(LESSON 1.1)**

63. How many pairs of numbers could be formed this way without repeating any numbers? **49 pairs**

64. Which whole numbers from 1 to 100 would not appear in any of the pairs? **50 and 100**

65. Based on the previous two exercises, what is the sum of the numbers from 1 to 100? **5050**

Look Beyond

66. Plot the data points from the table below on a coordinate plane. Use a straightedge to draw the line that best fits the points on the scatter plot.

x	0	3	6	9	12
y	14	20	26	32	38

67. What is the relationship of the points to the line? **The points lie on the line.**

Include this activity in your portfolio.

Leonard Fibonacci was a famous mathematician who lived in the Middle Ages. He is famous for a special sequence of numbers now called the Fibonacci sequence. The sequence starts as follows:

$$1, 1, 2, 3, 5, 8, \ldots$$

This sequence of numbers appears in many places in nature. Pine cones usually have 5 spiral arms in one direction and 8 in the other. The sneezewort plant has 1 main stem with 2 branches, 3 branches, 5 branches, and so on. Sunflowers have 5 spirals in one direction and 8 in the other in their seed head.

Find the next four Fibonacci numbers in the sequence. Copy and complete the table below.

Term number	1	2	3	4	5	6	7	8	9	10
Fibonacci number	1	1	2	3	5	8	?	?	?	?

Describe the pattern that you discover in the sequence.

66.

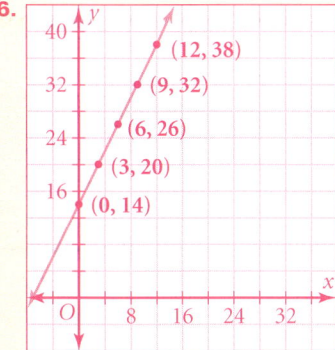

internetconnect

GO TO: go.hrw.com
KEYWORD: MA1 Fibonacci

To support the Portfolio Activity, students may visit the HRW Web site, where they can obtain additional information on Leonard Fibonacci and explore other occurrences of the Fibonacci sequence in nature.

LESSON 1.4   29

Prepare

NCTM Principles & Standards
1–3, 5, 6, 8–10

LESSON RESOURCES

- Practice 1.5
- Reteaching Master 1.5
- Enrichment Master 1.5
- Cooperative-Learning Activity 1.5
- Lesson Activity 1.5
- Problem Solving/Critical Thinking Master 1.5

QUICK WARM-UP

Find the next three terms of each sequence.

1. 11, 17, 23, 29, 35, . . .
   41, 47, 53

2. 76, 68, 60, 52, 44, . . .
   36, 28, 20

3. 1, 5, 13, 29, 61, . . .
   125, 253, 509

Find the values for y by substituting 2, 4, 6, 8, and 10 for x.

4. $y = 2x - 5$
   −1, 3, 7, 11, 15

5. $y = 3x + 1$
   7, 13, 19, 25, 31

Also on Quiz Transparency 1.5

Teach

Why Write the words *dependent* and *independent* on the board or overhead. Ask students to suggest meanings for each word. Have students give examples of real-world situations in which something is dependent or independent. For example, infants are dependent on their parents.

30 LESSON 1.5

1.5 Representing Linear Patterns

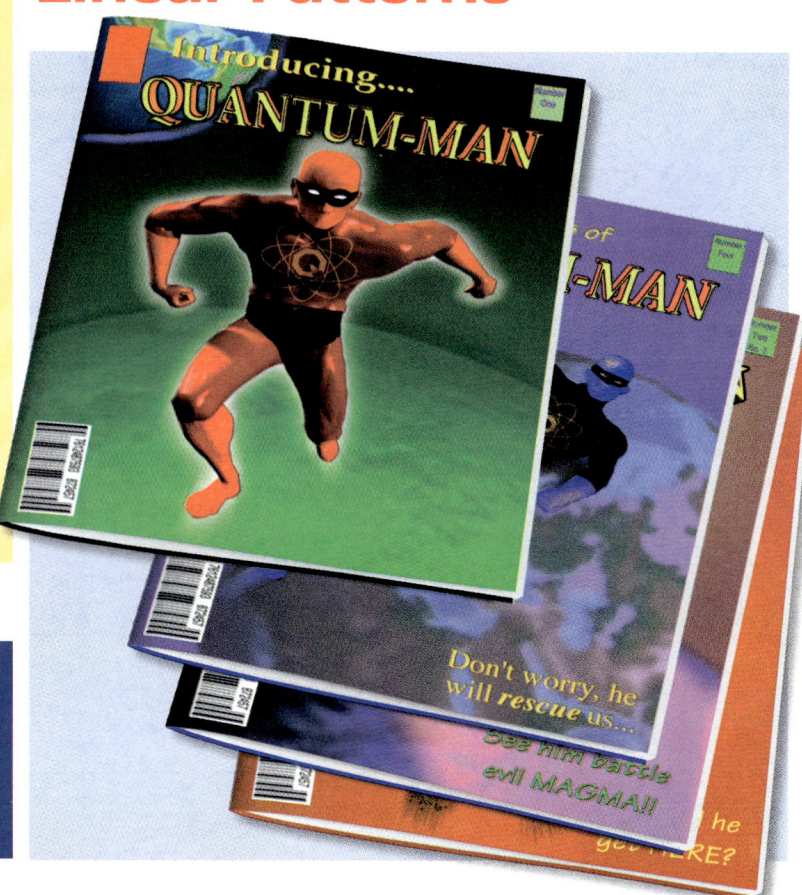

Objectives

- Represent linear patterns with equations.
- Represent linear equations with graphs.

Why Many situations, like the cost of membership in a comic book club, follow linear patterns. Linear equations can help you calculate solutions for problems that involve these situations.

Generalizing Data Patterns With Equations

The Acme Comix Club charges a $14 membership fee and $4 per comic book purchased. The table below represents this situation.

Number of comic books purchased	0	1	2	3	4	5
Total cost ($)	14	18	22	26	30	34

First differences   4   4   4   4   4

Since the data for total cost show constant first differences, the pattern is linear. The equation $c = 4b + 14$ represents this situation, where c represents the total cost and b represents the number of comic books purchased. Notice that the number 4 in the equation is the same as the first differences, and represents the price per comic book. The number 14 in the equation represents the initial cost to join the club.

CRITICAL THINKING Explain how the first differences are used to write an equation for total cost.

Alternative Teaching Strategy

USING COGNITIVE STRATEGIES Display the graph below on the board or overhead. Ask students to pick points on the graph and to write the coordinates in a table similar to the one at far right. Encourage students to list the x-values in ascending order.

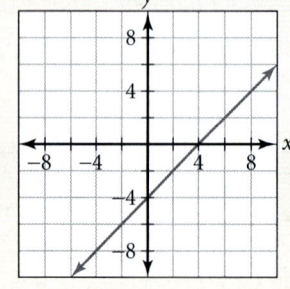

x	0	1	2	3	4	5	6
y	−4	−3	−2	−1	0	1	2

Lead students to find a pattern in the data from the table. Show students how to write an equation based on the pattern. Tell students that because the differences between the y-values in the table are 1 and because y is −4 when x is 0, the equation is $y = x - 4$. Explain to students that if the differences in the y-values were 2 and y is still −4 when x is 0, the equation would be $y = 2x - 4$.

EXAMPLE 1

APPLICATION
CONSUMER ECONOMICS

The costs associated with being a member of RR Comic Book Club are presented in the table below. **Find the first differences and write an equation to represent the data pattern.**

Number of comic books purchased	0	1	2	3	4	5
Total cost ($)	18	21	24	27	30	33

- SOLUTION

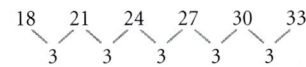

The first differences are 3. Therefore, the club charges $3 per book.

The cost for 0 books is $18. Therefore, the membership fee is $18.
The relationship is *total cost* = 3 • *number of comic books* + 18.

Use c for total cost and b for the number of comic books to write the equation.

$$c = 3b + 18$$

TRY THIS The table shows the costs associated with an audio CD club that charges a membership fee. Write an equation to represent the pattern in the data.

Number of CDs purchased	0	1	2	3	4	5
Total cost ($)	21	31	41	51	61	71

Activity
Modeling an Equation With a Graph

APPLICATION
TRAVEL

You will need: a graphics calculator or graph paper

A passenger ferry from Seattle, Washington, to Victoria, British Columbia, travels at a rate of 53 miles per hour. The distance the ferry travels in a given period of time can be represented by the following:

$$\text{distance} = \text{rate} \cdot \text{time}$$
$$d = 53t$$

1. Complete the table of values for the equation $d = 53t$.

Time in hours	0.5	1.0	1.5	2.0	2.5
Distance in miles	26.5	53.0	?	?	?

2. Graph each ordered pair from the table on a coordinate plane with time, t, on the horizontal axis and distance, d, on the vertical axis.

3. Connect the points. Describe the shape that is formed by connecting the points.

CHECKPOINT ✓ 4. Use the graph to predict how far the ferry would travel in 3 hours and in 4 hours.

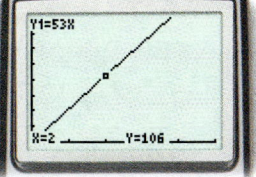

Use the graph and the trace feature on a graphics calculator.

Interdisciplinary Connection

BIOLOGY The heart-rate formula given in Example 3 is just one of many algebraic formulas that are used by exercise physiologists. If possible, invite an exercise physiologist to speak to your class. Ask students to research additional uses of algebra in this profession.

Inclusion Strategies

VISUAL STRATEGIES When creating a table of values for x and y, some students may benefit from an expanded table. For example, a table for $y = 5x - 1$, could be set up as shown below. Setting up the table in this way also gets students accustomed to seeing x and y in the same order as an ordered pair (x, y).

x	$y = 5x - 1$	y
1	$y = 5(1) - 1$	4
2	$y = 5(2) - 1$	9
3	$y = 5(3) - 1$	14

CRITICAL THINKING
The first differences, if constant, give the rate of change of the variable.

ADDITIONAL EXAMPLE 1

The costs associated with being a member of a CD club are presented in the table below. **Find the first differences and write an equation to represent the data pattern.**

Number of CDs purchased	Total cost ($)
0	25
1	34
2	43
3	52
4	61
5	70

The first differences are 9. The cost for zero CDs is $25. The equation is $t = 9c + 25$, where t is the total cost and c is the number of CDs.

TRY THIS
$c = 10n + 21$

Activity Notes

In this Activity, students use a formula for distance to complete a table and then graph the data from the table. Students use the graph to predict distances based on the time elapsed.

CHECKPOINT ✓
4. 159 miles in 3 hours; 212 miles in 4 hours

See Keystroke Guide, page 782.

LESSON 1.5 **31**

ADDITIONAL EXAMPLE 2

Jeremy drives at an average rate of 50 mph. **Write a linear equation to express the distance he travels in a given amount of time. Graph the equation for times from 0 hours to 10 hours.**

$d = 50t$, where d represents the distance in miles and t represents the time in hours

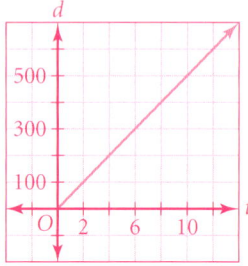

See Keystroke Guide, page 782.

Teaching Tip

TECHNOLOGY Be sure that students understand the special terms associated with graphs on their graphics calculators. On the TI-82 and TI-83 graphics calculators, pressing WINDOW displays the screen called the window editor. The numbers entered on this screen for **Xmin**, **Xmax**, **Ymin**, and **Ymax** determine the portion of the coordinate plane that the calculator displays. This is called the *viewing window*. The numbers entered for **Xscl** and **Yscl** (X scale and Y scale) determine the number of units represented by each tick mark on the x- and y-axes. To omit the tick marks, use **Xscl=0** and **Yscl=0**.

Linear Equations and Graphs

When you connect the points plotted in the previous activity, the graph of the equation $d = 53t$ is a straight line. For this reason, the equation is called a **linear equation**.

Drawing a line to connect a series of points does not automatically make every point on the line a solution to the problem. For example, it would make no sense to solve a distance problem for a negative time value, although you could certainly extend the line to include some negative values. As you solve problems, you must decide which kinds of answers are reasonable for the situation you are representing.

In the equation $d = 53t$, t is called the **independent variable**, and d is called the **dependent variable**. The value of d depends on the value chosen for t.

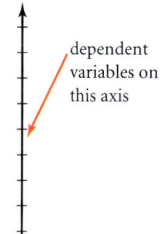

On a graph, the independent variable is represented on the horizontal axis and the dependent variable is represented on the vertical axis.

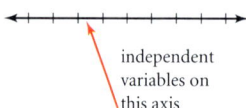

EXAMPLE 2

APPLICATION
TRAVEL

A jet airplane is flying at a rate of 280 miles per hour. **Write a linear equation to express the distance traveled in a given amount of time. Graph the equation for times from 0 to 10 hours.**

● **SOLUTION**

The distance traveled can be represented by the following relationship:
$$\text{distance} = \text{rate} \cdot \text{time}$$

The rate is 280 miles per hour. Let d represent the distance in miles and t represent the time in hours. The linear equation that expresses the distance traveled is as follows:
$$d = 280t$$

To find ordered pairs for the graph, make a table. Since t is the independent variable, choose values of t. Use the equation to determine the resulting values of d.

Time (hours), t	0	2	5	10
Distance (miles), d	0	560	1400	2800

Graph the ordered pairs. Let the horizontal axis represent the independent variable, t, and let the vertical axis represent the dependent variable, d. Connect the points to make a line.

Enrichment

Extend the study to include nonlinear graphs. Refer students to the rocket problem presented in Example 2 of Lesson 1.1. Tell them that the height of the rocket at a given time can be modeled by the *quadratic* equation $y = 224x - 16x^2$. Have students graph this equation on their graphics calculator with the following settings for the viewing window:

Xmin=0   Ymin=0
Xmax=20   Ymax=800
Xscl=1   Yscl=50

Have students use the trace feature of the calculator to identify several points on the graph. Ask them to explain what the coordinates of each point represent.

Have students compare their graph of the rocket data with the graph of the ferry data in the Activity on page 31. Ask them to identify how the graphs are alike and how they are different.

32 LESSON 1.5

You can also graph the equation on a graphics calculator. Remember that some calculators use **Y1** to represent the dependent variable, distance, and **X** to represent the independent variable, time.

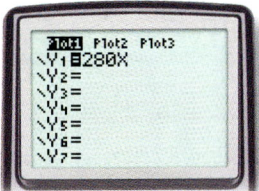
Enter **280X** after **Y1=**.

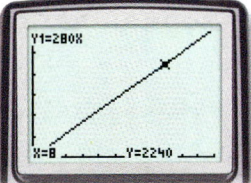
Set the window of the calculator as shown.

Press GRAPH.

TRY THIS A passenger train averages 54 mph. Write a linear equation that represents the distance traveled in a given time. Graph the equation for times from 0 to 15 hours.

TRY THIS

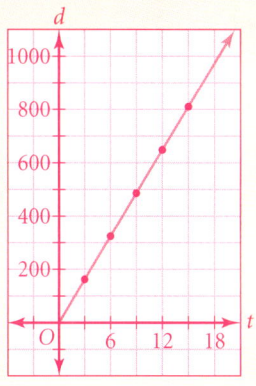

$d = 54t$

EXAMPLE 3

APPLICATION
HEALTH

Exercise physiologists suggest that a reasonable estimate for the maximum heart rate during exercise is no more than 220 beats per minute minus the person's age. **Represent the maximum heart rate by r and the age by a. Write a linear equation to express the maximum heart rate, r, in terms of age, a. Graph the equation for ages from 10 to 50.**

• **SOLUTION**

Since the rate, r, equals 220 beats per minute minus the age, a, write the equation $r = 220 - a$. This equation is a linear equation. The independent variable is a and the dependent variable is r.

Use the equation $r = 220 - a$ to create a table for the following values of a: 10, 20, 30, 40, and 50.

a	10	20	30	40	50
r	210	200	190	180	170

Let the horizontal axis represent age, a, and the vertical axis represent heart rate, r. Graph each ordered pair, and connect the points to draw a line.

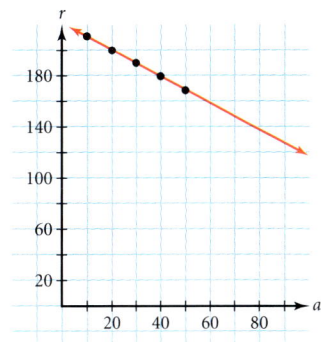

To graph the equation on a graphics calculator, enter **220 – X** after **Y1=** on a graphics calculator. Set the limits of the domain and range in the range window. Then press GRAPH.

Note: For some people the proper level of exercise should be lower. You should consult a doctor if there is any doubt about your exercise routine.

CRITICAL THINKING In Example 3, what values of a are reasonable for the problem?

ADDITIONAL EXAMPLE 3

During a flood, a river rose to a level that was 18 inches above normal. Then the waters started to recede at a rate of 1 inch per hour. Represent the number of hours that the water recedes by h and the river's level in inches above normal by l. **Write a linear equation to express the level, l, in terms of the hours, h. Graph the equation for times from 0 to 12 hours.** $l = 18 - h$

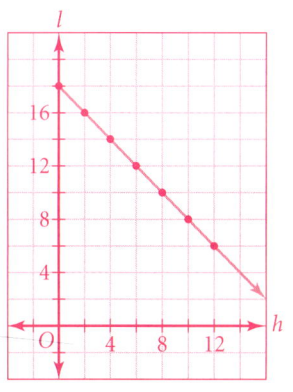

See Keystroke Guide, page 782.

CRITICAL THINKING
Answers may vary. Any value of a that makes sense as an age for a human doing exercise, say 2–100.

Reteaching the Lesson

COOPERATIVE LEARNING Have students work in groups of three. Give students the following exercise:

A train is traveling at a rate of 100 mph. Write a linear equation to express the distance traveled in a given amount of time. Create a table of values and graph the equation for times from 0 to 8 hours.

h, hours	0	1	2	3	4	5	6	7	8
d, miles	0	100	200	300	400	500	600	700	800

$d = 100h$

Assign one student to write an equation and create a table of values, one student to draw a graph, and the third student to use a graphics calculator to check the table of values and the graph.

Repeat the problem two times, rotating the roles of each group member and changing the rate of the train to 120 mph and 150 mph.

LESSON 1.5 **33**

Assess

Selected Answers

Exercises 5–7, 9–45 odd

ASSIGNMENT GUIDE

In Class	1–7
Core	8–22, 29–37 odd
Core Plus	8–28 even, 29–39
Review	40–46
Preview	47–51

✏ Extra Practice can be found beginning on page 738.

Answers to odd-numbered Extra Practice exercises can be found immediately after Selected Answers in the Pupil's Edition.

6. $d = 47t$

Time, t	0	1	2	3
Distance, d	0	47	94	141

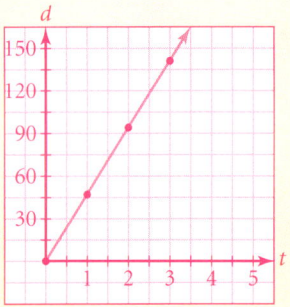

Practice

(Practice worksheet 1.5 Representing Linear Patterns — answers shown)

Find the first differences for each data set, and write an equation to represent the data pattern.

1.
0	1	2	3	4	5	6
13	20	27	34	41	48	55

First dif. 7, $y = 7x + 13$

2.
0	1	2	3	4	5	6
19	24	29	34	39	44	49

First dif. 5, $y = 5x + 19$

3.
0	1	2	3	4	5	6
0	4	8	12	16	20	24

First dif. 4, $y = 4x$

4.
0	1	2	3	4	5	6
4	23	42	61	80	99	118

First dif. 19, $y = 19x + 4$

5.
0	1	2	3	4	5	6
20	120	220	320	420	520	620

First dif. 100, $y = 100x + 20$

6.
0	1	2	3	4	5	6
396	362	328	294	260	226	192

First dif. –34, $y = 396 - 34x$

7.
x	1	2	3	4	5
y	40	80	120	160	200

8.
x	1	2	3	4	5
y	60	65	70	75	80

9.
x	1	2	3	4	5
y	33	52	71	90	109

10.
x	1	2	3	4	5
y	346	268	190	112	34

11.
x	1	2	3	4	5
y	23	28	33	38	43

12.
x	1	2	3	4	5
y	65	59	53	47	41

Make a table of values for each question below, using 1, 2, 3, 4, and 5 as values for x. Draw a graph for each equation by plotting points from your data set.

7. $y = 40x$ 8. $y = 55 + 5x$ 9. $y = 19x + 14$ 10. $y = 424 - 78x$
11. $y = 5x + 18$ 12. $y = 71 - 6x$ 13. $y = 175 - 20x$ 14. $y = 2000 - 250x$

Graphs for Exercises 7–14 are given in the answer key.

13.
x	1	2	3	4	5
y	155	135	115	95	75

14.
x	1	2	3	4	5
y	1750	1500	1250	1000	750

Suppose that the cost to order baseball tickets is $17 per ticket plus $2.50 handling charge per order (regardless of how many tickets are ordered).

15. How much does an order of 4 tickets cost? $70.50
16. How much does an order of 6 tickets cost? $104.50
17. Let t represent the number of tickets, and write an equation for the cost, c, of an order of tickets. $c = 17t + 2.50$

Exercises

Communicate

1. Explain what an independent variable is.
2. Explain what a dependent variable is.
3. Describe how first differences of number sequences relate to linear equations.
4. Describe the graph of a linear equation.

Guided Skills Practice

APPLICATIONS

5. TELECOMMUNICATIONS The costs associated with a particular cellular phone plan are shown below. Find the first differences and write an equation to represent the data pattern. *(EXAMPLE 1)*

Hours per month	0	1	2	3	4	5
Total cost ($)	25	37	49	61	73	85

First dif. 12
$c = 12h + 25$

6. TRANSPORTATION A hovercraft ferry crossing the English Channel averages 47 miles per hour. Write an equation, make a table, and draw a graph to show the distance the hovercraft travels for times from 0 to 3 hours. *(EXAMPLE 2)*

You can cross the English Channel by hovercraft in under 30 minutes.

7. CONTESTS The prize for a baking contest is $200. If a contestant spends $6 per batch on ingredients to practice a new recipe, write an equation to describe the profit, p, the contestant will make if he wins in terms of the number of practice batches, b, used to perfect the recipe. *(EXAMPLE 3)* $p = 200 - 6b$

Practice and Apply

Find the first differences for each data set, and write an equation to represent the data pattern.

First dif. 6, $y = 6x + 17$

8.
0	1	2	3	4	5	6
17	23	29	35	41	47	53

First dif. 6, $y = 6x + 25$

9.
0	1	2	3	4	5	6
25	31	37	43	49	55	61

First dif. 120, $y = 120x + 60$

10.
0	1	2	3	4	5	6
60	180	300	420	540	660	780

First dif. 3, $y = 3x$

11.
0	1	2	3	4	5	6
0	3	6	9	12	15	18

17.
x	1	2	3	4	5
y	19	26	33	40	47

18.
x	1	2	3	4	5
y	213	206	199	192	185

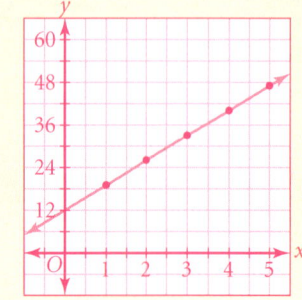

34   LESSON 1.5

12.

	0	1	2	3	4	5	6
	0	110	220	330	440	550	660

First dif. 110, $y = 110x$

13.

	0	1	2	3	4	5	6
	500	455	410	365	320	275	230

First dif. −45, $y = 500 − 45x$

14.

	0	1	2	3	4	5	6
	250	240	230	220	210	200	190

First dif. −10, $y = 250 − 10x$

15.

	0	1	2	3	4	5	6
	5	26	47	68	89	110	131

First dif. 21, $y = 21x + 5$

16.

	0	1	2	3	4	5	6
	201	181	161	141	121	101	81

First dif. −20, $y = 201 − 20x$

Make a table of values for each equation, using 1, 2, 3, 4, and 5 as values for *x*. Draw a graph for each equation by plotting points from your data set.

17. $y = 12 + 7x$ 18. $y = 220 − 7x$ 19. $y = 30x$ 20. $y = 12x$
21. $y = 300 − 2x$ 22. $y = 23 + 5x$ 23. $y = 15 + 7x$ 24. $y = 425 − 5x$
25. $y = 120 − 10x$ 26. $y = 65x$ 27. $y = 110x$ 28. $y = 53 + 11x$

CONNECTIONS

29. **COORDINATE GEOMETRY** Graph the data from the Acme Comix Club (page 30) and the RR Comic Book Club (Example 1, page 31) on the same coordinate grid. Connect the points with a line for each club. From the graph, find the point of intersection. What does this point represent?

30. **STATISTICS** Suppose that the average salary for a particular career begins at $30,000 for someone with no experience and increases by $2500 per year of experience.
    a. Complete a table that shows the average salary through 5 years of experience. Graph each ordered pair from the table.
    b. Use the graph to predict the average salary in the field after 10 years of experience. $55,000

APPLICATIONS

31. **TRAVEL** A paddleboat traveling on the Colorado River, in Austin, Texas, averages 14 miles per hour. Write an equation, make a table, and draw a graph to show the distance that the paddleboat travels for times from 0 to 5 hours.

32. **TRAVEL** A family plans a summer vacation to several national parks. They plan to average 58 miles per hour when they are traveling between the parks. Write an equation, make a table, and draw a graph to show the distance the family can travel for times from 0 to 8 hours. Use your graph to determine how far they could travel if they drove for 10 hours.

19.

x	1	2	3	4	5
y	30	60	90	120	150

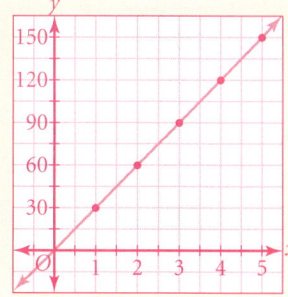

20.

x	1	2	3	4	5
y	12	24	36	48	60

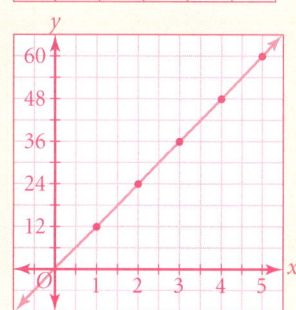

Exercises 47–50 extend the concepts explored in this lesson. Students must choose values for *x*, create a table, and graph the points to determine if the data are linear. Parabolas, such as the one they investigate in Exercise 51, will be studied in greater depth in Lesson 10.1.

33.

Number of CD's, n	1	2	3	4	5
Total cost including membership, c	37.25	48.50	59.75	71.00	82.25

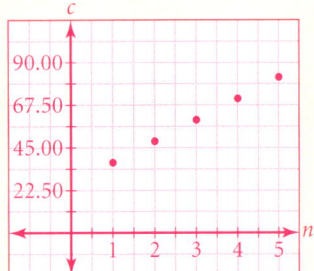

The equation is
$c = 11.25n + 26$.

34. a. distance = $3h$

b.

Hours	1	2	3	4	5
Distance	3	6	9	12	15

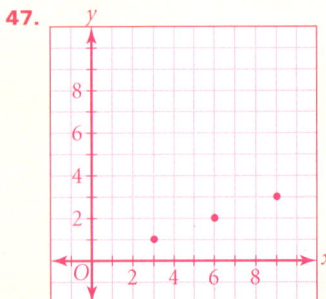

The graph is linear.

The answers to Exercises 49–51 can be found in Additional Answers beginning on page 810.

33. CONSUMER ECONOMICS Frank enjoys listening to music at work. The CD club that he is a member of charges $11.25 for each CD plus a $26.00 membership fee. Make a table of the costs for buying 1, 2, 3, 4, and 5 CDs from the club. Draw a graph and write an equation to represent the situation.

34. RECREATION Don walks at a rate of 3 miles per hour. You can determine the distance he walks by multiplying the rate by the number of hours he walks.
   a. Represent hours by *h*, and write an equation for the distance, *d*.
   b. Make a table and graph the equation for the distance that Don travels based on the number of hours that he walks.

35. HEALTH Using the equation given in Example 3, $r = 220 - a$, what should be the maximum heart rate for a 36-year-old person? **184 beats per minute**

36. ENTERTAINMENT How many $5 movie tickets can you buy with $32? **6**

37. FUND-RAISING How many $5 raffle tickets do you have to sell to raise $32? **7**

38. GIFTS If 5 people equally split the cost of a $32 birthday gift, how much should each pay? **$6.40**

CHALLENGE

39. Explain why the reasonable answers to Exercises 36, 37, and 38 are all different. **Exercises 36 and 37 deal with whole numbers of objects. Exercise 38 allows for a decimal answer.**

Look Back

40. What are the next three terms of the sequence 1, 1, 4, 10, 19, …? **(LESSON 1.1) 31, 46, 64**

41. How many differences are needed to reach a constant difference for the sequence 1, 2, 6, 15, 31, …? What is the constant difference for the sequence? **(LESSON 1.1) 3 differences, 2**

42. Make a table to solve the equation $7x + 5 = 40$. **(LESSON 1.2)** $x = 5$

Evaluate each expression. **(LESSON 1.3)**

43. $2 \cdot 14 \div 2 + 5$ **19**

44. $6 + 12 \div 6 - 4$ **4**

45. $4[(12 - 3) \cdot 2] \div 12$ **6**

46. $[3(4) - 6] - [(15 - 7) \div 4]$ **4**

Look Beyond

For each equation, make a table, plot the points, and tell whether the graph is linear.

47. $y = \frac{x}{3}$ **linear**

48. $y = \frac{3}{x}$ **not linear**

49. $x + y = 12$ **linear**

50. $x - y = 12$ **linear**

51. The graph of the path of a rocket is a curve called a parabola. The highest point of this curve is called the vertex. Graph the set of points below, and determine which point is the vertex. **(5, 25)**

x	0	1	2	3	4	5	6	7	8	9	10
y	0	9	16	21	24	25	24	21	16	9	0

47.

The graph is linear.

48.

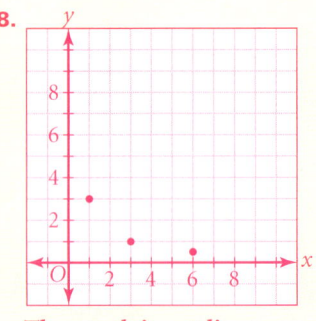

The graph is not linear.

1.6 Scatter Plots and Lines of Best Fit

Objectives
- Interpret data in a scatter plot.
- Find a line of best fit on a scatter plot by inspection.

Why Scatter plots provide a picture of the relationship between two sets of data, like those related to weather. Both scatter plots and lines of best fit can be used to predict relationships and future occurrences, such as tornado frequency.

APPLICATION
METEOROLOGY

Year	1950	1955	1960	1965	1970	1975	1980	1985	1990	1995
# of tornadoes	201	593	616	897	654	919	866	684	1133	1234

Scientists, such as meteorologists, often look for patterns and relationships between sets of numerical data. The frequency of tornadoes in a certain geographic area or during a specific period of time might show a pattern.

Tornadoes are devastating storms that can last from a few minutes to more than an hour and that often strike with little or no warning. Progress has been made through the study of tornadoes at places such as the Storm Prediction Center in Norman, Oklahoma. Although tornadoes cannot be prevented, advances are being made in tornado prediction. One area of interest is the number of tornadoes that have occurred during past years.

An effective way to see a relationship in data is to display the information as a scatter plot. A **scatter plot** shows how two variables relate to each other by showing how closely the data points cluster to a line.

To make a scatter plot, you need to
- have clearly defined variables that you can assign to the axes,
- form ordered pairs from your data, and
- locate each ordered pair on the scatter plot in the same way that you graph a point on a coordinate plane.

Alternative Teaching Strategy

TECHNOLOGY Have students use a graphics calculator to create a scatter plot and find the line of best fit for the tornado data in the lesson.

If you are using the TI-82 or TI-83 graphics calculator, press STAT to access the **STAT EDIT** menu. Choose **1:Edit...** Enter the years in **L1** and the corresponding numbers of tornadoes in **L2**.

Now press 2nd Y= to display the **STAT PLOTS** menu. Choose **1:Plot1...** On the screen that appears, select **On** and the scatter plot icon. Be sure that **L1** appears next to **Xlist:** and **L2** appears next to **Ylist:**. Select a type of mark for plotting the points. To display the scatter plot, press ZOOM **9:ZoomStat**.

To graph the line of best fit, use the following keystrokes:

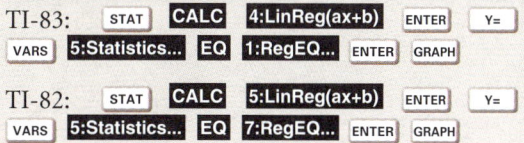

Prepare

NCTM PRINCIPLES & STANDARDS
1–3, 5–10

LESSON RESOURCES

- Practice 1.6
- Reteaching Master 1.6
- Enrichment Master 1.6
- Cooperative-Learning Activity 1.6
- Lesson Activity 1.6
- Problem Solving/Critical Thinking Master 1.6
- Teaching Transparency 4
- Teaching Transparency 5

QUICK WARM-UP

Graph each point on the same coordinate plane.

1. $A(15, 700)$
2. $B(10, 350)$
3. $C(25, 300)$

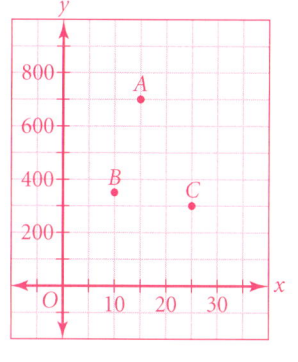

Also on Quiz Transparency 1.6

Teach

Why The relationship between two sets of data provides important information in many real-world situations. Have students make a list of situations in which they have seen two sets of data associated with one another. Ask them to describe any relationship between the sets of data.

LESSON 1.6   37

ADDITIONAL EXAMPLE 1

Match the scatter plots below of data related to characteristics of adults ages 25 to 75 with the statements.

A.
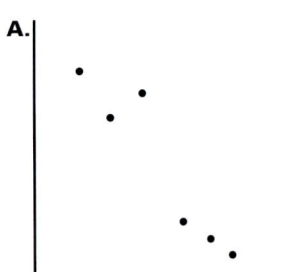

B.
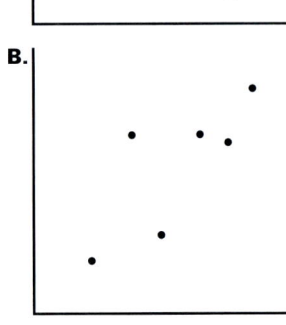

C.
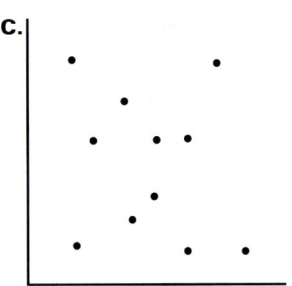

1. The variables studied are age and hair color. **C**
2. The variables studied are age and strength. **A**
3. The variables studied are age and weight. **B**

Use Teaching Transparency 4.

TRY THIS

Answers may vary. Sample answer: The variables *average daily temprature* and *number of people swimming* would likely show a strong positive correlation. The variables *average speed* and *traveling time* would likely show a strong negative correlation.

Correlation

Scatter plots provide a convenient way to determine whether a relationship exists between two variables. The relationship is called a **correlation**.

EXAMPLE 1 Compare the scatter plots of data related to characteristics of children from 0 to 18 years old with the statements below. **Match scatter plots a, b, and c with statements 1, 2, and 3.**

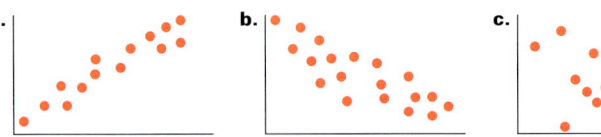

1. The variables studied are *age* and *eye color*.
2. The variables studied are *age* and *the time needed to run a fixed distance*.
3. The variables studied are *age* and *height*.

● **SOLUTION**

Scatter plot **a** shows a strong positive correlation. A **positive correlation** occurs when both variables increase. The correlation is strong because the data points are close to a straight line. As a child's age increases, his or her height increases, so this scatter plot would fit statement **3**.

Scatter plot **b** shows a strong negative correlation. A **negative correlation** occurs when one variable increases and the other variable decreases. As a child gets older and stronger, the time needed to run a fixed distance decreases, so this scatter plot would fit statement **2**.

Scatter plot **c** shows little or no correlation because the data points are randomly scattered. Eye color has no relation to age, so this scatter plot would fit statement **1**.

TRY THIS Think of two examples of related data. Describe the kind of correlation you expect from the data and why.

Activity
Working With Trends in Data

APPLICATION
ECOLOGY

TECHNOLOGY
GRAPHICS CALCULATOR

You will need: a graphics calculator, or graph paper and a pencil

Suppose that a certain wildlife reserve has an increase in its buffalo population. The data is shown below.

Year	0	2	4	6	8	10	12	14	16
Population	50	75	90	115	145	165	180	200	230
Differences		25	15	25	?	?	?	?	?

1. Use a graphics calculator or draw a scatter plot of the data.
2. Copy the table and complete the sequence of first differences.

Interdisciplinary Connection

SOCIAL SCIENCE Correlation statistics are widely used by social workers, psychologists, and psychiatrists. A trend in certain types of social behavior can be identified and analyzed by using scatter plots and lines of best fit. Have students research various sociological or psychological studies that have utilized these mathematical tools.

Enrichment

Have students research data about a subject that interests them and that can be represented by a scatter plot. Ask students to draw a scatter plot for their data with either graph paper or a graphics calculator and to describe any correlation they see.

3. A line of best fit is a line drawn on a scatter plot that shows the trend in the data. Looking at the pattern of differences, you might think that the population has increased by about 20 buffalo every 2 years or by about 10 every year. Begin trying to draw a line of best fit by using $y = 10x + 50$ as your equation.

4. Since that line is not the line of best fit, adjust the equation to better fit the data by using a guess-and-check strategy.

CHECKPOINT ✓ **5.** Use your equation to predict the population in the 10th year ($x = 10$). How close is it to the value that was given in the table? Use it to predict the population in the 20th year. You can also make a prediction by using the TABLE or TRACE feature of your graphics calculator.

Finding Lines of Best Fit

Scatter plots can be used to show trends in data. When the points in a scatter plot are represented by a trend line or a **line of best fit**, you can study the line to see how the data behave. You may have a basis to predict what the data might be for values not given.

EXAMPLE 2 Use the data from the Storm Prediction Center in Norman, Oklahoma, to find the line of best fit for the scatter plot of the number of storms in given years.

APPLICATION
METEOROLOGY

Describe the correlation shown in the scatter plot.

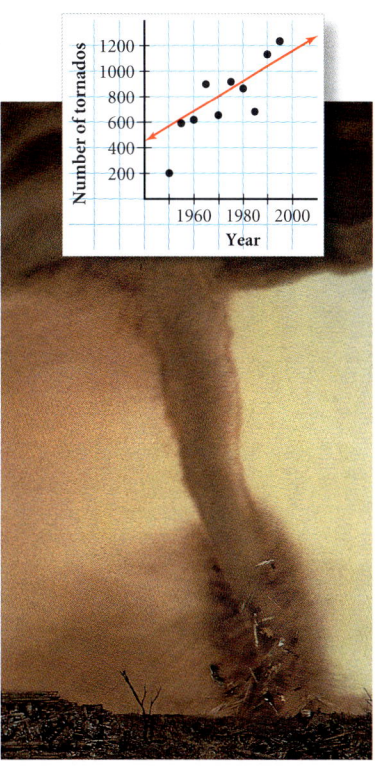

● **SOLUTION**

Plot the data as coordinates, (year, number of tornadoes), on a grid with the years on the horizontal axis and the number of tornadoes on the vertical axis.

Use the following steps to help you find the line of best fit:

- Use a straightedge, such as a clear ruler or a piece of uncooked spaghetti, to model the line.
- To fit the line to the points, choose your line so that it best matches the overall trend. The line does not have to pass through any of the points.

Since the data for both variables in the scatter plot are increasing, the correlation is positive.

Activity Notes

In Step 4 of the Activity, students may need help in guessing an equation that better fits the data. Starting with $y = 10x + 50$, they should graph a new equation with a different coefficient of x together with the scatter plot until they feel the line illustrates a better representation of the data. The regression equation is approximately $y = 11.04x + 50.56$.

CHECKPOINT ✓
5. Using the regression equation for the data, the prediction for the 10th year is ≈160, which is relatively close to the value in the table. The prediction for the 20th year is ≈270.

See Keystroke Guide, page 782.

ADDITIONAL EXAMPLE 2

Draw the line of best fit for the scatter plot.

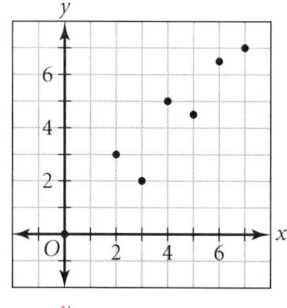

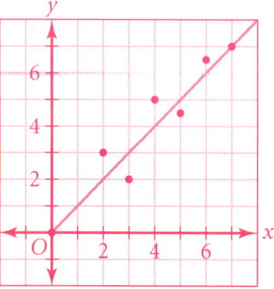

Use Teaching Transparency 5.

Inclusion Strategies

USING DISCUSSION Students who like to get exact answers may feel slightly uncomfortable with parts of this lesson. Discuss with students the fact that much of the data that occurs in the real world does not produce exact answers. Be sure students understand that scatter plots usually show a relationship between data sets that does not apply *exactly* to every pair of data values in the sets.

Reteaching the Lesson

USING COGNITIVE STRATEGIES Provide students with at least four graphs that show lines of best fit: two that show a positive correlation and two that show a negative correlation. Have each student create a scatter plot to go with each line. Then ask students how to draw a line of best fit for a data set that shows little or no correlation. Lead students to see that the line would be flat because a slant in either direction would indicate some correlation.

LESSON 1.6

TRY THIS

There may be 1400–1500 tornadoes reported.

CRITICAL THINKING

Answers may vary. Sample answer: The scales for the axes may be stretched or condensed, changing the steepness of the line. Steeper lines may indicate more dramatic relationships than lines that are less steep.

Assess

Selected Answers

Exercises 5–7, 9–23 odd

ASSIGNMENT GUIDE	
In Class	1–7
Core	8–18
Core Plus	8–18
Review	19–24
Preview	25, 26

✏ Extra Practice can be found beginning on page 738.

Answers to odd-numbered Extra Practice exercises can be found immediately after Selected Answers in the Pupil's Edition.

TRY THIS Use the line of best fit to predict how many tornadoes may be reported in the United States in 2015 if the trend continues.

There is usually only one line that *best* fits a set of data. A line of best fit should take all of the points into consideration.

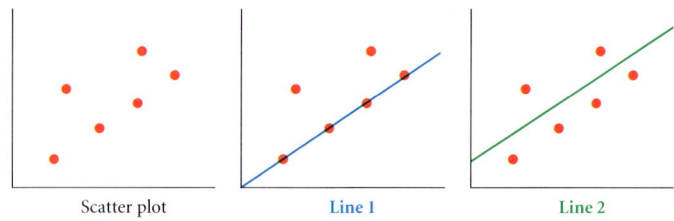

Scatter plot　　　Line 1　　　Line 2

Line 1 goes through several of the points, but it completely ignores the others.

Line 2 is closer to four points than it is to the other two, but it takes all of the points into account. Line 2 is a better fit, even though it doesn't go through any of the points.

CRITICAL THINKING Explain how the scales used for the axes of a scatter plot could affect the interpretation of the line of best fit.

Exercises

Communicate

1. Refer to the scatter plot in Example 2. Explain how to use the line of best fit to make a prediction.
2. Describe two data sets that have a strong negative correlation.
3. Describe two data sets that have a strong positive correlation.
4. Describe the correlation for the following scatter plots as strong positive, strong negative, or little to none. Explain the reason for your answer.

   a. 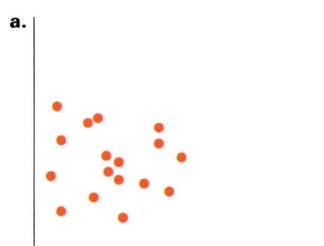　　b.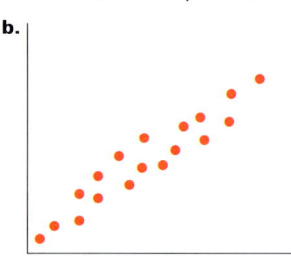

5.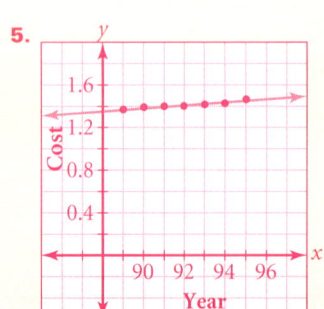

40　LESSON 1.6

● *Guided Skills Practice*

APPLICATION

ECONOMICS The table below shows the price of a half-gallon of whole milk from 1989 to 1995.

Year	1989	1990	1991	1992	1993	1994	1995
Cost	1.37	1.39	1.40	1.39	1.43	1.44	1.48

Source: Statistical Abstract of the U.S.

5. Plot the data in a scatter plot and draw a line of best fit. **(EXAMPLES 1 AND 2)**

6. What kind of correlation exists? **(EXAMPLE 1)** strong positive

7. Use the line of best fit you drew in Exercise 6 to estimate the cost of a half-gallon of whole milk in 1999. **(EXAMPLE 2)** approx. $1.51

● *Practice and Apply*

For each scatter plot, describe the correlation as strong positive, strong negative, or little to none.

8. little to none

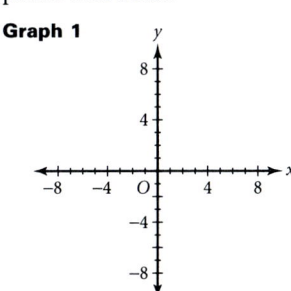

9. strong positive

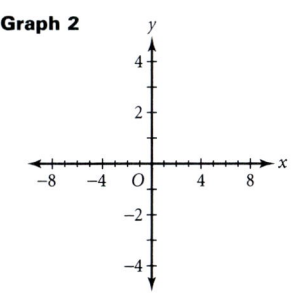

10. Draw the two coordinate axes shown below on your graph paper. Plot the same points, $(1, 2)$, $(2, 3)$, and $(3, 4)$, on each set of axes. Connect the points with a line.

Graph 1

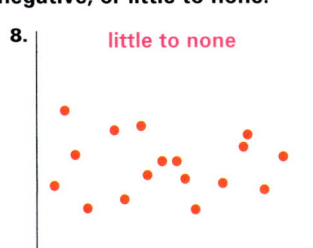

Graph 2

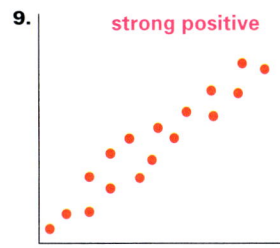

a. Which graph is steeper, graph 1 or graph 2?
b. What caused one graph to be steeper, even though the points are the same?
c. How does a change in scale on the axes affect the steepness of the line? (*Scale* refers to the distance between the units on an axis.)
d. Does the scale of the graph affect the correlation? Explain your answer.

10. a. Graph 2 is steeper than Graph 1.
b. The y-axis of Graph 2 ranges from −4 to 4, while all the other axes range from −8 to 8.
c. The scale on the y-axis of Graph 2 is "stretched out" compared to Graph 1. This makes the distance between the plotted points appear greater, and hence changes the steepness of the line connecting them.
d. No, the scale does not affect the correlation. The comparison of whether the variables are increasing or decreasing does not change when the scale changes.

Technology

Students can use a graphics calculator to create the scatter plots in Exercises 11–15 and 17.

Error Analysis

Watch for students who draw an inappropriate line of best fit because they believe the line must contain at least two of the points on the scatter plot. Stress the fact that the line of best fit does not necessarily pass through any of the points on the scatter plot.

internetconnect

GO TO: go.hrw.com
KEYWORD: MA1 Nose Count

In this activity, students examine a 10-year census and create scatter plots of various data. Then students make predictions and check their predictions against actual figures.

10. Graph 1

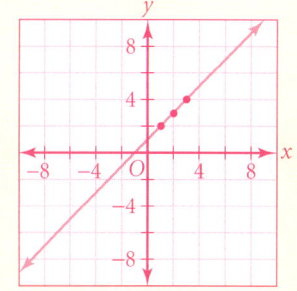

Graph 2

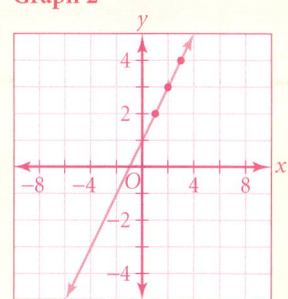

LESSON 1.6 **41**

11.

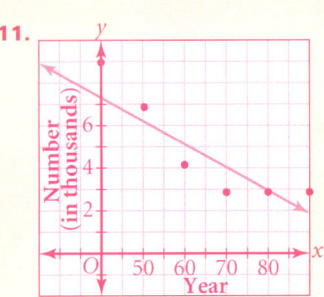

There is a strong negative correlation.

12.

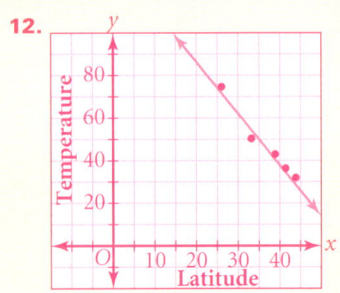

There is a strong negative correlation.

13.

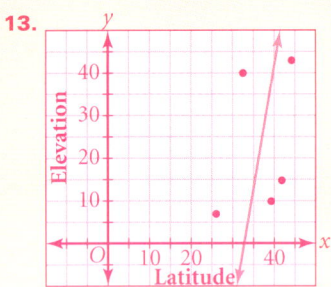

There is no correlation.

APPLICATIONS

11. DEMOGRAPHICS Make a scatter plot for the data below on the number of people working on farms in various years, and draw a line of best fit. Describe the correlation as strong positive, strong negative, or little to none.

Year	1940	1950	1960	1970	1980	1990
Number of farm workers in thousands	8995	6858	4132	2881	2818	2864

Source: *The World Almanac*, 1997

GEOGRAPHY Use the following data to answer Exercises 12–14:

City	North latitude	Elevation (in feet)	Maximum normal temp. for Jan. (in °F)
Miami, FL	26°	7	75
Charleston, SC	33°	40	50
Washington, DC	39°	10	43
Boston, MA	42°	15	36
Portland, ME	44°	43	31

12. Graph latitude and temperature, and draw a line of best fit. Describe the correlation.

13. Graph latitude and elevation, and draw a line of best fit. Describe the correlation.

14. Graph elevation and temperature, and draw a line of best fit. Describe the correlation.

15. SPORTS Make a scatter plot of the following data for the women's 400-meter run in the Olympics:

Year	1964	1968	1972	1976	1980	1984	1988	1992
Seconds	52.00	52.00	51.08	49.29	48.88	48.83	48.65	48.83

Source: *The World Almanac*, 1997

16. Describe the correlation for the scatter plot you made in Exercise 15.

APPLICATION

MUSIC As the length of a string such as a guitar string changes, the frequency, measured in vibrations per second (vps), changes according to data in the following table:

Length of string (cm)	40	60	80	100	120
Frequency (vps)	660	440	330	264	220

17. Make a scatter plot of the data.

18. As the length of the string increases, what happens to the frequency? Is the correlation between the length of a string and the frequency positive or negative?

18. The frequency decreases. There is a strong negative correlation between length and frequency.

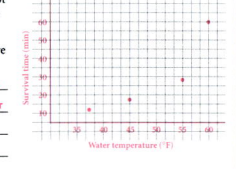

14.

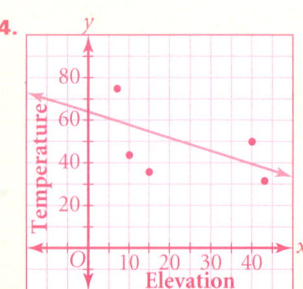

There is a slight negative correlation.

15.

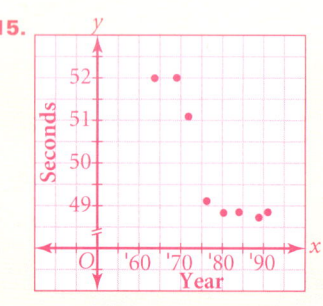

 **Look Back**

Perform the following calculations:

19. $\sqrt{144}$ 12
20. $\sqrt{7+2}$ 3
21. $\sqrt{170-1}$ 13
22. $2\sqrt{16}$ 8

23. What is the greatest common factor for 30, 60, and 15? 15

24. Pair the equivalent expressions, and explain the relationship between them.
    a. $2 \cdot 2 \cdot 2 \cdot 2$ b. $4(2)$ c. $2 + 2 + 2 + 2$ d. 2^4
    a and d, b and c

 Look Beyond

25. What values of x will make $2x^2 + 4 = 12$ true? $x = 2$ or $x = -2$
26. What values of n will make $42 - 2n^2 = 10$ true? $n = 4$ or $n = -4$

Include this activity in your portfolio.

Math test scores	60	40	80	20	65	55	100	90	85
Science test scores	70	35	90	50	65	40	95	85	90

The graph shows a median-fit line for a scatter plot that compares students' math and science test scores. The dashed red lines divide the data into three groups. The median points and median-fit line are indicated in blue.

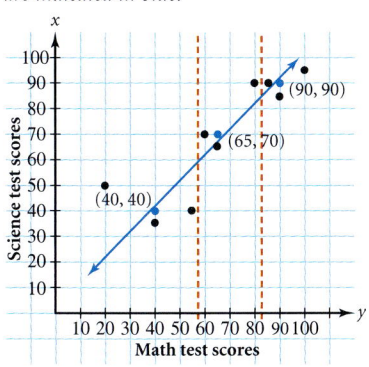

1. Find a set of real-world data similar to the tornado data in this lesson. You might look for information on the amount of precipitation in a particular region or on the occurrence of hurricanes, for example. Be sure that you have at least 15 points to plot.

2. Make a table and a scatter plot of your data. Describe the correlation.

3. Use two vertical lines to separate the data points on your scatter plot into 3 equal groups.

4. Within each group of data points, find the median of the x-coordinates and the median of the y-coordinates. The median is found by listing the numbers from least to greatest and picking the middle value. For example, if 1, 3, 5, 8, and 9 are the y-coordinates in one region, the median value is 5.

5. Use the median coordinates to create 3 ordered pairs, one for each region. For example, if the median x-coordinate for the first region is 5 and the median y-coordinate for the first region is 7, the ordered pair is (5, 7). Plot each of these points on your scatter plot.

6. Align a ruler with the outer two median points. Keeping the ruler parallel to the original position, slide the ruler one-third of the distance toward the inner median point. Draw the line. The result is a **median-fit line**, which is one type of *best-fit line*.

7. Make a prediction for a future year based on your data and the median-fit line.

WORKING ON THE CHAPTER PROJECT

You should now be able to complete Activities 3 and 4 of the Chapter Project on page 44.

16. There is a strong negative correlation between the year and the time.

17.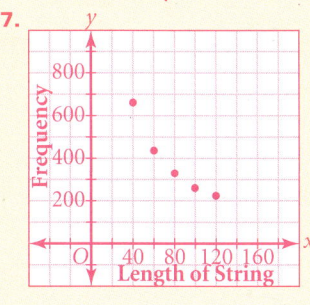

 **Look Beyond**

Exercises 25 and 26 require students to solve quadratic equations. This topic is covered in Chapter 10. At this point the students can solve the equations by using guess-and-check.

ALTERNATIVE
Assessment

Portfolio Activity

The Portfolio Activity can be used as preparation for the Chapter Project or as a separate activity. In this Portfolio Activity, students are to gather real-world information, make a table and scatter plot of the data, draw a median-fit line, and make a prediction for a future year based on the data. The use of a graphics calculator is recommended.

Answers to Portfolio Activities can be found in Additional Answers of the Teacher's Edition.

Student Technology Guide

NAME _____ CLASS _____ DATE _____

Student Technology Guide
1.6 Scatter Plots and Lines of Best Fit

A graphics calculator can be used to display a scatter plot. For example, the winning times for the women's 200-meter individual medley swimming event in the Olympics is shown in the table below.

Year	1968	1972	1984	1988	1992	1996
Time (sec)	144.7	143.07	132.64	132.59	131.59	133.93

To display the scatter plot, you first need to enter the data. Use keystrokes STAT EDIT ENTER and enter the years in list L1 and the time in list L2.

Then follow the steps shown below.

Choose the STAT PLOT menu by pressing 2nd Y=, and match the settings with the screen shown below.

Adjust the window values to see a good graph. A sample is shown below.

Press GRAPH. A sample graph is shown below. Add labels if the graph is copied onto paper.

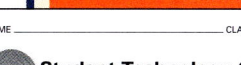

Make a scatter plot for each data set. Graphs are given in the answer key.

1. times for the men's 100-meter butterfly swimming event in the Olympics

Year	1976	1980	1984	1988	1992	1996
Seconds	54.35	54.92	53.08	53.0	52.32	52.27

Answers may vary.

2. the average number of hours of TV watched daily by a household

Year	1989	1990	1991	1992	1993	1994
Hours	6.92	6.93	7.1	7.3	7.4	7.3

Answers may vary.

3. the estimated number of on-line households in millions

Year	1995	1996	1997	1998	1999	2000
Number	15	23.4	34.0	45.2	56.7	66.6

Answers may vary.

4. the estimated world population in millions

Year	1950	1970	1990	1996	2020
Population	2556	3706	5282	5772	7600

Answers may vary.

Project

Focus
Throughout Chapter 1, students learn that algebra can be used to describe patterns. In this Chapter Project, they apply what they have learned to analyze patterns in several diverse areas, ranging from the golden rectangle of ancient times to the modern wave cheer.

Motivate
Before the students complete each activity, discuss it as a class.

For Activity 1, ask students if they have ever heard of the golden ratio. If so, ask them to share what they know about this ratio with the class. If possible, provide students with additional information about the use of golden rectangles by the ancient Greeks.

For Activity 2, ask students what they know about the history of the number π.

For Activity 3, ask students if they have ever done a wave cheer at a sporting event.

Activity 1
1. $\frac{5}{3} = 1.\overline{6}$, or approximately 1.667. This number is close to the golden ratio; it is different by about 5 one-hundredths.

Length	Width	Ratio
5	3	$\frac{5}{3} = 1.667$
8	5	$\frac{8}{5} = 1.6$
13	8	$\frac{13}{8} = 1.625$
21	13	$\frac{21}{13} = 1.615$
34	21	$\frac{34}{21} = 1.619$
55	34	$\frac{55}{34} = 1.618$

Sketch rectangles with these dimensions

As the Fibonacci numbers get larger, in centimeters or millimeters, the ratio $\frac{length}{width}$ gets closer to the golden ratio.

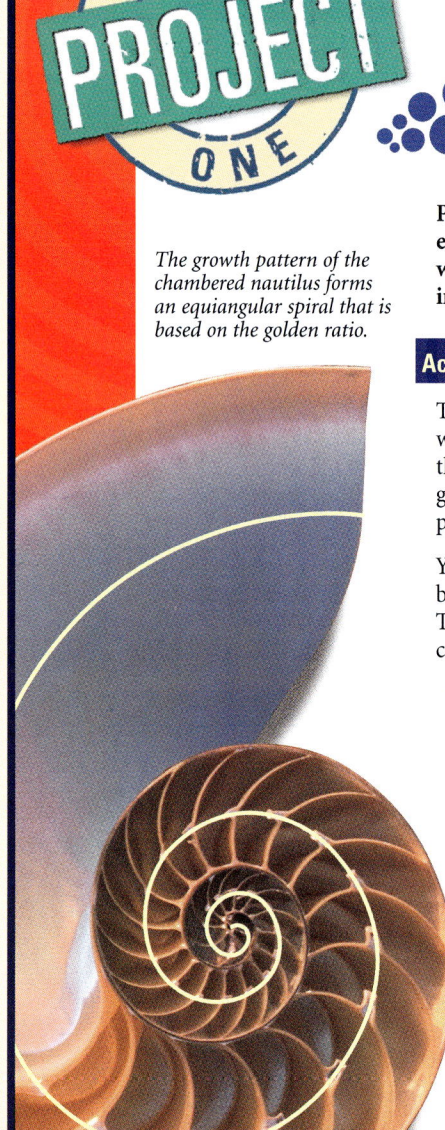

The growth pattern of the chambered nautilus forms an equiangular spiral that is based on the golden ratio.

CHAPTER PROJECT ONE

WORKING WITH Patterns in Data

People have been working with patterns to solve problems from the earliest evidence of mathematical thought. In this project, you will work with patterns to solve a variety of problems and to gain insight into the world of algebra.

Activity 1

The ancient Greeks were interested in constructing golden rectangles, which were considered pleasing to look at. The ratio of the length to the width of these rectangles (approximately 1.618) was called the golden ratio. Psychologists have found that consumers subconsciously prefer rectangular products that are close to golden rectangles.

You can use consecutive Fibonacci numbers to build rectangles that are close to golden rectangles. The rectangle at right is built by using the consecutive Fibonacci numbers 3 and 5.

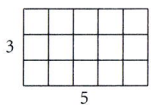

1. Use a calculator to write the ratio $\frac{5}{3}$ as a decimal by dividing, $5 \div 3$. How close is the ratio $\frac{5}{3}$ to the golden ratio?

   Use consecutive pairs of Fibonacci numbers to construct "almost" golden rectangles.

Length	Width	Ratio
5	3	$\frac{5}{3} = 1.667$
8	5	
13	8	
?	13	
?	?	

   - Complete the table by using consecutive Fibonacci numbers for width and length through a length of 55.
   - Sketch a rectangle with the length and width in centimeters or millimeters.
   - Compute the ratio $\frac{length}{width}$ as a decimal and compare it with the golden ratio.

2. Which of the following closely resemble golden rectangles?
   a. 3×5 index cards **b.** 5×7 photograph
   c. 5×8 photograph **d.** 8×10 photograph

3. Do you think that golden rectangles are the most pleasing to the eye?

4. Find five rectangles in the real world that have ratios close to the golden ratio.

2. a, c

3. Answers may vary.

4. Answers may vary. Find examples similar to choices in Exercise 2.

44 CHAPTER 1 PROJECT

Activity 2

1. Work with your class to find 20 circular objects. Measure their circumference, *c*, and diameter, *d*, and build a table of values.
2. For each circular object, compute the ratio $\frac{c}{d}$. Average the ratios.
3. Make a scatter plot of the data.
4. How do the number π and the equation $c = \pi d$ relate to this activity?

Activity 3

Simulate a wave cheer with members of your class.

1. To simulate the cheer, each person stands and sits in succession. Record the time it takes from start to finish and enter it in the table. (If there are not enough people for a wave, go back to the first person and repeat the wave until you reach the required number.)

People	4	8	12	16	20	P
Time	?	?	?	?	?	T = ?

2. Make a scatter plot of the data on graph paper. Draw a line of best fit for the data.
3. Use the line of best fit to answer these questions.
   a. How long would it take 60 people to complete the wave?
   b. How many people would be needed to complete the wave in 2 minutes?

Cooperative Learning

For **Activity 1**, have students work in groups of three. One student in each group can sketch and label the rectangles, another can use a calculator to compute the decimal equivalent for the ratio, and the third student can record the information in the table. All students in each group should work together to answer the last three questions.

Begin **Activity 2** with students working together as a class. Have each of 20 students contribute a set of measurements for Step 1. Then have students work in pairs to complete Steps 2–4.

For **Activity 3**, assign one student the task of keeping the time and another the task of recording the data. The rest of the class can participate in the wave. Then have students work in groups of two or three to complete Steps 2 and 3.

Discuss

After students have completed each activity, have them share their results with the class. Be sure students are satisfied with the results of one activity before moving on to the next. After they have completed all three activities, ask students to share what they have learned about the relationship between patterns and algebra.

Activity 2

1. Examples may include soup cans, dinner plates, bicycle tires, etc.
2. After computing each ratio, add the ratios and divide the total by the number of ratios, which should be 20.
3. A scatter plot can be produced from the data either by hand or by using a graphics calculator. Use the diameter, *d*, as the independent variable and the circumference, *c*, as the dependent variable.
4. The average calculated in Exercise 2 should be approximately equal to π. The line of best fit for the scatter plot in Exercise 3 should be approximately equal to the line whose equation is $y = \pi x$.

Activity 3

1. Students should perform the activity to gather data for the table. The last entry will be an approximate equation based on the pattern in the data.
2. The line of best fit should be approximately equal to the graph of the equation determined in Exercise 1.
3. a. Calculate the value using the equation for the line of best fit.
   b. Use the guess-and-check method to find the best value.

CHAPTER 1 PROJECT **45**

Chapter Review

Check students' graphs.
21. Yes, the points lie on a straight line.
22. No, the points do not lie on a straight line.
23. No, the points do not lie on a straight line.

Chapter Test, Form A

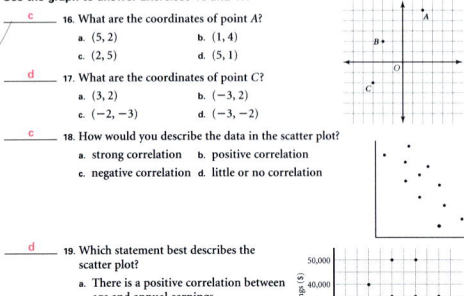

Chapter Review and Assessment

VOCABULARY

algebraic order of operations 19	inclusion symbols 19	positive correlation 38
algebraic expression 11	independent variable 32	quadrant 24
conjecture 5	line of best fit 39	scatter plot 37
constant difference 5	linear equation 32	solution 13
coordinates 24	median-fit line 43	variable 11
coordinate plane 24	negative correlation 38	x-axis 24
correlation 38	number sequence 5	y-axis 24
dependent variable 32	ordered pair 25	
equation 12	origin 25	

Key Skills & Exercises

LESSON 1.1

Key Skills

Extend a sequence by working backward from the differences.

Apply the method of differences until you get a constant. Then use the differences to find the next two terms.

10, 13, 19, 28, 40, …

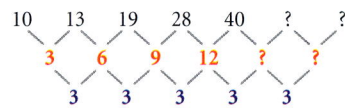

Work backward to find the next two terms.
40 + 15 = 55   55 + 18 = 73
The next two terms are 55 and 73.

LESSON 1.2

Key Skills

Evaluate expressions for given values of the variable.

Evaluate $2x + 3$ for $x = 2$ and $x = 3$.

$2(2) + 3$ $2(3) + 3$
$4 + 3$ $6 + 3$
7 9

Write an equation to represent a problem situation.

Write an equation for the cost of renting a car, using the variables c and d.
  The cost is $100 plus $20 per day.
    $c = 100 + 20d$

Exercises

Find the next two terms of each sequence.

1. 1, 5, 11, 19, 29, … **41, 55**
2. 1, 1, 6, 16, 31, … **51, 76**
3. 90, 70, 54, 42, 34, … **30, 30**
4. 1, 4, 7, 10, … **13, 16**
5. 8, 16, 26, 38, … **52, 68**
6. 100, 99, 97, 94, … **90, 85**
7. The fourth and fifth terms of a sequence are 21 and 32. If the second differences are a constant 3, what are the first five terms of the sequence? **6, 8, 13, 21, 32**

Exercises

Evaluate each expression when x is 1, 2, 3, and 4.

8. $3x$ **3, 6, 9, 12**
9. $5x + 2$ **7, 12, 17, 22**
10. $10 - x$ **0, 0, 7, 6**
11. $2x + 4$ **6, 8, 10, 12**
12. $x + 7$ **8, 9, 10, 11**
13. $4x - 5$ **−1, 3, 7, 11**
14. Mr. O'Neil is counting algebra books in the school bookroom. He counts 50 crates plus 7 additional books. Write an equation for this situation in terms of the total number of books, t, and the number of books in a crate, c. **$t = 50c + 7$**

24.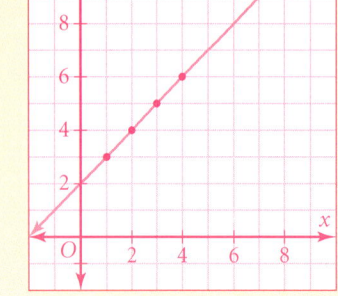

25.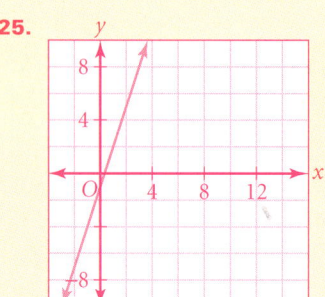

46   CHAPTER 1 REVIEW

LESSON 1.3
Key Skills

Use the algebraic order of operations to find the value of expressions.

To evaluate the expression $5(7 - 4) - 6^2 \div 3 + 1$, follow the algebraic order of operations: grouping symbols, exponents, multiplication/division, and addition/subtraction.

$$5(7 - 4) - 6^2 \div 3 + 1$$
$$5(3) - 6^2 \div 3 + 1$$
$$5(3) - 36 \div 3 + 1$$
$$15 - 12 + 1$$
$$4$$

Exercises

Evaluate.

15. $17 - 4 \cdot 3$ **5**
16. $32 - 24 \div 6 - 4$ **24**
17. $3 \cdot 4^2 - [24 \div (6 - 4)]$ **36**
18. $3(4 + 7 \cdot 2) - 100 \div 50$ **52**
19. $2[3(4 \cdot 5 + 2) - 1] - 4$ **126**
20. $(3^2 + 2^2) - 2(3^2 - 2^2)$ **3**

LESSON 1.4
Key Skills

Graph an equation by plotting points.

Make a table for the equation $y = 2x$ by substituting the values 1, 2, 3, 4, and 5 for x and finding the values for y.

x	1	2	3	4	5
y	2	4	6	8	10

Graph the ordered pairs on a coordinate plane and connect the points to make a line.

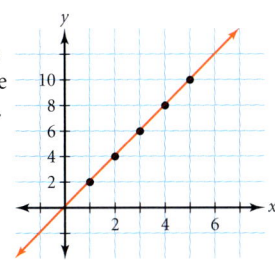

Exercises

Graph each list of ordered pairs. Tell whether the points lie on a straight line.

21. (1, 2), (2, 3), (3, 4)
22. (2, 5), (4, 2), (5, 2)
23. (1, 3), (1, 4), (3, 5)

Graph each equation by plotting points and connecting the points to make a line.

24. $y = x + 2$
25. $y = 3x - 1$
26. $y = x + 2$
27. $y = x$

LESSON 1.5
Key Skills

Write and graph linear equations representing problem situations.

The cost for membership in a CD club is shown in the table below.

Number of CDs purchased (n)	0	1	2	3	4
Cost in dollars (c)	10	21	32	43	54
		11	11	11	11

Exercises

28. A book club charges a membership fee in addition to the cost of each book purchased. Write an equation to represent the data shown in the table below, and then graph.

Number of books purchased (n)	0	1	2	3	4
Cost in dollars (c)	15	27	39	51	63

26.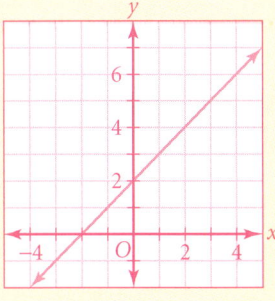

27.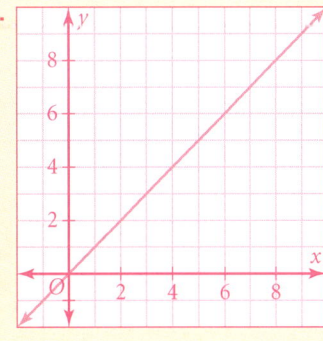

28. $c = 15 + 12n$

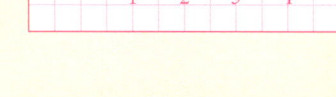

Chapter Test, Form B

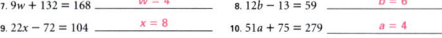

Chapter Assessment
Chapter 1, Form B, page 1

Find the next three numbers in each sequence.
1. 74, 65, 56, 47, ...   **38, 29, 20**
2. 19, 22, 18, 21, ...   **17, 20, 16**
3. 41, 48, 46, 53, ...   **51, 58, 56**
4. The first four terms of a sequence are 37, 41, 47, and 55. Find the first and second differences. What are the next three terms?
   first differences: 4, 6, 8; second difference: 2; next three terms: 65, 77, 91

Make a table to show the substitutions of 1, 2, 3, 4, 5 for the variable.

5. Find the values of $7z$ by substituting for z.

z	1	2	3	4	5
7z	7	14	21	28	35

6. Find the values of $3y$ by substituting for y.

y	1	2	3	4	5
3y	3	6	9	12	15

Use guess-and-check to solve each equation.
7. $9w + 132 = 168$ **w = 4**   8. $12b - 13 = 59$ **b = 6**
9. $22x - 72 = 104$ **x = 8**   10. $51a + 75 = 279$ **a = 4**

Evaluate each expression.
11. $6 \cdot 5 + 3 \cdot 2$ **36**   12. $9 \cdot 3 + 4 \div 3$ **28.333**
13. $0.7 \cdot 2.4 + 7$ **8.68**   14. $6.9 \cdot 4.2 + 6.7 \cdot 5$ **62.48**

For $a = 7$, $b = 5$, and $c = 2$, evaluate each expression.
15. $a^2 + b^2$ **74**   16. $a^2 + b - c$ **52**   17. $(a + c) \cdot b$ **45**

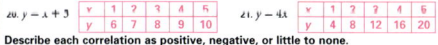

Chapter Assessment
Chapter 1, Form B, page 2

Graph the ordered pairs. State whether they lie on a straight line.

18. (1, 5), (3, 7), (4, 8)   **yes**   19. (22, 25), (23, 24), (24, 23)   **yes**

 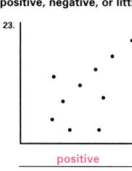

Find the values for y by substituting 1, 2, 3, 4, 5 for x. Make a table.

20. $y = x + 3$

x	1	2	3	4	5
y	6	7	8	9	10

21. $y = 4x$

x	1	2	3	4	5
y	4	8	12	16	20

Describe each correlation as positive, negative, or little to none.

22. little to none   23. positive   24. negative

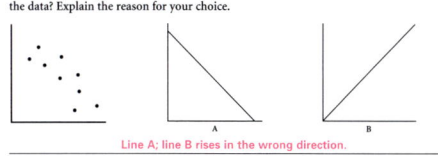

25. Given the scatter plot shown at the left below, which line better fits the data? Explain the reason for your choice.

   Line A; line B rises in the wrong direction.

CHAPTER 1 REVIEW   **47**

29. $d = 5h$

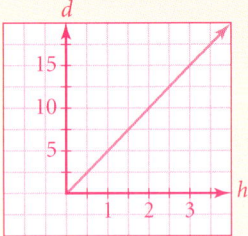

Since the data show constant first differences, the pattern is linear. The linear equation $c = 11n + 10$ represents the situation. Graph the equation by plotting points from the table and connecting them to form a line.

29. Marcia rides her bike at a rate of 5 miles per hour. Write an equation, make a table, and draw a graph to show the distance Marcia travels for times from 0 to 3 hours.

LESSON 1.6
Key Skills

Draw a scatter plot for data, and determine the correlation.

The table shows the change in a population over a period of 80 years.

Year	1900	1920	1940	1960	1980
People per square mile	21.5	29.9	37.2	50.6	64

Plot the ordered pairs from the table on a coordinate plane to form a scatter plot.

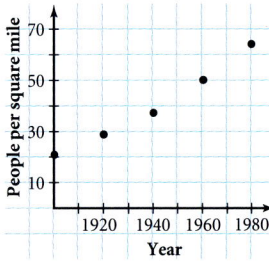

There is a strong positive correlation because both variables increase.

Exercises

The table below shows how the cost of food items that were $1 in 1982 has increased over the years.

Year	1982	1985	1988	1991
Cost	$1	$1.06	$1.18	$1.36

30. Make a scatter plot for the data.

31. Describe the correlation in the scatter plot from Exercise 30. **strong positive**

The table below shows the average daily high temperature at certain points on a mountain.

Elev. (ft)	3000	6000	9000	12,000
Temp. (°F)	75	65	52	40

32. Make a scatter plot for the data.

33. Describe the correlation in the scatter plot from Exercise 32. **strong negative**

Applications

34. ENTERTAINMENT If it costs $13 per person to enter an amusement park, how many people can get in for $98? **7 people with $7 left over**

35. ENTERTAINMENT Suppose that the cost to order concert tickets is $22 per ticket, plus a $5 handling charge per order (regardless of how many tickets are ordered).
  a. How much does an order of 4 tickets cost? **$93**
  b. How much does an order of 8 tickets cost? **$181**
  c. Let t represent the number of tickets, and write an equation for the cost, c, of an order of tickets. **$c = 22t + 5$**

30.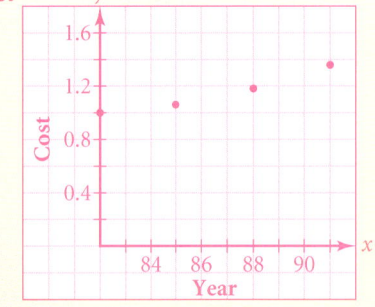

32.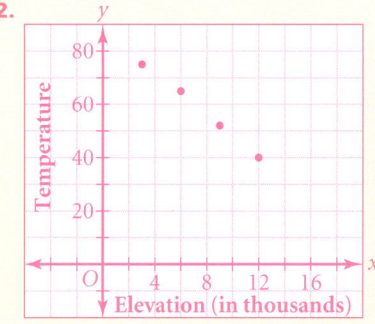

Alternative Assessment

Performance Assessment

1. Make a table to represent the following problem: Paul saved $1 the first month. Then each month for the next 6 months, Paul saved double the amount of money he saved the previous month.

   How much money did Paul have at the end of 7 months?

2. Ramos simplified the expression $8 - 5 \cdot 3 + 4^2$ and wrote 25 for the answer. Latisha said the correct answer is 9. Only one of the answers is correct. Which is correct and why?

3. Dawna, Teri, and David deliver newspapers each day. On any given day, Teri delivers 75 newspapers. The number of newspapers that Dawna delivers varies each day. David delivers twice as many newspapers as Dawna does. Let n represent the number of newspapers that Dawna delivers.
   a. Write an expression that represents the number of newspapers delivered by all three people.
   b. One day the three of them delivered 240 newspapers. Write an equation that represents the number of newspapers delivered by all three people.
   c. Use guess-and-check to find the number of papers delivered by Dawna on that day.

Portfolio Projects

1. Research Pascal's triangle. Make a drawing of the first six rows of Pascal's triangle. List as many patterns as possible that are found in the triangle.

2. Look up *variable* in the dictionary. How does the dictionary meaning differ from the meaning given on page 11? How are the meanings the same? Give examples of how the word *variable* is used for each meaning.

3. Cut out newspaper or magazine articles that include linear graphs or scatter plots. Explain how the variables for each graph are used to represent the data in the article.

Peer Assessment

Write a paragraph explaining how you would make a graph on a coordinate plane to represent the following situation:

   An electrician charges $45 for a house call and $18 for each hour on the job.

Exchange your paper with another student. Follow the directions in your partner's paragraph to make the graph. Create another problem situation with graphing and exchange papers again. Repeat this process until you and your partner are proficient at communicating how to make a graph.

---

internetconnect

The HRW Web site contains many resources to reinforce and expand your knowledge of how algebra is used to reveal patterns in data. This Web site also provides Internet links to other sites where you can find information and real-world data for use in research projects, reports, and activities that involve determining patterns in data. Visit the HRW Web site at **go.hrw.com**, and enter the keyword **MA1 CH1** to access the resources for this chapter.

Alternative Assessment

Portfolio Projects

1.
$$\begin{array}{c}1\\1\ \ 1\\1\ \ 2\ \ 1\\1\ \ 3\ \ 3\ \ 1\\1\ \ 4\ \ 6\ \ 4\ \ 1\\1\ \ 5\ \ 10\ \ 10\ \ 5\ \ 1\end{array}$$

   Every sequence of numbers on the diagonals of Pascal's triangle is a pattern.

2. Answers may vary. The dictionary meaning of the word *variable* is much broader than the mathematical definition, though it primarily means something changing. Example: The variable weather conditions made it difficult to make plans for the day.

3. Students will most likely find examples in the front sections of a newspaper, or in news-oriented magazines such as *Time* or *Newsweek*. Students should be able to identify the variables being compared in the data.

Performance Assessment

1. At the end of six months Paul had saved $1 + 2 + 4 + 8 + 16 + 32 = 63$ dollars.

2. Latisha is correct. Ramos reached the incorrect answer of 25 by performing the operations from left to right.

3. a. total number delivered = $75 + n + 2n$
   b. $240 = 75 + n + 2n$
   c. Dawna delivered 55 newspapers.

Cumulative Assessment

College Entrance Exam Practice

Multiple-Choice and Quantitative-Comparison Samples

The first half of the Cumulative Assessment contains two types of items found on standardized tests—multiple-choice questions and quantitative-comparison questions. Quantitative-comparison items emphasize the concepts of equality, inequality, and estimation.

Free Response Grid Samples

The second half of the Cumulative Assessment is a free-response section. This part of the Cumulative Assessment requires student-produced response items like those commonly found on college entrance exams. These questions require the use of machine-scored answer grids. You may wish to have students practice answering these items in preparation for standardized tests.

CHAPTER 1 CUMULATIVE ASSESSMENT

College Entrance Exam Practice

QUANTITATIVE COMPARISON For Questions 1–4 write
A if the quantity in column A is greater than the quantity in column B;
B if the quantity in column B is greater than the quantity in column A;
C if the two quantities are equal; or
D if the relationship cannot be determined from the information given.

	Column A	Column B	Answers	
1.	$\frac{1}{2} + \frac{3}{5}$	$\frac{1}{2} \cdot \frac{3}{5}$	Ⓐ Ⓑ Ⓒ Ⓓ	A
2.	The next term in the sequence 7, 16, 25, 34, 43, ...	The next term in the sequence 2, 4, 8, 16, 32, ...	Ⓐ Ⓑ Ⓒ Ⓓ [Lesson 1.1]	B
3.	$\sqrt{225}$	$3^2 + 2(7 - 4)$	Ⓐ Ⓑ Ⓒ Ⓓ [Lesson 1.3]	C
4.	$\frac{6}{10}$	$\left(\frac{6}{10}\right)^2$	Ⓐ Ⓑ Ⓒ Ⓓ	A

5. If Shelly drives at an average speed of 50 miles per hour, how many hours will it take her to drive 500 miles? *(LESSON 1.2)* c
   a. 6 hours b. 8 hours
   c. 10 hours d. more than 10 hours

6. Simplify the following expression:
   $(90 \div 3^2) - 9 + 5^2$ *(LESSON 1.3)* b
   a. 24 b. 26
   c. 98 d. 126

7. What number is missing from this pattern?
   232, 343, _____, 565, 676, 787 *(LESSON 1.1)*
   a. 424 b. 403 c
   c. 454 d. 554

If notebooks cost 59 cents, find the cost of the items below. *(LESSON 1.2)*

8. 2 notebooks **$1.18**   9. 5 notebooks **$2.95**

10. 12 notebooks **$7.08**

11. Write an equation to model this situation: How many notebooks can you buy for $14.75?
    14.75 = 0.59n

Alice sells memberships for an exercise club. Her weekly pay is $250 plus $20 for each membership sold. *(LESSONS 1.4 AND 1.5)*

12. Write an equation for Alice's weekly pay, P, in terms of the number of memberships she sells, m. **p = 250 + 20m**

13. Make a table to show Alice's weekly pay for 1, 2, 3, 4, and 5 memberships sold. **270, 290, 310, 330, 350**

14. Suppose that she earns $330 one week. How many memberships did she sell? **4**

15. Make a graph by using the data from your table.

16. Use your graph to find how many memberships Alice needs to sell in order to earn at least $380 in one week. **7**

15.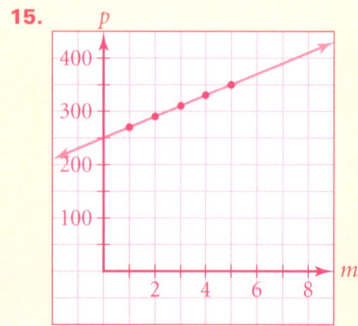

17. Make a table of the values and evaluate the expression $5x + 7$ by substituting 1, 2, 3, 4, and 5 for x. **(LESSON 1.2)** 12, 17, 22, 27, 32

18. Use the table of values in Exercise 17 to solve the equation $5x + 7 = 22$. $x = 3$

Write a list of factors for each number. Circle the prime factors.

19. 20 1, ②, 4, ⑤, 10, 20   **20.** 42 6, ⑦, 14, 21, 42   **21.** 76 1, ②, 4, ⑲, 38, 76
(with 1, ②, ③ shown for 19)

Determine whether each number is prime or composite.

22. 19 prime **23.** 30 composite **24.** 31 prime

Evaluate. (LESSON 1.3)

25. $\frac{3}{4} - \frac{3}{5}$ $\frac{3}{20}$

26. $1\frac{2}{3} - \frac{7}{8}$ $\frac{19}{24}$

27. $\frac{7}{15} \cdot \frac{3}{28}$ $\frac{1}{20}$

28. $\frac{1}{2} \div \frac{3}{5}$ $\frac{5}{6}$

29. $3 + 27 \div 3^2 - (7 - 5)$ 4

30. $(9 + 37) + 11$ 57   **31.** $20 \cdot (5 \cdot 19)$ 1900

Graph each list of ordered pairs and tell whether the points lie on a straight line. (LESSON 1.4)

32. (0, 1), (4, 9), (2, 5)   **33.** (1, 9), (8, 4), (4, 5)

Describe the correlation for each scatter plot below as strong positive, strong negative, or little to none. (LESSON 1.6) strong negative

34.

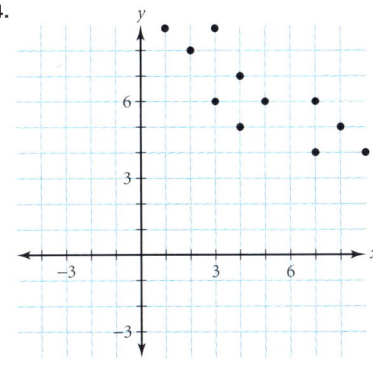

35. little or none

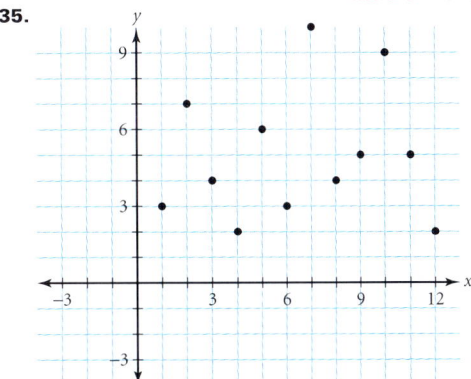

36. Make a scatter plot of this data for the amount of time spent studying and the score on a test the next day. **(LESSON 1.6)**

Time studying (in minutes)	15	12	20	30	25	40	60	50	55	45
Test score	65	68	75	80	79	85	97	80	88	73

37. Use a straightedge to estimate the line of best fit for the scatter plot from Exercise 36. **(LESSON 1.6)**

FREE-RESPONSE GRID
The following questions may be answered by using a free-response grid such as that commonly used by standardized test services.

Refer to the sequence 1, 2, 5, 10, 17, 26, … (LESSON 1.1)

38. Find the second differences. 2

39. Find the next term. 37

40. If the constant second difference of a sequence is 2, the first of the first differences is 5, and the first term is 12, find the fifth term of the sequence. **(LESSON 1.1)** 44

Find the next term of each sequence. (LESSON 1.1)

41. 1, 3, 7, 13, … 21   **42.** 99, 93, 86, 78, … 69

32.
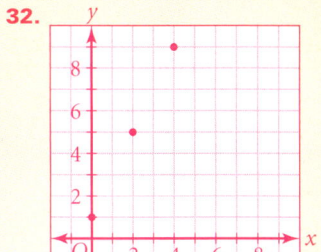
The points lie on a straight line.

33.
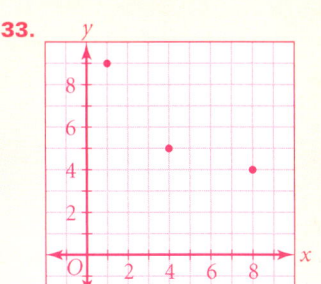
The points do not lie on a straight line.

36.
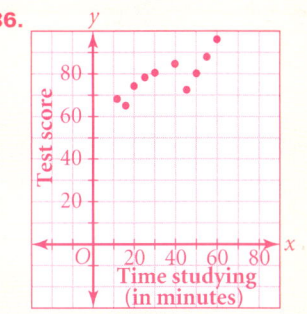

37.
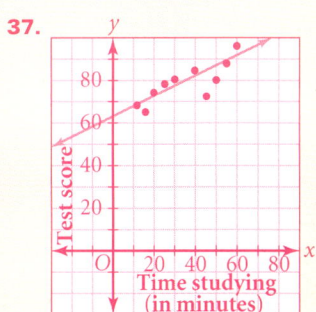

2 Operations in Algebra

CHAPTER PLANNING GUIDE

Lesson	2.1	2.2	2.3	2.4	2.5	2.6	2.7	Projects and Review
Pupil Edition Pages	54–59	60–66	67–72	73–79	80–88	89–93	96–103	94–95, 104–111
Extra Practice (Pupil Edition)	741	741	742	742	743	743	744	
Practice Workbook	7	8	9	10	11	12	13	
Reteaching Masters	13–14	15–16	17–18	19–20	21–22	23–24	25–26	
Enrichment Masters	7	8	9	10	11	12	13	
Cooperative-Learning Activities	7	8	9	10	11	12	13	
Lesson Activities	7	8	9	10	11	12	13	
Problem Solving/ Critical Thinking	7	8	9	10	11	12	13	
Student Study Guide	7	8	9	10	11	12	13	
Spanish Resources	7	8	9	10	11	12	13	
Student Technology Guide	7	8	9	10	11	12	13	
Assessment Resources	18	19	20	21	23	24	25	22, 26–31
Teaching Transparencies	6	7, 8, 9		10	11		12	
Quiz Transparencies	2.1	2.2	2.3	2.4	2.5	2.6	2.7	
Writing Activities for Your Portfolio								4–6
Tech Prep Masters								5–8
Long-Term Projects								5–8

LESSON PACING GUIDE

	2.1	2.2	2.3	2.4	2.5	2.6	2.7	Projects and Review
Traditional	2 days	2 days	2 days	2 days	1 day	1 day	2 days	2 days
Block	1 day	1 day	1 day	1 day	$\frac{1}{2}$ day	$\frac{1}{2}$ day	1 day	1 day
Two-Year	4 days	4 days	4 days	4 days	2 days	2 days	4 days	4 days

CONNECTIONS AND APPLICATIONS

Lesson	2.1	2.2	2.3	2.4	2.5	2.6	2.7	Review
Algebra	54–59	60–66	67–72	73–79	80–88	89–93	96–103	104–111
Geometry	59				87	93	102	108
Business and Economics	59	60, 65	72	73, 78	80, 82, 86, 87	93	102	
Statistics		65	71	78				105
Patterns in Data			71					
Life Skills			67				102	
Science and Technology	54	65	70, 72	77				104
Sports and Leisure					83, 88		99	105
Cultural Connection: Asia		60						

BLOCK-SCHEDULING GUIDE

Day	Lesson	Teacher Directed: Lesson Examples, Teaching Transparencies	Student Guided: Activity, Try This	Cooperative-Learning Activity, Lesson Activity, Student Technology Guide	Practice: Practice & Apply, Extra Practice, Practice Workbook	Assessment: Quiz, Mid-Chapter Assessment	Problem Solving, Reteaching
1	2.1	10 min	15 min	15 min	65 min	15 min	15 min
2	2.2	10 min	15 min	15 min	65 min	15 min	15 min
3	2.3	10 min	15 min	15 min	65 min	15 min	15 min
4	2.4	10 min	15 min	15 min	65 min	15 min	15 min
5	2.5	8 min	8 min	8 min	35 min	8 min	8 min
5	2.6	7 min	7 min	7 min	30 min	7 min	7 min
6	2.7	15 min	15 min	15 min	65 min	15 min	15 min
7	Assess.	50 min PE: Chapter Review	90 min PE: Chapter Project, Writing Activities	90 min Tech Prep Masters	65 min PE: Chapter Assessment, Test Generator	30 min Chap. Assess. (A or B), Alt. Assess. (A or B), Test Generator	

PE: Pupil's Edition

Alternative Assessment

The following are suggestions for an alternative assessment for students who may benefit from a different type of assessment than the regular chapter quizzes and the mid-chapter and end-of-chapter assessment materials. Many of the questions are open-ended, and students' answers will vary.

Performance Assessment

1. **a.** Summarize the rules by which you can add, subtract, multiply, or divide two real numbers.
   b. For various situations involving real numbers, give an example of how the rules for adding, subtracting, multiplying, and dividing real numbers are used to find sums, differences, products, and quotients.

2. When you deposit money into a bank account, you add to the current amount in the account. When you withdraw money, you subtract from the current amount.
   a. Explain how addition and subtraction of real numbers are used in banking.
   b. What does it mean when your account is *overdrawn*?

3. **a.** How can you tell whether two expressions have like terms? Give an example of two like terms and two unlike terms.
   b. Explain how like terms affect the adding or subtracting of two expressions.

Portfolio Project

Suggest that students perform the following project for inclusion in their portfolios.

1. **a.** Find examples of how mental math is used in everyday life.
   b. Suppose that Jamie wants to buy 3 tapes that cost $3.98 each. Explain how Jamie can calculate the total cost without tax by using a calculator.
   c. Give some examples from everyday life in which a calculator can help you make calculations and decisions.

---

internet connect

The table below identifies the pages in this chapter that contain technology information and support in the side columns.

Content Links	
Lesson Links	pages 72, 85, 91
Portfolio Links	pages 79, 88
Graphics Calculator Support	page 782

Resource Links
For information about teacher and parent resources as well as professional development help, visit **www.hrw.com/math**.

Technical Support HRW has assembled a team of dedicated technical and teaching professionals and a comprehensive service program to provide you with the support you need.

- The HRW Technical Support Line operates from 7 A.M. to 6 P.M. central time, Monday through Friday, at (800) 323-9239.
- The HRW Technical Support Center on the World Wide Web is available 24 hours a day, seven days a week, at **www.hrwtechsupport.com**.
- You can e-mail our Technical Support Center at **tsc@hrwtechsupport.com**.
- The Technical Support Center's fax-on-demand service at (800) 352-1680 offers solutions to common problems and answers to frequently asked questions.

Technology

Lesson Suggestions and Calculator Examples
(Keystrokes are based on a TI-83 calculator.)

Lesson 2.1 The Real Numbers and Absolute Value
In this lesson, students will be introduced to a calculator function.

absolute value of $x \rightarrow$ **abs(X)**

Students can access this function by using the **NUM** menu. To see the menu, press MATH and select NUM 1:abs(.

In this lesson, students will also learn about repeating decimals by performing simple division of integers.

Lesson 2.2 Adding Real Numbers
Students can add real numbers by noticing like and unlike signs and by using absolute values. With practice, students will begin to appreciate shortcuts in addition. For example, students will get the same sum for $5 + (-7)$ by using either key sequence below.

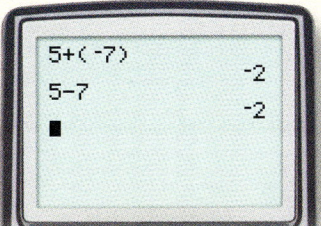

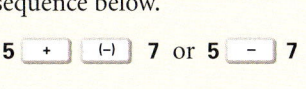

Lesson 2.3 Subtracting Real Numbers
This lesson is a companion lesson to Lesson 2.2. Students will gain practice in the subtraction of both whole numbers and mixed numbers. For example, ask for the difference $5 - \frac{7+1}{4}$. Ask students to contrast the two key sequences below.

5 − (7 + 1 ÷ 4)
5 − 7 + 1 ÷ 4

Lesson 2.4 Multiplying and Dividing Real Numbers
In this lesson, students will use a calculator to make conjectures about the rules for multiplication and division of real numbers. Have students perform the Activity on pages 73 and 74 with a calculator. To enhance the Activity, have students use a calculator to find each of the products below.

$2.3 \times (-5.1)$ $(-5.5) \times 7.3$ $10.2 \times (-0.5)$

In each case, students will obtain a negative answer and will learn that it was produced by a pair of numbers with unlike signs. In a similar way, you can extend the part of the Activity that deals with division.

Lesson 2.5 Properties and Mental Computation
In this lesson, students will learn when it is and is not necessary to use a calculator. The mathematical properties explained in this lesson will help students calculate some problems by using only mental math.

Consider offering students expressions like the ones below as well as those found in the exercises.

$2.3 \times 5.8 + 2.3 \times 4.2$ $2.3 \times 5.4 + 2.3 \times 4.2$

For which expression would you use a calculator to find the result? For which expression would you use mental math to find the result?

Lesson 2.6 Adding and Subtracting Expressions
In this lesson, students will use graphs to see how like terms combine. For example, consider $2x + 3x$ and $5x$. If these two expressions are equal for all values of x, then the graphs of $y = 2x + 3x$ and $y = 5x$ should be the same line. The calculator display at right below shows only one graph. This happens because the graphs are on top of one another, or $2x + 3x = 5x$.

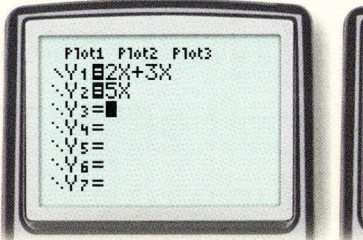 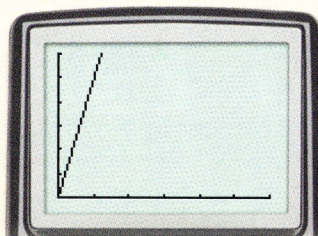

Lesson 2.7 Multiplying and Dividing Expressions
In this lesson, students will apply the Distributive Property to expressions in which x appears in two factors. Division by a number is shown as another application of the Distributive Property, where one factor is the reciprocal of the divisor.

Have students practice using x^2 and ^ . Practice problems could involve both positive and negative bases. Caution students to use (and) when the base is a fraction.

For further information, refer to the
- technology discussions in the lessons.
- lesson-related teacher's commentary in the side columns of this *Teacher's Edition*.
- lesson-related *Student Technology Guide* masters.
- *HRW Technology Handbook*.

Teaching Resources

Basic Skills Practice (2 pages per skill)

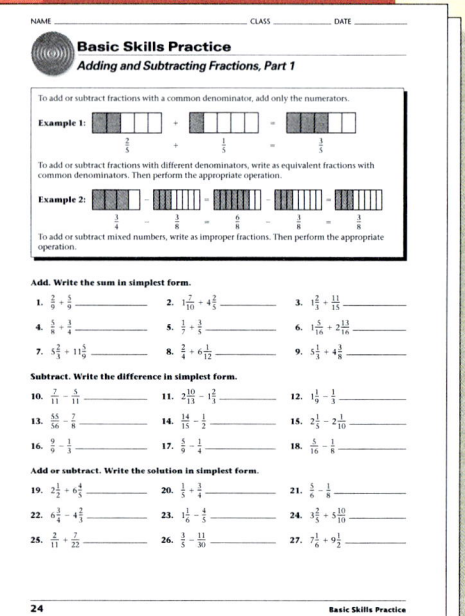

Reteaching (2 pages per lesson)

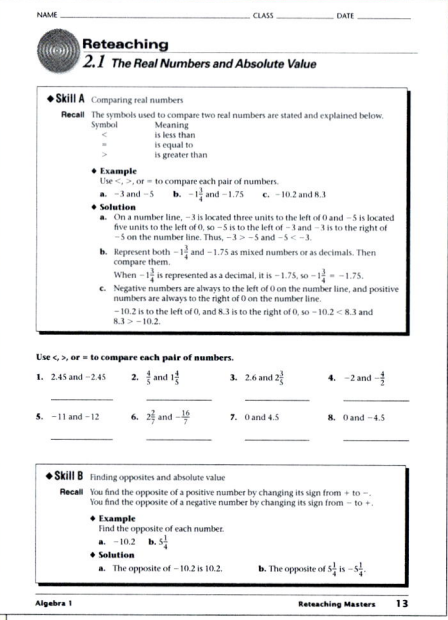

Enrichment (1 page per lesson)

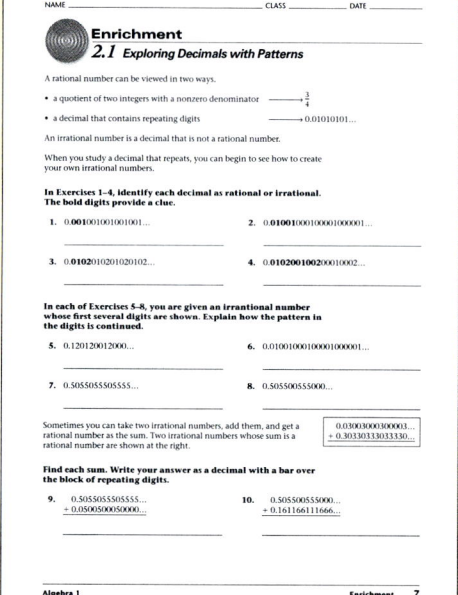

Cooperative-Learning Activity (1 page per lesson)

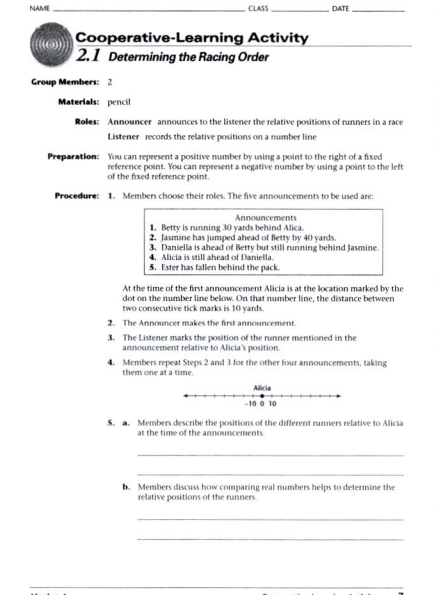

Lesson Activity (1 page per lesson)

Long-Term Project (4 pages per chapter)

Problem Solving/Critical Thinking
(1 page per lesson)

Writing Activities
(3 pages per chapter)

Tech Prep Masters
(4 pages per chapter)

Spanish Resources
(1 page per lesson)

Quizzes
(1 page per lesson)

Mid-Chapter Test
(1 page per chapter)

About Chapter 2

Background Information
In Chapter 2, students study operations with real numbers and the basic properties that govern these operations. Then they learn to use the properties to perform mental computations and to simplify algebraic expressions.

CHAPTER RESOURCES

- Block-Scheduling Handbook
- Writing Activities for Your Portfolio
- Tech Prep Masters
- Long-Term Project
- Assessment Resources:
  Mid-Chapter Assessment
  Chapter Assessments
  Alternative Assessments
- Test and Practice Generator
- Technology Handbook
- End-of-Course Exam Practice

Chapter Objectives

- Compare real numbers [2.1]
- Simplify expressions involving opposites and absolute value. [2.1]
- Use algebra tiles to model addition. [2.2]
- Add numbers with like signs. [2.2]
- Add numbers with unlike signs. [2.2]

Operations in Algebra

ONE OF THE FIRST BOOKS ABOUT ALGEBRA WAS written in Arabic by a ninth-century scientist named Muhammad ibn Muas al-Khwarizmi. The title of the book was shortened to *al-jabr*, now spelled *algebra*. The word *algebra* comes from part of the Arab title that means "equals can be added to both sides of an equation." Al-Khwarizmi used his *al-jabr* to help him in his scientific work in geography and astronomy.

Lessons

2.1 • The Real Numbers and Absolute Value

2.2 • Adding Real Numbers

2.3 • Subtracting Real Numbers

2.4 • Multiplying and Dividing Real Numbers

2.5 • Properties and Mental Computation

2.6 • Adding and Subtracting Expressions

2.7 • Multiplying and Dividing Expressions

Project Using a Spreadsheet

An astrolabe is a device that was used by early astronomers to study the positions of stars.

About the Photos
Astrolabes were highly developed and used by the Islamic World during the Middle Ages. The astrolabe pictured above is called a planespheric astrolabe. It was used to find the altitudes of nearby stars and later was adopted by mariners for navigation. Other uses for astrolabes include establishing the time of day or night and surveying applications, such as determining the height of a mountain.

Pages from an Arabic manuscript of al-Khwarizmi's algebra book

- Use algebra tiles to model subtraction. [2.3]
- Define subtraction in terms of addition. [2.3]
- Subtract numbers with like and unlike signs. [2.3]
- Multiply and divide positive and negative numbers. [2.4]
- Define the Properties of Zero. [2.4]
- State and apply the Commutative, Associative, and Distributive Properties. [2.5]
- Use the Commutative, Associative, and Distributive Properties to perform mental computations. [2.5]
- Use the Distributive Property to combine like terms. [2.6]
- Simplify expressions with several variables. [2.6]
- Multiply expressions containing variables. [2.7]
- Divide expressions containing variables. [2.7]

About the Chapter Project

A spreadsheet is a tool for organizing and analyzing numeric data that uses positive and negative numbers, expressions, and equations. The Chapter Project, *Using a Spreadsheet*, on page 104, is an introduction to the use of spreadsheets and provides an opportunity to apply the algebraic concepts covered in the chapter. After completing the Chapter Project, you will be able to do the following:

- Use a computer spreadsheet to analyze simple data sets.

About the Portfolio Activities

Throughout the chapter, you will be given opportunities to complete Portfolio Activities that are designed to support your work on the Chapter Project.

- The Portfolio Activity on page 66 describes how profit and loss are indicated on a spreadsheet.
- Average monthly temperature ranges for Calgary, Alberta, Canada, are the topic of the Portfolio Activity on page 79.
- In the Portfolio Activity on page 88, a spreadsheet is used to help describe windchill data.
- The Portfolio Activity on page 103 presents a bicycle-rental problem.

Additional Portfolio Activities appear at the end of Lessons 2.2, 2.4, 2.5, and 2.7. Each serves as preparation for the Chapter Project. The Portfolio Activities as well as the Chapter Project Activities are appropriate for inclusion in the students' portfolios. Students should be encouraged to include in their portfolios any other work in which they feel a sense of pride or a sense of accomplishment.

 internet connect

Internet connect features appearing throughout the chapter provide keywords for the HRW Web site that lead to links for math resources, tutorial assistance, references for student research, classroom activities, and all teaching resource pages. The HRW Web site also provides updates to the technology used in the text. Listed at right are the keywords for the Internet Connect activities referenced in this chapter. Refer to the side column on the page listed to read about the activity.

LESSON	KEYWORD	PAGE
2.3	MA1 U.S Jobs	72
2.4	MA1 TEMP Extremes	79
2.5	MA1 Napier's Bones	85
2.5	MA1 Windchill	88
2.6	MA1 Stocks	91

CHAPTER 2 **53**

Prepare

NCTM PRINCIPLES & STANDARDS 1–3, 6–10

LESSON RESOURCES

- Practice 2.1
- Reteaching Master 2.1
- Enrichment Master 2.1
- Cooperative-Learning Activity 2.1
- Lesson Activity 2.1
- Problem Solving/Critical Thinking Master 2.1
- Teaching Transparency 6

QUICK WARM-UP

Evaluate each expression.

1. $190 - 128 \div 4$ 158
2. $21 - 6 \cdot 3 + 4^2$ 19
3. $8 \cdot 9 + 7 \cdot 5$ 107
4. $2(5 + 8 - 6) + 11$ 25
5. $3^2 \cdot 4 + [9 - (2^2 - 1)]$ 42

Also on Quiz Transparency 2.1

Teach

Why On the board or overhead, make a list of words that are opposites. For example, give students the word *black*, and ask them to state its opposite, *white*. Other possible pairs are listed below.

on/off	in/out
begin/end	over/under
up/down	young/old

Tell students that just like words, numbers have opposites as well.

2.1 The Real Numbers and Absolute Value

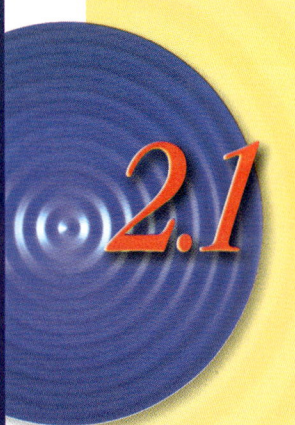

Objectives

- Compare real numbers.
- Simplify expressions involving opposites and absolute value.

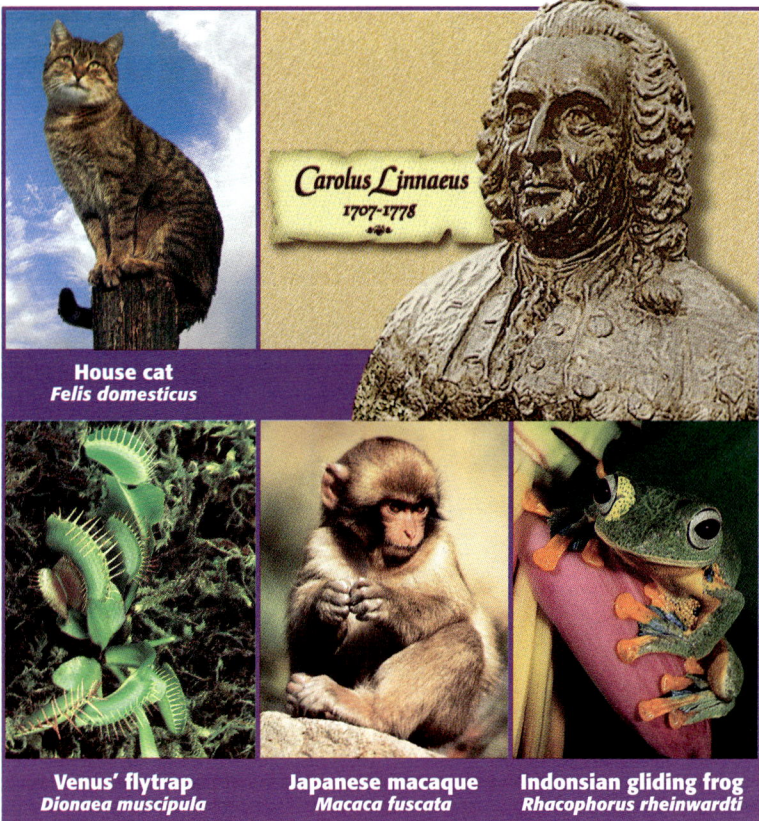

House cat *Felis domesticus*

Venus' flytrap *Dionaea muscipula*

Japanese macaque *Macaca fuscata*

Indonesian gliding frog *Rhacophorus rheinwardti*

Why Taxonomy deals with the naming and classification of plants and animals. Another system helps to classify numbers in mathematics. Understanding the different types of numbers and comparing numbers are important skills in many real-world situations.

APPLICATION
BIOLOGY

Scientists have classified millions of plants and animals by using a system called taxonomy that was developed in the mid-1700s by Swedish naturalist Carolus Linnaeus. This system groups living things by common characteristics in increasingly smaller and more specific categories. In mathematics numbers are classified in a similar, though much less complicated, system.

Sets of Numbers

Different sets of numbers can be represented by lists of their members. For example:

Natural or counting numbers: $N = 1, 2, 3, 4, \ldots$
Whole numbers: $W = 0, 1, 2, 3, 4, \ldots$
Integers: $I = \ldots -3, -2, -1, 0, 1, 2, 3, \ldots$

Recall that the three dots in a list indicate that the list continues indefinitely in a similar pattern.

Alternative Teaching Strategy

USING MODELS Use tape to mark off a number line on the floor of the classroom with numbers ranging from –10 to 10. Explain that the number line represents the entire set of real numbers. Then describe each subset of the real numbers, referring to the number line. Have a student demonstrate each subset by standing on a number that is representative of that subset. For opposites and absolute values, have two students start at zero and walk in opposite directions the same distance from zero.

The sets N, W, and I can also be shown on a number line. Remember that 0 is known as the origin and is neither positive nor negative. Numbers to the right of 0 are **positive**, and those to the left of 0 are **negative**.

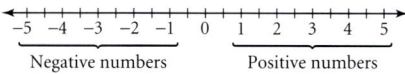

Although the labeled numbers are typically integers, there are many numbers on a number line that are not integers. Numbers like $\frac{2}{3}$, $-\frac{4}{15}$, and $\frac{23}{5}$ are included on a number line and are called **rational numbers**.

Definition of Rational Number

Any number that can be expressed in the form $\frac{a}{b}$, where a and b are integers and $b \neq 0$, is a rational number.

The set of rational numbers cannot be listed as the integers can be, but this set does include the integers since any integer, a, can be expressed as $\frac{a}{1}$.

The set of rational numbers has an important characteristic that distinguishes it from other types of numbers. Any rational number can be written as either a terminating decimal or a repeating decimal. Repeating decimals do not end: they continuously repeat a pattern of digits at some point to the right of the decimal. Terminating decimals simply end at a certain point.

Activity
Terminating and Repeating Decimals

1. Use division to write each rational number as a decimal rounded to the nearest thousandth.

   a. $\frac{4}{5}$ b. $\frac{4}{11}$ c. $\frac{5}{8}$ d. $\frac{7}{9}$ e. $\frac{5}{12}$ f. $\frac{3}{4}$

CHECKPOINT ✔ 2. Which numbers in Step 1 are terminating decimals? Change $\frac{5}{8}$ to a decimal by dividing. What happens during the division process that causes the decimal to terminate?

CHECKPOINT ✔ 3. Which numbers in Step 1 are repeating decimals? Change $\frac{4}{11}$ and $\frac{5}{12}$ to decimals by dividing. What happens during the division process that causes the decimals to repeat?

Use a bar above the repeating digits to indicate a repeating decimal. For example, $\frac{2}{11}$ is equal to 0.1818..., which is written $0.\overline{18}$.

 **Notes**

This Activity gives students an opportunity to discover why some fractions result in a repeating decimal and other fractions result in a terminating decimal.

CHECKPOINT ✔

2. $\frac{4}{5}, \frac{5}{8},$ and $\frac{3}{4}$ are terminating decimals. Answers may vary. Sample answer: During the division process, the 8 divides evenly into the remainder.

CHECKPOINT ✔

3. $\frac{4}{11}, \frac{7}{9}$ and $\frac{5}{12}$ are repeating decimals. Answers may vary. Sample answer: During the division process the same numbers are generated over and over.

Enrichment

Tell students to imagine that they have a bag of sticks with certain characteristics. Each stick is either a unit stick of a fixed length or a multiple of the unit stick. Ask students to explain if it is possible to construct a square and its diagonal from this set of sticks. Students should discover that since the diagonal of a unit square is $\sqrt{2}$, they will not be able to construct the diagonal from the given sticks.

CRITICAL THINKING
We need the equal sign because each point on the number line can be represented in many ways.

ADDITIONAL EXAMPLE ①

Insert an ordering symbol to make each statement true.

a. -3 __?__ 6 $<$

b. -8.5 __?__ -8.6 $>$

c. $\frac{4}{5}$ __?__ $\frac{16}{20}$ $=$

d. $7\frac{3}{7}$ __?__ 9 $<$

Numbers such as $\sqrt{7}$ and $\sqrt{11}$ are the square roots of numbers that are not perfect squares. You will study square roots, or radicals, in a later chapter. When evaluating these square roots, the results are nonrepeating decimals.

$$\sqrt{7} = 2.64575131\ldots$$
$$\sqrt{11} = 3.3166247\ldots$$

These numbers continue without repeating a pattern of digits. Since they are neither terminating nor repeating decimals, these numbers are not rational and cannot be expressed in the form $\frac{a}{b}$, where a and b are integers and $b \neq 0$. They are called **irrational numbers**.

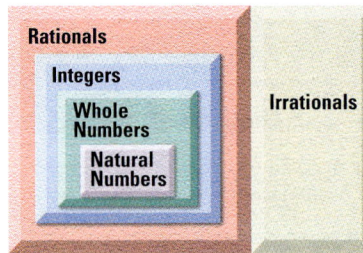

The real numbers

The set of rational numbers and the set of irrational numbers together form the set of **real numbers**. The Venn diagram at left shows how all of the sets of numbers discussed relate to one another.

Ordering Numbers

The number line represents the entire set of real numbers. Each point on the number line represents exactly one number. Each number in the set of real numbers corresponds with exactly one point on the number line. This allows the numbers on the number line to be ordered. Numbers on the number line increase in value from left to right.

The symbols used for ordering are as follows:

     $<$ less than $\leq$ less than or equal to
     $>$ greater than $\geq$ greater than or equal to
     $=$ equal to

CRITICAL THINKING   Since each point on the number line represents exactly one number, how can two numbers be equal to one another? That is, why do we need the equal sign for ordering numbers?

EXAMPLE ① Insert an ordering symbol to make each statement true.

a. 5 ___ -7 b. $\frac{1}{2}$ ___ $\frac{8}{16}$ c. -4.8 ___ -4.7 d. $4\frac{2}{3}$ ___ 6

• **SOLUTION**

a. 5 is to the right of -7 on a number line, so $5 > -7$.
b. $\frac{1}{2}$ and $\frac{8}{16}$ represent the same number, so $\frac{1}{2} = \frac{8}{16}$.
c. -4.8 is to the left of -4.7 on a number line, so $-4.8 < -4.7$.
d. $4\frac{2}{3}$ is to the left of 6 on a number line, so $4\frac{2}{3} < 6$.

Inclusion Strategies

ENGLISH LANGUAGE DEVELOPMENT Help students understand the terms *rational*, *irrational*, and *absolute value* by explaining the background of these terms. The term *rational* means that the number can be written as a ratio, or fraction. Conversely, *irrational* means that the number cannot be written as a ratio. Students need to understand that the absolute value of a number is always positive. Finding the absolute value of a negative number is like covering or mentally erasing the negative sign in front of a number.

TRY THIS Insert an ordering symbol to make each statement true.

a. −9 ___ −12   b. $\frac{1}{2}$ ___ $\frac{3}{5}$ c. −6.3 ___ 5   d. $\frac{4}{12}$ ___ $\frac{7}{21}$

TRY THIS
a. >   b. <
c. <   d. =

Opposites and Absolute Value

On the number line, two numbers that lie on opposite sides of 0 and are the same distance from 0 are **opposites**. The opposite of 2 is negative 2, written as −2. The opposite of a positive number is always a negative number.

−10 −9 −8 −7 −6 −5 −4 −3 −2 −1 0 1 2 3 4 5 6 7 8 9 10

As you can see from the number line, the opposite of −2 is +2. In fact, the opposite of a negative number is always a positive number. This is represented with symbols as follows:

$$-(-2) = 2$$

The opposite of zero is zero itself.

$$-0 = 0$$

ADDITIONAL EXAMPLE 2

Evaluate each expression.

a. −(−10)   10

b. $-\left(-\frac{3}{5}\right)$ $\frac{3}{5}$

c. −(8.6 − 1.5)   −7.1

EXAMPLE 2 Evaluate each expression.

a. $-\left(-\frac{2}{3}\right)$ b. −(0)   c. −(7.4 − 3.2)

● **SOLUTION**

a. The opposite of negative $\frac{2}{3}$ is $\frac{2}{3}$.

b. The opposite of 0 is 0.

c. The opposite of the difference between 7.4 and 3.2 is −4.2.

TRY THIS Evaluate.

a. $-\left(-\frac{4}{7}\right)$ b. −(−0)   c. −(6.2 − 2.7)

TRY THIS
a. $\frac{4}{7}$
b. 0
c. −3.5

Use Teaching Transparency 6.

Opposite numbers are helpful in understanding other important concepts in algebra, such as absolute value.

Definition of Absolute Value

The absolute value of a real number x is the distance from x to 0 on a number line. The symbol $|x|$ means the absolute value of x.

Notice that a number and its opposite have the same absolute value. The absolute values of 2 and −2 are the same, 2. Each is 2 units from 0.

The absolute value of a positive number or 0 is the same as the number itself. The absolute value of a negative number is the opposite of the number.

Reteaching the Lesson

USING MODELS On the board or overhead, draw a Venn diagram similar to the one on page 56. Have students draw the diagram on their own paper as you explain each set. For opposites and absolute value, have students construct a number line from −6 to 6 with half-inch spacing between the numbers. Students can confirm that opposites are the same distance from zero by folding the number line on itself with the crease at zero.

ADDITIONAL EXAMPLE 3

Simplify.

a. $|-8|$ b. $|16|$ c. $|7+2|$
   8   16   9

TRY THIS
a. 13.2
b. 5

Assess

Selected Answers
Exercises 4–15, 17–65 odd

ASSIGNMENT GUIDE

In Class	1–15
Core	17–61 odd
Core Plus	16–56 even, 57–61
Review	62–66
Preview	67

Extra Practice can be found beginning on page 738.

EXAMPLE 3 Simplify.

a. $|7|$ b. $|-14.2|$
c. $|0|$ d. $|9+4|$

SOLUTION

a. $|7| = 7$ b. $|-14.2| = 14.2$
c. $|0| = 0$ d. $|9+4| = |13| = 13$

TRY THIS Evaluate.
a. $|13.2|$ b. $|17-12|$

Exercises

Communicate

1. How are the numbers $\frac{2}{3}$ and $\frac{3}{5}$ different in terms of their decimal equivalents?
2. Explain why an integer is also a rational number.
3. What does absolute value tell you about a number? Why is the absolute value of 6 the same as the absolute value of –6?

Guided Skills Practice

Insert <, >, or = to make each statement true. (EXAMPLE 1)

4. 7 **>** 3   5. $2\frac{1}{2}$ **>** –3   6. –0.2 **<** –0.1   7. $\frac{1}{2}$ **>** $\frac{1}{3}$

Find the opposite of each expression. (EXAMPLE 2)

8. 2 **–2**   9. $-\frac{4}{5}$ **$\frac{4}{5}$**   10. 0.5 **–0.5**   11. 0 **0**

Evaluate. (EXAMPLE 3)

12. $|15|$ **15**   13. $\left|-12\frac{1}{2}\right|$ **$12\frac{1}{2}$**   14. $|2+6|$ **8**   15. $|6-7|$ **1**

Practice and Apply

Insert <, >, or = to make each statement true.

16. 0 **<** 5   17. $-4\frac{1}{4}$ **<** –4   18. –0.6 **<** 0.6   19. 5.4 **<** $5\frac{1}{2}$

20. –2 **<** 1   21. –7 **>** –9   22. $\frac{2}{3}$ **>** $-1\frac{1}{3}$ 23. 6.7 **<** $6\frac{3}{4}$

24. –4.62 **<** –4.6   25. 7.94 **<** 7.95   26. $-\frac{1}{5}$ **>** $-\frac{1}{4}$ 27. –8.2 **>** $-8\frac{1}{4}$

Practice

2.1 The Real Numbers and Absolute Value

Insert <, >, or = to make each statement true.

1. 5 **>** –10   2. 1.40 **=** 1.4   3. 19.5 **<** $19\frac{7}{10}$
4. $-9\frac{1}{3}$ **<** $-7\frac{5}{32}$ 5. 8.8 **>** 0   6. 12.2 **>** 12.02
7. 0 **>** –3   8. –18 **>** $-18\frac{1}{2}$ 9. –2.3 **=** $2\frac{3}{10}$
10. $0.\overline{3}$ **=** $\frac{1}{3}$ 11. 98.59 **<** 98.6   12. 2 **>** 0.667

Find the opposite of each number.

13. 418 **–418**   14. –4.8 **4.8**   15. 0.2 **–0.2**   16. $-\frac{3}{8}$ **$\frac{3}{8}$**
17. $\frac{n}{4}$ **$-\frac{n}{4}$**   18. 76 **–76**   19. –32 **32**   20. 1 **–1**
21. –19.5 **19.5**   22. $16\frac{2}{3}$ **$-16\frac{2}{3}$**   23. –x **x**   24. 1953 **–1953**

Find the absolute value of each number.

25. 17 **17**   26. –100 **100**   27. $29\frac{1}{5}$ **$29\frac{1}{5}$**   28. –4.12 **4.12**
29. $-13\frac{1}{2}$ **$13\frac{1}{2}$**   30. $\frac{3}{5}$ **$\frac{3}{5}$**   31. –22 **22**   32. 3.1416 **3.1416**
33. 85 **–85**   34. –52 **52**   35. 1971 **1971**   36. $-\frac{9}{1000}$ **$\frac{9}{1000}$**

Simplify each expression.

37. –(4 ÷ 2) **–2**   38. –(0.8 + 0.095) **–0.895**   39. –(–m) **m**   40. $-\left(\frac{13}{100} - \frac{3}{100}\right)$ **$-\frac{1}{10}$**
41. $|-1.8|$ **1.8**   42. $|-(-0.1)|$ **0.1**   43. $\left|\frac{21}{7}\right|$ **3**   44. $|4-4|$ **0**
45. $|-3| + |3|$ **6**   46. $-\left|\frac{2}{5}\right|$ **$-\frac{2}{5}$**   47. $|-4| \cdot |-4|$ **16**   48. $\left|8\frac{7}{8} - 8\right|$ **$\frac{7}{8}$**

---

58 LESSON 2.1

Find the opposite of each number.

28. 17   −17
29. −17   17
30. 0   0
31. (12.8 − 5)   −7.8
32. x   −x
33. −1.2   1.2
34. $\frac{4}{5}$ $-\frac{4}{5}$
35. $-\frac{13}{20}$ $\frac{13}{20}$

Find the absolute value of each number.

36. −6   6
37. 18   18
38. 0   0
39. −10.3   10.3
40. $\frac{9}{19}$ $\frac{9}{19}$
41. −3.8   3.8
42. 17.1   17.1
43. $-\frac{3}{4}$ $\frac{3}{4}$

44. Name a value of x such that
   a. −x is negative.
   b. −x is positive.
   c. −x is neither positive nor negative.

Simplify each expression.

45. −(−10)   10
46. −(6.8 − 4.9)   −1.9
47. −(3 · 8)   −24
48. −(17 − 17)   0
49. |3| 3
50. |−20| 20
51. $\left|\frac{20}{5}\right|$ 4
52. |−(−3)| 3
53. |6| + |−6| 12
54. $\left|-2\frac{1}{3}\right|$ $2\frac{1}{3}$
55. |−1| · |8| 8
56. −|16 − 10| −6

Determine whether each statement is true or false.

57. On a number line, x and −x are the same distance from 0.   **True**
58. 0 is neither a whole number nor an integer.   **False**
59. If x is a positive integer, then |x| = −x.   **False**
60. There are two real numbers whose absolute value is 6.   **True**

CONNECTION

61. **COORDINATE GEOMETRY** Determine the final point on a number line if you start at −7, move 3 spaces to the left, and then move $4\frac{1}{2}$ spaces to the right. Write an equation that models this situation. $-5\frac{1}{2}; -7 - 3 + 4\frac{1}{2} = -5\frac{1}{2}$

Look Back

Find the next two terms in each number sequence. *(LESSON 1.1)*

62. 6, 13, 22, 33, ...   46, 61
63. 1, 7, 23, 50, 89, ...   141, 207

APPLICATION

CONSUMER ECONOMICS A truck rental firm charges $25 plus 30 cents per mile to rent a moving truck. *(LESSON 1.2)*

64. Write an equation that shows the cost, c, of renting the truck in terms of the number of miles, m.   c = 25 + 0.3m
65. Make a table and find the rental cost for 25, 50, 75, 100, and 125 miles.
66. How many miles can you drive the truck if you have budgeted $58 for the rental?   **110 miles**

Look Beyond

67. Use a graphics calculator to plot the function y = ABS(x). Describe the graph.   **The graph is V-shaped. The point of the V is at the origin.**

---

44. a. Answers may vary.
   Sample answer: x = 7
   b. Answers may vary.
   Sample answer: x = −7
   c. x = 0

65.

m	25	50	75	100	125
c	$32.50	$40	$47.50	$55	$62.50

67.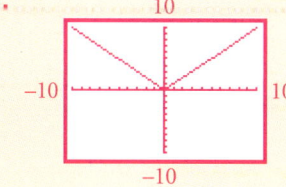

Prepare

NCTM PRINCIPLES & STANDARDS
1, 2, 5–10

LESSON RESOURCES

- Practice 2.2
- Reteaching Master 2.2
- Enrichment Master 2.2
- Cooperative-Learning Activity 2.2
- Lesson Activity 2.2
- Problem Solving/Critical Thinking Master 2.2
- Teaching Transparency 7
- Teaching Transparency 8
- Teaching Transparency 9

QUICK WARM-UP

Find each sum.

1. $868 + 75$ **943**
2. $\frac{5}{7} + \frac{1}{2}$ **$\frac{17}{14}$, or $1\frac{3}{14}$**
3. $7.2 + 6.9$ **14.1**

Find each difference.

4. $401 - 83$ **318**
5. $\frac{5}{6} - \frac{3}{10}$ **$\frac{8}{15}$**
6. $16 - 9.24$ **6.76**

Also on Quiz Transparency 2.2

Teach

Why Have students describe situations in which they have seen positive and negative numbers used. For example, students interested in sports may cite the use of positive and negative numbers when referring to gains and losses of yards in football and when recording the number of strokes above or below par in golf.

2.2 Adding Real Numbers

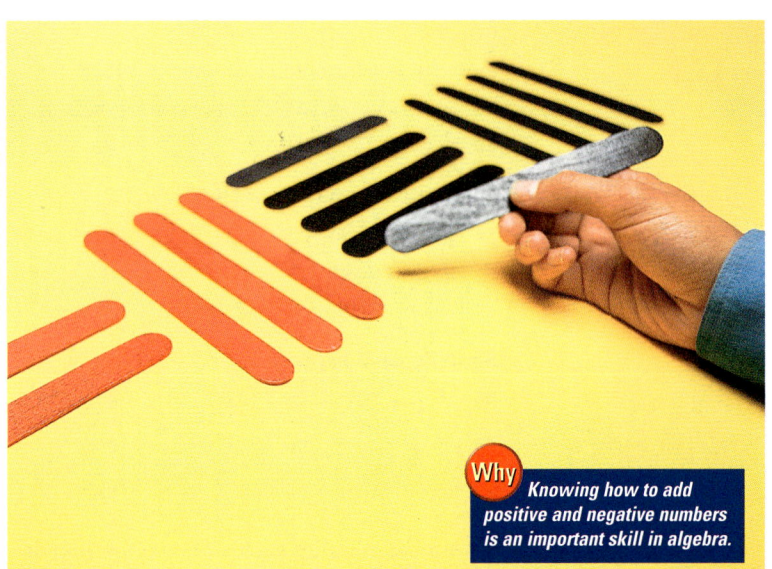

Objectives

- Use algebra tiles to model addition.
- Add numbers with like signs.
- Add numbers with unlike signs.

Why Knowing how to add positive and negative numbers is an important skill in algebra.

You can use sticks of different colors to practice an ancient Chinese method of addition.

APPLICATION
BUSINESS

CULTURAL CONNECTION: ASIA More than 2000 years ago, the Chinese used rods of different colors to represent positive and negative numbers. Consider the following problem:

A business transaction involved a loss of 23 monetary units (−23) and a gain of 54 monetary units (+54). What was the overall gain or loss?

To solve a problem like this, vertical rods were used to represent units or ones, and horizontal rods were used to represent tens. Rods of one color were positive (gains), and rods of another color were negative (losses).

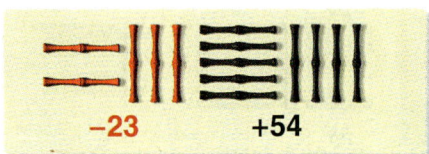

A gain and a loss of equal amounts do not change the total. Therefore, pairs of red and black rods can be eliminated from the layout because they have no effect on the total.

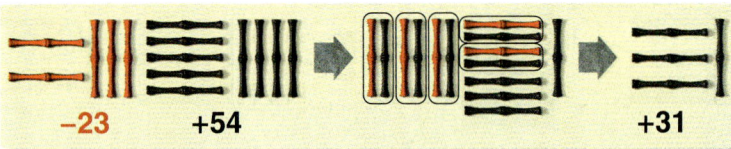

Alternative Teaching Strategy

USING MODELS Use a number line to model addition of real numbers. Begin by discussing the structure of a conventional number line, with positive numbers to the right of zero and negative numbers to the left. Then show students how addition can be modeled as a series of moves on this number line. A positive addend is represented by a move to the right, and a negative addend by a move to the left. Arrows are used to show the moves.

For example, model $5 + (-11)$ as follows:
Start at zero and move 5 units to the *right*.
Move 11 units *left* from this point.
End at −6.
Thus, $5 + (-11) = -6$.

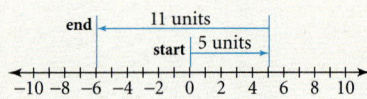

Have students draw a similar model for $-5 + 11$. Discuss how the models are alike and different.

Adding Integers

Like the rods in the Chinese mathematical example, algebra tiles can be used to model integer addition. One color of tile can represent the positive integers, and another color of tile can represent the negative integers.

To model the addition of two integers with like signs (both positive or both negative), represent each number with appropriately colored tiles. Then combine the tiles. Count the combined tiles and use the same sign (positive or negative) as the numbers you added.

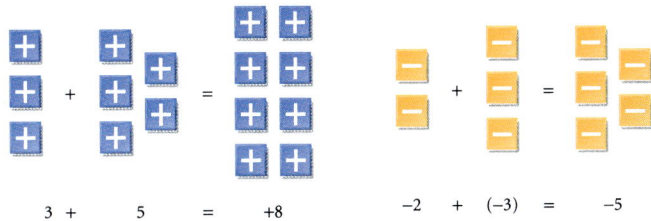

$$3 + 5 = +8 \qquad -2 + (-3) = -5$$

Modeling the addition of two integers with unlike signs (one positive and one negative) involves neutral pairs of tiles. A **neutral pair** consist of one positive and one negative tile. Since the value of a neutral pair is zero, they do not affect the total value of a group of tiles.

The value of a neutral pair is 0.

To model the addition of two integers with unlike signs, represent each number with appropriately colored tiles. Then combine the tiles. Remove all neutral pairs. The remaining tiles represent the sum. The color of the tiles indicates the sign of the sum.

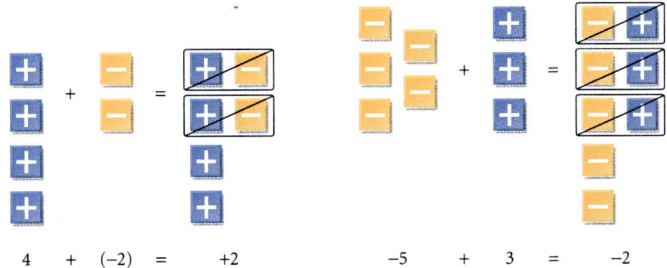

$$4 + (-2) = +2 \qquad -5 + 3 = -2$$

Teaching Tip

Be sure students understand the term *signed numbers*. The symbol − in front of a number indicates that its sign is negative. If there is no symbol in front of a nonzero number, its sign is understood to be positive. Emphasize the fact that the number 0 is neither positive nor negative.

Use Teaching Transparency 7.

Interdisciplinary Connection

HISTORY The plus and minus signs, + and −, first appeared in print in an arithmetic book published in Leipzig, Germany, in 1489 by Johann Widman. However, no one is certain of the origin of the signs. There is some evidence that they originally were warehouse markings used to indicate whether the contents of a barrel of goods exceeded or fell short of a standard weight.

Have students imagine that the standard weight of a barrel of goods is 16 ounces. Have them complete the table at right.

Deviation (ounces)	−2	−0.03	0	+0.5	+1.29
Actual weight (ounces)	14	15.97	16	16.5	17.29

The symbols + and − also indicate the operations of addition and subtraction. However, there have been other addition and subtraction symbols throughout history. Have students research these symbols and make a display of their findings.

LESSON 2.2

Activity Notes

Students generally feel comfortable adding numbers that have like signs. However, some students experience difficulty when the signs of the addends are not the same. This Activity provides all students with a concrete model that they can use to visualize the process. Students should be able to use the model as the basis for writing a formal rule for adding numbers with unlike signs. Be sure they understand that their rule should include the concept of absolute value.

Cooperative Learning

Have students work in groups of four to complete this Activity. Each group should choose one member to read the steps from the textbook, one to manipulate the tiles, one to keep a record of the tile arrangements, and another to write notes and summarize the group's results. All group members should be involved in a discussion of the Activity.

CHECKPOINT ✓
6. Find the difference of the absolute values, and use the sign of the number with the greater absolute value.

Teaching Tip

Have students write several original examples that illustrate each rule for addition in algebra. Ask them to share their examples with the class.

Use Teaching Transparency 8.

Activity: Modeling Addition—Unlike Signs

You will need: positive and negative tiles

1. Arrange 5 positive tiles and 8 negative tiles as shown.

2. Rearrange the tiles to form as many neutral pairs as you can. (One neutral pair has been circled.)

3. Remove all of the neutral pairs. How many tiles remain? Are they positive or negative? What is the answer to $5 + (-8)$?

4. Experiment by modeling other sums, such as $-10 + 6$ and $10 + (-3)$.

5. Look for a pattern in your results. For example, when the number of negative tiles is greater than the number of positive tiles, what is the sign of the answer?

CHECKPOINT ✓ 6. How can you add numbers with unlike signs without using algebra tiles? Write your own rule. (Hint: Use absolute value.)

Using Rules for Addition

By now, you may have already discovered the rules for addition in algebra. The rules use the definition of absolute value (Lesson 2.1).

Rules for Adding Two Signed Numbers

Addition of two numbers with like signs
1. Find the *sum* of the absolute values.
2. Use the sign common to both numbers.

Addition of two numbers with unlike signs
1. Find the *difference* of the absolute values.
2. Use the sign of the number with the greater absolute value.

Enrichment

Discuss with students how positive and negative numbers are used in the countdown for the launch of a space shuttle. Explain that a time of 30 seconds before liftoff is identified as *T minus 30*. A time of 30 seconds after liftoff is *T plus 30*. An expression such as *T minus 30 and counting* means that the liftoff is proceeding on schedule, while *T minus 30 and holding* indicates that the liftoff has been postponed.

Give students the following problem:

A maneuver that begins at *T* minus 20 requires 45 seconds for completion. At what time will the maneuver be completed? $-20 + 45 = 25$; *T* plus 25

Have students work in groups of three or four. Ask each group to write two other problems related to times before or after liftoff that can be solved by adding signed numbers. Have the groups share their problems with the entire class.

EXAMPLE 1 Find each sum.
a. $-2 + (-8)$
b. $15 + (-23)$
c. $14 + (-5)$

○ **SOLUTION**

a. Since the numbers have like signs (both negative), find the sum of their absolute values. Use the sign common to both numbers.

$|-2| = 2$ $|-8| = 8$ $2 + 8 = 10$ $-2 + (-8) = -10$

b. Since the numbers have unlike signs, find the difference of their absolute values. Use the sign of the number with the greater absolute value.

$|15| = 15$ $|-23| = 23$ $23 - 15 = 8$ $15 + (-23) = -8$

c. Since the numbers have unlike signs, find the difference of their absolute values. Use the sign of the number with the greater absolute value.

$|14| = 14$ $|-5| = 5$ $14 - 5 = 9$ $14 + (-5) = 9$

TRY THIS Find each sum.
a. $-3 + (-12)$
b. $17 + (-32)$
c. $-15 + 18$

The rules of algebraic addition apply to all real numbers.

EXAMPLE 2 Find each sum.
a. $-1.3 + 2.5$
b. $-\frac{1}{4} + \left(-1\frac{1}{4}\right)$
c. $-3.7 + (-8.6)$
d. $-4\frac{2}{3} + 5\frac{1}{3}$

○ **SOLUTION**

a. Since the numbers have unlike signs, find the difference of their absolute values. Use the sign of the number with the greater absolute value.

$|-1.3| = 1.3$ $|2.5| = 2.5$ $2.5 - 1.3 = 1.2$ $-1.3 + 2.5 = 1.2$

b. Since the numbers have like signs (both negative), find the sum of their absolute values. Use the sign common to both numbers.

$\left|-\frac{1}{4}\right| = \frac{1}{4}$ $\left|-1\frac{1}{4}\right| = 1\frac{1}{4}$ $\frac{1}{4} + 1\frac{1}{4} = 1\frac{2}{4} = 1\frac{1}{2}$ $-\frac{1}{4} + \left(-1\frac{1}{4}\right) = -1\frac{1}{2}$

c. Since the numbers have like signs, find the sum of their absolute values. Use the sign common to both numbers.

$|-3.7| = 3.7$ $|-8.6| = 8.6$ $3.7 + 8.6 = 12.3$ $-3.7 + (-8.6) = -12.3$

d. Since the numbers have unlike signs, find the difference of their absolute values. Use the sign of the number with the greater absolute value.

$\left|-4\frac{2}{3}\right| = 4\frac{2}{3}$ $\left|5\frac{1}{3}\right| = 5\frac{1}{3}$ $5\frac{1}{3} - 4\frac{2}{3} = \frac{2}{3}$ $-4\frac{2}{3} + 5\frac{1}{3} = \frac{2}{3}$

TRY THIS Find each sum.
a. $-9 + 3.7$
b. $7 + \left(-1\frac{3}{4}\right)$
c. $-6.3 + (-2.5)$
d. $4\frac{1}{5} + \left(-7\frac{2}{3}\right)$

ADDITIONAL EXAMPLE 1

Find each sum.
a. $-6 + (-4)$ -10
b. $-14 + 21$ 7
c. $18 + (-7)$ 11

TRY THIS
a. -15
b. -15
c. 3

Use Teaching Transparency 9.

Teaching Tip

Remind students that when a negative number follows a plus sign, they should enclose that number in parentheses. For example, they should write $-5 + (-3)$, not $-5 + -3$.

ADDITIONAL EXAMPLE 2

Find each sum.
a. $1.52 + (-3.98)$ -2.46
b. $-\frac{3}{11} + \frac{8}{11}$ $\frac{5}{11}$
c. $-4.5 + (-3.7)$ -8.2
d. $-6\frac{5}{8} + 2\frac{7}{8}$ $-3\frac{3}{4}$

TRY THIS
a. -5.3 b. $5\frac{1}{4}$
c. -8.8 d. $-3\frac{7}{15}$

Inclusion Strategies

ENGLISH LANGUAGE DEVELOPMENT In the English language, there are many pairs of words that commonly are associated with positive and negative numbers. For example, when directions are identified by signed numbers, *up and down* or *above and below* often are used to describe the positive and negative directions, respectively. Have students brainstorm to create a list of similar word pairs. Students who are familiar with other languages may enjoy an opportunity to translate the word pairs into those languages.

Reteaching the Lesson

USING PATTERNS On the board or overhead, write each group of addition equations below. Have students copy and complete the pattern of the sums. After students have finished, ask them to write a rule that describes the relationship.

$1 + 2 = 3$	$1 + (-2) = -1$	$-1 + 2 = 1$	$-1 + (-2) = -3$
$1 + 3 = 4$	$1 + (-3) = -2$	$-1 + 3 = 2$	$-1 + (-3) = -4$
$1 + 4 = 5$	$1 + (-4) = -3$	$-1 + 4 = 3$	$-1 + (-4) = -5$
$1 + 5 = ?$	$1 + (-5) = ?$	$-1 + 5 = ?$	$-1 + (-5) = ?$
$1 + 6 = ?$	$1 + (-6) = ?$	$-1 + 6 = ?$	$-1 + (-6) = ?$

LESSON 2.2

Assess

Selected Answers

Exercises 5–10, 11–89 odd

ASSIGNMENT GUIDE

In Class	1–10
Core	11–14, 19–60, 67, 69
Core Plus	12–66 even, 67–69
Review	70–90
Preview	91

✏ Extra Practice can be found beginning on page 738.

Technology

A calculator may be helpful for Exercises 19–69. You may wish to have students work the exercises with paper and pencil first and then use a calculator to verify their results.

Practice

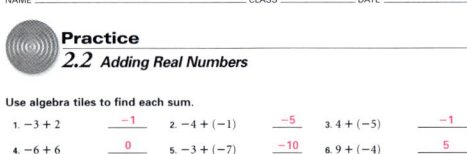

Practice
2.2 Adding Real Numbers

Use algebra tiles to find each sum.

1. $-3 + 2$ **−1**
2. $-4 + (-1)$ **−5**
3. $4 + (-5)$ **−1**
4. $-6 + 6$ **0**
5. $-3 + (-7)$ **−10**
6. $9 + (-4)$ **5**
7. $7 + (-4)$ **3**
8. $-7 + 9$ **2**
9. $-8 + 3$ **−5**
10. $10 + (-9)$ **1**
11. $-1 + (-9)$ **−10**
12. $(-6) + (-2)$ **−8**
13. $-4 + (-1) + 3$ **−2**
14. $5 + (-1) + (-2)$ **2**
15. $-3 + (-4) + 2$ **−5**

Find each sum.

16. $-35 + 40$ **5**
17. $15 + (-28)$ **−13**
18. $60 + (-18)$ **42**
19. $-17 + (-19)$ **−36**
20. $42 + (-56)$ **−14**
21. $-34 + 28$ **−6**
22. $59 + (-59)$ **0**
23. $-86 + 85$ **−1**
24. $-45 + (-45)$ **0**
25. $-68 + (-15)$ **−83**
26. $-3 + (-1) + (-2)$ **−6**
27. $4 + (-7) + (-4)$ **−7**
28. $-54 + 63 + (-20)$ **−11**
29. $-78 + (-78) + 50$ **50**
30. $-6 + (-42) + 24$ **−24**
31. $-24 + (-62) + (-11)$ **−97**
32. $-5 + |-4|$ **−1**
33. $|5| + |-4|$ **9**
34. $|-5| + |-4|$ **9**
35. $|-5| + |4| + (-9)$ **0**
36. $-48 + |-64| + (-32)$ **−16**
37. $-568 + (-43) + |-57|$ **−554**

Substitute 4 for a, -6 for b, and 3 for c. Evaluate each expression.

38. $a + (b + c)$ **1**
39. $a + |b + c|$ **7**
40. $(a + c) + b$ **1**
41. $(a + c) + |b|$ **13**
42. $|a + b| + c$ **5**
43. $|a + c| + b$ **1**
44. $|c| + b + a$ **1**
45. $a + |b| + |c|$ **13**

The Additive Identity and Additive Inverse

The sum of any number and 0 is equal to the number. Therefore, 0 is called the **identity element for addition**.

Identity Property for Addition

For all real numbers a, $a + 0 = a$ and $0 + a = a$.

Related to the identity element, 0, are **additive inverses**. Another name for additive inverse is *opposite*. The sum of two additive inverses is 0, the identity element.

$$-7 + 7 = 0 \qquad 4 + (-4) = 0 \qquad -3 + 3 = 0$$

Additive Inverse Property

For every real number a, there is exactly one real number $-a$ such that $a + (-a) = 0$ and $-a + a = 0$.

Exercises

Communicate

1. Describe how to add two integers that have the same sign.
2. Describe how to add two integers that have different signs.
3. Explain how to use algebra tiles to model the following sums:
   a. $-4 + (-5)$ **b.** $-3 + 6$
4. Explain how the Identity Property for Addition and the Additive Inverse Property are useful when modeling addition with algebra tiles.

Guided Skills Practice

Find each sum. (EXAMPLE 1)

5. $-3 + (-15)$ **−18**
6. $12 + (-7)$ **5**
7. $-24 + 37$ **13**

Find each sum. (EXAMPLE 2)

8. $-2.35 + (-1.76)$ **−4.11**
9. $-5.77 + 12.5$ **6.73**
10. $-5\frac{3}{4} + 1\frac{1}{4}$ **−4$\frac{1}{2}$**

64   LESSON 2.2

Practice and Apply

Use algebra tiles to find the following sums:

11. $-5 + (-2)$ **−7**
12. $-3 + 3$ **0**
13. $2 + (-6)$ **−4**
14. $-2 + 9$ **7**
15. $7 + (-3)$ **4**
16. $-2 + (-6)$ **−8**
17. $8 + (-5)$ **3**
18. $6 + (-6)$ **0**

Find each sum.

19. $-7 + (-15)$ **−22**
20. $-10 + 7$ **−3**
21. $23 + (-15)$ **8**
22. $28 + (-50)$ **−22**
23. $5.7 + (-3.2)$ **2.5**
24. $-17 + (-34)$ **−51**
25. $\frac{4}{17} + \left(-\frac{12}{17}\right)$ **$-\frac{8}{17}$**
26. $84 + (-18)$ **66**
27. $-29 + 14$ **−15**
28. $-43 + (-82)$ **−125**
29. $-308 + (-80)$ **−388**
30. $-56 + 107$ **51**
31. $-5 + (-5)$ **−10**
32. $-5 + 5$ **0**
33. $-13 + 20$ **7**
34. $13 + (-20)$ **−7**
35. $2.85 + 4.96$ **7.81**
36. $-5.3 + 1.4$ **−3.9**
37. $6.98 + (-0.2)$ **6.78**
38. $0.17 + (-3.45)$ **−3.28**
39. $-0.5 + 0.25$ **−0.25**
40. $0.26 + (-1.5)$ **−1.24**
41. $3.7 + (-0.7)$ **3**
42. $5.6 + (-3.2)$ **2.4**
43. $-8.27 + (-3.2)$ **−11.47**
44. $5.23 + (-2.7)$ **2.53**
45. $-3.4 + 7$ **3.6**
46. $-14 + 2.5$ **−11.5**
47. $-\frac{4}{15} + \left(-\frac{7}{15}\right)$ **$-\frac{11}{15}$**
48. $\frac{11}{25} + \left(-\frac{13}{25}\right)$ **$-\frac{2}{25}$**
49. $-\frac{1}{2} + \frac{1}{3}$ **$-\frac{1}{6}$**
50. $-\frac{3}{4} + \frac{1}{5}$ **$-\frac{11}{20}$**
51. $-1\frac{2}{5} + 3\frac{1}{4}$ **$1\frac{17}{20}$**
52. $2\frac{1}{4} + \left(-5\frac{2}{5}\right)$ **$-3\frac{3}{20}$**
53. $\frac{11}{3} + \left(-\frac{11}{3}\right)$ **0**
54. $-\frac{3}{2} + \left(-\frac{10}{7}\right)$ **$-2\frac{13}{14}$**

Let $a = 2$, $b = 3$, and $c = -5$. Evaluate each expression.

55. $a + b$ **5**
56. $a + (-b)$ **−1**
57. $-a + b$ **1**
58. $-a + (-b)$ **−5**
59. $a + c$ **−3**
60. $a + (-c)$ **7**
61. $-b + c$ **−8**
62. $-b + (-c)$ **2**
63. $a + b + c$ **0**
64. $(a + b) + (-a) + (-b)$ **0**
65. $-c + (-b) + (a) + (-b)$ **1**

CHALLENGE

66. $2(a^2 - b) + 3c - 5ab$ **−43**

CONNECTION

67. **STATISTICS** Suppose that a quarterback had the following plays in a football game: a pass for a gain of 7 yards, a sack for a loss of 2 yards, a sneak for a gain of 6 yards, and a sack for a loss of 13 yards.
    a. What was the total yardage gained or lost on these four plays? **−2**
    b. Suppose that the next two plays were a gain of 24 yards and then a loss of 17 yards. What was the total yardage for all six plays? **5**

APPLICATIONS

Glucose, $C_6H_{12}O_6$

68. **BANKING** Jeff opened a savings account with $57. Later, he deposited another $32. Finally, he made a withdrawal of $45 in order to buy some new CDs. What was the remaining balance in his savings account? **$44**

69. **CHEMISTRY** Protons have a charge of +1 and electrons have a charge of −1. The total charge of a particle is the sum of its proton charges and electron charges. Find the total charge of a particle with the following numbers of protons and electrons:
    a. 12 protons, 10 electrons **+2**
    b. 16 protons, 18 electrons **−2**
    c. 10 protons, 10 electrons **0**

Exercise 91 uses the relationship between addition and multiplication to preview multiplication of signed numbers. This concept will be studied in depth in Lesson 2.4.

ALTERNATIVE
Assessment

Portfolio Activity

The Portfolio Activity can be used as preparation for the Chapter Project or as a separate activity. In this Portfolio Activity, students see how positive and negative numbers are used to indicate profits and losses on a business spreadsheet.

Answers to Portfolio Activities can be found in Additional Answers of the Teacher's Edition.

83–86.

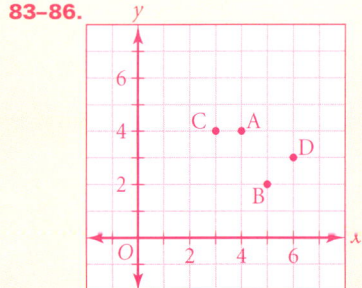

70. Find the sum $1 + 2 + 3 + \cdots + 40$. **(LESSON 1.1)** **820**

Find the next two terms in each number sequence. (LESSON 1.1)

71. 40, 37, 34, 31, . . . **28, 25**   **72.** −1, 3, 7, . . . **11, 15**   **73.** 1, 3, 6, 10, . . . **15, 21**

74. 5, 11, 18, 26, . . . **35, 45**   **75.** 2, 4, 8, 16, . . . **32, 64**   **76.** 2, 4, 7, 11, . . . **16, 22**

Evaluate. (LESSON 1.3)

77. $15 \cdot 2 - 7$ **23**   **78.** $[3(4-2)^2] + 7$ **19**   **79.** $12 + 3^2 + (9-6)$ **24**

80. $5^2 + 2^2 + 4^2$ **45**   **81.** $(7 + 3 \cdot 2) - 2^2$ **9**   **82.** $100 - 99 \div 33 \div 3$ **99**

Plot each point on the same coordinate plane. (LESSON 1.4)

83. $A(4, 4)$ **84.** $B(5, 2)$ **85.** $C(3, 4)$ **86.** $D(6, 3)$

Find the absolute value of each number. (LESSON 2.1)

87. 17 **17**   **88.** −1.2 **1.2**   **89.** $-\frac{13}{20}$ **$\frac{13}{20}$**   **90.** $(12.8 - 5)$ **7.8**

Look Beyond

91. You may recall from arithmetic that adding the same number repeatedly can be represented by multiplication and vice versa.

$$2 + 2 + 2 + 2 + 2 = 5(2) = 10$$

With this in mind, represent $7(-2)$ as an addition problem, and then find the value of $7(-2)$.

$(-2) + (-2) + (-2) + (-2) + (-2) + (-2) + (-2) = 7(-2) = -14$

Include this activity in your portfolio.

The table below comes from the quarterly spreadsheet that a business prints out. Study the spreadsheet and answer the questions below.

	Revenue	Expenses	Profit
January	25, 626	12, 810	12, 816
February	10, 887	14, 098	−3211
March	12, 978	12, 995	−17
TOTAL	49, 491	39, 903	9588

1. Explain how the numbers in the profit column are calculated.
2. Why are some of the numbers in the profit column negative?
3. Explain how 9588 in the profit column can be calculated in two different ways.

66 LESSON 2.2

2.3 Subtracting Real Numbers

Objectives
- Use algebra tiles to model subtraction.
- Define subtraction in terms of addition.
- Subtract numbers with like and unlike signs.

Why In algebra, the properties of the basic operations are defined in terms of addition and multiplication. You must understand how to convert subtraction to addition in order to use the properties when you solve equations.

When Mark uses his debit card, the bank subtracts, or debits, his account by the amount of his purchase.

**APPLICATION
PERSONAL FINANCE**

Suppose that Mark starts with $50 in his account and uses his debit card for a $20 purchase. This transaction can be represented in two ways.

Subtraction: $50 - 20$ Addition: $50 + (-20)$

According to either method, Mark will have $30 left.

$50 - 20 = 30$ $50 + (-20) = 30$

In algebra, it is often useful to think of subtraction as *adding the opposite* of the number that is being subtracted, as shown above. Let's first consider the more familiar idea of subtraction as "taking away."

Using algebra tiles, you can model a subtraction problem such as $6 - 2$ by beginning with 6 tiles and taking away 2 tiles.

How could you model $2 - 6$ by taking away tiles? The answer involves neutral pairs.

Prepare

NCTM PRINCIPLES & STANDARDS
1, 2, 5–10

LESSON RESOURCES
- Practice 2.3
- Reteaching Master 2.3
- Enrichment Master 2.3
- Cooperative-Learning Activity 2.3
- Lesson Activity 2.3
- Problem Solving/Critical Thinking Master 2.3

QUICK WARM-UP

Find the opposite of each number.

1. -9 **9**   2. 3.7 **-3.7**
3. 0 **0**   4. $-\frac{1}{3}$ **$\frac{1}{3}$**

Find each sum.

5. $-5 + 8$ **3**
6. $-13 + (-6)$ **-19**
7. $-\frac{3}{5} + \frac{2}{5}$ **$-\frac{1}{5}$**
8. $6.9 + (-4.28)$ **2.62**

Also on Quiz Transparency 2.3

Teach

Why The financial transaction described on this page is an application of subtraction that may be familiar to many students. There are many other real-world situations that involve subtraction. Write the subtraction expression $3 - (-2)$ on the board or overhead. Have students suggest real-world situations in which a person might want to compute this difference.
Sample: How much higher is a score of 3 than a score of –2?

Alternative Teaching Strategy

USING MODELS Temperature changes can be used to model subtraction of real numbers. A positive number represents an increase in temperature, while a negative number represents a decrease.

Have students consider a situation in which the temperature was 30°F at the end of the day and 10°F at the start of that same day. Ask them what number identifies the change in temperature. **20** Have them write a subtraction equation to represent the situation. **$30 - 10 = 20$**

Now have students consider a situation in which the temperature was 10°F at the end of the day and 30°F at the start of the same day. Ask them what number identifies the temperature change. **-20** Have them write a subtraction equation to represent the situation. **$10 - 30 = -20$**

Have students write similar subtraction equations to represent temperature changes from –10°F to 30°F and from 30°F to –10°F.
$30 - (-10) = 40; -10 - 30 = -40$

LESSON 2.3 **67**

ADDITIONAL EXAMPLE 1

Use algebra tiles to model the subtraction −4 − 1.

Start with 4 negative tiles.
Add 1 neutral pair.
Take away 1 positive tile.
There are 5 negative tiles left.
−4 − 1 = −5

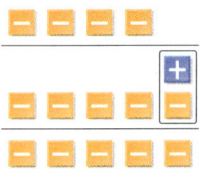

TRY THIS

After we take away 10 positive tiles, there will be 6 negative tiles left. 4 − 10 = −6

Activity Notes

The method students will use to complete this Activity is similar to the tile method shown in Example 1. Students should discover that after 3 positive tiles are taken away, the result is the same as if 3 negative tiles had been added. You may wish to have students model another subtraction expression, such as 2 − (−3), in which they can see that the result of taking away 3 negative tiles is the same as adding 3 positive tiles. This will support the conclusion that the result of subtracting a number is the same as adding its opposite.

CHECKPOINT ✔
3. adding 3 negative tiles

EXAMPLE 1

Use algebra tiles to model the subtraction 2 − 6.

● SOLUTION

Start with 2 positive tiles.

Since you want to subtract 6, you will need 4 more positive tiles than you already have. Add 4 neutral pairs (4 positive and 4 negative tiles). Remember that neutral pairs do not change the value.

Now you can subtract, or *take away*, 6 positive tiles. There will be 4 negative tiles left.
2 − 6 = −4

TRY THIS Use algebra tiles to model the subtraction 4 − 10.

Subtraction as Adding the Opposite

The bank account example on page 67 showed that a subtraction problem can be represented by an addition problem. The addition problem 50 + (−20) has the same answer as the subtraction problem 50 − 20.

Definition of Subtraction

For all real numbers *a* and *b*:

$$a - b = a + (-b)$$

Activity
Modeling the Definition of Subtraction

You will need: algebra tiles

1. Suppose that you want to *subtract* 3 positive tiles from some number that is represented by another group of tiles (not shown). No matter what the other number is, you can make sure that there will be enough tiles to subtract if you add 3 neutral pairs to the group.

2. After you have added the 3 neutral pairs, what will the next step be?

CHECKPOINT ✔ 3. What is the overall effect of adding 3 neutral pairs and removing 3 positive tiles?

4. State how the steps in this activity show that subtracting a given number is the same as adding the opposite of that number.

Interdisciplinary Connection

ACCOUNTING Accountants are financial professionals who use subtraction of real numbers in their daily work. Have students do research to find out more about how accountants use subtraction of real numbers. If possible, invite an accountant to talk to your class about his or her work.

Enrichment

Have students solve "missing number" problems such as the following:

3 − ? = 9	−6	−8 − ? = 8	−16
? − 7 = −5	2	0 − ? = 5	−5
9 − ? = −6	15	? − (−4) = 9	5

68 LESSON 2.3

EXAMPLE 2 Use the definition of subtraction to find each difference.

a. $-3 - (-4)$
b. $-5.6 - (-1.4)$
c. $-1\frac{2}{3} - 4\frac{1}{3}$
d. $5 - 2.3$

● SOLUTION

To subtract a number, find its opposite, and then add.

a. The opposite of -4 is 4.
$-3 - (-4) = -3 + 4 = 1$

b. The opposite of -1.4 is 1.4.
$-5.6 - (-1.4) = -5.6 + 1.4 = -4.2$

c. The opposite of $4\frac{1}{3}$ is $-4\frac{1}{3}$.
$-1\frac{2}{3} - 4\frac{1}{3} = -1\frac{2}{3} + \left(-4\frac{1}{3}\right) = -6$

d. The opposite of 2.3 is -2.3.
$5 - 2.3 = 5 + (-2.3) = 2.7$

TRY THIS Use the definition of subtraction to find each difference.

a. $-11 - \left(-5\frac{1}{2}\right)$
b. $-2.7 - 5.3$
c. $2 - 5$
d. $8.2 - 6$

Measuring Distance on a Number Line

Sometimes you will need to determine the distance between points on a number line. One method of doing this uses absolute value.

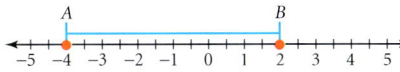

To find the distance between points A and B, subtract the values, and then find the absolute value of the difference.

The distance between -4 and 2 is
$|2 - (-4)| = |6| = 6$, or $|(-4) - 2| = |-6| = 6$.

In general, if you let a represent the value for point A and let b represent the value for point B, the distance between points A and B on a number line is $|b - a|$ or $|a - b|$.

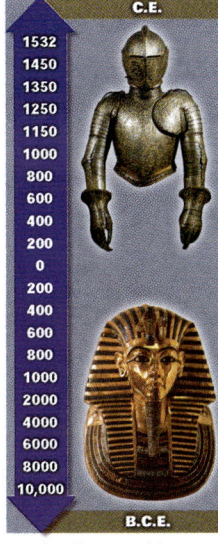

A timeline provides an easy visual reference for a sequence of dates.

EXAMPLE 3 What is the distance between -5 and 3 on a number line?

● SOLUTION

The distance between -5 and 3 is $|-5 - 3| = |-8| = 8$, or $|3 - (-5)| = |8| = 8$.

TRY THIS Find the distance between -2 and 13 on a number line.

Inclusion Strategies

ENGLISH LANGUAGE DEVELOPMENT Listen carefully as students read a subtraction expression such as $-4 - 5$. Be sure they interpret it as *negative four minus five*, not *minus four minus five*. Stress the fact that distinguishing between the sign of a number and the operation of subtraction is necessary in order to apply the definition of subtraction correctly.

Reteaching the Lesson

USING PATTERNS Give students the subtraction equations at right. Point out the pattern among the differences that have been calculated. Have them find the missing differences by extending the pattern. Using the completed subtraction equations, lead them to see that subtracting a number is the same as adding its opposite.

$4 - 7 = \square$
$4 - 6 = \square$
$4 - 5 = \square$
$4 - 4 = 0$
$4 - 3 = 1$
$4 - 2 = 2$
$4 - 1 = 3$
$4 - 0 = 4$
$4 - (-1) = \square$
$4 - (-2) = \square$
$4 - (-3) = \square$

ADDITIONAL EXAMPLE 2

Use the definition of subtraction to evaluate each difference.

a. $-10 - (-10)$ 0
b. $-5.8 - 4.8$ -10.6
c. $2\frac{1}{4} - \left(-1\frac{3}{4}\right)$ 4

TRY THIS

a. $-5\frac{1}{2}$
b. -8
c. -3
d. 2.2

ADDITIONAL EXAMPLE 3

Find the distance between -4 and -10 on a number line.

6

TRY THIS

15

Assess

Selected Answers
Exercises 6–15, 17–85 odd

ASSIGNMENT GUIDE

In Class	1–15
Core	16–27, 29–67 odd
Core Plus	16–46 even, 48–68
Review	69–85
Preview	86, 87

✐ Extra Practice can be found beginning on page 738.

Mid-Chapter Assessment for Lessons 2.1 through 2.3 can be found on page 21 of the *Assessment Resources*.

Closure

The sum of any two real numbers is a real number. To mathematicians, this means that the set of real numbers is *closed under addition*. The set of real numbers is also closed under subtraction. As a comparison, consider the whole numbers. Remember that the whole numbers consist of only the positive integers and zero. Are the whole numbers closed under addition and subtraction?

Adding two positive integers, adding a positive integer and zero, and adding zero and zero all result in a whole number. Therefore, the whole numbers are closed under addition.

$$2 + 4 = 6 \qquad 3 + 0 = 3 \qquad 0 + 0 = 0$$

The sums are all whole numbers.

What about subtraction? You cannot determine the value of a subtraction expression such as $4 - 8$ without using negative integers.

$$4 - 8 = -4 \qquad \text{Not a whole number}$$

Because the whole numbers do not include the negative integers, the whole numbers are not closed under subtraction.

Exercises

Communicate

1. How does the addition of neutral pairs of algebra tiles model a subtraction problem?

APPLICATION
2. **METEOROLOGY** The temperature is 5°F. Explain how to find the temperature if it dropped by 25°F.

3. Describe how you would solve the problem $-4 - 7$ by using algebra tiles.

4. What is meant by "adding the opposite"?

5. Explain why the real numbers are closed under subtraction.

Guided Skills Practice

Find each difference. *(EXAMPLES 1 AND 2)*

6. $-8 - 5$ **−13**   7. $-9 - (-11)$ **2**   8. $-5 - 14$ **−19**   9. $6 - (-8)$ **14**

APPLICATION
10. **NAVIGATION** A submarine at a depth of 98 feet descends another 89 feet. What is the new depth of the submarine? *(EXAMPLES 1 AND 2)* **−187**

Find the distance between each pair of points on a number line. *(EXAMPLE 3)*

11. $3, -7$ **10**   12. $-5, -6$ **1**   13. $13, -13$ **26**   14. $-3, 13$ **16**   15. $8, -6$ **14**

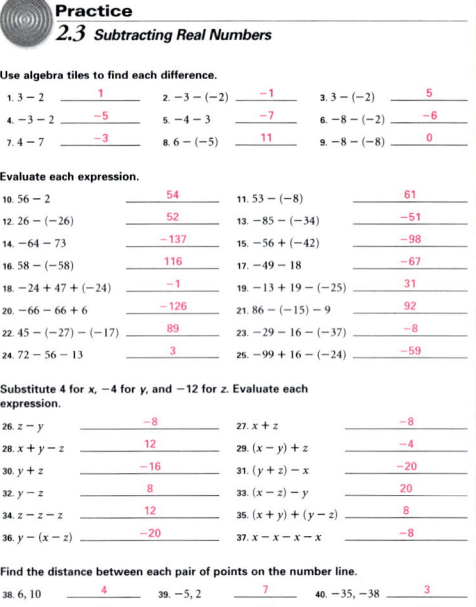

Practice 2.3 Subtracting Real Numbers

Use algebra tiles to find each difference.
1. $3 - 2$ **1**
2. $-3 - (-2)$ **−1**
3. $3 - (-2)$ **5**
4. $-3 - 2$ **−5**
5. $-4 - 3$ **−7**
6. $-8 - (-2)$ **−6**
7. $4 - 7$ **−3**
8. $6 - (-5)$ **11**
9. $-8 - (-8)$ **0**

Evaluate each expression.
10. $56 - 2$ **54**
11. $53 - (-8)$ **61**
12. $26 - (-26)$ **52**
13. $-85 - (-34)$ **−51**
14. $-64 - 73$ **−137**
15. $-56 + (-42)$ **−98**
16. $58 - (-58)$ **116**
17. $-49 - 18$ **−67**
18. $-24 + 47 + (-24)$ **−1**
19. $-13 + 19 - (-25)$ **31**
20. $-66 - 66 + 6$ **−126**
21. $86 - (-15) - 9$ **92**
22. $45 - (-27) - (-17)$ **89**
23. $-29 - 16 - (-37)$ **−8**
24. $72 - 56 - 13$ **3**
25. $-99 + 16 - (-24)$ **−59**

Substitute 4 for x, −4 for y, and −12 for z. Evaluate each expression.
26. $z - y$ **−8**
27. $x + z$ **−8**
28. $x + y - z$ **12**
29. $(x - y) + z$ **−4**
30. $y + z$ **−16**
31. $(y + z) - x$ **−20**
32. $y - z$ **8**
33. $(x - z) - y$ **20**
34. $z - z - z$ **12**
35. $(x + y) + (y - z)$ **8**
36. $y - (x - z)$ **−20**
37. $x - x - x - x$ **−8**

Find the distance between each pair of points on the number line.
38. $6, 10$ **4**
39. $-5, 2$ **7**
40. $-35, -38$ **3**
41. $-13, 26$ **39**
42. $-44, -29$ **15**
43. $-15, 73$ **88**

70 LESSON 2.3

Practice and Apply

Find each difference.

16. $12 - 7$ **5**
17. $-9 - 4$ **−13**
18. $-15 - (-12)$ **−3**
19. $11 - (-3)$ **14**
20. $-10 - 7$ **−17**
21. $24 - (-11)$ **35**
22. $-86 - (-92)$ **6**
23. $-57 - 14$ **−71**
24. $-117 - 82$ **−199**
25. $84.2 - (-12.5)$ **96.7**
26. $-65.4 - 32.8$ **−98.2**
27. $94 - (-16)$ **110**

Find each sum or difference.

28. $67 - 3$
29. $42 - (-9)$
30. $-10 - (-21)$
31. $-35 - 17$
32. $33 - (-33)$
33. $-78 + (-45)$
34. $990 - (-155)$
35. $-97 - 88$
36. $-43 + (-15)$
37. $-77 - (-77)$
38. $-108 + 118$
39. $85 - (-12)$
40. $-2.05 - 30.4$
41. $1.08 - (-6.79)$
42. $-0.012 - 0.65$
43. $64.5 - 65$
44. $\frac{1}{5} - \frac{4}{5}$
45. $\frac{3}{16} - \left(-\frac{7}{16}\right)$
46. $-\frac{1}{2} - \frac{1}{3}$
47. $\frac{3}{8} - \left(-\frac{3}{4}\right)$

28. 64
29. 51
30. 11
31. −52
32. 66
33. −123
34. 1145
35. −185
36. −58
37. 0
38. 10
39. 97
40. −32.45
41. 7.87
42. −0.662
43. −0.5
44. $-\frac{3}{5}$
45. $\frac{5}{8}$
46. $-\frac{5}{6}$
47. $\frac{9}{8}$

Let $x = 5$, $y = -3$, and $z = -10$. Evaluate each expression.

48. $x - y$ **8**
49. $-x - y$ **−2**
50. $-x - z$ **5**
51. $x - (-x)$ **10**
52. $y - z$ **7**
53. $-z - y$ **13**
54. $-z - (-y)$ **7**
55. $x - (-y)$ **2**

Find the distance between each pair of points on a number line.

56. $4, 9$ **5**
57. $-6, 15$ **21**
58. $-47, -23$ **24**
59. $-12, 74$ **86**
60. $-17, 5$ **22**
61. $12, 39$ **27**
62. $-3, 52$ **55**
63. $-86, -23$ **63**

CONNECTIONS

64. **STATISTICS** Susan, a marathon athlete, wants to improve her time in the Boston Marathon by 12 minutes this year.
    a. If her finish time last year was 2 hours and 7 minutes, what is her goal for this year? **1 hours 55 minutes**
    b. Suppose that at the race Susan manages to improve her time by 14 minutes over the previous year. What is her new finish time? **1 hours 53 minutes**

65. **PATTERNS IN DATA** Use constant differences to find the next three terms in each sequence.
    a. $-10, -7, -4, -1, \ldots$ **2, 5, 8**
    b. $60, 45, 30, 15, \ldots$ **0, −15, −30**
    c. $-20, -11, -3, 4, \ldots$ **10, 15, 19**

Technology

A calculator may be helpful for Exercises 16–55. You may wish to have students complete the exercises with paper and pencil first and then use a calculator to verify their answers.

Error Analysis

When subtracting integers, students usually remember to change the operation to addition, but they may forget to take the opposite of the second number. Encourage them to copy the subtraction problem exactly as given and then write the related addition problem immediately below it. To focus their attention on the two changes that must be made, have them draw an arrow between the operation signs and an arrow between the opposites, as shown below.

$$-8 - (-3)$$
$$\downarrow \quad \downarrow$$
$$-8 + 3$$

Student Technology Guide

2.3 Subtracting Real Numbers

You can use the sign change key [(-)] on a calculator to find the opposite of a number and help you find the difference of two integers. (It is important to note that the sign change key and the subtraction key are not the same. If you press the subtraction key [-] to enter a negative number or to find the opposite of a number, an error may occur.) To find the difference $-10 - (-6)$ with a calculator, you can use the key sequence below.

[(-)] 10 [-] [(-)] 6 [)] ENTER

The display will show −4. Thus, $-10 - (-6) = -4$.

Find each difference.

1. $-16 - 12$ **−28**
2. $-17 - (-11)$ **−6**
3. $18 - (-15)$ **33**
4. $19 - (-13)$ **32**
5. $-6 - (-16)$ **10**
6. $6 - (-16)$ **22**
7. $-25 - 18$ **−43**
8. $-16 - (-14)$ **−2**
9. $-420 - (-160)$ **−260**
10. $950 - (-675)$ **1625**
11. $0.38 - (-3.44)$ **3.82**
12. $-9.52 - (-9.52)$ **0**

The absolute-value function can be used to find the distance between two points on a number line. The subtraction key [-] is used to find the difference. For example, to find the distance between −8 and −5, calculate $|-8 - (-5)|$ by using the following keystrokes:

MATH NUM 1:abs([(-)] 8 [-] [(-)] 8 [)] ENTER

The display will show 3, so the distance between −8 and −5 is 3.

Find the distance between each pair of points on a number line.

13. $-9, 5$ **14**
14. $16, -14$ **30**
15. $-62, -43$ **19**
16. $-0.43, 1.9$ **2.33**

LESSON 2.3 **71**

Look Beyond

In this lesson, students learned how to find the distance between two points on a number line. This concept is extended in Exercises 86 and 87, in which students are given one point and a fixed distance and are asked to find a second point. Be sure they realize that there always are two points which satisfy the condition. Students will explore this concept in greater depth when they work with absolute-value equations in Lesson 6.5.

internetconnect

GO TO: go.hrw.com
KEYWORD: MA1 U.S. Jobs

In this activity, students research Web sites to find information on the current job market and use this information to predict which job fields are growing the fastest.

APPLICATIONS

66. INVENTORY The school cafeteria had 27 cases of juice before lunch. After lunch, there were 14 cases of juice left. How many cases of juice were used during lunch? **13 cases**

67. TEMPERATURE Theresa noticed that the thermostat in her freezer was set at 5°F. She lowered the thermostat by 7°F. Later, she lowered the thermostat another 2°F. At what temperature was the thermostat finally set? **−4°F**

68. BANKING Millie's checking account balance is $205 on the 13th of the month. She doesn't get paid again until the 1st of the following month. Millie has to pay a cable bill of $42.50 and a phone bill of $37.23.

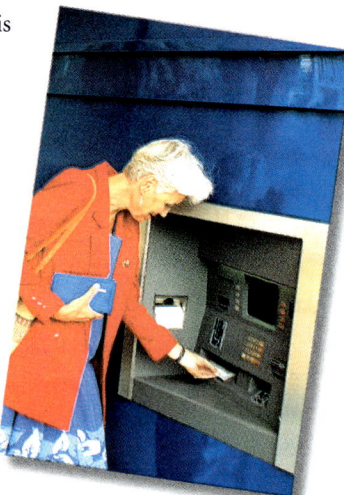

  a. Once these two bills are paid, how much money is left for Millie to spend this month? **$125.27**

CHALLENGE

  b. If Millie usually withdraws $20 for spending money every few days, how many times during the rest of the month can she do this if she has no other bills to pay? If she has withdrawn spending money twice and then withdraws $50 for a special occasion on the 21st of the month, how much money will Millie have left?
  6 times; $35.27

Look Back

Evaluate. *(LESSON 1.3)*

69. $36 - 12 \div 3 - 20$ **12**
70. $3 \cdot 5 + 7 \div 2$ **18.5**
71. $28 \div 2 \cdot 7 + 4$ **102**
72. $100 - 99 \div 3$ **67**
73. $(8^2 + 3 \cdot 2) - 16 \cdot 2$ **38**
74. $100 \div [5 - (2 \cdot 3 - 6)]$ **20**

Plot each set of ordered pairs on a coordinate plane and tell whether the points lie on a straight line. *(LESSON 1.4)*

75. (4, 11), (1, 5), (3, 9)   **76.** (1, −3), (7, 2), (5, 6)   **77.** (4, 9), (2, 8), (8, 11)
straight line   not a straight line   straight line

Find each sum. *(LESSON 2.2)*

78. $-18 + (-23)$ **−41**
79. $-95 + 78$ **−17**
80. $-31 + (-27)$ **−58**
81. $53 + (-29)$ **24**
82. $-3.2 + (-2.7)$ **5.9**
83. $5 + (-3.6)$ **1.4**
84. $-10.11 + (-3.9)$ **−14.01**
85. $3.7 + (-5.9)$ **−2.2**

Look Beyond

86. Let x represent an unknown point on a number line. If the distance between x and 4 on the number line is 7, what number(s) does x represent? (Hint: Draw a number line, and try different values of x.) **$x = 11$ or $x = -3$**

87. What number(s) does x represent if the distance between x and 4 is 10? **$x = -6$ or $x = 14$**

2.4 Multiplying and Dividing Real Numbers

Objectives
- Multiply and divide positive and negative numbers.
- Define the Properties of Zero.

Why In business, measures of gain and loss are reported with positive and negative numbers. Calculations with this data often involve multiplying and dividing. When calculating with positive and negative numbers, it is important to know what sign the result will have.

APPLICATION: BANKING

A college student is planning her budget based on her bank statements. She is studying the role that her monthly sources of income and her expenditures play in her final balance. Examine what would happen if she made the following changes:

Change	Effect	Representation	Result
Begin new part-time job	Add 2 deposits of $100	$(2)(100) = 200$	Increase of $200
Take bus to new job	Add 15 withdrawals of $1	$(15)(-1) = -15$	Decrease of $15
Quit old part-time job	Remove 2 deposits of $50	$(-2)(50) = -100$	Decrease of $100
Eat out less often	Remove 2 withdrawals of $15	$(-2)(-15) = 30$	Increase of $30

Activity
Multiplication and Division Patterns

Part I

1. Complete each pattern.

   a. $(2)(3) = 6$
   $(2)(2) = 4$
   $(2)(1) = 2$
   $(2)(0) = 0$
   $(2)(-1) = \underline{\ ?\ }$
   $(2)(-2) = \underline{\ ?\ }$
   $(2)(-3) = \underline{\ ?\ }$

   b. $(3)(3) = 9$
   $(2)(3) = 6$
   $(1)(3) = 3$
   $(0)(3) = 0$
   $(-1)(3) = \underline{\ ?\ }$
   $(-2)(3) = \underline{\ ?\ }$
   $(-3)(3) = \underline{\ ?\ }$

   c. $(3)(-3) = -9$
   $(2)(-3) = -6$
   $(1)(-3) = -3$
   $(0)(-3) = 0$
   $(-1)(-3) = \underline{\ ?\ }$
   $(-2)(-3) = \underline{\ ?\ }$
   $(-3)(-3) = \underline{\ ?\ }$

CHECKPOINT ✓

2. Complete each statement.

   a. $(+) \cdot (+) = \underline{\ +\ }$
   b. $(+) \cdot (-) = \underline{\ ?\ }$
   c. $(-) \cdot (+) = \underline{\ ?\ }$
   d. $(-) \cdot (-) = \underline{\ ?\ }$

Prepare

NCTM PRINCIPLES & STANDARDS
1, 2, 5, 6, 8–10

LESSON RESOURCES
- Practice 2.4
- Reteaching Master 2.4
- Enrichment Master 2.4
- Cooperative-Learning Activity 2.4
- Lesson Activity 2.4
- Problem Solving/Critical Thinking Master 2.4
- Teaching Transparency 10

QUICK WARM-UP

Find each product.

1. $(3.8)(4.1)$ **15.58**
2. $\frac{3}{5} \times \frac{1}{6}$ **$\frac{1}{10}$**

Find each quotient.

3. $6.8 \div 0.4$ **17**
4. $\frac{1}{10} \div \frac{1}{4}$ **$\frac{2}{5}$**

Also on Quiz Transparency 2.4

Teach

Why There are many real-world situations that involve multiplication and division of integers. Write the expressions -10×4 and $-10 \div 4$ on the board or overhead. Have students suggest real-world situations in which it might be necessary to compute this product or quotient. **Samples: What is the result if you have 4 losses of $10 each? What is the result if 4 people share a loss of $10 equally?**

☞ For Activity Notes and Checkpoint, see page 74.

Alternative Teaching Strategy

USING COGNITIVE STRATEGIES Preview the Distributive Property by showing students that $3[2 + (-2)]$ and $3(2) + 3(-2)$ are equal.

Ask students if they agree that $3(0) = 0$ is true. Then present them with the set of statements shown at right.

$3(0) = 0$
$3[2 + (-2)] = 0$
$3(2) + 3(-2) = 0$
$6 + 3(-2) = 0$

Point out the changes on the left side of the equal sign. Have students justify each change. Then ask what the value of $3(-2)$ must be if the last statement is to be true. **−6** Have students make a conjecture about the product of a positive and a negative number. **The product is negative.**

Now give students the set of statements shown at right. This time, lead them to see that the value of $(-3)(-2)$

$-3(0) = 0$
$-3[2 + (-2)] = 0$
$(-3)(2) + (-3)(-2) = 0$
$-6 + (-3)(-2) = 0$

must be 6 and to conjecture that the product of two negative numbers is positive.

LESSON 2.4   73

Activity Notes

Be sure that students complete Part I of the Activity before moving on to Part II. At the end of Part I, students should be able to describe in their own words the rules for multiplying two signed numbers. At the end of Part II, they should be able to write the rules for dividing two signed numbers.

CHECKPOINT ✓
2. a. +
   b. −
   c. −
   d. +

CHECKPOINT ✓
3. a. +
   b. −
   c. −
   d. +

Use Teaching Transparency 10.

Part II

1. Find each product. Examine all of the signs.
   a. (8)(7) = __?__   b. (6)(−5) = __?__
   a. (−4)(2) = __?__   b. (−8)(−1) = __?__

2. Multiplication is related to division. You can write related division facts for any multiplication fact. For example, the related division facts for 6 • 5 = 30 are 30 ÷ 5 = 6 and 30 ÷ 6 = 5. Write the related division facts for each of the multiplication facts in Step 1. Compare the signs in the related division and multiplication facts.

CHECKPOINT ✓
3. Complete each statement.
   a. (+) ÷ (+) = __?__   b. (+) ÷ (−) = __?__
   c. (−) ÷ (+) = __?__   d. (−) ÷ (−) = __?__

When multiplying or dividing signed numbers, follow the rules below.

Rules for Multiplying Two Signed Numbers

Multiplying Two Numbers With Like Signs
1. Find the product of the absolute values of the numbers.
2. Write the product as a positive number.

   (+) • (+) = (+)   (−) • (−) = (+)

Multiplying Two Numbers With Unlike Signs
1. Find the product of the absolute values of the numbers.
2. Write the product as a negative number.

   (+) • (−) = (−)   (−) • (+) = (−)

Rules for Dividing Two Signed Numbers

Dividing Two Numbers With Like Signs
1. Find the quotient of the absolute values of the numbers.
2. Write the quotient as a positive number.

   (+) ÷ (+) = (+)   (−) ÷ (−) = (+)

Dividing Two Numbers With Unlike Signs
1. Find the quotient of the absolute values of the numbers.
2. Write the quotient as a negative number.

   (+) ÷ (−) = (−)   (−) ÷ (+) = (−)

Interdisciplinary Connection

SOCIAL STUDIES Some economists analyze changes in the cost of living. They often use signed numbers to indicate amounts and rates of change. For example, the average cost of one gallon of regular leaded gasoline in the United States in 1980 was $1.191. An economist identified the change in cost from 1980 to 1988 as −$0.292. What was the average cost of one gallon of regular leaded gasoline in the United States in 1988? **$0.899** What signed number represents the average change per year in this period? **−$0.0365**

Enrichment

Have students solve "missing number" problems such as the following:

Problem	Answer
5(_?_)(−1)(−7) = −175	**−5**
(_?_ + 7) • 14 = 0	**−7**
6 ÷ _?_ + 4 = −2	**−1**
$\frac{5 \cdot \_?\_}{2} = -20$	**−8**
36 ÷ (−3) • (−| _?_ |) = 6	**0.5, −0.5**
| _?_ | • |−9| = −63	**no solution**

EXAMPLE 1 Find each product.
  a. $(9)(-4)$
  b. $(-4.5)(-2)$
  c. $(-5)(1.43)$

● **SOLUTION**
  a. $(9)(-4) = -36$ *The answer is negative.*
  b. $(-4.5)(-2) = 9$ *The answer is positive.*
  c. $(-5)(1.43) = -7.15$ *The answer is negative.*

TRY THIS Find each product.
  a. $(8)(-6)$
  b. $(-4)(-5.3)$
  c. $(-7)(2.6)$

EXAMPLE 2 Find each quotient.
  a. $\frac{9}{-3}$
  b. $(-4.5) \div (-1.5)$
  c. $\frac{-50}{25}$

● **SOLUTION**
  a. $(9) \div (-3) = -3$ *The answer is negative.*
  b. $(-4.5) \div (-1.5) = 3$ *The answer is positive.*
  c. $(-50) \div (25) = -2$ *The answer is negative.*

TRY THIS Find each quotient.
  a. $\frac{16}{-4}$
  b. $(-3.2) \div (-0.8)$
  c. $\frac{-25}{5}$

The Multiplicative Identity and Reciprocals

The product of any number and 1 is equal to the number. Therefore, 1 is called the **identity element for multiplication**.

Identity Property for Multiplication

For all real numbers a, $a \cdot 1 = a$ and $1 \cdot a = a$.

Related to the identity element, 1, are **multiplicative inverses**, or **reciprocals**. The product of two reciprocals is 1.

$7 \cdot \frac{1}{7} = 1$ $\frac{2}{3} \cdot \frac{3}{2} = 1$ $-\frac{4}{3} \cdot \left(-\frac{3}{4}\right) = 1$

Multiplicative Inverse Property

For every nonzero real number a, there is exactly one number $\frac{1}{a}$ such that
$a \cdot \frac{1}{a} = 1$ and $\frac{1}{a} \cdot a = 1$.

The number $\frac{1}{a}$ is called the reciprocal or multiplicative inverse of a.

ADDITIONAL EXAMPLE 1

Find each product.
  a. $(-15)(2)$ -30
  b. $(-6.1)(-9)$ 54.9
  c. $(8.3)(-4)$ -33.2

TRY THIS
a. -48
b. 21.2
c. -18.2

ADDITIONAL EXAMPLE 2

Find each quotient.
  a. $\frac{-36}{4}$ -9
  b. $-3.8 \div 19$ -0.2
  c. $\frac{-48}{-16}$ 3

TRY THIS
a. -4
b. 4
c. -5

Inclusion Strategies

VISUAL LEARNERS Some students will feel more comfortable working with a visual representation of the rules for multiplication and division. Have them make a chart like the one below in their notebooks.

$(+) \cdot (+) = (+)$ $(+) \div (+) = (+)$
$(-) \cdot (-) = (+)$ $(-) \div (-) = (+)$
$(+) \cdot (-) = (-)$ $(+) \div (-) = (-)$
$(-) \cdot (+) = (-)$ $(-) \div (+) = (-)$

Encourage these students to copy a given multiplication or division problem onto their paper and to insert a plus sign in front of each positive number. Tell them to use a colored highlighting marker to mark the signs of all the numbers involved and to mark the operation symbol. They can then match the highlighted signs and operation symbol with the appropriate "rule" in their chart.

LESSON 2.4

ADDITIONAL EXAMPLE 3

Find each quotient by using reciprocals.

a. $-54 \div 9$ $\quad -54 \cdot \frac{1}{9} = -6$

b. $\frac{-14}{-\frac{7}{8}}$ $\quad -14 \cdot \left(-\frac{8}{7}\right) = 16$

c. $\frac{11}{16} \div \left(-\frac{3}{4}\right)$

$\quad \frac{11}{16} \cdot \left(-\frac{4}{3}\right) = -\frac{11}{12}$

TRY THIS

a. 7

b. $-\frac{5}{27}$

c. $-\frac{5}{6}$

ADDITIONAL EXAMPLE 4

Find each product or quotient.

a. $\frac{1}{2}(0)$ $\quad 0$

b. $\frac{0}{-12}$ $\quad 0$

c. $6.05 \div 0$ $\quad$ undefined

Another way to divide numbers is to *multiply by reciprocals*. The quotient $12 \div (-3)$ can also be written as the product $12 \cdot \frac{1}{-3}$. The result is the same: $\frac{12}{-3}$, or -4.

EXAMPLE 3

Find each quotient by using reciprocals.

a. $24 \div (-8)$ b. $-4 \div \frac{8}{10}$ c. $\frac{4}{5} \div \frac{3}{10}$

SOLUTION

a. The reciprocal of -8 is $\frac{1}{-8}$.

$$24 \div (-8) = 24 \cdot \frac{1}{-8} = \frac{24}{-8} = -3$$

b. The reciprocal of $\frac{8}{10}$ is $\frac{10}{8}$.

$$-4 \div \frac{8}{10} = -4 \cdot \frac{10}{8} = \frac{-40}{8} = -5$$

c. The reciprocal of $\frac{3}{10}$ is $\frac{10}{3}$.

$$\frac{4}{5} \div \frac{3}{10} = \frac{4}{5} \cdot \frac{10}{3} = \frac{40}{15} = 2\frac{2}{3}$$

TRY THIS Find each quotient by using reciprocals.

a. $\frac{-49}{-7}$ b. $\frac{5}{9} \div (-3)$ c. $\frac{2}{3} \div \left(-\frac{4}{5}\right)$

Properties of Zero

Zero is the number that falls between positive numbers and negative numbers on a number line. However, *zero is neither positive nor negative*. There are some special rules for multiplying and dividing with zero.

Properties of Zero

Let a represent any real number.

1. The product of any real number and zero is zero.

   $a \cdot 0 = 0$ and $0 \cdot a = 0$

2. Zero divided by any nonzero real number is zero.

   $\frac{0}{a} = 0$, where $a \neq 0$

3. Division by zero is undefined. (Division by zero is not possible.)

Zero does not have a reciprocal because any number multiplied by 0 is 0.

EXAMPLE 4

Find each product or quotient.

a. $(0)(-4)$ b. $\frac{0}{-4}$ c. $4.5 \div 0$

Reteaching the Lesson

USING MANIPULATIVES Use tiles or counters that have two sides of different colors. Designate one color as positive and one as negative.

Have students work in pairs. Give each pair the same number of tiles. Have students separate the tiles into two piles. Ask them to turn all of the tiles in one pile to show the positive color and those in the other pile to show the negative color. Then have students separate each pile into groups that each contain the same number. Ask students to explain how this represents division by a positive number. Make a list of all the division problems they have modeled.

Now have students count out a given number of positive tiles. Have them repeat this four times, placing the tiles into one large pile as they count. Have students explain how this represents multiplication of two positive numbers. Repeat the activity with negative tiles. Then have students list all of the multiplication problems they can model with the set of tiles they have.

76 LESSON 2.4

● SOLUTION
a. $(0)(-4) = 0$ *The product of any number and zero is zero.*
b. $\frac{0}{-4} = 0$ *Zero divided by any nonzero number is zero.*
c. $4.5 \div 0$ is undefined.   *Division by zero is undefined.*

TRY THIS Find each product or quotient.
a. $\frac{0}{-8}$ b. $\frac{0}{27}$ c. $-45 \cdot 0$ d. $\frac{17}{0}$

CRITICAL THINKING Which one of the Properties of Zero applies to $0 \div 0$?

TECHNOLOGY
GRAPHICS CALCULATOR

Calculators will indicate an error when you try to divide by zero.

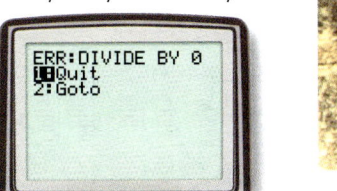

These are two symbols the Maya used for zero.

Exercises

Communicate

1. Explain the rule for multiplying 7 and -8.
2. Explain the rule for dividing -36 by -9.
3. Explain what a reciprocal is.
4. The multiplication equation for $\frac{6}{0} = n$ would be $0 \cdot n = 6$ or "zero times the number n is 6." What is the value of n? Why?
5. Explain how multiplying by the reciprocal in order to divide is similar to adding the opposite in order to subtract.

Guided Skills Practice

Find each product. *(EXAMPLE 1)*
6. $(-3)(-8)$ 24   7. $(-5)(-6)$ 30   8. $(-15)(9)$ -135 9. $(-3)(17)$ -51

Find each quotient. *(EXAMPLES 2 AND 3)*
10. $\frac{15}{-3}$ -5 11. $-42 \div 6$ -7 12. $\frac{-18}{-6}$ 3   13. $\frac{3}{7} \div \frac{6}{14}$ 1

Find each product or quotient. *(EXAMPLE 4)*
14. $7 \cdot 0$ 0   15. $0 \cdot (-9)$ 0   16. $\frac{-3}{0}$ undefined   17. $\frac{0}{3}$ 0

Teaching Tip

TECHNOLOGY If you are using the TI-83 graphics calculator, an attempt to divide by zero will result in the screen shown in the student text. If you are using the TI-82, the screen will appear as follows:

ERR:DIVIDE BY 0
1:Goto
2:Quit

On either calculator, selecting **Quit** simply returns you to the home screen. Selecting **Goto** returns you to the preceding screen, and the flashing cursor appears within the expression at the point where the division by zero was attempted.

TRY THIS
a. 0   b. 0
c. 0   d. undefined

CRITICAL THINKING
Division by zero is undefined.

Assess

Selected Answers
Exercises 6–17, 19–79 odd

ASSIGNMENT GUIDE	
In Class	1–17
Core	18–63, 65
Core Plus	18–58 even, 60–66
Review	67–80
Preview	81–83

✏ Extra Practice can be found beginning on on page 738.

Technology

A calculator may be helpful for Exercises 18–59 and 64–66. You may wish to have students complete the exercises with paper and pencil first and then use a calculator to verify their answers.

LESSON 2.4 **77**

Error Analysis

Students often work hastily and confuse the rules for addition of signed numbers with the rules for multiplication. Also, many student mistakenly read an expression like $(-3)(-8)$ as an addition problem. To focus students' attention on the correct operation and the appropriate set of rules, have them copy each expression and circle the operation symbol. In particular, encourage them to rewrite an expression such as $(-3)(-8)$ as $(-3) \times (-8)$ or as $(-3) \cdot (-8)$.

Look Beyond

Exercises 81–83 preview division of algebraic expressions. You may want to suggest that students rewrite each expression as the sum or difference of two expressions. In Exercise 81, for instance, have them write the following:

$$\frac{3w + 24}{-3} = \frac{3w}{-3} + \frac{24}{-3}$$

Students will learn more about dividing algebraic expressions in Lesson 2.7.

26. −23
27. −12
28. −5.76
29. 64
30. −5.35
31. −8
32. 9
33. 18.72
34. −1288
35. −2.1276
36. 6.35
37. $-1\frac{65}{96}$
38. −24.27
39. $-11\frac{1}{9}$

Practice and Apply

Find each product or quotient.

18. $(-4)(6)$ **−24**
19. $(5)(-7)$ **−35**
20. $(-14) \div (-2)$ **7**
21. $(12)(-11)$ **−132**
22. $27 \div (-3)$ **−9**
23. $(-17)(-3)$ **51**
24. $(-21)(-6)$ **126**
25. $(-17)(6)$ **−102**
26. $115 \div (-5)$
27. $(-84) \div 7$
28. $(-1.2)(4.8)$
29. $(-8)(-8)$
30. $(10.7)(-0.5)$
31. $8.8 \div (-1.1)$
32. $-2.7 \div (-0.3)$
33. $(-5.2)(-3.6)$
34. $(-23)(56)$
35. $(-0.591)(3.6)$
36. $(-220.345) \div (-34.7)$
37. $\left(\frac{7}{8}\right)\left(-\frac{23}{12}\right)$
38. $\frac{412.59}{-17}$
39. $\left(\frac{10}{3}\right) \div \left(-\frac{3}{10}\right)$

Write the reciprocal of each number.

40. -12 $-\frac{1}{12}$
41. 17 $\frac{1}{17}$
42. $\frac{6}{25}$ $\frac{25}{6}$
43. $\frac{4}{-9}$ $-\frac{9}{4}$

Evaluate.

44. $\frac{8}{-2}$ **−4**
45. $-22 \div (-1)$ **22**
46. $(-12)6$ **−72**
47. $-1.2 \div (-6)$ **0.2**
48. $\frac{3.4}{-17}$ **−0.2**
49. $(-5)(6)(-6)$ **180**
50. $-(-8)(-6)$ **−48**
51. $(-4)(-7)$ **28**
52. $\frac{-8+10}{-2}$ **−1**
53. $\frac{-4+7}{-6}$ $-\frac{1}{2}$
54. $\frac{-225}{-5}$ **45**
55. $(-47)(23)$ **−1081**
56. $\frac{-1208}{24}$ $-50\frac{1}{3}$
57. $\frac{(7)(-1)}{(-7)}$ **1**
58. $\frac{(-6)(-12)}{3}$ **24**
59. $96 \div \left(-\frac{2}{3}\right)$ **−144**

Kammi says, "Two negatives always make a positive." State whether this is true for the following operations:

60. addition **False**
61. subtraction **False**
62. multiplication **True**
63. division **True**

CONNECTION

a. 162 points; 54 points
b. 216 points
c. 24 points

64. **STATISTICS** Suppose that a football team scored 27 touchdowns in a 9-game season and made the two-point conversion each time.
   a. How many points came from touchdowns? from two-point conversions?
   b. How many total points did they score?
   c. If they scored the same number of points per game, how many points were scored in one game?

APPLICATION

65. **BANKING** Juan opened a savings account with a $20 deposit. He made 4 more deposits of $20 each and 5 withdrawals of $10 each.
   a. What was the amount of Juan's deposits? **$100**
   b. What was the amount of his withdrawals? **$50**
   c. If Juan wants to save another $200, how many $25 deposits must he make? **8 deposits**

66. **BUSINESS** A pet store owner is dividing a 50-pound sack of birdseed into smaller sacks to sell individually.
   a. How many $3\frac{1}{3}$-pound bags are there in one large sack? **15**
   b. How many $1\frac{2}{3}$-pound bags are there in one large sack? **30**

Practice

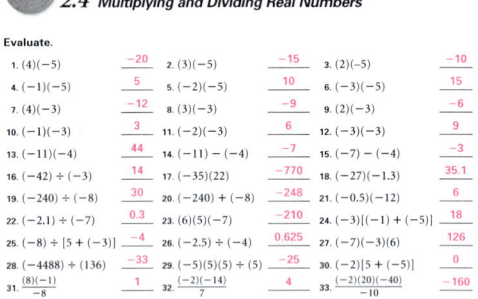

Practice 2.4 Multiplying and Dividing Real Numbers

Evaluate.

1. $(4)(-5)$ **−20**
2. $(3)(-5)$ **−15**
3. $(2)(-5)$ **−10**
4. $(-1)(-5)$ **5**
5. $(-2)(-5)$ **10**
6. $(-3)(-5)$ **15**
7. $(4)(-3)$ **−12**
8. $(3)(-3)$ **−9**
9. $(2)(-3)$ **−6**
10. $(-1)(-3)$ **3**
11. $(-2)(-3)$ **6**
12. $(-3)(-3)$ **9**
13. $(-11)(-4)$ **44**
14. $(-11) \div (-4)$ **−7**
15. $(-7) \div (-4)$ **−3**
16. $(-42) \div (-3)$ **14**
17. $(-35)(22)$ **−770**
18. $(-27)(-1.3)$ **35.1**
19. $(-240) \div (-8)$ **30**
20. $(-240) \div (-8)$ **−248**
21. $(-0.5)(-12)$ **6**
22. $(-2.1) \div (-7)$ **0.3**
23. $(6)(5)(-7)$ **−210**
24. $(-3)[(-1) + (-5)]$ **18**
25. $(-8) \div [5 + (-3)]$ **−4**
26. $(-2.5) \div (-4)$ **0.625**
27. $(-7)(-3)(6)$ **126**
28. $(-4488) \div (136)$ **−33**
29. $(-5)(5)(5) \div (5)$ **−25**
30. $(-2)[5 + (-5)]$ **0**
31. $\frac{(8)(-1)}{-8}$ **1**
32. $\frac{(-2)(-14)}{7}$ **4**
33. $\frac{(-2)(20)(-40)}{-10}$ **−160**

Tell whether each statement is true or false.

34. The product of two negative numbers is positive. **true**
35. The quotient of two negative numbers is positive. **true**
36. The average of a set of negative numbers is positive. **false**
37. The difference of two positive numbers is always positive. **false**
38. The sum of two positive numbers is positive. **true**

Stephanie opened a savings account with a $35 deposit. She made a total of 6 additional deposits of $15 each and withdrawals of $5, $10, and $15.

39. What is the total amount that Stephanie deposited in her account after her initial deposit? **$90**
40. What is the total amount that Stephanie withdrew from her account after her initial deposit? **$30**
41. What is the total amount currently in Stephanie's account? **$95**

Look Back

Find the second differences for each sequence. *(LESSON 1.1)*

67. 1, 3, 7, 13, 21, . . .   **2**   **68.** 17, 16, 14, 11, . . .   **−1**   **69.** −24, −20, −12, 0, . . .   **4**

Evaluate. *(LESSON 1.3)*

70. $14 \cdot 2 - 7$ **21**   **71.** $15 \div 3 - 2 \cdot 2$ **1**   **72.** $3^2 \cdot 2 + 14 \div 7$ **20**

Find each sum or difference. *(LESSONS 2.2 AND 2.3)*

73. $-23 + (-51)$ **−74**   **74.** $-17 - (-17)$ **0**   **75.** $-18 + (-83)$ **−101**   **76.** $-32 - (-5)$ **−27**

Let $q = 8$, $r = -2$, and $s = -5$. **Evaluate each expression.** *(LESSON 2.3)*

77. $q - r$ **10**   **78.** $-q - r$ **−6**   **79.** $-r - (-s)$ **−3**   **80.** $s - (-r)$ **−7**

Rewrite each expression in a simpler form.

81. $\dfrac{3w + 24}{-3}$ **−w − 8**   **82.** $\dfrac{-5y - 45}{5}$ **−y − 9**   **83.** $\dfrac{6x + 12}{2}$ **3x + 6**

Month	J	F	M	A	M	J	J	A	S	O	N	D
High	−7	−2	3	11	17	20	24	23	18	12	3	−2
Low	−15	−14	−9	−3	3	7	9	8	4	−1	−8	−13
Range	8	?	?	?	?	?	?	?	?	?	?	?

Include this activity in your portfolio.

Average high and low temperatures are part of the data recorded about a region's climate. The table above shows the average high and low temperatures in Calgary, Alberta, Canada, in degrees Celsius.

1. The range is the difference between the average high and the average low for a given period. Copy and complete the table by filling in the range data.

2. Use points on a number line or thermometer to show that the smallest range is in January.

3. Display the data as a double line graph with the average highs on one line and the average lows on the other line. (Note: The data do not represent linear equations.)

WORKING ON THE CHAPTER PROJECT

You should now be able to complete Activities 1 and 2 of the Chapter Project on page 104.

ALTERNATIVE Assessment

Portfolio Activity

The Portfolio Activity can be used as preparation for the Chapter Project or as a separate activity. In the Portfolio Activity on this page, students use subtraction of signed numbers to compute ranges within a set of data.

Answers to Portfolio Activities can be found in Additional Answers of the Teacher's Edition.

GO TO: go.hrw.com
KEYWORD: MA1 TEMP Extremes

To support the Portfolio Activity, students may visit the HRW Web site, where they can obtain information about monthly temperature extremes in the United States.

Student Technology Guide

NAME _____ CLASS _____ DATE _____

Student Technology Guide
2.4 *Multiplying and Dividing Real Numbers*

Suppose that you want to carry the set of instructions at right for the values of x below.

0, 5, 10, 15, 20, 25, and 30

   a. Start with x.
   b. Multiply by −2.
   c. Multiply by −4.5.
   d. Divide by −3.
   e. Divide by −4.

You could use a calculator and carry out the instructions for each value of x, one at a time. However, a spreadsheet can help you find the results for all of the values of x at once. Put the given values of x in the first column. Then use columns B, C, D, and E to calculate the results of performing one instruction after another. Column E will contain the final results.

	A	B	C	D	E
1	0				
2	5				
3	10				
4	15				
5	20				
6	25				
7	30				

• Enter = A1*(−2) into cell B1.
• Select Edit and then Fill Down.
• Enter = B1*(−4.5) into cell C1.
• Select Edit and then Fill Down.
• Enter = C1/(−3) into cell D1.
• Select Edit and then Fill Down.
• Enter = D1/(−4) into cell E1.
• Select Edit and then Fill Down.
The final values are given in column E.

You can represent the set of instructions above as $\dfrac{x(-2)(-4.5)}{(-3)(-4)}$.

Use a spreadsheet to find the final values for each set of instructions. Use 0, 10, 20, 30, 40, 50, and 60 for input values. Give answers to the nearest tenth. Then write each set of instructions as an expression.

1. a. Start with a number.
   b. Multiply by −2.4.
   c. Divide by −4.
   d. Multiply by 1.8.
   e. Divide by −2.   **0.0, −86.4, −172.8, −259.2, −345.6, −432.0, −518.4** $\dfrac{x(-2.4)(1.8)}{(-4)(-2)}$

2. a. Start with a number.
   b. Multiply the number by −5.
   c. Subtract 6.2.
   d. Multiply the result by −3.2
   e. Divide by −10.   **−2.0, −18.0, −34.0, −50.0, −66.0, −82.0, −98.0** $\dfrac{x(-5) - 6.2)(-3.2)}{(-10)}$

3. Write a set of instructions to evaluate the expression $\dfrac{x(-12)(3.8)}{(-6)(8)}$.   **Answers may vary.**
   Sample answer: 1. Start with a number. 2. Multiply the number by −12.
   3. Multiply the result by 3.8. 4. Divide the result by −6. 5. Divide by 8.

LESSON 2.4   79

Prepare

NCTM PRINCIPLES & STANDARDS 1, 2, 5–10

LESSON RESOURCES

- Practice 2.5
- Reteaching Master 2.5
- Enrichment Master 2.5
- Cooperative-Learning Activity 2.5
- Lesson Activity 2.5
- Problem Solving/Critical Thinking Master 2.5
- Teaching Transparency 11

QUICK WARM-UP

Evaluate.

1. $44 + (36 + 32)$ — 112
2. $3(4 + 2)$ — 18
3. $5 - (6 + 2)$ — -3
4. $2 \cdot (25 \cdot 3)$ — 150
5. $4(9.5)$ — 38

Also on Quiz Transparency 2.5

Teach

Why The ability to accurately perform mental computations is important in many everyday situations. The properties that allow expressions to be rewritten can be very helpful when performing these mental computations. Have students discuss their experiences with mental computations. Some situations they may cite are estimating the amount of a discount or determining the number of additional points needed to win a game.

2.5 Properties and Mental Computation

Objectives
- State and apply the Commutative, Associative, and Distributive Properties.
- Use the Commutative, Associative, and Distributive Properties to perform mental computations.

Why In algebra, it is necessary to rearrange and rewrite expressions in order to solve equations. The Commutative, Associative, and Distributive Properties enable you to do this. The properties are also useful when performing mental computations.

APPLICATION
INVENTORY

Felicia works at a river-tube rental company. Each morning she checks the inventory of tubes stored in three sheds. She counts 44, 27, and 56 tubes and quickly calculates the total, 127, mentally. How do you think Felicia used mental computation to calculate the sum $44 + 27 + 56$?

The Commutative Properties

Compare the sums in columns A and B. How are they alike and how are they different? Compare the products in columns C and D. How are they alike and how are they different?

Sums

Column A	Column B
$56 + 11 = 67$	$11 + 56 = 67$
$28 + 32 = 60$	$32 + 28 = 60$
$13 + 87 = 100$	$87 + 13 = 100$

Products

Column C	Column D
$6 \cdot 11 = 66$	$11 \cdot 6 = 66$
$8 \cdot 9 = 72$	$9 \cdot 8 = 72$
$10 \cdot 54 = 540$	$54 \cdot 10 = 540$

Columns A and B illustrate the **Commutative Property of Addition**. Changing the order of the numbers in an addition expression does not change the sum.

Columns C and D illustrate the **Commutative Property of Multiplication**. Changing the order of the numbers in a multiplication expression does not change the product.

Alternative Teaching Strategy

TECHNOLOGY Have students use their calculators to investigate the properties presented in this lesson. Give them sets of addition and multiplication problems such as the following:

For the Commutative Properties:

$15,907 + 29,489$ $\quad$ $15,907 \cdot 29,489$

$29,489 + 15,907$ $\quad$ $29,489 \cdot 15,907$

For the Associative Properties:

$189 + (908 + 86)$ $\quad$ $18 \cdot (719 \cdot 86)$

$(189 + 908) + 86$ $\quad$ $(18 \cdot 719) \cdot 86$

For the Distributive Property:

$68(58 + 45)$ $\quad$ $49(34 - 96)$

$68 \cdot 58 + 68 \cdot 45$ $\quad$ $49 \cdot 34 - 49 \cdot 96$

After they have completed each set of problems, ask them to make a conjecture about the property illustrated.

80 LESSON 2.5

Changing the order of two terms in an expression is known as *commuting the terms.*

Commutative Property of Addition

For all real numbers a and b:
$$a + b = b + a$$

Commutative Property of Multiplication

For all real numbers a and b:
$$a \cdot b = b \cdot a$$

CRITICAL THINKING Do you think that any two numbers can be divided in either order without changing the quotient? Explain.

The Associative Properties

Compare the sums in columns A and B. How are they alike and how are they different? Compare the products in columns C and D. How are they alike and how are they different?

Sums

Column A	Column B
$(12 + 8) + 15 = 35$	$12 + (8 + 15) = 35$
$(39 + 4) + 16 = 59$	$39 + (4 + 16) = 59$
$(24 + 36) + 7 = 67$	$24 + (36 + 7) = 67$

Products

Column C	Column D
$(2 \cdot 5) \cdot 16 = 160$	$2 \cdot (5 \cdot 16) = 160$
$(20 \cdot 5) \cdot 13 = 1300$	$20 \cdot (5 \cdot 13) = 1300$
$(3 \cdot 11) \cdot 10 = 330$	$3 \cdot (11 \cdot 10) = 330$

Columns A and B illustrate the **Associative Property of Addition**. Changing the grouping of the numbers in an addition expression does not change the sum.

Columns C and D illustrate the **Associative Property of Multiplication**. Changing the grouping of the numbers in a multiplication expression does not change the product.

The Associative Property of Addition

For all real numbers a, b, and c:
$$(a + b) + c = a + (b + c)$$

The Associative Property of Multiplication

For all real numbers a, b, and c:
$$(a \cdot b) \cdot c = a \cdot (b \cdot c)$$

CRITICAL THINKING
Answers may vary. Sample answer: No, the order of the numbers being divided does matter. For example, $6 \div 3 = 2$ but $3 \div 6 = \frac{1}{2}$.

Teaching Tip
Emphasize the fact that each of the Commutative and Associative Properties involves only one operation. In contrast, the Distributive Property involves two operations: multiplication and addition, or multiplication and subtraction.

Now give students a subtraction problem such as 701 − 125, and ask them to investigate why there is no Commutative Property for Subtraction. Ask them how they might rewrite 701 − 125 so that a Commutative Property can be applied to it. **701 + (−125)** Repeat this investigation with a division problem such as 296 ÷ 4. **Rewrite it as** $296 \cdot \frac{1}{4}$.

Interdisciplinary Connection

HOME ECONOMICS The recipe for a cake calls for $2\frac{1}{3}$ cups of flour. The home economics class is going to make 12 of these cakes for a charity bake sale. Show how to use properties to mentally compute the amount of flour needed for all of the cakes. **Use the Distributive Property; $12 \cdot 2\frac{1}{3} = 12 \cdot \left(2 + \frac{1}{3}\right) = 12 \cdot 2 + 12 \cdot \frac{1}{3} = 24 + 4 = 28$; they need 28 cups of flour.**

LESSON 2.5

ADDITIONAL EXAMPLE 1

April works at a neighborhood convenience store. Each morning she checks the stock of baked goods. One morning she counts 54 bagels, 39 muffins, and 26 pastries, and she mentally computes the total, 119. **Show how the Commutative and Associative Properties are used to regroup the terms to make the sum easier to compute mentally.**

$(54 + 39) + 26$
$= (39 + 54) + 26$
   Comm. Prop.
$= 39 + (54 + 26)$
   Assoc. Prop.
$= 39 + 80$
$= 119$

Use Teaching Transparency 11.

Activity Notes

Be sure students see that when performing mental computations of the type shown here, the goal is to rewrite one factor so that part of the resulting computation involves a multiplication by 10 or by a multiple of 10. Most people find it relatively easy to do this type of "tens multiplication" mentally. Students should recognize that this method of mental computation is valid because of the Distributive Property.

EXAMPLE 1

APPLICATION
INVENTORY

At the beginning of this lesson, Felicia used mental computation to calculate the number of tubes stored in three sheds. **Show how the Commutative and Associative Properties are used to regroup the numbers to make the sum easier to compute mentally.**

● **SOLUTION**

$(44 + 27) + 56$
$= (27 + 44) + 56$ *Commutative Property of Addition*
$= 27 + (44 + 56)$ *Associative Property of Addition*
$= 27 + 100$
$= 127$

The Distributive Property

The **Distributive Property** involves addition, subtraction, and multiplication. An important property of numbers is that multiplication can be *distributed* over addition and subtraction.

The Distributive Property works in two directions. To find the product $3(5 + 2)$, multiplication is distributed over addition.

$$3(5 + 2) = 3 \cdot 5 + 3 \cdot 2$$

When used in reverse, a common factor can be removed from the terms.

$$3 \cdot 5 + 3 \cdot 2 = 3(5 + 2)$$

The Distributive Property works similarly for multiplication over subtraction. For example:

$$4(7 - 3) = 4 \cdot 7 - 4 \cdot 3$$

And, in reverse:

$$4 \cdot 7 - 4 \cdot 3 = 4(7 - 3)$$

The Distributive Property of Multiplication Over Addition and Subtraction

For all real numbers a, b, and c:
$a(b + c) = ab + ac$ and $(b + c)a = ba + ca$
and
$a(b - c) = ab - ac$ and $(b - c)a = ba - ca$

You can use the Commutative, Associative, and Distributive Properties to rearrange expressions for quick mental computation.

Enrichment

Use properties to prove these statements.

1. $(5 \cdot x) \cdot (3 \cdot y) = (5 \cdot 3) \cdot (x \cdot y)$

$(5 \cdot x) \cdot (3 \cdot y) = 5[x \cdot (3 \cdot y)]$
   Assoc. Prop.
$= 5[(x \cdot 3) \cdot y]$
   Assoc. Prop.
$= 5[(3 \cdot x) \cdot y]$
   Comm. Prop.
$= (5 \cdot 3) \cdot (x \cdot y)$
   Assoc. Prop.

2. $(7 + 8) + (6 + 3) = (7 + 3) + (8 + 6)$

$(7 + 8) + (6 + 3) = 7 + [8 + (6 + 3)]$
   Assoc. Prop.
$= 7 + [(6 + 3) + 8]$
   Comm. Prop.
$= 7 + [(3 + 6) + 8]$
   Comm. Prop.
$= (7 + 3) + (6 + 8)$
   Assoc. Prop.
$= (7 + 3) + (8 + 6)$
   Comm. Prop.

Activity
Exploring Mental Computation

1. Examine the table below, reading each row from left to right. Copy and complete the table.

Column A	Column B	Column C	Column D
4(15)	4(10 + 5)	4 · 10 + 4 · 5	40 + 20 = 60
8(17)	8(10 + 7)	8 · 10 + 8 · 7	?
5(22)	5(10 + 10 + 2)	?	?
12(25)	(10 + 2)25	10 · 25 + 2 · 25	?
7(21)	7(20 + 1)	?	?
6(128)	?	?	?
8(452)	?	?	?
11(3651)	?	?	?

2. Describe the process shown in each row from left to right.

CHECKPOINT ✓ 3. Based on the process that you described in Step 2, explain how the Distributive Property can be used in the mental computation of products.

CRITICAL THINKING Explain how knowing multiplication facts, such as $3 \times 8 = 24$, by memory helps you to do mental computation. Describe the advantages of using mental computation instead of relying on a calculator.

EXAMPLE 2 John rents a river tube for 8 hours at $3.50 per hour. **Use the Distributive Property to find the total cost.**

APPLICATION
RECREATION

SOLUTION
The total cost is modeled by the expression 8(3.5). Use the Distributive Property to find the sum of two products.
$8(3.5) = 8(3) + 8(0.5)$
$= 24 + 4$
$= 28$
The total cost is $28.

TRY THIS Use the Distributive Property to find the cost of renting a river tube for 8 hours at $4.75 per hour. Find the cost of renting a river tube for 7 hours at $5.50 per hour.

CRITICAL THINKING Use the Distributive Property to explain why $-(4 - 7) = -4 + 7$ and $-(-5 + 6) = 5 - 6$.

Inclusion Strategies

VISUAL LEARNERS Provide visual reinforcement of the Distributive Property by drawing diagrams such as the ones below. Describe the diagrams as showing that the number outside the parentheses is "distributed" to each term inside the parentheses.

Distribute multiplication over addition:

$$a(b + c) = ab + ac$$
$$(b + c)a = ba + ca$$

Distribute multiplication over subtraction:

$$a(b - c) = ab - ac$$
$$(b - c)a = ba - ca$$

CHECKPOINT ✓
3. We can calculate the product of two numbers using mental computation by expanding one of the numbers and then using the Distributive Property.

CRITICAL THINKING
Answers may vary. Sample answer: Knowing multiplications facts means you do not have to rely on a calculator. Quickly calculating products is one advantage of using mental computation.

ADDITIONAL EXAMPLE 2

Tarek worked 30 hours last week. He earns $6.50 per hour. **Use the Distributive Property to calculate Tarek's total wages for last week.**

$30(6.5) = 30(6) + 30(0.5)$
$= 180 + 15$
$= 195$
Tarek's total wages were $195.

TRY THIS
The total cost is $38 and $38.50, respectively.

CRITICAL THINKING
$-(4 - 7) = -1(4 - 7)$
$= -1(4) - 1(-7)$
$= -4 + 7$

$-(-5 + 6) = -1(-5 + 6)$
$= -1(-5) - 1(6)$
$= 5 - 6$

LESSON 2.5

ADDITIONAL EXAMPLE 3

Find the opposite of each expression.

a. $-c + d$ $c - d$

b. $-a - 2$ $a + 2$

TRY THIS
a. $-a + b$
b. $x - 8$

ADDITIONAL EXAMPLE 4

Prove: $b + (a - b) = a$. Justify each step.

$b + (a - b) = (a - b) + b$
  Comm. Prop. of Add.

$= a + [(-b) + b]$
  Assoc. Prop. of Add.

$= a + 0$
  Additive Inverse Prop.

$= a$
  Additive Identity Prop.

TRY THIS
$a + (b + c) = (a + b) + c$
  Associative Property
$(a + b) + c = (b + a) + c$
  Commutative Property
$a + (b + c) = (b + a) + c$
  Transitive Property

Opposites of Numbers and Expressions

You can find the opposite of a number by multiplying the number by -1. For example, $-1 \cdot 8 = -8$, and $-1 \cdot -8 = 8$. You can find the opposite of an addition or subtraction expression in a similar way by using the Distributive Property.

EXAMPLE Find the opposite of each expression.
  a. $a + b$ b. $x - 8$

● **SOLUTION**

Multiply each expression by -1.

a. $-1(a + b) = (-1)(a) + (-1)(b) = -a + (-b) = -a - b$

b. $-1(x - 8) = (-1)(x) - (-1)(8) = -x - (-8) = -x + 8$

TRY THIS Find the opposite of each expression.
  a. $a - b$ b. $-x + 8$

Proving Statements

The table below summarizes the important properties of the real numbers discussed in this chapter.

THE PROPERTIES BELOW ARE TRUE FOR ALL REAL NUMBERS a, b, AND c.		
	Addition	**Multiplication**
Commutative	$a + b = b + a$	$ab = ba$
Associative	$(a + b) + c = a + (b + c)$	$(ab)c = a(bc)$
Identity	0 is the identity. $a + 0 = 0 + a = a$	1 is the identity. $a \cdot 1 = 1 \cdot a = a$
Inverse	$-a$ is the additive inverse of a.	For $a \neq 0$, $\frac{1}{a}$ is the multiplicative inverse of a.
Distributive	$a(b + c) = ab + ac$ and $(b + c)a = ba + ca$	

The following properties of equality are also true:

Properties of Equality

For all real numbers a, b, and c:

Reflexive Property $a = a$ (A number is equal to itself.)

Symmetric Property   If $a = b$, then $b = a$.

Transitive Property   If $a = b$ and $b = c$, then $a = c$.

Substitution Property   If $a = b$, then a can be replaced by b and b can be replaced by a.

Reteaching the Lesson

ENGLISH LANGUAGE DEVELOPMENT Help students internalize the meanings of the terms *commutative*, *associative*, and *distributive* by discussing everyday words related to them.

For example, note that the origins of the words *commutative* and *commute* are the same. Discuss what it means to commute to and from school or work. Ask how the meanings of the two words might be related. **Sample: Both involve exchanging places.**

Have students discuss what it means to *associate* with another person. Ask how this meaning might be related to the mathematical term *associative*. **Sample: Both involve grouping or joining together.**

Discuss what it means to *distribute* a set of papers to the class. Ask how this is similar to applying the *Distributive* Property in algebra. **Sample: Both involve giving the same thing to each member of a group.**

84 LESSON 2.5

All of the properties of real numbers can be used to prove mathematical statements. In mathematics, a statement that can be proved true is called a **theorem**. In statements of theorems, variables represent real numbers.

EXAMPLE 4 Prove: $(a + b) + (-a) = b$. Justify each step.

● SOLUTION

Proof

Statement	Reason
1. $(a + b) + (-a) = (b + a) + (-a)$	1. Commutative Prop. of Add.
2. $ = b + [a + (-a)]$	2. Associative Prop. of Add.
3. $ = b + 0$	3. Additive Inverse Prop.
4. $ = b$	4. Identity Prop. Add.
5. $(a + b) + (-a) = b$	5. Transitive Prop. of Equality

TRY THIS Prove: $a + (b + c) = (b + a) + c$.

EXAMPLE 5 Prove: If $a = b$, then $ac = bc$. Justify each step.

● SOLUTION

Proof

Statement	Reason
1. $a = b$	1. Given
2. $ac = ac$	2. Reflexive Prop. of Equality
3. $ac = bc$	3. Substitution Prop. of Equality

TRY THIS Prove: If $a = b$, then $ca = cb$.

Exercises

Communicate

1. Explain why it is helpful to know the properties discussed in this lesson.
2. Give examples of how you can use the Commutative and Associative Properties to help you compute mentally.
3. Explain how to use mental computation and the Distributive Property to find each product.
   a. $8(21)$
   b. $11(35)$
   c. $14(22)$
4. River tubes rent for $3.50 per hour. One weekend Bill and Tracy rent a tube for 5 hours on Saturday and 7 hours on Sunday. Explain how to use the Distributive Property to compute the total rental fee for the weekend.

ADDITIONAL EXAMPLE 5

Prove: If $a = b$, then $a + c = b + c$.

1. $a = b$ — Given
2. $a + c = a + c$ — Reflexive Prop. of Equality
3. $a + c = b + c$ — Substitution

TRY THIS

$a = b$ — Given
$ca = ca$ — Reflexive Property
$ca = cb$ — Substitution

Assess

Selected Answers
Exercises 5–11, 13–51 odd

ASSIGNMENT GUIDE

In Class	1–11
Core	12–41, 43, 45
Core Plus	12–42 even, 43–46
Review	47–52
Preview	53, 54

✐Extra Practice can be found beginning on page 738.

internetconnect

GO TO: go.hrw.com
KEYWORD: MA1 Napier's Bones

In this activity, students learn about a centuries-old calculating machine that shows how the multiplication algorithm can be performed mechanically.

Technology

A calculator may be helpful for Exercises 42–46. You may wish to have students complete the exercises with paper and pencil first and then use a calculator to verify their answers.

Error Analysis

When regrouping an expression such as $a - (b + c)$, students often forget to take the opposite of both terms in parentheses, and they give the answer as $(a - b) + c$. Encourage them to use the technique shown below. They should copy the given expression, rewrite it to show the multiplication by -1, and then insert arrows as a reminder that they must apply the Distributive Property.

$$a - (b + c)$$
$$= a + (-1)(b + c)$$

Guided Skills Practice

Complete each step, and name the property used. *(EXAMPLE 1)*

5. $(36 + 15) + 64$
   $= (15 + \underline{?}) + 64$ **36** Comm. Prop.
   $= 15 + (36 + \underline{?})$ **64** $\underline{?}$ Prop. **Assoc.**
   $= 15 + \underline{?}$ **100**
   $= \underline{?}$ **115**

6. $20 \cdot (72 \cdot 5)$
   $= 20 \cdot (\underline{?} \cdot 72)$ **5** $\underline{?}$ Prop. **Comm.**
   $= (\underline{?} \cdot 5) \cdot 72$ **20** $\underline{?}$ Prop. **Assoc.**
   $= \underline{?} \cdot \underline{?}$ **100, 72**
   $= \underline{?}$ **7200**

APPLICATION

7. **INCOME** Jarrel works 9 hours on Saturday and 6.5 hours on Sunday. He earns $12 per hour. Use the Distributive Property to compute Jarrel's total wages for the weekend. Remember to use the algebraic order of operations and to check your answer. *(EXAMPLE 2)* **$186**

Find the opposite of each expression. *(EXAMPLE 3)*

8. $x + y$ **$-x - y$**
9. $-a - b$ **$a + b$**
10. $x - y$ **$-x + y$**

11. Prove: $a + (b + c) = (a + c) + b$. *(EXAMPLES 4 AND 5)*

Practice and Apply

Complete each step, and name the property used.

12. $(24 + 27) + 56$
    $= (27 + \underline{?}) + 56$ **24** Comm. Prop.
    $= 27 + (24 + \underline{?})$ **56** $\underline{?}$ Prop. **Assoc.**
    $= 27 + \underline{?}$ **80**
    $= \underline{?}$ **107**

13. $25(7 + 4)$
    $= 25 \cdot \underline{?} + \underline{?} \cdot 4$ **7, 25** $\underline{?}$ Prop. **Distr.**
    $= \underline{?} + \underline{?}$ **175, 100**
    $= \underline{?}$ **275**

Use the Associative, Commutative, and Distributive Properties to find each sum or product. Show your work and name the properties used.

14. $(27 + 98) + 73$ **198**
15. $(45 \cdot 32) \cdot 0$ **0**
16. $(87 \cdot 5) \cdot 2$ **870**
17. $50 \cdot (118 \cdot 20)$ **118,000**
18. $(688 + 915) + 312$ **1915**
19. $(25 \cdot 78) \cdot 4$ **7800**
20. $2 \cdot (129 \cdot 5)$ **1290**
21. $(133 + 52) + 67$ **252**

Use the Distributive Property and mental computation to calculate each product.

22. $4 \cdot 28 = 4(20 + 8)$
    $= 4 \cdot 20 + 4 \cdot 8$
    $= 80 + 32$
    $112 = \underline{?}$

23. $40 \cdot 18 = 40(10 + 8)$
    10, 8 $= 40 \cdot \underline{?} + 40 \cdot \underline{?}$
    400, 320 $= \underline{?} + \underline{?}$
    $720 = \underline{?}$

24. $9 \cdot 680 = 9(600 + 80)$
    9, 9 $= \underline{?} \cdot 600 + \underline{?} \cdot 80$
    5400, 720 $= \underline{?} + \underline{?}$
    $6120 = \underline{?}$

25. $70 \cdot 540 = 70(500 + 40)$
    70, 500, 70, 40 $= \underline{?} \cdot \underline{?} + \underline{?} \cdot \underline{?}$
    35,000, 2800 $= \underline{?} + \underline{?}$
    $37,800 = \underline{?}$

11. $a + (b + c) = a + (c + b)$ **Commutative Property**
    $= (a + c) + b$ **Associative Property**

86 LESSON 2.5

Use the Distributive Property to rewrite each expression.

26. $9xy - 21xyz$ **$3xy(3 - 7z)$**
27. $rs + rq$ **$r(s + q)$**
28. $xy - wy$ **$y(x - w)$**
29. $4pq + pr$ **$p(4q + r)$**
30. $3de - 15df$ **$3d(e - 5f)$**
31. $35st + 20rs$ **$5s(7t + 4r)$**

Name the property illustrated. Be specific.

32. $32 + 17 = 17 + 32$ **Comm.**
33. $13 \cdot 21 - 13 \cdot 9 = 13(21 - 9)$ **Distr.**
34. $6(4.7 - 2) = 6(4.7) - 6(2)$ **Distr.**
35. $4(5x) = (4 \cdot 5)x$ **Assoc.**
36. $-8.2(2 + 5.3) = (2 + 5.3)(-8.2)$ **Comm.**
37. $(6 - 3)5 = 6 \cdot 5 - 3 \cdot 5$ **Distr.**

38. Prove: $(a + b) + c = (c + b) + a$. Justify each step.
39. Prove: $(ab)c = a(cb)$. Justify each step.
40. Prove: If $a = b$, then $a + c = b + c$. Justify each step.
41. Prove: If $a = b$, then $c + a = c + b$. Justify each step.

CONNECTION

42. **GEOMETRY** Use the Distributive Property to find the perimeter of the following rectangles:
    a. a rectangle with a length of 57 and a width of 43 **200**
    b. the total perimeter of 3 rectangles with a length of 167 and a width of 133 plus 3 rectangles with a length of 82 and a width of 68 **2700**

APPLICATIONS

43. **INCOME** Marcia earns $15 per hour tutoring math students. She tutored for 4 hours on Friday and for 6 hours on Saturday. Using two different methods, find the total amount Marcia earned for tutoring on Friday and Saturday. **$150**

44. **INCOME** Janie works 5 hours on Saturday and 4.5 hours on Sunday. She earns $12 per hour. Use the Distributive Property to compute Janie's total earnings for the weekend. Remember to use the algebraic order of operations and to check your answer. **$114**

45. **ACCOUNTING** At a music store, 876 CDs are sold for $12 each during a sale. Use the Distributive Property to compute the total value of the CDs sold during the sale. Show your work. **$10,512**

46. **INVENTORY** Sally owns a bakery. Each morning she has to count all of the baked goods that are delivered. How can Sally use mental computation to count the baked goods shown below? What is the total?

$(26 + 24) + (21 + 39) + 28$
$50 + 60 + 28$
138

38. $(a + b) + c$
    $= (b + a) + c$ Comm. Prop.
    $= b + (a + c)$ Assoc. Prop.
    $= b + (c + a)$ Comm. Prop.
    $= (b + c) + a$ Assoc. Prop.
    $= (c + b) + a$ Comm. Prop.

39. $(ab)c = a(bc)$ Assoc. Prop.
    $= a(cb)$ Comm. Prop.

40. $a = c$ Given
    $a + c = a + c$ Reflexive Property of Equality
    $a + c = b + c$ Substitution Property of Equality

41. $a = b$ Given
    $c + a = c + a$ Reflexive Property of Equality
    $c + a = c + b$ Substitution Property of Equality

Student Technology Guide

2.5 Properties and Mental Computation

You can use a spreadsheet to verify the Distributive Property. Enter values into column A, and then use the other columns to rewrite the expressions. Note that a computer uses an asterisk for multiplication and that it is not an optional symbol even if parentheses are used.

Example: Verify that $8*(10+15) = 8*10 + 8*15$ is true.

Enter the values 8, 10, and 15 into cells A1, A2, and A3, respectively.
Enter = A1*(A2+A3) in cell B1.
Enter = A1*A2+A1*A3 in cell C1.
The values should be the same.

The answer 200 is in both cells B1 and C1.

Use a spreadsheet to verify each equation. Then give the value of each expression.

1. $20*(40 + 12) = 20*40 + 20*12$ _____ 1040
2. $-4.5*(15 + 3.2) = -4.5*15 + (-4.5)*3.2$ _____ -81.9
3. $-3*(-12 + (-6)) = -3*(-12) + (-3)*(-6)$ _____ 54
4. $320*(1200 + 890) = 320*1200 + 320*890$ _____ 668,800
5. $125*(80 + 45) = 125*80 + 125*45$ _____ 15,625

The Distributive Property works in a similar way for multiplication over subtraction.

6. Let $a = 10$, $b = 18$, and $c = 15$. Describe the steps you would take to verify the Distributive Property, $a(b - c) = ab - ac$, with a spreadsheet. Give the contents of each cell. Enter the values 10, 18, and 15 into the cells of column A. Enter = A1 * (A2 − A3) in cell B1. Enter = A1 * A2 − A1 * A3 in cell C1. The values in cells B1 and C1 should be the same. The answer 30 is in both cells B1 and C1.

7. Use Exercise 6 with modification to verify that $12*(20 - 8) = 12*20 - 12*8$. Enter the values 12, 20, and 8 into the cells of column A. Enter = A1 * (A2 − A3) in cell B1. Enter = A1 * A2 − A1 * A3 in cell C1. The values in cells B1 and C1 should be the same.

LESSON 2.5 87

Exercises 53 and 54 preview the use of the Distributive Property when multiplying two binomials. Students will study this method in greater detail in Lesson 9.3.

ALTERNATIVE
Assessment
Portfolio Activity

The Portfolio Activity can be used as preparation for the Chapter Project or as a separate activity. The Portfolio Activity on this page requires students to recall their work with constant differences in Chapter 1. Here they apply constant differences to extend a sequence that involves negative numbers.

GO TO: go.hrw.com
KEYWORD: MA1 Windchill

To support the Portfolio Activity, students may visit the HRW Web site, where they can obtain additional information regarding windchill factors in various cities.

Make a table of values for *x* and *y* by substituting 1, 2, 3, 4, 5, and 10 for *x* in the equations below. Plot the resulting points on a coordinate plane. **(LESSON 1.5)**

47. $y = 3x$ **48.** $y = 2x - 1$

APPLICATION

49. ENTERTAINMENT A recording studio charges a fee of $100 plus $45 per hour of recording. Write an equation for the total charge, *c*, based on the number of hours, *h*, of recording. Make a table of charges for 1 through 8 hours. **(LESSON 1.5)**

Evaluate. **(LESSONS 2.2, 2.3 AND 2.4)**

50. $-5 + (-3)$ **−8** **51.** $8 - (-17)$ **25** **52.** $14 \cdot (-3)$ **−42**

Look Beyond

Use the Distributive Property to complete each exercise.

53. $25 \cdot 76 = (20 + 5) \cdot (70 + 6)$
$= 20 \cdot (70 + 6) + 5 \cdot (70 + 6)$
$= 20 \cdot 70 + 20 \cdot 6 + 5 \cdot 70 + 5 \cdot 6$
$= \underline{?} + \underline{?} + \underline{?} + \underline{?}$ **1400, 120, 350, 30**
$= \underline{?}$ **1900**

54. $26 \cdot 34 = (30 - 4) \cdot (30 + 4)$
$= (30 - 4) \cdot 30 + (30 - 4) \cdot 4$
$= 30 \cdot 30 - 4 \cdot 30 + 30 \cdot 4 - 4 \cdot 4$
$= \underline{?} - \underline{?} + \underline{?} - \underline{?}$ **900, 120, 120, 16**
$= \underline{?}$ **884**

Include this activity in your portfolio.

Windchill is a measure of the heat loss from exposed skin due to a combination of wind and low air temperature. The table below shows the windchill at an actual temperature of 25°F.

Windchill Temperatures at 25°F

Wind speed (mph)	15	20	25	30	35	40	45
Windchill (°F)	2	−3	−7	−10	?	?	?

1. Use a computer spreadsheet to compute the differences in the sequence of windchill numbers. Is the sequence linear?
2. Use differences to extend the sequence in the table. What do you observe?
3. Make a prediction for the windchill at wind speeds of 35, 40, and 45 mph.

47.

x	1	2	3	4	5	10
y	3	6	9	12	15	30

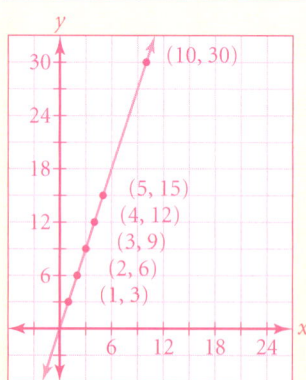

48.

x	1	2	3	4	5	10
y	1	3	5	7	9	19

49.

h	1	2	3	4	5	6	7	8
c	145	190	235	280	325	370	415	460

$c = 100 + 45h$

88 LESSON 2.5

2.6 Adding and Subtracting Expressions

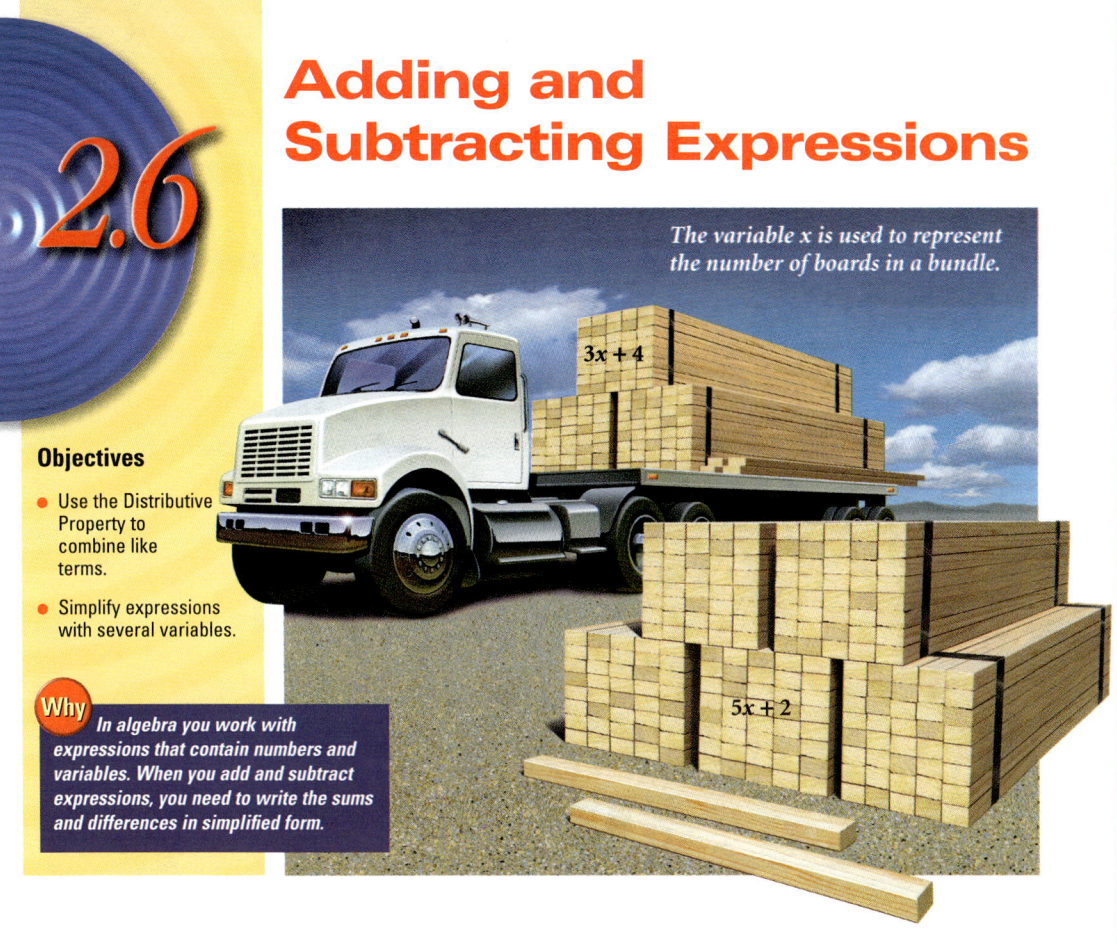

The variable x is used to represent the number of boards in a bundle.

3x + 4

5x + 2

Objectives
- Use the Distributive Property to combine like terms.
- Simplify expressions with several variables.

Why In algebra you work with expressions that contain numbers and variables. When you add and subtract expressions, you need to write the sums and differences in simplified form.

Combining Like Terms

The two lumber shipments in the illustration above are represented by the expressions $3x + 4$ and $5x + 2$. If you write out the sum of the two expressions, the result—a new expression—might be written as follows:

$(3x + 4) + (5x + 2)$ *sum of the shipments*

In this expression, the parts that are added are called **terms**. The terms $3x$ and $5x$ are **like terms** because they contain *the same form of the variable x*. The number part of the *x*-term, in this case 3 or 5, is called the **coefficient** of the *x*-term. The constants, 4 and 2, are also like terms.

The expression for the sum of the two shipments of lumber can be **simplified** by removing the parentheses and combining like terms. The terms for the bundles of lumber ($3x$ and $5x$) and the terms for the loose boards (4 and 2) are combined.

$8x + 6$ *simplified expression*

An expression is simplified when all like terms have been combined and all parentheses have been removed.

Alternative Teaching Strategy

USING MODELS Draw one rectangle with a length of *x* and width of 3 and a second rectangle with a length *x* and width of 5. Ask students to give an expression for the area of each rectangle.

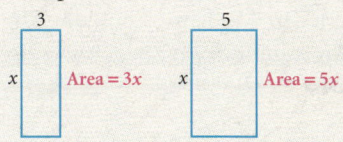

Now draw the two rectangles so they share a side, as shown at right. Ask students to give an expression for the area of the new rectangle.

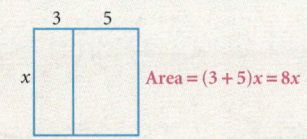

Point out that the rectangles model the addition of like terms. That is, the area of the new rectangle is the sum of the areas of the smaller rectangles, so $3x + 5x = 8x$. Be sure students see that the model would not work with unlike terms, such as $3x$ and $5y$. Then show a similar model for a subtraction problem such as $9x - 2x$.

Prepare

NCTM PRINCIPLES & STANDARDS 1–3, 6–10

LESSON RESOURCES
- Practice 2.6
- Reteaching Master 2.6
- Enrichment Master 2.6
- Cooperative-Learning Activity 2.6
- Lesson Activity 2.6
- Problem Solving/Critical Thinking Master 2.6

QUICK WARM-UP

Find each sum.
1. $12 + (-8)$ 4
2. $-6 + (-5)$ -11
3. $-11 + 7$ -4

Find each difference.
4. $4 - 9$ -5
5. $-7 - 8$ -15
6. $55 - (-20)$ 75

Also on Quiz Transparency 2.6

Teach

Why Algebraic expressions can be used to describe many real-world situations. One example of such a situation is given at the beginning of the lesson. Have students suggest other situations in which algebraic expressions might be used. Conclude by discussing situations in which it might be necessary to add or subtract these expressions.

LESSON 2.6   **89**

Activity Notes

As students work through the Activity, they should come to see the two types of tiles as concrete representations of unlike terms. Make sure they understand that a neutral pair must contain a positive tile and a negative tile *of the same type*. In Step 3, note that students are given a choice—they may model the subtraction by taking away 2 *x*-tiles and 1 unit tile, or they may apply the definition of subtraction and add 2 negative *x*-tiles and 1 negative unit tile. After completing the activity, they should appreciate why it is not permissible to combine unlike terms.

CHECKPOINT ✔

4.
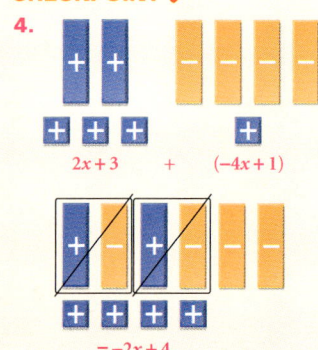

$2x + 3$ + $(-4x + 1)$

$= -2x + 4$

ADDITIONAL EXAMPLE 1

Use the Distributive Property to show that the following are true statements:

a. $11y + 7y = 18y$
$11y + 7y = (11 + 7)y$ Distributive Property
$(11 + 7)y = 18y$ Simplify.

b. $3m - 8m = -5m$
$3m - 8m = (3 - 8)m$ Distributive Property
$(3 - 8)m = -5m$ Simplify.

Activity
Exploring Combining Like Terms

You will need: algebra tiles

1. Add $3x + 4$ and $2x + 1$ by using algebra tiles.

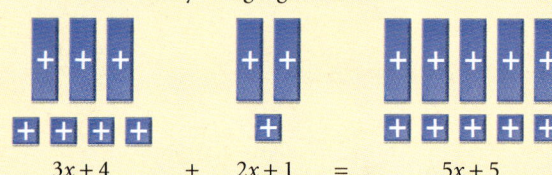

$3x + 4$ + $2x + 1$ = $5x + 5$

2. Add $4x + 3$ and $3x - 1$ by using algebra tiles. Recall the technique of removing neutral pairs.

$4x + 3$ + $3x - 1$ = $7x + 2$

3. Subtract $2x + 1$ from $3x + 4$. You can remove two *x*-tiles and 1 unit tile, or you can change the tiles representing $2x + 1$ to their opposites and add.

$(3x + 4) - (2x + 1) = (3x + 4) + (-2x - 1)$

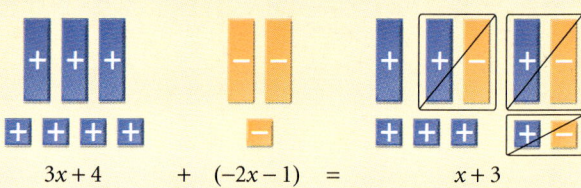

$3x + 4$ + $(-2x - 1)$ = $x + 3$

CHECKPOINT ✔

4. Subtract $4x - 1$ from $2x + 3$. You may want to add neutral pairs of *x*-tiles to the original expression.

Combining like terms is an application of the Distributive Property.

EXAMPLE 1 Use the Distributive Property to show that the following are true statements:
a. $2x + 3x = 5x$
b. $4x - 2x = 2x$

● **SOLUTION**

a. $2x + 3x = (2 + 3)x$ *Dist. Prop.*
   $(2 + 3)x = 5x$ *Simplify.*

b. $4x - 2x = (4 - 2)x$ *Dist. Prop.*
   $(4 - 2)x = 2x$ *Simplify.*

TRY THIS Use the Distributive Property to show that the following statements are true:
a. $7x + 2x = 9x$
b. $9x - 3x = 6x$

Enrichment

Have students write simplified expressions that represent the sum of an integer, *x*, and each of the following:

the next consecutive integer	$2x + 1$
the next two consecutive integers	$3x + 3$
the next three consecutive integers	$4x + 6$
the next four consecutive integers	$5x + 10$

Ask students to describe the number pattern that they see in the answers. Ask them to extend the pattern to find the sum of an integer, *x*, and the next eight consecutive integers. $9x + 36$

Inclusion Strategies

VISUAL LEARNERS Have students use colored pencils to indicate the like terms in algebraic expressions. For example, they can use a red pencil to write all of the *x*-terms and a blue pencil to write all of the constant terms. Students should then be aware that they can combine only the terms written in the same color.

Adding and Subtracting Expressions

To add two expressions, remove the parentheses, use the Commutative and Associative Properties to rearrange and group like terms, and simplify by combining like terms.

$(3x + 2) + (4x + 3)$	*Given*
$3x + 2 + 4x + 3$	*Remove parentheses.*
$(3x + 4x) + (2 + 3)$	*Rearrange and group like terms.*
$7x + 5$	*Combine like terms.*

You can subtract one number from another by adding its opposite. This is also true for expressions. To subtract one expression from another, change the subtracted expression to its opposite by changing the sign of each term in the expression. Then add the expressions.

$(2x + 3) - (4x - 1)$	*Given*
$(2x + 3) + (-4x + 1)$	*Definition of subtraction*
$2x + 3 - 4x + 1$	*Remove parentheses. (Note: + (-4x) becomes -4x by the definition of subtraction.)*
$2x - 4x + 3 + 1$	*Rearrange and group like terms.*
$-2x + 4$	*Combine like terms.*

EXAMPLE 2 Simplify.
  a. $(3x + 4) + (2x - 1)$
  b. $(4d - 2) - (5d - 3)$
  c. $(8x - 2y) + (5x + 6y)$
  d. $(5a + 3b) - (2a + 5b)$

● **SOLUTION**

a.
$(3x + 4) + (2x - 1)$	*Given*
$(3x + 2x) + (4 - 1)$	*Rearrange and group like terms.*
$5x + 3$	*Combine like terms.*

b.
$(4d - 2) - (5d - 3)$	*Given*
$(4d - 2) + (-5d + 3)$	*Definition of subtraction*
$(4d - 5d) + (-2 + 3)$	*Rearrange and group like terms.*
$-d + 1$	*Combine like terms.*

c.
$(8x - 2y) + (5x + 6y)$	*Given*
$(8x + 5x) + (-2y + 6y)$	*Rearrange and group like terms.*
$13x + 4y$	*Combine like terms.*

d.
$(5a + 3b) - (2a + 5b)$	*Given*
$(5a + 3b) + (-2a - 5b)$	*Definition of subtraction*
$(5a - 2a) + (3b - 5b)$	*Rearrange and group like terms.*
$3a - 2b$	*Combine like terms.*

TRY THIS Simplify.
  a. $(3x + 6) + (-4x + 5)$
  b. $(-4m + 3) - (-6m + 3)$
  c. $(4x - 3y) + (2x + 2y)$
  d. $(3m + 17n) - (5m + 15n)$

TRY THIS
a. $7x + 2x = (7 + 2)x$
   $= 9x$
b. $9x - 3x = (9 - 3)x$
   $= 6x$

ADDITIONAL EXAMPLE 2

Simplify.
a. $(4y - 5) + (-3y + 2)$
   $y - 3$
b. $(a + 9) - (-3a - 7)$
   $4a + 16$
c. $(3m - 8n) + (11n - 2m)$
   $m + 3n$
d. $(7x - 13y) - (8x + 6y)$
   $-x - 19y$

TRY THIS
a. $-x + 11$
b. $2m$
c. $6x - y$
d. $-2m + 2n$

GO TO: go.hrw.com
KEYWORD: MA1 Stocks

In this activity, students explore positive and negative numbers by examining gains and losses in the prices of specific stocks.

Reteaching the Lesson

USING MANIPULATIVES Have each student bring several dimes, nickels, and pennies to class. Tell students that they will use one dime to represent the variable d, one nickel to represent the variable n, and one penny to represent the constant 1. Heads will represent +, and tails will represent –. Work with students to model simple expressions like $2d$ (two dimes, heads up); $-3n$ (three nickels, tails up); 4 (four pennies, heads up); and $2n - 5$ (two nickels, heads up, and five pennies, tails up).

Now have students manipulate the coins to find sums and differences such as those below. Be sure they know that heads and tails of the same type of coin form a neutral pair. Tell them that they may model subtraction as taking away or as the addition of an opposite.

$(3d + 8) + (4d + 3)$	$7d + 11$
$(5n + 8) - (4n + 3)$	$n + 5$
$(5n - 6) + (-3n + 2)$	$2n - 4$
$(-4n + 3) - (-2n - 4)$	$-2n + 7$

LESSON 2.6

Assess

Selected Answers
Exercises 5–11, 13–63 odd

ASSIGNMENT GUIDE

In Class	1–11
Core	12–16, 21–40, 49, 51, 53
Core Plus	12–48 even, 49–54
Review	55–64
Preview	65

● Extra Practice can be found beginning on page 738.

Technology

A calculator may be helpful for exercises that involve decimals or larger numbers. You may wish to have students complete the exercises with paper and pencil first and then use a calculator to verify their answers.

Exercises

● Communicate

1. Define *like terms*. Give three examples of like terms.
2. Describe the steps involved in using the Distributive Property to add $3x$ and $8x$.
3. Explain what is meant by "adding the opposite" when subtracting expressions.
4. Explain how to simplify the following subtraction problem: $(9x + 3) - (4x - 1)$.

● Guided Skills Practice

Use the Distributive Property to show that the statements below are true. (EXAMPLE 1)

5. $-7x + 10x = 3x$ $\quad -7x + 10x = (-7 + 10)x = 3x$
6. $5h - 9h = -4h$ $\quad 5h - 9h = (5-9)h = -4h$
7. $3d + (-2d) = d$ $\quad 3d + (-2d) = (3-2)d = d$

Simplify. (EXAMPLE 2)

8. $(2x + 5) + (8x - 9)$ **$10x - 4$**
9. $(6c + 2) - (-3c + 7)$ **$9c - 5$**
10. $(5f - g) - (3f + g)$ **$2f - 2g$**
11. $(8x + 2y) + (8x - 2y)$ **$16x$**

● Practice and Apply

Use the Distributive Property to show that the following are true statements:

12. $3x + 4x = 7x$
13. $4m - 10m = -6m$
14. $5z + (-2z) = 3z$
15. $\frac{1}{3}p + \frac{2}{3}p = p$
16. $4.7d - 6.7d = -2d$
17. $11f - 15f = -4f$
18. $0.3r + 4.6r = 4.9r$
19. $-8x - 4x = -12x$
20. $\frac{1}{4}y - \frac{1}{2}y = -\frac{1}{4}y$

Simplify the following expressions:

21. $9x - 3x$ **$6x$**
22. $8y - 2y$ **$6y$**
23. $5c - (3 - 2c)$ **$7c - 3$**
24. $7d - (1 - d)$ **$8d - 1$**
25. $(7r + 2s) + (6r + 3s)$ **$13r + 5s$**
26. $(9k + 2j) + (11k - 2j)$ **$20k$**
27. $(2a - 1) - (-5a - 5)$ **$7a + 4$**
28. $(4m - 2n) + (-3m - 4n)$ **$m - 6n$**
29. $(5x + 7) - (2x - 7)$ **$3x + 14$**
30. $(-x + 2y) + (2x + 3y)$ **$x + 5y$**
31. $(3r - 7) + (17r - 6)$ **$20r - 13$**
32. $(-7f + 2) - (6f + 3)$ **$-13f - 1$**
33. $(5x - 1) + (-3x - 1)$ **$2x - 2$**
34. $(-4x - 2y) - (4x + 2y)$ **$-8x - 4y$**
35. $(3.7x + 2) - (1.7x + 3)$ **$2x - 1$**
36. $(-5x + 2y) - (3x + 2y)$ **$-8x$**
37. $(9v - 8w) - (8v - 9w)$ **$v + w$**
38. $(2q + 3) - (-4q + 5) + (6q - 7)$ **$12q - 9$**
39. $(9 + 4y) - (-1 + 8y) + (7 - y)$ **$17 - 5y$**
40. $(1.1a + 1.2b) + (2a - 0.8b)$ **$3.1a + 0.4b$**

---

12. $3x + 4x = (3 + 4)x$
    $= 7x$
13. $4m - 10m = (4 - 10)m$
    $= -6m$
14. $5z + (-2z) = (5 - 2)z$
    $= 3z$
15. $\frac{1}{3}p + \frac{2}{3}p = \left(\frac{1}{3} + \frac{2}{3}\right)p$
    $= p$
16. $4.7d - 6.7d = (4.7 - 6.7)d$
    $= -2d$
17. $11f - 15f = (11 - 15)f$
    $= -4f$
18. $0.3r + 4.6r = (0.3 + 4.6)r$
    $= 4.9r$
19. $-8x - 4x = (-8 - 4)x$
    $= -12x$
20. $\frac{1}{4}y - \frac{1}{2}y = \left(\frac{1}{4} - \frac{1}{2}\right)y$
    $= \left(\frac{1}{4} - \frac{2}{4}\right)y$
    $= -\frac{1}{4}y$

92 LESSON 2.6

41. $(5x + 5y) - (5x + 7y)$ $-2y$
42. $(-x - y) + (-x - y)$ $-2x - 2y$
43. $(3.5m - 2.5n) + (2.7m - 3.7n)$ $6.2m - 6.2n$
44. $(5p - 6r) - (p + r)$ $4p - 7r$
45. $(2x + 7y) + (2x + 7y)$ $4x + 14y$
46. $(515x + 755y) - (350x + 250y)$ $165x + 505y$

CHALLENGE
47. $\left(\frac{x}{2} + 1\right) - \left(\frac{x}{3} - 1\right)$ $\frac{1}{6}x + 2$
48. $\left(\frac{2m}{5} + \frac{1}{2}\right) + \left(\frac{m}{10} + \frac{5}{2}\right)$ $\frac{1}{2}m + 3$

CONNECTIONS
49. **GEOMETRY** Write an expression for the area of the whole rectangle. Explain how the Distributive Property can be used to find this area.

$xy + xz = x(y + z)$

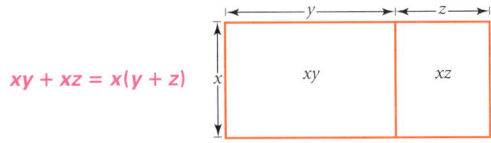

GEOMETRY Give an expression in simplified form for the perimeter of each figure.

50. 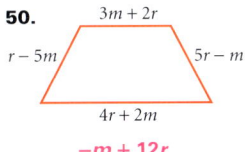 $-m + 12r$
51. 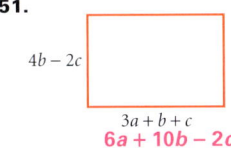 $6a + 10b - 2c$
52. 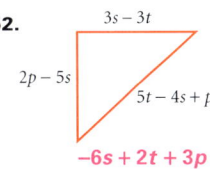 $-6s + 2t + 3p$

APPLICATIONS

53. **MANUFACTURING** A piece of fabric is cut into 12 equal lengths with 3 inches left over. Define a variable, and write an expression for the length of the original piece. $l = 12x + 3$

54. **INVENTORY** The school kitchen had 11 cases of juice plus 3 extra cans of juice. After lunch they had 6 cases of juice and no extra cans. How many cans of juice were sold during lunch? (Note: A case contains 24 cans.)
123 cans

Look Back

55. Find the next three terms in the sequence below. *(LESSON 1.1)*
    $10, 20, 40, 80, 160, \ldots$ **320, 640, 1280**

Place parentheses to make each equation true. *(LESSON 1.3)*

56. $28 \div 2 - 4 \cdot 1 = 10$
    none required
57. $16 \div 5 + 3 \div 2 = 1$
    $16 \div (5 + 3) \div 2 = 1$
58. $40 \cdot 2 + 10 \cdot 4 = 1680$
    $40 \cdot (2 + 10 \cdot 4) = 1680$

Evaluate. *(LESSONS 2.2 AND 2.3)*

59. $-2 + (-3)$ -5
60. $-7 + 6$ -1
61. $-5 - 5(-2)$ 5
62. $-10 - 5$ -15

Compute mentally. *(LESSON 2.5)*

63. $67 + 25 + 33$ **125**
64. $5 \cdot 11 \cdot 2$ **110**

Look Beyond

CONNECTION
65. **GEOMETRY** Draw a diagram to model the area represented by $(x + y)(x + y)$.

65.

	$(x + y)$	
	x	y
$(x+y)$ x	x^2	xy
y	xy	y^2

Error Analysis

When simplifying an expression such as $4x - x$, students sometimes just eliminate the variable and give the answer as 4. Review the tile model with these students, stressing that when 1 x-tile is removed from a group of 4 x-tiles, 3 x-tiles remain. To reinforce the model, encourage students to rewrite an expression like $4x - x$ as $4x - 1x$.

Look Beyond

In Exercise 65, students are asked to draw a diagram to model the product of two algebraic expressions. If they need a hint, ask them to recall the area formula for a rectangle. Students will examine this model in greater detail in Lesson 9.2.

Student Technology Guide

Student Technology Guide
2.6 Adding and Subtracting Expressions

You can use a graphics calculator to verify that $-8x + 12x = 4x$ is true. There are two approaches you can use. In either approach, begin by entering the expressions on the left and right of the equation, $-8x + 12x$ and $4x$, into the function list.

- If you graph the two resulting equations in the list, you will see that the graphs coincide. This means that $-8x + 12x$ and $4x$ have the same value for each value of x.
- If you make tables of values, you will see that the values of the two expressions are the same. This means that $-8x + 12x$ and $4x$ have the same value for each value of x.

Press **Y=**, and enter the left side of the equation into Y1 and the right side of the equation into Y2.

Press **GRAPH** to see that the graphs are identical. The two graphs lie on top of each other.

Press **2nd GRAPH** to check the table. The values of Y1 and Y2 are identical for each value of X.

Use a graphics calculator to verify each equation.

1. $3(x + 2) = 3x + 3 \cdot 2$
2. $-4(x + 6) = -4x + (-24)$
3. $6x - 9x = -3x$
4. $3x + (-8x) = -5x$
5. $(9 + 3x) - (-2 + 5x) = 11 + (-2x)$
6. $8x(36 + 12) = 288x + 96x$

7. Describe how a graphics calculator can be used to verify an equation.
Answers may vary. Sample answer: Plot both sides of the equation on the same coodinate axes and verify that they have the same graph. Then use the TABLE

8. A student wrote $3x + 4 + 2x = 9x$ by adding terms. Use graphing to feature of the show that $3x + 4 + 2x \ne 9x$ for all values of x. What is the correct sum? calculator to verify that the y-values of the two graphs are the same for each x-value.
$5x + 4$

9. Use a graphics calculator to determine whether $3x - 4x$ is equal to $-4x + 3x$. Which property supports your conclusion?
true; Commutative Property of Addition

Eyewitness Math

Focus

Since the sinking of the *Titanic*, many theories have been proposed to explain how and why the ship sank so quickly. Students use known relationships about the volume and weight of water to predict how long the gash made by the iceberg must have been in order for the ship to sink in less than 4 hours.

Motivate

Ask students if they have ever seen a movie or documentary about the sinking of the Titanic. Have students offer opinions on why they think the Titanic sank so quickly. Ask if they think that theories on the sinking of the Titanic can be proved or disproved using only mathematical calculations.

Have students read the two articles. Ask them if they have any questions that need clarification. Poll the class on whether they think that the theory presented is plausible. In other words, do they agree or disagree with the theory? Record the results of the poll.

Tell students that they will use known mathematical relationships to help them decide if the theory has merit.

ABANDON SHIP!
Titanic, Giant White Star Liner, Sinks After Collision With Iceberg on Her Maiden Voyage...

Richmond Times-Dispatch

NEW YORK, APRIL 15, 1912 — The greatest marine disaster in the history of the world occurred last night, when the *Titanic*, of the White Star Line, the biggest and finest of steamships, crammed herself against an iceberg and sank... in less than four hours.

Swinging from the westerly steamship lane at the Grand Banks of Newfoundland to take the direct run to this port, she hurled her giant bulk against an iceberg that rose from an immense field, drifted unseasonably from the Arctic. Running at high speed into that grim and silent enemy of seafarers, the shock crushed her bow... Through rent plates and timbers, water rushed...

When the *Titanic* plunged headlong against a wall of ice... her fate established that no modern steamship is unsinkable.

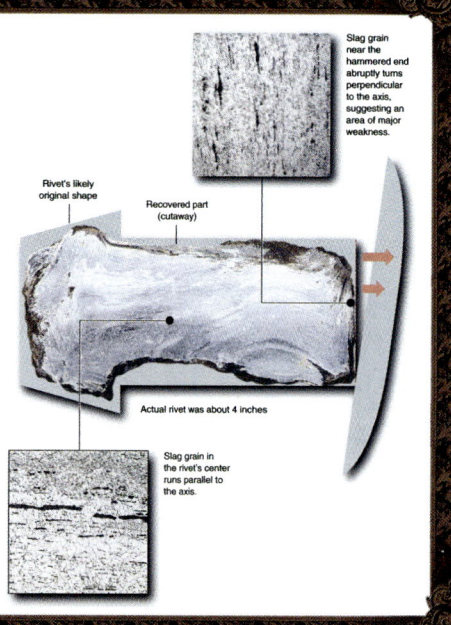

Slag grain near the hammered end abruptly turns perpendicular to the axis, suggesting an area of major weakness.

Rivet's likely original shape

Recovered part (cutaway)

Actual rivet was about 4 inches

Slag grain in the rivet's center runs parallel to the axis.

When the passenger ship Titanic sank only four days into its maiden voyage, it set the stage for years of debate and speculation. The design of the ship was thought to be so safe that it would be virtually impossible for it to sink. Since the night of April 14, 1912, survivors, scholars, and theorists have presented various explanations for what happened and how the "unsinkable" ship sank so quickly. Until the wreckage was discovered in 1985, there was no evidence to support or refute any of the theories

Faulty Rivets Emerge as Clues to *Titanic* Disaster

January 27, 1998 — Ever since the *Titanic* was discovered in the depths of the North Atlantic... her steel plates melting into rivers of rust, expeditions have repeatedly probed the hulk. And investigators armed with a growing body of evidence have been working to solve riddles posed by the opulent liner's sinking.

Now, after years of analysis and any number of false leads, experts say they have preliminary evidence suggesting that the *Titanic*, the biggest ship of her day, a dream of luxury come to life, may have been done in by structural weaknesses in some of her smallest and least glamorous parts: the rivets.

Experts, diving down nearly two and a half miles to peer through thick mud with sound waves, discovered that the *Titanic*'s bow had been pierced by six thin wounds.

The finding laid to rest the myth that the iceberg had sliced open a 300-foot gash in the ship's side and strengthened interest in the possibility of rivet failure.

The 46,000-ton *Titanic* was made of steel held together with some three million rivets. They secured both beams and plates. Suspicions of rivet failure have long haunted the famous disaster... But mysteries remained because the area in the lower bow damaged by the iceberg remained hidden under thick layers of mud.

Now new evidence suggests that they were defective and that the collision with the iceberg, rather than cutting a gash in the hull, caused the rivets to pop. An analysis shows that they had high concentrations of slag, a glassy material, making them brittle and prone to fractures. The hammered ends, or tails, of the analyzed rivets have vanished. They presumably popped off during the disaster.

Testimony given by witnesses to the tragedy was analyzed during the British Inquiry in 1912 by naval architect Edward Wilding. The evidence implied damage to the first through fifth and possibly the sixth of the ship's "watertight" compartments. This meant a possible total of at least 249 feet for the length of the damage.

Based on testimony regarding the times at which water reached various decks within the hull, Wilding concluded that the *Titanic* took on about 16,000 cubic feet of water in the first 40 minutes. Using that rate of entry, he calculated that a total area of 12 square feet must have been opened in the ship's hull by the iceberg.

by William J. Broad, New York Times

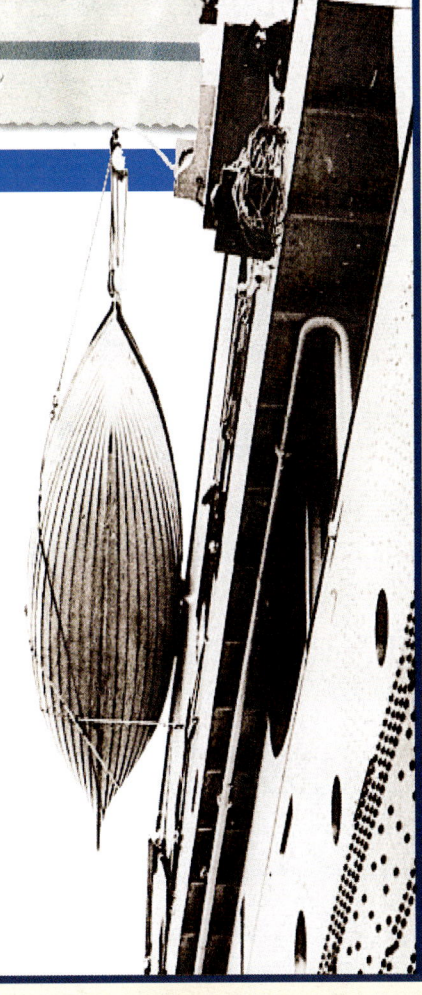

Cooperative Learning

1. If 1 cubic foot is equivalent to 7.48 gallons, calculate the number of gallons of water that the *Titanic* took on during the first 40 minutes after colliding with the iceberg.

2. If one gallon of water weighs 8.33 pounds, find the total added weight of the water that entered *Titanic* during the first 40 minutes.

3. Speculation about the length of the damage ranged as high as 300 feet. Assuming that the damage to the hull actually covered 200 feet in length, calculate the average width of the hole based on Wilding's estimate for the total area of the damage. Use the formula *area = length • width*.

4. Do you think that it is possible for a gash of that length to have the width that you calculated? How likely do you think it is that a continuous gash of this size would be created by an iceberg?

5. Given the information about the rivets and the six thin wounds from the article, what do you think is the most likely explanation for the damage to the *Titanic*?

6. **Project** Imagine that you are planning a new expedition to the wreckage of the *Titanic*. Write a brief report describing the purpose of your expedition and what information you would look for. Would you try to assess the damage more accurately through observations, collect pieces of evidence, or perhaps bring back artifacts and souvenirs? How would you justify the purpose of your mission in relation to its cost?

Cooperative Learning

Have students work in pairs. Instruct one student of each pair to complete Step 1 while the other completes Step 2. Be sure that partners share their answers before moving to the next step. Partners should work together to discuss and complete Steps 3–5.

Discuss

Bring the groups together to discuss their results. Lead the class in a debate on the validity of the proposed theory. Encourage students to use their calculations to support their arguments.

Ask students if their answer to the poll is still the same or if they have changed their opinion of the theory presented. Interested students can do research on other possible theories about the sinking of the *Titanic*.

---

1. **119,680 gallons**

2. **996,934.4 pounds (approx. 498 tons)**

3. **200w = 12; w = 0.06 feet (approx. $\frac{3}{4}$ inch)**

4. **The gash seems too narrow. A large iceberg would probably make a wider gash.**

5. **The rivets popped on some of the plates and created a series of shorter but much wider gaps in the hull.**

6. **Answers will vary.**

Prepare

NCTM PRINCIPLES & STANDARDS
1–3, 6, 8–10

LESSON RESOURCES

- Practice 2.7
- Reteaching Master 2.7
- Enrichment Master 2.7
- Cooperative-Learning Activity 2.7
- Lesson Activity 2.7
- Problem Solving/Critical Thinking Master 2.7
- Teaching Transparency 12

QUICK WARM-UP

Find each product.

1. $(-5)(18)$ −90
2. $(-9)(-7)$ 63

Find each quotient.

3. $16 \div (-8)$ −2
4. $\frac{-54}{-18}$ 3

Also on Quiz Transparency 2.7

Teach

Why Many real-world situations involve multiplication of algebraic expressions. Show students a simple 3-by-5 grid of square floor tiles such as the one below.

Lead them to see that if the length of one side of a tile is x inches, then the area of the grid is $(3x)(5x)$ square inches. Tell them that, in this lesson, they will learn to write a product such as $(3x)(5x)$ in a simpler form.

96   LESSON 2.7

2.7 Multiplying and Dividing Expressions

The number of tiles in a rectangular area can be found by multiplying the number of tiles along the length by the number of tiles along the width.

Objectives

- Multiply expressions containing variables.
- Divide expressions containing variables.

Why When solving equations, you often need to multiply or divide not only single numbers, but also larger expressions that may include variables.

Modeling Multiplication With Algebra Tiles

To model multiplication with algebra tiles, you can arrange the tiles in rectangular patterns (including squares). For example, the arrangement below models the multiplication $2 \cdot 3$.

The area of the rectangle is $2 \cdot 3$, or 6 unit tiles.
$2 \cdot 3 = 6$

You can also model more complicated multiplications, such as $(2 + 3) \cdot (4 + 2)$.

The area of the rectangle is $(2 + 3) \cdot (4 + 2)$, or 30 unit tiles.
$(2 + 3) \cdot (4 + 2) = 30$

Alternative Teaching Strategy

USING COGNITIVE STRATEGIES Divide the class into groups of three or four. Give them the following multiple-choice question:

$(4x)(3x) =$
a. $12x$ b. $7x$ c. $12x^2$ d. $7x^2$

Have students work together to determine the correct answer. Direct each group to write a convincing argument that supports their choice.
The correct choice is c.

After all groups have written their arguments, have one member of each group present to the class a summary of their argument. During these presentations, elicit the group's reasons, if any, for eliminating choices **a**, **b**, and **d**.

Repeat the activity with the following question:

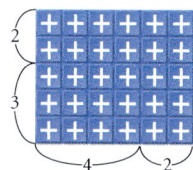

$\frac{8x^2}{2x} =$
a. $6x^2$ b. $6x$ c. $4x^2$ d. $4x$

The correct choice is d.

When *one of the factors* of a product contains a variable such as x, you will need to include x-tiles. The arrangement below models the product $2 \cdot 3x$.

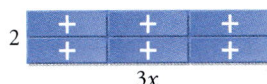

The area of the rectangle is
$2 \cdot 3x$, or 6 x-tiles.
$2 \cdot 3x = 6x$.

When *both of the factors* of a product contain a variable such as x, a new tile is needed to represent the product. Think of a tile that has the same height and length as the long side of an x-tile. The area of the tile is $x \cdot x$, which is written as x^2 and read as "x squared."

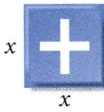

The area of the new tile is
$x \cdot x$, or x^2.

To model $2x \cdot 3x$, use x^2-tiles. Think of $2x \cdot 3x$ as the area of a rectangle with dimensions of $2x$ by $3x$.

The area of the rectangle is
$2x \cdot 3x$, or 6 x^2-tiles.
$2x \cdot 3x = 6x^2$

Exponents

In later work you will use variables that occur as factors more than twice, such as $x \cdot x \cdot x = x^3$ (read as "x cubed").

An expression such as x^2 or x^3 is called a **power** (in this case, a power of x). Other powers are illustrated in the table below.

Symbol	Read as	Mathematical meaning
3^1	3 to the first power	3
3^2	3 to the second power, or 3 squared	$3 \cdot 3$
3^3	3 to the third power, or 3 cubed	$3 \cdot 3 \cdot 3$
3^4	3 to the fourth power	$3 \cdot 3 \cdot 3 \cdot 3$
3^5	3 to the fifth power	$3 \cdot 3 \cdot 3 \cdot 3 \cdot 3$
y^4	y to the fourth power	$y \cdot y \cdot y \cdot y$
$5y^4$	5 multiplied by y to the fourth power	$5 \cdot y \cdot y \cdot y \cdot y$
y^n	y to the nth power	$y \cdot y \cdot y \cdot \ldots \cdot y$

In an expression such as x^n, x is called the **base** and n is called the exponent. The **exponent** is the number of times the base appears as a factor.

CRITICAL THINKING Describe the shape of the geometric object that could be used to represent x^3.

Teaching Tip

Encourage students to use algebra tiles as long as necessary until they fully understand how to multiply algebraic expressions. You may want to have them make drawings to record their work with algebra tiles.

CRITICAL THINKING
a cube

Interdisciplinary Connection

PHYSICS To find the distance in miles that a car travels, you multiply its speed in miles per hour by the time traveled in hours. Ask students to write an algebraic expression for the distance a car travels in x hours at an average speed of 55 miles per hour. **55x** Then ask them to write two different forms of an expression for the distance traveled when the time is increased by 2 hours.
$55(x + 2) = 55x + 110$

Multiplying and Dividing Expressions

To multiply expressions, you can use the definition of exponents along with the Commutative, Associative, and Distributive Properties.

EXAMPLE 1 Simplify. a. $2x(3x - 4)$ b. $(-2)(5a - 4)$

SOLUTION

a. $2x(3x - 4) = 2x(3x) - 2x(4)$ *Use the Distributive Property.*
$\qquad\qquad\quad = 6x^2 - 8x$

b. $(-2)(5a - 4) = (-2)(5a) - (-2)(4)$
$\qquad\qquad\quad\;\; = -10a + 8$

TRY THIS Simplify. a. $3x(-5x - 4)$ b. $(-4)(3x + 4)$

EXAMPLE 2 Simplify: $(5x + 3y - 7) - 3(2x - y)$.

SOLUTION

Use the definition of subtraction to rewrite the expression as an addition problem.

$(5x + 3y - 7) - 3(2x - y) = (5x + 3y - 7) + (-3)(2x - y)$

Distribute -3 over $(2x - y)$.

$(5x + 3y - 7) + (-3)(2x - y) = 5x + 3y - 7 + (-3)(2x) - (-3)(y)$
$\qquad\qquad\qquad\qquad\qquad\quad = 5x + 3y - 7 - 6x + 3y$
$\qquad\qquad\qquad\qquad\qquad\quad = (5x - 6x) + (3y + 3y) - 7$
$\qquad\qquad\qquad\qquad\qquad\quad = -x + 6y - 7$

TRY THIS Simplify: $(2b - 5c) - 4(3b + c)$.

EXAMPLE 3 Model the division problem $\frac{2x + 6}{2}$ with algebra tiles.

SOLUTION

$2x + 6$

Because you are dividing by 2, divide the tiles into two equal groups.

$x + 3$
$x + 3$

Since each set contains 1 x-tile and 3 unit-tiles, $\frac{2x + 6}{2} = x + 3$.

Enrichment

Have students work in pairs. Provide each pair with a bag containing six number cubes—three of one color to represent positive numbers and three of another color to represent negative numbers. One student in each pair should draw three cubes from the bag and roll them. The student should then use the numbers that come up to write an algebraic expression that can be simplified by using the Distributive Property. For example, if a student rolls -1, 3, and -4, the expression $(-4)(-x + 3)$ can be written. The other student should then simplify the expression. Have students reverse their roles and repeat the activity.

ADDITIONAL EXAMPLE 1

Simplify.

a. $8x(-5x - 9)$ $-40x^2 - 72x$

b. $(-3a - 4)(-5)$ $15a + 20$

TRY THIS

a. $-15x^2 - 12x$

b. $-12x - 16$

ADDITIONAL EXAMPLE 2

Simplify:
$(4a - b + 3) + 2(5a + 2b - 1)$.

$14a + 3b + 1$

TRY THIS

$-10b - 9c$

ADDITIONAL EXAMPLE 3

Model the division expression $\frac{3x + 6}{3}$ with tiles.

Start with 3 x-tiles and 6 unit tiles.
Divide the tiles into 3 groups.

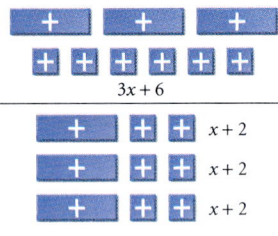
$3x + 6$

$x + 2$
$x + 2$
$x + 2$

$\frac{3x + 6}{3} = x + 2$

Use Teaching Transparency 12.

TRY THIS Use algebra tiles to model the division problem $\frac{8x-4}{4}$.

As the algebra-tile model suggests, when you divide an expression by a number, each term in the expression is divided by that number.

> **Dividing an Expression**
>
> For all real numbers a, b, and c, where $c \neq 0$:
>
> $$\frac{a+b}{c} = \frac{a}{c} + \frac{b}{c} \quad \text{and} \quad \frac{a-b}{c} = \frac{a}{c} - \frac{b}{c}$$

CRITICAL THINKING The division expression $\frac{a+b}{c}$ can be written as $\frac{1}{c}(a+b)$. Show how $\frac{1}{c}(a+b)$ gives the rule for dividing a sum.

EXAMPLE 4 Simplify $\frac{2x^2+6}{2}$.

● **SOLUTION**

There are two ways of working this problem. One way is to divide each term in the numerator by the divisor:

$$\frac{2x^2+6}{2} = \frac{2x^2}{2} + \frac{6}{2}$$
$$= x^2 + 3$$

Another way of solving the problem is to represent it as a multiplication problem and to use the Distributive Property.

$$\frac{2x^2+6}{2} = \frac{1}{2}(2x^2+6)$$
$$= \frac{1}{2}(2x^2) + \frac{1}{2}(6)$$
$$= x^2 + 3$$

EXAMPLE 5

APPLICATION
RECREATION

George is in charge of supplies for a mountain-climbing guide company. He is trying to locate carabiners for four climbing teams. George's preferred brand of carabiners are available at three outfitters as follows:

Outfitter A has 4 boxes plus 6 loose carabiners. Outfitter B has 6 boxes, but one box is missing 3 carabiners. Outfitter C has 10 boxes plus 5 loose carabiners.

Assuming each box contains an equal number of carabiners, write an expression describing the total number of carabiners available.

Write another expression describing the number of carabiners available for each team if the carabiners are divided equally among the teams.

Carabiners are used in mountain climbing to attach the climber to a rope.

TRY THIS

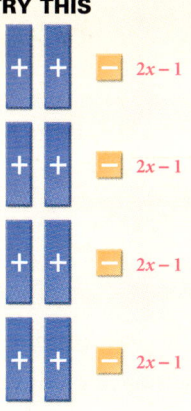

$$\frac{8x-4}{4} = 2x-1$$

CRITICAL THINKING
$$\frac{1}{c}(a+b) = \frac{1}{c} \cdot a + \frac{1}{c} \cdot b$$
$$= \frac{a}{c} + \frac{b}{c}$$
$$= \frac{a+b}{c}$$

ADDITIONAL EXAMPLE 4

Simplify $\frac{10x^2+5}{5}$.

$2x^2 + 1$

☞ For Additional Example 5, see page 100.

Inclusion Strategies

LINGUISTIC LEARNERS Tell students to imagine that they have been asked to produce two brief videos. One video will show how algebra tiles are used when multiplying expressions, and the second will show how they are used when dividing expressions. Tell students that the video may "star" either one or two people. Students should choose an appropriate multiplication problem and an appropriate division problem to feature in the video and then write the script that the actors will use to demonstrate the algebra-tile procedures.

Additional EXAMPLE 5

Haya is in charge of landscaping six large flower beds in a city park. She wants to plant tulip bulbs in each bed. The bulbs are available at four local garden suppliers, as indicated below.

Supplier A has 9 bags plus 17 loose bulbs.

Supplier B has 3 bags plus 15 loose bulbs.

Supplier C has 7 bags plus 5 loose bulbs.

Supplier D has 11 bags plus 5 loose bulbs.

Each bag contains the same number of tulip bulbs. Haya plans to divide the bulbs equally among the six flower beds.

Write an expression describing the total number of tulip bulbs available. $30t + 42$, where t represents the number of tulips in one bag

Write another expression describing the number of tulip bulbs available for each flower bed. $5t + 7$, where t represents the number of tulips in one bag

Additional EXAMPLE 6

Simplify $\frac{9 - 6x}{-3}$.

$-3 + 2x$

TRY THIS
a. $2x^2 - 3$
b. $-2x^2 + 3$

● **SOLUTION**

Using c to represent the number of carabiners in one box, the number available from each outfitter can be written as follows:

Outfitter A: $4c + 6$
Outfitter B: $6c - 3$
Outfitter C: $10c + 5$

The total available carabiners can be represented by adding the three expressions.

$$(4c + 6) + (6c - 3) + (10c + 5) = 4c + 6 + 6c - 3 + 10c + 5$$
$$= (4c + 6c + 10c) + (6 - 3 + 5)$$
$$= 20c + 8$$

To represent the number of carabiners available for each team, divide the $20c + 8$ by 4.

$$\frac{20c + 8}{4} = \frac{20c}{4} + \frac{8}{4}$$
$$= 5c + 2$$

EXAMPLE 6 Simplify $\frac{8x - 4}{-4}$.

● **SOLUTION**

Remember that because the quantity $8x - 4$ is to be divided by -4, each term in the quantity must be divided by -4.

$$\frac{8x - 4}{-4} = \frac{8x}{-4} - \frac{4}{-4}$$
$$= -2x - (-1)$$
$$= -2x + 1$$

Using the Distributive Property, multiply each term in the numerator by $\frac{1}{-4}$.

$$\frac{8x - 4}{-4} = \frac{8x}{-4} - \frac{4}{-4}$$
$$= \frac{1}{-4}(8x - 4)$$
$$= \left(\frac{1}{-4}\right)(8x) - \left(\frac{1}{-4}\right)(4)$$
$$= -2x + 1$$

TRY THIS Simplify each expression by using two different methods.

a. $\frac{6x^2 - 9}{3}$
b. $\frac{10x^2 - 15}{-5}$

Reteaching the Lesson

USING MANIPULATIVES Show students 4 paper cups, each of which contains 2 x-tiles. Write $2x$ on the side of each cup. Now spill all of the tiles into a single pile. Elicit from students that there are 8 x-tiles in the pile and that the expression $8x$ represents them. Lead students to see that the process they observed models the multiplication problem $(4)(2x) = 8x$. Repeat the process for the multiplication problem $3(x + 2)$ by placing 1 x-tile and 2 unit tiles in each of 3 cups. Ask students to verify that the product is $3x + 6$.

Now place 12 x-tiles into a pile. Tell students that you want to place the same number of tiles in each of 3 cups. Lead them to see that $12x$ represents the tiles in the pile, that $4x$ represents the tiles in each cup, and that the process models the division problem $\frac{12x}{3}$. Repeat the process for $\frac{12x + 6}{6}$, placing 12 x-tiles and 6 unit tiles in a pile and dividing them among 6 cups. Have students verify that the quotient this time is $2x + 1$.

Exercises

Communicate

1. Explain how to write an expression that shows Jan's income if she earns $6 per hour and works
   a. 4 hours.   b. 2.5 hours.   c. h hours.
2. Explain how to simplify $-2(4m + 5)$.
3. Describe how to simplify the expression $\frac{5p - 15}{-5}$.
4. Are all of the expressions below equivalent? If not, explain how they are different.
   a. $(4x - 6) \div 2$ b. $(4x - 6)(0.5)$ c. $\frac{4x - 6}{2}$ d. $\frac{-6 + 4x}{2}$
5. Explain why $\frac{2x + 6}{2}$ is not equivalent to $x + 6$.

Guided Skills Practice

Simplify. (EXAMPLE 1)

6. $3(4x - 2)$ $12x - 6$
7. $-3(5x + 2)$ $-15x - 6$
8. $-8(4 - x)$ $-32 + 8x$

Simplify. (EXAMPLE 2)

9. $(2x + 3y - 6) + 3(2x + y)$ $8x + 6y - 6$
10. $5(b + c) - 4(3b - 2c + 1)$ $-7b + 13c - 4$

Use algebra tiles to model each division problem. (EXAMPLE 3)

11. $\frac{2x + 8}{2}$
12. $\frac{4x^2 + 10}{2}$
13. $\frac{5y^2 + 20}{5}$

Simplify. (EXAMPLES 4, 5, AND 6)

14. $\frac{-6x + 10}{-2}$ $3x - 5$
15. $\frac{12x - 18}{-6}$ $-2x + 3$
16. $\frac{3x^2 - 21}{3}$ $x^2 - 7$

Practice and Apply

Simplify the following expressions. Use the Distributive Property if needed.

17. $2 \cdot 6x$ $12x$
18. $-6x \cdot 2$ $-12x$
19. $6x \cdot 2x$ $12x^2$
20. $12x \cdot 3x$ $36x^2$
21. $-66x \div 2$ $-33x$
22. $-12x \div 3$ $-4x$
23. $-1.2x \cdot 3x$ $-3.6x^2$
24. $-2(6x + 3)$ $-12x - 6$
25. $7x^2 - (3 - x^2)$ $8x^2 - 3$
26. $2x^2 - (4 - x^2)$ $3x^2 - 4$
27. $x^2 - 2(3 - x^2)$ $3x^2 - 6$
28. $3x \cdot 5 + 2x \cdot 2$ $19x$
29. $-2 \cdot 8x$ $-16x$
30. $-2(4x - 1)$ $-8x + 2$
31. $8x \cdot 2x$ $16x^2$
32. $-8x \div 2$ $-4x$
33. $-3(7x - 3)$ $-21x + 9$
34. $-21x \cdot 3x$ $-63x^2$
35. $21x \cdot 3x$ $63x^2$
36. $-21x \div 3$ $-7x$
37. $8x^2 - (2 - 5x^2)$ $13x^2 - 2$

Assess

Selected Answers

Exercises 6–16, 17–85 odd

ASSIGNMENT GUIDE

In Class	1–16
Core	17–37, 39–45 odd, 49–57 odd, 60
Core Plus	18–50 even, 47, 52–61
Review	62–86
Preview	87, 88

✎ Extra Practice can be found beginning on page 738.

Technology

A calculator may be helpful for performing the multiplication and division calculations in some of the exercises. You may wish to have students complete the exercises with paper and pencil first and then use a calculator to verify their answers.

13.

$5y^2 + 20$

$\frac{5y^2 + 20}{5} = y^2 + 4$

11.

$x + 4$

$x + 4$

$2x + 8$

$\frac{2x + 8}{2} = x + 4$

12.

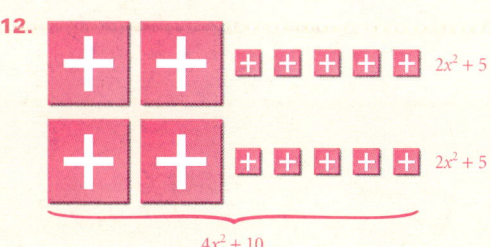

$2x^2 + 5$

$2x^2 + 5$

$4x^2 + 10$

$\frac{4x^2 + 10}{2} = 2x^2 + 5$

LESSON 2.7   101

Error Analysis

When simplifying an expression such as $6x \cdot 2x$, students frequently think of *like terms* and write either $8x$ or $12x$. Remind students that like terms are parts of an *addition* expression. To reinforce the fact that no addition is involved in an expression like this, suggest that they rewrite it as follows:

$$6x \cdot 2x = 6 \cdot x \cdot 2 \cdot x$$
$$= 6 \cdot 2 \cdot x \cdot x$$
$$= 12x^2$$

38. $8x^2 - 10(2 - 5x^2)$ **$58x^2 - 20$**

39. $\dfrac{11 - 33y}{11}$ **$1 - 3y$**

40. $\dfrac{-10x + 35}{5}$ **$-2x + 7$**

41. $\dfrac{-90x + 2.7}{-9}$ **$10x - 0.3$**

42. $\dfrac{5w + 15}{-5}$ **$-w - 3$**

43. $\dfrac{8m + 10}{2}$ **$4m + 5$**

44. $\dfrac{30x + 20}{-5}$ **$-6x - 4$**

45. $\dfrac{4y^2 + 4}{4}$ **$y^2 + 1$**

46. $\dfrac{27x - 18}{-9}$ **$-3x + 2$**

CHALLENGE

47. $\dfrac{30x^2 + 20x - 8}{-5}$ **$-6x^2 - 4x + 1.6$**

48. $\dfrac{36y - 30}{12}$ **$3y - 2.5$**

49. $\dfrac{49.5 + 10x}{10x}$ **$\dfrac{4.95}{x} + 1$**

50. Which of the expressions below are equivalent?
    a. $5(3x - 7)$ **b.** $5 \cdot (3x - 7)$ **c.** $(5)(7 - 3x)$ **d.** $5(3x + 7)$ **a, b**

51. Which of the expressions below are equivalent?
    a. $(6 - 12x) \div 6$ **b.** $(6 - 12x) \cdot \dfrac{1}{6}$ **c.** $\dfrac{6 - 12x}{6}$ **d.** $1 - 12x$ **a, b, and c**

CONNECTIONS

GEOMETRY The formula for the area of a rectangle is $A = lw$. Find the area for each rectangle by evaluating the formula for the given length and width.

52.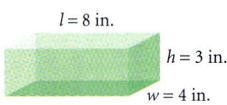
$l = 6$ cm, $w = 4$ cm
24 cm^2

53.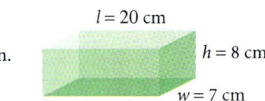
$l = 10$ in., $w = 3.5$ in.
35 in.^{2}

54.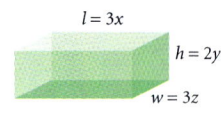
$l = 3x$, $w = 2x$
$6x^2$

GEOMETRY The formula for the volume of a rectangular prism is $V = lwh$. Find the area for each rectanglular prism by evaluating the formula for the given length, width, and height.

55. $l = 8$ in., $h = 3$ in., $w = 4$ in.
96 in.^{3}

56. $l = 20$ cm, $h = 8$ cm, $w = 7$ cm
1120 cm^3

57. $l = 3x$, $h = 2y$, $w = 3z$
$18xyz$

APPLICATIONS

58. **INCOME** Nicole has a part-time job three days after school and on Saturdays. If she earns $5.25 per hour, find Nicole's earnings for each of the following four days:
    a. 4 hours on Friday **$21.00**
    b. 7 hours on Saturday **$36.75**
    c. h hours on Monday **$5.25h$**
    d. $(h + 1)$ hours on Tuesday **$5.25(h + 1)$**

CHALLENGE

59. Write an expression in simplified form that shows the total number of hours that Nicole worked during the four days described in Exercise 58.
$12 + 2h$

60. **BUSINESS** A plumber charges $35 an hour plus a fixed service fee of $20 per job. Write an expression and determine how much the plumber would charge for a job that takes
    a. 1 hour. **$55**   **b.** 3 hours. **$125**   **c.** 7.5 hours. **$282.50**

61. **BUSINESS** On Sundays, the plumber described in Exercise 60 doubles his hourly rate. Write an expression and determine how much the plumber would charge on Sundays for a job that takes
    a. 1 hour. **$90**   **b.** 3 hours. **$230**   **c.** 7.5 hours. **$545**

Practice

2.7 Multiplying and Dividing Expressions

Simplify the following expressions. Use the Distributive Property if needed.

1. $2 \cdot 7x$ **$14x$**
2. $-4t \cdot 3$ **$-12t$**
3. $4a \cdot 3a$ **$12a^2$**
4. $-4(2d - 3)$ **$-8d + 12$**
5. $-2.3y \cdot y$ **$-2.3y^2$**
6. $\dfrac{-12w}{2w}$ **-6**
7. $7y \cdot 9y$ **$63y^2$**
8. $rt \cdot rk$ **r^2tk**
9. $-5(7 - 4e)$ **$-35 + 20e$**
10. $-36z \div (-4)$ **$9z$**
11. $\dfrac{-15x}{3}$ **$-5x$**
12. $-4(6y - 4)$ **$-24y + 16$**
13. $\dfrac{4g - 4r}{4}$ **$g - r$**
14. $-3(2 - 7m)$ **$-6 + 21m$**
15. $3x - (2x - 5)$ **$x + 5$**
16. $\dfrac{3 + 24r}{3}$ **$1 + 8r$**
17. $\dfrac{10 - 5x}{-5}$ **$-2 + x$**
18. $\dfrac{-18w + 12}{-6}$ **$3w - 2$**
19. $-y(c + d)$ **$-yc - yd$**
20. $\dfrac{32y + 24y}{4y}$ **14**

A computer consultant charges $50 per hour. How much would the consultant charge for

21. 3 hours? **$150**
22. 7.5 hours? **$375**
23. t hours? **$50t$**

A telephone company charges $40 per hour for repair work, plus a $25 service charge per job. How much would a customer be charged for a job that takes

24. 2 hours? **$105**
25. 3.5 hours? **$165**
26. t hours? **$40t + 25$**

George makes $8.00 an hour at his part-time job. Find his earnings for each of the following days:

27. 3 hours on Monday **$24**
28. 7 hours on Saturday **$56**

102 LESSON 2.7

Look Back

Evaluate. *(LESSONS 2.2, 2.3, AND 2.4)*

62. $3 + (-2)$ **1**   **63.** $1.4 - (-5)$ **6.4**   **64.** $-10 \cdot 21$ **−210**   **65.** $4.7 \cdot (-2)$ **−9.4**

66. $3.6 - (-3.6)$ **7.2**   **67.** $2.7 + (-5)$ **−2.3**   **68.** $1\frac{1}{2} - 2$ **$-\frac{1}{2}$**   **69.** $87 \div (-3)$ **−29**

70. $\frac{-8}{4}$ **−2**   **71.** $-150 \div (-15)$ **10**   **72.** $-15 \cdot 2.1$ **−31.5**   **73.** $3.7 - (-2.6)$ **6.3**

74. $-8.5 \div 5$ **−1.7**   **75.** $0.1 - 2.7$ **−2.6**   **76.** $15 \cdot (-3)$ **−45**   **77.** $86 + (-92)$ **−6**

78. $19 + (-2)$ **17**   **79.** $3\frac{1}{3} + \left(-1\frac{1}{2}\right)$ **$1\frac{5}{6}$**   **80.** $5\frac{1}{2} - (-5)$ **$10\frac{1}{2}$**   **81.** $-3.25 \div (-2.5)$ **1.3**

Simplify. *(LESSON 2.6)*

82. $(-3x + 7) + (2x - 1)$ **$-x + 6$**   **83.** $(5m - 2) + (5m - 2)$ **$10m - 4$**

84. $(-4x - 2y) - (-3x - 4y)$ **$-x + 2y$**   **85.** $(3x + 2y) + (3x - 2y) - (3x - 2y)$ **$3x + 2y$**

86. $(x^2 + 2y + 4) - (2x - 3y - 2)$ **$x^2 - 2x + 5y + 6$**

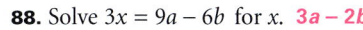

Look Beyond

87. Simplify: $\frac{2x^2 + 4x}{4x}$. **$\frac{1}{2}x + 1$**   **88.** Solve $3x = 9a - 6b$ for x. **$3a - 2b$**

Include this activity in your portfolio.

Visitors to a city park can rent bicycles at the following rates:
- Adult: $2 per hour plus $4
- Child: $1 per hour plus $3

1. Write an expression for the cost of renting an adult bicycle with H as the number of hours.

2. Write an expression for the cost of renting a child's bicycle with H as the number of hours.

3. Write two different expressions for renting 1 adult's bicycle and 1 child's bicycle for H hours.

4. Write two different expressions for renting 2 adult's bicycles and 3 children's bicycles for H hours.

WORKING ON THE CHAPTER PROJECT

You should now be able to complete Activity 3 of the Chapter Project on page 104.

In Exercise 87, students use their division skills to simplify a rational expression. They will learn more about this process in Lesson 11.3. In Exercise 88, students are given an equation that involves several variables, and they use division to solve for one of them. They will do further work with equations like this in Lesson 3.6.

ALTERNATIVE Assessment

Portfolio Activity

The Portfolio Activity can be used as preparation for the Chapter Project or as a separate activity. In the Portfolio Activity on this page, students create algebraic expressions that represent a real-world situation and use their addition skills to simplify the expressions.

Answers to Portfolio Activities can be found in Additional Answers of the Teacher's Edition.

Student Technology Guide

Project

Focus
Spreadsheets can be used to organize and manipulate data. Students will be asked to set up a spreadsheet, enter data into it, and use the spreadsheet's functions to analyze the data. They also will see how the spreadsheet can be used to make a graph of the data. Then they will compare spreadsheets.

Motivate
Explain to students that each cell in a spreadsheet has an address, or name. Provide students with a blank table that has 3 columns and 3 rows. Have them label the columns with the letters *A*, *B*, and *C* and label the rows with the numbers *1*, *2*, and *3*. Have them work in pairs to devise a way to use the letters and numbers to name each cell in the table. Ask each pair to share their ideas with the class.

Activity 1
1. 0
2. 3
3. –3
4. Row 1 = 0
   Row 2 = 0
   Row 3 = 0
   Column A = 0
   Column B = 0
   Column C = 0
   Diagonal 1 = 0
   Diagonal 2 = 0
5. They all equal 0.
6. Students should use the SUM command in a spreadsheet program to find that all of the sums are 0.

Chapter Project Two: Using a Spreadsheet

Computer spreadsheets can perform many computations quickly. Business people use spreadsheets to organize and analyze data.

The cells are identified by the column letter and row number, as shown.

	A	B	C
1	A1	B1	C1
2	A2	B2	C2
3	A3	B3	C3

Activity 1

Below is an example of a simple spreadsheet.

	A	B	C
1	1	–4	3
2	2	0	–2
3	–3	4	–1

1. What is the number in cell B2?
2. What is the number in cell C1?
3. What is the number in cell A3?
4. Use mental math to find the sum of each row, column, and diagonal.
5. What did you discover about all of the sums?
6. If you are using a spreadsheet on a computer, use the SUM command to find all of the row, column, and diagonal sums.
7. Adjust the values to create other number arrays in which the sums of the rows, columns, and diagonals are all the same (these are called "magic squares").

7. Students should create a magic square. For example,

2	–8	6
4	0	–4
–6	8	–2

is a magic square.

104 CHAPTER 2 PROJECT

Activity 2

The table below shows the average high and low temperature in Minneapolis, Minnesota, in degrees Fahrenheit.

	A	B	C	D	E	F	G	H	I	J	K	L	M
1	Month	J	F	M	A	M	J	J	A	S	O	N	D
2	Average high	21	27	39	57	69	79	84	81	71	59	41	26
3	Average low	3	9	23	36	48	58	63	60	50	39	25	10
4	Average range	?	?	?	?	?	?	?	?	?	?	?	?

1. Put the data into a spreadsheet. Compute the range for each month.
2. Use the spreadsheet to build a double line graph for the data. How does the double line graph display the monthly ranges?
3. Assume that the months are all the same length, and compute the average annual high, the average annual low, and the average annual range.
4. Write a paragraph describing the differences in the averages found in Step 3.

Activity 3

Visitors to a national park can rent equipment at the following rates:

Canoe $6 per hour plus $14
Kayak $5 per hour plus $12

Build a spreadsheet to determine the cost of renting the following combinations of canoes and kayaks for 1 through 10 hours:

	A	B	C	D	E	F	G	H	I
1	Combinations								
2	Number of canoes	1	1	1	1	2	2	2	2
3	Number of kayaks	1	2	3	4	1	2	3	4

Satellite photo of Minneapolis-St. Paul, Minnesota

Cooperative Learning

For **Activities 1–3**, have students work in pairs. When entering given data into a spreadsheet, one student of the pair can read the table entries aloud while the other types at the keyboard. Encourage the reader to give a specific location for each entry. For example, the reader might tell the student at the keyboard, "In cell A1, the entry is 1." Be sure students take turns assuming the role of reader.

The pairs of students should work together to investigate the software, learn how to create formulas, and locate any special functions they need. Be sure students know that, in spreadsheet usage, the word range generally refers to a span of cells. In **Activity 1**, for example, the *range* of cells that contains data is A1 through C3. To compute a statistical range, they most likely will need to create a subtraction formula that incorporates the spreadsheet's maximum and minimum functions.

Discuss

Bring the class together to discuss their answers to the questions. Have them suggest real-world situations in which spreadsheet tables might be useful. Have each pair of students choose one situation, research appropriate data, and create a spreadsheet.

Activity 2

1. 18, 18, 16, 21, 21, 21, 21, 21, 21, 20, 16, 16

2.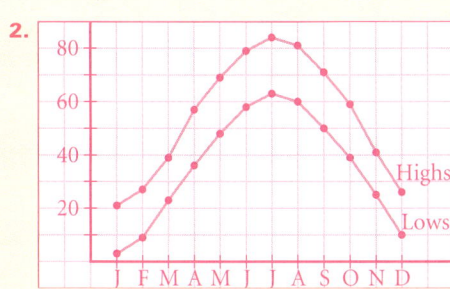

Month	J	F	M	A	M	J	J	A	S	O	N	D
Average high	21	27	39	57	69	79	84	81	71	59	41	26
Average low	3	9	23	36	48	58	63	60	50	39	25	10

The monthly ranges are visualized by the space between the two line graphs.

3. Average High = 54.50
   Average Low ≈ 35.33
   Average Range ≈ 19.17

4. Answers may vary.

Activity 3

Students should build the described spreadsheet. The cost for renting c canoes and k kayaks for h hours can be found using
$$c(6h + 14) + k(5h + 12)$$
for $h = 1$ to 10.

Chapter Review

Chapter 2 Review and Assessment

VOCABULARY

absolute value 57	integers 54	real numbers 56
additive inverse 64	irrational numbers 56	reciprocal 75
Associative Property 81	like terms 89	Reflexive Property 84
base 97	multiplicative inverse 75	simplified 89
coefficient 89	natural numbers 54	Substitution Property 84
Commutative Property 81	negative numbers 55	subtraction 68
Distributive Property 82	neutral pair 61	Symmetric Property 84
exponent 97	opposites 57	term 89
identity element for addition 64	positive numbers 55	theorem 85
	power 97	Transitive Property 84
identity element for multiplication 75	rational numbers 55	whole numbers 54

Key Skills & Exercises

LESSON 2.1

Key Skills

Compare values of real numbers.

Insert an ordering symbol to make the following statements true:

a. 8 ___ 5   **b.** −2 ___ 7

a. 8 is to the right of 5 on a number line, so 8 > 5.

b. 7 is to the right of −2 on a number line, so 7 > −2.

Find the absolute value of a number.

To find the absolute value of −7, find the distance it is from 0 on the number line.
$$|-7| = 7$$

Exercises

Insert an ordering symbol to make each statement true.

1. 12 **<** 15
2. −19 **<** 0.2
3. 23.4 **>** 23
4. $-1\frac{2}{3}$ **<** $\frac{23}{47}$

Find the absolute value of each number.

5. −73 **73**
6. 12 **12**
7. −15.2 **15.2**
8. $-\frac{16}{53}$ $\frac{16}{53}$
9. $\frac{23}{54}$ $\frac{23}{54}$
10. 0 **0**

LESSON 2.2

Key Skills

Add two or more real numbers.

When numbers have like signs, find the sum of their absolute values, and use the common sign. When numbers have unlike signs, find the difference of their absolute values and use the sign of the number with the greater absolute value.
$-4 + (-7) = -11$ $26 + (-9) = 17$ $-26 + 9 = -17$

Exercises

Find each sum.

11. −17 + 6 **−11**
12. 48 + (−15) **33**
13. −23 + (−25) + 3 **−45**
14. −39 + 68 **29**
15. 33 + (−55) **−22**
16. −214 + 214 **0**
17. 6 + (−7) + (−9) **−10**
18. −8 + 8 + (−12) **−12**

Chapter Test, Form A

NAME _____ CLASS _____ DATE _____

Chapter Assessment
Chapter 2, Form A, page 1

Write the letter that best answers the question or completes the statement.

b 1. Find the opposite of −(3 + 3.5).
   a. −6.5 b. 6.5 c. 0.5 d. −0.5

c 2. Add: −25 + 26 + |−18|.
   a. 32 b. −16 c. 19 d. 68

a 3. Juan has $63 in his savings account. He withdrew $16 one week and deposited $28 the next week. What is the balance in his account?
   a. $75 b. $107 c. $19 d. $51

c 4. Complete: −2 − (−7) = −2 + ?.
   a. −7 b. −9 c. 7 d. 9

b 5. Subtract: −3 − (−47).
   a. −50 b. 44 c. 50 d. −44

d 6. Evaluate (−64)(−110).
   a. −7040 b. 46 c. 174 d. 7040

d 7. On Sunday, the temperature was 82°F in Phoenix and −7°F in Minneapolis. How much warmer was it in Phoenix that day?
   a. 75°F b. −7°F c. 82°F d. 89°F

c 8. Evaluate 7b + 9 for b = $\frac{1}{2}$.
   a. 23 b. $3\frac{1}{2}$ c. $12\frac{1}{2}$ d. $16\frac{1}{2}$

b 9. Which expression is equivalent to (4)(6 − 2x)?
   a. 24 − 2x b. 24 − 8x c. 3(8 − 2x) d. 6(4 − 2x)

c 10. Sam charges $0.89 per pound for grapes. Which expression shows the cost for p pounds of grapes?
   a. 0.89 ÷ p b. 0.89 + p c. 0.89p d. 0.89(9p − 1)

NAME _____ CLASS _____ DATE _____

Chapter Assessment
Chapter 2, Form A, page 2

c 11. Add: (2b − 4) + (8b − 6).
   a. 10b + 2 b. 6b − 2 c. 10b − 10 d. −6b + 10

d 12. Add: (5e + 9f) + (2e − 4f).
   a. 7e + 13 b. 7e − 5f c. 3e + 5f d. 7e + 5f

d 13. Let a = −2 and b = 4. Find a + b.
   a. −6 b. −2 c. 4 d. 2

d 14. Subtract: (6b + 1) − (2b + 4).
   a. 3b − 3 b. 4b + 5 c. 8b + 5 d. 4b − 3

b 15. A recipe for clam chowder says to add water to the clam broth to get $2\frac{1}{3}$ cups of liquid. If there is $\frac{1}{2}$ cup of clam broth, how much water should be added?
   a. $\frac{1}{3}$ cup b. $1\frac{5}{6}$ cup c. $\frac{5}{6}$ cup d. $\frac{1}{2}$ cup

c 16. Identify the property illustrated by 4(3 + 2) = (4 · 3) + (4 · 2).
   a. Associative b. Commutative c. Distributive d. Transitive

b 17. Identify the property illustrated by $\frac{3}{4} + \frac{2}{3} = \frac{2}{3} + \frac{3}{4}$.
   a. Associative b. Commutative c. Distributive d. Transitive

a 18. Identify the property illustrated by 4 + (3 + 7) = (4 + 3) + 7.
   a. Associative b. Commutative c. Distributive d. Transitive

LESSON 2.3

Key Skills

Subtract real numbers.

To subtract –8, add its opposite.
–16 – (–8) = –16 + 8 = –8

To subtract 19, add its opposite.
–38 – 19 = –38 + (–19) = –57

Exercises

Find each difference.

19. 9 – (–15) **24**
20. 48 – (–48) **96**
21. –13 – 28 **–41**
22. 39 – (–18) **57**
23. –67 – (–42) **–25**
24. –23 – (–72) **49**

LESSON 2.4

Key Skills

Multiply and divide real numbers.

When multiplying or dividing numbers with like signs, the result is positive. When multiplying or dividing numbers with unlike signs, the result is negative.

a. (–14)(–3) = 42
b. (–72) ÷ (24) = –3
c. (–8)[(–7) + 7] = (–8)(0) = 0

Exercises

Find each product or quotient.

25. (–12)(–5) **60**
26. (–0.9)(3) **–2.7**
27. (54) ÷ (–18) **–3**
28. (–121) ÷ (–11) **11**
29. (–2)(–2)(–2)(–2) **16**
30. (–6)(–3) ÷ (–9) **–2**
31. 45[8 + (–8)] **0**
32. (–4)(4) ÷ (–1) **16**
33. (–5)(–5)(–1)(1) **–25**
34. (–3)(4 + 7) **–33**

LESSON 2.5

Key Skills

Identify number properties. Find a sum or product by using mental computation.

Commutative Property

$a + b = b + a \quad a \cdot b = b \cdot a$

To simplify (18 + 6) + 4:
(18 + 6) + 4 = 18 + (6 + 4)
= 18 + 10 = 28

Associative Property

$(a + b) + c = a + (b + c)$
$(a \cdot b) \cdot c = a \cdot (b \cdot c)$

To simplify (17 · 5) · 2:
(17 · 5) · 2 = 17 · (5 · 2)
= 17 · 10 = 170

Distributive Property

$a(b + c) = ab + ac$

To simplify 7(31):
7(31) = 7(30 + 1)
= 7(30) + 7(1)
= 210 + 7 = 217

Exercises

Use mental math to find each sum or product. Identify the property or properties that you use.

35. (27 + 8) + 12   **47, Assoc.**
36. (25 · 87) · 4   **8700, Assoc./Comm.**
37. (6.2 + 7.1) + 3.8   **17.1, Assoc./Comm.**
38. (63 · 20) · 5   **6300, Assoc.**
39. (3 + 10) + 7   **20, Assoc./Comm.**
40. (3 · 5) · 20   **300, Assoc.**
41. (8.7 + 21.1) + 1.3   **31.1, Assoc./Comm.**
42. (7 + 15) + 13   **35, Assoc./Comm.**

Complete each step, and name the property used.

43. (7 + 48) + 23   Given
    = (48 + _?_) + 23   **7** Commutative Property
    = 48 + (7 + _?_) **23**   _?_ Property **Assoc.**
    = 48 + _?_   **30**
    = _?_   **78**

44. 25 · (18 · 4)   Given
    = 25 · (_?_ · 18) **4**   _?_ Property **Comm.**
    = (_?_ · 4) · 18   **25** _?_ Property **Assoc.**
    = _?_ · _?_   **100, 18**
    = _?_   **1800**

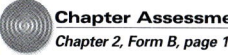

Chapter Assessment
Chapter 2, Form B, page 1

Find each sum.
1. (27) + (–35) **–8**
2. –64 + 61 **–3**
3. –3 + (–4) **–7**

Find each difference.
4. (38) – (–19) **57**
5. 62 – 83 **–21**
6. –13 – (–89) **76**

Use mental math to find each sum or product.
7. –30 + (–14) + 10 **–34**
8. (32 · 5) · 20 **3200**

Evaluate.
9. (–32)(–16) **512**
10. (–48) ÷ (–8) **6**

Evaluate $4p + 7$ for the following values of p:
11. 30 **127**
12. 5.5 **29**
13. $\frac{1}{3}$ **$8\frac{1}{3}$**

Simplify the following expressions. Use the Distributive Property if needed.
14. $3 \cdot 4y$ **12y**
15. $-4y \cdot 3$ **–12y**
16. $-12y \div (-3)$ **4y**
17. $\frac{10 - 22x}{10}$ **1 – 2.2x**
18. $\frac{-60x + 3.3}{6}$ **–10x + 0.55**
19. $24x \cdot 2x$ **$48x^2$**

20. In California, Mt. Whitney rises to 14,495 feet above sea level. Nearby, Death Valley sinks to 282 feet below sea level. What is the difference in elevation between these two points? **14,777 feet**
21. What property is illustrated by the equation $m + n = n + m$? **Commutative Property of Addition**
22. Evaluate $|-35|$. **35**

Simplify each expression.
23. $(3c + 7) + (4c - 3)$ **7c + 4**
24. $(9x + 2) - (3x + 7)$ **6x – 5**

Chapter Assessment
Chapter 2, Form B, page 2

Evaluate each expression for $a = 7$, $b = 5$, and $c = 2$.
25. $a + b$ **12**
26. $a + b - c$ **10**
27. $(a + c) \cdot b$ **45**

Find the opposite of each expression. Then simplify if possible.
28. $8y + 3y$ **–8y – 3y; –11y**
29. $3m - 4m$ **–3m + 4m; m**
30. $-b - c$ **b + c**

Perform the indicated operations.
31. $10x - 2x$ **8x**
32. $5d - (2 - d)$ **6d – 2**
33. $(5r + 2s) + (6r + 4s)$ **11r + 6s**
34. $(8v - 7w) - (7v - 2w)$ **v – 5w**

35. Make up a real-world situation that can be modeled by the equation $x + 19 = 45$.
Answers may vary. Sample answer: At noon, 19 students were sitting in the cafeteria. By 12:15 p.m., there were 45 students in the cafeteria. How many students entered the cafeteria during that time?

36. Ray bought 6 cartons of small bedding plants and 2 pots of full-grown plants. The store ran a sale the following week, so he purchased 3 additional cartons of the small plants and 4 pots of full-grown plants. Represent the cartons and pots algebraically, and determine the expression for the sum. **Let C represent the number of cartons and let P represent the number of pots. Then 6C + 2P + 3C + 4P = 9C + 6P.**

 Jeffrey opened a savings account with a $30 deposit. He made 2 more deposits of $26 and 2 withdrawals of $15 and $18.
37. What is the total amount currently in Jeffrey's account? **$49**
38. What is the cumulative change in Jeffrey's account since he opened it? **$33**

Simplify each expression.
39. $-4(5x + 2)$ **–20x – 8**
40. $(4x)(3) + (5x)(2)$ **22x**
41. $x^2 - (5 - 2x^2)$ **$3x^2 - 5$**

LESSON 2.6
Key Skills

Add and subtract expressions, and combine like terms.

When an expression in parentheses has a plus sign in front of it, drop the parentheses and combine like terms. When an expression in parentheses has a minus sign in front of it, change the minus sign to a plus sign and multiply the terms inside the parentheses by -1.

$(7x + 4y) - (6x - 5y)$	Given
$(7x + 4y) + (-6x + 5y)$	Definition of subtraction
$7x + (-6x) + 4y + 5y$	Rearrange terms.
$x + 9y$	Combine like terms.

Exercises

Simplify.

45. $(6a - 1) + (5a - 4)$ **$11a - 5$**
46. $(7 - t) + (3t + 4)$ **$2t + 11$**
47. $(8n - 4) - (6n - 3)$ **$2n - 1$**
48. $(6d + 3a) - (4d - 7a)$ **$2d + 10a$**
49. $\left(\frac{x}{3} - 2\right) + \left(\frac{x}{2} + 4\right)$ **$\frac{5}{6}x + 2$**
50. $(4a - 3b - c) - (4b - 7) + (9a + 5)$
    $13a - 7b - c + 12$
51. $(3x + 2) - (2x - 1)$ **$x + 3$**
52. $(7y - 3) + (6y + 2) - (3y - 7)$ **$10y + 6$**

LESSON 2.7
Key Skills

Multiply and divide expressions.

a. $-4(5x - 6)$ *Given*
    $= (-4)(5x) - (-4)(6)$ *Distributive Property*
    $= -20x + 24$ *Multiply.*

b. $(2a - 3b + 1) - 4(a + 6b - 3)$ *Given*
    $= 2a - 3b + 1 - 4a - 24b + 12$ *Dist. Prop.*
    $= -2a - 27b + 13$ *Combine like terms.*

c. $\frac{12 - 8y}{-4} = \frac{12}{-4} + \left(\frac{-8y}{-4}\right) = -3 + 2y$

Exercises

Simplify.

53. $3 \cdot 9x$ **$27x$**
54. $-33d \div 3$ **$-11d$**
55. $-2(7b - 2)$ **$-14b + 4$**
56. $-2.4f \cdot 2f$ **$-4.8f^2$**
57. $\frac{-30v + 3.6}{-3}$ **$10v - 1.2$**
58. $9r^2 - 8(4 - 3r^2)$ **$33r^2 - 32$**
59. $\frac{6(x + 1)}{3}$ **$2x + 2$**
60. $(5a^2 + 2a)(6m^2)$ **$30a^2m^2 + 12am^2$**
61. $\frac{-90w + 24w}{-3}$ **$22w$**
62. $\frac{3x^2 + 30}{3}$ **$x^2 + 10$**

Applications

GEOMETRY The formula for the perimeter, P, of a triangle with sides of length a, b, and c is $P = a + b + c$.

63. Use the formula to find P for $a = (x + 7)$, $b = (y - 8)$, and $c = (3x + 2y)$.   **$4x + 3y - 1$**
64. Use the formula to find P for $a = (x + 3)$, $b = (x + 7)$, and $c = (x - 10)$.   **$3x$**

Alternative Assessment

Performance Assessment

1. The temperature is 3°F at 2 A.M. During the next hour, the temperature falls 5°F.
   a. Use a diagram to show how a number line can be used to find the new temperature.
   b. Explain how algebra tiles can be used to find the new temperature.

2. Matlin knows that the product of five integers is negative. She told a classmate that an odd number of the integers must be negative. Is she right? Explain your answer.

3. $5 + 3 = 3 + 5$ illustrates the Commutative Property of Addition. Give an example that illustrates the Associative Property of Addition.

4. One side of a rectangle is represented by $2x$. The other side is represented by $x - 3$.
   a. Use a diagram to show how algebra tiles can be used to find the area of the rectangle.
   b. Explain how to use the Distributive Property to find the area of the rectangle.

5. For each number below, write three different problems that will result in that number as an answer. (Example: 17; $2(3 + 18) - 25 = 17$)
   a. 27   b. 301   c. 0

Portfolio Projects

1. Pick a stock from the business pages of your newspaper. Make a table to keep track of the daily changes in the price of the stock for two weeks. Explain how positive and negative numbers are used in your table.

2. Find the record high and low temperatures on January 15 and July 15 for the capital cities in three to five midwestern states. Use a number line to graph the data. Find the difference between the greatest record high and least record low.

3. Find the date of each of the following historical events in a reference work:
   a. death of Pericles in Athens
   b. signing the Declaration of Independence in Philadelphia
   c. beginning of the Han dynasty in China
   d. approximate birth of Euclid in Alexandria
   e. completion of the Taj Mahal in India

   On a time line, mark a zero to separate B.C.E. from C.E. Place the dates you found on this time line.

Peer Assessment

Complete this activity with a partner. Begin with the expression n. Choose another expression and an operation for your partner to perform. For example, add 5. After simplifying and writing the new expression, your partner should choose an operation and expression for you to perform. Check your partner's work and then perform the new operation and simplify.

Continue with this pattern until one partner makes a mistake. The partner who finds the mistake should write the correctly simplified expression in order to win the round. Otherwise the round is a tie. Begin a new round with n until 10 rounds have been completed.

---

internetconnect

The HRW Web site contains many resources to reinforce and expand your knowledge of the use of operations in algebra. The Web site also provides Internet links to other sites where you can find information and real-world data for use in research projects, reports, and activities that involve performing operations in algebra. Visit the HRW Web site at **go.hrw.com**, and enter the keyword **MA1 CH2**.

Alternative Assessment

Performance Assessment

1. a.

   The new temperature is -2°.

   b.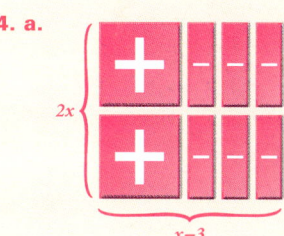

2. Matlin is right. Answers may vary. Sample answer: Since the product of two negative numbers is positive, there must be an odd number of negative integers to make the product negative.

3. Answers may vary. Sample answer:
   $(5 + 3) + 2 = 5 + (3 + 2)$

4. a.

   $2x \{$ [diagram of algebra tiles, 2 rows × 4 columns: first column + tiles, remaining three columns - tiles] $\}$ $x - 3$

   b. $2x(x - 3) = 2x(x) + 2x(-3)$
      $= 2x^2 - 6x$
   The area of the rectangle is $2x^2 - 6x$.

5. Answers may vary.
   a. Sample: $5(2 - 5) + 42$
   b. Sample: $19(5 + 7) + 7(8 + 3) - 4$
   c. Sample: $10(8 - 2) - 5(7 + 5)$

Portfolio Projects

1. Answers will vary depending on the stock chosen. A positive number means the stock price went up, and a negative number means the stock price went down.

2. Answers will vary depending on the cities chosen. Students can find this data in an almanac or on the Internet.

3. Students can find these dates at a library or on the Internet. Students can then plot the dates on a number line.

Cumulative Assessment

College Entrance Exam Practice

Multiple-Choice and Quantitative-Comparison Samples

The first half of the Cumulative Assessment contains two types of items found on standardized tests—multiple-choice questions and quantitative-comparison questions. Quantitative-comparison items emphasize the concepts of equality, inequality, and estimation.

Free Response Grid Samples

The second half of the Cumulative Assessment is a free-response section. This part of the Cumulative Assessment requires student-produced response items like those commonly found on college entrance exams. These questions require the use of machine-scored answer grids. You may wish to have students practice answering these items in preparation for standardized tests.

CHAPTERS 1–2 CUMULATIVE ASSESSMENT

College Entrance Exam Practice

QUANTITATIVE COMPARISON For Questions 1–4, write
A if the quantity in Column A is greater than the quantity in Column B;
B if the quantity in Column B is greater than the quantity in Column A;
C if the two quantities are equal; or
D if the relationship cannot be determined from the information given.

	Column A	Column B	Answers	
1.	$23 \cdot 43 + 23 \cdot 59$	$23 \cdot (43 + 59)$	A B C D [Lesson 2.5]	**C**
2.	$-17 - (-46)$	$-17 - 46$	A B C D [Lesson 2.3]	**A**
3.	$-113 \div 48$	$113 \div (-48)$	A B C D [Lesson 2.4]	**C**
4.	$\frac{2}{3}$	$\frac{5}{6}$	A B C D	**B**

5. Evaluate the expression $a + (-b)$ for $a = 2$ and $b = -5$. **(LESSON 2.2)** **b**
   a. -3 b. 7
   c. 3 d. none of these

6. Apply the method of finding differences to the sequence 4, 19, 44, 79, 124, . . . What is the value of the constant difference that you get? **(LESSON 1.1)** **b**
   a. 5 b. 10
   c. 15 d. 25

7. Which of the following is equivalent to $(8x - 4z) - (5z + x)$? **(LESSON 2.6)** **c**
   a. $8x - 4z - 5z + x$ b. $8x + x - 4z - 5z$
   c. $8x - 4z - 5z - x$ d. none of these

8. The following equation is an example of which property? **(LESSON 2.5)** **a**
   $7 + (11 + 4) = (7 + 11) + 4$
   a. Associative Property
   b. Commutative Property
   c. Distributive Property
   d. Addition Property

9. Which number is the opposite of -7? **(LESSON 2.1)** **b**
   a. 0 b. $+7$ c. -14 d. -1

10. Find $|-6|$. **(LESSON 2.1)** **a**
    a. $+6$ b. -6 c. 0 d. 12

11. Multiply: $2x(3x - 1)$. **(LESSON 2.7)** $6x^2 - 2x$

Simplify. **(LESSON 2.6)**

12. $(4x + 5y) + (x + 7y)$ $5x + 12y$

13. $(x + y + z) + (2w + 3y + 7)$ $x + 4y + z + 2w + 7$

14. $(2a - 3) - (a - 5)$ $a + 2$

15. $(4y + 9) - (7y - 2) + (13 - y)$ $-4y + 24$

16. $(3x + 7) - (2x - 2)$ $x + 9$

17. $(8x + 2y) + (3x + 5y)$ $11x + 7y$

18. $(17x - 2y + z) + (3x - y) + (2y - z)$ $20x - y$

19. $(8x - 3y) + (8x - 3y) - (2x + 2y)$ $14x - 8y$

What are the coordinates of the given points?
(LESSON 1.4)

20. A (1, 3)
21. B (10, 1)
22. C (7, 2)
23. D (5, 8)

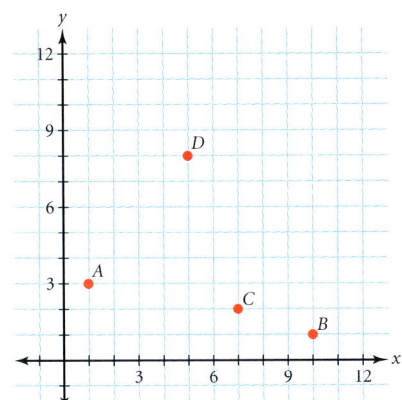

Find each sum or difference. (LESSONS 2.2 AND 2.3)

24. −39 + 68 **29**
25. 8 − 14 − 27 **−33**
26. −23 − (−72) **49**
27. −8 + 8 + (−12) **−12**

Make a table of values for each equation, using 1, 2, 3, 4, and 5 as values for x. Draw a graph for each equation by plotting points from your table. (LESSON 1.5)

28. $y = 2x + 3$
29. $y = x − 7$

30. The chart below shows how the cost of food items that were $3.00 in 1982 has increased over the years. Make a scatter plot, and use a straightedge to estimate the line of best fit.
(LESSON 1.6)

Year	1982	1985	1988	1991	1994	1997
Cost	3.00	3.18	3.54	3.54	4.08	4.95

31. Simplify: $(3b + 109) − (2b + 53)$. **b + 56**
(LESSON 2.6)

32. Simplify: $\frac{4x - 12}{2}$. (LESSON 2.7) **2x − 6**

33. Simplify: $(8x + 3y − 2z) + (4x − y + 5z)$.
(LESSON 2.6) **12x + 2y + 3z**

34. Evaluate: $(5 + 2^3) − (−3)(2 + 4)$. **31**
(LESSON 1.3)

35. Find the next three terms in the sequence: 1, 5, 11, 19, ... (LESSON 1.1) **29, 41, 55**

FREE-RESPONSE GRID
The following questions may be answered by using a free-response grid such as that commonly used by standardized test services.

36. Find the next term in the sequence: 7, 16, 25, 34, 43, ...
(LESSON 1.1) **52**

37. How many $6 movie tickets can you buy with $42? (LESSON 1.2) **7**

38. Evaluate: $(6 + 3^2) − 5(4 − 7)$. (LESSON 1.3) **30**

39. Find the opposite of −5. (LESSON 2.1) **5**

40. Evaluate: $8 + (−3.4)$. (LESSON 2.2) **4.6**

41. Evaluate: $-14 \div \left(-\frac{1}{2}\right)$. (LESSON 2.4) **28**

42. Let $q = 6$, $r = −3$, and $s = −2$. Evaluate: $q + r − s$. (LESSON 2.3) **5**

Find the value of each expression when the variable is 4. (LESSON 1.2)

43. $3p$ **12**
44. $6y − 7$ **17**
45. $2m + 13$ **21**
46. $−2s + 9$ **1**

47. Find the second difference for the sequence 1, 3, 8, 16, 27, ... (LESSON 1.1) **3**

48. Evaluate: $|−7|$. (LESSON 2.1) **7**

49. Evaluate: $\frac{(−3)(−7)}{42}$. (LESSON 2.4) **$\frac{1}{2}$**

28.
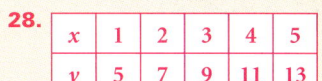
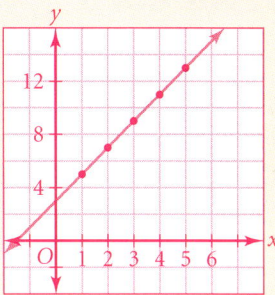

29.

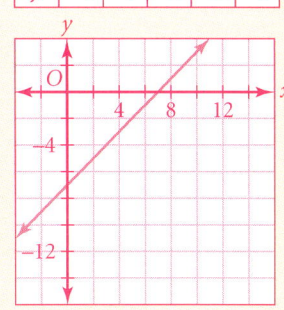

30.
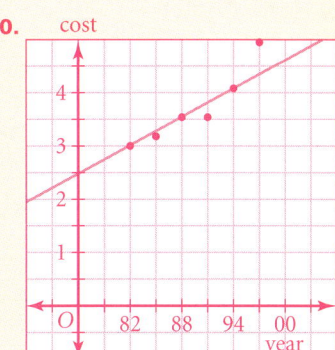

3 Equations

CHAPTER PLANNING GUIDE

Lesson	3.1	3.2	3.3	3.4	3.5	3.6	Projects and Review
Pupil Edition Pages	114–121	122–128	129–134	135–140	141–146	147–153	154–161
Extra Practice (Pupil Edition)	744	745	745	746	746	747	
Practice Workbook	14	15	16	17	18	19	
Reteaching Masters	27–28	29–30	31–32	33–34	35–36	37–38	
Enrichment Masters	14	15	16	17	18	19	
Cooperative-Learning Activities	14	15	16	17	18	19	
Lesson Activities	14	15	16	17	18	19	
Problem Solving/ Critical Thinking	14	15	16	17	18	19	
Student Study Guide	14	15	16	17	18	19	
Spanish Resources	14	15	16	17	18	19	
Student Technology Guide	14	15	16	17	18	19	
Assessment Resources	32	33	34	36	37	38	35, 39–44
Teaching Transparencies		13	14	15		16	
Quiz Transparencies	3.1	3.2	3.3	3.4	3.5	3.6	
Writing Activities for Your Portfolio							7–9
Tech Prep Masters							11–14
Long-Term Projects							9–12

LESSON PACING GUIDE

Traditional	2 days	2 days	2 days	1 day	1 day	2 days	2 days
Block	1 day	1 day	1 day	$\frac{1}{2}$ day	$\frac{1}{2}$ day	1 day	1 day
Two-Year	4 days	4 days	4 days	2 days	2 days	4 days	4 days

CONNECTIONS AND APPLICATIONS

Lesson	3.1	3.2	3.3	3.4	3.5	3.6	Review
Algebra	114–121	122–128	129–134	135–140	141–146	147–153	156–161
Geometry	119, 120	123, 126		139	145	149, 150, 152	
Business and Economics	120	127	129, 131, 133, 134	135, 136	141, 146	148	158
Statistics	117, 119			138, 139			
Life Skills	114, 120	122		139		152	
Science and Technology	118		132, 133	137, 140		147, 148	
Sports and Leisure	120	124, 127				152	
Cultural Connection: Africa	121		134				
Other	120	126, 127		140	146		

BLOCK-SCHEDULING GUIDE

Day	Lesson	Teacher Directed: Lesson Examples, Teaching Transparencies	Student Guided: Activity, Try This	Cooperative-Learning Activity, Lesson Activity, Student Technology Guide	Practice: Practice & Apply, Extra Practice, Practice Workbook	Assessment: Quiz, Mid-Chapter Assessment	Problem Solving, Reteaching
1	3.1	15 min	15 min	15 min	65 min	15 min	15 min
2	3.2	15 min	15 min	15 min	65 min	15 min	15 min
3	3.3	15 min	15 min	15 min	65 min	15 min	15 min
4	3.4	5 min	8 min	8 min	35 min	8 min	8 min
4	3.5	5 min	7 min	7 min	30 min	7 min	7 min
5	3.6	10 min	15 min	15 min	65 min	15 min	15 min
6	Assess.	50 min PE: Chapter Review	90 min PE: Chapter Project, Writing Activities	90 min Tech Prep Masters	65 min PE: Chapter Assessment, Test Generator	30 min Chap. Assess. (A or B), Alt. Assess. (A or B), Test Generator	

PE: Pupil's Edition

Alternative Assessment

The following are suggestions for an alternative assessment for students who may benefit from a different type of assessment than the regular chapter quizzes and the mid-chapter and end-of-chapter assessment materials. Many of the questions are open-ended, and students' answers will vary.

Performance Assessment

1. **a.** State the Properties of Equality.
   b. Give an example of an equation that can be solved in one step by using each of the properties stated in part **a**.
   c. Give an example of an equation that requires the use of two Properties of Equality. Show how the properties are used to solve your equation.

2. **a.** Define the Distributive Property.
   b. Give an example of an equation that requires the use of the Distributive Property. Show how the Distributive Property is used to solve your equation.
   c. A student wrote $3(2x - 1)$ as $6x - 1$. Explain why this is incorrect.

3. Write a verbal description and an algebraic statement for each word or phrase below. Identify the operation denoted by each.

   a. decreased by   **b.** sum
   c. difference   **d.** product
   e. quotient   **f.** ratio
   g. twice   **h.** remainder
   i. more than   **j.** per
   k. reduced by   **l.** consecutive integers

Portfolio Project

Suggest that students choose one of the following projects for inclusion in their portfolios.

1. Write the equation modeled by the algebra tiles below. Explain how to solve the equation by using the tiles, the Properties of Equality, a table, and a graph.

2. An equation that is true for every meaningful value of the variable is an *identity*. An equation that is true for some values of the variable but not others is a *conditional*. An equation that is false for all values of the variable is a *contradiction*. The graphs of each side of an identity are the same, the graphs of each side of a conditional have a point of intersection, and the graphs of each side of a contradiction never intersect. Classify each equation below as an identity, conditional, or contradiction.

   a. $3x + 4 = 5(x - 2)$ **b.** $x + y = y + x$
   c. $2x - 1 = 2x + 1$ **d.** $y(y - 3) = y^2 - 3y$
   e. $1 + \frac{3}{y} = \frac{y+3}{y}$ **f.** $2z - 5 = z + 7$

---

internet connect

The table below identifies the pages in this chapter that contain technology information and support in the side columns.

Content Links	
Lesson Links	page 127
Portfolio Links	page 121
Graphics Calculator Support	page 782

Resource Links

For information about teacher and parent resources as well as professional development help, visit **www.hrw.com/math**.

Technical Support HRW has assembled a team of dedicated technical and teaching professionals and a comprehensive service program to provide you with the support you need.

- The HRW Technical Support Line operates from 7 A.M. to 6 P.M. central time, Monday through Friday, at (800) 323-9239.
- The HRW Technical Support Center on the World Wide Web is available 24 hours a day, seven days a week, at **www.hrwtechsupport.com**.
- You can e-mail our Technical Support Center at **tsc@hrwtechsupport.com**.
- The Technical Support Center's fax-on-demand service at (800) 352-1680 offers solutions to common problems and answers to frequently asked questions.

Technology

Lesson Suggestions and Calculator Examples
(Keystrokes are based on a TI-83 calculator.)

Lesson 3.1 Solving Equations by Adding and Subtracting

This lesson focuses on solving an equation by using one step, either addition or subtraction. With practice, students can set up the solution to such an equation mentally and then use a calculator to complete the arithmetic involved.

Given	Press
$x + 2.6 = 10.1$ →	10.1 [-] 2.6
$x - 3.1 = 10.1$ →	10.1 [+] 3.1

Lesson 3.2 Solving Equations by Multiplying and Dividing

This lesson is the companion to Lesson 3.1. Continue to encourage students to set up the solution mentally before attempting to use a calculator. To challenge students who may become too sure of themselves, give the equation below.

$$\frac{3}{5}x = 12$$

Pressing 12 [÷] [( ] 3 [÷] 5 [ )] will give the correct solution. Pressing 12 [÷] 3 [÷] 5 will give an incorrect solution. Ask students to write the equation that would be solved by pressing 12 [÷] 3 [÷] 5.

Lesson 3.3 Solving Two-Step Equations

The technology discussion on page 132 illustrates a graphical approach to solving an equation in x. The procedure is as follows:

1. Set the left and the right sides of the equation equal to y.
2. Graph the two equations for y in terms of x.
3. The x-value of the point where the graphs intersect is the desired solution.

$$18 + 7x = 74 \rightarrow \begin{array}{l} y = 18 + 7x \\ y = 74 \end{array}$$

Remind students that window settings are important here. If the window does not show a point of intersection, they cannot conclude that there is no solution.

Lesson 3.4 Solving Multistep Equations

The technology discussion on page 137 uses the graphical approach to solving an equation in x described in Lesson 3.3 above.

Enter the left and right sides of the equation into **Y1** and **Y2**, respectively.

$$5x - 6 = -6x + 16 \rightarrow \begin{array}{l} \text{Y1=5X–6} \\ \text{Y2=–6X+16} \end{array}$$

Students can compare the y-values of both functions by creating a table as shown below.

After entering expressions for **Y1** and **Y2**, have students press [2nd] [WINDOW] to access the **TBLSET** menu and then enter the settings shown at right.

Press [2nd] [GRAPH] to view the y-values for the functions entered in **Y1** and **Y2**. Scroll down the table by using the arrow keys. Notice that the y-values are equal when $x = 2$.

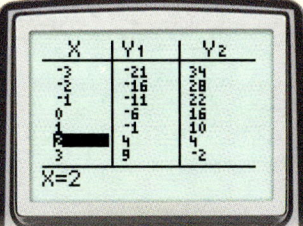

Lesson 3.5 Using the Distributive Property

The table shown on page 141 can be created on a TI-83 or TI-82 graphics calculator.

1. Enter the left and right sides of the equation into **Y1** and **Y2**, respectively.

2. Go to the **TBLSET** menu and enter the following values:
   TblStart=5 (or TblMin=5)
   ΔTbl=5
   Indpnt: **Auto** Ask
   Depend: **Auto** Ask

3. Press [2nd] [GRAPH] and find the x-value that gives a y-value of 48 in both the **Y1** and **Y2** columns.

Lesson 3.6 Using Formulas and Literal Equations

Students can enter formulas into the **Y=** editor and store values for the parameters of the formulas in alpha characters. For example, to find the area of a trapezoid with a height of 10 and bases of 12 and 15, enter the formula **Y1=.5*H(A+B)** into the **Y=** editor. Then press [2nd] [MODE] to exit the **Y=** editor. Now press **10** [STO▶] [ALPHA] [^] [ENTER] to store 10 in **H**. Store 12 in **A** and 15 in **B** by using a similar procedure. Press [VARS] and select **Y-VARS** **1: Function**. Highlight **1: Y1** and press [ENTER] twice to view the area of the given trapezoid.

For further information, refer to the

- technology discussions in the lessons.
- lesson-related teacher's commentary in the side columns of this *Teacher's Edition*.
- lesson-related *Student Technology Guide* masters.
- *HRW Technology Handbook*.

Teaching Resources

Basic Skills Practice
(2 pages per skill)

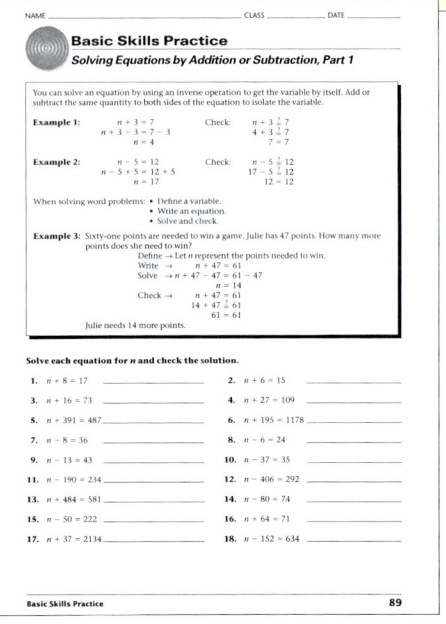

Reteaching
(2 pages per lesson)

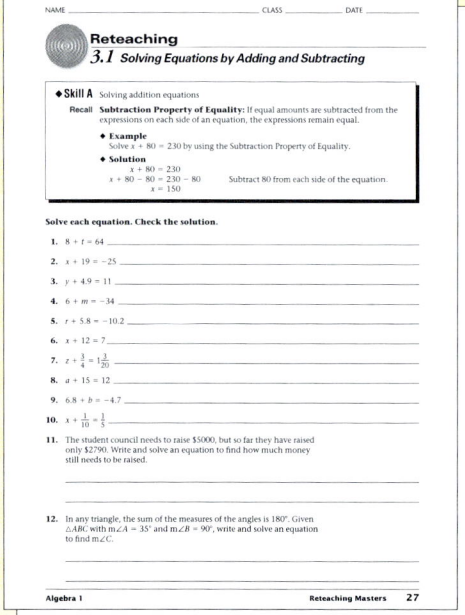

Enrichment
(1 page per lesson)

Cooperative-Learning Activity
(1 page per lesson)

Lesson Activity
(1 page per lesson)

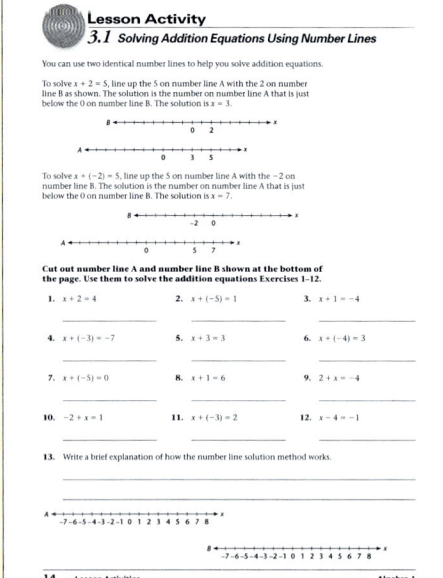

Long-Term Project
(4 pages per chapter)

About Chapter 3

Background Information

In Chapter 3, students learn to solve linear equations in one variable by applying the properties of equality. This is expanded to include two-step equations, multi-step equations, and equations that are solved by using the Distributive Property. The chapter concludes with a study of literal equations and formulas.

CHAPTER RESOURCES

- Block-Scheduling Handbook
- Writing Activities for Your Portfolio
- Tech Prep Masters
- Long-Term Project
- Assessment Resources:
   Mid-Chapter Assessment
   Chapter Assessments
   Alternative Assessments
- Test and Practice Generator
- Technology Handbook
- End-of-Course Exam Practice

Equations

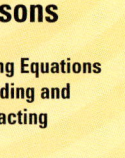

SOLVING EQUATIONS IS OFTEN NECESSARY WHEN working with problems that require a number for an answer. For example, in business, equations are used to determine bank interest and sales commissions. In sports, batting averages and other statistics are often determined by using equations. You can use an equation to determine how many miles per gallon of gasoline your car can be driven.

Lessons

3.1 • Solving Equations by Adding and Subtracting

3.2 • Solving Equations by Multiplying and Dividing

3.3 • Solving Two-Step Equations

3.4 • Solving Multistep Equations

3.5 • Using the Distributive Property

3.6 • Using Formulas and Literal Equations

Project
Track Team Schemes

interest = principal × rate × time

tip = 15% × total bill

About the Photos

The photos reflect many problems that may be familiar to students and that can be solved by using an equation. Annual rates for both loans and savings are determined by using an equation that includes factors such as interest rate, term, and principal amount. The number of miles per gallon that a car can be driven and the amount of tip for service at a restaurant can be modeled by basic multiplication and division equations. Students may be intrigued by the equation for determining a batting average. Extend the example about batting averages by comparing batting averages and percents (that is, a batting average of 0.300 means that the batter got a hit in 30% of his at bats).

Chapter Objectives

- Solve equations by using addition and subtraction. [3.1]
- Solve equations by using multiplication and division. [3.2]
- Write equations that represent real-world situations. [3.3]
- Solve two-step equations. [3.3]
- Write and solve multistep equations. [3.4]
- Use the Distributive Property to solve equations. [3.5]
- Solve real-world problems by using multistep equations. [3.5]
- Solve literal equations for a specific variable. [3.6]
- Use formulas to solve problems. [3.6]

About the Chapter Project

Throughout Chapter 3 you will be introduced to algebraic procedures that will expand your mathematical problem-solving skills. These skills include recognizing a pattern in a numeric relationship that can be modeled by an equation. In the Chapter Project, *Track Team Schemes*, on page 154 you will apply these skills to a fund-raising scenario. After completing the Chapter Project, you will be able to do the following:

- Write and solve multistep equations.

About the Portfolio Activities

Throughout the chapter, you will be given opportunities to complete Portfolio Activities that are designed to support your work on the Chapter Project.

- Egyptian equation-solving techniques are the topic of the Portfolio Activity on page 121.
- The Portfolio Activity on page 128 introduces a problem concerning the sale of T-shirts by a high school track team.
- The Portfolio Activity on page 140 extends the T-shirt sale problem from page 128.
- The Portfolio Activity on page 153 presents a famous problem about Greek mathematician Diophantus.

Portfolio Activities appear at the end of Lessons 3.1, 3.2, 3.4, and 3.6. Each serves as preparation for the Chapter Project. The Portfolio Activities as well as the Chapter Project Activities are appropriate for inclusion in the student's portfolio. Students should be encouraged to include in their portfolios any other work in which they feel a sense of pride or a sense of accomplishment.

 internetconnect

Internet Connect features appearing throughout the chapter provide keywords for the HRW Web site that lead to links for math resources, tutorial assistance, references for student research, classroom activities, and all teaching resource pages. The HRW Web site also provides updates to the technology used in the text. Listed at right are the keywords for the Internet Connect activities referenced in this chapter. Refer to the side column on the page listed to read about the activity.

LESSON	KEYWORD	PAGE
3.1	MA1 Egyptian	121
3.2	MA1 Oscar at Bat	127

Prepare

NCTM PRINCIPLES & STANDARDS
1–3, 6–10

LESSON RESOURCES

- Practice 3.1
- Reteaching Master 3.1
- Enrichment Master 3.1
- Cooperative-Learning Activity 3.1
- Lesson Activity 3.1
- Problem Solving/Critical Thinking Master 3.1

QUICK WARM-UP

Find each sum or difference.

1. $-10 + 4$ -6
2. $12 + (-12)$ 0
3. $14 - (-14)$ 28
4. $-8 - (-8)$ 0

Also on Quiz Transparency 3.1

Teach

Why Ask students to translate the equations $x + 4 = 7$ and $y - 5 = 9$ into English sentences. **A number x plus 4 is equal to 7. A number y minus 5 is equal to 9.** Have them give the solutions. **$x = 3; y = 14$** Point out that it is relatively easy to solve these equations by using mental math. Tell students that in this lesson, they will learn methods for solving equations which might not be as easy to solve mentally.

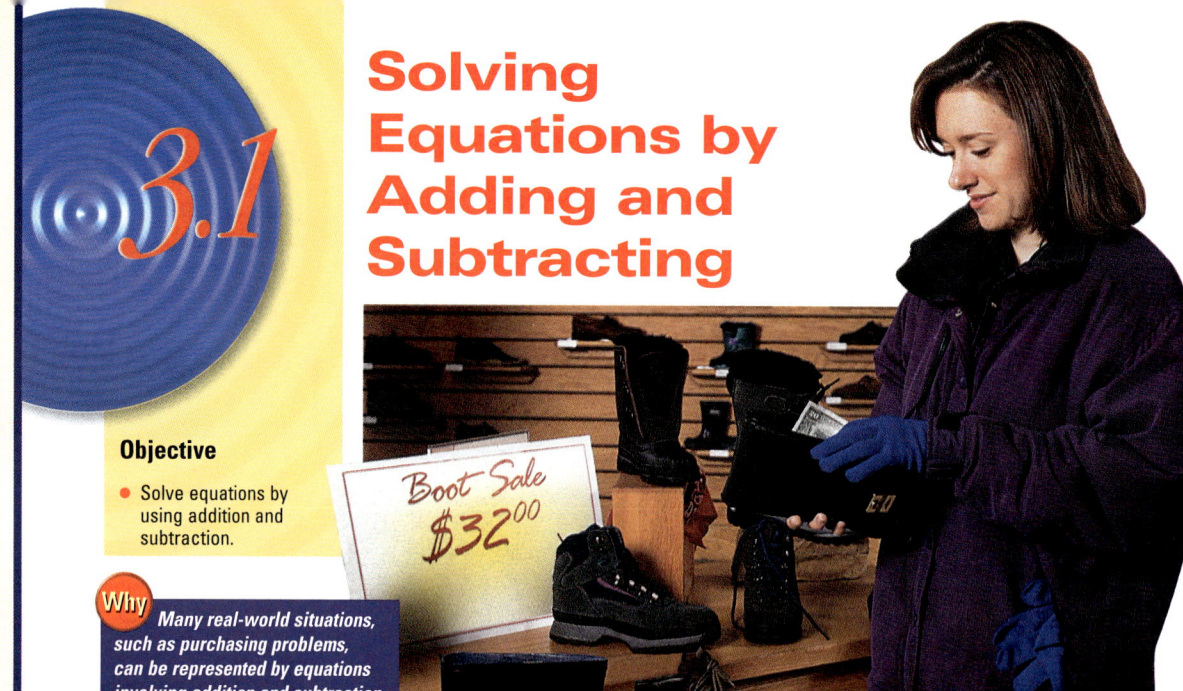

3.1 Solving Equations by Adding and Subtracting

Objective
- Solve equations by using addition and subtraction.

Why Many real-world situations, such as purchasing problems, can be represented by equations involving addition and subtraction.

Solving Equations by Using Subtraction

APPLICATION
PERSONAL FINANCE

Think about the simple problem below. How can it be modeled by a variable and an equation?

Riva has $20 and wants to purchase a pair of boots that cost $32. How much more money does she need to buy the boots?

The situation can be modeled by the following equation, where x represents the additional amount in dollars that Riva needs:

$$20 + x = 32, \text{ or } x + 20 = 32$$

Tiles can be used to represent the steps in solving the equation. Suppose that the x-tile represents an unknown number of unit tiles. The x-tile and the unit tiles represent the equation $x + 20 = 32$.

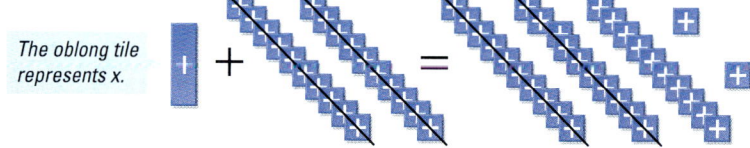

The oblong tile represents x.

Take away 20 unit tiles from each side of the equation. The quantities on either side of the equal sign remain equal. The x-tile is now isolated. Count the remaining tiles on the other side of the equal sign to find the value of x.

Subtracting the same number of tiles from each side of an equation models the Subtraction Property of Equality.

Alternative Teaching Strategy

USING VISUAL MODELS
Draw a balance scale like the one at right on the board or overhead. Remind students that a scale like this is balanced as long as there are equal weights on each side. If weight is added or subtracted from one side of a scale that is balanced, the same must be done to the other side in order to maintain the balance.

Explain that an equation such as $x + 7 = 9$ is balanced, like a scale, on either side of the equal sign. To isolate the variable, x, you subtract 7 from the left side. To maintain the balance, however, you must also subtract 7 from the right side. To provide visual reinforcement of the balance-scale model, show students how to write the solution of an equation in the vertical format shown at right.

$$\begin{array}{r} x + 7 = 9 \\ -7 = -7 \\ \hline x = 2 \end{array}$$

114 LESSON 3.1

Subtraction Property of Equality

For all real numbers a, b, and c, if $a = b$, then $a - c = b - c$.

Subtracting equal amounts from each side of an equation results in an equivalent equation.

EXAMPLE 1 Use the Subtraction Property of Equality to solve each equation. Check the solution.

　a. $x + 7 = 9$　　**b.** $8 + a = 2$　　**c.** $10.7 = 6.2 + m$

● **SOLUTION**

a. Subtract 7 from each side of the equation.

$$x + 7 = 9$$
$$x + 7 - 7 = 9 - 7$$
$$x = 2$$

b. Subtract 8 from each side of the equation.

$$8 + a = 2$$
$$8 + a - 8 = 2 - 8$$
$$a = -6$$

c. Subtract 6.2 from each side of the equation.

$$10.7 = 6.2 + m$$
$$10.7 - 6.2 = 6.2 + m - 6.2$$
$$4.5 = m$$
$$m = 4.5 \text{ Sym. Prop.}$$

To check the solution, substitute 2 for x in the original equation.

$$2 + 7 \stackrel{?}{=} 9$$
$$9 = 9 \text{ True}$$

To check the solution, substitute -6 for a in the original equation.

$$8 + (-6) \stackrel{?}{=} 2$$
$$2 = 2 \text{ True}$$

To check the solution, substitute 4.5 for m in the original equation.

$$10.7 \stackrel{?}{=} 6.2 + 4.5$$
$$10.7 = 10.7 \text{ True}$$

TRY THIS Solve each equation. Check the solution.

　a. $a + 5 = 4$　　**b.** $\frac{1}{2} + n = 11\frac{1}{2}$　　**c.** $27.2 = h + 45.6$

EXAMPLE 2 A school band needs new uniforms. So far they have raised $2344.10 toward a goal of $5000. **Write and solve an equation to find how much more money the band must raise.**

PROBLEM SOLVING

● **SOLUTION**

Organize the information.
Unknown: the amount still needed
Known: the amount already raised

Write a sentence with words and operation symbols.
the amount still needed + the amount already raised = the goal

Write an equation. Let x represent the amount still needed in dollars.
$x + 2344.10 = 5000$

Solve the equation.
$x + 2344.10 - 2344.10 = 5000 - 2344.10$ 　　*Subtr. Prop. of Eq.*
$x = 2655.90$

The band needs to raise $2655.90.

Interdisciplinary Connection

ECONOMICS When you finance a purchase with a lending institution, you often must make a *down payment* in cash. The amount of a down payment is usually calculated as a percent of the cost of the item you are buying. The amount of the loan is the difference between the down payment and the cost. This relationship can be modeled by the equation $L = C - D$, where L represents the amount of the loan, C represents the cost of the item, and D represents the down payment. Solve this equation for C, and then solve for D.
$C = L + D; D = C - L$

Inclusion Strategies

VISUAL LEARNERS Some students have difficulty understanding that the variable in an equation is a placeholder for an unknown value. These students may find it helpful to rewrite a given equation with an empty blank in place of the variable. Then the process of solving the equation becomes a search for a number to fill the blank. For instance, you might help students visualize the equation in part **a** of Example 1 as shown below.

$x + 7 = 9 \rightarrow \underline{} + 7 = 9 \rightarrow 2 + 7 = 9 \rightarrow x = 2$

Teaching Tip

Students learned the meanings of the terms *equation* and *solution to an equation* in Lesson 1.2. You may want to review these definitions at the beginning of this lesson.

ADDITIONAL EXAMPLE 1

Use the Subtraction Property of Equality to solve each equation. Check your solution.

a. $s + 5.3 = -9.7$　　$s = -15$

b. $-20 = -38 + b$　　$b = 18$

c. $\frac{5}{2} + x = \frac{1}{2}$　　$x = -2$

TRY THIS

a. $a = -1$

b. $n = 11$

c. $h = -18.4$

ADDITIONAL EXAMPLE 2

Marion wants to earn $150 to buy a new CD player. So far, he has earned $55.50 toward this goal. **Write and solve an equation to find how much more money Marion must earn to meet his goal.**
$x + 55.50 = 150$, where x represents the amount in dollars to be earned; Marion must still earn $94.50.

LESSON 3.1

Activity Notes

This Activity provides students with a concrete representation of the Addition Property of Equality. You may want to review the meaning of *neutral pair* with students before they proceed. Students should see that adding 2 unit tiles to each side creates 2 neutral pairs on the left side, which can then be removed. The result is that the *x*-tile has been isolated on one side of the equation.

Cooperative Learning

Have students do the Activity in pairs. One student should manipulate the tiles while the other makes a sketch of the tile arrangement at each stage. The pair should then work together to arrive at the symbolic representation in the Checkpoint.

CHECKPOINT ✓

3. $x - 2 + 2 = 8 + 2$
   $x = 10$

ADDITIONAL EXAMPLE 3

Use the Addition Property of Equality to solve each equation. Check your solution.

a. $t - 9.3 = 4.1$ $t = 13.4$

b. $w - \frac{1}{3} = -\frac{1}{3}$ $w = 0$

c. $-53 = n - 25$ $n = -28$

TRY THIS

a. $a = 12$

b. $n = -5$

c. $h = 6.8$

Solving Equations by Using Addition

Activity
Modeling the Addition Property of Equality

You will need: algebra tiles

Use algebra tiles to model the process of solving the equation $x - 2 = 8$.

1. Place 1 *x*-tile and 2 negative unit tiles on the left side of the work area and 8 unit tiles on the right side of the work area. How many positive unit tiles must be added to each side to isolate the *x*-tile?

2. Simplify the model by removing the neutral pairs. What is the solution?

CHECKPOINT ✓ 3. Fill in the blanks to show the process for arriving at the solution.

$$x - 2 + \underline{} = 8 + \underline{}$$
$$x = \underline{}$$

Equations involving subtraction are solved by using addition.

Addition Property of Equality

For all real numbers a, b, and c, if $a = b$, then $a + c = b + c$.

Adding equal amounts to each side of an equation results in an equivalent equation.

EXAMPLE 3

Use the Addition Property of Equality to solve each equation. Check the solution.

a. $x - 3 = 2$ b. $-12 = a - 8$ c. $m - 3.2 = 5.6$

● **SOLUTION**

a. Add 3 to each side of the equation.
   $x - 3 = 2$
   $x - 3 + 3 = 2 + 3$
   $x = 5$

b. Add 8 to each side of the equation.
   $-12 = a - 8$
   $-12 + 8 = a - 8 + 8$
   $-4 = a$
   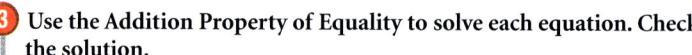
   $a = -4$ Sym. Prop.

c. Add 3.2 to each side of the equation.
   $m - 3.2 = 5.6$
   $m - 3.2 + 3.2 = 5.6 + 3.2$
   $m = 8.8$

To check the solution, substitute 5 for *x* in the original equation.
$5 - 3 \stackrel{?}{=} 2$
$2 = 2$ *True*

To check the solution, substitute −4 for *a* in the original equation.
$-12 \stackrel{?}{=} -4 - 8$
$-12 = -12$ *True*

To check the solution, substitute 8.8 for *m* in the original equation.
$8.8 - 3.2 \stackrel{?}{=} 5.6$
$5.6 = 5.6$ *True*

TRY THIS Solve each equation. Check the solution.

a. $a - 5 = 7$ b. $-8 = n - 3$ c. $h - 4.6 = 2.2$

Enrichment

Give students the equation shown at right, together with all of the steps of the solution. Point out that this is the equation in part **b** of Example 1, but this time several intermediate steps have been included in the solution. Have them justify each step.

After students have correctly justified each step, have them choose one of the equations given in Example 3 and write a similar solution, with justifications for each step.

Steps	Justifications
$8 + a = 2$	Given
$(8 + a) + (-8) = 2 + (-8)$	Add. Prop. of Equality
$8 + [a + (-8)] = 2 + (-8)$	Assoc. Prop. of Add.
$8 + [(-8) + a] = 2 + (-8)$	Comm. Prop. of Add.
$[8 + (-8)] + a = 2 + (-8)$	Assoc. Prop. of Add.
$0 + a = 2 + (-8)$	Add. Inverse Prop.
$a = 2 + (-8)$	Identity Prop. for Add.
$a = -6$	Simplify.

EXAMPLE 4 Solve each equation. Check the solution.

a. $5 - x = 7$
b. $13 - m = 7$

SOLUTION

a. $5 - x = 7$
$5 - x - 5 = 7 - 5$ Subtraction Property of Equality
$-x = 2$
$x = -2$ The opposite of x is 2; therefore, x is –2.

CHECK
$5 - (-2) \stackrel{?}{=} 7$
$7 = 7$ True

b. $13 - m = 7$
$13 - m - 13 = 7 - 13$ Subtraction Property of Equality
$-m = -6$
$m = 6$ The opposite of m is –6; therefore, m is –6.

CHECK
$13 - 6 \stackrel{?}{=} 7$
$7 = 7$ True

ADDITIONAL EXAMPLE 4

Solve each equation and check the solution.

a. $3 - x = 34$ $x = -31$
b. $5 - x = 17$ $x = -12$

ADDITIONAL EXAMPLE 5

After spending $45.90 at the mall, Leah had $23.50 remaining. **Write and solve an equation to find the amount Leah had originally.** $x - 45.90 = 23.50$, where x represents the original amount in dollars; Leah originally had $69.40.

EXAMPLE 5 In a dart game, the range of scores was 47 points, and the lowest score was 52. Write and solve an equation to find the highest score in this dart game.

CONNECTION STATISTICS

PROBLEM SOLVING

SOLUTION

ORGANIZE
The range is the difference between the highest and lowest scores.

range = highest score – lowest score

Because the highest score is unknown, let h represent this value.

WRITE
range = highest score – lowest score
$47 = h - 52$

SOLVE
$47 = h - 52$
$47 + 52 = h - 52 + 52$ Add. Prop. of Eq.
$99 = h$
$h = 99$ Symmetric Property of Equality

CHECK
Substitute 99 for h in the equation $47 = h - 52$.
Since $47 = 99 - 52$, the highest score was 99.

Math CONNECTION

STATISTICS The *range* of a set of data is the absolute value of the difference between the largest value and the smallest value. When comparing several sets of similar data, the range can be useful for indicating which set has the largest spread. When used alone, the range can sometimes be misleading.

Teaching Tip

TECHNOLOGY To use a calculator to find the solution of an equation, have students first write the solution in *calculator-ready* form. That is, they should isolate the variable on one side of the equal sign and write the required calculation on the other side. In part **c** of Example 3, for instance, have them write $m = 5.6 + 3.2$. Then, if using a TI-82 or TI-83 graphics calculator, they can obtain the solution by pressing **5.6** **+** **3.2** ENTER .

Reteaching the Lesson

USING LANGUAGE SKILLS Work with students to translate equations into verbal questions that incorporate the word *add* or *subtract*. For example, translate the equation $x + 3 = 8$ as "To what number do you add three to get eight?"

Now discuss the fact that addition and subtraction are inverse operations. Remind students that this means they can think of using one operation to "undo" the other. Refer back to the equation $x + 3 = 8$. Since 3 is *added* to the variable, you *subtract* 3 from each side to undo the operation.

Emphasize that the goal is to isolate the variable on one side of the equal sign.

Repeat the discussion, this time focusing on the equation $x - 3 = 8$. In this case, the translation is "From what number do you subtract three to get eight?" Since 3 is *subtracted* from the variable, you *add* 3 to each side to undo the operation.

LESSON 3.1 **117**

See Keystroke Guide, page 782.

Teaching Tip

TECHNOLOGY If you are using a TI-82 or TI-83 graphics calculator to create the graph shown on this page, press WINDOW and set the viewing window as follows:

Xmin=0   Ymin=0
Xmax=9.4 Ymax=15
Xscl=1   Yscl=1

Press Y= . In the **Y=** editor, enter **Y1=X+7** and **Y2=10**, as shown. Be sure to clear all other equations. Press GRAPH. To access the **CALCULATE** menu, press 2nd TRACE. Choose **5:intersect**, and then press ENTER ENTER ENTER.

To view the table, first press 2nd WINDOW and make these choices in **TABLE SETUP**.

TblStart=0 (or TblMin=0)
ΔTbl=1
Indpnt: **Auto** Ask
Depend: **Auto** Ask

Then press 2nd GRAPH.

TECHNOLOGY
GRAPHICS
CALCULATOR

The equation $x + 7 = 10$ is the linear equation $x + 7 = y$ when $y = 10$. You can find the solution to the equation by graphing the lines $y = x + 7$ and $y = 10$ and finding the x-value at the intersection point.

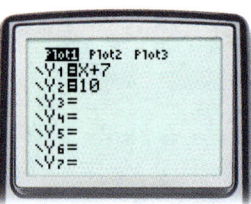
Enter the equations in the **Y=** window.

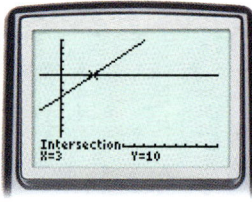
Find the point of intersection: $x = 3$ when $y = 10$.

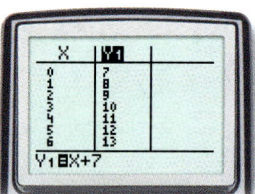
You can also observe values for $y = x + 7$ in the table window to confirm that $x = 3$ when $y = 10$.

Exercises

Communicate

1. Compare the Addition Property of Equality with the Subtraction Property of Equality. Explain how to decide which one to use to solve an equation.

2. Describe situations that can be modeled by the following equations:
   a. $x + 20 = 50$ **b.** $x - 20 = 50$

3. Explain how to solve each equation by using algebra tiles.
   a. $x + 6 = 10$ **b.** $x - 6 = 10$

4. Explain how to solve $10 = x + 7$ algebraically.

Guided Skills Practice

Solve each equation by using the Subtraction Property of Equality. Check the solution. *(EXAMPLE 1)*

5. $x + 7 = 31$ $x = 24$
6. $-7.2 = x + 3.5$ $x = -10.7$
7. $3\frac{1}{4} + m = 5\frac{3}{8}$ $m = 2\frac{1}{8}$

8. Jill has saved $137 toward the purchase of a mountain bike. The bike costs $547. Write and solve an equation to find out how much more money she needs. *(EXAMPLE 2)* $x + 137 = 547$; Jill needs $410.

Solve each equation by using the Addition Property of Equality. Check the solution. *(EXAMPLE 3)*

9. $y - 7.5 = -2.7$ $y = 4.8$
10. $32 = m - 3$ $m = 35$
11. $r - 4\frac{1}{3} = 1\frac{1}{2}$ $r = 5\frac{5}{6}$
12. $x - 7 = 10$ $x = 17$
13. $p - 10 = -15$ $p = -5$
14. $m - 3.5 = 7.5$ $m = 11$

118 LESSON 3.1

Solve each equation. Check the solution. *(EXAMPLE 4)*

15. $4 - m = 7$ $m = -3$ **16.** $10 - x = 2$ $x = 8$ **17.** $13 = 7 - x$ $x = -6$

18. The difference between the normal level and the 50-year flood level of a lake is 29 feet. If the normal level is 690 feet, what is the 50-year flood level? *(EXAMPLE 5)* The flood level is 719 feet.

Assess

Selected Answers
Exercises 5–18, 19–83 odd

ASSIGNMENT GUIDE

In Class	1–18
Core	19–42, 65–75 odd
Core Plus	20–62 even, 64–76
Review	77–84
Preview	85–87

Extra Practice can be found beginning on page 738.

Technology

A calculator may be helpful in Exercises 19–63. You may want to have students solve each equation with paper and pencil and then use a calculator to check their solutions.

Practice and Apply

Solve each equation. Check the solution.

19. $a - 16 = 15$ $a = 31$ **20.** $a - 4 = -10$ $a = -6$ **21.** $11 = t + 29$ $t = -18$
22. $m + 54 = 36$ $m = -18$ **23.** $r - 10 = -80$ $r = -70$ **24.** $g - 27 = 148$ $g = 175$
25. $b - 109 = 58$ $b = 167$ **26.** $24 = h - 53$ $h = 77$ **27.** $y + 37 = -110$
28. $396 = z + 256$ $z = 140$ **29.** $819 = g + 75$ $g = 744$ **30.** $1\frac{1}{4} = x + \frac{3}{4}$ $x = \frac{1}{2}$
31. $x + 9 = 6$ $x = -3$ **32.** $x - 10 = -4$ $x = 6$ **33.** $x + 6 = -4$ $x = -10$
34. $y + \frac{2}{3} = \frac{7}{9}$ $y = \frac{1}{9}$ **35.** $3 = g - 1.2$ $g = 4.2$ **36.** $b - 4.4 = 7$ $b = 11.4$
37. $c + 2.1 = 3.5$ $c = 1.4$ **38.** $d + 8.7 = 11.9$ **39.** $k - 2.1 = -6.3$
40. $j - 6.8 = -12.4$ **41.** $g - \frac{2}{6} = \frac{5}{6}$ $g = 1\frac{1}{6}$ **42.** $1\frac{2}{7} = x + \frac{4}{5}$ $x = \frac{17}{35}$
43. $y - 7.4 = 8.3$ $y = 15.7$ **44.** $-3.7 = y - 1.2$ **45.** $x + 4.2 = -6.1$
46. $5 + m = 10$ $m = 5$ **47.** $-8 + x = 4$ $x = 12$ **48.** $9 - x = 3$ $x = 6$
49. $-3 - p = 2$ $p = -5$ **50.** $6 - y = -3$ $y = 9$ **51.** $-4 - m = -3$ $m = -1$
52. $x + 6.2 = -5.3$ **53.** $456 = a - 529$ **54.** $d + 904 = -759$
55. $t + \frac{6}{7} = \frac{2}{3}$ $t = -\frac{4}{21}$ **56.** $t - \frac{3}{5} = \frac{1}{2}$ $t = 1\frac{1}{10}$ **57.** $k + 5.8 = -3.2$
58. $h + 9.1 = -5.3$ **59.** $n - \frac{7}{9} = \frac{1}{6}$ $n = \frac{17}{18}$ **60.** $-\frac{1}{3} = j + \frac{1}{5}$ $j = -\frac{8}{15}$
61. $m + 2.3 = 7$ $m = 4.7$ **62.** $15 = x - 4.6$ $x = 19.6$ **63.** $x + \frac{1}{2} = 1$ $x = \frac{1}{2}$

27. $y = -147$
38. $d = 3.2$
39. $k = -4.2$
40. $j = -5.6$
44. $y = -2.5$
45. $x = -10.3$
52. $x = -11.5$
53. $a = 985$
54. $d = -1663$
57. $k = -9.0$
58. $h = -14.4$

In Exercises 64–76, write an equation for each situation. Then solve the equation.

CONNECTIONS

64. STATISTICS The range of a set of scores is 28. If the highest score is 47, what is the lowest score?

65. GEOMETRY Supplementary angles are pairs of angles whose measures total 180°. Determine the measure of an angle that is supplementary to an angle with a measure of 92°.

66. GEOMETRY The measures of the interior angles of a triangle total 180°. Two angles of a triangle have a measure of 50° each. Find the measure of the third angle.

64. $28 = 47 - L$
The lowest score is 19.

65. $92 + s = 180$
The angle measure is 88°.

66. $50 + 50 + a = 180$
The measure of the third angle is 80°.

Error Analysis

When solving an equation like $x + 12 = 5$, some students may simply add the numbers they see, giving the solution as $x = 12 + 5 = 17$. Similarly, they may solve an equation like $z - 7 = 10$ by subtracting, giving the solution as $z = 10 - 7 = 3$. Remind students that the same operation must be performed on *each side* of the equal sign. Stress the importance of checking a proposed solution in the original equation.

67. $x + 30 = 70$
The measure of angle *SRT* is 40°.

68. $x + 18 = 70$
The measure of angle *SRT* is 52°.

69. $m + 53 = 67$
Lisa will have $14 left.

70. $s - 23 = 49$
Robert saved $72.

71. $s + 23 = 49$
Sandy saved $26.

72. $89 + y = 94$
The second-string running back ran 5 yd.

Practice

CONNECTIONS

67. GEOMETRY Angle *SRU* of triangle *RSU* measures 70°. Angle *TRU* measures 30°. Write and solve an equation to find the measure of angle *SRT*.

68. GEOMETRY Angle *SRU* of triangle *RSU* measures 70°. Angle *TRU* measures 18°. Write and solve an equation to find the measure of angle *SRT*.

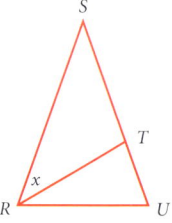

APPLICATIONS

69. PERSONAL FINANCE Lisa wants to buy a $53 computer program and has $67. How much will she have left after she buys the program?

70. PERSONAL FINANCE Robert has saved $23 more than he needs to buy a tennis racket that costs $49. How much has he saved?

71. PERSONAL FINANCE Sandy has saved $23 less than she needs to buy a tennis racket that costs $49. How much has she saved?

72. SPORTS The first- and second-string running backs on a football team ran for a total of 94 yards. If the first-string back ran 89 yards, how many yards did the second-string back run?

73. TRAVEL If the odometer on Sarah's car registered 23,580 when she finished a vacation trip of 149 miles, what did it register when she started the trip?

74. TRAVEL If the odometer on Sarah's car registered 23,580 before a vacation trip of 149 miles, what did it register when she finished her trip?

75. CALENDAR Carl sees a calendar which indicates that his birthday, December 21, is the 355th day of the year and that the current day, October 15, is the 288th day of the year. How many days away is his birthday?

76. CONSUMER ECONOMICS Sam bought two items that cost $12.98 and $14.95, plus tax. He got $0.39 in change from $30. How much was the tax?

73. $m + 149 = 23{,}580$
The odometer reading was 23,431 mi.

74. $m - 149 = 23{,}580$
The odometer reading was 23,729 mi.

75. $d + 288 = 355$
Carl's birthday is 67 days away.

76. $12.98 + 14.95 + t + 0.39 = 30.00$
The tax was $1.68.

Look Back

77. Find the sum $1 + 2 + 3 + \cdots + 12$. (LESSON 1.1) **78**

78. State the pattern and find the next three numbers in the sequence 6, 0, 12, 6, 18, 12, ... (LESSON 1.1) **24, 18, 30**

79. Suppose that a rocket took 23 seconds from takeoff to its return to Earth. Find how many seconds after takeoff the rocket reached its maximum height. (LESSON 1.1) **$t = 11.5$**

80. Find the perimeter of a square with sides that are 2.5 inches long. (LESSON 2.2) **10 inches**

81. If poster board costs $0.39 a piece, how many pieces of poster board can Shannon buy for $2.50? (LESSON 2.4) **6 poster boards**

Perform the indicated operations. (LESSON 2.6)

82. $(3x - 7) - (x - 4)$
    $2x - 3$
83. $(8 - 7y) + 2y$
    $8 - 5y$
84. $(n - 2) - (-3n + 4)$
    $4n - 6$

Look Beyond

Show how to solve these equations by using algebra tiles.

85. $3x = 18$
86. $2x = 12$
87. $2x + 5 = 11$

Include this activity in your portfolio.

Egyptian Equation Solving
About 4000 years ago, Egyptian mathematicians wrote textbooks to prepare students for government and business jobs. In 1858, a Scotsman named Rhind bought one of these textbooks, which had been copied on papyrus by a scribe named Ahmes. Part of this papyrus is shown in the photo at right.

Problem 70 of the Rhind papyrus asks, "What is the amount of meal in each loaf of bread if 2520 *ro* of meal is made into 100 loaves?" In the form of an equation, this is written as $100m = 2520$.

The Egyptians solved the equation by using the table method and by guess-and-check.

$100m = 2520$

m	1	10	20	5	$\frac{1}{5}$
$100m$	100	1000	2000	500	20

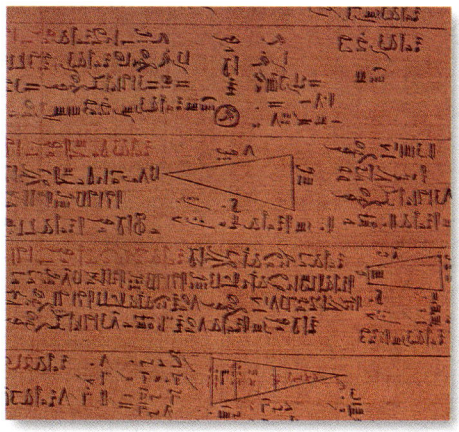

For the equation $100m = 2520$, they needed to multiply $100(20) + 100(5) + 100\left(\frac{1}{5}\right)$ on the left side to get 2520 on the right side. They found that m was $20 + 5 + \frac{1}{5}$, or $25\frac{1}{5}$ *ro*.

Solve these equations as the Egyptians might have.

1. $40m = 4800$
2. $60j = 930$
3. $24p = 252$
4. $10w = 2432$

Look Beyond

In Exercises 85–87, students investigate types of equations that they will study formally in Lesson 3.2 and in Lesson 3.3. At this point, however, students should be able to arrive at correct solutions by using mental math or algebra tiles.

ALTERNATIVE
Assessment

Portfolio Activity
The Portfolio Activity can be used as preparation for the Chapter Project or as a separate activity. In the Portfolio Activity on this page, students learn about a method used by the ancient Egyptians to solve certain equations of the form $ax = b$.

internetconnect

GO TO: go.hrw.com
KEYWORD: MA1 Egyptian

To support the Portfolio Activity, students may visit the HRW Web site, where they can examine a method for multiplying numbers used by ancient Egyptians.

Student Technology Guide

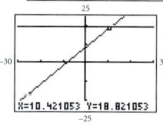

85. Model the left side of the equation with 3 *x*-tiles and the right side with 18 positive unit tiles. Separate the *x*-tiles on the left side into 3 groups, and separate the unit tiles on the right into 3 groups, one group for each of the *x*-tiles on the left side. There are 6 positive unit tiles in each group. $x = 6$

86. Model the left side of the equation with 2 *x*-tiles and the right side with 12 positive unit tiles. Separate the *x*-tiles on the left side into two groups, and separate the 12 positive unit tiles on the right side into 2 groups, one group for each of the *x*-tiles on the left side. There are 6 positive unit tiles in each group. $x = 6$

87. Model the left side of the equation with 2 *x*-tiles and 5 positive unit tiles and the right side with 11 positive unit tiles. Take away 5 positive unit tiles from each side. Separate the 6 positive unit tiles on the right side into 2 groups, one for each of the *x*-tiles on the right side. There are 3 positive unit tiles in each group. $x = 3$

LESSON 3.1

Prepare

NCTM Principles & Standards 1–3, 5–10

LESSON RESOURCES

- Practice 3.2
- Reteaching Master 3.2
- Enrichment Master 3.2
- Cooperative-Learning Activity 3.2
- Lesson Activity 3.2
- Problem Solving/Critical Thinking Master 3.2
- Teaching Transparency 13

QUICK WARM-UP

Find each product or quotient.

1. $(6)(-7)$ -42
2. $(-8)(-12)$ 96
3. $-148 \div (-4)$ 37
4. $-8 \div 16$ -0.5
5. $(-1.5)(0.3)$ -0.45

Also on Quiz Transparency 3.2

Teach

Why Equations are important tools in real-world problem solving. Give students the equations $12x = 102$ and $\frac{n}{12} = 8.5$. Have them suggest real-world situations that these equations might represent.
Samples: What is your wage per hour if you earn $102 for working 12 hours? What total pay for 12 hours of work gives you a wage of $8.50 per hour?

122 LESSON 3.2

3.2 Solving Equations by Multiplying and Dividing

Objective
- Solve equations by using multiplication and division.

Why Carpentry problems are one type of many real-world problems that can be modeled by equations involving multiplication and division.

APPLICATION
CARPENTRY

Alicia needs to purchase twenty 8-foot boards to complete the framing of a house. In order to compare costs, she needs to know the cost per foot. At one store she finds 8-foot boards for $2.40 each. What is the cost per foot?

Let x be the cost in dollars per foot. Then the equation $8x = 2.40$ models this situation.

How can you find the solution to this equation?

Activity
Finding Solutions to Multiplication Equations

1. Explain why the equation $8x = 2.40$ models the situation presented above.
2. Draw a picture of a board divided into eight 1-foot pieces. How does drawing a picture help you to solve the problem?

x								$8x$
1	2	3	4	5	6	7	8	
?								$2.40

3. Solve $8x = 2.40$ by using guess-and-check.
4. Examine the algebraic method for solving $8x = 2.40$ shown below. Why does this method work?

$$8x = 2.40$$
$$\frac{8x}{8} = \frac{2.40}{8} \quad \text{Divide both sides by 8.}$$
$$x = 0.30 \quad \text{Simplify.}$$

CHECKPOINT ✓ 5. Use the method from Step 4 to solve the following equations:

$8x = 3.20$ $10x = 25$

Alternative Teaching Strategy

USING MODELS Model the process of solving an equation such as $3x = 12$ by using algebra tiles. First place 3 x-tiles on the left side of the equal sign and 12 unit tiles on the right.

Now divide the tiles on each side of the equal sign into groups of equal size.

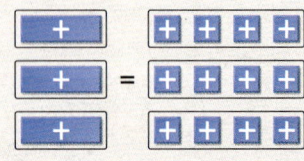

Each x-tile can be matched with a group of 4 unit tiles, so the solution is $x = 4$.

Have students create a similar model for the process of solving $4x = -12$. **Use 4 x-tiles and 12 negative unit tiles. Each x-tile can be matched with a group of 3 negative unit tiles, so $x = -3$.**

Solving Equations by Using Division

You can use the Division Property of Equality to solve equations such as the one presented at the beginning of the lesson.

> **The Division Property of Equality**
>
> For all real numbers a, b, and c, if $a = b$ and $c \neq 0$, then $\frac{a}{c} = \frac{b}{c}$.
>
> Dividing each side of an equation by equal amounts results in an equivalent equation.

EXAMPLE 1 Solve each equation. Check the solution.

a. $3x = -9$ b. $2 = 7a$ c. $-6m = -3.6$

● **SOLUTION**

a. Divide each side of the equation by 3.
$$3x = -9$$
$$\frac{3x}{3} = \frac{-9}{3}$$
$$x = -3$$

b. Divide each side of the equation by 7.
$$2 = 7a$$
$$\frac{2}{7} = \frac{7a}{7}$$
$$\frac{2}{7} = a$$
$$a = \frac{2}{7} \quad \text{Sym. Prop.}$$

c. Divide each side of the equation by -6.
$$-6m = -3.6$$
$$\frac{-6m}{-6} = \frac{-3.6}{-6}$$
$$m = 0.6$$

To check the solution, substitute -3 for x in the original equation.
$$3(-3) \stackrel{?}{=} -9$$
$$-9 = -9 \quad \text{True}$$

To check the solution, substitute $\frac{2}{7}$ for a in the original equation.
$$2 \stackrel{?}{=} 7\left(\frac{2}{7}\right)$$
$$2 = 2 \quad \text{True}$$

To check the solution, substitute 0.6 for m in the original equation.
$$-6(0.6) \stackrel{?}{=} -3.6$$
$$-3.6 = -3.6 \quad \text{True}$$

TRY THIS Solve each equation. Check the solution.

a. $-5x = 4.5$ b. $-30 = -6m$ c. $40 = 5p$

EXAMPLE 2

CONNECTION
GEOMETRY

In a regular pentagon, all of the interior angles are equal in measure. **The sum of the angle measures is 540°. Write and solve an equation to find the measure of each angle.**

● **SOLUTION**

The total number of angles is 5. The sum of the measures of the angles is 540°. Let x be the measure of each angle. Then $5x$ is the sum of the measures of all 5 angles and $5x = 540$.

Solve for x.
$$5x = 540$$
$$\frac{5x}{5} = \frac{540}{5} \quad \text{Div. Prop. of Equality}$$
$$x = 108 \quad \text{Simplify.}$$

CHECK
$$5(108) \stackrel{?}{=} 540$$
$$540 = 540 \quad \text{True}$$
Each angle has a measure of 108°.

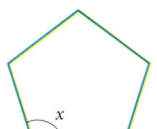

Activity Notes

In this Activity, students are given an equation that represents a real-world situation and are asked to solve it by using the guess-and-check method. Then they see how to obtain the same result by an algebraic method. Students should arrive at the conclusion that they can solve an equation of the form $ax = b$, where $a \neq 0$, by dividing each side of the equation by a.

CHECKPOINT ✔

5. $8x = 3.20$ $10x = 25$
$\frac{8x}{8} = \frac{3.20}{8}$ $\frac{10x}{10} = \frac{25}{10}$
$x = 0.40$ $x = 2.5$

ADDITIONAL EXAMPLE 1

Solve each equation. Check the solution.

a. $-5t = -15$ $t = 3$
b. $4y = 9$ $y = \frac{9}{4}$
c. $8p = -4.8$ $p = -0.6$

TRY THIS

a. $x = -0.9$
b. $m = 5$
c. $p = 8$

Use Teaching Transparency 13.

☞ For Additional Example 2 and Math Connection, see page 124.

Interdisciplinary Connection

PHYSICS According to Newton's second law of motion, the ratio of the force acting upon a body to the acceleration that the force produces is a constant that is called the *mass* of the body. This relationship is given as $\frac{F}{a} = m$, where F represents force, a represents acceleration, and m represents mass. Have students solve this equation for F and then for a. $F = ma$; $a = \frac{F}{m}$

Near the surface of Earth, every object experiences a constant acceleration due to gravity. The force that accelerates the object downward is its weight. Using Newton's second law of motion, written in the form $F = ma$, this relationship becomes $w = mg$, where w represents weight, m represents mass, and g represents acceleration due to gravity. Have students solve this equation for m and then for g. $m = \frac{w}{g}$; $g = \frac{w}{m}$

LESSON 3.2   **123**

ADDITIONAL EXAMPLE 2

In a regular octagon, all of the interior angles are equal in measure. The sum of the interior angle measures is 1080°. **Write and solve an equation to find the measure of each interior angle.**
$8x = 1080; x = 135$
The measure of each interior angle is 135°.

Math CONNECTION

GEOMETRY A *regular polygon* is a polygon whose angles are all congruent *and* whose sides are all congruent. The sum of the measures of the interior angles of any polygon with n sides is $(n-2)180°$. Thus, the measure of each interior angle is $\frac{(n-2)180°}{n}$.

Teaching Tip

You may need to remind students that the *coefficient* of a term is its numerical factor.

ADDITIONAL EXAMPLE 3

Solve each equation. Check the solution.

a. $\frac{n}{7} = -1.2$ $n = -8.4$

b. $5 = \frac{y}{-6}$ $y = -30$

c. $\frac{r}{-3} = -30$ $r = 90$

TRY THIS
a. $t = -16$
b. $d = -16.2$
c. $m = 4.5$

124   LESSON 3.2

Solving Equations by Using Multiplication

APPLICATION: SPORTS

To compute a batting average, the number of hits is divided by the number of times at bat.

$$\frac{\text{Hits}}{\text{Times at bat}} = \text{Average}$$

Suppose that Alex batted 150 times during his varsity baseball career. If his final batting average is 0.300, how many hits did he get?

Let H represent the number of hits. The problem can be modeled by the following division equation:
$$\frac{H}{150} = 0.300$$

In algebraic notation, $\frac{H}{150}$ is the same as $\frac{1}{150}(H)$. To solve the equation, multiply both sides of the equation by 150 because 150 is the reciprocal of $\frac{1}{150}$, the coefficient of H.

$$150\left(\frac{H}{150}\right) = 150(0.300)$$
$$H = 45$$

Alex had 45 hits in his varsity baseball career.

The Multiplication Property of Equality

For all real numbers a, b, and c, if $a = b$, then $ac = bc$.

Multiplying each side of an equation by equal amounts results in an equivalent equation.

EXAMPLE 3 Solve each equation. Check the solution.

a. $\frac{m}{5} = 1.3$ b. $-3 = \frac{x}{-4}$ c. $\frac{w}{5} = -10$

SOLUTION

a. Multiply each side of the equation by 5.
$$\frac{m}{5} = 1.3$$
$$(5)\frac{m}{5} = (5)1.3$$
$$m = 6.5$$

To check the solution, substitute -6.5 for m in in the original equation.
$$\frac{6.5}{5} \stackrel{?}{=} 1.3$$
$$1.3 = 1.3 \text{ True}$$

b. Multiply each side of the equation by -4.
$$-3 = \frac{x}{-4}$$
$$(-4)(-3) = (-4)\frac{x}{-4}$$
$$12 = x$$
$$x = 12$$

To check the solution, substitute 12 for x in the original equation.
$$-3 \stackrel{?}{=} \frac{12}{-4}$$
$$-3 = -3 \text{ True}$$

c. Multiply each side of the equation by 5.
$$\frac{w}{5} = -10$$
$$(5)\frac{w}{5} = (5)(-10)$$
$$w = -50$$

To check the solution, substitute -50 for w in the original equation.
$$\frac{-50}{5} \stackrel{?}{=} -10$$
$$-10 = -10 \text{ True}$$

TRY THIS Solve each equation. Check the solution.

a. $\frac{t}{-3.2} = 5$ b. $\frac{d}{3} = -5.4$ c. $3 = \frac{m}{1.5}$

Inclusion Strategies

USING SYMBOLS Some students may become confused when solving the equations in this lesson because familiar symbols for multiplication and division are not evident. You may want to give students a form that utilizes these symbols and show how the method of solving equations in this form parallels the methods used to solve the equations in Lesson 3.1.

$x + 3 = 21$ → $x + 3 - 3 = 21 - 3$
$x - 3 = 21$ → $x - 3 + 3 = 21 + 3$
$3x = 21$ → $3 \cdot x = 21$ → $3 \cdot x \div 3 = 21 \div 3$
$\frac{x}{3} = 21$ → $x \div 3 = 21$ → $x \div 3 \cdot 3 = 21 \cdot 3$

Enrichment

Tell students that equations with exactly the same solution are called *equivalent equations*. For example, $-5y = 20$ and $d + 3 = -1$ are equivalent because the solution of each is -4. Give students the equation $4r = -32$ and instruct them to find three equivalent equations: one of the form $x + a = b$, one of the form $x - a = b$, and one of the form $\frac{x}{a} = b$. **samples:** $n + 2 = -6$; $y - 5 = -13$; $\frac{a}{2} = -4$

Using Reciprocals

Recall that in order to divide by a nonzero number, you multiply by the reciprocal of that number. For example, $10 \div 5$ can be rewritten as $10 \times \frac{1}{5}$. The value of both expressions is the same, 2.

Note the following:

$$\frac{a}{b} \cdot \frac{b}{a} = \frac{ab}{ba}$$
$$= \frac{ab}{ab}$$
$$= 1$$

This tells you that the reciprocal of a rational number, $\frac{a}{b}$, is $\frac{b}{a}$. Reciprocals can be used to solve multiplication and division equations.

EXAMPLE 4 Solve each equation by using reciprocals. Check each solution.

a. $\frac{3}{5}y = 10$ b. $-\frac{1}{3} = 5x$ c. $\frac{2b}{3} = -10$

SOLUTION

a. The reciprocal of the coefficient of y is $\frac{5}{3}$. Multiply each side of the equation by $\frac{5}{3}$.

$\left(\frac{5}{3}\right)\frac{3}{5}y = \left(\frac{5}{3}\right)10$

$y = \frac{50}{3}$, or $16\frac{2}{3}$

b. The reciprocal of the coefficient of x is $\frac{1}{5}$. Multiply each side of the equation by $\frac{1}{5}$.

$\left(\frac{1}{5}\right)\left(-\frac{1}{3}\right) = \left(\frac{1}{5}\right)5x$

$-\frac{1}{15} = x$

$x = -\frac{1}{15}$ Sym. Prop.

c. The reciprocal of the coefficient of b is $\frac{3}{2}$. Multiply each side of the equation by $\frac{3}{2}$.

$\left(\frac{3}{2}\right)\frac{2b}{3} = \left(\frac{3}{2}\right)(-10)$

$b = -\frac{30}{2}$, or -15

To check the solution, substitute $\frac{50}{3}$ for y in the original equation.

$\frac{3}{5}\left(\frac{50}{3}\right) \stackrel{?}{=} 10$

$\frac{50}{5} \stackrel{?}{=} 10$

$10 = 10$ True

To check the solution, substitute $-\frac{1}{15}$ for x in the original equation.

$-\frac{1}{3} \stackrel{?}{=} 5\left(-\frac{1}{15}\right)$

$-\frac{1}{3} \stackrel{?}{=} -\frac{5}{15}$

$-\frac{1}{3} = -\frac{1}{3}$ True

To check the solution, substitute -15 for b in the original equation.

$\frac{2(-15)}{3} \stackrel{?}{=} -10$

$-\frac{30}{3} \stackrel{?}{=} -10$

$-10 = -10$ True

TRY THIS Solve by using reciprocals. Check the solution.

a. $4m = -\frac{3}{8}$ b. $\frac{2d}{5} = \frac{3}{2}$ c. $-\frac{4}{5} = 3x$

CRITICAL THINKING Explain how multiplying by a reciprocal instead of dividing is similar to adding the opposite instead of subtracting.

Reteaching the Lesson

USING LANGUAGE SKILLS Work with students to translate equations into verbal questions that incorporate the word *multiply* or *divide*. For example, translate the equation $8x = 40$ as "What number do you multiply by eight to get forty?"

Now discuss the fact that, just as addition and subtraction are inverse operations, so are multiplication and division. Remind students that this means they can think of using one operation to "undo" the other. Refer back to the equation $8x = 40$. Since the variable, x, is *multiplied* by 8, you *divide* each side by 8 to undo the operation. Emphasize that the goal is to isolate the variable on one side of the equal sign.

Repeat the discussion, this time focusing on the equation $\frac{x}{8} = 5$. In this case the translation is "What number do you divide by eight to get five?" Since the variable, x, is *divided* by 8, you *multiply* each side by 8 to undo the operation.

Teaching Tip

In the statement about the reciprocal of a number, be sure students note that you must divide by a nonzero number. Point out that if a real number, a, were the reciprocal of zero, then the product $0 \cdot a$ would have to equal 1. However, this contradicts the fact that the product of zero and any real number is zero. Therefore, zero has no reciprocal.

ADDITIONAL EXAMPLE 4

Solve each equation by using reciprocals. Check each solution.

a. $-\frac{9}{4}z = 10$

$z = -\frac{40}{9}$, or $-4\frac{4}{9}$

b. $-6s = \frac{1}{4}$ $s = -\frac{1}{24}$

c. $-\frac{3c}{4} = -15$ $c = 20$

TRY THIS

a. $m = -\frac{3}{32}$

b. $d = \frac{15}{4}$

c. $x = -\frac{4}{15}$

Teaching Tip

In part **c** of Example 4, some students may wonder why the reciprocal of the coefficient of b is $\frac{3}{2}$. You may want to show them the following justification:

$\frac{2b}{3} = \frac{2 \cdot b}{3 \cdot 1} = \frac{2}{3} \cdot \frac{b}{1} = \frac{2}{3} \cdot b$

The expression $\frac{2}{3} \cdot b$ is the same as $\frac{2}{3}b$. This means that the coefficient of b is $\frac{2}{3}$ and the reciprocal of the coefficient is $\frac{3}{2}$.

CRITICAL THINKING

Adding the opposite makes use of the additive inverse of a number. Multiplying by the reciprocal uses the multiplicative inverse of a number. These numbers have unique properties relative to the identity elements for multiplication and addition.

LESSON 3.2

ADDITIONAL EXAMPLE 5

Frank needs to make a 360-mile trip in 9 hours. **Write and solve an equation to determine what speed he should average.**

$9s = 360$; 40 mph

Assess

Selected Answers
Exercises 6–19, 21–71 odd

ASSIGNMENT GUIDE

In Class	1–19
Core	20–42, 53–59 odd
Core Plus	20–52 even, 53–59
Review	60–71
Preview	72–74

✏ **Extra Practice** can be found beginning on page 738.

EXAMPLE 5
APPLICATION
FOOD
PROBLEM SOLVING

An apple pie is cut into 6 pieces. Each piece contains a total of 18 grams of fat. **Write and solve an equation to determine how many grams of fat are in the whole pie.**

● **SOLUTION**

Write an equation. Let x be the number of grams of fat in the whole pie. Since the pie is cut into 6 equal pieces, the equation is as follows:

$$\frac{1}{6}x = 18$$

Solve for x.

$$\frac{1}{6}x = 18$$
$$(6)\left(\frac{1}{6}x\right) = (6)(18)$$
$$x = 108$$

The whole pie contains 108 grams of fat.

Exercises

● Communicate

Explain how to solve the following equations for x:

1. $592x = 812$
2. $5x = \frac{1}{10}$
3. $-21 = \frac{x}{12}$
4. $\frac{4x}{5} = \frac{1}{2}$

CONNECTION

5. **GEOMETRY** From the formula for the circumference of a circle, explain how to find the diameter, d, in terms of the circumference, C, and π.

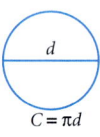
$C = \pi d$

● Guided Skills Practice

6. $x = -2$
7. $x = -20$
8. $x = \frac{1}{4}$
9. $m = -5$

Solve each equation. Check the solution. *(EXAMPLE 1)*

6. $-14x = 28$
7. $0.5x = -10$
8. $4 = 16x$
9. $-3m = 15$

CONNECTION

10. **GEOMETRY** The sum of the measures of the interior angles of a regular octagon is 940°. Write and solve an equation to find the measure of each angle. *(EXAMPLE 2)* $8a = 940$; $a = 117.5°$

12. $x = -6$
14. $m = 6.3$

Solve each equation. Check the solution. *(EXAMPLE 3)*

11. $\frac{x}{2} = -3$ $x = -6$
12. $\frac{x}{-5} = 1.2$
13. $4 = \frac{x}{2.1}$ $x = 8.4$
14. $\frac{m}{-3} = -2.1$

Practice

3.2 Solving Equations by Multiplying and Dividing

State the property needed to solve each equation. Then solve it.

1. $\frac{x}{36} = 6$ — Multiplication Property of Equality; $x = 216$
2. $5.75p = 46$ — Division Property of Equality; $p = 8$
3. $8 = -64y$ — Division Property of Equality; $y = -\frac{1}{8}$
4. $\frac{x}{3} = 3$ — Multiplication Property of Equality; $x = 9$
5. $6b = 54$ — Division Property of Equality; $b = 9$
6. $\frac{r}{-5} = 5$ — Multiplication Property of Equality; $r = -25$
7. $-10y = 5$ — Division Property of Equality; $y = -\frac{1}{2}$
8. $-16n = -128$ — Division Property of Equality; $n = 8$

Solve each equation.

9. $658b = 2632$ $b = 4$
10. $-4g = 36$ $g = -9$
11. $65x = 35$ $x = \frac{7}{13}$
12. $\frac{x}{3} = -24$ $x = -72$
13. $9y = 153$ $y = 17$
14. $\frac{c}{44} = -44$ $c = -1936$
15. $\frac{x}{0.07} = 5$ $x = 0.35$
16. $8t = -35$ $t = -4\frac{3}{8}$
17. $0.66p = 4.62$ $p = 7$
18. $-70g = 4200$ $g = -60$
19. $\frac{b}{20} = -20$ $b = -400$
20. $-86b = 43$ $b = -\frac{1}{2}$
21. $\frac{m}{-15} = 0$ $m = 0$
22. $\frac{x}{6} = -3$ $x = -18$
23. $4.4 = 2.2t$ $t = 2$
24. $16n = 320$ $n = 20$
25. $-4v = 68.8$ $v = -17.2$
26. $\frac{x}{-2} = -125$ $x = 250$
27. $-480 = -24z$ $z = 20$
28. $10d = \frac{3}{5}$ $d = \frac{3}{50}$
29. $0.25q = 25$ $q = 100$
30. $\frac{w}{0.5} = -15$ $w = -7.5$
31. $-1 = \frac{p}{-555}$ $p = 555$
32. $-495 = 99y$ $y = -5$

Solve each formula for the variable indicated.

33. $A = \frac{1}{2}bh$ for b $b = \frac{2A}{h}$
34. $z = wxy$ for w $w = \frac{z}{xy}$
35. $v = \frac{d}{t}$ for d $d = vt$

126 LESSON 3.2

Use reciprocals to solve each equation. Check your solution.
(EXAMPLES 4 AND 5)

15. $x = -12$
18. $m = -\frac{9}{4}$

15. $\frac{3x}{2} = -18$ 16. $5x = \frac{1}{4}$ $x = \frac{1}{20}$ 17. $\frac{x}{4} = \frac{3}{4}$ $x = 3$ 18. $\frac{2}{3}m = -\frac{3}{2}$

19. **FOOD** A chocolate cake is cut into 8 pieces. Each piece contains a total of 14 grams of fat. Write and solve an equation to determine how many grams of fat are in the whole cake. $\frac{c}{8} = 14$, $c = 112$

Practice and Apply

Solve each equation and check your solution.

20. $7x = 56$ $x = 8$
21. $-7m = -14$ $m = 2$
22. $-3x = 9$ $x = -3$
23. $5.6v = 7$ $v = 1.25$
24. $-13 = \frac{y}{3}$ $y = -39$
25. $\frac{b}{-9} = 6$ $b = -54$
26. $\frac{x}{27} = -26$ $x = -702$
27. $\frac{p}{-9} = 0.9$ $p = -8.1$
28. $84 = -12y$ $y = -7$
29. $111x = -888$ $x = -8$
30. $-3x = -4215$ $x = 1405$
31. $-4x = -3228$ $x = 807$
32. $\frac{x}{-7} = -1.4$ $x = 9.8$
33. $6 = \frac{x}{0.5}$ $x = 3$
34. $7 = -56w$ $w = -\frac{1}{8}$
35. $-3f = 15$ $f = -5$
36. $888x = 111$ $x = \frac{1}{8}$
37. $0.505 = 0.505x$ $x = 1$
38. $4b = -15$ $b = -\frac{15}{4}$
39. $2a = 1$ $a = \frac{1}{2}$
40. $\frac{p}{111} = -10$ $p = -1110$
41. $2a = 13$ $a = \frac{13}{2}$
42. $\frac{m}{-9} = 0$ $m = 0$
43. $\frac{b}{15} = 1$ $b = 15$
44. $\frac{2x}{3} = \frac{3}{4}$ $x = \frac{9}{8}$
45. $-\frac{7x}{5} = \frac{3}{10}$ $x = -\frac{3}{14}$
46. $-\frac{1}{4} = \frac{x}{2}$ $x = -\frac{1}{2}$
47. $\frac{-m}{4} = 2$ $m = -8$
48. $-4p = 16$ $p = -4$
49. $\frac{x}{-7} = 5$ $x = -35$
50. $-2w = 13$ $w = -\frac{13}{2}$
51. $\frac{w}{-5} = 10$ $w = -50$
52. $\frac{-x}{3} = 10$ $x = -30$

APPLICATIONS

BUSINESS Max is apprenticing as a carpenter. Write and solve an equation to describe each situation in Exercises 53–57.

53. If one roll of masking tape costs $1.15, how many rolls can Max buy with $5.75?

54. Max finds masking tape in packages of 4 rolls for $4.32. What is the cost of one roll in the package? Is the cost of a single roll of masking tape (from Exercise 53) more than the cost of one roll in a package of 4.

55. Max's employer has $18.00 to spend on tape measures for his crew. How many can he buy with that amount if each tape measure costs $4.50?

56. What is the price of one extension cord if Max can buy a package of 6 for $7.26?

57. Max buys a package of 4 AA-type batteries for $2.52. What is the price for each battery?

58. **TRAVEL** Natalie's family wants to make a 400-mile drive in 8 hours. Write and solve an equation to determine what speed they must average for the trip.

59. **TRAVEL** Maria wants to drive 320 miles in 8 hours. Write and solve an equation to determine what speed she must average for the trip.

53. Max can buy 5 rolls of tape.
54. One roll of tape costs $1.08; yes, it costs more to buy a single roll.
55. He can buy 4 tape measures.
56. One extension cord costs $1.21.
57. The price of one battery is 63 cents.
58. They must travel 50 mph.
59. Maria should drive 40 mph.

Technology

A calculator may be helpful in Exercises 20–59. You may want to have students solve each equation with paper and pencil and then use a calculator to check their solutions.

Error Analysis

When solving an equation that involves a negative coefficient, students may erroneously think of the negative sign as a subtraction sign and attempt to add the same number to each side of the equation. For instance, given $-4g = 20$, they may say that the solution is $g = 20 + 4 = 24$. Encourage students to rewrite the left side of the equation as a multiplication expression: $-4 \cdot g = 20$. It should then be clear that division is the operation required to find the solution.

internetconnect

GO TO: go.hrw.com
KEYWORD: MA1 Oscar at Bat

In this activity, students extend their averaging skills by exploring applications other than sports.

Student Technology Guide

Student Technology Guide
3.2 Solving Equations by Multiplying and Dividing

In the worked-out solutions below, you can see how easy it is to solve some equations in one step. Once you see the patterns, you can use a calculator to solve simple equations in just a few moments.

Multiplication equation
$2.35x = 10.2$
$\frac{2.35x}{2.35} = \frac{10.2}{2.35}$
Solve for x by dividing both sides by 2.35.

Division equation
$\frac{x}{2.35} = 10.2$
$2.35\left(\frac{x}{2.35}\right) = 10.2 \times 2.35$
Solve for x by multiplying both sides by 2.35.

You can carry out the thinking above by using the key sequences below.

10.2 [÷] 2.35 → 4.340425532
The value of x is about 4.3.

10.2 [×] 2.35 [ENTER] → 23.97
The value of x is exactly 23.97.

You can solve an equation like $-\frac{5x}{2} = 18.5$ as shown below.

THINK   ACT
$\left(-\frac{5}{2}\right)x = 18.5$ → 18.5 [×] [(-)] 5 [÷] [ENTER] → -7.4

Use a calculator to solve each equation.

1. $6x = 42$ 2. $-4x = -28$ 3. $-5x = 36$
   $x = 7$ $x = 7$ $x = -7.2$

4. $\frac{x}{-6} = 9$ 5. $\frac{x}{4.2} = 6.5$ 6. $-\frac{3x}{5} = \frac{9}{2}$
   $x = -54$ $x = 27.3$ $x = -7.5$

7. $1.5x = -9.0$ 8. $8 = -4x$ 9. $-\frac{3}{5}x = \frac{21}{25}$
   $x = -6$ $x = -2$ $x = 1.4$, or $1\frac{2}{5}$

To solve a multiplication equation involving mixed numbers, consider the following key sequence:

                              number on right side ÷ number on left side
$2.35x = 14\frac{5}{6}$ → [(] 14 [+] 5 [÷] 6 [)] [÷] 2.35 [ENTER] → 6.312056738

Use a calculator to solve each equation to the nearest tenth.

10. $6.5x = 42\frac{1}{4}$ 11. $-2.4x = 16\frac{3}{4}$ 12. $5.5x = 36\frac{1}{10}$
    6.5   about -7.0 about 6.6

LESSON 3.2 **127**

In Exercises 72–74, students investigate the process of solving equations in which a variable appears on both sides of the equal sign. They will learn to solve equations of this type in Lesson 3.3. At this point, remind students that their goal is to isolate the variable on one side of the equal sign. Ask them to consider ways that they might accomplish this.

ALTERNATIVE
Assessment

Portfolio Activity

The Portfolio Activity can be used as preparation for the Chapter Project or as a separate activity. In the Portfolio Activity on this page, students explore the concept of profit. They discover that they can analyze certain business situations by writing and solving a multistep equation.

Answers to Portfolio Activities can be found in Additional Answers of the Teacher's Edition.

60. Insert inclusion symbols to make $30 \cdot 7 - 4 \cdot 10 \div 2 \div 2 = 225$ true.
(**LESSON 1.3**) $30 \cdot (7 - 4) \cdot 10 \div 2 \div 2 = 225$

Evaluate each absolute value. (**LESSON 2.1**)

61. $|0|$ 0   **62.** $|-5|$ 5   **63.** $\left|-\frac{1}{6}\right|$ $\frac{1}{6}$ **64.** $|9|$ 9

Evaluate each expression for $a = -3$, $b = 2$, and $c = 0$. (**LESSON 2.4**)

65. $\frac{a + b \cdot c}{-3}$ 1   **66.** $\frac{a \cdot b}{b + c}$ -3 **67.** $\frac{a + b}{b + c}$ $-\frac{1}{2}$

68. Shawna started at ground level and walked down 13 steps. She noticed that she had forgotten her purse, so she walked 47 steps back up to her apartment. She then walked down 27 steps to meet her friend. How many steps from ground level is she? Show the numerical expression that models this situation. (**LESSONS 2.2 AND 2.3**) $7; -13 + 47 - 27$

Solve each equation. Check the solution. (**LESSON 3.1**)

69. $x + 5 = 7$ $x = 2$ **70.** $x - 2.4 = 5.7$ $x = 8.1$ **71.** $x + \frac{7}{5} = \frac{2}{3}$ $x = -\frac{11}{15}$

Solve each equation. Check the solution.

72. $3y = 10 + 5y$ $y = -5$ **73.** $4x + 0.5 = 5x$ $x = 0.5$ **74.** $x + 4 = 3x - 2$ $x = 3$

Include this activity in your portfolio.

The track team at Hale High School plans to sell T-shirts to raise money for new equipment. In order for the track team to make a profit, the income from selling the T-shirts must be greater than the cost of making the T-shirts. At Ted's Tees, printing costs are $0.80 per shirt and the cost for each T-shirt is $3.75. The shop also charges a $125 fee per order for the silk-screen design.

1. Based on the prices at Ted's Tees, determine the cost for ordering 100, 250, 500, and 750 T-shirts.

2. Determine the cost of a single printed T-shirt for orders of 100, 250, 500, and 750 T-shirts.

3. If the track team orders 250 T-shirts and wishes to make a $500 profit, how much should they charge for each T-shirt?

WORKING ON THE CHAPTER PROJECT

You should now be able to complete Activity 1 of the Chapter Project on page 154.

3.3 Solving Two-Step Equations

Objectives
- Write equations that represent real-world situations.
- Solve two-step equations.

Why The beach-towel problem presented below can be modeled by an equation that takes more than one step to solve.

APPLICATION
CONSUMER ECONOMICS

The beach hat costs $18 and the total bill, before tax, is $74. What is the cost of one beach towel?

This problem can be modeled with an equation. Let t represent the cost of one beach towel in dollars. Since there are 7 beach towels, the expression $7t$ represents the cost for all of the towels.

cost of hat + cost of towels = total bill
$$18 + 7t = 74$$

Solving the equation fot t requires two steps.

Activity
Using a Table

Solve the equation for the cost, t, of one beach towel by using a table. Complete the table through $10 for the cost of a single towel.

Cost of single towel ($)	Cost of hat + cost of towels, $18 + 7t$	Total ($)
0	$18 + 7(0)$	18
1	$18 + 7(1)$	25
2	?	?

1. Copy and complete the table above.

CHECKPOINT ✓ 2. Explain how to determine the cost of a single towel when the total bill is $74.

3. Describe other ways to solve the equation $18 + 7t = 74$ for t.

Alternative Teaching Strategy

USING VISUAL MODELS Students can use algebra tiles to solve two-step equations. Using algebra tiles for an overhead projector, set up the equation $2x - 5 = 1$. Students can do the same at their desks with their own tiles.

Solve by adding 5 positive tiles (blue) to both sides of the equation. Since the 5 negative tiles and 5 positive tiles cancel each other out, the resulting equation is $2x = 6$. Divide the 6 unit-tiles into two groups to show division by 2. The solution is $x = 3$.

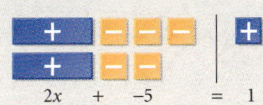

Prepare

NCTM PRINCIPLES & STANDARDS
1, 2, 6–10

LESSON RESOURCES
- Practice 3.3
- Reteaching Master 3.3
- Enrichment Master 3.3
- Cooperative-Learning Activity 3.3
- Lesson Activity 3.3
- Problem Solving/Critical Thinking Master 3.3
- Teaching Transparency 14

QUICK WARM-UP

Solve for x.

1. $3 + x = 11$		8
2. $x - 7 = 19$		26
3. $21 - x = 36$		−15
4. $3x = 18$		6
5. $9x = 45$		5
6. $6x = 15$		2.5
7. $\frac{x}{3} = 4$		12
8. $\frac{x}{15} = 3$		45

Also on Quiz Transparency 3.3

Teach

Why Ask students to choose a number, double it, and add 10. Then ask how many steps were used to find the answer. 2 Next, using problems like those in the Quick Warm-Up, review with students what they recently learned about solving one-step equations. Show them that a two-step equation involves a combination of two different operations, such as addition and multiplication.

☞ For Activity Notes and the answer to the Checkpoint, see page 130.

LESSON 3.3 **129**

Activity Notes

In this Activity, students use a table and the guess-and-check method to solve a two-step equation. Students should then observe the table and intuitively determine the steps for solving the given equation. They should discover that they should "undo" the operation of addition before multiplication to isolate the x.

Cooperative Learning

Have students do Activity 1 in pairs. They should both copy and complete the table on their own papers. Then they can compare answers and resolve any differences. Students should work together to answer Exercises 2 and 3.

CHECKPOINT ✔

2. Read the chart at $74. The cost of a single towel is $8 when the total bill is $74.

ADDITIONAL EXAMPLE 1

Solve $18 + 7t = 46$ in order to find the cost of one beach towel. $t = 4$

ADDITIONAL EXAMPLE 2

Solve.

a. $21 + 6x = -3$ $x = -4$

b. $2x - 7 = 13$ $x = 10$

TRY THIS

a. $y = 2$ b. $p = -22.5$

ADDITIONAL EXAMPLE 3

Solve.

a. $\frac{x}{5} + 2 = 6$ $x = 20$

b. $\frac{x}{10} - 6 = 2$ $x = 80$

You can solve $18 + 7t = 74$ algebraically. Solving this equation for t requires two steps.

EXAMPLE 1 Solve $18 + 7t = 74$ for t in order to find the cost of one beach towel.

● SOLUTION

$$18 + 7t = 74$$
$$18 + 7t - 18 = 74 - 18 \quad \text{Subtr. Prop. Eq.}$$
$$7t = 56 \quad \text{Simplify.}$$
$$\frac{7t}{7} = \frac{56}{7} \quad \text{Div. Prop. Eq.}$$
$$t = 8$$

CHECK
$$18 + 7(8) \stackrel{?}{=} 74$$
$$18 + 56 \stackrel{?}{=} 74$$
$$74 = 74 \quad \text{True}$$

One beach towel costs $8.

EXAMPLE 2 Solve.

a. $10x + 3 = 33$ b. $-47 = 3x - 50$

● SOLUTION

a. $10x + 3 = 33$
$$10x + 3 - 3 = 33 - 3 \quad \text{Subtr. Prop. Eq.}$$
$$10x = 30 \quad \text{Simplify.}$$
$$\frac{10x}{10} = \frac{30}{10} \quad \text{Div. Prop. Eq.}$$
$$x = 3$$

CHECK
$$10(3) + \stackrel{?}{=} 33$$
$$30 + 3 \stackrel{?}{=} 33$$
$$33 = 33 \quad \text{True}$$

b. $-47 = 3x - 50$
$$-47 + 50 = 3x - 50 + 50 \quad \text{Add. Prop. Eq}$$
$$3 = 3x \quad \text{Simplify.}$$
$$\frac{3}{3} = \frac{3x}{3} \quad \text{Div. Prop. Eq.}$$
$$1 = x$$

CHECK
$$-47 \stackrel{?}{=} 3(1) - 50$$
$$-47 \stackrel{?}{=} 3 - 50$$
$$-47 = -47 \quad \text{True}$$

TRY THIS Solve.

a. $15y + 31 = 61$ b. $-35 = 2p + 10$

EXAMPLE 3 Solve.

a. $\frac{m}{-5} + 3 = -2.5$ b. $5 = \frac{x}{3} - 14$

● SOLUTION

a. $\frac{m}{-5} + 3 = -25$
$$\frac{m}{5} + 3 - 3 = -2.5 - 3 \quad \text{Subtr. Prop. Eq.}$$
$$\frac{m}{-5} = -5.5 \quad \text{Simplify.}$$
$$-5\left(\frac{m}{-5}\right) = -5(-5.5) \quad \text{Mult. Prop. Eq.}$$
$$m = 27.5$$

CHECK
$$\frac{27.5}{-5} + 3 \stackrel{?}{=} -2.5$$
$$-5.5 + 3 \stackrel{?}{=} -2.5$$
$$-2.5 = -2.5 \quad \text{True}$$

Interdisciplinary Connection

HOME ECONOMICS Suppose that Maria's cellular phone company charges a $30 monthly fee and $0.40 a minute for each long-distance call. Write an expression for the monthly charges if Maria uses her phone for m minutes. $30 + 0.4m$ If Maria's bill is $52, how many minutes of long-distance calls did she make? **55 minutes**

Enrichment

Both of the following problems contain errors. Have students find the correct answer and describe the error in each problem.

a. $3x + 5 = 2$
$3x + 5 - 5 = 2 - 5$
$3x = 7$
$x = \frac{7}{3}$

b. $\frac{x}{5} - 4 = 6$
$\frac{x}{5} - 4 + 4 = 6 + 4$
$\frac{x}{5} = \frac{10}{5}$
$x = 2$

In Step 3 of **a**, $2 - 5 = -3$, not 7. The solution is $x = -1$. In Step 3 of **b**, 10 should be multiplied by 5, not divided by 5. The solution is $x = 50$.

b.
$5 = \frac{x}{3} - 14$
$5 + 14 = \frac{x}{3} - 14 + 14$ *Add. Prop. Eq.*
$19 = \frac{x}{3}$ *Simplify.*
$3(19) = 3\left(\frac{x}{3}\right)$ *Mult. Prop. Eq.*
$x = 57$

CHECK
$5 \stackrel{?}{=} \frac{57}{3} - 14$
$5 \stackrel{?}{=} 19 - 14$
$5 = 5$ *True*

Use Teaching Transparency 14.

TRY THIS
a. $m = -12$ b. $x = -52$

TRY THIS Solve.
a. $-4.2 = \frac{m}{2} + 1.8$
b. $\frac{x}{-4} - 3 = 10$

ADDITIONAL EXAMPLE 4

Erin buys two types of film, color and black-and-white. The color film costs $3 per roll and the black-and-white film normally costs $5 per roll. She buys 5 rolls of color film and 6 rolls of black-and-white film. Erin is expecting to pay $45 but is surprised to find that her total is $39. She realizes that the black-and-white film is on sale. **How much did she pay for each roll of black-and-white film? Write and solve an equation.**

$15 + 6b = 39$; $b = \$4$ per roll

EXAMPLE 4

APPLICATION
CONSUMER ECONOMICS

Jim bought 6 glasses and 6 mugs. Each mug cost $5.98 and each glass cost $2.99. The salesperson tells Jim that his total bill, before tax, is $60.60. Jim estimates that 1 mug and 1 glass would cost between $8 and $9, so 6 mugs and 6 glasses should cost between $48 and $54. Jim thinks that $60.60 is too high. Jim noticed that the correct price was scanned for the mugs. **What price did the scanner give for each of the glasses?**

SOLUTION

The cost for 6 mugs is $5.98 · 6, or $35.88.

Let *g* represent the cost of one glass. The cost of 6 glasses is represented by the expression 6*g*.

PROBLEM SOLVING   **Write an equation** that models the problem.

cost of mugs + cost of glasses = total bill
     35.88   +   6*g*   =   60.60

Solve for *g*.

$35.88 + 6g = 60.60$ *Given*
$35.88 - 35.88 + 6g = 60.60 - 35.88$ *Subtraction Property of Equality*
$6g = 24.72$ *Simplify.*
$\frac{6g}{6} = \frac{24.72}{6}$ *Division Property of Equality*
$g = 4.12$ *Simplify.*

Jim was charged $4.12 instead of $2.99 for each glass.

Teaching Tip

When solving two-step equations, some students cannot figure out which operation to "undo" first in order to isolate the variable. Review the order of operations. Explain that when "undoing" operations, they should use a backward order of operations. For instance, in the equation $2x + 5 = 20$, they should "undo" addition first because addition comes before multiplication in a *backward* order of operations.

Reteaching the Lesson

GUIDED ANALYSIS Write the following equation on the board or overhead: $3x + 5 = 14$. Working in pairs, have students describe how they would solve the equation. Answers will vary, but many students will say that they would subtract 5 from 14 and divide by 3. Show students that this answer is correct by demonstrating the steps in the proof shown at right.

$3x + 5 = 14$ *Given*
$3x + 5 - 5 = 14 - 5$ *Sub. Prop. of Equality*
$3x = 9$ *Simplify.*
$\frac{3x}{3} = \frac{9}{3}$ *Div. Prop. of Equality*
$x = 3$ *Simplify.*

After each group reads their answers, have them create rules for remembering which operation to perform first. Possible answers are to use a backwards order of operations, or to "undo" the operation of the number "farthest away from" the *x*.

LESSON 3.3   **131**

Teaching Tip

TECHNOLOGY If you are using a TI-82 or TI-83 graphics calculator in Example 4, choose the following window settings:

Xmin = –1   Ymin = –5
Xmax = 9.4   Ymax = 120
Xscl = 1   Yscl = 10

See Keystroke Guide, page 782.

Assess

Selected Answers
Exercises 6–12, 13–89 odd

ASSIGNMENT GUIDE

In Class	1–12
Core	13–21, 31–36, 46–57
Core Plus	22–57
Review	58–89
Preview	90–93

✏ Extra Practice can be found beginning on page 738.

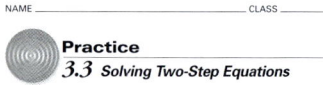

TECHNOLOGY GRAPHICS CALCULATOR

The solution to the beach towels and hat problem presented at the beginning of the lesson (page 129) can be represented by the intersection of two lines. The equations for the lines are based on the two sides of the original equation, $18 + 7t = 74$.

$y = 18 + 7x$ *Replace t with x.*
$y = 74$

Enter the equations in a graphics calculator and graph the lines. The solution for *x* is the value of the *x*-coordinate at the point of intersection. Use the trace or intersection feature to find the point of intersection.

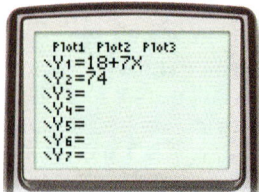 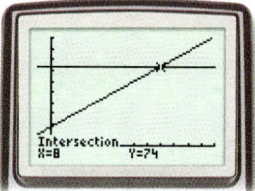

The value of the x-coordinate at the point of intersection is 8.

Exercises

Communicate

Explain the steps needed to solve the following equations:

1. $2x + 1 = 7$
2. $15 - 4m = 10$
3. $\frac{x}{24} + 33 = 84$
4. Explain how to use a table to solve an equation.
5. Explain how to use a graph to solve an equation.

Guided Skills Practice

Solve each equation. (EXAMPLES 1 AND 2)

6. $4p - 13 = 27$ $p = 10$
7. $-2 = 6x + 4$ $x = -1$
8. $15 + 2.5m = 70$ $m = 22$

Solve each equation. (EXAMPLE 3)

9. $\frac{x}{4} - 3 = 21$ $x = 96$
10. $-40 = \frac{g}{-3} - 20$ $g = 60$
11. $\frac{m}{-4} + 5.5 = -3.2$ $m = 34.8$

12. Leo bought 6 glasses and 6 mugs. Leo knows that he was charged $3.99 for each mug and that the total bill before tax was $65.88. Write and solve an equation to determine how much Leo was charged for each glass. (EXAMPLE 4) **$6.99**

Practice and Apply

Solve each equation.

13. $5x + 9 = 39$ $x = 6$
14. $9p + 11 = -7$ $p = -2$
15. $6 - 2d = 42$ $d = -18$
16. $9 - c = -13$ $c = 22$
17. $2m + 5 = 17$ $m = 6$
18. $9p + 20 = -7$ $p = -3$
19. $5x + 9 = 54$ $x = 9$
20. $3 + 2x = 21$ $x = 9$
21. $6 - 8d = -42$ $d = 6$
22. $9 - 14z = 51$ $z = -3$
23. $12 = 9x - 6$ $x = 2$
24. $16 = 5w - 9$ $w = 5$
25. $-4 - 11w = 18$
26. $-7 - 13y = 32$
27. $36 = -3y + 12$ $y = -8$
28. $4y + 3 = 13$ $y = 2.5$
29. $10m + 3.4 = 7$
30. $-3.7 = 2m + 5.1$
31. $127 = 2x + 17$ $x = 55$
32. $4m + 3 = 15$ $m = 3$
33. $-8m - 12 = 20$
34. $47 = 2f - 3$ $f = 25$
35. $5.2 + 1.3x = -1.3$
36. $8p - 15 = 87$
37. $\frac{x}{5} - 2 = 3.7$ $x = 28.5$
38. $4 - \frac{m}{2} = 10$ $m = -12$
39. $15 = \frac{a}{3} - 2$ $a = 51$
40. $10 = \frac{x}{4} + 5$ $x = 20$
41. $\frac{p}{3} - 2 = -56$
42. $45 + \frac{x}{3} = -20$
43. $\frac{z}{5} - 22 = -20$ $z = 10$
44. $\frac{v}{4} + 8 = 11$ $v = 12$
45. $\frac{d}{10} - 1 = -31$
46. $\frac{h}{12} - 5 = -17$
47. $-4 = \frac{x}{2} - 5$ $x = 2$
48. $-45 = \frac{m}{-2} + 4$ $m = 98$
49. $\frac{x}{-5} + 3.2 = 1.4$ $x = 9$
50. $1\frac{1}{2} + \frac{m}{-2} = 3\frac{1}{2}$
51. $\frac{x}{2} - 4\frac{1}{3} = -2\frac{1}{4}$ $x = 4\frac{1}{6}$

Answers:
25. $w = -2$
26. $y = -3$
29. $m = 0.36$
30. $m = -4.4$
33. $m = -4$
35. $x = -5$
36. $p = 12.75$
41. $p = -162$
42. $x = -195$
45. $d = -300$
46. $h = -144$
50. $m = -4$

APPLICATIONS

52. **ECOLOGY** The number of trees planted by the Alpine Nursery in April was 3 more than twice the number of trees planted by the nursery in March. If 71 trees were planted in April, write and solve an equation to find how many trees were planted in March.

53. **CONSUMER ECONOMICS** The $347 selling price for a stereo is $35 more than 3 times the wholesale cost. Write and solve an equation to find the wholesale cost.

54. **COMMUNICATION** A long-distance phone company charges $4.99 per month plus 10¢ per minute. Write and solve an equation to determine how many minutes were used in a month if the monthly bill is $16.99.

55. **CONSUMER ECONOMICS** An online music club sells compact discs for $15.95 each plus $2.95 shipping and handling per order. If Kara's total bill is $98.65, write and solve an equation to determine the number of compact discs Kara purchased.

$15.95 per disc, plus a shipping and handling charge of $2.95 per order.

52. $2m + 3 = 71$
    $m = 34$

53. $3c + 35 = 347$
    $c = 104$

54. $4.99 + 0.1m = 16.99$
    $m = 120$ minutes

55. $15.95d + 2.95 = 98.65$
    $d = 6$

Mid-Chapter Assessment for Lessons 3.1 through 3.3 can be found on page 35 of the *Assessment Resources*.

Technology

Students may find a calculator or computer software helpful in Exercises 13–57. You may wish to have students complete the exercises with paper and pencil first. They can then use a graphics calculator to graph the solutions or use a calculator or software to solve the equations.

Error Analysis

When solving equations such as $5x + 9 = 39$, students may have trouble deciding what operation to perform first to isolate the variable. Some may divide by 5 first and then subtract by 9. Remind them that they need to use a backward order of operations. Also to avoid errors, remind students that neatly writing the steps to an equation becomes more important as the equations become more complex. Have them check their solutions by looking for errors in each step.

Student Technology Guide

3.3 Solving Two-Step Equations

You can use the trace or intersection features of a graphics calculator to solve two-step equations by graphing. For example, you can solve the equation $3x - 4 = 5$ as follows:

1. Enter the expressions $3x - 4$ and 5 in the function list. Use the keystrokes [Y=] 3 [X,T,θ,n] [−] 4 [▼] 5.

2. Press [GRAPH]. Change the window dimensions, if necessary, so that the intersection point is displayed on the graph.

3. To use the trace feature, press [TRACE]. Use the arrow keys, [◄] and [►], to move to the point where the lines intersect. You can also use [▲] and [▼] to move between the two lines that you have graphed.

4. For more accurate results, you can use the intersection feature. To use the intersection feature, press [2nd] [TRACE] 5: intersect [ENTER] [ENTER] [ENTER]. The exact solution is $x = 3$.

Use a graphical approach to solve each equation.

1. $5x - 7 = 3$ $x = 2$
2. $-7x + 18 = 11$ $x = 1$
3. $2.3x - 5.7 = -1.79$ $x = 1.7$
4. $\frac{x}{8} + 5 = 11$ $x = 48$
5. $12 - \frac{x}{7} = 13$ $x = -7$
6. $\frac{x}{11} + 16 = 42$ $x = 286$

Often, the solution to a two-step equation is a fraction. To find the exact fraction that solves $7x + 3 = -2$, first use the intersection feature as described above. You should obtain $x \approx -0.714$. The intersection feature stores this value as the variable X. Now you can find the fractional solution by pressing [2nd] [MODE] [X,T,θ,n] [MATH] [MATH] 1:▶Frac [ENTER] [ENTER].

Use a graphical approach to solve each equation. Express each answer as a fraction.

7. $3x - 7 = -5$ $x = \frac{2}{3}$
8. $11x + 5 = 3$ $x = -\frac{2}{11}$
9. $16 - 5x = 13$ $x = \frac{3}{5}$
10. $18x - 4 = 17$ $x = \frac{7}{6}$
11. $52 - 22x = 17$ $x = \frac{35}{22}$
12. $8x + 11 = -14$ $x = -\frac{25}{8}$

Look Beyond

Exercises 90–93 look ahead to multistep equations, which will be covered in Lesson 3.4. In Exercises 91 and 93, the Distributive Property is used to solve the equations. This will be covered in Lesson 3.5.

56. $8s + 7.50 = 31.50$
$s = 3$
$3.00 per pair

57. $1.5h + 4 = 10$
$h = 4$
The basket holds 4 hekats.

APPLICATION

56. CONSUMER ECONOMICS Mike bought 8 pairs of socks and a T-shirt. His bill was $31.50. Write and solve an equation to determine the price for each pair of socks if the cost of the T-shirt was $7.50.

57. CULTURAL CONNECTION: AFRICA Write and solve an equation to find the answer to this problem from ancient Egypt, written about 3800 years ago. Fill a large basket $1\frac{1}{2}$ times. Then add 4 *hekats* (a *hekat* is about half of a bushel). The total is 10 *hekats*. How many *hekats* does the basket hold?

Look Back

Find each product. *(LESSON 2.4)*

58. $-4(-9)$ **36**   **59.** $-7(-8)$ **56**   **60.** $-13(7)$ **−91**   **61.** $6(-5)$ **−30**
62. $(-5)(-0.5)$ **2.5**   **63.** $(4.2)(-3)$ **−12.6**   **64.** $(-1.5)(-1.5)$ **2.25**   **65.** $(45)(-0.5)$ **−22.5**

Find each quotient. *(LESSON 2.4)*

66. $\frac{27}{-3}$ **−9**   **67.** $-45 \div 9$ **−5**   **68.** $\frac{-42}{-6}$ **7**   **69.** $-96 \div 12$ **−8**
70. $\frac{-30}{-6}$ **5**   **71.** $\frac{40}{-5}$ **−8**   **72.** $-10 \div 2$ **−5**   **73.** $\frac{-55}{11}$ **−5**

Solve each equation and check the solution. *(LESSON 3.1)*

74. $d = 31$
75. $g = 2$
76. $x = -926$
77. $h = 1$
84. $x = -85$
85. $y = 3$

74. $d + 23 = 54$ **75.** $g - 18 = -16$ **76.** $x + 73 = -853$ **77.** $h - \frac{7}{12} = \frac{5}{12}$
78. $x + 14 = 24$ **79.** $5 - m = 2$ **80.** $m + 17 = -3$ **81.** $p - 13 = -3.2$
$x = 10$ $m = 3$ $m = -20$ $p = 9.8$

Solve each equation and check the solution. *(LESSON 3.2)*

82. $3x = 18$ $x = 6$ **83.** $\frac{2}{3}x = 8$ $x = 12$ **84.** $\frac{x}{5} = -17$ **85.** $-17y = -51$
86. $-5x = 20$ **87.** $\frac{m}{-2} = 30$ **88.** $\frac{n}{5} = -15$ **89.** $-13p = 26$
$x = -4$ $m = -60$ $n = -75$ $p = -2$

Look Beyond

Solve each equation.

90. $2x + 3 = 4x - 5$ $x = 4$ **91.** $7(x - 2) = 4x$ $x = \frac{14}{3}$ **92.** $7x - 2 = x + 16$ $x = 3$
93. $3(x - 3) + 10 = 4x - 3x + 8$ $x = \frac{7}{2}$

3.4 Solving Multistep Equations

Why Solutions to real-world problems often have multiple steps. These steps involve various mathematical properties.

Objective
- Write and solve multistep equations.

APPLICATION: BUSINESS

Mateo is the office manager at a small company and regularly orders office supplies. Two different suppliers offer the same box of pens at the same price per box. However, supplier A will give a $2 discount on 4 boxes of pens. For the same total cost as supplier A, supplier B will sell 3 boxes of pens and charge $4 shipping.

Let x equal the price in dollars for a box of pens. The expression $4x - 2$ models the total cost for a purchase from supplier A and $3x + 4$ models the total cost for a purchase from supplier B. To find the price, x, of a single box of pens, set the expressions equal to each other and solve for x.

$$4x - 2 = 3x + 4$$

Activity: Modeling Equations

You will need: algebra tiles

1. Use algebra tiles to model the equation $4x - 2 = 3x + 4$.

$4x - 2 \qquad = \qquad 3x + 4$

2. How can you get all of the x-tiles on the left side? Do this with your algebra tiles.

3. Describe your model for Step 2 by completing the equation below.
$$4x - 2 = 3x + 4$$
$$4x - 2 - (?) = 3x + 4 - (?)$$
$$(?) - 2 = 4$$

CHECKPOINT ✓ 4. You now have an equation with a variable on only one side. Finish solving the equation by using tiles, and describe the step(s) you used.

Alternative Teaching Strategy

USING VISUAL MODELS Draw a balance scale on the board or overhead. Remind students that a scale like this is balanced when there are equal weights on each side. Tell students that you can model the equation $x + 6 = 2x + 4$ on the scale by placing a 6-ounce weight plus an unknown weight on one side and a 4-ounce weight and two unknown weights on the other side. Ask students what the unknown weight must be in order to balance the scale. **2 ounces** Show students that if they add or subtract an equal amount from each side of the scale, the scale remains balanced. Demonstrate this by subtracting one unknown weight and one 4-ounce weight from each side.

Prepare

NCTM PRINCIPLES & STANDARDS 1–3, 5–10

LESSON RESOURCES
- Practice 3.4
- Reteaching Master 3.4
- Enrichment Master 3.4
- Cooperative-Learning Activity 3.4
- Lesson Activity 3.4
- Problem Solving/Critical Thinking Master 3.4
- Teaching Transparency 15

QUICK WARM-UP

Solve for x.

1. $3 + 2x = 11$		4
2. $13x - 7 = 19$		2
3. $21 - 5x = 36$		−3
4. $3x + 6 = 18$		4
5. $9x + 8 = 45$		4.11
6. $6x + 17 = 28$		1.83

Also on Quiz Transparency 3.4

Teach

Why Ask students to name the properties of equality needed to solve $4m - 8 = 16$. (**Addition Property of Equality and Division Property of Equality**) Remind students that like terms can be added and subtracted, as in $4m - m = 3m$. Ask students what steps they think are necessary to isolate the variable m in the equation $4m - 8 = 16 + m$. **Subtract m from both sides, add 8 to both sides, and divide by 4 on both sides.**

☞ For Activity Notes and the answer to the Checkpoint, see page 136.

LESSON 3.4 **135**

Activity Notes

In this Activity, students use algebra tiles to solve multi-step equations. By using the tiles, they discover how to isolate the variable on one side of the equal sign. As they solve the equation with tiles, they should be able to complete the same steps written on paper.

Cooperative Learning

Students can work in groups of two or three, depending on the availability of algebra tiles for the classroom. While one or two students manipulate the tiles to solve the equation, the remaining student in the group can write the corresponding steps on paper.

CHECKPOINT ✔

4. Add 2 positive unit tiles to each side, and remove neutral pairs. This leaves 1 *x*-tile on the left side and 6 positive unit tiles on the right side. $x = 6$

ADDITIONAL EXAMPLE 1

Solve $2x + 5 = 6x + 4$. Check the solution. $x = 0.25$

ADDITIONAL EXAMPLE 2

Solve $-10z + 17 = 21 - 6z$. Check the solution. $z = -1$

TRY THIS
a. $x = 10$ b. $m = -3$

Use Teaching Transparency 15.

136   LESSON 3.4

Solving the equation presented on page 135 involves several steps, including isolating the variable on one side of the equation.

EXAMPLE 1   Solve $4x - 2 = 3x + 4$ to find the cost of one box of pens. Check the solution.

APPLICATION
BUSINESS

SOLUTION
Begin by subtracting $3x$ from both sides in order to isolate the variable on the left side of the equation.

$$4x - 2 = 3x + 4 \quad \textit{Given}$$
$$4x - 2 - 3x = 3x + 4 - 3x \quad \textit{Subtraction Property of Equality}$$
$$x - 2 = 4 \quad \textit{Simplify}$$
$$x - 2 + 2 = 4 + 2 \quad \textit{Addition Property of Equality}$$
$$x = 6 \quad \textit{Addition Property of Equality}$$

CHECK
$4(6) - 2 \stackrel{?}{=} (6) + 4$
$24 - 2 \stackrel{?}{=} 18 + 4$
$22 = 22$ *True*

The cost of one box of pens is $6.

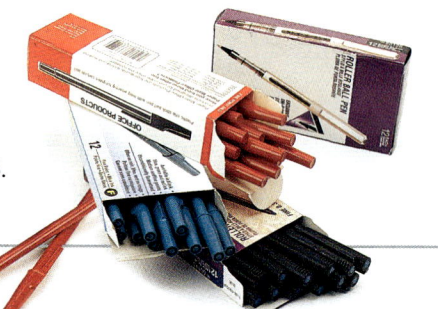

EXAMPLE 2   Solve $5x - 6 = -6x + 16$. Check the solution.

SOLUTION
Begin by adding $6x$ to each side.

$$5x - 6 = -6x + 16 \quad \textit{Given}$$
$$5x - 6 + 6x = -6x + 16 + 6x \quad \textit{Addition Property of Equality}$$
$$11x - 6 = 16 \quad \textit{Simplify.}$$
$$11x - 6 + 6 = 16 + 6 \quad \textit{Addition Property of Equality}$$
$$11x = 22 \quad \textit{Simplify.}$$
$$\frac{11x}{11} = \frac{22}{11} \quad \textit{Division Property of Equality}$$
$$x = 2$$

CHECK
$5(2) - 6 \stackrel{?}{=} -6(2) + 16$
$4 = 4$ *True*

TRY THIS   Solve. Check the solution.
a. $5x - 7 = 4x + 3$ b. $-3m + 14 = 5m + 38$

Interdisciplinary Connection

PHYSICS   A body is in *uniformly accelerated motion* when it moves in a straight line at a constant acceleration. The speed of such an object at the end of a specified time interval can be found by using the equation $v = v_o + at$. In this equation, v represents the final speed of the object, v_o represents its initial speed, a represents a constant acceleration, and t represents the time interval.

Suppose that a ball is positioned at the top of an incline. It is then set in motion at an initial speed of 2.0 meters per second. The ball rolls down the incline with a constant acceleration of 4.2 meters per second squared. Its final speed at the bottom of the incline is 23 meters per second. How many seconds did it take the ball to roll down the incline? **5 seconds**

TECHNOLOGY GRAPHICS CALCULATOR

The solution to the equation in Example 2 can be represented by the intersection of two lines. The solution for x is the value of the x-coordinate at the point of intersection. The two lines are represented by the following equations:

$$y = 5x - 6$$
$$y = -6x + 16$$

Enter the equations in a graphics calculator and graph the lines. Use the trace or intersection feature to find the point of intersection. The value of the x-coordinate at the point of intersection is 2.

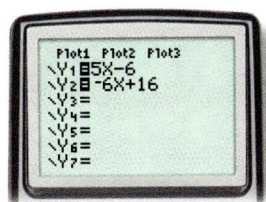
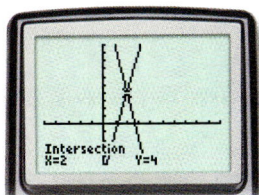

Teaching Tip

TECHNOLOGY If you are using a TI-82 or TI-83 graphics calculator in Example 2, choose the following window settings:

Xmin=-5   Ymin=-5
Xmax=10   Ymax=10
Xscl=1   Yscl=1

Press [Y=] to enter the **Y=** editor. Clear any existing equations and then enter **5x–6** next to **Y1=** and **–6x+16** next to **Y2=**. Press [GRAPH]. Press [2nd] [TRACE] to access the **CALCULATE** menu. Choose **5:intersect** and then press [ENTER] [ENTER] [ENTER].

See Keystroke Guide, page 782.

EXAMPLE 3 Solve $x - \frac{1}{4} = \frac{x}{3} + 7\frac{3}{4}$.

● **SOLUTION**

Clear the equation of fractions by changing the mixed number to a fraction and then multiplying both sides by the least common denominator, 12. Then solve for x.

$x - \frac{1}{4} = \frac{x}{3} + \frac{31}{4}$	Given
$12\left(x - \frac{1}{4}\right) = 12\left(\frac{x}{3} + \frac{31}{4}\right)$	Multiplication Property of Equality
$12x - 12\left(\frac{1}{4}\right) = 12\left(\frac{x}{3}\right) + 12\left(\frac{31}{4}\right)$	Distributive Property
$12x - 3 = 4x + 93$	Simplify.
$12x - 3 - 4x = 4x + 93 - 4x$	Subtraction Property of Equality
$8x - 3 = 93$	Simplify.
$8x - 3 + 3 = 93 + 3$	Addition Property of Equality
$8x = 96$	Simplify.
$\frac{8x}{8} = \frac{96}{8}$	Division Property of Equality
$x = 12$	Simplify.

Check the result by substituting 12 for x in the original equation.

$$12 - \frac{1}{4} \stackrel{?}{=} \frac{12}{3} + 7\frac{3}{4}$$
$$11\frac{3}{4} = 11\frac{3}{4} \quad \text{True}$$

TRY THIS Solve $x - 2 = \frac{x}{5} + 6$.

ADDITIONAL EXAMPLE 3

Solve $\frac{x}{8} + \frac{3}{4} = 6 + \frac{x}{2}$.

$x = -14$

TRY THIS
$x = 10$

Enrichment

Have students work in pairs. Give them the following set of equations to solve:

a. $\frac{x}{4} = 20$ **b.** $\frac{4}{x} = 20$ **c.** $\frac{4}{x+1} = 20$

The process of solving the equation in **a** should be familiar. Tell students to extend this process to devise a method for solving the equations in **b** and **c**. **a.** $x = 80$; **b.** $x = 0.2$; **c.** $x = -0.8$ Now have each student create a similar set of equations and write them on a sheet of paper. Partners should then exchange papers and solve each other's equations.

Inclusion Strategies

KINESTHETIC LEARNERS Kinesthetic learners may appreciate the opportunity to work more complicated problems on the board. Once they work the problems they could copy them onto their paper. The ability to stand, move around, and even talk out loud as they work could help them think as they find the solutions.

ADDITIONAL EXAMPLE 4

Michael's test scores are 87, 86, 91, and 90. **What does he need to score on a fifth test to have an average of 90?** 96

TRY THIS
$x = 40$

Assess

Selected Answers
Exercises 6–13, 15–49 odd

ASSIGNMENT GUIDE

In Class	1–13
Core	14–21, 28–31, 34–36
Core Plus	22–38
Review	39–50
Preview	51–54

Extra Practice can be found beginning on page 738.

EXAMPLE 4 Heather is studying for her third test in history. She scored 79 and 88 on her first two tests. **What score does Heather need on her third test so that her average will be exactly 85?**

SOLUTION

Let s represent the score Heather needs on her third test. The average of her three tests can be modeled by $\frac{79 + 88 + s}{3}$.

$$\frac{79 + 88 + s}{3} = 85 \qquad \text{Given}$$

$$\frac{167 + s}{3} = 85 \qquad \text{Simplify.}$$

$$3\left(\frac{167 + s}{3}\right) = 3(85) \qquad \text{Multiplication Property of Equality}$$

$$167 + s = 255 \qquad \text{Simplify.}$$

$$167 + s - 167 = 255 - 167 \qquad \text{Subtraction Property of Equality}$$

$$s = 88 \qquad \text{Simplify.}$$

Heather needs to score 88 on her third test to have an average of exactly 85.

CHECK

$$\frac{79 + 88 + 88}{3} \stackrel{?}{=} 85$$

$$85 = 85 \qquad \text{True}$$

TRY THIS Solve $\frac{90 + 80 + x}{3} = 70$.

Exercises

Communicate

Explain the steps needed to solve the following equations:

1. $23p + 57 = p + 984$
2. $5z - 5 = -2z + 19$
3. $3 - 2x = 4x + 15$
4. Explain how to solve $3x + 6 = 2x + 4$ by using a graph.

CONNECTION

5. **STATISTICS** Brian scored 85 on each of his first two algebra tests. Explain how to determine the score that Brian needs on his third test in order to have an average of exactly 90.

Guided Skills Practice

Solve each equation and check the solution. (EXAMPLES 1 AND 2)

6. $2b + 4 = -8$ $b = -6$
7. $3 - 7h = 10$ $h = -1$
8. $f = -6 + 2f$ $f = 6$
9. $4x - 1 = 2x + 5$ $x = 3$

Solve each equation and check the solution. (EXAMPLE 3)

10. $\frac{2}{3} - 3x = \frac{1}{3} + 2x$ $x = \frac{1}{15}$
11. $\frac{1}{6} + 3n = \frac{7}{24} + n$ $n = \frac{1}{16}$
12. $\frac{3a}{4} + \frac{7}{6} = -a + \frac{7}{24}$ $a = -\frac{1}{2}$

Reteaching the Lesson

USING MODELS Write the following equation on the board or overhead:

$$3w - 8 = 5w + 7$$

Ask students to describe how they would proceed in solving the equation. Elicit a step-by-step procedure from them until you arrive at the solution, $w = -7.5$. As each step is completed, work with students to create a general guideline that they can follow to solve any multistep equation. A sample set of guidelines is given at right.

How to Solve Multistep Equations
Step 1: Simplify by combining like terms.
Step 2: "Undo" any additions or subtractions.
Step 3: "Undo" any multiplications or divisions.

Have students use the guidelines they created to solve the following equations:

1. $6 = 2j - 2$ $j = 4$
2. $4r + 9 - r = 9r$ $r = 1.5$
3. $7p - 2 = 13 + 4p$ $p = 5$

Practice

3.4 Solving Multistep Equations

Solve each equation.

1. $4x + 7 = 3x + 18$ $x = 11$
2. $5y - 5 = 7y - 3$ $y = -1$
3. $4a - 6 = -2a + 14$ $a = 3\frac{1}{3}$
4. $4m - 5 = 3m + 7$ $m = 12$
5. $5x - 7 = 3x + 2$ $x = 4\frac{1}{2}$
6. $10y + 10 = 4 - 4y$ $y = -\frac{3}{7}$
7. $13 - 8v = 5v + 2$ $v = \frac{11}{13}$
8. $7 - 5y = 4y - 2$ $y = 1$
9. $2 + 3y = 4y - 1$ $y = 3$
10. $-7 - 3z = 8 + 2z$ $z = -3$
11. $7w - 19 = 5w - 5$ $w = 7$
12. $28 + 2a = 5a + 7$ $a = 7$
13. $5x + 32 = 8 - x$ $x = -4$
14. $m - 12 = 3m + 4$ $m = -8$
15. $2(x + 1) = 3x - 3$ $x = 5$
16. $5(3x + 5) = 4x - 8$ $x = -3$
17. $2r - 4 = 2(6 - 7r)$ $r = 1$
18. $8y - 3 = 5(2y + 1)$ $y = -4$
19. $2z - 5(z + 2) = -8 - 2z$ $z = -2$
20. $5t - 2(5 + 4t) = 3 + t - 8$ $t = -1\frac{1}{4}$
21. $15n + 25 = 2(n - 7)$ $n = -3$
22. $4y + 2 = 3(6 - 4y)$ $y = 1$
23. $2(3x - 1) = 3(x + 2)$ $x = 2\frac{2}{3}$
24. $9y - 8 + 4y = 7y + 16$ $y = 4$
25. $14d - 22 + 5d = 12d - 8$ $d = 2$
26. $23x + 34 = 23 - 12x + 7x$ $x = -\frac{11}{28}$
27. $29 - 3s = 23(2s - 3)$ $s = 2$
28. $12 - 5(2w - 13) = 3(2w - 5)$ $w = 5\frac{3}{4}$
29. $8 + 5(3q - 4) = 7(q - 12)$ $q = -9$
30. $2(y + 2) + y = 19 - (2y + 3)$ $y = 2.4$
31. $0.3w - 4 = 0.8 - 0.2w$ $w = 9.6$
32. $2.1z = 1.2z - 9$ $z = -10$
33. $12 + 2.1w = 1.3w$ $w = -15$
34. $3.5(j + 4) = 1.4(16 + j)$ $j = 4$
35. $4.5 - 1.9m = 20.1 - 2m$ $m = 156$
36. $x - 0.09 = 2.22 - 0.1x$ $x = 2.1$
37. $\frac{1}{2}x + 7 = \frac{3}{4}x - 4$ $x = 44$
38. $\frac{1}{4}y = \frac{2}{5}y - 1$ $y = 6\frac{2}{3}$
39. $\frac{1}{2}z = 3z - \frac{4}{5}$ $z = \frac{8}{25}$
40. $\frac{a}{2} - \frac{1}{3} = \frac{a}{3} - \frac{1}{2}$ $a = -1$
41. $2\left(\frac{1}{3}w + \frac{1}{4}\right) = 4 + \frac{1}{3}w$ $w = 10\frac{1}{2}$
42. $\frac{1}{4}(7 + 3r) = -\frac{1}{8}r$ $r = -2$

138 LESSON 3.4

CONNECTION

13. STATISTICS Steve is about to take his fifth math test. His scores on the first four tests were 84, 92, 71, and 94. What score does Steve need on his fifth test in order to have an average of exactly 85? **(EXAMPLE 4)** 84

Practice and Apply

Solve and check each equation.

14. $2x - 2 = 4x + 6$ $x = -4$
15. $3x + 5 = 2x + 2$ $x = -3$
16. $4x + 3 = 5x - 4$ $x = 7$
17. $2x - 5 = 4x - 1$ $x = -2$
18. $5x + 24 = 2x + 15$ $x = -3$
19. $5y - 10 = 14 - 3y$ $y = 3$
20. $12 - 6z = 10 - 5z$ $z = 2$
21. $5m - 7 = -6m - 29$ $m = -2$
22. $-10x + 3 = -3x + 12 - 4x$ $x = -3$
23. $6p - 12 = -4p + 18$ $p = 3$
24. $1.8x + 2.8 = 2.5x + 2.1$ $x = 1$
25. $2.6h + 18 = 2.4h + 22$ $h = 20$
26. $5h - 7 = 2h + 2$ $h = 3$
27. $4n + 1 = 12 + 5n$ $n = -11$
28. $1 - 3x = 2x + 8$ $x = -\frac{7}{5}$
29. $3a - 8 = \frac{a}{2} + 2$ $a = 4$
30. $\frac{w}{2} + 7 = \frac{w}{3} + 9$ $w = 12$
31. $6 - \frac{t}{4} = 8 + \frac{t}{2}$ $t = -\frac{8}{3}$

CHALLENGE

32. $x + \frac{5}{8} + \frac{3x}{4} = \frac{2}{3} + 5x$ $x = -\frac{1}{78}$

CONNECTIONS

33. GEOMETRY The area of a rectangle increased by 5 is the same as 3 less than twice the area of the rectangle. Write and solve an equation to determine the area of the rectangle.

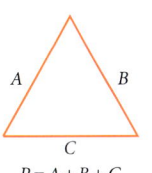

34. GEOMETRY The perimeter of a triangle increased by 3 is the same as 35 decreased by 7 times the perimeter. Write and solve an equation to determine the perimeter of the triangle.

$P = A + B + C$

35. STATISTICS In her first five basketball games, Rachel scored 27, 18, 27, 32, and 21 points. Write and solve an equation to determine how many points she must score in the sixth game in order to have an average of exactly 25 points per game.

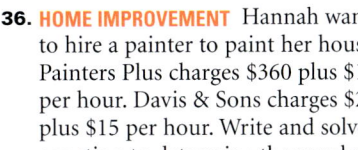

APPLICATION

36. HOME IMPROVEMENT Hannah wants to hire a painter to paint her house. Painters Plus charges $360 plus $12 per hour. Davis & Sons charges $279 plus $15 per hour. Write and solve an equation to determine the number of hours for which the two costs would be the same.

Technology
Students may find a calculator or computer software helpful in Exercises 14–38. You may wish to have students complete the exercises with paper and pencil first. They can then use a graphics calculator to graph the solutions or use a calculator or software to check their work.

Error Analysis
In an equation such as $3n + 10 = 2n$, many students will add $3n$ and $2n$ to arrive at $10 = 5n$. Remind them that they can combine like terms only within an expression on one side of the equal sign. To combine $3n$ and $2n$, remind them to subtract one of the terms from both sides of the equal sign.

33. $a + 5 = 2a - 3$; area = 8

34. $p + 3 = 35 - 7p$; perimeter = 4

35. $\dfrac{27 + 18 + 27 + 32 + 21 + x}{6} = 25$
  $x = 25$
Rachel must score 25 points in the sixth game.

36. $12x + 360 = 15x + 279$
  $x = 27$
The cost is the same at 27 hours.

LESSON 3.4 **139**

Look Beyond

In Exercises 51–54, students investigate some simple rational equations. They are asked to solve the equations first by graphing and then by devising a symbolic method. Rational equations will be studied in depth in Lesson 11.5.

ALTERNATIVE Assessment

Portfolio Activity

The Portfolio Activity can be used as preparation for the Chapter Project or as a separate activity. In the Portfolio Activity on this page, students use a multistep equation to identify a *break-even point*. The concept of the break-even point is important in determining the minimum amount of sales necessary for a business to make a profit.

Answers to Portfolio Activities can be found in Additional Answers of the Teacher's Edition.

APPLICATIONS

37. **TEMPERATURE** If a temperature is increased by 80 degrees, it is the same as 6 times the temperature. Write and solve an equation to determine the temperature.

38. **GARDENING** Jake and Sergio are planting tulip bulbs. Jake has planted 60 bulbs and is planting at a rate of 44 bulbs per hour. Sergio has planted 96 bulbs and is planting at a rate of 32 bulbs per hour. In how many hours will Jake and Sergio have planted the same number of bulbs? What number of bulbs will this be? Write and solve an equation to determine the answer.

Look Back

Solve each equation and check the solution. *(LESSON 3.1)*

39. $x - 3 = 4$ 40. $y + 17 = 2$ 41. $-3 = m + 12$ 42. $5 - x = 10$
     $x = 7$ $y = -15$ $m = -15$ $x = -5$

Solve each equation and check the solution. *(LESSON 3.2)*

43. $3m = 24$ 44. $\dfrac{x}{-3} = 5$ 45. $35 = -3.5x$ 46. $-8.2 = \dfrac{m}{-2}$
     $m = 8$ $x = -15$ $x = -10$ $m = 16.4$

Solve each equation and check the solution. *(LESSON 3.3)*

47. $4x + 3 = 20$ 48. $\dfrac{2}{3}x - 4 = 12$ 49. $-2.3 = 3y + 0.7$ 50. $15 = \dfrac{m}{8} + 12$
     $x = 4.25$ $x = 24$ $y = -1$ $m = 24$

Look Beyond

Solve $\dfrac{1}{x} = \dfrac{2}{x-2}$ by using each method.

51. graphing $x = -2$ 52. algebra $x = -2$

Solve $\dfrac{6}{x+2} = \dfrac{3}{x-2}$ by using each method.

53. graphing $x = 6$ 54. algebra $x = 6$

Include this activity in your portfolio.

Refer to the T-shirt problem in the Portfolio Activity on page 128. Suppose that the track team sells the T-shirts for $7.50.

1. If t is the number of T-shirts sold, explain why the equation $125 + 4.55t = 7.50t$ represents the point at which the income from the sales is equal to the cost of making the T-shirts.

2. Solve the equation in Step 1.

3. Suppose that the T-shirts sell for $10. What is the new value of t?

WORKING ON THE CHAPTER PROJECT

You should now be able to complete Activity 2 of the Chapter Project on page 154.

37. $6t = t + 80$
     $t = 16$
38. $60 + 44h = 96 + 32h$
     $h = 3$
     3 hours, 192 bulbs each

3.5 Using the Distributive Property

Why The Distributive Property can be very useful when solving algebra problems.

Objectives

- Use the Distributive Property to solve equations.
- Solve real-world problems by using multistep equations.

**APPLICATION
CONSUMER ECONOMICS**

Paul buys 3 booklets of bus passes at the student discount of $4 off each booklet for a total of $48. How can you find the original price for each of the booklets?

To find the original price of the bus-pass booklet, write an equation. Let x represent the original price of the booklet in dollars. Then $x - 4$ represents the discounted price that Paul paid for each booklet.

Since Paul bought 3 booklets, the following equation models the situation:

$$3(x - 4) = 48$$

Activity
Using a Table to Solve for x

You can solve the problem presented above by building a table of values for the equation.

Original price, x	$3(x-4)$	Total ($)
5	$3(5-4)$	3
10	$3(10-4)$	?
15	?	?
20	?	?
25	?	?

1. Copy and complete the table.
2. Explain how to determine the original cost of each booklet when the total is $48.

CHECKPOINT ✓ 3. Describe how to solve $3(x - 4) = 48$ by using the Distributive Property.

Alternative Teaching Strategy

USING VISUAL MODELS Students can use algebra tiles to model the Distributive Property. If tiles are not available, students may draw tiles on their paper. To model the equation $3(x - 4) = 12$, first remind students how to model the expression $x - 4$. Repeat this three times to model $3(x - 4)$.

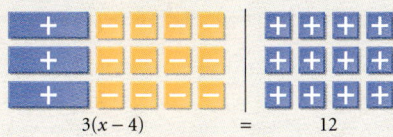

$3(x - 4) \qquad = \qquad 12$

Next combine the tiles on the left side by using the Distributive Property to get $3x - 12 = 12$. Solve the equation by adding 12 tiles to both sides of the equation and dividing by 3.

$x = 8$

The solution is $x = 8$. Ask students if they see any other ways to solve the equation. **Possible answer: First divide both sides by 3 to get $x - 4 = 4$. Add 4 to both sides to get $x = 8$.**

Prepare

NCTM PRINCIPLES & STANDARDS 1–3, 5–10

LESSON RESOURCES

- Practice 3.5
- Reteaching Master 3.5
- Enrichment Master 3.5
- Cooperative-Learning Activity 3.5
- Lesson Activity 3.5
- Problem Solving/Critical Thinking Master 3.5

QUICK WARM-UP

Solve for x.

1. $2 + 3x = 11$ 3
2. $6x - 7 = 11$ 3
3. $21 - 5x = 16$ 1
4. $\frac{x}{6} + 3 = 4$ 6
5. $\frac{x}{9} - 3 = 6$ 81

Also on Quiz Transparency 3.5

Teach

Why Play a game like the following with students: Subtract 7 from your age, double that number, add 14, and divide the sum by 2. The answer is the age of the student. Show students that this series of steps can be represented by $\frac{2(x-7) + 14}{2} = x$ and is true for any value of x. Explain that many real-world problems can be represented by expressions and equations that involve the Distributive Property.

☞ For Activity Notes and the answer to the Checkpoint, see page 142.

LESSON 3.5 **141**

Activity Notes

In this Activity, students use a table and the guess-and-check method to solve a multistep equation. Students should observe the table and intuitively determine the steps for solving the given equation. They should discover that they should use the Distributive Property first when solving the equation.

Cooperative Learning

Have students do the Activity in pairs. Both partners should copy and complete the table on their paper. Then they can compare answers and resolve any differences. Students should work together to complete Steps 2 and 3.

CHECKPOINT ✔

3. First distribute 3 over the expression $(x - 4)$. Add 12 to both sides and then divide both sides by 3 to get the result, $x = 20$.

ADDITIONAL EXAMPLE 1

Solve $4(x + 1) = 16$. $x = 3$

CRITICAL THINKING
First divide both sides by 3. Then add 4 to both sides to get $x = 20$.

ADDITIONAL EXAMPLE 2

Solve.

a. $2(6 + d) = -8$ $d = -10$

b. $5(t - 7) = 3$ $t = 7.6$

You can solve the problem presented on page 141 algebraically. Start by using the Distributive Property.

EXAMPLE 1 Solve the equation from page 141, $3(x - 4) = 48$, to find the original price of each of the booklets that Paul bought.

● **SOLUTION**

To solve an equation that has a variable inside parentheses, you can use the Distributive Property to remove the parentheses.

$3(x - 4) = 48$	*Given*
$3x - 12 = 48$	*Distributive Property*
$3x - 12 + 12 = 48 + 12$	*Addition Property of Equality*
$3x = 60$	*Simplify.*
$\frac{3x}{3} = \frac{60}{3}$	*Division Property of Equality*
$x = 20$	*Simplify.*

CHECK

$3(20 - 4) \stackrel{?}{=} 48$

$3(16) \stackrel{?}{=} 48$

$48 = 48$ *True*

The original price of each booklet is $20.

CRITICAL THINKING Explain how the equation in Example 1 could be solved algebraically without using the Distributive Property.

EXAMPLE 2 Solve each equation. Check the solution.

a. $3y - 8 - y = 6$ b. $2(x + 5) = -16$ c. $4x - 8(x + 1) = 8$

● **SOLUTION**

a.
$3y - 8 - y = 6$	*Given*
$2y - 8 = 6$	*Combine like terms.*
$2y - 8 + 8 = 6 + 8$	*Addition Property of Equality*
$2y = 14$	*Simplify.*
$\frac{2y}{2} = \frac{14}{2}$	*Division Property of Equality*
$y = 7$	*Simplify.*

CHECK

$3(7) - 8 - 7 \stackrel{?}{=} 6$ *True*

$6 = 6$

b.
$2(x + 5) = -16$	*Given*
$2x + 10 = -16$	*Distributive Property*
$2x + 10 - 10 = -16 - 10$	*Subtraction Property of Equality*
$2x = -26$	*Simplify.*
$\frac{2x}{2} = \frac{-26}{2}$	*Division Property of Equality*
$x = -13$	*Simplify.*

Inclusion Strategies

TECHNOLOGY Some students may be motivated to do their in-class assignments by getting the chance to use technology to check their work. After working a predetermined number of problems, allow students to check the problems by using a calculator or computer software. Students may graph the solutions on a graphics calculator or use a calculator or computer software that solves equations.

Enrichment

Both of the following problems contain errors. Have students find the correct answer and describe the error in each problem.

a. $4(x + 5) = 9$ b. $5(x + 6) = 7(x + 2)$
   $4x + 5 = 9$ $5x + 30 = 7x + 2$
   $4x = 4$ $-2x = -28$
   $x = 1$ $x = 14$

In Step 2 of part **a**, the distribution is not complete. It should be $4x + 20 = 9$. The solution is $x = -2.75$. In Step 2 of part **b**, the distribution is not complete. $7x + 2$ should be $7x + 14$. The solution is $x = 8$.

142 LESSON 3.5

CHECK
$$2(-13 + 5) \stackrel{?}{=} -16$$
$$-16 = -16 \quad \text{True}$$

c. $4x - 8(x + 1) = 8$ — *Given*
$4x - 8x - 8 = 8$ — *Distributive Property*
$-4x - 8 = 8$ — *Combine like terms.*
$-4x - 8 + 8 = 8 + 8$ — *Addition Property of Equality*
$-4x = 16$ — *Simplify.*
$\frac{-4x}{-4} = \frac{16}{-4}$ — *Division Property of Equality*
$x = -4$ — *Simplify.*

CHECK
$$4(-4) - 8(-4 + 1) \stackrel{?}{=} 8$$
$$8 = 8 \quad \text{True}$$

TRY THIS Solve each equation.
a. $4t + 7 - t = 19$ **b.** $5(p - 2) = -15$ **c.** $3m + 2(4m - 6) = 10$

Solving an equation sometimes results in a special situation, as illustrated in Example 3.

EXAMPLE 3 Solve each equation.
a. $2(x + 4) - 5 = 2x + 3$ **b.** $4m + 1 - m = 3m + 9$

SOLUTION
a. $2(x + 4) - 5 = 2x + 3$
$2x + 8 - 5 = 2x + 3$ — *Distributive Property*
$2x + 3 = 2x + 3$ — *Simplify.*
$2x + 3 - 2x = 2x + 3 - 2x$ — *Subtraction Property of Equality*
$3 = 3$ — *True*

All real numbers are solutions to this equation because the simplified form is always true.

b. $4m + 1 - m = 3m + 9$
$3m + 1 = 3m + 9$ — *Combine like terms.*
$3m + 1 - 3m = 3m + 9 - 3m$ — *Subtraction Property of Equality*
$1 = 9$ — *False*

There are no solutions to this equation because the simplified form is always false.

TRY THIS Solve each equation.
a. $3(y - 2) + 4 = 3y - 2$ **b.** $7s + 3 - 2s = 5s + 6$

TRY THIS
a. $t = 4$ **b.** $p = -1$ **c.** $m = 2$

ADDITIONAL EXAMPLE 3

Solve each equation.

a. $4m - 1 = 2(2m - 3) + 5$
all real numbers

b. $6(m + 2) = 4m + 7 + 2m$
no solution

Teaching Tip
Students may make errors because they do not complete the distribution, such as simplifying $2(x + 3)$ to $2x + 3$. These students could use tiles to help them see that $2(x + 3)$ is equal to $x + x + 3 + 3$, or $2x + 6$ and that $3(x + 3)$ is the same as $x + x + x + 3 + 3 + 3$, or $3x + 9$.

TRY THIS
a. all real numbers
b. no solution

Reteaching the Lesson

INVITING PARTICIPATION Playing the "pass it back" game is a fun way for students to review what they learned. In this game, the teacher arranges students' desks in rows of 4 or 5. The first person in each row is given an equation to solve. The solution must require at least 4 or 5 steps, depending on the number of people in each row. The first person performs only the first step and passes his paper to the person behind him. The second person performs the second step and passes the paper back, and so on. Ideally, the last person should provide the answer and run the paper up to the teacher. The first team to get the correct answer to the teacher wins. If additional steps are required, the last person in the row should give the paper to the person in front of him, and the team should continue passing it forward until all of the steps are complete and there is an answer.

ADDITIONAL EXAMPLE 4

Solve the equation
$5x - 2(3x + 7) = 7x + 12$.
$x = -3.25$

TRY THIS
$y = -5$

Assess

Selected Answers
Exercises 5–11, 13–83 odd

ASSIGNMENT GUIDE

In Class	1–11
Core	12–24, 33–47 odd
Core Plus	12–48 even
Review	49–83
Preview	84–87

✎ Extra Practice can be found beginning on page 738.

EXAMPLE 4

Solve $4x - 3(2x + 4) = 8x - 25$.

SOLUTION

$4x - 3(2x + 4) = 8x - 25$	Given
$4x - 3(2x) + (-3)(4) = 8x - 25$	Distributive Property
$4x - 6x - 12 = 8x - 25$	Simplify.
$-2x - 12 = 8x - 25$	Simplify.
$-2x - 12 - 8x = 8x - 25 - 8x$	Subtraction Property of Equality
$-10x - 12 = -25$	Simplify.
$-10x - 12 + 12 = -25 + 12$	Addition Property of Equality
$-10x = -13$	Simplify.
$\frac{-10x}{-10} = \frac{-13}{-10}$	Division Property of Equality
$x = 1.3$	Simplify.

Check by substituting 1.3 for x in the original equation.

$4x - 3(2x + 4) = 8x - 25$
$4(1.3) - 3[2(1.3) + 4] \stackrel{?}{=} 8(1.3) - 25$
$5.2 - 3(6.6) \stackrel{?}{=} 10.4 - 25$
$5.2 - 19.8 \stackrel{?}{=} 10.4 - 25$
$-14.6 = -1.46$ True

TRY THIS   Solve $4y - 7(y + 6) = 5y - 2$.

Exercises

Communicate

Explain the steps needed to solve the following equations:

1. $2(x - 4) = 11$ **2.** $3(x + 1) + x = 5x$ **3.** $4(x - 4) + 3 = 2x - 1$

4. Describe how to solve $4(x + 4) = 108$ by using a table.

Guided Skills Practice

Solve each equation. *(EXAMPLES 1 AND 2)*

5. $5s + 7 - 2s = 16$ $s = 3$ **6.** $4(f + 2) = 20$ $f = 3$ **7.** $5r - 3(r + 6) = 24$
$r = 21$

Solve each equation. *(EXAMPLE 3)*

8. All real numbers
9. No solution

8. $4(m + 3) + 7 = 4m + 19$ **9.** $3p - 7 + 2p = 5p + 12$

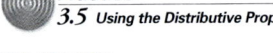

Practice
3.5 Using the Distributive Property

Solve each equation.

1. $2(a + 2) = 2$ $a = -1$
2. $3(b - 3) = 3$ $b = 4$
3. $4(c - 4) = 4$ $c = 5$
4. $5(5 - d) = -5$ $d = 6$
5. $6(6 - e) = 6$ $e = 5$
6. $-7(f + 7) = 7$ $f = -8$
7. $-8(8 - g) = 8$ $g = 9$
8. $1 + 4(2h + 1) = -35$ $h = -5$
9. $3(5i - 2) + 2 = 26$ $i = 2$
10. $7(2 - j) - 5 = 30$ $j = -3$
11. $-4 - 6(2 - k) = 8$ $k = 4$
12. $20 - 3(m - 1) = -4$ $m = 9$
13. $5(n + 2) = 3n$ $n = -5$
14. $-2(2p - 3) = 2p$ $p = 1$
15. $8(7 - 3q) = -17q$ $q = 8$
16. $3r = 2(7 + 5r)$ $r = -2$
17. $s + 3 = 3(11 - 3s)$ $s = 3$
18. $7t + 4 = 2(7t - 5)$ $t = 2$
19. $2u - 4 = -4(u - 5)$ $u = 4$
20. $2(v + 1) = 6v - 46$ $v = 12$
21. $2(2w + 2) = 3(2w - 2)$ $w = 5$
22. $-4(1 + x) = 5(1 - x)$ $x = 9$
23. $-6(y - 2) = 7(2 - y)$ $y = 2$
24. $9(z + 1) = -3(5 + z)$ $z = -2$
25. $1 + 2(a + 1) = 3(a + 4)$ $a = -9$
26. $5(b - 6) = 3 + 8(b + 9)$ $b = -35$
27. $2c - (c + 6) = 4(c - 2)$ $c = \frac{2}{3}$
28. $1 + 2(d + 1) = 3 + 4(d + 5)$ $d = -10$
29. $1 - 2(e - 1) = -3 - 4(3 - 5)$ $e = 7$
30. $3 - (2 - f) + 1 = 2(2 - f)$ $f = \frac{2}{3}$
31. $g + 5(g + 1) = 4(2g - 2) + 11g$ $g = 1$
32. $2(3h - 1) + 4h = 10(2 - 3h) + 38h$ $h = 11$
33. $h + 1 + 2(h + 1) = 3(h + 2) - h + 2$ $h = 5$
34. $3(i - 3) - 7(i + 3) = 4(2i - 3) - 8(2i + 3)$ $i = -\frac{3}{2}$
35. $9(2 - j) + 3(5 + 2j) = 2(7 - 2j) - 4(j - 1)$ $j = -3$
36. $34.8k + 0.2(k - 4) = 1.2 - 9(2 - 3k)$ $k = -2$
37. $0.1(2m + 3) - 4m = 1.1(2 - 3m) - 2.4$ $m = 1$
38. $0.5(4n - 4) = 1 + 3n$ $n = -3$

144   LESSON 3.5

Solve each equation. *(EXAMPLE 4)*

10. $3y + 7(2y - 4) = y - 12$ **y = 1**   **11.** $5(y + 4) - 3y = 4y + 12$ **y = 4**

Practice and Apply

Solve each equation.

12. $4n - 2 + 7n = 20$ **n = 2**
13. $3(r - 4) = 9$ **r = 7**
14. $6x - 4(2x + 1) = 12$ **x = –8**
15. $6(t + 7) - 20 = 6t$ **No solution**
16. $4x + 3 - 2x = 2x + 7$ **No solution**
17. $3y - 2(3y + 2) = 8$ **y = –4**
18. $3(x + 1) = 2x + 7$ **x = 4**
19. $3w - 1 - 4w = 4 - 2w$ **w = 5**
20. $8x - 5 = 4x + 4 - 2x$ **x = 1.5**
21. $15 - 3y = y + 13 + y$ **y = 0.4**
22. $4a - 4 = -2a + 14$ **a = 3**
23. $4m - 5 = 3m + 7$ **m = 12**
24. $2(x + 1) = 3x - 3$ **x = 5**
25. $2m - 4 = 2(6 - 7m)$ **m = 1**
26. $8y - 3 = 5(2y + 1)$ **y = –4**
27. $9y - 8 + 4y = 7y + 16$ **y = 4**
28. $0.3w - 4 = 0.8 - 0.2w$ **w = 9.6**
29. $5x + 32 = 8 - x$ **x = –4**
30. $4y + 2 = 3(6 - 4y)$ **y = 1**
31. $m - 12 = 3m + 4$ **m = –8**
32. $2(8y - 7) = 2(3 + 8y)$ **No solution**
33. $2(2x + 3) = 8x + 5$ **x = 0.25**
34. $3 - 2(y - 1) = 2 + 4y$ **y = 0.5**
35. $3(2r - 1) + 5 = 5(r + 1)$ **r = 3**
36. $5(x + 1) + 2 = 2x + 11$ **x = $\frac{4}{3}$**
37. $4z - (z + 6) = 3z - 4$ **No solution**

38. x = –11
39. y = –1
40. m = 0.75
41. All real numbers

38. $3x - 2(x + 6) = 4x - (x - 10)$
39. $8y - 4 + 3(y + 7) = 6y - 3(y - 3)$
40. $1.4m - 0.6(m - 2) = 2.4m$
41. $16x - 3(4x + 7) = 6x - (2x + 21)$
42. $6.3y = 5.2y - 1.1y + 12.1$ **y = 5.5**
43. $(w - 3) - 5(w + 7) = 10(w + 3) - (7w + 5)$ **w = –9**

CONNECTIONS

44. GEOMETRY The perimeters of the two rectangles at right are equal. Write and solve an equation to find the value of x and the common perimeter.
x = 15; perimeter = 90

45. GEOMETRY Write and solve an equation to determine the value of x for which the perimeter of the rectangle at right will be equal to the perimeter of the square.
x = 13

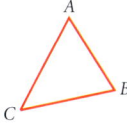

46. x = 9°;
m∠A = 70°,
m∠B = 53°,
m∠C = 57°

46. GEOMETRY The sum of the angle measures in a triangle is 180°. In triangle ABC at right, m∠A = 4x + 34, m∠B = 7x – 10, and m∠C = 5x + 12. Write and solve an equation to find the value of x and the measures of angles A, B, and C.

Technology

Students may find a calculator or computer software helpful in Exercises 12–48. You may wish to have students complete the exercises with paper and pencil first. They can then use a graphics calculator to graph the solutions or use a calculator or software to solve the equations.

Error Analysis

A common error when using the Distributive Property is failing to distribute a number over the whole expression. For instance, students may simplify $3(2x + 4)$ to $6x + 4$ instead of $6x + 12$. In other cases, students simply drop the parentheses: for instance $2(x + 5)$ may be written as $2x + 5$ instead of $2x + 10$. Remind students that they must multiply the number outside the parentheses by every term inside the parentheses.

Student Technology Guide

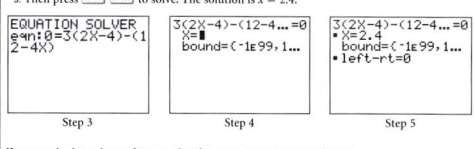

LESSON 3.5 **145**

Look Beyond

Exercises 84–87 look ahead to solving inequalities, which will be covered in Chapter 6.

47. $x + 2x + (2x + 7) = 327$
$x = 64$

Jill: 64 votes
Morgan: 128 votes
Trisha: 135 votes

68. $x = 3$
69. $x = 10$
70. $x = 11.2$
71. $x = 6.1$
72. $m = -4$
73. $r = -20$
74. $m = -7.2$
75. $x = -17$
76. $p = 3$
77. $x = 0.5$
78. $x = 1$
79. $x = 11$
84. $x > 1\frac{1}{3}$
85. $x > -8$
86. $x \le 8$
87. $y \ge 9$

APPLICATIONS

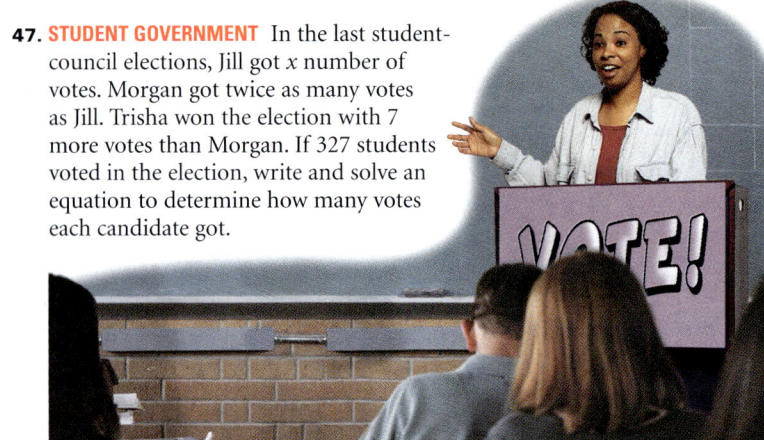

47. STUDENT GOVERNMENT In the last student-council elections, Jill got x number of votes. Morgan got twice as many votes as Jill. Trisha won the election with 7 more votes than Morgan. If 327 students voted in the election, write and solve an equation to determine how many votes each candidate got.

48. SMALL BUSINESS To reduce inventory, Millennium Games offers single-game CDs at a $5 discount. Roger buys 4 of these single-game CDs for $80. Write and solve an equation to determine the original price of a single-game CD.
$4(x - 5) = 80$; $x = \$25$

 Look Back

Determine the next three terms of each sequence. (LESSON 1.1)

49. 1, 3, 7, 13, … **50.** 50, 41, 32, 23, … **51.** $1, \frac{1}{2}, \frac{1}{4}, \frac{1}{8}, \ldots$ $\frac{1}{16}, \frac{1}{32}, \frac{1}{64}$
21, 31, 43   14, 5, −4

Add. (LESSON 2.2)

52. $-3 + (-7)$ −10   **53.** $4.6 + (-2.7)$ 1.9   **54.** $-1\frac{1}{2} + \left(-1\frac{1}{2}\right)$ −3   **55.** $3 + (-2.1)$ 0.9

Subtract. (LESSON 2.3)

56. $3 - (-7)$ 10   **57.** $3 - 4$ −1   **58.** $-1\frac{2}{3} - \left(-3\frac{1}{3}\right)$ $1\frac{2}{3}$ **59.** $0 - 3$ −3

Multiply. (LESSON 2.4)

60. $(3)(-2)$ −6   **61.** $(-2.7)(-2.3)$ 6.21   **62.** $(5)(-0.3)$ −1.5   **63.** $\left(-\frac{1}{2}\right)\left(-\frac{1}{2}\right)$ $\frac{1}{4}$

Divide. (LESSON 2.4)

64. $\frac{-30}{5}$ −6   **65.** $-\frac{1}{5} \div \frac{1}{5}$ −1   **66.** $-5.2 \div (-1.3)$ 4   **67.** $300 \div (-2.5)$ −120

Solve. (LESSONS 3.1 AND 3.2)

68. $3x = 9$ **69.** $\frac{x}{5} = 2$ **70.** $4 = x - 7.2$ **71.** $4 - x = -2.1$
72. $5m = 20$ **73.** $\frac{r}{-2} = 10$ **74.** $m + 4.7 = -2.5$ **75.** $-15 = x + 2$
76. $27p = 81$ **77.** $100x = 50$ **78.** $3 - x = 2$ **79.** $-7 = 4 - x$
80. $-5p = 35$ **81.** $202 = -101m$ **82.** $m + 7.6 = 14$ **83.** $p - 3.1 = -2.7$
$p = -7$ $m = -2$ $m = 6.4$ $p = 0.4$

Look Beyond

Solve each inequality and graph the solution on a number line.

84. $3x > 4$ **85.** $-4x < 32$ **86.** $x - 3 \le 5$ **87.** $2y - 7 \ge 11$

84. [number line with open circle at ~1.33, shaded right, marks at −4, −2, 0, 2, 4]

85. [number line with open circle at −8, shaded right, marks at −10, −8, −6, −4, −2]

86. [number line with closed circle at 8, shaded left, marks at 2, 4, 6, 8, 10]

87. [number line with closed circle at 9, shaded right, marks at 2, 4, 6, 8, 10]

3.6 Using Formulas and Literal Equations

Objectives
- Solve literal equations for a specific variable.
- Use formulas to solve problems.

Why Formulas are algebraic generalizations that can be applied to a variety of problems.

An equation that involves two or more variables is called a **literal equation**. A **formula** is a literal equation that states a rule for a relationship among quantities. For example, the formula $C = \frac{5}{9}(F - 32)$ relates degrees Fahrenheit to degrees Celsius. The variables in a formula are often related to the names of the quantities involved, such as F for Fahrenheit.

When working with formulas, the values for one or more of the variables are sometimes known, leaving the value of one variable to be determined. You can use the temperature conversion formula given above, for example, to easily convert from degrees Fahrenheit to degrees Celsius.

EXAMPLE ① Convert each temperature from degrees Fahrenheit to degrees Celsius by using the formula $C = \frac{5}{9}(F - 32)$.

APPLICATION
TEMPERATURE

a. 68°F   b. 32°F   c. 23°F

- **SOLUTION**

Substitute the given Fahrenheit temperature for F in the formula.

a. 68°F
$C = \frac{5}{9}(68 - 32)$
$C = \frac{5}{9}(36)$
$C = 20$
68°F = 20°C

b. 32°F
$C = \frac{5}{9}(32 - 32)$
$C = \frac{5}{9}(0)$
$C = 0$
32°F = 0°C

c. 23°F
$C = \frac{5}{9}(23 - 32)$
$C = \frac{5}{9}(-9)$
$C = -5$
23°F = −5°C

TRY THIS Convert each temperature from degrees Fahrenheit to degrees Celsius.

a. 50°F   b. 212°F   c. 104°F

Prepare

NCTM PRINCIPLES & STANDARDS
1–4, 6–10

LESSON RESOURCES
- Practice 3.6
- Reteaching Master 3.6
- Enrichment Master 3.6
- Cooperative-Learning Activity 3.6
- Lesson Activity 3.6
- Problem Solving/Critical Thinking Master 3.6
- Teaching Transparency 16

QUICK WARM-UP

Solve for x.

1. $2(x + 3) = 6$ $x = 0$
2. $3(x - 7) = 2x + 6$
   $x = 27$
3. $4(x + 3) = 5(2x + 1)$
   $x = \frac{7}{6}$
4. $7(2x - 4) = 14x - 28$
   all real numbers
5. $6(2x - 5) = 12x + 2$
   no solution

Also on Quiz Transparency 3.6

Teach

Why Help students to recall geometry formulas that they learned in earlier courses, such as the area and perimeter of a rectangle ($A = bh$ and $P = 2b + 2h$). Explain to students that formulas are used in many areas of daily life. Other formulas that students may recognize are $d = rt$, $a^2 + b^2 = c^2$, and $E = mc^2$.

☞ For Additional Example 1 and answers to Try This, see page 148.

Alternative Teaching Strategy

HANDS-ON STRATEGIES First show students how to solve literal equations. Then have students use geometry formulas to solve hands-on problems. Give students the formula for the volume of a rectangular solid, $V = lwh$. Working in groups of three, tell them to build a box out of cardboard with a length of 7 inches, a width of 5 inches, and a volume of 210 cubic inches. Students must solve the formula for h to find the height, 6 inches.

Next ask students to draw a rectangle whose area and perimeter are the same. $lw = 2l + 2w$, $l = \frac{2w}{(w-2)}$, $w = \frac{2l}{(l-2)}$ Sample dimensions are $w = 3$ and $l = 6$. Students can also use formulas for other types of solids, such as cylinders and pyramids. They can use materials such as cardboard, posterboard, clay, or Styrofoam to build the solids.

LESSON 3.6   **147**

Use Teaching Transparency 16.

ADDITIONAL EXAMPLE

Find the area of the trapezoid by using the formula $A = \frac{1}{2}h(b_1 + b_2)$, if $h = 5$, $b_1 = 6$, and $b_2 = 9$. $A = 37.5$

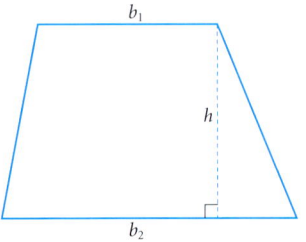

TRY THIS
a. $10°C$ b. $100°C$ c. $40°C$

ADDITIONAL EXAMPLE 2

Solve $d = k + s$ for s.
$s = d - k$

ADDITIONAL EXAMPLE 3

Solve $r = \frac{d}{t}$ for d.
$d = rt$

TRY THIS
a. $P = A - I$ b. $r = \frac{C}{2\pi}$

Solving for a Variable

With some literal equations and formulas you will need to solve the equation for one variable in terms of the other variables.

EXAMPLE The formula $A = P + I$ shows that the total amount of money, A, received from an investment equals the principal, P (the original amount of money invested), plus the interest, I. **Write a formula for the interest, I, in terms of the principal, P, and the total amount, A. In other words, solve the formula for I.**

APPLICATION
INVESTMENTS

● **SOLUTION**

To isolate I, subtract P from both sides of the equation.
$$A = P + I$$
$$A - P = P + I - P \quad \text{Subtraction Property of Equality}$$
$$A - P = I$$
$$I = A - P \quad \text{Symmetric Property of Equality}$$

EXAMPLE 3 *Density* is the mass of a material divided by its volume. The formula for density is $D = \frac{m}{V}$. **Find a formula for the mass, m, in terms of the density, D, and the volume, V.**

APPLICATION
PHYSICS

● **SOLUTION**

Multiply both sides of the equation by V.
$$D = \frac{m}{V}$$
$$V(D) = V\left(\frac{m}{V}\right) \quad \text{Multiplication Property of Equality}$$
$$VD = m \quad \text{Simplify.}$$
$$m = VD \quad \text{Symmetric Property of Equality}$$

TRY THIS Solve each formula for the indicated variable.
a. $A = P + I$, for P
b. $C = 2\pi r$, for r

EXAMPLE 4 Solve each equation for the indicated variable.
a. $y = mx + b$, for b
b. $3x + 4y = 12$, for y

● **SOLUTION**

a.
$$y = mx + b$$
$$y - mx = mx + b - mx \quad \text{Subtraction Property of Equality}$$
$$y - mx = b \quad \text{Simplify.}$$
$$b = y - mx \quad \text{Symmetric Property of Equality}$$

Inclusion Strategies

USING SYMBOLS When solving literal equations, writing the desired variable in red can help students remember which variable they are trying to isolate.

Enrichment

GUIDED RESEARCH Have students choose careers that interest them. Using the library or Internet, have students research that field and report any formulas that professionals in that field may use in their work. The following are some research ideas: electrical engineering and the use of Ohm's law, medicine and the use of Clark's rule for determining dosages (see Interdisciplinary Connection, page 149), accounting and the use of formulas for determining simple and compound interest, and law enforcement and the use of formulas to analyze automobile accident sites.

148 LESSON 3.6

b.
$$3x + 4y = 12$$
$$3x + 4y - 3x = 12 - 3x \quad \text{Subtraction Property of Equality}$$
$$4y = 12 - 3x \quad \text{Simplify.}$$
$$\frac{4y}{4} = \frac{12 - 3x}{4} \quad \text{Division Property of Equality}$$
$$y = 3 - \frac{3}{4}x \quad \text{Simplify.}$$
$$y = 3 + \left(-\frac{3}{4}x\right) \quad \text{Definition of subtraction}$$
$$y = -\frac{3}{4}x + 3 \quad \text{Commutative Property}$$

This is called the $y = mx + b$ form and will be useful in the study of linear equations.

ADDITIONAL EXAMPLE 4

Solve for the indicated variable.

a. $y = mx + b$ for m
$m = \frac{y - b}{x}$

b. $p = 6x + 2z$ for z
$z = \frac{p - 6x}{2}$

Working With Formulas

EXAMPLE 5

CONNECTION
GEOMETRY

Jayne has 396 linear feet of chicken wire and is putting a fence up on her small farm to fence in her chickens. She wants the length of the pen to be twice the width. **Find the dimensions of the pen.**

ADDITIONAL EXAMPLE 5

Louis has 480 feet of chicken wire and is building a pen. He wants the length of the pen to be 3 times the width. **Find the dimensions of the pen.** width = 60 feet, length = 180 feet

SOLUTION

PROBLEM SOLVING

Draw a diagram of the pen. Since the length is twice the width, label the sides w and $2w$.

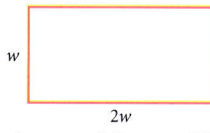

The length of fencing that Jayne has equals the perimeter of the pen. Use the formula for the perimeter of a rectangle, $P = 2l + 2w$, to write an equation for a perimeter of 396 feet.

$$P = 2l + 2w \quad \text{Perimeter formula}$$
$$396 = 2(2w) + 2w \quad \text{Substitute } 2w \text{ for } l.$$

Solve the equation to find the dimensions.

$$396 = 2(2w) + 2w$$
$$396 = 4w + 2w$$
$$396 = 6w$$
$$\frac{396}{6} = \frac{6w}{6} \quad \text{Division Property of Equality}$$
$$66 = w \quad \text{Simplify.}$$
$$w = 66 \quad \text{Symmetric Property of Equality}$$

The width is 66 feet, so the length $2w$, is $2(66)$, or 132 feet.

Math CONNECTION

GEOMETRY Examples 5, 6, and 7 use formulas for the perimeter of a rectangle ($P = 2l + 2w$), the area of a trapezoid ($A = \frac{1}{2}h(b_1 + b_2)$), and the circumference of a circle ($C = 2\pi r$). Other geometry formulas include the volume of a rectangular solid ($V = lwh$), the area of a triangle ($A = \frac{1}{2}bh$), and the volume of a cylinder ($V = \pi r^2 h$).

Interdisciplinary Connection

MEDICINE Clark's rule is one formula that is used to determine the correct dosage of medicine for children based on the dosage given to an adult: $C = A \times \frac{w}{150 \text{ lb}}$, where C is the child's dose, A is the adult dose, and w is the weight of the child.

a. The adult dosage of a certain medicine is 100 mg. Find the dosage, to the nearest milligram, for a child weighing 40 pounds.
27 mg

b. How much would a child have to weigh in order to receive half of an adult dosage? 75 lb (Note: This example is for instructional purposes only and is not meant to replace the advice of a physician.)

LESSON 3.6 **149**

ADDITIONAL EXAMPLE 6

The area of a trapezoid is 90 square inches. The height of the trapezoid is 12 inches. The longer base is 9 inches. Use the formula for the area of a trapezoid, $A = \frac{1}{2}h(b_1 + b_2)$, to find the length of the shorter base.

6 inches

ADDITIONAL EXAMPLE 7

Use the formula $C = 2\pi r$ to find the value of r for the given circumference. Use 3.14 for π.

a. $C = 28$ b. $C = 67$
   $r = 4.45$ $r = 10.67$

TRY THIS

a. $r = 5.41$ b. $r = 11.15$

EXAMPLE 6

CONNECTION: GEOMETRY

Jeffrey is building a tabletop in the shape of a trapezoid. The area of the tabletop is to be 18 square feet. The longer base of the trapezoid is 7 feet and the height of the trapezoid is 3.5 feet. **Use the formula for the area of a trapezoid, $A = \frac{1}{2}h(b_1 + b_2)$, to find the length of the shorter base.**

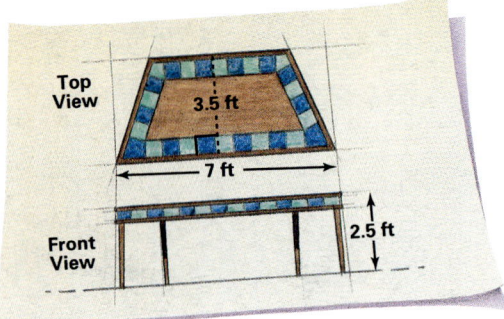

SOLUTION

Substitute 18 for A, 3.5 for h, and 7 for b_1 in the formula and solve for b_2, the length of the second base.

$A = \frac{1}{2}h(b_1 + b_2)$	
$18 = \frac{1}{2}(3.5)(7 + b_2)$	Substitute.
$18 = (1.75)(7 + b_2)$	Simplify.
$18 = (1.75)(7) + (1.75)(b_2)$	Distributive Property
$18 = 12.25 + 1.75b_2$	Simplify.
$18 - 12.25 = 12.25 + 1.75b_2 - 12.25$	Subtraction Property of Equality
$5.75 = 1.75b_2$	Simplify.
$\frac{5.75}{1.75} = \frac{1.75b_2}{1.75}$	Division Property of Equality
$3.2857 \approx b_2$	Simplify.
$b_2 \approx 3.2857$	The shorter base should be approximately 3.29 feet.

EXAMPLE 7

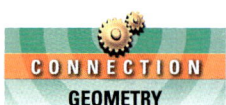
CONNECTION: GEOMETRY

The circumference of a circle C, can be determined from the formula $C = 2\pi r$, where r is the radius of the circle. **Use the formula to find the value of r for each circumference below. Use 3.14 as an approximation for π, and round your answers to the nearest hundredth.**

a. $C = 25$ b. $C = 12$

SOLUTION

a.
$C = 2\pi r$	
$25 = 2(3.14)r$	Substitute.
$25 = 6.28r$	Simplify.
$\frac{25}{6.28} = \frac{6.28r}{6.28}$	Div. Prop. of Eq.
$3.98 \approx 3.98$	Simplify.
$r \approx 3.98$	

b.
$C = 2\pi r$	
$12 = 2(3.14)r$	Substitute.
$12 = 6.28r$	Simplify.
$\frac{12}{6.28} = \frac{6.28r}{6.28}$	Div. Prop. of Eq.
$1.91 \approx r$	Simplify
$r \approx 1.91$	

TRY THIS   Use the formula $C = 2\pi r$ to find the value of r for each circumference below. Use 3.14 as an approximation for π, and round your answers to the nearest hundredth.

a. $C = 34$ b. $C = 70$

Reteaching the Lesson

GUIDED ANLYSIS Begin by reviewing how to solve one or two multistep equations and writing the corresponding property of equality next to each step. Replace the numbers in the equation with variables to show that the properties of equality hold true for both numbers and variables. Next have students solve simple one- or two-step literal equations, listing the corresponding properties of equality next to each step. When students understand, have them attempt literal equations of increasing difficulty. Students who have mastered the process could be partnered with students who are still struggling in order to tutor them during the class period.

Exercises

Communicate

1. Explain what a literal equation is.
2. Describe how formulas can be helpful in mathematics.
3. Explain how to solve each equation for the indicated variable.
   a. $C = \pi d$, for d b. $A + B = C$, for B c. $3m + 7 = y$, for m
4. Explain how to find the length of a rectangle when you know the perimeter and the width.

Guided Skills Practice

Convert each temperature from degrees Fahrenheit to degrees Celsius by using the formula $C = \frac{5}{9}(F - 32)$. *(EXAMPLE 1)*

5. 59°F **15°C**   6. 14°F **−10°C**   7. 149°F **65°C**

Solve for the indicated variable. *(EXAMPLES 2, 3, AND 4)*

8. $A - P = I$, for A 9. $y = mx$, for m 10. $a + bc = d$, for b

11. A rectangle has a length that is 3 times its width. Use the formula $P = 2l + 2w$ to determine the length and width if the perimeter is 32. *(EXAMPLE 5)* **$w = 4, l = 12$**

12. A trapezoid has an area of 22 square inches. The longer base measures 6 inches and the height is 4 inches. Use the formula $A = \frac{1}{2}h(b_1 + b_2)$ to find the length of the shorter base. *(EXAMPLE 6)* **$b = 5$**

13. The circumference of a circle is 30. Use the formula $C = 2\pi r$ to find the radius, r. Use 3.14 as an approximation for π, and round your answer to the nearest hundredth. *(EXAMPLE 7)* **$r \approx 4.78$**

Practice and Apply

Use the formula $C = \frac{5}{9}(F - 32)$ to convert each temperature from degrees Fahrenheit to degrees Celsius.

14. 23°F **−5°C**   15. 86°F **30°C**   16. 18°F **−7.8°C**   17. 125°F **51.7°C**

Solve each equation for the indicated variable.

18. $T + M = R$, for T 19. $M = T - R$, for T 20. $R + T = M$, for T
21. $a + b = c$, for b 22. $c = a - b$, for b 23. $a + b = -c$, for a
24. $a - b = c$, for a 25. $C = \pi d$, for d 26. $y = mx + 6$, for m
27. $A = lw$, for w 28. $3t = r$, for t 29. $5p + 9c = p$, for c
30. $ma = q$, for a 31. $ax + r = 7$, for x 32. $2a + 2b = c$, for b
33. $4y + 3x = 5$, for x 34. $3x + 7y = 2$, for y 35. $y = 3x + 3b$, for b

---

Assess

Selected Answers
Exercises 5–13, 15–65 odd

ASSIGNMENT GUIDE

In Class	1–13
Core	14–29, 36–41, 45–53 odd
Core Plus	14–17, 24–35, 42–52 even
Review	54–65
Preview	66, 67

✏ Extra Practice can be found beginning on page 738.

Answers:

8. $A = I + P$
9. $m = \frac{y}{x}$
10. $b = \frac{d - a}{c}$

18. $T = R - M$
19. $T = M + R$
20. $T = M - R$
21. $b = c - a$
22. $b = a - c$
23. $a = -b - c$
24. $a = c + b$
25. $d = \frac{C}{\pi}$
26. $m = \frac{y - 6}{x}$
27. $w = \frac{A}{l}$
28. $t = \frac{r}{3}$
29. $c = -\frac{4p}{9}$
30. $a = \frac{q}{m}$
31. $x = \frac{7 - r}{a}$
32. $b = \frac{c - 2a}{2}$
33. $x = \frac{5 - 4y}{3}$
34. $y = \frac{2 - 3x}{7}$
35. $b = \frac{y - 3x}{3}$

LESSON 3.6

Technology

Students may find calculators helpful in Exercises 14–17 and 36–44. When evaluating formulas, the arithmetic required may prove very time consuming without the use of a calculator.

Error Analysis

A student who cannot decide which operation to perform on a variable may benefit from writing the corresponding property of equality next to each step of a solution. Showing students how to solve an equation with numbers next to a literal equation can help them see that the properties hold true for numbers and variables.

CONNECTIONS

36. $r \approx 4.78$ inches
37. $r \approx 0.80$ centimeters
38. $r \approx 15.92$ meters
39. $r \approx 2.39$ meters
40. $r \approx 11.31$ inches
41. $r \approx 8.28$ feet

GEOMETRY Use the formula $C = 2\pi r$ to find the radius, r, for each circumference below. Use 3.14 as an approximation for π, and round your answers to the nearest hundredth.

36. $C = 30$ inches
37. $C = 5$ centimeters
38. $C = 100$ meters
39. $C = 15$ meters
40. $C = 71$ inches
41. $C = 52$ feet
42. $C = 3$ feet
    $r \approx 0.48$ feet
43. $C = 11$ inches
    $r \approx 1.75$ inches
44. $C = 200$ meters
    $r \approx 31.85$ meters

GEOMETRY Use the formula $A = \frac{1}{2}h(b_1 + b_2)$ to find the length of the missing base for trapezoids with the given dimensions. Round your answers to the nearest hundredth.

45. $A = 200, b_1 = 24, h = 10$ $b_2 = 16$
46. $A = 150, b_1 = 20, h = 9$ $b_2 = 13\frac{1}{3}$
47. $A = 28, b_1 = 8, h = 4$ $b_2 = 6$
48. $A = 50, b_1 = 10, h = 6$ $b_2 = 6\frac{2}{3}$
49. $A = 120, b_1 = 20, h = 10$ $b_2 = 4$
50. $A = 200, b_1 = 50, h = 5$ $b_2 = 30$

APPLICATIONS

51. **LANDSCAPING** Fred is putting a decorative border around a rectangular flower garden. The total perimeter of the garden is 200 feet and the length is 4 times the width. Use the formula $P = 2l + 2w$ to find the length and the width. $w = 20, l = 80$

52. **TRAVEL** You can estimate the time, t, in hours that it takes to fly a distance, d, in miles by using the formula $t = \frac{d}{500} + \frac{1}{2}$. For example, a flight of 1000 miles takes about 2.5 hours.

    a. Use the formula to estimate the time that it takes to fly 1300 miles. **3.1 hours**
    b. Solve the formula for d. $d = 500t - 250$
    c. Use the formula from **b** to find how many miles you can fly in 4 hours. $d = 1750$

53. **THEATRE** Alicia is painting a rectangular box as a part of the set for a school play. The box has a length of 3 feet, a width of 2 feet, and a height of 1 foot. Use the formula $A = 2(lw + hw + lh)$, where A is the total surface area, l is the length, w is the width, and h is the height, to find the total surface area of the box. **22 square feet**

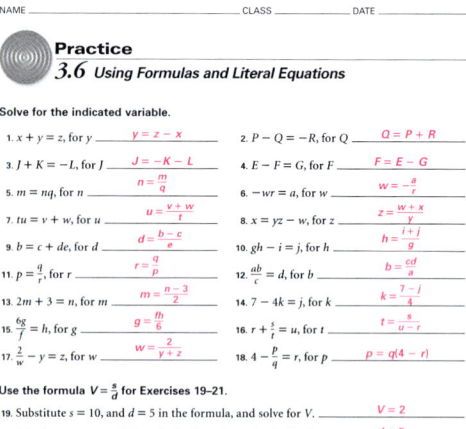

Look Back

Evaluate. (LESSON 2.3)

54. 89 − (−14) **103**   **55.** 400 − (−111) **511**   **56.** −674 − (−900) **226**

Evaluate. (LESSON 2.4)

57. (−3)(−3)(−1)(−1) **9**   **58.** (1)(−1)(−1)(−1) **−1**   **59.** $\frac{-22}{2}$ **−11**   **60.** $\frac{-16}{-4}$ **4**

Use the Distributive Property to simplify each expression. (LESSON 2.5)

61. $(3x − 2y + 1) − 3(x + 2y − 1)$
    −8y + 4

62. $3(a + b) − 2(a − b)$
    a + 5b

Solve each equation. (LESSON 3.3)

63. $3x − 2 = 19$
    x = 7

64. $−4x + 15 = −45$
    x = 15

65. $18x + 53 = 7x − 68$
    x = −11

Look Beyond

Sometimes inequalities are used to indicate values that make sense in an equation, such as in these examples.

If $y = \frac{1}{x}$, then $x \neq 0$.   If $y = \sqrt{x}$, then $x \geq 0$.

Complete each of the following:

66. If $y = \frac{1}{x-2}$, then $x \neq \underline{?}$. **2**

67. If $y = \sqrt{x + 3}$, then $x \geq \underline{?}$. **−3**

Look Beyond

Exercises 66 and 67 look ahead to solving rational equations, which is covered in Chapter 11, and solving radical equations, which is covered in Chapter 12.

ALTERNATIVE Assessment

Portfolio Activity

The Portfolio Activity can be used as preparation for the Chapter Project or as a separate activity. In the Portfolio Activity on this page, students write and solve an equation in order to solve the classic Diophantus problem.

Answers to Portfolio Activities can be found in Additional Answers of the Teacher's Edition.

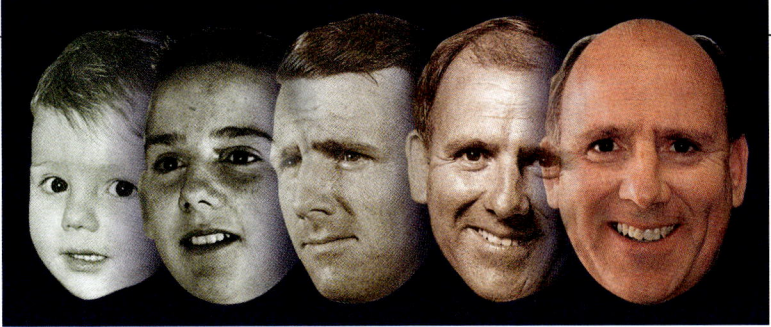

This series of photos shows the age progression for one man.

Include this activity in your portfolio.

Diophantus of Alexandria was a mathematician who lived in the third century C.E. He has been called the father of algebra. The following problem was written about Diophantus:

Diophantus lived $\frac{1}{6}$ of his life as a child, $\frac{1}{12}$ more as a youth, $\frac{1}{7}$ more before he married, and 5 years more before his son was born.

His son was alive for only $\frac{1}{2}$ Diophantus's life.

Diophantus then found solace in his studies for 4 years more, before he too died.

Write and solve an equation to determine how long Diophantus lived.

WORKING ON THE CHAPTER PROJECT

You should now be able to complete Activity 3 of the Chapter Project on page 154.

Student Technology Guide

Student Technology Guide
3.6 Using Formulas and Literal Equations

If you know the values of all but one variable in a literal equation, you can use the solver feature to find the value of the remaining variable. For example, if you have a rectangle whose width is 3 and whose perimeter is 18, you can use the formula $2l + 2w = P$, where $w = 3$ and $P = 18$, to find the length.

Rewrite the formula as $2l + 2w − P = 0$. Press MATH MATH 0:Solver ENTER ▲. You should see eqn:0= on your screen. Enter the equation by pressing 2 ALPHA) + 2 ALPHA - - ALPHA 8.

Press ENTER to display the "interactive solver editor." Enter the known values and align the cursor with the desired variable, L, by pressing ENTER 3 ENTER 18 ▲ ▲. Now press ALPHA ENTER to solve. The length is 6.

Given the indicated values, use a calculator to solve each equation for the indicated variable.

1. $A = \frac{1}{2}bh$, for b; $A = 12$, $h = 5$
   b = 4.8

2. $2x + 3y = z$, for x; $y = 3$, $z = 5$
   x = −2

3. $2w + 2l = P$, for w; $l = 15$, $P = 42$
   w = 6

4. $3x + 4y = 12$, for y; $x = −16$
   y = 15

5. $F = ma$, for a; $F = 96$, $m = 3$
   a = 32

6. $y = mx + b$, for m; $x = 3$, $y = 4$, $b = 1$
   m = 1

7. $2x + 3y = 5z$, for y; $x = 3$, $z = 6$
   y = 8

8. $V = IR$, for I; $V = 6$, $R = 12$
   $I = \frac{1}{2}$

9. $s = pq + r$, for r; $s = 15$, $p = 8$, $q = 4$
   r = −17

10. $C = 2\pi r$, for r; $C = 18$
    $r = \frac{9}{\pi}$

Project

Focus

Students apply their knowledge of equation solving to two situations related to a track team. First they are asked to investigate how different price structures and levels of sales affect the break-even point of a fund-raising venture. Then they turn their attention to determining the appropriate dimensions and suitable layout for a new track.

Motivate

Explain to students that the amount of money a business takes in is called its *revenue*. When the operating expenses for a given period of time are subtracted from the revenue for that same period, the result is the *profit* for the period. This relationship can be modeled by the equation $P = R - E$, where P represents profit, R represents revenue, and E represents operating expenses. Lead students in an analysis of this profit equation. Note that the amount of profit, P, is positive when $R > E$, that is, when revenue exceeds expenses. In contrast, when $R < E$, expenses exceed revenue and the amount of profit, P, is negative. In this case, the business incurs a *loss*. When $R = E$, the revenue and expenses are equal, resulting in a profit, P, of 0. This is called the *break-even point* of the business.

CHAPTER PROJECT THREE

TRACK TEAM Schemes

Help Track & Field... Buy a 'T'

The school board for Hale High School has approved funding for a new track. The Hale High School track team is raising money to buy new equipment for the new track and the track program at the school.

Activity 1

The Hale High School track team is selling T-shirts to raise money. The Portfolio Activity on page 136 included the following information about the cost of the T-shirts:

> Ted's Tees charges $3.75 per shirt, $0.80 to print each shirt with the design, and $125 per order for the silk-screen design.

1. Suppose that the track team wishes to make a profit of $1000 from the sale of T-shirts. Explore several pricing schemes that result in this level of profit by adjusting the amount that the track team charges for the shirts and the number of shirts that need to be sold.

2. Optional: Create a spreadsheet that shows the cost, income, and profit for each of the pricing schemes from Step 1.

Activity 2

1. For each pricing scheme that you explored in Activity 1, write an equation that represents how many T-shirts need to be sold to reach the break-even point, the point at which the cost and income are the same.

2. For each pricing scheme that you explored in Activity 1, write an equation that represents how many T-shirts need to be sold to make a profit of $1000.

Activity 1

1. Answers will vary. sample answers: 2500 shirts at $5; 207 shirts at $10

2. Check students' spreadsheets.

Activity 2

1. Answers will vary. Sample answers: For shirts selling at $5, $5x = 3.75x + 0.80x + 125$, breaking even at 278 shirts. For shirts selling at $10, $10x = 3.75x + 0.80x + 125$, breaking even at 23 shirts.

2. Answers will vary. Sample answers: At $5 per shirt, $5x - (3.75x + 0.80x + 125) = 1000$, $x = 2500$; 2500 shirts need to be sold. At $10 per shirt, $10x - (3.75x + 0.80x + 125) = 1000$, $x = 206.42$; 207 shirts need to be sold.

Activity 3

The Hale High School track coaches need to make a preliminary survey of the school property in order to determine the best site for the new track.

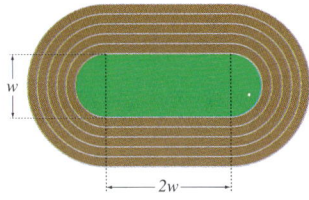

The inside lane of the track will enclose an area that is the shape of a rectangle with a semicircle at each end. The length of the rectangular part of the track should be twice the width.

The International Amateur Athletic Federation (IAAF) is the governing body of track and field. The IAAF rules specify that an outdoor running track should measure no less than 400 meters around the inside lane (perimeter). If available, a spreadsheet program is a useful for tool for completing this activity.

1. Write an equation for the perimeter of the track, using the information in the diagram at left.
2. Find the value of w which ensures that the inside lane of the track will produce a lap length of no less than 400 meters.
3. IAAF rules state that each lane should measure about 1.25 meters in width. Find the perimeter of the outer edge of the inside lane.
4. Suppose that the track is to have 6 lanes. Find the perimeter of the outer edge of each lane.
5. Assume that the inner edge of the inside lane is 400 meters. Explain how you could stagger the starting point for each lane to account for the differences in the lane perimeters.

Activity 3

1. $2w + 2w + \pi w = 400$
2. Using 3.14 for π, $2w + 2w + \pi w = 400$, so $w \approx 56.02$ meters.
3. $2w + 2w + \pi(w + 2.5)$; substitute 56.02 for w; $2(56.02) + 2(56.02) + \pi(56.02 + 2.5)$ ≈ 407.83 meters
4. lane 1 $\approx$ 407.83 meters

   lane 2 $\approx$ 415.68 meters

   lane 3 $\approx$ 423.53 meters

   lane 4 $\approx$ 431.38 meters

   lane 5 $\approx$ 439.23 meters

   lane 6 $\approx$ 447.08 meters

5. Stagger lane 2 approximately 7.85 meters ahead of lane 1. Stagger lanes 3–6 approximately 7.85 meters ahead of each previous lane.

Cooperative Learning

Have students work in groups of four. For **Activity 1**, each group member should try a different price and determine the number of T-shirts that the group would need to sell at that price in order to make a profit of $1000. Each student should also experiment with different numbers of T-shirts sold to investigate the price the group would have to charge for each T-shirt in order to reach the profit goal. After making a decision about the price, each student should complete **Activity 2** by using that price. After completing Activities 1 and 2, group members should compare results and arrive at a consensus about the best pricing plan.

For **Activity 3**, the group can work together to create an appropriate equation for the perimeter of the track and to determine the value of w. Then students can work in pairs for Steps 3 and 4. The group can then work together again in Step 5.

Discuss

Bring the class together so that the groups can share their results. Discuss reasons for similarities or differences among the T-shirt prices chosen by the groups. Ask for volunteers to explain why the starting point for each runner on the track must be staggered.

You might want to have students research the dimensions of their own school track and develop a method for locating suitable starting points.

Chapter Review

Chapter Review and Assessment

VOCABULARY

Addition Property of Equality 116	formula 147	Subtraction Property of Equality 115
Division Property of Equality 123	literal equation 147	
	Multiplication Property of Equality 124	

Key Skills & Exercises

LESSON 3.1

Key Skills

Solve equations by using addition and subtraction.

Solve $x + 15 = 11$.

$$x + 15 = 11$$
$$x + 15 - 15 = 11 - 15 \quad \text{Subtr. Prop. of Equality}$$
$$x = -4 \quad \text{Simplify.}$$

Check: $-4 + 15 = 11$ True

Solve $-8 = y - 14$.

$$-8 = y - 14$$
$$-8 + 14 = y - 14 + 14 \quad \text{Add. Prop. of Equality}$$
$$6 = y \quad \text{Simplify.}$$
$$y = 6$$

Check: $-8 = 6 - 14$ True

Exercises

Solve each equation. Check the solution.

1. $r + 26 = 16$ $r = -10$
2. $a + 1.5 = 3.6$ $a = 2.1$
3. $t + 7 = -5$ $t = -12$
4. $y - 13 = 12$ $y = 25$
5. $24 = x - 19$ $x = 43$
6. $m + \frac{1}{2} = \frac{5}{6}$ $m = \frac{1}{3}$
7. $h - \frac{1}{6} = \frac{2}{3}$ $h = \frac{5}{6}$
8. $k + 5 = 7$ $k = 2$
9. $m - 3 = 15$ $m = 18$
10. $x + 17 = 2$ $x = -15$

LESSON 3.2

Key Skills

Solve equations by using multiplication and division.

Solve $-9k = 108$.

$$-9k = 108 \quad \text{Given}$$
$$\frac{-9k}{-9} = \frac{108}{-9} \quad \text{Division Property of Equality}$$
$$k = -12 \quad \text{Simplify.}$$

Solve $\frac{w}{-5} = -2.2$.

$$\frac{w}{-5} = -2.2 \quad \text{Given}$$
$$-5\left(\frac{w}{-5}\right) = -5(-2.2) \quad \text{Mult. Prop. of Equality}$$
$$w = 11 \quad \text{Simplify.}$$

Exercises

Solve each equation.

11. $17x = -85$ $x = -5$
12. $-4g = -56$ $g = 14$
13. $-2.2h = 33$ $h = -15$
14. $24f = 150$ $f = 6.25$
15. $-8w = 0.5$ $w = -0.0625$
16. $\frac{y}{-2.4} = -10$ $y = 24$
17. $\frac{t}{5} = -24$ $t = -120$
18. $\frac{-c}{3} = 12$ $c = -36$
19. $\frac{w}{8} = -19$ $w = -152$
20. $\frac{x}{-3} = 0$ $x = 0$

Chapter Test, Form A

NAME _____ CLASS _____ DATE _____

Chapter Assessment
Chapter 3, Form A, page 1

Write the letter that best answers the question or completes the statement.

a   1. Solve $x + 0.54 = 2$.
    a. $x = 1.46$ b. $x = 2.54$ c. $x = 0.34$ d. $x = 0.46$

b   2. Solve $\frac{x}{9.2} = 8.1$.
    a. $x = 745.2$ b. $x = 74.52$ c. $x = 17.3$ d. $x = 73.52$

a   3. Solve $5x + 1.7 = 7x - 12.3$.
    a. $x = 7$ b. $x = 8$ c. $x = 5$ d. $x = 6$

b   4. Kendall wants to drive 280 miles in 7 hours. What speed in miles per hour should she average for the trip?
    a. 45   b. 40   c. 50   d. 55

d   5. Jerry pays $1.40 per gallon for premium gasoline. He spends $16.80 and wants to know how many gallons he bought. Which equation models this situation?
    a. $16.80 + x = 1.40$ b. $16.80 - 1.40 = x$
    c. $1.40 - x = 16.80$ d. $1.40x = 16.80$

d   6. Solve $3m - 17 = 19$.
    a. $m = 3$ b. $m = 36$ c. $m = 15$ d. $m = 12$

c   7. Solve $\frac{3x}{2} + 6 = 9$.
    a. $x = 10$ b. $x = -10$ c. $x = 2$ d. $x = -2$

a   8. Solve $6 - 6(x - 2) = 3(3x + 1)$.
    a. $x = 1$ b. $x = 0$ c. $x = -\frac{17}{15}$ d. $x = -\frac{1}{5}$

b   9. The cost of 10 pencils plus tax of $0.15 is $2.15. Which equation models the situation?
    a. $10(x + 0.15) = 2.15$ b. $10x + 0.15 = 2.15$
    c. $10x = 2.15 + 0.15$ d. $10 + 0.15x = 2.15$

NAME _____ CLASS _____ DATE _____

Chapter Assessment
Chapter 3, Form A, page 2

c   10. If $c - \frac{5}{8} = 6$, what is the value of c?
    a. $5\frac{3}{8}$ b. $-1\frac{7}{8}$ c. $6\frac{5}{8}$ d. $-5\frac{3}{8}$

d   11. Solve $2x + 4 = 84$.
    a. $x = -40$ b. $x = -37$ c. $x = 37$ d. $x = 40$

a   12. Solve $2(9 - 2y) = 42$.
    a. $y = -6$ b. $y = 15$ c. $y = -15$ d. $y = 6$

c   13. Solve $0.4r - 1.2 = 0.3r + 0.6$.
    a. $r = \frac{18}{7}$ b. $r = 6$ c. $r = 18$ d. not given

b   14. Which equation has no solution?
    a. $2x + 4 = 2(x + 2)$ b. $2x + 4 = 2(x + 1)$
    c. $3x + 5 = 2(x + 3)$ d. Neither a nor b has a solution.

c   15. Solve for P: $I = Prt$
    a. $P = I - rt$ b. $P = \frac{rt}{I}$ c. $P = \frac{I}{rt}$ d. $P = \frac{I \cdot r}{t}$

d   16. The price of a shirt has been reduced by $20. The cost of 10 shirts at the reduced price is $50. Which equation models the situation?
    a. $50(x - 20) = 10$ b. $50(x + 20) = 10$
    c. $10(x + 20) = 50$ d. $10(x - 20) = 50$

a   17. Solve for x: $y = mx + b$
    a. $x = \frac{y - b}{m}$ b. $x = \frac{b + y}{m}$ c. $x = \frac{y}{m} + b$ d. $x = y - \frac{b}{m}$

b   18. A triangle has an area of 72 square feet. Use the formula $A = \frac{1}{2}bh$ to find the height of the triangle if the base is 24 feet.
    a. 60 feet   b. 6 feet   c. 864 feet   d. 1.5 feet

LESSON 3.3

Key Skills

Solve two-step equations.

Solve $4x + 5 = 15$.

$$4x + 5 = 15$$
$$4x + 5 - 5 = 15 - 5 \quad \text{Sub. Prop. of Equal.}$$
$$4x = 10 \quad \text{Simplify.}$$
$$\frac{4x}{4} = \frac{10}{4} \quad \text{Div. Prop. of Equal.}$$
$$x = 2.5 \quad \text{Simplify.}$$

Solve $5 - \frac{x}{3} = 2$.

$$5 - \frac{x}{3} = 2$$
$$5 - \frac{x}{3} + 5 = 2 + 5 \quad \text{Add. Prop. of Equal.}$$
$$-\frac{x}{3} = 7 \quad \text{Simplify.}$$
$$-3\left(-\frac{x}{3}\right) = -3(7) \quad \text{Mult. Prop. of Equal.}$$
$$x = -21 \quad \text{Simplify.}$$

Exercises

Solve.

21. $3a + 7 = 31$ $a = 8$
22. $-2x + 10 = -4$ $x = 7$
23. $\frac{y}{5} + 4 = 40.5$ $y = 182.5$
24. $5x + 3 = -17$ $x = -4$
25. $-7 = 4 + \frac{x}{3}$ $x = -33$
26. $-8 = \frac{x}{2} - 5$ $x = -6$
27. $-3x - 2 = -4$ $x = \frac{2}{3}$
28. $5x + 17 = -17$ $x = -6.8$
29. $3 - 2x = 3$ $x = 0$
30. $\frac{x}{4} - 5 = 5$ $x = 40$
31. $4x + 3 = 23$ $x = 5$
32. $45 = 2w - 32$ $w = 38.5$
33. $60 = 5 - \frac{m}{7}$ $m = -385$

LESSON 3.4

Key Skills

Solve multistep equations.

Solve $5x + 7 = 2x - 2$.

$$5x + 7 = 2x - 2$$
$$5x + 7 - 2x = 2x - 2 - 2x \quad \text{Sub. Prop. of Equal.}$$
$$3x + 7 = -2 \quad \text{Simplify.}$$
$$3x + 7 - 7 = -2 - 7 \quad \text{Sub. Prop. of Equal.}$$
$$3x = -9 \quad \text{Simplify.}$$
$$\frac{3x}{3} = \frac{-9}{3} \quad \text{Div. Prop. of Equal.}$$
$$x = -3 \quad \text{Simplify.}$$

Exercises

Solve.

34. $n - 6 = 2n - 14$ $n = 8$
35. $2m + 5 = \frac{m}{5} - 4$ $m = -5$
36. $4x + 18 = x$ $x = -6$
37. $5m - 3 = 2m - 9$ $m = -2$
38. $15p + 4 = 5p$ $p = -0.4$
39. $\frac{m}{2} + 3 = \frac{m}{5} + 5$ $m = 6\frac{2}{3}$
40. $8x - 1.5 = 2x$ $x = 0.25$
41. $3f - 2 = 7$ $f = 3$
42. $2x + 3 = 3x + 5$ $x = -2$
43. $5m - 3 = 2$ $m = 1$
44. $-14x = 3x + 17$ $x = -1$
45. $-2x + 3 = x - 6$ $x = 3$
46. $8f + 7 = 15$ $f = 1$
47. $11m + 2 = 9m - 18$ $m = -10$

Chapter Test, Form B

NAME _____ CLASS _____ DATE _____

Chapter Assessment
Chapter 3, Form B, page 1

Write an equation or an inequality to represent each situation.

1. Paul has $10. Tom and Paul have $15 together.   Let x represent the amount of money Tom has. Then x + 10 = 15.
2. Melissa has 5 times as much money as Manuel. They have $60 together.   Let x represent the amount of money Manuel has. Then x + 5x = 60.
3. The cost of 8 CD's plus a sales tax of $1.34 is $105.26.   Let c represent the cost of one CD. Then 8c + 1.34 = 105.26.

Solve each equation.

4. $-4b = 48$ $b = -12$
5. $x + 9 = 23$ $x = 14$
6. $-3x = -1764$ $x = 588$
7. $\frac{x}{20} - 6 = 14$ $x = 400$
8. $\frac{c}{-3} = 0.3$ $c = -0.9$
9. $\frac{b}{-1} = -20$ $b = 20$
10. $5x - 2 = 13$ $x = 3$
11. $\frac{4}{5}y = 16$ $y = 20$
12. $\frac{-a}{5.2} = -7$ $a = 36.4$
13. $12x - (1 - 2x) = -29$ $x = -2$
14. $3x - 2(x + 3) = 4x - 7$ $x = \frac{1}{3}$
15. $-3(x + 2) = -18$ $x = 4$
16. $2.9m + 1.7 = 3.5 + 2.3m$ $m = 3$
17. $\frac{3}{4}y - 4 = 7 + \frac{1}{2}y$ $y = 44$
18. Tara wants to buy one set of towels that costs $25.00 for the set and 6 decorative guest towels. Her total purchase was $40 before taxes. Write and solve an equation to determine how much she was charged for each decorative towel. $6x + 25 = 40, 2.50

NAME _____ CLASS _____ DATE _____

Chapter Assessment
Chapter 3, Form B, page 2

19. Solve $0.4x + 3.91 = 21.05$. $x = 42.85$
20. Solve $x = 0.5x + 3.8$. $x = 7.6$
21. Solve $5x + 2(3x - 1) = x$. $x = \frac{1}{5}$
22. Solve $6x - 4(x + 1) = 2(x - 3) + 2$.   All real numbers are the solution to this equation.
23. Solve $4(x + 1) = 2(x - 3) - (1 - 2x)$.   There are no real number solutions to this equation.
24. Convert 41°F to Celsius by using the formula $C = \frac{5}{9}(F - 32)$.   5°C
25. Solve $r = s - t$ for t. $t = s - r$
26. Solve $ab - cd = 0$ for c. $c = \frac{ab}{d}$
27. The formula for the sum of the angles of the measures of a triangle is $a + b + c = 180$. Use the formula to find the measure of the third angle when the measures of two of the angles are 105° and 25°.
   50°
28. A trapezoid has an area of 105 square inches. The shorter base measures 8 inches and the height is 10 inches. Use the formula, $A = \frac{1}{2}h(b_1 + b_2)$ to find the length of the longer base.
   13 inches
27. The price of an automobile tire has been reduced by $15.00. The cost of a set of 4 tires at the reduced price is $600. Use an equation to find the original cost of a tire.
   $165

CHAPTER 3 REVIEW **157**

LESSON 3.5

Key Skills

Use the Distributive Property to solve multistep equations.

Solve $5x + 3(x - 3) + 7 = 2x - 1$.

$5x + 3(x - 3) + 7 = 2x - 1$	
$5x + 3x - 9 + 7 = 2x - 1$	Distributive Prop.
$8x - 2 = 2x - 1$	Combine like terms.
$8x - 2 - 2x = 2x - 1 - 2x$	Subtr. Prop. of Equal.
$6x - 2 = -1$	Simplify.
$6x - 2 + 2 = -1 + 2$	Add. Prop. of Equal.
$6x = -1$	Simplify.
$\frac{6x}{6} = \frac{-1}{6}$	Div. Prop. Equal.
$x = -\frac{1}{6}$	

Exercises

Solve.

48. $7x - 2(x + 6) = -2$ $x = 2$
49. $3x + 4(x - 2) = 7 + x$ $x = 2.5$
50. $5(x - 3) = 10$ $x = 5$
51. $6(2x + 7) = 102$ $x = 5$
52. $3 - 4x = 5(x + 6)$ $x = -3$
53. $x + 7 = 2(x - 1)$ $x = 9$
54. $5m - 3(m - 3) = 2(m + 3)$ No solution
55. $6(x + 2x + 1) = 12x$ $x = -1$

LESSON 3.6

Key Skills

Manipulate literal equations.

Solve $4m + s = 3p$ for m.

$4m + s = 3p$	
$4m + s - s = 3p - s$	Sub. Prop. of Equal.
$4m = 3p - s$	Simplify.
$\frac{4m}{4} = \frac{3p - s}{4}$	Div. Prop. of Equal.
$m = \frac{3p - s}{4}$	Simplify.

Use formulas to solve problems.

Use the formula $A = lw$ to find l if A is 40 and w is 4.

$A = lw$	
$40 = 4l$	Substitution
$\frac{40}{4} = \frac{4l}{4}$	Div. Prop. of Equal.
$10 = l$	Simplify.
$l = 10$	Sym. Prop.

Exercises

Solve for the indicated variable.

56. $A + B = C$, for A $A = C - B$
57. $y = mx + b$, for m $m = \frac{y - b}{x}$
58. $9x + 3y = 10$, for y $y = \frac{10 - 9x}{3}$
59. $ax + b = c$, for b $b = c - ax$
60. $T + \frac{S}{3} = R$, for S $S = 3(R - T)$

Use the formula $C = 2\pi r$ to find r for each circumference below. Use 3.14 as an approximation for π, and round your answers to the nearest hundredth.

61. $C = 100$ meters 15.92 meters
62. $C = 10$ centimeters 1.59 centimeters
63. $C = 15$ feet 2.39 feet
64. $C = 25$ centimeters 3.98 centimeters
65. $C = 28$ inches 4.46 inches

Applications

66. **CONSUMER ECONOMICS** Marsha needs to buy painting supplies. Paintbrushes cost $1.75 each and paint costs $2.45 per jar. Write an equation that shows how much will be spent, s, if she buys b paintbrushes and j jars of paint. If she has $125 to spend and she buys 22 paintbrushes, how many jars of paint can she buy?

    $s = 1.75b + 2.45j$
    $125 = 1.75(22) + 2.45j$
    $j \approx 35.31$
    Marsha can buy 35 jars of paint.

Alternative Assessment

Performance Assessment

1. The members of the sports club are raising money to buy new equipment that costs $345. So far they have raised $180. How much does the club still need to raise?
   a. Write an addition equation to model the situation.
   b. Explain how you can use algebra tiles to solve this addition equation.
   c. Write a subtraction equation that models the situation.
   d. Explain how you can use algebra tiles to solve this subtraction equation.
   e. Why do both methods result in the same answer?

2. Pick a number from 1 to 9. Add 3 to the number, and multiply the sum by 2. Write down the product. Subtract 6 from this number. Take one-half of this number. Use algebraic expressions and equations to show that the final answer is always the same as the original number.

3. Matt solved the equation $-5x = 25$ by using multiplication. Angie solved the same equation by using division and got the same result as Matt. Explain how Matt and Angie solved the same equation by two different methods.

4. Explain how to solve the equation $2x + 9 = 5x - 6$ with a graphics calculator.

Portfolio Projects

1. A squirrel and a dog weigh the same as 2 cats. A dog weighs as much as 2 squirrels and a cat. How many squirrels will balance a dog?

   Find a solution to the problem any way you can, then find an algebraic solution. Remember, if one quantity equals another, they can replace each other.

2. The Rhind papyrus is one of the earliest records of equation solving. Write a short history of the Rhind papyrus and explain how equations are used to solve problems in it.

Peer Assessment

Complete this activity with a partner. Begin by writing an equation that can be solved in one step. After solving the equation, your partner should write an equation for you to solve.

Check your partner's work, correct any errors, and then solve the new equation. Continue with this pattern until you have each solved 10 equations. Be sure to use equations involving addition, subtraction, multiplication, and division during this activity. Then repeat the entire activity with multistep equations.

---

internetconnect

The HRW Web site contains many resources to reinforce and expand your knowledge of equations. The Web site also provides Internet links to other sites where you can find information and real-world data for use in research projects, reports, and activities that involve equations. Visit the HRW Web site at **go.hrw.com**, and enter the keyword **MA1 CH3**.

Alternative Assessment

Performance Assessment

1. a. $x + 180 = 345$
   b. Use 180 tiles on the left and 345 tiles on the right. Subtract 180 tiles from both sides.
   c. $x = 345 - 180$
   d. Use 345 tiles. Subtract 180 tiles.
   e. You are subtracting 180 tiles for each method.

2. Let x represent the number. Follow these steps: $x + 3$; $2(x + 3) = 2x + 6$; $2x$; x; the result is the number you started with.

3. Dividing by -5 is the same as multiplying by $-\frac{1}{5}$.

4. Graph $y = 2x + 9$ and $y = 5x - 6$ on the same coordinate plane. The x-coordinate of the point of intersection is the solution.

Portfolio Projects

1. Answers will vary. Sample algebraic solution:
   Let $s =$ squirrels, $d =$ dogs, and $c =$ cats.
   $s + d = 2c$
   $2s + c = d$
   Substitute for d in the first equation.
   $s + (2s + c) = 2c$
   $c = 3s$
   Thus, one cat weighs the same as 3 squirrels.

   Substitute for c in the first equation.
   $s + d = 2(3s)$
   $d = 5s$
   Thus, a dog weighs the same as 5 squirrels.

2. Answers will vary.

Cumulative Assessment

College Entrance Exam Practice

Multiple-Choice and Quantitative-Comparison Samples

The first half of the Cumulative Assessment contains two types of items found on standardized tests—multiple-choice questions and quantitative-comparison questions. Quantitative-comparison items emphasize the concepts of equality, inequality, and estimation.

Free Response Grid Samples

The second half of the Cumulative Assessment is a free-response section. This part of the Cumulative Assessment requires student-produced response items like those commonly found on college entrance exams. These questions require the use of machine-scored answer grids. You may wish to have students practice answering these items in preparation for standardized tests.

CHAPTERS 1–3 CUMULATIVE ASSESSMENT

College Entrance Exam Practice

QUANTITATIVE COMPARISON For Questions 1–4, write
A if the quantity in Column A is greater than the quantity in Column B;
B if the quantity in Column B is greater than the quantity in Column A;
C if the two quantities are equal; or
D if the relationship cannot be determined from the information given.

	Column A	Column B	Answers	
1.	$2x + 1$ ($x = 6$)	$4x - 3$	Ⓐ Ⓑ Ⓒ Ⓓ [Lesson 1.3]	B
2.	$(-11) \cdot (121)$	$(121) \div (-11)$	Ⓐ Ⓑ Ⓒ Ⓓ [Lesson 2.4]	B
3.	$(3y + 2) + (2y - 5)$	$(2y + 6) - (y + 7)$	Ⓐ Ⓑ Ⓒ Ⓓ [Lesson 2.6]	D
4.	The solution to the equation $-\frac{3}{4}x = 3$	The solution to the equation $-\frac{4}{3}x = -4$	Ⓐ Ⓑ Ⓒ Ⓓ [Lesson 3.3]	B

5. Which expression is equal to 19? **(LESSON 1.4)** d
  a. $15 \cdot (6 + 4) - 57$
  b. $2^3 - 6 \cdot (3 + 2)$
  c. $4^2 - 5^2 \div 5 + 2(0)$
  d. $9 \div 3 \cdot 2^3 - 5$

6. What is the value of $-5\frac{5}{6} \cdot \frac{3}{5}$? **(LESSON 2.4)** c
  a. $-5\frac{1}{2}$
  b. $2\frac{1}{3}$
  c. $-3\frac{1}{2}$
  d. none of these

7. Which expression is *not* equivalent to $\frac{32 - 8x}{4}$? **(LESSON 2.7)** d
  a. $8 - 2x$
  b. $\frac{16 - 4x}{2}$
  c. $(32 - 8x) \div 4$
  d. none of these

8. Which expression is not equal to the others? **(LESSON 2.6)** c
  a. $3x - 2(x - 3)$
  b. $3x + 2(3 - x)$
  c. $3x - 2(x + 3)$
  d. $3x - 2(-3 + x)$

9. What is the solution to the equation $\frac{q}{139.2} = -58$? **(LESSON 3.2)** b
  a. -2.4
  b. -8073.6
  c. -0.417
  d. -7551.6

Find each sum. **(LESSON 2.2)**

10. $-25 + (-4)$ -29

11. $-36 + 6 + (-6)$ -36

12. $452 + (-452)$ 0

13. Solve the equation $-3w + 5 = -16$. **(LESSON 3.5)** $w = 7$

14. The perimeter of a rectangle is 52 and its width is 6 less than its length. Find the length and width of the rectangle. **(LESSON 3.6)** $l = 16$, $w = 10$

15. Solve: $6 - (3 + 2x) = -3x + (2x - 4)$. **(LESSON 3.5)** $x = 7$

16. Solve: $y + 4.5 = 6$. **(LESSON 3.1)** $y = 1.5$

17. Solve: $2(x + 7) = 15$. **(LESSON 3.5)** $x = \frac{1}{2}$

18. Solve: $3p - 2 + 2p = 8$. **(LESSON 3.3)** $p = 2$

Solve each equation. (LESSON 3.1)

19. $x + 7 = 6$ $x = -1$ **20.** $t - 9 = 11$ $t = 20$
21. $z - 8.4 = 1.25$ $z = 9.65$ **22.** $82 = 21 + r$ $r = 61$

Determine the next three terms of each sequence. (LESSON 1.1)

23. 1, 4, 8, 13, 19, …   26, 34, 43
24. 2, 4, 8, 14, 22, …   32, 44, 58
25. 10, 5, 0, −5, …   −10, −15, −20
26. 3, 6, 12, 24, …   48, 96, 192

27. Use the graph of Ellen's earnings below to determine how much she would earn if she worked 12 hours. (LESSON 1.5) $72

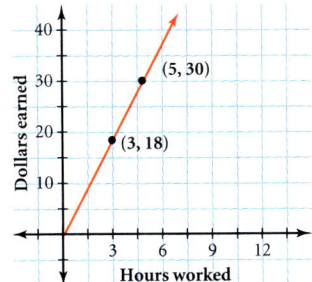

Solve each equation for the variable indicated. (LESSONS 3.1, 3.2, 3.3 AND 3.6)

28. $a + 7.5 = 3.2$, for a $a = -4.3$
29. $-6.5 = x + 1.7$, for x $x = -8.2$
30. $T = I + P$, for P $P = T - I$
31. $d = rt$, for r $r = \frac{d}{t}$
32. $3x + 7 = 2x - 6$, for x $x = -13$
33. $\frac{x}{4} + 2 = 3x + 35$, for x $x = -12$
34. $-5 + 2p = 3p$, for p $p = -5$

FREE-RESPONSE GRID
The following questions may be answered by using a free-response grid such as that commonly used by standardized test services.

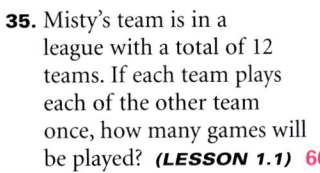

35. Misty's team is in a league with a total of 12 teams. If each team plays each of the other team once, how many games will be played? (LESSON 1.1)   66

36. Evaluate: $(-4)(4) \div (-1)$. (LESSON 2.4)   16

37. Use the method of differences to find the next term of the sequence 45, 42, 37, 30, 21, … (LESSON 1.1)   10

38. Use mental math to simplify $(45 + 23) + 107$. (LESSON 2.5)   175

Determine the x-coordinate of each point listed below. (LESSON 1.4)

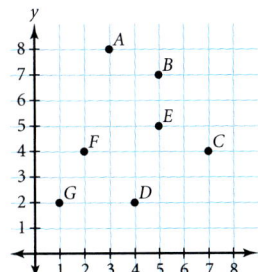

39. point A 3
40. point D 4
41. point F 2
42. point G 1

4 Proportional Reasoning and Statistics

CHAPTER PLANNING GUIDE

Lesson	4.1	4.2	4.3	4.4	4.5	4.6	Projects and Review
Pupil Edition Pages	164–170	171–177	180–185	186–192	193–200	201–207	178–179, 208–215
Extra Practice (Pupil Edition)	747	748	748	749	749	750	
Practice Workbook	20	21	22	23	24	25	
Reteaching Masters	39–40	41–42	43–44	45–46	47–48	49–50	
Enrichment Masters	20	21	22	23	24	25	
Cooperative-Learning Activities	20	21	22	23	24	25	
Lesson Activities	20	21	22	23	24	25	
Problem Solving/ Critical Thinking	20	21	22	23	24	25	
Student Study Guide	20	21	22	23	24	25	
Spanish Resources	20	21	22	23	24	25	
Student Technology Guide	20	21	22	23	24	25	
Assessment Resources	45	46	47	49	50	51	48, 52–57
Teaching Transparencies	17	18		19	20, 21, 22, 23	24	
Quiz Transparencies	4.1	4.2	4.3	4.4	4.5	4.6	
Writing Activities for Your Portfolio							10–12
Tech Prep Masters							15–18
Long-Term Projects							13–16

LESSON PACING GUIDE

Traditional	2 days	2 days	2 days	2 days	1 day	1 day	2 days
Block	1 day	1 day	1 day	1 day	$\frac{1}{2}$ day	$\frac{1}{2}$ day	1 day
Two-Year	4 days	4 days	4 days	4 days	2 days	2 days	4 days

CONNECTIONS AND APPLICATIONS

Lesson	4.1	4.2	4.3	4.4	4.5	4.6	Review
Algebra	164–170	171–177	180–185	186–192	193–200	201–207	210–215
Geometry	167, 168, 169, 170						
Business and Economics	167, 169				195		212
Statistics		177		190, 191	196, 198, 199	206	
Probability			184				
Life Skills	169	173, 174, 176, 177			199		212
Science and Technology	168	171, 175	185	191	200	203, 204, 206	
Sports and Leisure	164, 169		184	187, 189–192	197, 199	207	
Other		177			199		212

BLOCK-SCHEDULING GUIDE

Day	Lesson	Teacher Directed: Lesson Examples, Teaching Transparencies	Student Guided: Activity, Try This	Cooperative-Learning Activity, Lesson Activity, Student Technology Guide	Practice: Practice & Apply, Extra Practice, Practice Workbook	Assessment: Quiz, Mid-Chapter Assessment	Problem Solving, Reteaching
1	4.1	10 min	15 min	15 min	65 min	15 min	15 min
2	4.2	15 min	15 min	15 min	65 min	15 min	15 min
3	4.3	10 min	15 min	15 min	65 min	15 min	15 min
4	4.4	10 min	15 min	15 min	65 min	15 min	15 min
5	4.5	5 min	8 min	8 min	35 min	8 min	8 min
	4.6	5 min	7 min	7 min	30 min	7 min	7 min
6	Assess.	50 min PE: Chapter Review	90 min PE: Chapter Project, Writing Activities	90 min Tech Prep Masters	65 min PE: Chapter Assessment, Test Generator	30 min Chap. Assess. (A or B), Alt. Assess. (A or B), Test Generator	

PE: Pupil's Edition

Alternative Assessment

The following are suggestions for an alternative assessment for students who may benefit from a different type of assessment than the regular chapter quizzes and the mid-chapter and end-of-chapter assessment materials. Many of the questions are open-ended, and students' answers will vary.

Performance Assessment

1. Use the equation *percentage = base × rate* in the problems below.
   a. Write and solve three equations: one in which the percentage is unknown, one in which the rate is unknown, and one in which the base is unknown.
   b. Suppose that you buy an item in a store and that there is a sales tax. In this situation, what are the base, rate, and percentage? Give an example that shows you have identified each quantity correctly.

2. Suppose that you want to find the experimental probability of rolling 2 on a number cube. How could you use fractions to help find the probability?

3. a. Describe and illustrate a situation in which you could represent data with a circle graph.
   b. Describe and illustrate a situation in which you could represent data with a line graph or a bar graph.

Portfolio Project

Suggest that students choose one of the following projects for inclusion in their portfolios.

1. A *scale drawing* is a two-dimensional representation of an object. The drawing may be larger or smaller than the actual object. The scale relates lengths of the scale drawing to the lengths of the actual object.
   a. Choose a fairly simple object and sketch a scale drawing of it.
   b. Describe how proportion plays a role in your work.

2. Conduct a simple statistical survey, such as finding heights of plants that are the same age. Describe how statistical measures and graphs could help organize the data.

internetconnect

The table below identifies the pages in this chapter that contain technology information and support in the side columns.

Content Links	
Lesson Links	pages 175, 183, 192
Portfolio Links	pages 185, 207
Graphics Calculator Support	page 782

Resource Links

For information about teacher and parent resources as well as professional development help, visit **www.hrw.com/math**.

Technical Support HRW has assembled a team of dedicated technical and teaching professionals and a comprehensive service program to provide you with the support you need.

- The HRW Technical Support Line operates from 7 A.M. to 6 P.M. central time, Monday through Friday, at (800) 323-9239.
- The HRW Technical Support Center on the World Wide Web is available 24 hours a day, seven days a week, at **www.hrwtechsupport.com**.
- You can e-mail our Technical Support Center at **tsc@hrwtechsupport.com**.
- The Technical Support Center's fax-on-demand service at (800) 352-1680 offers solutions to common problems and answers to frequently asked questions.

Technology

Lesson Suggestions and Calculator Examples
(Keystrokes are based on a TI-83 calculator.)

Lesson 4.1 Using Proportional Reasoning
In this lesson, students will learn that there is more than one way to determine whether two ratios are equal. Have students use a calculator to evaluate $\frac{8}{10} \stackrel{?}{=} \frac{24}{30}$, as shown at right. Ask students how the third evaluation verifies that $\frac{8}{10} = \frac{24}{30}$ is true.

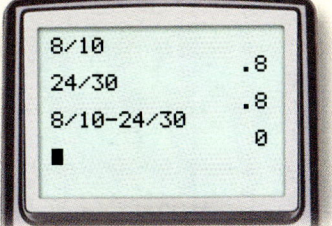

Lesson 4.2 Percent Problems
Have students consider the statement "20 is 80% of what number." Show them the two approaches listed below.

$$20 \times 80\% = x \qquad 20 = 80\% \times x$$

Then ask students which equation will give the correct answer. Elicit the response that the equation on the right is correct. Reason indicates that x must be greater than 20. Have students write the equation in calculator-ready form by converting the percent to a decimal and then isolating the x on one side of the equation. Check to see that students arrive at the correct answer, 25.

Lesson 4.3 Introduction to Probability
In this lesson, students will use their knowledge of fractions to solve a problem that deals with experimental probability. As an extension to the lesson, consider posing the following problem:

In an coin-toss experiment, heads appeared 48 times in 100 tosses. What can you say about the experimental probability if in the next 100 tosses, heads occur 46 times? 52 times?

The calculator display at right shows that when the number of heads decreases (increases), the experimental probability decreases (increases).

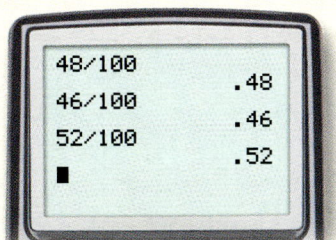

Lesson 4.4 Measures of Central Tendency
Help students learn to calculate statistical measures on the calculator.

1. To enter data into the list, press STAT, and select 1:Edit...
2. To calculate statistical measures for the list, press STAT, select CALC 1:1-Var Stats, and then press ENTER.

The calculator will display a list of statistical measures including the mean, $\bar{x}$, and the median, Med. The display will show other statistical measures that are not relevant here. On later occasions, students can discover the meaning of those other measures.

Lesson 4.5 Graphing Data
In this lesson, students will learn about line graphs, one type of statistical display.

Students will need an organized list of steps that they can follow in order to make the plot.

1. Enter the data into list 1 (**L1**) and list 2 (**L2**) by pressing STAT and selecting 1:Edit...

2. Press 2nd Y= ENTER to view the plot selection screen. Select the line graph.

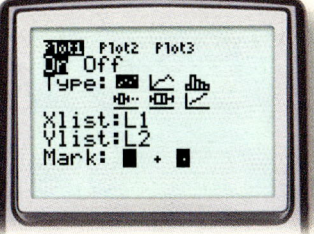

3. Press WINDOW to adjust the window settings and then display the graph by pressing GRAPH.

Have students make a line graph for each data set below.

(1, 2), (3, 5), (5, 8), (7, 11)   (1, 2), (3, 6), (5, 7), (7, 11)

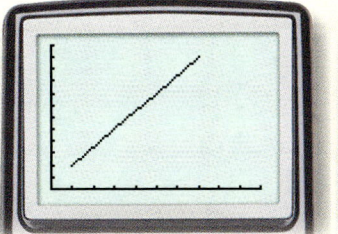 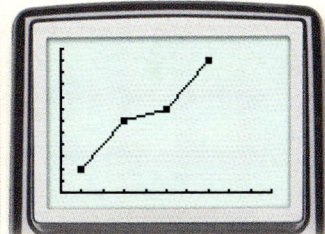

Discuss similarities and differences in the data sets.

Lesson 4.6 Other Data Displays
Take time to teach students the valuable skill of making a box-and-whisker plot. Press 2nd Y= ENTER in order to view the plot selection screen. Students will need to practice using the arrow keys to make the needed selections. (For an example of how to create a box-and-whisker plot, see page 786.)

For further information, refer to the

- technology discussions in the lessons.
- lesson-related teacher's commentary in the side columns of this *Teacher's Edition*.
- lesson-related *Student Technology Guide* masters.
- *HRW Technology Handbook*.

Teaching Resources

Basic Skills Practice
(2 pages per skill)

Reteaching
(2 pages per lesson)

Enrichment
(1 page per lesson)

Cooperative-Learning Activity
(1 page per lesson)

Lesson Activity
(1 page per lesson)

Long-Term Project
(4 pages per chapter)

About Chapter 4

Background Information

In Chapter 4, students study the familiar concepts of proportion and percent from an algebraic perspective. They also learn how these concepts are applied to basic principles of experimental probability and to fundamental techniques in the study of statistics. The statistics sequence begins with an examination of the most basic of statistical measures: mean, median, mode, and range. It then proceeds to a study of fundamental data displays: bar graphs, line graphs, circle graphs, stem-and-leaf plots, histograms, and box-and-whisker plots.

CHAPTER RESOURCES

- Block-Scheduling Handbook
- Writing Activities for Your Portfolio
- Tech Prep Masters
- Long-Term Project
- Assessment Resources
  Mid-Chapter Assessment
  Chapter Assessments
  Alternative Assessments
- Test and Practice Generator
- Technology Handbook
- End-of-Course Exam Practice

4 Proportional Reasoning and Statistics

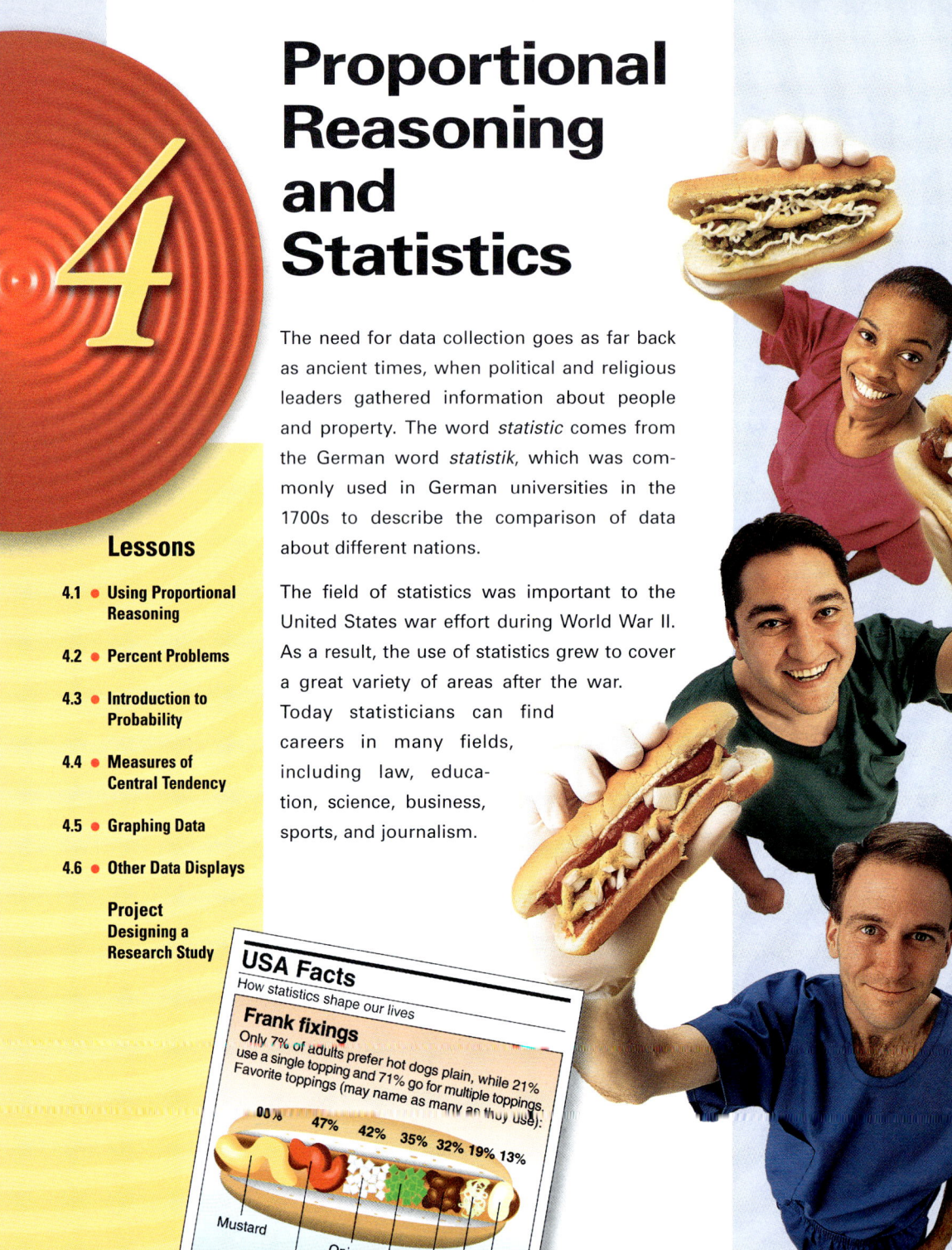

The need for data collection goes as far back as ancient times, when political and religious leaders gathered information about people and property. The word *statistic* comes from the German word *statistik*, which was commonly used in German universities in the 1700s to describe the comparison of data about different nations.

The field of statistics was important to the United States war effort during World War II. As a result, the use of statistics grew to cover a great variety of areas after the war. Today statisticians can find careers in many fields, including law, education, science, business, sports, and journalism.

Lessons

4.1 • Using Proportional Reasoning

4.2 • Percent Problems

4.3 • Introduction to Probability

4.4 • Measures of Central Tendency

4.5 • Graphing Data

4.6 • Other Data Displays

Project
Designing a Research Study

About the Photos

The photos show different statistics about everyday life. Tell students that they need to educate themselves about statistics because deceptive or false conclusions can be made to sound legitimate by using statistical language.

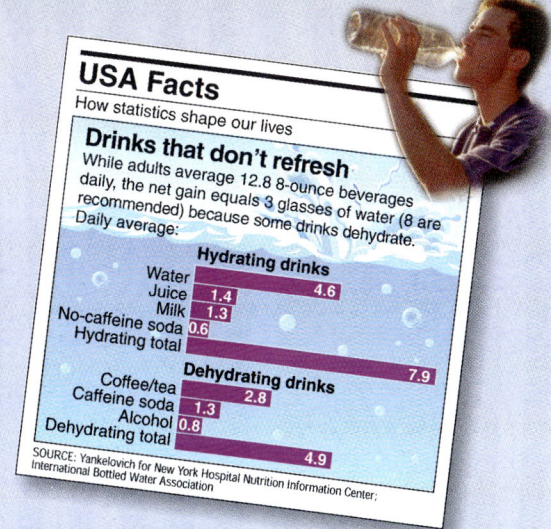

Central Park, New York City

Chapter Objectives

- Identify the means and extremes of a proportion. [4.1]
- Use proportions to solve problems. [4.1]
- Find equivalent fractions, decimals, and percents. [4.2]
- Solve problems involving percent. [4.2]
- Find the experimental probability that an event will occur. [4.3]
- Find the mean, median, mode, and range of a data set. [4.4]
- Represent data with frequency tables. [4.4]
- Interpret line graphs, bar graphs, and circle graphs. [4.5]
- Represent data with circle graphs. [4.5]
- Analyze graphs in order to find a misleading presentation of data. [4.5]
- Interpret stem-and-leaf plots, histograms, and box-and-whisker plots. [4.6]
- Represent data with stem-and-leaf plots, histograms, and box-and-whisker plots. [4.6]

About the Chapter Project

The Chapter Project, *Designing a Research Study*, on page 208, provides an opportunity for you to design your own research study, compile the results, and report on the results. After completing the Chapter Project, you will be able to do the following:

- Analyze data by using graphs, plots, and measures of central tendency.

About the Portfolio Activities

Throughout the chapter, you will be given opportunities to complete Portfolio Activities that are designed to support your work on the Chapter Project.

- Soft-drink preferences are the topic of the Portfolio Activity on page 170.
- Simulating a probability experiment is the topic of the Portfolio Activity on page 185.
- Graphing test scores is the topic of the Portfolio Activity on page 207.

Portfolio Activities appear at the end of Lessons 4.1, 4.3, and 4.6. Each serves as preparation for the Chapter Project. The Portfolio Activities and the Chapter Project are appropriate for inclusion in the student's portfolio. Students should be encouraged to include in their portfolios any other work in which they feel a sense of pride or a sense of accomplishment.

internet connect

Internet Connect features appearing throughout the chapter provide keywords for the HRW Web site that lead to links for math resources, tutorial assistance, references for student research, classroom activities, and all teaching resource pages. The HRW Web site also provides updates to the technology used in the text. Listed at right are the keywords for the Internet Connect activities referenced in this chapter. Refer to the side column on the page listed to read about the activity.

LESSON	KEYWORD	PAGE
4.2	MA1 Rural Exodus	175
4.3	MA1 Simpson's Paradox	183
4.3	MA1 Coin Toss	185
4.4	MA1 Voting Schemes	192
4.6	MA1 Endangered Species	207

CHAPTER 4 **163**

Prepare

NCTM PRINCIPLES & STANDARDS 1–3, 5–10

LESSON RESOURCES

- Practice 4.1
- Reteaching Master 4.1
- Enrichment Master 4.1
- Cooperative-Learning Activity 4.1
- Lesson Activity 4.1
- Problem Solving/Critical Thinking Master 4.1
- Teaching Transparency 17

QUICK WARM-UP

Write each as a fraction in simplest form.

1. $\frac{4}{16}$ $\frac{1}{4}$ 2. $\frac{28}{35}$ $\frac{4}{5}$
3. $\frac{48}{64}$ $\frac{3}{4}$ 4. $\frac{72}{44}$ $\frac{18}{11}$

Solve each equation.

5. $6k = 96$ $k = 16$
6. $54 = 3n$ $n = 18$
7. $32r = 4$ $r = 0.125$
8. $0.5y = 15$ $y = 30$

Also on Quiz Transparency 4.1

Teach

Why Students may not realize how often they use proportional reasoning in their everyday lives. Lead a class discussion in which students make a list of situations that involve proportional reasoning. The list probably will include situations such as decreasing or increasing the yield of a recipe, reading a map scale, and building a model airplane. Point out that in this lesson students will learn how algebra can be used as a tool to solve proportion problems.

164 LESSON 4.1

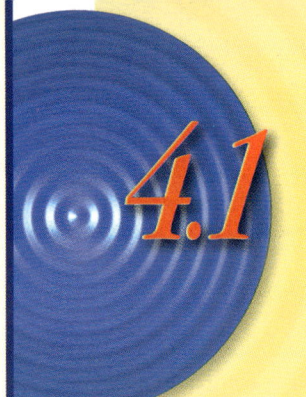

4.1 Using Proportional Reasoning

Objectives

- Identify the means and extremes of a proportion.
- Use proportions to solve problems.

Why Proportional reasoning is an important tool in many professions, including science, engineering, and business. Proportions are also common in everyday activities such as cooking and gardening.

Proportions

APPLICATION SPORTS

In the first round of a timed free-throw contest, Ben made 28 out of 40 shots and Alicia made 21 out of 30 shots. Who was more accurate?

Ben	Alicia
$\frac{28}{40} = \frac{7}{10}$	$\frac{21}{30} = \frac{7}{10}$

Because both fractions simplify to $\frac{7}{10}$, $\frac{28}{40}$ is equal to $\frac{21}{30}$. The contestants were equally accurate. Fractions, such as $\frac{7}{10}$ and $\frac{15}{5}$, are *ratios*. A **ratio** is a comparison of two quantities that uses division.

The equation $\frac{28}{40} = \frac{21}{30}$ is called a **proportion**. Any statement that two ratios are equal, $\frac{a}{b} = \frac{c}{d}$, is a proportion. You can use proportions to solve many types of problems.

Alternative Teaching Strategy

USING COGNITIVE STRATEGIES Some students intuitively see a "mental math" alternative for solving certain proportions. Here is an example.

$\frac{5}{2} = \frac{n}{6}$ 6 is equal to 2 • 3, so n must be equal to 5 • 3, or 15.

Remind students of the Identity Property for Multiplication: The product of any number and 1 is equal to the number. Show how this property and the Substitution Property together justify the method shown above.

$\frac{5}{2} = \frac{5}{2} \cdot 1 = \frac{5}{2} \cdot \frac{3}{3} = \frac{15}{6}$! $n = 15$

Now have students apply this method to the proportions below.

1. $\frac{9}{4} = \frac{36}{r}$ 36 is equal to 9 • 4, so r is equal to 4 • 4, or 16.
2. $\frac{7}{3} = \frac{42}{a}$ 42 is equal to 7 • 6, so a is equal to 3 • 6, or 18.
3. $\frac{8}{9} = \frac{12}{z}$ 12 is equal to 8 • 1.5, so z is equal to 9 • 1.5, or 13.5.

Means and Extremes

A proportion such as $\frac{a}{b} = \frac{c}{d}$ can also be written in the form $a:b = c:d$. The inside terms, b and c, are called the **means**. The outside terms, a and d, are called the **extremes**. The product of the means and the product of the extremes can be used to solve proportion problems.

Activity
Exploring Relationships in Proportions

1. For each proportion, identify the means and the extremes.
   a. $\frac{1}{2} = \frac{5}{10}$ b. $\frac{3}{4} = \frac{9}{12}$
   c. $\frac{15}{25} = \frac{30}{50}$ d. $\frac{4}{6} = \frac{6}{9}$

2. For each proportion in Step 1, find the product of the means and the product of the extremes. (For example, in part **a**, find $2 \cdot 5$ and $1 \cdot 10$.)

3. Describe what you observe about the relationship between the product of the means and the product of the extremes.

4. For any proportion $\frac{a}{b} = \frac{c}{d}$, where $b \neq 0$ and $d \neq 0$, ad and bc are called cross products. Explain how cross products and products of the means and extremes are related.

CHECKPOINT ✓

5. Explain how you can use the cross products to solve the following proportion for x:

$$\frac{x}{4} = \frac{42}{24}$$

In a true proportion, the product of the means and the product of the extremes, also called **cross products**, are equal. Because of this relationship, it is possible to use cross products to determine whether a proportion is true and to solve for a missing value in a proportion.

Cross Products Property

For all real numbers a, b, c, and d, where $b \neq 0$ and $d \neq 0$:

If $\frac{a}{b} = \frac{c}{d}$, then $a \cdot d = b \cdot c$.

Interdisciplinary Connection

CHEMISTRY To obtain a mixture needed for a laboratory experiment, 11 ounces of distilled water must be added for every 5 ounces of a chemical solution. Suppose that you need a total of 3 quarts of the mixture. How many ounces of distilled water do you need? How many ounces of the chemical solution? (Hint: 1 quart = 32 ounces)
66 ounces; 30 ounces

Inclusion Strategies

AUDITORY/LINGUISTIC LEARNERS Students with a more verbal orientation may look at the proportion $\frac{a}{b} = \frac{c}{d}$ and classify it as "just another equation." To reinforce the fact that it represents a very special type of reasoning, encourage them to always read a proportion as "*a* is to *b* as *c* is to *d*." You might also lead a discussion in which they compare the process of solving a proportion to the process of completing a linguistic analogy such as "*big* is to *small* as *tall* is to ? ."

Activity Notes

In Steps 1–4 of the Activity, students will discover that in a true proportion, the product of the means is equal to the product of the extremes. After they complete Step 4, you may wish to give them one or more proportions that *clearly are not* true, such as $\frac{2}{3} = \frac{4}{5}$, and demonstrate that the cross products are not equal. In Step 5, students are asked to consider how cross products can be used to find an unknown term of a proportion.

Cooperative Learning

You may wish to have students work on the Activity in pairs. In Step 2, one student in each pair can be the *means calculator*, while the other is the *extremes calculator*. Each student in the pair should write the appropriate multiplication expressions and products in a column on a sheet of paper, as shown here.

Means	Extremes
$2 \cdot 5 = 10$	$1 \cdot 10 = 10$
$4 \cdot 9 = 36$	$3 \cdot 12 = 36$
$25 \cdot 30 = 750$	$15 \cdot 50 = 750$
$6 \cdot 6 = 36$	$4 \cdot 9 = 36$

When the partners compare their results, the two lists of identical products should provide visual reinforcement of the Cross-Products Property.

CHECKPOINT ✓

5. Set the cross products equal to each other and then solve for x.
   $x \cdot 24 = 4 \cdot 42$
   $x = 7$

ADDITIONAL EXAMPLE 1

Use cross products to determine whether each proportion is true.

a. $\frac{6}{27} = \frac{2}{9}$

$6 \cdot 9 = 54, 27 \cdot 2 = 54$; true

b. $\frac{4}{3} = \frac{6}{5}$

$4 \cdot 5 = 20, 3 \cdot 6 = 18$; false

ADDITIONAL EXAMPLE 2

Solve each proportion.

a. $\frac{12}{5} = \frac{15}{r}$ $r = 6.25$

b. $\frac{6}{y} = \frac{4}{82}$ $y = 123$

c. $\frac{9.2}{23} = \frac{a}{5}$ $a = 2$

d. $\frac{d}{0.8} = \frac{6}{0.16}$ $d = 30$

TRY THIS
a. $x = 7$
b. $n = 10$

CRITICAL THINKING
Multiply both sides of the equation by the least common multiple of the denominators to solve for x.

Use Teaching Transparency 17.

Teaching Tip

If students have difficulty remembering which terms of a proportion are the means and which are the extremes, encourage them to write the proportion in the form $a:b = c:d$. In this form, it is easier to see that the terms a and d are in the "extreme" positions.

166 LESSON 4.1

EXAMPLE 1

Use cross products to determine whether each proportion is true.

a. $\frac{2}{3} \stackrel{?}{=} \frac{3}{4}$

b. $\frac{3}{5} \stackrel{?}{=} \frac{15}{25}$

● SOLUTION

a. Find the cross products.
$2 \cdot 4 = 8 \quad 3 \cdot 3 = 9 \quad 8 \neq 9$

The proportion is not true.

$\frac{2}{3} \neq \frac{3}{4}$

b. Find the cross products.
$3 \cdot 25 = 75 \quad 5 \cdot 15 = 75 \quad 75 = 75$

The proportion is true.

$\frac{3}{5} = \frac{15}{25}$

When one term of a proportion is unknown, such as in $\frac{3}{7} = \frac{x}{14}$, you can use cross products to find the value of the variable that makes a true proportion. First find the cross products and set them equal to each other.

$$7 \cdot x = 3 \cdot 14$$
$$7x = 42$$

Then solve for the variable.

$$7x = 42$$
$$\frac{7x}{7} = \frac{42}{7}$$
$$x = 6$$

EXAMPLE 2

Solve each proportion.

a. $\frac{15}{m} = \frac{10}{9}$

b. $\frac{5}{60} = \frac{3}{x}$

c. $\frac{x}{8} = \frac{7.5}{20}$

d. $\frac{9}{3} = \frac{m}{5}$

● SOLUTION

a. $\frac{15}{m} = \frac{10}{9}$

$15 \cdot 9 = 10m$
$135 = 10m$
$\frac{135}{10} = \frac{10m}{10}$
$13.5 = m$
$m = 13.5$

b. $\frac{5}{60} = \frac{3}{x}$

$5x = 180$
$\frac{5x}{5} = \frac{180}{5}$
$x = 36$

c. $\frac{x}{8} = \frac{7.5}{20}$

$20x = 60$
$\frac{20x}{20} = \frac{60}{20}$
$x = 3$

d. $\frac{9}{3} = \frac{m}{5}$

$45 = 3m$
$\frac{45}{3} = \frac{3m}{3}$
$15 = m$
$m = 15$

TRY THIS Solve each proportion. a. $\frac{x}{3.5} = \frac{5}{2.5}$ b. $\frac{13}{2} = \frac{65}{n}$

CRITICAL THINKING What steps would be required to solve each proportion in Example 2 without using cross products?

Enrichment

Justify each step in this proof, for all numbers a and c and all nonzero numbers b and d.

$\frac{a}{b} = \frac{c}{d}$ Given

$\left(\frac{a}{b}\right)(bd) = \left(\frac{c}{d}\right)(bd)$ Mult. Prop. of Equality

$\left(\frac{a}{b}\right)(bd) = (bd)\left(\frac{c}{d}\right)$ Comm. Prop. of Mult.

$\left(\frac{a}{b} \cdot b\right)d = b\left(d \cdot \frac{c}{d}\right)$ Assoc. Prop. of Mult.

$(a \cdot 1)d = b(1 \cdot c)$ Mult. Inverse Property

$ad = bc$ Comm. Prop. of Mult

Now have students prove the statements below. Assume that a, b, c, and d represent nonzero real numbers.

1. If $\frac{a}{b} = \frac{c}{d}$, then $\frac{a}{c} = \frac{b}{d}$.

$ad = bc$; $\frac{ad}{cd} = \frac{bc}{cd}$; $\frac{a}{c} = \frac{b}{d}$

2. If $\frac{a}{b} = \frac{c}{d}$, then $\frac{b}{a} = \frac{d}{c}$.

$ad = bc$; $\frac{ad}{ac} = \frac{bc}{ac}$; $\frac{d}{c} = \frac{b}{a}$; $\frac{b}{a} = \frac{d}{c}$

3. If $\frac{a}{b} = \frac{c}{d}$, then $\frac{a+b}{b} = \frac{c+d}{d}$.

$\frac{a}{b} + \frac{b}{b} = \frac{c}{d} + \frac{d}{d}$; $\frac{a+b}{b} = \frac{c+d}{d}$

EXAMPLE 3

CONNECTION
GEOMETRY

Given: rectangle ABCD ~ rectangle WXYZ. **Find s.**

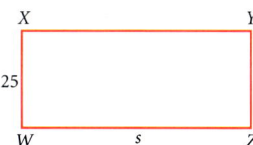

● SOLUTION

Because the rectangles are similar, the corresponding sides are proportional.
Set up a proportion. $\frac{9}{25} = \frac{27}{s}$

Solve the proportion for s. $675 = 9s$
$75 = s$

The length is 75.

TRY THIS

Given: triangle MNO ~ triangle UVW.
Find x.

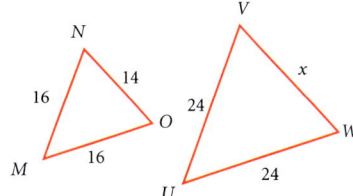

EXAMPLE 4

APPLICATION
INVENTORY

The goal of the Bartlesville City Library is to have 2 books in stock for every 5 residents. The most recent census shows that the population is 12,455. **Find the total number of books that the library should have in order to meet the goal.**

● SOLUTION

PROBLEM SOLVING

First **make a table** to organize the information.

Number of books	2	n
Number of residents	5	12,455

Write the proportion shown in the table. $\frac{2}{5} = \frac{n}{12,455}$

Then use cross products to solve the proportion.
$5n = 2 \cdot 12{,}455$
$5n = 24{,}910$
$\frac{5n}{5} = \frac{24{,}910}{5}$
$n = 4982$

The library should have 4982 books in order to meet its goal.

Reteaching the Lesson

USING COGNITIVE STRATEGIES Give students the following problem: A recipe that yields 24 muffins requires 2 cups of flour. How much flour is needed to make 40 of these muffins?

Remind students that a ratio compares two quantities. Ask what quantities are being compared.
cups of flour, number of muffins

Now show them how to use these answers to set up a proportion as shown at right.

$$\begin{array}{c}\text{Recipe} \quad \text{Actual}\end{array}$$
$$\frac{\text{cups of flour}}{\text{no. of muffins}} \to \frac{\Box}{\Box} = \frac{\Box}{\Box} \leftarrow \frac{\text{cups of flour}}{\text{no. of muffins}}$$

Help them use the set-up to solve the problem.

$\frac{2}{24} = \frac{n}{40}$; $n = 3\frac{1}{3}$; **it takes $3\frac{1}{3}$ cups of flour to make 40 muffins.**

ADDITIONAL EXAMPLE 3

In the figure below, $\triangle GHJ \sim \triangle KLM$. **What is the unknown length, z?**

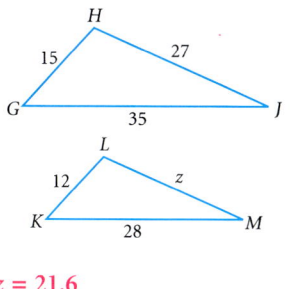

$z = 21.6$

Math CONNECTION

GEOMETRY Similar figures are geometric figures that have the same shape, but not necessarily the same size. If the similar figures are polygons, then their vertices can be paired in such a way that corresponding angles are congruent and corresponding sides are proportional. The vertices of the figures are always named in the order of the correspondence. Be sure students know that you read the symbol ~ as "is similar to."

TRY THIS
$x = 21$

ADDITIONAL EXAMPLE 4

Jenny hiked 14 miles in 3 hours. **If she continues to hike at this pace, how long will it take her to hike 35 miles?**

7.5 hours

LESSON 4.1

Assess

Selected Answers
Exercises 5–12, 13–63 odd

ASSIGNMENT GUIDE

In Class	1–12
Core	13–18, 25–39, 47–51 odd
Core Plus	14–44 even, 46–51
Review	52–64
Preview	65, 66

 Extra Practice can be found beginning on page 738.

Exercises

Communicate

1. What is a proportion? Give an example.
2. Explain how to find the means and the extremes of a proportion.
3. Make a statement about the relationship between the means and the extremes of a proportion.
4. Describe two ways to solve for n in the proportion $\frac{n}{36} = \frac{2}{3}$.

Guided Skills Practice

Determine whether each proportion is true. *(EXAMPLE 1)*

5. $\frac{3}{4} \stackrel{?}{=} \frac{6}{8}$ **True**
6. $\frac{12}{19} \stackrel{?}{=} \frac{5}{7}$ **Not true**
7. $\frac{4}{21} \stackrel{?}{=} \frac{16}{84}$ **True**

Solve each proportion. *(EXAMPLE 2)*

8. $\frac{5}{7} = \frac{x}{21}$ **15**
9. $\frac{10}{p} = \frac{85}{153}$ **18**
10. $\frac{12}{14} = \frac{84}{m}$ **98**

CONNECTION

11. **GEOMETRY** Rectangle *QRST* and rectangle *LMNO* are similar. Write and solve a proportion to find the missing length, *b*. *(EXAMPLE 3)*

$b = 18\frac{2}{3}$

APPLICATION

12. **BIOLOGY** To keep the proper balance between fish species A and fish species B in an experimental fish pond, a wildlife biologist needs to keep the ratio of species A to species B at 2 to 3. Write and solve a proportion to find how many fish of species B are needed if there are 256 fish of species A. *(EXAMPLE 4)*
384 of species B

Practice and Apply

Determine whether each proportion is true.

13. $\frac{15}{9} \stackrel{?}{=} \frac{35}{21}$ **True**
14. $\frac{12}{9} \stackrel{?}{=} \frac{18}{12}$ **Not true**
15. $\frac{56}{24} \stackrel{?}{=} \frac{49}{21}$ **True**
16. $\frac{27}{21} \stackrel{?}{=} \frac{35}{28}$ **Not true**
17. $\frac{18}{8} \stackrel{?}{=} \frac{108}{48}$ **True**
18. $\frac{3}{5} \stackrel{?}{=} \frac{81}{135}$ **True**
19. $\frac{3}{13} \stackrel{?}{=} \frac{10}{65}$ **Not true**
20. $\frac{12}{20} \stackrel{?}{=} \frac{27}{45}$ **True**
21. $\frac{8}{10} \stackrel{?}{=} \frac{24}{30}$ **True**
22. $\frac{12}{8} \stackrel{?}{=} \frac{48}{34}$ **Not true**
23. $\frac{24}{3} \stackrel{?}{=} \frac{72}{12}$ **Not true**
24. $\frac{12}{16} \stackrel{?}{=} \frac{60}{80}$ **True**

168 LESSON 4.1

Solve each proportion. Round to the nearest hundredth if necessary.

25. $\frac{27}{18} = \frac{42}{n}$ 28
26. $\frac{38}{19} = \frac{n}{20}$ 40
27. $\frac{42}{28} = \frac{36}{n}$ 24
28. $\frac{n}{48} = \frac{72}{96}$ 36
29. $\frac{21.5}{x} = \frac{64.5}{18}$ 6
30. $\frac{x}{37.2} = \frac{16}{24.8}$ 24
31. $\frac{30.8}{112} = \frac{y}{10}$ 2.75
32. $\frac{t}{25} = \frac{471}{15}$ 785
33. $\frac{1.2}{1.5} = \frac{8}{m}$ 10
34. $\frac{72}{34} = \frac{9}{g}$ 4.25
35. $\frac{7}{15} = \frac{f}{48}$ 22.4
36. $\frac{4}{m} = \frac{1.6}{22}$ 55
37. $\frac{21}{36} = \frac{4.8}{d}$ 8.23
38. $\frac{21}{56} = \frac{x}{7.2}$ 2.7
39. $\frac{p}{12} = \frac{21}{63}$ 4
40. $\frac{45}{18} = \frac{3}{k}$ 1.2
41. $\frac{22}{36} = \frac{x}{198}$ 121
42. $\frac{60}{21} = \frac{20}{s}$ 7
43. $\frac{52}{13} = \frac{8}{q}$ 2
44. $\frac{y}{16} = \frac{12}{4}$ 48
45. $\frac{74}{p} = \frac{92}{46}$ 37

9.6 ft 46. Suppose that a 6-foot tall person casts a shadow that is 10 feet long. A nearby statue casts a 16-foot shadow. How tall is the statue?

19.5 ft 47. A 52-foot building casts a shadow that is 48 feet long. At the same time, the shadow cast by a nearby statue is 18 feet long. How tall is the statue?

CONNECTION

48. **GEOMETRY** Triangle *ABC* is similar to triangle *MNP*. Find *s*.

s = 2.52 in.

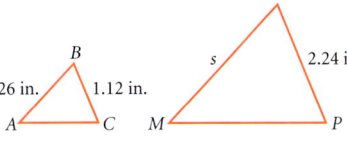

APPLICATIONS

6.4 lbs 49. **CONSUMER ECONOMICS** A fertilizer manufacturer recommends 4 pounds of fertilizer for every 1500 square feet of lawn. How many pounds of fertilizer are needed to cover 2400 square feet of lawn?

about 27.4 gallons 50. **TRAVEL** Matthew drove his car 425 miles on 18.2 gallons of gasoline. At this rate, how many gallons of gasoline will Matthew need to travel 640 miles? Give your answer to the nearest tenth of a gallon.

51. **COOKING** A recipe for chili for 12 people calls for 3 pounds of ground beef, 8 tomatoes, and various spices.
   a. How many pounds of ground beef are needed to make
   12.5 lbs   enough chili for 50 people?
   b. How many pounds of ground beef are needed to make enough chili
   16.25 lbs   for 65 people?

Look Beyond

In Exercises 65 and 66, students preview a important fact about similar polygons: If the ratio of the corresponding sides of two similar polygons is 1:a, then the ratio of the areas of the polygons is 1:a^2. Students will study this relationship in greater depth in a future geometry course.

ALTERNATIVE
Assessment

Portfolio Activity

The Portfolio Activity can be used as preparation for the Chapter Project or as a separate activity. In this Portfolio Activity, students investigate the way a bar graph is used to organize and display data.

Answers to Portfolio Activities can be found in Additional Answers of the Teacher's Edition.

52. Subtract: $\frac{1}{12} - \frac{2}{30}$. (**LESSON 2.3**) $\frac{1}{60}$

Simplify each expression. (**LESSONS 2.6 AND 2.7**)

53. $3 \cdot 5x$ **15x**

54. $-2y \cdot 3y$ **$-6y^2$**

55. $7(x-3) - (5-4x)$ **$11x - 26$**

56. $4s + 7 - 2s - 5$ **$2s + 2$**

57. $4t \cdot (-3t)$ **$-12t^2$**

58. $\frac{4y+2}{2}$ **$2y + 1$**

Solve each equation. (**LESSON 3.2**)

59. $\frac{w}{-12} = 13$ **-156**

60. $-18 = \frac{x}{12}$ **-216**

61. $100x = 216$ **2.16**

62. $3p = -18$ **-6**

63. $-4y = 124$ **-31**

64. $5t = 123$ **24.6**

CONNECTION

65. GEOMETRY The ratio of both the lengths and the widths of the rectangles is 1:3. Find the ratio of the areas. $\frac{1}{9}$

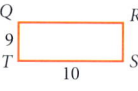

66. For the rectangles in Exercise 65, does the ratio of the areas equal the ratio of the lengths and the widths? Why or why not? No; ratio of area is $\frac{1}{9}$, other ratios are $\frac{1}{3}$

Include this activity in your portfolio.

Horacha used a computer to build a bar graph for the soft-drink preferences reported in a survey of her science class.

1. 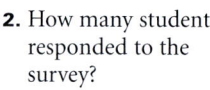 Which soft drink was the most popular?

2. 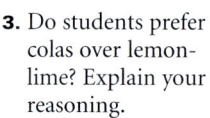 How many students responded to the survey?

3. 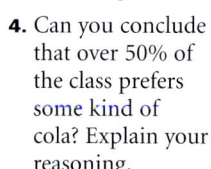 Do students prefer colas over lemon-lime? Explain your reasoning.

4. Can you conclude that over 50% of the class prefers some kind of cola? Explain your reasoning.

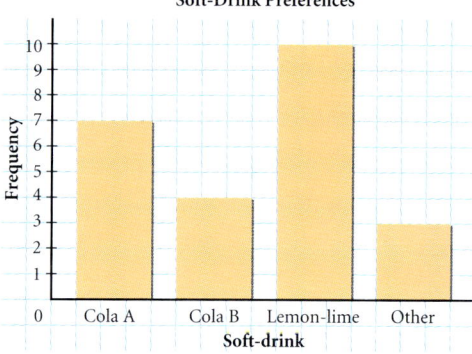

5. How many soft-drink varieties do you think are included in the "other" category? Why?

WORKING ON THE CHAPTER PROJECT

You should now be able to complete Activities 1 and 2 of the Chapter Project on page 208.

4.2 Percent Problems

Objectives
- Find equivalent fractions, decimals, and percents.
- Solve problems involving percent.

Why Percents are often found in newspapers, magazines, and other media. They are used in tables, charts, graphs, and advertisements to describe situations like discounts, taxes, and weather.

Fractions, Decimals, and Percents

APPLICATION
METEOROLOGY

A **percent** is a *ratio* that compares a number with 100. For example, 50% is the same as $\frac{50}{100}$, or $\frac{1}{2}$, and 100% is the same as $\frac{100}{100}$, or 1.

Relative humidity is a common example of percent. For example, at 30°C, 1 cubic meter of air can hold no more than 26 grams of water. At this amount, the relative humidity is 100%. At other amounts, ratios are used to determine the relative humidity. If 1 cubic meter of air at 30°C contains 6.5 grams of water, the relative humidity is 25% because $\frac{6.5}{26} = \frac{1}{4}$, or 25%.

When solving these types of problems, it is helpful to know how to write percents as decimals and as fractions. Sometimes you may wish to write the fractions in simplest terms.

$60\% = \frac{60}{100} = 0.60$ $\qquad\qquad\qquad \frac{60}{100} = \frac{3}{5}$

Prepare

NCTM PRINCIPLES & STANDARDS
1, 2, 5, 6, 8–10

LESSON RESOURCES
- Practice 4.2
- Reteaching Master 4.2
- Enrichment Master 4.2
- Cooperative-Learning Activity 4.2
- Lesson Activity 4.2
- Problem Solving/Critical Thinking Master 4.2
- Teaching Transparency 18

QUICK WARM-UP

Write as a decimal.
1. $\frac{4}{5}$ 0.8   2. $1\frac{3}{8}$ 1.375
3. $\frac{7}{100}$ 0.07   4. $\frac{117}{100}$ 1.17

Solve.
5. $0.4y = 50$ 6. $25t = 15$
   $y = 125$ $t = 0.6$
7. $\frac{3}{100} = \frac{s}{75}$ 8. $\frac{12}{100} = \frac{18}{a}$
   $s = 2.25$ $a = 150$

Also on Quiz Transparency 4.2

Teach

Why Taxes, prices of sale items, and salary increases are just a few of the many quantities that are calculated with percents. Have students make a list of references to percent that they have seen in the past 24 hours. Examples may include grades, class rankings, octane ratings of gasoline, and interest rates. Point out that this lesson will show how algebra can be used as a tool to solve percent problems.

Alternative Teaching Strategy

USING COGNITIVE STRATEGIES Remind students that a fraction names part of a whole. For instance, $\frac{3}{8}$ can be considered 3 parts out of 8.

Any percent can be written as a fraction with denominator 100, so 55% is equal to $\frac{55}{100}$. This means it can be considered 55 parts out of 100.

Now show students how they can use the concept of parts and wholes to set up and solve a proportion for any percent problem. Use examples like those shown at right.

What number is 20% of 50?

$\frac{\text{part}}{\text{whole}} \leftarrow \frac{20}{100} = \frac{n}{50} \rightarrow \frac{\text{part}}{\text{whole}}$ $n = 10$

What percent of 60 is 15?

$\frac{\text{part}}{\text{whole}} \leftarrow \frac{n}{100} = \frac{15}{60} \rightarrow \frac{\text{part}}{\text{whole}}$ $n = 25$

40% of what number is 10?

$\frac{\text{part}}{\text{whole}} \leftarrow \frac{40}{100} = \frac{10}{n} \rightarrow \frac{\text{part}}{\text{whole}}$ $n = 25$

LESSON 4.2   **171**

ADDITIONAL EXAMPLE 1

Write each percent as a decimal and as a fraction or mixed number in simplest terms.

a. 80%   0.8; $\frac{4}{5}$

b. 450%   4.5; $\frac{9}{2}$

c. 9.5%   0.095; $\frac{19}{200}$

TRY THIS

a. 0.4; $\frac{2}{5}$

b. 1.75; $1\frac{3}{4}$

c. 0.025; $\frac{1}{40}$

Use Teaching Transparency 18.

Teaching Tip

In previous courses, students probably were encouraged to memorize the most commonly used fraction-percent equivalents. You may want to quiz them to make sure they recall these.

$\frac{1}{8} = 12\frac{1}{2}\%$ $\frac{1}{4} = 25\%$

$\frac{3}{8} = 37\frac{1}{2}\%$ $\frac{1}{2} = 50\%$

$\frac{5}{8} = 62\frac{1}{2}\%$ $\frac{3}{4} = 75\%$

$\frac{7}{8} = 87\frac{1}{2}\%$ $\frac{1}{1} = 100\%$

$\frac{1}{5} = 20\%$ $\frac{2}{5} = 40\%$

$\frac{3}{5} = 60\%$ $\frac{4}{5} = 80\%$

$\frac{1}{6} = 16\frac{2}{3}\%$ $\frac{1}{3} = 33\frac{1}{3}\%$

$\frac{2}{3} = 66\frac{2}{3}\%$ $\frac{5}{6} = 83\frac{1}{3}\%$

EXAMPLE 1

Write each percent as a decimal and as a fraction or mixed number in simplest terms.

a. 75%   b. 110%   c. 4.4%

SOLUTION

a. $75\% = \frac{75}{100} = 0.75$ $\frac{75}{100} = \frac{3}{4}$

b. $110\% = \frac{110}{100} = 1.10$ $\frac{110}{100} = \frac{11}{10}$, or $1\frac{1}{10}$

c. $4.4\% = \frac{4.4}{100} = \frac{44}{1000} = 0.044$ $\frac{4.4}{100} = \frac{44}{1000} = \frac{11}{250}$

TRY THIS Write each percent as a decimal and as a fraction or mixed number in simplest terms.

a. 40%   b. 175%   c. 2.5%

In a percent statement such as "25% of 40 is 10," 25 is the **percent rate**, 40 is known as the **base**, and 10 is the **percentage**. If you know two of the parts in a percent problem, there are different methods to find the third part.

The Proportion Method

You can use proportions to solve percent problems. A percent bar can help you to visualize the proportion.

1. The percentage is unknown. 25% of 40 is x.

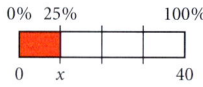

$\frac{25}{100} = \frac{x}{40} \rightarrow 100x = 1000 \rightarrow x = 10$

2. The base is unknown. 25% of x is 10.

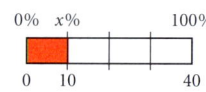

$\frac{25}{100} = \frac{10}{x} \rightarrow 25x = 1000 \rightarrow x = 40$

3. The percent rate is unknown. $x\%$ of 40 is 10.

$\frac{x}{100} = \frac{10}{40} \rightarrow 40x = 1000 \rightarrow x = 25$

The Equation Method

You can set up and solve an equation for the unknown part of a percent problem if the other two parts are known. Write percents as decimals when using them in this type of equation.

1. The percentage is unknown.

25% of 40 is x.
$0.25 \cdot 40 = x$
$10 = x$

2. The base is unknown.

25% of x is 10.
$0.25 \cdot x = 10$
$x = 40$

3. The percent rate is unknown.

x of 40 is 10.
$x \cdot 40 = 10$
$x = 0.25$, or 25%

Interdisciplinary Connection

BUSINESS Sometimes business partners do not share the ownership of a business equally. In such a case, they generally share the profits of the business proportionally rather than equally. The proportion is determined by the percent of the business that each owns.

Alvarez, Brown, and Chow are partners in a business that generated $500,000 in profits. You know that Alvarez owns 40% of the business and that Chow has received $175,000 of the profits. Find the amount of the profits that Alvarez and Brown should receive and the percent of the business owned by Brown and Chow.

profits: Alvarez should receive $200,000.
          Brown should receive $125,000.

percents owned: Brown owns 25%.
                     Chow owns 35%.

EXAMPLE 2

APPLICATION
DISCOUNTS

The sophomore class is sponsoring a trip to see a play. Student tickets normally cost $8. If at least 20 people buy tickets as a group, there is a 30% discount. **How much will each ticket cost at the discounted price?**

The play "Julius Caesar" was written by William Shakespeare between 1599 and 1600.

● **SOLUTION**

Proportion Method
You can visualize the problem by using a percent bar. The base, or regular cost, $8, is the length of the bar. Because the tickets are 30% off, the discounted price is 70% of the original price (100% − 30% = 70%). Shade 70% of the bar, and use x to represent the percentage, or discounted price.

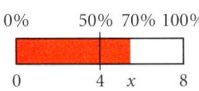

Write the proportion $\frac{70}{100} = \frac{x}{8}$ from the percent bar.

Solve the proportion. $\frac{70}{100} = \frac{x}{8} \rightarrow 100x = 560 \rightarrow x = 5.60$

The discounted price is $5.60.

Equation Method
As in the proportion method, use x to represent the discounted price.

The discounted price, x, is 70% of the full price, $8.

Model this statement with an equation and solve for x.

$x = 0.70 \cdot 8$

$x = 5.60$

The discounted price is $5.60.

TRY THIS Find each answer by using the proportion method or the equation method.
a. What percent of 80 is 15? b. Find 115% of 200.
c. 45 is 40% of what number?

EXAMPLE 3

APPLICATION
DISCOUNTS

A VCR with a sale price of $239.40 is advertised as 40% off. **What was the original price?**

● **SOLUTION**

Let w represent the original price, or base, and note that $239.40 is 60% of the original price (100% − 40% = 60%). Model this information with an equation.

$$239.40 = 0.60w$$

Apply the Division Property of Equality and simplify.
$$\frac{239.40}{0.60} = \frac{0.60w}{0.60}$$
$$399 = w$$

The original price was $399.

Inclusion Strategies

VISUAL LEARNERS
Give students 10-by-10 grids to make visual models of percents. For example, have them shade one row or column and then write a fraction, a decimal, and a percent that represent the shading, as shown at right. Repeat the activity for several other grid areas.

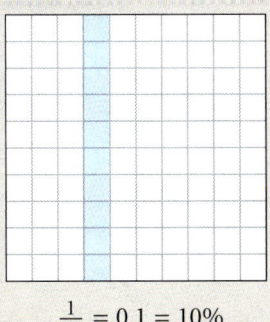

$\frac{1}{10} = 0.1 = 10\%$

Enrichment

Have students discuss these problems in groups.

A retailer buys an item at cost and marks it up 50% to set the selling price. Later the retailer discounts the selling price 50%. Is the new price equal to the original cost? Explain. **No; $0.5(1.5x) = 0.75x$; it is 75% of the original cost.**

The original selling price of an item is discounted 30%. Then this discounted price is further discounted 30%. Is the new price 40% of the original selling price? Explain. **No; $0.7(0.7x) = 0.49x$; it is 49% of the original price.**

ADDITIONAL EXAMPLE 2

The price of a videotape is marked as $14.97. At the cash register, the clerk tells you that you will be charged an additional 7.5% in sales tax. **What is the total price including tax?** **$16.09**

TRY THIS
a. 18.75%
b. 230
c. 112.5

ADDITIONAL EXAMPLE 3

Mario bought a coat on sale for $29.40. This sale price was advertised as 60% of the original price. **What was the original price?** **$49**

Teaching Tip

TECHNOLOGY If students are using scientific calculators, there may be a percent key that can be used in percent calculations. On the TI-34 scientific calculator, for instance, you can find 18% of 64 by using this key sequence.

18 [2nd] [=] [×] 64 [=]

Key sequences vary among calculators, so students should consult the manuals to find the appropriate sequence for their calculator. Most graphics calculators do *not* have a special percent key.

LESSON 4.2 **173**

ADDITIONAL EXAMPLE 4

Anne is hosting a party at a skating rink. The regular cost of a student admission to the rink is $5.75. However, the manager told Anne that the cost of each student admission will be discounted to $4.60 if there are at least 25 students at the party. **What percent of the regular cost of admission is this discounted cost?** 80%

TRY THIS
6.5%

ADDITIONAL EXAMPLE 5

A company is expanding its building space to accommodate an increase in the number of employees. The floor space in the building was 10,500 square feet before the expansion. Now the floor space is 14,000 square feet. **Find the percent of increase in the amount of floor space.**

$33\frac{1}{3}\%$

TRY THIS
25%

EXAMPLE 4 Most states have a sales tax for certain items. Linda paid $47.74 for an item priced at $45.25. **Find the sales tax rate to the nearest tenth of a percent.**

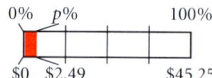
APPLICATION
TAXES

● **SOLUTION**

First find the amount of sales tax. Subtract the price of the item from the total amount paid.

$$\$47.74 - \$45.25 = \$2.49$$

The amount of sales tax was $2.49. You can use a percent bar to show the proportion. Let p represent the percent rate.

Write the proportion from the percent bar.

$$\frac{p}{100} = \frac{2.49}{45.25}$$

Solve the proportion.

$p \approx 5.5$ The sales tax rate is 5.5%.

TRY THIS Find the sales tax rate to the nearest tenth of a percent if the total paid for an item priced at $50.75 was $54.05.

Many types of problems involve finding a percent of increase or percent of decrease after a change has occurred. It is important to remember when working these types of problems that the original amount is always used as the base.

EXAMPLE 5 A family is adding additional rooms to their house. The house originally had 1500 square feet of floor space. After the additions are completed, the house will have 1800 square feet of floor space. **Find the percent of increase in floor space.**

● **SOLUTION**

First find the amount of the increase in floor space. Find the difference between the amounts of floor space in square feet before and after the additions.

$$1800 - 1500 = 300$$

The amount of increase in floor space is 300 square feet. Thus, some percent of 1500 is 300. Let p represent the percent rate in decimal form. You can model the problem with an equation.

$$p \cdot 1500 = 300, \text{ or } 1500p = 300$$

Solve for p and convert the decimal to a percent.

$$\frac{1500p}{1500} = \frac{300}{1500} \rightarrow p = 0.20$$

The percent of increase in floor space is 20%.

TRY THIS Suppose that the amount of floor space in Example 5 had been increased from 1500 square feet to 1875 square feet. Find the percent of increase.

Reteaching the Lesson

USING SYMBOLS Suggest that students think of the relationship among the base, rate, and percentage in terms of the following formula:

percentage = base · rate
p = b · r
$p = br$

As the first step of answering a percent question, tell them to identify the values of p, b, and r. They can then substitute these values into the formula and solve the resulting equation. Demonstrate the process with the examples at right.

1. What is 3.5% of 22? $r = 0.035, b = 22, p = \underline{\ ?\ }$
   $p = br \rightarrow$ $p = 22(0.035)$
   $p = 0.77$

2. 18 is 45% of what number $p = 18, r = 0.45, b = \underline{\ ?\ }$
   $p = br \rightarrow$ $18 = b(0.45)$
   $40 = b$

3. What percent of 44 is 16.5? $b = 44, p = 16.5, r = \underline{\ ?\ }$
   $p = br \rightarrow$ $16.5 = 44r$
   $0.375 = r$
   $37.5\% = r$

EXAMPLE 6 The price of a car that originally sold for $17,000 has been reduced to $14,450. **Find the percent of decrease in the price of the car.**

● **SOLUTION**

First find the amount of the decrease in the price. Find the difference between the original price and the reduced price.

$$17{,}000 - 14{,}450 = 2550$$

The amount of decrease in the price is $2550. Thus, some percent of 17,000 is 2550. Let p represent the percent rate in decimal form. You can model the problem with an equation.

$$p \cdot 17{,}000 = 2550, \text{ or } 17{,}000p = 2550$$

Solve for p and convert the decimal to a percent.

$$\frac{17{,}000p}{17{,}000} = \frac{2550}{17{,}000} \rightarrow p = 0.15$$

The percent of decrease is 15%.

TRY THIS Suppose that the price of the car in Example 6 had been reduced from $17,000 to $13,430. Find the percent of decrease in the price of the car.

Exercises

Communicate

1. Explain the procedure for changing a percent to a decimal.
2. Explain how a percent bar helps to model a percent problem.
3. What are the advantages of using the proportion method to solve percent problems?
4. What are the advantages of using the equation method to solve percent problems?

Guided Skills Practice

Write each percent as a decimal and as a fraction or mixed number in simplest terms. *(EXAMPLE 1)*

5. 34% 0.34; $\frac{17}{50}$
6. 20% 0.2; $\frac{1}{5}$
7. 130% 1.3; $1\frac{3}{10}$
8. 7.5% 0.075; $\frac{3}{40}$

Find each answer. *(EXAMPLE 2)*

9. 70% of 150 is what number? 105
10. 14% of 50 is what number? 7

APPLICATION

11. **HEALTH** Tim has reduced his cholesterol level by 20% by dieting and exercising. If his original level was 238, what is his new cholesterol level? *(EXAMPLE 2)* 190.4

ADDITIONAL EXAMPLE 6

The price of a 3-bedroom, 2-bath house was originally listed as $125,000. The real estate company that owns the property is having difficulty selling the house, so the price has been reduced to $97,500. **Find the percent of decrease in the price of the house.** 22%

TRY THIS 21%

Assess

Selected Answers
Exercises 5–22, 23–87 odd

ASSIGNMENT GUIDE

In Class	1–22
Core	23–75 odd
Core Plus	24–70 even, 71–76
Review	77–87
Preview	88

✎ Extra Practice can be found beginning on page 738.

internetconnect

GO TO: go.hrw.com
KEYWORD: MA1 Rural Exodus

In this activity, students gather data from the U.S Census Bureau on urban and rural populations to discover how the population has shifted. This activity illustrates the importance of looking at percentages of a population rather than the population size.

LESSON 4.2 175

Technology

A calculator may be helpful for Exercises 43–76. You may want to have students complete the exercises with paper and pencil first and then use the calculator to verify their answers.

43. $x = 28$

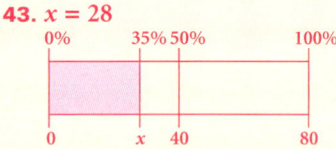

44. $x = 90$

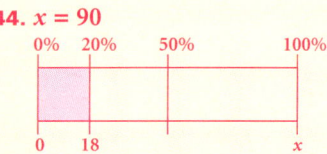

Find each answer. *(EXAMPLE 3)*

12. 17 is 50% of what number? **34**
13. 25 is 40% of what number? **62.5**

Find each answer. *(EXAMPLE 4)*

14. What percent of 50 is 5? **10%**
15. What percent of 350 is 70? **20%**

APPLICATION

16. **TAXES** Jim paid $25.85 to purchase a shirt that cost $23.50. What was the tax rate, rounded to the nearest tenth of a percent? *(EXAMPLE 4)* **10%**

A yard originally had a perimeter of 200 meters. Find the percent of increase for each new perimeter below. *(EXAMPLE 5)*

17. 240 meters **20%**
18. 230 meters **15%**
19. 280 meters **40%**

A car originally cost $30,000. Find the percent of decrease for each new price below. *(EXAMPLE 6)*

20. $24,000 **20%**
21. $27,500 **$8\frac{1}{3}$%**
22. $16,500 **45%**

Practice and Apply

Write each percent as a decimal.

23. 55% **0.55**
24. 1.2% **0.012**
25. 8% **0.08**
26. 45.3% **0.453**
27. 0.5% **0.005**
28. 73% **0.73**
29. 186% **1.86**
30. 24% **0.24**
31. 4.5% **0.045**
32. 17.2% **0.172**

Write each percent as a fraction or mixed number in lowest terms.

33. 47% **$\frac{47}{100}$**
34. 18% **$\frac{9}{50}$**
35. 3.5% **$\frac{7}{200}$**
36. 0.01% **$\frac{1}{10,000}$**
37. 125% **$1\frac{1}{4}$**
38. 52% **$\frac{13}{25}$**
39. 38% **$\frac{19}{50}$**
40. 0.8% **$\frac{1}{125}$**
41. 210% **$2\frac{1}{10}$**
42. 7.3% **$\frac{73}{1000}$**

Draw a percent bar for each problem, and use a proportion to find the solution.

43. Find 35% of 80.
44. 18 is 20% of what number?
45. What percent of 40 is 60?
46. What percent of 20 is 50?
47. 48 is 40% of what number?
48. Find 45% of 120.

Find each answer.

49. Find 40% of 50. **20**
50. Find 200% of 50. **100**
51. 2 is what percent of 100? **2%**
52. Find 45% of 80. **36**
53. 30 is 60% of what number? **50**
54. What percent of 80 is 10? **12.5%**
55. Find 180% of 40. **72**
56. 8 is 20% of what number? **40**
57. Find 30% of 120. **36**
58. Find 50% of 124. **62**
59. What percent of 80 is 60? **75%**
60. Find 175% of 92. **161**
61. 27 is 60% of what number? **45**
62. What percent of 35 is 105? **300%**

A house originally had 1700 square feet of floor space. Find the percent of increase for each floor space given below.

63. 1955 square feet **15%**
64. 1870 square feet **10%**
65. 1785 square feet **5%**
66. 2057 square feet **21%**

45. $x = 150$

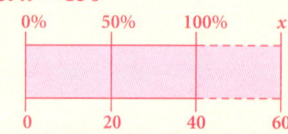

46. $x = 250$

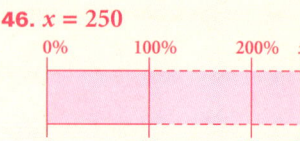

47. $x = 120$

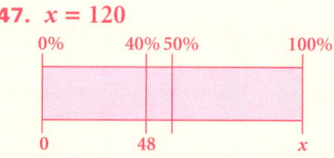

48. $x = 54$

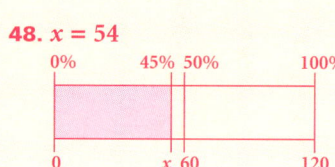

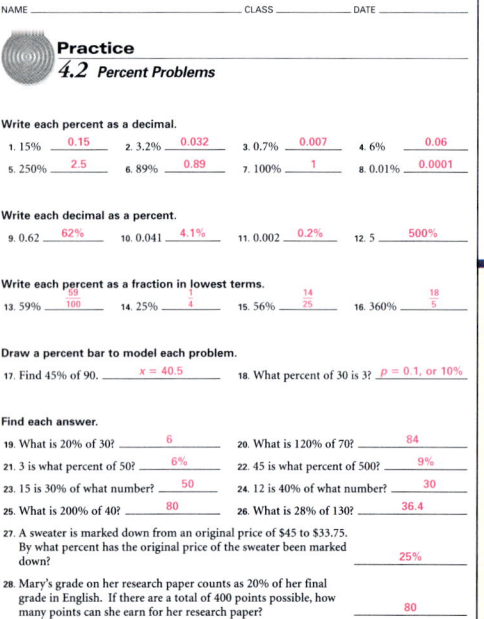

176 LESSON 4.2

A car originally cost $15,000. Find the percent of decrease for each new price.

67. $14,250 **5%**   **68.** $13,050 **13%**   **69.** $13,650 **9%**   **70.** $12,300 **18%**

71. A teacher says that 40% of your grade will be based on homework. If there is a total of 500 points possible toward your grade, how many points are possible for homework? **200 points**

CONNECTION

72. STATISTICS The 1996 presidential election had the following results:

Candidate	Popular vote	Electoral vote
Clinton (Democrat)	45,590,703	379
Dole (Republican)	37,816,307	159
Perot (Independent)	7,866,284	0

a. $C \approx 50\%$; $D \approx 41\%$; $P \approx 9\%$

b. $C \approx 70\%$; $D \approx 30\%$; $P \approx 0\%$

c. $C \approx 36\%$; $D \approx 33\%$; $P \approx 31\%$

a. What percent of the popular vote did each candidate get?
b. What percent of the electoral vote did each candidate get?
c. If 12,300,000 Clinton voters and 8,100,000 Dole voters voted for Perot, what would each candidate's percentage of the popular vote have been?

CHALLENGE

APPLICATIONS

73. DISCOUNTS A book is marked down from an original price of $20 to $15. By what percent has the book been marked down? **25%**

74. DISCOUNTS A coat is marked for sale at $160, which is 40% off the original price. What was the original price to the nearest dollar? **$267**

75. TAXES If there is a 5% sales tax on the sale price of the coat in Exercise 74, what is the total amount that you would have to pay? **$168**

76. POLITICAL SCIENCE The newspaper reports that only 42% of the registered voters in a small town voted in an election. If 11,960 people voted, how many registered voters are there? **about 28,476**

Look Back

Evaluate. *(LESSONS 2.2, 2.3, AND 2.4)*

77. $-3 + 7$ **4**   **78.** $(-3)(7)$ **−21**   **79.** $-3 - (-7)$ **4**   **80.** $(-3)(-7)$ **21**

Simplify. *(LESSON 2.6)*

81. $(4x + 7) + (9x - 9)$ **$13x - 2$**   **82.** $(3x^2 + 2x) + (3x^2 - 2x)$ **$6x^2$**

83. $(-6c + 5) - (-3c + 8)$ **$-3c - 3$**   **84.** $(-y + 2z) + (-3y + 3z)$ **$-4y + 5z$**

Simplify. *(LESSON 2.7)*

85. $\dfrac{4x + 18x^2}{2}$ **$2x + 9x^2$**   **86.** $\dfrac{15f^2 + 35}{-5}$ **$-3f^2 - 7$**   **87.** $\dfrac{3x^2 - 21}{3}$ **$x^2 - 7$**

Look Beyond

88. Suppose that a $1000 investment grows by 8% each year. Complete the table below.

2, ≈ $1166.40;

3, ≈ $1259.71;

4, ≈ $1360.49

Years	0	1	2	3	4
Value ($)	1000	1080	?	?	?

Error Analysis

In calculating a percent increase or decrease, students often mistakenly use the final amount as the base. If students experience this difficulty, suggest that they write the following equations at a prominent location in their notebooks and refer to them whenever they do this type of calculation:

$$\dfrac{\text{percent of}}{\text{increase}} = \dfrac{\text{amount of increase}}{\text{original amount}}$$

$$\dfrac{\text{percent of}}{\text{decrease}} = \dfrac{\text{amount of decrease}}{\text{original amount}}$$

Look Beyond

In Exercise 88, students use a series of percent multiplications to explore the value of an investment that earns *compound interest*. In Lesson 8.7, students will learn how to solve problems like this more quickly by using an exponential function.

Student Technology Guide

4.2 Percent Problems

Many scientific calculators have a % function you can use to calculate percentages such as 42% of 70. On calculators that do not have a % function, use division by 100.

On a calculator that has a % function, enter: 70 [×] 42 [2nd] [·] [·]

On a calculator that does not have a % function, you enter: 70 [×] 42 [÷] 100 [ENTER]

By either set of key strokes, 42% of 70 = 29.4.

Find each answer.

1. 18% of 90   **16.2**   2. 30% of 140   **42**
3. 125% of 40   **50**   4. 0.5% of 1050   **5.25**

To find what percent one number is of another number, write down an equation. Then solve the equation. What percent is 40 of 320?

$40 = x \text{ percent} \times 320$

$x \text{ percent} = \dfrac{40}{320}$

On a calculator that has a % function, enter: 40 [÷] 320 [2nd] [·] [·]

On a calculator that does not have a % function, enter: 40 [÷] 320 [×] 100 [ENTER]

By either set of key strokes, 40 is 12.5% of 320.

Find each answer.

5. What percent of 80 is 24?   **30%**   6. What percent of 250 is 45?   **18%**
7. What percent of 40 is 50?   **125%**   8. What percent of 36 is 180?   **500%**

You can also use a calculator to find the unknown base in a percent problem. For example, 80 is 40% of what number?

$80 = 40\% \times x$

$x = \dfrac{80}{40\%}$

On a calculator that has a % function, enter: 80 [÷] 40 [2nd] [·] [·]

On a calculator that does not have a % function, enter: 80 [÷] [(] 40 [÷] 100 [)] [ENTER]

Find each answer.

9. 234 is 45% of what number?   **520**   10. 28 is 70% of what number?   **40**
11. 120 is 150% of what number?   **80**   12. 12.5 is 20% of what number?   **62.5**

LESSON 4.2 **177**

Eyewitness Math

Focus

This article describes the controversy over removing the grizzly bear from the endangered species list and explains that the true grizzly population is hard to ascertain. Students simulate the *tag-and-recapture* technique described in the article by taking objects from a paper bag at random and using the sample to estimate the entire "population" of objects within the bag.

Motivate

As students read the article, have them consider its main points. Ask the following questions:

- What was the estimated number of grizzlies when they were placed on the Endangered Species List in 1975?
- What claim about the grizzly population is made by federal officials?
- What do environmentalists say about the grizzly population?

Discuss with students the difficulties involved in attempting to count every grizzly. For example, it would be difficult to ensure that each animal is counted only once. Any type of count would be likely to disturb the animals, possibly disrupting their feeding and mating patterns. It may also be simply impossible to find them all.

EYEWITNESS MATH

BARELY ENOUGH GRIZZLIES?

Counting Big Bears
Government Wants Them Off the Endangered List; How Many is Enough?

By Marj Charlier–Staff Reporter of the Wall Street Journal

KALISPELL, Mont.—The solitary monarch of the wild, the grizzly rules the rugged high country. Scientists call it *Ursus arctus horribilus*, and the name suggests the terror it inspires. Fearless and unpredictable, the bear can kill with a single rake of a claw.

Man is the grizzly's only natural enemy, but the clash almost proved fatal to the species. Unrestricted hunting and development of the animal's habitat decimated the bear population over the last century. By 1975 when the grizzly was placed on the federal government's Endangered Species List, only 1,000 were thought to be alive.

Now, federal officials say the grizzly population has rebounded, though they concede there is no hard evidence to support that conclusion, and they are proposing to take the bear off the list.

A Death Sentence?

Many environmentalists, however, believe that even if the number of bears has increased, the grizzly population is still too small to strip the animal of federal protection. Such a plan, they say, would be a death sentence for the species. "If the grizzly bear is delisted, it will be strangled by development and go extinct quite promptly," says Lance Olsen, president of the nonprofit Great Bear Foundation.

Population Unknown

Complicating the issue is the difficulty of determining how many grizzlies roam the Rockies and their foothills. Everyone concedes that the 1,000 figure for 1975 was only a guess. Attempts to take a bear census in the years since then have proved difficult, and no one knows how many grizzlies there are now.

Richard Mace, a wildlife biologist for the state of Montana, knows how tough it is to count grizzlies. Displaying a battered steel box with a jagged rip down one side, he explains that the box once held a tree-mounted camera that was rigged to snap photos when it sensed body heat from a large animal. Biologists had placed some 43 cameras in a section of Montana's Swan Mountains, hoping to get a more accurate bear count.

Apparently, Mr. Mace says, when the camera flashed, the grizzly slashed. Its two-inch long claws ripped through the steel casing, which is a bit thicker than a coffee can. The bear dug out the camera and chewed it to pieces.

Cooperative Learning

Counting animals in the wild is not easy. How do you know if you have counted some more than once? How can you possibly find them all? Wildlife managers sometimes use a technique called *tag and recapture*. To see how the tag-and-recapture method works, you can try it—with objects instead of animals.

You will need:
- a batch of several hundred objects (macaroni, beans, crumpled pieces of paper) in a paper bag or other container
- a marker (or some other way to *tag*, or mark, the objects)

1. Follow steps **a–e** to estimate the number of objects in your batch.
   a. Remove a handful of objects from the batch and mark them. Record how many you have marked.
   b. Return the marked objects to the batch. Mix thoroughly.
   c. Take 30 objects from the batch at random. Count the number of marked objects in your sample. Copy the chart and fill in the first row. Repeat for each group member.

Sample	Total number of objects in sample (N_s)	Number of marked objects in sample (M_s)
1.	30	
2.	30	

   d. How do you think the ratio of marked objects to total number of objects in your samples compares with the ratio of marked objects to total number of objects in the whole batch?
   e. Use the equation below to estimate the size of the whole batch. (Average your data for M_s and N_s.)

$$\frac{M_b}{N_b} = \frac{M_s}{N_s}$$

   M_b is the number of **marked objects** in the batch.
   N_b is the **total number of objects** in the batch.
   M_s is the number of **marked objects** in the sample.
   N_s is the **total number of objects** in the sample.

   f. Check your estimate by counting the actual number of objects in the batch. Within what percent of the actual number was your estimate?

2. Describe how the tag-and-recapture method could be used to estimate an animal population.

3. Do you think the tag-and-recapture method would work for a grizzly bear population? Why or why not?

Cooperative Learning

Have students work in pairs. Give each pair a bag of objects that you have prepared in advance. One student in each pair should collect the first sample, and the other should record the data. They then can reverse roles and collect a second sample.

After the data have been recorded, suggest that each student perform the required calculations individually. They can then compare their results and work together to resolve any differences.

Discuss

After the groups have completed the experiment and answered all the questions, bring the class together for a discussion of the results. Ask whether they think their estimates are reasonable compared with the actual number of objects in the bag. If they feel their estimates are *not* reasonable, have them suggest ways to make better estimates. A logical conclusion may be that it is necessary to increase the size of the samples.

Be sure that students understand how the tag-and-recapture method is used with animals in the wild. Ask if they think this is a practical method for estimating a population of wild animals.

You may want to suggest that students research the current status of the grizzly population and find information about other animals that have been on the endangered species list.

1. Answers will vary.

2. Tag and recapture provides a ratio for a sample that can be extended to the entire population.

3. Answers will vary.

Prepare

NCTM PRINCIPLES & STANDARDS
1, 2, 5, 6, 8–10

LESSON RESOURCES

- Practice 4.3
- Reteaching Master 4.3
- Enrichment Master 4.3
- Cooperative-Learning Activity 4.3
- Lesson Activity 4.3
- Problem Solving/Critical Thinking Master 4.3

QUICK WARM-UP

Write each fraction as a decimal and as a percent.

1. $\frac{3}{4}$ 0.75; 75%
2. $\frac{7}{10}$ 0.7; 70%
3. $\frac{23}{100}$ 0.23; 23%
4. $\frac{1}{8}$ 0.125; 12.5%
5. $\frac{2}{9}$ $0.\overline{2}$; $22\frac{2}{9}\% \approx 22.2\%$
6. $\frac{125}{300}$ $0.41\overline{6}$; $41\frac{2}{3}\% \approx 41.7\%$

Also on Quiz Transparency 4.3

Teach

Why What are the chances that a certain event will happen? The branch of mathematics called *probability* provides a way to answer that question. Invite students to make a list of everyday situations that involve chance. Examples might include the chances of drawing a winning card in a card game or the likelihood that a baseball player will get a hit in the next at bat.

4.3 Introduction to Probability

Objective
- Find the experimental probability that an event will occur.

Why Experimental probability is the basis for many conclusions and predictions, including the daily weather forecast and which sports teams are expected to win. The experiments that you perform in this lesson will help you understand more about experimental probability.

The percent chances of rain that weather forecasters use are based on previous situations with similar circumstances.

The **probability** of an event is the likelihood that the event will occur.

If the weather forecaster says that the probability of rain is 0%, this means that she is certain that it will not rain. If she says that the probability of rain is 100%, this means that she is certain that it will rain.

Probability is measured on a scale from 0% to 100%, or from 0 to 1.

```
Impossible    Less likely      More likely      Certain
|—————————————————————————————————————————————|
0%                                              100%
```

How can you find the probability that something will happen? One way is by *observing* the way that things happen in the real world. For example, meteorologists study weather data such as temperature, pressure, and humidity. Information about past weather patterns is compared with current patterns to predict the chance of rain each day.

You can conduct a simpler experiment. Consider the following question: If you roll a pair of number cubes repeatedly, how often would you expect the pair to show a sum of 7? One way to answer the question is to make observations about trials of your own—as in the following activity.

Alternative Teaching Strategy

USING TECHNOLOGY Set up a graphics calculator to simulate 50 rolls of a number cube. If you are using a TI-83 graphics calculator, press MATH ▶ ▶ ▶ and choose **5:randInt(** to access the random-integer generator. Set the range of the integers by pressing **1** , **6**).

If you are using a TI-82 graphics calculator, press MATH ▶ and choose **4:int**. Then press (**6** MATH ▶ ▶ ▶ and choose **1:rand**. Press + **1**).

On either calculator, you can now press ENTER 50 times to simulate 50 rolls of a number cube.

Explain to the class the meaning of the terms *event, experiment, trial,* and *outcome*. Discuss the concept of *experimental probability*.

Have students record the results and use them to calculate the experimental probability of the following outcomes:

1. 4
2. 4 or more
3. an even number
4. a prime number

Answers will vary.

180   LESSON 4.3

Activity Two Probability Experiments

You will need: a coin and two number cubes

You may want to work in groups of two or more.

Part I

1. Roll a pair of number cubes 20 times and record the number of times that a sum of 7 occurs.
2. Compare your results with the results of other groups in your class.
3. Find a percent for the entire class by dividing the total number of sums of 7 that occurred by the total number of rolls for the whole class.

CHECKPOINT ✓
4. Are your group and class results what you expected? Explain.

Part II

1. Toss a coin 20 times and record how many times it comes up heads. Divide this number by 20 to find the percent of heads that occurred.
2. Compare your results with the results of other groups in your class.
3. Find a percent for the entire class by dividing the total number of heads that came up by the total number of tosses for the whole class.

CHECKPOINT ✓
4. Are your group and class results what you expected? Explain.

Experimental Probability

A probability experiment consists of a number of *trials*, such as the tossing of a coin or the rolling of a pair of number cubes. For any given trial, the result, or outcome, may or may not be favorable. That is, it may or may not be the event that you are interested in studying. For example, if you are interested in the probability that a tossed coin comes up heads, then heads is a **favorable outcome**. In some instances you may be interested in counting unfavorable outcomes to find the probability that an event does not occur.

Activity Notes

Students most likely will discover that results vary considerably from group to group. In Part 1, each group should find the experimental probability of rolling a 7 to be close to $\frac{1}{6}$. In Part 2, they should find the experimental probability of tossing heads to be close to $\frac{1}{2}$. The probabilities should be even closer to $\frac{1}{9}$ and $\frac{1}{2}$ when the results from the entire class are pooled.

CHECKPOINT ✓
Part I
4. Answers may vary. Students should expect answers to be $\frac{1}{6}$ or about 17%.

Part II
4. Answers may vary. Students should expect answers to be about 50%.

Teaching Tip

The probabilities that students calculate in this lesson are based solely on the results of observations that arise from experiments. That is why this type of probability is given the name *experimental probability*. In Chapter 13, students will learn that it is possible to calculate a *theoretical probability* without experimentation.

Interdisciplinary Connection

BIOLOGY A type of plant called the Japanese four-o'clock has flowers that are red, white, or pink. If two red four-o'clocks are crossed, the offspring are red. If two white four-o'clocks are crossed, the offspring are white. However, if two pink four-o'clocks are crossed, the offspring are sometimes pink, sometimes red, and sometimes white.

Suppose that two pink four-o'clocks are crossed and produce 150 offspring: 81 pink, 36 red, and 33 white. Find the experimental probability of each color. **pink: 54%; red: 24%; white: 22%**

Inclusion Strategies

HANDS-ON STRATEGIES The topic of experimental probability is tailor-made for students who learn best through hands-on involvement. Students should have ready access to a supply of number cubes and disks with two distinct sides (to emulate coins). Encourage them to replicate some of the experiments described in the exercises and to compare their results with the stated results. Students also may enjoy conducting similar experiments with items for which all outcomes are not equally likely. Two such items that can be obtained easily are paper cups and thumbtacks.

LESSON 4.3 **181**

ADDITIONAL EXAMPLE 1

Two number cubes were rolled 200 times. A number greater than 9 appeared as a sum 28 times. **Based on this experiment, what is the experimental probability that two number cubes will show a sum that is greater than 9 when rolled?**

$\frac{28}{200} = 0.14$, or 14%

TRY THIS

$\frac{29}{100} = 0.29$, or 29%

ADDITIONAL EXAMPLE 2

Two coins were tossed 100 times. Two tails appeared 19 times. **Based on this experiment, what is the experimental probability that two tails will appear when these coins are tossed again?**

$\frac{19}{100} = 0.19$, or 19%

TRY THIS

$\frac{52}{200} = 0.26$, or 26%

The number of favorable outcomes divided by the total number of trials is called the **experimental probability** of the event.

Experimental Probability

For a sufficient number of trials, the experimental probability of an event, $P(E)$, is given by the following formula:

$$P(E) = \frac{\text{(number of favorable outcomes)}}{\text{(total number of trials)}}$$

Assume that all examples and exercises in this section refer only to fair number cubes and coins.

EXAMPLE 1 Two number cubes were rolled 100 times. An even number appeared as a sum 47 times. **Based on this experiment, what is the experimental probability that two rolled number cubes will show a sum that is an even number?**

● **SOLUTION**
There were 100 trials and 47 successful outcomes. Thus, in this case, the experimental probability of a successful outcome (showing a sum that is an even number) is as follows:
$\frac{47}{100} = 0.47$, or 47%

TRY THIS Two number cubes were rolled 100 times. A sum greater than 8 appeared 29 times. Based on this experiment, what is the experimental probability that two rolled number cubes will show a sum greater than 8?

EXAMPLE 2 Two coins were tossed 200 times. Two heads appeared 48 times. **Based on this experiment, what is the experimental probability that two flipped coins will show two heads?**

● **SOLUTION**
There were 200 trials and 48 successful outcomes. Thus, in this case, the experimental probability of a successful outcome (showing two heads) is as follows:
$\frac{48}{200} = 0.24$, or 24%

TRY THIS Two coins were tossed 200 times. Two tails appeared 52 times. Based on this experiment, what is the experimental probability of two flipped coins showing two tails?

Enrichment

A coin that is minted and distributed by a government is usually a *fair* coin. This means that on any given toss of the coin, both sides are equally likely to come up.

Have students create an "unfair" coin by firmly taping together a nickel and a penny. They should be sure that one heads is showing and one tails is showing. Then have them design and conduct an experiment to find the experimental probability of heads and of tails for this coin.

Reteaching the Lesson

INVITING PARTICIPATION Give each student a large circle with its center clearly marked. Tell them to divide it in as many sections as they want and to distinguish the sections with different numbers or colors. To make a pointer, have them put the point of a pencil in the bend of a paper clip or hairpin and hold the pencil point firmly at the center of the circle. Have them spin the pointer 30 times and record the results. Ask them to find the experimental probability for each section of the circle.

Exercises

Communicate

1. Explain what is meant by *experimental probability*.
2. Describe an experiment that could be used to find the experimental probability of getting at least 3 heads in 4 tosses of a coin.
3. Is it possible for two separate groups to conduct the same experiment to determine an experimental probability and get different results? Explain or give an example.
4. Is it possible for someone to conduct the same probability experiment twice and get different results? Explain or give an example.
5. Explain why the number of trials performed in a probability experiment can affect the outcome.

Guided Skills Practice

Two number cubes were rolled 100 times. Based on the results below, find the experimental probability of each outcome. *(EXAMPLE 1)*

6. An odd number appeared as a sum 53 times. **0.53 or 53%**
7. A sum of less than 4 appeared 7 times. **0.07 or 7%**
8. At least one 6 appeared 29 times. **0.29 or 29%**

Two coins were flipped 200 times. Based on the results below, find the experimental probability of each outcome. *(EXAMPLE 2)*

9. Two heads appeared 46 times. **0.23 or 23%**
10. Two tails appeared 51 times. **0.255 or 25.5%**
11. One heads and one tails appeared 106 times. **0.53 or 53%**

Practice and Apply

Two number cubes were rolled 100 times. Based on the results below, find the experimental probability of each outcome.

12. A sum of less than 5 appeared 18 times. **0.18 or 18%**
13. An even number appeared on at least one cube 76 times. **0.76 or 76%**
14. A sum of 9 appeared 13 times. **0.13 or 13%**
15. A sum of 7 or 11 appeared 24 times. **0.24 or 24%**
16. An even number appeared on both cubes 21 times. **0.21 or 21%**
17. Two 6s appeared 3 times. **0.03 or 3%**

Assess

Selected Answers
Exercises 6–11, 13–41 odd

ASSIGNMENT GUIDE

In Class	1–11
Core	13–33 odd
Core Plus	12–34 even
Review	35–41
Preview	42, 43

✎ Extra Practice can be found beginning on page 738.

Mid-Chapter Assessment for Lessons 4.1 through 4.3 can be found on page 48 of the *Assessment Resources*.

Technology

A calculator may be helpful for Exercises 12–34. You may want to have students complete the exercises with paper and pencil first and then use the calculator to verify their answers.

internetconnect

GO TO: go.hrw.com
KEYWORD: MA1 Simpson's Paradox

In this activity, students use the definition of probabilty to find probabilities for two subsets of data. Then students find the same probabilities after combining the two sets, and they discover a puzzling situation. This activity illustrates a phenomenon called "Simpson's paradox."

Error Analysis

Students sometimes mistakenly report the number of favorable outcomes as the probability of the event in question. Remind them that a probability is a ratio and that the ratio must be equal to a number from 0 to 1 inclusive. They may find it helpful to write the following inequality at a prominent place in their notebooks.

$$0 \text{ (0\%)} \leq \text{probability of an event} \leq 1 \text{ (100\%)}$$

Three coins were tossed 200 times. Based on the results below, find the experimental probability of each outcome.

18. At least one coin came up tails 178 times. **0.89 or 89%**
19. At least one coin came up heads 173 times. **0.865 or 86.5%**
20. Exactly two coins came up heads 84 times. **0.42 or 42%**
21. Exactly two coins came up tails 67 times. **0.335 or 33.5%**
22. All three coins came up tails 27 times. **0.135 or 13.5%**
23. All three coins came up heads 22 times. **0.11 or 11%**

Two coins were tossed 20 times with the following results (H represents heads and T represents tails):

Trial	1	2	3	4	5	6	7	8	9	10	11	12	13	14	15	16	17	18	19	20
Coin 1	H	H	H	H	T	H	T	T	H	H	H	T	T	H	T	T	T	H	H	H
Coin 2	T	H	H	T	T	T	T	H	T	T	T	H	H	H	H	H	H	H	T	H

Based on the data above, find the following experimental probabilities:

24. Both coins came up the same. **0.35 or 35%**
25. Both coins are heads. **0.25 or 25%**
26. At least one coin is heads. **0.9 or 90%**
27. Neither coin is heads. **0.1 or 10%**

Conduct 20 trials of your own to find the following experimental probabilities. Report your results in a table.

28. The sum of two rolls of a number cube is 9. **Answers may vary.**
29. Three tosses of a coin will result in at least 2 heads. **Answers may vary.**

CHALLENGE

30. To determine an experimental probability, Fred conducts 15 trials and Ted conducts 16 trials. Is it possible for them to arrive at the same experimental probability? Explain. **Yes. If Fred gets $\frac{15}{15}$ and Ted gets $\frac{16}{16}$, or if they both get a probability of 1.**

CONNECTION

PROBABILITY The result of a school experiment to find the probability of tossing three heads in a row was 11.5%. Twenty groups participated in the experiment.

31. Suppose each group performed 30 trials. How many favorable outcomes occurred? unfavorable outcomes? **favorable: 69; unfavorable: 531**

32. Suppose there were 46 favorable outcomes total. How many trials did the twenty groups perform? **20 trials**

APPLICATIONS

33. **GAMES** Many board games offer an extra turn if a player rolls doubles (both cubes showing the same number). Conduct several experiments with two number cubes to determine the experimental probability of rolling doubles for different numbers of trials. Complete a table for 5, 10, 15, 20, 25, and 30 trials. Show the number of doubles and the experimental probability for each number of trials. Compare your results with others in your class. **Answers will vary.**

34. **GAMES** Combine all of your trials from Exercise 33, and count all of the doubles. From those totals, find the overall experimental probability. Do you think this is a reliable estimate of the probability of rolling doubles during a game? Explain. **Yes.**

34. **Answers may vary. Since you probably will not roll the number cubes in an actual game more than the number of trials performed, the results are most likely reliable.**

Practice

NAME _____ CLASS _____ DATE _____

Practice
4.3 Introduction to Probability

Two number cubes were rolled 150 times. Find each experimental probability.

1. a sum less than 4 appeared 15 times _____ **10%**
2. a sum of 7 appeared 11 times _____ **7.3**
3. a sum of 6 or greater appeared 85 times _____ **56.6**

Two coins were tossed 10 times with the outcomes shown in the table below.

Trial	1	2	3	4	5	6	7	8	9	10
Coin 1	H	H	T	T	H	T	T	H	H	T
Coin 2	T	T	T	H	T	H	H	T	H	T

Use the data above to find each experimental probability.

4. At least one coin shows tails. _____ **90%**
5. Both coins show the same side of the coin. _____ **30%**
6. Neither coin shows heads. _____ **20%**

A survey was conducted to find out how students get to school. The results of the survey are shown in the table below.

Method of transportation	School bus	Car	Bicycle	Walk
Number of students	87	71	25	45

Use the the data above to find each experimental probability.

7. A student rides a bus to school. _____ **≈ 38.2%**
8. A student walks to school. _____ **≈ 19.7%**
9. A student rides to school in a car or in a bus. _____ **≈ 69.3%**
10. A student rides a bicycle or walks to school. _____ **≈ 30.7%**
11. A student does not walk to school. _____ **≈ 80.3%**

Look Back

Use guess-and-check to solve each equation. (LESSON 1.2)

35. $-1 = 4 - x$ **5**   **36.** $2x + 1 = 1$ **0**

Simplify. (LESSONS 2.2, 2.3, AND 2.4)

37. $-17 + (-12)$ **−29**   **38.** $(-3) \cdot (-13) + (-41)$ **−2**   **39.** $\frac{-4}{12} - \left(\frac{3}{-6}\right)$ $\frac{1}{6}$

Solve each equation for t. (LESSON 3.5)

40. $7(t - r) = 2r - 3(4r + t)$ $-\frac{3}{10}r$ **41.** $\frac{3(4t - 2)}{2} = \frac{3}{2} - (5 - t)$ $-\frac{1}{10}$

Look Beyond

42. How many different pairs of numbers can be generated by one red cube and one green cube if the faces of each cube are numbered from 1 to 6? **36**

43. How many of the pairs from Exercise 42 have a sum of 3? **2 pairs**

PORTFOLIO ACTIVITY

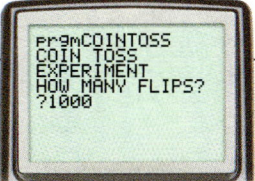
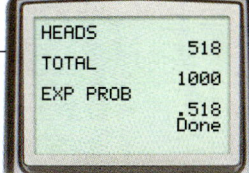

Include this activity in your portfolio.

You can use a graphics calculator to simulate a probability experiment. With a calculator, you can quickly conduct experiments with many more trials than you could by hand. The following program uses a random-number generator to simulate tossing a coin a given number of times:

Disp "COIN TOSS":Disp "EXPERIMENT"
Disp "HOW MANY FLIPS?"
Input N
0 → H
For(X,1,N)
int(2rand) → T
If T=0:H+1 → H
End
Disp "HEADS",H,"TOTAL",N
Disp "EXP PROB",H/N

1. What experimental probability is shown on the display for a 1000-trial experiment above?

2. Use the program and a graphics calculator to complete the table.

Number of trials	Heads	Experimental probability
4		
10		
20		
30		

3. Compare your results with those of others in your class.

4. Combine all of the trials and find the total number of heads. What is the overall experimental probability?

WORKING ON THE CHAPTER PROJECT

You should now be able to complete Activities 3 and 4 of the Chapter Project on page 208.

 internetconnect

GO TO: go.hrw.com
KEYWORD: MA1 Coin Toss

To support the Portfolio Activity, students may visit the HRW Web site, where they can explore more simulated experiments that use coins and number cubes.

LESSON 4.3 **185**

Prepare

NCTM PRINCIPLES & STANDARDS
1, 2, 5, 6, 8–10

LESSON RESOURCES

- Practice 4.4
- Reteaching Master 4.4
- Enrichment Master 4.4
- Cooperative-Learning Activity 4.4
- Lesson Activity 4.4
- Problem Solving/Critical Thinking Master 4.4
- Teaching Transparency 19

QUICK WARM-UP

Arrange each set of numbers from least to greatest.

1. 7, 19, 1, 14, 4, 9, 15, 5
   1, 4, 5, 7, 9, 14, 15, 19

2. 1.1, 0.25, 7.8, 2.5, 11.1, 0.5
   0.25, 0.5, 1.1, 2.5, 7.8, 11.1

Find each difference.

3. 9.2 – 5.8   **3.4**
4. 16 – 5.1   **10.9**

Also on Quiz Transparency 4.4

Teach

Why Write these two sets of numbers on the board or overhead.

49, 81, 88, 82, 85
73, 73, 73, 81, 85

Have students guess the average of each set. Then tell them that each average is 77. Ask if they think the number 77 is "representative" of each set. Have them suggest other numbers that might be more representative.

4.4 Measures of Central Tendency

Objectives

- Find the mean, median, mode, and range of a data set.
- Represent data with frequency tables.

Why Sometimes sets of numbers can be confusing. Statistics are used to present all kinds of information in a simplified form. It is important to understand different types of statistics and what they tell you about a particular set of data.

Descriptive Statistics

Statistics can help you understand and describe raw data. For example, the *mean*, *median*, and *mode* of a data set each tell you something about the *central tendency* of the data—that is, how the data tend to cluster around a value. The *range* provides information about the *spread* of the data.

Definitions: Mean, Median, Mode, and Range

The **mean** of a data set is the quotient when the sum of all of the elements is divided by the total number of elements.

The **median** of a data set is the middle number in the set when the elements are placed in numerical order. If there are an even number of elements, the median is the average of the two middle numbers.

The **mode** of a data set is the element, if any, that occurs most often. There may be no mode, one mode, or several modes.

The **range** of a data set is the difference between the greatest and least values in the set.

Alternative Teaching Strategy

USING VISUAL MODELS On the board or overhead, draw a horizontal line. Draw tick marks on the line and label them with the integers from zero to ten.

Poll students about the amount of time per week they spend studying math, rounded to the nearest hour. As each responds, place an × above the corresponding integer. You will now have a *line plot* like the one shown at the right.

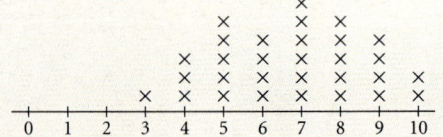

Ask which number has the most ×s. Explain that this is the *mode* of the data. Ask which × is in the middle, and note that the number below is the *median*. To find the *mean*, have students multiply each number by the number of ×s above it, add the products, and then divide by the total number of ×s.

186 LESSON 4.4

EXAMPLE 1

APPLICATION: SPORTS

Find the mean, median, mode, and range of the batting averages on page 186 for five National League players: 0.372, 0.366, 0.362, 0.333, and 0.327.

- **SOLUTION**

To find the *mean*, add the 5 batting averages, and then divide the sum by 5.
0.372 + 0.366 + 0.362 + 0.333 + 0.327 = 1.76
1.76 ÷ 5 = 0.352
The mean is **0.352**.

To find the *median*, arrange the five batting averages in ascending order. Because the number of averages is odd, take the middle value.
0.327, 0.333, 0.362, 0.366, 0.372
The median is **0.362**, the middle value.

To find the *mode*, look for a value that occurs more often than any other. Because each batting average occurs just once, there is no mode for the set.

To find the *range*, subtract the lowest value from the highest value in the set.
0.372 − 0.327 = 0.045
The range is **0.045**.

EXAMPLE 2

Find the mean, median, mode, and range for each set of data.
  a. 3, 5, 8, 5, 8, 7, 16, 10, 2, 7, 1, 5, 6, 1
  b. 160, 100, 150, 125, 125, 150, 114

- **SOLUTION**

a. To find the mean, add the numbers, and then divide by 14 (the number of numbers in the set of data).

3 + 5 + 8 + 5 + 8 + 7 + 16 + 10 + 2 + 7 + 1 + 5 + 6 + 1 = 84
84 ÷ 14 = 6   The mean is 6.

To find the median and mode, arrange the data in ascending order.

1, 1, 2, 3, 5, 5, **5, 6,** 7, 7, 8, 8, 10, 16

The median is the middle value. Since there is an even number of numbers, 14, the middle value falls between the seventh and eighth numbers, 5 and 6. Add and divide by 2 in order to find the median.
5 + 6 = 11   11 ÷ 2 = 5.5   The median is 5.5.

The mode is the value that occurs most frequently. There are three 5s. The mode is 5.

To find the range, subtract the lowest from the highest value in a set.
16 − 1 = 15   The range is 15.

b. To find the mean, add the numbers, and then divide by 7 (the number of numbers in the set of data).
160 + 100 + 150 + 125 + 125 + 150 + 114
924 ÷ 7 = 132   The mean is 132.

To find the median and the mode, arrange the data in ascending order.
100, 114, 125, 125, 150, 150, 160

ADDITIONAL EXAMPLE 1

The data below are mileage ratings in miles per gallon for nine new cars.
   22.0, 17.0, 33.0, 33.0, 19.5, 29.0, 32.0, 24.5, 33.0

Find the mean, median, mode, and range of the data.

mean: 27 miles per gallon
median: 29 miles per gallon
mode: 33 miles per gallon
range: 16 miles per gallon

ADDITIONAL EXAMPLE 2

Suppose that Sara has test grades of 81, 99, 85, and 88. She has one test left and she would like to have a test average of 90. **What must Sara score on the final test in order to meet her goal?**   97

Use Teaching Transparency 19.

Teaching Tip

TECHNOLOGY If you are using a TI-82 or TI-83 graphics calculator, you may wish to use its mean function. First access the **LISTMATH** menu as follows:

TI-83: [2nd] [STAT] [▶] [▶]

TI-82: [2nd] [STAT] [▶]

Choose **3:Mean(**. Now input your data. The numbers 6, 8, and 9 are used here to illustrate the procedure.

[2nd] [( ] 6 [,] 8 [,] 9 [2nd] [ )] [ )]

Press [ENTER] to see the mean.

There also is a median function. To use it, follow the same steps as for the mean, but choose **4:Median(** from the **LISTMATH** menu.

Interdisciplinary Connection

BUSINESS When ordering shoes, the manager of a shoe store must decide how many of each size to stock. It is important that there is enough stock to meet the demand for the most popular sizes. That is, the manager needs to identify the mode(s) of shoe sizes among potential customers.

Have students gather data about the shoe sizes of all students in the class. Instruct them to make two frequency tables, one for girls' shoe sizes and one for boys'. Have them determine the most common girls' and boys' shoe sizes in the class.

Inclusion Strategies

TACTILE LEARNERS Punch holes in a strip of cardboard and mark it as shown.

| 20 | 21 | 22 | 23 | 24 | 25 | 26 | 27 | 28 | 29 | 30 |

Loop a piece of string about 1 foot long through the top hole. Tape the string to the edge of a desk so that the strip is suspended. Give students 12 paper clips. Have them hang all the paper clips in the holes in such a way that the strip remains level. Lead them to see that the paper clips represent a set of data with mean 25.

LESSON 4.4   **187**

Teaching Tip

When finding a median, students tend to arrange the data in increasing numerical order. Make sure they are aware that it also is acceptable to arrange the data in decreasing order.

TRY THIS
mean: 0.3324
median: 0.329
mode: 0.328
range: 0.019

ADDITIONAL EXAMPLE 3

One day, several students in Ms. McCrossin's homeroom were discussing lengths of last names. They made the table below showing the number of letters in the last names of all the students in the homeroom. **Find the mode, median, mean, and range of their data.**

Number of letters	Frequency
4	||
5	|||||
6	||||
7	||||| ||
8	||
9	
10	|
11	|

mode: 7 letters
median: 6.5 letters
mean: 6.5 letters
range: 7 letters

The median is the middle value. Since there is an odd number of numbers, 7, the middle value is the fourth number.
$$\text{The median is } 125.$$

The mode is the value that occurs most frequently. There are two 125s and two 150s. Therefore, this set of data has two modes, 125 and 150.
To find the range, subtract the lowest from the highest value in the set.
$$160 - 100 = 60 \qquad \text{The range is } 60.$$

TRY THIS Give the mean, median, mode, and range of the batting averages on page 186 for five American League players.

EXAMPLE 3 Suppose that Dale has test scores of 75, 95, 82, and 77. He has one test remaining and his goal is to have a test average of exactly 85. **What must Dale score on the final test in order to meet his goal?**

SOLUTION
Valuable information about the data set can be found without any calculations. First, order the data.
$$\{75, 77, 82, 95\}$$

The highest and lowest scores are both 10 points from 85, so their average is 85, and they can be disregarded. The other two scores are both lower than 85, which means that the new score must to be higher than 85 in order to balance these two scores.

The exact score can be calculated algebraically by solving a related equation.
$$\frac{75 + 77 + 82 + 95 + x}{5} = 85$$
$$329 + x = 425$$
$$x = 96$$

Dale must earn a 96 on his final test in order to have an average of 85 for his test grades.

Frequency Tables

A **frequency table** is a method of displaying data in which tally marks are used to show how often each element of the set occurs. Instead of listing each occurrence of a number, a tally mark is placed next to the number each time it occurs. The following frequency table shows the scores on a 100-point quiz. How many scores are represented in the table?

Score	55	60	65	70	75	80	85	90	95	100
Frequency	|	|	||	|||	||||	|||	||||	||||	||	|

Enrichment

Give students the following alternative method for finding a mean:

1. Choose a number that appears to be close to the mean. Call it the trial mean.
2. Subtract the trial mean from each data value and record the differences.
3. Find the mean of the differences.
4. Add the mean of the difference to the trial mean.

The result is the true mean of the data set.

Have students apply this method to the data below.
$$85 \ 94 \ 88 \ 83 \ 91 \ 90 \ 83 \ 92 \ 83 \ 96$$

trial mean: 90 (sample)
differences: −5, 4, −2, −7, 1, 0, −7, 2, −7, 6
mean of differences: −15 ÷ 10 = −1.5
true mean: 90 + (−1.5) = 88.5

Have students find the mean conventionally to verify this method. Suggest that they practice this method with other data sets. Challenge them to find an algebraic justification for this method.

Activity
Using Frequency Tables

Use the frequency table on page 188 to complete this activity.

1. What is the mode of the scores? Explain how a frequency table makes it easy to find the mode.
2. What is the median of the scores? Explain how to find the median from the frequency table.
3. Without computing, estimate the mean. Then list all of the scores and compute the mean.

CHECKPOINT ✓ 4. How can a frequency table make it easier to compute the mean, median, and mode of a set of data?

EXAMPLE 4
APPLICATION: SPORTS

Cathy has been swimming in the 50-meter butterfly event since she started high school. During that time, she has kept track of what place she finished in the event at each meet she attended. **Find the mode, median, mean, and range of the places finished, using the data in the table below.**

Place	1st	2nd	3rd	4th	5th	6th	7th	8th	9th	10th
Frequency	IIII	III	IIII	IIII I	II	IIII	IIII	III	III	I

SOLUTION

To find the mode, find the value in the frequency table with the most tally marks. Fourth place has five marks, so the mode of the places finished is 4th.

To find the median, first count the tally marks in the frequency table. There are 33 marks in this table. Next count to find the location of the center tally mark, in this case the 17th mark. That mark is found at 5th place, so the median of the places finished is 5th.

To find the mean, first multiply each place finished by the number of tally marks below it. Then add these products and divide by the total number of tally marks, 33.

$4 + 6 + 12 + 20 + 10 + 24 + 28 + 24 + 27 + 10 = 165$

$165 \div 33 = 5$

The mean of the places finished is 5th.

The range is the difference between the highest and lowest places finished, in this case, $10 - 1 = 9$. Thus, the range of the places finished is 9.

CRITICAL THINKING

Mean, median, and mode each tell you something different about a set of data. Which statistic do you think is the best average for the places that Cathy finished in the 50-meter butterfly event? Which statistic do you think should be used to find the average of your grades? the average number of people in a family?

Activity Notes

The Activity guides students in an exploration of a frequency table. Students should discover that a frequency table can reduce the time it takes to determine the mean, median, and mode of a set of data.

Cooperative Learning

You may wish to have students do the Activity in groups of three. One student can be identified as the advocate for the mode, one for the median, and one for the mean. The group can work together to complete Steps 1–4. Each student should then write an argument explaining why a frequency table is or is not "helpful" in finding their assigned measures of central tendency.

CHECKPOINT ✓

4. A frequency table is a way to organize data, therefore making it easier to work with.

CRITICAL THINKING

Answers may vary. sample answers: mode; mean; median

Reteaching the Lesson

USING VISUAL MODELS Have students work in pairs. Give each pair 70 pennies. Tell them to make stacks of 8, 12, 5, 11, 8, 9, and 17 pennies, in that order, and to record the order. Then have them rearrange the stacks of pennies in order from shortest to tallest and record the new order. **5, 8, 8, 9, 11, 12, 17**

Now ask the following questions:

What is the difference in height between the tallest and shortest stacks? **12**

What statistic does this represent? **the range**

What is the height of the middle stack? **9**

What statistic does this represent? **the median**

What height is most common? **8**

What statistic does this represent? **the mode**

Have students rearrange the coins so that each stack has the same number. Ask them to identify the number of coins in each stack. **10** Ask what statistic this represents. **the mean**

LESSON 4.4

Assess

Selected Answers
Exercises 5–13, 15–69 odd

ASSIGNMENT GUIDE	
In Class	1–13
Core	15–57 odd
Core Plus	14–52 even, 54–58
Review	59–69
Preview	70

✏ Extra Practice can be found beginning on page 738.

Exercises

Communicate

1. Explain the difference between the mean and the median of a set of data.

2. Describe how you would find the median of the following set of data:
6, 8, 4, 9, 3, 4, 2, 5, 4, 0, 7

CONNECTION

3. **STATISTICS** A census report says that the average American family has 1.5 children. What measure of central tendency was most likely used? What problems do you think occur with statistics like this?

4. Explain how to determine the mean of the data set displayed in the frequency table below.

Grade	70	75	80	85	90	95	100
Frequency								||| |||| ||| |||| ||||

Guided Skills Practice

Find the mean, median, mode, and range for each set of data. (EXAMPLES 1 AND 2)

5. 2, 5, 8, 4, 2, 8, 3, 7, 6
 5; 5; 2 and 8; 6
6. 12, 29, 14, 8, 36, 26, 15
 20; 15; none; 28
7. 5, 9, 8, 10, 7, 3
 7; 7.5; none; 7
8. 0.25, 0.91, 0.65, 0.63, 0.82, 0.73, 0.21
 0.6; 0.65; none; 0.7
9. 8.3, 2.5, 1.7, 7.6, 9, 5.2, 3.1, 4
 5.175; 4.6; none; 7.3
10. 2, 5, 3, 6, 12, 9, 1, 4, 2, 4
 4.8; 4; 2 and 4; 11

APPLICATION

SPORTS Use the frequency table below, which compares the number of runs per game scored by a softball team, for Exercises 11–13.

Total runs scored	0	1	2	3	4	5	6	7	8
Frequency	||	||||	|||	|||||	|||	|||	||	||||		

11. How many games are represented in the frequency table? *(EXAMPLE 2)*
 27 games
12. Find the mean number of runs scored by the softball team.
 (EXAMPLE 2) **3.7**
13. Find the median, mode, and range for the runs scored per game.
 (EXAMPLE 2) **3, 3, 8**

Practice and Apply

Find the mean of each set of data.

14. 18, 15, 12, 14, 16 **15**
15. 5, 10, 12, 13, 10 **10**
16. 18, 24, 16, 13, 27, 22 **20**
17. 87, 86, 92, 95, 75 **87**
18. 45, 54, 53, 47, 38, 63 **50**
19. 91, 46, 85, 96, 78, 84 **80**

Practice

4.4 Measures of Central Tendency

Find the mean, median, mode(s), and range for each data set.

1. 13, 13, 10, 8, 7, 6, 4, 5
 mean **8.25** median **7.5** mode(s) **13** range **8**
2. 20, 30, 35, 24, 36, 47, 48
 mean **≈ 34.29** median **35** mode(s) **none** range **28**
3. 2, 5, 4, 1, 6, 7, 4, 3, 2, 1
 mean **3.5** median **3.5** mode(s) **1, 2, 4** range **6**
4. 130, 140, 135, 125, 160, 175
 mean **≈ 144.17** median **137.5** mode(s) **none** range **50**
5. 16, 18, 39, 200, 31, 39
 mean **≈ 57.17** median **35** mode(s) **39** range **184**

The Sleep Shop conducted a survey to determine the average number of hours that people sleep at night. The results are shown at right. Use this data for Exercises 6–12.

Number of Hours Spent Sleeping at Night
5 8 6 7 4
9 8 7 5 9
8 10 7 7 8
6 8 7 7 8
9 8 7 5 9
10 7 8 8 6

6. Make a frequency table for the data.

Hours	4	5	6	7	8	9	10
Frequency	1	2	3	7½	7½	7½	2

Find the measures of central tendency for the data.

7. mean **7.4**   8. median **8**   9. mode **8**   10. range **6**

11. Which measure of central tendency do you think gives the best indication of the number of hours the "typical" person spends sleeping each night? Explain. **The mode is the best indication of the typical person. According to the survey, most people sleep 8 hours (the mode) each night.**

12. Suppose that another person was surveyed and said that he spends 3 hours sleeping at night. How would this affect the mean, median, mode, and range? **The new measures of central tendency are as follows: mean: 7.2, median: 8, mode: 8, range: 7**

190 LESSON 4.4

24. 15
25. 10
26. 20
27. 87
28. 50
29. 84.5
30. 2.9
31. 114
32. 0.45
33. 52

34. 6
35. 8
36. 14
37. 20
38. 25
39. 50
40. 2.8
41. 25
42. 0.56
43. 55

20. 1.4, 2.6, 3.2, 1.8, 4.2, 3.3 **2.75**
21. 123, 114, 98, 102, 115 **110.4**
22. 0.25, 0.68, 0.12, 0.45, 0.5 **0.4**
23. 45, 57, 78, 23, 59, 43, 52 **51**

24–33. Find the median of each set of data in Exercises 14–23.

34–43. Find the range for each set of data in Exercises 14–23.

Find the mode, if any, of each set of data.

44. 5, 2, 6, 8, 2, 3, 1, 3, 2, 4, 7, 9 **2**
45. 16, 23, 19, 6, 15, 17, 18, 12, 19, 23, 19 **19**
46. 50, 48, 50, 55, 48, 57, 56 **48 and 50**
47. 97, 101, 98, 103, 101, 98, 97, 107, 98 **98**
48. 91, 87, 98, 97, 91, 100, 94 **91**

49–53. Find the mean, median, and range of each data set in Exercises 44–48.

CONNECTION

49. $4\frac{1}{3}$; 3.5; 8
50. 17; 18; 17
51. 52; 50; 9
52. 100; 98; 10
53. 94; 94; 13

APPLICATIONS

54. STATISTICS A class had the following test scores:
85, 100, 60, 81, 80, 75, 70, 90, 85, 65, 85, 75, 100, 75, 70, 95, 80, 60, 90, 70, 60, 85, 55, 60, 65, 65, 65, 70, 73, 75, 80, 81, 85, 90, 95

a. Make a frequency table.
b. Find the mean. **77**
c. Find the median. **75**
d. Find the mode. **85**

55. Suppose that Frank has test grades of 84, 92, 89, and 93. He has one test left, and his goal is to have a test average of 90. What must Frank score on the final test in order to meet his goal? **92**

56. SPORTS The basketball coach at the high school made the following list of the points scored by each member of the freshman basketball team:
26, 18, 4, 18, 18, 5, 6, 11, 4, 9, and 8.

a. Make a frequency table for the data.
b. Find the median number of points scored. **9**
c. Find the mean of the data. **11.$\overline{54}$**
d. What is the mode? **18**
e. Which measure of central tendency would you use to give the best impression of the players' abilities? Explain your choice. **mode; it has the highest value.**

57. BIOLOGY A biologist studying red-tailed hawks counts the eggs in their nests. Four nests have 1 egg, 65 nests have 2 eggs, and 5 nests have 3 eggs. He concludes that red-tailed hawks usually have 2 eggs at a time in the nest. Which measure of central tendency did the biologist most likely use? **mode**

The red-tailed hawk, found in North America, frequently preys on rodents.

Technology

A calculator is a useful tool for finding the mean of virtually any set of data. You may want to encourage students to use their calculators for Exercises 49–58, which involve greater numbers of data values.

Error Analysis

Many students confuse the meanings of the terms *mean*, *median*, and *mode*. Suggest some word associations such as the ones below to help them remember the differences.

- The *median* of a highway is the strip in the middle of two opposing lanes of traffic.
- A fashion *mode* is the current, most common style.

Student Technology Guide

4.4 Measures of Central Tendency

You can use the statistics features of a graphics calculator to find some measures of central tendency, including the mean. For example, to find the mean of the 10 quiz scores shown at right, first enter the scores into list L1 of the calculator.

32, 28, 18, 15, 25
30, 15, 22, 23, 18

The first step in statistical analysis is to enter the data.

STAT EDIT 1:Edit 32 ENTER 28 ENTER 18 ENTER 15 ENTER
25 ENTER 30 ENTER 15 ENTER 22 ENTER 23 ENTER 18 ENTER

Once the data is entered, find the mean as follows:

STAT CALC 1:1-Var Stats ENTER

Read the mean, denoted by $\bar{x}$, as 22.6. The average of the 10 quiz scores is 22.6 points.

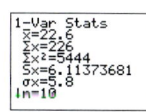

You can also use the calculator to sort the quiz-score data.

STAT EDIT 2:SortA(ENTER 2nd 1 ENTER STAT EDIT 1:Edit ▼

Once the data is sorted, you can scan the sorted list to find the mode(s). The modes are 15 and 18.

In this data set of 10 values, the median is the average of the fifth and sixth scores. From the sorted list, they are 22 and 23.

$$\text{median} = \frac{22 + 23}{2} = 22.5$$

Find the mean, median, and mode(s) of each set of data. Round your answers to the nearest tenth if necessary.

1. 18, 32, 14, 25, 37, 21, 28, 32, 26
   mean: 25.9, median: 26, mode: 32
2. 27, 34, 21, 17, 25, 19, 10, 51, 21, 18
   mean: 24.3, median: 21, mode: 21
3. 125, 118, 121, 152, 153, 137, 139, 155, 161
   mean: 140.1, median: 139, mode: none
4. 1.5, 1.8, 2.3, 1.8, 3.2, 3.0, 2.3, 4.1, 5.2
   mean: 2.8, median: 2.3, mode: 1.8 and 2.3
5. 1.2, 1.3, 1.0, 1.0, 1.0, 1.4, 1.4, 0.9, 0.9
   mean: 1.1, median: 1.0, mode: 1.0
6. 22.5, 20.1, 22.5, 23.6, 30.0, 29.6, 20.5, 26.4
   mean: 24.4, median: 23.05, mode: 22.5
7. 3.45, 3.44, 3.44, 3.55, 3.52, 3.53, 3.52
   mean: 3.5, median: 78, mode: 3.44 and 3.52
8. 89, 65, 65, 82, 84, 99, 92, 90, 66, 68, 74, 78, 75
   mean: 79, median: 23.05, mode: 65

54. a.

Test scores	55	60	65	70	73	75	80	81	85	90	95	100
Frequency	I	IIII	IIII	IIII	I	IIII	III	II	IIII	III	II	II

56. a.

Points scored	4	5	6	8	9	11	18	26
Frequency	II	I	I	I	I	I	III	I

LESSON 4.4 **191**

Exercise 70 previews the use of circle graphs as a means of displaying data. This topic will be studied in detail in Lesson 4.5. If students have access to a statistical graphing utility, they can use it to make the circle graph in part **b** of Exercise 70. If they are making the graph manually, they will need a compass, protractor, and straightedge. Remind them that one revolution around the center of a circle spans 360°.

internetconnect

GO TO: go.hrw.com
KEYWORD: MA1 Voting Schemes

In this activity, students examine different voting methods to discover how small differences in the computation of votes can change the outcome of an election.

APPLICATION

58. **RECREATION** The table at right shows the average number of hours spent watching television for certain age groups.
   a. Find the mean number of hours.
   b. Find the median number of hours.
   c. What is the mode of the data?
   d. What is the range of the data?

CHALLENGE

   e. Suppose that the number of hours per week spent watching television for females aged 25–54 is 32. Would this change the mean, median, mode, or range? Explain.

Age group (in years)	Number of hours per week spent watching TV
Children 2–5	24
Children 6–12	20
Female 13–17	21
Female 18–24	25
Female 25–54	30
Female 55+	44
Male 13–17	21
Male 18–24	22
Male 25–54	28
Male 55+	38

 Look Back

59. Write an equation to describe the following situation and solve it by using guess-and-check: If tickets for a movie cost $7 each, how many can you buy with $91? **(LESSON 1.2)**

Find each sum. (LESSON 2.1)

60. $-12 + (-7)$ 61. $23.4 + 142.7$ 62. $-302 + (-128) + 23$

Solve each equation. Check your solution. (LESSONS 3.2 AND 3.3)

63. $2a + 3 = 27$ 64. $413y = 1239$ 65. $\frac{x}{-16} = 23$

66. $3x - 14 = 88$ 67. $4.3b + 12.7 = 29.9$

Solve for the variable in each proportion. (LESSON 4.1)

68. $\frac{5}{7} = \frac{d}{56}$ 69. $\frac{20}{q} = \frac{45}{36}$

 Look Beyond

70. Use the frequency table you created in Exercise 54.
   a. What percent of the students scored 60 or below? from 61 to 70? from 71 to 80? from 81 to 90? from 91 to 100?
   b. Use your answer from part **a** to make a circle graph, or pie chart, for the data.

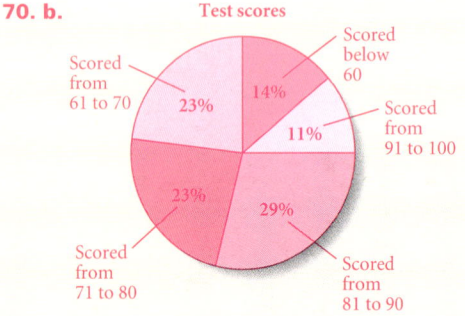

4.5 Graphing Data

Objectives
- Interpret line graphs, bar graphs, and circle graphs.
- Represent data with circle graphs.
- Analyze graphs in order to find a misleading presentation of data.

Why Graphs are used to present large amounts of information in ways that are easy to understand. Although graphs can be very informative, they can sometimes be misleading. It is important that you learn how to interpret and analyze them.

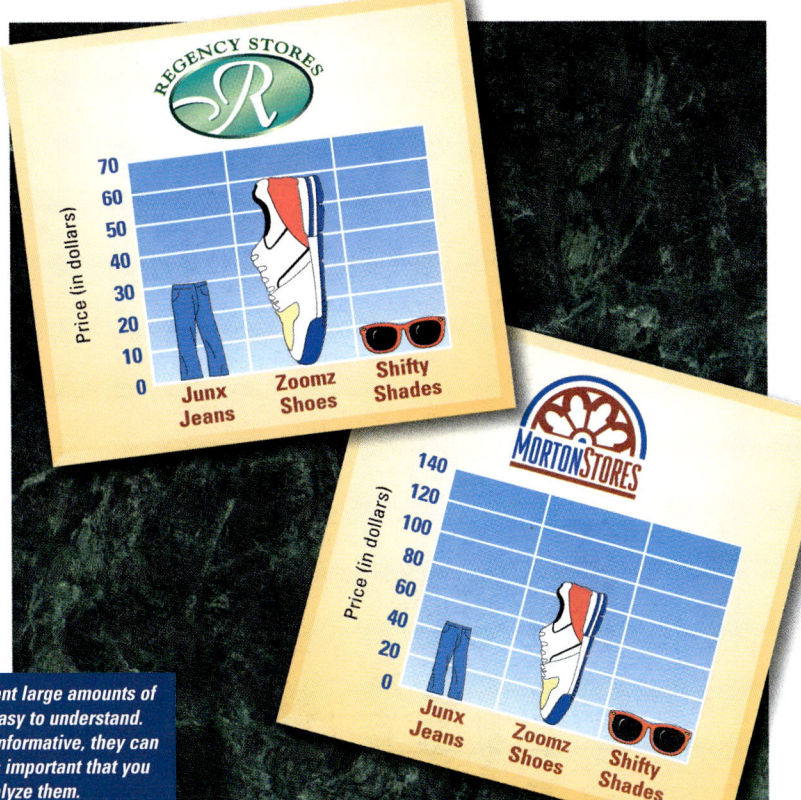

Common Types of Graphs

Graphs are very useful for interpreting large amounts of data at a glance. Three common forms of graphs are bar graphs, line graphs, and circle graphs. A circle graph is sometimes called a pie chart because of its resemblance to a pie that has been sliced.

Bar graphs use rectangular bars or objects to represent data. They can be used to compare individual years or other distinct items, as shown for the stores above.

A **line graph** is a type of graph that uses line segments between known data points to show changes that have occurred over time. An example of a line graph is shown above. By looking at the general trend, or pattern, of this graph, can you make a reasonable guess about the stock market in the future?

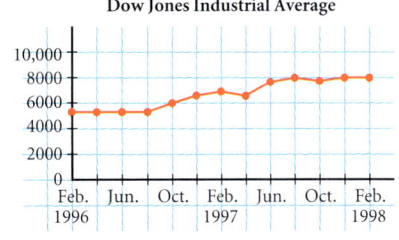

Alternative Teaching Strategy

USING TECHNOLOGY Provide students with computer software that has a statistical graphing component. Have them input these data.

Number of Supporters						
	Apr.	May	Jun.	Jul.	Aug.	Sept.
Ruiz	420	450	430	490	470	480
Chen	470	480	460	450	450	440

Tell them to apply the software's line graph style to the data.

Point out that the table of data shows the number of supporters of two candidates in an upcoming election. Ask students how altering the scale of the graph affects one's impression of the data. Elicit the fact that a scale with larger increments leads one to believe that the amount of support has changed more dramatically over time than it actually has.

Provide students with other tables of data that are suitable for exploring the bar graph and circle graph styles of the software.

Prepare

NCTM PRINCIPLES & STANDARDS
1–3, 5–10

LESSON RESOURCES
- Practice 4.5
- Reteaching Master 4.5
- Enrichment Master 4.5
- Cooperative-Learning Activity 4.5
- Lesson Activity 4.5
- Problem Solving/Critical Thinking Master 4.5
- Teaching Transparency 20
- Teaching Transparency 21
- Teaching Transparency 22
- Teaching Transparency 23

QUICK WARM-UP

Write each as a decimal. Round to the nearest hundredth, if necessary.

1. 68%   **0.68**   2. 3.2%   **0.032**
3. $\frac{37}{50}$ **0.74**   4. $\frac{37}{150}$ **≈0.25**

Also on Quiz Transparency 4.5

Teach

Why Ask students if they have ever heard the saying, "A picture is worth a thousand words." Tell them that graphs are pictures that can convey a great deal of information. Ask students to describe different types of graphs that they have seen recently.

LESSON 4.5   193

Activity Notes

In this Activity, students read and analyze two bar graphs. They will see how two graphs that appear very similar at first glance can, in fact, convey very different information. They should arrive at the conclusion that if two graphs are to be used to compare two sets of data, it is imperative that the graphs be drawn with the same scale.

CHECKPOINT ✓

3. The difference in the vertical scales of the graphs makes it possible to misinterpret the information. Use the same vertical scale.

Teaching Tip

Be sure students are aware that they can read only *approximations* of data from bar graphs and line graphs of the type shown in this lesson. When reading a number from one of these graphs, they should take care to preface it with the word *about* or *approximately*.

Use Teaching Transparency 20.

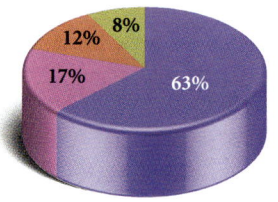
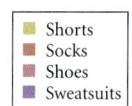
- Shorts
- Socks
- Shoes
- Sweatsuits

In a **circle graph**, or pie chart, data are represented as parts of a circle. This is most effective for looking at parts or percents of a whole. For example, a sporting goods store that needs to know what percent of its sales are from its shoes, shorts, socks, and sweatsuits, might use a circle graph like the one at left.

If the store had $1,000,000 in sales last year, the approximate dollar amount for each of its items is the percent given for the item multiplied by $1,000,000. The dollar amount of shoe sales, which accounted for 17% of the store's sales, is given by the following:

$$0.17 \cdot 1{,}000{,}000 = 170{,}000$$

Thus, the total amount of sales from shoes was $170,000.

Unfortunately, graphs are sometimes used to intentionally mislead consumers. This is why it is important to be able to read and interpret graphs in a knowledgeable manner.

Activity
Misleading Graphs

The bar graphs at left show the sales (in millions of dollars) of two retail companies for five consecutive years.

1. What is the approximate amount of sales for the Regency stores in 1997? What is the approximate amount of sales for the Morton stores in 1997? Which company's bar graph has a longer bar to represent sales in 1997? Which company actually had greater sales in 1997?

2. What is the approximate amount of sales for the Regency stores in 2000? What is the approximate amount of sales for the Morton stores in 2000? Which company's bar graph has a shorter bar to represent sales in 2000? Which company actually had lower sales in 2000?

CHECKPOINT ✓

3. What features of the graphs make it possible to misinterpret the information? How might the information be presented to make a more accurate visual comparison?

4. Redraw the graphs to present a more accurate visual comparison.

Interdisciplinary Connection

SOCIAL STUDIES In many ways, a government is a business. As with a business, the amount of money a government takes in is called its *revenue*. The circle graph at right shows the expected sources of revenue for the United States government for 1999. The total amount of revenue expected for 1999 is $1743 billion. Find the amount that will come from individual income taxes and the amount that will come from corporate income taxes.

Federal Revenues by Source, 1999

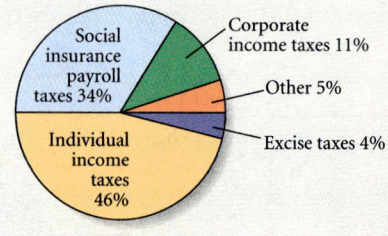

Individual income taxes: ≈ $802 billion
Corporate income taxes: ≈ $192 billion

194 LESSON 4.5

Working With Graphs

APPLICATION: ECONOMICS

A line graph can be used to make predictions based on current trends.

From the graph above, you can tell in what year the unemployment rate was the greatest. You can also find the year of the lowest unemployment rate, 1945, and several years when the rate was the same, 1955, 1960, 1975, and 1990. The graph also clearly shows the trend of increasing unemployment from 1930 to 1935 and sharply decreasing unemployment from 1935 to 1945.

CRITICAL THINKING

What historical event might account for the rise in the unemployment rate from 1930 to 1935? What event would account for the decline in the unemployment rate from 1935 to 1945?

EXAMPLE 1

APPLICATION: INVESTMENTS

Don invested $100 each in four companies in 1995. The line graph at right shows the value of the stock over several years. **Find the approximate values of Don's MRI and MMart stock in 1999. Construct a bar graph that shows the approximate value of Don's WMB stock from 1995 to 1999.**

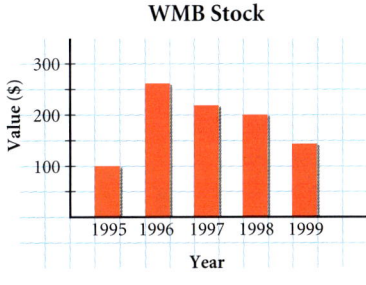

- **SOLUTION**

MRI: The graph shows that the value of Don's MRI stock was about $290 in 1999.

MMart: The graph shows that the value of Don's MMart stock was about $140 in 1999.

WMB: The approximate value of Don's WMB stock was $100 in 1995, $260 in 1996, $220 in 1997, $200 in 1998, and $140 in 1999. The bar graph for this information is shown at right.

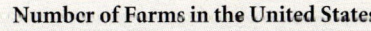

Inclusion Strategies

LINGUISTIC LEARNERS Virtually every graph tells a story. Have students search through newspapers and magazines and select a graph that interests them. Then have them write what they perceive to be the "story" of the graph.

For example, you might give students the graph at right. Tell them to imagine a person born in 1890 who lived on a farm. Have them write an essay in which they describe what the graph tells them about changes this person might have experienced from 1890 through 1990.

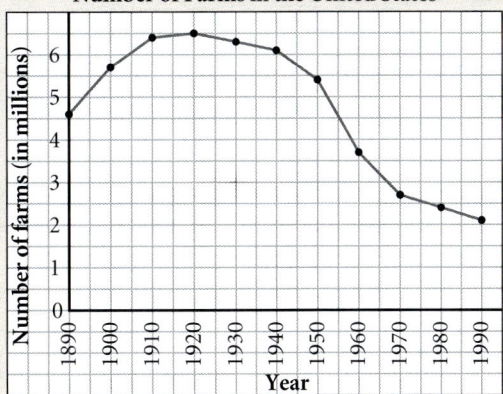

CRITICAL THINKING
the stock market crash of 1929; World War II

Use Teaching Transparency 21.
Use Teaching Transparency 22.

ADDITIONAL EXAMPLE 1

Belynda invested $200 in each of two stocks in 1994. The line graph below shows the value of the stocks over several years. **Find the approximate values of her WEB1 and UNI stocks at the end of 1999. Construct a bar graph that shows the value of Belynda's WEB1 stock at the end of each year from 1994 to 1999.**

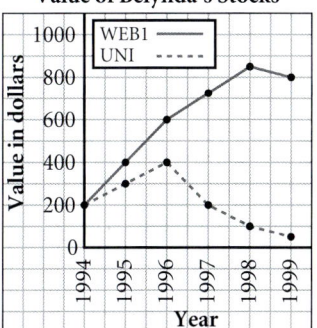

WEB1: about $800
UNI: about $50

LESSON 4.5 195

TRY THIS
1997 MRI = $340
1999 CBC = $160

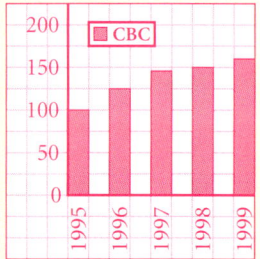

ADDITIONAL EXAMPLE 2

The circle graph below shows the sources of the profit for a manufacturer of personal-care products in a recent year. The manufacturer's total profits for this year were $520 million. **Find the amount of profit from medical products, from hair-care products, and from skin-care products.**

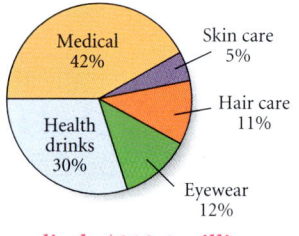

medical: $218.4 million
hair care: $57.2 million
skin care: $26 million

Math CONNECTION

STATISTICS Circle graphs display data as parts of a whole. For this reason, circle graphs are usually labeled with percents.

TRY THIS
investments: $2560
clothing: $1920
miscellaneous: $7040

TRY THIS Find the approximate value of Don's MRI stock in 1997 and of Don's CBC stock in 1999. Construct a bar graph that shows the approximate value of Don's CBC stock from 1995 to 1999.

While line graphs and bar graphs are useful for presenting data and analyzing trends, a circle graph is a better method for representing how different elements in a whole quantity are distributed.

EXAMPLE 2 The circle graph at right shows how the average American family spends its money. **Find how much money, on average, a family with an annual income of $32,000 spends on housing, transportation, and food.**

American Family Spending
Investments 8%
Clothing 6%
Miscellaneous 22%
Food 15%
Transportation 18%
Housing 31%

● **SOLUTION**
Housing: Multiply 31% by $32,000.
$$0.31 \cdot 32{,}000 = 9920$$
The average amount that an American family with an annual income of $32,000 spends on housing is $9920.

Transportation: Multiply 18% by $32,000.
$$0.18 \cdot 32{,}000 = 5760$$
The average amount that an American family with an annual income of $32,000 spends on transportation is $5760.

Food: Multiply 15% by $32,000.
$$0.15 \cdot 32{,}000 = 4800$$
The average amount that an American family with an annual income of $32,000 spends on food is $4800.

TRY THIS Find how much money, on average, a family with an annual income of $32,000 spends on investments, clothing, and miscellaneous expenses.

Enrichment

Ask students to bring in recent newspapers and magazines. Divide the class into groups of 4 or 5. Instruct students to find examples of as many different kinds of graphs as they can. Encourage them to search not only for simple line graphs, bar graphs, and circle graphs, but also for multiple line graphs, stacked bar graphs, double circle graphs, and so on. Have the groups share their graphs with the class explaining what each represents. Ask them to identify any graphs they feel are misleading and to explain their reasoning.

Reading circle graphs is a useful skill. There may be times when you need to create a circle graph. In some cases, you will need to use a protractor to construct your graph.

EXAMPLE 3 Use the information in the table to make a circle graph for the number of cats in each breed that are entered in the Feline Fanciers cat show.

APPLICATION
CATS

Breed	Persian	Abyssinian	Balinese	Manx	Siamese	Total
Number of cats	99	30	66	41	106	342

• **SOLUTION**

Find the percent for each category.

There are 99 Persian cats in the show. The total number of cats entered in the show is 342.

$$\frac{99}{342} \approx 0.29, \text{ or approximately } 29\%$$

The Persian breed accounts for 29% of the cats entered in the show.

Find the number of degrees that should be used to represent this category in the graph. The total number of degrees in a circle graph is 360.

$$0.29 \cdot 360° = 104.4°$$

Continue with the remaining categories. The original table can be extended to include columns for the percent and the number of degrees.

Breed	Number of cats	Percent	Number of degrees
Persian	99	$\frac{99}{342} \approx 0.29 \approx 29\%$	$0.29 \cdot 360° \approx 104.4°$
Abyssinian	30	$\frac{30}{342} \approx 0.09 \approx 9\%$	$0.09 \cdot 360° \approx 32.4°$
Balinese	66	$\frac{66}{342} \approx 0.19 \approx 19\%$	$0.19 \cdot 360° \approx 68.4°$
Manx	41	$\frac{41}{342} \approx 0.12 \approx 12\%$	$0.12 \cdot 360° \approx 43.2°$
Siamese	106	$\frac{106}{342} \approx 0.31 \approx 31\%$	$0.31 \cdot 360° \approx 111.6°$
Total	342	100%	360°

Abyssinian cats are known for their quiet personalities and slender bodies.

Use a compass to construct the circle, and mark the center of the circle with a point. Then use a protractor to draw angles with the indicated measures. It may be necessary to round your angle measures to the nearest whole number.

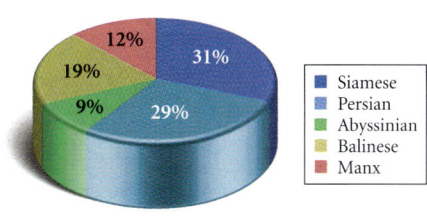

Cat Show Entries

Use Teaching Transparency 23.

ADDITIONAL EXAMPLE 3

Use the information in the table below to make a circle graph for the number of dogs in each breed that are entered in a dog show.

Breed	Number of dogs
Golden retriever	90
Doberman pinscher	146
Great Dane	21
German shepherd	63
Irish setter	44
Total	364

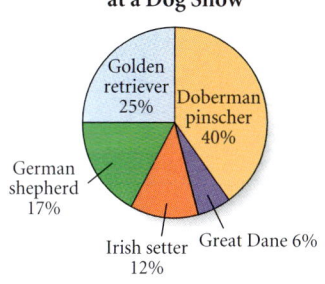

Breeds of Dogs at a Dog Show

Reteaching the Lesson

COOPERATIVE LEARNING Have students work in pairs. Give them the bar graph shown at right or one similar to it. Have each student write five questions about the graph. Now have the partners exchange questions. Each partner should answer the other's questions. Then the partners should check each other's answers. Have them work together to resolve any differences in their perceptions of the graph.

Repeat the activity two more times, using a suitable line graph and circle graph.

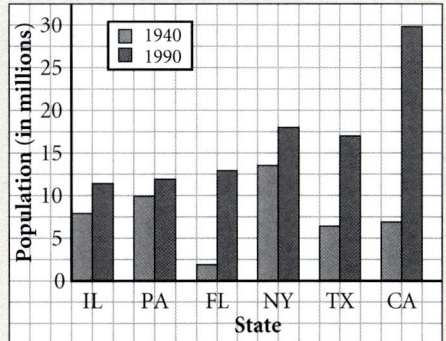

LESSON 4.5

TRY THIS

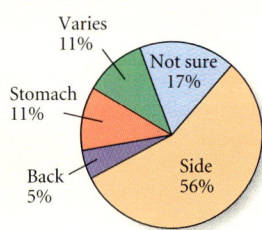

TRY THIS A survey of 500 Americans was conducted to find the position in which they fall asleep. Make a circle graph that shows the percent of people in each category.

Sleeping position	Side	Back	Stomach	Varies	Not sure
Number	280	25	55	55	85

Exercises

Assess

Selected Answers
Exercises 5–8, 9–35 odd

ASSIGNMENT GUIDE

In Class	1–8
Core	9–23
Core Plus	9–24
Review	25–36
Preview	37

✏ Extra Practice can be found beginning on page 738.

Communicate

1. Why are graphs important statistical tools?
2. Explain how graphs can be misleading.
3. Name and describe three different types of graphs that can be used to display data.
4. Explain the difference between a bar graph and a line graph.

Guided Skills Practice

The graphs below are based on the 1990 United States census.

CONNECTION

Northeast: 88,550,000
West: 60,720,000
South: 53,130,000
Midwest: 50,600,000

5. **STATISTICS** Based on the bar graph, what is the smallest population group in the United States? What is the second smallest group? **(EXAMPLE 1)** under 5; 65 and over

6. Based on the circle graph, how many people lived in each region listed if the total population in the United States in 1990 was approximately 253,000,000? **(EXAMPLE 2)**

7. How many people would you expect to have been living in the Northeast in 1995 if the distribution stayed the same and the total population that year was estimated at 265,000,000? **(EXAMPLE 2)** 92,750,000

8. Suppose that you asked 90 people whether or not they liked a certain movie. Thirty people told you that they did not like the movie, 45 people said that they did like it, and 15 people were unsure whether they liked it. Make a circle graph that displays the data. **(EXAMPLE 3)**

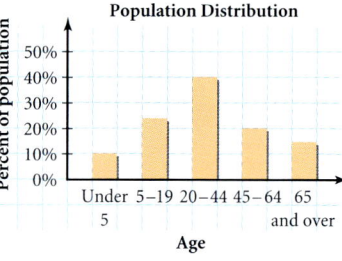

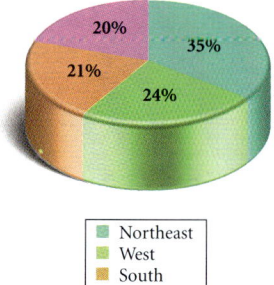

■ Northeast
■ West
■ South
■ Midwest

Practice

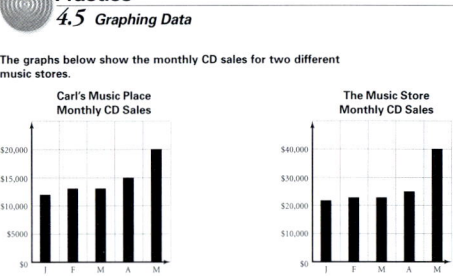

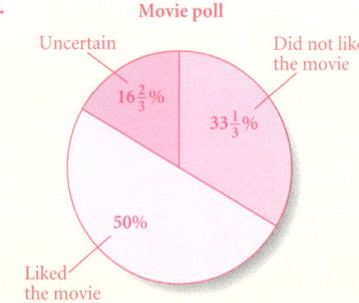

8.

198 LESSON 4.5

Practice and Apply

CONNECTION

STATISTICS These graphs show the price, in dollars, for the stock of two companies. Use the graphs for Exercises 9–11.

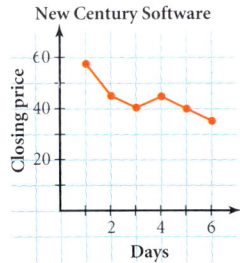
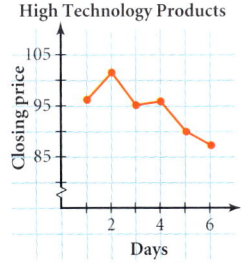

9. Which company had the highest one-day stock price for the week?
   High Technology Products
10. Which company had the lowest one-day stock price during the week?
    New Century Software
11. Why might it be misleading to display these two graphs together?
    The scales of the two graphs are different.

APPLICATIONS

GEOGRAPHY Use the graph of average high temperatures for Exercises 12–14.

Madrid; July 12. Which city had the highest temperature? In which month did this temperature occur?

Moscow; January 13. Which city had the lowest temperature? In which month did this temperature occur?

Moscow; ≈ 60° 14. Which city had the greatest increase in temperature from January to July? What was the approximate amount of the increase?

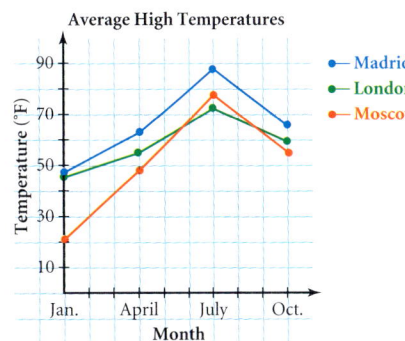

15. **DOGS** Use the information below to make a circle graph for the number of dogs in each breed that are in a Canine Companions program.

Breed	Golden Retriever	Yellow Lab	Collie	Cocker Spaniel	Dalmatian	Total
Number of dogs	105	73	24	47	91	340

FIRE PREVENTION Use the circle graph about the causes of house fires for Exercises 16–19.

16. If there were 957,000 fires in one year, how many of them were caused by smoking?
    95,700
17. If there were 957,000 fires in one year, how many were caused by faulty electrical wiring?
    248,820
18. According to this graph, what is the most likely cause of a house fire?
    Electrical wiring
19. Based on these statistics, what steps could you take to reduce fire hazards in your home? **Answers may vary.**

Causes of House Fires
- 26% Electrical wiring
- 18% Cooking
- 5% Unknown
- 5% Arson
- 10% Smoking
- 14% Open flames
- 22% Children playing

15.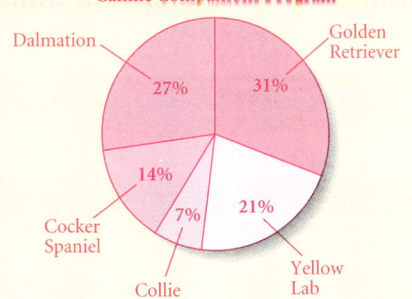
Canine Companions Program
- Golden Retriever 31%
- Yellow Lab 21%
- Collie 7%
- Cocker Spaniel 14%
- Dalmation 27%

Technology

A calculator may be helpful in performing the calculations in Exercises 6, 7, 16, 17, and 24.

Error Analysis

Students may have difficulty reading bar graphs when the top of a bar falls between grid lines. Suggest that they use a ruler or the edge of a piece of paper to visually determine where the top of the bar would intersect a grid line if it were extended. This technique also is helpful in determining the coordinates of points on a line graph.

Student Technology Guide

4.5 Graphing Data

The spreadsheet at right shows the number of votes received by each candidate in a recent mayoral election.

A graph is often useful as an aid to help you visualize and analyze data such as this. Two common forms of graphs are circle and bar graphs. You can make each of these graphs by using a spreadsheet's graphics options.

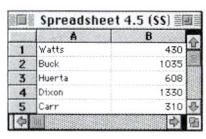

Choose the type of graph you want to create. A circle graph, also called a pie chart, is shown. You may need to state that the categories come from column A and the numerical data comes from column B. Complete the steps needed to display the graph.

1. Suppose that Dixon really received 100 votes more than what is recorded in the spreadsheet and each of the other candidates received 25 fewer votes. Modify the entries in the spreadsheet, and then construct a new circle graph. How do the new percents reflect this change in the data?
   Watts, 11%; Buck, 27%; Huerta, 16%;
   Dixon, 38%; Carr, 8%
   See answer section for the circle graph.

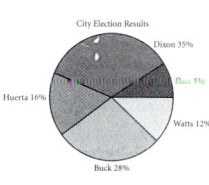

Next choose a bar graph, also called a column graph. Specify column A as the x-axis and column B as the y-axis. Graph. Add labels so that the graph is easier to analyze.

2. What kind of data is displayed in both graphs?
   the distribution of votes in an election

3. How do the graphs differ in what they show?
   The circle graph shows percents while the bar graph shows actual numbers of votes.

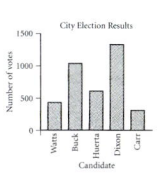

In Exercise 37, students use the formula $C = 2\pi r$ to find the circumferences of two circles. Students may recall this formula from work in previous courses, and they will study it in greater depth if they take a course in geometry. Be sure students are aware that the labeled measures are the radii of the circles. Remind them that the diameter of a circle is twice its radius.

24. No, orthopedic injuries and strokes increased by about 31%, head injuries by about 67%, and spinal cord injuries and other injuries by about 50%.

APPLICATION

HEALTH The bar graph shown here is in the annual report for a rehabilitation hospital. Use the graph for Exercises 20–24.

20. Which category had the greatest increase in the number of patients from 1998 to 1999?

21. Which category had the smallest increase in the number of patients from 1998 to 1999?

22. How many patients in 1999 were diagnosed with an orthopedic injury? a spinal cord injury?

23. What was the increase in the number of patients diagnosed with a stroke from 1998 to 1999?

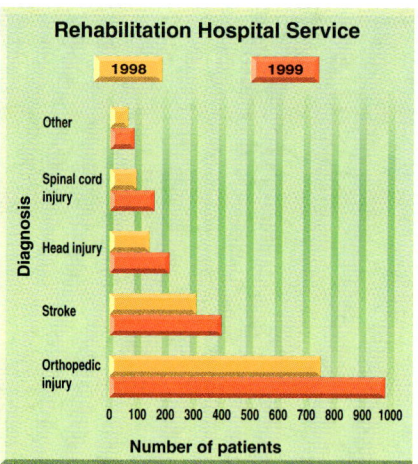

CHALLENGE

24. Was the percent increase in the number of patients from 1998 to 1999 the same for each type of injury diagnosed at the rehabilitation hospital? Explain your reasoning.

 Look Back

Simplify. *(LESSONS 2.5 AND 2.6)*

25. $(3x - 7) + (5x + 2)$

26. $4(7s - 3) - 8(s + 2)$

Solve each equation. Check your answers. *(LESSONS 3.4 AND 3.5)*

27. $4t + 7 = 23 + 2t$

28. $3(5h - 12) = 9$

29. $2x + 3(x + 7) = 41$

30. $4y + 17 = 3y - 26$

31. $4(m + 7) + 3m - 5 = 2m + 23$

32. $3(x + 5) = 5(x + 3)$

Find each answer. *(LESSON 4.2)*

33. What is 35% of 160?

34. 14 is what percent of 200?

Find the mean, median, mode, and range for each set of data. *(LESSON 4.4)*

35. 4, 5, 18, 12, 6, 4, 10, 15

36. 100, 87, 92, 94, 90, 100, 88

Look Beyond

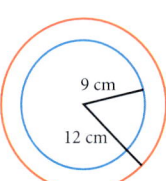

37. The formula for the circumference of a circle is $C = 2\pi r$, where r is the radius of the circle. Use this formula to find the sum of the circumferences of the two circles shown at left.

4.6 Other Data Displays

Gray whales, which live in the northern Pacific Ocean, were recently taken off the endangered species list. Some scientists collect data such as the length of adult gray whales.

Objectives

- Interpret stem-and-leaf plots, histograms, and box-and-whisker plots.
- Represent data with stem-and-leaf plots, histograms, and box-and-whisker plots.

Why *Displays such as stem-and-leaf plots, histograms, and box-and-whisker plots provide a way to study large sets of data more easily and arrive at conclusions more quickly. This process is helpful in fields like science, medicine, and technology.*

Prepare

NCTM PRINCIPLES & STANDARDS 1–3, 5, 6, 8–10

LESSON RESOURCES

- Practice 4.6
- Reteaching Master 4.6
- Enrichment Master 4.6
- Cooperative-Learning Activities 4.6
- Lesson Activity 4.6
- Problem Solving/Critical Thinking Master 4.6
- Teaching Transparency 24

QUICK WARM-UP

Find the mean, median, mode(s), and range for the set of data below.

Number	Frequency
11					
12			
16					
19		
22			

mean: 15; median: 16; modes: 11, 16; range: 11

Also on Quiz Transparency 4.6

Stem-and-Leaf Plots

Length of Adult Gray Whales (in meters)			
12.7	13.5	14.0	14.2
12.8	13.5	14.0	14.3
12.8	13.6	14.0	14.3
12.8	13.6	14.1	14.3
12.9	13.6	14.1	14.4
13.0	13.7	14.1	14.4
13.1	13.7	14.1	14.4
13.1	13.7	14.1	14.5
13.2	13.8	14.1	14.5
13.3	13.8	14.1	14.5
13.4	13.9	14.2	14.6
13.4	13.9	14.2	15.2

A **stem-and-leaf plot** displays data by dividing each element into two parts. The stem-and-leaf plot below gives the lengths of several adult whales. The numbers to the left of the vertical line are called **stems** and represent the first part of each element. The numbers to the right of the line are called **leaves** and frequently represent the last digit of each element.

Stem-and-Leaf Plot	
Stems	**Leaves**
12	7, 8, 8, 8, 9
13	0, 1, 1, 2, 3, 4, 4, 5, 5, 6, 6, 6, 7, 7, 7, 8, 8, 9, 9
14	0, 0, 0, 1, 1, 1, 1, 1, 1, 1, 2, 2, 2, 3, 3, 3, 4, 4, 4, 5, 5, 5, 6
15	2

To record data in a stem-and-leaf plot, first examine the data set to determine what the stems and leaves will be. In the stem-and-leaf plot above, the stems are whole numbers, and the leaves are tenths.

If the data set consists of numbers from 15 to 55, the stems can represent the tens digits, and the leaves can represent the ones digits.

Teach

Why Tell students to imagine a large set of data, such as a set of 50 test scores. Ask them to consider when it might be important to display the data so that each score is known, and when it might only be necessary to see the data grouped in intervals.

Samples: It may be necessary to know the value of each score in order to compute additional statistics. A grouped display may be all that is needed to see general characteristics of the data.

Alternative Teaching Strategy

USING TECHNOLOGY Students can use the histogram feature of a graphics calculator to examine the frequency of items in a set of data.

If they are using the TI-82 or TI-83 graphics calculator, tell them to first clear any existing data from **LIST 1**. Then have them enter the data from Example 1 on page 202 into **LIST 1**. To do this, they should press [STAT], choose **1:Edit...**, and enter the values under **L1**.

Now have them press [2nd] [Y=] and choose **1:Plot1**. Tell them to select **On**, and then press [▼] [▶] [▶] [ENTER] to choose the histogram icon. Then they should press [Y=] and clear any equations from the **Y=** editor.

Have them press [WINDOW] and make these settings.

Xmin=5   Xmax=16   Xscl=1
Ymin=0   Ymax=12   Yscl=1

Now they can press [GRAPH] to view the histogram. If they then press [TRACE], they can use the [◀] and [▶] keys to display the size of each interval (min=) and the frequency in that interval (n=).

LESSON 4.6 **201**

ADDITIONAL EXAMPLE 1

Construct a stem-and-leaf plot for the data below.

Rainfall (in inches)				
2.3	2.1	3.3	1.5	2.2
2.2	2.2	2.1	2.4	1.1
3.1	2.8	2.2	3.6	2.0
1.6	1.9	4.7	3.0	1.9
2.1	2.2	2.7	2.8	2.1

Rainfall (in inches)	
Stems	Leaves
1	1, 5, 6, 9, 9
2	0, 1, 1, 1, 1, 2, 2, 2, 2, 2, 3, 4, 7, 8, 8
3	0, 1, 3, 6
4	7

TRY THIS

Stems	Leaves
6	4
7	6, 9, 9
8	0, 0, 1, 6
9	1, 6, 7, 9

Teaching Tip

To demonstrate the relationship between stem-and-leaf plots and histograms, give students a set of data. Tell them to make a stem-and-leaf plot and a histogram of the data on separate sheets of paper. Have them give the paper with the histogram a quarter-turn clockwise and compare the two data displays.

EXAMPLE 1

Breathing Intervals (in minutes)			
5.6	8.6	10.3	11.3
14.8	14.7	13.6	12.2
6.5	15.5	8.6	13.8
8.4	11.0	13.5	15.8
15.6	12.6	10.5	6.7
7.9	9.6	10.8	11.7
15.6	11.4	9.2	6.8
7.1	9.2	10.5	11.4
16.2	14.0	12.9	11.5
12.3	13.8	10.1	14.8
13.7	14.2	12.6	8.5
10.1	12.7	12.2	11.4
12.6	13.9	13.8	14.7
12.2	13.7	14.2	11.4
10.7	13.9	12.7	16.7
13.9	15.6	7.3	

Gray whales breathe every 5 to 15 minutes. They can, however, go as long as 40 minutes without breathing. A scientist studying a group of whales recorded the whales' breathing intervals (in minutes). **Construct a stem-and-leaf plot of the data in the table at left.**

- **SOLUTION**

To construct the stem-and-leaf plot, first determine the stems and leaves. The numbers range from 5.6 to 16.7, so let the stems be whole numbers, from 5 to 16, and let the leaves be tenths. Then record the leaves next to the appropriate stems. Arrange the leaves in each row from least to greatest.

Breathing Intervals (in minutes)	
Stems	Leaves
5	6
6	5, 7, 8
7	1, 3, 9
8	4, 5, 6, 6
9	2, 2, 6
10	1, 1, 3, 5, 5, 7, 8
11	0, 3, 4, 4, 4, 4, 5, 7
12	2, 2, 2, 3, 6, 6, 6, 7, 7, 9
13	5, 6, 7, 7, 8, 8, 8, 9, 9, 9
14	0, 2, 2, 7, 7, 8, 8
15	5, 6, 6, 6, 8
16	2, 7

TRY THIS Make a stem-and-leaf plot for the following data:
97, 86, 64, 79, 91, 81, 80, 99, 80, 96, 79, 76.

Histograms

A **histogram** is a bar graph that has no space between the bars and that shows the frequency of data. You can construct a histogram very easily from the data in a stem-and-leaf plot. Refer to the stem-and-leaf plot for the lengths of gray whales on page 201.

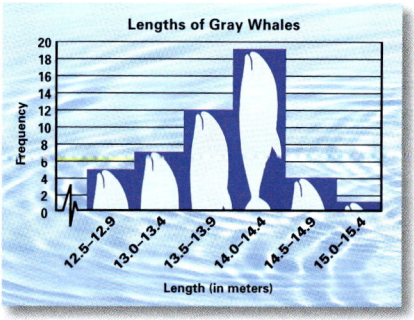

The lengths are listed along the horizontal axis of the histogram, and the frequencies of the numbers in the data set are listed along the vertical axis. Notice that the lengths are listed on the horizontal axis in equal intervals, such as from 12.5 to 12.9, instead of by each individual measurement.

Interdisciplinary Connection

SCIENCE Todd and Tessa are working on a science project. Part of the project requires them to collect and record the daily high and low temperatures in their city over a 30-day period. Below are the data they collected.

Highs: 82, 85, 79, 81, 78, 79, 79, 65, 66, 68, 74, 75, 61, 65, 65, 72, 74, 78, 66, 67, 58, 58, 60, 62, 48, 55, 57, 60, 58, 59

Lows: 65, 66, 62, 62, 60, 62, 63, 48, 48, 50, 60, 61, 51, 50, 49, 52, 58, 61, 45, 46, 39, 38, 42, 45, 29, 35, 36, 39, 36, 39

Make box-and-whisker plots to display both the high temperatures and the low temperatures.

Key Points:
highs: 48, 60, 66, 78, 85
lows: 29, 39, 49.5, 61, 66

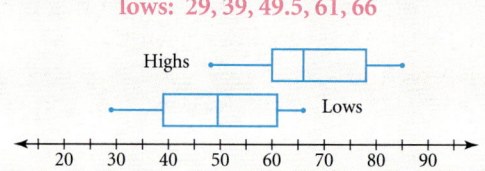

202 LESSON 4.6

Box-and-Whisker Plots

Box-and-whisker plots, also called box plots, show how data is distributed by using the median, the upper and lower quartiles, and the greatest and least values of the data.

Interpreting a Box-and-Whisker Plot

The grades for two of Ms. Garth's classes are listed below.

Student	1	2	3	4	5	6	7	8	9	10	11	12	13
Class 1	60	80	65	85	80	75	75	100	70	90	90	80	75
Class 2	70	85	75	85	70	85	85	90	100	90	95	85	85

1. Order the data for class 1 from least to greatest, and find the median grade.
2. Calculate the median grade for the first 6 grades of class 1.
3. Calculate the median grade for the last 6 grades of class 1.
4. The data for class 1 is displayed below in a box-and-whisker plot. What value is represented by the vertical line in the middle of the box?

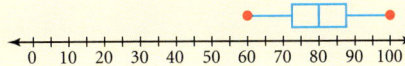

5. The vertical line at 72.5 represents the lower quartile and the vertical line at 87.5 represents the upper quartile. Explain how these values compare with the lower and upper medians you computed in Steps 2 and 3.
6. Explain in your own words what the dots at the far left and right of the horizontal line represent.

CHECKPOINT ✓ 7. Sketch a box-and-whisker plot of the data for class 2.

EXAMPLE 2

APPLICATION
BIOLOGY

Use the stem-and-leaf plot of the length of gray whales (in meters) to construct a box-and-whisker plot.

Length of Gray Whales (in meters)	
Stems	Leaves
12	7, 8, 8, 8, 9
13	0, 1, 1, 2, 3, 4, 4, 5, 5, 6, 6, 6, 7, 7, 7, 8, 8, 9, 9
14	0, 0, 0, 1, 1, 1, 1, 1, 1, 2, 2, 2, 3, 3, 3, 4, 4, 4, 5, 5, 5, 6
15	2

Inclusion Strategies

VISUAL LEARNERS The concepts of stem-and-leaf plots and box-and-whisker plots are rich in visual imagery. Students with drawing ability may enjoy creating imaginative instructional posters that relate stem-and-leaf plots to plants and box-and-whisker plots to cats. If you wish, you can have a contest. Display all of the posters in the classroom and have the class vote on the most original poster.

Enrichment

Give students this box-and-whisker plot.

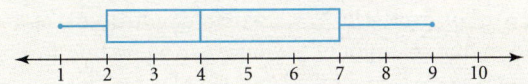

Ask them to identify the percent of the data that falls within the intervals below.
1. $2 \leq n \leq 7$ 50%   2. $4 \leq n \leq 7$ 25%
3. $1 \leq n \leq 7$ 75%   4. $1 \leq n \leq 9$ 100%

Now have them identify different intervals that correspond to 25%, 50%, and 75%.

25%: $1 \leq n \leq 2$, $2 \leq n \leq 4$, or $7 \leq n \leq 9$
50%: $1 \leq n \leq 4$ or $4 \leq n \leq 9$; 75%: $2 \leq n \leq 9$

Activity Notes

In this Activity, students investigate the technique for drawing a box-and-whisker plot. Students should discover that only five key numbers are needed to draw a box-and-whisker plot: the median of the data, the lower quartile, the upper quartile, the least value, and the greatest value. Tell students that these five statistics are so significant that they sometimes are called a *five-number summary* of a data set.

Cooperative Learning

You may wish to have students do the Activity in pairs. In Step 7, one student can be responsible for calculating the required statistics, while the other draws the box-and-whisker plot.

CHECKPOINT ✓

7.
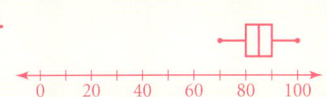

Use Teaching Transparency 24.

ADDITIONAL EXAMPLE 2

Use the table of amounts of rainfall from Additional Example 1 to construct a box-and-whisker plot.

median: 2.2
lower quartile: 2.05
upper quartile: 2.8
least value: 1.1
greatest value: 4.7

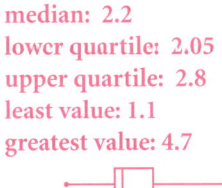

See Keystroke Guide, page 782.

LESSON 4.6 **203**

Teaching Tip

TECHNOLOGY If you are using the TI-82 or TI-83 graphics calculator, you can use the procedure below to make the box-and-whisker plot shown on Example 2.

First clear any existing data from **LIST 1** in the **STAT LIST** editor. Press `STAT`, choose `4:ClrList`, and press `2nd` `1` `ENTER`.

Now enter your data into **LIST 1**. Press `STAT` and choose `1:Edit...`. Enter the data values under **L1**, pressing `ENTER` after each one.

Next, press `2nd` `Y=`. Choose `1:Plot 1...`. In the window that appears, choose `On`. Then choose the *boxplot* icon.

Press `Y=` and clear any existing equations from the `Y=` editor. Then press `WINDOW` and enter the settings below. (Ignore the other settings.)

  Xscl=1   Ymin=0   Yscl=0

To see display, press `ZOOM` and choose `9:ZoomStat`. If you now press `TRACE`, you can use the ◄ and ► keys to find the upper and lower quartiles, the median, and the greatest and least values.

TRY THIS
a. 30   b. 80
c. 65   d. 45
e. 55

● **SOLUTION**

First draw a number line. Use points to mark the least value, 12.7, and greatest value, 15.2. Identify the median with a vertical mark above the number line.

The median is 13.95, the mean of the 24th and 25th ordered values.

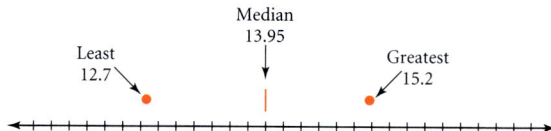

Find the median of all the measurements in the lower half, below 13.95. This measurement, called the **lower quartile**, is 13.45.

Find the median of all the measurements in the upper half, above 13.95. This measurement, called the **upper quartile**, is 14.2.

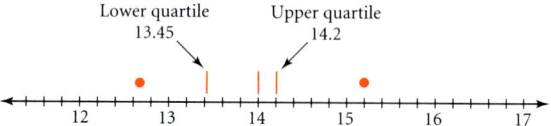

Identify the upper and lower quartiles with vertical marks above the number line as you did for the median.

Draw a rectangular box from the lower quartile to the upper quartile. Complete the box-and-whisker plot by drawing the line segments from each end of the box to the points for the least and greatest values.

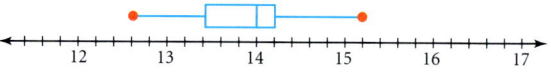

TRY THIS For the data set represented by the box-and-whisker plot below, give the values of the following:

  a. the lowest score   b. the highest score   c. the upper quartile
  d. the lower quartile   e. the median

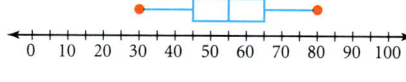

TECHNOLOGY GRAPHICS CALCULATOR

Once data are entered into a graphics calculator, they can easily be represented in a number of ways. The box-and-whisker plot from Example 2 is shown below on a graphics calculator display.

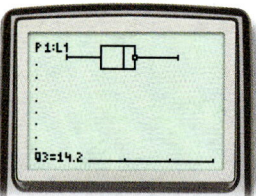

Reteaching the Lesson

GUIDED ANALYSIS Write a set of data on the board or overhead. Work with students to make a stem-and-leaf plot of the data. As you do this, lead them to develop a set of numbered steps that outline a general procedure for making a stem-and-leaf plot for any set of data. Be sure they copy the steps into their notebooks. A sample set of steps is given at right.

After students feel comfortable with the procedure for stem-and leaf plots, repeat the activity for box-and-whisker plots.

How to Make a Stem-and-Leaf Plot

1. Make a blank table with two columns.
2. Identify the stems for the data.
3. Write the stems in order from least to greatest in the first column of the table. Write the word *Stems* at the top of the column.
4. Write the word *Leaves* at the top of the second column of the table.
5. Identify the leaves for the data and write each leaf to the right of the corresponding stem.
6. Rearrange the leaves for each stem in order from least to greatest.

204 LESSON 4.6

Exercises

Communicate

1. Explain how to construct a stem-and-leaf plot.
2. Explain how to make a histogram from a stem-and-leaf plot.
3. Explain how to determine the median and the range from a box-and-whisker plot.
4. Explain how to find the upper and lower quartiles from a box-and-whisker plot.

Guided Skills Practice

5. Construct a stem-and-leaf plot of the following data: 12.1, 10.3, 12.6, 15.2, 11.5, 14.4, 14.3, 13.9, 17.3, 14.1, 13.7, 16.6, 15.5, 13.8, 10.7, 16.7, 15.2, 13.8, 12.1, 15.2, 12.2, 13.4. **(EXAMPLE 1)**
6. Construct a box-and-whisker plot of the data from Exercise 5. **(EXAMPLE 2)**

Practice and Apply

Construct a stem-and-leaf plot for each set of data.

7. 14.8, 12.2, 20.3, 18.7, 16.4, 15.7, 17.2, 15.9, 12.5, 18.9, 16.6, 16.3, 15.1, 19.1, 13.3, 16.4
8. 47, 42, 39, 55, 68, 61, 14, 53, 61, 36, 53, 48, 59, 23, 44, 37, 62, 11, 36, 41, 44, 48, 53
9. 144, 179, 136, 183, 153, 147, 127, 142, 136, 176, 148, 153, 125, 127, 128, 133, 135, 138, 139, 141, 153, 154, 159, 162, 175, 184
10. 17, 12, 22, 38, 84, 76, 37, 28, 11, 46, 83, 78, 65, 32, 48, 56, 13, 82, 11, 13, 14, 23, 32, 41, 47, 49, 73
11. 1.2, 5.6, 8.4, 6.5, 9.7, 6.3, 2.5, 4.6, 4.8, 6.8, 6.7, 2.5, 3.4, 7.8, 9.1, 5.2, 8.4, 9.2, 2.7
12. 13, 25, 29, 38, 27, 26, 12, 24, 25, 27, 26, 28, 16, 18, 32, 24, 34, 29, 21, 20, 19, 37
13. 7, 12, 18, 9, 5, 23, 5, 16, 19, 18, 4, 8, 26, 29, 5, 6, 7, 19, 10, 12, 17
14. 12, 11, 16, 18, 20, 21, 28, 34, 19, 16, 14, 17, 12, 18, 14, 16, 24
15. 26.4, 26.1, 27.6, 29.5, 29.4, 23.1, 25.8, 29.6, 24.7, 23.6, 21.4, 20.8, 29.4, 21.6, 23.9
16. 328, 349, 336, 307, 378, 355, 322, 374, 371, 337, 314, 407, 303, 335, 339, 359, 366, 393

17–26. Construct a box-and-whisker plot for each set of data in Exercises 7–16.

Assess

Selected Answers
Exercises 5–6, 7–53 odd

ASSIGNMENT GUIDE

In Class	1–6
Core	7–11, 17–21, 27–43 odd
Core Plus	8–26 even, 28–44 even
Review	45–54
Preview	55–57

✎ Extra Practice can be found beginning on page 738.

5. Stems: ones
Leaves: tenths

Stems	Leaves
10	3, 7
11	5
12	1, 1, 2, 6
13	4, 7, 8, 8, 9
14	1, 3, 4
15	2, 2, 2, 5
16	6, 7
17	3

6. 10.3 12.2 13.85 15.2 17.3
(box-and-whisker plot on number line 10–18)

7. Stems: ones
Leaves: tenths

Stems	Leaves
12	2, 5
13	3
14	8
15	1, 7, 9
16	3, 4, 4, 6
17	2
18	7, 9
19	1
20	3

8. Stems: tens
Leaves: ones

Stems	Leaves
1	1, 4
2	3
3	6, 6, 7, 9
4	1, 2, 4, 4, 7, 8, 8
5	3, 3, 3, 5, 9
6	1, 1, 2, 8

9. Stems: tens
Leaves: ones

Stems	Leaves
12	5, 7, 7, 8
13	3, 5, 6, 6, 8, 9
14	1, 2, 4, 7, 8
15	3, 3, 3, 4, 9
16	2
17	5, 6, 9
18	3, 4

The answers to Exercises 10–26 can be found in Additional Answers beginning on page 810.

LESSON 4.6 **205**

Technology

A calculator may be helpful in performing some of the calculations in Exercises 27–44.

Error Analysis

When a set of data contains an odd number of items, students sometimes mistakenly include the median when calculating the upper and lower quartiles. Suggest that after they identify the median in the list of data, they put a large × through it as a reminder that it should not be used again.

Look Beyond

In Exercises 55–57, students generate data, graph the corresponding points on a coordinate plane, and find the slope of the line determined by the points. Students will study these concepts in Chapter 5.

The answers to Exercises 32, 37, and 39 can be found in Additional Answers beginning on page 810.

Range and median stay the same. The mode would be 59 and 60. The mean would be slightly less.

STATISTICS The stem-and-leaf plot below shows the number of runs credited to the top players in a baseball league. The stems are tens digits, and the leaves are ones digits.

27. Another player had 60 runs. If his data were added to the stem-and-leaf plot, would the range, mean, median, or mode be affected? If so, how?

28. What percent of the players scored more than 80 runs? Round your answer to the nearest tenth of a percent. **17.2%**

29. What percent of the players scored fewer than 60 runs? Round your answer to the nearest tenth of a percent. **44.8%**

30. Construct a histogram of the data.

31. Construct a box-and-whisker plot of the data.

Number of Runs	
Stems	Leaves
3	5, 5
4	0, 2, 2
5	2, 2, 5, 6, 7, 9, 9, 9
6	0, 0, 1, 2, 3, 5
7	4, 6, 7, 9, 9
8	1, 1, 8, 9
9	—
10	4

APPLICATIONS

BIOLOGY A scientist recorded the duration of humpback whale songs. Use the data below for Exercises 32–38.

32. Use the data to make a stem-and-leaf plot.

33. What is the range of the data? **5.9**

34. Find the mean and median of the data. **17.0125; 16.9**

35. Is there a mode? If so, what is it? **17.2 and 19.3**

All 3 are so close that it does not matter.

36. What measure of central tendency do you think best answers this question? Why?

37. Construct a histogram of the data.

38. Construct a box-and-whisker plot of the data.

Duration of Humpback Whale Songs (in minutes)			
16.6	19.9	17.2	18.9
14.5	18.1	15.6	17.8
15.3	16.3	17.2	16.2
19.7	14.1	15.5	19.3
20.0	17.4	18.2	16.1
19.3	15.0	14.9	15.2

BIOLOGY One well-known type of dolphin is the bottle-nosed dolphin, which can weigh up to 600 pounds. Use the data below for Exercises 39–44.

39. Use the data to make a stem-and-leaf plot.

40. What is the range of the data? **340**

41. What is the median of the data? **497**

42. What is the mean of the data? **474.05**

43. Is there a mode? If so, what is it? **Yes; 497**

44. What measure of central tendency do you think best represents the average weight in this data set? **474.05**

Weight of Bottle-nose Dolphins (in pounds)				
345	574	249	497	588
567	586	383	484	491
275	499	478	343	549
399	512	589	576	497

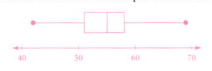

Practice 4.6 Other Data Displays

In the table at right are the ages of the first 42 presidents of the United States when they were sworn into office.

57	61	57	57	58	57
61	54	68	51	49	64
50	48	65	52	56	46
54	49	50	47	55	55
54	42	51	56	55	51
54	51	60	62	43	55
56	61	52	69	64	46

1. In the space below, make a stem-and-leaf plot of the data.

Stems	Leaves
4	2, 3, 6, 6, 7, 8, 9, 9
5	0, 0, 1, 1, 1, 1, 2, 2, 4, 4, 4, 5, 5, 5, 6, 6, 6, 7, 7, 7, 8
6	0, 1, 1, 1, 2, 4, 5, 8, 9

2. What is the range of the data? **27**
3. What is the median of the data? **55**
4. What are the lower and upper quartiles for this data? **51 and 58**
5. What is the mean of the data? **approximately 54.8**
6. What is the mode of the data? **The modes are 51, 54, 55, and 57.**
7. What is the average age of a president of the United States when he is sworn into office? What measure of central tendency do you think best answers this question? Why? **The mean is generally considered to represent the average, so the average age of a president taking office is about 54.8 years. In some cases the median or the mode may be a better estimate. In this case, the median and the mode are very close, so either would be a good estimate.**
8. In the space below, construct a box-and-whisker plot for this data.

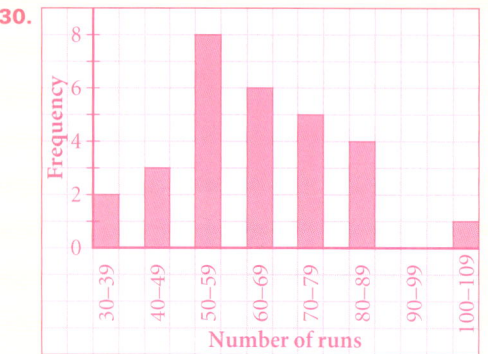

30.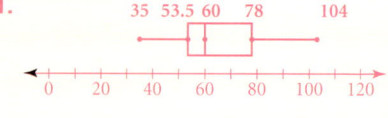

31. 35 53.5 60   78   104

38. 15.4 16.9 18.55
   14.1   20.0

206 LESSON 4.6

 **Look Back**

Find each product. *(LESSON 2.4)*

45. $8 \cdot 1\frac{1}{4}$ **10**   **46.** $\frac{2}{3} \cdot 2\frac{3}{5}$ $\frac{26}{15}$ or $1\frac{11}{15}$ **47.** $6\frac{1}{5} \cdot 3\frac{3}{10}$ $\frac{1023}{50}$ or $20\frac{23}{50}$

Simplify. *(LESSONS 2.6 AND 2.7)*

48. $2(3x + 7) - x$ **$5x + 14$**   **49.** $5y - 7 + 2y + 3$ **$7y - 4$**   **50.** $\frac{3x + 18}{x + 6} - 2x + 4$ **$-2x + 7$**

51. Solve $4y - 3 = 6y - 2$ for y. *(LESSON 3.4)* $y = -\frac{1}{2}$

Write each percent as a fraction or mixed number in simplest terms. *(LESSON 4.2)*

52. 80% $\frac{4}{5}$ **53.** 38% $\frac{19}{50}$ **54.** 3.5% $\frac{7}{200}$

 Look Beyond

APPLICATION

TRAVEL Mary leaves the city in her car at a steady rate of 40 miles per hour. The distance that she drives away from the city depends on the number of hours that she drives. Her distance from the city equals the rate times the number of hours.

55. Write an equation to describe the distance that Mary drives in terms of the number of hours that she drives. $d = 40t$

56. Find the distance she would travel in 1, 2, 3, 4, and 5 hours and plot the points on a graph. Connect the points.

57. Describe the graph. How far up does the graph move for every hour that it moves over?

The graph moves up 40 for every 1 that it moves to the right. The slope is 40.

Include this activity in your portfolio.

The grades on a test in a history class are listed in the table below.

Student	1	2	3	4	5	6	7	8	9	10	11	12	13
Grade	60	80	65	85	80	75	75	100	70	90	90	80	75

1. Represent the data in three different types of displays.

2. Explain why you chose each of the three different types of displays.

3. Which display is the best way to show this data and why?

WORKING ON THE CHAPTER PROJECT

You should now be able to complete Activities 5 and 6 of the Chapter Project on page 208.

56.

d	$40t$
40	40(1) = 40
80	40(2) = 80
120	40(3) = 120
160	40(4) = 160
200	40(5) = 200

ALTERNATIVE Assessment

Portfolio Activity

The Portfolio Activity can be used as preparation for the Chapter Project or as a separate activity. In the Portfolio Activity on this page, students are given a set of data and must display it in three different ways. They are then asked to defend their choice of data displays and to identify which one seems best suited to the data.

Answers to Portfolio Activities can be found in Additional Answers of the Teacher's Edition.

 internetconnect

GO TO: go.hrw.com
KEYWORD: MA1 Endangered Species

To support the Portfolio Activity, students may visit the HRW Web site, where they can examine population data on the California condor and mortality data on the Florida manatee, both of which are endangered species.

Student Technology Guide

Student Technology Guide
4.6 Other Data Displays

On some graphics calculators, you can create a box-and-whisker plot from data that you enter. The calculator also gives the lower quartile, median, and upper quartile. To construct a box-and-whisker plot for the data at right, follow these steps.

84, 92, 77, 65, 81
82, 84, 68, 95, 78
91, 99, 62, 77, 78
91, 90, 91, 68, 85

Begin by entering the data.

STAT EDIT 1:Edit 84 ENTER 92 ENTER 77 ENTER, and so on until all of the data is entered into L1.

Then follow these steps.

1. Press 2nd Y=. When the screen looks like the one below, press ENTER.

2. Choose a box-and-whisker plot by matching the screen below.

3. Press WINDOW. Choose the settings shown below.

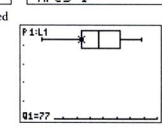

To find the lower quartile, median, and upper quartile, proceed as follows:

Press GRAPH and TRACE to see the screen to the right. This display shows that the lower quartile is 77. Use the right arrow key to display the median, 83. Use the right arrow key again to display the upper quartile, 91.

Construct a box-and-whisker plot for each set of data. Find the lower quartile, median, and upper quartile.

1. 14.5, 15.8, 21.2, 19.4, 18.6, 20.0, 15.6, 17.1, 19.9, 13.9, 19.0, 18.7, 18.9, 17.1, 18.5, 18.3, 13.8
lower quartile: 15.7; median: 18.5; upper quartile: 19.2

2. 320, 328, 398, 356, 376, 381, 410, 415, 455, 318, 398, 376, 356, 399, 300, 402, 410, 329
lower quartile: 329; median: 378.5; upper quartile: 402

3. 1.2, 1.5, 1.5, 1.6, 1.1, 1.3, 1.4, 1.5, 1.6, 1.5, 1.9, 2.0, 2.2, 4.0, 5.0
lower quartile: 1.4; median: 1.5; upper quartile: 1.6

LESSON 4.6

Project

Focus

In this project, students choose a topic of research and make a hypothesis. Then they design and conduct a study that they believe will prove or disprove the hypothesis. Using the concepts they studied in this chapter, they organize and display the data they collected and prepare an analysis of the results.

Motivate

As part of their work, scientists, marketing experts, political experts, and many other professionals make hypotheses. These hypotheses must then be proven true or false. The facts and information that students see in books, newspapers, and magazines were researched, tested, organized, and analyzed. This project gives students an opportunity to see what is involved in this process.

CHAPTER PROJECT FOUR

Designing a Research Study

Many types of questions can be investigated through research studies. In this project, you will choose a question and answer it experimentally by collecting and analyzing data.

What time of day is a polar bear most active?

Activity 1
Choose a Topic

Decide on a topic that you would like to research. State the question that your study will be designed to answer.

Activity 2
Make a Hypothesis

Write a statement to describe the answer you think you will find in your research. Explain why you think this will be the outcome.

Activity 1

Students should give a specific topic, as in the example.

Activity 2

Students should give a specific answer and explanation. Example: A polar bear is most active between 9 A.M. and 11 A.M. The polar bear has plenty of rest after sleeping all night, and it has just eaten breakfast, so it has energy.

Activity 3

Questions or descriptions should be as specific as possible. Example: How many times per day do polar bears eat? At approximately what times of the day do polar bears eat? Do polar bears nap after eating?

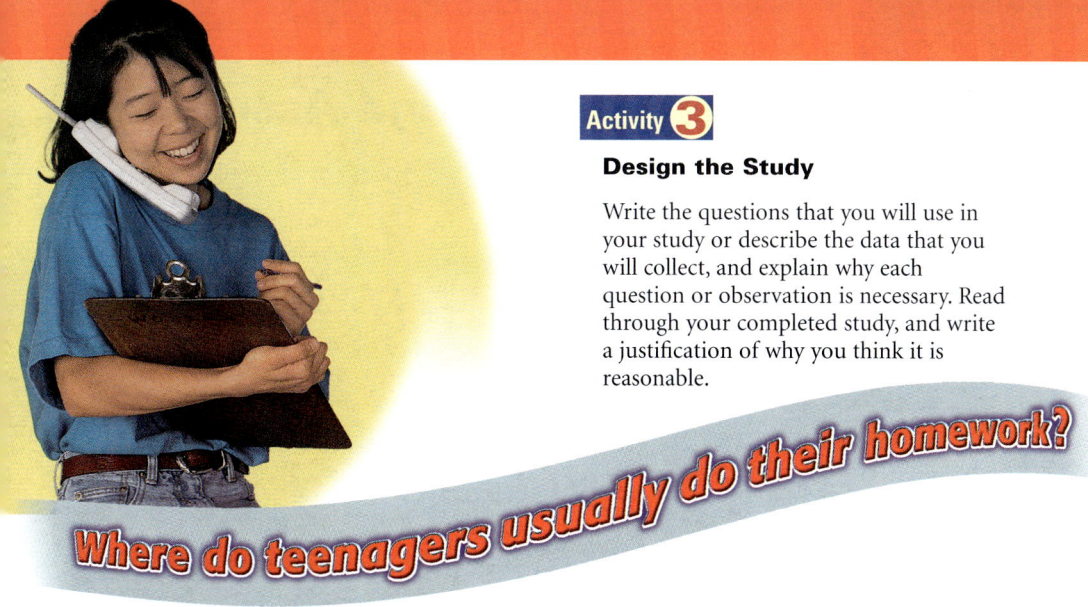

Design the Study

Write the questions that you will use in your study or describe the data that you will collect, and explain why each question or observation is necessary. Read through your completed study, and write a justification of why you think it is reasonable.

Where do teenagers usually do their homework?

Conduct the Study

Collect as large a sample for your study as you can manage. Be sure that you collect enough data (at least 10 values) so that you will be able to accurately test your hypothesis.

Display the Data

Summarize the results of your study in a table. Then use one of the methods you studied in this chapter (bar graph, line graph, circle graph, histogram, stem-and-leaf plot, or box-and-whisker plot) to display your data. You might also want to use a scatter plot.

Is a person's height related to his or her running speed?

Analyze the Results

Explain how the data you gathered supports or contradicts the hypothesis you made in Activity 2.

Cooperative Learning

If you wish, you may have students do the project in small groups or in pairs. Each group should pick a topic and proceed with the activities described. This will allow students to draw on multiple experiences and points of view when designing the study. A group effort generally makes it possible to gather a large amount of data in a shorter period of time.

Discuss

As students are choosing their topics and making hypotheses, discuss with them the reasonableness of their hypotheses. Students may be able to share insights which indicate that a given hypothesis cannot be tested in a practical manner.

While designing and conducting the studies, bring the class together periodically to discuss any problems they have encountered. Have students suggest possible ways to resolve the problems.

After students have completed the project, discuss the displays and analyses of the data. Ask whether all students feel confident that the analyses are valid. If some students express doubts, ask them to explain.

Activity 4
Students should use as many resources as necessary, such as the Internet, almanacs, and actual surveys if possible.

Activity 5
Students should use an appropriate method to display data. For example, a circle graph may not be an appropriate choice to display the times of day that polar bears are most active.

Activity 6
Students should refer to the hypothesis made in Activity 2 in their explanations.

Chapter Review

Chapter Review and Assessment

VOCABULARY

bar graph 193	leaves 201	probability 180
base 172	line graph 193	proportion 164
box-and-whisker plot 203	lower quartile 212	proportion method 172
circle graph 194	mean 186	range 186
cross product 165	means 165	ratio 164
equation method 172	median 186	stem 201
experimental probability ... 182	mode 186	stem-and-leaf plot 201
extremes 165	percent 171	trials 181
frequency table 188	percent rate 172	upper quartile 204
histogram 202	percentage 172	

Key Skills & Exercises

LESSON 4.1

Key Skills

Solve a proportion.

Solve $\frac{n}{32} = \frac{3}{8}$.

Method A

$\frac{n}{32} = \frac{3}{8}$

$32 \cdot \frac{n}{32} = 32 \cdot \frac{3}{8}$

$n = \frac{96}{8} = 12$

Method B

$\frac{n}{32} = \frac{3}{8}$

$32 \cdot 3 = 8 \cdot n$

$96 = 8n$

$12 = n$

Exercises

Solve each proportion.

1. $\frac{n}{6} = \frac{12}{9}$ **8**
2. $\frac{13}{n} = \frac{39}{27}$ **9**
3. $\frac{11.4}{8} = \frac{45.6}{x}$ **32**
4. $\frac{16}{41} = \frac{x}{820}$ **320**
5. $\frac{x}{5} = \frac{3}{15}$ **1**
6. $\frac{n}{2} = \frac{18}{3}$ **12**
7. $\frac{4}{5} = \frac{3}{x}$ **3.75**
8. $\frac{15}{3} = \frac{m}{30}$ **150**

LESSON 4.2

Key Skills

Solve percent problems.

Find 30% of 15.

Use the proportion method.

$\frac{30}{100} = \frac{x}{15}$

$100x = 450$

$\frac{100x}{100} = \frac{450}{100}$

$x = 4.5$

30% of 15 is 4.5.

What percent of 50 is 75?

Use the equation method.

$x \cdot 50 = 75$

$\frac{x \cdot 50}{50} = \frac{75}{50}$

$x = 1.5 = 150\%$

75 is 150% of 50.

Exercises

Find each answer.

9. What is 55% of 60? **33**
10. 28 is 70% of what number? **40**
11. What is 200% of 40? **80**
12. What is 35% of 140? **49**
13. What percent of 90 is 40.5? **45%**
14. 4 is what percent of 50? **8%**
15. 10 is what percent of 40? **25%**
16. 47% of 250 is what number? **117.50**
17. 31 is 50% of what number? **62**
18. 2 is 200% of what number? **1**
19. 87% of 1500 is what number? **1305**
20. What is 14% of 350? **49**

Chapter Test, Form A

Chapter Assessment
Chapter 4, Form A, page 1

Write the letter that best answers the question or completes the statement.

b 1. Which of the following is a true proportion?
  a. $\frac{7}{8} = \frac{8}{7}$ b. $\frac{16}{28} = \frac{36}{63}$ c. $\frac{4}{7} = \frac{12}{28}$ d. $\frac{18}{45} = \frac{3}{15}$

c 2. Solve the proportion $\frac{5}{8} = \frac{15}{c}$.
  a. 45 b. 3 c. 24 d. 16

a 3. What is 3.687 written as a percent?
  a. 368.7% b. 0.03687% b. 36.76% d. 0.3687%

b 4. What is 37% of 50?
  a. 0.185 b. 18.5 c. 185 d. 1.85

d 5. 15 is what percent of 45?
  a. $66\frac{2}{3}\%$ b. 30% b. 300% d. $33\frac{1}{3}\%$

d 6. 40% of what number is 20?
  a. 0.05 b. 0.5 b. 5 d. 50

b 7. The cost of a jacket is increased from $75 to $80. What is the percent of increase?
  a. 87.5% b. $6\frac{2}{3}\%$ b. 6.25% d. $66\frac{2}{3}\%$

A number cube is rolled 10 times. Use the table of results to find each probability.

Trial	1	2	3	4	5	6	7	8	9	10
Number rolled	4	3	6	6	1	5	2	6	2	5

a 8. rolling 1 on the number cube
  a. 10% b. 20% c. 40% d. 50%

b 9. rolling 5 on the number cube
  a. 10% b. 20% c. 40% d. 50%

Chapter Assessment
Chapter 4, Form A, page 2

Write the letter that best answers each question or completes each statement.

The bar graph shows the percent of 18-year-olds with high school diplomas for the given years.

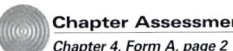

b 10. What year had the highest percent of 18-year-olds with high school diplomas?
  a. 1975 b. 1970
  c. 1985 d. not given

c 11. Between what consecutive years did the greatest increase or decrease in the percent of 18-year-olds with diplomas occur?
  a. 1955 and 1960 b. 1970 and 1975
  c. 1950 and 1955 d. not given

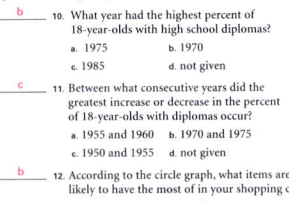

b 12. According to the circle graph, what items are you likely to have the most of in your shopping cart?
  a. meat b. produce
  c. canned goods d. not given

a 13. If you have 50 items in your cart, how many would probably be seafood items?
  a. 5 b. 10
  c. 20 d. none

Use the frequency table to answer Exercises 14–16.

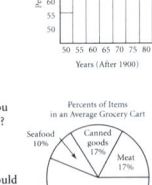

a 14. What is the mean of the data?
  a. 80.4 b. 80
  c. 82 d. not given

d 15. What is the mode of the data?
  a. 78 b. 89
  c. 95 d. not given

c 16. What is the median of the data?
  a. 79 b. 80.4
  c. 80 d. not given

210 CHAPTER 4 REVIEW

LESSON 4.3

Key Skills

Find the experimental probability for a particular event.

Two number cubes were tossed 100 times. A sum greater than 8 appeared 34 times. Find the experimental probability of rolling a sum greater than 8.

There were 100 trials and 34 successful outcomes. Thus, in this case, the experimental probability of a successful outcome (rolling a sum that is greater than 8) is as follows:

$$\frac{34}{100} = 0.34 = 34\%$$

Exercises

Two number cubes were tossed 150 times. Based on the results given below, find the experimental probability for each outcome.

21. A sum greater than 6 appeared 67 times. ≈ **44.7%**
22. A pair of 3s appeared 4 times. ≈ **2.7%**
23. A sum of 11 appeared 11 times. ≈ **7.3%**
24. At least one 4 appeared 23 times. ≈ **15.3%**

LESSON 4.4

Key Skills

Find the mean, median, mode, and range for a set of data.

Lucy received the following grades on her mathematics tests during a one-month period: 80, 75, 80, 95, and 100.

mean: $\frac{80 + 75 + 80 + 95 + 100}{5} = 86$

median: Arrange the grades in order: 75, 80, 80, 95, 100. The median is the middle grade, which is 80.

mode: The grade that appears most often is 80.

range: The range is 100 − 75, or 25.

Exercises

Find the mean, median, mode, and range for each set of data.

25. 6, 8, 8, 11, 7   **8; 8; 8; 5**
26. 5, 7, 10, 13, 21, 3   **9.83; 8.5; none; 18**
27. 55, 85, 96, 102, 135, 85, 55, 96, 55, 206   **97; 90.5; 55; 151**
28. 87, 65, 32, 103, 59   **69.2; 65; none; 71**
29. 0.352, 0.376, 0.385, 0.368, 0.395   **0.3752; 0.376; none; 0.043**
30. 1.5, 1.6, 1.8, 1.1, 1.4   **1.48; 1.5; none; 0.7**
31. 8, 10, 12, 24, 36, 47, 8, 12, 8   **18.3̄ ; 12; 8; 39**

LESSON 4.5

Key Skills

Interpret and construct bar graphs, line graphs, and circle graphs.

The circle graph below shows annual operating expenses for a car. The total expenses were $2148, and $720 was spent on gasoline. The angle of the section for gasoline is calculated as follows:

$$\frac{720}{\text{total cost}} \cdot 360° = \frac{720}{2148} \cdot 360° \approx 120.7°$$

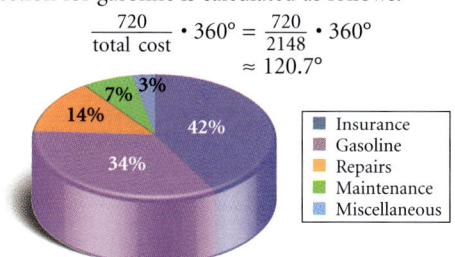

Exercises

32. If the total amount of the operating expenses were $2500, what amount would have been spent on insurance? Use the circle graph at left. **$1050**

33. Brenda used her car-expense record to budget her expenses for the next year. Make a circle graph of the budget shown below.

Gasoline	$800
Maintenance	$200
Repairs	$200
Insurance	$900
Miscellaneous	$100

33.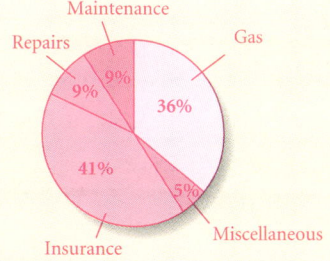
Car-Expense Budget

Chapter Test, Form B

Chapter Assessment
Chapter 4, Form B, page 1

Solve each proportion.

1. $\frac{4}{5} = \frac{8}{p}$ **p = 10**
2. $\frac{q}{16} = \frac{3}{4}$ **q = 12**
3. $\frac{6}{r} = \frac{15}{40}$ **r = 16**

4. Roberto earned $6.50 by mowing lawns in his neighborhood for 2 hours. How much will he earn by mowing lawns for 5 hours?   **$16.25**

Write each percent as a decimal.

5. 72%   **0.72**
6. 7.5%   **0.075**
7. 112%   **1.12**

Write each percent as a fraction in lowest terms.

8. 4.5%   **$\frac{9}{200}$**
9. 20%   **$\frac{1}{5}$**
10. 104%   **$1\frac{1}{25}$**

Find each answer.

11. What is 35% of 200?   **70**
12. 40 is what percent of 25?   **160%**
13. 25 is what percent of 125?   **20%**
14. Find 115% of 80.   **92**
15. What number is 70% of 15?   **10.5**
16. What is 2.5% of 120?   **3**
17. The sales tax is 7.5%. If you buy $80 worth of merchandise, how much is your total bill?   **$86**

Two number cubes were rolled 10 times. Use the table of results to find the experimental probability of each outcome.

Trial	1	2	3	4	5	6	7	8	9	10
Cube 1	6	5	4	1	1	5	3	6	4	6
Cube 2	4	3	4	3	2	5	2	2	3	1

18. 5 is rolled on the first cube.   **20%**
19. 2 is rolled on the second cube.   **30%**
20. Doubles are rolled.   **20%**
21. Even numbers are rolled on both cubes.   **30%**

Chapter Assessment
Chapter 4, Form B, page 2

The line graph shows the typing speed of individuals who practice different amounts of time during the week.

22. What is the typing speed of the person who practices 9 hours per week?   **50 wpm**
23. How many hours did a person practice to achieve a typing speed of 75 words per minute (wpm)?   **12 hours**
24. What is the increase in typing speed from a person who practices 2 hours a day to a person who practices 11 hours a day?   **45 wpm**
25. What does this graph tell you about typing speed in terms of hours of practice?   **The more hours of practice, the faster the typing speed**

26. Find the mean, median, mode, and range for the following data:
21, 19, 18, 17, 22, 44
mean = 23.5; median = 20; is no mode; range = 27

27. If another value of 18 were added to the list of numbers above, how would it affect the measures of central tendency?
The mean would decrease to 22.7. The median would change to 19.5. The mode would be 18. The range would not be affected.

28. Create a stem-and-leaf plot for the following data:
14 20 31 40 55 16
27 35 16 33 27 27
26 32 24 23 46 21

Stem	Leaves
1	4 6 6
2	0 1 3 4 6 7 7 7
3	1 2 3 5
4	0 6
5	5

29. Using the stem-and-leaf plot from Exercise 28, find the mean, median, mode, and range of the data.
mean = 28.5; median = 27; mode = 27; range = 41

30. Find the upper and lower quartiles for the stem-and-leaf plot from Exercise 28.
upper quartile: 33; lower quartile: 21

31. Using the results from Exercises 29 and 30, make a box-and-whisker plot.

38. Stems: tens
    Leaves: ones

Stems	Leaves
5	0, 8
6	0, 5, 8
7	4, 5, 6, 8, 9
8	0, 1, 4, 5, 5, 8
9	2, 2, 3, 8, 9

39.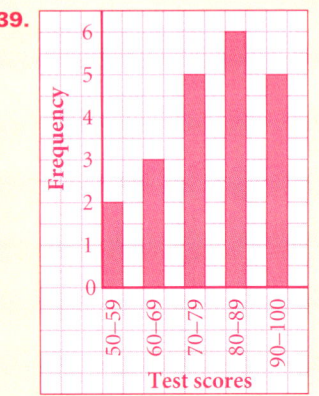

40.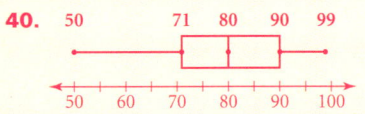

LESSON 4.6

Key Skills

Use data to construct stem-and-leaf plots, histograms, and box-and-whisker plots.

Employees in a company are surveyed to find out their ages. A stem-and-leaf plot, histogram, and box-and-whisker plot are each made from the data: 23, 18, 46, 23, 51, 27, 34, 39, 31, and 56.

Stems	Leaves
1	8
2	3, 3, 7
3	1, 4, 9
4	6
5	1, 6

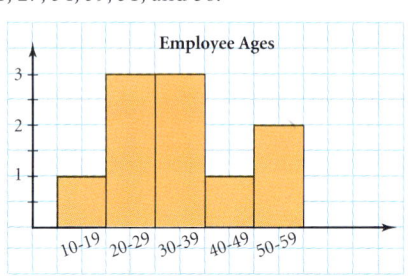

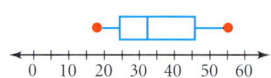

Exercises

For the data set at left, identify

34. the lower quartile.
35. the upper quartile.
36. the median.
37. the range.

Display the following data for test scores as indicated below:

76, 78, 92, 65, 98, 80, 92, 60, 85, 85, 99, 50, 74, 75, 81, 93, 58, 79, 84, 68, 88

38. in a stem-and-leaf plot
39. in a histogram
40. in a box-and-whisker plot

Applications

41. **EDUCATION** The tuition costs at five universities are $11,800; $14,500; $12,500; $17,000; and $18,600. Find the mean, median, mode, and range of the tuition costs.

42. **CONSUMER ECONOMICS** In late December, the McGroty family budgeted the following amounts for essential expenses during the new year: food, $6500; housing, $10,300; clothing, $2500; transportation, $7000; health and personal care, $1800; and miscellaneous, $2000. Draw a circle graph to display the data.

43. **PATTERNS IN DATA** Typing speeds on an exam for Ms. Huxley's Typing I class were: 24, 32, 19, 22, 35, 25, 42, 36, 21, 32, 24, 34, 31, 21, 18, 15, 28, 38, 22, 36, and 27. Ms Huxley's Typing II class had speeds of 32, 45, 44, 42, 53, 62, 27, 36, 44, 52, 38, 47, 26, 38, and 48. Construct a box-and-whisker plot for each of these data sets, and compare the results.

42.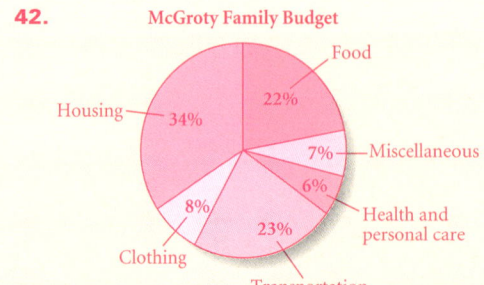

43.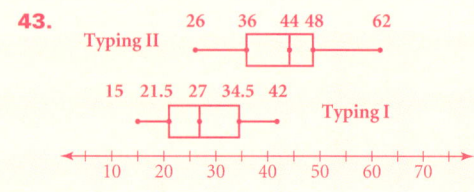

212 CHAPTER 4 REVIEW

Alternative Assessment

Performance Assessment

1. Use the proportion $\frac{n}{16} = \frac{15}{24}$ to answer each of the following:
   a. What are the means and the extremes of the proportion?
   b. What is the product of the means? What is the product of the extremes?
   c. What is the relationship between the products found in part **b**?
   d. Solve for *n*. What is the result?
   e. Explain how to solve a proportion with one unknown.

2. A major discount store offers a 20% discount on all merchandise. The original price of a sweater is $50.
   a. How can you use a percent bar to find the discount?
   b. How can you use a proportion to find the discount?
   c. How can you use an equation to find the discount?
   d. You know that a shirt at the same store is on sale for $40. How can you use an equation to find the original price of the shirt? Write an equation and solve.

3. Your five test scores are 70, 82, 90, 86, and 78. Your teacher will record either the mean or median as your final score. Which would you choose? Explain why.

Portfolio Projects

1. Find a statistical graph in a newspaper that can be interpreted in more than one way. Explain how you would use the graph to convince someone that your interpretation is the correct one.

2. Use a tape measure to find the length of your ring finger, hand, arm, etc. Use a ratio to compare each of these lengths with the length of your hand. Make a table to organize your data. Note any interesting data. Compare your findings with others in your class.

3. A bag has 25 chips numbered from 1 to 25. You reach into the bag and remove two chips. Design an experiment to find the experimental probability that the sum of the numbers you draw is less than 10.

Peer Assessment

Research data that can be represented by a graph, such as company profits, stock values, average rainfall amounts, average salaries, and so on. Using the data, create a graph that is deliberately misleading. Have another student study your graph and explain how the graph is misleading.

internetconnect

The HRW Web site contains many resources to reinforce and expand your knowledge of proportional reasoning and statistics. This Web site also provides Internet links to other sites where you can find information and real-world data for use in research projects, reports, and activities that involve proportional reasoning. Visit the HRW Web site at **go.hrw.com**, and enter the keyword **MA1 CH4** to access the resources for this chapter.

Alternative Assessment

Performance Assessment

1. a. means = 16 and 15
      extremes = *n* and 24
   b. product of means = 240
      product of extremes = 24*n*
   c. They are equivalent.
   d. *n* = 10
   e. Answers may vary. Sample: Find the product of the means and the product of the extremes. Set these products equal to each other. Solve the equation for the variable.

2. a. Answers may vary. Sample: The percent bar can help you visualize each component to set up a proportion.
   b. Let the variable (*D*) equal the discount amount. Set up a proportion as follows:
      $\frac{20}{100} = \frac{D}{50}$
      Solve for *D*.
   c. Let the variable (*D*) equal the discount amount. Set up an equation as follows:
      $D = 0.20 \cdot 50$
      Solve for *D*.
   d. If the shirt is 20% off, then you are paying 80% of the original price.
      Let *p* = original price.
      $0.8p = 40$
      $p = 50$

3. Choose the median since it is greater.

Portfolio Projects

Answers may vary. Check students' responses.

Cumulative Assessment

College Entrance Exam Practice

Multiple-Choice and Quantitative-Comparison Samples

The first half of the Cumulative Assessment contains two types of items found on standardized tests—multiple-choice questions and quantitative-comparison questions. Quantitative comparison items emphasize the concepts of equalities, inequalities, and estimation.

Free Response Grid Samples

The second half of the Cumulative Assessment is a free-response section. A portion of this part of the Cumulative Assessment consists of student-produced response items commonly found on college entrance exams. These questions require the use of machine-scored answer grids. You may wish to have students practice answering these items in preparation for standardized tests.

CHAPTERS 1-4 CUMULATIVE ASSESSMENT

College Entrance Exam Practice

QUANTITATIVE COMPARISON For Questions 1–6, write
A if the quantity in Column A is greater than the quantity in Column B;
B if the quantity in Column B is greater than the quantity in Column A;
C if the two quantities are equal; or
D if the relationship cannot be determined from the information given.

	Column A	Column B	Answers
1.	The number of $14 CDs you can buy for $86	The number of $12 CDs you can buy for $75	Ⓐ Ⓑ Ⓒ Ⓓ [Lesson 1.2]
2.	$|-12.4|$	$|12.4|$	Ⓐ Ⓑ Ⓒ Ⓓ [Lesson 2.1]
3.	$15 \cdot (6+4) - 7$	$2^3 - 6 \cdot (3+2)$	Ⓐ Ⓑ Ⓒ Ⓓ [Lesson 1.3]
4.	$-18 + (-3)$	$-17 + (-3)$	Ⓐ Ⓑ Ⓒ Ⓓ [Lessons 2.1 and 2.2]
5.	The solution to $v + 10 = 7$	The solution to $v - 6 = 5$	Ⓐ Ⓑ Ⓒ Ⓓ [Lesson 3.1]
6.	$(-15)(-4) \div (10)$	$(-15)(-4) \div (-10)$	Ⓐ Ⓑ Ⓒ Ⓓ [Lesson 2.4]

7. The correlation between two variables is described as strong positive. What is the relationship between the two variables? **(LESSON 1.6)**
   a. As one variable increases, the other also tends to increase.
   b. As one variable increases, you cannot tell if the other tends to increase or decrease.
   c. As one variable increases, the other tends to decrease.

8. Which sequence shows a linear relationship? **(LESSON 1.5)**
   a. 60, 30, 15, 7.5, 3.75, ...
   b. 6, 12, 24, 48, 96, ...
   c. 6, 9, 13.5, 20.25, 30.375, ...
   d. 64, 58, 52, 46, ...

9. What is the value of $\frac{5^2 + 11}{9 \cdot 3 + 9}$? **(LESSON 2.4)**
   a. 36
   b. 0
   c. 1
   d. $\frac{4}{9}$

10. What percent is equivalent to 0.5? **(LESSON 4.2)**
    a. 0.5%
    b. 50%
    c. 5%
    d. 0.05%

11. Jamie made 80% of his free throws in his game last night. If he attempted 15 free throws, how many free throws did he make? **(LESSON 4.2)**

12. What is the solution to the equation $\frac{q}{139.2} = -58$? (**LESSON 3.2**) **b**
    a. −2.4 b. −8073.6
    c. −0.417 d. −7551.6

13. Construct a stem-and-leaf plot for the following data: 12, 18, 26, 24, 29, 32, 37, 16, 14, 12, 29, 27, 34, 30, and 11. (**LESSON 4.6**)

14. Is 18t a variable, an expression, or an equation? Explain your answer. (**LESSON 1.2**)
    Expression

What are the coordinates of the given points? (**LESSON 1.4**)

15. X **(−2, 2)**
16. Y **(1, −4)**
17. Z **(−3, −1)**

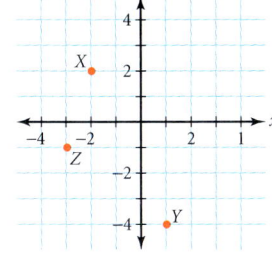

18. Solve the proportion $\frac{g}{12} = \frac{-3}{20}$. (**LESSON 4.1**)
    −1.8
19. What is the reciprocal of $-\frac{1}{9}$? (**LESSON 2.4**)
    −9

Simplify each expression. (**LESSONS 2.6 AND 2.7**)

20. $(9w - 5) - (3w - 4)$ **6w − 1**
21. $(-54s) \div (6s)$ **−9**
22. $4x^2 - 4(3x^2 + 7)$ **−8x² − 28**
23. $16x^2 - 8x^2$ **8x²**
24. Solve the proportion $\frac{15}{n} = \frac{20}{32}$. (**LESSON 4.1**) **24**
25. Solve $2y + 4.5 = 6y - 1.5$. (**LESSON 3.4**) **y = 1.5**
26. What is the solution to the proportion $\frac{c}{16} = \frac{3}{8}$? (**LESSON 4.1**) **6**

Solve each equation. (**LESSONS 3.1, 3.2, AND 3.3**)

27. $x - 3 = 14$ **17**
28. $y + 4 = 23$ **19**
29. $3x = -39$ **−13**
30. $6p = 24$ **4**
31. $3d - 15 = 30$ **15**
32. Evaluate $\frac{(-5)(-10)}{(-2)(5)}$. (**LESSON 2.4**) **−5**
33. Find the opposite of the expression $3x - 7$. (**LESSON 2.5**) **−3x + 7**
34. Complete each step, and name the property used. (**LESSON 2.5**)

    $(74 + 23) + 26$
    $= (23 + ?) + 26$ **74** Commutative
    $= 23 + (74 + ?)$ **26** Associative
    $= 23 + ?$ **100** addition
    $= ?$ **123** addition

FREE-RESPONSE GRID
The following questions may be answered by using a free-response grid such as that commonly used by standardized test services.

35. What is the solution to the equation $3r + 2(r - 4) = 12$? (**LESSON 3.5**) **4**

36. Find the mean of the following grades: 96, 83, 92, 79, 94, and 87. (**LESSON 4.3**) **88.5**

37. Hannah got 84% of the questions correct on a test. If there were 50 questions on the test, how many did she answer correctly? (**LESSON 4.2**) **42**

38. Two number cubes were rolled 100 times. A sum greater than 7 appeared 29 times. Based on this experiment, what is the experimental probability, expressed as a decimal, that a sum greater than 7 will appear on two rolled cubes? (**LESSON 4.4**) **0.29**

13. Stem: tens
    Leaves: ones

Stems	Leaves
1	1, 2, 2, 4, 6, 8
2	4, 6, 7, 9, 9
3	0, 2, 4, 7

5 Linear Functions

CHAPTER PLANNING GUIDE

Lesson	5.1	5.2	5.3	5.4	5.5	5.6	Projects and Review
Pupil Edition Pages	218–225	226–233	236–243	244–251	252–257	258–263	234–235, 264–273
Extra Practice (Pupil Edition)	750	751	751	752	752	753	
Practice Workbook	26	27	28	29	30	31	
Reteaching Masters	51–52	53–54	55–56	57–58	59–60	61–62	
Enrichment Masters	26	27	28	29	30	31	
Cooperative-Learning Activities	26	27	28	29	30	31	
Lesson Activities	26	27	28	29	30	31	
Problem Solving/Critical Thinking	26	27	28	29	30	31	
Student Study Guide	26	27	28	29	30	31	
Spanish Resources	26	27	28	29	30	31	
Student Technology Guide	26	27	28	29	30	31	
Assessment Resources	58	59	60	62	63	64	61, 65–70
Teaching Transparencies		25	26		27, 28, 29	30	
Quiz Transparencies	5.1	5.2	5.3	5.4	5.5	5.6	
Writing Activities for Your Portfolio							13–15
Tech Prep Masters							21–24
Long-Term Projects							17–20

LESSON PACING GUIDE

Traditional	2 days	1 day	1 day	2 days	2 days	2 days	2 days
Block	1 day	$\frac{1}{2}$ day	$\frac{1}{2}$ day	1 day	1 day	1 day	1 day
Two-Year	4 days	2 days	2 days	4 days	4 days	4 days	4 days

CONNECTIONS AND APPLICATIONS

Lesson	5.1	5.2	5.3	5.4	5.5	5.6	Review
Algebra	218–225	226–233	236–243	244–251	252–257	258–263	266–273
Geometry	224	232	241	251	257	259, 262	
Business and Economics			240, 241, 242				270
Life Skills	218		241, 242	251			
Science and Technology	221, 225	228, 229, 232	236, 237, 239, 240, 242	246, 251	254	258, 262	270
Sports and Leisure	221, 225	227	242	244, 245, 246			270
Cultural Connection: Africa				251			
Cultural Connection: Americas		233					
Other	222, 225	232	237, 240	251	253, 257		270

BLOCK-SCHEDULING GUIDE

Day	Lesson	Teacher Directed: Lesson Examples, Teaching Transparencies	Student Guided: Activity, Try This	Cooperative-Learning Activity, Lesson Activity, Student Technology Guide	Practice: Practice & Apply, Extra Practice, Practice Workbook	Assessment: Quiz, Mid-Chapter Assessment	Problem Solving, Reteaching
1	5.1	10 min	15 min	15 min	65 min	15 min	15 min
2	5.2	8 min	8 min	8 min	35 min	8 min	8 min
	5.3	7 min	7 min	7 min	30 min	7 min	7 min
3	5.4	10 min	15 min	15 min	65 min	15 min	15 min
4	5.5	15 min	15 min	15 min	65 min	15 min	15 min
5	5.6	10 min	15 min	15 min	65 min	15 min	15 min
6	Assess.	50 min PE: Chapter Review	90 min PE: Chapter Project, Writing Activities	90 min Tech Prep Masters	65 min PE: Chapter Assessment, Test Generator	30 min Chap. Assess. (A or B), Alt. Assess. (A or B), Test Generator	

PE: Pupil's Edition

Alternative Assessment

The following are suggestions for an alternative assessment for students who may benefit from a different type of assessment than the regular chapter quizzes and the mid-chapter and end-of-chapter assessment materials. Many of the questions are open-ended, and students' answers will vary.

Performance Assessment

1. A motorist drives at a constant speed of 35 miles per hour. The equation $d = 35t$ represents distance, d, in miles traveled in t hours.
   a. How is the rate of change in this situation related to the slope of the graph of $d = 35t$?
   b. How can you determine whether distance traveled varies directly as time in the equation?

2. a. Write an equation for each line described below.
   line a: has slope of -3 and y-intercept of 2.5
   line b: contains $(-2, 5)$ and $(3, -4)$
   line c: has slope of 5 and contains $(3, 7)$
   b. What information do you need to know in order to find an equation for a line? Summarize the various forms that the equation can take.

3. a. On graph paper, draw two nonvertical lines that are perpendicular and pass through points with integer coordinates.
   b. Explain how your drawing helps verify that nonvertical perpendicular lines have slopes that are negative reciprocals of one another.

Portfolio Project

Suggest that students choose one of the following projects for inclusion in their portfolios.

1. a. Describe a real-world example of direct variation and use equations, tables, and graphs to explain the meaning of *direct variation*.
   b. Give a real-world example of a relationship that is not a direct-variation relationship in order to help clarify the definition of direct variation.

2. Explain how to use parallel and perpendicular lines to draw a parallelogram on the coordinate plane and how to write equations to represent the lines containing the sides. Draw illustrations to justify your answer.

internetconnect

The table below identifies the pages in this chapter that contain technology information and support in the side columns.

Content Links	
Lesson Links	pages 227, 240, 257
Portfolio Links	pages 233, 243
Graphics Calculator Support	page 782

Resource Links

For information about teacher and parent resources as well as professional development help, visit **www.hrw.com/math**.

Technical Support HRW has assembled a team of dedicated technical and teaching professionals and a comprehensive service program to provide you with the support you need.

- The HRW Technical Support Line operates from 7 A.M. to 6 P.M. central time, Monday through Friday, at (800) 323-9239.
- The HRW Technical Support Center on the World Wide Web is available 24 hours a day, seven days a week, at **www.hrwtechsupport.com**.
- You can e-mail our Technical Support Center at **tsc@hrwtechsupport.com**.
- The Technical Support Center's fax-on-demand service at (800) 352-1680 offers solutions to common problems and answers to frequently asked questions.

Technology

Lesson Suggestions and Calculator Examples
(Keystrokes are based on a TI-83 calculator.)

Lesson 5.1 Linear Functions and Graphs
In this lesson, students will create and examine scatter plots to determine whether lists of ordered pairs represent functions. The graphics calculator displays below show the ordered pairs (3, 2), (5, 6), (7, 10), and (9, 14) in tabular and graphical form.

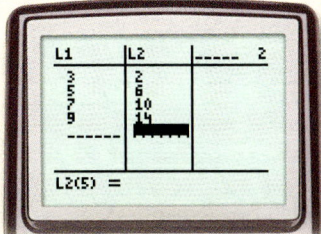 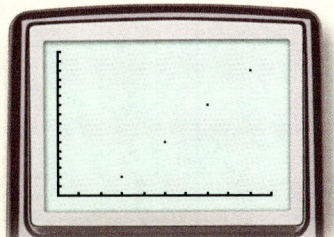

Discuss with students that because no two points lie on the same vertical line, the set of ordered pairs determines a function.

Lesson 5.2 Defining Slope
In this lesson, students will learn how to find the slope of a line, when two points are given, by using either point first. The display at right shows two calculations of the slope of the line containing (−2, 3) and (5, −4).

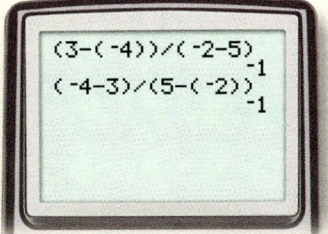

Tell students that the two choices for entering the coordinates are restricted to choosing one point as (x_2, y_2) and subtracting the x- and y-coordinates in the same order.

Lesson 5.3 Rate of Change and Direct Variation
An alternative definition of direct variation states the following: If y varies directly as x and (x_1, y_1) and (x_2, y_2) satisfy the relationship, then $\frac{y_1}{x_1} = \frac{y_2}{x_2}$. This definition allows students to turn direct-variation problems into proportion problems and then use the calculator to evaluate the proportion as described in Lesson 4.1. Ask students how the constant of variation k can be used in the definition above. Students should answer that $\frac{y_1}{x_1} = k$ and $k = \frac{y_2}{x_2}$.

Lesson 5.4 The Slope-Intercept Form
In this lesson, students will explore how different values of m and b affect the graph of the equation $y = mx + b$. Have students graph each set of equations (shown at the top of the next column) on the same set of coordinate axes. Make sure students understand how changes to the values of m and b affect the graphs.

 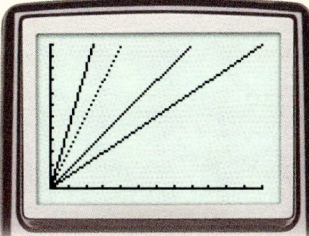

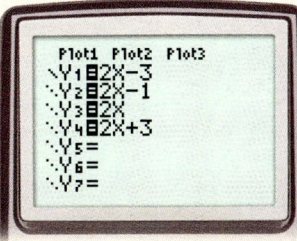 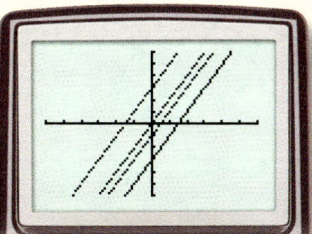

Lesson 5.5 The Standard and Point-Slope Forms
In order to use a graphics calculator to graph an equation, students must solve the equation for y. However, the equation does not need to be simplified before it is entered into the calculator. For example, to graph $5x + 4y = 20$, students can write the following:

$$5x + 4y = 20 \rightarrow y = \frac{20 - 5x}{4}$$

Then they can enter the fractional expression by using parentheses to group the numerator. Explain that a graph-ready expression may help reduce the possibility of error.

Lesson 5.6 Parallel and Perpendicular Lines
In this lesson, students will establish the relationships between the slopes of parallel and perpendicular lines by graphing the equations below.

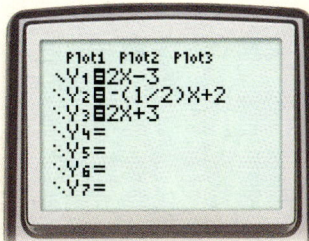 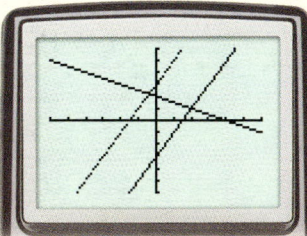

For further information, refer to the
- technology discussions in the lessons.
- lesson-related teacher's commentary in the side columns of this *Teacher's Edition*.
- lesson-related *Student Technology Guide* masters.
- *HRW Technology Handbook*.

Teaching Resources

Basic Skills Practice
(2 pages per skill)

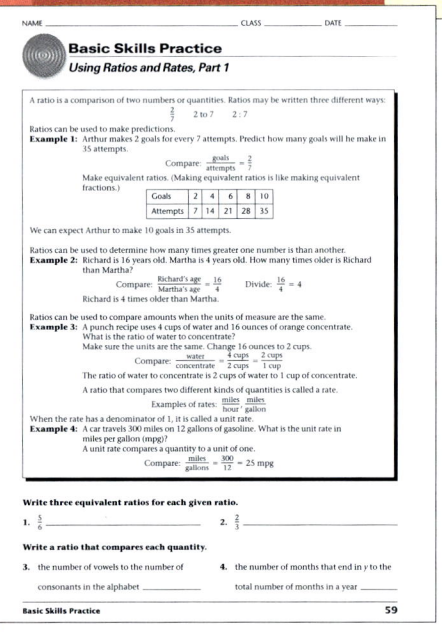

Reteaching
(2 pages per lesson)

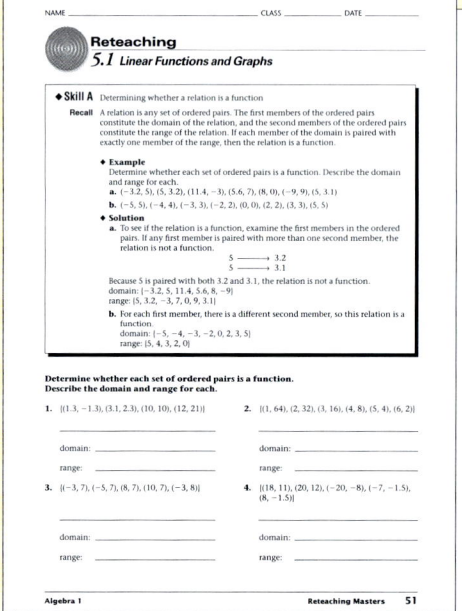

Enrichment
(1 page per lesson)

Cooperative-Learning Activity
(1 page per lesson)

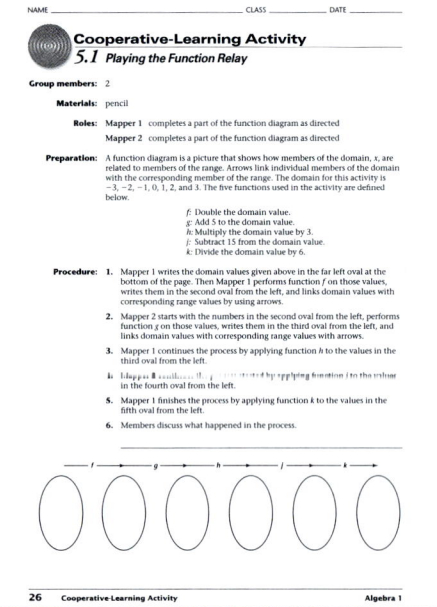

Lesson Activity
(1 page per lesson)

Long-Term Project
(4 pages per chapter)

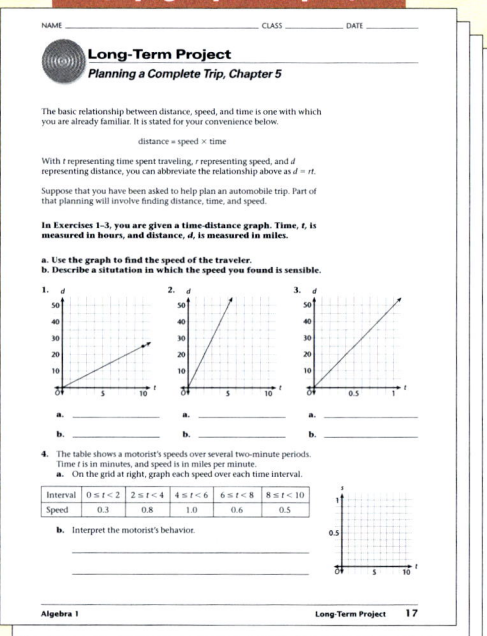

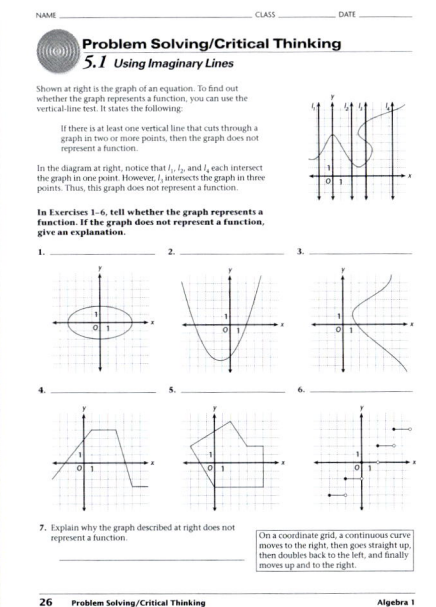

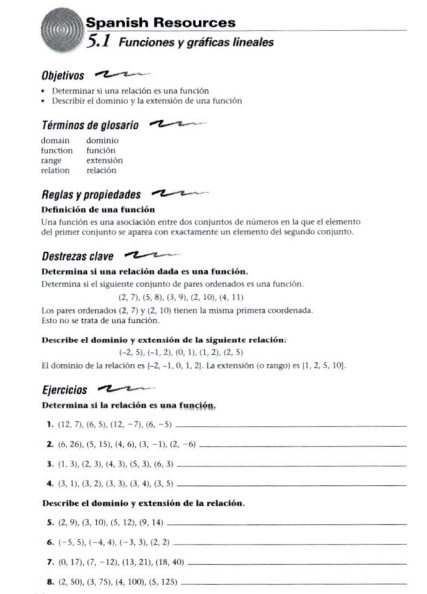

CHAPTER 5 INTERLEAF **216F**

About Chapter 5

Background Information
In Chapter 5, students are introduced to the concept of a *function* and the special category of *linear functions*. They observe that the graph of a linear function is a line, and they learn to read a wealth of information from the slope of this line. They then investigate several forms of writing an equation for the line. The chapter concludes with a study of parallel and perpendicular lines.

CHAPTER RESOURCES
- Block-Scheduling Handbook
- Writing Activities for Your Portfolio
- Tech Prep Masters
- Long-Term Project
- Assessment Resources:
  Mid-Chapter Assessment
  Chapter Assessments
  Alternative Assessments
- Test and Practice Generator
- Technology Handbook
- End-of-Course Exam Practice

Chapter Objectives
- Determine whether a relation is a function. [5.1]
- Describe the domain and range of a function. [5.1]
- Calculate the slope of a line by using the rise and run. [5.2]
- Calculate the slope of a line from the ratio of the differences in *x*- and *y*-coordinates. [5.2]

Linear Functions

5

Lessons

5.1 • Linear Functions and Graphs

5.2 • Defining Slope

5.3 • Rate of Change and Direct Variation

5.4 • The Slope-Intercept Form

5.5 • The Standard and Point-Slope Forms

5.6 • Parallel and Perpendicular Lines

Project Modeling Trends in Data

THE GREAT PYRAMID OF GIZEH IN EGYPT WAS built over a 30-year period around 2900 B.C.E. The builders used blocks averaging 2.5 tons, but some weighed as much as 54 tons. The blocks were transported from a quarry many miles away. The pyramid has a square base, and the error in constructing the right angles of the base was only 1 part in 14,000.

In order to build the pyramids, the builders needed to solve problems about slope. One problem from the Rhind papyrus asks how to find the *seked* (a form of slope) of a pyramid with a base of 2520 *palms* (618 feet) on each side and a height of 250 *cubits* (429 feet).

The pyramid at the entrance to the Louvre in Paris, France, provides a contrast to the older architecture of the main building.

About the Photos
This controversial steel-and-glass pyramid, which crowns the entrance to the Louvre in Paris, was designed by the American architect I.M. Pei. Ask students how they think mathematics may have been used in the building of the pyramid.

The strong lines in these high-rise buildings show the importance of linear functions in architecture.

Pyramids, Gizeh, Egypt

- Find the rate of change from a graph. [5.3]
- Solve and graph direct-variation equations. [5.3]
- Define and explain the components of the slope-intercept form of a linear equation. [5.4]
- Use the slope-intercept form of a linear equation. [5.4]
- Define and use the standard form for a linear equation. [5.5]
- Define and use the point-slope form for a linear equation. [5.5]
- Identify parallel lines and perpendicular lines by comparing their slopes. [5.6]
- Write equations of lines that are parallel and perpendicular to given lines. [5.6]

About the Chapter Project

Algebra provides the power to model real-world data. Beyond merely representing data on a graph or chart, algebra makes it possible to use *functions* to model data. In the Chapter Project, *Modeling Trends in Data*, on page 264, you will learn how to model data with linear equations by using linear regression. After completing the Chapter Project, you will be able to do the following:

- Draw a scatter plot of real-world data and determine the equation of the line of best fit or regression line.
- Make predictions about trends in data based on regression lines.

About the Portfolio Activities

Throughout the chapter, you will be given opportunities to complete Portfolio Activities that are designed to support your work on the Chapter Project.

- Relating measures of diameter and measures of circumference is the topic of the Portfolio Activity on page 233.
- Recognizing real-world data that can be modeled with a graph is the topic of the Portfolio Activity on page 243.
- Correlations between sets of data are the topic of the Portfolio Activity on page 263.

Portfolio Activities appear at the end of Lessons 5.2, 5.3, and 5.6. Each serves as preparation for the Chapter Project. The Portfolio Activities as well as the Chapter Project Activities are appropriate for inclusion in the student's portfolio. Students should be encouraged to include in their portfolios any other work in which they feel a sense of pride or a sense of accomplishment.

internet connect

Internet connect features appearing throughout the chapter provide keywords for the HRW Web site that lead to links for math resources, tutorial assistance, references for student research, classroom activities, and all teaching resource pages. The HRW Web site also provides updates to the technology used in the text. Listed at right are the keywords for the Internet Connect activities referenced in this chapter. Refer to the side column on the page listed to read about the activity.

LESSON	KEYWORD	PAGE
5.2	MA1 Marathon	227
5.2	MA1 Golden Ratio	233
5.3	MA1 Import Export	240
5.3	MA1 Ordered Pairs	243
5.5	MA1 Check Digits	257

Prepare

NCTM PRINCIPLES & STANDARDS
1–3, 5, 6, 8–10

LESSON RESOURCES

- Practice 5.1
- Reteaching Master 5.1
- Enrichment Master 5.1
- Cooperative-Learning Activity 5.1
- Lesson Activity 5.1
- Problem Solving/Critical Thinking Master 5.1

QUICK WARM-UP

Graph each point on a coordinate plane.

1. $A(3, 3)$ 2. $B(6, 2)$
3. $C(1, 0)$ 4. $D(0, 5)$
5. $E(7, 4)$ 6. $F(4, 7)$

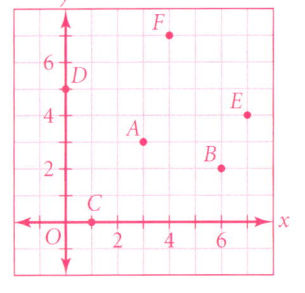

Also on Quiz Transparency 5.1

Teach

Why Remind students that there are many times when one quantity depends on another. Have them make a list of such situations. **Samples: Distance traveled depends on time traveled; the amount of change received depends on the amount given to the cashier.** Tell students that, mathematically, it is said that one of these quantities *is a function of* the other.

Linear Functions and Graphs

Objectives

- Determine whether a relation is a function.
- Describe the domain and range of a function.

Why Linear functions are the most basic algebraic functions. Many real-world problems can be modeled by linear functions.

Many people keep track of their height at different ages.

APPLICATION
HEALTH

Sets of paired numbers have a special meaning in algebra. Consider the following tables:

Table 1

	Age (years)	Height (inches)
Chad	14	68
Twan	11	64
Joshua	13	65
Sam	16	72
Sarah	15	64
Paul	14	67
Mike	15	66
Crystal	12	62

This table shows a pairing of ages with heights.

Table 2

Joshua's age (years)	Joshua's height (inches)
11	63
12	64
13	65
14	70
15	72
16	72

This table shows the age and height data for one student.

Each table presents two sets of paired numbers. A **relation** is a pairing between two sets of numbers. The two sets of numbers in Table 2 represent a special relation called a *function*.

Alternative Teaching Strategy

USING DEFINITIONS Discuss with students the definitions of relation, domain, and range as presented on pages 218 and 219. Then give them the following alternative definition of a function:

A function is a relation in which each element of the domain is paired with exactly one element of the range. Each member of the range is paired with at least one member of the domain.

Write the following domains and ranges on the board or overhead:

1. domain: {–2, 1, 0, 5}   range: {–6, –1, 0, 3}
2. domain: {–2, 1, 0, 5}   range: {–6, –1, 0}
3. domain: {–2, 1}   range: {–6, –1, 0}

Ask students to write two functions for each domain and range. Discuss their answers.
samples:
1. {(–2, –6), (1, –1), (0, 0), (5, 3)}
   {(–2, 3), (1, 0), (0, –1), (5, –6)}
2. {(–2, –6), (1, –1), (0, 0), (5, 0)}
   {(–2, –6), (1, –6), (0, –1), (5, 0)}
3. impossible

Activity
Exploring Relations and Functions

1. Write the ordered pairs from each table on page 218. The first few have been done for you.

   Table 1: (14, 68), (11, 64), …   Table 2: (11, 63), (12, 64), …

2. Study the first coordinates of the set of ordered pairs from Table 1. Are any of the first coordinates repeated?

3. Study the first coordinates of the set of ordered pairs from Table 2. Are any of the first coordinates repeated?

4. Study the second coordinates from Table 1. Are any of the second coordinates repeated?

5. Study the second coordinates from Table 2. Are any of the second coordinates repeated?

CHECKPOINT ✓ 6. Both tables represent relations. The second table represents a function. Based on your observations from Steps 2–5, what makes a relation a function?

Activity Notes

In this Activity, students will compare a relation and a function. Focus their attention on the relationship between a given age and the corresponding height. In Table 1, they should notice that there are instances in which more than one height corresponds to a given age. In Table 2, they should notice that exactly one height corresponds to a given age.

CHECKPOINT ✓
6. The second coordinates may be repeated, but none of the first coordinates may be repeated.

Relations and Functions

A relation is a pairing between two sets of numbers. A convenient way to represent a relation is as a set of ordered pairs. When writing a set of numbers or ordered pairs, use braces to enclose the set. The set of ordered pairs that represents the relation in Table 1 on page 218 is {(14, 68), (11, 64), (13, 65), (16, 72), (15, 64) (14, 67) (15, 66), (12, 62)}. The first coordinates in the set of ordered pairs are the **domain** of the relation, and the second coordinates are the **range** of the relation. The domain of the relation in Table 1 is {11, 12, 13, 14, 15, 16}. The range of the relation is {62, 64, 65, 66, 67, 68, 72}.

A function is a relation that meets a special condition.

Definition of a Function

A **function** is a pairing between two sets of numbers in which each element of the first set is paired with exactly one element of the second set.

Teaching Tip

You may want to review the language of sets with students. Remind them that a *set* is a collection of objects. Each object is called a *member*, or *element*, of the set. For some students, this will be the first time they have seen the bracket notation, { }, used as a means of enclosing the members of a set.

Table 2 on page 218 represents a function. The set of ordered pairs from Table 2 is {(11, 63), (12, 64), (13, 65), (14, 70), (15, 72), (16, 72)}.

Some students may be the same age but different heights as shown in Table 1 on page 218.

Notice that no two ordered pairs have the same first coordinate. In a function, some of the second coordinates may be the same, but all of the first coordinates must be different.

Interdisciplinary Connection

BUSINESS A salesperson whose earnings depend on an amount sold is said to work on *commission*. The commission usually is determined as a percent of the amount sold. This percent is called the *commission rate*. There are some salespeople who earn a *base salary* plus a commission.

A company has advertised a job opening for a salesperson. The base salary is $250 per week. In addition, the advertisement includes the table at right, which shows the total that can be earned each week depending on sales.

Amount of sales during the week	Total weekly earnings
$1000	$300
$2000	$350
$3000	$400
$4000	$450
$5000	$500

Write an equation for the situation.

$y = 250 + 0.05x$, where y represents the total weekly earnings and x represents the amount of sales during the week

LESSON 5.1

Teaching Tip

Although domains and ranges often are listed in increasing order, be sure students are aware that the order does not matter. Also, if two ordered pairs have the same first coordinate or same second coordinate, tell students that the number is only listed once in the domain or range.

ADDITIONAL EXAMPLE 1

Determine whether each set of ordered pairs is a function. Describe the domain and range for each.

a. $\{(1, 5), (3, 4), (5, 2)\}$
   This is a function.
   domain: $\{1, 3, 5\}$
   range: $\{2, 4, 5\}$

b. $\{(2, 8), (4, 1), (1, 2), (4, 7)\}$
   This is not a function.
   domain: $\{1, 2, 4\}$
   range: $\{1, 2, 7, 8\}$

TRY THIS

Answers may vary. Sample: A set of ordered pairs that is a function is $\{(0, 1), (1, 2), (2, 3), (3, 4)\}$. A set of ordered pairs that is not a function is $\{(1, 5), (2, 6), (1, 7), (3, 8)\}$.

ADDITIONAL EXAMPLE 2

Complete each ordered pair so that it is a solution to the equation $3x - 2y = 12$.

a. $(2, \underline{?})$ -3 b. $(-2, \underline{?})$ -9
c. $(\underline{?}, 3)$ 6 d. $(\underline{?}, -6)$ 0

A function has a domain and a range just as any relation does. The domain of the function from Table 2 on page 218 is $\{11, 12, 13, 14, 15, 16\}$. The range is $\{63, 64, 65, 70, 72\}$.

EXAMPLE 1 Determine whether each set of ordered pairs is a function. Describe the domain and range for each.

a. $\{(3,4), (3, 6), (5, 14), (7, 14)\}$ b. $\{(4, 12), (5, 18), (7, 12), (8, 19)\}$

SOLUTION

a. The ordered pairs (3, 4) and (3, 6) have the same first coordinate. This is not a function. The domain is $\{3, 5, 7\}$, and the range is $\{4, 6, 14\}$.

b. Since none of the ordered pairs have the same first coordinate, this is a function. The domain is $\{4, 5, 7, 8\}$, and the range is $\{12, 18, 19\}$.

TRY THIS Write a set of four ordered pairs that is a function and a set of four ordered pairs that is not a function.

Linear Functions

Three common methods are used to describe functions—tables that list ordered pairs, function rules, and graphs.

Since a function is a special pairing between two sets of numbers, a convenient way to provide a rule for a function is to use an equation in two variables such as those shown below.

$$y = 5x \qquad d = 5t \qquad 3x + 2y = 10 \qquad -10p = 2q + 1$$

An equation in two variables can have infinitely many solutions. Solutions to an equation in two variables can be represented by ordered pairs of numbers. The set of all ordered pairs that are solutions to a given equation together make up the function. The equation *describes* the function. Functions and equations in two variables can be linear, as presented in this lesson, or not linear, as presented in later lessons.

EXAMPLE 2 Which point(s) lie(s) on the graph of the equations $2x + y = 9$?

a. $P(1, 8)$ b. $Q\left(\frac{5}{2}, 4\right)$ c. $R(0, 9)$ d. $S(4, 0)$

SOLUTION

a. Substitute 1 for x and 8 for y.
$(2)(1) + 8 \stackrel{?}{=} 9 \rightarrow 2 + 8 \stackrel{?}{=} 9 \rightarrow 10 = 9$ False

b. Substitute $\frac{5}{2}$ for x and 4 for y.
$(2)\left(\frac{5}{2}\right) + 4 \stackrel{?}{=} 9 \rightarrow 5 + 4 \stackrel{?}{=} 9 \rightarrow 9 = 9$ True

c. Substitute 0 for x and 9 for y.
$(2)(0) + 9 \stackrel{?}{=} 9 \rightarrow 0 + 9 \stackrel{?}{=} 9 \rightarrow 9 = 9$ True

d. Substitute 4 for x and 0 for y.
$(2)(4) + 0 \stackrel{?}{=} 9 \rightarrow 8 + 0 \stackrel{?}{=} 9 \rightarrow 8 = 9$ False

Inclusion Strategies

VISUAL LEARNERS For students who are visually oriented, a picture called a *mapping diagram* can help determine whether a relation is a function. For instance, the first mapping diagram at right shows the relation $\{(3, 4), (3, 6), (5, 14), (7, 14)\}$ from part **a** of Example 1. It clearly shows that two ordered pairs, (3, 4) and (3 6), have the same first coordinate. Therefore, this relation is not a function.

In contrast, the second mapping diagram below shows the relation $\{(4, 12), (5, 18), (7, 12), (8, 19)\}$ from part **b** of Example 1. This time it is clear that no two ordered pairs have the same first coordinate, so this relation is a function.

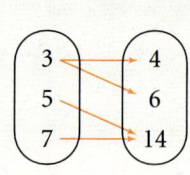
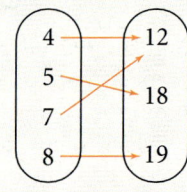

TRY THIS Which point(s) lie(s) on the graph of $3x + y = 15$?
  a. $W(3, 5)$ b. $X(4, 3)$ c. $Y(0, 1.5)$ d. $Z(5, 0)$

TRY THIS
a. false   b. true
c. false   d. true

A table of values can help you to determine a rule for a linear function.

ADDITIONAL EXAMPLE 3

The table below shows the amount that a company charges for bicycle rental for up to 8 hours. **Write an equation for the function. Write a set of ordered pairs for the function. Identify the domain and range values. Then graph the function.**

EXAMPLE 3

APPLICATION
RECREATION

The table below shows the amount that a company charges for a raft rental for up to 10 hours. An initial deposit is included in the amounts shown. **Write an equation for the function. Write a set of ordered pairs for the function. Identify the domain and range values.**

Time in hours, x	1	2	3	4	5	6	7	8	9	10
Rental cost in dollars, y	15	20	25	30	35	40	45	50	55	60

Time (hours), x	Rental cost (dollars), y
1	12
2	16
3	20
4	24
5	28
6	32
7	36
8	40

• **SOLUTION**

PROBLEM SOLVING

Use the differences in the rental cost to **determine a pattern**. For each hour, the cost increases by $5. The cost for 1 hour is $15, which includes the deposit. If the rental fee is $5 per hour, then the deposit is $10. Renting a raft costs $10 plus $5 per hour. The equation is as follows:

$y = 5x + 10$

The set of ordered pairs is {(1, 15), (2, 20), (3, 25), (4, 30), (5, 35), (6, 40), (7, 45), (8, 50), (9, 55), (10, 60)}.

The domain is {1, 2, 3, 4, 5, 6, 7, 8, 9, 10}.

The range is {15, 20, 25, 30, 35, 40, 45, 50, 55, 60}.

The domain and range do not include amounts that are not whole numbers because the rental fee is based on hourly rates and the domain of the function is restricted to hours from 1 to 10 only.

equation: $y = 4x + 8$
ordered pairs: {(1, 12), (2, 16), (3, 20), (4, 24), (5, 28), (6, 32), (7, 36), (8, 40)}
domain: {1, 2, 3, 4, 5, 6, 7, 8}
range: {12, 16, 20, 24, 28, 32, 36, 40}

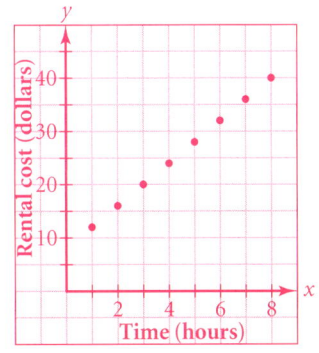

TECHNOLOGY
GRAPHICS CALCULATOR

On some calculators you can list the ordered pairs and plot the corresponding points to show the graph of the function in Example 3.

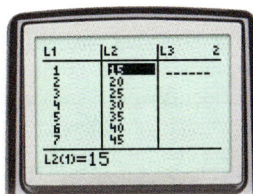 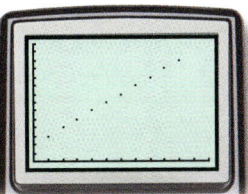

CRITICAL THINKING

Does a charge of $10 for 0 hours make sense for the problem in Example 3? Why or why not?

See Keystroke Guide, page 782.

CRITICAL THINKING
Answers may vary. Sample: No, if a customer does not rent a raft, then he or she should not pay a fee. If a customer rents for less than 1 hour they would be charged the flat fee of $10.

Enrichment

The actual time of a rental is usually rounded to the cost associated with the nearest hour that the rate is based upon. Ask students how much they would be charged for the raft rental in Example 2 for each of the times below.

4 h	$30	4 h 5 min	$35
4 h 30 min	$35	4 h 50 min	$35
4 h 59 min	$35	5 h	$35

Have students draw a graph for all times from 0 hours to 10 hours. **See graph at right.**

This type of function is called a *step function*. Ask why the function is given that name. **The graph has the appearance of a set of steps.**

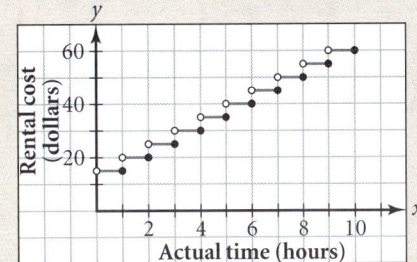

LESSON 5.1 **221**

TRY THIS

equation: $y = 3.5x + 15$

ordered pairs: {(1, 18.50), (2, 22), (3, 25.50), (4, 29), (5, 32.50), (6, 36), (7, 39.50), (8, 43), (9, 46.50), (10, 50)}

domain: {1, 2, 3, 4, 5, 6, 7, 8, 9, 10}
range: {18.50, 22, 25.50, 29, 32.50, 36, 39.50, 43, 46.50, 50}

ADDITIONAL EXAMPLE 4

The graph shows the distance that a car travels over time. **Identify the dependent and independent variables, and describe the domain and range. Write an equation for the linear function.**

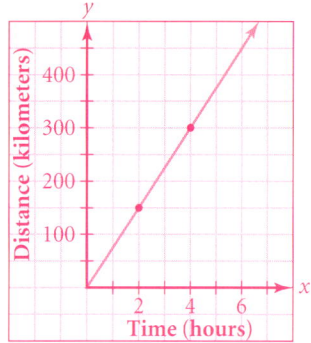

The dependent variable is distance, and the independent variable is time. Both the domain and range consist of all real numbers greater than or equal to 0. An equation for the function is $y = 75x$.

TRY THIS

The dependent variable is distance, and the independent variable is time. Both the domain and range consist of all real numbers greater than or equal to 0. An equation for the function is $y = 50x$.

TRY THIS The table below shows the amount that a company charges for a raft rental for up to 10 hours. An initial deposit is included in the amounts shown. Write an equation for the function. Write a set of ordered pairs for the function. Identify the domain and range values.

Time in hours, x	1	2	3	4	5	6	7	8	9	10
Rental cost in dollars, y	18.50	22	25.50	29	32.50	36	39.50	43	46.50	50

Graphs can also be used to represent functions. This method is especially helpful for functions with infinite domains that are impossible to list in a table.

You can read some of the ordered pairs from the graph of a linear function and use these ordered pairs to write an equation.

EXAMPLE 4

APPLICATION
TRANSPORTATION

The graph shows the distance that a plane travels over time. **Identify the dependent and independent variables, and describe the domain and range. Write an equation for the linear function.**

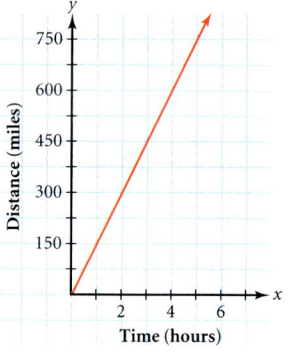

● **SOLUTION**

The dependent variable is distance. The independent variable is time. Both the domain values and the range values are the set of all real numbers greater than or equal to 0.

Build a table of values to find the equation.

Time in hours, x	0	1	2	3	4	5
Distance in miles, y	0	150	300	450	600	750

Use the differences between the y-values to determine a pattern. The y-values start with 0 and increase consistently by 150. An equation that represents this function is $y = 150x$.

TRY THIS The graph shows the distance that a tour bus is from its home base over time. Identify the dependent and independent variables and describe the domain and range. Write an equation for the linear function.

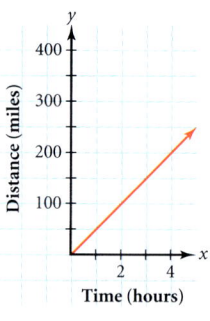

Reteaching the Lesson

USING LANGUAGE Start with a function table like the one at right. Ask students to state how y is related to x. *y is three more than x* Then show them how to translate their words into an equation for a function.

x	y
−4	−1
−2	1
1	4
5	8
7	10

y is three more than x
↓ ↓ ↓ ↓
y = 3 + x

Now repeat the activity with the tables below.

x	y
−3	−9
−2	−6
−1	−3
2	6
4	12

x	y
−5	−11
−1	−3
0	−1
2	3
6	11

x	y
−4	−2
−2	−1
0	0
6	3
10	5

x	y
−8	1
−6	2
−2	4
0	5
10	10

$y = 3x$ $y = 2x − 1$ $y = 0.5x$ $y = 0.5x + 5$

222 LESSON 5.1

Another way to express the equation for the function in Example 4 is to use **function notation**, $f(x) = 150x$. This is read "f of x equals $150x$." Notice that "$f(x)$" replaces "y." Any letters may be used to represent the variables, but the independent variable must appear within the parentheses and in the expression on the right side of the equation. You will study the use of function notation further in Chapter 14.

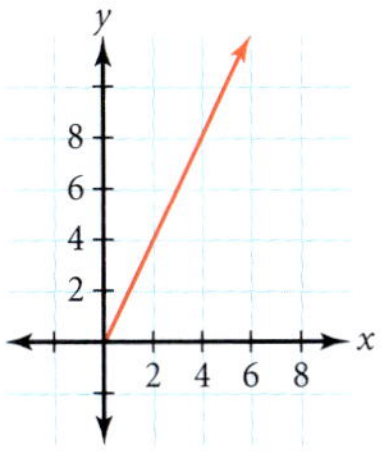

Assess

Selected Answers
Exercises 5–12, 13–45 odd

ASSIGNMENT GUIDE

In Class	1–12
Core	13–29, 31, 33
Core Plus	13–34
Review	35–45
Preview	46, 47

Extra Practice can be found beginning on page 738.

Technology
A graphics calculator may be helpful for Exercises 12, 30, and 31. You might suggest that students graph their equations on a calculator and determine whether the calculator graph matches the given graph.

Exercises

Communicate

1. Study each set of ordered pairs and determine whether each is a function. Explain your reasoning. Describe the domain and range for each.
   a. {(4, 5), (5, 4), (3, 4), (1, 5)}
   b. {(−2, 3), (2, −3), (4, 2), (2, −4)}
   c. {(1, 3), (2, 4), (3, 5), (4, 6), (5, 7)}
   d. {(1, 4), (1, 5), (1, 6), (1, 7)}

2. Explain what makes a function different from any other relation.

3. How do you know that a point lies on a line given the equation of the line?

4. Describe how to find an equation from the graph of a linear function.

Guided Skills Practice

5. D: {3, 4, 9}
   R: {−5, −2, 5}
   Function

6. D: {1, 2, 3, 4}
   R: {1}
   Function

11. $y = 0.5x + 2$
    D: {1, 2, 3, 4, 5, 6}
    R: {2.5, 3, 3.5, 4, 4.5, 5}

Describe the domain and range of each relation below. Determine whether each relation is a function. *(EXAMPLE 1)*

5. {(9, −5), (4, 5), (3, −2)}
6. {(1, 1), (2, 1), (3, 1), (4, 1)}

Complete each ordered pair so that it is a solution to $4x + y = 21$. *(EXAMPLE 2)*

7. (5, ?) (5, 1)
8. (?, 9) (3, 9)
9. (0, ?) (0, 21)
10. (?, 0) $(5\frac{1}{4}, 0)$

11. The table shows per minute rates for a cellular phone call, including a $2 fixed fee per call. Use the table to write an equation for this function. Identify the domain and the range shown in the table. *(EXAMPLE 3)*

Minutes of use	1	2	3	4	5	6
Rate in dollars	2.50	3.00	3.50	4.00	4.50	5.00

12. The graph shows a linear function. Make a table of values, and determine an equation. *(EXAMPLE 4)*
    x: 0, 2, 4
    y: 0, 4, 8
    $y = 2x$

LESSON 5.1   223

Error Analysis

Students frequently confuse the terms *domain* and *range*. Encourage them to use alphabetical order to match the terms with the proper coordinates. That is, *D* appears before *R* in the alphabet, so the domain is the set of first coordinates. The letter *R* appears after *D*, so the range is the set of second coordinates.

34.
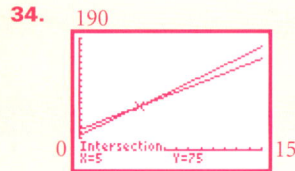

The CD club in Exercise 32 is a better deal for buying less than 5 CDs. The CD club in Exercise 33 is a better deal for buying more than 5 CDs. The cost is exactly the same for buying exactly 5 CDs.

Practice and Apply

13. D: {−2, 5.7, 9.2}; R: {−3.4, 3.4, 4}; Function

14. D: {0, 3, 4}; R: {0, 3, 4}; Not a function

15. D: {−3, −1, 0, 1, 3}; R: {−2, −1, 4}; Function

16. D: {2}; R: {1, 1000}; Not a function

17. D: {3, 4, 5, 9, 12}; R: {3, 4, 5, 9, 12}; Function

For each relation, describe the domain and range. Determine whether the relation is a function. Explain your reasoning.

13. {(9.2, 3.4), (5.7, −3.4), (−2, 4)}

14. {(0, 3), (0, 4), (4, 0), (3, 0)}

15. {(−1, −2), (1, −2), (3, −1), (−3, −2), (0, 4)}

16. {(2, 1), (2, 1000)}

17. {(9, 9), (3, 3), (4, 4), (5, 5), (12, 12)}

Complete each ordered pair so that it is a solution to $2x - y = 14$.

18. (8, ?) **(8, 2)** **19.** (10, ?) **(10, 6)** **20.** (0, ?) **(0, −14)** **21.** (?, 0) **(7, 0)**

22. (5, ?) **(5, −4)** **23.** (−5, ?) **(−5, −24)** **24.** (3, ?) **(3, −8)** **25.** (?, 3) **($8\frac{1}{2}$, 3)**

26. (?, 6) **(10, 6)** **27.** (?, −4) **(5, −4)** **28.** (?, −7) **($3\frac{1}{2}$, −7)** **29.** (?, 10) **(12, 10)**

CONNECTIONS

30. COORDINATE GEOMETRY Game Menagerie offers a deal for frequent video-game players. The graph shows the cost of playing from 0 to 30 video games. Make a table of values and write an equation for the situation.

x: 0, 10, 20, 30
y: 5, 7.5, 10, 12.5
$y = \frac{1}{4}x + 5$

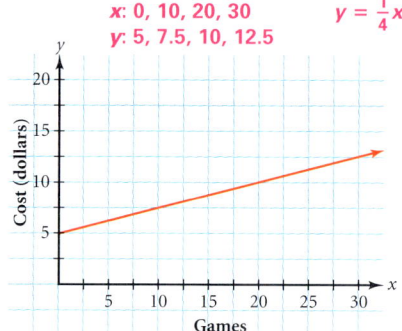

x: 0, 1, 2, 3, 4
y: 0, 5, 10, 15, 20
$y = 5t$
D: $x \geq 0$
R: $y \geq 0$

31. COORDINATE GEOMETRY The graph shows how far a hiker is from the starting point of a 30-mile trail. Make a table of values, write an equation for the graph, and describe the domain and range.

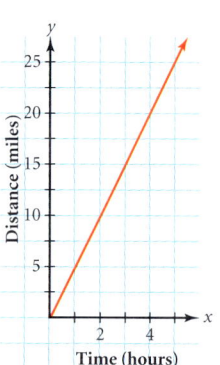

35.

x	1	2	3	4	5	10
y	3	5	7	9	11	21

36.

x	1	2	3	4	5	10
y	4	9	14	19	24	49

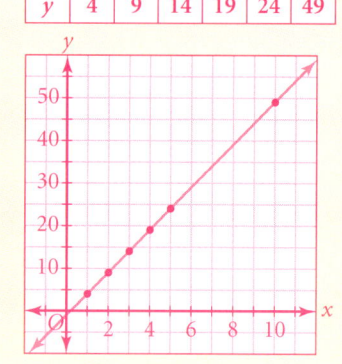

APPLICATIONS

32. ENTERTAINMENT The table shows the total costs associated with purchases from an audio CD club, including an initial membership fee of $25. Write an equation for this situation. **y = 10x + 25**

CDs purchased	0	2	4	6	8	10	12	14
Cost	$25	$45	$65	$85	$105	$125	$145	$165

33. ENTERTAINMENT The table shows the total costs associated with purchases from another audio CD club, including an initial membership fee of $35. Write an equation for this situation. **y = 8x + 35**

CDs purchased	0	2	4	6	8	10	12	14
Cost	$35	$51	$67	$83	$99	$115	$131	$147

34. TECHNOLOGY Use a graphics calculator to graph the equations from Exercises 32 and 33. Compare the two options. Which CD club offers the better deal and why?

Look Back

For Exercises 35 and 36, build a table of values by substituting 1, 2, 3, 4, 5, and 10 for *x*. Sketch a graph on graph paper. **(LESSON 1.5)**

35. $y = 2x + 1$ **36.** $y = 5x - 1$

APPLICATION

DEMOGRAPHICS In a small community, the percent of adults with college degrees has steadily increased, as shown in the table. **(LESSON 1.6)**

Year	1955	1960	1965	1970	1975	1980	1985	1990	1995	2000
Percent	12	15	22	25	35	48	52	52	55	60

37. Make a scatter plot of the data.

38. Draw a line of best fit to predict the percent in the years 2005 and 2010.

39. Write an equation for your line of best fit. **sample: y = 1.2x + 12**

40. Is the line a good predictor over the next 50 years? Why or why not?

Use mental computation to evaluate each expression. **(LESSON 2.5)**

41. $300 - 196$ **104**   **42.** $10 \cdot 30$ **300**   **43.** $\frac{480}{16}$ **30**   **44.** $1000 \cdot 1000$ **1,000,000**

45. Find the area of a square with a side length of 7 centimeters. **49 cm²**

Look Beyond

x	1	2	3	5	6	4	7	8
y	1	4	9	25	36	16	49	64

46. Study the table above. Is the relation linear? **No**

47. Write an equation for the data in the table above. Graph the equation, and describe the shape of the graph.

37.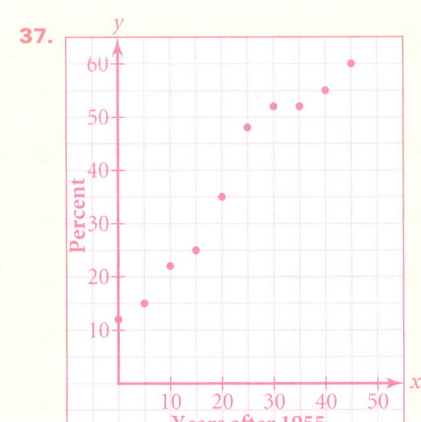
Years after 1955

38.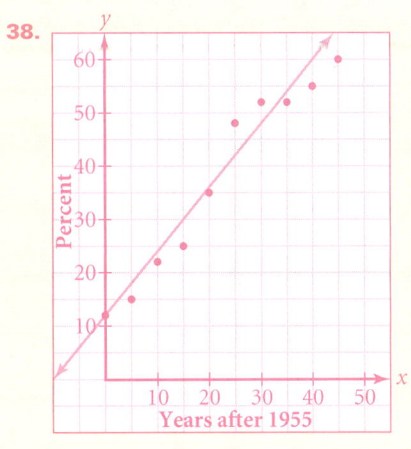
Years after 1955

Look Beyond

In Exercises 46 and 47, students investigate a relation that is not linear. They may discover that it is not linear by graphing the ordered pairs, or they may notice that the first differences in the table of data are not constant. The relation is a quadratic function, which students will study in detail in Chapter 10.

40. No, the line eventually predicts percentages greater than 100, which is impossible.

47. $y = x^2$

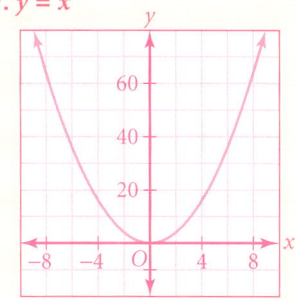

The graph is curved like a "U," with its lowest point at the origin.

Student Technology Guide

Student Technology Guide
5.1 Linear Functions and Graphs

When you are given a list of ordered pairs or a table of values, you can tell graphically if a function is determined by creating a scatter plot. If any two points in the scatter plot lie above the same vertical line, then the relation is not a function.

Example: Plot the ordered pairs (3, 1), (4, 2), (5, 4), (3, 4), (7, 5), and (10, 6). Use the graph to tell whether the set of ordered pairs represents a function.
- Prepare for data entry.
  Clear all functions from the function list. The following sequence will clear lists L1 and L2 in preparation for new data entry.
  Press STAT 4:ClrList 2nd 1 ENTER STAT 4:ClrList 2nd 2 ENTER
- Enter the data.
  Press STAT 1:Edit ENTER.
  Using the arrow keys, highlight the entry in row 1 of list L1. To enter the x-values: Press 3 ENTER 4 ENTER ... 7 ENTER 10 ENTER
  Using the arrow keys, highlight the entry in row 1 of list L2. To enter the y-values: Press 1 ENTER 2 ENTER ... 5 ENTER 6 ENTER
- Make the necessary range settings.
  Press WINDOW. Choose 0 ≤ x ≤ 12 and 0 ≤ y ≤ 8.
- Make the scatter plot.
  Press 2nd Y= 1:Plot 1...On ENTER On
  Type: GRAPH.

The scatter plot shows point (3, 4) lying above point (3, 1). The given list of ordered pairs does not represent a function.

Make a scatter plot for each table of values or list of ordered pairs. Use the graph to tell whether the relation is a function.

1.
x	1	2	3	4	5	6
y	1	1	1	1	1	1

function

2.
x	1.5	2.5	3.5	6.5	4.5	1.5
y	1.5	5.0	6.0	6.5	4.0	3.0

not a function

3. (2, 1), (3, 2), (7, 6), (5, 4), (4, 3), (6, 5)   function

4. (2, 3), (4, 4), (2, 5), (2, 4), (3, 5), (5, 6)   not a function

5. (2, 1), (1, 2), (2, 3), (4, 6), (7, 2)   not a function

6. (3, 3), (4, 4), (5, 5), (6, 6), (7, 7)   function

Prepare

NCTM PRINCIPLES & STANDARDS 1–10

LESSON RESOURCES

- Practice 5.2
- Reteaching Master 5.2
- Enrichment Master 5.2
- Cooperative-Learning Activity 5.2
- Lesson Activity 5.2
- Problem Solving/Critical Thinking Master 5.2
- Teaching Transparency 25

QUICK WARM-UP

Evaluate.

1. $4 - 9$ −5
2. $6 - (-4)$ 10
3. $\frac{-8-4}{9-5}$ −3
4. $\frac{-5-(-5)}{1-(-1)}$ 0

Also on Quiz Transparency 5.2

Teach

Why Discuss with students their understanding of the word *slope*. Have them identify situations in which the term is used. **samples: ski slope, the slope of a roof** Ask them why it might be important to assign a specific number to a slope.

Defining Slope

5.2

Objectives

- Calculate the slope of a line by using the rise and run.
- Calculate the slope of a line from the ratio of the differences in *x*- and *y*-coordinates.

Why By using the information in the diagram, you can find the slope, or steepness, of the hill for each of the hikers.

Rich: 24 cm, 60 cm   Kara: 18 cm, 45 cm

Finding Slope From Rise and Run

Slope is the ratio of vertical rise to horizontal run. Although the two hikers shown above take different-sized steps, the slope of the hill is the same for both of them.

In algebra, slope is defined in terms of the coordinate plane. Real-world information must be translated in order to be studied accurately.

The hikers' strides uphill can be represented by triangles that show the vertical **rise** and horizontal **run** of each stride. Placing the triangles on a coordinate plane provides a way to determine the slopes of the lines that represent the hikers' paths up the hill.

Alternative Teaching Strategy

USING MANIPULATIVES Place a cardboard box on top of a desk or table. Attach a ruler to the box in a vertical position, with its zero mark visible and pointing downward. Place another ruler on the horizontal surface of the table or desk, with its zero mark next to the zero mark of the vertical ruler.

Hold a piece of string with one end at 4 on the vertical ruler and the other end at 4 on the horizontal ruler. Lead students to see that the ratio $\frac{\text{vertical distance from } 0}{\text{horizontal distance from } 0}$ is equal to 1. Ask them to describe another way to hold the string that gives a ratio of 1. Elicit several answers, and demonstrate each. Be sure students notice that in each case, the "slant" of the string is the same.

Repeat the activity to demonstrate several cases of the ratio 2 and the ratio $\frac{1}{2}$. Tell students that this ratio, which characterizes the amount of "slant," is called *slope*.

226   LESSON 5.2

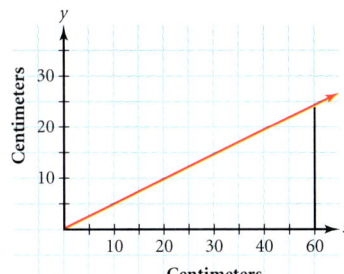
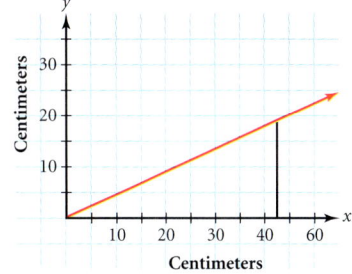

EXAMPLE ① Find the slopes of the lines that represent each hiker's path up the hill.

APPLICATION
RECREATION

○ **SOLUTION**

To calculate the slope, find the ratio of rise to run for each hiker. The horizontal distance of Rich's step (the run) is 60 centimeters, and the vertical distance (the rise) is 24 centimeters. The run for Kara's step is 45 centimeters and the rise is 18 centimeters.

Rich

slope = $\frac{\text{rise}}{\text{run}} = \frac{24}{60} = \frac{2}{5}$, or 0.4

The slope is $\frac{2}{5}$, or 0.4.

Kara

slope = $\frac{\text{rise}}{\text{run}} = \frac{18}{45} = \frac{2}{5}$, or 0.4

TRY THIS Find the slope of a line that represents the uphill path of a hiker if the rise is 12 inches and the run is 24 inches.

A graph that represents hiking downhill shows a line that is falling from left to right. The rise is negative in this case.

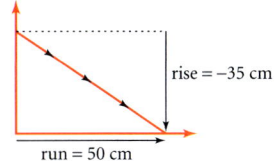

slope = $\frac{\text{rise}}{\text{run}} = \frac{-35}{50} = -\frac{7}{10} = -0.7$

On a coordinate plane, lines with **positive slope** go up as you move from left to right. Lines with **negative slope** go down as you move from left to right.

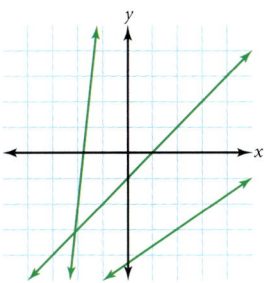
Lines with positive slope

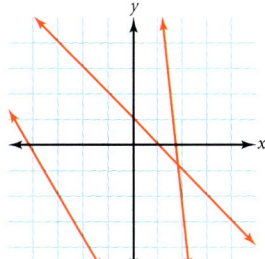
Lines with negative slope

ADDITIONAL
EXAMPLE ①

Yara and Nidal are hiking up a hill. The run for Yara's step is 32 inches, and the rise is 8 inches. The run for Nidal's step is 36 inches, and the rise is 9 inches. **Find the slope of the lines representing each hikers pathway up the hill.** $\frac{1}{4}$

TRY THIS
$\frac{1}{2}$

internet connect

GO TO: go.hrw.com
KEYWORD: MA1 Marathon

In this activity, students use the definition of slope to determine the steepness, or gradient, of a road or highway. Then students explore different gradients found along the Boston Marathon course.

Now hold one end of the string firmly at 2 on the horizontal ruler. Gradually move the other end up the vertical ruler, pausing several times to have students calculate the slope. Lead them to observe that as the "slant" of the string increases, so does the slope.

Enrichment

Replace each ? with a coordinate so that the line containing the two points has the given slope.

1. (1, 1), (? , −2); slope = 1   −2
2. (10, ?), (13, 7); slope = $\frac{2}{3}$ 5
3. (−3, 4), (? , 6); undefined slope   −3
4. (12, 2), (? , 16); slope = $-\frac{7}{5}$ 2
5. (−3, 9), (5, ?); slope = 0   9

Have students create five similar problems and then exchange problems with a partner.

LESSON 5.2 **227**

Use Teaching Transparency 25.

ADDITIONAL EXAMPLE 2

The coordinate plane below shows the position of a ladder resting against the side of a building. **Find the slope of the ladder.**

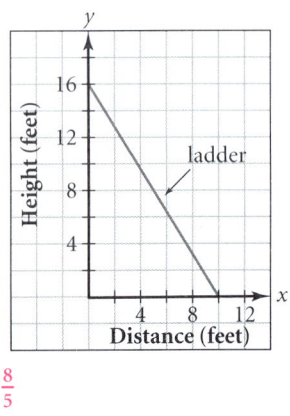

$-\frac{8}{5}$

Teaching Tip

When seeing the definition of slope for the first time, students might not know how to read the numbers that distinguish the coordinates of the two points. Tell them that the numbers are called *subscripts* and that y_2 is read as "y sub two," y_1 as "y sub one," and so on. Note that this is easier to say than "the second y-coordinate," "the first y-coordinate," and so on.

ADDITIONAL EXAMPLE 3

Find the slope of the line containing the points $P(-2, 4)$ and $Q(5, -6)$.

$-\frac{10}{7}$

EXAMPLE 2

APPLICATION: PHYSICS

Maya and her class are studying the effects of friction. In an experiment, they lean a meter stick against a wall. The object of the experiment is to find the steepness of the meter stick when it first starts to slide. They observe that it does not slide when the top of the meter stick is 80 centimeters from the tabletop and the bottom is 60 centimeters from the wall. When the steepness is any less, the meter stick slides.

The students recorded this information in a graph. The y-axis represents the wall, and the x-axis represents the tabletop. **From Maya's perspective, what is the slope of the meter stick just before it begins to slide?**

● **SOLUTION**

To find the slope from Maya's perspective, use the points where the line representing the meter stick touches the x- and y-axes. The run is the number of units to the right (60), and the rise is the number of units downward (−80).

$$\text{slope} = \frac{\text{rise}}{\text{run}} = \frac{-80}{60} = -\frac{4}{3}$$

You can find the slope of a line if you know the coordinates of two points on the line.

Definition of Slope

Given two points with coordinates (x_1, y_1) and (x_2, y_2), the formula for the slope, m, of the line containing the points is

$$m = \frac{y_2 - y_1}{x_2 - x_1}.$$

EXAMPLE 3 Find the slope of the line containing the points $M(-2, -6)$ and $N(3, 5)$.

● **SOLUTION**

$\text{slope} = \frac{y_2 - y_1}{x_2 - x_1}$

Choose M as the first point, with coordinates (x_1, y_1). Let N be the second point, with coordinates (x_2, y_2). Subtract the x-coordinates in the same order as the y-coordinates.

$$\text{slope} = \frac{5 - (-6)}{3 - (-2)} = \frac{11}{5}$$

Inclusion Strategies

TACTILE LEARNERS Use a geoboard with the pegs arranged in a square array to model a coordinate plane. Position two rubber bands to represent the axes.

Students can manipulate other rubber bands to find the slope of a line that contains two given points. To do this, they should first loop a rubber band around the pegs that represent the points and then loop two other rubber bands in positions that represent the rise and run. They can now determine the slope by using the slope ratio.

In the figure at left, for example, a rubber band has been looped around the pegs that represent $(-2, 4)$ and $(3, -4)$. To move from the first point to the second, a rubber band is stretched 8 units down, and then another rubber band is stretched 5 units right. The slope is $\frac{-8}{5}$, or $-\frac{8}{5}$.

TRY THIS   Find the slope of the line containing the points $P(2, -3)$ and $Q(-2, 15)$.

CRITICAL THINKING   What is the effect on the value found for the slope if the *x*-coordinates are not subtracted in the same order as the *y*-coordinates?

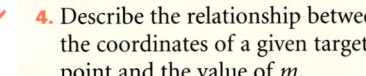

Fitting a Line to a Point

You will need: graph paper or a graphics calculator

1. Graph the point $A(3, 6)$ on a coordinate plane.

2. Use the linear equation $y = mx$, and substitute values for m until you find a line that intersects point A. Note that m represents the slope of the line.

3. Repeat the process for points B through F: $B(2, 8)$, $C(6, 3)$, $D(3, -6)$, $E(2, -8)$, $F(4, 7)$.

CHECKPOINT ✓

4. Describe the relationship between the coordinates of a given target point and the value of m.

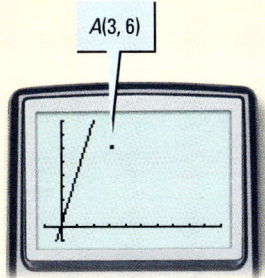

The calculator shows point A and the line $y = 4x$.

Horizontal and Vertical Lines

Horizontal and vertical lines are special cases that are important to the study of slope.

EXAMPLE 4 Find the slope of the horizontal line shown at right.

● **SOLUTION**

For a horizontal line, the *y*-coordinate of every point on the line is the same.

Choose two points on the line, and use the slope formula.

For the points $(3, 2)$ and $(-2, 2)$:

slope $= \frac{2-2}{-2-3} = \frac{0}{-5} = 0$

The slope of the horizontal line is 0.

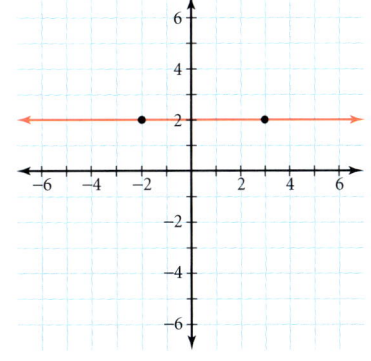

Reteaching the Lesson

WORKING BACKWARD   Have students draw a set of coordinate axes on graph paper. Give them a positive slope, such as 3, and have them draw any line with that slope. If necessary, remind them to use the ratio $\frac{\text{rise}}{\text{run}}$ to establish the slope.

Now ask them to pick any two points on the line and write the coordinates. Have them substitute those coordinates into the slope ratio $\frac{y_2 - y_1}{x_2 - x_1}$. Tell them to simplify the result. If the simplified ratio is not equal to the given slope, work with them to find the source of the error.

Repeat the activity with a positive fractional slope, such as $\frac{5}{2}$, a negative slope, a slope of 0, and an undefined slope.

TRY THIS

$-\frac{9}{2}$

CRITICAL THINKING

The slope turns out to be the opposite of what it should be.

Activity Notes

In this Activity, students explore slopes of lines that contain the origin. They should find that, for any point $A(x, y)$ on such a line, the ratio $\frac{y}{x}$ is equal to the slope of the line. Stress that this relationship does not hold for lines that do *not* contain the origin.

CHECKPOINT ✓

4. Answers may vary. Sample answer: The linear equation in the form $y = mx$ that goes through point (a, b) has the slope $m = \frac{b}{a}$.

Teaching Tip

TECHNOLOGY   If you are using a TI-82 or TI-83 graphics calculator, you can graph the point $A(3, 6)$ as follows:

First clear all equations from the **Y=** editor. Then press **WINDOW** and make these settings.

    Xmin=-5   Ymin=-10
    Xmax=10   Ymax=10
    Xscl=1   Yscl=1

Now clear any data from **LIST 1** and **LIST 2**. Then press **STAT** and choose **1:EDIT…**. Enter **3** under **L1** and **6** under **L2**. Then press **2nd** **Y=** **ENTER** and turn **Plot 1 ON**. Press **GRAPH**. The other points can be graphed in a similar manner.

To test an equation, press **Y=** and enter the equation next to **Y1=**. Then press **GRAPH**. When testing a different equation, be sure to first clear **Y1**.

See Keystroke Guide, page 782.

☞ For Additional Example 4, see page 230.

LESSON 5.2

Teaching Tip

Students often confuse the meanings of *horizontal* and *vertical*. Remind them that a horizontal line has the same orientation as the *horizon*.

ADDITIONAL EXAMPLE 4

Use the coordinates of point R and point S to find the slope of the horizontal line shown below.

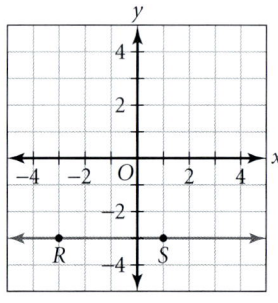

$\frac{-3-(-3)}{-3-1} = \frac{0}{-4} = 0$

The slope of the horizontal line is 0.

ADDITIONAL EXAMPLE 5

Use the coordinates of point J and point K to find the slope of the vertical line shown below.

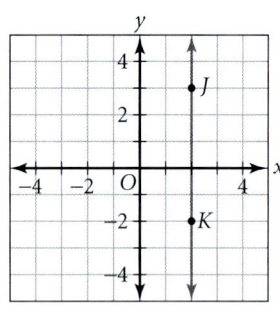

$\frac{3-(-2)}{2-2} = \frac{5}{0}$

The slope of the vertical line is undefined.

TRY THIS

a. The slope is undefined.
b. The slope is 0.

230 LESSON 5.2

EXAMPLE 5

Find the slope of the vertical line shown at right.

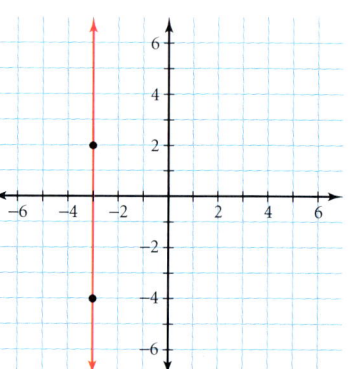

SOLUTION

For a vertical line, the *x*-coordinate of every point on the line is the same.

Choose two points on the line, and use the slope formula.

For the points $(-3, -4)$ and $(-3, 2)$:

slope = $\frac{2-(-4)}{-3-(-3)} = \frac{6}{0}$ *Dividing by 0 is not possible.*

The slope of the vertical line is *undefined* because division by 0 is impossible. Therefore, the slope formula cannot be applied to vertical lines.

The Slopes of Horizontal and Vertical Lines

The slope of a **horizontal line** is 0.

The slope of a **vertical line** is undefined.

TRY THIS Plot each pair of points and draw the line that contains them. Find the slope of each line.

  a. (5, 6) and (5, −1)   **b.** (−1, 4) and (5, 4)

Exercises

Communicate

1. Explain how to draw a line that has a rise of 4 and a run of 3 and that passes through the origin.

2. Describe the slope of a line with a negative rise and a positive run.

3. Why is the slope of a vertical line undefined?

4. Suppose that *s* is positive. Compare a line with a slope of *s* to a line with a slope of −*s*. Describe how the lines are different.

5. Points R(5, −3) and N(9, −4) are on line *k*. Explain how to find the slope of line *k*.

● **Guided Skills Practice**

What is the slope of each line? *(EXAMPLES 1 AND 2)*

6. $-\frac{6}{7}$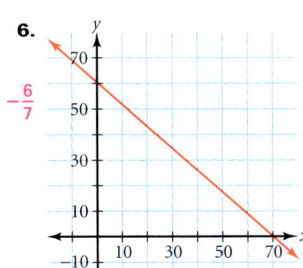

7. $\frac{1}{2}$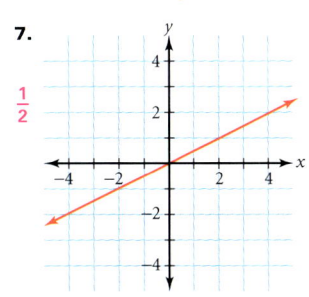

8. Find the slope of a line containing the points $(-1, 7)$ and $(8, -3)$. $-\frac{10}{9}$
   (EXAMPLE 3)

9. Plot the points $(-3, 3)$ and $(6, 3)$, and draw the line that contains them. Describe the line. What is its slope? *(EXAMPLE 4)* horizontal; 0

10. Plot the points $(4, 7)$ and $(4, 3)$, and draw the line that contains them. Describe the line. What is its slope? *(EXAMPLE 5)* vertical; undefined

● **Practice and Apply**

Find the slope of the lines graphed below.

11. 1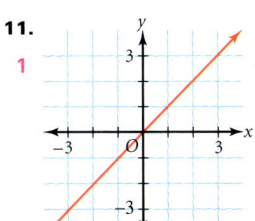

12. $-\frac{1}{3}$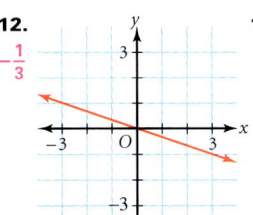

13. 0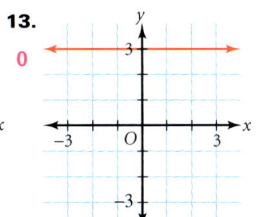

14. 2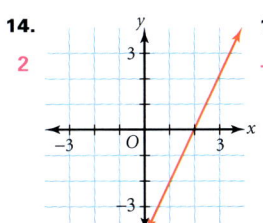

15. -2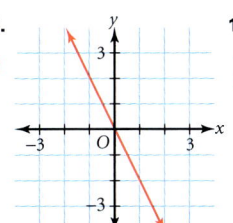

16. $\frac{1}{2}$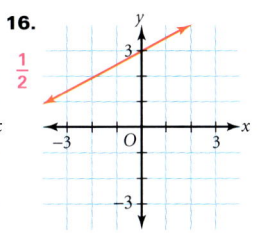

17. und.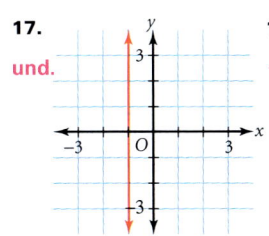

18. -1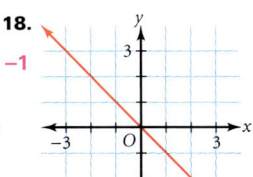

19. 2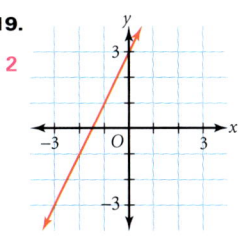

9.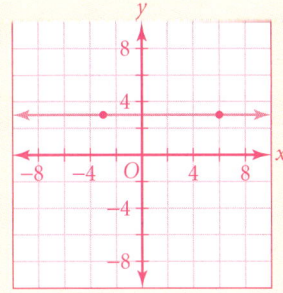
horizontal; 0

10.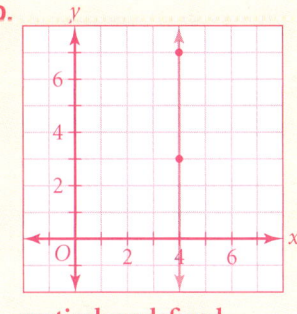
vertical; undefined

Assess

Selected Answers
Exercises 6–10, 11–61 odd

ASSIGNMENT GUIDE

In Class	1–10
Core	11–25, 32–37, 45–49 odd
Core Plus	12–44 even, 45–50
Review	51–61
Preview	62, 63

✎ Extra Practice can be found beginning on page 738.

Technology

A calculator may be helpful for Exercises 32–43. You may want students to first find the slope with paper and pencil and then use a calculator to verify their answers. You might suggest that they write the simplified slope as well as a calculator-ready form. For example, in Exercise 32, they might write the slope as 0.25, but also write $(6 - 4) \div (9 - 1)$ next to it.

LESSON 5.2 **231**

Error Analysis

Students often calculate a slope incorrectly because they have subtracted the *y*-coordinates and *x*-coordinates in a different order. To overcome this difficulty, suggest that they write the coordinates of the points on paper by using one color for *x*-coordinates and another for *y*-coordinates. They can then use the same colors to write the subtraction problem, making sure that the colors appear in the same order in each part of the problem.

Look Beyond

Exercises 62 and 63 preview the slope-intercept form for the equation of a line. They also explore the relationship between the slopes of parallel lines and of perpendicular lines. Students will study these topics in greater depth in Lessons 5.4 and 5.6.

Find the slope for each rise and run.

20. rise: 1, run: 5 $\frac{1}{5}$
21. rise: 3, run: 10 $\frac{3}{10}$
22. rise: −1, run: 7 $-\frac{1}{7}$
23. rise: −6, run: 3 −2
24. rise: 5, run: −3 $-\frac{5}{3}$
25. rise: 5, run: −2 $-\frac{5}{2}$
26. rise: −9, run: −5 $\frac{9}{5}$
27. rise: −4, run: −3 $\frac{4}{3}$
28. rise: −8 + 3, run: 4 + 6 $-\frac{1}{2}$
29. rise: 5 − 6, run: 4 − 7 $\frac{1}{3}$
30. rise: $-4\frac{1}{5}$, run: 3 $-\frac{7}{5}$
31. rise: $3\frac{1}{2}$, run: $2\frac{1}{2}$ $\frac{7}{5}$

Find the slope of the line that contains each pair of points.

32. $M(9, 6), N(1, 4)$ $\frac{1}{4}$
33. $M(23, 1), N(2, 6)$ $-\frac{5}{21}$
34. $M(2, 24), N(2, 9)$ und.
35. $M(8, 10), N(8, 7)$ und.
36. $M(-2, 2), N(4, -4)$ −1
37. $M(10, 4), N(7, 4)$ 0
38. $M(-9, 16), N(-11, 16)$ 0
39. $M(2, 7), N(0, 0)$ $\frac{7}{2}$
40. $M(7, -4), N(7, 8)$ und.
41. $M(3\frac{1}{5}, 8\frac{9}{10}), N(8\frac{3}{10}, 7\frac{1}{5})$ $-\frac{1}{3}$
42. $M(23\frac{1}{5}, 29), N(1\frac{4}{5}, 8\frac{1}{5})$ $\frac{104}{107}$
43. $M(-3\frac{1}{10}, 3\frac{1}{5}), N(3\frac{1}{10}, -3\frac{2}{5})$ $-\frac{33}{31}$

CHALLENGE

44. A line passes through the origin and the point (c, d). What is the slope of the line? $\frac{d}{c}$

45. A line passes through the point (a, b) and the point (c, d). What is the slope of the line? $\frac{d-b}{c-a}$ or $\frac{b-d}{a-c}$

CONNECTION

46. **GEOMETRY** Find or draw at least four pictures of houses or buildings that have roofs with different slopes. Use a ruler to find the slope from the rise and run.

APPLICATIONS

47. **PHYSICS** If the temperature of a beverage changes from 200°F at a time of 0 minutes to 150°F at 10 minutes, what is the rate of change? In other words, what is the slope of the line connecting the points (0, 200) and (10, 150)? −5

≈ 0.018 or 96 feet per mile.

48. **GEOGRAPHY** Denver's elevation is 1 mile (5280 feet). The elevation of Loveland Pass, 70 miles west of Denver, is 11,990 feet. What is the average slope of the land from Denver to Loveland Pass? (Use 70 miles to approximate the run)

46. Answers may vary. Students should show roofs with different steepness, and calculate slope by measuring $\frac{rise}{run}$.

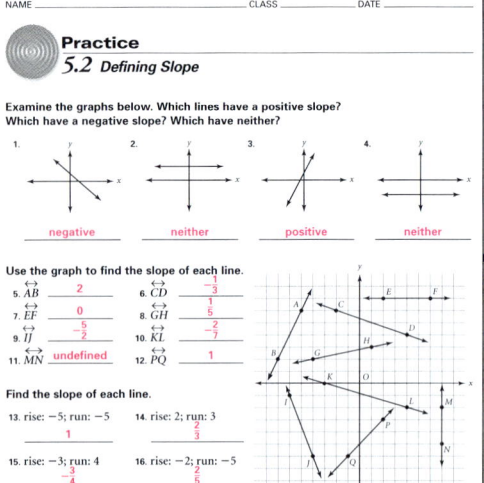

232 LESSON 5.2

CULTURAL CONNECTION: AMERICAS Mexican pyramids are somewhat different in design from Egyptian pyramids. Mexican pyramids have stairways on each side.

49. If the height (rise) of the stairway is 27 meters and the length (run) is 25 meters, what is the slope of the stairway? $\frac{27}{25} = 1.08$

50. The average rise of a step on the pyramid is 29.7 centimeters, and the average run is 27.5 centimeters. Find the slope of a step, and compare it with the slope of the stairway.
1.08; The slope of a step is the same as the slope of the stairway.

There are 91 steps in each of the 4 stairways on the Mayan-Toltec pyramid at Chichén Itzá, in the Yucatán.

Look Back

Evaluate. *(LESSON 2.1)*

51. $|-7|$ 7   **52.** $|50|$ 50   **53.** $-|29|$ −29   **54.** $-|-99|$ −99

55. If there are 8 people contributing to a charity, how much would each person have to contribute to raise $98? *(LESSON 2.4)* $12.25 per person

Solve for x. *(LESSON 3.3, 3.4, AND 3.6)*

56. $3x + 16 = 19$ $x = 1$ **57.** $28 = -4 + 4x$ $x = 8$ **58.** $4x + 5 = 2x + 5$ $x = 0$

59. $-2x + 10 = 5x$ **60.** $a + x = 3x + b$ **61.** $2x - 7 = a + 8$

$x = \frac{10}{7}$ $x = \frac{a-b}{2}$ $x = \frac{a+15}{2}$

Look Beyond

62. Graph $y = 4x$, $y = 4x - 2$, and $y = 4x + 3$ on the same axes. Find the slope of each line. What do you notice about the lines?

63. Graph $y = 3x - 2$ and $y = -\frac{1}{3}x - 2$ on the same axes. Find the slope of each line. What do you notice about the lines?

Include this activity in your portfolio.

Use a strip of paper or cloth tape measure to measure the diameter and circumference of several round objects. Record the measurements in a table. Then graph the data with the diameters on the x-axis and the circumferences on the y-axis. Fit the data with a straight line. How well does the line fit the data? Find the slope of the line. From this data and graph, what conjectures can you make about the relationship between diameter and circumference?

WORKING ON THE CHAPTER PROJECT
You should now be able to complete Activity 1 of the Chapter Project on page 264.

62.
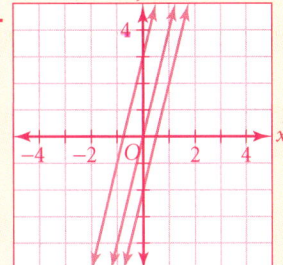

The slope for all three lines is 4.
The lines are parallel; they never intersect.

63.
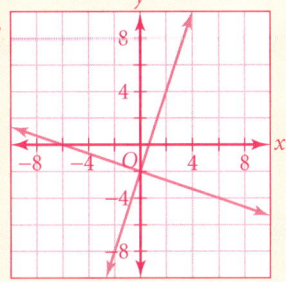

The slope for $y = 3x - 2$ is 3.
The slope for $y = -\frac{1}{3}x - 2$ is $-\frac{1}{3}$.
The lines are perpendicular; they intersect at $(0, -2)$.

ALTERNATIVE Assessment

Portfolio Activity

The Portfolio Activity can be used as preparation for the Chapter Project or as a separate activity. In the Portfolio Activity on this page, students measure the circumference and diameter of several circular objects and then graph the results. They should observe that the relationship between circumference and diameter is linear.

Answers to Portfolio Activities can be found in Additional Answers of the Teacher's Edition.

internetconnect

GO TO: go.hrw.com
KEYWORD: MA1 Golden Ratio

To support the Portfolio Activity, students may visit the HRW Web site, where they can utilize the Fibonacci sequence to find the golden ratio and explore some of its many real-world applications.

Student Technology Guide

5.2 Defining Slope

You can use a calculator to find the slope of the line containing two given points. Be sure to enter the coordinates in the correct order.

Example: Find the slope of the line containing $M(-3, 4)$ and $N(5, 7)$. Choose M as the first point, with coordinates (x_1, y_1). Choose N as the second point, with coordinates (x_2, y_2).

slope $= \frac{y_2 - y_1}{x_2 - x_1} = \frac{7 - 4}{5 - (-3)} = (7 - 4) \div (5 + 3)$

Press (7 − 4) ÷ (5 + 3) ENTER.

The slope is 0.375. To display this result as a fraction, press MATH MATH 1:▶Frac ENTER. The equivalent fraction is $\frac{3}{8}$.

Example: Find the slope of the line containing $M(3, 7)$ and $N(3, -2)$.

slope $= \frac{y_2 - y_1}{x_2 - x_1} = \frac{-2 - 7}{3 - 3} = (-2 - 7) \div (3 - 3)$

Press (− 2 − 7) ÷ (3 − 3) ENTER.

The calculator reports an error because division by zero is not possible. The line is vertical, and its slope is undefined.

Press ENTER to go on to another calculation.

Use a calculator to find the slope of the line that contains each pair of points. If the slope is not an integer, give your answer as a fraction. If the slope is undefined, write *undefined*.

1. $M(8, 1)$, $N(2, 6)$ $-\frac{5}{6}$
2. $M(-4, 6)$, $N(-4, 3)$ undefined
3. $M(2, -4)$, $N(5, 8)$ 4
4. $M(12, 17)$, $N(-5, 13)$ $\frac{4}{17}$
5. $M(11, -3)$, $N(8, 5)$ $-\frac{8}{3}$
6. $M(-4, 17)$, $N(3, 15)$ $-\frac{2}{7}$
7. $M(2.6, 3.9)$, $N(5.2, -1.3)$ -2
8. $M(7.4, 3.1)$, $N(7.4, 9.5)$ undefined
9. $M(-6.8, 3.1)$, $N(-4.3, 10.6)$ 3

Eyewitness Math

Focus

An architect's mission to make stairways safer sparks a debate about the balance between safety and cost. Students explore the mathematics of stairway design in order to take part in the debate.

Motivate

Before students read the articles, be sure they understand the terms *tread* and *riser*, which may be unfamiliar.

After reading the first article, discuss the issues raised. Ask questions such as the following:

- Why does the architect want building codes changed? **He feels new specifications will ensure a greater degree of safety on stairs.**
- Why does the National Association of Home Builders want building codes to remain unchanged? **They believe a change will increase the cost of building a house.**

Discuss the fact that a balance between safety and cost is necessary because it may not always be practical to make something completely safe. Point out that there may be a point of diminishing returns. That is, in trying to make something a little safer, the costs may be increased to the point where the item is too expensive for the average person to afford it.

Tell students that the second article describes the mathematics of stair design in greater detail. After reading this article, students should have a greater understanding of the issues underlying the debate.

EYEWITNESS MATH

Watch Your Step!

Cutaway diagram of steps of stairway
TREAD RISER

Atlanta Architect Steps Up Quest for Safe Staircases

Numerous Accidents, Deaths Push Researcher into Career

ATLANTA— ... Every year, 1 million Americans seek medical treatment for falls on staircases. About 50,000 are hospitalized and 4,000 die.

The United States isn't unique. In Japan, more people die on stairs than in fires, architect John Templer said. Yet stair safety has largely been ignored. Templer's interest began as a student at Columbia University, when someone asked him why people were always falling down the steps outside Lincoln Center. Templer discovered there was no research on stair safety. Nobody even counted falls.

Curious, he visited Lincoln Center with his family, and his sister-in-law tripped on the steps. Templer called his mentor to propose the topic as a thesis and learned that the man had just broken his leg falling down a stairwell of a subway station.

A career was born.

"All stairs are dangerous; it's a matter of degree," Templer said. "There are ways to mitigate that danger, if we could get that message to people."

Stairs evolved from ladders and first were used for defense. Narrow winding staircases, for instance, hampered intruders. Europeans wrought stairs into works of art, building grand palace staircases that gradually steepened, forcing visitors into a slow, stately pace as they approached royalty.

Most stairs now have 9-inch treads and $8\frac{1}{2}$-inch risers, a size determined around 1850, Templer said. But people today have bigger feet that hang over the edges of stairs, throwing them off balance, he said.

Stairs also are too high, Templer concluded after experiments in which he forced volunteers to trip on collapsible stairs. They were harnessed so they didn't tumble all the way to the floor, but Templer used videos to simulate how they would have landed.

He wants building codes revised for stairs with 11-inch treads and 7-inch risers. His proposal prompted a lobbying blitz from the National Association of Home Builders, which contends that larger stairs would add at least 150 square feet and $1,500 in costs to a house.

NAHB's Richard Meyer dismissed Templer's work, saying people fall when stairs are improperly lighted, have loose carpeting, or have objects placed in the way.

That's true, too, Templer said. But he said his experiments, funded by the National Science Foundation, prove stair shape is a large problem.

By Lauran Neergaard
—THE ASSOCIATED PRESS

Count Your Way to Stair Success

The riser height (rise) and the tread width (run) determine how comfortable the stairs will be to use. If the rise and run are too great, the stairs will strain your legs and be hard to climb. If the rise and run are too small, you may whack your toe on the back of each step.

Over the years, carpenters have determined that tread width times riser height should equal somewhere between 72 and 75 when the measurements are in inches.

On the main stair, the maximum rise should be no more than $8\frac{1}{4}$ inches and the minimum run should be no less than 9 inches.

To determine how many steps, or treads you need, measure from the top of the finished floor on the lower level to the top of the finished floor on the upper level.

To figure the rise and run in a house with 8-foot ceilings, for instance, start by figuring the total vertical rise. By the time you add floor joists, subfloor, and finish floor, the total is about 105 inches.

A standard number of treads in a stair between first and second floors is 14. One hundred-five divided by 14 equals $7\frac{1}{2}$. That means the distance from the top of each step to the top of the next step will be $7\frac{1}{2}$ inches.

With a riser height of $7\frac{1}{2}$ inches, tread width (run) should be at least 9 inches. Ten inches is a more comfortable run; when you multiply $7\frac{1}{2}$ inches by 10 inches, you get 75—within the conventional *ratio* of 72 to 75. With fourteen 10-inch treads, the total run of the stairs will be 140 inches. In other words, the entire stair will be 105 inches tall and 140 inches deep.

You can alter the rise and run to some extent. If you used 15 risers instead of 14, for instance, the rise would be 7 inches, and the tread width would be $10\frac{1}{2}$ inches (7 times $10\frac{1}{2}$ equals 73.5, within the rise and run guidelines).

By Karol V. Menzie and Randy Johnson
—BALTIMORE SUN

Cooperative Learning

1. In the news article above, is the term *ratio* used correctly? Explain.
2. Find the angle of each of the following staircases. Draw the triangle and measure the angle.
   a. The 105-inch-by-140-inch staircase described in the article above.
   b. The same staircase changed as described in the article to include 15 risers instead of 14.
   c. The same staircase changed to meet Templer's preference for 7-inch risers and 11-inch treads.
3. Which of the three staircases in Step 2 would take up the most room? Explain.
4. Suppose that you did not want the building code for stairs to change. Write an argument (with numbers and a diagram) to show that larger stairs would add at least 150 square feet to a typical house. Make up your own estimates of the size of a typical house.
5. Measure the rise and run of stairs in your school or home. Do they match the information in the two news articles about standard stairs?

---

1. No; 72 to 75 is actually the range of a product (tread width × riser height). In the subsequent paragraph, the article refers to this product more appropriately as the "rise and run guidelines."
2. a. ≈ 37°
   b. ≈ 34°
   c. ≈ 32.5°
3. Stairway c; because it has the smallest slope, it will extend farthest into the room.
4. Answers may vary. For example, stairway c is about 2 feet longer than stairway a. To make up for that lost space, the room would need to be about 2 feet longer. In a house 40 feet wide, that's an additional 80 square feet. Since placing the wall 2 feet further out also affects the second story, that's another 80 additional square feet, for a total extra area of 160 square feet.
5. Answers may vary.

Cooperative Learning

Have students work in pairs. In Step 1, partners can share their perceptions of the use of the word *ratio*. For Steps 2 and 3, suggest that the partners first answer the questions individually and then get together to compare results and resolve any discrepancies.

In Step 4, partners can first work together to brainstorm ideas and then work independently to formulate individual arguments.

For Step 5, each partner should measure a different set of stairs. The partners can then work together to determine whether each set of stairs meets the criteria outlined in the article.

Discuss

Bring the class together to share results. Have several students present the arguments they prepared for Step 4. Discuss similarities and differences among the arguments. See if it is possible for the entire class to arrive at a consensus.

Have students share their measurements of stairs in the school or in their homes.

Prepare

NCTM PRINCIPLES & STANDARDS 1–6, 8–10

LESSON RESOURCES

- Practice 5.3
- Reteaching Master 5.3
- Enrichment Master 5.3
- Cooperative-Learning Activity 5.3
- Lesson Activity 5.3
- Problem Solving/Critical Thinking Master 5.3
- Teaching Transparency 26

QUICK WARM-UP

Solve each equation.

1. $15t = 3$ $t = 0.2$
2. $42 = 0.6m$ $m = 70$
3. $(12)(9) = d$ $d = 108$
4. $r = (8.4)(1.5)$ $r = 12.6$

Also on Quiz Transparency 5.3

Teach

Why The graph of a function shows the relationship between two variables. The rate of change of the graph is the slope of the graph at a given point. The rate of change can describe things such as speed (miles per hour) and pay rate (dollars per hour).

5.3 Rate of Change and Direct Variation

When aloft, commercial planes maintain a fairly constant speed. This makes it possible to model their motion during a flight with a linear graph.

Objective
- Find the rate of change from a graph.
- Solve and graph direct-variation equations.

Why Rate of change and direct variation are two concepts directly related to slope, which has many real-world applications.

Rate of Change

APPLICATION
AVIATION

The graph at right shows the relationship between the variables for a plane moving at a constant rate. Time is represented by the horizontal axis, and distance traveled is represented by the vertical axis.

The **rate of change**, or speed, of the plane can be determined from the graph. After 20 minutes, the plane has traveled 200 miles. Use a formula for rate of change as follows:

$$\text{rate of change} = \frac{\text{change in distance}}{\text{change in time}} = \frac{200}{20} = \frac{10}{1}$$

The plane travels 10 miles for every 1 minute that it travels.

Notice that the rate of change was determined by calculating the rise, 200 miles, over the run, 20 minutes. In fact, the slope of a linear graph is the same as the rate of change of the linear function that it represents.

Alternative Teaching Strategy

USING TABLES Have students complete the following tables.

$y = 2x$		
x	y	$\frac{y}{x}$
4	8	2
5	10	2
8	16	2
10	20	2
20	40	2

$y = 2x + 10$		
x	y	$\frac{y}{x}$
4	18	4.5
5	20	4
8	26	3.25
10	30	3
20	50	2.5

Ask students how the ratios in the third column of each table differ. Elicit the response that the ratios in the first table are constant, while the ratios in the second table are decreasing. Have students write another equation that they think will behave in the same way as $y = 2x$. Tell them to make a table of values to test their conjecture. Discuss the results. Students should observe that any equation of the form $y = kx$ will yield a constant ratio that is called the *constant of variation*. Note that this constant of variation is represented by the value of k in the equation.

236 LESSON 5.3

Some other examples of rates of change include the following:

60 miles per hour
1.5 children per marriage
1 mile per minute

4.5 police officers per 1000 people
1 birth per 4 hours
20 cases of lung cancer per 1000 smokers

EXAMPLE 1

**APPLICATION
TRANSPORTATION**

The graph shows the distance a car travels at a constant speed. **What is the speed of the car?**

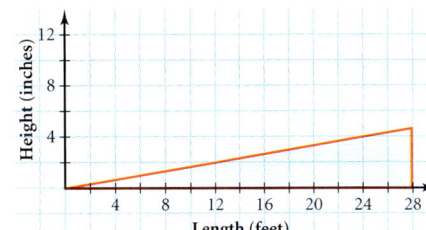

● **SOLUTION**

Select two points, such as the ones shown on the graph. Use the slope formula.

Let (x_1, y_1) be $(1.0, 60)$.
Let (x_2, y_2) be $(2.0, 120)$.

$$\frac{\text{change in distance}}{\text{change in time}} = \frac{(y_2 - y_1)}{(x_2 - x_1)} = \frac{120 - 60}{2.0 - 1.0} = \frac{60 \text{ miles}}{1.0 \text{ hour}}$$

The car travels 60 miles for every 1 hour it travels.

TRY THIS

Find the rate of change of the ramp represented in the drawing at right. How many inches does the ramp rise for every foot of length?

EXAMPLE 2

**TECHNOLOGY
GRAPHICS
CALCULATOR**

The graph shown at right was made from data collected by a motion detector. In the experiment, a student walked in a straight line away from the motion detector. **Describe the movement of the student by identifying the rates of change shown by the graph.**

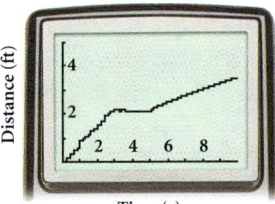

Time in seconds is indicated on the horizontal axis and distance in feet is indicated on the vertical axis.

● **SOLUTION**

The graph of the data can be divided into three segments. Using the formula rate of change = $\frac{\text{change in distance}}{\text{change in time}}$, the rate of change for the first segment is $\frac{2}{3}$ feet per second. The middle segment is a horizontal line, so the rate of change is 0 feet per second. The rate of change for the last segment in the graph is approximately $\frac{1}{3}$ foot per second. Thus, the student walked away from the motion detector at $\frac{2}{3}$ feet per second for 3 seconds. Then the student stood still for 2 seconds. Finally, the student walked away from the motion detector at $\frac{1}{3}$ foot per second for 4 seconds.

Some calculators can be used with equipment like a motion detector to collect data for analysis.

Interdisciplinary Connection

GEOGRAPHY Suppose that the scale on a map of northeastern Africa is 1 inch : 200 miles. Write an equation that relates the number of miles of actual distance, m, to the number of inches of map distance, i. **$m = 200i$** What is the constant of variation? **200 miles per inch** What actual distance is represented by $2\frac{3}{8}$ inches of map distance? **475 miles** What map distance would represent an actual distance of 350 miles? **$1\frac{3}{4}$ inches**

ADDITIONAL EXAMPLE 1

The graph below shows the distance a certain train travels at a constant speed. **What is the speed of the train?**

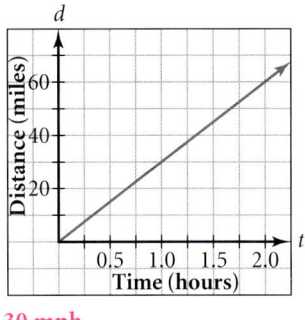

30 mph

TRY THIS

$\frac{\text{change in height}}{\text{change in length}} = \frac{4}{24} = \frac{1}{6} \approx 0.17$

The ramp rises about 0.17 inches for every foot.

ADDITIONAL EXAMPLE 2

Describe the movement of a person by identifying the rates of change shown on the graph below.

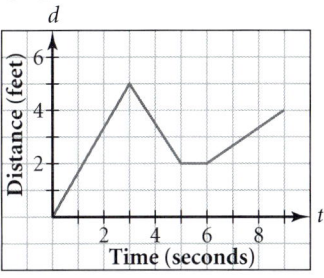

The rate of change of the 4 segments, in feet per second, is $\frac{5}{3}$, $-\frac{3}{2}$, 0, and $\frac{2}{3}$ respectively. Thus, the person walked forward at $1\frac{2}{3}$ feet per second for 3 seconds, then walked backward at $1\frac{1}{2}$ feet per second for 2 seconds, stood still for 1 second, and then moved forward again at $\frac{2}{3}$ feet per second for 3 seconds.

☞ For a Teaching Tip about the use of technology, see page 238.

LESSON 5.3

Teaching Tip

TECHNOLOGY A motion detector is a device that can be connected to a graphics calculator to record changes in motion. The device measures the distance that an object is from the detector. This distance is transmitted to the graphics calculator. When properly programmed, the graphics calculator graphs the distances. Stress to students that a positive slope indicates motion away from the detector; a slope of zero indicates no change in distance, which is a lack of motion; and a negative slope indicates motion toward the detector.

ADDITIONAL EXAMPLE 3

If y varies directly as x and $y = 8$ when $x = 4$, find the constant of variation and write an equation of direct variation.

$k = 2; y = 2x$

TRY THIS
$k = \frac{10}{2} = 5; y = 5x$

ADDITIONAL EXAMPLE 4

If y varies directly as x and $y = 36$ when $x = 9$, find x when $y = 48$.

$x = 12$

TRY THIS
$\frac{35}{7} = \frac{y}{84}; y = 420$

Direct Variation

A Japanese high-speed train passes near Mount Fuji.

One type of rate of change is direct variation. An example of a rate of change that is a direct variation is average speed. For example, using t for time in minutes and d for distance in miles, the equation for a high-speed train traveling at 3 miles per minute can be written as follows:

$$d = 3t$$

Notice that 3 is a constant. In terms of direct variation, you can state the equation as follows:

"The distance traveled by a high-speed train moving at a constant speed of 3 miles per minute varies directly as the time traveled."

Direct Variation

If y varies directly as x, then $y = kx$, or $\frac{y}{x} = k$, where k is the **constant of variation**.

EXAMPLE 3 If y varies directly as x and $y = 4$ when $x = 0.5$, find the constant of variation and write an equation of direct variation.

● **SOLUTION**

Since $\frac{y}{x} = k$, substitute 4 for y and 0.5 for x to find the constant of variation.

$$\frac{4}{0.5} = k \qquad 8 = k \qquad k = 8$$

Use the constant of variation, k, to write an equation of direct variation. Substitute 8 for k.

$$y = 8x$$

TRY THIS If y varies directly as x and $y = 10$ when $x = 2$, find the constant of variation and write an equation of direct variation.

EXAMPLE 4 If y varies directly as x and $y = 27$ when $x = 6$, find x when $y = 45$.

● **SOLUTION**

In a direct variation, all values of $\frac{y}{x}$ must equal some constant, k. Therefore, all values of $\frac{y}{x}$ must be equal. You can set up and solve a proportion to find x.

$$\frac{27}{6} = \frac{45}{x}$$
$$27x = 270$$
$$x = 10$$

TRY THIS If y varies directly as x and $y = 35$ when $x = 7$, find y when $x = 84$.

Inclusion Strategies

TACTILE LEARNERS Provide students with a ruler; a measuring cup of water, marked in ounces; and a cylindrical glass container, such as a cylindrical drinking glass. Tell them to pour 1 ounce of water into the container at a time. After each ounce, they should record the height of the water in the container. Have them explain how the experiment relates to direct variation, identify the constant of variation, and graph the function. **The height of the water in the container varies directly as the number of ounces of water. Constants and graphs will vary.**

Enrichment

Have students work in groups of three or four to brainstorm a list of relationships that can be modeled by direct-variation equations. Urge them to consider relationships they have encountered in other school subjects. Challenge them to find a relationship in which the constant of variation is negative. Then have them write a direct-variation equation for each relationship on the list, taking care to define each variable and to identify the constant of variation.

238 LESSON 5.3

EXAMPLE 5

APPLICATION
PHYSICS

Hooke's law relates the distance a spring stretches to the amount of force applied to the spring. Hooke's law is expressed mathematically as follows:

$$F = kd$$

In the equation, the variable d is the distance of the stretch in meters and the variable F is the amount of the force in Newtons. (A force of one Newton is equivalent to the weight of one medium-sized apple.) The letter k is a constant, meaning that k has a definite value that does not change or vary in the equation for a particular spring. The actual value of k for a given spring is determined by experiment.

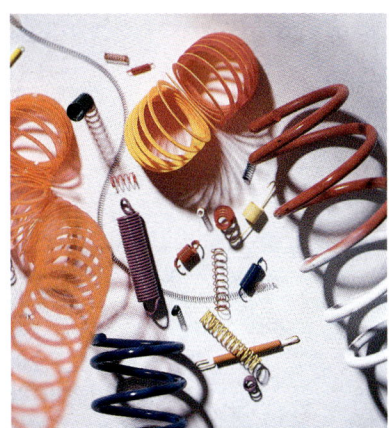

A force of 20 Newtons stretches a spring a distance of 0.5 meter. **Find the constant, k, for the spring.**

SOLUTION

Substitute the values for F, the force applied, and d, the distance of the stretch, into the equation for Hooke's law. Then solve for k.

$F = kd$
$20 = k \cdot (0.5)$
$k = 20 \div 0.5$
$k = 40$

The spring constant is 40.

TRY THIS A spring with a spring constant, k, of 28 is stretched a distance of 0.25 meter by a lead weight. Find the force in Newtons exerted by the lead weight.

Activity
Exploring the Graphs of Direct Variations

You will need: graph paper

1. Complete the table of values below for the equation $y = 0.5x$.

x	0	1	2	3	4	5
y	?	?	?	?	?	?

2. Use the points in the table to draw a graph of the direct variation.
3. What is the shape of the graph? Does it pass through the origin?
4. Select two points on the graph and use them to find the slope of the graph.

CHECKPOINT  5. Repeat Steps 1–4 with several different constants for the direct variation. Make conjectures about the shape of the graph, whether it passes through the origin, and its slope.

The graph of a direct variation is a straight line that passes through the origin. The constant of variation is the slope of the graph.

Reteaching the Lesson

GUIDED ANALYSIS Write the following on the board or overhead:

Suppose that y varies directly as x.
If $y = 24$ when $x = 6$, find y when $x = -4$.
If $y = 24$ when $x = 6$, find x when $y = -4$.

Ask students to compare the two problems. Ask how they are alike and how they are different. As students respond, urge them to use language such as *direct variation* and *constant of variation*. Now work with them to solve each problem, taking care to write the solutions side by side. Then compare the solutions. $y = -16; x = -1$

ADDITIONAL EXAMPLE 5

A force of 4 Newtons stretches a spring a distance of 0.25 meters. **Find the constant, k, for the spring.**
16

TRY THIS
$F = kd = 28(0.25) = 7$ Newtons

Use Teaching Transparency 26.

Activity Notes

In this Activity, students investigate the graph of a direct-variation equation. Students should arrive at the conclusion that the graph of any direct-variation equation is a line that passes through the origin. When the equation is written in the form $y = kx$, k is not only the constant of variation, but also the slope of the graph.

Cooperative Learning

You may wish to have students do the Activity in pairs. Suggest that each student do Steps 1–4 individually, and then have partners compare and discuss results. For Step 5, have partners assign each other a constant of variation. Suggest that they again compare results and then work together to formulate a conjecture.

CHECKPOINT ✓
5. Answers may vary. All graphs should be lines through the origin with slopes equal to the constants of variation.

LESSON 5.3

Assess

Selected Answers
Exercises 5–10, 11–53 odd

ASSIGNMENT GUIDE

In Class	1–10
Core	11–31 odd, 32–34
Core Plus	12–26 even, 27–36
Review	37–53
Preview	54

 Extra Practice can be found beginning on page 738.

Mid-Chapter Assessment for Lessons 5.1 through 5.3 can be found on page 61 of the *Assessment Resources*.

---

internetconnect

GO TO: go.hrw.com
KEYWORD: MA1 Import Export

In this activity, students study slope as a rate of change to investigate the change in imports and exports of semi-conductors between the United States and Eastern Asia.

Exercises

Communicate

1. Allen is riding his mountain bike over three different types of terrain. Examine the graph. When is he riding uphill? downhill? on level ground?

2. Describe the slope of a hill in terms of a rate of change.

3. For a direct-variation equation, explain how the slope of the graph and the constant of variation are related.

4. **FUNDRAISING** At the homecoming football game, the senior class makes $1.50 for every program sold. Describe the direct variation for the amount of money the senior class will make from selling football programs.

Guided Skills Practice

5. The graph at right shows the cost of renting videos. What is the rate of change, or cost per video?
   (EXAMPLE 1) $3 per video

APPLICATIONS

INCOME The graph below represents an employee's daily wages as a function of the number of hours that he works in a day. **(EXAMPLE 2)**

6. Use the graph to find the employee's hourly rate of pay if he works no more than 8 hours per day. **$10 per hour**

7. Use the graph to find his hourly rate of pay after he has worked 8 hours in a given day (overtime pay). **$15 per hour**

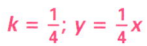 $k = \frac{1}{4}$; $y = \frac{1}{4}x$

8. If y varies directly as x and $y = 7$ when $x = 28$, find the constant of variation and write an equation of direct variation. **(EXAMPLE 3)**

9. If y varies directly as x and $y = 30$ when $x = 15$, find x when $y = 50$.
   (EXAMPLE 4) $x = 25$

10. **PHYSICS** A force of 8 Newtons stretches a spring a distance of 0.25 meter. Find the constant, k, for the spring. **(EXAMPLE 5)** $k = 32$

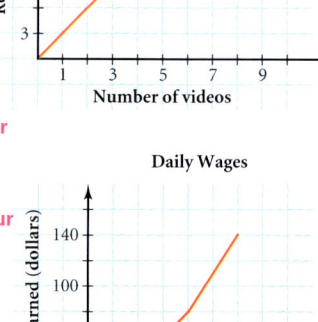

240 LESSON 5.3

Practice and Apply

APPLICATIONS

11. **RENTALS** The graph at right shows the monthly cost of renting videos, including an initial membership fee. Find the rate of change, or cost per video. **$2.50 per video**

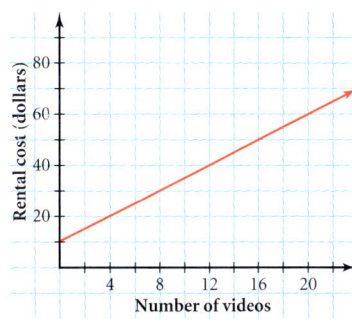

12. **INCOME** The graph at right shows Amy's wages as a function of the number hours she works in a week. Use the graph to find Amy's hourly pay rate if she works no more than 40 hours in a week and after she has worked more than 40 hours in a given week (overtime pay).
$12 per hour to 40
$18 per hour after 40

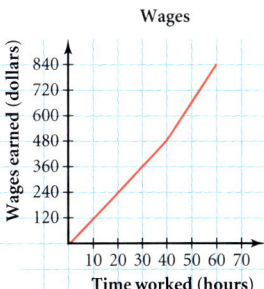

For the following values of x and y, y varies directly as x. Find the constant of variation, and write an equation of direct variation.

13. $k = 7$; $y = 7x$
14. $k = 0.2$; $y = 0.2x$
15. $k = 0.625$; $y = 0.625x$
16. $k = 0.6$; $y = 0.6x$
17. $k = 11$; $y = 11x$
18. $k = -0.4$; $y = -0.4x$
19. $k = 0.3$; $y = 0.3x$
20. $k = 4$; $y = 4x$
21. $k = 7$; $y = 7x$
22. $k = 0.25$; $y = 0.25x$

13. y is 14 when x is 2.
14. y is 2.6 when x is 13.
15. y is 5 when x is 8.
16. y is 3 when x is 5.
17. y is 132 when x is 12.
18. y is -2 when x is 5.
19. y is 4.5 when x is 15.
20. y is 84 when x is 21.
21. y is 84 when x is 12.
22. y is 2 when x is 8.

For the following values of x and y, y varies directly as x. Find the missing value.

23. y is 14 when x is 2. Find x when y is 21. **$x = 3$**
24. y is 5 when x is 8. Find y when x is 28. **$y = 17.5$**
25. y is 56 when x is 7. Find x when y is 8. **$x = 1$**
26. y is 27 when x is 3. Find x when y is 4.5. **$x = 0.5$**

CONNECTIONS

27. **GEOMETRY** The perimeter, p, of a square varies directly as the length, l, of a side. Write an equation of direct variation that relates the two variables. **$p = 4l$**

28. **GEOMETRY** If the base area of a cylindrical container is kept constant, the volume of the container varies directly as the height. The table shows the volume and height for various cylinders with the same base area. Find the constant of variation. Then find the volume if the height is 12 centimeters.

Height (cm)	3	4	5	6
Volume (cm³)	27	36	45	54

$k = 9$; $V = 108$ cm³

Technology
A calculator may be helpful for verifying answers in many of the exercises. You may want to suggest that students solve the exercises on paper, but also write their results in calculator-ready form.

Error Analysis
When solving direct-variation problems, students often identify the constant of variation correctly, but then automatically multiply the given value, whether it is x or y, by the constant. For example, consider the problem "If y varies directly as x and $y = 6$ when $x = 12$, find x when $y = 16$." Many students easily identify 0.5 as the constant of variation. However, they may proceed to write $x = (0.5)(16) = 8$. Stress the importance of writing the general equation $y = kx$ at each step and writing the known values—x, y, or k—directly beneath it to determine whether multiplication or division is required.

LESSON 5.3 **241**

Exercise 54 previews the graphs of a quadratic function, a reciprocal function, and a square-root function. Students will study the graphs of these functions in detail in Chapters 10 and 12.

29. $y = 8x$

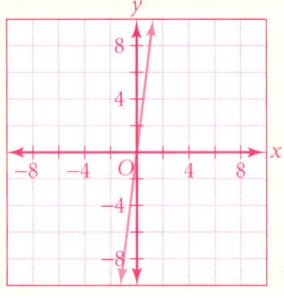

30. $f = 2000m$

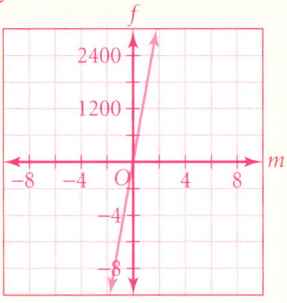

APPLICATIONS

29. INCOME Suppose that the hourly wage at a job is $8. Let *x* represent the number of hours, and let *y* represent the total amount earned. Write an equation of direct variation, and graph the equation.

30. AVIATION An airplane ascends at a rate of 2000 feet per minute. Write an equation of direct variation, and graph the equation.

31. HEALTH Suppose that you are walking at 4 miles per hour. Let *x* represent the time you spend walking, and let *y* represent the distance you walk. Write an equation of direct variation, and graph the equation.

RECREATION On the second day of a hike, two hikers start out from different camps at the same time. The graph shows the hikers' progress on day two.

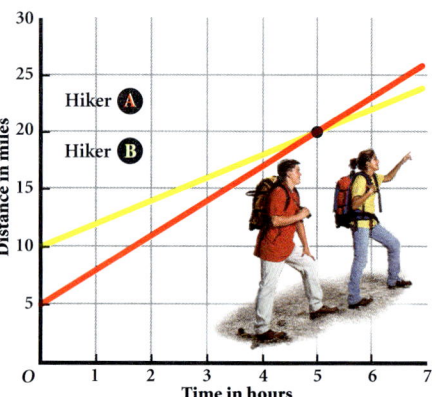

3 miles per hour **32.** Find the rate of travel on day two for hiker A.

2 miles per hour **33.** Find the rate of travel on day two for hiker B.

34. After how many hours on day two will the hikers have traveled the same total distance? **5 hours**

CHALLENGE

35. After how many hours on day two will hiker A be 3 miles ahead of hiker B in total distance? **After 8 hours**

APPLICATION

36. PHYSICS Find a sturdy rubber band or spring and suspend it so that you can measure its length without any weight on it. Find several weights that the rubber band or spring will hold. Hang the weights on the rubber band or spring, and measure the distance of the stretch. Make a table of the weights and the distances of the stretches. Make a scatter plot, and find the line of best fit. Use a graphics calculator, if available, to find the equation of the line of best fit. Write a description of your findings.

Look Back

Graph each point on a coordinate plane. (LESSON 1.4)

37. $A(-2, 3)$ **38.** $B(0, 8)$ **39.** $C(-5, -1)$

Make a table for each equation and find values of *y* by substituting 1, 2, 3, 4, and 5 for *x*. (LESSON 1.5)

40. $y = 7x$ **41.** $y = \frac{1}{2}x - 3$

42. A ski boat crosses a lake at an average speed of 35 mph. Write an equation, make a table, and draw a graph for the distance the boat travels for times from 0 to 3 hours. (LESSON 1.5)

Use the Distributive Property to show that the equations below are true. (LESSON 2.5)

43. $-7x + 11x = 4x$ **44.** $8h - 9h = -h$ **45.** $3d + 5d = 8d$

Practice

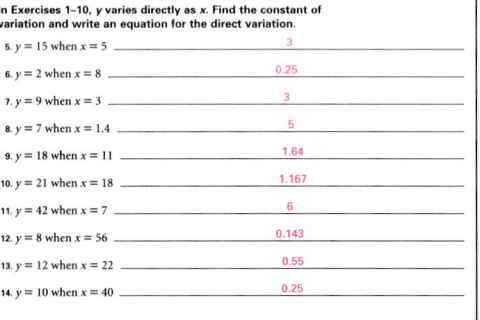

31. $y = 4x$

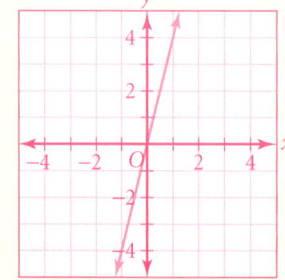

36. Answers may vary. Students should find that the line of best fit is of the form $y = kx$, where *k* is the constant of the rubber band or spring.

37–39.

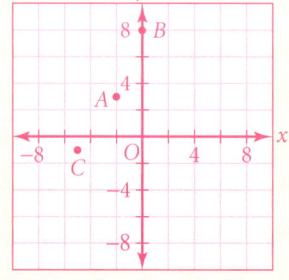

The answers to Exercises 40–45 and 54c can be found in Additional Answers beginning on page 810.

Simplify. *(LESSON 2.5)*

46. $(3x + 7) + (4x - 9)$ **$7x - 2$**
47. $(9c + 3) - (-8c + 2)$ **$17c + 1$**
48. $(-3d - 12) - (-2d + 4)$ **$-d - 16$**
49. $(6h - 2) - (4h + 3)$ **$2h - 5$**
50. $(-5t + 2) + (t - 12)$ **$-4t - 10$**
51. $(f + 3) - (6f - 45)$ **$-5f + 48$**
52. $(2a + b - c) + (a - 3b + 5c)$
    $3a - 2b + 4c$
53. $(b + 9f - 2g) - (5b - 4f + 12g)$
    $-4b + 13f - 14g$

54. Graph each function below on a coordinate plane. Let the horizontal axis represent x and the vertical axis represent $f(x)$. Describe each graph. Are the graphs linear?

    a. $f(x) = 2x^2$ **b.** $f(x) = \frac{2}{x}, x \neq 0$ **c.** $f(x) = \sqrt{x}, x \geq 0$

A rain gauge is used to determine rainfall amounts.

Include this activity in your portfolio.

1. Describe three sets of real-world data that can be modeled as ordered pairs on a coordinate plane (for example, regional rainfall amounts per month).

2. Pick one of your sets of data from Step 1 and draw a graph. Connect the data points on your graph. Describe the shape of the line connecting the points. Is it straight, curved, or jagged? (Recall that drawing a line to connect points is a tool to help visualize the relationship and does not mean that the relationship holds true for every point on the line.)

3. Using the data you graphed in Step 2, redraw the graph without connecting the data points—this creates a scatter plot of your data.

4. Use a clear ruler to draw a line of best fit for your data.

5. Make a prediction for a data point that is not on your graph. For example, if you chose to model monthly rainfall totals, predict the rainfall for the next month.

6. Explain why you are or are not able to make a prediction from your graph. Do you have confidence in your prediction? Why or why not?

WORKING ON THE CHAPTER PROJECT

You should now be able to complete Activity 2 of the Chapter Project on page 264.

54. a.
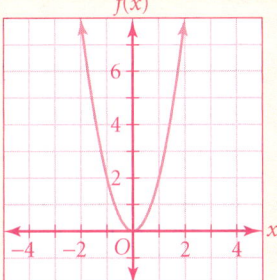

The graph is not linear. It looks like a U, with the lowest point at the origin.

b.
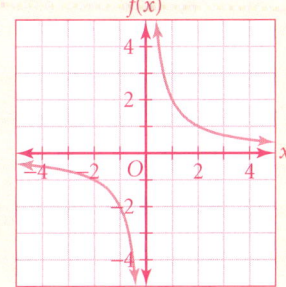

The graph is not linear. It has two separate pieces, both curved, and never touches either axis.

LESSON 5.3 243

Prepare

NCTM PRINCIPLES & STANDARDS 1–6, 8–10

LESSON RESOURCES

- Practice 5.4
- Reteaching Master 5.4
- Enrichment Master 5.4
- Cooperative-Learning Activity 5.4
- Lesson Activity 5.4
- Problem Solving/Critical Thinking Master 5.4

QUICK WARM-UP

Find the slope of the line that contains each pair of points.

1. $A(3, 6), B(1, 12)$ −3
2. $C(0, −2), D(−6, −2)$ 0

Complete each ordered pair so that it is a solution of the equation $y = −2x + 3$.

3. $(2, \underline{?})$ −1   4. $(0, \underline{?})$ 3
5. $(\underline{?}, 11)$ −4   6. $(\underline{?}, 0)$ 1.5

Also on Quiz Transparency 5.4

Teach

Why Ask students if they need to know every characteristic of an object before it can be identified. Elicit examples from students. For example, all that is needed to identify a specific house is an address. Similarly, only two key characteristics are needed to identify an equation of a line.

244   LESSON 5.4

5.4 The Slope-Intercept Form

Objectives
- Define and explain the components of the slope-intercept form of a linear equation.
- Use the slope-intercept form of a linear equation.

Why The slope-intercept form allows you to write an equation for a line when you can identify only the slope and y-intercept.

APPLICATION
SPORTS

PROBLEM SOLVING

Kim, a skater, pays an initial $50 for club fees at the rink, plus $3 for every hour that she practices. She uses a graph like the one at right to show the pattern of her skating expenses.

A graph representing what Kim must pay for her practice time does *not* pass through the origin. This is because she has to pay an initial $50 in club fees each month before she can practice.

Examine the data. Let x be the number of hours Kim practices. Let y be the total cost for her skating expenses for the month. **Make a table** of x- and y-values from the data.

Time in hours, x	0	1	2	3	4	5	6	7	8	9	10
Cost in dollars, y	50	53	56	59	62	65	68	71	74	77	80

                3   3   3   3   3   3   3   3   3   3

Notice that the first differences are constant. Constant first differences indicate a linear pattern. Therefore, an equation that represents Kim's practice expenses is linear. In this case, the total cost is a function of the number of hours that Kim practices. She uses a time card that keeps precise track of her practice times so that the cost can also be calculated for fractional parts of hours.

To find the slope of the graph that models Kim's expenses, pick any two points from the table. For any points that you choose, the following is true:

$$\text{slope} = m = \frac{\text{change in } y}{\text{change in } x} = \frac{3}{1} = 3$$

Alternative Teaching Strategy

USING FUNCTION TABLES Give students the graph at right. For each line, have them choose five points and make a table of ordered pairs. (At right is a possible table for points on line p.) For each table, have them write a function of the form $y = mx + b$. Ask students to make a conjecture about the relationship between the function and the graph. Lead them to see that the coefficient of x is the slope of the line and that the constant is the y-intercept.

line p	
x	y
−5	−6
−4	−4
−2	0
0	4
1	5

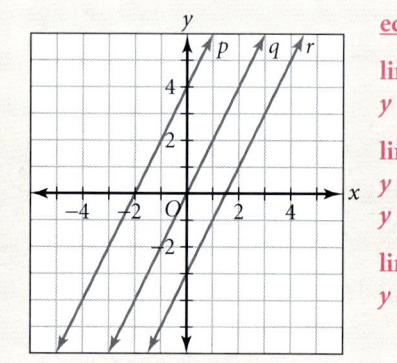

equations

line p:
$y = 2x + 4$

line q:
$y = 2x$, or
$y = 2x + 0$

line r:
$y = 2x − 3$

The slope represents the fact that Kim must pay $3 for every hour she uses the skating rink.

The table also shows that when x is 0, y is 50. Thus, (0, 50) is on the graph. The y-value of the point where the line crosses the y-axis is called the **y-intercept**. In this case, the y-intercept of 50 represents the initial $50 fee that Kim must pay.

In the study of linear equations, the y-intercept is usually represented by b and the slope by m. The graph of an equation of the form $y = mx + b$ has a slope of m and crosses the y-axis at b. If you substitute 3 for m and 50 for b, an equation that models the cost of Kim's practice becomes $y = 3x + 50$.

Slope-Intercept Form

The slope-intercept form for a line with a slope of m and a y-intercept of b is

$$y = mx + b.$$

The ordered pairs (x, y) that are solutions to the equation $y = mx + b$ represent a function. For any such function, the set of independent-variable values, in this case x, is the domain. The set of dependent-variable values, in this case y, is the range. You will have to consider what values are reasonable for the domain and range of any particular problem.

EXAMPLE 1

APPLICATION
SPORTS

The equation of the line that represents Kim's monthly skating costs is $y = 3x + 50$. **Use the slope-intercept form to construct the graph of the equation.**

● **SOLUTION**

The form is $y = mx + b$. Since b is 50, measure 50 units up from the origin on the y-axis and graph a point.

From that point, measure the rise and then the run of the slope to locate a second point.

The slope, m, is 3. You can use any equivalent of $\frac{3}{1}$ to measure the rise and run. The y-axis is marked in intervals of 10 units, so use $\frac{30}{10}$. From the initial point (0, 50), move 10 units to the right and then 30 units up. Graph the second point. Draw a line through the points.

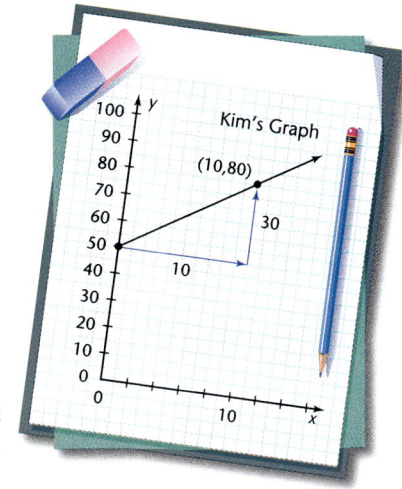

ADDITIONAL EXAMPLE 1

A cylindrical storage silo is filled with grain to a height of 5 feet. The silo is then filled with more grain so that the height rises at a constant rate of 2 feet per minute. The equation $y = 2x + 5$ expresses the height in feet of the grain, y, after the silo has been filling for x minutes. **Use the slope-intercept form to graph the equation.**

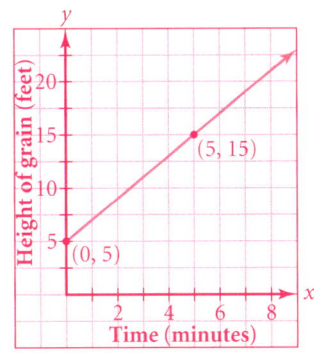

Interdisciplinary Connection

SOCIAL STUDIES Often it is possible to use a linear equation to model the growth of a population over a relatively short period of time.

The graph at right shows the growth in population of a small town over the 10-year period after a major electronics firm relocated to the town. Write an equation in slope-intercept form for the graph. $y = 2000x + 15{,}000$ What was the population at the start of the 10-year period? **15,000** What was the average rate of change for the population? **an increase of 2000 people per year**

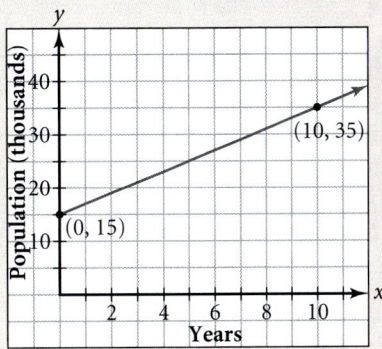

LESSON 5.4

TRY THIS

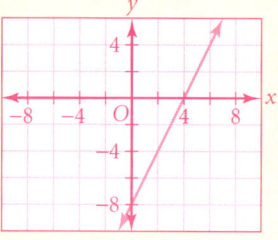

CRITICAL THINKING

The line represents cost, which will not be negative. A realistic maximum number of hours would be about 250.

Activity Notes

In this Activity, students use a graphics calculator to explore the graphs of several equations of the form $y = 2x + b$. They should observe that all of the lines have the same slope, 2, but that the differing values of b determine different y-intercepts.

You may wish to extend the Activity by having students graph several different equations of the form $y = mx + 2$. In this case, they will observe that all of the lines have the same y-intercept, 2, but that the differing values of m result in different slopes.

CHECKPOINT ✓
5. It raises or lowers the graph along the y-axis.

See Keystroke Guide, page 782.

246 LESSON 5.4

TRY THIS Use the slope-intercept form to construct the graph of the equation $y = 2x - 8$.

CRITICAL THINKING Why do you think the line in Kim's graph on page 245 does not extend past the y-axis? What do you think is a realistic maximum value for the domain? In other words, what is a realistic maximum number of hours of skating practice for one month?

Activity
Exploring Graphs of Linear Functions

TECHNOLOGY
GRAPHICS CALCULATOR

You will need: graph paper or a graphics calculator

Graph the following linear functions on the same axes:

$y = 2x$ $y = 2x + 3$ $y = 2x + 5$ $y = 2x - 3$

1. How are the graphs of these lines alike?
2. How are the graphs of these lines different?
3. Predict what the graph of $y = 2x - 4$ looks like.
4. Check your prediction by graphing.

CHECKPOINT ✓ 5. Explain how changing the value of b in an equation of the form $y = mx + b$ affects the graph of the function.

From Two Points to an Equation

When you know two points on a line, you can determine the equation for that line. First calculate the slope, m, by using the slope formula. Then calculate b by using the slope-intercept form and one of the points.

Recall from Lesson 5.2 that for two points with coordinates (x_1, y_1) and (x_2, y_2), the formula for the slope is as follows:

$$m = \frac{\text{change in } y}{\text{change in } x} = \frac{\text{difference in } y}{\text{difference in } x} = \frac{y_2 - y_1}{x_2 - x_1}$$

EXAMPLE Kim's graph shows that after 7 hours of practice the total cost for her skating is $71. It also shows that after 5 hours of practice the total cost is $65. **Write an equation in slope-intercept form for the line that models this situation.**

APPLICATION
SPORTS

● **SOLUTION**
You can represent the data as points by writing the time and cost as ordered pairs, $(5, 65)$ and $(7, 71)$. To write the equation of the line from these two points, substitute the values into the slope formula.

$$m = \frac{y_2 - y_1}{x_2 - x_1}$$

$$m = \frac{71 - 65}{7 - 5} = \frac{6}{2} = 3$$

Substitute 3 for m in $y = mx + b$.

$$y = 3x + b$$

Inclusion Strategies

VISUAL LEARNERS Have students use color coding to emphasize the connection between the slope-intercept form of an equation and the graph of the equation. Suggest that they rewrite a given equation with the coefficient of x in one color and the constant in a different color. Then, when graphing the equation, they can use the "constant color" to graph the y-intercept and the "coefficient color" to indicate the slope.

Then choose either point, and substitute the values for x and y into the equation. If you use the point $(5, 65)$, substitute 5 for x and 65 for y. Then solve for b.

$$y = 3x + b$$
$$65 = 3(5) + b$$
$$65 = 15 + b$$
$$50 = b$$

Now substitute 3 for m and 50 for b in $y = mx + b$. The equation in slope-intercept form for the line is $y = 3x + 50$.

TRY THIS Write an equation in slope-intercept form for the line containing the points $(3, 3)$ and $(5, 7)$.

CRITICAL THINKING Explain why the slope-intercept form is a helpful way to write a linear equation. Is $3x - 4y = 7$ a linear equation? Why or why not?

Finding Intercepts

The slope-intercept form makes it very easy to find the y-intercept since it is given by b in the equation $y = mx + b$. You can also use this form to find the **x-intercept**, which is the x-coordinate of the point where the line crosses the x-axis. To do this, substitute 0 for y in the equation. Then solve for x. The solution is the x-intercept.

EXAMPLE 3 Identify the x- and y-intercepts of each line.
  a. $y = -3x + 12$ **b.** $y = 2x - 18$

● **SOLUTION**

a. The y-intercept is 12.

To find the x-intercept, substitute 0 for y in the equation. Then solve.

$$y = -3x + 12$$
$$0 = -3x + 12$$
$$3x = 12$$
$$x = 4$$

The x-intercept is 4.

b. The y-intercept is -18.

To find the x-intercept, substitute 0 for y in the equation. Then solve.

$$y = 2x - 18$$
$$0 = 2x - 18$$
$$18 = 2x$$
$$9 = x$$

The x-intercept is 9.

TRY THIS Identify the x- and y-intercepts of each line.
  a. $y = -4x + 8$ **b.** $y = 3x - 15$

Teaching Tip

TECHNOLOGY If you are using a TI-82 or TI-83 graphics calculator, it is possible to enter all of the equations in the Activity by using a single expression. First press [Y=] and clear any existing equations. Then enter the following next to **Y1=**:

2 [X,T,θ,n] + 2nd (0 ,
3 , 5 , (-) 3 2nd)

To see the graphs, press [ZOOM] and select **6: ZStandard**.

ADDITIONAL EXAMPLE 2

Write an equation in slope-intercept form for the line that passes through the points $(-4, 7)$ and $(10, 0)$.
$y = -\frac{1}{2}x + 5$

TRY THIS
$y = 2x - 3$

CRITICAL THINKING
The slope-intercept form provides the information necessary to sketch a line quickly. $3x - 4y = 7$ is a linear equation. It can be rewritten as $y = \frac{3}{4}x - \frac{7}{4}$.

ADDITIONAL EXAMPLE 3

Identify the y-intercept of each line.

a. $y = 3x - 16$ -16

b. $y = 78x + 157$ 157

TRY THIS
a. The y-intercept is 8.
   The x-intercept is 2.

b. The y-intercept is -15.
   The x-intercept is 5.

Enrichment

Give students the following set of equations:

$x = 2y + 2$ $x = -2y + 2$
$x = 4y + 2$ $x = \frac{1}{2}y + 2$

Have them graph the equations on the same set of coordinate axes. (Encourage them to first solve for y.) Ask what all of the lines have in common. Elicit the fact that all pass through the point $(2, 0)$.

Now repeat the activity with this set of equations.

$x = 2y + 2$ $x = 2y - 2$
$x = 2y - 6$ $x = 2y$

This time elicit the fact that all have a slope of $\frac{1}{2}$.

Have students make a conjecture about what information can be read from an equation of the form $x = ny + a$. Sample: The variable n is the inverse of the slope, and a is the x-intercept. Have them write other equations of the form $x = ny + a$ to test their conjecture. Students might enjoy creating an original name for this form of equation, such as "inverse-slope/x-intercept form."

LESSON 5.4 **247**

ADDITIONAL EXAMPLE 4

Write an equation for each line.

a.
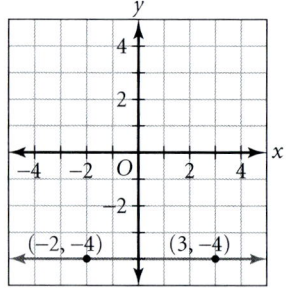
$y = -4$

b.
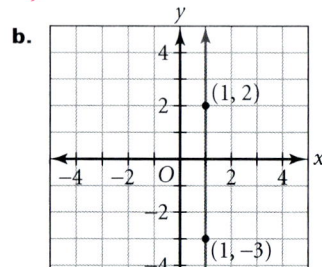
$x = 1$

TRY THIS
a. $y = -3$
b. $x = 3$

Horizontal and Vertical Lines

The intercepts make it very easy to write equations for both horizontal and vertical lines. Since horizontal lines have a slope of 0, their equations are of the form $y = b$. Vertical lines have undefined slope (Lesson 5.2), but an equation can still be written by using the x-intercept. If the x-intercept is a, then the equation for a vertical line is $x = a$.

Equations of Horizontal and Vertical Lines

The equation of a horizontal line is $y = b$, where b is the y-intercept.

The equation of a vertical line is $x = a$, where a is the x-intercept.

EXAMPLE 4 Write an equation for each line.

a. 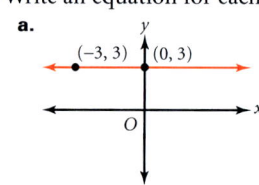 b.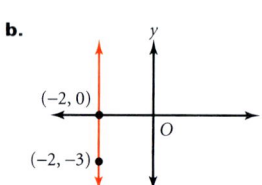

● **SOLUTION**

a. Notice that the line is horizontal. The y-intercept is 3. Substitute into the equation for a horizontal line.
$y = 3$

b. Notice that the line is vertical. The x-intercept is −2. Substitute into the equation for a vertical line.
$x = -2$

TRY THIS

Write an equation for each line.

a. 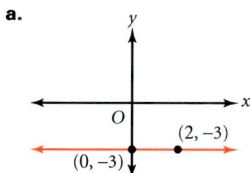 b.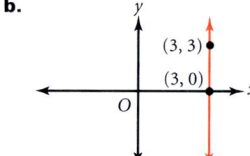

Reteaching the Lesson

COOPERATIVE LEARNING Give students the following "What's my equation?" puzzles:

I have a slope of 2. My y-intercept is −7. What's my equation? $y = 2x - 7$

I pass through (−1, 5). I pass through the origin. What's my equation? $y = -5x$

I pass through (6, −2). I pass through (3, 0). What's my equation? $y = -\frac{2}{3}x + 2$

Have students work in pairs to solve the puzzles. If a certain difficulty seems to arise in several groups, discuss it with the entire class.

Now have each student create a set of five similar "What's my equation?" puzzles on a sheet of paper. Have partners exchange papers and solve each other's puzzles. When both have finished solving, have them check the answers and work together to resolve any difficulties.

248 LESSON 5.4

Exercises

Communicate

1. Line *l* passes through the origin and the point $A(3, 6)$. Explain how to write an equation for line *l*.
2. If the slope of a line is h and it passes through the origin, what is an equation of the line?
3. Describe how you can determine an equation of the function graphed at right.
4. How does changing the value of b affect the graph of $y = mx + b$?
5. How does changing the value of m affect the graph of $y = mx + b$?

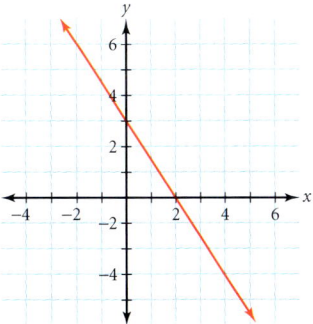

Guided Skills Practice

Graph each equation below by using the slope and *y*-intercept. (EXAMPLE 1)

6. $y = -4x - 1$
7. $y = \frac{3}{5}x - 2$

8. Write an equation for a line passing through the points $(4, 5)$ and $(6, -2)$. (EXAMPLE 2) $y = \frac{-7}{2}x + 19$

9. Identify the *x*- and *y*-intercepts of the line $y = -x + 7$. (EXAMPLE 3)
   x-intercept is 7; *y*-intercept is 7

Write an equation for each line. (EXAMPLE 4)

10. a horizontal line that passes through the point $(6, 0)$ $y = 0$
11. a vertical line that passes through the point $(0, 0)$ $x = 0$

Practice and Apply

Identify the *x*- and *y*-intercepts of each line.

12. $y = 4x + 5$
13. $y = 8x - 1$
14. $y = -3x + 5$
15. $y = -2x + 13$
16. $y = 17x - 4$
17. $y = -5x - 9$

12. $-\frac{5}{4}$, 5
13. $\frac{1}{8}$, -1
14. $\frac{5}{3}$, 5
15. $\frac{13}{2}$, 13
16. $\frac{4}{17}$, -4
17. $-\frac{9}{5}$, -9

Graph each equation by using the slope and *y*-intercept.

18. $y = -5x + 7$
19. $y = \frac{4}{5}x - 3$
20. $y = -\frac{3}{2}x + 8$
21. $y = 0.5x - 4$
22. $y = -3x + 2.5$
23. $y = -x - 15$

20.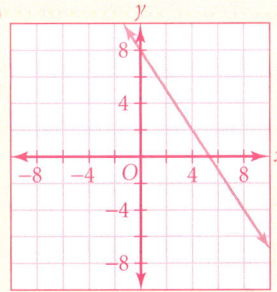

21.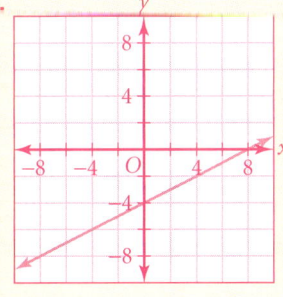

The answers to Exercises 22 and 23 can be found on page 250.

Assess

Selected Answers
Exercises 6–11, 13–73 odd

ASSIGNMENT GUIDE

In Class	1–11
Core	13–61 odd
Core Plus	12–60 even, 62–68
Review	69–74
Preview	75

 Extra Practice can be found beginning on page 738.

6.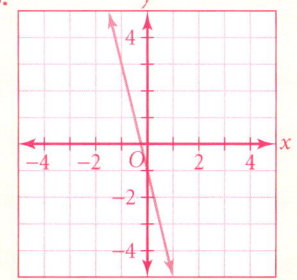

7.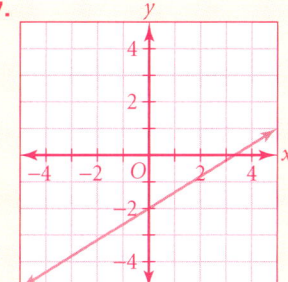

18.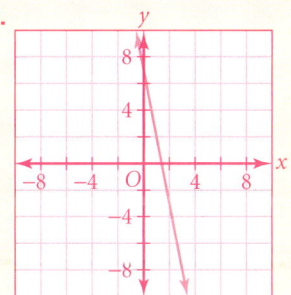

19.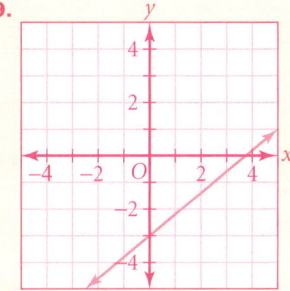

LESSON 5.4   **249**

Technology

You may want students to use a graphics calculator to check their work in Exercises 18–66. If an exercise calls for students to graph an equation, suggest that they first use graph paper and then use a calculator. They can now compare the graphs to see if they are alike.

In exercises that require students to write an equation, they can graph the proposed equation with the graphics calculator and determine whether the calculator graph has the required characteristics. For example, the calculator's **TRACE** feature will allow them to determine whether a given point lies on the line.

24. $y = \frac{-6}{5}x + 6$
25. $y = \frac{3}{2}x - \frac{1}{2}$
26. $y = -x + 8$
27. $y = \frac{4}{11}x - \frac{6}{11}$
28. $y = \frac{-4}{13}x - \frac{20}{13}$
29. $y = \frac{-4}{11}x - \frac{49}{11}$
30. $y = \frac{-3}{2}x - 9$
31. $y = x$
32. $y = \frac{-4}{3}x - \frac{1}{3}$

50. $m = -5; y = 0$
51. $m = -5; y = 3$
52. $m = 0; y = 7$
   horizontal line through (0, 7)
53. m, und.; no y-int.
   vertical line through (7, 0)
54. $m = -1; y = -3$
55. $m = 2; y = -8$
56. $m = \frac{1}{4}; y = 10$
57. $m = \frac{4}{5}; y = 9$
58. $m = -1; y = 7$
59. $m = \frac{1}{8}; y = 3$
60. $m = \frac{-2}{3}; y = 0$
61. $m = 3; y = -1$

Write an equation in slope-intercept form for the line that contains each pair of points.

24. (0, 6), (5, 0)
25. (3, 4), (−1, −2)
26. (1, 7), (5, 3)
27. (7, 2), (−4, −2)
28. (−5, 0), (8, −4)
29. (7, −7), (−4, −3)
30. (−4, −3), (−2, −6)
31. (6, 6), (−2, −2)
32. (−1, 1), (5, −7)

Write an equation in slope-intercept form for each line graphed below.

33. $y = -x$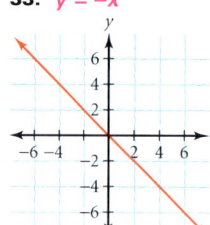
34. $y = \frac{2}{3}x$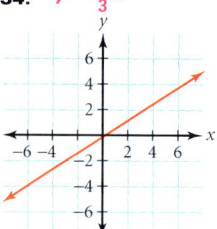
35. $y = \frac{1}{2}x - 2$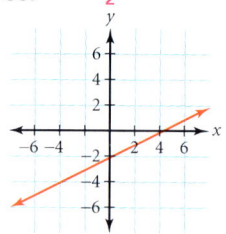

36. $y = \frac{-3}{2}x + 4$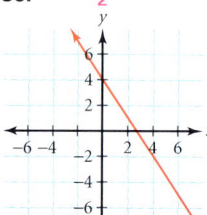
37. $y = \frac{3}{2}x + 4$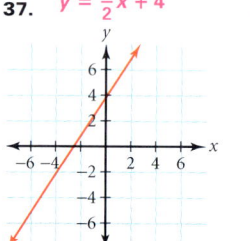
38. $y = -x - 4$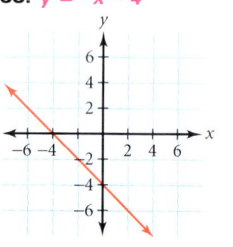

Write an equation for the line that contains each pair of points.

39. (3, 4), (3, −3) $x = 3$
40. (−4, 1), (6, 1) $y = 1$
41. (2, −5), (10, −5) $y = -5$
42. (5, −3), (5, 1) $x = 5$

Write an equation in slope-intercept form for the line that fits each description below.

43. contains the origin and has a slope of −2 $y = -2x$
44. contains the origin and has a slope of $\frac{2}{3}$ $y = \frac{2}{3}x$
45. crosses the y-axis at −1 and has a slope of 5 $y = 5x - 1$
46. contains the origin and (4, 5) $y = \frac{5}{4}x$
47. contains (−3, 2) and (6, 6) $y = \frac{4}{9}x + \frac{10}{3}$
48. contains (0, −2) and (7, 1) $y = \frac{3}{7}x - 2$
49. contains (6, 8) and (−3, 4) $y = \frac{4}{9}x + \frac{16}{3}$

Without graphing, describe what the graph of each equation looks like. Include information about the slope and y-intercept.

50. $y = -5x$
51. $y = -5x + 3$
52. $y = 7$
53. $x = 7$
54. $y = -x - 3$
55. $y = 2x - 8$
56. $y = \frac{1}{4}x + 10$
57. $y = \frac{4}{5}x + 9$
58. $y = 7 - x$
59. $y = 3 + \frac{1}{8}x$
60. $y = \frac{-2x}{3}$
61. $y = \frac{6x}{2} - 1$

22.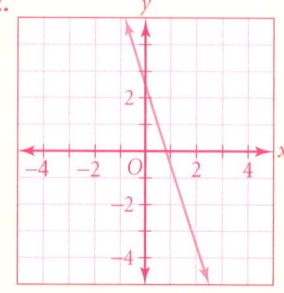
23.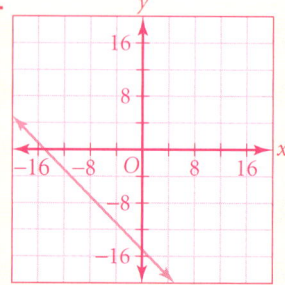

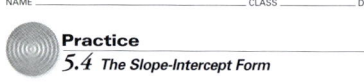

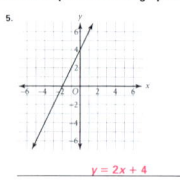
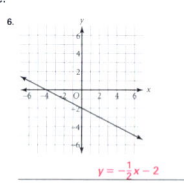

250 LESSON 5.4

CONNECTION

COORDINATE GEOMETRY Match each equation with the appropriate description.

62. $x + y = 9$ **d**
63. $xy = 9$ **e**
64. $x = 9$ **b**
65. $y = 9$ **a**
66. $y = 9x$ **c**

a. a horizontal line 9 units above the origin
b. a vertical line 9 units to the right of the origin
c. a line through the origin with a slope of 9
d. a line with a slope of –1 and a y-intercept of 9
e. something other than a straight line

APPLICATIONS

67. **ECOLOGY** Suppose that the water level of a river is 34 feet and that it is receding at a rate of 0.5 foot per day. Write an equation for the water level, L, after d days. Graph the equation. In how many days will the water level be 26 feet? $L = -0.5d + 34$; 16 days

CHALLENGE

68. **BUDGET** Saul's father is thinking of buying his son a six-month movie pass for $40. With the pass matinees cost $1.00. If matinees are normally $3.50 each, how many times must Saul attend in order for it to benefit his father to buy the pass? **17 times or more**

Look Back

APPLICATION

FUND-RAISING For the following numbers of people, how much will each person have to contribute to reach a goal of $250? *(LESSON 3.2)*

69. 25 people **$10**
70. 15 people **$16.67**
71. 20 people **$12.50**
72. 150 people **$1.67**

73. Solve $3a + 9d = 7$ for d. *(LESSON 3.6)* $d = \dfrac{7 - 3a}{9}$

74. Dan earns 15% on each $20 box of candies he sells. How many boxes will he have to sell to earn at least $100? *(LESSON 4.2)*
    Dan must sell at least 34 boxes.

Look Beyond

75. **CULTURAL CONNECTION: AFRICA** The Rhind papyrus that describes the mathematics of building the ancient pyramids of Egypt refers to a *seked*. A seked can be stated mathematically as follows:

$$\text{seked} = \frac{\text{horizontal distance in palms}}{\text{vertical distance in cubits}}$$

Consider a pyramid with a base of 2520 palms on each side and a height of 250 cubits. Scribes took half of the length of the base as the run and the height as the rise. They calculated the seked to be $\dfrac{1260}{250}$. Compare the calculation of the seked with the calculation of the slope for the same pyramid. Explain the relationship between the seked and slope.

67. $L = -0.5d + 34$; 16 days

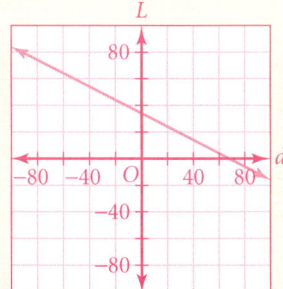

75. The seked is $\dfrac{\text{run}}{\text{rise}}$. The slope is $\dfrac{\text{rise}}{\text{run}}$. The seked uses different units for rise and run.

Error Analysis

When finding an equation for the line that passes through two points, students often find a correct slope, but then err in the process of solving for b. Remind them that the second point can serve as a check on their work. That is, if the coordinates of the second point do not satisfy the equation they wrote, they know they must go back and find some error in their work.

Look Beyond

In Exercise 75, students explore the *seked*, a ratio that was used by the ancient Egyptians when building pyramids. Students are asked to compare this ratio to the slope ratio.

Student Technology Guide

5.4 The Slope-Intercept Form

Using a graphics calculator, you can graph the line with a given slope and y-intercept. Then you can use the graph to obtain other information about the line.

Example: A line has a slope of $\tfrac{3}{4}$ and a y-intercept of $\tfrac{9}{4}$.

Use a graph to identify the point where the graph crosses the x-axis. You can enter the equation directly as shown below.

Press [Y=] [(] 3 [÷] 4 [)] [X,T,θ,n] [+] 9 [÷] 4.

Alternately, you can mentally change the fractions into decimals.

$\tfrac{3}{4} = 0.75$ and $\tfrac{9}{4} = 2.25$ ⟶ Press [Y=] .75 [X,T,θ,n] [+] 2.25.

When [GRAPH] is pressed, either method will produce a correct graph as shown at the right below.

It is often necessary to adjust the viewing window settings of the graphics calculator. If your graph does not match the one shown, you can press [WINDOW] and then set Xmin = −4.7, Xmax = 4.7, Xscl = 1, Ymin = −3.1, Ymax = 3.1, and Yscl = 1.

The graph shows that the line crosses the x-axis at about (−3, 0). Usually it is best to select a window that shows where the graph crosses the x-axis and where it crosses the y-axis. Often, you will want to select a window such that one unit on the x-axis is the same length as one unit on the y-axis. This occurs when Ymax − Ymin is about $\tfrac{2}{3}$(Xmax − Xmin).

Each line described below crosses the x-axis at an integer value of x. Use a graph to identify the point where the graph crosses the x-axis.

1. slope of 2, y-intercept of 8 **(−4, 0)**
2. slope of $\tfrac{1}{4}$, y-intercept of −3 **(12, 0)**
3. slope of $-\tfrac{3}{2}$, y-intercept of $\tfrac{15}{2}$ **(5, 0)**
4. slope of $\tfrac{7}{5}$, y-intercept of 14 **(−10, 0)**
5. slope of $\tfrac{4}{3}$, y-intercept of $\tfrac{8}{3}$ **(−2, 0)**
6. slope of $-\tfrac{3}{10}$, y-intercept of 6 **(20, 0)**
7. slope of $\tfrac{3}{8}$, y-intercept of $-\tfrac{9}{8}$ **(3, 0)**
8. slope of $\tfrac{1}{7}$, y-intercept of 3 **(−21, 0)**

Prepare

NCTM PRINCIPLES & STANDARDS 1–10

LESSON RESOURCES

- Practice 5.5
- Reteaching Master 5.5
- Enrichment Master 5.5
- Cooperative-Learning Activity 5.5
- Lesson Activity 5.5
- Problem Solving/Critical Thinking Master 5.5
- Teaching Transparency 27
- Teaching Transparency 28
- Teaching Transparency 29

QUICK WARM-UP

Solve for the indicated variable.

1. $t = r + s$ for s
   $s = t - r$

2. $-5g = h$ for g
   $g = \frac{h}{-5}$

3. $q = mn - p$ for m
   $m = \frac{p + q}{n}$

4. $4x + 2y = z$ for y
   $y = \frac{z - 4x}{2}$

Also on Quiz Transparency 5.5

Teach

Why Write 0.25, $\frac{1}{4}$, and 25% on the board or overhead. Ask students how the three expressions are related. Elicit from them the fact that they are different forms of the same number. Have them discuss situations in which it would be better to use one form rather than another. Point out that, in the same way, there are different forms for equations of lines.

5.5 The Standard and Point-Slope Forms

Wynton Marsalis studied at the Juilliard School of Music and played trumpet with the New Orleans Philharmonic Orchestra before gaining fame as a jazz musician.

Objectives

- Define and use the standard form for a linear equation.
- Define and use the point-slope form for a linear equation.

Why The equation of a line can be written in different forms. If you know which form to use, you can often save yourself time and effort.

Wynton Marsalis was born in New Orleans on October 16, 1961. You can write this date in many ways. Each way has its advantages. The long form is easy to read. The short form is quick and easy to write. The computer form is easy to sort by year, month, or day.

October 16, 1961	10-16-61	19611016
long form	**short form**	**computer form**

You have been using the slope-intercept form of the equation for a line, $y = mx + b$, to solve problems. As for dates, there are other forms for the equation of a line. Each form has advantages.

Standard Form

Consider an equation written in the form $Ax + By = C$. After values for A, B, and C are determined, a solution to this equation is any ordered pair of numbers (x, y) that makes the equation true. When these ordered pairs are graphed, they form a straight line.

Standard Form of a Linear Equation

A linear equation in the form $Ax + By = C$ is in standard form when
- A, B, and C are real numbers, and
- A and B are not both zero.

Alternative Teaching Strategy

GUIDED ANALYSIS Write these equations on the board or overhead.

$y = 2x - 4$ $\quad 2x = y + 4$ $\quad 2x - y = 4$
$\frac{1}{2}y = x - 2$ $\quad x = \frac{1}{2}y + 2$ $\quad x - \frac{1}{2}y = 2$

Tell students that two linear equations are *equivalent* if the graph of each is the same line. Ask which equations in the list above are equivalent. Elicit from them the fact that all of the given equations are equivalent to each other. This means that, for any given line, its equation can be written in many different forms.

Now write the following:
- $Ax + By = C$
- A, B, and C are real numbers.
- A and B are not both zero.

Ask which equations in the given list fit this description. Lead them to identify the equations in the third column of the list. Tell them that the form of these two equations is considered a *standard form of a linear equation*. Display a graph of the equation. Then demonstrate how, using either of the two standard forms, it is relatively easy to identify both the x- and y-intercepts of the graph.

252 LESSON 5.5

EXAMPLE 1 Write each linear equation in standard form.

a. $y = -2x + 3$ b. $x = -13y + 4$ c. $\frac{3}{4}x - 2 = 3y$

● **SOLUTION**

a. $y = -2x + 3$
   $2x + y = 3$

b. $x = -13y + 4$
   $x + 13y = 4$

c. $\frac{3}{4}x - 2 = 3y$
   $\frac{3}{4}x - 2 - 3y = 0$
   $\frac{3}{4}x - 3y = 2$

This is acceptable as standard form. It could also be written as $\frac{3}{4}x + (-3y) = 2$.

TRY THIS Write each linear equation in standard form.

a. $y = -4x - 12$ b. $2y - 5 = 3x$ c. $5.7x - 5.3 = 4.3y$

EXAMPLE 2

APPLICATION
FUND-RAISING

Jackie is in charge of selling tickets for the school jazz concert at $2.00 for student tickets and $4.00 for adult tickets. She hopes that the total ticket sales will be about $600 in order to cover expenses and make a modest profit. **Write a linear equation in standard form to model this situation, and then graph the equation.**

● **SOLUTION**

Let x represent the number of adult tickets.
Let y represent the number of student tickets.

An equation in standard form that models the problem is $4x + 2y = 600$.

To graph an equation in standard form, substitute 0 for x, and solve for y. This gives the y-intercept—the y-coordinate of the point where the line crosses the y-axis. Then substitute 0 for y and solve for x. This gives the x-intercept—the x-coordinate of the point where the line crosses the x-axis.

Start with the equation $4x + 2y = 600$.

First let $x = 0$.

$4x + 2y = 600$
$0 + 2y = 600$
$y = 300$

The y-intercept is 300.

Then let $y = 0$.

$4x + 2y = 600$
$4x + 0 = 600$
$x = 150$

The x-intercept is 150.

Draw the line that connects the two intercepts. The coordinates of any point on this line will solve the equation. Recall though that only whole numbers make sense for the domain and range in this situation. Why?

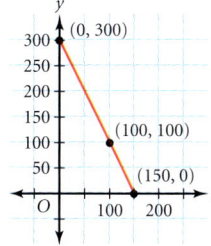

Use Teaching Transparency 27.

ADDITIONAL EXAMPLE 1

Write each linear equation in standard form.

a. $y = -3x + 1$
   $3x + y = 1$

b. $5y - 2 = 4x$
   $4x - 5y = -2$

c. $\frac{4}{5}x = 2y - 3$
   $\frac{4}{5}x - 2y = -3$

TRY THIS

a. $4x + y = -12$
b. $3x - 2y = -5$
c. $5.7x - 4.3y = 5.3$

Teaching Tip

The situation described in Example 2 is an example of a *linear combination* problem. The standard form of a linear equation is especially helpful in solving this type of problem.

ADDITIONAL EXAMPLE 2

A donation of $300 was made to the Classical Music Club for the purpose of building their collection of CDs. At the local music store, some classical CDs cost $10 and others cost $15. The club wants to know how many of each type they can buy with the $300. (Do not consider the sales tax.) **Write an equation in standard form to model this situation.**
$10x + 15y = 300$, where x represents the number of $10 CDs and y represents the number of $15 CDs

---

Interdisciplinary Connection

BUSINESS The conditions that limit the activities of a business are called business *constraints*. Constraints are frequently expressed as a linear equation in standard form.

Shannon is starting a small business called Shannon's Shirts. Right now she plans to stock just T-shirts and sweatshirts. The supplier she has chosen will charge her $2.50 for each T-shirt and $4.50 for each sweatshirt. Shannon has $1800 to purchase the shirts. Write an equation to model the number of T-shirts and sweatshirts she can buy from the supplier. **$2.5x + 4.5y = 1800$, where x represents the number of T-shirts and y represents the number of sweatshirts**

If Shannon buys no sweatshirts, how many T-shirts can she buy? **720** If Shannon buys no T-shirts, how many sweatshirts can she buy? **400** Find two possible combinations of T-shirts and sweatshirts that she can buy. **samples: 360 T-shirts, 200 sweatshirts; 180 T-shirts, 300 sweatshirts**

LESSON 5.5 253

Teaching Tip

TECHNOLOGY If you are entering the equation given here into a TI-82 or TI-83 graphics calculator, press `Y=` to enter the **Y=** editor. Clear any existing equations by pressing ▼ to place the cursor into the line that contains the equation and then pressing `CLEAR`.

Now press ▲ as many times as necessary to move the cursor to the right of **Y1=**. Then press `(-)` `2` `X,T,θ,n` `+` `300` `ENTER`.

Next press `WINDOW` and set the window as follows:

Xmin=–10   Ymin=–10
Xmax=200   Ymax=350
Xscl=50   Yscl=50

To view the graph, press `GRAPH`.

See Keystroke Guide, page 782.

CRITICAL THINKING
The result is $y = -\frac{A}{B}x + \frac{C}{B}$, which is a linear equation in slope-intercept form.

Use Teaching Transparency 28.

ADDITIONAL EXAMPLE 3

A line with a slope of –5 contains the point (2, 3). **Write an equation in point-slope form for the line.**
$y - 3 = -5(x - 2)$

TECHNOLOGY GRAPHICS CALCULATOR

To enter the equation from Example 2 in a graphics calculator, first rewrite the equation in slope-intercept form, $y = mx + b$.

$4x + 2y = 600$
$2y = 600 - 4x$
$y = 300 - 2x$
$y = -2x + 300$

The equation can now be entered into the calculator by using the **Y=** button.

CRITICAL THINKING   If the equation of a line is written in standard form, $Ax + By = C$, the slope is $-A/B$ and the y-intercept is C/B. Explain why.

Point-Slope Form

If you know the slope of a line and the coordinates of one of its points, you can write a third form of an equation for a line called the **point-slope form**.

The point-slope form is derived from the slope formula, $m = \frac{y_2 - y_1}{x_2 - x_1}$.

$$m = \frac{y_2 - y_1}{x_2 - x_1}$$

$$m(x_2 - x_1) = \frac{y_2 - y_1}{x_2 - x_1}(x_2 - x_1)$$

$$m(x_2 - x_1) = y_2 - y_1$$

If you replace the point (x_2, y_2) in the equation with the notation for any point on the line, (x, y), the result is as follows:

$$y - y_1 = m(x - x_1)$$

Point-Slope Form

The form $y - y_1 = m(x - x_1)$ is the point-slope form for the equation of a line. The coordinates x_1 and y_1 are taken from a given point, (x_1, y_1), and m is the slope.

EXAMPLE 3 A line with a slope of 3 contains the point (2, –7). **Write an equation in point-slope form for the line.**

• **SOLUTION**
Let $m = 3$, $x_1 = 2$, and $y_1 = -7$. Substitute the given values into the point-slope form of the equation.

$$y - y_1 = m(x - x_1)$$
$$y - (-7) = 3(x - 2)$$

Inclusion Strategies

LINGUISTIC LEARNERS Students who are verbal may perceive the *Summary of the Forms for the Equation of a Line* as a maze of symbols. Suggest that they write a description of each form in their own words. For instance, they might describe standard form as the form in which the x- and y-terms are on one side of the equal sign and the constant term is on the other side. For each form, encourage them to include a sentence or two that explains what information can be read from the equation and the types of situations in which the form is useful.

Enrichment

Have students investigate the following situations concerning the form $Ax + By = C$:

1. What is the effect when you hold A and B constant and vary C? **The slope remains the same. The y-intercept varies.**
2. What is the effect when you hold B and C constant and vary A? **The y-intercept remains the same. The slope varies.**
3. What is the effect when you keep A and C constant and vary B? **Both the slope and the y-intercept vary.**

TRY THIS A line with a slope of –4 contains the point (4, 2). Write an equation in point-slope form for the line.

TRY THIS
$y - 2 = -4(x - 4)$

EXAMPLE 4 Write $y - 149 = 9(x - 16)$ in each form.
  a. slope-intercept form b. standard form

SOLUTION

a. Begin with the original equation and solve for y.
$$y - 149 = 9(x - 16)$$
$$y - 149 = 9x - 144$$
$$y = 9x + 5$$

b. Then change the slope-intercept form to standard form, $Ax + By = C$.
$$y = 9x + 5$$
$$y - 9x = 5$$
$$-9x + y = 5$$
$$9x - y = -5$$

ADDITIONAL EXAMPLE 4

Write $y + 20 = 7(x - 10)$ in each form.

a. slope-intercept form
$y = 7x - 90$

b. standard form
$7x - y = 90$

TRY THIS Write $y - 23 = 5(x - 4)$ in each form.
  a. slope-intercept form b. standard form

TRY THIS
a. $y = 5x + 3$
b. $5x - y = -3$

EXAMPLE 5 Write an equation in point-slope form for a line that contains the points (–1, 10) and (5, 8).

SOLUTION

The point-slope form simplifies the way this problem is solved. Let (x_1, y_1) be (–1, 10) and (x_2, y_2) be (5, 8). Find the slope.

$$\text{slope} = \frac{y_2 - y_1}{x_2 - x_1} = \frac{8 - 10}{5 - (-1)} = \frac{-2}{6} = -\frac{1}{3}$$

Use the point (–1, 10) for (x_1, y_1). Substitute the slope and the coordinates of the point into the equation $y - y_1 = m(x - x_1)$.

$$y - 10 = -\frac{1}{3}[x - (-1)]$$

ADDITIONAL EXAMPLE 5

Write an equation in point-slope form for a line that contains the points (3, –6) and (–5, 2).
$y - (-6) = -1(x - 3)$, or
$y + 6 = -(x - 3)$

TRY THIS Write an equation in point-slope form for the line that contains the points (5, 65) and (7, 71).

TRY THIS
$y - 65 = 3(x - 5)$ or
$y - 71 = 3(x - 7)$

Use Teaching Transparency 29.

SUMMARY OF THE FORMS FOR THE EQUATION OF A LINE

Name	Form	Example
Slope-intercept	$y = mx + b$	$y = -3x + 5$
Standard	$Ax + By = C$	$3x + y = 5$
Point-slope	$y_1 - y = m(x_1 - x)$	$y - 2 = -3(x - 1)$

Reteaching the Lesson

COOPERATIVE LEARNING Have students work in groups of four, seated in a circular arrangement. Each student should write a slope and the coordinates of a point on a clean sheet of paper. The student should also write the point-slope form of the equation of the line determined by that slope and point.

Have students pass their paper to the person seated to the right. The students should now write the slope-intercept form of the equation on the paper they received.

Have students pass the papers to the right again. This time they should write the standard form of the equation.

Have them pass the papers to the right again. Now each student should graph the equation. After all graphs are complete, have students discuss which form of the equation they used to draw the graph. Have them explain the reasons for their choices.

LESSON 5.5

Assess

Selected Answers
Exercises 5–16, 17–61 odd

ASSIGNMENT GUIDE

In Class	1–16
Core	17–28, 29–47 odd
Core Plus	18–48 even
Review	49–61
Preview	62, 63

✎ Extra Practice can be found beginning on page 738.

Technology

You might suggest that students use a graphics calculator to check their answers in Exercises 32–40.

Exercises

Communicate

1. Explain how to write the equation $5y - 2 = -3x$ in standard form.
2. Tell how you would identify the intercepts for the graph of the equation $3x + 6y = 18$.
3. Explain how to rewrite $x - 3y = 9$ in slope-intercept form.
4. How would you use the point-slope formula to write the equation of a line containing the points $(-2, 4)$ and $(4, -8)$?

Guided Skills Practice

Write each equation in standard form. *(EXAMPLES 1 AND 2)*

5. $y = 3x + 7$ 6. $2y = 3x - 4$ 7. $3x = -7y - 17$
   $3x - y = -7$ $3x - 2y = 4$ $3x + 7y = -17$

Write an equation in point-slope form for the line with each slope that contains the given point. *(EXAMPLE 3)*

8. slope of 2; (3, 4)   9. slope of -2; $(-3, 4)$ 10. slope of $\frac{1}{3}$; $(3, -4)$

Write the equations below in slope-intercept form and in standard form. *(EXAMPLE 4)*

11. $y - 50 = 8(x - 4)$ 12. $3y = 9x + 15$ 13. $y = 10(-4x + 3)$

Write an equation in point-slope form for the line that contains each pair of points. *(EXAMPLE 5)*

14. $(5, -2), (-2, 5)$ 15. $(-3, 3), (-4, 4)$ 16. $(3, 2), (-3, -2)$
    $y - (-2) = -1(x - 5)$ $y - 3 = -1(x - (-3))$ $y - 2 = \frac{2}{3}(x - 3)$

Practice and Apply

Write each equation in standard form.

17. $4x = -3y + 24$ 18. $7y = -5x - 35$ 19. $6x + 4y + 12 = 0$
20. $2x = 4y$ 21. $6x - 8 = 2y + 6$ 22. $x = \frac{2}{3}y + 6$
23. $7x + 14y = 3x - 10$ 24. $5 = y - x$ 25. $7x + 2y + 14 = 0$
26. $2x - 4 = 3y + 6$ 27. $6 = x - y$ 28. $3 + 9y = -x - 12$
    $2x - 3y = 10$ $x - y = 6$ $x + 9y = -15$

Write an equation in point-slope form for the line that has the given slope and that contains the given point.

29. slope 3, (2, 6)   30. slope -4, $(-3, 1)$ 31. slope $\frac{1}{5}$, $(-4, -2)$
    $y - 6 = 3(x - 2)$ $y - 1 = -4(x - (-3))$ $y - (-2) = \frac{1}{5}(x - (-4))$

Find the x- and y-intercepts for the graph of each equation.

32. $x + y = 10$ 33. $3x - 2y = 12$ 34. $5x + 4y = 12$
35. $4x - 5y = 20$ 36. $2x - 7y = 14$ 37. $9x + y = 18$

Answers (side column):

8. $y - 4 = 2(x - 3)$
9. $y - 4 = -2(x - (-3))$
10. $y - (-4) = \frac{1}{3}(x - 3)$
11. $y = 8x + 18$
     $8x - y = -18$
12. $y = 3x + 5$
     $3x - y = -5$
13. $y = -40x + 30$
     $40x + y = 30$
17. $4x + 3y = 24$
18. $5x + 7y = -35$
19. $6x + 4y = -12$
20. $2x - 4y = 0$
21. $6x - 2y = 14$
22. $3x - 2y = 18$
23. $4x + 14y = -10$
24. $x - y = -5$
25. $7x + 2y = -14$
32. $x = 10; y = 10$
33. $x = 4; y = -6$
34. $x = \frac{12}{5}; y = 3$
35. $x = 5; y = -4$
36. $x = 7; y = -2$
37. $x = 2; y = 18$

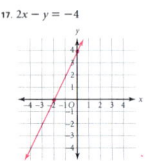 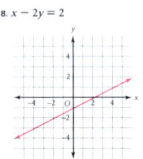

Practice

5.5 The Standard and Point-Slope Forms

Write each equation in standard form.

1. $2x = -5y + 11$ —— $2x + 5y = 11$ 2. $3y = -x - 20$ —— $x + 3y = -20$
3. $4x - 7y + 15 = 0$ —— $4x - 7y = -15$ 4. $9x = 6y$ —— $9x - 6y = 0$
5. $2x + 10 = 3y - 1$ —— $2x - 3y = -11$ 6. $2x = \frac{1}{2}y + 3$ —— $4x - y = 6$

Find the x- and y-intercepts for each equation.

7. $x + y = 5$ —— x-intercept: 5; y-intercept: 5   8. $3x + 5y = 15$ —— x-intercept: 5; y-intercept: 3
9. $4x - 3y = 12$ —— x-intercept: 3; y-intercept: -4 10. $x - 3y = 6$ —— x-intercept: 6; y-intercept: -2
11. $x - y = -3$ —— x-intercept: -3; y-intercept: 3   12. $4x = -5y$ —— x-intercept: 0; y-intercept: 0
13. $2x + y = 1$ —— x-intercept: $\frac{1}{2}$; y-intercept: 1   14. $x = \frac{2}{3}y$ —— x-intercept: 0; y-intercept: 0
15. $\frac{x}{4} - y = 2$ —— x-intercept: 8; y-intercept: -2 16. $x = -6y - 2$ —— x-intercept: -2; y-intercept: $-\frac{1}{3}$

Use intercepts to graph each equation.

17. $2x - y = -4$ 18. $x - 2y = 2$

Write an equation in standard form for each line.

19. through (4, 5) and with a slope of 1 —— $x - y = -1$
20. crosses the x-axis at $x = -3$ and the y-axis at $y = 6$ —— $2x - y = -6$
21. through (1, 6) and with a slope of 2 —— $2x - y = -4$
22. through (3, 7) and (0, -2) —— $3x - y = 2$
23. through (1, 5) and $(-3, 1)$ —— $x - y = -4$

256 LESSON 5.5

Write an equation for the line that contains each pair of points.

38. (7, −2), (7, 8)   **39.** (0, 3), (−2, −1)   **40.** (−7, 5), (16, 5)
$x = 7$ $y = 2x + 3$ $y = 5$

41. The equation of a given line is $6x + 2y = 40$. What is the slope of the line?
−3

42. Compare the graphs of $4a + 2s = 588$ and $2a + s = 294$. What do you find?
The graphs are identical.

CHALLENGE
CONNECTION

COORDINATE GEOMETRY The equation $x = 4$ cannot be written in slope-intercept form because the slope is undefined. It can, however, be written in standard form as $1x + 0y = 4$. Copy and complete the table below. Write the equivalent forms of each equation or "undefined slope" when appropriate.

43. undefined slope;
$1x + 0y = 1$

44. $y = 0x + 4$;
$0x + 1y = 4$

45. $y = -x + 5$;
$1x + 1y = 5$

46. $y = 4x + 0$;
$4x - 1y = 0$

47. $y = \frac{1}{4}x + 0$;
$1x - 4y = 0$

	Given	Slope-intercept form	Standard form
43.	$x = 1$	?	?
44.	$y = 4$	?	?
45.	$x + y = 5$	?	?
46.	$y = 4x$	?	?
47.	$x = 4y$	?	?

APPLICATION

48. FUND-RAISING Tickets to a school play are $3 for students and $5 for adults. Write an equation in standard form which shows that the total sales for s student tickets and a adult tickets is $700. Graph the equation with adult tickets on the horizontal axis and student tickets on the vertical axis.

Anne Frank 1929–1945

Look Back

49. Describe a scatter plot with a strong negative correlation. **(LESSON 1.6)**
as the *x*-value increases, the *y*-value decreases

Evaluate each expression for $x = 1$, $y = 1$, and $z = 2$.
(LESSONS 2.2 AND 2.3)

50. $x^2 + y + z^2$ 6   **51.** $x - y + z$ 2   **52.** $x + y - z$ 0   **53.** $-(x + y + z)$ −4

Solve each equation. **(LESSON 3.2)**

54. $-5y = 30$ **55.** $3x = 420$ **56.** $\frac{y}{9} = 36$ **57.** $\frac{x}{2} = 108$
$y = -6$ $x = 140$ $y = 324$ $x = 216$

Find the slope of the line containing the origin and the given point. Then give the equation of the line containing the origin and the given point. **(LESSON 5.2)**

58. $A(3, 6)$ **59.** $B(2, 8)$ **60.** $C(6, 3)$ **61.** $D(-5, -7)$
$2; y = 2x$ $4; y = 4x$ $\frac{1}{2}; y = \frac{1}{2}x$ $\frac{7}{5}; y = \frac{7}{5}x$

Look Beyond

62. Will two distinct vertical lines ever intersect? **No**

63. What kind of angle is formed when a horizontal and a vertical line intersect? **Right angle**

Error Analysis

When changing equations from one form to another, students often make errors in the signs of the terms. One possible remedy is to first rewrite any subtractions as additions. For example, given the equation $4x - 2y = 8$, suggest that students write $4x + (-2y) = 8$.

Look Beyond

In Exercises 62 and 63, remind students that the situations are described by the terms *parallel lines* and *perpendicular lines*, respectively. Students will study equations of parallel and perpendicular lines in Lesson 5.6.

internetconnect

GO TO: go.hrw.com
KEYWORD: MA1 Check Digits

In this activity, students use simple arithmetic to explore *check digits*, numbers inserted in UPC codes to verify data read by a scanner.

Student Technology Guide

42.
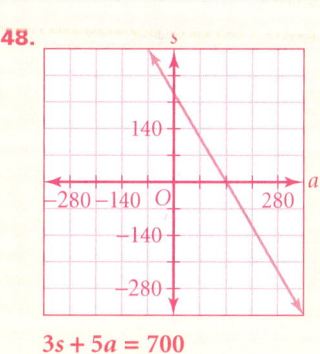

48.
$3s + 5a = 700$

LESSON 5.5   **257**

Prepare

NCTM PRINCIPLES & STANDARDS 1–6, 8–10

LESSON RESOURCES

- Practice 5.6
- Reteaching Master 5.6
- Enrichment Master 5.6
- Cooperative-Learning Activity 5.6
- Lesson Activity 5.6
- Problem Solving/Critical Thinking Master 5.6
- Teaching Transparency 30

QUICK WARM-UP

Find each product.

1. $(3)\left(\frac{1}{3}\right)$ **1**

2. $\left(\frac{4}{7}\right)\left(-\frac{7}{4}\right)$ **−1**

Identify the slope and the *y*-intercept of each line. Solve for *y* if necessary.

3. $y = -2x + 5$
   slope: −2; *y*-intercept: 5

4. $2x + 2y = 5$
   slope: −1; *y*-intercept: 2.5

Also on Quiz Transparency 5.6

Teach

Why Discuss with students their understanding of the terms *parallel* and *perpendicular*. Review the formal definitions given in the text. Ask them to suggest situations in which it might be important to construct lines that are perfectly parallel or perpendicular. **samples: carpentry, building railroad tracks**

5.6 Parallel and Perpendicular Lines

	°F	°C	K
Water boils	212	100	373
Water freezes	32	0	273
Absolute zero	−459	−273	0

Objectives

- Identify parallel lines and perpendicular lines by comparing their slopes.
- Write equations of lines that are parallel and perpendicular to given lines.

Why Identifying parallel and perpendicular lines by comparing slopes is an important step toward recognizing relationships between lines without graphing.

Water looks different at very cold temperatures, like those found near an iceberg, and at very hot temperatures, like those found in a geyser.

APPLICATION
PHYSICS

The table above compares temperatures for three scales—Fahrenheit, Celsius, and Kelvin.

The formulas for converting from Celsius and Kelvin to Fahrenheit are $F = \frac{9}{5}C + 32$ (Celsius to Fahrenheit) and $F = \frac{9}{5}K - 460$ (Kelvin to Fahrenheit). The two equations can be rewritten by using *y* for *F* and *x* for *C* and *K*; then they can be graphed on the same coordinate plane.

$y = \frac{9}{5}x + 32$

$y = \frac{9}{5}x - 460$

Notice that the two lines are parallel. Both lines have the same slope, $\frac{9}{5}$.

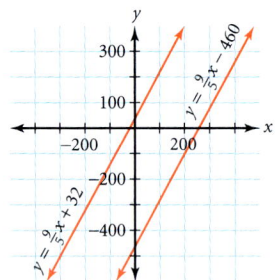

Alternative Teaching Strategy

USING PATTERNS Give students the table at right. Point out the fact that both of the lines given have the same slope. Now have students complete the table. After this has been done, ask if y_1 will ever equal y_2 for a given value of *x*. Lead them to see that y_1 will always be 3 greater than y_2. This means that the two lines will never have a point in common. That is, they will never intersect. Remind students that two lines in a plane that do not intersect are called *parallel lines*. Tell them that any two lines with the same slope are parallel.

x	$y_1 = 4x + 1$	$y_2 = 4x - 2$
−3	−11	−14
−2	−7	−10
−1	−3	−6
0	1	−2
1	5	2
2	9	6
3	13	10

Parallel Lines

If two different lines have the same slope, the lines are **parallel**.
If two nonvertical lines are parallel, they have the same slope.
Two parallel, vertical lines have undefined slopes.

EXAMPLE 1 Write an equation in slope-intercept form of a line parallel to $y = 3x - 7$ with a y-intercept of 4.

● **SOLUTION**

The slope of $y = 3x - 7$ is 3; a line parallel to this line has a slope of 3.

The y-intercept will be 4, so the equation of the new line is $y = 3x + 4$.

TRY THIS Write an equation in slope-intercept form of a line parallel to $y = 0.5x + 5$ with a y-intercept of -2.

In geometry, **perpendicular lines** form right angles. In the next activity, you will discover the algebraic relationship between the slopes of two lines that are perpendicular.

Activity
Exploring Slopes and Perpendicular Lines

You will need: graph paper and a sheet of clear plastic or tracing paper

On a piece of graph paper with the x- and y-axes marked, draw a line that has a run of 4 units and a rise of 3 units, as shown below. Make a copy on the plastic or tracing paper. Rotate the copy of the graph 90 degrees clockwise, as shown. When one graph is placed on top of the other, you can see the relationship between the two lines.

The lines are perpendicular because the line on the second graph was rotated 90 degrees from the line on the first graph.

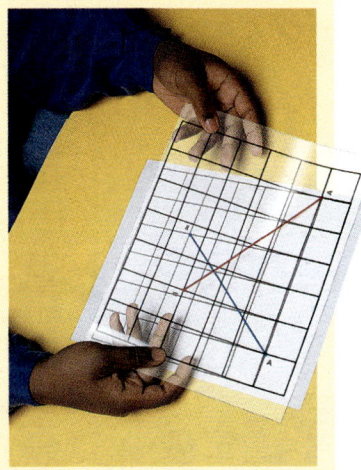

1. What is the slope of the original line?
2. Find the slope of the rotated line. Count the squares to find the rise and the run.
3. What is the sign for the slope of the rotated line?

CHECKPOINT ✓ 4. What is the relationship between the slope of the rotated line and the slope of the original line?

ADDITIONAL EXAMPLE 1

Write an equation in slope-intercept form of a line parallel to $y = -x + 7$ with a y-intercept of -5.
$y = -x - 5$

TRY THIS
$y = 0.5x - 2$

Math CONNECTION

GEOMETRY In the definition of parallel lines, it is important to stress that the lines are in the same plane. If two lines do not intersect and are not in the same plane, they are called *skew lines*. In the definition of perpendicular lines, be sure students understand that a right angle is an angle whose measure is 90°.

Activity Notes

Be sure students understand how to rotate the paper 90°. If they do not already know, tell them that the measure of a complete rotation is 360°. This means that a rotation of 90° is one-quarter of a complete rotation.

Students should discover that the product of the slopes of the original segment and the rotated segment is -1. You may wish to have them try the same experiment with one or more additional segments to convince themselves that this relationship always holds true.

CHECKPOINT ✓
4. They are negative reciprocals of each other.

Enrichment

Give students these definitions from geometry: A *quadrilateral* is a figure that has four sides. A *parallelogram* is a quadrilateral in which both pairs of opposite sides are parallel. A *rectangle* is a parallelogram with four right angles.

Have students find equations for four lines that form each figure: a parallelogram, a rectangle, and a quadrilateral that is not a parallelogram.

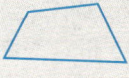 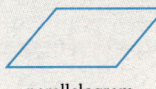 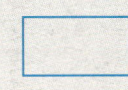

quadrilateral   parallelogram   rectangle

LESSON 5.6 **259**

Use Teaching Transparency 30.

ADDITIONAL EXAMPLE 2

Write an equation in slope-intercept form of a line that has a y-intercept of -5 and that is perpendicular to the line $y = -2x + 3$.

$y = \frac{1}{2}x - 5$

TRY THIS

$y = -\frac{1}{4}x + 6$

ADDITIONAL EXAMPLE 3

Write an equation in point-slope form for the line that contains the point $(4, -3)$ and that is perpendicular to the line $-10x + 2y = 8$.

$y + 3 = -\frac{1}{5}(x - 4)$

TRY THIS

$y - (-2) = -\frac{1}{2}(x - 3)$

Consider the equations $y = 2x + 3$ and $y = -\frac{1}{2}x - 1$. The graph shows that the two lines are perpendicular to each other.

Notice that the slopes of the two lines are *negative reciprocals* of each other.

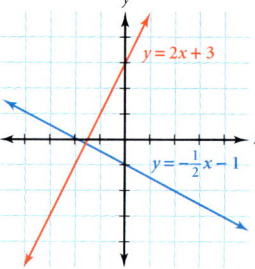

Perpendicular Lines

If the slopes of two lines are m and $-\frac{1}{m}$, the lines are perpendicular.

EXAMPLE 2 Write an equation in slope-intercept form of a line that has a y-intercept of 4 and that is perpendicular to the line $y = 3x + 2$.

- **SOLUTION**

The slope of the perpendicular line is the negative reciprocal of 3, or $-\frac{1}{3}$, and the y-intercept is 4. The slope-intercept form of the equation is $y = -\frac{1}{3}x + 4$.

TRY THIS Write an equation in slope-intercept form of a line that has a y-intercept of 6 and that is perpendicular to $y = 4x + 2$.

EXAMPLE 3 Write an equation in point-slope form for the line that contains the point $(4, 5)$ and that is perpendicular to the line $2x + 3y = 7$.

- **SOLUTION**

First solve the given equation for y in order to find the slope-intercept form of the equation.

$$y = -\frac{2}{3}x + \frac{7}{3}$$

For a line with a slope of $-\frac{2}{3}$, any line with a slope of $\frac{3}{2}$ is perpendicular to that line. Since the perpendicular line also contains the point $(4, 5)$, you can substitute the coordinates and slope into the point-slope form, $y - y_1 = m(x - x_1)$.

$$y - 5 = \frac{3}{2}(x - 4)$$

TRY THIS Write an equation in point-slope form for the line that contains the point $(3, -2)$ and that is perpendicular to the line $4x - 2y = -6$.

Inclusion Strategies

TACTILE LEARNERS When they are asked to write an equation for a line parallel or perpendicular to a given line, suggest that students first graph the given line on a sheet of graph paper. Warn them to make the line heavy enough to be seen through a second layer of paper. Then, after writing a proposed equation for the parallel or perpendicular line, tell them to graph that line on a separate sheet of graph paper. If they now position the second sheet of paper over the first, they should be able to determine whether the proposed equation meets the given condition.

Reteaching the Lesson

COOPERATIVE LEARNING Have students work in pairs. Have each student write an equation in standard form for a line and give the coordinates of a point not on the line. The partners should exchange papers and find an equation of a line parallel to and a line perpendicular to the given line through the given point. Encourage them to use graphics calculators to graph the lines and verify their results.

Exercises

Communicate

1. Explain how to write an equation for a line parallel to $y = 4x + 3$.
2. The slope of a line is $\frac{2}{3}$. Explain how to find the slope of a line that is perpendicular to that line.
3. How would you find the slope of a line perpendicular to the line $y = \frac{1}{3}x + 2$?
4. Describe how to find an equation for a line perpendicular to $y = 4x + 3$.

Guided Skills Practice

Write an equation in slope-intercept form for the line parallel to each line below and with a y-intercept of 5. *(EXAMPLE 1)*

5. $y = 2x + 3$ 6. $y = -3x$ 7. $4y = x$ 8. $-y = -6x + 2$

Write an equation in slope-intercept form for a line perpendicular to each line below and with a y-intercept of 4. *(EXAMPLE 2)*

9. $y = 3x - 3$ 10. $y = -3x$ 11. $5y = x$ 12. $-6y = x$

Write an equation in point-slope form for the line that contains the point (4, 5) and is perpendicular to the graph of the equation. *(EXAMPLE 3)*

13. $2x + 3y = 4$ 14. $x - 3y = 8$ 15. $-2x - 8y = 16$

Practice and Apply

Find the slope of each line.

16. $y = 4x + 10$ **4**
17. $3x + y = 7$ **−3**
18. $10 = -5x + 2y$ **$\frac{5}{2}$**
19. $4x - 3y = 12$ **$\frac{4}{3}$**
20. $y = \frac{1}{3}x - 3$ **$\frac{1}{3}$**
21. $3x - y = 7$ **3**
22. $2x - y = 14$ **2**
23. $3x + 2y = 51$ **$-\frac{3}{2}$**
24. $13 = 20x - 5y$ **4**
25. $3y = -4x + 2$ **$-\frac{4}{3}$**
26. $\frac{2}{3}x + 6y = 1$ **$-\frac{1}{9}$**
27. $4x - \frac{1}{4}y = 8$ **16**

Answers (left column):
5. $y = 2x + 5$
6. $y = -3x + 5$
7. $y = \frac{1}{4}x + 5$
8. $y = 6x + 5$
9. $y = \frac{-1}{3}x + 4$
10. $y = \frac{1}{3}x + 4$
11. $y = -5x + 4$
12. $y = 6x + 4$
13. $y - 5 = \frac{3}{2}(x - 4)$
14. $y - 5 = -3(x - 4)$
15. $y - 5 = 4(x - 4)$

Assess

Selected Answers
Exercises 5–15, 17–83 odd

ASSIGNMENT GUIDE

In Class	1–15
Core	17–65 odd
Core Plus	16–60 even, 61–67
Review	68–84
Preview	85, 86

✎ Extra Practice can be found beginning on page 738.

Technology
You might suggest that students use a graphics calculator to check their answers in Exercises 40–57.

Error Analysis
Watch for students who simply read the coefficient of x as the slope, regardless of the form of equation they are given. You might reinforce the need to analyze the equation by working with them to create a visual such as the one below, which they can copy into their notebooks.

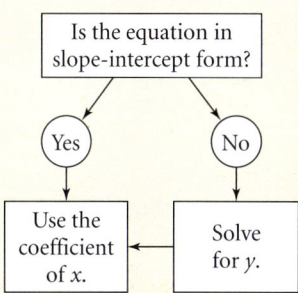

Finding Slope From an Equation

LESSON 5.6 **261**

56. Answers may vary. Sample:

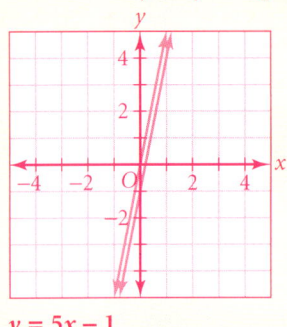

$y = 5x - 1$

57. Answers may vary. Sample:

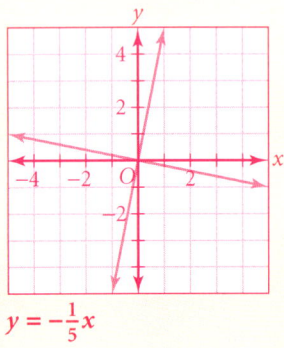

$y = -\frac{1}{5}x$

Find the slope of a line that is perpendicular to the following lines:

28. $y = -\frac{1}{3}x + 10$ **3**
29. $-\frac{1}{2}x - y = 20$ **2**
30. $13 = -x + y$ **−1**
31. $3x + 12y = 12$ **4**
32. $y = 5x + 10$ $-\frac{1}{5}$
33. $3x + y = 2$ $\frac{1}{3}$
34. $20 = -5x + 2y$ $-\frac{2}{5}$
35. $4x - 4y = 12$ **−1**
36. $2y = 5x + 11$ $-\frac{2}{5}$
37. $-4x + 8y = 17$ **−2**
38. $12x - 3y = 10$ $-\frac{1}{4}$
39. $4y = 20x - 3$ $-\frac{1}{5}$

Write an equation in a standard form for a line containing the point (2, 3) and parallel to each of the following lines:

40. $x + y = 1$
41. $3x = 7y + 2$
42. $y = 2x - 3$
43. $3y = 2x$
44. $7x - 2y = 10$
45. $11 = 3y + 2x$

Write an equation in slope-intercept form for each line according to the given information.

	Contains	Is parallel to
46.	(3, −5)	$5x − 2y = 10$
47.	(−2, 7)	$y = 3x − 4$
48.	(2, 4)	$y = 7$
49.	(2, −4)	$y = 3x − 4$
50.	(−1, 4)	$y = 2x + 5$

	Contains	Is perpendicular to
51.	(3, −5)	$5x + 2y = 10$
52.	(2, 7)	$y = 3x − 4$
53.	(2, −4)	$y = 7$
54.	(2, −4)	$3x + y = 5$
55.	(−1, 4)	$y = 2x − 5$

Graph the line $y = 5x$.

56. Sketch a line that is parallel to $y = 5x$. Write its equation.

57. Sketch a line that is perpendicular to $y = 5x$. Write its equation.

What is the slope of a line that is

58. parallel to a horizontal line?   **0**
59. perpendicular to a horizontal line?   **Undefined**
60. parallel to a vertical line?   **Undefined**
61. perpendicular to a vertical line?   **0**

CONNECTIONS

GEOMETRY Write the equations of four lines that intersect to form a square whose sides are

62. *parallel* to the axes.
63. *not parallel* to the axes.

64. GEOMETRY If one side of a square is contained by the line $y = \frac{3}{4}x + 5$, write equations of lines that could contain the other 3 sides.

APPLICATION

65. PHYSICS Write the formulas for converting from Fahrenheit, F, to Celsius, C, and to Kelvin, K. Rewrite them by using x for F and y for K and C. Graph the two equations on the same coordinate plane.

66. What is the relationship between the two lines in Exercise 65? Find the slope of these two lines.

CHALLENGE

67. What relationship do these two lines have to the lines formed by the equations for converting from Celsius and Kelvin to Fahrenheit?

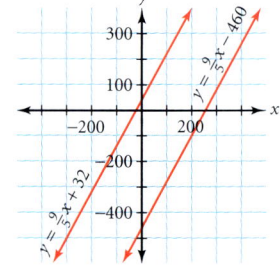

The equation $y = \frac{9}{5}x + 32$ converts Celsius to Fahrenheit. $y = \frac{9}{5}x - 460$ converts Kelvin to Fahrenheit.

40. $x + y = 5$
41. $3x - 7y = -15$
42. $2x - y = 1$
43. $2x - 3y = -5$
44. $7x - 2y = 8$
45. $2x + 3y = 13$

46. $y = \frac{5}{2}x - \frac{25}{2}$
47. $y = 3x + 13$
48. $y = 4$
49. $y = 3x - 10$
50. $y = 2x + 6$
51. $y = \frac{2}{5}x - \frac{31}{5}$
52. $y = \frac{-1}{3}x + \frac{23}{3}$
53. $x = 2$
54. $y = \frac{1}{3}x - \frac{14}{3}$
55. $y = -\frac{1}{2}x + \frac{7}{2}$

66. Each line has a slope of $\frac{5}{9}$. They are parallel.

67. These lines are less steep.

62. Answers may vary. Sample:
$x = 4$ $\quad$ $x = -4$
$y = 4$ $\quad$ $y = -4$

63. Answers may vary. Sample:
$y = x$ $\quad$ $y = x + 3$
$y = -x$ $\quad$ $y = -x + 3$

64. Answers may vary. Sample:
$y = \frac{3}{4}x - 5$ $\quad$ $y = -\frac{4}{3}x + 5$
$y = -\frac{4}{3}x - \frac{10}{3}$

65. $y = \frac{5}{9}x - \frac{160}{9}$
$y = \frac{5}{9}x + \frac{2300}{9}$

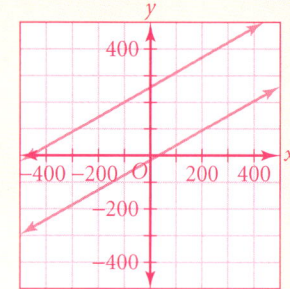

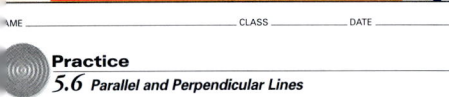

262 LESSON 5.6

 **Look Back**

68. Place parentheses to make $2 \cdot 7 + 35 \div 7 - 10 = 2$ true. *(LESSON 1.3)*
$2 \cdot (7 + 35) \div 7 - 10 = 2$

Evaluate each expression. *(LESSONS 2.2 AND 2.3)*

69. $-4 + (-3) + 1$ -6
70. $-2 + 3 + (-7) + 3$ -3
71. $-12 + 4 - (-4)$ -4
72. $-3 + (-11) - (-8)$ -6
73. $-5 + (-2) - 13$ -20
74. $8 - (-5) + 23$ 36
75. $15 + (-11) - 6$ -2
76. $26 + (-16) - (-8)$ 18

Simplify. *(LESSON 2.6)*

77. $5x^2 + 7y$
78. $3x + 7y + 4$
79. $5x^2 + 10xy + 2x$
80. $5x + 10y + 5$
81. $15x^2 + 8x + 8$
82. $4xy + 2x + 4y$
83. $13x^2 + 5xy + 2x$
84. $4y^2 + 15y + 6xy$

77. $2x^2 + 3y + 4y + 3x^2$
78. $3x + 2 + 4y + 2 + 3y$
79. $2x + 3xy + 5x^2 + 7xy$
80. $3x + 4y + 2x + 5 + 6y$
81. $4x^2 + 5x + 8 + 11x^2 + 3x$
82. $3xy + 2x + 4y + xy$
83. $9x^2 + 5xy + 2x + 4x^2$
84. $4y^2 + 12y + 6xy + 3y$

 **Look Beyond**

How many ordered pairs will satisfy two linear equations simultaneously if the equations have graphs

85. that are parallel? Zero
86. that are perpendicular? One

Include this activity in your portfolio.

One type of scientific research involves determining correlations between sets of data. For example, for many years researchers have studied the correlation between high speeds and automobile accident rates. Use current newspapers or other media to find reports that claim a correlation between sets of data. For each report, answer the following questions in your portfolio:

- Are you convinced that the claimed correlation actually exists? Why or why not?
- What evidence is used to support the claimed correlation?
- Does the report use graphs to represent the correlation? If so, describe whether or not the graph accurately represents the claimed correlation. If a graph is not used, explain how a graph might help or hinder the report's claim.

WORKING ON THE CHAPTER PROJECT

You should now be able to complete Activity 3 of the Chapter Project on page 264.

 **Look Beyond**

Exercises 85 and 86 preview systems of equations. Students will study this topic in greater detail in Chapter 7.

ALTERNATIVE
Assessment

Portfolio Activity

The Portfolio Activity can be used as preparation for the Chapter Project or as a separate activity. In the Portfolio Activity on this page, students research and analyze sets of data for which there is a purported correlation.

Answers to Portfolio Activities can be found in Additional Answers of the Teacher's Edition.

Student Technology Guide

NAME _____ CLASS _____ DATE _____

Student Technology Guide
5.6 *Parallel and Perpendicular Lines*

A graphics calculator is useful for determining whether two lines are parallel or perpendicular. However, it is a good idea to check your work with algebraic techniques because the results shown on a graphics calculator can be misleading.

Example: Tell whether the lines $y = 0.5x + 2$ and $y = 0.55x - 1$ are parallel, perpendicular, or neither.
The lines may appear to be parallel, but the slopes, 0.5 and 0.55, are not quite the same. The lines are neither parallel nor perpendicular.

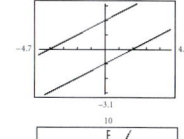

Example: Tell whether the lines $y = 3x + 1$ and $y = -\frac{1}{3}x - 2$ are parallel, perpendicular, or neither.
If you look at the graph in the viewing window shown at right, lines do not appear to be parallel or perpendicular. But the slopes, 3 and $-\frac{1}{3}$, are negative reciprocals. How can this be?

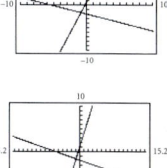

When viewing perpendicular lines, it is important to use a square viewing window—that is, a viewing window in which one unit on the x-axis is the same length as one unit on the y-axis. On many graphics calculators, you can convert any viewing window to a square window by pressing ZOOM 5:ZSquare ENTER. This converts the window above to the one at right.
The lines $y = 3x + 1$ and $y = -\frac{1}{3}x - 2$ graphed at the right are perpendicular.

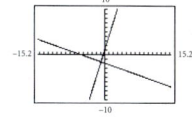

Tell whether the lines in each pair are parallel, perpendicular, or neither.

1. $y = \frac{2}{3}x - 1$, $y = \frac{2}{3}x + 4$ parallel
2. $y = \frac{5}{2}x - 3$, $y = \frac{2}{5}x + 4$ neither
3. $y = \frac{3}{8}x + 2$, $y = -\frac{8}{3}x - 1$ perpendicular
4. $9x + 4y = 18$, $2x + y = 5$ neither
5. $7x - 3y = 4$, $3x + 7y = 8$ perpendicular
6. $2x + 5y = 10$, $4x + 10y = 3$ parallel

LESSON 5.6 **263**

Project

Focus

The average speeds at the Indianapolis 500 auto race over the years can be modeled by a linear equation. Students are asked to make a scatter plot of the speeds for selected years and approximate a line of best fit. Then they research a different set of data and find a linear equation that models it.

Motivate

Discuss the data for the Indianapolis 500 auto race. Ask students what accounts for the obvious increase in speed over the years. Encourage students who are familiar with automobile racing to share some of their technical knowledge. Students who are familiar with the Indianapolis 500 can share their knowledge of the history of that race.

Ask students what they envision for the future. Do they think the speeds will increase steadily? Do they think there will come a time when the speeds will level off? Or do they think that, at some future time, some technological advance will cause a dramatic increase in speeds?

Activity 1

1. (graph shows years on the *x*-axis, and speed in mph on the *y*-axis)

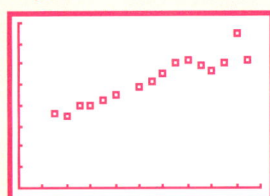

2. Answers will vary. Students' lines should approximate $y = 1.23x + 73.09$, shown below.

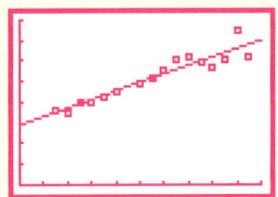

CHAPTER PROJECT FIVE

Modeling Trends IN DATA

The scoreboard shows the average speed at the Indianapolis 500 from 1915 to 1995 in 5-year intervals.

Over the last 100 years, technological improvements have significantly increased the top speeds for automobiles. Speeds at races, such as the Indianapolis 500, have steadily increased. This increase in speed over time can be approximately modeled by a linear equation. In this project, you will use your knowledge of linear equations to model real-world data.

Activity 1

1. Use the data from the scoreboard above to create a scatter plot with years on the horizontal axis and speeds on the vertical axis.

2. Use a clear ruler to draw a line of best fit on your scatter plot.

3. Find the slope and *y*-intercept of your line of best fit, and write the equation for the line.

4. Based on your line of best fit, estimate the speeds for 2000, 2005, and the year of the most recent Indianapolis 500. Use an almanac or other information source to find the actual speeds for the years that you estimated. What might explain the difference between your estimates and the actual speeds?

3. The slope is approximately 1.2 and the *y*-intercept is approximately 73.

4. For 2000, the prediction is 175.644. For 2005, the prediction is 180.771. Answers about differences between estimates and actual speeds will vary. sample answers: rule changes that result in slower race times, technology that allows a sudden jump in the average speed upwards

Activity 2

1.

L1	L2	L3 2
15	89.84	------
20	88.16	
25	101.13	
30	100.45	
35	106.24	
40	114.28	
50	124	

L2 ={89.84,88.16...

Linear equations can be useful for representing real-world data even when the data does not fit exactly on a line. The line of best fit you drew for Activity 1 is an example of a line used to model a trend in data. The line of best fit is also called a regression line (the data tends to "regress" to the line). Regression lines can help you to analyze trends in data.

Activity 2 (Optional, requires the use of a graphics calculator)

Graphics calculator technology allows you to be more precise with the scatter plot and regression line. The calculator will automatically calculate the slope and the y-intercept of the regression line.

1. Select the **Stat** key of your calculator. Place the information from the table into a two-variable (x, y) data table.

2. Locate and select **Reg** or **LinReg** for the linear regression. Here, the slope for the line of best fit is a, and the y-intercept is b.

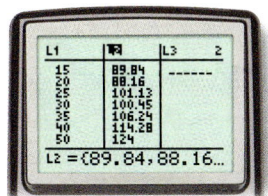

3. The slope, represented by a on the calculator screen, is the rate of increase in the average speed, about 1.03. Use the slope and the y-intercept (represented by b) to write the equation for the regression line (line of best fit). Graph the regression line by entering the equation after **Y=** on your calculator.

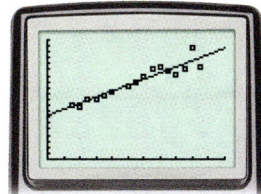

Activity 3

Choose one of the topics listed at right or one of your own that can be modeled by linear regression. You will need to do research in the library, on the Internet, or by conducting your own experiment in order to find data for your topic. Write a report about your data. In your report, include a scatter plot of your data showing the regression line and the equation of the regression line. Discuss the type of correlation represented by your data. Make predictions for data points that are not shown on your scatter plot, and discuss why the actual data may vary. Explain why linear regression is or is not a good way to model your data.

Suggested topics:
- air temperature and the number of cricket chirps per minute
- foot length and hat size
- commuter travel time and the distance from the destination
- number of advertisements and sales
- years of experience and salary

Cooperative Learning

Have students work in pairs. In **Activity 1**, each student should draw the scatter plot and determine a line of best fit. The partners can then compare lines of best fit to see how closely they match. Stress the fact that when using this technique, the lines of best fit are approximations. That is, there is no one "right answer" for the line of best fit. To research the data, one partner can do library research while the other searches the Internet.

In **Activity 2**, the students can work together to determine the correct calculator procedure for finding the regression equation. Be sure students understand that the calculator uses a well-defined algorithm to find the regression equation, so this time there will be just one "right answer."

In **Activity 3**, suggest that the students reverse their roles from Activity 1. Partners can share the data and analyze it together. However, you may want each partner to prepare an independent report of the results.

Discuss

Bring the class together to discuss the results of Activity 1. Have each group report their predictions for speeds in future races. Ask students to explain any differences among the predictions.

Have each group report the results of their research from Activity 3. Give other students the opportunity to ask questions about the results.

Activity 3
Student solutions should model Activity 1 or Activity 2.

2.

3.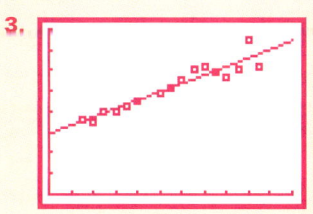

Chapter Review

5 Chapter Review and Assessment

VOCABULARY

constant of variation 238	parallel 259	run 226
direct variation 238	perpendicular 259	slope 228
domain 219	point-slope form 254	slope-intercept form 245
function 219	range 219	standard form 252
function notation 223	rate of change 236	vertical line 230
Hooke's law 239	relation 218	x-intercept 247
horizontal line 230	rise 226	y-intercept 245

Key Skills & Exercises

LESSON 5.1

Key Skills

Determine whether a given relation is a function.

Determine whether the set of ordered pairs below is a function.

$$\{(4, 5), (5, 7), (7, 12), (4, 9)\}$$

The ordered pairs (4, 5) and (4, 9) have the same first coordinate. This is not a function.

Describe the domain and range of the following relations:

$$\{(2, 3), (7, 2), (8, 4), (9, 13), (6, 12)\}$$

The domain is {2, 6, 7, 8, 9}.
The range is {2, 3, 4, 12, 13}.

Exercises

Determine whether each relation is a function.

1. {(7, 12), (6, 9), (7, 8), (4, 11)} Not
2. {(5, 5), (10, 7), (7, 15), (4, 9)} Function
3. {(8, 5), (5, 3), (8, 12), (1, 5)} Not
4. {(6, 5), (3, 7), (7, 9), (8, 9)} Function

Describe the domain and range for each relation.

5. {(8, 5), (2, 5), (3, 12), (6, 5)}
6. {(5, 5), (7, 6), (12, 7), (13, 24)}
7. {(2, 5), (4, 7), (1, 9), (8, 5)}
8. {(34, 4), (4, 15), (10, 4), (9, 17)}

5. D: {2, 3, 6, 8}; R: {5, 12}
6. D: {5, 7, 12, 13}; R: {5, 6, 7, 24}
7. D: {1, 2, 4, 8}; R: {5, 7, 9}
8. D: {4, 9, 10, 34}; R: {4, 15, 17}

LESSON 5.2

Key Skills

Find the slope of a line from the rise and run.

If the rise between two points on a line is 6 and the run between the same two points is 2, find the slope of the line.

$$\text{slope} = m = \frac{6}{2} = 3$$

Exercises

Find the slope of a line with the given rise and run.

9. rise of 4 run of 8 $\frac{1}{2}$
10. rise of 16 run of 4   4
11. rise of 12 run of 18 $\frac{2}{3}$
12. rise of 8 run of 12 $\frac{2}{3}$

Chapter Test, Form A

Chapter Assessment
Chapter 5, Form A, page 1

Write the letter that best answers the question or completes the statement.

a 1. What is the slope of a line that passes through the origin and the point (−2, 1)?
   a. $-\frac{1}{2}$ b. 2   c. $\frac{1}{2}$ d. −2

d 2. What is the slope of a line with a rise and a run of 4?
   a. $\frac{4}{5}$ b. $-\frac{4}{5}$ c. $\frac{5}{4}$ d. 1

c 3. An equation of a line that has a slope of 0 and contains the point (0, −2) is
   a. $x = 2$.   b. $x = -2$.   c. $y = -2$.   d. $y = x - 2$.

c 4. In a graph of a linear function, the change in y is 3. The change in x is 9. The rate of change is
   a. 3.   b. 9.   c. $\frac{1}{3}$.   d. $\frac{1}{9}$.

b 5. The slope of the line containing the points (2, −4) and (3, −7) is
   a. −11.   b. −3.   c. $-\frac{1}{3}$.   d. 3.

d 6. Write the equation $y = \frac{1}{2}x + 5$ in standard form.
   a. $x - 2y = 10$ b. $x - 2y - 10 = 0$ c. $-x + 2y = 10$ d. $x - 2y = -10$

a 7. Which line is parallel to $y = \frac{1}{4}x + 2$?
   a. $y = \frac{1}{4}x + 5$ b. $y = -4x + 5$ c. $y = -\frac{1}{4}x + 5$ d. $y = 4x + 5$

c 8. If y varies directly as x and y = 8 when x = 3, find an equation of direct variation.
   a. $y = \frac{3}{8}x$ b. $y = -\frac{8}{3}x$ c. $y = \frac{8}{3}x$ d. $y = 8x$

Chapter Assessment
Chapter 5, Form A, page 2

c 9. If y varies directly as x and $y = \frac{2}{3}$ when $x = \frac{1}{2}$, find an equation of direct variation.
   a. $y = \frac{3}{4}x$ b. $y = -\frac{4}{3}x$ c. $y = \frac{4}{3}x$ d. $y = -\frac{3}{4}x$

b 10. What are the slope and y-intercept of the equation $3x - 4y = 4$?
   a. slope: −3; y-intercept: −4   b. slope: $\frac{3}{4}$; y-intercept: −1
   c. slope: 3; y-intercept: 4   d. slope: $\frac{3}{4}$; y-intercept: −4

a 11. What is the slope of a line perpendicular to $x - 3y = 4$?
   a. −3   b. 3   c. $\frac{1}{3}$ d. $-\frac{1}{3}$

c 12. Which line contains the point (0, 2) and is perpendicular to $y = \frac{1}{4}x + 2$?
   a. $y = \frac{1}{4}x + 1$ b. $y = -\frac{1}{4}x + 2$ c. $y = -4x + 2$ d. $y = 4x + 2$

b 13. Which point is on the graph of the equation $3x + 6y = 27$?
   a. (0, 9)   b. (5, 2)   c. (1, −5)   d. (−5, 7)

b 14. An equation for the line that contains (2, 4) and has an undefined slope is
   a. $y = 2x + 4$.   b. $x = 2$.   c. $y = 2$.   d. $y = 0$.

c 15. An equation for the line that crosses the x-axis at $x = -1$ and the y-axis at $y = 3$ is
   a. $3x + y = 3$.   b. $3x = y + 3$.   c. $y = 3x + 3$.   d. $y + 3x = -3$.

b 16. Which equation represents a horizontal line?
   a. $y = x$ b. $y = -2$ c. $x = 0$ d. $y = 2x + 1$

266 CHAPTER 5 REVIEW

Find the slope of a line from two points.

Find the slope of a line containing the points $A(-2, 4)$ and $B(3, 5)$.

slope $= m = \dfrac{\text{difference in } y\text{-values}}{\text{difference in } x\text{-values}} = \dfrac{5-4}{3-(-2)} = \dfrac{1}{5}$

Find the slope of the line containing the given points.

13. $A(-3, 2)$, $B(2, 3)$ $\dfrac{1}{5}$

14. $A(2, 5)$, $B(4, 1)$ -2

15. $A(-5, 4)$, $B(1, 4)$ 0

LESSON 5.3
Key Skills

Find the rate of change from a graph.

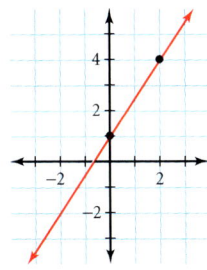

To find the rate of change from the linear graph, find the slope of the line.

Use the slope formula and points $(0, 1)$ and $(2, 4)$.

$m = \dfrac{y_2 - y_1}{x_2 - x_1} = \dfrac{4 - 1}{2 - 0} = \dfrac{3}{2}$

The slope of a line is referred to as the rate of change because the slope is the rate that the y-values change compared to the rate that the x-values change.

Exercises

Find the rate of change for each linear graph.

16. 1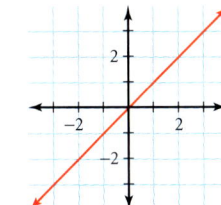

17. $-\dfrac{1}{2}$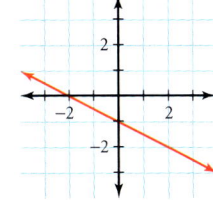

18. -3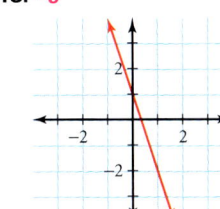

19. $\dfrac{3}{2}$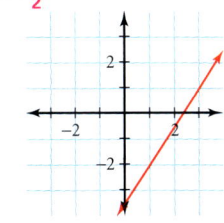

20. -2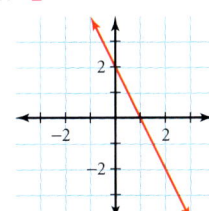

21. 2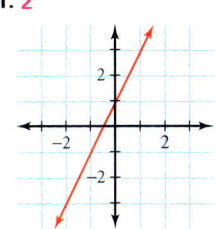

22. $\dfrac{1}{2}$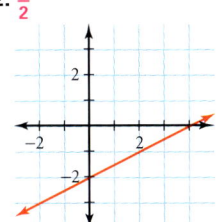

23. $\dfrac{1}{2}$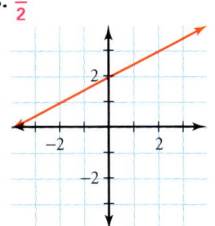

Chapter Test, Form B

NAME _____ CLASS _____ DATE _____

Chapter Assessment
Chapter 5, Form B, page 1

Each pair of points is on a line. Find the slope of each line.
1. $M(0, 1), N(1, 1)$ __0__   2. $M\left(4, 2\tfrac{1}{2}\right), N(0, 2)$ __$\tfrac{1}{8}$__

Find an equation of a line that contains the origin and the given point.
3. $(2, -2)$ __$y = -x$__   4. $(3, 4)$ __$y = \tfrac{4}{3}x$__

Suppose that y varies directly as x. Write an equation for the direct variation.
5. when $x = 2, y = 8$ __$y = 4x$__   6. $y = 3.6$ when $x = 6$ __$y = \tfrac{3}{5}x$, or $y = 0.6x$__

Find the y-intercept of the graph of each equation.
7. $y = -3x$ __0__   8. $3x - 6y = 12$ __-2__

Write an equation for the specified line.
9. slope: 4; y-intercept: -3 __$y = 4x - 3$__   10. contains $(2, -9)$; slope: -1 __$y = -x - 7$__
11. contains $(3, -1)$ and $(6, -3)$ __$y = -\tfrac{2}{3}x + 1$__   12. contains $(6, -12)$; slope: -2 __$y = -2x$__

Marta worked 6 hours and earned $51.
13. Write a direct-variation equation that models the situation. __$y = \$8.50x$__
14. At the same rate, how much will Marta earn in 9 hours? __$\$76.50$__
15. Write the equation $y + 1 = \tfrac{1}{2}(x - 6)$ in standard form. __$x - 2y = 8$__
16. Write the equation $8x + 4y = 24$ in slope-intercept form. __$y = -2x + 6$__
17. Robert ran 6 miles in 72 minutes. What was his speed in miles per hour? __5 miles per hour__

Write an equation for each line described below. Write your answer in slope-intercept form.
18. crosses the x-axis at $x = -2$ and the y-axis at $y = -1$ __$y = -\tfrac{1}{2}x - 1$__
19. contains the point $(2, 4)$ and is perpendicular to the graph of $y = \tfrac{1}{3}x + 1$ __$y = -3x + 10$__
20. contains the point $(-1, 0)$ and is parallel to the graph of $2x + 3y = 6$ __$y = -\tfrac{2}{3}x - \tfrac{2}{3}$__

NAME _____ CLASS _____ DATE _____

Chapter Assessment
Chapter 5, Form B, page 2

21. Graph the line $y = 3x$. Then sketch the graph of a line that is parallel to $y = 3x$ and write its equation.

22. Graph the line $y = \tfrac{3}{4}x + 2$. Then sketch the graph of a line that is perpendicular to $y = \tfrac{3}{4}x + 2$ and write its equation.

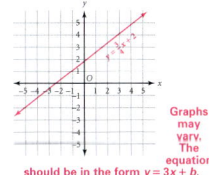 Graphs may vary. The equation should be in the form $y = 3x + b$.

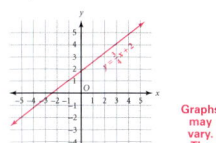 Graphs may vary. The equation should be in the form $y = -\tfrac{4}{3}x + b$.

23. Draw the graph of the line with an x-intercept of $\tfrac{4}{3}$ and y-intercept of $\tfrac{4}{3}$.

24. Graph the equation $x - 4y = 2$ by finding the intercepts.

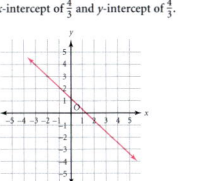

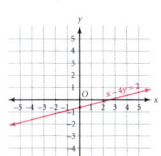

CHAPTER 5 REVIEW **267**

47.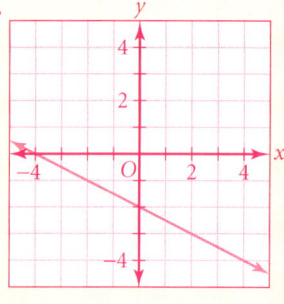

48.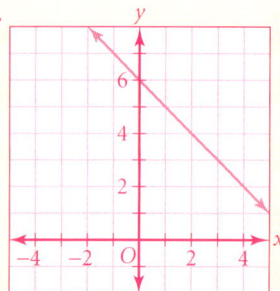

49.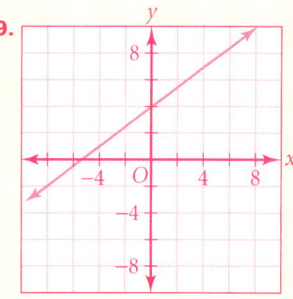

50.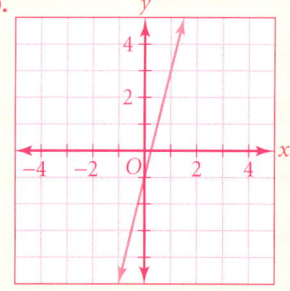

Solve problems of direct variation.

To find the constant of variation for a spring by using Hooke's law, substitute values for the force and distance and solve for the constant, k. A force of 8 Newtons stretches a spring 3 meters.

$$F = kd$$
$$8 = k(3)$$
$$k = \frac{8}{3}$$
$$k \approx 2.7$$

If y varies directly as x and $y = 24$ when $x = 6$, find the constant of variation, k, and write an equation of direct variation.

Substitute 24 for y and 6 for x in $\frac{y}{x} = k$ to find the constant of variation, k.

$$\frac{24}{6} = k \qquad k = 4$$

Use the constant of variation, k, to write an equation of direct variation. Substitute 4 for k.

$$\frac{y}{x} = k$$
$$\frac{y}{x} = 4$$
$$y = 4x$$

Use Hooke's law, $F = kd$, to find the constant, k, for each spring described below.

24. force: 2 Newtons; stretch: 0.6 meter $3\frac{1}{3}$
25. force: 10 Newtons; stretch: 0.4 meter 25
26. force: 10 Newtons; stretch: 1.0 meter 10
27. force: 5 Newtons; stretch: 0.1 meter 50
28. force: 20 Newtons; stretch: 0.8 meter 25

If y varies directly as x for the values of x and y below, find the constant of variation, k, and write an equation of direct variation.

29. $y = 4, x = 10$ $k = \frac{2}{5}; y = \frac{2}{5}x$
30. $y = 16, x = 8$ $k = 2; y = 2x$
31. $y = -5, x = 15$ $k = \frac{-1}{3}; y = \frac{-1}{3}x$
32. $y = -12, x = 18$ $k = \frac{-2}{3}; y = \frac{-2}{3}x$
33. $y = -4, x = -16$ $k = \frac{1}{4}; y = \frac{1}{4}x$
34. $y = 2, x = 2$ $k = 1; y = x$
35. $y = 4, x = -12$ $k = \frac{-1}{3}; y = \frac{-1}{3}x$
36. $y = 6, x = 9$ $k = \frac{2}{3}; y = \frac{2}{3}x$

LESSON 5.4

Key Skills

Find an equation in slope-intercept form for a line.

Write an equation for the line
a. with a slope of 3 and y-intercept of 5.
b. passing through $(-2, 3)$ and $(-1, 5)$.

a. Substitute 3 for m and 5 for b.
$$y = mx + b$$
$$y = 3x + 5$$

b. $m = \frac{5-3}{-1-(-2)} = 2$

Substitute 2 for m, -2 for x, and 3 for y in the equation $y = mx + b$. Then solve for b.
$$3 = 2(-2) + b$$
$$3 = -4 + b$$
$$7 = b$$

Exercises

Write an equation in slope-intercept form for each line described below.

37. slope of 2 and y-intercept of 1 $y = 2x + 1$
38. slope of $-\frac{1}{2}$ and y-intercept of 5 $y = -\frac{1}{2}x + 5$
39. containing $(0, -4)$ with a slope of 8 $y = 8x - 4$
40. containing $(4, 2)$ and $(5, -4)$ $y = -6x + 26$
41. containing $(2, -1)$ and $(-3, -1)$ $y = -1$
42. containing $(-6, 0)$ and $(0, -2)$ $y = -\frac{1}{3}x - 2$
43. slope of -3 and y-intercept of 6 $y = -3x + 6$
44. containing $(4, 6)$ and $(-2, 0)$ $y = x + 2$
45. slope of 5 and y-intercept of -2 $y = 5x - 2$
46. containing $(2, 5)$ and $(-6, 9)$ $y = -\frac{1}{2}x + 6$

51.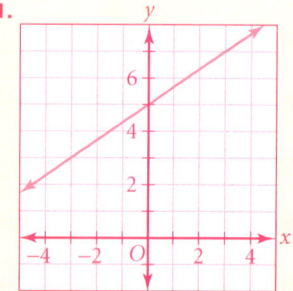

Substitute 2 for *m* and 7 for *b*.

$$y = mx + b$$
$$y = 2x + 7$$

Graph an equation in slope-intercept form.

Graph the equation $y = \frac{1}{2}x + 5$.

The slope is $\frac{1}{2}$ and the *y*-intercept is 5.

Locate (0, 5) and move 2 units to the right and 1 unit up from that point. Graph this point. Draw the line containing the points.

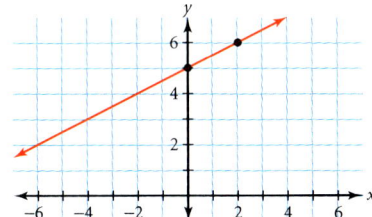

Graph each equation.

47. $y = -\frac{1}{2}x - 2$
48. $y = -x + 6$
49. $y = \frac{3}{4}x + 4$
50. $y = 4x - 1$
51. $y = \frac{2}{3}x + 5$
52. $y = -\frac{5}{4}x + 5$
53. $y = \frac{1}{2}x + 6$
54. $y = x + 4$
55. $y = -2x - 3$
56. $y = -4x + 1$

54.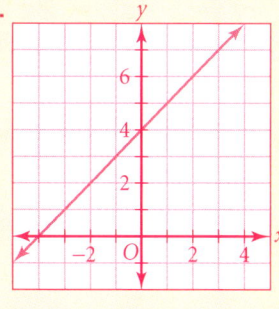

55.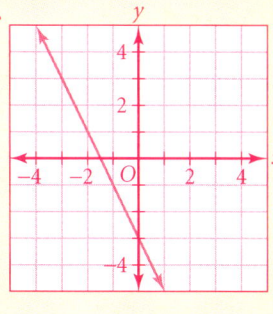

56.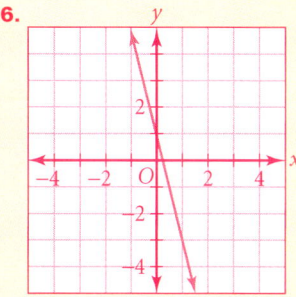

LESSON 5.5

Key Skills

Write equations in standard form.

Write $y + 10 = -8x - 22$ in standard form.

$$y + 10 = -8x - 22$$
$$8x + y + 10 = -22$$
$$8x + y = -32$$

Write equations in point-slope form.

Write an equation for the line containing the points (4, 5) and (2, 1).

$$m = \frac{1-5}{2-4} = \frac{-4}{-2} = 2$$

Use the point (4, 5) for (x_1, y_1).

$$y - 5 = 2(x - 4)$$

Exercises

Write the following equations in standard form:

57. $y + 9 = 4x - 8$
    $4x - y = 17$
58. $y - 4 = -x + 1$
    $x + y = 5$
59. $y - 13 = 2x + 4$
    $2x - y = -17$
60. $3x + y + 6 = 9$
    $3x + y = 3$

Write an equation in point-slope form for the following lines:

61. with a slope of 2 and a *y*-intercept of -1
    $y - (-1) = 2(x - 0)$
62. containing (0, 4) and (1, 2)
    $y - 2 = -2(x - 1)$ or $y - 4 = -2(x - 0)$
63. containing (3, 5) with a slope of 4
    $y - 5 = 4(x - 3)$
64. crossing the *x*-axis at $x = -2$ and the *y*-axis at $y = 1$ $y - 1 = \frac{1}{2}(x - 0)$ or $y - 0 = \frac{1}{2}(x - (-2))$
65. with a slope of 3 and *y*-intercept of -3
    $y - (-3) = 3(x - 0)$
66. containing $(-2, 4)$ with a slope of -2
    $y - 4 = -2(x - (-2))$
67. crossing the *x*-axis at $x = 3$ and the *y*-axis at $y = 6$ $y - 0 = -2(x - 3)$ or $y - 6 = -2(x - 0)$
68. with a slope of -4 and *y*-intercept of -1
    $y - (-1) = -4(x - 0)$

52.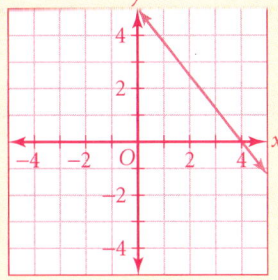

53.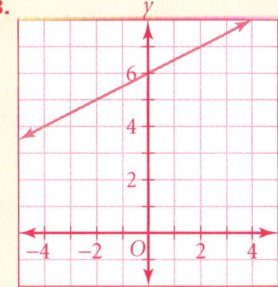

78. Answers may vary. Sample: The rate of change doesn't change, but its numerical value does because the units of measurement are different.

LESSON 5.6
Key Skills

Recognize and write equations of lines parallel and perpendicular to other lines.

Write an equation in slope-intercept form for the line containing the point (0, 4) and

a. parallel to the line $2x + 3y = 5$.
b. perpendicular to the line $2x + 3y = 5$.
Find the slope of the line $2x + 3y = 5$ by writing the equation in slope-intercept form.

$$y = -\frac{2}{3}x + \frac{5}{3}$$

a. The slope of the parallel line is $-\frac{2}{3}$. The y-intercept is 4. The equation of the parallel line is $y = -\frac{2}{3}x + 4$.

b. The slope of the perpendicular line is the negative reciprocal of $-\frac{2}{3}$, or $\frac{3}{2}$. The equation is $y = \frac{3}{2}x + 4$.

Exercises

Find the slopes of lines that are parallel and perpendicular to the following lines:

69. $y = \frac{2}{3}x + 4$ parallel: $\frac{2}{3}$; perp.: $-\frac{3}{2}$
70. $y = -7x - 1$ parallel: -7; perp.: $\frac{1}{7}$

Write an equation for the line that contains the point (1, 5) and is

71. parallel to $2x + y = -1$. $y - 5 = -2(x - 1)$
72. perpendicular to $2x + y = -1$. $y - 5 = \frac{1}{2}(x - 1)$

Applications

73. GEOGRAPHY The elevation of Houston is 120 feet. The elevation of El Paso is 4000 feet. Suppose that the distance between the two cities is a straight line of 760 miles. Traveling from Houston to El Paso, what is the average slope of the land? **0.00097 or 5 feet per mile**

TRAVEL Marie had driven 110 miles when she stopped at a gas station. After she left the station, she kept track of the hours she drove. Based on an average of 50 miles per hour for the rest of the trip, write an equation to find the total distance she drove. Determine the total distance traveled if she drove the following number of hours after leaving the gas station: **Use $d = 110 + 50t$**

74. 1.6 hours **190 mi.**   **75.** 2.9 hours **255 mi.**   **76.** 3.3 hours **275 mi.**

77. METEOROLOGY If the air temperature is 76°F at 8 A.M. and 92°F at 4 P.M., find the rate of change of temperature per hour.
2 degrees per hour

78. METEOROLOGY Would the rate of change in Exercise 77 be different if you used the Celsius temperature scale and converted hours to minutes? Write a paragraph explaining your answer.

79. INCOME A person's rate is the amount that he or she earns per hour. If you earn $500 for a 40-hour work week, what is your pay rate? **$12.50 per hour**

Alternative Assessment

Performance Assessment

1. Many electricians use a fixed-cost equation to determine how much to charge for a house call. For example, an electrician might charge $45 for coming to your house and $25 for each hour of service. The fixed-cost equation is $y = 45 + 25x$.
   a. How are the slope and y-intercept related to the equation?
   b. Explain why the fixed-cost equation is a linear function.
   c. What is the rate of change in the fixed-cost equation? How is the rate of change related to the slope of the line represented by the equation? Why?

2. Explain how to graph an equation of the form $y = mx + b$.

3. Give an example of a direct-variation problem.
   a. Set up an equation to find the constant of variation.
   b. Use the constant of variation to solve the problem.

4. Write the equation $2x - y = 8$ in each of the following forms:
   a. slope-intercept form
   b. standard form
   c. point-slope form

Portfolio Projects

1. Use a diagram to show how slope is used to construct stairs.

2. Explain how to find a line of best fit on a graphic calculator. How is the slope of the line of best fit related to the correlation shown on a scatter plot.

3. Hooke's law is one example of a direct variation formula. Find at least two other examples of direct variation formulas used in science and write a report that illustrates their use.

4. Write a paragraph explaining the conditions needed for the graph of $Ax + By + C = 0$ to be parallel to the x-axis or y-axis. Include a graph for each condition.

Peer Assessment

Complete this activity with a partner. Begin by giving the coordinates of two points. Your partner should write an equation in slope-intercept form for the line containing the two points and then choose two different points.

Check your partner's work, correct any errors, and then write an equation for the line containing the two new points. Continue with this pattern until you have each written 10 equations. Then take turns rewriting each of the 20 equations in either standard or point-slope form.

internetconnect

The HRW Web site contains many resources to reinforce and expand your knowledge of linear functions. The Web site also provides Internet links to other sites where you can find information and real-world data for use in research projects, reports, and activities that involve a constant rate of change. Visit the HRW Web site at **go.hrw.com**, and enter the keyword **MA1 CH5** to access the resources for this chapter.

Alternative Assessment

Performance Assessment

1. a. The slope represents the hourly rate charged. The y-intercept represents the fixed charge for the visit.
   b. If the equation is written as $y = 25x + 45$, it is in the form $y = mx + b$.
   c. The rate of change is $25 per hour. The slope of the line represents the same information, namely how cost is changing over time.

2. b represents the y-intercept of the line. Travel along the y-axis until the value of b is reached. Draw a point to represent b. m, defined as $\frac{rise}{run}$, is the slope of the line. Use the slope to find another point by starting at point b, moving vertically the distance of the rise, and then moving horizontally the distance of the run. This process gives the second point. Connect the two points with a straight line.

3. The cost (c) of apples varies directly as the number of apples (n) purchased. 6 pounds of apples cost $3. How much would 15 pounds cost?
   a. $c = kn$
      $3 = k6$
      $k = \frac{1}{2}$
      So, $c = \frac{1}{2}n$.
   b. $c = \frac{1}{2}n$
      $c = \frac{1}{2}(15) = \$7.50$

4. a. $y = 2x - 8$
   b. $2x - y = 8$
   c. $y - (-8) = 2(x - 0)$

Portfolio Projects

1.

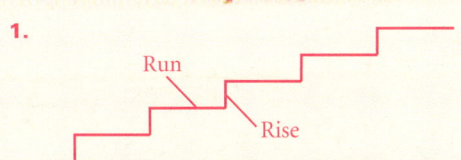

2. Answers may vary. Most graphics calculators have a linear regression feature. If the scatter plot has a positive correlation, the slope of the line of best fit is positive. If the scatter plot has a negative correlation, the slope of the line of best fit is negative.

3. Answers may vary. Students may research examples such as isothermal work.

4. Answers may vary. If $A = 0$ but $B \neq 0$, the line will be parallel to the x-axis. If $B = 0$ but $A \neq 0$, the line will be parallel to the y-axis.

$A = 0, B \neq 0, y = -\frac{C}{B}$

$A \neq 0, B = 0, x = -\frac{C}{A}$

Cumulative Assessment

College Entrance Exam Practice

Multiple-Choice and Quantitative-Comparison Samples

The first half of the Cumulative Assessment contains two types of items found on standardized tests—multiple-choice questions and quantitative-comparison questions. Quantitative-comparison items emphasize the concepts of equality, inequality, and estimation.

Free-Response Grid Samples

The second half of the Cumulative Assessment is a free-response section. This part of the Cumulative Assessment requires student-produced response items like those commonly found on college entrance exams. These questions require the use of machine-scored answer grids. You may wish to have students practice answering these items in preparation for standardized tests.

CHAPTERS 1-5 CUMULATIVE ASSESSMENT

College Entrance Exam Practice

QUANTITATIVE COMPARISON For Questions 1–6, write
A if the quantity in Column A is greater than the quantity in Column B;
B if the quantity in Column B is greater than the quantity in Column A;
C if the two quantities are equal; or
D if the relationship cannot be determined from the information given.

	Column A	Column B	Answers	
1.	The solution to the equation $4x + 7 = -13$	The solution to the equation $-3x + 11 = 29$	Ⓐ Ⓑ Ⓒ Ⓓ [Lesson 3.3]	A
2.	The slope of the line containing (7, 2) and (8, 7)	The slope of the line containing (−1, −3) and (2, 9)	Ⓐ Ⓑ Ⓒ Ⓓ [Lesson 5.2]	A
3.	$(-21)(-4)$	$(-7)(12)$	Ⓐ Ⓑ Ⓒ Ⓓ [Lesson 2.4]	A
4.	First differences of the sequence 2, 7, 12, 17, . . .	First differences of the sequence 1, 5, 14, 28, . . .	Ⓐ Ⓑ Ⓒ Ⓓ [Lesson 1.1]	D
5.	20% of 120	35% of 70	Ⓐ Ⓑ Ⓒ Ⓓ [Lesson 4.2]	B
6.	The slope of a horizontal line	The slope of a line containing (0, 1) and (9, 1)	Ⓐ Ⓑ Ⓒ Ⓓ [Lesson 5.2]	C

7. What is the solution to the equation $3p - 173 = 220$? **(LESSON 3.3)** b
   a. 78
   b. 131
   c. 15.6
   d. none of these

8. What number is missing from the following sequence? **(LESSON 1.1)** c
   12, 36, ___, 84, 108, 132
   a. 52
   b. 70
   c. 60
   d. 44

9. Evaluate the expression $-s + 2t$ for $s = -3$ and $t = 7$. **(LESSON 2.4)** b
   a. −7
   b. 17
   c. 11
   d. none of these

Solve each proportion. **(LESSON 4.1)**

10. $\frac{x}{7} = \frac{52}{13}$
    $x = 28$

11. $\frac{1.2}{y} = \frac{2.3}{13.8}$
    $y = 7.2$

12. $\frac{63}{102} = \frac{p}{64}$
    $p = 39\frac{9}{17}$

Write an equation for the line containing the origin and the points below. **(LESSON 5.2)**

13. (2, 1)
14. (−3, 6)
15. (4, −12)
16. (−1, −13)

17. Solve $-4w + 17 = 65$ for w. **(LESSON 3.3)**

13. $y = \frac{1}{2}x$
14. $y = -2x$
15. $y = -3x$
16. $y = 13x$
17. $w = -12$

272 CHAPTERS 1-5 CUMULATIVE ASSESSMENT

Simplify. *(LESSON 1.3)*

18. $4 \cdot 3 + 12 \div 2$ **18**
19. $[14(8 \div 4) + 7] \div 5$ **7**

Find each answer. *(LESSON 4.2)*

20. What is 15% of 110? **16.5**
21. 105 is 70% of what number? **150**
22. 12 is what percent of 50? **24%**

If pencils cost 29 cents, find the cost of the following: *(LESSON 1.2)*

23. 3 pencils **$0.87**
24. 5 pencils **$1.45**
25. 14 pencils **$4.06**
26. Write an equation to model this situation: How many pencils can you buy for $8.70? *(LESSON 1.2)* **$8.70 = 0.29p$**
27. Write an equation in standard form for the line graphed below. *(LESSON 5.5)*
    $0.5x + y = 0.75$

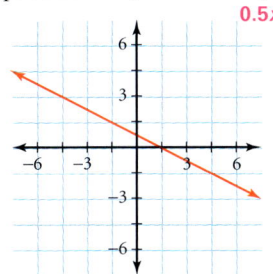

Write an equation in slope-intercept form for each line described below. *(LESSON 5.5)*

28. slope $= -3$, y-intercept $= 4$ **$y = -3x + 4$**
29. slope $= 7$, containing $(0, -2)$ **$y = 7x - 2$**
30. containing $(0, 2)$ and $(2, 5)$ **$y = \frac{3}{2}x + 2$**
31. Solve the equation $y - 7 = 24$. *(LESSON 3.1)*
    $y = 31$

Find the mean, median, mode, and range for each set of data. *(LESSON 4.4)*

32. $\{8, 11, 7, 9, 5\}$ **8; 8; none; 6**
33. $\{98, 91, 87, 100, 88\}$ **92.8; 91; none; 13**
34. $\{0.21, 0.17, 0.28, 0.14\}$ **0.2; 0.19; none; 0.14**
35. Determine the slope of the line represented by $2x + 4y = 1$. *(LESSON 5.5)* **$-\frac{1}{2}$**
36. Solve: $3x - 5 = 13$. *(LESSON 3.3)* **$x = 6$**
37. Solve: $4 + x = 6$. *(LESSON 3.1)* **$x = 2$**
38. Solve: $2x - 5 = 7$. *(LESSON 3.3)* **$x = 6$**
39. Solve: $x - 5 = 13$. *(LESSON 3.1)* **$x = 18$**
40. Solve: $\frac{38}{14} = \frac{a}{7}$. *(LESSON 4.1)* **$a = 19$**
41. Find 60% of 180. *(LESSON 4.2)* **108**
42. Determine the next three terms of the following sequence: 1, 4, 8, 13, 19, … *(LESSON 1.1)* **26, 34, 43**
43. Evaluate: $2 - [(3 \cdot 4 - 6 \cdot 7) \cdot 3 - 4]$. *(LESSON 1.3)* **96**
44. Find 30% of 180. *(LESSON 4.2)* **54**

FREE-RESPONSE GRID

The following questions may be answered by using a free-response grid such as that commonly used by standardized test services.

45. Find the slope of a line passing through $A(4, 5)$ and $B(-2, 2)$. *(LESSON 5.2)* $\frac{1}{2}$
46. Simplify: $5(7 + 13) - 40 \div 8$. *(LESSON 1.3)* **95**
47. What is 55% of 20? *(LESSON 4.2)* **11**
48. Solve the equation $\frac{t}{3.5} = 14.2$. *(LESSON 3.2)*
    $t = 49.7$

6 Inequalities and Absolute Value

CHAPTER PLANNING GUIDE

Lesson	6.1	6.2	6.3	6.4	6.5	Projects and Review
Pupil Edition Pages	276–281	282–288	289–293	294–299	300–307	308–315
Extra Practice (Pupil Edition)	753	754	754	755	755	
Practice Workbook	32	33	34	35	36	
Reteaching Masters	63–64	65–66	67–68	69–70	71–72	
Enrichment Masters	32	33	34	35	36	
Cooperative-Learning Activities	32	33	34	35	36	
Lesson Activities	32	33	34	35	36	
Problem Solving/ Critical Thinking	32	33	34	35	36	
Student Study Guide	32	33	34	35	36	
Spanish Resources	32	33	34	35	36	
Student Technology Guide	32	33	34	35	36	
Assessment Resources	71	72	73	75	76	74, 77–82
Teaching Transparencies	31				32	
Quiz Transparencies	6.1	6.2	6.3	6.4	6.5	
Writing Activities for Your Portfolio						16–18
Tech Prep Masters						25–28
Long-Term Projects						21–24

LESSON PACING GUIDE

	6.1	6.2	6.3	6.4	6.5	Projects and Review
Traditional	1 day	1 day	2 days	1 day	1 day	2 days
Block	$\frac{1}{2}$ day	$\frac{1}{2}$ day	1 day	$\frac{1}{2}$ day	$\frac{1}{2}$ day	1 day
Two-Year	2 days	2 days	4 days	2 days	2 days	4 days

CONNECTIONS AND APPLICATIONS

Lesson	6.1	6.2	6.3	6.4	6.5	Review
Algebra	276–281	282–288	289–293	294–299	300–307	310–315
Geometry		287		296	304	
Business and Economics	277, 279, 280	285, 286, 287	293		300, 307	312
Statistics				294		
Maximum/Minimum					301	
Transformations				298		
Life Skills					306	
Science and Technology	281		290, 293	299		
Sports and Leisure		287		299	302	312
Other	280		289		307	313

BLOCK-SCHEDULING GUIDE

Day	Lesson	Teacher Directed: Lesson Examples, Teaching Transparencies	Student Guided: Activity, Try This	Cooperative-Learning Activity, Lesson Activity, Student Technology Guide	Practice: Practice & Apply, Extra Practice, Practice Workbook	Assessment: Quiz, Mid-Chapter Assessment	Problem Solving, Reteaching
1	6.1	8 min	8 min	8 min	35 min	8 min	8 min
	6.2	7 min	7 min	7 min	30 min	7 min	7 min
2	6.3	15 min	15 min	15 min	65 min	15 min	15 min
3	6.4	5 min	8 min	8 min	35 min	8 min	8 min
	6.5	5 min	7 min	7 min	30 min	7 min	7 min
4	Assess.	50 min PE: Chapter Review	90 min PE: Chapter Project, Writing Activities	90 min Tech Prep Masters	65 min PE: Chapter Assessment, Test Generator	30 min Chap. Assess. (A or B), Alt. Assess. (A or B), Test Generator	

PE: Pupil's Edition

Alternative Assessment

The following are suggestions for an alternative assessment for students who may benefit from a different type of assessment than the regular chapter quizzes and the mid-chapter and end-of-chapter assessment materials. Many of the questions are open-ended, and students' answers will vary.

Performance Assessment

1. **a.** Define the word *inequality*.
   b. Describe a real-world situation in which a meaningful inequality could be solved by the Subtraction Property of Inequality. Write an inequality to model your situation and then solve it.
   c. Describe a real-world situation in which a meaningful inequality could be solved by the Multiplication Property of Inequality. Write an inequality to model your situation and then solve it.

2. **a.** Define the term *absolute value*.
   b. Explain how to solve an absolute-value equation. Write an absolute-value equation and solve it.
   c. Explain how to solve an absolute-value inequality. Write an absolute-value inequality and solve it.

Portfolio Project

Suggest that students choose one of the following projects for inclusion in their portfolios.

1. Compare and contrast solving an equation with solving an inequality. Give examples to justify your response.

2. The *triangle inequality* states that the length of one side of a triangle is less than the sum of the lengths of the other two sides.
   a. Draw several triangles and measure their lengths.
   b. Do you believe the triangle inequality? Explain.

3. Describe a real-world situation which uses a range of numbers. Write an absolute-value inequality to model the situation.

internetconnect

The table below identifies the pages in this chapter that contain technology information and support in the side columns.

Content Links	
Lesson Links	page 291
Portfolio Links	page 288
Graphics Calculator Support	page 782

Resource Links

For information about teacher and parent resources as well as professional development help, visit **www.hrw.com/math**.

Technical Support HRW has assembled a team of dedicated technical and teaching professionals and a comprehensive service program to provide you with the support you need.

- The HRW Technical Support Line operates from 7 A.M. to 6 P.M. central time, Monday through Friday, at (800) 323-9239.
- The HRW Technical Support Center on the World Wide Web is available 24 hours a day, seven days a week, at **www.hrwtechsupport.com**.
- You can e-mail our Technical Support Center at **tsc@hrwtechsupport.com**.
- The Technical Support Center's fax-on-demand service at (800) 352-1680 offers solutions to common problems and answers to frequently asked questions.

Technology

Lesson Suggestions and Calculator Examples
(Keystrokes are based on a TI-83 calculator.)

Lesson 6.1 Solving Inequalities
This lesson presents the opportunity to introduce more calculator capabilities. To access an inequality symbol, have students press [2nd] [MATH], highlight **TEST**, choose an inequality symbol from the list below, and then press [ENTER].

The graphical representation of $x \geq 3$ is shown below. Have students practice graphing inequalities.

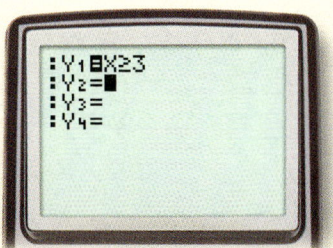 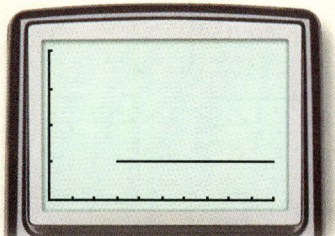

Lesson 6.2 Multistep Inequalities
The technology discussion on page 285 shows how a table can be used to decide which phone plan is less expensive. The strategy is as follows:

1. Enter the relevant expressions into the **Y=** editor.

2. Set up and create two side-by-side tables of values from the **TABLE SETUP** menu.

3. Compare the values in the table to find the solution.

Lesson 6.3 Compound Inequalities
In this lesson, students will build on graphing simple inequalities and will learn how to graph a pair of inequalities linked by AND or OR. Students can use the calculator to graph compound inequalities by using the **LOGIC** menu, which can be accessed by pressing [2nd] [MATH] [▶]. Then have them select **1:and** or **2:or**. For example, if students used this procedure to graph $x \geq 3$ AND $x \leq 9$, they would see the screens shown below.

 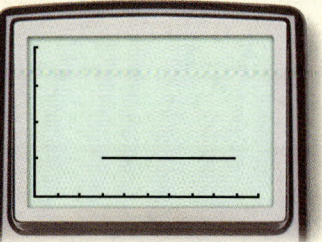

Lesson 6.4 Absolute-Value Functions
This lesson introduces the parent absolute-value function and many of its variations. Students can graph absolute-value functions on a TI-83 graphics calculator by pressing [MATH] [▶] [ENTER] to access the absolute-value feature. If using a TI-82, they should press [2nd] [x^{-1}]. Have students graph several absolute-value functions so that they can observe how changes to the equations affect the graphs.

Lesson 6.5 Absolute-Value Equations and Inequalities
In this lesson, students will combine the concepts of absolute value and inequality. Have students graph $|x - 3| \leq 6$ by using the following key sequence:

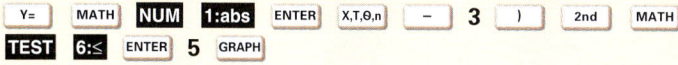

In Example 5, students can verify the solution by entering the original equation next to **Y1=** and then entering the solution, **X<4 or X>8**, next to **Y2=**. If the graphs of **Y1** and **Y2** coincide, then the solution is correct. Students can use this method to check their work when completing the exercises.

For further information, refer to the

- technology discussions in the lessons.
- lesson-related teacher's commentary in the side columns of this *Teacher's Edition*.
- lesson-related *Student Technology Guide* masters.
- *HRW Technology Handbook*.

Teaching Resources

Basic Skills Practice
(2 pages per skill)

Reteaching
(2 pages per lesson)

Enrichment
(1 page per lesson)

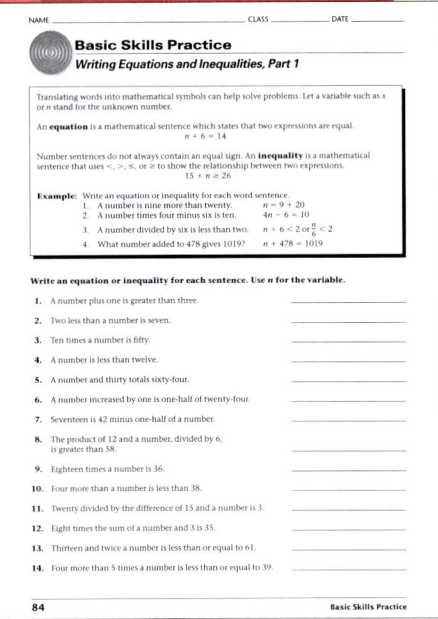

Cooperative-Learning Activity
(1 page per lesson)

Lesson Activity
(1 page per lesson)

Long-Term Project
(4 pages per chapter)

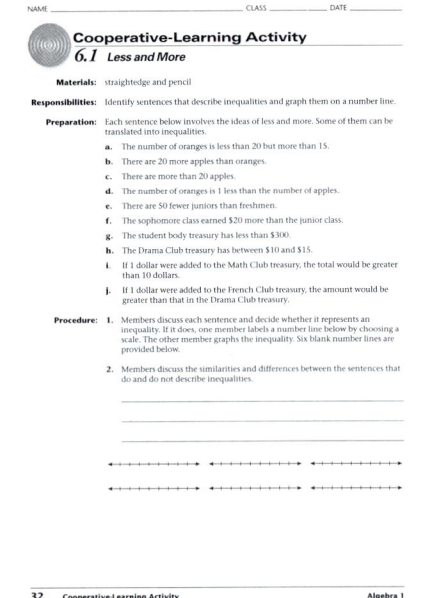

Problem Solving/Critical Thinking
(1 page per lesson)

Writing Activities
(3 pages per chapter)

Tech Prep Masters
(4 pages per chapter)

Spanish Resources
(1 page per lesson)

Quizzes
(1 page per lesson)

Mid-Chapter Test
(1 page per chapter)

CHAPTER 6 INTERLEAF 274F

About Chapter 6

Background Information

In Chapter 6, students learn the properties of inequality and use them to solve linear inequalities in one variable. Students learn to graph the solutions to linear inequalities on a number line. Students then explore the absolute-value function and its features. The chapter concludes by combining the concepts of absolute value and inequalities.

CHAPTER RESOURCES

- Block-Scheduling Handbook
- Writing Activities for Your Portfolio
- Tech Prep Masters
- Long-Term Project
- Assessment Resources:
  Mid-Chapter Assessment
  Chapter Assessments
  Alternative Assessments
- Test and Practice Generator
- Technology Handbook
- End-of-Course Exam Practice

Chapter Objectives

- State and use symbols of inequality. [**6.1**]
- Solve inequalities that involve addition and subtraction. [**6.1**]
- State and apply the Multiplication and Division Properties of Inequality. [**6.2**]
- Solve multistep inequalities in one variable. [**6.2**]

Inequalities and Absolute Value

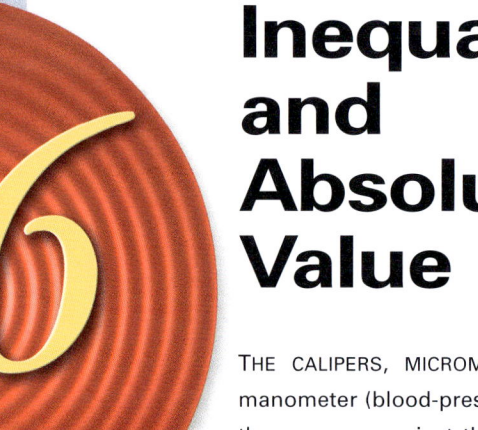

Lessons

6.1 • Solving Inequalities

6.2 • Multistep Inequalities

6.3 • Compound Inequalities

6.4 • Absolute-Value Functions

6.5 • Absolute-Value Equations and Inequalities

Project
Planning a Snowboarding Trip

THE CALIPERS, MICROMETER, AND SPHYGMOmanometer (blood-pressure gauge) shown on these pages are just three of the many measuring instruments used daily in work environments. But measurements are never absolute. Whatever unit is chosen as being accurate enough for the work at hand, a smaller unit is always possible. For example, centimeters or millimeters are the smallest units found on a common ruler, but physicists frequently use nanometers (one-billionth of a meter) or picometers (one-trillionth of a meter).

Because of the impossibility of making absolute measurements, there is always a *margin of error* in the size of manufactured products. Quality control standards are determined on the basis of how large a difference from the standard can be allowed without affecting the performance of the product. Similarly, statistical probability standards determine the margin of error that is acceptable in opinion polls and scientific studies.

Inequalities and absolute value can be used to express margins of error, and in this chapter, you will learn the meaning of these concepts and methods for applying them to a variety of situations.

About the Photos

Measuring instruments like those pictured above always have a margin of error. In this chapter, students will learn how inequalities and absolute value can be used to express margins of error.

The micrometer can be used to measure relatively small distances such as the diameter of a human hair.

A sphygmomanometer is used to measure blood pressure.

Precision instruments, such as assembly line robots, need to work within a very narrow margin of error.

- Graph the solution sets of compound inequalities. [**6.3**]
- Solve compound inequalities. [**6.3**]
- Explore features of the absolute-value function. [**6.4**]
- Explore basic transformations of the absolute-value function. [**6.4**]
- Solve absolute-value equations. [**6.5**]
- Solve absolute-value inequalities, and express the solution as a range of values on a number line. [**6.5**]

PORTFOLIO ACTIVITIES

Portfolio Activities appear at the end of Lessons 6.1 and 6.2. Each serves as preparation for the Chapter Project. The Portfolio Activities as well as the Chapter Project Activities are appropriate for inclusion in the student's portfolio. Students should be encouraged to include in their portfolios any other work in which they feel a sense of pride or a sense of accomplishment.

About the Chapter Project

Different algebraic functions can be used to model different situations. In the Chapter Project, *Planning a Snowboarding Trip*, on page 308, you will investigate the use of a *step function* to model a problem situation. After completing the Chapter Project, you will be able to do the following:

- Recognize the relationship between pairs of numbers as a step function.
- Graph a step function.
- Recognize, write, and graph a *greatest integer* function (an important type of step function).

About the Portfolio Activities

Throughout the chapter, you will be given opportunities to complete Portfolio Activities that are designed to support your work on the Chapter Project.

- Piecewise functions are the topic of the Portfolio Activity on page 281.
- Absolute-value functions are the topic of the Portfolio Activity on page 288.

internet connect

Internet Connect features appearing throughout the chapter provide keywords for the HRW Web site that lead to links for math resources, tutorial assistance, references for student research, classroom activities, and all teaching resource pages. The HRW Web site also provides updates to the technology used in the text. Listed at right are the keywords for the Internet Connect activities referenced in this chapter. Refer to the side column on the page listed to read about the activity.

LESSON	KEYWORD	PAGE
6.2	MA1 Puzzles	288
6.3	MA1 Hurricane Force	291

CHAPTER 6 **275**

Prepare

NCTM PRINCIPLES & STANDARDS 1–4, 6–10

LESSON RESOURCES

- Practice 6.1
- Reteaching Master 6.1
- Enrichment Master 6.1
- Cooperative-Learning Activity 6.1
- Lesson Activity 6.1
- Problem Solving/Critical Thinking Master 6.1
- Teaching Transparency 31

QUICK WARM-UP

Write the number that corresponds to each lettered point on the number line.

```
   F   C     A  D E   B
◄──●───●─────●──●─●───●──►
  -2  -1     0  1     2
```

1. A 0
2. B 2
3. C -1
4. D $\frac{2}{3}$
5. E $1\frac{1}{3}$
6. F $-1\frac{2}{3}$

Also on Quiz Transparency 6.1

Teach

Why Work with students to make a list of real-world situations in which relationships between quantities are described by phrases such as *no more than*, *at least*, *greater than*, and *at most*.
Samples: You must be *at least* a certain height to get on some rides at amusement parks. You may drive on this road at a speed *no greater than* 35 miles per hour.

6.1 Solving Inequalities

Why Many real-world problems, such as the price of the sandwiches pictured, involve finding a range of possibilities instead of one specific answer. Inequalities can be used to model this type of problem.

Objectives

- State and use symbols of inequality.
- Solve inequalities that involve addition and subtraction.

Inequalities

An **inequality** is a statement that two expressions are not equal. The sentence "Lunch will cost more than $2" can be represented by the inequality $L > 2$. The solution to an inequality is the set of numbers that make the inequality true. In this case, the range of values in the solution includes any amount greater than $2.

Several inequalities are used in algebra.

Statements of Inequality	
a is less than b.	$a < b$
a is greater than b.	$a > b$
a is less than or equal to b (or a is at most b).	$a \leq b$
a is greater than or equal to b (or a is at least b).	$a \geq b$
a is greater than b and less than c.	$b < a < c$
a is greater than or equal to b and less than or equal to c.	$b \leq a \leq c$
a is not equal to b.	$a \neq b$

Alternative Teaching Strategy

USING TECHNOLOGY Have students use the table feature of a graphics calculator to explore the inequality $x + 4 < 11$. If you are using a TI-82 or TI-83 graphics calculator, have students press `Y=` to enter the **Y=** editor. Tell them to clear any existing equations and then enter **X+4** next to **Y1=**.

Then instruct students to press `2nd` `WINDOW` and make these choices in the **TABLE SETUP** screen.
    TblStart=0 (or TblMin=0)   Indpnt: **Auto** Ask
    ΔTbl=1   Depend: **Auto** Ask
Have them press `2nd` `GRAPH` to see the table.

Have students use ▼ to scroll down the table until they arrive at 11 in the **Y1** column. Ask them to name the corresponding value of x. **7** Lead them to see that 7 is the value of x for which $x + 4 = 11$ is true. Ask them to name some values of x for which $x + 4 < 11$ is true. **6, 5, 4, 3, 2, …** Point out that these are just a few of the many *solutions* of $x + 4 < 11$. Repeat the activity for ΔTbl=.25. This time, $x + 4 < 11$ is true when x equals 6.75, 6.5, 6.25, 6, 5.75, 5.5, … Ask students to describe all possible solutions of $x + 4 < 11$. **all real numbers less than 7**

276   LESSON 6.1

EXAMPLE 1

APPLICATION
CONSUMER ECONOMICS

Michael can spend at most $3.10 for lunch. He buys a hamburger and a drink for $2.15. **Write an inequality that models how much Michael can spend on dessert and stay within his spending limit.**

- **SOLUTION**

Let d be the amount Michael can spend on dessert. The total amount he can spend is $d + 2.15$. Michael's situation can be modeled by the inequality $d + 2.15 \leq 3.10$, which is read as "d plus 2.15 is less than or equal to 3.10."

TRY THIS Trish has only $3.75 to spend for lunch. She buys a cheeseburger for $2.50. Write an inequality that models how much Trish can spend on dessert and stay within her spending limit.

Properties of Inequality

The inequality $d + 2.15 \leq 3.10$ can be thought of as the combination of the following two statements:

$$d + 2.15 < 3.10 \text{ or } d + 2.15 = 3.10$$

The equation $d + 2.15 = 3.10$ is true when $d = 0.95$. Notice that the inequality $d + 2.15 < 3.10$ is true when $d < 0.95$. This suggests that properties used to solve inequalities are similar to those used to solve equations.

Addition and Subtraction Properties of Inequality

Let a, b, and c be real numbers.

If $a < b$, then $a + c < b + c$.
If $a > b$, then $a + c > b + c$.

If $a < b$, then $a - c < b - c$.
If $a > b$, then $a - c > b - c$.

ADDITIONAL EXAMPLE 1

Anne can spend at most $15.00 when she goes to see a movie. She has to spend $1.25 each way for a subway ride, and the movie ticket is $7.00. **Write an inequality that models how much Anne can spend on refreshments and stay within her spending limit.** $r + 9.50 \leq 15.00$, where r is the amount she can spend on refreshments

TRY THIS

Let d represent the amount Trish can spend on another item. The total amount Trish can spend is $3.75, so her situation can be modeled by the inequality $d + 2.50 \leq 3.75$.

Teaching Tip

Students sometimes have difficulty remembering whether the statement $a \leq b$ is interpreted as $a < b$ or $a = b$ or as $a < b$ and $a = b$. Give them a statement such as $m \leq 5$ and point out that it is not possible for a number to be equal to 5 *and* less than 5 at the same time. Thus, the number must be either equal to 5 *or* less than 5.

Interdisciplinary Connection

HEALTH The recommended daily calorie intake for male and female teenagers ages 15 to 18 is 2800 and 2100, respectively. On a certain day, Maria and José each consumed 600 calories at breakfast and 1100 calories at lunch. Write and solve inequalities that describe the number of calories each can consume at dinner and stay within the recommended limits. $1700 + m \leq 2100$, where m is the number of calories Maria can consume at dinner; $m \leq 400$; $1700 + j \leq 2800$, where j is the number of calories José can consume at dinner; $j \leq 1100$

Inclusion Strategies

ENGLISH LANGUAGE DEVELOPMENT If students have language difficulties or if English is their second language, it may be difficult for them to perceive distinctions between several pairs of words and phrases used in this lesson: *dot* versus *open circle*; *shading right* versus *shading left*; *included* versus *excluded*; *at most* versus *at least*; and so on. You may find it helpful to conduct a class discussion in which students share with each other their strategies for remembering the distinctions.

LESSON 6.1 **277**

ADDITIONAL EXAMPLE 2

Solve the inequality. Check the solution.

$2.5 + j \leq 4.8$ $j \leq 2.3$

TRY THIS

$c < -18$

Use Teaching Transparency 31.

ADDITIONAL EXAMPLE 3

Graph the solution to $h + 8 \geq 4$.

$h \geq -4$

TRY THIS

$t < -10$

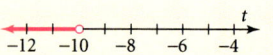

EXAMPLE 2 Solve the inequality $m - 8 \geq 2$. Check the solution.

○ **SOLUTION**

$m - 8 \geq 2$
$m - 8 + 8 \geq 2 + 8$ Addition Property of Inequality
$m \geq 10$ Simplify.

To check, substitute values near 10 into the original inequality to see if the result is true.

Substitute 10 for m.
$10 - 8 \stackrel{?}{\geq} 2$
$2 \geq 2$ True

Substitute 11 for m.
$11 - 8 \stackrel{?}{\geq} 72$
$3 \geq 2$ True

TRY THIS Solve the inequality $c + 7 < -11$.

Number Lines and Inequalities

The solution to an inequality is the set of all numbers that make the statement true. The solution set might be listed, described, or graphed.

An inequality is graphed as an interval, a ray, or a line. Included endpoints are shown as dots, •, while endpoints that are not included are shown as open circles, ○.

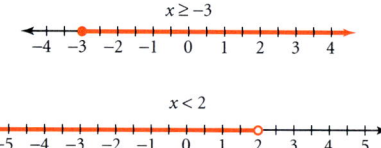

EXAMPLE 3 Graph the solution to $5 > x - 3$.

○ **SOLUTION**

First solve the inequality for x.
$5 > x - 3$ Given
$5 + 3 > x - 3 + 3$ Addition Property of Inequality
$8 > x$ Simplify.
$x < 8$ It is customary to put the variable to the left of the inequality sign in the final answer.

To graph the result on a number line, shade all points to the left of 8 on the number line. This includes all points whose coordinates are less than 8. To indicate that 8 is *not included*, put an *open circle* at 8.

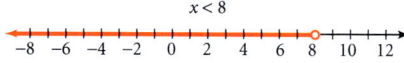

TRY THIS Graph the solution to $t + 7 < -3$.

Enrichment

Let a, b, c, and d be real numbers. Ask students if the following statement is true or false: If $a = b$ and $c = d$, then $a + c = b + d$. **true** If they have difficulty understanding why it is true, have them substitute values for a, b, c, and d. Then repeat the activity with this statement: If $a < b$ and $c < d$, then $a + c < b + d$. **true**

Now give them this open statement: If $a < b$ and $c > d$, then __?__. Have them determine ways they might fill in the blank to make a true statement. **Samples:** $a + d < b + c, a - c < b - d$

Reteaching the Lesson

USING MANIPULATIVES Write $x + 5 \leq 7$ on the board or overhead. Directly below it, write $x + 5 = 7$. Have students use tiles to solve the equation. $x = 2$ Point out that they can use the same steps to solve the inequality. Ask them what the solution of $x + 5 \leq 7$ is. $x \leq 2$

Now give them the inequality $x \geq 6$. Tell them to work backward by adding and subtracting tiles to model inequalities for which $x \geq 6$ is the solution. **Samples:** $x + 3 \geq 9, x - 4 \geq 2$

278   LESSON 6.1

EXAMPLE 4 Write an inequality that describes the points graphed on each number line.

a.

b.

● **SOLUTION**

a. This number line shows a ray that includes all real numbers greater than −4. Since the endpoint at −4 is an open circle, −4 is not included in the solution set. The inequality that describes this graph is $x > -4$.

b. This number line shows a ray that includes all real numbers less than or equal to 3. Since the endpoint at 3 is shaded, 3 is included in the solution set. The inequality that describes this graph is $x \leq 3$.

TRY THIS Write an inequality that describes the points graphed on each number line.

a.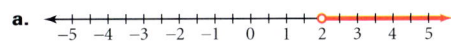

b.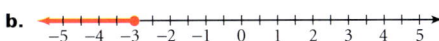

ADDITIONAL EXAMPLE 4

Write an inequality that describes the points graphed on each number line.

a.

$x \geq -3$

b.

$x < 5$

TRY THIS

a. $x > 2$

b. $x \leq -3$

Assess

Selected Answers
Exercises 5–14, 15–55 odd

ASSIGNMENT GUIDE

In Class	1–14
Core	15–26, 36–38, 41–47 odd
Core Plus	16–34 even, 36–47
Review	48–55
Preview	56–61

✎ Extra Practice can be found beginning on page 738.

Exercises

Communicate

1. How are the Addition and Subtraction Properties of Inequality similar to the Addition and Subtraction Properties of Equality?

2. What steps are necessary to solve the inequality $x - 4 \leq 1$? Name the property you would use for each step.

3. Explain how to graph the solution to $x + 3 < 7$ on a number line.

4. How would you write an inequality that describes the points graphed on the number line below?

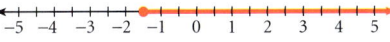

Guided Skills Practice

APPLICATION

$18 + s \leq 27$

5. **CONSUMER ECONOMICS** Jerry has only $27 for school supplies. He spends $18 on a backpack. Write an inequality that models how much more Jerry can spend on school supplies and stay within his limit. **(EXAMPLE 1)**

LESSON 6.1 **279**

Technology

In Exercises 15–35, a calculator may be helpful for checking the solution of the related equation. If students arrive at a solution such as $x > -2$, suggest that they use a calculator to verify that -2 is the solution of the related equation.

Error Analysis

Some students have difficulty with a result such as $-5 < t$ and may graph it incorrectly as all numbers less than -5. Suggest that they print results like this on a sheet of paper and then flip the paper over from right to left. When they hold the paper in front of a light source, they can read the inequality in a more familiar form, with the variable on the left. Although the numbers and letters will appear backward, the direction of the inequality will be correct.

---

6. $b \leq -12$
7. $x > 14$
8. $m \geq 6$
9. $t > -5$
10. $x \geq 7$
11. $y < -5$
12. $x \geq 20$
13. $x < -9$

Solve each inequality and graph each solution on a number line. *(EXAMPLES 2 AND 3)*

6. $b + 1 \leq -11$ 7. $x - 9 > 5$ 8. $m - 14 \geq -8$ 9. $-2 < t + 3$
10. $x - 3 \geq 4$ 11. $y + 7 < 2$ 12. $3 \leq x - 17$ 13. $-5 > 4 + x$

14. Write an inequality that describes the points graphed on the number line below. *(EXAMPLE 4)*

 $x < 4$

Practice and Apply

Solve each inequality.

15. $x + 8 > -1$ $x > -9$
16. $x - 6 \leq 7$ $x \leq 13$
17. $x - 7 > 18$ $x > 25$
18. $b + 4 \leq 18$ $b \leq 14$
19. $67 \geq y + 28$ $y \leq 39$
20. $a + 18 > -3$ $a > -21$
21. $m - 9 \leq 40$ $m \leq 49$
22. $a - 7 > 2.3$ $a > 9.3$
23. $x + 0.04 > 0.6$ $x > 0.56$
24. $x - 0.1 < 8$ $x < 8.1$
25. $x + \frac{3}{4} \geq \frac{1}{2}$ $x \geq -\frac{1}{4}$
26. $x + \frac{3}{4} < 1$ $x < \frac{1}{4}$
27. $v - 3 \leq 17$ $v \leq 20$
28. $17 > b - 5$ $b < 22$
29. $t - 76 \leq 50$ $t \leq 126$
30. $h + 15 \geq 5$ $h \geq -10$
31. $v + 6.2 \geq 8.1$ $v \geq 1.9$
32. $m - 2.2 < -12.2$ $m < -10$
33. $x + \frac{2}{3} \leq \frac{5}{9}$ $x \leq -\frac{1}{9}$
34. $t - \frac{7}{4} \leq \frac{3}{4}$ $t \leq \frac{5}{2}$
35. $c - \frac{3}{7} \geq \frac{6}{13}$ $c \geq \frac{81}{91}$

Write an inequality that describes the set of points graphed on each number line.

36. $x \geq -1$
37. $x \leq 3.5$
38. $x > -2$

CHALLENGES

39. If 8 more than twice a number is less than 6 more than the number, what can the number be? **any number less than -2**

40. If 3 less than 4 times a number is greater than or equal to 3 less than 7 times the number, what is the largest the number can be? **0**

APPLICATIONS

41. **AUDITORIUM SEATING** A school auditorium can seat 450 people for graduation. The graduates will use 74 seats. Write and solve an inequality to describe the number of additional people who can be seated in the auditorium.
$x + 74 \leq 450$ $x \leq 376$

42. **INVESTMENTS** A bank promises that the interest rate on a certain savings account will never drop below 3.5%. Write an inequality to represent this situation. $r \geq 0.035$

---

6. $b \leq -12$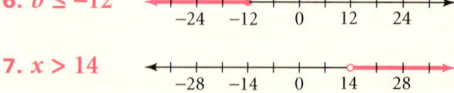
7. $x > 14$
8. $m \geq 6$
9. $t > -5$
10. $x \geq 7$
11. $y < -5$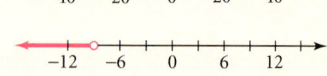
12. $x \geq 20$
13. $x < -9$

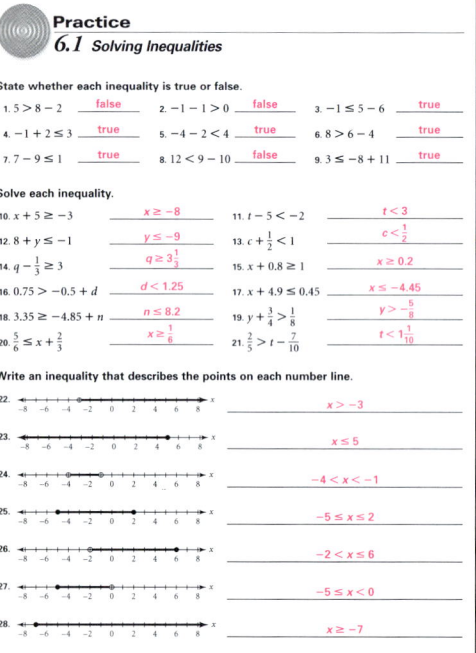

280 LESSON 6.1

APPLICATION

TEMPERATURE Write an inequality or an equation in terms of *t* to represent each situation.

43. a temperature of at least 40°F $t \geq 40$
44. a temperature of no more than 54°F $t \leq 54$
45. a temperature above 34°F $34 < t$
46. a temperature of 78°F $t = 78$
47. Let *t* represent an initial temperature reading. After a rise of 5°F, the temperature was between 70°F and 80°F inclusive. What inequality represents this situation? $70 \leq t + 5 \leq 80$ or $65 \leq t \leq 75$

 Look Back

48. The points on a scatter plot lie near a line that rises from left to right. Name the type of correlation. **(LESSON 1.6)** positive
49. The points on a scatter plot lie near a line that falls from left to right. Name the type of correlation. **(LESSON 1.6)** negative

Evaluate. (LESSON 2.1)

50. $|-4.5|$ 4.5
51. $|-9|$ 9
52. $|-3 + 4|$ 1

Solve. (LESSONS 3.1 AND 3.3)

53. $x + 7 = 10$ $x = 3$
54. $15 - x = 25$ $x = -10$
55. $3 - 2x = 17$ $x = -7$

 Look Beyond

Solve for *x*.

56. $bx + c = a$ $x = \frac{a-c}{b}$
57. $tx - 7 = r$ $x = \frac{r+7}{t}$
58. $Ax + Bx = C$ $x = \frac{C}{A+B}$

Find three values of *x* that make each inequality true and three values of *x* that make each inequality false.

59. $2x < 8$
60. $4x + 5 \leq 16$
61. $8x - 3 > 33$

Include this activity in your portfolio.

Some functions have graphs that can be assembled from sections of other graphs. This type of function is called a *piecewise function*. The following function is a piecewise function:

$$y = \begin{cases} -x & \text{if } x < 0 \\ x & \text{if } x \geq 0 \end{cases}$$

The table below has been completed for *x*-values of −2 and 0.5. Because $x < 0$ for $x = -2$, the equation $y = -x$ is used to determine *y*, and $y = -(-2) = 2$. Because $x \geq 0$ for $x = 0.5$, the equation $y = x$ is used to determine *y*, and $y = 0.5$. Complete the table for the remaining *x*-values.

x	−4	−3	−2	−1	−0.5	0	0.5	1	2	3	4
y	? 4	? 3	2	? 1	?	? 0	0.5	? 1	? 2	? 3	? 4

0.5

59. Answers may vary. True values will be less than 4; false values will be 4 or greater.

60. Answers may vary. True values will be less than or equal to $\frac{11}{4}$. False values will be greater than $\frac{11}{4}$.

61. Answers may vary. True values will be greater than $\frac{36}{8}$, or $4\frac{1}{2}$. All others will be false.

In Exercises 56–61, students investigate equations and inequalities that require multiple steps to solve. Students will do further work of this type in Lesson 6.2.

ALTERNATIVE
Assessment

Portfolio Activity

The Portfolio Activity can be used as preparation for the Chapter Project or as a separate activity. In the Portfolio Activity on this page, students are introduced to the piecewise function that is equivalent to the absolute-value function. Students then complete a table of values for the function, which they will use for the Portfolio Activity on page 288.

Answers to Portfolio Activities can be found in Additional Answers of the Teacher's Edition.

Student Technology Guide

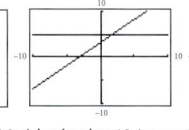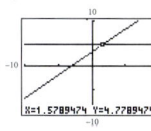

LESSON 6.1 **281**

Prepare

NCTM PRINCIPLES & STANDARDS 1–10

LESSON RESOURCES

- Practice 6.2
- Reteaching Master 6.2
- Enrichment Master 6.2
- Cooperative-Learning Activity 6.2
- Lesson Activity 6.2
- Problem Solving/Critical Thinking Master 6.2

QUICK WARM-UP

Solve each equation. Check your solution.

1. $-1.6t = -5$ $t = 3.125$
2. $b - 3.4 = -6$ $b = -2.6$
3. $4a + 6 = 9 + 8a$
              $a = -0.75$
4. $\frac{z}{3} - 3 = -12$ $z = -27$

Also on Quiz Transparency 6.2

Teach

Why Write these inequalities on the board or overhead:
- $x + 3 \geq 4 + 3$
- $x + (-3) \geq 4 + (-3)$
- $3x \geq 3(4)$
- $(-3)x \geq (-3)(4)$

Have students use substitution to determine which of these inequalities has the solution $x \geq 4$. They should discover that all but the last one have the solution $x \geq 4$. Tell them that, in this lesson, they will learn why the last inequality is different from the others.

Multistep Inequalities

6.2

By sending a signal to the lander and performing a series of mathematical calculations, scientists were able to pinpoint Pathfinder's landing site on Mars to within 5 meters of its actual location.

Objectives
- State and apply the Multiplication and Division Properties of Inequality.
- Solve multistep inequalities in one variable.

Why Many real-world situations are solved not by finding a single answer, but by finding a minimum, a maximum, or a range of answers that satisfy the conditions. For problems of this type, you often need to solve an inequality.

Solving inequalities sometimes involves multiplying and dividing. The activity and the list of properties that follow will help you to understand how to use multiplication and division to solve inequalities.

Activity
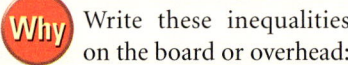

Multiplying and Dividing Inequalities

1. Copy and complete the table. Notice that in each row of the table $A < B$.

A	B	$2A$	$2B$	$-2A$	$-2B$	$\frac{A}{2}$	$\frac{B}{2}$	$\frac{A}{-2}$	$\frac{B}{-2}$
2	4	4	8	−4	−8	1	2	−1	−2
−2	2	?	?	?	?	?	?	?	?
−4	6	?	?	?	?	?	?	?	?
−6	−2	?	?	?	?	?	?	?	?
−8	−6	?	?	?	?	?	?	?	?

Alternative Teaching Strategy

USING TECHNOLOGY Have students use the table feature of a graphics calculator to explore the inequality $2x < 10$. If you are using a TI-82 or TI-83 graphics calculator, have students press [Y=] to enter the **Y=** editor. Tell them to clear any existing equations and enter **2X** next to **Y1=**.

Then instruct students to press [2nd] [WINDOW] and make these choices in the **TABLE SETUP** screen.
    TblStart=0 (or TblMin=0)   Indpnt: Auto Ask
    ΔTbl=.5   Depend: Auto Ask
Have them press [2nd] [GRAPH] to see the table.

Have students use [▼] to scroll down the table until they arrive at **10** in the **Y1** column. Ask them to name the corresponding value of x. **5** Ask students to examine the table and describe all values of x for which $2x < 10$ is true. $x < 5$

Repeat the activity for $-2x < 10$. Be sure students enter **−2X** next to **Y1=**. This time they will need to use [▲] to locate **10** in the **Y1** column. Ask them to name the corresponding value of x. **−5** Have them examine the table and describe all values of x for which $-2x < 10$ is true. $x > -5$

282   LESSON 6.2

2. **a.** Compare 2A and 2B in each row. Write your findings as an inequality.
   b. Compare −2A and −2B in each row. Write your findings as an inequality.
3. **a.** Compare $\frac{A}{2}$ and $\frac{B}{2}$ in each row. Write your findings as an inequality.
   b. Compare $\frac{A}{-2}$ and $\frac{B}{-2}$ in each row. Write your findings as an inequality.

CHECKPOINT ✓
4. **a.** Write a rule to explain what you must do to an inequality when you multiply or divide each side by the same positive number.
   b. Write a rule to explain what you must do to an inequality when you multiply or divide each side by the same negative number.

Multiplication and Division Properties of Inequalities

Let a, b, and c be nonzero real numbers.

For $c > 0$:

$\quad$ If $a < b$, then $ac < bc$ and $\frac{a}{c} < \frac{b}{c}$.

$\quad$ If $a > b$, then $ac > bc$ and $\frac{a}{c} > \frac{b}{c}$.

For $c < 0$:

$\quad$ If $a < b$, then $ac > bc$ and $\frac{a}{c} > \frac{b}{c}$.

$\quad$ If $a > b$, then $ac < bc$ and $\frac{a}{c} < \frac{b}{c}$.

These properties are the same for inequalities involving ≤ and ≥.

Example 1 illustrates the use of multiplication and division by positive numbers in solving inequalities.

EXAMPLE ① Solve each inequality.

a. $3x > 6$ **b.** $16 \leq 2y$
c. $-25 \geq 5x$ **d.** $\frac{a}{3} \geq 4$

● **SOLUTION**

a. $3x > 6$
$\frac{3x}{3} > \frac{6}{3}$
$x > 2$

b. $16 \leq 2y$
$\frac{16}{2} \leq \frac{2y}{2}$
$8 \leq y$
$y \geq 8$

c. $-25 \geq 5x$
$\frac{-25}{5} \geq \frac{5x}{5}$
$-5 \geq x$
$x \leq -5$

d. $\frac{a}{3} \geq 4$
$3\left(\frac{a}{3}\right) \geq 3(4)$
$a \geq 12$

TRY THIS Solve.
a. $6y > 12$ **b.** $18 < 3x$
c. $5m \leq -35$ **d.** $\frac{x}{7} \geq -6$

Interdisciplinary Connection

ACCOUNTING The total cost of producing a product can be modeled by the equation $c = un + f$, where c represents the total cost, u represents the unit cost of the product, n represents the number of units, and f represents fixed costs. A manufacturer wants the total cost per month for a calculator to be no more than $400,000. Fixed costs per month are $110,000, and the unit cost is $25. Write an inequality for this situation. Find the maximum number of calculators that can be manufactured in one month under these conditions.
$25n + 110,000 \leq 400,000; n \leq 11,600$

Enrichment

Replace ? with >, <, or = so that each statement is true for all real-number values of the variables.

1. If $a < b$ and $c < 0$, then ac ? bc.   >
2. If $g < h$ and $k > 0$, then gk ? hk.   <
3. If $x < y$ and $z = 0$, then xz ? yz.   =
4. If $s < 0$ and $t < 0$, then st ? 0.   >
5. If $a < b$, then $-a$? $-b$.   >
6. If $p < q$ and $r < 0$, then $p + r$? $q + r$.   <
7. If $j < k$, then k ? j.   >
8. If $x < y$ and $y < 0$, then xy ? y^2.   >
   (Hint: $y^2 = y \cdot y$)

Activity Notes

In this Activity, students generate a table of values that provides a visual demonstration of the Multiplication Property of Inequality and the Division Property of Inequality. They should arrive at the conclusions in the summary box on this page.

Cooperative Learning

You may wish to have students do the Activity in pairs, with one partner completing the columns for 2A, 2B, $\frac{A}{2}$, and $\frac{B}{2}$, while the other partner completes the columns for −2A, −2B, $\frac{A}{-2}$, and $\frac{B}{-2}$. The first partner should then answer part **a** of Steps 2–4, while the second partner answers part **b**. The partners can then compare their results.

CHECKPOINT ✓
4. **a.** Treat the inequality sign the same as an equal sign; it doesn't change.
   b. Reverse the inequality sign.

ADDITIONAL EXAMPLE ①

Solve each inequality.

a. $4d < 16$ $d < 4$
b. $56 \leq 8y$ $y \geq 7$
c. $11p \leq -121$ $p \leq -11$
d. $-5 > \frac{a}{9}$ $a < -45$

TRY THIS
a. $y > 2$ **b.** $x > 6$
c. $m \leq -7$ **d.** $x \geq -42$

LESSON 6.2

ADDITIONAL EXAMPLE 2

Solve each inequality.

a. $-18 < 9h$ $h > -2$
b. $-6w \geq 36$ $w \leq -6$
c. $-28 \geq -14b$ $b \geq 2$
d. $\frac{m}{8} < -5$ $m < -40$

TRY THIS
a. $x > -12$ b. $x < -6$
c. $x \geq -2$ d. $x \geq 972$

ADDITIONAL EXAMPLE 3

Solve each inequality and graph each solution on a number line.

a. $4s - (3 + 5s) \leq 17$

     $s \geq -20$

b. $3x - 2(x + 4) \leq 3x - 8$

     $x \geq 0$

TRY THIS
a. $m \geq \frac{11}{3}$
b. $p \geq \frac{7}{4}$

Example 2 illustrates the use of multiplication or division by negative numbers in solving inequalities.

EXAMPLE 2 Solve each inequality.

a. $-x > 9$ b. $-2m < 10$
c. $15 \leq -5x$ d. $-\frac{r}{7} \geq -3$

● **SOLUTION**

Either multiplying or dividing by –1 will change –x to x.

a. $-x > 9$
    $-1(-x) < -1(9)$
    $x < -9$

b. $-2m < 10$
    $\frac{-2m}{-2} > \frac{10}{-2}$
    $m > -5$

c. $15 \leq -5x$
    $\frac{15}{-5} \geq \frac{-5x}{-5}$
    $-3 \geq x$
    $x \leq -3$

d. $-\frac{r}{7} \geq -3$
    $-7\left(-\frac{r}{7}\right) \leq -7(-3)$
    $r \leq 21$

TRY THIS   Solve.

a. $-x < 12$ b. $-4x > 24$ c. $-16x \leq 32$ d. $\frac{x}{-9} \leq -108$

Example 3 illustrates multi-step inequalities, where multiplication (or division) and addition (or subtraction) are both required in the solutions.

EXAMPLE 3 Solve each inequality.

a. $3x - (2 + 6x) \leq 28$
b. $2x - 6 \geq 3x - 2$

● **SOLUTION**

a. $3x - (2 + 6x) \leq 28$ *Given*
    $3x - 2 - 6x \leq 28$ *Distributive Property*
    $-3x - 2 \leq 28$ *Simplify.*
    $-3x - 2 + 2 \leq 28 + 2$ *Addition Property of Inequality*
    $-3x \leq 30$ *Simplify.*
    $\frac{-3x}{-3} \geq \frac{30}{-3}$ *Division Property of Inequality*
    $x \geq -10$

b. When the variable appears on both sides of the inequality, you can often avoid inequality-sign changes by choosing the operation(s) which will result in a positive coefficient for the variable.

    $2x - 6 \geq 3x - 2$
    $2x - 2x - 6 \geq 3x - 2x - 2$
    $-6 \geq x - 2$
    $-6 + 2 \geq x$
    $-4 \geq x$
    $x \leq -4$

By choosing to subtract 2x in the second step, x has a positive coefficient in the the third step.

TRY THIS   Solve each inequality.

a. $2m + 7(m - 1) \geq 26$
b. $3p + 2 \leq 7p - 5$

Inclusion Strategies

KINESTHETIC LEARNERS Use a "human number line" to demonstrate the meaning of an inequality such as $-a \geq 2$. Draw a large number line on a roll of paper with the integers from −8 to 8 spaced about 1 foot apart. Place it on the floor. Assign the numbers −2, −3.5, −6, and −7.75 to four students. Have each student write his or her number on a sheet of paper. Ask the four students to stand at the point on the number line that corresponds to their number and to hold up the papers so that the numbers are visible to the class. Lead the class to see that each number is a solution to $a \leq -2$.

Now select four other students. Assign each to be the "opposite" of one of the students standing on the number line. Tell the "opposite students" to write their assigned opposite (2, 3.5, 6, or 7.75) on a sheet of paper. Then ask these students to stand at the corresponding point on the number line, holding up their papers so that the numbers are visible to the class. Lead the class to see that each of these numbers is a solution to the inequality $-a \geq 2$.

EXAMPLE 4

APPLICATION
CONSUMER ECONOMICS

Eastern Cellular offers two monthly phone plans. **For what range of minutes would it cost less to use plan 1?**

Eastern Cellular Phone Company
Monthly Rates
Plan 1: $25 plus 50¢ per minute
Plan 2: 70¢ per minute

SOLUTION

Method A
First write an inequality to represent the problem.

Plan 1: $25 + 0.5m$
Plan 2: $0.7m$

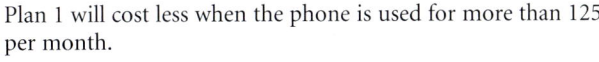

Solve the inequality.

$25 + 0.5m < 0.7m$	
$25 + 0.5m - 0.5m < 0.7m - 0.5m$	Subtraction Property of Inequality
$25 < 0.2m$	Simplify.
$\frac{25}{0.2} < \frac{0.2m}{0.2}$	Division Property of Inequality
$125 < m$	Simplify.
$m > 125$	

Plan 1 will cost less when the phone is used for more than 125 minutes per month.

TECHNOLOGY
GRAPHICS CALCULATOR

Method B
You can use the table feature on a graphics calculator. Enter the expression for plan 1 as **Y₁** and the expression for plan 2 as **Y₂**. Then key in the table feature.

The table shows that the costs are equal at 125 minutes. Plan 1 costs less when the number of minutes is greater than 125.

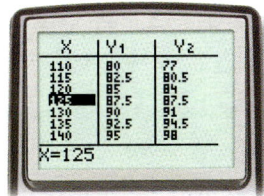

TRY THIS
Jennifer, the manager of a fast-food restaurant, can choose one of two salary options.

Option 1: $2000 a month
Option 2: $500 a month plus $10 dollars per hour

For what range of hours would option 2 pay more?

Reteaching the Lesson

COOPERATIVE LEARNING Have students work in groups of two or three. Give each group the two sets of inequalities at right. Tell them to match each inequality in column I with its solution in column II.

Now have each group work together to create similar sets of eight inequalities and solutions. Have groups of students exchange sets of inequalities and match them with the correct solution.

I	II	
1. $3z - 6 < -9$	**a.** $z > 1$	1. d
2. $3z - 6 > 9$	**b.** $z < 1$	2. e
3. $3z + 6 > -9$	**c.** $z > -1$	3. g
4. $3z + 6 > 9$	**d.** $z < -1$	4. a
5. $-3z - 6 > 9$	**e.** $z > 5$	5. h
6. $-3z + 6 > -9$	**f.** $z < 5$	6. f
7. $-3z - 6 > -9$	**g.** $z > -5$	7. b
8. $-3z + 6 < 9$	**h.** $z < -5$	8. c

ADDITIONAL EXAMPLE 4

A company reimburses its sales personnel for work-related travel. The salesperson can choose one of two options. With option A, the salesperson receives a fixed amount of $150 per month. With option B, the salesperson receives $45 per month plus 35¢ per mile traveled. **Under what conditions is option B more beneficial to the salesperson?**

Option B is more beneficial when the salesperson travels more than 300 miles per month.

See Keystroke Guide, page 782.

Teaching Tip

TECHNOLOGY If you are using a TI-82 or TI-83 graphics calculator, you can create the table shown on this page by following these steps. Press [Y=]. In the **Y=** editor, first clear any existing equations. Then enter **25 + .5X** next to **Y₁=** and **.7X** next to **Y₂=**. Now press [2nd] [WINDOW], and make these choices on the **TABLE SETUP** screen that appears.

TblStart=100 (or TblMin=100)
ΔTbl=5
Indpnt: **Auto** Ask
Depend: **Auto** Ask

Press [2nd] [GRAPH] to see the table. Use [▼] to scroll down the table until you arrive at the place where the values in the **Y₁** and **Y₂** columns are equal.

TRY THIS
Let h represent the number of hours Jennifer works.
Option 1: 2000
Option 2: $500 + 10h$
$500 + 10h > 2000$
$h > 150$
Option 2 pays more if Jennifer works more than 150 hours each month.

LESSON 6.2 285

Assess

Selected Answers
Exercises 8–18, 19–91 odd

ASSIGNMENT GUIDE

In Class	1–18
Core	19–65 odd, 67, 68
Core Plus	20–66 even, 67–71
Review	72–91
Preview	92–95

✏ Extra Practice can be found beginning on page 738.

Technology
In Exercises 19–66, a calculator may be helpful for checking the solution of the related equation.

Practice

Practice
6.2 Multistep Inequalities

Write an inequality that corresponds to each statement.
1. x is less than y. — $x < y$
2. W is greater than B. — $W > B$
3. a is less than or equal to 10. — $a \leq 10$
4. x is greater than or equal to 100. — $x \geq 100$
5. r is positive. — $r > 0$
6. q is nonnegative. — $q \geq 0$
7. M cannot equal 0. — $M \neq 0$; $M < 0$ or $M > 0$
8. V is between 4.5 and 4.6 inclusive. — $4.5 \leq V \leq 4.6$

Tell whether each statement is true or false.
9. $6.9 \geq 6.9$ — true
10. $9.66 > 9.606$ — true
11. $10.2 < 10.02$ — false
12. $\frac{1}{2} \leq \frac{1}{3}$ — false
13. $8.91 > 8.901$ — true
14. $0 > -1$ — true
15. $\frac{1}{7} < \frac{1}{5}$ — true
16. $-5 < -2$ — true
17. $-6 \geq -4$ — false

Solve each inequality.
18. $x + 3 < 4$ — $x < 1$
19. $m - 10 \geq 50$ — $m \geq 60$
20. $T - 5 \leq 8$ — $T \leq 13$
21. $7 - N \geq 16$ — $N \leq -9$
22. $6x < -12$ — $x < -2$
23. $x + 12 > 10$ — $x > -2$
24. $3.5b > -7$ — $b > -2$
25. $2x \geq 6$ — $x \geq 3$
26. $\frac{n}{5} > 20$ — $n > 100$
27. $\frac{r}{-2} \leq 7$ — $r \geq -14$
28. $-5t \leq 45$ — $t \geq -9$
29. $\frac{x}{4} + 7 < 10$ — $x < 12$
30. $\frac{-x}{5} + 4 \geq -1$ — $x \leq 25$
31. $6x - 2 \leq 4$ — $x \leq 1$
32. $x + 3 < 8 - x$ — $x < 2\frac{1}{2}$
33. $x - 1 < -5$ — $x < -4$
34. $-5x \geq 2x - 6$ — $x \leq \frac{6}{7}$
35. $7 - x \leq 3$ — $x \geq 4$
36. $16 + \frac{m}{2} < 9$ — $m < -14$
37. $14 > 4 - \frac{j}{3}$ — $j > -30$

Exercises

Communicate

Describe the steps needed to solve each inequality.
1. $x + 1 > 4$
2. $x - 3 \leq 13$
3. $-3p < 12$
4. $-4x - 2 \geq 2x + 3$
5. How do the Multiplication and Division Properties of Inequality differ from the Multiplication and Division Properties of Equality?
6. Why is $x < -2$ not the solution to the inequality $-3x < 6$?
7. Demonstrate why the inequality symbol must be reversed when both sides of $a < b$ are multiplied by -1. Use both a numerical example and a number line to support your reasoning.

Guided Skills Practice

Solve each inequality. (EXAMPLE 1)
8. $8x \leq 48$ $x \leq 6$
9. $3x < 48$ $x < 16$
10. $\frac{x}{5} \geq -4$ $x \geq -20$
11. $\frac{d}{7} \geq -9$ $d \geq -63$

Solve each inequality. (EXAMPLE 2)
12. $-3b > 27$ $b < -9$
13. $-4x < -16$ $x > 4$
14. $-\frac{x}{3} \geq 12$ $x \leq -36$
15. $-\frac{x}{7} < -3$ $x > 21$

Solve each inequality. Graph each solution on a number line. (EXAMPLE 3)
16. $2x - (4x + 2) \leq 38$ $x \geq -20$
17. $-18y - 4 > 50$ $x < -3$

APPLICATION
18. **CONSUMER ECONOMICS** A digital telephone service provider offers its customers two monthly service plans.

Plan A: $30 + $0.50 per minute   Plan B: $10 + $0.75 per minute
For what range of minutes does plan B cost less? (EXAMPLE 4)
for between 0 and 80 minutes/month

Practice and Apply

Solve each inequality.
19. $2h > 32$ $h > 16$
20. $8x \leq -56$ $x \leq -7$
21. $5y \leq 22$ $y \leq \frac{22}{5}$
22. $-12b > 3$ $b < -\frac{1}{4}$
23. $\frac{y}{6} > 17$ $y > 102$
24. $\frac{z}{4} \leq -8$ $z \leq -32$
25. $\frac{x}{8} < 1$ $x < 8$
26. $\frac{u}{-3} \geq 21$ $u \leq -63$
27. $-3y \geq 48$ $y \leq -16$
28. $-6z \leq 96$ $z \geq -16$
29. $-\frac{1}{5}m \geq 8$ $m \leq -40$
30. $-\frac{g}{14} \leq 7$ $g \geq -98$
31. $2x + 8 \geq 11$ $x \geq \frac{3}{2}$
32. $4x - 11 < 20$ $x < \frac{31}{4}$
33. $3x - 7 > 18$ $x > \frac{25}{3}$
34. $2.4z + 9.6 \geq 21.3$ $z \geq 4.875$
35. $8 - h > 9$ $h < -1$
36. $6 - x > -1$ $x < 7$
37. $-5b + 4 \leq 18$ $b \geq -\frac{14}{5}$
38. $-3a - 7 > 2.3$ $a < -3.1$
39. $6y + 7 \geq 2y + 28$ $y \geq \frac{21}{4}$
40. $1.3b - 9.2 < 7.7$ $b < 13$
41. $4m + 1 < 5m - 7$ $m > 8$
42. $10.8 \leq 2c - 11.3$ $c \geq 11.05$
43. $38.25 > -2q + 29.5$ $q > -4.375$
44. $\frac{-d}{3} + 4 < 10$ $d > -18$
45. $\frac{-f}{5} - 1 < 3$ $f > -20$

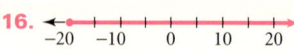

16. [number line with closed circle at -20, shaded right]
17. [number line with open circle at -3, shaded left]

286   LESSON 6.2

58. $b \geq -\frac{4}{3}$
59. $y \geq 7$
60. $r > -\frac{2}{9}$
61. $q < 5.81$
62. $h < 1.125$
63. $p \leq 4$
64. $n > \frac{-27}{2}$
65. $d > \frac{28}{3}$
66. $m \geq -56$

CONNECTION

APPLICATIONS

$550 + 6.50g \leq 1200$

$g \leq 100; 100$

Solve each inequality.

46. $15 - \frac{t}{4} > 10$ $t < 20$
47. $\frac{x}{4} - 2 \leq 13$ $x \leq 60$
48. $\frac{5}{7}x + 27 \leq 12$ $x \leq -21$
49. $2x + 5 \leq -3$ $x \leq -4$
50. $-3t > 8$ $t < -\frac{8}{3}$
51. $5c - 9 > 20$ $c > \frac{29}{5}$
52. $3(m + 7) \leq 28$ $m \leq \frac{7}{3}$
53. $4p + 7 - p \leq 31$ $p \leq 8$
54. $7m - 2(m - 1) > -48$ $m > -10$
55. $3p + (2 - 3p) > 28$ **No solution**
56. $y + 5 + 5y \geq -7$ $y \geq -2$
57. $2(m - 3) + 8m \leq 11$ $m \leq \frac{17}{10}$
58. $-3b + 12 \leq 16$
59. $4y + 5 \geq y + 26$
60. $-6r - 10 < 3r - 8$
61. $3.3q - 8.4 < 10.8$
62. $-5.6h + 8.6 > 2.3$
63. $\frac{p}{7} - \frac{1}{14} \leq \frac{1}{2}$
64. $-\frac{2}{3}n + 12 < 21$
65. $\frac{1}{4}d - \frac{3}{9} > 2$
66. $-\frac{7}{8}m - 9 \leq 40$

67. **GEOMETRY** The length, l, of a rectangle is at least 5 centimeters more than the width, w.

   a. Express the statement above as an inequality. $l \geq 5 + w$

   b. Write an inequality for the statement above when the width is 20 centimeters. $l \geq 25$

68. **SMALL BUSINESS** Brian charges $15 plus $5.50 per hour to mow yards. Greg charges $12 plus $6.25 per hour.

   a. Write two equations to show how much Brian and Greg charge, c, in terms of the number of hours, h, that they work. **Brian: $c = 5.50h + 15$; Greg: $c = 6.25h + 12$**

   b. When is Brian's charge greater than Greg's charge? **from 0 to 4 hours**

   c. When are Brian's and Greg's charges equal? **4 hours**

69. **TRAVEL** Jeff's car averages 18 miles per gallon of gasoline. What is the greatest number of gallons of gasoline that he will need if he travels no more than 450 miles? **25 gallons**

70. **THEATER** Members of the drama club are putting on a play. They need to spend $560 for costumes, sets, and programs. They plan to sell tickets for $7 each. How many tickets must be sold in order for the drama club to make a profit? **81 tickets**

71. **SMALL BUSINESS** A catering business specializes in catering wedding receptions. They charge $550 for setting up the buffet and an additional $6.50 per guest.

   a. Mr. and Mrs. Hiroshige want to spend no more than $1200 on the catering for their daughter's wedding reception. Write an inequality in terms of the number of guests, g, that they can invite to the reception.

   b. Solve the inequality that you wrote for part **a**. What is the maximum number of guests that can be invited to the reception?

Error Analysis

Students often forget to change the direction of the inequality symbol when multiplying and dividing by negative numbers. Encourage them to develop the habit of checking values at and near the boundary of the solution set. For instance, if their proposed solution is $y < 5$, they should verify that

- 5 is the solution of the related equation;
- 4 is a solution of the inequality; and
- 6 is *not* a solution of the inequality.

Student Technology Guide

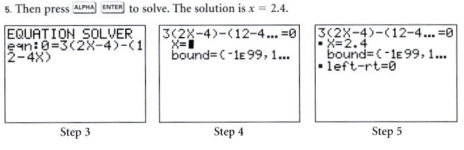

Student Technology Guide
6.2 Multistep Inequalities

LESSON 6.2 **287**

 **Look Beyond**

In Exercises 92–95, students preview multistep inequalities with two variables. They will encounter two-variable inequalities again in Lesson 7.5.

ALTERNATIVE
Assessment

Portfolio Activity

The Portfolio Activity can be used as preparation for the Chapter Project or as a separate activity. In this Portfolio Activity, students are introduced to the absolute-value function and its graph. They will need to use the table of values they completed for the piecewise function in the Portfolio Activity on page 281. Students will learn more about the absolute-value function in Lesson 6.4.

Answers to Portfolio Activities can be found in Additional Answers of the Teacher's Edition.

internet connect

GO TO: go.hrw.com
KEYWORD: MA1 Puzzles

To support the Portfolio Activity, students may visit the HRW Web site, where they can examine more puzzles and discover how they work mathematically.

 **Look Back**

Evaluate. *(LESSON 2.3)*

72. $76 - (-43)$ **119**   **73.** $-111 - 400$ **−511**   **74.** $12 - 546$ **−534**
75. $80 - (-31)$ **111**   **76.** $-1 - 99$ **−100**   **77.** $-31 - (-50)$ **19**

Evaluate. *(LESSON 2.4)*

78. $(-3)(4)(-3)$ **36**   **79.** $(-2)(-2)(-2)(-2)$ **16**   **80.** $(5)(-3)(-3)(5)$ **225**
81. $(-4)(-4)(-4)$ **−64**   **82.** $(-2)(5)(-3)$ **30**   **83.** $(-25)(2)(-2)(-25)$ **−2500**

Simplify each expression. *(LESSONS 2.6 AND 2.7)*

84. $3x + 2y + 5x + 7y$ **$8x + 9y$**   **85.** $3(x + 2y) + 4(2x + y)$ **$11x + 10y$**

Solve each equation. *(LESSON 3.3)*

86. $2x + 7 = 15$ **$x = 4$**   **87.** $65 = 3x - 7$ **$x = 24$**   **88.** $6y + 17 = 32$ **$y = \frac{5}{2}$**
89. $3m - 10 = 23$ **$m = 11$**   **90.** $130 = 4p - 30$ **$p = 40$**   **91.** $8x + 7 = -17$ **$x = -3$**

 **Look Beyond**

92. $y \leq x + \frac{1}{5}$; yes
93. $y \geq 1$; no
94. $y < 17x - 1$; no
95. $y < x$; no

Solve each inequality for *y*. Then state whether the point (0, 0) satisfies the inequality.

92. $5(y + 2x) \leq 15x + 1$ **93.** $y - 6x - \frac{1}{2}(y - 3x) + 1 \leq \frac{3}{2}(y - 3x)$
94. $4x - 2(y + 1) > -30x$ **95.** $y - 2x + 1 < 1 - x$

Include this activity in your portfolio.

Refer to the table of values for the piecewise function in the Portfolio Activity on page 281.

1. Plot the points from the table on a coordinate plane. Sketch the graph of the function by connecting the points.

2. Describe the shape of the graph. What is the equation of the line to the left of the *y*-axis? What is the equation of the line to the right of the *y*-axis?

The function you graphed is a well-known and useful function in mathematics. It is called the *absolute-value function*. The notation

$$\begin{cases} -x \text{ if } x < 0 \\ x \text{ if } x \geq 0 \end{cases}$$ is generally replaced by the symbol $|x|$. You will learn more

about the absolute-value function in Lesson 6.4.

WORKING ON THE CHAPTER PROJECT

You should now be able to complete Activities 1 and 2 of the Chapter Project on page 308.

6.3 Compound Inequalities

Objectives
- Graph the solution sets of compound inequalities.
- Solve compound inequalities.

Swimming pools require maintenance to ensure that they stay clean and free of bacteria. Adding chlorine to a pool is one way to ensure that bacteria do not have a chance to grow.

Why Many real-world situations have a range of values that are acceptable. Compound inequalities describe these types of limited ranges.

When two inequalities are combined into one statement, the result is called a **compound inequality**. The words *and* and *or* are used to describe the relationship between the two parts of the inequality.

EXAMPLE ① The United States standard for safe levels of chlorine in swimming pools is at least 1 part per million and no greater than 3 parts per million. **Write and graph a compound inequality to describe this situation.**

APPLICATION
POOL MAINTENANCE

SOLUTION

Let c be the level of chlorine in parts per million.

The inequality representing at least 1 part per million is $c \geq 1$. The inequality representing no more than 3 parts per million is $c \leq 3$.

Both $c \geq 1$ AND $c \leq 3$ must be true. The two inequalities can be combined into one statement: $1 \leq c \leq 3$.

The graph of the compound inequality is the *intersection* region of the two parts of the inequality.

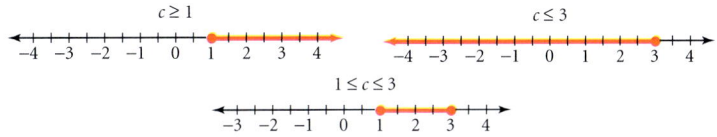

The graph of the solution set shows the set of values that make both inequalities true.

Prepare

NCTM PRINCIPLES & STANDARDS 1–4, 6–10

LESSON RESOURCES
- Practice 6.3
- Reteaching Master 6.3
- Enrichment Master 6.3
- Cooperative-Learning Activity 6.3
- Lesson Activity 6.3
- Problem Solving/Critical Thinking Master 6.3

QUICK WARM-UP

Solve each inequality.

1. $16 + h < 25$ $h < 9$
2. $4m \geq 22$ $m \geq 5.5$

Solve each inequality and graph.

3. $6x - 10 \geq 14$ $x \geq 4$

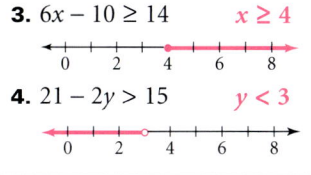

4. $21 - 2y > 15$ $y < 3$

Also on Quiz Transparency 6.3

Teach

Why Give students these two sentences: The dog is black. The cat is white. Ask students how the two sentences could be rewritten as only one sentence. Students may suggest using the words *and*, *or*, and *but* to connect the sentences. Explain to students that the new sentence formed is called a compound sentence. Compound sentences, or statements, are also used in algebra.

☞ For Additional Example 1, see page 290.

Alternative Teaching Strategy

USING VISUAL MODELS Have students use the area above a number line as a scratch pad to draw the graph for each case of a compound inequality. Students can then decide which part of the graph for each case belongs on the graph of the solution. This technique is shown below for the solution to part **a** of Example 3 on page 291.

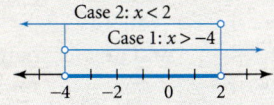

Inclusion strategies

VISUAL LEARNERS Many students can successfully graph each case of a compound inequality separately but may have difficulty graphing both cases on the same number line. Encourage students to use different colored pencils to graph each case of a compound inequality. Tell students that when AND is used in the compound inequality, they should include in their solution the parts that are *common to both* of the colored graphs. When OR is used in the compound inequality, tell students they should include all areas of each colored graph.

LESSON 6.3 **289**

ADDITIONAL EXAMPLE 1

Alan bought an aquarium and some fish at the pet store. He was told that his fish required water with at least 6 but no more than 10 parts per million of dissolved oxygen. **Write and graph a compound inequality to describe the situation.**

$6 \leq x \leq 10$

ADDITIONAL EXAMPLE 2

On certain highways, it is illegal for a car to drive less than 45 or more than 65 miles per hour. **Write and graph a compound inequality to describe the speeds that are prohibited.**

$x < 45$ OR $x > 65$

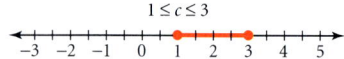

Notice that in Example 1 that $c \geq 1$ AND $c \geq 3$. A compound inequality involving AND has a solution region that represents an **intersection**, or overlapping, of the solution regions for the separate parts of the inequality. This type of compound inequality is called a **conjunction**.

$1 \leq c \leq 3$

Both $c \geq 1$ AND $c \leq 3$ must be true to make the conjunction true.

Example 2 illustrates the other type of compound inequality, a *disjunction*.

EXAMPLE 2

APPLICATION
AIR TRAVEL

In a certain restricted air space, airplanes must fly below 10,000 feet or above 15,000 feet. **Write and graph a compound inequality to describe this situation.**

● **SOLUTION**

Let a be the altitude in feet.

The inequality that represents flying below 10,000 feet is $a < 10,000$. The inequality that represents flying above 15,000 feet is $a > 15,000$.

Either $a < 10,000$ OR $a > 15,000$ can be true. The two inequalities can be combined into one statement using the word OR: $a < 10,000$ OR $a > 15,000$.

The graph of the compound inequality includes the regions on the opposite sides of the two endpoints determined by the individual inequalities.

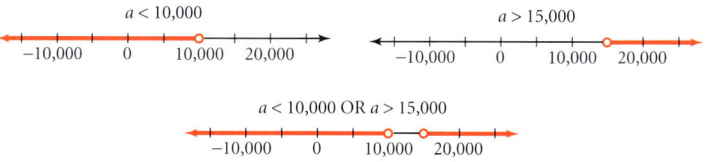

The graph of the solution set shows the set of values that make either one inequality or the other true.

In Example 2, either $a \leq 10,000$ OR $a \leq 15,000$. A compound inequality involving OR has solution regions that are the **union**, or the total, of the solution regions of the separate parts of the inequality. This type of compound inequality is a **disjunction**.

If either $a < 10,000$ OR $a > 15,000$ is true, then the disjunction is true.

Enrichment

Draw the four graphs shown below on the board or overhead. Have students write the compound inequality that represents each graph.

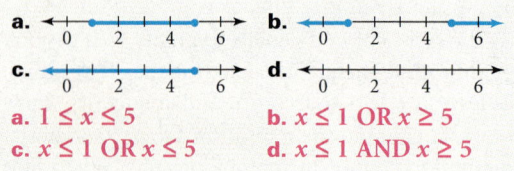

a. $1 \leq x \leq 5$
b. $x \leq 1$ OR $x \geq 5$
c. $x \leq 1$ OR $x \leq 5$
d. $x \leq 1$ AND $x \geq 5$

Reteaching the Lesson

INVITING PARTICIPATION Before class starts, write the integers from −10 to 10 and some simple fractions on separate sheets of paper and ensure that every student will have a paper. Give each student one paper, and instruct the class to make a human number line at the front of the classroom to represent each of the following inequalities:
• numbers greater than 3
• numbers less than or equal to −4
• numbers less than 4 and greater than −4
• numbers less than or equal to −7 OR numbers greater than or equal to 5

EXAMPLE 3 Graph each compound inequality.
 a. $-4 < x$ AND $x < 2$
 b. $x < 2$ OR $x \geq 5$

SOLUTION

a. Notice that the acceptable values of x must be both greater than -4 AND less than 2. This conjuction may also be written as $-4 < x < 2$.

Therefore, the solution is the set of values between -4 and 2.

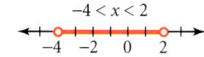

b. The acceptable values for this compound inequality may be either less than 2 OR greater than or equal to 5.

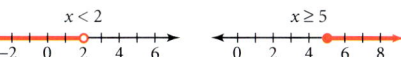

In this case the solution consists of the values that lie on opposite sides of the endpoints and that include the endpoint at 5.

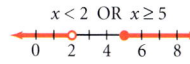

TRY THIS Graph each compound inequality.
 a. $q < 5$ OR $q > 16$
 b. $-3 \leq y \leq 6$

Example 4 illustrates how to solve a multi-step conjuction.

EXAMPLE 4 Solve and graph $-5 \leq 3x + 7 \leq 10$.

SOLUTION

Solve each inequality separately and then combine the results to form a compound inequality.

$$-5 \leq 3x + 7 \qquad\qquad 3x + 7 \leq 10$$
$$-12 \leq 3x \qquad\qquad\qquad 3x \leq 3$$
$$-4 \leq x \qquad\qquad\qquad\quad x \leq 1$$

The solution to the compound inequality is region where both inequalities are true and where the graphs overlap.

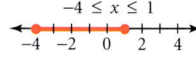

The solution is $-4 \leq x \leq 1$.

TRY THIS Solve and graph $-6 < 2x + 4 \leq 10$.

GO TO: go.hrw.com
KEYWORD: MA1 Hurricane Force

In this activity, students examine scales used for measuring the strength of hurricanes and tornadoes in order to categorize some given storms.

ADDITIONAL EXAMPLE 3

Graph each compound inequality.

a. $3 < x < 7$

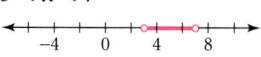

b. $x > 6$ OR $x \leq -1$

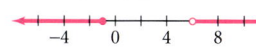

TRY THIS

a.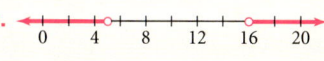

b.

ADDITIONAL EXAMPLE 4

Solve and graph $-2 \leq 5x + 8 \leq 18$.

$-2 \leq x \leq 2$

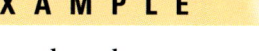

TRY THIS

$-5 < x \leq 3$

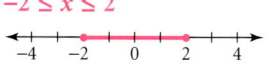

ADDITIONAL EXAMPLE 5

Solve and graph $6x \leq -30$ OR $4x + 9 \geq -3$.

$x \leq -5$ OR $x \geq -3$

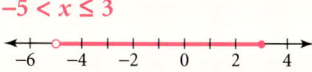

TRY THIS

$x \leq 3$ OR $x > 10$

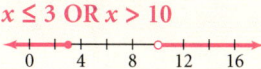

LESSON 6.3 **291**

ADDITIONAL EXAMPLE 6

Solve and graph $7x + 8 \leq 43$ OR $x - 16 > -13$.

$x \leq 5$ OR $x > 3$

Assess

Selected Answers
Exercises 5–18, 19–43 odd

ASSIGNMENT GUIDE

In Class	1–18
Core	19–40
Core Plus	19–40
Review	41–44
Preview	45, 46

✏ Extra Practice can be found beginning on page 738.

Mid-Chapter Assessment for Lessons 6.1 through 6.3 can be found on page 74 of the *Assessment Resources*.

Practice

EXAMPLE 5

Solve and graph $3x \leq -12$ OR $2x + 7 \geq 5$.

● SOLUTION

Simply solve each inequality separately, keeping the word OR between them, and then graph the resulting statements.

$3x \leq -12$	OR	$2x + 7 \geq 5$
$x \leq -4$		$2x \geq -2$
		$x \geq -1$

$x \leq -4$ OR $x \geq -1$

The resulting graph is constructed as follows:

$x \leq -4$ OR $x \geq -1$
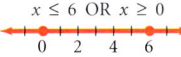

TRY THIS Solve and graph $5x \leq 15$ OR $3x - 7 > 23$.

EXAMPLE 6

Solve and graph $2x + 1 \leq 13$ OR $x - 5 \geq -5$.

● SOLUTION

$2x + 1 \leq 13$	OR	$x - 5 \geq -5$
$2x \leq 12$		$x \geq 0$
$x \leq 6$		

Since OR is used, the solution region includes all real numbers and is not confined to the intersection of the two individual solution sets.

The graph of the solution set is constructed as follows:

$x \leq 6$ OR $x \geq 0$

Exercises

Communicate

1. Describe how to write $y > -6$ AND $y < 4$ as a compound inequality.

2. Describe how to solve the compound inequality $-2 < 2x < 4$.

3. **a.** Give a real-world situation that can be described by a compound inequality.

   b. Give the compound inequality that describes the situation from Exercise 3.

4. Describe the differences between the following compound inequalities:

   a. $-4 < x \leq 5$ **b.** $-4 < x$ OR $x \leq 5$

5.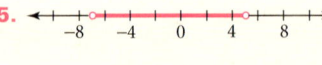

6.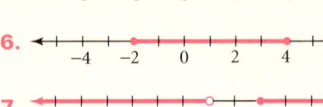

7.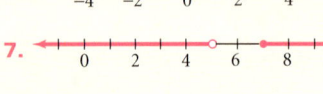

8.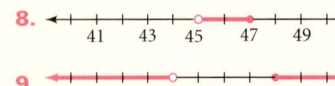

9.

10.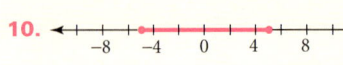

11.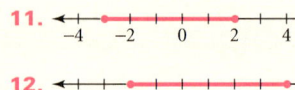

12.

13.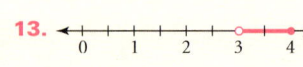

14.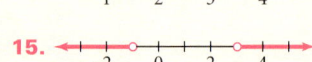

15.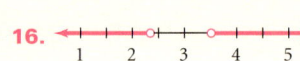

16.

292 LESSON 6.3

Guided Skills Practice

Graph each compound inequality. *(EXAMPLES 1, 2, AND 3)*

5. $-7 < x < 5$
6. $-2 \leq y \leq 4$
7. $q < 5$ OR $q \geq 7$
8. $45 < z \leq 47$
9. $v < 4$ OR $v \geq 12$
10. $-5 \leq x \leq 5$

11. $-3 \leq x \leq 2$
12. $-2 \leq x \leq 4$
13. $3 < x \leq 4$
14. $x \leq 3$ or $x > 4$
15. $x > 3$ or $x < -1$
16. $x > \frac{7}{2}$ or $x < \frac{7}{3}$
17. All real numbers
18. All real numbers

Solve and graph each compound inequality. *(EXAMPLE 4)*

11. $-7 \leq 3x + 2 \leq 8$
12. $-2 \leq 4x + 6 \leq 22$
13. $8 < 3x - 1 \leq 11$

Solve and graph each compound inequality. *(EXAMPLE 5)*

14. $2x \leq 6$ OR $3x > 12$
15. $7x > 21$ OR $2x < -2$
16. $2x > 7$ OR $3x < 7$

Solve and graph each compound inequality. *(EXAMPLE 6)*

17. $3x - 4 \leq 11$ OR $2x + 5 > 5$
18. $2x - 4 > -2$ OR $2x - 6 < -2$

Practice and Apply

Graph each compound inequality.

19. $2 < x < 5$
20. $x > 4$ OR $x < 2$
21. $-1 \leq x < 7$
22. $z < 4$ OR $z \geq 7$
23. $-4 < n \leq 6$
24. $x > -14$ OR $x < -18$
25. $-6 < v \leq 3$
26. $x < 4$ OR $x \geq 6$
27. $3.5 < t \leq 7.5$
28. $y > -4.5$ OR $y < -6.2$
29. $t < 6$ OR $t \geq 8.4$
30. $-4.3 < t \leq 6.9$

31. $-9 \leq x < 3$
32. $13 \leq z < 15$
33. All real numbers
34. $-4 < x \leq 9$
35. $1 < x < 7$
36. $x < -8$ or $x \geq -6$
37. $-\frac{5}{2} < x < \frac{7}{2}$
38. All real numbers

Solve and graph the following compound inequalities:

31. $-3 \leq 2x + 15 < 21$
32. $5 \leq z - 8 < 7$
33. $2x < 14$ OR $3x \geq 18$
34. $-14 < 2x - 6 \leq 12$
35. $-5 < 2x - 7 < 7$
36. $-2x > 16$ OR $-3x \leq 18$
37. $-5 < 2x < 7$
38. $-2x < 14$ OR $3x < -12$

APPLICATIONS

39. **WEATHER** The high temperature on a certain day was 98°F and the low was 78°F. If t represents the temperature, write a compound inequality that describes the range of temperatures on this day. $78 \leq t \leq 98$

40. **SALES** If a store makes a profit of $3 on each T-shirt that it sells, how many T-shirts must it sell to make a profit of at least $400?
    at least 134 T-shirts

Look Back

41. Find the sum $1 + 2 + 3 + 4 + 5 + 6$ by using a geometric dot pattern. *(LESSON 1.1)* 21

42. If each of 6 teams played each of the other teams just once in a tournament, how many games would there be? *(LESSON 1.1)* **15 games**

43. Evaluate: $4^2 \cdot 16[2 + 3(2 \cdot 4 + 7)] - 26$. *(LESSON 1.3)* **12,006**

44. Simplify: $(7 - 3y) - (8 + 2y)$. *(LESSON 2.6)* $-1 - 5y$

Look Beyond

45. Solve $2x + 3y > 18$ for y. $y > \frac{18 - 2x}{3}$
46. When x is 4, find the values of y that make the inequality in Exercise 45 true. $y > \frac{10}{3}$

17.
18.
19.
20.
21.
22.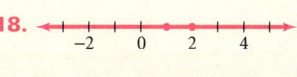

23.
24.
25.
26.
27.

The answers to Exercises 28–38 can be found in Additional Answers beginning on page 810.

Technology

In Exercises 31–38, a calculator may be helpful for checking the solutions of the related equations. For example, if students arrive at a solution such as $x > 2$ and $x < 5$, suggest that they use a calculator to verify that 3 is a solution of both inequalities.

Error Analysis

Students often get the graphs of compound inequalities confused. Remind students that if AND is used in a compound inequality, then the graph contains only the area that is common to both cases of the compound inequality. If OR is used in a compound inequality, the graph contains the combined graphs of each case of the compound inequality.

Look Beyond

In Exercises 45 and 46, students investigate an inequality with two variables. These types of inequalities will be studied in detail in Lesson 7.5.

Student Technology Guide

LESSON 6.3 **293**

Prepare

NCTM Principles & Standards 1–10

LESSON RESOURCES

- Practice 6.4
- Reteaching Master 6.4
- Enrichment Master 6.4
- Cooperative-Learning Activity 6.4
- Lesson Activity 6.4
- Problem Solving/Critical Thinking Master 6.4
- Teaching Transparency 32

QUICK WARM-UP

Evaluate.

1. $|-11|$ 11
2. $|2 - 6|$ 4
3. $|6 - 2|$ 4

Find the distance between each pair of points on a number line.

5. 1, 7   6
6. −2, 3   5

Also on Quiz Transparency 6.4

Teach

Why Students have learned how to solve inequalities and graph them on a number line. Tell students that they will now explore quantities in which the distance from zero, not the direction, is important.

Math
CONNECTION

STATISTICS The absolute error, in this case, is the amount by which each student overestimated or underestimated the time elapsed.

294   LESSON 6.4

6.4 Absolute-Value Functions

Objectives

- Explore features of the absolute-value function.
- Explore basic transformations of the absolute-value function.

Why The absolute-value function allows you to consider the magnitude of a number without regard for its sign. This is useful in various kinds of measurements, including statistics.

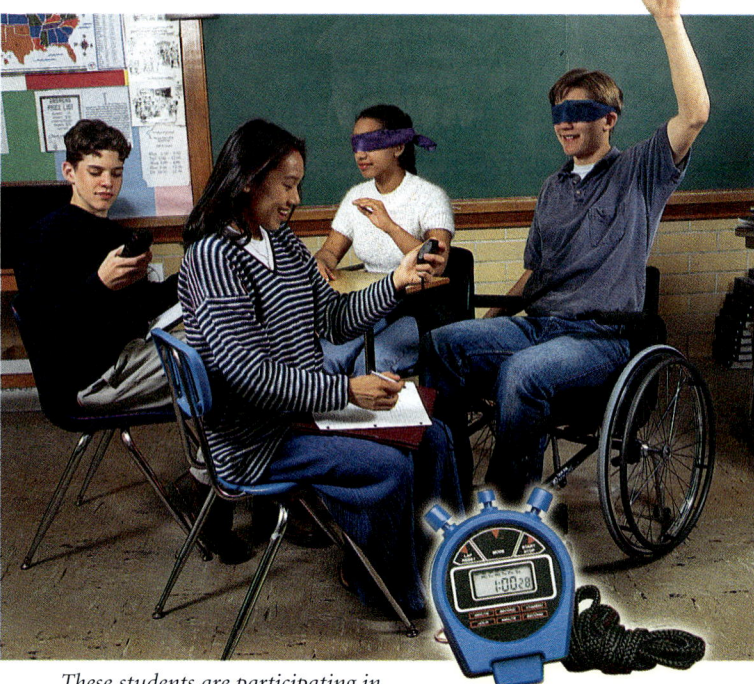

These students are participating in an experiment in which some of the students estimate elapsed time.

CONNECTION
STATISTICS

The students in the experiment shown above work in pairs. Half of the students are blindfolded and then told to raise a hand. They are told to put their hand down when they think that 1 minute has passed. The other half of the students act as timers. Later they switch roles.

Data from the first group is entered in a spreadsheet with the students' names in column A and their estimates of the time elapsed in column B.

TECHNOLOGY
SPREADSHEET

Compute the error *in seconds* in column C by subtracting 60 from each student's time.

In column C, a negative number means the estimate was *under* 1 minute. The absolute error in column D shows only the *amount* of the error.

	A	B	C	D
1	Student	Time	Error	Abs. Error
2	Tricia	49	−11	11
3	Keira	59	−1	1
4	Tom	51	−9	9
5	Louise	65	5	5
6	James	68	8	8
7	Sakeenah	77	17	17
8	Hong	66	6	6
9	Louis	54	−6	6
10	Mary	67	7	7
11	Maria	46	−14	14
12	Marcus	62	2	2
13	Shamar	73	13	13
14	Lois	61	1	1
15	Dianne	53	−7	7
16	Suzanne	64	4	4

Alternative Teaching Strategy

USING VISUAL MODELS Have students build a table of values for $y = x$ and for $y = |x|$. Lead them to see that the y-values of the two equations are the same when x is positive and are opposites when x is negative. Have the students graph each equation.

x	$y = x$	$y = |x|$
−4	−4	4
−2	−2	2
0	0	0
2	2	2
4	4	4

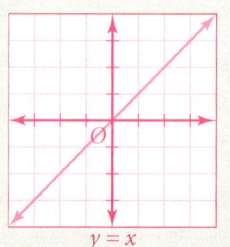
$y = x$

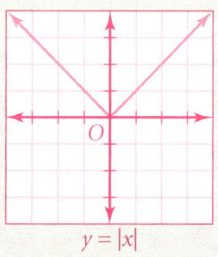
$y = |x|$

Point out that the left side of the graph of $y = x$ is flipped over the x-axis to form the graph of $y = |x|$. Ask students how the table of values for the two equations helps to explain the "flipping" of the left side of the $y = x$ graph.

For example, row 10 shows that Mary thought 1 minute had elapsed after 67 seconds. Her error was 67 − 60 = 7, indicating that her answer was 7 seconds *over* 1 minute. Row 15 shows that Diane's time was 53 seconds, so her error was 53 − 60 = −7, or 7 seconds *under* 1 minute.

Look carefully at column D in the spreadsheet. Compare the numbers in column C with those in column D. What is the sign of each number in column D? How does it compare with the sign of the corresponding number in column C?

The **absolute value** of a number can be shown on a number line, as you learned in Lesson 2.1. Recall that the absolute value of a number is its distance from zero.

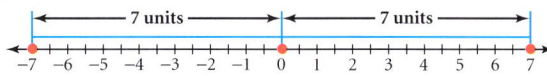

7 is 7 units from 0, so the absolute value of 7 is 7.

Thus, $|7| = 7$.

−7 is 7 units from 0, so the absolute value of −7 is 7.

Thus, $|-7| = 7$.

The geometric illustration leads to the following algebraic interpretation: The absolute value of a number is the number itself when the number is positive or zero. The absolute value of a number is the opposite of the number when the number is negative.

Algebraic Definition of Absolute Value

For all real numbers a:

$|a| = a$, for $a \geq 0$
$|a| = -a$, for $a < 0$

EXAMPLE 1 Find each of the following absolute values:
  a. $|8 − 2|$ **b.** $|2 − 8|$ **c.** $|3 − 3|$

● **SOLUTION**

a. $|8 − 2| = |6|$
      $= 6$

b. $|2 − 8| = |-6|$
      $= 6$

c. $|3 − 3| = |0|$
      $= 0$

TRY THIS Find each of the following absolute values:
  a. $|5 − 14|$ **b.** $|12 − 2|$

A function written in the form $y = |x|$ or $y = \text{ABS}(x)$ is an example of an **absolute-value function**.

Teaching Tip

Some students confuse taking the absolute value of a number with taking the opposite of a number. For example, some students may think that because $|-5| = 5$, then $|5| = -5$. Encourage students to think of absolute-value as taking only the magnitude of a number into account, not its sign.

Teaching Tip

TECHNOLOGY If you are using a TI-83 graphics calculator, you can access its absolute-value function by pressing MATH ▶ ENTER. On the TI-82, the keystrokes are 2nd x^{-1}.

ADDITIONAL EXAMPLE 1

Find each of the following absolute values:
  a. $|7 − 9|$ 2
  b. $|9 − 7|$ 2
  c. $|5 − 6|$ 1

TRY THIS
a. 9
b. 10

Interdisciplinary Connection

BIOLOGY Approximately one in four American adults has high blood pressure, or hypertension. Blood pressure is the force that blood exerts on the walls of the blood vessels. When the heart contracts and is pumping blood, blood pressure is at its greatest. This is called the systolic pressure. When the heart is at rest, blood pressure falls. This is called the diastolic pressure. A normal systolic level is less than 130 mm Hg, and a normal diastolic level is less than 85 mm Hg. Borderline levels for systolic and diastolic pressure are 130–140 mm Hg and 85–90 mm Hg, respectively. Have students write absolute-value expressions to describe the borderline levels for both pressures. In addition, have students research normal blood cholesterol levels, write absolute-value expressions to describe those levels, and report their findings to the class.

Let s represent systolic pressure and and let d represent diastolic pressure. The borderline levels can be modeled by $|s − 135| \leq 5$ and $|d − 87.5| \leq 2.5$.

LESSON 6.4

ADDITIONAL EXAMPLE 2

a. Find the domain and the range of $y = |x - 1|$.

domain: all real numbers; range: all nonnegative real numbers

b. Graph the function $y = |x - 1|$ for x-values from −3 to 3.

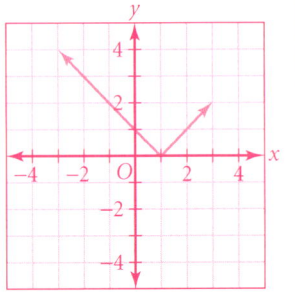

Teaching Tip

TECHNOLOGY If you are using a TI-82 or TI-83 graphics calculator, you can create the table shown in Example 2 by following these steps. Press [Y=]. Clear any existing equations and then press [MATH] [▶] [ENTER] [X,T,θ,n]. Now press [2nd] [WINDOW] and make the following choices in **TABLE SETUP**:

TblStart=−3 (or TblMin=−3)
ΔTbl=1
Indpnt: **Auto** Ask
Depend: **Auto** Ask

Now press [2nd] [GRAPH] to view the table. Press [GRAPH] to see the graph.

Math CONNECTION

COORDINATE GEOMETRY In Example 2, point out to students that the range is readily apparent from the graph because the function never dips below the x-axis.

TRY THIS
Check students' graphs.
a. domain: all real numbers; range: all nonnegative real numbers
b. domain: all real numbers; range: all nonnegative real numbers

296 LESSON 6.4

EXAMPLE

a. Find the domain and the range of the absolute-value function $y = |x|$.
b. Graph the function $y = |x|$ for x-values from −3 to 3.

● **SOLUTION**

a. Since you can find the absolute value of any number, the domain is all real numbers. Since $|0| = 0$ and $|x|$ is positive otherwise, the range is all non-negative real numbers, or $y \geq 0$.

b. Make a table of ordered pairs. For every value of x in the table, write the absolute value of x as the y-value. For example, if x is −2, then y is $|-2|$ or 2.

x	−3	−2	−1	0	1	2	3
y	3	2	1	0	1	2	3

CONNECTION
COORDINATE GEOMETRY

Plot the ordered pairs, and then connect the points because the domain includes all real numbers.

The graph of the absolute-value function has a distinctive **V** shape. The tip of the **V** is at the origin, (0, 0).

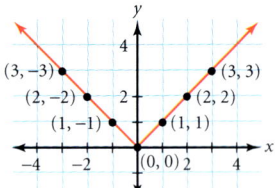

TRY THIS Find the domain and range of each function. Then graph each function.
a. $y = 2|x|$
b. $y = |x - 2|$

Activity
Manipulating the Absolute-Value Function

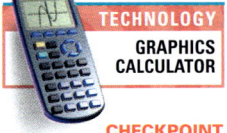
TECHNOLOGY
GRAPHICS CALCULATOR

You will need: a graphics calculator or graph paper

1. Graph the functions $y = |x|$ and $y = |x - 3|$ together on a same coordinate plane.

CHECKPOINT ✓ 2. What effect does the 3 have on the graph?

3. Graph the functions $y = |x|$ and $y = |x| - 3$ on the same coordinate plane.

CHECKPOINT ✓ 4. What effect does the 3 have on the graph?

5. Graph the functions $y = |x|$ and $y = -|x|$ together on a coordinate plane.

CHECKPOINT ✓ 6. What effect does the negative sign have on the graph?

The function $y = |x|$ is called the absolute-value **parent function** because it is the most basic absolute-value function.

The graph of a parent function changes if the function rule is changed in some way. There are certain kinds of changes called **transformations**. (Transformations will be presented in detail in Chapter 14.) Two types of transformations are illustrated in Example 3.

A **translation** shifts the parent-function graph horizontally or vertically.

Inclusion Strategies

INVITING PARTICIPATION Create a large coordinate plane on the floor. Have one student start at the origin and walk x units to the right and y units up. Have the second student mirror the first student's path by walking x units to the *left* (or $-x$ units) and y units up. Explain to the class that both students are the same distance from the origin. Have another pair of students repeat the exercise. Point out the V-shape that the students have made. Tell them that they have just modeled the absolute-value function.

Enrichment

Replace each ? with >, <, or = so that each statement is true.

1. $|-4|$? $|4|$ =
2. $|-2|$? $|2|$ >
3. -7 ? $-|7|$ =
4. -9 ? $-|-9|$ =
5. $-|-1|$? 1 <

Write the expression without using the absolute-value symbol.

6. $|-9|$ 9
7. $|4 - 3|$ $(4 - 3)$
8. $|-1| - |-2|$ $(1 - 2)$
9. $-3 - |-7|$ $(-3 - 7)$

A **reflection** creates a mirror image of the parent-function graph across the **line of reflection**. If the line of reflection is the *x*-axis, then each point (a, b) on the original graph becomes $(a, -b)$ on the reflection.

EXAMPLE ③ Graph $y = |x|$ and the given transformation function on the same coordinate plane. Identify the type of transformation.
a. $y = |x + 7|$
b. $y = -|x|$

● **SOLUTION**

a. Substituting $x + 7$ for the x in the parent function causes its graph to shift 7 units to the left. This shift is a (horizontal) translation.

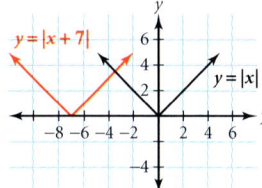

b. Multiplying the parent function by -1 causes the parent-function graph to reflect across the *x*-axis. Therefore, this transformation is a reflection.

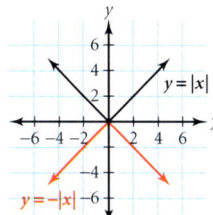

CRITICAL THINKING Normally, a reflection of the parent graph across the *y*-axis is caused by substituting $-x$ for x in the parent function. What is the result of doing this in the absolute-value function? Explain.

Exercises

● *Communicate*

1. When guessing how long it takes for a minute to elapse, Andy guessed 51 seconds and Abi guessed 68 seconds. Explain how to compare the amount of their errors.
2. Give an example of a value of *a* for which $|a| = -a$, if possible. Explain.
3. Describe the graph of an absolute-value function.
4. Can the value of $|x|$ ever be negative? Explain.
5. Can the value of $|x|$ ever be zero? Explain.

Reteaching the Lesson

WORKING BACKWARD Draw the graph of $y = |x|$ on the board or overhead. Have students build a table of values based on the graph. Now ask students to describe a rule that would explain the relationship between the *x*-values and the *y*-values.

Activity Notes

In this Activity, students explore the relationship between variations in the equation $y = |x|$ and variations in its graph.

CHECKPOINT ✓
2. The 3 shifts the graph 3 units to the right.
4. The 3 shifts the graph 3 units down.
6. It creates a *reflection* of the graph across the *x*-axis.

See Keystoke Guide, page 782.

ADDITIONAL EXAMPLE ③

Graph $y = |x|$ and the given transformation on the same coordinate plane. Identify the type of transformation.
a. $y = |x| - 2$
b. $y = -3|x|$

a. shifted 2 units down; vertical transformation

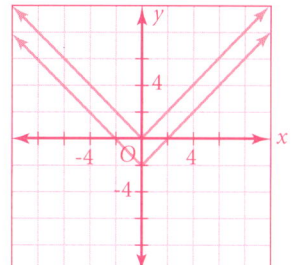

b. compressed and reflected across the *x*-axis; reflection

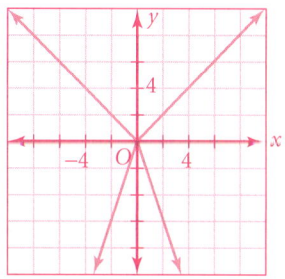

CRITICAL THINKING

The graph does not change because the left half of the absolute-value parent function is the reflection of the right half by definition.

LESSON 6.4

Assess

Selected Answers
Exercises 6–11, 13–83 odd

ASSIGNMENT GUIDE

In Class	1–11
Core	13–59 odd, 62
Core Plus	12–60 even, 61–64
Review	65–83
Preview	84

✏ Extra Practice can be found beginning on page 738.

Technology

A graphics calculator may be helpful for creating some of the graphs in the exercises. You may want students to first draw the graphs on graph paper and then use a graphics calculator to verify their results.

Practice

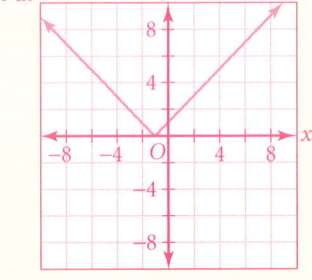

Guided Skills Practice

Find each absolute-value. *(EXAMPLE 1)*

6. $|5 - 12|$ **7**   7. $|13 - 12|$ **1**   8. $|-3 - 3|$ **6**   9. $|4 - 4|$ **0**

10. Find the domain and range of each function, and then graph the function. *(EXAMPLE 2)*

   d: all reals   **d: all reals**
   a. $y = |x + 1|$ r: all non-negative   b. $y = |x| + 1$ r: all reals ≥ 1

11. Graph $y = |x|$ and each of the following functions on the same coordinate plane. Identify the type(s) of transformation(s) of the graph of $y = |x|$ that occurs with each of the new functions. *(EXAMPLE 3)*

   a. $y = |x + 1|$ b. $y = |x| + 1$ c. $y = -|x| + 1$

Practice and Apply

Find the absolute value (ABS) of each number.

12. 17 **17**   13. −33 **33**   14. −7.11 **7.11**   15. 8.67 **8.67**
16. $\frac{4}{3}$ **$\frac{4}{3}$**   17. −2.5 **2.5**   18. $-3\frac{5}{11}$ **$3\frac{5}{11}$**   19. 79.2 **79.2**

Evaluate.

20. $|13 - 24|$ **11**   21. $|0 - 3|$ **3**   22. $|4 - 12|$ **8**   23. $|1 - 11|$ **10**
24. $|0 - (-3)|$ **3**   25. $|1 - 27|$ **26**   26. $|11 - 3|$ **8**   27. $|-14 - (-14)|$ **0**
28. $|-13 + 13|$ **0**   29. $|-5 - 2|$ **7**   30. $|5 - (-3)|$ **8**   31. $|-11 - 11|$ **22**
32. $|-5 + (-5)|$ **10**   33. $|0 - 5|$ **5**   34. $|5 - 10|$ **5**   35. $|-5 - 10|$ **15**

36–47 Domain: All real numbers
36. $y \geq 0$ 37. $y \geq 0$
38. $y \geq 2$ 39. $y \geq -4$
40. $y \leq 0$ 41. $y \leq 0$
42. $y \leq 2$ 43. $y \leq -4$
44. $y \geq 0$ 45. $y \geq 0$
46. $y \geq -1$ 47. $y \geq 0$

Find the domain and range of each function.

36. $y = |x + 4|$ 37. $y = |x - 5|$ 38. $y = |x| + 2$ 39. $y = |x| - 4$
40. $y = -|x + 4|$ 41. $y = -|x - 5|$ 42. $y = -|x| + 2$ 43. $y = -|x| - 4$
44. $y = 4|x|$ 45. $y = \frac{1}{2}|x|$ 46. $y = 4|x| - 1$ 47. $y = 4|x - 1|$

48–55. For Exercises 48–55, refer to the functions given in Exercises 36–43. Graph $y = |x|$ and the given function on the same coordinate plane. Identify the transformations of the absolute-value parent function.

Find $|a - b|$ and $|b - a|$ for each of the following:

56. $a = 5, b = 3$ **2**   57. $a = 5, b = -3$ **8**   58. $a = -5, b = 3$ **8**   59. $a = 3, b = 5$ **2**

60. Based on your answers for Exercises 56–59, what conclusions can you make about $|a - b|$ and $|b - a|$? **They are equal**

CHALLENGE

61. **TRANSFORMATIONS** Write a function for each transformation of $y = |x|$ described below.

   a. The graph of $y = |x|$ is translated 5 units to the left and then 2 units down. **$y = |x + 5| - 2$**

   b. The graph of $y = |x|$ is translated 5 units up and then reflected across the x-axis. **$y = -(|x| + 5)$**

   c. The graph of $y = |x|$ is reflected across the x-axis and then translated 2 units up. **$y = -|x| + 2$**

10. a. 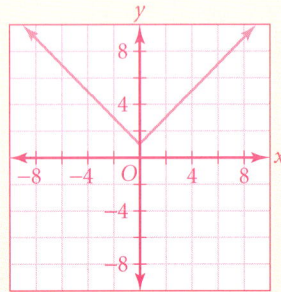   b.

The answers to Exercise 11 can be found in Additional Answers beginning on page 810.

298 LESSON 6.4

APPLICATIONS

62. CHEMISTRY Four students reported their findings about how much sodium was in a salt compound. If the actual amount was 8.4 milligrams, find the error and the absolute error for each student.
  a. Alisha reported 8.2 milligrams. −0.2; 0.2
  b. Najee reported 9.0 milligrams. 0.6; 0.6
  c. Nell reported 8.1 milligrams. −0.3; 0.3
  d. Paula reported 8.4 milligrams. 0; 0

TRAVEL An express train leaves from Houston traveling north to St. Louis at an average speed of 80 miles per hour. After about 3 hours the train passes through Dallas. If t is the travel time in hours from Houston, the distance between the train and Dallas is given by $y = 80|3 − t|$.

63. After 2 hours, how far has the train traveled? Where is the train at this time in relation to Dallas? 160 miles; 80 miles south of Dallas

64. After 4 hours, how far has the train traveled? Where is the train at this time in relation to Dallas? 320 miles; 80 miles north of Dallas

Look Back

65. Identify the next three numbers in the sequence below. *(LESSON 1.1)*
 2, 6, 10, 14, 18, . . . 22, 26, 30

66. If Miguel has $2.00, how many bookmarks can he buy if the bookmarks cost $0.38 each? *(LESSON 2.4)* 5 bookmarks

Evaluate. *(LESSONS 2.2, 2.3, AND 2.4)*

67. $-3 + 4$ 1 **68.** $-3(0.3)$ −0.9 **69.** $-15 − (−15)$ 0 **70.** $60 \div -3$ −20
71. $(-10)\left(\frac{1}{2}\right)$ −5 **72.** $-1.4 − (−3)$ 1.6 **73.** $4\left(-1\frac{1}{2}\right)$ −6 **74.** $(-3.2) \div 4$ −0.8

75. Graph the functions below on the same coordinate plane and describe the similarities and differences. *(LESSON 5.4)*
  a. $y = x + 2$ b. $y = \frac{2}{3}x − 1$ c. $y = x$

Write each linear equation in slope-intercept form. *(LESSON 5.4)*

76. $3x + 2y = 1$ **77.** $4x = 2y$ $y = 2x$ **78.** $4y = 0$ $y = 0$ **79.** $2x − 2y = 17$ $y = x − \frac{17}{2}$

Solve each inequality and graph the solution on a number line. *(LESSON 6.1)*

80. $x + 7 \leq 3$ **81.** $x − 3 \geq 2$ **82.** $x + 15 \leq -1$ **83.** $x − 3 > 4$
 $x \leq -4$ $x \geq 5$ $x < -16$ $x > 7$

Look Beyond

84. On a number line, $\sqrt{x^2}$ can be thought of as the distance from the origin, 0, to the point x. In the coordinate plane, $\sqrt{x^2 + y^2}$ can be thought of as the distance from the origin (0, 0) to the point (x, y). Use $\sqrt{x^2 + y^2}$ to find the distance from the origin to each of the following points:
  a. (3, 4) 5   b. (12, 5) 13   c. (−8, 6) 10   d. (−15, −20) 25

Look Beyond

In Exercise 84, students learn the formula for finding the distance between the origin and a point on the coordinate plane. Students will study the distance formula in more depth in Chapter 12.

Error Analysis

Make sure that students' absolute-value graphs have two opposite slopes. Remind them that the two halves of the graph should be mirror images.

75. Graphs **a** and **c** have the same slope, 1, but different y-intercepts, 2 and 0. Graph **b** has a different slope from the others, $\frac{2}{3}$, and a different y-intercept, −1. The x-intercepts are all different: −2, $\frac{3}{2}$, and 0, respectively.

---

48. horizontal translation 4 units left

49. horizontal translation 5 units right

50. vertical translation 2 units up

51. vertical translation 4 units down

52. horizontal translation 4 units left, then reflection

53. horizontal translation 5 units right, then reflection

54. reflection, then vertical translation 2 units up

55. reflection, then vertical translation 4 units down

The graphs for Exercises 48–55 can be found in Additional Answers beginning on page 810.

Student Technology Guide

6.4 Absolute-Value Functions

You can use a graphics calculator to evaluate absolute values.
To find $|-18|$, enter MATH NUM 1:abs(ENTER (-) 18) ENTER.
To find $|a − b|$ and $|b − a|$ when $a = −5$ and $b = 6$, use the following keystrokes:
(-) 5 STO► ALPHA MATH ALPHA . 6 STO► ALPHA MATRX ALPHA
. MATH NUM 1:abs(ENTER ALPHA MATH − ALPHA MATRX)
ENTER MATH NUM 1:abs(ENTER ALPHA MATRX − ALPHA MATH
) ENTER

```
abs(-18)
              18
-5→A:6→B:abs(A-B
)
              11
abs(A-B)
              11
```

Use a graphics calculator to find the absolute value of each number.

1. 5	2. −21	3. 3.7	4. 0
5	21	3.7	0

Use a graphics calculator to evaluate.

5. $\|17 − 5\|$	6. $\|-6 − (−3)\|$	7. $\|8 − 12\|$	8. $\|-4.6 + 8.6\|$
12	3	4	4

Use a graphics calculator to find $|a − b|$ and $|b − a|$ for each of the following:

9. $a = 4, b = -3$	10. $a = -8, b = -5$	11. $a = 16, b = 21$	12. $a = -6, b = 5$
$\|a − b\|$ 7	$\|a − b\|$ 3	$\|a − b\|$ 5	$\|a − b\|$ 11
$\|b − a\|$ 7	$\|b − a\|$ 3	$\|b − a\|$ 5	$\|b − a\|$ 11

A graphics calculator also can be used to graph absolute-value functions.
To graph $y = |x + 2| − 1$, first set the window dimensions. Absolute-value functions look best in a square window, such as the one shown. Enter the function as Y1 by pressing Y=
MATH NUM 1:abs(ENTER X,T,θ,n + 2) − 1.
Then press GRAPH.

Graph each function.

13. $y = |x − 1| + 2$ 14. $y = |x + 3| − 3$ 15. $y = 2 − |x|$ 16. $y = |x − 2| − 2$

Prepare

NCTM PRINCIPLES & STANDARDS 1–10

LESSON RESOURCES

- Practice 6.5
- Reteaching Master 6.5
- Enrichment Master 6.5
- Cooperative-Learning Activity 6.5
- Lesson Activity 6.5
- Problem Solving/Critical Thinking Master 6.5
- Teaching Transparency 32

QUICK WARM-UP

Evaluate each expression.

1. $|-6|$ **6**   2. $-|-2|$ **−2**
3. $-|5.1|$ **−5.1**   4. $|0|$ **0**

Simplify each expression.

5. $-(y+3)$ **$-y-3$**
6. $-(2x-9)$ **$-2x+9$**

Also on Quiz Transparency 6.5

Teach

Why Ask students if they have ever tried on two pairs of the same size and style of shoe and had one pair fit while the other was too tight. Ask students to explain why this is a common occurrence and to share with the class their own experiences with variation in products. Explain to students that a margin of error is often allowed when manufacturing an item. Point out that absolute value can be used to represent the difference between an actual measurement and a manufacturer's quality-control standard of measurement.

6.5 Absolute-Value Equations and Inequalities

A company manufactures a special gear for a car. If the gear is made too large or too small, it will not turn smoothly and may wear out quickly.

Objectives

- Solve absolute-value equations.
- Solve absolute-value inequalities, and express the solution as a range of values on a number line.

Why Absolute-value expressions are often used to describe differences from a given value, such as tolerances in manufacturing.

Error and Absolute Value

APPLICATION
MANUFACTURING

A gear is designed with a specification of 3.50 centimeters for its diameter. Its actual diameter must be within 0.01 centimeter of 3.50 centimeters in order for the gear to work properly. This can be expressed as 3.50 ± 0.01.

The difference between the actual measure and the specified measure is called the **error**. For example, a gear with an actual measure of 3.48 centimeters would have the following error:

$$3.48 - 3.50 = -0.02 \qquad \text{Error}$$

The **absolute error** is the absolute value of the difference between the actual measure and the specified measure.

$$|3.48 - 3.50| = |-0.02| = 0.02 \qquad \text{Absolute error}$$

The greatest and least acceptable values for the diameter of the gear are the values for which the absolute error is 0.01. Letting d represent the actual diameter of the gear, you can use the following **absolute-value equation** to find the greatest and least acceptable values:

$$|d - 3.50| = 0.01$$

Because of the definition of absolute value, *the equation will have two solutions*.

Alternative Teaching Strategy

USING VISUAL MODELS Remind students that the absolute value of a number is related to distance on a number line. Tell them that they can interpret absolute-value equations as follows:

$|n| = a$ → n is exactly a units from 0.
$|n - b| = a$ → n is exactly a units from b or n and b are exactly a units away from each other.

Use the set of examples at right to develop this concept and extend it to the solutions of absolute-value inequalities.

Find all numbers exactly 3 units from 0.

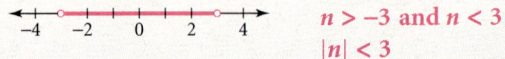

$n = -3$ or $n = 3$; $|n| = 3$

Find all numbers less than 3 units from 0.

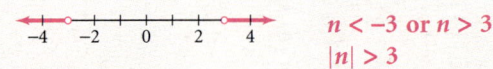

$n > -3$ and $n < 3$; $|n| < 3$

Find all numbers more than 3 units from 0.

$n < -3$ or $n > 3$; $|n| > 3$

300 LESSON 6.5

Recall the definition of the absolute value of a number that you learned in Lesson 6.4: $|a| = a$ for $a \geq 0$ and $|a| = -a$ for $a < 0$. If there is an expression inside the absolute-value sign, it is usually helpful to restate the definition as follows:

For $a \geq 0$, $|x| = a$ is equivalent to $x = a$ or $x = -a$.

EXAMPLE 1 Solve the absolute-value equation for the gear problem.
$$|d - 3.50| = 0.01$$

CONNECTION
MAXIMUM/MINIMUM

● SOLUTION

To solve an absolute-value equation, you must consider two cases.

CASE 1
Consider the quantity within the absolute-value symbols to be positive.

$|d - 3.50| = 0.01$
$d - 3.50 = 0.01$
$d = 3.51$

$|3.51 - 3.50| = 0.01$ *True*

CASE 2
Consider the quantity within the absolute-value symbols to be negative.

$|d - 3.50| = 0.01$
$d - 3.50 = -0.01$
$d = -0.01 + 3.50$
$d = 3.49$

$|3.49 - 3.50| = |-0.01| = 0.01$ *True*

The solution shows that the greatest and least allowable lengths for the gear diameter are 3.51 centimeters and 3.49 centimeters, respectively.

TRY THIS Solve $|x - 10| = 4.5$.

EXAMPLE 2 Solve $|3x - 2| = 10$.

● SOLUTION

CASE 1
Consider the quantity within the absolute-value symbols to be positive.

$|3x - 2| = 10$
$3x - 2 = 10$
$3x = 12$
$x = 4$

$|3(4) - 2| \stackrel{?}{=} 10$
$|12 - 2| \stackrel{?}{=} 10$
$|10| = 10$ *True*

CASE 2
Consider the quantity within the absolute-value symbols to be negative.

$|3x - 2| = 10$
$3x - 2 = -10$
$3x = -8$
$x = -\frac{8}{3}$

$|3(-\frac{8}{3}) - 2| \stackrel{?}{=} 10$
$|-8 - 2| \stackrel{?}{=} 10$
$|-10| = 10$ *True*

TRY THIS Solve $|2x - 4| = 8$.

CRITICAL THINKING
a. How can you tell immediately that $|5x - 2| = -12$ has no solution?
b. What apparent solutions do you get if you use the method in Example 2?

Find all numbers exactly 3 units from 2.

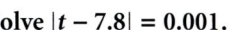

$n = -1$ or $n = 5$;
$|n - 2| = 3$

Find all numbers less than 3 units from 2.

$n > -1$ and $n < 5$;
$|n - 2| < 3$

Find all numbers more than 3 units from 2.

$n < -1$ or $n > 5$;
$|n - 2| > 3$

Interdisciplinary Connection

CHEMISTRY The pH of a substance is a measure of how acidic or alkaline it is. A pH scale extends from 0 to 14. Solutions in the interval from 0 to 2 inclusive are highly acidic and are considered to be strong acids. Solutions in the interval from 12 to 14 inclusive are highly alkaline and are considered to be strong bases. Write an absolute-value inequality that describes the pH of a strong acid or a strong base relative to a neutral pH of 7.

$|x - 7| \geq 5$, where x represents the pH of the substance

Teaching Tip

Students learned the meaning of the term *absolute value* in Lesson 2.1. You may want to review the definition of this term at the beginning of this lesson.

ADDITIONAL EXAMPLE 1

Solve $|t - 7.8| = 0.001$.

$t = 7.801$ or $t = 7.799$

Math CONNECTION

MAXIMUM/MINIMUM The solution of an absolute-value equation gives a range of acceptable values for a problem. Students will study the concept of maximum and minimum in more depth in future courses such as precalculus and calculus.

TRY THIS
$x = 14.5$ or $x = 5.5$

ADDITIONAL EXAMPLE 2

Solve $|2z + 21| = 33$.

$z = 6$ or $z = -27$

TRY THIS
$x = 6$ or $x = -2$

CRITICAL THINKING
a. Absolute-value expressions can never equal a negative value.
b. $x = -2$ or $x = \frac{14}{5}$

LESSON 6.5

ADDITIONAL EXAMPLE 3

A tool-and-die worker receives an order for a bolt whose diameter must be within $\frac{1}{64}$ inch of $\frac{3}{8}$ inch. **Use an absolute-value inequality to find the range of acceptable diameters.**

$|x - \frac{3}{8}| \leq \frac{1}{64}$; all diameters between $\frac{23}{64}$ inch and $\frac{25}{64}$ inch inclusive, or $\frac{23}{64} \leq x \leq \frac{25}{64}$

Absolute Value and Inequalities

In the gear problem, the acceptable values for d fall within the boundary values of 3.49 centimeters and 3.51 centimeters. To find the acceptable values, you can use an **absolute-value inequality**. In this case, the inequality $|d - 3.50| \leq 0.01$ expresses the requirement that the actual diameter of a gear must fall within 0.01 centimeter of 3.50 centimeters. A similar inequality is solved in Example 3.

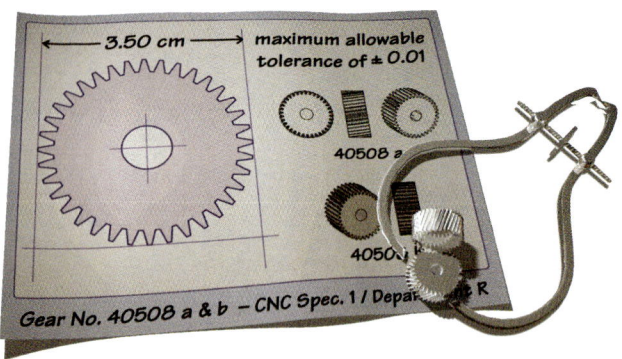

EXAMPLE 3

APPLICATION
CRAFTS

The square metal frame for a piece of stained glass will fit properly into the space in a block if the error in the actual length of the sides is within 0.15 centimeter of 48 centimeters. **Use an absolute-value inequality to find the boundary values.**

● **SOLUTION**

The absolute error can be expressed as $|x - 48.00|$, where x is the actual measurement of the square frame. The allowable absolute error is no greater than 0.15 centimeter. Since the absolute error must be less than or equal to 0.15 centimeter, use an inequality.

Solve $|x - 48.00| \leq 0.15$.

CASE 1	CASE 2
Find the upper boundary.	Find the lower boundary.
$\|x - 48.00\| \leq 0.15$	$\|x - 48.00\| \leq 0.15$
$x - 48.00 \leq 0.15$	$x - 48.00 \geq -0.15$
$x \leq 48.15$	$x \geq 47.85$

The allowable lengths for the sides of the square frame are all of the measures between 47.85 and 48.15 centimeters inclusive, or $47.85 \leq x \leq 48.15$.

Inclusion Strategies

VISUAL LEARNERS Many students successfully solve both cases of an absolute-value inequality and then have difficulty understanding how to graph both cases on the same number line. Suggest that they first graph the solution to each case above the number line and then use these separate graphs to decide what the final graph should be. This technique is shown at right for the solution to Example 4 on page 303.

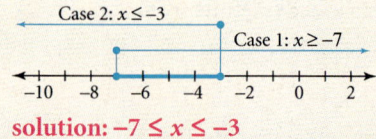

solution: $-7 \leq x \leq -3$

EXAMPLE 4 Solve $|x - (-5)| \leq 2$.

● **SOLUTION**

Solve $-2 \leq x - (-5) \leq 2$ by solving each part of the inequality seperately.

CASE 1
$-2 \leq x - (-5)$
$-2 \leq x + 5$
$-7 \leq x$

CASE 2
$x - (-5) \leq 2$
$x + 5 \leq 2$
$x \leq -3$

Then combine the solutions in the form of a conjunction.
$$-7 \leq x \leq -3$$

Notice that when an absolute-value expression is in the form $|x + 5|$, $+5$ is a shortened form of $-(-5)$. Therefore, when you graph the solution of the inequality, -5 will be the midpoint of the solution region.

TRY THIS Solve $|x + 6| \leq 4$.

CRITICAL THINKING What is the total length of the solution interval for $|x - a| \leq b$?

EXAMPLE 5 Solve $|x - 6| > 2$.

● **SOLUTION**

Consider two cases.

CASE 1
The quantity $x - 6$ is positive.
$|x - 6| > 2$
$x - 6 > 2$
$x > 8$

CASE 2
The quantity $x - 6$ is negative.
$|x - 6| > 2$
$-(x - 6) > 2$
$-x + 6 > 2$
$-x > -4$
$x < 4$

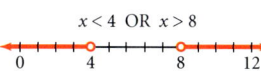

Check by substituting numbers from the solution region into the original inequality.

TRY THIS Solve and graph each absolute-value inequality.
a. $|x + 7| > 3$
b. $|x - 5| \geq 2$

CRITICAL THINKING How can you tell immediately that there is no solution for $|x - 5| < -1$?

Use Teaching Transparency 32.

ADDITIONAL EXAMPLE 4

Solve $|x - (-4)| < 5$.

$-9 < x < 1$

TRY THIS
$-10 \leq x \leq -2$

CRITICAL THINKING
$2b$

ADDITIONAL EXAMPLE 5

Solve $|x - 8| > 3$.

$x > 11$ or $x < 5$

TRY THIS
a. $x > -4$ or $x < -10$

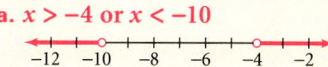

b. $x \geq 7$ and $x \leq 3$

CRITICAL THINKING
Because an absolute value cannot be a negative value, $|x - 5|$ cannot be less than -1.

Enrichment

Decide whether each statement is *always*, *sometimes*, or *never* true. If the answer is *sometimes*, describe the conditions that are necessary for the statement to be true.

1. $|x + y| = |y + x|$ always

2. $|x + y| = |x| + |y|$ sometimes; true when x and y are both negative or both positive

3. $|x - y| = |y - x|$ always

4. $|x - y| = |x| - |y|$ sometimes; true when x and y are both positive and $x > y$, when x and y are both negative and $x < y$, or when $y = 0$

5. $|xy| = |yx|$ always

6. $|xy| = |x| \cdot |y|$ always

7. $|x \div y| = |y \div x|$ sometimes; true when x and y are both nonzero and $x = y$

8. $|x \div y| = |x| \div |y|$ sometimes; true when $y \neq 0$

LESSON 6.5   **303**

Math
CONNECTION

COORDINATE GEOMETRY Fundamental concepts related to the coordinate plane were presented in Lesson 1.4. You may want to review these with students before they proceed with the Activity.

 **Notes**

You may want to begin the Activity by having students graph just $y = 6$. Be sure they understand that the graph is a horizontal line 6 units above the x-axis. After they add $y = |4x - 2|$, lead them to focus on x-values. They should discover that the line $y = 6$ divides the coordinate plane into three regions that they can use to visualize the solutions of $|4x - 2| = 6$, $|4x - 2| < 6$, and $|4x - 2| > 6$.

CHECKPOINT ✓
4. a. $x = -3$ and $x = 2$
 b. $-3 < x < 2$
 c. $x < -3$ or $x > 2$

See Keystroke Guide, page 782.

Teaching Tip

TECHNOLOGY If you are using a TI-82 or TI-83 graphics calculator for the Activity, press [Y=]. In the **Y=** editor, clear any existing equations and then enter the following next to **Y1=**:

TI-83: [MATH] [▶] [ENTER] 4 [X,T,θ,n] − 2)

TI-82: [2nd] [x^{-1}] (4 [X,T,θ] − 2)

Place the cursor **Y2=** and enter 6. Press [GRAPH]. To find the first point of intersection, press [2nd] [TRACE]. Choose **5:intersect**, and press [ENTER] [ENTER] [ENTER]. To find the second point, press [2nd] [TRACE] and choose **5:intersect** again. Press [ENTER] [ENTER], and then press [▶] until the cursor is near the second point. Then press [ENTER].

304 LESSON 6.5

Graphing Absolute-Value Equations

CONNECTION
COORDINATE GEOMETRY

You can use graphs of absolute-value and linear equations in two variables to illustrate one-variable absolute-value equations and inequalities. A graphics calculator will also give you solutions using this method, as you can discover in the following activity.

Activity
Graphing to Find Solutions

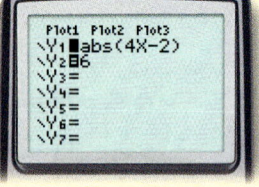

TECHNOLOGY
GRAPHICS CALCULATOR

You will need: a graphics calculator or graph paper

1. Graph $y = |4x - 2|$ and $y = 6$ on the same coordinate plane.

 If you are using a graphics calculator, press **graph** to see the graph on the calculator screen.

 To set the window, use
 Xmin = −10, Xmax = 10,
 Ymin = −5, and Ymax = 10.

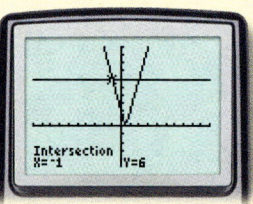

2. Explain how you can use the graph to find the solutions to the equation $|4x - 2| = 6$.

3. Explain how you can use the graph to find the solution to the inequalities $|4x - 2| < 6$ and $|4x - 2| > 6$.

CHECKPOINT ✓ 4. Use a graphics calculator to solve each of the following:
 a. $|2x + 1| = 5$ b. $|2x + 1| < 5$ c. $|2x + 1| > 5$

 **6** Use graphing to solve: a. $|x + 2| = 6$ b. $|x + 2| > 6$

● **SOLUTION**

Step 1
Graph the absolute-value function $y = |x + 2|$.

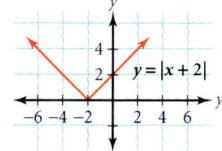

Step 2
Graph the constant function $y = 6$ on the same coordinate plane.

a. The x-values of the two intersection points of the graphs will be the solutions to $|x + 2| = 6$, $x = 4$ or $x = -8$.

b. All of the x-values of the linear graph on the outside of each intersection point will be the solutions to $|x + 2| > 6$, $x > 4$ or $x < -8$.

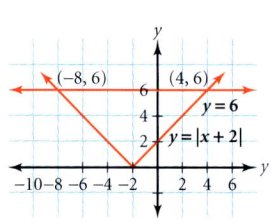

Reteaching the Lesson

USING VISUAL MODELS Draw a large number line on a roll of paper with the integers from −8 to 8 spaced about 1 foot apart. Place it on the floor. Have two students stand back-to-back at 0. Have one student walk to the number 5 and the other to −5. Point out that although one student walked in the negative direction and the other in the positive direction, each walked the same distance, 5 units. Lead the class to see that the students have modeled the solutions to $|x| = 5$: $x = 5$ or $x = -5$. Emphasize that the number used to describe distance is always nonnegative.

Repeat the activity with the students beginning at 2 rather than 0. Tell them to walk 5 units in opposite directions. They should arrive at 7 and −3. Lead the class to see that the students have modeled the solutions to $|x - 2| = 5$: $x = 7$ or $x = -3$.

Now have two students start at −3 and walk 5 units in opposite directions. They should arrive at 2 and −8. This time the students have modeled the solutions to $|x - (-3)| = 5$, which is equivalent to $|x + 3| = 5$. The solutions are $x = 2$ or $x = -8$.

Exercises

Communicate

1. Compare the solutions of $4 = |x + 5|$ with the solution of $4 \geq |x + 5|$.
2. **a.** Explain the meaning of the specification 45 ± 0.001 centimeters.
   b. Explain how to write an absolute-value expression to represent 45 ± 0.001 centimeters.
3. Why must you consider two cases when solving absolute-value equations and inequalities?
4. Describe how to use an absolute-value inequality to represent all of the values on a number line that are within 3 units of -7.

Guided Skills Practice

Solve each equation. *(EXAMPLE 1)*

5. $|x - 5.5| = 0.5$
   $x = 5$ or $x = 6$
6. $|x - 2| = 5$
   $x = 7$ or $x = -3$
7. $|a - 3.3| = 7$
   $a = 10.3$ or $a = -3.7$

Solve each equation. *(EXAMPLE 2)*

8. $|5x - 14.5| = 2.5$
   $x = 3.4$ or $x = 2.4$
9. $|3z + 5| = 16$
   $z = -7$ or $z = 3\frac{2}{3}$
10. $|2v + 6| = 8$
    $v = -7$ or $v = 1$

11. What numbers are less than or equal to 4 units from 10 on a number line? Write and solve an inequality that represents this situation. *(EXAMPLE 3)*

Solve each inequality. *(EXAMPLE 4)*

12. $|5x - 14.5| \leq 2.5$
13. $|3x + 2.5| \leq 23.5$
14. $|4x + 5.5| \leq 22.5$

15. Solve $|x - 10| > 3$. *(EXAMPLE 5)* $x < 7$ or $x > 13$

Solve by graphing on a coordinate plane. *(EXAMPLE 6)*

16. $|8x + 4| \geq 20$
17. $|3x - 6| < 3$

Practice and Apply

Solve each equation if possible. Check your answers.

18. $|x - 5| = 3$
19. $|x - 1| = 6$
20. $|x - 2| = 4$
21. $|x - 8| = 5$
22. $|5x - 1| = 4$
23. $|2x + 4| = 7$
24. $|4x + 5| = -1$
    no solution
25. $|-1 + x| = -3$
    no solution
26. $|7x + 4| = 18$
    $x = 2$ or $x = -3\frac{1}{7}$

Solve each inequality. Check your answers.

27. $|x - 3| < 7$
28. $|x - 12| < 6$
29. $|x + 4| > 8$
30. $|x + 9| > 1$
31. $|x - 8| \leq -4$
32. $|x - 5| \leq -2$
33. $|x - 2| > 6$
34. $|x - 2| \leq 10$
35. $|x - 16| \leq 2$
36. $|x + 1| < 5$
37. $|x + 4| > 2$
38. $|x + 5| \leq 35$

Side column answers:

11. $|x - 10| \leq 4$; $6 \leq x \leq 14$
12. $2.4 \leq x \leq 3.4$
13. $-8\frac{2}{3} \leq x \leq 7$
14. $-7 \leq x \leq 4.25$
16. $x \geq 2$ or $x \leq -3$
17. $1 < x < 3$
18. $x = 2$ or $x = 8$
19. $x = 7$ or $x = -5$
20. $x = 6$ or $x = -2$
21. $x = 3$ or $x = 13$
22. $x = 1$ or $x = -\frac{3}{5}$
23. $x = 1\frac{1}{2}$ or $x = -5\frac{1}{2}$
27. $-4 < x < 10$
28. $6 < x < 18$
29. $x > 4$ or $x < -12$
30. $x < -10$ or $x > -8$
31. no solution
32. no solution
33. $x < -4$ or $x > 8$
34. $-8 \leq x \leq 12$
35. $14 \leq x \leq 18$
36. $-6 < x < 4$
37. $x < -6$ or $x > -2$
38. $-40 \leq x \leq 30$

16.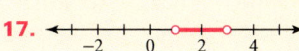
17. (number line graph)

ADDITIONAL EXAMPLE 6

Use graphing to solve:
a. $|x - 1| = 3$
b. $|x - 1| > 3$

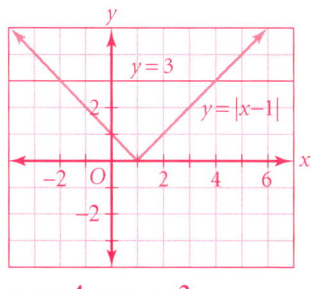

a. $x = 4$ or $x = -2$
b. $x > 4$ or $x < -2$

Assess

Selected Answers
Exercises 5–17, 19–63 odd

ASSIGNMENT GUIDE

In Class	1–17
Core	19–63 odd
Core Plus	18–60 even, 61–63
Review	64–78
Preview	79

✏ Extra Practice can be found beginning on page 738.

LESSON 6.5

Technology

In Exercises 18–54, a calculator may be helpful for checking the solutions of the equations or of related equations. If students are using the TI-83 graphics calculator, they can access the absolute-value function by pressing [MATH] ▶ [ENTER]. On the TI-82, the keystrokes are [2nd] [x^{-1}].

39. $d > -0.667$ or $d < -4$
40. $b > 6.5$ or $b < -1.5$
41. $0.286 \leq y \leq 0.857$
42. $0.111 \leq w \leq 1.444$
43. $x \geq 1.625$ or $x \leq -0.125$
44. $t \geq 1.111$ or $t \leq -0.222$
45. $-7.125 < x < 4.875$
46. $-1.5 < h < -0.214$
47. $0.4 < x < 2.8$
48. $-3.4 \leq d \leq -1.6$
49. $y > 2.34$ or $y < 0.66$
50. $x > 73.25$ or $x < 66.75$
51. $x > 0.404$ or $x < -0.404$
52. $x > 0.023$ or $x < 0.01$
53. $x > 7300$ or $x < 3100$
54. $x > 60.4$ or $x < 59.6$

Solve. Round your answers, if necessary, to the nearest thousandth.

39. $|3d + 7| > 5$
40. $|2b - 5| > 8$
41. $|4 - 7y| \leq 2$
42. $|7 - 9w| \leq 6$
43. $\left|\frac{3}{4} - x\right| \geq \frac{7}{8}$
44. $\left|\frac{4}{9} - t\right| \geq \frac{2}{3}$
45. $\left|\frac{2}{3}x + \frac{3}{4}\right| < 4$
46. $\left|\frac{7}{9}h + \frac{2}{3}\right| < \frac{1}{2}$
47. $\left|\frac{1}{2}x - \frac{4}{5}\right| < 0.6$
48. $\left|\frac{1}{3}d + \frac{5}{6}\right| \leq 0.3$
49. $|0.5y - 0.75| > 0.42$
50. $|0.002x - 0.14| > 0.0065$
51. $|0.001 - 5.2x| > 2.1$
52. $|0.005 - 0.3x| > 0.002$
53. $|5.2 - 0.001x| > 2.1$
54. $|0.3 - 0.005x| > 0.002$

55. The distance between x and -2 on a number line is 7.
   a. Draw a number-line diagram to illustrate the given statement.
   b. Translate the given statement into an absolute-value equation.
      $|x - (-2)| = 7$

Refer to the following inequality for Exercises 56–59: $|x - 3| < 4$.

56. Translate the absolute-value inequality into a sentence that begins with "The distance between." The distance between x and 3 is less than 4 units.
57. Graph the solution to the inequality on a number line.
58. Describe the solution in words.
59. Give five specific numbers that satisfy the inequality.

CHALLENGE

60. Write all the ways in which an absolute-value inequality can be written to represent all of the values within (and including) 3 units of -7. (Hint: Remember that $|a - b| = |b - a|$.)

APPLICATION

$|w - 168| = 3;$
$w = 171$ or $w = 165$

$|w - 168| \leq 3;$
$165 \leq w \leq 171$

61. **HEALTH** According to a height and weight chart, Mark's ideal wrestling weight is 168 pounds. Mark will be satisfied if he is no more than 3 pounds from his ideal weight.

   a. Draw a number-line diagram to illustrate the given situation. Identify the boundary values and acceptable weights.
   b. Write an absolute-value equation to describe the boundary values. Show the algebraic solution.
   c. Write an absolute-value inequality to describe the acceptable weights. Show the algebraic solution.

Practice

6.5 Absolute-Value Equations and Inequalities

Find the values of x that solve each absolute-value equation. Check your answers.

1. $|x + 2| = 5$ $x = 3$ or $x = -7$
2. $|x + 6| = 7$ $x = 1$ or $x = -13$
3. $|x - 7| = 4$ $x = 11$ or $x = 3$
4. $|x - 3| = 5$ $x = 8$ or $x = -2$
5. $|4x - 2| = 6$ $x = 2$ or $x = -1$
6. $|3x + 5| = 11$ $x = 2$ or $x = -5\frac{1}{3}$
7. $|-4 + x| = 7$ $x = 11$ or $x = -3$
8. $|x - 2.75| = 0.05$ $x = 2.8$ or $x = 2.7$

Find the values of x that solve each absolute-value inequality. Graph each answer on the number line provided. Check your answers.

9. $|x + 2| > 7$ $x > 5$ or $x < -9$
10. $|x + 1| \leq 8$ $-9 \leq x \leq 7$
11. $|-2 - x| \geq 4$ $x \leq -6$ or $x \geq 2$
12. $|x + 1| \geq 4$ $x \geq 3$ or $x \leq -5$
13. $|x - 3| > 2$ $x > 5$ or $x < 1$
14. $|4 - x| \geq 5$ $x \leq -1$ or $x \geq 9$
15. $|x + 2| > 2$ $x > 0$ or $x < -4$
16. $|x - 5| \leq 1$ $4 \leq x \leq 6$
17. $|x + 2| < 2$ $-4 < x < 0$

---

55. a. $x?$ ... $x?$ (number line from -9 to 5, marking -9 and 5)
b. $|x - (-2)| = 7; x = 5$ or $x = -9$

57. (number line showing open circles at -1 and 7)

58. x is between (but does not include) 7 and -1.

59. 2, 0, 5, 6, 3

60. $|x - (-7)| \leq 3$, $|x + 7| \leq 3$, $|-7 - x| \leq 3$
Either the first or the last inequality can be directly related to the graph. (The second inequality is an algebraic simplification of the first.)

61. a. (number line showing 163, 165, 167, 169, 171, 173 with marks at 165 and 171)
boundary values: 165, 171; acceptable weights: all weights between and including 165 and 171

306 LESSON 6.5

APPLICATIONS

62. MANUFACTURING A valve for the space shuttle must be within 0.001 millimeter of 5 millimeters in diameter.

  a. Write an absolute-value equation to describe the boundary values for the given situation. Solve your equation.

  b. What are the acceptable diameters for the valve? **between 4.999 and 5.001 mm, inclusive**

  c. Describe these acceptable values with an absolute-value inequality. $|d - 5| \leq 0.001$

63. GOVERNMENT A polling organization reports that 52% of registered voters preferred candidate A. The polling technique used has a **margin of error** of ±3%, meaning that the results are considered to be accurate within a range of 3% on either side of the reported figure.

a. upper: 55%; lower: 49%

b. Let *b* represent the percentage of voters favoring candidate **B**. $|b - 52| \leq 3$

c. Yes, given the error of ±3%, candidate A could be preferred by as few as 49% of the voters. This percentage does not represent a majority.

  a. What are the upper and lower boundaries for the actual percent of voters who support candidate A?

  b. Show how to represent the upper and lower boundaries, B, using an absolute-value equation.

  c. Is it possible that candidate A is not actually preferred by the majority of voters? Explain.

Error Analysis

When solving equations of the form $-(x + a) = b$, students often forget to take the opposite of both terms in parentheses and write $-x + a = b$. Encourage them to rewrite the equation to show a multiplication by -1 and to insert arrows as a reminder that they must apply the Distributive Property.

$$-(x + a) = b$$
$$(-1)(x + a) = b$$
$$-x - a = b$$

Look Beyond

Students have learned to solve equations where one side is an absolute value. In Exercise 79, students are asked to extend this skill to include equations where both sides are absolute values.

Look Back

Evaluate each expression for $p = 4$, $q = -1$, and $r = -2$. (LESSON 2.1)

64. $pqr - q$ **9**   **65.** $\dfrac{pq}{r}$ **2**   **66.** $\dfrac{pqr}{q} + pqr$ **0**

Simplify. (LESSON 2.5)

67. $-5(8c + 3)$ **$-40c - 15$**   **68.** $9(7b + 2)$ **$63b + 18$**   **69.** $-4(-5k + 8)$ **$20k - 32$**

Solve each equation. (LESSONS 3.1, 3.2, AND 3.4)

70. $x + 7 = 4$ $x = -3$ **71.** $2x + 3 = 3x + 1$ $x = 2$ **72.** $m - 7 = 2m + 3$ $m = -10$

73. $\dfrac{y}{7} = 2$ $y = 14$ **74.** $3x = -21$ $x = -7$ **75.** $-14p = -28$ $p = 2$

Solve each inequality. (LESSON 6.2)

76. $4x + 5 < 25$ $x < 5$ **77.** $6y - 10 > 5$ $y > \dfrac{5}{2}$ **78.** $9m - 8 < 4 + 8m$ $m < 12$

Look Beyond

79. Solve for *x*.

$x = 5$ or $x = -5$ $x = a$ or $x = -a$ $x = 1$ or $x = -\dfrac{3}{7}$

  a. $|x| = |-5|$ b. $|x| = |a|$ c. $|4x + 1| = |3x + 2|$

62. a. $|d - 5| = 0.001$; $d = 5.001$ or $d = 4.999$

Student Technology Guide

NAME _____ CLASS _____ DATE _____

Student Technology Guide
6.5 Absolute-Value Equations and Inequalities

The logic feature of a graphics calculator can be used to solve an absolute-value inequality. For example, to solve $|x - 4| > 3$, enter the following key sequence:

[Y=] [MATH] [NUM] 1:abs([ENTER] [X,T,θ,n] [−] 4 [)] [2nd] [MATH] [TEST] 3: [ENTER] 3 [ENTER]

Press [WINDOW]. Choose ranges such as Xmin = −10, Xmax = 10, Ymin = −2, and Ymax = 2. Press [GRAPH]. The endpoints of the two horizontal line segments on the display give you the clue to the solution.

To locate the endpoints, press [TRACE]. Those values of *x* are 1 and 7. Since the inequality involves >, the solution has two parts, $x < 1$ or $x > 7$.

Use a graphics calculator to solve each absolute-value inequality.

1. $|x + 2| > 6$ 2. $|x + 8| > 2$ 3. $|x - 3| > 1$
   $x > 4$ or $x < -8$ $x < -10$ or $x > -6$ $x < 2$ or $x > 4$

4. $|x - 2| > 3$ 5. $|x + 1| > 2.5$ 6. $|x - 3.2| > 1.6$
   $x < 1$ or $x > 5$ $x < -3.5$ or $x > 1.5$ $x < 1.6$ and $x > 4.8$

You can also use a graphics calculator to solve $|x - 4| \leq 3$. The key sequence is similar to the one above. However, you use ≤ rather than >.

Notice the single line segment on the display. As before, locate the endpoints of it to help find the solution to the given inequality. Since the inequality involves ≤, the endpoints of the segment are part of the solution. The solution to $|x - 4| \leq 3$ is $1 \leq x \leq 7$.

Use a graphics calculator to solve each absolute-value inequality.

7. $|x - 12| \leq 4$ 8. $|x - 4| < 5$ 9. $|x + 21| < 13$
   $8 \leq x \leq 16$ $-1 \leq x \leq 9$ $-34 \leq x \leq -8$

LESSON 6.5 **307**

Project

Focus

A *step function* is a piecewise function that consists of different constant range values for different intervals of the domain of the function. In this Chapter Project, students must complete a table of values and a graph for two step functions. The graph can be used to visually describe the situations, and can provide valuable insight into the relationship.

Motivate

Work with students to brainstorm a list of situations or pricing systems that use rounding to calculate the price. Lead them to see that post offices and long distance companies both use rounding to determine prices. Tell students that these situations can be modeled by step functions.

CHAPTER PROJECT SIX

planning a SNOWBOARDING TRIP

Mr. Line coaches a snowboarding team and teaches algebra in Bellevue, Washington. He is preparing to take his snowboarding team to a halfpipe competition and asks his algebra class to determine the number of vans the team will need for the trip and the amount of money the team will need to pay the entry fee for the competition. In this project, you will imagine that you are in Mr. Line's class and use step functions to determine the number of vans and the amount money the team will need.

Activity 1

The Vans

The snowboarding team normally has 17 members, but several members have been playing too many video games after school and may have to stay home to catch up on their homework.

1. Let n represent the number of team members going to the competition and let v represent the number of vans the team will need. Given that each van seats 6 team members, complete the table at right.

2. Plot the pairs of numbers from the table on a coordinate plane. The independent variable is n and the dependent variable is v. You can see that the value of v stays the same until the value of n reaches a certain number. Then the value of v jumps. Because the graph looks like a flight of steps, this type of function is known as a step function.

3. Suppose that 4 students have to stay home and do their homework. Describe how Mr. Line could use the graph that you created to determine the number of vans the team will need.

n	v
1	?
2	?
3	?
4	?
5	?
6	?
7	?
8	?
9	?
10	?
11	?
12	?
13	?
14	?
15	?
16	?
17	?

Activity 1

1. 1, 1, 1, 1, 1, 1, 2, 2, 2, 2, 2, 2, 3, 3, 3, 3, 3

3. Answers may vary.

2.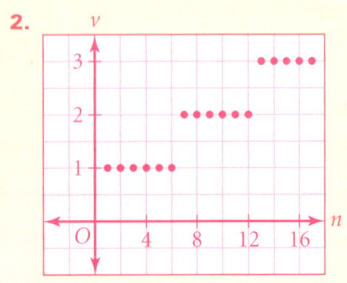

Cooperative Learning

Have students work in pairs. For **Activity 1**, one student should complete the table in Step 1, and the other student should draw the graph in Step 2. They should discuss and complete Step 3 together.

For **Activity 2**, partners should exchange roles. One student should complete a table of values, and the other should graph the equation in Step 1. Have the partners discuss and complete Step 2 together.

Discuss

Bring the class together to discuss their answers to the questions. Have them suggest real-world situations that can be modeled by step functions. Have each pair of students choose one situation, research appropriate data, and graph a step function for the data.

Activity 2

The Entry Fee

The greatest-integer function is a step function that can be used to model many situations. The greatest-integer function rounds a given number down to the nearest integer. For example, INT(2) = 2, INT(2.9) = 2, INT$\left(\frac{7}{3}\right)$ = 2, and INT(−2.1) = −3. It is important to note that the greatest-integer function always rounds *down*, and not *up*, to the nearest integer.

The entry fee for the competition depends on the number of team members competing. Let n represent the number of team members competing and let f represent the entry fee in dollars. The entry fee is given by the equation $f = 10\left[\text{INT}\left(\frac{n}{5}\right) + 1\right]$.

1. Make a table of values and graph the given equation above for integer values of n such that $1 \leq n \leq 17$.
2. Refer to the equation above and your graph to describe the pricing structure for the entry fee.

Activity 2

1. a.

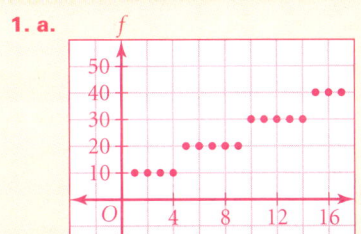

2. The fee is $10 for the first 4 entrants. After that, the fee increases by $10 for every 5 additional entrants.

n	1	2	3	4	5	6	7	8	9	10	11	12	13	14	15	16	17
f	10	10	10	10	20	20	20	20	20	30	30	30	30	30	40	40	40

Chapter Review

Chapter Review and Assessment

VOCABULARY

absolute error 300	Division Property of Inequality 283	Subtraction Property of Inequality 277
absolute-value equation ... 300	error 300	reflection 297
absolute-value function 295	inequality 276	transformations 296
absolute-value inequality .. 302	intersection 290	translation 296
Addition Property of Inequality 277	line of reflection 297	union 290
compound inequality 289	Multiplication Property of Inequality 283	
conjunction 290	parent function 296	
disjunction 290		

Chapter Test, Form A

Chapter Assessment
Chapter 6, Form A, page 1

Write the letter that best answers the question or completes the statement.

b 1. Which inequality corresponds to the statement "b is less than or equal to 1"?
  a. $b \geq 1$ b. $b \leq 1$ c. $b > 1$ d. $b < 1$

d 2. Which inequality is true?
  a. $0 \leq -7$ b. $-5 > 1$ c. $1.5 < 1.2$ d. $3.90 > 3.09$

b 3. Which sentence represents $n < 4$ on the number line?
  a. Draw a shaded circle at 4. Shade the numbers to the right of 4.
  b. Draw an unshaded circle at 4. Shade the numbers to the left of 4.
  c. Draw a shaded circle at 4. Shade the numbers to the right of 4.
  d. Draw a shaded circle at 4. Shade the numbers to the left of 4.

c 4. Burt wants to buy a pair of shoes that cost $49.95. He also wants to buy a T-shirt, but he cannot spend more than $60. Which inequality models this situation?
  a. $49.95 - x > 60$ b. $49.95 + x \geq 60$ c. $x + 49.95 \leq 60$ d. $60 + x \geq 49.95$

d 5. Solve $|x - 3.45| = 0.25$.
  a. $x = 3$ and $x = 3.7$ b. $x = -3$ and $x = -3.7$
  c. $x = 3.2$ d. $x = 3.2$ and $x = 3.7$

a 6. Solve $\frac{m}{10} - 16 < 40$.
  a. $m < 560$ b. $m > 560$ c. $m < 240$ d. $m > 240$

a 7. Solve $-8p < -56$.
  a. $p > 7$ b. $p < 7$ c. $p > -7$ d. $p < -7$

b 8. Solve $10 \leq 8 - x$.
  a. $x \leq 2$ b. $x \leq -2$ c. $x \geq 2$ d. $x \geq -2$

a 9. Solve $12 + 5x > 7x - 12$.
  a. $x < 12$ b. $x < -12$ c. $x > 12$ d. $x < -12$

Chapter Assessment
Chapter 6, Form A, page 2

c 10. Which inequality represents the graph?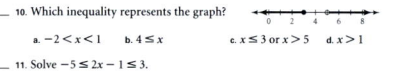
  a. $-2 < x < 1$ b. $4 \leq x$ c. $x \leq 3$ or $x > 5$ d. $x > 1$

d 11. Solve $-5 \leq 2x - 1 \leq 3$.
  a. $-3 \leq x \leq 2$ b. $x \leq -2$ or $x \geq 2$ c. $x \leq -3$ or $x \geq 2$ d. $-2 \leq x \leq 2$

b 12. A company is producing gauges for a contractor. The overhead charges are $500. The production costs need to be kept under $1000. How many gauges can the company produce at $10 each and keep the costs under $1000?
  a. under 100 b. under 50 c. over 100 d. over 50

d 13. Which equation represents the graph?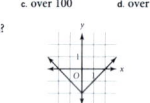
  a. $y = |x + 2|$ b. $y = |x - 2|$
  c. $y = |x| + 2$ d. $y = |x| - 2$

a 14. What is the domain and range of the function $y = -|x|$?
  a. Domain: all real numbers; range: all negative numbers and 0
  b. Domain: all negative numbers and 0; range: all real numbers
  c. Domain: all real numbers; range: all positive numbers and 0
  d. Domain: all positive numbers and 0; range: all real numbers

b 15. Evaluate $|-20 - 20|$.
  a. -40 b. 40 c. 0 d. not here

c 16. Solve $|x + 1| = 5$.
  a. $x = 6$ and $x = -4$ b. $x = 6$ and $x = 4$
  c. $x = -6$ and $x = 4$ d. $x = -6$ and $x = -4$

c 17. Solve $|x - 6| \geq 14$.
  a. $-8 \leq x \leq 20$ b. $x \geq 20$ or $x \leq -8$ c. $x \geq 20$ or $x \leq -8$ d. not given

a 18. Solve $|x + 8| \geq 1$.
  a. $-9 < x < -7$ b. $-9 > x > -7$ c. $7 < x < 8$ d. no solution

Key Skills & Exercises

LESSON 6.1
Key Skills

Solve inequalities, and graph the solution on the number line.

For inequalities involving addition and subtraction, solve as you would an equation. Then graph the solution on a number line.

$$5t - 6 \leq 4t + 1$$
$$5t - 4t - 6 \leq 4t - 4t + 1$$
$$t - 6 \leq 1$$
$$t - 6 + 6 \leq 1 + 6$$
$$t \leq 7$$

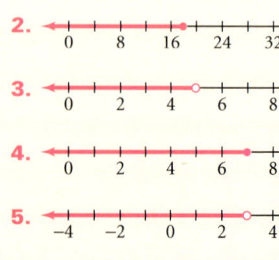

Exercises

Solve each inequality, and graph the solution on a number line.

1. $x + 5 > 10$ $x > 5$
2. $n - 15 \leq 3$ $n \leq 18$
3. $g + 6 < 11$ $g < 5$
4. $11 \geq b + 4$ $b \leq 7$
5. $y + 0.09 < 3.09$ $y < 3$
6. $d - \frac{2}{3} \geq \frac{3}{9}$ $d \geq 1$
7. $\frac{4}{5} \leq x - \frac{1}{5}$ $x \geq 1$
8. $x - 3.2 \leq 2.7$ $x \leq 5.9$
9. $x + 3.5 \leq 7$ $x \leq 3.5$
10. $17 \leq x - 2.6$ $x \geq 19.6$
11. $m + 3 > 0$ $m > -3$
12. $m - 3.6 < 6$ $m < 9.6$

LESSON 6.2
Key Skills

Solve multistep inequalities.

Solve $14 - x \leq -6$.

$14 - x \leq -6$	Given
$14 - 14 - x \leq -6 - 14$	Subtr. Prop. of Inequality
$-x \leq -20$	Simplify.
$-1(-x) \geq -1(-20)$	Mult. Prop. of Inequality
$x \geq 20$	Simplify.

Remember that if each side of an inequality is multiplied or divided by a negative number, the inequality symbol is reversed.

Exercises

Solve.

13. $2x + 4 < 6$ $x < 1$
14. $8 - 3y \geq 2$ $y \leq 2$
15. $5r - 12 > 48$ $r > 12$
16. $\frac{-p}{8} \leq -3$ $p \geq 24$
17. $7 < \frac{-b}{4}$ $b < -28$
18. $t + 3 \geq 9 - t$ $t \geq 3$
19. $11m - 17 > 13m$ $m > \frac{30}{11}$
20. $5y + 2 > 7$ $y > 1$
21. $8x + 7 \leq 2x + 2$ $x \leq -\frac{5}{6}$
22. $4x + 1 > 7x - 14$ $x < 5$
23. $k + \frac{1}{2} < \frac{k}{4} + 2$ $k < 2$
24. $\frac{3}{5}k - 4 \leq \frac{1}{5}k + 7$ $k \leq 27.5$
25. $-\frac{3}{5}k + 4 \leq -\frac{1}{5}k$ $k \geq 10$

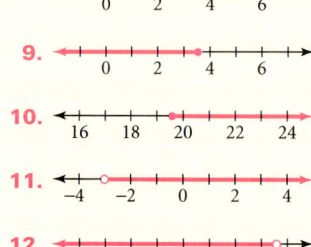

310   CHAPTER 6 REVIEW

LESSON 6.3
Key Skills

Solve and graph compound inequalities.

Solve each inequality separately and then combine the result into a compound inequality.

Conjuction: Solve and graph $-4 \leq 2x - 6 < 12$.

$-4 \leq 2x - 6$ AND $2x - 6 < 12$
$1 \leq x$ AND $x < 9$
$1 \leq x < 9$

Combine the graphs of the two inequalities in order to graph the compound inequality.

Disjunction: Solve $3x < -6$ OR $5x + 2 > -3$. Then graph the solution on a number line.

$3x < -6$ $\quad\quad\quad$ $5x + 2 > 3$
$\quad\quad\quad\quad\quad\quad\quad\quad$ $5x > -5$

$x < -2$ $\quad$ OR $\quad$ $x > -1$

Exercises

Solve each inequality, if possible. Then graph the solution on a number line.

26. $-5 \leq 4x \leq 12$ $\quad -\frac{5}{4} \leq x \leq 3$
27. $3 < 2x + 1 \leq 9$ $\quad 1 < x \leq 4$
28. $-13 < 3x + 5 \leq 26$ $\quad -6 < x \leq 7$
29. $12 \leq -4y + 8 < 16$ $\quad -2 < y \leq -1$
30. $-5 \geq x + 2 \geq -7$ $\quad -7 \geq x \geq -9$
31. $-5 \leq 2 - x \leq 7$ $\quad -5 \leq x \leq 7$
32. $3x + 2 > 11$ OR $-(3x - 2) > 4$ $\quad x > 3$ or $x < -\frac{2}{3}$
33. $2x < 12$ OR $3x \geq 27$ $\quad x < 6$ or $x \geq 9$
34. $-3x \geq 12$ OR $2x > 6$ $\quad x \leq -4$ or $x > 3$
35. $3a + 2 \geq -2$ OR $-4a + 1 \geq -3$ **All real numbers**
36. $x + 3 \geq 4$ OR $3x + 7 \leq 22$ **All real numbers**

LESSON 6.4
Key Skills

Find the absolute value of an expression.

Find each of the following:

a. $|7 - 3|$ $\quad$ b. $|3 - 7|$

a. $|7 - 3| = |4|$
    $= 4$

b. $|3 - 7| = |-4|$
    $= 4$

Describe the domain and range of an absolute-value function.

Describe the domain and range of $y = 3|x|$.

Since you can find the absolute value of any number, the domain is all real numbers. When $x = 0$, $3|x| = 0$. When x has any other value, the value of $3|x|$ will be a positive number. Therefore, the range is all non-negative numbers, or $y \geq 0$.

Exercises

Evalutate each of the following:

37. $|4 - 6|$ **2**
38. $|3 - 10|$ **7**
39. $|14 - 18|$ **4**
40. $|18 - 14|$ **4**
41. $|3 - 20|$ **17**
42. $|13 - 6|$ **7**
43. $|17 - 3|$ **14**
44. $|3 - 17|$ **14**

Describe the domain and range of each absolute-value function.

45. $y = 5|x|$ $y \geq 0$
46. $y = \frac{1}{2}|x|$ $y \geq 0$
47. $y = |x - 2|$ $y \geq 0$
48. $y = |x - 3|$ $y \geq 0$
49. $y = |x| - 2$ $y \geq -2$
50. $y = |x| + 3$ $y \geq 3$
51. $y = -|x|$ $y \leq 0$
52. $y = |x| - 4$ $y \geq -4$
53. $y = 4|x| + 3$ $y \geq 3$
54. $y = -4|x| + 3$ $y \leq 3$

45–54 Domain: All real numbers

26.
27.
28.
29.
30.
31.

32.
33.
34.
35.
36.

LESSON 6.1

Key Skills

Solve absolute-value equations.

Solve $|x - 2| = 3$.

Case 1	Case 2
$x - 2 = 3$	$x - 2 = -3$
$x = 5$	$x = -1$

The solution is $x = 5$ or $x = -1$.

Solve absolute-value equations.

Conjuction: Solve $|x + 5| < 7$.

Case 1	Case 2
$x + 5 > -7$	$x + 5 < 7$
$x > -12$	$x < 2$

The solution is $-12 < x < 2$.

Disconjuction: Solve $|3x + 5| > 7$.

Case 1	Case 2
$3x + 5 > 7$	$3x + 5 > -7$
$3x > 3$	$3x > -12$
$x > \frac{2}{3}$	$x > -4$

The solution is $x < -4$ or $x > \frac{2}{3}$.

Exercises

Solve, if possible.

55. $|n - 4| = 8$ $n = 12$ or $n = -4$
56. $|3t + 2| = 6$ $t = \frac{4}{3}$ or $t = -\frac{8}{3}$
57. $|2d - 5| = 15$ $d = 10$ or $d = -5$
58. $|x + 7| = 3$ $x = -4$ or $x = -10$
59. $|3x + 2| = 4$ $x = \frac{2}{3}$ or $x = -2$
60. $|y - 7| = 16$ $y = 23$ or $y = -9$
61. $|k - 3| \leq 1$ $2 \leq k \leq 4$
62. $|h - 6| < 2$ $4 < h < 8$
63. $|2 - m| \leq 2$ $0 \leq m \leq 4$
64. $|7 - x| \leq 4$ $3 \leq x \leq 11$
65. $|x + 2| \leq -1$ no solution
66. $|3x + 3| < 0$ no solution
67. $|2r + 4| \geq 6$ $r \leq -5$ or $r \geq 1$
68. $|p - 13| \geq 3$ $p \leq 10$ or $p \geq 16$
69. $5 \leq |3 + 2m|$ $m \leq -4$ or $m \geq 1$
70. $6 \leq |6x + 3|$ $x \geq \frac{1}{2}$ or $x \leq -\frac{3}{2}$
71. $|x - 7| \geq 0$ All real numbers
72. $|x - 7| > -5$ All real numbers

Applications

73. **CONSUMER ECONOMICS** Tamara paid $5.45 for a lunch. She had less than $5 left after paying her lunch bill. Write and solve an inequality representing the amount of money that Tamara could have had before she paid her lunch bill. $x - 5.45 < 5$

74. **BANKING** Randy had some money in his savings account. After he deposited $50, there was at least $180 in his account. Write and solve an inequality to determine the least amount that could have been in the account before his deposit. $x + 50 \geq 180$

75. **SAVINGS** The cost of a VCR is at least $280. Rhonda has saved $90. Write and solve an inequality to determine the least amount that Rhonda still needs to save in order to buy a VCR. $x + 90 \geq 280$

76. **HOBBIES** Rita has $15 to spend on new baseball cards. She decides to buy a card that costs $3. Write and solve an inequality to find the greatest amount that she can spend on other baseball cards. $x + 3 \leq 15$

Alternative Assessment

Performance Assessment

1. Use a number line to model the following inequalities:
   a. $x > -5$
   b. $x < 3$
   c. $x \leq 14$
   d. $x \leq -2$

2. Draw a number line to represent each situation.
   a. Luke is at least 16 years old.
   b. The tank will hold 20 gallons at most.
   c. You must be 18 or older to vote.
   d. Children under 12 eat free.
   e. Buy a minimum of 6 tickets.
   f. Today's maximum temperature was 80°.
   g. The test scores fell between 72 and 95, inclusive.

3. Write an inequality to represent each situation.
   a. all numbers greater than 4
   b. all numbers that are at most 10
   c. all numbers that are at least −4
   d. the solution to $x - 5 \leq 0$

Portfolio Projects

1. For each inequality below, translate the inequality into words. (For example, the translation for $x \leq 2$ might be "x is less than or equal to 2.")
   a. $x \leq 5$
   b. $x > 3$
   c. $3 \leq x \leq 15$
   d. $-2 < x \leq 5$

2. **DEMOGRAPHICS** Accuracy rates for opinion polls are often reported with a margin of error of plus or minus a certain number of percentage points. For example, a poll might have a margin of error of ±3%. Find three examples of polls reported in some media source.
   a. Describe the margin of error and the absolute margin of error for each poll.
   b. Take the reported percentages of one of the individual question results from each poll and notate the result and its margin of error as an absolute value inequality.
   c. Does the stated margin of error allow for a possible reordering of the results? (For example, could the answer reported to be most popular actually be less popular than another if the polling error was actually 3%?)

3. Describe the steps for solving the inequality $3x + 6 > 2x + 10$. Describe the solution set and graph the solution set.

Peer Assessment

Complete this activity with a partner. Begin by writing an inequality that can be solved in one step. After solving the inequality, your partner should write an inequality for you to solve.

Check your partner's work, correct any errors, and then solve the new inequality. Continue until you have each solved 10 inequalities. Be sure to use inequalities involving addition, subtraction, multiplication, and division. Try to include inequalities that involve conjunctions with no solutions and disjunctions whose solution is all real numbers.

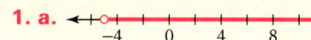

The HRW Web site contains many resources to reinforce and expand your knowledge of equations. The Web site also provides Internet links to other sites where you can find information and real-world data for use in research projects, reports, and activities that involve interpreting inequalities. Visit the HRW Web site at go.hrw.com, and enter the keyword **MA1 CH6**.

Alternative Assessment

Performance Assessment

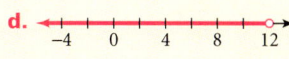

3. a. $x > 4$
   b. $x \leq 10$
   c. $x \geq -4$
   d. $x \leq 5$

Portfolio Projects

1. a. x is less than or equal to 5
   b. x is greater than 3
   c. x is greater than or equal to 3 and less than or equal to 15
   d. x is greater than −2 and less than or equal to 5

2. a. Answers will vary.
   b. The margin of error should be of the form ±x%. The absolute margin of error should be of the form x%.
   c. Answers will vary. The margin of error allows for a possible reordering if the reported difference in the responses of any two choices is less than the reported margin of error. Sample answer: Those who favor cola A:45%; those who favor cola B:43%; reported margin of error: ±4%. Because the range of the actual preference for cola A is 41%–49% and for cola B is 39%–47%, cola B could actually be preferred by more people.

3. Subtract 6 from both sides. Subtract $2x$ from both sides. The result is $x > 4$.

Cumulative Assessment

College Entrance Exam Practice

CHAPTERS 1-6 CUMULATIVE ASSESSMENT
College Entrance Exam Practice

Multiple-Choice and Quantitative-Comparison Samples

The first half of the Cumulative Assessment contains two types of items found on standardized tests—multiple-choice questions and quantitative-comparison questions. Quantitative-comparison items emphasize the concepts of equality, inequality, and estimation.

Free Response Grid Samples

The second half of the Cumulative Assessment is a free-response section. This part of the Cumulative Assessment requires student-produced response items like those commonly found on college entrance exams. These questions require the use of machine-scored answer grids. You may wish to have students practice answering these items in preparation for standardized tests.

QUANTITATIVE COMPARISON For Questions 1–4 write
A if the quantity in Column A is greater than the quantity in Column B;
B if the quantity in Column B is greater than the quantity in Column A
C if the two quantities are equal; or
D if the relationship cannot be determined from the information given

	Column A	Column B	Answers	
1.	$-18 + 4$	$-12 + (-13)$	Ⓐ Ⓑ Ⓒ Ⓓ [Lesson 2.2]	A
2.	0.45	$\frac{45}{100}$	Ⓐ Ⓑ Ⓒ Ⓓ [Lesson 4.2]	C
3.	The smallest integer solution to $x + 5 \geq 3$	The greatest integer solution to $x - 7 < -4$	Ⓐ Ⓑ Ⓒ Ⓓ [Lesson 6.1]	B
4.	The slope of the line passing through $(2, 3)$ and $(4, 7)$	The slope of the line $3x + 5y = 12$	Ⓐ Ⓑ Ⓒ Ⓓ [Lesson 5.5]	A

5. Which expression is equal to 2? **(LESSON 1.3)**
   a. $3 \cdot 4 + 2 \div 5$ b. $2^7 \div 2^6$ **b**
   c. $4(3 - 2) \div 0$ d. $2^4 - 2^3$

6. What is the solution to the equation $23y = 115$? **(LESSON 3.2)**
   a. $y = 45$ b. $y = 5$ **b**
   c. $y = 192$ d. $y = 15$

7. What is the solution to the inequality $-2x + 3 \leq 7$? **(LESSON 6.2)**
   a. $x \leq 2$ b. $x \leq 5$ **c**
   c. $x \geq -2$ d. $x > -2$

8. Which of the following is the standard form of the linear equation $y + 4 = 2x - 7$? **(LESSON 5.5)**
   a. $y = 2x - 11$ b. $2x + y = -11$ **d**
   c. $-2x + y = 11$ d. $2x - y = 11$

9. What is the solution to the equation $\frac{m}{2.3} = 5$? **(LESSON 3.2)**
   a. $m = 1.5$ b. $m = -1.25$ **d**
   c. $m = 1.15$ d. $m = 11.5$

Find each sum or difference. **(LESSONS 2.2 AND 2.3)**

10. $-25 + 35$ **10**
11. $-36 + 7 + (-23)$ **−52**
12. $323 + (-233)$ **90**
13. $690 - (-235)$ **925**
14. $-34 - (-34)$ **0**
15. $45 + (-23)$ **22**
16. $-3.4 + (-2.34)$ **−5.74**
17. $7.8 - 19.2$ **−11.4**
18. Solve: $y + 6.5 = 7$. **(LESSON 3.1)** $y = 0.5$
19. Solve: $3x + 7 = x - 4$. **(LESSON 3.4)** $x = -\frac{11}{2}$
20. Solve: $3(y + 4) - 23 = 5y$. **(LESSON 3.5)** $y = -\frac{11}{2}$
21. Solve: $y + 7 < 3$. **(LESSON 6.1)** $y < -4$
22. Solve: $-2x - 3 \geq 4x + 7$. **(LESSON 6.2)** $x \leq -\frac{5}{3}$
23. What percent of 40 is 10? **(LESSON 4.2)**   25%

314   CHAPTERS 1–6 CUMULATIVE ASSESSMENT

24. 24 is what percent of 72? *(LESSON 4.2)* $33\frac{1}{3}\%$

25. 30% of 150 is what number? *(LESSON 4.2)* 45

26. Find the next three terms of the sequence 5, 10, 20, 35, . . . *(LESSON 1.1)* 55, 80, 110

27. Simplify $(7x + 2) - (3x - 2y + 2)$. *(LESSON 2.6)* $4x + 2y$

Find the slope of the line passing through the given points. *(LESSON 5.2)*

28. (3, 5) and (4, 6) 1

29. (9, −10) and (3, −2) $-\frac{4}{3}$

30. (5, 5) and (−5, 5) 0

31. (3, −4) and (4, −50) −46

32. (9, 0) and (0, 8) $-\frac{8}{9}$

Write each linear equation in slope-intercept form. *(LESSON 5.4)*

33. $3y - 2x = 14$ $y = \frac{2}{3}x + 14$

34. $y - 3x = 2$ $y = 3x + 2$

35. $4x = 3y + 2$ $y = \frac{4}{3}x - \frac{2}{3}$

Write an equation in standard form for a line passing through the given points. *(LESSON 5.5)*

36. (4, 6) and (3, 7) $x + y = 10$

37. (−2, −3) and (3, 2) $x - y = 1$

38. (−3, −3) and (3, 3) $x - y = 0$

39. (−2, 0) and (−3, −1) $x - y = -2$

40. (3, −5) and (7, −3) $x - 2y = 13$

Find the mean, median, and mode for each set of numbers. *(LESSON 4.4)*

41. {1, 3, 3, 4, 5, 6, 6, 7, 8, 9, 10}

42. {300, 320, 120, 125, 126, 129}

43. {12, 12, 14, 14, 15, 11, 10, 10, 10, 15, 12}

44. {1, 2, 3, 4, 5, 5, 6, 5, 4, 3, 10}

Simplify each expression. *(LESSON 2.6)*

45. $(9w - 5) - (3w - 4)$ $6w - 1$

46. $4x^2 - 4(3x^2 + 7)$ $-8x^2 - 28$

47. $(9m - 5n) + (-3m - 4n)$ $6m - 9n$

48. $(3x + 7y) - (4x + 5y)$ $-x + 2y$

Solve each proportion. *(LESSON 4.1)*

49. $\frac{x}{5} = \frac{10}{2}$ 25

50. $\frac{3}{m} = \frac{27}{45}$ 5

51. $\frac{2}{5} = \frac{m}{27}$ $10\frac{4}{5}$

FREE-RESPONSE GRID
The following questions may be answered by using a free-response grid such as that commonly used by standardized test services.

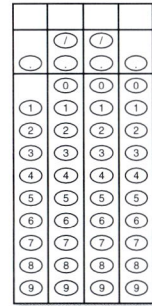

52. Simplify $4^2 \div 8 + 5(8 - 2) \cdot 2$. *(LESSON 1.3)* 62

53. Solve: $\frac{5}{8} - x = \frac{1}{2}$. *(LESSON 3.3)* $\frac{1}{8}$

54. Maurice has scores of 88, 90, and 80 on three tests. What score must he make on his next test to have a mean score of exactly 85? *(LESSON 4.4)* 82

55. Find the slope of a line that passes through the origin and (4, 2). *(LESSON 5.2)* $\frac{1}{2}$

41. mean: $5\frac{7}{11}$; median: 6; modes: 3, 6

42. mean: $186\frac{2}{3}$; median: $127\frac{1}{2}$; mode: none

43. mean: $12\frac{3}{11}$; median: 12; modes: 10, 12

44. mean: $4\frac{4}{11}$; median: 4; mode: 5

7 Systems of Equations and Inequalities

CHAPTER PLANNING GUIDE

Lesson	7.1	7.2	7.3	7.4	7.5	7.6	Projects and Review
Pupil Edition Pages	318–325	326–330	331–337	338–344	345–352	353–359	360–367
Extra Practice (Pupil Edition)	756	756	757	757	758	758	
Practice Workbook	37	38	39	40	41	42	
Reteaching Masters	73–74	75–76	77–78	79–80	81–82	83–84	
Enrichment Masters	37	38	39	40	41	42	
Cooperative-Learning Activities	37	38	39	40	41	42	
Lesson Activities	37	38	39	40	41	42	
Problem Solving/ Critical Thinking	37	38	39	40	41	42	
Student Study Guide	37	38	39	40	41	42	
Spanish Resources	37	38	39	40	41	42	
Student Technology Guide	37	38	39	40	41	42	
Assessment Resources	83	84	85	87	88	89	86, 90–95
Teaching Transparencies	33, 34	35		36, 37	38	39	
Quiz Transparencies	7.1	7.2	7.3	7.4	7.5	7.6	
Writing Activities for Your Portfolio							19–21
Tech Prep Masters							31–34
Long-Term Projects							25–28

LESSON PACING GUIDE

	7.1	7.2	7.3	7.4	7.5	7.6	Projects and Review
Traditional	1 day	1 day	2 days	2 days	2 days	2 days	2 days
Block	$\frac{1}{2}$ day	$\frac{1}{2}$ day	1 day	1 day	1 day	1 day	1 day
Two-Year	2 days	2 days	4 days	4 days	4 days	4 days	4 days

CONNECTIONS AND APPLICATIONS

Lesson	7.1	7.2	7.3	7.4	7.5	7.6	Review
Algebra	318–325	326–330	331–337	338–344	345–352	353–359	360–367
Geometry		329, 330	335	343	351		364
Business and Economics	318	328	336	344	345, 349	359	
Statistics	324						
Number Theory						359	
Life Skills			331, 333		351		
Science and Technology	319–325		337	338, 339, 341	346	356, 358, 359	
Sports and Leisure	322	326	330, 336		350, 351	358	312
Cultural Connection: Asia		330					
Cultural Connection: Africa			337				
Other		330		340	352		

BLOCK-SCHEDULING GUIDE

Day	Lesson	Teacher Directed: Lesson Examples, Teaching Transparencies	Student Guided: Activity, Try This	Cooperative-Learning Activity, Lesson Activity, Student Technology Guide	Practice: Practice & Apply, Extra Practice, Practice Workbook	Assessment: Quiz, Mid-Chapter Assessment	Problem Solving, Reteaching
1	7.1	5 min	8 min	8 min	35 min	8 min	8 min
	7.2	5 min	7 min	7 min	30 min	7 min	7 min
2	7.3	10 min	15 min	15 min	65 min	15 min	15 min
3	7.4	10 min	15 min	15 min	65 min	15 min	15 min
4	7.5	10 min	15 min	15 min	65 min	15 min	15 min
5	7.6	15 min	20 min	15 min	65 min	15 min	15 min
7	Assess.	50 min PE: Chapter Review	90 min PE: Chapter Project, Writing Activities	90 min Tech Prep Masters	65 min PE: Chapter Assessment, Test Generator	30 min Chap. Assess. (A or B), Alt. Assess. (A or B), Test Generator	

PE: Pupil's Edition

Alternative Assessment

The following are suggestions for an alternative assessment for students who may benefit from a different type of assessment than the regular chapter quizzes and the mid-chapter and end-of-chapter assessment materials. Many of the questions are open-ended, and students' answers will vary.

Performance Assessment

1. **a.** Solve each system by any method. Justify your choice.
   $$\begin{cases} 2x + 3y = 18 \\ -5x + 4y = 20 \end{cases} \quad \begin{cases} 2x + y = 5 \\ x = \frac{1}{2}y \end{cases} \quad \begin{cases} y = -2.4x + 3 \\ y = 1.5x - 3 \end{cases}$$
   b. State the definition of an *inconsistent system*. Are any of the systems above inconsistent?

2. Write a detailed solution to the system $\begin{cases} x + 2y \leq 6 \\ y \leq -2x + 5 \end{cases}$, where $x \geq 0$ and $y \geq 0$. Sketch the solution region on a coordinate plane.

3. A canoeist paddles upstream to a point 15 miles away. The trip takes 2.5 hours. The trip downstream takes 1.75 hours. What is the speed of the current and the canoeist's speed in still water?
   a. Make a sketch and write a system of equations to represent the situation. Then solve the system.
   b. Explain how to solve the problem if the distance is 20 miles.

Portfolio Project

Suggest that students choose one of the following projects for inclusion in their portfolios.

1. Use a system of linear equations in x and y to make a triangle in the coordinate plane.
   a. Write two linear equations in x and y that have a unique solution.
   b. Explain how to add an equation in x and y to the pair of equations that you wrote in part **a** in order to get a triangle. How does an inconsistent system help you solve the problem?

2. Suppose that you are challenged with the following puzzle: Choose a two-digit number whose digits add up to 8 and whose tens digit is 2 more than its units digit. What is the number?
   a. Explain why the challenger will always guess the correct number.
   b. Create different digit puzzles, each with only one answer. Challenge friends with them.

---

internetconnect

The table below identifies the pages in this chapter that contain technology information and support in the side columns.

Content Links

Lesson Links	pages 319, 330, 334
Portfolio Links	page 325
Graphics Calculator Support	page 782

Resource Links

For information about teacher and parent resources as well as professional development help, visit **www.hrw.com/math**.

Technical Support HRW has assembled a team of dedicated technical and teaching professionals and a comprehensive service program to provide you with the support you need.

- The HRW Technical Support Line operates from 7 A.M. to 6 P.M. central time, Monday through Friday, at (800) 323-9239.
- The HRW Technical Support Center on the World Wide Web is available 24 hours a day, seven days a week, at **www.hrwtechsupport.com**.
- You can e-mail our Technical Support Center at **tsc@hrwtechsupport.com**.
- The Technical Support Center's fax-on-demand service at (800) 352-1680 offers solutions to common problems and answers to frequently asked questions.

Technology

Lesson Suggestions and Calculator Examples
(Keystrokes are based on a TI-83 calculator.)

Lesson 7.1 Graphing Systems of Equations
In this lesson, students will learn that a system of two equations in x and y may be satisfied by

1. exactly one ordered pair,
2. no ordered pair, or
3. infinitely many ordered pairs.

Have students graph $y = x$. Then have students graph other equations in x and y: ones that intersect $y = x$ at only one point, ones that are parallel to $y = x$, and ones that have the same graph as $y = x$. Continue the experiment until students see the various possibilities.

Show students that a solution found by graphing often is approximate and, therefore, does not exactly satisfy either given equation.

Lesson 7.2 The Substitution Method
In this lesson, students will learn the algebraic strategy of transforming a pair of equations in two variables into a single equation in one variable. Then they can use the previously learned skills to find the value of the variable. Students can use a calculator to check their answers.

Consider the system below.
$$\begin{cases} 15x - 5y = 30 \\ y = 2x + 3 \end{cases}$$

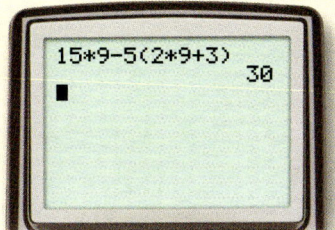

The solution is (9, 21). To check the solution, have students evaluate the expression $15(9) - 5[2(9) + 3]$ as shown at right. The display indicates that the solution for x is correct. Evaluate the expression $2(9) + 3$ to check the solution for y.

Lesson 7.3 The Elimination Method
In this chapter, students will learn three different methods for solving a system of equations in two variables. Once students see that a system can be solved easily by one method, they realize that the choice of a different method may be inefficient. You may want to have students set their calculators aside until they have given convincing evidence that a calculator is really the best choice as a method of solution.

Lesson 7.4 Consistent and Inconsistent Systems
Students may ask whether there is an algebraic criterion to determine whether a system has a unique solution. Consider giving students the statement, shown at the top of the next column, about the solutions of systems of equations.

Given $\begin{cases} ax + by = c \\ dx + ey = f \end{cases}$, where $ae - bd \neq 0$, the system has exactly one solution.

To apply this criterion, the equations must be in standard form. For example, the system in Example 1 on page 339 can be written as $\begin{cases} x + y = 4 \\ -3x + y = -6 \end{cases}$. Ask students to identify a, b, c, and d and then apply the criterion stated above. They should see that exactly one solution is indicated.

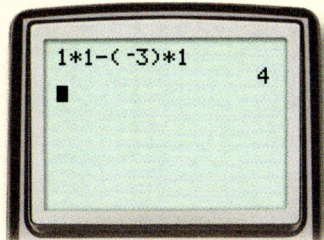

Lesson 7.5 Systems of Inequalities
The solution to a linear inequality in x and y is very different from the graph of an equation in x and y. This lesson allows students to explore calculator capabilities that they have not experienced before.

Consider $\begin{cases} y \geq -\frac{3}{4}x \\ y \leq x - 1 \end{cases}$. Students will need to tell the calculator that it should shade above the first boundary line and below the second one.

1. Enter **–3X/4** as **Y1**, and **X–1** as **Y2**.
2. Use the arrow keys and ENTER as often as needed to display ◤ to the left of **Y1** and ◣ to the left of **Y2**.

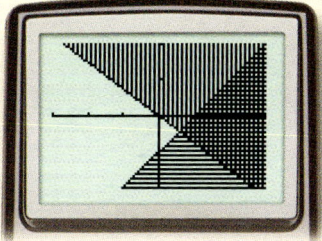

Lesson 7.6 Classic Puzzles in Two Variables
In this lesson, students will use what they learned about solving systems of equations in two variables. Challenge students with the chemical-solution example on page 356. Use the calculator editing feature to solve related problems.

For further information, refer to the

- technology discussions in the lessons.
- lesson-related teacher's commentary in the side columns of this *Teacher's Edition*.
- lesson-related *Student Technology Guide* masters.
- *HRW Technology Handbook*.

Teaching Resources

Basic Skills Practice
(2 pages per skill)

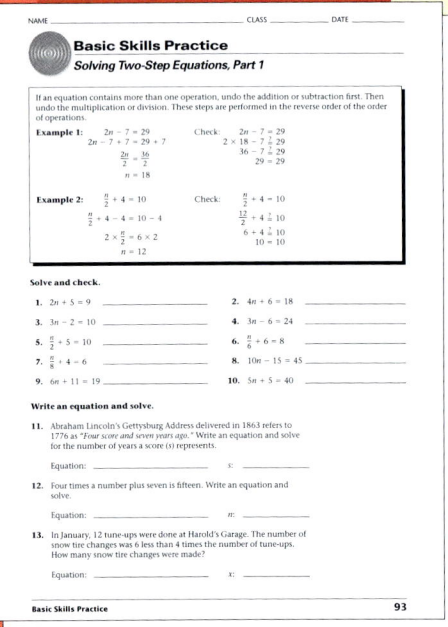

Reteaching
(2 pages per lesson)

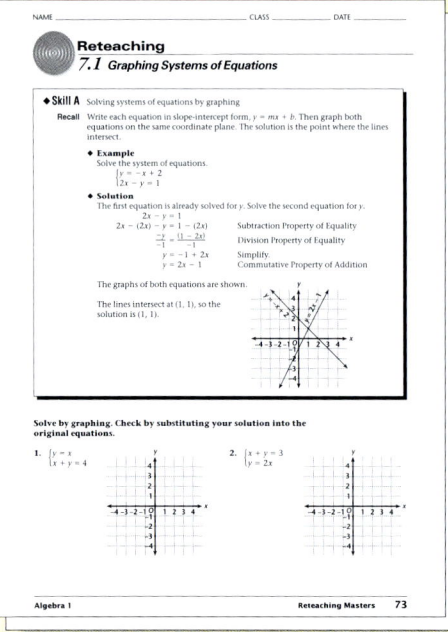

Enrichment
(1 page per lesson)

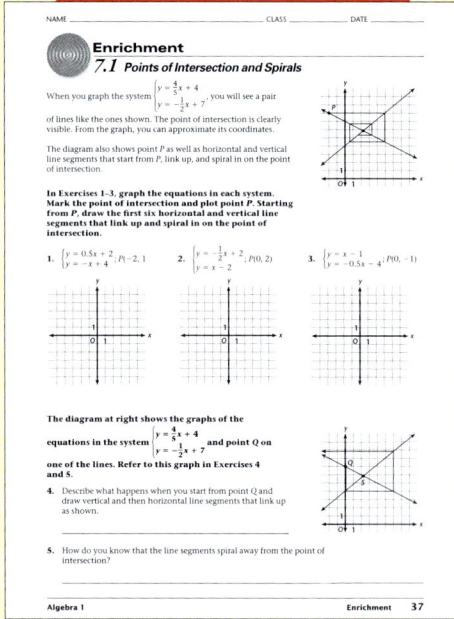

Cooperative-Learning Activity
(1 page per lesson)

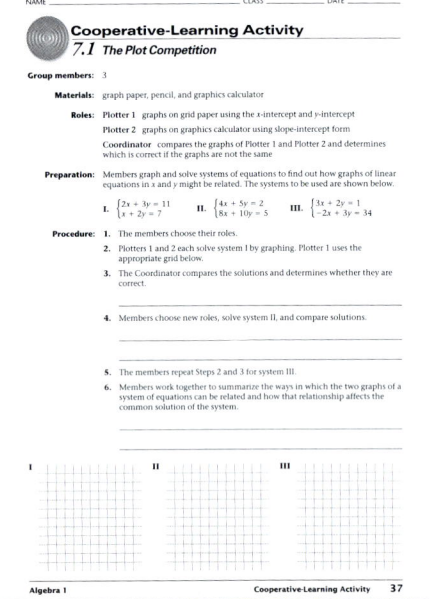

Lesson Activity
(1 page per lesson)

Long-Term Project
(4 pages per chapter)

316E CHAPTER 7 INTERLEAF

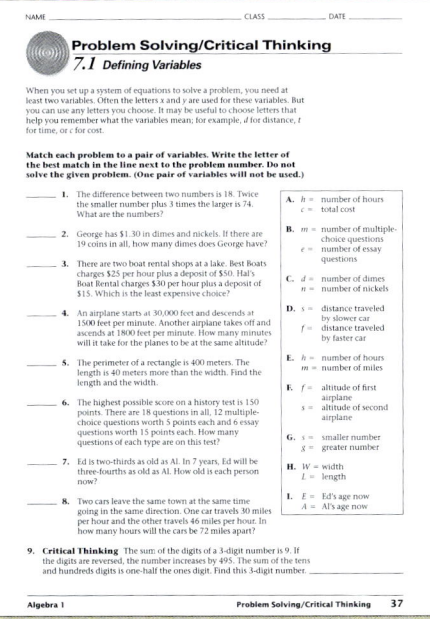
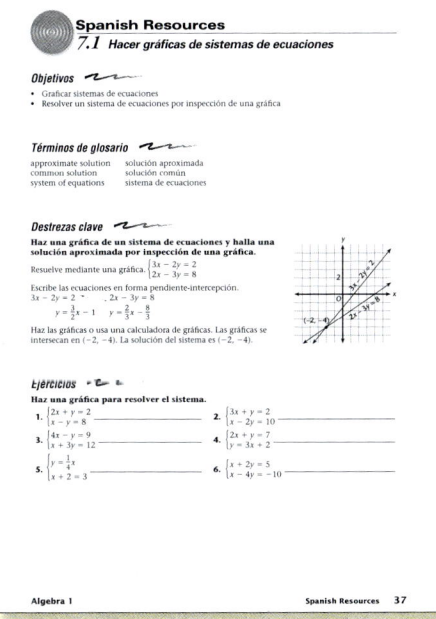

About Chapter 7

Background Information

In Chapter 7, students learn three techniques for solving systems of linear equations in two variables: graphing, substitution, and elimination. They learn to recognize when such a system has exactly one solution, when it has infinitely many solutions, and when it has no solution. Their experience with systems of linear equations is then used as the foundation for the study of systems of linear inequalities. The final lesson of the chapter examines how systems can be used to solve some classic mathematical puzzles.

CHAPTER RESOURCES

- Block-Scheduling Handbook
- Writing Activities for Your Portfolio
- Tech Prep Masters
- Long-Term Project
- Assessment Resources:
  Mid-Chapter Assessment
  Chapter Assessments
  Alternative Assessments
- Test and Practice Generator
- Technology Handbook
- End-of-Course Exam Practice

Systems of Equations and Inequalities

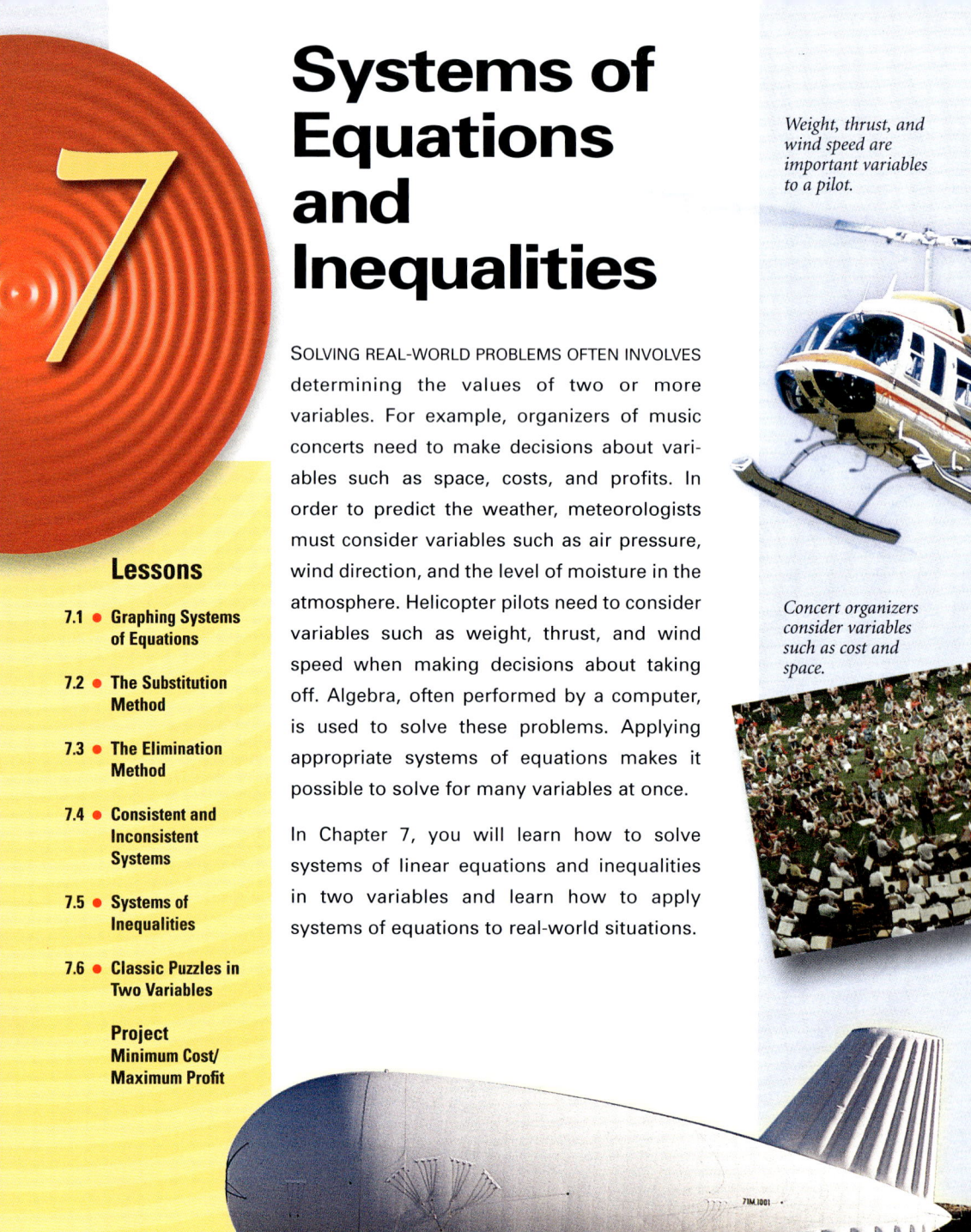

Weight, thrust, and wind speed are important variables to a pilot.

Concert organizers consider variables such as cost and space.

Lessons

7.1 ● Graphing Systems of Equations

7.2 ● The Substitution Method

7.3 ● The Elimination Method

7.4 ● Consistent and Inconsistent Systems

7.5 ● Systems of Inequalities

7.6 ● Classic Puzzles in Two Variables

Project
Minimum Cost/
Maximum Profit

SOLVING REAL-WORLD PROBLEMS OFTEN INVOLVES determining the values of two or more variables. For example, organizers of music concerts need to make decisions about variables such as space, costs, and profits. In order to predict the weather, meteorologists must consider variables such as air pressure, wind direction, and the level of moisture in the atmosphere. Helicopter pilots need to consider variables such as weight, thrust, and wind speed when making decisions about taking off. Algebra, often performed by a computer, is used to solve these problems. Applying appropriate systems of equations makes it possible to solve for many variables at once.

In Chapter 7, you will learn how to solve systems of linear equations and inequalities in two variables and learn how to apply systems of equations to real-world situations.

About the Photos

Systems of equations and inequalities are used to solve real-world problems in diverse situations. Some of the many applications of systems that students will examine throughout this chapter are illustrated in the photographs on these pages. Students will see how systems are used to solve problems about entertainment, speeds of race cars, navigation of airplanes, and the current and wind speeds encountered while boating. Depending on past experience, students may already be familiar with some of these situations.

Meteorologists use systems of equations to make sense out of weather data.

About the Chapter Project

Writing and solving equations and systems of equations enable you to solve some very complex problems—graphically and algebraically. In the Chapter Project, *Minimum Cost/Maximum Profit*, on page 360 you will learn how a manager of a camping supply store can maximize profit when selling trail mix to backpackers. After completing the Chapter Project, you will be able to do the following:

- Write and solve systems of equations and inequalities.
- Solve an optimization problem.

About the Portfolio Activities

Throughout the chapter, you will be given opportunities to complete Portfolio Activities that are designed to support your work on the Chapter Project.

- Baseball and basketball players' salaries are the topic of the Portfolio Activity on page 325.
- Finding the correct amount to charge for tickets to a magic show is the topic of the Portfolio Activity on page 337.
- Deciding the number of color pages for a high-school yearbook is the topic of the Portfolio Activity on page 352.

Chapter Objectives

- Graph systems of equations. [7.1]
- Solve a system of equations by inspecting a graph. [7.1]
- Find an exact solution to a system of linear equations by using the substitution method. [7.2]
- Use the elimination method to solve a system of equations. [7.3]
- Choose an appropriate method to solve a system of equations. [7.3]
- Identify consistent and inconsistent systems of equations. [7.4]
- Identify dependent and independent systems of equations. [7.4]
- Graph the solution to a linear inequality. [7.5]
- Graph the solution to a system of linear inequalities. [7.5]
- Solve traditional math puzzles in two variables. [7.6]

Portfolio Activities appear at the end of Lessons 7.1, 7.3, and 7.5. Each serves as preparation for the Chapter Project. The Portfolio Activities as well as the Chapter Project Activities are appropriate for inclusion in the student's portfolio. Students should be encouraged to include in their portfolios any other work in which they feel a sense of pride or a sense of accomplishment.

internet connect

Internet Connect features appearing throughout the chapter provide keywords for the HRW Web site that lead to links for math resources, tutorial assistance, references for student research, classroom activities, and all teaching resource pages. The HRW Web site also provides updates to the technology used in the text. Listed at right are the keywords for the Internet Connect activities referenced in this chapter. Refer to the side column on the page listed to read about the activity.

LESSON	KEYWORD	PAGE
7.1	MA1 Athletes	325
7.2	MA1 House Gender	330
7.3	MA1 Sports Opera	334

CHAPTER 7 **317**

Prepare

NCTM PRINCIPLES & STANDARDS 1–3, 5, 6, 8–10

LESSON RESOURCES

- Practice 7.1
- Reteaching Master 7.1
- Enrichment Master 7.1
- Cooperative-Learning Activity 7.1
- Lesson Activity 7.1
- Problem Solving/Critical Thinking Master 7.1
- Teaching Transparency 33
- Teaching Transparency 34

QUICK WARM-UP

Solve for y.

1. $4x - y = 7$
   $y = 4x - 7$
2. $x + 2y = 8$
   $y = -0.5x + 4$
3. $\frac{2}{5}y - 8x = 3$
   $y = 20x + 7.5$
4. $4 - 6y = 9x$
   $y = -\frac{3}{2}x + \frac{2}{3}$

Also on Quiz Transparency 7.1

Teach

Why Graphs of systems of linear equations can show trends over a period of time and are often used in business to convey this information. Such graphs often model situations that involve two sets of conditions. Have students discuss situations that might involve two conditions, such as how supply and demand of a product are affected by different market conditions.

7.1 Graphing Systems of Equations

Objectives
- Graph systems of equations.
- Solve a system of equations by inspecting a graph.

Why Graphing systems of equations in two unknowns makes it possible to solve for more than one variable at the same time. You may encounter situations in which the graph of a system of equations can be used to help make the decision that is best for you.

Newspaper advertisements provide the Smiths with information about costs.

Modeling Options With Equations

APPLICATION
CONSUMER ECONOMICS

The Smith family is trying to decide between installing cable TV or purchasing a VCR and renting tapes. The Smiths need to compare the cost of each option over several months in order to make an appropriate decision.

Cable TV: One-time installation charge of $45, and $34 per month
VCR: Purchase a VCR for $185 and rent 8 tapes per month at $2.50 per tape, which is $20 per month

One way to compare the two options is to create a graph. The graph will show when both options will cost the same. First model each option by using an equation.

Let m represent the number of months.
Let c represent the total costs for an option.

Cable TV cost = cost for m months + installation

$$c = 34m + 45$$

VCR and tape cost = tape rental for m months + VCR

$$c = 20m + 185$$

Alternative Teaching Strategy

INVITING PARTICIPATION Display a set of coordinate axes on the board or overhead. Graph the point $A(3, 1)$. Ask students for the equations of two lines that pass through the point. Graph the students' lines.

Now write the equations of the lines in system form. For example, if they identify the lines $x + y = 4$ and $x - y = 2$, write the system at right.

$$\begin{cases} x + y = 4 \\ x - y = 2 \end{cases}$$

Ask students to explain the relationship between the coordinates $(3, 1)$ and the equations in the system. Elicit the response that $(3, 1)$ is a solution of each equation. Point out the fact that the lines intersect at exactly one point, which is $A(3, 1)$. This means that $(3, 1)$ is, in fact, the *only* solution common to both equations.

Tell students that the two equations taken together are called a *system of linear equations* and that $(3, 1)$ is the *solution to the system*.

318 LESSON 7.1

Graph each equation on the same coordinate plane, with months on the horizontal axis and cost on the vertical axis.

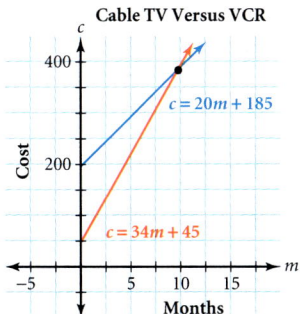

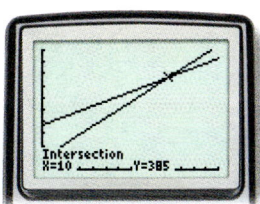

On a graphics calculator, use the intersection feature to find the point of intersection.

The lines representing the two options intersect at (10, 385). This means that at 10 months the cost of each option is $385. The graph shows that each cost continues to increase, but the cost of cable TV increases at a faster rate. After 10 months, the VCR is a better choice.

Two equations in two variables are called a **system of equations**. A **solution** to a system of equations in two variables is an ordered pair of numbers that is a solution to each equation in the system. The solution to the system in the problem about the Smith family is (10, 385). The solution to a system of linear equations can be found by graphing.

EXAMPLE 1 Solve by graphing. $\begin{cases} y = 3x + 1 \\ y = -x + 5 \end{cases}$

The brace { is used to indicate that the equations form a system.

● SOLUTION

TECHNOLOGY GRAPHICS CALCULATOR

Use a graphics calculator or graph paper to graph both equations on the same coordinate plane.

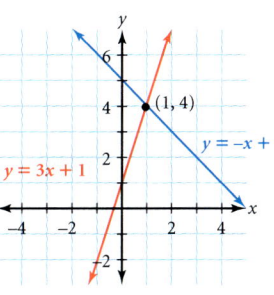

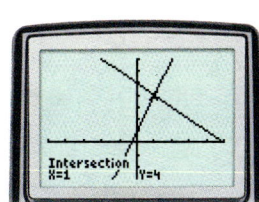

From the graph, the solution is (1, 4).

Check the solution by substituting 1 for x and 4 for y in each equation.

$y = 3x + 1$ $\qquad$ $y = -x + 5$
$4 \stackrel{?}{=} 3(1) + 1$ $\qquad$ $4 \stackrel{?}{=} -(1) + 5$
$4 = 4$ $\quad$ True $\qquad$ $4 = 4$ $\quad$ True

Because the substitution is true for each equation, (1, 4) is the solution.

Teaching Tip

TECHNOLOGY If you are using a TI-82 or TI-83 graphics calculator, you can display the point of intersection of two lines with the following procedure: First use the **Y=** editor to enter the equations as **Y1** and **Y2**. Press GRAPH. Press 2nd TRACE to access the **CALCULATE** menu. Choose **5:intersect** and then press ENTER ENTER ENTER.

Use Teaching Transparencies 33 and 34.

ADDITIONAL EXAMPLE 1

Solve by graphing.

$\begin{cases} y = 5x - 4 \\ y = -2x + 3 \end{cases}$

(1, 1)

See Keystroke Guide, page 782.

Teaching Tip

TECHNOLOGY On TI-82 and TI-83 graphics calculators, the **TRACE** feature allows you to approximate the coordinates for the point of intersection of two lines. After you graph the lines, press TRACE. Then use ◄ or ► to move the cursor along one line to the point where it meets the other. The coordinates of the location of the cursor are displayed at the bottom of the screen.

Interdisciplinary Connection

HOME ECONOMICS Write a system of equations that models this problem. Then use the graph of the system to solve the problem.

A recipe for punch calls for two-thirds as much orange juice as ginger ale. For an upcoming reception, the students in a home economics class need to make 6 gallons, or 24 quarts, of this punch. How many quarts of ginger ale and orange juice do they need?

Let x represent the number of quarts of ginger ale, and let y represent the number of quarts of orange juice.

$\begin{cases} x + y = 24 \\ y = \frac{2}{3}x \end{cases}$

They will need 14.4 quarts of ginger ale and 9.6 quarts of orange juice.

LESSON 7.1

TRY THIS
(−1, 0)

ADDITIONAL EXAMPLE 2

Solve by graphing.
$\begin{cases} -7x + y = -13 \\ x - 5y = -3 \end{cases}$
(2, 1)

See Keystroke Guide, page 782.

Teaching Tip

TECHNOLOGY If you are using a TI-82 or TI-83 graphics calculator, you can locate the solution of a system of two equations by using the **TABLE** feature. First use the **Y=** editor to enter the equations as **Y1** and **Y2**. Press `2nd` `WINDOW` to access the **TABLE SETUP** screen, and choose the following settings:

TblStart=0 (or TblMin=0)
ΔTbl=1
Indpnt: Auto Ask
Depend: Auto Ask

Press `2nd` `GRAPH` to display the table. Then use ▲ and ▼ to locate the place in the table where **Y1** and **Y2** are equal, or nearly equal.

TRY THIS
a. (1, 0)   b. (−1, −3)

CRITICAL THINKING
No, there is no common solution. The lines are parallel.

TRY THIS   Solve the system by graphing. Check your answer. $\begin{cases} y = 2x + 2 \\ y = -x - 1 \end{cases}$

In order to graph a system of equations, you may need to change each equation to slope-intercept form. Whether you are using a graphics calculator or graph paper, the slope-intercept form is often easier to use.

EXAMPLE 2   Solve by graphing. $\begin{cases} 3x + y = 4 \\ x - 2y = 6 \end{cases}$

● **SOLUTION**

Change each equation to slope-intercept form, $y = mx + b$, by solving for y in terms of x.

$3x + y = 4$ $x - 2y = 6$
$y = -3x + 4$ $-2y = -x + 6$
                          $y = \frac{1}{2}x - 3$

TECHNOLOGY GRAPHICS CALCULATOR

Graph both equations on the same coordinate plane to find a common solution.

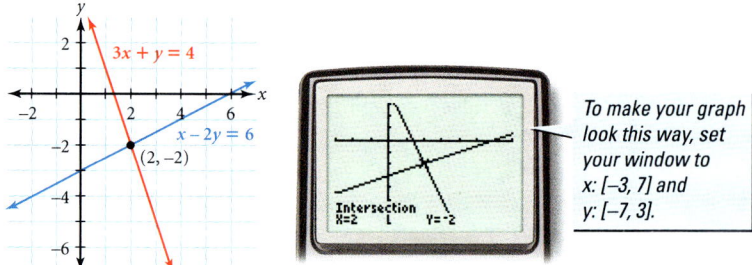

To make your graph look this way, set your window to x: [−3, 7] and y: [−7, 3].

The solution is (2, −2).

Check the solution by substituting 2 for x and −2 for y in each equation.

$3x + y = 4$ $x - 2y = 6$
$3(2) + (-2) \stackrel{?}{=} 4$ $2 - 2(-2) \stackrel{?}{=} 6$
$4 = 4$ True $6 = 6$ True

TRY THIS   Solve each system below by graphing. Check your answers.

a. $\begin{cases} y + 2x = 2 \\ y + x = 1 \end{cases}$ b. $\begin{cases} y = 3x \\ y = -4x - 7 \end{cases}$

CRITICAL THINKING   Consider the following system: $\begin{cases} y = -2x + 3 \\ y = -2x + 10 \end{cases}$

Does the graph of this system show a common solution? Why or why not?

Inclusion Strategies

VISUAL LEARNERS   Have students write each equation in a system with a pencil of a different color. Then tell them to graph each equation in the corresponding color. Suggest that they use a third color to draw the point of intersection and to write the coordinates of that point.

Activity

Exploring Approximate Solutions

You will need: graph paper and a graphics calculator

1. Graph this system on graph paper. $\begin{cases} 4x + 3y = 6 \\ 2y = x + 2 \end{cases}$

2. Estimate the solution from your graph. Check your estimated solution by substituting your values for x and y into the equations. How accurate are your estimates?

3. Graph the system from Step 1 on a graphics calculator. Use the trace feature or intersection feature to find the solution to the system.

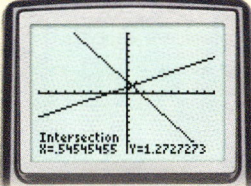

4. Test the solution from the calculator by substituting the values for x and y into the equations. Note that you can use the **Frac** feature on some calculators to find fractions for decimal values.

CHECKPOINT ✓

5. Discuss the advantages and disadvantages of the paper-and-pencil and calculator methods for solving systems of equations.

Approximate Solution

It is sometimes difficult to find an exact solution from a hand-drawn graph of a system of equations. A reasonable estimate for a point of intersection is an **approximate solution**.

EXAMPLE ❸ Find an approximate solution by graphing. $\begin{cases} y = -0.2x + 1 \\ y = 2x - 7 \end{cases}$

● **SOLUTION**

An approximate solution for this system is (3.7, 0.3).

Check the solution by substituting **3.7** for x and **0.3** for y in the original equations.

$y = -0.2x + 1$ $y = 2x - 7$
$0.3 \stackrel{?}{=} -0.2(3.7) + 1$ $0.3 \stackrel{?}{=} 2(3.7) - 7$
$0.3 \stackrel{?}{=} -0.74 + 1$ $0.3 \stackrel{?}{=} 7.4 - 7$
$0.3 \approx 0.26$ $0.3 \approx 0.4$

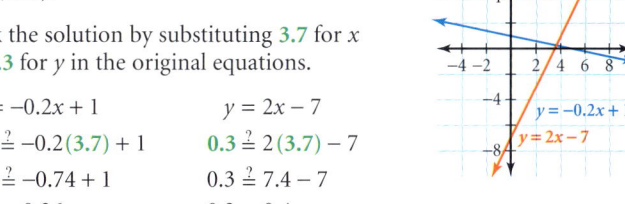

A check shows that the approximate solution is reasonable to the nearest tenth.

You can also check your approximate solution with a graphics calculator.

Enrichment

An increasing number of American households are choosing to obtain their television reception through a satellite dish rather than a cable. Have students consider the following situation: One company markets satellite service for an initial fee of $300, which includes the purchase of the dish and installation, and then charges $20 per month for programming. A cable television company in the same area requires an initial fee of $14 for installation and then charges $31 per month thereafter. Ask students to solve a system of equations to compare the costs of the two services.

For the first 25 months, the total amount invested in cable service would be less than the amount invested in satellite service. At 26 months, the costs would be the same. Thereafter, the cost of the satellite service would be less.

Have students research the costs of cable and satellite service in their own area, including the initial fees and monthly charges for each. Then have them determine which option would be less expensive after 1 year, 2 years, and 5 years.

Activity Notes

In this Activity, students graph a system of equations with paper and pencil and with a graphics calculator. Then they compare the two methods. They should come to realize that each method has its advantages and disadvantages and that each is a valuable part of their "toolbox" of graphing skills.

☞ For a tip about displaying the coordinates as fractions on a graphics calculator, see page 322.

See Keystroke Guide, page 782.

Cooperative Learning

You might want students to do the Activity in pairs. One student in the pair can solve the system with paper and pencil, while the other uses a graphics calculator. The students can then compare their answers and the methods used to obtain them. If you wish, give the students a second system to solve. Have them switch methods and repeat the Activity.

CHECKPOINT ✓

5. Answers may vary. Sample: Solving by paper and pencil may be better for seeing the actual intersection, but finding the exact value can be difficult. Solving with a graphics calculator is faster, but intersection points may be rounded and the actual intersection may be difficult to see.

ADDITIONAL EXAMPLE ❸

Find an approximate solution by graphing.

$\begin{cases} y = -0.4x - 4 \\ y = 3x + 5 \end{cases}$

Answers may vary. Sample answer: about $(-2.5, -3)$

LESSON 7.1 **321**

Teaching Tip

TECHNOLOGY If you are using a TI-82 or TI-83 graphics calculator for the Activity on page 321, use this procedure to find the fractions for the coordinates of the point of intersection.

First graph the equations and use the **CALCULATE** menu to find the point of intersection. Then press [2nd] [MODE] to return to the home screen. To bring the *x*-coordinate to the home screen, press [ALPHA] [STO▶] [ENTER]. Then press [2nd] [(-)] [MATH] and select **1:▶Frac**. Press [ENTER] to see the fraction. To recall the *y*-coordinate, press [ALPHA] [1] [ENTER].

TRY THIS

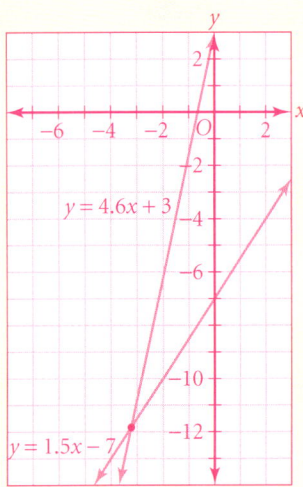

about (−3, −12)

ADDITIONAL EXAMPLE 4

Several books are on sale at a bookstore. Some of these books cost $3, while others cost $5. One day last week, 80 of these books were sold. The total amount of the sales was $300. **How many of each type of book were sold?**

50 books at $3 and 30 books at $5

TRY THIS   Find an approximate solution by graphing. $\begin{cases} y = 4.6x + 3 \\ y = 1.5x - 7 \end{cases}$

Using Systems to Solve Problems

EXAMPLE ④

**APPLICATION
HOBBIES**

A birding club holds an annual photography contest among its members. After a set time limit in a particular park, contestants receive 4 points for photos of songbirds and 20 points for photos of birds of prey. Last year's winner had a total of 200 points from 38 photos of individual birds. **How many of each type of bird did the winner photograph?**

• **SOLUTION**

PROBLEM SOLVING   First define the two variables. Then write equations to model the problem.

ORGANIZE
Let *x* represent the number of 4-point songbird photos.
Let *y* represent the number of 20-point bird-of-prey photos.

WRITE

points for songbirds	+	points for birds of prey	= 200 points
$4x$	+	$20y$	= 200

number of 4-point photos	+	number of 20-point photos	= 38 photos
x	+	y	= 38

SOLVE
Change the equations to slope-intercept form.

$4x + 20y = 200$
$\quad 20y = -4x + 200$
$\quad\quad y = -\frac{1}{5}x + 10$

$x + y = 38$
$\quad y = -x + 38$

Graph the two equations on the same coordinate plane.

The solution is (35, 3).

An owl is an example of a bird of prey.

The winner photographed 35 songbirds and 3 birds of prey.

CHECK

$35 + 3 \stackrel{?}{=} 38$
$38 = 38$ True

$4(35) + 20(3) \stackrel{?}{=} 200$
$140 + 60 \stackrel{?}{=} 200$
$200 = 200$ True

Reteaching the Lesson

COOPERATIVE LEARNING   Have students work in groups of three. Tell them to draw any two lines on a coordinate plane and determine how the lines are related, that is, whether or not they intersect.

If the lines do not intersect, be sure students understand that the system has no solution.

If the lines intersect, have the students find the coordinates of the point of intersection. Be sure they understand that the point of intersection is the only point that lies on both lines.

Now tell students to write the equation of each line. They should then substitute the coordinates of the point of intersection into each equation. The result should be two true statements. Stress that the coordinates of the point are the solution to the system.

Have students repeat this activity with several different pairs of lines.

Exercises

Communicate

Exercises 1 and 2 refer to the following system of equations: $\begin{cases} 2x - 3y = 4 \\ x + 4y = -9 \end{cases}$

1. How do you write each equation in slope-intercept form? Why is this form used to find values for the variables?

2. Explain how to find the solution for the system from a graph.

3. How do you graph the system at right? Explain how to make a reasonable estimate of the solution by inspecting the graph. Why is it important to check your estimate? $\begin{cases} x + y = 3 \\ x - y = 4 \end{cases}$

Guided Skills Practice

Solve by graphing. (EXAMPLES 1 AND 2)

4. $\begin{cases} -2x + y = 1 \\ y = -x + 4 \end{cases}$
5. $\begin{cases} y + 2x = 0 \\ 2y = -x - 6 \end{cases}$
6. $\begin{cases} 2x + 3y = -12 \\ 4x - 4y = 4 \end{cases}$

Graph each system and find an approximate solution for each system by examining the graphs. Round your approximations to the nearest tenth. (EXAMPLE 3)

7. $\begin{cases} \frac{1}{2}x - y = 2 \\ y = -\frac{2}{3}x \end{cases}$
8. $\begin{cases} 2y - x = 6 \\ 3x + y = -5 \end{cases}$

9. Larry has $4 in nickels and quarters. If there are 36 coins in all, how many quarters does Larry have? (EXAMPLE 4) $n = 25, q = 11$

Practice and Apply

Solve by graphing. Round approximate solutions to the nearest tenth. Check by substituting the approximate solutions into the original equations.

10. $\begin{cases} y = -\frac{1}{3}x + 2 \\ y = 2x + 12 \end{cases}$
11. $\begin{cases} y = \frac{3}{4}x + 3 \\ y = \frac{2}{3}x + 1 \end{cases}$
12. $\begin{cases} y = x \\ y = -x \end{cases}$

13. $\begin{cases} x + 7 = y \\ -4x + 2 = y \end{cases}$
14. $\begin{cases} a = 400 - 2b \\ a = b + 100 \end{cases}$
15. $\begin{cases} 3y = 4x - 2 \\ y = 2x - 2 \end{cases}$

16. $\begin{cases} x - 2y = 10 \\ 3y = 30 - 2x \end{cases}$
17. $\begin{cases} x = 2 \\ 2y = 4x + 2 \end{cases}$
18. $\begin{cases} -7m + 14n = -21 \\ 3m + 15 = 2n \end{cases}$

19. $\begin{cases} 9x - y = -7 \\ 6y - 2x = 15 \end{cases}$
20. $\begin{cases} x + 2 = y + 13 \\ 5 - x = y - 4 \end{cases}$
21. $\begin{cases} x - 3 = y - 3 \\ 2x + 3y = 10 \end{cases}$

Assess

Selected Answers
Exercises 4–9, 11–55 odd

ASSIGNMENT GUIDE

In Class	1–9
Core	11–23 odd, 28–31, 39–43 odd
Core Plus	10–26 even, 28–34, 35–43 odd
Review	44–55
Preview	56–59

✏ Extra Practice can be found beginning on page 738.

4.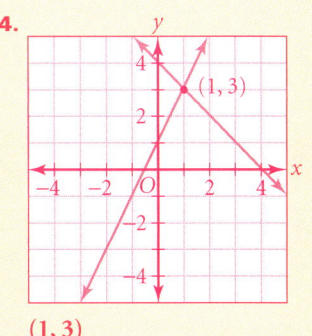
(1, 3)

5.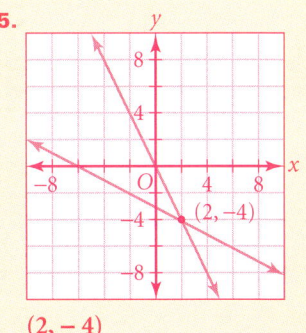
(2, −4)

6.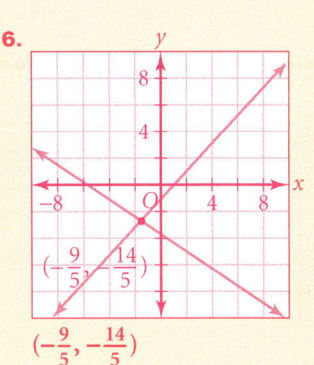
$\left(-\frac{9}{5}, -\frac{14}{5}\right)$

7.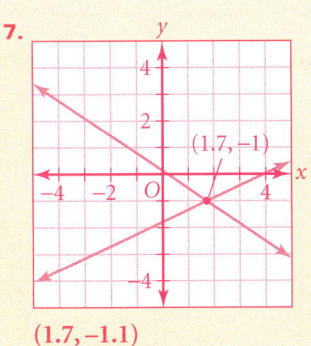
(1.7, −1.1)

8.
(−2.3, 1.9)

The answers to Exercises 10–21 can be found in Additional Answers beginning on page 810.

LESSON 7.1

Technology

A graphics calculator can be used to solve any system in the exercises on pages 323–325. You may wish to have students first use paper and pencil and then use a graphics calculator to check their solutions.

Error Analysis

When solving a system with paper and pencil, students' results may be inaccurate because they did not position the lines properly on the coordinate plane. Encourage students to draw the graphs as accurately as possible and then check their results by substituting the coordinates of the point of intersection into each equation in the system.

22.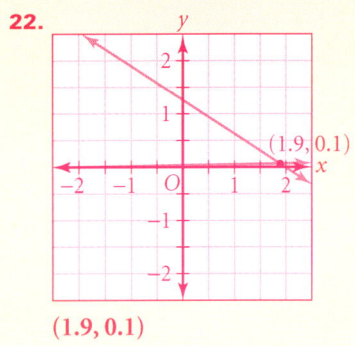
(1.9, 0.1)

22. $\begin{cases} 25y - 0.4x = 0.8 \\ 0.5x + 0.75y = 1 \end{cases}$
23. $\begin{cases} 3c + 2d = -6 \\ -3c + 2d = 6 \end{cases}$
24. $\begin{cases} 2x + 10y = -5 \\ 6x + 4y = 2 \end{cases}$
25. $\begin{cases} 5x + 6y = 14 \\ 3x + 5y = 7 \end{cases}$
26. $\begin{cases} 3x + 5y = 12 \\ 7x - 5y = 8 \end{cases}$
27. $\begin{cases} 7x - 9y = 13 \\ \frac{3}{4}y = \frac{1}{2}x + \frac{4}{3} \end{cases}$

Determine whether the point (2, 10) is a solution for each system of equations.

28. $\begin{cases} y = 2x - 4 \\ y = x + 8 \end{cases}$ **no**
29. $\begin{cases} y = -x + 12 \\ x = -y + 16 \end{cases}$ **no**
30. $\begin{cases} y = x + 8 \\ y = -3x + 16 \end{cases}$ **yes**
31. $\begin{cases} x + 3y = 6 \\ -6 = 2x + 12 \end{cases}$ **no**

32. Graph the equations in Exercises 28–31 to check your answers.
33. Find the point of intersection of $y = x - 8$ and $y = -2x + 6$. $\left(4\frac{2}{3}, -3\frac{1}{3}\right)$
34. Use a graph to find the approximate point of intersection of $y = -x + 12$ and $y = 2x - 4$, giving values to the nearest tenth. Estimate the fractional value of these decimals. Check your estimate by substituting into the original equations. **(5.3, 6.7) or $\left(5\frac{1}{3}, 6\frac{2}{3}\right)$**

Refer to points A(3,5), B(4, −1), and C(9,3). Lines AB and AC meet at point A.

35. Write an equation for line AB. **$y = -6x + 23$**
36. Write an equation for line AC. **$y = -\frac{1}{3}x + 6$**
37. Check your equations by graphing.
38. Use the coordinates of the point of intersection to check your equations. **The point of intersection is (3, 5).**

CONNECTIONS

STATISTICS Use the following data to solve Exercises 39–42:

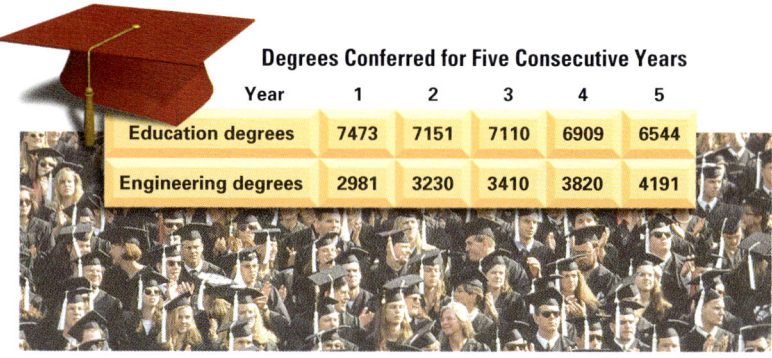

Degrees Conferred for Five Consecutive Years

Year	1	2	3	4	5
Education degrees	7473	7151	7110	6909	6544
Engineering degrees	2981	3230	3410	3820	4191

39. **TECHNOLOGY** Use a graphics calculator to enter the data for education degrees as ordered pairs (years, degrees) in a table. Use your calculator to show a scatter plot of the data.

Education: $y = -210x + 7667.4$

40. **TECHNOLOGY** Use a graphics calculator to enter the data for engineering degrees as ordered pairs (years, degrees) in a table. Use your calculator to show a scatter plot of the data.

Engineering: $y = 301x + 2623.4$

41. **TECHNOLOGY** Use your graphics calculator to find the equation for the line of best fit for each scatter plot in Exercises 39 and 40.

23.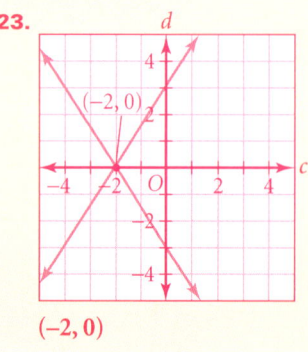
(−2, 0)

24.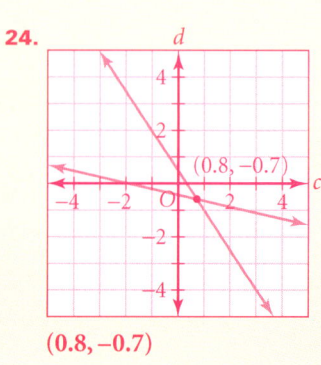
(0.8, −0.7)

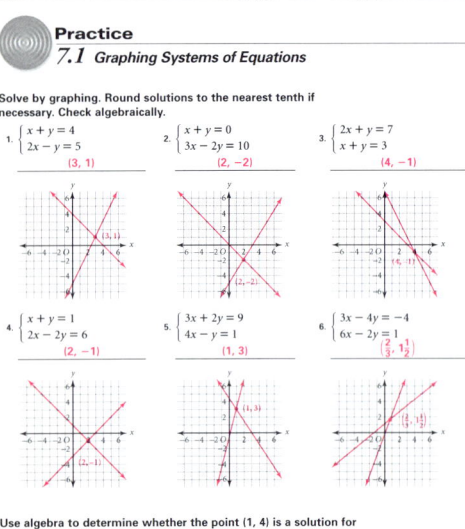

324 LESSON 7.1

42. Find the approximate point of intersection for the two lines you found in Exercise 41. Based on the data in the table, about how many years beyond the given data will it be before the number of education degrees and the number of engineering degrees conferred are the same? **about 5 years beyond the given data**

APPLICATION

43. **AVIATION** A plane initially at 28,000 feet is descending at a rate of 1800 feet per minute, and a second plane initially at 1500 feet is climbing at a rate of 2300 feet per minute. Write and graph a system of equations to determine how many minutes it will take for the planes to be at the same altitude. **≈ 6.5 minutes**

Look Back

44. Place parentheses and brackets to make $21 \div 3 - 4 + 0 + 6 \cdot (-3) = -39$ true. **(LESSON 1.3)** $21 \div (3 - 4) + 0 + 6 \cdot (-3) = -39$

45. If data points are randomly scattered, what type of correlation exists? **(LESSON 1.6) little or no correlation**

Evaluate. (LESSONS 2.2 AND 2.3)

46. $-3 + \left(-\frac{1}{4}\right)$ $-3\frac{1}{4}$ **47.** $1.5 - (-1.5)$ **3**   **48.** $3\frac{1}{2} - 5\frac{2}{3}$ $-2\frac{1}{6}$

Evaluate. (LESSON 2.4)

49. $(-3)(2)$ **−6**   **50.** $(-1.7)(-2.5)$ **4.25**   **51.** $4\left(-2\frac{1}{3}\right)$ $-9\frac{1}{3}$

Simplify each expression. (LESSON 2.6)

52. $-4x + 2x$ **−2x**   **53.** $3y + (-2x) - 3y$ **−2x**   **54.** $x - y + x$ **2x − y**

55. Solve: $3(5 - 2x) - (8 - 6x) = -9 + 2(3x + 4) - 10$. **(LESSON 3.5)** $x = 3$

Look Beyond

Substitute each value of x into the equation $-2x + 4y = 12$, and solve for y.

56. $x = -6$ **57.** $x = 8y$ **58.** $x = y - 3$ **59.** $x = 3y + 1$
$y = 0$ $y = -1$ $y = 3$ $y = -7$

Include this activity in your portfolio.

From 1980 to 1990, there was a shift in salaries for basketball players and baseball players. The graph reflects the average salaries for the players.

- Estimate the year that the salaries were equal and estimate the salary.
- Estimate the *y*-intercept of each graph and use it with the point of intersection to determine the equation of each line.
- Extend the graph to predict the salaries for 2000.
- Find the current salary data and the data for 2000 in an almanac or on-line, and compare these data with your results.
- Find data for football salaries in an almanac or on-line, and compare them with the baseball and basketball salaries.

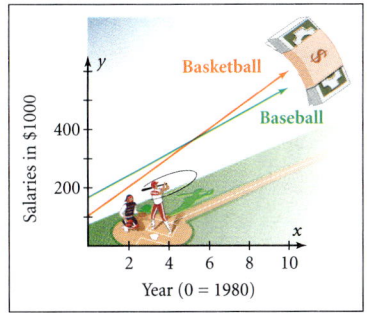

Look Beyond

Exercises 56–59 preview the substitution method for solving systems of equations. Students will learn this method in Lesson 7.2.

ALTERNATIVE Assessment

Portfolio Activity

The Portfolio Activity can be used as preparation for the Chapter Project or as a separate activity. In the Portfolio Activity on this page, students solve a system with the given graph. Then they approximate the equations in the system and use the equations to make predictions.

internetconnect

GO TO: go.hrw.com
KEYWORD: MA1 Athletes

To support the Portfolio Activity, students may visit the HRW Web site, where they can obtain more data that relates to the salaries of professional athletes.

Student Technology Guide

7.1 Graphing Systems of Equations

You can use a graphics calculator to solve a system of linear equations such as $\begin{cases} 3x + y = 4 \\ x - 2y = 6 \end{cases}$.

First you need to solve each equation for *y*.

$3x + y = 4$ $x - 2y = 6$
$y = 4 - 3x$ $y = \frac{x - 6}{2}$

The equations can be entered as shown. Notice that there is no need to write the slope-intercept form.

Now press GRAPH. You can use TRACE and the arrow keys, ◄ and ►, to locate the intersection point. If you define the window like the one shown, you can often avoid the messy decimals that appear while tracing. The solution appears to be $(2, -2)$. You can verify that this is the exact solution by substituting into the original equations.

Many graphics calculators offer an intersect feature that easily gives an approximation of the solution.

To find the solution of $\begin{cases} y = \frac{2}{5}x + 4 \\ y = -\frac{2}{7}x + 2 \end{cases}$ graph each equation. Then press 2nd TRACE 5:intersect ENTER ENTER ENTER.

Note that the values are now stored as the variables *x* and *y*. You can convert them to fractions as follows:

2nd MODE X,T,θ,n MATH MATH ►Frac ENTER ENTER ALPHA
1 MATH MATH ►Frac ENTER ENTER

The solution is $\left(-\frac{35}{12}, \frac{17}{6}\right)$.

Use a graphics calculator to solve each system by graphing. Write the solutions as fractions.

1. $\begin{cases} y = 3x - 5 \\ y = \frac{2}{3}x + 1 \end{cases}$ $\left(\frac{18}{7}, \frac{19}{7}\right)$ **2.** $\begin{cases} y = \frac{1}{4}x - 2 \\ y = 6x + 3 \end{cases}$ $\left(-\frac{20}{23}, -\frac{51}{23}\right)$ **3.** $\begin{cases} 2x - 3y = 8 \\ 5x + 2y = 4 \end{cases}$ $\left(\frac{28}{19}, -\frac{32}{19}\right)$

25.

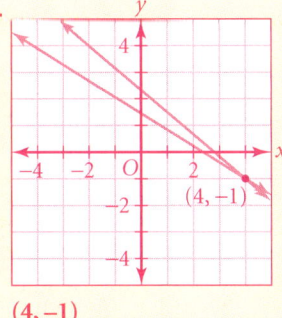

$(4, -1)$

26.

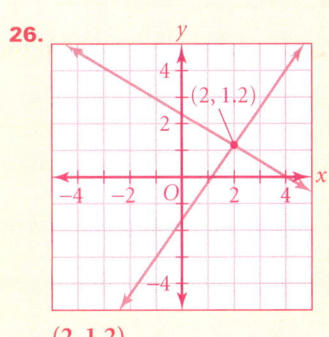

$(2, 1.2)$

The answers to Exercises 27, 32, 37, and 39–43 can be found in Additional Answers beginning on page 810.

LESSON 7.1 **325**

Prepare

NCTM PRINCIPLES & STANDARDS
1–3, 5, 6, 8–10

LESSON RESOURCES

- Practice 7.2
- Reteaching Master 7.2
- Enrichment Master 7.2
- Cooperative-Learning Activity 7.2
- Lesson Activity 7.2
- Problem Solving/Critical Thinking Master 7.2
- Teaching Transparency 35

QUICK WARM-UP

Simplify.

1. $4r + 3(r + 5)$ $7r + 15$
2. $z - 2(z - 2)$ $-z + 4$
3. $5(4p - 7) - 2p$ $18p - 35$

Solve for y.

4. $2x + y = 5$ $y = -2x + 5$
5. $6x - 2y = 12$ $y = 3x - 6$

Also on Quiz Transparency 7.2

Teach

Why In sports, coaches often *substitute* one player with another who plays the same position. Ask students to suggest other situations in which substitutions are made. Tell them that you also can make substitutions in equations. That is, you can replace a variable in an equation with an equivalent expression.

7.2 The Substitution Method

Objective
- Find an exact solution to a system of linear equations by using the substitution method.

Why Systems of equations often require exact solutions. In order to make a profit, the owner of a small business at the auto race track must know the exact amount to charge for each of her two products.

The 12 Hours at Sebring automobile race is an endurance test. Two or more drivers for each racing team take turns driving in the race, which covers more than 1800 miles.

You can find an exact solution to a system of linear equations without graphing. One method of doing this is to use substitution.

Activity
Exploring Substitution

APPLICATION
AUTO RACING

Two drivers on a team at the 12 Hours at Sebring automobile race drive for a total of 157 laps. Driver two drives 21 laps less than driver one. How many laps did each driver drive?

1. First use equations to model the problem.
   Let x represent the number of laps for driver one.
   Let y represent the number of laps for driver two.
   Then $\begin{cases} x + y = 157 \\ y = x - 21 \end{cases}$ is the system that models the problem.

PROBLEM SOLVING

2. Use the **guess-and-check** method to find the values for x and y that solve the system.

3. Look at the second equation in the system, $y = x - 21$. How can this information about y be used in the first equation?

4. Because $y = x - 21$, substitute $x - 21$ in place of y in the first equation. Solve the new equation for x.

5. Substitute this value of x into the second equation to find y.

CHECKPOINT ✓

6. Compare the values for x and y with the answer you found by using guess-and-check. Are they the same?

Alternative Teaching Strategy

USING MODELS Begin by giving students the system at right. $\begin{cases} y = x - 3 \\ 4x + 2y = 6 \end{cases}$
Point out that they can model y by modeling $x - 3$, as shown.

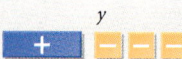

Ask students how they might model $4x + 2y = 6$ with tiles. They probably will note that there are no y-tiles. Tell them that they can substitute 2 "$x - 3$" sets in place of $2y$ as shown at right.

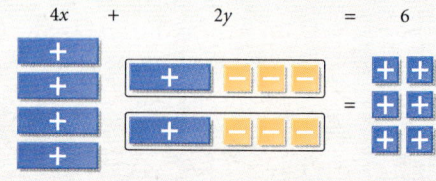

Have students manipulate these tiles to arrive at the solution, $x = 2$. Now return to the tiles that represent y and replace the x-tile with 2 unit tiles.

Thus, $y = -1$. The solution to the system is $(2, -1)$.

326 LESSON 7.2

If you know the value of one variable in a system of equations, you can find the solution for the system by substituting the known value of the variable into one of the equations. This method is called **substitution**.

EXAMPLE 1 Solve by using substitution. $\begin{cases} 8x + 2y = 19 \\ x = 3 \end{cases}$

• **SOLUTION**

Since $x = 3$, you can substitute the value of x into the first equation.

$$8(3) + 2y = 19$$

Solve the resulting equation for y.

$$24 + 2y = 19$$
$$2y = -5$$
$$y = -2.5$$

The solution is $(3, -2.5)$.

Check the solution by substituting the values for x and y into the first equation.

$$8(3) + 2(-2.5) \stackrel{?}{=} 19$$
$$24 + (-5) \stackrel{?}{=} 19$$
$$19 = 19 \quad \text{True}$$

EXAMPLE 2 Solve by using substitution. $\begin{cases} 15x - 5y = 30 \\ y = 2x + 3 \end{cases}$

• **SOLUTION**

Substitute $2x + 3$ for y into the first equation, and solve for x.

$$15x - 5(2x + 3) = 30$$
$$15x - 10x - 15 = 30$$
$$5x - 15 = 30$$
$$5x = 45$$
$$x = 9$$

Substitute 9 for x into the equation $y = 2x + 3$, and solve for y.

$$y = 2(9) + 3$$
$$y = 18 + 3$$
$$y = 21$$

The solution is $(9, 21)$.

Check the solution by substituting the values for x and y into the original equations.

$$15(9) - 5(21) \stackrel{?}{=} 30 \qquad 21 \stackrel{?}{=} 2(9) + 3$$
$$135 - 105 \stackrel{?}{=} 30 \qquad 21 \stackrel{?}{=} 18 + 3$$
$$30 = 30 \quad \text{True} \qquad 21 = 21 \quad \text{True}$$

TRY THIS Solve each system by using substitution.

a. $\begin{cases} 2x + 5y = 14 \\ y = 5 \end{cases}$ b. $\begin{cases} -3x + 2y = 31 \\ x = 0.5y + 6 \end{cases}$

Interdisciplinary Connection

CONSUMER EDUCATION At StarCars Rentals, you can rent a car for $35 per day plus 10¢ per mile driven. At URentEm Car Rentals, you can rent a car for $20 per day plus 25¢ per mile driven. For what number of miles driven will the cost of a StarCars rental be the same as the cost of a URentEm rental? **100 miles**

At which establishment would the cost be less if you planned to drive 40 miles? 200 miles? **URentEm; StarCars**

Inclusion Strategies

LINGUISTIC LEARNERS Have students write a general procedure for using the substitution method to solve systems of equations. A sample is given.

Solving a System by Substitution

1. Solve for one variable in one of the equations.
2. Substitute into the second equation.
3. Solve to find the value of the other variable.
4. Substitute into the first equation.
5. Solve to find the value of the first variable.
6. Write the solution as an ordered pair.
7. Check the ordered pair in both equations.

Activity Notes

This Activity gives students the opportunity to explore the substitution method of solving systems of equations. After completing this Activity, students should see that a systematic method for solving two equations in two unknowns, such as the substitution method, can be more efficient and less time-consuming than the guess-and-check method.

CHECKPOINT ✓
6. They are the same.

ADDITIONAL EXAMPLE 1

Solve by using substitution.

$\begin{cases} 10x + 2y = 10 \\ y = -10 \end{cases}$

$(3, -10)$

ADDITIONAL EXAMPLE 2

Solve by using substitution.

$\begin{cases} 10x + 2y = 10 \\ y = 3x - 3 \end{cases}$

$(1, 0)$

TRY THIS

a. $\left(-\frac{11}{2}, 5\right)$
b. $(55, 98)$

LESSON 7.2

ADDITIONAL EXAMPLE 3

Solve by using substitution.
$\begin{cases} 4x - 3y = 5 \\ 6x + y = 2 \end{cases}$

$(0.5, -1)$

TRY THIS

$\left(\dfrac{41}{20}, \dfrac{13}{20}\right)$

CRITICAL THINKING

The coefficient of y is 1, so solving for this variable is faster.

Use Teaching Transparency 35.

ADDITIONAL EXAMPLE 4

Sam sells T-shirts at a baseball park. He has 50 of last year's T-shirts and 200 of this year's T-shirts in stock. He knows that his customers will pay $5 more for this year's T-shirt. He needs to make a total of $3750 from T-shirt sales. **How much should he charge for each type?**
last year's T-shirt: $11
this year's T-shirt: $16

EXAMPLE 3 Solve by using substitution. $\begin{cases} 3x + y = 4 \\ 5x - 7y = 11 \end{cases}$

● **SOLUTION**

To use substitution in this example, solve the first equation for y.

$3x + y = 4$
$3x + y - 3x = 4 - 3x$
$y = 4 - 3x$

> Choose the equation that is easier to solve.

Substitute $4 - 3x$ for y in the second equation and solve for x.

$5x - 7y = 11$
$5x - 7(4 - 3x) = 11$
$5x - 28 + 21x = 11$
$26x - 28 = 11$
$26x = 39$
$x = 1.5$

Substitute 1.5 for x in the first equation and solve for y.

$3(1.5) + y = 4$
$4.5 + y = 4$
$y = -0.5$

The solution is $(1.5, -0.5)$. Check the solution by substituting the values for x and y into the original equations.

TRY THIS Solve by using substitution. $\begin{cases} 6x - 2y = 11 \\ x + 3y = 4 \end{cases}$

CRITICAL THINKING In Example 3, what is the advantage in choosing to solve for y in the first equation?

EXAMPLE 4 Pam sells racing team caps at her concession stand at an automobile race track. She has 100 of last year's caps and 300 of this year's caps in stock. Pam knows from past experience that her customers will pay $7 more for this year's cap than for last year's. She has calculated that she needs to make $5300 in sales of the caps. **Assuming that she will sell all of the caps, how much should Pam charge for each type of cap?**

APPLICATION
SMALL BUSINESS

● **SOLUTION**

Define the variables.
Let m represent the cost of this year's cap.
Let n represent the cost of last year's cap.

Write a system of linear equations. $\begin{cases} 300m + 100n = 5300 \\ m = n + 7 \end{cases}$

Substitute $n + 7$ for m in the first equation, and solve for n.

$300(n + 7) + 100n = 5300$
$300n + 2100 + 100n = 5300$
$400n + 2100 = 5300$
$400n = 3200$
$n = 8$

Substitute 8 for n in the second equation, and solve for m.

$m = 8 + 7$
$m = 15$

The solution is $(15, 8)$.

Pam should charge $15 for this year's caps and $8 for last year's caps.

Enrichment

Compare the two systems at right. How are they alike? How are they different?

$\begin{cases} x + y = 3 \\ y = 2x \\ y = x + 1 \end{cases}$ $\begin{cases} x + y = 3 \\ y = 2x \\ y = x - 1 \end{cases}$

Sample: Each is a system of three linear equations in two variables. The solution to the first system is (1, 2). There is no common solution to all three equations in the second system, so this system has no solution.

Have students create a set of systems similar to these. **Students' original systems will vary.**

Reteaching the Lesson

COOPERATIVE LEARNING Give students the system of equations at right. Have students work in pairs. Instruct one student of each pair to solve the first equation for y and the other student to solve the second equation for y. Then have both solve the system independently by using substitution. Each should arrive at the solution $(1, -2)$. Have them compare their work and discuss whether it matters which substitution is made. Discuss with the class which substitution they think is more efficient.

$\begin{cases} 2x - 4y = 10 \\ y + 5x = 3 \end{cases}$

Practice

Practice
7.2 The Substitution Method

Solve and check each system by using the substitution method.

1. $\begin{cases} y = 2x \\ 2x + y = -12 \end{cases}$ $(-3, -6)$
2. $\begin{cases} y = x + 3 \\ 3x + y = 11 \end{cases}$ $(2, 5)$
3. $\begin{cases} y = 2x + 1 \\ x + 3y = 31 \end{cases}$ $(4, 9)$
4. $\begin{cases} x + y = 3 \\ 4x - 2y = 18 \end{cases}$ $(4, -1)$
5. $\begin{cases} 2x - 3y = -25 \\ 3x + y = 1 \end{cases}$ $(-2, 7)$
6. $\begin{cases} x - 2y = -3 \\ 4x - 3y = 8 \end{cases}$ $(5, 4)$
7. $\begin{cases} 2x + 5y = -7 \\ 3x - y = -2 \end{cases}$ $(-1, -1)$
8. $\begin{cases} 2x - y = -11 \\ 3x - 6y = 6 \end{cases}$ $(-8, -5)$
9. $\begin{cases} x - y = 4 \\ 2x - 3y = 6 \end{cases}$ $(6, 2)$
10. $\begin{cases} 3x + y = -3 \\ x - 3y = 11 \end{cases}$ $(0.2, -3.6)$
11. $\begin{cases} x - y = 10 \\ 2x + 3y = 5 \end{cases}$ $(7, -3)$
12. $\begin{cases} 2x + y = 2 \\ 4x - 2y = -4 \end{cases}$ $(0, 2)$
13. $\begin{cases} x = y - 4.2 \\ 2x - 3y = -9 \end{cases}$ $(-3.6, 0.6)$
14. $\begin{cases} -2x - y = 4 \\ x + y = -3 \end{cases}$ $(-1, -2)$
15. $\begin{cases} 4x - y = -2 \\ -8x + y = 3 \end{cases}$ $(-0.25, 1)$
16. $\begin{cases} 2x - 2y = 2 \\ 3x + y = -9 \end{cases}$ $(-2, -3)$

Graph each system and estimate the solution. Then use the substitution method to get an exact solution.

17. $\begin{cases} y = 2x \\ 2x + y = 7 \end{cases}$ exact solution: $\left(1\dfrac{3}{4}, 3\dfrac{1}{2}\right)$
18. $\begin{cases} x + y = 2 \\ x - 2y = 0 \end{cases}$ exact solution: $\left(1\dfrac{1}{3}, \dfrac{2}{3}\right)$

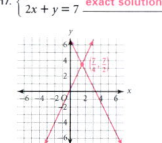
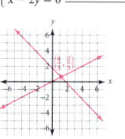

328 LESSON 7.2

Exercises

Communicate

1. If you know $y = 42$, explain how to use substitution to solve $y = 2x + 8$ for x.

2. Given the equations $-4x + y = 2$ and $2x + 3y = 34$, tell how to find an expression to substitute and how to find the solution.

3. Explain how to use substitution to solve this system. $\begin{cases} x - 2y = 8 \\ 2x + 3y = 23 \end{cases}$

Guided Skills Practice

Solve by using substitution. Check your answers.
(EXAMPLES 1, 2, AND 3)

4. $\begin{cases} 5x = 3y + 12 \\ x = 5 \end{cases}$ $(5, 4\frac{1}{3})$

5. $\begin{cases} 3x - 2y = 2 \\ y = 2x + 8 \end{cases}$ $(-18, -28)$

6. $\begin{cases} 5x - y = 1 \\ 3x + y = 1 \end{cases}$ $(0.25, 0.25)$

7. $\begin{cases} 2x + y = 1 \\ 10x = 4y + 2 \end{cases}$ $\left(\frac{1}{3}, \frac{1}{3}\right)$

Write and solve a system of equations for the problem below. Check your answer. (EXAMPLE 4)

8. The sum of two numbers is 27. The larger number is 3 more than the smaller number. Find the two numbers. **12 and 15**

Practice and Apply

Solve by using substitution, and check your answers.

9. $\begin{cases} 2x + 8y = 1 \\ x = 2y \end{cases}$ $\left(\frac{1}{6}, \frac{1}{12}\right)$

10. $\begin{cases} x = 7 \\ 2x + y = 5 \end{cases}$ $(7, -9)$

11. $\begin{cases} 3x + y = 5 \\ 2x - y = 10 \end{cases}$ $(3, -4)$

12. $\begin{cases} y = 5 - x \\ 1 = 4x + 3y \end{cases}$ $(-14, 19)$

13. $\begin{cases} 2x + y = -92 \\ 2x + 2y = -98 \end{cases}$ $(-43, -6)$

14. $\begin{cases} 4x + 3y = 13 \\ x + y = 4 \end{cases}$ $(1, 3)$

15. $\begin{cases} 6y = x + 18 \\ 2y - x = 6 \end{cases}$ $(0, 3)$

16. $\begin{cases} 2x + y = 1 \\ 10x - 4y = 2 \end{cases}$ $\left(\frac{1}{3}, \frac{1}{3}\right)$

17. $\begin{cases} 2x + 3y = 7 \\ x + 4y = 9 \end{cases}$ $(0.2, 2.2)$

18. $\begin{cases} 4x - y = 15 \\ -2x + 3y = 12 \end{cases}$ $(5.7, 7.8)$

19. $\begin{cases} 2y + x = 4 \\ y - x = -7 \end{cases}$ $(6, -1)$

20. $\begin{cases} 4y - x = 4 \\ y + x = 6 \end{cases}$ $(4, 2)$

21. $\begin{cases} y = x - 3 \\ x + y = 5 \end{cases}$ $(4, 1)$

22. $\begin{cases} 2x + 3y = 21 \\ -3x - 6y = -24 \end{cases}$ $(18, -5)$

23. $\begin{cases} 5x - 7y = 31 \\ -4x + 2y = -14 \end{cases}$ $(2, -3)$

24. $\begin{cases} 3x + y = 21 \\ 10x + 5y = 65 \end{cases}$ $(8, -3)$

25. $\begin{cases} -3y = 9x + 24 \\ 6y + 2x = 32 \end{cases}$ $(-5, 7)$

26. $\begin{cases} 12x + 4y = 22 \\ 3x - 8y = -10 \end{cases}$ $\left(1\frac{7}{27}, 1\frac{13}{18}\right)$

27. $\begin{cases} 11x + 4y = -17 \\ -6x + y = 22 \end{cases}$ $(-3, 4)$

28. $\begin{cases} -5x + 7y = -41 \\ 7x + y = 25 \end{cases}$ $(4, -3)$

CONNECTION

29. **GEOMETRY** Find the dimensions of the rectangle shown at right if its length is twice its width.

$W = 34\frac{2}{3}$ $L = 69\frac{1}{3}$

Perimeter = 208

Assess

Selected Answers
Exercises 4–8, 9–43 odd

ASSIGNMENT GUIDE

In Class	1–8
Core	9–23, 29–33 odd
Core Plus	10–28 even, 29–34
Review	35–43
Preview	44, 45

✎ Extra Practice can be found beginning on page 738.

Student Technology Guide

NAME _____ CLASS _____ DATE _____

Student Technology Guide
7.2 The Substitution Method

A graphics calculator makes it easy to solve systems of linear equations by using substitution.

Example: Solve the system $\begin{cases} 15x - 5y = 30 \\ y = 2x + 3 \end{cases}$.

Since the second equation is already solved for y, substitute $2x + 3$ for y in the first equation to get $15x - 5(2x + 3) = 30$. There is no need to simplify further. To find x, let Y1 represent the left side of this single equation and let Y2 be 30, as shown at left below.

 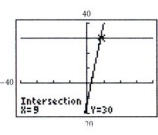

Press `GRAPH` `TRACE`. Or, use the intersect feature.
`2nd` `TRACE` `5: intersect` `ENTER` `ENTER`
The use of the intersect feature is shown at right above.
To find y, recall that $y = 2x + 3$. Since the value of x is stored as the variable X, you can press 2 `X,T,θ,n` `+` 3 `ENTER`.
The solution is (9, 21).

Here are a few things to notice about the example above.

- There is no need to write down the expression $15x - 5(2x + 3)$. You can simply think of $15x - 5y$, but replace y with $(2x + 3)$ when you get to it.
- It is important to use an appropriate viewing window. Since you graphed Y2 = 30, you needed to make sure that 30 is included between Ymin and Ymax.
- Finally, notice that the intersect feature gives X = 9 and Y = 30. However, since the equation you were solving involved only x, this is *not* the value of y you were looking for; another step is needed to find y.

Use a calculator to solve by using substitution.

1. $\begin{cases} 7x - 3y = -18 \\ y = 5x - 2 \end{cases}$ (3, 13)

2. $\begin{cases} 3x + 5y = -101 \\ y = -4x + 7 \end{cases}$ (8, -25)

3. $\begin{cases} 4x + 6y = -36 \\ x = 3y + 9 \end{cases}$ (-3, -4)

LESSON 7.2

Technology

A graphics calculator may be helpful for Exercises 9–34. You may want to have students first complete the exercises with paper and pencil and then use the calculator to verify their answers.

Error Analysis

Students often will arrive at an incorrect solution to a system because they fail to simplify properly after making a substitution. Remind them to put parentheses around the expression that is substituted and to check that they have applied the Distributive Property correctly if it is used.

Exercises 44 and 45 extend the concept of substitution in systems of linear equations in three variables. Be sure students are aware that a solution to such a system is an *ordered triple* of numbers, (x, y, z). Students will study systems like these if they take a more advanced course in algebra.

---

internetconnect

GO TO: go.hrw.com
KEYWORD: MA1 House Gender

In this activity, students collect gender data on members of the House of Representatives. Students then use this data to develop systems of equations that can be used to predict future numbers.

CONNECTION

30. GEOMETRY In the triangle shown, the sum of the measures of angle A and angle B is 90°. If angle A is 30° less than twice angle B, then find the degree measure for each angle. $m\angle A = 50°$, $m\angle B = 40°$

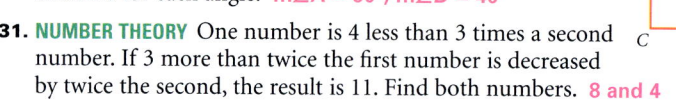

31. NUMBER THEORY One number is 4 less than 3 times a second number. If 3 more than twice the first number is decreased by twice the second, the result is 11. Find both numbers. **8 and 4**

Write and solve a system of equations for each problem.

APPLICATIONS

32. RECREATION A hot-air balloon is rising from the ground at a rate of 4 feet every second. Another balloon at 756 feet is descending at a rate of 3 feet every second. In how many seconds will the two balloons be at the same altitude? **108 seconds**

33. FUND-RAISING At the Boy Scout "all you can eat" spaghetti dinner, the troop served 210 people and raised $935. Adult meals cost $6.00 and child meals cost $3.50.
  a. Write an equation for the *total amount raised* from the adult and child dinners served. $6a + 3.5c = 935$
  b. Write an equation for the *total number* of adult and child dinners served.
  c. Solve the system of equations from parts **a** and **b**. How many adult and child dinners were served? b. $a + c = 210$ c. **80 adults, 130 children**

34. CULTURAL CONNECTION: ASIA A Chinese problem states that several people combined their money to buy a farm tool to share. If each person paid 8 coins, they had 3 coins too many. If each paid 7 coins, they were 4 coins short. How many people were there, and what was the price in coins of the farm tool? **7 people, 53 coins**

 Look Back

APPLICATION

35. RECREATION In a foot race, Sam finished 20 feet in front of Joe. Joe was 5 feet behind Mark, and Mark was 10 feet behind Rob. Tom was 15 feet ahead of Rob. In what order did they finish? *(LESSON 2.2)*
Tom, Sam, Rob, Mark, Joe

Solve each equation for *x*. *(LESSON 3.2)*

36. $\frac{x}{15} = 3$ $x = 45$ **37.** $\frac{3}{x} = 15$ $x = \frac{1}{5}$ **38.** $\frac{15}{x} = 3$ $x = 5$ **39.** $\frac{x}{3} = 15$ $x = 45$

40. The number 12.6 is 42% of what number? *(LESSON 4.2)* **30**

Tell whether the following equations represent lines that are parallel, perpendicular, or neither: *(LESSON 5.6)*

41. $\begin{cases} 3y = 2x - 15 \\ 3x + 2y = 24 \end{cases}$ **42.** $\begin{cases} 2y = x - 12 \\ 2y - x = 12 \end{cases}$ **43.** $\begin{cases} y = x - 1 \\ y = -x + 3 \end{cases}$
    perpendicular   **parallel**   **perpendicular**

 Look Beyond

Examine the following systems. Use substitution to solve for *x* and *y*.

44. $\begin{cases} x + 2y + 3z = 8 \\ y + 2z = 3 \\ z = 2 \end{cases}$ $x = 4$, $y = -1$

45. $\begin{cases} 2x + 3y + 5z = 44 \\ 2y - 6z = 4 \\ z = 4 \end{cases}$ $x = -9$, $y = 14$

7.3 The Elimination Method

Jason drove 125 miles on the 2-day trip. He drove 350 miles on the 4-day trip.

Objectives

• Use the elimination method to solve a system of equations.

• Choose an appropriate method to solve a system of equations.

Why The elimination method is another technique that can provide a quick solution to complex systems of equations such as the one presented in the car-rental problem.

**APPLICATION
RENTAL FEES**

To find the daily fee and the per-mile cost at Airport Rent-A-Car, you can set up and solve a system of equations.

First define the variables.
  Let d represent the cost per day.
  Let m represent the cost per mile.

Use the information from the rental bills to set up a system of equations.

$$\begin{cases} 2d + 125m = 95.75 \\ 4d + 350m = 226.50 \end{cases}$$ represents Bill No. 001
represents Bill No. 002

Both the graphing method and the substitution method provide options for finding the solution to this system. However, neither of these methods would be easy to apply to this system. In this lesson you will develop a method for solving a system like this one.

Alternative Teaching Strategy

GUIDED ANALYSIS Review the Addition and Subtraction Properties of Equality.
  If $a = b$, then $a + c = b + c$.
  If $a = b$, then $a - c = b - c$.
Then remind students of these related forms.
  If $a = b$ and $c = d$, then $a + c = b + d$.
  If $a = b$ and $c = d$, then $a - c = b - d$.

Now give students the following systems:

$$\begin{cases} 2x + y = 9 \\ x - y = 6 \end{cases} \quad \begin{cases} 2x + y = 9 \\ x + y = 6 \end{cases}$$

Lead students in a comparison of the two systems. Then work with them to apply the properties and find the solutions. **(5, –1); (3, 3)**

Now discuss with students how the Multiplication Property of Equality must be applied to solve systems such as the ones given below.

$$\begin{cases} 2x + y = 9 \\ x + 2y = 6 \end{cases} \quad \begin{cases} 2x + 3y = 9 \\ 3x + 2y = 6 \end{cases}$$

Work with students to find the solutions. **(4, 1); (0, 3)**

Prepare

NCTM PRINCIPLES & STANDARDS
1–3, 5, 6, 8–10

LESSON RESOURCES

• Practice 7.3
• Reteaching Master 7.3
• Enrichment Master 7.3
• Cooperative-Learning Activity 7.3
• Lesson Activity 7.3
• Problem Solving/Critical Thinking Master 7.3

QUICK WARM-UP

Simplify.

1. $(6x - y) + (-7x + y)$ **$-x$**
2. $-4(x + 3y)$ **$-4x - 12y$**

Solve each system.

3. $\begin{cases} 5x - 2y = 3 \\ y = 2x \end{cases}$ **(3, 6)**

4. $\begin{cases} 2x + y = -1 \\ y = -3x - 3 \end{cases}$ **(–2, 3)**

Also on Quiz Transparency 7.3

Teach

Why Ask students what is meant by an elimination tournament in sports. Be sure they understand that it is a series of games played among a set of participants, generally to determine a single champion, and that some participants are eliminated at each stage. Tell students that in this lesson, they will learn how to solve a system of equations by a process of eliminating terms until just one remains.

LESSON 7.3   **331**

Activity Notes

This Activity introduces students to the elimination method of solving systems of equations. After completing the Activity, students should discover that when the coefficients of terms containing the same variable are opposites, adding the equations eliminates that variable.

CHECKPOINT ✓

5. Answers may vary. Sample answer: Opposites can be used to eliminate one (or more) variables by combining equations. When only one variable remains, its value can be found by simplification and then substituted to solve for the other variable(s).

ADDITIONAL EXAMPLE 1

Solve by elimination.

$\begin{cases} 4x + 5y = 6 \\ 3x - 5y = 8 \end{cases}$

$\left(2, -\frac{2}{5}\right)$

Activity
Exploring Using Opposites

Consider this system of equations: $\begin{cases} 3x + 2y = 7 \\ 5x - 2y = 9 \end{cases}$

1. Which terms of the equations are opposites?
2. Use the Addition Property of Equality to combine the two equations into one equation (add $3x$ and $5x$, $2y$ and $-2y$, and 7 and 9). How many variables does the resulting equation have?
3. Solve the resulting equation for x, and then solve for y by substituting the value for x into one of the original equations.
4. Check the solution in both equations of the system.

CHECKPOINT ✓ 5. Explain how opposites can be used to solve a system of equations.

The Elimination Method

In the activity, you used the **elimination method** to solve a system of equations. This method uses opposites to eliminate one of the variables.

EXAMPLE 1 Solve by elimination. $\begin{cases} 3x + 4y = 7 \\ 2x - 4y = 13 \end{cases}$

- **SOLUTION**

Use the Addition Property of Equality to combine the equations, and solve the resulting equation for x.

$3x + 4y = 7$
$\underline{2x - 4y = 13}$
$5x + 0 = 20$
$5x = 20$
$x = 4$

$4y$ and $-4y$ are opposites.

To find y, substitute 4 for x in one of the original equations.

$3x + 4y = 7$
$3(4) + 4y = 7$
$12 + 4y = 7$
$4y = -5$
$y = -1.25$

The solution is $(4, -1.25)$.

Check by substituting **4** for x and **−1.25** for y in the original equations.

$3(4) + 4(-1.25) = 7$ $2(4) - 4(-1.25) = 13$
$12 + (-5) = 7$ $8 - (-5) = 13$
$7 = 7$ *True* $13 = 13$ *True*

In the system of equations, notice that the y-terms are opposites. These types of systems are the easiest to solve by using the elimination method.

Sometimes it is necessary to multiply one or both equations by a number in order to produce opposites that will eliminate each other. This process

Interdisciplinary Connection

BUSINESS A company wants to advertise its product in a certain magazine. The magazine offers two ad plans for a 12-issue run.

Plan A: 6 full-page ads, 6 half-page ads, $2550
Plan B: 8 full-page ads, 4 half-page ads, $2800

The company's business manager needs to find the costs of one full-page ad and one half-page ad in order to compare them with the costs at other magazines. Use a system of equations to find the costs. **full-page ad: $275; half-page ad: $150**

Inclusion Strategies

VISUAL STRATEGIES Show students how to use a notation like the following to keep track of their multipliers.

$\begin{cases} 4x + 2y = 8 \\ 5x + 6y = 3 \end{cases} \begin{array}{l} \xrightarrow{(5)} \\ \xrightarrow{(-4)} \end{array} \begin{array}{l} 20x + 10y = 40 \\ \underline{-20x - 24y = -12} \\ -14y = 28 \\ y = -2 \end{array}$

After students have multiplied, encourage them to maintain the vertical alignment of the terms that contain the same variables and of the constants, as shown above.

is easiest when a variable in one of the equations has a coefficient of 1. However, it can be applied to more complicated systems such as the one solved in Example 2.

EXAMPLE 2 Solve by elimination. $\begin{cases} 2x + 3y = 1 \\ 5x + 7y = 3 \end{cases}$

● **SOLUTION**

Multiply the first equation by 5 and the second equation by −2 in order to produce opposites that will eliminate each other.

$$\begin{cases} (5)2x + (5)3y = (5)1 \\ (-2)5x + (-2)7y = (-2)3 \end{cases} \Rightarrow \begin{cases} 10x + 15y = 5 \\ -10x - 14y = -6 \end{cases}$$

Use the Addition Property of Equality to combine the equations, and solve the resulting equation for y.

$$\begin{array}{r} 10x + 15y = 5 \\ -10x - 14y = -6 \\ \hline y = -1 \end{array}$$

To find x, substitute -1 for y in one of the original equations.

$$2x + 3y = 1$$
$$2x + 3(-1) = 1$$
$$2x - 3 = 1$$
$$2x = 4$$
$$x = 2$$

The solution is $(2, -1)$.

Check by substituting **2** for x and **−1** for y in the original equations.

$$\begin{array}{ll} 2(2) + 3(-1) = 1 & 5(2) + 7(-1) = 3 \\ 4 + (-3) = 1 & 10 + (-7) = 3 \\ 1 = 1 \quad \text{True} & 3 = 3 \quad \text{True} \end{array}$$

EXAMPLE 3 Use elimination to solve the car-rental problem from page 331.

$$\begin{cases} 2d + 125m = 95.75 \\ 4d + 350m = 226.50 \end{cases}$$

APPLICATION
RENTAL FEES

● **SOLUTION**

Multiply each term of the first equation by −2.

$$\begin{cases} (-2)2d + (-2)125m = (-2)95.75 \\ 4d + 350m = 226.50 \end{cases}$$

Use the Addition Property of Equality to combine the equations, and solve the resulting equation for m.

$$\begin{array}{r} -4d + (-250m) = -191.50 \\ 4d + 350m = 226.50 \\ \hline 100m = 35 \\ m = 0.35 \end{array}$$

Then substitute 0.35 for m in the first equation to find d.

$$2d + 125(0.35) = 95.75$$
$$2d + 43.75 = 95.75$$
$$2d = 52$$
$$d = 26$$

The solution is $(26, 0.35)$. A check shows that the solution is correct. The car was rented for $26 a day plus $0.35 per mile.

ADDITIONAL EXAMPLE 2

Solve by elimination.

$\begin{cases} 3x - 4y = 10 \\ 5x + 7y = 3 \end{cases}$

$(2, -1)$

ADDITIONAL EXAMPLE 3

A company ordered new bookcases and file cabinets, which arrived in two shipments. One shipment contained 6 bookcases and 11 file cabinets and was accompanied by an invoice for $956. A second shipment contained 9 bookcases and 5 file cabinets and was accompanied by an invoice for $698. **Write a system of equations to represent the situation. Then use the system to find the cost of each bookcase and each file cabinet.**

Let b represent the cost in dollars of one bookcase. Let c represent the cost in dollars of one file cabinet.

$\begin{cases} 6b + 11c = 956 \\ 9b + 5c = 698 \end{cases}$

The cost of one bookcase is $42, and the cost of one file cabinet is $64.

Enrichment

Have students work in pairs. Each student should write three systems of equations: one that the student thinks is best solved by graphing, one that is best solved by substitution, and one that is best solved by elimination. Each should be identified with the appropriate direction line: *solve by graphing, solve by substitution,* or *solve by elimination.* Have partners exchange papers and solve each other's systems. They can then discuss whether the indicated method was appropriate.

Reteaching the Lesson

USING COGNITIVE STRATEGIES Give students the system at right. Point out that to solve this system by elimination, each equation must be multiplied by a different number. Ask students to identify sensible choices for the multipliers. Elicit the responses 4 and −6, −4 and 6, 7 and −2, and −2 and 7. Divide the class into four groups. Have each group solve the system by using a different pair of multipliers. Discuss the results with the class. **The solution is (−1, 4).**

$\begin{cases} 6x + 2y = 2 \\ 4x + 7y = 24 \end{cases}$

TRY THIS

a. (3, –1)
b. (4, 3)

Assess

Selected Answers
Exercises 10–13, 15–63 odd

ASSIGNMENT GUIDE

In Class	1–13
Core	15–33 odd, 37–39, 41–47 odd
Core Plus	14–34 even, 35–48
Review	49–64
Preview	65

 Extra Practice can be found beginning on page 738.

Mid-Chapter Assessment for Lessons 7.1 through 7.3 can be found on page 86 of the Assessment Resources.

internetconnect

GO TO: go.hrw.com
KEYWORD: MA1 Sports Opera

In this activity, students collect data on the spending habits of consumers and use the data to compare consumer spending on spectator sports and cultural performances. Students also use the data to predict spending in the future.

TRY THIS Solve by elimination. a. $\begin{cases} 2x - y = 7 \\ 5x + 4y = 11 \end{cases}$ b. $\begin{cases} 3a - 2b = 6 \\ 5a + 7b = 41 \end{cases}$

Choosing a Method

You have learned several methods for solving systems of linear equations. What method should you choose? There may be more than one good choice for solving a system.

SUMMARY OF METHODS FOR SOLVING SYSTEMS		
Example	Suggested method	Why
$\begin{cases} 6x + y = 10 \\ y = 5 \end{cases}$	Substitution	The value of y is known and can be easily substituted into the other equation.
$\begin{cases} 2x - 5y = -20 \\ 4x + 5y = 14 \end{cases}$	Elimination	$5y$ and $-5y$ are opposites and are easily eliminated.
$\begin{cases} 9a - 2b = -11 \\ 8a - 4b = 25 \end{cases}$	Elimination	b can be easily eliminated by multiplying the first equation by -2.
$\begin{cases} 324p + 456t = 225 \\ 178p - 245t = 150 \end{cases}$	Graph on a graphics calculator	The coefficients are large numbers, so other methods may be cumbersome.

Communicate

In the following systems, which terms are opposites? Explain how you would solve each system.

1. $\begin{cases} x + 7y = 13 \\ x - 7y = 5 \end{cases}$
2. $\begin{cases} 2x - 3y = 8 \\ 5x + 3y = 20 \end{cases}$
3. $\begin{cases} 2a + b = 6 \\ -2a - 3b = 8 \end{cases}$

Describe the steps necessary to solve each system by using the elimination method.

4. $\begin{cases} 2x + 3y = 9 \\ 3x + 6y = 7 \end{cases}$
5. $\begin{cases} 2x - 5y = 1 \\ 3x - 4y = -2 \end{cases}$
6. $\begin{cases} 9a + 2b = 2 \\ 21a + 6b = 4 \end{cases}$

Choose a method for solving each system, and explain your decision.

7. $\begin{cases} 3x + 2y = 5 \\ 5x - 2y = 7 \end{cases}$
8. $\begin{cases} 4x + 3y = 15 \\ 3x + y = 10 \end{cases}$
9. $\begin{cases} 5.6x + 8.77y = 31.17 \\ 7.3x - 2.34y = 26.86 \end{cases}$

334 LESSON 7.3

Guided Skills Practice

Solve each system by elimination, and check your solution.

10. $\begin{cases} 3x + 2y = 5 \\ 5x - 2y = 7 \end{cases}$ (EXAMPLE 1) (1.5, 0.25)

11. $\begin{cases} 4m + 3n = 13 \\ 2m - 4n = 1 \end{cases}$ (EXAMPLE 2) (2.5, 1)

12. $\begin{cases} 2j - 2k = 4 \\ 3j + 5k = -10 \end{cases}$ (EXAMPLE 3) (0, −2)

13. $\begin{cases} 2x + 3y = 1 \\ -3x - 4y = 0 \end{cases}$ (EXAMPLE 3) (−4, 3)

Practice and Apply

Solve each system of equations by elimination, and check your solution.

14. $\begin{cases} -x + 2y = 12 \\ x + 6y = 20 \end{cases}$ (−4, 4)

15. $\begin{cases} 2p + 3q = 18 \\ 5p - q = 11 \end{cases}$ (3, 4)

16. $\begin{cases} -4x + 3y = -1 \\ 8x + 6y = 10 \end{cases}$ $\left(\dfrac{3}{4}, \dfrac{2}{3}\right)$

17. $\begin{cases} 2m - 3n = 5 \\ 5m - 3n = 11 \end{cases}$ $\left(2, -\dfrac{1}{3}\right)$

18. $\begin{cases} 6x - 5y = 3 \\ -12x + 8y = 5 \end{cases}$ $\left(-\dfrac{49}{12}, -\dfrac{11}{2}\right)$

19. $\begin{cases} 4s + 3t = 6 \\ -2s + 6t = 7 \end{cases}$ $\left(\dfrac{1}{2}, \dfrac{4}{3}\right)$

Solve each system of equations by using any method.

20. $\begin{cases} 2m = 2 - 9n \\ 21n = 4 - 6m \end{cases}$ $\left(-\dfrac{1}{2}, \dfrac{1}{3}\right)$

21. $\begin{cases} -x - 7 = 3y \\ 6y = 2x - 14 \end{cases}$ $\left(0, -\dfrac{7}{3}\right)$

22. $\begin{cases} \dfrac{2}{3}x = \dfrac{2}{3} - \dfrac{1}{6}y \\ y = 3x - 12 \end{cases}$ $\left(\dfrac{16}{7}, -\dfrac{36}{7}\right)$

23. $\begin{cases} 0.6a = 3.2b + 4.6 \\ 2.9b = 0.3a + 4.8 \end{cases}$ approx. (36.8, 5.5)

24. $\begin{cases} p = 1.5b + 4 \\ 0.8p + 0.4b = 0 \end{cases}$ (−2, 1)

25. $\begin{cases} 2x = 3y - 12 \\ \dfrac{1}{3}x = 4y + 5 \end{cases}$ (−9, −2)

26. $\begin{cases} 2x + 3y = 2 \\ 3x - 4y = 16 \end{cases}$ $\left(\dfrac{56}{17}, -\dfrac{78}{51}\right)$

27. $\begin{cases} 2x - 5y = -14 \\ -7x + 4y = -5 \end{cases}$ (3, 4)

28. $\begin{cases} 3x - 2y = -26 \\ 5x + 3y = 9 \end{cases}$ $\left(-\dfrac{60}{19}, \dfrac{157}{19}\right)$

29. $\begin{cases} 11x + 2y = -8 \\ 8x + 3y = -5 \end{cases}$ $\left(-\dfrac{14}{17}, \dfrac{9}{17}\right)$

30. $\begin{cases} -7x - 3y = 10 \\ 2x + 2y = -8 \end{cases}$ $\left(\dfrac{1}{2}, -\dfrac{9}{2}\right)$

31. $\begin{cases} 3x - 2y = 2 \\ 4x - 7y = 33 \end{cases}$ (−4, −7)

32. $\begin{cases} 2x + 4y = 40 \\ 7x - 3y = 4 \end{cases}$ (4, 8)

33. $\begin{cases} 13x - 3y = -50 \\ 12x + 5y = 16 \end{cases}$ (−2, 8)

34. $\begin{cases} 2x - 7y = 20 \\ 5x + 8y = -1 \end{cases}$ (3, −2)

CHALLENGE

35. $\begin{cases} \dfrac{2}{3}x - \dfrac{3}{5}y = -\dfrac{17}{15} \\ \dfrac{8}{5}x - \dfrac{7}{6}y = -\dfrac{3}{10} \end{cases}$ $\left(\dfrac{257}{41}, \dfrac{363}{41}\right)$

CONNECTION

36. **GEOMETRY** The perimeter of a rectangle is 24, and its length is 3 times its width. Find the length and width of this rectangle. L = 9, W = 3

Use the equations $y = x - 2$ and $y = 2x$ for Exercises 37–39.

37. Find the common solution by graphing. (−2, −4)

38. Solve the system algebraically. (−2, −4)

39. What method did you use in Exercise 38 and why?

39. Answers may vary. Sample answer: Substitution, since both equations are solved for *y*.

Technology

A graphics calculator may be helpful for verifying answers in Exercises 14–19.

In Exercises 20–34, students may identify the graphics calculator as their method of choice. You may want to have them first use a different method to solve these systems and then use the graphics calculator to verify their answers.

Error Analysis

When subtracting one equation in a system from another, students often subtract only the term to be eliminated and add the remaining terms. For example, given the system $\begin{cases} 6x + 2y = 5 \\ 3x + 2y = 11 \end{cases}$, they may perform the subtraction mentally and write $9x = 16$. Encourage them to rewrite the entire system, showing the result after one of the equations is multiplied by −1. For the given system, they might write $\begin{cases} 6x + 2y = 5 \\ -3x - 2y = -11 \end{cases}$.

LESSON 7.3

56.
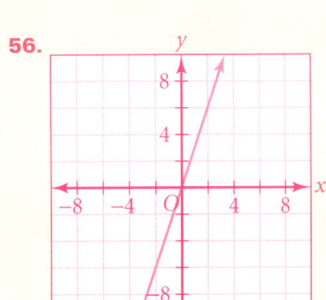
slope = 3

57.
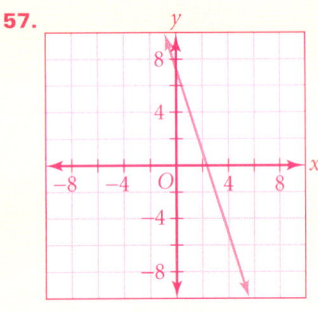
slope = –3

58.
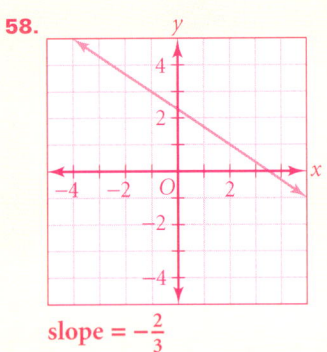
slope = $-\frac{2}{3}$

For Exercises 40–48 write a system of equations for each problem, and choose the best method to solve each system. Solve and check your answers.

APPLICATIONS

40. CONSUMER MATH Mrs. Jones is celebrating Gauss's birthday by having a pizza party for her two algebra classes. She orders 3 pizzas and 3 bottles of soda for $23.34 for her 2nd period class and 4 pizzas and 6 bottles of soda for $32.70 for her 7th period class. How much does each pizza and each bottle of soda cost? **pizza: $6.99   soda: $0.79**

41. INCOME As a parking attendant, John earns a fixed salary for the first 15 hours he works each week and then additional pay for any time over 15 hours. During the first week, John worked 25 hours and earned $240; the second week, he worked 22.5 hours and earned $213.75. What is John's weekly salary and overtime pay per hour? **$135 per week, $10.50 per hour overtime**

42. INVESTMENTS Mr. Moore sells his tractor for $6000. He makes two investments, one at a bank paying 5% interest per year and the rest in stocks yielding 9% interest per year. If he earns a total of $380 in interest in the first year, how much is invested at each rate? **$4000 at 5% $2000 at 9%**

43. BUSINESS At a local music store, single-recording tapes cost $6.99 and concert tapes cost $10.99. At the end of the day a total of 25 tapes were sold. If the sales for the day total $230.75 for single-recording tapes and concert tapes, find the number of each type of tape sold. **11 singles   14 concert tapes**

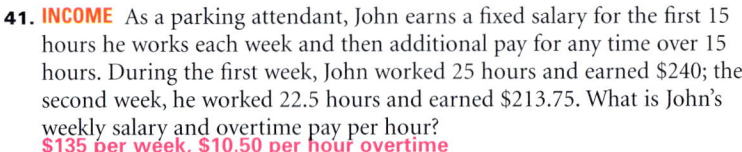

Some people invest the money they get from selling an item like a tractor.

44. FINANCE To rent an apartment, a one-time deposit is frequently required with the first month's rent. Roberto paid $900 the first month and a total of $6950 during the first year. Find the monthly rent and the amount of the deposit. **rent: $550, deposit: $350**

45. RECREATION The Shamrock Inn offers two holiday weekend specials. The first is 2 nights and 4 meals for $205, while the other is 3 nights and 8 meals for $342.50. At these rates, what is the room rate and the cost of a meal? **rooms: $67.50   meals: $17.50**

46. CONSUMER MATH Shopping at Super Sale Days, Martha buys her children 3 shirts and 2 pairs of pants for $85.50. She returns during the sale and buys 4 more shirts and 3 more pairs of pants for $123. What is the sale price of the shirts and pants? **shirts: $10.50   pants: $27**

47. BUSINESS At a local bakery, single-crust pies sell for $5.99 and double-crust pies sell for $9.99. The total number of pies sold on Monday was 25. If the sales total for Monday was $189.75 for these two types of pies, find the number of each type sold. **15 pies at $5.99   10 pies at $9.99**

48. SALES The owner of a health-food store sells a mixture of seeds and nuts. The seeds cost 20¢ per ounce, and the nuts cost 30¢ per ounce. Forty-two ounces of the mixture has a value of $11.00. How many ounces of seeds and nuts did the owner use in the mix? **16 ounces of seeds, 26 ounces of nuts**

59.
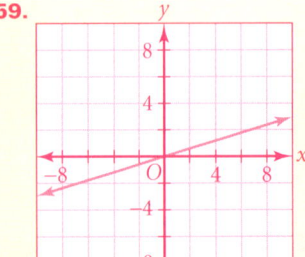
slope = $\frac{2}{7}$

60.
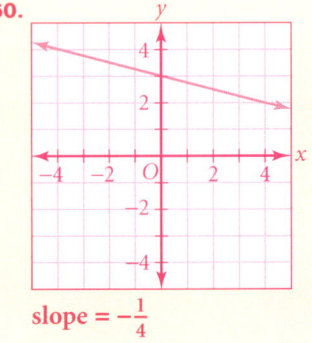
slope = $-\frac{1}{4}$

Look Back

Find the slope of each line. *(LESSON 5.4)*

49. $7x - 3y = 22$ $\frac{7}{3}$ **50.** $3x + 2y = 6$ $-\frac{3}{2}$ **51.** $y = 2x$ **2**

52. CULTURAL CONNECTION: AFRICA The following problem is in the Rhind papyrus by Ahmes: A bag contains equal weights of gold, silver, and lead purchased for 84 *sha'ty*. What is the amount (in *deben* weight) of each metal if the price for 1 *deben* weight is 6 *sha'ty* for silver, 12 *sha'ty* for gold, and 5 *sha'ty* for lead? *(LESSON 3.3)* $3\frac{15}{23}$ **deben**

Solve each equation for the variable. *(LESSONS 3.3 AND 3.4)*

53. $-5 = -x + 7$ **x = 12**   **54.** $3x - 6 = 2x + 1$ **x = 7**   **55.** $\frac{1}{2}x + 3 = 2$ **x = –2**

Draw a graph and give the slope for each linear equation. *(LESSON 5.5)*

56. $y = 3x$ **57.** $y = -3x + 7$ **58.** $3y + 2x = 7$ **59.** $7y = 2x$

60. $2x + 8y = 24$ **61.** $-7x = 3y + 9$ **62.** $-3y = 7x + 15$ **63.** $18x + 15y = 45$

64. Write the equation of the line that passes through the point $(-3, 8)$ and is parallel to $y = 0.8x - 7$. *(LESSON 5.6)* **y = 0.8x + 10.4**

Look Beyond

65. TECHNOLOGY Graph the lines represented by the equations below on the same coordinate plane and describe the graph. Use a graphics calculator if one is available.

a. $2x - 3y = 6$ **b.** $4x - 6y = 18$

The lines are parallel.

Include this activity in your portfolio.

An entertainment promotor for a magic show must make $161,000 from ticket sales in order to cover the costs and make a reasonable profit. The auditorium has 3000 seats, with 1800 seats on the lower floor and 1200 balcony seats. Attendees will pay $20 more for the lower-floor seats than for the balcony seats.

- Write a system of equations with *f* for the floor-seat ticket price and *b* for the balcony-seat ticket price.
- Solve the system graphically.
- Solve the system algebraically by using either substitution or elimination.
- What should the entertainment promoter charge for each type of ticket?
- Suppose that the show will take place at a large arena with 6000 balcony seats and 7000 lower-floor seats. Ticket sales must total $500,000. You may choose the price difference between the tickets. Determine appropriate ticket prices.

WORKING ON THE CHAPTER PROJECT

You should now be able to complete Activity 1 of the Chapter Project on page 360.

61.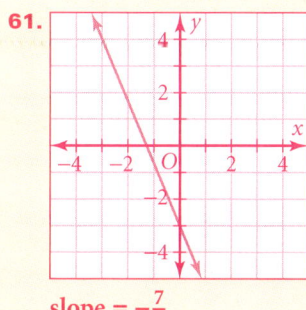
slope = $-\frac{7}{3}$

62.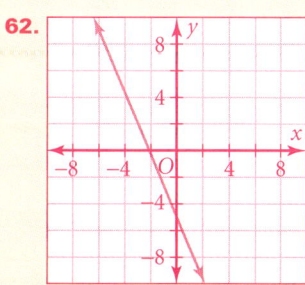
slope = $-\frac{7}{3}$

The answers to Exercises 63 and 65 can be found in Additional Answers beginning on page 810.

Look Beyond

Exercise 65 previews systems of equations that are inconsistent, meaning they have no solution. Students will study consistent and inconsistent systems in greater depth in Lesson 7.4.

ALTERNATIVE
Assessment

Portfolio Activity

The Portfolio Activity can be used as preparation for the Chapter Project or as a separate activity. In the Portfolio Activity on this page, students use systems of linear equations to investigate a problem involving ticket sales.

Answers to Portfolio Activities can be found in Additional Answers of the Teacher's Edition.

Student Technology Guide

Student Technology Guide

7.3 The Elimination Method

Suppose that you want to write formulas for x and y, given the system $\begin{cases} ax + by = e \\ cx + dy = f \end{cases}$.

To solve for x, multiply the first equation by d and the second equation by $-b$.

$adx + bdy = de$
$-bcx - bdy = -bf$
$\overline{adx - bcx = de - bf}$
$(ad - bc)x = de - bf$
$x = \frac{de - bf}{ad - bc}$

To solve for y multiply the first equation by c and the second equation by $-a$.

$acx + bcy = ce$
$-acx - ady = -af$
$\overline{bcy - ady = ce - af}$
$(bc - ad)y = ce - af$
$y = \frac{ce - af}{bc - ad}$

If you program these formulas into your graphics calculator, you will be able to automatically solve any system of two linear equations that are given in standard form.

- Press [PRGM] [NEW] **1: Create New** [ENTER],
- To name the program SOLVESYS, press [2nd] [ALPHA] [LN] [7] [ )] [6] [SIN] [LN] [1] [LN] [ENTER].
- Press [PRGM] [I/O] **2: Prompt** [ENTER] [ALPHA] [MATH] [,] [ALPHA] [MATRX] [,] [ALPHA] [SIN] [,] [ALPHA] [PRGM] [,] [ALPHA] [x⁻¹] [,] [ALPHA] [COS] [ENTER]
- Continue using [ALPHA] to enter (DE − BF)/(AD − BC), and then press [STO▶] [X,T,θ,n] [ENTER].
- Enter the formula (CE − AF)/(BC − AD), and then press [STO▶] [ALPHA] [1] [ENTER].
- Finally, press [2nd] [{] [X,T,θ,n] [,] [ALPHA] [1] [2nd] [}] [MATH] [MATH] **1: ▶Frac** [ENTER]

Your program should appear as shown at right. Press [2nd] [MODE] to exit programming mode.

Example: Solve the system $\begin{cases} 3x + 4y = 8 \\ 2x - 5y = -3 \end{cases}$.

Press [PRGM] [EXEC] **1: SOLVESYS** [ENTER] [ENTER] 3 [ENTER] [ENTER] 8 [ENTER] 2 [ENTER] [(-)] 5 [ENTER] [(-)] 3 [ENTER]

The solution is $(\frac{28}{23}, \frac{25}{23})$.

```
PROGRAM:SOLVESYS
:Prompt A,B,E,C,
D,F
:(DE-BF)/(AD-BC)
→X
:(CE-AF)/(BC-AD)
→Y
:{X,Y}▶Frac
```

Use the calculator program to solve each system of equations.

1. $\begin{cases} 5x - y = 18 \\ 3x + 2y = 7 \end{cases}$ $(\frac{43}{13}, -\frac{19}{13})$ 2. $\begin{cases} 8x + 7y = -5 \\ 3x + 9y = 11 \end{cases}$ $(-\frac{122}{51}, \frac{103}{51})$ 3. $\begin{cases} 4x - 6y = 12 \\ 5x + 47 = -13 \end{cases}$ $(-\frac{15}{23}, -\frac{56}{23})$

4. $\begin{cases} x + 7y = -25 \\ -2x + 3y = -18 \end{cases}$ **(3, –4)**   5. $\begin{cases} 9x - 2y = 4 \\ -7x + 5y = -3 \end{cases}$ $(\frac{14}{31}, \frac{1}{31})$ 6. $\begin{cases} 4x + 7y = -6 \\ 2x + 11y = 3 \end{cases}$ $(-\frac{29}{10}, \frac{4}{5})$

Prepare

NCTM PRINCIPLES & STANDARDS 1–3, 5–10

LESSON RESOURCES

- Practice 7.4
- Reteaching Master 7.4
- Enrichment Master 7.4
- Cooperative-Learning Activity 7.4
- Lesson Activity 7.4
- Problem Solving/Critical Thinking Master 7.4
- Teaching Transparency 36
- Teaching Transparency 37

QUICK WARM-UP

Solve.

1. $5a + (7 - 2a) = 4a - 1$
   $a = 8$

2. $9z - 2(2z - 3) = 6z + 6 - z$
   all real numbers

3. $-4n - 2 + 3n = 8 - n + 6$
   no real solutions

Also on Quiz Transparency 7.4

Teach

Why Place two strands of uncooked spaghetti on the stage of an overhead projector. Tell students to imagine that the strands are lines that extend to infinity at each end. Manipulate the strands as you lead students to see that there are exactly three ways the lines can relate to each other if they remain in the same plane—they can intersect at one point, be parallel, or coincide.

7.4 Consistent and Inconsistent Systems

Objectives

- Identify consistent and inconsistent systems of equations.
- Identify dependent and independent systems of equations.

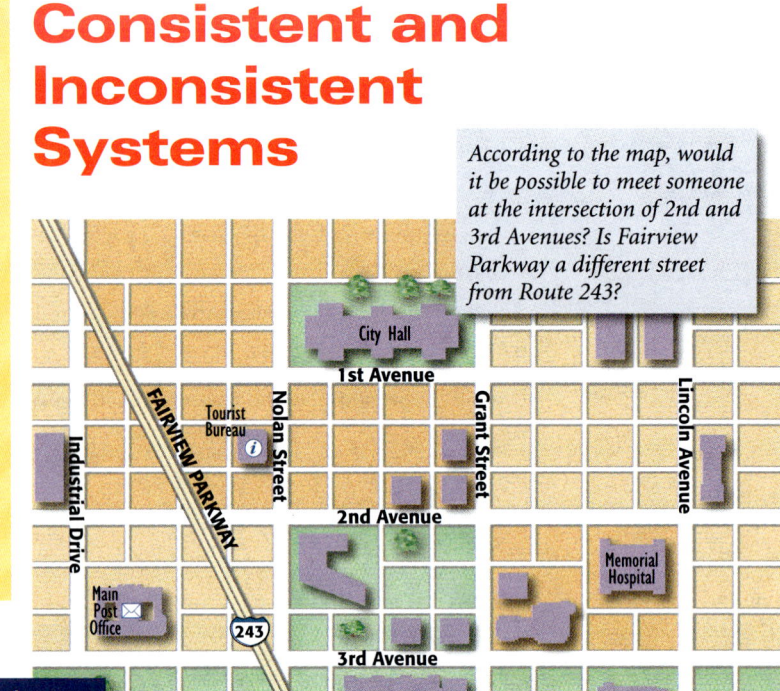

According to the map, would it be possible to meet someone at the intersection of 2nd and 3rd Avenues? Is Fairview Parkway a different street from Route 243?

Why Some real-world situations may have no solutions and some may have many solutions. A clear understanding of consistent and inconsistent systems can help you model and better understand these real-world situations.

The point of intersection for the graphs of the equations of a system is similar to the intersection of streets on the map above. However, some systems of equations are like the streets that do not intersect. These have no solution. Some systems of equations are like the street with two names. These have an infinite number of solutions.

Activity
Exploring No Solution and Many Solutions

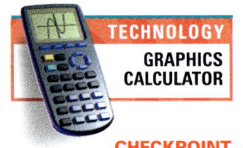

TECHNOLOGY
GRAPHICS CALCULATOR

You will need: a graphics calculator or graph paper

1. Use a graphics calculator or graph paper to graph this system.
   $\begin{cases} y = -3x + 4 \\ y = -3x + 8 \end{cases}$
   a. Do the lines intersect?

   CHECKPOINT ✓ **b.** Is there a solution for this system? Why or why not?

2. Use a graphics calculator or graph paper to graph this system.
   $\begin{cases} y = 2x - 3 \\ 3y = 6x - 9 \end{cases}$
   a. Describe the graph of this system.
   b. Do the lines intersect?

   CHECKPOINT ✓ **c.** Is there a solution for this system? Why or why not?

Alternative Teaching Strategy

INVITING PARTICIPATION Have students graph $x + y = 3$. Tell them to draw any line that intersects the graph at the point $A(-2, 5)$ and to write the equation of that line. They should then use the two equations to form a system. Now tell them to solve their system algebraically. They should arrive at $(-2, 5)$ as the solution.

Now repeat the activity. This time, however, tell students to draw any line parallel to $x + y = 3$. After they have attempted to solve their systems, ask them to describe what is different from the first solution. They should note that all of the variable terms were eliminated and that the remaining constants form a false statement. Tell them this indicates a system that is *inconsistent*. Discuss the definition of this term.

Repeat the activity again, this time asking students to write an equation whose graph is exactly the same as the graph of $x + y = 3$. Now the algebraic solution of their systems should yield a true statement. Tell them that this indicates a *dependent* system, and discuss the definition.

338 LESSON 7.4

Inconsistent and Consistent Systems

A system of equations may have one unique solution, infinitely many solutions, or no solution at all. Systems that have one or many solutions are called **consistent**. Systems with no solution are called **inconsistent**.

CONSISTENT SYSTEMS		INCONSISTENT SYSTEMS
One unique solution	Infinitely many solutions	No solution
$\begin{cases} x + 2y = 3 \\ 2x - y = 1 \end{cases}$	$\begin{cases} x + y = 3 \\ 3x + 3y = 9 \end{cases}$	$\begin{cases} -2x + y = -4 \\ -2x + y = 3 \end{cases}$
The lines intersect.	The lines are the same.	The lines are parallel.

EXAMPLE 1 Determine whether the system is consistent or inconsistent by solving it algebraically and graphically.

$$\begin{cases} y = 4 - x \\ y = 3x - 6 \end{cases}$$

SOLUTION

Method 1
Solve the system algebraically by substituting $3x - 6$ for y in the first equation.

$3x - 6 = 4 - x$ $y = 4 - 2.5$
$4x = 10$ $y = 1.5$
$x = 2.5$

The solution is $(2.5, 1.5)$.
The system is consistent.

Method 2
Draw a graph.

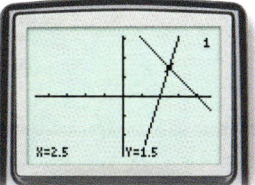

The system is consistent.

EXAMPLE 2 Determine whether the system is consistent or inconsistent by solving it algebraically and graphically.

$$\begin{cases} 2x - y = -1 \\ 4x - 2y = 4 \end{cases}$$

SOLUTION

Method 1
Use elimination to solve the system algebraically. Multiply the first equation by -2.

$\begin{cases} (-2)2x - (-2)y = (-2)(-1) \\ 4x - 2y = 4 \end{cases}$
$\Downarrow$
$\quad -4x + 2y = 2$
$\quad \underline{4x - 2y = 4}$
$\qquad \quad 0 \neq 6$

There is no ordered pair that satisfies the original equations. The system is inconsistent.

Method 2
Draw a graph.

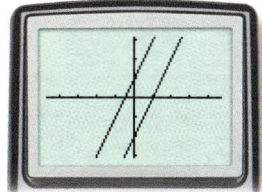

The system is inconsistent.

Inclusion Strategies

VISUAL LEARNERS Have students make a concept map like the one shown at right. Encourage them to add any other information that helps them to remember the different types of solutions, such as information about relationships between the y-intercepts or slopes.

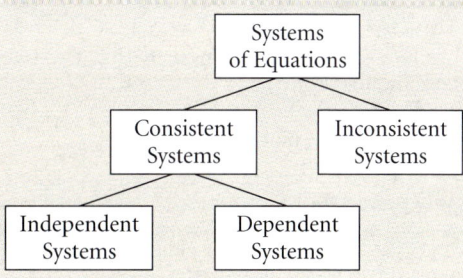

Activity Notes

This Activity sets the stage for the work with inconsistent and dependent systems that follows. Students should observe that it is possible for a system to have no solution or infinitely many solutions.

See Keystroke Guide, page 782.

CHECKPOINT ✓
1. b. No, the lines are parallel.
2. c. Yes, there are infinitely many solutions because the two lines are the same.

Use Teaching Transparency 36.

ADDITIONAL EXAMPLE 1

Determine whether the system is consistent or inconsistent by solving it algebraically and graphically.

$$\begin{cases} y = 5 - x \\ y = 2x - 7 \end{cases}$$

The solution is $(4, 1)$. The system is consistent.

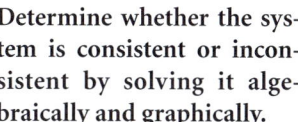

See Keystroke Guide, page 782.

ADDITIONAL EXAMPLE 2

Determine whether the system is consistent or inconsistent by solving it algebraically and graphically.

$$\begin{cases} 3x - 6y = -12 \\ x - 2y = 4 \end{cases}$$

The system is inconsistent.

☞ For a tip about the use of technology in Examples 1 and 2, see page 340.

LESSON 7.4

Teaching Tip

TECHNOLOGY If you are using a TI-82 or TI-83 graphics calculator, you can obtain the graphs shown in the solutions of Examples 1 and 2 by entering the equations as **Y1** and **Y2** in the **Y=** editor, pressing ZOOM, and choosing **4:ZDecimal**.

In Example 1, you can use either **TRACE** or the **CALCULATE** menu to identify the point of intersection. In Example 2, if you try to use the intersection option of the **CALCULATE** menu, you will get an error message. However, note that this error message does not always mean that the lines are parallel. You will receive the same message if the lines intersect at a point outside the viewing window.

TRY THIS
a. The system is inconsistent.
b. The system is consistent.

Math CONNECTION

PATTERNS IN DATA Students may need to review scatter plots and lines of best fit, which were covered in Lesson 1.6.

ADDITIONAL EXAMPLE 3

Based on the trend in the chart below, will the life expectancy of men and women in the United States ever be the same?

Life Expectancy in Years at Birth (United States)		
Year	Men	Women
1988	71.4	78.3
1989	71.7	78.5
1990	71.8	78.8
1991	72.0	78.9
1992	72.3	79.1

☞ The solution appears at the top of page 341.

340 LESSON 7.4

TRY THIS Determine whether each system is consistent or inconsistent by solving it algebraically and graphically.

a. $\begin{cases} y = 2x - 1 \\ y = 2x + 6 \end{cases}$
b. $\begin{cases} 2x - y = -1 \\ 4x + 2y = 7 \end{cases}$

Definition of Consistent System

A system of equations is consistent if it has one or infinitely many solutions.

Definition of Inconsistent System

A system of equations is inconsistent if it has no solution.

EXAMPLE 3

APPLICATION
DEMOGRAPHICS

Based on the trend in the chart, will the average ages when first married for men and women ever be the same?

Average Age When First Married

1970	1975	1980	1985	1990
20.6 22.5	20.8 22.7	21.8 23.6	23.0 24.8	24.0 25.9

[Source: Statistical Abstract of the United States, 1996]

The data shows that the average age when first married increased from 1970 to 1990 for both men and women.

CONNECTION
PATTERNS IN DATA

The average age when first married could also be studied for other time periods.

SOLUTION

The data in the chart can be compared by making two scatter plots on the same coordinate plane. The initial year, 1970, is assigned a value of 0.

Use a graphics calculator to determine the equation for a line of best fit for each set of data. With coefficients rounded to the nearest hundredth, the equations form the following system:

$\begin{cases} y = 0.18x + 20.24 & \text{Women's data} \\ y = 0.18x + 22.12 & \text{Men's data} \end{cases}$

The slope is the same in both equations of the system. The lines are parallel. A graph of the lines of best fit confirms that the system is inconsistent.

If the trend continues, the average ages when first married for men and women will not converge.

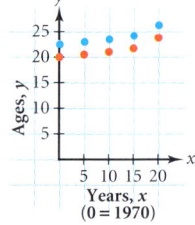

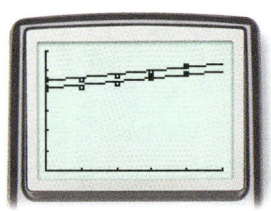

The calculator screen shows the lines of best fit for the data.

Enrichment

Give students the equation $3x - 4y = 8$. Have them write a second equation such that the two equations together form

a. an inconsistent system.
b. a dependent system.
c. an independent system.

Challenge students to write a generalization that describes all equations that will form each type of system.

Answers will vary. Sample generalizations:

a. any equation that can be written in the form $y = \frac{3}{4}x + b$, where b is any real number except -2

b. any equation that can be written in the form $3ax - 4ay = 8a$, where a is any real number except 0

c. any equation that can be written in the form $y = mx + b$, where $m \neq \frac{3}{4}$, or any equation that can be written in the form $x = a$

Independent and Dependent Systems

Consistent systems of equations can be divided further into two categories—independent and dependent.

An **independent system** has only one solution. One unique ordered pair, (x,y), satisfies both equations.

➡ $\begin{cases} x + 2y = 3 \\ 2x - y = 1 \end{cases}$ ➡

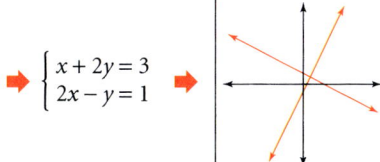

The graph shows that the lines intersect at one unique point.

A **dependent system** has infinitely many solutions. Every ordered pair that is a solution of the first equation is also a solution of the second equation.

➡ $\begin{cases} x - y = 2 \\ 2x - 2y = 4 \end{cases}$ ➡

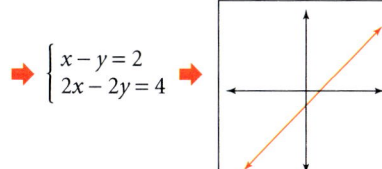

EXAMPLE 4 Determine whether the system is dependent or independent by solving it algebraically and graphically.

$\begin{cases} y = 2x - 1 \\ 3x - 1.5y = 1.5 \end{cases}$

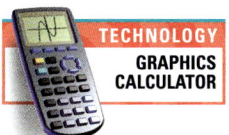
TECHNOLOGY
GRAPHICS CALCULATOR

• **SOLUTION**

Method 1

Solve the system algebraically by substituting $2x - 1$ for y in the second equation.

$3x - 1.5(2x - 1) = 1.5$
$3x - 3x + 1.5 = 1.5$
$1.5 = 1.5$

Since the substitution resulted in the expression $1.5 = 1.5$, the equation is true for any value of x. This means that all of the ordered pairs on the line are solutions to both equations. Thus, the system is dependent.

Method 2

Solve the system graphically by drawing a graph or by using a graphics calculator.

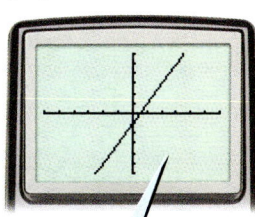

The graph shows that the equations represent the same line.

CRITICAL THINKING Use the equation $y = -3x + 4$ with different equations to create systems that are consistent and independent, consistent and dependent, and inconsistent.

Reteaching the Lesson

USING TABLES Write the following systems on the board or overhead:

1. $\begin{cases} y = 5x + 2 \\ 2y = 10x + 4 \end{cases}$
2. $\begin{cases} 3y = 9x - 6 \\ y = 3x + 2 \end{cases}$
3. $\begin{cases} 3y = 9x - 6 \\ 4y = 20x - 8 \end{cases}$
4. $\begin{cases} 3y = 9x - 6 \\ 4y = 20x - 16 \end{cases}$

Tell students to rewrite each equation in slope-intercept form. Have them graph each system and record observations about the slopes and y-intercepts in a chart like the one at right.

	Slope	y-intercept	Graph	Type of system	No. of solutions
1.	same	same	same line	consistent dependent	infinitely many
2.	same	diff.	parallel	inconsistent	0
3.	diff.	same	intersecting	consistent independent	1
4.	diff.	diff.	intersecting	consistent independent	1

Let x represent the year, with 1988 assigned a value of 0. Let y represent the life expectancy in years. Then a system representing lines of best fit for the data is

$\begin{cases} y = 0.2x + 71.4 \\ y = 0.2x + 78.3 \end{cases}$

If the trend continues, the life expectancy for men and women will never be the same.

Use Teaching Transparency 37.

ADDITIONAL EXAMPLE 4

Determine whether the system is dependent or independent by solving it algebraically and graphically.

$\begin{cases} y = 4x - 3 \\ 10x - 2.5y = 7.5 \end{cases}$

The system is dependent.

See Keystroke Guide, page 782.

Teaching Tip

TECHNOLOGY If you are using a TI-82 or TI-83 graphics calculator, you can obtain the graph shown in the solution of Example 4 as follows: First press WINDOW and make the following settings:

Xmin=-6   Ymin=-6
Xmax=6   Ymax=6
Xscl=1   Yscl=1

Then press Y= and enter the equations as Y1 and Y2 in the Y= editor. Press GRAPH to view the graph.

☞ For a sample answer to Critical Thinking, see page 342.

CRITICAL THINKING

Answers may vary. Sample:

$\begin{cases} y = -3x + 4 \\ y = \frac{1}{3}x + 4 \end{cases}$ consistent and independent

$\begin{cases} y = -3x + 4 \\ 2y = -6x + 8 \end{cases}$ consistent and dependent

$\begin{cases} y = -3x + 4 \\ y = -3x - 7 \end{cases}$ inconsistent

Assess

Selected Answers
Exercises 6–10, 11–63 odd

ASSIGNMENT GUIDE

In Class	1–10
Core	11–31 odd, 32–37
Core Plus	12–28 even, 29–41
Review	42–64
Preview	65

✏ Extra Practice can be found beginning on page 738.

Exercises

Communicate

Explain how to identify each system of equations as inconsistent, dependent, or independent.

1. $\begin{cases} y = 2x - 3 \\ 3y = 6x - 9 \end{cases}$
2. $\begin{cases} y = -3x + 2 \\ y = 2x + 2 \end{cases}$
3. $\begin{cases} x + y = 4 \\ x + y = 5 \end{cases}$

4. Explain how to write an equation that would form a dependent system with the equation $y = 2x + 3$.

5. Explain how to write an equation that would form an inconsistent system with the equation $y = -3x + 4$.

Guided Skills Practice

Determine whether each system is consistent or inconsistent by solving it graphically and algebraically. **(EXAMPLES 1, 2, AND 3)**

6. $\begin{cases} y = -4x + 6 \\ 2y = -8x - 8 \end{cases}$ incons.
7. $\begin{cases} c + d = 7 \\ 2c + 3d = 7 \end{cases}$ cons.
8. $\begin{cases} y = 0.34x + 0.45 \\ y = 0.34x + 4.5 \end{cases}$ incons.

9. Use the equation $y = 4x + 3$ with another equation to create a system that is dependent. **(EXAMPLE 4)**

10. Use the equation $y = -2x + 7$ with another equation to create a system that is independent. **(EXAMPLE 4)**

Practice and Apply

Solve each system algebraically. Identify each system as consistent and dependent, consistent and independent, or inconsistent.

11. $\begin{cases} x + y = 7 \\ x + y = -5 \end{cases}$ no sol. incons.
12. $\begin{cases} y = \frac{1}{2}x + 9 \\ 2y - x = 1 \end{cases}$ no sol. incons.
13. $\begin{cases} 3b = 2a - 24 \\ 4b = 3a - 3 \end{cases}$ (−87, −66) cons. ind.

14. $\begin{cases} y = -2x - 4 \\ 2x + y = 6 \end{cases}$ no sol. incons.
15. $\begin{cases} 2f + 3g = 11 \\ f - g = -7 \end{cases}$ (−2, 5) cons. ind.
16. $\begin{cases} 4x = y + 5 \\ 6x + 4y = -9 \end{cases}$ $\left(\frac{1}{2}, -3\right)$ cons. ind.

17. $\begin{cases} 4r + t = 8 \\ t = 4 - 2r \end{cases}$ (2, 0) cons. ind.
18. $\begin{cases} y = \frac{3}{2}x - 4 \\ 2y - 8 = 3x \end{cases}$ no sol. incons.
19. $\begin{cases} 3p = 3m - 6 \\ p = m + 2 \end{cases}$ no sol. incons.

20. $\begin{cases} 2y + x = 8 \\ y = 2x + 4 \end{cases}$ (0, 4) cons. ind.
21. $\begin{cases} k + 6h = 8 \\ k = -6h + 8 \end{cases}$ inf. sol. cons. dep.
22. $\begin{cases} x - 5y = 10 \\ -5 = -x + 6 \end{cases}$ $\left(11, \frac{1}{5}\right)$ cons. ind.

23. $\begin{cases} 8x + 3y = 7 \\ 3x + 2y = 7 \end{cases}$ (−1, 5) cons. ind.
24. $\begin{cases} 2x - \frac{3}{4}y = 17 \\ -3x - \frac{5}{8}y = -11 \end{cases}$ $\left(\frac{151}{28}, -\frac{58}{7}\right)$ cons. ind.
25. $\begin{cases} -6x + 8y = -16 \\ -2x - 3y = -28 \end{cases}$ (8, 4) cons. ind.

26. $\begin{cases} 8x - 3y = -42 \\ \frac{2}{3}x + \frac{1}{3}y = 0 \end{cases}$ (−3, 6) cons. ind.
27. $\begin{cases} \frac{3}{5}x - \frac{8}{3}y = -105 \\ \frac{2}{3}x - \frac{4}{9}y = -26 \end{cases}$ (−15, 36) cons. ind.
28. $\begin{cases} \frac{5}{7}x - \frac{2}{3}y = 6 \\ \frac{3}{14}x - 2y = -9 \end{cases}$ (14, 6) cons. ind.

9. Answers may vary. Sample answer:
$\begin{cases} y = 4x + 3 \\ 2y = 8x + 6 \end{cases}$

10. Answers may vary. Sample answer:
$\begin{cases} y = -2x + 7 \\ y = \frac{1}{2}x + 7 \end{cases}$

Practice

7.4 Consistent and Inconsistent Systems

Solve each system algebraically.

1. $\begin{cases} x + y = 5 \\ -2x - 2y = -10 \end{cases}$ infinitely many solutions
2. $\begin{cases} x + 2y = 6 \\ x + 2y = -4 \end{cases}$ no solution
3. $\begin{cases} 2x - y = 1 \\ x + y = 5 \end{cases}$ (2, 3)
4. $\begin{cases} 3x + y = 5 \\ 3x + y = -2 \end{cases}$ no solution
5. $\begin{cases} 2y = -x - 4 \\ 2x = -4y - 8 \end{cases}$ infinitely many solutions
6. $\begin{cases} 4x - 2y = -2 \\ y = 2x + 1 \end{cases}$ no solution
7. $\begin{cases} 3x - 2y = 6 \\ -6x + 4y = -12 \end{cases}$ infinitely many solutions
8. $\begin{cases} x - y = 2 \\ -x + y = -2 \end{cases}$ infinitely many solutions
9. $\begin{cases} 4x + y = 10 \\ y = -4x + 5 \end{cases}$ no solution
10. $\begin{cases} 2x + y = 3 \\ 4x = 6 - y \end{cases}$ (1.5, 0)

Determine whether each system is dependent, independent, or inconsistent.

11. $\begin{cases} 2x + y = 8 \\ y = -2x + 8 \end{cases}$ dependent
12. $\begin{cases} 3x - y = 3 \\ y = 3x - 9 \end{cases}$ inconsistent
13. $\begin{cases} y + 4 = -5x \\ y = 6x - 7 \end{cases}$ independent
14. $\begin{cases} 3x - 2y = -5 \\ 4y = 6x + 10 \end{cases}$ dependent
15. $\begin{cases} 6x - 2y = -10 \\ y = 3x + 2 \end{cases}$ inconsistent
16. $\begin{cases} y + 6x = 18 \\ y = -7x + 2 \end{cases}$ independent
17. $\begin{cases} y + 4x = 3 \\ 2y = 8x + 6 \end{cases}$ independent
18. $\begin{cases} x - y = 3 \\ 2y = 2x + 6 \end{cases}$ inconsistent
19. $\begin{cases} y = 3x + 1 \\ 3x = 1 - y \end{cases}$ independent
20. $\begin{cases} 3x = y + 2 \\ 3y = x + 2 \end{cases}$ independent
21. $\begin{cases} 4 = x - y \\ 8 + 2y = 2x \end{cases}$ dependent
22. $\begin{cases} x + 5y = 10 \\ y = 5x - 10 \end{cases}$ independent
23. $\begin{cases} 4x - 2y = 1 \\ 2x - 4y = -1 \end{cases}$ independent
24. $\begin{cases} 5x - 2y = 4 \\ 5x - 4 = 2y \end{cases}$ dependent

342 LESSON 7.4

CONNECTIONS

29. The equations for lines TU and UV form a dependent system. If the slope of line UV is -3 and the points $T(-3, p)$ and $U(5, 2-p)$ are given, find the value of p. **$p = 13$**

30. **NUMBER THEORY** One number is 24 more than another number. If the sum of the numbers is 260, what are the numbers? **118 and 142**

31. **GEOMETRY** The sum of the measures of two angles is 180°. What is the measure of each angle if one angle is 30° more than twice the other? **50° and 130°**

COORDINATE GEOMETRY Use the graph below to answer Exercises 32–37.

32. Write equations for lines AB and CD. $\overleftrightarrow{AB}$: $y = -\frac{3}{2}x + 10$ $\overleftrightarrow{CD}$: $y = -\frac{3}{2}x + 3$

33. What kind of system do lines AB and CD represent? Explain your answer. **inconsistent**

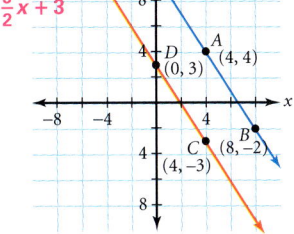

34. Write the equation of a line that forms an independent system with line AB and that contains point A. Graph your line.

35. Write the equation of a line that forms a dependent system with line CD and that contains point D. Graph your line.

36. What is the point of intersection of $y = 6$ and line AB? of $y = 6$ and line CD? $\left(\frac{8}{3}, 6\right)$; $(-2, 6)$

37. Write the equation of a line that forms an inconsistent system with $y = 6$.

38. **GEOMETRY** Triangle ABC below has a perimeter of 21, while rectangle $PQRS$ has a perimeter of 48. Write a system of equations for the perimeters of the figures, and find the dimensions of the rectangle.

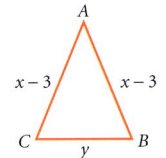 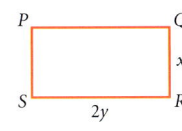 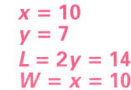

$x = 10$
$y = 7$
$L = 2y = 14$
$W = x = 10$

APPLICATION

Vegiland:
$y = 500x - 949{,}930$

Megaveggies:
$y = 500x - 946{,}250$

SALES The chart shows the increase in total sales of two small produce companies from 1980 to 2000.

39. Write a system of equations that can be used to predict the total sales, y, in a given year, x, for each company.

40. If the trend continues, will the sales of Vegiland ever equal the sales of Megaveggies? If so, when? **Never, the system is inconsistent.**

	Vegiland Sales	Megaveggies Sales
1980	$40,070	$43,750
1985	$42,570	$46,250
1990	$45,070	$48,750
1995	$47,570	$51,250
2000	$50,070	$53,750

Technology

A graphics calculator may be helpful for Exercises 11–28. You may want to have students first complete the exercises with paper and pencil and then use the calculator to verify their answers.

Error Analysis

When students arrive at a result such as $5 = 5$, they may assign that value to one of the variables and proceed to solve the system on that basis. Point out that in a statement like $5 = 5$, both variables have been eliminated. Remind them that this means there is a "special solution." In this case, it indicates that there are infinitely many solutions.

34. Answers may vary. Sample answer: $y = 4$

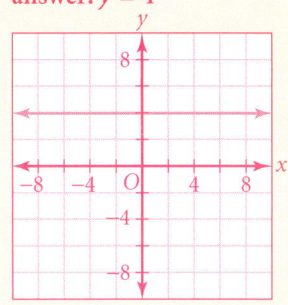

35. Answers may vary. Sample answer: $2y = -3x + 6$

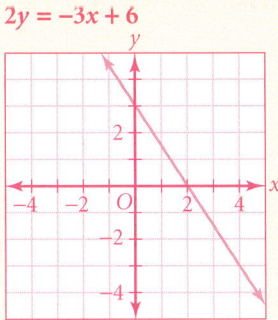

37. Answers may vary. Sample answer: $y = 8$

Student Technology Guide

7.4 Consistent and Inconsistent Systems

Suppose that you want to find out whether each system below is consistent and dependent, consistent and independent, or inconsistent.

a. $\begin{cases} y = 2x + 1 \\ y = 2x + 3 \end{cases}$ b. $\begin{cases} y = -\frac{1}{3}x + 1 \\ y = 2x + 3 \end{cases}$ c. $\begin{cases} y = -x + 1 \\ y = -\frac{1}{2}x + 6 \end{cases}$

a. Press Y= 2 X,T,θ,n + 1 ENTER 2 X,T,θ,n + 3 GRAPH. The lines are parallel so the system is inconsistent.

b. Note that this system is the same as the previous one, except that the 2 in the first equation has been changed to $-\frac{1}{3}$.

To edit Y1, press Y= 2nd DEL (-) 1 ÷ 3 X,T,θ,n + 1 GRAPH.
The lines intersect at a point, so the system is consistent and independent.

c. As in part b, enter the new system by editing the previous one. Press Y= ▶ DEL DEL DEL DEL DEL ENTER 2nd DEL (-) 1 ÷ ▶ ▶ ▶ 6 GRAPH.
You might get a graph like the one at left below, suggesting dependence. However, pressing TRACE and the ▲ or ▼ button shows that the y-value changes. For a better view, change to a viewing window such as that shown at right below. The system is consistent and independent.

Use a graphics calculator to identify each system as consistent and dependent, consistent and independent, or inconsistent.

1. $\begin{cases} y = \frac{1}{4}x - 3 \\ y = \frac{1}{4}x + 25 \end{cases}$ 2. $\begin{cases} y = x - 2 \\ y = 2x + 4 \end{cases}$ 3. $\begin{cases} y = \frac{1}{3}x + 4 \\ y = \frac{x + 12}{3} \end{cases}$

inconsistent consistent and independent consistent and dependent

In Exercise 65, students must rewrite the formula $d = rt$ as $t = \frac{d}{r}$ and then use this new formula to calculate two different times. Time-and-distance problems will be explored in greater depth when students study rational equations in Chapter 11.

APPLICATION

41. INCOME Suppose that you are offered the choice of two different jobs. The first has a starting salary of $2500 per month with a $750-per-month expense account and an annual bonus of $800. The other job pays $3250 per month and has two semiannual bonuses of $350 each. Explain which of the two jobs you would take and why.
Choose Job 1. It pays $100 more per year.

Look Back

Find the value of each expression when $a = -1$, $b = -3$, and $c = 4$.
(LESSON 2.4)

42. abc **12**
43. $-abc$ **−12**
44. $-(a + b)$ **4**
45. $-a + bc$ **−11**
46. $a^2 + b^2$ **10**
47. $a + b + c$ **0**
48. $-(a + b + c)$ **0**

Solve for c. **(LESSON 3.3 AND 3.6)**

49. $2c - 5 = 15$ **10**
50. $-3c + 4 = -14$ **6**
51. $4c + 7 = -5$ **−3**
52. $ac + b = d$ $\frac{d-b}{a}$

53. $3d - a$
53. $\frac{a+c}{d} = 3$
54. $ad - c = -b$
55. $-b(a - c) = d$
56. $c(b - d) = a$

54. $b + ad$

55. $\frac{d + ab}{b}$

56. $\frac{a}{b-d}$

57. Graph the equation $2x + 5y = 7$ and give the slope and y-intercept.
(LESSON 5.4) slope $= -\frac{2}{5}$, y-intercept $= \frac{7}{5}$

58. Write $3x + 2y = 6x + 9$ in standard form. **(LESSON 5.5)** $3x - 2y = -9$

Solve each system of equations. **(LESSONS 7.2 AND 7.3)** $\left(-\frac{26}{7}, -\frac{8}{7}\right)$

59. $\begin{cases} 2x + 3y = 6 \\ x + y = 4 \end{cases}$ **(6, −2)**
60. $\begin{cases} -t + 3v = 9 \\ t - v = 12 \end{cases}$ $\left(\frac{45}{2}, \frac{21}{2}\right)$
61. $\begin{cases} -4p - 8q = 24 \\ p - 5q = 2 \end{cases}$

62. $\begin{cases} x + y = 3 \\ 2x = 7y + 7 \end{cases}$
63. $\begin{cases} y = 2x + 1 \\ x = 2y + 1 \end{cases}$ **(−1, −1)**
64. $\begin{cases} 2y + 3x = 7 \\ -3y + 2x = -3 \end{cases}$

$\left(\frac{28}{9}, -\frac{1}{9}\right)$ $\left(\frac{15}{13}, \frac{23}{13}\right)$

Look Beyond

65. To get to work, Charles takes Interstate 87 for 15 miles at 60 miles per hour and then takes Central Avenue for 12 miles at 40 mph. How many minutes does it take Charles to get to work? **33 minutes**

57.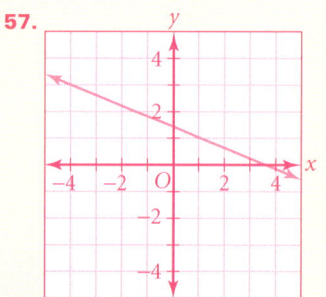

7.5 Systems of Inequalities

Objectives
- Graph the solution to a linear inequality.
- Graph the solution to a system of linear inequalities.

Why Systems of inequalities are especially useful in discovering ranges of possibilities or, in some cases, for choosing the best option. For example, the producers of a stage play can use systems of inequalities to prevent cost from exceeding the budget.

Shakespeare's last play, *The Tempest*, is a favorite among costume and set designers because of the creative challenges and opportunities it presents. Costumes and sets are two of the major expenses in a production of the play.

Prepare

NCTM PRINCIPLES & STANDARDS
1–3, 5–10

LESSON RESOURCES
- Practice 7.5
- Reteaching Master 7.5
- Enrichment Master 7.5
- Cooperative-Learning Activity 7.5
- Lesson Activity 7.5
- Problem Solving/Critical Thinking Master 7.5
- Teaching Transparency 38

QUICK WARM-UP

Solve each inequality. Graph the solution on a number line.

1. $-3s \geq 3$ $s \leq -1$
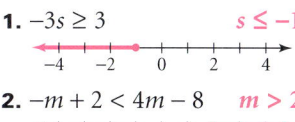

2. $-m + 2 < 4m - 8$ $m > 2$

3. $b + 4 < 2$ or $6 < 2b$
$b < -2$ or $b > 3$
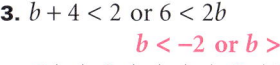

4. $-9 \leq 2q - 1 < 5$
$-4 \leq q < 3$
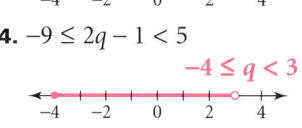

Also on Quiz Transparency 7.5

Two-Variable Inequalities

APPLICATION
FINANCE

The amount of money spent by a theater company on costumes and sets for its production of *The Tempest* must not exceed its budget of $5000. If x represents the amount of money for costumes and y the amount for sets, the situation can be represented by the following inequality:

$$x + y \leq 5000$$

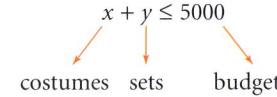
costumes   sets   budget

Notice that the form of this inequality is like a *linear equation* with the equal sign replaced by an inequality symbol. Therefore, this inequality is a **linear inequality**.

The inequalities you studied in Chapter 6 were graphed on number lines because they had just one variable. Inequalities with two variables are graphed on a coordinate plane.

Teach

Why Give students the following situation:

Ed has $100 to spend at a sale. Tapes cost $3 apiece. CDs cost $9 apiece. Ed can buy x tapes and y CDs.

Ask students for possible values of x and y, and list them as ordered pairs, (x, y). Point out that the ordered pairs are solutions to the inequality $3x + 9y \leq 100$.

Alternative Teaching Strategy

USING VISUAL MODELS Use this activity as a means of introducing systems of inequalities after students have learned how to graph linear inequalities.

Have students work in pairs. Provide each pair with two transparencies, each showing an identical blank coordinate plane. Assign each pair of students a system of linear inequalities in two variables.

Have each partner graph a different inequality from the assigned system. Each should use a marker of a different color to shade the graph. They should then place their transparencies together, matching the axes, in order to form the solution to the system.

Repeat this activity as many times as necessary for students to grasp the concept of graphing systems of inequalities. Be sure to include systems in which both boundary lines are dashed and systems in which just one boundary line is dashed.

LESSON 7.5 **345**

Activity Notes

Any line separates a plane into three regions—the line itself and two *half-planes*. In this Activity, students discover that graphing a linear inequality involves graphing a linear equation that serves as a *boundary* and then deciding which half-plane contains points whose coordinates satisfy the inequality.

See Keystroke Guide, page 782.

CHECKPOINT ✔

2. *S* is over the budget and *R* is under the budget. Any point above the line represents spending over the budget, while any point under the line represents spending under the budget.

Teaching Tip

TECHNOLOGY If you are using a TI-83 graphics calculator, you can graph linear inequalities by using the *graph style* feature of the **Y=** editor. For instance, to graph the inequality in the Activity, press WINDOW and make these settings:

Xmin=−1000   Ymin=−1000
Xmin=6000   Ymax=6000
Xscl=1000   Yscl=1000

Then go to the **Y=** editor and enter **5000−X** for **Y1**. Press ◄ to move the cursor to the graph style icon to the left of the equal sign. Press ENTER ENTER ENTER to select "shade below" for the graph style. Then press GRAPH.

Activity
Modeling a Linear Inequality With a Graph

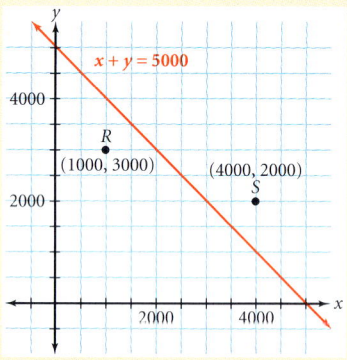

You will need: a graphics calculator or graph paper

Refer to the theater problem from the beginning of the lesson. An inequality that models the situation is $x + y \leq 5000$.

A graph of the inequality will show the possible solutions to the problem.

1. Replace the inequality sign with an equal sign. Then graph the resulting equation as shown at right. (Because the amount of money spent on costumes and sets is positive, draw the line only in the first quadrant.)

   The line that you graphed represents all combinations of costume and set expenses that add up to *exactly* $5000.

CHECKPOINT ✔

2. Look at points *R* and *S*. Which one goes over budget? Which one goes under? What can you conjecture about other points that are above or below the line? Test your conjecture for a few randomly selected points.

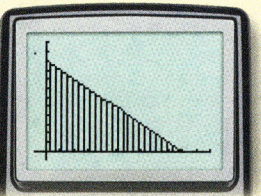

3. Shade the part of the plane where the expenses are either on or under the budget of $5000. (Because the amount of money spent on costumes and sets is positive, shade only in the first quadrant.) *The graph shows where the inequality is true.* The shaded part of your graph should include the straight line, which is known as a boundary line, because the expenses can equal $5000.

Graphing Linear Inequalities in Two Variables

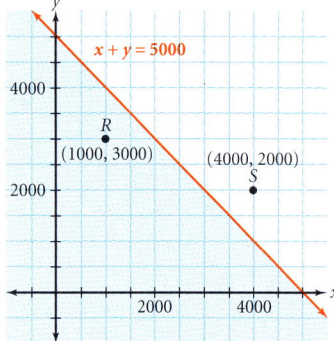

The graph at left shows the solution to the inequality $x + y \leq 5000$. The equation $x + y = 5000$ forms a **boundary line** that divides the coordinate plane into two half-planes. The boundary line is included in the solution because the inequality symbol, $\leq$, contains the equal sign. The shaded area and the boundary line contain all the ordered pairs that make the inequality true.

When the inequality symbol is $\leq$ or $\geq$, the points on the boundary line are included in the solution, and the line is solid.

When the inequality symbol is $<$ or $>$, the points on the boundary line are not included in the solution, and the line is dashed.

Interdisciplinary Connection

BUSINESS Invite a member of your local business community to talk to your class about the ways that systems of inequalities are used in his or her work. Ask your guest to bring data that students can use to write problems that involve systems of inequalities. Encourage students to take detailed notes during the presentation. Then have students work in pairs to write several problems about the data presented by your guest.

EXAMPLE 1 Graph the inequality $x - 2y < 4$.

SOLUTION

Graph the boundary line by using the *equation* $x - 2y = 4$. Because the inequality symbol $<$ does not include the equal sign, use a dashed line for the boundary.

Solve the inequality in terms of positive y, with y on the left side.

$$x - 2y < 4$$
$$-2y < 4 - x$$
$$-y < 2 - 0.5x$$
$$y > -2 + 0.5x$$

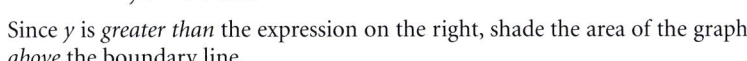

Since y is *greater than* the expression on the right, shade the area of the graph *above* the boundary line.

Check the solution by selecting a point in the shaded region to see if it satisfies the inequality. For the point (0, 0):

$$x - 2y < 4$$
$$0 - 2(0) < 4$$
$$0 < 4 \quad \text{True}$$

> You can choose any point in the region, but (0, 0) is an easy one to check.

The solution is correct.

EXAMPLE 2 Graph the inequality $2x - y \geq -3$.

SOLUTION

Graph the boundary line by using the equation $2x - y = -3$. Because the inequality symbol $\geq$ includes the equal sign, use a solid line for the boundary.

Solve the inequality in terms of y, with y on the left side.

$$2x - y \geq -3$$
$$-y \geq -3 - 2x$$
$$y \leq 3 + 2x$$

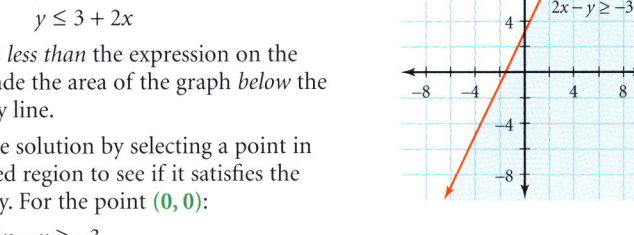

Since y is *less than* the expression on the right, shade the area of the graph *below* the boundary line.

Check the solution by selecting a point in the shaded region to see if it satisfies the inequality. For the point (0, 0):

$$2x - y \geq -3$$
$$2(0) - 0 \geq -3$$
$$0 > -3 \quad \text{True}$$

The solution is correct.

TRY THIS Graph each inequality. **a.** $3y < 2x + 7$ **b.** $y \geq 0.5x - 3$

Inclusion Strategies

ENGLISH LANGUAGE DEVELOPMENT Review translations of words into mathematical symbols. For example, remind students that the symbol $\leq$ can represent the phrase *no more than* and the symbol $\geq$ can represent *at least*. Make a list of all the phrases students can think of that are associated with the symbols $>$, $<$, $\geq$, and $\leq$. Have them write several statements that incorporate the phrases. Then ask students to write their statements using mathematical symbols.

VISUAL LEARNERS Suggest that students use a different color to graph each inequality in a system. The intersection of the two graphs will then be easier to identify.

ADDITIONAL EXAMPLE 1

Graph the inequality $2x + 3y \geq 6$.

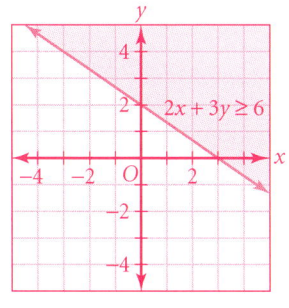

ADDITIONAL EXAMPLE 2

Graph the inequality $3x - y > 2$.

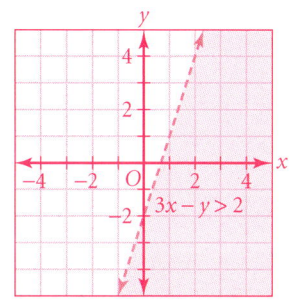

TRY THIS

a.
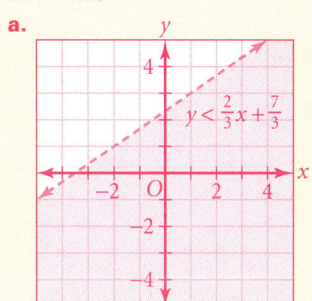

b.
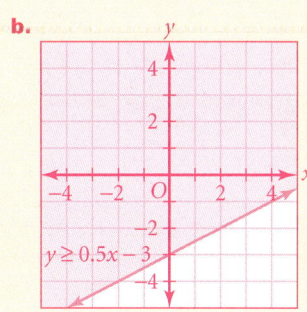

LESSON 7.5

Use Teaching Transparency 38.

ADDITIONAL EXAMPLE 3

Solve by graphing.
$$\begin{cases} y \geq \tfrac{3}{4}x - 3 \\ y < -2x + 1 \end{cases}$$

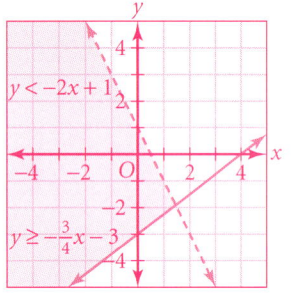

TRY THIS

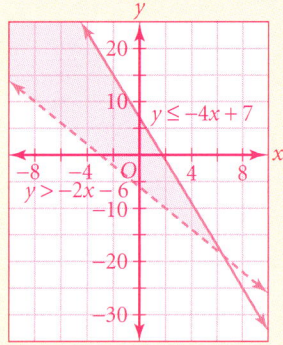

Teaching Tip

TECHNOLOGY If you are using a TI-82 graphics calculator, you can use its **DRAW** feature to graph systems of linear inequalities in a limited way. For the system in Example 3, first clear all equations from the **Y=** editor. Press `2nd` `PRGM` to access the **DRAW** menu. Choose `7:Shade(`. To the right of the parenthesis, enter `–0.5X–1, X+3, 2)`. Then press `ENTER` to display the graph. Notice that the correct region is shaded. However, it is not possible to draw a dashed boundary line at $y = x + 3$.

Systems of Linear Inequalities

A *system of linear inequalities* is like a system of linear equations, but with inequalities rather than equations. The solution of a system of linear inequalities is the *intersection* of the solutions of each inequality. Every point in the intersection region satisfies the system.

EXAMPLE 3 Solve by graphing. $\begin{cases} y \leq -\tfrac{1}{2}x - 1 \\ y > x + 3 \end{cases}$

- **SOLUTION**

  Graph each inequality, and combine the graphs on one coordinate plane.

  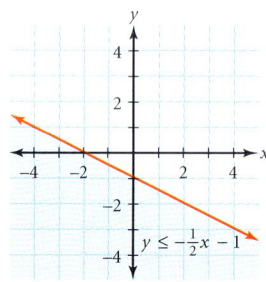 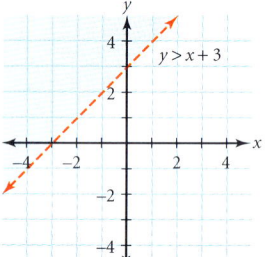

  The solution of the system is the intersection of the two shaded regions in the combined graph at right, including the part of the solid boundary line that the intersection region touches.

  Try a point from the intersection region, such as (–4, 0), to test in both inequalities.

  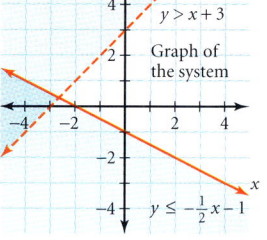

  $2y \leq -\tfrac{1}{2}x - 1$ $\quad y > x + 3$
  $0 \leq -\tfrac{1}{2}(-4) - 1$ $\quad 0 > -4 + 3$
  $0 \leq 1$ True $\quad 0 > -1$ True

TRY THIS Solve by graphing. $\begin{cases} y \leq -4x + 7 \\ y > -2x - 6 \end{cases}$

EXAMPLE 4 During the summer, Kara wants to earn at least $126 per week by working part-time. She can make $6 an hour doing yard work and $7 an hour working at the library. She can work no more than 30 hours each week. **Write a system of inequalities to represent Kara's situation, and use it to find all of the combinations of hours and jobs that she can work.**

Enrichment

Give students several systems of three or four inequalities to graph. Two possible systems and their graphs are shown below. Have students write and graph the solution of several systems of their own. Then have students exchange papers with a partner and solve each other's systems.

1. $\begin{cases} x \geq 0 \\ y \geq 0 \\ x + y \leq 4 \end{cases}$

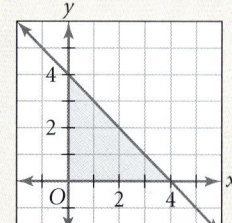

2. $\begin{cases} x \geq 0 \\ y \leq 0 \\ x \leq 3 \\ y > x - 4 \end{cases}$

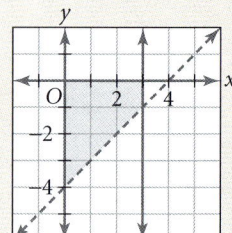

348 LESSON 7.5

PROBLEM SOLVING

APPLICATION
INCOME

Kara works part of the time at the library and part of the time doing yard work.

● **SOLUTION**

Define the variables.
Let m represent the number of hours doing yard work.
Let l represent the number of hours working at the library.
Write and graph inequalities that satisfy the conditions in the problem.

Kara can work no more than 30 hours per week.

$$m + l \leq 30$$

The shaded area shows the possible number of hours that Kara can work at each job.

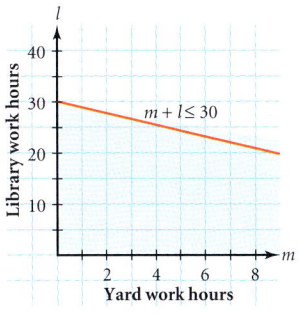

Kara wants to earn at least $126 per week.

$$6m + 7l \geq 126$$

The shaded area shows the possible number of hours that Kara can work to make at least $126.

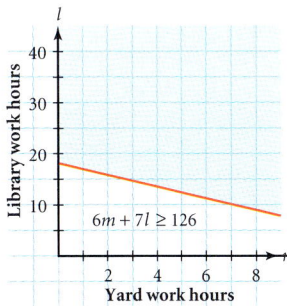

Combine the graphs of the inequalities on a single grid.

The intersection of the two shaded regions (the darker region) is the solution set of the system of inequalities.

You can check the solution by testing a point inside the intersection region. For example, the point (4, 20) is in the region.

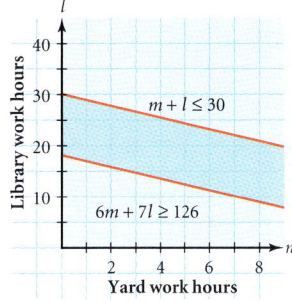

The number of hours worked is $4 + 20$, or 24, which is within Kara's 30-hour maximum. The amount earned is $4(7) + 20(6)$, or $148, which meets her earnings requirement.

CRITICAL THINKING

Describe reasonable domains and ranges for the variables in the system from Example 4. Why should only points in the first quadrant be considered as a solution?

ADDITIONAL EXAMPLE 4

Megan has at most $1500 to invest. She plans to invest some of the money in a long-term CD at 6% and some of it in a short-term CD at 3%. She wants to earn at least $75 in interest per year. **Write a system of inequalities to represent the situation, and use it to find all of the combinations of investments that Megan can make with these two CDs.**

Let l represent the amount in dollars invested in the long-term CD, and let s represent the amount in dollars invested in the short-term CD.

$$\begin{cases} 0.06l + 0.03s \geq 75 \\ l + s \leq 1500 \end{cases}$$

The shaded region of this graph shows the combinations of investments that Megan can make.

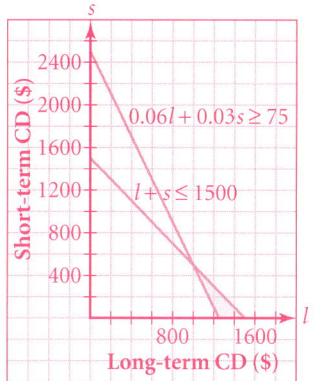

CRITICAL THINKING
The variables m and l can only have values between 0 and 30, with $m + l \leq 30$. Since neither m nor l will be negative, it is only necessary to consider the first quadrant.

Reteaching the Lesson

COOPERATIVE LEARNING Have students work in groups of four. Tell the students in each group to count off so that each is assigned a number from 1 to 4.

Now give the groups a system of inequalities. Have student 1 graph the boundary line for the first inequality. Student 1 should then pass the graph to student 2, who graphs the correct half-plane for the first inequality. Have student 3 and student 4 graph the second inequality in a similar manner.

Give students several different systems to graph in this way, assigning different roles to students 1, 2, 3, and 4 each time.

Assess

Selected Answers

Exercises 7–15, 17–73 odd

ASSIGNMENT GUIDE

In Class	1–15
Core	16–25, 32–46, 56–58
Core Plus	16–54 even, 56–59
Review	60–73
Preview	74–76

✏ Extra Practice can be found beginning on page 738.

7.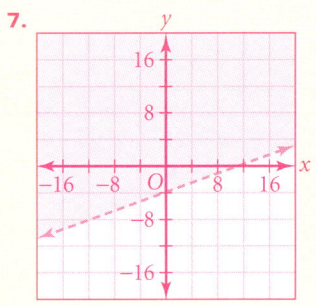

Practice

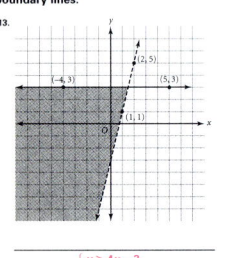
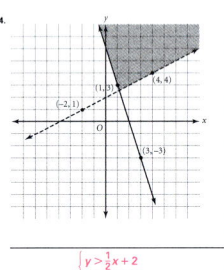

Exercises

Communicate

1. Explain how you would decide whether to draw a solid or a dashed line for the boundary for each of the following:
   a. $y \geq x - 3$
   b. $y < x - 3$

2. How do you choose the area to shade for the inequality $x + 2y < 2$? What is a good point to test in the solution region?

Discuss how to graph the solution of the following systems of inequalities. Describe the process for determining the shaded region that satisfies both inequalities in each system.

3. $\begin{cases} y < 1 \\ x < 1 \end{cases}$
4. $\begin{cases} 8x + 4y > 12 \\ y < 3 \end{cases}$
5. $\begin{cases} x \leq 3 \\ x - 2y \geq 2 \end{cases}$
6. $\begin{cases} y - 3 > x \\ y + 2x > 4 \end{cases}$

Guided Skills Practice

Graph each inequality. (EXAMPLES 1 AND 2)

7. $x - 3y < 12$
8. $4x - y \leq 7$
9. $-2y > 3x - 1$
10. $-3r - 5t \geq 10$

Solve by graphing. (EXAMPLE 3)

11. $\begin{cases} 3x + y > 3 \\ 4x + 3y \leq 9 \end{cases}$
12. $\begin{cases} y > \frac{3}{4}x - 2 \\ -x + 6y \leq 12 \end{cases}$
13. $\begin{cases} -2c + 6 < d \\ c - d < 5 \end{cases}$
14. $\begin{cases} x > -2 \\ 5y - 3x \geq -20 \end{cases}$

APPLICATION

15. **GARDENING** Marta is planning to add at least 30 new plants to her garden. She wants to add perennials ($4.00 each) and annuals ($1.50 each) and spend no more than $100. Write a system of inequalities to represent the situation and solve by graphing. (EXAMPLE 4)

Practice and Apply

Graph each inequality.

16. $y \geq \frac{1}{2}x + 2$
17. $-2x + 3y < -15$
18. $4y + x > -1$
19. $2x - 7y \leq -14$
20. $6y \leq -4x - 24$
21. $10x - 15y \leq 30$
22. $7x - y < 3$
23. $y < 2x$
24. $3x + 5y \leq 25$
25. $5x - 4y < -24$
26. $2m - 7n \geq 14$
27. $-3r - 5t < 10$
28. $k > \frac{5}{7}h - 9$
29. $8c - 2f > 10$
30. $4m - 3p < 9$
31. $j - k < -1$

Solve by graphing.

32. $\begin{cases} 2x - 3y > 6 \\ 5x + 4y < 12 \end{cases}$
33. $\begin{cases} x - 4y \leq 12 \\ 4y + x \leq 12 \end{cases}$
34. $\begin{cases} y < x - 5 \\ y \leq 3 \end{cases}$
35. $\begin{cases} 2x + y \leq 4 \\ 2y + x \geq 8 \end{cases}$
36. $\begin{cases} 5y < 2x - 5 \\ 4x + 3y \leq 9 \end{cases}$
37. $\begin{cases} 4y < 3x + 8 \\ y \leq 1 \end{cases}$

8.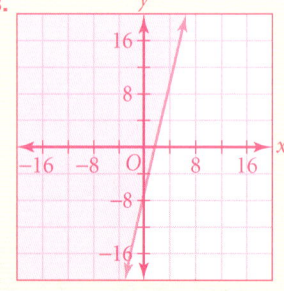

9.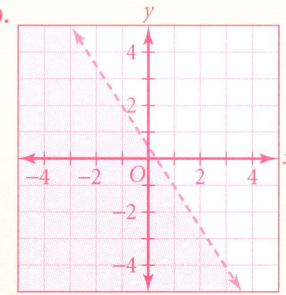

350 LESSON 7.5

38. $\begin{cases} x \geq 4 \\ 0 < y \end{cases}$
39. $\begin{cases} x - 2y < 8 \\ x + y \geq 5 \end{cases}$
40. $\begin{cases} 3x - y > -2 \\ x - y > -1 \end{cases}$

41. $\begin{cases} 2y \leq -x + 6 \\ -3x + 4y < -8 \end{cases}$
42. $\begin{cases} -5a + b < -1 \\ -a + 5b > -5 \end{cases}$
43. $\begin{cases} 4w - z \geq 2 \\ 3w + z \geq 7 \end{cases}$

44. $\begin{cases} y < x \\ y > -x \end{cases}$
45. $\begin{cases} 7q - 5r < 30 \\ 5q + 7r \leq 42 \end{cases}$
46. $\begin{cases} 5c > 4b \\ c \leq 3 \end{cases}$

47. $\begin{cases} t < 5 \\ -3t - 7v > 14 \end{cases}$
48. $\begin{cases} z - x > 3 \\ z + x < 3 \end{cases}$
49. $\begin{cases} y - 2x \leq 2 \\ y + 2x \geq 2 \end{cases}$

50. $\begin{cases} 3m - 2n > 0 \\ 2m + 3n < -3 \end{cases}$
51. $\begin{cases} 8x + 2y \geq 1 \\ x - 3y < 4 \end{cases}$
52. $\begin{cases} 2l + 3m < 4 \\ -\frac{1}{2}l + \frac{1}{3}m > -5 \end{cases}$

53. $\begin{cases} x - 2y < 2 \\ 3x + \frac{1}{2}y \leq -2 \end{cases}$
54. $\begin{cases} w + v < 2 \\ v - w \geq -3 \end{cases}$
55. $\begin{cases} 8x + 5y \leq -2 \\ -5x \leq -3y + 2 \end{cases}$

Technology

A graphics calculator may be helpful for Exercises 16–58. You may wish to have students first solve the systems with paper and pencil and then use a graphics calculator to check their solutions.

Error Analysis

When graphing linear inequalities, students often use the wrong type of boundary line or shade the wrong half-plane. Stress the importance of paying careful attention to the inequality symbol in order to determine whether the boundary line should be dashed or solid. Remind them that they should substitute values from each half-plane into the inequality to verify which half-plane contains the points that satisfy it.

CONNECTION

56. **GEOMETRY** Write the system of inequalities defined by each perimeter described below, and graph the common solution. Make sure that the solution represents real-world possibilities.

Rectangle ABCD's perimeter is at most 30 centimeters, while rectangle PQRS's perimeter is at least 12 centimeters.

APPLICATIONS

57. **SPORTS** In the Lincoln High School basketball game, Troy made only 2-point field goals and 1-point free throws. He made no more than 16 points in the game. Find the combinations of field goals and free throws Troy could have made.

58. **ACADEMICS** On most days Anna spends no more than 2 hours on her math and science homework. If math always takes about twice as long as science, what is the maximum time that Anna usually spends on science?

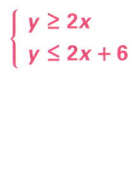 $\frac{2}{3}$ hours

38.
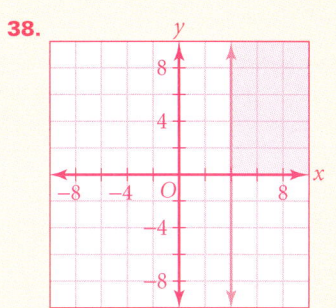

CHALLENGE

59. Write the system of inequalities that represents the shaded region. Use the points shown to find equations for the boundary lines.

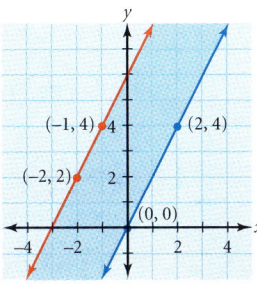

$\begin{cases} y \geq 2x \\ y \leq 2x + 6 \end{cases}$

39.

40.

Student Technology Guide

7.5 Systems of Inequalities

You can use a graphics calculator to graph a system of inequalities such as $\begin{cases} y \geq -\frac{3}{4}x \\ y \leq x - 1 \end{cases}$

Method 1: Press Y=. Enter −3X/4 as Y1 and X − 1 as Y2. Use the arrow keys and ENTER as often as needed to display ◣ to the left of Y1 and ◤ to the left of Y2. Press GRAPH. The overlapping region indicates the solution to the system.

Method 2: Press Y= and use the arrow key ▼ and CLEAR to remove any expressions. Then press 2nd PRGM ENTER − 3 X,T,θ,n + 4 , X,T,θ,n − 1) ENTER.

This tells the calculator to shade above the line $y = -\frac{3}{4}x$ and below the line $y = x - 1$.

Note: If you solve several problems using this method, you will need to press 2nd PRGM Draw 1: Clr Draw after each problem.

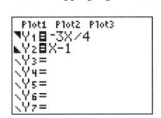 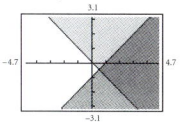

Use a graphics calculator to graph each system of linear inequalities. Describe the solution region.

See answer section for graphs and sample descriptions.

1. $\begin{cases} y \leq 2x - 1 \\ y \leq -\frac{2}{3}x + 1 \end{cases}$
2. $\begin{cases} y \geq 3\frac{3}{2}x - 2 \\ y \leq \frac{1}{2}x - 1 \end{cases}$
3. $\begin{cases} 3x - 4y \geq -3 \\ 3x - 4y \leq 2 \end{cases}$

4. $\begin{cases} 4x + 7y \leq 28 \\ x \geq 0 \\ y \geq 0 \end{cases}$
5. $\begin{cases} y \geq -2 \\ y \leq x + 4 \\ y \leq -2x + 4 \end{cases}$
6. $\begin{cases} 2x - y \leq -3 \\ x - y \leq -1 \\ 3x + y \geq 0 \end{cases}$

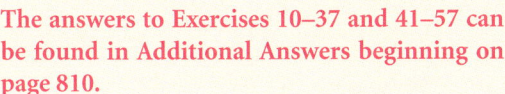
The answers to Exercises 10–37 and 41–57 can be found in Additional Answers beginning on page 810.

LESSON 7.5 351

Students are already familiar with the equation $y = mx + b$. In Exercises 74–76, students are introduced to an alternative form of the equation. This form, $f(x) = mx + b$, is called function notation. Functions will be studied in Chapters 10 and 11.

ALTERNATIVE Assessment

Portfolio Activity

The Portfolio Activity can be used as preparation for the Chapter Project or as a separate activity. In the Portfolio Activity on this page, students use guess-and-check to find three solutions to a system of inequalities and then they graph the system to verify their solutions.

Answers to Portfolio Activities can be found in Additional Answers of the Teacher's Edition.

60. Find the first differences for the following sequence: 5, 17, 37, 65, 101, ...
(LESSON 1.1) **12, 20, 28, 36**

61. Find the second differences for the following sequence: 5, 17, 37, 65, 101, ...
(LESSON 1.1) **a constant 8**

62. Simplify. $-8 - 2\left[3 + 6\left(5 - \frac{4}{2}\right) + 7\right]$ *(LESSON 1.3)* **−64**

63. FUND-RAISING Suppose that 20 businesses contribute equally at a charity benefit. If the total raised was $1200, how much did each business contribute? *(LESSON 2.4)* **$60**

Simplify. *(LESSON 2.6)*

64. $2x^2 + 4 + 3y - x^2 + 2$
$x^2 + 3y + 6$

65. $3a + a + z + 4a + 2b^2 - b$
$8a + 2b^2 - b + z$

Use the slopes to determine whether each system represents lines that are parallel, perpendicular, or neither. *(LESSON 5.6)*

66. $\begin{cases} x + y = 7 \\ x + 3y = 9 \end{cases}$ neither

67. $\begin{cases} -2x + y = -5 \\ -2x - y = 5 \end{cases}$ neither

68. $\begin{cases} -x + 2y = 6 \\ -x + 2y = -3 \end{cases}$ parallel

69. $\begin{cases} x - 2y = 4 \\ 2x + y = 1 \end{cases}$ perpendicular

70. $\begin{cases} 3x = 2y \\ 3x - 2y = 0 \end{cases}$ parallel

71. $\begin{cases} x - y = 4 \\ y + x = 2 \end{cases}$ perpendicular

72. $\begin{cases} 3x = 4 \\ 3y = 2 \end{cases}$ perpendicular

73. $\begin{cases} 8x + 2y = 5 \\ x + y = -4 \end{cases}$ neither

Look Beyond

For $f(x) = 2x + 4$, evaluate the following:

74. $f(3)$ **10**

75. $f(-2)$ **0**

76. $f\left(-\frac{1}{2}\right)$ **3**

Include the following activity in your portfolio:

The high-school yearbook staff is deciding how many color pages to include in the yearbook. In order to provide the appropriate visual appeal, the yearbook staff has decided that the number of black-and-white pages must be no more than twice the number of color pages. Due to cost constraints, there can be no more than 40 color pages in the yearbook.

- Use guess-and-check to determine three different combinations of pages that will work.
- Write a system of inequalities that models the situation.
- Solve the system graphically. Show that your three different combinations lie inside the solution region.
- Explain what the intersection region represents.

WORKING ON THE CHAPTER PROJECT

You should now be able to complete Activities 2 and 3 of the Chapter Project on page 360.

7.6 Classic Puzzles in Two Variables

Objective
• Solve traditional math puzzles in two variables.

Math phobic's nightmare

A math phobic is a person who is afraid of math.

Why Traditional math puzzles may be a way of testing and sharpening your math skills. The concepts you learn from them may help you solve problems or make discoveries. Einstein used concepts from wind and current problems to investigate and make discoveries about the speed of light.

PROBLEM SOLVING

Traditional math puzzles like the one the angel in the cartoon is presenting may indeed seem frightening. To someone who doesn't understand them, they may seem almost impossible. But there are specific ways to approach problems of this type, and a study of them will sharpen your algebra skills.

Age Puzzles

The key to solving age puzzles is knowing how to represent a person's age in the future or the past.

If a person's age now is x, then 5 years from now the person's age will be $x + 5$. Five years ago, the person's age was $x - 5$.

Once you understand this part, you will need to set up a system of two equations in two unknowns. You already know how to do this.

Alternative Teaching Strategy

USING TECHNOLOGY You can use the **TABLE** or **TRACE** feature of a graphics calculator to find solutions to any of the puzzles studied in this lesson. For instance, if you are using a TI-82 or TI-83 graphics calculator to solve the puzzle in Example 1, press [Y=] and enter **X+32** next to **Y1** and **5X+16** next to **Y2**.

To use the **TABLE** feature, press [2nd] [WINDOW] and make the following settings:

TblStart=0 (or **TblMin=0**)   **Indpnt:** Auto Ask
∆Tbl=1   **Depend:** Auto Ask

Press [2nd] [GRAPH]. Then use [▼] to locate the place where **Y1** and **Y2** are equal, at $x = 4$.

To use the **TRACE** feature, press [WINDOW] and set the viewing window as follows:

Xmin=0   Ymin=0
Xmax=10   Ymax=60
Xscl=1   Yscl=5

Press [GRAPH] and then [2nd] [TRACE]. Choose **5:intersect**, then press [ENTER] [ENTER] [ENTER].

Prepare

NCTM Principles & Standards 1, 2, 5, 6, 8–10

LESSON RESOURCES

• Practice 7.6
• Reteaching Master 7.6
• Enrichment Master 7.6
• Cooperative-Learning Activity 7.6
• Lesson Activity 7.6
• Problem Solving/Critical Thinking Master 7.6
• Teaching Transparency 39

QUICK WARM-UP

Write an equation to model each situation.

1. A person drove 150 miles in 3 hours at an average rate of r miles per hour.
   $3r = 150$

2. An 8-ounce mixture of nuts consists of p ounces of peanuts and c ounces of cashews. $8 = p + c$

3. A collection of q quarters is worth $11.75.
   $0.25q = 11.75$

4. In a certain number, the tens digit, t, is 3 less than the units digit, u.
   $t = u - 3$

Also on Quiz Transparency 7.6

Teach

Why Ask students to describe some puzzles that they have worked on recently. Make a list on the board or overhead. Have students describe strategies that they used to solve the puzzles. Explain that some classic mathematical puzzles can be solved by using an equation in two variables.

LESSON 7.6   **353**

ADDITIONAL EXAMPLE 1

A mother is 20 years older than her daughter. In 7 years the mother will be 3 times as old as her daughter. **How old is each now?**

At the present time, the mother is 23 years old, and her daughter is 3 years old.

TRY THIS

June is 14 years old, and her brother is 10 years old.

ADDITIONAL EXAMPLE 2

A plane leaves Washington, D.C., and heads toward Los Angeles, which is 2300 miles away. The plane, flying against the wind, takes 5 hours to reach Los Angeles. After refueling, the plane returns to Washington, D.C., traveling with the wind, in 4 hours. **Find the speed of the wind and the speed of the plane with no wind.**

The speed of the wind is 57.5 miles per hour. The speed of the plane with no wind is 517.5 miles per hour.

EXAMPLE 1

A father is 32 years older than his son. In 4 years the father will be 5 times as old as his son. **How old is each now?**

- **SOLUTION**

Write two equations describing the age of the father and son at two different times. Let f represent the father's present age. Let s represent the son's present age. You are given the following information:

$$f = s + 32$$

In 4 years the father will be 5 times the son's age. Add 4 years to each age and express this information in an equation.

$$f + 4 = 5(s + 4)$$

Solving the system $\begin{cases} f = s + 32 \\ f + 4 = 5(s + 4) \end{cases}$ gives $f = 36$ and $s = 4$.

At the present time, the father is 36 years old, and his son is 4 years old.

TRY THIS

June is 4 years older than her brother. Six years ago, she was twice as old as he was. How old is each now?

Wind and Current Puzzles

Wind and current puzzles have a common idea. When a plane is traveling with the wind (a tailwind), the speed of the wind is added to the speed of the plane with no wind. When traveling against the wind (a headwind), the speed of the wind is subtracted from the plane's speed with no wind.

EXAMPLE 2

APPLICATION
AVIATION

A plane leaves New York City and heads for Chicago, which is 750 miles away. The plane, flying against the wind, takes 2.5 hours to reach Chicago. After refueling, the plane returns to New York City, traveling with the wind, in 2 hours. **Find the speed of the wind and the speed of the plane with no wind.**

- **SOLUTION**

Define the variables. Put the information in a table.

Let x represent the speed of the plane with no wind.
Let y represent the speed of the wind.

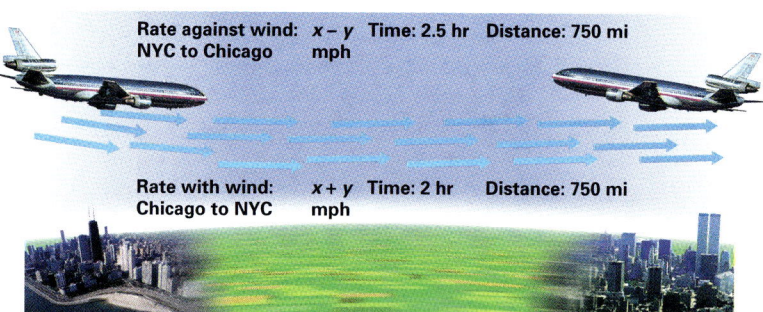

Rate against wind: $x - y$ mph Time: 2.5 hr Distance: 750 mi
NYC to Chicago

Rate with wind: $x + y$ mph Time: 2 hr Distance: 750 mi
Chicago to NYC

Interdisciplinary Connection

ACCOUNTING One of the duties of a vending machine operator is to collect the change from each machine and record the number of each type of coin for accounting purposes. Due to a malfunction, one machine accepted only nickels and quarters. An automatic counter on the machine indicated the total number of coins, 601, and their total value, $76.85. How many of each type of coin were in the machine?

There were 367 nickels and 234 quarters.

354 LESSON 7.6

To write the system of equations, recall the relationship between rate, time, and distance.

$$\text{rate} \times \text{time} = \text{distance}$$

$$\begin{cases}(x-y)(2.5) = 750 \\ (x+y)(2) = 750\end{cases} \Rightarrow \begin{cases}x - y = 300 \\ x + y = 375\end{cases}$$

Solving the system gives $x = 337.5$ and $y = 37.5$.

The speed of the plane with no wind is 337.5 miles per hour. The speed of the wind is 37.5 miles per hour.

TRY THIS

A riverboat traveling with the current takes 2.5 hours to make a trip of 45 miles. The return trip, against the current, takes 2.75 hours. Find the speed of the current and the speed of the boat in still water.

Number-Digit Puzzles

In most number-digit problems the strategy is to write the value of a number in expanded form. You can write 52 as $5(10) + 2$. If you reverse the digits in 52, you get 25, which you can write as $2(10) + 5$.

EXAMPLE 3 The sum of the digits of a two-digit number is 7. The original two-digit number is 3 less than 4 times the number with its digits reversed. **Find the original two-digit number.**

● **SOLUTION**

Let t represent the tens digit and u represent the units digit of the original number.

Original number: $10t + u$

Number with digits reversed: $10u + t$

You are also given $t + u = 7$.

The original two-digit number is 3 less than 4 times the number with its digits reversed.

$$10t + u = 4(10u + t) - 3$$

Solve the system: $\begin{cases} t + u = 7 \\ 10t + u = 4(10u + t) - 3 \end{cases}$

Solving the system gives $t = 6$ and $u = 1$.

If t is 6 and u is 1, then the original number is $10(6) + 1$, or 61. The number 61 is a two-digit number that is 3 less than 4 times itself with its digits reversed. Since $4[10(1) + 6] - 3 = 61$, the solution is correct.

TRY THIS

A two-digit number whose tens digit is 2 more than the units digit is 3 more than 6 times the sum of its digits. Find the original number.

TRY THIS

The speed of the current is $\frac{9}{11}$ miles per hour. The speed of the boat in still water is $17\frac{2}{11}$ miles per hour.

ADDITIONAL EXAMPLE 3

The sum of the digits of a two-digit number is 8. The original two-digit number is 17 less than 2 times the number with its digits reversed. **Find the original two-digit number.** 53

TRY THIS

The original number is 75.

Inclusion Strategies

VISUAL LEARNERS All of the puzzles in this lesson can be approached by organizing the given information in a table such as those used in Examples 4 and 5. Students with a visual learning style may find these tables helpful. A table like the one below, containing information from Example 2, can be used for any wind or current problem.

	rate	×	time	= distance
With the wind	$x + y$		2	750
Against the wind	$x - y$		2.5	750

Enrichment

Have students create puzzles of the type studied in the lesson. Alert them to the fact that it generally is easier to do this by working backward. That is, they should start with an answer and then develop the conditions needed to arrive at that answer.

When creating a coin puzzle, for example, first choose the kind and number of coins, such as 18 nickels and 24 quarters. Then the first condition can be *There are 42 coins with a total value of $6.90*. A second condition might be *There are 6 more quarters than nickels*.

LESSON 7.6

ADDITIONAL EXAMPLE 4

In a coin bank there are 275 dimes and nickels worth a total of $21.85. **How many dimes and how many nickels are in the bank?**
There are 162 dimes and 113 nickels in the coin bank.

TRY THIS
There are 18 quarters and 30 nickels.

Use Teaching Transparency 39.

ADDITIONAL EXAMPLE 5

A 5% acid solution is mixed with a 9% acid solution. **How many ounces of each solution are needed to obtain 40 ounces of a 6% acid solution?**
30 ounces of the 5% acid solution and 10 ounces of the 9% acid solution

Coin Puzzles

Classic coin problems can usually be set up by using a system of equations—one involving the number of coins and the other involving the monetary value of the coins in cents.

EXAMPLE 4 In a coin bank there are 250 dimes and quarters worth a total of $39.25. **Find how many dimes, d, and how many quarters, q, are in the bank.**

○ **SOLUTION**
Create a table to organize the information.

	Quarters	Dimes	Total
Number of coins	q	d	250 coins
Value in cents	$25q$	$10d$	3925¢

From the table, write two equations based on the totals.

$$\begin{cases} q + d = 250 \\ 25q + 10d = 3925 \end{cases}$$

By solving the system, you find that $q = 95$ and $d = 155$.
The coin bank has 95 quarters and 155 dimes.

TRY THIS In the math department water-cooler fund there are 48 quarters and nickels in all, with a total value of $6.00. Find how many quarters and how many nickels are in the fund.

Chemical-Solution Puzzles

Chemical-solution puzzles involve mixing solutions, typically acids, of different strengths. The strengths of the solutions are expressed as percents. The strategy is to keep track of the total amount of acid, in ounces or grams, in the original solutions and in the final solution.

EXAMPLE 5 A 1.5% acid solution is mixed with a 4% acid solution. How many ounces of each solution are needed to obtain 60 ounces of a 2.5% acid solution?

APPLICATION
CHEMISTRY

○ **SOLUTION**
Draw a diagram and make a table to help solve this problem.

Reteaching the Lesson

GUIDED ANALYSIS Write the name of each type of classic puzzle on the board or overhead. Work with students to make a "profile" of each type of puzzle. The profile should contain identifying characteristics, commonly used variables, a blank form of any table used, hints for writing equations, and any other information students have found to be useful. Have students copy each profile. Then give them a new example of each type of puzzle. Students should work in pairs to solve each puzzle by using the information in the profile.

Define the variables.

Let x represent the number of ounces of the 1.5% acid solution.

Let y represent the number of ounces of the 4% acid solution.

Place the information in a table to help organize the facts.

	First solution	Second solution	Third solution
Amount of solution	x	y	60 oz
Percent acid	1.5%	4%	2.5%
Amount of acid	$0.015x$	$0.04y$	$0.025(60)$

From the table, write two equations. $\begin{cases} x + y = 60 \\ 0.015x + 0.04y = 0.025(60) \end{cases}$

Solving the system gives the solution (36, 24).

To get the required 60 ounces of a 2.5% acid solution, mix 36 ounces of the 1.5% acid solution with 24 ounces of the 4% acid solution.

TRY THIS How many ounces of a 2% acid solution should be mixed with a 4% acid solution to produce 20 ounces of a 3.6% acid solution?

TRY THIS

Mix 4 ounces of the 2% acid solution with 16 ounces of the 4% acid solution.

Assess

Selected Answers
Exercises 5–9, 11–31 odd

ASSIGNMENT GUIDE	
In Class	1–9
Core	10–21
Core Plus	10–21
Review	22–31
Preview	32, 33

✎ Extra Practice can be found beginning on page 738.

Exercises

Communicate

1. Discuss the problem-solving strategy you would use to solve the following problem, and explain how to set up the equations:

   You are running a chemistry experiment that requires you mix a 25% alcohol solution with a 5% alcohol solution. This mixture produces 20 liters of a solution that contains 2.6 liters of pure alcohol. How many liters of the 25% alcohol solution must you use?

Migratory birds sometimes fly several thousand miles to their winter nesting sites.

A bird can fly with the wind 3 times as fast as it can fly against the wind.

2. How can you represent the rate of the bird flying in still air (no wind) and the rate of the wind?

3. How can you represent the rate of the bird flying with the wind and the rate against the wind?

4. Discuss how the rate of the bird flying with the wind compares with the rate of the bird flying in still air.

LESSON 7.6

Technology

A graphics calculator may be helpful in Exercises 10–21. You may wish to have students first write and solve the systems on paper and pencil and then use the graphics calculator to check their solutions.

Error Analysis

When solving coin and mixture puzzles, many students have difficulty working with the decimals involved. Suggest that they multiply each side of an equation with decimals by a multiple of 10 to clear the decimals. For example, if the equation is $0.05x + 0.2y = 0.16$, multiplying each side by 100 will result in the equation $5x + 20y = 16$.

Guided Skills Practice

APPLICATIONS

5. Gabriel is 16 years older than Sam. In 4 years Gabriel will be twice as old as Sam. What are their ages now? **(EXAMPLE 1)** Gabriel is 28, Sam is 12

6. RECREATION Chris can row a boat a distance of 20 miles in 2.5 hours when he is rowing with the current. It takes him 3 hours to row the same distance when he is rowing against the current. What is the rate of the current? **(EXAMPLE 2)** $\frac{2}{3}$ mph

7. The sum of the digits of a two-digit number is 8. If 16 is added to the original number, the result is 3 times the original number with its digits reversed. Find the original number. **(EXAMPLE 3)** 62

8. When Jim cleaned out the reflecting pool at the library, he found 20 nickels and quarters. The collection of nickels and quarters totaled $2.60. How many quarters did Jim find? **(EXAMPLE 4)** 8 quarters

9. CHEMISTRY Janet is mixing a 15% glucose solution with a 35% glucose solution. This mixture produces 35 liters of a solution that contains 6 liters of pure glucose (100% solution). How many liters of each solution is Janet using in the mixture? **(EXAMPLE 5)** 31.25 liters of the 15% solution
3.75 liters of the 35% solution

Practice and Apply

APPLICATION

10. RECREATION Catherine is camping along a river. It takes her 1.5 hours to paddle her canoe 6 miles upstream from her campsite. Catherine turns her canoe around and returns 6 miles downstream to her campsite in exactly 1 hour. What is the rate of the river's current and the rate of Catherine's paddling in still water?

canoe = 5 mph
current = 1 mph

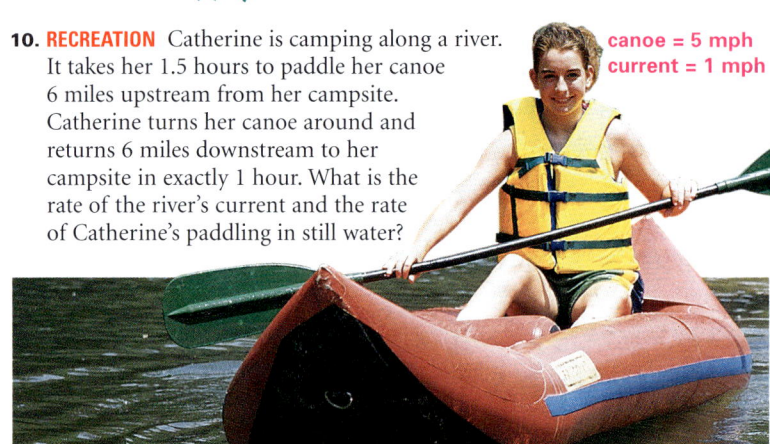

11. The coin box of a vending machine contains 6 times as many quarters as dimes. If the total amount of money in quarters and dimes is $28.80, how many quarters and how many dimes are in the box?
18 dimes, 108 quarters

12. Find the two-digit number whose tens digit is 4 less than its units digit if the original number is 2 more than 3 times the sum of the digits. 26

13. In 15 years, Maya will be twice as old as David is now. In 15 years, David will be as old as Maya will be 10 years from now. How old are they now?
Maya is 25, David is 20

14. The sum of Anjie's age and Carol's age is 66. Carol's age is 2 more than 3 times Anjie's age. How old are Carol and Anjie?
Carol is 50, Anjie is 16

15. Wymon is 20 years older than Sabrina. In 8 years, Wymon will be twice as old as Sabrina. How old is each now? Wymon is 32, Sabrina is 12

Practice

7.6 Classic Puzzles in Two Variables

1. The tens digit of a 2-digit number is twice the units digit. The sum of the digits is 12. Find the original number. **84**

2. Olga has 5 times as many dimes as nickels. If she has $3.30, how many of each coin does she have? **30 dimes and 6 nickels**

3. Milk that is 4% butterfat is mixed with milk that is 1% butterfat to obtain 18 gallons of milk that is 2% butterfat. How many gallons of each type of milk are needed? **6 gallons of 4% milk and 12 gallons of 1% milk**

4. A banker invested a portion of $10,000 at 5% interest per year and the remainder at 4% interest per year. If the banker received $470 in interest after 1 year, how much was invested at each rate? **$7000 at 5% interest and $3000 at 4% interest**

5. The sum of Nora's age and her grandmother's age is 71. Four times Nora's age is 6 less than her grandmother's age. Find their ages. **Nora: 13; grandmother: 58**

6. Sailing with the current, a boat takes 3 hours to travel 48 miles. The return trip, against the current, takes 4 hours. Find the average speed of the boat for the entire trip and the speed of the current. **boat: 14 miles per hour; current: 2 miles per hour**

7. A coin bank contains nickels, dimes, and quarters totaling $5.45. If there are twice as many quarters as dimes and 11 more nickels than quarters, how many of each coin are in the bank? **25 nickels, 7 dimes, and 14 quarters**

8. Find the two-digit number whose tens digit is 3 less than the units digit. The original number is 6 more than 4 times the sum of the digits. **58**

9. A daughter is 28 years younger than her father. In 5 years, the father will be 3 times as old as his daughter. How old is each now? **daughter: 9; father: 37**

10. A chemist has a solution that is 20% peroxide and another solution that is 70% peroxide. She wishes to make 100 L of a solution that is 35% peroxide. How much of each solution should she use? **70 liters of 20% solution and 30 liters of 70% solution**

358 LESSON 7.6

CONNECTION

16. A coin bank contains $17.00 in pennies and nickels. If there are 1140 coins in the bank, how many of the coins are nickels? **140 nickels**

17. **NUMBER THEORY** The sum of the digits of a positive two-digit number is 13. If twice the first digit is 1 less than the second digit, what is the two-digit number? **49**

APPLICATIONS

18. **CHEMISTRY** How many ounces of a 20% salt solution should be mixed with a 10% salt solution to produce 45 ounces of a 14% salt solution? **18 ounces of the 20% solution, 27 ounces of the 10% solution**

19. **CHEMISTRY** A 4% salt solution is mixed with a 16% salt solution. How many milliliters of each solution are needed to obtain 600 milliliters of a 10% solution? **300 ml of each solution**

20. **AVIATION** With a tailwind, a jet flew 2000 miles in 4 hours, but the return trip against the same wind required 5 hours. Find the jet's speed and the wind speed. **jet = 450 mph, wind = 50 mph**

21. **BUSINESS** A grocery store sells a mixture of peanuts and raisins for $1.75 per pound. If peanuts cost $1.25 per pound and raisins cost $2.75 per pound, what amount of raisins and peanuts go into 1 pound of the mixture?

$\frac{2}{3}$ **pound of peanuts**

$\frac{1}{3}$ **pound of raisins**

Look Back

Evaluate each expression for $x = 2$, $y = 1$, and $z = 1$. *(LESSON 1.3)*

22. $2x + 4y - 3z$ **5**   23. $\frac{2x-z}{5}$ **$\frac{3}{5}$**   24. $\frac{x+y}{z}$ **3**   25. $3xyz$ **6**

26. Simplify. $(29 + 1) \div 3 \cdot 2 - 4$ *(LESSON 1.3)* **16**

Simplify. *(LESSON 2.7)*

27. $-[2(a + b)]$ **$-2a - 2b$**   28. $-[-11(x - y)]$ **$11x - 11y$**   29. $2(x - y) + 3[-2(x + y)]$ **$-4x - 8y$**

Find the slope of each line. *(LESSON 5.5)*

30. $3x + 4y = 12$ **$-\frac{3}{4}$**   31. $5y = 10x - 20$ **2**

Look Beyond

Find each square root.

32. $\sqrt{\frac{25}{625}}$ **$\frac{1}{5}$**   33. $\sqrt{a^2}$ **a if $a \geq 0$; $-a$ if $a < 0$**

Look Beyond

In Exercises 32 and 33, students investigate square roots. You may need to review the meaning of a square root. Point out that the value of the expression $\sqrt{a}$ is always nonnegative. Students will study square roots and their properties in Chapter 12.

Student Technology Guide

NAME _____ CLASS _____ DATE _____

Student Technology Guide
7.6 Classic Puzzles in Two Variables

In a coin bank there are 250 dimes and quarters worth a total of $39.25. How many dimes, d, and quarters, q, are in the bank? To answer this question, you can solve the system shown at right.
$\begin{cases} q + d = 250 \\ 25q + 10d = 3925 \end{cases}$

Step 1: Choose one variable and solve each equation for the chosen variable.
$\begin{cases} q + d = 250 \\ 25q + 10d = 3925 \end{cases} \rightarrow \begin{cases} q = 250 - d \\ q = (3925 - 10d)/25 \end{cases}$

Step 2: Let y represent the variable you chose, and let x represent the other variable. In this case, $x = d$ and $y = q$. Press [Y=] 250 [−] [X,T,θ,n] [ENTER] [( ] 3925 [−] 10 [X,T,θ,n] [ )] [÷] 25 [GRAPH]. Adjust the window settings as necessary.

Step 3: Using [TRACE] or the calculator's intersect feature you can determine that the lines intersect at (155, 95). Since x represents d, the number of dimes, and y represents q, the number of quarters, you can see that there are 155 dimes and 95 quarters.

Now suppose that you want to solve a related problem in which there are only 220 dimes and quarters worth a total of $39.25. The only thing that has changed is the total number of coins. You can edit your definition of Y1 by pressing [Y=] [▶] 2. Then solve as before. The intersection point is (105, 115), so now there are 105 dimes and 115 quarters.

Use a graphics calculator to solve each puzzle.

1. A collection of nickels and pennies is worth a total of $8.33. If there are 297 coins in all, how many of each type of coin is in the collection? **134 nickels and 163 pennies**

2. In one day, a carnival booth collected a total of $565 in $1 and $5 bills. If there were 297 bills in all, how many of each type of bill were there? **230 $1 bills and 67 $5 bills**

3. Five years ago, Kevin was 3 times as old as Tami. Now, he is only twice as old. How old are they now? **Kevin is 20, and Tami is 10.**

4. In 7 years, Nicole will be twice as old as Rolando. Now, she is 3 times as old. How old are they now? **Nicole is 21, and Rolando is 7.**

5. Pat prepared 25 liters of a 26% sugar solution by mixing together a 20% solution and a 30% solution. How much of each solution did Pat use? **10 liters of 20% solution and 15 liters of 30% solution**

Project

Focus

The problem of how to minimize cost and maximize profit can be solved with a system of linear inequalities. Students will be asked to explain the meaning of various systems of equations and inequalities. Then they will graph the systems and use the graphs to find the optimal way to package trail mix.

Motivate

If any students have gone backpacking, have them share their knowledge about limits for the weight of a backpack. Ask them to relate their experiences in finding ways to keep the weight below those limits. Discuss the importance of considering the amounts of certain nutrients in the foods that are packed.

Have students read the problem described at the top of this page. Ask them to propose possible solutions to the problem. After discussing their solutions, explain that they can apply the skills they acquired in this chapter in an efficient method called *linear programming*.

CHAPTER PROJECT SEVEN

Minimum Cost Maximum Profit

The manager of a camping supply store wants to make a trail mix of nuts and raisins to provide all the required nutrients at the lowest cost. The chart shows the nutritional information and cost per ounce.

	Calories/oz	Fat/oz	Cost/oz
Nuts	150	13 g	$0.40
Raisins	90	0 g	$0.10

Backpackers usually have a weight limit for their pack, so each package of trail mix will contain no more than 18 ounces. The snack should supply at least 1800 Calories, with no more than 117 grams of fat. How many ounces each of raisins and nuts should each package of trail mix contain to minimize the costs?

This optimization problem can be solved by using a method called linear programming. The activities below lead to the solution produced by this method. Let n represent the number of ounces of nuts. Let r represent the number of ounces of raisins.

Activity 1

1. The equation $n + r = 18$ represents that the total weight of the nuts and raisins is 18 ounces. The equation $150n + 90r = 1800$ represents that there is a total of 1800 calories in the mix.

2. Solve the first equation for n, and substitute into the second equation. $r = 15; n = 3$

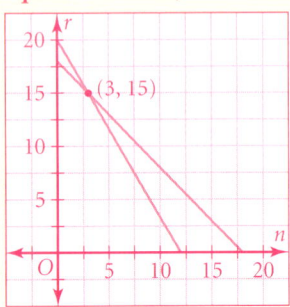

There are 3 oz of nuts and 15 oz of raisins in the mix.

3. There is no fat in raisins.

4. The equation $13n = 117$ represents that there is a total of 117g of fat in the mix. The other equations are the same as Exercise 1.

5. $\begin{cases} 13n = 117 \\ 150n + 90r = 1800 \end{cases}$

    Solve the first equation for n and substitute into the second equation.
    $n = 9, r = 5$

    $\begin{cases} n + r = 18 \\ 13n = 117 \end{cases}$

Solve the second equation for n and substitute into the first equation.
$n = 9, r = 9$

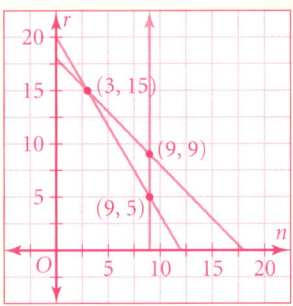

360 CHAPTER 7 PROJECT

Activity 1

1. Explain what the two equations in the following system represent:
$$\begin{cases} n + r = 18 \\ 150n + 90r = 1800 \end{cases}$$
2. Solve the system by substitution and by graphing.
3. Why can you represent the total amount of fat by the expression $13n$?
4. Explain what the two equations in each system below represent.
$$\begin{cases} 13n = 117 \\ 150n + 90r = 1800 \end{cases} \qquad \begin{cases} n + r = 18 \\ 13n = 117 \end{cases}$$
5. Solve each system by substitution and by graphing. Graph both systems on the same coordinate plane.

Activity 2

1. Explain what each inequality in the following system represents:
$$\begin{cases} n + r \le 18 \\ 150n + 90r \ge 1800 \\ 13n \le 117 \end{cases}$$
2. Graph each inequality on the coordinate plane you used in Activity 1.
3. What does the common area (intersection region) represent in the problem?
4. Pick four different points in the intersection region and show that each one satisfies all of the conditions.

Activity 3

1. Nuts cost $0.40 per ounce and raisins cost $0.10 per ounce. Complete the following exercises:
   a. Write an expression for the total cost of the nuts.
   b. Write an expression for the total cost of the raisins.
   c. Write an equation for the combined cost of nuts and raisins. This is known as the *optimization equation*.
2. Use graphing technology to graph each of the three *constraint inequalities* from Step 1 of Activity 2. Shade the region that represents the solution.
3. Explain what the vertices of the triangle represent in the problem.
   $A(3, 15) \qquad B(9, 9) \qquad C(9, 5)$
4. Substitute the coordinates of each vertex of the shaded region into the optimization equation that you wrote in part **c** of Step 1. Find the solution to the optimization equation. The equation with the lowest cost is the minimum.
5. How many ounces of each ingredient should be used per package to keep the costs at a minimum?

The shaded region, $\triangle ABC$, represents the solution set. This region, determined by the linear inequalities containing the possible solutions, is called the feasibility region.

Cooperative Learning

Have students do the Chapter Project in pairs. In **Activity 1**, one student of each pair can solve the first system by substitution while the other solves it by graphing. Then they can reverse their roles for the systems in Step 4.

In **Activity 2**, be sure students are aware that the inequalities in Step 1 are related to the equations from Activity 1. One student in each pair can graph the inequalities, and the other can check points in the solution region.

In **Activity 3**, each student should work independently to write the expressions in Step 1. Partners can then compare the expressions they wrote, resolve any differences, and do the remaining steps together.

Discuss

After students have completed all three activities, bring the entire class together to create a step-by-step description of the linear programming method that was used to solve the problem. Ask students to suggest other situations in which they think linear programming would be an appropriate problem-solving method. Make a list of the situations on the board or overhead. Have each pair select one of the situations listed and then write and solve a linear programming problem related to it.

Activity 2

1. The inequality $n + r \le 18$ represents that the total weight of the nuts and raisins mix can be no more than 18 ounces. The inequality $150n + 90r \ge 1800$ represents that the mix should supply at least 1800 calories. The inequality $13n \le 117$ represents that there should be no more than 117 grams of fat in the mix.

2.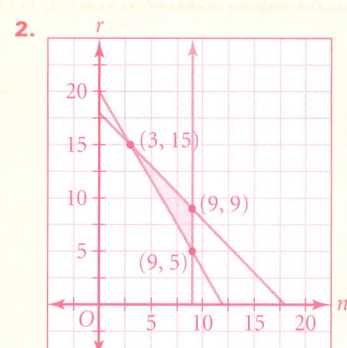

3. This region represents all possible combinations that satisfy all three requirements for weight, calories, and fat.

4. Answers may vary. Sample answers:
   $(8, 8), (7, 9), (6, 10), (5, 12)$

The answers to Activity 3 can be found in Additional Answers beginning on page 810.

CHAPTER 7 PROJECT **361**

Chapter Review

Chapter 7 Review and Assessment

VOCABULARY

approximate solution 321	dependent system 341	linear inequalities 345
boundary line 346	elimination method 332	solution 319
common solution 319	inconsistent system 339	substitution method 327
consistent system 339	independent system 341	system of equations 319

Key Skills & Exercises

LESSON 7.1

Key Skills

Graph a system of equations and estimate a solution from a graph by inspection.

Solve by graphing. $\begin{cases} x + 3y = 12 \\ x - y = 4 \end{cases}$

Write the equations in slope-intercept form.

$x + 3y = 12$ $x - y = 4$
$y = -\frac{x}{3} + 4$ $y = x - 4$

Graph the lines on graph paper or on a graphics calculator. The solution is the point of intersection, (6, 2). Check algebraically.

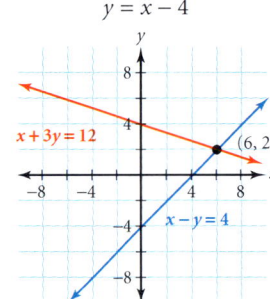

Exercises

Solve each system by graphing.

1. $\begin{cases} x - y = 6 \\ x + 2y = -9 \end{cases}$ 2. $\begin{cases} y = -x \\ y = 2x \end{cases}$

3. $\begin{cases} x + 6y = 3 \\ 3x + y = -8 \end{cases}$ 4. $\begin{cases} 3x + y = 6 \\ y + 2 = x \end{cases}$

5. $\begin{cases} 8x + 6y = 2 \\ x + 2y = -3 \end{cases}$ 6. $\begin{cases} -x + 2y = 1 \\ x - 2y = 3 \end{cases}$

7. $\begin{cases} y = 3x + 7 \\ y = 2x - 6 \end{cases}$ 8. $\begin{cases} y = -3x + 1 \\ y = -2x \end{cases}$

LESSON 7.2

Key Skills

Find an exact solution to a system of equations by using the substitution method.

Solve by using the substitution method. $\begin{cases} x + 4y = 1 \\ 2x + 7y = 6 \end{cases}$

Solve $x + 4y = 1$ for x to get $x = 1 - 4y$. Then substitute the expression $1 - 4y$ for x in the equation $2x + 7y = 6$. Solve for y.

$2x + 7y = 6$
$2(1 - 4y) + 7y = 6$
$2 - 8y + 7y = 6$
$2 - y = 6$
$-y = 4$
$y = -4$

Exercises

Solve each system of linear equations by using substitution.

9. $\begin{cases} 2x + y = 1 \\ x + y = 2 \end{cases}$ **(-1, 3)**   10. $\begin{cases} x - y = 6 \\ 2x - 4y = 28 \end{cases}$ **(-2, -8)**

11. $\begin{cases} x + 2y = 1 \\ 2x - 8y = -1 \end{cases}$ **$\left(\frac{1}{2}, \frac{1}{4}\right)$**   12. $\begin{cases} 4x = 3y + 44 \\ x + y = -3 \end{cases}$ **(5, -8)**

13. $\begin{cases} y = \frac{1}{2}x + 2 \\ 3x + y = 7 \end{cases}$ **$\left(\frac{10}{7}, \frac{19}{7}\right)$**   14. $\begin{cases} \frac{3}{5}x - 2y = 5 \\ x = \frac{4}{5}y + 2 \end{cases}$ **$\left(0, -\frac{5}{2}\right)$**

15. $\begin{cases} 4m + 5n = 6 \\ -3m - 2n = -4 \end{cases}$ **$\left(\frac{8}{7}, \frac{2}{7}\right)$**   16. $\begin{cases} 8a - \frac{1}{2}b = \frac{3}{4} \\ b = \frac{1}{2}a + 7 \end{cases}$ **$\left(\frac{17}{31}, \frac{451}{62}\right)$**

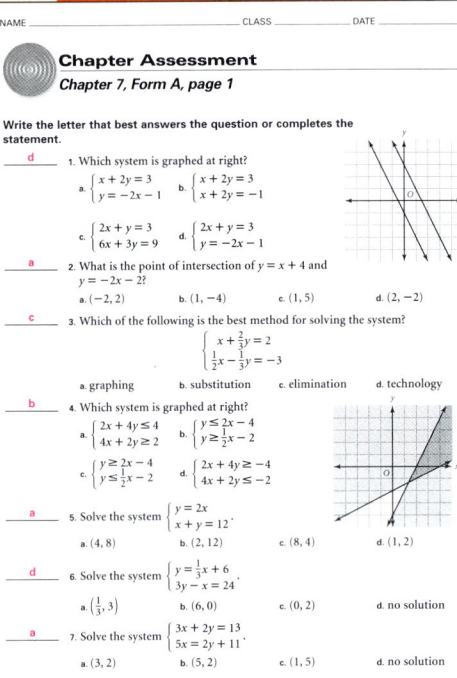

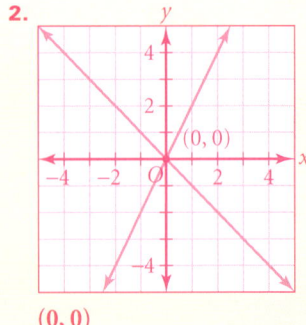

2. (0, 0)

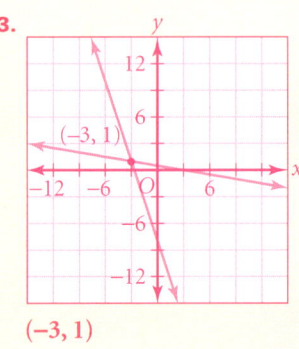

3. (-3, 1)

362   CHAPTER 7 REVIEW

Now substitute –4 for y in one of the original equations. Solve for x.

$x + 4(-4) = 1$
$x - 16 = 1$
$x = 17$

The solution is $(17, -4)$.

LESSON 7.3
Key Skills

Find an exact solution to a system of equations by using the elimination method.

Solve by using elimination. $\begin{cases} x + y = -3 \\ 2x - 3y = -11 \end{cases}$

Use the Multiplication Property of Equality to get opposite coefficients for one of the variables in the equations.

$\begin{cases} 3(x+y) = 3(-3) \\ 2x - 3y = -11 \end{cases} \Rightarrow \begin{cases} 3x + 3y = -9 \\ 2x - 3y = -11 \end{cases}$

Add the expressions on corresponding sides of the equations, and solve for the remaining variable.

$3x + 3y = -9$
$2x - 3y = -11$
$5x = -20$
$x = -4$

Substitute –4 for x in $x + y = 3$. Solve for y.

$-4 + y = -3$
$y = 1$

The solution is $(-4, 1)$.

Exercises

Solve each system of linear equations by using elimination.

17. $\begin{cases} 4x + 5y = 3 \\ 2x + 5y = -11 \end{cases}$ **(7, –5)**

18. $\begin{cases} 2x + 3y = -6 \\ -5x - 9y = 14 \end{cases}$ **$\left(-4, \frac{2}{3}\right)$**

19. $\begin{cases} 0.5x + y = 0 \\ 0.9x - 0.2y = -2 \end{cases}$ **(–2, 1)**

20. $\begin{cases} -2x + 4y = 12 \\ 3x - 2y = -10 \end{cases}$ **(–2, 2)**

21. $\begin{cases} 0.7m + 3n = 2.5 \\ 0.7m + 2n = 3 \end{cases}$ $\left(\frac{40}{7}, -\frac{1}{2}\right)$

22. $\begin{cases} 5x + 7y = 4 \\ 10x - 2y = -2 \end{cases}$

23. $\begin{cases} 4x - 2y = 3 \\ x + 3y = 1 \end{cases}$ $\left(\frac{11}{14}, \frac{1}{14}\right)$

24. $\begin{cases} -7x - 3y = 3 \\ -5x + 6y = 2 \end{cases}$ $\left(-\frac{8}{19}, -\frac{1}{57}\right)$

22. $\left(-\frac{3}{40}, \frac{5}{8}\right)$

LESSON 7.4
Key Skills

Identify consistent and inconsistent systems of equations. Identify consistent systems as dependent or independent.

a. $\begin{cases} y = 2x + 6 \\ y = 2x - 5 \end{cases}$

The two equations have the same slope. This is an *inconsistent* system.

b. $\begin{cases} 5x + y = 3 \\ 10x + 2y = 6 \end{cases}$

The two equations are both the same equation: $y = -5x + 3$. This is a *consistent, dependent* system.

c. $\begin{cases} y = \frac{1}{2}x + 2 \\ y = 3x + 7 \end{cases}$

These two equations do not have the same slope. This is a *consistent, independent* system.

Exercises

Identify each system of equations as independent, dependent, or inconsistent.

25. $\begin{cases} 3x - 2y = 7 \\ 4y = -14 + 6x \end{cases}$ **dep.**

26. $\begin{cases} 4y = 2x + 20 \\ 3x - y = -20 \end{cases}$ **indep.**

27. $\begin{cases} 2x + 5y = 7 \\ 3y = 2x + 17 \end{cases}$ **indep.**

28. $\begin{cases} x + y = 7 \\ 28 - 2y = 2x \end{cases}$ **incons.**

29. $\begin{cases} 3x + 6y = 2 \\ 6x + 6y = 4 \end{cases}$ **indep.**

30. $\begin{cases} 2x = y - 3 \\ y = -2x + y \end{cases}$ **indep.**

4.

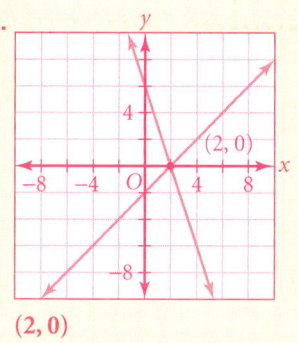

(2, 0)

5.

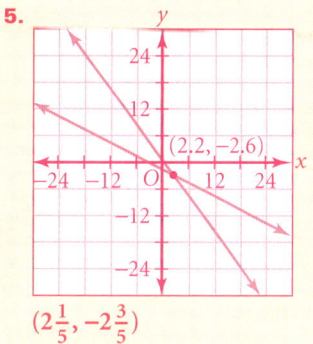

$\left(2\frac{1}{5}, -2\frac{3}{5}\right)$

The answers to Exercises 7 and 8 can be found in Additional Answers beginning on page 810.

6.

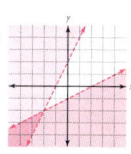

no solution

Chapter Test, Form B

NAME _____ CLASS _____ DATE _____

Chapter Assessment
Chapter 7, Form B, page 1

1. Explain how to write an equation that would form a dependent system with the equation $2x - y = 5$.
   Multiply each term of the equation $2x - y = 5$ by the same number.

2. Describe the boundary line for the graph of $x + 3y > 4$.
   The boundary is the dashed line $x + 3y = 4$.

Graph each system on the grid provided to find a solution. Check your answer algebraically.

3. $\begin{cases} 2x + 4y = -4 \\ y = x + 2 \end{cases}$

4. $\begin{cases} 3x - y = 3 \\ \frac{1}{3}y = x - 1 \end{cases}$

(–2, 0) — **Infinite number of solutions**

Solve by using the best method for each system. Check your solutions.

5. $\begin{cases} 2x + 3y = 7 \\ x + 2y = 4 \end{cases}$ **(2, 1)**

6. $\begin{cases} 4x - y = 11 \\ 2x + y = 1 \end{cases}$ **(2, –3)**

7. $\begin{cases} 3x + 2y = 1 \\ 6x + 5y = 4 \end{cases}$ **(–1, 2)**

8. $\begin{cases} 3x - 3y = 18 \\ x + y = 3 \end{cases}$ **(4.5, –1.5)**

9. $\begin{cases} x - 4y = 5 \\ 2x + 4y = -2 \end{cases}$ **(1, –1)**

10. $\begin{cases} 6x - 2y = 6 \\ 3x = y + 2 \end{cases}$ **no solution**

11. The sum of two numbers is 48. If the larger number is 6 less than twice the smaller number, find the two numbers. **30 and 18**

12. The perimeter of a rectangle is 64 meters. If the length is 2 more than twice the width, find the dimensions of the rectangle. **width: 10 meters; length: 22 meters**

NAME _____ CLASS _____ DATE _____

Chapter Assessment
Chapter 7, Form B, page 2

Identify each system as *consistent and independent*, *consistent and dependent*, or *inconsistent*.

13. $\begin{cases} 3x + 2y = 6 \\ 9x + 6y = 18 \end{cases}$ **consistent and dependent**

14. $\begin{cases} y + 2x = 10 \\ y - 2x = 6 \end{cases}$ **consistent and independent**

15. $\begin{cases} y - 3x = 1 \\ y = 3x - 1 \end{cases}$ **inconsistent**

Answer the following questions by referring to the graph:

16. Write the equation of each line:
    $\overleftrightarrow{AB}$ **$x - y = -4$**
    $\overleftrightarrow{CD}$ **$2x + y = -2$**

17. Write the equation of a line that would form an inconsistent system with line AB.
    any equation with a slope of 1 and a y-intercept not equal to 4

18. Write the equation of a line that would form a dependent system with line CD.
    any equation whose coefficients are a multiple of those in $2x + y = -2$

Graph the solution set of each system on the grid provided.

19. $\begin{cases} y \leq x + 3 \\ y \geq x + 2 \end{cases}$

20. $\begin{cases} y - 2x > 2 \\ 2y - x < -4 \end{cases}$

21. A parking meter contains $8.25 in dimes and quarters. If the number of quarters is 9 more than twice the number of dimes, how many of each coin is in the parking meter? **10 dimes and 29 quarters**

22. The sum of the ages of Jon and his brother is 40. Jon's age is 8 more than his brother's age. Find the age of each. **John is 24, and his brother is 16.**

23. Traveling with the wind, a plane takes $2\frac{1}{2}$ hours to fly a distance of 1500 miles. The return flight against the wind takes 3 hours. Find the speed of the plane and the speed of the wind. **The speed of the plane is 550 miles per hour, and the rate of wind is 50 miles per hour.**

CHAPTER 7 REVIEW **363**

31.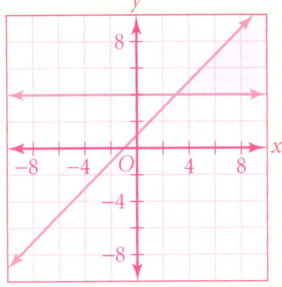

32.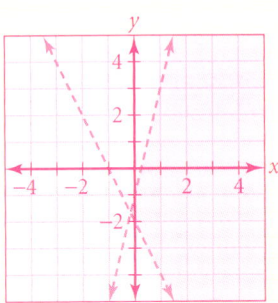

33.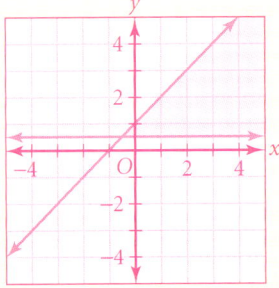

34.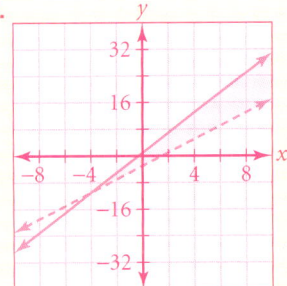

35.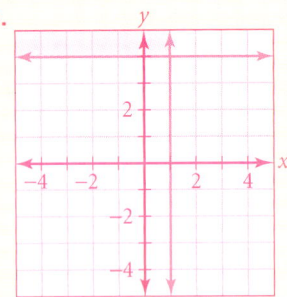

LESSON 7.5
Key Skills

Graph the solution to a system of linear inequalities.

Graph the solution set for this system:
$$\begin{cases} y < -x - 1 \\ y \leq x + 1 \end{cases}$$

Graph $y < -x - 1$ with a dashed boundary line and graph $y \leq x + 1$ with a solid boundary line. In both inequalities, y is less than the expression on the right, so shade the area of the graph below each boundary line. The solution is the region of the plane where the shaded regions overlap.

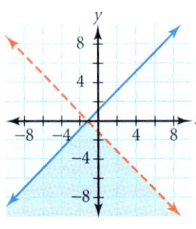

Exercises

Solve each system of linear inequalities by graphing.

31. $\begin{cases} y \geq 4 \\ y \leq x + 1 \end{cases}$ 32. $\begin{cases} 4x - y > 1 \\ 2x + y > -2 \end{cases}$

33. $\begin{cases} y - 1 \leq x \\ 4y \geq 2 \end{cases}$ 34. $\begin{cases} 2x - y < 3 \\ 2y - 2 \leq 6x \end{cases}$

35. $\begin{cases} x \leq 1 \\ y \geq 4 \end{cases}$ 36. $\begin{cases} -x < y + 7 \\ y \geq x - 7 \end{cases}$

LESSON 7.6
Key Skills

Solve traditional math puzzles.

Mr. Carver invested $5000, part at 5% interest and the rest at 3.9% interest. If he earned $233.50 in interest the first year, how much did he invest at each rate?

Make a table to organize the information.

Amount	x	y	$5000
Interest	$0.05x$	$0.039y$	$233.50

From the table, write two equations. $\begin{cases} x + y = 5000 \\ 0.05x + 0.039y = 233.50 \end{cases}$

The solution to the system is (3500, 1500). Mr. Carver invested $3500 at 5% and $1500 at 3.9%.

Exercises

Use a system of equations to solve each problem.

37. Samantha has a total of 43 quarters and nickels in her purse. If her change is worth $7.75, how many nickels and how many quarters does Samantha have? **15 nickels, 28 quarters**

38. A chemist wants to combine a solution that is 30% glycerin with pure (100%) glycerin. He wants to produce 42 milliliters of a solution that is 50% glycerin. How much of the 30% solution and how much of the pure glycerin should the chemist use?
30 ml of 30% solution
12 ml of 100% solution

Applications

39. **CHEMISTRY** How many ounces of a 25% acid solution should be mixed with a 50% acid solution to produce 100 ounces of a 40% acid solution? **40 ounces**

40. **GEOMETRY** The sum of the measures of angles x, y, and z of triangle XYZ is 180°. Angle x has a measure of 60°. The measure of angle y is 15° more than twice the measure of the angle z. What are the measures of angles y and z? **$y = 85°, z = 35°$**

36.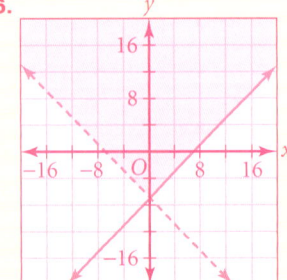

7 Alternative Assessment

Performance Assessment

1. Compare the strengths and weaknesses of the graphing, substitution, and elimination methods for solving systems of equations. Summarize your results in a chart.

2. The steps for solving a system of equations are listed below, but an error is present. Find and describe the error. Then find the correct solution.

   System: $\begin{cases} y = 5x - 2 \\ y - 2 = 5x \end{cases}$ 1st step: $\begin{cases} 5x + y = -2 \\ -5x + y = 2 \end{cases}$

   2nd step: $2y = 0$, $y = 0$ Solution: $\left(-\frac{2}{5}, 0\right)$

3. Suppose that you are the owner of a regional delivery service and that you are contemplating the purchase of 5 new delivery trucks. You have narrowed the possible choices to two models, and you know the following information:

	Model A	Model B
Purchase price	$35,000	$45,000
Expense	$0.40/mile	$0.38/mile
Miles driven per year	20,000	20,000
Useful life	8 years	12 years

   Make a graph to illustrate the options. Decide which model to purchase and justify your decision.

Portfolio Projects

1. Write a paragraph describing how to use a graphics calculator to solve a system of linear equations.

2. Draw the graph of a system of equations that is dependent, of one that is independent, and of one that is inconsistent. Describe a way to determine whether a system is dependent, independent, or inconsistent by comparing the slopes and the y-intercepts.

3. Research the cost of cable TV and the cost of a VCR and tape rental. Model each option with a linear equation and form a system (similar to the problem on pages 318–319 in Lesson 7.1). Draw a graph of the system, and describe which is a better option and why.

Peer Assessment

Complete this activity with a partner. Begin by writing a system of equations. You may want to model your system after some of the exercises from this chapter, or use exercises that you have not yet been assigned. Your partner should choose a method, solve the system, and then write a system for you to solve.

Check your partner's work, correct any errors, and then solve the new system. Continue with this pattern until you have each solved 10 systems of equations. During this activity, try to include systems that can be most easily solved by using each of the methods that you learned in this chapter, graphing, substitution, and elimination.

internetconnect

The HRW Web site contains many resources to reinforce and expand your knowledge of systems of equation and inequalities. This Web site also provides Internet links to other sites where you can find information and real-world data for use in research projects, reports, and activities that involve solving systems of equations. Visit the HRW Web site at go.hrw.com, and enter the keyword **MA1 CH7**.

Performance Assessment

1. Answers may vary. Sample answer:

Method	Strengths	Weaknesses
Graphing	Gives the visual interpretation of the solution. Good for tedious numbers.	Difficult to locate exact solutions.
Substitution	Easy to solve if one equation is already solved for one of the variables.	Can be tedious if equations are in standard form or the numbers are cumbersome.
Elimination	Easy to solve if both equations are in the same form.	May require up to 2 extra steps to set up the coefficients for elimination.

2. The error is Step 1. Step 1 should be
   $\begin{cases} -5x + y = -2 \\ -5x + y = 2 \end{cases}$

   The graphs of these equations have the same slope but different y-intercepts which makes them parallel. Therefore there is no solution.

3. Check students' graphs. Choose model A since it will cost less over the life of both trucks.

Alternative Assessment

Portfolio Projects

1. Answers may vary. Students may include details about the intersect function, the trace function, or the importance of an appropriate viewing window.

2. Answers may vary. Sample answers:

   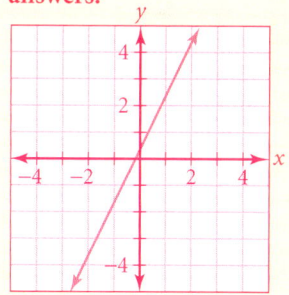

   dependent

   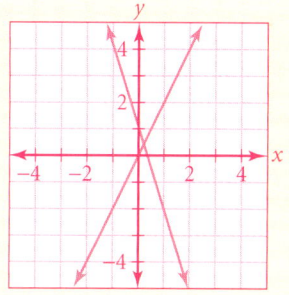

   independent

   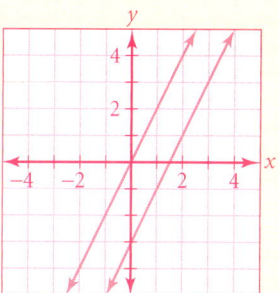

   inconsistent

   In a dependent system, one equation is a constant multiple of the other. In an independent system, the equations have different y intercepts and different slopes (which are not constant multiples of each other). In an inconsistent system, the equations have the same slope but different y-intercepts.

3. Answers may vary. Other examples are nutrition analysis and agricultural planning.

Cumulative Assessment

College Entrance Exam Practice

Multiple-Choice and Quantitative-Comparison Samples

The first half of the Cumulative Assessment contains two types of items found on standardized tests—multiple-choice questions and quantitative-comparison questions. Quantitative-comparison items emphasize the concepts of equality, inequality, and estimation.

Free-Response Grid Samples

The second half of the Cumulative Assessment is a free-response section. This part of the Cumulative Assessment requires student-produced response items like those commonly found on college entrance exams. These questions require the use of machine-scored answer grids. You may wish to have students practice answering these items in preparation for standardized tests.

CHAPTERS 1–7 CUMULATIVE ASSESSMENT

College Entrance Exam Practice

QUANTITATIVE COMPARISON For Questions 1–5, write
A if the quantity in Column A is greater than the quantity in Column B;
B if the quantity in Column B is greater than the quantity in Column A;
C if the two quantities are equal; or
D if the relationship cannot be determined from the information given.

	Column A	Column B	Answers	
1.	x where $\begin{cases} 3x + 2y = 6 \\ x + y = 0 \end{cases}$	y	Ⓐ Ⓑ Ⓒ Ⓓ [Lesson 7.2]	**A**
2.	0.37	$\frac{37}{10}$	Ⓐ Ⓑ Ⓒ Ⓓ [Lesson 4.2]	**B**
3.	The eighth term of the sequence 2, 5, 8, 11, 14, …	The eighth term of the sequence 32, 30, 28, 26, 24, …	Ⓐ Ⓑ Ⓒ Ⓓ [Lesson 1.1]	**A**
4.	The solution to $x + 3 \geq 4$	The solution to $x - 7 < 10$	Ⓐ Ⓑ Ⓒ Ⓓ [Lesson 6.1]	**D**
5.	The slope of the line passing through the points (1, 2) and (0, 3)	The slope of the line $4x + 4y = 12$	Ⓐ Ⓑ Ⓒ Ⓓ [Lesson 5.5]	**C**

6. Which expression is the solution to the equation $-4g = -64$? **(LESSON 3.2)** d
  a. $\frac{-64}{4}$
  b. $(-64)(-4)$
  c. -16
  d. 16

7. What is the reciprocal of $\frac{1}{6}$? **(LESSON 2.4)** c
  a. 1.6
  b. -1.6
  c. 6
  d. -6

8. What are the first three terms of a sequence if the fourth and fifth terms are 27 and 39 and the second differences are a constant 2? **(LESSON 1.1)** b
  a. 21, 23, 25
  b. 3, 9, 17
  c. 6, 8, 10
  d. 13, 15, 17

9. If a basketball league has 8 teams and each team plays each of the other teams once, how many games will be played? **(LESSON 1.1)** a
  a. 28
  b. 36
  c. 72
  d. 56

10. What is $|8.6|$? **(LESSON 2.1)** c
  a. 8
  b. -8
  c. 8.6
  d. -8.6

11. Which point is on the line $3x - 8y = 1$? **(LESSON 5.1)** c
  a. $M(7, -4)$
  b. $N(-2, -3)$
  c. $O(3, 1)$
  d. $P(4, 2)$

12. Which number line shows the solution set to the inequality $x - 6 > -4$? (**LESSON 6.1**) **d**

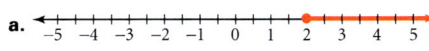

13. What is the solution to $\frac{3}{5}p = -3$?
    (**LESSON 3.2**) **c**
    a. $-\frac{9}{5}$ b. $-\frac{5}{9}$
    c. -5 d. $-\frac{1}{5}$

14. What is 65% of 120? (**LESSON 4.2**) **a**
    a. 78 b. 185
    c. 7800 d. 780

15. Which of the following lines is vertical?
    (**LESSON 5.2**) **b**
    a. $y = -5$ b. $x = 3$
    c. $y = -3x + 1$ d. $y = 2x - 2$

Evaluate. (**LESSON 6.4**)

16. $|-2.2|$ **2.2** 17. ABS(7) **7** 18. ABS(−9.6) **9.6**

Find each sum. (**LESSON 2.2**)

19. $-25 + (-4)$ **−29** 20. $-36 + 6 + (-6)$ **−36**
21. $452 + (-452)$ **0**

Solve each equation. (**LESSON 3.3**)

22. $3t + \frac{1}{6} = \frac{2}{3}$ $t = \frac{1}{6}$ 23. $2a - 5 = 11$ $a = 8$
24. $\frac{r}{4} + 6 = -4$ $r = -40$

25. What is the y-intercept of the line that is parallel to the line $2x + 3y = 4$ and contains the point $(3, -1)$? (**LESSON 5.6**) **1**

26. Solve $4 - 3t \geq 19$. Show the solution set on a number line. (**LESSON 6.2**)

27. Simplify: $\frac{15 - 36x}{3}$. (**LESSON 2.7**) **5 − 12x**

28. Write an equation for the line passing through the points $(7, -7)$ and $(-2, 4)$. (**LESSON 5.4**)
    $y = -\frac{11}{9}x + \frac{14}{9}$

29. Mark has 3 nickels, 5 dimes, and 7 quarters in his pocket. He chooses one coin. What is the probability that the coin is a dime?
    (**LESSON 4.3**) $\frac{1}{3}$

30. The perimeter of a rectangle is 52 and its width is 6 less than its length. Find the length and width. (**LESSON 7.2**) **L = 16, W = 10**

Solve each system. (**LESSONS 7.1, 7.2, AND 7.3**)

31. $\begin{cases} y = 7x \\ 2x - y = -10 \end{cases}$ **(2, 14)** 32. $\begin{cases} 5x + 2y = 8 \\ x + y = 1 \end{cases}$ **(2, −1)**

FREE-RESPONSE GRID
The following questions may be answered by using a free-response grid such as that commonly used by standardized test services:

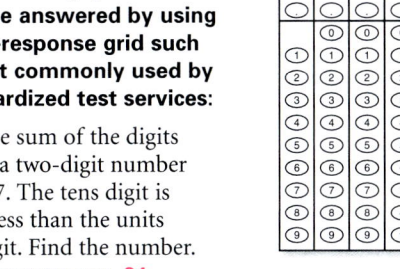

33. The sum of the digits of a two-digit number is 7. The tens digit is 1 less than the units digit. Find the number.
    (**LESSON 7.6**) **34**

34. What is the solution to the proportion $\frac{c}{16} = \frac{3}{8}$? (**LESSON 4.1**) **6**

35. A suit is marked down from an original price of $75 to $63.75. By what percent of the original price has the suit been marked down?
    (**LESSON 4.2**) **15%**

36. Simplify: $4^2 \div 8 + 5(8 - 2) \cdot 2$. (**LESSON 1.3**) **62**

37. How much would each of 16 people need to contribute toward a gift that costs $136?
    (**LESSON 2.4**) **$8.50**

38. Solve $7(p + 4) = 49$. (**LESSON 3.5**) **3**

39. Find the slope of the line that passes through the origin and $(4, 2)$. (**LESSON 5.2**) $\frac{1}{2}$

40. 16 is 20% of what number? (**LESSON 4.2**) **80**

---

26. $t \leq -5$

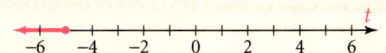

8 Exponents and Exponential Functions

CHAPTER PLANNING GUIDE

Lesson	8.1	8.2	8.3	8.4	8.5	8.6	8.7	Projects and Review
Pupil Edition Pages	370–376	377–382	383–389	390–395	398–403	404–408	409–415	396–397, 416–423
Extra Practice (Pupil Edition)	759	759	760	760	761	761	762	
Practice Workbook	43	44	45	46	47	48	49	
Reteaching Masters	85–86	87–88	89–90	91–92	93–94	95–96	97–98	
Enrichment Masters	43	44	45	46	47	48	49	
Cooperative-Learning Activities	43	44	45	46	47	48	49	
Lesson Activities	43	44	45	46	47	48	49	
Problem Solving/Critical Thinking	43	44	45	46	47	48	49	
Student Study Guide	43	44	45	46	47	48	49	
Spanish Resources	43	44	45	46	47	48	49	
Student Technology Guide	43	44	45	46	47	48	49	
Assessment Resources	96	97	98	99	101	102	103	100, 104–109
Teaching Transparencies	40			41			42	
Quiz Transparencies	8.1	8.2	8.3	8.4	8.5	8.6	8.7	
Writing Activities for Your Portfolio								22–24
Tech Prep Masters								35–38
Long-Term Projects								29–32

LESSON PACING GUIDE

Traditional	2 days	2 days	2 days	1 day	1 day	2 days	2 days	2 days
Block	1 day	1 day	1 day	$\frac{1}{2}$ day	$\frac{1}{2}$ day	1 day	1 day	1 day
Two-Year	4 days	4 days	4 days	2 days	2 days	4 days	4 days	4 days

CONNECTIONS AND APPLICATIONS

Lesson	8.1	8.2	8.3	8.4	8.5	8.6	8.7	Review
Algebra	370–376	377–382	383–389	390–395	398–403	404–408	409–415	418–423
Geometry	373–375	380, 382	385, 388		403			
Business and Economics			389		402		410, 411, 413, 414	420
Number Theory				395				
Probability				394			414	
Science and Technology	372, 374, 375		383, 388	394	398–403	408	409, 412, 413–415	420
Sports and Leisure		382						
Cultural Connection: Europe				395		408		
Other					402	404, 405, 407, 408	412, 413, 414	

BLOCK-SCHEDULING GUIDE

Day	Lesson	Teacher Directed: Lesson Examples, Teaching Transparencies	Student Guided: Activity, Try This	Cooperative-Learning Activity, Lesson Activity, Student Technology Guide	Practice: Practice & Apply, Extra Practice, Practice Workbook	Assessment: Quiz, Mid-Chapter Assessment	Problem Solving, Reteaching
1	8.1	10 min	15 min	15 min	65 min	15 min	15 min
2	8.2	10 min	15 min	15 min	65 min	15 min	15 min
3	8.3	10 min	15 min	15 min	65 min	15 min	15 min
4	8.4	10 min	15 min	15 min	65 min	15 min	15 min
5	8.5	15 min	15 min	15 min	65 min	15 min	15 min
6	8.6	10 min	15 min	15 min	65 min	15 min	15 min
7	8.7	15 min	15 min	15 min	65 min	15 min	15 min
8	Assess.	50 min PE: Chapter Review	90 min PE: Chapter Project, Writing Activities	90 min Tech Prep Masters	65 min PE: Chapter Assessment, Test Generator	30 min Chap. Assess. (A or B), Alt. Assess. (A or B), Test Generator	

PE: Pupil's Edition

Alternative Assessment

The following are suggestions for an alternative assessment for students who may benefit from a different type of assessment than the regular chapter quizzes and the mid-chapter and end-of-chapter assessment materials. Many of the questions are open-ended, and students' answers will vary.

Performance Assessment

1. You are given the following tasks:
   - Multiply two powers with the same base.
   - Divide two powers with the same base.
   - Raise a monomial in x to a power.

   a. Using examples, explain how to carry out each task. State any laws of exponents that you would use to complete the task.

   b. Simplify $\frac{(2x^3)(3x)^3}{(5x)^2}$ by using some combination of the laws of exponents. Give a detailed solution with a justification for each step.

2. a. The equation $A = P(1 + r)^t$ can represent exponential growth or decay. Explain when it represents growth and when it represents decay. Give real-world examples to support your explanation.

   b. If money grows at a rate of 5% per year compounded annually, make a table to show how that money grows over the first 20 years.

Portfolio Project

Suggest that students choose one of the following projects for inclusion in their portfolios.

1. Hermaine says that when you multiply powers of 2, divide powers of 2, or raise a power of 2 to an integer power, you get another power of 2. Do you believe Hermaine? Use examples to test this claim and support your response.

2. Mr. Knox told Zelda that an object worth $1000 was changing in value by 3% per year. Zelda asked herself whether 3% was a rate of growth or a rate of decay. Write a report that shows what happens in each case. How important is it to report whether a rate is one of growth or decay? What is the difference in value after 10 years given each interpretation?

---

internetconnect

The table below identifies the pages in this chapter that contain technology information and support in the side columns.

Content Links	
Lesson Links	pages 374, 402, 408
Portfolio Links	pages 376, 395, 415
Graphics Calculator Support	page 782

Resource Links

For information about teacher and parent resources as well as professional development help, visit **www.hrw.com/math**.

Technical Support HRW has assembled a team of dedicated technical and teaching professionals and a comprehensive service program to provide you with the support you need.

- The HRW Technical Support Line operates from 7 A.M. to 6 P.M. central time, Monday through Friday, at (800) 323-9239.
- The HRW Technical Support Center on the World Wide Web is available 24 hours a day, seven days a week, at **www.hrwtechsupport.com**.
- You can e-mail our Technical Support Center at **tsc@hrwtechsupport.com**.
- The Technical Support Center's fax-on-demand service at (800) 352-1680 offers solutions to common problems and answers to frequently asked questions.

Technology

Lesson Suggestions and Calculator Examples
(Keystrokes are based on a TI-83 calculator.)

Lesson 8.1 Laws of Exponents: Multiplying Monomials
Have students evaluate the expressions below.

$$2^3 \times 2^4 - 2^7$$
$$2^1 \times 2^3 - 2^4$$
$$2^3 \times 2^3 - 2^6$$

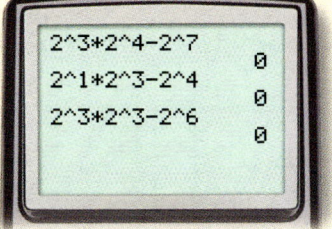

The evaluations are shown in the display at right. The 0 result tells students that the two parts of each subtraction are equal. This exercise can help students understand the Product-of-Powers Property.

Lesson 8.2 Laws of Exponents: Powers and Products
This lesson is a continuation of Lesson 8.1. Have students evaluate the expressions below.

$$(2^3)^2 - 2^6$$
$$(2^2)^2 - 2^4$$
$$(2^1)^5 - 2^5$$

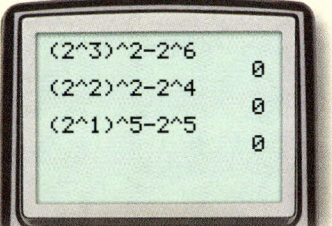

Notice that although all the exponents and parentheses are not needed in order to evaluate, their presence helps students see the pattern.

Lesson 8.3 Laws of Exponents: Dividing Monomials
Students can enhance their knowledge about the laws of exponents by using a calculator. Continue the exploration technique suggested for Lessons 8.1 and 8.2. This time, however, have students create expressions to represent the Quotient-of-Powers Property, which is stated below.

If x is a nonzero number and m and n are positive integers with $m > n$, then $\frac{x^m}{x^n} = x^{(m-n)}$.

The following expressions and calculator display at right resemble what students may suggest. Have students use a graphics calculator to solve.

$$2^4 \div 2^3 - 2^{(4-3)}$$
$$2^6 \div 2^2 - 2^{(6-2)}$$
$$2^8 \div 2^3 - 2^{(8-3)}$$

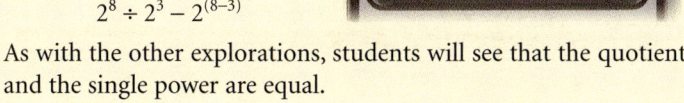

As with the other explorations, students will see that the quotient and the single power are equal.

Lesson 8.4 Negative and Zero Exponents
In this lesson, students will extend the laws of exponents to a larger domain. Continue the exploration technique suggested for Lessons 8.1 through 8.3 with zero and negative integers.

Lesson 8.5 Scientific Notation
This lesson will teach students how to use the **EE** function of a graphics calculator. This can be accessed by pressing [2nd] [,]. For example, to enter 1.49×10^8 into the calculator in scientific notation, press **1.49** [2nd] [,] **8** [ENTER].

The display at right shows the quotient $\frac{1.49 \times 10^8}{2.98 \times 10^5}$. Notice that the result is shown in decimal form.

Lesson 8.6 Exponential Functions
When you introduce students to a new class of function, it is helpful to contrast the new function with functions that the students have already seen. Consider having students graph the three functions below.

$$f(x) = 2x \qquad g(x) = x^2 \qquad h(x) = 2^x$$

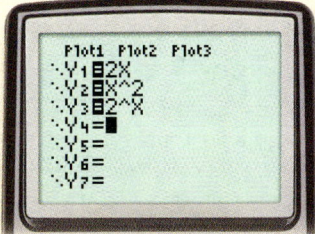
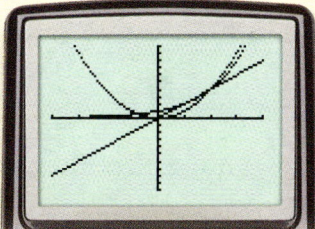

Ask students to compare the graph of the new function, $h(x) = 2^x$, with the graphs of the others.

Lesson 8.7 Applications of Exponential Functions
In the equation $A = P(1 + r)^t$, A and P represent amounts of the same type and t represents elapsed time. The equation can be applied to money, population, and other problems involving exponential growth or decay. Encourage students to examine such models by using equations, tables, and graphs.

For further information, refer to the
- technology discussions in the lessons.
- lesson-related teacher's commentary in the side columns of this *Teacher's Edition*.
- lesson-related *Student Technology Guide* masters.
- *HRW Technology Handbook*.

Teaching Resources

Basic Skills Practice
(2 pages per skill)

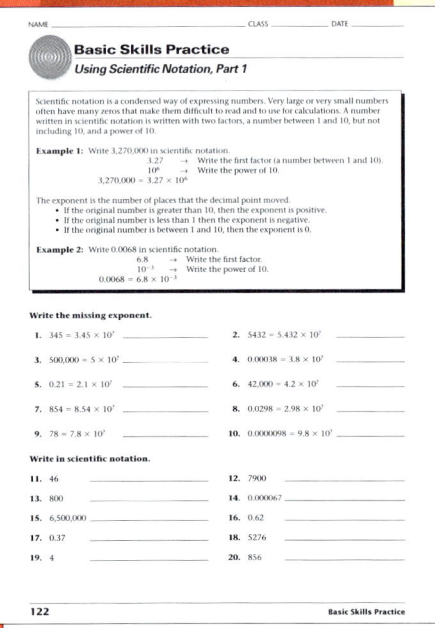

Reteaching
(2 pages per lesson)

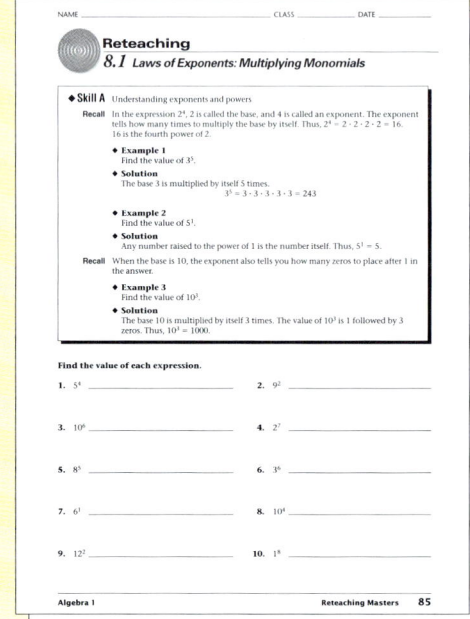

Enrichment
(1 page per lesson)

Cooperative-Learning Activity
(1 page per lesson)

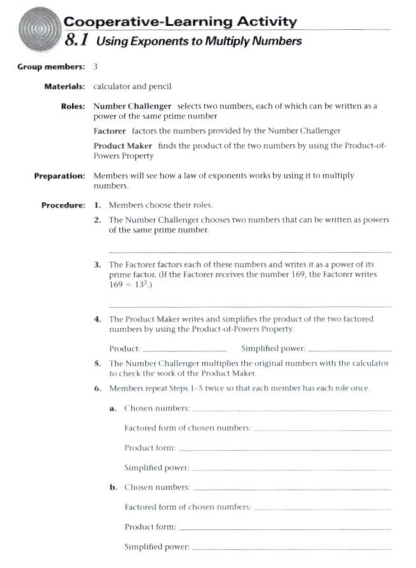

Lesson Activity
(1 page per lesson)

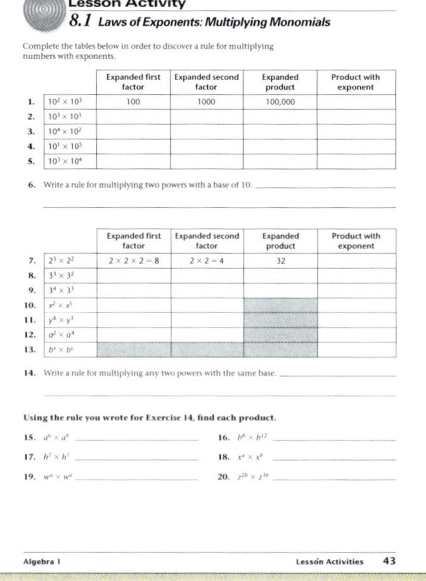

Long-Term Project
(4 pages per chapter)

Problem Solving/Critical Thinking
(1 page per lesson)

Writing Activities
(3 pages per chapter)

Tech Prep Masters
(4 pages per chapter)

Spanish Resources
(1 page per lesson)

Quizzes
(1 page per lesson)

Mid-Chapter Test
(1 page per chapter)

CHAPTER 8 INTERLEAF 368F

About Chapter 8

Background Information

In Chapter 8, students study exponential expressions. They learn the laws of exponents and see how these laws are used to simplify expressions containing positive, negative, and zero exponents. They also learn how to find products and quotients of monomials. The chapter ends with an introduction to exponential functions.

CHAPTER RESOURCES

- Block-Scheduling Handbook
- Writing Activities for Your Portfolio
- Tech Prep Masters
- Long-Term Project
- Assessment Resources:
   Mid-Chapter Assessment
   Chapter Assessments
   Alternative Assessments
- Test and Practice Generator
- Technology Handbook
- End-of-Course Exam Practice

Chapter Objectives

- Define *exponents* and *powers*. [8.1]
- Find products of powers. [8.1]
- Simplify products of monomials. [8.1]
- Find the power of a power. [8.2]
- Find the power of a product. [8.2]

Exponents and Exponential Functions

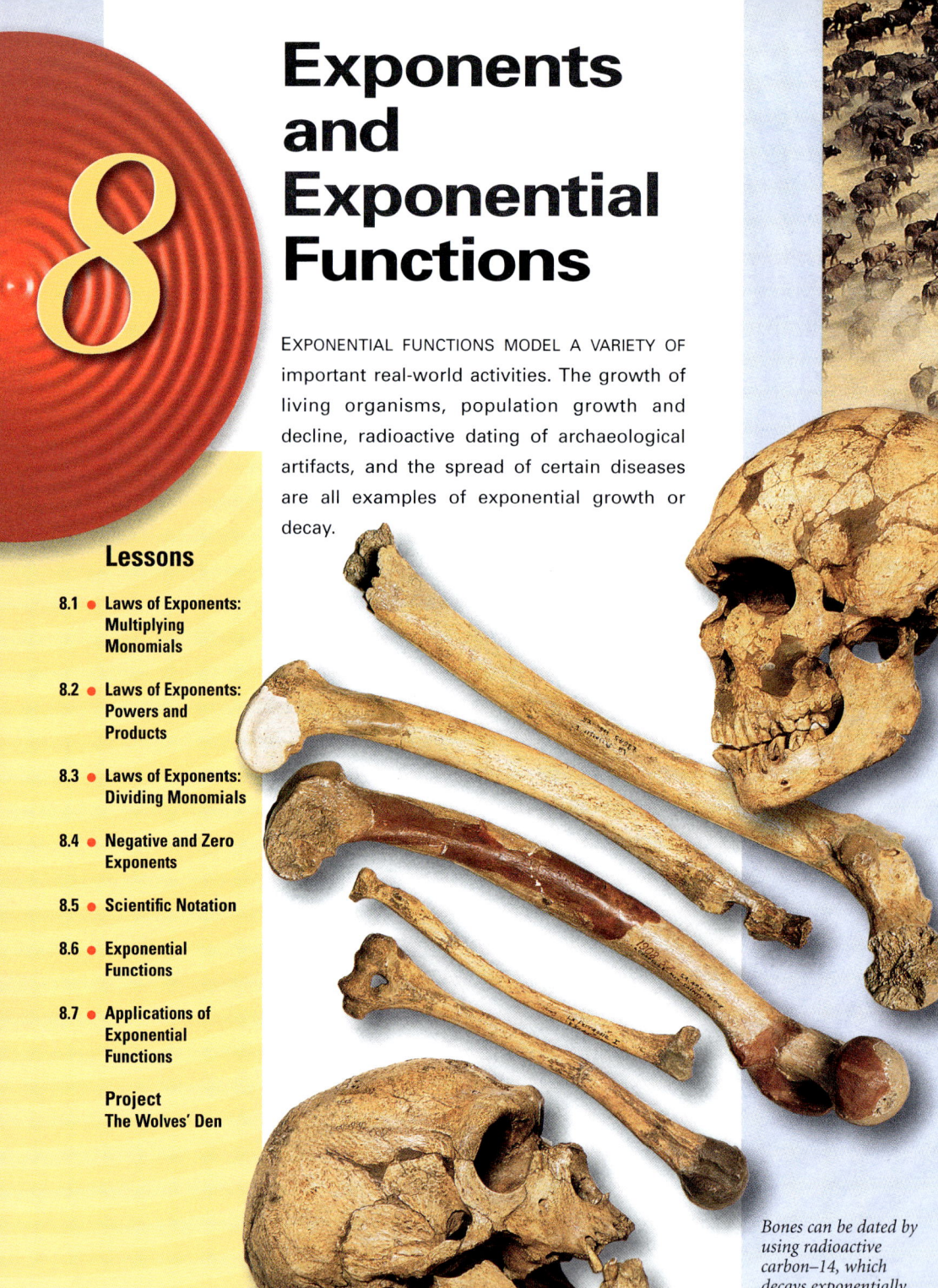

EXPONENTIAL FUNCTIONS MODEL A VARIETY OF important real-world activities. The growth of living organisms, population growth and decline, radioactive dating of archaeological artifacts, and the spread of certain diseases are all examples of exponential growth or decay.

Lessons

8.1 • Laws of Exponents: Multiplying Monomials

8.2 • Laws of Exponents: Powers and Products

8.3 • Laws of Exponents: Dividing Monomials

8.4 • Negative and Zero Exponents

8.5 • Scientific Notation

8.6 • Exponential Functions

8.7 • Applications of Exponential Functions

Project
The Wolves' Den

Bones can be dated by using radioactive carbon–14, which decays exponentially.

About the Photos

Carbon-14 is a rare form of carbon found in living things. Because the rate of decay of carbon-14 is constant, scientists are able to approximate the age of organic materials by measuring the amount of carbon-14 remaining in an organism. The half-life of carbon, which is the time it takes for one-half of the radioactive atoms to decay, is 5730 years. Thus, a sample that initially contained 24 grams of carbon-14 will have 12 grams of carbon-14 left after 5730 years and 6 grams of carbon-14 left after 11,460 years.

The explosive, exponential growth of certain bacteria can cause the rapid spread of a disease.

- Simplify quotients of powers. [**8.3**]
- Simplify powers of fractions. [**8.3**]
- Understand the concepts of negative and zero exponents. [**8.4**]
- Simplify expressions containing negative and zero exponents. [**8.4**]
- Recognize the need for special notation in scientific calculations. [**8.5**]
- Perform computations involving scientific notation. [**8.5**]
- Understand exponential functions and how they are used. [**8.6**]
- Recognize differences between graphs of exponential functions with different bases. [**8.6**]
- Use exponential functions to model applications that include growth and decay in different contexts. [**8.7**]

About the Chapter Project

You have already seen how algebra makes it possible to use *functions* to model data. In the Chapter Project, *The Wolves' Den*, on page 416, you will compare linear and exponential models of data and decide which works better for the population that you study. After completing the Chapter Project, you will be able to do the following:

- Determine whether trends in given data sets are linear or exponential.

About the Portfolio Activities

Throughout the chapter, you will be given opportunities to complete Portfolio Activities that are designed to support your work on the Chapter Project.

- Comparing linear and exponential growth is the topic of the Portfolio Activity on page 376.
- Population growth is the topic of the Portfolio Activity on page 395.
- Compound interest is the topic of the Portfolio Activity on page 415.

Portfolio Activities appear at the end of Lessons 8.1, 8.4, and 8.7. Each serves as preparation for the Chapter Project. The Portfolio Activities as well as the Chapter Project Activities are appropriate for inclusion in the student's portfolio. Students should be encouraged to include in their portfolios any other work in which they feel a sense of pride or a sense of accomplishment.

internetconnect

Internet connect features appearing throughout the chapter provide keywords for the HRW Web site that lead to links for math resources, tutorial assistance, references for student research, classroom activities, and all teaching resource pages. The HRW Web site also provides updates to the technology used in the text. Listed at right are the keywords for the Internet Connect activities referenced in this chapter. Refer to the side column on the page listed to read about the activity.

LESSON	KEYWORD	PAGE
8.1	MA1 Number Sieves	374
8.1	MA1 Doubling	376
8.4	MA1 Centenarians	395
8.5	MA1 Scientific Notation	402
8.6	MA1 Snowflake	408
8.7	MA1 Compound Interest	415

CHAPTER 8 **369**

Prepare

NCTM Principles & Standards 1–6, 8–10

LESSON RESOURCES

- Practice 8.1
- Reteaching Master 8.1
- Enrichment Master 8.1
- Cooperative-Learning Activity 8.1
- Lesson Activity 8.1
- Problem Solving/Critical Thinking Master 8.1
- Teaching Transparency 40

QUICK WARM-UP

Evaluate.

1. $2 \cdot 2 \cdot 2 \cdot 2$	16
2. $(2)(3)(4)$	24
3. $(2)(-3)(4)$	-24
4. $(-2)(-3)(-4)$	-24
5. $(-2)(3)(-4)$	24
6. $(-14)(-7)$	98

Also on Quiz Transparency 8.1

Teach

Why Ask students what is meant by a *property* of an operation. Lead them to see that a property is a rule that governs the operation. Have them give several examples. Then review the meaning of *exponent*. Point out that an exponent is a mathematical notation that has a special meaning. In this lesson and the lessons that follow, they will learn that exponents have their own special set of properties, called *laws of exponents*.

8.1 Laws of Exponents: Multiplying Monomials

Objectives
- Define exponents and powers.
- Find products of powers.
- Simplify products of monomials.

Why Real-world calculations frequently involve multiplying numbers by themselves. The area of a square, for example, requires that you multiply the length of a side by itself.

The expression 2^{1000} represents 2 to the thousandth power. It is the product of a thousand 2s multiplied together. The number 2 is called the **base**, and 1000 is the **exponent**. Recall that a base and exponent together are called a power.

An exponent tells how many times a number is used as a factor.

Exponents

For all real numbers x and all positive integers m,

$$x^m = \underbrace{x \cdot x \cdot x \cdot \cdots \cdot x}_{m \text{ factors}}.$$

Notice that when $m = 1$, $x^m = x^1 = x$.

EXAMPLE ① Evaluate.
    a. 2^8 **b.** 5^3 **c.** 3^4 **d.** 6^1

SOLUTION
a. $2^8 = 2 \cdot 2 \cdot 2 \cdot 2 \cdot 2 \cdot 2 \cdot 2 \cdot 2 = 256$
b. $5^3 = 5 \cdot 5 \cdot 5 = 125$
c. $3^4 = 3 \cdot 3 \cdot 3 \cdot 3 = 81$
d. $6^1 = 6$

Alternative Teaching Strategy

USING PATTERNS Have students complete the following table:

$10^1 \cdot 10^1 = 10 \cdot 10 = 100 = 10^?$ 2
$10^1 \cdot 10^2 = 10 \cdot 100 = 1000 = 10^?$ 3
$10^1 \cdot 10^3 = 10 \cdot 1000 = 10{,}000 = 10^?$ 4
$10^1 \cdot 10^4 = 10 \cdot 10{,}000 = 100{,}000 = 10^?$ 5
$10^2 \cdot 10^2 = 100 \cdot 100 = 10{,}000 = 10^?$ 4
$10^2 \cdot 10^3 = 100 \cdot 1000 = 100{,}000 = 10^?$ 5
$10^2 \cdot 10^4 = 100 \cdot 10{,}000 = 1{,}000{,}000 = 10^?$ 6

Tell students to study the patterns they see in the table. Ask them to make a conjecture about the product of two powers that have the same base. **When you multiply two powers that have the same base, you can add the exponents.**

Have students test their conjecture with the products below. Tell them to show the answer both as a power of the given base and in simplified form. Encourage them to check their results with a calculator.

1. $2^4 \cdot 2^5$ $2^9 = 512$ **2.** $5^2 \cdot 5^4$ $5^6 = 15{,}625$
3. $7^3 \cdot 7$ $7^4 = 2401$ **4.** $3^2 \cdot 3^4 \cdot 3$ $3^7 = 2187$

370 LESSON 8.1

TRY THIS Evaluate.
a. 2^6 b. 4^3 c. 3^3 d. 8^1

Product of Powers

There is a simple and convenient way to multiply powers with the same base. In the activity that follows, you will discover this multiplication rule yourself.

Activity: Multiplying Powers

1. Examine the example below.

   In decimal notation: $1000 \cdot 10 = 10{,}000$
   In exponential notation: $10^3 \cdot 10^1 = 10^4$
                             product form / simplified form

   Complete the missing numbers as shown above.
   a. $1000 \cdot 100 = \underline{\ ?\ }$
      $10^3 \cdot 10^2 = \underline{\ ?\ }$
   b. $\underline{\ ?\ } \cdot \underline{\ ?\ } = \underline{\ ?\ }$
      $10^3 \cdot 10^3 = \underline{\ ?\ }$

2. Complete the table below for $10^4 \cdot 10^2$. Write the missing product in simplified form.

Decimal notation	$10{,}000 \cdot 100 = 1{,}000{,}000$
Exponential notation	$10^4 \cdot 10^2 = \underline{\ ?\ }$

3. Write each term as a product of factors. Then use the definition of exponents and powers to write each product below as a single power.
   a. $2^3 \cdot 2^2$ b. $5^1 \cdot 5^3$ c. $10^2 \cdot 10^4$ d. $3^2 \cdot 3^2$

CHECKPOINT ✓
4. For any positive number x and any positive integers m and n, how would you simplify $x^m x^n$? Fill in the box: $x^m x^n = x^\square$.

In the previous activity you may have discovered a property that lets you simplify the product of two powers with the same base. The property tells you to keep the base the same and add the exponents. For example:

$$x^3 \cdot x^5 = x^{3+5} = x^8$$

Product-of-Powers Property

For all nonzero real numbers x and all integers m and n,

$$x^m \cdot x^n = x^{m+n}.$$

Interdisciplinary Connection

CONSUMER ECONOMICS Many people save for the future by annually depositing a fixed amount of money into an interest-earning account. To find the total amount, A, accumulated in such an account, you can use the formula $A = D\left[\frac{(1+r)^n - 1}{r}\right]$, where D represents the fixed amount of each deposit, r represents the annual interest rate, and n is the number of years.

A family wants to prepare for the cost of their child's college education. At the beginning of each year, they plan to deposit $2000 into an account that earns 8% annual interest, compounded annually. How much money will be in the account at the end of the 10th year? **$28,973.12**

ADDITIONAL EXAMPLE 1

Evaluate.
a. 4^3 b. 7^2 c. 5^1 d. 2^4
   64   49   5   16

TRY THIS
a. 64   b. 64
c. 27   d. 8

Use Teaching Transparency 40.

Activity Notes

In this Activity, students develop an understanding of the Product-of-Powers Property through investigation. In Steps 1 and 2, they perform calculations with powers of ten. Students may remember how to find these products by counting zeros. In Step 3, they apply the definition of exponents to expand each factor and then use the definition again to write the product as a single power. Students should conclude that when multiplying powers with the same base, you can add the exponents.

Cooperative Learning

You may wish to have students do the Activity in pairs. In Step 3, have each student complete all the calculations. Then partners can exchange papers and check each other's work. Suggest that each student create two similar exercises. Have partners exchange exercises, find the products, and check each other's work.

CHECKPOINT ✓
4. $m + n$

LESSON 8.1   371

ADDITIONAL EXAMPLE 2

Simplify.

a. $3^2 \cdot 3^3$ $3^5 = 243$

b. $6^2 \cdot 6$ $6^3 = 216$

c. $j^3 \cdot j^4$ j^7

d. $7^a \cdot 7^b$ 7^{a+b}

TRY THIS

a. 3^6 b. 8^4

c. z^9 d. a^{z+3}

ADDITIONAL EXAMPLE 3

Suppose that a colony of bacteria doubles in size every hour. The colony contains 500 bacteria at noon. **How many bacteria will the colony contain at 2 P.M. and at 4 P.M. of the same day?**
2000; 8000

CRITICAL THINKING

a. Solve $x + 3 = 7$ or $2^x \cdot 8 = 128$.

b. Solve $x + 4 = 6$ or $2^x \cdot 16 = 64$.

Teaching Tip

TECHNOLOGY If you are using a TI-82 or TI-83 graphics calculator, you use the [^] key to enter an exponent. For example, you enter 5^2 by pressing 5 [^] 2 [ENTER]. Caution students that when using exponents, they may need to use parentheses. Note, for example, that the key sequence 2 [^] (2 + 2) [ENTER] returns **16**, but the sequence 2 [^] 2 + 2 [ENTER] returns **6**.

EXAMPLE 2 Simplify.

a. $2^3 \cdot 2^4$ b. $8 \cdot 8^3$ c. $y^2 \cdot y^5$ d. $5^m \cdot 5^p$

● **SOLUTION**

a. $2^3 \cdot 2^4 = 2^{3+4} = 2^7 = 128$ b. $8 \cdot 8^3 = 8^{1+3} = 8^4 = 4096$

c. $y^2 \cdot y^5 = y^{2+5} = y^7$ d. $5^m \cdot 5^p = 5^{m+p}$

TRY THIS Simplify.

a. $3^5 \cdot 3^1$ b. $8^2 \cdot 8^2$ c. $z^6 \cdot z^3$ d. $a^z \cdot a^3$

EXAMPLE 3

**APPLICATION
BIOLOGY**

Suppose that a colony of bacteria doubles in size every hour. If the colony contains 1000 bacteria at noon, how many bacteria will the colony contain at 3 P.M. and at 5 P.M. of the same day?

● **SOLUTION**

At 3 P.M. there will be $1000 \cdot 2^3$, or 8000, bacteria. At 5 P.M., two hours later, the bacteria will double two more times. There will be $(1000 \cdot 2^3) \cdot 2^2 = 1000 \cdot 2^{3+2} = 1000 \cdot 2^5$, or 32,000, bacteria in the colony.

CRITICAL THINKING Describe two different ways of solving the following equations for x:

a. $2^x \cdot 2^3 = 2^7$ b. $2^x \cdot 2^4 = 2^6$

**TECHNOLOGY
SCIENTIFIC CALCULATOR**

You can use a graphics calculator to evaluate large numerical products involving exponents. Check the calculator you use to find out how to perform these computations. Most calculators use the [y^x] or [^] key to indicate an exponent. To evaluate the product $8^3 \cdot 8^5$, you might use the following keystrokes:

8 [^] 3 [×] 8 [^] 5 [ENTER]

or 8 [y^x] 3 [×] 8 [y^x] 5 [=]

Inclusion Strategies

AUDITORY LEARNERS Some students become confused when they hear a single exponential expression being read in different ways. Write x^2 and x^3 on the board or overhead and work with the class to make a list like this.

x^2	x^3
x to the second power	x to the third power
the second power of x	the third power of x
x squared	x cubed
the square of x	the cube of x

Enrichment

Numbers of the form $2^p - 1$, where p is a prime number, are called *Mersenne numbers* after French mathematician Marin Mersenne (1588–1648). When a Mersenne number is also a prime number, it is called a *Mersenne prime*. Find the first five Mersenne numbers and determine whether they are Mersenne primes.

3, prime; 7, prime; 31, prime; 127, prime; 2047, not prime

372 LESSON 8.1

Multiplying Monomials

The Product-of-Powers Property can be used to find the product of more complex expressions such as $5a^2b$ and $-2ab^3$. Expressions like $5a^2b$ and $-2ab^3$ are called *monomials*.

Definition of Monomial

A **monomial** is an algebraic expression that is either a constant, a variable, or a product of a constant and one or more variables. The constant is called the **coefficient**.

To simplify a product of monomials, use the following steps:

1. Remove the parentheses and use the Commutative and Associative Properties of Multiplication to rearrange the terms. Group the constants together, and then group like variables together.
2. Simplify by using the Product-of-Powers Property where appropriate.

EXAMPLE 4 Simplify.
  a. $(5t)(-30t^2)$
  b. $(-4a^2b)(-ac^2)(3b^2c^2)$

● SOLUTION
  a. $(5t)(-30t^2) = (5)(-30)(t)(t^2) = -150t^3$
  b. $(-4a^2b)(-ac^2)(3b^2c^2) = (-4)(-1)(3)(a^2)(a)(b)(b^2)(c^2)(c^2) = 12a^3b^3c^4$

TRY THIS Simplify.
  a. $(3m^2)(60mp^2)$
  b. $(8xz)(-10y)(-2yz^2)$

EXAMPLE 5 The volume, V, of a right rectangular prism can be found by using the formula $V = lwh$, where l is the length, w is the width, and h is the height. Suppose that a prism has a length of $2xy$, a width of $3xy$, and a height of $6xyz$. **Find the volume.**

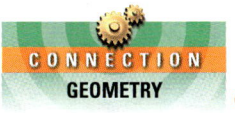

● SOLUTION
$V = lwh$
$= (2xy)(3xy)(6xyz)$
$= (2)(3)(6)(x)(x)(x)(y)(y)(y)(z)$
$= 36x^3y^3z$

CRITICAL THINKING Explain how to use the Product-of-Powers Property to simplify each of the following products:
  a. $x^{3c} \cdot x^{4c}$
  b. $(x+c)^d \cdot (x+c)^{2d}$

Reteaching the Lesson

WORKING BACKWARD To focus students' attention on the addition of the exponents in the Product-of-Powers Property, give them exercises that require them to work backward. For example, write the following on a board or overhead:

$$y^\square \cdot y^2 = y^{10}$$

Ask what number must be placed in the box in order for the exponent on the right side of the equal sign to be 10. Elicit the response 8. Urge students to verbalize reasons for their choice.

Now give them this problem.

$$y^\square \cdot y^\square = y^{10}$$

Work with them to make a complete list of positive integer choices for the missing exponents. 1, 9; 2, 8; 3, 7; 4, 6; 5, 5

Now have them work in pairs to create their own missing-exponent problems by using the form below.

$$(x^\square z^\square)(x^\square z^\square) = (x^\square z^\square)$$

ADDITIONAL EXAMPLE 4

Simplify.
  a. $(-2c^2)(3c^3)$
     $-6c^5$
  b. $(3xy^2)(-5x^2y)(y^3z)$
     $-15x^3y^6z$

TRY THIS
  a. $180m^3p^2$
  b. $160xy^2z^3$

ADDITIONAL EXAMPLE 5

The length of a right rectangular prism is $2a^2b$, its width is $3ab^2$, and its height is $3a^2bc$. **Write an expression for its volume.** $18a^5b^4c$

Math CONNECTION

GEOMETRY A *right prism* is a three-dimensional figure whose two parallel bases are bounded by congruent polygons and whose other faces are rectangular. If the bases also are rectangular, then the prism is a *right rectangular prism*.

CRITICAL THINKING
  a. The base of each term is x, so use the Product-of-Powers Property to add the exponents. $3c + 4c = 7c$, so $x^{3c} \cdot x^{4c} = x^{7c}$.
  b. The base of each term is $x + c$, so use the Product-of-Powers Property to add the exponents. $d + 2d = 3d$, so $(x+c)^d \cdot (x+c)^{2d} = (x+c)^{3d}$.

LESSON 8.1

Assess

Selected Answers
Exercises 6–17, 19–73 odd

ASSIGNMENT GUIDE

In Class	1–17
Core	18–63 odd
Core Plus	18–58 even, 60–64
Review	65–73
Preview	74–78

✐ Extra Practice can be found beginning on page 738.

Exercises

Communicate

1. Use 10^{12} to explain the relationship between the exponent in a power of 10 and the number of zeros in the decimal notation.
2. How can you tell how many times the base is used as a factor when 3^6 is rewritten in factored form?
3. Explain why you add the exponents when multiplying $7^3 \cdot 7^5$.
4. Explain how you could verify that the Product-of-Powers Property is true for $2^3 \cdot 2^4$.
5. The product $x \cdot x$ is equal to x^2. Explain how this illustrates the Product-of-Powers Property.

Guided Skills Practice

Evaluate. (EXAMPLE 1)

6. 5^4 **625**
7. 2^3 **8**
8. 3^4 **81**
9. 11^1 **11**

Simplify each product. (EXAMPLE 2)

10. $2^5 \cdot 2^4$ **512**
11. $f^5 \cdot f^9$ **f^{14}**
12. $t^4 \cdot t^4$ **t^8**
13. $x^t \cdot x^r$ **x^{t+r}**

APPLICATION

14. **BIOLOGY** Suppose that a bacterial colony doubles in size every half-hour. If the colony contains 2000 bacteria at 10 A.M., how many bacteria will the colony contain at 1 P.M. and at 2 P.M.? (EXAMPLE 3) **128,000; 512,000**

Simplify each product. (EXAMPLE 4)

15. $(8p^3)(40m^7p^6)$ **$320m^7p^9$**
16. $(-4x^3z^2)(-6y^5)(-y^3z^7)$ **$-24x^3y^8z^9$**

Exponents can be used to model bacterial growth rates.

CONNECTION

17. **GEOMETRY** The volume, V, of a right rectangular prism can be found by using the formula $V = lwh$, where l is the length, w is the width, and h is the height. Suppose that a prism has a length of $3rt$, a width of $5rt$, and a height of $2rty$. Find the volume. (EXAMPLE 5) **$30r^3t^3y$**

Practice and Apply

Find the value of each expression.

18. 4^5 **1024**
19. 2^6 **64**
20. 3^6 **729**
21. 5^3 **125**
22. 10^4 **10,000**
23. 10^7 **10,000,000**
24. 100^3 **1,000,000**
25. 6^4 **1296**
26. 8^3 **512**
27. 3^7 **2187**
28. 10^8 **100,000,000**
29. 9^4 **6561**
30. 2^8 **256**
31. 2^4 **16**
32. 2^5 **32**
33. 2^2 **4**

internetconnect

GO TO: go.hrw.com
KEYWORD: MA1 Number Sieves

In this activity, students use number sieves to generate sequences of numbers with special properties. A number sieve is a series of tests that filters out unwanted values. Number sieves have been used by mathematicians for hundreds of years.

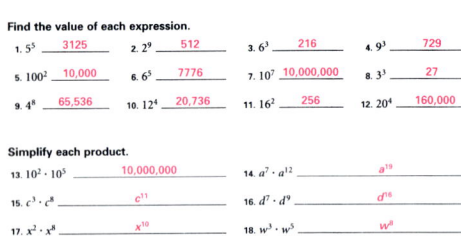

374 LESSON 8.1

Simplify each product. Leave the product in exponent form.

34. $10^3 \cdot 10^4$ **10^7**   35. $10^3 \cdot 10^5$ **10^8**   36. $2^3 \cdot 2^4$ **2^7**   37. $3^4 \cdot 3^2$ **3^6**

38. $10^6 \cdot 10^2$ **10^8**   39. $10^4 \cdot 10^8$ **10^{12}**   40. $4^3 \cdot 4^5$ **4^8**   41. $6^2 \cdot 6^4$ **6^6**

42. Use your answers for Exercises 32 and 33 to evaluate $2^5 \cdot 2^2$. **128**

43. Express the result of Exercise 42 as a single power of 2. **2^7**

Simplify each product.

44. $(6x^2)(4x^2y^3)$ **$24x^4y^3$**   45. $(3x^3z^2)(-6y^5)$ **$-18x^3y^5z^2$**

46. $(5p^3)(-m^8p^2)$ **$-5m^8p^5$**   47. $(-7m^5n^4)(8n^3)$ **$-56m^5n^7$**

48. $(10g^3j^8v^6)(11gj^8)$ **$110g^4j^{16}v^6$**   49. $(-13b^7)(2b^2f^4)$ **$-26b^9f^4$**

50. $(4f^9h^3)(-5f^6)(-3h^2)$ **$60f^{15}h^5$**   51. $(3x^ay^bz^c)(-y^fz^g)$ **$-3x^ay^{b+f}z^{c+g}$**

Some of the following expressions can be written as a whole number power of 10; others cannot. If possible, rewrite the given expression as a whole number power of 10. If not possible, write the expression in decimal notation. Then simplify.

52. $10^4 \cdot 10^2$ **10^6**   53. $10^4 + 10^2$ **10,100**   54. $10^4 - 10^2$ **9,900**   55. 10^{2+3} **10^5**

TECHNOLOGY For exponents, 2^3 is entered as 2 [^] 3 on some calculators and as 2 [x^y] 3 on others. Evaluate 2^3. Familiarize yourself with the way the calculator you are using handles exponents. Then use the calculator to evaluate the following:

56. 2^{20} **1,048,576**   57. 3^{10} **59,049**   58. 0.5^4 **0.0625**   59. 0.8^5 **0.32768**

CONNECTION

60. **GEOMETRY** The area, A, of a trapezoid can be found by using the formula $A = \frac{1}{2}(b_1 + b_2)h$, where b_1 is one base, b_2 is the other base, and h is the height. Suppose that a trapezoid has a base of $5x$, another base of $9x$, and a height of $8xy$. Find the area. **$56x^2y$**

APPLICATION

BIOLOGY Food chains explain why even small organisms affect much larger ones. The following is an approximation of how a food chain works:
- A whale might eat 10 seals per day.
- A seal might eat 100 pounds of fish per day.
- Each pound of fish might need 10 pounds of zooplankton per day.
- Each pound of zooplankton might need 10 pounds of algae per day.

Express the following as a power of 10:

61. the number of seals needed to sustain a whale **10^1**

62. the number of pounds of fish needed to sustain a whale **10^3**

63. the number of pounds of zooplankton needed to sustain a whale **10^4**

64. the number of pounds of algae needed to sustain a whale **10^5**

Technology

A calculator may be helpful for Exercises 18–33. You may want to have students first do the exercises with pencil and paper and then use a calculator to verify their results. Exercises 56–59 give students an opportunity to use a calculator to solve more difficult problems involving exponents.

Error Analysis

Students often multiply exponents when they should add them. For instance, some will write $a^3 \cdot a^4 = a^{12}$. Encourage students to apply the definition of an exponent.

$$\underbrace{\underbrace{a \cdot a \cdot a}_{a^3} \cdot \underbrace{a \cdot a \cdot a \cdot a}_{a^4}}_{a^7}$$

Student Technology Guide

NAME _____ CLASS _____ DATE _____

Student Technology Guide

8.1 Laws of Exponents: Multiplying Monomials

The Product-of-Powers Property tells how to simplify a product of powers with the same base. To simplify the product, add the exponents and keep the base the same.

To verify the Product-of-Powers Property, you can use a calculator to compare $3^2 \cdot 3^5$ and 3^7. Notice that the screen at right shows two different ways to enter 3^2. They are 3 [^] 2 and 3 [x^2].

You can use the Product-of-Powers Property to simplify an expression before you find its value. This will reduce the number of calculator keystrokes you must use to find the value of the expression.

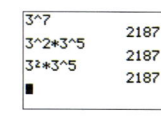

```
3^7              2187
3^2*3^5          2187
3²*3^5           2187
```

Write each product as a single power. Then use a calculator to find the value of each expression.

1. $2^3 \cdot 2^4$ **2^7; 128**   2. $10^5 \cdot 10^4$ **10^9; 1,000,000,000**   3. $5^5 \cdot 5^8$ **5^{13}; 1,220,703,125**   4. $3^7 \cdot 3^7$ **3^{14}; 4,782,969**

5. $10^3 \cdot 10^2$ **10^5; 100,000**   6. $4^5 \cdot 4^4$ **4^9; 262,144**   7. $7^6 \cdot 7^3$ **7^9; 40,353,607**   8. $6^3 \cdot 6^3$ **6^6; 46,656**

If an expression has a negative base, use parentheses to group the sign with the base. Use the sign change key, [(-)], to enter the negative number.
For example, to find the value of $(-3)^4$, use the following key sequence:

[(] [(-)] 3 [)] [^] 4 [ENTER] → 81

Use a calculator to find the value of each expression.

9. $(-4)^2$ **16**   10. $(-4)^3$ **-64**   11. $(-4)^4$ **256**   12. $(-4)^5$ **-1024**

13. $(-2)^2 \cdot (-2)^3$ **-32**   14. $(-5)^4 \cdot (-5)^2$ **15,625**   15. $(-8)^3 \cdot (-8)^3$ **262,144**   16. $(-4)^6 \cdot (-4)$ **$-16,384$**

17. Suppose that you want to evaluate $(-2)^4$ with a calculator. Explain why it is important to use parentheses when you enter this expression.
Answers will vary. Sample answer: If parentheses are not used, the calculator will evaluate the power of the positive base first and then take the opposite. This give an answer of -16, but the correct answer is 16.

LESSON 8.1 **375**

In Exercises 74–78, students use a calculator to discover that the result of raising a number to the power $\frac{1}{2}$ is the same as taking the square root of the number. Students will study *rational exponents* such as this if they take a more advanced course in algebra.

ALTERNATIVE Assessment

Portfolio Activity

The Portfolio Activity can be used as preparation for the Chapter Project or as a separate activity. In this Portfolio Activity, students compare the effects of repeatedly adding a given number with those of repeatedly multiplying by a given number. They discover that the products increase much more rapidly than the sums.

---

internetconnect

GO TO: go.hrw.com
KEYWORD: MA1 Doubling

To support the Portfolio Activity, students may visit the HRW Web site, where they can study exponential growth by examining various problems involving doubling.

Look Back

Find each answer. (LESSON 4.2)

65. 25 is 40% of what number? **62.5**   **66.** 16 is 20% of what number? **80**

67. What is 30% of 180? **54**   **68.** What is 35% of 160? **56**

Graph each function by making a table and plotting points. (LESSON 6.4)

69. $y = |x + 6|$ **70.** $y = |x - 5|$ **71.** $y = |3x|$

Solve each system of equations. (LESSON 7.3)

72. $\begin{cases} y = 2x + 11 \\ 3x = 29 - 5y \end{cases}$ **(–2, 7)**   **73.** $\begin{cases} 2x - 3y = 5 \\ -6x + 7y = -9 \end{cases}$ **(–2, –3)**

 Look Beyond

Use a calculator to evaluate the following:

74. $4^{0.5}$ **2**   **75.** $9^{0.5}$ **3**   **76.** $81^{0.5}$ **9**   **77.** $100^{0.5}$ **10**

78. What is the value of $b^{0.5}$, where b is any positive number? $\sqrt{b}$

Prize A! Start with $100. Repeatedly add $100 to the previous day's total.

Prize B! Start with $0.01. Repeatedly multiply the previous day's total by 2.

Include this activity in your portfolio.

After playing a certain game, the winner has 10 seconds to decide between the two prizes described above. For both prizes, the winner gets the amount of money calculated for the 20th day.

1. Make a table for the first four days of each prize.
2. After four days, which prize would you pick?
3. Do you think prize B will ever exceed prize A?
4. Continue the table for each prize through 20 days.
5. Did prize B ever exceed prize A?
6. Extend the table to 30 days. How much is each prize worth?

WORKING ON THE CHAPTER PROJECT

You should now be able to complete Activity 1 of the Chapter Project on page 416.

69.

x	y
–9	3
–8	2
–7	1
–6	0
–5	1
–4	2
–3	3

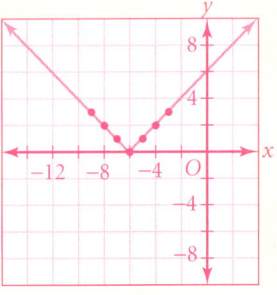

70.

x	y
2	3
3	2
4	1
5	0
6	1
7	2
8	3

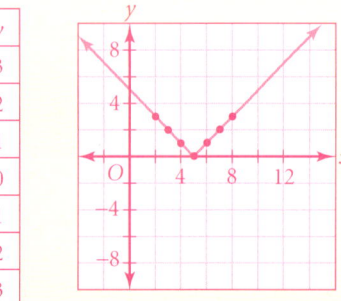

71.

x	y
–3	9
–2	6
–1	3
0	0
1	3
2	6
3	9

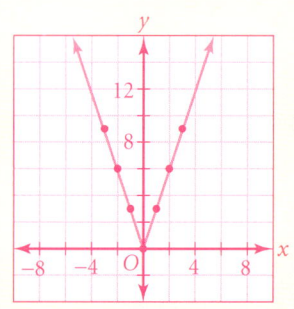

8.2 Laws of Exponents: Powers and Products

Objectives
- Find the power of a power.
- Find the power of a product.

Why Variables with exponents are commonly found in real-world problems such as those involving area and volume.

The volume of a sphere such as this hot-air balloon can be found with the formula $V = \frac{4}{3}\pi r^3$. If you substitute an expression for the value of the radius, r, the whole expression that is substituted must be cubed.

In Lesson 8.1 you learned about products of powers—that is, multiplying powers. Using the Product-of-Powers Property (page 371), you can simplify expressions by adding exponents to multiply powers.

$$a^2 \cdot a^3 = a^{2+3} = a^5$$
$$x^5 \cdot x^6 = x^{5+6} = x^{11}$$

In algebra you will sometimes need to raise *a power to a power*. For example, you might need to raise a^2 to the third power. This would be written as $(a^2)^3$. Do you think you should add the exponents in this case? You will examine this question in the following activity.

Alternative Teaching Strategy

USING TECHNOLOGY Give students the following sets of expressions to simplify on their calculators. Remind them to use the parenthesis keys and the ^ key in the correct order.

1. $(2^3)(2^3)$ $(2^3)^2$ 2^6
2. $(3^2)(3^2)$ $(3^2)^2$ 3^4
3. $(3^3)(3^3)$ $(3^3)^2$ 3^6
4. $(2^2)(2^2)(2^2)$ $(2^2)^3$ 2^6
5. $(2^3)(2^3)(2^3)$ $(2^3)^3$ 2^9

Be sure that students record their results. Then ask them to examine the results and make a conjecture about the exponents when a power such as 8^3 is raised to another power, as in the expression $(8^3)^4$. Elicit the response that it is permissible to multiply the exponents to obtain a simpler expression: $(8^3)^4 = 8^{3 \cdot 4} = 8^{12}$. Encourage them to use their calculators to verify this conclusion. Then give students the expression $(x^m)^n$ and ask them to generalize their findings. Lead them to see that $(x^m)^n$ can be written in a simplified form as x^{mn}.

Prepare

NCTM PRINCIPLES & STANDARDS
1–4, 6–10

LESSON RESOURCES
- Practice 8.2
- Reteaching Master 8.2
- Enrichment Master 8.2
- Cooperative-Learning Activity 8.2
- Lesson Activity 8.2
- Problem Solving/Critical Thinking Master 8.2

QUICK WARM-UP

Evaluate each expression.

1. a^2 for $a = 7$ **49**
2. $\frac{1}{b^2}$ for $b = 3$ **$\frac{1}{9}$**
3. $c^2 d$ for $c = 4$ and $d = 5$ **80**
4. $(fg)^2$ for $f = 4$ and $g = 2$ **64**
5. $h(j + k)^2$ for $h = 2$, $j = 3$, and $k = 1$ **32**

Also on Quiz Transparency 8.2

Teach

Why Draw a square on the board or overhead. Ask students to state a formula that can be used to find the area, A, of the square. Elicit the response $A = s \cdot s$, or $A = s^2$, where s represents the length of one side of the square. Ask them to give an expression for the area of the square when the length of one side is 3, 10, x, and x^5. Elicit the responses 3^2, 10^2, x^2, and $(x^5)^2$, respectively. Tell them that in this lesson, they will learn how to write an expression such as $(x^5)^2$ in a simpler form.

LESSON 8.2   **377**

Activity Notes

In this Activity, students discover the Power-of-a-Power Property. In Step 1, students will see that the exponents are not added. That is, $(a^2)^3$ simplifies not to a^5, but rather to a^6. In Step 4, students are asked to write the Power-of-a-Power Property in their own words and to provide a justification for it.

Cooperative Learning

You may wish to have students do the Activity in groups of four. They can work together in Steps 1 and 2. In Step 3, group members can first work individually and then work together to compare results. In Step 4, each student should again work individually to write a rule. Then the group can discuss all the proposed rules and come to a consensus on a single rule that they feel is best.

CHECKPOINT ✔

4. Answers may vary. Sample answer: When raising a power to a power, multiply the exponents. The rule is true because repeated addition of exponents can be translated to multiplication.

ADDITIONAL EXAMPLE 1

Simplify each expression and find the value when possible.

a. $(3^2)^4$ 6561
b. $(2^5)^2$ 1024
c. $(r^4)^2$ r^8
d. $(a^n)^4$ a^{4n}

Activity: Raising a Power to a Power

1. Study the following:
The expression $(a^2)^3$ means a^2 raised to the third power. By the definition of an exponent, a^2 is used as a factor 3 times.
$$a^2 \cdot a^2 \cdot a^2 = (a \cdot a) \cdot (a \cdot a) \cdot (a \cdot a)$$
$$= a \cdot a \cdot a \cdot a \cdot a \cdot a$$
$$= a^6$$

As you can see, the result is *not* a^5, so the exponents are not added. Does the correct answer suggest a rule for raising a power to a power? Make a guess and keep it in mind as you do Steps 2 and 3.

2. How would you simplify the expression $(x^5)^2$? Following the pattern in Step 1, expand this expression and show how to simplify it. Does your result agree with the rule you guessed in Step 1?

3. Write down a power of a power of your own choosing. Then repeat the process in Step 2.

CHECKPOINT ✔ **4.** Write a rule for raising a power to a power. In your own words, explain why you think it is true.

In the activity, you may have discovered the following rule: To raise a power to a power, multiply the exponents.

Power-of-a-Power Property

For all nonzero real numbers x and all integers m and n,
$$(x^m)^n = x^{mn}.$$

EXAMPLE 1

Simplify and find the value of each expression when possible.

a. $(2^3)^4$ **b.** $(10^3)^2$ **c.** $(p^2)^5$ **d.** $(x^m)^2$

SOLUTION

a. $(2^3)^4 = 2^{3 \cdot 4}$ **b.** $(10^3)^2 = 10^{3 \cdot 2}$
$ = 2^{12}$ $ = 10^6$
$ = 4096$ $ = 1{,}000{,}000$

c. $(p^2)^5 = p^{2 \cdot 5}$ **d.** $(x^m)^2 = x^{m \cdot 2}$
$ = p^{10}$ $ = x^{2m}$

TRY THIS Simplify and find the value of each expression when possible.
a. $(2^6)^2$ **b.** $(10^4)^3$ **c.** $(y^3)^5$ **d.** $(m^3)^x$

Interdisciplinary Connection

GRAPHIC ARTS A student is planning a modern art piece that will feature collages of material covering the surface of a small square and the surface of a large square. The length of one side of the large square is to be 3 times the length of one side of the small square. The student needs to determine the cost of materials. In order to experiment with the effect of different sizes on the cost, the student decides to first use the variable s to represent the length of one side of the small square.

1. Write an expression for the area of each.
a. the small square s^2
b. the large square $(3s)^2$, or $9s^2$
c. both squares $s^2 + 9s^2$

2. Find the amount of material that will be needed for both squares when one side of the small square is the given length.
a. 2 feet   40 square feet
b. 3 feet   90 square feet
c. 4 feet   160 square feet

378 LESSON 8.2

Raising a Monomial to a Power

Sometimes an exponential expression has a monomial for a base, as in the examples below.

$$(xy)^2 \qquad (xy^2)^2 \qquad (\pi r^2)^4$$

You can simplify expressions like these by using the definition of exponents.

EXAMPLE 2 Simplify $(xy^2)^3$ by using the definition of exponents.

SOLUTION

The exponent outside the parentheses, 3, shows that the monomial factor, (xy^2), is used as a factor 3 times. Remove the parentheses, and simplify by regrouping and multiplying.

$$(xy^2)^3 = (xy^2)(xy^2)(xy^2)$$
$$= (x \cdot x \cdot x)(y^2 \cdot y^2 \cdot y^2)$$
$$= x^3 y^6$$

TRY THIS Simplify each expression by using the definition of exponents.
a. $(xy)^2$ b. $(xy^2)^2$ c. $(\pi r^2)^4$

Power-of-a-Product Property

For all nonzero real numbers x and y, and all integers m and n,
$$(xy)^n = x^n y^n.$$

The Power-of-a-Product Property can be extended to include any number of terms inside the parentheses. For example:

$$(xyz)^n = x^n y^n z^n$$

EXAMPLE 3 Verify the Power-of-a-Product Property by finding the value of each expression in two different ways.
a. $(2 \cdot 3)^3$ b. $(2 \cdot 5)^5$

SOLUTION
a. $(2 \cdot 3)^3 = 2^3 \cdot 3^3 = 8 \cdot 27 = 216$ or $(2 \cdot 3)^3 = 6^3 = 216$
b. $(2 \cdot 5)^5 = 2^5 \cdot 5^5 = 32 \cdot 3125 = 100,000$ or $(2 \cdot 5)^5 = 10^5 = 100,000$

TRY THIS Verify the Power-of-a-Product Property by finding the value of each expression in two different ways.
a. $(3 \cdot 3)^2$ b. $(2 \cdot 5)^6$

Inclusion Strategies

TACTILE LEARNERS Have students build cubes out of unit cubes to understand why $(2x)^3$ must be equal to $2^3 x^3$, or $8x^3$. Start with one unit cube and point out that its volume is 1 cubic unit. Tell them to build a cube whose edges are double in length. They should build a $2 \times 2 \times 2$ cube and observe that its volume is 8 times 1 cubic unit, or 8 cubic units. Now have them build a $3 \times 3 \times 3$ cube and determine that its volume is 27 cubic units. This time, when they double the length of each edge, the volume of the resulting $6 \times 6 \times 6$ cube is 8 times 27 cubic units, or 216 cubic units.

Enrichment

Have students work in pairs to solve these "missing exponent" problems.

1. $(2^2)^? = 2^{12}$ 6
2. $(3^?)^4 = 81$ 1
3. $(x^4 y^?)^3 = x^{12} y^6$ 2
4. $(a^2 b^3)^? = a^6 b^9$ 3
5. $(p^q)^? = p^{5q}$ 5
6. $(y^m)^? = y^{2mn}$ $2n$
7. $(4^2)^? = 2^8$ 2
8. $(9^?)^3 = 3^{12}$ 2
9. $(25^6)^? = 5^{12}$ 1
10. $(8^?)^4 = 2^{24}$ 2

Then have each student make up five original problems like the ones above and write them on a clean sheet of paper. Partners should exchange papers and solve each other's problems.

TRY THIS
a. 2^{12}, or 4096
b. 10^{12}, or 1,000,000,000,000
c. y^{15}
d. m^{3x}

ADDITIONAL EXAMPLE 2

Simplify $(a^3 b)^2$ by using the definition of exponents.

$(a^3 b)^2 = (a^3 b)(a^3 b)$
$= (a^3 \cdot a^3)(b \cdot b)$
$= a^6 b^2$

TRY THIS
a. $x^2 y^2$
b. $x^2 y^4$
c. $\pi^4 r^8$

ADDITIONAL EXAMPLE 3

Find the value of each expression in two different ways.

a. $(3 \cdot 4)^2$
$(3 \cdot 4)^2 = 3^2 \cdot 4^2$
$= 9 \cdot 16$
$= 144$
$(3 \cdot 4)^2 = 12^2$
$= 144$

b. $(4 \cdot 2)^3$
$(4 \cdot 2)^3 = 4^3 \cdot 2^3$
$= 64 \cdot 8$
$= 512$
$(4 \cdot 2)^3 = 8^3$
$= 512$

TRY THIS
a. $(3 \cdot 3)^2 = 3^2 \cdot 3^2 = 81$
$(3 \cdot 3)^2 = 9^2 = 81$
b. $(2 \cdot 5)^6 = 2^6 \cdot 5^6 = 1,000,000$
$(2 \cdot 5)^6 = 10^6 = 1,000,000$

ADDITIONAL EXAMPLE 4

Simplify by using the Power-of-a-Product Property.

a. $(cd^3)^2$ c^2d^6

b. $(x^my^2z)^4$ $x^{4m}y^8z^4$

TRY THIS

a. x^2y^6

b. $x^6y^6z^6$

c. $a^nb^nc^nd^n$

ADDITIONAL EXAMPLE 5

The volume, V, of a cube is given by the formula $V = e^3$, where e is the length of one edge of the cube. Suppose that the length of one edge is multiplied by a factor of 4. **Write a simplified expression for the new volume of the cube.** $64e^3$

Practice

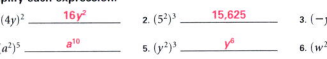

8.2 Laws of Exponents: Powers and Products

Simplify each expression.

1. $(4y)^2$ __$16y^2$__   2. $(5^2)^3$ __$15{,}625$__   3. $(-y^5)^4$ __y^{20}__
4. $(a^2)^5$ __a^{10}__   5. $(y^2)^3$ __y^6__   6. $(w^2)^2$ __w^4__
7. $(w^4)^6$ __w^{24}__   8. $(-8c^5)^2$ __$64c^{10}$__   9. $(-3h^9)^3$ __$-27h^{27}$__
10. $(-y^4d^6)^8$ __$y^{32}d^{48}$__   11. $(-c^5h^6)^3$ __$-c^{15}h^{18}$__   12. $(-15h^9k^7)^2$ __$225h^{18}k^{14}$__
13. $(k^9)^5(k^3)^2$ __k^{51}__   14. $(3y^6)^2(x^5y^2z)$ __$9x^5y^{14}z$__
15. $(4h^3)^2(-2g^3h)^3$ __$-128g^9h^9$__   16. $(14a^4b^6)^2(a^6b^3)^7$ __$196a^{50}b^{33}$__

Evaluate each monomial for $x = 5$, $y = -1$, and $z = -4$.

17. y^4 __1__   18. $3x^3$ __375__   19. $2y^2$ __2__
20. z^2 __16__   21. $(yz)^2$ __16__   22. $(yx)^2$ __25__
23. x^2z^2 __400__   24. y^x __-1__   25. $-y^x$ __1__

26. What is the area of a square if each edge of the square has a length of $3a^5$?
__$9a^{10}$__

27. What is the area of a rectangle if one side has a length of $12x^3$ and the other side has a length of $6x^2$?
__$72x^5$__

Find the volume of the cube for each edge length, e.

28. $e = 5y^4$ __$125y^{12}$__

29. $e = 3x^7y^5$ __$27x^{21}y^{15}$__

380 LESSON 8.2

EXAMPLE 4

Simplify.

a. $(x^2y)^3$ b. $(ab^2c^n)^5$

SOLUTION

a. $(x^2y)^3 = x^{2 \cdot 3}y^3 = x^6y^3$ b. $(ab^2c^n)^5 = a^5b^{2 \cdot 5}c^{n \cdot 5} = a^5b^{10}c^{5n}$

TRY THIS Simplify.

a. $(xy^3)^2$ b. $(x^2y^2z^2)^3$ c. $(abcd)^n$

EXAMPLE 5

CONNECTION GEOMETRY

The volume, V, of a sphere is given by the formula $V = \frac{4}{3}\pi r^3$. Suppose that the radius, r, of a sphere is equal to $6s^2$, where s is the radius of a smaller sphere. **Find the volume of the larger sphere in terms of the radius of the smaller sphere.**

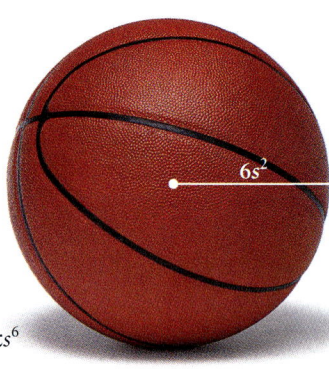

SOLUTION

$V = \frac{4}{3}\pi(6s^2)^3 = \frac{4}{3}\pi(216s^6) = 288\pi s^6$

Powers of –1

Recall that $(-1) \cdot (-1) = 1$; that is, $(-1)^2 = 1$. However, $(-1) \cdot (-1) \cdot (-1) = -1$. Thus, $(-1)^3 = -1$. This pattern suggests the following rule:

> **Powers of –1**
>
> Even powers of –1 are equal to 1.
> Odd powers of –1 are equal to –1.

EXAMPLE 6

Simplify.

a. $(-t)^5$ b. $-t^4$ c. $(-5x)^3$

SOLUTION

a. $(-t)^5 = (-1 \cdot t)^5 = (-1)^5 \cdot (t^5) = -1 \cdot t^5 = -t^5$
b. $-t^4$ is already in simplest form. The exponent applies only to t.
c. $(-5x)^3 = (-5)^3x^3 = -125x^3$

TRY THIS Simplify.

a. $(-y)^4$ b. $-y^{100}$ c. $(-3y)^4$

Reteaching the Lesson

VISUAL STRATEGIES Review with students the definition of *exponent*: In the expression x^m, the exponent, m, indicates how many times the base, x, is used as a factor. Use the visual device of the "boxes" in the display at the right to show that the Power-of-a-Power Property and Power-of-a-Product Property are really just applications of the definition. Examine the expressions beneath the boxes and stress the fact that the question is, How many times is x (or y) used as a factor? Note that the answer to this question is the exponent of x (or y) in the simplified expression.

$\boxed{x}^3 = \boxed{x} \cdot \boxed{x} \cdot \boxed{x}$
$\phantom{\boxed{x}^3} = x \cdot x \cdot x = x^3$

$\boxed{x^2}^3 = \boxed{x^2} \cdot \boxed{x^2} \cdot \boxed{x^2}$
$\phantom{\boxed{x^2}^3} = x \cdot x \cdot x \cdot x \cdot x \cdot x = x^6$

$\boxed{xy}^3 = \boxed{xy} \cdot \boxed{xy} \cdot \boxed{xy}$
$\phantom{\boxed{xy}^3} = x \cdot y \cdot x \cdot y \cdot x \cdot y = x^3y^3$

Exercises

Communicate

1. Explain the difference between a^2a^3 and $(a^2)^3$.
2. Explain why $(x^a)^b = (x^b)^a$ is true.
3. Explain why you multiply the exponents when you raise a power to a power.
4. Explain why $(-1)^2 > (-1)^3$ is true.
5. Explain why $(a^s)^t = a^{st}$ and $(xy)^n = x^n y^n$ are true.

Guided Skills Practice

Simplify and find the value of each expression when possible. (EXAMPLE 1)

6. $(3^5)^2$ **59,049**
7. $(10^3)^2$ **1,000,000**
8. $(b^4)^9$ b^{36}
9. $(z^7)^y$ z^{7y}

Simplify each expression by using the definition of exponents. (EXAMPLE 2)

10. $(ab)^8$ $a^8 b^8$
11. $(x^2y^4)^3$ $x^6 y^{12}$
12. $(c^2h^7)^4$ $c^8 h^{28}$
13. $(m^2n^5)^3$ $m^6 n^{15}$

Verify the Power-of-a-Product Property by finding the value of each expression in two different ways. (EXAMPLE 3)

14. $(5 \cdot 5)^3$ **15,625**
15. $(3 \cdot 10)^7$ **21,870,000,000**
16. $(4 \cdot 3)^2$ **144**
17. $(5 \cdot 11)^2$ **3025**

Simplify by using the Power-of-a-Product Property. (EXAMPLE 4)

18. $(x^2y^3)^5$ $x^{10}y^{15}$
19. $(a^4b^4c^4)^{11}$ $a^{44}b^{44}c^{44}$
20. $(qrs)^t$ $q^t r^t s^t$
21. $(x^2y^5)^3$ $x^6 y^{15}$

Simplify. (EXAMPLE 5)

22. $(-6)^3$ -216
23. $(-b)^{10}$ b^{10}
24. $(-5k)^3$ $-125k^3$
25. $(-1)^{1001}$ -1

Practice and Apply

Simplify each of the following:

26. $(3x)^2$ $9x^2$
27. $(6^2)^3$ **46,656**
28. $(-x^2)^3$ $-x^6$
29. $(10^2)^3$ **1,000,000**
30. $(2x^4)^3$ $8x^{12}$
31. $(-3b^2)^5$ $-243b^{10}$
32. $(-2r^3)^2$ $4r^6$
33. $(-3x^6)^5$ $-243x^{30}$
34. $(-10m^4)^3$
35. $(5j^2k^3)^4$
36. $(-6xy^4z^5)^2$
37. $2(3a^2)^3$
38. $10(-5b^5)^2$
39. $(8n^2p)^3$
40. $(7j^2)^5$
41. $(-ab^5)(a^3)^2$
42. $(-v^4w^3)^2(-v^3)^4$
43. $(5f^3t^7)^3(d^4f^2)^5$
44. $(x^2)(y^2)^5$
45. $(3x^2)^2(12y^3)^3$
46. $(y^2)^4(x^3y^5)^6$
47. $(-4b^7e^3)^2(-h^3)^3$
48. $(-x^3)(-y^2)^3$
49. $(xy)^2(x^2y^2)^2$
50. $\left(\dfrac{1}{3x^2y}\right)^3$
51. $\left(\dfrac{3x^2}{2y^2}\right)^5$
52. $\left(\dfrac{8x^2}{2x^2}\right)^2$
53. $\left(\dfrac{-1}{5x^3y^5}\right)^{11}$

CHALLENGE

Answers

34. $-1000m^{12}$
35. $625j^8k^{12}$
36. $36x^2y^8z^{10}$
37. $54a^6$
38. $250b^{10}$
39. $512n^6p^3$
40. $343j^{6}$
41. $-a^7b^5$
42. $v^{20}w^6$
43. $125d^{20}f^{19}t^{21}$
44. x^2y^4
45. $15,552x^4y^9$
46. $x^{18}y^{38}$
47. $-16b^{14}e^6h^9$
48. x^3y^6
49. x^6y^6
50. $\dfrac{1}{27x^6y^3}$
51. $\dfrac{243x^{10}}{32y^{10}}$
52. 16
53. $\dfrac{-1}{48,828,125x^{33}y^{55}}$

Math CONNECTION

GEOMETRY In general, when the radius of a sphere is multiplied by a factor of x, its volume is multiplied by a factor of x^3. In Example 5, the radius s is being multiplied by a factor of $6s$ to give the new radius of $6s^2$, so the original volume, $\frac{4}{3}\pi s^3$, will be multiplied by a factor of $(6s)^3$, or $216s^3$. Note that $\frac{4}{3}\pi s^3 \cdot 216s^3$ is equal to $288\pi s^6$.

ADDITIONAL EXAMPLE 6

Simplify by using the Power-of-a-Product Property.

a. $(-2y)^4$ $16y^4$

b. $(-6h)^3$ $-216h^3$

c. $-10m^6$ $-10m^6$

TRY THIS

a. y^4

b. $-y^{100}$

c. $81y^4$

Student Technology Guide

8.2 Laws of Exponents: Powers and Products

An expression such as $(3^4)^5$ involves a power raised to a power.
To evaluate $(3^4)^5$ with a graphics calculator, use the key sequence:
[1] 3 [^] 4 [)] [^] 5 [ENTER]

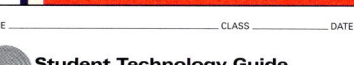

Alternatively, you can use the Power-of-a-Power Property to simplify $(3^4)^5$ as 3^{20} before you use the calculator.
3 [^] 20 [ENTER]

Example: Simplify $(4^3)^2$. Then find the value of this expression.
Using the Power-of-a-Power Property, $(4^3)^2 = 4^{2 \cdot 3} = 4^6$.
Press 4 [^] 6 [ENTER] ⟶ 4096

Evaluate each expression. Write fractional answers as needed.

1. $(2^2)^4$ $2^8 = 256$
2. $(5^2)^5$ $5^{10} = 9,765,625$
3. $(3^4)^3$ $3^{12} = 531,441$
4. $\left[\left(\dfrac{1}{2}\right)^2\right]^3$ $\dfrac{1}{64}$
5. $[(-3)^3]^3$ $-19,683$
6. $\left[\left(\dfrac{1}{2}\right)^3\right]^3$ $\dfrac{1}{512}$

The Power-of-a-Product Property says that if x and y are any numbers and n is a positive integer, then $(xy)^n = x^n y^n$. You can use a graphics calculator to verify this property.

Example: Verify the Power-of-a-Product Property by evaluating $(3 \cdot 4)^2$ in two different ways.
Using the Power-of-a-Product Property, $(3 \cdot 4)^2 = 3^2 \cdot 4^2$.
[1] 3 [×] 4 [)] [^] 2 [ENTER] ⟶ 144
[1] 3 [^] 2 [×] [1] 4 [^] 2 [)] [ENTER] ⟶ 144

Verify the Power-of-a-Product Property by finding the value of each expression in two different ways. Write the value of the expression.

7. $(3 \cdot 2)^4$ **1296**
8. $(5 \cdot 2)^3$ **1000**
9. $(7 \cdot 9)^5$ **992,436,543**

LESSON 8.2 **381**

Assess

Selected Answers

Exercises 6–25, 27–69 odd

ASSIGNMENT GUIDE

In Class	1–25
Core	27–49 odd, 55–65 odd
Core Plus	26–48 even, 50–65
Review	66–69
Preview	70, 71

✎ Extra Practice can be found beginning on page 738.

Technology

A calculator may be helpful for performing the calculations in some of the exercises. You might want students to first do the exercises with paper and pencil and then use a calculator to check their answers.

Error Analysis

Students commonly forget to apply an exponent outside parentheses to a constant factor with the parentheses. For example, when given $(2x^2)^3$, they may write the simplified form as $2x^6$. Suggest that they rewrite such an expression so that every factor within parentheses has a visible exponent, as in $(2^1x^2)^3$. This will serve as a reminder to consider the exponent of each factor in the simplified expression.

Look Beyond

In Exercises 70 and 71, students explore some interesting and unexpected properties of sums of two squares. Encourage students to employ guess-and-check as a problem-solving strategy.

Evaluate each monomial for $a = 10$.

54. $2a^3$ **2000**   **55.** $(2a)^3$ **8000**   **56.** $-2a^2$ **–200**   **57.** $(-2a)^2$ **400**

Evaluate each of the following:

58. $(-1)^1$ **–1**   **59.** $(-1)^2$ **1**   **60.** $(-1)^3$ **–1**

61. $(-1)^4$ **1**   **62.** $(-1)^5$ **–1**   **63.** $(-1)^6$ **1**

CONNECTION

64. GEOMETRY Find the volume of the cube if each edge, e, is doubled. **$8e^3$**

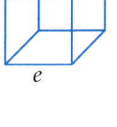

APPLICATION

65. RECREATION A summer camp needs to build a field for a new sport invented by one of its employees. The field needs to be rectangular and to have a length three times as long and a width twice as the long as the side of an existing square field. Use s to represent the length of the sides of the square field. Find the area of the new field. **$6s^2$**

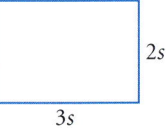

Look Back

66. Distance, d, varies directly as time, t. If a car travels 195 miles in 3 hours, determine the constant, k, and write an equation for the direct variation. How far will the car travel in 4 hours? **(LESSON 5.3)** $k = 65$; $d = 65t$; **260 miles**

CONNECTION

67. COORDINATE GEOMETRY A line passes through the points $P(14, -3)$ and $Q(-5, 27)$. Write an equation for this line and an equation for a line perpendicular to this line. **(LESSON 5.6)**

67. $y - (-3) = -\frac{30}{19}(x - 14)$
or $y - 27 = -\frac{30}{19}(x + 5)$;
$y - (-3) = \frac{19}{30}(x - 14)$
or $y - 27 = \frac{19}{30}(x + 5)$

Solve for x. (LESSON 6.5)

68. $|x - 3| \leq 7$ **$-4 \leq x \leq 10$**

69. $|2x + 4| > 8$ **$x < -6$ or $x > 2$**

Look Beyond

70. Show that $1^2 + 8^2 = 4^2 + 7^2$. A consequence of this is $14^2 + 87^2 = 41^2 + 78^2$. How were the numbers 14, 87, 41, and 78 chosen?

71. Show that $14^2 + 87^2 = 41^2 + 78^2$ is true. Is $17^2 + 84^2 = 71^2 + 48^2$ true? How were the numbers 17, 84, 71, and 48 chosen?

70. $1^2 + 8^2 = 4^2 + 7^2$
$1 + 64 = 16 + 49$
$65 = 65$

The numbers 14, 87, 41, and 78 were chosen by pairing the first numbers and second numbers on each side of the equal sign.

71. $14^2 + 87^2 = 41^2 + 78^2$
$196 + 7569 = 1681 + 6084$
$7765 = 7765$

$17^2 + 84^2 = 71^2 + 48^2$
$289 + 7056 = 5041 + 2304$
$7345 = 7345$

The numbers 17, 84, 71 and 48 were chosen by pairing the outer numbers and inner numbers from the equation $1^2 + 8^2 = 4^2 + 7^2$.

8.3 Laws of Exponents: Dividing Monomials

Objectives
- Simplify quotients of powers.
- Simplify powers of fractions.

Why The study of ratios, which is important in sciences such as biology and physics, requires an understanding of quotients, which may involve exponents.

APPLICATION
BIOLOGY

A plant that has a large volume in comparison to its surface area will hold heat and moisture well. Do you think the ratio between volume and surface area is greater for small or large plants?

Quotients of Powers

In a quotient of powers, one power is divided by another. You can simplify a quotient of powers by using the definition of powers and exponents. You can begin by expressing the powers in terms of their factors.

$$\frac{a^6}{a^4} = \frac{a \cdot a \cdot a \cdot a \cdot a \cdot a}{a \cdot a \cdot a \cdot a} = a \cdot a = a^2$$

In the activity that follows, you will use discover a rule for simplifying quotients of powers.

Prepare

NCTM PRINCIPLES & STANDARDS 1–4, 6–10

LESSON RESOURCES
- Practice 8.3
- Reteaching Master 8.3
- Enrichment Master 8.3
- Cooperative-Learning Activity 8.3
- Lesson Activity 8.3
- Problem Solving/Critical Thinking Master 8.3

QUICK WARM-UP

Evaluate.
1. $54 \div (-6)$ -9
2. $-24 \div (-4)$ 6
3. 2^5 32
4. 5^3 125

Simplify.
5. $(m^3)^6$ m^{18}
6. $(y^r)^4$ y^{4r}
7. $(3a^2)^5$ $243a^{10}$
8. $(p^2q^3)^2$ p^4q^6

Also on Quiz Transparency 8.3

Teach

Why Remind students that subtraction of real numbers is the inverse of addition and that division is the inverse of multiplication. Remind students that to *multiply* powers with the same base, they learned that they could *add* exponents. Ask them how they think the exponents will be handled when they *divide* powers with the same bases. **They will be subtracted.**

Alternative Teaching Strategy

USING COGNITIVE STRATEGIES Have students work in groups of three or four. Give them the following multiple-choice question:

$\frac{12m^{12}}{4m^4} = \underline{\ ?\ }$

a. $3m^3$ **b.** $8m^8$ **c.** $3m^8$ **d.** $8m^4$

Have students work together to determine the correct answer. Direct each group to write a convincing argument that supports their choice.
The correct choice is c.

After all groups have written their arguments, have one member of each group present to the class a summary of their argument. During these presentations, elicit the group's reasons, if any, for eliminating choices **a**, **b**, and **d**.

Repeat the activity with the following question:

$\frac{-10t^4}{-2t} = \underline{\ ?\ }$

a. $-5t^4$ **b.** $5t^3$ **c.** $-8t^4$ **d.** $8t^3$

The correct choice is b.

LESSON 8.3   **383**

Activity Notes

In this Activity, students investigate the Quotient-of-Powers property. They should arrive at the conclusion that when you divide powers with the same base, you can subtract the exponent of the divisor from the exponent of the dividend.

Cooperative Learning

You may wish to have students do the Activity in pairs. In Step 2, have each student complete all the calculations. Then partners can compare their answers.

CHECKPOINT ✔

5. Subtract the exponents.
$\frac{2^5}{2^3} = 2^{5-3} = 2^2 = 4$

ADDITIONAL EXAMPLE 1

Simplify.

a. $\frac{3^5}{3^2}$ **b.** $\frac{2^{12}}{2^5}$

$3^3 = 27$ $2^7 = 128$

TRY THIS

a. $10^8 = 100,000,000$

b. $2^5 = 32$

c. $3^3 = 27$

ADDITIONAL EXAMPLE 2

Simplify.

a. $\frac{r^9}{r^7}$ **b.** $\frac{n^6}{n}$ **c.** $\frac{h^a}{h^b}$

r^2 n^5 h^{a-b}

TRY THIS

a. m^{n-p}

b. y^9

c. 10^y

384 LESSON 8.3

Activity

Discovering a Rule for Quotients of Powers

1. Use the table to write the quotient $\frac{10^6}{10^2}$ as a power of 10.

Form	Numerator	Denominator	Quotient
Decimal	1,000,000	100	10,000
Exponential	10^6	10^2	?

2. Make a similar table for each of the following:

a. $\frac{10^5}{10^2}$ **b.** $\frac{2^6}{2^2}$ **c.** $\frac{3^4}{3^3}$

3. Guess the exponential notation for $\frac{10^5}{10^1}$. Make a table to check your guess.

4. For any positive number a and any positive integers m and n, where $m > n$, what is the simplified form for $\frac{a^m}{a^n}$?

CHECKPOINT ✔ **5.** Explain how to simplify $\frac{2^5}{2^3}$.

Quotient-of-Powers Property

For all nonzero real numbers x and all integers m and n, where $m > n$,

$$\frac{x^m}{x^n} = x^{m-n}.$$

EXAMPLE 1 Simplify.

a. $\frac{2^6}{2^4}$ **b.** $\frac{10^9}{10^2}$

● **SOLUTION**

a. $\frac{2^6}{2^4} = 2^{6-4} = 2^2 = 4$ **b.** $\frac{10^9}{10^2} = 10^{9-2} = 10^7 = 10,000,000$

TRY THIS Simplify.

a. $\frac{10^{12}}{10^4}$ **b.** $\frac{2^{10}}{2^5}$ **c.** $\frac{3^{10}}{3^7}$

EXAMPLE 2 Simplify. Assume that the conditions of the property are satisfied.

a. $\frac{x^a}{x}$ **b.** $\frac{x^{a+b}}{x^c}$ **c.** $\frac{x^{m+1}}{x}$

● **SOLUTION**

a. $\frac{x^a}{x} = \frac{x^a}{x^1} = x^{a-1}$ **b.** $\frac{x^{a+b}}{x^c} = x^{a+b-c}$ **c.** $\frac{x^{m+1}}{x} = \frac{x^{m+1}}{x^1} = x^{m+1-1} = x^m$

TRY THIS Simplify.

a. $\frac{m^n}{m^p}$ **b.** $\frac{y^{10}}{y}$ **c.** $\frac{10^{y+1}}{10}$

Inclusion Strategies

VISUAL LEARNERS Use colored pencils to show how the exponents are written when the Quotient-of-Powers Property is applied. Write the exponent in the numerator in one color and the exponent in the denominator in another color. Continue using the same colors for the numbers as you demonstrate how to find the exponent in the quotient by subtraction.

Quotients of Monomials

To find quotients of monomials, treat the coefficients separately, and use the Quotient-of-Powers Property to simplify powers with the same base.

EXAMPLE 3 Simplify. Assume that the conditions of the property are satisfied.

a. $\dfrac{-4x^2y^5}{2xy^3}$ b. $\dfrac{c^4b}{c^2a}$

SOLUTION

Simplify any numerical coefficients. Then subtract the exponents of the powers in the denominator from the exponents of the powers with the same base in the numerator.

a. $\dfrac{-4x^2y^5}{2xy^3} = \left(\dfrac{-4}{2}\right)(x^{2-1})(y^{5-3}) = -2xy^2$

b. $\dfrac{c^4b}{c^2a} = c^{4-2}\left(\dfrac{b}{a}\right) = \dfrac{c^2b}{a}$

TRY THIS Simplify. Assume that the conditions of the property are satisfied.

a. $\dfrac{6ab^2}{-2ab}$ b. $\dfrac{(5y^2)(-81t^4)}{45t^3}$

EXAMPLE 4

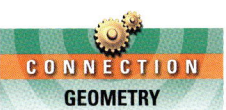
CONNECTION
GEOMETRY

The volume, *V*, of a sphere can be found with the formula $V = \frac{4}{3}\pi r^3$. The surface area, *S*, of a sphere can be found with the formula $S = 4\pi r^2$. **Find the ratio of the volume of a sphere to its surface area. What happens to this ratio as the radius of the sphere increases?**

SOLUTION

Find the ratio by dividing the expression for the volume, *V*, by the expression for the surface area, *S*.

$$\dfrac{V}{S} = \dfrac{\frac{4}{3}\pi r^3}{4\pi r^2} = \left(\dfrac{\frac{4}{3}}{4}\right)\left(\dfrac{\pi}{\pi}\right)\left(\dfrac{r^3}{r^2}\right) = \dfrac{1}{3}r$$

By substituting different values for *r*, you will see that the value of the ratio increases as the value of *r* increases.

ADDITIONAL EXAMPLE 3

Simplify.

a. $\dfrac{-3x^3y^4}{-2x^2y^2}$ b. $\dfrac{ab^2c^3}{ac^2}$

$\dfrac{3}{2}xy^2$ b^2c

TRY THIS

a. $-3b$

b. $-9ty^2$

ADDITIONAL EXAMPLE 4

The volume, *V*, of a cube can be found with the formula $V = e^3$, where *e* represents the length of one edge of the cube. Its surface area, *S*, can be found with the formula $S = 6e^2$. **Find the ratio of the volume of a cube to its surface area. As the length of an edge of the cube increases, does the value of this ratio increase, decrease, or stay the same?**

$\dfrac{e}{6}$; increases

Math CONNECTION

GEOMETRY Students may need to be reminded of the difference between volume and surface area of a three-dimensional figure. The volume is the number of *cubic units* that can be contained in its interior without overlapping. The surface area is the number of *square units* that can be contained on all of its surfaces without overlapping.

Enrichment

Have students work in pairs to solve these "missing exponent" problems.

1. $\dfrac{2^5}{2^?} = 2^3$ 2
2. $\dfrac{5^?}{5^2} = 5^{10}$ 12
3. $\dfrac{a^6}{a^?} = a^2$ 4
4. $\dfrac{w^?}{w} = w^6$ 7
5. $\dfrac{ab^?}{ab^2} = b^5$ 7
6. $\dfrac{x^4y^?z^5}{x^2y^4z} = x^2z^4$ 4
7. $\dfrac{k^n}{k^?} = k^2$ $n-2$
8. $\dfrac{g^?}{g^x} = g^x$ $2x$

Then have each student make up five original problems like the ones given and write them on a clean sheet of paper. Partners should exchange papers and solve each other's problems.

LESSON 8.3 **385**

CRITICAL THINKING

Answers may vary. Sample answer: A large object will lose more heat or moisture than a small object of the same shape. When the shapes are the same, a larger volume corresponds to a larger surface area.

ADDITIONAL EXAMPLE 5

Simplify each expression.

a. $\left(\dfrac{1}{3}\right)^3$ $\dfrac{1}{27}$

b. $\left(\dfrac{8}{2}\right)^3$ 64

c. $\left(\dfrac{c^2}{d^3}\right)^4$ $\dfrac{c^8}{d^{12}}$

TRY THIS

a. $\dfrac{16}{81}$

b. $\dfrac{216}{8}$, or 27

c. $\dfrac{x^{3z}}{2^{yz}}$

CRITICAL THINKING Evaporation and heat loss take place at the surface of an object. Do you think that a large or a small object of the same shape will lose less heat and moisture? Explain how this relates to the ratio of volume to surface area.

The anole at left has a smaller ratio of volume to surface area than the Komodo dragon above.

Powers of Fractions

In order to multiply a fraction by a fraction, recall that you must find the product of the numerators and the product of the denominators.

$$\dfrac{3}{4} \times \dfrac{7}{8} = \dfrac{3 \times 7}{4 \times 8} = \dfrac{21}{32}$$

This procedure leads to the rule for finding a power of a fraction.

$$\left(\dfrac{a}{b}\right)^n = \underbrace{\dfrac{a}{b} \cdot \dfrac{a}{b} \cdot \dfrac{a}{b} \cdot \cdots \cdot \dfrac{a}{b}}_{n \text{ factors}} = \dfrac{a \cdot a \cdot a \cdot \cdots \cdot a}{b \cdot b \cdot b \cdot \cdots \cdot b} = \dfrac{a^n}{b^n}$$

Power-of-a-Fraction Property

For all real numbers a, b, and n, where $n \geq 0$ and $b \neq 0$

$$\left(\dfrac{a}{b}\right)^n = \dfrac{a^n}{b^n}.$$

EXAMPLE 5 Simplify each expression.

a. $\left(\dfrac{3}{4}\right)^2$ b. $\left(\dfrac{10}{5}\right)^4$ c. $\left(\dfrac{x^n}{y^n}\right)^2$, where $n > 0$

SOLUTION

a. $\left(\dfrac{3}{4}\right)^2 = \dfrac{3^2}{4^2} = \dfrac{9}{16}$

b. $\left(\dfrac{10}{5}\right)^4 = \dfrac{10^4}{5^4} = \dfrac{10{,}000}{625} = 16$ or $\left(\dfrac{10}{5}\right)^4 = 2^4 = 16$

c. $\left(\dfrac{x^n}{y^n}\right)^2 = \dfrac{(x^n)^2}{(y^n)^2} = \dfrac{x^{2n}}{y^{2n}}$

TRY THIS Simplify each expression.

a. $\left(\dfrac{2}{3}\right)^4$ b. $\left(\dfrac{6}{2}\right)^3$ c. $\left(\dfrac{x^3}{2^y}\right)^z$, where $y > 0$ and $z > 0$

Reteaching the Lesson

COOPERATIVE LEARNING Have students work in groups of three or four. Have them create a summary of all the laws of exponents they learned in Lessons 8.1 through 8.3 by filling out a chart like the one shown at right.

	Rule (words)	Example (numbers)	Example (variables)
Product of Powers			
Quotient of Powers			
Power of a Product			
Power of a Quotient			
Power of a Power			

386   LESSON 8.3

Exercises

Communicate

1. Explain how to simplify $\frac{2^5}{2^4}$.
2. Explain why you cannot use the same procedure to simplify $\frac{2^5}{2^4}$ and $\frac{5^2}{4^2}$.
3. Use $2^3 \cdot 2^2$ and $\frac{2^3}{2^2}$ to compare the Product-of-Powers Property with the Quotient-of-Powers Property.
4. Explain why $\left(\frac{f}{g}\right)^6 = \frac{f^6}{g^6}$ is true.

Guided Skills Practice

Use the Quotient-of-Powers Property to simplify each quotient. Then find the value of the result. *(EXAMPLE 1)*

5. $\frac{10^9}{10^3}$ 1,000,000
6. $\frac{3^{12}}{3^9}$ 27
7. $\frac{7^{10}}{7^7}$ 343

Use the Quotient-of-Powers Property to simplify each quotient. Assume that the conditions of the property are satisfied. *(EXAMPLE 2)*

8. $\frac{a^c}{a^d}$ a^{c-d}
9. $\frac{d^{12}}{d}$ d^{11}
10. $\frac{m^{y+1}}{m}$ m^y

Simplify. Assume that the conditions of the Quotient-of-Powers Property are satisfied. *(EXAMPLES 3 AND 4)*

11. $\frac{-8x^3z^7}{2x^2yz}$ $\frac{-4xz^6}{y}$
12. $\frac{h^3k^7}{h^2k^4}$ hk^3
13. $\frac{p^2r^4s^5}{r^3s^2}$ p^2rs^3

Simplify each expression. *(EXAMPLE 5)*

14. $\left(\frac{5}{8}\right)^2$ $\frac{25}{64}$
15. $\left(\frac{3}{4}\right)^4$ $\frac{81}{256}$
16. $\left(\frac{a^3}{4^t}\right)^n$ $\frac{a^{3n}}{4^{tn}}$

Practice and Apply

Use the Quotient-of-Powers Property to simplify each quotient. Then find the value of the result.

17. $\frac{10^{11}}{10^6}$ 100,000
18. $\frac{4^{11}}{4^7}$ 256
19. $\frac{6^{12}}{6^4}$ 1,679,616
20. $\frac{4^3}{4^2}$ 4
21. $\frac{5^2}{5}$ 5
22. $\frac{10^5}{10^2}$ 1000
23. $\frac{3^5}{3^3}$ 9
24. $\frac{11^{42}}{11^{39}}$ 1331
25. $\frac{10^{15}}{10^5}$ 10,000,000,000
26. $\frac{7^6}{7^2}$ 2401
27. $\frac{8^3}{8}$ 64
28. $\frac{2^{16}}{2^{14}}$ 4

Assess

Selected Answers
Exercises 5–16, 17–87 odd

ASSIGNMENT GUIDE

In Class	1–16
Core	17–81 odd
Core Plus	18–80 even, 81–83
Review	84–88
Preview	89, 90

✏ Extra Practice can be found beginning on page 738.

LESSON 8.3 **387**

Technology

A calculator may be helpful in performing some of the calculations involved in the exercises. You might want students to do the exercises first with paper and pencil and then use the calculator to check their answers.

Error Analysis

Students sometimes divide the exponents when they should subtract. For instance, some will simplify $\frac{m^6}{m^2}$ as m^3. Encourage them to apply the definition of an exponent.

$$\frac{m^6}{m^2} = \frac{m \cdot m \cdot m \cdot m \cdot m \cdot m}{m \cdot m} = m^4$$

Simplify each expression. Assume that the conditions of the Quotient-of-Powers Property are met.

29. $\left(\frac{a}{b}\right)^3$ $\frac{a^3}{b^3}$
30. $\left(\frac{10^x}{y^3}\right)^2$ $\frac{10^{2x}}{y^6}$
31. $\left(\frac{5x}{y^6}\right)^3$ $\frac{125x^3}{y^{18}}$
32. $\left(\frac{16x^2}{4y^5}\right)^3$ $\frac{64x^6}{y^{15}}$

33. $\left(\frac{a^4}{b^2}\right)^3$ $\frac{a^{12}}{b^6}$
34. $\left(\frac{10g^7}{h^3}\right)^6$ **34.** $\frac{1,000,000g^{42}}{h^{18}}$
35. $\left(\frac{9d^4}{e^6}\right)^2$ $\frac{81d^8}{e^{12}}$
36. $\left(\frac{8w^7}{16}\right)^3$ $\frac{w^{21}}{8}$

37. $\left[\frac{c^5p^4}{(cp)^2}\right]^3$ c^9p^6
38. $\left(\frac{6x^5}{24x^4}\right)^2$ $\frac{x^2}{16}$
39. $\left(\frac{2w^5}{3f^4}\right)^5$ $\frac{32w^{25}}{243f^{20}}$
40. $\left(\frac{4y^7}{40y^2}\right)^2$ $\frac{y^{10}}{100}$

41. $\left(\frac{7p^q}{f}\right)^d$ $\frac{7^d p^{qd}}{f^d}$
42. $\left(\frac{ac^5}{b^3}\right)^z$ $\frac{a^z c^{5z}}{b^{3z}}$
43. $\left(\frac{wx^t}{w}\right)^t$ x^{t^2}
44. $\left(\frac{c^2a^3}{b^x}\right)^{xy}$ $\frac{c^{2xy}a^{3xy}}{b^{x^2y}}$

Find each quotient. Assume that the conditions of the Quotient-of-Powers Property are met.

45. $\frac{8r^5}{4r^3}$ $2r^2$
46. $\frac{70x^4}{7x^3}$ $10x$
47. $\frac{-2a^5}{4a^2}$ $-\frac{a^3}{2}$
48. $\frac{-3z^5}{27z^3}$ $-\frac{z^2}{9}$

49. $\frac{-p^6}{10p^3}$ $-\frac{p^3}{10}$
50. $\frac{-2a^2b^7}{-5ab}$ $\frac{2ab^6}{5}$
51. $\frac{-x^2y^5}{-xy^2}$ xy^3
52. $\frac{49a^7b^2}{7a^5b}$ $7a^2b$

53. $\frac{48a^4b^2}{-1.2ac^5}$ $-\frac{40a^3b^2}{c^5}$
54. $\frac{0.08r^{12}}{0.004r^3}$ $20r^9$
55. $\frac{4.38u^2v^{10}w^5}{0.1w^2v^3}$ **55.** $43.8u^2v^7w^3$
56. $\frac{x^4y^3z^2}{x^3y^2z}$ xyz

57. $\frac{-r^4}{-5r^3}$ $\frac{r}{5}$
58. $\frac{-24a^5b^{14}}{-36a^3b}$ $\frac{2a^2b^{13}}{3}$
59. $\frac{-c^5d^7}{-c^4y^7}$ $\frac{cd^7}{y^7}$
60. $\frac{-x^3y^7}{-x^3y^4}$ y^3

61. $\frac{5.6d^{15}e^{13}}{-7(de)^{12}}$ **61.** $-0.8d^3e$
62. $\frac{0.27p^4}{0.09p^3}$ $3p$
63. $\frac{156q^5(r^2s)^7}{12q^2s^3}$ $13q^3r^{14}s^4$
64. $\frac{128a^4(bc^2)^3}{32abc}$ $4a^3b^2c^5$

TECHNOLOGY Calculators can be used to evaluate expressions involving monomials.

Store 10 for A and 2 for B in the memory of a calculator. Use the calculator to evaluate each of the following. Round each answer to the nearest hundredth if necessary.

65. $\frac{A^2}{B^3}$ 12.5
66. $\frac{1}{A^2+B^3}$ 0.01
67. $\frac{A^2}{B^3+B}$ 10
68. $\frac{3}{A^2B-B^3}$ 0.02

69. $\frac{AB^2}{(AB)^2}$ 0.1
70. $\frac{A^2+B^2}{A^2B^2}$ 0.26
71. $\frac{-A^2}{(-A)^2}$ -1
72. $\frac{B^A}{A^B}$ 10.24

Use a calculator to evaluate each quotient for $A = 9.8$ and $B = 2.1$. Round each answer to the nearest hundredth if necessary.

73. $\frac{A^2}{B^3}$ 10.37
74. $\frac{1}{A^2+B^3}$ 0.01
75. $\frac{A^2}{B^3+B}$ 8.45
76. $\frac{3}{A^2B-B^3}$ 0.02

77. $\frac{AB^2}{(AB)^2}$ 0.1
78. $\frac{A^2+B^2}{A^2B^2}$ 0.24
79. $\frac{-A^2}{(-A)^2}$ -1
80. $\frac{B^A}{A^B}$ 11.92

CONNECTION

81. **GEOMETRY** Find the ratio of the volume of a sphere to the area of a circle with the same radius as the sphere.
Area of circle: πr^2
Volume of sphere: $\frac{4}{3}\pi r^3$ $\frac{4}{3}r$

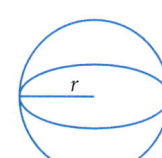

NAME _____ CLASS _____ DATE _____

Practice
8.3 Laws of Exponents: Dividing Monomials

Use the Quotient-of-Powers Property to simplify each quotient. Then find the value of the result.

1. $\frac{10^6}{10^2}$ $10,000$
2. $\frac{4^{17}}{4^{14}}$ 64
3. $\frac{6^8}{6^5}$ 7776
4. $\frac{9^{210}}{9^{207}}$ 729
5. $\frac{2^{r+1}}{2^r}$ 2
6. $\frac{8^{y+3}}{8^y}$ 512

Simplify each expression. Assume that the conditions of the Quotient-of-Powers Property are met.

7. $\left(\frac{x}{y}\right)^6$ $\frac{x^6}{y^6}$
8. $\left(\frac{5c}{d^3}\right)^2$ $\frac{25c^2}{d^6}$
9. $\left(\frac{4d^3}{c^5}\right)^3$ $\frac{64d^9}{c^{15}}$
10. $\left(\frac{3w}{g^{10}}\right)^3$ $\frac{243w^3}{g^{30}}$
11. $\left(\frac{-4s^6}{t^3r^5}\right)^3$ $\frac{-64s^{18}}{t^9r^{15}}$
12. $\left(\frac{-2d^{11}f^6}{c^{18}}\right)^2$ $\frac{4d^{22}f^{12}}{c^{36}}$
13. $\left(\frac{30d}{5d^4}\right)^5$ $\frac{7776}{d^{15}}$
14. $\left(\frac{-24t^6}{8t^3}\right)^3$ $-27t^9$
15. $\left(\frac{d^{11}f^{16}}{d^6f^6}\right)^2$ $d^{10}f^{20}$
16. $\frac{6t^3}{2t}$ $3t^2$
17. $\frac{-40s^6}{20s^3}$ $-2s^3$
18. $\frac{21d^{18}e^5}{7d^{11}e^3}$ $3d^7e^2$
19. $\frac{-16w^7r^2}{-4wr}$ $4w^6r$
20. $\frac{a^2b^3c^5}{-a^2b^7c^4}$ $-a^3b^2c$
21. $\frac{4.2x^4y^{14}}{0.6x^9y^9}$ $\frac{7y^5}{x^5}$

Evaluate each quotient given $x = 2$, $y = -2$, and $z = 10$.

22. $\frac{x^3}{x}$ 4
23. $\frac{y^4}{y}$ -8
24. $\frac{x^3y}{xy^3}$ 1
25. $\frac{z^2x^2y}{zxy^2}$ -1000
26. $\frac{(yz)^2}{z}$ 40
27. $\frac{y^3(3zx)^2}{9x^3}$ -400
28. $\frac{z^{x+1}}{z^x}$ 10
29. $\frac{z^{x+x}}{z^{x+3}}$ 1000
30. $\left(\frac{xz}{y}\right)^3$ -1000

APPLICATIONS

82. ECONOMICS The 1996 population of the United States was 265 million people. The *Washington Post* reports that the national debt for 1996 was 3.7 trillion dollars. To find the amount of debt per person, divide the debt by the population. Simplify the following expression:

$$\frac{3.7 \times 10^{12}}{2.65 \times 10^{8}} \quad \$13{,}962.26$$

CHALLENGE

83. PACKAGING A box manufacturer has the option of producing hat boxes as right rectangular prisms or as cylinders. If the hats are y inches high and x^2 inches wide at the widest part of the brim, write a ratio of the surface area of the smallest possible prism to the surface area of the smallest possible cylinder. Simplify the ratio as much as possible. $\frac{4}{\pi}$

Look Back

Solve each equation for n. (LESSONS 3.4 AND 3.5)

84. $(4n - 20) - n = n + 12$ **n = 16**   **85.** $9(n + k) = 5n + 17k + 32$ **n = 2k + 8**

Solve each system. (LESSONS 7.2 AND 7.3)

86. $\begin{cases} 5x - 3y = 32 \\ 2x = 12.8 \end{cases}$ **(6.4, 0)**   **87.** $\begin{cases} 7x - 3y = 2 \\ 2x - y = -5 \end{cases}$ **(17, 39)**

88. A canoer takes 2.5 hours to travel 7.5 miles upstream. The return trip takes 2 hours. Find the speed of the current and the speed of the canoe in still water. (LESSON 7.6) **canoe 3.375 mph**
current 0.375 mph

Look Beyond

89. Find the ratio of the area of the square to the area of the circle.

$\frac{4}{\pi}$

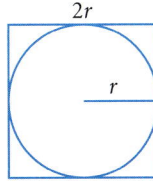

90. Notice that $r^2 \div r^2 = 1$, where $r \neq 0$. Does this suggest a definition of a zero exponent? Explain.

90. Yes, $\frac{r^2}{r^2} = r^{2-2} = r^0$.

Also, $\frac{r^2}{r^2} = 1$.

Therefore, $r^0 = 1$.

Look Beyond

In Exercise 89, students use a ratio to compare the area of a square with the area of a circle inscribed in it. This ratio is applied in geometric probability, which students will study if they take a course in geometry or a second course in algebra. In Exercise 90, students investigate the meaning of a zero exponent, which they will study in Lesson 8.4.

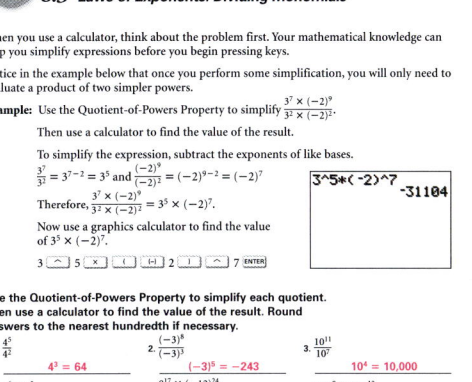

LESSON 8.3 **389**

Prepare

NCTM PRINCIPLES & STANDARDS 1, 2, 5–10

LESSON RESOURCES

- Practice 8.4
- Reteaching Master 8.4
- Enrichment Master 8.4
- Cooperative-Learning Activity 8.4
- Lesson Activity 8.4
- Problem Solving/Critical Thinking Master 8.4
- Teaching Transparency 41

QUICK WARM-UP

Evaluate.

1. $3^2 \cdot 3^4$ **729**   2. $2^6 \cdot 2$ **128**
3. $\frac{9^{13}}{9^{11}}$ **81**   4. $\frac{2^{12}}{2^6}$ **64**

Simplify.

5. $k^4 \cdot k^8$ k^{12} 6. $(-2t)^3$ $-8t^3$
7. $\frac{a^4 b^7}{ab^6}$ $a^3 b$ 8. $\frac{8v^8}{2v^2}$ $4v^6$

Also on Quiz Transparency 8.4

Teach

Why Ask students if they know the size of an atom. Encourage them to guess an actual measure. Then tell them that scientists have found the diameter of a typical atom to be about 0.0000000001 meter. Point out that it is easy to make a mistake in writing a number like this. However, when you use negative exponents, the measure can be written more concisely as 10^{-10} meter.

8.4 Negative and Zero Exponents

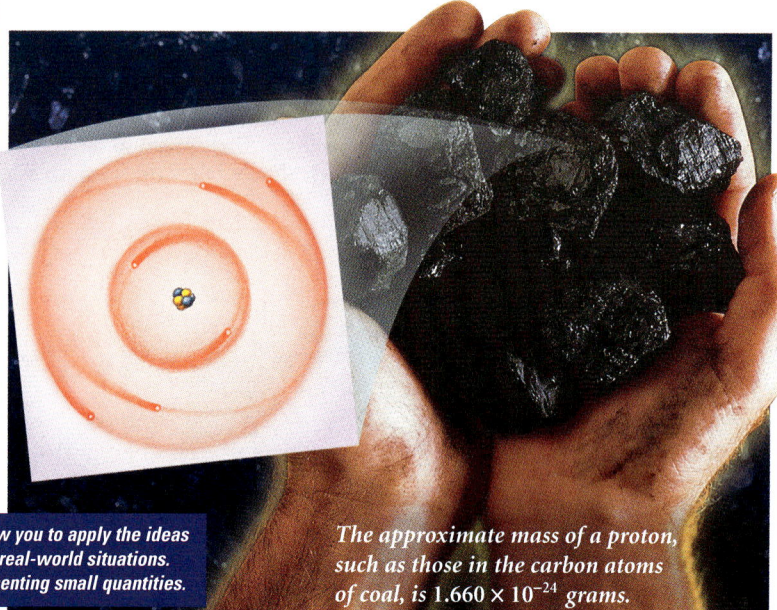

Objectives

- Understand the concepts of negative and zero exponents.
- Simplify expressions containing negative and zero exponents.

Why Negative exponents allow you to apply the ideas that you have learned to more real-world situations. They are also useful for representing small quantities.

The approximate mass of a proton, such as those in the carbon atoms of coal, is 1.660×10^{-24} grams.

The expression 10^1 is defined to be 10. How would you define 10^{-1}?

When the Quotient-of-Powers Property is applied to an expression such as $\frac{10^4}{10^3}$, the result is $\frac{10^4}{10^3} = 10^{4-3} = 10^1$.

If the Quotient-of-Powers Property is applied when the exponent in the numerator is less than the exponent in the denominator, the result is a negative exponent. For example, $\frac{10^3}{10^4} = 10^{3-4} = 10^{-1}$.

If you write the powers as factors, the result is $\frac{1}{10}$.

The result above indicates that 10^{-1} can be defined as $\frac{1}{10}$.

Definition of Negative Exponent

For all nonzero real numbers x and all integers n,

$$x^{-n} = \frac{1}{x^n}.$$

You can use this definition to simplify expressions with negative exponents. All the properties of exponents presented in Lessons 8.1 through 8.3 also apply to negative and zero exponents.

Alternative Teaching Strategy

USING PATTERNS Have students examine this table and note the pattern.

$$10^4 = 10 \cdot 10 \cdot 10 \cdot 10 = 10{,}000$$
$$10^3 = 10 \cdot 10 \cdot 10 = 1000$$
$$10^2 = 10 \cdot 10 = 100$$
$$10^1 = 10 = 10$$

Ask how the next row would appear if the pattern were continued. Then write the next row and discuss the meaning of a zero exponent.

$$10^0 = 1 = 1$$

Now ask how the next row would appear. Write it and discuss the meaning of –1 as an exponent.

$$10^{-1} = \frac{1}{10} = 0.1$$

Finally, have students write the next two rows.

$$10^{-2} = \frac{1}{10 \cdot 10} = 0.01$$
$$10^{-3} = \frac{1}{10 \cdot 10 \cdot 10} = 0.001$$

Be sure they understand that a negative exponent represents the reciprocal of a number. Emphasize that it is does *not* represent a negative number.

390   LESSON 8.4

EXAMPLE 1

Evaluate each expression.

a. $2^{-3} \cdot 2^2$ b. $\dfrac{10^3}{10^{-1}}$ c. $\dfrac{7^5}{7^7}$

● **SOLUTION**

a. Use the Product-of-Powers Property. Add the exponents and simplify.
$$2^{-3} \cdot 2^2 = 2^{-3+2} = 2^{-1} = \tfrac{1}{2^1} = \tfrac{1}{2}$$

b. Use the Quotient-of-Powers Property. Subtract the exponents and simplify.
$$\tfrac{10^3}{10^{-1}} = 10^{3-(-1)} = 10^4 = 10{,}000$$

c. Use the Quotient-of-Powers Property. Subtract the exponents and simplify.
$$\tfrac{7^5}{7^7} = 7^{5-7} = 7^{-2} = \tfrac{1}{7^2} = \tfrac{1}{49}$$

TRY THIS   Evaluate each expression.

a. $\dfrac{10^5}{10^7}$ b. $3^{-5} \cdot 3^8$

CRITICAL THINKING   Consider a^n and a^{-n}. How are these two expressions related?

Activity

Defining x^0

1. Evaluate each expression.

   a. $\dfrac{10^5}{10^5}$ b. $\dfrac{2^3}{2^3}$ c. $\dfrac{5^4}{5^4}$

2. Use the Quotient-of-Powers Property to simplify each expression in Step 1.

CHECKPOINT ✓

3. Based on Steps 1 and 2, what are the numerical values of 10^0, 2^0, and 5^0?

4. Why would you define x^0 to be 1 for any nonzero number x?

Zero as an Exponent

For any nonzero number x, $x^0 = 1$.

EXAMPLE 2

Simplify and write each expression with positive exponents only.

a. $c^{-4} \cdot c^4$ b. $\dfrac{d^4 f^3}{d^6 f^3}$

● **SOLUTION**

a. $c^{-4} \cdot c^4 = c^{-4+4} = c^0 = 1$ b. $\dfrac{d^4 f^3}{d^6 f^3} = (d^{4-6})(f^{3-3}) = d^{-2} f^0 = d^{-2} = \dfrac{1}{d^2}$

Interdisciplinary Connection

PHYSICS Some scientists estimate that the distance from Earth to the edge of the observable universe is about 10^{26} meters. The diameter of a typical atom is about 10^{-10} meter. How much larger is the distance to the edge of the observable universe compared with the diameter of an atom? **about 10^{36} times the diameter** How much smaller is the diameter of an atom compared with the distance to the edge of the observable universe? **about 10^{-36} times the distance**

Inclusion Strategies

TACTILE LEARNERS Give students several squares of paper, such as patty paper. Have them successively double the squares to show positive integral powers of 2. Have them successively cut the squares in half to show negative integral powers of 2. Then the original square represents 2^0, or 1.

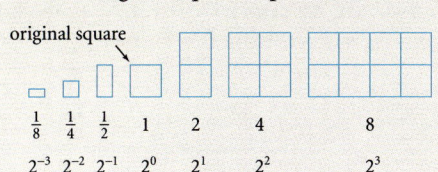

ADDITIONAL EXAMPLE 1

Evaluate each expression.

a. $3^5 \cdot 3^{-7}$ $\dfrac{1}{9}$

b. $\dfrac{5^2}{5^{-2}}$ 625

c. $\dfrac{4^4}{4^7}$ $\dfrac{1}{64}$

TRY THIS

a. $10^{-2} = \dfrac{1}{100}$

b. $3^3 = 27$

CRITICAL THINKING

They are reciprocals of each other.

Activity Notes

In this Activity, students use the Quotient-of-Powers Property to investigate the meaning of zero as an exponent. They should find that any nonzero number raised to the zero power is equal to 1.

CHECKPOINT ✓

3. 1

ADDITIONAL EXAMPLE 2

Simplify and write each expression with positive exponents only.

a. $\dfrac{r^4}{r^6}$ $\dfrac{1}{r^2}$ b. $\dfrac{m^5 n}{m^2 n^9}$ $\dfrac{m^3}{n^8}$

LESSON 8.4

TRY THIS

a. $b^0 = 1$

b. $\dfrac{y}{x}$

ADDITIONAL EXAMPLE 3

Simplify and write each expression with positive exponents only.

a. $5d^{-3}$ $\dfrac{5}{d^3}$

b. $\dfrac{p^{-1}}{q^{-5}}$ $\dfrac{q^5}{p}$

c. $\dfrac{-20a^{-2}b^4}{5a^{-3}b^{-1}}$ $-4ab^5$

TRY THIS

a. $\dfrac{b}{5a^3}$

b. $-\dfrac{4}{x^3}$

CRITICAL THINKING

$\left(\dfrac{a}{b}\right)^{-n} = \dfrac{a^{-n}}{b^{-n}} = \dfrac{\frac{1}{a^n}}{\frac{1}{b^n}} = \dfrac{1}{a^n} \cdot \dfrac{b^n}{1} = \dfrac{b^n}{a^n} = \left(\dfrac{b}{a}\right)^n$

Use Teaching Transparency 41.

Teaching Tip

The expression 0^0 is undefined. If students wonder why this is true, tell them to consider that 0^0 would be equal to 0^{n-n}. But $0^{n-n} = \dfrac{0^n}{0^n}$, which is undefined.

TRY THIS Simplify and write each expression with positive exponents only.

a. $\dfrac{b^6 \cdot b^{-2}}{b^4}$

b. $\dfrac{x^2 y^3}{x^3 y^2}$

Expressions containing variables with negative and zero exponents can be simplified by using the properties of exponents.

EXAMPLE 3 Simplify and write each expression with positive exponents only.

a. $-3y^{-2}$

b. $\dfrac{4c^4 d^5}{12c^4 d^8}$

c. $\dfrac{m^2}{n^{-3}}$

● **SOLUTION**

a. $-3y^{-2} = (-3)(y^{-2}) = \dfrac{-3}{y^2}$

b. $\dfrac{4c^4 d^5}{12c^4 d^8} = \dfrac{1}{3}(c^{4-4})(d^{5-8}) = \dfrac{1}{3}c^0 d^{-3} = \dfrac{1}{3d^3}$

c. $\dfrac{m^2}{n^{-3}} = m^2 n^{-(-3)} = m^2 n^3$

TRY THIS Simplify and write each expression with positive exponents only.

a. $\dfrac{5a^2 b^3}{25a^5 b^2}$

b. $-4x^{-3}$

CRITICAL THINKING If a and b are nonzero numbers and n is any integer, show that $\left(\dfrac{a}{b}\right)^{-n} = \left(\dfrac{b}{a}\right)^n$ is true.

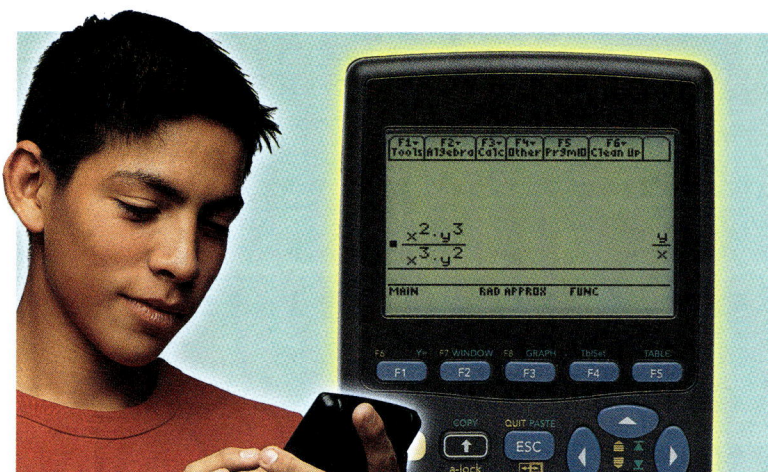

Some calculators can perform algebraic manipulations such as simplifying rational expressions.

SUMMARY OF THE PROPERTIES OF EXPONENTS

Let x and y be any number and let m and n be integers.

Product of Powers	Quotient of Powers	Power of a Power	Power of a Product	Power of a Fraction
$x^m \cdot x^n = x^{m+n}$	$\dfrac{x^m}{x^n} = x^{m-n}$	$(x^m)^n = x^{mn}$	$(xy)^m = x^m y^m$	$\left(\dfrac{x}{y}\right)^n = \dfrac{x^n}{x^n}$

Enrichment

Challenge students to use four 4s to write expressions with values from 1 to 10. They may use exponents as well as other operations. **Answers will vary.**

samples: $4^{-4} \cdot 4^4 = 1$; $\dfrac{4}{4} + \dfrac{4}{4} = 2$; $\dfrac{4+4+4}{4} = 3$;

$\dfrac{4-4}{4} + 4 = 4$; $4^{4-4} + 4 = 5$; $\dfrac{4+4}{4} + 4 = 6$;

$4 + 4 - \dfrac{4}{4} = 7$; $\dfrac{4 \cdot 4}{4} + 4 = 8$; $4 + 4 + \dfrac{4}{4} = 9$;

$\dfrac{4}{\sqrt{4}} + 4\sqrt{4} = 10$

Reteaching the Lesson

INVITING PARTICIPATION Tell each student to write an expression similar to those in Examples 2 and 3. Now have the students count off by fours.

Ask all of the "1s" to write their expressions on the board. Then have the "2s" go to the board and simplify the expressions.

Now the "2s" should write their expressions on the board, and the "3s" should simplify. Continue in this manner, matching "3s" and "4s" and then "4s" and "1s" until all students have participated.

Exercises

Communicate

1. Explain how to show that $5^{-2} = \frac{1}{5^2}$ makes sense.
2. Why can $5^2 \cdot 4^{-3}$ not be simplified by using the properties of exponents?
3. Explain why $(3a)^{-2}$ does not equal $\frac{3}{a^2}$.
4. What is the meaning of a negative exponent?
5. What is the meaning of a zero exponent?

Guided Skills Practice

Evaluate each expression. *(EXAMPLE 1)*

6. $7^{-5} \cdot 7^2$ $\frac{1}{343}$
7. $\frac{10^7}{10^{11}}$ $\frac{1}{10,000}$
8. $\frac{6^8}{6^2}$ $46,656$
9. $\frac{4^5}{4^7}$ $\frac{1}{16}$

Simplify and write each expression with positive exponents only. *(EXAMPLE 2)*

10. $\frac{a^5 \cdot a^{-5}}{a^8}$ $\frac{1}{a^8}$
11. $\frac{x^6 y^7}{x^6 y^6}$ y
12. $\frac{r^{-5} s^2}{r^{-3} s^2}$ $\frac{1}{r^2}$
13. $\frac{a^{-3} b^3}{a^3 b^5}$ $\frac{1}{a^6 b^2}$

Simplify and write each expression with positive exponents only. *(EXAMPLE 3)*

14. $-5y^{-6}$ $-\frac{5}{y^6}$
15. $\frac{3c^6 d^3}{36 c^4 d^8}$ $\frac{c^2}{12 d^5}$
16. $\frac{j^3}{h^{-4}}$ $j^3 h^4$
17. $\frac{64 a^7 b^5}{25 a b^{10}}$ $\frac{64 a^6}{25 b^5}$

Practice and Apply

Evaluate each expression.

18. $6^4 \cdot 6^{-2}$ 36
19. $\frac{11^8}{11^{13}}$ $\frac{1}{161,051}$
20. $\frac{9^{12}}{9^3}$ $387,420,489$
21. $\frac{7^5}{7^{12}}$ $\frac{1}{823,543}$
22. $8^{-6} \cdot 8^3$ $\frac{1}{512}$
23. $\frac{13^8}{13^{15}}$ $\frac{1}{62,748,517}$
24. $\frac{5^3}{5^{10}}$ $\frac{1}{78,125}$
25. $3^4 \cdot 3^{-6}$ $\frac{1}{9}$

Simplify and write each expression with positive exponents only.

26. $\frac{f^7 \cdot f^{-7}}{f^2}$ $\frac{1}{f^2}$
27. $\frac{a^5 b^8}{a^5 b^3}$ b^5
28. $\frac{s^{-7} t^2}{s^{-3} t^3}$ $\frac{1}{s^4 t}$
29. $\frac{3x^4 y^8}{27 x^6 y^4}$ $\frac{y^4}{9 x^2}$
30. $-6p^{-7}$ $-\frac{6}{p^7}$
31. $\frac{4x^8 y^4}{24 x^3 y^{11}}$ $\frac{x^5}{6 y^7}$
32. $\frac{r^5}{s^{-12}}$ $r^5 s^{12}$
33. $\frac{8 a^4 b^7 c^{-4}}{3 a^6 b^{-6} c^{-4}}$ $\frac{8 b^{13}}{3 a^2}$

Evaluate each expression.

34. -3^2 -9
35. $(-3)^2$ 9
36. 3^{-2} $\frac{1}{9}$
37. $-(3^{-2})$ $-\frac{1}{9}$

38. Copy the table and fill in the remaining entries by continuing the pattern.

Decimal notation	10,000	1000	100	10	?	?	?
Exponential notation	10^4	10^3	10^2	10^1	?	?	?

Answers: 1, 0.1, 0.01 ; 10^0, 10^{-1}, 10^{-2}

Assess

Selected Answers
Exercises 6–17, 19–69 odd

ASSIGNMENT GUIDE

In Class	1–17
Core	19–59 odd
Core Plus	18–58 even, 60–63
Review	64–70
Preview	71

✐ Extra Practice can be found beginning on page 738.

Mid-Chapter Assessment for Lessons 8.1 through 8.4 can be found on page 100 of the *Assessment Resources*.

LESSON 8.4   **393**

Technology

In Exercises 52–57, students are asked to consider when a calculator is and is not a reasonable tool for evaluating expressions involving exponents. Exercises 58 and 59 provide them with an opportunity to examine situations that produce an error message on calculators.

Error Analysis

Given an expression such as $\frac{m^3}{m^5}$, students often do a quick mental "computation" and write m^2. Encourage them to always write an intermediate step that shows the subtraction of the exponents in the proper order.

$$\frac{m^3}{m^5} = m^{3-5} = m^{-2} = \frac{1}{m^2}$$

---

39. The corresponding numbers 10,000 and 10^4 in the table on the previous page are equal. Are the other corresponding numbers that you wrote equal? **yes**

Write each of the following without negative or zero exponents:

40. 2^{-3} $\frac{1}{2^3}$ **41.** 10^{-5} $\frac{1}{10^5}$ **42.** a^3b^{-2} $\frac{a^3}{b^2}$

43. $c^{-4}d^3$ $\frac{d^3}{c^4}$ **44.** $v^0w^2y^{-1}$ $\frac{w^2}{y}$ **45.** $(a^2b^{-7})^0$ **1**

46. r^6r^{-2} r^4 **47.** $-t^{-1}t^{-2}$ $-\frac{1}{t^3}$ **48.** $\frac{m^2}{m^{-3}}$ m^5

49. $\frac{2a^{-5}}{a^{-6}}$ $2a$ **50.** $\frac{(2a^3)(10a^5)}{4a^{-1}}$ $5a^9$ **51.** $\frac{b^{-2}b^4}{b^{-3}b^4}$ b

TECHNOLOGY Tell whether each expression is easier to evaluate on a calculator or mentally. Explain why you think so. Then evaluate each expression.

52. $\frac{2.56^7}{2.56^6}$ **2.56**   **53.** $\frac{2.56^6}{2.56^7}$ **0.390625**

54. 0^7 **0**   **55.** 7^0 **1**

56. $(2.992 \times 9.554)^0$ **1**   **57.** $(19.43 \times 0)^{18}$ **0**

58. TECHNOLOGY What happens when you try to compute $2^{2^{36}}$ on a calculator?

59. TECHNOLOGY What happens when you try to compute 0^0 on a calculator?

CONNECTION

CHALLENGE

PROBABILITY The probability, p, of correctly guessing a multiple-choice question with four answer choices is $\frac{1}{4}$, and the probability, q, of guessing incorrectly is $\frac{3}{4}$. Expressions for finding the probability of getting a certain number of answers correct on a test with 5 multiple-choice questions are shown in the chart below.

No. correct	Probability	Value
0	p^0q^5	$\left(\frac{1}{4}\right)^0\left(\frac{3}{4}\right)^5 = 0.237$
1	$5p^1q^4$	? **0.396**
2	$10p^2q^3$	? **0.264**
3	$10p^3q^2$	? **0.088**
4	$5p^4q^1$	? **0.015**
5	p^5q^0	? **0.001**

60. Find the missing values in this chart, to the nearest thousandth.

61. Which number of correct answers is most likely? **1**

62. Which number of correct answers is least likely? **5**

63. Find the sum of all six probability values in the table. The sum of all probabilities should equal to 1. If the sum is not 1, explain the discrepancy.

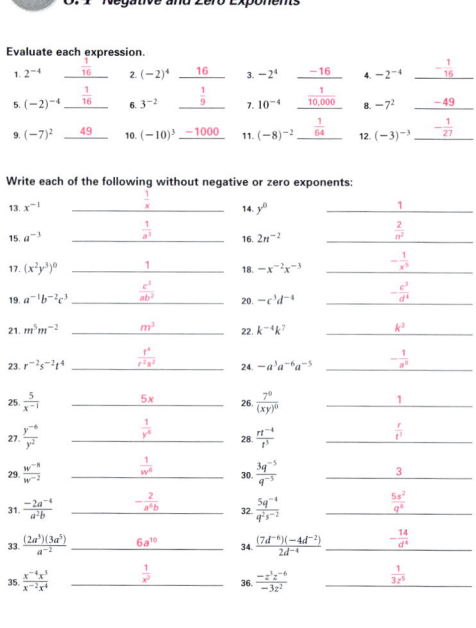

58. You get an error message: "overflow." The number is too large to display.

59. You get an error message because 0^0 is undefined.

63. 1.001 is the sum. Rounding errors explain the discrepancy.

 **Look Back**

64. What are the next three terms of the sequence 8, 11, 16, 23, 32, ... ? **(LESSON 1.1)** 43, 56, 71

Solve each equation for *y*. How are the equations related?
(LESSON 3.6)

65. $5(3y) = 3y - 24$ $y = -2$ **66.** $aby = by - c$ $y = \frac{-c}{ab - b}$

67. What is the slope of a vertical line? How is it different from the slope of a horizontal line? **(LESSON 5.2)**

68. Draw a number line showing the values of *x* that solve the inequality $3x - 6 > -4x + 12$. Describe the solution set of this inequality in words. **(LESSON 6.2)**

Let $A = 3 \cdot 3 \cdot 3 \cdot 3 \cdot 3 \cdot x \cdot x \cdot x \cdot x \cdot x \cdot x \cdot y \cdot y \cdot y \cdot y$ and
$B = 2 \cdot 2 \cdot 2 \cdot 2 \cdot 3 \cdot 3 \cdot 3 \cdot x \cdot x \cdot x \cdot x \cdot y$.

69. Write *A* and *B* in exponential form. **(LESSON 8.1)**

70. Write the product $A \cdot B$ in the original form and in exponential form. **(LESSON 8.1)**

 Look Beyond

Pierre de Fermat
1601-1665

CULTURAL CONNECTION: EUROPE Pierre de Fermat (1601–1665) was a French mathematician who helped develop number theory and probability. He studied numbers of the form $2^{2^n} + 1$, which are now called Fermat numbers.

71. NUMBER THEORY Fermat conjectured that each of the numbers of this form is prime. Though an unusually gifted mathematician, he was wrong in thinking that the sixth number, $2^{2^5} + 1$, is prime; in fact, it has a factor of 641. Use a calculator to find the other factor. Later research has suggested that *no* Fermat number with $n > 5$ is prime. Even great mathematicians can make mistakes. 6,700,417

Include this activity in your portfolio.

In 1995 it was reported that there were 60,000 centenarians (people 100 or more years old) in the United States. The researcher predicted a 7% annual growth rate for this age group and claimed that the number of centenarians would reach 232,000 by the year 2015.

1. Verify the prediction numerically.
2. Write a function to model the problem. Use your calculator to determine when the number of centenarians will reach 1,000,000.

WORKING ON THE CHAPTER PROJECT
You should now be able to complete Activities 2 and 3 of the Chapter Project on page 416.

67. The slope of a vertical line is undefined. The slope of a horizontal line is 0.

68. $x > \frac{18}{7}$

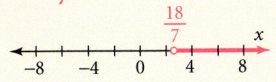

69. $A = 3^5 x^6 y^4$
$B = 2^4 3^3 x^4 y$

70. $A \cdot B = 2 \cdot 2 \cdot 2 \cdot 2 \cdot 3 \cdot 3 \cdot 3 \cdot 3 \cdot 3 \cdot 3 \cdot 3 \cdot 3 \cdot x \cdot x \cdot x \cdot x \cdot x \cdot x \cdot x \cdot x \cdot x \cdot x \cdot y \cdot y \cdot y \cdot y \cdot y$
$A \cdot B = 3^5 x^6 y^4 \cdot 2^4 3^3 x^4 y = 2^4 3^8 x^{10} y^5$

Eyewitness Math

Focus

A 1990 study of card shuffling reported in the *New York Times* raises the notion of matching a mathematical model to a physical reality and brings up some surprising results about the disadvantages of perfection. In answering a series of questions about the situation, students discover that an explanation for the dilemma lies in a familiar exponential expression.

Motivate

Before students read the article, discuss the questions at the top of the page. As students read, consider pausing after the first seven paragraphs to be sure they understand the story. Ask questions such as the following:

- What did the mathematicians discover?
- Why was the problem complicated?
- Why does the article mention "analyzing speech patterns," which seems to have nothing to do with shuffling cards?
- What is the dovetail, or riffle, shuffle?

After reading the article, go over *Getting Lost in the Shuffle.* Discuss the diagram, which illustrates one imperfect shuffle for a "deck" of 13 cards. Discuss the meaning of the graph. Note that the curve labeled *1st shuffle* shows that after only one shuffle, the first card is much more likely to remain near the front of the deck. In looking at the curves for successive shuffles, help students see how a flatter curve means the card is not much more likely to be in one place than another.

EYEWITNESS MATH

ALL MIXED UP?

You are playing a card game and it's your turn to deal. How many times do you shuffle the deck? Is that enough to be sure the cards are thoroughly mixed? Is there such a thing as shuffling too much? Read the article below to learn how two mathematicians, one of them a magician as well, found some answers that surprised shufflers around the world.

In Shuffling Cards, 7 Is Winning Number

It takes just seven ordinary, imperfect shuffles to mix a deck of cards thoroughly, researchers have found. Fewer are not enough and more do not significantly improve the mixing.

The mathematical proof, discovered after studies of results from elaborate computer calculations and careful observation of card games, confirms the intuition of many gamblers, bridge enthusiasts and casual players that most shuffling is inadequate.

The finding has implications for everyone who plays cards and everyone, from casino operators to magicians, who has a stake in knowing whether a shuffle is random . . .

No one expected that the shuffling problem would have a simple answer, said Dr. Dave Bayer, a mathematician and computer scientist at Columbia who is co-author of the recent discovery. Other problems in statistics, like analyzing speech patterns to identify speakers, might be amenable to similar approaches, he said . . .

Dr. Persi Diaconis, a mathematician and statistician at Harvard University who is another author of the discovery, said the methods used are already helping mathematicians analyze problems in abstract mathematics that have nothing to do with shuffling or with any known real-world phenomena.

Dr. Diaconis, who is also a magician, has invented several card tricks and has been carefully watching casino dealers and casual card players shuffle for the past 20 years. The researchers studied the dovetail or riffle shuffle, in which the deck of cards is cut in half and the two halves are riffled together. They said this was the most commonly used method of shuffling cards. But, Dr. Diaconis said it produces a card order that is "far from random." . . .

Dr. Diaconis began working with Dr. Jim Reeds at Bell Laboratories and showed that a deck is perfectly mixed if it is shuffled between 5 and 20 times.

Next, Dr. Diaconis worked with Dr. Aldous and showed that it takes 5 to 12 shuffles to perfectly mix a deck. But, said Dr. Diaconis, "nobody in practice shuffles 12 times," adding, "We needed some new ideas."

In the meantime, he also worked on "perfect shuffles," those that exactly interlace the cards. Almost no one, except a magician can do perfect shuffles every time. But Dr. Diaconis showed several years ago that if a person actually does perfect shuffles, the cards would never be thoroughly mixed. He derived a mathematical proof showing that if a deck is perfectly shuffled eight times, the cards will be in the same order as they were before the shuffling.

To find out how many ordinary shuffles were necessary to mix a deck, Dr. Diaconis and Dr. Bayer watched players shuffle. He also watched Las Vegas dealers to see how perfectly they would interlace the cards they shuffled . . .

The researchers did extensive simulations of shuffling on a computer. To get the proof, the researchers looked at a lot of shuffles, guessed that the answer was seven, and finally proved it by finding an abstract way to describe what happens when cards are shuffled.

"When you take an honest description of something realistic and try to write it out in mathematics, usually it's a mess," Dr. Diaconis said. "We were lucky that the formula fit the real problem. That is just miraculous, somehow."

By Gina Kolata

Getting Lost in the Shuffle

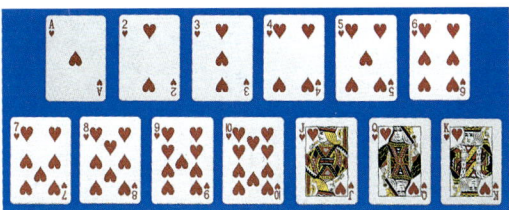

Even after a deck of cards is cut and shuffled, fragments of the original arrangement remain. In this example, the hearts are cut into two sequences: ace through 6, and 7 through king. Then the sequences are shuffled.

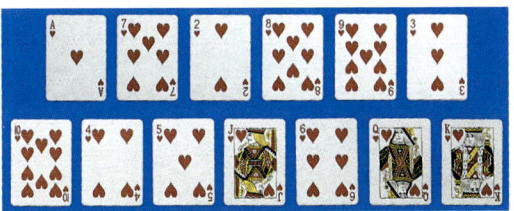

The cards are no longer ordered, but the sequence ace through 6 remains, with cards from the other fragment interspersed, also in sequence. The larger number of cards in play, the more shuffles are required to mix them.

Curved lines show the odds that the first card in a deck will occupy any other position in the deck after shuffling from one to seven times. After one shuffle, for example, the first card is very likely to be one of the first few cards in the deck and very unlikely to be even five or six cards back. After four shuffles, it is still far more likely to be at the beginning of the deck than at the end. Only after seven shuffles does the card have about the same odds of being in any given position.

Cooperative Learning

1. Based on the graph at right, after 4 shuffles, how many times more likely is it for the original 1st card to be in the 1st position than in the 35th position? Explain.

2. a. What is the difference between a perfect shuffle and an ordinary shuffle?
   b. Is the shuffle in the diagram above perfect? Explain.

3. Use a diagram or model to find out how many perfect shuffles it takes for a deck to return to its original order if the deck consists of

   a. 2^2 cards.   b. 2^3 cards.   c. 2^4 cards.

4. a. How many perfect shuffles do you think it would take for a deck of 32 cards to return to its original order? Why?
   b. Use a diagram or model to check your prediction. What did you find?

5. Would a machine that always makes perfect shuffles be useful when you play a card game? Why or why not?

Cooperative Learning

Have students work in groups of three. Each group will need either a standard deck of playing cards or 32 index cards numbered 1 through 32.

Have students do Steps 1 and 2 together. The question in Step 1 will help determine whether all students understand the graph. For Step 2, ask that one student in each group volunteer to demonstrate for the group a perfect shuffle of an 8-card deck, using cards numbered 1 through 8.

In Step 3, each student in the group can do one part. The group can then come together to share results and resolve any difficulties. The group should do Steps 4 and 5 together.

Discuss

Have students make a brief presentation of their results to the entire class. Ask students for their reactions to what they learned. Were they surprised to discover that a process which may have seemed random could, in fact, be very structured?

Interested students can do further research into the ways that the solution techniques used here can be applied to other problems in statistics.

---

1. about 6 or 7 times greater (although other estimates may be reasonable, as well); the probability of the card still being first is about 0.07, and the probability of it being 35th is about 0.01.

2. a. In a perfect shuffle, the cards from the two halves of the deck alternate: you get one card from one half, one from the other half, then one from the first half, and so on. In an ordinary shuffle, you may get two cards from one half, then three from the other, then one from the first, and so on.

   b. No; the first four cards alternate, but then the 3 of hearts should come before the 9 of hearts. Also, the 4 and 5 of hearts are together; they should be separated by a card from the right half of the deck.

3. a. 2
   b. 3
   c. 4

4. a. 5; because 32 is 2^5, which continues the pattern in Activity 3
   b. It does take 5 perfect shuffles for the deck of 32 cards to return to its original order.

5. No; because the machine would not mix the cards randomly, and, with a certain number of shuffles, it might not mix them at all

Prepare

NCTM Principles & Standards 1–4, 6, 8–10

LESSON RESOURCES

- Practice 8.5
- Reteaching Master 8.5
- Enrichment Master 8.5
- Cooperative-Learning Activity 8.5
- Lesson Activity 8.5
- Problem Solving/Critical Thinking Master 8.5

QUICK WARM-UP

Evaluate.

1. $5^5 \cdot 5^{-1}$
   625
2. $10^4 \cdot 10^{-6}$
   $\frac{1}{100}$, or 0.01
3. $\frac{10^7}{10^3}$
   10,000
4. $\frac{3^5}{3^7}$
   $\frac{1}{9}$
5. (3.2)(100)
   320
6. (3.2)(0.01)
   0.032

Also on Quiz Transparency 8.5

Teach

Why Ask students if they know the meaning of *nanosecond*. **one billionth of a second** Show them how this can be written as 0.000000001 second. Work with them to rewrite it as a power of 10. 10^{-9} **second** Now ask how you could write the time for a computer operation that requires 7 nanoseconds. 7×10^{-9} **second** Tell them that this form of the number is called *scientific notation*.

8.5 Scientific Notation

Objectives

- Recognize the need for special notation in scientific calculations.
- Perform computations involving scientific notation.

Why The study of the universe involves measurements of very large and very small quantities. Scientific notation provides an easier way to do calculations with very large and very small numbers both in scientific fields and in other fields.

APPLICATION
ASTRONOMY

Betelgeuse is one of the stars in the constellation Orion. The approximate distance between Earth and Betelgeuse is 6,000,000,000,000,000 miles. This number, with all its zeros, is hard to read and to use in calculations. A special way to express large and small numbers in a compact form is **scientific notation**. In scientific notation, the distance between Earth and Betelgeuse is written as 6×10^{15} miles.

A number written in scientific notation is a product with two factors—a number from 1 to 10, including 1 but not including 10, and a power of 10. The first factor includes only significant digits (0s are included only if they lie between two other digits) and is often rounded to simplify the notation. Therefore, all numbers written in scientific notation are considered approximations.

Your knowledge of the powers of 10 will help you to write numbers in scientific notation.

Alternative Teaching Strategy

USING TECHNOLOGY Have students enter these multiplications into their calculator and record the results exactly as they appear on the screen.

863 × 10	8630
863 × 100	86300
863 × 1000	863000
863 × 10,000	8630000
863 × 100,000	86300000
863 × 1,000,000	863000000
863 × 10,000,000	8630000000
863 × 100,000,000	8.63E10

Tell students to observe the pattern. Lead them to see that 8.63E10 must represent 86,300,000,000. Now have them enter 8.63×10^{10}. They should see 8.63E10 displayed, so $8.63 \times 10^{10} = 86,300,000,000$. The expression, 8.63×10^{10} is called *scientific notation*, and 86,300,000,000 is called *decimal form*. Point out that 8.63E10 is "shorthand" for scientific notation.

Repeat the activity with multiplications from 863×0.1 through 863×0.0001. This time, lead students to see that $8.63 \times 10^{-4} = 0.000863$.

398 LESSON 8.5

Activity

Exploring Powers of 10

1. Complete the table for powers of 10.

Value	1000	100	10	1	$\frac{1}{10}$	$\frac{1}{100}$	$\frac{1}{1000}$
Power	$10^{?}$	$10^{?}$	$10^{?}$	$10^{?}$	$10^{?}$	10^{-2}	$10^{?}$

2. Copy and complete the table below.

Decimal	Expanded form	Scientific notation
1200	1.2×1000	$1.2 \times 10^{?}$
120	1.2×100	$1.2 \times 10^{?}$
12	1.2×10	$1.2 \times 10^{?}$
1.2	1.2×1	$1.2 \times 10^{?}$
0.12	$1.2 \times \frac{1}{10}$	$1.2 \times 10^{?}$
0.012	$1.2 \times \frac{1}{100}$	$1.2 \times 10^{?}$
0.0012	$1.2 \times \frac{1}{1000}$	$1.2 \times 10^{?}$

CHECKPOINT ✓ 3. Explain how to write each of the following in scientific notation:
a. 0.000012   b. 120,000

EXAMPLE 1

APPLICATION
ASTRONOMY

The distance between Earth and Proxima Centauri, the nearest star other than the sun, is about 24,000,000,000,000 miles—read as 24 trillion miles. **Write this number in scientific notation.**

← Proxima Centauri

● **SOLUTION**

Place the decimal point between the 2 and the 4 (the first factor must be between 1 and 10), and count the number of digits it moved.

$$24{,}000{,}000{,}000{,}000.$$

Because the decimal point was moved 13 places to the left, use 10^{13} as the other factor to maintain equality with the original number, 24 trillion.

$$24{,}000{,}000{,}000{,}000 = 2.4 \times 10^{13}$$

TRY THIS   Write 875,000 in scientific notation.

Interdisciplinary Connection

PHYSICS The structure of an atom can be described in terms of protons, neutrons, and electrons. The protons and neutrons make up the core of the atom, called its nucleus. The mass of a proton is about 1.7×10^{-27} kilogram, and the mass of a neutron is about 1.7×10^{-27} kilogram. Surrounding the nucleus are the electrons, each having a mass of about 9.1×10^{-31} kilogram. Which has the greater mass: an electron or a proton? How many times greater is its mass? **proton; about 2000 times greater**

Inclusion Strategies

TACTILE LEARNERS Obtain a supply of 3-inch-by-5-inch index cards. Make several sets of cards, each of which contains at least 12 cards with the digit zero, at least three cards with each digit from 1 through 9, a card with a decimal point, and a card with ×. Have students work in groups of three or four. Give each group a set of cards, several numbers in scientific notation, and several numbers in decimal form. Have them manipulate the cards to display the decimal form of each number in scientific notation and vice versa.

Activity Notes

In this Activity, students investigate the meaning of scientific notation. They first write the given numbers as products involving powers of ten in decimal form. Then they translate the powers to the exponential form. In Step 3, they are asked to apply and explain the process in two additional situations.

Cooperative Learning

You may wish to have students do the Activity in pairs. Have them work together to complete Step 1. For the table in Step 2, have one student identify the missing exponent and have the other check the work. Then have partners switch roles as they proceed from row to row. In Step 3, the partners should first work independently and then compare their answers.

CHECKPOINT ✓

3. a. $0.000012 = 1.2 \times \frac{1}{100{,}000}$
$ = 1.2 \times 10^{-5}$

b. $120{,}000 = 1.2 \times 100{,}000$
$\phantom{120{,}000} = 1.2 \times 10^{5}$

ADDITIONAL EXAMPLE 1

The mean distance between the sun and Pluto is about 3,700,000,000 miles. **Write this distance in scientific notation.** 3.7×10^{9} miles

TRY THIS
8.75×10^{5}

LESSON 8.5

ADDITIONAL EXAMPLE 2

When the planet Saturn is closest to Earth, the distance from Earth to Saturn is about 7.5×10^8 miles. Suppose that you traveled at a speed of 1000 miles per hour. **How long would it take to get from Earth to Saturn?**
at least 750,000 hours, or at least 86 years

TRY THIS
3×10^{24}

ADDITIONAL EXAMPLE 3

It has been estimated that the diameter of a typical atom is between 0.0000000106 and 0.000000054 centimeter. **Write these lengths in scientific notation.**
1.06×10^{-8} centimeter
5.4×10^{-8} centimeter

TRY THIS
4.02×10^{-8}

ADDITIONAL EXAMPLE 4

Express the product and the quotient in scientific notation.

a. $(2 \times 10^{-2})(7 \times 10^8)$
1.4×10^7

b. $\frac{3.1 \times 10^3}{4 \times 10^{10}}$
7.75×10^{-8}

TRY THIS
a. 3.9×10^2
b. 6×10^1

EXAMPLE 2

APPLICATION: ASTRONOMY

Betelgeuse is 6×10^{15} miles from Earth. **If you traveled at a speed of 1000 miles per hour, how long would it take to get from Earth to Betelgeuse?**

SOLUTION

Use the properties of exponents and scientific notation.

$$(6 \times 10^{15}) \div 1000 = \frac{6 \times 10^{15}}{10^3} = 6 \times 10^{15-3} = 6 \times 10^{12}$$

The flight would take about 6 trillion hours, or about 685 million years!

TRY THIS Find $3 \times 10^{28} \div 10^4$.

EXAMPLE 3

APPLICATION: PHYSICS

A typical X ray has a wavelength of about 0.000000000125 meter. **Write this number in scientific notation.**

SOLUTION

Since the first factor must be a number between 1 and 10, place the decimal after the 1. Because the decimal point moves 10 places to the right, use 10^{-10} as the other factor to maintain equality with the original number, 0.000000000125.

$$0.000000000125 = 1.25 \times 10^{-10}$$

TRY THIS Write 0.0000000402 in scientific notation.

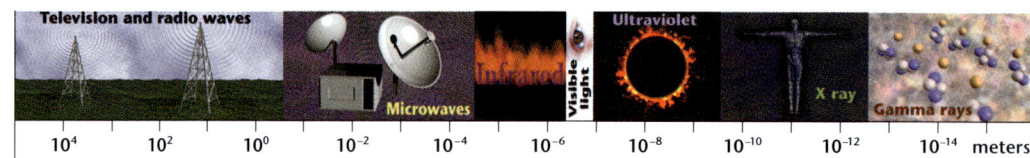

Answers to multiplication and division problems involving scientific notations should always be rounded to the least number of significant digits in the original problem.

EXAMPLE 4

Express the product and quotient in scientific notation.

a. $(3 \times 10^3)(4 \times 10^{-5})$
b. $\frac{2.5 \times 10^6}{5 \times 10^2}$

SOLUTION

a. $(3 \times 10^3)(4 \times 10^{-5}) = (3 \times 4)(10^3 \times 10^{-5})$
$= 12 \times 10^{-2}$
$= (1.2 \times 10^1) \times 10^{-2}$
$= 1.2 \times (10^1 \times 10^{-2})$
$= 1.2 \times 10^{-1}$

b. $\frac{2.5 \times 10^6}{5 \times 10^2} = \left(\frac{2.5}{5}\right)\left(\frac{10^6}{10^2}\right)$
$= 0.5 \times 10^4$
$= (5 \times 10^{-1}) \times 10^4$
$= 5 \times (10^{-1} \times 10^4)$
$= 5 \times 10^3$

TRY THIS Express the product and quotient in scientific notation.
a. $(2.5 \times 10^5)(1.56 \times 10^{-3})$
b. $\frac{3.6 \times 10^5}{6 \times 10^3}$

Enrichment

Have students work in pairs to solve these "missing exponent" problems.

1. $(9 \times 10^2)(4 \times 10^?) = 3.6 \times 10^8$ 5
2. $(5 \times 10^?)(3 \times 10^8) = 1.5 \times 10^6$ -3
3. $(2.5 \times 10^?)(4 \times 10^{-11}) = 1 \times 10^{-16}$ -6
4. $(3.8 \times 10^p)(7.5 \times 10^q) = 2.85 \times 10^?$ $p + q + 1$
5. $(2 \times 10^m)(8.1 \times 10^?) = 1.62 \times 10^{m+n}$ $n - 1$

Then have each student make up five original problems like the ones above and write them on a clean sheet of paper. Partners should exchange papers and solve each other's problems.

Reteaching the Lesson

GUIDED ANALYSIS Work with students to outline a general procedure like the one below for writing a number in scientific notation.

1. Move the decimal point until you have a number a such that $-1 \le a < 10$.
2. Count the number of places, n, that the decimal point was moved.
3. If the move was to the *left*, write $a \times 10^n$. If the move was to the *right*, write $a \times 10^{-n}$.

Then outline a similar procedure for writing a number in decimal form.

EXAMPLE 5

TECHNOLOGY — SCIENTIFIC CALCULATOR

The speed of light is about 2.98×10^5 kilometers per second. **If the sun is about 1.49×10^8 kilometers from Earth, how long does it take sunlight to reach Earth?**

● **SOLUTION**

Use the [EE] key.
1.49 [EE] 8 [÷] 2.98 [EE] 5 [=] 500

Or use the [EXP] key.
1.49 [EXP] 8 [÷] 2.98 [EXP] 5 [=] 500

Sunlight takes about 500 seconds to reach Earth.

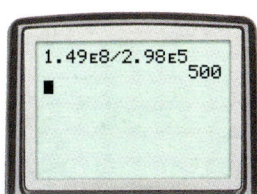

```
1.49E8/2.98E5
                500
■
```

Notice that on the calculator screen above, "E8" represents "10^8" and "E5" represents "10^5". This is just an easy way to display a number in scientific notation on a calculator or computer. The two forms are interchangeable.

Exercises

● **Communicate**

1. Why is it helpful to be able to write a number such as 23,000,000,000 in scientific notation?

2. Explain what moving the decimal point 8 places to the left does to the value of a number. What power of 10 must you multiply by in order to compensate for this movement of the decimal point?

3. Explain what moving the decimal point 12 places to the right does to the value of a number. What power of 10 must you multiply by in order to compensate for this movement of the decimal point?

● **Guided Skills Practice**

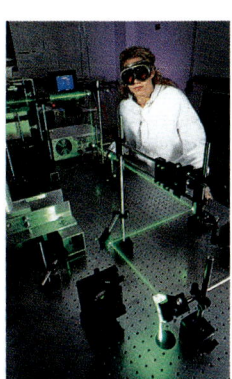

A laser is a focused beam of light.

Write each number in scientific notation. *(EXAMPLE 1)*

4. 463,000,000 4.63×10^8
5. 25,000 2.5×10^4
6. 367,700,000,000 3.677×10^{11}

Express each result in scientific notation. *(EXAMPLE 2)*

7. Find $6 \times 10^{14} \div 10^2$. 6×10^{12}
8. Find $3 \times 10^{13} \div 10^5$. 3×10^8

Write each number in scientific notation. *(EXAMPLE 3)*

9. 0.000597 5.97×10^{-4}
10. 0.000000165 1.65×10^{-7}
11. 0.00000457 4.57×10^{-6}

Express each result in scientific notation. *(EXAMPLE 4)*

12. $(3.5 \times 10^3)(2.67 \times 10^{-5})$ 9.345×10^{-2}
13. $\dfrac{2.4 \times 10^4}{6 \times 10^1}$ 4×10^2

14. The speed of light is about 2.98×10^5 kilometers per second. If a certain star is about 2.36×10^{14} kilometers from Earth, how long does it take the star's light to reach Earth? *(EXAMPLE 5)* **about 25.1 years**

ADDITIONAL EXAMPLE 5

The speed of light is about 1.86×10^5 miles per second. When Pluto is closest to the sun, the distance from the sun to Pluto is approximately 2.76×10^9 miles. **How long does it take sunlight to reach Pluto?** **at least 14,839 seconds, or at least 4.12 hours**

Teaching Tip

TECHNOLOGY If you are using a TI-82 or TI-83 graphics calculator, you can do the calculation in Example 5 by entering **1.49** [2nd] [,] 8 [÷] 2.98 [2nd] [,] 5 [ENTER].

Assess

Selected Answers
Exercises 4–14, 15–73 odd

ASSIGNMENT GUIDE

In Class	1–14
Core	15–53 odd, 59–68
Core Plus	16–44 even, 46–68
Review	69–73
Preview	74, 75

✎ Extra Practice can be found beginning on page 738.

LESSON 8.5

Technology

A calculator may be helpful for Exercises 32–45. You may wish to have students first complete the exercises with paper and pencil and then use a calculator to verify their answers.

Error Analysis

Students often think that any number written in the form $a \times 10^n$ is in scientific notation. For instance, they may arrive at a result such as 12.5×10^6 and leave the answer in this form. Stress that the first factor must be at least 1 and less than 10.

internetconnect

GO TO: go.hrw.com
KEYWORD: MA1 Scientific Notation

In this activity, students apply their knowledge of scientific notation to examine some of the smallest and largest numbers found in the field of cosmology, the study of the beginning, development, and fate of the universe.

Practice and Apply

Write each number in scientific notation.

15. 2,000,000 2×10^6
16. 8,000,000,000,000 8×10^{12}
17. 340,000 3.4×10^5
18. 58,000 5.8×10^4
19. 0.00008 8×10^{-5}
20. 0.0000005 5×10^{-7}
21. 0.000234 2.34×10^{-4}
22. 0.000000082 8.2×10^{-8}
23. 1,350,000,000 1.35×10^9

Write each number in decimal notation.

24. 3×10^4 30,000
25. 4×10^8 400,000,000
26. 6.7×10^{10} 67,000,000,000
27. 9.01×10^5 901,000
28. 4×10^{-7} 0.0000004
29. 5×10^{-9} 0.000000005
30. 8.8×10^{-12} 0.0000000000088
31. 7.2×10^{-10} 0.00000000072

Perform the following computations. Write your answers in scientific notation.

32. $(2 \times 10^4)(3 \times 10^5)$ 6×10^9
33. $(6 \times 10^8)(1 \times 10^6)$ 6×10^{14}
34. $(8 \times 10^6)(2 \times 10^{10})$ 1.6×10^{17}
35. $(9 \times 10^6)(7 \times 10^6)$ 6.3×10^{13}
36. $(9 \times 10^6) + (7 \times 10^6)$ 1.6×10^7
37. $(9 \times 10^6) - (7 \times 10^6)$ 2×10^6
38. $(8.2 \times 10^6)(3.1 \times 10^6)$ 2.542×10^{13}
39. $(1.9 \times 10^8)(2 \times 10^{10})$ 3.8×10^{18}
40. $\dfrac{(2 \times 10^6)(9 \times 10^4)}{6 \times 10^4}$ 3×10^6
41. $(2 \times 10^{10})(6 \times 10^7)(4 \times 10^{-8})$ 4.8×10^{10}
42. $\dfrac{6 \times 10^5}{2 \times 10^2}$ 3×10^3
43. $\dfrac{9 \times 10^8}{2 \times 10^4}$ 4.5×10^4
44. $\dfrac{3 \times 10^4}{6 \times 10^7}$ 5×10^{-4}
45. $\dfrac{16 \times 10^5}{2 \times 10^{12}}$ 8×10^{-7}

APPLICATIONS

SOCIAL STUDIES The following are some large numbers that appeared in studies of different aspects of American culture. Write each number in scientific notation.

46. 125,000: the number of passengers bumped on United States airlines each year due to overbooking 1.25×10^5
47. $5 trillion: the total value of outstanding stock in the stock market 5×10^{12}
48. 5.4 million: the number of American businesses owned by women 5.4×10^6
49. 3.7 million: the number of square feet in the Pentagon, the world's largest office building 3.7×10^6

ECONOMICS The national debt for the United States in 1992 was listed as $4064.6 billion.

50. Write 4064.6 billion in scientific notation. 4.0646×10^{12}
51. This number can be expressed as about $4 \times 10^?$. 4×10^{12}
52. Assuming that the population of the United States was about 250 million in 1992, how much was the debt per person?   about $16,260
53. Write your age at your last birthday in seconds. Use scientific notation.
54. **ASTRONOMY** The distance from Earth to the moon is about 248,000 miles. Write the distance in miles from Earth to the moon in scientific notation.
2.48×10^5 miles

53. Answers may vary. Sample answer: 4.41504×10^8 seconds

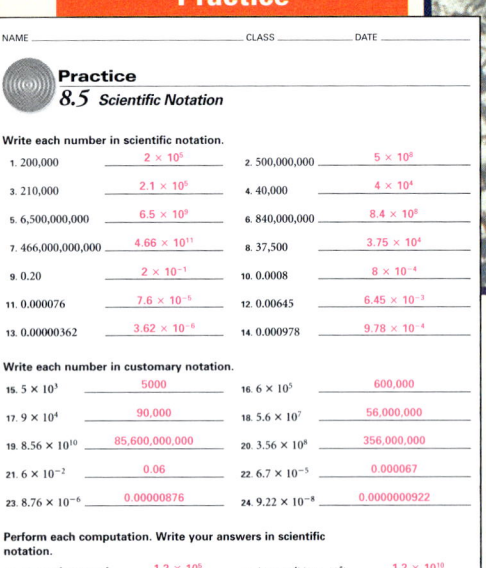

402 LESSON 8.5

APPLICATIONS

ASTRONOMY The distance that light travels in one year, 5.87×10^{12} miles, is called a light-year.

55. Write this number in decimal notation. **5,870,000,000,000**

56. If the distance from Betelgeuse to Earth is 6×10^{15} miles, how many years does it take light to travel from that star to Earth? **1.02×10^3 years**

57. If the distance from Proxima Centauri to Earth is 2.14×10^{13} miles, how many years does it take light to travel from Proxima Centauri to Earth? **3.6 years**

58. **ASTRONOMY** According to Nobel Prize–winning physicist Leon Lederman, the universe is 10^{18} seconds old. How many years is this? Write your answer in words. **approximately 31.7 billion years**

TECHNOLOGY Write the following numbers in decimal notation:

59. 2.3 E 04 **23,000**
60. 5.6 E 03 **5600**
61. 7.22 E –03 **0.00722**
62. 1.01 E –04 **0.000101**
63. –2.8 E 02 **–280**
64. –9.303 E –04 **–0.0009303**

65. What is the largest number that can be shown on your calculator without using scientific notation?

66. What is the largest number that can be shown on your calculator by using scientific notation?

67. What is the smallest positive number that can be shown on your calculator without using scientific notation?

68. What is the smallest positive number that can be shown on your calculator by using scientific notation?

Look Back

69. Find the value of the expression. $3 + (-4) - (-9) - [-8 + 2(3-5)]$ **(LESSONS 2.2 AND 2.3) 20**

70. Find the product. $(5)(-3)[-2(4)](-12)$ **(LESSON 2.4) –1440**

71. Simplify. $(6x + 4y - 8) - 5(-7x + y)$ **(LESSON 2.6) $41x - y - 8$**

72. **COORDINATE GEOMETRY** Given points $A(3, 7)$ and $B(-3, 12)$, find the slope of the line joining the points. **(LESSON 5.2) $-\frac{5}{6}$**

73. Graph the system of inequalities below and describe the solution region. **(LESSON 7.5)**
$$\begin{cases} x - y > 5 \\ -3x + y \leq 9 \end{cases}$$

Look Beyond

TECHNOLOGY Use the constant feature of your calculator to write the given terms of the following sequences in decimal notation:

74. $0.9^1, 0.9^2, 0.9^3, 0.9^4, 0.9^5, \ldots$ **0.9, 0.81, 0.729, 0.6561, 0.59049**

75. $1.1^1, 1.1^2, 1.1^3, 1.1^4, 1.1^5, \ldots$ **1.1, 1.21, 1.331, 1.4641, 1.61051**

Look Beyond

In Exercises 74 and 75, students investigate some sequences that give rise to exponential functions. They will study exponential functions in Lesson 8.6.

If you are using a TI-82 or TI-83 graphics calculator, you can evaluate the terms of the sequence in Exercise 74 by entering **0.9** ENTER ✕ **0.9** and then pressing ENTER four more times. The sequence in Exercise 75 can be evaluated in a similar manner.

---

65. Answers may vary. Sample answer: 9,999,999,999

66. Answers may vary. Sample answer: $9.999999999 \times 10^{99}$

67. Answers may vary. Sample answer: 0.001

68. Answers may vary. Sample answer: 1.0×10^{-99}

73.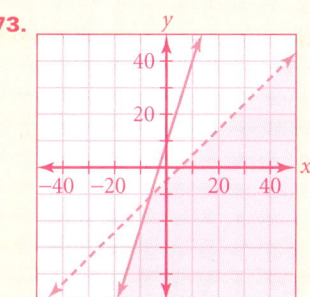

The region describing the solution is simultaneously all points to the right of $y < x - 5$, excluding the points to the right of $y < x - 5$ and to the left of $y \leq 3x + 9$.

Student Technology Guide

8.5 Scientific Notation

Most calculators automatically display very large or very small numbers in scientific notation. You can also choose to enter very large or very small numbers in scientific notation.

The Andromeda Galaxy, one of the Milky Way's nearest neighbors, is about 19,000,000,000,000,000,000 kilometers away. Written in scientific notation, $19{,}000{,}000{,}000{,}000{,}000{,}000 = 1.9 \times 10^{19}$.

To enter 1.9×10^{19}, press
1.9 2nd , 19 ENTER.

Notice that the "E" on the calculator screen represents "times 10 to the ____ power."

You can use properties of exponents to simplify problems involving scientific notation and then use a calculator to find the answer. If you want the calculator to provide answers in scientific notation, carry out the following key sequence before you begin your work:

MODE Sci ENTER

Example: The nearest star to Earth, other than the Sun, is Proxima Centauri, 2.4×10^{13} miles away. If a spaceship traveled at 80,000 miles per day, how long would it take to reach Proxima Centauri?

Rewrite the problem as $\frac{2.4 \times 10^{13}}{8.0 \times 10^4}$. Use the Quotient-of-Powers Property to simplify $\frac{2.4 \times 10^{13}}{8.0 \times 10^4}$ as $\frac{2.4 \times 10^9}{8.0}$.

Press 2.4 2nd , 9 ÷ 8.0 ENTER.

It would take 3×10^8 days (about 820,000 years!) to reach Proxima Centauri.

Perform the following computations. Write your answers in scientific notation.

1. $(2 \times 10^2)(3 \times 10^3)$ **6×10^5**
2. $(5 \times 10^7)(1 \times 10^4)$ **5×10^{11}**
3. $(3.8 \times 10^4)(2.1 \times 10^{-6})$ **7.98×10^{-2}**
4. $\frac{6 \times 10^7}{3 \times 10^4}$ **2×10^3**
5. $\frac{7.8 \times 10^{11}}{6.5 \times 10^5}$ **1.2×10^6**
6. $\frac{8.5 \times 10^6}{3.4 \times 10^{15}}$ **2.5×10^{-9}**

7. If a spaceship traveled at a speed of 200,000 kilometers per day, how long would it take to get from Earth to the Andromeda Galaxy? **9.5×10^{13} days**

Prepare

NCTM PRINCIPLES & STANDARDS 1–6, 8–10

LESSON RESOURCES

- Practice 8.6
- Reteaching Master 8.6
- Enrichment Master 8.6
- Cooperative-Learning Activity 8.6
- Lesson Activity 8.6
- Problem Solving/Critical Thinking Master 8.6

QUICK WARM-UP

Evaluate.

1. 3^0 **1**
2. 3^1 **3**
3. 3^2 **9**
4. 3^3 **27**
5. 3^4 **81**
6. 3^5 **243**
7. $(0.5)^0$ **1**
8. $(0.5)^1$ **0.5**
9. $(0.5)^2$ **0.25**
10. $(0.5)^3$ **0.125**
11. $(0.5)^4$ **0.0625**
12. $(0.5)^5$ **0.03125**

Also on Quiz Transparency 8.6

Teach

Why Every person has 2 biological parents and 4 biological grandparents. Ask students how many great-grandparents a person has. **8** How many great-great-grandparents does a person have? **16** Tell students that this pattern is an example of an *exponential function*. The number of grandparents is said to *increase exponentially*.

8.6 Exponential Functions

Objectives
- Understand exponential functions and how they are used.
- Recognize differences between graphs of exponential functions with different bases.

Why Exponential functions model problems of growth and decline in populations, financial trends, and scientific experiments.

You have studied linear functions in which values in the range change by a fixed *amount*. The graphs of these functions form straight lines. **Exponential functions** apply to situations in which values in the range change instead by a fixed *rate*. The graphs of exponential functions do not form straight lines.

For example, $y = 2^x$ is an exponential function. The base, 2, is the number that is repeatedly multiplied. The exponent, x, represents the number of times that the base, 2, occurs in the multiplications. The function is graphed at right.

APPLICATION
DEMOGRAPHICS

According to the headline above, how many people are added to the population in one year? How many are added in one minute?

$1.52\% \times 5.85$ billion $= 0.0152 \times 5,850,000,000$
$= 88,920,000$ people per year
$\approx 243,616$ people per day
$\approx 10,151$ people per hour
≈ 169 people per minute

In the time it took you to read this page, the population of the world probably increased by several dozen people.

CRITICAL THINKING How was the increase for one day obtained from the increase for one year?

Alternative Teaching Strategy

USING TECHNOLOGY Have students use a graphics calculator to graph the familiar linear equation $y = 2x$ and the exponential equation $y = 2^x$ simultaneously. Have them compare and contrast what they see.

If they are using a TI-82 or TI-83 graphics calculator, tell them to press **Y=**. They should first clear all equations from the **Y=** editor and then enter **2X** next to **Y1=** and **2^X** next to **Y2=**.

Now tell them to press **WINDOW** and set the viewing window as follows:

Xmin=–10 Xmax=10 Xscl=1
Ymin=–2 Ymax=20 Yscl=2

Then tell them to press **GRAPH**. Ask them how the graphs are alike and how they are different.

To get a closer look at the behavior of the two graphs in the first quadrant, tell students to press **WINDOW** and reset the viewing window as follows:

Xmin=0 Xmax=3 Xscl=1
Ymin=0 Ymax=5 Yscl=1

Activity
Population Growth

1. Complete the table below or build a spreadsheet to estimate the increase in the world's population from 1997 to 2007. Start with the 1997 population of 5.85 billion. Assuming that the population increases by 1.52% each year, the new population each year will be 101.52% of the previous year's population, so use 1.0152 as a multiplier.

Year	Process	Exponential notation	Population (in billions)
1997	5.85	$5.85(1.0152)^0$	5.85
1998	5.85(1.0152)	$5.85(1.0152)^1$	5.94
1999	5.85(1.0152)(1.0152)	$5.85(1.0152)^2$	6.03
2000	5.85(1.0152)(1.0152)(1.0152)	$5.85(1.0152)^3$	6.12
...	...	...	...

2. At this rate, when will the population reach 20 billion?

CHECKPOINT ✔ 3. Examine the pattern in the table in Step 1. What is the formula for finding the population if you know the original amount and rate of growth?

General Growth Formula

Let P be the amount after t years at a yearly growth rate of r, where r is expressed as a decimal. If the original amount is A, then the formula is as follows:

$$P = A(1 + r)^t$$

This formula works for population, compound interest, and many other situations in which growth takes place exponentially.

EXAMPLE 1

APPLICATION
DEMOGRAPHICS

In 1992, India had an estimated population of about 886 million and was growing at a yearly rate of 1.9%. **At this rate, by how many people will the population increase in 10 years?**

TECHNOLOGY
SCIENTIFIC CALCULATOR

SOLUTION

Use the general growth formula. If you have a calculator, use the y^x or $\wedge$ key. You may also use a constant multiplier of 1.019 in your computation.

$$P = A(1 + r)^t$$
$$= 886(1.019)^{10}$$
$$\approx 1069$$

The increase is $1069 - 886 = 183$, or about 183 million people.

Interdisciplinary Connection

BIOLOGY The number of bacteria in a certain culture doubles every half-hour. At 10:30 A.M., there were 2000 bacteria in the culture. At what time were there 1000 bacteria? At what time will there be 100,000 bacteria?
10:00 A.M. of that day; between 1:00 P.M. and 1:30 P.M. of that day

Inclusion Strategies

TACTILE/KINESTHETIC LEARNERS Show students a ream (500 sheets) of paper. Place one sheet from the ream on a table. Ask one student to come up and double the amount, making a stack. Ask a second student to come up and double the stack. Continue with four more students, recording the number of sheets at each stage. Ask how many more students would have to come up for all 500 sheets to be used. 3 Ask how many sheets of paper would be needed if every student in the class were to participate. 2^n, where n is the number of students

---

CRITICAL THINKING
by dividing by the number of days in one year

Activity Notes

In this Activity, students look for a pattern in an example of exponential growth, and they use the pattern to derive a formula that models the growth. They should observe that after t years of growth, the population is equal to the original population multiplied by the yearly factor t times.

Cooperative Learning

If you wish, students can do the Activity in pairs. In Step 1, one partner can operate a calculator, while the other dictates the multiplications and records the results. Both should work together to complete Steps 2 and 3.

CHECKPOINT ✔

3. new population = (original population) × $(1 + \text{rate})^{\text{no. of years}}$

ADDITIONAL EXAMPLE 1

When Christopher was born, his parents put $1000 into a savings account. The amount of money in the account is guaranteed to increase at a yearly rate of 4.5%. Assume that no additional deposits are made to the account and no withdrawals are made. **By what amount will the account increase in 18 years?**
$2208.48

LESSON 8.6

TRY THIS
approximately 18,690,000

ADDITIONAL EXAMPLE 2

Make a table of values and graph the function $y = 3^x$.

x	y
−3	0.0̄3̄7̄
−2	0.1̄
−1	0.3̄
0	1
1	3
2	9
3	27

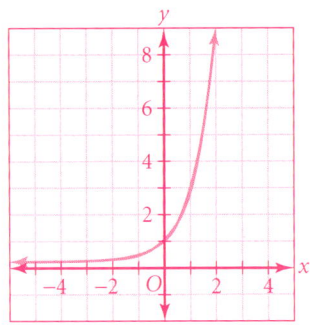

TRY THIS Suppose that the population of a country was 200,000,000 at the last census. If the population grows at an annual rate of 1.5%, by how many people will the population increase in the next 6 years?

EXAMPLE 2 Make a table of values and graph the function $y = 2^x$.

SOLUTION

x	−3	−2	−1	0	1	2	3
2^x	0.125	0.25	0.5	1	2	4	8

The graph of the function $y = 2^x$ shows the characteristics of an exponential function.

The graph remains above the x-axis for all values of x. When x is 0, the function equals 1. After crossing the y-axis, the function increases rapidly.

Notice that, as with linear functions, each value in the domain, x, has a unique value in the range, y.

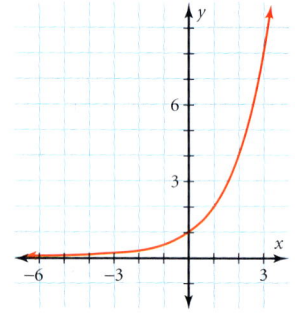

EXAMPLE 3 Graph $y = 3^x$, $y = 4^x$, and $y = \left(\frac{1}{2}\right)^x$ on the same axes. Explain how the graphs are similar and how they are different.

SOLUTION

You can graph each function on a graphics calculator, or you can make a table of values and graph the functions on paper. To simplify your calculations, use integer values for x.

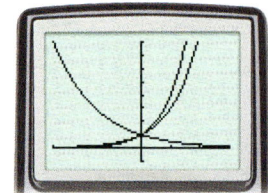

x	−3	−2	−1	0	1	2	3
3^x	0.04	0.11	0.33	1	3	9	27
4^x	0.0156	0.0625	0.25	1	4	16	64
$\left(\frac{1}{2}\right)^x$	8	4	2	1	$\frac{1}{2}$	$\frac{1}{4}$	$\frac{1}{8}$

The graphs of the first two functions are similar in that each rises from left to right. The third function does the opposite. Each graph crosses the y-axis at $y = 1$. Each gets closer and closer to the x-axis but never crosses it.

The first two functions are different in that the graph of $y = 4^x$ rises faster than the graph of $y = 3^x$ for values of x greater than 0.

CRITICAL THINKING Consider the graphs of $y = 3^x$ and $y = 4^x$. For values of x less than 0, which graph rises faster? Explain your reasoning.

Enrichment

Use a graphics calculator to graph $y = 2^x$ and $y = x^2$ simultaneously. Compare the graphs.

1. How do they seem alike? **Answers will vary.**
2. How do they seem different? **Answers will vary.**
3. Where do the graphs intersect? **(2, 4), (4, 16)**
4. What is the meaning of the point(s) of intersection? **These points indicate the two instances in which $2^x = x^2$: when $x = 2$ and when $x = 4$.**

Reteaching the Lesson

USING TABLES Have students make the table below to track the growth of $1000 invested at a yearly rate of 5% and compare it with the result from the general growth formula.

Year	Amt. at start	Interest (5%)	Amt. at end
1	$1000.00	$50.00	$1050.00
2	$1050.00	$52.50	$1102.50
3	$1102.50	$55.13	$1157.63
4	$1157.63	$57.88	$1215.51

$P = 1000(1 + 0.05)^4 = \$1215.51$

Practice

8.6 Exponential Functions

Graph each of the following:

1. $y = 7^x$
2. $y = 0.1^x$
3. $y = 1.5^x$

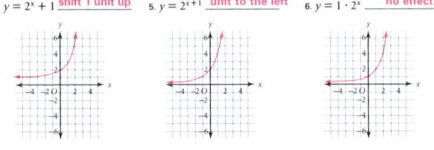

Graph each function. In each case, describe the effect of 1 in the transformation of the parent function.

4. $y = 2^x + 1$ **shift 1 unit up**
5. $y = 2^{x+1}$ **shift 1 unit to the left**
6. $y = 1 \cdot 2^x$ **no effect**

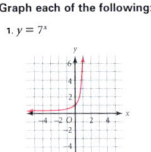 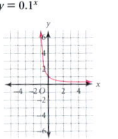 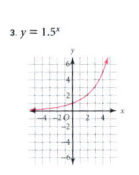

If $f(x) = \left(\frac{1}{2}\right)^x$, find the following values for the function:

7. $f(2)$ __$\frac{1}{4}$__
8. $f(0)$ __1__
9. $f(-1)$ __2__
10. $f(-2)$ __4__
11. $f(-3)$ __8__
12. $f(-4)$ __16__
13. $f(4)$ __$\frac{1}{16}$__
14. $f(3)$ __$\frac{1}{8}$__

The population of a city is about 200,000 and is growing at a rate of about 0.6% per year.

15. What multiplier is used to find the new population each year? __1.006__
16. Estimate the population 4 years from now. __about 204,843__

406 LESSON 8.6

Exercises

Communicate

1. Describe how an exponential function changes as the base increases.
2. Explain how population and money earning interest can both be represented by the formula $P = 1000(1.08)^t$.
3. Explain how you would determine whether the graph of $y = x^4$ represents an exponential function.
4. Explain how you would determine whether the graph of $y = 4^x$ represents an exponential function.

Guided Skills Practice

APPLICATION

5. **DEMOGRAPHICS** In 1995, the population of Zimbabwe was about 11 million and was growing at a rate of 1.9% per year. At this rate, by how many people will this population increase by 2005? **(EXAMPLE 1)**
 ≈ 2,278,057 people
6. Make a table of values for $y = 5^x$ and $y = \left(\frac{1}{5}\right)^x$ and graph the functions on the same axes. Explain how the graphs are similar and how they are different. **(EXAMPLES 2 AND 3)**

Harare is the capital of Zimbabwe, a country in Africa.

Practice and Apply

Graph each of the following:

7. $y = 2^x$
8. $y = 4^x$
9. $y = 7^x$
10. $y = 12^x$
11. $y = 8^x$
12. $y = 10^x$
13. $y = 1^x$
14. $y = 6^x$
15. $y = 0.3^x$
16. $y = \left(\frac{1}{2}\right)^x$
17. $y = \left(\frac{1}{10}\right)^x$
18. $y = \left(\frac{1}{5}\right)^x$
19. $y = 0.4^x$
20. $y = \left(\frac{1}{4}\right)^x$
21. $y = \left(\frac{2}{3}\right)^x$
22. $y = 0.8^x$
23. $y = \left(\frac{3}{4}\right)^x$
24. $y = \left(\frac{4}{3}\right)^x$
25. $y = \left(\frac{2}{5}\right)^x$
26. $y = \left(\frac{5}{2}\right)^x$
27. $y = \left(\frac{1}{3}\right)^x$
28. $y = 3^x$
29. $y = \left(\frac{3}{7}\right)^x$
30. $y = \left(\frac{7}{3}\right)^x$

APPLICATION

DEMOGRAPHICS The population of Spain was about 39 million in 1992 and was growing at a rate of about 0.3% per year.

31. What multiplier is used to find the new population each year? **1.003**
32. Use this information to estimate the population in 1993. ≈ **39,117,000**
33. Estimate the population in the year 2000. ≈ **39,945,887**

ADDITIONAL EXAMPLE 3

Graph $y = 5^x$ and $y = 10^x$ on the same axes. Explain how the graphs are similar and how they are different.

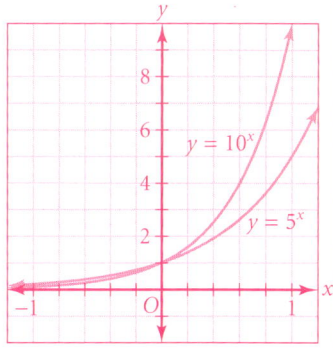

The graphs are similar in that each rises from left to right, each crosses the y-axis at 1, and each gets closer to the x-axis as x decreases. They are different in that the graph of $y = 10^x$ rises faster than the graph of $y = 5^x$ for values of x greater than 0.

CRITICAL THINKING

For values of x less than zero, $y = 3^x$ rises faster than $y = 4^x$. This happens because $4^x < 3^x$ for $x < 0$; that is, $\left(\frac{1}{4}\right)^x < \left(\frac{1}{3}\right)^x$ for $x > 0$.

Student Technology Guide

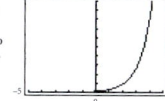

---

6. $y = 5^x$ $\qquad$ $y = \left(\frac{1}{5}\right)^x$

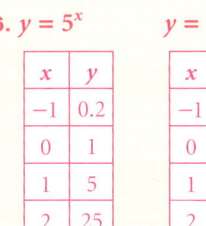

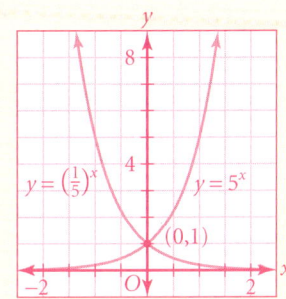

The graphs are mirror images of each other. As x gets larger, one increases and the other decreases. They both pass through the point (0, 1).

The answers to Exercises 7–30 can be found in Additional Answers beginning on page 810.

LESSON 8.6 **407**

Assess

Selected Answers
Exercises 5–6, 7–49 odd

ASSIGNMENT GUIDE

In Class	1–6
Core	7–10, 19–22, 27–30, 31–39 odd
Core Plus	11–18, 23–26, 31–40
Review	41–50
Preview	51–53

✏ Extra Practice can be found beginning on page 738.

Technology

For Exercises 7–30, you may want students to first draw the graphs on graph paper and then use a graphics calculator to verify their results. You probably will want students to use a calculator for Exercises 31–40.

Error Analysis

When applying the general growth formula, students often use r^t rather than $(1 + r)^t$. Remind them to use common sense. That is, for *growth* to occur, the growth factor must be greater than 1.

Look Beyond

In Exercises 51–53, students investigate variations of a simple rational function. Rational functions will be studied in greater detail in Chapter 11.

internet connect
GO TO: go.hrw.com
KEYWORD: MA1 Snowflake

In this activity, students learn about exponential growth by examining a figure called the Koch snowflake.

408 LESSON 8.6

APPLICATION

DEMOGRAPHICS The population of Italy was about 58 million in 1995 and was growing at a rate of about 0.1% per year. The population of France was about 57.8 million in 1995 and was growing at a rate of about 0.4% per year. Use this information to estimate the following populations:

The Spanish Steps of Rome and the Arc de Triomphe are historical monuments that attract tourists from all over the world.

34. Italy in 1996 ≈ **58,058,000**
35. Italy in 1998 ≈ **58,174,174**
36. Italy in 2000 ≈ **58,290,581**
37. France in 1996 ≈ **58,031,200**
38. France in 1998 ≈ **58,496,378**
39. France in 2000 ≈ **58,965,285**
40. In what year did the population of France surpass that of Italy? **1997**

Look Back

41. Find the slope of the line containing the points (−1, 4) and (2, −3). (**LESSON 5.2**) $-\frac{7}{3}$

APPLICATION

42. PHYSICS A spring with a constant, k, of 17 is stretched a distance of 0.6 meters. Find the amount of force used to stretch the spring. (**LESSON 5.3**) **10.2 Newtons**

Graph each equation by using the slope and *y*-intercept. (**LESSON 5.4**)

43. $y = -3x - 2$ **44.** $y = \frac{2}{5}x - 4$ **45.** $y = -\frac{1}{5}x + 7$

Solve each system of equations. (**LESSON 7.3**)

46. $\begin{cases} 2t + 3w = -4 \\ -2t + w = -1 \end{cases}$ $t = -\frac{1}{8}$ $w = -\frac{5}{4}$

47. $\begin{cases} 2x + 2y = 12 \\ -15x + 3y = 9 \end{cases}$ $x = \frac{1}{2}$ $y = \frac{11}{2}$

48. $\begin{cases} 5k + 4m = 10 \\ 5k - 2m = 10 \end{cases}$ $k = 2$ $m = 0$

Simplify by using the order of operations and properties of exponents. (**LESSONS 8.1 AND 8.2**)

49. $(x^4 \cdot x^2)^2 + (y^3 \cdot y^2)^3$ $x^{12} + y^{15}$

50. $(a^2b^4)^2(a^2b)^3$ $a^{10}b^{11}$

Look Beyond

Maria Gaetana Agnesi 1718-1799

CULTURAL CONNECTION: EUROPE Maria Gaetana Agnesi was born in Italy in 1718. She is considered to be the first woman in modern times to achieve a reputation for mathematical works. In her best known work, written in 1748, she discusses a curve that is now known as the Agnesi curve. This curve is represented by the function $f(x) = \frac{a^3}{x^2 + a^2}$. Let the domain of f be $\{-3, -2, -1, 0, 1, 2, 3\}$.

51. Graph the equation for $a = 1$.
52. Graph the equation for $a = 5$.
53. How do these two graphs compare?

---

43. slope = −3; *y*-intercept = −2

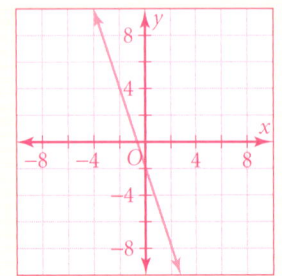

44. slope = $\frac{2}{5}$; *y*-intercept = −4

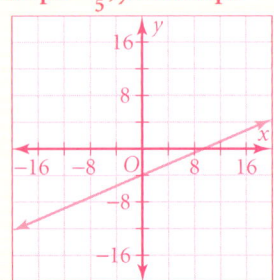

The answers to Exercises 45 and 51–53 can be found in Additional Answers beginning on page 810.

8.7 Applications of Exponential Functions

Objective
- Use exponential functions to model applications that include growth and decay in different contexts.

Artifacts such as these can be dated by using the carbon-14 method.

Why How much money should be invested to reach a target amount? How old is a wooden bowl found at an archaelogical site? These are two of the many questions that can be answered with the use of exponential functions.

Years	0	5700	11,400	17,100	22,800	28,500	34,200
Fraction of carbon-14 remaining	$1 = \left(\frac{1}{2}\right)^0$	$\frac{1}{2} = \left(\frac{1}{2}\right)^1$	$\frac{1}{4} = \left(\frac{1}{2}\right)^2$	$\frac{1}{8} = \left(\frac{1}{2}\right)^3$	$\frac{1}{16} = \left(\frac{1}{2}\right)^4$	$\frac{1}{32} = \left(\frac{1}{2}\right)^5$	$\frac{1}{64} = \left(\frac{1}{2}\right)^6$

APPLICATION
ARCHAEOLOGY

Exponential functions can be used to determine the age of fossils and archaeological finds. This method is based on the fact that the amount of radioactive carbon in an organic object decreases at a fixed rate over time.

Carbon-14 dating is a reliable method of determining the age of organic objects up to 40,000 years old. Carbon-14 is a radioactive form of carbon that has a half-life of 5700 years. This means that half of the carbon-14 is converted to nonradioactive carbon every 5700 years. For example, it can be concluded that an organic object with half as much carbon-14 as its living counterpart died 5700 years ago. Every 5700 years, half of the remaining amount of carbon-14 decays.

Prepare

NCTM Principles & Standards 1–6, 8–10

LESSON RESOURCES

- Practice 8.7
- Reteaching Master 8.7
- Enrichment Master 8.7
- Cooperative-Learning Activity 8.7
- Lesson Activity 8.7
- Problem Solving/Critical Thinking Master 8.7
- Teaching Transparency 42

QUICK WARM-UP

Evaluate each expression.

1. x^3 for $x = 4$ **64**
2. 3^x for $x = 4$ **81**
3. $3y^z$ for $y = 2$ and $z = 4$ **48**
4. $7(1 + r)^2$ for $r = 5$ **252**
5. $4(1 + r)^t$ for $r = 2$ and $t = 6$ **2916**

Also on Quiz Transparency 8.7

Teach

Why Ask students to identify some reasons for establishing a long-term savings goal. Elicit responses such as saving money to pay for college, to buy an automobile, and to put a down payment on a house. Tell them to imagine that the goal is to have $5000 in a savings account 4 years from now. Point out that exponential functions can help determine how much money must be placed in the account now to reach that goal.

Alternative Teaching Strategy

INVITING PARTICIPATION Have the class agree on a savings goal to investigate. For example, a goal might be to have $5000 saved as a down payment on an automobile 3 years from now.

Tell students to suppose that a savings account is guaranteed to increase at a yearly rate of 4%. Ask them to guess the amount they would need to deposit in this account now in order to meet the goal 3 years from now. They should assume that no additional deposits will be made to the account and that no withdrawals will be made.

Use the general growth formula, $P = A(1 + r)^t$, substituting 0.04 for r and 3 for t. Have students use their guesses as values of A and solve for P. Test several guesses until the class agrees that they have arrived at a reasonable value of A.

Point out that they have been using guess-and-check as a problem-solving strategy. Ask if anyone sees a more efficient way to solve the problem. Elicit the response that a solution can be obtained by solving $5000 = A(1 + 0.04)^3$ for A. The amount of the initial deposit should be $4444.98.

LESSON 8.7 **409**

Use Teaching Transparency 42.

ADDITIONAL EXAMPLE 1

Use the table on page 409 to estimate the age of an object that has 24% of its original carbon-14 remaining.

a little more than 11,400 years old

TRY THIS
22% is slightly less than $\frac{1}{4}$, so the object is a little more than 11,400 years old. A reasonable estimate is about 12,000 years.

ADDITIONAL EXAMPLE 2

Suppose that your family wants to make a down payment of $10,000 on a house 4 years from now. They plan to deposit money into an account that is guaranteed to increase at a yearly rate of 7%. What amount do they need to deposit now in order to meet their goal? Assume that no additional deposits will be made to the account and that no withdrawals are made.

about $7628.95

TRY THIS
$3000 = A(1 + 0.04)^4$
$A \approx \$2565$

EXAMPLE 1 Use the table on the previous page to estimate the age of an object that has 10% of its original carbon-14 remaining.

● **SOLUTION**
Since 10%, or $\frac{1}{10}$, is between $\frac{1}{8}$ and $\frac{1}{16}$, the age is between 17,100 and 22,800 years. A reasonable estimate is about 19,000 years old.

TRY THIS Use the table to estimate the age of an object that has 22% of its original carbon-14 remaining.

Financial calculations are another common use of exponential functions.

EXAMPLE 2 Suppose that you wish to have $1000 for a down payment on a car and that you want to buy the car when you graduate in 3 years. The current interest rate for a savings account is 5% compounded annually. **How much do you need to deposit in a savings account now in order to have enough to buy the car in 3 years?**

APPLICATION
PERSONAL FINANCE

● **SOLUTION**
PROBLEM SOLVING
Use a formula. The general growth formula, $P = A(1 + r)^t$, can be used to find the value of an investment earning compound interest. Let P represent the amount of money after t years. Let r represent the annual interest rate (expressed as a decimal). Substitute the values into the formula, and solve for A, the original amount.

$$P = A(1 + r)^t$$
$$1000 = A(1.05)^3$$
$$1000 = A(1.157625)$$
$$\frac{1000}{1.157625} = A$$
$$863.84 \approx A$$

You will need to deposit about $863.84.

TRY THIS Suppose that Maria's parents want to have $3000 for tuition when she goes to college in 4 years. The current interest rate for a savings account is 4% compounded annually. How much do Maria's parents need to deposit in a savings account now in order to have $3000 in 4 years?

Interdisciplinary Connection

PSYCHOLOGY Suppose that you memorize a list of 100 words. However, for each week that passes, you forget one-sixth of the words you knew the week before. Then the number of words, K, you still know after n weeks can be modeled by $K = 100\left(\frac{5}{6}\right)^n$. How many words will you still know after 2 weeks? At what point will you still know only about 25 words? about 69 words; between 7 weeks and 8 weeks

Suppose that for each week that passes, you forget only one-ninth of the words you knew the week before. Write an equation that can model the number of words, K, that you still know after n weeks. $K = 100\left(\frac{8}{9}\right)^n$

EXAMPLE 3

APPLICATION
INVESTMENTS

The value of a sculpture has been growing at a rate of 5% per year for 4 years. **If the sculpture is worth $14,586.08 now, what was its value 4 years ago?**

SOLUTION

Use the general growth formula. Since you are trying to find the value 4 years ago, substitute −4 for t.

$P = A(1 + r)^t$
$= 14{,}586.08(1.05)^{-4}$
$\approx 14{,}586.08(0.8227)$
$\approx 12{,}000$

The value 4 years ago was about $12,000.

TRY THIS

An art museum recently sold a painting for $4,500,000. The museum directors claim that its value has been growing at a rate of 7% per year for the last 50 years. What was the painting worth 50 years ago?

EXAMPLE 4

Ginny has determined that she answers an individual step correctly in an algebra problem 90% of the time. She does 100 problems, each with 2 steps.
a. In how many problems can she expect to answer both steps correctly?
b. Calculate how many she can expect to answer correctly if each problem has 3 steps.

SOLUTION

a. In 100 problems she can expect to get the first step correct 90 times. Of these 90 times, she should get the second step correct 90% of the time. Because 90% of 90 is 81, she should get both steps correct 81 times out of 100. Notice that the answer is equal to $100(0.90)^2$.
b. From the solution in part **a**, you know that in 100 problems she should have the first 2 steps correct 81 times. She should also get the third step correct 90% of these times, or about 73 times. Notice that the answer is equal to $(0.90)^3$.

TRY THIS

Gavin gets a given step correct in an algebra problem 92% of the time. He does 100 problems, each with 3 steps. Calculate the number of problems in which he should expect to get all 3 steps correct.

In this lesson you have seen several examples of *exponential growth* and *exponential decay*. The general formula for exponential change gives the amount, P, when A is the initial amount, r is the rate of change expressed in decimal form, and t is the time expressed in years:

$$P = A(1 + r)^t$$

For **exponential growth**, r is positive. For **exponential decay**, r is negative.

ADDITIONAL EXAMPLE 3

The value of a painting is now $150,000. Suppose that its value has been growing at a rate of 8% per year for 5 years. **What was the value of the painting 5 years ago?**
about $102,087.48

TRY THIS
$P = 4{,}500{,}000(1 + 0.07)^{-50}$
$P \approx \$152{,}765$

ADDITIONAL EXAMPLE 4

Lionel has determined that he answers an individual step correctly in an algebra problem 95% of the time. He does 100 problems, each with 4 steps. **Calculate the probability that Lionel will correctly answer all 4 steps in a given problem.**
about 81%

TRY THIS
$100(0.92)^3 \approx 78$ problems

Inclusion Strategies

TACTILE LEARNERS Have students work in groups to perform an experiment that models radioactive decay. Give each group 50 pennies inside a large cup with a lid. Tell them to shake the cup and then spill the pennies onto a flat surface. Now they should remove the pennies that show tails and record the number of heads. Have them repeat the process with the remaining pennies until there are no coins remaining. Tell them to make a scatter plot of their data. On the same set of axes, they should then graph the equation $y = 50(0.05)^x$. Discuss the results.

Enrichment

Often "annual interest" actually is compounded more frequently than once per year. When $p\%$ annual interest is compounded n times per year, $\frac{p}{n}\%$ interest is paid in each payment period. Have students find the amount in an account after 5 years if $10,000 is deposited at 4.8% annual interest compounded as follows:

1. annually $10{,}000(1.048)^5$; $12,641.73
2. semiannually $10{,}000(1.024)^{10}$; $12,676.51
3. quarterly $10{,}000(1.012)^{20}$; $12,694.34
4. monthly $10{,}000(1.004)^{60}$; $12,706.41

LESSON 8.7

ADDITIONAL EXAMPLE 5

Suppose that the population of a city is 200,000 in the year 2000 and is decreasing at a rate of 2% per year. **When will the population of this city be less than 175,000? Solve the problem by using the following methods:**

a. a graph

The city's population will be less than 175,000 in about $6\frac{3}{4}$ years.

b. a table

Year	Population
0	200,000
1	196,000
2	192,080
3	188,238
4	184,474
5	180,784
6	177,168
7	173,625

The city's population will be less than 175,000 between 6 and 7 years after the year 2000.

See Keystroke Guide, page 782.

EXAMPLE 5

APPLICATION
DEMOGRAPHICS

The population of a city in the United States was 152,494 in 2000 and is decreasing at a rate of about 1% per year. When will the city's population fall below 140,000? **Solve the problem by using the following methods:**
  a. a graph
  b. a table

● **SOLUTION**

Each year the population will be 100% − 1%, or 99%, of what it was the previous year, so the multiplier is 0.99. The population x years after 2000 is given by $y = 152,494(0.99)^x$.

PROBLEM SOLVING
TECHNOLOGY
GRAPHICS CALCULATOR

a. Draw a graph. Graph **Y1**= $152,494(0.99)^x$ and **Y2**= 140,000, and find the point of intersection. The value of y is 140,000 when x is approximately 8.5. Therefore, the city's population will fall below 140,000 in about $8\frac{1}{2}$ years.

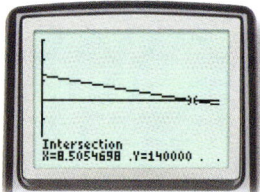

b. Make a table. The population decreases by a factor of 0.99 each year. Use the constant feature on a calculator to generate a table of values for the population. Continue the table until you reach the year in which the value falls below 140,000.

No. of Years	Population
x	$y = 152,494(0.99)^x$
0	152,494
1	150,969
2	149,459
3	147,965
4	146,485
5	145,020
6	143,570
7	142,134
8	140,713
9	139,306

According to the table, the population falls below 140,000 between 8 and 9 years after 2000.

TRY THIS The population of a small city in the United States was 16,225 in 2000 and was increasing at a rate of about 2% per year. When will the city's population be above 17,000? Solve the problem by using the following methods:
  a. a graph
  b. a table

Reteaching the Lesson

GUIDED ANALYSIS Remind students of the general growth formula: $P = A(1 + r)^t$. Point out that they have worked with three types of problems related to the formula, depending on which quantity is the unknown. Review each type of problem in relation to compound interest.

1. P is unknown.
An account earns 6% annual interest, compounded annually. $2000 is deposited. Find the amount in the account after 4 years.
$P = 2000(1 + 0.06)^4$; $2524.95

2. A is unknown.
An account earns 6% annual interest, compounded annually. The goal is to have $2000 in the account after 4 years. What amount must be deposited now?
$2000 = A(1 + 0.06)^4$; $1584.19

3. t is unknown.
An account earns 6% annual interest, compounded annually. $2000 is deposited. When will the amount in the account be $3000?
in about 7 years

412 LESSON 8.7

Exercises

Communicate

1. Describe three real-world situations that can be modeled by exponential functions.
2. Explain the formula for exponential growth, $P = A(1 + r)^t$.
3. Explain the formula for exponential decay, $P = A(1 - r)^t$.
4. Explain why r is subtracted from 1 in the formula for exponential decay.
5. Explain what is meant by the term *half-life* in relation to a radioactive substance.

Guided Skills Practice

APPLICATIONS

6. **ARCHAEOLOGY** Use the table on page 409 to estimate the age of an object that has 25% of its original carbon-14 remaining. *(EXAMPLE 1)* **11,400 years old**

7. Suppose that David wants to have $2500 saved at the end of 6 years. The current interest rate for a savings account is 3.5% compounded annually. How much will David need to invest in a savings account now in order to have $2500 in 6 years? *(EXAMPLE 2)* **$2033.75**

8. **INVESTMENTS** Mike sold an antique car for $32,000. The car's value has been growing at a rate of 7% per year for the last 25 years. What was the car valued at 25 years ago? *(EXAMPLE 3)* **$5895.97**

9. Sally gets a given step in an algebra problem correct 85% of the time. She does 100 problems, each with 5 steps. Calculate the number of problems in which she should expect to get all 5 steps correct. *(EXAMPLE 4)* **44 problems**

10. **DEMOGRAPHICS** The population of a city in the United States was 160,525 in 1990 and was decreasing at a rate of about 1.5% per year. When will the city's population be below 155,000? *(EXAMPLE 5)* **1993**

Henry Ford employed about 10 people in 1903, when the first Model A was made in Detroit.

Practice and Apply

Each digit of a repeating decimal such as $0.33333\overline{3} = \frac{1}{3}$ can be described in terms of an exponential function. Evaluate each expression.

11. $3\left(\frac{1}{10}\right)^1$ **0.3**
12. $3\left(\frac{1}{10}\right)^2$ **0.03**
13. $3\left(\frac{1}{10}\right)^3$ **0.003**
14. $3\left(\frac{1}{10}\right)^4$ **0.0003**

Using the table on page 409, estimate the age of an object that has the following amounts of its original carbon-14 remaining.

15. 5% **approx. 24,000 years**
16. 28% **approx. 10,500 years**
17. 50% **approx. 5,700 years**
18. 100% **0 years**

TRY THIS

a.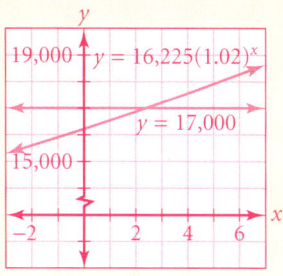

in about 2.4 years, during the year 2003

b.
Year	Population
0	16,225
1	16,550
2	16,880
3	17,218

Assess

Selected Answers
Exercises 6–10, 11–49 odd

ASSIGNMENT GUIDE

In Class	1–10
Core	11–18, 19–39 odd
Core Plus	12–18 even, 19–39
Review	40–49
Preview	50

✎ Extra Practice can be found beginning on page 738.

LESSON 8.7

Technology

A calculator will be helpful for performing the calculations required in many of the exercises. You may wish to have students use graphics calculators for Exercises 23–39.

Error Analysis

In problems involving exponential decay, students often substitute the given percent as the value of r, forgetting to first subtract it from 100%. In Exercises 32–35, for example, they may incorrectly use 0.02 as the value of r rather than 0.98.

23.
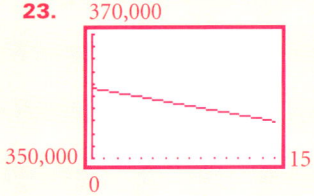

CONNECTION

19. **PROBABILITY** If the value of x in 2^x was chosen from the list 1, 2, 3, 4, and 5 and each value came up twice in 10 trials, find the probability that 2^x is greater than 10. $\frac{2}{5}$

20. **PROBABILITY** Repeat Exercise 19 if the value of x was chosen from the list 1, 2, 3, . . ., 100 and each value came up 3 times in 300 trials. $\frac{97}{100}$

APPLICATIONS

ARCHAEOLOGY Use the table on page 409 to estimate the age of an object that has

21. 25% of its original carbon-14 remaining. **approx. 11,400 years old**

22. 1% of its original carbon-14 remaining. **approx. 37,500 years old**

23. **TECHNOLOGY** The population of a metropolitan area was 361,280 in 1998 and was decreasing at a rate of about 0.1% per year. Draw a graph that shows the population over the next 15 years.

DEMOGRAPHICS Use the graph from Exercise 23 to predict the population of the city in the following years:

24. 2000 ≈ **360,558** 25. 2003 ≈ **359,477** 26. 2005 ≈ **358,759** 27. 2010 ≈ **356,968**

INVESTMENTS An investment has been growing in value at a rate of 8% per year and now has a value of $8200. Find the value of the investment

28. in 5 years. **$12,048.49**
29. in 10 years. **$17,703.18**
30. 5 years ago. **$5,580.78**
31. 10 years ago. **$3,798.19**

INVESTMENTS An investment is losing value at a rate of 2% per year and now has a value of $94,000. Find the value of the investment

32. in 5 years. **$84,968.55**
33. in 10 years. **$76,804.84**
34. 5 years ago. **$103,991.41**
35. 10 years ago. **$115,044.83**

36. **ARCHAEOLOGY** Use the table on page 409 to estimate the age of an object that has 58% of its original carbon-14 remaining. **approx. 5200 years old**

Paint used in cave paintings can be dated by using the carbon-14 method.

37. **INVESTMENTS** Suppose that Carl wants to have $1800 saved at the end of 3 years. The current interest rate for a savings account is 6.5% compounded annually. How much will Carl need to invest in a savings account now in order to have $1800 in 3 years? **$1490.13**

38. **INVESTMENTS** Marla invested $3000 in a new company. The value of her investment has been growing at a rate of 7.5% per year for the last 5 years. What is the value of her investment now? If the value of the investment continues to grow at this rate, what will be the value of her investment in 5 years? **$4306.89; $6183.09**

39. **DEMOGRAPHICS** The population of a city is 100,000 and is increasing at a rate of about 2.5% per year. When will the city's population be above 125,000? **in 10 years**

40. $y = 18x$
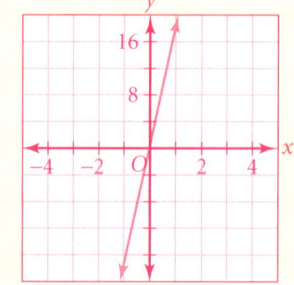

44.
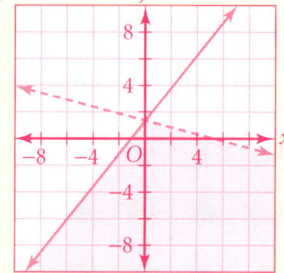

Practice

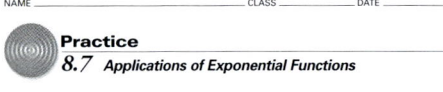

Practice
8.7 Applications of Exponential Functions

From 1964 to 1992 the percent of female drivers in the United States rose steadily. The percent increase is modeled by $y = 40(1.03)^x$, where x is the number of years, beginning with $x = 0$ in 1964, and y is the percent increase.

1. Use the model to estimate the percent increase of female drivers in the United States in 1975. **55.4%**

2. Estimate the percent increase of female drivers in the United States in the year 2000. **116%**

The "Mendelssohn" Stradivarius violin was estimated to be worth approximately $1,700,000 in 1990. The violin is expected to increase in value by approximately 7.5% each year.

3. Use a calculator to complete the table.

Number of years	0	1	2	3	4	5
Value (in millions)	1.7	1.83	1.96	2.11	2.27	2.44

4. Find an equation to model the growth in value. **$(1.7 \times 10^6)(1.075)^x$**

5. Graph the function from Exercise 4 by using a graphics calculator. **Check students' graphs.**

6. Estimate the value of the violin in the year 2010. **7.22×10^6**

The International Telecommunication Union estimated 130,000,000 telephone subscribers in the United States in 1991. The number of telephone subscribers is estimated to increase 8% each year.

7. Find a function to model this growth in subscribers. **$(1.3 \times 10^8)(1.08)^x$**

8. Use the formula to estimate the number of subscribers in the year 2001. **2.81×10^8**

9. Graph this function by using a graphics calculator. **Check students' graphs.**

10. Estimate the number of subscribers in the year 2011. **6.06×10^8**

414 LESSON 8.7

Look Back

40. Suppose that the hourly wage at a job is $18. Let x represent the number of hours, and let y represent the total wages earned. Write an equation of direct variation and graph it. **(LESSON 5.3)**

Solve each system by using any method. **(LESSONS 7.1, 7.2, AND 7.3)**

41. $\begin{cases} 4x + 2y = 15 \\ 6x - y = 3 \end{cases}$ $x = \frac{21}{16}$ $y = \frac{39}{8}$

42. $\begin{cases} 12x + 3y = 36 \\ 2x - 6y = -20 \end{cases}$ $x = 2$ $y = 4$

43. $\begin{cases} 1.6x + 0.7y = 2.8 \\ 8x - 5.6y = 1.7 \end{cases}$ $x \approx 1.159$ $y \approx 1.352$

Use graphing to solve each system of inequalities. **(LESSON 7.5)**

44. $\begin{cases} 2x + 8y < 10 \\ 4x - 3y \geq -4 \end{cases}$

45. $\begin{cases} 2b + 6d > -4 \\ b - 5d \geq 6 \end{cases}$

46. $\begin{cases} 3x - 2y \geq 5 \\ 3x + 2y \leq -4 \end{cases}$

Graph each function. **(LESSON 8.6)**

47. $y = 0.5^x$

48. $y = \left(\frac{1}{3}\right)^x$

49. $y = \left(\frac{1}{5}\right)^x$

Look Beyond

50. **TECHNOLOGY** A number that is often used to describe the growth of invested money is e, which is about 2.718. Find the e^x key on your calculator. Write the first eight digits of e. (Use the e^x key and use 1 for x.) Then draw the graph of $y = e^x$. $e \approx 2.7182818$

Include this activity in your portfolio.

The table below shows the growth of $5000 earning a fixed rate of interest compounded annually.

1. Find the rate by looking at successive quotients.
2. Write a function for the relationship.
3. Calculate the amount through year 10.
4. When will the amount meet or exceed $10,000? $15,000?

Year	Amount	Quotients
0	5000	
1	5375	$\frac{5375}{5000} = ?$
2	5778.13	?
3	6211.48	?
…	…	…
x	y	?

WORKING ON THE CHAPTER PROJECT

You should now be able to complete Activity 4 of the Chapter Project on page 416.

45.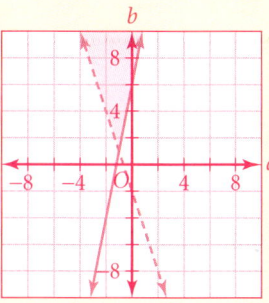

46.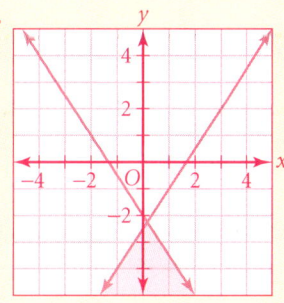

The answers to Exercises 47–50 can be found in Additional Answers beginning on page 810.

Project

Focus
Equations can be used to model the growth of animal populations and to make predictions about that growth. Students will be asked to study a table of data for a wolf population, graph it, and approximate a line of best fit. Then they will model the same data with an exponential function and compare the models.

Motivate
The subject of endangered species appears frequently in the news. If this has happened recently, ask students to bring to class relevant articles from newspapers or magazines. Ask them to share any other knowledge of wild animal populations they may have acquired through science courses or through personal interest.

Ask students to share their perceptions of the growth of animal populations. Do they think that, in general, such growth occurs at a constant rate?

Ask what factors affect the growth of an animal population. Elicit answers such as availability of food, the introduction of new predators into a territory, and human actions that alter the environment.

CHAPTER PROJECT EIGHT: THE WOLVES' DEN

An endangered species is one with so few individuals left that the species may become extinct. Wolves are among the species protected by the Endangered Species Act. Because the number of wolves in the United States was reduced and even eliminated in some areas, a program of re-introduction was started to rebuild the population under controlled circumstances. Re-introduction takes place when animals are set free to start a new population in areas where the species used to live. There are many factors that have to be considered when attempting such a large project with such far-reaching consequences. If the population grows too quickly, the resources available for the animals will be exhausted. However, if the population does not grow as quickly as expected, researchers must figure out what factors are causing the problem. The U.S. Fish and Wildlife Service is responsible for tracking and managing the re-introduction of wolves in various areas around the country.

Activity 1
In 1999, there should be about 115 wolves in the population.

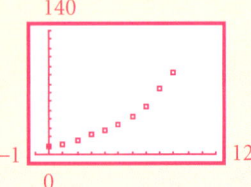

Using the linear regression capability of a graphics calculator, the line of best fit is $y = 8\frac{26}{33}x - \frac{41}{55}$.

When $x = 10$,
$y = 8\frac{26}{33}(10) - \frac{41}{55}$
$= \frac{290}{33}(10) - \frac{41}{55}$
$= 87\frac{2}{15}$

≈ 87 wolves in 1999, which is considerably less than the prediction

Activity 1

The table below shows the population of wolves living in a hypothetical wildlife preserve. In this simulation, 10 wolves were re-introduced to the area in 1989.

Year	1989	1990	1991	1992	1993	1994	1995	1996	1997	1998
Population	10	12	16	23	27	34	43	55	75	93

Predict the population of wolves in the wildlife preserve for the current year by looking at the table. Then make a scatter plot on your graphics calculator. Does the relationship among the data look linear? Write an equation for a line of best fit. Then use it to check your prediction.

You can use the idea of percent or growth rate to model population data with exponential functions. You can find the growth rate for each year by computing the ratio $\frac{\text{new population}}{\text{previous population}}$. The multiplier from 1989 to 1990 is $\frac{12}{10} = 1.2$, or 120%. This indicates a 20% percent increase over the first year.

Activity 2

Use the model $y = 10(1.2)^x$ to build a table and graph and see if a 20% growth rate models the data well. Try other growth rates to refine your model. Observe the changes in the population for different rates. Make a prediction about the population over the next 10 years.

Activity 3

Find data on the actual wolf population in a particular area or on another population that interests you. You can use your county, state, or city population or a wildlife population. Use a graphics calculator to find a reasonable line that fits the data and to make predictions about the population over next 10 years. Next, determine a reasonable rate of growth or decay. Use a graphics calculator to find a reasonable exponential function that fits the data (you may need to try several growth rates) and to make predictions about the population over next 10 years.

Activity 4

Compare the two models from Activity 3 and explain which one you think is a better predictor.

Cooperative Learning

Students can do the Chapter Project in groups of three. For **Activity 1**, one student can read the data aloud, another can enter it into the calculator, and the third can check that the data has been entered correctly. The group can then work together to decide on an appropriate linear model.

In **Activity 2**, the group can first brainstorm three possible growth rates. Each student should make a table and graph for one of the rates. Then the whole group can compare and discuss the results.

To obtain the data for **Activity 3**, group members can pursue different avenues of research. For example, one student can do library research, one can search the Internet, and the third can contact a government agency that oversees wildlife populations. Group members can then share the data.

Discuss

When all groups have completed the Chapter Project, have each group make a brief presentation of their findings. Discuss whether, in general, the exponential model was the best model of growth.

Have the class analyze the "big picture" in your area. Do local wildlife populations seem relatively stable? Is a particular type of animal population dwindling? Is this a national or worldwide trend, or is it local? What are the reasons for it? If the species is threatened or endangered, what can be done to reverse the trend?

Activity 2

x	0	1	2	3	4	5	6	7	8	9
$y = 10(1.2)^x$	10	12	14	17	21	25	30	36	43	52

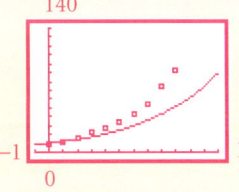

This model does not increase as quickly as the actual data.

Try $y = 10(1.3)^x$.

x	0	1	2	3	4	5	6	7	8	9
$y = 10(1.3)^x$	10	13	17	22	29	37	48	63	82	106

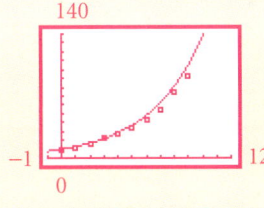

This model is much closer to the actual data, but is a little higher.

Activity 3

Students may find data in an almanac, on the Internet, or by calling wildlife reserves. Students should follow the steps described in Activities 1 and 2 to create their models.

Activity 4

Students will most likely find the exponential models to be the best predictors.

Chapter 8 Chapter Review and Assessment

VOCABULARY

base 370	monomial 373	Powers of –1 380
coefficient 373	negative exponent 390	Product-of-Powers Property 371
exponent 370	Power-of-a-Fraction Property 386	Quotient-of-Powers Property 384
exponential decay 411	Power-of-a-Power Property 378	scientific notation 398
exponential function 404	Power-of-a-Product Property 379	zero exponent 391
exponential growth 411		
general growth formula ... 405		

Key Skills & Exercises

LESSON 8.1

Key Skills

Simplify the product of monomials containing exponents.

Simplify $(4a^2b)(-3a^3b^2)$.

Multiply the constants. Rearrange the terms to group the like variables together.

$(4a^2b)(-3a^3b^2) = -12(a^2)(a^3)(b)(b^2)$

Simplify by using the Product-of-Powers Property. Add the exponents of like variables.

$-12(a^2)(a^3)(b)(b^2) = -12a^5b^3$

Exercises

Simplify each product.

1. $(5a^2)(4a^3)$ **$20a^5$**
2. $(3m^2)(-2m^2n)$ **$-6m^4n$**
3. $(7s^4t^3)(4t)$ **$28s^4t^4$**
4. $(-p^2)(6pq^2)$ **$-6p^3q^2$**
5. $(4m^3)(3mn^2)$ **$12m^4n^2$**
6. $(2a^2bc^2)(5b^2c)$ **$10a^2b^3c^3$**
7. $(-4xy)(-3x^2z)(2y^2z^2)$ **$24x^3y^3z^3$**
8. $(2x^ay^b)(-3x^my)$ **$-6x^{a+m}y^{b+1}$**

LESSON 8.2

Key Skills

Use the properties of exponents to simplify an expression.

Simplify $(x^4)(x^3y^2)^2$.

Use the Power-of-a-Power Property and the Power-of-a-Product Property to simplify $(x^3y^2)^2$.

$(x^4)(x^3y^2)^2 = (x^4)(x^6y^4)$

Use the Product-of-Powers Property to simplify $(x^4)(x^6y^4)$.

$(x^4)(x^6y^4) = x^{10}y^4$

Exercises

Simplify each expression.

9. $(c^4)^2$ **c^8**
10. $(s^3)^a$ **s^{3a}**
11. $(4x^2y)^3$ **$64x^6y^3$**
12. $2(r^2)^5$ **$2r^{10}$**
13. $(-p^2)^3(p)$ **$-p^7$**
14. $(-xy^2)^6$ **x^6y^{12}**
15. $(-3mn^2)(2m^3)^3$ **$-24m^{10}n^2$**
16. $(5b^3c^2)^4(2d^4)$ **$1250b^{12}c^8d^4$**

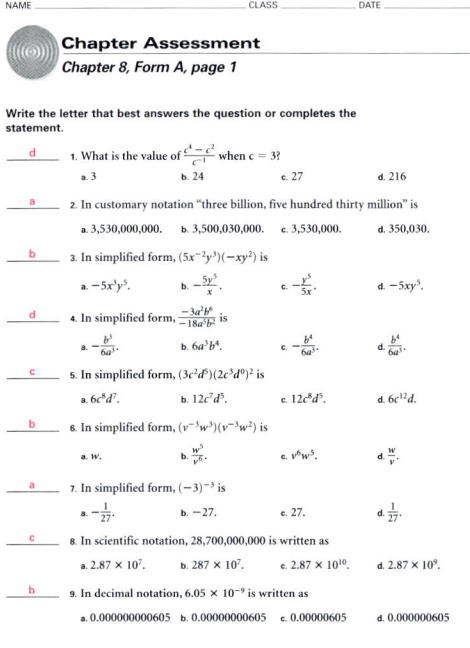

418 CHAPTER 8 REVIEW

LESSON 8.3

Key Skills

Use the properties of exponents to simplify expressions containing fractions.

Simplify. **a.** $\dfrac{6x^2y^3}{2xy^2}$ **b.** $\left(\dfrac{4c^5}{5d^3}\right)^2$

a. Simplify by using the Quotient-of-Powers Property. First simplify any numerical coefficients. Then subtract the exponents in the denominator from the exponents in the numerator.

$$\dfrac{6x^2y^5}{2xy^2} = 3xy^3$$

b. Simplify by using the Power-of-a-Fraction Property.

$$\left(\dfrac{4c^5}{5d^3}\right)^2 = \dfrac{(4c^5)^2}{(5d^3)^2} = \dfrac{16c^{10}}{25d^6}$$

Exercises

Simplify each expression.

17. $\left(\dfrac{x}{y}\right)^4$ $\dfrac{x^4}{y^4}$ **18.** $\left(\dfrac{2a^2}{b}\right)^3$ $\dfrac{8a^6}{b^3}$

19. $\left(\dfrac{4w^2}{6w^4}\right)^3$ $\dfrac{8}{27w^6}$ **20.** $\left(\dfrac{12fg^2}{6g}\right)^2$ $4f^2g^2$

21. $\left(\dfrac{y^4}{5xy^2}\right)^3$ $\dfrac{y^6}{125x^3}$ **22.** $\left(\dfrac{7p^2q}{21q}\right)^2$ $\dfrac{p^4}{9}$

23. $\left[\dfrac{-15(ab)^2}{5a^2b}\right]^5$ $-243b^5$ **24.** $\left(\dfrac{st^2}{s^4}\right)^r$ $\dfrac{t^{2r}}{s^{3r}}$

LESSON 8.4

Key Skills

Simplify expressions containing negative and zero exponents.

Simplify $\dfrac{-8a^3c^5}{2a^4b^3c^2}$ and write the expression with positive exponents only.

Use the properties of exponents and the definition of negative exponents.

$$\dfrac{-8a^3c^5}{2a^4b^3c^2} = -4(a^{3-4})(b^{-3})(c^{5-2}) = -4a^{-1}b^{-3}c^3 = \dfrac{-4c^3}{ab^3}$$

Exercises

Simplify each expression. Write the expression with positive exponents only.

25. 3^{-2} $\dfrac{1}{3^2} = \dfrac{1}{9}$ **26.** a^2b^{-3} $\dfrac{a^2}{b^3}$

27. $a^0b^{-2}c^3$ $\dfrac{c^3}{b^2}$ **28.** $5x^{-3}y^2$ $\dfrac{5y^2}{x^3}$

29. $\dfrac{12p^{-3}}{4p}$ $\dfrac{3}{p^4}$ **30.** $\left(\dfrac{4a^{-2}}{3b^{-3}}\right)^3$ $\dfrac{64b^9}{27a^6}$

31. $\dfrac{(15t^{-4})(t^3)}{-3t^{-2}}$ $-5t$ **32.** $(8b^2)(2b^{-8})(2b^6)$ 32

LESSON 8.5

Key Skills

Perform computations involving scientific notation.

Write the product $(3 \times 10^3)(4 \times 10^5)$ in scientific notation.

Rearrange the factors: $(3 \times 4)(10^3 \times 10^5)$

Multiply: 12×10^8

Rewrite the first factor as a decimal between 1 and 10: 1.2×10^1

Complete the simplification: $12 \times 10^8 = (1.2 \times 10^1)(10^8) = 1.2 \times 10^9$

Exercises

Write each number in scientific notation.

33. 5,900,000 **34.** 0.0000075 7.5×10^{-6}
5.9×10^6

Perform each computation. Write the answer in scientific notation.

35. $(3 \times 10^2)(5 \times 10^5)$ 1.5×10^8

36. $(2.1 \times 10^5)(3 \times 10^{-3})$ 6.3×10^2

37. $(8 \times 10^2) + (2 \times 10^2)$ 1×10^3

38. $(9 \times 10^5) - (3 \times 10^5)$ 6×10^5

39. $\dfrac{9 \times 10^7}{3 \times 10^4}$ 3×10^3

40. $\dfrac{8 \times 10^4}{2 \times 10^{-2}}$ 4×10^6

45.
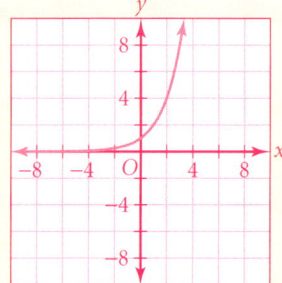

46.
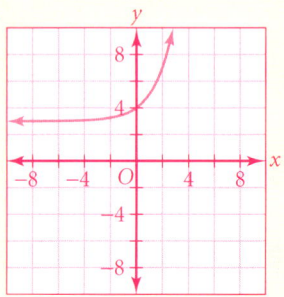

47.
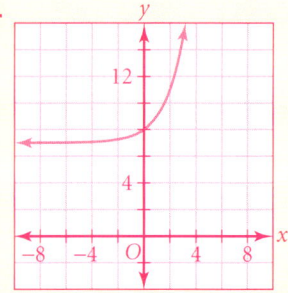

48.
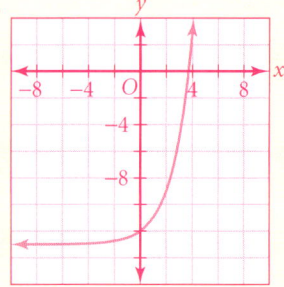

49.
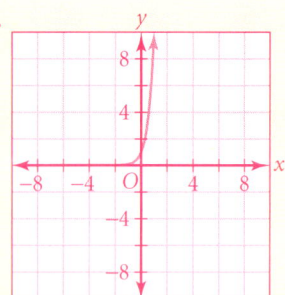

LESSON 8.6
Key Skills

Understand and graph exponential functions.

The current population of a city is 120,000. It is estimated that the population of city will grow 1.5% annually. What will the city's population be in 10 years?

Use the general growth formula, $P = A(1 + r)^t$, where A is the original amount and P is the amount after t years at a yearly growth rate of r, expressed as a decimal.

$P = 120,000(1.015)^{10} = 139,264.899$

In 10 years, the population of the city will be about 139,265.

Graph the exponential function $y = 4^x$.

You can make a table or use a graphics calculator. Notice that the graph remains above the x-axis for all values of x.

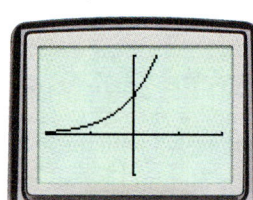

Exercises

The population of New Zealand was about 3.3 million in 1991 and was growing at the rate of about 0.8% per year. Estimate the population

41. in the year 2000. ≈ 3,545,347

42. in the year 2003. ≈ 3,631,118

43. in the year 2010. ≈ 3,839,406

44. in the year 2020. ≈ 4,157,855

Graph each exponential function.

45. $y = 2^x$

46. $y = 2^x + 3$

47. $y = 2^x + 7$

48. $y = 2^x - 13$

49. $y = 10^x$

LESSON 8.7
Key Skills

Use exponential functions to model applications.

An investment has been losing money at a rate of 3% per year and now has a value of $6500. What was the value of the investment 5 years ago?

Since you need to find the value 5 years ago, use the general growth formula with $t = -5$.

$P = A(1 + r)^t = 6500(1 - 0.03)^{-5} \approx \$7569.28$

Exercises

An investment is growing at a rate of 5.5% per year and now has a value of $6300. Find the value of the investment

50. in 5 years. $8,233.85

51. 5 years ago. $4820.35

52. in 10 years. $10,761.31

53. 10 years ago. $3,688.21

Applications

54. ASTRONOMY The sun has a diameter of about 864,000 miles. Write this number in scientific notation. 8.64×10^5

55. FINANCE The value of a famous painting is increasing at the rate of 6% per year and has a value of $800,000 in the year 2000. Find the value of the painting in the year 2005. $1,070,580.46

8 Alternative Assessment

Performance Assessment

1. Use the definition of exponents to simplify $(3ab^3)(2a^2b)$.
2. Explain how the Power-of-a-Product Property is used to simplify $(2cd^2)^3$. What is the simplified expression?
3. Explain why $-(5^3) = (-5)^3$ but $-(5^4) \neq (-5)^4$.
4. Explain how to
   a. write 10,760,000 in scientific notation.
   b. write 0.0000528 in scientific notation.
5. Explain how to use the general growth formula to find the current value of an investment made several years ago. Include the variables you need to know as well as the steps for finding the current value.

Portfolio Projects

1. List the properties of exponents. Give an example of each of property and how it is used to simplify an expression.
2. Begin with the fact that $3^3 = 27$ and $3^2 = 9$. Make a table and write a paragraph to convince a classmate that $3^0 = 1$.
3. Scientific notation is used in the study of atoms. Research several subatomic particles and explain how scientific notation is used to record information about them.
4. Find an example of a population and an example of compound interest that can be used with the general growth formula.

5. Find several very large or very small numbers used in a newspaper or magazine article. Rewrite the numbers in the article by using scientific notation.
6. The capacity of computer chips is said to double every 18 months. Write a report on how the power of 2 is used in computer chip development.

Peer Assessment

Complete this activity with a partner. Begin by writing an expression that can be simplified by using the laws of exponents. Your partner should simplify the expression and then write a new expression for you to simplify.

Check your partner's work, correct any errors, and then simplify the new expression. Continue with this pattern until you have each simplified 20 expressions. Be sure to use expressions involving multiplication, division, negative exponents, powers of powers, and products of powers during this activity.

Repeat the entire activity by writing numbers in standard notation and having your partner rewrite them in scientific notation. Remember to use very large and very small numbers for this part of the activity.

---

internetconnect

The HRW Web site contains many resources to reinforce and expand your knowledge of exponents and exponential functions. The Web site also provides Internet links to other sites where you can find information and real-world data for use in research projects, reports, and activities that involve scientific notation or exponential growth. Visit the HRW Web site at **go.hrw.com** and enter the keyword **MA1 CH8**.

Alternative Assessment

Performance Assessment

1. $6a^3b^4$

2. The Product-of-Powers Property says that if the bases are the same, add the exponents to simplify. This gives the following:
$(2)(2)(2)(c^{1+1+1})(d^{2+2+2})$
$= 8c^3d^6$

3. $(-5)^3 = (-1)^3(5)^3 = -(5^3)$, but $(-5)^4 = (-1)^4(5)^4 = 5^4$.
-1 raised to an odd power is negative, but -1 raised to an even power is positive.

4. a. Move the decimal point 7 places to the left, between the one and the zero. $10,760,000 = 1.076 \times 10^7$
   b. Move the decimal point 5 places to the right, between the five and the two. $0.0000528 = 5.28 \times 10^{-5}$

5. In $P = A(1 + r)^t$,
P = current value
A = original value
r = rate of growth
t = time in years
Substitute in the appropriate values for A, r, and t and evaluate to find P.

Portfolio Projects

1. Examples may vary. Samples:
   Product-of-Powers Property:
   $a^4 \cdot a^3 = a^{4+3} = a^7$
   Power-of-a-Product Property:
   $(ab)^3 = a^3b^3$
   Quotient-of-Powers Property: $\frac{a^5}{a^2} = a^{5-2} = a^3$
   Power-of-a-Fraction Property: $\left(\frac{a}{b}\right)^7 = \frac{a^7}{b^7}$
   Power-of-a-Power Property:
   $(a^3)^6 = a^{3 \cdot 6} = a^{18}$

2. Answers may vary. Sample: $\frac{3^3}{3^3} = \frac{27}{27} = 1$, but by the Quotient-of-Powers Property, $\frac{3^3}{3^3} = 3^{3-3} = 3^0$, so $3^0 = 1$.

3^{-3}	3^{-2}	3^{-1}	3^0	3^1	3^2	3^3
$\frac{1}{27}$	$\frac{1}{9}$	$\frac{1}{3}$	1	3	9	27

3. Answers may vary. Students may cite examples of ways in which scientific notation is used to describe mass of particles, lifetime, and energy.

4. Answers may vary. Students may use examples of populations found in almanacs and may get rates of interest from a local bank.

5. Answers may vary.

6. Answers may vary. Students may summarize the history of computer chip development with respect to doubling capacity, and may make predictions for future developments.

Cumulative Assessment

College Entrance Exam Practice

Multiple-Choice and Quantitative-Comparison Samples

The first half of the Cumulative Assessment contains two types of items found on standardized tests—multiple-choice questions and quantitative-comparison questions. Quantitative-comparison items emphasize the concepts of equality, inequality, and estimation.

Free-Response Grid Samples

The second half of the Cumulative Assessment is a free-response section. A portion of this part of the Cumulative Assessment consists of student-produced response items commonly found on college entrance exams. These questions require the use of machine-scored answer grids. You may wish to have students practice answering these items in preparation for standardized tests.

CHAPTERS 1–8 CUMULATIVE ASSESSMENT

College Entrance Exam Practice

QUANTITATIVE COMPARISON For Questions 1–5 write
A if the quantity in Column A is greater than the quantity in Column B;
B if the quantity in Column B is greater than the quantity in Column A;
C if the two quantities are equal; or
D if the relationship cannot be determined from the information given.

	Column A	Column B	Answers	
1.	$6 + 5^2 \cdot 2$	$(3+7)^2 \div 2$	Ⓐ Ⓑ Ⓒ Ⓓ [Lesson 1.3]	A
2.	The solution to $\frac{x}{5} = 1.2$	The solution to $\frac{x}{10} = 2.4$	Ⓐ Ⓑ Ⓒ Ⓓ [Lesson 3.2]	B
3.	$(2^4)^3$	$2^3 \cdot 2^4$	Ⓐ Ⓑ Ⓒ Ⓓ [Lesson 8.2]	A
4.	The slope of the line passing through $(1, 3)$ and $(4, 5)$	The slope of the line passing through $(-1, 3)$ and $(4, 5)$	Ⓐ Ⓑ Ⓒ Ⓓ [Lesson 5.2]	A
5.	110% of 85	85% of 110	Ⓐ Ⓑ Ⓒ Ⓓ [Lesson 4.2]	C

6. Let $x = 3$ and $y = -4$. Simplify $x - 8 - y$. **(LESSON 2.3)** d
   a. 15   b. −9   c. 1   d. −1

7. Solve the system. **(LESSON 7.2)** d
   $\begin{cases} 2x - 4 = -12 \\ y = x + 3 \end{cases}$
   a. (3, 0)   b. (0, 3)   c. (−3, 0)   d. (−4, −1)

8. Solve the equation $5x + 5 = 10$. **(LESSON 3.3)** b
   a. −1   b. 1   c. 3   d. −3

9. Which equation models the following situation: If each CD costs $12, how many CDs can you buy for $45? **(LESSON 1.2)** b
   a. $x = 12 + 45$ b. $x = \frac{45}{12}$
   c. $x = (45)(12)$ d. $x = 45 - 12$

10. Solve the system for x. **(LESSON 7.2)** d
    $\begin{cases} y = 1 - 4x \\ 3y = 2x + 2 \end{cases}$
    a. 14   b. $-\frac{1}{14}$ c. −14   d. $\frac{1}{14}$

11. Find the median of the data set: 45, 36, 17, 24, and 33. **(LESSON 4.4)** a
    a. 33   b. 31   c. 28   d. 36

12. Find the slope of the line passing through $A(2, 9)$ and $B(4, -3)$. **(LESSON 5.2)** d
    a. 6   b. $-\frac{1}{6}$ c. $\frac{1}{6}$ d. −6

13. Evaluate $y = 3^x$ for $x = 3$. **(LESSON 8.6)** c
    a. 9   b. 3   c. 27   d. 1

14. Evaluate $3x^2 + 2y$ for $x = 3$ and $y = 2$. **(LESSON 1.3)** c
    a. 24   b. 77   c. 31   d. 14

15. Solve $6x - 3 \leq 12$. (**LESSON 6.2**) **b**
    a. $x \leq 1.5$ b. $x \leq 2.5$ c. $x \leq -2.5$ d. $x < 2.5$
16. Simplify: $\frac{15x - 6}{3}$. (**LESSON 2.7**) **$5x - 2$**
17. Solve $x - \frac{2}{3} = 1\frac{1}{3}$. (**LESSON 3.1**) **2**
18. Find the range for the data set: 45, 36, 17, 24, and 33. (**LESSON 4.4**) **28**
19. In the equation $y = 5x + 3$, which variable is the independent variable? (**LESSON 1.5**) **x**
20. Find the mean of the data set: 28, 36, 34, 29, and 31. (**LESSON 4.4**) **31.6**
21. Solve the system. $\begin{cases} x + y = 5 \\ y + 1 = x \end{cases}$ **$x = 3, y = 2$** (**LESSON 7.3**)
22. Find the next two terms in the sequence: 5, 6, 8, 11, ... (**LESSON 1.1**) **15, 20**
23. Write the equation of the line passing through $(0, 3)$ and $(-1, 0)$. (**LESSON 5.5**) **$y = 3x + 3$**
24. Simplify $\frac{5x^2 - 10}{-5}$. (**LESSON 2.7**) **$-x^2 + 2$**
25. Simplify $(3x + 4) - (8x - 12)$. (**LESSON 2.6**) **$-5x + 16$**
26. Find the perimeter of a rectangle if $l = 3a + 2b$ and $w = 4a - 3b$. (**LESSON 2.6**) **$14a - 2b$**
27. Solve for x and graph the solution on a number line: $x - 7 \geq -4$. (**LESSON 6.1**) **$x \geq 3$**
28. Solve for a and graph the solution on a number line: $3a - 5 \leq 16$. (**LESSON 6.2**) **$a \leq 7$**
29. Simplify $\frac{16x + 4}{-2}$. (**LESSON 2.7**) **$-8x - 2$**
30. Solve $|x - 4| = 7$. (**LESSON 6.5**) **$x = -3$ or $x = 11$**
31. Solve $|a + 4| \leq 12$. (**LESSON 6.5**) **$-16 \leq a \leq 8$**
32. Write an equation for the line passing through the points $(3, 5)$ and $(5, 13)$. (**LESSON 5.5**) **$y - 13 = 4(x - 5)$ or $y - 5 = 4(x - 3)$**
33. Write the equation for the line that passes through the point $(5, 3)$ and has a slope of -2. (**LESSON 5.5**) **$y - 3 = -2(x - 5)$**
34. A new car being sold at $30,000 is advertised as "25% off." Find the original price of the car. (**LESSON 4.2**) **$40,000**

35. Solve the system of equations by elimination, and check your solution. (**LESSON 7.3**)
    $\begin{cases} 4x - 8y = 12 \\ 6x + 10y = -10 \end{cases}$ **$x = \frac{5}{11}$, $y = -\frac{112}{88}$**
36. Solve the system by graphing. (**LESSON 7.5**)
    $\begin{cases} 2x - 4y \leq 4 \\ -3x - 6y > 6 \end{cases}$
37. The sum of the digits of a positive two-digit number is 10. If twice the first digit is 2 more than the second digit, what is the two-digit number? (**LESSON 7.6**) **46**
38. Simplify: $\frac{24a^7 b^3}{48a^9}$. (**LESSON 8.3**) **$\frac{b^3}{2a^2}$**

FREE-RESPONSE GRID
The following questions may be answered by using a free-response grid such as that commonly used by standardized test services.

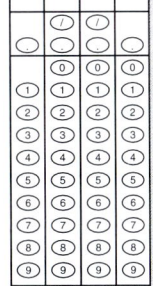

39. Evaluate $\frac{x+y}{2} + \frac{x-y}{2}$ for $x = 8$ and $y = 2$. (**LESSON 1.3**) **8**
40. Evaluate $3x^2 + 4y^2$ for $x = 0.5$ and $y = 1$. (**LESSON 1.3**) **4.75**
41. Solve $\frac{18}{a} = \frac{21}{14}$. (**LESSON 4.1**) **$a = 12$**
42. Simplify $\frac{2^3 \cdot 2^5}{2^4}$. (**LESSON 8.3**) **16**
43. A weight of 6 pounds stretches a spring a distance of 12 inches. Find the constant, k, for the spring. (**LESSON 5.3**) **$k = 0.5$**
44. Find the sum of the integers from 1 to 55. (**LESSON 1.1**) **1540**
45. Solve $4a + 6 = -2a + 12$ for a. (**LESSON 3.4**) **$a = 1$**
46. Solve for x: $\frac{x}{12} = \frac{9}{36}$. (**LESSON 4.1**) **$x = 3$**
47. Find 30% of 180. (**LESSON 4.2**) **54**

36. **(0, −1)**

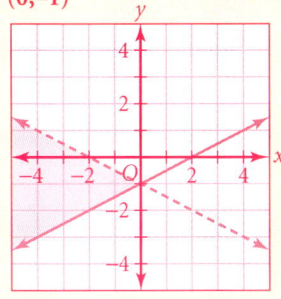

27. $x \geq 3$

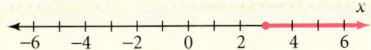

28. $a \leq 7$

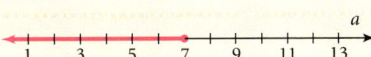

9 Polynomials and Factoring

CHAPTER PLANNING GUIDE

Lesson	9.1	9.2	9.3	9.4	9.5	9.6	9.7	9.8	Projects and Review
Pupil Edition Pages	426–431	432–437	438–442	443–447	448–451	452–457	458–463	464–469	470–477
Extra Practice (Pupil Edition)	762	763	763	764	764	765	765	766	
Practice Workbook	50	51	52	53	54	55	56	57	
Reteaching Masters	99–100	101–102	103–104	105–106	107–108	109–110	111–112	113–114	
Enrichment Masters	50	51	52	53	54	55	56	57	
Cooperative-Learning Activities	50	51	52	53	54	55	56	57	
Lesson Activities	50	51	52	53	54	55	56	57	
Problem Solving/Critical Thinking	50	51	52	53	54	55	56	57	
Student Study Guide	50	51	52	53	54	55	56	57	
Spanish Resources	50	51	52	53	54	55	56	57	
Student Technology Guide	50	51	52	53	54	55	56	57	
Assessment Resources	110	111	112	113	115	116	117	118	114, 119–124
Teaching Transparencies	43	44, 45	46						
Quiz Transparencies	9.1	9.2	9.3	9.4	9.5	9.6	9.7	9.8	
Writing Activities for Your Portfolio									25–27
Tech Prep Masters									41–44
Long-Term Projects									33–36

LESSON PACING GUIDE

Traditional	2 days	2 days	2 days	2 days	1 day	1 day	1 day	1 day	2 days
Block	1 day	1 day	1 day	1 day	$\frac{1}{2}$ day	$\frac{1}{2}$ day	$\frac{1}{2}$ day	$\frac{1}{2}$ day	1 day
Two-Year	4 days	4 days	4 days	4 days	2 days	2 days	2 days	2 days	4 days

CONNECTIONS AND APPLICATIONS

Lesson	9.1	9.2	9.3	9.4	9.5	9.6	9.7	9.8	Review
Algebra	426–431	432–437	438–442	443–447	448–451	452–457	458–463	464–469	472–477
Geometry	428	436	441, 442	446	449, 451	452, 456	463		474
Probability								469	
Business and Economics	426		440, 442	443, 444, 445, 446		456		468	474
Life Skills		436					463		
Science and Technology		437		444, 445, 446, 447	451			464, 465, 467, 468	
Sports and Leisure				447			458		
Other	431					457			
Cultural Connection: Africa						456			
Cultural Connection: Americas							463		
Cultural Connection: Europe and Asia									470

BLOCK-SCHEDULING GUIDE

Day	Lesson	Teacher Directed: Lesson Examples, Teaching Transparencies	Student Guided: Activity, Try This	Cooperative-Learning Activity, Lesson Activity, Student Technology Guide	Practice: Practice & Apply, Extra Practice, Practice Workbook	Assessment: Quiz, Mid-Chapter Assessment	Problem Solving, Reteaching
1	9.1	10 min	15 min	15 min	65 min	15 min	15 min
2	9.2	10 min	15 min	15 min	65 min	15 min	15 min
3	9.3	10 min	15 min	15 min	65 min	15 min	15 min
4	9.4	10 min	15 min	15 min	65 min	15 min	15 min
5	9.5	5 min	8 min	8 min	35 min	8 min	8 min
5	9.6	5 min	7 min	7 min	30 min	7 min	7 min
6	9.7	8 min	8 min	8 min	35 min	8 min	8 min
6	9.8	7 min	7 min	7 min	30 min	7 min	7 min
7	Assess.	50 min PE: Chapter Review	90 min PE: Chapter Project, Writing Activities	90 min Tech Prep Masters	65 min PE: Chapter Assessment, Test Generator	30 min Chap. Assess. (A or B), Alt. Assess. (A or B), Test Generator	

PE: Pupil's Edition

Alternative Assessment

The following are suggestions for an alternative assessment for students who may benefit from a different type of assessment than the regular chapter quizzes and the mid-chapter and end-of-chapter assessment materials. Many of the questions are open-ended, and students' answers will vary.

Performance Assessment

1. **a.** Simplify $(5x - 3) + (3x - 1)$, $(5x - 3) - (3x - 1)$, and $(5x - 3)(3x - 1)$. Explain how the Distributive Property is used to find the sum, difference, and product.
   b. A rectangle has a length of $2a - 1$ and a width of $3a + 2$. What operation would you use to find a single expression for area? Write that expression and find the area if $a = 10$.

2. **a.** A linear function is a polynomial function. Describe the graph of a linear function.
   b. A quadratic function is a polynomial function. Describe the graph of a quadratic function.

3. **a.** Solve each equation below.
   $x^2 - 4 = 0$ $x^2 + 4x + 4 = 0$ $x^2 + 5x + 4 = 0$
   b. How are the equations and solution processes similar and how are they different?

Portfolio Project

Suggest that students choose one of the following projects for inclusion in their portfolios.

1. **a.** Represent $x^2 + x + 1$ by using algebra tiles. Can the representation be arranged in a rectangular region?
   b. What does your response to part **a** tell you about factoring $x^2 + x + 1$?
   c. Explain how you can use algebra tiles to determine whether a quadratic expression can be factored.

2. The area of a square is given by the expression $(x + 1)^2$.
   a. Find x such that the area of the square is 100 square units.
   b. Describe the procedure for finding x when the side length of a square is given by $x + a$ and the area is b.

internetconnect

The table below identifies the pages in this chapter that contain technology information and support in the side columns.

Content Links	
Lesson Links	pages 431, 446, 467
Portfolio Links	pages 437, 447, 469
Graphics Calculator Support	page 782

Resource Links

For information about teacher and parent resources as well as professional development help, visit **www.hrw.com/math**.

Technical Support HRW has assembled a team of dedicated technical and teaching professionals and a comprehensive service program to provide you with the support you need.

- The HRW Technical Support Line operates from 7 A.M. to 6 P.M. central time, Monday through Friday, at (800) 323-9239.
- The HRW Technical Support Center on the World Wide Web is available 24 hours a day, seven days a week, at **www.hrwtechsupport.com**.
- You can e-mail our Technical Support Center at **tsc@hrwtechsupport.com**.
- The Technical Support Center's fax-on-demand service at (800) 352-1680 offers solutions to common problems and answers to frequently asked questions.

Technology

Lesson Suggestions and Calculator Examples
(Keystrokes are based on a TI-83 calculator.)

Lesson 9.1 Adding and Subtracting Polynomials
If you have access to a TI-92 calculator, enter the expressions shown in the Activity on page 427 into the calculator. For example, to find the difference in Step 4, use the following keystrokes:

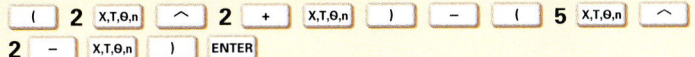

A similar procedure can be used to solve other polynomial addition and subtraction problems in this lesson.

Lesson 9.2 Modeling Polynomial Multiplication
Although this lesson emphasizes the use of algebra tiles, you can also use a graphics calculator to help students see that the product of two linear binomials in x gives an expression whose graph is a parabola, or a graph involving x^2. Enter the two products shown below as **Y1** and **Y2**, graph, and discuss the results with the class.

1. $(x-2)(x+3)$ **2.** $(2x+3)(x-4)$

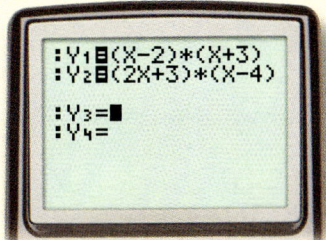 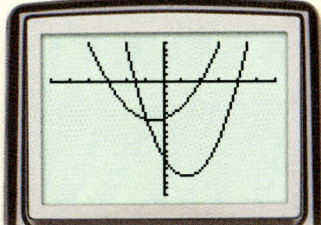

Lesson 9.3 Multiplying Binomials
Have students graph $y = (x+3)(x-4)$, identify the x-intercepts, and compare the intercepts with the numbers in the product. The display below can be part of the discussion. It suggests that one x-intercept is -3. Ask students to generalize and find the x-intercepts of the graph of $y = (x+a)(x-b)$.

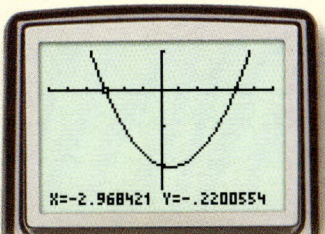

Lesson 9.4 Polynomial Functions
Surface area or volume problems may result in a third-degree polynomial equation. This type of equation does not have a straight or **U**-shaped graph. Have students use the graphics calculator to explore the graphs of different polynomial equations. Two such graphs are shown in the diagrams at the top of the next column.

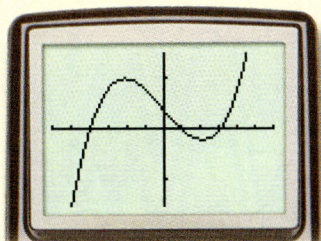 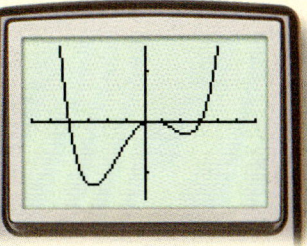

Lesson 9.5 Common Factors
The first step in factoring a polynomial expression is to find the GCF of the numerical coefficients. On the TI-83, the GCF function is named **gcd**, which stands for greatest common divisor. You can access this function by pressing MATH ▶ and selecting **9:gcd(**. The calculator display at right shows that the GCF of 15 and 12 is 3.

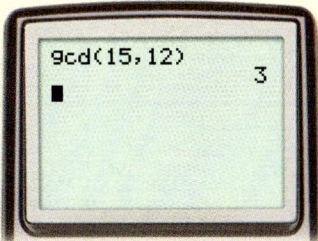

Lesson 9.6 Factoring Special Polynomials
After factoring problems by using pencil and paper, consider a calculator enrichment problem. How are the x-intercepts of $y = x^2 - a^2$, where a is a fixed positive number, related to one another?

Lesson 9.7 Factoring Quadratic Trinomials
One reason for having students factor quadratic trinomials on their own is to help students see that trial and error plays an important role in mathematical work. If a calculator does the factoring for them, they do not learn to appreciate the need for trial and error. However, you may want to show students how to factor a quadratic trinomial on the TI-92. To do this, press **F2**, select **2:factor(**, enter the expression, and press ENTER.

Lesson 9.8 Solving Equations by Factoring
Discuss with students two other approaches to solving the problem shown on the bottom of page 465.

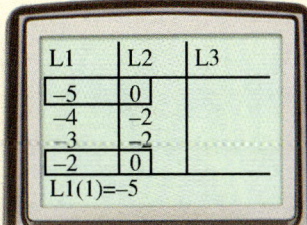 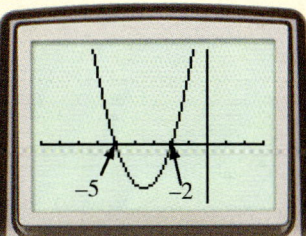

For further information, refer to the
- technology discussions in the lessons.
- lesson-related teacher's commentary in the side columns of this *Teacher's Edition*.
- lesson-related *Student Technology Guide* masters.
- *HRW Technology Handbook*.

Teaching Resources

Basic Skills Practice
(2 pages per skill)

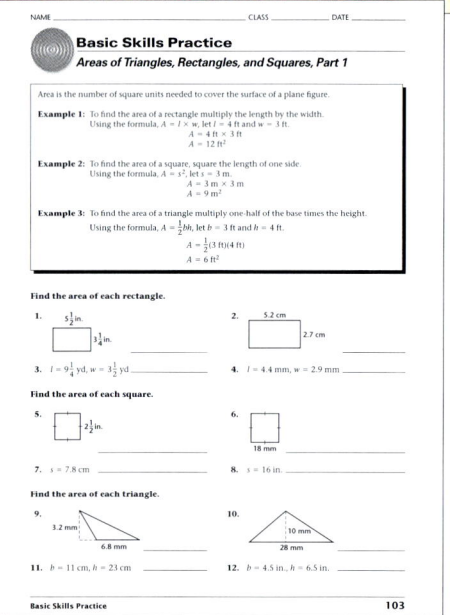

Reteaching
(2 pages per lesson)

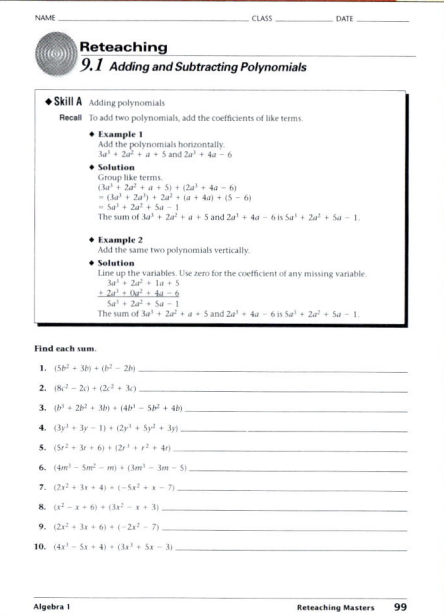

Enrichment
(1 page per lesson)

Cooperative-Learning Activity
(1 page per lesson)

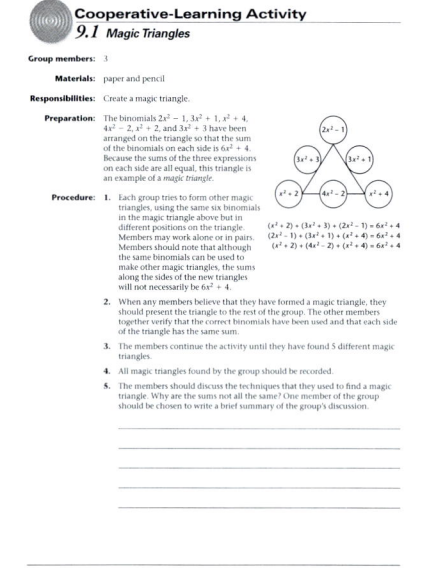

Lesson Activity
(1 page per lesson)

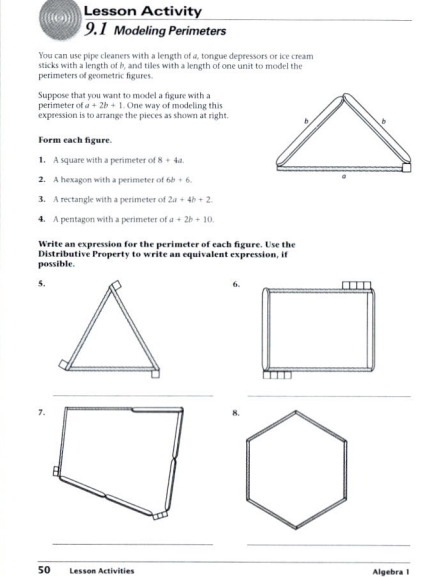

Long-Term Project
(4 pages per chapter)

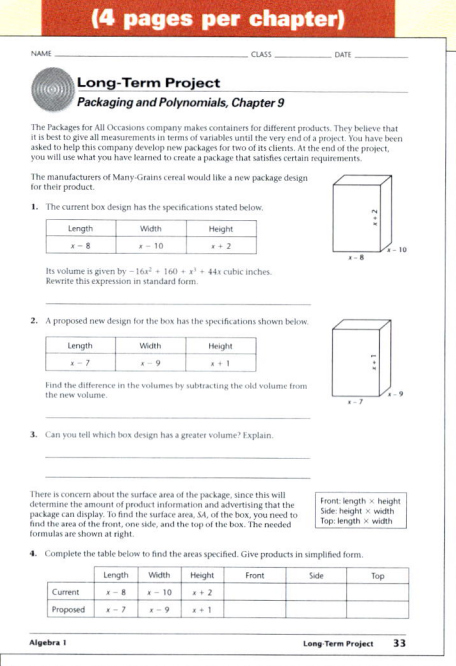

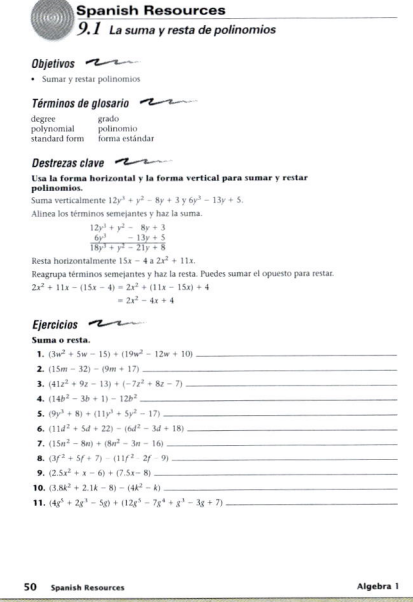

About Chapter 9

Background Information

In Chapter 9, students are introduced to polynomials. They learn to add, subtract, and multiply polynomials, and they investigate some simple polynomial functions. Then they learn to factor, with an emphasis on the greatest common factor and on special patterns involved in factoring quadratic trinomials. In the final lesson of the chapter, students apply their experiences with factoring to the solution of quadratic equations.

CHAPTER RESOURCES

- Block-Scheduling Handbook
- Writing Activities for Your Portfolio
- Tech Prep Masters
- Long-Term Project
- Lesson Activity Masters
- Assessment Resources:
  Mid-Chapter Assessment
  Chapter Assessments
  Alternative Assessments
- Test and Practice Generator
- Technology Handbook
- End-of-Course Exam Practice

Polynomials and Factoring

POLYNOMIALS ARE USEFUL FOR REPRESENTING area and volume functions, problems involving gravity, consumer problems, and many other types of real-world situations.

Lessons

- 9.1 Adding and Subtracting Polynomials
- 9.2 Modeling Polynomial Multiplication
- 9.3 Multiplying Binomials
- 9.4 Polynomial Functions
- 9.5 Common Factors
- 9.6 Factoring Special Polynomials
- 9.7 Factoring Quadratic Trinomials
- 9.8 Solving Equations by Factoring

Project Powers, Pascal, and Probability

About the Photos

One way to approximate the volume of a hot air balloon is to imagine the balloon being sliced into a stack of circles or ovals. The volume would then equal the average area of the ovals multiplied by the height.

The volume and surface area of a hot-air balloon can be modeled using polynomial functions.

Chapter Objectives

- Add and subtract polynomials. [**9.1**]
- Use algebra tiles to model the products of binomials. [**9.2**]
- Mentally simplify special products of binomials. [**9.2**]
- Find products of binomials by using the Distributive Property. [**9.3**]
- Find products of binomials by using the FOIL method. [**9.3**]
- Define polynomial functions. [**9.4**]
- Solve problems involving polynomial functions. [**9.4**]
- Factor a polynomial by using the greatest common factor. [**9.5**]
- Factor a polynomial by using a binomial factor. [**9.5**]
- Factor perfect-square trinomials. [**9.6**]
- Factor the difference of two squares. [**9.6**]
- Factor quadratic trinomials by using algebra tiles. [**9.7**]
- Factor quadratic trinomials by using guess-and-check methods. [**9.7**]
- Solve equations by factoring. [**9.8**]

About the Chapter Project

You have already seen how algebra makes it possible to use functions in many different types of situations. In the Chapter Project, *Powers, Pascal, and Probability*, on page 470, you will explore connections between Pascal's triangle and polynomial functions that can be used to simplify calculations. After completing the Chapter Project, you will be able to do the following:

- Apply Pascal's triangle to a probability problem.

About the Portfolio Activities

Throughout the chapter, you will be given opportunities to complete Portfolio Activities that are designed to support your work on the Chapter Project.

- Forming Pascal's triangle is the topic of the Portfolio Activity on page 437.
- Modeling triangular numbers with polynomial functions is the topic of the Portfolio Activity on page 447.
- Modeling number sequences with polynomial functions is the topic of the Portfolio Activity on page 457.
- Modeling rectangular numbers with polynomial functions is the topic of the Portfolio Activity on page 470.

Portfolio Activities appear at the end of Lessons 9.2, 9.4, 9.6, and 9.8. Each serves as preparation for the Chapter Project. The Portfolio Activities as well as the Chapter Project Activities are appropriate for inclusion in the student's portfolio. Students should be encouraged to include in their portfolios any other work in which they feel a sense of pride or a sense of accomplishment.

 internet connect

Internet Connect features appearing throughout the chapter provide keywords for the HRW Web site that lead to links for math resources, tutorial assistance, references for student research, classroom activities, and all teaching resource pages. The HRW Web site also provides updates to the technology used in the text. Listed at right are the keywords for the Internet Connect activities referenced in this chapter. Refer to the side column on the page listed to read about the activity.

LESSON	KEYWORD	PAGE
9.1	MA1 Poly Ops	431
9.2	MA1 Pascal	437
9.4	MA1 Regression	446
9.4	MA1 Figurate	447
9.8	MA1 Zeros	467
9.8	MA1 Rectangular	469

Prepare

NCTM Principles & Standards
1–3, 6, 8–10

LESSON RESOURCES

- Practice 9.1
- Reteaching Master 9.1
- Enrichment Master 9.1
- Cooperative-Learning Activity 9.1
- Lesson Activity 9.1
- Problem Solving/Critical Thinking Master 9.1
- Teaching Transparency 43

QUICK WARM-UP

Simplify.

1. $-9b + 8b$ $-b$
2. $6y - y$ $5y$
3. $4m + (7 - m)$ $3m + 7$
4. $c - (3c + 1)$ $-2c - 1$
5. $(3r + 4) + (-2r + 8)$ $r + 12$
6. $(-2g + 1) - (2g + 1)$ $-4g$
7. $(5j - 7k) - (4j - 6k)$ $j - k$

Also on Quiz Transparency 9.1

Teach

Why Work with students to make a list of words that begin with the prefixes *mono-*, *bi-*, *tri-*, and *poly-*. Discuss the meaning of each word. Remind students of the meaning of the word *monomial*, which they learned in Chapter 8. Ask them to make a conjecture about the meanings of the words *binomial*, *trinomial*, and *polynomial*.

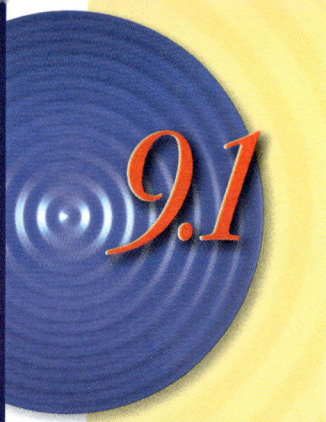

9.1 Adding and Subtracting Polynomials

Objective
- Add and subtract polynomials.

Why The arc of a football, the distance traveled by an accelerating car, and volume of an object can be modeled by polynomials.

The volume of the box on the left is $(x + 2)(x + 2)(x + 2)$, or $x^3 + 6x^2 + 12x + 8$.

APPLICATION
PACKAGING

A **polynomial** is a monomial, or a sum or difference of monomials. The expression for the volume of the box above, $x^3 + 6x^2 + 12x + 8$, is an example of a polynomial in one variable, x.

The **degree** of a polynomial in one variable is determined by the exponent with the greatest value within the polynomial. The degree of $9 - 4x^2$ is 2.

The terms of a polynomial may appear in any order. However, in **standard form**, the terms of a polynomial are ordered from left to right in descending order, which means from the greatest exponent to the least.

EXAMPLE ❶ Write $9 + x - 4x^2$ in standard form.

● **SOLUTION**

Reorder the terms of the polynomial in descending order according to the exponents of the variables.

$$9 + x - 4x^2 = 9 + x + (-4x^2)$$
$$= -4x^2 + x + 9$$

Some polynomial expressions have special names that are determined either by their degree or by the number of terms, as illustrated in the table.

Alternative Teaching Strategy

CONNECTING TO REAL LIFE Have students recall the general growth formula. Ask them to write an expression for the amount of money in an account after 4 years if $50 is deposited and the yearly growth rate is 8%. Elicit the response $50(1.08)^4$.

Now ask them to consider what happens if $50 is the original amount, but an additional $50 is deposited at the beginning of each successive year. Lead them to see that the total in the account after 4 years would be the found as follows:

$$50(1.08)^4 + 50(1.08)^3 + 50(1.08)^2 + 50(1.08)$$

↑	↑	↑	↑
amount deposited in year 1	amount deposited in year 2	amount deposited in year 3	amount deposited in year 4

For $x = 1.08$, the total can be represented by $50x^4 + 50x^3 + 50x^2 + 50x$. Point out that a variable expression like this is useful because it allows you to find the amount after 4 years for different values of x. Tell students that this expression is an example of a *polynomial*.

426   LESSON 9.1

Polynomial	# of terms	Name by # of terms	Degree	Name by degree
12	1	monomial	0	constant
$8x$	1	monomial	1	linear
$4x^2 + 3$	2	binomial	2	quadratic
$5x^3 + x^2$	2	binomial	3	cubic
$3x^2 - 4x + 6$	3	trinomial	2	quadratic
$3x^4 - 4x^3 + 6x^2 - 7$	4	polynomial	4	quartic

ADDITIONAL EXAMPLE 1

Write $9 - 3m^2 - m^3 + 2m$ in standard form.

$-m^3 - 3m^2 + 2m + 9$

Use Teaching Transparency 43.

Activity
Polynomial Addition and Subtraction

You will need: algebra tiles

1. Use algebra tiles to model this sum of binomials: $(2x + 2) + (3x - 3)$.

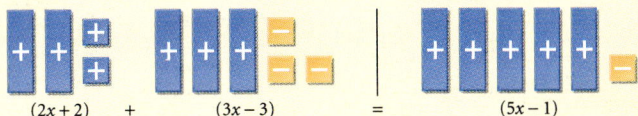

$(2x + 2) \quad + \quad (3x - 3) \quad = \quad (5x - 1)$

2. Check your operations with polynomials by substituting several values for x to verify the following:

$$(2x + 2) + (3x - 3) = 5x - 1$$

3. Use algebra tiles to simplify the polynomial expression $(2x + 2) - (3x - 1)$, and complete the following:

$$(2x + 2) - (3x - 1) = ?$$

Show that the equation is true by substituting -3, 4, and 10 for x.

CHECKPOINT ✓ 4. Explain how to simplify the polynomial addition expression $(2x^2 + x) + (5x^2 - x)$. Explain how to simplify the polynomial subtraction expression $(2x^2 + x) - (5x^2 - x)$.

Activity Notes

This Activity provides students with a concrete model for adding and subtracting polynomials. Remind students that once they have grouped like tiles, any neutral pairs may be removed. You may want to have students model other addition and subtraction expressions, such as $(x - 1) + (4x + 5)$ and $(4x - 2) - (6 - 3x)$.

After completing the Activity, students should recognize that to add polynomials, you combine like terms. To subtract polynomials, you take the opposite of each term of the polynomial being subtracted and then add.

CHECKPOINT ✓

4. Group like terms and add. Change subtraction to addition of the opposite, group like terms, and add.

Adding Polynomials

Polynomials can be added in vertical or horizontal form. In vertical form, align the like terms and add.

$$\begin{aligned} x^2 + x + 1 \\ \underline{2x^2 + 3x + 2} \\ 3x^2 + 4x + 3 \end{aligned}$$

In horizontal form, regroup like terms and add.

$$(x^2 + x + 1) + (2x^2 + 3x + 2) = (x^2 + 2x^2) + (x + 3x) + (1 + 2)$$
$$= 3x^2 + 4x + 3$$

Inclusion Strategies

VISUAL LEARNERS Have students rewrite an addition or subtraction of polynomials by using colored pencils to indicate like terms. Encourage them to use placeholders for "absent" terms. For example, they might write the following:

$(x^3 + 2x - 1) + (x^2 - 3)$
$= (x^3 + 0x^2 + 2x - 1) + (0x^3 + x^2 + 0x - 3)$

↑ ↑ ↑ ↑ ↑ ↑ ↑ ↑
green red blue black green red blue black

Enrichment

Have students work in pairs to solve these "missing polynomial" problems.

1. $(r^3 + r^2 - 5) + (\underline{\ ?\ }) = r^2 + 2r - 7$
   $-r^3 + 2r - 2$

2. $(2a^2 + a) - (\underline{\ ?\ }) = -a^3 + a^2 + 3a - 1$
   $a^3 + a^2 - 2a + 1$

3. $(\underline{\ ?\ }) - (2k^4 - 2k^2 - 3) = k^4 + k^3 - 2k^2 + k$
   $3k^4 + k^3 - 4k^2 + k - 3$

Then have each student make up three original problems like the ones given. Partners should trade papers and solve each other's problems.

Teaching Tip

Some students believe that x and x^2 must be like terms. Algebra tiles provide a concrete way to demonstrate why this is incorrect. Remind students that like terms contain the same variables and that the variables must be raised to the same power.

ADDITIONAL EXAMPLE 2

Find the sum $(3d^2 + 5d - 1) + (-4d^2 - 5d + 2)$ by using

a. the vertical form.
$$3d^2 + 5d - 1$$
$$\underline{-4d^2 - 5d + 2}$$
$$-d^2 + 1$$

b. the horizontal form.
$(3d^2 + 5d - 1) + (-4d^2 - 5d + 2)$
$= (3d^2 - 4d^2) + (5d - 5d) + (-1 + 2)$
$= -d^2 + 0 + 1$
$= -d^2 + 1$

TRY THIS
$x^2 - x + 4$

ADDITIONAL EXAMPLE 3

Write a polynomial expression for the perimeter of each polygon.

a.
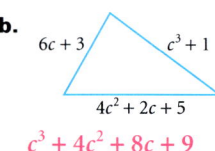
$8z^2 - 10z + 14$

b.
$c^3 + 4c^2 + 8c + 9$

Math CONNECTION

GEOMETRY A *polygon* is a plane figure formed by three or more segments joined at their endpoints, with none of the segments lying on the same line. Each segment is called a *side* of the polygon. The *perimeter* of a polygon is the sum of the lengths of its sides.

TRY THIS
$2x^3 + 6x^2 + 6x$

428 LESSON 9.1

EXAMPLE 2 Find the sum $(2x^2 - 3x + 5) + (4x^2 + 7x - 2)$ by using
a. the vertical form.
b. the horizontal form.

● **SOLUTION**

a. Align like terms and add.
$$2x^2 - 3x + 5$$
$$\underline{4x^2 + 7x - 2}$$
$$6x^2 + 4x + 3$$

b. Regroup like terms and add.
$(2x^2 - 3x + 5) + (4x^2 + 7x - 2) = (2x^2 + 4x^2) + (-3x + 7x) + (5 - 2)$
$ = 6x^2 + 4x + 3$

TRY THIS Find the sum $(3x^2 + 5x) + (4 - 6x - 2x^2)$ by using the vertical form and the horizontal form.

EXAMPLE 3 Write a polynomial expression for the perimeter of each polygon.

CONNECTION GEOMETRY

a.

b.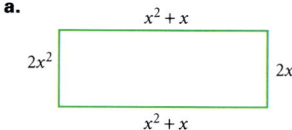

● **SOLUTION**

Vertical form

a.
$$x^2 + x$$
$$2x^2$$
$$2x^2$$
$$\underline{x^2 + x}$$
$$6x^2 + 2x$$

b.
$$a^3 + 2a$$
$$a + 1$$
$$\underline{2a^3 + a + 3}$$
$$3a^3 + 4a + 4$$

Horizontal form

a. $(x^2 + x) + 2x^2 + 2x^2 + (x^2 + x) = x^2 + x^2 + 2x^2 + 2x^2 + x + x$
$ = 6x^2 + 2x$

b. $(a^3 + 2a) + (a + 1) + (2a^3 + a + 3) = a^3 + 2a^3 + 2a + a + a + 1 + 3$
$ = 3a^3 + 4a + 4$

TRY THIS Write a polynomial expression for the perimeter of the polygon shown at right.

Reteaching the Lesson

COOPERATIVE LEARNING Have students work in groups of three. Write these terms on the board or overhead.

$5n^2 \quad -2 \quad n \quad 6 \quad -n^2 \quad -4n$

Tell students that they are to choose from among these terms to write an addition of two binomials whose sum is a binomial. Each group member's addition should have a different sum.

three possible additions: $(5n^2 + n) + (-n^2 - 4n) = 4n^2 - 3n$; $(5n^2 - 2) + (-n^2 + 6) = 4n^2 + 4$; $(n - 2) + (-4n + 6) = -3n + 4$

Repeat the activity for the conditions below.

• Write an addition of two binomials whose sum is a trinomial.
sample: $(5n^2 + n) + (-n^2 + 6) = 4n^2 + n + 6$

• Write a subtraction of two binomials whose difference is a binomial.
sample: $(5n^2 + n) - (-n^2 - 4n) = 6n^2 + 5n$

• Write a subtraction of two binomials whose difference is a trinomial.
sample: $(n + 6) - (-n^2 - 4n) = n^2 + 5n + 6$

Subtracting Polynomials

Recall that subtraction can be modeled by adding the opposite.

EXAMPLE 4 Find the difference.
  a. $(3x^2 - 2x + 8) - (x^2 - 4)$
  b. $(4x^2 + 3x + 2) - (2x^2 - 3x + 7)$

● **SOLUTION**

a. The binomial $x^2 - 4$ can be written as $x^2 + 0x - 4$, where $0x$ is a "placeholder" for the missing x-term. Remember to change signs when subtracting.

Vertical form
$$\begin{array}{r} 3x^2 - 2x + 8 \\ -(x^2 + 0x - 4) \\ \hline 2x^2 - 2x + 12 \end{array}$$

Horizontal form
$$\begin{aligned} (3x^2 - 2x + 8) - (x^2 - 4) &= 3x^2 - 2x + 8 - x^2 + 4 \\ &= (3x^2 - x^2) - 2x + (8 + 4) \\ &= 2x^2 - 2x + 12 \end{aligned}$$

b. **Vertical form**
$$\begin{array}{r} 4x^2 + 3x + 2 \\ -(2x^2 - 3x + 7) \\ \hline 2x^2 + 6x - 5 \end{array}$$

Horizontal form
$$\begin{aligned} (4x^2 + 3x + 2) - (2x^2 - 3x + 7) &= 4x^2 + 3x + 2 - 2x^2 + 3x - 7 \\ &= (4x^2 - 2x^2) + (3x + 3x) + (2 - 7) \\ &= 2x^2 + 6x - 5 \end{aligned}$$

TRY THIS Subtract $x^2 + 2x - 4$ from $3x^2 - 2x + 3$.

ADDITIONAL EXAMPLE 4

Find the difference.
$(10p^2 + 5) - (7p^2 - 2p + 3)$

$3p^2 + 2p + 2$

TRY THIS
$2x^2 - 4x + 7$

Teaching Tip

TECHNOLOGY If you are using a TI-92 to find the difference in Example 4a, use these keystrokes.

(3X ^ 2 − 2X +
8) − (X ^ 2 −
4) ENTER

Press CLEAR, and then use these keystrokes for part **b**.

(4X ^ 2 + 3X +
2) − (2X ^
2 − 3X + 7) ENTER

Assess

Selected Answers
Exercises 6–16, 17–61 odd

ASSIGNMENT GUIDE	
In Class	1–16
Core	17–46, 50–54
Core Plus	19–46 even, 47–54
Review	55–61
Preview	62

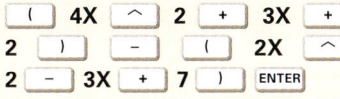 Extra Practice can be found beginning on page 738.

Exercises

Communicate

1. Explain how to use algebra tiles to represent $5x^2 - 2x + 3$.
2. How do you determine the degree of a polynomial?
3. Classify the polynomial $3x^3 + 4x^2 - 7$ in two different ways.
4. Explain how to find $(3b^3 - 2b + 1) - (b^3 + b - 3)$.
5. Use $5x^2 - 2x + 3x^4 - 6$ to explain how to write a polynomial in standard form.

LESSON 9.1 **429**

Technology

A calculator may be helpful in performing some of the additions and subtractions involved in the exercises. You may want to have students first do the exercises with paper and pencil and then use a calculator to check their answers.

Error Analysis

When subtracting polynomials, students often forget to change the sign of each term in the second polynomial. For instance, given $(z^2 + 2) - (3z^2 - z + 1)$, they may write $z^2 + 2 - 3z^2 - z + 1$. Suggest that they rewrite the subtraction to show a multiplication by –1 and then insert arrows as a reminder that they must apply the Distributive Property.

$(z^2 + 2) - (3z^2 - z + 1)$

$= (z^2 + 2) + (-1)(3z^2 - z + 1)$

17.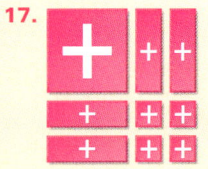

Practice

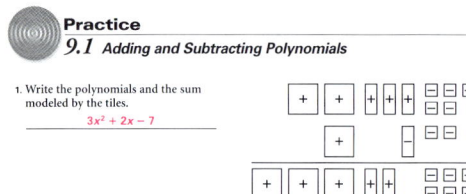

Guided Skills Practice

Write each polynomial in standard form. (EXAMPLE 1)

6. $x + 3x^3 - 2$
   $3x^3 + x - 2$
7. $15 + 2x - 3x^2$
   $-3x^2 + 2x + 15$
8. $3x^4 + 23 - 2x + 2x^2$
   $3x^4 + 2x^2 - 2x + 23$

Find each sum by using the horizontal form and the vertical form. (EXAMPLE 2)

9. $(3x^3 + 2x - 5) + (x^3 - 4)$ $\quad 4x^3 + 2x - 9$
10. $(15x^2 + 5x) + (3x^2 - 3)$ $\quad 18x^2 + 5x - 3$
11. $(12x^3 + 7x + 2) + (4x^2 + 3x - 6)$
    $12x^3 + 4x^2 + 10x - 4$
12. $(10x^4 - 3x^2 + 2) + (3x^3 + 2x^2 - 13)$
    $10x^4 + 3x^3 - x^2 - 11$

Write a polynomial expression for the perimeter of each polygon shown below. (EXAMPLE 3)

13. 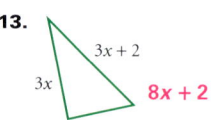 $8x + 2$

14. 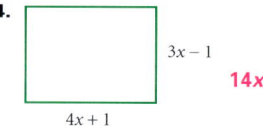 $14x$

Find each difference by using the horizontal form and the vertical form. (EXAMPLE 4)

15. $(12x^2 + 5x + 11) - (10x^2 + 3x + 2)$ $\quad 2x^2 + 2x + 9$
16. $(3x^4 + 2x^2) - (2x^4 + 3)$ $\quad x^4 + 2x^2 - 3$

Practice and Apply

17. Draw an algebra-tile model of $x^2 + 4x + 4$.
18. Draw an algebra-tile model of $(3x^2 + 6) + (2x^2 - 1)$.

Write each polynomial in standard form.

19. $6 + c + c^3$ 20. $5x^3 - 1 + 5x^4 + 5x^2$ 21. $10 + p^7$
    $c^3 + c + 6$ $5x^4 + 5x^3 + 5x^2 - 1$ $p^7 + 10$

Name each polynomial by the number of terms and by degree.

22. $4r + 1$ 23. $x^3 + x^4 + x - 1$ 24. $y + y^3$
    Binomial; linear   Polynomial; quartic   Binomial; cubic
25. $3x + 1$ 26. $8x^2 - 1$ 27. $8x^2 - 2x + 3$
    Binomial; linear   Binomial; quadratic   Trinomial; quadratic

Give an example of each type of polynomial. (28–30. Answers may vary. Sample answer:)

28. quadratic trinomial   29. linear binomial   30. cubic monomial
    $x^2 + x + 1$ $x + 5$ x^3

Find the sum or difference by using the vertical form.

31. $(3x^2 + 4x^4 - x + 1) + (3x^4 + x^2 - 6)$ $\;7x^4 + 4x^2 - x - 5$
32. $(2y^3 + y^2 + 1) + (3y^3 - y^2 + 2)$ $\;5y^3 + 3$
33. $(4r^4 + r^3 - 6) + (r^3 + r^2)$ $\;4r^4 + 2r^3 + r^2 - 6$
34. $(2c - 3) + (c^2 + c + 4)$ $\;c^2 + 3c + 1$
35. $(x^3 + x^2 + 7) - (x^2 + x)$ $\;x^3 - x + 7$
36. $(4y^2 - y + 6) - (3y^2 - 4)$ $\;y^2 - y + 10$
37. $(5c^3 + 10c + 5) - (4c^3 - c^2 - 1)$ $\;c^3 + c^2 + 10c + 6$
38. $(x^3 - x + 4) - (8x^3)$ $\;-7x^3 - x + 4$

Find the sum or difference by using the horizontal form.

39. $(y^3 - 4) + (y^2 - 2)$ $\;y^3 + y^2 - 6$
40. $(x^3 + 2x - 1) + (3x^2 + 4)$ $\;x^3 + 3x^2 + 2x + 3$
41. $(3s^2 + 7s - 6) + (s^3 + s^2 - s - 1)$ $\;s^3 + 4s^2 + 6s - 7$
42. $(w^3 + w - 2) + (4w^3 - 7w + 2)$ $\;5w^3 - 6w$
43. $(y^2 + 3y + 2) - (3y - 2)$ $\;y^2 + 4$
44. $(3x^2 - 2x + 10) - (2x^2 + 4x - 6)$ $\;x^2 - 6x + 16$
45. $(3x^2 - 5x + 3) - (2x^2 - x - 4)$ $\;x^2 - 4x + 7$
46. $(2x^2 + 5x) - (x^2 - 3)$ $\;x^2 + 5x + 3$

18.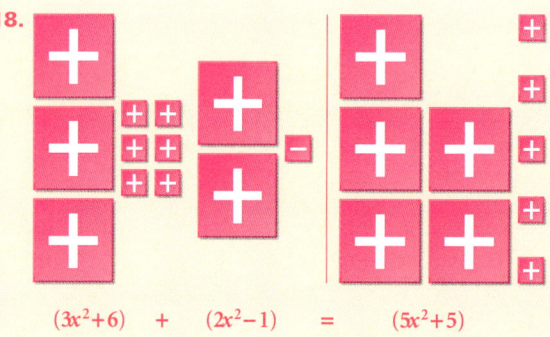

$(3x^2 + 6) \quad + \quad (2x^2 - 1) \quad = \quad (5x^2 + 5)$

430   LESSON 9.1

Simplify. Express all answers in standard form.

47. $(1 - 4x - x^4) + (x - 3x^2 + 9)$ $-x^4 - 3x^2 - 3x + 10$
48. $(5 - 3x - 1.4x^2) - (13.7x - 62 + 5.6x^2)$ $-7x^2 - 16.7x + 67$
49. $(4x^5 - 3x^3 - 3x - 5) - (x - 2x^2 + 3x + 5)$ $4x^5 - 3x^3 + 2x^2 - 7x - 10$

C O N N E C T I O N

GEOMETRY Exercises 50–52 refer to the dimensions of the figure at left.

50. Find the polynomial expression that represents the perimeter of the large rectangle. $16x + 12$
51. Find the polynomial expression that represents the perimeter of the small rectangle. $10x + 10$
52. Find the difference between the perimeters of the two rectangles. $6x + 2$

A P P L I C A T I O N

INTERIOR DECORATING Tina is estimating the amount of wallpaper border trim that she will need for a rectangular room. The trim pieces that she plans to use are precut x feet long. She estimates that one length of the room will take 3 pieces plus an extra 4 feet of trim and that one width of the room will take 2 pieces plus an extra 3 feet of trim.

53. Write a polynomial expression for the length of the walls to be trimmed and one for the width.   width: $2x + 3$; length: $3x + 4$
54. What is the total amount of border trim, in terms of x, needed for all four walls?
$10x + 14$ feet

Look Back

Multiply. *(LESSON 3.5)*

55. $6x^2 - 24x$

55. $6(x^2 - 4x)$ 56. $8(m^2 + 6m)$ $8m^2 + 48m$

57. Solve $\frac{6x - 12x + 18}{3} = 1$ for x. *(LESSON 3.4)* $\frac{5}{2} = 2\frac{1}{2}$

58. Solve $-2(a + 3) = 5 - 6(2a - 7)$ for a. *(LESSON 3.5)* 5.3

59. Solve this system of equations. $\begin{cases} 6x = 4 - 2y \\ 12x - 4y = 16 \end{cases}$ *(LESSON 7.3)*
$(1, -1)$

Write each number in scientific notation. *(LESSON 8.5)*

60. $7{,}100{,}000$ 7.1×10^6 61. $8{,}900{,}000{,}000$ 8.9×10^9

Look Beyond

62. Use algebra tiles to model a square whose area can be represented by $x^2 + 6x + 9$. What is the length of a side of the square?

62.

The length of one side is $x + 3$.

Look Beyond

In Exercise 62, students investigate a *perfect-square trinomial*. Special case polynomials such as perfect-square trinomials and differences of two squares will be studied in detail in Lesson 9.6.

internetconnect

GO TO: go.hrw.com
KEYWORD: MA1 Poly Ops

In this activity, students explore some of the costs associated with a specific business to illustrate the use of operations on polynomials.

Student Technology Guide

NAME _____ CLASS _____ DATE _____

Student Technology Guide
9.1 Adding and Subtracting Polynomials

When you add or subtract polynomials, you combine like terms. Using a spreadsheet can help you keep track of like terms.

Example: Simplify $(x^4 + 7x^2 - 6x + 2) - (3x^3 - 2x^2 + 17) + (2x^4 - 8x^3 + 2x + 1)$.
- In cells A2, A3, and A4, list the polynomials you are adding or subtracting. Be sure to include a negative sign for any subtracted polynomial.
- In cells B1 through F1, list the types of terms in the polynomials. Use the symbol ^ to indicate an exponent.
- Enter the coefficient of each term in the appropriate cell. Remember to *change the sign* of each coefficient when one polynomial is subtracted from another. Enter a zero if there is no such term and a 1 for terms with no written coefficient.
- In cell B5, enter =SUM(B2:B4). Similarly, enter formulas for the sums in cells C5, D5, E5, and F5.

	A	B	C	D	E	F
1		x^4	x^3	x^2	x	const.
2	x^4 + 7x^2 - 6x + 2	1	0	7	-6	2
3	-(3x^3 - 2x^2 + 17)	0	-3	2	0	-17
4	2x^4 - 8x^3 + 2x + 1	2	-8	0	2	1
5		3	-3	9	-4	-14

Write the answer, $3x^4 - 3x^3 + 9x^2 - 4x - 14$.

Simplify. Express all answers in standard form.

1. $(x^2 - 7x + 5) + (3x^2 + 4x - 8)$
   $4x^2 - 3x - 3$
2. $(2a^3 - a + 3) - (4a^2 + 3a + 1)$
   $2a^3 - 4a^2 - 4a + 2$
3. $(4c^4 - 2c + 8) - (5c^3 - 3c^2 - 9c)$
   $4c^4 - 5c^3 + 3c^2 + 7c + 8$
4. $(1.7x^3 - 2.1x + 7.7) - (4.4x^2 + 7.9x)$
   $1.7x^3 - 4.4x^2 - 10x + 7.7$

On a graphics calculator, such as the TI-92, you can enter polynomials in the same form in which they appear on paper. The calculator will give the correct sum or difference even if the terms of the polynomials are not written in ascending or descending order of exponents.

The display at right shows the difference $2 + 3x^2 + 5x - (x^2 + 4)$. Notice that the calculator displays the answer as a polynomial whose terms are arranged in descending order.

Prepare

NCTM PRINCIPLES & STANDARDS
1, 2, 6, 8–10

LESSON RESOURCES

- Practice 9.2
- Reteaching Master 9.2
- Enrichment Master 9.2
- Cooperative-Learning Activity 9.2
- Lesson Activity 9.2
- Problem Solving/Critical Thinking Master 9.2
- Teaching Transparency 44
- Teaching Transparency 45

QUICK WARM-UP

Simplify.

1. $(-7)(-2)$ 14
2. $(-3)(4t)$ $-12t$
3. $3(s+5)$ $3s+15$
4. $4(-n+2)$ $-4n+8$
5. $4-(h+2)$ $-h+2$

Also on Quiz Transparency 9.2

Teach

Why Ask students how they might direct someone who needs to travel from the school to a location some distance away. Elicit several possible responses, such as making a map, writing a set of instructions, or simply giving verbal instructions. Tell students that in this lesson, they will be directed to multiply polynomials in several different ways: by making a model, by writing a set of algebraic steps, and by multiplying mentally.

9.2 Modeling Polynomial Multiplication

Objectives
- Use algebra tiles to model the products of binomials.
- Mentally simplify special products of binomials.

Why Algebra tiles provide a physical representation of polynomial multiplication. This provides another way of thinking about the Distributive Property.

Algebra tiles can be used as a model for multiplication. The model is related to the fact that the area of a rectangle is equal to the product of the length and the width.

In the product $(x+1)(2x+2)$, two binomials are multiplied. You can model this product by using algebra tiles, which will show you how to find the product.

Activity
Exploring Multiplication of Polynomials

You will need: algebra tiles (or use paper-and-pencil sketches)

Part I: Finding Products

1. The model below shows how $x+1$ can be multiplied by 2. The result is $2x+2$. This is an example of which property?

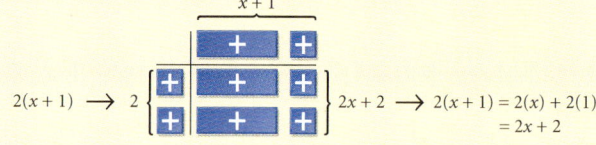

Alternative Teaching Strategy

USING VISUAL MODELS Have students use graph paper models to verify polynomial multiplications for different values of the variable. For instance, the figures below model $x(x+3) = x^2 + 3x$ for $x=4$ and $x=5$.

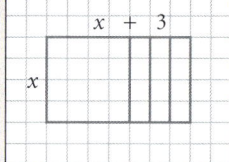
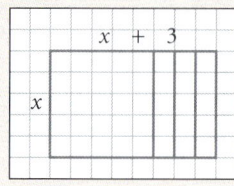

For $x=4$:
$4(4+3) = 28$
$4^2 + 3(4) = 28$
$4(4+3) = 4^2 + 3(4)$

For $x=5$:
$5(5+3) = 40$
$5^2 + 3(5) = 40$
$5(5+3) = 5^2 + 3(5)$

432   LESSON 9.2

2. What two factors are multiplied in the model at right? What is the product shown by the tiles?

3. Check the product in Step 2 by substituting 10 for x.

CHECKPOINT ✓ 4. Use tiles to find the product $(2x + 2)(x + 4)$, and check your answer by substituting a value for x.

Part II: Introducing Negative Numbers

1. The tiles shown at right model the product of two binomials. Remove as many neutral pairs from the product as you can. What is the simplified product?

$(x + 2)(x - 3) = $ ___?___

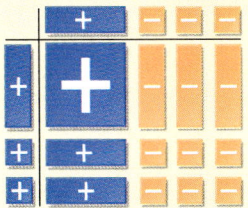

2. Check the product by substituting a value, such as 10, for x.

CHECKPOINT ✓ 3. Use tiles to find the product $(x + 3)(2x - 4)$, and check your answer by substituting a value for x.

Part III: Special Products

1. Use tiles to find the following products:

$(x + 2)(x - 2)$ $(x + 3)(x - 3)$

What do these two products have in common? Use your observations to find the product $(x + 5)(x - 5)$ without using tiles.

2. Use tiles to find the following products:

$(x + 2)(x + 2)$ $(x + 3)(x + 3)$

What do these two products have in common? Use your observations to find the product $(x + 5)(x + 5)$ without using tiles.

CHECKPOINT ✓ 3. Use tiles to find the following products:

$(x - 2)(x - 2)$ $(x - 3)(x - 3)$

What do these two products have in common? Use your observations to find the product $(x - 5)(x - 5)$ without using tiles.

CRITICAL THINKING Given the algebra tiles that model a second-degree polynomial, will you always be able to form a rectangle with the tiles? Explain.

Interdisciplinary Connection

AGRICULTURE A square field that is 100 feet long and 100 feet wide can yield 75 bushels of a certain crop. Suppose that the length of the field is increased by a certain amount, but the width is decreased by that same amount. Will the yield remain the same? Explain. **No. The yield is directly proportional to the area. The area of the original field is 10,000 square feet. After increasing the length by x feet and decreasing the width by x feet, the area becomes $(100 + x)(100 - x)$ square feet, or $(10,000 - x^2)$ square feet.**

Inclusion Strategies

LINGUISTIC LEARNERS Some students may see the statements of the three special products on page 434 as a maze of symbols that look virtually identical. Suggest that they copy each statement as given in the text and then follow it immediately with a sentence or two that describes the special product in their own words.

Use Teaching Transparency 44.

Activity Notes

This Activity provides students with a visual model for multiplying polynomials. In Parts 1 and 2, students use algebra tiles to multiply a binomial by a monomial and to multiply two binomials. In Part 3, they explore multiplications that yield two types of special products: the difference of two squares and a perfect-square trinomial. After completing this Activity, students should be able to use tiles to model multiplications of polynomials. They should also be able to find special products without the aid of the tiles.

CHECKPOINT ✓
4. $(2x + 2)(x + 4) = 2x^2 + 10x + 8$
Check:
$[2(3) + 2][(3) + 4] \stackrel{?}{=}$
$2(3)^2 + 10(3) + 8$
$(8)(7) \stackrel{?}{=} 18 + 30 + 8$
$56 = 56$

CHECKPOINT ✓
3. $(x + 3)(2x - 4) = 2x^2 + 2x - 12$
Check:
$[(4) + 3][2(4) - 4] \stackrel{?}{=}$
$2(4)^2 + 2(4) - 12$
$(7)(4) \stackrel{?}{=} 32 + 8 - 12$
$28 = 28$

CHECKPOINT ✓
3. $(x - 2)(x - 2) = x^2 - 4x + 4$
$(x - 3)(x - 3) = x^2 - 6x + 9$
Each simplified product is of the form $x^2 - 2ax + a^2$.
$(x - 5)(x - 5) = x^2 - 10x + 25$

☞ For illustrations of the algebra tiles, see Additional Answers.

CRITICAL THINKING
No, not all second-degree polynomials can be factored. A clever student might suggest that it could be done if the tiles could be cut into parts to represent fractions of tiles.

LESSON 9.2 **433**

ADDITIONAL EXAMPLE 1

Use algebra tiles to find each product. Check your results by substituting a value for x in each expression.

a. $(x+4)(x+1)$

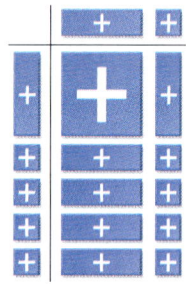

$(x+4)(x+1) = x^2 + 5x + 4$

b. $(x+4)(x-1)$

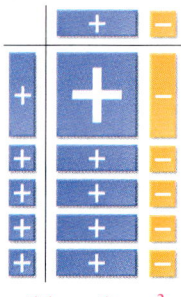

$(x+4)(x-1) = x^2 + 3x - 4$

Use Teaching Transparency 45.

ADDITIONAL EXAMPLE 2

Find the following products without using tiles:

a. $(2x+3)(2x-3)$
   $4x^2 - 9$

b. $(3x+1)(3x+1)$
   $9x^2 + 6x + 1$

c. $(x-10)(x-10)$
   $x^2 - 20x + 100$

EXAMPLE 1

Use algebra tiles to find each product. Check your results by substituting a value for x in each expression.

a. $(x+2)(x+4)$
b. $(x-2)(x+4)$

SOLUTION

a. $(x+2)(x+4) = x^2 + 6x + 8$
b. $(x-2)(x+4) = x^2 + 2x - 8$

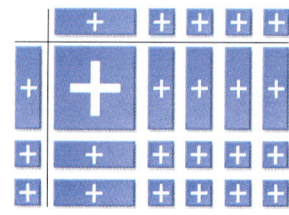

Let $x = 5$, for example.
$(5+2)(5+4) \stackrel{?}{=} 5^2 + 6(5) + 8$
$63 = 63$

Let $x = 3$, for example.
$(3-2)(3+4) \stackrel{?}{=} 3^2 + 2(3) - 8$
$7 = 7$

Special Products

For any real numbers a and b and for any expressions involving real numbers and variables:

$$(a+b)(a-b) = a^2 - b^2$$
$$(a+b)(a+b) = (a+b)^2 = a^2 + 2ab + b^2$$
$$(a-b)(a-b) = (a-b)^2 = a^2 - 2ab + b^2$$

EXAMPLE 2

Find the following products without using tiles:

a. $(x-6)(x+6)$
b. $(2x+6)(2x+6)$
c. $(2x-3)(2x-3)$

SOLUTION

a. Use the rule $(a+b)(a-b) = a^2 - b^2$, where $a = x$ and $b = 6$.
$$(x-6)(x+6) = x^2 - 6^2 = x^2 - 36$$

b. Use the rule $(a+b)^2 = a^2 + 2ab + b^2$, where $a = 2x$ and $b = 6$.
$$(2x+6)^2 = (2x)^2 + 2(2x)(6) + 6^2 = 4x^2 + 24x + 36$$

c. Use the rule $(a-b)^2 = a^2 - 2ab + b^2$, where $a = 2x$ and $b = 3$.
$$(2x-3)^2 = (2x)^2 - 2(2x)(3) + 3^2 = 4x^2 - 12x + 9$$

Enrichment

Describe the conditions under which each statement is true.

1. $(a+b)^2 = a^2 + b^2$
   a and b are real numbers such that $a = 0$ or $b = 0$.

2. $(a+b)^2 > a^2 + b^2$
   a and b are real numbers such that $a > 0$ and $b > 0$ or $a < 0$ and $b < 0$.

3. $(a+b)^2 < a^2 + b^2$
   a and b are real numbers such that $a > 0$ and $b < 0$ or $a < 0$ and $b > 0$.

Reteaching the Lesson

COOPERATIVE LEARNING Have students work in pairs. Each student should take a clean sheet of paper and draw an algebra-tile model for the product of two binomials. The student should record the multiplication and the product on a separate sheet of paper.

Now have partners trade the papers on which they drew their models. Each should write the multiplication and product for the other's model. When both are finished, they should check each other's answers and resolve any discrepancies.

Exercises

Communicate

1. Explain how you can use algebra tiles to simplify each product below.
   a. $(x+2)(x+3)$ **b.** $(x-4)(x-4)$ **c.** $(x+6)(x+6)$
2. Give an example of a product represented by an algebra-tile model in which you would remove neutral pairs in order to simplify it.
3. Explain how you can multiply $(x+7)(x-7)$ mentally.
4. Use an algebra-tile model to show why $(x+4)^2$ does not equal x^2+16.

Guided Skills Practice

5. Use algebra tiles to find each product. Check your results by substituting a value for x in each expression. **(EXAMPLE 1)**
   a. $(x+1)(2x+3)$ **b.** $(2x+1)(3x-3)$ **c.** $(x+2)(x-4)$
6. Find each product without using tiles. **(EXAMPLE 2)**
   a. $(x+5)(x-5)$ **b.** $(3x+2)(3x+2)$ **c.** $(2x-8)(2x-8)$
   x^2-25 $9x^2+12x+4$ $4x^2-32x+64$

Practice and Apply

Write a product of binomial factors for each model. Use the model to find the product. Check your results by substituting a value for x into each expression.

7. 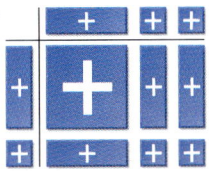 $(x+2)(x+1)$ 8. 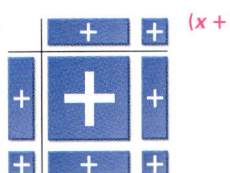 $(x+1)(x+1)$

9. 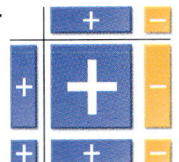 $(x-1)(x+1)$ 10.  $(x-3)(x-1)$

11. $(x-4)(x+2)$
12. $(2x-1)(x+2)$

11. 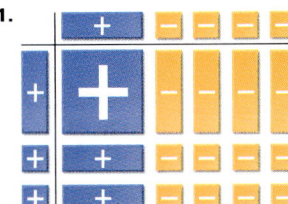 12.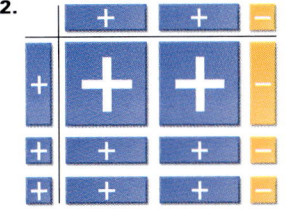

Assess

Selected Answers
Exercises 5–6, 7–59 odd

ASSIGNMENT GUIDE

In Class	1–6
Core	7–45 odd, 48–51
Core Plus	8–46 even, 47–51
Review	52–60
Preview	61–64

✎ Extra Practice can be found beginning on page 738.

5. a.
$2x^2+5x+3$

b.
$6x^2-3x-3$

c.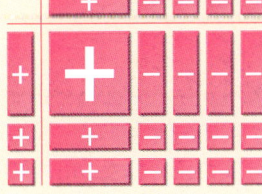
x^2-2x-8

LESSON 9.2

Technology

A calculator may be helpful in performing some of the calculations involved in the exercises. You may want to have students first do the exercises with paper and pencil and then use a calculator to check their answers.

Error Analysis

When squaring a binomial, students will often simply square each term. For instance, they may write $(x + 3)^2 = x^2 + 9$ and $(x - 5)^2 = x^2 - 25$. Suggest that they first write the multiplication in expanded form:

$$(x + 3)^2 = (x + 3)(x + 3)$$
$$(x - 5)^2 = (x - 5)(x - 5)$$

Encourage them to check their answers by substituting nonzero values for the variable(s).

13.

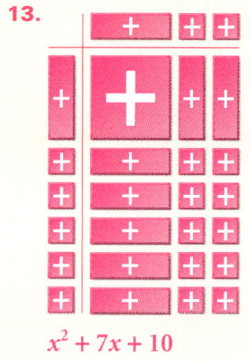

$x^2 + 7x + 10$

22. $4x^2 - 4$
23. $25x^2 - 1$
24. $x^2 - 16$
25. $x^2 + 2x + 1$
26. $x^2 - 36$
27. $9x^2 - 24x + 16$
28. $x^2 + 16x + 64$
29. $x^2 - 14x + 49$
30. $25x^2 + 90x + 81$
31. $4x + 8$
32. $12x + 42$
33. $5x + 50$
34. $4x - 16$
35. $3x + 24$
36. $-2x + 6$
37. $24x^2 - 32x$
38. $-3x^3 - 12x^2$
39. $x - x^2$
40. $x^2 - 2x$
41. $-x^2 + 5x$
42. $-4x^2 + 12x$

CHALLENGE

CONNECTION

APPLICATION

Model each product with tiles, and give the simplified product. Check by substituting a value for x.

13. $(x + 5)(x + 2)$
14. $(x + 2)(x - 1)$
15. $(2x + 1)(x + 1)$
16. $(x + 5)(x - 2)$
17. $(x + 2)(x + 1)$
18. $(3x + 5)(x - 7)$
19. $(x - 2)(4x - 3)$
20. $(x + 5)(x - 7)$
21. $(x - 5)(2x + 12)$

Find each product by using the rules for special products.

22. $(2x + 2)(2x - 2)$
23. $(5x - 1)(5x + 1)$
24. $(x + 4)(x - 4)$
25. $(x + 1)^2$
26. $(x + 6)(x - 6)$
27. $(3x - 4)^2$
28. $(x + 8)^2$
29. $(x - 7)^2$
30. $(5x + 9)^2$

Use the Distributive Property to find each product.

31. $4(x + 2)$
32. $6(2x + 7)$
33. $5(x + 10)$
34. $4(x - 4)$
35. $3(x + 8)$
36. $-2(x - 3)$
37. $8(3x^2 - 4x)$
38. $-3x(x^2 + 4x)$
39. $x(1 - x)$
40. $-x(2 - x)$
41. $x(-x + 5)$
42. $4x(3 - x)$

Determine whether each statement is true by substituting 10 and another value for x.

43. $x(2x) = 2x^2$ **True**
44. $x(2x + 1) = 2x^2 + 2$ **False**
45. $x(2x - 1) = 2x^2 - x$ **True**
46. $(x + 3)(x + 1) = 2x^2 + 3x + 3$ **False**

47. Explain why the statement $(x - 10)(x + 3) = (x - 10)(x + 4)$ is true for $x = 10$ but is false for $x \neq 10$. **When $x = 10$, $x - 10 = 0$. Thus the product is zero on both sides. When $x \neq 10$, the products will be different.**

GEOMETRY A landscaper is designing Mary's backyard. Mary wants a rectangular flower bed and lawn in her backyard, as shown below.

48. Write an expression for the area of the flower bed. $10(x)\,yd^2$
49. Write an expression for the area of the lawn. $(20)(10)\,yd^2$
50. Write an expression for the area of the backyard. $10(x + 20)\,yd^2$

51. **DISCOUNTS** A grocery store has pieces of dinnerware for sale in the housewares department. Glasses are $1.50 each, plates are $2.50 each, cups are $1.25 each, and saucers are $1.00 each. A 25% discount is given for purchases over $10.00. Suppose that a customer buys x glasses and y plates and receives the discount. Write an expression for the amount of the discount that the customer receives. $0.375x + 0.625y$

14.

$x^2 + x - 2$

The illustrations of the algebra tiles for Exercises 15–21 can be found in Additional Answers beginning on page 810.

15. $2x^2 + 3x + 1$
16. $x^2 + 3x - 10$
17. $x^2 + 3x + 2$
18. $3x^2 - 16x - 35$
19. $4x^2 - 11x + 6$
20. $x^2 - 2x - 35$
21. $2x^2 + 2x - 60$

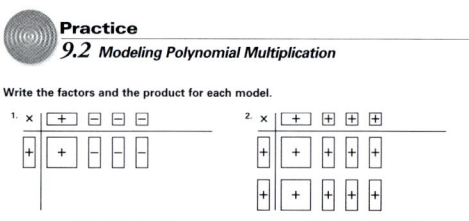

436 LESSON 9.2

Look Back

Solve each equation or inequality. (LESSON 6.5)

52. $|x - 2| = 5$
53. $|2x - 1| = -1$
54. $|x| \leq 5$ $-5 \leq x \leq 5$

52. $x = 7$ or $x = -3$
53. No solution

55. Solve this system of linear equations by graphing. (LESSON 7.1)
$\begin{cases} 6y = 8x - 2 \\ 2x + 3y = 2 \end{cases}$ $\left(\frac{1}{2}, \frac{1}{3}\right)$

56. Solve this system of linear equations by substitution. (LESSON 7.2)
$\begin{cases} x + y = 7 \\ 4x + 2y = 10 \end{cases}$ $(-2, 9)$

57. Solve this system of linear equations by elimination. (LESSON 7.3)
$\begin{cases} -4x + 3y = -1 \\ -8x - 6y = -10 \end{cases}$ $\left(\frac{3}{4}, \frac{2}{3}\right)$

Find each sum or difference. (LESSON 9.1)

58. $(x^2 + 2x) + (x + 1)$
    $x^2 + 3x + 1$
59. $(x^2 + 2x) - (x + 1)$
    $x^2 + x - 1$
60. $(2x^2 + x) - (2x^2 - x)$
    $2x$

Look Beyond

The zeros of the polynomial function $y = x^2 - 4$ are 2 and -2 because these values result in a value of 0 for y when they are substituted for x. Find the zeros of the following functions:

61. $y = x^2 - 9$ 3 and -3
62. $y = x^2 - 16$ 4 and -4
63. $y = x^2 - 25$ 5 and -5

64. **TECHNOLOGY** Graph the functions in Exercises 61–63. How can you identify the zeros on the graphs? Explain your reasoning.

Include this activity in your portfolio.

The first four rows of Pascal's triangle are shown below.

Row 0	1	Sum = 1
Row 1	1 1	Sum = 2
Row 2	1 2 1	Sum = ?
Row 3	1 3 3 1	Sum = ?

1. Complete rows 4, 5, and 6.
2. Compute the sum for each row.
3. What should be the sums for rows 7, 8, and 9?

WORKING ON THE CHAPTER PROJECT

You should now be able to complete Activity 1 of the Chapter Project on page 470.

Blaise Pascal 1623–1662

55.

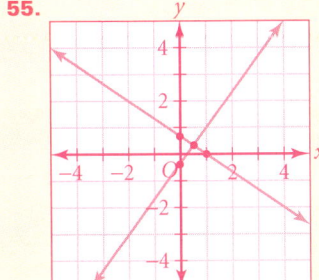

64. See Additional Answers for the graphs. The zeros are where the function crosses the x-axis, or where $y = 0$.

Prepare

NCTM PRINCIPLES & STANDARDS 1–4, 6, 8–10

LESSON RESOURCES

- Practice 9.3
- Reteaching Master 9.3
- Enrichment Master 9.3
- Cooperative-Learning Activity 9.3
- Lesson Activity 9.3
- Problem Solving/Critical Thinking Master 9.3
- Teaching Transparency 46

QUICK WARM-UP

Simplify.

1. $(6z)(4z)$ $24z^2$
2. $-2(w + 5)$ $-2w - 10$
3. $6q - 9 - 5q$ $q - 9$
4. $(k^2 - 9k) + (k^2 + 9)$ $2k^2 - 9k + 9$
5. $(d^2 - 4d) + (4d - 1)$ $d^2 - 1$

Also on Quiz Transparency 9.3

Teach

Why Explain to students that an *acronym* is a word formed from the initial letters of a series of words or from parts of the words. For example, *radar* is an acronym for *radio detecting and ranging*. Work with students to make a list of other familiar acronyms and the words from which they are taken. Then add *FOIL* to the list. Tell students that in today's lesson, they will learn how this new acronym may help them to multiply binomials.

9.3 Multiplying Binomials

Objectives
- Find products of binomials by using the Distributive Property.
- Find products of binomials by using the FOIL method.

Why The Distributive Property can be used to multiply a monomial and a binomial. It can also be used to multiply two binomials.

Mathematical software can find products like the one shown on the computer screen. To understand what such a product means and how the computer calculates it, you should first learn to work with products without the aid of a computer or calculator.

Using the Distributive Property

The Distributive Property can be used to show that the following is true:

$$(x + 3)(x + 2) = x^2 + 5x + 6$$

$(x + 3)(x + 2) = (x + 3)x + (x + 3)2$	*Distribute the first binomial to each term of the second binomial.*
$= x^2 + 3x + 2x + 6$	*Simplify by using the Distributive Property.*
$= x^2 + 5x + 6$	*Add like terms.*

EXAMPLE 1 Use the Distributive Property to show that $(x + 4)(x - 5) = x^2 - x - 20$ is true.

Alternative Teaching Strategy

USING MODELS Remind students that the formula $A = lw$ gives the area, A, of a rectangle with length l and width w. Show them how to use this fact to model a multiplication of polynomials.

On the board or overhead, draw the rectangle shown at right. Ask students to use the formula above to write an expression for the area.

$A = (x + 3)(x + 1)$

Divide each dimension of the rectangle into parts. Ask students to write an expression for the area of each small rectangle. By adding the four expressions, you can show that the equation $(x + 3)(x + 1) = x^2 + 4x + 3$ is true.

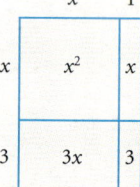

438   LESSON 9.3

● **SOLUTION**

$(x + 4)(x - 5)$ *Given*
$= (x + 4)x + (x + 4)(-5)$ *Distributive Property*
$= x^2 + 4x - 5x - 20$
$= x^2 - x - 20$

TRY THIS Use the Distributive Property to show that $(x - 1)(x + 9) = x^2 + 8x - 9$ is true.

Activity
The Distributive Property and Algebra Tiles

You will need: algebra tiles (or paper and pencil for sketching)

1. The algebra-tile model at right represents the product $(x + 3)(x + 2)$. Explain the relationship between the tile model and the use of the Distributive Property to find the same product on the previous page.

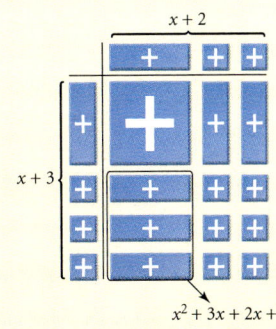

$x^2 + 3x + 2x + 6$

2. Simplify the product $(x + 4)(x + 5)$ by using the Distributive Property. Then model the product by using algebra tiles. Explain the relationship between the tile model and the appropriate use of the Distributive Property in the multiplication.

CHECKPOINT ✓ 3. Write a procedure for quickly finding a product like the one in Step 2.

4. Repeat Step 2 for the product $(x + 4)(x - 5)$. Does your procedure from Step 2 work for this product, or does it need revising? Explain.

The FOIL Method

A popular mnemonic (memory) device for multiplying two binomials mentally, which you may already know, is called the **FOIL method**.

- Multiply the **F**irst terms.
- Multiply the **O**utside terms.
- Multiply the **I**nside terms. Add the outside and inside products.
- Multiply the **L**ast terms.

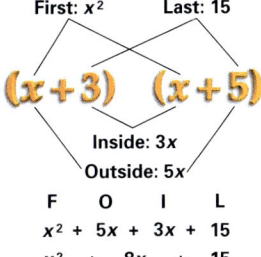

First: x^2 Last: 15
Inside: $3x$
Outside: $5x$

F O I L
$x^2 + 5x + 3x + 15$
$x^2 + 8x + 15$

Interdisciplinary Connection

CONSUMER ECONOMICS A rectangular garden is 30 feet wide and 40 feet long. It is to be surrounded by a path of uniform width. Let p represent the width in feet of the path. Write a simplified expression for the total area of the garden and the path and for the area of the path alone.
$4p^2 + 140p + 1200; 4p^2 + 140p$

Suppose that the path will be covered with gravel to a depth of 3 inches, and that the gravel costs $2 per cubic foot. What is total cost of the gravel if the path is to be 2 feet wide? 4 feet wide? $148; $312

Inclusion Strategies

VISUAL LEARNERS Suggest that students rewrite a given multiplication of binomials by using a pencil of a different color for each binomial. They should then use the same colors when applying the Distributive Property. For instance, this is how they might write $(c + 4)(c + 5)$.

$\underbrace{(c + 4)}_{red} \underbrace{(c + 5)}_{blue} = \underbrace{(c + 4)}_{red} \underbrace{c}_{blue} + \underbrace{(c + 4)}_{red} \underbrace{5}_{blue}$

ADDITIONAL EXAMPLE 1

Use the Distributive Property to show that
$(x + 3)(x - 8) = x^2 - 5x - 24$
is true.

$(x + 3)(x - 8)$
$= (x + 3)x - (x + 3)8$
$= x^2 + 3x - 8x - 24$
$= x^2 - 5x - 24$

TRY THIS
$(x - 1)(x + 9)$
$= (x - 1)x + (x - 1)9$
$= x^2 - x + 9x - 9$
$= x^2 + 8x - 9$

Activity Notes

This Activity leads students to see the relationship between applying the Distributive Property and using algebra tiles to find a product of binomials. Students should note that the tiles representing $(x + 3)(x + 2)$ can be split into two groups: one representing $(x + 3)x$ and the other representing $(x + 3)2$.

CHECKPOINT ✓
3. Sample answer: Multiply the first terms of each factor, add the product of the first term of the first factor and the second term of the second factor, add the product of the second term of the first factor and the first term of the second factor, and add the product of the second terms of each factor.

Use Teaching Transparency 46.

LESSON 9.3 **439**

ADDITIONAL EXAMPLE 2

Use the FOIL method to find each product.

a. $(2s + 1)(5s + 3)$
   $10s^2 + 11s + 3$

b. $(4g - 7)(2g + 3)$
   $8g^2 - 2g - 21$

c. $(3m - 5)(3m + 5)$
   $9m^2 - 25$

CRITICAL THINKING
$(a - b)(a + b)$
$= a^2 + ab - ab - b^2$
$= a^2 - b^2$

ADDITIONAL EXAMPLE 3

A rectangular poster is twice as long as it is wide. It is mounted on a piece of cardboard with a 3-inch border on all sides. The area of the border alone is 306 square inches. **What are the length and width of the poster?**
30 inches, 15 inches

EXAMPLE 2

Use the FOIL method to find each product.

a. $(3x + 2)(5x + 1)$ b. $(x + 9)(2x - 4)$ c. $(5m - 2)(5m + 2)$

● SOLUTION

a. $(3x + 2)(5x + 1) = 15x^2 + 3x + 10x + 2$
   $= 15x^2 + 13x + 2$

b. $(x + 9)(2x - 4) = 2x^2 - 4x + 18x - 36$
   $= 2x^2 + 14x - 36$

c. $(5m - 2)(5m + 2) = 25m^2 + 10m - 10m - 4$
   $= 25m^2 - 4$

CRITICAL THINKING In Lesson 9.2 you learned a rule for simplifying products like the one in part **c** of Example 2. Use the FOIL method to explain why that rule works.

EXAMPLE 3

Cruz's Frame Shop makes a mat by cutting out the inside of a rectangular mat board. **Use the measurements from the diagram at right to find the length and width of the original mat if the area of the mat board is 148 square inches.**

Creating a mat to frame a picture requires precise cutting.

● SOLUTION

Find the areas of the outer and inner rectangles formed by the mat.

Area of the outer rectangle	Area of the inner rectangle
$(2x - 1)(x + 6)$	$(x + 2)(2x - 5)$
$2x^2 + 12x - x - 6$	$2x^2 - 5x + 4x - 10$
$2x^2 + 11x - 6$	$2x^2 - x - 10$

Substitute each area into the formula below and then solve for x.

Area of outer rectangle − Area of inner rectangle = Area of the mat
$2x^2 + 11x - 6$ − $(2x^2 - x - 10)$ = 148
$2x^2 + 11x - 6$ − $2x^2 + x + 10$ = 148
$12x + 4$ = 148
$12x$ = 144
x = 12

The length of the original rectangle was $2x - 1$. Substitute 12 for x.

$$2x - 1 = 2(12) - 1 = 23$$

The width was $x + 6$. Substitute 12 for x.

$$x + 6 = 12 + 6 = 18$$

The original mat measured 23 inches by 18 inches.

Practice

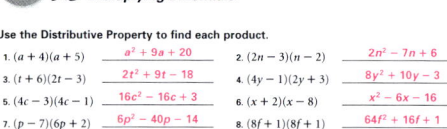

Enrichment

Have students simplify these expressions by applying the Distributive Property.

1. $(a + 1)(a^2 + a + 1)$ $a^3 + 2a^2 + 2a + 1$
2. $(a + 1)(a^2 - a + 1)$ $a^3 + 1$
3. $(a + 1)(a^2 + a - 1)$ $a^3 + 2a^2 - 1$
4. $(a + 1)(a^2 - a - 1)$ $a^3 - 2a - 1$
5. $(a - 1)(a^2 + a + 1)$ $a^3 - 1$
6. $(a - 1)(a^2 - a + 1)$ $a^3 - 2a^2 + 2a - 1$
7. $(a - 1)(a^2 + a - 1)$ $a^3 - 2a + 1$
8. $(a - 1)(a^2 - a - 1)$ $a^3 - 2a^2 + 1$

Reteaching the Lesson

USING COGNITIVE STRATEGIES Show how to multiply two binomials by using the familiar vertical format for multiplying two numbers.

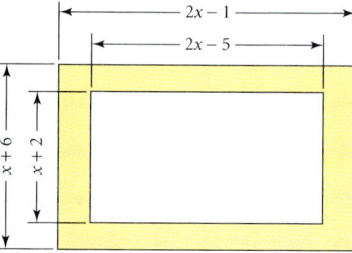

Have students practice this method for parts **a**, **b**, and **c** of Example 2.

440   LESSON 9.3

Exercises

Communicate

1. Show how to use algebra tiles to model each product.
   a. $(x+1)(x+2)$ b. $(x-1)(x-2)$

2. Show how to use the Distributive Property to model each product.
   a. $(x+1)(x+2)$ b. $(x-1)(x-2)$

3. Describe how to find $(2x+3)(x-4)$ by using the FOIL method.

4. Use algebra tiles to model the expressions $(x+2)^2$ and $x^2 + 2^2$. How are the two expressions different?

Guided Skills Practice

5. Use the Distributive Property to show that $(x+5)(x-3) = x^2 + 2x - 15$ is true. **(EXAMPLE 1)**

6. Use the FOIL method to find each product. **(EXAMPLE 2)**
   a. $(x+3)(x+6)$ b. $(a-4)(a-9)$ c. $(m-3)(m+9)$

7. **GEOMETRY** The area of the shaded part of the figure is 156. Find the length and width of the outer rectangle. **(EXAMPLE 3)**
   length: 24; width: 23

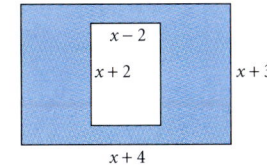

Practice and Apply

Use the Distributive Property to find each product.

8. $(x+2)(x+5)$ 9. $(a-3)(a+4)$ 10. $(b-1)(b-3)$
11. $(y+3)(y-2)$ 12. $(c+5)(c+5)$ 13. $(d+5)(d-5)$
14. $(2y)(y+3)$ $2y^2 + 6y$ 15. $(w^2)(w+1)$ $w^3 + w^2$ 16. $(3b)(b+3)$ $3b^2 + 9b$

Use the FOIL method to find each product.

17. $(3y-2)(y-1)$ 18. $(5p+3)(p+1)$ 19. $(2q-1)(2q+1)$
20. $(2x+5)(2x-3)$ 21. $(4m+1)(5m-3)$ 22. $(2w-9)(3w-8)$
23. $(3x-5)(3x-5)$ 24. $(7s+2)(2s-3)$ 25. $\left(y-\frac{1}{2}\right)\left(y+\frac{1}{2}\right)$
26. $\left(y-\frac{1}{3}\right)\left(y+\frac{5}{9}\right)$ 27. $(c^2+1)(c^2+2)$ 28. $(2a^2+3)(2a^2+3)$
29. $(a+c)(a+2c)$ 30. $(p+q)(p+q)$ 31. $(a^2+b)(a^2-b)$
32. $(x^2+y)(x+y)$ 33. $(c+d)(2c+d)$ 34. $(1.2m+5)(0.8m-4)$

Selected Answers

6. a. $x^2 + 9x + 18$
   b. $a^2 - 13a + 36$
   c. $m^2 + 6m - 27$

8. $x^2 + 7x + 10$
9. $a^2 + a - 12$
10. $b^2 - 4b + 3$
11. $y^2 + y - 6$
12. $c^2 + 10c + 25$
13. $d^2 - 25$
17. $3y^2 - 5y + 2$
18. $5p^2 + 8p + 3$
19. $4q^2 - 1$
20. $4x^2 + 4x - 15$
21. $20m^2 - 7m - 3$
22. $6w^2 - 43w + 72$
23. $9x^2 - 30x + 25$
24. $14s^2 - 17s - 6$
25. $y^2 - \frac{1}{4}$
26. $y^2 + \frac{2}{9}y - \frac{5}{27}$
27. $c^4 + 3c^2 + 2$
28. $4a^4 + 12a^2 + 9$
29. $a^2 + 3ac + 2c^2$
30. $p^2 + 2pq + q^2$
31. $a^4 - b^2$
32. $x^3 + x^2y + xy + y^2$
33. $2c^2 + 3cd + d^2$
34. $0.96m^2 - 0.8m - 20$

5. $(x+5)(x-3) = (x+5)(x) + (x+5)(-3)$
   $= x^2 + 5x - 3x - 15$
   $= x^2 + 2x - 15$

Assess

Selected Answers
Exercises 5–7, 9–51 odd

ASSIGNMENT GUIDE

In Class	1–7
Core	9–37 odd
Core Plus	8–34 even, 35–38
Review	39–52
Preview	53–55

✎ Extra Practice can be found beginning on page 738.

Student Technology Guide

9.3 Multiplying Binomials

If two polynomial functions have identical graphs, the functions are equivalent. You can use this idea and a graphics calculator to check your answer when multiplying binomials.

Example: Determine whether $(x+3)(x-2)$ is equal to $x^2 + x - 6$.
- Press [Y=].
- For Y₁, enter (X,T,θ,n + 3) (X,T,θ,n − 2).
- For Y₂, enter X,T,θ,n x^2 + X,T,θ,n − 6.
- Press [GRAPH].

It looks as if the calculator is displaying only one graph. However, there are actually two identical graphs, which means that the two expressions are equivalent.

To confirm this, you can press [TRACE] and use the up ▲ and down ▼ arrow keys to toggle between the graphs. Although the equation at the upper left of the screen changes, the trace point does not move.

Find each product. Check your answer by graphing.

1. $(x+5)(x+3)$ 2. $(x-1)(x+4)$ 3. $(x-2)(2x-7)$ 4. $(3x+1)(2x-3)$
   $x^2 + 8x + 15$ $x^2 + 3x - 4$ $2x^2 - 11x + 14$ $6x^2 - 7x - 3$

Some calculators, such as the TI-92, can multiply expressions involving variables. To find the product $(6x-4)(3x+12)$, you can use the following key sequence:

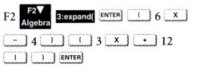

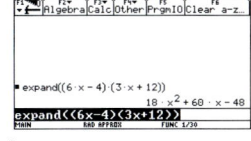

Use a graphics calculator to find each product.

5. $(x+9)(x-5)$ 6. $(2x+7)(2x-7)$ 7. $(4x+1)(9x+1)$ 8. $\left(\frac{2}{3}x-3\right)\left(\frac{1}{5}x+7\right)$
   $x^2 + 4x - 45$ $4x^2 - 49$ $36x^2 + 13x + 1$ $\frac{2}{15}x^2 + 4\frac{1}{15}x - 21$

LESSON 9.3 441

Technology

A calculator may be helpful in performing some of the calculations in the exercises. You may want to have students first do the exercises with paper and pencil and then use a calculator to check their answers.

Error Analysis

When multiplying two binomials, many students will multiply only the first and last terms. For instance, given $(t+4)(t-7)$, students may write $t^2 - 28$. Remind them that the four letters in the mnemonic *FOIL* indicate that there must be four multiplications. Encourage them to check their answers by substituting one or two nonzero values for the variable(s).

In Exercises 53–55, students investigate the relationship between the factored form of a quadratic expression and the *x*-intercepts of the related quadratic function. You may wish to have students use a graphics calculator for these exercises. The zeros of quadratic functions will be discussed in greater detail in Lesson 9.8.

CONNECTIONS

36. The area of the square is larger, by 1 square unit

37. The area of the square is larger, by 1 square unit

APPLICATION

42. $\frac{1}{2}$; parallel

43. $\frac{1}{2}$, -2; perpendicular

44. undefined, 0; perpendicular

46. (4, 1)
47. No solution
48. $\left(\frac{55}{13}, \frac{-37}{13}\right)$

53. $x = -3$ and $x = 2$
54. $x = -2$
55. $x = -2$ and $x = 2$

---

35. **GEOMETRY** Find the area in terms of *y* of a square rug whose sides are $y + 6$ units long. $y^2 + 12y + 36$ square units

36. **GEOMETRY** Which has the larger area, a square with sides that are $x + 1$ units long or a rectangle with a length of $x + 2$ units and a width of *x* units? What is the difference in areas?

37. **GEOMETRY** Which has a greater area, a square with sides that are $x - 1$ units long or a rectangle with a length of *x* units and a width of $x - 2$ units? What is the difference in areas?

38. **CONSTRUCTION** An architect is designing a rectangular pavilion that has a length of $x + 8$ units and a width of $x - 4$ units. Draw a diagram and label the dimensions. Write an expression representing the area covered by the pavilion.

Look Back

Find the slope of each line. *(LESSON 5.4)*

39. $4x + 3y = 12$ $-\frac{4}{3}$
40. $y = 4x$ 4
41. $3y - 7x = 10$ $\frac{7}{3}$

State the slope of each line, and indicate whether each pair of lines is parallel, perpendicular, or neither. *(LESSON 5.6)*

42. $2y = x + 6$
    $2y = x - 3$
43. $x - 2y = 4$
    $2x + y = 1$
44. $x = 7$
    $y = 7$

45. Write the equation of the line through $(-2, -1)$ and parallel to $y = -2x - 4$. *(LESSON 5.6)* $y = -2x - 5$

Solve each system of equations by graphing. Check your solutions by using substitution. *(LESSON. 7.1)*

46. $\begin{cases} 2x + y = 9 \\ x - 2y = 2 \end{cases}$
47. $\begin{cases} x + y = 4 \\ x + y = 10 \end{cases}$
48. $\begin{cases} 3x + 2y = 7 \\ x + 5y = -10 \end{cases}$

49. Find the two-digit number whose tens digit is 2 less than its units digit if the original number is 4 times the sum of the digits. *(LESSON 7.6)* 24

Simplify each expression. *(LESSON 8.3)*

50. $\left(\frac{x^4}{y^2}\right)^3$ $\frac{x^{12}}{y^6}$
51. $\left(\frac{c^3 b^4}{b^2}\right)^2$ $c^6 b^4$
52. $\left[\frac{x^3 y^5}{(x^2)^3}\right]^4$ $\frac{y^{20}}{x^{12}}$

Look Beyond

At what point does the graph of each function cross the *x*-axis?

53. $y = (x + 3)(x - 2)$
54. $y = (x + 2)^2$
55. $y = x^2 - 4$

38.

$A = (x + 8)(x - 4)$
$A = x^2 + 4x - 32$

442 LESSON 9.3

9.4 Polynomial Functions

Objectives
- Define polynomial functions.
- Solve problems involving polynomial functions.

Why Surface area and volume formulas are two examples of polynomial functions that can be used in real-world situations such as manufacturing.

Shari, the production manager at Power Wash Inc., wants to change the size of the Super Suds box.

You have already worked with several types of functions, including linear functions. Linear functions are a specific type of **polynomial function**, which can be used to help solve a wide variety of problems.

Polynomial Function

A *polynomial function* is a function that consists of a monomial, or the sum or difference of two or more monomials.

APPLICATION
PACKAGING

Consider the following problem: Power Wash Incorporated manufactures a variety of laundry detergents. The production manager notices that the production costs of Super Suds laundry detergent are rising. Instead of raising prices, the production manager considers lowering the production costs by changing the size of the packaging.

The Super Suds box currently has a base with a length of 20 centimeters and a width of 10 centimeters. The height of the current box is 30 centimeters.

Alternative Teaching Strategy

USING TECHNOLOGY Students can use spreadsheet software to test the dimensions for the box in the Activity on page 444. A partial spreadsheet is shown at right. It can be created as follows:

Enter **0** in cell **B2**. Enter **=B2+.1** in cell **B3**.
Highlight cells **B3–B100** and choose **FILL DOWN**.
Enter **=2*B2** in cell **A2**.
Highlight cells **A2–A100** and choose **FILL DOWN**.
Enter **30** in cell **C2**.
Highlight cells **C2–C100** and choose **FILL DOWN**.
Enter **=A2*B2*C2** in cell **D2**.

Highlight cells **D2–D100** and choose **FILL DOWN**.
Enter **=A2*B2+A2*C2+B2*C2** in cell **E2**.
Highlight cells **E2–E100** and choose **FILL DOWN**.

Super Suds Box				
A	B	C	D	E
Length	Width	Height	Volume	Surface
0.00	0.00	30.00	0.00	0.00
0.20	0.10	30.00	0.60	18.04
0.40	0.20	30.00	2.40	36.16
0.60	0.30	30.00	5.40	54.36
0.80	0.40	30.00	9.60	72.64
1.00	0.50	30.00	15.00	91.00

Prepare

NCTM PRINCIPLES & STANDARDS 1–6, 8–10

LESSON RESOURCES
- Practice 9.4
- Reteaching Master 9.4
- Enrichment Master 9.4
- Cooperative-Learning Activity 9.4
- Lesson Activity 9.4
- Problem Solving/Critical Thinking Master 9.4

QUICK WARM-UP

Complete the table for each function.

1.

$y = 2x + 1$	
x	y
−3	−5
0	1
1	3
7	15

2.

$y = 7 − x$	
x	y
−6	13
−4	11
2	5
8	−1

Also on Quiz Transparency 9.4

Teach

Why Ask students why a manufacturer might want to change the volume or surface area of a product's packaging. Responses might include decreasing the cost of the packaging, making the package more visually appealing, making the package easier to hold, or offering the product in an increased variety of sizes. Tell students that a polynomial function can be used to evaluate the effects of changing the volume or surface area of an object.

LESSON 9.4 **443**

ADDITIONAL EXAMPLE 1

A box is shaped like a right rectangular prism. Its length is 28 centimeters, its width is 11 centimeters, and its height is 35 centimeters. **Find the volume of the box.**
10,780 cubic centimeters

TRY THIS
$V = 9375$ cubic centimeters

Activity Notes

In this Activity, students investigate the effect on the volume and surface area of a box as the width of the base is varied. After completing the Activity, students should appreciate that a small change in the width can cause a significant change in the volume and surface area.

CHECKPOINT ✓
4. As the width increases, the volume and surface area both increase. A width of 10 centimeters gives the original volume and surface area of the box.

ADDITIONAL EXAMPLE 2

A container of oatmeal is shaped like a cylinder with radius 2 inches and height 7 inches. The manufacturer wants to increase the volume of the container 25% by increasing the height, but maintaining the same radius. **Find the volume of the original container, the new volume, and the height of the new container.**

orig. vol.: ≈88.0 cubic inches;
new vol.: ≈110.0 cubic inches;
new height: ≈8.75 inches

See Keystroke Guide, page 782.

444 LESSON 9.4

EXAMPLE 1

APPLICATION: PACKAGING

Find the volume of the current Super Suds box.

◯ SOLUTION

First write an equation that expresses the volume of the Super Suds box.

You can use the formula $V = Bh$, where B is the area of the base and h is the height, to find the volume, V, of the Super Suds box. You know that $B = lw$, so $V = lwh$.

You can substitute the current dimensions of the box into this formula to find the volume.

$$V = lwh$$
$$V = 20 \cdot 10 \cdot 30$$
$$V = 6000$$

The volume of the current Super Suds box is 6000 cubic centimeters.

TRY THIS Suppose that a box has a length of 25 centimeters, a width of 15 centimeters, and a height of 25 centimeters. Find the volume of the box.

Activity
Exploring Volume and Surface Area

You will need: graphics calculator (optional)

1. Find the surface area of the original Super Suds box, using the dimensions given on the previous page.

2. If the production manager wants to reduce the volume of the original box by 10%, what would the new volume be? If the production manager wants to reduce the surface area of the original box by 10%, what would the new surface area be?

3. The production manager decides to reduce the size of the rectangular base, keeping the same height and keeping the length twice the width. Explain why you can use the following functions, where x is the width of the base, to represent the volume and surface area of the Super Suds box.

$$V = 2x \cdot x \cdot 30 \qquad S = 2(2x \cdot x) + 2(2x \cdot 30) + 2(x \cdot 30)$$
$$= 60x^2 \qquad\qquad\quad = 4x^2 + 120x + 60x$$
$$\qquad\qquad\qquad\qquad\quad = 4x^2 + 180x$$

CHECKPOINT ✓
4. Examine the table at right, where X is width, Y1 is volume, and Y2 is surface area. Does changing the width of the base have a similar effect on both the volume and the surface area? What width gives the original volume of the box? the original surface area?

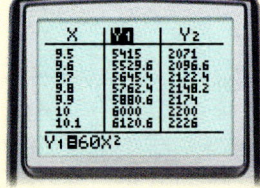

5. Suppose that the production manager decides to reduce the width of the base to 9.3 centimeters. What would be the volume and surface area of the new Super Suds box? What percent reduction in volume is this? What percent reduction in surface area is this?

Inclusion Strategies

TACTILE LEARNERS Some students will perceive the tables of numbers generated on these pages as little more than number-crunching. To give some meaning to the numbers, provide students with poster board or other heavy paper and have them create models of the original boxes and the new boxes with the altered dimensions.

EXAMPLE ❷

APPLICATION
PACKAGING

The manufacturer of a powdered drink mix wants to increase the volume of the container by 20% by increasing the height. The current container is a cylinder with a radius of 6 centimeters and a height of 10 centimeters. **Find the volume of the original container, the new volume, and the height of the new container.**

● **SOLUTION**

The formula for volume of a cylinder is $V = Bh$, or $V = \pi r^2 h$. First find the volume of the original container.

$$V = \pi r^2 h$$
$$= \pi \cdot 6^2 \cdot 10$$
$$= 360\pi$$
$$\approx 1131$$

The volume of the original container is approximately 1131 cubic centimeters.

Find the new volume after a 20% increase.

$$\text{New volume} = 1.20 \cdot 1131$$
$$\approx 1357.2$$

The volume after a 20% increase would be approximately 1357.2 cubic centimeters.

You can use a graphics calculator to make a table for the function $V = 36\pi h$, where h is the height of the container. Substitute x for h and enter $y = 36\pi x$ into your graphics calculator.

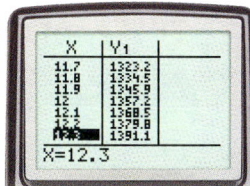

The height that increases the volume of the container by approximately 20% is 12 centimeters.

TRY THIS

Suppose that the container in Example 2 currently has a radius of 8 centimeters and a height of 10 centimeters and that the manufacturer wants to increase the volume by 25%. Find the volume of the original container, the new volume, and the height of the new container.

An **identity** is an equation that is true for all values of the variable or variables.

EXAMPLE ❸

TECHNOLOGY
GRAPHICS CALCULATOR

Show that the equations $y = (x + 2)^3$ and $y = x^3 + 6x^2 + 12x + 8$ are equivalent for integer values of x from −3 to 3. That is, show that $(x + 2)^3 = x^3 + 6x^2 + 12x + 8$ is an identity.

● **SOLUTION**

Build a table of values for the functions. Enter Y1 = (X+2)^3 and Y2 = X^3 + 6X^2 + 12X + 8 into a graphics calculator, and compare the values in the table.

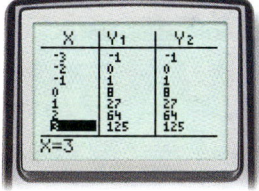

Since the columns for Y1 and Y2 have the same values, the functions are equivalent for integer values from −3 to 3.

Enrichment

Provide students with several empty product containers that are shaped like right rectangular prisms and cylinders. Have students work in groups of three or four. Tell each group to choose one container that will be considered their original container. Have them design a set of three containers—the original container, one with a volume 20% less than the original, and one with a volume 20% greater than the original—by varying only one dimension of the original container.

Reteaching the Lesson

USING VISUAL STRATEGIES

Review the definition of function. Ask students to name the types of functions they have studied so far in this book. Work with them to determine the relationships among these functions by creating a schematic like the one at right.

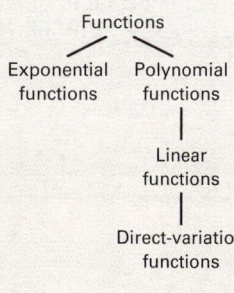

TRY THIS
original volume = 640π, or $\approx$ 2011 cubic centimeters, new volume = 800π, or $\approx$ 2513 cubic centimeters, new height = $12\frac{1}{2}$ centimeters

ADDITIONAL
EXAMPLE ❸

Show that $f(x) = (x + 3)^3$ and $g(x) = x^3 + 9x^2 + 27x + 27$ name the same function over the domain {−3, −2, −1, 0, 1, 2, 3}. That is, show that $(x + 3)^3 = x^3 + 9x^2 + 27x + 27$ is true for integer values of x from −3 to 3 inclusive.

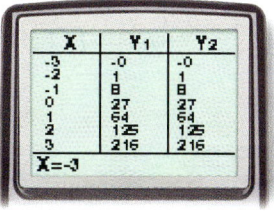

See Keystroke Guide, page 782.

Teaching Tip

TECHNOLOGY If you are using a TI-82 or TI-83 graphics calculator, you can create the table shown in Example 3 as follows: Press [Y=] and enter (X+2)^3 next to Y1, and X^3+6X^2+12X+8 next to Y2. Then press [2nd] [WINDOW] to access the **TABLE SETUP** screen, and choose the settings shown below:

TblStart= −3 (or TblMin= −3)
ΔTbl=1
Indpnt: **Auto** Ask
Depend: **Auto** Ask

Press [2nd] [GRAPH] to display the table.

LESSON 9.4 **445**

Assess

Selected Answers
Exercises 4–6, 7–33 odd

ASSIGNMENT GUIDE

In Class	1–6
Core	7–28
Core Plus	7–28
Review	29–34
Preview	35–37

✎ Extra Practice can be found beginning on page 738.

Mid-Chapter Assessment for Lessons 9.1 through 9.4 can be found on page 114 of the *Assessment Resources.*

Technology
A graphics calculator may be helpful in creating the tables for Exercises 7–18 and 23–28.

Error Analysis
Students who have difficulty with Exercises 7–18 may be evaluating the expressions incorrectly. It may help to review the order of operations with them.

Practice

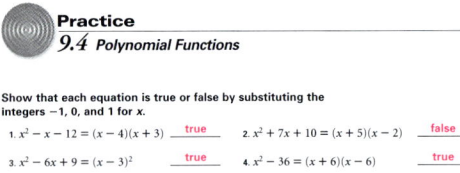

Practice
9.4 Polynomial Functions

Show that each equation is true or false by substituting the integers −1, 0, and 1 for x.
1. $x^2 − x − 12 = (x − 4)(x + 3)$ ____true____
2. $x^2 + 7x + 10 = (x + 5)(x − 2)$ ____false____
3. $x^2 − 6x + 9 = (x − 3)^2$ ____true____
4. $x^2 − 36 = (x + 6)(x − 6)$ ____true____
5. $x^2 + 5x + 25 = (x + 5)^2$ ____false____
6. $4x^2 − 16t^2 = (2x − 4t)(2x + 4t)$ ____false____

Write a function for the indicated measurement of each geometric solid.
7. the surface area of a cube with an edge $3x$ centimeters long
   $S = 54x^2$
8. the volume of a cube with an edge x inches long
   $V = x^3$
9. the surface area of a rectangular solid with a base length of 5 inches and a base width of 3 inches
   $S = 16x + 30$
10. the volume of a rectangular solid with a base length of 4.5 centimeters and a base width of 3.2 centimeters
    $V = 14.4x$
11. the surface area of a cylinder with a height of 7 inches and a radius of x inches
    $S = 2\pi x^2 + 14\pi x$, or $S = 2\pi x(x + 7)$
12. the volume of a cylinder with a height of 1.3 meters and a diameter of x meters
    $V = 0.325\pi x^2$

446 LESSON 9.4

Exercises

Communicate

1. Define a polynomial function and give two examples.
2. Explain how to write the functions for the volume and surface area of a cube with an edge length of x.
3. Explain how to use a graphics calculator table to show that $x^2 − 4 = (x + 2)(x − 2)$ is true for integer values of x from −3 to 3.

Guided Skills Practice

4. Suppose that the base of a box has a length of 20 centimeters and a width of 15 centimeters and that the box is 25 centimeters high. Find the volume of the box. **(EXAMPLE 1)** 7500 cm³

APPLICATION

5. original volume ≈ 6283 cm³; new volume ≈ 7540 cm³; new height ≈ 24 cm

5. **PACKAGING** The manufacturer of a powdered drink mix wants to increase the volume of the container by 20% by increasing the height. The current container is a cylinder with a radius of 10 centimeters and a height of 20 centimeters. Find the volume of the original container, the new volume, and the height of the new container. **(EXAMPLE 2)**

6. **TECHNOLOGY** Use a graphics calculator to show that the functions $y = x^3 + 9x^2 + 27x + 27$ and $y = (x + 3)^3$ are equivalent for integer values of x from −3 to 3. That is, show that $(x + 3)^3 = x^3 + 9x^2 + 27x + 27$ is an identity. **(EXAMPLE 3)**

Practice and Apply

Create a table to verify that each equation is an identity by substituting integers from −3 to 3 for x.

7. $x^2 + 5x + 6 = (x + 2)(x + 3)$
8. $(x − 4)(x − 2) = x^2 − 6x + 8$
9. $x^2 − 5x + 6 = (x − 2)(x − 3)$
10. $(x + 7)(x − 1) = x^2 + 6x − 7$
11. $x^2 − 9 = (x − 3)(x + 3)$
12. $(x − 6)(x + 4) = x^2 − 2x − 24$
13. $x^3 − 8 = (x − 2)(x^2 + 2x + 4)$
14. $(x + 5)(x + 3) = x^2 + 8x + 15$
15. $x^3 + 12x^2 + 48x + 64 = (x + 4)^3$
16. $(x − 6)(x − 7) = x^2 − 13x + 42$
17. $x^2 − 6x + 9 = (x − 3)^2$
18. $(x + 5)^2 = x^2 + 10x + 25$

CONNECTION

GEOMETRY Write the indicated equation for each geometric solid.

19. the volume of a cube with an edge of $2x$ centimeters $V = (2x)^3 = 8x^3$ cm³
20. the surface area of a cube with an edge of x centimeters long $S = 6x^2$ cm²
21. the surface area of a rectangular solid with a base length of 7 inches, a base width of 3 inches, and a height of x inches $S = (20x + 42)$ cm²

22. $V = 5\pi x^2$ m³

22. the volume of a cylinder with a height of 5 meters and a radius of x meters

 internetconnect

GO TO: go.hrw.com
KEYWORD: MA1 Regression

In this activity, students match data points to polynomial curves. Students should discover that the higher the degree of the polynomial, the better the fit of the curve to the data points.

The answers to Exercises 6–18 and 37 can be found in Additional Answers beginning on page 810.

APPLICATION

SPORTS Sports balls come in different sizes. Complete the table for each radius given. Round your answers to the nearest tenth.

	Radius r	Circumference $C = 2\pi r$	Surface area $S = 4\pi r^2$	Volume $V = \frac{4\pi r^3}{3}$
23.	18 inches	113.1 in	4071.5 in²	24,429.0 in³
24.	8 inches	50.3 in	804.2 in²	2144.7 in³
25.	5 inches	31.4 in	314 in²	523.6 in³
26.	2 inches	12.6 in	50.3 in²	33.5 in³
27.	1 inch	6.3 in	12.6 in²	4.2 in³
28.	0.5 inch	3.1 in	3.1 in²	0.5 in³

Look Back

Simplify. (LESSONS 2.2, 2.3, AND 2.4)

29. $-2(-3) + 4(-8)$
    −26
30. $-11(7) - (-12)(-3)$
    −113
31. $22(-2) + (-9)(-7)$
    19

Write an equation for the line containing each pair of points.
(LESSON 5.5)

32. $(2, 3), (-2, 3)$
    $y = 3$
33. $(0, -1), (4, 3)$
    $y = x - 1$
34. $(-1, 1), (-2, 2)$
    $y = -x$

Look Beyond

TECHNOLOGY Use a graphics calculator to graph each set of polynomial functions on the same coordinate plane. Describe the similarities and differences.

35. $y = x$, $y = x^3$, $y = x^5$
36. $y = x^2$, $y = x^4$, $y = x^6$
37. $y = x^2 - 2$, $y = x^3 - 2x$, $y = x^4 - 2x^2$

PORTFOLIO ACTIVITY

Include this activity in your portfolio.

The sequence of triangular numbers, 1, 3, 6, 10, . . ., can be represented by $\frac{n(n+1)}{2}$, where n is the term number.

1. Draw the next two models in the sequence below. Use the models below to show that the triangular numbers can also be obtained from the polynomial function $y = \frac{x^2 + x}{2}$, where x is the term number.

Term number	1	2	3	4	5
	•	•• ••	••• ••• •••	?	?

2. Verify the identity $\frac{x^2 + x}{2} = \frac{x(x+1)}{2}$ by building a table of values for the two functions $y = \frac{x^2 + x}{2}$ and $y = \frac{x(x+1)}{2}$.

Look Beyond

In Exercises 35–37, students analyze the graphs of some simple polynomial functions. Students will learn more about graphs like these if they take a more advanced course in algebra.

ALTERNATIVE Assessment

Portfolio Activity

The Portfolio Activity can be used as preparation for the Chapter Project or as a separate activity. In the Portfolio Activity on this page, students investigate the polynomial function that defines the *triangular numbers*.

internetconnect

GO TO: go.hrw.com
KEYWORD: MA1 Figurate

To support the Portfolio Activity, students may visit the HRW Web site, where they can obtain additional information about triangular numbers and their relationship to other figurate numbers.

35.

Answers may vary. Sample answer: $y = x$ is linear, but $y = x^3$ and $y = x^5$ are not linear. All three graphs increase in value from left to right and pass through (−1, −1), (0, 0), and (1, 1).

36.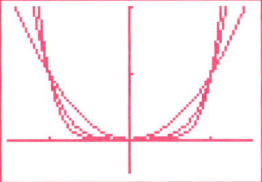

Answers may vary. Sample answer: Each graph has a slightly different curvature. All three graphs open upward, are U-shaped, and pass through (−1, 1), (0, 0), and (1, 1).

Student Technology Guide

Student Technology Guide
9.4 Polynomial Functions

When you use a graphics calculator to graph a polynomial function, you may need to change the viewing window to see the most important parts of the graph.

The graph of a quadratic function is shaped like a "**U**" or an upside-down "**U**" with sloping sides. This curve is called a *parabola*. An important point on the parabola is its turning point, or *vertex*, at the tip of the "**U**." The points where the graph crosses the *x*-axis or *y*-axis are also important.

Example: Graph $y = x^2 + 8x + 5$. Give a viewing window that shows the vertex and any points where the graph crosses the *x*-axis or *y*-axis.

Press [Y=] [X,T,θ,n] [x²] [+] [8] [X,T,θ,n] [+] [5] [GRAPH].
Ranges from −9 to 9 for *x* and from −6 to 6 for *y* give the graph shown at right. Although you can see the points where the graph crosses the axes, the vertex is not visible—it is below the bottom of this viewing window.

To see the vertex, change the viewing window. Press [WINDOW] and enter new values. Here, Ymin has been decreased to −12 to extend the window downward.

The new window shows the vertex.

The settings for the viewing window that shows the vertex and all the points where the graph crosses the axes is Xmin = −9, Xmax = 9, Ymin = −12, and Ymax = 6. (Note that there are many possible viewing windows that show all these points.)

Graph each quadratic function. Choose a viewing window that shows the vertex and any points where the graph crosses the *x*-axis or *y*-axis. Give the viewing window you choose.

Note: For each problem, the smallest integer settings for the viewing window are given. Other answers are possible.

1. $y = x^2 - 6x + 5$
   $0 \le x \le 5$ and $-4 \le y \le 5$
2. $y = x^2 + 4x - 4$
   $-5 \le x \le 1$ and $-8 \le y \le 0$
3. $y = x^2 + 10x + 8$
   $-10 \le x \le 0$ and $-17 \le y \le 8$
4. $y = -4x^2 + 16x + 28$
   $-2 \le x \le 6$ and $0 \le y \le 44$
5. $y = x^2 - 16x + 55$
   $0 \le x \le 11$ and $-9 \le y \le 55$
6. $y = -2x^2 + 13x + 16$
   $-2 \le x \le 8$ and $0 \le y \le 38$

Prepare

NCTM PRINCIPLES & STANDARDS
1–4, 6, 8–10

LESSON RESOURCES

- Practice 9.5
- Reteaching Master 9.5
- Enrichment Master 9.5
- Cooperative-Learning Activity 9.5
- Lesson Activity 9.5
- Problem Solving/Critical Thinking Master 9.5

QUICK WARM-UP

Find the greatest common factor (GCF) of each set of numbers.

1. 6, 15   3
2. 16, 24, 60   4
3. 3, 6, 14, 28   1

Multiply.

4. $4(h^2 - 5)$ $4h^2 - 20$
5. $2b(b^2 - 9b)$ $2b^3 - 18b^2$

Also on Quiz Transparency 9.5

Teach

Why Ask students to describe instances in which one process will "undo" another. For instance, they may recall that addition undoes subtraction and vice versa. They also may have used word-processing or graphics software in which there is an "undo" option in a menu. Tell students that in this lesson, they will see how to undo the multiplication of two binomials by the process of factoring.

9.5 Common Factors

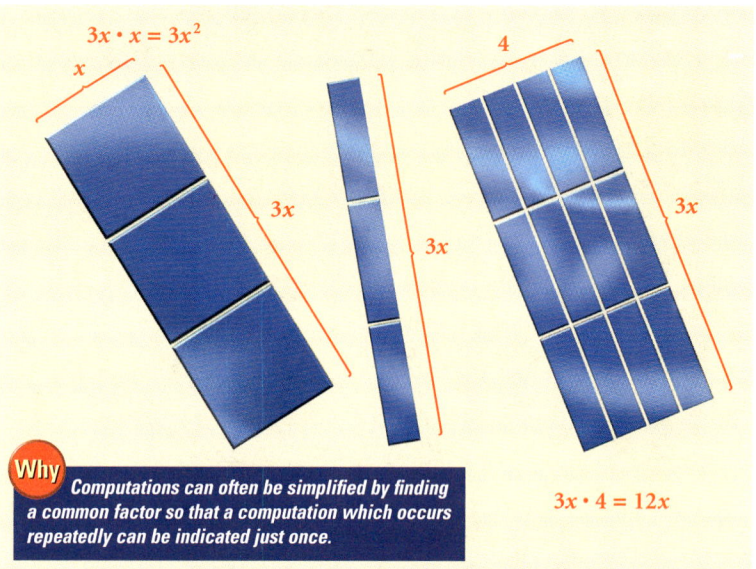

Objectives

- Factor a polynomial by using the greatest common factor.
- Factor a polynomial by using a binomial factor.

Why Computations can often be simplified by finding a common factor so that a computation which occurs repeatedly can be indicated just once.

When two numbers are multiplied, they form a product. Each number is called a **factor** of that product. When you are asked to factor a polynomial such as $3x^2 + 12x$, the first step is to examine the terms for the *greatest common factor*, or GCF. When you write $3x^2 + 12x$ as $(3x)x + (3x)4$, you can see that the two terms have $3x$ as the greatest common factor. Using the Distributive Property, you can then write the following:

$$3x^2 + 12x = 3x(x + 4)$$

common monomial factor

E X A M P L E **1** Factor each polynomial.
 a. $5am - 5an$
 b. $5x^3 - 3y^2$
 c. $2c^4 - 4c^3 + 6c^2$

● **SOLUTION**

Find the greatest common factor, or GCF, of the terms. If possible, use the Distributive Property to rewrite the polynomial.

 a. The GCF is $5a$. $5am - 5an = 5a(m - n)$
 b. The GCF is 1. $5x^3 - 3y^2$ cannot be factored using the method described above.
 c. The GCF is $2c^2$. $2c^4 - 4c^3 + 6c^2 = 2c^2(c^2 - 2c + 3)$

To check each factorization, multiply the two factors to find the original expression.

Alternative Teaching Strategy

USING MODELS Have students count out 2 x^2-tiles and 6 x-tiles. Tell them to arrange the tiles into a rectangle. Lead them to the arrangement shown at right. Point out that the width of the rectangle is $2x$ and the length is $x + 3$. Thus, they have shown that $x^2 + 6x$ can be rewritten as $2x(x + 3)$. That is, they have factored $x^2 + 6x$. Use this activity as an introduction to a discussion of factoring.

Inclusion Strategies

VISUAL LEARNERS Show students how to use visual cues to identify common factors. First write a polynomial in an expanded form, using the prime factorization of the coefficients. Then use a different geometric figure to identify each common factor. For instance, this is how you might factor $10x^2z - 25xz + 15xz^2$ as $5xz(2x - 5 + 3z)$.

$2 \cdot 5 \cdot x \cdot x \cdot z - 5 \cdot 5 \cdot x \cdot z + 3 \cdot 5 \cdot x \cdot z \cdot z$

$2 \cdot \boxed{5} \cdot \boxed{x} \cdot x \cdot \boxed{z} - \boxed{5} \cdot 5 \cdot \boxed{x} \cdot \boxed{z} + 3 \cdot \boxed{5} \cdot \boxed{x} \cdot \boxed{z} \cdot z$

$\boxed{5}\,\boxed{x}\,\boxed{z}(2 \cdot x - 5 + 3 \cdot z)$

448 LESSON 9.5

TRY THIS Factor each polynomial.
  a. $6ab + 3a$
  b. $5x^3 + 10x^2 - 20x$
  c. $2x^2 - 3y^2$

Some polynomials cannot be factored using the methods described in this lesson. Additional methods for factoring polynomials will be covered in later lessons.

A polynomial may have a common factor that is a binomial.

EXAMPLE 2 Factor $r(t + 1) + s(t + 1)$.

SOLUTION

Notice that $r(t + 1) + s(t + 1)$ contains the **common binomial factor** $t + 1$. Use the Distributive Property to write the following:

$$r(t + 1) + s(t + 1) = (r + s)(t + 1)$$

You can visualize this factoring procedure with a geometric model. The model shows the same total area in two different arrangements.

CONNECTION GEOMETRY

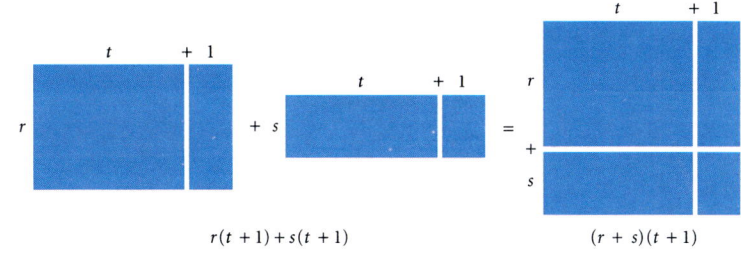

$r(t+1) + s(t+1)$ $(r + s)(t + 1)$

TRY THIS Factor $a(x - 3) + 6(x - 3)$.

Sometimes, polynomials can be factored by *grouping* terms.

EXAMPLE 3 Factor $x^2 + x + 2x + 2$ by grouping.

SOLUTION

There are several ways to factor this polynomial by grouping. One way is to group the terms that have a common number or variable as a factor. Treat $x^2 + x$ as one expression, and treat $2x + 2$ as another expression.

$x^2 + x + 2x + 2 = (x^2 + x) + (2x + 2)$ *Group terms.*
$\qquad\qquad\qquad = x(x + 1) + 2(x + 1)$ *Factor the GCF from each group.*
$\qquad\qquad\qquad = (x + 2)(x + 1)$ *Use the Distributive Property.*

To check, multiply the factors.

TRY THIS Factor $ax + bx + ay + by$ by grouping.

Enrichment

Replace each ? with the binomial that will make the statement true. Assume that n is a positive integer.

1. $y^{n+1} + y^n = y^n(\underline{\ ?\ })$ $\quad y + 1$
2. $t^{2n} + t^n = t^n(\underline{\ ?\ })$ $\quad t^n + 1$
3. $r^n + r^{n-1} = r^{n-1}(\underline{\ ?\ })$ $\quad r + 1$
4. $a^n b + ab^n = ab(\underline{\ ?\ })$ $\quad a^{n-1} + b^{n-1}$
5. $p^n q^{3n} + p^n q^{2n} = p^n q^n(\underline{\ ?\ })$ $\quad q^{2n} + q^n$

Tell students to create two original problems like the ones above. Have students exchange and solve each other's problems.

Reteaching the Lesson

COOPERATIVE LEARNING Have students work in pairs. Each student should write a multiplication of a trinomial in x by a monomial in x and then write the product. For instance, a student might write $4x(x^2 + 2x - 3)$ as the multiplication and $4x^3 + 8x^2 - 12x$ as the product. Tell students to write the product on a clean sheet of paper. Partners should exchange the papers and factor the other's product.

Repeat the activity for a multiplication of a binomial in m and n by a monomial in m and n.

ADDITIONAL EXAMPLE 1

Factor each polynomial over the integers.

a. $4y^3 - 16y^4$ $\quad 4y^3(1 - 4y)$
b. $6p + 15p^2 - 9pq$
  $\quad 3p(2 + 5p - 3q)$
c. $8r^4 + 17s^4$ $\quad$ prime

TRY THIS
a. $3a(2b + 1)$
b. $5x(x^2 + 2x - 4)$
c. $2x^2 - 3y^2$ is prime.

ADDITIONAL EXAMPLE 2

Factor $a(b + 4) + c(b + 4)$.
$(a + c)(b + 4)$

Math CONNECTION

GEOMETRY Students use a rectangular geometric model to visualize the multiplication of two binomials. Remind students that the area of a rectangle is found by multiplying its length times its width.

TRY THIS
$(a + 6)(x - 3)$

ADDITIONAL EXAMPLE 3

Factor $mn + mp + 5n + 5p$ by grouping.
$(m + 5)(n + p)$

TRY THIS
$ax + bx + ay + by$
$= (ax + bx) + (ay + by)$
$= (a + b)x + (a + b)y$
$= (a + b)(x + y)$

LESSON 9.5 **449**

Assess

Selected Answers
Exercises 6–12, 13–67 odd

ASSIGNMENT GUIDE

In Class	1–12
Core	13–39 odd, 43–48, 55–60
Core Plus	14–58 even, 59, 60
Review	61–67
Preview	68

✎ Extra Practice can be found beginning on page 738.

Exercises

Communicate

1. Describe what a factor of a polynomial is.
2. What property of mathematics is used when you factor by using the GCF?
3. Explain how to find the GCF.
4. Once you have found the GCF of the following numbers and expressions, how do you find the remaining factor?
   a. 60 and 150
   b. x^3y^5 and x^5y^2
   c. $25(x+y)$ and $39(x+y)$
5. Explain how to group terms to factor the polynomial $5y^2 + 5y + 2y + 2$.

Guided Skills Practice

Factor each polynomial. *(EXAMPLE 1)*

6. $10bc + 5b$ $5b(2c+1)$
7. $3x^2 - 4y$ Does not factor
8. $4x^2y + 12xy^2 - 20xy$ $4xy(x+3y-5)$

Factor. *(EXAMPLE 2)*

9. $x(y+1) - 2(y+1)$ $(x-2)(y+1)$
10. $m(n^2+3) + 4(n^2+3)$ $(m+4)(n^2+3)$

Factor by grouping. *(EXAMPLE 3)*

11. $m^2 + m - 4m - 4$ $(m-4)(m+1)$
12. $x^2 - 2ax + 3ax - 6a^2$ $(x+3a)(x-2a)$

Practice and Apply

Factor each polynomial.

13. $4x - 16$ $4(x-4)$
14. $5r^2 + 10r$ $5r(r+2)$
15. $2n^2 + 4$ $2(n^2+2)$
16. $5m^2 - 35$ $5(m^2-7)$
17. $3p^2 + 12p$ $3p(p+4)$
18. $4f^2 + 4f$ $4f(f+1)$
19. $2x^2 - 4$ $2(x^2-2)$
20. $5n^2 - 10$ $5(n^2-2)$
21. $3x^2 + 6x$ $3x(x+2)$
22. $x^9 - x^2$ $x^2(x^7-1)$
23. $k^5 + k^2$ $k^2(k^3+1)$
24. $4a^8 - 20a^6 + 8a^4$
25. $4x^2 + 2x - 6$
26. $7x^2 - 28x - 14$
27. $27y^3 + 18y^2 - 81y$
28. $4n^3 - 16n^2 + 8n$
29. $100 + 25s^5 - 50s$
30. $2p^4r - 8p^3r^2 + 16p^2r^3$
31. $3m^3 - 9m^2 + 3m$
32. $90 + 15a^5 - 45a$
33. $2x^3y - 8x^2y^2 + 17xy^3$
34. $4x^4 - 24x^2$
35. $7y^3 - 21y^2 + 14y$
36. $42r - 14r^3$
37. $3x^2 + 21x^4$
38. $8a^4 + 4a^3 - 12a^2$
39. $2t^3 - 130t^5$

CHALLENGE

40. $x^{n+3} + x^n$
41. $9w^{2n} + 21w^{2n+1}$
42. $x^{3n+21} + 2x^{2n+14}$

Answers (side column)

24. $4a^4(a^4 - 5a^2 + 2)$
25. $2(2x^2 + x - 3)$
26. $7(x^2 - 4x - 2)$
27. $9y(3y^2 + 2y - 9)$
28. $4n(n^2 - 4n + 2)$
29. $25(4 + s^5 - 2s)$
30. $2p^2r(p^2 - 4pr + 8r^2)$
31. $3m(m^2 - 3m + 1)$
32. $15(6 + a^5 - 3a)$
33. $xy(2x^2 - 8xy + 17y^2)$
34. $4x^2(x^2 - 6)$
35. $7y(y^2 - 3y + 2)$
36. $14r(3 - r^2)$
37. $3x^2(1 + 7x^2)$
38. $4a^2(2a^2 + a - 3)$
39. $2t^3(1 - 65t^2)$
40. $x^n(x^3 + 1)$
41. $3w^{2n}(3 + 7w)$
42. $x^{2n+14}(x^{n+7} + 2)$

Practice

9.5 Common Factors

Identify each polynomial as prime or not prime.

1. $9x^2 + 24x + 15$ not prime
2. $9a^2 + 24a$ not prime
3. $9m^2 + 16$ prime
4. $x^2 + 5x - 6$ not prime
5. $2x^2 + 2x + 4$ not prime
6. $4x^2 + 25y^2$ prime

Factor each polynomial by using the GCF.

7. $3x^2 - 9x$ $3x(x-3)$
8. $6r^3 + 3r^2 - 7r$ $r(6r^2 + 3r - 7)$
9. $6w^4 + 3w^2 - 9$ $3(2w^4 + w^2 - 3)$
10. $3q^7 + 6q^5 + 9q$ $3q(q + 2q^4 + 3)$
11. $18b^3 - 36b^2 - 9$ $9(2b^3 - 4b^2 - 1)$
12. $3v^6 - 9v^4 + 147v^2$ $3v^2(v^4 - 3v^2 + 49)$
13. $xy - 2x^2y + xy^3$ $xy(1 - 2x + y^2)$
14. $10wz^2 - 5wz + 15w^2z$ $5wz(2z - 1 + 3w)$

Write each as the product of two binomials.

15. $y(y+1) + 2(y+1)$ $(y+2)(y+1)$
16. $3(c+d) - a(c+d)$ $(3-a)(c+d)$
17. $2(x+w) - z(x+w)$ $(2-z)(x+w)$
18. $6(x-2) + y(x-2)$ $(6+y)(x-2)$
19. $a(a-b) - b(a-b)$ $(a-b)(a-b)$
20. $3x(x+4) - 2(x+4)$ $(3x-2)(x+4)$
21. $4(x-2y) + x(x-2y)$ $(4+x)(x-2y)$
22. $z(4-w) + y(4-w)$ $(z+y)(4-w)$
23. $7x(x-1) + (x-1)$ $(7x+1)(x-1)$
24. $mn(r+1) - (r+1)$ $(mn-1)(r+1)$

Factor by grouping.

25. $4ax - bx + 4ay - by$ $(4a-b)(x+y)$
26. $2pq^2 + 4pq - 2q - 4$ $2(pq-1)(q+2)$
27. $3a + 3 - a^2 - a$ $(a+1)(3-a)$
28. $6m^3 + 4m - 9m^2 - 6$ $(3m^2+2)(2m-3)$
29. $3m - 12 + m^3 - 4m^2$ $(m-4)(m^2+3)$
30. $ab^2 + 5a - 6b^2 - 30$ $(a-6)(b^2+5)$
31. $2cd - c - 6d + 3$ $(2d-1)(c-3)$
32. $4 - 2x + 6 - 3x$ $5(2-x)$
33. $2xz + 2yz + x + y$ $(x+y)(2z+1)$
34. $mx + 5m + nx + 5n$ $(x+5)(m+n)$
35. $xy + 2y + x + 2$ $(x+2)(y+1)$
36. $3x^2 + 3xy - 2xy - 2y^2$ $(3x-2y)(x+y)$

LESSON 9.5

43. $(x + 1)(x + 2)$
44. $(y + 3)(5 - x)$
45. $(x + y)(a + b)$
46. $(4 + p)(3q - 4)$
47. $(x - 1)(x + 2)$
48. $(x - 4)(r + t)$
49. $(a - 3)(5a + 4)$
50. $(w + 4)(2w - 3)$
51. $(x - 2)(2 - x)$
56. $(3a - 2)(4b - 5)$

Write each polynomial as the product of two binomials.

43. $x(x + 1) + 2(x + 1)$
44. $5(y + 3) - x(y + 3)$
45. $a(x + y) + b(x + y)$
46. $(4 + p)3q - 4(4 + p)$
47. $x(x - 1) + 2(x - 1)$
48. $r(x - 4) + t(x - 4)$
49. $5a(a - 3) + 4(a - 3)$
50. $2w(w + 4) - 3(w + 4)$
51. $2(x - 2) + x(2 - x)$
52. $8(y - 1) - x(y - 1)$ $(y - 1)(8 - x)$
53. $2r(r - s)^2 - 3(r - s)^2$ $(r - s)^2(2r - 3)$
54. $3(m + 7) - n(m + 7)$ $(m + 7)(3 - n)$

Factor by grouping.

55. $2x + 2y + ax + ay$ $(x + y)(2 + a)$
56. $12ab - 15a - 8b + 10$
57. $3(x + y) + 12(x + y)$ $15(x + y)$
58. $x^2 + 3x + 4x + 12$ $(x + 3)(x + 4)$

CONNECTION

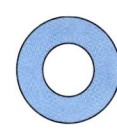

*The shaded area between two concentric circles is called an **annulus**.*

GEOMETRY Donuts and bagels are in the shape of an annulus. The formula for the area of the annulus is $A = \pi R^2 - \pi r^2$, where R is the radius of the larger circle and r is the radius of the smaller circle.

59. Write the formula for the area of an annulus in factored form. $\pi(R^2 - r^2)$

60. Use the factored form to find the area of an annulus formed by concentric circles with radii of 8 and 5. $39\pi \approx 122.5$

Look Back

61. Evaluate $8 - 5(2 - 6) - 3^2$. **(LESSON 1.3)** 19

62. Solve the formula $A = \frac{1}{2}h(b_1 + b_2)$ for h. **(LESSON 3.6)** $h = \frac{2A}{b_1 + b_2}$

63. Solve for x: $3(x - 5) - 2(x + 8) = 7x + 3$. **(LESSON 3.5)** $-\frac{17}{3}$

64. Graph $x - 4y = 8$. **(LESSON 5.1)**

65. Solve. $\begin{cases} 4x = 11 + 15y \\ 6x + 5y = 0 \end{cases}$ **(LESSON 7.3)** $\left(\frac{1}{2}, -\frac{3}{5}\right)$

66. Exponential, because the first differences are not constant.

66. Is the sequence 3, 6, 12, 24, 48, ... representative of a linear relationship or an exponential relationship? Why? **(LESSON 8.6)**

67. Suppose that you want to save $6000 to buy a car when you graduate in 3 years. The current interest rate on a certificate of deposit (CD) is 8% compounded annually. How much will you need to invest in a CD now to buy the car in 3 years? **(LESSON 8.7)** $\approx \$4763$

Look Beyond

APPLICATION

68. **PHYSICS** A certain projectile is fired vertically into the air at an initial velocity of 320 feet per second. Its motion can be modeled by the equation $h = -16t^2 + 320t$, where h is the height at time t. Graph the equation, and find t when $h = 1200$.

64.
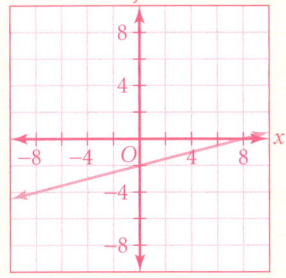

68.
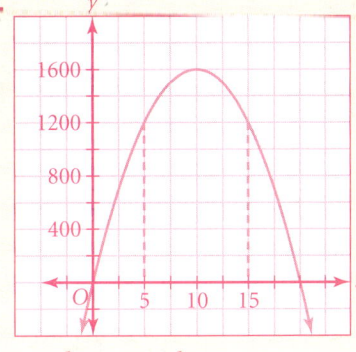

5 and 15 seconds

Technology

A calculator may be helpful in performing some of the multiplications involved when checking factorizations in Exercises 19–39.

Error Analysis

When the greatest common factor of the terms of a polynomial is itself one of the terms, students may forget to include 1 in the factored form. For example, they may factor $2a^2 + 4a + 2$ as $2(a^2 + 2a)$. Encourage students to check their factorizations by multiplying.

Look Beyond

In Exercise 68, students investigate a graph that is a parabola. They will study parabolas in greater detail in Lesson 10.1.

Student Technology Guide

NAME _____ CLASS _____ DATE _____

Student Technology Guide
9.5 Common Factors

A graphics calculator allows you to find the greatest common denominator of two fractions. You can use this feature to find the greatest common factor of several whole numbers.

To find the greatest common factor of 148 and 96, press the following keys:

[MATH] [NUM] [9:gcd] 148 [,] 96 [)] [ENTER]

Notice that the left parenthesis is provided automatically but you must press the key to enter the right parenthesis.

gcd(148,96)   4

The GCF of 148 and 96 is 4.

Find the GCF of each pair of numbers.

1. 16 and 44   2. 12 and 78   3. 144 and 168   4. 105 and 84
   4   6   24   21

When you factor a polynomial, finding the GCF of the coefficients can be the most difficult part of the problem. By using the process shown above, you can use a graphics calculator to find the GCF of the coefficients.

Example: Factor $72x^3y^7 + 48x^2y^9$.

The powers of x are x^3 and x^2, so the greatest common factor includes x^2. The powers of y are y^7 and y^9, so the greatest common factor includes y^7.

To find the greatest common factor of 72 and 48, press [MATH] [NUM] [9:gcd] 72 [,] 48 [)] [ENTER].

gcd(72,48)   24
72/24   3
48/24   2

The greatest common factor of the coefficients is 24.

x^3 and x^2 $\xrightarrow{gcd}$ x^2 y^7 and y^9 $\xrightarrow{gcd}$ y^7

The greatest common factor of the two terms is $24x^2y^7$. The factored expression is $24x^2y^7(3x + 2y^2)$.

Factor each polynomial by using the GCF.

5. $24x^3 + 8x^7$ 6. $5c^3 + 15c^2$ 7. $24y^4 - 32y^{11}$ 8. $27a^7 + 105a$
   $8x^3(3 + x^4)$ $5c^2(c + 3)$ $8y^4(3 - 4y^7)$ $3a(9a^6 + 35)$

9. $6x^2y^9 + 4x^2y^5$ 10. $56x^3 - 42x^7$ 11. $16m^5n + 12m^4n^3$ 12. $165a^5b^8 - 225a^2b^{12}$
   $2x^2y^5(3y^4 + 2)$ $14x^3(4 - 3x^4)$ $4m^4n(4m + 3n^2)$ $15a^2b^8(11a^3 - 15b^4)$

Prepare

NCTM PRINCIPLES & STANDARDS 1–10

LESSON RESOURCES

- Practice **9.6**
- Reteaching Master **9.6**
- Enrichment Master **9.6**
- Cooperative-Learning Activity **9.6**
- Lesson Activity **9.6**
- Problem Solving/Critical Thinking Master **9.6**

QUICK WARM-UP

Multiply.

1. $4(j + 6)$ $4j + 24$
2. $(n - 9)(n + 9)$ $n^2 - 81$
3. $(2t + 5)(2t + 5)$
                  $4t^2 + 20t + 25$

Factor.

4. $3k^2 + 21$ $3(k^2 + 7)$
5. $2y^2 - 15y$ $y(2y - 15)$
6. $6c^3 + 9c^2$ $3c^2(2c + 3)$

Also on Quiz Transparency 9.6

Teach

Why Ask students to name some items that have a pattern. Some of the many items they may list are fabrics, wall papers, the petals of a flower, and the leaves of a tree. Then ask them to cite instances in which knowing a pattern may help to solve a real-world problem. Tell them that in this lesson, they will see how recognizing a pattern can help in factoring a polynomial.

9.6 Factoring Special Polynomials

Objectives

- Factor perfect-square trinomials.
- Factor the difference of two squares.

Why Perfect-square trinomials and the difference of two squares are polynomials that can be easily factored by applying simple rules.

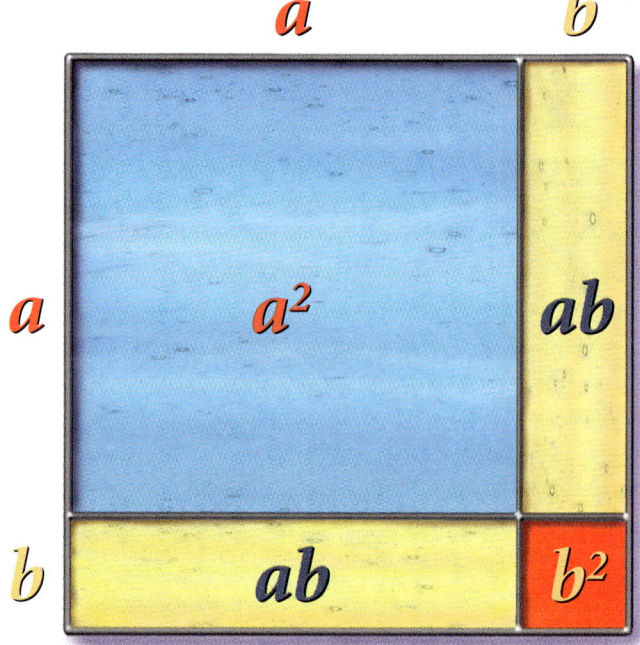

$$(a+b)^2 = a^2 + 2ab + b^2$$

Perfect-Square Trinomials

CONNECTION GEOMETRY

You can see from the figure that the area of the large square is both $(a + b)^2$ and $a^2 + 2ab + b^2$. This figure illustrates why the product $a^2 + 2ab + b^2$ is called a **perfect-square trinomial.**

The same pattern is used to square the expression $a - b$. Square the first term, add twice the product of the terms, and add the square of the last term, noting that the middle term of the result has a minus sign.

$$(a - b)^2 = a^2 - 2ab + b^2$$

To see if a trinomial such as $4x^2 + 12x + 9$ is a perfect square, use this test.

- Is the first term, $4x^2$, a perfect square? Yes: $(2x)^2$
- Is the last term, 9, a perfect square? Yes: 3^2
- Is the middle term, $12x$, twice the product of the square roots of the first and last terms? Yes: $2(\sqrt{4x^2})(\sqrt{9}) = 2(2x)(3) = 12x$

Thus, $4x^2 + 12x + 9$ is a perfect-square trinomial.

Alternative Teaching Strategy

USING MODELS Have students count out 1 x^2-tile, 6 x-tiles, and 9 unit tiles. Tell them to make a rectangle. Lead them to the arrangement shown at right. Point out that both the length and width are $x + 3$, so the rectangle is a square. The trinomial that the tiles represent, $x^2 + 6x + 9$, is called a *perfect-square trinomial*. Its factorization is $(x + 3)^2$.

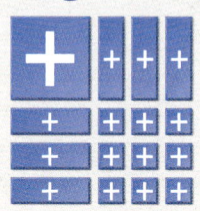

Now have students put aside all tiles except for 1 x^2-tile. Tell them to add as many x-tiles and unit-tiles as necessary to model a different perfect-square trinomial. Ask them to write the trinomial and its factorization. Likely responses are 1 x^2-tile, 2 x-tiles, and 1 unit-tile, or $x^2 + 2x + 1 = (x + 1)^2$, and 1 x^2-tile, 4 x-tiles, and 4 unit tiles, or $x^2 + 4x + 4 = (x + 2)^2$. Repeat the activity, this time starting out with 4 x^2-tiles, 4 x-tiles, and 1 unit tile. Be sure to lead students to model $4x^2 + 4x + 1 = (2x + 1)^2$.

452 LESSON 9.6

EXAMPLE 1 State whether each expression is a perfect-square trinomial. Explain.
   a. $x^2 + 8x + 16$ b. $3x^2 + 16x + 16$ c. $x^2 + 3x + 9$

SOLUTION

a. Yes; the first and last terms are perfect squares and the middle term is twice the product of the square roots of the first and last terms.

b. No; the first term, $3x^2$, is not a perfect square.

c. No; the middle term is not twice the product of the square roots of the first and last terms.

Rule for Factoring a Perfect-Square Trinomial

For all numbers a and b:

$a^2 + 2ab + b^2 = (a+b)(a+b) = (a+b)^2$, and
$a^2 - 2ab + b^2 = (a-b)(a-b) = (a-b)^2$

EXAMPLE 2 Factor each expression.   a. $x^2 - 10x + 25$ b. $49y^4 + 14y^2 + 1$

SOLUTION

a. Since the expression is a perfect-square trinomial, apply the rule. Using the square roots of the first and last terms write $x^2 - 10x + 25 = (x-5)^2$.

b. Since the expression is a perfect-square trinomial, apply the rule. Using the square roots of the first and last terms write $49y^4 + 14y^2 + 1 = (7y^2 + 1)^2$.

TRY THIS Factor each expression.   a. $9s^2 + 24s + 16$ b. $64a^4 - 16a^2b + b^2$

The Difference of Two Squares

When you multiply $(a+b)(a-b)$, the product is $a^2 - b^2$. This product is called the **difference of two squares**.

$(a+b)(a-b) = a(a-b) + b(a-b)$
$= a^2 - ab + ab - b^2$
$= a^2 - b^2$

EXAMPLE 3 State whether each expression is a difference of two squares. Explain.
   a. $4a^2 - 25$ b. $9x^2 - 15$ c. $16r^2 + 49$

SOLUTION

a. Yes; the expression involves subtraction, $4a^2$ is the square of $2a$, and 25 is the square of 5.
b. No; 15 is not a perfect square.
c. No; it is the sum of two squares.

Math CONNECTION

GEOMETRY Remind students that the formula for the area, A, of a square is $A = s^2$, where s represents the length of one side.

ADDITIONAL EXAMPLE 1

State whether each expression is a perfect-square trinomial. Explain.

a. $25r^2 + 30r + 3$
No; the last term, 3, is not a perfect square.

b. $x^2 - 12xy + 36y^2$
Yes; $36y^2 = (6y)^2$, and $12xy = 2(x)(6y)$.

c. $4a^2 - 18a + 81$
No; $4a^2 = (2a)^2$ and $81 = 9^2$, but $18a \neq 2(2a)(9)$.

ADDITIONAL EXAMPLE 2

Factor each expression.

a. $x^2 + 24x + 144$
$(x+12)^2$

b. $100m^2 - 140m + 49$
$(10m-7)^2$

TRY THIS
a. $(3s+4)^2$ b. $(8a^2 - b)^2$

ADDITIONAL EXAMPLE 3

State whether each expression is a difference of two squares. Explain.

a. $4y^2 + 49$
No; it is the sum of two squares.

b. $64 - 5a^4$
No; $5a^4$ is not a perfect square.

c. $81g^2 - 16$
Yes; $81g^2 = (9g)^2$, $16 = 4^2$, and the expression involves subtraction.

Interdisciplinary Connection

GRAPHIC ARTS The students in a graphics art class are creating square collages. Each collage is to be mounted on a board so that there is a 3-inch border around the collage. One student said that the area of the board needed can be calculated using the expression $s^2 + 36s + 36$, where s represents the length of one side of the collage. Ask students whether this is correct. Have them explain their responses. If they feel it is incorrect, tell them to write a correct expression for the area of the board. incorrect; $s^2 + 12s + 36$

Inclusion Strategies

TACTILE LEARNERS Have students make a cardboard model that demonstrates how the figure at right proves the identity $a^2 - 2ab + b^2 = (a-b)^2$.

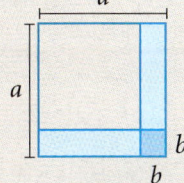

Start with: Take away: Put back:

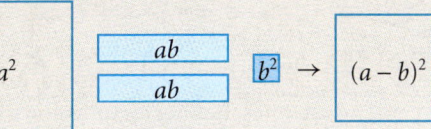

LESSON 9.6

Activity Notes

In this Activity, students first focus on several multiplication expressions of the form $(a + b)(a - b)$. They should recall from their work in Lesson 9.2 that the product takes the form $a^2 - b^2$. At the end of the Activity, they should realize that they can use the relationship to work backwards. That is, given a binomial of the form $a^2 - b^2$, they can factor it as $(a + b)(a - b)$.

CHECKPOINT ✔
3. $a^2 - b^2 = (a + b)(a - b)$

ADDITIONAL EXAMPLE 4

Factor each expression.

a. $m^2 - 121$
 $(m + 11)(m - 11)$

b. $9r^2 - 100s^2$
 $(3r + 10s)(3r - 10s)$

TRY THIS
$(5w + 9)(5w - 9)$

CRITICAL THINKING
$a^8 - b^8 = (a^4)^2 - (b^4)^2$, so it can be factored as the difference of two squares: $(a^4 + b^4)(a^4 - b^4)$. However, $a^4 - b^4 = (a^2)^2 - (b^2)^2$ can be factored as the difference of two squares as well: $(a^2 + b^2)(a^2 - b^2)$, and so on. The final factorization is $a^8 - b^8 = (a^4 + b^4)(a^2 + b^2)(a + b)(a - b)$.

Activity

Patterns in Differences of Two Squares

1. Multiply the following pairs of binomials. Describe the pattern that you observe.
   a. $(x + 2)(x - 2)$
   b. $(x + 3)(x - 3)$
   c. $(x + 4)(x - 4)$
   d. $(x + 5)(x - 5)$

2. Use the pattern from Step 1 to factor the following differences of two squares:
   a. $x^2 - 1$
   b. $x^2 - 36$
   c. $x^2 - 64$
   d. $x^2 - 100$

CHECKPOINT ✔
3. Based on the patterns from Steps 1 and 2, write a rule for factoring binomials of the form $a^2 - b^2$.

There is a useful rule for factoring differences of two squares.

Rule for Factoring a Difference of Two Squares

For all numbers a and b: $a^2 - b^2 = (a + b)(a - b)$

EXAMPLE 4 Factor each expression.
a. $x^2 - 4$
b. $36a^2 - 49b^2$

● **SOLUTION**

a. Apply the rule for factoring a difference of two squares.
$$x^2 - 4 = (x + 2)(x - 2)$$

b. Apply the rule for factoring a difference of two squares.
$$36a^2 - 49b^2 = (6a)^2 - (7b)^2$$
$$= (6a + 7b)(6a - 7b)$$

TRY THIS Factor $25w^2 - 81$.

CRITICAL THINKING Explain why $a^8 - b^8$ has more than two factors. What are all of the factors?

EXAMPLE 5 Find $17 \cdot 13$ by using the difference of two squares.

● **SOLUTION**
Think of $17 \cdot 13$ as $(15 + 2)(15 - 2)$.
The product is $15^2 - 2^2 = 225 - 4 = 221$.

TRY THIS Find $34 \cdot 26$ by using the difference of two squares.

Enrichment

Have students work in pairs to factor the following expressions:

1. $(k + 2)^2 - 1$ $(k + 3)(k + 1)$
2. $(d + 2)^2 - 4$ $d(d + 4)$
3. $(z + 3)^2 - z^2$ $3(2z + 3)$
4. $(m + 2)^2 - (n + 2)^2$ $(m - n)(m + n + 4)$
5. $(a + b)^2 - (c + d)^2$
     $(a + b + c + d)(a + b - c - d)$

Then have each student write three similar expressions to be factored. Partners should exchange papers and factor each other's expressions.

Reteaching the Lesson

COGNITIVE STRATEGIES Give students several trinomials such as those below. For each, ask them to explain why it is *not* a perfect-square trinomial. Then tell them to make a perfect-square trinomial by changing just one term in the expression.

1. $t^2 + 32t + 64$ Sample: Change 32 to 16.
2. $h^2 - 30h + 60$ Sample: Change 60 to 225.
3. $4w^2 + 28w - 49$ Sample: Change − to +.
4. $16q^2 + 6q + 1$ Sample: Change 16 to 9.

Repeat the activity, using expressions that can be converted to a difference of two squares.

454   LESSON 9.6

Exercises

Communicate

1. Describe the process for finding the factors of the perfect-square trinomial $x^2 + 20x + 100$.
2. Explain how to factor $4x^2 - 12x + 9$.
3. Explain how to factor this difference of two squares: $p^2 - 121$.

Guided Skills Practice

State whether each expression is a perfect-square trinomial. Explain. (EXAMPLE 1)

4. $x^2 + 16x + 64$
5. $4b^2 - 32bc + 64c^2$
6. $3x^2 + 2x - 7$

Factor. (EXAMPLE 2)

7. $x^2 + 10x + 25$ $(x + 5)^2$
8. $36a^2 - 60a + 25$ $(6a - 5)^2$
9. $16x^2 + 8xy + y^2$ $(4x + y)^2$

State whether each expression is a difference of two squares. Explain. (EXAMPLE 3)

10. $9q^2 - 16r^2$
11. $4k^4 - 25l^2$
12. $25x^2 - 5y^2$

Factor. (EXAMPLE 4)

13. $a^2 - b^2$ $(a + b)(a - b)$
14. $36x^2 - 49y^2$ $(6x + 7y)(6x - 7y)$
15. $4m^2 - 144n^2$ $(2m + 12n)(2m - 12n)$

Find each product by using the difference of two squares. (EXAMPLE 5)

16. $47 \cdot 53$ $(50 - 3)(50 + 3) = 2491$
17. $97 \cdot 103$ $(100 - 3)(100 + 3) = 9991$

Practice and Apply

Factor each polynomial completely.

18. $y^2 + 8y + 16$ $(y + 4)^2$
19. $x^2 + 4x + 4$ $(x + 2)^2$
20. $r^2 - 18r + 81$ $(r - 9)^2$
21. $x^2 - 4$ $(x + 2)(x - 2)$
22. $36d^2 + 12d + 1$
23. $4t^2 - 1$
24. $81 - 4m^2$
25. $25x^2 - 9$
26. $y^2 - 100$
27. $4x^2 - 20x + 25$
28. $100 - 36q^2$
29. $16c^2 - 25$
30. $p^2 - q^2$
31. $9c^2 - 4d^2$
32. $16x^2 + 72xy + 81y^2$
33. $9a^2 - 12a + 4$
34. $49x^2 - 42xy + 9y^2$
35. $a^2x^2 + 2axb + b^2$
36. $4m^2 + 4mn + n^2$
37. $81a^4 - 9b^2$
38. $x^4 - y^4$
39. $x^2(25 - x^2) - 4(25 - x^2)$
40. $(x - 1)x^2 - 2x(x - 1) + (x - 1)$
41. $(3x + 5)(x^2 - 3) - (3x + 5)$
42. $(x^2 - y^2)(x^2 + 2xy) + (x^2 - y^2)(y^2)$

Answers:

10. Yes, both terms are perfect squares.
11. Yes, both terms are perfect squares.
12. No, $5y^2$ is not a perfect square.
22. $(6d + 1)^2$
23. $(2t + 1)(2t - 1)$
24. $(9 + 2m)(9 - 2m)$
25. $(5x + 3)(5x - 3)$
26. $(y + 10)(y - 10)$
27. $(2x - 5)^2$
28. $(10 + 6q)(10 - 6q)$
29. $(4c + 5)(4c - 5)$
30. $(p + q)(p - q)$
31. $(3c + 2d)(3c - 2d)$
32. $(4x + 9y)^2$
33. $(3a - 2)^2$
34. $(7x - 3y)^2$
35. $(ax + b)^2$
36. $(2m + n)^2$
37. $(9a^2 + 3b)(9a^2 - 3b)$
38. $(x^2 + y^2)(x + y)(x - y)$
39. $(x + 2)(x - 2)(5 + x)(5 - x)$
40. $(x - 1)^3$
41. $(3x + 5)(x + 2)(x - 2)$
42. $(x - y)(x + y)^3$

CHALLENGE

4. Yes; explanations may vary. Sample explanation: The first and last terms are perfect squares, and the middle term is twice the product of the square roots of the first and last terms.

5. Yes; explanations may vary. Sample explanation: The first and last terms are perfect squares, and the middle term is the opposite of twice the product of the square roots of the first and last terms.

6. No; explanations may vary. Sample explanation: The first and last terms are not perfect squares.

ADDITIONAL EXAMPLE 5

Find $39 \cdot 41$ by using the difference of two squares.

$39 \cdot 41 = (40 - 1)(40 + 1)$
$= 40^2 - 1^2$
$= 1600 - 1$
$= 1599$

TRY THIS

$(30 + 4)(30 - 4) = 30^2 - 4^2$
$= 900 - 16$
$= 884$

Assess

Selected Answers
Exercises 4–17, 19–63 odd

ASSIGNMENT GUIDE

In Class	1–17
Core	18–38, 43–57 odd
Core Plus	18–48 even, 49–58
Review	59–63
Preview	64

Extra Practice can be found beginning on page 738.

Technology

A calculator may be helpful in performing some of the multiplications involved in checking factorizations in Exercises 18–48.

Error Analysis

Students often confuse taking the square root of a number with taking half of a number. For instance, they may look at an expression like $z^2 - 16$ and quickly factor it as $(z + 8)(z - 8)$ rather than as $(z + 4)(z - 4)$. Remind students to always check their factorizations by multiplying.

Find the missing term in each perfect-square trinomial.

43. $x^2 - 14x + ?$ **49**
44. $16y^2 + ? + 9$ **24y**
45. $25a^2 + 60a + ?$ **36**
46. $9x^2 + ? + 25$ **30x**
47. $x^2 - 12x + ?$ **36**
48. $? - 36y + 81$ **4y²**

49. **CULTURAL CONNECTION: AFRICA**
Abu Kamil, known as the "Egyptian calculator," used geometric models to solve problems around the year 900 C.E. You can make replicas of his models with algebra tiles or paper rectangles.

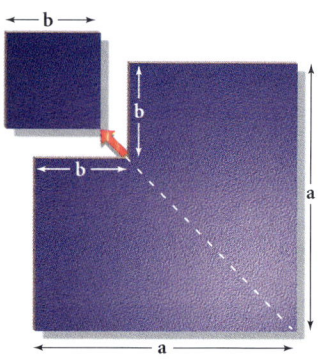

Use a sheet of paper to create a model of $a^2 - b^2$ as shown. Remove the b^2 section from the corner. Then cut along the dotted lines, and move the pieces to form a rectangle. Explain how this represents the factorization of $a^2 - b^2$.

CONNECTIONS

GEOMETRY The area of a square is represented by $n^2 - 12n + 36$.

50. Find the length of each side of the square. **n − 6**
51. Find the perimeter of the square. **4n − 24**

52. **GEOMETRY** A small circle with a radius of r units is drawn within a large circle with a radius of R units, as shown at left. What is the area of the large circle? What is the area of the shaded part? Factor the expression for the shaded area. (Hint: First factor out the common monomial factor, π. Then factor the binomial.) **Large circle: πR^2; Shaded area: $\pi R^2 - \pi r^2$; Shaded area, factored: $\pi(R + r)(R - r)$**

APPLICATION

CONSTRUCTION A construction firm is designing a fountain to be built in several locations in varying sizes. The diagram below shows the dimensions of the proposed fountain.

53. What is the area of the top of one of the bases, including the top surface of the water? **$x^2 + 8x + 16$**

54. $x^2 - 2x + 1$

54. What is the area of one of the square water surfaces in the center?

55. What is the area of the top of one of the stone surfaces surrounding the water? **10x + 15**

56. Factor the polynomial that represents the area of the top of one of the stone surfaces. **5(2x + 3)**

57. Draw a rectangle by using the two factors you found in Exercise 56 for the dimensions.

CHALLENGE

58. Show that the area of this rectangle equals the area of the stone surface.

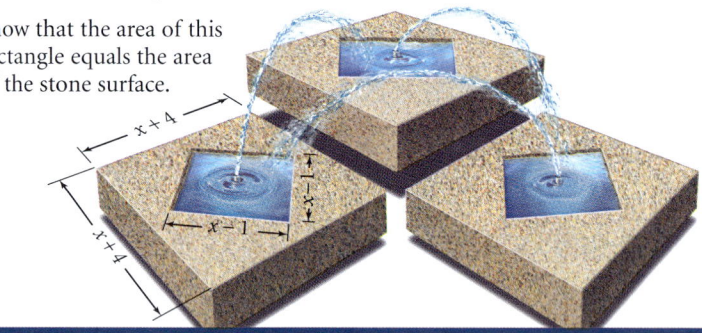

Practice

9.6 Factoring Special Polynomials

Use the generalization of a perfect-square trinomial or the difference of two squares to find each product.

1. $(x - 1)^2$ $x^2 - 2x + 1$
2. $(2x + 3)^2$ $4x^2 + 12x + 9$
3. $(a + 5)^2$ $a^2 + 10a + 25$
4. $(3g + 1)^2$ $9g^2 + 6g + 1$
5. $(2a - 5)^2$ $4a^2 - 20a + 25$
6. $(x - y)^2$ $x^2 - 2xy + y^2$
7. $(x + 2y)^2$ $x^2 + 4xy + 4y^2$
8. $(3m - n)^2$ $9m^2 - 6mn + n^2$
9. $(x - 3)(x + 3)$ $x^2 - 9$
10. $(4y + 1)(4y - 1)$ $16y^2 - 1$
11. $(5x + 2)(5x - 2)$ $25x^2 - 4$
12. $(6r - s)(6r + s)$ $36r^2 - s^2$

Factor each polynomial completely.

13. $x^2 - 25$ $(x + 5)(x - 5)$
14. $y^2 - 49$ $(y - 7)(y + 7)$
15. $2x^2 - 2$ $2(x + 1)(x - 1)$
16. $4y^2 - 16$ $4(y - 2)(y + 2)$
17. $16y^2 - 25$ $(4y - 5)(4y + 5)$
18. $121 - x^2$ $(11 - x)(11 + x)$
19. $9 - 16y^2$ $(3 - 4y)(3 + 4y)$
20. $x^2 - 6x + 9$ $(x - 3)^2$
21. $x^2 + 10x + 25$ $(x + 5)^2$
22. $y^2 - 16y + 64$ $(y - 8)^2$
23. $z^2 - 20z + 100$ $(z - 10)^2$
24. $64x^2 - 48xy + 9y^2$ $(8x - 3y)^2$
25. $25 - 10a + a^2$ $(5 - a)^2$
26. $4m^2 + 4m + 1$ $(2m + 1)^2$
27. $81r^2s^2 - 100t^4q^4$ see below
28. $16x^2 + 24x + 9$ $(4x + 3)^2$
29. $x^2 + 2xy + y^2$ $(x + y)^2$
30. $4a^2 + 4ab + b^2$ $(2a + b)^2$
31. $y^2 - x^2$ $(y - x)(y + x)$
32. $25a^2 - 1$ $(5a - 1)(5a + 1)$
33. $1 - x^2y^4$ $(1 - xy^2)(1 + xy^2)$
34. $144c^2 - 120cd + 25d^2$ $(12c - 5d)^2$
35. $6a^2 - 216b^2$ $6(a - 6b)(a + 6b)$
36. $49x - xy^2$ $x(7 - y)(7 + y)$
37. $12x^4 - 12$ $12(x^2 + 1)(x - 1)(x + 1)$
38. $b^4 - 2b^2 + b^2 - 2$ $(b^2 + 1)(b^2 - 2)$
39. $4c^2 - 24c + 36$ $4(c - 3)^2$
40. $4m^4n^2 + 4m^3n^2 + m^2n^2$ $m^2n^2(2m + 1)^2$
27. $(9rs - 10t^2q^2)(9rs + 10t^2q^2)$

49.

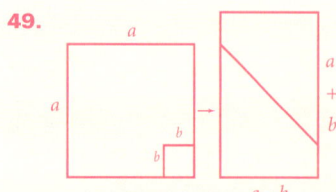

Answers may vary. Sample answer: The rectangle formed has dimensions $(a + b)(a - b)$; therefore, by comparing areas, $a^2 - b^2 = (a + b)(a - b)$.

57.

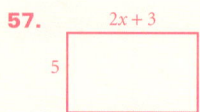

58.
$x^2 + 8x + 16$ area of large square
$-\ x^2 - 2x + 1$ area of water square
$\overline{10x + 15}$ area of stone surface

Area of rectangle $= 5(2x + 3)$
$= 10x + 15$, which is also the area of the stone surface

 **Look Back**

CONNECTION

59. STATISTICS Mark has scores of 78, 83, and 92 on his first three tests. What must his score be on the next test in order to have an average of 87? **(LESSON 4.4)** 95

60. Find the slope of a line that is parallel to the line represented by the equation $3y = 5 - 4x$. **(LESSON 5.6)** $-\frac{4}{3}$

61. Find the slope of a line that is perpendicular to the line you found in Exercise 61. **(LESSON 5.6)** $\frac{3}{4}$

62. Graph the solution of $|x| < 4$ on a number line. **(LESSON 6.5)**

63. Graph this system of inequalities. $\begin{cases} y < x - 5 \\ y \geq 3 \end{cases}$ **(LESSON 7.5)**

 Look Beyond

64. Given the trinomial $x^2 + x - 42$, what are possible pairs of factors for -42? Which factors result in the sum of 1? What does the sign before the last term of the trinomial determine?

1
3

2
4

3
5
15

3   8   15

Include this activity in your portfolio.

1. Find the next three terms in the sequence 0, 3, 8, 15, . . . Compare this sequence with the sequence of square numbers, 1, 4, 9, 16, . . . Write a polynomial function to describe each sequence.

2. Draw the next two models in the sequence shown above. Label the dimensions.

3. Use the model to explain why $x^2 - 1 = (x + 1)(x - 1)$.

4. Verify the identity $x^2 - 1 = (x + 1)(x - 1)$ by building a table for the two functions, $y = x^2 - 1$ and $y = (x + 1)(x - 1)$, using integer values from -3 to 3 for x.

WORKING ON THE CHAPTER PROJECT

You should now be able to complete Activities 2 and 3 of the Chapter Project on page 470.

 **Look Beyond**

Exercise 64 previews a method for factoring any quadratic trinomial. Students will learn more about this method in Lesson 9.7.

ALTERNATIVE Assessment

Portfolio Activity

The Portfolio Activity can be used as preparation for the Chapter Project or as a separate activity. In the Portfolio Activity on this page, students investigate a geometric pattern that models the identity $x^2 - 1 = (x + 1)(x - 1)$.

Answers to Portfolio Activities can be found in Additional Answers of the Teacher's Edition.

62.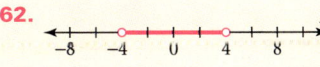

63.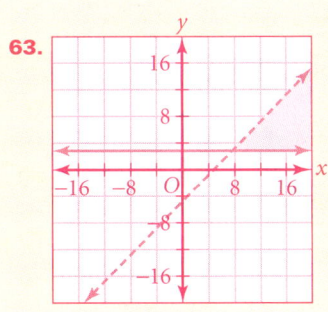

64. (−1, 42), (1, −42), (−2, 21), (2, −21), (−3, 14), (3, −14), (−6, 7), (6, −7); (−6, 7); answers may vary. Sample answer: The sign before the last term determines whether the signs of the factors will be the same or different.

Student Technology Guide

9.6 Factoring Special Polynomials

In Lesson 9.6, you saw special factoring patterns for differences of squares and perfect-square trinomials. You can use a graphics calculator to see whether a quadratic expression is one of these special polynomials.

Set a quadratic expression equal to y and graph the resulting function.

- If the graph touches the x-axis at two points that are an equal distance from the origin, the expression is a *difference of two squares*.
- If the graph touches the x-axis at exactly one point, the expression is a *perfect-square trinomial*.

Difference of two squares: $y = x^2 - 4$ Perfect-square trinomial: $y = x^2 + 6x + 9$

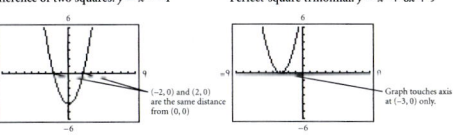

If the graph does neither of these things, the expression is neither a difference of two squares nor a perfect-square trinomial.

Example: State whether $4x^2 - 12x + 9$ is a perfect-square trinomial, a difference of two squares, or neither.

First, graph $y = 4x^2 - 12x + 9$. Press Y= 4 X,T,θ,n x^2 − 12 X,T,θ,n + 9.

Look at the graph. It appears to touch the x-axis at only one point, near (1.5, 0). This means that $4x^2 - 12x + 9$ is a perfect-square trinomial.

To be more confident that the graph actually touches the axis, use TRACE to move as close as you can to that point and then press ZOOM 2: Zoom In ENTER ENTER to examine the graph more closely.

State whether each polynomial is a perfect-square trinomial, a difference of two squares, or neither.

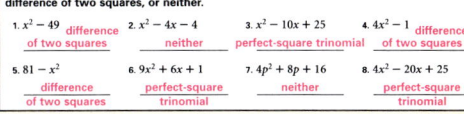

LESSON 9.6 **457**

Prepare

NCTM PRINCIPLES & STANDARDS 1–4, 6–10

LESSON RESOURCES

- Practice 9.7
- Reteaching Master 9.7
- Enrichment Master 9.7
- Cooperative-Learning Activity 9.7
- Lesson Activity 9.7
- Problem Solving/Critical Thinking Master 9.7

QUICK WARM-UP

Factor each polynomial by using the GCF.

1. $24v + 16$ $8(3v + 2)$
2. $6t^2 - 18t$ $6t(t - 3)$
3. $8a^4 + 16a^2 + 10$ $2(4a^4 + 8a^2 + 5)$

Factor each polynomial.

4. $n^2 - 144$ $(n + 12)(n - 12)$
5. $16z^2 - 40z + 25$ $(4z - 5)^2$

Also on Quiz Transparency 9.7

Teach

Why Ask students to describe real-world situations that they might approach by working backward. A sample situation might be working backward from the time school starts to plan a time to get up, a time to eat breakfast, a time to leave the house, and so on. Tell students that in this lesson, they will learn how they can factor certain types of polynomials by working backward.

9.7 Factoring Quadratic Trinomials

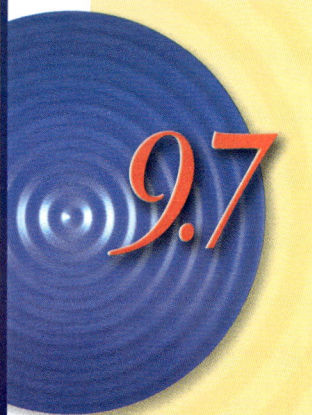

Objectives

- Factor quadratic trinomials by using algebra tiles.
- Factor quadratic trinomials by using guess-and-check methods.

Why If you examine certain products carefully, you will discover patterns that enable you to recognize the factors. This will simplify your work with real-world problems involving trinomials.

APPLICATION FINE ARTS

If a rectangular area is covered by 30 uniform squares, what could the length and width of the area be? To answer this question, you would need to find pairs of numbers that are factors of 30.

Factoring With Tiles

Suppose that the area of a rectangle is represented by the quadratic trinomial $x^2 + 6x + 8$. How can you find expressions for its length and width? One way to find the length and width is to use algebra tiles. If the tiles can be arranged to form an appropriate rectangle, the length and width can be determined.

Start with tiles that model $x^2 + 6x + 8$.

$x^2 + 6x + 8$

Arrange the tiles in a rectangle. Count the tiles representing the length and the width.

The length is represented by $x + 4$, and the width is represented by $x + 2$. To check, multiply the factors.

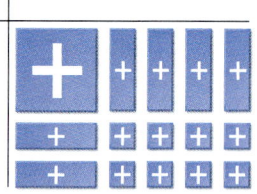

Alternative Teaching Strategy

USING PATTERNS AND TABLES Have students find these products. Then discuss the results.

$(y + 2)(y + 18)$	$y^2 + 20y + 36$
$(y + 3)(y + 12)$	$y^2 + 15y + 36$
$(y + 4)(y + 9)$	$y^2 + 13y + 36$
$(y + 6)(y + 6)$	$y^2 + 12y + 36$

Lead students to see that the constant in each product is 36 because each pair of binomials contains factors of 36. Note that the y-terms of the products differ, with the coefficient being determined by the sum of the factors of 36.

Now write $c^2 + 3c - 10$ on the board or overhead. Tell students that to factor this trinomial, they must find two factors of -10 whose sum is 3. Work with them to make a table like the one at right. The factors that satisfy the conditions are -2 and 5, so $c^2 + 3c - 10 = (c - 2)(c + 5)$.

Factors of –10		Sum of Factors
–1	10	9
–2	5	3
2	–5	–3
1	–10	–9

458   LESSON 9.7

You may need to add neutral pairs when finding factors with algebra tiles. For example, examine the tiles for the quadratic trinomial $x^2 + 2x - 8$.

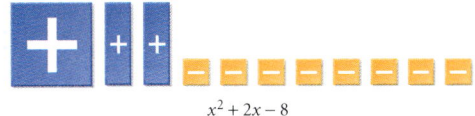

$x^2 + 2x - 8$

Place the variable tiles first. Then try a 2 × 4 arrangement of the negative unit tiles.

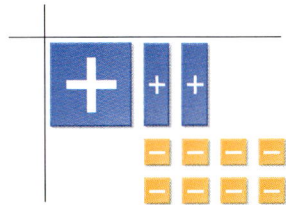

Complete the rectangle by adding 2 positive and 2 negative x-tiles.

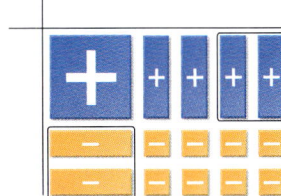

Finally, determine the factor tiles.

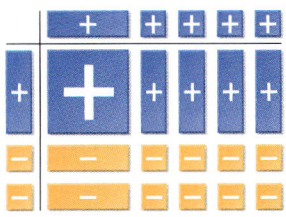

Thus, the factors of $x^2 + 2x - 8$ are $(x - 2)$ and $(x + 4)$. To check, find the product.

$$(x - 2)(x + 4) = x^2 + 4x - 2x - 8$$
$$= x^2 + 2x - 8$$

CRITICAL THINKING Sometimes you may need to try more than one arrangement to find the correct pattern. Suppose, for example, that your first attempt in the problem above was to arrange the negative unit tiles in a 1 × 8 pattern, as shown below.

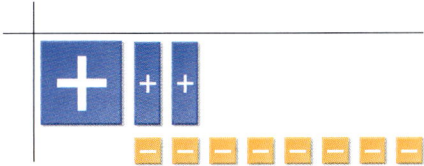

Explain why it is impossible to complete an appropriate rectangle with this arrangement.

TRY THIS Explain how to use tiles to factor $x^2 - 2x - 8$. Check by multiplying the factors and simplifying.

Inclusion Strategies

VISUAL LEARNERS Some students will benefit from drawing an area diagram to verify a factorization. For instance, the diagram at right gives a visual confirmation that the following is true:

$$4p^2 + 4p - 15 = (2p + 5)(2p - 3)$$

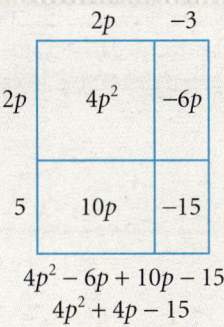

$4p^2 - 6p + 10p - 15$
$4p^2 + 4p - 15$

Teaching Tip

You may need to remind students of the meaning of *quadratic trinomial*.

CRITICAL THINKING
You can only add neutral pairs, which is impossible in this arrangement.

TRY THIS
Start with tiles that model $x^2 - 2x - 8$. Try to place the tiles in a rectangle, adding neutral pairs if necessary. Add 2 positive and 2 negative x-tiles. The factors are $(x - 4)$ and $(x + 2)$.

Check:
$(x - 4)(x + 2) = x^2 + 2x - 4x - 8$
$ = x^2 - 2x - 8$

LESSON 9.7

ADDITIONAL EXAMPLE 1

Factor $x^2 + 12x + 35$.

$(x + 5)(x + 7)$

CRITICAL THINKING

The constant term of a trinomial is the product of the constant terms of the factors. The product of a positive number and a negative number is negative, so the only other possibilities are that they are both positive or both negative.

ADDITIONAL EXAMPLE 2

Factor $x^2 - 12x + 35$.

$(x - 5)(x - 7)$

TRY THIS

a. $(x + 4)(x + 10)$

b. $(x - 2)(x - 8)$

CRITICAL THINKING

The trinomials are the same except for the sign of the middle term. The factors in Example 2 have the opposite sign of the factors in Example 1.

Factoring by Working Backward

Because factoring is related to multiplication, a quadratic trinomial can be factored by working backward with the FOIL method.

EXAMPLE 1 Factor $x^2 + 6x + 8$.

SOLUTION

PROBLEM SOLVING

Work backward. Because the constant term of the trinomial is positive, the constant terms in the binomial factors must be both positive or both negative.

$(x + \_\_\_)(x + \_\_\_)$ or $(x - \_\_\_)(x - \_\_\_)$

Because the coefficient of the middle term is positive, the constant terms in the binomial factors must both be positive. Find a pair of numbers that have a product of 8 and a sum of 6.

Test 2 and 4.

$(x + 2)(x + 4) = x^2 + 4x + 2x + 8$
$= x^2 + 6x + 8$ Correct

Therefore, the factors are $(x + 2)$ and $(x + 4)$.

CRITICAL THINKING

If the constant term of a trinomial is positive, explain why the constant terms in the binomial factors must be both positive or both negative.

EXAMPLE 2 Factor $x^2 - 6x + 8$.

SOLUTION

PROBLEM SOLVING

Work backward. Because the constant term of the trinomial is positive, the constant terms in the binomial factors must be both positive or both negative.

$(x + \_\_\_)(x + \_\_\_)$ or $(x - \_\_\_)(x - \_\_\_)$.

Because the coefficient of the middle term is negative, the constant terms in both binomial factors must be negative. Find a pair of numbers that have a product of 8 and a sum of −6.

You can see that −2 and −4 give the correct result.

$(x - 2)(x - 4) = x^2 - 2x - 4x + 8$
$= x^2 - 6x + 8$ Correct

Therefore, the factors are $(x - 2)$ and $(x - 4)$.

TRY THIS Factor each trinomial.
     a. $x^2 + 14x + 40$ b. $x^2 - 10x + 16$

CRITICAL THINKING Compare Examples 1 and 2. Explain how the trinomials are related. How are the binomial factors related?

Enrichment

Give students the trinomial $x^2 + 6x + n$. Tell them to find three positive and three negative values of n that will make the trinomial factorable over the integers.

samples:
$n = 5: x^2 + 6x + 5 = (x + 1)(x + 5)$
$n = 8: x^2 + 6x + 8 = (x + 2)(x + 4)$
$n = 9: x^2 + 6x + 9 = (x + 3)(x + 3)$
$n = -7: x^2 + 6x - 7 = (x - 1)(x + 7)$
$n = -16: x^2 + 6x - 16 = (x - 2)(x + 8)$
$n = -27: x^2 + 6x - 27 = (x - 3)(x + 9)$

Repeat the activity for the trinomial $ax^2 + 7x + b$. This time, tell students to find any six pairs of values of a and b, where $a \ne 1$, that will make the trinomial factorable over the integers.

samples:
$a = 6, b = 1: 6x^2 + 7x + 1 = (6x + 1)(x + 1)$
$a = 5, b = 2: 5x^2 + 7x + 2 = (5x + 2)(x + 1)$
$a = 4, b = 3: 4x^2 + 7x + 3 = (4x + 3)(x + 1)$
$a = 3, b = 4: 3x^2 + 7x + 4 = (3x + 4)(x + 1)$
$a = 2, b = 5: 2x^2 + 7x + 5 = (2x + 5)(x + 1)$
$a = 8, b = -1: 8x^2 + 7x - 1 = (8x - 1)(x + 1)$

EXAMPLE 3 Factor $x^2 + 2x - 8$.

● SOLUTION

PROBLEM SOLVING Work backward. Because the constant term of the trinomial is negative, one factor will contain $(x - \_\_\_)$, and the other will contain $(x + \_\_\_)$.

Because the coefficient of the middle term is positive, the positive factor must be larger than the absolute value of the negative factor. Find a pair of numbers that have a product of –8 and a sum of 2.

You can see that –2 and 4 give the correct result.

$(x - 2)(x + 4) = x^2 + 4x - 2x + 8$
$ = x^2 + 2x - 8$ Correct

Thus, the factors are $(x - 2)$ and $(x + 4)$.

CRITICAL THINKING For the trinomial in Example 3, explain why the positive factor must be larger than the absolute value of the negative factor.

EXAMPLE 4 Factor $x^2 - 2x - 8$.

● SOLUTION

PROBLEM SOLVING Work backward. Because the constant term of the trinomial is negative, one factor will contain $(x - \_\_\_)$, and the other will contain $(x + \_\_\_)$.

Because the coefficient of the middle term is negative, the positive constant term must be smaller than the absolute value of the negative constant term. Find a pair of numbers that have a product of –8 and a sum of –2.

You can see that 2 and –4 give the correct result.

$(x + 2)(x - 4) = x^2 - 4x + 2x - 8$
$ = x^2 - 2x - 8$ Correct

Thus, the factors are $(x + 2)$ and $(x - 4)$.

TRY THIS Factor each trinomial.
   a. $x^2 + 9x + 18$ **b.** $x^2 - 10x + 21$ **c.** $x^2 + 4x - 21$ **d.** $x^2 - x - 20$

CRITICAL THINKING Compare Examples 3 and 4. Explain how the trinomials are related. How are the binomials related?

Sometimes the terms of a trinomial has a common monomial factor. To write the trinomial as a product, it may be convenient to factor out the common monomial first.

EXAMPLE 5 Write $2x^3 + 16x^2 + 24x$ as a product.

● SOLUTION

$2x^3 + 16x^2 + 24x = 2x(x^2 + 8x + 12)$
$ = 2x(x + 2)(x + 6)$

ADDITIONAL EXAMPLE 3

Factor $x^2 + 2x - 15$.

$(x + 5)(x - 3)$

CRITICAL THINKING
The middle term is positive.

ADDITIONAL EXAMPLE 4

Factor $x^2 - 2x - 15$.

$(x - 5)(x + 3)$

TRY THIS
a. $(x + 3)(x + 6)$
b. $(x - 3)(x - 7)$
c. $(x - 3)(x + 7)$
d. $(x + 4)(x - 5)$

CRITICAL THINKING
The trinomials are the same except for the sign of the middle term. The constant terms of the binomial factors are opposites.

ADDITIONAL EXAMPLE 5

Write $3x^3 + 27x^2 + 42x$ as a product.

$3x(x + 2)(x + 7)$

Reteaching the Lesson

COOPERATIVE LEARNING Have students work in groups of four. Each student should write three products of binomials, one in each of the following forms, on a sheet of paper:

$(x + \underline{\ ?\ })(x + \underline{\ ?\ })$
$(x - \underline{\ ?\ })(x - \underline{\ ?\ })$
$(x + \underline{\ ?\ })(x - \underline{\ ?\ })$

Tell students to multiply their binomials and then write the products on three separate index cards. Each student should then give the index cards to the student to the right, and all group members should factor the polynomials in the set of cards they receive. Tell them to record the factorizations on a sheet of paper. Continue passing the cards until all group members have worked with each set of cards. Students should then compare their factorizations and resolve any discrepancies.

Repeat the activity for binomial multiplications of each form below.

$(\underline{\ ?\ }x + \underline{\ ?\ })(\underline{\ ?\ }x + \underline{\ ?\ })$
$(\underline{\ ?\ }x - \underline{\ ?\ })(\underline{\ ?\ }x - \underline{\ ?\ })$
$(\underline{\ ?\ }x + \underline{\ ?\ })(\underline{\ ?\ }x - \underline{\ ?\ })$

Teaching Tip

TECHNOLOGY If you are using a TI-92 to factor Exercise 5, first press [F2] and select [2:factor(]. Enter X ^ 2 + 7X + 12, and then press [ENTER]. Follow the same procedure for the other expressions.

Assess

Selected Answers
Exercises 5–9, 11–45 odd

ASSIGNMENT GUIDE

In Class	1–9
Core	10–37, 40
Core Plus	10–40
Review	41–46
Preview	47

✐ Extra Practice can be found beginning on page 738.

SUMMARY OF FACTORING METHODS

Method	Example polynomial	Factored form
Use the GCF	$4x^2 + 12x$	$4x(x + 3)$
Factor by grouping	$x^2 + x + 4x + 4$	$(x + 4)(x + 1)$
Perfect-square trinomial pattern	$x^2 + 8x + 16$ $x^2 - 8x + 16$	$(x + 4)(x + 4)$ $(x - 4)(x - 4)$
Difference of squares pattern	$x^2 - 25$	$(x + 5)(x - 5)$
Working backward	$x^2 + 3x + 2$	$(x + 2)(x + 1)$

Exercises

● Communicate

1. Explain how to use algebra tiles to factor each polynomial.
   a. $x^2 - 5x + 4$ **b.** $x^2 - 4x - 12$ **c.** $x^2 + 6x + 9$

2. Write each trinomial in factored form. Tell why the signs in the binomial factors are the same or opposite.
   a. $x^2 + x - 6$ **b.** $x^2 - 7x + 10$ **c.** $x^2 + 2x - 15$

3. Explain how to use the guess-and-check method to factor $x^2 - 5x - 24$.

4. If the third term of a trinomial is 36, write the possible factor pairs.

● Guided Skills Practice

5. Factor $x^2 + 7x + 12$. **(EXAMPLE 1)** $(x + 3)(x + 4)$
6. Factor $x^2 - 7x + 12$. **(EXAMPLE 2)** $(x - 3)(x - 4)$
7. Factor $x^2 + x - 12$. **(EXAMPLE 3)** $(x - 3)(x + 4)$
8. Factor $x^2 - x - 12$. **(EXAMPLE 4)** $(x + 3)(x - 4)$
9. Write $3x^3 + 18x^2 + 27x$ as a product. **(EXAMPLE 5)** $(3x)(x + 3)^2$

● Practice and Apply

Use algebra tiles to factor each of the following trinomials:

10. $x^2 + 5x + 6$ 11. $x^2 - x - 12$ 12. $x^2 - 6x + 5$ 13. $x^2 - 6x + 9$

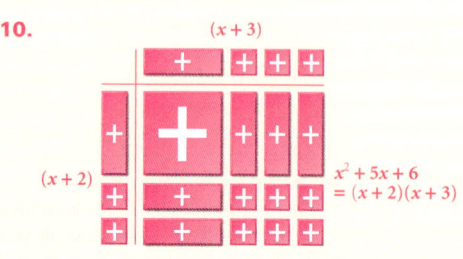

10. $x^2 + 5x + 6 = (x + 2)(x + 3)$

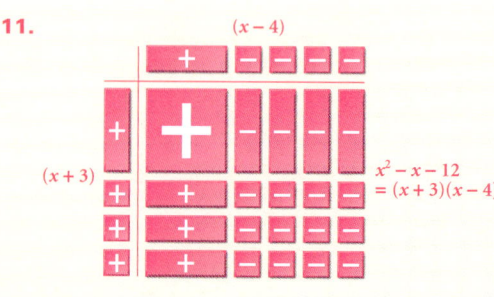

11. $x^2 - x - 12 = (x + 3)(x - 4)$

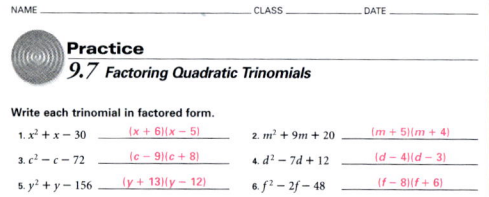

Practice 9.7 Factoring Quadratic Trinomials

Write each trinomial in factored form.

1. $x^2 + x - 30$ $(x + 6)(x - 5)$
2. $m^2 + 9m + 20$ $(m + 5)(m + 4)$
3. $c^2 - c - 72$ $(c - 9)(c + 8)$
4. $d^2 - 7d + 12$ $(d - 4)(d - 3)$
5. $y^2 + y - 156$ $(y + 13)(y - 12)$
6. $f^2 - 2f - 48$ $(f - 8)(f + 6)$

For each polynomial, write all of the factor pairs of the third term, and then circle the pair that would successfully factor the polynomial.

7. $n^2 - 8n + 15$ 1, 15; -1, -15; 3, 5; ⦶-3, -5⦵
8. $t^2 - 121$ -1, 121; 1, -121; ⦶-11, 11⦵
9. $s^2 + 5s + 4$ ⦶-4, -1⦵; 4, 1; 2, 2; -2, -2
10. $q^2 - 2q - 35$ -1, 35; 1, -35; -5, 7; ⦶5, -7⦵

Write each trinomial as a product of its factors. Use factoring patterns, graphing, or algebra tiles to assist you in your work.

11. $g^2 - 3g - 40$ $(g - 8)(g + 5)$
12. $h^2 + 6h - 40$ $(h + 10)(h - 4)$
13. $j^2 + 22j + 40$ $(j + 20)(j + 2)$
14. $k^2 - 39k - 40$ $(k - 40)(k + 1)$
15. $x^2 - x - 12$ $(x + 3)(x - 4)$
16. $y^2 - 7y - 18$ $(y - 9)(y + 2)$
17. $a^2 - 9a + 14$ $(a - 7)(a - 2)$
18. $x^2 - 5x - 6$ $(x - 6)(x + 1)$
19. $x^2 - 8x + 15$ $(x - 3)(x - 5)$
20. $p^2 + 18p + 45$ $(p + 3)(p + 15)$
21. $x^2 - 9x - 36$ $(x + 3)(x - 12)$
22. $x^2 - x - 42$ $(x + 6)(x - 7)$
23. $x^2 - 81$ $(x + 9)(x - 9)$
24. $x^2 + 17x + 60$ $(x + 5)(x + 12)$
25. $x^2 + 4x - 12$ $(x - 2)(x + 6)$
26. $x^2 + 14x - 32$ $(x - 2)(x + 16)$
27. $x^2 + 12x + 35$ $(x + 5)(x + 7)$
28. $x^2 - x - 72$ $(x - 9)(x + 8)$
29. $x^2 + 17x + 72$ $(x + 9)(x + 8)$
30. $x^2 - 17x + 72$ $(x - 9)(x - 8)$
31. $x^2 + x - 72$ $(x + 9)(x - 8)$
32. $x^3 + 7x^2 + 12x$ $x(x + 3)(x + 4)$
33. $2x^2 + 6x + 3$ $(x + 3)(2x + 1)$
34. $6x^2 + 8x - 30$ $(3x - 5)(2x + 6)$
35. $5x^2 + 7xy + 2y^2$ $(x + y)(5x + 2y)$
36. $14x^2 + 23xy + 3y^2$ $(2x + 3y)(7x + y)$

462 LESSON 9.7

14. $(a + 5)(a - 7)$
15. $(p - 2)(p + 6)$
16. $(y - 2)(y - 3)$
17. $(b + 3)(b - 8)$
18. $(n - 2)(n - 9)$
19. $(z - 4)(z + 5)$
22. $(x + 4)(x - 7)$
23. $(s - 3)(s - 21)$
24. $(x - 4)(x + 7)$
26. $4xy(x - 1)(x - 4)$
27. $6y(y - 1)(y - 2)$
29. $(a + 3)(a - 1)(a + 6)$
31. $(x + 6)(x - 6)(x + 2)$
32. $-(x^2 + 2)(x - 2)(x + 2)$
37. $(a + b)(2x + y)$
38. $\pm 13, \pm 8, \pm 7$
39. $\pm 11, \pm 4, \pm 1$

CHALLENGE

CONNECTION

Factor each trinomial. If a trinomial cannot be factored, write *prime*.

14. $a^2 - 2a - 35$
15. $p^2 + 4p - 12$
16. $y^2 - 5y + 6$
17. $b^2 - 5b - 24$
18. $n^2 - 11n + 18$
19. $z^2 + z - 20$
20. $x^2 + 5x + 12$ Prime
21. $a^2 - 2a + 7$ Prime
22. $x^2 - 3x - 28$
23. $s^2 - 24s + 63$
24. $x^2 - 3x - 28$
25. $m^2 - 3m + 17$ Prime

Factor each polynomial. In some cases you will need to factor out a constant monomial term.

26. $4x^3y - 20x^2y + 16xy$
27. $6y^3 - 18y^2 + 12y$
28. $x^2 - 18x + 81$ $(x - 9)^2$
29. $(a + 3)(a^2 + 5a) - 6(a + 3)$
30. $5x^3 - 50x^2 + 45x$ $5x(x - 1)(x - 9)$
31. $x^3 + 2x^2 - 36x - 72$
32. $-x^4 + 2x^2 + 8$
33. $64p^4 - 16$ $16(2p^2 + 1)(2p^2 - 1)$
34. $z^2 - 5z - 36$ $(z + 4)(z - 9)$
35. $x^2 - 2x + 1$ $(x - 1)^2$
36. $125x^2y - 5x^4$ $5x^2(25y - x^2)$
37. $2ax + ay + 2bx + by$

38. Find all possible values of b such that $x^2 + bx + 12$ can be factored.
39. Find all possible values of b such that $x^2 + bx - 12$ can be factored.
40. **GEOMETRY** What are the dimensions of a quadrilateral with an area of $49x^2 - 64$? Could the figure be a square? a rectangle?
(7x + 8) units by (7x - 8) units; The sides are not equal, thus it could not be a square, however it could be a rectangle.

 Look Back

Solve. *(LESSONS 3.1 AND 3.2)*

41. $\frac{3}{4}a = 163$ $217\frac{1}{3}$
42. $\frac{z}{-8} = \frac{11}{12}$ $-7\frac{1}{3}$
43. $w - \frac{7}{9} = 93$ $93\frac{7}{9}$

APPLICATION

44. **SALES TAX** A freshman class is donating a tree to a local children's hospital. If sales tax is 5.5%, what is the total price of a tree marked at $139? *(LESSON 4.2)* **$146.65**

45. A line with a slope of -2 passes through point $A(4, -1)$. Write the equation of the line that passes through this point and that is perpendicular to the original line. *(LESSON 5.6)* $y = \frac{1}{2}x - 3$

46. Represent the inequality $-4 < x < 5$ on a number line. *(LESSON 6.3)*

 Look Beyond

CULTURAL CONNECTION Americas de Padilla, a Guatemalan mathematician, wrote this problem 250 years ago.

47. Find two numbers, given that the second is 3 times the first. If you multiply the first number by the second number and the second number by 4, the sum of the products is 420.
The numbers are 10 and 30 or −14 and −42.

Technology

A calculator may be helpful in performing some of the multiplications involved in checking factorizations in Exercises 14–37.

Error Analysis

When factoring a trinomial of the form $x^2 + bx - c$, students often switch the signs in the binomial factors. For instance, when factoring $z^2 + z - 30$, they may write $(z + 5)(z - 6)$. Remind them to check both the product *and the sum* of the factors.

 Look Beyond

In Exercise 47, the solution of a classic problem provides a preview of solving equations by factoring. Students will learn this method of solving equations in Lesson 9.8.

46.

Student Technology Guide

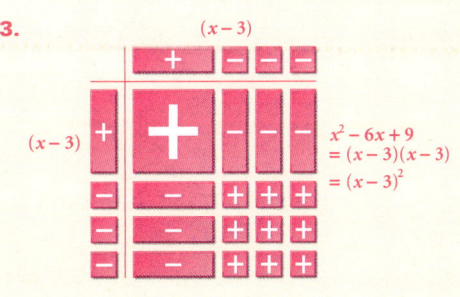

12.
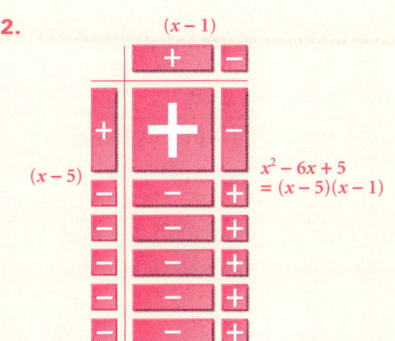
$(x - 1)$
$(x - 5)$
$x^2 - 6x + 5$
$= (x - 5)(x - 1)$

13.
$(x - 3)$
$(x - 3)$
$x^2 - 6x + 9$
$= (x - 3)(x - 3)$
$= (x - 3)^2$

LESSON 9.7 **463**

Prepare

NCTM PRINCIPLES & STANDARDS
1–6, 8–10

LESSON RESOURCES

- Practice 9.8
- Reteaching Master 9.8
- Enrichment Master 9.8
- Cooperative-Learning Activity 9.8
- Lesson Activity 9.8
- Problem Solving/Critical Thinking Master 9.8

QUICK WARM-UP

Factor each polynomial.

1. $u^2 + 6u + 8$ $(u + 2)(u + 4)$
2. $k^2 + 4k + 4$ $(k + 2)^2$
3. $b^2 - b - 12$ $(b + 3)(b - 4)$

Solve each equation.

4. $8 - j = 18$ $j = -10$
5. $-3c = -42$ $c = 14$
6. $2(x + 10) = 0$ $x = -10$

Also on Quiz Transparency 9.8

Teach

Why Tell students to imagine that a ball is thrown upward. Ask them to describe the height of the ball over time. Elicit the response that the height increases quickly at first and then more and more slowly until it reaches its greatest height. Next the height decreases slowly and then more and more quickly. Tell students that the height of the ball can be modeled by a polynomial function.

464   LESSON 9.8

9.8 Solving Equations by Factoring

Objective
- Solve equations by factoring.

Why Factoring can be used to help solve equations that model real-world problems.

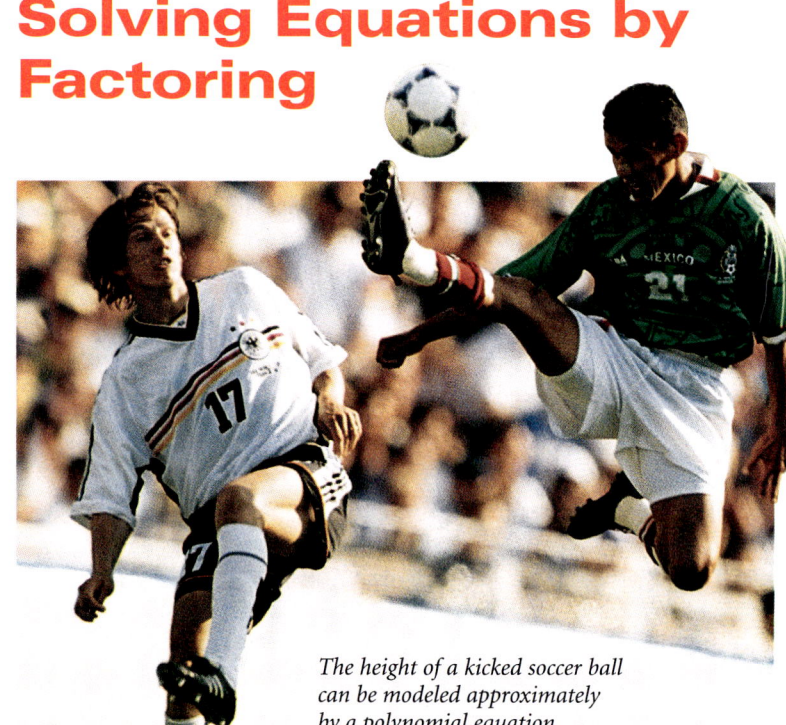

The height of a kicked soccer ball can be modeled approximately by a polynomial equation.

APPLICATION
PHYSICS

The height of a soccer ball kicked with an initial upward velocity of 48 feet per second can be approximated by the equation $y = -16x^2 + 48x$, where y is the height in feet and x is the time in seconds. You can use factoring to find the values of x when y is 0.

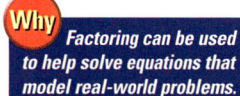

Exploring the Zeros of a Function

You will need: graphics calculator (optional)

1. Complete the table by using the equation from the soccer-ball problem, $y = -16x^2 + 48x$.

Time, x	0	1	2	3
Height, y	?	?	?	?

2. When is the soccer ball on the ground (at a height of 0)?
3. Write the right side of the equation in factored form.
4. What values of x make one of the factors equal to 0?

CHECKPOINT ✓
5. Explain how you can use the factors to find the time when the soccer ball hits the ground.
6. Use a graphics calculator or graph paper to graph the equation. How can you tell from the graph when the soccer ball is on the ground?

Alternative Teaching Strategy

USING COGNITIVE STRATEGIES Write these equations on the board or overhead.

$3(x - 4) = 0$ $3(x + 9) = 0$

Work with students to solve the equations by using properties of operations and of equality.

$3(x - 4) = 0$	$3(x + 9) = 0$	
$3x - 12 = 0$	$3x + 27 = 0$	Dist. Prop.
$3x = 12$	$3x = -27$	Add. Prop.
$x = 4$	$x = -9$	Dist. Prop.

Ask students to compare the solutions of the equations with the given equations. Elicit the observation that the solution of the first equation is the same as the solution of $x - 4 = 0$ and that the solution of the second equation is the same as the solution of $x + 9 = 0$. Use these observations as an introduction to the Zero Product Property. Then work with students to apply the property to solve these equations.

$x(x - 5) = 0$	$x = 0$ or $x = 5$
$(x - 2)(x + 7) = 0$	$x = 2$ or $x = -7$
$x^2 + 4x - 45 = 0$	$x = -9$ or $x = 5$

The Zeros of a Function

A function such as $y = (x - 2)(x - 4)$ is a *polynomial function* because the product on the right simplifies to a polynomial, $x^2 - 6x + 8$. (You can verify this by using algebra tiles.) What happens when you substitute the numbers 2 and 4 for x in the function?

$y = (x - 2)(x - 4) = (2 - 2)(2 - 4) = (0)(-2) = 0$ Substitute 2 for x.
$y = (x - 2)(x - 4) = (4 - 2)(4 - 4) = (2)(0) = 0$ Substitute 4 for x.

The numbers 2 and 4 are called the *zeros* of the function $y = x^2 - 6x + 8$ because the value of the function is 0 when these numbers are substituted for the independent variable, x.

When a polynomial function is written as a product of binomials, you can easily find the zeros of the polynomial function by using the Zero Product Property.

Zero Product Property
If a and b are real numbers such that $ab = 0$, then $a = 0$ or $b = 0$.

EXAMPLE 1 Find the zeros of the polynomial function $y = (x + 2)(x - 3)$.

- **SOLUTION**

The function will equal 0 when either factor equals 0. Set each factor equal to 0 and solve for x.

First factor	Second factor
$(x + 2) = 0$	$(x - 3) = 0$
$x = -2$	$x = 3$

The zeros of the polynomial function are -2 and 3.

Check by substituting these values into the original polynomial function.

$y = (x + 2)(x - 3) = (-2 + 2)(-2 - 3) = (0)(-5) = 0$ Substitute –2 for x.
$y = (x + 2)(x - 3) = (3 + 2)(3 - 3) = (5)(0) = 0$ Substitute 3 for x.

Since the polynomial function equals 0 in each case, the solutions are correct.

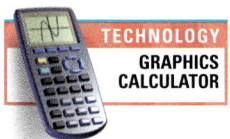
TECHNOLOGY GRAPHICS CALCULATOR

Use the table feature of your graphics calculator to build a table of values for the function $y = (x + 5)(x + 2)$.

Look for x-values that give a y-value of 0. You can see that the zeros of the function $y = (x + 5)(x + 2)$ are -5 and -2.

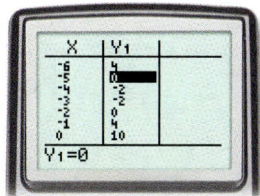

Interdisciplinary Connection

PHYSICS Scientific measurements are often taken in metric units. When t is the elapsed time in seconds, the equation $h = -4.9t^2 + vt$ gives the height, h, in meters of a projectile that is thrown upward from ground level with an initial velocity of v meters per second. Suppose that a projectile is thrown upward at a speed of 98 meters per second. After how many seconds will it hit the ground?
20 seconds

Inclusion Strategies

VISUAL LEARNERS Be sure that students are not interpreting the graph in the Activity as a picture of the path of the ball. Emphasize this with a demonstration such as the following: Toss a very light ball directly upward over a table. It will probably fall back to the table in about two seconds. Work with students to approximate a graph that plots time in seconds on the x-axis and height in feet on the y-axis. They should see that although the path of the ball lies entirely along a vertical line, the time-height graph will be the familiar shape of a parabola.

Activity Notes

In this Activity, students investigate a quadratic function. They see how to use three techniques—a table, a graph, and factoring—to determine the two values of x for which the value of the function is zero. After completing this Activity, students should have an intuitive understanding of the meaning of *zero of a function*.

See Keystroke Guide, page 782.

CHECKPOINT ✓
5. Each value of x that makes one of the factors equal to 0 is a time when the soccer ball is on the ground. Obviously, the soccer ball starts on the ground, which corresponds to $x = 0$.

ADDITIONAL EXAMPLE 1

Find the zeros of the polynomial function below.
$f(x) = (x - 6)(x + 5)$
6, –5

Teaching Tip

TECHNOLOGY If you are using a TI-82 or TI-83 graphics calculator, you can obtain the table shown at left as follows: Press Y= and enter **(X+5)(X+2)** next to Y1. Then press 2nd WINDOW and make the choices shown below in the **TABLE SETUP** screen.

TblStart=–6 (or TblMin=–6)
ΔTbl=1
Indpnt: Auto Ask
Depend: Auto Ask

Then press 2nd GRAPH to display the table.

See Keystroke Guide, page 782.

LESSON 9.8   **465**

Teaching Tip

You may wish to tell students that the shape of the graph of a quadratic equation is called a *parabola*. This term will be presented formally in Chapter 10.

ADDITIONAL EXAMPLE 2

Solve by factoring:
$-8x^2 - 40x = 0$.
$x = 0$ or $x = -5$; the solutions are 0 and 5.

See Keystroke Guide, page 782.

TRY THIS
$x = 0$ or $x = 10$

ADDITIONAL EXAMPLE 3

Solve by factoring.

a. $x^2 + 12x + 27 = 0$
$x = -3$ or $x = -9$; the solutions are -3 and -9.

b. $2x^2 + 5x - 12 = 0$
$x = -4$ or $x = \frac{3}{2}$; the solutions are -4 and $\frac{3}{2}$.

Using Factoring to Solve Equations

The solution to an equation like $-16x^2 + 96x = 0$ can be found by writing the equation in factored form and considering the values that make either factor equal to 0. The Zero Product Property allows you to solve many polynomial equations by factoring.

EXAMPLE 2 Solve by factoring: $-16x^2 + 96x = 0$.

TECHNOLOGY
GRAPHICS CALCULATOR

SOLUTION

Write the equation in factored form.

$-16x^2 + 96x = 0$
$-16x(x - 6) = 0$

The equation is true when $-16x = 0$ or $x - 6 = 0$. Solve each factor for x.

$-16x = 0$ $x - 6 = 0$
$x = 0$ $x = 6$
$x = 0$ or $x = 6$ The solutions are 0 and 6.

You can check your solutions by using the table and graph features on some graphics calculators.

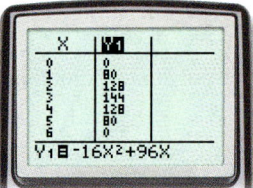 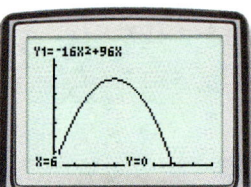

The table and the graph show that $y = 0$ when $x = 0$ or $x = 6$.

TRY THIS Solve by factoring: $-10x^2 + 100x = 0$.

EXAMPLE 3 Solve by factoring.

a. $x^2 + 5x - 24 = 0$ b. $4x^2 + x - 5 = 0$

SOLUTION

a. Write the equation in factored form.

$x^2 + 5x - 24 = 0$
$(x + 8)(x - 3) = 0$

The equation is true when $x + 8 = 0$ or $x - 3 = 0$.

Solve each factor for x.

$x + 8 = 0$ $x - 3 = 0$
$x = -8$ $x = 3$
$x = -8$ or $x = 3$ The solutions are -8 and 3.

Enrichment

Have students work in pairs. Tell them to write a quadratic equation that has the given pair of solutions. The equation should be written in the form $ax^2 + bx + c = 0$, where a, b, and c are integers.

1. 2, 1 — sample: $x^2 - 3x + 2 = 0$
2. $-7, -3$ — sample: $x^2 + 10x + 21 = 0$
3. $-4, 5$ — sample: $x^2 - x - 20 = 0$
4. $0, -8$ — sample: $x^2 + 8x = 0$
5. $\frac{1}{2}, 4$ — sample: $2x^2 - 9x + 4 = 0$
6. $-\frac{2}{5}, \frac{1}{3}$ — sample: $15x^2 + x - 2 = 0$

Reteaching the Lesson

GUIDED ANALYSIS Work with students to outline a step-by-step procedure for solving quadratic equations by factoring. Apply the procedure to two or three sample equations, and refine the procedure as necessary. The following is one possible procedure:

1. Rewrite the quadratic equation, if necessary, so that one side is zero.
2. Factor the nonzero side into linear factors.
3. Set each linear factor equal to zero.
4. Solve each linear equation.

b. Write the equation in factored form.

$$4x^2 + x - 5 = 0$$
$$(4x + 5)(x - 1) = 0$$

The equation is true when $4x + 5 = 0$ or $x - 1 = 0$.

Solve each factor for x.

$$4x + 5 = 0 \qquad x - 1 = 0$$
$$4x = -5 \qquad x = 1$$
$$x = -\tfrac{5}{4}$$

$x = -\tfrac{5}{4}$ or $x = 1$ The solutions are $-\tfrac{5}{4}$ and 1.

TRY THIS Solve by factoring.
a. $x^2 + 6x - 16 = 0$ **b.** $2x^2 + 5x - 3 = 0$

TRY THIS
a. $x = -8$ or $x = 2$
b. $x = \tfrac{1}{2}$ or $x = -3$

In Example 3, part **b**, you found that the solutions to the equation $4x^2 + x - 5 = 0$ are $-\tfrac{5}{4}$ and 1. Thus, $-\tfrac{5}{4}$ and 1 are called the zeros of the polynomial function $y = 4x^2 + x - 5$. You can find the zeros of a polynomial function by using special menu commands on many graphics calculators.

 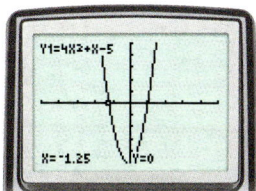

Teaching Tip

TECHNOLOGY If you are using a TI-82 or TI-83 graphics calculator, you can obtain the screens shown at left as follows: Press [Y=] and enter **4X^2+X–5** next to **Y1**. Then press [WINDOW] and make these settings.

Xmin=–5   Ymin=–5
Xmax= 5   Ymax= 5
Xscl= 1   Yscl= 1

Press [GRAPH] to view the graph.

Now press [2nd] [TRACE] and choose **2:**. Use [◄] to move the cursor above the x-axis. Press [ENTER]. Then use [►] to move the cursor below the x-axis, and press [ENTER] [ENTER].

Use a similar procedure to locate the second zero.

See Keystroke Guide, page 782.

Assess

Selected Answers
Exercises 6–14, 15–49 odd

Exercises

Communicate

1. Explain what it means to solve an equation by factoring. Give two examples.
2. What are the zeros of a polynomial function?
3. Describe a polynomial function whose zeros are 2 and 3.
4. Is it possible for a polynomial function to have no real zeros? Explain.
5. The height of a projectile over time can be modeled by a polynomial function. Explain what the zeros of this function would represent.

ASSIGNMENT GUIDE	
In Class	1–14
Core	15–40
Core Plus	15–41
Review	42–50
Preview	51–53

Extra Practice can be found beginning on page 738.

internetconnect
GO TO: go.hrw.com
KEYWORD: MA1 Zeros

In this activity, students discover some properties of higher degree polynomials by exploring a Web site that has an interactive cubic polynomial graphing utility and a Web site that approximates solutions for third- and fourth-degree polynomials.

LESSON 9.8

Technology

A graphics calculator may be helpful for Exercises 21–35. You may wish to have students first solve with paper and pencil and then use the calculator to check their answers.

Error Analysis

When solving a quadratic equation by factoring, students may factor correctly and then incorrectly identify the signs of the solutions. For example, given $x^2 + 6x - 16 = 0$, they may write $(x-2)(x+8) = 0$, which is correct, but then give the solutions as -2 and 8. Remind them of the importance of checking all solutions in the original equation.

36.

Width	Area
100	2500
75	3750
25	2500

37. 3750 ft^2; the largest pen possible will be a square with a side length of 62.5 ft; area = 62.5^2 = 3906.25 ft^2.

6. $x = -5$ and $x = -2$
7. $x = -2$ and $x = 10$
8. $x = -11$ and $x = -13$
9. $x = 0$ or $x = \frac{1}{5}$
10. $x = 0$ or $x = \frac{1}{4}$
11. $x = 0$ or $x = 2$
15. $x = -5$ and $x = 2$
16. $x = -15$ and $x = 7$
17. $x = -6$
21. $x = 3$
22. $x = -1$ or $x = 7$
23. $x = 1$ or $x = 9$
24. $x = 3$ or $x = 6$
25. $x = 3$ or $x = -6$
26. $x = \frac{1}{2}$ or $x = 1$
27. $x = -\frac{1}{2}$ or $x = 1$
28. $x = -4$ or $x = 9$
29. $x = 3$ or $x = 12$
30. $x = -2$ or $x = 18$
31. $x = -3$ or $x = 12$
32. $x = 3$ or $x = -4$
33. $x = 3$ or $x = 4$
34. $x = -3$ or $x = -4$
35. $x = -3$ or $x = 4$

APPLICATION

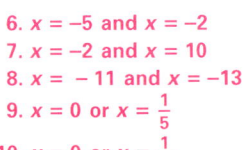

Guided Skills Practice

Identify the zeros of each function. *(EXAMPLE 1)*

6. $y = (x+5)(x+2)$
7. $y = (x+2)(x-10)$
8. $y = (x+11)(x+13)$

Solve by factoring. *(EXAMPLE 2)*

9. $-25x^2 + 5x = 0$
10. $12x^2 - 3x = 0$
11. $-15x^2 + 30x = 0$

Solve by factoring. *(EXAMPLE 3)*

12. $x^2 + 7x + 10 = 0$
    $x = -2$ or $x = -5$
13. $x^2 - 7x + 10 = 0$
    $x = 2$ or $x = 5$
14. $2x^2 + x - 10 = 0$
    $x = 2$ or $x = -\frac{5}{2}$

Practice and Apply

Identify the zeros of each function.

15. $y = (x+5)(x-2)$
16. $y = (x+15)(x-7)$
17. $y = (x+6)(x+6)$
18. $y = (x+4)(x-3)$
    $x = -4$ and $x = 3$
19. $y = (x+10)(x-3)$
    $x = -10$ and $x = 3$
20. $y = (2x+2)(2x-2)$
    $x = -1$ and $x = 1$

Solve by factoring.

21. $x^2 - 6x + 9 = 0$
22. $x^2 - 6x - 7 = 0$
23. $x^2 - 10x + 9 = 0$
24. $x^2 - 9x + 18 = 0$
25. $x^2 + 3x - 18 = 0$
26. $2x^2 - 3x + 1 = 0$
27. $2x^2 - x - 1 = 0$
28. $x^2 - 5x - 36 = 0$
29. $x^2 - 15x + 36 = 0$
30. $x^2 - 16x = 36$
31. $x^2 - 9x = 36$
32. $x^2 + x = 12$
33. $x^2 - 7x = -12$
34. $x^2 + 7x = -12$
35. $x^2 - x = 12$

CONSTRUCTION Alice has 250 feet of fencing. She decides to use it to build a pen. Use this information in Exercises 36–41.

36. Copy and complete the table of values to show the length, width, and area of each possible pen.

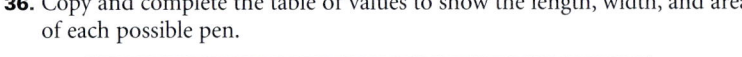

Fixed perimeter	Length	Width	Area
250	25	?	?
250	50	?	?
250	100	?	?

37. Which pen listed in the table has the largest area? Use other dimensions to find the pen that has the largest possible area.

38. **TECHNOLOGY** Use your graphics calculator and the equation Y1=X(125–X) to build a table of values, where Y1 is the area and X is the length. Explain why this equation can be used to build the table above.

39. Find the zeros by solving the equation $0 = x(125 - x)$. What do the zeros mean in this problem? $x = 0$ and $x = 125$. These are the values of the length that will make the area equal zero.

40. Graph the equation in Exercise 39. Find the dimensions that maximize the area of the pen.

CHALLENGE

41. Solve the equation $2500 = x(125-x)$ by factoring. What are the solutions of the equation? What does the equation $2500 = x(125-x)$ mean in this problem?

Practice

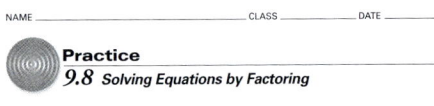

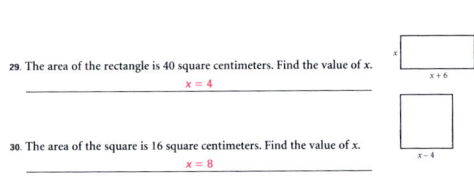

38.

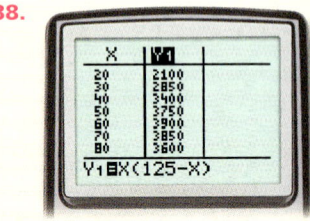

Area = $l \times w$
     = $l(125 - l)$

This is the same equation as $y = x(125 - x)$, where y is the area and x is the length.

40.

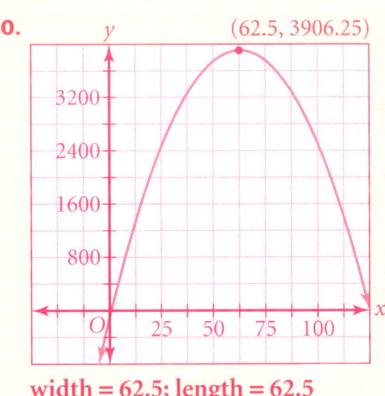

width = 62.5; length = 62.5

468 LESSON 9.8

43. $r = \frac{m}{4t}$
44. $r = \frac{4t}{s}$
45. $x = 1.2$
46. $x = \frac{16}{3} = 5\frac{1}{3}$
47. $x \geq -6$
50. $\frac{3}{25}$

Look Back

Solve each equation for r. (LESSON 3.6)

42. $C = 2\pi r$ $r = \frac{C}{2\pi}$
43. $m = 4rt$
44. $rs = 4t$

Solve each equation or inequality for x. (LESSONS 3.6 AND 6.2)

45. $2.5x - 2 = 1$
46. $5x = 2x + 16$
47. $x - 1 \leq 2x + 5$

PROBABILITY Two coins were tossed 50 times. Two heads appeared 23 times. Two tails appeared 21 times. Find the experimental probability of each outcome below. (LESSON 4.3)

48. two heads $\frac{23}{50}$
49. two tails $\frac{21}{50}$
50. one heads and one tails

Look Beyond

Solve each equation by graphing.

51. $x^2 + 5x - 6 = 0$
52. $x^2 + x - 1 = 0$
53. $2x^2 - 2x + 3 = 0$

PORTFOLIO ACTIVITY

Include this activity in your portfolio.

The first three numbers in the sequence of rectangular numbers are modeled at right.

(1×2=2, 2×3=6, 3×4=12)

1. Draw the next two rectangular numbers by using the pattern shown.
2. Use the pattern to find the first 10 rectangular numbers.
3. Use the dimensions of each rectangle to write a function for the sequence.
4. Draw the next two rectangular numbers by using the model at right. Use this model to write a polynomial function for the sequence.

   (1 + 1, 4 + 2, 9 + 3)

5. Write an identity by setting the two functions from Steps 3 and 4 equal to each other. Use the table feature on a graphics calculator to verify the identity.
6. What is the relationship between the rectangular numbers and the triangular numbers? (Hint: Recall the Portfolio Activity on page 413.)

WORKING ON THE CHAPTER PROJECT

You should now be able to complete Activity 4 of the Chapter Project on page 470.

---

41. $x = 25$ or $x = 100$; x represents the width, $(125 - x)$ represents the length, and 2500 represents the area.

51.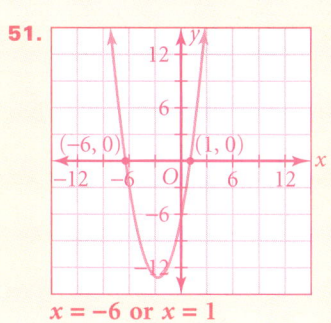
$x = -6$ or $x = 1$

52.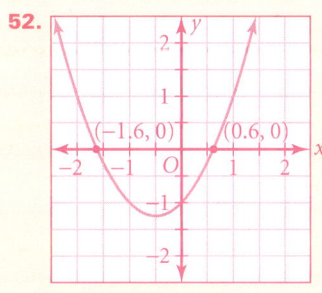
$x \approx -1.6$ or $x \approx 0.6$

53. no real solution; see Additional Answers for the graph.

Look Beyond

In Exercises 51–53, students use graphs of quadratic equations to identify solutions. Students will learn about the graphs of quadratic equations in Chapter 10.

ALTERNATIVE Assessment

Portfolio Activity

The Portfolio Activity can be used as preparation for the Chapter Project or as a separate activity. In the Portfolio Activity on this page, students investigate the polynomial function that defines the *rectangular numbers*.

internet connect

GO TO: go.hrw.com
KEYWORD: MA1 Rectangular

To support the Portfolio Activity, students may visit the HRW Web site, where they can obtain additional information about rectangular numbers, also called *pronic* numbers.

Student Technology Guide

Student Technology Guide
9.8 Solving Equations by Factoring

There are several ways that a graphics calculator can help you find the zeros of a polynomial function. Here you will see two different methods for finding the zeros of $y = 2x^2 - 11x + 12$.

Making a table: To find the zeros of a function, you can use the table feature of a graphics calculator and search for the x-value(s) that correspond to a y-value of 0.

- Press [Y=] 2 [X,T,θ,n] [x²] [−] 11 [X,T,θ,n] [+] 12 to enter the function. You can set the amount of change between x-values in the table. For this problem, it is reasonable to start with increments of 1. Press [2nd] [WINDOW] and set ΔTbl to 1.
- Press [2nd] [GRAPH]. Use the up [▲] and down [▼] arrows to look for the x-values that correspond to a y-value of 0. You can see that $x = 4$ is one solution to this equation. This looks as if it is the only zero for this function. Notice, however, that the y-value *changes sign* between $x = 1$ and $x = 2$. This suggests that there is another zero between these x-values.
- To search for a zero between $x = 1$ and $x = 2$, press [2nd] [WINDOW]. Set TblStart to 1 and ΔTbl to 0.1. Press [2nd] [GRAPH] again. You can see that $y = 0$ when $x = 1.5$.

The zeros of $y = 2x^2 - 11x + 12$ are $x = 1.5$ and $x = 4$.

Graphing: You can also search for zeros by graphing a function and looking for the points where the graph crosses the x-axis. (Remember that if a point is on the x-axis, its y-coordinate is 0.)

You have already entered $y = 2x^2 - 11x + 12$, so you can graph this function by pressing [GRAPH]. Press [TRACE] and use the left [◄] and right [►] arrow keys to move to a point where $y = 0$. For a closer look at the graph, press [ZOOM] 2:Zoom In [ENTER] [ENTER]. You may want to trace and zoom several times.

The trace cursor in the screen at right is near the zero at $x = 1.5$. You can also see the zero at $x = 4$.

Use either a graph or a table to identify the zeros of each function. Use the other method to check your answer.

1. $y = x^2 + 2x - 8$ 2. $y = 6x^2 - 18x$ 3. $y = 2x^2 + 3x - 5$ 4. $y = 5x^2 + 27x + 5$
   −4 and 2   0 and 3   −2.5 and 1   −4 and −1.4

LESSON 9.8   469

Project

Focus

The number pattern known as Pascal's triangle, after French mathematician Blaise Pascal (1623–1662), was in fact known to the ancient Chinese as early as 1100 C.E. As students continue their study of mathematics, they will learn that a wealth of information is encoded in the pattern. Here they see how the coefficients of the *expansion* of $(a + b)^n$ can be read from the rows of the triangle. Then they perform a simple coin-toss experiment and discover how the probabilities of the outcomes are related to the expansion.

Motivate

Before students do each activity, discuss it as a class.

For Activity 1, ask students if they have ever heard of Pascal's triangle. If so, have them share what they know about the properties of the triangle.

For Activity 2, ask students to guess what the outcome of the experiment might be.

For Activity 3, ask students if they have seen or used tree diagrams in other classes. If so, have them relate what the tree diagrams represented and how they were used.

CHAPTER PROJECT NINE

POWERS, PASCAL, AND PROBABILITY

CULTURAL CONNECTION: EUROPE AND ASIA If a great mathematical discovery were made today, the discoverer would be discussed on the evening news. Throughout history, it has been much more difficult to share information about important mathematical ideas. Pascal's triangle, although known to the ancient Chinese for centuries, was discovered independently by French mathematician Blaise Pascal.

Activity 1

Look at the different powers of the binomial $(a + b)$, and compare the expanded form with Pascal's triangle.

Power	Expanded Form	Pascal's Triangle
$(a + b)^0$	1	1
$(a + b)^1 =$	$1a + 1b$	1 1
$(a + b)^2 =$	$1a^2 + 2ab + 1b^2$	1 2 1
$(a + b)^3 =$	$1a^3 + 3a^2b + 3ab^2 + 1b^3$	1 3 3 1
$(a + b)^4 =$	$1a^4 + 4a^3b + 6a^2b^2 + 4ab^3 + 1b^4$	1 4 6 4 1
$(a + b)^5 =$	$1a^5 + 5a^4b + 10a^3b^2 + 10a^2b^3 + 5ab^4 + 1b^5$	1 5 10 10 5 1

1. Start with the expanded form of $(a + b)^2$, and multiply it by $a + b$. What is the result?
2. Do the same with the expanded forms of $(a + b)^3$ and $(a + b)^4$. What is the result?
3. How does the number of terms in each row of the expanded form compare with the exponent in the power of the binomial?
4. How many terms would you expect to be in the expansion of $(a + b)^7$?
5. Recalling from the Portfolio Activity on page 437 that new elements in the triangle are created by adding pairs of elements from a previous row, extend the triangle through row 10.
6. Write the coefficients of each term in the expansion of $(a + b)^7$.
7. Look at the terms in the expanded forms. Describe the pattern in the exponents as you read them from left to right in each row. What happens to the exponents of a? What happens to the exponents of b?
8. Write the complete algebraic expansion of $(a + b)^7$ without multiplying.

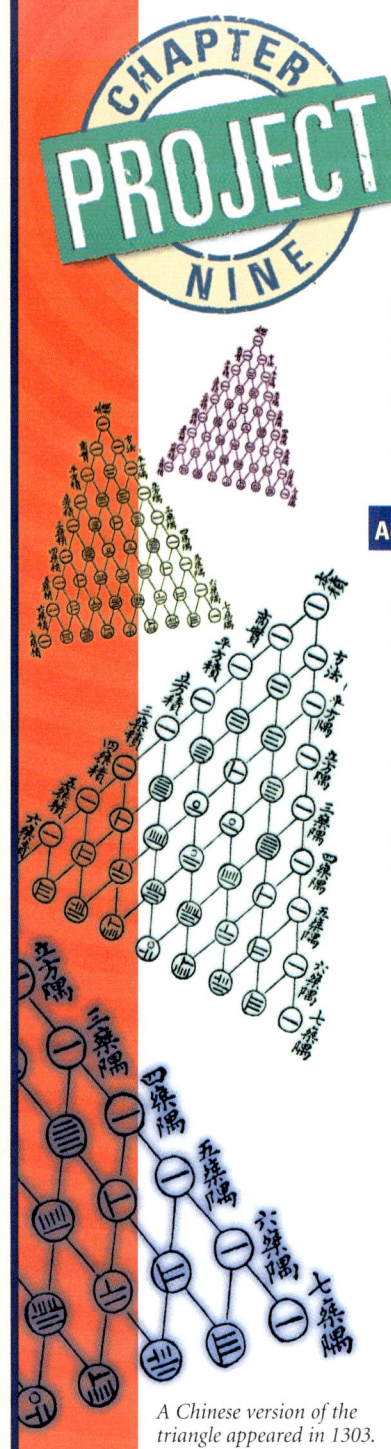
A Chinese version of the triangle appeared in 1303.

Activity 1

1. $(a + b)^2(a + b)$
  $= (a^2 + 2ab + b^2)(a + b)$
  $= a^3 + 3a^2b + 3ab^2 + b^3$
  $= (a + b)^3$

2. $(a + b)^3(a + b)$
  $= (a^3 + 3a^2b + 3ab^2 + b^3)(a + b)$
  $= a^4 + 4a^3b + 6a^2b^2 + 4ab^3 + b^4$
  $= (a + b)^4$
  $(a + b)^4(a + b)$
  $= (a + b)^4(a + b)$
  $= (a^4 + 4a^3b + 6a^2b^2 + 4ab^3 + b^4)(a + b)$
  $= a^5 + 5a^4b + 10a^3b^2 + 10a^2b^3 + 5ab^4 + b^5$
  $= (a + b)^5$

3. The number of terms is one more than the power of the binomial.

4. eight terms

5. See Additional Answers.

6. 1 7 21 35 35 21 7 1

7. The exponents of a decrease by 1 in each term from left to right, and the exponents of b increase by 1 in each term.

8. $a^7 + 7a^6b + 21a^5b^2 + 35a^4b^3 + 35a^3b^4 + 21a^2b^5 + 7ab^6 + b^7$

Activity 2

Answers may vary.

Activity 2

1. Copy the table.

	3 heads	2 heads, 1 tails	1 heads, 2 tails	3 tails
Tally marks	?	?	?	?
Totals	?	?	?	?
Totals ÷ 4	?	?	?	?

2. Toss three coins a total of 32 times. Record the results in the table.

3. Divide each total by 4, and round the answer to the nearest whole number. What are the results for each column?

Activity 3

A tree diagram is a method for considering all possible outcomes of an experiment.

1. Look at the tree diagram shown at right for the coin tosses from Activity 2.
2. How many ways are there to toss 3 heads?
3. How many ways are there to toss 3 tails?
4. How many ways are there to toss 2 heads and 1 tails?
5. How many ways are there to toss 1 heads and 2 tails?
6. Assume that each outcome in the diagram occurred once in 8 trials. What is the probability of each outcome?

Activity 4

1. Write the expanded form of $(h + t)^3$.
2. The probability of getting heads when tossing a coin is $\frac{1}{2}$. The probability of getting tails when tossing a coin is $\frac{1}{2}$. Let h and t each equal $\frac{1}{2}$, and find the value of *each term* in the expanded form of $(h + t)^3$.
3. How are the values of each term related to
   a. the probabilities you calculated in Activity 3?
   b. the results you obtained in Activity 2?
4. If you tossed 5 coins at a time, what would be the probability of getting 5 heads? 4 heads and 1 tails?

Cooperative Learning

Have students work in pairs. For **Activity 1**, partners should first do Steps 1 and 2 individually. They should then come together to compare their results, resolve any discrepancies, and complete the remaining steps.

For **Activity 2**, one student can toss the coin while the other records the results.

Partners can work together to answer the questions in **Activity 3**. In **Activity 4**, the partners should begin by answering the questions individually and then work together to reach an agreement on the answers.

Discuss

After students have completed all four Activities, bring the class together to discuss the results. Remind students that the probabilities calculated in Activity 2 were experimental probabilities, which are familiar to them from Chapter 4. In Activities 3 and 4 they worked with *theoretical probabilities*, which are based solely on ratios of possible outcomes. They will learn more about theoretical probabilities in Chapter 13.

You may wish to have students do further research on the history of Pascal's triangle and other information that can be read from it.

Activity 3

2. 1   3. 1
4. 3   5. 3

6. The probability of 3 heads is $\frac{1}{8}$; the probability of 3 tails is $\frac{1}{8}$; the probability of 2 heads and 1 tail is $\frac{3}{8}$; the probability of 1 head and 2 tails is $\frac{3}{8}$.

Activity 4

1. $h^3 + 3h^2t + 3ht^2 + t^3$

2. $h^3 = \left(\frac{1}{2}\right)^3 = \frac{1}{8}$

   $3h^2t = 3\left(\frac{1}{2}\right)^2\left(\frac{1}{2}\right) = \frac{3}{8}$

   $3ht^2 = 3\left(\frac{1}{2}\right)\left(\frac{1}{2}\right)^2 = \frac{3}{8}$

   $t^3 = \left(\frac{1}{2}\right)^3 = \frac{1}{8}$

3. a. The probability of 3 heads = $h^3 = \frac{1}{8}$; the probability of 3 tails = $t^3 = \frac{1}{8}$; the probability of 2 heads and 1 tail = $3h^2t = \frac{3}{8}$; the probability of 1 head and 2 tails = $3ht^2 = \frac{3}{8}$.

   b. Answers may vary. Sample answer: $8h^3$ = the number of times 3 heads occurred, divided by 4; $8t^3$ = the number of times 3 tails occurred, divided by 4; $8(3h^2t) + 1$ = the number of times 2 heads and 1 tail occurred, divided by 4; $8(3ht^2) - 1$ = the number of times 1 head and 2 tails occurred, divided by 4.

4. The probability of getting 5 heads would be $h^5 = \left(\frac{1}{2}\right)^5 = \frac{1}{32}$. The probability of getting 4 heads and 1 tail would be $5h^4t = 5\left(\frac{1}{2}\right)^4\left(\frac{1}{2}\right) = \frac{5}{32}$.

CHAPTER 9 PROJECT **471**

Chapter Review

Chapter Review and Assessment

VOCABULARY

common binomial factor . . . 449	FOIL method 439	special products 434
common monomial factor . . 448	identity 445	standard form 426
degree 426	perfect-square trinomial . . . 452	Zero Product Property 465
difference of two squares . . 453	polynomial 426	
factor 448	polynomial function 443	

Key Skills & Exercises

LESSON 9.1

Key Skills

Use the horizontal form and the vertical form to add and subtract polynomials.

Add $4x^2 + 6x$ and $2x^2 - 8x - 3$ vertically. Align the like terms and add.

$$\begin{array}{r} 4x^2 + 6x \\ +\ 2x^2 - 8x - 3 \\ \hline 6x^2 - 2x - 3 \end{array}$$

Subtract $6b^2 + 4b - 8$ from $10h^2 - 13$ horizontally. Remember that you can subtract by adding the opposite.

$(10b^2 - 13) - (6b^2 + 4b - 8) =$
$10b^2 - 13 - 6b^2 - 4b + 8 = 4b^2 - 4b - 5$

Exercises

Add or subtract.

1. $(3x^2 - 4x + 2) + (2x^2 + 3x - 2)$ $5x^2 - x$
2. $(c^3 + 4c^2 + 6) + (c^2 + 3c - 5)$ $c^3 + 5c^2 + 3c + 1$
3. $(8d^2 - d) - (2d^2 + 4d - 5)$ $6d^2 - 5d + 5$
4. $(w^3 - 3w + 9) - (8w^3)$ $-7w^3 - 3w + 9$
5. $(10m^2 - m + 4) - (2m^2 + m)$ $8m^2 - 2m + 4$
6. $(7c + 3) + (3c^2 - 7c - 2)$ $3c^2 + 1$
7. $(8x^2 + x) - (2x^2 - 3x)$ $6x^2 + 4x$
8. $(5x^3 + 2x^2 - x) + (5x^3 + 3x^2 - 2)$
9. $(7t^5 + 2t^3 - t^2) - (3t^5 - 4t^4 + 3t^2)$

8. $10x^3 + 5x^2 - x - 2$
9. $4t^5 + 4t^4 + 2t^3 - 4t^2$

LESSON 9.2

Key Skills

Use the Distributive Property to find the product of a monomial and a binomial.

Find the product: $8(x + 3)$.

$8(x + 3) = (8 \cdot x) + (8 \cdot 3) = 8x + 24$

Use the rules for special products to find the product of two binomials.

Find the product: $(2a + 3)^2$.

Use the special product rule

$(2a + 3)^2 = (2a)^2 + 2(2a)(3) + (3)^2 = 4a^2 + 12a + 9$

Exercises

Find each product.

10. $5(x - 5)$ $5x - 25$
11. $4y(y + 2)$ $4y^2 + 8y$
12. $-x(2x - 3)$ $-2x^2 + 3x$
13. $(x + 7)(x - 7)$ $x^2 - 49$
14. $(x - 4)(x + 4)$ $x^2 - 16$
15. $(x + 1)(x + 1)$ $x^2 + 2x + 1$
16. $(5x - 2)^2$ $25x^2 - 20x + 4$
17. $(3x - 8)(3x + 8)$ $9x^2 - 64$

Chapter Test, Form A

NAME _____ CLASS _____ DATE _____

Chapter Assessment
Chapter 9, Form A, page 1

Write the letter that best answers the question or completes the statement.

d 1. The sum of $3r^4 - 7r^2 + r - 9$ and $r^4 + 5r^3 - 2r^2 + 1$ is
   a. $4r^4 - 12r^2 + 3r - 10$
   b. $4r^4 - 2r^2 - r - 8$
   c. $4r^4 + 5r^3 - 5r^2 + r - 8$
   d. $4r^4 + 5r^3 - 9r^2 + r - 8$

b 2. The difference when $x^3 - 5x^2 + 7$ is subtracted from $x^5 - 2x^3 + 3x^2 - 3$ is:
   a. $x^5 - x^3 - 3x^2 + 4$
   b. $x^5 - 3x^3 + 8x^2 - 10$
   c. $-x^5 + x^3 - 8x^2 + 10$
   d. $x^5 - x^3 - 2x^2 + 4$

a 3. Which expression is modeled by the algebra tiles?
   a. $4x^2 - 2x + 1$
   b. $(2x - 1)^2$
   c. $(2x - 1)(2x + 1)$
   d. $4x^2 + 2x - 1$

In Exercises 4–7, simplify each expression.

a 4. $6y^2(3y^2 - 1) =$
   a. $18y^4 - 6y^2$
   b. $9y^4 - 6y$
   c. $18y^4 - 1$
   d. $18y^3 - 6y^2$

c 5. $(4p - 3)^2 =$
   a. $16p^2 - 6p + 9$
   b. $16p^2 - 12p + 9$
   c. $16p^2 - 24p + 9$
   d. $16p^2 - 12p - 9$

d 6. $(5t - 3)(2t + 2) =$
   a. $10t^2 - 16t - 6$
   b. $10t - t - 6$
   c. $15t^2 + 4t + 4$
   d. $10t^2 + 4t - 6$

a 7. $(3x - 2)(3x + 2) =$
   a. $9x^2 - 4$
   b. $3x^2 - 6x - 4$
   c. $9x^2 - 12x - 4$
   d. $9x^2 - 2x - 4$

b 8. What is the GCF of $3a^4b^2c$ and a^2b^3c?
   a. $3a^2b^3c$
   b. a^2b^2c
   c. $3a^4b^3c$
   d. a^2b^3c

NAME _____ CLASS _____ DATE _____

Chapter Assessment
Chapter 9, Form A, page 2

a 9. What is the factored form of $2a^2b^3 + 8a^3b^2 - 6a^2b$?
   a. $2a^2b(b^2 + 4ab - 3)$
   b. $2ab(ab^2 + 4ab - 3a)$
   c. $2a^2b^3(8a^3b^2 - 6a^2b)$
   d. $2a^2b^2(b + 4a - 3)$

d 10. What are the solutions to the equation $x^2 - 5x + 6 = 0$?
   a. -2 and 3
   b. 2 and -3
   c. -2 and -3
   d. 2 and 3

In Exercises 11–14, write the letter that represents the factored form of each expression.

d 11. $9x^2 - 36x + 36$
   a. $9(x + 2)(x - 2)$
   b. $3(x + 3)(3x + 4)$
   c. $9(x - 4)(x + 1)$
   d. $9(x - 2)^2$

c 12. $25a^2 - 9b^4$
   a. $16(a + b^2)(a - b^2)$
   b. $25a^2(1 - 3b^2)(1 + 3b^2)$
   c. $(5a - 3b^2)(5a + 3b^2)$
   d. $(5a - 3b^2)^2$

b 13. $6x^2 + x - 12$
   a. $2(3x + 2)(x - 3)$
   b. $(3x - 4)(2x + 3)$
   c. $(3x + 4)(2x - 3)$
   d. $6(x + 1)(x - 2)$

b 14. $2y(y - z)^2 - 5(y - z)^2$
   a. $(2y - 5)(y - z)$
   b. $(2y - 5)(y - z)^2$
   c. $(2y - 5)(y + z)^2$
   d. $(2y - 5)(2y^2 + z)^2$

a 15. Which expression represents the shaded region in the figure at right?
   a. $x^2 - 25$
   b. $(x - 5)^2$
   c. $x^2 + 25$
   d. $(x + 5)^2$

c 16. The perimeter of a rectangle with a length of $y + 5$ units and a width of $y - 2$ units can be represented by
   a. $y^2 + 3y - 10$
   b. $2y + 3$
   c. $4y + 6$
   d. $y^2 + 7y - 10$

472 CHAPTER 9 REVIEW

LESSON 9.3
Key Skills

Multiply two binomials by using the Distributive Property or the FOIL method.

Use the Distributive Property to find $(x+6)(x+2)$.

$(x+6)(x+2) = (x+6)x + (x+6)2$
$= x^2 + 6x + 2x + 12$
$= 2x^2 + 8x + 12$

Use the FOIL method to find $(2x+3)(x-1)$.
$(2x+3)(x-1) = 2x^2 - 2x + 3x - 3 = 2x^2 + x - 3$

Exercises

Use the Distributive Property to find each product.

18. $(y+9)(y-2)$
 $y^2 + 7y - 18$
19. $(2p-9)(p+5)$
 $2p^2 + p - 45$

Use the FOIL method to find each product.

20. $(x+3)(x-4)$
21. $(5d-8)(d-1)$
22. $(4w+3z)(w+z)$
23. $(3m+5)(m+5)$

20. $x^2 - x - 12$
21. $5d^2 - 13d + 8$
22. $4w^2 + 7wz + 3z^2$
23. $3m^2 + 20m + 25$

LESSON 9.4
Key Skills

Use polynomial functions to model the volume and surface area of solid figures.

The volume of a cylinder that is 10 inches high is $V = 10B$, where B is the area of the circular base. The area of a circle with a radius of r is $A = \pi r^2$, so the volume of the cylinder is modeled by the function $V = 10\pi r^2$.

Use an identity to express a polynomial as the product of other polynomials.

The equation $(x-1)^2 = x^2 - 2x + 1$ is an identity (true for all values of x). For example, if $x = -3$, then the left side, $(-3-1)^2$, is $(-4)^2$, or 16. The right side, $(-3^2) - 2(-3) + 1$, is $9 + 6 + 1$, or 16 also.

Exercises

24. The formula for the volume of a cone is $V = \frac{1}{3}Bh$, where B is the area of the circular base. Write a polynomial function that represents the volume of a cone with a height of 9 feet and a base that has a radius of r.
 $V = 3\pi r^2$

25. The equation $(x+2)(x-2) = x^2 - 4$ is an identity. Show that each side of the equation has the same value when x is replaced by each of the following numbers: $-3, -2, -1, 0, 1, 2,$ and 3.

LESSON 9.5
Key Skills

Factor a polynomial by using the GCF or by grouping.

Factor $8m^6 + 4m^4 - 2m^2$.
The GCF is $2m^2$. Factor $2m^2$ from each term.
$8m^6 + 4m^4 - 2m^2 = 2m^2(4m^4 + 2m^2 - 1)$

Factor $a^2 + 3a - 4a + 4$ by grouping.
Group the terms with common factors and use the Distributive Property.
$a^2 + 3a + 4a + 4 = (a^2 + 3a) + (4a + 12)$
$= a(a+3) + 4(a+3)$
$= (a+4)(a+3)$

Exercises

Factor each polynomial by using the GCF.

26. $x^3 + 3x^2$ $x^2(x+3)$
27. $b^4 + 15b^3 + 5b$ $b(b^3 + 15b^2 + 5)$
28. $24m^5 + 16m^4 - 8m^3$ $8m^3(3m^2 + 2m - 1)$
29. $2a^4y - 4a^3y^2 + 8a^2y$ $2a^2y(a^2 - 2ay + 4)$

Factor each polynomial by grouping.

30. $a(x+1) + b(x+1)$ $(a+b)(x+1)$
31. $x^2 + 2x + 4x + 8$ $(x+4)(x+2)$
32. $s^2 - s + 4s - 4$ $(s+4)(s-1)$
33. $2m^2 + 3m - 10m - 15$ $(m-5)(2m+3)$

25.

x	$(x+2)(x-2)$	$x^2 - 4$
-3	5	5
-2	0	0
-1	-3	-3
0	-4	-4
1	-3	-3
2	0	0
3	5	5

Chapter Test, Form B

Chapter Assessment
Chapter 9, Form B, page 1

1. Identify $9x^3 + 1$ by degree and by the number of terms. __cubic binomial__
2. Explain why $3x + 4y$ is considered prime.
 There are exactly two factors, 1 and $3x + 4y$.
3. Give an example of a perfect-square trinomial.
 Answers may vary. Sample answer: $x^2 + 6x + 9$
4. If you wrote $x^2 - 8x + 12$ in factored form, would the signs in the monomial factors be the same or opposite? Why?
 Same; the sign of the last term is positive.
5. If you wrote $x^2 - 8x + 12$ in factored form, would the signs in the monomial factors be positive or negative? Why?
 Negative; the sign of the middle term is negative.

Add. Express all answers in standard form.

6. $3y^3 - 8y^2 + 7y - 9$ and $9y^3 + 7y^2 - 5y - 6$ $12y^3 - y^2 + 2y - 15$
7. $w^2 + 12w - 6w^3 + 3$ and $5w^3 - w^2 + 3w - 11$ $-w^3 + 15w - 8$
8. $5a^4 - 9a^2 + 7a + 15$ and $a^4 + 7a^2 - 2a^3 - 8$ $6a^4 - 2a^3 - 2a^2 + 7a + 7$

Subtract. Express all answers in standard form.

9. $x^3 + 3x^2 - 5x$ from $4x^3 - 9x^2 + 12$ $3x^3 - 12x^2 + 5x + 12$
10. $6b^3 - 2b^2 + 7$ from $5b^3 - 11b^2 + 3b - 6$ $-b^3 - 9b^2 + 3b - 13$
11. $8d^4 + 5d - 3d^2 + 1$ from $13d^4 - 5d^2 + 3d^3 - 1$ $5d^4 + 3d^3 - 2d^2 - 5d - 2$

Simplify. Express all answers in standard form.

12. $(3 - 2x + x^3) + (3x - 5x^3 + 8)$ $-4x^3 + x + 11$
13. $(7 - 9y^2 + 3y^3) - (3y^2 + 8 - 9y^3)$ $12y^3 - 12y^2 - 1$
14. Express the area of the triangle shown at right in terms of x.
 $3x^2 - 5x - 12$

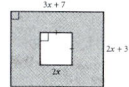

Chapter Assessment
Chapter 9, Form B, page 2

Find each product.

15. $3w(5w^2 - 1)$ $15w^3 - 3w$
16. $(4a - 2b^2)(4a + 2b^2)$ $16a^2 - 4b^4$
17. $(x+3)^2$ $x^2 + 6x + 9$
18. $(2x-3)^2$ $4x^2 - 12x + 9$
19. $(d+5)(d-3)$ $d^2 + 2d - 15$
20. $(3t-8)(2t+3)$ $6t^2 - 7t - 24$

Factor each polynomial completely.

21. $9r^2 - 4$ $(3r-2)(3r+2)$
22. $z^2 + 7z - 18$ $(z+9)(z-2)$
23. $m^2 - 10m + 25$ $(m-5)^2$
24. $7a^3 + 28a^2 - 35a$ $7a(a+5)(a-1)$
25. $y^2 - 3y - 18$ $(y-6)(y+3)$
26. $b(b-1) + 5(b-1)$ $(b+5)(b-1)$

Solve each equation.

27. $x^2 + 3x - 4 = 0$ $x = -4$ and $x = 1$
28. $x^2 - 4 = 0$ $x = -2$ and $x = 2$
29. $x^2 - 8x + 15 = 0$ $x = 3$ and $x = 5$
30. $x^2 - 2x + 1 = 0$ $x = 1$

31. Express the perimeter of the rectangle shown at right in terms of x.
 $10x + 20$
32. Express the area of the rectangle shown at right in terms of x.
 $6x^2 + 23x + 21$
33. Express the area of the shaded region in terms of x. $2x^2 + 23x + 21$

The area of a square is represented by $4p^2 - 12p + 9$.

34. Express the length of each side in terms of p. $2p - 3$
35. Express the perimeter of the square in terms of p. $8p - 12$

LESSON 9.6
Key Skills

Use rules to factor special polynomials.

Factor. **a.** $x^2 - 6x + 9$ **b.** $x^2 - 9$

a. $x^2 - 6x + 9$ is a perfect-square trinomial.
$$x^2 - 6x + 9 = (x - 3)^2$$

b. $x^2 - 9$ is a difference of two squares.
$$x^2 - 9 = (x + 3)(x - 3)$$

Exercises

Factor each polynomial completely.

34. $x^2 - 25$ *(x + 5)(x − 5)*

35. $4x^2 + 4x + 1$ *(2x + 1)²*

36. $9x^2 - 6x + 1$ *(3x − 1)²*

37. $3s^2 - 6s + 3$ *3(s − 1)²*

LESSON 9.7
Key Skills

Factor a quadratic trinomial by working backward.

Factor $x^2 + 7x + 12$ by using guess-and-check.

Since the constant term and the middle term are positive, the constant terms in both factors must be positive, and the binomial factors can be written in the form $(x + \_\_\_)(x + \_\_\_)$. The positive constant terms must have a product of 12 and a sum of 7. Thus, $x^2 + 7x + 12 = (x + 3)(x + 4)$.

Exercises

Factor each trinomial. If a trinomial cannot be factored, write *not factorable*.

38. $c^2 + 8c + 15$ *(c + 3)(c + 5)*

39. $m^2 - 2m - 8$ *(m + 2)(m − 4)*

40. $t^2 + 8t + 12$ *(t + 2)(t + 6)*

41. $r^2 + r - 56$ *(r − 7)(r + 8)*

42. $a^2 - 15x + 48$ *Prime*

43. $b^2 + 11b + 28$ *(b + 4)(b + 7)*

LESSON 9.8
Key Skills

Solve an equation by factoring.

Solve $x^2 + x - 12 = 0$ by factoring.

Write the equation in factored form.
$(x + 4)(x - 3) = 0$

Set each factor equal to 0 and solve for x.

$x + 4 = 0 \qquad x - 3 = 0$
$x = -4 \qquad x = 3$

The solutions are −4 and 3.

Exercises

Solve by factoring.

44. $x^2 - 8x + 12 = 0$ *x = 2 or x = 6*

45. $x^2 + 2x - 15 = 0$ *x = 3 or x = −5*

46. $x^2 - 5x - 24 = 0$ *x = −3 or x = 8*

47. $x^2 + 7x + 12 = 0$ *x = −3 or x = −4*

48. $x^2 - 14x + 40 = 0$ *x = 4 or x = 10*

Applications

49. GEOMETRY The length of a rectangle is represented by $x + 1$. Its width is represented by $x - 3$. If the area of the rectangle is 117 square centimeters, what are the dimensions of the rectangle? *13 cm, 9 cm*

50. LANDSCAPING A rectangular garden is designed so that its length is twice its width. The number of feet in the perimeter of the garden is equal to the number of square feet in its area. What are the dimensions of the garden? *3 feet, 6 feet*

9 Alternative Assessment

Performance Assessment

1. Use an algebra-tile drawing to simplify $(3x - 1) + (4 - 2x) + (x + 3)$. What is the result?

2. Use an algebra-tile drawing to find the product of $(x + 2)$ and $(x + 1)$. What is the result?

3. A quilt is 48 inches by 60 inches. A border x inches wide is to be added to the edge of the quilt. Draw a diagram and explain how to write an expression for the area of the border.

4. Use an algebra-tile drawing to find the factors of $x^2 - x - 2$. What are the factors?

5. Use the FOIL method to explain why $2x^2 + 2x + 3$ cannot be factored.

6. Solve the following problem. First draw a picture. Then explain how you solved the problem.

    The length of a rectangle is 4 meters longer than its width. The area of the rectangle is 96 square meters. Find the rectangle's perimeter.

Portfolio Projects

1. The words *polygon*, *monomial*, *binomial*, and *trinomial* include prefixes that indicate the number of terms in that type of algebraic expression. Make a list of words that start with each prefix, and use a dictionary to write the definition of each word.

2. Draw a geometric model to show that $(a - b)(a + b) = a^2 - b^2$ is true.

3. Write a paragraph explaining how to use graph paper to model $(x + 3)(x + 5)$. Let each square on the graph paper be 1 square unit.

4. Design a poster that illustrates the FOIL method for multiplying two binomials. Use geometric symbols such as triangles and squares for the variables and constants. Write a set of rules for using the illustration to multiply $(2x + 5)$ and $(x - 3)$.

5. Write a paragraph explaining how you know when a trinomial is factored completely.

Peer Assessment

Complete this activity with a partner. Begin by writing an expression involving adding, subtracting, or multiplying polynomials. For example, $(x + 2)(x - 3)$ involves multiplication. After completing the specified operation, your partner should write a similar expression for you to simplify.

Check your partner's work, correct any errors, and then simplify the new expression. Continue with this pattern until you have each simplified 20 expressions. Be sure to use different operations.

Repeat the entire activity by writing polynomials that can be factored and having your partner simplify each polynomial. Remember to use polynomials that can be factored by using the patterns you have learned for this part of the activity.

internetconnect

The HRW Web site contains many resources to reinforce and expand your knowledge of polynomials and factoring. The Web site also provides Internet links to other sites where you can find information and real-world data for use in research projects, reports, and activities that involve polynomial functions. Visit the HRW Web site at **go.hrw.com**, and enter the keyword **MA1 CH9**.

Alternative Assessment

Performance Assessment

1.
   $2x + 6$

2.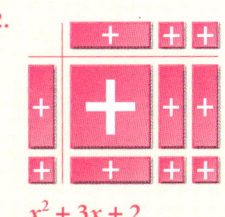
   $x^2 + 3x + 2$

3.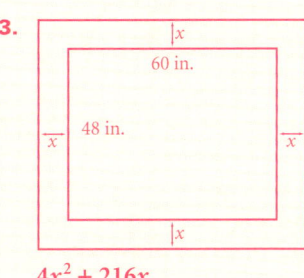
   $4x^2 + 216x$

4.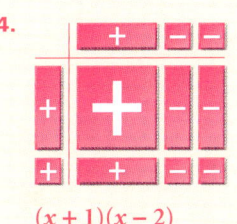
   $(x + 1)(x - 2)$

5. There are no combinations of the factors $\pm 3, \pm 1$ and $\pm 2, \pm 1$ so that the sum of the outer products and the inner products has a coefficient of 2.

6.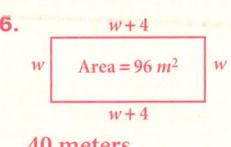
   40 meters

Portfolio Projects

1. Answers may vary. Sample words: polygamy, monotone, bicycle, and tripod

2. Sample answer: Start with a square with side lengths of a, subtract a square with side lengths of b from one corner, cut the remainder in half and rearrange into a rectangle.

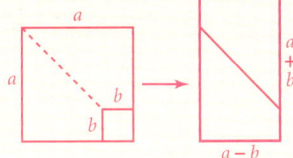

3. Answers may vary. Students should include the solution $x^2 + 8x + 15$ in their answers.

4. Students should include an explanation of what the letters in FOIL represent. The solution $2x^2 - x - 15$ should be displayed.

5. Answers may vary. Sample answer: A trinomial is factored completely when all possible methods of factoring have been exhausted: finding common factors, reversing the FOIL method, grouping, factoring special polynomials such as the difference of squares, and finally the guess-and-check method.

Cumulative Assessment

College Entrance Exam Practice

CHAPTERS 1-9 CUMULATIVE ASSESSMENT
College Entrance Exam Practice

Multiple-Choice and Quantitative-Comparison Samples

The first half of the Cumulative Assessment contains two types of items found on standardized tests—multiple-choice questions and quantitative-comparison questions. Quantitative comparison items emphasize the concepts of equalities, inequalities, and estimation.

Free Response Grid Samples

The second half of the Cumulative Assessment is a free-response section. This part of the Cumulative Assessment requires student-produced response items like those commonly found on college entrance exams. These questions require the use of machine-scored answer grids. You may wish to have students practice answering these items in preparation for standardized tests.

QUANTITATIVE COMPARISON For Questions 1–5, write
A if the quantity in Column A is greater than the quantity in Column B;
B if the quantity in Column B is greater than the quantity in Column A;
C if the two quantities are equal; or
D if the relationship cannot be determined from the information given.

	Column A	Column B	Answers	
1.	reciprocal of -2	reciprocal of -3	Ⓐ Ⓑ Ⓒ Ⓓ [Lesson 2.4]	B
2.	$-18 + 4$	$-10 + (-4)$	Ⓐ Ⓑ Ⓒ Ⓓ [Lesson 2.2]	C
3.	$|-5.2|$	$|4.9|$	Ⓐ Ⓑ Ⓒ Ⓓ [Lesson 2.1]	A
4.	45% of 16	60% of 12	Ⓐ Ⓑ Ⓒ Ⓓ [Lesson 4.2]	C
5.	$2 \cdot 3^2$	$3 \cdot 2^2$	Ⓐ Ⓑ Ⓒ Ⓓ [Lesson 8.1]	A

6. What are the next three terms in the sequence 6, 12, 24, 48, ... ? **(LESSON 1.1)** b
   a. 72, 96, 120
   b. 96, 192, 384
   c. 86, 162, 240
   d. 50, 52, 54

7. Solve: $16y = -120$. **(LESSON 3.2)** d
   a. 1920
   b. -1920
   c. 7.5
   d. -7.5

8. Find the equation of the line passing through $(-1, 2)$ and parallel to $y = 2x - 1$.
   (LESSON 5.6) c
   a. $y = -\frac{x}{2} + 4$
   b. $y = \frac{x}{2} + 2$
   c. $y = 2x + 4$
   d. $y = -x - 1$

9. Solve. $\begin{cases} 2x + 3y = 9 \\ x - 4y = -23 \end{cases}$ **(LESSON 7.3)** b
   a. $(3, 5)$
   b. $(-3, 5)$
   c. $(3, -5)$
   d. $(-3, -5)$

10. A television is marked at $198. What is the total cost of the television if the sales tax is 7.5%? **(LESSON 4.2)** a
    a. $212.85
    b. $14.85
    c. $183.15
    d. $205.50

11. Simplify: $\frac{4x^3y^4}{2x^5y^2}$. **(LESSON 8.3)** c
    a. $2x^8y^6$
    b. $2x^{-2}y^{-2}$
    c. $2x^{-2}y^2$
    d. $2x^2y^2$

12. Simplify: $(2x - 1)(x + 3)$. **(LESSON 9.3)** c
    a. $2x^2 - 3$
    b. $2x^2 + 5x + 3$
    c. $2x^2 + 5x - 3$
    d. $2x^2 - 5x + 3$

13. Solve: $2(x + 2) = 12$. **(LESSON 3.5)** a
    a. 4
    b. 22
    c. -4
    d. 26

14. Solve: $|2x + 3| = 7$. *(LESSON 6.5)* **c**
  a. 5, −3   b. 4, −4
  c. 2, −5   d. 3, −6

15. Which number is not a solution of $|2x + 1| < 5$? *(LESSON 6.5)* **d**
  a. −2   b. 0
  c. 1   d. −3

16. Do the points $A(-2, 2)$, $B(1, 4)$, and $C(-4, 0)$ lie on a straight line? *(LESSON 5.5)* **No**

17. What is the slope of the line represented by $x = -4$? *(LESSON 5.4)* **Undefined**

18. Solve: $4 - 3t \geq 19$. *(LESSON 6.2)* **$t \leq -5$**

19. Find the opposite of $-(a - b)$. *(LESSON 2.1)* **$a - b$**

20. A jacket has been marked down from an original price of $85 to $59.50. By what percent of the original price has the jacket been marked down? *(LESSON 4.2)* **30%**

21. The perimeter of a rectangle is 52 inches. Its width is 6 inches less than its length. Find the dimensions of the rectangle. *(LESSON 3.3)* **length = 16 in, width = 10 in**

22. Write 0.0000025 in scientific notation. *(LESSON 8.5)* **2.5×10^{-6}**

23. Factor: $5xy^2 + 10x^2y^2 - 5xy$. *(LESSON 9.7)* **$5xy(y + 2xy - 1)$**

24. What is the missing number? *(LESSON 1.1)*
  0 1 2 3 4
  1 2 5 10 ?   **17**

25. Simplify: $(3x^2 + 2x + 1) - (x^2 - 5x + 3)$. *(LESSON 9.1)* **$2x^2 + 7x - 2$**

26. Simplify: $\frac{8x^2y^3}{2xy}$. *(LESSON 8.3)* **$4xy^2$**

27. Solve the system. $\begin{cases} 2x + y = 24 \\ y + 3 = x \end{cases}$ *(LESSON 7.2)* **(9, 6)**

28. Evaluate: $\left[\left(\frac{x^3y^4}{y^2}\right)^3 \left(\frac{xy^2}{x^2}\right)^4\right]^2$ *(LESSON 8.3)* **$x^{10}y^{28}$**

29. Write an equation for the line passing through the points $(7, -8)$ and $(-4, -4)$. *(LESSON 5.5)*
  $y - (-8) = -\frac{4}{11}(x - 7)$ or $y - (-4) = -\frac{4}{11}(x - (-4))$

FREE-RESPONSE GRID
The following questions may be answered by using a free-response grid such as that commonly used by standardized test services.

30. A city is growing at a rate of 7% per year. What multiplier is used to find the new population each year? *(LESSON 8.6)* **1.07**

31. Simplify: $4^2 \div 8 + 5(8 - 2) \cdot 2$. *(LESSON 1.3)* **62**

32. Find the slope of the line that passes through the origin and $(4, 2)$. *(LESSON 5.4)* **$\frac{1}{2}$**

33. Solve: $7(p + 4) = 49$. *(LESSON 3.5)* **$p = 3$**

34. Simplify: $54 - 68 + |80|$. *(LESSON 2.1)* **66**

35. Evaluate $\frac{x+y}{3} + \frac{x-y}{2}$ for $x = 14$ and $y = 10$. *(LESSON 1.3)* **10**

36. Solve the system for x. $\begin{cases} y = 3 + 2x \\ 2y = 5x + 4 \end{cases}$ *(LESSON 7.2)* **(2, 7)**

37. Find the slope of the line passing through the points $(-1, 3)$ and $(4, 5)$. *(LESSON 5.5)* **$\frac{2}{5}$**

38. Solve for x: $\frac{x}{12} = 13$. *(LESSON 3.2)* **156**

39. Find the next term in the sequence: 3, 5, 8, 13, 21, 33 . . . *(LESSON 1.1)* **50**

40. Evaluate $x^2 + 3y$ for $x = 4$ and $y = 0.5$. *(LESSON 1.3)* **17.5**

41. Solve for d: $\frac{15}{d} = \frac{5}{2}$. *(LESSON 4.1)* **$d = 6$**

42. What percent of 80 is 20? *(LESSON 4.2)* **25%**

43. Give the slope of a line that is perpendicular to $2x - 3y = 15$. *(LESSON 5.6)* **$-\frac{3}{2}$**

44. Evaluate: $2^3 - 3^4$. *(LESSON 1.3)* **−73**

45. Solve: $3(p - 5) = 30$. *(LESSON 3.5)* **$p = 15$**

10 Quadratic Functions

CHAPTER PLANNING GUIDE

Lesson	10.1	10.2	10.3	10.4	10.5	10.6	Projects and Review
Pupil Edition Pages	480–485	486–491	492–497	498–503	506–510	511–515	504–505, 516–523
Extra Practice (Pupil Edition)	766	767	767	768	768	769	
Practice Workbook	58	59	60	61	62	63	
Reteaching Masters	115–116	117–118	119–120	121–122	123–124	125–126	
Enrichment Masters	58	59	60	61	62	63	
Cooperative-Learning Activities	58	59	60	61	62	63	
Lesson Activities	58	59	60	61	62	63	
Problem Solving/ Critical Thinking	58	59	60	61	62	63	
Student Study Guide	58	59	60	61	62	63	
Spanish Resources	58	59	60	61	62	63	
Student Technology Guide	58	59	60	61	62	63	
Assessment Resources	125	126	127	129	130	131	128, 132–137
Teaching Transparencies	47, 48			49	50, 51		
Quiz Transparencies	10.1	10.2	10.3	10.4	10.5	10.6	
Writing Activities for Your Portfolio							28–30
Tech Prep Masters							45–48
Long-Term Projects							37–40

LESSON PACING GUIDE

	10.1	10.2	10.3	10.4	10.5	10.6	Projects and Review
Traditional	2 days	2 days	1 day	1 day	2 days	2 days	2 days
Block	1 day	1 day	$\frac{1}{2}$ day	$\frac{1}{2}$ day	1 day	1 day	1 day
Two-Year	4 days	4 days	2 days	2 days	4 days	4 days	4 days

CONNECTIONS AND APPLICATIONS

Lesson	10.1	10.2	10.3	10.4	10.5	10.6	Review
Algebra	480–485	486–491	492–497	498–503	506–510	511–515	518–523
Geometry		489–490		500, 503	510		520
Maximum/Minimum						511	
Number Theory	485						
Business and Economics					510	511	
Science and Technology	480, 482, 483, 485	486, 487, 488, 489, 490	497	501		513, 515	
Sports and Leisure				503			520
Other				503			
Cultural Connection: Africa					506		

BLOCK-SCHEDULING GUIDE

Day	Lesson	Teacher Directed: Lesson Examples, Teaching Transparencies	Student Guided: Activity, Try This	Cooperative-Learning Activity, Lesson Activity, Student Technology Guide	Practice: Practice & Apply, Extra Practice, Practice Workbook	Assessment: Quiz, Mid-Chapter Assessment	Problem Solving, Reteaching
1	10.1	10 min	15 min	15 min	65 min	15 min	15 min
2	10.2	10 min	15 min	15 min	65 min	15 min	15 min
3	10.3	5 min	8 min	8 min	35 min	8 min	8 min
3	10.4	5 min	7 min	7 min	30 min	7 min	7 min
4	10.5	10 min	15 min	15 min	65 min	15 min	15 min
5	10.6	10 min	15 min	15 min	65 min	15 min	15 min
6	Assess.	50 min PE: Chapter Review	90 min PE: Chapter Project, Writing Activities	90 min Tech Prep Masters	65 min PE: Chapter Assessment, Test Generator	30 min Chap. Assess. (A or B), Alt. Assess. (A or B), Test Generator	

PE: Pupil's Edition

Alternative Assessment

The following are suggestions for an alternative assessment for students who may benefit from a different type of assessment than the regular chapter quizzes and the mid-chapter and end-of-chapter assessment materials. Many of the questions are open-ended, and students' answers will vary.

Performance Assessment

1. **a.** For each equation below, choose a method of solution and solve the equation, if possible.

   $2x^2 = 32$ $x^2 - 16 = 0$
   $(x-1)^2 = 16$ $x^2 + 8x + 16 = 0$
   $x^2 + 16 = 0$ $x^2 + 6x - 16 = 0$

   b. Explain how you can recognize patterns in equations and use pattern recognition to help choose a method of solution.

2. **a.** State the quadratic formula.

   b. Explain how to use the quadratic formula to find the roots of quadratic equations. Then find the roots of several examples.

Portfolio Project

Suggest that students choose one of the following projects for inclusion in their portfolios.

1. Let $x^2 + bx + c = 0$.
   a. Explain how to solve the equation for $b = 0$ and $c \neq 0$ and for $b \neq 0$ and $c = 0$.
   b. Use graphs to show how the given equation can have no real roots, one real root, or two real roots. For each situation, state a sample equation.

2. You can make a "smile" on the coordinate plane by graphing two parabolas opening upward and shading the region between them.
   a. Using quadratic inequalities, sketch a "smile" in the coordinate plane. State the inequalities you used.
   b. Using quadratic inequalities, sketch other patterns in the coordinate plane.

internetconnect

The table below identifies the pages in this chapter that contain technology information and support in the side columns.

Content Links	
Lesson Links	pages 484, 503, 507
Portfolio Links	pages 497
Graphics Calculator Support	page 782

Resource Links

For information about teacher and parent resources as well as professional development help, visit **www.hrw.com/math**.

Technical Support HRW has assembled a team of dedicated technical and teaching professionals and a comprehensive service program to provide you with the support you need.

- The HRW Technical Support Line operates from 7 A.M. to 6 P.M. central time, Monday through Friday, at (800) 323-9239.
- The HRW Technical Support Center on the World Wide Web is available 24 hours a day, seven days a week, at **www.hrwtechsupport.com**.
- You can e-mail our Technical Support Center at **tsc@hrwtechsupport.com**.
- The Technical Support Center's fax-on-demand service at (800) 352-1680 offers solutions to common problems and answers to frequently asked questions.

Technology

Lesson Suggestions and Calculator Examples
(Keystrokes are based on a TI-83 calculator.)

Lesson 10.1 Graphing Parabolas
In this lesson, students will learn about the parent function of a parabola and transformations on it. In the Activity on page 481, students engage in a graphing exploration in which they study the function $y = x^2$ and vertical and horizontal translations of its graph.

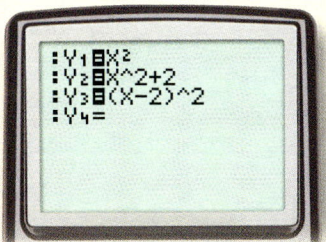 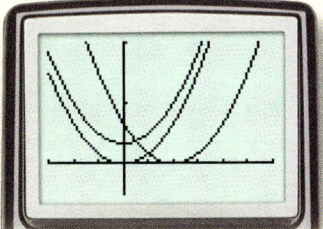

Students should see that when a parabola with a minimum is translated, the graph still has a minimum. Furthermore, the axis of symmetry is unchanged when a vertical translation is performed. However, under a horizontal translation, the axis of symmetry moves to the right or left accordingly.

10.2 Solving Equations by Using Square Roots
This lesson introduces square roots. Students can access the square root function on the TI-83 by pressing `2nd` `x²`. (Point out that pressing `2nd` before a key accesses what is written above the key face.) For example, to solve $3x^2 = 16$ on the TI-83, write $x = \sqrt{\frac{16}{3}}$, and press `2nd` `x²` `16` `÷` `3` `)` `ENTER`.

Lesson 10.3 Completing the Square
Point out to students that one purpose of completing the square is to write a quadratic function in vertex form. Using this form, the coordinates of the vertex and an equation for the axis of symmetry can be determined. Point out to students that they can use the graphics calculator to perform the reverse process; that is, they can graph the quadratic equation, identify the coordinates of the vertex, and use them to write the vertex form of the equation. The vertex form will then give the completion of the square.

For example, from the displays below, students can see the function $y = x^2 - 2x + 3$ and its graph. Tracing suggests that the coordinates of the vertex are (1, 2). Students should verify this by using

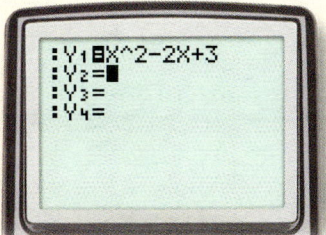 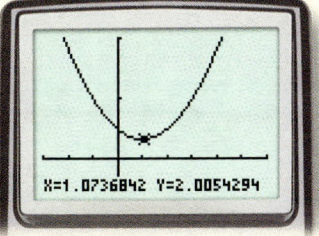

substitution. Now students can write the equation in vertex form, $y = (x - 1)^2 + 2$. This expression is exactly what students would get if they were to complete the square by using tiles or algebra.

Lesson 10.4 Solving Equations of the Form $x^2 + bx + c = 0$
Example 3 on page 500 provides an opportunity to investigate a twist on a problem. Find x such that $\begin{cases} y = x^2 + 3x - 15 \\ y = -5 \end{cases}$. This system cannot be solved by the algebraic methods taught in Chapter 7 because one of the equations in the system is nonlinear. However, the graphical approach taught in Chapter 7 may be applicable here. The display below shows the graphical approach. Tracing indicates that -5 may be a solution.

This problem may also present the opportunity to introduce the calculator's intersect feature, which can be accessed by pressing `2nd` `TRACE` and selecting `5:intersect`.

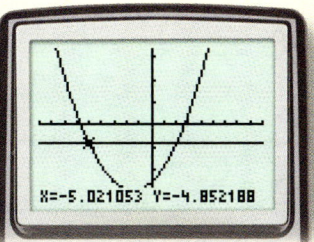

Lesson 10.5 The Quadratic Formula
Have students use the quadratic equation format to solve for the first solution on the calculator. After they obtain the first solution, they can press `2nd` `ENTER` to recall the expression already evaluated and use `◄` to change + to − in order to get the second solution.

Lesson 10.6 Graphing Quadratic Inequalities
In this lesson, students extend what they know about graphing linear inequalities to quadratic inequalities.

To graph $y \geq x^2$, enter x^2, place the cursor at the far left of the function line and press `ENTER` until ◣ appears. (This symbol tells the calculator to shade the region above the graph.) The resulting graph is shown above.

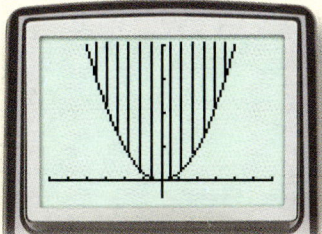

For further information, refer to the

- technology discussions in the lessons.
- lesson-related teacher's commentary in the side columns of this *Teacher's Edition*.
- lesson-related *Student Technology Guide* masters.
- *HRW Technology Handbook*.

Teaching Resources

Basic Skills Practice
(2 pages per skill)

Reteaching
(2 pages per lesson)

Enrichment
(1 page per lesson)

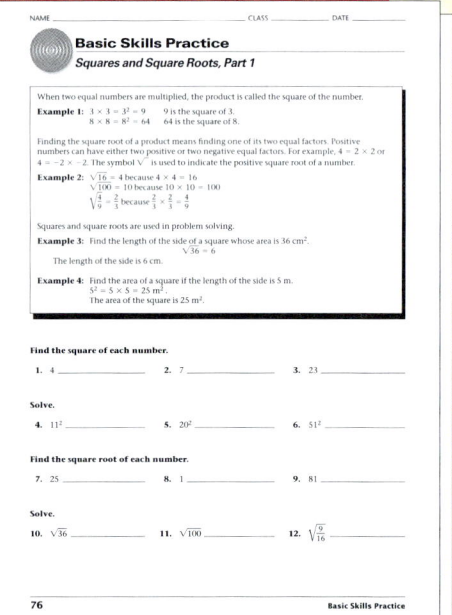

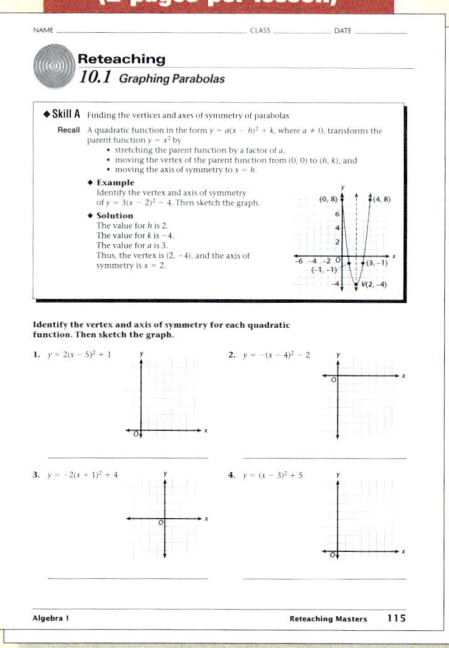

Cooperative-Learning Activity
(1 page per lesson)

Lesson Activity
(1 page per lesson)

Long-Term Project
(4 pages per chapter)

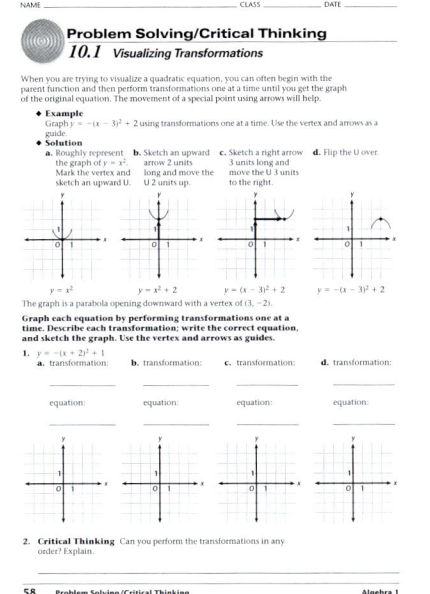

About Chapter 10

Background Information

Students were introduced to quadratic equations in Chapter 9, where they learned how to solve by factoring. Chapter 10 begins with a study of the graph of a quadratic equation, the *parabola*, and then moves on to present four other methods of solution: using square roots, graphing, completing the square, and using the quadratic formula. In the final lesson of the chapter, the technique for solving quadratic equations is extended to include quadratic inequalities.

CHAPTER RESOURCES

- Block-Scheduling Handbook
- Writing Activities for Your Portfolio
- Tech Prep Masters
- Long-Term Project
- Assessment Resources:
  Mid-Chapter Assessment
  Chapter Assessments
  Alternative Assessments
- Test and Practice Generator
- Technology Handbook
- End-of-Course Exam Practice

Chapter Objectives

- Discover how adding a constant to the parent function $y = x^2$ affects the graph of the function. [10.1]
- Use the zeros of a quadratic function to find the vertex of the graph of the function. [10.1]

Quadratic Functions

10

YOU HAVE SEEN THAT THE HEIGHT OF A PROJECTILE can be modeled by a quadratic function. Often the height versus time of the flight of a projectile will resemble a parabola, the graph of a quadratic function. Stopping distances for various vehicles can also be modeled by quadratic functions. In this chapter you will learn different methods for analyzing quadratic functions and quadratic equations.

The height of a projectile, such as a baseball, can be modeled by using a quadratic function.

Lessons

10.1 • Graphing Parabolas

10.2 • Solving Equations by Using Square Roots

10.3 • Completing the Square

10.4 • Solving Equations of the Form $x^2 + bx + c = 0$

10.5 • The Quadratic Formula

10.6 • Graphing Quadratic Inequalities

Project
What's the Difference?

Quadratic functions can sometimes be used to model depreciation rates.

About the Photos

Quadratic functions are widely used to model problems in math and science. Many pheonomena in the physical world can be described by quadratic functions. For example, in a game of tennis, there is a quadratic relationship between the height of the ball above the ground after a bounce and the amount of time that has elapsed since the bounce.

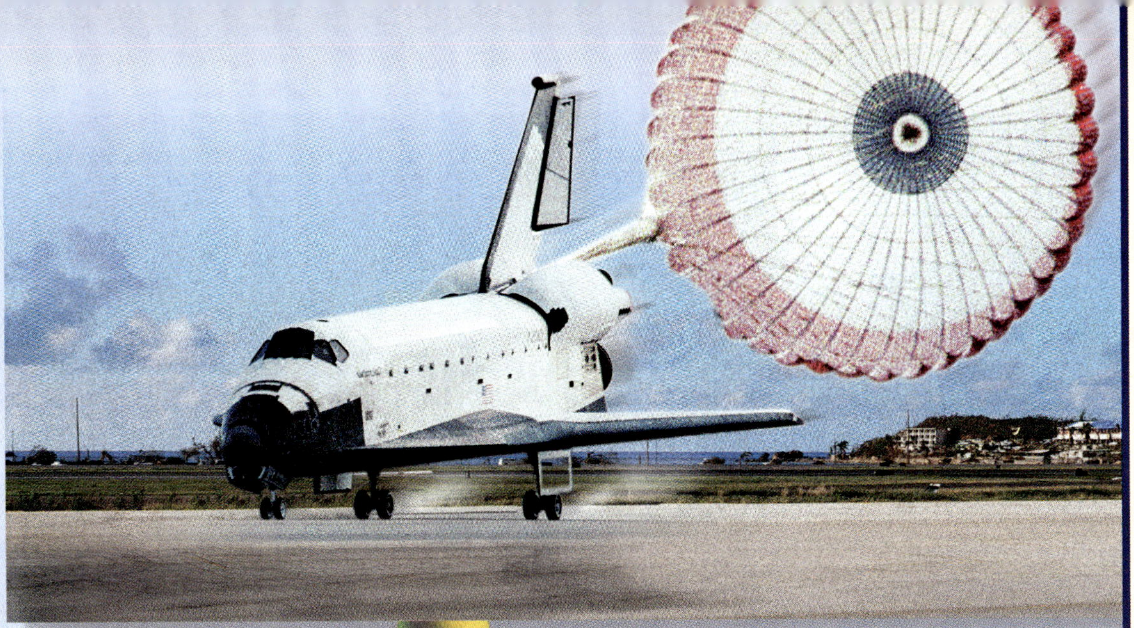

A quadratic function can be used to model braking distance.

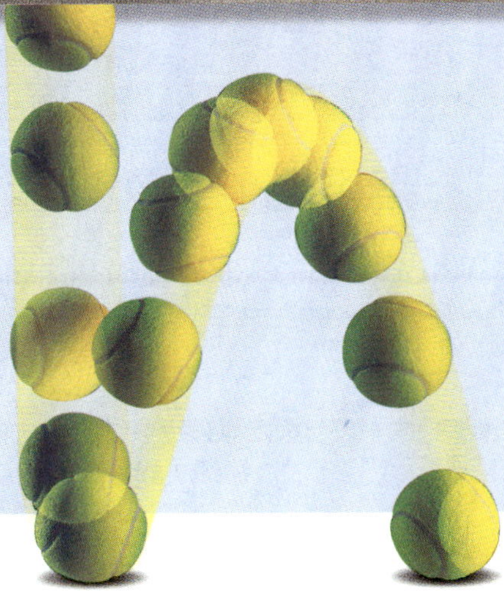

- Solve equations of the form $ax^2 = k$. [**10.2**]
- Solve equations of the form $x^2 = k$ where x is replaced by an algebraic expression. [**10.2**]
- Form a perfect-square trinomial from a given quadratic binomial. [**10.3**]
- Write a given quadratic function in vertex form. [**10.3**]
- Solve quadratic equations by completing the square or by factoring. [**10.4**]
- Use the quadratic formula to find solutions to quadratic equations. [**10.5**]
- Use the quadratic formula to find the zeros of a quadratic function. [**10.5**]
- Evaluate the discriminant to determine how many real roots a quadratic equation has and whether it can be factored over the integers. [**10.5**]
- Solve and graph quadratic inequalities and test solution regions. [**10.6**]

Portfolio Activities appear at the end of Lessons 10.2, 10.3, and 10.6. Each serves as preparation for the Chapter Project. The Portfolio Activities as well as the Chapter Project Activities are appropriate for inclusion in the student's portfolio. Students should be encouraged to include in their portfolios any other work in which they feel a sense of pride or a sense of accomplishment.

About the Chapter Project

You have seen many circumstances in which quadratic functions can be used to model real-world situations. In the Chapter Project, *What's the Difference*, on page 516, you will explore connections between quadratic functions and the differences of a sequence. After completing the Chapter Project you will be able to do the following:

- Use the method of finite differences to determine a function rule.

About the Portfolio Activities

Throughout the chapter, you will be given opportunities to complete Portfolio Activities that are designed to support your work on the Chapter Project.

- Using a motion detector to study a falling object is the topic of the Portfolio Activity on page 491.
- The general function for a falling object is the topic of the Portfolio Activity on page 497.
- Relating anxiety level and performance ratings is the topic of the Portfolio Activity on page 515.

internet connect

Internet connect features appearing throughout the chapter provide keywords for the HRW Web site that lead to links for math resources, tutorial assistance, references for student research, classroom activities, and all teaching resource pages. The HRW Web site also provides updates to the technology used in the text. Listed at right are the keywords for the Internet Connect activities referenced in this chapter. Refer to the side column on the page listed to read about the activity.

LESSON	KEYWORD	PAGE
10.1	MA1 Parabolas	484
10.3	MA1 Gravity	497
10.4	MA1 Motion Equations	503
10.5	MA1 Quadratic Formula	507

CHAPTER 10

Prepare

NCTM PRINCIPLES & STANDARDS
1–3, 6, 8–10

LESSON RESOURCES

- Practice 10.1
- Reteaching Master 10.1
- Enrichment Master 10.1
- Cooperative-Learning Activity 10.1
- Lesson Activity 10.1
- Problem Solving/Critical Thinking Master 10.1
- Teaching Transparency 47
- Teaching Transparency 48

QUICK WARM-UP

Solve by factoring.

1. $x^2 + 4x - 12 = 0$
   $x = -6$ or $x = 2$

2. $x^2 - x - 30 = 0$
   $x = -5$ or $x = 6$

3. $x^2 - 8x + 16 = 0$
   $x = 4$

4. $x^2 - 1 = 0$
   $x = -1$ or $x = 1$

5. $x^2 - x = 0$
   $x = 0$ or $x = 1$

Also on Quiz Transparency 10.1

Teach

Why The graph of a quadratic function is a curve called a *parabola*. To visualize a parabola, tell students to think of the path taken by a football when a field goal is attempted. The highest point of the path is the *vertex* of the parabola. Ask students to describe other real-world situations that involve parabolic paths.

Graphing Parabolas
10.1

Objectives

- Discover how adding a constant to the parent function $y = x^2$ affects the graph of the function.

- Use the zeros of a quadratic function to find the vertex of the graph of the function.

Why Fireworks displays require careful planning in order to be safe and successful. Parabolas can model the height of a fireworks shell over time as well as other types of motion.

APPLICATION
PHYSICS

The displays that you see during fireworks celebrations are typically launched at carefully timed intervals from mortars. The height of a shell after it is launched can be modeled by a quadratic function.

Definition of Quadratic Function

A quadratic function is a function of the form $y = ax^2 + bx + c$ where a, b, and c are real numbers and $a \neq 0$.

Because it is the simplest quadratic function, the function $y = x^2$ is the parent function of the class of quadratic functions.

The graph of the parent function $y = x^2$ is shown at right. The shape is called a **parabola**. The lowest point on the parabola, which is the **minimum value** of the function, is the point $(0, 0)$. This point is the **vertex** of the parabola. The function $y = -x^2$ has a maximum value and is discussed on page 481.

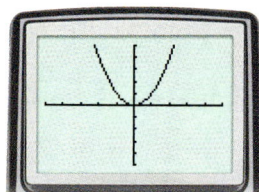

You can study the effect of adding a constant to the quadratic parent function in the following activity.

Alternative Teaching Strategy

CONNECTING TO PRIOR KNOWLEDGE Write these equations on the board or overhead.
$y = 2x \quad y = 2^x \quad y = x^2$
Have students graph the equations on three different sets of coordinate axes.

When students have completed their work, lead them in a discussion of the graphs. Ask how the graphs are alike and how they are different. Elicit observations about the distinctive ∪-shape of the graph of $y = x^2$. Tell students that this shape is called a *parabola*.

Ask students to write a different equation whose graph they think might be a parabola. Have them graph their equations. Discuss the results. Lead students to see that any equation that can be written in the form $y = ax^2 + bx + c$, where $a \neq 0$, will have a graph that is a parabola. Tell them that any function that can be defined by an equation of this type is called a *quadratic function*.

480 LESSON 10.1

Activity
Transformations of $y = x^2$

You will need: pencil and graph paper, or a graphics calculator

1. Build a table of values and make a graph for each of these functions:
   $y = x^2$ $y = x^2 + 1$ $y = x^2 - 1$

2. What effect does the addition or subtraction in Step 1 have on the graph of the parent function?

3. Build a table of values and make a graph for each of these functions:
   $y = x^2$ $y = (x + 2)^2$ $y = (x - 2)^2$

CHECKPOINT ✓
4. What effect does the addition or subtraction in Step 3 have on the graph of the parent function?

5. To observe the effect of combining Steps 1 and 3, build a table of values and make a graph for each of these functions:
   $y = x^2$ $y = (x - 2)^2 + 1$ $y = (x - 2)^2 - 1$ $y = (x + 2)^2 + 1$

CHECKPOINT ✓
6. What effect do addition and subtraction in Step 5 have on the graph of the parent function?

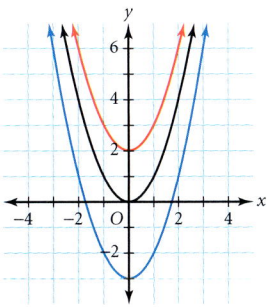

The functions $y = x^2 + 2$ and $y = x^2 - 3$ are **vertical translations** of the parent function $y = x^2$. The parabola is moved *up* or *down*.

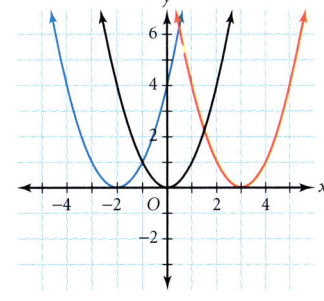

The functions $y = (x + 2)^2$ and $y = (x - 3)^2$ are **horizontal translations** of the parent function. The parabola is moved *left* or *right*.

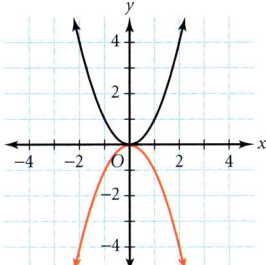

The function $y = -x^2$ is a **reflection**, or mirror image, of the parent function. The vertex, $(0, 0)$, now corresponds to the **maximum** value of the function $y = -x^2$.

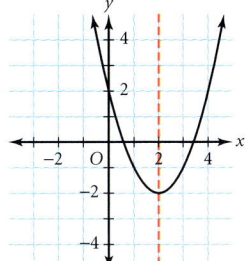

In a parabola, a vertical line drawn through the vertex is called the **axis of symmetry**. The half of the parabola on one side of the axis is a reflection of the half on the other side.

Activity Notes

The *quadratic parent function* is defined by the equation $y = x^2$. In this Activity, students explore the effect on the graph of the parent function as the equation is altered. You may want to suggest that they focus their attention on the change in the position of the vertex. In Step 1, they discover that the form $y = x^2 + k$ results in a vertical translation of k units. In Step 3, they see how the form $y = (x - h)^2$ yields a horizontal translation of h units. In Step 5, they observe that the form $y = (x - h)^2 + k$ causes both a vertical and a horizontal translation.

Cooperative Learning

You may wish to have students do the Activity in pairs. One student can do Steps 1 and 2, while the other does Steps 3 and 4. Partners can then discuss their results and make a conjecture about what will happen in Step 5. Have them work individually to test their conjecture in Steps 5 and 6.

CHECKPOINT ✓
4. Adding 2 shifts the graph 2 units to the left, and subtracting 2 shifts the graph 2 units to the right.

CHECKPOINT ✓
6. If the operation inside the squared term is subtraction, the graph is shifted to the right, and if the operation is addition, the graph is shifted to the left. If a number is added to the squared term, the graph is shifted up, and if a number is subtracted from the squared term, the graph is shifted down.

Interdisciplinary Connection

PHYSICS When an object is dropped from a height of a feet, its height, h, in feet above the ground after an elapsed time of t seconds is given by the equation $h = -16t^2 + a$, where $a > 0$. How is the graph of $h = -16t^2 + a$ related to the graph of $h = -16t^2$? **It is a parabola of the same shape that is translated a units upward.**

How long does it take an object to reach the ground if it is dropped from a height of 16 feet? 64 feet? **1 second; 2 seconds**

Suppose that you want it to take 3 seconds for an object to fall to the ground. From what height must it be dropped? (Hint: Translate the graph upward until the x-intercept is 3.) **144 feet** Assume that the object is dropped from this height. Write an equation to model its height, h, in feet above the ground after t seconds. $h = -16t^2 + 144$

LESSON 10.1 **481**

Teaching Tip

Be sure students understand that the sign of *a* has no effect on the size or shape of the parabola; the sign of *a* determines whether the parabola opens upward or downward.

Use Teaching Transparency 47.

ADDITIONAL EXAMPLE 1

Identify the vertex and the axis of symmetry for the graph of each equation. Check by graphing.

a. $y = (x + 5)^2 - 4$
vertex: $(-5, -4)$
axis of symmetry: $x = -5$

b. $y = -3(x - 1)^2 + 2$
vertex: $(1, 2)$
axis of symmetry: $x = 1$

TRY THIS

a. vertex: $(1, -2)$
axis of symmetry: $x = 1$

b. vertex: $(-3, 1)$
axis of symmetry: $x = -3$

Teaching Tip

TECHNOLOGY If you are using a TI-82 or TI-83 graphics calculator to create one of the graphs in Example 1, enter the equation next to **Y1** in the **Y=** editor. Then press ZOOM and choose **6: ZStandard**.

Use Teaching Transparency 48.

The Vertex Form

The Vertex Form of a Quadratic Function

The **vertex form** of the quadratic function is $y = a(x - h)^2 + k$, where (h, k) is the vertex. The line $x = h$ is the axis of symmetry. When *a* is positive, the vertex is the lowest point on the parabola. When *a* is negative, the vertex is the highest point on the parabola.

EXAMPLE ① Identify the vertex and the axis of symmetry for the graph of each equation. Check by graphing.

a. $y = 2(x - 3)^2 - 8$
b. $y = -(x + 3)^2 + 4$

● **SOLUTION**

a. Vertex: $(3, -8)$
Axis of symmetry: $x = 3$
Since *a* is positive, the vertex is the lowest point on the parabola.

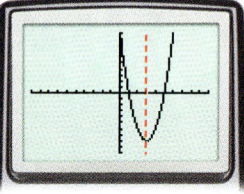

b. Vertex: $(-3, 4)$
Axis of symmetry: $x = -3$
Since *a* is negative, the vertex is the highest point on the parabola.

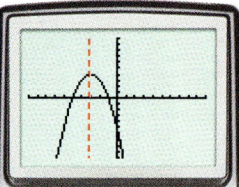

TRY THIS Identify the vertex and the axis of symmetry for each quadratic function.
a. $y = (x - 1)^2 - 2$
b. $y = -(x + 3)^2 + 1$

Using Zeros to Find the Vertex

In Chapter 9 you learned how to find the zeros of quadratic functions by factoring. For example, the zeros of the quadratic function $y = x^2 - 8x + 7$ are 1 and 7.

A parabola is symmetric. One half of the parabola is the mirror image of the other half. The axis of symmetry passes through a point midway between the zeros.

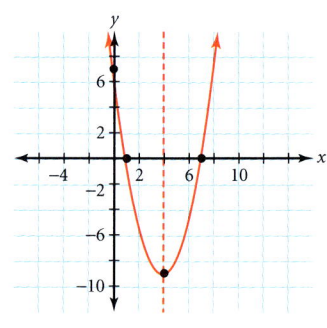

The zeros can be used to find the vertex of the parabola.

Inclusion Strategies

TACTILE LEARNERS Have students create a "master graph" of the parent quadratic function on a sheet of graph paper. Then, to graph an equation of the form $y = (x - h)^2 + k$, tell them to first draw an identical set of axes on another sheet of graph paper, being sure that the markings are dark enough to be seen through the master graph sheet. Have them use the equation to determine the translation and then perform the translation by sliding the master graph over the axes.

Enrichment

Have students investigate the graph of the equation $x = y^2$. Ask questions such as these.

1. What is the shape of the graph? **parabola**
2. How is the graph of $x = y^2$ related to the graph of $y = x^2$? **It is the graph of $y = x^2$ rotated a quarter-turn to the right.**
3. How is the graph of $x = y^2 + 3$ related to the graph of $x = y^2$? **translation 3 units to the right**
4. How is the graph of $x = (y - 3)^2$ related to the graph of $x = y^2$? **translation 3 units upward**

EXAMPLE 2

Use the zeros to find the vertex of the graph of $y = x^2 - 6x + 5$. Then write the vertex form of the equation.

SOLUTION

Find the zeros by setting the function equal to 0. Start by factoring.

$$x^2 - 6x + 5 = 0$$
$$(x - 1)(x - 5) = 0$$

Set each binomial of the factored form equal to 0 and solve for x.

$$x - 1 = 0 \quad \text{or} \quad x - 5 = 0$$
$$x = 1 \quad\quad\quad\quad x = 5$$

The zeros of the function are 1 and 5.

Because the parabola is symmetrical, the vertex is halfway between the zeros. To find the x-coordinate of the vertex, find the midpoint, or average, of the x-coordinates of the zeros.

$$x = \frac{5+1}{2} = 3$$

Substitute for x in the function to find the y-coordinate of the vertex.

$$y = x^2 - 6x + 5$$
$$= 3^2 - 6(3) + 5$$
$$= -4$$

Thus, the vertex is $(3, -4)$. Write the equation in vertex form.

$$y = (x - 3)^2 - 4$$

TRY THIS Use the zeros to find the vertex and the vertex form of the quadratic function $y = x^2 - 7x + 10$.

EXAMPLE 3

APPLICATION
PHYSICS

A fireworks shell is fired from a mortar. Its height from 0 to 14 seconds is modeled by the function $y = -16(x - 7)^2 + 784$, where x is the time in seconds and y is the height in feet. **If the shell falls to the ground without exploding, how many seconds will it take for the shell to hit the ground?**

SOLUTION

In the form given, the vertex, the axis of symmetry, and the maximum value can be read directly from the function.

function: $y = -16(x - 7)^2 + 784$
vertex: $(7, 784)$
axis of symmetry: $x = 7$
maximum value: 784

Sketch a graph of the function by using this information. The zeros are 0 and 14. Thus, the shell will hit the ground after 14 seconds of flight.

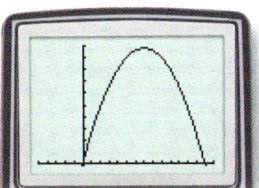

ADDITIONAL EXAMPLE 2

Use the zeros to find the vertex of the graph of $y = x^2 - 6x + 8$. Then write the vertex form of the equation.
vertex: $(3, -1)$
vertex form: $y = (x - 3)^2 - 1$

TRY THIS
vertex: $(3.5, -2.25)$
vertex form: $y = (x - 3.5)^2 - 2.25$

ADDITIONAL EXAMPLE 3

A fireworks shell is fired from a mortar, and its height is given by the function $y = -16(x - 3)^2 + 144$, where y is the height in feet and x is the elapsed time in seconds. If the shell falls to the ground without exploding, how many seconds will it take for the shell to hit the ground?
The shell will hit the ground after 6 seconds.

Teaching Tip

TECHNOLOGY If you are using a TI-82 or TI-83 graphics calculator to create the graph in Example 3, first enter **−16(X−7)^2+784** next to **Y1** in the **Y=** editor. Then press **WINDOW** and make these settings.

Xmin= −5   Ymin=0
Xmax=15   Ymax=800
Xscl=1   Yscl=80

Press **GRAPH** to view the graph.

Reteaching the Lesson

USING VISUAL STRATEGIES
Show students the graph at right. Tell them that the curve is a parabola and that an equation for it is $y = x^2$. Remind them that this is a graph of the quadratic parent function.

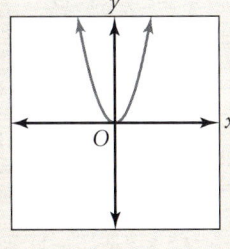

Now give them several transformations of the curve, one at a time. Two possibilities are shown at right. Do *not* show any units on the axes.

Have students write possible equations for each transformation. Ask for volunteers to share their equations, and discuss their results.

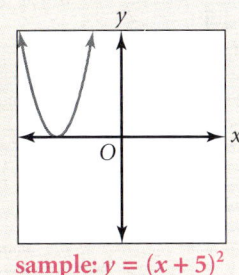

sample: $y = (x + 5)^2$

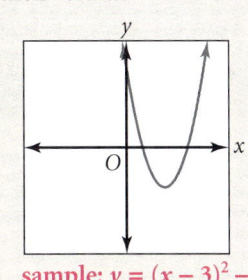

sample: $y = (x - 3)^2 - 3$

LESSON 10.1   483

Assess

Selected Answers

Exercises 5–10, 11–53 odd

ASSIGNMENT GUIDE

In Class	1–10
Core	11–16, 20–25, 32–37, 42–46
Core Plus	12–40 even, 41–46
Review	47–54
Preview	55

✏ Extra Practice can be found beginning on page 738.

Technology

A graphics calculator may be helpful for Exercises 11–40. You may wish to have students first do the exercises with paper and pencil and then use the calculator to check their answers.

Practice

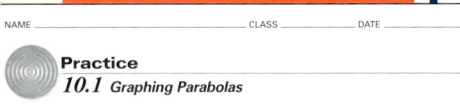
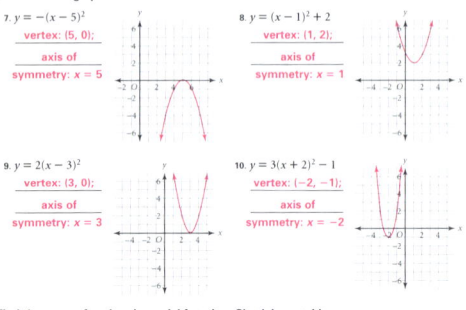

Exercises

Communicate

1. Discuss how the graph of the quadratic function $y = x^2 - 8$ differs from the graph of the parent function $y = x^2$.
2. Discuss how the graph of the quadratic function $y = (x - 8)^2$ differs from the graph of the parent function $y = x^2$.
3. Explain how to find the vertex of $y = 2(x - 3)^2 - 8$.
4. Explain how to find the axis of symmetry of $y = 2(x - 3)^2 - 8$.

Guided Skills Practice

Identify the vertex and the axis of symmetry for each quadratic function. *(EXAMPLE 1)*

5. $y = (x + 6)^2 - 4$
   (–6, –4); x = –6
6. $y = (x - 3)^2 - 8$
   (3, –8); x = 3
7. $y = -4(x + 3)^2 + 6$
   (–3, 6); x = –3

Use the zeros to find the vertex and the vertex form of each quadratic function. *(EXAMPLE 2)*

8. $y = x^2 - 12x + 20$
   (6, –16);
   $y = (x - 6)^2 - 16$
9. $y = x^2 - 3x - 10$
   $(\frac{3}{2}, \frac{-49}{4})$;
   $y = (x - \frac{3}{2})^2 - \frac{49}{4}$
10. $y = x^2 + 11x + 18$
    $(\frac{-11}{2}, \frac{-49}{4})$;
    $y = (x + \frac{11}{2})^2 - \frac{49}{4}$

Practice and Apply

Compare the graphs of the following functions to the graph of $y = x^2$. Describe the vertical and horizontal translations of each vertex.

11. $y = (x - 2)^2 + 3$
12. $y = 3(x - 5)^2 - 2$
13. $y = -(x - 2)^2 + 1$
14. $y = -7(x + 6)^2 - 2$
15. $y = -7(x - 3)^2 - 2$
16. $y = 5(x + 4)^2 - 7$
17. $y = \frac{1}{3}(x + 4)^2 - 5$
18. $y = \frac{1}{4}(x + 7)^2 + 14$
19. $y = 5(x + 2)^2 - 8$

Find the vertex and axis of symmetry for the graph of each function, and then sketch the graph.

20. $y = -2(x + 4)^2 - 3$
21. $y = \frac{1}{2}(x - 2)^2 + 3$
22. $y = (x - 3)^2 - 7$
23. $y = -(x + 2)^2 - 4$
24. $y = -4(x - 3)^2 - 8$
25. $y = \frac{1}{2}(x - 4)^2 + 5$
26. $y = 3(x + 6)^2 - 3$
27. $y = (x + 5)^2 - 7$
28. $y = (x - 2)^2 + 9$
29. $y = -3(x - 5)^2 + 2$
30. $y = -(x + 3)^2 - 2$
31. $y = 2(x + 5)^2 + 7$

Use factoring to find the zeros of each function.

32. $y = x^2 + 8x - 9$ **1, –9**
33. $y = x^2 - 20x + 100$ **10**
34. $y = x^2 - x - 72$ **–8, 9**
35. $y = x^2 + 6x - 7$ **1, –7**
36. $y = x^2 + 4x - 5$ **1, –5**
37. $y = x^2 + 2x - 24$ **4, –6**
38. $y = x^2 + 18x + 81$ **–9**
39. $y = x^2 + 2x - 63$ **7, –9**
40. $y = x^2 - 5x + 6$ **2, 3**

GO TO: go.hrw.com
KEYWORD: MA1 Parabolas

In this activity, students investigate the properties of parabolas with an interactive Web site that allows students to change the constants in the general equation of a parabola and gauge the effects.

11. The graph is shifted 2 units to the right and 3 units up.

12. The graph is shifted 5 units to the right and 2 units down.

13. The graph is shifted 2 units to the right and 1 unit up.

14. The graph is shifted 6 units to the left and 2 units down.

15. The graph is shifted 3 units to the right and 2 units down.

C O N N E C T I O N

41. NUMBER THEORY Use a quadratic equation to find two numbers whose difference is 5 and whose product is 24. *–3 and –8, or 3 and 8*

A P P L I C A T I O N

PHYSICS The graph at right represents the relationship between the time, in seconds, when a projectile is propelled vertically into the air and its height.

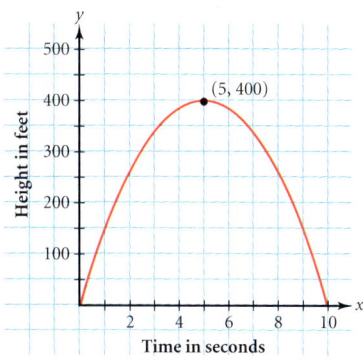

42. What is the maximum height reached by the projectile? *400 feet*

43. How long does it take for the projectile to reach its maximum height? *5 seconds*

44. How long does it take for the projectile to return to the ground? *10 seconds*

45. What is the axis of symmetry of the graph? *x = 5*

46. PHYSICS An arrow is shot upward at an initial velocity of 40 meters per second. Use the function $h = 40t - 5t^2$ to find the height in meters after 5 seconds. Round your answers to the nearest tenth. *75.0 m*

The longbow was used by soldiers in medieval times.

Look Back

The expression $2(x - 3)^2 + 1$ involves multiplication, an operation inside parentheses, an exponent, and addition. *(LESSON 1.3)*

47. Which of these operations should be done first? *subtraction inside parentheses*

48. Which should be done second? *exponent*

49. Which should be done third? *multiplication*

Solve. *(LESSON 7.2)*

50. $\begin{cases} x = y \\ 2x + 3y = 2 \end{cases}$ $\left(\frac{2}{5}, \frac{2}{5}\right)$

51. $\begin{cases} y = 4 \\ 2x = 3y \end{cases}$ *(6, 4)*

52. Find the two-digit number whose tens digit is 4 less than its units digit if the original number is 2 more than 3 times the sum of the digits. *26* *(LESSON 7.6)*

Factor. *(LESSON 9.6)*

53. $x^2 + 10x + 25$ $(x + 5)^2$

54. $x^2 - 14x + 49$ $(x - 7)^2$

Look Beyond

A P P L I C A T I O N

55. ECOLOGY A student group hopes to collect 100,000 cans to recycle. Find how much space the cans will occupy if each can has a circumference of 21 centimeters and a height of 12.5 centimeters. Express your answer in cubic centimeters and in cubic meters (1,000,000 cm³ = 1 m³).

≈ 43,867,081 cm³ or ≈ 43.9 m³

Error Analysis

When reading the coordinates of the vertex from an equation in vertex form, students often reverse the sign of the *x*-coordinate. For example, given $y = (x - 3)^2 - 7$, they identify the vertex as $(-3, -7)$. Encourage them to check by substituting the coordinates into the equation. If the resulting statement is false, the vertex coordinates are incorrect.

Look Beyond

Exercise 55 is a nonroutine problem. You may want to encourage students to either draw a diagram or act it out. Note that extensions of this problem can provide opportunities for open-ended problem solving. For example, students might determine the dimensions and number of boxes needed to pack the cans.

The answers to Exercises 21–31 can be found in Additional Answers beginning on page 810.

Student Technology Guide

Student Technology Guide
10.1 Graphing Parabolas

The graph of a quadratic function, such as $y = x^2 - 6x + 5$, is called a parabola and has a turning point called the vertex. The *axis of symmetry* of a parabola passes through its vertex. If you know the coordinates of the vertex, it is easy to identify the axis of symmetry. You can use a graphics calculator to find the vertex and axis of symmetry of a parabola. In order to do this, you need to have a complete view of the parabola.

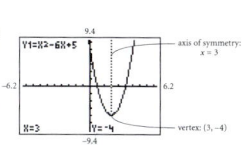

axis of symmetry: $x = 3$
vertex: (3, –4)

Example: Identify the vertex and the axis of symmetry for $y = x^2 + 4x - 5$.

- Enter and graph the equation.

If you cannot see the vertex, press WINDOW, change Xmin, Xmax, Ymin, or Ymax, and press GRAPH again. Changing Xmin to –18 and Ymin to –12 works well here.

- Trace to the vertex of the parabola by pressing TRACE and using the arrow keys to move along the graph. The vertex appears to be near (–2, –9).

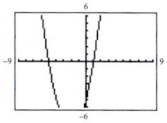 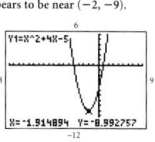

When a parabola crosses the *x*-axis in two points, you can locate the two *x*-intercepts and find their average in order to find the *x*-coordinate of the vertex of the parabola.

Identify the vertex and the axis of symmetry for each quadratic function.

1. $y = x^2 + 4x + 2$
   vertex: (–2, –2); axis of symmetry: $x = -2$
2. $y = x^2 - 8x - 3$
   vertex: (4, –19); axis of symmetry: $x = 4$
3. $y = 2x^2 + 6x - 9$
   vertex: (–1.5, –13.5); axis of symmetry: $x = -1.5$
4. $y = -x^2 - 10x + 9$
   vertex: (–5, 34); axis of symmetry: $x = -5$

---

16. The graph is shifted 4 units to the left and 7 units down.

17. The graph is shifted 4 units to the left and 5 units down.

18. The graph is shifted 7 units to the left and 14 units up.

19. The graph is shifted 2 units to the left and 8 units down.

20. vertex: (–4, –3)
axis of symmetry: $x = -4$

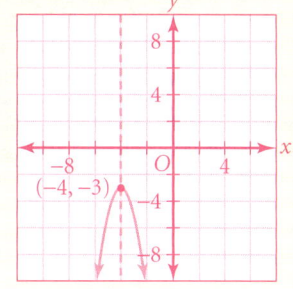

Prepare

NCTM PRINCIPLES & STANDARDS 1–3, 5–10

LESSON RESOURCES

- Practice 10.2
- Reteaching Master 10.2
- Enrichment Master 10.2
- Cooperative-Learning Activity 10.2
- Lesson Activity 10.2
- Problem Solving/Critical Thinking Master 10.2

QUICK WARM-UP

Find each product.

1. $(-5)(-5)$ **25**
2. $(11)(11)$ **121**

Solve each equation.

3. $2x - 5 = 11$ ***x = 8***
4. $-3y + 12 = 24$ ***y = −4***
5. $-16t + 320 = 0$ ***t = 20***

Also on Quiz Transparency 10.2

Teach

Why When an object is in free fall, the speed at which it falls increases at a rate of 32 feet per second per second. This rate is called the *acceleration of gravity*. The object's height, h, in feet above the ground after falling for t seconds is given by the equation $h = -16t^2 + h_0$, where h_0 is the height in feet from which the object was dropped. Ask students to suggest ways that this equation might be used in real-world situations.

10.2 Solving Equations by Using Square Roots

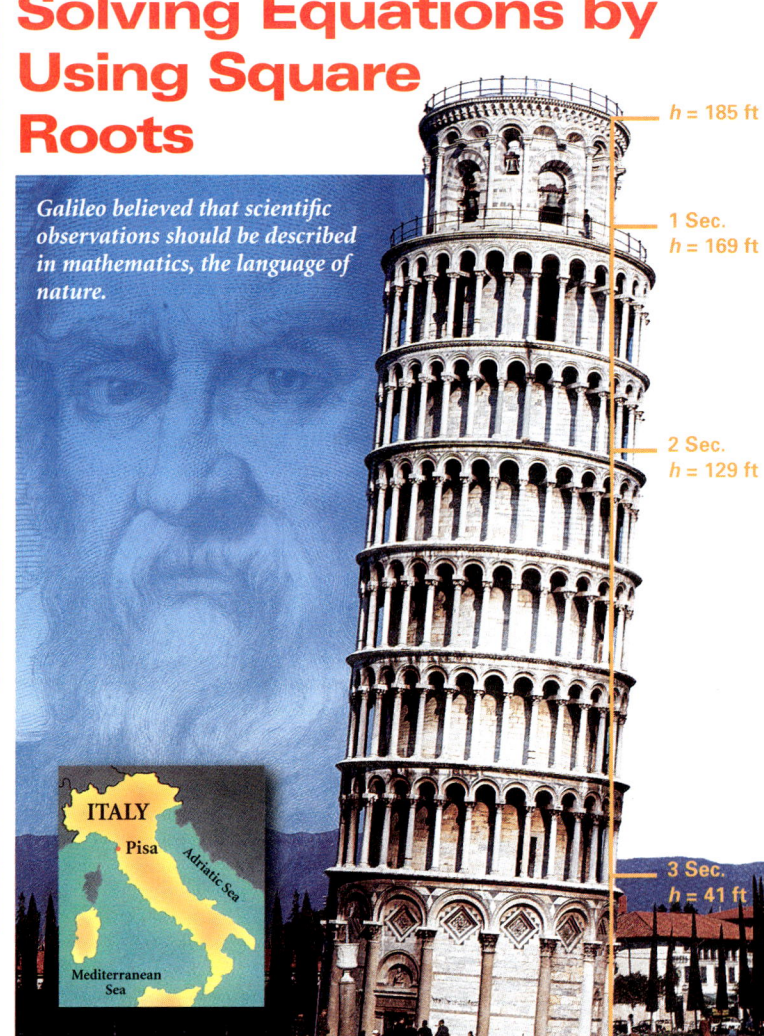

Galileo believed that scientific observations should be described in mathematics, the language of nature.

Objectives

- Solve equations of the form $ax^2 = k$.
- Solve equations of the form $x^2 = k$ where x is replaced by an algebraic expression.

Why The solutions to quadratic equations can give you answers to problems in the real world, such as the height of a falling object.

APPLICATION
PHYSICS

Galileo Galilei, born in Pisa, Italy, in 1564, discovered that objects fall with the same acceleration, regardless of the mass of the object. In a famous experiment, he hypothesized that if a cannon ball and a small stone were dropped from the Leaning Tower of Pisa, both would hit the ground at the same time.

Suppose that an object is dropped from the top of the Leaning Tower of Pisa, 185 feet above the ground. The object's height in feet, h, after t seconds is given by the following function:

$$h = -16t^2 + 185$$

How long will it take for the object to hit the ground? You will investigate this question in the following activity.

Alternative Teaching Strategy

USING TECHNOLOGY Write $x - 2 = 4$ on the board or overhead. If you are using TI-82 or TI-83 graphics calculators, have students press [Y=] and enter **X–2** next to **Y1** and **4** next to **Y2**. Then tell them to press [ZOOM] and choose **6: ZStandard**. Note that the two graphs intersect at one point. Lead them to see that the x-coordinate of this point is the value of x for which $x - 2 = 4$. To find the point of intersection, have them press [2nd] [TRACE], choose **5: intersect**, then press [ENTER] [ENTER] [ENTER]. The x-coordinate is 6, so the solution to $x - 2 = 4$ is $x = 6$.

Now display the equation $x^2 = 4$. Have students press [Y=] and then [CLEAR] and enter **X^2** next to **Y1**. Now have them press [GRAPH]. Discuss the display with them, noting that this time the graph of $y = 4$ intersects the graph of $y = x^2$ at *two* points. Have them use the **CALCULATE** menu to identify the x-coordinates of these points, which are 2 and –2. (To locate the second point, they will need to use the [◄] key to move the cursor to the left of the y-axis before pressing [ENTER] [ENTER] [ENTER].) Lead them to see that the solution to $x^2 = 4$ is $x = 2$ or $x = -2$.

Tables for Falling Objects

1. Build a table of values for the function $h = -16t^2 + 185$.

Time, t	0	1	2	3	4	5
Height, h	?	?	?	?	?	?

2. The value of h is 0 when the object hits the ground. Use your table to find the approximate value of t when $h = 0$.

3. Refine your estimate from Step 2 by building a new table with increments of 0.1 for t near the value that you found in Step 2.

CHECKPOINT ✓ 4. Refine your estimate further by building a table with increments of 0.01 for t near the value that you found in Step 3.

Using Square Roots

The equation $x^2 = 9$ can be stated as the question, "What number multiplied by itself is 9?" There are two such numbers, 3 and -3. Each of these is called a **square root** of 9.

The *positive* square root of a number is called the **principal square root**. The principal square root of a number is indicated by the **radical sign**, $\sqrt{}$. In general, if $x^2 = k$ and $k > 0$, then $x = \sqrt{k}$ or $x = -\sqrt{k}$. You can write this as $\pm\sqrt{k}$, which is read as "plus or minus the square root of k."

Square roots are the solutions to equations of the form $x^2 = k$, as you will see in the following example.

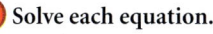

Solve each equation.

  a. $x^2 = 25$ b. $x^2 = 10$ c. $x^2 = \frac{9}{4}$

TECHNOLOGY
SCIENTIFIC CALCULATOR

• SOLUTION

a. The solution to $x^2 = 25$ is $x = \sqrt{25}$ or $x = -\sqrt{25}$. Thus, $x = 5$ or $x = -5$.

b. The solution to $x^2 = 10$ is $x = \sqrt{10}$ or $x = -\sqrt{10}$. There is no exact rational number answer, but an approximation can be found by using the $\sqrt{}$ key on your calculator.

$$\boxed{\sqrt{}}\ 10\ \boxed{=}\ \ 3.16227766$$

Thus, $x \approx 3.16$ or $x \approx -3.16$.

c. The solution to $x^2 = \frac{9}{4}$ is $x = \sqrt{\frac{9}{4}}$ or $x = -\sqrt{\frac{9}{4}}$. Thus, $x = \frac{3}{2}$ or $x = -\frac{3}{2}$.

TRY THIS Solve each equation.

  a. $x^2 = 100$ b. $x^2 = \frac{25}{4}$ c. $x^2 = \frac{16}{49}$

Interdisciplinary Connection

PHYSICS The motion of a pendulum from one end of its swing to the opposite end and back to its starting point is called an *oscillation* of the pendulum. The time required for one oscillation is the *period* of the pendulum. For small swings, the period, T, in seconds of a pendulum l meters long is given by the formula $T = 2\pi\sqrt{\frac{l}{9.8}}$. Find the period of a pendulum that is 1.5 meters long. Round to the nearest hundredth of a second. **about 2.46 seconds**

Inclusion Strategies

VISUAL LEARNERS Some students will benefit from working with a concrete representation of the principal square root of a number. Have them use graph paper to draw squares with area 1 square unit, 4 square units, 9 square units, and 16 square units. Ask them to give the length of one side of each square. **1 unit; 2 units; 3 units; 4 units** Explain that the area of each square represents a number and that the length of the side of the square represents the principal square root of that number.

Activity Notes

In this Activity, students use a series of tables to approximate the solution to the equation $h = -16t^2 + 185$ for $h = 0$ and $t > 0$. They should understand that this gives the number of seconds it takes for an object to fall to the ground when dropped from a height of 185 feet.

Cooperative Learning

You may wish to have students do the Activity in pairs. Have them work together to complete the table and find the approximation in Step 2. For Step 3, one partner can build the table, and the other can make the approximation. They can then reverse this procedure in Step 4.

CHECKPOINT ✓

4.
Time, t	Height, h
3.39	1.1264
3.40	0.04
3.41	-1.05
3.42	-2.142

ADDITIONAL EXAMPLE

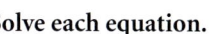

Solve each equation.

a. $x^2 = 36$
  $x = -6$ or $x = 6$

b. $x^2 = 15$
  $x = -\sqrt{15}$ or $x = \sqrt{15}$

c. $x^2 = \frac{36}{25}$
  $x = -\frac{6}{5}$ or $x = \frac{6}{5}$

TRY THIS
a. $x = \pm 10$
b. $x = \pm\frac{5}{2}$
c. $x = \pm\frac{4}{7}$

LESSON 10.2

Teaching Tip

TECHNOLOGY If you are using a TI-82 or TI-83 graphics calculator, you access the square root function by pressing [2nd] [x^2].

ADDITIONAL EXAMPLE 2

Find the exact time that it takes for an object to fall from a tower that is 320 feet tall.

$\sqrt{20}$ seconds

ADDITIONAL EXAMPLE 3

Solve $(x + 4)^2 - 25 = 0$.

$x = -9$ or $x = 1$

TRY THIS
a. $t = \pm \frac{12}{5}$
b. $x = 13$ or $x = -5$

The results of Example 1 lead to a generalization for finding the solution to a quadratic equation of the form $ax^2 = k$.

> **Solving $ax^2 = k$, Where $\frac{k}{a}$ Is Nonnegative**
>
> An equation of the form $ax^2 = k$, where $\frac{k}{a} \geq 0$ and $a \neq 0$, has the following solutions:
>
> $$x = \sqrt{\frac{k}{a}} \text{ or } x = -\sqrt{\frac{k}{a}}$$

EXAMPLE 2

APPLICATION
PHYSICS

Find the exact time that it takes for an object to fall from the top of the Leaning Tower of Pisa.

SOLUTION

Use the equation obtained by setting the function from the beginning of the lesson equal to 0.

$-16t^2 + 185 = 0$
$-16t^2 = -185$
$t^2 = \frac{-185}{-16} = \frac{185}{16}$
$t = \pm\sqrt{\frac{185}{16}} = \pm\frac{\sqrt{185}}{\sqrt{16}} = \pm\frac{\sqrt{185}}{4}$

The *exact* time is $\frac{\sqrt{185}}{4}$ seconds. This value can be approximated to the nearest hundredth as 3.40 seconds.

Although $-\frac{\sqrt{185}}{4}$ is a solution to the equation, it is not reasonable for this particular problem because time is not measured in negative units.

In the next example, an *expression* takes the place of x in an equation of the form $ax^2 = k$.

EXAMPLE 3 Solve the equation $(x - 2)^2 - 9 = 0$.

SOLUTION

$(x - 2)^2 - 9 = 0$ The solutions are $2 + 3$, or 5, and
$(x - 2)^2 = 9$ $2 - 3$, or -1. Check each solution
$x - 2 = \pm\sqrt{9}$ in the original equation.
$x = 2 \pm 3$

TRY THIS Solve each equation.
a. $25t^2 - 144 = 0$
b. $(x - 4)^2 - 81 = 0$

Enrichment

Tell students to write two different quadratic equations that have the solutions listed below. The equations should be in the form $ax^2 = k$ or $a(x - h)^2 = k$. Samples are given below.

1. $-4, 4$ $x^2 = 16, 2x^2 = 32$
2. $2, -6$ $(x + 2)^2 = 16, 2(x + 2)^2 = 32$
3. $-\sqrt{2}, \sqrt{2}$ $x^2 = 2, 2x^2 = 4$
4. only 2 $(x - 2)^2 = 0, 2(x - 2)^2 = 0$

Reteaching the Lesson

GUIDED ANALYSIS Work with students to outline a procedure for using square roots to solve a quadratic equation. A sample set of steps is given below.

1. Rewrite the equation, if necessary, to obtain an equation of the form $a(x - h)^2 = k$.
2. Apply the Division Property of Equality to obtain an equation of the form $(x - h)^2 = \frac{k}{a}$.
3. Take the square root of each side and solve for x.

CRITICAL THINKING  In Example 3, the expression $(x - 2)$ takes the place of x in an equation of the form $ax^2 = k$. What numbers represent a and k?

CONNECTION
COORDINATE GEOMETRY

For the function $y = (x - 2)^2 - 9$, the vertex of the graph is $(2, -9)$ and the axis of symmetry is $x = 2$. Because the coefficient of the quadratic term is positive, the parabola opens upward and has a minimum value. For $y = 0$, the solutions are 5 and -1 (see Example 3). In other words, the graph of the function crosses the x-axis at $(5, 0)$ and $(-1, 0)$.

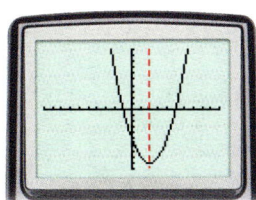

CRITICAL THINKING
$a = 1, k = 9$

Math
CONNECTION

COORDINATE GEOMETRY Point out to students that the solutions to $(x - 2)^2 - 9 = 0$ obtained by using square roots are consistent with those obtained by analyzing the graph of $y = (x - 2)^2 - 9$, as was done in Lesson 10.1.

Exercises

Assess

Selected Answers
Exercises 6–12, 13–71 odd

Communicate

1. The function $h = -16t^2 + 300$ represents the height in feet, h, of a falling body in terms of time, t, in seconds. Explain how to make a table of values to find the point(s) where the function crosses the horizontal axis.

2. Explain how to find the square root of 100. Explain why you use the $\pm$ sign.

3. Explain how to solve the following equations:
   a. $x^2 = 64$ **b.** $x^2 = 8$ **c.** $x^2 = \frac{16}{100}$

4. Discuss how to solve each equation by using square roots.
   a. $(x + 3)^2 - 25 = 0$ **b.** $(x - 8)^2 = 2$

5. Explain how to make a rough sketch of the graph of $y = (x - 3)^2 - 4$ by using the vertex, axis of symmetry, and zeros.

ASSIGNMENT GUIDE

In Class	1–12
Core	13–57 odd
Core Plus	14–50 even, 52–59
Review	60–71
Preview	72, 73

✎ Extra Practice can be found beginning on page 738.

Guided Skills Practice

Solve each equation. *(EXAMPLE 1)*

6. $x^2 = 64$ **±8**
7. $x^2 = \frac{16}{25}$ **±$\frac{4}{5}$**
8. $x^2 = 17$ **≈ ±4.12**

APPLICATIONS

9. **PHYSICS** Find the time, t, in seconds it takes for an object to fall 600 feet by using the function $h = -16t^2 + 600$. *(EXAMPLE 2)* **about 6.12 seconds**

10. **PHYSICS** Find the time, t, in seconds it takes for an object to fall 800 feet by using the function $h = -16t^2 + 800$. *(EXAMPLE 2)* **about 7.07 seconds**

11. Solve the equation $(x + 2)^2 - 144 = 0$. *(EXAMPLE 3)* **$x = 10$ or -14**

12. Solve the equation $(x - 3)^2 - 400 = 0$. *(EXAMPLE 3)* **$x = 23$ or -17**

LESSON 10.2

Technology

Students will need a calculator to do the approximations required in Exercises 17–48, 58, and 59.

Error Analysis

When solving an equation of the form $x^2 = k$, students often forget the negative solution. Remind them that the only time an equation of this type has exactly one solution is when $k = 0$. When $k > 0$, they must always look for two solutions.

49. vertex: (4, −3)
axis of symmetry: $x = 4$
zeros: 5.73 and 2.27

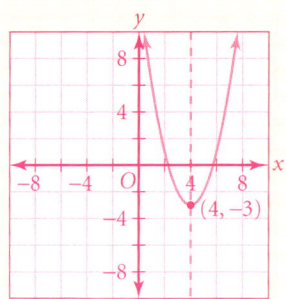

Practice and Apply

Find each positive square root. Round answers to the nearest hundredth when necessary.

13. $\sqrt{121}$ 11   **14.** $\sqrt{144}$ 12   **15.** $\sqrt{625}$ 25   **16.** $\sqrt{36}$ 6

17. $\sqrt{44}$ 6.63   **18.** $\sqrt{90}$ 9.49   **19.** $\sqrt{88}$ 9.38   **20.** $\sqrt{19}$ 4.36

Solve each equation. Round answers to the nearest hundredth when necessary.

21. $x^2 = 25$ **22.** $x^2 = 169$ **23.** $x^2 = 81$ **24.** $x^2 = 625$

25. $x^2 = 12$ **26.** $x^2 = 24$ **27.** $x^2 = 18$ **28.** $x^2 = 54$

29. $x^2 = \frac{25}{81}$ **30.** $x^2 = \frac{49}{121}$ **31.** $x^2 = \frac{36}{49}$ **32.** $x^2 = \frac{4}{100}$

33. $x^2 = 32$ **34.** $3x^2 = 135$ **35.** $2x^2 = 56$ **36.** $x^2 = 63$

37. $(x+4)^2 - 25 = 0$ **38.** $(x-5)^2 - 9 = 0$ **39.** $(x+1)^2 - 1 = 0$

40. $(x-2)^2 - 6 = 0$ **41.** $(x+7)^2 - 5 = 0$ **42.** $6(x-3)^2 - 12 = 0$

43. $(x+3)^2 = 36$ **44.** $(x-2)^2 = 144$ **45.** $(x-8)^2 = 81$

46. $(x-1)^2 = 11$ **47.** $(x+5)^2 = 10$ **48.** $7(x+6)^2 = 105$

21. $x = \pm 5$
22. $x = \pm 13$
23. $x = \pm 9$
24. $x = \pm 25$
25. $x = \pm 3.46$
26. $x = \pm 4.90$
27. $x = \pm 4.24$
28. $x = \pm 7.35$
29. $x = \pm \frac{5}{9}$
30. $x = \pm \frac{7}{11}$
31. $x = \pm \frac{6}{7}$
32. $x = \pm \frac{1}{5}$
33. $x = \pm 5.66$
34. $x = \pm 6.71$
35. $x = \pm 5.29$
36. $x = \pm 7.94$
37. $x = 1$ or $x = -9$
38. $x = 8$ or $x = 2$
39. $x = 0$ or $x = -2$
40. $x = 4.45$ or $x = -0.45$
41. $x = -4.76$ or $x = -9.24$
42. $x = 4.41$ or $x = 1.59$
43. $x = 3$ or $x = -9$
44. $x = 14$ or $x = -10$
45. $x = 17$ or $x = -1$
46. $x = 4.32$ or $x = -2.32$
47. $x = -1.84$ or $x = -8.16$
48. $x = -2.13$ or $x = -9.87$

Find the vertex, axis of symmetry, and zeros of each function. Then sketch the graph of each.

49. $y = (x-4)^2 - 3$ **50.** $y = (x+2)^2 - 1$ **51.** $y = (x-4)^2 - 2$

Refer to the function $y = (x+4)^2 - 4$ to complete Exercises 52–55.

52. Find the vertex. **(−4, −4)**

53. Find the axis of symmetry. $x = -4$

54. Find the zeros. **−2 and −6**

55. Sketch the graph.

CHALLENGE

56. Explain why $x^2 + 100 = 0$ has no real solutions.

CONNECTION

57. GEOMETRY Use the formula $S = 4\pi r^2$ to find the radius of a sphere with a surface area of 90 square meters. (Use 3.14 for π.) $r \approx 2.68$ meters

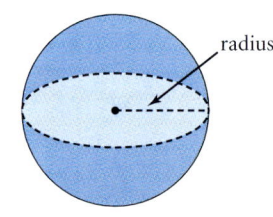

radius

APPLICATION

PHYSICS An object is dropped from a point 700 feet above the ground. Its height in feet, h, after t seconds is given by the function $h = -16t^2 + 700$. Find both approximate and exact answers for Exercises 58 and 59.

58. How long will it take for the object to reach the ground? ≈ **6.61 s**

59. How long will it take for the object to reach a height of 100 feet? ≈ **6.12 s**

Practice

Practice
10.2 Solving Equations of the Form $ax^2 = k$

Find each square root. Round answers to the nearest hundredth.
1. $\sqrt{1}$ 1   2. $\sqrt{81}$ 9   3. $\sqrt{16}$ 4   4. $\sqrt{225}$ 15
5. $\sqrt{30}$ 5.48   6. $\sqrt{18}$ 4.24   7. $\sqrt{110}$ 10.49   8. $\sqrt{55}$ 7.42

Solve each equation for x. Round answers to the nearest hundredth.
9. $x^2 = 16$ 4 and −4   10. $x^2 = 900$ 30 and −30
11. $x^2 = 75$ 8.66 and −8.66   12. $x^2 = \frac{1}{4}$ $\frac{1}{2}$ and $-\frac{1}{2}$
13. $x^2 = \frac{4}{49}$ $\frac{2}{7}$ and $-\frac{2}{7}$ 14. $x^2 = \frac{9}{25}$ $\frac{3}{5}$ and $-\frac{3}{5}$
15. $(x-2)^2 = 16$ 6 and −2   16. $(x+2)^2 = 16$ 2 and −6
17. $x^2 - 4 = 0$ 2 and −2   18. $4 = (x+3)^2$ −1 and −5
19. $-(x+3)^2 + 4 = 0$ −1 and −5   20. $(x-2)^2 - 36 = 0$ 8 and −4
21. $(x-3)^2 = 18$ 7.24 and −1.24   22. $x^2 + 1 = 17$ 4 and −4
23. $x^2 + 2 = 5$ 1.73 and −1.73   24. $(x+5)^2 = 7$ −2.35 and −7.65

Find the vertex, axis of symmetry, and zeros of each function. Sketch the graph.
25. $f(x) = (x-1)^2 - 1$ vertex: (1, −1); axis of symmetry: $x = 1$; zeros: 0 and 2
26. $g(x) = (x+2)^2 - 4$ vertex: (−2, −4); axis of symmetry: $x = -2$; zeros: 0 and −4

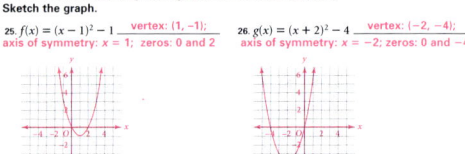

50. vertex: (−2, −1)
axis of symmetry: $x = -2$
zeros: −1 and −3

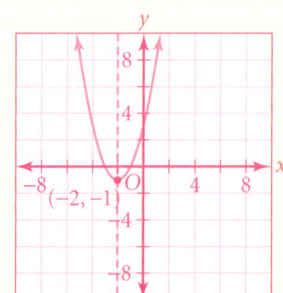

51. vertex: (4, −2)
axis of symmetry: $x = 4$
zeros: 2.59 and 5.41

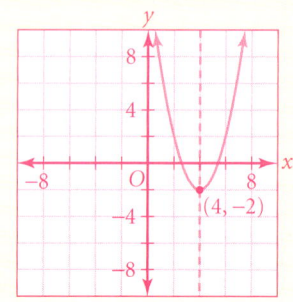

Look Back

Find each slope. (LESSON 5.4)

60. $y = 2x + 3$ **2**
61. $2x + 3y = 4$ $-\frac{2}{3}$
62. $x = y$ **1**
63. $4x + 7 = 5y$ $\frac{4}{5}$
64. $y - 2x = 3$ **2**
65. $y = 10$ **0**

66. Write an equation in slope-intercept form for a line that contains $(2, -4)$ and is perpendicular to the line $2x - y = 5$. (LESSON 5.6) $y = -\frac{1}{2}x - 3$

Simplify each expression. (LESSONS 8.2 AND 8.3)

67. $\frac{-6x^2y^2}{2x}$ $-3xy^2$
68. $\frac{b^3c^4}{bc}$ b^2c^3
69. $(p^2)^4(2a^2b^3)^2$ $4a^4b^6p^8$

Add or subtract each polynomial. (LESSON 9.1)

70. $(3x^2 - 2x + 1) + (2x^2 + 4x + 6)$
$5x^2 + 2x + 7$

71. $(x^2 + 2x - 1) - (2x^2 - 5x + 7)$
$-x^2 + 7x - 8$

Look Beyond

72. Evaluate $\sqrt{5^2 - 4 \cdot 1 \cdot 2}$. ≈ 4.12
73. Evaluate $\sqrt{6^2 - 4(1)(9)}$. **0**

Include this activity in your portfolio.

Use a motion detector to investigate a falling object.

1. Drop an object from 10 feet. Use the motion detector to find the object's height above the ground over time.
2. Use the time and height data to compute second differences in the height values. What are the second differences?
3. Make a scatter plot of the data. Write an equation of the form $y = -16x^2 + h$, where h is the initial height, to fit the data.

55.
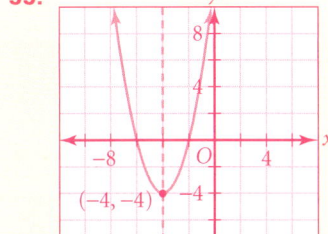

56. Answers may vary. Sample answer: $x^2 + 100$ has no real solutions because otherwise $x^2 = -100$, which is not true for any real number. The square of any real number can never be negative.

Look Beyond

Exercises 72 and 73 preview radical expressions of the type that students will encounter when applying the quadratic formula in Lesson 10.5. Be sure students understand that the radical sign is an inclusion symbol in the order of operations.

ALTERNATIVE Assessment

Portfolio Activity

The Portfolio Activity can be used as preparation for the Chapter Project or as a separate activity. In the Portfolio Activity on this page, students investigate a falling object, make a scatter plot of the time and height data, and fit a quadratic model to it.

Answers to Portfolio Activities can be found in Additional Answers of the Teacher's Edition.

Student Technology Guide

NAME _____ CLASS _____ DATE _____

Student Technology Guide
10.2 Solving Equations of the Form $ax^2 = k$

You can use a calculator to help solve equations of the form $ax^2 = k$.

Use the following steps:
• Isolate the x^2 term.
• In $ax^2 = k$, $\frac{k}{a}$ must be positive. If not, the equation has no real-number solutions.
• Use a calculator to take the square root of the constant term. Use $\pm$ to account for both solutions.

Example: Solve $-16x^2 = -248$.

$\frac{-16}{-16}x^2 = \frac{-248}{-16}$ Divide both sides by -16.

$x^2 = \frac{31}{2}$ Simplify. Note that the constant term is positive.

$x = \pm\sqrt{\frac{31}{2}}$ Take the square root of each side. $\sqrt{(31/2)}$ 3.937003937

Use a calculator to find a decimal approximation.
Press [2nd] [x^2] (31 ÷ 2) [ENTER].
The solutions are $x \approx \pm 3.94$.

Solve each equation, if possible. Round answers to the nearest hundredth when necessary. If the equation has no solution, write *no solution*.

1. $x^2 = 75$ ± 8.66
2. $6x^2 = 18$ ± 1.73
3. $-7x^2 = -192$ ± 5.24
4. $-9x^2 = 1989$ no solution

Example: The Eiffel Tower is 984 feet tall. By solving the equation $0 = 984 - 16t^2$, you will find the time it takes an object to fall from the top of the Eiffel Tower to the ground.

Solving the equation $0 = 984 - 16t^2$ for t gives the equation $t = \sqrt{\frac{123}{2}}$.

• Press [2nd] [x^2] (123 ÷ 2) [ENTER].
The object would take about 7.84 seconds to fall to the ground.

5. The John Hancock Center in Chicago is 1127 feet tall. About how long would it take an object to fall from the top of the John Hancock Center to the ground?   about 8.39 seconds

LESSON 10.2   **491**

Prepare

NCTM PRINCIPLES & STANDARDS
1–3, 5, 6, 8–10

LESSON RESOURCES

- Practice 10.3
- Reteaching Master 10.3
- Enrichment Master 10.3
- Cooperative-Learning Activity 10.3
- Lesson Activity 10.3
- Problem Solving/Critical Thinking Master 10.3

QUICK WARM-UP

Evaluate each expression.

1. $\left(\frac{6}{2}\right)^2$ 9   2. $\left(-\frac{10}{2}\right)^2$ 25

3. $\left(\frac{7}{2}\right)^2$ $\frac{49}{4}$ 4. $\left(-\frac{1}{2}\right)^2$ $\frac{1}{4}$

Factor each expression.

5. $x^2 + 2x + 1$ $(x+1)^2$

6. $x^2 + 8x + 16$ $(x+4)^2$

7. $x^2 - 10x + 25$ $(x-5)^2$

Also on Quiz Transparency 10.3

Teach

Why Write $3(x-2) + 5 = -16$ on the board or overhead. Remind students how they can use properties of equality to *transform* the equation into the form $x = -5$. Point out that with this transformed equation, it is easy to identify the solution. Tell them that today they will learn how to transform an equation of the form $y = x^2 + bx + c$ into vertex form. Ask them to suggest reasons why the vertex form might be more useful.

492   LESSON 10.3

Completing the Square

10.3

Objectives

- Form a perfect-square trinomial from a given quadratic binomial.

- Write a given quadratic function in vertex form.

Why If you write the quadratic function $y = x^2 + bx + c$ in the form $y = (x - h)^2 + k$, you can find the vertex and the axis of symmetry of the parabola. A technique called completing the square will enable you to do this. This technique simplifies graphing and is helpful for complicated real-world problems.

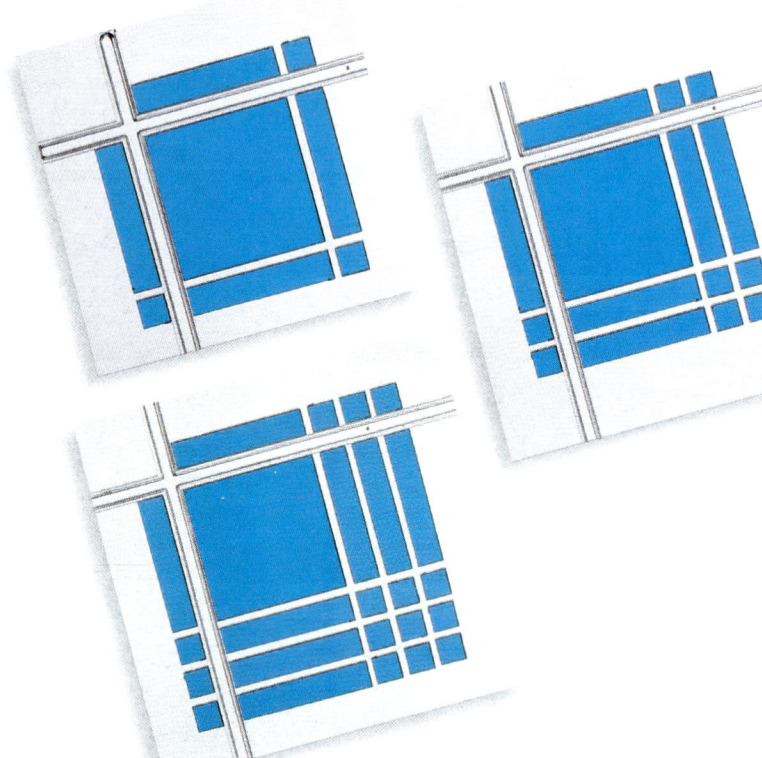

The algebra tiles shown above represent the factorization of a perfect-square trinomial. Do you see why?

You can represent $x^2 + 4x + 4$ with tiles, as shown in the illustration above. Notice that the tiles form a complete square. Recall from Lesson 9.6 that an expression such as $x^2 + 4x + 4$ is called a *perfect-square trinomial* because it can be written as $(x + 2)^2$, the *square of a binomial*.

These tiles represent the binomial $x^2 + 6x$. Can you form a square by rearranging them and adding a certain number of 1-tiles? How many 1-tiles do you need to add?

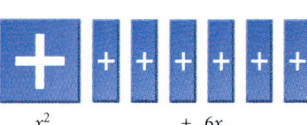

Alternative Teaching Strategy

USING COGNITIVE STRATEGIES Review with students the meaning of a *perfect-square trinomial*. Have them give a few examples. Then ask them to expand the general forms $(x + a)^2$ and $(x - a)^2$.
$x^2 + 2ax + a^2, x^2 - 2ax + a^2$

Now give students the expression $x^2 + 12x$. Point out that it is a binomial. Ask what one thing can be done to change it to a perfect-square trinomial. Elicit the response that an appropriate constant must be added. Then show students the method at right for identifying that constant.

$$\begin{array}{ccc} x^2 + \boxed{2a}\,x + a^2 & & 2a = 12 \\ x^2 + \boxed{12}\,x + \,? & \to & a = 6 \\ \downarrow \quad \downarrow \quad \downarrow & & a^2 = 36 \\ x^2 + 12x + 36 & & \end{array}$$

Repeat the procedure to complete the square given the binomial $x^2 + 11x$. $x^2 + 11x + \frac{121}{4}$

Activity
Completing the Square With Tiles

You will need: algebra tiles

1. Arrange tiles as shown at right to represent $x^2 + 6x$.

2. Fill in the empty space in your new arrangement by adding 1-tiles so that a complete square is formed. How many 1-tiles did you add?

3. Your completed square represents a perfect-square trinomial. What is the trinomial? Rewrite the trinomial as the square of a binomial.

4. Repeat Steps 1–3 for the tile arrangement at right.

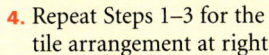
x^2 $+ 8x$

CHECKPOINT ✓ 5. Try to think of a rule for the number of 1-tiles you added in each case. Write a formula or a procedure for the rule.

In the activity you may have discovered the following rule:

Completing the Square

To make a perfect-square trinomial from a binomial of the form $x^2 + bx$, add the square of half of the coefficient of the x-term, $\left(\frac{b}{2}\right)^2$.

$$x^2 + bx + \left(\frac{b}{2}\right)^2$$

In addition to being faster than using algebra tiles, the rule for completing the square can be used for cases in which algebra tiles cannot be used, such as with large numbers or fractions.

EXAMPLE 1 Complete the square by finding the value that makes a perfect-square trinomial when added to the binomial $x^2 + 5x$.

- **SOLUTION**

The coefficient of the x-term is 5. Take half of this number and square it.

$$\left(\frac{5}{2}\right)^2 = \frac{25}{4} = 6\frac{1}{4}$$

Add $6\frac{1}{4}$ to the original expression, $x^2 + 5x$, to make a perfect-square trinomial, $x^2 + 5x + 6\frac{1}{4}$. This can also be written as $\left(x + \frac{5}{2}\right)^2$.

Inclusion Strategies

VISUAL/TACTILE LEARNERS Students who rely on the visual representation provided by algebra tiles may falter when encountering the situation in Example 1, which cannot be represented by the tiles. Tell them to imagine that the tiles can be cut into halves and fourths, and draw a diagram such as the one shown at right. Students who need the experience of actually manipulating tiles can make themselves a set of paper tiles that includes a few $\frac{1}{2}x$-tiles, $\frac{1}{2}$-unit tiles, and $\frac{1}{4}$-unit tiles.

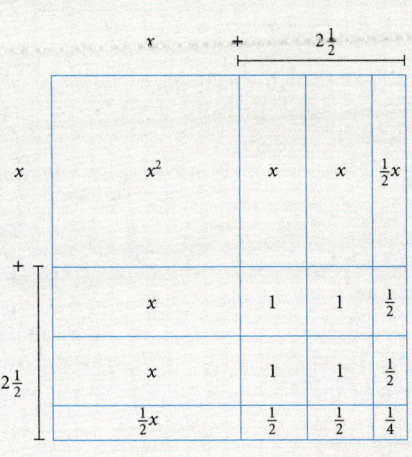

Activity Notes

In this Activity, students investigate a concrete model for the process of completing the square. They begin with 1 x^2-tile and a number of x-tiles, and they are asked to determine how many unit tiles are needed to complete the square. They should arrive at the generalization that the required number of unit tiles will always be the square of half the number of x-tiles.

Cooperative Learning

Students can do the Activity in pairs. In Steps 1–3, one student can manipulate the tiles while the other makes a sketch to record the arrangement. The students can switch roles for Step 4 and then work together to write the generalization in Step 5.

CHECKPOINT ✓
5. Add the square of one-half the number of x-tiles.

ADDITIONAL EXAMPLE 1

Complete the square by finding the value that makes a perfect-square trinomial when added to the binomial $x^2 - 9x$.

$x^2 - 9x + \left(\frac{9}{2}\right)^2 =$

$x^2 - 9x + \frac{81}{4}$, or $\left(x - \frac{9}{2}\right)^2$

LESSON 10.3

TRY THIS

$\left(\frac{20}{2}\right)^2 = 100$

ADDITIONAL EXAMPLE 2

Complete the square for $x^2 - 16x$.

$x^2 - 16x + 64$, or $(x - 8)^2$

TRY THIS Find the value that makes a perfect-square trinomial when added to the binomial $x^2 + 20x$.

The rule for completing the square also works when the coefficient of the x-term is negative, as the next example illustrates.

EXAMPLE 2 Complete the square for $x^2 - 10x$.

○ **SOLUTION**

The coefficient of the x-term is -10. Take one-half this number and square it.

$$\left(\frac{b}{2}\right)^2 = \left(\frac{-10}{2}\right)^2 = (-5)^2 = 25$$

Add 25 to the original expression, $x^2 - 10x$, to make a perfect-square trinomial, $x^2 - 10x + 25$, or $(x - 5)^2$.

The diagram below represents the right side of the equation $y = x^2 + 6x + 5$. It is easy to see that this is not a perfect-square trinomial because the tiles do not form a square.

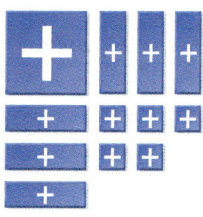

$y = x^2 + 6x + 5$

By adding neutral pairs of 1-tiles, you can create an arrangement that represents the same quadratic function—but includes a perfect square. The new arrangement suggests a way to rewrite the original function.

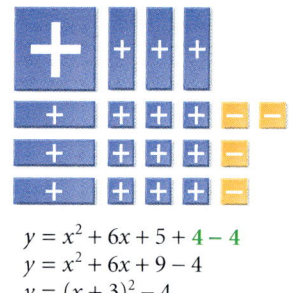

$y = x^2 + 6x + 5 + 4 - 4$
$y = x^2 + 6x + 9 - 4$
$y = (x + 3)^2 - 4$

The vertex (h, k) of the function is $(-3, -4)$. (Recall Lesson 10.1.)

The algebra-tile model above suggests the following procedure for converting a quadratic function in which $a = 1$ to vertex form. (You will learn in a later lesson what to do for functions in which $a \neq 1$.)

Enrichment

Have students work from a table of values to graph $x = y^2 + 2y + 3$. Ask them to describe the graph. **a parabola that opens to the right** Now tell them to rewrite the right side of the equation to complete the square. $x = (y + 1)^2 + 2$ Ask them what characteristic of the graph they can read from the new equation. $x = (y + 1)^2 + 2 \rightarrow x = [y - (-1)]^2 + 2$; **the vertex of the parabola is (2, −1).** Have them make a conjecture about the vertex of the parabola with the equation $x = (y - b)^2 + a$ and then test their conjecture. **The vertex is (a, b).**

Reteaching the Lesson

COOPERATIVE LEARNING Have students work in pairs. Each student should write an equation of the form $y = x^2 + bx + c$ such that $x^2 + bx + c$ is *not* a perfect-square trinomial. Then have students rewrite the equation in vertex form and record the answer on a separate sheet of paper.

Have partners exchange their original equations. Each should write the vertex form of the other's equation. When they are done, they should compare answers and resolve any disagreements.

Converting to Vertex Form

1. Start with the quadratic function. $y = x^2 + 6x + 5$
2. Group the x^2-terms and x-terms. $y = (x^2 + 6x) + 5$
3. From the coefficient of the x-term, find the number you need to add to make a perfect-square trinomial. $\frac{1}{2} \cdot 6 = 3; \; 3^2 = 9$
4. Complete the square by adding the number to the expression inside the parentheses. Then subtract the number from the constant term outside the parentheses. $y = (x^2 + 6x + 9) + 5 - 9$
5. Simplify. $y = (x^2 + 6x + 9) - 4$
6. Write in the form $y = (x - h)^2 + k$. $y = (x + 3)^2 - 4$

EXAMPLE 3

a. Rewrite $y = x^2 - 8x + 7$ in the form $y = (x - h)^2 + k$.
b. Find the vertex of the parabola represented by this function.

● **SOLUTION**

a. Follow the steps for converting to vertex form.

1. $y = x^2 - 8x + 7$
2. $y = (x^2 - 8x) + 7$
3. $\frac{1}{2}(-8) = -4; \; (-4)^2 = 16$
4. $y = (x^2 - 8x + 16) + 7 - 16$
5. $y = (x^2 - 8x + 16) - 9$
6. $y = (x - 4)^2 - 9$

b. The vertex of the parabola is $(4, -9)$.

ADDITIONAL EXAMPLE 3

a. Rewrite $y = x^2 + 2x + 5$ in the form $y = (x - h)^2 + k$.
$y = (x + 1)^2 + 4$

b. Find the vertex of the parabola whose equation is given in part a.
$(-1, 4)$

Assess

Selected Answers
Exercises 6–8, 9–73 odd

ASSIGNMENT GUIDE

In Class	1–8
Core	9–12, 17–34, 43–57, 65–68
Core Plus	18–42 even, 43–68
Review	69–74
Preview	75–77

✏ Extra Practice can be found beginning on page 738.

Technology
A graphics calculator may be helpful for Exercises 52–68. You may wish to have students do the exercises first with paper and pencil and then use the calculator to check their results.

Exercises

● *Communicate*

1. Describe how to use algebra tiles to complete the square for $x^2 + 4x$.
2. Describe how to use algebra to complete the square for $x^2 - 7x$.
3. How do you rewrite $y = x^2 + 10x + 25 - 25$ in the form $y = (x - h)^2 + k$?
4. Describe how to find the vertex of $y = x^2 + 2x$.
5. Explain how to rewrite $y = x^2 - 10x + 11$ in the form $y = (x - h)^2 + k$. How can you find the vertex from this form of the equation?

LESSON 10.3

Error Analysis

When completing the square, students often forget to take half of the coefficient of the linear term. For example, given $x^2 + 3x$, they may write $x^2 + 3x + 9$. Encourage them to write the factored form and multiply to check their work. In this case, if they write $(x + 3)^2$ for the factored form and then multiply, they should discover that the result is $x^2 + 6x + 9$ rather than $x^2 + 3x + 9$.

9.
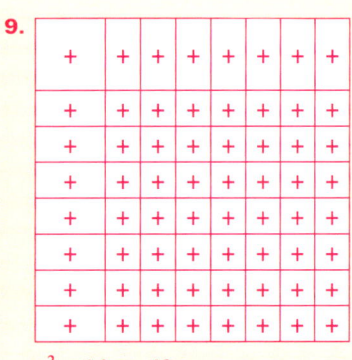
$x^2 + 14x + 49$

Guided Skills Practice

6. Find the value that makes a perfect-square trinomial when added to the binomial $x^2 + 19x$. **(EXAMPLE 1)** $90\frac{1}{4}$

7. Find the value that makes a perfect-square trinomial when added to the binomial $x^2 - 3x$. **(EXAMPLE 2)** $2\frac{1}{4}$

8. Rewrite $y = x^2 + 6x + 18$ in the form $y = (x - h)^2 + k$. **(EXAMPLE 3)**
$y = (x + 3)^2 + 9$

Practice and Apply

Use algebra tiles to complete the square for each binomial. Draw a sketch of your algebra-tile model. Write each perfect-square trinomial in the form $x^2 + bx + c$.

9. $x^2 + 14x$ 10. $x^2 - 14x$ 11. $x^2 + 8x$ 12. $x^2 - 8x$
13. $x^2 + 4x$ 14. $x^2 - 10x$ 15. $x^2 + 12x$ 16. $x^2 + 10x$

Rewrite each function in the form $y = (x - h)^2 + k$.

17. $y = x^2 + 8x + 16 - 16$ 18. $y = x^2 - 4x + 4 - 4$
19. $y = x^2 - 10x + 25 - 25$ 20. $y = x^2 + 14x + 49 - 49$
21. $y = x^2 - 16x + 64 - 64$ 22. $y = x^2 + 20x + 100 - 100$

Complete the square by using Example 2 as a model.

23. $x^2 + 6x$ 24. $x^2 - 2x$ 25. $x^2 + 12x$ 26. $x^2 - 12x$
27. $x^2 + 7x$ 28. $x^2 - 10x$ 29. $x^2 + 15x$ 30. $x^2 - 5x$
31. $x^2 + 16x$ 32. $x^2 + 20x$ 33. $x^2 - 9x$ 34. $x^2 + 40x$
35. $x^2 - 36x$ 36. $x^2 + 44x$ 37. $x^2 - 17x$ 38. $x^2 + 30x$
39. $x^2 - 13x$ 40. $x^2 + 13x$ 41. $x^2 + 22x$ 42. $x^2 - 44x$

Refer to the function $y = x^2 + 10x$ to complete Exercises 43–45.

43. Complete the square and rewrite the function in the form $y = (x - h)^2 + k$.
$y = (x + 5)^2 - 25$
44. Find the vertex of the function. $(-5, -25)$
45. Find the maximum or minimum value of the function. -25

Find the vertex of the parabolas represented by the following functions:

46. $y = x^2 + 3$ **(0, 3)**   47. $y = x^2 - 4$ **(0, −4)**   48. $y = -x^2 + 2$ **(0, 2)**
49. $y = x^2 - 1$ **(0, −1)**   50. $y = -x^2 + 6$ **(0, 6)**   51. $y = x^2 - 5$ **(0, −5)**

Rewrite each function in the form $y = (x - h)^2 + k$ by using Example 3 as a model. Find the vertex of each.

52. $y = x^2 + 3$
53. $y = x^2 - 4$
54. $y = x^2 + 2$
55. $y = x^2 - 1$
56. $y = 6 + x^2$
57. $y = -5 + x^2$
58. $y = x^2 + 4x - 1$
59. $y = x^2 - 2x - 3$
60. $y = x^2 + 10x - 12$
61. $y = x^2 - x + 2$
62. $y = -5 - 6x + x^2$
63. $y = x^2 + \frac{1}{3}x - 3$

CHALLENGE

64. $3y = 3x^2 - 24x + 18$ $y = (x - 4)^2 - 10; (4, -10)$

Answers:
17. $y = (x + 4)^2 - 16$
18. $y = (x - 2)^2 - 4$
19. $y = (x - 5)^2 - 25$
20. $y = (x + 7)^2 - 49$
21. $y = (x - 8)^2 - 64$
22. $y = (x + 10)^2 - 100$
23. $(x + 3)^2$ 24. $(x - 1)^2$
25. $(x + 6)^2$ 26. $(x - 6)^2$
27. $(x + \frac{7}{2})^2$ 28. $(x - 5)^2$
29. $(x + \frac{15}{2})^2$ 30. $(x - \frac{5}{2})^2$
31. $(x + 8)^2$ 32. $(x + 10)^2$
33. $(x - \frac{9}{2})^2$ 34. $(x + 20)^2$
35. $(x - 18)^2$ 36. $(x + 22)^2$
37. $(x - \frac{17}{2})^2$ 38. $(x + 15)^2$
39. $(x - \frac{13}{2})^2$ 40. $(x + \frac{13}{2})^2$
41. $(x + 11)^2$ 42. $(x - 22)^2$
52. $y = (x - 0)^2 + 3; (0, 3)$
53. $y = (x - 0)^2 - 4; (0, -4)$
54. $y = (x - 0)^2 + 2; (0, 2)$
55. $y = (x - 0)^2 - 1; (0, -1)$
56. $y = (x - 0)^2 + 6; (0, 6)$
57. $y = (x - 0)^2 - 5; (0, -5)$
58. $y = (x + 2)^2 - 5; (-2, -5)$
59. $y = (x - 1)^2 - 4; (1, -4)$
60. $y = (x + 5)^2 - 37; (-5, -37)$
61. $y = (x - \frac{1}{2})^2 + \frac{7}{4}; (\frac{1}{2}, \frac{7}{4})$
62. $y = (x - 3)^2 - 14; (3, -14)$
63. $y = (x + \frac{1}{6})^2 - 3\frac{1}{36}; (\frac{-1}{6}, -3\frac{1}{36})$

10.
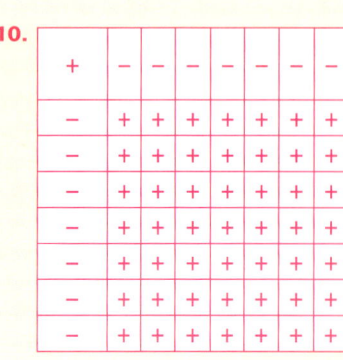
$x^2 - 14x + 49$

11.
$x^2 + 8x + 16$

12.
$x^2 - 8x + 16$

The answers to Exercises 13–16 can be found in Additional Answers beginning on page 810.

Practice

Practice
10.3 Completing the Square

Use algebra tiles to complete the square for each binomial. Draw a sketch of your algebra-tile model. Write each perfect-square trinomial in the form $x^2 + bx + c$.

1. $x^2 - 6x$ ___ $x^2 - 6x + 9$ 2. $x^2 + 2x$ ___ $x^2 + 2x + 1$
3. $x^2 - 4x$ ___ $x^2 - 4x + 4$ 4. $x^2 + 10x$ ___ $x^2 + 10x + 25$

Complete the square.

5. $x^2 + 8x$ ___ $x^2 + 8x + 16$ 6. $x^2 - 8x$ ___ $x^2 - 8x + 16$
7. $x^2 + 11x$ ___ $x^2 + 11x + \frac{121}{4}$ 8. $x^2 - 11x$ ___ $x^2 - 11x + \frac{121}{4}$

Rewrite each function in the form $y = (x - h)^2 + k$.

9. $y = x^2 - 4x + 4 - 4$ ___ $(x - 2)^2 - 4$ 10. $y = x^2 + 4x + 4 - 4$ ___ $(x + 2)^2 - 4$
11. $y = x^2 + 18x + 81 - 81$ ___ $(x + 9)^2 - 81$ 12. $y = x^2 - 18x + 81 - 81$ ___ $(x - 9)^2 - 81$
13. $y = x^2 - 24x + 144 - 144$ ___ $(x - 12)^2 - 144$ 14. $y = x^2 + 22x + 121 - 121$ ___ $(x + 11)^2 - 121$

Refer to the equation $y = x^2 + 6x$ for Exercises 15 and 16.

15. Complete the square and rewrite in the form $y = (x - h)^2 + k$. ___ $(x + 3)^2 - 9$
16. Find the vertex and maximum (or minimum) value. ___ vertex: $(-3, -9)$; minimum value: -9

Rewrite each equation in the form $y = (x - h)^2 + k$. Find the vertex of each quadratic function.

17. $f(x) = x^2 - 3$ ___ $(0, -3)$ 18. $f(x) = x^2 - 2$ ___ $(0, -2)$
19. $f(x) = x^2 + 2x - 8$ ___ $f(x) = (x + 1)^2 - 9; (-1, -9)$ 20. $f(x) = x^2 - 4x + 3$ ___ $f(x) = (x - 2)^2 - 1; (2, -1)$

496   LESSON 10.3

APPLICATION

PHYSICS For a certain projectile, the relationship between time in seconds, t, and height in feet, h, is given by the function $h = -16t^2 + 192t$.

65. Graph the function.

66. Complete the square to find the vertex of the parabola represented by the function. **(6, 576)**

67. What is the maximum height reached by the projectile? **576 feet**

68. How long does it take for the projectile to reach its maximum height? **6 seconds**

Some volcanoes eject incandescent lava as high as 470 meters.

Look Back

69. Graph each line on the same coordinate plane. *(LESSON 5.4)*
   a. $y = -\frac{1}{3}x + 5$ b. $4x - 2y = 2$ c. $x = 2y$

Solve each system algebraically. Check by graphing.
(LESSONS 7.2 AND 7.3)

70. $\begin{cases} x - 3y = 3 \\ 2x - y = -4 \end{cases}$ **(−3, −2)**

71. $\begin{cases} x - 2y = 0 \\ x + y = 3 \end{cases}$ **(2, 1)**

Write each expression without negative exponents. *(LESSON 8.4)*

72. $a^2 b^{-3}$ $\dfrac{a^2}{b^3}$

73. $\dfrac{m^5}{m^{-2}}$ m^7

74. $\dfrac{2n^{-6}}{n^{-4}}$ $\dfrac{2}{n^2}$

Look Beyond

75. The height in feet, h, of a baseball t seconds after it is hit is given by the function $h = -16t^2 + 100t$. Use the following methods to find t when the ball is at a height of 100 feet: **5 or 1.25 seconds**
   a. graphing   b. algebra

Complete the square for each of the following:

76. $2x^2 + 8x + 1$ $2(x + 2)^2 - 7$

77. $3x^2 + 5x - 4$ $3(x + \frac{5}{6})^2 - 6\frac{1}{12}$

Include this activity in your portfolio.

1. Build a table for the falling-object function $y = -16x^2 + 1000$ for x-values from 1 to 6.

2. Take successive differences of the y-values until they are constant. What is the constant?

3. The general form of the falling object function is $y = -32\left(\frac{1}{2}\right)x^2 - h$.

   The acceleration due to gravity is −32 feet per second squared. Use your data from the Portfolio Activity on page 491 to find the force of gravity from your experimental data.

Look Beyond

In Exercises 75–77, students are asked to solve a quadratic equation and to complete the square when the coefficient of the quadratic term is not 1. Students will learn more about these skills in Lesson 10.5.

ALTERNATIVE Assessment

Portfolio Activity

The Portfolio Activity can be used as preparation for the Chapter Project or as a separate activity. In the Portfolio Activity on this page, students find that the constant second differences of the falling-object function are determined by the acceleration due to gravity.

internetconnect

GO TO: go.hrw.com
KEYWORD: MA1 Gravity

To support the Portfolio Activity, students may visit the HRW Web site, where they can explore different methods for measuring gravity and learn how weight is related to gravity.

Student Technology Guide

NAME _____ CLASS _____ DATE _____

Student Technology Guide
10.3 Completing the Square

In Lesson 10.3, you learned how to find the vertex of a quadratic equation by first writing the equation in vertex form, $y = (x - h)^2 + k$, and then reading the coordinates of the vertex, (h, k). By using a graphics calculator, you can reverse this process—find the coordinates of the vertex in order to write the vertex form of the equation.

Example: Find the coordinates of the vertex of $y = x^2 - 6x + 7$. Then rewrite the equation in the form $y = (x - h)^2 + k$.

- Enter and graph the equation as shown at right. Y= X,T,θ,n x^2 − 6 X,T,θ,n + 7 GRAPH

- As shown in the diagram at right, the vertex appears to be (3, −2). But can you be sure the coordinates are really integers?

- Check that the coordinates are integers by forcing the trace cursor to go to $x = 3$. To do this, press 3. The 3 appears at the bottom left of the screen. Then press ENTER. The trace point moves to (3, −2), as shown at right. This suggests that (3, −2) truly is the vertex of the parabola.

The vertex has coordinates (3, −2). Thus, $h = 3$ and $k = −2$. The vertex form of $y = x^2 − 6x + 7$ is $y = (x − 3)^2 − 2$.

When searching for a vertex, you may prefer to trace as close as you can to the vertex and press ZOOM 2: Zoom In ENTER ENTER to zoom in on the trace point. You can repeat this process as needed until you are satisfied with the accuracy of the coordinates.

Find the coordinates of the vertex of each parabola. You may find it helpful to zoom. Round coordinates to the nearest hundredth if necessary. Then rewrite each function in the form $y = (x − h)^2 + k$.

1. $y = x^2 + 4x + 7$
   vertex: (−2, 3);
   $y = (x + 2)^2 + 3$

2. $y = x^2 − 9$
   vertex: (0, −9);
   $y = (x + 0)^2 − 9$

3. $y = −x^2 + 10x − 15$
   vertex: (5, 10);
   $y = (x − 5)^2 + 10$

4. $y = x^2 + 3x − 5$
   vertex: (−1.5, −7.25);
   $y = (x + 1.5)^2 − 7.25$

5. $y = −x^2 − 7x + 9$
   vertex: (−3.5, 21.25);
   $y = (x + 3.5)^2 + 21.25$

6. $y = x^2 − 3.6x + 1.51$
   vertex: (1.8, −1.73);
   $y = (x − 1.8)^2 − 1.73$

65.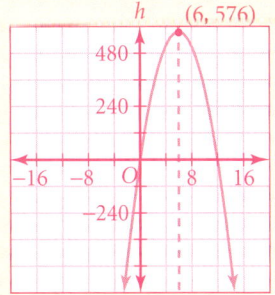

69.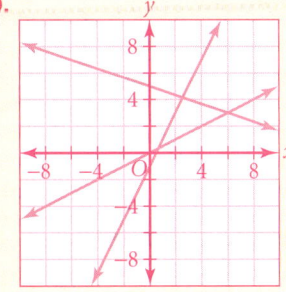

Prepare

NCTM PRINCIPLES & STANDARDS 1–3, 6–10

LESSON RESOURCES

- Practice 10.4
- Reteaching Master 10.4
- Enrichment Master 10.4
- Cooperative-Learning Activity 10.4
- Lesson Activity 10.4
- Problem Solving/Critical Thinking Master 10.4
- Teaching Transparency 49

QUICK WARM-UP

Solve each equation.

1. $(x - 3)^2 - 16 = 0$
   $x = -1$ or $x = 7$

2. $(x + 4)^2 - 25 = 0$
   $x = 1$ or $x = -9$

Factor each expression.

3. $x^2 + x - 6$
   $(x + 3)(x - 2)$

4. $x^2 - 8x + 12$
   $(x - 2)(x - 6)$

Also on Quiz Transparency 10.4

Teach

 Why Write the following equations on the board or overhead:

$y = x^2 + bx + c$ $y = 0$

Point out that the first is an equation for a quadratic function, and the second is an equation for the *x*-axis. Ask students what would be determined by solving $x^2 + bx + c = 0$. **the point where the quadratic and the *x*-axis intersect, which are the zeros of the quadratic function**

10.4 Solving Equations of the Form $x^2 + bx + c = 0$

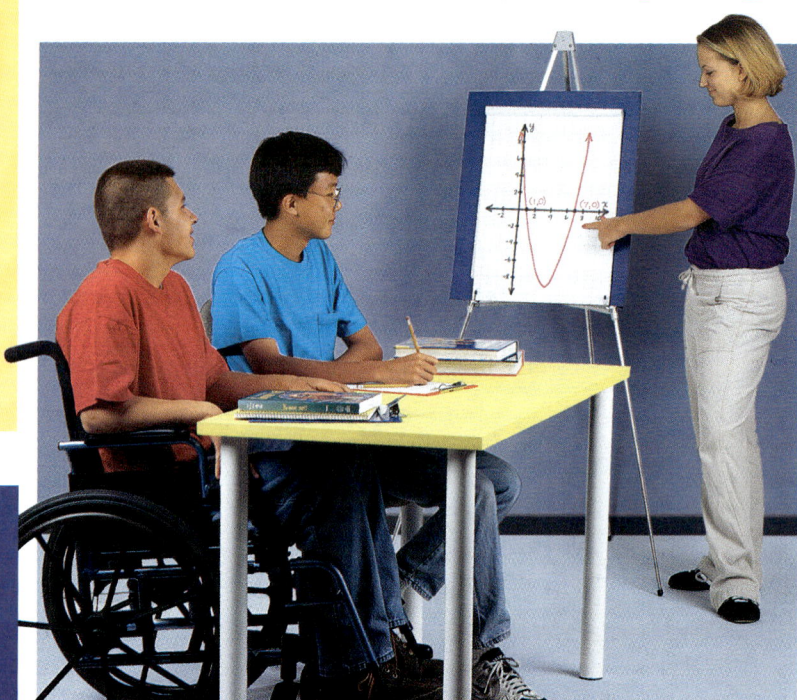

Objective
- Solve quadratic equations by completing the square or by factoring.

Why You have solved equations of the form $x^2 = k$. Equations of the form $x^2 + bx + c = 0$ can also be solved by completing the square. This type of equation can represent many different real-world situations.

In Lesson 10.3 you found the vertex of $y = x^2 - 8x + 7$ by completing the square. When this function is rewritten in the form $y = (x - 4)^2 - 9$, the vertex, (4, −9), can be found by inspection. To find the zeros of this function, substitute 0 for *y*, and rewrite the equation in the form $n^2 = k$, where *n* is an expression involving *x*.

$$(x - 4)^2 - 9 = 0$$
$$(x - 4)^2 = 9$$
$$x - 4 = \pm 3$$

$x - 4 = 3$ or $x - 4 = -3$
$x = 7$ $x = 1$

The solutions are 7 and 1. These two solutions indicate the points where the graph of $y = x^2 - 8x + 7$ intersects the *x*-axis. Substitute 7 and 1 for *x* in the original function to check.

$$y = 7^2 - 8(7) + 7 = 49 - 56 + 7 = 0$$
$$y = 1^2 - 8(1) + 7 = 1 - 8 + 7 = 0$$

The zeros of the function are (7, 0) and (1, 0). These are the points where the parabola intersects the *x*-axis.

CRITICAL THINKING Why are the zeros of the function $y = x^2 - 8x + 6$ and the solutions to the equation $x^2 - 8x + 6 = 0$ the same?

Alternative Teaching Strategy

USING TECHNOLOGY Give students the equation $y = x^2 - 8x + 7$ and have them use a graphics calculator to find the zeros of the related function.

If they are using TI-82 or TI-83 calculators, tell them to begin by pressing `Y=` and entering `X^2−8x+7` next to **Y1**. To view the graph, they can press `ZOOM` and choose **6: ZStandard**.

Now have them press `2nd` `TRACE` and choose **2: zero** (or **2: root**). Tell them to use `◄` to move the cursor above the *x*-axis, and then press `ENTER`.

Now they should use `►` to move the cursor below the *x*-axis and press `ENTER` `ENTER`. The display will show the first zero as **X=1**.

Lead students in a similar procedure near the second *x*-intercept, placing the cursor first below the *x*-axis and then above and pressing `ENTER` each time. The display will show **X=7**.

Discuss the zeros with students. Ask how they might have obtained the zeros without using the calculator. Elicit the response that they could factor $x^2 - 8x + 7$ and solve $(x - 1)(x - 7) = 0$.

EXAMPLE 1 Find the zeros of $y = x^2 - 6x + 7$.

● **SOLUTION**

Replace y with 0, complete the square, and solve.

$$(x^2 - 6x) + 7 = 0$$
$$(x^2 - 6x) + 7 + 9 - 9 = 0$$
$$(x^2 - 6x + 9) + 7 - 9 = 0$$
$$(x^2 - 6x + 9) - 2 = 0$$
$$(x - 3)^2 - 2 = 0$$
$$(x - 3)^2 = 2$$

$$x - 3 = \sqrt{2} \quad \text{or} \quad x - 3 = -\sqrt{2}$$
$$x = 3 + \sqrt{2} \quad\quad\quad\quad x = 3 - \sqrt{2}$$

The zeros are $3 + \sqrt{2}$ and $3 - \sqrt{2}$.

TRY THIS Find the zeros of $y = x^2 - 7x + 10$.

When a quadratic can be factored, you can quickly solve the quadratic equation. For example, $x^2 - 6x + 8$ can be factored as $(x-4)(x-2)$, so $x^2 - 6x + 8 = 0$ is the same as $(x-4)(x-2) = 0$.

The Zero Product Property (Lesson 9.8) can be used to solve $(x-4)(x-2) = 0$.

$$x - 4 = 0 \quad \text{or} \quad x - 2 = 0$$
$$x = 4 \quad\quad\quad\quad x = 2$$

The solutions are 4 and 2.

EXAMPLE 2 Solve $x^2 + x = 2$ by
a. completing the square.
b. factoring.

● **SOLUTION**

a.
$$x^2 + x = 2$$
$$x^2 + x + \left(\frac{1}{2}\right)^2 = 2 + \left(\frac{1}{2}\right)^2$$
$$\left(x + \frac{1}{2}\right)^2 = \frac{9}{4}$$
$$x + \frac{1}{2} = \frac{3}{2} \quad \text{or} \quad x + \frac{1}{2} = -\frac{3}{2}$$
$$x = 1 \quad\quad\quad\quad x = -2$$

b.
$$x^2 + x = 2$$
$$x^2 + x - 2 = 0$$
$$(x + 2)(x - 1) = 0$$
$$x + 2 = 0 \quad \text{or} \quad x - 1 = 0$$
$$x = -2 \quad\quad\quad\quad x = 1$$

The solutions are 1 and −2.

CRITICAL THINKING Sometimes both completing the square and factoring could be used to solve a quadratic equation of the form $x^2 + bx + c = 0$. How would you decide which method to use?

CRITICAL THINKING
By definition, zeros are the values of x that make the function equal to zero.

ADDITIONAL EXAMPLE 1

Find the zeros of the function $y = x^2 - 4x + 1$.

$2 + \sqrt{3}, 2 - \sqrt{3}$

TRY THIS
The zeros are 2 and 5.

ADDITIONAL EXAMPLE 2

Solve $x^2 + 2x = 3$ by

a. completing the square.
$x^2 + 2x + 1 = 3 + 1$
$(x + 1)^2 = 4$
$x = -3$ or $x = 1$

b. factoring.
$x^2 + 2x - 3 = 0$
$(x + 3)(x - 1) = 0$
$x = -3$ or $x = 1$

CRITICAL THINKING
Answers may vary. Sample answer: If the value of c is a perfect square, the expression may be easily factored. Otherwise, try completing the square.

Interdisciplinary Connection

INDUSTRIAL ARTS A student wants to make a planter from a single square sheet of metal. Four congruent squares will be cut from the corners of the sheet, and then the sides will be folded up to make an open-top box with a square base. The height of the planter is to be 8 inches, and its volume is to be 288 cubic inches. What should be the size of the original sheet of metal? It should be a square with side lengths of 22 inches.

Inclusion Strategies

VISUAL LEARNERS Suggest that students keep track of the process of completing the square by color-coding. For instance, this is how they might color-code the second line of the solution of Example 2.

$$\underbrace{(x^2 - 6x)}_{\text{blue}} + \underbrace{7}_{\text{red}} + \underbrace{9}_{\text{blue}} - \underbrace{9}_{\text{red}} = \underbrace{0}_{\text{red}}$$

They should then carry these colors through all related calculations in the rest of the solution.

Use Teaching Transparency 49.

ADDITIONAL EXAMPLE 3

Let $y = x^2 - 6x + 12$. **Find the value(s) of x when y is 4.**
$x = 2$ or $x = 4$

See Keystroke Guide, page 782.

Teaching Tip

TECHNOLOGY If you are using a TI-82 or TI-83 graphics calculator, you can obtain the graph shown in the solution for Example 3 by pressing [WINDOW] and making the following settings:

Xmin=–9.4	Ymin=–20
Xmax=9.4	Ymax=10
Xscl=1	Yscl=1

Then press [Y=] and enter X^2+3X–15 next to Y1 and –5 next to Y2. Press [GRAPH] to view the graph.

TRY THIS
The solutions are 2 and 6.

Activity Notes

In this Activity, students first solve a problem related to an open-top box by making and investigating models of the box. Then they see how a quadratic equation can be used to solve the problem. Be sure students understand that only one of the two solutions to the equation is a meaningful solution.

EXAMPLE 3 Let $y = x^2 + 3x - 15$. Find the value(s) of x when y is –5.

● **SOLUTION**

Method A
Graph $y = x^2 + 3x - 15$ and $y = -5$ on the same set of axes. Use a graphics calculator if you have one. Find the point(s) of intersection. You can see that $y = -5$ when $x = 2$ or $x = -5$.

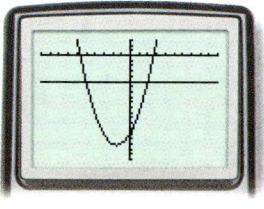

Method B
$x^2 + 3x - 15 = -5$
$x^2 + 3x - 10 = 0$ *Addition Property of Equality*
$(x - 2)(x + 5) = 0$ *Factor.*
$x - 2 = 0$ or $x + 5 = 0$ *Zero Product Property*
$x = 2$ $x = -5$

The solutions are 2 and –5.

TRY THIS Let $y = x^2 - 8x + 22$. Find the value(s) of x when y is 10.

CONNECTION
GEOMETRY

Activity
Package Design

You will need: paper, a ruler, and scissors

A box with a square base and no top is to be made by cutting 2-inch squares from each corner of a square piece of paper and folding up the sides of the remaining piece. The volume of the box is to be 8 cubic inches. What should be the dimensions of the original piece of paper?

Part I
1. Start with an 8-inch square piece of paper. Cut out 2-inch squares from each corner.
2. Fold the paper into a box with an open top, and measure the length, width, and height. Calculate the volume ($V = lwh$). Is the volume more or less than 8 cubic inches?
3. Find the correct size for the original piece of paper by trying 7-inch, 6-inch, and 5-inch squares.

Enrichment
Complete the square to find the zeros of each function. Then find the sum of the zeros.

1. $y = x^2 - 2x - 6$ $1 + \sqrt{7}, 1 - \sqrt{7}; 2$
2. $y = x^2 + 4x - 7$ $-2 + \sqrt{11}, -2 - \sqrt{11}; -4$
3. $y = x^2 - 6x + 3$ $3 + \sqrt{6}, 3 - \sqrt{6}; 6$
4. $y = x^2 + 8x + 11$ $-4 + \sqrt{5}, -4 - \sqrt{5}; -8$

5. Write a conjecture about the sum of the zeros of a function in the form $y = x^2 + bx + c$.
   Sample: The sum of the zeros is $-b$.

Part II

1. Let x be the length of the side of the square piece of paper. Explain why $(x - 4)$ represents both the length and width of the box and why 2 is the height.

2. Write an expression for the volume of the box.

CHECKPOINT ✓

3. Set the expression equal to 8, and solve for x in order to find the size of the original square. (Hint: To solve for x, divide both sides by 2, and then write $(x - 4)^2$ for $(x - 4)(x - 4)$.)

CRITICAL THINKING

Explain why $x = 2$ is not a meaningful solution for the box problem in the preceding activity.

CHECKPOINT ✓

3. $x = 6$ or $x = 2$

The original square should be a 6-inch square because a 2-inch square does not make sense for this problem.

CRITICAL THINKING

x is the length of the side of the piece of paper, and squares with 2-inch sides must be cut from each corner. This is impossible if the original piece of paper is only 2 inches on each side.

EXAMPLE ❹ Find the points where the graph of $y = x^2 - 10x + 21$ intersects the graph of $y = x - 3$.

TECHNOLOGY
GRAPHICS CALCULATOR

● **SOLUTION**

Method A

Graph the functions, and find the points of intersection.

The graphs intersect at $(8, 5)$ and $(3, 0)$.

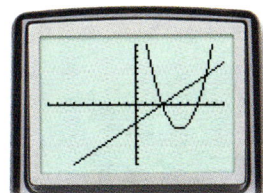

Method B

Use substitution to solve the following system: $\begin{cases} y = x^2 - 10x + 21 \\ y = x - 3 \end{cases}$

Substitute $x - 3$ from the second equation for y in the first equation and simplify.

$$x^2 - 10x + 21 = x - 3$$
$$x^2 - 10x + 21 - (x - 3) = x - 3 - (x - 3)$$
$$x^2 - 11x + 24 = 0$$

Solve the new quadratic equation by factoring.

$$(x - 8)(x - 3) = 0$$

$x - 8 = 0$ or $x - 3 = 0$
$x = 8$ $\qquad$ $x = 3$

Substitute 8 and 3 for x in the equation $y = x - 3$.

$y = x - 3$ $\qquad$ $y = x - 3$
$y = 8 - 3$ $\qquad$ $y = 3 - 3$
$y = 5$ $\qquad$ $y = 0$

The graphs intersect at $(8, 5)$ and $(3, 0)$.

TRY THIS Find the points where the graph of $y = -2x + 4$ intersects the graph of $y = x^2 - 4x + 1$.

ADDITIONAL EXAMPLE ❹

Find the points where the graphs of $y = x^2 - 4x + 11$ and $y = x + 5$ intersect.

$(2, 7)$ and $(3, 8)$

Teaching Tip

TECHNOLOGY If you are using a TI-82 or TI-83 graphics calculator, you can obtain the graph shown in Example 4 as follows: Press [Y=] and enter **X^2–10X+21** next to **Y1** and **X–3** next to **Y2**. Be sure to clear all other equations. Then press [ZOOM] and choose **6: ZStandard**. If you wish, you can also use the **CALCULATE** menu to locate the points of intersection.

TRY THIS

The graphs intersect at $(-1, 6)$ and $(3, -2)$.

Reteaching the Lesson

GUIDED ANALYSIS Display figure 1 below on the board or overhead. Ask students what points P and Q represent. **the zeros of a quadratic function** Ask how they would find the zeros.

1.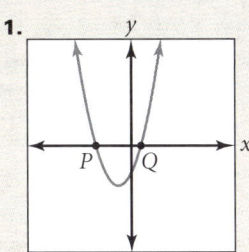
2.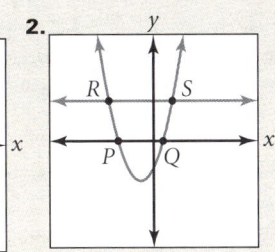

Solve $x^2 + bx + c = 0$. Now tell students that an equation for the function is $y = x^2 + 2x - 2$. Work with them to find the zeros. $-1 + \sqrt{3}, -1 - \sqrt{3}$

Next, show students figure 2. Ask how it is different from figure 1. **A line with the equation $y = n$ has been added.** Ask students how to find the x-coordinates of points R and S. **Solve the equation $x^2 + bx + c = n$.** Now tell them that equations for the line and parabola are $y = 3$ and $y = x^2 + 2x - 2$. Have them solve $x^2 + 2x - 2 = 3$.
$x = -1 + \sqrt{6}$ or $x = -1 - \sqrt{6}$

LESSON 10.4 **501**

Assess

Selected Answers
Exercises 6–15, 17–61 odd

ASSIGNMENT GUIDE

In Class	1–15
Core	16–45, 47–53 odd
Core Plus	16–50 even, 51–54
Review	55–62
Preview	63

● Extra Practice can be found beginning on page 738.

Technology

A graphics calculator may be helpful for Exercises 16–49. You may wish to have students first do the exercises with paper and pencil and then use the calculator to check their answers.

Practice

NAME _____ CLASS _____ DATE _____

Practice
10.4 Solving Equations of the Form $x^2 + bx + c = 0$

Find the zeros of each function. Graph to check.
1. $y = x^2 - 13x + 42$ ___**6 and 7**___
2. $y = x^2 + 13x + 42$ ___**−7 and −6**___
3. $y = x^2 - x - 42$ ___**−6 and 7**___
4. $y = x^2 + 7x + 12$ ___**−4 and −3**___
5. $y = x^2 - x - 12$ ___**−3 and 4**___
6. $y = x^2 + x - 12$ ___**−4 and 3**___

Solve each equation by factoring.
7. $x^2 - 4x - 60 = 0$ ___**−6 and 10**___
8. $x^2 - 10x + 21 = 0$ ___**3 and 7**___
9. $x^2 + 5x + 4 = 0$ ___**−4 and −1**___
10. $x^2 + 2x - 24 = 0$ ___**−6 and 4**___
11. $x^2 - x - 6 = 0$ ___**−2 and 3**___
12. $x^2 - 5x + 4 = 0$ ___**1 and 4**___
13. $x^2 + 11x + 30 = 0$ ___**−6 and −5**___
14. $x^2 + 2x - 8 = 0$ ___**−4 and 2**___
15. $x^2 + 6x + 9 = 0$ ___**−3**___
16. $x^2 - 2x + 1 = 0$ ___**1**___
17. $x^2 + x - 30 = 0$ ___**−6 and 5**___
18. $x^2 + 5x - 14 = 0$ ___**−7 and 2**___

Solve each equation by completing the square.
19. $x^2 + 5x + 6 = 0$ ___**−3 and −2**___
20. $x^2 + 3x + 2 = 0$ ___**−2 and −1**___
21. $x^2 + x - 12 = 0$ ___**−4 and 3**___
22. $x^2 + 7x + 10 = 0$ ___**−5 and −2**___
23. $x^2 - 2x = 3$ ___**−1 and 3**___
24. $x^2 + 4x = 5$ ___**−5 and 1**___
25. $x^2 + 6 = 7x$ ___**1 and 6**___
26. $x^2 + 24 = 11x$ ___**3 and 8**___

Find the point(s) where the graphs intersect. Graph to check.
27. $\begin{cases} y = 4 \\ y = x^2 \end{cases}$ ___**(2, 4) and (−2, 4)**___
28. $\begin{cases} y = -x - 5 \\ y = x^2 + 3x - 10 \end{cases}$ ___**(1, −6) and (−5, 0)**___
29. $\begin{cases} y = 7x - 35 \\ y = x^2 - 3x - 10 \end{cases}$ ___**(5, 0)**___
30. $\begin{cases} y = x - 2 \\ y = x^2 - 7x + 10 \end{cases}$ ___**(2, 0) and (6, 4)**___

502 LESSON 10.4

Exercises

● Communicate

1. How are the x-intercepts of the graph of $y = x^2 - 12x + 36$ related to the zeros of the function?
2. Explain how to find the zeros of $y = x^2 - 6x + 8$ by completing the square.
3. Describe how to solve $x^2 - 2x = 15$ by completing the square.
4. Graph $y = x^2 - 7x + 12$. Explain how to find the value of x when y is 2.
5. Describe two methods of finding the points where the graph of $y = x - 1$ intersects the graph of $y = x^2 - 3x + 2$.

● Guided Skills Practice

Find the zeros of each function. *(EXAMPLE 1)*
6. $y = x^2 - 7x - 8$ **−1 and 8**
7. $y = x^2 + 4x + 2$ **−2 ± √2**
8. $y = x^2 - 8x - 8$ **4 ± 2√6**
9. Solve $x^2 + 12x = -27$ by completing the square and by factoring. *(EXAMPLE 2)* **−3 or −9**

For $y = x^2 - 8x + 15$, find the value(s) of x for each value of y below. *(EXAMPLE 3)*
10. 0 **3 or 5**
11. 3 **6 or 2**
12. −1 **4**
13. 8 **7 or 1**

Find the points where the graphs of the two functions intersect. *(EXAMPLE 4)*
14. $y = x^2 - 9x + 25$ and $y = 3x - 11$ **(6, 7)**
15. $y = x^2 + 4x - 12$ and $y = 9x + 2$ **(−2, −16) or (7, 65)**

● Practice and Apply

Find the zeros of each function. Graph to check.
16. $y = x^2 - 2x - 8$ **−2 or 4**
17. $y = x^2 + 6x + 5$ **−1 or −5**
18. $y = x^2 - 4x + 4$ **2**

Solve each equation by factoring.
19. $x^2 - 2x - 3 = 0$ **−1 or 3**
20. $x^2 + 4x - 5 = 0$ **1 or −5**
21. $x^2 + 7x + 12 = 0$ **−4 or −3**
22. $x^2 - 10x + 24 = 0$ **4 or 6**
23. $x^2 - 3x = 10$ **−2 or 5**
24. $x^2 - 8x = -15$ **5 or 3**
25. $x^2 - 6x + 9 = 0$ **3**
26. $x^2 + 10x = -25$ **−5**
27. $x^2 - 2x + 1 = 0$ **1**

Solve by completing the square. Round your answers to the nearest hundredth when necessary.
28. $x^2 - 2x - 15 = 0$ **5 or −3**
29. $x^2 + 4x - 5 = 0$ **1 or −5**
30. $x^2 - x - 20 = 0$ **5 or −4**
31. $x^2 + 2x - 6 = 0$ **1.65 or −3.65**
32. $x^2 - 4x = 12$ **6 or −2**
33. $x^2 + x - 6 = 0$ **2 or −3**
34. $x^2 + 2x = 5$ **1.45 or −3.45**
35. $x^2 + 4x = 1$ **0.24 or −4.24**
36. $x^2 + 8x + 13 = 0$ **−2.27 or −5.73**

16.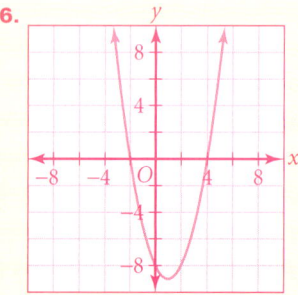

17.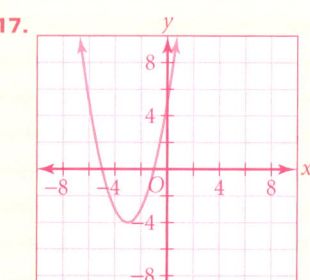

37. 0 or −10
38. 4 or 6
39. 2.30 or −1.30
40. 2 or −6
41. 1.19 or −4.19
42. 3.30 or −0.30
43. 5.54 or −0.54
44. 0 or 1
45. 2.58 or −6.58
46. (3, 9) and (−3, 9)
47. (−1, 4) and (3, 4)
48. (2, 1)
49. (0, 3) and (5, 8)

CONNECTION

APPLICATION

CHALLENGE

Solve each equation by factoring or by completing the square.

37. $b^2 + 10b = 0$
38. $r^2 − 10r + 24 = 0$
39. $x^2 − x − 3 = 0$
40. $x^2 + 4x − 12 = 0$
41. $s^2 + 3s − 5 = 0$
42. $t^2 − 3t − 1 = 0$
43. $p^2 − 5p − 3 = 0$
44. $x^2 − x = 0$
45. $q^2 + 4q − 17 = 0$

Find the points where the graphs of each system intersect. Graph to check.

46. $\begin{cases} y = 9 \\ y = x^2 \end{cases}$
47. $\begin{cases} y = 4 \\ y = x^2 − 2x + 1 \end{cases}$
48. $\begin{cases} y = x − 1 \\ y = x^2 − 3x + 3 \end{cases}$
49. $\begin{cases} y = x + 3 \\ y = x^2 − 4x + 3 \end{cases}$

50. Find two consecutive even integers whose product is 224.   **14 and 16 or −16 and −14**

51. **GEOMETRY** The length of a rectangular flag is 4 yards longer than the width. The area of the flag is 140 square yards. Find the length and the width.
width = 10 yds, length = 14 yds

The flag that flew over Fort McHenry inspired Francis Scott Key to write "The Star Spangled Banner".

52. **PHYSICS** A boat fires an emergency flare that travels upward at an initial velocity of 25 meters per second. To find the height in meters, h, of the flare at t seconds, use the function $h = −5t^2 + 25t$. After how many seconds will the flare be at a height of 10 meters? Round your answer(s) to the nearest tenth.   **4.6 seconds or 0.4 seconds**

53. **PHOTOGRAPHY** The perimeter of a rectangular photograph is 80 centimeters. Find the dimensions of the photograph if the length is 4 centimeters more than the width and the area is 396 square centimeters.   **width = 18 cm, length = 22 cm**

54. **MARCHING BAND** A marching band uses a rectangular marching formation with 8 columns and 10 rows. An equal number of columns and rows were added to the band's marching formation when 40 new members joined at the start of the school year. How many rows and columns were added?
2 rows and 2 columns

Look Back

Solve each system of equations. *(LESSONS 7.2 AND 7.3)*

55. $\begin{cases} x − 2y = 1 \\ 4x + 2y = −1 \end{cases}$ $(0, −\frac{1}{2})$
56. $\begin{cases} 5x − 2y = 11 \\ 3x + 5y = 19 \end{cases}$ $(3, 2)$

Write each number in standard notation. *(LESSON 8.5)*

57. 4×10^4 **40,000**
58. 6.5×10^7 **65,000,000**
59. 9.6×10^{-5} **0.000096**

Write each polynomial in factored form. *(LESSON 9.5)*

60. $2b^2 + 6b − 36$
    2(b + 6)(b − 3)
61. $6w(w^2 − w)$
    6w²(w − 1)
62. $8p^4 − 16p$
    8p(p³ − 2)

Look Beyond

63. Substitute values for a and b to determine whether the following statements are true:
    a. $\sqrt{a + b} = \sqrt{a} + \sqrt{b}$ **Not always true**
    b. $\sqrt{a \cdot b} = \sqrt{a} \cdot \sqrt{b}$ **True**

18.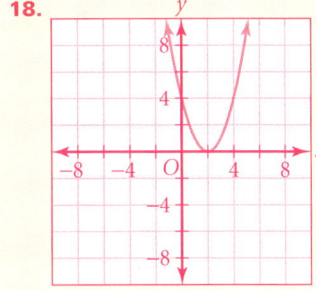

Error Analysis

When solving an equation such as $x^2 − 3x = 9$ by factoring, students may begin the solution by writing $x(x − 3) = 9$. This may lead to erroneous solutions such as 0 and 3. Remind students that they are applying the Zero-Product Property, which requires the equation to be written with one side equal to zero.

Look Beyond

Exercise 63 previews the simplification of radical expressions, which students will study in depth in Chapter 12.

internetconnect

GO TO: go.hrw.com
KEYWORD: MA1 Motion Equations

In this activity, students access Web sites with animations and other graphics to discover how equations of motion can illustrate properties of polynomials.

Student Technology Guide

NAME _____ CLASS _____ DATE _____

Student Technology Guide
10.4 Solving Equations of the Form $x^2 + bx + c = 0$

In Lesson 10.4, you solved equations of the form $x^2 + bx + c = 0$ by completing the square and factoring. You can use a graphics calculator to solve equations where $x^2 + bx + c$ is equal to any real number.

Example: Let $y = x^2 + 6x + 5$. Find the value(s) of x when y is −3.

- Enter $y = x^2 + 6x + 5$ as Y1 by pressing [Y=] [X,T,θ,n] [x²] [+] 6 [X,T,θ,n] [+] 5.
- Then enter $y = −3$ as Y2 by pressing [Y=] [(-)] 3.
- Press [GRAPH] to see the graphs of the equations.
- Use [TRACE] and the left and right arrow keys to trace along the parabola to one of the points where the graphs intersect. The first point appears to be at $x = −2$. You can use the intersect feature of the calculator to confirm this.
- With the trace point near the point of intersection as shown, press [2nd] [TRACE] [5: intersect] [ENTER] [ENTER] [ENTER]. This confirms that the intersection point is at $x = −2$.
- There appears to be another point of intersection at $x = −4$. Check by tracing near that point and then using the intersect feature to confirm that $y = −3$ when $x = −4$.

For $y = x^2 + 6x + 5$, $y = −3$ when $x = −4$ or $x = −2$.

Notice that you can easily investigate other y-values for this function by pressing [Y=] and changing Y2.

For each function, find the value(s) of x when $y = −1$ and when $y = 4$. If necessary, round answers to the nearest hundredth.

1. $y = x^2 + 4x + 3$
   $y = −1: x = −2;$
   $y = 4: −4.24$ and 0.24
2. $y = x^2 − 6x + 7$
   $y = −1;$ 2 and 4;
   $y = 4: 0.55$ and 5.45
3. $y = −x^2 − 4x + 6$
   $y = −1: −5.32$ and $1.32;$
   $y = 4: −4.45$ and 0.45
4. $y = x^2 − 4$
   $y = −1: −1.73$ and
   $1.73; y = 4: −2.82$ and 2.82
5. $y = −2x^2 + 4x + 4$
   $y = −1: −0.87$ and
   $2.87; y = 4: 0$ and 2
6. $y = x^2 + 4x + 4$
   $y = −1:$ no solution;
   $y = 4: −4$ and 0

LESSON 10.4 **503**

Eyewitness Math

Focus

An article describing the unprecedented midair rescue of an unconscious skydiver includes data that does not stand up to scrutiny. Students use equations of motion to determine why the numbers in the story do not make sense and to make a conjecture about numbers that would be more plausible.

Motivate

After students read the article, ask how they think the numbers were calculated, particularly the altitudes cited. Note that although skydivers do wear altimeters to measure altitude, it is doubtful that Robertson would look at an altimeter in the midst of a life-and-death crisis. The altitudes were most likely based on estimates made by the skydivers and by observers on the ground.

Discuss the diagrams at the bottom of page 504, making the following points:

- When a skydiver first exits the airplane, most of his or her motion is horizontal. Then air resistance slows the horizontal speed, while gravity increases the vertical speed. That is why the skydiver's motion becomes increasingly vertical.

- Before reaching terminal velocity, the motion of a skydiver is complex. The force of air resistance changes constantly, which causes the acceleration to change constantly. Once the skydiver reaches terminal velocity, the acceleration is constant, so it is much easier to quantify at that point.

EYEWITNESS MATH

RESCUE AT 2000 FEET

Robertson explains how he saved a life at 2,000 ft.

After jumping out of a plane, the force of gravity causes the diver to accelerate downward.

As the sky diver's speed increases, the force of air resistance (or *drag*) plays a larger role.

Eventually the two forces balance, and the sky diver falls at a steady rate, called the *terminal velocity*.

A MIRACULOUS SKY RESCUE

The jump began as a routine skydiving exercise, part of a convention of 420 parachutists sponsored by Skydive, Arizona, but it quickly turned into a test of nerve, instinct, and courage ... Moments after he went out the open hatch of a four-engine DC-4 airplane near Coolidge, Arizona, sky diver Gregory Robertson, 35, could see that Debbie Williams, 31, a fellow parachutist with a modest 50 jumps to her credit, was in big trouble. Instead of "floating" in the proper stretched-out position parallel to the earth, Williams was tumbling like a rag doll. In attempting to join three other divers in a hand-holding ring formation, she had slammed into the backpack of another chutist and was knocked unconscious.

From his instructor's position above the other divers, Robinson reacted with instincts that had been honed by 1700 jumps during time away from his job as an AT&T engineer in Phoenix. He straightened into a vertical dart, arms pinned to his body, ankles crossed, head aimed at the ground in what chutists call a "no-lift" dive, and plummeted toward Williams ... *At 3500 feet, about 10 seconds before impact, Robertson caught up with Williams*, almost hitting her but slowing his own descent by assuming the open-body, froglike position. He angled the unconscious sky diver so her chute could open readily and *at 2000 feet, with some 6 seconds left, yanked the ripcord* on her emergency chute, then pulled his own ripcord. The two sky divers floated to the ground. Williams, a fifth-grade teacher from Post, Texas, landed on her back, suffering a skull fracture and a perforated kidney—but alive. In the history of recreational skydiving, there has never been such a daring rescue in anyone's recollection.

Time, May 4, 1987

Does the article give you a complete picture of the rescue? How can you tell if the numbers in the article are accurate?

You can use what you know about distance, time, and speed to check the facts in the article and to get a fuller sense of what took place during this amazing feat. First, study these diagrams to get an idea of some of the forces that affect sky divers in free fall.

504 EYEWITNESS MATH

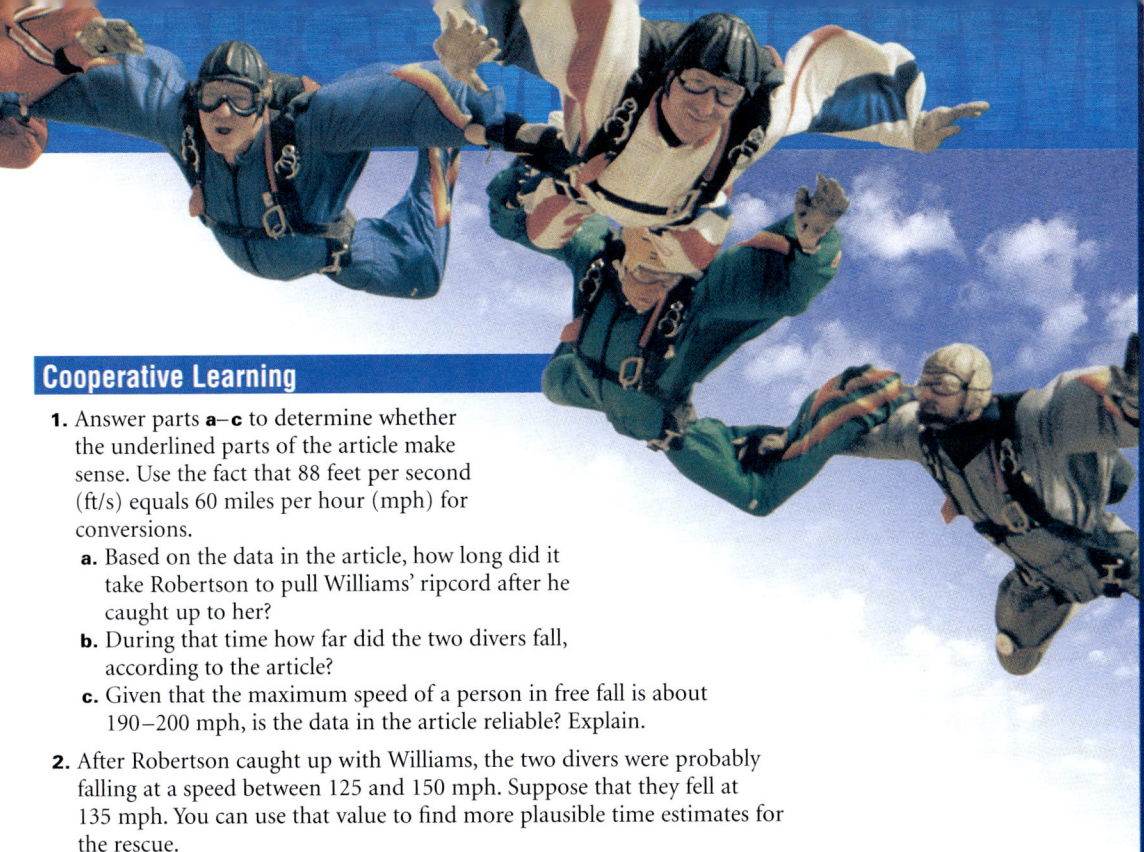

Cooperative Learning

1. Answer parts **a–c** to determine whether the underlined parts of the article make sense. Use the fact that 88 feet per second (ft/s) equals 60 miles per hour (mph) for conversions.
   a. Based on the data in the article, how long did it take Robertson to pull Williams' ripcord after he caught up to her?
   b. During that time how far did the two divers fall, according to the article?
   c. Given that the maximum speed of a person in free fall is about 190–200 mph, is the data in the article reliable? Explain.

2. After Robertson caught up with Williams, the two divers were probably falling at a speed between 125 and 150 mph. Suppose that they fell at 135 mph. You can use that value to find more plausible time estimates for the rescue.
   a. About how many seconds from impact would the two sky divers be at 3500 feet? at 2000 feet?
   b. Suppose that Robertson did catch up to Williams at 3500 feet and did pull her ripcord at 2000 feet. About how much time would have elapsed between those two events if the divers were falling at 135 mph?

3. While Robertson was catching up to Williams, he changed his body position to speed up and then slow down. Such motion is difficult to calculate exactly, but you can make estimates for this part of the rescue by using a single average speed for Robertson. Assume that Robertson was at 8500 feet and Williams was at 8300 feet when Robertson started trying to catch up.
   a. Write an equation for t in terms of v_r and v_w, where v_r is Robertson's average speed in ft/s while catching up to Williams, v_w is Williams' speed in ft/s, and t is the time in seconds from the time when Robertson starts to go after Williams to the time when he reaches her. (Hint: Compare the distances the two divers fell during the time in question.)
   b. Use your equation to find t if $v_r = 206$ (about 140 mph) and $v_w = 198$ (about 135 mph).
   c. Based on your results from part **b**, at what altitude would Robertson catch up to Williams?

4. **Project** Imagine that you are a journalist writing about the rescue. Use your data from Steps 2 and 3 to write a paragraph that will give your readers a reasonably accurate sense of the event.

---

1. a. 4 seconds
   b. 1500 feet
   c. no; 1500 feet/4 seconds = 375 feet/second or 256 mph

2. a. 135 mph = 198 feet/second; time to impact at 3500 feet: $\frac{3500}{198}$ or 17.7 seconds; time to impact at 2000 feet: $\frac{2000}{198}$ or 10.1 seconds
   b. 17.7 − 10.1 = 7.6 seconds

3. a. $t = \dfrac{200}{(v_r - v_w)}$
   b. $t = \dfrac{(200)}{(206 - 198)} = \dfrac{200}{8} = 25$
   c. 3350 feet

4. Answers will vary, but numbers used should be consistent both with figures given in the lesson and with each other. For example, speeds, times, and distances should satisfy basic equations of motion.

Prepare

NCTM PRINCIPLES & STANDARDS 1–3, 6–10

LESSON RESOURCES

- Practice 10.5
- Reteaching Master 10.5
- Enrichment Master 10.5
- Cooperative-Learning Activity 10.5
- Lesson Activity 10.5
- Problem Solving/Critical Thinking Master 10.5
- Teaching Transparency 50
- Teaching Transparency 51

QUICK WARM-UP

Find each square root.

1. $\sqrt{64}$ **8**
2. $\pm\sqrt{144}$ **±12**

Evaluate each expression.

3. $6^2 - 4(1)(3)$ **24**
4. $2^2 - 4(1)(-3)$ **16**
5. $\dfrac{-6 + \sqrt{6^2 - 4(1)(5)}}{2(1)}$ **−1**

Also on Quiz Transparency 10.5

Teach

Why Ask students to describe real-world problems that can be solved in two or more ways. As an example, point out that the total of three payments of $19.97 can be found using the sum 19.97 + 19.97 + 19.97 or the product 3(19.97). Tell students that, similarly, there are several methods for solving quadratic equations.

10.5 The Quadratic Formula

Objectives

- Use the quadratic formula to find solutions to quadratic equations.
- Use the quadratic formula to find the zeros of a quadratic function.
- Evaluate the discriminant to determine how many real roots a quadratic equation has and whether it can be factored.

Why The height of an object thrown into the air, such as a javelin, can be modeled by a quadratic function. The zeros of the quadratic function tell when the object is at ground level.

Deriving the Quadratic Formula

CULTURAL CONNECTION: ASIA Since early Babylonian times, scholars have known how to solve the standard quadratic equation $ax^2 + bx + c = 0$ by using a formula. To see how the quadratic formula is developed, follow the steps as the equations $2x^2 + 12x + 4 = 0$ and $ax^2 + bx + c = 0$ are solved below.

1. Divide each side of each equation by the coefficient of x^2.

$$2x^2 + 12x + 4 = 0 \qquad ax^2 + bx + c = 0$$
$$x^2 + 6x + 2 = 0 \qquad x^2 + \frac{b}{a}x + \frac{c}{a} = 0$$

2. Subtract the constant term from each side of each equation.

$$x^2 + 6x = -2 \qquad x^2 + \frac{b}{a}x = -\frac{c}{a}$$

3. Complete the square. Then simplify.

$$x^2 + 6x + 9 = -2 + 9 \qquad x^2 + \frac{b}{a}x + \left(\frac{b}{2a}\right)^2 = -\frac{c}{a} + \left(\frac{b}{2a}\right)^2$$
$$(x+3)^2 = 7 \qquad\qquad \left(x + \frac{b}{2a}\right)^2 = -\frac{c}{a} + \frac{b^2}{4a^2}$$
$$\qquad\qquad\qquad\qquad \left(x + \frac{b}{2a}\right)^2 = \frac{b^2 - 4ac}{4a^2}$$

Alternative Teaching Strategy

COOPERATIVE LEARNING Display $x^2 - 6x + 8 = 0$ on the board or overhead. Review with the class the three methods they have learned for solving this type of equation: graphing, factoring, and completing the square.

Have students work in groups of three. Each student in the group should solve the equation by using a method different from the other two students. Have the students in each group compare their answers and resolve any differences. All should agree that the solution is $x = 2$ or $x = 4$.

Display $x^2 - 6x + 4 = 0$. Have each group work in the same manner to solve the equation. Discuss the results. Note that the student who chooses factoring can not solve this equation, because $x^2 - 6x + 4$ cannot be factored over the integers.

Have a volunteer show how to solve the equation by completing the square. Tell the class that this method can be used to find the solution to any quadratic equation. Using the presentation on page 506, show them how completing the square leads to the quadratic formula.

4. Take the square root of each side. Then simplify.

$$x + 3 = \pm\sqrt{7}$$
$$x = 3 \pm \sqrt{7}$$

$$x + \frac{b}{2a} = \pm\sqrt{\frac{b^2 - 4ac}{4a^2}}$$
$$x = \frac{-b \pm \sqrt{b^2 - 4ac}}{2a}$$

The equation above and to the right is known as the **quadratic formula**.

The Quadratic Formula

For $ax^2 + bx + c = 0$, where $a \neq 0$:

$$x = \frac{-b \pm \sqrt{b^2 - 4ac}}{2a}$$

EXAMPLE 1 Use the quadratic formula to solve $x^2 - 10 + 3x = 0$.

• **SOLUTION**

Rewrite $x^2 - 10 + 3x = 0$ in standard form, $x^2 + 3x - 10 = 0$. Identify the coefficients and the constant: $a = 1$, $b = 3$, and $c = -10$. Substitute these values into the quadratic formula. Then simplify.

$$x = \frac{-3 \pm \sqrt{(3)^2 - 4(1)(-10)}}{2(1)}$$
$$= \frac{-3 \pm \sqrt{9 + 40}}{2}$$

$$x = \frac{-3 + 7}{2} \quad \text{or} \quad x = \frac{-3 - 7}{2}$$
$$x = \frac{4}{2} = 2 \quad\quad\quad x = \frac{-10}{2} = -5$$

The solutions are 2 and −5. Check by substitution.

TRY THIS Use the quadratic formula to solve $x^2 - 12 + 4x = 0$.

You now have five different methods for solving a quadratic equation.
1. taking square roots
2. factoring
3. completing the square
4. using the quadratic formula
5. graphing

EXAMPLE 2 Find the zeros of $y = 3x^2 + 2x - 4$.

• **SOLUTION**

Solve $0 = 3x^2 + 2x - 4$. Use the quadratic formula, with $a = 3$, $b = 2$, and $c = -4$.

$$x = \frac{-2 \pm \sqrt{4 - (-48)}}{6}$$
$$= \frac{-2 \pm \sqrt{52}}{6}$$

A calculator shows that $x \approx 0.87$ or $x \approx -1.54$. Check by substitution.

Interdisciplinary Connection

BUSINESS The amount of money that a business takes in from the sales of goods or services is called its *revenue*. When the expenses for a given period of time are subtracted from the revenue for that same period, the result is the *profit* for the period. A profit of 0 indicates the *break-even* point.

For x units of goods sold, a certain company uses these functions to project revenue, R, and costs, C.

$$R(x) = 8x^2 + 10x - 240 \quad\quad C(x) = 6x^2 + 62x$$

What is the break-even point? **30 units**

Inclusion Strategies

INVITING PARTICIPATION Some students become anxious when confronted with several different methods of approaching a problem. Lead a class discussion in which students share their perceptions of the advantages and disadvantages of the different methods for solving quadratic equations. Encourage them to share any special techniques they have developed. For example, some may offer the observation that even though graphing does not always yield exact solutions, it always helps them to get solutions that are "in the ballpark."

Use Teaching Transparency 50.

ADDITIONAL EXAMPLE 1

Use the quadratic formula to solve $x^2 - 3x - 28 = 0$.
$x = -4$ or $x = 7$

TRY THIS
The solutions are −6 and 2.

Teaching Tip

Be sure students are aware that the ± sign in the quadratic formula indicates that there are two expressions to evaluate: $\frac{-b + \sqrt{b^2 - 4ac}}{2a}$ and $\frac{-b - \sqrt{b^2 - 4ac}}{2a}$.

ADDITIONAL EXAMPLE 2

Find the zeros of $y = 2x^2 - 9x + 5$.
$x = \frac{9 \pm \sqrt{41}}{4}$;
$x \approx 0.65$ or $x \approx 3.85$

internetconnect

GO TO: go.hrw.com
KEYWORD: MA1 Quadratic Formula

In this activity, students access a brief animation demonstrating the derivation of the quadratic formula and then explore an interactive utility capable of solving any quadratic equation. Students use this interactive graphing program to investigate the properties of quadratic functions.

Use Teaching Transparency 51.

ADDITIONAL EXAMPLE 3

Find the number of real solutions to each equation.

a. $6x^2 - 4x = 5$
   two real solutions

b. $6x^2 = 9x - 4$
   no real solutions

CRITICAL THINKING

If the discriminant is a perfect square, then the results of the quadratic formula are either intergers or rational numbers that can be expressed as the quotients of integers. From these results, it is easy to work backward and determine what the factors of the quadratic expression should be.

ADDITIONAL EXAMPLE 4

Factor $6x^2 - x - 2$.
$(2x + 1)(3x - 2)$

The Discriminant

The discriminant of a quadratic equation allows you to determine how many real-number solutions the quadratic equation has.

The Discriminant

Given the quadratic equation $ax^2 + bx + c = 0$, where $a \neq 0$, the **discriminant** is defined as $b^2 - 4ac$.

- If the value of the discriminant is less than 0, the quadratic equation has no real solutions.
- If the value of the discriminant is 0, the quadratic equation has exactly one real solution.
- If the value of the discriminant is greater than 0, the quadratic equation has two real solutions.

EXAMPLE 3 Find the number of real solutions to each quadratic equation.

a. $3x^2 - 2x + 1 = 0$ b. $4x = 4x^2 + 1$

● **SOLUTION**

a. For $a = 3$, $b = -2$, and $c = 1$, the discriminant $b^2 - 4ac$ has a value of -8. Therefore, $3x^2 - 2x + 1 = 0$ has no real solutions.

b. Write the equation in standard form, $4x^2 - 4x + 1 = 0$. For $a = 4$, $b = -4$, and $c = 1$, the discriminant $b^2 - 4ac$ has a value of 0. Therefore, $4x = 4x^2 + 1$ has one real solution.

If the discriminant is a perfect square, the quadratic equation can be factored. You can use the quadratic formula to find the factors.

CRITICAL THINKING Examine the quadratic formula and explain why a quadratic equation can be factored if the discriminant is a perfect square.

EXAMPLE 4 Determine whether $6x^2 + 23x + 20$ can be factored and, if possible, determine the factored form.

● **SOLUTION**

For $a = 6$, $b = 23$, and $c = 20$, the discriminant $b^2 - 4ac$ has a value of 49, which is a perfect square.

Use the quadratic formula to find the factors.

$$x = \frac{-23 \pm \sqrt{49}}{12} = \frac{-23 \pm 7}{12}$$

Simplify. $x = -\frac{30}{12} = -\frac{5}{2}$ or $x = -\frac{16}{12} = -\frac{4}{3}$

Practice

NAME _____ CLASS _____ DATE _____

Practice
10.5 The Quadratic Formula

Identify a, b, and c for each quadratic equation.

1. $x^2 + 3x + 2 = 0$ $a = 1, b = 3, c = 2$
2. $x^2 - 2x + 1 = 0$ $a = 1, b = -2, c = 1$
3. $x^2 + 7x - 3 = 0$ $a = 1, b = 7, c = -3$
4. $x^2 - 5x + 3 = 0$ $a = 1, b = -5, c = 3$
5. $3x^2 - 4 = 0$ $a = 3, b = 0, c = -4$
6. $2x^2 + 15x = 0$ $a = 2, b = 15, c = 0$
7. $10x^2 + 1 + 6x = 0$ $a = 10, b = 6, c = 1$
8. $7x + x^2 - 2 = 0$ $a = 1, b = 7, c = -2$
9. $x^2 = 36$ $a = 1, b = 0, c = -36$
10. $x^2 + 36 = 12x$ $a = 1, b = -12, c = 36$
11. $x^2 - 6x = 0$ $a = 1, b = -6, c = 0$
12. $3x^2 = 5x$ $a = 3, b = -5, c = 0$

Find the value of the discriminant and determine the number of solutions for each equation.

13. $4x^2 + 3x + 1 = 0$ -7; no solutions
14. $x^2 + 5x + 1 = 0$ 21; two solutions
15. $x^2 + 2x - 6 = 0$ 28; two solutions
16. $2x^2 + 2x + 4 = 0$ -28; no solutions
17. $x^2 - 5x + 1 = 0$ 21; two solutions
18. $7 - x^2 + 6x = 0$ 64; two solutions
19. $x^2 - 4x + 4 = 0$ 0; one solution
20. $3 - 4x + 2x^2 = 0$ -8; no solutions

Use the quadratic formula to solve each equation. Give answers to the nearest hundredth. Check by substitution.

21. $3x^2 - 4x - 2 = 0$ -0.39 and 1.72
22. $2x^2 - 5x - 4 = 0$ -0.64 and 3.14
23. $3x^2 - 8 = 0$ -1.63 and 1.63
24. $6x^2 - 5x + 1 = 0$ 0.5 and 0.33
25. $x^2 - 8x + 8 = 0$ 1.17 and 6.83
26. $x^2 - 5 = 0$ -2.24 and 2.24

Choose any method to solve each quadratic equation. Give answers to the nearest hundredth.

27. $x^2 - x - 2 = 0$ -1 and 2
28. $x^2 - 7x + 12 = 0$ 3 and 4
29. $x^2 + 3x + 2 = 0$ -2 and -1
30. $x^2 - 25 = 0$ -5 and 5
31. $3x^2 + 2x - 5 = 0$ -1.67 and 1
32. $x^2 + 3x + 1 = 0$ -0.38 and -2.62
33. $2x^2 - 10x + 8 = 0$ 1 and 4
34. $x^2 + 5x + 4 = 0$ -4 and -1

Enrichment

1. Consider a quadratic equation of the form $nx^2 + nx + n = 0$, where $n \neq 0$. Explain why such an equation has no real solutions.
   For $n \neq 0$, $n^2 - 4(n)(n) < 0$.

2. Consider a quadratic equation of the form $nx^2 + 2nx + 3n = 0$, where $n \neq 0$. Does such an equation have any real solutions? Explain.
   No; for $n \neq 0$, $(2n)^2 - 4(n)(3n) < 0$.

3. Create an original problem of this type. Write the solution, and then trade problems with a partner. **Answers will vary.**

Reteaching the Lesson

VISUAL STRATEGIES On the board or overhead, write the equations using one color for a and 2, another for b and 5, and a third for c and 1.

$ax^2 + bx + c = 0$ $2x^2 + 5x + 1 = 0$

Now continue the color-coding in a display of the general quadratic formula and its application to the equation from above.

$$x = \frac{-b \pm \sqrt{b^2 - 4ac}}{2a} \qquad x = \frac{-5 \pm \sqrt{5^2 - 4(2)(1)}}{2(2)}$$

Work with students to simplify the expression, arriving at $x \approx -2.28$ or $x \approx -0.22$.

Work backward to find the factors.

$$x = -\frac{5}{2} \qquad x = -\frac{4}{3} \qquad \textit{Given}$$
$$2x = -5 \qquad 3x = -4 \qquad \textit{Mult. Prop. of Equality}$$
$$2x + 5 = 0 \qquad 3x + 4 = 0 \qquad \textit{Add. Prop. of Equality}$$

The factored form of $6x^2 + 23x + 20$ is $(2x + 5)(3x + 4)$.

CULTURAL CONNECTION: EUROPE In the sixteenth century, Italian mathematician Girolamo Cardano developed a way to work with square roots of negative numbers. Cardano defined $\sqrt{-1} \cdot \sqrt{-1} = -1$. Seventeenth century French mathematician Rene Descartes defined the **imaginary unit**, i, such that $i = \sqrt{-1}$ and $i^2 = -1$. **Imaginary numbers** have the form $i\sqrt{a}$, where a is any positive real number.

The imaginary numbers and the real numbers together form the set of **complex numbers**. These numbers can be expressed in the form $a + bi$, where a and b are real numbers. Quadratic equations with no real-number solutions often have solutions that are complex numbers. You will study complex numbers in later mathematics courses.

Assess

Selected Answers
Exercises 4–15, 17–53 odd

ASSIGNMENT GUIDE

In Class	1–15
Core	17–43 odd
Core Plus	16–44 even
Review	45–53
Preview	54

✎ Extra Practice can be found beginning on page 738.

Technology
Students will need a calculator to do the approximations required in some of the exercises. A graphics calculator may be helpful for checking answers in Exercises 49–51.

Exercises

Communicate

1. Discuss how to identify the coefficients and constant (a, b, and c) in the equation $2n^2 + 3n = 7$ in order to use the quadratic formula.

2. Explain how to find the discriminant for $2t^2 + 3t - 7 = 0$.

3. Discuss how to use the discriminant to determine the number of real solutions to each equation.

   a. $2x^2 - x - 2 = 0$ **b.** $x^2 - 2x = -7$

Guided Skills Practice

Use the quadratic formula to solve each equation. (EXAMPLE 1)

4. $x^2 - 7x - 8 = 0$
   $x = 8$ or $x = -1$
5. $2x^2 + 4x + 2 = 0$
   $x = -1$
6. $4x^2 + 4x - 5 = 0$
   $x \approx 0.72$ or $x \approx -1.72$

Find the zeros of each function. (EXAMPLE 2)

7. $y = 2x^2 - 8x - 8$
   4.83 or –0.83
8. $y = 3x^2 - 2x + 5$
   no real solutions
9. $y = x^2 + 7x - 3$
   0.41 or –7.41

Find the number of real solutions for each equation. (EXAMPLE 3)

10. $3x^2 + 6x + 3 = 0$ **1**
11. $x^2 - 2x + 3 = 0$ **none**
12. $-2x^2 + 7x = 5$ **2**

13. $2(3x - 7)(x + 1)$
14. $-3(x + 1)(5x + 2)$
15. $(7x - 6)(2x + 7)$

Factor each expression. (EXAMPLE 4)

13. $6x^2 - 8x - 14$
14. $-15x^2 - 21x - 6$
15. $14x^2 + 37x - 42$

Student Technology Guide

Student Technology Guide
10.5 The Quadratic Formula

You can use the quadratic formula, $x = \frac{-b \pm \sqrt{b^2 - 4ac}}{2a}$, to solve a quadratic equation of the form $ax^2 + bx + c = 0$. A calculator can help you evaluate the formula.

Example: Use the quadratic formula to solve $-4x + 7 = 2x^2$. Give answers to the nearest hundredth.
- Rewrite $-4x + 7 = 2x^2$ in standard form by subtracting $2x^2$ from each side. This gives $-2x^2 - 4x + 7 = 0$.
- Identify a, b, and c: $a = -2$, $b = -4$, and $c = 7$.
- Substitute a, b, and c into the quadratic formula.
  $x = \frac{-(-4) \pm \sqrt{(-4)^2 - 4(-2)(7)}}{2(-2)}$
- Simplify some of the expression mentally to get
  $x = \frac{4 \pm \sqrt{16 + 56}}{-4}$.
- Use a calculator to evaluate the expression below.
  $\frac{4 - \sqrt{16 + 56}}{-4}$

Press: (4 + 2nd x^2 16 + 56)) ÷ (-) 4 ENTER.

- Edit the expression in the calculator so that you can evaluate the following expression:
  $\frac{4 - \sqrt{16 + 56}}{-4}$

```
(4+√(16+56))/-4
      -3.121320344
(4-√(16+56))/-4
       1.121320344
```

Press 2nd ENTER and change the first plus sign to a minus sign. Press ENTER.

The solutions to $-4x + 7 = 2x^2$ are –3.12 and 1.12.

Use the quadratic formula to solve each equation. Give answers to the nearest hundredth when necessary.

1. $x^2 - 5x + 6 = 0$ 2 and 3
2. $x^2 + 9x + 14 = 0$ –7 and –2
3. $3x^2 - 12 = -4x$ –2.77 and 1.44
4. $x^2 = 14 + 9x$ –1.35 and 10.35
5. $-4x^2 + 6x = 2$ 0.5 and 1
6. $-x^2 = -8x + 16$ $x = 4$

LESSON 10.5 **509**

Error Analysis

Students may incorrectly identify the values of a, b, and c to be substituted into the quadratic formula when the given equation is not in standard form, $ax^2 + bx + c = 0$. Encourage them to rewrite the given equation in standard form and to write the values of the variables ($a = \underline{?}$, $b = \underline{?}$, $c = \underline{?}$) before they apply the formula.

In Exercise 54, students are asked to make a conjecture about the graph of a quadratic inequality. They will study this topic further in Lesson 10.6.

49.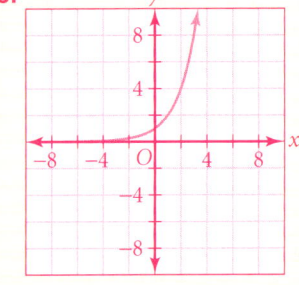

50.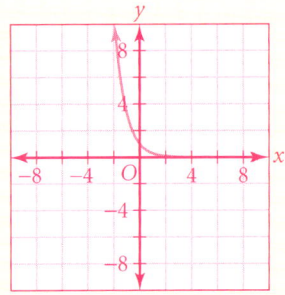

51.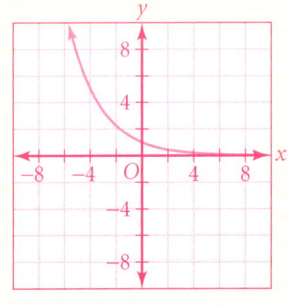

31. $2(x - 5)(x - 2)$
32. $2(x + 2)(x + 4)$
33. $4(x - 1)(x + 3)$
34. $2(x - 8)(x - 1)$
35. $(x - 1)(2x - 1)$
36. $3(x - 4)(x + 3)$
37. 2 or −6
38. ±3.61
39. $\frac{1}{2}$ or −3
40. −2 or −5
41. 4 or $-\frac{3}{2}$
42. 0.44 or −3.77

CONNECTION

APPLICATION

 **Practice and Apply**

Find the value of the discriminant and determine the number of real solutions for each equation.

16. $x^2 - x - 3 = 0$ 13; 2
17. $x^2 + 2x - 8 = 0$ 36; 2
18. $x^2 + 8x + 13 = 0$ 12; 2
19. $z^2 + 4z - 21 = 0$ 100; 2
20. $2y^2 - 8y = 8$ 128; 2
21. $8y^2 - 2 = 0$ 64; 2

Use the quadratic formula to solve each equation. Round your answers to the nearest hundredth when necessary. Check by substitution.

22. $a^2 - 4a - 21 = 0$ 7 or −3
23. $t^2 + 6t - 16 = 0$ 2 or −8
24. $m^2 + 4m - 5 = 0$ 1 or −5
25. $w^2 - 4w = 0$ 4 or 0
26. $x^2 + 9x = 0$ 0 or −9
27. $x^2 - 9 = 0$ 3 or −3
28. $x^2 + 2x = 0$ 0 or −2
29. $3m^2 = 2m + 1$ 1 or $-\frac{1}{3}$
30. $-x^2 + 6x - 9 = 0$ 3

Factor each expression by using the quadratic formula.

31. $2x^2 - 14x + 20$
32. $2x^2 + 12x + 16$
33. $4x^2 + 8x - 12$
34. $2x^2 - 18x + 16$
35. $2x^2 - 3x + 1$
36. $3x^2 - 3x - 36$

Choose any method to solve each quadratic equation. Round your answers to the nearest hundredth when necessary.

37. $x^2 + 4x - 12 = 0$
38. $x^2 - 13 = 0$
39. $2x^2 + 5x - 3 = 0$
40. $x^2 + 7x + 10 = 0$
41. $2a^2 - 5a - 12 = 0$
42. $3x^2 + 10x - 5 = 0$

43. **GEOMETRY** The seats in a theater are arranged in parallel rows that form a rectangular region. The number of seats in each row of the theater is 16 fewer than the number of rows. How many seats are in each row of a 1161-seat theater? 27

44. **ACCOUNTING** To approximate the profit per day for her business, Mrs. Howe uses the formula $p = -x^2 + 50x - 350$. The profit, p, depends on the number of cases, x, of decorator napkins that are sold. How many cases of napkins must she sell to break even ($p = 0$)? to make the maximum profit? Find the maximum profit. ≈ 8.42 or ≈ 41.58 cases; 25 cases; $275

 **Look Back**

Write each number in scientific notation. *(LESSON 8.5)*

45. 32,000,000 3.2×10^7
46. 67,000 6.7×10^4
47. 0.00654 6.54×10^{-3}
48. 0.00000091 9.1×10^{-7}

Graph. *(LESSON 8.6)*

49. $y = 2^x$
50. $y = 0.3^x$
51. $y = \left(\frac{2}{3}\right)^x$

Multiply. *(LESSON 9.3)*

52. $(x - 5)(x + 3)$ $x^2 - 2x - 15$
53. $(2x + 4)(2x - 2)$ $4x^2 + 4x - 8$

Look Beyond

54. From what you know about graphing quadratic equations and linear inequalities, describe what you think the graph of $y < x^2 + 3x + 2$ looks like.

54. Answers may vary. Sample answer: A parabola with the points outside shaded.

10.6 Graphing Quadratic Inequalities

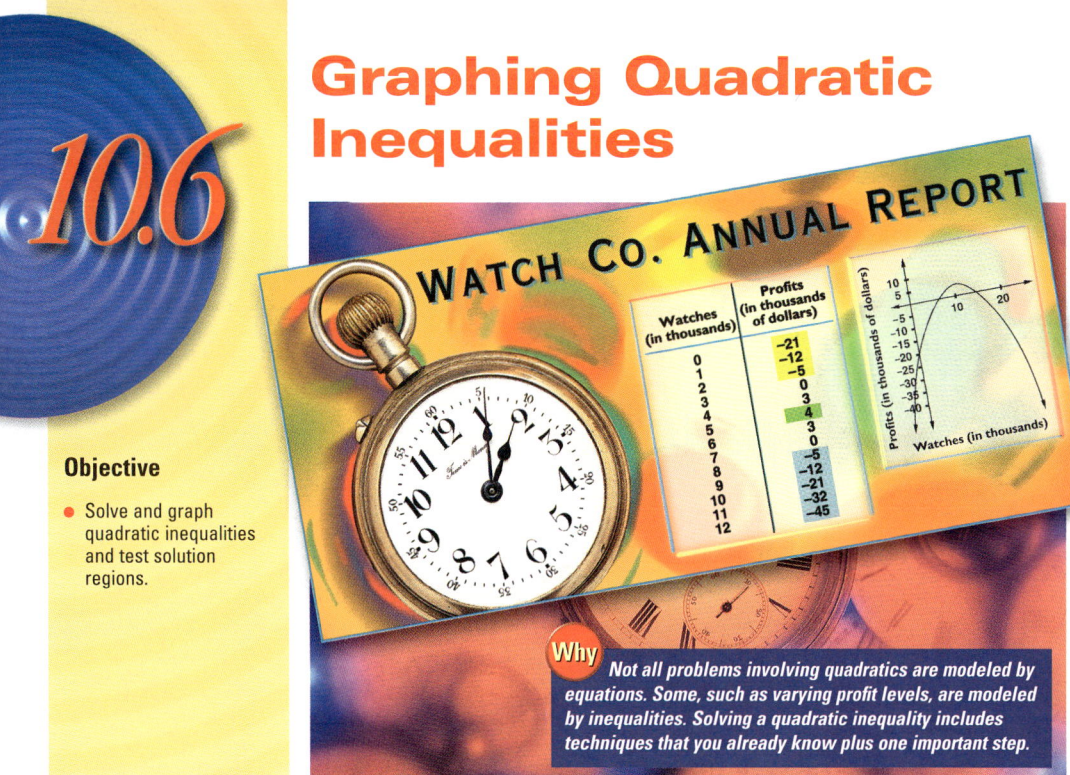

Objective
- Solve and graph quadratic inequalities and test solution regions.

Why Not all problems involving quadratics are modeled by equations. Some, such as varying profit levels, are modeled by inequalities. Solving a quadratic inequality includes techniques that you already know plus one important step.

**APPLICATION
BUSINESS**

To prepare the annual report, the president of a watch company gathers data about the number of watches produced per week and the corresponding profit. He uses the model $p = -x^2 + 10x - 21$, in which p represents profit in thousands of dollars and x represents the number of watches in thousands produced per week.

Production and Profit

**CONNECTION
MAXIMUM/MINIMUM**

Use the table of values and the graph of the profit function above to answer each question.

1. At what levels of production is the profit equal to 0?
2. At what levels of production is the profit greater than 0?
3. At what levels of production is the profit less than 0?
4. How many watches should be produced to maximize the profit?

CHECKPOINT ✓ 5. According to the profit function, what is the maximum profit that the company can make?

In the activity above, you solved a quadratic equation and two **quadratic inequalities**.

$-x^2 + 10x - 21 = 0$ $-x^2 + 10x - 21 > 0$ $-x^2 + 10x - 21 < 0$
Quadratic equation   Quadratic inequality   Quadratic inequality

---

Alternative Teaching Strategy

CONNECTING TO PRIOR KNOWLEDGE Have students graph the equations $y = 2x$ and $y = x^2$ on two sets of coordinate axes, placed side-by-side. Check their work. Then write the following inequalities on the board or overhead:

$y \geq 2x$ $\qquad\qquad$ $y \geq x^2$

Remind students that the first statement is called a *linear inequality*. Discuss with them the procedure they learned for graphing a statement of this type. Then tell them to adjust the graph of $y = 2x$ so that it becomes the graph of $y \geq 2x$.

Now ask them what name they think is given to a statement like $y \geq x^2$. Elicit the response *quadratic inequality*. Ask them to adjust the graph of $y = x^2$ in a way that they believe makes it the graph of $y \geq x^2$. Discuss the results. Have them make corrections if necessary.

Conclude by asking students how the graphs would be different for the following pairs of statements.

$y > 2x$ and $y > x^2$ $\qquad$ $y < 2x$ and $y < x^2$

Prepare

NCTM PRINCIPLES & STANDARDS
1–3, 5–10

LESSON RESOURCES

- Practice 10.6
- Reteaching Master 10.6
- Enrichment Master 10.6
- Cooperative-Learning Activity 10.6
- Lesson Activity 10.6
- Problem Solving/Critical Thinking Master 10.6

QUICK WARM-UP

Solve each equation.

1. $x^2 - 2x - 15 = 0$
   $x = -3$ or $x = 5$

2. $x^2 + 3x - 28 = 0$
   $x = -7$ or $x = 4$

Tell whether the given ordered pair satisfies the given inequality.

3. $(1, 4); y > -2x + 3$ $\qquad$ yes
4. $(-2, 3); y \leq -3x + 4$ $\qquad$ yes
5. $(-1, -2); 5x - y \geq 7$ $\qquad$ no

Also on Quiz Transparency 10.6

Teach

Why Remind students that the height of a ball thrown upward can be modeled by a quadratic equation. Ask them for instances in which it might be important to consider heights greater than or less than the height of the ball. Tell them that these heights can be modeled by a *quadratic inequality*.

☞ For Activity Notes and the answer to the Checkpoint, see page 512.

LESSON 10.6   **511**

Activity Notes

In this Activity, students use a quadratic function to explore the relationship between levels of production and profit. They should discover the levels of production that result in no profit, a positive profit, and a loss.

CHECKPOINT ✔
5. $4000

Math CONNECTION

MAXIMUM/MINIMUM The concepts of maximum and minimum are important to the study of calculus. In later mathematics courses, students will learn that a line tangent to a parabola at its maximum or minimum point has a slope of zero.

ADDITIONAL EXAMPLE 1

Solve $x^2 + 3x - 10 \leq 0$.

$-5 \leq x \leq 2$

TRY THIS
$x \leq -4$ or $x \geq 3$

CRITICAL THINKING
Sample answer: Like Method A in Example 1, graph the appropriate quadratic function and find the x-values that satisfy the inequality.

EXAMPLE 1 Solve $x^2 - 2x - 15 > 0$.

● **SOLUTION**

Method A
Graph $y = x^2 - 2x - 15$, and find the values of x for which values of y are greater than 0.

$x < -3$ or $x > 5$

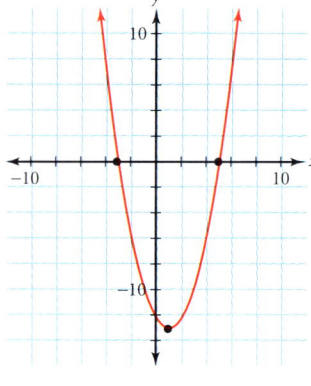

Method B
Solve the related quadratic equation, $x^2 - 2x - 15 = 0$.

$$x^2 - 2x - 15 = 0$$
$$(x - 5)(x + 3) = 0$$
$$x - 5 = 0 \quad \text{or} \quad x + 3 = 0$$
$$x = 5 \qquad\qquad x = -3$$

Use the x-axis as a number line, and plot the solutions. Since the inequality symbol is >, use open circles at −3 and 5.

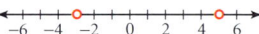

The solutions divide the x-axis into three intervals, to the left, between, and to the right of the two points. Test each interval. Choose a number from each interval, and substitute it for x in the original inequality.

Left of −3	Between −3 and 5	Right of 5
$x^2 - 2x - 15 > 0$	$x^2 - 2x - 15 > 0$	$x^2 - 2x - 15 > 0$
$(-5)^2 - 2(-5) - 15 \stackrel{?}{>} 0$	$(0)^2 - 2(0) - 15 \stackrel{?}{>} 0$	$(6)^2 - 2(6) - 15 \stackrel{?}{>} 0$
$25 + 10 - 15 \stackrel{?}{>} 0$	$0 - 0 - 15 \stackrel{?}{>} 0$	$36 - 12 - 15 \stackrel{?}{>} 0$
$20 > 0$ *True*	$-15 > 0$ *False*	$9 > 0$ *True*
Solution interval	No	Solution interval

Graph the solutions.

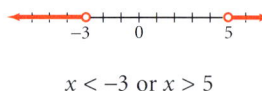

$x < -3$ or $x > 5$

TRY THIS Solve $x^2 + x - 12 \geq 0$.

CRITICAL THINKING Explain how you can determine which intervals to shade without testing numbers from all of the intervals.

Interdisciplinary Connection

DRIVER EDUCATION The distance required for an automobile to stop depends primarily on the speed at which it is traveling, the condition of the road, and the reaction time of the driver. Assuming a particular type of road surface and a reaction time of 0.1 second, the stopping distance in feet can be modeled by $y = 0.05x^2 + 1.1x$, where x is the speed in miles per hour. Use a graph to determine at which speeds the stopping distance is less than 60 feet (approximately four car-lengths). Round to the nearest mile per hour.
between 0 miles per hour and 25 miles per hour

Inclusion Strategies

KINESTHETIC LEARNERS Use masking tape to make a set of coordinate axes on the floor of the classroom. Use chalk to draw grid lines and a scale. Have students use string to model the graph of $y = 2x^2 - 8x$. Now have one student stand on a point inside the parabola. Discuss with students how the coordinates of this point represent a solution to $y < 2x^2 - 8x$. Have other students stand on other points inside the parabola and test whether they satisfy the inequality. Then repeat the activity for the inequality $y > 2x^2 - 8x$ and points *outside* the parabola.

EXAMPLE 2

Graph $y > -x^2 + 1$.

SOLUTION

Graph $y = -x^2 + 1$. Draw the parabola with a dashed curve to show that points on the curve $y = -x^2 + 1$ are *not* included in the solution.

To graph $y > -x^2 + 1$, shade the points outside the parabola.

As a check, test points inside and outside the parabola.

Test (**0, 0**), inside.   Test (**0, 3**), outside.

$y > -x^2 + 1$ $y > x^2 + 1$
$0 \overset{?}{>} 0 + 1$ $3 \overset{?}{>} 0 + 1$
$0 > 1$ False $3 > 1$ True

The graph of $y > -x^2 + 1$ is the set of points outside the parabola.

TRY THIS   Graph $y \geq x^2 - 7$.

EXAMPLE 3

APPLICATION
OPTICS

A cross section of the parabolic light reflector shown at right is described by the equation $y = \frac{1}{14}x^2 + \frac{7}{2}$. The bulb inside the reflector is located at the point (0, 7). Light from the bulb bounces off the reflector in parallel rays. This allows a flashlight to direct the light from the bulb in a narrow beam. **Determine which inequality below indicates the region in which the bulb is located.**

a. $y > \frac{1}{14}x^2 + \frac{7}{2}$

b. $y < \frac{1}{14}x^2 + \frac{7}{2}$

To determine which region satisfies the inequality, choose a point above the parabola for part **a**, and a point below the parabola for part **b**.

SOLUTION

a. $y > \frac{1}{14}x^2 + \frac{7}{2}$ **b.** $y < \frac{1}{14}x^2 + \frac{7}{2}$

Test (**0, 7**).   Test (**0, 7**).

$7 \overset{?}{>} \frac{1}{14}(0) + \frac{7}{2}$ $7 \overset{?}{<} \frac{1}{14}(0) + \frac{7}{2}$

$7 > \frac{7}{2}$ True $7 < \frac{7}{2}$ False

The solution to the inequality $y > \frac{1}{14}x^2 + \frac{7}{2}$ includes the point where the bulb is located.

A flashlight has a parabolic reflector to direct the beam of light.

ADDITIONAL EXAMPLE 2

Graph $y \leq -x^2 + 3$.

TRY THIS

ADDITIONAL EXAMPLE 3

The graph of $y = x^2 - 5x - 6$ is a parabola. It separates the coordinate plane into three regions: the points on the parabola, the points in the interior of the parabola, and the points in the exterior of the parabola. The point $(-2, -1)$ lies in the exterior region. **Determine which inequality below defines this region.**

a. $y > x^2 - 5x - 6$

b. $y < x^2 - 5x - 6$

The correct choice is **b.**

Enrichment

Have students work in pairs. Tell them to write four quadratic inequalities that have the given solution.

1. $-2 \leq x \leq 2$ any inequality that can be written in the form $a(x^2 - 4) \leq 0$, where $a > 0$, or $a(x^2 - 4) \geq 0$, where $a < 0$
2. $x < -3$ or $x > 1$ any inequality that can be written in the form $a(x - 1)(x + 3) > 0$, where $a > 0$, or $a(x - 1)(x + 3) < 0$, where $a < 0$

Challenge students to generalize their results.

Reteaching the Lesson

USING TABLES Have students graph $y = x^2$. Tell them to choose several points *inside* the parabola and make a table like the one at right. Ask them how y and x^2 are related. $y > x^2$ Point out that the region inside the parabola is the graph of $y > x^2$. Now have them make a table of points *outside* the parabola. Lead them to see that this region is the graph of $y < x^2$. Repeat the activity for $y = -2x^2$.

(x, y)	y	x²
(1, 2)	2	1
(2, 5)	5	4
(−1, 4)	4	1
(−2, 7)	7	4

LESSON 10.6   **513**

Assess

Selected Answers

Exercises 5–10, 11–45 odd

ASSIGNMENT GUIDE

In Class	1–10
Core	11–18, 25–33 odd, 36–38
Core Plus	12–34 even, 35–38
Review	39–46
Preview	47

✏ Extra Practice can be found beginning on page 738.

Technology

Exercise 38 is well-suited for the use of graphing technology. A graphics calculator may also be helpful for Exercises 11–35. You may wish to have students do these exercises first with paper and pencil and then use the calculator to check their answers.

Practice

NAME _____ CLASS _____ DATE _____

Practice
10.6 Graphing Quadratic Inequalities

Solve each quadratic inequality by using the Zero Product Property. Graph the solution on a number line.

1. $x^2 - 3x - 40 \geq 0$ $x \leq -5$ or $x \geq 8$
2. $x^2 + 9x + 14 < 0$ $-7 < x < -2$
3. $x^2 + 2x + 1 \geq 0$ all real numbers
4. $x^2 - x - 30 < 0$ $-5 < x < 6$
5. $x^2 - x - 12 > 0$ $x \leq -3$ or $x \geq 4$
6. $3x^2 + 3x + 2 < 0$ $-2 < x < -1$

Graph each quadratic inequality on the grid provided. Shade the solution region.

7. $y > x^2 - 4$
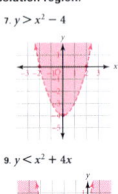

8. $y \geq -x^2 + 4$
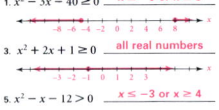

9. $y < x^2 + 4x$
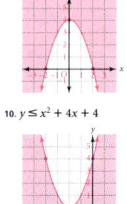

10. $y \leq x^2 + 4x + 4$
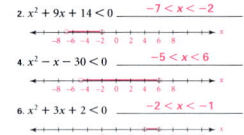

Exercises

Communicate

1. Explain how to solve $x^2 - 9x + 18 > 0$ by using the Zero Product Property.
2. Describe how to use a number line to graph and check the solution to the inequality from Exercise 1.
3. Explain how to determine whether the boundary line for the graph of each inequality is solid or dashed.
   a. $y < x^2 - 3$ **b.** $y \leq x^2 - 3$ **c.** $y \geq x^2 - 3$
4. Describe how to graph each inequality. Explain two ways to decide whether to shade inside or outside the parabola for each solution.
   a. $y < x^2 - 3$ **b.** $y \leq x^2 - 3$ **c.** $y \geq x^2 - 3$

Guided Skills Practice

Solve each inequality. *(EXAMPLE 1)*

5. $x^2 + 6x + 9 > 0$
   $x > -3$ or $x < -3$
6. $x^2 - 4 < 0$
   $-2 < x < 2$
7. $x^2 + 7x + 12 > 0$
   $x < -4$ or $x > -3$

Graph each inequality. Shade the solution region. *(EXAMPLES 2 AND 3)*

8. $y > -x^2 + 4$
9. $y < x^2 - 4x + 4$
10. $y < x^2 + 3x + 2$

Practice and Apply

Solve each quadratic inequality by using the Zero Product Property. Graph each solution on a number line.

11. $x^2 - 6x + 8 > 0$
12. $x^2 + 3x + 2 < 0$
13. $x^2 - 9x + 14 \leq 0$
14. $x^2 + 2x - 15 > 0$
15. $x^2 - 7x + 6 < 0$
16. $x^2 - 8x - 20 < 0$
17. $x^2 - 8x > -15$
18. $x^2 - 3x \leq 18$
19. $x^2 - 11x \geq -30$
20. $x^2 - 4x - 12 \geq 0$
21. $x^2 + 5x > 14$
22. $x^2 + 10x + 24 \geq 0$

Graph each quadratic inequality. Shade the solution region.

23. $y \geq x^2$
24. $y \geq x^2 + 4$
25. $y < -x^2$
26. $y > x^2$
27. $y > x^2 - 2x$
28. $y \geq 1 - x^2$
29. $y > -2x^2$
30. $y \leq x^2 - 3x$
31. $y \leq x^2 + 2x - 8$
32. $y \geq x^2 - 2x + 1$
33. $y \leq x^2 - 4x + 4$
34. $y \leq x^2 - 5x - 6$

CHALLENGE

35. Graph the solution region for $\begin{cases} y \geq x^2 - 9 \\ y \leq -x^2 + 9 \end{cases}$.

8.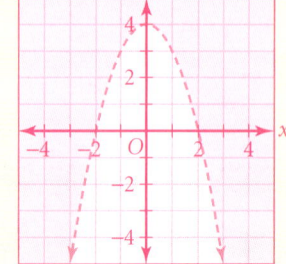

9.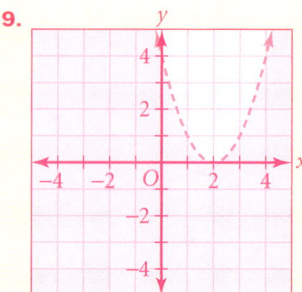

514 LESSON 10.6

APPLICATION

PHYSICS A projectile is fired vertically into the air. Its motion is described by $h = -16t^2 + 320t$, where h is the projectile's height (in feet) after t seconds. Use a quadratic inequality to find the answer to each question.

36. During what time interval(s) will the height of the projectile be below 1024 feet? *t < 4 or t > 16*

37. During what time interval(s) will the height of the projectile be above 1024 feet? *4 < t < 16*

38. TECHNOLOGY Use graphing technology to graph $y = -16x^2 + 320x$. Use this graph to check your answers to Exercises 36 and 37.

Look Back

Find each product. *(LESSON 8.2)*

39. $(-ab^2c)(a^2bc^3)$ *–a³b³c⁴*
40. $(3p^2q^3r^4)(-2pqr)$ *–6p³q⁴r⁵*

Write each number in scientific notation. *(LESSON 8.5)*

41. 0.825 *8.25 × 10⁻¹*
42. 0.000001 *1.0 × 10⁻⁶*
43. 0.0000074 *7.4 × 10⁻⁶*

Factor each polynomial. *(LESSON 9.5)*

44. $12y^2 - 2y$ *2y(6y – 1)*
45. $2ax + 6x + ab + 3b$ *(2x + b)(a + 3)*
46. $4x^2 - 24x + 32$ *4(x – 2)(x – 4)*

Look Beyond

47. Three identical rectangles are stacked to form a square. If the perimeter of each rectangle is 24 inches, what is the area of the square? *81 in²*

Include this activity in your portfolio.

A scientist measures a rat's performance rating on a task after measuring various anxiety levels in the rat. The data is listed in the chart below.

Anxiety level	2	4	6	8	10	12	14
Performance rating	31.5	36	38.5	39	37.5	34	28.5

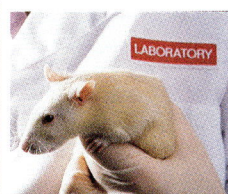

1. Make a scatter plot of the data. Use x for anxiety level and y for performance rating.
2. Use differences to determine the anxiety level when the performance rating is 0.
3. On a graphics calculator, use quadratic regression to find the equation that best fits the data.
4. Find the anxiety level that maximizes the performance rating.
5. Use the equation to find anxiety levels for performance ratings of at least 35.

WORKING ON THE CHAPTER PROJECT

You should now be able to complete the Chapter Project on page 516.

Error Analysis

Students often shade the incorrect region when graphing quadratic inequalities. Remind them to substitute values from each region into the inequality to verify that the correct region was shaded.

Look Beyond

Exercise 47 is a nonroutine problem. Suggest that students draw a diagram or use guess-and-check to solve.

ALTERNATIVE
Assessment

Portfolio Activity

The Portfolio Activity can be used as preparation for the Chapter Project or as a separate activity. In the Portfolio Activity on this page, students use a graphics calculator to find a quadratic regression equation for a given set of data.

Answers to Portfolio Activities can be found in Additional Answers of the Teacher's Edition.

Student Technology Guide

NAME _____ CLASS _____ DATE _____

Student Technology Guide
10.6 Graphing Quadratic Inequalities

You can use a graphics calculator to graph a quadratic inequality.

Example: Graph the quadratic inequality $y < x^2 - 7x + 8$. Tell whether the boundary line should be solid or dashed.

- Enter the equation for the boundary line of the inequality. Here, the boundary line is $y = x^2 - 7x + 8$. Enter this equation as Y1 by pressing [Y=] [X,T,θ,n] [x²] [−] 7 [X,T,θ,n] [+] 8.

Use the left arrow [◄] key to move the cursor to the left of the Y1 label. Since this inequality is satisfied by y-values *less than* $x^2 - 7x + 8$, press [ENTER] until you see a flashing triangle pointing down and to the left. Refer to the diagram at the left below. (For an inequality involving > or ≥, the triangle should point up and to the right.)

- Graph the inequality by pressing [GRAPH]. The result for $y < x^2 - 7x + 8$ is shown at right below.

 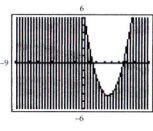

- A graphics calculator always shows a solid boundary line with the shading feature. You will need to look at the inequality to decide whether points on the boundary line satisfy the inequality. In this example, the inequality involves <, so the boundary should be dashed.

Graph each quadratic inequality. Tell whether the boundary line should be dashed or solid. See answer section for graphs.

1. $y \geq x^2 - 3x + 2$ *solid*
2. $y < 2x^2 + 9x + 5$ *dashed*
3. $y \geq -x^2 - 5x - 8$ *solid*
4. $y \leq \frac{1}{3}x^2 - 2x + 2$ *solid*
5. $y > -8x^2 - 2x + 4$ *dashed*
6. $y \geq 0.04x^2$ *solid*

10.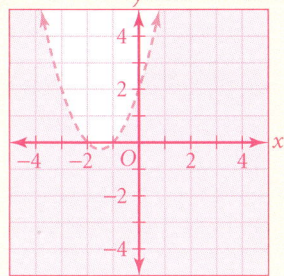

11. $x < 2$ or $x > 4$

12. $-2 < x < -1$

13. $2 \leq x \leq 7$

14. $x < -5$ or $x > 3$

15. $1 < x < 6$

The answers to Exercises 16–35 and 38 can be found in Additional Answers beginning on page 810.

LESSON 10.6 **515**

Project

Focus

When the second differences of a sequence are constant, the set of ordered pairs formed by the natural numbers and the corresponding terms of the sequence represents a quadratic function. Students will learn how to use the method of *finite differences* to write a rule for this function.

Motivate

On the board or overhead, show a diagram that represents 10 bowling pins in the triangular arrangement used in bowling. Point out that there is a "total" of 1 pin in the first row, a total of 3 pins after the second row is added, a total of 6 pins after the third row is added, and a total of 10 pins after the fourth row is added. Write the sequence 1, 3, 6, 10.

Now have students calculate the differences between the totals and write them in order. They should arrive at 2, 3, and 4. Tell them that these are called the *first differences* of the original sequence.

Next, have students find the differences between the first differences. They should arrive at 1 and 1. Tell them that these are the *second differences* of the original sequence.

Have the class compare and discuss the sequence and the sets of first and second differences. Be sure they observe that the second differences are constant.

What's the Difference?

There are many sequences that are generated by a function which you can identify.

Consider the square numbers. If n represents the dots on a side, the sequence shows the total number of dots in a square pattern.

n	1	2	3	4	5	6	7	...	n
$A(n)$	1	4	9	16	25	36	49	...	n^2

A function that can be used to produce this sequence is $A(n) = n^2$. The variable n is usually used in place of x when working with sequences. The variable n represents the number of the term in the sequence. The function rule instructs you to square the number of the term in order to produce the value of the term.

How do you determine the function rule? One method is called **finite differences**. If a sequence eventually produces a constant difference, this method can provide a way to find the function rule for that sequence.

Examine another sequence: 0, 5, 12, 21, 32, 45, 60. You want to find a function rule that can be used to generate any term of this sequence. Begin by finding differences until they are constant.

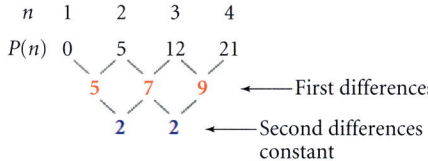

If the sequence produces constant second differences, you can expect a quadratic function. Write the quadratic function in the form of a polynomial function, $P(n) = an^2 + bn + c$. Evaluate the function for values of n from 1 to at least 3. Then make a table of the terms and differences, using the expressions you got by evaluating the general quadratic polynomial.

$P(n) = an^2 + bn + c$
$P(1) = a(1)^2 + b(1) + c = a + b + c$
$P(2) = a(2)^2 + b(2) + c = 4a + 2b + c$
$P(3) = a(3)^2 + b(3) + c = 9a + 3b + c$
$P(4) = a(4)^2 + b(4) + c = 16a + 4b + c$

These function rules correspond to the values of the terms in the actual sequence.

n	1	2	3	4
$P(n)$	$a+b+c$	$4a+2b+c$	$9a+3b+c$	$16a+4b+c$

$3a+b \quad 5a+b \quad 7a+b$ ← First differences

$2a \quad\quad 2a$ ← Second differences constant

Set the corresponding terms and expressions in each layout equal to each other and solve for a, b, and c.

Term 1	First difference	Second difference
$a+b+c=0$	$3a+b=5$	$2a=2$

Use substitution to find the values of a, b, and c.

Use $2a=2$ to solve for a. $a=1$

Use $3a+b=5$ and the value $3(1)+b=5$
for a to solve for b. $b=2$

Use $a+b+c=0$ and the values $1+2+c=0$
for a and b to solve for c. $c=-3$

Finally, replace a, b, and c in the quadratic polynomial, $P(n)=an^2+bn+c$, with the values you found, $a=1$, $b=2$, and $c=-3$.

$$P(n)=n^2+2n-3$$

Check your rule by substituting values for n to see if it generates the appropriate sequence.

$$0, 5, 12, 21, 32, 45, 60, \ldots$$

 **Find a Quadratic Function**

Form groups of 2 to 5 students, and choose a leader. The leader begins the activity by choosing a quadratic function with integer coefficients and generating a sequence of five numbers. The leader then shows only the sequence to the other students in the group. The group then tries to find the quadratic function that generated the sequence.

 **Find the Quadratic Function for Hexagonal Numbers**

The design of the nested hexagons in the building shown at right can be represented by the sequence of hexagonal numbers. Let n represent the number of dots per side. The terms of the sequence represent the total number of dots in each figure. Write the first four terms of the sequence of hexagonal numbers. Then use finite differences to identify the quadratic function that generates the sequence.

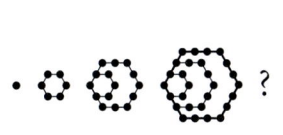

Cooperative Learning
Have students work in groups of two to five, as suggested in **Activity 1**. Begin **Activity 2** by having each group member work individually to identify the function. Each group can then come together to compare answers, resolve any differences, and arrive at a consensus on the function for hexagonal numbers.

Discuss
Bring the class together to discuss their results. Have students suggest other sequences that might be modeled by a quadratic function.

Tell students that the method of finite differences is a topic studied in a field of mathematics called *discrete mathematics*. Have them research the topics encompassed by discrete mathematics and how they are used in business and industry.

Activity 1
Answers may vary. Check students' work.

Activity 2
The first terms of the sequence of hexagonal numbers are 1, 6, 15, and 28. The sequence is generated by the quadratic function $P(n)=2n^2-n$.

Chapter Review

Chapter Review and Assessment

VOCABULARY

axis of symmetry 481	maximum value 481	radical 487
complete the square 493	minimum value 480	reflection 481
complex numbers 509	parabola 480	square root 487
discriminant 508	principal square root 487	vertex 480
horizontal translation 481	quadratic formula 507	vertex form 482
imaginary numbers 509	quadratic function 480	vertical translation 481
imaginary unit 509	quadratic inequality 511	

Key Skills & Exercises

LESSON 10.1

Key Skills

Find the vertex, axis of symmetry, and zeros of a quadratic function.

Find the vertex, axis of symmetry, and zeros of $y = (x-1)^2 - 4$ and sketch the graph.

The function $y = (x-1)^2 - 4$ is in vertex form, $y = a(x-h)^2 + k$, where (h, k) is the vertex, $x = h$ is the line of symmetry, and the zeros are the solution to $(x-1)^2 - 4 = 0$.

Thus, the vertex is (1, −4), axis of symmetry is $x = 1$.

$y = (x-1)^2 - 4$
$= x^2 - 2x - 3$
$= (x-3)(x+1)$

The zeros are 3 and −1.

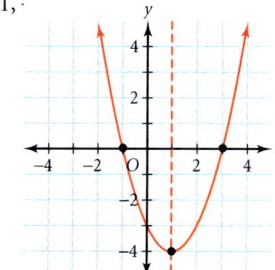

Exercises

Find the vertex and axis of symmetry for the graph of each function. Sketch each graph.

1. $y = (x+1)^2 - 4$
2. $y = 2(x-3)^2 - 2$

Factor each quadratic function to find the zeros.

3. $y = x^2 - 4x + 4$ $x = 2$
4. $y = x^2 - 3x - 10$ $x = -2$ or $x = 5$
5. $y = x^2 + 8x + 16$ $x = -4$
6. $y = x^2 - 25$ $x = -5$ or $x = 5$
7. $y = 2x^2 - 2x - 24$ $x = -3$ or $x = 4$
8. $y = 2x^2 - 6x$ $x = 0$ or $x = 3$

LESSON 10.2

Key Skills

Solve equations of the form $ax^2 = k$.

Solve the equation $(x-1)^2 - 16 = 0$.

Since $(x-1)^2 - 16 = 0$, $(x-1)^2 = 16$. The solutions for an equation of the form $ax^2 = k$ are $x = \sqrt{\frac{k}{a}}$ or $x = -\sqrt{\frac{k}{a}}$. For $(x-1)^2 = 16$, $x - 1 = 4$ or $x - 1 = -4$. Thus, $x = 5$ or $x = -3$. The solutions are 5 and −3.

Exercises

Solve each equation. Round answers to the nearest hundredth when necessary.

9. $x^2 = 25$ $x = \pm 5$
10. $x^2 = \frac{9}{144}$ $x = \pm\frac{1}{4}$
11. $x^2 - 4 = 0$ $x = \pm 2$
12. $x^2 = 12$ $x \approx \pm 3.46$
13. $(x+1)^2 - 4 = 0$ $x = 1$ or $x = -3$
14. $(x+7)^2 - 81 = 0$ $x = 2$ or $x = -16$
15. $(x-3)^2 - 3 = 0$ $x = 4.73$ or $x = 1.27$
16. $(x+5)^2 - 16 = 0$ $x = -1$ or $x = -9$

1. vertex: $(-1, -4)$
   axis of symmetry: $x = -1$

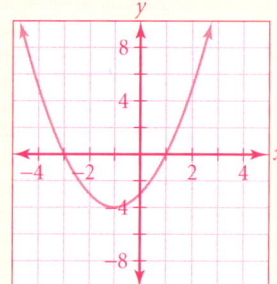

2. vertex: $(3, -2)$
   axis of symmetry: $x = 3$

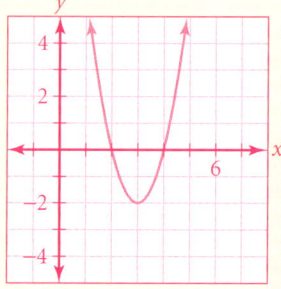

Chapter Test, Form A

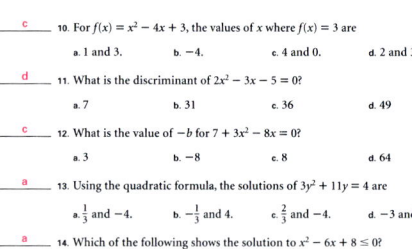

NAME _____ CLASS _____ DATE _____

Chapter Assessment
Chapter 10, Form A, page 1

Write the letter that best answers the question or completes the statement.

___d___ 1. What is the vertex of the graph of $y = \frac{1}{2}(x+6)^2 - 2$?
   a. (3, −2)   b. (−3, −1)   c. (6, −2)   d. (−6, −2)

___a___ 2. The axis of symmetry of the graph of $y = -(x-3)^2 + 5$ is
   a. $x = 3$.   b. $x = -3$.   c. $x = 5$.   d. $x = -5$.

___c___ 3. What is the factored form of $x^2 - 3x - 28$?
   a. $(x+7)(x-4)$ b. $(x-3)(x+28)$
   c. $(x-7)(x+4)$ d. $(x-14)(x+2)$

___c___ 4. The solutions to $(x+3)^2 - 25 = 0$ are
   a. −3 and 25.   b. −3 and 5.
   c. −8 and 2.   d. 8 and −2.

___b___ 5. Which of the following shows $x^2 - 6x + 5 = 0$ in the form $(x-h)^2 + k = 0$?
   a. $(x-6)^2 + 5 = 0$ b. $(x-3)^2 + (-4)$
   c. $(x-3)^2 + 11 = 0$ d. $(x-6)^2 = 0$

___a___ 6. The graph of $y = x^2 - 2x - 3$ intersects the x-axis at
   a. (3, 0) and (−1, 0).   b. (0, 3) and (0, −1).
   c. (−3, 0) and (1, 0).   d. (3, 0) and (1, 0).

___d___ 7. The solutions to $x^2 - 6x - 27 = 0$ are
   a. −9 and 3.   b. 6 and −3.   c. −6 and 3.   d. −3 and 9.

___c___ 8. The zeros of $y = x^2 + 8x + 7$ are
   a. 1 and 7.   b. 7 and 8.
   c. −7 and −1.   d. −7 and −8.

___c___ 9. Which of the following would complete the square for $x^2 + x$?
   a. $\frac{1}{2}$.   b. $-\frac{1}{2}$.   c. $\frac{1}{4}$.   d. $-\frac{1}{4}$.

NAME _____ CLASS _____ DATE _____

Chapter Assessment
Chapter 10, Form A, page 2

___c___ 10. For $f(x) = x^2 - 4x + 3$, the values of x where $f(x) = 3$ are
   a. 1 and 3.   b. −4.   c. 4 and 0.   d. 2 and 3.

___d___ 11. What is the discriminant of $2x^2 - 3x - 5 = 0$?
   a. 7   b. 31   c. 36   d. 49

___c___ 12. What is the value of $-b$ for $7 + 3x^2 - 8x = 0$?
   a. 3   b. −8   c. 8   d. 64

___a___ 13. Using the quadratic formula, the solutions of $3y^2 + 11y = 4$ are
   a. $\frac{1}{3}$ and −4.   b. $-\frac{1}{3}$ and 4.   c. $\frac{2}{3}$ and −4.   d. −3 and 4.

___a___ 14. Which of the following shows the solution to $x^2 - 6x + 8 \le 0$?

___c___ 15. Which inequality is graphed at right?
   a. $y \ge x^2 + 4x + 3$
   b. $y > x^2 + 4x + 3$
   c. $y < x^2 + 4x + 3$
   d. $y \le x^2 + 4x + 3$

___c___ 16. The profit function $p = -x^2 + 18x - 56$ describes a relationship between x, the number of calculators produced (in thousands), and p, the profit (in thousands of dollars). What is the break-even point for the company?
   a. between 4000 and 14,000
   b. less than 4000 or greater than 14,000
   c. 4000 or 14,000
   d. 14,000 only

518 CHAPTER 10 REVIEW

LESSON 10.3

Key Skills

Find the vertex of a parabola by completing the square.

Write $y = x^2 + 4x + 3$ in vertex form by completing the square. Find the vertex.

To complete the square, add and subtract the number that makes $x^2 + 4x + 3$ a perfect-square trinomial. To find this number take half the coefficient of x and square it. Write the function in vertex form, $y = a(x - h)^2 - k$.

$x^2 + 4x = (x^2 + 4x + 4) + 3 - 4$
$\quad\quad\quad\quad = (x + 2)^2 - 1$

The vertex is $(-2, -1)$.

Exercises

Find the number that completes the square.

17. $x^2 + 8x$ 16
18. $x^2 - 3x$ $\frac{9}{4}$

Rewrite each function in the form $y = a(x - h)^2 - k$. Find each vertex.

19. $y = x^2 - 8x + 18$ $y = (x - 4)^2 + 2; (4, 2)$
20. $y = x^2 - 14x + 50$ $y = (x - 7)^2 + 1; (7, 1)$
21. $y = x^2 + 12x + 16$ $y = (x + 6)^2 - 20; (-6, -20)$
22. $y = x^2 - 2x + 5$ $y = (x - 1)^2 + 4; (1, 4)$
23. $y = x^2 - 5x + 2$ $y = (x - \frac{5}{2})^2 - \frac{17}{4}; (\frac{5}{2}, \frac{-17}{4})$

LESSON 10.4

Key Skills

Solve a quadratic equation by completing the square.

Solve $x^2 - 4x + 3 = 0$ by completing the square. Find the solutions of the equation.

$\begin{array}{ll} x^2 - 4x + 3 = 0 & x - 2 = 1 \text{ or } x - 2 = -1 \\ x^2 - 4x = -3 & x = 3 \quad\quad\quad x = 1 \\ x^2 - 4x + 4 = -3 + 4 & \\ (x - 2)^2 = 1 & \end{array}$

The solutions to $x^2 - 4x + 3 = 0$ are 3 and 1.

Exercises

Find the solutions of each quadratic equation by factoring or completing the square.

24. $x^2 + 2x = 24$ $x = 4$ or $x = -6$
25. $x^2 + 5x + 6 = 0$ $x = -2$ or $x = -3$
26. $x^2 - 8x + 16 = 0$ $x = 4$
27. $x^2 - 2x - 35 = 0$ $x = -5$ or $x = 7$
28. $x^2 + 6x - 12 = 0$ $x \approx 1.58$ or $x \approx -7.58$

LESSON 10.5

Key Skills

Use the quadratic formula to solve quadratic equations and functions.

Solve $x^2 + 3x - 14 = 0$. What does the discriminant tell you about the solutions to the quadratic equation?

In the quadratic equation above, $a = 1$, $b = 3$, and $c = -14$. Substitute these values in the quadratic formula, $x = \frac{-b \pm \sqrt{b^2 - 4ac}}{2a}$, and simplify.

$x = \frac{-3 \pm \sqrt{9 - (4 \cdot 1)(-14)}}{2 \cdot 1}$

$\quad = \frac{-3 \pm \sqrt{65}}{2}$

Since the discriminant, $b^2 - 4ac = 65$, is positive, the equation has two real solutions, which are approximately 2.53 and −5.53.

Exercises

Find the value of the discriminant and determine the number of real solutions for each equation.

29. $x^2 + 3x - 6 = 0$ 33; 2 solutions
30. $5x^2 - 6x + 2 = 0$ −4; no real solutions

Use the quadratic formula to solve each quadratic equation. Round your solutions to the nearest hundredth when necessary.

31. $x^2 - 2x - 9 = 0$ $x \approx -2.16$ or $x \approx 4.16$
32. $4x^2 - 5x - 4 = 0$ $x \approx 1.80$ or $x \approx -0.55$
33. $4x^2 - 4x + 1 = 0$ $\frac{1}{2}$
34. $6x^2 - 7x - 3 = 0$ $x = \frac{3}{2}$ or $x = \frac{-1}{3}$
35. $8x^2 + 2x - 1 = 0$ $x = \frac{1}{4}$ or $x = -\frac{1}{2}$

CHAPTER 10 REVIEW   **519**

36. $x \leq -4$ or $x \geq -2$

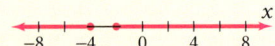

37. $-2 < x < 5$

38. $x < -6$ or $x > -3$

39. $-5 \leq x \leq 3$

40. $x \leq -4$ or $x \geq 7$

41.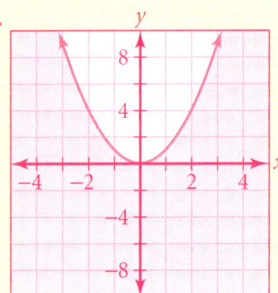

42.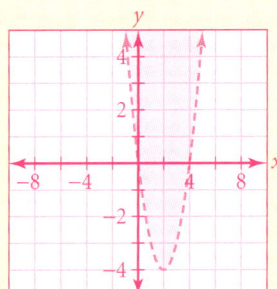

43.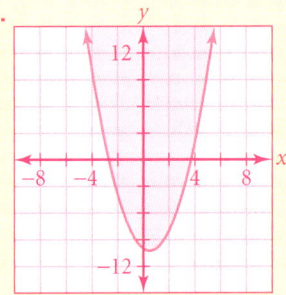

LESSON 10.6
Key Skills

Solve and graph quadratic inequalities.

Solve and graph $x^2 + 3x - 4 > 0$.

To solve $x^2 + 3x - 4 > 0$, find the solutions for $x^2 + 3x - 4 = 0$. Since $x^2 + 3x - 4 = (x+4)(x-1)$, the solutions for the equation are -4 and 1. Since the numbers less than -4 or greater than 1 produce a true inequality, the solutions for $x^2 + 3x - 4 > 0$ are $x < -4$ or $x > 1$.

Graph $y \leq x^2 + 2$.

Graph $y \leq x^2 + 2$. Draw a solid parabola to show that the solutions include the graph of $y = x^2 + 2$. Since the inequality is $\leq$, shade the points below the parabola.

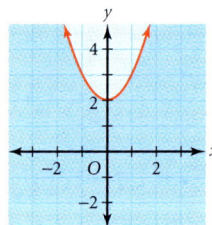

Exercises

Solve each quadratic inequality and graph the solution on a number line.

36. $x^2 + 6x + 8 \geq 0$
37. $x^2 - 3x - 10 < 0$
38. $x^2 + 9x + 18 > 0$
39. $x^2 + 2x - 15 \leq 0$
40. $x^2 - 3x - 28 \geq 0$

Graph each inequality.

41. $y \leq x^2$
42. $y > x^2 - 4x$
43. $y \geq x^2 - x - 10$
44. $y > x^2 + 5x + 6$
45. $y < x^2 + 4x - 12$

Applications

SPORTS The graph at right represents a relationship between the number of yards a football is thrown and its height above the ground.

46. What is the maximum height reached by the football? ≈ 12.5 feet

47. How far does the football travel before it reaches its maximum height? ≈ 15 yards

48. If a receiver catches the ball 30 yards from where the ball was thrown, what is the height of the ball when it is caught? ≈ 5 feet

49. **GEOMETRY** A piece of wrapping paper has an area of 480 square inches. A square piece of paper is cut from the wrapping paper. If the remaining piece of paper has an area of 455 square inches, what is the length of each side of the square piece? 5 in.

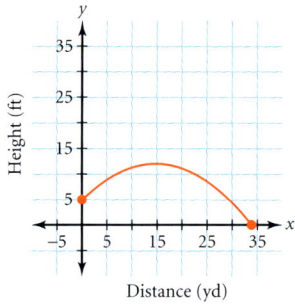

44.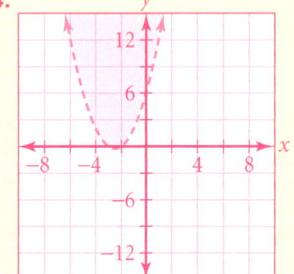

45.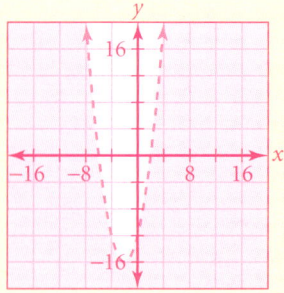

10 Alternative Assessment

Performance Assessment

1. The graph of $y = -(x - 2)^2 - 3$ is a parabola. How can you use the function to find
   a. the vertex of the parabola?
   b. the axis of symmetry of the parabola?
   c. whether the parabola opens up or down?
   d. the maximum or minimum value of the function?

2. Explain how to find the solution to $(2x - 1)^2 = 25$.

3. Use an algebra-tile drawing to complete the square for $x^2 + 8x$. Write the result as a perfect-square trinomial.

4. Explain how to write $y = x^2 + 2x + 1$ in vertex form. Find the following:
   a. the vertex of the parabola
   b. the axis of symmetry of the parabola
   c. the x-intercepts of the parabola

5. a. What is the relationship between the quadratic formula and the discriminant?
   b. What is the relationship between the discriminant and the number of real solutions of a quadratic equation?

6. Write an equation for a quadratic function and draw its graph. Write an equation for and graph on the same coordinate plane
   • a line that intersects the parabola at two points.
   • a line that does not intersect the parabola.

7. Explain how to use substitution to solve the following system of equations:
   $$\begin{cases} y = 2x \\ y = x^2 + 2x + 1 \end{cases}$$

Portfolio Projects

1. Make a table to illustrate how a change in a, b, or c effects the graph of $ax^2 + bx + c$.

2. In the seventeeth century, Galileo designed and experiment to drop two objects of different masses from the Leaning Tower of Pisa. Write a report of Galileo's experiment and how it relates to a quadratic function.

3. Design a poster to show how the solutions to a quadratic equation for the path h depend on the value of the discriminant h equation.

4. The height of an object thrown into the air is modeled by the quadratic function $h = -16t^2 + 64t$, where h is the height of the object and t is time in seconds. Use a graph to describe the time when the height of the object is less than 50 feet.

Peer Assessment

Complete this activity with a partner. Begin by giving your partner a quadratic equation. Your partner should first determine if the equation has real solutions and then find the zeros if they exist.

Check your partner's work and correct any errors. Continue this pattern until you have each solved 10 quadratic equations.

internet connect

The HRW Web site contains many resources to reinforce and expand your knowledge of quadratic functions. The Web site also provides Internet links to other sites where you can find information and real-world data for use in research projects, reports, and activities that involve parabolas and their transformations. Visit the HRW Web site at **go.hrw.com**, and enter the keyword **MA1 CH10**.

Alternative Assessment

Performance Assessment

1. a. $(2, -3)$
   b. $x = 2$
   c. down
   d. -3

2. Take the square root of both sides to get $2x - 1 = \pm 5$. Solve to get $x = 3$ or $x = -2$.

3. $x^2 + 8x + 16$

4. Rewrite the equation as $y = (x + 1)^2 + 0$.
   a. $(-1, 0)$
   b. $x = -1$
   c. -1

5. a. The discriminant is the radicand of the quadratic equation.
   b. The value of the discriminant determines the number of real solutions for the function.

6. Answers may vary.

7. Substitute $2x$ for y into the second equation.
   $2x = x^2 + 2x + 1$
   $x^2 = -1$; there are no real solutions.

1. Answers may vary. Sample answer:

	direction facing	as value increases	as value decreases
$a > 0$	parabola faces upward	parabola shrinks	parabola stretches
$a < 0$	parabola faces downward	parabola shrinks	parabola stretches
b	no effect	parabola shifts left	parabola shifts right
c	no effect	parabola moves up	parabola moves down

2. Students' reports should include that Galileo recognized projectile motion to be parabolic. He also asserted that, regardless of weight or mass, all objects fall at the same rate where there is no air resistance (in a vacuum).

3. Student posters should include the following:

value of the discriminant	number of real solutions
greater than zero	2
equal to zero	1
less than zero	none

4.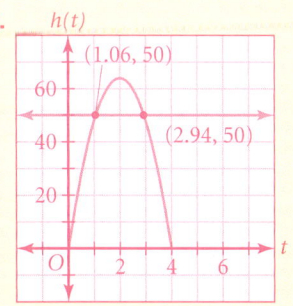

The height is less than 50 feet between 0 and 1.06, seconds, and after 2.94 seconds.

Cumulative Assessment

College Entrance Exam Practice

Multiple-Choice and Quantitative-Comparison Samples

The first half of the Cumulative Assessment contains two types of items found on standardized tests—multiple-choice questions and quantitative-comparison questions. Quantitative-comparison items emphasize the concepts of equality, inequality, and estimation.

Free Response Grid Samples

The second half of the Cumulative Assessment is a free-response section. This part of the Cumulative Assessment requires student-produced response items like those commonly found on college entrance exams. These questions require the use of machine-scored answer grids. You may wish to have students practice answering these items in preparation for standardized tests.

CHAPTERS 1–10 CUMULATIVE ASSESSMENT

College Entrance Exam Practice

QUANTITATIVE COMPARISON For questions 1–5 write
A if the quantity in Column A is greater than the quantity in Column B;
B if the quantity in Column B is greater than the quantity in Column A;
C if the quantities are equal; or
D if the relationship cannot be determined from the information given.

	Column A	Column B	Answers	
1.	$3c(c+1)$ for $c = -2$	$3c^2 + 3c$ for $c = -2$	Ⓐ Ⓑ Ⓒ Ⓓ [Lesson 9.2]	C
2.	slope of $2x + y = 9$	slope of $5x + y = 3$	Ⓐ Ⓑ Ⓒ Ⓓ [Lesson 5.4]	A
3.	perimeter of a square	area of a square	Ⓐ Ⓑ Ⓒ Ⓓ [Lesson 3.6]	D
4.	$(5+5)^2$	$(5^2 + 5^2)$	Ⓐ Ⓑ Ⓒ Ⓓ [Lesson 8.2]	A
5.	$\frac{-6}{3}$	$\frac{-6}{-3}$	Ⓐ Ⓑ Ⓒ Ⓓ [Lesson 2.4]	B

6. Find the vertex for $y = 3(x - 5)^2 + 8$. **(LESSON 10.1)** c
   a. $(-5, 8)$ b. $(5, -8)$
   c. $(5, 8)$ d. $(-5, 8)$

7. A shirt is marked down to $24. The rate of the discount is 20%. What was the original price of the shirt? **(LESSON 4.2)** a
   a. $30   b. $19.20
   c. $4.80   d. $48

8. Simplify: $8 + 6 \cdot 4 \div 2 + 2$. **(LESSON 1.3)** c
   a. 192   b. 9
   c. 22   d. 14

9. Subtract: $5x^3 + x - 1 - (x^2 + x + 3)$.
   (LESSON 9.1) a
   a. $5x^3 - x^2 - 4$ b. $4x^3 - x^2 - 4$
   c. $5x^3 + 2x - 4$ d. $5x - 3$

10. A coin is tossed two times, resulting in one heads, and one tails. What is the experimental probability of one tails? **(LESSON 4.3)** c
    a. $\frac{1}{4}$ b. $\frac{3}{4}$
    c. $\frac{1}{2}$ d. 1

11. Find the slope of the line containing $A(3, 2)$ and $B(3, -4)$. **(LESSON 5.2)** d
    a. 0   b. 1
    c. -6 d. undefined

12. Evaluate $x - (y - z)$ for $x = 2$, $y = -3$, and $z = 4$. **(LESSON 2.3)** a
    a. 9   b. -5
    c. 1   d. 3

13. Solve: $(x + 3)^2 - 16 = 0$. **(LESSON 10.2)** b
    a. -4 and 4   b. -7 and 1
    c. 7 and 1   d. -1 and 7

14. Simplify: $4^{-3} \cdot 4^3$. **(LESSON 8.4)** d
    a. 4   b. 0
    c. 12   d. 1

15. Solve. $\begin{cases} x + y = 5 \\ 2x - 3y = 0 \end{cases}$ (LESSON 7.2) **b**
   a. (2, 3) b. (3, 2)
   c. (−3, −2) d. (−2, −3)

16. Factor: $4x^2 - 4x + 1$. (LESSON 9.7) $(2x - 1)^2$

17. Solve: $36 = t - 24$. (LESSON 3.1) $t = 60$

18. Solve. $\begin{cases} y = 2x + 3 \\ y = -3x + 2 \end{cases}$ (LESSON 7.3) $\left(-\frac{1}{5}, \frac{13}{5}\right)$

19. Solve. $\begin{cases} 2x - 2y = -2 \\ 3x + 3y = 9 \end{cases}$ (LESSON 7.3) (1, 2)

20. Solve: $|x - 8| = 12$. (LESSON 6.5) 20 or −4

21. Write $3y = 5 - 2x$ in standard form. (LESSON 5.5) $2x + 3y = 5$

22. Solve: $\frac{n}{6} = \frac{15}{18}$. (LESSON 4.1) $n = 5$

23. Solve: $2x + 9 \leq 23$. (LESSON 6.2) $x \leq 7$

24. A city has a population of 3 million people. It is estimated that the population will increase at a yearly rate of 1.5%. At this rate, find the estimated population of the city in 5 years. (LESSON 8.7) 3,231,852

25. The third and fourth terms of a sequence are 15 and 23. If the second differences are a constant 2, what are the first seven terms of the sequence? (LESSON 1.1) 5, 9, 15, 23, 33, 45, 59

26. Simplify: $\frac{-100a^8b^{10}}{-20a^5b^9}$. (LESSON 8.3) $5a^3b$

27. Write an equation for a line that passes through the point (−1, 0) and is perpendicular to a line that has a slope of −2. (LESSON 5.6) $y = \frac{1}{2}(x + 1)$

28. If tickets to a concert cost $24 plus a $7 handling charge per order, regardless of how many tickets are ordered, how many tickets could you order for $199? (LESSON 3.3) 8 tickets

29. Factor $4x^2 - 25$. (LESSON 9.6) $(2x + 5)(2x - 5)$

FREE-RESPONSE GRID
The following questions may be answered by using a free-response grid such as that commonly used by standardized test services.

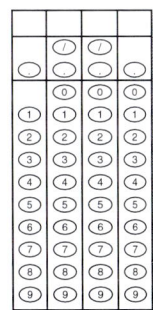

30. Todd scored 86 on his first test and 100 on his second test. What score does Todd need on his third test to have a 92 average for the three tests? (LESSON 4.4) 90

31. The height of an object is expressed by the function $y = -16(x - 5)^2 + 345$, where x is the time in seconds and y is the height in feet. Find the maximum height of the object. (LESSON 10.1) 345 ft

32. Add: $-7.5 + (-1.8)$. (LESSON 2.2) −9.3

33. The cost of 3 lunches and 4 dinners is $60. The cost of 5 lunches and 2 dinners is $44. Find the cost of one dinner. (LESSON 7.3) $12

34. If y varies directly as x and $y = 35$ when $x = 7$, find x when $y = 55$. (LESSON 5.3) $x = 11$

35. Find the mean of 48, 62, 38, and 52. (LESSON 4.4) 50

36. Simplify: $(3x^2y^3)^0$. (LESSON 8.4) 1

37. Find the constant difference for the sequence 1, 2, 6, 13, 23, … (LESSON 1.1) The second differences are a constant 3.

38. Solve: $\frac{2x}{3} = 12$. (LESSON 3.3) $x = 18$

39. Find the value of the discriminant for $x^2 - 6x + 9 = 0$. (LESSON 10.5) 0

40. How many real solutions are there for the equation in Exercise 39? 1 real solution

11 Rational Functions

CHAPTER PLANNING GUIDE

Lesson	11.1	11.2	11.3	11.4	11.5	11.6	Projects and Review
Pupil Edition Pages	526–531	532–537	538–543	546–551	552–557	558–565	544–545, 566–573
Extra Practice (Pupil Edition)	769	770	770	771	771	772	
Practice Workbook	64	65	66	67	68	69	
Reteaching Masters	127–128	129–130	131–132	133–134	135–136	137–138	
Enrichment Masters	64	65	66	67	68	69	
Cooperative-Learning Activities	64	65	66	67	68	69	
Lesson Activities	64	65	66	67	68	69	
Problem Solving/ Critical Thinking	64	65	66	67	68	69	
Student Study Guide	64	65	66	67	68	69	
Spanish Resources	64	65	66	67	68	69	
Student Technology Guide	64	65	66	67	68	69	
Assessment Resources	138	139	140	142	143	144	141, 145–150
Teaching Transparencies		52				53	
Quiz Transparencies	11.1	11.2	11.3	11.4	11.5	11.6	
Writing Activities for Your Portfolio							31–33
Tech Prep Masters							51–54
Long-Term Projects							41–44

LESSON PACING GUIDE

	11.1	11.2	11.3	11.4	11.5	11.6	Projects and Review
Traditional	1 day	1 day	2 days	2 days	2 days	2 days	2 days
Block	$\frac{1}{2}$ day	$\frac{1}{2}$ day	1 day	1 day	1 day	1 day	1 day
Two-Year	2 days	2 days	4 days	4 days	4 days	4 days	4 days

CONNECTIONS AND APPLICATIONS

Lesson	11.1	11.2	11.3	11.4	11.5	11.6	Review
Algebra	526–531	532–537	538–543	546–551	552–557	558–565	568–573
Geometry	530			541, 542			570
Statistics/Probability						565	
Coordinate Geometry, Transformations		536, 537		550	556		
Business and Economics	531	536			554, 557		
Science and Technology	528, 531	533, 535	542	551	554		
Sports and Leisure	531			546	556	565	570
Other	531		543				
Cultural Connection: Africa	527						
Cultural Connection: Americas		537					

BLOCK-SCHEDULING GUIDE

Day	Lesson	Teacher Directed: Lesson Examples, Teaching Transparencies	Student Guided: Activity, Try This	Cooperative-Learning Activity, Lesson Activity, Student Technology Guide	Practice: Practice & Apply, Extra Practice, Practice Workbook	Assessment: Quiz, Mid-Chapter Assessment	Problem Solving, Reteaching
1	11.1	5 min	8 min	8 min	35 min	8 min	8 min
	11.2	5 min	7 min	7 min	30 min	7 min	7 min
2	11.3	10 min	15 min	15 min	65 min	15 min	15 min
3	11.4	15 min	15 min	15 min	65 min	15 min	15 min
4	11.5	10 min	15 min	15 min	65 min	15 min	15 min
5	11.6	15 min	15 min	15 min	65 min	15 min	15 min
6	Assess.	50 min PE: Chapter Review	90 min PE: Chapter Project, Writing Activities	90 min Tech Prep Masters	65 min PE: Chapter Assessment, Test Generator	30 min Chap. Assess. (A or B), Alt. Assess. (A or B), Test Generator	

PE: Pupil's Edition

Alternative Assessment

The following are suggestions for an alternative assessment for students who may benefit from a different type of assessment than the regular chapter quizzes and the mid-chapter and end-of-chapter assessment materials. Many of the questions are open-ended, and students' answers will vary.

Performance Assessment

1. **a.** What does it mean for variables x and y to be in an inverse-variation relationship? Write two inverse-variation relationships using distance, speed, and elapsed time.
   b. Graph one of the inverse-variation relationships you wrote in Part **a**. Describe how the graph could be useful to a motorist.

2. **a.** Explain why $\frac{1}{x+1} = \frac{x-1}{8}$ is a rational equation that is also a proportion.
   b. Solve the equation in Part **a** by using algebra. Are all the solutions found true in the given equation? How do you know?
   c. Solve the equation in Part **a** by graphing. Compare and contrast the solution processes used in parts **b** and **c**.

Portfolio Project

Suggest that students choose one of the following projects for inclusion in their portfolios.

1. Research real-world examples of inverse variation, such as Boyle's law. What quantities are involved in this law and how do they interact?

2. The equation $\frac{t}{80} + \frac{t}{120} = 1$ tells how long it will take for two people to complete a job when one person takes 80 minutes and the other takes 120 minutes to complete the job alone.
   a. How does your knowledge of operations on rational expressions and solving a rational equation help you find t?
   b. Generalize your work in part **a** to find t when you are given other times for two workers working alone.

3. **a.** Suppose that $a \neq 0$. Is $a = \frac{1}{\frac{1}{a}}$ always true? Test to find out. Prove your conjecture by using algebra.
   b. Write an equation involving real numbers a and b that you believe to be true for all a and b. Prove your conjecture by using algebra.

internetconnect

The table below identifies the pages in this chapter that contain technology information and support in the side columns.

Content Links	
Lesson Links	pages 531, 551, 555
Portfolio Links	pages 543, 557
Graphics Calculator Support	page 782

Resource Links

For information about teacher and parent resources as well as professional development help, visit **www.hrw.com/math**.

Technical Support HRW has assembled a team of dedicated technical and teaching professionals and a comprehensive service program to provide you with the support you need.

- The HRW Technical Support Line operates from 7 A.M. to 6 P.M. central time, Monday through Friday, at (800) 323-9239.
- The HRW Technical Support Center on the World Wide Web is available 24 hours a day, seven days a week, at **www.hrwtechsupport.com**.
- You can e-mail our Technical Support Center at **tsc@hrwtechsupport.com**.
- The Technical Support Center's fax-on-demand service at (800) 352-1680 offers solutions to common problems and answers to frequently asked questions.

Technology

Lesson Suggestions and Calculator Examples

(Keystrokes are based on a TI-83 calculator.)

Lesson 11.1 Inverse Variation

In Lesson 5.3, students studied direct variation and saw that the graphical representation is a line. Use the graphics calculator here to show students that equations of the form $xy = k$, where $k > 0$, have curved graphs, as shown below.

 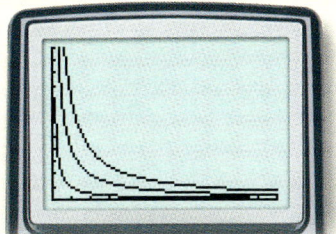

Students should see that each of the curves resulting from $y = \frac{k}{x}$ approach but do not touch the axes.

Lesson 11.2 Rational Expressions and Functions

There are two technology goals for you to consider in this lesson. The first goal is to help students see that rational expressions are not always defined for all real numbers. A table exploration can often show this clearly.

The graphics calculator table at right shows the evaluation of $\frac{x+1}{x^2-x-6}$ for a selected set of x-values. Ask students what the table of values indicates.

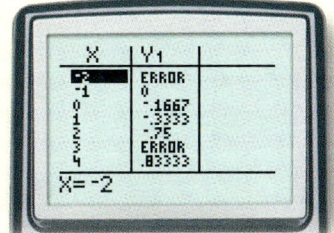

The second goal is to locate from the graph the x-value that gives a specified y-value. To accomplish this:

1. Use [Y=] to enter the function.
2. Use [WINDOW] to adjust the window settings.
3. Use [GRAPH] to display the graph.
4. Use [TRACE] to read the coordinates.

Lesson 11.3 Simplifying Rational Expressions

In this lesson, students will learn the algebraic counterpart to reducing fractions. As an enrichment, pose the following:

When does $\frac{\text{quadratic expression in } x}{\text{linear expression in } x}$ equal a linear expression in x?

Lesson 11.4 Operations With Rational Expressions

As in Lesson 11.3, the objective is to learn the algebraic counterpart to adding, subtracting, multiplying, and dividing rational expressions. Students should do this work with pencil and paper. You might use technology for enrichment. Ask students to find two simple rational expressions whose sum is $\frac{1}{x}$. Students can use the calculator to verify conjectures.

Lesson 11.5 Solving Rational Equations

The graphical solution approach has the advantage that students can see how many solutions are to be found. The displays below accompany the discussion on page 553.

 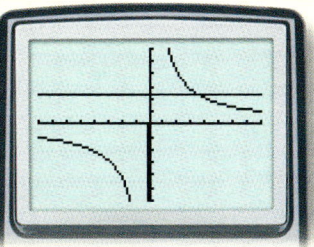

Lesson 11.6 Proof in Algebra

Ask students to use the calculator to test the truth of the following statement:

If $a > 0$ and $b > 0$, $-(a + b) < 0$.

The display at right shows a test for various values of a and b. In each case, the result is negative, so the statement appears to be true.

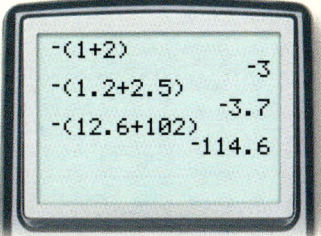

However, students need to understand that although an example is sufficient to *disprove* a statement, examples alone are not sufficient to *prove* a statement.

For further information, refer to the

- technology discussions in the lessons.
- lesson-related teacher's commentary in the side columns of this *Teacher's Edition*.
- lesson-related *Student Technology Guide* masters.
- *HRW Technology Handbook*.

Teaching Resources

Basic Skills Practice
(2 pages per skill)

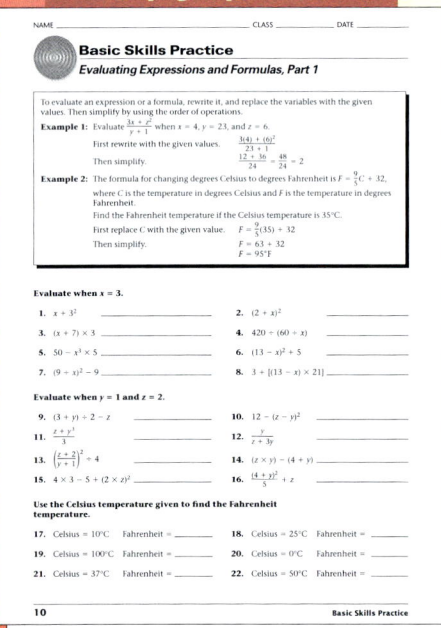

Reteaching
(2 pages per lesson)

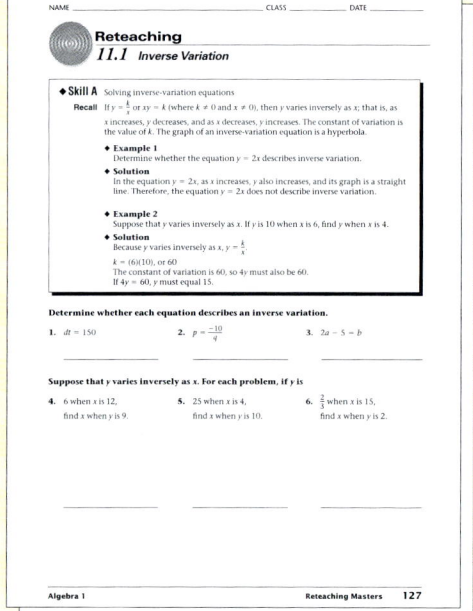

Enrichment
(1 page per lesson)

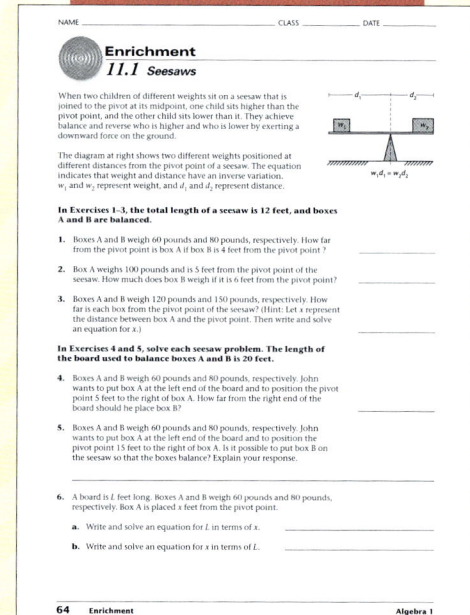

Cooperative-Learning Activity
(1 page per lesson)

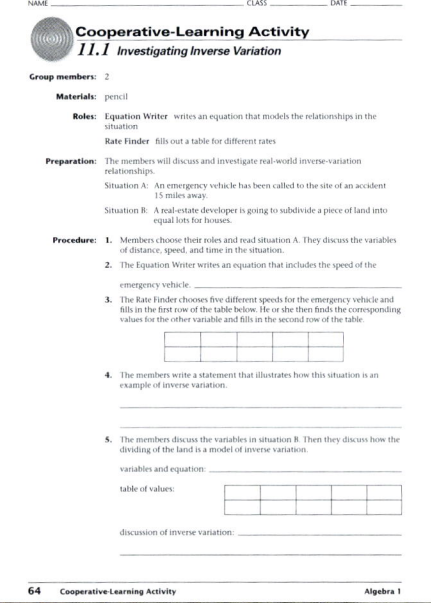

Lesson Activity
(1 page per lesson)

Long-Term Project
(4 pages per chapter)

Problem Solving/Critical Thinking
(1 page per lesson)

Writing Activities
(3 pages per chapter)

Tech Prep Masters
(4 pages per chapter)

Spanish Resources
(1 page per lesson)

Quizzes
(1 page per lesson)

Mid-Chapter Test
(1 page per chapter)

About Chapter 11

Background Information

In Chapter 11, an introductory lesson on inverse variation serves as the gateway to a study of rational expressions and rational functions. Students learn to simplify rational expressions, to perform basic operations with them, and to solve rational equations. The chapter concludes with a study of the absolute-value and integer functions and another look at algebraic proof. **This chapter requires graphics calculators for graphing rational functions.**

CHAPTER RESOURCES

- Block-Scheduling Handbook
- Writing Activities for Your Portfolio
- Tech Prep Masters
- Long-Term Project
- Assessment Resources:
   Mid-Chapter Assessment
   Chapter Assessments
   Alternative Assessments
- Test and Practice Generator
- Technology Handbook
- End-of-Course Exam Practice

Chapter Objectives

- Define and use two different forms of inverse variation to study real-world situations. [11.1]
- Define and illustrate the use of rational expressions and functions. [11.2]
- Factor the numerator and denominator to simplify rational expressions. [11.3]

Rational Functions

11

HAVE YOU EVER NOTICED THAT IT TAKES LONGER TO row upstream than downstream? Have you ever used a lever to lift a heavy object? Do you know what a gear ratio is?

The speed of a current affects the time spent rowing. The distance from the balance point affects the mechanical advantage of a lever. The gear ratio affects how fast you travel when riding a bicycle.

Rational expressions and functions can be used to model these types of relationships. A rational expression is a fraction that includes two algebraic expressions.

Lessons

11.1 • Inverse Variation

11.2 • Rational Expressions and Functions

11.3 • Simplifying Rational Expressions

11.4 • Operations With Rational Expressions

11.5 • Solving Rational Equations

11.6 • Proof in Algebra

Project Designing a Park

You can make a bicycle easier or more difficult to pedal by adjusting the gear ratios.

Rowing with oars is an example of using levers to propel an object.

About the Photos

Students can use objects that are familiar to them, such as bycicle gears, to help visualize the application of ratios and proportions. Allow students to briefly discuss what they know about bicycle gears.

The images in the photos depict several examples of situations in which rational expressions and functions can be used to model relationships.

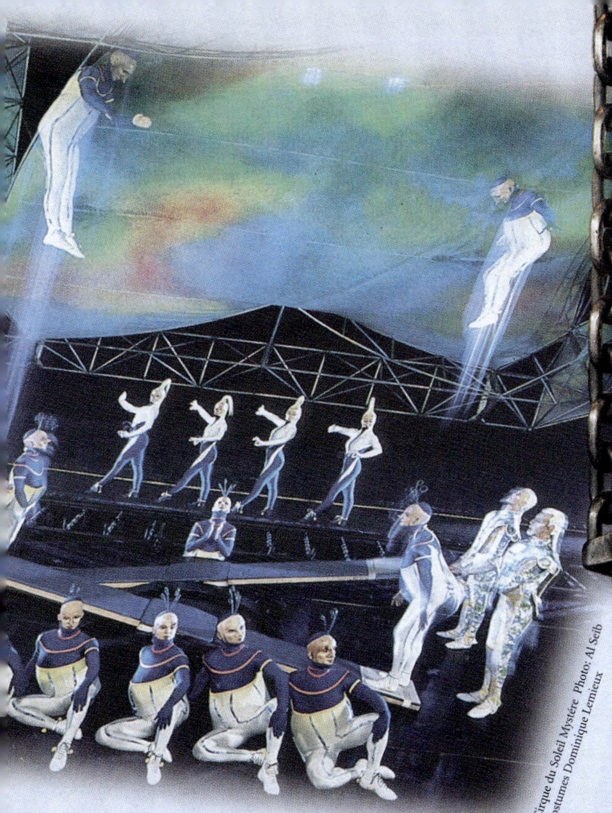

The ratio between the distances from the ends of the board to the balance point affects the height of the acrobats.

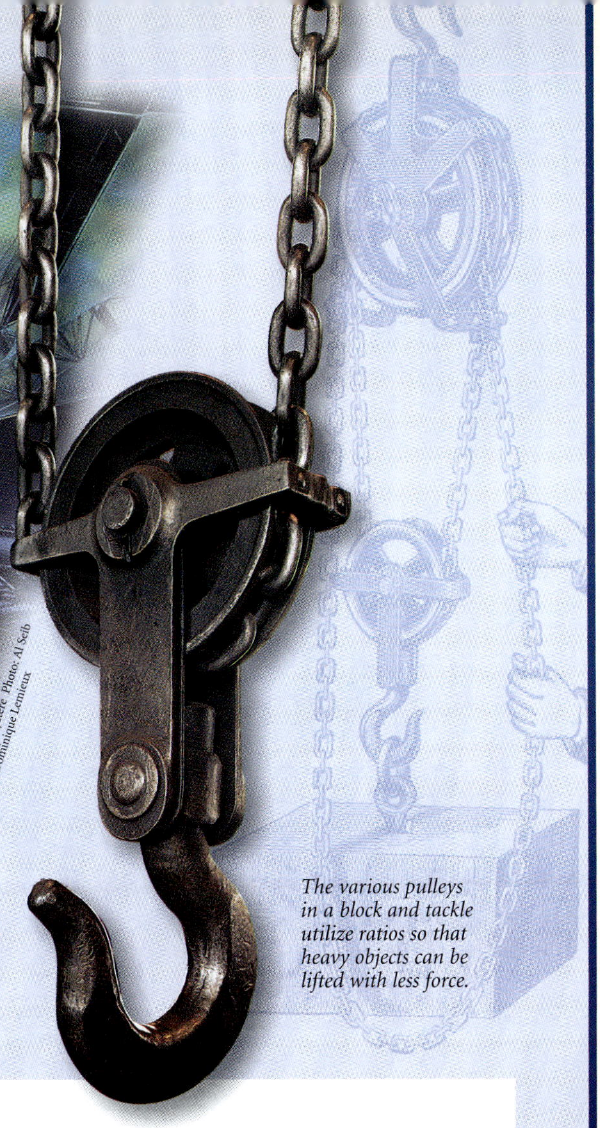

The various pulleys in a block and tackle utilize ratios so that heavy objects can be lifted with less force.

CHAPTER PORTFOLIO ACTIVITIES PROJECT

About the Chapter Project

Many different types of situations can be modeled by rational expressions and functions. In the Chapter Project, *Designing a Park*, on page 566 you will use rational expressions and functions to calculate costs and make design decisions. After completing the Chapter Project, you will be able to do the following:

- Use equations containing rational expressions to solve a real-world problem.

About the Portfolio Activities

Throughout the chapter, you will be given opportunities to complete Portfolio Activities that are designed to support your work on the Chapter Project.

- Balancing a lever is the topic of the Portfolio Activity on page 543.
- The relationship between gears on a bicycle is the topic of the Portfolio Activity on page 557.

- State the restrictions on the variable of a simplified rational expression. **[11.3]**
- Extend simplification techniques to other algebraic fractions. **[11.3]**
- Perform operations with rational expressions in a real-world example. **[11.4]**
- Add, subtract, multiply, and divide rational expressions. **[11.4]**
- Solve rational equations by using the common-denominator method. **[11.5]**
- Solve rational equations by graphing. **[11.5]**
- Define the parts of a conditional statement. **[11.6]**
- Define *converse.* **[11.6]**
- Define *proof.* **[11.6]**
- Prove theorems stated in conditional form. **[11.6]**

CHAPTER PORTFOLIO ACTIVITIES PROJECT

Portfolio Activities appear at the end of Lessons 11.3 and 11.5. Each serves as preparation for the Chapter Project. The Portfolio Activities as well as the Chapter Project Activities are appropriate for inclusion in the student's portfolio. Students should be encouraged to include in their portfolios any other work in which they feel a sense of pride or a sense of accomplishment.

internet connect

Internet connect features appearing throughout the chapter provide keywords for the HRW Web site that lead to links for math resources, tutorial assistance, references for student research, classroom activities, and all teaching resource pages. The HRW Web site also provides updates to the technology used in the text. Listed at right are the keywords for the Internet Connect activities referenced in this chapter. Refer to the side column on the page listed to read about the activity.

LESSON	KEYWORD	PAGE
11.1	MA1 Gear Ratios	531
11.3	MA1 Lever	543
11.4	MA1 Coaster Math	551
11.5	MA1 Rational Functions	555
11.5	MA1 Gears	557

Prepare

NCTM PRINCIPLES & STANDARDS
1–6, 8–10

LESSON RESOURCES

- Practice 11.1
- Reteaching Master 11.1
- Enrichment Master 11.1
- Cooperative-Learning Activity 11.1
- Lesson Activity 11.1
- Problem Solving/Critical Thinking Master 11.1

QUICK WARM-UP

In each case, y varies directly as x.

1. If y is 18 when x is 3, find y when x is 36.   **216**

2. If y is 24 when x is 5, find y when x is 6.   **28.8**

3. If y is 45 when x is 9, find x when y is 90.   **18**

4. If y is 10 when x is 12, find x when y is 42.   **50.4**

5. If y is 3.8 when x is 4, find x when y is 13.3.   **14**

Also on Quiz Transparency 11.1

Teach

Why Have students consider a vehicle traveling at a constant speed: the greater the time elapsed, the greater the distance traveled. Remind them that this illustrates a direct variation. Now have them consider a vehicle that must travel a fixed distance: the greater the constant speed, the less the time elapsed. Tell students that this situation illustrates *inverse variation*. Ask them to suggest other situations that they think involve inverse variation.

526   LESSON 11.1

11.1 Inverse Variation

Objective
- Define and use two different forms of inverse variation to study real-world situations.

Why Recall from Chapter 5 that direct variation describes a linear relationship between two variables. Inverse variation describes a nonlinear relationship between two variables. This type of relationship models many real-world situations, such as comparing the time required to complete a project with the number of people working.

Environmental projects such as tree planting are often made possible by the work of volunteer organizations. As the number of volunteers increases, the time it takes to plant a given number of trees decreases.

In *inverse variation*, as one quantity increases, another quantity decreases. In the activity that follows, you will explore an example of this type of variation.

Activity
Exploring Inverse Variation

A volunteer organization has 5000 trees to plant in a particular area. It is estimated that a team of workers can plant 10 trees per hour.

1. How long would it take 1 team to plant all of the trees?
2. How long would it take 50 teams to plant all of the trees?
3. How long would it take 100 teams to plant all of the trees?

CHECKPOINT ✓ 4. Write a function to represent the time in hours, t, that it takes x teams to plant 5000 trees.

Alternative Teaching Strategy

HANDS-ON STRATEGIES Have students work in groups of three. Give each group a set of 24 unit tiles. Have one student in each group build as many rectangles as possible using all 24 tiles, while a second student makes sketches of the rectangles. Some sample rectangles are shown below. The third student should record the data in a table like the one shown at right.

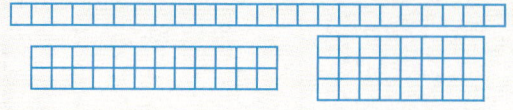

Length	Width	Area
1	24	24
2	12	24
3	8	24
4	6	24
6	4	24
8	3	24
12	2	24
24	1	24

Have each group work together to write an equation for the width, w, of each rectangle in terms of the length, l, and the area, 24.

$$w = \frac{24}{l}$$

Tell students that this is an *inverse-variation* equation and that 24 is the *constant of variation*.

Inverse Variation

In an inverse variation, y is said to vary inversely as x. This is written as $y = \frac{k}{x}$, or $xy = k$, where k is the constant of variation, $k \neq 0$, and $x \neq 0$.

EXAMPLE 1 If y varies inversely as x and $y = 5$ when $x = 15$, write an equation that shows how y is related to x.

SOLUTION

Use the first form of an inverse variation and the information given to find the constant of variation, k.

$y = \frac{k}{x}$ *Equation for inverse variation*

$5 = \frac{k}{15}$ *Substitute the given information.*

$k = 75$

Thus, the equation is $y = \frac{75}{x}$.

TRY THIS If y varies inversely as x and $y = 4$ when $x = 12$, write an equation that shows how y is related to x.

CULTURAL CONNECTION: AFRICA One of the earliest references to inverse variation comes from an ancient Egyptian papyrus that describes how bread was used as a medium of exchange in some Egyptian cities. The value of bread was based on the number of *hekats* of wheat used to make the bread. A rating system was created to assign values to different types of bread. As you can see from the table below, a *lower* rating number meant a *higher* value. This is an example of inverse variation. The table shows the value of 1000 loaves of bread with various ratings.

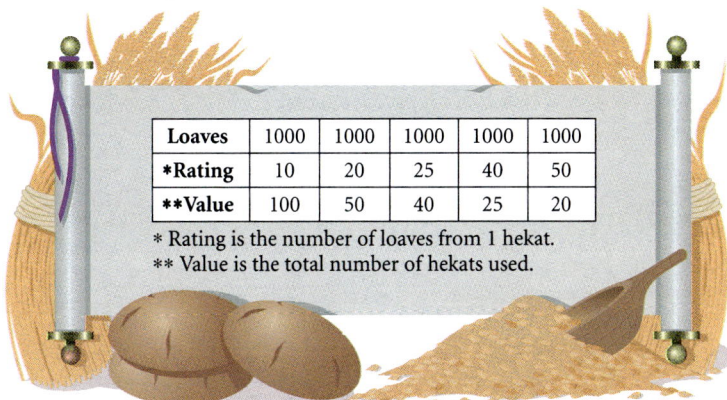

Loaves	1000	1000	1000	1000	1000
*Rating	10	20	25	40	50
**Value	100	50	40	25	20

* Rating is the number of loaves from 1 hekat.
** Value is the total number of hekats used.

Notice that the product of the rating and the value is always 1000. Thus, the relationship between the quantities in the table is an example of inverse variation of the form $xy = k$, or in this case, rating • value = 1000.

Activity Notes

In this Activity, students investigate a real-world situation that exemplifies an inverse-variation relationship. They should observe that the total number of trees is a constant and that the total time required varies inversely as the number of teams involved.

CHECKPOINT ✓

4. $t = \frac{500}{x}$

ADDITIONAL EXAMPLE 1

If y varies inversely as x and $y = 7$ when $x = 13$, write an equation that shows how y is related to x. Find the constant of variation.

$y = \frac{91}{x}$; $k = 91$

TRY THIS

$y = \frac{48}{x}$; $k = 48$

Interdisciplinary Connection

PHYSICS According to Ohm's law, the electrical *current* passing through a metal conductor varies inversely as the *resistance* of the conductor. The constant of variation is the *potential difference* of the conductor. The potential difference, usually measured in volts, also is called the *voltage*.

In a certain conductor, when the resistance is 1.2 ohms, the current is 5 amperes. Find the current when the resistance is 0.8 ohms. **7.5 amperes** What is the voltage of this conductor? **6 volts**

Inclusion Strategies

KINESTHETIC LEARNERS Students can participate in a demonstration of inverse variation. Have one student make 120 marks while another times the job. Then have two students each make 60 marks simultaneously, while another times them. Repeat with teams of three, four, five, six, eight, ten, and twelve students sharing the job of making 120 marks and others timing them. All students should keep track of the team sizes and the times. Then work with the class to model the data with a graph and an equation.

LESSON 11.1 **527**

ADDITIONAL EXAMPLE 2

Using the information in the table below, write an equation that shows the relationship between the variables x and y. If the relationship is an inverse variation, find the constant of variation, k.

x	y
1	36
2	18
3	12
4	9
5	7.2
6	6

$xy = 36$, or $y = \frac{36}{x}$; $k = 36$

TRY THIS

$pq = 60$, or $q = \frac{60}{p}$

EXAMPLE 2

APPLICATION: PHYSICS

The physics student at right is investigating the properties of levers. On one side of the lever, a 30-gram mass is fixed 10 centimeters from the fulcrum. On the other side of the lever, distances from the fulcrum are marked in 5-centimeter intervals along the lever arm. The student is using the guess-and-check method to find the mass required at each distance to balance the 30-gram mass. The results of the investigation are shown in the table below.

Distance of mass from fulcrum	5	10	15	20	25
Mass required to balance 30 grams	60	30	20	15	12

a. Examine the products of the corresponding distances and masses. Does there seem to be an inverse variation between these variables? Explain.

b. Write an equation that relates the variables. If an inverse variation exists, find the constant of variation, k.

c. Graph your equation.

SOLUTION

a. The table seems to illustrate an inverse variation because the product of distance and mass is always 300.

b. The inverse variation is represented by the following:

   distance from fulcrum • mass = k

   For 20 centimeters and 15 grams:

   $20 \cdot 15 = 300$

   Thus, the constant of variation, k, is 300.

   An equation that represents this inverse variation is $d \cdot m = 300$.

c. If $d \cdot m = 300$, then $m = \frac{300}{d}$.
Use the information in the table to plot points contained in the graph. Sketch the graph using the points as a guide.

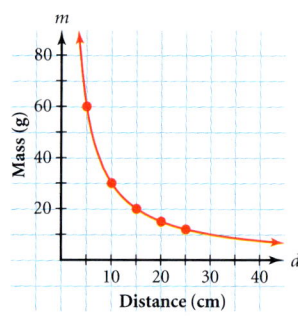

TRY THIS

Using the information in the table below, write an equation that shows the relationship between the variables p and q.

p	3	4	5	6
q	20	15	12	10

Enrichment

When a quantity varies directly as one quantity and inversely as another, the relationship is called a *combined variation*. Suppose that y varies directly as x and inversely as z.

1. When $x = 4$ and $z = 6$, $y = 12$.
   Find y when $x = 9$ and $z = 36$.
   (Hint: Use the equation $y = \frac{kx}{z}$.) $y = 4.5$

2. When $x = 6$ and $z = 15$, $y = 1.6$.
   Find z when $x = 8$ and $y = 10$. $z = 3.2$

Reteaching the Lesson

USING ALGORITHMS Give students the following *Product Rule for Inverse Variations*:

If (x_1, y_1) and (x_2, y_2) are solutions of the inverse variation equation $y = \frac{k}{x}$, then $x_1 y_1 = x_2 y_2$.

Show them how to apply this rule to inverse-variation problems such as the following:

If y varies inversely as x and $y = 4$ when $x = 3.5$, find y when $x = 2.5$.

$x_1 y_1 = x_2 y_2 \rightarrow (3.5)(4) = (2.5)(y_2) \rightarrow y_2 = 5.6$

528 LESSON 11.1

EXAMPLE 3

APPLICATION
INVESTMENT

The amount of time, t, in years that it takes to double a given amount of money with interest compounded annually can be approximated by the function $t = \frac{72}{r}$, where r is the annual interest rate. In this application, percent is not converted to decimal form. For example, if the annual interest rate is 7%, then $r = 7$. In business, this function is usually referred to as the **rule of 72**.

a. About how long does it take to double your money at 6% interest compounded annually?

b. Graph the function for the rule of 72.

● **SOLUTION**

a. To use the rule of 72, substitute 6 for r: $t = \frac{72}{6} = 12$. It takes about 12 years to double your money at 6% interest.

b. Choose values for r. Make a table and sketch the graph from the points.

Rule of 72

r	t
4	18
6	12
8	9
12	6

TRY THIS Use the rule of 72 to estimate how long it takes to double your money at 10% interest compounded annually.

ADDITIONAL EXAMPLE 3

Use the rule of 72 to estimate how long it takes to double your money if it is invested in an account that earns 4.5% annual interest, compounded annually.

about 16 years

TRY THIS
about 7.2 years

Assess

Selected Answers
Exercises 5–14, 15–39 odd

ASSIGNMENT GUIDE

In Class	1–14
Core	15–28, 29–35 odd
Core Plus	15–36
Review	37–40
Preview	41

✎ Extra Practice can be found beginning on page 738.

Exercises

● **Communicate**

1. Write two equivalent equations that express an inverse variation when y varies inversely as x.

2. Explain what an inverse variation between rate and time means. Give a real-world example of an inverse variation involving these variables.

3. If y is 3 when x is 8 and y varies inversely as x, explain how to find y when x is 2.

4. Explain how to use the rule of 72 to estimate how long it would take to double your money at 4% interest compounded annually.

LESSON 11.1 **529**

Technology

A calculator may be helpful for Exercises 23–36. You may wish to have students first complete the exercises with paper and pencil, writing their answers both in simplified form and in calculator-ready form. They can then use a calculator to verify that their simplified answers are correct.

Error Analysis

Some students have difficulty when an inverse-variation equation involves a variable in a denominator. In Exercise 23, for instance, they may arrive at $12 = \frac{48}{x}$ and then be unsure how to proceed. If this occurs, suggest that they write the equation as $\frac{12}{1} = \frac{48}{x}$ and use the Cross-Products Property to obtain the more "comfortable" equation $12x = 48$.

Guided Skills Practice

If y varies inversely as x, write an equation that shows how y is related to x for each situation below. *(EXAMPLE 1)*

5. y is 12 when x is 60. $y = \frac{720}{x}$

6. y is 3 when x is 4. $y = \frac{12}{x}$

7. y is 8 when x is 4. $y = \frac{32}{x}$

Use the information in each table of values to write an equation that shows the relationship between the variables. Tell whether the relationship is an inverse variation. *(EXAMPLE 2)*

8.

p	2	3	4
q	12	8	6

$pq = 24$, inv. var.

9.

p	4	15	20	60
q	30	8	6	2

$pq = 120$, inv. var.

10.

p	2	3	4	5
q	10	9	8	7

$p + q = 12$, not inv. var.

11.

p	5	15	25	75
q	45	15	9	3

$pq = 225$, inv. var.

Use the rule of 72 to estimate the amount of time that it takes to double a given amount of money at each interest rate, compounded annually. *(EXAMPLE 3)*

12. 8% 9 years   **13.** 4% 18 years   **14.** 12% 6 years

Practice and Apply

Determine whether each equation represents an inverse variation.

15. $rt = 400$ Yes   **16.** $y = \frac{-28}{x}$ Yes   **17.** $y = 10 - x$ No   **18.** $\frac{n}{5} = \frac{3}{m}$ Yes

19. $\frac{x}{y} = \frac{1}{2}$ No   **20.** $x = 10y$ No   **21.** $a = \frac{42}{b}$ Yes   **22.** $r = t$ No

For Exercises 23–28, y varies inversely as x.

23. If y is 8 when x is 6, find x when y is 12. 4

24. If y is 9 when x is 12, find y when x is 36. 3

25. If y is 3 when x is 32, find x when y is 4. 24

26. If y is 3 when x is −8, find x when y is −4. 6

27. If y is $\frac{3}{5}$ when x is −60, find y when x is 2. −18

28. If y is $\frac{3}{4}$ when x is 12, find y when x is 27. $\frac{1}{3}$

CONNECTIONS

29. GEOMETRY If the area of a triangle is constant, the length of the base varies inversely as the height. When the base is 22 centimeters, the height is 36 centimeters. What should the length of the base be when the height is 24 centimeters in order to maintain constant area? 33 cm

30. GEOMETRY The area of a rectangle is 36 square centimeters. What is the length of a rectangle with the same area and a width of 3 centimeters? Given that the area remains constant, explain why the formula for the area of a rectangle is an inverse variation.

30. 12 centimeters; The area of this rectangle is a model of inverse variation because $l \times w = k$, where k is 36.

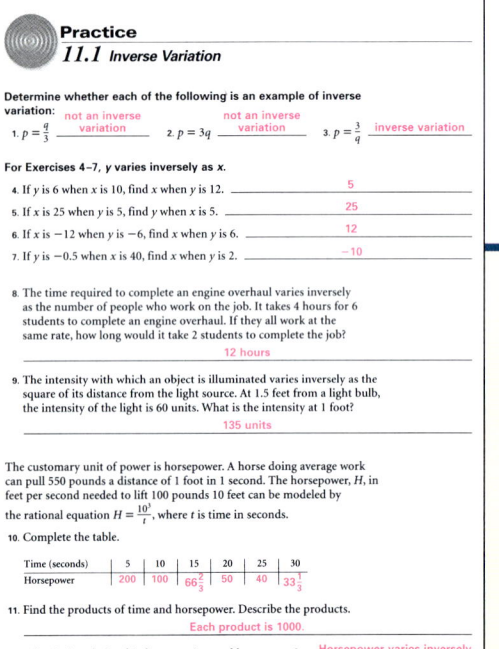

530   LESSON 11.1

APPLICATIONS

31. **MECHANICS** The speed at which a gear revolves varies inversely as the number of teeth on the gear. If a gear with 16 teeth revolves at a speed of 500 revolutions per minute, at what speed should a gear with 20 teeth revolve? **400 rpm**

32. **INVESTMENTS** Harold invests $500 of his 4-H prize money at 5% interest compounded annually. Use the rule of 72 to determine how long will it take him to double his money. **Approx. 15 years**

TRAVEL Time varies inversely as the average rate of speed over a given distance.

33. Mike travels 6 hours at an average rate of 80 kilometers per hour. Find the time required to travel the same distance if Mike's average rate were 90 kilometers per hour. **5 hrs. 20 min.**

34. A jet takes about 2.7 hours to fly from New York to Paris, averaging 2200 kilometers per hour. How long would it take a plane traveling at an average speed of 1760 kilometers per hour to make the same trip? **3.375 hr.**

35. **PHYSICS** The frequency of a radio wave varies inversely as its wavelength. If a 200-meter wave has a frequency of 3000 kilocycles, what is the wavelength of a wave that has a frequency of 2000 kilocycles? **300 m**

36. **MUSIC** A harp string vibrates to produce sound. The number of vibrations varies inversely as the length of the string. If a string that is 28 centimeters long vibrates 518 times per second, how long is a string that vibrates 370 times per second? **39.2 cm**

Look Back

37. The first three terms of a sequence are 1, 5, and 12. The second difference is a constant 3. What are the next three terms of the sequence? *(LESSON 1.1)* **22, 35, 51**

38. A picture of an award-winning cat is surrounded by a frame with a width of $3x$ and a length of $4x$. What is the perimeter of the frame in terms of x? *(LESSON 1.2)* **14x**

39. The total cost for an order of computer discs, including the $1.95 handling charge, is $57.60. The discs are sold in boxes of 10 and are on sale for $7.95 per box. How many discs are in the order? *(LESSON 3.3)* **70 discs**

40. What are the coordinates of the vertex of $y = -x^2 - 4x$? *(LESSON 10.1)* **(−2, 4)**

Look Beyond

41. The intensity, I, of light varies inversely as the square of the distance, d^2, from the light source. If I is 30 units when d is 6 meters, find I when d is 3 meters. **120 units**

Look Beyond

Exercise 41 extends the concept of inverse variation to a situation in which one quantity varies inversely as the *square* of another. You may need to help students set up this variation equation, using either $y = \frac{k}{x^2}$ or $x^2 y = k$. Students will study this type of variation in greater depth if they take a more advanced course in algebra.

internetconnect

GO TO: go.hrw.com
KEYWORD: MA1 Gear Ratios

In this activity, students learn more about inverse variation by investigating gears of different sizes.

Student Technology Guide

NAME _____ CLASS _____ DATE _____

Student Technology Guide
11.1 Inverse Variation

In Lesson 11.1, you saw that two quantities are inversely related if their product is a nonzero constant. You can use a graphics calculator to determine if quantities are inversely related.

Example: Tell whether the data in the table below represents an inverse variation. If it does, write an equation that shows the relationship between x and y.

x	15	24	28	35	45
y	840	525	450	360	280

- First, enter the data. Press STAT 1:Edit.. ENTER. Enter the x-values into L1 and then press ENTER.
- Use the right arrow ▶ key to move to L2. Enter the y-values into L2 and then press ENTER.
- Use the right arrow ▶ key to move to L3. Use the up arrow ▲ key to move to the L3 label, as shown at left below. Press 2nd 1 × 2nd 2 to make each number in L3 become the product of the corresponding numbers in L1 and L2. Press ENTER.
- If all numbers in L3 are equal, the data represents an inverse variation, and the number in L3 is the constant of variation.

The data represents an inverse variation relationship whose equation is $xy = 12{,}600$.

Tell whether each data set represents an inverse variation. If it does, write an equation that shows the relationship between the variables.

1.
x	10	25	40	50
y	40	30	20	10

does not represent an inverse variation

2.
x	120	180	240	540
y	18	12	9	4

inverse variation, $xy = 2160$

Prepare

NCTM PRINCIPLES & STANDARDS
1–3, 5, 6, 8–10

LESSON RESOURCES

- Practice 11.2
- Reteaching Master 11.2
- Enrichment Master 11.2
- Cooperative-Learning Activity 11.2
- Lesson Activity 11.2
- Problem Solving/Critical Thinking Master 11.2
- Teaching Transparency 52

QUICK WARM-UP

Solve each equation.

1. $4y - 15 = 0$ $y = 3.75$
2. $5c^2 - 20c = 0$
                $c = 0$ or $c = 4$
3. $m^2 - 2m + 1 = 0$ $m = 1$
4. $a^2 - 7a - 18 = 0$
               $a = -2$ or $a = 9$

Also on Quiz Transparency 11.2

Teach

Why Display the expression $\frac{100}{n}$ on the board or overhead. Ask students to describe some situations that the expression might represent. Samples: the amount received by each of n people who share a $100 prize equally; the time required to travel 100 miles at a constant speed of n miles per hour Tell students that this is an example of a rational expression.

CRITICAL THINKING

Sample answer: They can be thought of as ratios of polynomials.

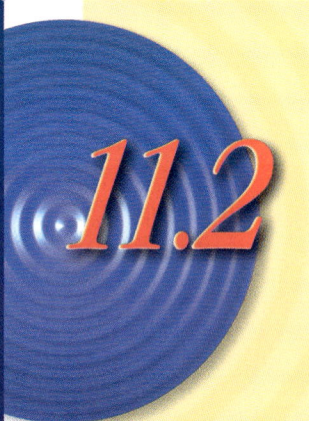

11.2 Rational Expressions and Functions

Objective
- Define and illustrate the use of rational expressions and functions.

Why *Rational functions provide convenient ways of expressing and solving some interesting and complex problems, such as calculating the average cost of pottery lessons.*

A small pottery studio gives pottery lessons. The charge includes a $55 initial fee plus $4.50 for each lesson.

A **rational expression** is an expression that can be written as a quotient of two polynomials.

Rational Expressions

If P and Q are polynomials and $Q \neq 0$, then an expression of the form $\frac{P}{Q}$ is a rational expression.

You can create a rational expression to represent the average cost of a pottery lesson.

If a student takes n pottery lessons, the total cost of the lessons will be $55 + 4.50n$. The average cost of a lesson can be found by dividing the total cost by the number of lessons taken. The result is the following rational expression:

$$\frac{55 + 4.50n}{n}$$

Notice that both the numerator and the denominator are first degree polynomials in the variable n.

CRITICAL THINKING   Why do you think such expressions are called *rational*?

Alternative Teaching Strategy

CONNECTING TO PRIOR KNOWLEDGE Review with students the meaning of the domain of a function. Ask them to give some examples of rules for functions whose domain could be the set of all real numbers. Lead them to responses such as $f(x) = 2x + 1$ and $g(x) = x^2$.

Now ask if they think it is possible for there to be a function whose domain could include any real number except 0. Elicit answers such as $f(x) = \frac{1}{x}$, $g(x) = \frac{1}{2x}$, and $h(x) = \frac{1}{x^2}$.

Continue with questions such as the one below.

Is it possible for there to be a function whose domain could include any real number except the following?

2   sample: $f(x) = \frac{1}{x - 2}$

0 or 2   sample: $g(x) = \frac{1}{x(x - 2)} = \frac{1}{x^2 - 2x}$

–1 or 2   sample: $h(x) = \frac{1}{(x + 1)(x - 2)} = \frac{1}{x^2 - x - 2}$

532 LESSON 11.2

Rational Functions

A function of the form $y = \frac{P}{Q}$, where $\frac{P}{Q}$ is a rational expression, is a **rational function**. The average cost per pottery lesson is as follows:

$$A = \frac{55 + 4.50n}{n}$$

In the following activity you will explore the domain and range of the rational function for the average cost of a pottery lesson shown above.

Activity
Graphing a Rational Function

You will need: graph paper or a graphics calculator

TECHNOLOGY
GRAPHICS CALCULATOR

1. Copy the table below and use your calculator to fill in the missing values.

Number of lessons, n	Average cost of a lesson, $A = \frac{55 + 4.50n}{n}$	Cost
1	$A = \frac{55 + 4.50(1)}{1}$	59.50
2	?	?
3	?	?
4	?	?
5	?	?
10	?	?
15	?	?
20	?	?
30	?	?
40	?	?

2. Does the function seem to be defined for all positive values of n? What happens when $n = 0$? Explain.

3. Negative and non-integer values of n would have no meaning for the pottery-lesson example, but is the function defined for negative and non-integer values of n? Complete the table for two negative and two non-integer values of n.

4. Graph the function with your graphics calculator, or sketch the graph from the data in your table. (If you use your table data, you may want to complete the table for additional values of n.)

CHECKPOINT ✓
5. Describe the shape of your graph near the vertical line $n = 0$. As n gets larger, what happens to A?

Interdisciplinary Connection

HEALTH Suppose that the normal adult dose of a certain medicine is 500 milligrams. Using a method called *Young's rule*, you can calculate the appropriate dose of this medicine for a child by using the function $f(x) = \frac{500x}{x + 12}$, where x represents the child's age in years. Use this function to calculate the appropriate dose of the medicine for an 8-year-old child. **200 milligrams** What do you think is a reasonable domain for this function in the given situation? **possible answer: $2 \leq x \leq 18$**

Inclusion Strategies

ENGLISH LANGUAGE DEVELOPMENT Many students who have language difficulties or for whom English is a second language may have trouble understanding some of the words used in this lesson, such as *asymptote, trivial, non-trivial,* and *rational*. Have these students write each word and its definition on a separate index card. Encourage them to keep the index cards handy when they do their assignments so that they can easily refer to the definitions.

 Notes

In this Activity, students investigate a rational function by using a table of values and a graph. They should discover that the values of the function *approach* 0, but will never actually include 0. The graph approaches the *y*-axis from both the positive and negative directions, but never touches the *y*-axis.

Cooperative Learning

Have students do the Activity in pairs. In Step 1, have one student operate the calculator while the other fills in the table. Partners should switch roles when extending the table in Step 3.

CHECKPOINT ✓
5. As the graph approaches $n = 0$ from either side, it sharply turns away from the *x*-axis and steadily approaches, but never reaches, the vertical line $n = 0$. As n gets larger, $A(n)$ gets closer to the value 4.5.

See Keystroke Guide, page 782.

Teaching Tip

TECHNOLOGY If you are using a TI-82 or TI-83 graphics calculator for the Activity, first press `Y=` and clear any existing equations. Enter **(55+4.50X)/X** next to **Y1**. Then press `2nd` `WINDOW` and make these choices in **TABLE SETUP**.

Indpnt: Auto [Ask]
Depend: [Auto] Ask

Press `2nd` `GRAPH`, and you will see an empty table. Enter the values for *n* in the **X** column, pressing `ENTER` after each. To view the graph, press `WINDOW` and make these settings:

Xmin= –47   Ymin= –31
Xmax=47   Ymax=31
Xscl=5   Yscl=5

Then press `GRAPH`.

LESSON 11.2   **533**

ADDITIONAL EXAMPLE 1

What is the domain of each rational function below?

a. $y = \frac{3}{2x}$

all real numbers except 0

b. $y = \frac{x+2}{x(x-2)}$

all real numbers except 0 and 2

c. $y = \frac{x+5}{x^2+x-20}$

all real numbers except −5 and 4

TRY THIS

a. all real numbers except −1
b. all real numbers except −1 and −2

CRITICAL THINKING

Answers may vary. Sample answers are given below.

trivial: $y = \frac{2x}{3}, y = \frac{14x}{7}$;

non-trivial: $y = \frac{17}{x}, y = \frac{3}{4x}$;

Non-trivial functions have a variable in the denominator.

ADDITIONAL EXAMPLE 2

a. Write the function $y = \frac{3x^2 + 2x}{x^2}$ in simplest terms and graph.

$y = 3 + \frac{2}{x}$

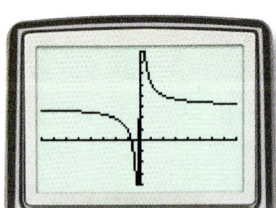

Xmin = −8   Ymin = −4
Xmax = 8   Ymax = 8
Xscl = 1   Yscl = 1

b. Use the graph of the function from part a to find the value of x for which the value of the function is 7. 0.5

534 LESSON 11.2

You may have seen in the activity that a situation may restrict the values that make sense for the domain of a particular function. In general, however, the domain of a rational function consists of all real-numbers that give a nonzero denominator. At each value that makes the denominator zero, the function is undefined, and there is either a vertical asymptote or hole. A **vertical asymptote** is a vertical line that the graph approaches but never crosses. A **hole** is an omitted point in the graph.

EXAMPLE 1

What is the domain of each of the following rational functions?

a. $y = \frac{1}{x}$ b. $y = \frac{x^2 - 4}{x - 2}$ c. $n = \frac{m+2}{m^2 - 5m + 6}$

SOLUTION

a. $y = \frac{1}{x}$ is undefined at $x = 0$. Thus, the domain of the function includes all real numbers except 0.

b. $y = \frac{x^2 - 4}{x - 2}$ is undefined when $x - 2 = 0$. Thus, the domain of the function includes all real numbers except 2.

c. $n = \frac{m+2}{m^2 - 5m + 6}$ is undefined when $m^2 - 5m + 6 = 0$.
Since $m^2 - 5m + 6$ can be factored to $(m-2)(m-3)$, the function is undefined when $m = 2$ or $m = 3$. Thus, the domain of the function is all real numbers except 2 and 3.

TRY THIS What is the domain of each of the following rational functions?

a. $y = \frac{1}{x+1}$ b. $y = \frac{x^2 + 0.5}{x^2 + 3x + 2}$

Non-trivial Rational Functions

The function $y = \frac{1}{x}$, graphed at right, is the simplest **non-trivial** rational function. A function like $y = \frac{x}{2}$, whose denominator is a constant, is a **trivial** rational function because it can be written as $y = \frac{1}{2}x$, which is a polynomial.

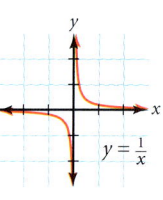

CRITICAL THINKING Name several examples of trivial and non-trivial functions. What are some differences that distinguish them?

EXAMPLE 2

a. Rewrite the function for the average price of a pottery lesson (page 533), $A = \frac{55 + 4.50n}{n}$, in simplest terms and graph.

b. Use the graph of the function from part a to determine how many lessons a person has taken if the average price of the lessons was $5.60.

Enrichment

Write a rational expression in two variables, a and b, for which the given values are excluded. **Sample answers are given.**

1. $a \neq 0$, $\frac{1}{ab}$
   $b \neq 0$

2. $a \neq 0$, $\frac{1}{ab - 3a}$
   $b \neq 3$

3. $a \neq 4$ $\frac{b}{a-4}$

4. $a \neq -2$, $\frac{1}{ab - 4a + 2b - 8}$
   $b \neq 4$

5. $a \neq b$ $\frac{a}{a-b}$

6. $a \neq -b$ $\frac{a}{a+b}$

Reteaching the Lesson

COOPERATIVE LEARNING In pairs, students should write one *non-trivial* rational expression in terms of x that meets the given condition. Partners should check each other's work. **Sample answers are given.**

There is exactly one excluded value of x. $\frac{1}{x+2}, x \neq -2$

There are exactly two excluded values of x. $\frac{1}{x^2 - 3x + 2}, x \neq 1, x \neq 2$

There are no excluded values of x. $\frac{1}{x^2 + 1}$

SOLUTION

a. Rewrite A by applying the definition for adding fractions and simplifying.

$$A = \frac{55 + 4.50n}{n} = \frac{55}{n} + \frac{4.50n}{n} = \frac{55}{n} + 4.5$$

In order to use your graphics calculator to graph the function, you may need to let $A = y$ and $n = x$.

b. Use the graph of the function to find the x-value of the function for which $y = 5.60$.

The x-value of 50 represents 50 lessons. This is the number of lessons for which the average cost is $5.60.

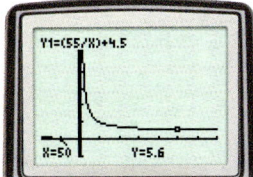

TECHNOLOGY GRAPHICS CALCULATOR

TRY THIS

a. Rewrite the function $y = \frac{83 - 44.3x}{x}$ in simplest terms and graph.

b. Use the graph of the function from part **a** to approximate the value of x for which the function equals -40.

Exercises

Communicate

1. Define the terms *rational expression* and *rational function*. Give two examples of each.
2. Describe how to determine if a rational expression is undefined for any values of x.
3. Discuss how to express the function $y = \frac{12x - 7}{x}$ in simplest form.

Guided Skills Practice

Identify the domain of each function. (EXAMPLE 1)

4. $y = \frac{2}{x-3}$ **5.** $y = \frac{x}{x^2 - 5x}$ **6.** $y = \frac{x^2 - 16}{x + 4}$

Rewrite each function in simplest terms and graph. (EXAMPLE 2)

7. $y = \frac{2 - 3x}{x}$ **8.** $y = \frac{5x + 7x^2}{x^2}$ **9.** $y = \frac{3}{x} - 5$

---

4. domain: all real numbers except 3

5. domain: all real numbers except 0 and 5

6. domain: all real numbers except −4

7. $y = \frac{2}{x} - 3$

8. $y = \frac{5}{x} + 7$

The answer to Exercise 9 can be found in Additional Answers beginning on page 810.

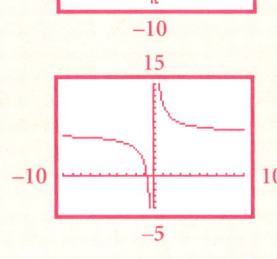

TRY THIS

a. $y = \frac{83}{x} - 44.3$

b. $x \approx 19.15$

Use Teaching Transparency 52.

Teaching Tip

TECHNOLOGY If you are using a TI-82 or TI-83 graphics calculator for Example 2, first press [Y=] and clear any existing equations. Enter **55/X+4.5** next to **Y1**. Press [WINDOW] and make the following settings:

Xmin=−47   Ymin=−31
Xmax=47   Ymax=31
Xscl=5   Yscl=5

Press [GRAPH]. Then press [TRACE] and use the left and right arrow keys to place the cursor at the point on the graph where $y = 5.6$

Assess

Selected Answers
Exercises 4–9, 11–61 odd

ASSIGNMENT GUIDE

In Class	1–9
Core	10–23, 27–32, 37–53 odd
Core Plus	10–50 even, 51–53
Review	54–62
Preview	63, 64

✎ Extra Practice can be found beginning on page 738.

LESSON 11.2 **535**

Technology

A graphics calculator may be helpful for some of the exercises. Depending on window settings, graphics calculators sometimes include asymptotes as part of a graph.

Error Analysis

When specifying the domain of a rational function, some students will exclude values of the variable for which the value of the *numerator* is 0. For example, given $f(x) = \frac{x}{x+1}$, they may write $x \neq 0$ and $x \neq -1$. Remind them that when the numerator of a fraction is 0, the value of the fraction is 0. In the case of the function given above, $\frac{0}{0+1} = \frac{0}{1} = 0$, so 0 is not an excluded value of x.

36. $x \neq 0$

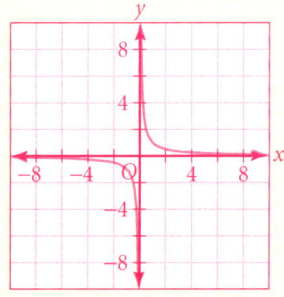

12. $m = \frac{1}{2}$ or $m = 0$
13. $x = 1$ or $x = -5$
14. $x = 2$ or $x = 1$
22. $-3\frac{2}{3}$, undef.
23. undef., $4\frac{2}{3}$
25. $-3\frac{6}{7}$, -9
26. -9, $-7\frac{1}{5}$

45. $y = \frac{4}{x} + 6$
46. $y = \frac{1}{x} - \frac{1}{3}$
47. $y = \frac{1}{2x} - 5$
48. $y = \frac{1}{x} - 1$
49. $y = \frac{9}{x} - 2$
50. $y = \frac{1}{2x} - \frac{9}{2}$

Practice and Apply

For what values of the variable is each rational expression undefined?

10. $\frac{3x+9}{x}$ $x = 0$ 11. $\frac{6}{y-2}$ $y = 2$ 12. $\frac{2m-5}{6m^2-3m}$ 13. $\frac{x^2+2x-3}{x^2+4x-5}$

14. $\frac{x^2-6x+8}{x^2-3x+2}$ 15. $\frac{3x^2+2x}{x^2}$ $x = 0$ 16. $\frac{4z-5}{3z+4}$ $z = -\frac{4}{3}$ 17. $\frac{2l^2+4l}{2l+5}$ $l = -\frac{5}{2}$
$x = 2$ or $x = 1$

Evaluate each rational function or expression for $x = 1$ and $x = -2$. Write *undefined* if appropriate.

18. $y = \frac{1}{x-5}$ $-\frac{1}{4}, -\frac{1}{7}$ 19. $y = \frac{2}{x}$ $2, -1$ 20. $y = \frac{-1}{x} + 3$ $2, 3\frac{1}{2}$

21. $y = \frac{1}{2x+3}$ $\frac{1}{5}, -1$ 22. $y = \frac{4}{x+2} - 5$ 23. $y = \frac{-2}{x-1} + 4$

24. $y = \frac{3}{x+1} + 2$ $3\frac{1}{2}, -1$ 25. $y = \frac{-6}{2x+5} - 3$ 26. $y = \frac{2}{3x-4} - 7$

27. $\frac{5x-1}{x}$ $4, 5.5$ 28. $\frac{4x}{x-1}$ undef., $\frac{8}{3}$ 29. $\frac{x^2-1}{x^2-4}$ 0, undef.

30. $\frac{x^2+2x}{x^2+x+2}$ $\frac{3}{4}, 0$ 31. $\frac{x^2-3}{x^2+4}$ $-\frac{2}{5}, \frac{1}{8}$ 32. $\frac{x^2-3x+4}{x^2+4x+4}$ $\frac{2}{9}$, undef.

33. $\frac{x^2-5}{x^2-4}$ $\frac{4}{3}$, undef. 34. $\frac{3x(x+2)}{x^2-1}$ undef., 0 35. $\frac{4x^2+2x-1}{x^2-2x+1}$ undef., $\frac{11}{9}$

Graph each rational function. List any values of x for which the function is undefined.

36. $y = \frac{1}{x}$ 37. $y = \frac{1}{x-2}$ 38. $y = \frac{1}{2(x-2)}$

39. $y = \frac{3}{2(x-2)}$ 40. $y = \frac{-3}{2(x-2)}$ 41. $y = \frac{-3}{2(x-2)} + 1$

42. $y = \frac{3}{x-1}$ 43. $y = \frac{2}{x-3} + 4$ 44. $y = \frac{-2}{x-4} + 3$

Rewrite each function in simplest terms.

45. $y = \frac{20+30x}{5x}$ 46. $y = \frac{3-x}{3x}$ 47. $y = \frac{x-10x^2}{2x^2}$

48. $y = \frac{x^{100}-x^{101}}{x^{101}}$ 49. $y = \frac{9}{x} - 2$ 50. $y = \frac{1-9x}{2x}$

CONNECTION

51. **COORDINATE GEOMETRY** Graph the trivial function $y = \frac{x+2}{4}$ and the non-trivial function $y = \frac{4}{x+2}$ on the same coordinate grid. List any values for which either function is undefined. List the coordinates of any points of intersection.

APPLICATION

CONSUMER MATH Ms. Kratochvil is financing her new car with a loan of $23,000. At a simple interest rate of 6.5%, the total amount to be repaid is $24,495. Her monthly payment, P, in dollars can be modeled by the rational function $P = \frac{24,495}{m}$, where m is the term of the loan in months.

52. Use graphing technology to graph the rational function P.

53. If Ms. Kratochvil can spend no more than $250 per month on her car payment, what is the shortest possible term for her loan? **98 months**

37. $x \neq 2$

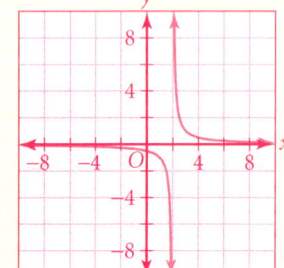

38. $x \neq 2$

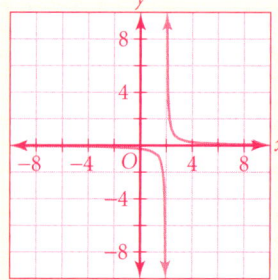

The answers to Exercises 39–44 and 52 can be found in Additional Answers beginning on page 810.

536 LESSON 11.2

 **Look Back**

54. Suppose that your class ordered 500 mugs with all the names of the class members. In the first week, 356 mugs were sold. What percent of the mugs was sold the first week? **(LESSON 4.2)** 71.2%

55. Write an equation for the line that contains the points (2, 1) and (–3, 5). **(LESSON 5.4)** $y - 1 = -\frac{4}{5}(x - 2)$ or $y - 5 = -\frac{4}{5}(x + 3)$

Rewrite each quadratic function in the form $y = (x - h)^2 + k$. (LESSON 10.3)

56. $y = x^2 - 4x + 6$
57. $y = x^2 + 14x + 39$
58. $y = x^2 - 10x + 32$
59. $y = x^2 + 4x + 5$
60. $y = x^2 - 2x - 3$
61. $y = x^2 + 6x - 8$

62. Solve $x^2 - 20 = -x$. **(LESSON 10.4)** $x = -5$ or $x = 4$

 Look Beyond

Benjamin Banneker, 1731–1806

63. **CULTURAL CONNECTION: AMERICAS** Benjamin Banneker, an early African American astronomer, is famous for the almanacs he created from 1791 to 1797. Banneker also enjoyed solving mathematical problems such as the one above, which appeared in his journal. The ladders shown in the drawing are each 60 feet long. The height of one windowsill is 37 feet and the height of the other is 23 feet. If the two ladders meet at the bottom as shown, what is the width to the nearest hundredth of a foot of the street that runs between the two buildings? Use the Pythagorean Theorem (page 592) to solve Banneker's problem. **102.65 ft.**

CONNECTION

64. **TRANSFORMATIONS** Graph the functions $y = x$, $y = x - 2$, and $y = 2x$ on the same coordinate plane. Describe the similarities and differences between the functions.

 Look Beyond

Exercise 63 previews the Pythagorean Theorem. Students will learn about this theorem and some of its applications in Lesson 12.3. Exercise 64 previews transformations, which will be studied further in Chapter 14.

64.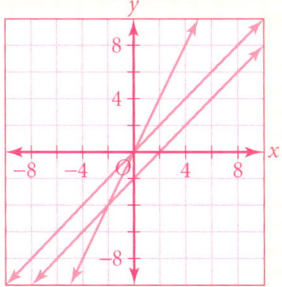

$y = x$ is the parent function. The graph of the line shifts 2 units to the right for $y = x - 2$. The slope becomes steeper for $y = 2x$; $y = x$ and $y = x - 2$ are parallel; $y = x$ and $y = 2x$ pass through the origin.

Student Technology Guide

NAME _____ CLASS _____ DATE _____

Student Technology Guide
11.2 Rational Expressions and Functions

You have seen graphs of the rational parent function $f(x) = \frac{1}{x}$. You can use a graphics calculator to graph other rational functions.

Example: Graph the rational function $f(x) = \frac{x+3}{x-2}$.
- Enter the function. Press [Y=] [( ] [X,T,θ,n] [+] 3 [ )] [÷] [( ] [X,T,θ,n] [–] 2 [ )]. Notice that you need to use parentheses to group the numerator and the denominator.
- Press [GRAPH]. The graph has two separate curved branches and is undefined at $x = 2$.

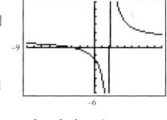

Note: The vertical line is *not* part of the graph. It is there because the calculator is (incorrectly) connecting a point just to the left of $x = 2$ to a point just to the right of $x = 2$.

Graph each rational function. List any values for which the function is undefined. See answer key for graphs.

1. $f(x) = \frac{x-1}{x+3}$ 2. $g(x) = \frac{1}{x-2} + 3$ 3. $f(x) = \frac{3}{2x-6} - 1$ 4. $h(x) = \frac{-2}{x+5}$

undefined at $x = -3$ undefined at $x = 2$ undefined at $x = 3$ undefined at $x = -5$

Example: Describe the transformations applied to the graph of $f(x) = \frac{1}{x}$ to create the graph of $g(x) = \frac{2}{x+6}$.
- Press [Y=] 1 [÷] [X,T,θ,n] to enter $\frac{1}{x}$ as Y1. Move to Y2 and press 2 [÷] [( ] [X,T,θ,n] [+] 6 [ )]. Then press [GRAPH].
- Compare the graphs. The graph of g is a vertical stretch and a leftward shift of the graph of f. Looking at $g(x) = \frac{2}{x+6}$, you can tell that the vertical stretch has a scale factor of 2 and that the leftward shift is 6 units.

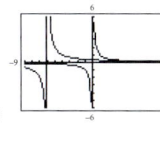

Describe the transformations applied to the graph of $f(x) = \frac{1}{x}$ to create the graph of the following rational functions. List any values for which the function is undefined.

5. $f(x) = \frac{3}{x+3}$ vertical stretch with a scale factor of 3, translated 3 units to the left, undefined at $x = -3$

6. $h(x) = -\frac{2}{x} + 3$ vertical flip, vertical stretch with scale factor 2, translated 3 units up, undefined at $x = 0$

51.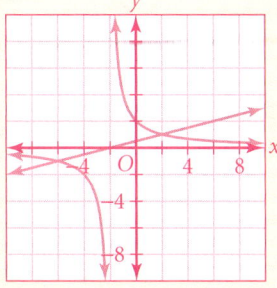

y is undefined at $x = -2$; points of intersection: $(-6, -1)$ and $(2, 1)$.

56. $y = (x - 2)^2 + 2$
57. $y = (x + 7)^2 - 10$
58. $y = (x - 5)^2 + 7$
59. $y = (x + 2)^2 + 1$
60. $y = (x - 1)^2 - 4$
61. $y = (x + 3)^2 - 17$

Prepare

NCTM PRINCIPLES & STANDARDS 1–10

LESSON RESOURCES

- Practice 11.3
- Reteaching Master 11.3
- Enrichment Master 11.3
- Cooperative-Learning Activity 11.3
- Lesson Activity 11.3
- Problem Solving/Critical Thinking Master 11.3

QUICK WARM-UP

Simplify. Write answers with positive exponents only.

1. $\frac{w^4}{w}$ w^3 2. $\frac{3x^2y^5}{15xy^2}$ $\frac{xy^3}{5}$

3. $\frac{a}{a^7}$ $\frac{1}{a^6}$ 4. $\frac{24m^6n^4}{8m^2n^7}$ $\frac{3m^4}{n^3}$

Factor completely.

5. $5k - 35$ $5(k - 7)$

6. $r^2 + 2r - 24$
      $(r + 6)(r - 4)$

7. $x^2 - 1$ $(x + 1)(x - 1)$

8. $3b^2 - 42b + 72$
      $3(b - 2)(b - 12)$

Also on Quiz Transparency 11.3

Teach

Why Ask students why it might be important to simplify a rational number such as $\frac{322}{805}$. Then ask them how they might go about simplifying it. Lead them to a process of factoring and then simplifying. In this case, the result is $\frac{2 \cdot 7 \cdot 23}{5 \cdot 7 \cdot 23}$, or $\frac{2}{5}$. Tell them that the process of simplifying rational expressions is similar.

11.3 Simplifying Rational Expressions

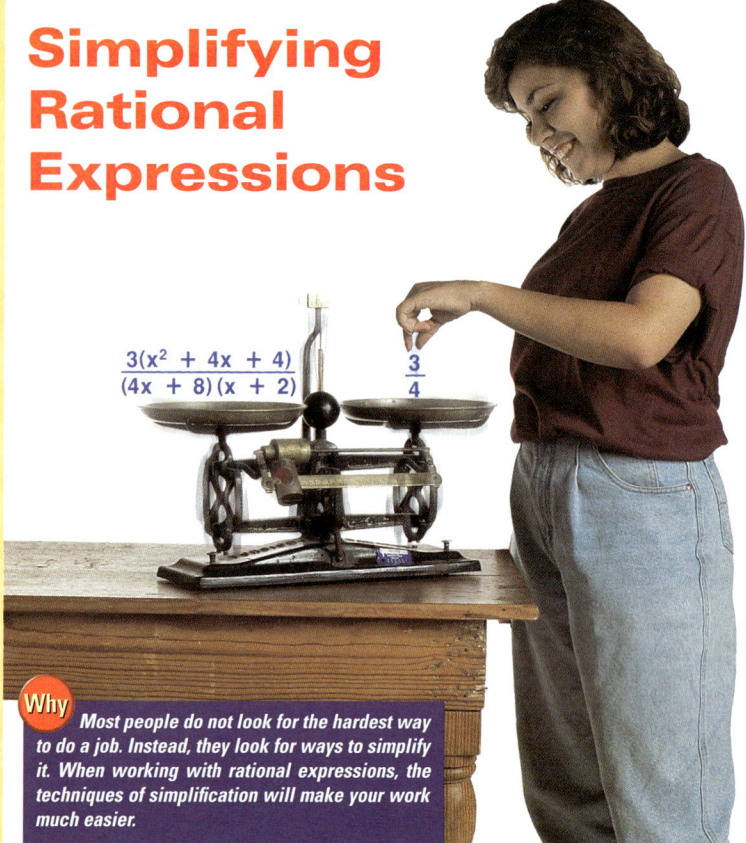

Objectives

- Factor the numerator and denominator to simplify rational expressions.
- State the restrictions on the variable of a simplified rational expression.
- Extend simplification techniques to other algebraic fractions.

Why Most people do not look for the hardest way to do a job. Instead, they look for ways to simplify it. When working with rational expressions, the techniques of simplification will make your work much easier.

Which rational expression is easier to evaluate for $x = 5$?

$$\frac{15x^4 - 120x^3 + 260x^2 - 120x + 245}{3x^5 - 24x^4 + 52x^3 - 24x^2 + 49x} \quad \text{or} \quad \frac{5}{x}$$

Although the two rational expressions are equivalent, the second expression is much easier to evaluate. You can often simplify a rational expression, making it easier to evaluate and graph.

Activity
Exploring Rational Functions

You will need: a graphics calculator or graph paper

Consider the equations $y = \frac{3x + 6}{3x}$ and $y = \frac{x + 2}{x}$.

1. Evaluate both equations for $x = -2$, $x = 5$, and $x = 10$.

CHECKPOINT ✓ 2. What happens if you try to evaluate both equations for $x = 0$?

3. Graph both equations.

CHECKPOINT ✓ 4. Make a hypothesis about the relationship between $y = \frac{3x + 6}{3x}$ and $y = \frac{x + 2}{x}$.

Alternative Teaching Strategy

USING COGNITIVE STRATEGIES Have students work in groups of three or four. Write the following problem on the board or overhead:

$$\frac{2y}{2y + 10} = \underline{?}$$

a. $\frac{1}{10}$ **b.** 10   **c.** $\frac{1}{5}$ **d.** $\frac{y}{y + 5}$, $y \neq -5$

Have groups work together to choose the correct answer to the question. **The correct answer is d.**

Now direct each group to write a convincing argument that supports their answer. After all groups have written their arguments, discuss the results with the class. Elicit from the students their reasons, if any, for eliminating choices **a**, **b**, and **c**. Then repeat the activity with the following problem:

$$\frac{z^2 - 2z - 3}{z^2 - 1} = \underline{?}$$

a. $2z + 3$ **b.** $\frac{z - 3}{z - 1}$ **c.** 3   **d.** $\frac{z - 3}{z - 1}$

$z \neq -1$, $z \neq -1$, $z \neq -1$, $z \neq 1$

$z \neq 1$ $z \neq 1$ $z \neq 1$ $z \neq 3$

The correct choice is b.

Recall the steps you would follow to rewrite $\frac{18}{24}$ in simplest form.

1. First write the numerator and denominator as products of prime factors.
$$\frac{18}{24} = \frac{2 \cdot 3 \cdot 3}{2 \cdot 2 \cdot 2 \cdot 3}$$

2. Separate out the fractions equal to one, and simplify.
$$\frac{2}{2} \cdot \frac{3}{3} \cdot \frac{3}{2 \cdot 2} = \frac{3}{4}$$

The same procedure is used to rewrite a rational expression in simplest form. A rational expression is in simplest form when the numerator and denominator have no common factors other than 1 or −1.

EXAMPLE 1 Simplify $\frac{5a+10}{5a}$. State any restrictions on the variable.

● **SOLUTION**

$\frac{5a+10}{5a} = \frac{5(a+2)}{5a}$ *Factor where possible.*

$= \frac{5}{5} \cdot \frac{a+2}{a}$ *Rewrite to show fractions equal to 1.*

$= \frac{a+2}{a}$, where $a \neq 0$ *Simplify.*

TRY THIS   Simplify $\frac{10-5x}{5x}$. State any restrictions on the variable.

Steps to Simplify Rational Expressions

1. Factor the numerator and denominator.
2. Express each common-factor pair in the numerator and denominator as 1.
3. Simplify and state any restrictions on the variable.

Be sure that only common factors are eliminated when you simplify. Notice the common factors in the examples below.

Expression	Common factor	Simplified form
$\frac{x(x+3)}{2x}$	x	$\frac{x+3}{2}$, $x \neq 0$
$\frac{(x+4)}{x}$	none	cannot be simplified further
$\frac{2x(x+1)}{4x^2}$	$2x$	$\frac{x+1}{2x}$, $x \neq 0$
$\frac{2x^2+3}{x}$	none	cannot be simplified further

Inclusion Strategies

VISUAL LEARNERS Some students will benefit from a display like the one below. Drawing a large 1 around the pairs of common factors gives visual emphasis to the importance of Step 2 in the *Steps to Simplify Rational Expressions* that appear on this page.

$$\frac{a^2 - 3a - 10}{a^2 - 6a + 5} = \frac{(a-5)(a+2)}{(a-5)(a-1)} = \frac{(a+2)}{(a-1)}$$

 **Notes**

In this Activity, students analyze two equations involving rational expressions. After graphing the equations and evaluating them for different values of *x*, students should discover that the equations are equivalent.

Cooperative Learning

You may wish to have students do the Activity in groups of three. Each group member can do one of the substitutions in Step 1, and then the group can share results. Each student should work individually to sketch the graphs in Step 3. The group can then compare graphs and work together to resolve any discrepancies.

CHECKPOINT ✔

2. The value for *y* is undefined when $x = 0$.

4. The functions are the same.

See Keystroke Guide, page 782.

ADDITIONAL EXAMPLE 1

Simplify $\frac{4m-20}{8m}$. State any restrictions on the variable.

$\frac{m-5}{2m}$, where $m \neq 0$

TRY THIS

$\frac{2-x}{x}$, where $x \neq 0$

LESSON 11.3   **539**

ADDITIONAL EXAMPLE 2

Simplify $\frac{7s-28}{s^2+3s-28}$. State any restrictions on the variable.

$\frac{7}{s+7}$, where $s \neq -7$ and $s \neq 4$

TRY THIS

$\frac{1}{x+3}$, where $x \neq -2$ and $x \neq -3$

CRITICAL THINKING

The original expression had a denominator of $(x+3)(x-2)$. For $x = -3$ and $x = 2$, the expression is undefined.

ADDITIONAL EXAMPLE 3

Simplify $\frac{a^2-b^2}{a^2+2ab+b^2}$. State any restrictions on the variables.

$\frac{a-b}{a+b}$, where $a \neq -b$

TRY THIS

$\frac{x-5}{-x+6}$, where $x \neq 6$ and $x \neq -6$

CRITICAL THINKING

While both graphs are straight lines, the graph of $y = \frac{x^2-36}{x+6}$ does not contain the point $(-6, -12)$.

ADDITIONAL EXAMPLE 4

A right cone has a radius of r, a height of h, and a slant height of l. **Find the ratio of its volume, $\frac{1}{3}\pi r^2 h$, to its surface area, $\pi r^2 + \pi r l$.**

$\frac{rh}{3(r+l)}$

When you simplify an expression, the simplified result must carry the *same restrictions on the variable* as the original expression.

EXAMPLE 2 Simplify $\frac{3x-6}{x^2+x-6}$. State any restrictions on the variable.

- **SOLUTION**

$\frac{3x-6}{x^2+x-6} = \frac{3(x-2)}{(x+3)(x-2)}$ *Factor the numerator and denominator.*

$= \frac{x-2}{x-2} \cdot \frac{3}{x+3}$ *Rewrite to show fractions equal to 1.*

$= \frac{3}{x+3}$ *Simplify.*

Since the denominator of the original expression, $x^2 + x - 6$, equals 0 when $x = 2$ or $x = -3$, the restricted values of the variable are 2 and -3.

TRY THIS Simplify $\frac{x+2}{x^2+5x+6}$. State any restrictions on the variable.

CRITICAL THINKING The simplified expression in Example 2 above would have just one restriction, $x \neq -3$. Why must the restriction $x \neq 2$ be added?

EXAMPLE 3 Simplify $\frac{x^2+x-20}{16-x^2}$. State any restrictions on the variable.

- **SOLUTION**

$\frac{x^2+x-20}{16-x^2} = \frac{x^2+x-20}{(-1)(x^2-16)}$ *Factor -1 from the denominator.*

$= \frac{(x+5)(x-4)}{(-1)(x+4)(x-4)}$ *Factor numerator and denominator.*

$= \frac{x+5}{(-1)(x+4)}$ *Simplify.*

$= \frac{x+5}{-x-4}$, where $x \neq 4$ and $x \neq -4$

TRY THIS Simplify $\frac{x^2+x-30}{36-x^2}$. State any restrictions on the variable.

CRITICAL THINKING How do the graphs of $y = \frac{x^2-36}{x+6}$ and $y = x-6$ differ?

Other Algebraic Fractions

In the examples you have studied so far in this lesson, both the numerator and denominator have been polynomials with just a single variable. However, the techniques you have learned can be extended to other algebraic fractions.

Enrichment

Have students work in pairs. Tell them to solve each equation for x in terms of a and to state any restrictions on a.

1. $x(a-2) = a^2 + 6a - 16$ $x = a + 8$, where $a \neq 2$
2. $ax + 3x = a^2 - 9$ $x = a - 3$, where $a \neq -3$
3. $ax - x = a^2 - 6a + 5$ $x = a - 5$, where $a \neq 1$
4. $a^2 + 5a - 4a = ax$ $x = a + 1$, where $a \neq 0$
5. $a^2 - 1 = ax + x$ $x = a - 1$, where $a \neq -1$

Reteaching the Lesson

WORKING BACKWARD Ask students to write a rational expression in terms of m that simplifies to $m - 3$. Tell them to state any restrictions on m.

Samples: $\frac{2m-6}{2}$, no restrictions; $\frac{m^2-4m+3}{m-1}$, where $m \neq 1$

Ask for volunteers to write their expressions on the board. Discuss the expressions with the class. Repeat the activity, this time asking students to write a rational expression in terms of m that simplifies to $\frac{m+2}{m-1}$. Sample: $\frac{2m+4}{2m-2}$

EXAMPLE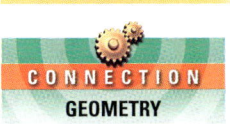

4 Find the ratio of the volume of a right circular cylinder with a radius of r and a height of h to its surface area. Simplify if possible.

● **SOLUTION**

PROBLEM SOLVING

Use a formula. The formula for the volume of a right circular cylinder is $V = \pi r^2 h$. The formula for the surface area of a right circular cylinder is $S = 2\pi r^2 + 2\pi rh$.

$$\frac{\text{volume}}{\text{surface area}} = \frac{\pi r^2 h}{2\pi r^2 + 2\pi rh}$$

$$= \frac{\pi r^2 h}{2\pi r(r + h)}$$

$$= \frac{\pi r}{\pi r} \cdot \frac{rh}{2(r + h)}$$

$$= \frac{rh}{2(r + h)}, \text{ where } r \neq -h \text{ and } r \neq 0$$

Therefore, the ratio $\frac{\text{volume}}{\text{surface area}}$ is $\frac{rh}{2(r + h)}$.

TRY THIS Find the ratio of the surface area of a cube, $6s^2$, to its volume, s^3. Simplify if possible.

CRITICAL THINKING In the final statement of the solution in Example 4, no restrictions on the variable were mentioned. Explain.

Math CONNECTION

GEOMETRY A *cylinder* is a three-dimensional figure that consists of two parallel congruent circular *bases* and a *lateral surface* that connects the boundaries of the bases. If the segment drawn between the centers of the bases is perpendicular to them, then the cylinder is a *right cylinder*.

TRY THIS

$\frac{6}{s}$

CRITICAL THINKING

A restriction would occur if a value for r or h made the denominator, $2(r + h)$, equal to 0. This could only happen if $r = h = 0$ or $r = -h$. Neither of these conditions is possible for a cylinder.

Exercises

Assess

Selected Answers
Exercises 6–17, 19–51 odd

● *Communicate*

1. When is a rational expression undefined?
2. Write a rational expression that is undefined for $x = 7$.
3. Discuss how to find the restrictions on a variable in a rational expression.
4. Explain what is meant by a common factor.
5. Describe the process that is used to write $\frac{x + 1}{x^2 + 2x + 1}$ in simplest form.

ASSIGNMENT GUIDE	
In Class	1–17
Core	18–45
Core Plus	18–47
Review	48–52
Preview	53

✐ Extra Practice can be found beginning on page 738.

Mid-Chapter Assessment for Lessons 11.1 through 11.3 can be found on page 141 of the *Assessment Resources*.

LESSON 11.3 **541**

Technology

You may wish to have students first do Exercises 30–45 with paper and pencil and then use the calculator to simultaneously graph the related rational functions for the given expression and for the simplified expression. If the graphs coincide, the simplification is correct.

Error Analysis

When simplifying rational expressions, students often attempt to divide *terms* rather than factors. For instance, given $\frac{2x+6}{2x}$, they may write 6 as the simplified form. Remind them that they are looking for *factors* that are common to the numerator and denominator. Urge them to check their results by substituting values for the variable into both the original and simplified expressions.

6. $2(b-3)$
7. $\frac{x+4}{2x}, x \neq 0$
8. $\frac{5n-3}{2}, n \neq 0$
9. $\frac{3}{2(x+2)}, x \neq -2$

Guided Skills Practice

Simplify each expression and state any restrictions on the variables.
(EXAMPLES 1, 2, 3, AND 4)

6. $\frac{6b-18}{3}$
7. $\frac{4x+16}{8x}$
8. $\frac{30n^2-18n}{12n}$
9. $\frac{24}{16x+32}$
10. $\frac{14ab}{49ab+98b}$
11. $\frac{2x-10}{x-5}$
12. $\frac{a-6}{a^2-6a}$
13. $\frac{x^2-y^2}{x^2+2xy+y^2}$
14. $\frac{x^2-2x+1}{2x-2}$
15. $\frac{x+4}{x^2+x-12}$
16. $\frac{y^2-49}{y^2+4y-21}$
17. $\frac{m^2+4m}{m^2+m-12}$

Practice and Apply

For what value(s) of the variable is each rational expression undefined?

18. $\frac{10x}{x-3}$ $x=3$
19. $\frac{10}{5-y}$ $y=5$
20. $\frac{r-6}{r}$ $r=0$
21. $\frac{7p}{p-4}$ $p=4$
22. $\frac{k-3}{3-k}$ $k=3$
23. $\frac{(a+3)(a+4)}{(a-3)(a+4)}$ $a=3, -4$
24. $\frac{c-4}{2c-10}$ $c=5$
25. $\frac{3}{y(y^2-5y+6)}$ $y=0, 3, 2$

Name the common factors of the numerator and denominator.

26. $\frac{9}{12}$ 3
27. $\frac{3(x+4)}{6x}$ 3
28. $\frac{x-y}{(x+y)(x-y)}$ $x-y$
29. $\frac{r+3}{r^2+5r+6}$ $r+3$

Simplify each expression and state any restrictions on the variables.

30. $\frac{16(x+1)}{30(x+2)}$
31. $\frac{3(a+b)}{6(a-b)}$
32. $\frac{4(c+2)}{10(2+c)}$
33. $\frac{3(x+y)(x-y)}{6(x+y)}$
34. $\frac{6m+9}{6}$
35. $\frac{7t+21}{t+3}$
36. $\frac{12+8x}{4x}$
37. $\frac{3d^2+d}{3d+1}$
38. $\frac{b+2}{b^2-4}$
39. $\frac{x-2}{x^2+2x-8}$
40. $\frac{-(a+1)}{a^2+8a+7}$
41. $\frac{4-k}{k^2-k-12}$
42. $\frac{c^2-9}{3c+9}$
43. $\frac{3n-12}{n^2-7n+12}$
44. $\frac{y^2+2y-3}{y^2+7y+12}$
45. $\frac{a^2-b^2}{(a+b)^2}$

CONNECTION

46. **GEOMETRY** The area of a certain rectangle is represented by $x^2+7x+12$ and its length is represented by $x+4$. Find the width in terms of x. $x+3$

APPLICATION

47. **BIOLOGY** Two equal populations of animals, one a predator and the other its prey, were released into the wild. Five years later a survey showed that the predator population was 3 times the original predator population, while the prey population was 8 times the original prey population. Then 12 more predators and 32 more prey were released into the area. What is the ratio of predator to prey after the release? Write your answer in simplest form. $\frac{3}{8}$

10. $\frac{2a}{7(a+2)}, a \neq -2$ and $b \neq 0$
11. $2, x \neq 5$
12. $\frac{1}{a}, a \neq 0, 6$
13. $\frac{(x-y)}{(x+y)}, x \neq -y$
14. $\frac{x-1}{2}, x \neq 1$
15. $\frac{1}{x-3}, x \neq -4, 3$
16. $\frac{y-7}{y-3}, y \neq -7, 3$
17. $\frac{m}{m-3}, m \neq -4, 3$

30. $\frac{8(x+1)}{15(x+2)}, x \neq -2$
31. $\frac{a+b}{2(a-b)}, a \neq b$
32. $\frac{2}{5}, c \neq -2$
33. $\frac{x-y}{2}, x \neq -y$
34. $\frac{2m+3}{2}$
35. $7, t \neq -3$
36. $\frac{3+2x}{x}, x \neq 0$
37. $d, d \neq -\frac{1}{3}$

542 LESSON 11.3

 **Look Back**

APPLICATION

48. **DISCOUNTS** Chin-Lyn buys a new sweater at a 20% discount and pays $23.60, before tax. What was the original price of the sweater? **(LESSON 4.2)** $29.50

49. Write an equation for the line that contains the points $\left(\frac{2}{3}, -2\right)$ and $\left(-\frac{5}{6}, 3\frac{1}{3}\right)$. **(LESSON 5.4)** $y = -\frac{32}{9}x + \frac{10}{27}$

50. Multiply: $(x+2)(x-2)$. **(LESSON 9.3)** $x^2 - 4$

51. Write the equation of the parabola whose vertex is (2, 3) and whose graph passes through the point (0, −1). **(LESSON 10.1)** $y = -(x-2)^2 + 3$

52. Which of the following parabolas open downward? **(LESSON 10.1)**
    a. $y = 3x^2 - 8$ b. $y = 9 - x^2$ c. $y = -x^2 + 4$ d. $y = \frac{3}{4}x^2 - 6$
    b and c

 **Look Beyond**

53. Eight squares are the same size, but each is a different color. The squares are stacked on top of each other as shown at right. Identify the order of the squares numerically from top to bottom. The top square is number 1

Include this activity in your portfolio.

Levers can be found in many real-world settings, from playground equipment to scissors in the office and pliers in the garage. When used correctly, levers are useful because the amount of effort required to apply a large force decreases as the lever arm length increases. To see how this works, perform the experiment described in Example 2 on page 528.

1. Use three different fixed masses. Place the balancing masses at five different points along the lever arm for each fixed mass.

2. For each fixed mass, record your data in a table similar to the one shown in the example.

3. Graph your data from each table on separate coordinate planes, and write a short report that describes your results. Be sure to include a function for each fixed mass that describes the relationship between the balanced mass and the distance from the fulcrum.

 **Look Beyond**

Exercise 53 is a nonroutine problem. You might suggest acting out the situation as an appropriate problem-solving strategy.

ALTERNATIVE
Assessment

Portfolio Activity

The Portfolio Activity can be used as preparation for the Chapter Project or as a separate activity. In the Portfolio Activity on this page, students investigate a lever and explore the relationship between mass and distance when the lever is balanced.

 internetconnect

GO TO: go.hrw.com
KEYWORD: MA1 Lever

To support the Portfolio Activity, students may visit the HRW Web site, where they can obtain additional information about levers and experiment with a simple lever.

Student Technology Guide

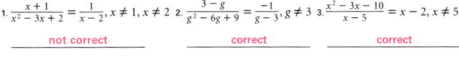

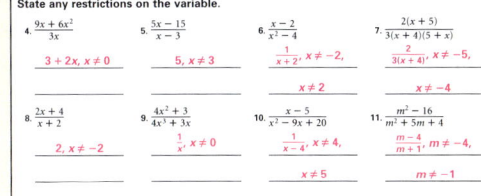

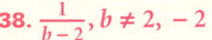

38. $\frac{1}{b-2}, b \neq 2, -2$

39. $\frac{1}{x+4}, x \neq 2, -4$

40. $\frac{-1}{a+7}, a \neq -1, -7$

41. $\frac{-1}{k+3}, k \neq 4, -3$

42. $\frac{c-3}{3}, c \neq -3$

43. $\frac{3}{n-3}, n \neq 3, 4$

44. $\frac{y-1}{y+4}, y \neq -3, -4$

45. $\frac{a-b}{a+b}, a \neq -b$

53. The top-left square is number 2. The squares are numbered counterclockwise, ending with number 8 at the top right.

LESSON 11.3 **543**

Eyewitness Math

Focus

An article that discusses some aspects of the debate over testing people for high iron levels raises an important question: Do the benefits of being tested for high iron levels outweigh the possible psychological trauma of a false positive result? Students use a hypothetical case to model the probabilities that a cancer test will give a true positive or a false positive. Then they determine the likelihood that a person actually has cancer if the test is positive.

Motivate

Before students read the article, ask them if they think that a 98% chance of success is reasonable in most situations. Discuss with students any situations in which a 98% chance of success would be unacceptable. For instance, would it be acceptable for the brakes on an automobile to function properly only 98% of the time?

THE EYEWITNESS MATH

HOW WORRIED SHOULD YOU BE?

NOW, THE TESTING QUESTION
U.S. NEWS & WORLD REPORT, SEPT. 21, 1992

Concerns about iron's role in heart disease, if ultimately confirmed, surely will touch off a debate over iron tests. Who should be tested? And should testing be a matter of public-health policy?...

The virtue of such broad screening would be to help ferret out many of the estimated 32 million Americans who carry a faulty gene that prompts their body to harbor too much iron. There is no test for the gene itself, but anyone with a high blood level of iron is a suspect. Siblings and children would be candidates for testing as well, since the condition is inherited ...

Two widely used blood tests for iron generally cost from $25 to $75 each. The first measures ferritin, a key iron-storing protein ...

The results should always be discussed with a doctor, since some physicians take recent studies seriously enough to worry about even moderately elevated ferritin levels. Moreover, the test might bear repeating. Erroneously high readings are relatively common.

Although people submit to medical tests because they want to find out if something is wrong, the results may cause them to worry. Read the following excerpt about an imaginary case and see how you would react.

Assume that there is a test for cancer which is 98 percent accurate; i.e., if someone has cancer, the test will be positive 98 percent of the time, and if one doesn't have it, the test will be negative 98 percent of the time. Assume further that .5 percent—one out of two hundred people—actually have cancer. Now imagine that you've taken the test and that your doctor somberly informs you that you've tested positive. The question is: How depressed should you be?

To find out whether your reaction would be justified, you need to look at the numbers more closely.

---

1. a. 0.5% of 10,000 = (0.005)(10,000) = 50

b. 98% of 50 = (0.98)(50) = 49

c. 9950; since 0.5% of the 10,000 are assumed to have cancer, then the remaining 99.5% can be assumed to not have it: (0.995)(10,000) = 9950. Or, since 50 out of 10,000 have cancer, 10,000 − 50, or 9950, do not.

d. 199; the test gives correct results 98% of the time and therefore gives incorrect results 2% of the time. (9950)(0.02) = 199

e. 248; 49 true positives plus 199 false positives

f. $P = \frac{49}{248}$, or about 0.2. You would have about a 20% chance of actually having cancer.

g. Answers may vary. For those who originally calculated there would be a 98% chance of having cancer, the 20% figure should be surprising. Though worried still, they should be much less worried with a 20% chance than with a 98% chance.

544 EYEWITNESS MATH

Cooperative Learning

1. Use the data in the previous excerpt to complete steps a–g. Explain each answer.

   a. If 10,000 people are tested, how many would you expect to actually have cancer?

   b. How many of those people that actually have cancer would you expect to test positive? (Such cases are called *true positives*.)

   c. If 10,000 people are tested, how many would you expect not to have cancer?

   d. How many of those people that do not have cancer would you expect to test positive? (Such cases are called *false positives*.)

   e. Add your results from parts **b** and **d** to find the total number of people expected to test positive.

   f. What is the probability, P, that if you tested positive, you actually have cancer?

   g. How does your answer from part **f** compare with your response to the question, How worried should you be?

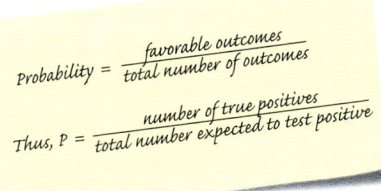

2. You used fictional data in Step 1. To explore what happens with different sets of data, you can find a general formula for P, the probability that if you tested positive, you actually have cancer. Use the following variables:

   N = the number of people tested
   a = the accuracy of the test (If the test is 98% accurate, then $a = 0.98$.)
   r = the portion of people tested that actually have what's being tested for (In Step 1, $r = 0.005$.)

   a. Write a formula for the number of true positives. (Hint: Look back at parts **a** and **b** of Step 1.)

   b. Write a formula for the number of false positives. (Hint: Look back at parts **c** and **d** of Step 1.)

   c. Write a formula for P. (Hint: Look back at parts **e** and **f** of Step 1.)

3. Look at your formula from part c of Step 2.

   a. Find P when $r = 1$. Explain why that result makes sense.

   b. For the situation in Activity 1, suppose that $r = 0.5$ (instead of $r = 0.005$). Would the results in part **f** of Step 1 have been as surprising? Explain.

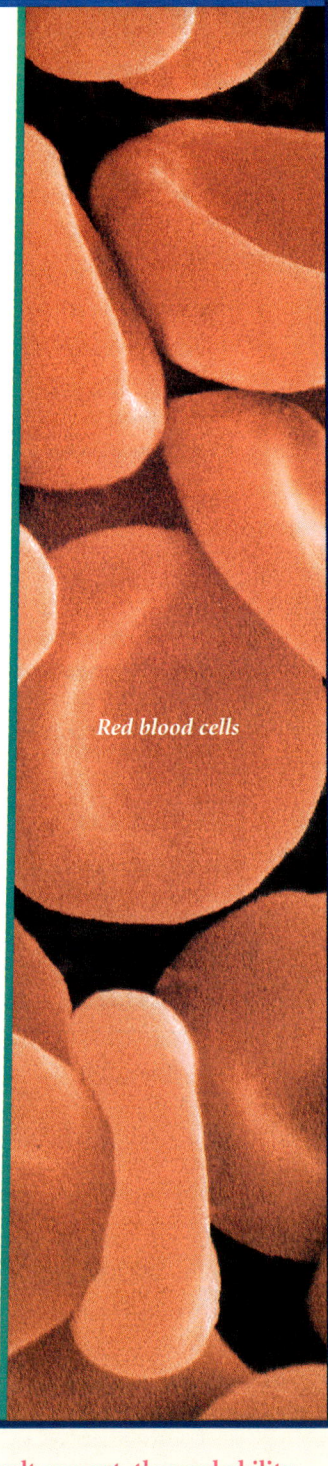

Red blood cells

Cooperative Learning

You may wish to have students do the activities in pairs. In parts **a–c** of **Activity 1**, one partner can operate a calculator while the other records the results. Partners can then switch roles for parts **d–f**. The partners should work together to complete part **g**.

In **Activity 2**, have one student of each pair do part **a** while the other does part **b**. The partners can then work together to write a formula for P in part **c**.

Have the partners do **Activity 3** together.

Discuss

Bring the class back together to discuss the findings of each pair of students. Discuss with students how they would feel if they tested positive in the situation described. Ask questions such as these:

- Would testing positive affect your lifestyle?
- Would the stress of getting a false positive override the benefit of knowing that you might have cancer?

---

2. a. number of true positives = arN
   b. number of false positives =
      $(1 - a)(N - rN)$, or $(1 - a)(1 - r)N$
   c. $P = \dfrac{arN}{(1-a)(1-r)N + arN}$
      $= \dfrac{ar}{(1-a)(1-r) + ar}$

3. a. $P = \dfrac{a(1)}{(1-a)(1-1) + a(1)}$
      $= \dfrac{a}{(1-a)(0) + a} = \dfrac{a}{a} = 1$

   The result makes sense because if $r = 1$, then everyone tested has the disease. So, no matter what result you get, the probability that you have the disease is 1.

   b. $P = \dfrac{ar}{(1-a)(1-r) + ar}$
      $= \dfrac{(0.98)(0.5)}{(1-0.98)(1-0.5) + (0.98)(0.5)}$
      $= \dfrac{0.49}{0.50} = 0.98$

   No, because the probability would have turned out to be the same as the test accuracy.

Prepare

NCTM PRINCIPLES & STANDARDS
1–6, 8–10

LESSON RESOURCES

- Practice 11.4
- Reteaching Master 11.4
- Enrichment Master 11.4
- Cooperative-Learning Activity 11.4
- Lesson Activity 11.4
- Problem Solving/Critical Thinking Master 11.4

QUICK WARM-UP

Evaluate. Write answers in simplest form.

1. $\frac{7}{12} + \frac{1}{12}$ $\frac{2}{3}$
2. $\frac{1}{2} + \frac{2}{3}$ $\frac{7}{6}$, or $1\frac{1}{6}$
3. $\frac{9}{10} - \frac{3}{4}$ $\frac{3}{20}$
4. $\frac{2}{5} \cdot \frac{5}{6} \cdot \frac{4}{5}$ $\frac{4}{15}$
5. $\frac{3}{8} \div \frac{2}{3}$ $\frac{9}{16}$
6. $\frac{3}{5} \div 6$ $\frac{1}{10}$

Also on Quiz Transparency 11.4

Teach

Why Remind students of the water-current puzzles they solved by using systems of equations in Lesson 7.6. Ask them to recall how a current affects the progress of a person or object traveling through the water at a constant speed. Tell them that in this lesson, they will see how current problems are related to rational expressions.

11.4 Operations With Rational Expressions

Objectives

- Perform operations with rational expressions in a real-world example.
- Add, subtract, multiply, and divide rational expressions.

Why Calculating travel times with and against a current is a problem that can be solved by using rational expressions. In solving real-world problems involving rational expressions, you will often need to perform arithmetic operations such as addition, subtraction, multiplication, and division.

APPLICATION
RECREATION

At a rowing club's annual picnic, the "Current Challenge" is a popular event. The challenge requires participants to paddle a kayak up and back down a 1-mile portion of the river. The above kayaker's average rate in still water is 4 times the speed of the river's current. You can use this information to study travel times for different river trips. As you will see, a single rational expression can be used to represent a kayaker's round-trip time.

You can write an equation for the total time it takes to paddle upstream and downstream.

$$\text{total time} = \text{time } \underbrace{\text{upstream}}_{\text{against current}} + \text{time } \underbrace{\text{downstream}}_{\text{with current}}$$

You can rewrite this equation for the total time in terms of the distance and rate.

Alternative Teaching Strategy

CONNECTING TO PRIOR KNOWLEDGE Review with students the addition of fractions.

$$\frac{1}{10} + \frac{1}{15} = \frac{1}{2 \cdot 5} + \frac{1}{3 \cdot 5}$$

$$\frac{3}{2 \cdot 3 \cdot 5} + \frac{2}{2 \cdot 3 \cdot 5} = \frac{3}{30} + \frac{2}{30}$$

$$= \frac{5}{30}$$

$$= \frac{1}{6}$$

Factor the denominators.

Rewrite, using the LCD.

Add the numerators.

Simplify.

Now show how the same steps can be used in the addition of rational expressions.

$$\frac{1}{2x+2} + \frac{1}{3x+3} = \frac{1}{2(x+1)} + \frac{1}{3(x+1)}$$

$$\frac{3}{2 \cdot 3 \cdot (x+1)} + \frac{2}{2 \cdot 3 \cdot (x+1)} = \frac{3}{6(x+1)} + \frac{2}{6(x+1)}$$

$$= \frac{5}{6(x+1)}$$

$$= \frac{5}{6x+6}, x \neq -1$$

546 LESSON 11.4

Since distance = rate • time, time = $\frac{\text{distance}}{\text{rate}}$. Thus, you can rewrite the equation for total time as follows:

$$\text{total time} = \frac{\text{distance}}{\text{rate against current}} + \frac{\text{distance}}{\text{rate with current}}$$

Let x represent the rate of the current. The rate of the kayaker in the illustration on the previous page is 4x. The distance each way is 1 mile. Substituting this information into the above equation gives the following equation:

$$\text{total time} = \frac{1 \text{ mile}}{4x \text{ mph} - x \text{ mph}} + \frac{1 \text{ mile}}{4x \text{ mph} + x \text{ mph}}$$

By combining like terms in the denominator of each rational expression, you obtain the following equation:

$$t = \frac{1}{3x} + \frac{1}{5x}, \text{ where } x \neq 0$$

To find a single rational expression for $\frac{1}{3x} + \frac{1}{5x}$, add them in the same way that you would add fractions with unlike denominators. Compare the methods for adding fractions and rational expressions.

Fractions	Rational expressions	Simplifying fractions and expressions
$\frac{1}{30} + \frac{1}{50}$	$\frac{1}{3x} + \frac{1}{5x}$	Given
$\frac{5}{5} \cdot \frac{1}{30} + \frac{3}{3} \cdot \frac{1}{50}$	$\frac{5}{5} \cdot \frac{1}{3x} + \frac{3}{3} \cdot \frac{1}{5x}$	Find a common denominator.
$\frac{5}{150} + \frac{3}{150}$	$\frac{5}{15x} + \frac{3}{15x}$	Change to common denominators.
$\frac{8}{150}$	$\frac{8}{15x}$	Simplify.

Thus, the total time for the kayaker in the illustration is $\frac{8}{15x}$, where $x \neq 0$.

CRITICAL THINKING

If you knew the time it took the kayaker to complete the trip, how could you find the rate of the current?

Multiplying two rational expressions is relatively simple. Just multiply the numerators and multiply the denominators.

EXAMPLE ❶ Multiply: $\frac{x-2}{x+3} \cdot \frac{x+3}{x-5}$. State any restrictions on the variable.

● **SOLUTION**

$\frac{x-2}{x+3} \cdot \frac{x+3}{x-5} = \frac{(x-2)(x+3)}{(x+3)(x-5)}$ *Multiply the numerators and the denominators.*

$= \frac{x-2}{x-5} \cdot \frac{x+3}{x+3}$ *Separate fractions that equal 1.*

$= \frac{x-2}{x-5}$, where $x \neq 5$ and $x \neq -3$ *Simplify.*

The restrictions on the variable must include the restrictions on the original expressions.

CRITICAL THINKING
Substitute the time for t and solve for x, which is the rate of the current. The kayaker's rate in still water is 4x, or $\frac{32}{15t}$.

ADDITIONAL EXAMPLE ❶

Multiply: $\frac{2n+2}{5n+10} \cdot \frac{n+2}{2n-2}$. State any restrictions on the variable.

$\frac{n+1}{5(n-1)}$, or $\frac{n+1}{5n-5}$, where $n \neq -2$ and $n \neq 1$

Inclusion Strategies

Use a similar procedure for introducing subtraction, multiplication, and division. The following are some examples:

$\frac{2}{3} - \frac{1}{4}; \frac{c}{c^2-2c+1} - \frac{1}{c-1}$ $\frac{5}{12}; \frac{1}{c^2-2c+1}, c \neq 1$

$\frac{1}{3} \cdot \frac{3}{5}; \frac{t-1}{t+3} \cdot \frac{t+3}{t+5}$ $\frac{1}{5}; \frac{t-1}{t+5}, t \neq -3, t \neq -5$

$\frac{3}{4} \div \frac{2}{3}; \frac{z-2}{z+2} \div \frac{z-1}{z+2}$ $\frac{9}{8}; \frac{z-2}{z-1}, z \neq -2, z \neq 1$

LINGUISTIC LEARNERS Some students will feel more comfortable if they outline a step-by-step procedure that they can use for each of the four operations with rational expressions. A sample procedure for multiplication is shown below.

1. Factor all numerators and denominators.
2. Divide numerators and denominators by any common factors.
3. Multiply numerators and then denominators.
4. Simplify numerators and denominators.

ADDITIONAL EXAMPLE 2

Multiply: $\frac{t-6}{t+6} \cdot \frac{4t}{9} \cdot \frac{t+6}{10t^2}$. State any restrictions on the variable.

$\frac{2(t-6)}{45t}$, or $\frac{2t-12}{45t}$, where $t \neq -6$ and $t \neq 0$

TRY THIS

$\frac{5}{2}, x \neq 0$

ADDITIONAL EXAMPLE 3

Divide: $\frac{c+5}{c-8} \div \frac{c-5}{c-8}$. State any restrictions on the variable.

$\frac{c+5}{c-5}$, where $c \neq 5$ and $c \neq 8$

TRY THIS

a. $\frac{x+5}{x-3}$, where $x \neq 2$ and $x \neq 3$
b. $\frac{1}{x+7}$, where $x \neq -7$ and $x \neq 2$

ADDITIONAL EXAMPLE 4

Add: $\frac{z}{z-9} + \frac{2}{z+1}$. State any restrictions on the variable.

$\frac{z^2+3z-18}{z^2-8z-9}$, where $z \neq 9$ and $z \neq -1$

EXAMPLE 2 Multiply: $\frac{x+2}{6x^2} \cdot \frac{x-3}{x+2} \cdot \frac{3x}{5}$. State any restrictions on the variable.

SOLUTION

$\frac{x+2}{6x^2} \cdot \frac{x-3}{x+2} \cdot \frac{3x}{5} = \frac{(x+2)(x-3)(3x)}{(6x^2)(x+2)(5)}$ *Multiply the numerators and the denominators.*

$= \frac{x-3}{(2x)(5)} \cdot \frac{x+2}{x+2} \cdot \frac{3x}{3x}$ *Separate fractions that equal 1.*

$= \frac{x-3}{10x}$, where $x \neq 0$ and $x \neq -2$ *Simplify.*

The restrictions on the variable must include the restrictions on the original expressions.

TRY THIS Multiply: $\frac{7}{x} \cdot \frac{x^2}{14} \cdot \frac{5}{x}$. State any restrictions on the variable.

Recall that division by a fraction is equivalent to multiplication by the reciprocal of the fraction. The same is true for rational expressions.

EXAMPLE 3 Divide: $\frac{x-2}{x} \div \frac{x-2}{x-3}$. State any restrictions on the variable.

SOLUTION

Invert the divisor and multiply:

$\frac{x-2}{x} \div \frac{x-2}{x-3} = \frac{x-2}{x} \cdot \frac{x-3}{x-2}$

$= \frac{x-3}{x}$

The restrictions are $x \neq 0$, $x \neq 3$, and $x \neq 2$. The restrictions must include $x \neq 2$ because the divisor, $\frac{x-2}{x-3}$, is zero when $x = 2$.

TRY THIS Divide the following expressions, and state any restrictions on the variable.

a. $\frac{x+5}{x-2} \div \frac{x-3}{x-2}$ b. $\frac{x-2}{x+7} \div (x-2)$

To add or subtract rational expressions, you will need to find a common denominator. To do this, you will need to use multiplication.

EXAMPLE 4 Add: $\frac{x}{x+2} + \frac{x}{x-1}$. State any restrictions on the variable.

SOLUTION

Multiply each expression by the appropriate equivalent of 1. The resulting expressions must have the same denominator.

$\frac{x}{x+2} + \frac{x}{x-1} = \frac{x-1}{x-1} \cdot \frac{x}{x+2} + \frac{x+2}{x+2} \cdot \frac{x}{x-1}$

Once the expressions have the same denominator, add the numerators.

$\frac{(x-1)x}{(x-1)(x+2)} + \frac{(x+2)x}{(x-1)(x+2)} = \frac{(x^2-x)+(x^2+2x)}{(x-1)(x+2)}$

$= \frac{2x^2+x}{x^2+x-2}$, where $x \neq 1$ and $x \neq -2$

Enrichment

Have students simplify these *continued fractions*. Remind them that the fraction bar is an inclusion symbol. Challenge them to create and simplify their own continued fractions.

1. $\cfrac{1}{2+\cfrac{1}{2+\frac{1}{2}}}$ $\frac{5}{12}$

2. $3 + \cfrac{1}{3+\cfrac{1}{1+\frac{1}{4}}}$ $\frac{62}{19}$

3. $\cfrac{1}{x+\cfrac{1}{x+\frac{1}{x}}}$ $\frac{x^2+1}{x^3+2x}$, where $x \neq 0$

Reteaching the Lesson

COOPERATIVE LEARNING Have students work in pairs. Tell each student to write one addition, one subtraction, one multiplication, and one division involving rational expressions, on separate index cards. Tell them to write each expression with just one variable. For each expression, they should perform the indicated operation and write the simplified sum, difference, product, or quotient on the back of the card. Remind them to include any restrictions on the variable. Have partners trade cards and simplify each other's expressions.

TRY THIS Add: $\frac{2x}{x+3} + \frac{x-1}{x-4}$. State any restrictions on the variable.

TRY THIS
$\frac{3(x^2 - 2x - 1)}{(x+3)(x-4)}$, where $x \neq -3$ and $x \neq 4$

CRITICAL THINKING How do you choose the appropriate equivalents for 1 in an addition problem like the one in Example 4?

CRITICAL THINKING
Decide what factor you have to multiply the denominator of each fraction by in order to get the least common denominator.

EXAMPLE 5 Subtract: $\frac{3a}{a^2 - 4} - \frac{2}{a+2}$. State any restrictions on the variable.

SOLUTION

Since $a^2 - 4 = (a+2)(a-2)$, multiply the second expression by $\frac{a-2}{a-2}$. Then subtract.

$$\frac{3a}{a^2-4} - \frac{2}{a+2} = \frac{3a}{(a+2)(a-2)} - \frac{2(a-2)}{(a+2)(a-2)}$$
$$= \frac{3a - 2a + 4}{(a+2)(a-2)}$$
$$= \frac{a+4}{a^2-4}, \text{ where } a \neq 2 \text{ and } a \neq -2$$

ADDITIONAL EXAMPLE 5

Subtract: $\frac{3a^2 - 9}{a^2 + 6a + 9} - \frac{a-3}{a+3}$. State any restrictions on the variable.

$\frac{2a^2}{a^2 + 6a + 9}$, where $a \neq -3$

TRY THIS Add or subtract the expressions below.

a. $\frac{x}{x^2-1} + \frac{2}{x+1}$ b. $\frac{4x-1}{8x} - \frac{2}{x}$

TRY THIS
a. $\frac{3x-2}{x^2-1}$, where $x \neq -1$ and $x \neq 1$
b. $\frac{4x-17}{8x}$, where $x \neq 0$

When completing operations with rational expressions that involve several steps, it is important to work carefully in order to avoid errors. You can check your work by substituting a number in the place of the variable in the initial and final expressions. The values of the two expressions should equal one another.

Exercises

Communicate

Refer to the expressions $\frac{x}{x+1}$ and $\frac{3}{x}$ for Exercises 1–5.

1. Explain how to find a common denominator for these two rational expressions.
2. Describe how to find the sum of these rational expressions.
3. Describe how to find the difference of the first expression subtracted from the second expression.
4. Describe how to find the product of these rational expressions.
5. Describe how to find the quotient of the first expression divided by the second expression.

Assess

Selected Answers
Exercises 6–20, 21–57 odd

ASSIGNMENT GUIDE

In Class	1–20
Core	21–51 odd
Core Plus	22–50 even, 51–54
Review	55–58
Preview	59–61

✐ Extra Practice can be found beginning on page 738.

LESSON 11.4

Technology

A graphics calculator may be helpful for some of the exercises. For Exercises 21–50, students can simultaneously graph on a calculator the related functions for the given expression and for the simplified expression. If the graphs coincide, the simplification is correct.

Error Analysis

In a subtraction of rational expressions, students may neglect to subtract *each term* in the numerator of the second fraction. Suggest that they enclose the numerator of the second fraction in parentheses as a reminder that they must distribute −1 over each term inside the parentheses.

6. $\frac{x+3}{x-4}, x \neq 4, 5$

7. $\frac{k-4}{k+8}, k \neq -8, -7$

8. $\frac{3(k-4)}{(k+2)}, k \neq -2, 2$

9. $\frac{4(x+4)}{3(x+1)}, x \neq -1, 2$

10. $\frac{2(k+1)}{3(2k+1)}, k \neq -6, -\frac{1}{2}$

11. $\frac{y+3}{2y}, y \neq 0, 9$

Guided Skills Practice

Multiply, and state the restrictions on the variable.
(EXAMPLES 1 AND 2)

6. $\frac{x+3}{x-5} \cdot \frac{x-5}{x-4}$

7. $\frac{k-4}{k+7} \cdot \frac{k+7}{k+8}$

8. $\frac{3k-6}{2k+4} \cdot \frac{2k-8}{k-2}$

Divide, and state the restrictions on the variable. **(EXAMPLE 3)**

9. $\frac{x+4}{3(x-2)} \div \frac{x+1}{4(x-2)}$

10. $\frac{k+1}{3(k+6)} \div \frac{2k+1}{2(k+6)}$

11. $\frac{y+3}{2y-18} \div \frac{y}{y-9}$

Simplify, and state the restrictions on the variable.
(EXAMPLES 4 AND 5)

12. $\frac{x}{x^2-1} + \frac{2}{x+1}$

13. $\frac{x}{x^2-9} + \frac{4}{x-3}$

14. $\frac{k}{k^2-4k+4} + \frac{5}{k-2}$

15. $\frac{z}{z^2+6z+9} + \frac{3}{z+3}$

16. $\frac{y}{y^2-8y+16} + \frac{3}{y-4}$

17. $\frac{x+4}{x-1} - \frac{x-3}{x-1}$

18. $\frac{x}{x+3} - \frac{x}{x-2}$

19. $\frac{x}{x+5} - \frac{x}{x-4}$

20. $\frac{a}{a-1} - \frac{a}{a+1}$

Practice and Apply

Perform the indicated operations. Simplify, and state the restrictions on the variable.

21. $\frac{5}{3x} + \frac{2}{3x}$

22. $\frac{8}{x+1} - \frac{5}{x+1}$

23. $\frac{2}{y+4} \div \frac{5}{y+4}$

24. $\frac{5x+4}{x-2} - \frac{7+3x}{x-2}$

25. $\frac{2}{a} + \frac{3}{a+1}$

26. $\frac{7}{3t} - \frac{8}{2t}$

27. $\frac{5}{2r} + \frac{3}{r}$

28. $\frac{5}{p} \div \frac{3}{p^2}$

29. $\frac{-2}{x+1} + \frac{3}{2(x+1)}$

30. $x - \frac{x-4}{x+4}$

31. $\frac{-3-d}{d-1} + 2$

32. $\frac{1}{b} + \frac{3}{b^2}$

33. $\frac{5}{3+n} \cdot \frac{3+n}{2+n}$

34. $\frac{x+2}{x(x+1)} \cdot \frac{x^2}{(x+2)(x+3)}$

35. $\frac{q^2-1}{q^2} \cdot \frac{q}{q+1}$

36. $\frac{1}{x+1} \cdot \frac{2}{x}$

37. $\frac{3}{x-1} \cdot \frac{5}{x}$

38. $\frac{y}{y-4} - \frac{1}{y}$

39. $\frac{-2}{x+1} + \frac{3}{x}$

40. $\frac{a-2}{a+1} + \frac{5}{a+3}$

41. $\frac{m+5}{m+2} \cdot \frac{m-3}{m-1}$

42. $\frac{2r}{(r+5)(r+1)} - \frac{r}{r+5}$

43. $\frac{3}{b^2+b-6} - \frac{5}{b-2}$

44. $\frac{4}{y-2} + \frac{5y}{y^2-4y+4}$

45. $\frac{2}{y-5} - \frac{5y-3}{y^2+y-30}$

46. $\frac{x}{x-1} - \frac{1}{x-1}$

47. $\frac{2}{x^2-9} - \frac{1}{2x+6}$

48. $\frac{13}{3x+12} + \frac{2x}{x+4}$

49. $\frac{x+5}{3x-21} \div \frac{x}{x-7}$

50. $\frac{4p}{p^2-2p-3} \cdot \frac{p-3}{p+1}$

CONNECTION

51. **COORDINATE GEOMETRY** On the same coordinate plane, sketch the graphs of $y = x$ and $y = (x-4)(x+4)$. Make a table of values for each function. Then sketch the graph of $y = x + (x-4)(x+4)$ on the same coordinate plane by using the values in your tables and looking at the two graphs that you drew. You may want to use a graphics calculator to check your work.

Practice
11.4 Operations With Rational Expressions

Perform the indicated operations. Simplify and state the restrictions on the variables.

1. $\frac{3}{5z} + \frac{4}{5z}$ — $\frac{7}{5z}, z \neq 0$

2. $\frac{6}{x-3} + \frac{2}{x-3}$ — $\frac{8}{x-3}, x \neq 3$

3. $\frac{5a}{a+b} - \frac{2a}{a+b}$ — $\frac{3a}{a+b}, a \neq -b$

4. $\frac{1}{x-1} - \frac{1}{x-1}$ — $1, x \neq 1$

5. $\frac{1}{p} - \frac{2}{q}$ — $\frac{q-2p}{pq}, p$ and $q \neq 0$

6. $\frac{c}{3d} + \frac{d}{3c}$ — $\frac{c^2+d^2}{3cd}, c$ and $d \neq 0$

7. $\frac{4}{mn} \cdot \frac{m}{2}$ — $\frac{2}{n}, m$ and $n \neq 0$

8. $\frac{y^2}{6} \cdot \frac{12}{y^3}$ — $\frac{2}{y}, y \neq 0$

9. $\frac{x^3}{6} + \frac{x^2}{12}$ — $\frac{2x^3+x^2}{12}$

10. $\frac{a^2(a-3)}{3-a}$ — $-a^2, a \neq 3$

11. $\frac{a^3}{(a+1)^2} + \frac{a}{(a+1)^2}$ — $\frac{a^3+a}{(a+1)^2}, a \neq -1$

12. $\frac{t^2+2t}{t^2-4} \cdot \frac{3t-6}{t}$ — $3, t \neq -2, 0,$ or 2

13. $\frac{y^2+5y+6}{y^2} \cdot \frac{y}{y+2}$ — $\frac{y+3}{y}, y \neq -2$ or 0

14. $\frac{3x}{3x-2} - \frac{2}{3x-2}$ — $1, x \neq \frac{2}{3}$

15. $\frac{m^2-n^2}{m^2n^2} \cdot \frac{mn}{1}$ — $\frac{m^2-n^2}{mn}, m$ and $n \neq 0$

16. $\frac{1}{x+1} + \frac{x}{x^2-1}$ — $\frac{2x-1}{x^2-1}, x \neq \pm 1$

17. $\frac{3x^2}{1-x^2} \cdot \frac{2-x}{1-x^2}$ — $\frac{3x-2}{1-x^2}, x \neq \pm 1$

18. $\frac{8b-40}{8} \cdot \frac{b+5}{b^2-25}$ — $1, b \neq \pm 5$

19. $\frac{6}{c^2-3c} \cdot \frac{2c^2-2c-12}{6c+12}$ — $\frac{2}{c}, c \neq -2, 0,$ or 3

20. $\frac{x}{x+y} - \frac{-3xy}{(x+y)^2}$ — $\frac{x^2+4xy}{(x+y)^2}, y \neq -x$

21. $4 + \frac{3x}{x^2-6}$ — $\frac{4x^2+3x-24}{x^2-6}, x \neq \pm\sqrt{6}$

22. $\frac{3}{z-1} + \frac{z}{1-z}$ — $\frac{z-3}{1-z}, z \neq 1$

23. $\frac{4}{x-3} \cdot \frac{1}{x}$ — $\frac{4}{x(x-3)}, x \neq 0$ or 3

24. $\frac{n+4}{16} \cdot \frac{12}{n^2-16}$ — $\frac{3}{4(n-4)}, n \neq \pm 4$

25. $\frac{2}{x+5} + \frac{3x}{x^2+4x-5}$ — $\frac{5x-2}{x^2+4x-5}, x \neq -5$ or 1

26. $\frac{3d}{d^2-9} + \frac{d}{3+d}$ — $\frac{d^2}{d^2-9}, d \neq \pm 3$

---

12. $\frac{3x-2}{x^2-1}, x \neq -1, 1$

13. $\frac{5x+12}{x^2-9}, x \neq -3, 3$

14. $\frac{6k-10}{k^2-4k+4}, k \neq 2$

15. $\frac{4z+9}{z^2+6z+9}, z \neq -3$

16. $\frac{4y-12}{y^2-8y+16}, y \neq 4$

17. $\frac{7}{x-1}, x \neq 1$

18. $\frac{-5x}{x^2+x-6}, x \neq -3, 2$

19. $\frac{-9x}{x^2+x-20}, x \neq -5, 4$

20. $\frac{2a}{a^2-1}, a \neq -1, 1$

21. $\frac{7}{3x}, x \neq 0$

22. $\frac{3}{x+1}, x \neq -1$

23. $\frac{2}{5}, y \neq -4$

24. $\frac{2x-3}{x-2}, x \neq 2$

25. $\frac{5a+2}{a^2+a}, a \neq -1, 0$

26. $\frac{-5}{3t}, t \neq 0$

27. $\frac{11}{2r}, r \neq 0$

550 LESSON 11.4

APPLICATION

ECOLOGY Students are creating a plan for an exhibit of a bay ecosystem. They determine that the water in the bay exhibit should be 4% salt by mass concentration. The water in the tank for a sea exhibit is 6% salt by mass concentration. The students' plan calls for them to dilute 750 kilograms of water taken from the sea tank by adding enough pure water to get the 4% salt concentration needed for the bay tank. The concentration can be determined by using the following equation:

$$\text{concentration} = \frac{\text{salt concentration of sea water} \cdot \text{mass of sea water}}{\text{mass of sea water} + \text{mass of pure water}}$$

(Note: Because pure water is 0% salt, it is not necessary to account for its concentration in the numerator of the expression.)

52. Write a rational function to represent the resulting concentration when w kilograms of pure water are added. $c = \frac{45}{750 + w}$

53. Graph the function from Exercise 52.

54. How much pure water needs to be added to 750 kilograms of sea water in order to produce a salt concentration of 4.0%? **375 kg**

Look Back

55. State the pattern and identify the next three terms of the sequence below. **(LESSON 1.1)** Second differences are +2; 39, 54, 71

$$-1, 6, 15, 26, \underline{}, \underline{}, \underline{}$$

56. Solve $0.3(x - 90) = 0.7(2x - 70)$. **(LESSON 3.5)** $x = 20$

57. Graph the solution to $\begin{cases} y < -2x + 6 \\ y \leq 3x - 4 \end{cases}$. **(LESSON 7.5)**

58. Express 25,000,000,000 in scientific notation. **(LESSON 8.5)** 2.5×10^{10}

Look Beyond

A proportion is the equality of two ratios, such as $\frac{5}{10} = \frac{2}{4}$. Each of these ratios equals $\frac{1}{2}$. This simplified fraction is called the constant of the proportion. Find the value of the variable in each proportion below, and state the constant of the proportion.

59. $\frac{26}{39} = \frac{x}{9}$ $x = 6; \frac{2}{3}$ **60.** $\frac{14}{x} = \frac{49}{28}$ $x = 8; \frac{7}{4}$ **61.** $\frac{15}{40} = \frac{27}{x}$ $x = 72; \frac{3}{8}$

Look Beyond

In Exercises 59–61, students find the constant of a proportion. You may wish to point out that for a proportion of the form $\frac{y_1}{x_1} = \frac{y_2}{x_2}$, the relationship between y and x is a direct variation, and the constant of the proportion is the constant of variation.

internetconnect

GO TO: go.hrw.com
KEYWORD: MA1 Coaster Math

In this activity, students learn more about direct- and inverse-proportion equations by investigating the dynamics of roller coasters and other amusement rides.

53.

57.

The answers to Exercises 28–51 can be found in Additional Answers beginning on page 810.

Student Technology Guide

11.4 Operations With Rational Expressions

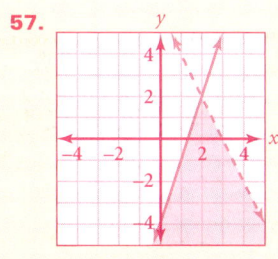

LESSON 11.4 **551**

Prepare

NCTM PRINCIPLES & STANDARDS
1–6, 8–10

LESSON RESOURCES

- Practice 11.5
- Reteaching Master 11.5
- Enrichment Master 11.5
- Cooperative-Learning Activity 11.5
- Lesson Activity 11.5
- Problem Solving/Critical Thinking Master 11.5

QUICK WARM-UP

Solve each equation.

1. $9g - 16 = 38$ $g = 6$
2. $\frac{k}{4} + 9 = 3$ $k = -24$
3. $2n^2 - 12n + 18 = 0$ $n = 3$
4. $t^2 + 3t = 10$ $t = 2$ or $t = -5$
5. Simplify $\frac{x^2 - 4}{6x} \cdot \frac{x}{x + 2}$. State any restrictions on the variable.

$\frac{x - 2}{6}$, where $x \neq 0$ and $x \neq -2$

Also on Quiz Transparency 11.5

Teach

 Why Display these equations.

$\frac{5m}{2} - \frac{m}{2} = 8$ $\frac{2}{5m} - \frac{2}{m} = 8$

Ask students how they are alike and how they are different. Ask how they would solve the first one. Work with them to arrive at the solution, $m = 4$. Explain that the second equation is called a *rational equation*. In this lesson, they will learn how to solve this type of equation.

552   LESSON 11.5

11.5 Solving Rational Equations

Objectives

- Solve rational equations by using the common-denominator method.
- Solve rational equations by graphing.

Why Many quantities, such as time, work, and average cost, can be modeled by rational equations.

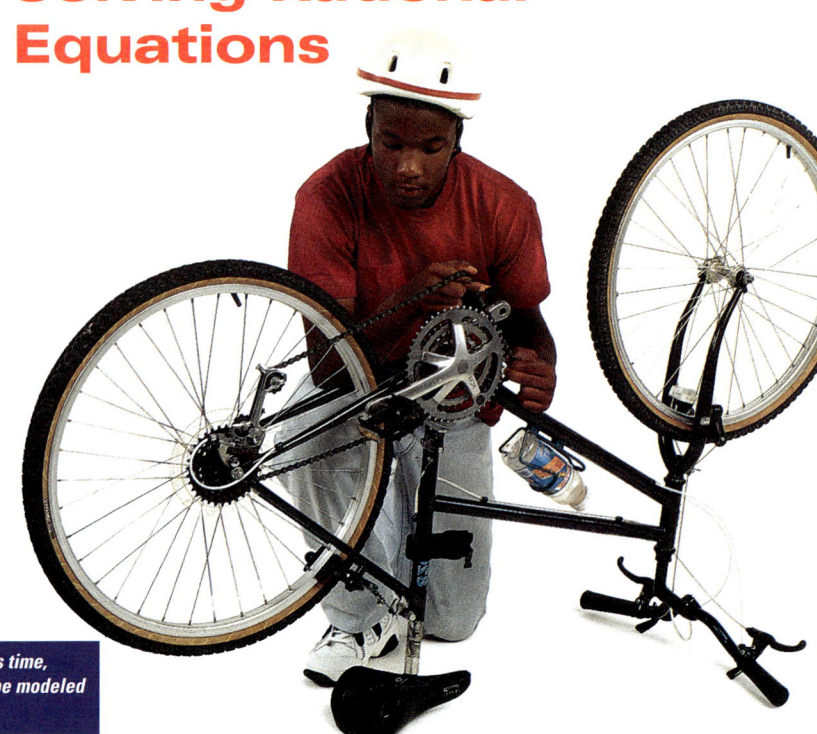

A cyclist left home and rode his bike 7 miles before his chain broke; then he walked home. During his return trip, the cyclist estimated that he bikes 5 times faster than he walks. If the entire trip took him 2 hours and 24 minutes, what are his walking and riding speeds? You can approach the problem by guessing and checking.

Activity
Guessing and Checking to Solve a Problem

1. It took the cyclist 2 hours and 24 minutes to bike 7 miles and then walk home. Convert the total time to a decimal.
2. Copy and complete the table below for a walking speed of 2 mph.

Walking speed (mph)	Walking time: $\frac{7 \text{ miles}}{\text{walking speed}}$	Riding speed: $5 \cdot$ walking speed	Riding time: $\frac{7 \text{ miles}}{\text{riding speed}}$	Total time (hours)
1.0	$7 \div 1 = 7$	$5 \cdot 1 = 5$	$7 \div 5 = 1.4$	8.4
2.0	?	?	?	?

CHECKPOINT ✓   3. Use the table and the guess-and-check method to find the walking speed that results in a total time of 2 hours and 24 minutes. What is the riding speed?

Alternative Teaching Strategy

CONNECTING TO PRIOR KNOWLEDGE Review with students the Cross-Products Property of proportions. Then display the equation from Example 1. Ask if it is possible to transform this into a proportion. Elicit the response that this can be accomplished by writing the left side as one fraction. Lead them through the process.

$$\frac{6}{x-1} + 2 = \frac{12}{x^2 - 1}$$

$$\frac{6}{x-1} + \frac{2(x-1)}{(x-1)} = \frac{12}{x^2 - 1}$$

$$\frac{6 + 2(x-1)}{x-1} = \frac{12}{x^2 - 1}$$

$$\frac{2x + 4}{x-1} = \frac{12}{x^2 - 1}$$

Show students how they can now apply the Cross-Products Property.

$$(2x + 4)(x^2 - 1) = (x - 1)12$$

Then work with them to arrive at the solution, $x = -4$ or $x = 1$.

You can also determine the cyclist's walking and riding speeds by solving an equation. The time for each part of the trip can be represented by the expression $\frac{\text{distance}}{\text{rate}}$. The total time can be represented by the equation $\frac{\text{riding distance}}{\text{riding speed}} + \frac{\text{walking distance}}{\text{walking speed}} = \text{total time}$. If x is the cyclist's walking speed, then $5x$ is his riding speed. You can rewrite the equation as follows:

$$\frac{7 \text{ miles}}{5x \text{ mph}} + \frac{7 \text{ miles}}{x \text{ mph}} = 2 \text{ hours and } 24 \text{ minutes, or } 2.4 \text{ hours}$$

Therefore, the equation that models the problem is $\frac{7}{5x} + \frac{7}{x} = 2.4$, where $x \neq 0$. The equation can be solved by two methods.

Common-Denominator Method

Multiply all of the terms in the original equation by the lowest common denominator to clear the equation of fractions. Using the lowest common denominator saves extra steps when solving the equation.

$$\frac{7}{5x} + \frac{7}{x} = 2.4$$
$$5x\left(\frac{7}{5x} + \frac{7}{x}\right) = 5x(2.4) \quad \text{Multiply both sides by } 5x\text{, the least common denominator.}$$
$$7 + 35 = 12x$$
$$12x = 42$$
$$x = 3.5$$

Thus, the cyclist walks at a rate of 3.5 mph. His biking speed is 5 times his walking speed, or 17.5 mph.

Graphing Method

Graph $y = \frac{7}{5x} + \frac{7}{x}$ and $y = 2.4$. You will find that the intersection is (3.5, 2.4).

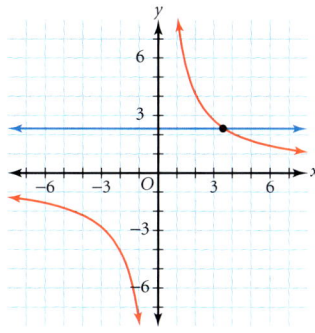

Since x represents walking speed, the cyclist walks at a rate of 3.5 mph. As in the common-denominator method, his biking speed is 5 times his walking speed, or 17.5 miles per hour.

CRITICAL THINKING

What is the real-world domain for the equation that describes the cyclist's total time? For what value(s) of the variable are the rational expressions in the equation undefined?

Activity Notes

In this Activity, students are given a real-world problem and are led to find a solution by making a table and using the guess-and-check method. As the lesson then progresses, students will see how the same problem can be solved more efficiently by using a rational equation.

Cooperative Learning

You may wish to have students do the Activity in pairs. In Step 1, partners can work individually to convert the total time to a decimal and then discuss their answers and come to an agreement. In Step 2, each partner can fill in the table. The partners can then compare results and do Step 3 together.

CHECKPOINT ✔
3. walking speed = 3.5 mph
   riding speed = 17.5 mph

CRITICAL THINKING
The real-world domain is $x > 0$. The expression is undefined for $x = 0$.

Interdisciplinary Connection

CHEMISTRY According to Boyle's law, the volume and pressure of a gas are related by the equation $\frac{V_1}{V_2} = \frac{P_2}{P_1}$, where V_1 and P_1 represent the initial volume and pressure and V_2 and P_2 represent the final volume and pressure. When the volume of a sample of hydrogen gas was decreased by 30 cubic meters, the pressure rose from 150 kilopascals to 200 kilopascals. What were the initial and final volumes of this hydrogen sample? **120 cubic meters; 90 cubic meters**

Inclusion Strategies

VISUAL LEARNERS With many problems involving rational equations, numbers and expressions can be organized in a table. Students with a visual style of learning may find this helpful. For instance, a table like the one shown below can be used for most distance problems. In this case, the information displayed is taken from the Activity.

	rate	× time	= distance
walking	x	$\frac{7}{x}$	7
riding	$5x$	$\frac{7}{5x}$	7

LESSON 11.5 **553**

ADDITIONAL EXAMPLE 1

Solve $\frac{2x+7}{x+2} = \frac{5}{x^2-4} - 2$ by the common-denominator method and by the graphing method.

$x = -3$ and $x = 2.25$

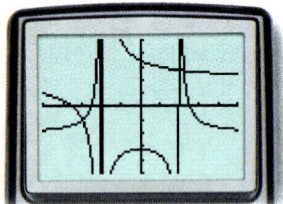

Xmin = –5   Ymin = –5
Xmax = 5   Ymax = 5
Xscl = 1   Yscl = 1

See Keystroke Guide, page 782.

TRY THIS
$x \approx -5.34$ or $x \approx 0.84$

ADDITIONAL EXAMPLE 2

An experienced house painter can paint a small house in 12 hours. The painter's apprentice can complete the same job in 20 hours. Suppose that the painter and the apprentice work together. **How long will it take them to paint the house?** 7.5 hours

Teaching Tip

After learning to solve rational equations, some students become confused and try to make the denominators "disappear" when simplifying rational expressions. Remind students that the Multiplication Property of Equality applies to *equations*. An expression can be multiplied only by an expression that is equivalent to 1.

554 LESSON 11.5

EXAMPLE 1 Solve $\frac{6}{x-1} + 2 = \frac{12}{x^2-1}$ by using the common-denominator method and the graphing method.

SOLUTION

Common-Denominator Method

$\frac{6}{x-1} + 2 = \frac{12}{(x+1)(x-1)}$ *Factor the denominator of the second expression.*

$(x+1)(x-1)\left(\frac{6}{x-1} + 2\right) = (x+1)(x-1)\left[\frac{12}{(x+1)(x-1)}\right]$ *Multiply both sides by the least common denominator.*

$6(x+1) + 2(x+1)(x-1) = 12$

$6x + 6 + 2x^2 - 2 = 12$

$2x^2 + 6x + 4 = 12$

$2x^2 + 6x - 8 = 0$

$2(x+4)(x-1) = 0$

$x + 4 = 0$ or $x - 1 = 0$ *Zero Product Property*
$x = -4$ $x = 1$

It is necessary to always check the original equation for restrictions on the variable. In this case, $x \neq -1$ and $x \neq 1$. Since 1 is excluded, $x = -4$ is the only solution.

TECHNOLOGY GRAPHICS CALCULATOR

Graphing Method

Graph $y = \frac{6}{x-1} + 2$ and $y = \frac{12}{x^2-1}$. Be sure that you enter the equations correctly; for example, the first equation should be entered as $6 \div (x-1) + 2$. The two equations have a common asymptote at $x = 1$ and intersect at $x = -4$.

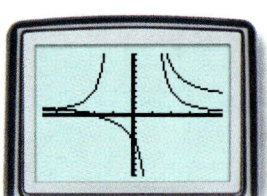

The solution is $x = -4$.

TRY THIS Solve $\frac{7}{x-2} + 2 = \frac{18}{(x-2)(x+3)}$ by using the common-denominator method and the graphing method. (Hint: Be careful to use parentheses properly when entering the expressions into your graphics calculator.)

EXAMPLE 2 Max and his younger brother Carl are earning extra money by delivering papers. Max can deliver all of the papers in 80 minutes when he works alone. Carl can deliver them in 2 hours when he works alone. **If Max and Carl work together, how many minutes will it take them to finish this job?**

SOLUTION

To solve the problem, write an expression for each brother's rate of work. Then, by multiplying by the time in minutes, t, that each brother works, you obtain the *fractional part* of the job that each brother completes. (Notice that when they work together, each brother works the same length of time although they work at different rates.) The finished job is equal to the sum of all the fractional parts, which represents 1 complete job.

Enrichment

Solve each rational inequality. (Hint: If $\frac{a}{b} > 0$, then either $a > 0$ and $b > 0$, or $a < 0$ and $b < 0$.)

1. $\frac{x}{x+1} > 0$
   $x < -1$ or $x > 0$

2. $\frac{x+1}{x-1} > 0$
   $x < -1$ or $x > 1$

3. $\frac{x+1}{x} < 0$
   $-1 < x < 0$

4. $\frac{x-2}{x+3} > 0$
   $x > 2$ or $x < -3$

Challenge students to generalize their results for inequalities of the form $\frac{x+m}{x+n} > 0$ and $\frac{x+m}{x+n} < 0$.

Reteaching the Lesson

USING VISUAL STRATEGIES Display the equation $\frac{2}{r-1} - 2 = \frac{3}{r}$. Direct students to first rewrite the equation as follows:

$\frac{2}{r-1} \cdot \boxed{} - \frac{2}{1} \cdot \boxed{} = \frac{3}{r} \cdot \boxed{}$

Lead them to identify $r(r-1)$ as the LCD of all the fractions. Show them how to fill in the boxes:

$\frac{2}{r-1} \cdot \boxed{\frac{r}{r}} - \frac{2}{1} \cdot \boxed{\frac{r(r-1)}{r(r-1)}} = \frac{3}{r} \cdot \boxed{\frac{(r-1)}{(r-1)}}$

The solutions are $r = -1$ and $r = 1.5$.

PROBLEM SOLVING

Make a table.

Fractional part of the job done by Max	+	Fractional part of the job done by Carl	=	The complete job, or 1 whole unit
$\left(\begin{array}{c}\text{rate of}\\\text{Max's work}\end{array}\right) \cdot \left(\begin{array}{c}\text{time}\\\text{worked}\end{array}\right)$	+	$\left(\begin{array}{c}\text{rate of}\\\text{Carl's work}\end{array}\right) \cdot \left(\begin{array}{c}\text{time}\\\text{worked}\end{array}\right)$	=	1
$\frac{1}{80} \cdot t$	+	$\frac{1}{120} \cdot t$	=	1
$\frac{t}{80}$	+	$\frac{t}{120}$	=	1

Solve $\frac{t}{80} + \frac{t}{120} = 1$.

Multiply each term by the lowest common denominator.

$$240 \cdot \frac{t}{80} + 240 \cdot \frac{t}{120} = 240 \cdot 1$$
$$3t + 2t = 240$$
$$5t = 240$$
$$t = 48$$

Working together, it will take Max and Carl 48 minutes to deliver all of the papers.

TRY THIS Suppose that Max can deliver all of the papers in 2 hours and Carl can deliver all of the papers in 3 hours. If Max and Carl work together, how many minutes will it take them to deliver all of the papers?

Exercises

Communicate

1. Explain the significance of restricted values of the variable in a rational equation.
2. How is the real-world domain of a rational equation determined in applications?

In Exercises 3–5, refer to $\frac{3}{c} - \frac{2}{c-1} = \frac{6}{c}$.

3. Describe how to identify the lowest common denominator.
4. Explain how to solve the rational equation by graphing.
5. How would you determine the values of the variable for which the equation is undefined?

TRY THIS

72 minutes, or 1 hour and 12 minutes

Assess

Selected Answers

Exercises 6–10, 11–39 odd

ASSIGNMENT GUIDE

In Class	1–10
Core	11–29 odd, 30, 33, 34
Core Plus	12–28 even, 29–34
Review	35–39
Preview	40

 Extra Practice can be found beginning on page 738.

Technology

A graphics calculator may be helpful for Exercises 20–28. You may wish to have students first do the exercises with paper and pencil and then use the calculator to verify their results.

internetconnect

GO TO: go.hrw.com
KEYWORD: MA1 Rational Functions

In this activity, students access an interactive web site and use rational functions to solve some simple problems in optics.

LESSON 11.5 **555**

Error Analysis

Students often forget to multiply a constant term of a rational equation by the LCD. Suggest that as soon as they copy an equation onto their paper, they should put parentheses around each side of the equation. This will serve as a reminder to distribute the LCD to *each term* within the parentheses.

Look Beyond

Exercise 40 is a nonroutine problem. Students may observe that Antonio advances a total of 2 steps every 5 seconds, and they may conclude that it will take him 250 seconds to reach the 100th step. You may want to suggest that they act out Antonio's progress as he nears the 100th step.

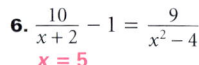

Guided Skills Practice

Solve each equation by using the common-denominator method and the graphing method. *(EXAMPLE 1)*

6. $\frac{10}{x+2} - 1 = \frac{9}{x^2 - 4}$ $x = 5$

7. $\frac{7}{x+1} - 1 = \frac{-20}{x^2 - 1}$ $x \approx 8.62$ or $x \approx -1.62$

8. $\frac{3}{x-2} - 3 = \frac{24}{x^2 - 4}$ No real sol.

Solve each problem below by using rational equations. *(EXAMPLE 2)*

9. Dan paints a house in 6 days. It takes Joan 9 days to paint the same house. How long will it take them to paint the house if they work together? **3.6 days**

10. Sue and Ed can clean their house in 7 hours. It takes Sue 15 hours to do the job alone. How long would it take Ed to do the job alone? **$13\frac{1}{8}$ hours**

11. $x = \frac{11}{12}$
12. $x = 6.25$ or $x = 1$
13. $x = 3.2$
14. $x = -0.7$
15. No sol.
16. $x = 2$
17. $y = \frac{1}{2}$
18. $x = 2$
19. $x = 3$
20. $x = 6$
21. $x = -1$
22. $x = -5$ or $x = 4$
23. $x = 2$
24. No sol.
25. No sol.
26. $x = 3$
27. $x = -1$ or $x = -\frac{1}{3}$
28. $x = -1$ or $x = -\frac{1}{3}$

Practice and Apply

Solve each rational equation by using the lowest common denominator.

11. $\frac{x-1}{x} + \frac{7}{3x} = \frac{9}{4x}$

12. $\frac{10}{2x} + \frac{4}{x-5} = 4$

13. $\frac{x-3}{x-4} = \frac{x-5}{x+4}$

14. $\frac{x+3}{x} + \frac{6}{5x} = \frac{7}{2x}$

15. $\frac{x}{x-3} = 5 + \frac{x}{x-3}$

16. $\frac{x+3}{x^2 - 9} - \frac{6}{x-3} = 5$

17. $\frac{4}{y-2} + \frac{5}{y+1} = \frac{1}{y+1}$

18. $\frac{10}{x+3} - \frac{3}{5} = \frac{10x+1}{3x+9}$

19. $\frac{3}{x-2} - \frac{6}{x^2 - 2x} = 1$

Solve each rational equation by graphing.

20. $\frac{2}{x} + \frac{1}{3} = \frac{4}{x}$

21. $\frac{1}{2} + \frac{1}{x} = \frac{1}{2x}$

22. $\frac{20-x}{x} = x$

23. $\frac{12x-7}{x} = \frac{17}{x}$

24. $\frac{3}{7x + x^2} = \frac{9}{x^2 + 7x}$

25. $\frac{x-3}{x-4} = \frac{x-5}{x-4}$

26. $\frac{3}{x-2} - \frac{6}{x^2 - 2x} = 1$

27. $\frac{1}{x-2} + \frac{16}{x^2 + x - 6} = -3$

28. $\frac{1}{x-2} + 3 = \frac{-16}{x^2 + x - 6}$

29. The sum of two numbers is 56. If the larger number is divided by the smaller, the quotient is 1 with a remainder of 16. Find the two numbers. **20 and 36**

CONNECTION

30. **COORDINATE GEOMETRY** Find any points of intersection for the graphs of the function $y = \frac{1}{x} + 3$ and the function $y = \frac{x-4}{x+2}$.
   (−5.31, 2.81) and (−0.19, −2.31)

APPLICATION

RECREATION Marisa pedals her paddleboat downstream at a still-water speed of 20 meters per minute for 500 meters before turning around and pedaling upstream to where she started. If Marisa maintains the same still-water speed, she will pedal only 300 meters upstream in the same time it took her to pedal 500 meters downstream.

31. downstream: $\frac{500}{20+c}$

    upstream: $\frac{300}{20-c}$

31. Write rational expressions that represent the time it takes Marisa to pedal downstream and upstream in the current.

32. What is the speed of the current? **5 meters per minute**

Practice

11.5 Solving Rational Equations

Solve the following rational equations by finding the lowest common denominator:

1. $\frac{2}{y} + 3 = \frac{29}{y}$ $y = 9$
2. $\frac{z}{2} - 7 = \frac{z}{3}$ $z = 42$
3. $\frac{1}{x} + \frac{3}{x} = \frac{5}{4}$ $x = \frac{16}{5}$
4. $\frac{2x}{3} - \frac{x}{2} = 5$ $x = 30$
5. $\frac{3-5c}{8} + \frac{3c-6}{5} = 1$ $c = -73$
6. $\frac{1}{4} - \frac{x-3}{8x} = 0$ $x = -3$
7. $\frac{1}{x+6} = \frac{2}{x+9} - \frac{1}{x+3}$ $x = -5$
8. $\frac{y-1}{y} - 2 = \frac{-y}{y+1}$ $y = -\frac{1}{2}$
9. $\frac{4b+5}{2b+1} = \frac{6b+1}{3b-1}$ $b = 2$
10. $\frac{x}{x^2 - 3} = \frac{5}{x+4}$ $x = -\frac{3}{2}$ and $x = \frac{5}{2}$
11. $\frac{6}{a} - 1 = a$ $a = -3$ and $a = 2$
12. $\frac{y}{y-5} + \frac{y}{y-5} = 3$ $y = 15$
13. $\frac{3x-1}{4} = \frac{x-5}{5} - 2$ $x = -5$
14. $\frac{4w}{3w-2} + \frac{2w}{3w+2} = 2$ $w = -2$
15. $\frac{2}{x} + \frac{1}{4} = \frac{11}{12}$ $x = 3$
16. $\frac{12}{x^2 - 16} - 3 = \frac{24}{x-4}$ $x = -6$ and $x = -2$
17. $\frac{1}{r^2 - 1} = \frac{2}{r^2 + r - 2}$ $r = 0$
18. $\frac{m}{m^2 - 1} + \frac{2}{m+1} = \frac{1}{2m-2}$ no solution
19. $\frac{t}{t-5} = \frac{1}{t+5} - \frac{17}{25-t^2}$ $t = -6$ and $t = 2$
20. $2 - \frac{3}{y^2 + 5y + 6} = \frac{y+3}{y+2}$ $y = -4$ and $y = 0$
21. $\frac{q-3}{4+q} = \frac{q-3}{7}$ $q = 3$
22. $\frac{g}{g-2} = 1 + \frac{g+7}{g+2}$ $g = -6$ and $g = 3$

Solve the following rational equations by graphing:

23. $\frac{3x}{5} = \frac{x}{2} - 1$ $x = -10$
24. $\frac{x}{4} + \frac{x-3}{3} = 6$ $x = 12$
25. $\frac{1}{x+2} + \frac{1}{x-3} = \frac{2}{x}$ $x = -12$
26. $\frac{2x-6}{3x+1} = \frac{x-1}{4x-4}$ $x = 5$

556 LESSON 11.5

SMALL BUSINESS At a car body shop, Kayla needs 5 hours working alone to paint a car. It takes Emily 7 hours to paint the same car.

Kayla: $\frac{x}{5}$

Emily: $\frac{x}{7}$

33. Write rational expressions that represent the fractional parts of the work that Kayla and Emily complete when painting the car together.

34. Write and solve a rational equation to find the time it would take them to paint the car if they work together. $\frac{x}{5} + \frac{x}{7} = 1$; 2 hours 55 minutes

Look Back

35. Solve: $x - 4 = -\frac{3}{4}(x + 2)$. **(LESSON 3.5)** $\frac{10}{7}$

36. Solve: $|2x - 7| \leq 7$. **(LESSON 6.5)** $0 \leq x \leq 7$

37. Solve: $\begin{cases} x + 5y = 9 \\ 3x - 2y = 10 \end{cases}$. **(LESSON 7.3)** (4, 1)

38. Multiply: $(3x - 8)(5x + 2)$. **(LESSON 9.3)** $15x^2 - 34x - 16$

39. Identify the vertex of $y = -2(x + 1)^2 + 7$. **(LESSON 10.1)** (−1, 7)

Look Beyond

40. As a way to vary his exercise, Antonio climbs 6 steps in 3 seconds and then carefully descends 4 steps in 2 seconds. Continuing this sequence of climbing and descending, how long will it take him to reach the 100th step? **250 seconds**

Include this activity in your portfolio.

The relationship between the number of teeth on the pedal gear and the number of teeth on the wheel gear of a bicycle affects the speed and effort of pedaling. For example, if the wheel gear has 16 teeth and the pedal gear has 54 teeth, then the gear ratio is $\frac{16}{54} \approx 0.296$. The reciprocal is $\frac{54}{16} = 3.375$. The reciprocal 3.375 represents the number of turns of the wheel for every full turn of the pedal.

1. Assume for this activity that the circumference of a bicycle tire is 2.19 meters. How far will a rider travel for each full turn of the pedal if the gear ratio is 0.25?

2. The wheel gear could have 13, 14, 15, 16, 18, 21, or 24 teeth and the pedal gear could have 49 or 54 teeth. Write a formula for the possible gear ratios, r, based on the number of teeth on the wheel gear, w, and the number of teeth on the pedal gear, p. Find the ratios and reciprocals for each possible combination.

3. Suppose that a bicycle rider makes 60 full turns of the pedal per minute. Choose two possible gear combinations and calculate the rate in miles per hour for each. (Note that 1 kilometer is approximately equal to 0.62 miles.)

WORKING ON THE CHAPTER PROJECT

You should now be able to complete the Chapter Project on page 566.

ALTERNATIVE Assessment

Portfolio Activitiy

The Portfolio Activity can be used as preparation for the Chapter Project or as a separate activity. In the Portfolio Activity on this page, students calculate distances traveled by a bicycle rider given different gear ratios. They also write functions and find ratios for various combinations of the pedal and wheel gears.

internetconnect

 GO TO: go.hrw.com
KEYWORD: MA1 Gears

To support the Portfolio Activity, students may visit the HRW Web site, where they can learn about the importance of gears in the design of cars and bicycles.

Answers to Portfolio Activities can be found in Additional Answers of the Teacher's Edition.

Student Technology Guide

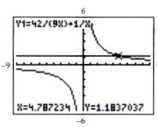
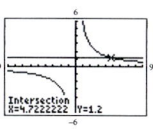

LESSON 11.5 **557**

Prepare

NCTM PRINCIPLES & STANDARDS
1, 2, 5–10

LESSON RESOURCES

- Practice 11.6
- Reteaching Master 11.6
- Enrichment Master 11.6
- Cooperative-Learning Activity 11.6
- Lesson Activity 11.6
- Problem Solving/Critical Thinking Master 11.6
- Teaching Transparency 53

QUICK WARM-UP

Name the property illustrated by each statement. Assume that all variables represent real numbers.

1. $3(10 + 7) = 3(10) + 3(7)$
   Distributive Property

2. $(k + 7)(0) = 0$
   Mult. Property of Zero

3. If $a + 6 = 5$, then $a + 6 + (-6) = 5 + (-6)$.
   Addition Property of Equality

4. If $r = 6 + 8$ and $6 + 8 = 14$, then $r = 14$.
   Transitive Property of Equality

Also on Quiz Transparency 11.6

11.6 Proof in Algebra

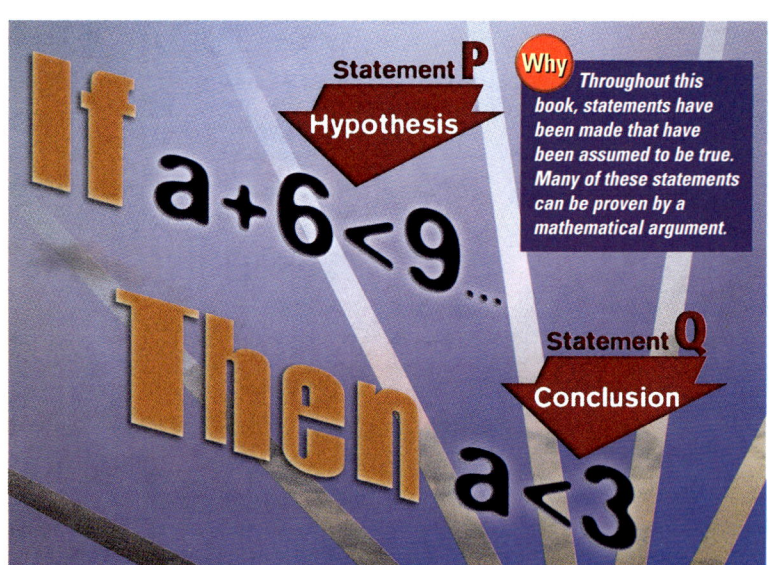

Objectives
- Define the parts of a conditional statement.
- Define *converse*.
- Define *proof*.
- Prove theorems stated in conditional form.

Why Throughout this book, statements have been made that have been assumed to be true. Many of these statements can be proven by a mathematical argument.

Theorems as "If-*p*-then-*q*" Statements

Mathematical statements can often be written as "if-*p*-then-*q*" statements, also called **conditional statements**. Once such statements are proven, they are called **theorems**.

In a conditional statement, the *p* portion (following the word *if*) is called the **hypothesis**. The *q* portion (following the word *then*) is called the **conclusion**. A **proof** is a series of statements in a logical sequence which demonstrate that the conclusion is true *whenever the hypothesis is true*.

In a proof, the hypothesis is assumed to be true. Each step in a proof is based on established *definitions*, *axioms* or *postulates* (basic statements that are accepted as true without proof), or previously proven theorems. Study the following logical argument, which is actually a proof of a very simple conditional statement:

Prove: If $a + 6 < 9$, then $a < 3$.

Statement	Reason
$a + 6 < 9$	Hypothesis or given
$a + 6 - 6 < 9 - 6$	Subtraction Property of Inequality
$a < 3$	Simplified form

Therefore, if $a + 6 < 9$, then $a < 3$. Notice the importance of the hypothesis. The conclusion, $a < 3$, taken by itself cannot be proven mathematically.

If you interchange the hypothesis and the conclusion, the new statement is called the **converse** of the original statement. "If *q*, then *p*" is the converse of "If *p*, then *q*." What is the converse of the conditional statement "If $a + 6 < 9$, then $a < 3$?" Is the converse true?

Teach

Why Ask students where they have seen the word *proof* used before. An example might be a court case in which lawyers assemble evidence as proof of a person's guilt or innocence. Discuss with students what is needed to prove that a statement is true in mathematics.

Alternative Teaching Strategy

CONNECTING TO PRIOR KNOWLEDGE Give students the equation $a + 7 = 5$ and review with them the progression of steps below.

$a + 7 = 5$	Given
$(a + 7) + (-7) = 5 + (-7)$	Add. Prop. of Eq.
$a + [7 + (-7)] = 5 + (-7)$	Assoc. Prop. of Add.
$a + 0 = 5 + (-7)$	Add. Inverse Prop.
$a = 5 + (-7)$	Add. Identity Prop.
$a = -7 + 5$	Comm. Prop. of Add.

Tell students they have proved this statement: If $a + 7 = 5$, then $a = -7 + 5$.

Define the terms *conditional statement*, *hypothesis*, and *conclusion*.

Now display the statement below and go through a similar set of steps and justifications.

If $a + b = c$, then $a = -b + c$.

Tell the students that they have now demonstrated that the statement is true for any values of *a*, *b*, and *c*. Explain that this type of general statement is called a *theorem*. Define the word *proof*, and discuss the concept of a *conjecture*.

558 LESSON 11.6

Mathematical theorems often start out as **conjectures**, or statements that seem to be true but have not yet been proven.

There are two types of reasoning that are used in mathematics: inductive and deductive. Both are necessary parts of mathematical thinking.

Inductive reasoning is the process of reasoning that a general principle is true because the special cases that you have seen are true. For example, you might conclude that all of the food from a particular restaurant is bad because you did not like the meal that you ate there. Inductive reasoning is not logically valid and does not constitute mathematical proof except in the special method called "mathematical induction." However, inductive reasoning is often used to formulate conjectures.

Deductive reasoning is the process of concluding that a special case of a proven general principle is true. For example, you might conclude that the sum of the interior angles of a particular triangle is 180 degrees because you know that the sum of the angles in any triangle is always 180 degrees. Deductive reasoning is logically valid and is the fundamental method by which mathematical facts are shown to be true.

Activity: Inductive Reasoning

1. Consider the equation $\frac{a}{b} = \frac{c}{d}$. Find nonzero values for a, b, c, and d that make the equation true. Substitute the same values for a, b, c, and d in each equation below.

   a. $\frac{b}{a} = \frac{d}{c}$ b. $\frac{a+b}{c} = \frac{c+b}{d}$ c. $\frac{a+b}{b} = \frac{c+d}{d}$

   d. $\frac{a}{c} = \frac{b}{d}$ e. $\frac{a-b}{b} = \frac{c-b}{d}$ f. $\frac{a}{b} = \frac{a+c}{b+d}$

2. Which of the equations in Step 1 are true for your values of a, b, c, and d?

3. Write if-then statements relating the original equation, $\frac{a}{b} = \frac{c}{d}$, with each equation in Step 1.

CHECKPOINT ✔ 4. Are your conditional statements conjectures or theorems? Explain.

In the activity, you may have made the following conjecture:

If $\frac{a}{b} = \frac{c}{d}$, then $\frac{a+b}{b} = \frac{c+d}{d}$.

This conjecture can be proven by adding 1 to each side of the equation in the hypothesis.

Statement	Reason
$\frac{a}{b} = \frac{c}{d}$	Hypothesis or given
$\frac{a}{b} + 1 = \frac{c}{d} + 1$	Addition Property of Equality
$\frac{a}{b} + \frac{b}{b} = \frac{c}{d} + \frac{d}{d}$	Substitution: $\frac{b}{b} = 1$ and $\frac{d}{d} = 1$
$\frac{a+b}{b} = \frac{c+d}{d}$	Definition of addition

Interdisciplinary Connection

COMPUTER SCIENCE Computer programmers use *if-then* statements to create *loops* in programs and to direct the program to a different *branch* of the program. When the condition following the word *if* is true, the computer executes the command following the word *then*. When the condition is false, the program simply proceeds to the next step. Give students the following excerpts from BASIC programs, and ask them to identify the *if* condition that must be true and the command that the computer will then execute upon finding it true.

150 IF V>=0 THEN LET S=S+V

Condition: The value of V must be positive or 0.
Command: The value of V will be added to the current value of S to create a new value of S.

170 IF V<=0 THEN 200
180 LET P=P+1
190 LET S=S+V
200 PRINT "ACCEPTABLE"

Condition: The value of V must be negative or 0.
Command: The word ACCEPTABLE will be printed.

Teaching Tip

Inductive reasoning often is described as "particular to general." That is, by using several particular cases, you make a general observation. In contrast, deductive reasoning is described as "general to particular." That is, by using several general statements, you make a particular observation.

Activity Notes

In this Activity, students write a set of true proportions and use inductive reasoning to make conjectures about relationships among the terms of any true proportion. After completing the Activity, students should understand the process of inductive reasoning that leads to a conjecture. The text that follows the Activity shows how deductive reasoning can be used to prove one of the conjectures.

Cooperative Learning

You may wish to have students do the Activity in groups of three. In Step 1, the group should first agree on the values of a, b, c, and d that will be used. Each group member can then test the values in two of the six statements. The group should share their results and work together to complete Steps 2 and 3.

CHECKPOINT ✔

4. Conjectures; they have not been proven yet.

Teaching Tip

Students should recognize the conjecture in Example 1 as the Cross-Products Property of proportions, which they studied in Lesson 4.1.

ADDITIONAL EXAMPLE 1

Prove the following conjecture: For all nonzero real numbers a, b, c, and d, if $\frac{a}{b} = \frac{c}{d}$, then $\frac{b}{a} = \frac{d}{c}$.

Statement	Reason
$\frac{a}{b} = \frac{c}{d}$	Given
$ad = bc$	Cross-Products Property
$\frac{ad}{ac} = \frac{bc}{ac}$	Div. Prop. of Eq.
$\frac{d}{c} = \frac{b}{a}$	Simplified form
$\frac{b}{a} = \frac{d}{c}$	Reflex. Prop. of Equality

ADDITIONAL EXAMPLE 2

Prove the following conjecture: If k is an odd number, then $k + 1$ is even.

Statement	Reason
k is an odd number.	Given
$k = 2n + 1$, where n is an integer	Definition of odd number
$k + 1 = (2n + 1) + 1$	Addition Prop. of Eq.
$k + 1 = 2n + (1 + 1)$	Assoc. Prop. of Add.
$k + 1 = 2n + 2$	Simplify.
$k + 1 = 2(n + 1)$	Distributive Property
$n + 1$ is an integer.	Definition of integer
$k + 1$ is an even number.	Definition of even number

EXAMPLE 1 Prove the following conjecture:

If $\frac{a}{b} = \frac{c}{d}$, then $ad = bc$.

● **SOLUTION**

Statement	Reason
$\frac{a}{b} = \frac{c}{d}$	Hypothesis or given
$bd\left(\frac{a}{b}\right) = bd\left(\frac{c}{d}\right)$	Multiplication Property of Equality
$da = bc$	Simplified form
$ad = bc$	Commutative Property of Multiplication

Therefore, if $\frac{a}{b} = \frac{c}{d}$, then $ad = bc$.

The definitions of even and odd numbers can also be used in proofs.

Definitions of Even and Odd Numbers

An even number is a number of the form $2n$, where n is an integer.

An odd number is a number of the form $2n + 1$, where n is an integer.

EXAMPLE 2 Prove the following conjecture:

If two numbers are both odd, then their sum is even.

● **SOLUTION**

Statement	Reason
a and b are two odd numbers.	Hypothesis or given
$a = 2k + 1$ and $b = 2p + 1$, where k and p are integers	Definition of odd numbers
$a + b = (2k + 1) + (2p + 1)$	Addition of real numbers
$ = 2k + 2p + 2$	Simplified form
$ = 2(k + p + 1)$	Distributive Property

Since $k + p + 1$ is an integer, $2(k + p + 1)$ is in the form $2n$, where n is an integer. Thus, $2(k + p + 1)$ is an even number.

Therefore, if two numbers are both odd, then their sum is even.

Inclusion Strategies

LINGUISTIC LEARNERS In Examples 1, 2, and 3, proofs are presented in a form called a *two-column proof*. Students whose learning style is more verbal sometimes find this form to be intimidating. These students may be more comfortable using a *paragraph proof*. This form generally contains the same steps as in a two-column proof, but the steps are linked in a style that more closely resembles a dialogue. For example, the proof from Example 2 is given in paragraph form at right.

From the given statement, you know that k is an integer, and from the definition of an integer, you know that $k + 1$ also is an integer. Now think of $k + 1$ as n. From the definition of even numbers, when n is an integer, $2n$ is an even number. So $2(k + 1)$ is an even number. By the Distributive Property, $2(k + 1)$ is equal to $2k + 2$. This means that, by the Substitution Property, $2k + 2$ is an even number. Therefore, if k is an integer, then $2k + 2$ is an even number.

EXAMPLE 3 Prove the following conjecture:

If k is an integer, then $2k + 2$ is an even number.

● SOLUTION

Statement	Reason
k is an integer.	Hypothesis or given
$k + 1$ is an integer.	Definition of integer
$2(k + 1)$ is even.	Definition of even numbers *Think of $k + 1$ as n.*
$2k + 2$ is even.	Distributive Property

Therefore, if k is an integer, then $2k + 2$ is an even number.

In addition to the sum shown in Example 2 on the previous page, it is easy to prove that the following are true:

odd + even = even + odd = odd   even + even = even
odd • even = even • odd = even   odd • odd = odd
even • even = even

Using these facts, you can see that a quadratic trinomial with all odd coefficients, such as $3x^2 + 25x + 15$, cannot be factored into linear terms with integer coefficients.

Suppose that a trinomial like $3x^2 + 25x + 15$ has the factors $(ax + b)$ and $(cx + d)$, where a, b, c, and d are integers. Then the products ac and bd, in this case 3 and 15, must both be odd. Therefore, a, b, c, and d must all be odd.

The middle term, in this case 25, is the sum of ad and bc, both of which must be odd since all four factors (a, b, c, and d) are odd. But the sum of two odd numbers will always be even and will never result in an odd middle coefficient. Thus, it is impossible to find integer values of a, b, c, and d that will work.

These facts about sums and products can also be used to solve simpler problems.

EXAMPLE 4 Find four even integers whose sum is 101, or explain why it is impossible.

● SOLUTION

The sum of any two even numbers will always be an even number, so adding four even integers will also result in an even integer. Thus, it is not possible to add four even integers and get an odd number such as 101 as the sum.

TRY THIS Find three odd integers whose product is 100, or explain why it is impossible.

Most mathematical statements can be proven true or shown to be false with a **counterexample**. There are some cases in which the statement is sometimes true and sometimes false, often depending on the value of a variable within the statement. It is important for you to be able to recognize these kinds of statements because statements that are only sometimes true cannot be used as reasons in proofs. However, they can often be rewritten with further specifications that will make them either always true or never true.

Enrichment

Have students create two true conditional statements for which the converse is false. One statement should reflect an everyday situation, and the other should be mathematical in nature. **Samples: If a person lives in Texas, then that person lives in the United States. If a and b are even numbers, then $a + b$ is an even number.** Now have students create two true conditional statements for which the converse is true. **Samples: If today is Monday, then today is the second day of the week. If $a - b = c$, then $a = b + c$.**

Tell students that when a conditional and its converse are both true, you can combine them into one true statement, called a *biconditional statement*, using the words *if and only if*. Challenge them to write several biconditional statements. **Samples: Today is Monday if and only if today is the second day of the week. $a - b = c$ if and only if $a = b + c$.**

ADDITIONAL EXAMPLE 3

Prove the following conjecture: The sum of two even numbers is an even number.

Statement	Reason
a and b are even numbers.	Hypothesis or given
Let $a = 2p$ and $b = 2q$, where p and q are integers.	Definition of even numbers
$a + b = 2p + 2q$	Substitution Property
$a + b = 2(p + q)$	Distributive Property
$p + q$ is an integer.	Closure
$a + b$ is an even number.	Definition of even numbers

ADDITIONAL EXAMPLE 4

Find three even numbers whose product is 149, or tell why it is impossible.

The product of two even numbers is an even number. When multiplying three even numbers, the product of any two of the numbers will be even. When that product is then multiplied by the third even number, the new product will also be even. Therefore, it is not possible to multiply three even numbers and get an odd number such as 149 as the product.

TRY THIS

Since the product of two odds is odd, the product of three odds would be odd also. Since 100 is even, there is no combination of 3 odd integers whose product is 100.

LESSON 11.6

ADDITIONAL EXAMPLE 5

Let a and b represent real numbers. Tell whether each statement is *always true*, *sometimes true*, or *never true*.

a. $2(a + b) = 2a + 2b$
   always true

b. $2(a + b) = 2a + b$
   sometimes true

c. $2(a + 1) = 2a + 1$
   never true

TRY THIS
a. This statement is true only when $a = 0$.
b. This statement is sometimes true.
c. This statement is always true.

Use Teaching Transparency 53.

CRITICAL THINKING
No, it is not possible. A counterexample is sufficient to disprove a statement, but examples alone are not enough to prove a statement.

EXAMPLE 5 Tell whether each statement is always true, sometimes true, or never true.

a. $a^2b^2 = (ab)^2$
b. $-b^2 = (-b)^2$
c. $|a + 2| = |a| + 2$

SOLUTION

a. This statement is always true based on the laws of exponents that you studied in Chapter 8.

b. This statement is sometimes true, depending on the value of b. If b is 0, then the equation is true. The equation is false for all other values of b.

c. This statement is sometimes true, depending on the value of a. If a is a positive number or 0, then the equation is true. But if a is negative, the equation is false. For example, $|3 + 2| = |3| + 2$, but $|-3 + 2| \neq |-3| + 2$.

TRY THIS Tell whether each statement is always true, sometimes true, or never true.

a. $a - b = -a + (-b)$
b. $ab = |a| \cdot |b|$
c. If $|x| < a$, then $-a < x < a$.

You have already studied many properties of real numbers, including the Associative, Commutative, and Distributive Properties. Another property that is valuable for studying sets of numbers is the **Closure Property**.

A set of numbers is *closed* under addition if the sum of any two numbers in the set is also in the set. Since the sum of any two integers is also an integer, the set of integers is closed under addition. The Closure Property applies in the same way to any set of numbers and any mathematical operation.

Closure Property

A set of numbers is said to be closed, or to have closure, under a given operation if the result of the operation on any two numbers in the set is also in the set.

Only one case in which the result is not in the set, a counterexample, is required to show that a set is not closed. For example, the set {0, 1, 2} is not closed under addition because $1 + 2 = 3$ and 3 is not in the set.

CRITICAL THINKING Is it possible to prove that a set is closed by using several examples? Why or why not?

EXAMPLE 6 Determine whether each set is closed under the given operation.

a. whole numbers; multiplication
b. {−1, 0, 1}; multiplication
c. whole numbers; subtraction
d. integers; subtraction

Reteaching the Lesson

COOPERATIVE LEARNING Write the five statements in the proof below on a sheet of paper. *Do not include the reasons.* Be sure to leave some space between each of the statements.

Statements	Reasons
$(a - b) + b$	Given
$= [a + (-b)] + b$	Definition of subtr.
$= a + [(-b) + b]$	Assoc. Prop. of Add.
$= a + 0$	Add. Inverse Prop.
$= a$	Add. Identity Prop.

Make several copies of the proof. Cut each copy into five strips, with one statement on each strip, and place the strips in an envelope, making sure that the strips are *not* in the original order.

Have students work in pairs. Give each pair one envelope. Tell them that there is a scrambled proof in the envelope and that their job is to reassemble it in the proper order. Tell them that they also must write the reason for each step. After they have done this, ask them to write the statement that was proved. $(a - b) + b = a$

LESSON 11.6

SOLUTION

a. Since the product of two whole numbers is always a whole number, the set of whole numbers is closed under multiplication.

b. Since the product of any two numbers in the set is also in the set, $\{-1, 0, 1\}$ is closed under multiplication.

c. Since $4 - 8 = -4$, the set of whole numbers is not closed under subtraction.

d. Since the difference of any two integers is always an integer, the set of integers is closed under subtraction.

TRY THIS Determine whether the set of integers is closed under division.

The Closure Property is used in the study of *fields*. The set of real numbers is probably the best-known field. The set of real numbers has the properties shown below for addition and multiplication. Any set that has the properties shown in the table below for any two operations is a **field**.

Property	Addition	Multiplication
Associative	$(a + b) + c = a + (b + c)$	$(a \cdot b) \cdot c = a \cdot (b \cdot c)$
Closure	$a + b$ is in the set.	$a \cdot b$ is in the set.
Commutative	$a + b = b + a$	$a \cdot b = b \cdot a$
Distributive	$a \cdot (b + c) = (a \cdot b) + (a \cdot c)$	
Identity	A unique identity element, 0, exists such that $a + 0 = 0 + a = a$.	A unique identity element, 1, exists such that $a \cdot 1 = 1 \cdot a = a$.
Inverse	For any a, there is an inverse, $-a$, such that $a + (-a) = -a + a = 0$.	For any $a \neq 0$, there is an inverse, $\frac{1}{a}$, such that $a \cdot \frac{1}{a} = \frac{1}{a} \cdot a = 1$.

CRITICAL THINKING Is the set $\{-1, 0, 1\}$ a field? Explain.

Exercises

Communicate

1. Name three types of reasons that can be used in a proof.
2. Identify the part of an if-then statement that is assumed to be true. What part has to be proven?
3. Explain what is meant by the converse of a conditional statement.

Error Analysis

In determining whether a set is closed under a given operation, students sometimes forget to check what happens when one element of the set is used twice. For instance, they may say that $\{-2, 0, 2\}$ is closed under addition because they have overlooked the sums $-2 + (-2) = -4$ and $2 + 2 = 4$. In a finite set such as this, you may want to suggest that students list all possible combinations of the elements.

4.
	$\frac{2}{3} = \frac{x}{9}$	Given
	$(3 \cdot 9)\frac{2}{3} = (3 \cdot 9)\frac{x}{9}$	Mult. Prop. of Eq.
	$18 = 3x$	Simplify.
	$\frac{18}{3} = \frac{3x}{3}$	Div. Prop. of Eq.
	$6 = x$	Simplify.

Therefore, if $\frac{2}{3} = \frac{x}{9}$, then $x = 6$.

The answers to Exercises 7–9 and 21–24 can be found in Additional Answers beginning on page 810.

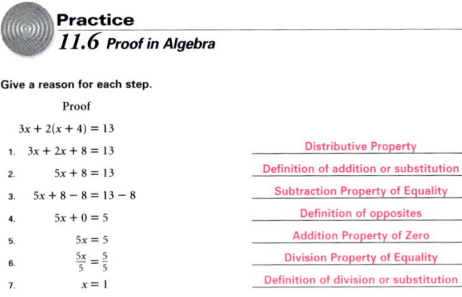

Guided Skills Practice

4. Prove: If $\frac{2}{3} = \frac{x}{9}$, then $x = 6$. **(EXAMPLE 1)**
5. Prove the following conjecture: If an even number is added to an odd number, then the sum is an odd number. **(EXAMPLE 2)**
6. Prove the following conjecture: If k is an integer, then $4k + 4$ is an even integer. **(EXAMPLE 3)**
7. Find four odd integers whose product is 100, or explain why it is impossible. **(EXAMPLE 4)**
8. Is the statement $ab = |a| \cdot |b|$ always true, sometimes true, or never true? **(EXAMPLE 5)**
9. Determine whether the set of integers is closed under multiplication. **(EXAMPLE 6)**

Practice and Apply

State a reason for each step in the proof of the following conjecture: If $3m - 4 = 0$, then $m = \frac{4}{3}$.

	Statement	Reason
	$3m - 4 = 0$	Hypothesis or given
10.	$3m - 4 + 4 = 0 + 4$	Addition Prop. Eq.
11.	$3m + 0 = 0 + 4$	Inverse Prop. Addition
12.	$3m = 4$	Identity Prop. Addition
13.	$\frac{3m}{3} = \frac{4}{3}$	Div. Prop. Eq.
14.	$1 \cdot m = \frac{4}{3}$	Equiv. form of 1
15.	$m = \frac{4}{3}$	Identity Prop. Mult.

State a reason for each step in the proof of the following conjecture: If $x = (a + b) + (-a)$, then $x = b$.

	Statement	Reason
16.	$x = (a + b) + (-a)$	Given
17.	$= (b + a) + (-a)$	Comm. Addition
18.	$= b + [a + (-a)]$	Assoc. Addition
19.	$= b + 0$	Inverse Prop. Addition
20.	$= b$	Identity Prop. Addition

For Exercises 21–26, write a proof for each conjecture. Give a reason for each step. Let all variables represent real numbers.

21. If $y = 5x - 3x$, then $y = 2x$.
22. If $n = (ax + b) + ay$, then $n = a(x + y) + b$.
23. If $a = b$, then $a + c = b + c$.
24. If a is an even number, then its square, a^2, is an even number.

5.
a is an even number	Given
b is an odd number	Given
$a = 2k$	Def. of even numbers
$b = 2p + 1$	Def. of odd numbers
$a + b = 2k + 2p + 1$	Add. of real numbers
$= 2(k + p) + 1$	Distributive Prop.

$2(k + p) + 1$ is in the form $2n + 1$, where n is equal to $k + p$ and $k + p$ is an integer. Thus, $2(k + p) + 1$ is an odd number.

6.
k is an integer	Given
$k + 1$ is an integer	Definition of integer
$2(k + 1)$ is even	Definition of even numbers
$2k + 2$ is even	Distributive Property
$2(2k + 2)$ is even	Definition of even numbers
$4k + 4$ is even	Distributive Property

Therefore, if k is an integer, $4k + 4$ is an even integer.

564 LESSON 11.6

25. The expressions $a - b$ and $b - a$ are opposites. (Hint: Prove that $(a - b) + (b - a) = 0$ is true. Then refer to the definition of opposites.)

26. If $x = \frac{a}{b} + \frac{c}{d}$, then $x = \frac{da + bc}{bd}$.

27. Find three odd integers whose sum is 50, or explain why it is impossible.

28. Find five even integers whose sum is 105, or explain why it is impossible.

29. Sometimes true
30. Sometimes true
31. Always true
32. Never true

For Exercises 29–32, tell whether each statement is always true, sometimes true, or never true.

29. $(-x)^n = x^n$ 30. $(a)(-a) = -1$ 31. $|x^2| = x^2$ 32. $a \cdot \frac{1}{a} = -1$

33. Determine whether the set $\{-2, -1, 0, 1, 2\}$ is closed under subtraction.

APPLICATION

34. **CONTESTS** With each purchase, a fast-food restaurant gives out a card with a number on it. Anyone with cards whose numbers add up to 100 wins a major prize. A group of students collects cards with the following numbers: 3, 3, 3, 6, 12, 15, 24, 42, 54, 60, 72, 84. Use the methods learned in this lesson to prove that it is impossible to form a sum of 100 from the numbers on these cards.

Look Back

35. Solve $x = 16x + 45$. **(LESSON 3.3)** $x = -3$

CONNECTION

36. **STATISTICS** What is the average length in centimeters of the hand spans (from the tip of the thumb to the tip of the little finger) in this sample: 16.2, 18.5, 23.0, 21.2, 22.9, 21.1, 20.6, 19.2? **(LESSON 4.4)** 20.3 cm

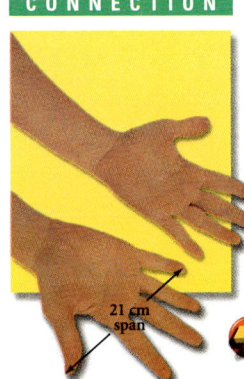

37. Write the equation of the line that is perpendicular to $2x - 3y = 9$ and contains the point $(-6, 12)$. **(LESSON 5.6)** $y = -\frac{3}{2}x + 3$

38. Solve: $x^2 - 8 = 2x$. **(LESSON 10.4)** $x = 4$ or $x = -2$

Look Beyond

39. What is wrong with the following proof, which gives the conclusion that $1 = 0$?

$$\text{Let } a = b$$
$$a - b = 0$$
$$\frac{a - b}{a - b} = \frac{0}{a - b}$$
$$1 = 0$$

Look Beyond

In Exercise 39, students analyze a proof that appears at first glance to make a logical argument in support of an illogical conclusion. If students have difficulty finding the flaw in the proof, suggest that they look for a logical contradiction between two steps.

The answers to Exercises 25 and 26 can be found in Additional Answers beginning on page 810.

27. The sum of 2 odd integers is even. The sum of that even integer and an odd integer is odd. Therefore, the sum of three odd integers must be odd and cannot be 50.

28. The sum of any number of even integers is even, so the sum of 5 even integers must be even and cannot be 105.

33. The set $\{-2, -1, 0, 1, 2\}$ is not closed under subtraction. For example, $(-2) - (2) = -4$, which is not in the set.

34. All of the numbers are multiples of 3. The sum of any of the numbers must be a multiple of 3, and therefore cannot add up to 100, because 100 is not a multiple of 3.

39. If $a = b$, then $a - b = 0$. The proof divides by $a - b$, which means it is dividing by 0, which is invalid.

Student Technology Guide

11.6 Proof in Algebra

You can use the IF function on a spreadsheet to check whether a conditional statement is true for the numbers in the spreadsheet cells. The spreadsheet can give different responses depending on whether the condition is true or false. A spreadsheet cannot prove a statement, but it can give evidence that the statement is true.

Example: Use a spreadsheet to test whether the statement below is true.
The sum of two consecutive integers is an odd integer.

The strategy is to generate several random pairs of consecutive integers and to see if their sums are always odd.

- Enter =INT(100*RAND()) into cell A1. RAND() gives a random number between 0 and 0.99. Multiplying by 100 gives a number between 0 and 99.99. The INT function removes the decimal part, leaving a number between 0 and 99.
- Enter =A1+1 into cell B1. This makes the number in B1 the integer after A1.
- Enter =IF(INT((A1+B1)/2)=(A1+B1)/2, "even," "odd") into cell C3. This instruction means the following:
 INT((A1+B1)/2) takes the sum of the integers, divides by 2, and removes any decimal part. (A1+B1)/2 takes the sum and divides it by 2.
 The IF function checks whether these numbers are equal. If the sum is even, then there is no decimal part when it is divided by two, so INT((A1+B1)/2) = (A1+B1)/2. If the sum is odd, (A1+B1)/2 will end in 0.5.
 When the IF condition is true, the word *even* is displayed. When the condition is false, the word *odd* is displayed.
- Click in cell A1 and drag to highlight all the cells from A1 to C10. Use the Fill Down command to fill these cells with the formulas in cells A1 through C1.

All entries in column C display the word *odd*. This suggests that the statement "The sum of two consecutive integers is an odd integer." is true. (It does not *prove* that the statement is true because it is not possible to test all integers.)

Although you used integers from 0 to 99 in this example, you may use any range of integers to test as long as the range is not too small.

	A	B	C
1	99	100	odd
2	63	64	odd
3	17	18	odd
4	78	79	odd
5	6	7	odd
6	1	2	odd
7	30	31	odd
8	25	26	odd
9	21	22	odd
10	0	1	odd

Use a spreadsheet to test whether the statement below is true.
The product of two consecutive integers is an even integer.

Does the statement appear to be true? Give the IF statement you used to test the statement.

The statement is true. IF statements may vary. Sample statement:
=IF(INT((A1*B1)/2)=(A1*B1)/2, "even," "odd")

LESSON 11.6 **565**

Project

Focus

The design of a new city park can be influenced by the space available and by the costs associated with each component of the park. Students will be asked to determine areas for parts of a proposed city park and to write rational functions that model the costs associated with each part. They will also make a scale drawing of the park that includes all of the features outlined in the Project.

Motivate

Before students begin the Project, have them make a list of some of the features they think should be included in a city park. Ask them to suggest reasons why some features may be omitted.

Explain to students that when city planners design a park, they must make trade-offs in order to stay within certain constraints. For instance, a winding road would give greater accessibility to more areas of the park, but would add to the cost.

CHAPTER PROJECT ELEVEN
designing a PARK

Suppose that your city has set aside a square piece of land with an area of 1 square mile on which to build a park. You are on the committee to design the park. Your committee has decided that the basic design will include a road, three parking lots, and a jogging trail. The road will run diagonally across the park. The parking lots will be at each end of this road and in the center of the park. The jogging trail will be constructed around the perimeter of the park.

Activity 1

The Basics

1. **a.** Find the length of the road. If the road is 50 feet wide, what is its area? Assume that the road forms a rectangle.
   b. Suppose that materials for the road cost $100 per square foot of area plus an additional $20 per foot of length. If the city requires an increase of 5 feet in the planned width of the road, calculate the new area and the total cost of the widened road.
   c. Using the cost information from part **b.**, write a rational function to determine the total cost per square foot based on the length and width of the road.

2. **a.** Find the length of the jogging trail. Suppose that the trail is 10 feet wide. Find the area of the jogging trail.
   b. Suppose that the jogging trail costs $20 per square foot of area plus $2 per foot for the outer perimeter. After an initial meeting, the committee wants to increase the width of the trail by 2 feet in order to accommodate more people. Calculate the new area and the total cost for the widened trail.
   c. Using the cost information from part **b.**, write a rational function to determine the total cost per square foot based on the length and width of the trail.

Activity 1

1. **a.** road length ≈ 7467 ft
   road area = 373,350 sq ft

   b. new road area = 410,685 sq ft
   road cost = $41,217,840

   c. l = length
   w = width
   cost = $100lw + 20l$

2. **a.** trail length = 21,120 ft
   trail area = 210,800 sq ft

   b. new trail area = 252,864 sq ft
   trail cost = $5,099,520

   c. l = length
   w = width
   cost = $20[4w(l - w)] + 2l$

Cooperative Learning

Students can do this Project in pairs. For **Activity 1**, one partner can do Step 1 while the other does Step 2. Partners should then share their results.

In **Activity 2**, the pair can work together to develop the table of dimensions for Step 1. Both students should then work individually to write and graph the function in Step 2. They should then compare results, resolve any differences, and do Step 3 together.

In **Activity 3**, have one student of each pair complete Step 1 while the other finds the area required in Step 2. The partners should share the research tasks in Step 2. They can then work together to make a scale drawing of the park.

Activity 2
Adding On

The first facility that the committee wants to build on the land is a recreation building. Based on successful parks in other areas, the committee determines that the recreation building needs to have an area of 14,400 square feet. In order to simplify the design and keep costs as low as possible, the building will be rectangular in shape and have the smallest possible perimeter.

1. Make a table of possible dimensions for the building. Use widths of 20, 40, 60, . . . , 200. Which dimensions have the smallest perimeter?
2. Write and graph a rational function for the perimeter, p, in terms of the width, x.
3. Which dimensions are possible if you want to keep the perimeter under 525 feet?

Activity 3
The Final Design

Suppose that after presenting the basic design to the city and getting feedback, the committee decides to add a pool, a baseball stadium, and six playground areas to the park. The pool will be Olympic size; the baseball stadium will seat 1000 people; and each playground area will be 80 yards by 120 yards.

1. Each of the three parking lots has dimensions of $\frac{1}{8}$ mile by $\frac{1}{8}$ mile. Find the area of the land that is not used by the parking lots, the jogging trail, the road, or the recreation facility.
2. Determine the total area of the six playgrounds. Do some research to estimate the areas of the pool and baseball, and determine how much land will remain undeveloped after all of the features have been added.
3. Create a complete scale drawing or model of the park that includes all of the features discussed by the committee.

Discuss

Bring the class together to discuss their results. Ask students how confident they would be about submitting their plan and the projected costs to a city council.

Discuss with students some factors that might cause the project to be more expensive. Ask them if they are aware of any projects in their community that have gone over budget. If so, discuss possible reasons why this occurred and whether the contractor could have foreseen the difficulties.

Activity 2

1.

Width	Length	Perimeter
20	720	1480
40	360	800
60	240	600
80	180	520
100	144	488
120	120	480
140	102.9	485.8
160	90	500
180	80	520
200	72	544

120ft × 120ft has the smallest perimeter.

2. $p = 2\left(\frac{14{,}400}{x} + x\right)$

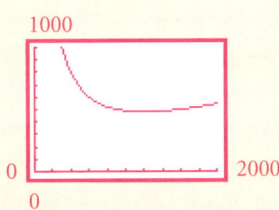

3. 80 ft × 180 ft, 100 ft × 144 ft, 120 ft × 120 ft, 140 ft × 102.9 ft, 160 ft × 90 ft

Activity 3

1. 25,893,651 sq ft

2. playgrounds: 518,400 sq ft
   Pool and stadium answers may vary. sample answers:
   pool: 11,500 sq ft
   stadium: 320,000 sq ft

3. Answers may vary.

Chapter Review

9. $y = \dfrac{-2}{x} + 4, x \neq 0$

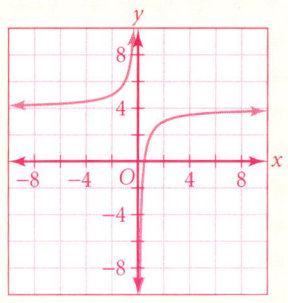

Chapter Test, Form A

Chapter Assessment
Chapter 11, Form A, page 1

Write the letter that best answers the question or completes the statement.

c   1. Which rational expression is undefined for $x = -1$?
    a. $\dfrac{x+1}{x^2-3}$ b. $\dfrac{x+3}{x^2+1}$ c. $\dfrac{x+3}{x+1}$ d. $\dfrac{x+1}{x-1}$

b   2. Which function is graphed at right?
    a. $f(x) = \dfrac{1}{x+1} + 1$
    b. $f(x) = \dfrac{1}{x-1} + 1$
    c. $f(x) = \dfrac{1}{x+1} - 1$
    d. $f(x) = \dfrac{1}{x-1} - 1$

d   3. The value of ABS $(-6.5) -$ ABS (6) is
    a. -0.5 b. -12.5 c. 12.5 d. 0.5

a   4. Given that y varies inversely as x, if y is $\frac{1}{2}$ when x is 20, what is the value of x when y is 5?
    a. 2   b. $\frac{1}{4}$ c. 4   d. $2\frac{1}{2}$

c   5. Time varies inversely as the average speed over a given distance. When Aaron travels 3 hours, his average speed is 75 km/h. What is Aaron's speed when his time is 5 hours?
    a. 15 km/h   b. 25 km/h   c. 45 km/h   d. 225 km/h

a   6. When simplified, $\dfrac{15a^2b^3}{20ab^5}$ is equal to
    a. $\dfrac{3a}{4b^2}$.   b. $\dfrac{a}{4b}$.   c. $4ab$.   d. $\dfrac{a^3b^8}{4}$.

c   7. When simplified, $\dfrac{3x^2-6x}{x^2-3x+2}$ is equal to
    a. $\dfrac{3}{x-1}$.   b. $\dfrac{x-2}{x-1}$.   c. $\dfrac{3x}{x-1}$.   d. 3.

c   8. What is the sum $\dfrac{5n}{3} + \dfrac{3}{2n}$?
    a. $\dfrac{5n+3}{6n}$.   b. $\dfrac{5n+3}{5n}$.   c. $\dfrac{10n^2+9}{6n}$.   d. $\dfrac{10n^2+9}{6n^2}$.

b   9. The difference $y - \dfrac{y-2}{y+2}$ is equal to
    a. $\dfrac{y^2-y-2}{y+2}$.   b. $\dfrac{y^2+y+2}{y+2}$.   c. $y - 1$.   d. $y + 1$.

Chapter Assessment
Chapter 11, Form A, page 2

b   10. The product $\dfrac{7a^3b}{2c} \cdot \dfrac{c^3}{21a^2b^3}$ is equal to
    a. $\dfrac{2a^2c^2}{3b^2}$.   b. $\dfrac{ac^2}{6b^2}$.   c. $\dfrac{a^3b^4c^4}{42}$.   d. $\dfrac{a^3c^3}{b^3}$.

a   11. What is the product of $\dfrac{x^2-3x+2}{x^2-1}$ and $\dfrac{x^2+x}{3x}$?
    a. $\dfrac{x-2}{3}$ b. $\dfrac{x-2}{3x}$ c. $\dfrac{x(x-2)}{3}$ d. $\dfrac{x-2}{3(x+1)}$

d   12. The quotient $\dfrac{p^3}{2qr^2} \div \dfrac{-p^2r}{4q}$ is
    a. $-\dfrac{p^5}{8p^2r}$.   b. $-\dfrac{2p}{r}$.   c. $-\dfrac{2}{pr}$.   d. $-\dfrac{2p}{r^3}$.

c   13. The quotient $\dfrac{w^2-6w+9}{w^2+5w+6} \div \dfrac{4w-12}{w+2}$ is
    a. $\dfrac{w-3}{4w-12}$.   b. $\dfrac{w-3}{w+3}$.   c. $\dfrac{w-3}{4w+12}$.   d. $\dfrac{w+3}{w-3}$.

b   14. For which values of k is $\dfrac{3k}{2k+1} + \dfrac{4}{k} = 0$ true?
    a. $k = \frac{2}{3}, k = 2$ b. $k = -\frac{2}{3}, k = -2$
    c. $k = \frac{3}{4}, k = -4$ d. $k = -\frac{3}{4}, k = 4$

b   15. The solution of the equation $\dfrac{2}{x-1} - \dfrac{1}{2} = -\dfrac{5}{2}$ is
    a. 2.   b. 0.   c. -2.   d. 1.

c   16. The difference between two numbers is 21. If the larger number is divided by the smaller, the quotient is 2 with a remainder of 6. What are the two numbers?
    a. 21 and 42   b. 3 and 19   c. 15 and 36   d. 17 and 38

d   17. If $\dfrac{m}{5} = \dfrac{m+2}{3}$, then
    a. $m = 5$.   b. $m = \frac{5}{4}$.   c. $m = -1$.   d. $m = -5$.

a   18. The value of INT(-4.8) is
    a. -5.   b. 5.   c. -4.   d. 4.

d   19. Using the definition of an odd number, which of the following shows that $4k + 17$ is an odd number?
    a. $4k + 17 = 2(2k + 8)$ b. $4k + 17 = 2(2k) + 17$
    c. $4k + 17 = 4k + 16 + 1$ d. $4k + 17 = 2(2k + 8) + 1$

568   CHAPTER 11 REVIEW

---

11 Chapter Review and Assessment

VOCABULARY

Closure Property 562	field 563	rational expression 532
conclusion 558	hole 534	rational function 533
conditional statement 558	hypothesis 558	rule of 72 529
conjecture 559	inductive reasoning 559	theorem 558
converse 558	inverse variation 526	trivial rational function 534
counterexample 561	non-trivial rational function 534	vertical asymptote 534
deductive reasoning 559	proof 558	

Key Skills & Exercises

LESSON 11.1
Key Skills

Identify and use inverse variation.

If y varies inversely as x and y is 10 when x is 3, the constant of variation is found by solving the equation $10 = \dfrac{k}{3}$. Thus, $k = 30$.

To find x when $y = 5$, use the constant of variation.

$$5 = \dfrac{30}{x} \qquad 5x = 30 \qquad x = 6$$

Exercises

1. If x varies inversely as y and $x = 75$ when $y = 20$, find y when $x = 25$. **y = 60**

2. If p varies inversely as q and $q = 36$ when $p = 15$, find p when $q = 24$. **p = 22.5**

3. If m varies inversely as n and $m = 0.5$ when $n = 5$, find m when $n = 50$. **m = 0.05**

4. If y varies inversely with x and if y is $\frac{1}{2}$ when x is 20, find x when y is 5. **x = 2**

LESSON 11.2
Key Skills

Evaluate rational functions. List the values of the variable for which functions are undefined.

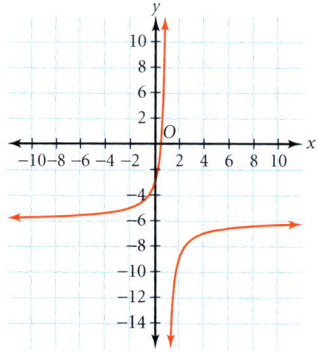

The function $y = \dfrac{-3}{x-1} - 6$ is graphed above. The graph has a vertical asymptote at $x = 1$. The function is undefined when $x - 1 = 0$, or $x = 1$.

Exercises

Evaluate each rational expression for $x = -1$ and $x = 2$. Write "undefined" if appropriate.

5. $\dfrac{4}{x}$ **−4, 2**
6. $\dfrac{3}{x+1}$ **undef., 1**
7. $\dfrac{-12}{2x-4}$ **2, undef.**
8. $\dfrac{-12}{x^2-x-2}$ **undef., undef.**

Graph each of the following functions. List any values of the variable for which the function is undefined.

9. $y = \dfrac{-2}{x} + 4$

10. $y = \dfrac{-3}{x+4} - 1$

11. $y = \dfrac{2}{x-2} + 5$

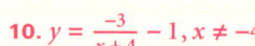

10. $y = \dfrac{-3}{x+4} - 1, x \neq -4$

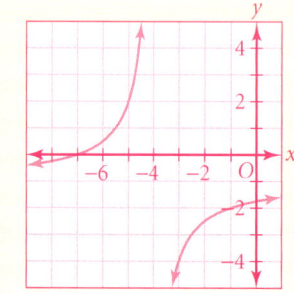

11. $y = \dfrac{2}{x-2} + 5, x \neq 2$

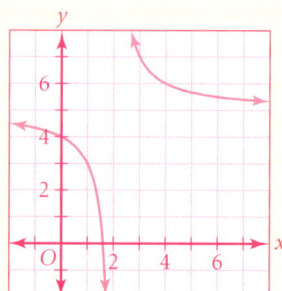

LESSON 11.3

Key Skills

Simplify rational expressions.

To simplify $\frac{12(x+y)}{3x^2 - 3y^2}$, factor the numerator and denominator, remove any common factors, and list any restrictions for the original expression.

$$\frac{12(x+y)}{3x^2 - 3y^2} = \frac{(3)(4)(x+y)}{3(x+y)(x-y)} = \frac{4}{x-y}, x \neq \pm y$$

Exercises

Simplify. List any restrictions.

12. $\frac{2x(x+3)}{8x^2}$

13. $\frac{8(x+1)}{10(x+1)^2}$

14. $\frac{2x^2 + 6x}{6x - 30}$

15. $\frac{2x^2 - x - 1}{1 - x^2}$

LESSON 11.4

Key Skills

Add, subtract, multiply, and divide rational expressions.

To add $\frac{1}{4x} + \frac{3}{5x^2}$, rewrite the expression with common denominators and add. List any restrictions for the original expressions.

$$\frac{1}{4x} + \frac{3}{5x^2} = \frac{1(5x)}{4x(5x)} + \frac{3(4)}{5x^2(4)}$$
$$= \frac{5x}{20x^2} + \frac{12}{20x^2} = \frac{5x + 12}{20x^2}, x \neq 0$$

To multiply $\frac{x+y}{5x} \cdot \frac{3x}{x^2 - y^2}$, multiply the numerators and multiply the denominators. Simplify, and list any restrictions for the original expressions.

$$\frac{x+y}{5x} \cdot \frac{3x}{x^2 - y^2} = \frac{(x+y)(3x)}{5x(x^2 - y^2)}$$
$$= \frac{(x+y)(3x)}{5x(x+y)(x-y)}$$
$$= \frac{3}{5(x-y)} \; x \neq 0, \text{ and } x \neq +y$$

To divide $\frac{x^2 - 4}{x - 5} \div \frac{x + 2}{x - 5}$, invert the divisor and multiply. List any restrictions for the original expressions. Also, restrict any values for the variable that make the divisor zero.

$$\frac{x^2 - 4}{x - 5} \div \frac{x + 2}{x - 5} = \frac{x^2 - 4}{x - 5} \cdot \frac{x - 5}{x + 2}$$
$$= \frac{(x+2)(x-2)(x-5)}{(x-5)(x+2)}$$
$$= x - 2, x \neq 5 \text{ and } x \neq -2$$

Exercises

Perform the indicated operations and simplify. List any restrictions.

16. $\frac{8mt}{3s} \cdot \frac{s^2}{16t}$

17. $\frac{2}{5x} + \frac{5}{10y}$

18. $\frac{16}{y^2 - 16} - \frac{2}{y + 4}$

19. $\frac{t^2 - 9}{6} \div \frac{3 - t}{9}$

20. $\frac{3}{4x} + \frac{5}{2}$

21. $\frac{a^2 - 2a - 15}{6} \cdot \frac{4}{3a + 9}$

22. $\frac{5x}{6} - \frac{3}{2x}$

23. $\frac{5r^2 q}{6q} \div \frac{3r}{q^2 r}$

24. $\frac{1}{2} + \frac{10}{5x}$

25. $\frac{x^2 + x - 12}{x - 3} \cdot \frac{x^2 - 25}{x^2 - x - 20}$

---

12. $\frac{x+3}{4x}, x \neq 0$

13. $\frac{4}{5(x+1)}, x \neq -1$

14. $\frac{x^2 + 3x}{3x - 15}, x \neq 5$

15. $\frac{2x+1}{-1-x}, x \neq -1, 1$

16. $\frac{ms}{6}, s \neq 0, t \neq 0$

17. $\frac{4y + 5x}{10xy}, x \neq 0, y \neq 0$

18. $\frac{24 - 2y}{y^2 - 16}, y \neq -4, 4$

19. $\frac{-3t - 9}{2}, t \neq 3$

20. $\frac{3 + 10x}{4x}, x \neq 0$

21. $\frac{2(a-5)}{9}, a \neq -3$

22. $\frac{5x^2 - 9}{6x}, x \neq 0$

23. $\frac{5r^2 q^2}{18}, q \neq 0, r \neq 0$

24. $\frac{x+4}{2x}, x \neq 0$

25. $x + 5, x \neq -4, 3, 5$

Chapter Test, Form B

NAME _____ CLASS _____ DATE _____

Chapter Assessment
Chapter 11, Form B, page 1

1. What condition causes a rational expression to be undefined? *A rational expression is undefined when the denominator is equal to 0.*

2. Explain how the expression $x - 3$ is related to $3 - x$. *$3 - x$ is the opposite of $x - 3$.*

3. In the statement "If $a + 3 > 8$, then $a > 5$," what part is assumed to be true? *$a + 3 > 8$*

Graph each rational function on the grid provided.

4. $y = \frac{1}{x} + 2$

5. $y = -\frac{1}{x+1}$

For Exercises 6–8, y varies inversely as x.

6. If y is 8 when x is 3, find x when y is 4. *$x = 6$*
7. If x is 12 when y is 5, find y when x is 15. *$y = 4$*
8. If y is 16 when x is $\frac{1}{2}$, find x when y is 1. *$x = 8$*
9. The frequency of a vibrating string is inversely proportional to its length. A violin string that is 10 inches long vibrates at a frequency of 512 cycles per second. Find the frequency of an 8-inch string. *640 cycles per second*

Simplify each expression.

10. $\frac{14x^4 y^2}{21x^3 y^3}$ *$\frac{2x}{3y}$* 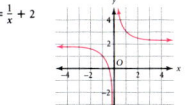 11. $\frac{2a^3 - 6a^2}{2a^2 - 18}$ *$\frac{a^2}{a+3}$*

12. $\frac{(m-1)^2}{(1-m)^2}$ *1* 13. $\frac{3k + 12}{16 - k^2}$ *$\frac{3}{4-k}$*

14. $\frac{5n - 10}{n^2 + 3n - 10}$ *$\frac{5}{n+5}$* 15. $\frac{b^2 - 2b - 15}{b^2 - 6b + 5}$ *$\frac{b+3}{b-1}$*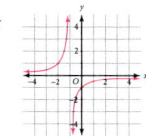

NAME _____ CLASS _____ DATE _____

Chapter Assessment
Chapter 11, Form B, page 2

Perform the indicated operations.

16. $\frac{5}{2x} + \frac{7}{6x^2}$ *$\frac{15x + 7}{6x^2}$* 17. $\frac{1}{5a} - \frac{4}{7a^2}$ *$\frac{7a - 20}{35a^2}$*

18. $\frac{3m^5}{4n^2} \cdot \frac{12n}{15m^2}$ *$\frac{3m^3}{5n}$* 19. $\frac{18p^3}{5q^4} \div \frac{9p}{q^3}$ *$\frac{2p^2 q}{5}$*

20. $\frac{y^2}{2y+3} + 3$ *$\frac{(y+3)^2}{2y+3}$* 21. $\frac{b^2 - 4b}{b - 1} - \frac{3}{1-b}$ *$b - 3$*

22. $\frac{2k^2 - 7k + 3}{2k^2 - 4k} \cdot \frac{k^2 - k - 2}{6k - 3}$ *$\frac{k^2 - 2k - 3}{6k}$*

23. $\frac{(z-3)^2}{z+1} \div \frac{z^2 - 9}{3z^2 + 3z}$ *$\frac{3z(z-3)}{z+3}$*

Solve algebraically.

24. $\frac{5}{x} + \frac{1}{2} = \frac{3}{x}$ *$x = -4$* 25. $\frac{x+2}{x} + \frac{2}{3x} = 5$ *$x = \frac{2}{3}$*

26. $\frac{x+1}{x^2 - 4} + \frac{7}{x+2} = -3$ *$n = 1$* 27. $\frac{5}{x-3} + 1 = \frac{1}{x^2 - 4x + 3}$ *$x = -3$ and $x = 2$*

28. Use the grid provided to solve $-\frac{2}{x} + \frac{1}{2} = 3$ by graphing. *$x = -\frac{4}{5}$*

Evaluate.

29. ABS (-5.8) *5.8* 30. INT $\left(3\frac{2}{3}\right)$ *3*

31. Jill and her younger sister Samantha mow lawns to earn extra money. If Jill can mow a lawn in 45 minutes and Samantha can mow a lawn in $1\frac{1}{2}$ hours, how long will it take if they work together? *30 minutes*

32. Write the hypothesis for the converse of the statement "If the length of the side of a square is $x + 3$, then the perimeter is $4x + 12$." *The perimeter is $4x + 12$.*

26. $y = 5$

27. $z = \frac{8}{7}$

28. $x = 3$ or $x = -2$

29. $w = -3$

30.
$2a + 3 = 5$	Given
$2a + 3 - 3 = 5 - 3$	Sub. Prop. of Eq.
$2a + 0 = 2$	Simplify.
$2a = 2$	Add. Prop. of Zero
$\frac{2a}{2} = \frac{2}{2}$	Div. Prop. of Eq.
$a = 1$	Simplify.

31.
$x = (a - b) - (b - a)$	Given
$= a - b - b + a$	Dist. Prop.
$= a + [-b + (-b)] + a$	Def. of Sub.
$= a + (-2b) + a$	Simplify.
$= a + a + (-2b)$	Comm. Prop. of Add.
$= 2a + (-2b)$	Simplify.
$= 2a - 2b$	Def. of Sub.

32.
$3x - 9 = 6$	Given
$\frac{3x}{3} - \frac{9}{3} = \frac{6}{3}$	Div. Prop. of Eq.
$x - 3 = 2$	Simplify.
$x - 3 + 3 = 2 + 3$	Add. Prop. of Eq.
$x = 5$	Simplify.

LESSON 11.5

Key Skills

Solve rational equations by the common-denominator method.

To solve $\frac{2}{x-1} + \frac{3}{x+1} = 1$, multiply both sides of the equation by the common denominator, $x^2 - 1$.

$(x^2 - 1)\left(\frac{2}{x-1} + \frac{3}{x+1}\right) = (x^2 - 1)1$

$2(x + 1) + 3(x - 1) = x^2 - 1$

$2x + 2 + 3x - 3 = x^2 - 1$

$x^2 - 5x = 0$

$x(x - 5) = 0$

$x = 0$ or $x = 5$

Exercises

Solve each equation by using the common-denominator method.

26. $\frac{5}{2y} - \frac{3}{10} = \frac{1}{y}$

27. $\frac{z+2}{2z} + \frac{z+3}{z} = 5$

28. $\frac{6-x}{x} = \frac{4}{x+1}$

29. $\frac{5}{w+6} - \frac{2}{w} = \frac{9w+6}{w^2+6w}$

LESSON 11.6

Key Skills

Prove algebraic statements.

To prove that if $x = a + 2(b + a)$, then $x = 3a + 2b$, list each step and give the reason next to the step.

Statement	Reason
1. $x = a + 2(b + a)$	Hypothesis or given
2. $= a + 2b + 2a$	Distributive Property
3. $= 3a + 2a + 2b$	Commutative Property
4. $= 3a + 2b$	Simplified form

Exercises

Prove each of the following:

30. If $2a + 3 = 5$, then $a = 1$.

31. If $x = (a - b) - (b - a)$, then $x = 2a - 2b$.

32. If $3x - 9 = 6$, then $x = 5$.

33. If $n = (2x + y) - (x - 3y)$, then $n = x + 4y$.

34. If $-6a - 5 = -95$, then $a = 15$.

Applications

35. **FITNESS** Millie joins a fitness club. There is a $300 fee to join and a monthly charge of $20. Write a rational expression to represent Millie's average cost per month for belonging to the fitness club. $\frac{300 + 20x}{x}, x \neq 0$

36. **GEOMETRY** The area ($\frac{1}{2}$ · base · height) of a certain triangle is given by $x^2 - 2x - 15$ and its height is given by $x - 5$. Write a simplified expression for the length of its base. $2x + 6, x \neq 5$

37. **TRAVEL** A trip of 189 miles can take 1 hour less if Gerald increases the average speed of his car by 12 miles per hour. What was the original average speed of the car? 42 mph

38. **GEOMETRY** The area (length · width) of certain rectangle is represented by $6x^2 + x - 35$. Its length is represented by $2x + 5$. Find the width of the rectangle. $3x - 7, x \neq -\frac{5}{2}$

33.
$n = (2x + y) - (x - 3y)$	Given
$= 2x + y - x + 3y$	Dist. Prop.
$= 2x - x + y + 3y$	Comm. Prop. of Add.
$= x + 4y$	Simplify.

34.
$-6a - 5 = -95$	Given
$-6a - 5 + 5 = -95 + 5$	Add. Prop. of Eq.
$-6a = -90$	Simplify.
$\frac{-6a}{-6} = \frac{-90}{-6}$	Div. Prop. of Eq.
$a = 15$	Simplify.

11 Alternative Assessment

Performance Assessment

1. Suppose y varies inversely as x and $y = 9$ when $x = 8$.
   a. How do you find the constant of variation?
   b. Find the value of x when $y = 12$.
2. a. Explain how to simplify $\frac{3}{x+1} + \frac{2}{x}$.
   b. What is the sum?
   c. What are the restrictions on the variable?
3. Let $\frac{6}{x-1} + \frac{3}{x+1} = 5, x \neq 1, x \neq -1$.
   a. Solve the equation by the graphing method.
   b. Solve the equation by the common denominator method.
   c. Which method do you prefer? Why?
4. Why is it necessary to use a common denominator to add two rational expressions, but not to find the product of two rational expressions?

Portfolio Projects

1. You will need 36 algebra tiles.
   a. Form as many rectangles as possible using all 36 tiles for each rectangle.
   b. Make a table of the lengths, l, and widths, w.
   c. Write and graph a function that relates the length to the width. What happens to the width as the length decreases?
   d. Compare your graph with the graph of $f(x) = \frac{1}{x}$.

2. For stringed instruments, the length of the string varies inversely as the frequency of its vibrations. Do some research to find out how the length of the string affects the note that is played on the instrument.

3. The golden ratio and golden rectangle are used in art and architecture. Write a report on how the golden rectangle was used by artists in the sixteenth century.

4. The greatest common factor and the least common multiple are terms used in number theory. Write a report about the use of these two terms in modular arithmetic.

Peer Assessment

Complete this activity with a partner. Begin by writing a rational expression that can be factored. For example, $4x^2 + 2x$ can be factored and rewritten as $2x(x + 1)$. After rewriting the expression in factored form, your partner should write an expression for you to factor.

Check your partner's work, correct any errors, and then factor the expression. Continue this pattern until you have each factored 20 expressions. Be sure to use expressions involving addition, subtraction, multiplication, division, parentheses, and exponents during this activity.

internetconnect

The HRW Web site contains many resources to reinforce and expand your knowledge of rational expressions and equations. This Web site also provides Internet links to other sites where you can find information and real-world data for use in research projects, reports, and activities that involve determining patterns in data. Visit the HRW Web site at **go.hrw.com**, and enter the keyword **MA1 CH11** to access the resources for this chapter.

Portfolio Projects

1. a. Students will form various rectangles and squares with the algebra tiles.
   b. Tables may vary. One possible table:

l	w
1	36
2	18
3	12
4	9
6	6
9	4
12	3
18	2
36	1

   c. $w = \frac{36}{l}$

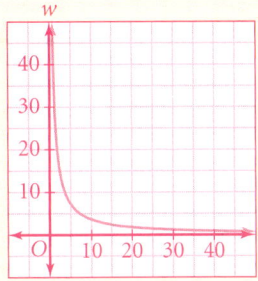

The width increases as the length decreases.

   d. The graph of $f(x) = \frac{1}{x}$ has the same shape, but is closer to the axes.

Alternative Assessment

Performance Assessment

1. a. Solve the equation $y = \frac{k}{x}$ for k; $k = 72$
   b. $x = 6$
2. a. Multiply the numerator and denominator by the factor required to result in the least common denominator.
   b. $\frac{5x+2}{x^2+x}$
   c. $x \neq -1, 0$
3. a.

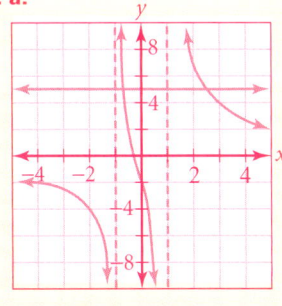

   b. $x \approx -0.65$ or $x \approx 2.45$
   c. Answers may vary.
4. Answers may vary. Sample answer: When rational expressions are multiplied together, the denominators are multiplied no matter what they are.

2. Answers may vary, but should mention that placing a finger on the string of an instrument shortens the portion of the string that is vibrating, and produces a higher frequency note.

3. Answers may vary, but might mention artwork by Leonardo da Vinci and Georges Seurat.

4. Answers may vary.

Cumulative Assessment

College Entrance Exam Practice

CHAPTERS 1–11 CUMULATIVE ASSESSMENT

College Entrance Exam Practice

Multiple-Choice and Quantitative-Comparison Samples

The first half of the Cumulative Assessment contains two types of items found on standardized tests—multiple-choice questions and quantitative-comparison questions. Quantitative-comparison items emphasize the concepts of equality, inequality, and estimation.

Free Response Grid Samples

The second half of the Cumulative Assessment is a free-response section. This part of the Cumulative Assessment requires student-produced response items like those commonly found on college entrance exams. These questions require the use of machine-scored answer grids. You may wish to have students practice answering these items in preparation for standardized tests.

QUANTITATIVE COMPARISON For Questions 1–5, write
A if the quantity in Column A is greater than the quantity in Column B;
B if the quantity in Column B is greater than the quantity in Column A;
C if the two quantities are equal; or
D if the relationship cannot be determined from the information given.

	Column A	Column B	Answers	
1.	$(-6)(9)$	$(6)(-9)$	Ⓐ Ⓑ Ⓒ Ⓓ [Lesson 2.4]	C
2.	x Given: $\begin{cases} x+y=3 \\ x-y=1 \end{cases}$	y	Ⓐ Ⓑ Ⓒ Ⓓ [Lesson 7.3]	A
3.	slope of a line that rises left to right	slope of a line that is horizontal	Ⓐ Ⓑ Ⓒ Ⓓ [Lesson 5.2]	A
4.	70% of x	90% of x	Ⓐ Ⓑ Ⓒ Ⓓ [Lesson 4.2]	D
5.	$\dfrac{5 \times 10^{-2}}{7 \times 10^{-5}}$	$\dfrac{5 \times 10^{2}}{7 \times 10^{5}}$	Ⓐ Ⓑ Ⓒ Ⓓ [Lesson 8.5]	A

6. Solve: $\begin{cases} 2x + y = 1 \\ x - y = -7 \end{cases}$ **(LESSON 7.3)** b
   - a. $(5, -2)$
   - b. $(-2, 5)$
   - c. $(2, -5)$
   - d. $(5, 2)$

7. What is the factored form of $4x^2 - 25$? **(LESSON 9.6)** b
   - a. $(2x - 5)^2$
   - b. $(2x + 5)(2x - 5)$
   - c. $(2x + 5)^2$
   - d. cannot be factored

8. Evaluate $x - (z - y)$ for $x = -4$, $y = 2$, and $z = -3$. **(LESSON 2.3)** c
   - a. -9
   - b. -5
   - c. 1
   - d. -1

9. Which equation represents the axis of symmetry for the quadratic function $y = (x - 2)^2 - 3$? **(LESSON 10.1)** c
   - a. $x = 3$
   - b. $x = -3$
   - c. $x = 2$
   - d. $x = -2$

10. The length of each side of a square is 3 cm. What is the square's perimeter? **(LESSON 2.4)** b
    - a. 15 cm
    - b. 12 cm
    - c. 3 cm
    - d. 9 cm

11. Solve: $\dfrac{n}{12} = \dfrac{12}{18}$. **(LESSON 4.1)** b
    - a. -2
    - b. 8
    - c. 1
    - d. -1

12. Simplify: $\dfrac{x^2 - 1}{x - 1}$. **(LESSON 11.3)** b
    - a. $x^2 + 1$
    - b. $x + 1$
    - c. $x - 1$
    - d. $x^2 - 1$

13. Solve: $-3x - 5 > -11$. **(LESSON 6.2)** d
    - a. $x > 2$
    - b. $x > -2$
    - c. $x < 6$
    - d. $x < 2$

14. Simplify: $(5x^2y)(2xy^{-1})^2$. (LESSON 8.4) d
    a. $10x^4y^{-1}$ b. $20x^4y$
    c. $20x^4y^2$ d. $20x^4y^{-1}$
15. Evaluate $3(x+1)^2 + 1$ for $x = 2$. (LESSON 1.3)
    28
16. Which number must be added to $x^2 + 4x$ to complete the square? (LESSON 10.3) 4
17. Write an equation for the line passing through the origin and $(3, -4)$. (LESSON 5.5) $y = -\frac{4}{3}x$
18. Tickets to a concert cost $24 per person, plus a $7 handling charge per order. How many tickets can be ordered for $199? (LESSON 3.3) 8
19. Simplify: $\frac{-100a^8b^{10}}{-20a^5b^9}$. (LESSON 8.3) $5a^3b$
20. Vernon scored 86 on his first two tests, 90 on his third test, and 78 on his fourth test. Find the median test score. (LESSON 4.4) 86
21. Write an equation for the line with a slope of -2 passing through $(-1, 0)$. (LESSON 5.5) $y = -2x - 2$
22. Factor: $3a(2c + d)^2 - 5(2c + d)^2$. (LESSON 9.5) $(3a - 5)(2c + d)^2$
23. Solve: $\frac{2}{x+1} = \frac{x}{6}$. (LESSON 11.5) $x = 3$ or $x = -4$
24. Solve. $\begin{cases} 3x + y = -2 \\ 2x + 3y = 8 \end{cases}$ (LESSON 7.3) $(-2, 4)$
25. Simplify: $(3x^2y^3)^0$. (LESSON 8.4) 1
26. Solve: $|x - 5| = 17$. (LESSON 6.5) $x = -12$ or $x = 22$
27. Solve: $\frac{7}{9} = \frac{x}{27}$. (LESSON 3.2) $x = 21$
28. Find 60% of 240. (LESSON 4.2) 144
29. 16 is 20% of what number? (LESSON 4.2) 80
30. Solve: $(x + 3)^2 = 25$. (LESSON 10.2) $x = -8$ or $x = 2$
31. If you invest $2000 in a CD that earns 5% interest, compounded annually, how much will the CD be worth in 5 years? (LESSON 8.7)
    $ 2552.56
32. Write an equation in slope-intercept form for the line that contains the point $(3, -7)$ and is perpendicular to $5x + 2y = 10$. (LESSON 5.6)
    $y = \frac{2}{5}x - \frac{41}{5}$

33. Solve: $x^2 - 3x - 18 = 0$. (LESSON 9.8) $x = 6$ or $x = -3$
34. Solve. $\begin{cases} 3x - 4y = -18 \\ 2x + 6y = 14 \end{cases}$ (LESSON 7.3) $(-2, 3)$
35. Solve. $\begin{cases} x = y \\ 2x + 3y = 7 \end{cases}$ (LESSON 7.2) $\left(\frac{7}{5}, \frac{7}{5}\right)$

FREE-RESPONSE GRID
The following questions may be answered by using a free-response grid such as that commonly used by standardized test services.

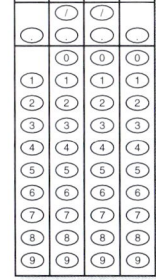

36. Evaluate $2x^2 + 7x - 20$ for $x = 5$. (LESSON 1.3) 65
37. Solve: $\frac{2x}{3} = 12$. (LESSON 3.2) $x = 18$
38. What is the slope of the line whose equation is $y = 2x + 5$? (LESSON 5.4) 2
39. Tim must score a total of at least 270 points on his tests to earn an A for the semester. On the first test, he scored an 85 and on the second test he scored an 89. What is the lowest score that Tim can receive on the third and final test and still earn an A? (LESSON 6.1) 96
40. Find the slope of a line that passes through the points $(4, 3)$ and $(-2, -3)$. (LESSON 5.2) 1
41. Solve: $3x^2 - 6x = -3$. (LESSON 9.8) $x = 1$
42. Solve: $\frac{x}{7} = \frac{12}{42}$. (LESSON 3.2) $x = 2$
43. Find the slope of the line represented by $2x + 3y = 5$. (LESSON 5.5) $-\frac{2}{3}$
44. Solve: $\frac{5}{11} = \frac{15}{w}$. (LESSON 4.1) $w = 33$
45. Find 15% of 260. (LESSON 4.2) 39

12 Radicals and Coordinate Geometry

CHAPTER PLANNING GUIDE

Lesson	12.1	12.2	12.3	12.4	12.5	12.6	12.7	12.8	Projects and Review
Pupil Edition Pages	576–582	583–590	591–597	598–605	606–612	613–620	621–627	628–635	636–645
Extra Practice (Pupil Edition)	772	773	773	774	774	775	775	776	
Practice Workbook	70	71	72	73	74	75	76	77	
Reteaching Masters	139–140	141–142	143–144	145–146	147–148	149–150	151–152	153–154	
Enrichment Masters	70	71	72	73	74	75	76	77	
Cooperative-Learning Activities	70	71	72	73	74	75	76	77	
Lesson Activities	70	71	72	73	74	75	76	77	
Problem Solving/ Critical Thinking	70	71	72	73	74	75	76	77	
Student Study Guide	70	71	72	73	74	75	76	77	
Spanish Resources	70	71	72	73	74	75	76	77	
Student Technology Guide	70	71	72	73	74	75	76	77	
Assessment Resources	151	152	153	155	156	157	158	159	154, 160–165
Teaching Transparencies	54	55, 56	57–59	60		61		62, 63	
Quiz Transparencies	12.1	12.2	12.3	12.4	12.5	12.6	12.7	12.8	
Writing Activities for Your Portfolio									34–36
Tech Prep Masters									55–58
Long-Term Projects									45–48

LESSON PACING GUIDE

	12.1	12.2	12.3	12.4	12.5	12.6	12.7	12.8	Projects and Review
Traditional	2 days	2 days	2 days	2 days	2 days	1 day	1 day	2 days	2 days
Block	1 day	1 day	1 day	1 day	1 day	$\frac{1}{2}$ day	$\frac{1}{2}$ day	1 day	1 day
Two-Year	4 days	4 days	4 days	4 days	4 days	2 days	2 days	4 days	4 days

CONNECTIONS AND APPLICATIONS

Lesson	12.1	12.2	12.3	12.4	12.5	12.6	12.7	12.8	Review
Algebra	576–582	583–590	591–597	598–605	606–612	613–620	621–627	628–635	636–645
Geometry		582	592, 594, 596, 597	599, 602	606	619	625		
Statistics/Probability								634, 635	
Coordinate Geometry, Trigonometry				604		616, 618, 620	624, 627		
Business and Economics			597			619	626		642
Science and Technology		584, 585, 586, 589		604, 605	609	613, 615, 616, 617, 619, 620	622, 623, 624	635	642
Sports and Leisure			594, 596			617, 620	621, 623, 626		
Other		582	596	598				628	642
Cultural Connection: Africa		58							
Cultural Connection: Americas				605					

BLOCK-SCHEDULING GUIDE

Day	Lesson	Teacher Directed: Lesson Examples, Teaching Transparencies	Student Guided: Activity, Try This	Cooperative-Learning Activity, Lesson Activity, Student Technology Guide	Practice: Practice & Apply, Extra Practice, Practice Workbook	Assessment: Quiz, Mid-Chapter Assessment	Problem Solving, Reteaching
1	12.1	15 min	15 min	15 min	65 min	15 min	15 min
2	12.2	15 min	20 min	15 min	65 min	15 min	15 min
3	12.3	10 min	15 min	15 min	65 min	15 min	15 min
4	12.4	10 min	15 min	15 min	65 min	15 min	15 min
5	12.5	10 min	15 min	15 min	65 min	15 min	15 min
6	12.6	5 min	8 min	8 min	35 min	8 min	8 min
6	12.7	5 min	7 min	7 min	30 min	7 min	7 min
7	12.8	15 min	15 min	15 min	65 min	15 min	15 min
8	Assess.	50 min PE: Chapter Review	90 min PE: Chapter Project, Writing Activities	90 min Tech Prep Masters	65 min PE: Chapter Assessment, Test Generator	30 min Chap. Assess. (A or B), Alt. Assess. (A or B), Test Generator	

PE: Pupil's Edition

Alternative Assessment

The following are suggestions for an alternative assessment for students who may benefit from a different type of assessment than the regular chapter quizzes and the mid-chapter and end-of-chapter assessment materials. Many of the questions are open-ended, and students' answers will vary.

Performance Assessment

1. a. Solve $\sqrt{4x-3} = 5$, giving a detailed solution. Explain how solving a linear equation helps you find x.
   b. Solve $\sqrt{4x-3} = x$, giving a detailed solution. Explain how solving a quadratic equation helps you find x.

2. A circle has a radius of 5 and its center at $C(3, 2)$.
   a. Write an equation for this circle.
   b. Write an equation for the circle translated 3 units to the left and 2 units down.

3. $\triangle ABC$ has vertices at $A(0, 0)$, $B(3, 5)$, and $C(3, 0)$.
   a. How do you know $\triangle ABC$ is a right triangle?
   b. Find the lengths of the sides of $\triangle ABC$.
   c. Approximate the measure of each angle to the nearest whole number of degrees.

Portfolio Project

Suggest that students choose one of the following projects for inclusion in their portfolios.

1. Consider numbers of the form $a + b\sqrt{2}$, where a and b are real numbers.
   a. Describe the results when you add, subtract, and multiply numbers of this form.
   b. How would you solve the equation below?
   $x + (a + b\sqrt{2}) = c + d\sqrt{2}$, where a, b, c, and d are real numbers

2. The diagram below shows a plot of land. Find the perimeter and area of the region. Justify your reasoning and results.

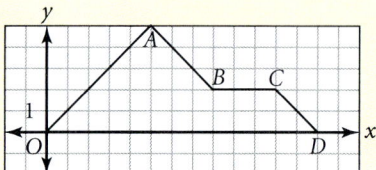

---

internetconnect

The table below identifies the pages in this chapter that contain technology information and support in the side columns.

Content Links

Lesson Links	pages 578, 587, 617, 624
Portfolio Links	pages 605, 627
Graphics Calculator Support	page 782

Resource Links

For information about teacher and parent resources as well as professional development help, visit **www.hrw.com/math**.

Technical Support HRW has assembled a team of dedicated technical and teaching professionals and a comprehensive service program to provide you with the support you need.

- The HRW Technical Support Line operates from 7 A.M. to 6 P.M. central time, Monday through Friday, at (800) 323-9239.
- The HRW Technical Support Center on the World Wide Web is available 24 hours a day, seven days a week, at **www.hrwtechsupport.com**.
- You can e-mail our Technical Support Center at **tsc@hrwtechsupport.com**.
- The Technical Support Center's fax-on-demand service at (800) 352-1680 offers solutions to common problems and answers to frequently asked questions.

Technology

Lesson Suggestions and Calculator Examples

(Keystrokes are based on a TI-83 calculator.)

Lesson 12.1 Operations With Radicals
In this lesson, students learn that if $a \geq 0$ and $b \geq 0$, $\sqrt{ab} = \sqrt{a}\sqrt{b}$ and $\sqrt{\dfrac{a}{b}} = \dfrac{\sqrt{a}}{\sqrt{b}}$ ($b \neq 0$).

To test these rules, have students choose positive numbers and evaluate expressions as shown at right.

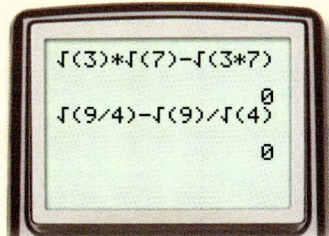

Lesson 12.2 Square-Root Functions and Radical Equations
Even though students have gained a great deal of practice with the graphics calculator, students still need to be attentive to the way in which they enter equations. The displays below show what happens when students fail to press) . Note that the calculator places all that follows the initial parentheses under the radicand unless the students press) after they enter the expression.

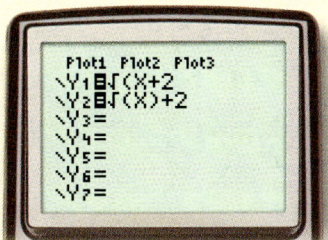

Lesson 12.3 The Pythagorean Theorem
A calculator can help students run tests. For example, can the numbers 3, 4, and 5 be the side lengths of a right triangle? Can 6, 7, and 8 be side lengths?

The display at right shows tests using the converse of the Pythagorean Theorem. The numbers 3, 4, and 5 give a right triangle, but the numbers 6, 7, and 8 do not. Ask students how the display shows these results.

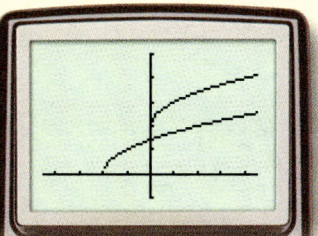

Lesson 12.4 The Distance Formula
Have students apply the distance formula to a pair of points without regard to the order in which x-coordinates are entered and y-coordinates are entered. The calculator gives the same result. Ask students why.

Lesson 12.5 Geometric Properties
Show students that in order to graph $x^2 + y^2 = 16$ on a graphics calculator they must separate it into two radical equations, $y = \sqrt{16 - x^2}$ and $y = -\sqrt{16 - x^2}$, and graph them in a square window.

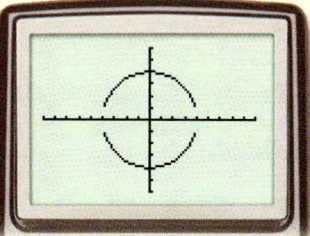

Lesson 12.6 The Tangent Function
To work with the tangent function on a graphics calculator, students need to use degree measurement. To select degree mode, press MODE, highlight Degree, and press ENTER. In $\triangle ABC$, m$\angle A = 35°$ and $BC = 4$. To find AC students should write $\tan 35° = \dfrac{4}{AC}$ and evaluate the expression for AC, as shown at right.

Lesson 12.7 The Sine and Cosine Functions
Again, students need to be sure the calculator is in degree mode when working with sine and cosine functions. To find AB in $\triangle ABC$ discussed in 12.6 above, students should write $\sin 35° = \dfrac{4}{AB}$ and evaluate the expression for AB.

Lesson 12.8 Introduction to Matrices
In this lesson, students will learn how to perform operations on matrices. Students can proceed as follows.

1. Enter the matrix or matrices, including dimensions.
2. Select the needed matrix or matrices by name.
3. Press the keys that give the desired operations.

Let $A = \begin{bmatrix} -2 & 0 \\ 3 & 5 \end{bmatrix}$ and $B = \begin{bmatrix} 0 & 4 \\ -2 & 1 \end{bmatrix}$.

The displays below show addition, subtraction, scalar multiplication, and matrix multiplication involving A and B.

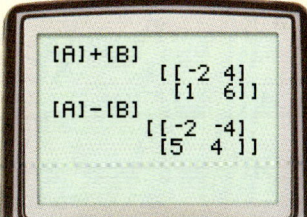
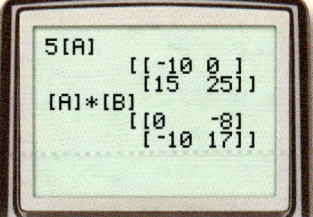

For further information, refer to the

- technology discussions in the lessons.
- lesson-related teacher's commentary in the side columns of this *Teacher's Edition*.
- lesson-related *Student Technology Guide* masters.
- *HRW Technology Handbook*.

Teaching Resources

Basic Skills Practice
(2 pages per skill)

Reteaching
(2 pages per lesson)

Enrichment
(1 page per lesson)

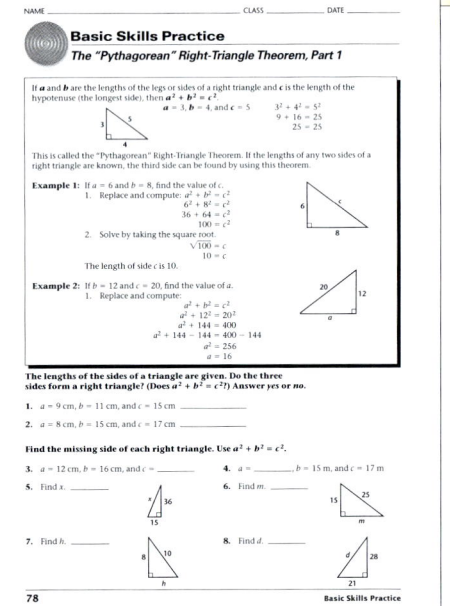

Cooperative-Learning Activity
(1 page per lesson)

Lesson Activity
(1 page per lesson)

Long-Term Project
(4 pages per chapter)

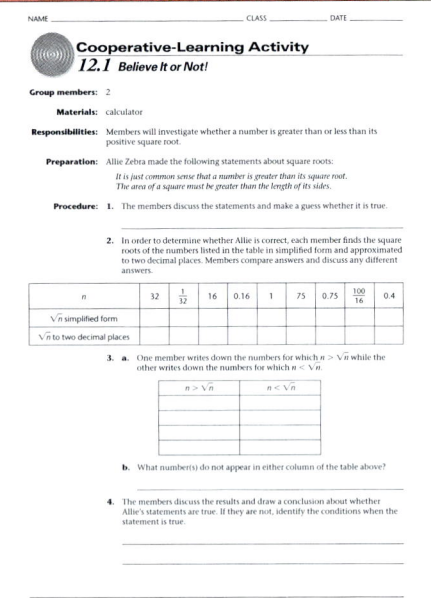

Problem Solving/Critical Thinking
(1 page per lesson)

Writing Activities
(3 pages per chapter)

Tech Prep Masters
(4 pages per chapter)

Spanish Resources
(1 page per lesson)

Quizzes
(1 page per lesson)

Mid-Chapter Test
(1 page per chapter)

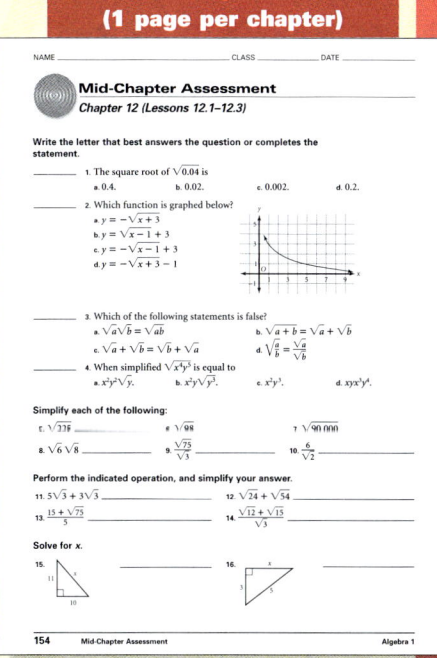

About Chapter 12

Background Information

In this chapter, students extend the study of radicals that began in Chapter 9. They learn to simplify radical expressions, to perform operations with radicals, and to solve radical equations. These skills then serve as the gateway to a study of selected geometric topics, including the Pythagorean Theorem, the distance formula, and the sine, cosine, and tangent ratios. The chapter concludes with a brief introduction to matrices.

CHAPTER RESOURCES

- Block-Scheduling Handbook
- Writing Activities for Your Portfolio
- Tech Prep Masters
- Long-Term Project
- Assessment Resources:
  Mid-Chapter Assessment
  Chapter Assessments
  Alternative Assessments
- Test and Practice Generator
- Technology Handbook
- End-of-Course Exam Practice

Chapter Objectives

- Identify or estimate square roots. [12.1]
- Define and write square roots in simplest radical form. [12.1]
- Perform mathematical operations with radicals. [12.1]
- Identify the graph of $y = \sqrt{x}$. [12.2]
- Solve equations containing radicals. [12.2]
- Solve equations by using radicals. [12.2]
- Find a side length of a right triangle given the lengths of its other two sides. [12.3]

12 Radicals, Functions & Coordinate Geometry

Lessons

- 12.1 Operations With Radicals
- 12.2 Square-Root Functions and Radical Equations
- 12.3 The Pythagorean Theorem
- 12.4 The Distance Formula
- 12.5 Geometric Properties
- 12.6 The Tangent Function
- 12.7 The Sine and Cosine Functions
- 12.8 Introduction to Matrices

Project Working the Angles

MONUMENTS SUCH AS THOSE PICTURED HERE are great accomplishments that can be even more fully appreciated with an understanding of the mathematical properties demonstrated by their construction. In order to understand the relationships between the various distances and angles used to construct such monuments, it is necessary to understand radicals, or square roots.

Radicals play an important role in algebra and geometry because they are essential to both the quadratic formula and the distance formula. In this chapter, you will learn techniques for solving equations that contain radicals and study several geometric properties and trigonometric functions that relate to radicals. You will have opportunities to apply these techniques to real-world situations ranging from construction to aviation.

Easter Island statues

Taj Mahal, India

About the Photos

The fields of architecture, art, and geometry often overlap. Angles and dimensions play a key role in the design of monuments like those pictured above.

The Taj Mahal, located in Agra, India, is built in perfect symmetry and looks equally graceful from any angle or direction. The interior of the tall minarets and tomb of the Taj Mahal have illusionary effects that can be attributed to its geometry.

For example, the lettering of the Qur'an verses around the archways appears to be uniform despite the varying heights of the lines of text.

The Great Wall of China is a triumph of engineering and determined planning. Begun in 221 B.C.E., it is one of the largest construction projects ever carried out, stretching about 4000 miles. The height of the wall ranges from 15 to 50 feet and its base width is between 15 and 30 feet.

The Great Wall of China

Stonehenge, England

- Apply the Pythagorean Theorem to real-world problems. [12.3]
- Use the distance formula to find the distance between two points in a coordinate plane. [12.4]
- Determine whether a triangle is a right triangle. [12.4]
- Apply the midpoint formula. [12.4]
- Define and use the equation of a circle. [12.5]
- Use the coordinate plane to investigate the diagonals of a rectangle and the midsegment of a triangle. [12.5]
- Identify and use the tangent ratio in a right triangle. [12.6]
- Find unknown side and angle measures in right triangles. [12.6]
- Define the sine and cosine ratios in a right triangle. [12.7]
- Find unknown side and angle measures in right triangles. [12.7]
- Understand the sine and cosine ratios as functions. [12.7]
- Determine the dimensions and addresses of a matrix. [12.8]
- Determine whether two matrices are equal. [12.8]
- Add, subtract, and multiply matrices. [12.8]
- Determine the multiplicative identity of a matrix. [12.8]

About the Chapter Project

Solving equations with radicals is an important skill that allows you to solve complex problems involving distance and various geometric and trigonometric properties. In the Chapter Project, *Working the Angles*, on page 636, you will learn how to build a hypsometer, use it to measure angles of elevation and depression, and calculate heights and distances from your observations.

After completing the Chapter Project, you will be able to do the following:

- Apply the inverse tangent function to a real-world situation.

About the Portfolio Activities

Throughout the chapter, you will be given opportunities to complete Portfolio Activities that are designed to support your work on the Chapter Project.

- Examining and graphing the action of pendulums is the topic of the Portfolio Activity on page 590.
- The "Wheel of Theodorus" is the topic of the Portfolio Activity on page 605.
- Golden rectangles are the topic of the Portfolio Activity on page 627.

Portfolio Activities appear at the end of Lessons 12.2, 12.4, and 12.7. Each serves as preparation for the Chapter Project. The Portfolio Activities as well as the Chapter Project Activities are appropriate for inclusion in the student's portfolio. Students should be encouraged to include in their portfolios any other work in which they feel a sense of pride or a sense of accomplishment.

internet connect

Internet Connect features appearing throughout the chapter provide keywords for the HRW Web site that lead to links for math resources, tutorial assistance, references for student research, classroom activities, and all teaching resource pages. The HRW Web site also provides updates to the technology used in the text. Listed at right are the keywords for the Internet Connect activities referenced in this chapter. Refer to the side column on the page listed to read about the activity.

LESSON	KEYWORD	PAGE
12.1	MA1 Continued Fractions	578
12.2	MA1 Planets	587
12.4	MA1 Theodorus	605
12.6	MA1 Right Triangles	617
12.7	MA1 Trig Functions	624
12.7	MA1 Golden	627

CHAPTER 12 **575**

Prepare

NCTM Principles & Standards 1–10

LESSON RESOURCES

- Practice 12.1
- Reteaching Master 12.1
- Enrichment Master 12.1
- Cooperative-Learning Activity 12.1
- Lesson Activity 12.1
- Problem Solving/Critical Thinking Master 12.1
- Teaching Transparency 54

QUICK WARM-UP

Evaluate.

1. 14^2 **196**   2. $(-5)^2$ **25**
3. $\sqrt{81}$ **9**   4. $-\sqrt{1}$ **−1**

Simplify.

5. $y \cdot y$ y^2
6. $p(p-2)$ $p^2 - 2p$
7. $(k+3)(k-3)$ $k^2 - 9$
8. $(n-4)(n-1)$ $n^2 - 5n + 4$

Also on Quiz Transparency 12.1

Teach

Why The concept of *square root* arises whenever you are given the area of a square and need to find the length of its sides. Ask students to suggest situations in which this skill might be needed. Tell them that in this lesson, they will learn new methods for evaluating square roots as well as techniques for performing operations with them.

12.1 Operations With Radicals

Why Radicals often appear when computing statistics or evaluating formulas in physics, but they can also be found in everyday activities like tiling.

Objectives

- Identify or estimate square roots.
- Define and write square roots in simplest radical form.
- Perform mathematical operations with radicals.

Estimating Square Roots

The blue square has an area of 12. How can the length of a side of the blue square be determined from the area?

Recall that the area of a square is found by squaring the length of a side.
$$Area = s^2$$

Thus, the square root of the area is the length of a side.

To estimate $\sqrt{12}$, find the two nearest perfect-square numbers, one that is less than 12 and another that is greater than 12—the numbers 9 and 16.

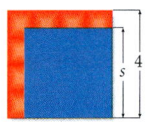 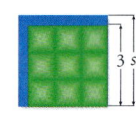 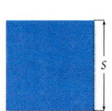

The area of the square, 12, is a little closer to 9 than it is to 16, so a good estimate for $\sqrt{12}$ is somewhere between 3.4 and 3.5. Multiply to refine the estimate to the nearest hundredth.

$$3.45^2 = 11.9025$$
$$3.46^2 = 11.9716$$
$$3.47^2 = 12.0409$$
$$3.48^2 = 12.1104$$

Thus, a reasonable estimate for $\sqrt{12}$ is 3.46. Recall from Lesson 10.2 that the principal square root, indicated by the radical sign, is always positive.

Alternative Teaching Strategy

USING COGNITIVE STRATEGIES   Have the students work in groups of three or four. On the board or overhead, write this multiple-choice question.

$7\sqrt{3} + \sqrt{3} = \underline{}$

a. 10   **b.** $7\sqrt{6}$ **c.** $8\sqrt{3}$ **d.** $8\sqrt{6}$ **e.** 21

Each group should work together to determine the correct answer.   **They should arrive at choice c.**

Instruct each group to write a convincing argument that supports their choice. Discuss the results with the entire class. Have one member of each group present a summary of the argument. During these presentations, elicit their reasons, if any, for eliminating the other choices.

Repeat the activity with the following question:

$(7\sqrt{3})(\sqrt{3}) = \underline{}$

a. 10   **b.** $7\sqrt{6}$ **c.** $8\sqrt{3}$ **d.** $8\sqrt{6}$ **e.** 21

The correct choice is e.

576   LESSON 12.1

Definition of Square Root

If a is a number greater than or equal to zero, $\sqrt{a}$ represents the principal, or positive, square root of a and $-\sqrt{a}$ represent the negative square root of a. The square roots of a have the following property:

$$\sqrt{a} \cdot \sqrt{a} = a \qquad (-\sqrt{a})(-\sqrt{a}) = a$$

You were introduced to the radical sign, or just *radical*, in Lesson 10.2. The number or expression under the radical is the **radicand**. You also learned in 10.2 that the solutions to $x^2 = a$ are $x = \sqrt{a}$ or $x = -\sqrt{a}$. This is often abbreviated as $x = \pm\sqrt{a}$, "x equals plus or minus the square root of a."

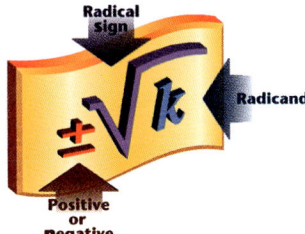

Use Teaching Transparency 54.

ADDITIONAL EXAMPLE 1

Evaluate each square root.
a. $\sqrt{100}$ 10
b. $-\sqrt{225}$ -15
c. $\sqrt{11}$ ≈ 3.32
d. $\pm\sqrt{121}$ $-11, 11$

TRY THIS
a. 14
b. -15

EXAMPLE 1 Evaluate each square root.
a. $\sqrt{25}$ b. $-\sqrt{144}$ c. $\sqrt{7}$ d. $\pm\sqrt{9}$

SOLUTION

a. $\sqrt{25} = 5$ because $5 \cdot 5 = 25$ and $\sqrt{25}$ means the principal, or positive, square root.
b. $-\sqrt{144} = -12$ because $-(\sqrt{144}) = -(12) = -12$.
c. $\sqrt{7}$ is not a perfect square; 4 and 9 are perfect squares, so $\sqrt{7}$ must be between 2 and 3. Multiplication will confirm that 2.65 is a good estimate to the nearest hundredth.
d. $\pm\sqrt{9} = 3$ or $\pm\sqrt{9} = -3$ because both the positive and negative roots are indicated.

TRY THIS Evaluate each square root.
a. $\sqrt{196}$ b. $-\sqrt{225}$

ADDITIONAL EXAMPLE 2

Simplify.
a. $7\sqrt{3} - 9\sqrt{3}$ $-2\sqrt{3}$
b. $6 - 4\sqrt{2} - 9 + 3\sqrt{2}$ $-3 - \sqrt{2}$
c. $4\sqrt{5} - \sqrt{7} + 2\sqrt{7} + 9\sqrt{5}$ $13\sqrt{5} + \sqrt{7}$
d. $m\sqrt{n} - \sqrt{n}$, for $n \geq 0$ $(m - 1)\sqrt{n}$

Simplifying Radicals and Radical Expressions

In order to add or subtract expressions that contain radicals, radicands must be equal. Terms that contain $\sqrt{5}$, for example, can be added by applying the Distributive Property.

$$2\sqrt{5} + 4\sqrt{5} = (2 + 4)\sqrt{5} = 6\sqrt{5}$$

EXAMPLE 2 Simplify.
a. $5\sqrt{6} - 2\sqrt{6}$ b. $5 + 6\sqrt{7} - 2\sqrt{7} - 3$
c. $8\sqrt{3} + 6\sqrt{2} - \sqrt{3} + 2\sqrt{2}$ d. $a\sqrt{x} + b\sqrt{x}$

Teaching Tip

TECHNOLOGY If you are using a TI-82 or TI-83 graphics calculator, you access its square-root function by pressing [2nd] [x^2].

Interdisciplinary Connection

GEOGRAPHY The formula for estimating the distance, d, in miles to the horizon from a height of h feet can be written as $d \approx \sqrt{\frac{2}{3}h}$. Rewrite this formula so that the denominator of the expression on the right side is rationalized. $d \approx \frac{\sqrt{6h}}{3}$

Use the formula to estimate the distance to the horizon from each height given at right. When appropriate, give answers in simplified radical form. Then use a calculator to approximate the value to the nearest tenth of a mile.

a. 24 feet ≈ 4 miles
b. 500 feet $\approx \frac{10\sqrt{30}}{3}$ miles, or ≈ 18.3 miles
c. 45,000 feet $\approx 100\sqrt{3}$ miles, or ≈ 173.2 miles
d. 5 miles (1 mile = 5280 feet) $\approx 40\sqrt{11}$ miles, or ≈ 132.7 miles

LESSON 12.1

TRY THIS

a. $3\sqrt{6}$
b. $-a - \sqrt{b}$

Activity Notes

In this Activity, students investigate these two key statements involving square roots.

$\sqrt{ab} = \sqrt{a} \cdot \sqrt{b}$
$\sqrt{a+b} = \sqrt{a} + \sqrt{b}$

They should arrive at the conclusion that the first statement is always true. As the lesson progresses, they will learn that this statement illustrates the Multiplication Property of Square Roots. The second statement, however, is *not* always true.

Cooperative Learning

You may wish to have students do the Activity in pairs. One partner can do Steps 1 and 2, while the other partner does Steps 3 and 4. The partners can then compare and discuss their results.

CHECKPOINT ✔

3. $\sqrt{9+16} = \sqrt{25} = 5$
$\sqrt{9} + \sqrt{16} = 3 + 4 = 7$

internetconnect

GO TO: go.hrw.com
KEYWORD: MA1 Continued Fractions

This activity presents continued fractions as a way of approximating radicals. They may also prove to be a source of continued investigation for the students and a jumping-off point for other topics.

578   LESSON 12.1

● **SOLUTION**

a. Use the Distributive Property to combine like terms. In this case, $\sqrt{6}$ is the common factor. Then simplify.
$5\sqrt{6} - 2\sqrt{6} = (5-2)\sqrt{6} = 3\sqrt{6}$

b. Rearrange in order to combine like terms.
$5 + 6\sqrt{7} - 2\sqrt{7} - 3 = (5-3) + (6-2)\sqrt{7} = 2 + 4\sqrt{7}$

c. Rearrange in order to combine like terms.
$8\sqrt{3} + 6\sqrt{2} - \sqrt{3} + 2\sqrt{2} = (8-1)\sqrt{3} + (6+2)\sqrt{2} = 7\sqrt{3} + 8\sqrt{2}$

d. Variables are treated in a similar way.
$a\sqrt{x} + b\sqrt{x} = (a+b)\sqrt{x}$

TRY THIS   Simplify.   a. $4\sqrt{6} - \sqrt{6}$ b. $a + 2\sqrt{b} - 2a - 3\sqrt{b}$

Activity
Operations and Radical Expressions

You can use your understanding of radicals to perform mathematical operations.

1. Compute $\sqrt{9 \cdot 16}$ in two different ways:
   a. $\sqrt{9 \cdot 16} = \sqrt{144} = ?$
   b. $\sqrt{9 \cdot 16} = \sqrt{9} \cdot \sqrt{16} = ?$
2. Give three examples illustrating that $\sqrt{ab} = \sqrt{a}\sqrt{b}$ is true.

CHECKPOINT ✔ 3. Compute $\sqrt{9+16}$ and $\sqrt{9} + \sqrt{16}$.

4. Give three examples illustrating that $\sqrt{a+b} = \sqrt{a} + \sqrt{b}$ is *not* always true.

Multiplication Property of Square Roots

For all numbers a and b, where $a \geq 0$ and $b \geq 0$:

$\sqrt{ab} = \sqrt{a}\sqrt{b}$

Radical expressions that are simplified are easier to manipulate algebraically. A square-root expression is in **simplest radical form** when all the following conditions are met:

1. No factor of the radicand is a perfect square other than 1.
2. The radicand contains no fractions.
3. No radical appears in the denominator of a fraction.

Inclusion Strategies

LINGUISTIC LEARNERS   Students with a verbal learning style may perceive this lesson as a maze of symbols. These students may find it helpful to write a step-by-step procedure in their own words for each key skill of the lesson. For instance, the following is a sample procedure that they might write for multiplying radical expressions:

1. Apply the Distributive Property, if necessary.
2. Within each term, multiply the rational numbers and multiply the radicals.
3. Write each term in simplest radical form.

Enrichment

If n is a positive integer and a and b are real numbers such that $b^n = a$, then b is called an *nth root* of a. If n is odd, there is exactly one real *n*th root of a, denoted $\sqrt[n]{a}$. If n is even and $a > 0$, there are two real *n*th roots of a. The positive *n*th root is denoted $\sqrt[n]{a}$, and the negative *n*th root is denoted $-\sqrt[n]{a}$. If n is even and $a < 0$, there are no real *n*th roots of a. Have students simplify these *n*th roots.

1. $\sqrt[3]{64}$ 2. $\sqrt[3]{-64}$ 3. $\sqrt[4]{81}$ 4. $\sqrt[4]{-81}$
   4   −4   3   not real

5. $\sqrt[3]{r^6 s^4}$ $r^2 s \sqrt[3]{s}$ 6. $\sqrt[4]{32 j^7 k^8}$ $2jk^2 \sqrt[4]{2j^3}$

EXAMPLE 3 Write in simplest radical form.

a. $\sqrt{12}$ b. $\sqrt{400}$ c. $\sqrt{x^2}$ d. $\sqrt{a^5 b^{10}}$

SOLUTION

Look for perfect-square factors, and apply the Multiplication Property of Square Roots. Then find the square roots of the perfect squares. Leave any factor that is not a perfect square in radical form.

a. $\sqrt{12} = \sqrt{4 \cdot 3} = \sqrt{4} \cdot \sqrt{3} = 2\sqrt{3}$
b. $\sqrt{400} = \sqrt{4 \cdot 100} = \sqrt{4} \cdot \sqrt{100} = 2 \cdot 10 = 20$

Because the radical sign designates the principle square root, the value of $\sqrt{x^2}$ must be positive. Use the absolute value sign to indicate this when the exponent of a variable in the radical is **even** and the simplified exponent outside of the radical is **odd**.

c. $\sqrt{x^2} = |x^1| = |x|$
d. $\sqrt{a^5 b^{10}} = \sqrt{(a^2)^2 \cdot a \cdot (b^5)^2} = a^2 |b^5| \sqrt{a}$

TRY THIS   Simplify $\sqrt{72 m^2 n^5}$.

CRITICAL THINKING   Why is the absolute-value sign not necessary for the following?

a. $\sqrt{x^3} = x\sqrt{x}$ b. $\sqrt{x^4} = x^2$

EXAMPLE 4 Simplify.

a. $\left(5\sqrt{3}\right)^2$ b. $\sqrt{3}\sqrt{6}$ c. $\sqrt{2}\left(6 + \sqrt{12}\right)$ d. $\left(3 - \sqrt{2}\right)\left(4 + \sqrt{2}\right)$

SOLUTION

a. Recall from the properties of exponents that the second power means the product of two identical factors. Rearrange the factors and multiply.
$\left(5\sqrt{3}\right)^2 = \left(5\sqrt{3}\right)\left(5\sqrt{3}\right) = (5 \cdot 5)\left(\sqrt{3}\sqrt{3}\right) = 25 \cdot 3 = 75$

b. The Multiplication Property of Square Roots allows you to multiply separate radicals. You can then factor the new product in a different way and simplify.
$\sqrt{3}\sqrt{6} = \sqrt{3 \cdot 6} = \sqrt{18} = \sqrt{9 \cdot 2} = \sqrt{9}\sqrt{2} = 3\sqrt{2}$

c. Use the Distributive Property to multiply, factor, and simplify the result. Notice that the radical is usually written last in each term.
$\sqrt{2}\left(6 + \sqrt{12}\right) = \sqrt{2} \cdot 6 + \sqrt{2} \cdot \sqrt{12} = 6\sqrt{2} + \sqrt{2 \cdot 12} = 6\sqrt{2} + \sqrt{24}$
$6\sqrt{2} + \sqrt{24} = 6\sqrt{2} + \sqrt{4 \cdot 6} = 6\sqrt{2} + 2\sqrt{6}$

d. To multiply differences and sums, multiply the two binomials by using the FOIL method.
$\left(3 - \sqrt{2}\right)\left(4 + \sqrt{2}\right) = 12 + 3\sqrt{2} - 4\sqrt{2} - 2 = 10 - \sqrt{2}$

TRY THIS   Simplify.   a. $\left(2\sqrt{7}\right)^2$ b. $\sqrt{2}\left(4 - \sqrt{8}\right)$

Reteaching the Lesson

USING VISUAL STRATEGIES   Show students how to simplify radical expressions by writing out all of the factors of the radicand and identifying pairs of factors. Lead them through this method by using the simplification of $\sqrt{72 a^3 b^4}$, which is shown at right. Then give students these practice exercises:

1. $\sqrt{225 r^2}$ 2. $\sqrt{98 m^3}$ 3. $\sqrt{150 p^5 q^2}$

   15r   **$7m\sqrt{2m}$**   **$5p^2 q\sqrt{6p}$**

$\sqrt{72 a^3 b^4}$
$= \sqrt{(2 \cdot 2) \cdot 2 \cdot (3 \cdot 3) \cdot (a \cdot a) \cdot a \cdot (b \cdot b) \cdot (b \cdot b)}$
$= \quad 2 \cdot \sqrt{2} \cdot 3 \quad \cdot \quad a \cdot \sqrt{a} \cdot b \cdot b$
$= 2 \cdot 3 \cdot a \cdot b \cdot b \cdot \sqrt{2} \cdot \sqrt{a}$
$= 6ab^2 \sqrt{2a}$

ADDITIONAL EXAMPLE 3

Write in simplest radical form.

a. $\sqrt{18}$ $3\sqrt{2}$

b. $\sqrt{3600}$ 60

c. $\sqrt{y^6}$ $|y^3|$

d. $\sqrt{r^6 s^3}$, for $s \geq 0$ $|r^3| s\sqrt{s}$

TRY THIS

$6|m| n^2 \sqrt{2n}$

CRITICAL THINKING

a. Within the real numbers when x^3 is under the square root sign, it is assumed to have a positive value. Therefore, the value of x must also be positive.

b. The values of x^4 and x^2 are always positive.

ADDITIONAL EXAMPLE 4

Simplify.

a. $\left(2\sqrt{3}\right)^2$ 12

b. $\left(\sqrt{8}\right)\left(\sqrt{14}\right)$ $4\sqrt{7}$

c. $\sqrt{3}\left(\sqrt{6} - 5\right)$
    $3\sqrt{2} - 5\sqrt{3}$

d. $\left(2 + \sqrt{7}\right)\left(2 - \sqrt{7}\right)$ -3

TRY THIS

a. 28
b. $4\sqrt{2} - 4$

Teaching Tip

Students can be confused about the use of absolute value signs when simplifying even roots that contain variables. Sometimes an even root of an even power of a variable simplifies to an odd power. Students should be reminded that the simplified odd power of the variable should be enclosed in an absolute-value sign to indicate the principal root.

LESSON 12.1   **579**

ADDITIONAL EXAMPLE 5

Simplify.

a. $\sqrt{\frac{9}{121}}$ $\frac{3}{11}$

b. $\sqrt{\frac{5}{16}}$ $\frac{\sqrt{5}}{4}$

c. $\sqrt{\frac{4}{7}}$ $\frac{2\sqrt{7}}{7}$

d. $\sqrt{\frac{x^5}{y^2z^4}}$, $x \geq 0$, $y > 0$, and $z \neq 0$ $\frac{x^2\sqrt{x}}{yz^2}$

TRY THIS

a. $\frac{2}{5}$

b. $\frac{2\sqrt{3}}{3}$

c. $\frac{|b|\sqrt{ac}}{c}$

Division Property of Square Roots

For all numbers $a \geq 0$ and $b > 0$:

$$\sqrt{\frac{a}{b}} = \frac{\sqrt{a}}{\sqrt{b}}$$

EXAMPLE 5 Simplify.

a. $\sqrt{\frac{16}{25}}$ b. $\sqrt{\frac{7}{16}}$ c. $\sqrt{\frac{a^2b^3}{c^2}}$ d. $\sqrt{\frac{9}{5}}$

● **SOLUTION**

Rewrite each square root by using the Division Property of Square Roots. Then simplify the numerator and denominator separately.

a. $\sqrt{\frac{16}{25}} = \frac{\sqrt{16}}{\sqrt{25}} = \frac{4}{5}$ b. $\sqrt{\frac{7}{16}} = \frac{\sqrt{7}}{\sqrt{16}} = \frac{\sqrt{7}}{4}$

c. $\sqrt{\frac{a^2b^3}{c^2}} = \frac{\sqrt{a^2b^2b}}{\sqrt{c^2}} = \frac{|a|b\sqrt{b}}{|c|}$ d. $\sqrt{\frac{9}{5}} = \frac{\sqrt{9}}{\sqrt{5}} = \frac{3}{\sqrt{5}}$

A radical remains in the denominator of the answer for part **d**. Change this expression to simplest radical form by multiplying by $\frac{\sqrt{5}}{\sqrt{5}}$, which is equivalent to multiplying by 1:

$$\frac{3}{\sqrt{5}} \cdot \frac{\sqrt{5}}{\sqrt{5}} = \frac{3\sqrt{5}}{\sqrt{5}\sqrt{5}} = \frac{3\sqrt{5}}{5}$$

Removing a radical expression from the denominator is known as **rationalizing the denominator**.

TRY THIS Simplify.   a. $\sqrt{\frac{4}{25}}$ b. $\sqrt{\frac{4}{3}}$ c. $\sqrt{\frac{ab^2}{c}}$

Exercises

● *Communicate*

1. Describe how you can use graph paper to determine the square root of 16, a perfect square.

2. Describe how you can use graph paper to estimate the square root of 19, a number that is not a perfect square.

3. Explain how to estimate $\sqrt{7}$ without a calculator or graph paper.

4. How is factoring used to simplify expressions that contain radicals such as $5\sqrt{90x^3y^4}$?

5. What is *simplest radical form*?

Practice

Practice
12.1 Operations With Radicals

Simplify each radical expression by factoring.

1. $\sqrt{144}$ **12**   2. $\sqrt{72}$ **$6\sqrt{2}$**
3. $\sqrt{288}$ **$12\sqrt{2}$**   4. $\sqrt{4000}$ **$20\sqrt{10}$**
5. $\sqrt{3264}$ **$8\sqrt{51}$**   6. $\sqrt{8775}$ **$15\sqrt{39}$**

Express each radical expression in simplest radical form.

7. $\sqrt{8}\sqrt{14}$ **$4\sqrt{7}$**   8. $\sqrt{3}\sqrt{27}$ **9**
9. $\sqrt{12}\sqrt{6}$ **$6\sqrt{2}$**   10. $\frac{\sqrt{72}}{\sqrt{6}}$ **$2\sqrt{3}$**
11. $\frac{\sqrt{162}}{\sqrt{3}}$ **$3\sqrt{6}$**   12. $\frac{\sqrt{500}}{\sqrt{50}}$ **$\sqrt{10}$**

Simplify each radical expression. Assume that all variables are non-negative and that all denominators are nonzero.

13. $\sqrt{m^6n}$ **$m^3\sqrt{n}$**   14. $\sqrt{a^8b^6}$ **a^4b^3**
15. $\sqrt{\frac{x^3}{y^3}}$ **$\frac{x\sqrt{x}}{y^2}$**   16. $\sqrt{\frac{x^5}{y^9}}$ **$\frac{x^2\sqrt{x}}{y^4}\sqrt{\frac{x}{y}}$**

If possible, perform each indicated operation and simplify your answer.

17. $\sqrt{12} + 3\sqrt{3}$ **$5\sqrt{3}$**   18. $3\sqrt{27} - 5\sqrt{3}$ **$4\sqrt{3}$**
19. $\frac{\sqrt{1}+\sqrt{25}}{\sqrt{2}}$ **$3\sqrt{2}$**   20. $(5 + \sqrt{5}) + (2 - \sqrt{3})$ **$7 + \sqrt{5} - \sqrt{3}$**

Simplify each radical expression.

21. $(2\sqrt{3})^2$ **12**   22. $2(\sqrt{3} + \sqrt{12})$ **$6\sqrt{3}$**
23. $(\sqrt{12} + 2)(\sqrt{12} - 2)$ **8**   24. $\sqrt{3}(2 + \sqrt{12})$ **$2\sqrt{3} + 6$**
25. $(\sqrt{3} + 2)^2$ **$7 + 4\sqrt{3}$**   26. $(\sqrt{3} + \sqrt{12})^2$ **27**

580 LESSON 12.1

Guided Skills Practice

Evaluate each square root. *(EXAMPLE 1)*

6. $\sqrt{36}$ **6**   7. $-\sqrt{64}$ **−8**   8. $\pm\sqrt{81}$ **±9**   9. $-\sqrt{121}$ **−11**

Simplify. *(EXAMPLE 2)*

10. $8\sqrt{3} - 6\sqrt{3}$ **$2\sqrt{3}$**   11. $9 + 3\sqrt{7} - 5\sqrt{7} + 4$ **$13 - 2\sqrt{7}$**

Express in simplest radical form. *(EXAMPLE 3)*

12. $\sqrt{32}$ **$4\sqrt{2}$**   13. $\sqrt{x^2y^7}$ **$|x|y^3\sqrt{y}$**   14. $\sqrt{27x^6}$ **$3|x^3|\sqrt{3}$**   15. $\sqrt{19a^7b^3}$ **$a^3b\sqrt{19ab}$**

Simplify. *(EXAMPLE 4)*

16. $(7\sqrt{11})^2$ **539**   17. $\sqrt{2}\sqrt{10}$ **$2\sqrt{5}$**   18. $(2 - \sqrt{3})(5 + \sqrt{3})$ **$7 - 3\sqrt{3}$**

Simplify. *(EXAMPLE 5)*

19. $\sqrt{\frac{9}{4}}$ **$\frac{3}{2}$**   20. $\sqrt{\frac{6}{49}}$ **$\frac{\sqrt{6}}{7}$**   21. $\sqrt{\frac{225}{18}}$ **$\frac{5\sqrt{2}}{2}$**   22. $\sqrt{\frac{x^7y^{14}}{z^3}}$ **$\frac{x^3|y^7|\sqrt{xz}}{z^2}$**

Practice and Apply

Find each square root. If the square root is irrational, approximate the value to the nearest hundredth.

23. $\sqrt{225}$ **15**   24. $-\sqrt{169}$ **−13**   25. $\pm\sqrt{11}$ **±3.32**   26. $\sqrt{\frac{4}{9}}$ **$\frac{2}{3}$**   27. $-\sqrt{40}$ **−6.32**

28. $-\sqrt{27}$ **−5.20**   29. $\sqrt{1000}$ **31.62**   30. $\sqrt{10{,}000}$ **100**   31. $-\sqrt{0.04}$ **−0.2**   32. $\sqrt{0.059}$ **0.24**

Simplify each radical by factoring.

33. $\sqrt{49}$ **7**   34. $\sqrt{196}$ **14**   35. $\sqrt{576}$ **24**   36. $\sqrt{3600}$ **60**   37. $\sqrt{192}$ **$8\sqrt{3}$**

38. $\sqrt{75}$ **$5\sqrt{3}$**   39. $\sqrt{98}$ **$7\sqrt{2}$**   40. $\sqrt{1620}$ **$18\sqrt{5}$**   41. $\sqrt{264}$ **$2\sqrt{66}$**   42. $\sqrt{648}$ **$18\sqrt{2}$**

Decide whether each statement is true or false. Assume that $a > 0$ and $b > 0$.

43. $\sqrt{a+b} = \sqrt{a} + \sqrt{b}$ **false**   44. $\sqrt{ab} = \sqrt{a}\sqrt{b}$ **true**   45. $\sqrt{\frac{a}{b}} = \frac{\sqrt{a}}{\sqrt{b}}, b \neq 0$ **true**

Express in simplest radical form.

46. $\sqrt{3}\sqrt{12}$ **6**   47. $\sqrt{8}\sqrt{18}$ **12**   48. $\sqrt{48}\sqrt{3}$ **12**   49. $\sqrt{54}\sqrt{6}$ **18**

50. $\sqrt{\frac{64}{16}}$ **2**   51. $\sqrt{\frac{96}{2}}$ **$4\sqrt{3}$**   52. $\frac{\sqrt{50}}{\sqrt{8}}$ **$\frac{5}{2}$**   53. $\frac{\sqrt{150}}{\sqrt{6}}$ **5**

54. $\sqrt{5}\sqrt{15}$ **$5\sqrt{3}$**   55. $\sqrt{98}\sqrt{14}$ **$14\sqrt{7}$**   56. $\sqrt{\frac{56}{8}}$ **$\sqrt{7}$**   57. $\frac{\sqrt{96}}{\sqrt{8}}$ **$2\sqrt{3}$**

Simplify each of the following. Assume that all variables are nonnegative and that all denominators are nonzero.

58. $\sqrt{a^4b^6}$ **$a^2|b^3|$**   59. $\sqrt{x^8y^9}$ **$x^4y^4\sqrt{y}$**   60. $\sqrt{\frac{p^9}{q^{10}}}$ **$\frac{p^4\sqrt{p}}{|q^5|}$**   61. $\sqrt{\frac{x^3}{y^6}}$ **$\frac{x\sqrt{x}}{|y^3|}$**

If possible, perform the indicated operations, and simplify your answer.

62. $3\sqrt{5} + 4\sqrt{5}$ **$7\sqrt{5}$**   63. $4\sqrt{5} + 2\sqrt{5} - 5\sqrt{5}$ **$\sqrt{5}$**   64. $\sqrt{6} + 2\sqrt{3} - \sqrt{6}$ **$2\sqrt{3}$**

65. $(4 + \sqrt{3}) + (1 - \sqrt{2})$ **$5 + \sqrt{3} - \sqrt{2}$**   66. $\frac{6 + \sqrt{18}}{3}$ **$2 + \sqrt{2}$**   67. $\frac{\sqrt{15} + \sqrt{10}}{\sqrt{5}}$ **$\sqrt{3} + \sqrt{2}$**

Assess

Selected Answers
Exercises 6–22, 23–89 odd

ASSIGNMENT GUIDE

In Class	1–22
Core	23–75 odd, 80–83
Core Plus	24–76 even, 77–83
Review	84–89
Preview	90–96

Extra Practice can be found beginning on page 738.

Student Technology Guide

12.1 Operations with Radicals

Although some square roots simplify to "nice" numbers, many are irrational numbers. You can use a calculator to find decimal approximations of square roots.

Example: Find each square root. If the square root is irrational, estimate its value to the nearest hundredth.

a. $\sqrt{21^2 + 28^2}$ b. $-\sqrt{\frac{5}{7}}$

On most calculators, $\sqrt{x}$ and x^2 are on the same key.

a. Press 2nd x^2 . Notice that some calculators automatically give you a left parenthesis. Then press 21 x^2 + 28 x^2 and press) to close the parentheses. Finally, press ENTER.

$\sqrt{21^2 + 28^2} = 35$

b. Press (-) 2nd x^2 5 ÷ 7) ENTER.

To the nearest hundredth, $-\sqrt{\frac{5}{7}} = -0.85$.

Find each square root. If the square root is irrational, estimate its value to the nearest hundredth.

1. $\sqrt{75}$ **8.66**   2. $\sqrt{5^2 + 12^2}$ **13**   3. $\sqrt{\frac{11}{121}}$ **0.30**   4. $\sqrt{0.015}$ **0.12**

The formula $D = \sqrt{1.5A}$, where D is distance in miles and A is altitude in feet, gives the distance to the horizon.

Example: How far is the horizon from a hot-air balloon at an altitude of 500 feet?

Press 2nd x^2 1.5 × 500 ENTER.

The horizon is about 27.39 miles away.

Find the distance to the horizon for each location. Give your answer to the nearest hundredth.

5. top of the Empire State Building, at a height of 1250 feet **43.30 miles**

6. top of Mauna Kea, at an altitude of 13,796 feet **143.85 miles**

Technology

Students will need a calculator to find the approximations required in Exercises 23–32. In Exercises 33–42, 46–57, and 62–79, they can check their answers by using a calculator to evaluate both the given expression and their simplification of it.

Error Analysis

When rationalizing a denominator, some students square both the numerator and the denominator. For example, given $\frac{3}{\sqrt{2}}$, they may write $\frac{3}{\sqrt{2}} \cdot \frac{3}{\sqrt{2}}$. Suggest that they draw a rectangle around the factor they are introducing to verify that it is equal to 1.

$$\frac{3}{\sqrt{2}} \cdot \boxed{\frac{\sqrt{2}}{\sqrt{2}}} = \frac{3\sqrt{2}}{\sqrt{2}\sqrt{2}} = \frac{3\sqrt{2}}{2}$$

Look Beyond

In Exercises 90 through 96, students learn how to express roots as fractional exponents. Explain to students that if n is even, a must be a positive real number because expressions such as $\sqrt[2]{-4}$ are not defined within the set of real numbers. However, if n is odd, a may be negative. For example, $\sqrt[3]{-1} = -1$ because $(-1)^3 = -1$.

---

73. $6\sqrt{5} - 5\sqrt{3}$
74. $6\sqrt{6} + 6\sqrt{3}$
76. $-5 - 2\sqrt{3}$

CHALLENGE
CONNECTION

APPLICATIONS

77. $7\sqrt{3} + 12$
78. $134\sqrt{3} + 96$
79. $21\sqrt{5} - 40$
80. ≈ 15.81 m
81. 12 in.
82. ≈ 5.29 mi

90. 5
91. 2
92. 4
93. 10
94. x^3
95. $x^{\frac{5}{2}} y^{\frac{3}{2}}$
96. $x^{\frac{37}{2}} y^{\frac{19}{2}}$

Simplify.

68. $(3\sqrt{5})^2$ **45**
69. $(4\sqrt{25})^2$ **400**
70. $\sqrt{12}\sqrt{6}$ **$6\sqrt{2}$**
71. $\sqrt{72}\sqrt{32}$ **48**
72. $3(\sqrt{5} + 9)$ **$3\sqrt{5} + 27$**
73. $\sqrt{5}(6 - \sqrt{15})$
74. $\sqrt{6}(6 + \sqrt{18})$
75. $(\sqrt{5} - 2)(\sqrt{5} + 2)$ **1**
76. $(\sqrt{3} - 4)(\sqrt{3} + 2)$
77. $\sqrt{3}(\sqrt{3} + 2)^2$
78. $\sqrt{12}(\sqrt{3} + 8)^2$
79. $\sqrt{5}(\sqrt{5} - 4)^2$

GEOMETRY Determine the side length of a square with each given area.

80. 250 square meters
81. 144 square inches
82. 28 square miles

83. **LANDSCAPING** The Smiths have a yard in the shape of a square. If the area of the yard is 676 square feet, what is the length of each side? **26 ft**

Look Back

Simplify. (LESSONS 8.2 AND 8.3)

84. $(-a^2b^2)^3(a^4b)^2$ $-a^{14}b^8$
85. $\frac{x^5y^7}{x^2y^3}$ x^3y^4
86. $\left(\frac{20x^3}{-4x^2}\right)^3$ $-125x^3$

Find each product. (LESSON 9.3)

87. $(2x - 4)(2x - 4)$
    $4x^2 - 16x + 16$
88. $(3a + 5)(2a - 6)$
    $6a^2 - 8a - 30$
89. $(6b + 1)(3b - 1)$
    $18b^2 - 3b - 1$

Look Beyond

Roots and Fractional Exponents

The radical sign can be used to indicate other roots of numbers. For example:

$\sqrt[3]{27}$ means "the principal third (or cubed) root of 27"

$\sqrt[4]{16}$ means "the principal fourth root of 16"

$\sqrt[3]{27} = 3$, since $3^3 = 27$ $\sqrt[4]{16} = 2$, since $2^4 = 16$

Evaluate.

90. $\sqrt[3]{125}$
91. $\sqrt[5]{32}$
92. $\sqrt[4]{256}$
93. $\sqrt[5]{100000}$

Fractional exponents can also be used to indicate roots. **For any real numbers $a \geq 0$ and integer $n \geq 2$, $a^{\frac{1}{n}} = \sqrt[n]{a}$.** For example:

$\sqrt{5} = 5^{\frac{1}{2}}$ $\sqrt[4]{7} = 7^{\frac{1}{4}}$ $\sqrt[3]{a^2} = a^{\frac{2}{3}}$

The laws of exponents can be used to simplify expressions with fractional exponents. For example:

$\left(x^{\frac{1}{2}}\right)^5 \left(x^{\frac{3}{2}}\right) = \left(x^{\frac{5}{2}}\right)\left(x^{\frac{3}{2}}\right) = x^{\frac{8}{2}} = x^4$ $\left(a^{\frac{1}{3}}\right)\left(a^{\frac{1}{3}}\right)\left(a^{\frac{1}{3}}\right) = \left(a^{\frac{1}{3}}\right)^3 = a^1 = a$

Simplify.

94. $\left(x^{\frac{1}{3}}\right)^4 (x^5)^{\frac{1}{3}}$
95. $(xy)^{\frac{1}{2}}\left(x^{\frac{1}{3}}\right)^6 \left(y^{\frac{1}{2}}\right)^2$
96. $\left(x^3y^{\frac{3}{2}}\right)^6 (xy)^{\frac{1}{2}}$

Enrichment

To reinforce the idea that fractional exponents indicate roots, show students the following:

For $a > 0$, $\sqrt{a} \cdot \sqrt{a} = (\sqrt{a})^2 = a$, which can also be shown as $(a^{\frac{1}{2}}) \cdot (a^{\frac{1}{2}}) = (a^{\frac{1}{2}})^2 = a^{\frac{1}{2} \cdot 2} = a^1 = a$. Similarly, $\sqrt[3]{a} \cdot \sqrt[3]{a} \cdot \sqrt[3]{a} = (\sqrt[3]{a})^3 = a$, which can be written as $(a^{\frac{1}{3}}) \cdot (a^{\frac{1}{3}}) \cdot (a^{\frac{1}{3}}) = (a^{\frac{1}{3}})^3 = a^1 = a$. Continue this pattern with $a^{\frac{1}{4}}$ and $a^{\frac{1}{5}}$.

If time permits, you may want to extend the study of fractional exponents to include negative fractional exponents. Students should already be familiar with negative exponents from Lesson 8.4. Give students the following definition: For any positive real number a and any positive integer n, $a^{-\frac{1}{n}} = \frac{1}{\sqrt[n]{a}}$. To reinforce the idea, show students the following:

$\frac{1}{\sqrt{a}} \cdot \frac{1}{\sqrt{a}} = \left(\frac{1}{\sqrt{a}}\right)^2 = \frac{1}{a}$, similarly $(a^{-\frac{1}{2}}) \cdot (a^{-\frac{1}{2}}) = (a^{-\frac{1}{2}})^2 = a^{-\frac{1}{2} \cdot 2} = a^{-1} = \frac{1}{a}$.

12.2 Square-Root Functions and Radical Equations

Objectives
- Solve equations containing radicals.
- Solve equations by using radicals.

Why Equations that contain radicals are often used in science to model natural phenomena. Scientists and students also use radicals to solve equations that contain squared numbers and variables. Radicals provide a necessary tool to solve many problems that occur in science and other areas.

The Square-Root Function

The function $y = \sqrt{x}$ is the simplest **square-root function**. Because $\sqrt{x}$ means the positive, or principal, square root of x, the range of $y = \sqrt{x}$ is limited to 0 and all positive real numbers. The domain is also limited to 0 and all positive real numbers because square roots of negative numbers are not real numbers.

The function $y = \sqrt{x}$ is graphed below. The graph of the function lies entirely in the first quadrant because the domain and range are both limited to 0 and all the positive real numbers.

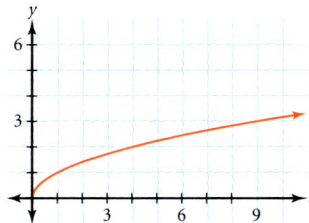

CRITICAL THINKING Describe similarities and differences between the graph of a square-root function and the graph of a quadratic function. Explain how the two functions are related.

Alternative Teaching Strategy

CONNECTING TO PRIOR KNOWLEDGE Write the equation $(x + 1)^2 = 4$ on the board or overhead. Remind students that this is a quadratic equation. Review with them the method of solving a quadratic equation by using square roots.

$$(x + 1)^2 = 4$$
$$\sqrt{(x+1)^2} = \pm\sqrt{4} \quad \leftarrow \text{Take the square root of each side.}$$
$$x + 1 = +\sqrt{4} \quad \text{or} \quad x + 1 = -\sqrt{4}$$
$$x + 1 = 2 \quad \text{or} \quad x + 1 = -2$$
$$x = 1 \quad \text{or} \quad x = -3$$

For non-negative numbers, taking the square root is the inverse of squaring. Thus, when an equation involves square roots, you can solve by squaring. Discuss the following example:

$$\sqrt{x + 1} = 4$$
$$(\sqrt{x+1})^2 = 4^2 \quad \leftarrow \text{Square each side.}$$
$$x + 1 = 16$$
$$x = 15$$

Then extend the method to solve $\sqrt{x + 12} = x$.
$x = 4$

Prepare

NCTM PRINCIPLES & STANDARDS 1–10

LESSON RESOURCES
- Practice 12.2
- Reteaching Master 12.2
- Enrichment Master 12.2
- Cooperative-Learning Activity 12.2
- Lesson Activity 12.2
- Problem Solving/Critical Thinking Master 12.2
- Teaching Transparency 55
- Teaching Transparency 56

QUICK WARM-UP

Solve each equation.

1. $2m + 9 = 6$ $m = -1.5$
2. $s^2 = 36$ $s = -6$ or $s = 6$
3. $(a + 2)^2 = 9$
             $a = -5$ or $a = 1$
4. $g^2 + 4g = 0$
             $g = 0$ or $g = -4$
5. $u^2 + 4u = 5$
             $u = -5$ or $u = 1$

Also on Quiz Transparency 12.2

Teach

Why Ask students if they know what makes one pendulum swing faster than another. They may be surprised to learn that for small swings, the time required for a pendulum to make one swing is determined by its length. Tell them that, in this lesson, they will see how a pendulum's motion can be defined by an equation involving a radical.

☞ For the answer to Critical Thinking, see page 584.

LESSON 12.2 **583**

Use Teaching Transparency 55.

CRITICAL THINKING
The square-root function looks like half a parabola with a horizontal orientation instead of a vertical orientation. The square-root function has an inverse operation relationship with the quadratic function, which is a squaring function.

Teaching Tip
Students may have a hard time determining when a square-root symbol represents the principal (positive) root and when it represents both the positive and negative roots. Remind them that when the symbol is actually present in the text, it represents one number, the principal (or positive) root. For example, $\sqrt{9}$ represents 3. If students introduce the symbol by taking the square root of 9 while solving a problem, then $\sqrt{9} = \pm 3$ because they are looking for *all* possible solutions, or all numbers that equal 9 when squared.

Use Teaching Transparency 56.

Solving Radical Equations

The equation relating the length of a pendulum to the number of swings per minute is one example of a radical equation.

APPLICATION
PHYSICS

In a physics class, groups of students are given six strings of different lengths, each tied to a washer. The students time one complete swing of each pendulum and then put all of their data together, make a table, and draw a graph of the data. The point (0, 0) represents the time for one complete swing of a pendulum with a length of 0.

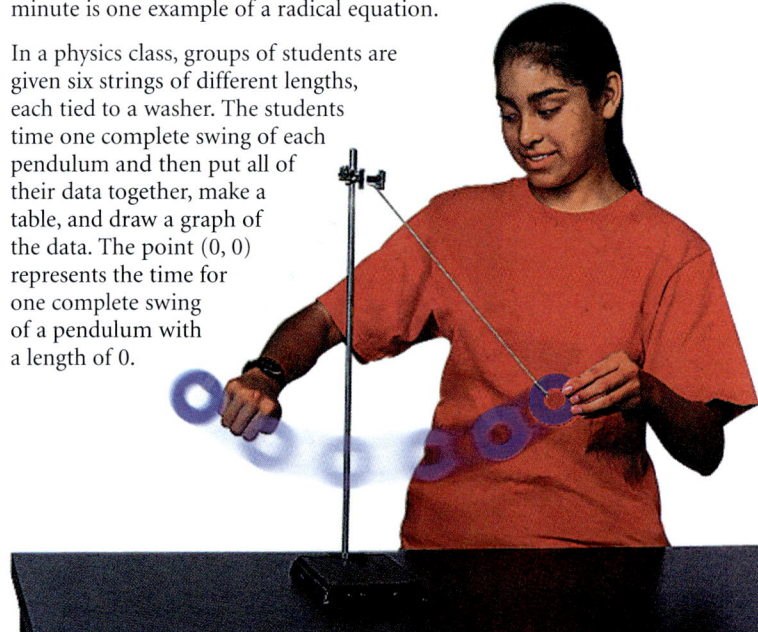

Let *l* represent the length of the pendulum in centimeters and *t* represent the time of one complete swing in seconds. Examine the data in the table and on the graph.

l	t
0	0
6	0.5
25	1.0
58	1.5
100	2.0
155	2.5

What conclusions can be drawn from the information in the graph? How can you use the graph to predict the time that it will take for a string of any length to make one complete swing?

Notice that the curve created by connecting the points resembles a transformation of the parent function $y = \sqrt{x}$.

The equation that models the motion of a pendulum is $t = 2\pi\sqrt{\dfrac{l}{g}}$. The variable *t* is the time in seconds, *l* is the length in centimeters, and *g* is the acceleration due to gravity (980 centimeters per second squared). This formula can be used to calculate the time that a pendulum of a given length should take to make one complete swing.

Interdisciplinary Connection

PHYSICS The formula $t = \sqrt{\dfrac{2d}{g}}$ gives the time, *t*, in seconds that it takes an object at rest to fall a distance of *d* meters when the acceleration due to gravity is *g* meters per second squared. On the moon, an object dropped from a height of 3.2 meters would reach the surface in 2 seconds. What is the acceleration due to gravity on the moon? **1.6 meters per second squared**

EXAMPLE 1

APPLICATION
PHYSICS

a. Determine the time in seconds that it takes a 75-centimeter pendulum to make one complete swing.
b. Determine the length in centimeters of a pendulum that takes 3 seconds to make one complete swing.

● **SOLUTION**

a. Substitute the values into the pendulum formula.

$$t = 2\pi\sqrt{\frac{75}{980}}$$

Use a calculator to help in the calculation.

$$t \approx 1.74$$

One swing of a 75-centimeter pendulum takes about 1.7 seconds.

b. Substitute the known information into the formula, and solve for l.

$3 = 2\pi\sqrt{\frac{l}{980}}$ *Given*

$3\sqrt{980} = 2\pi\sqrt{l}$ *Multiply both sides by $\sqrt{980}$.*

$\frac{3\sqrt{980}}{2\pi} = \sqrt{l}$ *Divide both sides by 2π.*

$\left(\frac{3\sqrt{980}}{2\pi}\right)^2 = l$ *Use the definition of square root.*

$223.4 \approx l$ *Use a calculator.*

The length of the pendulum is about 223.4 centimeters.

TRY THIS

a. Find the time in seconds that it takes a 50-centimeter pendulum to make one complete swing.
b. Find the length in centimeters of a pendulum that takes 5 seconds to make one complete swing.

A Foucault pendulum

EXAMPLE 2 Solve $\sqrt{x+2} = 3$.

● **SOLUTION**

The square-root and squaring operations are inverses of each other and can be used to solve equations.

$\sqrt{x+2} = 3$
$\left(\sqrt{x+2}\right)^2 = 3^2$ *Square both sides of the equation.*
$x + 2 = 9$
$x = 7$

Check by substituting. Because $\sqrt{7+2} = \sqrt{9} = 3$, the solution is 7.

TRY THIS Solve $\sqrt{2x-3} = 4$

ADDITIONAL EXAMPLE 1

a. Determine the time in seconds that it takes a 120-centimeter pendulum to make one complete swing.
about 2.2 seconds

b. Determine the length in centimeters of a pendulum that takes 1.5 seconds to make one complete swing.
about 55.9 centimeters

TRY THIS
a. about 1.42 seconds
b. about 620.6 centimeters

ADDITIONAL EXAMPLE 2

Solve $\sqrt{x-3} = 3$.
$x = 12$

See Keystroke Guide, page 782.

TRY THIS
$x = 9.5$

Inclusion Strategies

VISUAL LEARNERS Students may wonder why the range of the square-root function includes only principal square roots. It may be helpful for them to make a picture of the result if negative square roots were included. Tell students to graph both $y = \sqrt{x}$ and $y = -\sqrt{x}$ for nonnegative, real-number values of x. Their graphs should look like the one at right. Ask them to explain why this is not the graph of a function. Lead them to see that the graph fails the vertical-line test. That is, there are infinitely many vertical lines that intersect the graph more than once.

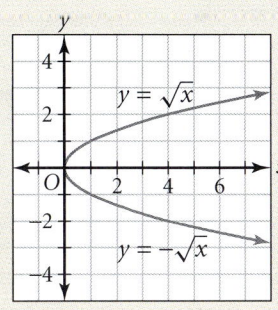

LESSON 12.2 **585**

Teaching Tip

TECHNOLOGY If you are using a TI-82 or TI-83 graphics calculator to create the graph at right, enter the equations as **Y1** and **Y2**. Then press ZOOM and select 6: ZStandard.

To find the point of intersection quickly, you can press 2nd TRACE to access the **CALCULATE** menu. Choose , and then press ENTER ENTER ENTER.

ADDITIONAL EXAMPLE 3

Solve $\sqrt{4x+5} = x$.

$x = 5$

TRY THIS

$x = 4$

ADDITIONAL EXAMPLE 4

Solve each equation for x.

a. $x^2 = 72$
   $x = 6\sqrt{2}$ or $x = -6\sqrt{2}$

b. $x^2 = 8^2 + 15^2$
   $x = 17$ or $x = -17$

c. $x^2 = y^2 - z^2$
   $x = \sqrt{y^2 - z^2}$ or
   $x = -\sqrt{y^2 - z^2}$

TECHNOLOGY GRAPHICS CALCULATOR

You can also solve the equation with a graphics calculator. Set an appropriate window. Enter the left side of the equation in the graphics calculator as $Y_1 = \sqrt{x+2}$. Then enter the right side of the equation as $Y_2 = 3$.

The calculator will graph both equations in the same window. You can then use the trace feature to find the point(s) of intersection. The point of intersection is (7, 3), so $x = 7$.

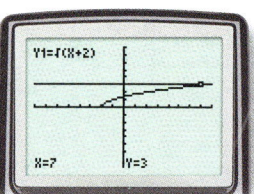

EXAMPLE 3 Solve $\sqrt{x+6} = x$.

SOLUTION

Notice that the variable appears on both sides of the equation.

$\sqrt{x+6} = x$
$(\sqrt{x+6})^2 = x^2$ *Square both sides of the equation.*
$x + 6 = x^2$
$0 = x^2 - x - 6$ *Form a quadratic equation equal to 0.*
$0 = (x-3)(x+2)$ *Factor.*
$x - 3 = 0$ or $x + 2 = 0$ *Set each factor equal to zero.*

The possible solutions are 3 and −2. You must check each possible solution by substituting it into the original equation to determine whether it is true or an *extraneous* solution.

CHECK

For $x = 3$, $\sqrt{3+6} = \sqrt{9} = 3$, so 3 is a solution.

For $x = -2$, $\sqrt{-2+6} = \sqrt{4} \neq -2$. Because $\sqrt{4}$ is defined as the positive square root of 4, −2 does not satisfy the original equation. It is an **extraneous solution**.

TRY THIS Solve $\sqrt{3x+4} = x$ for x, and check. Verify by graphing.

You can solve the equation $x^2 = 225$ by using the definition on page 577. Recall that the square root may have two solutions because the value of x can be positive or negative.

$x^2 = 225$
$x = \pm\sqrt{225}$
$x = \pm 15$

EXAMPLE 4 Solve each equation for x.

a. $x^2 = 150$ b. $x^2 = 4^2 + 3^2$ c. $x^2 = y^2 + z^2$

SOLUTION

a. $x^2 = 150$
   $x = \pm\sqrt{150} = \pm\sqrt{25 \cdot 6} = \pm 5\sqrt{6}$
   $x = 5\sqrt{6}$ or $x = -5\sqrt{6}$

Enrichment

Tell students that a number m is the geometric mean of two numbers a and b if $\frac{a}{m} = \frac{m}{b}$. Ask them to solve the proportion for m. They should arrive at $m = \sqrt{ab}$. Then have them work in pairs to complete the following set of exercises:

1. Find the geometric mean of 9 and 16.   **12**

2. The geometric mean of 20 and a number x is 30. Find x.   **45**

3. The geometric mean of two positive numbers is 8. One number is 4 times the other. Find the numbers.   **4 and 16**

4. The geometric mean of two positive numbers is 15. One number is 2 less than 3 times the other. Find the numbers.   **9 and 25**

Now have each student write two similar exercises related to a geometric mean. Partners should trade papers and work each other's exercises.

b.
$$x^2 = 4^2 + 3^2$$
$$x = \pm\sqrt{4^2 + 3^2} = \pm\sqrt{25} = \pm 5$$
$$x = 5 \quad \text{or} \quad x = -5$$

c. For an equation with variables, follow the same procedure that you use with numbers. Remember to include everything on the right side of the equation under one radical. Consider both positive and negative values as possible solutions.
$$x^2 = y^2 + z^2$$
$$x = \pm\sqrt{y^2 + z^2}$$
$$x = \sqrt{y^2 + z^2} \quad \text{or} \quad x = -\sqrt{y^2 + z^2}$$

TRY THIS Solve the following equations for x:
a. $x^2 = 200$ **b.** $x^2 = 5^2 + 12^2$ **c.** $x^2 = y^2 + z^3$

EXAMPLE 5 The area of a circular flower garden is 23 square yards. **What is the radius of the garden?**

● **SOLUTION**

The equation for the area of a circle is $A = \pi r^2$. Substitute 23 for A, and solve for r.

Since $\pi r^2 = 23$ and $r^2 = \frac{23}{\pi}$,
$$r = \pm\sqrt{\frac{23}{\pi}} \approx \pm\sqrt{\frac{23}{3.14}} \approx \pm\sqrt{7.325} \approx \pm 2.706.$$
Only the positive square root is meaningful in this case. The radius is approximately 2.7 yards.

EXAMPLE 6 Solve the equation $x^2 - 8x + 16 = 36$ by using radicals.

● **SOLUTION**

The trinomial $x^2 - 8x + 16$ is a perfect square. When it is factored, it can be written as $(x - 4)^2$. Write the equation in this form, and solve.
$$(x - 4)^2 = 36$$
$$x - 4 = \pm\sqrt{36}$$
$$x - 4 = \pm 6$$
$$x = 10 \quad \text{or} \quad x = -2$$

Since x can be any number, you must consider both positive and negative values.

Check by substitution to confirm the values are actual solutions to the original equation.

TRY THIS Solve the equation $x^2 + 10x + 25 = 49$ by using radicals.

Reteaching the Lesson

WORKING BACKWARDS Display the first graph at right on the board or overhead. Ask students to give equations for the line and curve that are graphed. $y = 2, y = \sqrt{x + 3}$ Ask them to give coordinates for the point of intersection. $(1, 2)$ Lead them to see that the x-coordinate, 1, is the solution to $\sqrt{x + 3} = 2$. Work with them to arrive at this solution algebraically. Repeat the activity with the second graph at right. This time, students should be able to arrive at 1 as the solution to $\sqrt{x + 3} = x + 1$.

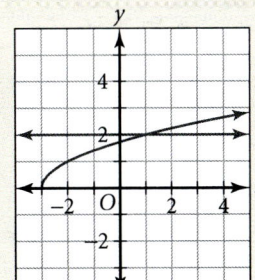
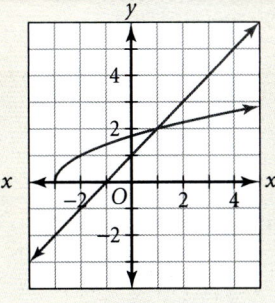

TRY THIS
a. $x = \pm 10\sqrt{2}$
b. $x = \pm 13$
c. $x = \pm\sqrt{y^2 + z^3}$

ADDITIONAL EXAMPLE 5

The area of a circular flower bed is 120 square feet. **What is the diameter of the bed?**

about 12.4 feet

ADDITIONAL EXAMPLE 6

Solve $x^2 + 12x + 36 = 1$ by using radicals.

$x = -7$ or $x = -5$

TRY THIS
$x = 2$ or $x = -12$

internetconnect

GO TO: go.hrw.com
KEYWORD: MA1 Planets

This activity introduces students to the work of the great scientists Johannes Kepler and Isaac Newton, and it gives students a chance to apply what they have learned about operations with radicals as well as direct and inverse variation. Students may also access other interesting Internet sites related to physics.

LESSON 12.2 **587**

Assess

Selected Answers
Exercises 5–28, 29–67 odd

ASSIGNMENT GUIDE

In Class	1–28
Core	29–59 odd
Core Plus	30–56 even, 58–61
Review	62–67
Preview	68–71

● Extra Practice can be found beginning on page 738.

Technology

A calculator with a square-root function is needed to obtain approximate solutions for Exercises 58 and 61b. A graphics calculator may be helpful for Exercises 54–57.

Practice

Practice
12.2 Square-Root Functions and Radical Equations

Solve each equation algebraically. Be sure to check your solution.

1. $\sqrt{x-7}=3$ $x=16$
2. $\sqrt{7-x}=3$ $x=-2$
3. $\sqrt{-x-7}=3$ $x=-16$
4. $\sqrt{x-3}=x$ no solution
5. $\sqrt{2x+3}=\sqrt{x+7}$ $x=4$
6. $\sqrt{2x+3}=5$ $x=11$
7. $\sqrt{5+x}=3$ $x=4$
8. $\sqrt{11-10x}=9$ $x=-7$
9. $\sqrt{3x-2}=-1$ no solution
10. $\sqrt{x-2}=3$ $x=11$
11. $\sqrt{2x+1}=x-1$ $x=4$
12. $\sqrt{2x+15}=x$ $x=5$

Use graphing technology or sketch the graphs on your graph paper to solve each equation, if possible.

13. $4\sqrt{x}=-x-3$ no solution
14. $\sqrt{x}+0.1=1.0$ $x=0.81$

Solve each equation, if possible. If not possible, explain why.

15. $5x^2=125$ $x=\pm 5$
16. $x^2-1=120$ $x=\pm 11$
17. $4x^2=120$ $x=\pm\sqrt{30}$
18. $\sqrt{x}=-49$ Not possible; a square root cannot be negative.
19. $x^2+169=0$ Not possible; a square root cannot be negative.
20. $\sqrt{x+20}=5$ Not possible; a square root cannot be negative.
21. $\sqrt{2x}=50$ $x=1250$
22. $\sqrt{x-1}+6=2$ Not possible; a square root cannot be negative.
23. $\sqrt{x+4}=1$ $x=-3$

588   LESSON 12.2

Exercises

● Communicate

1. State the domain of the function $y=\sqrt{x}$. Explain the reason for the restriction on the domain values.
2. Discuss which values of a make $\sqrt{a^2}=a$ true.
3. Explain why it is helpful to know the general form of an equation like the one that models the motion of a pendulum.
4. Explain why it is important to check all solutions found for an equation such as $\sqrt{6x+7}=x$.

● Guided Skills Practice

The regular rhythm of a pendulum helps some clocks to keep time.

Find the time in seconds it takes a pendulum with the given length to make one complete swing. Round your answers to the nearest hundredth. *(EXAMPLE 1)*

5. 60 cm ≈ **1.55 sec**
6. 150 cm ≈ **2.46 sec**
7. 100 cm ≈ **2.01 sec**
8. 45 cm ≈ **1.35 sec**

Find the length in centimeters of a pendulum that takes the given amount of time to make one complete swing. Round your answers to the nearest hundredth. *(EXAMPLE 1)*

9. 2 seconds ≈ **99.29 cm**
10. 2.5 seconds ≈ **155.15 cm**
11. 3.5 seconds ≈ **304.09 cm**
12. 4 seconds ≈ **397.18 cm**

Solve each equation. *(EXAMPLE 2)*

13. $\sqrt{x-2}=3$ **11**
14. $\sqrt{x+3}=1$ **–2**
15. $\sqrt{x-1}=2$ **5**
16. $\sqrt{x+5}=3$ **4**

Solve each equation, if possible. If there is no solution, write *no solution*. *(EXAMPLE 3)*

17. $\sqrt{x-1}=x$ **no sol.**
18. $\sqrt{4x-4}=x$ **2**
19. $\sqrt{3x-2}=x$ **1, 2**
20. $\sqrt{2x+24}=x$ **6**

Solve each equation for *x*. Give answers in simplified form. *(EXAMPLE 4)*

21. $x^2=800$
22. $x^2=250$
23. $x^2=8^2+6^2$
24. $x^2=v^2-t^2$

25. Find the radius of a circular flower garden that has an area of 18 square yards. *(EXAMPLE 5)*   **≈ 2.39 yards**

Solve each equation by using radicals. *(EXAMPLE 6)*

26. $x^2-8x+16=25$
27. $x^2+8x+16=4$
28. $x^2+4x+4=16$

21. $\pm 20\sqrt{2}$
22. $\pm 5\sqrt{10}$
23. ± 10
24. $\pm\sqrt{v^2-t^2}$
26. 9, –1
27. –2, –6
28. –6, 2

Practice and Apply

Solve each equation algebraically. Be sure to check your solution(s).

29. $\sqrt{x-5} = 2$ **9**
30. $\sqrt{x+7} = 5$ **18**
31. $\sqrt{2x} = 6$ **18**
32. $\sqrt{10-x} = 3$ **1**
33. $\sqrt{2x+9} = 7$ **20**
34. $\sqrt{2x-1} = 4$ **$\frac{17}{2}$**
35. $\sqrt{x+2} = x$ **2**
36. $\sqrt{6-x} = x$ **2**
37. $\sqrt{5x-6} = x$ **3, 2**
38. $\sqrt{x-1} = x-7$ **10**
39. $\sqrt{x+3} = x+1$ **1**
40. $\sqrt{2x+6} = x-1$ **5**
41. $\sqrt{x^2+3x-6} = x$ **2**
42. $\sqrt{x-1} = x-1$ **2, 1**
43. $\sqrt{x^2+5x+11} = x+3$ **2**

Solve each equation and write the solution(s) in simplest form.

44. $x^2 = 90$ **$\pm 3\sqrt{10}$**
45. $4x^2 = 7$ **$\pm \frac{\sqrt{7}}{2}$**
46. $3x^2 - 27 = 0$ **± 3**
47. $2x^2 = 48$ **$\pm 2\sqrt{6}$**
48. $x^2 - 8x + 16 = 0$ **4**
49. $x^2 - 12x + 36 = 0$ **6**

Solve each equation, if possible. If not possible, explain why.

53. Not possible; the square of a number must be positive.

50. $x^2 = 9$ **± 3**
51. $\sqrt{x} = 9$ **81**
52. $|x| = 9$ **± 9**
53. $x^2 = -9$

Graph each side of the equation on the same coordinate plane. Solve the equations that have solutions by finding the points of intersection of the graphs.

54. $\sqrt{x+12} = x$ **4**
55. $\sqrt{x-2} = x$ **no solution**
56. $\sqrt{x} = \frac{1}{3}x + \frac{2}{3}$ **1, 4**

57. a. domain: $x \leq 4$
       range: all reals
    b. domain: $x \leq 0$
       range: all reals

57. Find three ordered pairs for each function. Plot the ordered pairs for each pair of functions on the same coordinate axes. Sketch the graphs, and state the domain and range of each pair of graphs.

a. $y = \sqrt{4-x}$ and $y = -\sqrt{4-x}$ b. $y = \sqrt{-x}+3$ and $y = -\sqrt{-x}+3$

APPLICATIONS

PHYSICS The motion of a pendulum can be modeled by $t = 2\pi\sqrt{\frac{l}{32}}$, where l is the length of the pendulum in feet and t is the number of seconds required for one complete swing.

58. The mechanism of a grandfather clock is based on the motion of a pendulum. About how long should a pendulum be so that the time required for one complete swing is 1 second? **≈ 0.81 ft**

59. If the time required for one complete swing is doubled, by what value is the length multiplied? **4**

60. The time required for one complete swing of a pendulum is multiplied by c. By what value is the corresponding length multiplied? Generalize.
    When t is multiplied by c, l is multiplied by c^2.

61. **PHYSICS** The formula for kinetic energy is $E = \frac{1}{2}mv^2$, where E is the kinetic energy in joules, m is the mass of the object in kilograms, and v is the velocity of the object in meters per second.
    a. Solve the formula for v. Write in simplest radical form. $v = \frac{\sqrt{2Em}}{m}$
    b. If a baseball with a mass of 0.14 kilograms has 50 joules of kinetic energy, what is its velocity? **≈ 26.73 m/sec**

57. a.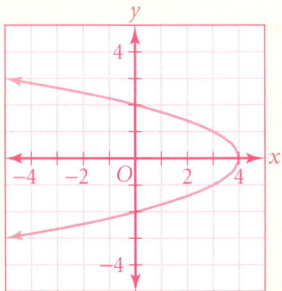

    b.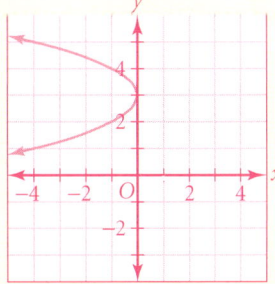

Error Analysis

Students sometimes forget that checking solutions to radical equations is a necessary part of the solution process. Even when they check, they may forget that $\sqrt{x}$ indicates only the positive square root of x. For instance, when checking the possible solutions in Example 4, they may believe that $\sqrt{4} = -2$ is a true statement. You may suggest that they use graphing as a second means of checking the solutions.

Look Beyond

In Exercise 68, students identify the coordinates of the midpoint of a segment on the coordinate plane. They will learn how to do this algebraically in Lesson 12.5. At this time, they should be able to find the midpoint by inspecting the graph of the segment. In Exercises 69–71, students preview translations, which they will learn more about in Chapter 14.

Student Technology Guide

12.2 Square-Root Functions and Radical Equations

You can use a graphics calculator to solve radical equations by graphing.

Example: Solve $\sqrt{2x-5} = 1$. If necessary, approximate the value of x to the nearest hundredth.

- Set up the problem as an intersection of two functions, $y = \sqrt{2x-5}$ and $y = 1$. To enter $y = $ as Y1, press [Y=] [2nd] [x^2]. 2 [X,T,R] [-] 5 [)]. Enter 1 as Y2.
- Press [GRAPH] [TRACE] and trace as closely to the intersection point as you can. The solution seems to be near $x = 3$.

Is the solution $x = 3$, or is it just close to 3? Here are two ways the calculator can help you decide.

Use the intersect feature. With the trace cursor near the intersection point, press [2nd] [TRACE]. 5: intersect [ENTER] [ENTER] [ENTER].

Press 3 [ENTER]. This moves the trace cursor to the point where $x = 3$. (Be sure you are tracing the graph of the square-root function and not the graph of the line.)

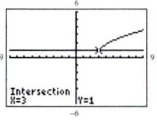

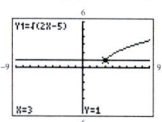

Using either method, you can see that $x = 3$ is the solution to $\sqrt{2x-5} = 1$.

Solve each equation. If necessary, estimate the value of x to the nearest hundredth.

1. $\sqrt{x+2} = 2$ $x = 2$
2. $\sqrt{2x-4} = -3$ $x = 0.5$
3. $\sqrt{3x+6} = 5$ $x \approx 6.33$
4. $\sqrt{4x+6} = x$ $x \approx 5.16$

5. In the Example, the trace cursor for $y = \sqrt{2x-5}$ does not appear unless $x \geq 2.5$. Why is this?
   Answers may vary. Sample answer: The graph does not exist for $x < 2.5$ because $2x - 5$ is negative, and so $\sqrt{2x-5}$ is not a real number for these x-values.

LESSON 12.2 589

Alternative Assessment

Portfolio Activity

The Portfolio Activity can be used as preparation for the Chapter Project or as a separate activity. In the Portfolio Activity on this page, students perform a pendulum experiment, graph the data, and compare their graph with the graph of the function that models the motion of a pendulum under ideal conditions.

Teaching Tip

The following is for the indicated steps of the Portfolio Activity:

2. The nature of the graph in Step 3 will be easier for students to see when the differences in the lengths of the strings are greater. Fairly accurate readings can be found by using the stopwatch function on many digital watches.

3–4. The logarithmic nature of the graph will be more or less clear depending on the scales chosen for the x- and y-axes.

5. If a stopwatch is used, it should be possible for the measured time and the prediction from the model to correspond to within approximately $\frac{1}{10}$ of a second.

65.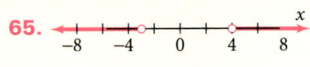
66.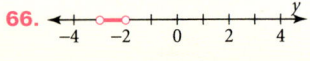
67.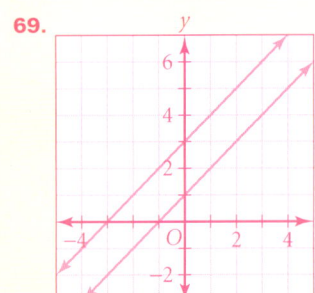

Look Back

62. Solve $-2x + 3 < 11$. (**LESSON 6.2**) $x > -4$

63. Factor $4x^4 - 16y^4$ completely. (**LESSON 9.6**) $4(x^2 - 2y^2)(x^2 + 2y^2)$

64. What values of the discriminant indicate that a quadratic equation has no real solution? Give an example. (**LESSON 10.5**)

64. When the value of the discriminant is less than 0; sample example: $x^2 + x + 1 = 0$ has a discriminant value of –3 and has no real solutions.

Solve and graph each inequality on a number line. (**LESSON 10.6**)

65. $x^2 - x - 12 > 0$
    $x > 4$ or $x < -3$
66. $y^2 + 5y + 6 < 0$
    $-3 < y < -2$
67. $a^2 - 6a - 16 < 0$
    $-2 < a < 8$

Look Beyond

68. What is the midpoint of the segment connecting the points (3, 2) and (9, 8)? (Hint: It may be helpful to accurately plot the points and connect them with a line.) (6, 5)

Graph each pair of functions on the same set of axes. Describe the relationship between the two graphs.

69. $y = x + 3$
    $y = (x - 2) + 3$
    parallel lines; second line shifted 2 units to right

70. $y = x^2 + 3$
    $y = (x - 2)^2 + 3$
    parabolas; second parabola shifted 2 units right

71. $y = |x| + 3$
    $y = |x - 2| + 3$
    v-shaped graphs; second graph shifted 2 units right

PORTFOLIO ACTIVITY

Include this activity in your portfolio.

Ideally, a graph of the length of a pendulum versus the time for one complete swing of the pendulum will be a transformation of the graph of the square-root parent function, $y = \sqrt{x}$, where x is the length of the string and y is the time for one complete swing of the pendulum.

1. Tie a piece of string to an object, such as a washer, and tie the other end to a fixed point where the washer can swing freely. Measure and record the length of the string (in centimeters) and time (in seconds) for ten complete swings of the pendulum. Divide the total time by 10 to get the average time of one swing.

2. Repeat Step 1 for six different string lengths.

3. Using the horizontal axis for the length of the string, l, and the vertical axis for the time for one complete swing, t, plot your data.

4. Does your graph appear to be a transformation of a square-root function?

5. The function that models the motion of a pendulum is $t = 2\pi\sqrt{\frac{l}{980}}$. Does this function model your data? If not, what are some possible reasons for the discrepancy?

69.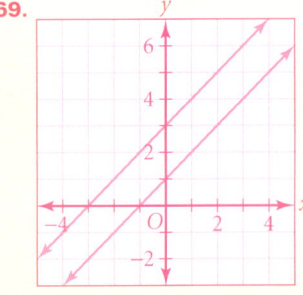
70.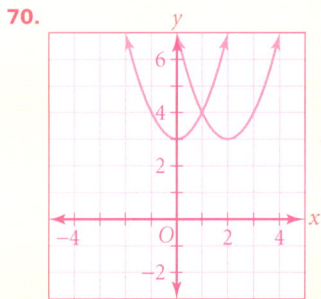
71.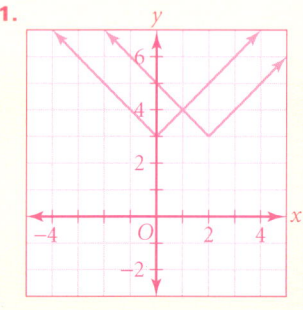

12.3 The Pythagorean Theorem

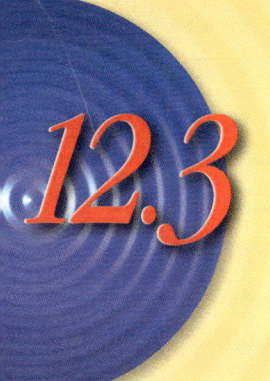

Right Triangle ABC

Objectives
- Find a side length of a right triangle given the lengths of its other two sides.
- Apply the Pythagorean Theorem to real-world problems.

Why Just as the Parthenon is a familiar and significant landmark, the Pythagorean Theorem is one of the most familiar and significant theorems in mathematics. To apply the theorem, you will need to use your knowledge of squares and square roots.

The Babylonian tablet lists numbers that show that the Babylonians knew about the relationship among the sides of right triangles many years before Pythagoras.

Prepare

NCTM PRINCIPLES & STANDARDS 1–4, 6–10

LESSON RESOURCES
- Practice 12.3
- Reteaching Master 12.3
- Enrichment Master 12.3
- Cooperative-Learning Activity 12.3
- Lesson Activity 12.3
- Problem Solving/Critical Thinking Master 12.3
- Teaching Transparency 57
- Teaching Transparency 58
- Teaching Transparency 59

QUICK WARM-UP

Solve each equation for x.

1. $x^2 = 49$ $x = \pm 7$
2. $x^2 = 48$ $x = \pm 4\sqrt{3}$
3. $x^2 + 144 = 169$ $x = \pm 5$
4. $x^2 + 3^2 = 5^2$ $x = \pm 4$
5. $x^2 = y^2 + z^2$ $x = \pm\sqrt{y^2 + z^2}$
6. $x^2 + y^2 = z^2$ $x = \pm\sqrt{z^2 - y^2}$

Also on Quiz Transparency 12.3

Recall from your earlier mathematics classes that a *right triangle* is one that has one 90° (or *right*) angle. The famous relationship among the sides of right triangles that you will study in this lesson has long been associated with the ancient Greeks. It is now known, however, that knowledge of this relationship goes back much further than the Greeks.

The conventional labeling of the sides and angles of a right triangle is shown at right.

The angles are labeled with capital letters. The right angle is commonly labeled *C*.

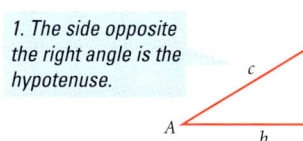

1. The side opposite the right angle is the hypotenuse.

2. The 90° angle is indicated by a square sign.

The side opposite an angle is labeled in lowercase with the same letter as the angle. In the diagram, side *c* is the **hypotenuse**. Sides *a* and *b* are called **legs**.

The ancient Babylonians, Chinese, and Hindus all knew of the right-triangle relationship that is known as the Pythagorean Theorem. European culture gave this name to the relationship in honor of Pythagoras, who was a famous Greek philosopher and mathematician of the sixth century B.C.E.

Teach

Why Tell students that the Pythagorean Theorem will allow them to find the length of any side of a right triangle given the lengths of the other two sides. Show students a sketch of a ladder resting against the side of a house. Ask them to suggest ways that the Pythagorean Theorem might be useful in this situation.

Use Teaching Transparency 57.

Alternative Teaching Strategy

USING TECHNOLOGY Use geometry software to create a dynamic demonstration of the Pythagorean Theorem. A sample sketch is shown at right. To make a sketch like this, construct $\overleftrightarrow{AC}$ and $\overleftrightarrow{CB}$ as perpendicular lines. Draw points *A* and *B* by choosing **Point on Object** from the **Construct** menu. For the list of measures, use the **Calculate** menu. To demonstrate the theorem, click and drag point *A* or point *B*. Ask students to describe the relationship between the changing values of $AC^2 + BC^2$ and AB^2.

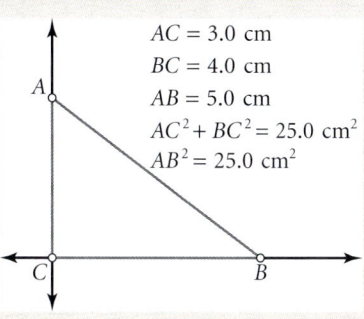

$AC = 3.0$ cm
$BC = 4.0$ cm
$AB = 5.0$ cm
$AC^2 + BC^2 = 25.0$ cm^2
$AB^2 = 25.0$ cm^2

LESSON 12.3 **591**

Activity Notes

In this Activity, students are led to make a conjecture about the *converse* of the Pythagorean Theorem. In Steps 1 and 2, they should note that the Pythagorean Theorem holds true for any right triangle they draw. In Step 3, they draw triangles that are *not* right, and they should discover that the Pythagorean Theorem does not apply.

Cooperative Learning

You may wish to have students do the Activity in pairs. In Steps 1, 2, and 3, each partner can draw two triangles and investigate the relationship among the sides. The partners can then discuss their results and work together to complete Step 4.

CHECKPOINT ✔

2. $c^2 = a^2 + b^2$; this agrees with the theorem.

Use Teaching Transparency 58.

Math CONNECTION

GEOMETRY Be sure students understand that the figure shown illustrates the Pythagorean Theorem only for the single case of a right triangle with sides of 3 units, 4 units, and 5 units. A proof of the theorem that utilizes an area model like this must demonstrate that the relationship holds true for all right triangles.

The Pythagorean Theorem

If given a right triangle with legs of length a and b and hypotenuse of length c, then $a^2 + b^2 = c^2$.

Activity
Exploring Right Triangles

You will need: large-grid graph paper or calculator

1. On a piece of graph paper, draw right triangles with the side lengths given in the table. (Make the legs, a and b, of the triangle horizontal and vertical. You can use a strip of graph paper as a ruler to measure the hypotenuse, c.)

Side a	Side b	Side c	a^2	b^2	c^2	a^2+b^2
3	4	5	?	?	?	?
5	12	13	?	?	?	?
16	30	34	?	?	?	?
13	84	85	?	?	?	?

CHECKPOINT ✔

2. Compare the entries under c^2 with those under $a^2 + b^2$. Explain how the relationships between them support the theorem given on the previous page.

3. Do you think that the theorem works for triangles other than right triangles? Test this idea by drawing and measuring some nonright triangles.

4. The converse of a theorem stated in if-then form is a new statement in which the phrases following the words *if* and *then* are interchanged. Write the converse of the Pythagorean theorem, which should begin with "If, in a triangle with sides a, b, and c, $c^2 = a^2 + b^2$, then ..." Based on Step 3, do you think the converse of the Pythagorean Theorem is true?

CONNECTION
GEOMETRY

Over the years, illustrations like the one at right have been used to show that the sum of the squares of the legs of a right triangle is equal to the square of the hypotenuse. Notice that the combined areas of the red and blue squares are equal to the area of the purple square.

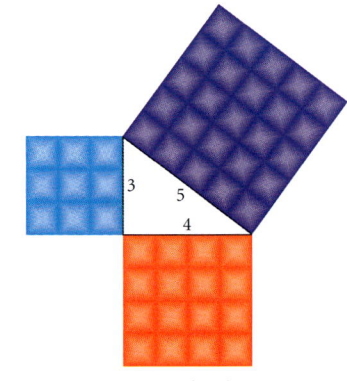

$3^2 + 4^2 = 5^2$

Interdisciplinary Connection

PHYSICS *Thermal expansion* is an increase in the length or volume of an object due to a change in temperature. It is a major reason why structures such as pipelines are built in sections with flexible joints that allow for expansion. Suppose that a 500-foot pipeline were laid with no joints. Heat could cause it to expand and buckle, as shown below. Find h, the height at the point of the buckle. **about 77.5 inches, or about 6.5 feet**

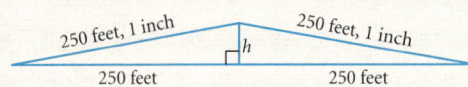

Inclusion Strategies

TACTILE LEARNERS Have students use drinking straws to model right triangles with sides defined by several common Pythagorean triples. For instance, have them cut straws to lengths of 3 centimeters, 4 centimeters, and 5 centimeters. They can then pull a length of string through the three pieces and join the ends of the string together to form the triangle. To obtain lengths greater than the length of one straw, two straws can be taped together. If students wish, they can use several different triangles to form a "Pythagorean mobile" for the classroom.

592 LESSON 12.3

EXAMPLE 1

Find the length of sides x, y, and z.

a.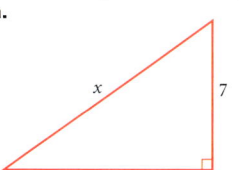
b.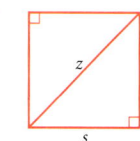
c. (square with diagonal z and sides s)

SOLUTION

a. Since x is the side opposite the right angle, x is the hypotenuse.
$$x^2 = 7^2 + 10^2$$
$$x^2 = 149$$
$$x = \sqrt{149} \approx 12.21$$

Since x represents the length of a side, it must be positive.

b. The lengths of the legs are 4 and 6.
$$y^2 = 4^2 + 6^2$$
$$y^2 = 52$$
$$y = \sqrt{52} \approx 7.21$$

c. The diagonal of the square is the hypotenuse of a right triangle. Each leg has length s.
$$z^2 = s^2 + s^2$$
$$z^2 = 2s^2$$
$$z = \sqrt{2s^2}$$
$$z = s\sqrt{2}$$

Thus, the formula for the diagonal, d, of a square with side length s is $d = s\sqrt{2}$.

Colored water is contained in the squares attached to the sides of a right triangle. As the device rotates, the water flows from the two squares opposite the hypotenuse, filling the square on the hypotenuse.

TRY THIS Find the length of sides x, y, and z.

a.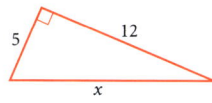
b.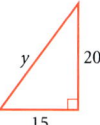
c.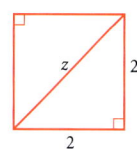

CRITICAL THINKING

You have seen that the Pythagorean Theorem can be used to find the length of the hypotenuse of a right triangle. Could it also be used to find the length of a leg? What information would be necessary?

ADDITIONAL EXAMPLE 1

Find the unknown length.

a.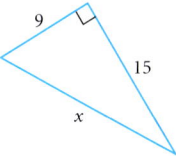

$x = \sqrt{306} \approx 17.49$

b. (right triangle with legs 10 and 24, hypotenuse y)

$y = \sqrt{676} = 26$

c.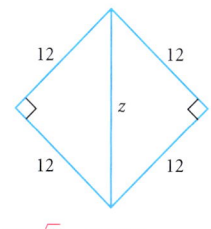

$z = 12\sqrt{2} \approx 16.97$

TRY THIS

a. $x = 13$
b. $y = 25$
c. $z = 2\sqrt{2}$

CRITICAL THINKING

The Pythagorean Theorem could be used to find the length of a leg of a right triangle given the lengths of the hypotenuse and the other leg.

Teaching Tip

Remind students that the hypotenuse of a right triangle is opposite the right angle, and it is the longest side of a right triangle.

Enrichment

Have students work in groups of three or four. Give them the figure at right. Ask them to show how it can be used to prove the Pythagorean Theorem.

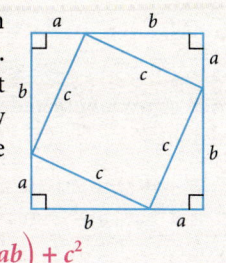

$$(a+b)^2 = 4\left(\frac{1}{2}ab\right) + c^2$$
$$a^2 + 2ab + b^2 = 2ab + c^2$$
$$a^2 + 2ab + b^2 - 2ab = 2ab + c^2 - 2ab$$
$$a^2 + b^2 = c^2$$

After students have done the first proof, have them repeat the activity for the figure at right. In this case, they may need help to recognize that the length of one side of the small square is $b - a$.

$$4\left(\frac{1}{2}ab\right) + (b-a)^2 = c^2$$
$$2ab + (b^2 - 2ab + a^2) = c^2$$
$$a^2 + b^2 = c^2$$

ADDITIONAL EXAMPLE 2

An airplane departs from an airport and flies 48 kilometers due north and then 20 kilometers due east. **At that point, how far is the airplane from the airport?**

52 kilometers

TRY THIS
60 units

CRITICAL THINKING
They are also the side lengths of right triangles.

Use Teaching Transparency 59.

ADDITIONAL EXAMPLE 3

The slant height of a regular square pyramid is 17 feet, and the height is 15 feet. **Find the length of one side of the base of the pyramid.**

16 feet

Math CONNECTION

GEOMETRY A *regular pyramid* is a three-dimensional figure that has one *base* bounded by a regular polygon, a point called the *vertex* that is not in the same plane as the base, and congruent triangular *faces* that connect the vertex to the sides of the base. The base of a regular square pyramid is bounded by a square. The height of one of the triangular faces is the *slant height* of the pyramid.

TRY THIS
about 57.2 meters

594 LESSON 12.3

EXAMPLE 2

APPLICATION: RECREATION

Your boat is traveling due north at 20 miles per hour. A friend's boat left at the same time from the same location and headed due west at 15 miles per hour. After an hour you get a call from your friend who tells you that he has just stopped because of engine trouble. **How far must you travel to reach your friend?**

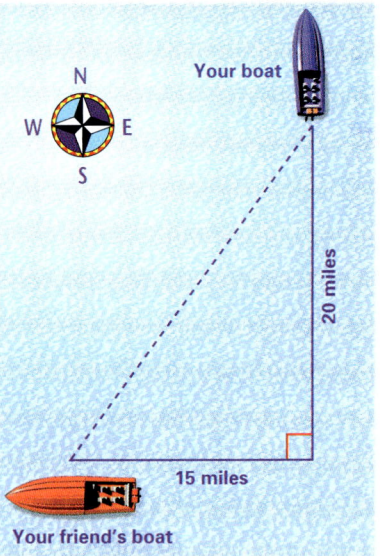

● **SOLUTION**

The legs of the triangle in the diagram are 15 miles and 20 miles long. Apply the Pythagorean Theorem to the triangle.

$$c^2 = a^2 + b^2$$
$$c^2 = 15^2 + 20^2$$
$$c^2 = 625$$
$$c = \sqrt{625} = 25$$

Since the answer represents a distance, only the positive root is used. You will need to travel 25 miles.

TRY THIS Find the hypotenuse of a right triangle whose legs are 36 and 48 units long.

CRITICAL THINKING What do you notice about multiples of the side lengths of a 3-4-5 right triangle (lengths of 6, 8, and 10, for example)?

EXAMPLE 3

The height of a pyramid with a square base can be found from the measurements of the base and the slant height. **Find the height of a pyramid whose square base measures 40 meters on each side and whose slant height is 52 meters.**

CONNECTION: GEOMETRY

● **SOLUTION**

Solve the triangle in the diagram below. Notice that the measure of one leg of the triangle is equal to half the length of a side of the base.

Apply the Pythagorean Theorem to △ABC with AB = 52 and $AC = \frac{1}{2} \cdot 40 = 20$.

$$52^2 = BC^2 + 20^2$$
$$BC^2 = 52^2 - 20^2$$
$$BC^2 = 2304$$
$$BC = \sqrt{2304} = 48$$

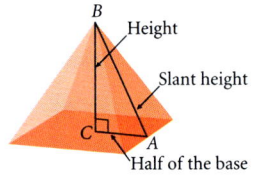

The height of the pyramid is 48 meters.

TRY THIS Find the height of a pyramid whose square base measures 36 meters on each side and whose slant height is 60 meters.

Reteaching the Lesson

USING MANIPULATIVES

Give each student a geoboard or provide several grids of dots that simulate the face of a geoboard. Have students work in groups of two or three.

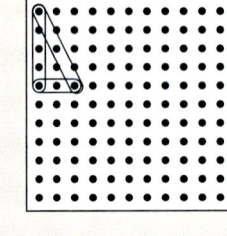

Tell them that they are to use right triangles to create segments of the following lengths on their geoboard:

1. $\sqrt{2}$ 1, 1
2. $\sqrt{5}$ 1, 2
3. $\sqrt{10}$ 1, 3
4. $\sqrt{13}$ 2, 3
5. $3\sqrt{2}$ 3, 3
6. $4\sqrt{2}$ 4, 4

Each segment created is the hypotenuse of a right triangle.

Now tell students to find the lengths of all segments that can be created in a 16-pin square of the geoboard. Have them list the lengths from least to greatest. Remind them to include horizontal and vertical segments. 1, $\sqrt{2}$, 2, $\sqrt{5}$, $2\sqrt{2}$, 3, $\sqrt{10}$, $\sqrt{13}$, 4, $\sqrt{17}$, $3\sqrt{2}$, $2\sqrt{5}$, 5, $4\sqrt{2}$

Exercises

Pythagoras
580-500 B.C.E.

Communicate

1. Identify the parts of a right triangle.
2. State the Pythagorean Theorem in your own words.
3. Explain how to find the hypotenuse of a right triangle if you know the lengths of the legs.
4. Describe how to rewrite the basic equation in the Pythagorean Theorem to determine the length of one of the legs.
5. Explain how the Pythagorean Theorem can be used to find the length of the diagonal of a square when given the length of its sides.

Guided Skills Practice

Find the length of sides x, y, and z in simplified form. (EXAMPLE 1)

6.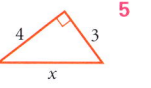
7.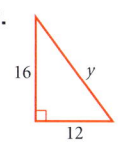
8.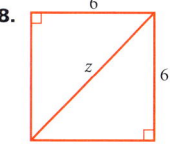

9. Calculate the length of the hypotenuse of a right triangle in simplified form whose legs are 6 and 10 units long. (EXAMPLE 2) $2\sqrt{34}$ units

10. Determine the height of a pyramid whose square base measures 30 meters on each side and whose slant height is 25 meters. (EXAMPLE 3) 20 m

Practice and Apply

Identify the hypotenuse and legs in each right triangle, and label them appropriately.

11. hyp = $\overline{AB}$ = c
    leg = $\overline{CB}$ = a
    leg = $\overline{AC}$ = b

12. hyp = $\overline{XZ}$ = y
    leg = $\overline{YZ}$ = x
    leg = $\overline{XY}$ = z

13. hyp = $\overline{ST}$ = r
    leg = s = $\overline{RT}$
    leg = t = $\overline{RS}$

18. 21.9
19. 24.3

11.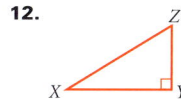
12.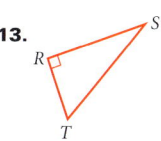
13.

For Exercises 14–21, the tables give the lengths of various sides of right triangles. Copy and complete the tables. Use a calculator and round each answer to the nearest tenth.

	Leg	Leg	Hypotenuse
14.	24	45	51 ?
16.	15 ?	8	17
18.	12	?	25
20.	30 ?	40	50

	Leg	Leg	Hypotenuse
15.	10	24 ?	26
17.	5	9	10.3 ?
19.	?	6	25
21.	0.75	1 ?	1.25

Assess

Selected Answers
Exercises 6–10, 11–39 odd

ASSIGNMENT GUIDE

In Class	1–10
Core	11–31 odd
Core Plus	12–32 even
Review	33–39
Preview	40, 41

✏ Extra Practice can be found beginning on page 738.

Mid-Chapter Assessment for Lessons 12.1 through 12.3 can be found on page 154 of the *Assessment Resources*.

Technology

Students will need a calculator with a square-root function for the approximations required in Exercises 14–21 and 30–32. The calculator also may be helpful for checking their answers to Exercises 22–27.

LESSON 12.3 **595**

Error Analysis

When applying the Pythagorean Theorem, some students will routinely add the squares of the two given lengths, without checking whether one is the length of the hypotenuse. Suggest that they begin each exercise involving the Pythagorean Theorem by writing the formula $a^2 + b^2 = c^2$ and determining if the value of c is unknown or if a given length should be substituted for c.

CONNECTION

28. a. $a = \sqrt{n^2 - \left(\frac{n}{2}\right)^2}$
    $= \frac{|n|\sqrt{3}}{2}$

Decide whether each set of numbers can represent the side lengths of a right triangle.

22. 3, 9, 7 **no**
23. 10, 6, 8 **yes**
24. 5, 9, 11 **no**
25. $\sqrt{5}, \sqrt{6}, \sqrt{11}$ **yes**
26. 4, $4\sqrt{3}$, 8 **yes**
27. $\sqrt{3}, \sqrt{4}, \sqrt{5}$ **no**

28. **GEOMETRY** Recall from your previous math classes that the *altitude* of a triangle forms a 90° angle with the base. The altitude of an *equilateral* triangle divides the triangle into 2 right triangles of the same size and shape. In the triangle at right, a is the altitude.
    a. Find the length of the altitude, a, in terms of the length of the sides, n.
    b. Copy and complete the table below. Simplify each answer.

Equilateral Triangles								
Side	4	6	8	10	12	20	n	? **34**
Half of a side	2	? **3**	? **4**	? **5**	? **6**	? **10**	? $\frac{n}{2}$	? **17**
Altitude	?	?	?	?	?	?	?	$17\sqrt{3}$

$2\sqrt{3}$ $3\sqrt{3}$ $4\sqrt{3}$ $5\sqrt{3}$ $6\sqrt{3}$ $10\sqrt{3}$ $\frac{|n|\sqrt{3}}{2}$

APPLICATIONS

LANDSCAPING A garden in the shape of a right triangle is represented by the figure at right.

29. What is the length of the third side of the garden? **≈ 9.8 ft**
30. How many feet of fencing must be bought to enclose the garden? (Assume that fencing is sold by the foot.) **34 ft**
31. If the cost of fencing is $4.98 per foot, how much will it cost to enclose the garden? **$169.32**

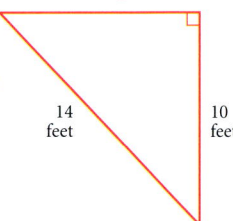

14 feet 10 feet

32. **SPORTS** A baseball diamond is a square with sides of 90 feet. To the nearest foot, how long is a throw from third base to first base? **≈ 127 ft**

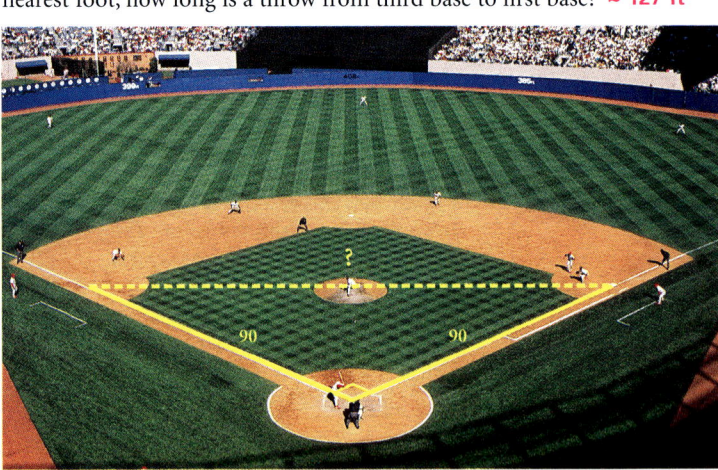

Practice

12.3 The Pythagorean Theorem

NAME _____ CLASS _____ DATE _____

Complete the table. Use a calculator and round each answer to the nearest tenth.

	Leg	Leg	Hypotenuse
1.	21	**72**	75
3.	6.5	7.2	**9.7**
5.	13	**84**	85
7.	**16**	63	65
9.	28	45	**53**
11.	48	**64**	80
13.	**0.6**	0.63	0.87

	Leg	Leg	Hypotenuse
2.	**18**	80	82
4.	20	21	29
6.	33	44	**55**
8.	**36**	48	60
10.	10	24	**26**
12.	$\frac{1}{5}$	$\frac{4}{15}$	$\frac{1}{3}$
14.	0.036	0.077	**0.085**

Solve for x.

15.   **15**
16.   **60**
17. 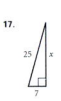 **24**

18. A picket fence has a gate that is 48 inches wide and 62 inches high. Find the length, to the nearest inch, of a diagonal brace for the gate. **78 inches**

19. The base of a 15-foot ladder is placed 9 feet from a wall. If the ladder leans against the wall, how high will it reach? **12 feet**

20. Cynthia drove 56 miles due south and 90 miles due west. If she drives in a straight line back to the point where she started, how far will she drive? **106 miles**

596 LESSON 12.3

 **Look Back**

APPLICATION

INCOME A student is paid a weekly salary of $20 plus $2.50 for every lamp that he sells. Use this information in Exercises 33–36. **(LESSON 1.2)**

33. Copy and complete the table.

Number of lamps sold	0	10	20	30	40	50
Week's pay	a. ?	b. ?	c. ?	d. ?	e. ?	f. ?

$20 $45 $70 $95 $120 $145

34. Write an equation that describes w, one week's pay, in terms of l, the number of lamps sold. $w = 20 + 2.50l$

35. If the student sells 29 lamps in a given week, what will his week's pay be? **$92.50**

36. If the student needs to earn $149 in a given week, how many lamps must he sell? **52**

Decide whether the given pairs of lines are parallel, perpendicular, or neither. (LESSON 5.6)

37. $y = \frac{1}{2}x + 3$ **neither**
    $y = 4x + 3$

38. $y = 3x - 4$ **perp.**
    $y = -\frac{1}{3}x + 2$

39. $-2x + y = 8$ **parallel**
    $-6x + 3y = 15$

 Look Beyond

CONNECTION

GEOMETRY A **Pythagorean triple** is a set of three positive integers that satisfy the Pythagorean Theorem. For example, {3, 4, 5} is a Pythagorean triple. The three expressions listed below can be used to create Pythagorean triples in a three-step process.

40. Answers may vary. Samples:
{5, 12, 13},
{7, 24, 25},
{8, 15, 17},
{9, 40, 41},
{20, 21, 29},
{11, 60, 61}

If p and q are integers and $p > q > 0$, then the following expressions give the elements of a Pythagorean triple:

$$\{p^2 - q^2,\ 2pq,\ p^2 + q^2\}$$

For $p = 2$ and $q = 1$:

$$p^2 - q^2 = 2^2 - 1^2 = 3$$
$$2pq = 2 \cdot 2 \cdot 1 = 4$$
$$p^2 + q^2 = 2^2 + 1^2 = 5$$

Thus, {3, 4, 5} is a Pythagorean triple.

41. Because there are infinitely many integers, the three expressions will generate infinitely many Pyth. triples.

40. Using the expressions above, make a list of at least 6 different Pythagorean triples.

41. Explain why the three expressions give infinitely many Pythagorean triples.

 **Look Beyond**

In Exercises 40 and 41, students work with algebraic expressions that can be used to generate Pythagorean triples. Students will investigate this topic further if they take a course in geometry.

Student Technology Guide

 Student Technology Guide

12.3 The Pythagorean Theorem

A calculator is a helpful tool when solving problems. However, it is a good idea to do the simpler math yourself in order to reduce the chances of pressing an incorrect key.

Example: The hypotenuse of a right triangle is 4 feet long. One leg is 2 feet long. To the nearest tenth of a foot, how long is the other leg?

- Use the Pythagorean Theorem, with $a = 2$ and $c = 4$. $c^2 = a^2 + b^2$, so $4^2 = 2^2 + b^2$
- Solve for b. $4^2 - 2^2 = b^2$, so $\sqrt{4^2 - 2^2} = b$
- Simplify. $\sqrt{4^2 - 2^2} = \sqrt{16 - 4} = \sqrt{12}$
- Press [2nd] [x^2] 12 [ENTER]. (The calculator may give you a left parenthesis. When finding the square root of a single number, you do not need to add the right parenthesis.)

√(12
         3.464101615

The triangle's other leg is about 3.5 feet long.

Find each missing length for right triangle ABC with hypotenuse c. Give your answer to the nearest tenth.

1. $a = 1, b = 3$ — **3.2**
2. $a = 4$ meters, $c = 6$ meters — **4.5 meters**
3. $b = 9$ inches, $c = 10$ inches — **≈ 4.4 inches**

Example: Samuel and Maria are riding their bicycles home from school. Maria rides east and Samuel rides north. How far away from each other are they when
a. Samuel has gone 45 yards and Maria has gone 55 yards?
b. Samuel has gone 95 yards and Maria has gone 92 yards?

In both cases, $d^2 = m^2 + s^2$, so $d = \sqrt{m^2 + s^2}$.

a. Press [2nd] [x^2] 45 [x^2] [+] 55 [x^2] [)] [ENTER].
Maria and Samuel are about 71.1 yards apart.

b. Press [2nd] [ENTER]. Use the left arrow [◄] key to move to 45 and type a 9 over the 4. Then move to 55 and replace it with 92. Press [ENTER]. Maria and Samuel are now about 132.2 yards apart.

√(45²+55²)
         71.06335202
√(95²+92²)
         132.2459829

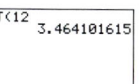

4. To the nearest tenth of a yard, how far away from each other are Maria and Samuel when
a. Samuel has gone 112 yards and Maria has gone 123 yards? **166.4 yards**
b. Samuel has gone 237 yards and Maria has gone 224 yards? **326.1 yards**

LESSON 12.3 597

Prepare

NCTM PRINCIPLES & STANDARDS 1–10

LESSON RESOURCES

- Practice 12.4
- Reteaching Master 12.4
- Enrichment Master 12.4
- Cooperative-Learning Activity 12.4
- Lesson Activity 12.4
- Problem Solving/Critical Thinking Master 12.4
- Teaching Transparency 60

QUICK WARM-UP

In a right triangle, a and b are the lengths of the legs and c is the length of the hypotenuse. **Find each unknown length. Give exact answers.**

1. $a = 6, b = 8, c = \underline{\ ?\ }$ 10
2. $a = \sqrt{2}, b = \sqrt{7}, c = \underline{\ ?\ }$ 3
3. $a = 1, b = 7, c = \underline{\ ?\ }$ $5\sqrt{2}$
4. $a = 15, c = 17, b = \underline{\ ?\ }$ 8
5. $b = \sqrt{7}, c = 4, a = \underline{\ ?\ }$ 3
6. $a = 1, c = 5, b = \underline{\ ?\ }$ $2\sqrt{6}$

Also on Quiz Transparency 12.4

Teach

Why Ask students how distances between cities or other locations can be related to coordinates. Among other observations, they might note that coordinate grids often are superimposed on city maps and that a scale is given. Similarly, lines of latitude and longitude, with known distances between them, provide coordinates for any point on Earth.

12.4 The Distance Formula

Objectives

- Use the distance formula to find the distance between two points in a coordinate plane.
- Determine whether a triangle is a right triangle.
- Apply the midpoint formula.

Why In an emergency situation, rescue personnel must be able to quickly determine how long it will take to reach an accident scene. The ability to calculate the distance between two points is of critical importance not only to pilots, but also to sailors, surveyors, architects, astronomers, and drafting professionals.

**APPLICATION
EMERGENCY SERVICES**

A medical center learns of an auto accident located 7 miles east and 11 miles north of the center. If the medical center's helicopter is now located 1 mile east and 3 miles north of the center, what is the shortest distance the helicopter can travel to get to the scene of the accident?

The *distance* between two points, A and B, is the *length* of $\overline{AB}$, which is the segment with A and B as endpoints.

The shortest distance between the helicopter and the accident is HA, which is the length of the hypotenuse of right triangle HVA. $\triangle HVA$ can be visualized by picturing $\overline{HV}$ as an east-west segment and $\overline{VA}$ as a north-south segment.

Alternative Teaching Strategy

USING MANIPULATIVES   Give students the figure at right together with a "ruler" of 16 adjacent grid squares, as shown below. Have them use the ruler to measure each segment in the figure to the nearest half-unit. Ask them how they can tell without measuring that $EF = JK$, but $EF \neq LM$. Use their observations to initiate a discussion of the Pythagorean Theorem and the distance formula. Have them use the distance formula to calculate the exact length of all the segments.

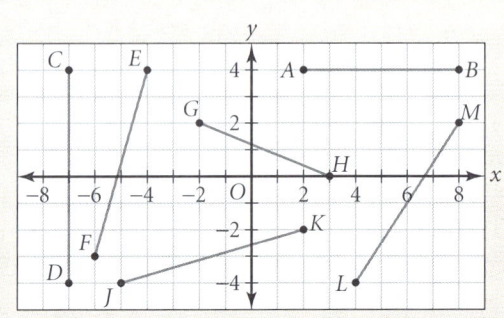

$AB = 6, EF = \sqrt{53} \approx 7.3, JK = \sqrt{53} \approx 7.3,$
$CD = 8, GH = \sqrt{29} \approx 5.4, LM = 2\sqrt{13} \approx 7.2$

598   LESSON 12.4

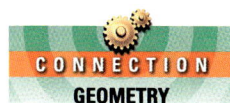

CONNECTION
GEOMETRY

You can find the length of $\overline{HV}$ by subtracting the first coordinates of V and H.

$$\text{length of } \overline{HV} = 7 - 1 = 6$$

You can find the length of $\overline{AV}$ by subtracting the second coordinates of A and V.

$$\text{length of } \overline{AV} = 11 - 3 = 8$$

Now use the Pythagorean Theorem.

$$c^2 = a^2 + b^2$$
$$(\text{length of } \overline{AH})^2 = 6^2 + 8^2$$
$$(\text{length of } \overline{AH})^2 = 100$$
$$\text{length of } \overline{AH} = \sqrt{100} = 10$$

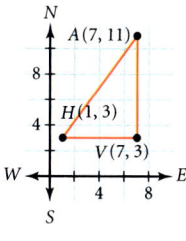

The shortest distance that the medical center's helicopter can travel is 10 miles.

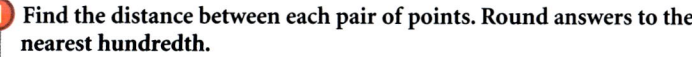

Distance Formula

The distance, d, between two points $A(x_1, y_1)$ and $B(x_2, y_2)$ is given by

$$d = \sqrt{(x_2 - x_1)^2 + (y_2 - y_1)^2}.$$

EXAMPLE 1 Find the distance between each pair of points. Round answers to the nearest hundredth.

a. (2, −3) and (8, 2)   **b.** (−4, −3) and (2, 0)

• **SOLUTION**
Use the distance formula.

a. $d = \sqrt{(x_2 - x_1)^2 + (y_2 - y_1)^2}$
$d = \sqrt{(8 - 2)^2 + [2 - (-3)]^2}$
$d = \sqrt{6^2 + 5^2}$
$d = \sqrt{36 + 25}$
$d = \sqrt{61}$
$d \approx 7.81$

b. $d = \sqrt{(x_2 - x_1)^2 + (y_2 - y_1)^2}$
$d = \sqrt{[2 - (-4)]^2 + [0 - (-3)]^2}$
$d = \sqrt{6^2 + 3^2}$
$d = \sqrt{36 + 9}$
$d = \sqrt{45}$
$d \approx 6.71$

TRY THIS Find the distance between (4, 5) and (−5, 7).

Interdisciplinary Connection

GEOGRAPHY A square grid is superimposed on the map of a city, with each unit on the grid representing one-quarter of a mile of actual distance. Two points of interest in the city have map coordinates of (3, 8) and (12, 2). What is the actual distance between them, to the nearest quarter of a mile? about $2\frac{3}{4}$ miles

Inclusion Strategies

TACTILE LEARNERS Have students use geoboards with pegs arranged in a square array to model distance problems on a coordinate plane. The figure at right, for example, shows a model for finding the distance between points with coordinates (−2, 5) and (5, −4).

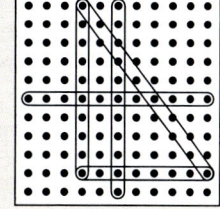

Math
CONNECTION

GEOMETRY Point out to students that when two points in a plane lie on a horizontal line, the distance between them is simply the absolute value of the difference of their x-coordinates. Similarly, the distance between two points that lie on a vertical line is the absolute value of the difference of their y-coordinates.

Use Teaching Transparency 60.

ADDITIONAL
EXAMPLE 1

Find the distance between each pair of points.

a. $J(11, 7)$ and $K(5, −1)$ 10

b. $P(−3, 6)$ and $Q(−2, 8)$
$\sqrt{5}$, or ≈ 2.24

TRY THIS
$d \approx 9.22$

LESSON 12.4   **599**

The Converse of the Pythagorean Theorem

Recall from Lesson 11.6 that when the *if* and *then* portions of a theorem are interchanged, the new statement is called the *converse* of the given theorem.

Converse of the Pythagorean Theorem

If a, b, and c are the lengths of the sides of a triangle and $a^2 + b^2 = c^2$, where $c > a$ and $c > b$, then the triangle is a right triangle.

CRITICAL THINKING Is the converse of an if-then statement always true? If not, give an example.

EXAMPLE 2 Given vertices $P(1, 2)$, $Q(3, -1)$, and $R(-5, -2)$, determine whether triangle PQR is a right triangle.

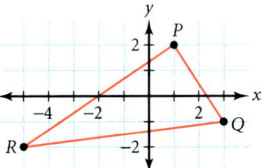

● **SOLUTION**

Draw triangle PQR on graph paper.

Use the distance formula to find the lengths of the sides.

$PR = \sqrt{(-5-1)^2 + (-2-2)^2}$
$= \sqrt{(-6)^2 + (-4)^2} = \sqrt{52}$

$PQ = \sqrt{(3-1)^2 + (-1-2)^2}$
$= \sqrt{2^2 + (-3)^2} = \sqrt{13}$

$RQ = \sqrt{[3-(-5)]^2 + [-1-(-2)]^2}$
$= \sqrt{8^2 + 1^2} = \sqrt{65}$

Now use the converse of the Pythagorean Theorem to test whether the three lengths can be the sides of a right triangle.

Is $(\sqrt{52})^2 + (\sqrt{13})^2 = (\sqrt{65})^2$ true?

Since $52 + 13 = 65$, the triangle is a right triangle.

TRY THIS Given vertices $P(-1, -2)$, $Q(5, 2)$, and $R(-3, 1)$, determine whether triangle PQR is a right triangle.

CRITICAL THINKING How can you use the slopes of PR and PQ to show that triangle PQR in Example 2 is a right triangle?

Enrichment

Have students work in pairs. First pose this question to them: If point P is in the same plane as $\overline{AB}$ and is 5 units from point A and 5 units from point B, is point P necessarily the midpoint of $\overline{AB}$? Lead them to see that the answer is no. Point P can be anywhere on the line that is perpendicular to $\overline{AB}$ at its midpoint. Tell them that this line contains all the points in the plane that are *equidistant* from points A and B.

Now give students the exercises at right. Have them find an equation for the line in the same plane as the given points that contains all points equidistant from them.

1. $A(-3, 10)$, $B(5, 10)$ $x = 1$
2. $C(10, -3)$, $D(10, 5)$ $y = 1$
3. $E(6, 0)$, $F(0, 6)$ $y = x$
4. $G(6, 2)$, $H(0, 8)$ $y = x + 2$
5. $J(-4, 0)$, $K(0, 4)$ $y = -x$
6. $L(4, 0)$, $M(0, 2)$ $y = 2x - 3$

CRITICAL THINKING
No; answers may vary. Sample answer: "If you are in Chicago, then you are in Illinois" is a true statement. The converse, "If you are in Illinois, then you are in Chicago" is not a true statement.

ADDITIONAL EXAMPLE 2

Given vertices $D(-3, -4)$, $E(2, -2)$, and $F(0, 3)$, determine whether $\triangle DEF$ is a right triangle.

$DE = \sqrt{(-3-2)^2 + [-4-(-2)]^2}$
$= \sqrt{29}$

$EF = \sqrt{(2-0)^2 + (-2-3)^2}$
$= \sqrt{29}$

$DF = \sqrt{(-3-0)^2 + (-4-3)^2}$
$= \sqrt{58}$

$DE^2 + EF^2 \stackrel{?}{=} DF^2$
$(\sqrt{29})^2 + (\sqrt{29})^2 \stackrel{?}{=} (\sqrt{58})^2$
$58 = 58$ True

Yes, $\triangle DEF$ is a right triangle.

Teaching Tip

You may want to remind students that a *vertex* of a triangle, or of any polygon, is a point where two sides intersect. The plural of *vertex* is *vertices*.

TRY THIS
$\triangle PQR$ is a right triangle.

CRITICAL THINKING
If the slopes of the two legs are negative reciprocals, then the legs are perpendicular.

The Midpoint Formula

 Deriving the Midpoint Formula

You will need: graph paper

1. Plot the following pairs of points on grid paper. Guess the coordinates of the midpoint, M, of $\overline{PQ}$ for each pair.
   a. $P(1, 6)$, $Q(7, 10)$ b. $P(-2, 3)$, $Q(6, 7)$ c. $P(-4, -1)$, $Q(8, 7)$

2. Use the distance formula to verify that your proposed midpoints are the midpoints of the given segments.

3. What rule can be applied to the x-coordinates of the endpoints of a segment to obtain the x-coordinate of the midpoint of the segment?

4. What rule can be applied to the y-coordinates of the endpoints of a segment to obtain the y-coordinate of the midpoint of the segment?

CHECKPOINT ✔ 5. Apply your rules from Steps 3 and 4 to calculate the coordinates of the midpoint of $\overline{AB}$, for points $A(12, 63)$ and $B(43, 20)$.

6. Use the distance formula to verify your answer.

Midpoint Formula

Given $\overline{PQ}$, where $P(x_1, y_1)$ and $Q(x_2, y_2)$, the coordinates of the midpoint, M, of $\overline{PQ}$, are $\left(\dfrac{x_1 + x_2}{2}, \dfrac{y_1 + y_2}{2}\right)$.

EXAMPLE 3 The streets of a certain city are laid out like a coordinate plane, with City Hall at the origin. Jacques lives 3 blocks east and 2 blocks north of City Hall. His friend Alise lives 11 blocks east and 8 blocks north of City Hall. They want to meet at the point midway between their homes. **At what point should they meet?**

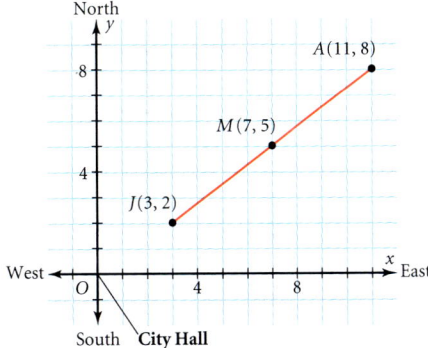

Activity Notes

In this Activity, students first guess the coordinates of the midpoints of three segments in a coordinate plane and then use the distance formula to verify that their guesses are correct. Then they generalize their results to create a formula for finding the coordinates of the midpoint of any segment given the coordinates of its endpoints.

CHECKPOINT ✔
5. $(27.5, 41.5)$

Teaching Tip

Remind students that a *midpoint* is the point that divides a segment into two congruent segments. Be sure they understand that for point M to be the midpoint of a given segment, the distance from M to one endpoint of the segment must equal the distance from M to the other endpoint.

ADDITIONAL EXAMPLE 3

Refer to the city described in Example 3. Alan lives 5 blocks west and 4 blocks south of City Hall, and his friend Mona lives 7 blocks east and 2 blocks north of City Hall. They want to meet midway between their homes. **At what location should they meet?**

The location midway between their homes is identified by the coordinates $(1, -1)$, which is 1 block east and 1 block south of City Hall.

Reteaching the Lesson

COOPERATIVE LEARNING Have students work in groups of four. Within each group, one student should be the "quadrant keeper" for quadrant I, another for quadrant II, another for quadrant III, and another for quadrant IV.

Write $A(1, 2)$ on the board or overhead. Tell the groups that they are to find four points that are exactly 5 units from point A, one in each quadrant. The only rule is that no point can lie on the same horizontal or vertical line as point A. Each quadrant keeper should then use the distance formula to verify that the distance from the point in his or her quadrant to point A is actually 5 units. **sample answer: quadrant I, $B(4, 6)$; quadrant II, $C(-2, 6)$; quadrant III, $D(-2, -2)$; quadrant IV, $E(4, -2)$**

Repeat the activity for the point $P(-2, 1)$. This time, have students find a point in each of the four quadrants that is exactly $\sqrt{13}$ units from point P. **sample answer: quadrant I, $Q(1, 3)$; quadrant II, $R(-5, 3)$; quadrant III, $S(-4, -2)$; quadrant IV, $T(1, -1)$**

LESSON 12.4 **601**

ADDITIONAL EXAMPLE 4

The center of a circle is at $R(-3, -5)$. One endpoint of a diameter of this circle is at $S(2, -3)$. **What are the coordinates of point T, the other endpoint of this diameter?**

$T(-8, -7)$

Math CONNECTION

GEOMETRY Remind students that a *diameter* of a circle is a segment that has its endpoints on the circle and that passes through the center of the circle. Since all points on a circle are equidistant from the center, the center of the circle is the midpoint of any diameter. Point out that a circle has infinitely many diameters.

TRY THIS

$B(-6, -7)$

Assess

Selected Answers
Exercises 6–20, 21–69 odd

ASSIGNMENT GUIDE

In Class	1–20
Core	21–53 odd, 57
Core Plus	22–54 even, 55–59
Review	60–69
Preview	70–74

✎ Extra Practice can be found beginning on page 738.

● **SOLUTION**

Plot $J(3, 2)$ and $A(11, 8)$ on a coordinate plane, as shown in the figure. To find the point midway between J and A, use the midpoint formula, which gives the "average" coordinates, written as $(\bar{x}, \bar{y})$.

$\bar{x} = \frac{x_1 + x_2}{2} = \frac{3 + 11}{2} = \frac{14}{2} = 7$

$\bar{y} = \frac{y_1 + y_2}{2} = \frac{2 + 8}{2} = \frac{10}{2} = 5$

They should meet at $M(7, 5)$, the midpoint of $\overline{JA}$.

EXAMPLE 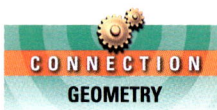 The center of a circle is $M(3, 4)$. **If one endpoint of a diameter is $A(-4, 6)$, what is the other endpoint, $B(x_2, y_2)$?**

CONNECTION
GEOMETRY

● **SOLUTION**

Substitute the coordinates of M and A into the midpoint formula, and solve.

$\frac{x_1 + x_2}{2} = \bar{x}$ $\frac{y_1 + y_2}{2} = \bar{y}$

$\frac{-4 + x_2}{2} = 3$ $\frac{6 + y_2}{2} = 4$

$-4 + x_2 = 6$ $6 + y_2 = 8$

$x_2 = 10$ $y_2 = 2$

The other endpoint is $B(10, 2)$.

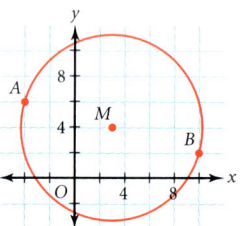

TRY THIS If the center of a circle is $M(-1, -2)$ and one endpoint of a diameter is $A(4, 3)$, what is the other endpoint, $B(x_2, y_2)$?

Exercises

● **Communicate**

1. Explain, in your own words, the difference between points and coordinates.
2. Discuss the relationship between the Pythagorean Theorem and the distance formula.
3. Explain how to find the distance between the points (2, 5) and (5, 7) in a coordinate plane.
4. Describe how to find the midpoint of the segment connecting the points from Exercise 3.
5. Explain what the converse of an if-then statement is.

36.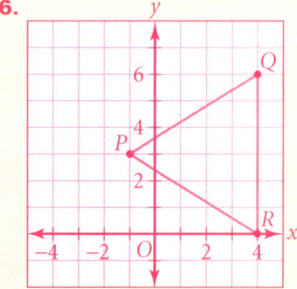

37.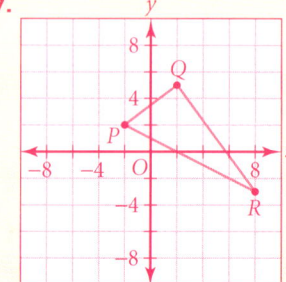

Guided Skills Practice

Find the distance between each pair of points. *(EXAMPLE 1)*

6. $A(2, 3)$, $B(5, 7)$ **5**
7. $C(-5, 6)$, $D(7, 1)$ **13**
8. $E(-3, -2)$, $F(5, -8)$ **10**

Determine whether each set of points represents the vertices of a right triangle. *(EXAMPLE 2)*

9. $A(3, 6)$, $B(1, 2)$, $C(-3, 4)$ **yes**
10. $D(2, 3)$, $E(-2, 9)$, $F(-5, 7)$ **yes**
11. $G(4, -2)$, $H(-1, 0)$, $I(6, -2)$ **no**
12. $J(6, 4)$, $K(3, 0)$, $L(2, 7)$ **yes**

What is the midpoint of a segment with the following endpoints? *(EXAMPLE 3)*

13. $A(5, -1)$, $B(-3, 7)$ **(1, 3)**
14. $J(-3, -2)$, $K(4, -6)$ $\left(\frac{1}{2}, -4\right)$
15. $M(-18, 13)$, $N(44, -13)$ **(13, 0)**
16. $O(17, -13)$, $P(14, 12)$ $\left(15\frac{1}{2}, -\frac{1}{2}\right)$

Given the coordinates of the midpoint, M, and one endpoint, E, of a segment, find the other endpoint, $F(x_2, y_2)$. *(EXAMPLE 4)*

17. $M(4, -3)$, $E(7, -7)$ **(1, 1)**
18. $M(-6, -2)$, $E(4, 8)$ **(−16, −12)**
19. $M\left(\frac{3}{2}, \frac{7}{2}\right)$, $E(4, 2)$ **(−1, 5)**
20. $M\left(\frac{9}{2}, \frac{1}{2}\right)$, $E(-4, 3)$ **(13, −2)**

Practice and Apply

Find the distance between each pair of points. Round answers to the nearest hundredth.

21. $A(4, 7)$, $B(1, 3)$ **5**
22. $P(5, 6)$, $Q(17, 11)$ **13**
23. $R(-5, -2)$, $S(-9, 3)$
24. $T(5, -2)$, $U(-6, 3)$
25. $P(-3, 8)$, $Q(4, 8)$ **7**
26. $J(5, -3)$, $K(5, 6)$ **9**
27. $F(7, 2)$, $G(-1, -1)$
28. $M(3, -2)$, $N(5, 4)$
29. $S(5, 6)$, $T(4, -3)$ **9.06**
30. $K(4, 2)$, $L(7, 1)$ **3.16**
31. $C(-1, -5)$, $D(3, -3)$
32. $Y(-4, 2)$, $Z(10, 6)$
33. $P(15, 2)$, $Q(-1, 2)$
34. $T(5, 7)$, $U(5, -10)$
35. $V(4, 8)$, $W(-7, 12)$

For Exercises 36–41, plot △PQR on graph paper. Decide which of the following terms apply to △PQR: scalene (no sides equal), isosceles (two sides equal), equilateral (three sides equal), or right.

36. $P(-1, 3)$, $Q(4, 6)$, $R(4, 0)$
37. $P(-2, 2)$, $Q(2, 5)$, $R(8, -3)$
38. $P(1, -1)$, $Q(-3, 2)$, $R(-3, -4)$
39. $P(5, 2)$, $Q(1, 1)$, $R(6, -2)$
40. $P(-4, -2)$, $Q(0, -2)$, $R(-2, 0)$
41. $P(1, -4)$, $Q(2, 1)$, $R(8, 0)$

Plot each segment in a coordinate plane, and identify the coordinates of the midpoint of the segment.

42. $\overline{AB}$, where $A(2, 5)$ and $B(9, 5)$
43. $\overline{EF}$, where $E(-4, -1)$ and $F(-2, -7)$
44. $\overline{GH}$, where $G(-5, 6)$ and $H(-1, -2)$
45. $\overline{IJ}$, where $I(4, 3)$ and $J(9, -4)$

In Exercises 46–47, the locations of two points are given relative to the origin, (0, 0). Calculate the coordinates of the midpoint of the segment connecting each point and the origin.

46. A is 2 units east and 3 units north of the origin. $\left(1, \frac{3}{2}\right)$
47. C is 2 units west and 5 units south of origin. $\left(-1, -\frac{5}{2}\right)$

---

23. 6.40
24. 12.08
27. 8.54
28. 6.32
31. 4.47
32. 14.56
33. 16
34. 17
35. 11.70
36. isosceles
37. scalene right
38. isosceles
39. isosceles right
40. isosceles right
41. scalene
42. (5.5, 5)
43. (−3, −4)
44. (−3, 2)
45. (6.5, −0.5)

Technology

Students will need a calculator with a square-root function for the approximations required in Exercises 21–35 and 56–59. The calculator also may be helpful for verifying their calculations in Exercises 42–53.

Error Analysis

When using the midpoint formula, students sometimes subtract rather than add the coordinates. Suggest that they plot the given points and the proposed midpoint to visually determine whether they have applied the midpoint formula correctly.

40.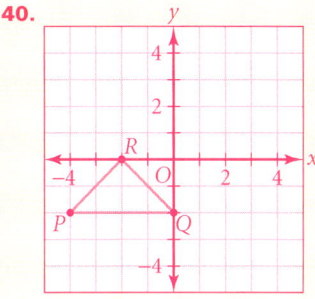

41.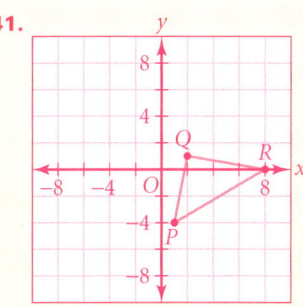

42.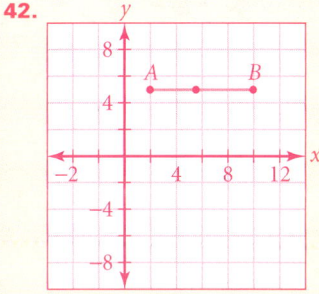

43.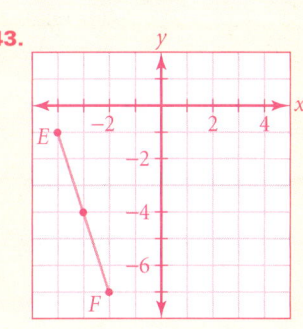

38.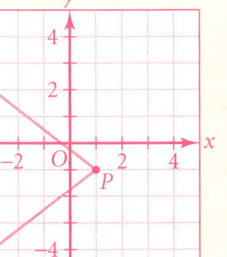

39.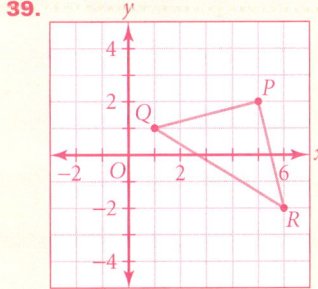

The answers to Exercises 44 and 45 can be found in Additional Answers beginning on page 810.

Look Beyond

In Exercises 70–73, students use their calculators to investigate the meaning of $\frac{1}{2}$ as an exponent. They will learn more about this and other *rational exponents* if they take a more advanced course in algebra.

54. a.
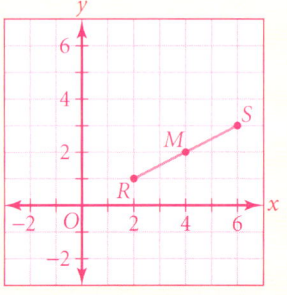

b. The slope of $\overline{RS}$ is $\frac{1}{2}$.

c. A line perpendicular to $\overline{RS}$ has a slope of –2.

d.
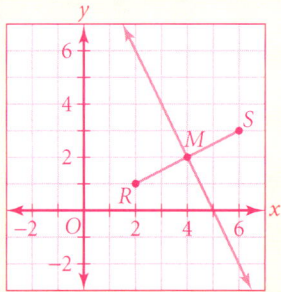

48. Use the midpoint formula to determine the coordinates of the midpoint of $\overline{PQ}$, where $P(10, 7)$ and $Q(2, 3)$. **(6, 5)**

49. If the coordinates of P are (a, b) and the coordinates of Q are (c, d), what are the coordinates of the midpoint of $\overline{PQ}$? $\left(\frac{a+c}{2}, \frac{b+d}{2}\right)$

The midpoint of $\overline{PQ}$ is M. Calculate the missing coordinates.

50. $P(2, -5), Q(6, 7), M(?, ?)$ **(4, 1)**   **51.** $P(4, 8), Q(?, ?), M(10, 7)$ **(16, 6)**

52. $P(?, ?), Q(6, -2), M(9, 4)$ **(12, 10)**   **53.** $P(3, ?), Q(?, 5), M(2, 8)$ **(3, 11)** and **(1, 5)**

CONNECTION

54. COORDINATE GEOMETRY Draw $\overline{RS}$, where $R(2, 1)$ and $S(6, 3)$, on a coordinate plane.

  a. Determine the coordinates of M, the midpoint of $\overline{RS}$, and plot M. **(4, 2)**
  b. Identify the slope of $\overline{RS}$. $\frac{1}{2}$
  c. What is the slope of any line perpendicular to $\overline{RS}$? **–2**
  d. Determine an equation for k, the line that is perpendicular to $\overline{RS}$ and contains point M. Draw line k on your graph. Line k is called the *perpendicular bisector* of $\overline{RS}$ because it is perpendicular to $\overline{RS}$ and contains the midpoint of $\overline{RS}$. **$k: y = -2x + 10$**
  e. Choose any two points on k other than M. Name them A and B. Compute the distance from A to R and from A to S. Compute the distance from B to R and from B to S. **Answers will vary. Should show that $AR = AS$ and $BR = BS$.**

CHALLENGE

55. Use right triangles and the Pythagorean Theorem to show that $(4, 7)$ is the midpoint of the segment connecting the points $(1, 3)$ and $(7, 11)$.

CONNECTION

COORDINATE GEOMETRY For Exercises 56–58, state the coordinates of Q, and calculate the length of the hypotenuse of each right triangle.

56. 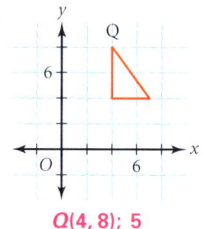   **57.** 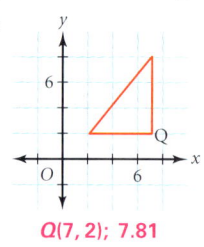   **58.**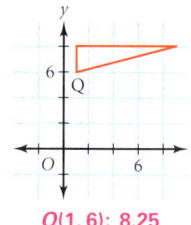

$Q(4, 8); 5$ $Q(7, 2); 7.81$ $Q(1, 6); 8.25$

APPLICATION

59. AVIATION A rescue helicopter located 3 miles west and 2 miles north of a Coast Guard communications center must respond to a ship located 5 miles east and 7 miles south of the center. What distance must the helicopter travel to reach the ship? **≈ 12.04 mi**

 Practice

12.4 The Distance Formula

For each figure, find the coordinates of Q and the lengths of the three sides of the triangle. Decide whether the triangle is scalene, isosceles, equilateral, or right.

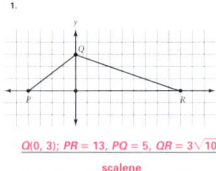

1. $Q(0, 3); PR = 13, PQ = 5, QR = 3\sqrt{10}$; scalene
2. $Q(3, 6); PR = 2, PQ = \sqrt{10}, QR = \sqrt{10}$; isosceles

Find the distance between the two given points.

3. $A(-2, 8), B(10, -1)$ ___15___
4. $M(17, 14), N(20, 10)$ ___5___
5. $X(-7, -1), Y(-6, -1)$ ___1___
6. $D(-3, 0), E(3, 2)$ ___$2\sqrt{10}$___

Find the coordinates of the midpoint of the segment.

7. $\overline{AB}$, with $A(-3, 5)$ and $B(3, 7)$ ___(0, 6)___
8. $\overline{MN}$, with $M(-4, 3)$ and $N(3, -5)$ ___$(-\frac{1}{2}, -1)$___
9. $\overline{XY}$, with $X(-4, 5)$ and $Y(-2, 1)$ ___(-3, 3)___
10. $\overline{CD}$, with $C(6, -4)$ and $D(4, 6)$ ___(5, 1)___
11. $\overline{FG}$, with $F(7, -1)$ and $F(5, -2)$ ___$(6, -\frac{3}{2})$___

The midpoint of $\overline{PQ}$ is M. Find the missing coordinates.

12. $P(52, -5), Q(16, 19), M(\underline{34, 7}\ )$
13. $P(6, -2), Q(\underline{12, 10}\ ), M(9, 4)$
14. $P(\underline{16, 6}\ ), Q(4, 8), M(10, 7)$
15. $P(-2, 25), Q(\underline{4, 11}\ ), M(1, 18)$

55.
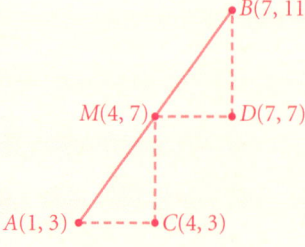

$(AC)^2 + (CM)^2 = (AM)^2$
$3^2 + 4^2 = (AM)^2$
$9 + 16 = (AM)^2$
$\sqrt{25} = AM$
$5 = AM$

$(MD)^2 + (BD)^2 = (MB)^2$
$3^2 + 4^2 = (MB)^2$
$9 + 16 = (MB)^2$
$\sqrt{25} = MB$
$5 = MB$

Since $AM = MB$, M is the midpoint of $\overline{AB}$.

Look Back

Factor each trinomial. *(LESSON 9.7)*

60. $y^2 + 35y + 300$ **(y + 20)(y + 15)**   **61.** $x^2 + 30x + 216$ **(x + 12)(x + 18)**

Find the coordinates of the vertex of each parabola by inspection.
(LESSON 10.1)

62. $y = (x - 2)^2 + 3$ **(2, 3)**   **63.** $y = (x + 3)^2 - 1$ **(−3, −1)**

64. $y = -2(x - 1)^2 - 4$ **(1, −4)**   **65.** $y = -(x + 4)^2 + 1$ **(−4, 1)**

66. Use algebra to find the x-intercept(s) of the graph of $y = x^2 - 6x - 10$. Give the exact solution(s). *(LESSON 10.5)* $3 \pm \sqrt{19}$

Use properties of radicals to simplify each expression. Use a calculator to check your answers. *(LESSON 12.1)*

67. $\sqrt{3}(\sqrt{12} - \sqrt{75})$ **−9**   **68.** $\dfrac{\sqrt{75}}{\sqrt{3}}$ **5**   **69.** $\dfrac{\sqrt{36} + \sqrt{81}}{\sqrt{9}}$ **5**

Look Beyond

TECHNOLOGY Use a scientific calculator to evaluate each expression.

70. $9^{\frac{1}{2}}$ **3**   **71.** $16^{\frac{1}{2}}$ **4**   **72.** $39^{\frac{1}{2}} \approx 6.24$ **73.** $2^{\frac{1}{2}} \approx 1.41$

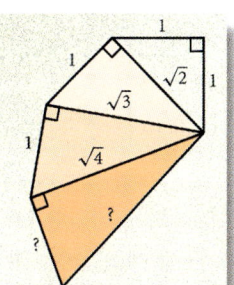

Include this activity in your portfolio.

CULTURAL CONNECTION: AFRICA The figure at left is known as the "Wheel of Theodorus" and is named after a Greek-speaking mathematician who lived in Alexandria, Egypt, in the mid-fifth century B.C.E.

Notice that the hypotenuse of each triangle acts as one of the legs of the next tringle in the spiral.

1. Continue the spiral by building each new right triangle on the hypotenuse of the previous one. Check the side lengths of each triangle with the Pythagorean Theorem. Add 4 new triangles.

2. Make a table relating the number of a given triangle in the spiral to the length of its hypotenuse.

3. Predict the length of the hypotenuse of the 10th triangle in the spiral.

4. Describe in terms of n the length of the hypotenuse of the nth right triangle in the spiral.

WORKING ON THE CHAPTER PROJECT

You should now be able to complete Activities 1 and 2 of the Chapter Project beginning on page 636.

ALTERNATIVE Assessment

Portfolio Activity

The Portfolio Activity can be used as preparation for the Chapter Project or as a separate activity. In the Portfolio Activity on this page, students use simple geometric techniques to construct segments with lengths of $\sqrt{2}$, $\sqrt{3}$, $\sqrt{4}$, $\sqrt{5}$, and so on. Then they analyze their constructions and generalize the pattern.

internet connect

GO TO: go.hrw.com
KEYWORD: MA1 Theodorus

To support the Portfolio Activity, students may visit the HRW Web site where they can obtain additional information about Theodorus of Cyrene and the golden spiral.

Student Technology Guide

NAME _____ CLASS _____ DATE _____

Student Technology Guide
12.4 The Distance Formula

If you know the coordinates of a triangle's vertices, you can use the distance formula and the Converse of the Pythagorean Theorem to decide whether the triangle is a right triangle. A calculator can simplify the process.
There is a convenient form of the Converse of the Pythagorean Theorem that you can use on a calculator. If c is the longest side of triangle ABC, then ABC is a right triangle if $c^2 - a^2 - b^2 = 0$.

Example: Given vertices $P(12, 14)$, $Q(-3, 24)$, and $R(-13, 6)$, determine whether triangle PQR is a right triangle.

- Use the distance formula to calculate the square of PQ. (Remember, the Converse of the Pythagorean Theorem involves the squares of the side lengths.)
$PQ^2 = (23 - 12)^2 + (-4 - 14)^2 = 11^2 + (-18)^2$
Press 11 x^2 + ((-) 18) x^2 ENTER.
$PQ^2 = 445$

$11^2 + (-18)^2$ 445
$(-36)^2 + 10^2$ 1396
$(-25)^2 + (-8)^2$ 689

- Calculate the squares of QR and PR.
$QR^2 = (-13 - 23)^2 + (6 - (-4))^2 = (-36)^2 + 10^2$. Press ((-) 36) x^2 + 10 x^2 ENTER. $QR^2 = 1396$. $PR^2 = (-13 - 12)^2 + (6 - 14)^2 = (-25)^2 + (-8)^2$. Press ((-) 25) x^2 + ((-) 8) x^2 ENTER. $PR^2 = 689$.
- Use the Converse of the Pythagorean Theorem. Subtract the squares of the lengths of the two shorter sides (here, PQ and PR) from the square of the length of the longest side (QR). Press 1396 − 445 − 689 ENTER.
Since $QR^2 - PQ^2 - PR^2 \neq 0$, PQR is *not* a right triangle.

Determine whether each set of points are the vertices of a right triangle.

1. $G(1, 7)$, $H(15, -6)$, $I(22, 3)$ 2. $P(-8, -13)$, $Q(4, -4)$, $R(16, -20)$ 3. $T(19, 1)$, $U(6, -7)$, $V(9, 12)$

not a right triangle   **right triangle**   **not a right triangle**

4. You are building a bookshelf and want to make sure the shelf is perpendicular to the back. You measure the shelf, the back, and the diagonal support. The shelf is 18 inches long; the back is 22 inches tall; and the support is 27 inches long.
Is the shelf perpendicular to the back? Explain your reasoning.
The shelf is not perpendicular to the back. Sample reason: $27^2 = 729$, $18^2 = 324$, and $22^2 = 484$. $729 - 324 - 484 \neq 0$, so the triangle in the diagram is not a right triangle.

Prepare

NCTM PRINCIPLES & STANDARDS
1–4, 6–10

LESSON RESOURCES

- Practice 12.5
- Reteaching Master 12.5
- Enrichment Master 12.5
- Cooperative-Learning Activity 12.5
- Lesson Activity 12.5
- Problem Solving/Critical Thinking Master 12.5

QUICK WARM-UP

Find the distance between each pair of points.

1. $R(2, 4)$, $S(2, -7)$ **11**
2. $M(-3, -6)$, $N(-1, -6)$ **2**
3. $O(0, 0)$, $P(-4, 3)$ **5**
4. $A(3, 5)$, $B(-6, 2)$ **≈ 9.49**

Find the coordinates of the midpoint of the segment with the given endpoints.

5. $J(1, 7)$, $K(3, 11)$ **(2, 9)**
6. $C(-8, -2)$, $D(2, 8)$ **(−3, 3)**

Also on Quiz Transparency 12.5

Teach

Why Tell students that the equation of a circle can be used to identify all points in a coordinate plane that are a given distance from a given point. Have them brainstorm real-world situations in which this might be useful. **sample answers: identifying the farthest places that a television or radio signal can be received; finding the farthest cities to which a plane can fly on a given amount of fuel**

Geometric Properties

12.5

The sandstone blocks of Stonehenge, an ancient monument in England, once formed a circle with a 160-foot radius.

Objectives

- Define and use the equation of a circle.
- Use the coordinate plane to investigate the diagonals of a rectangle and the midsegment of a triangle.

Why Studying a monument like Stonehenge can be difficult because of its size and the restrictions protecting the site. By representing objects such as Stonehenge as geometric figures on a coordinate plane, algebra can be used to study their basic geometric properties.

The Equation of a Circle

CONNECTION
GEOMETRY

A **circle** is the set of all points in a plane that are the same distance from a given point called the **center** of the circle. The distance from the center of the circle to any point on the circle is the **radius** of the circle.

You can represent the outer circle of stones at Stonehenge by using an equation for a circle with its center at (0, 0) and a radius of 160. You will need to use the distance formula (Lesson 11.4).

If a point (x, y) is on the circle, then its distance, d, from (0, 0) must be 160.

$$d = \sqrt{(x-0)^2 + (y-0)^2} = 160$$
$$\sqrt{x^2 + y^2} = 160$$
$$x^2 + y^2 = 160^2$$
$$x^2 + y^2 = 25{,}600$$

The equation $x^2 + y^2 = 25{,}600$ represents all possible points on the circle. If you replace the radius, 160, in the Stonehenge example with the variable r, the resulting equation is the *general equation for a circle centered at the origin*, (0, 0):

$$x^2 + y^2 = r^2$$

Alternative Teaching Strategy

HANDS-ON STRATEGIES Have students draw a set of coordinate axes on graph paper and use a compass to construct a circle with a radius of 10 units centered at the origin. Ask students to complete the table at right. Then ask them to make a conjecture about the coordinates of all the points on this circle. Elicit the fact that if a point with coordinates (x, y) is on this circle, then $x^2 + y^2 = 100$, or $x^2 + y^2 = 10^2$. Discuss why this makes sense in light of the distance formula. Ask them to generalize the conjecture for any circle with its center at the origin and a radius of r. $x^2 + y^2 = r^2$

(x, y)	$x^2 + y^2$	Is the point on the circle?
(0, 10)	100	yes
(−10, 0)	100	yes
(6, −8)	100	yes
(−7, −5)	74	no
(−8, −6)	100	yes
(9, 6)	117	no

606 LESSON 12.5

EXAMPLE  Use the distance formula to find the equation of a circle with a radius of 3 and its center at (1, 2).

SOLUTION

If a point (x, y) is on the circle, then its distance, d, from the center, $(1, 2)$, must be 3.

$$\sqrt{(x-1)^2 + (y-2)^2} = 3$$
$$(x-1)^2 + (y-2)^2 = 9$$

The method in the example above leads to the following equation:

The Equation of a Circle

The equation of a circle with a radius of r and its center at (h, k) is

$$(x - h)^2 + (y - k)^2 = r^2.$$

TRY THIS Find the equation of a circle with the given radius, r, and the given center, C.
a. $r = 1$, $C(0, 0)$
b. $r = 5$, $C(-2, 5)$

Activity
The Diagonals of a Rectangle

You will need: graph paper

1. Use the distance formula to find the length of each diagonal of rectangle $ABCO$.

2. Draw two other rectangles and find the lengths of their diagonals. In each case, place one vertex at the origin and two of the sides along the axes.

3. Make a conjecture about the diagonals of a rectangle.

CHECKPOINT ✓ 4. A rectangle is shown at right. Notice that the length of the rectangle is a and the width is b. Write an expression for the length of the diagonals of a rectangle with length a and width b.

5. Your result from Step 4 will be true for any rectangle with length a and width b. Explain how this result proves your conjecture from Step 3 about the diagonals of a rectangle that is positioned as shown.

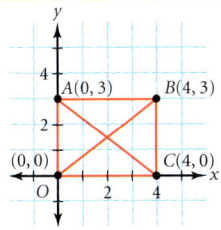

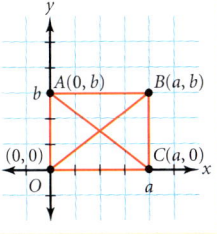

In the exercise set at the end of this lesson, you will be asked to discover and demonstrate other properties of rectangles.

Interdisciplinary Connection

ART The circle is the basis of many intricate patterns created by artists. State the equations of the circles that form the design shown below.

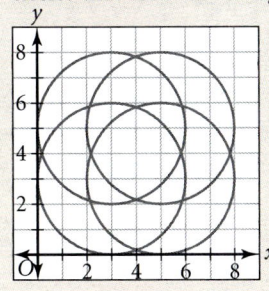

$(x - 3)^2 + (y - 3)^2 = 9$
$(x - 3)^2 + (y - 5)^2 = 9$
$(x - 5)^2 + (y - 3)^2 = 9$
$(x - 5)^2 + (y - 5)^2 = 9$

Inclusion Strategies

INVITING PARTICIPATION Students who have worked with geometry software might enjoy creating dynamic drawings to illustrate the Triangle Midsegment Theorem and to demonstrate the equality of the diagonals of a rectangle. If you have the equipment to project a computer screen for viewing, allow one or more of the students to present their drawings to the entire class. Otherwise, load the electronic files that contain their drawings onto individual computers, and have the class view and work with them in small groups.

Math
CONNECTION

GEOMETRY Be sure students are aware that the center of a circle lies in the same plane as the circle. If you consider a line through the center of a circle that is perpendicular to the plane of the circle, the points on the circle also are equidistant from any given point on that line.

ADDITIONAL
EXAMPLE 1

Use the distance formula to find the equation of a circle with a radius of 4 and its center at $C(-2, 3)$.

$(x + 2)^2 + (y - 3)^2 = 16$

TRY THIS
a. $x^2 + y^2 = 1$
b. $(x + 2)^2 + (y - 5)^2 = 25$

Activity Notes

In this Activity, students draw several rectangles and use the distance formula to investigate the lengths of the diagonals. They should arrive at the conclusion that the diagonals of a rectangle are equal in length.

Cooperative Learning

You may wish to have students do the Activity in pairs. Each pair can begin by doing Steps 1–3 together. Then have the entire class discuss their findings, resolve any discrepancies, and complete Steps 4–5.

CHECKPOINT ✓
4. $\sqrt{a^2 + b^2}$

LESSON 12.5

Math
CONNECTION

GEOMETRY Students should be aware that all triangles have three midsegments and that the theorem stated here applies to each of them.

ADDITIONAL EXAMPLE 2

Test the Triangle Midsegment Theorem for the triangle shown below.

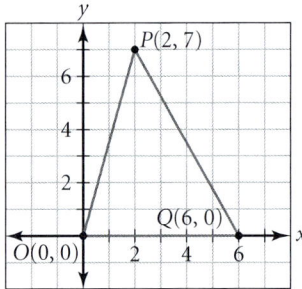

1. Begin by finding the midpoints, M and N, of sides $\overline{OP}$ and $\overline{PQ}$.
   $M(1, 3.5), N(4, 3.5)$

2. Find the slope of midsegment $\overline{MN}$ and the slope of side $\overline{OQ}$.
   slope of $\overline{MN} = 0$
   slope of $\overline{OQ} = 0$

3. Find the length of midsegment $\overline{MN}$ and the length of side $\overline{OQ}$.
   $MN = 3; OQ = 6$

The Triangle Midsegment Theorem holds true for midsegment $\overline{MN}$ and side $\overline{OQ}$. The slope of each is 0, so they are parallel and $MN = \frac{1}{2}OQ$.

Segments and Midpoints

CONNECTION
GEOMETRY

You can use the midpoint formula to test an important geometry theorem, the Triangle Midsegment Theorem.

Triangle Midsegment Theorem

The segment joining the midpoints of two sides of a triangle is parallel to the third side of the triangle and is half the length of the third side.

EXAMPLE 2

Test the Triangle Midsegment Theorem for the triangle shown at right.

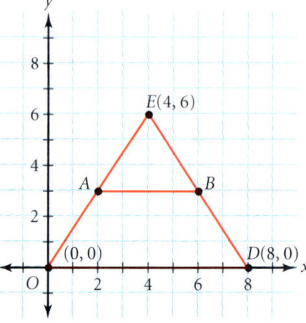

- **SOLUTION**

1. Begin by finding the midpoints, A and B, of sides $\overline{OE}$ and $\overline{ED}$ of the triangle.

Midpoint A

x-coordinate: $x = \frac{0+4}{2} = 2$

y-coordinate: $y = \frac{0+6}{2} = 3$

Thus, the midpoint is (2, 3).

Midpoint B

x-coordinate: $x = \frac{4+8}{2} = 6$

y-coordinate: $y = \frac{6+0}{2} = 3$

Thus, the midpoint is (6, 3).

2. Determine the slope of the **midsegment**, which connects the midpoints of two sides of the triangle, and the slope of the third side.

Slope of midsegment

$m = \frac{3-3}{6-2} = \frac{0}{4} = 0$

Slope of third side

$m = \frac{0-0}{8-0} = \frac{0}{8} = 0$

The slope is 0 in each case, so the lines are parallel, as stated in the theorem.

3. Calculate the length of the midsegment and of the third side of the triangle.

Length of midsegment

$d = 6 - 2 = 4$

Length of third side

$d = 8 - 0 = 8$

The length of the midsegment is one-half of the length of the third side, as stated in the theorem.

Enrichment

An *ellipse* is the set of all points P in a plane such that the sum of the distances from P to two fixed points is a constant. These fixed points are called the *foci* of the ellipse. For the ellipse at right, the foci are the points labeled F_1 and F_2. Give students the figure and have them calculate $AF_1 + AF_2$, $BF_1 + BF_2$, $CF_1 + CF_2$, and $DF_1 + DF_2$. They should get a constant sum of 10. Then have them use this fact and expressions for the distance from any point P to each focus to write an equation for the ellipse.

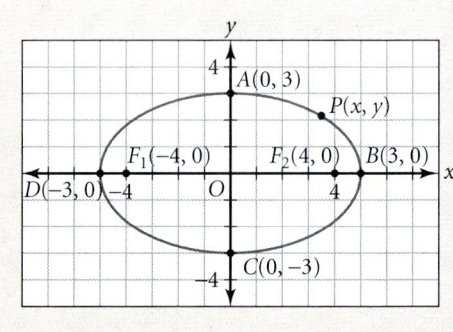

$\sqrt{(x+4)^2 + y^2} + \sqrt{(x-4)^2 + y^2} = 10$

The Focus and Directrix of a Parabola

One of the interesting properties of a parabola is that it consists of the set of all points equidistant from a given point, known as the **focus**, and a given line, known as the **directrix**.

EXAMPLE 3

APPLICATION: ENGINEERING

Engineers plan to locate a maintenance facility at a point that is equidistant from a TV tower and a road. The TV tower is 2 miles from the road. The diagram below shows several possible locations for the facility. **Write an equation that describes all possible locations of the maintenance facility.**

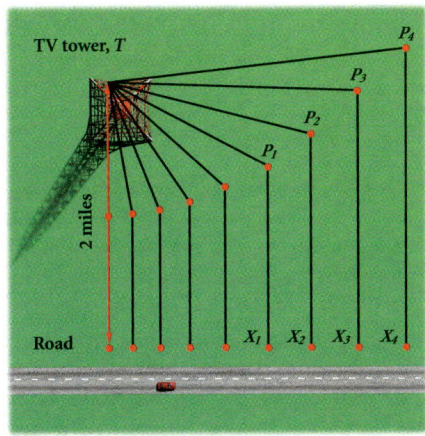

- SOLUTION

PROBLEM SOLVING

Make a diagram. Use a coordinate grid to model the problem. Let $T(0, 2)$ be the TV tower. Let $R(x, 0)$ be a point on the x-axis, which represents the road. Finally, let $P(x, y)$ be any point equidistant from points T and R.

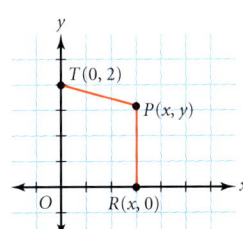

$$PR = PT$$
$$\sqrt{(x-x)^2 + (y-0)^2} = \sqrt{(x-0)^2 + (y-2)^2}$$
$$\sqrt{y^2} = \sqrt{x^2 + (y-2)^2}$$
$$y^2 = x^2 + (y-2)^2$$
$$y^2 = x^2 + y^2 - 4y + 4$$
$$4y = x^2 + 4$$
$$y = \tfrac{1}{4}x^2 + 1$$

The equation that describes all possible locations of the maintenance facility is $y = \tfrac{1}{4}x^2 + 1$. All possible locations of the facility, $P(x, y)$, lie on a parabola, with the TV tower representing the focus and the road representing the directrix.

TRY THIS Write an equation that describes all possible locations for another facility that are equidistant from a road and a radio tower that is 3 miles from the road.

ADDITIONAL EXAMPLE 3

Write an equation that describes the set of all points in the coordinate plane that are equidistant from the point $R(-4, 2)$ and the x-axis.

$y = \tfrac{1}{4}x^2 + 2x + 5$

TRY THIS

$y = \tfrac{1}{6}x^2 + \tfrac{3}{2}$

Reteaching the Lesson

USING VISUAL STRATEGIES
Give students the graph at right. Do *not* show units on the axes. Tell students that the figure is a circle with its center at the origin. Ask them to write a possible equation for the circle. Have several volunteers share their responses. **sample:** $x^2 + y^2 = 4$ Discuss what all the correct responses have in common.

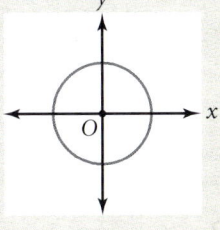

Repeat the activity with several transformations of the circle, such as those shown below.

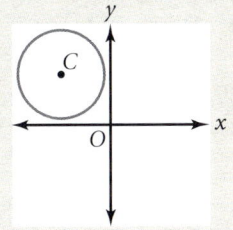

sample:
$(x + 4)^2 + (y - 4)^2 = 9$

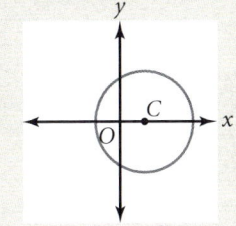

sample:
$(x - 1)^2 + y^2 = 9$

LESSON 12.5

Assess

Selected Answers

Exercises 5–7, 9–57 odd

ASSIGNMENT GUIDE

In Class	1–7
Core	9–27 odd, 29–35, 38–41
Core Plus	8–26 even, 28–41
Review	42–57
Preview	58, 59

✏ Extra Practice can be found beginning on page 738.

Technology

A graphics calculator may be helpful in Exercises 8–28. You may want to have students first do the exercises with paper and pencil and then use the calculator to verify their answers.

Practice

Practice
12.5 Geometric Properties

Find an equation of a circle with its center at the origin and with the given radius.

1. radius of 9 $x^2 + y^2 = 81$ 2. radius of 4 $x^2 + y^2 = 16$

Sketch each circle on the grid provided. Then find the equation of the circle.

3. center at (4, 0) and radius of 3   4. center at (−4, 0) and radius of 3

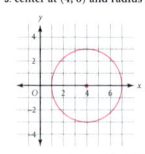 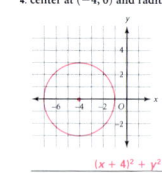

$(x − 4)^2 + y^2 = 9$ $(x + 4)^2 + y^2 = 9$

5. center at (0, 4) and radius of 3   6. center at (0, −4) and radius of 3

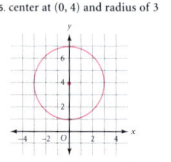 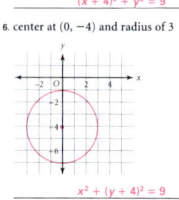

$x^2 + (y − 4)^2 = 9$ $x^2 + (y + 4)^2 = 9$

Identify the center and the radius of each circle.

7. $(x − 1)^2 + y^2 = 4$ 8. $(x + 1)^2 + (y − 2)^2 = 4$
   center at (1, 0); radius of 2   center at (−1, 2); radius of 2

Exercises

Communicate

1. Explain how the distance formula can be used to find the set of all points equidistant from a given point.
2. How can algebra help you to solve geometry problems? Include examples in your answer.
3. State the Triangle Midpoint Theorem in your own words. Why is it convenient to locate one vertex of a triangle at the origin when proving the theorem?
4. Define the focus and directrix of a parabola in your own words.

Guided Skills Practice

5. Find the equation of a circle with the given radius, r, and the given center, C. **(EXAMPLE 1)**
   a. $r = 3$, $C(0, 0)$ $x^2 + y^2 = 9$ **b.** $r = 4$, $C(−3, 4)$ $(x + 3)^2 + (y − 4)^2 = 16$

6. Test the Triangle Midsegment Theorem for the triangle shown at right. **(EXAMPLE 2)**

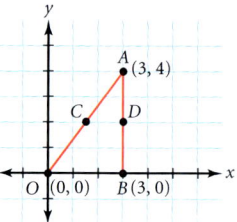

7. Write an equation for all points that are equidistant from a given point, (2, 3), and a given line, the x-axis. **(EXAMPLE 3)** $y = \frac{1}{6}x^2 - \frac{2}{3}x + \frac{13}{6}$

12. $(x − 1)^2 + (y − 1)^2 = 9$
13. $(x + 1)^2 + (y − 4)^2 = 16$
14. $(x − 4)^2 + (y + 2)^2 = 9$
15. $(x + 3)^2 + (y + 6)^2 = 16$
16. c : (1, 3)
    r : 3
17. c : (−1, −3)
    r : 3
18. c : (5, −3)
    r : 25
19. c : (−6, −4)
    r : 10
20. c : (1.5, 3.5)
    r : $\frac{25}{6}$
21. c : (−1, −4)
    r : $\frac{2}{5}$

Practice and Apply

Write the equation of a circle with its center at the origin and the given radius.

8. radius = 6 $x^2 + y^2 = 36$ 9. radius = 1.2 $x^2 + y^2 = 1.44$
10. radius = $\sqrt{3}$ $x^2 + y^2 = 3$ 11. radius = $5\sqrt{3}$ $x^2 + y^2 = 75$

Find the equation of a circle with the given radius, r, and the given center, C.

12. $r = 3$, $C(1, 1)$ 13. $r = 4$, $C(−1, 4)$
14. $r = 3$, $C(4, −2)$ 15. $r = 4$, $C(−3, −6)$

From each equation of a circle, give the center and the radius.

16. $(x − 1)^2 + (y − 3)^2 = 9$ 17. $(x + 1)^2 + (y + 3)^2 = 9$
18. $(x − 5)^2 + (y + 3)^2 = 625$ 19. $(x + 6)^2 + (y + 4)^2 = 100$
20. $(x − 1.5)^2 + (y − 3.5)^2 = \frac{625}{36}$ 21. $(x + 1)^2 + (y + 4)^2 = \frac{4}{25}$

---

6. midpoint of hypotenuse: C(1.5, 2)
midpoint of vertical leg: D(3, 2)
slope of midsegment $\overline{CD} = 0$,
slope of base $\overline{OB} = 0$
The slope is 0 in each case, therefore the lines are parallel.
length of midsegment $\overline{CD} = 1.5$,
length of base $\overline{OB} = 3$
The length of the midsegment is one-half the length of the third side.

Write an equation for each circle. C represents each center.

22.
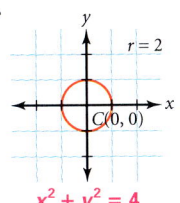
$x^2 + y^2 = 4$

23.
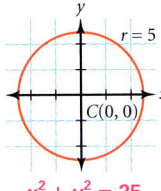
$x^2 + y^2 = 25$

24.
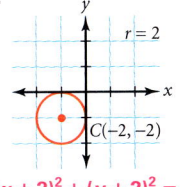
$(x + 2)^2 + (y + 2)^2 = 4$

25.
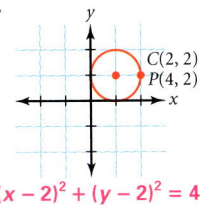
$(x - 2)^2 + (y - 2)^2 = 4$

26.
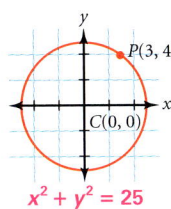
$x^2 + y^2 = 25$

27.
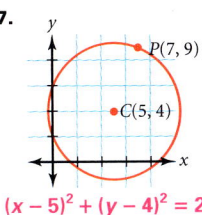
$(x - 5)^2 + (y - 4)^2 = 29$

Error Analysis

When writing the equation of a circle, students often will identify the radius correctly and then forget to square it. Stress the importance of checking the equation by substituting the coordinates of at least two points on the circle for x and y and determining if the result is a true statement.

28. a. Copy and complete the table of values below.

Values that satisfy the equation $x^2 + y^2 = 100$							
x	0	6	−6	8	−8	10	−10
y	±10	±8	±8?	±6?	±6?	0?	0?

b. Plot the values on a coordinate plane.
c. What shape is the graph? **circle**
d. What are the x-intercepts of the graph? **+10, −10**
e. What are the y-intercepts of the graph? **+10, −10**

Plot triangle PQR on graph paper, where P(2, 1), Q(4, 7), and R(12, 3). Refer to your drawing for Exercises 29–35.

29. Identify the coordinates of M, the midpoint of $\overline{PQ}$, and plot M. **(3, 4)**

30. Find the coordinates of N, the midpoint of $\overline{PR}$, and plot N. **(7, 2)**

31. Draw $\overline{MN}$, and give the length of $\overline{MN}$ in simplest terms. **$2\sqrt{5}$**

32. Find the length of $\overline{QR}$. **$4\sqrt{5}$**

33. Compare the lengths of $\overline{MN}$ and $\overline{QR}$.

34. Find the slopes of $\overline{MN}$ and $\overline{QR}$.

35. What do the slopes tell you about $\overline{MN}$ and $\overline{QR}$?

36. Use the distance formula to write the equation for all points $P(x, y)$ such that the sum of the distances from P to $A(-8, 0)$ and from P to $B(8, 0)$ is 20.

37. Write the equation for all points $P(x, y)$ such that the sum of the distances from P to $A(0, -4)$ and from P to $B(0, 4)$ is 10.

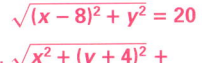

33. $QR = 2MN$
34. Slopes are both $-\frac{1}{2}$.
35. $\overline{MN}$ and $\overline{QR}$ are parallel.
36. $\sqrt{(x + 8)^2 + y^2} + \sqrt{(x - 8)^2 + y^2} = 20$
37. $\sqrt{x^2 + (y + 4)^2} + \sqrt{x^2 + (y - 4)^2} = 10$

CHALLENGES

28. h.
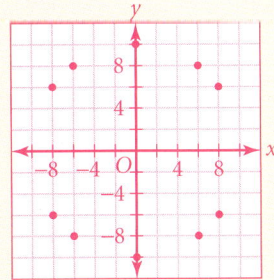

29–35.
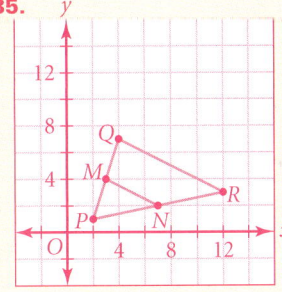

Student Technology Guide

12.5 Geometric Properties

If you draw a circle on a coordinate plane, you will notice that there are *two* y-values for most x-values. Because of this, you need to use a special method to graph a circle on a graphics calculator.

Before following the steps below, press ZOOM 6:ZStandard. This gives you the standard viewing window with x- and y-values from −10 to 10.

Example: Graph the circle $x^2 + y^2 = 25$.

- To enter this equation, you must solve it for y.
  $y^2 = 25 - x^2$
  $y = \pm\sqrt{25 - x^2}$
  Note that there are *two* equations, $y = \sqrt{25 - x^2}$ and $y = -\sqrt{25 - x^2}$.
- Enter the first equation as Y1. Press Y= 2nd x^2 25 − X,T,θ,n x^2) .
- Enter the second equation as Y2. Since the left side of the second equation is the opposite of the left side of the first, you can do this by moving to Y2 and pressing (−) VARS Y-VARS 1:Function... ENTER 1:Y₁ ENTER .
- Press GRAPH. As seen at left below, the graph looks like an ellipse not a circle. This is because, in this viewing window, the units on the x-axis are longer than those on the y-axis.

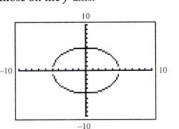
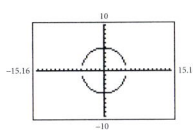

- To get an undistorted graph, as shown at right above, press ZOOM 5:ZSquare ENTER. (Or press WINDOW and change Xmin, Xmax, Ymin, and Ymax so that the range for the x-values is about 1.5 times the range for the y-values.)

Graph each circle on a graphics calculator. For each circle, give the two equations you used. *See answer section for graphs.*

1. $x^2 + y^2 = 9$
   $y = \sqrt{9 - x^2}$ and
   $y = -\sqrt{9 - x^2}$

2. $x^2 + y^2 = 36$
   $y = \sqrt{36 - x^2}$ and
   $y = -\sqrt{36 - x^2}$

3. $x^2 + y^2 = 10$
   $y = \sqrt{100 - x^2}$ and
   $y = -\sqrt{100 - x^2}$

Look Beyond

In Exercises 58 and 59, students use coordinates to prove the following two theorems of Euclidean geometry:

- The diagonals of a square are perpendicular.
- The diagonals of a rectangle bisect each other.

Students will investigate these theorems in greater depth if they take a course in geometry.

56.

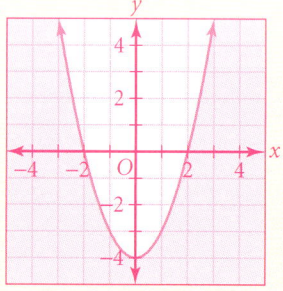

57.

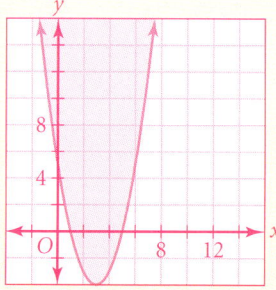

58. The figure is a square because the lengths of all four sides are equal.
slope of diagonal 1:
$\frac{0-a}{a-0} = \frac{-a}{a} = -1$

slope of diagonal 2:
$\frac{a-0}{a-0} = \frac{a}{a} = 1$

Since the slopes are negative reciprocals, the diagonals are perpendicular.
length of diagonal 1:
$\sqrt{(0-a)^2 + (a-0)^2} = \sqrt{a^2 + a^2} = a\sqrt{2}$

length of diagonal 2:
$\sqrt{(a-0)^2 + (a-0)^2} = \sqrt{a^2 + a^2} = a\sqrt{2}$

The lengths of the diagonals are equal.

38. (2, 1), yes

39. Diagonals bisect each other at their midpoints.

40. The slopes are $\frac{1}{2}$ and $-\frac{1}{2}$; not perpendicular.

41. The diagonals of a square are perpendicular.

59.

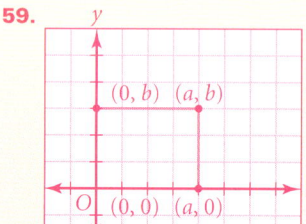

For Exercises 38–41, refer to the graph below.

38. Find the midpoint of each diagonal. Are the midpoints the same or not?

39. In this lesson you may have discovered that the diagonals of a rectangle have equal lengths. Make a new conjecture about the midpoints of the diagonals based on your answer to Exercise 38.

40. Find the slope of each diagonal. What is the relationship between the slopes? Are the diagonals perpendicular?

41. Construct a graph of a square (a rectangle with all sides of equal length). What conjecture can you make about the diagonals of a square?

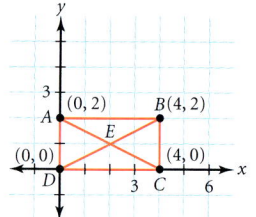

Look Back

State whether each expression correctly represents the distance between 2 and 5 on a number line. (LESSON 2.1)

42. $2 + 5$ no
43. $2 - 5$ no
44. $5 + 2$ no
45. $5 - 2$ yes
46. $|2 + 5|$ no
47. $|2 - 5|$ yes
48. $|5 + 2|$ no
49. $|5 - 2|$ yes

Write each number in scientific notation. (LESSON 8.5)

50. 2,300,000 2.3×10^6
51. 0.00000125 1.25×10^{-6}

Simplify. (LESSON 9.1)

52. $(x^2 + 3x + 5) + (7x^2 - 5x - 10)$
$8x^2 - 2x - 5$

53. $(8b^2 - 15) - (2b^2 + b + 1)$
$6b^2 - b - 16$

Find the value of the discriminant for each quadratic equation. (LESSON 10.5)

54. $3x^2 - 6x + 3 = 0$ 0
55. $2x^2 + 3x - 2 = 0$ 25

Graph each quadratic inequality. (LESSON 10.6)

56. $y < x^2 - 4$
57. $y > x^2 - 6x + 5$

Look Beyond

58. Explain why the figure at right is a square for any positive value of a. Prove your conjecture from Exercise 41 about the diagonals of a square.

59. Graph a rectangle with length a and width b. Prove that the diagonals bisect each other.

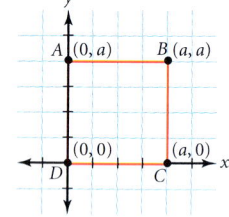

midpoint of diagonal 1:
$\left(\frac{a+0}{2}, \frac{0+b}{2}\right) = \left(\frac{a}{2}, \frac{b}{2}\right)$

midpoint of diagonal 2:
$\left(\frac{a+0}{2}, \frac{b+0}{2}\right) = \left(\frac{a}{2}, \frac{b}{2}\right)$

Since the midpoints of the two diagonals are the same point, the diagonals bisect each other at the point $\left(\frac{a}{2}, \frac{b}{2}\right)$.

12.6 The Tangent Function

Objectives

- Identify and use the tangent ratio in a right triangle.
- Find unknown side and angle measures in right triangles.

Why A pilot must be able to determine not only the distance to the destination, but also the exact direction. By using the Pythagorean Theorem and the tangent function, you can determine both side lengths and angle measures in right triangles.

APPLICATION
AVIATION

The direction in which a plane is traveling is referred to by pilots as the *heading* of the plane. The heading is an angle from 0 to 360 degrees, measured clockwise from the magnetic north.

A pilot at point *A* is 300 miles due south of a town at point *C*. A town at point *B* is 200 miles due east of point *C*. Assuming that there is no wind, in what direction must the pilot head to reach point *B* by a straight path?

To answer the question, you will need to find the measure of angle *A*. Examine the ratio given with respect to angle *A*. On a right triangle this ratio is called the **tangent** of angle *A*.

$$\text{tangent } A = \tan A = \frac{BC}{AC} = \frac{200}{300} = \frac{2}{3}$$

Prepare

NCTM PRINCIPLES & STANDARDS
1–4, 6–10

LESSON RESOURCES

- Practice 12.6
- Reteaching Master 12.6
- Enrichment Master 12.6
- Cooperative-Learning Activity 12.6
- Lesson Activity 12.6
- Problem Solving/Critical Thinking Master 12.6
- Teaching Transparency 61

QUICK WARM-UP

Write as a decimal.

1. $\frac{5}{8}$ **0.625**   2. $\frac{5.2}{13}$ **0.4**

Solve each equation.

3. $20 = \frac{w}{4}$ 4. $20 = \frac{4}{a}$
   $w = 80$ $a = 0.2$

Given $\triangle ABC \sim \triangle XYZ$, $AB = 6$, $AC = 12$, $BC = 9$, and $XY = 3$, find each length.

5. *XZ* **6**   6. *YZ* **4.5**

Also on Quiz Transparency 12.6

Teach

Why When you use a ruler or a car's odometer to measure a distance, you are making a *direct measurement*. Ask students to name some distances that are difficult to measure directly. Elicit responses such as the height of a building or the distance across a river. Tell them that the *tangent ratio* of a right triangle is often used to find these types of distances indirectly.

Alternative Teaching Strategy

USING TECHNOLOGY Use geometry software to create a dynamic demonstration of the tangent ratio. A sample sketch is shown at right. To make a sketch like this, construct $\overleftrightarrow{AC}$ and $\overleftrightarrow{AZ}$ as shown, draw points *C* and *C'* by choosing **Point on Object** from the **Construct** menu, and construct $\overleftrightarrow{BC}$ and $\overleftrightarrow{B'C'}$ by choosing **Perpendicular Line** from the **Construct** menu. For the list of measures, use the **Calculate** menu. Then click and drag point *Z* to demonstrate the tangent. Ask students to observe the values of the ratios.

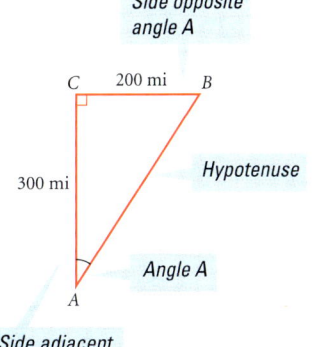

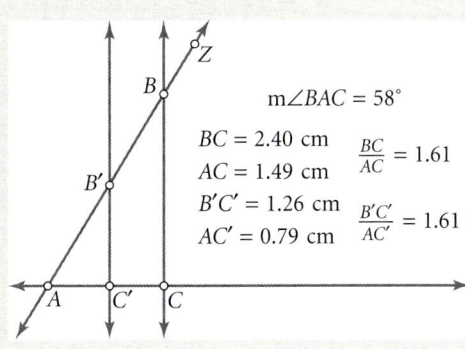

LESSON 12.6 **613**

Use Teaching Transparency 61.

Activity Notes

In this Activity, students investigate the tangent ratio in three similar right triangles. They should arrive at the conclusion that the value of the tangent ratio depends only on the measure of the acute angle. That is, no matter how large or small the right triangle, if one acute angle has a given measure, the tangent of that angle will always have the same value.

Cooperative Learning

You may wish to have students do the Activity in groups of three. Each student in a group should do Steps 1–3 individually, with each student using a different measure for ∠A. The group should then compare results and complete Step 4 together.

CHECKPOINT ✔

4. The tangent ratio does not depend on the size of the triangle, but only on the measurement of the angle. This is because similar triangles are proportional.

Teaching Tip

TECHNOLOGY Some scientific calculators require the user to enter the angle measure first and then press the trig ratio, while others, such as TI graphics calculators, require the user to press the trig ratio first and then enter the angle measure.

The Tangent Ratio

On a right triangle:

$$\text{tangent of angle } A = \tan A = \frac{\text{length of the leg opposite angle } A}{\text{length of the leg adjacent to angle } A}$$

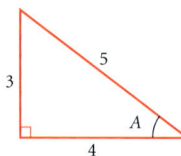

$\tan A = \frac{3}{4}$

The tangent ratio is one of the fundamental ratios of the branch of mathematics known as **trigonometry** (from Greek words meaning "triangle measure").

Before returning to the plane problem and calculating the proper heading, consider the following question: Do you think the tangent of an angle depends on the size of the triangle it is a part of? In the activity that follows, you will investigate this question.

Activity: The Tangent Ratio in Similar Triangles

You will need: graph paper, a centimeter ruler, and a calculator

$$\tan A = \frac{opp}{adj}$$

1. On a piece of graph paper, draw triangle ABC as shown at right. Identify side BC as the side *opposite* angle A, side AC as the side *adjacent* to angle A, and side AB as the *hypotenuse*.

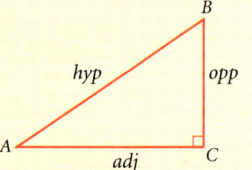

2. Draw segments B'C' and B"C" perpendicular to side AC. Each new segment is part of a smaller triangle that contains angle A. What leg is opposite angle A in triangle AB'C'? What leg is adjacent to angle A in triangle AB'C'?

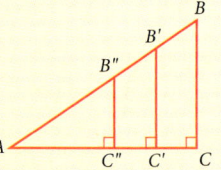

3. For each of the triangles containing angle A, measure the legs opposite and adjacent to angle A in millimeters. Use your calculator to find the tangent ratio for angle A in each triangle, rounding your answers to the nearest tenth. Keep a record of your results.

CHECKPOINT ✔ 4. Based on your approximate measurements and rounded results, does the tangent ratio for an angle depend on the size of the triangle in which the angle appears? Explain.

Interdisciplinary Connection

GEOGRAPHY The city of San Francisco is known for the steepness of its streets. The steepest street in the city is the stretch of Filbert Street between Hyde and Leavenworth, where it rises 1 foot for every 1.63 feet of horizontal distance. To the nearest tenth of a degree, what angle does the street form with the horizontal? **about 31.5°**

As you may have conjectured in the activity, the tangent ratio of an angle is independent of the size of the triangle in which the angle appears. In fact, the tangent ratio is a function of the angle measure. For any angle, there is exactly one value for the tangent of the angle. The domain of the tangent function is the set of angle measures, and the range is the set of numbers known as tangents.

The value of the tangent function increases as the angle measure increases from 0° to 90°, as the illustrations below indicate. (The side adjacent to angle X is 10 units in all of the triangles.)

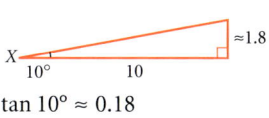
tan 10° ≈ 0.18

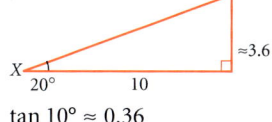
tan 10° ≈ 0.36

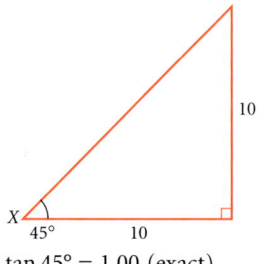
tan 45° = 1.00 (exact)

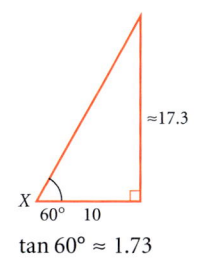
tan 60° ≈ 1.73

TECHNOLOGY
SCIENTIFIC CALCULATOR

You can use a calculator to find the tangent of an angle by using the [TAN] key. Because you are using degrees to measure the angles, be sure the calculator mode is set for degrees. For example, to find tan 45°, press [TAN] 45 [ENTER] (or [=]). The calculator will display the value 1.

In the problem of finding the angle for the plane's heading, the tangent of the angle is known. The problem is to find an angle between 0° and 90° whose tangent is $\frac{2}{3}$. The tangent function constantly increases over the interval from 0° to 90° so you can be sure that there is just one angle in the interval whose tangent has this value. "The angle whose tangent is $\frac{2}{3}$" is written as $\tan^{-1}\left(\frac{2}{3}\right)$, which is read as "the inverse tangent of $\frac{2}{3}$." You can use your calculator to find the angle by using the [TAN⁻¹] key or [INV] [TAN] keys.

The plane's direction, or heading, should be approximately 33.7°.

A precision flying team breaks from a formation.

Inclusion Strategies

LINGUISTIC LEARNERS Students who are uncomfortable with the symbolism of mathematics may become confused by the use of −1 in the expression $\tan^{-1} A$. Thinking that −1 is an exponent, they may interpret the expression as $\frac{1}{\tan A}$. To focus their attention on the difference, suggest that they write an essay with a title such as "The Adventures of −1." In this essay, they should describe the different roles they have seen −1 assume in their study of algebra. Be sure they address −1 as a real number, as an exponent, and as a notation for the inverse of a function.

Teaching Tip

TECHNOLOGY If you are using a TI-82 or TI-83 graphics calculator, you can check the mode setting by pressing [MODE]. If the third line of the display shows **Degree** highlighted, the calculator is in degree mode. If **Radian** is highlighted, you can change the setting by pressing [▼] [▼] [▶] [ENTER]. Then press [2nd] [MODE] to return to the home screen. To access the tangent function, press [TAN], and for the inverse-tangent function, press [2nd] [TAN].

LESSON 12.6 **615**

Math CONNECTION

TRIGONOMETRY The term *acute angle* may be unfamiliar to students. Be sure they understand that an acute angle is an angle whose measure is less than 90°.

ADDITIONAL EXAMPLE 1

Find the tangent of the acute angles in each right triangle. Then find the measures of the angles by using the inverse tangent function of your calculator. Round the angle measures to the nearest degree.

a.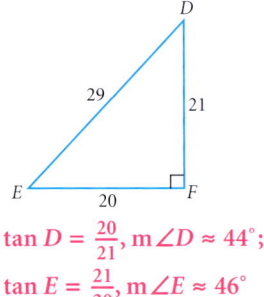

$\tan D = \frac{20}{21}, m\angle D \approx 44°$;
$\tan E = \frac{21}{20}, m\angle E \approx 46°$

b.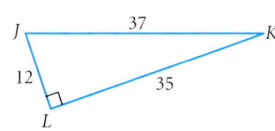

$\tan J = \frac{35}{12}, m\angle J \approx 71°$;
$\tan K = \frac{12}{35}, m\angle K \approx 19°$

TRY THIS

a. $\tan N = \frac{24}{7}, m\angle N \approx 74°$;
$\tan O = \frac{7}{24}, m\angle O \approx 16°$

b. $\tan P = \frac{16}{12}, m\angle P \approx 53°$;
$\tan R = \frac{12}{16}, m\angle R \approx 37°$

EXAMPLE 1

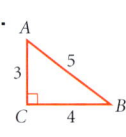

CONNECTION TRIGONOMETRY

Find the tangent of the acute angles in each triangle. Then find the measures of the angles by using the TAN⁻¹ key on your calculator. Round your answers for the angle measures to the nearest degree.

a.

b. (triangle YZX with Y, hypotenuse 13, ZY=5, ZX=12)

SOLUTION

TECHNOLOGY SCIENTIFIC CALCULATOR

a. $\tan A = \frac{\text{opposite}}{\text{adjacent}} = \frac{4}{3}$

$A = \tan^{-1}\left(\frac{4}{3}\right) \approx 53°$

$\tan B = \frac{\text{opposite}}{\text{adjacent}} = \frac{3}{4}$

$B = \tan^{-1}\left(\frac{3}{4}\right) \approx 37°$

b. $\tan Y = \frac{\text{opposite}}{\text{adjacent}} = \frac{12}{5}$

$Y = \tan^{-1}\left(\frac{12}{5}\right) \approx 67°$

$\tan X = \frac{\text{opposite}}{\text{adjacent}} = \frac{5}{12}$

$X = \tan^{-1}\left(\frac{5}{12}\right) \approx 23°$

TRY THIS Find the tangent of the acute angles in each triangle below. Then find the measures of the angles by using the TAN⁻¹ key or INV TAN keys on your calculator. Round your answers for the angle measures to the nearest degree.

a.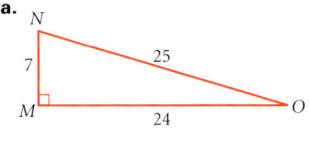

b.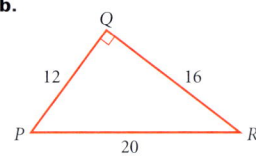

EXAMPLE 2

APPLICATION AVIATION

A plane climbs steadily to 30,000 feet. **If this altitude is reached when the airplane has covered a horizontal distance of 35 miles, what is its angle of ascent?**

SOLUTION

Draw a diagram to model the problem. Since 1 mile = 5280 feet, convert 35 miles to feet by multiplying by 5280.

TECHNOLOGY SCIENTIFIC CALCULATOR

$\tan A = \frac{30,000}{184,800}$

$A = \tan^{-1}\left(\frac{30,000}{184,800}\right) \approx 9°$

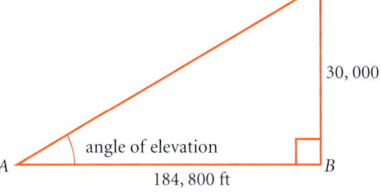

The airplane's angle of ascent is about 9°.

Enrichment

Have students find the measures of the angles that the line with the given equation forms with the *x*-axis and with the *y*-axis. If students need a hint to get started, tell them to consider how the slope of the line is related to the tangent ratio.

1. $y = x$ 45°; 45°
2. $y = 2x$ ≈63°; ≈27°
3. $y = 0.5x$ ≈27°; ≈63°
4. $y = 3x$ ≈72°; ≈18°
5. $y = 3x + 2$ ≈72°; ≈18°
6. $y = mx + b$ $\tan^{-1} m$; $90° - \tan^{-1} m$

Now have students write an equation for a line that satisfies each set of conditions below.

1. forms an angle of *p* degrees with the *x*-axis and passes through the point (0, *b*)
$y = (\tan p)x + b$

2. forms an angle of *q* degrees with the *y*-axis and passes through the point (0, *b*)
$y = [\tan(90 - q)]x + b$, or $y = \left(\frac{1}{\tan q}\right)x + b$

TRY THIS A 1500-foot building casts a 2000-foot shadow. What angle does the ray of sunlight shown form with the ground?

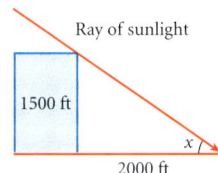

EXAMPLE ③

APPLICATION
RECREATION

A group of skateboarders wants to build a ramp with an angle of 12°. **What should be the rise for every 10 meters of run in order to achieve a 12° angle?**

• **SOLUTION**

Draw a diagram, and write the tangent ratio for the information that you have.

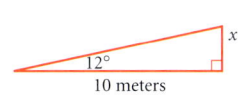

TECHNOLOGY
SCIENTIFIC CALCULATOR

Solve for x to find the rise. $\tan 12° = \frac{x}{10}$
$$x = 10 \tan 12°$$
$$x \approx 2.126$$

The rise should be about 2.1 meters for every 10 meters of run.

TRY THIS A park ranger on a 200-foot tower spots a wolf at an angle of depression of 4°. How far is the wolf from the base of the tower?

An angle of depression is an angle between a horizontal line and a line of sight to a point below.

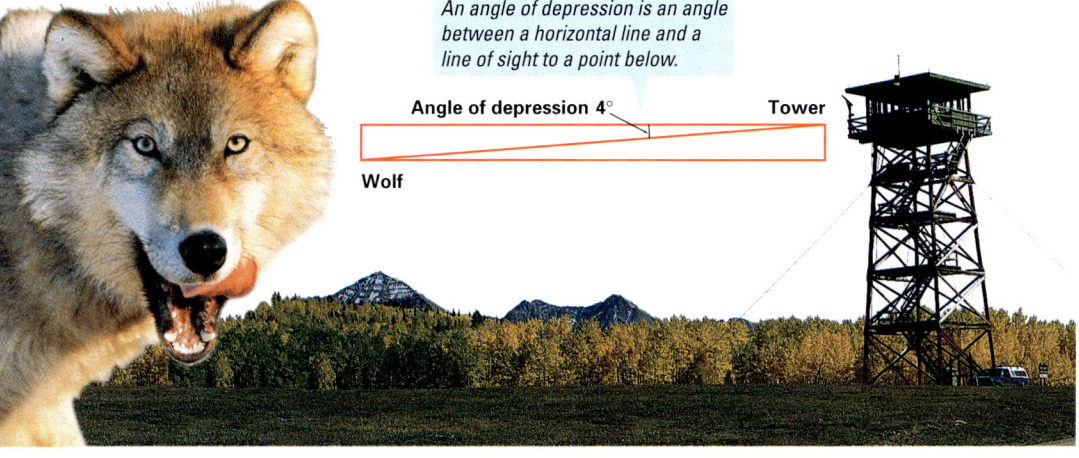

ADDITIONAL EXAMPLE ②

A ladder rests against a wall so that its top is 14 feet above ground level. The foot of the ladder is 5 feet from the base of the wall. **What angle does the ladder form with the ground?**   about 70°

TRY THIS
The angle of the ray of sunlight with the ground is about 37°.

ADDITIONAL EXAMPLE ③

After takeoff from an airport, an airplane's angle of ascent is 10°. The airplane climbs to an altitude of 10,000 feet. **At that point, what is the land distance between the airplane and the airport?**

about 56,713 feet, or about 10.7 miles

TRY THIS
The wolf is about 2860 feet from the tower.

internetconnect

GO TO: go.hrw.com
KEYWORD: MA1 Right Triangles

Finding the sides and angles of right triangles is one of the most common and useful applications at this level. This activity challenges students to use the methods from the text to solve slightly more complicated problems.

Reteaching the Lesson

COOPERATIVE LEARNING Have students work in groups of three or four. Each student should have graph paper, a protractor, and a straightedge.

Review with students the meaning of the tangent ratio. Then instruct students to draw right triangle ABC on the graph paper with $\angle C$ as the right angle and $\tan A$ equal to $\frac{3}{5}$. Within each group, students should draw a triangle that is a different size from the triangles of the other students. They should then use their protractors to measure $\angle A$.

When all students have finished, discuss the results with the class. Write several sets of measures on the board or overhead to stress the fact that the ratio $\frac{BC}{AC}$ is always equal to $\frac{3}{5}$ and that the measure of $\angle A$ is always approximately 31°.

Repeat this activity for several other tangent ratios. Be sure to give students some of the ratios in decimal form.

LESSON 12.6 **617**

Assess

Selected Answers

Exercises 4–6, 7–37 odd

ASSIGNMENT GUIDE

In Class	1–6
Core	7–25, 27, 29
Core Plus	7–30
Review	31–38
Preview	39, 40

✏ Extra Practice can be found beginning on page 738.

Technology

A calculator with tangent and inverse-tangent functions is needed for Exercises 7–30.

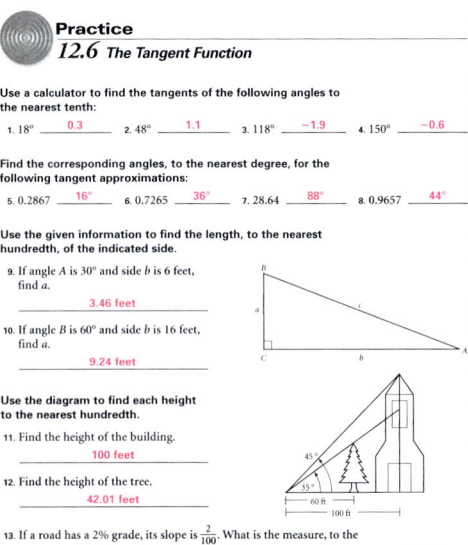

Exercises

● **Communicate**

1. Explain how the tangent function is related to the slope of a line.

2. How can you find the length, a, of the side opposite angle A in a right triangle when you know the measure of angle A and the length, b, of the other leg?

3. How can you find the measure of an acute angle in a right triangle when you know the lengths of the legs opposite and adjacent to that angle?

● **Guided Skills Practice**

CONNECTION

4. **TRIGONOMETRY** Find the tangent of each acute angle in the triangles below. Then find the measure of the angles by using the 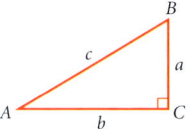 key of your calculator. Round answers to the nearest degree. *(EXAMPLE 1)*

   a.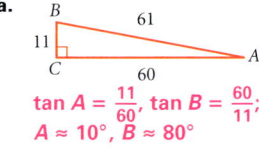

   $\tan A = \frac{11}{60}$, $\tan B = \frac{60}{11}$; $A \approx 10°$, $B \approx 80°$

   b.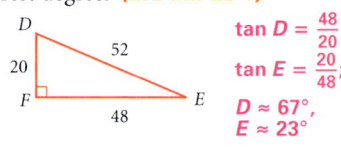

   $\tan D = \frac{48}{20}$, $\tan E = \frac{20}{48}$; $D \approx 67°$, $E \approx 23°$

5. The sun casts a 400-foot shadow from a 1000-foot building. What angle does the ray of sunlight shown form with the ground? *(EXAMPLE 2)* ≈ 68°

   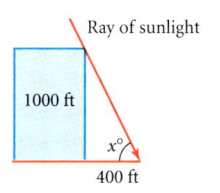

6. A park ranger on a 300-foot tower spots a wolf at an angle of depression of 6°. How far is the wolf from the base of the tower? *(EXAMPLE 3)* ≈ 2854 ft

● **Practice and Apply**

Use a calculator to find the tangent of each angle to the nearest ten-thousandth.

7. 45° **1**   8. 30° **0.5774**   9. 60° **1.7321**   10. 0° **0**

11. 38° **0.7813**   12. 57° **1.5399**   13. 89° **57.2900**   14. 89.9° **572.9572**

Find the measure of the angle whose tangent is given. Give each angle measure to the nearest whole number of degrees.

15. 1.7321 **60°**   16. 1.1918 **50°**   17. 0.2679 **15°**   18. 3.7321 **75°**

618 LESSON 12.6

CONNECTIONS

GEOMETRY Use the diagram at right for Exercises 19–24. Round your answers to the nearest hundredth.

19. If m∠A = 30° and b = 12 meters, find a. **6.93 m**
20. If m∠B = 60° and a = 10 feet, find b. **17.32 ft**
21. If a is 3 centimeters and tan A is 0.75, what is b? **4 cm**
22. If b is 23 feet and tan A is 1.0, what is a? **23 ft**
23. If a is 6 inches and m∠A ≈ 31°, what is b? **9.99 in.**
24. If m∠A ≈ 35° and b is 20 millimeters, what is a? **14 mm**

APPLICATIONS

25. **CONSTRUCTION** A home buyer wants the roof of his house to have an angle of 20°, as shown. How high does the roof rise above the horizontal line shown if the width of the roof is 28 meters? **≈ 5.1 m**

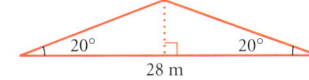

26. **ECOLOGY** A park ranger on a 200-foot tower spots the wildlife listed below. Determine the distances from the tower for the given angles of depression.

Animal	Angle of depression	Distance
bear	3°	3816 ? ft
raccoon	15°	746 ? ft
fox	20°	549 ? ft
moose	2°	5727 ? ft

APPLICATION

27. **ASTRONOMY** A science class conducts an experiment to determine the angle of elevation of the sun at different times during the day by measuring the shadow cast by a 28-foot goalpost. Copy the table shown below. Compute the angle of elevation from ground level for each shadow length. (The angle of elevation is the angle of a line of sight to a point *above* a horizontal line.)

Shadow length in feet	80	48	13	8	2	4
Angle of elevation	?	?	?	?	?	?

≈19° ≈30° ≈65° ≈74° ≈86° ≈82°

Error Analysis

Students sometimes incorrectly identify the hypotenuse as the adjacent side. Stress the fact that the tangent is the ratio of the lengths of two *legs*. Point out to students that when calculating angle measures, they can use a fundamental property of triangles to check their answers. That is, the sum of the measures of the angles of a triangle is 180°, so the sum of the measures of the acute angles of a right triangle must be 90°.

Student Technology Guide

Student Technology Guide

12.6 The Tangent Function

The tangent ratio is useful for solving many real-world problems. Using a calculator, you can quickly find the tangent of an angle or find the measure of an angle if you know its tangent.

Before following the steps below, press MODE and check that "Degree" is highlighted. If it is not, use the arrow keys to move to "Degree" and press ENTER.

Example: The angle of elevation from an airport, A, to a jet, J, is 7°. Find the altitude of the jet when it is 4000 meters from the airport. Round your answer to the nearest meter.

- Use the diagram to write an equation.
  $\tan 7° = \frac{opposite}{adjacent} = \frac{a}{4000}$
- Solve for a. $4000 \tan 7° = a$
- Use a calculator to find the answer.
  Press 4000 TAN 7 ENTER.

The jet's altitude is about 491 meters.

You can use the inverse tangent to solve for an unknown angle measure.

Example: The jet is at a distance of 12,000 meters from the airport and at an altitude of 2200 meters. To the nearest degree, what is the angle of elevation from the airport to the plane?

- Use the diagram from the first example to write an equation. $\tan A = \frac{opposite}{adjacent} = \frac{2200}{12,000}$
- Solve for angle A by taking the inverse tangent of each side.
  $A = \tan^{-1}\left(\frac{2200}{12,000}\right)$
- Use a calculator to find the answer. Press
  2nd TAN 2200 ÷ 12000 ENTER.

The angle of elevation is about 10°.

1. A flagpole that is f feet tall casts a shadow s feet long, as shown above. If F = 32° and s = 22 feet, how tall is the flagpole? Round your answer to the nearest foot. **f ≈ 14 feet**

2. Use your answer from Exercise 1 to find the angle of the sun's rays, F, when s = 10 feet. Round your answer to the nearest degree. **F ≈ 54°**

LESSON 12.6

In Exercises 39 and 40, students investigate the sine and cosine ratios. They will study these ratios in greater depth in Lesson 12.7. Exercise 40 introduces them to a basic trigonometric identity. They will learn more about trigonometric identities if they take a more advanced course in algebra.

APPLICATIONS

28. **HOBBIES** A model airplane and a car begin traveling from the same point. The airplane stays directly above the car. After 15 seconds, the car has traveled one-tenth of a mile, and the angle of elevation of the airplane is 30°. What is the altitude of the model airplane? ≈ **0.06 mi or 317 ft**

29. **CIVIL ENGINEERING** If a road has a 7% grade, it rises 7 feet per 100 feet of run. Thus, its slope is $\frac{7}{100}$. What is the angle of the road? ≈ **4°**

30. **CIVIL ENGINEERING** A water tower is 600 feet from an observer. A line of sight to the top of the water tank forms an angle of 28° with the ground. A line of sight to the bottom of the tank forms an angle of 26°. What is the distance from the bottom of the water tank to the top of the water tank? **26.39 ft**

Look Back

31. Solve $4x - 5 = 7(x - 3) + 2$ for x. *(LESSON 3.5)* $\frac{14}{3}$

32. State whether the quadratic equation $2x^2 + x - 15 = 0$ has any real solutions. If it does, use the quadratic formula to determine the solution(s). *(LESSON 10.5)* $\frac{5}{2}, -3$

33. Solve $x^2 - 5x + 6 \geq 0$. *(LESSON 10.6)* $x \geq 3$ or $x \leq 2$

34. Graph the solution to $x^2 - 16 < 0$. *(LESSON 10.6)* $-4 < x < 4$

35. Points $P(4, 9)$ and $Q(10, 5)$ are plotted on a coordinate plane. Find the coordinates of M, the midpoint of $\overline{PQ}$. *(LESSON 12.4)* (7, 7)

Using the information given in Exercise 35, find the length of each segment in simplified form. *(LESSON 12.4)*

36. $\overline{PQ}$ $2\sqrt{13}$ 37. $\overline{PM}$ $\sqrt{13}$ 38. $\overline{MQ}$ $\sqrt{13}$

Look Beyond

CONNECTION

TRIGONOMETRY Like the tangent (tan) ratio, the *sine* (sin) and *cosine* (cos) ratios are basic trigonometric functions. They provide additional tools for solving problems by using indirect measure.

40. $(\sin x)^2 + (\cos x)^2$
$= \left(\frac{b}{c}\right)^2 + \left(\frac{a}{c}\right)^2$
$= \frac{b^2}{c^2} + \frac{a^2}{c^2}$
$= \frac{b^2 + a^2}{c^2}$
$= \frac{c^2}{c^2} = 1$

$\tan x = \frac{b}{a} \qquad \sin x = \frac{b}{c} \qquad \cos x = \frac{a}{c}$

39. If the tangent of angle x is $\frac{3}{4}$, what are the sine and the cosine of x? (Hint: Use the Pythagorean Theorem to find c.) $\sin(x) = \frac{3}{5}, \cos(x) = \frac{4}{5}$

40. Show why $(\sin x)^2 + (\cos x)^2 = 1$ is true for any right triangle.

34.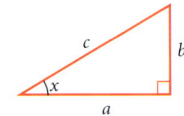

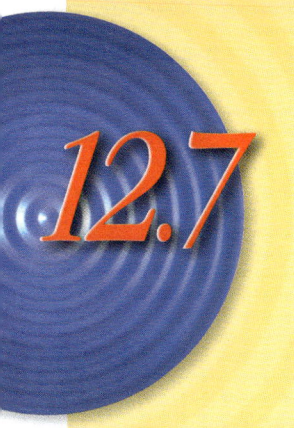

12.7 The Sine and Cosine Functions

Objectives

- Define the sine and cosine ratios in a right triangle.
- Find unknown side and angle measures in right triangles.

Why Directly measuring the steepness of something like a water slide or ramp can be difficult. The sine and cosine functions, like the tangent function, can be used in indirect measurement.

APPLICATION
RECREATION

As you go down the water slide, it feels almost like you are dropping straight down. How steep is the slide? You can use trigonometry to find an answer from the information in the diagram.

Recall the tangent ratio from Lesson 12.6.

$$\tan A = \frac{\text{length of the leg opposite } \angle A}{\text{length of the leg adjacent to } \angle A}$$

Two other important ratios, the *sine* and the *cosine*, are also based on the sides of a right triangle.

Prepare

NCTM PRINCIPLES & STANDARDS
1–4, 6–10

LESSON RESOURCES

- Practice 12.7
- Reteaching Master 12.7
- Enrichment Master 12.7
- Cooperative-Learning
  Activity 12.7
- Lesson Activity 12.7
- Problem Solving/Critical
  Thinking Master 12.7

QUICK WARM-UP

Use a calculator to evaluate the following:

1. tan 57° to the nearest ten-thousandth ≈ **1.5399**

2. $\tan^{-1}(0.8)$ to the nearest degree ≈ **39°**

In △ABC, m∠C = 90°, AC = 24, and BC = 10. Find the following:

3. tan A $\frac{5}{12}$ 4. tan B $\frac{12}{5}$

5. m∠A ≈ **23°**   6. m∠B ≈ **67°**

Also on Quiz Transparency 12.7

Teach

Why Give students the following statement:

You can use the tangent ratio when you know the measure of one side of a right triangle and the measure of one angle other than the right angle.

Ask them if this is a true statement. Lead them to see that it is true only if the side mentioned is a leg. Tell them that if the side is the hypotenuse, you must use one of two other ratios, called the *sine* ratio and the *cosine* ratio.

Alternative Teaching Strategy

USING TECHNOLOGY Use geometry software to demonstrate the sine ratio. A sample sketch is shown at right. To make the sketch, construct $\overleftrightarrow{AC}$ and $\overleftrightarrow{AZ}$ as shown, draw points C and C' by choosing **Point on Object** from the **Construct** menu, and construct $\overleftrightarrow{BC}$ and $\overleftrightarrow{B'C'}$ by choosing **Perpendicular Line** from the **Construct** menu. Use the **Calculate** menu to obtain the measures. To demonstrate the sine, click and drag point , and instruct students to observe the values of the ratios. Adjust the measures and repeat for the cosine ratio.

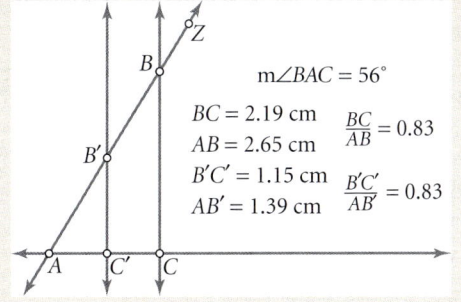

m∠BAC = 56°
BC = 2.19 cm $\frac{BC}{AB} = 0.83$
AB = 2.65 cm
B'C' = 1.15 cm $\frac{B'C'}{AB'} = 0.83$
AB' = 1.39 cm

LESSON 12.7 **621**

Activity Notes

In this Activity, students investigate the sine and cosine ratios in a right triangle as the length of the hypotenuse is held constant and the measures of the acute angles are varied. Students should discover that as the measure of an acute angle increases from 0° to 90°, the sine of the angle increases and the cosine of the angle decreases.

Cooperative Learning

You may wish to have students do the Activity in pairs. One partner can be the "sine calculator" who does Step 1, while the other is the "cosine calculator" who does Step 2. The partners can then compare results and complete Step 3 together.

CHECKPOINT ✔

3. As the measure of an angle increases from 0° to 90°, the value of the sine increases from 0 to 1 and the value of the cosine decreases from 1 to 0.

Teaching Tip

TECHNOLOGY If you are using a TI-82 or TI-83 graphics calculator, remember that you can check the mode setting by pressing [MODE]. If the third line shows **Degree** highlighted, the calculator is in degree mode. If Radian is highlighted, you can change the setting by pressing [▼] [▼] [▶] [ENTER]. Then press [2nd] [MODE] to return to the home screen.

To access the sine and cosine functions, press [SIN] or [COS].

TRY THIS

cos 65° ≈ 0.423
sin 25° ≈ 0.423

622 LESSON 12.7

Sine and Cosine

$$\text{sine of } \angle A = \sin A = \frac{\text{length of the leg opposite } \angle A}{\text{length of hypotenuse}} = \frac{opp}{hyp}$$

$$\text{cosine of } \angle A = \cos A = \frac{\text{length of the leg adjacent to } \angle A}{\text{length of hypotenuse}} = \frac{adj}{hyp}$$

Activity
Exploring Sines and Cosines

You will need: graph paper and a scientific calculator

1. In the illustrations below, the hypotenuse is the same in each triangle. What happens to the value of sin A as the measure of angle A increases? Does it increase or decrease? Write a conjecture about the sine of 0° and the sine of 90°.

$$\sin A = \frac{opp}{hyp}$$

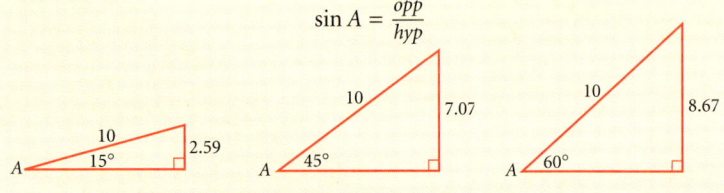

2. In the illustrations below, the hypotenuse is the same in each triangle. What happens to the value of cos A as the measure of angle A increases? Does it increase or decrease? Write a conjecture about the cosine of 0° and the cosine of 90°.

$$\cos A = \frac{adj}{hyp}$$

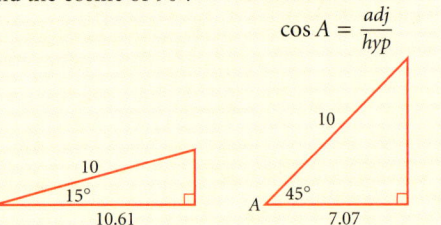
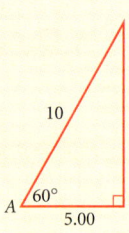

CHECKPOINT ✔ 3. Write a statement contrasting the behavior of the sine and cosine ratios as the measure of an angle increases from 0° to 90°.

TECHNOLOGY / SCIENTIFIC CALCULATOR

You can use your calculator to find the sine or cosine of an angle. Because the angle measure is given in degrees, your calculator must be in degree mode. To determine sin 10°, use the [SIN] key.

[SIN] 10 [ENTER] ≈ 0.174

TRY THIS Use the [COS] and [SIN] keys on your calculator to find cos 65° and sin 25°.

Interdisciplinary Connection

PHYSICS When a projectile is fired at an angle, there are two *components* to its velocity—an upward component and a forward component. Use the diagram to find the upward and forward components of a projectile fired with a velocity of 40 meters per second at an angle of 30° with the horizontal.

upward: 20 m/s; forward: ≈ 34.6 m/s

Inclusion Strategies

LINGUISTIC LEARNERS Show students how to use the mnemonic SOH-CAH-TOA as a means of remembering the three trigonometric ratios they are studying.

$$\sin = \frac{opposite\ leg}{hypotenuse}, \text{ or } s = \frac{o}{h}$$

$$\cos = \frac{adjacent\ leg}{hypotenuse}, \text{ or } c = \frac{a}{h}$$

$$\tan = \frac{opposite\ leg}{adjacent\ leg}, \text{ or } t = \frac{o}{a}$$

EXAMPLE ❶

APPLICATION
RECREATION

A parasail is towed behind a boat. The rope that is attached to the parasail is 35 feet long. The parasail is at an angle of 30° with the surface of the water. **Approximately how high above the water is the parasailer?** (The rope is connected to the boat at a point 1 foot above the surface of the water.)

● **SOLUTION**

Use the sine ratio to find b, the length of the side opposite the 30° angle.

$$\sin 30° = \frac{\text{opposite}}{\text{hypotenuse}}$$

$$\sin 30° = \frac{b}{35}$$

$$b = 35(\sin 30°)$$

$$b = 17.5 \quad \text{Use a calculator.}$$

Since the rope is connected to the boat 1 foot above the water, add 1 to 17.5. Thus, the parasail is 18.5 feet above the surface of the water.

TRY THIS Find the length of side a in the diagram at right.

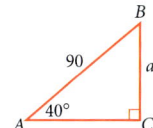

EXAMPLE ❷ Suppose that the horizontal distance of a proposed water slide is 115 feet and that the slide is 200 feet long. **If the maximum allowable angle of the slide is 35°, does the slide described meet this requirement?**

● **SOLUTION**

$$\cos B = \frac{\text{adjacent}}{\text{hypotenuse}} = \frac{115}{200}$$

$$B = \cos^{-1}\left(\frac{115}{200}\right) \quad (\text{Use the } \boxed{\cos^{-1}} \text{ key.})$$

$$B \approx 55°$$

The ramp described would be too steep.

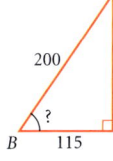

TRY THIS Find the measure of angle M in the diagram at right.

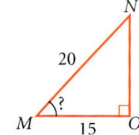

ADDITIONAL
EXAMPLE ❶

Find the length of side $\overline{PQ}$ in $\triangle PQR$ below.

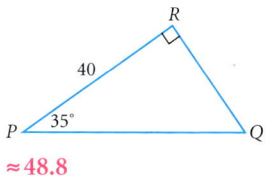

≈ 48.8

TRY THIS
≈ 57.9

ADDITIONAL
EXAMPLE ❷

Find the measure of $\angle Y$ in $\triangle XYZ$ below.

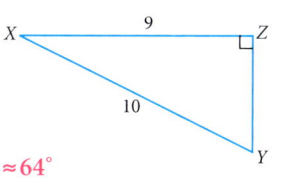

$\approx 64°$

TRY THIS
$\approx 41°$

Teaching Tip

TECHNOLOGY If you are using a TI-82 or TI-83 graphics calculator, you can access the inverse sine and inverse cosine functions by pressing [2nd] [SIN] and [2nd] [COS], respectively.

Enrichment

Have students refer to the figure at right to determine whether each statement below is true or false.

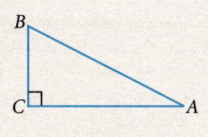

1. $\sin A = \cos B$ **true**
2. $\cos A = \sin B$ **true**
3. $\tan A = \frac{\cos A}{\sin A}$ **false**
4. $\tan B = \frac{\sin B}{\cos B}$ **true**
5. $\sin A = \frac{1}{\cos B}$ **false**
6. $\tan B = \frac{1}{\tan A}$ **true**
7. $\sin A = (\cos A)(\tan A)$ **true**
8. $\cos B = (\sin B)(\tan A)$ **true**

Reteaching the Lesson

WORKING BACKWARD Have students work in pairs. Each student should draw a right-triangle diagram in which an unknown angle measure can be found by using the labeled information and the sine ratio. Partners should trade papers and find the unknown angle measure in each other's diagram. Then have students draw and trade diagrams again, this time using the sine ratio to find an unknown length.

Repeat this activity for the cosine and tangent ratios.

LESSON 12.7

Assess

Selected Answers
Exercises 5–7, 9–75 odd

ASSIGNMENT GUIDE

In Class	1–7
Core	9–59 odd
Core Plus	8–60 even, 61
Review	62–75
Preview	76–80

✎ Extra Practice can be found beginning on page 738.

Technology
A calculator with trigonometric functions and their inverses is needed for Exercises 8–61.

Exercises

Communicate

1. Explain how to find the sine of an angle in a right triangle.
2. Explain how to find the cosine of an angle in a right triangle.
3. Explain how to find the length of a ramp that makes a 4° angle up to a 3-inch curb.
4. What are the maximum and minimum values of the sine and cosine ratios in a right triangle?

Guided Skills Practice

CONNECTION

5. **TRIGONOMETRY** Find the length of side s in the diagram at right.
   (EXAMPLE 1) ≈ **37.6**

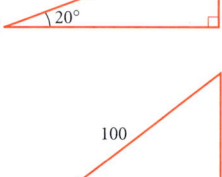

6. Find the measure of angle X in the diagram at right.
   (EXAMPLE 2) ≈ **36.9°**

7. **RECREATION** Suppose that the horizontal distance of a proposed rollercoaster drop is 125 feet and that the track is 200 feet long. If the maximum allowable angle of the ride is 45°, does the ride described meet this requirement? (Example 2)
The angle of the ride is about 51°, which is greater than the maximum allowable angle of 45°.

Practice and Apply

Approximate each sine and cosine. Give your answers to the nearest ten-thousandth.

8. sin 24° **0.4067**
9. cos 32° **0.8480**
10. cos 88° **0.0349**
11. sin 18° **0.3090**
12. sin 66.8° **0.9191**
13. cos 84.2° **0.1011**
14. cos 27.7° **0.8854**
15. sin 48.6° **0.7501**

---

internetconnect

GO TO: go.hrw.com
KEYWORD: MA1 Trig Functions

The three trigonometric functions defined in the text allow the student to solve many elementary applied problems. In this activity, they will deal with trig functions—primarily the tangent function—on a qualitative basis while discovering some of their properties.

624 LESSON 12.7

Find the acute angle measure for each approximate sine value below. Give your answers to the nearest tenth of a degree.

16. 0.56   **34.1°**   **17.** 0.892   **63.1°**   **18.** 0.129   **7.4°**   **19.** 0.759   **49.4°**
20. 0.5   **30.0°**   **21.** 0.707   **45.0°**   **22.** 0.563   **34.3°**   **23.** 0.445   **26.4°**
24. 0.26   **15.1°**   **25.** 0.972   **76.4°**   **26.** 0.0   **0°**   **27.** 1.0   **90°**

Find the acute angle measure for each approximate cosine value below. Give your answers to the nearest tenth of a degree.

28. 0.126   **82.8°**   **29.** 0.5   **60.0°**   **30.** 0.886   **27.6°**   **31.** 0.707   **45.0°**
32. 0.99   **8.1°**   **33.** 0.81   **35.9°**   **34.** 0.54   **57.3°**   **35.** 0.78   **38.7°**
36. 0.612   **52.3°**   **37.** 0.643   **50.0°**   **38.** 0.0   **90°**   **39.** 1.0   **0°**

For each triangle, state whether the sine or cosine function should be used to find the value of x. Then find x to the nearest tenth.

40. **sin; $x \approx 48.6°$**   **41.**   **42.**

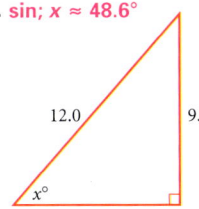
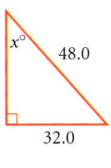
sin; $x \approx 41.8°$
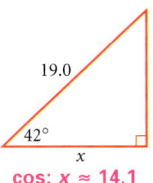
cos; $x \approx 14.1$

43.   **44.**   **45.**

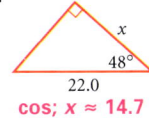
cos; $x \approx 14.7$
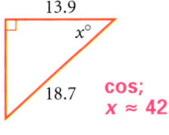
cos; $x \approx 42.0°$
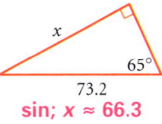
sin; $x \approx 66.3$

For Exercises 46–55, use right triangle ABC at right. Find the indicated side length or angle measure to the nearest tenth.

46. $b = 7$, $c = 10$, m$\angle B = $ __?__ $\approx$ **44.4°**
47. $b = 6$, $c = 13$, m$\angle A = $ __?__ $\approx$ **62.5°**
48. m$\angle B = 35°$, $c = 65$, $b = $ __?__ $\approx$ **37.3**
49. m$\angle B = 51°$, $c = 22$, $a = $ __?__ $\approx$ **13.8**
50. $b = 154$, $a = 25$, m$\angle B = $ __?__ $\approx$ **80.8°**
51. $a = 27$, $c = 36$, m$\angle B = $ __?__ $\approx$ **41.4°**
52. m$\angle B = 41°$, $c = 6$, $b = $ __?__ $\approx$ **3.9**
53. m$\angle B = 30°$, $b = 11$, $c = $ __?__ **22**
54. $a = 7$, $c = 14$, m$\angle B = $ __?__ $\approx$ **60°**
55. m$\angle B = 45°$, $c = 4$, $b = $ __?__ $\approx$ **2.8**

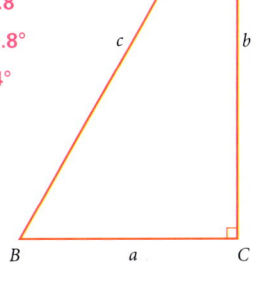

CONNECTION

56. GEOMETRY If m$\angle B = 60°$ and $c = 100$ meters, find the area of triangle ABC. **2165 sq m**

Error Analysis

Students sometimes misinterpret a diagram like the one at the bottom of page 621, believing that the *leg opposite* an angle is always to the right and that the *leg adjacent* is always at the bottom. Stress the fact that these terms are relative. It may be helpful to give them a page of right triangles in different orientations and ask them to identify the opposite leg and adjacent leg for each acute angle.

Look Beyond

In Exercises 76 and 77, students calculate areas of geometric figures by using trigonometric ratios to find the required measures. Students will explore such problems in greater detail if they take a course in geometry.

Exercises 78–80 give students an opportunity to investigate the graphs of the sine and cosine functions. Students will work with these graphs more extensively if they take a more advanced course in algebra.

APPLICATIONS

CONSTRUCTION A wheelchair ramp is to have an angle of 4.5° with the ground. The deck at the top of the ramp is 18 inches above ground level. Use this information for Exercises 57 and 58.

57. How long should the ramp be? ≈ 229.4 in.
58. How far from the deck should the ramp begin? ≈ 228.7 in.
59. **CONSTRUCTION** A 12-foot ladder is leaning against a building. The ladder makes a 75° angle with the ground. How far is the base of the ladder from the base of the building? ≈ 3.1 ft
60. **CIVIL ENGINEERING** A guy wire attached to the top of a tower makes an angle of 25° with the tower. If the wire is anchored to the ground 30 feet from the tower, how long is the wire? ≈ 70.98 ft

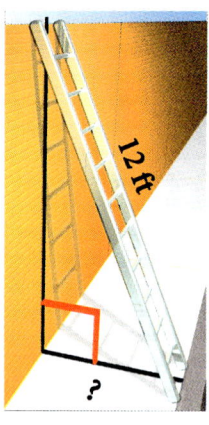

CHALLENGE

61. **AVIATION** An airplane traveling at 500 mph is descending from 30,000 feet to 20,000 feet at an angle of depression of 3°. How long, in seconds, does the descent take? ≈ 4.34 min, or ≈ 260 sec

Look Back

Solve each equation. Round your answers to the nearest hundredth. **(LESSON 10.2)**

62. $x^2 = 36$ ±6
63. $x^2 = 144$ ±12
64. $(x + 4)^2 - 36 = 0$ –10, 2
65. $(x - 1)^2 = 11$ ≈ 4.32, ≈ –2.32

Complete the square. **(LESSON 10.3)**

66. $x^2 + 10x$ $x^2 + 10x + 25 = (x + 5)^2$
67. $x^2 - 6x$ $x^2 - 6x + 9 = (x - 3)^2$
68. $x^2 + 9x$ $x^2 + 9x + \frac{81}{4} = \left(x + \frac{9}{2}\right)^2$
69. $x^2 - 7x$ $x^2 - 7x + \frac{49}{4} = \left(x - \frac{7}{2}\right)^2$

Use the quadratic formula to solve each equation. Round your answers to the nearest hundredth. **(LESSON 10.5)**

70. $t^2 + 6t - 22 = 0$ ≈ 2.57, –8.57
71. $h^2 + 6h + 5 = 0$ –5, –1

Find x for each triangle. Round your answers to the nearest tenth. **(LESSON 12.6)**

72. ≈ 20.2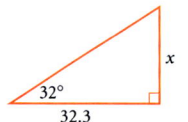
73. ≈ 6.5
74. ≈ 63.2°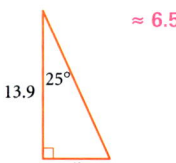
75. ≈ 63.5°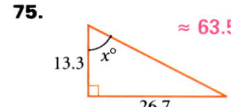

Practice
12.7 The Sine and Cosine Functions

Evaluate each expression to the nearest ten-thousandth.

1. sin 65° 0.9063
2. sin 85° 0.9962
3. cos 32° 0.8480
4. cos 15° 0.9659
5. sin 12° 0.2079
6. sin 50° 0.7660
7. cos 39° 0.7771
8. cos 89° 0.0175
9. sin 76° 0.9703
10. sin 82° 0.9903
11. cos 49° 0.6561
12. cos 61° 0.4848

For Exercises 13–20, use right triangle ABC shown at right. Find the indicated angle measure or length. Round your answer to the nearest tenth.

13. $b = 6, c = 8, m\angle B =$ 48.6°
14. $a = 8, c = 15, m\angle B =$ 57.8°
15. $b = 6, c = 9, m\angle A =$ 48.2°
16. $a = 8, c = 14, m\angle B =$ 55.2°
17. $b = 7, c = 10, m\angle B =$ 44.4°
18. $m\angle B = 40°, c = 30, b =$ 19.3
19. $m\angle B = 52°, c = 18, a =$ 11.1
20. $m\angle A = 46°, c = 9, b =$ 6.3

What angle measure has each approximate sine value? Round each answer to the nearest tenth of a degree.

21. 0.5992 36.8°
22. 0.8387 57.0°
23. 0.0872 5.0°
24. 0.2419 14.0°
25. 0.1139 6.5°
26. 0.7408 47.8°

What angle measure has each approximate cosine value? Round each answer to the nearest tenth of a degree.

27. 0.3746 68.0°
28. 0.8988 26.0°
29. 0.0087 89.5°
30. 0.9885 8.7°
31. 0.5490 56.7°
32. 0.0523 87.0°

Look Beyond

CONNECTIONS

TRIGONOMETRY Use trigonometry to find the area of each figure.

76. ≈ 19.3 sq units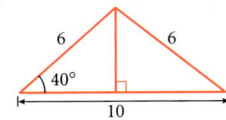

77. ≈ 40 sq units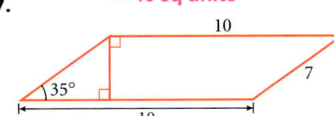

78. Use a calculator to complete the table, and sketch a graph of $y = \sin x$.

x	0°	30°	60°	90°	120°	150°	180°	210°	240°	270°	300°	330°
$\sin x$	?	?	?	?	?	?	?	?	?	?	?	?
	0	0.5	0.87	1	0.87	0.5	0	−0.5	−0.87	−1	−0.87	−0.5

79. Use a calculator to complete the table, and sketch a graph of $y = \cos x$.

x	0°	30°	60°	90°	120°	150°	180°	210°	240°	270°	300°	330°
$\cos x$	?	?	?	?	?	?	?	?	?	?	?	?
	1	0.87	0.5	0	−0.5	−0.87	−1	−0.87	−0.5	0	0.5	0.87

80. Compare the graphs you sketched in Exercises 78 and 79. How are they different? How are they similar?

The graphs are the same shape. The sine graph is shifted 90° to the right of the cosine graph.

Portfolio Activity

Include this activity in your portfolio.

The golden ratio was used in ancient civilizations to form "golden" rectangles. Golden rectangles are pleasing to the eye. A golden rectangle has the property that the ratio of the width to the length is equal to the ratio of the length to the length plus the width:

$$\frac{W}{L} = \frac{L}{L + W}$$

This television screen has dimensions of a golden rectangle.

You can compute the golden ratio by considering a rectangle with a width of 1.

1. Find the golden ratio by substituting 1 for W and solving for L, using the quadratic formula.

$$\frac{1}{L} = \frac{L}{L + 1}$$

2. Explain why a 3 × 5 card is close to being a golden rectangle.

3. Find at least three other real-world rectangles that are close to being golden rectangles.

WORKING ON THE CHAPTER PROJECT

You should now be able to complete Activities 3 and 4 of the Chapter Project beginning on page 636.

78.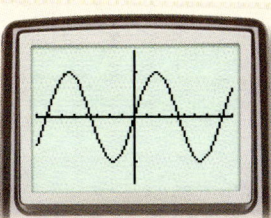

79.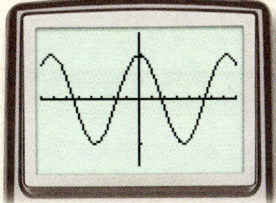

Prepare

NCTM PRINCIPLES & STANDARDS
1, 2, 5–10

LESSON RESOURCES

- Practice 12.8
- Reteaching Master 12.8
- Enrichment Master 12.8
- Cooperative-Learning Activity 12.8
- Lesson Activity 12.8
- Problem Solving/Critical Thinking Master 12.8
- Teaching Transparency 62
- Teaching Transparency 63

QUICK WARM-UP

Evaluate.

1. $-15 + 6$ -9
2. $5 + (-3.7)$ 1.3
3. $4 - 6$ -2
4. $-3 - (-2.1)$ -0.9
5. $(-12)(4)$ -48
6. $(-9)(-7)$ 63
7. $2(-5) + (-1)(0)$ -10
8. $(-3)(-8) + (-9)(2)$ 6

Also on Quiz Transparency 12.8

Teach

Why Matrices provide an important way to organize information. Once information is organized in a matrix, it is easy to manipulate. Have students discuss ways in which they have organized information in the past, both in their study of mathematics and in everyday situations. Their answers may include tables, lists, and tree diagrams.

12.8 Introduction to Matrices

Objectives

- Determine the dimensions and addresses of a matrix.
- Determine whether two matrices are equal.
- Add, subtract, and multiply matrices.
- Determine the multiplicative identity of a matrix.

Why Matrices are an easy way of combining information from different tables.

Costs associated with car ownership can be studied by using matrices.

Organizing Data by Using Matrices

APPLICATION
TRANSPORTATION

The table shows the costs associated with owning and operating a car for the given years.

Item	Unit	1980	1993
Cost per mile	Cents/mile	27.95	45.14
Gas and oil	Cents/mile	7.62	6.00
Maintenance	Cents/mile	5.86	2.40
Tires	Cents/mile	1.12	0.90
Additional costs (insurance, license and registration, depreciation, and finance charges)	Dollars/year	2033	4486

[Source: Statistical Abstract of the United States, 1996]

Data that are organized in a table are easy to read and manipulate. In mathematics, data can often be arranged in a special table of rows and columns enclosed in a set of brackets, [ ]. This arrangement is called a **matrix**.

A matrix is usually named with a capital letter. Matrix C contains the data from the table above.

$$C = \begin{bmatrix} 27.95 & 45.14 \\ 7.62 & 6.00 \\ 5.86 & 2.40 \\ 1.12 & 0.90 \\ 2033 & 4486 \end{bmatrix}$$

Alternative Teaching Strategy

USING TECHNOLOGY Have students use a graphics calculator to investigate matrix addition and subtraction. Use the matrices below.

$$A = \begin{bmatrix} -3 & 2 \\ 6 & -7 \end{bmatrix} \quad B = \begin{bmatrix} -4 & -1 \\ -8 & 9 \end{bmatrix}$$

If you are using a TI-82 or TI-83 graphics calculator, use the following key sequence to access the matrix editor and to enter matrix A:

MATRX ▶ ▶ ENTER 2 ENTER 2 ENTER (−) 3 ENTER 2 ENTER 6 ENTER (−) 7 ENTER

Then enter matrix B as follows:

(−) 1 ENTER (−) 8 ENTER 9 ENTER

To quit the matrix editor, press 2nd MODE. Then use the following key sequence to calculate $A + B$:

The result will be displayed as $\begin{bmatrix} -7 & 1 \\ -2 & 2 \end{bmatrix}$. Discuss how this sum was calculated.

Matrix C has 5 rows and 2 columns. The number of rows by the number of columns gives the row-by-column dimensions of the matrix. The dimensions of C are 5 by 2. This can be written as 5×2 or $C_{5 \times 2}$.

The information that appears in each position of the matrix is called an **entry**. Each entry in a matrix can be located by its **matrix address**. The entry at the address c_{32}, 2.4, is the entry at row 3 and column 2 of matrix C.

Addresses for matrix C

$$\begin{array}{c} \text{Column Column} \\ 1 \quad\quad 2 \end{array}$$

$$\begin{array}{c} \text{Row 1} \\ \text{Row 2} \\ \text{Row 3} \\ \text{Row 4} \\ \text{Row 5} \end{array} \begin{bmatrix} c_{11} & c_{12} \\ c_{21} & c_{22} \\ c_{31} & c_{32} \\ c_{41} & c_{42} \\ c_{51} & c_{52} \end{bmatrix}$$

The dimensions are 5×2.

A matrix is called a **square matrix** if its row-by-column dimensions are equal.

Matrix Equality

Two matrices are equal when their dimensions are the same and their corresponding entries are equal.

$$U = \begin{bmatrix} 1 & 2 & 3 \\ 4 & 5 & 6 \end{bmatrix} \quad V = \begin{bmatrix} 1 & 2 & 3 \\ 4 & 5 & 6 \end{bmatrix} \quad W = \begin{bmatrix} 1 & 4 \\ 2 & 5 \\ 3 & 6 \end{bmatrix} \quad X = \begin{bmatrix} 1 & 3 & 5 \\ 2 & 4 & 6 \end{bmatrix}$$

Matrix U and matrix V have the same dimensions and their corresponding entries are equal, so $U = V$. Matrix V and matrix W do not have the same dimensions, so $V \neq W$. Matrix U and matrix X have the same dimensions, but corresponding entries are not equal, so $U \neq X$.

EXAMPLE 1 Are matrices S and T equal? Explain.

$$S = \begin{bmatrix} 2^2 & (12-3) \\ \sqrt{49} & 11 \\ -3(9) & \frac{24}{3} \end{bmatrix} \quad T = \begin{bmatrix} 2(2) & \sqrt{81} \\ 7 & \frac{44}{4} \\ -(30-3) & 2^3 \end{bmatrix}$$

SOLUTION

Think of the characteristics that make two matrices equal.

The two matrices must have the same dimensions. In this example, both matrices have dimensions of 3 by 2.

The entries in both matrices simplify to $\begin{bmatrix} 4 & 9 \\ 7 & 11 \\ -27 & 8 \end{bmatrix}$, so the

corresponding entries in each matrix are equal. Thus, matrices S and T are equal.

Use the following key sequence to calculate $A - B$:

[MATRX] [ENTER] [−] [MATRX] [▼] [ENTER] [ENTER]

The result will be displayed as $\begin{bmatrix} 1 & 3 \\ 14 & -16 \end{bmatrix}$.

Discuss how this difference was calculated. If you wish, you can lead students in similar investigations to introduce the multiplication of a matrix by a scalar and by another matrix. Then give them a 2×2 matrix and challenge them to find the identity matrix for it.

Inclusion Strategies

VISUAL LEARNERS Have students use colored pencils to write 2×2 matrices. Have them use blue to write the entries in row 1, column 1; red to write the entries in row 1, column 2; green to write the entries in row 2, column 1; and purple to write the entries in row 2, column 2. To add or subtract the matrices, students should add or subtract the numbers written in the same color.

Use Teaching Transparency 62.

ADDITIONAL EXAMPLE 1

Are matrices A and B below equal? Explain.

$$A = \begin{bmatrix} 0.06 & \sqrt{64} \\ 1-11 & \frac{48}{6} \end{bmatrix}$$

$$B = \begin{bmatrix} \frac{3}{50} & 2^3 \\ (-1)(10) & 12 \end{bmatrix}$$

The matrices have the same dimensions, but not all of the corresponding entries are equal: $\frac{48}{6} \neq 12$. Thus, matrices A and B are not equal.

Teaching Tip

TECHNOLOGY When using a TI-83 or TI-82 graphics calculator, remind students that they must set the dimensions of the matrix before they input the entries of the matrix. To set the dimensions of matrix A, press [MATRX], select **Edit**, and choose **1:[A]**. The dimensions are entered at the top of the screen to the right of the name of the matrix. For example, **Matrix [A] 2 × 3** will create matrix A with 2 rows and 3 columns.

LESSON 12.8

Use Teaching Transparency 63.

ADDITIONAL EXAMPLE 2

Add the following matrices.

a. $R = \begin{bmatrix} 5 & 1 \\ 2 & 8 \end{bmatrix}$ $S = \begin{bmatrix} 4 & 6 \\ 9 & 0 \end{bmatrix}$

$\begin{bmatrix} 9 & 7 \\ 11 & 8 \end{bmatrix}$

b. $J = \begin{bmatrix} -4 & 3 & 7 & -5 & 0 \end{bmatrix}$

$K = \begin{bmatrix} 2 & -1 & -8 & 5 & 3 \end{bmatrix}$

$\begin{bmatrix} -2 & 2 & -1 & 0 & 3 \end{bmatrix}$

TRY THIS

$\begin{bmatrix} 8.82 \\ 14.05 \end{bmatrix}$

ADDITIONAL EXAMPLE 3

Let $X = \begin{bmatrix} 3 & 5 \\ 0 & -1 \\ -4 & 6 \\ 8 & 1 \end{bmatrix}$ and

$Y = \begin{bmatrix} 1 & 4 \\ -3 & 1 \\ -2 & -1 \\ 2 & 5 \end{bmatrix}$.

Perform the following operations:

a. $X + Y$ b. $X - Y$

$\begin{bmatrix} 4 & 9 \\ -3 & 0 \\ -6 & 5 \\ 10 & 6 \end{bmatrix}$ $\begin{bmatrix} 2 & 1 \\ 3 & -2 \\ -2 & 7 \\ 6 & -4 \end{bmatrix}$

Adding and Subtracting Matrices

A matrix is an effective tool for storing data in a way that makes sense visually. But what if the data changes? You may need to add or subtract matrices.

To add or subtract matrices, both matrices *must have the same dimensions*. Then add or subtract corresponding entries to create a new matrix.

EXAMPLE 2 Add the following matrices:

$A = \begin{bmatrix} 3 \\ 7 \\ 9 \end{bmatrix}$ $B = \begin{bmatrix} 11 \\ 15 \\ 16 \end{bmatrix}$

• **SOLUTION**

$A + B = \begin{bmatrix} 3 \\ 7 \\ 9 \end{bmatrix} + \begin{bmatrix} 11 \\ 15 \\ 16 \end{bmatrix} = \begin{bmatrix} 3 + 11 \\ 7 + 15 \\ 9 + 16 \end{bmatrix} = \begin{bmatrix} 14 \\ 22 \\ 25 \end{bmatrix}$

TRY THIS Add the following matrices: $A = \begin{bmatrix} 1.5 \\ 4.35 \end{bmatrix}$ $B = \begin{bmatrix} 7.32 \\ 9.7 \end{bmatrix}$

Addition of Matrices

If A and B are matrices with the same dimensions, then the matrix $(A + B)$ has the same dimensions with elements that are the sum of the corresponding elements of A and B.

$\begin{bmatrix} a & b & c \\ d & e & f \end{bmatrix} + \begin{bmatrix} g & h & i \\ j & k & l \end{bmatrix} = \begin{bmatrix} a+g & b+h & c+i \\ d+j & e+k & f+l \end{bmatrix}$

EXAMPLE 3 Let $A = \begin{bmatrix} 1 & 5 & 7 \\ 9 & 8 & 2 \end{bmatrix}$ and $B = \begin{bmatrix} 4 & 3 & 6 \\ 2 & 1 & 5 \end{bmatrix}$. Perform the following operations:

a. $A + B$ b. $B - A$

• **SOLUTION**

a. $A + B = \begin{bmatrix} 1 & 5 & 7 \\ 9 & 8 & 2 \end{bmatrix} + \begin{bmatrix} 4 & 3 & 6 \\ 2 & 1 & 5 \end{bmatrix} = \begin{bmatrix} 1+4 & 5+3 & 7+6 \\ 9+2 & 8+1 & 2+5 \end{bmatrix} = \begin{bmatrix} 5 & 8 & 13 \\ 11 & 9 & 7 \end{bmatrix}$

b. $B - A = \begin{bmatrix} 4 & 3 & 6 \\ 2 & 1 & 5 \end{bmatrix} - \begin{bmatrix} 1 & 5 & 7 \\ 9 & 8 & 2 \end{bmatrix} = \begin{bmatrix} 4-1 & 3-5 & 6-7 \\ 2-9 & 1-8 & 5-2 \end{bmatrix} = \begin{bmatrix} 3 & -2 & -1 \\ -7 & -7 & 3 \end{bmatrix}$

Enrichment

On a coordinate plane, have students draw $\overline{AB}$ with endpoints at $A(-2, 4)$ and $B(5, 3)$. Tell them that this segment can be represented by the matrix $\begin{bmatrix} -2 & 5 \\ 4 & 3 \end{bmatrix}$. Be sure that they notice how the coordinates are positioned in the matrix. Now have them find the sum of this matrix and the matrix $\begin{bmatrix} 2 & 2 \\ 6 & 6 \end{bmatrix}$. Tell them to draw the segment represented by the new matrix. $\begin{bmatrix} 0 & 7 \\ 10 & 9 \end{bmatrix}$

Ask students to describe the relationship between the two segments. Elicit from them that the new segment is 6 units above and 2 units to the right of the original segment, the segments are congruent, and the segments are parallel. That is, the new segment is a translation of the original segment. Have each student create a new *translation matrix* of the form $\begin{bmatrix} a & a \\ b & b \end{bmatrix}$. Have them perform the translation on $\overline{AB}$ and write the matrix addition problem that represents the translation.

LESSON 12.8

Scalar Multiplication of a Matrix

To multiply a matrix by a number, simply multiply each entry in the matrix by the number and write the product in the appropriate address.

$$3\begin{bmatrix} 16 & 18 \\ 14 & 22 \end{bmatrix} = \begin{bmatrix} (3)(16) & (3)(18) \\ (3)(14) & (3)(22) \end{bmatrix} = \begin{bmatrix} 48 & 54 \\ 42 & 66 \end{bmatrix}$$

This type of multiplication is called **scalar multiplication**. The **scalar** is the number multiplied by each entry in the matrix. In this case, the scalar is 3.

EXAMPLE 4
Perform the scalar multiplication.

$$5\begin{bmatrix} 3 & -2 \\ 4 & 5 \end{bmatrix}$$

● **SOLUTION**

Multiply each entry by 5.

$$5\begin{bmatrix} 3 & -2 \\ 4 & 5 \end{bmatrix} = \begin{bmatrix} (5)(3) & (5)(-2) \\ (5)(4) & (5)(5) \end{bmatrix} = \begin{bmatrix} 15 & -10 \\ 20 & 25 \end{bmatrix}$$

Multiplying Matrices

Multiplying two matrices creates a new matrix called the **product matrix**. In order for two matrices to be compatible for multiplication, the number of columns in the matrix on the left must match the number of rows in the matrix on the right.

To multiply, first multiply each entry in row 1 of the left matrix by the corresponding entry in column 1 of the right matrix. Then add the products and place the sum at the address p_{11} in a product matrix, P.

Repeat the procedure for the second row and second column, placing the sum in p_{21} of P, and so on.

$$A = \begin{bmatrix} a & d \\ b & e \\ c & f \end{bmatrix} \quad B = \begin{bmatrix} p & r \\ q & s \end{bmatrix} \quad A \times B = \begin{bmatrix} ap+dq & ar+ds \\ bp+eq & br+es \\ cp+fq & cr+fs \end{bmatrix}$$

As you proceed, the address of the sum will reflect the row number of the values used from A and the column number of the values used from B.

$$P = \begin{bmatrix} p_{11} & p_{12} \\ p_{21} & p_{22} \\ p_{31} & p_{32} \end{bmatrix}$$

CHECKPOINT ✔ What is the size of the product matrix when A is a 3×5 matrix and B is a 5×1 matrix?

Reteaching the Lesson

USING COGNITIVE STRATEGIES Write the following on the board or overhead:

$$A + B = \begin{bmatrix} -8+(-1) & -3+5 \\ 2+(-7) & -4+0 \end{bmatrix}$$

Tell students that the first addend in each entry comes from matrix A and the second addend comes from matrix B. Have them re-create matrices A and B and then write the simplified sum.

$$A = \begin{bmatrix} -8 & -3 \\ 2 & -4 \end{bmatrix} \quad B = \begin{bmatrix} -1 & 5 \\ -7 & 0 \end{bmatrix}$$

$$A + B = \begin{bmatrix} -9 & 2 \\ -5 & -4 \end{bmatrix}$$

Repeat the activity for subtraction and multiplication of matrices A and B.

ADDITIONAL EXAMPLE 4

Perform each scalar multiplication below.

a. $4\begin{bmatrix} 3 & 0 & -2 \\ 5 & -2 & 1 \end{bmatrix}$

$$\begin{bmatrix} 12 & 0 & -8 \\ 20 & -8 & 4 \end{bmatrix}$$

b. $-1\begin{bmatrix} -3 & 0 \\ 5 & -2 \\ 6 & -4 \\ -1 & 1 \end{bmatrix}$

$$\begin{bmatrix} 3 & 0 \\ -5 & 2 \\ -6 & 4 \\ 1 & -1 \end{bmatrix}$$

Teaching Tip

Show students the diagram below to help them identify which matrices can be multiplied and the dimension of the resulting product matrix.

$$\underbrace{a \times b \quad \text{times} \quad b \times c}_{\text{dimension of product}}$$
$$\overbrace{\phantom{a \times b \quad \text{times} \quad b \times c}}^{\text{must match}}$$

If the dimensions of two matrices are $a \times b$ and $b \times c$, then the matrices can be multiplied because the number of columns of the left matrix factor is equal to the number of rows of the right matrix factor, and the product matrix has dimension $a \times c$. If the inner dimensions shown in the diagram (the number of columns of the left matrix factor and the number of rows of the right matrix factor) are not equal, then the matrices cannot be multiplied, and we say the multiplication is undefined.

CHECKPOINT ✔

3×1

LESSON 12.8

ADDITIONAL EXAMPLE 5

Find each product. If multiplication is not possible, explain why.

a. $\begin{bmatrix} -1 & 0 & 1 \end{bmatrix} \begin{bmatrix} 2 \\ -5 \\ 1 \end{bmatrix}$

[−1]

b. $\begin{bmatrix} -1 & 0 & 1 \end{bmatrix} \begin{bmatrix} 6 & -3 \\ -2 & 0 \\ 4 & 5 \end{bmatrix}$

[−2 8]

c. $\begin{bmatrix} -1 & 0 & 1 \end{bmatrix} \begin{bmatrix} -2 & 4 & 1 \\ 3 & -5 & 0 \end{bmatrix}$

The matrices are not compatible for multiplication because the number of columns in the matrix on the left does not match the number of rows in the matrix on the right.

TRY THIS

AB does not exist because the number of columns of *A* does not equal the number of rows of *B*.

BA = [6 14]

ADDITIONAL EXAMPLE 6

Find the identity matrix for
$\begin{bmatrix} -1 & 4 & 1 & -2 \\ 3 & -6 & 0 & 2 \\ -4 & -5 & 2 & -3 \\ 5 & 0 & 1 & -1 \end{bmatrix}.$

$\begin{bmatrix} 1 & 0 & 0 & 0 \\ 0 & 1 & 0 & 0 \\ 0 & 0 & 1 & 0 \\ 0 & 0 & 0 & 1 \end{bmatrix}$

TRY THIS

$\begin{bmatrix} 1 & 0 & 0 & 0 \\ 0 & 1 & 0 & 0 \\ 0 & 0 & 1 & 0 \\ 0 & 0 & 0 & 1 \end{bmatrix}$

EXAMPLE 5 Find the product. If multiplication is not possible, explain why.

$\begin{bmatrix} 4 & -8.6 \\ -7 & 6 \end{bmatrix} \begin{bmatrix} -1 & -2 \\ 3 & 4.2 \end{bmatrix}$

SOLUTION

The matrices are compatible for multiplication. The number of columns in the matrix on the left matches the number of rows in the matrix on the right.

$\begin{bmatrix} 4 & -8.6 \\ -7 & 6 \end{bmatrix} \begin{bmatrix} -1 & -2 \\ 3 & 4.2 \end{bmatrix} = \begin{bmatrix} (4)(-1)+(-8.6)(3) & (4)(-2)+(-8.6)(4.2) \\ (-7)(-1)+(6)(3) & (-7)(-2)+(6)(4.2) \end{bmatrix}$

$= \begin{bmatrix} -29.8 & -44.12 \\ 25 & 39.2 \end{bmatrix}$

TRY THIS Find the products *AB* and *BA*, if multiplication is possible.

$A = \begin{bmatrix} 3 & 7 \end{bmatrix} \quad B = \begin{bmatrix} 2 \end{bmatrix}$

One special matrix is the **identity matrix** for multiplication. The identity matrix is always a square matrix. The number of rows equals the number of columns. Each entry on the **main diagonal** is 1; all other entries are 0.

$$\text{identity matrix} = \begin{bmatrix} 1 & 0 \\ 0 & 1 \end{bmatrix}$$

Any matrix multiplied by its identity matrix leaves that matrix unchanged.

EXAMPLE 6 What is the identity matrix for $B = \begin{bmatrix} 3 & 2 & 1 \\ 2 & 3 & 1 \\ 4 & 4 & 5 \end{bmatrix}$?

SOLUTION

Since *B* has dimensions of 3 × 3, the identity matrix for multiplication must be a 3 × 3 matrix.

The identity matrix for $B = \begin{bmatrix} 3 & 2 & 1 \\ 2 & 3 & 1 \\ 4 & 4 & 5 \end{bmatrix}$ is $I = \begin{bmatrix} 1 & 0 & 0 \\ 0 & 1 & 0 \\ 0 & 0 & 1 \end{bmatrix}$.

TRY THIS What is the identity matrix for $D = \begin{bmatrix} 3 & 2 & 1 & 0 \\ 5 & 1 & 2 & 6 \\ 4 & 3 & 1 & 7 \\ 4 & 6 & 5 & 1 \end{bmatrix}$?

Exercises

Communicate

1. Create a 3×5 matrix, A, and enter each address. Describe how the addresses are notated.
2. Is matrix addition commutative? Explain.
3. Is matrix multiplication commutative? Explain.

Guided Skills Practice

4. Determine which two of the matrices below are equivalent. **(EXAMPLE 1)**

$$H = \begin{bmatrix} 3 & -4 & 4(5) \\ 1 & 0 & 2 \end{bmatrix} \quad I = \begin{bmatrix} 3 & -4 & 20 \\ 1 & 0 & 12 \end{bmatrix} \quad J = \begin{bmatrix} (2+1) & -4 & 2(10) \\ 1 & 0 & 4-2 \end{bmatrix}$$

Add or subtract as indicated. **(EXAMPLES 2 AND 3)**

5. $\begin{bmatrix} 7 & 9 \\ -4 & 5 \end{bmatrix} + \begin{bmatrix} 18 & 4 \\ -3 & -16 \end{bmatrix}$

6. $\begin{bmatrix} -5 & 6 \\ 12 & 8 \end{bmatrix} - \begin{bmatrix} 4 & -19 \\ 17 & 7 \end{bmatrix}$

7. $\begin{bmatrix} 4 & -6 & 12 \\ -3 & 9 & 8 \\ 2 & 16 & -3 \end{bmatrix} - \begin{bmatrix} 1 & -2 & 3 \\ 27 & 8 & 7 \\ 5 & -3 & 14 \end{bmatrix}$

Perform each scalar multiplication. **(EXAMPLE 4)**

8. $5 \begin{bmatrix} 6 & -4 \\ -3 & 5 \end{bmatrix}$

9. $-2 \begin{bmatrix} 3 & 9 \\ -6 & -4 \end{bmatrix}$

10. $3 \begin{bmatrix} 2 & -5 & 1 \\ 12 & -3 & -8 \\ 14 & 10 & 4 \end{bmatrix}$

Find each product matrix, if it exists. **(EXAMPLE 5)**

11. $\begin{bmatrix} -6 & 4 \\ -23 & 3 \end{bmatrix} \begin{bmatrix} 9 & -4 \\ 6 & 2 \end{bmatrix}$

12. $\begin{bmatrix} 13 & 12 \\ -4 & 3 \end{bmatrix} \begin{bmatrix} 3 & 9 \\ -6 & -4 \end{bmatrix}$

13. $\begin{bmatrix} 3 & 8 & 7 \\ -3 & 14 & -6 \\ 8 & 3 & -7 \end{bmatrix} \begin{bmatrix} 2 & 5 \\ -7 & 9 \end{bmatrix}$

14. $\begin{bmatrix} 4 \\ 2 \\ 9 \end{bmatrix} \begin{bmatrix} 7 \\ 7 \\ 4 \end{bmatrix}$

Find the identity matrix for each matrix below. **(EXAMPLE 6)**

15. $\begin{bmatrix} 18 & 4 \\ -5 & -6 \end{bmatrix}$ $\begin{bmatrix} 1 & 0 \\ 0 & 1 \end{bmatrix}$

16. $\begin{bmatrix} 3 & 2 & 1 \\ 2 & 3 & 1 \\ 4 & 4 & 5 \end{bmatrix}$ $\begin{bmatrix} 1 & 0 & 0 \\ 0 & 1 & 0 \\ 0 & 0 & 1 \end{bmatrix}$

Answers:

4. H and J

5. $\begin{bmatrix} 25 & 13 \\ -7 & -11 \end{bmatrix}$

6. $\begin{bmatrix} -9 & 25 \\ -5 & 1 \end{bmatrix}$

7. $\begin{bmatrix} 3 & -4 & 9 \\ -30 & 1 & 1 \\ -3 & 19 & -17 \end{bmatrix}$

8. $\begin{bmatrix} 30 & -20 \\ -15 & 25 \end{bmatrix}$

9. $\begin{bmatrix} -6 & -18 \\ 12 & 8 \end{bmatrix}$

10. $\begin{bmatrix} 6 & -15 & 3 \\ 36 & -9 & -24 \\ 42 & 30 & 12 \end{bmatrix}$

11. $\begin{bmatrix} -30 & 32 \\ -189 & 98 \end{bmatrix}$

12. $\begin{bmatrix} -33 & 69 \\ -30 & -48 \end{bmatrix}$

13. does not exist

14. does not exist

Assess

Selected Answers
Exercises 4–16, 17–57 odd

ASSIGNMENT GUIDE

In Class	1–16
Core	17–41 odd
Core Plus	20–40 even
Review	42–58
Preview	59, 60

✏ Extra Practice can be found beginning on page 738.

Technology

A calculator may be helpful for performing the computations in many of the exercises. If students have access to a graphics calculator, they can use its matrix editor. You may wish to have students complete the exercises with paper and pencil first and then use a calculator to verify their answers.

LESSON 12.8

Error Analysis

When giving the dimensions of a matrix or the address of an entry, students may write the rows and columns in the incorrect order. Suggest that they think about locating a seat in a stadium or in a theater. That is, a person usually identifies the row first and then locates the seat within that row. Point out that the elements of a matrix are identified in a similar fashion: first the row and then the column.

32. $\begin{bmatrix} 96 & 165 \\ -47 & -112 \end{bmatrix}$

33. $\begin{bmatrix} 102 \\ 87 \\ -53 \end{bmatrix}$

34. $\begin{bmatrix} -6 \\ 20 \end{bmatrix}$

35. $a = -10$
    $b = 7$
    $c = 15$
    $d = \frac{1}{3}$
    $g = 75$
    $k = \frac{12}{5}$

36. 744 girls; 756 boys

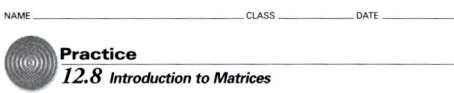

Practice and Apply

Perform the indicated matrix operations. If a solution is not possible, explain why.

17. $\begin{bmatrix} -3 & 4 \\ -4 & 0 \\ 2 & -5 \end{bmatrix} + \begin{bmatrix} 0 & -1 \\ -6 & 9 \\ -7 & 5 \end{bmatrix}$ 18. $\begin{bmatrix} 0.4 & -1.5 & 0.9 \\ 2.6 & 6.9 & 3.7 \end{bmatrix} + \begin{bmatrix} -4.7 & 2.6 & 6.9 \\ -7.3 & 9.8 & -5.5 \end{bmatrix}$

19. $\begin{bmatrix} -25 & 32 & 14 \\ 36 & -42 & -45 \\ -71 & 65 & 29 \end{bmatrix} - \begin{bmatrix} 16 & -34 & -55 \\ 21 & 11 & 22 \\ -43 & -67 & -44 \end{bmatrix} + \begin{bmatrix} 57 & 79 & 64 \\ -38 & -22 & -48 \\ -56 & 88 & 26 \end{bmatrix}$

20. $\begin{bmatrix} -1.4 & 4.3 & -9.6 \\ 15.8 & 1 & -3.5 \end{bmatrix} - \begin{bmatrix} 6.2 & 3.2 \\ -9.6 & 7.1 \\ 2.6 & -8.5 \end{bmatrix}$ 21. $\begin{bmatrix} 1 \\ 2 \\ 3 \end{bmatrix} - \begin{bmatrix} 3 \\ 2 \\ 1 \end{bmatrix}$

22. $\begin{bmatrix} 4 & 2 \\ 6 & -7 \\ 3 & 9 \end{bmatrix} + \begin{bmatrix} -5 & 7 \\ -3 & 9 \\ 3 & -6 \end{bmatrix} - \begin{bmatrix} -11 & 3 \\ 8 & -15 \\ -7 & -2 \end{bmatrix}$ 23. $\begin{bmatrix} -2 & 3 \\ 3 & -2 \end{bmatrix} - \begin{bmatrix} 4 & 6 \\ 4 & 6 \end{bmatrix}$

24. $\begin{bmatrix} -3 \\ -2 \\ -5 \end{bmatrix} - \begin{bmatrix} 1 & 2 & 3 \end{bmatrix}$ 25. $\begin{bmatrix} 3 & 2 \\ 6 & 4 \\ 9 & 6 \end{bmatrix} - \begin{bmatrix} -5 & 3 \\ 2 & -2 \\ -1 & 7 \end{bmatrix}$ 26. $5\begin{bmatrix} 3 & -2 \\ 7 & 8 \end{bmatrix}$

27. $-3\begin{bmatrix} 5 & 17 & -3 \\ 12 & -8 & 2 \\ -7 & 14 & 12 \end{bmatrix}$ 28. $-7\begin{bmatrix} 5 & -3 \\ 4 & -12 \\ 8 & 6 \end{bmatrix}$ 29. $\begin{bmatrix} 4 & -6 \\ 12 & 8 \end{bmatrix}\begin{bmatrix} 3 & -8 \\ 4 & 9 \end{bmatrix}$

30. $\begin{bmatrix} 3 & -7 & 18 \end{bmatrix}\begin{bmatrix} 4 \\ 13 \\ 9 \end{bmatrix}$ 31. $\begin{bmatrix} 3 & -8 \\ 1 & -3 \\ 4 & 16 \end{bmatrix}\begin{bmatrix} 8 & -6 & 9 \\ 12 & 4 & -2 \end{bmatrix}$

32. $\begin{bmatrix} 3 & 12 \\ -4 & -7 \end{bmatrix}\begin{bmatrix} -4 & 7 \\ 9 & 12 \end{bmatrix}$ 33. $\begin{bmatrix} 4 & 12 & 9 \\ -3 & -4 & 6 \\ 3 & 2 & -4 \end{bmatrix}\begin{bmatrix} 3 \\ -3 \\ 14 \end{bmatrix}$ 34. $\begin{bmatrix} 3 & -12 \\ 2 & 4 \end{bmatrix}\begin{bmatrix} 6 \\ 2 \end{bmatrix}$

Matrices M and N are equal. Find the values of a, b, c, d, g, and k.

35. $M = \begin{bmatrix} 2(a+4) & 77 & \frac{1}{3}c \\ -5d-1 & 30 & -\frac{1}{2}k \end{bmatrix}$ $N = \begin{bmatrix} -12 & 11b & 5 \\ -(3-d) & 0.4g & \frac{3}{4}k - 3 \end{bmatrix}$

STATISTICS In the Woodlake public school system, there are two junior high schools, Glenn and Kelly. The enrollment data for the electives music (Mu), art (Ar), technology (Te), and health (He) appear in the matrices below.

Glenn Middle School

	Mu	Ar	Te	He
Boys	447	199	514	389
Girls	498	352	432	399

Kelly Middle School

	Mu	Ar	Te	He
Boys	387	276	489	367
Girls	505	392	387	437

36. How many girls are enrolled in art? How many boys are enrolled in health?

37. Create a matrix to show the total enrollment in these electives.

38. How many students are enrolled in technology this year?

Answers (red):

17. $\begin{bmatrix} -3 & 3 \\ -10 & 9 \\ -5 & 0 \end{bmatrix}$

18. $\begin{bmatrix} -4.3 & 1.1 & 7.8 \\ -4.7 & 16.7 & -1.8 \end{bmatrix}$

19. $\begin{bmatrix} 16 & 145 & 133 \\ -23 & -75 & -115 \\ -84 & 220 & 99 \end{bmatrix}$

20. no solution; matrices have different dimensions

21. $\begin{bmatrix} -2 \\ 0 \\ 2 \end{bmatrix}$

22. $\begin{bmatrix} 10 & 6 \\ -5 & 17 \\ 13 & 5 \end{bmatrix}$

23. $\begin{bmatrix} -6 & -3 \\ -1 & -8 \end{bmatrix}$

24. no solution; matrices have different dimensions

25. $\begin{bmatrix} 8 & -1 \\ 4 & 6 \\ 10 & -1 \end{bmatrix}$

26. $\begin{bmatrix} 15 & -10 \\ 35 & 40 \end{bmatrix}$

27. $\begin{bmatrix} -15 & -51 & 9 \\ -36 & 24 & -6 \\ 21 & -42 & -36 \end{bmatrix}$

28. $\begin{bmatrix} -35 & 21 \\ -28 & 84 \\ -56 & -42 \end{bmatrix}$

29. $\begin{bmatrix} -12 & -86 \\ 68 & -24 \end{bmatrix}$

30. $\begin{bmatrix} 83 \end{bmatrix}$

31. $\begin{bmatrix} -72 & -50 & 43 \\ -28 & -18 & 15 \\ 224 & 40 & 4 \end{bmatrix}$

37. | | Mu | Ar | Te | He |
|---|---|---|---|---|
| Boys | 834 | 475 | 1003 | 756 |
| Girls | 1003 | 744 | 819 | 836 |

Total enrollment $\begin{bmatrix} 1837 & 1219 & 1822 & 1592 \end{bmatrix}$

38. 1822 students

634 LESSON 12.8

APPLICATION

SPORTS Suppose that you keep all the statistics for the girls' basketball team in a matrix, a portion of which is shown below. You record 3-point baskets, 2-point baskets, and 1-point free throws. The matrices show the statistics for the regular season and the playoff series. Use these matrices to complete Exercises 39–41.

Baskets Scored in Regular Season

	3 pts	2 pts	1 pt
Alesia	3	15	12
Jennel	1	20	16
Betsy	3	17	13
Katie	4	6	9
Nancy	0	14	18

Baskets Scored in Playoff Series

	3 pts	2 pts	1 pt
Alesia	2	10	3
Jennel	1	2	4
Betsy	2	3	2
Katie	1	6	3
Nancy	0	3	4

39. Create matrices to show the total points scored by each player in the regular season and during the playoff series.

40. Add the two matrices from Exercise 39 to create a new matrix.

41. How many total points were scored by the team during the regular season and the playoff series combined? **327**

 **Look Beyond**

Exercises 59 and 60 provide a preview of *theoretical probability*, which students will study in Lesson 13.1. Suggest that students think back to experimental probability, which they studied in Chapter 4.

Look Back

Find each sum or difference. (LESSONS 2.2 AND 2.3)

42. $-8 + 6$ **−2**   **43.** $-7 - (-3)$ **−4**   **44.** $8 + (-4)$ **4**   **45.** $-5 + (-6)$ **−11**

Match the name of the property with the equation that best illustrates it. (LESSON 2.5)

46. $7 + (11 + 4) = (7 + 11) + 4$ **b**   a. Commutative Property for Addition
47. $4 + 9 = 9 + 4$ **a**   b. Associative Property for Addition
48. $12 + 18 = 6(2 + 3)$ **c**   c. Distributive Property

Solve each equation. (LESSON 3.1)

49. $x + 1.4 = -5.6$ **−7**   **50.** $6 = c - (-6)$ **0**   **51.** $m + (-47) = 31$ **78**

Find the slope of the line that contains each pair of points. (LESSON 5.2)

52. $M(2, 5), N(-6, 7)$ $-\frac{1}{4}$ **53.** $M(7, 4), N(-12, 3)$ $\frac{1}{19}$

Solve each equation by factoring. (LESSON 9.8)

54. $x^2 + x - 12 = 0$ **55.** $x^2 - 2x - 35 = 0$ **56.** $x^2 - 49 = 0$
   −4, 3   **7, −5**   **−7, 7**

Use the quadratic formula to find exact answers. (LESSON 10.5)

57. $2x^2 + 12x + 14 = 0$ **58.** $3x^2 - 3x - 36 = 0$
   $-3 \pm \sqrt{2}$ **4, −3**

Look Beyond

CONNECTION

59. PROBABILITY If you flipped a coin 100 times, how many times would you expect it to land heads up? **about 50**

60. If you rolled a 6-sided number cube 100 times, how many times would you expect a 3 or a 5 to appear? **about 33**

---

39. Regular Season

Alesia	51
Jennel	59
Betsy	56
Katie	33
Nancy	46

Playoff Series

Alesia	29
Jennel	11
Betsy	14
Katie	18
Nancy	10

40. Total Points

Alesia	80
Jennel	70
Betsy	70
Katie	51
Nancy	56

Student Technology Guide

12.8 Introduction to Matrices

Many graphics calculators allow you to enter matrices and perform operations on them. To enter a matrix into a graphics calculator, you need to name the matrix, state its dimensions, and then enter its entries in order row by row.

To enter $\begin{bmatrix} -2 & 0 \\ 3 & 5 \end{bmatrix}$, follow the key sequence shown below.

[MATRX] [EDIT] [1:A] 2 [ENTER] 2 [ENTER]
(-) 2 [ENTER] 0 [ENTER] 3 [ENTER] 5 [ENTER]

Let $A = \begin{bmatrix} -2 & 0 \\ 3 & 5 \end{bmatrix}$ and $B = \begin{bmatrix} 0 & 4 \\ -2 & 1 \end{bmatrix}$.

Example: Find $A + B$ and $A - B$.
• Enter matrices A and B as explained above.
• Press [MATRX] [NAMES] [1:A] + [MATRX] [NAMES] [2:B] [ENTER] [ENTER].
• Press [2nd] [ENTER], change + to −, then press [ENTER].

Example: Find $5A$ and AB.
• Enter matrix A as explained above.
• Press 5 [MATRX] [NAMES] [1:A] [ENTER] [ENTER].
• Enter matrices A and B as explained above.
• Press [MATRX] [NAMES] [1:A] × [MATRX] [NAMES] [2:B] [ENTER] [ENTER].

Note: Check the dimensions of the matrices you want to multiply to be sure that the product exists. If the product does not exist, you will get an error message.

Perform the following matrix operations. If a solution is not possible, explain why.

1. $\begin{bmatrix} 900 & 1200 \\ 1650 & 400 \\ 600 & 100 \end{bmatrix} + \begin{bmatrix} 1100 & 800 \\ 550 & 375 \\ 450 & 900 \end{bmatrix}$ 2. $\begin{bmatrix} 120 & -110 & 95 \\ 108 & 72 & -76 \\ 45 & -32 & -12 \end{bmatrix} - \begin{bmatrix} 90 & 20 & -38 \\ -92 & 48 & 0 \\ -24 & 9 & -21 \end{bmatrix}$

See answer section for solutions.   See answer section for solutions.

3. $\begin{bmatrix} -3 & 3.5 \\ 0 & 12 \end{bmatrix} - \begin{bmatrix} -1.2 & 0 \\ 5 & 3.5 \end{bmatrix}$ 4. $-\frac{3}{4} \begin{bmatrix} 4 & -10 & 5 \\ 12.4 & 0 & 1 \end{bmatrix}$ 5. $\begin{bmatrix} 2.5 & 6 \\ -6 & 11 \end{bmatrix} \begin{bmatrix} 6 & -6 & 5 \\ -5 & 1 & -1 \end{bmatrix}$

See answer section for solutions.   See answer section for solutions.   See answer section for solutions.

Project

Focus

Hypsography is the scientific study of Earth's topologic configuration above sea level, concentrating on the measurement and mapping of land elevations. An instrument used to measure these elevations is called a *hypsometer*. In this Chapter Project, students use a washer, string, and protractor to build a simple hypsometer. Then they measure angles of elevation and depression and use their results to make indirect measurements of height.

Motivate

Work with students to brainstorm a list of nearby objects whose heights are already known or can be measured directly with some ease. Then make a second list of objects whose heights are not known and cannot be measured directly. Ask students if they know of any techniques for measuring unknown heights. Tell them that in this Chapter Project, they will learn how to make a simple device that can be used to measure heights indirectly by applying the mathematics they learned in this chapter.

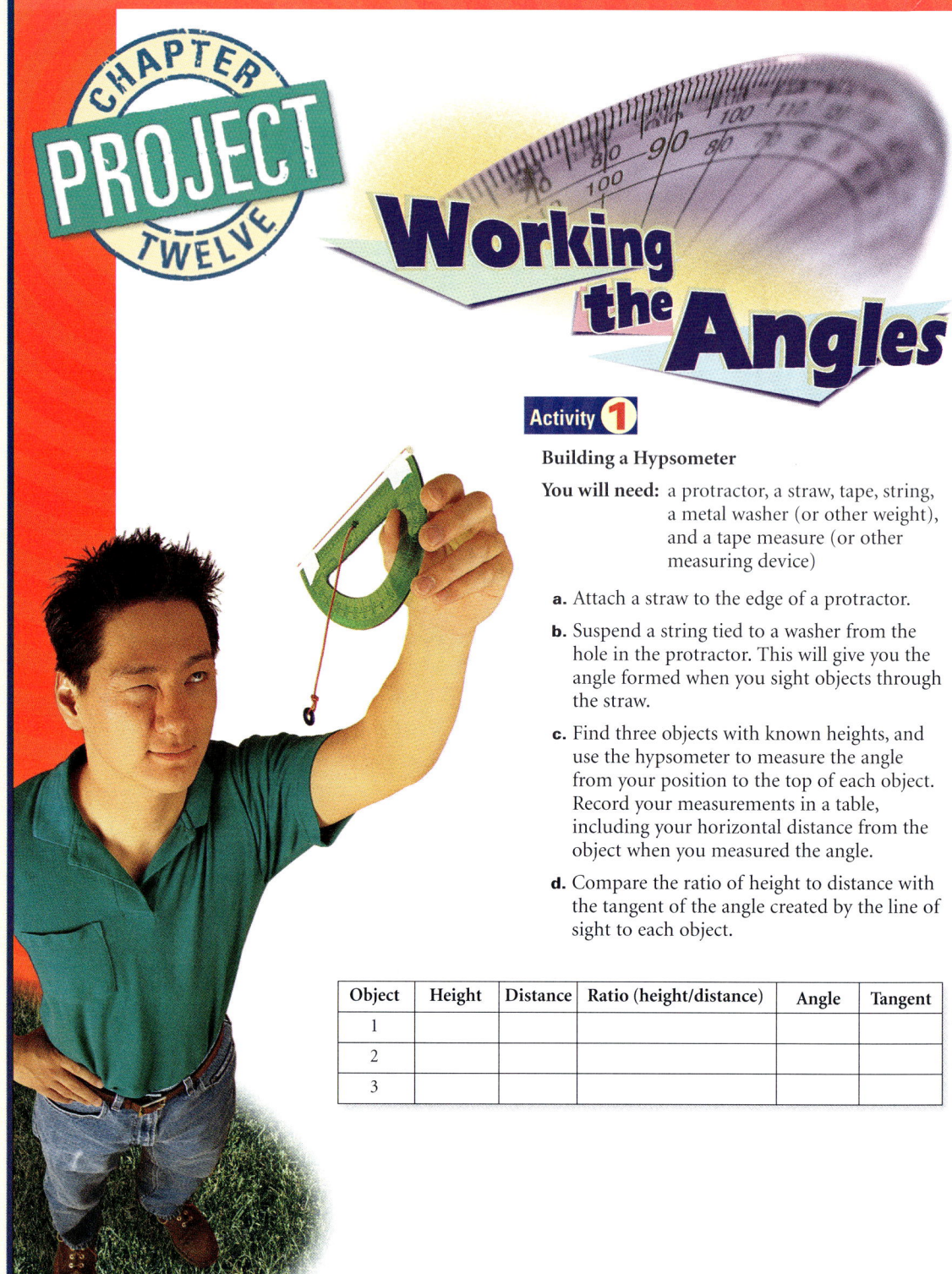

CHAPTER PROJECT TWELVE

Working the Angles

Activity 1

Building a Hypsometer

You will need: a protractor, a straw, tape, string, a metal washer (or other weight), and a tape measure (or other measuring device)

a. Attach a straw to the edge of a protractor.

b. Suspend a string tied to a washer from the hole in the protractor. This will give you the angle formed when you sight objects through the straw.

c. Find three objects with known heights, and use the hypsometer to measure the angle from your position to the top of each object. Record your measurements in a table, including your horizontal distance from the object when you measured the angle.

d. Compare the ratio of height to distance with the tangent of the angle created by the line of sight to each object.

Object	Height	Distance	Ratio (height/distance)	Angle	Tangent
1					
2					
3					

Activity 1

a–b. Students should gather the materials listed and construct a hypsometer as described.

c. Students should choose three objects and use their hyposmeters to take measurements. Data should be collected in the table given.

d. Students should observe that the ratio height to distance is equal to the tangent of the angle.

Activity 2

Measuring Up

a. Use your hypsometer to measure the *angle of elevation* of three objects with unknown heights.

b. Measure the horizontal distance between the hypsometer and the objects.

c. Record your measurements in a table. Use the $\tan^{-1}$ function on your calculator to compute the height.

Object	Distance	Angle	Height
1	?	?	?
2	?	?	?
3	?	?	?

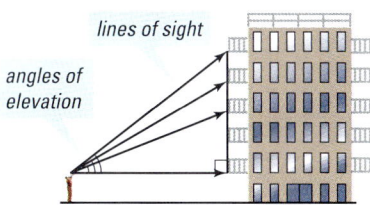

Activity 3

Measuring Down

a. Use your hypsometer to measure the angle between the building and your line of sight from three unknown heights to a specific object on the ground. Call these angles A, A', and A''. (The unknown heights can be different floors of a building.)

b. Measure the horizontal distance on the ground between the object and a point beneath the unknown heights.

c. Use the $\tan^{-1}$ function on your calculator and the formula $\tan A = \frac{d}{h}$, where d is distance and h is height, to compute the heights.

Object	Distance	Angle	Height
1	?	?	?
2	?	?	?
3	?	?	?

Cooperative Learning

Have students work in groups of three. In **Activity 1**, the group can work together to make the hypsometer and prepare a table for recording results. Then each student can use the hypsometer to take one of the three sets of measurements. Each student should record his or her results in the table. The group can then work together to complete the table and discuss the results.

Activities 2 and **3** can be conducted in a similar fashion, with each group member taking and recording one set of measurements. The group can then complete the table together.

Discuss

After all groups have completed the three activities, bring the class together for a discussion of the results. If two or more groups calculated the height of the same object, have them compare results to see if they agree. If their results differ, discuss possible reasons for the differences. Have students assess the accuracy of their hypsometers. Ask them to suggest real-world situations in which it might be useful to have a hypsometer available for measuring a height.

Activity 2

a–b. Students should choose 3 objects above them and use their hypsometers to take measurements.

c. Students should calculate the height of the three objects by using the table given.

Activity 3

a–b. Students should choose an object below them and use their hypsometer to take measurements from 3 different heights.

c. Students should calculate the height of the viewing position by using the inverse tangent function and record data in the table given.

Chapter Review and Assessment

VOCABULARY

addition of matrices 630	identity matrix	rationalizing the
angle of depression 617	for multiplication 632	denominator 580
center 606	legs 591	right angle 591
circle 606	main diagonal 632	right triangle 591
Converse of the	matrix 628	scalar 631
Pythagorean Theorem 600	matrix address 629	scalar multiplication 631
cosine 622	matrix equality 629	simplest radical form 578
directrix 609	midpoint formula 601	sine 622
distance formula 599	midsegment 608	square matrix 629
Division Property	Multiplication Property	square root 577
of Square Roots 580	of Square Roots 578	square-root function 583
entry 629	product matrix 631	tangent 613
equation of a circle 607	Pythagorean Theorem 591	tangent ratio 614
extraneous solution 586	Pythagorean triple 597	Triangle Midpoint
focus 609	radicand 577	Theorem 608
hypotenuse 591	radius 606	trigonometry 614

Key Skills & Exercises

LESSON 12.1

Key Skills

Simplify radical expressions, and estimate square roots.

To express $\sqrt{12} + 2\sqrt{3}$ in simplest radical form, first use the Multiplication Property of Square Roots.

$$\sqrt{12} + 5\sqrt{3} = \sqrt{4 \cdot 3} + 5\sqrt{3}$$
$$= \sqrt{4} \cdot \sqrt{3} + 5\sqrt{3}$$
$$= 2\sqrt{3} + 5\sqrt{3} = 7\sqrt{3}$$
$$\approx 12.12$$

Exercises

Find the value of each square root. If the square root is irrational, find the value to the nearest hundredth.

1. $\sqrt{20}$ **4.47**   2. $\sqrt{\frac{9}{16}}$ **$\frac{3}{4}$, or 0.75**

Simplify each radical expression.

3. $\sqrt{a^2b^7}$ **$|a|b^3\sqrt{b}$**   4. $\sqrt{2} + 3\sqrt{7} - 3\sqrt{2}$
5. $(2\sqrt{3})^2$ **12**   6. $\sqrt{3}(2 - \sqrt{12})$
7. $(\sqrt{5} - 6)(\sqrt{5} + 6)$ **−31**   8. $(\sqrt{17} - 8)(\sqrt{17} + 4)$

4. $-2\sqrt{2} + 3\sqrt{7}$ 6. $2\sqrt{3} - 6$ 8. $-15 - 4\sqrt{17}$

LESSON 12.2

Key Skills

Solve radical equations, and use radicals to solve equations.

To solve $\sqrt{x + 2} = 5$, square both sides of the equation. $(\sqrt{x + 2})^2 = 5^2$

$$x + 2 = 25, x = 23$$

To solve $2x^2 = 36$, use the definition of square root. $2x^2 = 36, x^2 = 18$

$$x = \pm\sqrt{18} = \pm\sqrt{9 \cdot 2} = \pm 3\sqrt{2}$$

Exercises

Solve each equation. Give exact answers. Remember to check your solutions.

9. $\sqrt{x - 7} = 2$ **11**   10. $\sqrt{3x + 4} = 1$ **−1**
11. $\sqrt{x^2 + 6x - 1} = x + 4$ 12. $\sqrt{x^2 - 2x + 1} = x - 5$
13. $x^2 = 40$ **$\pm 2\sqrt{10}$**   14. $2x^2 - 32 = 0$ **± 4**
15. $5x^2 + 14 = 139$ **± 5**   16. $x = \sqrt{x + 12}$ **4**
11. no solution   12. no solution

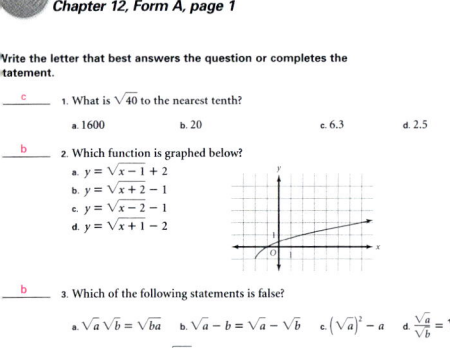

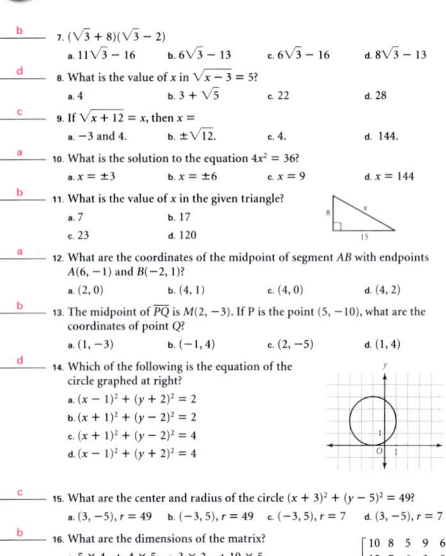

638 CHAPTER 12 REVIEW

LESSON 12.3

Key Skills

Use the Pythagorean Theorem to find the length of a missing side of a right triangle.

The Pythagorean Theorem, $a^2 + b^2 = c^2$, can be used to find the length of the hypotenuse of the right triangle.

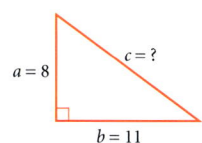

$a^2 + b^2 = c^2$
$8^2 + 11^2 = c^2$
$64 + 121 = c^2$
$185 = c^2$
$c = \pm\sqrt{185} \approx \pm 13.60$

Since the length of a segment must be positive, the length of the hypotenuse is approximately 13.6 meters.

Exercises

Use the Pythagorean Theorem to find the length of the missing side of each right triangle.

17.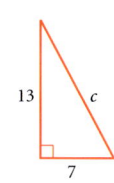
$c = \sqrt{218}$, or ≈ 14.76

18.
$a = 12\sqrt{2}$, or ≈ 16.97

19. Find the hypotenuse of a right triangle whose legs have lengths of 12 and 5. **13**

20. Can a triangle with side lengths of 2, $\sqrt{3}$, and $\sqrt{5}$ form a right triangle? Explain. **no**

21. Can a triangle with side lengths of $\sqrt{5}$, 2, and 3 form a right triangle? Explain. **yes**

LESSON 12.4

Key Skills

Use the distance formula to find the distance between two points, and use the midpoint formula to find the midpoint of a segment.

To find the distance between (2, 3) and (−4, −1), use the distance formula.

$d = \sqrt{(x_2 - x_1)^2 + (y_2 - y_1)^2}$
$= \sqrt{(-4 - 2)^2 + (-1 - 3)^2}$
$= \sqrt{36 + 16} = \sqrt{52} = 2\sqrt{13} \approx 7.21$

To find the midpoint of a segment whose endpoints are (−2, 6) and (3, 4), use the midpoint formula, $\left(\dfrac{x_1 + x_2}{2}, \dfrac{y_1 + y_2}{2}\right)$.

$\left(\dfrac{x_1 + x_2}{2}, \dfrac{y_1 + y_2}{2}\right) \rightarrow \left(\dfrac{-2 + 3}{2}, \dfrac{6 + 4}{2}\right) \rightarrow \left(\dfrac{1}{2}, 5\right)$

Notice that the midpoint formula simply finds the average of the x-coordinates and the average of the y-coordinates.

Exercises

Find the distance between the two points.

22. $A(0, 3)$ and $B(2, 8)$ ≈ 5.39
23. $X(-1, 5)$ and $Y(3, -8)$ ≈ 13.60
24. $M(5, 2)$ and $N(3, 6)$ ≈ 4.47
25. $S(6, 4)$ and $T(9, 0)$ **5**

Find the midpoint of each segment with the given endpoints.

26. $M(5, 2)$ and $N(3, 6)$ **(4, 4)**
27. $P(-3, 4)$ and $Q(-7, -1)$ **(−5, 1.5)**
28. $R(8, -2)$ and $S(-1, 4)$ **(3.5, 1)**
29. $M(-4, 3)$ and $N(5, -2)$ $\left(\dfrac{1}{2}, \dfrac{1}{2}\right)$
30. $R(5, -7)$ and $S(-8, -12)$ $\left(-\dfrac{3}{2}, -\dfrac{19}{2}\right)$
31. $Q(3, 7)$ and $R(-8, -8)$ $\left(-\dfrac{5}{2}, -\dfrac{1}{2}\right)$
32. $A(7, -4)$ and $B(8, 4)$ $\left(\dfrac{15}{2}, 0\right)$
33. $A(a, b)$ and $B(3a, 2b)$ $\left(2a, \dfrac{3}{2}b\right)$

20. A triangle with side lengths of 2, $\sqrt{3}$, and $\sqrt{5}$ cannot be a right triangle because it would contradict the Pythagorean Theorem: $(2)^2 + \left(\sqrt{3}\right)^2 \neq \left(\sqrt{5}\right)^2$.

21. A triangle with side lengths of $\sqrt{5}$, 2, and 3 is a right triangle because it satisfies the Pythagorean Theorem: $(2)^2 + \left(\sqrt{5}\right)^2 = (3)^2$.

Chapter Test, Form B

NAME _____ CLASS _____ DATE _____

Chapter Assessment
Chapter 12, Form B, page 1

1. For what values of x is $\sqrt{-x + 5}$ defined? __ $x \leq 5$ __

Find each square root. If the square root is irrational, find the value to the nearest hundredth.

2. $\sqrt{81}$ __9__   3. $-\sqrt{225}$ __−15__   4. $\sqrt{\dfrac{4}{16}}$ __0.5__   5. $\sqrt{32}$ __5.66__

6. Find the length of the side of a square whose area is 49 square centimeters. __7 centimeters__

Graph each function on the grid provided. Give the domain and range.

7. $y = 2\sqrt{x - 1}$

8. $y = \sqrt{x - 3}$

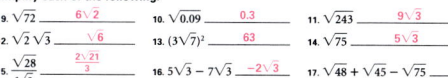

domain: $x \geq 1$; range: $y \geq 0$ domain: $x \geq 3$; range: $y \geq 0$

Simplify each of the following:

9. $\sqrt{72}$ __$6\sqrt{2}$__   10. $\sqrt{0.09}$ __0.3__   11. $\sqrt{243}$ __$9\sqrt{3}$__
12. $\sqrt{2}\sqrt{3}$ __$\sqrt{6}$__   13. $(3\sqrt{7})^2$ __63__   14. $\sqrt{75}$ __$5\sqrt{3}$__
15. $\dfrac{\sqrt{28}}{\sqrt{3}}$ __$\dfrac{2\sqrt{21}}{3}$__   16. $5\sqrt{3} - 7\sqrt{3}$ __$-2\sqrt{3}$__   17. $\sqrt{48} + \sqrt{45} - \sqrt{75}$ __$3\sqrt{5} - \sqrt{3}$__

Solve each equation algebraically.

18. $\sqrt{x - 3} = 7$ __$x = 52$__   19. $\sqrt{3x} = 9$ __$x = 27$__   20. $\sqrt{2x + 5} = 3$ __$x = 2$__

Use matrix A for Exercises 21 and 22.

21. Give the entry at A_{24}. __9__ $A = \begin{bmatrix} 12 & 7 & -1 & -2 \\ 5 & 0 & 3 & 9 \end{bmatrix}$

22. Give the dimensions of matrix A. __2×4__

NAME _____ CLASS _____ DATE _____

Chapter Assessment
Chapter 12, Form B, page 2

Solve for x.

23. $x = 20$

24. $x = 15$

Find the distance between each pair of points.

25. $A(-2, 3), B(6, -3)$ __10__   26. $P(0, -2), Q(3, 4)$ __$3\sqrt{5}$__

27. Given $P(-1, 8), Q(-1, 5), R(3, 5)$, determine whether $\triangle PQR$ is a scalene, isosceles, equilateral, or right triangle. __right triangle__

The midpoint of $\overline{CD}$ is M. Calculate the missing coordinates.

28. $C(3, 6), D(7, 2), M(\underline{5}, \underline{4})$ 29. $C(-3, 1), D(5, -5), M(\underline{1}, \underline{-2})$
30. $C(4, 1), D(\underline{2}, \underline{3}), M(3, 2)$ 31. $C(\underline{8}, \underline{-2}), D(2, 4), M(5, 1)$

From each equation of a circle, give the center and radius.

32. $x^2 + y^2 = 5$ __center (0, 0); radius $\sqrt{5}$__   33. $(x - 3)^2 + (y + 1)^2 = 49$ __center (3, −1); radius 7__

34. Graph $(x - 2)^2 + y^2 = 9$ below.

Use the information in the given triangle ABC to find the indicated length or angle measure. Give lengths to the nearest hundredth and degree measures to the nearest whole number.

35.

36.

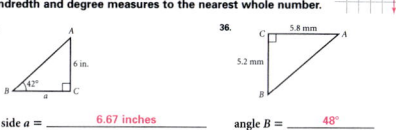

side $a = $ __6.67 inches__   angle $B = $ __48°__

LESSON 12.5
Key Skills

Find the equation of a circle whose center and radius are given.

To find the equation of a circle with its center at $(2, -3)$ and a radius of 5, use the equation of a circle, $(x - h)^2 + (y - k)^2 = r^2$, where (h, k) is the center and r is the radius. Thus, the equation is $(x - 2)^2 + (y + 3)^2 = 25$.

Exercises

Find the equation of each circle with the given center and radius.

34. center at $(0, 0)$ and radius of 2 $x^2 + y^2 = 4$

35. center at $(-2, 5)$ and radius of 5 $(x + 2)^2 + (y - 5)^2 = 25$

36. 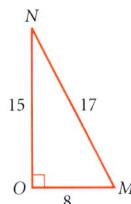 $(x - 3)^2 + (y - 7)^2 = 16$

37. 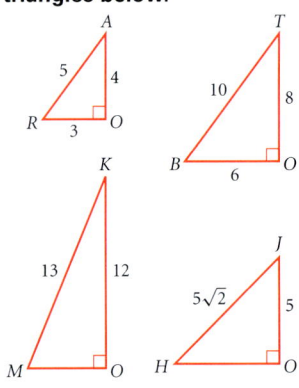 $(x + 3)^2 + (y + 2)^2 = 36$

From each equation of a circle, give the center and radius.

38. $(x - 2)^2 + (y - 3)^2 = 36$ $(2, 3); 6$

39. $(x + 0.25)^2 + y^2 = 6.25$ $(-0.25, 0); 2.5$

LESSON 12.6
Key Skills

Find the tangent of an acute angle of a right triangle.

The tangent of an acute angle in a right triangle is the ratio of the length of its opposite side to the length of its adjacent side.

The tangent of angle M is $\frac{opp}{adj} = \frac{15}{8}$. The tangent of angle N is $\frac{opp}{adj} = \frac{8}{15}$.

Exercises

Find the indicated tangent ratio for the right triangles below.

40. $\tan M$ $\frac{12}{5}$
41. $\tan K$ $\frac{5}{12}$
42. $\tan A$ $\frac{3}{4}$
43. $\tan B$ $\frac{8}{6}$, or $\frac{4}{3}$
44. $\tan R$ $\frac{4}{3}$
45. $\tan T$ $\frac{6}{8}$, or $\frac{3}{4}$
46. $\tan H$ $\frac{5}{5}$, or 1
47. $\tan J$ $\frac{5}{5}$, or 1

LESSON 12.7

Key Skills

Find the sine and cosine of an acute angle of a right triangle.

The sine of an angle in a right triangle is the ratio of the length of the side opposite the angle to the length of the hypotenuse of the triangle.

The cosine of an angle in a right triangle is the ratio of the length of the side adjacent to the angle to the length of the hypotenuse of the triangle.

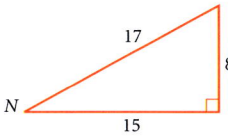

The sine of angle N is $\sin N = \frac{opp}{hyp} = \frac{8}{17}$.

The cosine of angle N is $\cos N = \frac{adj}{hyp} = \frac{15}{17}$.

If you know the sine or cosine of an angle, the degree measure can be found by using the $\sin^{-1}$ or $\cos^{-1}$ function on your calculator.

The measure of angle N can be found by using either $\sin^{-1}$ or $\cos^{-1}$.

$\sin^{-1}(N) = \sin^{-1}\left(\frac{8}{17}\right) = 28.07°$

$\cos^{-1}(N) = \cos^{-1} = \cos^{-1}\left(\frac{15}{17}\right) = 28.07°$

Exercises

Find the indicated sine or cosine to the nearest hundredth.

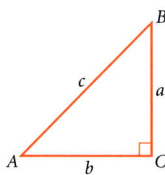

48. $a = 5, c = 9, \cos\angle B = ?$ ≈ **0.56**
49. $b = 6, c = 12, \cos\angle A = ?$ **0.5**
50. $a = 1, c = 2, \sin\angle A = ?$ **0.5**
51. $b = 8, c = 15, \sin\angle B = ?$ ≈ **0.53**
52. $a = 5, c = 14, \cos\angle B = ?$ ≈ **0.36**
53. $a = 6, c = 10, \cos\angle B = ?$ **0.6**
54. $b = 3, c = 18, \sin\angle B = ?$ ≈ **0.17**
55. $a = 8, c = 15, \sin\angle A = ?$ ≈ **0.53**

Find the degree measure of each angle to the nearest hundredth.

56. $b = 7, c = 24, m\angle A = ?$ ≈ **73.04°**
57. $a = 14, c = 27, m\angle B = ?$ ≈ **58.77°**
58. $a = 3, c = 9, m\angle A = ?$ ≈ **19.47°**
59. $a = 9, c = 20, m\angle A = ?$ ≈ **26.74°**

LESSON 12.8

Key Skills

Add and subtract matrices.

In order to add or subtract matrices, the matrices must have the same dimensions. Then the corresponding entries from each matrix are added or subtracted.

$\begin{bmatrix} 5 & 7 \\ -3 & -9 \end{bmatrix} - \begin{bmatrix} 9 & 3 \\ -4 & -8 \end{bmatrix} = \begin{bmatrix} 5-9 & 7-3 \\ -3-(-4) & -9-(-8) \end{bmatrix}$

$= \begin{bmatrix} -4 & 4 \\ 1 & -1 \end{bmatrix}$

Exercises

Perform the indicated matrix operations.

60. $\begin{bmatrix} -2 & 8 \\ -12 & 4 \end{bmatrix} + \begin{bmatrix} 4 & 6 \\ -1 & -7 \end{bmatrix}$ $\begin{bmatrix} 2 & 14 \\ -13 & -3 \end{bmatrix}$

61. $\begin{bmatrix} 2 & 14 \\ 3 & -10 \end{bmatrix} + \begin{bmatrix} -4 & 1 \\ -3 & 8 \end{bmatrix}$ $\begin{bmatrix} -2 & 15 \\ 0 & -2 \end{bmatrix}$

62. $\begin{bmatrix} 5 & -1 \\ 3 & 8 \end{bmatrix} - \begin{bmatrix} 2 & 4 \\ 7 & -1 \end{bmatrix}$ $\begin{bmatrix} 3 & -5 \\ -4 & 9 \end{bmatrix}$

63. $\begin{bmatrix} 11 & 14 \\ 2 & 16 \end{bmatrix} - \begin{bmatrix} 4 & 3 \\ -2 & -14 \end{bmatrix}$ $\begin{bmatrix} 7 & 11 \\ 4 & 30 \end{bmatrix}$

64. $\begin{bmatrix} 253 & 786 \\ 212 & 1452 \end{bmatrix} - \begin{bmatrix} 221 & 753 \\ 819 & 1221 \end{bmatrix}$ $\begin{bmatrix} 32 & 33 \\ -607 & 231 \end{bmatrix}$

Multiply a matrix by a scalar and by another matrix.

To multiply a matrix by a scalar, multiply each element in the matrix by the scalar.

$$-7\begin{bmatrix} 4 & -8 \\ -4 & 7 \end{bmatrix} = \begin{bmatrix} -7(4) & -7(-8) \\ -7(-4) & -7(7) \end{bmatrix}$$
$$= \begin{bmatrix} -28 & 64 \\ 28 & -49 \end{bmatrix}$$

To multiply a matrix by another matrix, the number of columns in the matrix on the left must match the number of rows in the matrix on the right. To multiply, find the product of each entry in a row of the matrix on the left with the corresponding column entry in the matrix on the right. Then add the products.

$$\begin{bmatrix} -6 & 4 \\ -3 & 8 \end{bmatrix}\begin{bmatrix} 8 & -2 \\ 4 & -1 \end{bmatrix}$$
$$= \begin{bmatrix} -6(8)+4(4) & -6(-2)+4(-1) \\ -3(8)+8(4) & (-3)(-2)+8(-1) \end{bmatrix}$$
$$= \begin{bmatrix} -32 & 8 \\ 8 & -2 \end{bmatrix}$$

Perform the indicated matrix operations.

65. $4\begin{bmatrix} -3 & 12 \\ 24 & -8 \end{bmatrix}$ $\begin{bmatrix} -12 & 48 \\ 96 & -32 \end{bmatrix}$

66. $-12\begin{bmatrix} 4 & -12 \\ -14 & -11 \end{bmatrix}$ $\begin{bmatrix} -48 & 144 \\ 168 & 132 \end{bmatrix}$

67. $23\begin{bmatrix} 14 & 2 \\ 13 & -6 \end{bmatrix}$ $\begin{bmatrix} 322 & 46 \\ 299 & -138 \end{bmatrix}$

68. $\begin{bmatrix} 4 & 3 \\ 8 & 7 \end{bmatrix}\begin{bmatrix} 6 & 3 \\ 8 & 9 \end{bmatrix}$ $\begin{bmatrix} 48 & 39 \\ 104 & 87 \end{bmatrix}$

69. $\begin{bmatrix} -3 & 8 \\ -14 & 7 \end{bmatrix}\begin{bmatrix} -2 & -8 \\ 3 & -6 \end{bmatrix}$ $\begin{bmatrix} 30 & -24 \\ 49 & 70 \end{bmatrix}$

70. $\begin{bmatrix} -6 & -2 \\ 4 & 8 \end{bmatrix}\begin{bmatrix} -3 & -8 \\ 9 & 10 \end{bmatrix}$ $\begin{bmatrix} 0 & 28 \\ 60 & 48 \end{bmatrix}$

71. $\begin{bmatrix} 1 & 0 \\ 0 & 1 \end{bmatrix}\begin{bmatrix} 8 & 9 \\ 12 & 356 \end{bmatrix}$ $\begin{bmatrix} 8 & 9 \\ 12 & 356 \end{bmatrix}$

72. $\begin{bmatrix} 1 & 0 \\ 0 & 1 \end{bmatrix}\begin{bmatrix} 856 & 231 \\ -876 & 329 \end{bmatrix}$ $\begin{bmatrix} 856 & 231 \\ -876 & 329 \end{bmatrix}$

73. $\begin{bmatrix} 845 & -765 \\ -156 & -943 \end{bmatrix}\begin{bmatrix} 1 & 0 \\ 0 & 1 \end{bmatrix}$ $\begin{bmatrix} 845 & -765 \\ -156 & -943 \end{bmatrix}$

Applications

74. **CONSTRUCTION** A ladder is leaning against a house. It forms an angle of 55° with the house. If the ladder touches the house at a point 12 feet from the ground, how far is the bottom of the ladder from the bottom of the house? ≈ 17.1 ft

75. **SPORTS** At the end of a ski lift, riders are 120 feet above their starting point. The ski lift travels a horizontal distance of 500 feet.
    a. How long is the cable that carries the ski lift? ≈ 514.2 ft
    b. What angle does the cable make with the ground at the bottom of the lift? ≈ 13.5°

76. **TRANSPORTATION** Suppose you are driving up a hill whose total height is 2000 feet. Your odometer shows that you drove a total of 6 miles to reach the top. What horizontal distance have you traveled? (Hint: 1 mile = 5280 feet) ≈ 31,617 ft, or 5.99 mi

Alternative Assessment

Performance Assessment

1. A teacher received a student paper showing the following steps:
   $\sqrt{18} = \sqrt{9+9} = \sqrt{9} + \sqrt{9} = 3 + 3 = 6$
   a. Explain why the answer is not correct.
   b. Explain how to find the correct answer.

2. On a coordinate grid, a ranger tower is located at the origin, (0, 0). Two fires are spotted at $A(-1, 2)$ and $B(3, 4)$.
   a. Find the distance between the tower and each fire.
   b. Find the midpoint between the two fires.

3. Explain how to use the distance formula to find the equation of a circle with its center at (0, 0) and a radius of 6.

Portfolio Projects

1. The time it takes a pendulum to make a complete swing can be represented by a radical function.
   a. Research the origins of the pendulum clock.
   b. What is the relationship between the time it takes a pendulum to complete one full swing and its length?
   c. Why is the relationship you found in part **b** a radical function?

2. During the floods in the Nile Valley, property lines were often washed out. Find out how Egyptian farmers used "rope stretching" to re-establish their property lines. Write a paragraph relating this method to the Pythagorean Theorem.

3. Find out how construction workers use a carpenter's square when they build a house. Write an instruction booklet that explains how to use a carpenter's square to draw right angles.

Peer Assessment

Complete this activity with a partner. Begin by writing a radical expression that can be simplified. After simplifying the radical, your partner should write a radical expression for you to simplify.

Check your partner's work, correct any errors, and then simplify the new radical expression. Continue until you have each simplified 20 expressions. Be sure to use expressions involving multiplication and division inside the radical, multiplication and division of different radicals, powers of powers, and products of powers during this activity.

internetconnect

The HRW Web site contains many resources to reinforce and expand your knowledge of radical functions and coordinate geometry. The Web site also provides Internet links to other sites where you can find information and real-world data for use in research projects, reports, and activities that involve radical functions and coordinate geometry. Visit the HRW Web site at **go.hrw.com**, and enter the keyword **MA1 CH12**.

Alternative Assessment

Performance Assessment

1. a. The sum of square roots is not equal to the square root of a sum.
   b. Factor out a perfect square.
   c. $3\sqrt{2}$

2. a. $OA \approx 2.2$
      $OB = 5$
   b. $(\overline{x}, \overline{y}) = (1, 3)$

3. A circle of radius 6 is the set of points 6 units from its center, (0, 0).
   $\sqrt{(x-0)^2 + (y-0)^2} = 6$

Portfolio Projects

1. a. Students should research the history of the pendulum.
   b. $P = 2\pi\sqrt{\dfrac{l}{g}}$, where P = period, l = length, and g = gravity.
   c. It has a square root in it.

2. Students should research the Egyptian rope stretching method, and relate it to the Pythagorean Theorem.

3. Students should investigate the use of a carpenter's square to write their instruction booklets.

Cumulative Assessment

College Entrance Exam Practice

CHAPTERS 1–12 CUMULATIVE ASSESSMENT
College Entrance Exam Practice

Multiple-Choice and Quantitive-Comparison Samples

The first half of the Cumulative Assessment contains two types of items found on standardized tests—multiple-choice questions and quantitative-comparison questions. Quantitative-comparison items emphasize the concepts of equality, inequality, and estimation.

Free Response Grid Samples

The second half of the Cumulative Assessment is a free-response section. This part of the Cumulative Assessment requires student-produced response items like those commonly found on college entrance exams. These questions require the use of machine-scored answer grids. You may wish to have students practice answering these items in preparation for standardized tests.

QUANTITATIVE COMPARISON For Questions 1–5 write:
A if the quantity in Column A is greater than the quantity in Column B;
B if the quantity in Column B is greater than the quantity in Column A;
C if the quantities are equal; or
D if the relationship cannot be determined from the information given.

	Column A	Column B	Answers
1. C	The mean of 16, 26, 21, 23, 19	The median of 16, 26, 21, 23, 19	Ⓐ Ⓑ Ⓒ Ⓓ [Lesson 4.4]
2. A	The solution to $\frac{6}{n} = \frac{9}{15}$	The solution to $\frac{n}{15} = \frac{9}{25}$	Ⓐ Ⓑ Ⓒ Ⓓ [Lesson 4.1]
3. B	The solution to $5(x+4) = 25$	The solution to $4(x-3) = 16$	Ⓐ Ⓑ Ⓒ Ⓓ [Lesson 3.5]
4. B	$(2+\sqrt{3})(2-\sqrt{3})$	8	Ⓐ Ⓑ Ⓒ Ⓓ [Lesson 12.1]
5. A	$\frac{8b-2}{-6}$, where $b=4$	$3(b-6)$, where $b=4$	Ⓐ Ⓑ Ⓒ Ⓓ [Lesson 2.4]

6. If y varies directly as x and y is 3 when x is 8, what is the constant of variation? **(LESSON 5.3)** d
   a. $\frac{8}{3}$ b. 3
   c. 8 d. $\frac{3}{8}$

7. Let $f(x) = 2^x$. Find $f(x)$ when $x = 3$. **(LESSON 8.6)** d
   a. 6 b. $\frac{1}{6}$
   c. 4 d. 8

8. Simplify: $(x-y) - (y+x)$. **(LESSON 2.3)** c
   a. $2x + 2y$ b. $2x - 2y$
   c. $-2y$ d. $2y - 2x$

9. Solve: $\begin{cases} y = x + 3 \\ 2x - 3y = -12 \end{cases}$ **(LESSON 7.2)** c
   a. $(-3, -6)$ b. $(3, -6)$
   c. $(3, 6)$ d. $(-3, 6)$

10. Solve: $|x + 5| < 5$. **(LESSON 6.5)** a
    a. $-10 < x < 0$ b. $-5 < x < 5$
    c. $-10 < x < 5$ d. $-10 < x < 10$

11. Find 30% of 50. **(LESSON 4.2)** d
    a. 1.5 b. 150
    c. 1500 d. 15

12. Find the slope of the line connecting the points $M(3, -5)$ and $N(-6, 14)$. **(LESSON 5.2)** c
    a. $\frac{7}{3}$ b. $\frac{19}{9}$
    c. $-\frac{19}{9}$ d. $\frac{9}{19}$

13. Find the equation of the line passing through the points $(3, 4)$ and $(-1, -2)$. **(LESSON 5.5)** d
    a. $y = 3x - 1$ b. $y = \frac{2}{3}x - 2$
    c. $y = \frac{3}{2}x - 1$ d. $y = \frac{3}{2}x - \frac{1}{2}$

14. Multiply: $(2x + 4)(2x - 4)$. **(LESSON 9.3)** a
    a. $4x^2 - 16$ b. $x^2 - 4$
    c. $2x^2 - 16x - 16$ d. $2x^2 + 16x - 16$

15. What is the slope of a line perpendicular to $2x + 4y = 5$? **(LESSON 5.6)** a
    a. 2   b. $\frac{1}{2}$
    c. -2 d. $-\frac{1}{2}$

16. Find the distance between $A(0, 1)$ and $B(4, 5)$. **(LESSON 12.4)** d
    a. 16   b. $8\sqrt{2}$
    c. 2   d. $4\sqrt{2}$

17. Solve: $5x + 8 \leq 3x$. **(LESSON 6.2)** a
    a. $x \leq -4$ b. $x \leq 4$
    c. $x \geq 4$ d. $x \geq -4$

18. Factor: $2x^2 + 12x + 18$. **(LESSON 9.6)** d
    a. $(2x + 3)^2$ b. $(2x + 3)(2x - 3)$
    c. $(x + 3)^2$ d. $2(x + 3)^2$

19. Let $P = \begin{bmatrix} 3 & 0 \\ -5 & 1 \end{bmatrix}$ and $Q = \begin{bmatrix} 5 & -1 \\ -9 & -1 \end{bmatrix} \cdot \begin{bmatrix} -2 & 1 \\ 4 & 2 \end{bmatrix}$
    Find $P - Q$. **(LESSON 12.8)**

20. Solve: $9x^2 = 1$. **(LESSON 10.2)** $\pm\frac{1}{3}$

21. List the domain of the following relation: $\{(2, -3), (2, 3), (4, 5), (4, -5)\}$. **(LESSON 5.1)** {2, 4}

22. Solve: $\frac{1.2}{7.5} = \frac{n}{2.5}$. **(LESSON 4.1)** 0.4

23. Find the midpoint of $\overline{MN}$, where $M = (3, -4)$ and $N = (7, 8)$. **(LESSON 12.4)** (5, 2)

24. What is the y-intercept of the graph of $3x - 2y = 6$? **(LESSON 5.5)** -3

25. Factor: $6x^2 + 9x + 3$. **(LESSON 9.6)** $3(2x + 1)(x + 1)$

26. Mary has a solution that is 60% alcohol and another that is 20% alcohol. How much of each should she use to make 100 milliliters of a solution that is 52% alcohol? **(LESSON 7.6)** 80 ml of 60%, 20 ml of 20%

27. Simplify: $8 - (3 - 5) - 4$. **(LESSON 2.3)** 6

28. Solve: $x + \frac{1}{2} = \frac{x}{2} + 3$. **(LESSON 3.4)** 5

FREE-RESPONSE GRID
The following questions may be answered by using a free-response grid such as that commonly used by standardized test services.

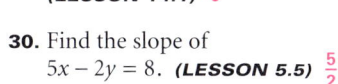

29. Let $g(x) = x^2 - 5x + 6$. Evaluate $g(3)$. **(LESSON 14.1)** 0

30. Find the slope of $5x - 2y = 8$. **(LESSON 5.5)** $\frac{5}{2}$

31. One month, Latoya spent $300 more than one-third of her usual budget. If Latoya spent $390 that month, how much is Latoya's usual budget? **(LESSON 3.3)** $270

32. Simplify: $5 - |-3| - 4$. **(LESSON 2.2)** -2

33. Simplify: $\left(\frac{2}{3}\right)^{-3}$. **(LESSON 8.4)** $\frac{27}{8}$

34. Name the next term in the sequence 8, 13, 21, 34, 55, … **(LESSON 1.1)** 89

35. Solve $3x - 5 = 16$ for x. **(LESSON 3.3)** 7

36. What number is 45% of 60? **(LESSON 4.2)** 27

37. Find the slope of the line containing the points (3, 4) and (6, 10). **(LESSON 5.2)** 2

38. Find the first differences of the sequence 2, 8, 14, 20, 26, … **(LESSON 1.1)** 6

39. Name the constant differences of the sequence 20, 21, 26, 35, 48, … **(LESSON 1.1)** 4

40. 9 is what percent of 60? **(LESSON 4.2)** 15%

41. Evaluate: $\sqrt{121}$. **(LESSON 12.1)** 11

42. What is the length of the hypotenuse of a right triangle with legs of 5 and 12 units? **(LESSON 12.3)** 13

43. Find the distance between the points $M(5, -6)$ and $N(-7, -1)$. **(LESSON 12.4)** 13

13 Probability

CHAPTER PLANNING GUIDE

Lesson	13.1	13.2	13.3	13.4	13.5	Projects and Review
Pupil Edition Pages	648–653	654–660	661–666	667–673	676–681	674–675, 682–689
Extra Practice (Pupil Edition)	776	777	777	778	778	
Practice Workbook	78	79	80	81	82	
Reteaching Masters	155–156	157–158	159–160	161–162	163–164	
Enrichment Masters	78	79	80	81	82	
Cooperative-Learning Activities	78	79	80	81	82	
Lesson Activities	78	79	80	81	82	
Problem Solving/ Critical Thinking	78	79	80	81	82	
Student Study Guide	78	79	80	81	82	
Spanish Resources	78	79	80	81	82	
Student Technology Guide	78	79	80	81	82	
Assessment Resources	166	167	168	170	171	169, 172–177
Teaching Transparencies	64			65	66	
Quiz Transparencies	13.1	13.2	13.3	13.4	13.5	
Writing Activities for Your Portfolio						37–39
Tech Prep Masters						61–64
Long-Term Projects						49–52

LESSON PACING GUIDE

Traditional	2 days	2 days	2 days	1 day	1 day	2 days
Block	1 day	1 day	1 day	$\frac{1}{2}$ day	$\frac{1}{2}$ day	1 day
Two-Year	4 days	4 days	4 days	2 days	2 days	4 days

CONNECTIONS AND APPLICATIONS

Lesson	13.1	13.2	13.3	13.4	13.5	Review
Algebra	648–653	654–660	661–666	667–673	676–681	682–689
Geometry	653	660	665, 666		681	
Statistics		658, 659, 660				
Business and Economics		660	663, 665			686
Life Skills		654	661, 662, 666	667, 669, 673		
Science and Technology	652		665, 666		676, 678, 680	
Sports and Leisure	648	657, 660	665	672, 673	676, 677, 680	
Other	652					
Cultural Connection: Americas	652					

BLOCK-SCHEDULING GUIDE

Day	Lesson	Teacher Directed: Lesson Examples, Teaching Transparencies	Student Guided: Activity, Try This	Cooperative-Learning Activity, Lesson Activity, Student Technology Guide	Practice: Practice & Apply, Extra Practice, Practice Workbook	Assessment: Quiz, Mid-Chapter Assessment	Problem Solving, Reteaching
1	13.1	10 min	15 min	15 min	65 min	15 min	15 min
2	13.2	10 min	15 min	15 min	65 min	15 min	15 min
3	13.3	10 min	15 min	15 min	65 min	15 min	15 min
4	13.4	5 min	8 min	8 min	35 min	8 min	8 min
4	13.5	5 min	7 min	7 min	30 min	7 min	7 min
5	Assess.	50 min PE: Chapter Review	90 min PE: Chapter Project, Writing Activities	90 min Tech Prep Masters	65 min PE: Chapter Assessment, Test Generator	30 min Chap. Assess. (A or B), Alt. Assess. (A or B), Test Generator	

PE: Pupil's Edition

Alternative Assessment

The following are suggestions for an alternative assessment for students who may benefit from a different type of assessment than the regular chapter quizzes and the mid-chapter and end-of-chapter assessment materials. Many of the questions are open-ended, and students' answers will vary.

Performance Assessment

1. Jar A contains 5 red and 7 green marbles and jar B contains 5 red, 1 white, and 14 blue marbles.
   a. Explain how to find the theoretical probability of selecting a red marble from jar A. Find that probability.
   b. How would you use your answer to part **a** to help find the probability of selecting a red marble from jar A AND another red marble from jar A, assuming that the first marble was not replaced.
   c. Explain how to find the probability of selecting a red marble from jar A AND a red marble from jar B. What probability facts are involved in finding the answer?

2. How does the Fundamental Counting Principle help you select one hat from 5 different hats, one pair of shoes from 4 different pairs of shoes, and one jacket from 4 different jackets?

Portfolio Project

Suggest that students choose one of the following projects for inclusion in their portfolios.

1. A trucker needs to drive from point A to point B by using one of 3 available routes, then from point B to point C by using one of 2 different routes, and finally from point C to point D by using one of 5 different routes.
   a. Illustrate the trucker's choices.
   b. Find the number of possible ways the trucker can get from one point to the location two points away.
   c. The trucker does not want so many choices. Explain how the number of choices might be reduced. Then find how many choices remain.
   d. Make a presentation that shows how the Fundamental Counting Principle can be used to deal with trucking route problems.

2. Suppose that the probabilities that a motorist will turn right, go straight, or turn left when reaching a certain intersection are equal. Design and carry out a simulation for the action of the next 15 motorists reaching the intersection.

internet connect

The table below identifies the pages in this chapter that contain technology information and support in the side columns.

Content Links	
Lesson Links	pages 657, 664, 666
Portfolio Links	pages 653, 666
Graphics Calculator Support	page 782

Resource Links

For information about teacher and parent resources as well as professional development help, visit **www.hrw.com/math**.

Technical Support HRW has assembled a team of dedicated technical and teaching professionals and a comprehensive service program to provide you with the support you need.

- The HRW Technical Support Line operates from 7 A.M. to 6 P.M. central time, Monday through Friday, at (800) 323-9239.
- The HRW Technical Support Center on the World Wide Web is available 24 hours a day, seven days a week, at **www.hrwtechsupport.com**.
- You can e-mail our Technical Support Center at **tsc@hrwtechsupport.com**.
- The Technical Support Center's fax-on-demand service at (800) 352-1680 offers solutions to common problems and answers to frequently asked questions.

Technology

Lesson Suggestions and Calculator Examples
(Keystrokes are based on a TI-83 calculator.)

Lesson 13.1 Theoretical Probability
In this lesson, students will apply their knowledge of fractions to probability problems. Students will use the calculator to enter a fraction and receive the answer as a decimal, which they may convert to a percent, if necessary. For example, if the probability of an event is $\frac{3}{28}$, students should enter the fraction, mentally round the result to the nearest hundredth, 0.11, and then change the decimal to a percent, 11%. The calculator display above shows students a shortcut for this process.

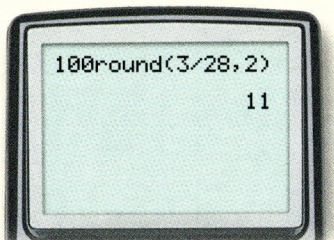

Lesson 13.2 Counting the Elements of Sets
In this lesson, you will have the opportunity to discuss with students the simplification that can be done before they use a calculator to carry out the needed evaluations. Consider, for example, Example 3 on page 658. The needed probability is represented by $\frac{12}{52} + \frac{13}{52} - \frac{3}{52}$. All the denominators are the same, so students can simply evaluate $\frac{12+13-3}{52}$, as shown at right.

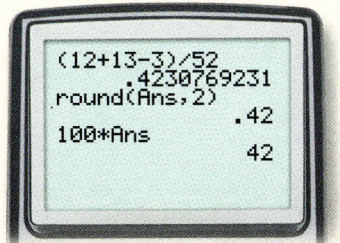

Lesson 13.3 The Fundamental Counting Principle
Consider an extension to this lesson that introduces students to the notion of a factorial. Pose the following problem:

In how many ways can Tom select 5 letters from {a, b, c, d, e} and arrange them in order without repeating a letter?

There are $5\times4\times3\times2\times1$ ways of doing it by the Fundamental Counting Principle. The product $5\times4\times3\times2\times1$ is abbreviated 5!. The calculator display above shows the calculation of this product by using 5 MATH PRB 4:!.

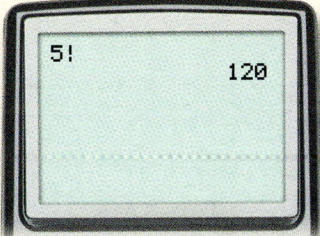

Lesson 13.4 Independent Events
In this lesson, students will have the opportunity to decide whether events are independent or dependent. Consider posing these problems.

1. A jar contains 5 red marbles and 7 green marbles. A marble is chosen at random, then replaced. A second marble is drawn. What is the probability of choosing 2 red marbles?

2. A jar contains 5 red marbles and 7 green marbles. A marble is chosen at random but not replaced. A second marble is drawn. What is the probability of choosing 2 red marbles?

The solutions are shown at right. Notice that the results are not the same. The probability of 2 red marbles is less when the first marble is not replaced. Discuss why this makes sense.

Lesson 13.5 Simulations
To carry out a simulation on a graphics calculator, students will need to become familiar with its random number generator, which can be accessed by pressing MATH and selecting PRB 5:randInt(. This feature will generate random integers suitable for a simulation of a coin toss or a roll of a number cube. You might also use a spreadsheet. The spreadsheet shown below simulates 60 coin tosses with 1 representing the event "heads" and 0 representing the event "tails." Cell A1 contains **=INT(2*RAND())** and cell F11 contains the formula **=SUM(A1:F10)/60**, which gives the experimental probability of heads in 60 tosses.

	A	B	C	D	E	F
1	1	0	0	1	0	0
2	1	0	1	1	0	0
3	1	0	0	1	0	0
4	0	1	0	1	0	0
5	0	1	0	0	1	0
6	1	0	0	0	1	1
7	1	0	1	0	0	0
8	0	0	0	0	0	1
9	0	1	0	0	1	0
10	1	1	0	0	1	1
11					P(H)	0.383

For further information, refer to the

- technology discussions in the lessons.
- lesson-related teacher's commentary in the side columns of this *Teacher's Edition*.
- lesson-related *Student Technology Guide* masters.
- *HRW Technology Handbook*.

Teaching Resources

Basic Skills Practice
(2 pages per skill)

Reteaching
(2 pages per lesson)

Enrichment
(1 page per lesson)

Cooperative-Learning Activity
(1 page per lesson)

Lesson Activity
(1 page per lesson)

Long-Term Project
(4 pages per chapter)

646E CHAPTER 13 INTERLEAF

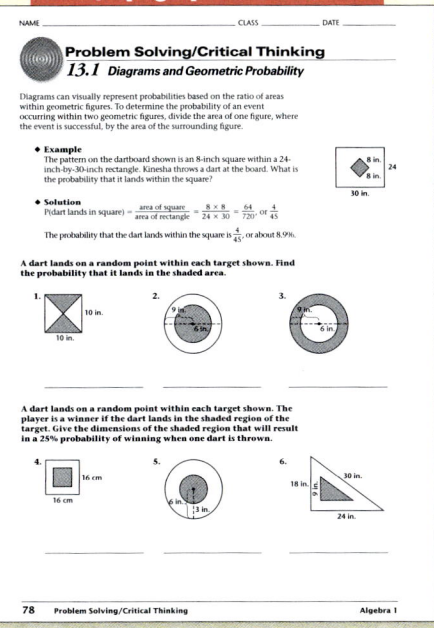

About Chapter 13

Background Information
Students learned about experimental probability in Chapter 4. In this chapter, they move on to the concept of theoretical probability. They use counting principles to determine the number of elements in a set and then use set sizes as a basis for calculating probabilities. The chapter concludes with a look at probabilities derived from simulations.

CHAPTER RESOURCES

- Block-Scheduling Handbook
- Writing Activities for Your Portfolio
- Tech Prep Masters
- Long-Term Project
- Assessment Resources:
   Mid-Chapter Assessment
   Chapter Assessments
   Alternative Assessments
- Test and Practice Generator
- Technology Handbook
- End-of-Course Exam Practice

Probability

PROBABILITY TOUCHES YOUR LIFE IN MANY WAYS. Games, insurance, matters of health and diet, and countless other situations involve an element of chance. Statistics makes extensive use of probability to study, predict, and draw conclusions from data. Sometimes probabilities are found experimentally by looking at data. Other times, probabilities are found theoretically by considering all the possibilities. In this chapter, you can learn about both kinds of probability and explore examples of probability in action.

Lessons

- 13.1 Theoretical Probability
- 13.2 Counting the Elements of Sets
- 13.3 The Fundamental Counting Principle
- 13.4 Independent Events
- 13.5 Simulations

Project
Happy Birthday!

About the Photos
The photos reflect several events that have a very low probability of occurring. The postage stamp pictured above was printed with the airplane upside down. Printing errors on postage stamps are rare, so this stamp is a valuable collector's item. Other rare events shown here are a four-leaf clover, triplets, an albino cow, an albino squirrel, snow in the desert, and a coin toss in which all of the coins turned up heads. Ask students if they can think of other events that have a very small probability of occurring.

The pictures on these pages illustrate rare events such as the birth of triplets, coins coming up all heads, and misprinted postage stamps.

About the Chapter Project

The study of probability provides the power to predict the outcome of chance events over a series of trials. In the Chapter Project, *Happy Birthday!*, on page 682, you will perform an experiment that will allow you to predict the chances that two people in a group will have the same birthday or birth month. After completing the Chapter Project, you will be able to do the following:

- Perform a probability experiment.

About the Portfolio Activity

Throughout the chapter, you will be given opportunities to complete Portfolio Activities that are designed to support your work on the Chapter Project.

- Theoretical probability is the topic of the Portfolio Activity on page 653.
- Experimental probability is the topic of the Portfolio Activity on page 666.
- The results of a three-game series between two evenly matched teams is the topic of the Portfolio Activity on page 673.
- A comparison of the probabilities of independent and dependent events is the topic of the Portfolio Activity on page 681.

Chapter Objectives

- List or describe the sample space of an experiment. [13.1]
- Find the theoretical probability of a favorable outcome. [13.1]
- Find the union and intersection of sets. [13.2]
- Count the elements of sets. [13.2]
- Apply the Addition of Probabilities Principle. [13.2]
- Use tree diagrams to count the number of choices that can be made from sets. [13.3]
- Use the Fundamental Counting Principle to count the number of choices that can be made from sets. [13.3]
- Find the probability of independent events. [13.4]
- Design and perform simulations to find experimental probabilities. [13.5]

Portfolio Activities appear at the end of Lessons 13.1, 13.3, 13.4, and 13.5. Each serves as preparation for the Chapter Project. The Portfolio Activities as well as the Chapter Project Activities are appropriate for inclusion in the student's portfolio. Students should be encouraged to include in their portfolios any other work in which they feel a sense of pride or a sense of accomplishment.

internet connect

Internet Connect features appearing throughout the chapter provide keywords for the HRW Web site that lead to links for math resources, tutorial assistance, references for student research, classroom activities, and all teaching resource pages. The HRW Web site also provides updates to the technology used in the text. Listed at right are the keywords for the Internet Connect activities referenced in this chapter. Refer to the side column on the page listed to read about the activity.

LESSON	KEYWORD	PAGE
13.1	MA1 Blaise	653
13.2	MA1 Fair Polyhedra	657
13.2	MA1 Bells	664
13.3	MA1 Probability	666
13.5	MA1 Simulation	679

CHAPTER 13 **647**

Prepare

NCTM PRINCIPLES & STANDARDS
1–3, 5, 6, 8–10

LESSON RESOURCES

- Practice 13.1
- Reteaching Master 13.1
- Enrichment Master 13.1
- Cooperative-Learning Activity 13.1
- Lesson Activity 13.1
- Problem Solving/Critical Thinking Master 13.1
- Teaching Transparency 64

QUICK WARM-UP

Write as a percent.

1. $\frac{15}{30}$ **50%** 2. $\frac{30}{48}$ **62.5%**

3. Two number cubes were rolled 120 times. A sum of 7 appeared 18 times. Based on these results, what is the experimental probability that a sum of 7 will appear on the next roll of these number cubes? $\frac{3}{20} = 15\%$

Also on Quiz Transparency 13.1

Teach

Why Review with students the meaning of experimental probability. Tell them that for certain events, it is not necessary to actually conduct an experiment in order to cite a reasonable probability. Ask them to suggest some events for which this might be true, such as tossing a fair coin and rolling a fair number cube. Tell them that in this lesson, they will learn how to find the *theoretical probability* of such events.

13.1 Theoretical Probability

Objectives

- List or describe the sample space of an experiment.
- Find the theoretical probability of a favorable outcome.

Why Theoretical probability is a mathematical model that is helpful in making predictions. Exact models can be found without actually doing experiments.

The coin toss before a football game is considered a fair way to determine initial possession because heads or tails are equally likely.

APPLICATION
SPORTS

The result of a coin toss is used to determine which team first gets the ball in a football game. Whether the coin lands heads or tails is an interesting question of probablility.

Consider the following probability experiment:

Toss a coin one time and see whether it lands heads or tails.

For this experiment, which is very simple, there are two possible *outcomes*— heads or tails. The outcomes are said to be **equally likely outcomes** because a tossed coin is just as likely to be heads as tails.

In the long run, a tossed coin tends to come up heads half of the time and tails half of the time (recall Lesson 4.3). Thus, the probability of getting heads is said to be $\frac{1}{2}$, and the probability of getting tails is said to be $\frac{1}{2}$. In function notation:

$$P(\text{heads}) = \frac{1}{2} \quad \text{and} \quad P(\text{tails}) = \frac{1}{2}$$

The sum of the probabilities is 1.

Alternative Teaching Strategy

INVITING PARTICIPATION Create two or three different spinners. Place one on a desk or table, hidden from the view of the class. Have one student spin it a number of times, calling out the result each time. The other class members should record the results. After several spins, ask the class to guess what the spinner looks like.

If the class guesses correctly, repeat the activity with a different spinner. If they do not guess correctly, have the student continue spinning until the class arrives at the correct conclusion.

After completing the activity with all of the spinners you made, lead the class in a discussion of theoretical probability. Point out that they were eventually able to describe the appearance of each spinner because its appearance dictates the theoretical outcome of any experiment involving it. For instance, if a spinner is divided into three equal regions, the theoretical probability that any one region comes up on a given spin is one-third. The experimental results of this spinner should be very close to that.

Similarly, for a single roll of a number cube there are six possible outcomes— 1, 2, 3, 4, 5, and 6—and each of them is equally likely.

$P(1) = \frac{1}{6}$ $P(2) = \frac{1}{6}$ $P(3) = \frac{1}{6}$
$P(4) = \frac{1}{6}$ $P(5) = \frac{1}{6}$ $P(6) = \frac{1}{6}$

The sum of the probabilities is again 1.

The set of all of possible outcomes of a probability experiment is called the **sample space** of the experiment. For the experiment of tossing a coin once, a sample space is {heads, tails} or {H, T}. For rolling a number cube once, a sample space is {1, 2, 3, 4, 5, 6}.

EXAMPLE ❶ Find the sample space for the experiment of tossing a coin twice.

● SOLUTION

There are four different possible results from tossing a coin twice.

First Toss	Second Toss
H	H
H	T
T	H
T	T

You can write these in set notation as {HH, HT, TH, TT}.

TRY THIS Find the sample space for the experiment of tossing a coin three times.

In a probability experiment, to say that an outcome is "favorable" does not mean that it is good or bad. A **favorable outcome** is the outcome you are looking for when you do the experiment.

EXAMPLE ❷ A coin is tossed twice. You are interested in the probability that both tosses will be heads or both will be tails. **What are the favorable outcomes in the sample space for the experiment? How many favorable outcomes are there?**

● SOLUTION

The sample space for the experiment is {**HH**, HT, TH, **TT**}. (See Example 1.) The outcomes in which the two tosses are both heads or both tails (the favorable outcomes) are shown in red. There are two favorable outcomes.

TRY THIS A coin is tossed three times. You are interested in the probability that two or more of the tosses will be heads. What is the sample space for the experiment? What are the favorable outcomes? How many favorable outcomes are there?

Interdisciplinary Connection

PSYCHOLOGY The cards at the right are used in a test for extrasensory perception. To take the test, you face the other way while the tester concentrates on one shape. You must then name the shape. Assuming that you know the four choices, what is the probability of guessing the shape correctly if the test is done once? if the test is done twice? $\frac{1}{4}; \frac{1}{16}$

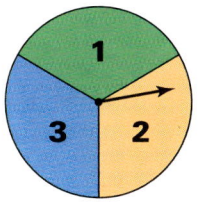

Inclusion Strategies

TACTILE LEARNERS Some students may find it helpful to model a sample space with the objects described. For example, to understand the sample space for two tosses of a coin, have them use eight pennies to model all possible ways that the penny could land on the two tosses. If they place the pennies on a sheet of paper, they can write the H and T representations directly beneath them.

H H   H T   T H   T T

ADDITIONAL EXAMPLE ❶

Find the sample space for two spins of this spinner.

{(1, 1), (1, 2), (1, 3), (2, 1), (2, 2), (2, 3), (3, 1), (3, 2), (3, 3)}

TRY THIS
There are eight different possible results for tossing a coin three times.
{HHH, HHT, HTH, HTT, THH, THT, TTH, TTT}

ADDITIONAL EXAMPLE ❷

A fair coin is tossed twice. You are interested in the probability that at least one of the tosses will come up heads. **What are the favorable outcomes in the sample space for this experiment? How many favorable outcomes are there?**

The set of favorable outcomes is {HH, HT, TH}. There are three favorable outcomes.

TRY THIS
The sample space for the experiment is {HHH, HHT, HTH, HTT, THH, THT, TTH, TTT}. The outcomes in which two or more of the tosses are heads are the favorable outcomes. They are {HHH, HHT, HTH, THH}. There are four favorable outcomes.

Use Teaching Transparency 64.

ADDITIONAL EXAMPLE 3

A number cube is rolled twice. **What is the probability that the sum will be 10?**

$P(\text{sum of } 10) = \frac{1}{12}$

TRY THIS

The sample space is {HHH, HHT, HTH, HTT, THH, THT, TTH, TTT}. The outcomes in which one or two of the tosses are heads are {HHT, HTH, HTT, THH, THT, TTH}.

$P(1 \text{ OR } 2 \text{ heads}) = \frac{6}{8} = \frac{3}{4}$

ADDITIONAL EXAMPLE 4

Which has the greater probability: 10 chances out of 12, or 25 chances out of 30?

neither, $\left(\frac{10}{12} = \frac{25}{30} = \frac{5}{6}\right)$

TRY THIS

$\frac{17}{71} > \frac{19}{91}$, so 17 chances out of 71 has a greater probability.

To find the theoretical probability of something happening, you first need to define the sample space and the favorable outcomes.

> **Theoretical Probability**
>
> Let n be the number of elements in a sample space in which each element is equally likely to occur. Let E be an event for which there are f favorable outcomes in the sample space.
>
> $$P(E) = \frac{f}{n}$$

EXAMPLE 3 A number cube is rolled twice. **What is the probability that the sum of the rolls will be 9?**

● **SOLUTION**

The sample space for the experiment is {(1, 1), (1, 2), (1, 3), (1, 4), (1, 5), (1, 6), (2, 1), (2, 2), (2, 3), (2, 4), (2, 5), (2, 6), (3, 1), (3, 2), (3, 3), (3, 4), (3, 5), **(3, 6)**, (4, 1), (4, 2), (4, 3), (4, 4), **(4, 5)**, (4, 6), (5, 1), (5, 2), (5, 3), **(5, 4)**, (5, 5), (5, 6), (6, 1), (6, 2), **(6, 3)**, (6, 4), (6, 5), (6, 6)}

Out of 36 possibilities, there are 4 favorable outcomes. The probability is $P(\text{sum of } 9) = \frac{4}{36} = \frac{1}{9}$.

TRY THIS A coin is tossed three times. What is the probability that only one or only two of the tosses will be heads?

EXAMPLE 4 On the first day of class, a teacher puts a sticker under 3 of the 20 chairs in her classroom. A second teacher puts a sticker under 4 of the 28 chairs in her classroom. Students who sit in the chairs with stickers will receive a $5 gift certificate to the school cafeteria. **In which classroom is the probability of receiving a gift certificate greater?**

● **SOLUTION**

The sample space for the first teacher consists of 20 chairs, and there are 3 favorable outcomes. The sample space for the second teacher consists of 28 chairs, and there are 4 favorable outcomes.

$P(\text{receiving certificate in classroom } 1) = \frac{3}{20}$, or 15%

$P(\text{receiving certificate in classroom } 2) = \frac{4}{28}$, or about 14%

Thus, a student in the first classroom has a slightly greater probability of receiving a gift certificate than a student in the second classroom.

TRY THIS Which has the greater probability: 19 chances out of 91 or 17 chances out of 71?

Enrichment

Another way to describe a probability is to calculate *odds*, which are defined as follows:

$$\text{odds of an event} = \frac{\text{number of favorable outcomes}}{\text{number of unfavorable outcomes}}$$

For instance, if a number cube is rolled, the odds in favor of rolling a 4 are given as *1 to 5*. Find the odds listed below for one roll of a fair number cube.

1. odds in favor of an even number   **3 to 3**
2. odds in favor of a number greater than 5   **1 to 5**
3. odds against a number greater than 5   **5 to 1**

Reteaching the Lesson

WORKING BACKWARD Display a list of the sample space for two rolls of a number cube. Give students each probability below. Tell them to describe an event that has the given probability. **Sample responses are given.**

1. $\frac{1}{36}$ **sum of 2**
2. $\frac{1}{18}$ **sum of 3**
3. $\frac{1}{9}$ **sum of 9**
4. $\frac{1}{6}$ **two equal numbers**
5. 0 **sum of 1**
6. 1 **sum > 1**

Exercises

Communicate

1. Based on his record so far this season the probability that Nat will make a successful free throw is 40%. Is this a theoretical probability or an experimental probability (recall Lesson 4.3)? Explain.
2. In a two-player game, 4 coins are tossed. The first player wins if the coins come up all heads or all tails. Otherwise the second player wins. Does each player have an equally likely chance of winning? Explain.
3. When finding the theoretical probability of rolling a certain number on a number cube, why must each possibility be equally likely?

Guided Skills Practice

Find the sample space for each experiment. (EXAMPLE 1)

4. rolling a number cube twice
5. tossing a coin four times

Find the number of *favorable* outcomes in the sample space for each experiment. (EXAMPLE 2)

6. A number cube is rolled twice. The sum of the rolls is 7. **6**
7. A coin is tossed three times. All three tosses are heads. **1**

Find the probability of each outcome. (EXAMPLE 3)

8. the probability that two rolls of a number cube will have a sum of 7 $\frac{1}{6}$
9. the probability that three tosses of a coin will be three heads $\frac{1}{8}$

Which has the greater probability? (EXAMPLE 4)

10. 12 chances out of 53 or 18 chances out of 74 **18 chances out of 74**
11. 34 chances out of 71 or 53 chances out of 117 **34 chances out of 71**

Practice and Apply

An integer between 1 and 25, inclusive, is drawn at random. Find each probability.

12. The integer is even. $\frac{12}{25}$
13. The integer is odd. $\frac{13}{25}$
14. The integer is a multiple of 3. $\frac{8}{25}$
15. The integer is a multiple of 5. $\frac{1}{5}$

Assess

Selected Answers

Exercises 4–11, 13–35 odd

ASSIGNMENT GUIDE

In Class	1–11
Core	12–23, 25–29
Core Plus	12–29
Review	30–35
Preview	36

✎ Extra Practice can be found beginning on page 738.

Technology

If students have access to a calculator that can display fractions, it may be helpful for simplifying some of the answers to the exercises. You may wish to have students do the exercises with paper and pencil first and then use the calculator to verify their answers.

5.

First toss	Second toss	Third toss	Fourth toss
H	H	H	H
H	H	H	T
H	H	T	H
H	H	T	T
H	T	H	H
H	T	H	T
H	T	T	H
H	T	T	T
T	H	H	H
T	H	H	T
T	H	T	H
T	H	T	T
T	T	H	H
T	T	H	T
T	T	T	H
T	T	T	T

---

4. {1 – 1, 1 – 2, 1 – 3, 1 – 4, 1 – 5, 1 – 6,
2 – 1, 2 – 2, 2 – 3, 2 – 4, 2 – 5, 2 – 6,
3 – 1, 3 – 2, 3 – 3, 3 – 4, 3 – 5, 3 – 6,
4 – 1, 4 – 2, 4 – 3, 4 – 4, 4 – 5, 4 – 6,
5 – 1, 5 – 2, 5 – 3, 5 – 4, 5 – 5, 5 – 6,
6 – 1, 6 – 2, 6 – 3, 6 – 4, 6 – 5, 6 – 6}

LESSON 13.1

Error Analysis

Students sometimes incorrectly identify the sample space for an experiment. For instance, given the experiment of two rolls of a number cube, they may identify the sample space simply as {1, 2, 3, 4, 5, 6}. Remind them that if two actions are taken, the sample space must consist of *pairs*. Similarly, if three actions are taken, the sample space must consist of *triples*.

Find the probability of each outcome.

16. the probability that two rolls of a number cube will have a sum of 3 $\frac{1}{18}$
17. the probability that two tosses of a coin will be one heads and one tails $\frac{1}{2}$
18. the probability that two rolls of a number cube will have a sum of 11 $\frac{1}{18}$
19. the probability that three tosses of a coin will be at least two heads $\frac{1}{2}$

Suppose that you select a letter of the English alphabet at random. Find each probability.

20. The letter is *r*. $\frac{1}{26}$ 21. The letter is a vowel. $\frac{5}{26}$
22. The letter is a consonant. $\frac{21}{26}$ 23. The letter is in the word *mathematics*. $\frac{4}{13}$

CHALLENGE

24. John is trying to solve the equation $2x - 1 = 15$ on an algebra quiz. He doesn't remember how to do it, but he reasons as follows: The answer, *x*, is multiplied by 2 and comes out to be around 15. He assumes that the answer must be a positive one-digit number, and he makes a guess. Find the probability that he guesses the correct solution. $\frac{1}{9}$

APPLICATIONS

25. **GEOGRAPHY** Find the probability that the name of a state chosen at random from the United States begins with the letter *M*. $\frac{4}{25}$

26. **ASTRONOMY** Find the probability that a planet chosen at random in our solar system is closer to the Sun than Earth is. (Hint: Of the 9 planets, only Mercury and Venus are closer to the Sun than Earth is.) $\frac{2}{9}$

27. **LANGUAGE ARTS** List all possible arrangements of the letters in the word *tap*. Find the probability that an arrangement chosen at random is a word. $\frac{1}{2}$

CULTURAL CONNECTION: AMERICAS The Blackfoot people of Montana played a game of chance in which they threw 4 decorated game pieces made from buffalo ribs. Three were blank on one side, but the fourth, called the chief, was engraved on both sides like a coin with heads and tails. The highest score was 6 points, for a throw of 3 blanks and the chief landing heads up. For 3 blanks and the chief reversed, the score was 3 points.

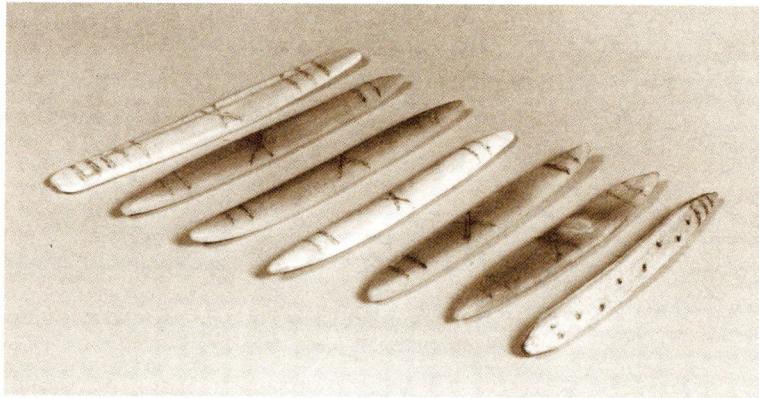

28. What is the probability of throwing 3 blanks and the chief landing heads up? $\frac{1}{16}$

29. What is the probability of throwing 3 blanks and the chief reversed? $\frac{1}{16}$

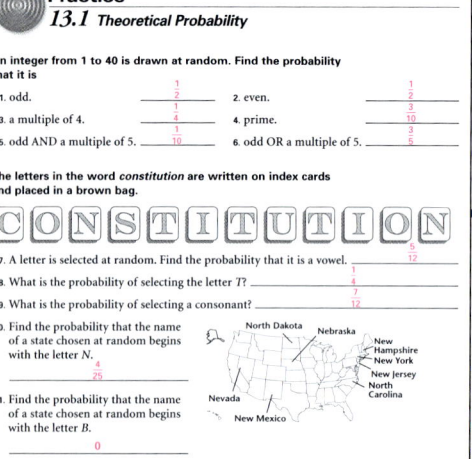

652   LESSON 13.1

Look Back

30. Solve the equation $y = mx + b$ for m. (LESSON 3.6) $m = \frac{y-b}{x}$

31. An item that costs $53 totals $56.18 with sales tax. Use the proportion method to find the rate of the sales tax. (LESSON 4.2) $\frac{3.18}{53.00} = \frac{x}{100}$, 6%

Karl's typing speeds on 5 tests were 19, 21, 26, 21, and 27 words per minute.

32. What is the mode of Karl's typing speeds? (LESSON 4.4) 21

33. What is the median of Karl's typing speeds? (LESSON 4.4) 21

34. Given $-6 \leq n - 4 < 10$, graph the solution to the inequality on a number line for values of n that are integers. (LESSON 6.3)

35. Draw the graph of the function $y = \frac{1}{x}$ for the interval $\frac{1}{4} < x \leq 4$. What is the range of the function? (LESSON 11.2)

Look Beyond

CONNECTION

36. **GEOMETRY** A dart is thrown at a circular target. Assuming that it is equally likely to land anywhere on the target, what is the probability that it hits the bull's-eye, the red region in the center? (The measurements indicate the diameter of each circular region.) $\frac{1}{16}$

Include this activity in your portfolio.

A game is played by rolling two number cubes. Player A wins if the sum is 2, 3, 4, 5, 10, 11, or 12. Player B wins if the sum is 6, 7, 8, or 9.

Sum	2	3	4	5	6	7	8	9	10	11	12
# of ways to Roll Sum	1	2	?	?	?	6	5	?	?	?	1

1. Complete the table above.
2. How many favorable sums are possible for player A?
3. How many favorable sums are possible for player B?
4. Find the theoretical probability of player A winning the game.
5. Find the theoretical probability of player B winning the game.
6. Which player has the better chance of winning? Explain.

Look Beyond

In Exercise 36, students explore a problem involving *geometric probability*. They will learn more about this topic if they take a course in geometry or a more advanced course in algebra.

ALTERNATIVE Assessment

Portfolio Activity

The Portfolio Activity can be used as preparation for the Chapter Project or as a separate activity. In the Portfolio Activity on this page, students examine the rules for a game to determine which player has the greater chance of winning.

internetconnect

GO TO: go.hrw.com
KEYWORD: MA1 Blaise

To support the Portfolio Activity, students may visit the HRW Web site where they can obtain additional information about Blaise Pascal, his work in probability theory, and some interesting problems in probability.

Student Technology Guide

34.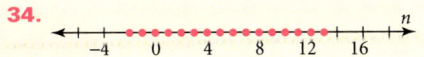

35.

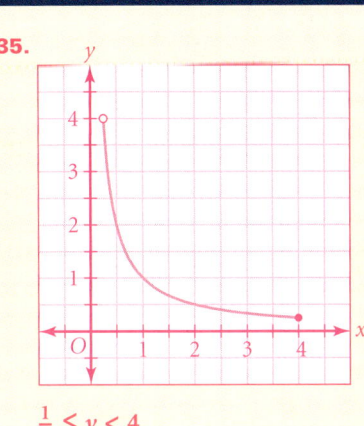

$\frac{1}{4} \leq y < 4$

Prepare

NCTM PRINCIPLES & STANDARDS
1–3, 5–10

LESSON RESOURCES

- Practice 13.2
- Reteaching Master 13.2
- Enrichment Master 13.2
- Cooperative-Learning Activity 13.2
- Lesson Activity 13.2
- Problem Solving/Critical Thinking Master 13.2

QUICK WARM-UP

A fair number cube is rolled once. **Find the probability of each event below.**

1. 1 $\frac{1}{6}$
2. an even number $\frac{1}{2}$
3. 4 $\frac{1}{6}$
4. a prime number $\frac{1}{2}$

A fair number cube is rolled two times. **Find the probability of each event below.**

5. two 3s $\frac{1}{36}$
6. a sum of 5 $\frac{1}{9}$

Also on Quiz Transparency 13.2

Teach

Why Tell students that any collection of data arranged for easy retrieval is called a *database*. Ask them to name some databases that they use regularly. Examples might include the telephone book, an address book, and an itemized list of baseball card collection. Ask students to describe how they use databases like these.

13.2 Counting the Elements of Sets

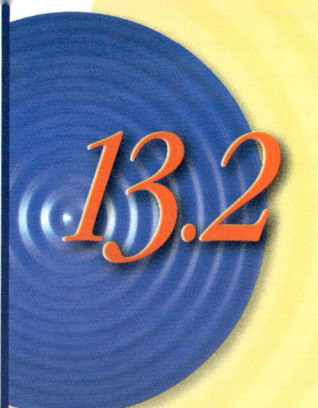

Objectives

- Find the union and intersection of sets.
- Count the elements of sets.
- Apply the Addition of Probabilities Principle.

Why In many situations, you need to count the number of favorable outcomes that occur in subsets of favorable outcomes.

APPLICATION
INTERPERSONAL SKILLS

Maria keeps a database of addresses of her friends and relatives. The user of a database can list records and fields in various ways. Maria will need these capabilities in order to write thank you notes for her Quinceañera gifts.

A database user, such as Maria in the photo, may want to create a number of different lists. For example, Maria can print out a list of all the people in her database who are her relatives.

	A	B	C	D	E	F	G
1	Last name	First name	Street	City	State	Zip	Relative?
2	Sanchez	Richardo	1823 Putnam	Santa Fe	NM	87010	Yes
3	Gutierrez	Alfred	3201 Swan's Crossing	Victoria	TX	77901	Yes
4	Lopez	Pauline	1573 Piedmont	Austin	TX	78746	Yes
5	Ortiz	Mary	809 Millwood	Pensacola	FL	32514	Yes
6	Hays	Laura	2337 Sneed	Edinburg	TX	78539	Yes
7	Trimble	Alicia	408 Elm Creek	Berea	OH	44017	Yes

She can also print out a list of people in her database who live in her home state, Texas.

	A	B	C	D	E	F	G
1	Last name	First name	Street	City	State	Zip	Relative?
2	Gutierrez	Alfred	3201 Swan's Crossing	Victoria	TX	77901	Yes
3	Lopez	Pauline	1573 Piedmont	Austin	TX	78746	Yes
4	Hays	Laura	2337 Sneed	Edinburg	TX	78539	Yes
5	Vasquez	David	784 Fairfield Dr	San Antonio	TX	78034	No
6	Gomez	George	2305 Main	San Antonio	TX	78042	No

If she wishes, she can also print out a list of people in her database who are her relatives AND who live in Texas.

	A	B	C	D	E	F	G
1	Last name	First name	Street	City	State	Zip	Relative?
2	Gutierrez	Alfred	3201 Swan's Crossing	Victoria	TX	77901	Yes
3	Lopez	Pauline	1573 Piedmont	Austin	TX	78746	Yes
4	Hays	Laura	2337 Sneed	Edinburg	TX	78539	Yes

In computer software and in logic, the word AND is a **logical operator**. To understand the logical operator AND, Venn diagrams are useful.

Alternative Teaching Strategy

USING TECHNOLOGY The day before you plan to teach this lesson, have students in your class agree on data they would like to gather about themselves. One possible topic might be the types of pets that students have, with categories such as *dog, cat, bird, fish, rabbit,* and *other*. Another possibility might be categories of hobbies. Have students write their name and the pertinent data on a sheet of paper. Collect the papers, and compile the data into one list.

On the day of the lesson, give students the list of data and have them make a database with appropriate software. Show them how to use the **Sort** function of the software to identify all students who fit a certain category, such as *students who have dogs*.

Now have them suggest categories that intersect, such as *students who have dogs* and *students who have cats*. Have them gather the data for the two categories. Show them how to make a Venn diagram to represent the data.

654 LESSON 13.2

The "AND" Relationship in Venn Diagrams

In a **Venn diagram**, circles are used to represent sets. When circles overlap, the overlapping region is known as the *intersection* of the sets.

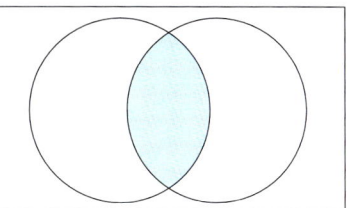

The intersection of the sets represented by the circles is the shaded region.

The **intersection** of two sets consists of the elements that are common to both sets. The logical relationship AND represents the intersection of two sets.

The intersection of sets is shown by the symbol ∩. The intersection of sets A and B is written as A ∩ B.

Compare this Venn diagram with Maria's three separate printouts on the previous page.

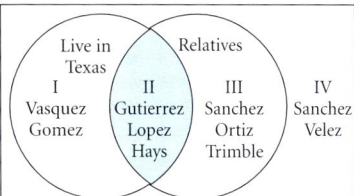

The "OR" Relationship in Venn Diagrams

Maria can print out a list of people in her database who live in Texas OR who are relatives, as shown.

	A	B	C	D	E	F	G
1	Last name	First name	Street	City	State	Zip	Relative?
2	Sanchez	Richardo	1823 Putnam	Santa Fe	NM	87010	Yes
3	Gutierrez	Alfred	3201 Swan's Crossing	Victoria	TX	77901	Yes
4	Lopez	Pauline	1573 Piedmont	Austin	TX	78746	Yes
5	Ortiz	Mary	809 Millwood	Pensacola	FL	32514	Yes
6	Hays	Laura	2337 Sneed	Edinburg	TX	78539	Yes
7	Vasquez	David	784 Fairfield Dr	San Antonio	TX	78034	No
8	Trimble	Alicia	408 Elm Creek	Berea	OH	44017	Yes
9	Gomez	George	2305 Main	San Antonio	Tx	78042	No

The logical relationship OR represents the *union* of sets.

The **union** of two sets consists of all elements of both sets. In a Venn diagram of the union of two sets, the whole circles are shaded.

The union of two sets is shown by the symbol ∪. The union of sets A and B is written as A ∪ B.

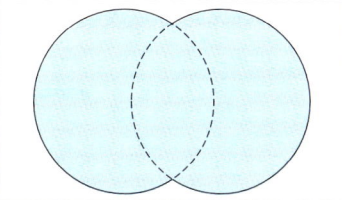

Union of sets with some elements in common

Interdisciplinary Connection

SOCIAL STUDIES Demographers who analyze polls and voting records must know how to count elements of sets that intersect. For example, suppose that 100 people voted in a local election. The voters included 45 men and 55 women. In the final count, 70 people voted in favor of the issue, 32 of whom were men. Draw a Venn diagram to illustrate this situation.

1. How many women voted against the issue? **17**

2. How many of the 100 voters were men OR voted against the issue? **62**

Inclusion Strategies

KINESTHETIC LEARNERS In a large open space, use string to make a large Venn diagram like those shown in this lesson. Identify two attributes of class members, such as *boys* and students *wearing a white shirt*. Choose eight students and ask them to stand at the appropriate places in the Venn diagram. Have the class count and record the number of students in each region. Discuss the principle for counting the elements of sets. Then have students suggest other sets of attributes that can be used to illustrate the principle.

Activity Notes

In this Activity, students take a look at the meaning of Venn diagrams by listing the individual elements of given sets in the appropriate regions. Since the elements are taken from a database of names, be sure that students make the diagrams large enough to write in the names. Students should arrive at the conclusion that to calculate the total number of elements in two sets, you add the number of elements in the two sets, but you must then subtract the number of elements that appear in *both* sets.

Cooperative Learning

You may wish to have students do the Activity in pairs. For Steps 1–4, one student can prepare the Venn diagram for *Texas* and *Ohio*, while the other prepares the diagram for *Ohio* and *Relative*. They can switch assignments to prepare the diagrams for Step 5. The partners should work together to answer all of the questions.

CHECKPOINT ✔

6. To find the number of people who are relatives OR live in Ohio, add the number of people who are relatives, 6, to the number who live in Ohio, 1, and subtract the number of those who overlap, 1.
$6 + 1 - 1 = 6$

	A	B	C	D	E	F	G
1	Last name	First name	Street	City	State	Zip	Relative?
2	Sanchez	Richardo	1823 Putnam	Santa Fe	NM	87010	Yes
3	Gutierrez	Alfred	3201 Swan's Crossing	Victoria	TX	77341	Yes
4	Lopez	Pauline	1573 Piedmont	Austin	TX	78746	Yes
5	Ortiz	Mary	809 Millwood	Pensacola	FL	32514	Yes
6	Hays	Laura	2337 Sneed	Edinburg	TX	78539	Yes
7	Shwartz	Ricky	489 King Blvd.	Baton Rouge	LA	70422	No
8	Velez	Henry	6097 Nash	Oklahoma City	OK	73114	No
9	Vasquez	David	784 Fairfield	San Antonio	TX	78034	No
10	Trimble	Alicia	408 Elm Creek	Berea	OH	44017	Yes
11	Gomez	George	2305 Main	San Antonio	Tx	78042	No

Use the database printout to complete the activity below.

Activity

Venn Diagrams for AND and OR

Make two large copies of the Venn diagram at right. The circles in one of your diagrams should be labeled Texas and Ohio. The other diagram should have labels for Ohio and Relative. Fill in each diagram with last names from Maria's database placed in the appropriate places. Then answer the following questions:

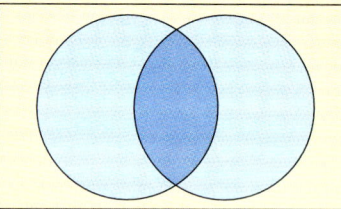

1. How many people in the database live in Ohio? in Texas?
2. How many people in the database live in Ohio AND Texas? in Ohio OR Texas?
3. How many people in the database are relatives of Maria?
4. How many people in the database live in Ohio OR are relatives of Maria? live in Ohio AND are relatives of Maria?
5. Draw new diagrams and shade the portions that represent people in the database who
   a. live in Ohio OR Texas.
   b. live in Ohio AND Texas.
   c. live in Ohio AND are relatives of Maria.
   d. live in Ohio OR are relatives of Maria.

CHECKPOINT ✔

6. Write a rule for finding the number of items in one set OR the other. (Hint: Use the number of people who live in Ohio, the number who are relatives, and the number who are relatives OR who live in Ohio in your answer.)

You may have noticed that it is important to not count the elements in the intersection of the circles twice.

Counting Elements of Sets

Suppose that there are m elements in set M and n elements in set N; then the total number of elements in the two sets is $m + n - t$, where t is the number of elements in the intersection of M and N.

Enrichment

Give students the Venn diagram at right, and have them use symbols to write an expression that describes each numbered region in terms of the sets.

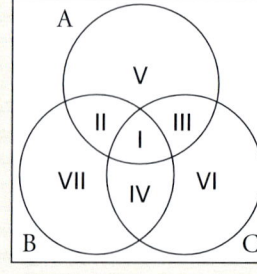

I:	$A \cap B \cap C$
II:	$A \cap B - (A \cap B \cap C)$
III:	$A \cap C - (A \cap B \cap C)$
IV:	$B \cap C - (A \cap B \cap C)$
V:	$A - (A \cap B) - (A \cap C) + (A \cap B \cap C)$
VI:	$C - (A \cap C) - (B \cap C) + (A \cap B \cap C)$
VII:	$B - (A \cap B) - (B \cap C) + (A \cap B \cap C)$

EXAMPLE 1

APPLICATION: GAMES

In a card game called Crazy Eights, if the queen of hearts is showing, you can play only a queen, a heart, or an eight. **How many different cards can be played in this situation, assuming that all cards except the queen of hearts are available?**

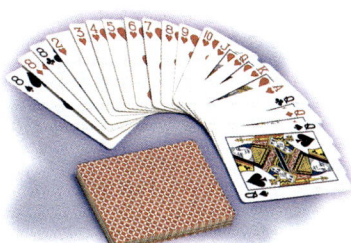

● **SOLUTION**

Do not count the queen of hearts (it has already been played). There are 3 other queens, 12 other hearts, and 4 eights—with the eight of hearts being counted twice. Thus, there are $3 + 12 + 4 - 1$, or 18, cards that can be played.

TRY THIS   How many cards in a deck of playing cards are spades OR aces?

EXAMPLE 2

If you draw a card at random from a complete deck of 52 playing cards, what is the probability that you will draw a king? What is the probability that you will draw a queen? What is the probability that you will draw a king OR a queen?

● **SOLUTION**

There are 4 kings and 4 queens in the deck.

$$P(\text{king}) = \frac{4}{52} = \frac{1}{13} \quad \text{and} \quad P(\text{queen}) = \frac{4}{52} = \frac{1}{13}$$

Since there are 8 cards in the set of favorable outcomes for drawing a king or a queen, the probability is as follows:

$$P(\text{king OR queen}) = \frac{8}{52} = \frac{2}{13}$$

TRY THIS   If you draw a card at random from a complete deck of 52 playing cards, what is the probability that you will draw a spade OR a club?

Example 2 suggests that you can add the probabilities of two outcomes to find the probability that one OR the other will occur.

$$P(\text{king OR queen}) = P(\text{king}) + P(\text{queen}) = \frac{4}{52} + \frac{4}{52} = \frac{8}{52}, \text{ or } \frac{2}{13}$$

When the two outcome sets from a sample space do not overlap, they are said to be **disjoint**.

Addition of Probabilities Principle

If M and N are *intersecting* sets of outcomes in the same sample space, then
$$P(M \text{ OR } N) = P(M) + P(N) - P(M \cap N).$$

If M and N are *disjoint* sets of outcomes in the same sample space, then
$$P(M \text{ OR } N) = P(M) + P(N).$$

ADDITIONAL EXAMPLE 1

How many cards in a standard deck of 52 playing cards are hearts OR kings?

16 cards

TRY THIS

13 cards are spades and 4 cards are aces, but 1 card is both and should not be counted twice. 16 cards are spades OR aces.

ADDITIONAL EXAMPLE 2

A bag contains 3 blue marbles, 4 red marbles, and 5 green marbles. You reach into the bag and randomly draw a marble.

a. What is the probability that you will draw a blue marble? $\frac{1}{4}$

b. What is the probability that you will draw a red marble? $\frac{1}{3}$

c. What is the probability that you will draw a blue marble OR a red marble? $\frac{7}{12}$

TRY THIS

$P(\text{spade OR club}) = \frac{26}{52} = \frac{1}{2}$

Reteaching the Lesson

USING VISUAL STRATEGIES Obtain one yellow and one blue transparency sheet for use on an overhead projector. Cut a large circle from each. Place the yellow circle on the overhead. As students observe, use a washable marker to write the names of all the girls in the class on the left side of the circle. Then place the blue circle on the overhead. This time, write the names of all students in the class who have a certain characteristic, such as *walked to school this morning*, on the right side of the circle. Count the names in each circle.

Now overlap the circles, yellow on the left and blue on the right. The overlapping section will be green. Have students identify any names that appear in both circles. Erase those names from both the yellow and blue circles and write them in the green section. Explain that these names are the intersection of the two sets. Count the names in each region and use the numbers to illustrate the principle for counting elements of sets that appears on page 656.

---

 internetconnect

GO TO: go.hrw.com
KEYWORD: MA1 Fair Polyhedra

Students are now familiar with number cubes. In this activity, they will discover fair number solids with more sides. Then they will calculate the probabilities of with more than six outcomes.

LESSON 13.2

ADDITIONAL EXAMPLE 3

What is the probability of drawing a red face card OR a heart from a standard deck of 52 playing cards? (Note: Hearts and diamonds are red cards. Clubs and spades are black cards.) $\frac{4}{13}$

TRY THIS

$P(\text{odd-numbered spade OR ace}) = \frac{5+3}{52} \approx 0.15$

EXAMPLE 3

What is the probability of drawing a face card OR a spade from a deck of playing cards? (Note: The face cards are the 4 kings, 4 queens, and 4 jacks. There are 13 spades.)

SOLUTION

There are 12 face cards and 13 spades. Three of the 13 spades are also face cards, so there are 3 cards in the intersection of the two sets. Let F be the set of face cards and S the set of spades.

$$P(F \text{ OR } S) = P(F) + P(S) - P(F \cap S) = \frac{12}{52} + \frac{13}{52} - \frac{3}{52} = \frac{22}{52} = \frac{11}{26} \approx 0.42$$

TRY THIS

What is the probability of drawing an odd-numbered spade OR an ace from a deck of playing cards? (Note: The numbered cards are ace through 10 in each of the four suits; an ace counts as a one.)

SUMMARY OF RELATIONSHIPS BETWEEN TWO SETS

Intersection	Union	Disjoint

Exercises

Communicate

1. Explain some ways that a database can be used to organize sets of data.
2. In the database at the beginning of this lesson, name a person other than Richardo Sanchez who is in the intersection of the set of people who are relatives and the set of people who do not live in Texas.
3. In the Crazy Eights situation (Example 1), explain how the eight of hearts is counted twice.

CONNECTION

STATISTICS A survey of student participation in school music programs had the results shown in the table below. Explain how to find each of the following:

	Participated	Did not participate
Boys	43	59
Girls	49	57

4. How many students are girls? **106**
5. How many students participated? **92**
6. How many students are girls AND participated? **49**
7. How many students are girls OR participated? **149**

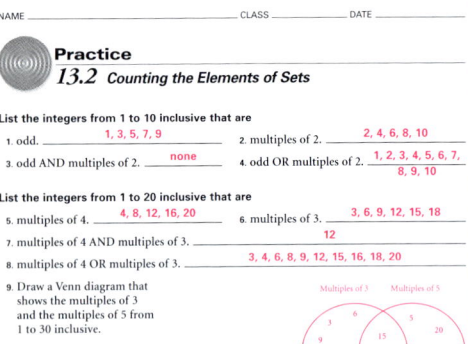

Practice

13.2 Counting the Elements of Sets

List the integers from 1 to 10 inclusive that are
1. odd. **1, 3, 5, 7, 9**
2. multiples of 2. **2, 4, 6, 8, 10**
3. odd AND multiples of 2. **none**
4. odd OR multiples of 2. **1, 2, 3, 4, 5, 6, 7, 8, 9, 10**

List the integers from 1 to 20 inclusive that are
5. multiples of 4. **4, 8, 12, 16, 20**
6. multiples of 3. **3, 6, 9, 12, 15, 18**
7. multiples of 4 AND multiples of 3. **12**
8. multiples of 4 OR multiples of 3. **3, 4, 6, 8, 9, 12, 15, 16, 18, 20**
9. Draw a Venn diagram that shows the multiples of 3 and the multiples of 5 from 1 to 30 inclusive.
10. Which numbers are multiples of 3 AND of 5? **15, 30**
11. How is this shown in the Venn diagram? **the intersection of both circles**

A marketing representative gave supermarket customers a sample taste of a new soft drink. The results are shown in the table.

	Liked drink	Disliked drink	Total
Men	16	14	30
Women	19	10	29
Total	35	24	59

12. How many customers are men? **30**
13. How many customers disliked the new soft drink? **24**
14. How many customers are men AND disliked the soft drink? **14**
15. How many customers are men OR disliked the soft drink? **40**

● *Guided Skills Practice*

In a deck of 52 playing cards, how many are the type listed below?
(EXAMPLE 1)

8. a queen OR a heart 16
9. a heart OR red 26
10. a seven OR an eight 8

If you draw a card at random from a complete deck of 52 playing cards, what is the probability that you will draw the cards below?
(EXAMPLE 2)

11. a two OR a club $\frac{4}{13}$
12. a spade OR an even-numbered card $\frac{7}{13}$
13. a jack OR an eight $\frac{2}{13}$

If you draw a card at random from a complete deck of 52 playing cards, what is the probability that you will draw the cards below?
(EXAMPLE 3)

14. an even numbered card OR a red card $\frac{9}{13}$
15. a face card OR a black card (Note: There are 12 face cards: 4 kings, 4 queens, and 4 jacks.) $\frac{8}{13}$
16. a spade OR a nine $\frac{4}{13}$

Assess

Selected Answers
Exercises 8–16, 17–43 odd

ASSIGNMENT GUIDE

In Class	1–16
Core	17–34
Core Plus	17–38
Review	39–44
Preview	45

✎ Extra Practice can be found beginning on page 738.

● *Practice and Apply*

List the integers from 1 to 10 inclusive that are

17. even. {2, 4, 6, 8, 10}
18. multiples of 3. {3, 6, 9}
19. even AND multiples of 3. {6}
20. even OR multiples of 3. {2, 3, 4, 6, 8, 9, 10}

CONNECTION

STATISTICS Mark took a class survey to get student opinions about a new rule concerning students driving to school. Use the table to at right answer the questions below.

	Favor rule	Oppose rule	Total
Boys	4	9	13
Girls	7	10	17
Total	11	19	30

21. How many of those surveyed are boys? 13
22. How many of those surveyed are girls AND oppose the rule? 10
23. How many of those surveyed are girls OR oppose the rule? 26
24. How many of those surveyed are boys AND favor the rule? 4
25. How many of those surveyed are boys OR favor the rule? 20

List the integers from 1 to 20 inclusive that are multiples of the following numbers:

26. 5 {5, 10, 15, 20}
27. 3 {3, 6, 9, 12, 15, 18}
28. 5 AND 3 {15}
29. 5 OR 3 {3, 5, 6, 9, 10, 12, 15, 18, 20}
30. People waiting to hear a concert are given numbers. At one point, the people with tickets numbered from 100 to 150 inclusive are allowed to enter. How many people are allowed to enter? 51

Student Technology Guide

13.2 *Counting the Elements of Sets*

When using the Addition Counting Principle, you often add and subtract fractions with the same denominator. You can use a calculator to simplify this process.

Example: What is the probability of drawing a heart or a king from a deck of playing cards?
- There are 13 hearts and 4 kings. One of the kings is a heart. Let H be the set of hearts and K be the set of kings.
  $P(H \text{ or } K) = P(H) + P(K) - P(H \cap K) = \frac{13}{52} + \frac{4}{52} - \frac{1}{52}$
- Use a calculator to find the probability. Press (13 + 4 − 1) ÷ 52 ENTER.
  $P(H \text{ or } K) \approx 0.31$

If you draw a card at random from a deck of 52 playing cards, what is the probability that you will draw the following? Give your answers as decimals rounded to the nearest hundredth.

1. a face card or a 6 0.31
2. a club or a face card 0.42
3. a heart or an even card 0.54

Example: The results of a student survey are shown at right. What is the probability that a student is a boy or brings lunch?

	Boys	Girls	Total
Brings lunch	80	67	147
Buys lunch	32	48	80
Total	112	115	

- Let B be the set of boys and R be the set of students who bring lunch.
  $P(B \text{ or } R) = P(B) + P(R) - P(B \cap R) = \frac{112}{227} + \frac{147}{227} - \frac{80}{227}$
- Use a calculator to find the probability.
  Press (112 + 147 − 80) ÷ 227 ENTER.
  $P(B \text{ or } R) \approx 0.79$

Use the information in the table above to find each probability. Give your answers as decimals rounded to the nearest hundredth.

4. the student is a girl or brings lunch 0.86
5. the student is a boy or buys lunch 0.70

Technology

A calculator may be helpful for the additions and subtractions required in some of the exercises, particularly Exercises 33 and 34. You may wish to have students do the exercises with paper and pencil first and then use the calculator to verify their answers.

Error Analysis

When counting the elements in a union of two sets, students commonly forget to subtract the number of elements in the intersection. Encourage them to use a Venn diagram to illustrate a given situation. If necessary, they should actually show the individual elements of the sets, as was done in the Activity on page 656.

In Exercise 45, students investigate a situation in which a probability can be found by using the *complement* of a set. Students will learn more about this technique in Lesson 13.4.

CONNECTIONS

STATISTICS In a class of 28 students, 17 have brown eyes, and 13 have black hair. This includes 10 students who have both.

31. Show this information in a Venn diagram.

32. How many students have either brown eyes OR black hair? **20**

STATISTICS A survey of student participation in team sports had the results shown in the table below.

	Participated	Did not participate
Boys	124	239
Girls	111	250

33. How many students are girls AND participated? **111**

34. How many students are girls OR participated? **485**

CHALLENGE

35. If it takes 15 minutes to saw a log into 3 pieces, how long does it take to saw it into 4 pieces? (Hint: The answer is not 20.) **22.5 minutes**

APPLICATION

GAMES Tell how many of the following are in an ordinary deck of 52 cards:

36. aces **4**   **37.** red cards **26**   **38.** cards that are red OR aces **28**

Look Back

Simplify. *(LESSON 1.3)*

39. $-[-3 + (-7)] + 23 - 2(7 - 5)$ **29**   **40.** $4(a + 2b) - 5(b - 3a)$ **19a + 3b**

CONNECTION

41. GEOMETRY The volume of a cone is $V = \frac{1}{3}\pi r^2 h$. Solve the formula for h. *(LESSON 3.6)* $\frac{3V}{\pi r^2} = h$

Solve.

42. $3(x + 5) - 23 = 2x - 47$ *(LESSON 3.5)* $x = -39$

43. $\frac{x}{3} = -3x + 5$ *(LESSON 3.4)* $x = \frac{3}{2}$

APPLICATION

44. INVESTMENTS Chou deposits $150 in a bank that pays 5% interest per year, compounded annually. How much money will she have in this account after 5 years? *(LESSON 8.7)* **about $191.44**

 Look Beyond

45. Two coins are tossed. Find the probability that neither one will be heads. $\frac{1}{4}$

31.

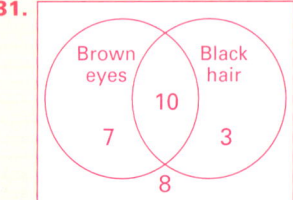

13.3 The Fundamental Counting Principle

Objectives

- Use tree diagrams to count the number of choices that can be made from sets.
- Use the Fundamental Counting Principle to count the number of choices that can be made from sets.

Why Using multiplication to count the number of choices you can make is extremely helpful in a variety of situations ranging from games to important decisions.

Abi is scheduling her classes for next year. After she registers for her required courses, she may choose one elective from cluster 1 and one from cluster 2.

APPLICATION
SCHEDULING

In certain situations you need to count the number of combined choices that are possible when one choice comes from one set of things and one choice comes from another set of things.

EXAMPLE ① Use a tree diagram to find the number of elective pairs (from the illustration above) that are possible for Abi.

- **SOLUTION**

  1. Make branches for band, orchestra, and chorus. A dot indicates a decision point.

     To each branch, add home economics and woodworking.

  2. Count the total number of choices by following all of the possible pathways, starting with the first decision point. There are 6 possible sets of choices.

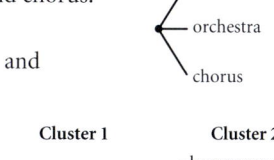

Alternative Teaching Strategy

USING TABLES As an alternative to tree diagrams, show students how to organize and count choices in a table.

Suppose that an event consists of flipping a coin two times. Work with students to create the table of possible outcomes shown at right.

	H	T
H	H, H	H, T
T	T, H	T, T

Now suppose that a third stage, such as rolling a number cube, is added to the event. Students can create a new table by using the four outcomes from the preceding table.

	1	2	3	4	5	6
H, H	H, H, 1	H, H, 2	H, H, 3	H, H, 4	H, H, 5	H, H, 6
H, T	H, T, 1	H, T, 2	H, T, 3	H, T, 4	H, T, 5	H, T, 6
T, H	T, H, 1	T, H, 2	T, H, 3	T, H, 4	T, H, 5	T, H, 6
T, T	T, T, 1	T, T, 2	T, T, 3	T, T, 4	T, T, 5	T, T, 6

Prepare

NCTM PRINCIPLES & STANDARDS
1–3, 5–10

LESSON RESOURCES

- Practice 13.3
- Reteaching Master 13.3
- Enrichment Master 13.3
- Cooperative-Learning Activity 13.3
- Lesson Activity 13.3
- Problem Solving/Critical Thinking Master 13.3

QUICK WARM-UP

How many outcomes are in the sample space for each experiment?

1. flipping a coin once — **2**
2. flipping a coin twice — **4**
3. flipping a coin three times — **8**
4. rolling a number cube once — **6**
5. rolling a number cube twice — **36**

Also on Quiz Transparency 13.3

Teach

Why Knowing how to find a total number of choices is important in many situations. For example, there are several cities where more than one area code is needed because the number of telephone customers is greater than the number of possible combinations of 7-digit phone numbers. Ask students to suggest other situations in which this type of planning might be necessary.

☞ For Additional Example 1, see page 662.

LESSON 13.3 **661**

ADDITIONAL EXAMPLE 1

A restaurant serves four kinds of sandwiches: turkey, roast beef, chicken, and vegetarian. With each sandwich, the restaurant offers a free beverage: coffee, tea, or milk. **Use a tree diagram to find the number of sandwich-and-beverage combinations that the restaurant offers.**

There are 12 combinations.

TRY THIS
There are 12 meal combinations.

Activity Notes

In this Activity, students count choices in a familiar situation: How many outfits can you make from a given number of items in your wardrobe? In Steps 1 and 2, students may need to use a tree diagram. By the time they arrive at Step 4, however, they should begin to see that the number of possible outfits can be counted by multiplying the number of choices for each of the two categories of clothing.

CHECKPOINT ✓
3. (# of shirts) • (# of ties) = (# of combinations)

ADDITIONAL EXAMPLE 2

A certain style of T-shirt is manufactured in 4 sizes. Each size is available in 9 different colors. **How many different size-and-color combinations must a clothing store stock in order to carry the complete line of these T-shirts?**

The store must stock 36 combinations.

TRY THIS A diner at a cafe may select 1 of 6 main dishes and 1 of 2 desserts. Use a tree diagram to find how many different meals are possible.

You can use multiplication to find the number of sets of choices that are possible as illustrated by the following activity.

Activity — Sets of Choices

APPLICATION CLOTHING

Michael has 3 dress shirts (blue, grey, and white) and 2 ties (black and blue). In the steps that follow, you will find how many different shirt-and-tie combinations he can make from the 3 shirts and 2 ties.

1. Suppose that Michael chooses the blue shirt. How many choices are possible for the tie?
2. How many choices for a shirt-and-tie combination are available to Michael?

CHECKPOINT ✓ 3. How might your answer to Step 2 involve multiplication?

4. Repeat this activity for 5 shirts and 3 ties.

In the activity, you may have discovered the following principle:

Fundamental Counting Principle

If there are m ways to make a first choice and n ways to make a second choice, then there are $m \cdot n$ ways to make a first choice and a second choice.

EXAMPLE 2 If you choose 1 boy out of 12 boys and 1 girl out of 13 girls, how many ways can this be done?

● **SOLUTION**
Since there are 12 boys, there are 12 ways to make the first choice. Thus, $m = 12$. Since there are 13 girls, there are 13 ways to make the second choice. Thus, $n = 13$. In all there are $m \cdot n = 12 \cdot 13 = 156$ ways to make a first and a second choice.

TRY THIS A car buyer can choose one of three different models of a new car. Five different colors are available for each. How many ways can the buyer make the choices? Relate your answer to the Fundamental Counting Principle.

Interdisciplinary Connection

CIVICS Suppose that a state numbers license plates according to this system: A letter of the alphabet is followed by a four-digit number from 1000 through 9999, inclusive, which is then followed by a letter of the alphabet. How many different license plates can this state issue?
26 • 9 • 10 • 10 • 10 • 26 = 6,084,000

Have students research the basic numbering system used for license plates in your area and determine how many possible plates can be issued.

Inclusion Strategies

TACTILE LEARNERS Some students will benefit from using concrete objects to model all possible choices in a given situation. Let the types of materials that are available dictate the situations you give them. For example, if you have a supply of number cubes and pennies, ask students to model all possible outcomes when a roll of a number cube is followed by a flip of a coin. Have them place each combination of objects on a sheet of paper and write a description of the combination next to the objects.

The Fundamental Counting Principle can be extended to more than two choices, as the next example illustrates.

EXAMPLE 3

APPLICATION
INVENTORY

A store stocks 20 styles of shoes. Each style has 4 sizes and 3 colors. **How many pairs of shoes must the store carry in order to have one pair of each choice of size, style, and color?**

○ **SOLUTION**

For each of the 20 styles there are 4 sizes. This means there are 20 • 4, or 80, choices of style and size.

For each of these 80 choices, there are 3 colors. Thus, there are 80 • 3, or 240, choices in all.

You can also perform all of the multiplications at once:

20 • 4 • 3 = 240

TRY THIS

Tammy is selecting colored pens. The first characteristic is the point, which may be either felt tip or ball-point. The second characteristic is the color—red, blue, green, or black. The third characteristic is the brand—Able or Blakely. How many possible selections of a pen can Tammy make?

EXAMPLE 4

A deck of playing cards has 26 red cards and 26 black cards. There are 4 aces —2 red and 2 black. Two decks of cards are shuffled separately. A card is drawn from each deck. **How many ways are there of drawing a black card from the first deck AND an ace from the second deck?**

○ **SOLUTION**

There are 26 ways of drawing a black card from the first deck, and 4 ways of drawing an ace from the second deck. By the Multiplication Counting Principle, there are 26 • 4, or 104, ways to draw a black card from one deck AND an ace from a second deck.

TRY THIS

There are 3 • 5, or 15, ways to choose a car.

ADDITIONAL EXAMPLE 3

A print shop offers custom-made personalized stationary. The customer can choose white, pink, or blue paper. A border around the edge can be either classic or modern in style. The border can be printed in either black, blue, brown, red, purple, or green ink. **How many choices of stationary are possible?**

There are 36 choices in all.

TRY THIS

There are 2 • 4 • 2, or 16, possible selections.

ADDITIONAL EXAMPLE 4

Two standard decks of cards are shuffled separately. Then a card is drawn from each deck. **In how many ways can you draw an ace from the first deck and a heart from the second deck?**

There are 52 ways to draw the cards described.

Enrichment

Have students work in groups of three or four. Give each group several restaurant menus. Tell them to choose one menu and decide on categories that would make up a typical meal at the restaurant. For example, the categories might be *appetizer, entrée, dessert,* and *beverage*. After a group has agreed on a menu and the categories, they should work together to find the number of possible orders. They might also consider how the number of orders is affected if a diner does not order an item from one or more of the categories.

Reteaching the Lesson

WORKING BACKWARD Have students write and solve a problem that could be modeled by the tree diagram at right. **Sample: If you have white, blue, and tan shirts and brown and black pairs of pants, you can make six different outfits.** Students can then make their own tree diagram, trade with a partner, and write a problem for each other's diagrams.

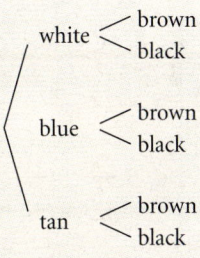

LESSON 13.3 663

TRY THIS
There are 26 • 12, or 312, ways.

CRITICAL THINKING
The results would be different if both cards were drawn from one deck. If you draw one of the black aces, you could not draw it again on the second draw. However, the result would be the same if one deck were used with replacement because the second draw would also be from a complete deck.

internet connect

GO TO: go.hrw.com
KEYWORD: MA1 Bells

Combinations and permutations are basic tools for solving probability and other mathematical problems. In this activity, the terms are defined and explored in the context of English Bell Change Ringing. Students access a site where they will learn about this art, hear sound samples, and see how sample spaces grow rapidly as more elements are added.

Practice

NAME _____ CLASS _____ DATE _____

Practice
13.3 The Fundamental Counting Principle

Jerry has 4 T-shirts, 2 pairs of jeans, and 3 pairs of shoes that can be worn together in any combination.

1. How many ways are there for Jerry to choose an outfit? **24**
2. How many outfits can include a blue T-shirt, assuming that it is one of the 4 T-shirts? **6**
3. How many outfits can include a pair of black jeans, assuming that it is one of the 2 pairs of jeans? **12**
4. How many outfits can include a pair of brown shoes, assuming that it is one of the 3 pairs of shoes? **8**

How many ways are there to arrange the letters in each of the following words?

5. cat **6**
6. math **24**
7. sport **120**
8. problem **5040**

9. The Camachos are decorating their kitchen. There are 7 different wallpaper patterns from which to choose. There are 5 borders that coordinate with each wallpaper pattern. How many combinations are possible? **35**

10. Members of the pep club are selling carnations. They are selling red, pink, or white carnations in packages of 2, 4, or 6. If all carnations in a package must be the same color, how many choices are offered? **9**

11. A popular car model comes in 5 different colors, with a standard or automatic transmission, and with or without a sports package. How many different types of cars are available in this model? **20**

12. How many ways are there to name the given line? **6**   A   B   C

13. Name all of these ways.
    AB, AC, BC,
    BA, CA, CB

TRY THIS A card is drawn from a deck of playing cards. Then it is replaced. A second card is drawn. How many ways are there of drawing a red card followed by a face card? (The face cards are the 4 kings, 4 queens, and 4 jacks. For each kind of face card, 2 are red, and 2 are black.)

CRITICAL THINKING In Example 4, two different decks were used. Would the result be the same or different if both cards were drawn from one deck? Would the result be the same or different if one deck were used and the first card chosen was *replaced* before drawing the second card?

Exercises

● Communicate

1. Explain the difference between these two sentences:
   • I am going to the mall AND I am going to study.
   • I am going to the mall OR I am going to study.

2. Suppose that you are playing a game in which the first step is to flip a coin. If the coin lands heads up, you will then roll a number cube. If the coin lands tails up, you will choose a card from a stack of 10. Explain how to find the number of branches on a tree diagram representing this situation.

3. Two cards are drawn from an ordinary deck of 52 cards. The first card is replaced before the second card is drawn. Explain how to find the number of possible ways to draw a 3 followed by a red card.

● Guided Skills Practice

Use a tree diagram to find the combined number of choices that are possible in each situation. *(EXAMPLE 1)*

4. choosing a food item and a drink from a menu with 6 food and 4 drink choices **24**

5. flipping a coin and then rolling a number cube **12**

Use the Fundamental Counting Principle to find the number of choices that are possible for each situation. *(EXAMPLE 2)*

6. A pasta dish is offered with 2 choices of pasta and 4 choices of sauces. **8**

7. A businessman has 6 ties that will go with 3 shirts. **18**

Use the Fundamental Counting Principle to find the number of possibilities for each situation. *(EXAMPLE 3)*

8. A car has the following options: 2 different types of engines, 2 body styles, 6 colors, 2 seat fabrics, and 3 interior colors. How many choices are there? **144**

9. A store stocks 10 styles of shorts. Each style has 5 sizes and 6 colors. How many pairs of shorts must the store carry to have at least one pair of each size, style, and color? **300**

664   LESSON 13.3

Use the Fundamental Counting Principle to answer each question. Two decks of playing cards are shuffled separately. A card is drawn from each deck. **(EXAMPLE 4)**

10. How many ways are there of drawing a king from the first deck AND a spade from the second deck? **52**

11. How many ways are there of drawing a red card from the first deck AND a face card from the second deck? **312**

Practice and Apply

12. In how many ways can Leanna choose 1 CD AND 1 tape from a collection of 4 CDs and 5 tapes? **20**

13. Juliana is trying to choose 1 CD AND 1 tape from a collection of 2 CDs and 3 tapes that she likes. In how many ways can she do this? **6**

14. A menu contains 4 appetizers, 3 salads, 5 main courses, and 6 desserts. How many ways are there to choose one of each? **360**

CONNECTION

15. **GEOMETRY** How many ways are there to name the triangle shown at right? **6**

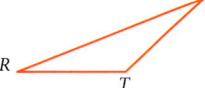

APPLICATIONS

16. **CONSUMER ECONOMICS** The Rahalls are ordering a new car. They can choose from 9 exterior colors. For each of these, there are 4 interior colors. How many versions are possible? **36**

17. **GAMES** At the start of a game of chess, white moves first and has 20 possible moves. For each of these possibilities, black has 20 possible countermoves. How many possibilities are there for the first round? **400**

18. **TELECOMMUNICATIONS** How many different 7-digit phone numbers are possible if the first 3 digits are 555? Assume that 555-0000 is allowed. **10,000**

19. **TELECOMMUNICATIONS** How many different 7-digit phone numbers are possible? Assume that phone numbers cannot start with 0. **9,000,000**

LANGUAGE ARTS A word that is obtained by rearranging the letters of another word is called an *anagram*. For example, *art* is an anagram of *rat*.

20. Find another anagram of *rat*. **tar**

21. How many ways are there to arrange the letters of the word *rat*, including non-words, using each letter exactly once? **6**

22. Write three anagrams of *stake*.
   samples: steak, takes, skate

23. How many ways are there to arrange the six letters of the word *foster*? **720**

24. Find two anagrams of *foster*. **samples: forest, softer**

Assess

Selected Answers
Exercises 4–11, 13–31 odd

ASSIGNMENT GUIDE

In Class	1–11
Core	12–24
Core Plus	12–24
Review	25–31
Preview	32–35

✎ Extra Practice can be found beginning on page 738.

Technology

A calculator may be helpful for doing the multiplications required in some of the exercises. You may wish to have students do the exercises with paper and pencil first and then use the calculator to verify their answers.

Student Technology Guide

NAME _____ CLASS _____ DATE _____

Student Technology Guide
13.3 The Fundamental Counting Principle

In Lesson 13.3, you used the Multiplication Counting Principle to find the total number of ways a series of choices can be made. A calculator can help you do such calculations quickly.

When several factors are the same, use exponents to simplify the problem.

Example: The license plates in a certain state have 3 letters followed by 3 numbers. How many different license plates are possible?

```
26^3*10^3
      17576000
```

• There are 26 possibilities for each letter (A through Z), and 10 possibilities for each digit (0 through 10). The total number of license plates is: $26 \cdot 26 \cdot 26 \cdot 10 \cdot 10 \cdot 10 = 26^3 \cdot 10^3$.
• Use the exponent key to enter this expression into the calculator. Press 26 [^] 3 [×] 10 [^] 3 [ENTER].
  There are 17,576,000 possible license plates.

1. Suppose that the license plates in another state consist of 4 digits followed by 2 letters. How many different license plates are possible?
   6,760,000

An exclamation point has a special mathematical meaning; 5!, read "5 factorial," means, "multiply all the whole numbers from 5 down to 1." So, $5! = 5 \cdot 4 \cdot 3 \cdot 2 \cdot 1 = 120$.

Example: How many batting orders are possible for 9 players on a softball team?

```
9!
      362880
```

• There are 9 players who can bat first. Once that choice is made, there are only 8 choices for the second player, 7 for the third player, and so on. The number of possible batting orders is $9 \cdot 8 \cdot 7 \cdot 6 \cdot 5 \cdot 4 \cdot 3 \cdot 2 \cdot 1 = 9!$.
  Press 9 [MATH] [PRB] 4 [ENTER] [ENTER].
  There are 362,880 possible batting orders for the 9 players.

Evaluate each of the following.

2. 4! **24**
3. 6! **720**
4. 12! **479,001,600**
5. 17! **≈3.56 × 10¹⁴**

6. A choir plans to sing 8 songs at a concert. In how many different orders could they sing the songs?
   40,320

LESSON 13.3   665

Error Analysis

Some students may become confused and add rather than multiply when applying the Fundamental Counting Principle. You might suggest that these students continue to use tree diagrams or tables to help them visualize all possible choices.

In Exercises 32–35, students see how the Fundamental Counting Principle helps in calculating a probability. Exercise 35 involves the probability of independent events, which will be studied in greater depth in Lesson 13.4.

ALTERNATIVE
Assessment

Portfolio Activity

The Portfolio Activity can be used as preparation for the Chapter Project or as a separate activity. In the Portfolio Activity on this page, students design and perform a probability experiment with number cubes and then compare the experimental probabilities that arise from their results with the theoretical probabilities.

Answers to Portfolio Activities can be found in Additional Answers of the Teacher's Edition.

25. Hexagonal numbers form a sequence: 1, 6, 15, 28, 45, 66, 91, . . . Find the next two hexagonal numbers. **(LESSON 1.1)** 120 and 153

Solve. **(LESSON 3.3)**

26. $\frac{3}{2}x = 6$ $x = 4$ **27.** $\frac{x}{-4} = -\frac{5}{8}$ $x = \frac{5}{2}$ **28.** $\frac{x}{5} = 10$ $x = 50$

CONNECTION

29. **COORDINATE GEOMETRY** Points $P(7, -4)$ and $Q(-2, 5)$ are on a line. What is the slope of that line? **(LESSON 5.2)** −1

30. If two lines are parallel, how are their slopes related? **(LESSON 5.6)** slopes are equal

APPLICATION

31. **CHEMISTRY** How many milliliters of a 9% acid solution should be mixed with 450 milliliters of a 1.6% acid solution to produce a 5% acid solution? **(LESSON 7.6)** 382.5 milliliters

APPLICATION

ACADEMICS Find the probability that Abi from page 661 chooses the following:

32. band from cluster 1 $\frac{1}{3}$ **33.** woodworking from cluster 2 $\frac{1}{2}$

34. band OR woodworking $\frac{2}{3}$ **35.** band AND woodworking $\frac{1}{6}$

Include this activity in your portfolio.

Design and conduct an experiment with two number cubes to find the probability of getting the following sums as outcomes:

1. 2 OR 12
2. 1
3. 7 OR 11
4. 6 OR 8

Perform at least 100 trials, and list each outcome as either favorable (the sum indicated) or unfavorable (not the sum indicated).

Calculate the theoretical probabilities of the four outcomes listed at left and compare with the results of your experiment.

WORKING ON THE CHAPTER PROJECT

You should now be able to complete Activity 1 of the Chapter Project on page 682.

internetconnect

GO TO: go.hrw.com
KEYWORD: MA1 Probability

To support the Portfolio Activity, students may visit the HRW Web site, where they can obtain additional information about concepts and problems in probability theory and games of chance.

13.4 Independent Events

Objective
* Find the probability of independent events.

Why There are many times when you may want to know the probability that more than one event occurs.

APPLICATION
SCHEDULING

In Lesson 13.3, Abi selects one elective from cluster 1 (band, orchestra, or chorus) and one elective from cluster 2 (home economics or woodworking). If all of her choices are equally likely, how can you find the probability that she chooses band AND woodworking?

Modeling Independent Events

You can model the sample space of Abi's choices geometrically. Draw two vertical lines to divide the model into three equal regions, one for each elective in cluster 1.

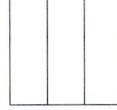

Then draw a horizontal line to divide the model into two equal regions, one for each elective in cluster 2.

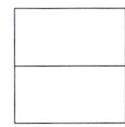

Thus, as you saw in Lesson 13.3, there are 6 possibilities. That is, the sample space for the Abi's decision has 6 elements, one of which is band AND woodworking. The probability of her selecting band and woodworking is as follows:

$$P(\text{band AND woodworking}) = \frac{1}{6}$$

 ← Woodworking

Alternative Teaching Strategy

USING COGNITIVE STRATEGIES You may wish to enhance students' understanding of independent events by providing a definition of *dependent events*: Two events are dependent if the occurrence of the first event affects the probability of the occurrence of the second. As each set of independent events in the lesson is encountered, ask students what, if anything, might change the events described to dependent events.

In the case of Abi's choice of electives, for instance, ask students how the situation changes if one elective in cluster 2 is only offered in the same time period as an elective in cluster 1. Similarly, in Example 2, discuss the effect if John were depending on Debbie to give him a ride to the lunch date.

In contrast, note that Example 1 involves two events that always are independent. If two fair number cubes are rolled, the result of one cube can never affect the result of the other.

Prepare

NCTM PRINCIPLES & STANDARDS
1–3, 5–10

LESSON RESOURCES
* Practice 13.4
* Reteaching Master 13.4
* Enrichment Master 13.4
* Cooperative-Learning Activity 13.4
* Lesson Activity 13.4
* Problem Solving/Critical Thinking Master 13.4
* Teaching Transparency 65

QUICK WARM-UP

A card is drawn from a standard deck of 52 playing cards. **Find the probability of each event.**

1. 5 $\frac{1}{13}$ 2. heart $\frac{1}{4}$
3. 5 OR heart $\frac{4}{13}$ 4. 5 AND hearts $\frac{1}{52}$

A fair number cube is rolled once. **Find the probability of each event.**

5. 1 $\frac{1}{6}$ 6. even $\frac{1}{2}$
7. 1 OR even $\frac{2}{3}$ 8. 2 OR even $\frac{1}{2}$

Also on Quiz Transparency 13.4

Teach

Why Many situations involve events that do not affect each other. For example, if you flip a coin twice, the outcome of the first flip has no effect on the outcome of the second. Tell students that events like these are called *independent events*. Have them suggest other situations that involve independent events.

LESSON 13.4 **667**

Activity Notes

In this Activity, students further investigate the situation that was used to introduce the lesson. Another choice is added to each cluster of electives, and students answer questions relating to the new choices. In the process, they consider when it is appropriate to multiply probabilities and when it is appropriate to add. They should come to associate addition with the logical connector OR and to associate multiplication with AND.

Cooperative Learning

You may wish to have student do the Activity in pairs. Partners can work together on Steps 1 and 2. Then one partner can do Steps 3 and 4, while the other does Steps 5 and 6. They can then work together again to compare results and complete Step 7.

CHECKPOINT ✓

7. Multiplication is appropriate when dealing with intersections, which are denoted by AND. Addition is appropriate when dealing with unions, which are denoted by OR.

Teaching Tip

If two events are mutually exclusive and both have nonzero probabilities, then they are not independent. For example, suppose that the probability of rain is 20% and the probability of a clear sky is 80%. Rain and a clear sky are mutually exclusive events with nonzero probabilities. They are *not* independent events because $P(\text{rain and clear sky}) = 0$ and $P(\text{rain}) \cdot P(\text{clear sky}) = (0.8)(0.2) = 0.16$, so $P(\text{rain and clear sky}) \neq P(\text{rain}) \cdot P(\text{clear sky})$.

Now consider each of the choices separately. The probability of choosing woodworking is $\frac{1}{2}$, and the probability of choosing band is $\frac{1}{3}$.

Notice that $\frac{1}{2} \times \frac{1}{3} = \frac{1}{6}$.

This suggests that you can multiply the probabilities of the individual events to find the probability of both of them occurring. This is in fact true whenever two events, such as Abi's choices, are *independent*.

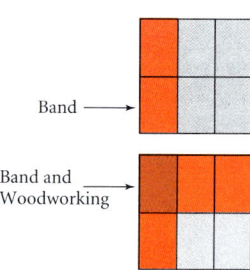

Independent Events

Two events are independent if the occurrence of one does not affect the probability of the occurrence of the other.

Abi's choices are independent because her choice from cluster 1 does not affect her choice from cluster 2.

Probabilities of Independent Events

If A and B are independent events, then $P(A \text{ AND } B) = P(A) \cdot P(B)$.

Activity: Using a Model

1. Extend the model of Abi's choices in the situation above by adding dance to cluster 1 and auto shop to cluster 2.
2. Assuming that Abi's choices are all equally likely, how many equally likely possibilities are there?
3. What is the probability that she will choose band? woodworking? band AND woodworking?
4. In Step 3, should you add or multiply the first two probabilities to find the third probability?
5. What is the probability that she will choose chorus? dance? chorus OR dance?
6. In Step 5, should you add or multiply the first two probabilities to find the third probability?

CHECKPOINT ✓ 7. When is multiplication appropriate in the questions above? When is addition appropriate in the questions above? (Relate your answer to the use of the logical relationships OR and AND.)

Interdisciplinary Connection

GENETICS Identical twins come from the same egg and are the same gender. Fraternal twins come from different eggs and can be different in gender. Assume that the theoretical probability that twins will be fraternal is $\frac{2}{3}$ and that the theoretical probability that a baby will be female is $\frac{1}{2}$. Find the following probabilities:

1. P(twins will be fraternal *and* both twins will be male) $\frac{2}{3} \cdot \frac{1}{4} = \frac{1}{6}$
2. P(twins will be identical *or* both twins will be female) $\frac{1}{3} + \frac{1}{4} - \frac{1}{12} = \frac{1}{2}$

Inclusion Strategies

INVITING PARTICIPATION The situations described in the lesson and the exercises provide several opportunities for students with limited attention spans to be actively involved. For example, you can introduce the lesson by having the class act out the choosing of electives described on page 667. A discussion of the results can lead to listing the sample space and comparing the experimental results with the theoretical probability of each choice. Similarly, students can work in small groups to act out the selection of cards in Examples 3 and 4.

668 LESSON 13.4

EXAMPLE 1 Todd rolls a red and a green number cube. **What is the probability that the red cube shows an odd number AND the green cube shows a number greater than or equal to 5?**

- **SOLUTION**

Because the result on one cube does not affect the result on the other, the events are independent. Thus, the probabilities can be multiplied.

Three of the six numbers on a cube are odd.
$$P(\text{rolling an odd number on the red cube}) = \frac{3}{6} = \frac{1}{2}$$

Two of the six numbers on a cube are greater than or equal to 5.
$$P(\text{rolling a 5 or greater on the green cube}) = \frac{2}{6} = \frac{1}{3}$$

To find the probability of both happening, multiply the individual probabilities. $\quad \frac{1}{2} \times \frac{1}{3} = \frac{1}{6}$

TRY THIS A red and a green number cube are rolled. Find the probability that the red cube shows a number that is greater or equal to 5 AND the green cube shows a number less than or equal to 2.

EXAMPLE 2

APPLICATION
SCHEDULING

John and Debbie have trouble being punctual. John is late 40% of the time, and Debbie is late 20% of the time. **Assuming that they are just as likely to be late on any particular occasion, what is the probability that both will be on time for a lunch date?**

- **SOLUTION**

On graph paper, divide a 10-by-10 square into two parts with a vertical line, as shown in the first figure. This shows a 60% probability that John is on time. Now divide the square into two parts with a horizontal line, as shown in the second figure. This shows an 80% probability that Debbie is on time.

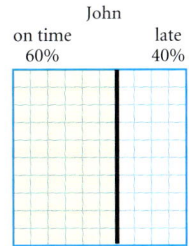 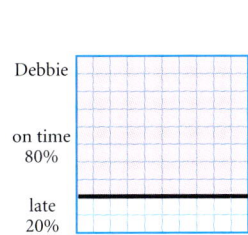 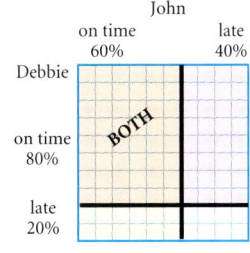

When the squares for John and Debbie are combined, the lines divide the 100-block square into four regions. The region marked BOTH represents the probability that both are on time. It contains 8 · 6, or 48, out of 10 · 10, or 100 squares. The probability that both John and Debbie are on time is $\frac{48}{100}$, or 48%. Notice that $\frac{48}{100}$ is the product of $\frac{6}{10}$ and $\frac{8}{10}$.

TRY THIS Find the probability that Mark and Tracy will both be on time if Mark is on time 40% of the time and Tracy is on time 70% of the time.

ADDITIONAL EXAMPLE 1

Two fair number cubes, one red and one green, are rolled at the same time. **What is the probability that the red cube shows an even number AND the green cube shows a prime number?** $\frac{1}{4}$

TRY THIS
The probability is $\frac{1}{9}$.

Use Teaching Transparency 65.

ADDITIONAL EXAMPLE 2

In a close softball game, Crystal is at bat, and Karen is the next to bat. Crystal gets on base in 65% of her at bats. Karen gets on base in 40% of her at bats. **What is the probability that both girls will get on base?** 26%

TRY THIS
The probability is 28%.

Teaching Tip

See the Internet Connect feature listed on page 657 for an Internet site that identifies many polyhedral shapes.

Enrichment

The figure at right is a *regular octahedron*. It has 8 congruent triangular faces. When you roll a fair number cube that has this shape, each face is equally likely to come up.

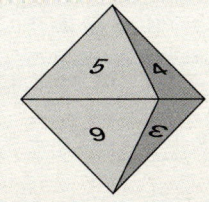

Have students work in small groups. Each group should write 10 probability questions related to rolling one or more of these number cubes.

Groups can then trade questions, answer them, and check each other's answers.

You can extend this activity to include three other types of regular number cubes.

tetrahedron: 4 triangular faces
dodecahedron: 12 pentagonal faces
icosahedron: 20 triangular faces

Number cubes with these shapes, as well as others, are available from commercial suppliers.

LESSON 13.4 **669**

ADDITIONAL EXAMPLE 3

Two fair number cubes are rolled. **Find the probability that both cubes show a number less than 3.**
$\frac{1}{9}$

TRY THIS

The probability is $\frac{1}{169}$.

ADDITIONAL EXAMPLE 4

Two fair number cubes are rolled. **Find the probability that at least one cube shows a number greater than 2.**
$\frac{8}{9}$

TRY THIS

Method 1:

$P(\text{sum of 3}) + P(\text{sum of 4}) + P(\text{sum of 5}) + P(\text{sum of 6}) + P(\text{sum of 7}) + P(\text{sum of 8}) + P(\text{sum of 9}) + P(\text{sum of 10}) + P(\text{sum of 11}) + P(\text{sum of 12}) = \frac{35}{36}$

Method 2:

$1 - P(\text{sum of 2}) = \frac{35}{36}$

EXAMPLE 3 Sarah draws a card, replaces it, and draws another card from a regular 52-card deck. **Find the probability that the 2 cards she draws are black. In a regular deck, half of the cards are black.**

SOLUTION

There are 52 ways of drawing the first card. After Sarah replaces the first card, there are 52 ways of drawing the second card. Since she draws a first card and a second card, use the Fundamental Counting Principle. There are 52 • 52, or 2704, equally likely ways of drawing the two cards when the first card is replaced.

A successful event occurs if Sarah draws two black cards. There are 26 ways to draw the first black card. After the first black card is replaced, there are 26 ways to draw the second black card. Apply the Fundamental Counting Principle. There are 26 • 26, or 676, ways to draw two black cards.

The probability of drawing two black cards is $\frac{676}{2704}$, or 25%.

TRY THIS Thelma draws a card from a regular 52-card deck. After replacing it, she draws another card from the same deck. Find the probability that the two cards she draws are aces.

Finding Probabilities With the Complement

Finding the probability of certain events can often require several calculations if the probability is found directly. In these cases, it is often easier to count the ways that something does *not* happen.

EXAMPLE 4 Two cards are drawn with replacement from an ordinary deck of 52 cards. **Find the probability that at least one of them is red.**

SOLUTION

The event *at least one card is red* includes every event that can happen when two cards are drawn *except* the event of getting two black cards. Mathematically this can be stated as $P(\text{at least one red}) = 1 - P(\text{two black cards})$.

In the previous example, we found that the probability of drawing two black cards is $\frac{676}{2704}$. Therefore, $P(\text{at least one red}) = 1 - \frac{676}{2704} = \frac{2028}{2704}$, or 75%.

TRY THIS If two number cubes are rolled, what is the probability that the sum is greater than or equal to 3? Solve the problem two different ways.

When you find the number of ways that an event can occur by considering the number of ways that it does *not* occur, you are using the **complement** of the set of ways that it can occur.

Reteaching the Lesson

USING COGNITIVE STRATEGIES Write the following on the board or overhead:

There are 6 red balls and 2 blue balls in a bag. Draw one ball, replace it, and then draw another.

Have students identify the sample space for this experiment. {(R, R), (R, B), (B, R), (B, B)}

Give students the matching exercises at right. Have them work in pairs to make the matches. Tell them that one choice will be used twice.

1. $P(R, R)$ 2. $P(R, B)$ 3. $P(B, R)$ 4. $P(B, B)$
5. $P(\text{at least one R})$ 6. $P(\text{at least one B})$
7. $P(R, R) + P(R, B) + P(B, R) + P(B, B)$

a. 1   b. $\frac{15}{16}$ c. $\frac{9}{16}$ d. $\frac{7}{16}$ e. $\frac{3}{16}$ f. $\frac{1}{16}$

1. c   2. e   3. e   4. f   5. b   6. d   7. a

Now have each pair of students think of a similar experiment, identify the sample space, and write a similar set of exercises. Pairs can trade sets of exercises, make the matches, and then check each other's answers.

Exercises

Communicate

1. Explain what it means for two events to be independent.
2. Name two ways to represent the probabilities of independent events with a diagram.
3. Explain why you use the Fundamental Counting Principle when determining the probabilities of independent events.
4. Explain how to find a probability by using the complement.

Guided Skills Practice

A red number cube and a green number cube are rolled together. Find each probability. *(EXAMPLE 1)*

5. Both numbers are even. $\frac{1}{4}$
6. Both numbers are greater than 3. $\frac{1}{4}$
7. The green cube is even AND the red cube is less than 4. $\frac{1}{4}$

Use area models to find each probability below. *(EXAMPLE 2)*

8. Melissa is interviewing for a job in the suburb of another city and decides that the best way to get there is to fly to the city and take a train to the suburb. If the plane is on time 90% of the time and the train is on time 70% of the time, what is the probability that both will be on time? **63%**

9. When commuting by train, a passenger notices that his first train is on time 90% of the time and his connecting train is on time 80% of the time. What is the probability that both will be on time? **72%**

Suppose that you draw two cards with replacement from a regular deck of playing cards. Find each probability. *(EXAMPLE 3)*

10. Both cards are kings. $\frac{1}{169}$
11. Both cards are spades. $\frac{1}{16}$
12. Both cards are red aces. $\frac{1}{676}$

Two cards are drawn with replacement from an ordinary deck of 52 cards. Find each probability. *(EXAMPLE 4)*

13. At least one card is not a king. $\frac{168}{169}$
14. At least one card is not a spade. $\frac{15}{16}$
15. At least one card is not a red ace. $\frac{675}{676}$

Assess

Selected Answers
Exercises 5–15, 17–41 odd

ASSIGNMENT GUIDE

In Class	1–15
Core	16–21, 23–34
Core Plus	16–34
Review	35–42
Preview	43

✎ Extra Practice can be found beginning on page 738.

LESSON 13.4 **671**

Error Analysis

Students sometimes identify a complement incorrectly when the choices involved are numerical. For example, if the event is two number cubes that each show a number greater than 3, students may identify the complement as two number cubes that each show a number less than 3. Have students list the sample space, and demonstrate that the two events are not complements because there is an element of the sample space (3) that is not included in either event. The complement of the event that each of two number cubes shows a number greater than 3 is that at least one of the number cubes shows a number that is less than or equal to 3.

Look Beyond

Exercise 43 touches on the concept of *expected value*. Students will learn more about this topic if they take a more advanced course in algebra.

Practice

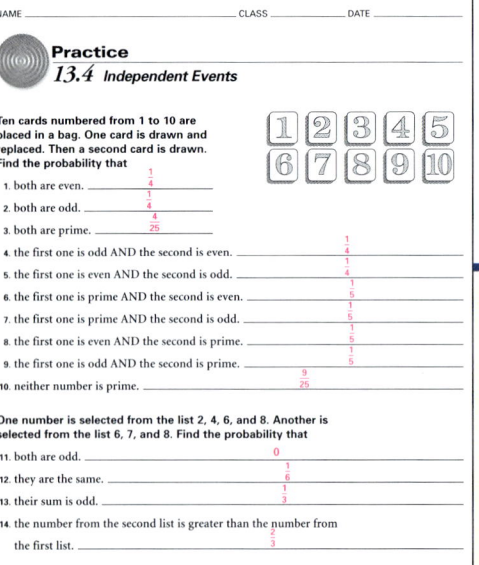

Practice and Apply

A coin is tossed twice. Tell whether the events in each pair below are dependent or independent.

16. The first toss is heads, and the second toss is heads. **Ind.**
17. The first toss is heads, and the second toss is tails. **Ind.**
18. The first toss is tails, and the second toss is heads. **Ind.**

Use the grid below to find the following probabilities:

19. Event A occurs. **80%**
20. Event B occurs. **70%**

CHALLENGE

21. Event A occurs AND event B occurs. **56%**
22. Event A occurs OR event B occurs. **94%**

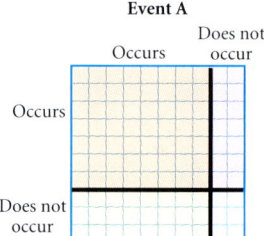

APPLICATIONS

GAMES One number is selected from the list {1, 3, 5, 7}. Another is selected from the list {5, 6, 7}. Find each probability.

23. The two numbers selected are even. **0**
24. The two numbers selected are the same. $\frac{1}{6}$
25. The sum of the two numbers selected is even. $\frac{2}{3}$
26. The number selected from the second list is greater than the number selected from the first list. $\frac{2}{3}$

GAMES Five chips numbered 1 through 5 are placed in a bag. A chip is drawn and replaced. Then a second chip is drawn. Find the following probabilities:

27. Both chips are even. $\frac{4}{25}$ 28. Both chips are odd. $\frac{9}{25}$
29. The first chip is even AND the second chip is odd. $\frac{6}{25}$
30. One chip is even AND the other is odd. $\frac{12}{25}$

RECREATION John and Jim are each hiking from Wapitu Falls to Songbird Lake on the trails shown on the map. Assume that each trail is equally likely to be chosen.

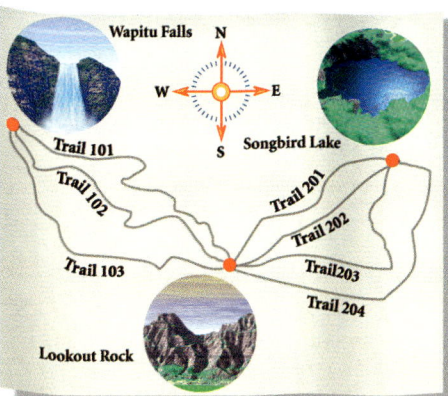

31. Find the probability that John takes trail 101 from Wapitu Falls to Lookout Rock. $\frac{1}{3}$
32. Find the probability that John takes trail 201 from Lookout Rock to Songbird Lake. $\frac{1}{4}$
33. Find the probability that John takes trail 101 from Wapitu Falls to Lookout Rock AND takes trail 201 from Lookout Rock to Songbird Lake. $\frac{1}{12}$
34. Find the probability that John takes trail 101 from Wapitu Falls to Lookout Rock OR he takes trail 201 from Lookout Rock to Songbird Lake (or both). $\frac{1}{2}$

Look Back

35. Find the product: $-6(2)\left(\dfrac{5}{-3}\right)$. *(LESSON 2.4)* 20

Find the slope of each line. *(LESSON 5.4)*

36. $y = 4x$ 4   **37.** $y = 4x - 3$ 4   **38.** $y = -4x$ −4   **39.** $y = \dfrac{x}{4}$ $\dfrac{1}{4}$

40. Tim leaves home on his bike and rides at a rate of 12 miles per hour. Two hours after he leaves, Tim's mother remembers that Tim has a dentist's appointment. She sets out after him in her car at 40 miles per hour. How long does it take her to catch up to him? *(LESSON 7.6)*
0.857 hour or about 51.4 minutes

APPLICATIONS

41. ACADEMICS At one school, 48 students signed up to take Spanish, and 23 signed up to take French. If there were 12 students that signed up for both, how many students total signed up for these foreign languages?
(LESSON 13.2) **59 students**

42. TRAVEL There are 3 highways from Birchville to Pine Springs. From Pine Springs to Clearwater, there are 5 roads. How many possible ways are there to travel from Birchville to Clearwater? *(LESSON 13.3)* **15 routes**

Look Beyond

APPLICATION

43. CONTESTS Two radio stations are each having a contest. At the first station, the prize is $1000, and your probability of winning is $\dfrac{1}{100{,}000}$. At the second station, the prize is $25, and your probability of winning is $\dfrac{1}{2000}$. Which contest provides a better return for the listener's time? Compare the probabilities of winning each contest. (Hint: What would you expect to happen if you played each contest 100,000 times?)
The second station's contest has a better chance of yielding money.

Include this activity in your portfolio.

Design and conduct an experiment for the following situation: Two evenly matched teams play a three-game series. Perform at least 100 trials and list the experimental probability of the outcomes below.

Team 1 wins the following number of games:
  a. exactly 1 game   **b.** exactly 2 games   **c.** 3 games

Team 2 wins the following number of games:
  a. exactly 1 game   **b.** exactly 2 games   **c.** 3 games

Calculate the theoretical probabilities of each outcome above and compare them with the results of your experiment.

ALTERNATIVE Assessment

Portfolio Activity

The Portfolio Activity can be used as preparation for the Chapter Project or as a separate activity. In the Portfolio Activity on this page, students design and perform a simulation of an experiment and use the results to calculate experimental probabilities. Then they compare the experimental and theoretical probabilities.

Answers to Portfolio Activities can be found in Additional Answers of the Teacher's Edition.

Student Technology Guide

NAME _____ CLASS _____ DATE _____

Student Technology Guide
13.4 Independent Events

The probability of two independent events is the product of the probabilities of the individual events. You can use a calculator with fraction capabilities to express answers to these problems in different ways.

Example: Dana draws a card from an ordinary deck of 52 cards. She replaces it and then draws another card. What is the probability that she draws a face card and then an ace? Express the answer as a decimal rounded to the nearest hundredth and as a fraction.

- $P(\text{face card}) = \dfrac{12}{52}$ and $P(\text{ace}) = \dfrac{4}{52}$. The probability of a face card, then an ace is $\dfrac{12}{52} \times \dfrac{4}{52}$.
- To express the answer as a decimal, press [ ]
  12 [÷] 52 [ ] [×] [ ] 4 [÷] 52 [ ] ENTER.
- To express the answer as a fraction, press [2nd] [ENTER], then press [MATH] [▶Frac] [ENTER] [ENTER].

```
(12/52)*(4/52)
      .0177514793
(12/52)*(4/52)▶F
rac
            3/169
```

The probability of drawing a face card and then an ace is $\dfrac{3}{169} \approx 0.02$.

Find each probability. Give your answers as fractions and as decimals rounded to the nearest hundredth.

1. the probability that Dana draws a red card, replaces it, and then draws a face card $\dfrac{3}{26} \approx 0.12$

2. the probability that Dana draws an ace, replaces it, and then draws a red face card $\dfrac{3}{338} \approx 0.01$

You can use *expected value* to analyze a situation in which events have different values. The expected value is found by multiplying the probability of each event by its value and then adding the products.

Example: In a carnival game, players have a 5% chance of winning a $10 stuffed animal, a 24% chance of winning a $4 spinning top, and a 71% chance of winning a $0.75 ball. Find the expected value.

- Multiply the probability of each event (in decimal form) by its value, and then add the results. The expected value of the game is: $0.05 \times 10 + 0.24 \times 4 + 0.71 \times 0.75$.
  The expected value of the game is about $1.99.

3. Suppose that an oil well has a 7% chance of making a $1,000,000 profit, a 35% chance of making a $360,000 profit, and a 58% chance of losing $150,000. What is the expected value of the oil well?   **$109,000**

Eyewitness Math

Focus

A study of the shooting records of basketball players challenges the widely held belief that a player sometimes develops "hot hands." Students work with some of the researchers' raw data to get a feel for the way the study was conducted. Then they are asked to form their own opinion about the concept of hot hands.

Motivate

Discuss hot streaks and cold streaks in sports. Have students who are familiar with basketball explain what is meant by "hot hands." Ask if they know of any professional players who are currently considered to have hot hands.

Have students read the article. Ask them if they have any questions that need clarification. Poll the class on their opinion: Do they believe there is such a thing as hot hands, or do they think shots are made or missed randomly? Record the results of the poll.

Tell students that they will analyze some of the actual data from the study to reach a more informed opinion. Point out that the set of data used by the researchers actually included many more players and games than that given here.

Be sure students understand how to interpret the numbers. Point out that a space between groups of numbers indicates the end of one game and the start of another. Stress that a string of four 1s in a single game counts as three instances of a hit followed by a hit. Also point out that the first shot in a game is not counted because the preceding shot was in a different game.

THE EYEWITNESS MATH

HOT HANDS OR HOOP-LA?

'Hot Hands' Phenomenon: A Myth?

The gulf between science and sports may never loom wider than in the case of hot hands. Those who play, coach, or otherwise follow basketball believe almost universally that a player who has successfully made his last shot or last few shots—a player with hot hands—is more likely to make his next shot. An exhaustive statistical analysis led by a Stanford University psychologist examined thousands of shots in actual games and found otherwise: the probability of a successful shot does not depend at all on the shots that came before.

To the psychologist, Amos Tversky, the discrepancy between reality and belief highlights the extraordinary differences between events that are random and events that people perceive as random. When events come in clusters and streaks, people look for explanations; they refuse to believe they are random, even though clusters and streaks do occur in random data.

To test the theory, researchers got the records of every shot taken from the field by the Philadelphia 76ers over one and a half seasons. When they looked at every sequence of two shots by the same player—hit-hit, hit-miss, miss-hit, or miss-miss—they found that a hit followed by a miss was actually a tiny bit likelier than a hit followed by a hit.

They also looked at sequences of more than two shots. Again, the number of long successful streaks was no greater than would have been expected in a random set of data, with every event independent of its predecessor.

The New York Times, April 19, 1988

Researchers' Survey of 100 Basketball Fans

Does a player have a better chance of making a shot after having just made his last two shots than he does after having just missed his last two shots?

Is it important for players to pass the ball to someone who has just made several (2, 3, or 4) shots in a row?

DISCUSS Do you agree with the majority of basketball fans surveyed? Why or why not?

---

1. a.

Player A

After a hit	
no. of times next shot is made	no. of times next shot is missed
17	19

Player B

After a hit	
no. of times next shot is made	no. of times next shot is missed
16	18

b. Player A—hit followed by a hit: 17 of 36 or about 47%; hit followed by a miss: 19 of 36 or about 53%

Player B—hit followed by a hit: 16 of 34 or about 47%; hit followed by a miss: 18 of 34 or about 53%

c. about 50% for each because whether the player just hit or missed, the chances of making the next shot would be equal

Cooperative Learning

To form your own opinion about the hot-hands study, it helps to know how the researchers tabulated their data. For simplicity, you will use a portion of their data, which shows field-goal attempts by two players for several games. Both players usually make about half of their shots from the field.

A	11100010 01110111010010 000011101 0100000101 010101001110 1001111 01001010001 011110110
B	10101000100 11100000110011 0101010111001111 111001001101 0001101010 10010001111

(1 = hit, 0 = miss, space = new game)

1. First, look at what happens after a shot is made within each game.
   a. Copy and complete Table 1 below, using the data for either player A or player B.
   b. What percent of the time was a hit followed by another hit? What percent of the time was a hit followed by a miss?
   c. What would you expect your results in part **b** to be if there were no such thing as *hot hands*, that is, if the shots were hit or missed just by chance? Explain. (Assume that the player makes 50% of his shots on average.)

Table 1:   After a hit

No. of times next shot is made	No. of times next shot is missed

Table 2:   After a miss

No. of times next shot is missed	No. of times next shot is made

2. Now look at what happens after a shot is missed within each game.
   a. Copy and complete Table 2. Use the same player you used for Table 1.
   b. What percent of the time was a miss followed by a hit? What percent of the time was a miss followed by another miss?
   c. What would you expect your results in part **b** to be if the shots were hit or missed just by chance?
3. Do you think there is such a thing as hot hands in basketball? Why or why not? How does your belief affect the strategy you would use if you were coaching a basketball team?

Cooperative Learning

Have students work in pairs. In Steps 1 and 2, one partner can work with the data for player A, while the other works with the data for player B. The partners can do parts **a** and **b** individually and then compare results and do part **c** together.

To answer part **c**, you might suggest that students perform a series of coin tosses. Using heads to represent a hit and tails to represent a miss, the result for several tosses is a set of random data that simulates what would happen if the hits or misses were chance occurrences.

Partners should discuss Step 3 together. Then each student should write an individual response to the questions.

Discuss

Bring the groups together to discuss their results. After the discussion, repeat the opinion poll. Compare these results to the results of the earlier poll. Discuss whether there was any significant change.

As a follow-up, have students collect their own shooting data by watching an NBA, college, or high school team play. They can compare these data to random data generated by coin tosses. Be sure students understand that the coin-toss data are based on the assumption that a player has a 50% chance of success on any given shot.

---

2. a.

Player A

After a miss	
no. of times next shot is missed	no. of times next shot is made
15	21

Player B

After a miss	
no. of times next shot is missed	no. of times next shot is made
16	18

b. Player A—miss followed by miss: 15 of 36 or about 42%; miss followed by a hit: 21 of 36 or about 58%

Player B—miss followed by miss: 16 of 34 or about 47%; miss followed by hit: 18 of 34 or about 53%

c. about 50% for each because whether the player just hit or missed, the chances of making the next shot would be equal

3. Answers may vary. Sample answer: The study showed that the shooting records of players is about what you would get if they hit or missed at random. A good coaching strategy is to get the ball to the player with the highest overall shooting percentage.

EYEWITNESS MATH 675

Prepare

NCTM PRINCIPLES & STANDARDS 1–3, 5–10

LESSON RESOURCES

- Practice 13.5
- Reteaching Master 13.5
- Enrichment Master 13.5
- Cooperative-Learning Activity 13.5
- Lesson Activity 13.5
- Problem Solving/Critical Thinking Master 13.5
- Teaching Transparency 66

QUICK WARM-UP

A number cube was rolled 120 times. **Based on the results in the table, find the experimental probability of each event below.**

Event	Occurrences
1	18
2	21
3	17
4	20
5	23
6	21

1. 4 $\frac{1}{6}$ **2.** 1 $\frac{3}{20}$

3. even $\frac{31}{60}$ **4.** odd $\frac{29}{60}$

5. 1 OR 4 $\frac{19}{60}$ **6.** 4 OR even $\frac{31}{60}$

Also on Quiz Transparency 13.5

Teach

Why Many computer applications are designed to *simulate* actual experiences for the purpose of either work or entertainment. Ask students to describe computer simulations that are familiar to them. Examples might include action games and flight simulators.

13.5 Simulations

Objective
- Design and perform simulations to find experimental probabilities.

Why Experiments that use lists of random numbers can be used to simulate outcomes in order to find experimental probabilities. Simulations are used to study games, sports, weather, political elections, and many other situations.

APPLICATION
SPORTS

Amy usually makes 50% of her field-goal attempts in games.

In a certain game, Amy made four consecutive baskets. How often do you think Amy makes four baskets in a row in a game in which she takes 20 shots? If you had the actual shot-by-shot records of Amy's performance, you could see how often this happens. Even without the records, it is possible to find the **experimental probability** by using a *simulation*. A **simulation** is an experiment with mathematical characteristics that are similar to the actual event.

Suppose that Amy usually makes exactly half of the shots she takes in a game. If Amy takes 20 shots, a simulation can be designed to find the experimental probability that she will make four successful shots in a row.

TECHNOLOGY
GRAPHICS CALCULATOR

Outcomes are random if all possible outcomes are equally likely. A graphics calculator can be used to generate random numbers. (See the Calculator Keystroke Guide on page 782.)

1. Set up your calculator to randomly generate 0s and 1s.
2. Let 1 represent a successful shot, and let 0 represent a missed shot.
3. Each randomly generated number represents a shot. Generate 20 random numbers (these 20 numbers are considered one **trial**), and write them down in order.
4. Repeat Step 3 at least 10 times. A larger number of simulations should result in an experimental probability that is closer to the theoretical probability.
5. Count the number of trials in which a sequence of at least four consecutive 1s appears, and divide this result by the number of trials.

Alternative Teaching Strategy

USING TECHNOLOGY Use the random-number function of your spreadsheet software to create trials for the Activity on page 677. Let 1 represent a success, and let 0 represent a miss. In cell **A1**, enter **=INT(10*RAND()/5)**. Then highlight cells **A1** through **A20** and choose **Fill Down** from the **Edit** menu. This will randomly generate 20 0s or 1s, which represent one trial. To generate a greater number of trials, such as 12, highlight cells A1 through **A20** and the corresponding cells in 11 columns to the right. Choose **Fill Right**.

Interdisciplinary Connection

HEALTH Approximately 45% of the population of the United States has type O blood. This is an important statistic because anyone in need of a transfusion can receive red blood cells from someone with type O blood.

Suppose that two blood donors at a local clinic are chosen at random. Design and perform a simulation, and then give the experimental probability that at least one of the donors has type O blood. **Simulations and probabilities will vary.**

676 LESSON 13.5

Trial	Shot									
	1	2	3	4	5	6	7	8	9	10
1	0	0	1	0	0	0	0	1	0	0
2	1	1	1	0	0	0	1	1	1	1
3	0	1	1	0	1	1	1	1	0	1
4	0	1	0	1	1	0	0	1	0	0
5	0	1	0	1	1	0	0	1	1	0
6	0	0	0	1	1	0	0	0	1	1
7	1	1	1	0	0	1	0	1	0	1
8	0	0	0	0	0	0	1	1	1	0
9	1	1	0	1	1	0	0	0	1	1
10	1	1	1	0	0	1	0	1	0	1

Of the 10 trials shown, there were at least four consecutive 1s recorded in 2 of the trials (which ones?). The experimental probability, in this case, that Amy will make four baskets in a row is $\frac{2}{10}$. Note that this is a relatively small number of trials, so this experimental probability may be significantly different from the theoretical probability.

When designing a simulation of an actual event to approximate an experimental probability, you must do the following:

1. Choose a random-number generator, such as a coin toss, number-cube roll, calculator, computer, or other method, that can be used to simulate the situation. Decide how to perform the experiment so that it simulates one trial.
2. Perform a large number of trials, and record the results.
3. Use the number of favorable outcomes, f, and the number of trials, t, to compute the experimental probability: $P = \frac{f}{t}$.

Activity
Basketball Simulation

APPLICATION
SPORTS

You will need: a coin

A coin can be used to simulate the basketball situation previously described. Let heads represent a success, and tails represent a miss.

1. Toss a coin 20 times, and record the sequence of heads and tails. This represents one game.
2. Does your sequence contain at least four consecutive heads?
3. Does your trial represent making four baskets in a row or not?
4. Conduct several trials, each representing one game with 20 attempted shots. Divide the total number of four consecutive heads by the total number of trials.

CHECKPOINT ✓ 5. What is the experimental probability of four consecutive baskets?

Enrichment

Show students how to use simulation methods to estimate the area of an irregular shape. Give them the figure at right. Note that the area of the square is 100 square units and that the curve is part of the graph of $y = 0.5x^2$. The goal is to find the area of the shaded region. Have students randomly generate integral coordinates for 20 points on or inside the square. Instruct them to find the percent of those points that lie in the shaded region, multiply by 100 square units, and record the result. Repeat this several times. The average of the results is an estimate of the area of the shaded region.

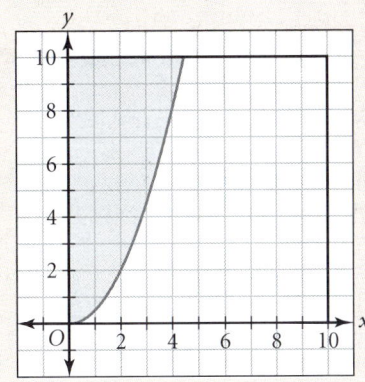

The area of the shaded region is ≈29.8 square units. Students' estimates will vary.

Teaching Tip
Random numbers were originally generated by equipment similar to that used in gambling casinos. For this reason, simulation methods that utilize random numbers are often called *Monte Carlo methods*, after the famous Monte Carlo casino that is located in the European principality of Monaco.

Activity Notes
In this Activity, students use a coin toss to simulate the shots made by a basketball player who has 50% chance of success for any shot. As students investigate the likelihood that this player will have four successful shots in a row, they will see that the occurrence of this event varies from simulation to simulation. However, they will learn that they can calculate an experimental probability by averaging the results of several simulations.

Cooperative Learning
You may wish to have students do the Activity in groups of three. For each trial, one student can toss the coin, one can record the data, and the third can count the occurrences of four consecutive successes. Encourage students to change roles from trial to trial. The group can then work together to calculate the experimental probability.

CHECKPOINT ✓
5. Answers may vary. Students should calculate the experimental probability of four consecutive baskets from the data gathered in Step 4.

LESSON 13.5 677

ADDITIONAL EXAMPLE 1

Refer to the weather report described in Example 1. The table below shows a second simulation of 20 trials. Let 1, 2, 3, and 4 represent rain. Let any other number represent no rain. **According to the results below, on how many weekends did it rain on at least one day?**

Trial	1st number	2nd number
1	2	6
2	5	8
3	1	4
4	10	3
5	8	10
6	3	4
7	1	10
8	2	1
9	6	9
10	10	3
11	3	2
12	1	8
13	1	5
14	4	10
15	1	9
16	7	3
17	10	8
18	4	3
19	10	10
20	1	2

15 weekends out of 20, or 75% of the weekends

TRY THIS
Students' answers should be modeled on Example 1. Results should be close to (0.6)(0.6) = 0.36, or 36%.

CRITICAL THINKING
The larger the number of trials, the more accurate the experimental probability should be.

Use Teaching Transparency 66.

EXAMPLE 1
APPLICATION
METEOROLOGY

When watching weather reports to plan outdoor activities for the weekend, the *chance of rain* is often mentioned. This is given as a percent. The given percent is simply an *experimental probability*. The weather conditions that exist have resulted, in the past, in rain the given percent of the time. Suppose that the weather report is for a 40% chance of rain on Saturday and a 40% chance of rain on Sunday. **Create a simulation and discuss the results. According to your results, on how many weekends did it rain at least one day?**

SOLUTION

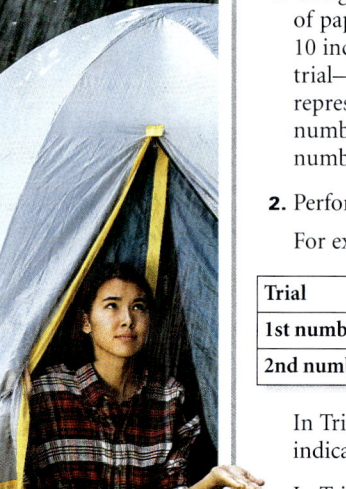

1. Using a random-number generator such as a graphics calculator or slips of paper numbered from 1 to 10, generate pairs of numbers from 1 to 10 inclusive. Each pair of numbers that you generate will represent one trial—the first number represents Saturday and the second number represents Sunday. Because the numbers 1 through 4 are 40% of the numbers 1 through 10, let 1, 2, 3, and 4 represent rain. Let any other number represent no rain.

2. Perform a large number of trials, and record the results.

   For example, the following table shows a simulation of 20 trials:

Trial	1	2	3	4	5	6	7	8	9	10	11	12	13	14	15	16	17	18	19	20
1st number	3	5	10	6	3	9	8	9	4	8	6	2	4	8	10	7	1	7	7	5
2nd number	10	1	9	10	6	10	5	5	2	6	5	1	3	2	4	7	2	7	8	10

In Trial 1, the number 3 indicates rain on Saturday and the number 10 indicates no rain on Sunday.

In Trial 2, the number 5 indicates no rain on Saturday and the number 1 represents rain on Sunday.

In Trial 9, the 4 and the 2 represent rain on both Saturday and Sunday.

In Trial 11, the 6 and the 5 represent no rain on both Saturday and Sunday.

According to the results, rain occurred on at least one day of the weekend 9 out of 20 times, or 45% of the time.

TRY THIS Suppose that the weather report gives a 60% chance of rain on Saturday and a 60% chance of rain on Sunday. Create a simulation and discuss the results. According to your results, what is the probability that it will rain on both days?

CRITICAL THINKING Explain how the number of trials affects the accuracy of the probability of an event determined by experiment.

Computer spreadsheets are powerful tools for many applications. One useful feature of spreadsheets is their ability to use formulas to perform many calculations with a single set of instructions. To set up a random number generator for Example 1 in a spreadsheet, you would perform the following steps:

1. Select a cell and enter the formula = Int(10*Rand())+1. This generates a random number from 1 to 10 inclusive.

Inclusion Strategies

INVITING PARTICIPATION Simulations provide an opportunity to actively involve students who believe that mathematics has no relevance to their lives. Encourage them to find a "success statistic" relating to an area of interest. For example, if the student has a favorite video game, the game might keep a record of the student's wins and losses. Have them create a question to investigate, such as "What is the probability that I will win three of the next five games I play?" They can then design an appropriate simulation and calculate the experimental probability.

Reteaching the Lesson

GUIDED ANALYSIS Have students work in groups. Tell them to suppose that there is a 50% chance of rain today and a 50% chance of rain tomorrow. Assign each group one method: coin, number cube, spinner, slips of paper, or random-number generator. Have each group plan and carry out a simulation to find experimental probability that it will rain both days. Bring the class together to share results. Discuss how their simulations would be different if the chance of rain were 30% each day. Note that the coin and number cube would no longer be viable simulation methods.

678 LESSON 13.5

2. Copy that cell, select a row or column of cells, and use the "Paste Special" feature from the Edit menu. You will select "Paste Formulas" from the options given.

3. The selected cells will then automatically fill with randomly generated numbers from 1 to 10 inclusive.

	A	B	C	D	E	F
1	2	2	4	8	2	3
2	3	1	6	7	7	4
3	3	1	5	7	7	5
4	4	1	9	5	5	8
5	7	2	10	10	7	8
6	4	2	4	4	3	5
7	2	9	7	10	2	5
8	4	1	6	4	10	3
9	5	1	6	8	6	7
10	9	3	5	1	7	4
11						

Exercises

Communicate

1. What are the three steps for designing a simulation?

2. Assume that the chance of making a shot in a basketball game is 50%. Give three different ways to simulate making or missing a shot.

3. Suppose that you want to simulate a 30% chance of rain. Tell how to do this by using slips of paper numbered from 1 to 10 inclusive.

4. Name three things that could be simulated by tossing a coin one time.

5. Tell how you might simulate the situation of correctly guessing the answer to a multiple-choice question with 5 possible responses.

6. Suppose that 90% of the flights for an airline are on time. Explain how to design a simulation to find the experimental probability that three consecutive flights are on time.

Scene from the movie "Singing in the Rain"

Guided Skills Practice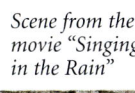

Use coin tosses to simulate a 50% chance of rain for three days. Perform 20 trials and combine your results with those of your classmates to answer the questions below. (EXAMPLE 1)

7. For the entire class, what was the percent of trials that had rain on all three days?

8. For the entire class, what was the percent of trials that had no rain?

---

7. The class should combine their results to calculate the percent of trials that had rain on all three days.

8. The class should combine their results to calculate the percent of trials that had no rain.

 internetconnect

GO TO: go.hrw.com
KEYWORD: MA1 Simulation

In this activity, students will access a site with a simple random-number generator to simulate traffic patterns at an intersection. They will then analyze the data and make a decision based on their results. Finally, they are asked to think of other ways to model the problem.

Assess

Selected Answers
Exercises 7–8, 9–27 odd

ASSIGNMENT GUIDE

In Class	1–8
Core	9–22
Core Plus	9–22
Review	23–27
Preview	28

Extra Practice can be found beginning on page 738.

Technology
If students have access to graphics calculators or spreadsheet software, you may wish to have them use this technology to perform the simulations in Exercises 19–22.

Error Analysis
Students often set up a random-number simulation correctly, but become confused when interpreting the results. Suggest that they make their tables of results big enough both the numbers and the interpretations. For instance, given the table in Example 1, they can write *rain* next to numbers from 1 through 4 and *no rain* next to the other numbers.

LESSON 13.5 **679**

9. 2 successful shots out of 10

10. Out of the 10 trials, 5 had results of more than 50% success.

11. INT(RAND()*2)

12. did not make the shot; made the shot

13. Let the numbers 1 to 8 represent the occurrence of rain, and the numbers 9 and 10 represent no rain. Generate 2 random numbers, and let the results represent the weather on day 1 and day 2, respectively. Repeat for 10 trials. Divide the number of trials where rain occurred on at least one day by 10. The quotient is the experimental probability.

Practice and Apply

APPLICATIONS

SPORTS Refer to the data for the basketball simulation on page 677.

9. How many baskets did Amy make in the first trial?

10. In how many trials did Amy make baskets for more than 50% of her shots?

11. What formula is used to generate random numbers in this spreadsheet?

12. What does a value of 0 mean in this chart? What does a value of 1 mean in this chart?

METEOROLOGY Suppose that you want to simulate an 80% chance of rain on each of two consecutive days in order to find the probability of there being rain on at least one of the two days. Explain how you would perform the simulation by using the following methods:

13. generating random numbers from 1 to 10 inclusive

14. using slips of paper

Describe one way to simulate the random selection of the following:

15. a day of the week
16. a day of the year
17. 1 student out of 24 students
18. 1 of 8 team captains

Design a simulation for each situation in Exercises 19–22, and give the results of your simulations.

19. Pick two random numbers from 1 to 100 inclusive. Give the experimental probability that both numbers are less than or equal to 20.

20. A person has a 30% chance of correctly guessing a number. What is the experimental probability that a person will correctly guess the number on 2 out of 5 tries?

21. Design a simulation and then give the experimental probability that a family with 4 children has 2 boys and 2 girls, assuming that the births of boys and girls are equally likely.

22. Design a simulation and then give the experimental probability that a student guesses correctly on all 3 questions of a true-false quiz.

"Guess your weight" is a game of chance that is popular at some carnivals.

14. Write numbers on five slips of paper. Let the numbers 1 to 4 represent the occurrence of rain, and let 5 represent no rain. Draw a slip twice (with replacement), with each draw representing 1 day. Record the result. Repeat for 10 trials. Divide the number of trials where rain occurred on at least one day by 10. The quotient is the experimental probability.

15. Use a calculator to generate numbers from 1 to 7; use the command INT(RAND*7)+1.

16. Use a calculator to generate numbers from 1 to 365; use the command INT(RAND*365)+1.

17. Use a calculator to generate numbers from 1 to 24; use the command INT(RAND*24)+1.

18. Use a calculator to generate numbers from 1 to 8; use the command INT(RAND*8)+1.

Look Back

23. Write an expression that will generate the *n*th term of the following sequence: 4, 8, 12, 16, 20, 24, ... (**LESSON 1.1**) $t = 4n$, where $n = 1, 2, 3, 4, ...$

24. Write an equation to represent the length, *l*, of a piece of fabric whose width, *w*, is 74 centimeters less than the length. (**LESSON 1.2**) $l = w + 74$

25. Solve Exercise 24 for the width, *w*, if the length, *l*, is 86 centimeters. (**LESSON 3.1**) **12 centimeters**

CONNECTION

26. **COORDINATE GEOMETRY** A line passes through the origin and the point $A(3, -2)$. Write an equation for the line that passes through point *A* and is perpendicular to the original line. (**LESSON 5.6**) $3x - 2y = 13$

27. Find each product. Are the products equal? (**LESSON 12.8**)

$$\begin{bmatrix} -1 & 3 \\ 5 & 2 \end{bmatrix} \begin{bmatrix} 2 & -4 \\ 1 & -3 \end{bmatrix} \qquad \begin{bmatrix} 2 & -4 \\ 1 & -3 \end{bmatrix} \begin{bmatrix} -1 & 3 \\ 5 & 2 \end{bmatrix}$$

Look Beyond

28. Draw the graphs of $y = x^2$ and $y = 2^x$, and explain the differences between them.

Include this activity in your portfolio.

A bag contains 12 marbles: 6 red (R), 4 green (G), and 2 yellow (Y). Two marbles are randomly drawn. Design a simulation and find the experimental probabilities of the events below. Copy and complete the table.

	First marble returned (independent events)	First marble *not* returned (dependent events)
$P(R, \text{then } R)$	?	?
$P(R, \text{then } G)$	?	?
$P(R, \text{then } Y)$	?	?
$P(G, \text{then } R)$	?	?
$P(G, \text{then } G)$	?	?
$P(G, \text{then } Y)$	?	?
$P(Y, \text{then } R)$	?	?
$P(Y, \text{then } G)$	?	?
$P(Y, \text{then } Y)$	?	?
SUM	?	?

Write a short summary that describes your results. Describe and explain differences in the results when returning the marble and when not returning the marble.

WORKING ON THE CHAPTER PROJECT
You should now be able to complete Activities 2 and 3 of the Chapter Project on page 682.

19. Use a calculator to generate numbers from 1 to 100; use the command INT(RAND*100)+1. Generate 2 random numbers; record the result. This represents 1 trial. Repeat for 20 trials. Divide the number of trials where both numbers are less than or equal to 20 by 20. The quotient is the experimental probability. Answers may vary.

20. Use a calculator to generate numbers from 1 to 10; use the command INT(RAND*10) + 1. Let 1, 2, or 3 represent a win. Generate 5 random numbers; record the result. This represents 1 trial. Repeat for 10 trials. Divide the number of trials where there are 2 wins by 10. The quotient is the experimental probability. Answers may vary.

The answers to Exercises 21, 22, 27, and 28 can be found in Additional Answers beginning on page 810.

In Exercise 28, students investigate the difference between the graph of $y = x^2$ and of $y = 2^x$. Students will revisit the topic of graphing functions in Lesson 14.1.

ALTERNATIVE
Assessment

Portfolio Activity

The Portfolio Activity can be used as preparation for the Chapter Project or as a separate activity. In the Portfolio Activity on this page, students find experimental probabilities for a set of independent events and for a set of dependent events related to drawing colored marbles from a bag. Students summarize and analyze their results, and then explain the differences of the probabilities for the independent and dependent events.

Student Technology Guide

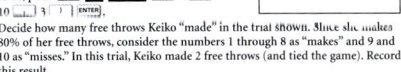

Student Technology Guide
13.5 Simulations

A simulation is a probability experiment that models a real-world event. You can use a graphics calculator to perform a simulation.

Example: At the end of a close basketball game, Keiko is fouled while attempting a three-point shot and is awarded 3 free throws. Keiko usually makes 80% of her free throws, and her team is behind by 2 points. Find the probabilities that (1) Keiko's team wins the game, (2) Keiko's team ties the game (and it goes into overtime), and (3) Keiko's team loses the game.

Notice that Keiko's team wins if she makes all 3 free throws, ties if she makes any 2 of the 3, and loses if she makes fewer than 2.

- To generate a set of 3 random integers from 1 to 10, press MATH PRB 5:randInt ENTER 1 , 10 , 3) ENTER.
- Decide how many free throws Keiko "made" in the trial shown. Since she makes 80% of her free throws, consider the numbers 1 through 8 as "makes" and 9 and 10 as "misses." In this trial, Keiko made 2 free throws (and tied the game). Record this result.
- Press ENTER to generate more trials. Record each result.
- Calculate the probabilities. For the 6 trials shown, the experimental probabilities are:

$P(\text{win}) = \frac{3}{6} = 50\%$
$P(\text{tie}) = \frac{3}{6} = 50\%$
$P(\text{loss}) = \frac{0}{6} = 0\%$

After conducting 50 trials, the probabilities are $P(\text{tie}) = \frac{25}{50} = 50\%$, and $P(\text{loss}) = \frac{2}{50} = 4\%$.

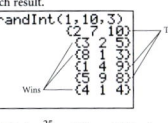

Use a graphics calculator to simulate 25 trials for each situation. Give the results of each simulation.

1. Use the situation in the example, but change Keiko's successful free-throw percent to 60%. **Answers may vary.** Theoretical probabilities are: $P(\text{win}) = 21.6\%$, $P(\text{tie}) = 43.2\%$, and $P(\text{loss}) = 35.2\%$.

2. On average, 30% of the September days in Portland, Oregon, are rainy. For a 4-day period, find the experimental probabilities that it rains (1) all 4 days, (2) 2 or more days, (3) no days. **Answers may vary.** Theoretical probabilities are: $P(4 \text{ rainy days}) = 0.81\%$, $P(2 \text{ or more rainy days}) = 34.83\%$, and $P(\text{no rainy days}) = 24.01\%$.

LESSON 13.5 **681**

Project

Focus

How many people must be present in order for it to be likely that two people in the group have the same birthday? In this Project, students use the probability concepts and simulation techniques they learned in this chapter to explore this question and others related to it. They will predict the number of people with the same birth month or birthday in a group of a given size. Then they will predict the number of people required in order to have a given probability of a match.

Motivate

Ask students if they think that any of their classmates were born in the same month that they were. Discuss the fact that if the class size is 13 or greater, at least two students *must* have the same birth month. Ask their opinions about the probability of repeated birth months in groups of smaller sizes, such as 2, 6, and 8.

Now repeat the discussion, this time asking if they think that any of their classmates were born on the same day of the year that they were. Discuss the fact that to be certain that two people in the class have the same birthday, there would need to be 367 students in the class. Ask how many students they think must be in the class for there to be a 50% probability that two will have the same birthday. Tell students that this project will provide a mathematical means of determining the number needed to meet this condition.

CHAPTER PROJECT THIRTEEN — Happy Birthday!

Activity 1

Matching Birth Months

How many of your friends or family were born in your birth month? on your birthday? In this experiment you will explore the probability of matching birth months and birthdays.

1. Do you think all birthdays are equally likely? Explain your reasoning.

2. Estimate the following probability: If you choose 4 people at random, at least 2 will have the same birth month.

3. Estimate the following probability: If you choose 6 people at random, at least 2 will have the same birth month.

4. Devise a method for randomly picking 6 students from your class and execute it.

5. Check whether 2 of the students chosen have the same birth month.

6. Repeat this experiment 10 times. Compute the experimental probability 2 people having the same birth month in a group of 6 people.

7. How does the experimental probability compare with your prediction from Step 3?

If you assume that all birth months are equally likely, you can use a coin and number cube to conduct this experiment. The sample space is shown below.

J	F	M	A	M	J	J	A	S	O	N	D
H1	H2	H3	H4	H5	H6	T1	T2	T3	T4	T5	T6

Or use a calculator by numbering the months 1–12.

For the TI-83, the keystrokes for randomly generating the integers from 1 to 12 inclusive are given below.

7. Use a coin and number cube, or a calculator, to repeat the experiment for 10 trials. Compute the experimental probability of 2 people having the same birth month in a group of 6 people. Combine the trials for the whole class, and record the experimental probability for the class data.

8. Andrea said, "Since there are 12 months and I am choosing 6 people at random there is a 50% probability that I will get at least 2 with the same birth month." Is she right?

9. What is the probability of 2 people having the same birth months in a group of 13 people chosen at random?

Activity 1

1. Yes; reasoning may vary.
2. List students' estimates.
3. List students' estimates.
4. Have the class randomly pick 6 students and record matching birth months.
5. Have the class repeat Step 4 ten times. Calculate $P(match) = \frac{f}{10}$.
6. Discuss results and compare to predictions from Step 3.
7. Have each student conduct 10 trials and combine results. Calculate the experimental probability.
8. no
9. 100%

Activity 2

Matching Birthdays

1. What do you think is the probability 2 people having the same birthday in a group of 35 people? Check the birthdays of students in your class for matches.

   You can use a graphics calculator or spreadsheet software to simulate picking 35 people at random and looking for at least one birthday match. One trial is shown below. Working with a partner, one person can generate the data, and the other can record the birthdays generated by crossing out values on a chart. Two strikes through a single date will represent a match.

   TECHNOLOGY
   GRAPHICS CALCULATOR

   For the TI-83, the keystrokes for randomly generating the integers from 1 to 365 are:

2. Explain why the function INT(365 rand) + 1 is used to generate random birthdays.

3. Perform 10 trials of the experiment. Compute the experimental probability of a birthday match in a group of 35 people.

4. Collect the data from your class, and compute the experimental probability of a match. Does the experimental probability agree with your guess from Step 1?

Activity 3

In Activity 2 you may have discovered that probabilities are often difficult to estimate. You will now explore the birthday problem in a different way.

1. Simulate the birthday problem for 35 people at least 100 times, either by performing all of the simulations yourself or by combining data from the entire class.

2. Find the number of matches for trials 1–5, 6–10, 11–15, and so on. Create a histogram to display the frequency distribution. It will show the frequency of the matches for each block of 5 numbers.

3. How many people do you think you need in order to have
   a. a 50% chance of a match?
   b. a 90% chance of a match?
   c. a 99% chance of a match?

Cooperative Learning

Have students do these activities in pairs.

Since several questions require students to know the birthdays of their classmates, you may want to collect this data in advance. List the students' names and birthdays on a sheet of paper and provide each pair of students with a copy of the list.

In **Activity 1**, partners can work together to answer the questions. In Step 5, each partner should perform five of the required trials.

In **Activity 2**, one partner can generate the data for five trials in Step 3, while the other records the data. The partners can then switch roles for the other five trials.

In **Activity 3**, have the partners work together to create the histogram in Step 2. Each student should do Step 3 individually. Then the partners can work together again to compare and discuss their answers.

Discuss

Bring the class together after Activity 1 and after Activity 2 to discuss their findings.

After Activity 3, have the pairs display their histograms, and discuss with the class the similarities and differences among them. For each part of Step 3, first poll students about their answers. Then conduct a discussion to see if the class can arrive at a consensus about the number of people required for each probability.

Activity 2

1. Poll the class for birthdays and look for matches.

2. The rand function generates a number between 0 and 1. Multiplying by 365, taking the integer component, and adding 1 produces a number between 1 and 365, inclusive, which can represent all possible birthdays in the year.

3. Record data from 10 trials and compute the experimental probability.

4. Compute the experimental probability based on data from the class.

Activity 3

1. Answers may vary.
2. Answers may vary.
3. Answers may vary.

Chapter Review

13 Chapter Review and Assessment

VOCABULARY

Addition of Probabilities Principle 657	experimental probability ... 676	sample space 649
AND 654	favorable outcome 649	simulation 676
complement 670	Fundamental Counting Principle 662	theoretical probability 650
counting elements of sets .. 656	independent events 668	tree diagram 661
disjoint 657	intersection 655	trial 676
equally likely outcomes 648	OR 654	union 655
		Venn diagram 655

Key Skills & Exercises

LESSON 13.1

Key Skills

Find the theoretical probability of a favorable outcome.

Suppose that you roll a number cube once. The sample space is 1, 2, 3, 4, 5, and 6. The probability that you will roll a number greater than 4 is the ratio of the favorable outcomes (f), 5 and 6, to the number of elements in the sample space (n), 6.

$$P(E) = \frac{f}{n}$$
$$P(\text{greater than } 4) = \frac{2}{6} = \frac{1}{3}$$

Exercises

A chip is drawn from a bag containing chips numbered from −10 to 10 inclusive. Find the probability that the number drawn is

1. negative. $\frac{10}{21}$
2. even. $\frac{11}{21}$
3. equal to 0. $\frac{1}{21}$
4. less than or equal to −5. $\frac{2}{7}$

A number cube is rolled once. Find the probability that you roll

5. an even number. $\frac{1}{2}$
6. a prime number. $\frac{1}{2}$
7. a number not evenly divisible by 3. $\frac{2}{3}$
8. a number between 2 and 6 inclusive. $\frac{5}{6}$

LESSON 13.2

Key Skills

Use the Addition of Probabilities Principle.

To find the probability of drawing an ace or a ten in an ordinary deck of 52 playing cards, use the Addition of Probabilities Principle.

There are 4 aces in the deck. There are 4 tens in the deck.

$$P(\text{ace}) = \frac{4}{52} = \frac{1}{13} \qquad P(\text{ten}) = \frac{4}{52} = \frac{1}{13}$$

Use the Addition of Probabilities Principle.

$$P(\text{ace OR ten}) = P(\text{ace}) + P(\text{ten}) = \frac{4}{52} + \frac{4}{52}$$
$$= \frac{8}{52}, \text{ or } \frac{2}{13}$$

Exercises

9. In a music class, 11 students play the clarinet and 15 play the flute, including 8 who play both. How many students are in the class? **18**

10. In King High School, there are 245 freshmen in science classes and 238 in math classes. If 150 freshmen are in science AND math classes, how many are in math OR science? **333**

In an ordinary deck of cards,

11. how many cards are black OR face cards? **32**

12. how many cards are red OR numbered cards? **44**

If you toss a number cube, what is the probability that you will toss

13. a 4 OR a number greater than 3? $\frac{1}{2}$

14. an odd OR a prime number? $\frac{2}{3}$

Chapter Test, Form A

NAME _____ CLASS _____ DATE _____

Chapter Assessment
Chapter 13, Form A, page 1

Write the letter that best answers the question or completes the statement.

___a___ 1. A coin is tossed three times. You are interested in the probability that exactly two tosses will be tails. What is the number of favorable outcomes?
   a. 3   b. 4   c. 2   d. 8

___b___ 2. A box contains 12 tickets numbered 1 through 12. One ticket is drawn at random. What is the probability that the number on the ticket is a multiple of 3?
   a. $\frac{1}{4}$ b. $\frac{1}{3}$ c. $\frac{1}{2}$ d. $\frac{1}{12}$

___b___ 3. Which simulation is the best method for finding the experimental probability of a 60% chance of snow?
   a. Toss a coin 60 times and record the sequence of heads or or tails.
   b. Put 6 slips of paper marked *snow* and 4 slips of paper marked *no snow* into a bag and draw them at random.
   c. Roll a number cube 100 times and record how many times a 6 is rolled.
   d. Generate a sequence of random integers from 1 to 40.

___c___ 4. To find the probability when you know the size of the sample space and the number of favorable outcomes, which of the following steps should you take?
   a. Add the number of favorable outcomes and the size of the sample space.
   b. Multiply the number of favorable outcomes by the size of the sample space.
   c. Divide the number of favorable outcomes by the size of the sample space.
   d. Divide the size of the sample space by the number of favorable outcomes.

___a___ 5. The table below shows the results of a simulation in which one number cube was rolled 10 times. What is the experimental probability of getting a 5 on one roll?

Trials	1	2	3	4	5	6	7	8	9	10
Number	4	3	6	6	1	5	2	6	2	5

   a. $\frac{1}{5}$ b. 5   c. 1   d. $\frac{1}{6}$

NAME _____ CLASS _____ DATE _____

Chapter Assessment
Chapter 13, Form A, page 2

___d___ 6. Which integers from 1 to 20 inclusive are multiples of 4 AND multiples of 8?
   a. 4, 8, 16   b. 4, 8, 12, 16, 20
   c. 2, 4, 6, 8, 10   d. 8, 16

___b___ 7. A plane seats 3 passengers on each side of the aisle. Passengers seated in rows 20 through 25 were asked to board the plane first. Assuming that each seat was booked, how many passengers was this?
   a. 18   b. 36   c. 30   d. 45

___a___ 8. An athletic store stocks 15 kinds of basketball shoes in 6 sizes. Each kind of shoe comes in 3 ankle heights—low-, mid-, and high-tops. How many pairs of basketball shoes should the manager order if he wants to carry one of each combination?
   a. 270   b. 24   c. 90   d. 51

___c___ 9. Points X, Y, Z, A, and B lie on the same line. How many ways are there to name the line by using 2 letters?
   a. 5   b. 4
   c. 20   d. 25

___d___ 10. An integer from 1 to 30 inclusive is drawn at random. Find the probability that the number drawn is a multiple of 6.
   a. $\frac{1}{5}$ b. 5   c. 6   d. $\frac{1}{6}$

___c___ 11. An integer from 1 to 20 inclusive is drawn at random. What is the probability the number drawn is even AND a multiple of 4?
   a. 4   b. 5   c. $\frac{1}{4}$ d. $\frac{1}{5}$

___a___ 12. What is the probability that when two number cubes are rolled, the sum is less than 8?
   a. $\frac{1}{4}$ b. $\frac{7}{36}$ c. $\frac{1}{6}$ d. $\frac{2}{3}$

___c___ 13. What is the probability for event B shown in the grid pattern at right?
   a. 4   b. $\frac{8}{36}$
   c. $\frac{1}{9}$ d. $\frac{1}{3}$

A	B
C	D

684 CHAPTER 13 REVIEW

LESSON 13.3

Key Skills

Use the Fundamental Counting Principle.

To find the number of T-shirts stocked by a craft store that stocks 5 styles, 4 sizes, and 10 different colors, use the Fundamental Counting Principle, which states that if there are *m* ways to make a first choice and *n* ways to make a second choice, then there are $m \cdot n$ ways to make a first choice and a second choice. Thus, there are $5 \cdot 4 \cdot 10$, or 200, T-shirts stocked by the craft store.

Exercises

15. How many ways are there to arrange A, B, C, and D? **24**

16. A dress shop stocks 6 styles of blouses. Each blouse comes in 7 sizes and 5 colors. How many different selections are possible? **210**

17. If you use only the digits 1–9, how many 3-digit numbers can you write? **729**

18. How many different results can you get from flipping a coin and then rolling a number cube? **12**

Two decks of ordinary playing cards are shuffled separately. A card is drawn from each deck. How many ways are there to draw

19. a queen from the first deck AND a face card from the second deck? **48**

20. a face card from the first deck AND a red card from the second deck? **312**

LESSON 13.4

Key Skills

Find the probability of two independent events.

There are 2 red marbles and 4 blue marbles in a bag. You draw a marble, replace it, and then draw another marble. The probability of selecting a red marble and a blue marble is $\frac{2}{6} \cdot \frac{4}{6} = \frac{8}{36} = \frac{2}{9}$.

Exercises

There are 10 cards numbered 1 through 10 in a box. One card is drawn and then replaced. Then another card is drawn. Find the probability that

21. the first card is 4 and the second is 6. $\frac{1}{100}$

22. both cards are less than 5. $\frac{4}{25}$

23. both cards are multiples of 3. $\frac{9}{100}$

24. both cards are the same. $\frac{1}{10}$

There are 5 blue marbles and 3 red marbles in a bag. One marble is drawn and replaced. A second marble is drawn. Find the probability that

25. the first marble is red and the second is blue. $\frac{15}{64}$

26. both marbles are red. $\frac{9}{64}$

27. both marbles are blue. $\frac{25}{64}$

CHAPTER 13 REVIEW 685

28. Answers may vary. Sample answer: Use a calculator and the command INT(RAND*2) to generate the random numbers 0 and 1. Let 0 represent an incorrect response and 1 represent a correct response. Generate 10 numbers; record the results. This represents 1 trial. Repeat for 20 trials. Divide the number of trials where all 10 responses were correct by 20. The quotient represents the experimental probability.

29. Answers may vary. Sample answer: Let heads represent a correct response and tails represent an incorrect response. Flip a coin 10 times; record the results. This represents 1 trial. Repeat for 20 trials. Divide the number of trials where all 10 responses were correct by 20. The quotient represents the experimental probability.

LESSON 13.5

Key Skills

Use a simulation to find experimental probability.

To simulate randomly choosing the correct response to a multiple-choice question with 5 possibilities, use a number cube to simulate the experiment. Let 1, 2, 3, 4, or 5 represent the possible responses, with 1 representing a correct response. Roll again if the result is a 6. One roll represents one trial. The following chart indicates a possible simulation:

Trial	1	2	3	4	5	6	7	8	9	10
Roll	1	4	2	1	3	3	1	4	3	5

Exercises

Suppose that you want to simulate answering all of the questions on a 10-question true-false quiz correctly. Design a simulation using

28. random numbers.

29. coins.

Applications

MARKETING The Hamburger Hut is giving away free food during their grand opening. They are distributing 1000 cards—100 are for a free hamburger, 150 are for free fries, and 750 are for a free soft drink.

If one card is drawn, what is the probability of winning

30. a hamburger? $\frac{1}{10}$

31. a soft drink? $\frac{3}{4}$

If two cards are drawn with replacement, what is the probability of winning

32. a hamburger OR soft drink? $\frac{31}{40}$

33. a hamburger AND soft drink? $\frac{3}{40}$

34. **FINANCE** Tim, Mary, and Paul have applied for a loan to buy a used car. The probability that Tim's application is approved is 75%. The probability that Mary's application is approved is 80%. The probability that Paul's application is approved is 62.5%. What is the probability that all three loans will be approved? 37.5%

35. **FINANCE** In Exercise 34, what is the probability that only Mary's application will be approved? 7.5%

13 Alternative Assessment

Performance Assessment

1. Suppose that you roll a number cube and toss a coin.
   a. List the sample space for the experiment as a set of ordered pairs.
   b. What is the probability of rolling a 3?
   c. What is the probability of getting tails?

2. Use a Venn diagram and whole numbers to illustrate
   a. the AND relationship between two sets.
   b. the OR relationship between two sets.
   c. the intersection of two disjoint sets.

3. Explain how to find the number of elements in set A OR set B when
   a. A and B have five elements in common.
   b. A and B have no elements in common.

4. A jar contains 5 blue marbles, 3 green marbles, and 2 yellow marbles.
   a. One marble is drawn and returned to the jar. A second marble is drawn. What is the probability that the first marble is blue and the second marble is green?
   b. One marble is drawn but not replaced. A second marble is drawn. What is the probability that the first marble is blue and the second marble is green?

5. How do you find the probability that at least one of two independent events will occur by using the complement of a set?

6. Explain how to use slips of paper numbered 1 through 10 to simulate a 20% chance of rain.

Portfolio Projects

1. Write a paragraph about the terms used in probability. Include the terms: sample space, outcomes, successful, experimental, theoretical, experiment, random, and fair.

2. Find how probability is used to set the odds that an event will or will not occur. Pick three states that have lottery contests. Compare the odds of winning the big prize.

3. The theory of probability began with a problem posed to two famous mathematicians of the seventeenth century. Write a paragraph naming the mathematicians and discussing the problem they were to solve.

Peer Assessment

Complete this activity with a partner. Begin by describing a situation that can be modeled by a simulation, such as the probability of rain over a period of days. Have your partner design a simulation and conduct 20 trials. Based on the results, determine the probability of the event you described.

Then have your partner describe a situation for which you design and conduct a simulation. Your partner should then determine the probability of the event based on your results. Continue this pattern until each of you has conducted 5 simulations.

internetconnect

The HRW Web site contains many resources to reinforce and expand your knowledge of probability. The Web site also provides Internet links to other sites where you can find information and real-world data for use in research projects, reports, and activities that involve probability. Visit the HRW Web site at go.hrw.com, and enter the keyword **MA1 CH13.**

Alternative Assessment

Performance Assessment

1. a. {(1, H), (1, T), (2, H), (2, T), (3, H), (3, T), (4, H), (4, T), (5, H), (5, T), (6, H), (6, T)}
   b. $\frac{1}{6}$ c. $\frac{1}{2}$

2 a. {1, 2, 3, 4} AND {2, 4, 6, 8}

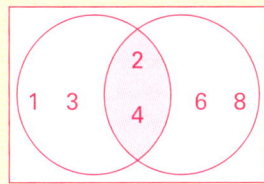

b. {1, 2, 3, 4} OR {2, 4, 6, 8}

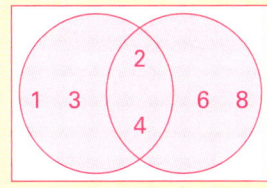

c. {1, 2, 3, 4} ∩ {5, 6, 7, 8}

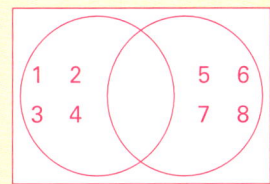

3. Let a be number of elements in A. Let b be the number of elements in B. Let x be the number of elements in (A or B).
   a. $x = a + b - 5$
   b. $x = a + b$

4. a. $\frac{3}{20}$ b. $\frac{1}{6}$

5. Subtract the number of ways each event could fail to occur from the total possible outcomes. The difference represents the ways each event could occur. Divide each difference by the total number of outcomes to find the probability of each event occurring. The probability that the two events will occur is the product of these two probabilities.

6. Let the numbers 1 and 2 represent the occurrence of rain, and let the numbers 3, 4, 5, 6, 7, 8, 9, and 10 represent no rain.

Portfolio Projects

1. Students should include the following terms:
   sample space: the set of all possible *outcomes* of an *experiment*
   outcomes: the results of an *experiment*
   favorable: the *outcome* you are looking for in an *experiment*
   experimental: the actual *outcome* of an *experiment*
   theoretical: the predicted *outcome* of an *experiment*, assuming that each element of the sample space is equally likely to occur, known as *random* or *fair*

2. Have students discuss the individual and combined results.

3. Students should mention the correspondence between Pascal and Fermat about this type of question: A player has 8 tries in which to throw a one on a number cube. If the game is interrupted after 3 unsuccessful throws, how should the players be rewarded?

Cumulative Assessment

College Entrance Exam Practice

Multiple-Choice and Quantitative-Comparison Samples

The first half of the Cumulative Assessment contains two types of items found on standardized tests—multiple-choice questions and quantitative-comparison questions. Quantitative-comparison items emphasize the concepts of equality, inequality, and estimation.

Free Response Grid Samples

The second half of the Cumulative Assessment is a free-response section. This part of the Cumulative Assessment requires student-produced response items like those commonly found on college entrance exams. These questions require the use of machine-scored answer grids. You may wish to have students practice answering these items in preparation for the standardized tests.

CHAPTERS 1–13 CUMULATIVE ASSESSMENT

College Entrance Exam Practice

QUANTITATIVE COMPARISON For questions 1–5 write
A if the quantity in Column A is greater than the quantity in Column B;
B if the quantity in Column B is greater than the quantity in Column A;
C if the quantities are equal; or
D if the relationship cannot be determined from the information given.

	Column A	Column B	Answers	
1.	The number of $15 tickets you can buy for $90	The number of $12 tickets you can buy for $84	(A) (B) (C) (D) [Lesson 3.2]	B
2.	$(-18)(-4)$	$-18(-2)$	(A) (B) (C) (D) [Lesson 2.4]	B
3.	$\frac{2\sqrt{2}}{3}$	$\frac{2\sqrt{10}}{5}$	(A) (B) (C) (D) [Lesson 12.1]	B
4.	$(9 \times 10^6)(7 \times 10^6)$	$(1.9 \times 10^8)(3 \times 10^4)$	(A) (B) (C) (D) [Lesson 8.5]	A
5.	The sine of angle A	The cosine of angle A	(A) (B) (C) (D) [Lesson 12.7]	D

6. Find $|-3.6| - |4.5|$. **(LESSON 2.1)** b
   a. 0.9
   b. −0.9
   c. 8.1
   d. −8.1

7. Solve: $2x^2 = 288$. **(LESSON 10.2)** c
   a. 12
   b. 72
   c. ±12
   d. ±72

8. What is the slope of a line parallel to $2x + 3y = -5$? **(LESSON 5.6)** b
   a. 2
   b. $-\frac{2}{3}$
   c. $\frac{3}{2}$
   d. 3

9. What is the probability of choosing a blue chip from a bag containing 4 blue chips and 3 red chips? **(LESSON 13.1)** c
   a. $\frac{3}{4}$
   b. $\frac{4}{3}$
   c. $\frac{4}{7}$
   d. $\frac{3}{7}$

10. Simplify: $3\sqrt{2} - \sqrt{5} + 6\sqrt{2} + 9\sqrt{5}$. **(LESSON 12.1)** d
    a. $9\sqrt{2} + 9$
    b. $-3\sqrt{2} + 8\sqrt{5}$
    c. $-3\sqrt{2} + 9$
    d. $9\sqrt{2} + 8\sqrt{5}$

11. Find 65% of 20. **(LESSON 4.2)** c
    a. 1300
    b. 0.0325
    c. 13
    d. 3.25

12. Which spreadsheet command generates a random number from the list 3, 4, 5? **(LESSON 13.5)** d
    a. INT(RAND()*5)+5
    b. INT(RAND()*3)+5
    c. INT(RAND()*5)+3
    d. INT(RAND()*3)+3

13. Which function has a graph that is identical to that of $s = t^2$? **(LESSON 10.1)** a
    a. $y = x^2$
    b. $y = x$
    c. $y = |x|$
    d. $y = \frac{1}{x}$

14. Simplify: $\frac{x^2 + 5x + 6}{x + 3}$. **(LESSON 11.3)** a
    a. $x + 2, x \neq -3$
    b. $x + 2, x \neq 3$
    c. -2
    d. $-2, -3$

15. What is the simplified form of $(-2m^2n^4)^2$? **(LESSON 8.2)** d
    a. $-4m^4n^6$
    b. $4m^4n^6$
    c. $-4m^4n^8$
    d. $4m^4n^8$

16. There are 48 students in the choir and 50 students in the orchestra. If 15 students are in both the choir and orchestra, what is the total number of students in the choir and the orchestra? **(LESSON 13.2)** 83

17. Write an inequality that describes the points graphed on the number line. **(LESSON 6.1)** $a > 2$

18. If x varies inversely as y and $x = 5$ when $y = 7$, find x when $y = 10$. **(LESSON 11.1)** 3.5

19. Write the common denominator for $\frac{x}{x+3} + \frac{2}{x-3}$. **(LESSON 11.4)** $x^2 - 9$

20. Find two numbers whose sum is 32 and whose product is 255. **(LESSON 7.6)** 15 and 17

21. Factor: $4x^2 - 16$. **(LESSON 9.6)** $4(x-2)(x+2)$

22. Solve: $(x+1)^2 - 4 = 0$. **(LESSON 10.2)** $x = -3, 1$

23. The first three terms of a sequence are 4, 7, and 12. The second differences are a constant 2. Find the next three terms. **(LESSON 1.1)** 19, 28, 39

24. Simplify: $3w(4w + 5) - (w + 10)$. **(LESSON 9.2)** $12w^2 + 14w - 10$

25. Write and solve an equation for this situation: A computer is on sale for $1800. The sale price is $\frac{2}{3}$ of the regular price. Find the regular price of the computer. **(LESSON 3.2)** $2700

26. Evaluate: $\frac{-15}{-3}$. **(LESSON 2.4)** 5

27. Solve: $13 - 3x = -8$. **(LESSON 3.3)** $x = 7$

28. Solve: $|5x + 14| < 11$. **(LESSON 6.5)** $-5 < x < -\frac{3}{5}$

29. What percent of 160 is 60? **(LESSON 4.2)** 37.5%

30. What is the probability of getting an even number when rolling a number cube? **(LESSON 4.3)** $\frac{1}{2}$

31. What is the slope of the line containing the points $A(8, 4)$ and $B(-3, 10)$? **(LESSON 5.2)** $-\frac{6}{11}$

32. A force of 5 Newtons stretches a spring 35 centimeters. Find the constant, k, for the spring. **(LESSON 5.3)** $\frac{100}{7}$ N/m, or about 14.3 N/m

33. Solve the system of equations. **(LESSON 7.3)** $p = 3, q = 4$ $\begin{cases} 4p + 6q = 36 \\ 10p - 2q = 22 \end{cases}$

FREE-RESPONSE GRID

The following questions may be answered by using a free-response grid such as that commonly used by standardized test services.

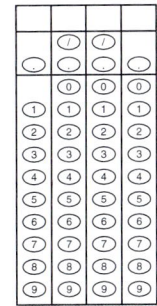

34. Evaluate $2x - (3y + 4)$ for $x = -3$ and $y = -2$. **(LESSON 1.3)** -4

35. How many ways can you arrange 6 different letters? **(LESSON 13.3)** 720

36. A large truck has a load capacity of 9 tons more than a smaller truck. The ratio of their load capacities is 5 to 2. Find the load capacity of the smaller truck. **(LESSON 4.1)** 6 tons

37. Evaluate $x^2 - 4x + 7$ for $x = 3$. **(LESSON 10.1)** 4

38. Find the slope of the line containing the points $M(-1, -2)$ and $N(3, 4)$. **(LESSON 5.2)** $\frac{3}{2}$

14 Functions and Transformations

CHAPTER PLANNING GUIDE

Lesson	14.1	14.2	14.3	14.4	14.5	Projects and Review
Pupil Edition Pages	692–699	700–706	707–713	714–719	722–727	720–721, 728–735
Extra Practice (Pupil Edition)	779	779	780	780	781	
Practice Workbook	83	84	85	86	87	
Reteaching Masters	165–166	167–168	169–170	171–172	173–174	
Enrichment Masters	83	84	85	86	87	
Cooperative-Learning Activities	83	84	85	86	87	
Lesson Activities	83	84	85	86	87	
Problem Solving/ Critical Thinking	83	84	85	86	87	
Student Study Guide	83	84	85	86	87	
Spanish Resources	83	84	85	86	87	
Student Technology Guide	83	84	85	86	87	
Assessment Resources	178	179	180	182	183	181, 184–189
Teaching Transparencies	67	68, 69		70	71	
Quiz Transparencies	14.1	14.2	14.3	14.4	14.5	
Writing Activities for Your Portfolio						40–42
Tech Prep Masters						65–68
Long-Term Projects						53–56

LESSON PACING GUIDE

Traditional	1 day	1 day	2 days	2 days	2 day	2 days
Block	$\frac{1}{2}$ day	$\frac{1}{2}$ day	1 day	1 day	1 day	1 day
Two-Year	2 days	2 days	4 days	4 days	4 days	4 days

CONNECTIONS AND APPLICATIONS

Lesson	14.1	14.2	14.3	14.4	14.5	Review
Algebra	692–699	700–706	707–713	714–719	722–727	728–735
Geometry	699			719	726	
Statistics		706	712	719		
Business and Economics		706	713		726	732
Science and Technology	699	700	707, 708, 709	718		732
Sports and Leisure	692, 699		713		722, 726	
Other		704				
Cultural Connection: Asia				718		

BLOCK-SCHEDULING GUIDE

Day	Lesson	Teacher Directed: Lesson Examples, Teaching Transparencies	Student Guided: Activity, Try This	Cooperative-Learning Activity, Lesson Activity, Student Technology Guide	Practice: Practice & Apply, Extra Practice, Practice Workbook	Assessment: Quiz, Mid-Chapter Assessment	Problem Solving, Reteaching
1	14.1	5 min	8 min	8 min	35 min	8 min	8 min
	14.2	5 min	7 min	7 min	30 min	7 min	7 min
2	14.3	10 min	15 min	15 min	65 min	15 min	15 min
3	14.4	10 min	15 min	15 min	65 min	15 min	15 min
4	14.5	10 min	15 min	15 min	65 min	15 min	15 min
5	Assess.	50 min PF: Chapter Review	90 min PE: Chapter Project, Writing Activities	90 min Tech Prep Masters	65 min PE: Chapter Assessment, Test Generator	30 min Chap. Assess. (A or B), Alt. Assess. (A or B), Test Generator	

PE: Pupil's Edition

Alternative Assessment

The following are suggestions for an alternative assessment for students who may benefit from a different type of assessment than the regular chapter quizzes and the mid-chapter and end-of-chapter assessment materials. Many of these questions are open-ended, and students' answers will vary.

Performance Assessment

1. Let $f(x) = x^2$.
   a. Graph the function that results from each transformation below taken one at a time.
      i. translation 2 units to the right
      ii. translation 3 units down
      iii. vertical stretch by a factor of 2
      iv. reflection across the x-axis
   b. Explain how your work in part **a** helps you graph $g(x) = -2(x - 2)^2 - 3$.

2. a. One pencil costs $0.20. Write an equation for the cost, c, in dollars of p pencils.
   b. How do you know the equation represents a function?
   c. Explain why it is important that the equation represents a function.
   d. What would happen if cost were not a function?

Portfolio Project

Suggest that students choose one of the following two projects for inclusion in their portfolios.

1. The equation $c = 0.33 + 0.29[w - 1]$ gives the cost, c, of mailing a letter that weighs w ounces. The first ounce costs $0.33 and each additional ounce or fraction thereof costs $0.29.
   a. Research another real-world situation in which the cost of the first unit is a certain amount and the cost of each additional unit or fraction thereof is a different amount.
   b. What is the parent function and how are transformations involved?

2. Your task is to design a logo on the coordinate plane that is based on a parent function, such as $y = |x|$, and transformations of it.
   a. Design and sketch such a logo by using transformations you have learned.
   b. Suppose that you may also use a reflection across the y-axis produced by changing (x, y) to $(-x, y)$. Make a second logo using transformations you learned and reflection across the y-axis.

internet connect

The table below identifies the pages in this chapter that contain technology information and support in the side columns.

Content Links	
Lesson Links	pages 696, 697, 717
Graphics Calculator Support	page 782

Resource Links

For information about teacher and parent resources as well as professional development help, visit **www.hrw.com/math**.

Technical Support HRW has assembled a team of dedicated technical and teaching professionals and a comprehensive service program to provide you with the support you need.

- The HRW Technical Support Line operates from 7 A.M. to 6 P.M. central time, Monday through Friday, at (800) 323-9239.
- The HRW Technical Support Center on the World Wide Web is available 24 hours a day, seven days a week, at **www.hrwtechsupport.com**.
- You can e-mail our Technical Support Center at **tsc@hrwtechsupport.com**.
- The Technical Support Center's fax-on-demand service at (800) 352-1680 offers solutions to common problems and answers to frequently asked questions.

Technology

Lesson Suggestions and Calculator Examples
(Keystrokes are based on a TI-83 calculator.)

Lesson 14.1 Graphing Functions and Relations
This lesson will reinforce and extend Lesson 5.1 on functions and graphs. Refer to the lesson suggestions and calculator examples presented for that lesson. The graphics calculator displays below show the ordered pairs (3, 2), (5, 6), (7, 10), and (9, 14) in tabular and in graphical form.

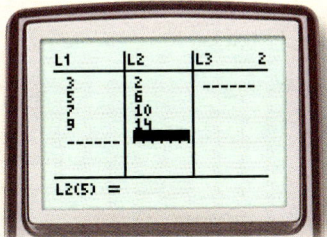
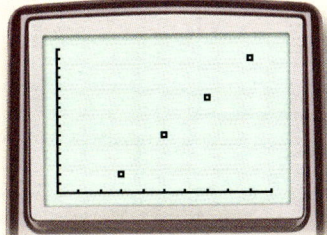

If students imagine using the vertical-line test, they will see that no two points in the graph lie along the same vertical line. Thus, the set determines a function.

Lesson 14.2 Translations
The displays below show the parent function $y = x^2$ translated 0, 2, and 4 units up and 0, 2, and 4 units to the left.

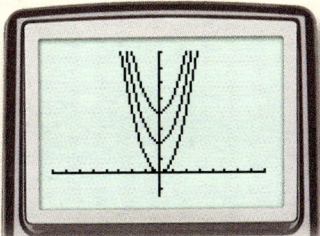
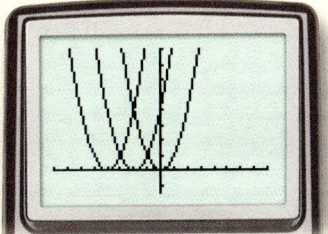

Have students practice choosing a parent function, predicting the effect of a particular translation type, amount, and direction, and then using the graphics calculator to verify their predictions.

Lesson 14.3 Stretches and Compressions
The displays below show the effect of scale factors 1, 0.5, and 2 on the parent functions $y = |x|$ and $y = x^2$.

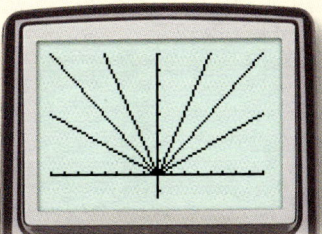
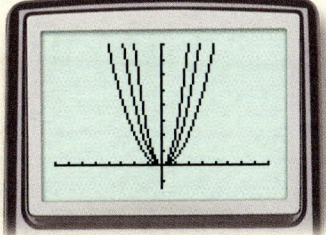

Have students practice choosing a parent function, predicting the effect of a particular stretch, and then using the graphics calculator to verify their predictions.

Lesson 14.4 Reflections
It is not difficult for students to believe that when a graph is reflected across the *x*-axis, the equation related to the graph becomes the opposite of the original equation. Apply what students learned about reflections to another concept, such as slope.

The display at right shows the graphs of two linear functions, with each being the reflection of the other across the *x*-axis. Ask students how the slopes are related and how the *y*-intercepts are related. Ask students for equations for the lines to verify their conjecture algebraically.

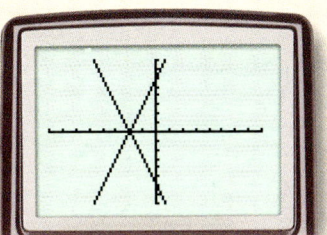

Lesson 14.5 Combining Transformations
Ask students to graph the parent function $y = |x|$, then translate it 2 units to the right, and then horizontally compress it by a factor of 3. These transformations combine to give the graph of $y = 3|x - 2|$. The display at right shows the graphs.

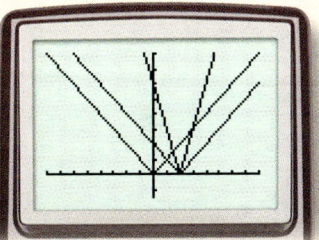

For further information, refer to the

- technology discussions in the lessons.
- lesson-related teacher's commentary in the side columns of this *Teacher's Edition*.
- lesson-related *Student Technology Guide* masters.
- *HRW Technology Handbook*.

Teaching Resources

Basic Skills Practice
(2 pages per skill)

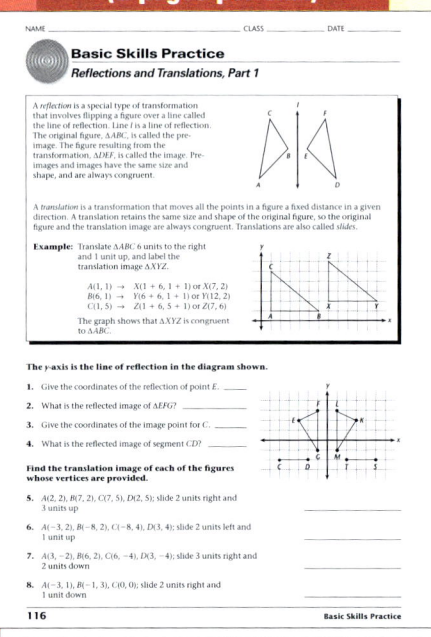

Reteaching
(2 pages per lesson)

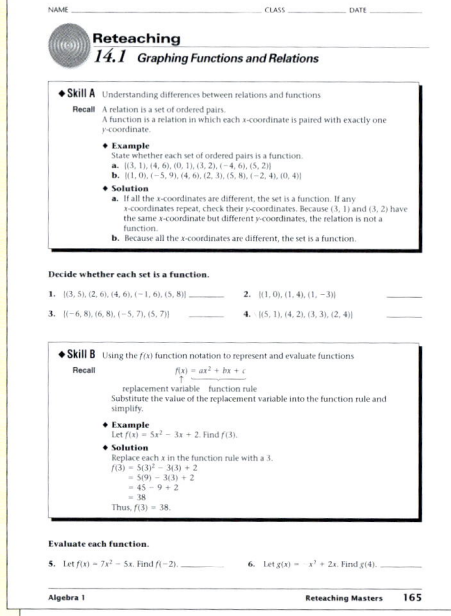

Enrichment
(1 page per lesson)

Cooperative-Learning Activity
(1 page per lesson)

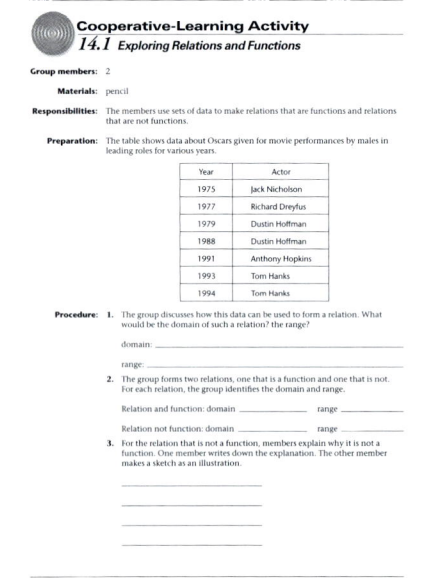

Lesson Activity
(1 page per lesson)

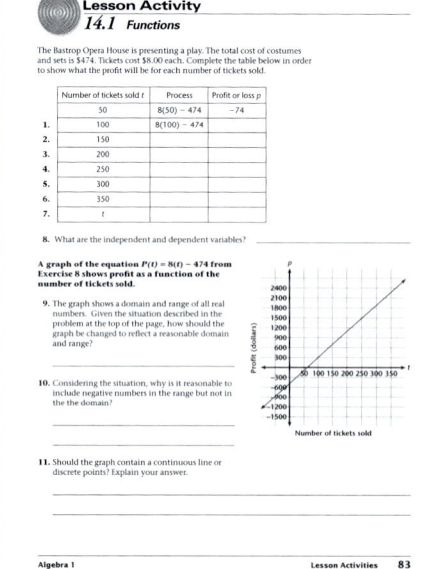

Long-Term Project
(4 pages per chapter)

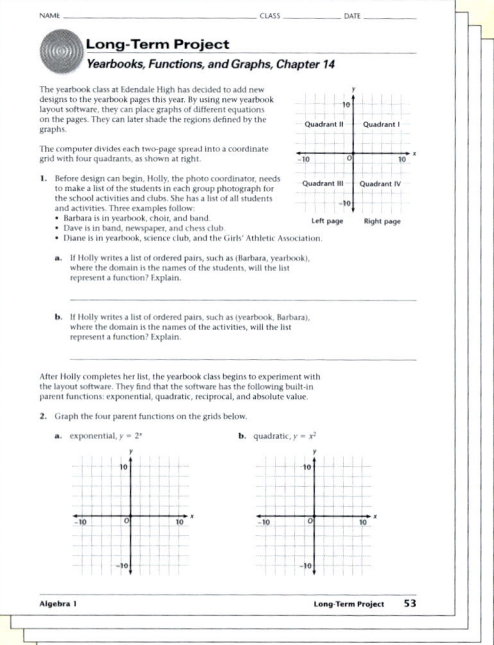

CHAPTER 14 INTERLEAF

About Chapter 14

Background Information

Students already have learned some characteristics of linear, quadratic, polynomial, and exponential functions. In Chapter 14, they take a closer look at the graphs of these and other types of functions. They learn to recognize the graphs of the parent functions, and they examine how the graph of a parent is affected by three types of transformations: vertical stretches, vertical reflections, and translations.

CHAPTER RESOURCES

- Block-Scheduling Handbook
- Writing Activities for Your Portfolio
- Tech Prep Masters
- Long-Term Project
- Assessment Resources:
  Mid-Chapter Assessments
  Chapter Assessments
  Alternative Assessments
- Test and Practice Generator
- Technology Handbook
- End-of-Course Exam Practice

Chapter Objectives

- Use models to understand functions and relations. [14.1]
- Evaluate functions by using function rules. [14.1]
- Identify the functions of some important families of functions. [14.1]
- Describe how changes to the rule of a function correspond to the translation of its graph. [14.2]

Functions and Transformations

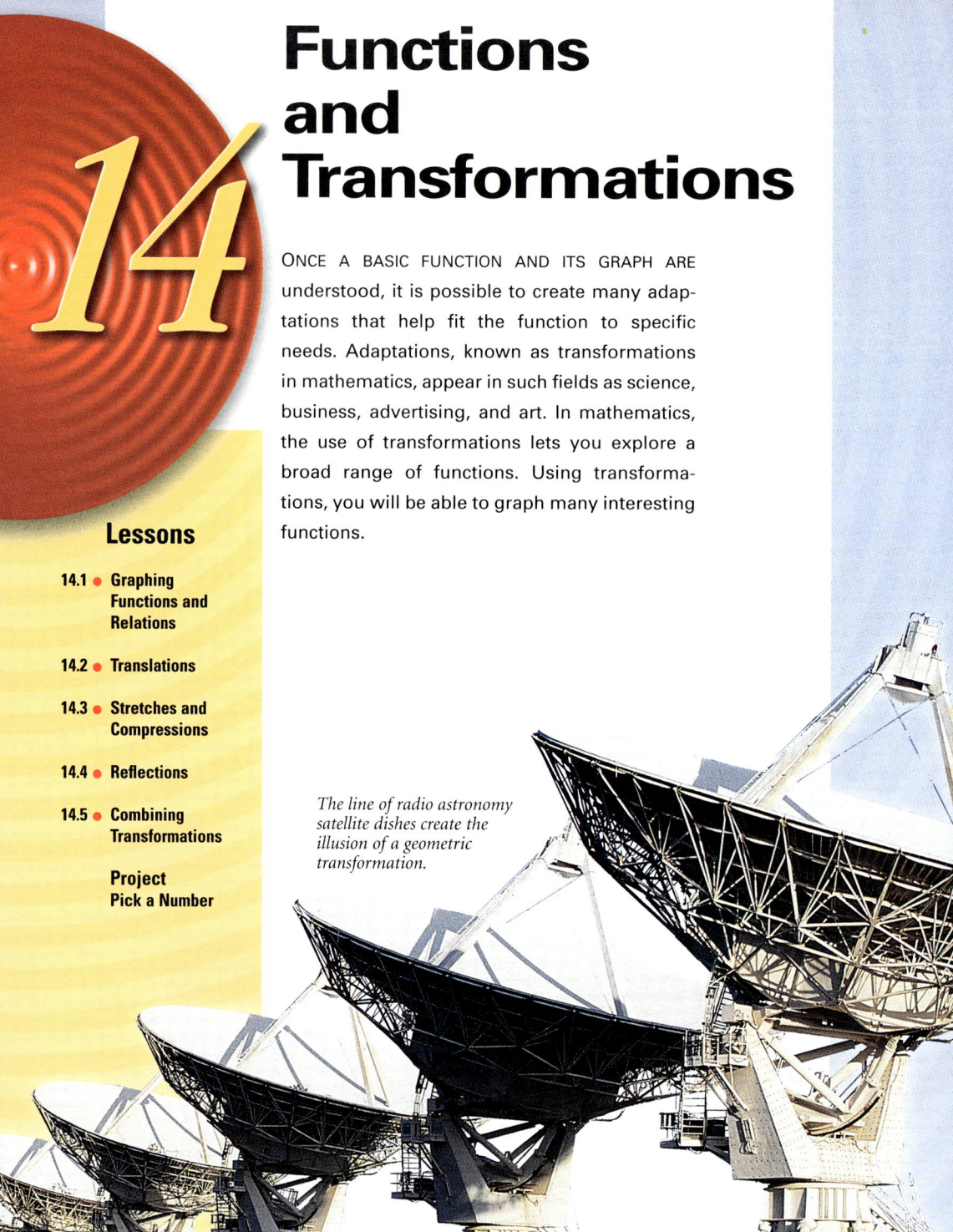

ONCE A BASIC FUNCTION AND ITS GRAPH ARE understood, it is possible to create many adaptations that help fit the function to specific needs. Adaptations, known as transformations in mathematics, appear in such fields as science, business, advertising, and art. In mathematics, the use of transformations lets you explore a broad range of functions. Using transformations, you will be able to graph many interesting functions.

Lessons

14.1 Graphing Functions and Relations

14.2 Translations

14.3 Stretches and Compressions

14.4 Reflections

14.5 Combining Transformations

Project Pick a Number

The line of radio astronomy satellite dishes create the illusion of a geometric transformation.

About the Photos

Have students look at the photographs on these pages and identify as many transformations as possible. Remind students of the different types of transformations—slides, reflections, and stretches. Have students discuss other occurrences of transformations, for example, a person on roller skates, a reflection in a pond, or putting a picture on clay and then stretching the clay.

Patterns in architecture, agriculture, and industry model geometric transformations.

- Describe how changes to the rule of a function stretch or compress its graph. [14.3]
- Describe how a change to the rule of a function corresponds to a reflection of its graph. [14.4]
- Study a real-world application of transformed functions. [14.5]
- Graph functions that involve more than one transformation. [14.5]

Portfolio Activities appear at the end of Lessons 14.4 and 14.5. Each serves as preparation for the Chapter Project. The Portfolio Activities as well as the Chapter Project Activities are appropriate for inclusion in the student's portfolio. Students should be encouraged to include in their portfolios any other work in which they feel a sense of pride or a sense of accomplishment.

About the Chapter Project

Algebra provides the power to simplify many seemingly complex situations through the use of functions. In the Chapter Project, *Pick a Number*, on page 728, you will investigate the relationship between number games and composite functions. After completing the Chapter Project you will be able to do the following:

- Define and apply the concept of composite functions.

About the Portfolio Activities

Throughout the chapter, you will be given opportunities to complete Portfolio Activities that are designed to support your work on the Chapter Project.

- Recognizing transformed functions is the topic of the Portfolio Activity on page 719.
- Comparing flying time and mileage between cities is the topic of the Portfolio Activity on page 727.

internet connect

Internet Connect features appearing throughout the chapter provide keywords for the HRW Web site that lead to links for math resources, tutorial assistance, references for student research, classroom activities, and all teaching resource pages. The HRW Web site also provides updates to the technology used in the text. Listed at right are the keywords for the Internet Connect activities referenced in this chapter. Refer to the side column on the page listed to read about the activity.

LESSON	KEYWORD	PAGE
14.1	MA1 Weather	696
14.1	MA1 Graphs	697
14.4	MA1 Friezes	717

CHAPTER 14 **691**

Prepare

NCTM PRINCIPLES & STANDARDS 1–3, 6, 8–10

LESSON RESOURCES

- Practice 14.1
- Reteaching Master 14.1
- Enrichment Master 14.1
- Cooperative-Learning Activity 14.1
- Lesson Activity 14.1
- Problem Solving/Critical Thinking Master 14.1
- Teaching Transparency 67

QUICK WARM-UP

Evaluate each expression.

1. $r - 8$, for $r = -3$ -11
2. $|j| - 8$, for $j = -3$ -5
3. $m^2 - 8$, for $m = -3$ 1
4. $(y + 2)^2$, for $y = -10$ 64
5. $6p^2$, for $p = 2$ 24

Also on Quiz Transparency 14.1

Teach

Why Remind students that when one quantity depends on another, it is said that the first quantity is a function of the other. Write these statements on the board or overhead.

The cost of a long-distance telephone call is a function of ? .
The cost of a long-distance telephone call is not a function of ? .

Ask students to name several ways that the blanks can be filled in to make the statements true. **samples: the time of day, the temperature**

14.1 Graphing Functions and Relations

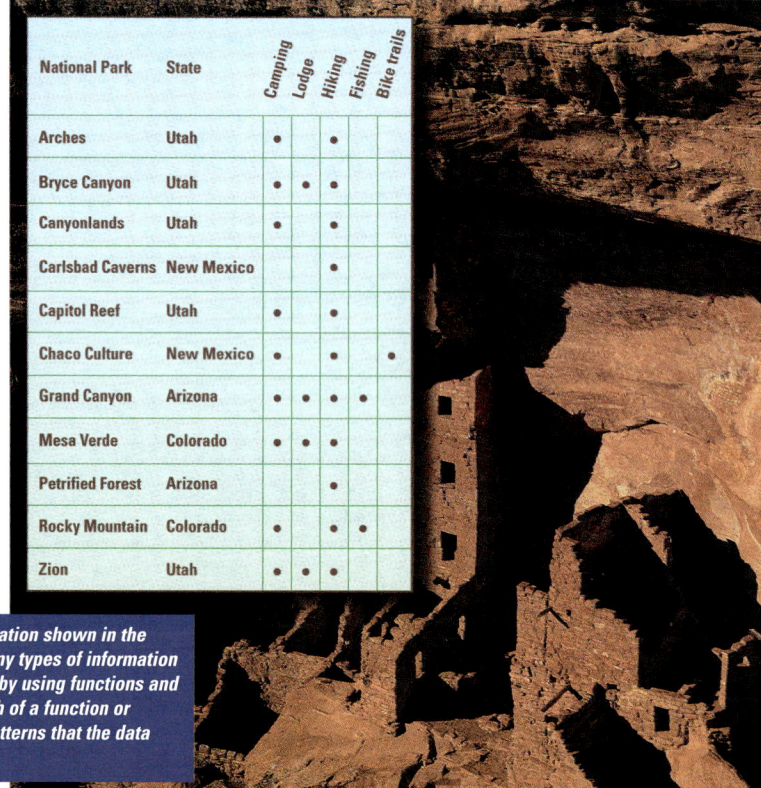

Objectives

- Use models to understand functions and relations.
- Evaluate functions by using function rules.
- Identify the parent functions of some important families of functions.

Why The national park information shown in the chart is an example of the many types of information that can be better understood by using functions and relations as models. The graph of a function or relation often shows visual patterns that the data alone would not reveal.

Mesa Verde National Park covers 52,122 acres in southwest Colorado. It contains some of the best-preserved cliff dwellings in the United States.

Recall that a relation is a set of ordered pairs. The first elements of the ordered pairs form a set called the *domain*, and the second elements of the ordered pairs form a set called the *range*. A function is a special type of relation in which each element of the domain is paired with exactly one element of the range.

EXAMPLE 1

APPLICATION
RECREATION

Consider each relation below. **Draw a model of each and tell whether the relation is a function.**

a. {(Canyonlands, camping), (Capitol Reef, hiking), (Bryce Canyon, lodge), (Grand Canyon, fishing)}
b. {(Arches, hiking), (Chaco Culture, bike trails), (Zion, lodge), (Petrified Forest, hiking)}
c. {(Rocky Mountain, fishing), (Mesa Verde, lodge), (Carlsbad Caverns, hiking), (Mesa Verde, camping)}

Alternative Teaching Strategy

USING TECHNOLOGY Give students the function $f(x) = x^2 - 3x + 1$. If they are using TI-82 or TI-83 graphics calculators, tell them to press [Y=], clear any existing equations, and enter X^2–3X+1 next to **Y1**. To view the graph of the function when the domain is the set of all real numbers, have them press [ZOOM] and select **6: ZStandard**.

Now, to visualize the vertical-line test, have students press [2nd] [PRGM] and select **4: Vertical** from the **DRAW** menu. They can now use the [◄] and [►] keys to move a vertical line left and right.

To analyze finite domains, they can use the table feature. For instance, to find the range of f over the domain {–3, –2, –1, 0, 1, 2, 3}, have students press [2nd] [WINDOW] and make these choices in the **TABLE SETUP** menu.

    Indpnt: Auto **Ask**
    Depend: **Auto** Ask

Next, have the students press [2nd] [GRAPH] and enter the domain values in the **X** column, pressing [ENTER] after each. The values in the **Y1** column indicate that the range is {–1, 1, 5, 11, 19}.

692 LESSON 14.1

● **SOLUTION**

Represent each relation by listing the elements of the domain and the range in two seperate columns. Connect the elements of the ordered pairs with lines.

a.
```
Bryce Canyon      camping
Canyonlands       fishing
Capitol Reef      hiking
Grand Canyon      lodge
```

The relation in **a** is a function because each element in the first set is paired with exactly one element in the second set.

b.
```
Arches            bike trails
Chaco Culture     hiking
Petrified Forest  lodge
Zion
```

The relation in **b** is a function even though one element in the second set is paired with two different elements in the first set.

c.

The relation in **c** is not a function because an element of the first set is paired with more than one element in the second set.

Graphs of Functions and Relations

Relations can be graphed. Once a relation is graphed, the **vertical-line test** can be used to determine whether the relation is a function.

Activity
Vertical-Line Test

You will need: graph paper

1. Draw a model of each relation below.
   a. {(2, 3), (3, 5), (8, 6), (7, 9)}   b. {(2, 3), (0, 4), (−9, 5), (8, 3)}
   c. {(2, 3), (2, 9), (3, 5), (8, 3)}   d. {(2, 3), (−2, 3), (2, −3), (−2, −3)}
2. Determine whether or not each relation in Step 1 is a function.
3. Graph each relation from Step 1 on a separate coordinate grid.
4. For each relation, try to draw a vertical line that intersects the graph in more than one place.

CHECKPOINT ✓ 5. What do you notice about the graphs of the relations that are functions compared with those of the other relations? Explain how the vertical-line test distinguishes functions from other relations.

As you may have discovered in the activity, if an element of the domain is paired with more than one element of the range, a vertical line can be drawn through both points on a graph. Therefore, a relation is a function if it is impossible to draw a vertical line that intersects the graph of the relation more than once.

To find a specific value of the function, such as $f(22)$, have students press [2nd] [MODE] to go to the home screen. To store 22 as the value of x, have them press 22 [STO▶] [ALPHA] [STO▶] [ENTER]. Now, to evaluate f for this value of x, they can use these key sequences.
   TI-83: [VARS] [▶] [ENTER] [ENTER] [ENTER]
   TI-82: [2nd] [VARS] [ENTER] [ENTER] [ENTER]
The number displayed on the screen, 419, is $f(22)$.

Interdisciplinary Connection

PHYSICS A brick falls from the top of scaffolding that is 40 meters high. The distance that the brick falls is a function of the time that it has been falling. This function can be defined by the rule $f(t) = 4.9t^2$, where t is the time in seconds and $f(t)$ is the distance in meters. Evaluate $f(2.5)$ and explain its meaning in the context of the situation.
$f(2.5) = 30.625$; after 2.5 seconds, the distance the brick has fallen is 30.625 meters.

ADDITIONAL EXAMPLE 1

Draw a model for each relation and tell whether it is a function. Give a reason for each answer.

a. {(−3, −2), (−1, 4), (6, 8)}

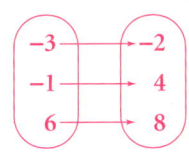

This is a function. Each element of the domain is paired with exactly one element of the range.

b. {(−3, −2), (−1, 4), (6, −2)}

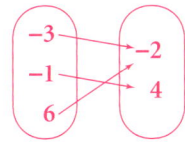

This is a function. Each element of the domain is paired with exactly one element of the range.

c. {(−3, −2), (−1, 4), (−1, 8)}

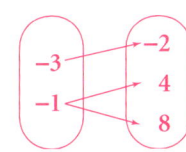

This is not a function. One element of the domain, −1, is paired with two different elements of the range, 4 and 8.

Activity Notes

This Activity gives students a visual method for determining whether a given relation is a function. After completing the Activity, students should realize that if a vertical line intersects the graph of a relation at more than one point, then the relation is not a function.

☞ For Cooperative Learning notes and an answer to the Checkpoint, see page 694.

LESSON 14.1   **693**

Cooperative Learning

Have students do the Activity in groups of four. Each group member can choose a different one of the relations given in Step 1 and work with it in Steps 2, 3, and 4. The group can then come together to compare results and discuss and complete Step 5.

CHECKPOINT ✔

5. A vertical line intersects the graph of a function only once but may intersect the graph of a relation more than once. If an element of the domain is paired with more than one element of the range, a vertical line can be drawn through more than one point on the graph of the relation.

ADDITIONAL EXAMPLE 2

Which of the following graphs represent functions?

a.
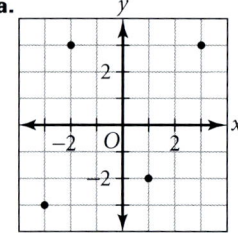

b.
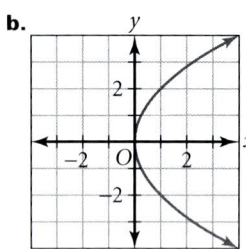

c.
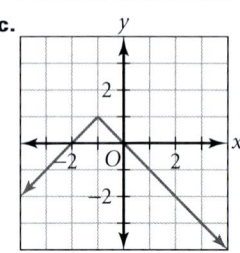

Graphs **a** and **c** represent functions.

EXAMPLE 2

Which of the following graphs represent functions?

a.
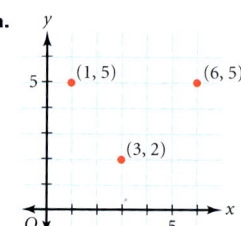

b.
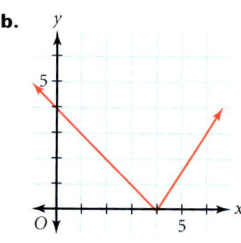

c.
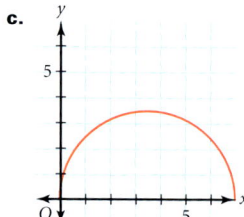

d.
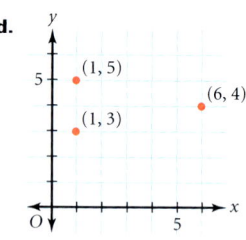

e.
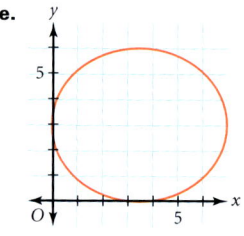

f.
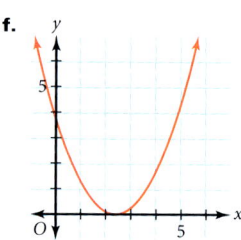

● SOLUTION

It is impossible to draw a vertical line that will intersect graphs **a**, **b**, **c**, and **f** more than once, so these graphs represent functions. Neither graph **d** nor graph **e** represents a function because the vertical line $x = 1$ intersects graph **d** twice and the vertical line $x = 2$, for example, intersects graph **e** twice.

Function Notation and Function Rules

The statement $y = x^2$ represents a quadratic function. Another way to express the same function is to use **function notation**. In function notation, the dependent variable, y, is written as $f(x)$—or as $g(x)$, $h(x)$, etc. Thus, $y = x^2$ would be written as $f(x) = x^2$, $g(x) = x^2$, $h(x) = x^2$, etc. Notice that the independent variable is placed inside the parentheses.

The statement $f(x) = x^2$ is read as "f of x equals x squared." The expression on the right side of the statement is called the **function rule**. To evaluate the function for a given value of x, replace x in the function rule with that value. For example, if $x = 3$, then $f(3) = 3^2 = 9$.

Inclusion Strategies

TACTILE/KINESTHETIC LEARNERS On the board or overhead, draw several graphs such as those in Example 2. Have students use a yard stick or meter stick to simulate the vertical-line test, which determines whether each graph represents a function. Have them move the stick to different locations on the graph to show that if the graph does represent a function, a vertical line that intersects the graph at no more than one point could be drawn anywhere on the graph. Similarly, if the stick touches more than one point on the graph, then explain that the graph does not represent a function.

The elements of the domain of a function are sometimes called the *input* of the function rule. The elements of the range are sometimes called the *output* of the function rule.

input
(an element from ➡ function rule ➡ output
the domain)   (an element from
                                        the range)

You can think of a function rule as a machine that accepts domain values and returns range values.

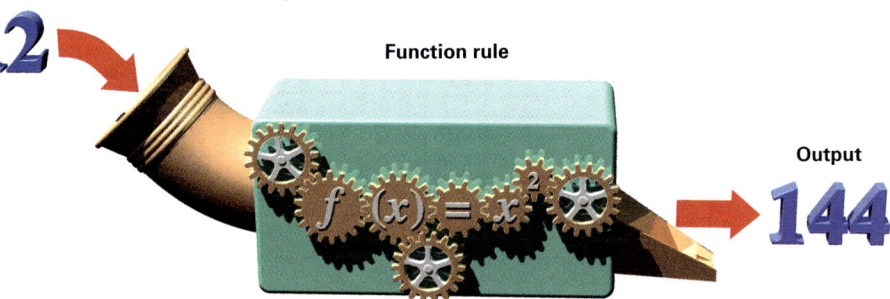

It is important to consider what domain and range values are appropriate for a particular situation. If no particular domain is given, the domain is understood to be all real numbers for which the function rule is defined.

EXAMPLE 3 If $h(x) = x^2 - 2x + 1$, what is $h(4)$?

● **SOLUTION**

The notation $h(4)$ indicates that you substitute 4 for x in the function rule $x^2 - 2x + 1$. Next, simplify to find the value of the function when x is 4.

$h(x) = x^2 - 2x + 1$
$h(4) = 4^2 - 2(4) + 1$
$h(4) = 9$

TRY THIS If $g(x) = 2x^2 - 3x - 2$, what is $g(3)$?

EXAMPLE 4 Consider the function $f(x) = x^2$, where the domain of the function consists of the integers from −2 to 2 inclusive. **Make a list of the ordered pairs of the function and draw a model of the function.**

● **SOLUTION**

Use the function rule x^2 to find the range values for each of the domain values. For example, for the domain value −2, the range value is $(-2)^2$, or 4. Thus, (−2, 4) is an ordered pair of the function. The complete list of ordered pairs of the function is as follows:

$\{(-2, 4), (-1, 1), (0, 0), (1, 1), (2, 4)\}$

Use the list of ordered pairs to draw the function model.

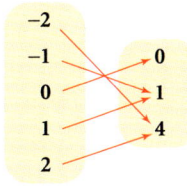

Use Teaching Transparency 54.

ADDITIONAL EXAMPLE 3

If $h(x) = x^2 + 4x - 7$, what is $h(-3)$? −10

TRY THIS

7

ADDITIONAL EXAMPLE 4

Consider the function defined by $h(x) = x^2 + 2x - 1$, where the domain consists of the integers from −4 to 2, inclusive. **Make a list of the ordered pairs of the function and draw a model of the function.**

$\{(-4, 7), (-3, 2), (-2, -1), (-1, -2), (0, -1), (1, 2), (2, 7)\}$

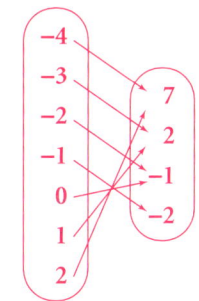

Enrichment

Tell students that not every equation in two variables defines a function. Give them the equations at right and ask them to use ordered pairs to explain why each relation represented by the equations is *not* a function. In each case, they should consider x to be the dependent variable, and they should assume that the domain is the set of all real numbers.

In each case, a single element of the domain can be paired with two elements of the range. Sample ordered pairs are given.

1. $x = y^2$ (9, 3), (9, −3)
2. $x = |y|$ (4, 4), (4, −4)
3. $|x| + |y| = 3$ (1, 2), (1, −2)
4. $x^2 + y^2 = 25$ (3, 4), (3, −4)
5. $x^4 + y^4 = 2$ (1, 1), (1, −1)

Challenge students to create two other equations in two variables that do not represent functions.

LESSON 14.1

TRY THIS

{(−3, 11), (−2, 6), (−1, 3), (0, 2), (1, 3), (2, 6), (3, 11)}

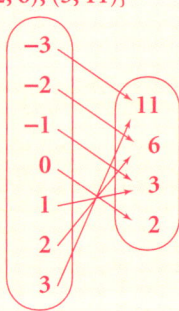

ADDITIONAL EXAMPLE

Identify the parent function for each of the following functions:

a. $y = |x - 2|$
 $y = |x|$

b. $y = -x^2 - 7x + 12$
 $y = x^2$

c. $y = \frac{-1}{x - 5}$
 $y = \frac{1}{x}$

---

internetconnect

GO TO: go.hrw.com
KEYWORD: MA1 Weather

In this activity, students can obtain information about relations that arise in real-world weather data.

TRY THIS Consider the function $f(x) = x^2 + 2$, where the domain of the function consists of the integers from −3 to 3 inclusive. Make a list of the ordered pairs of the function and draw a model of the function.

Parent Functions

In nature, a mother or a father may have many offspring. In mathematics, a **parent function** can be transformed into many—in fact, *infinitely* many—new functions. A *family* of functions is comprised of these new functions and the parent function. Every function can be classified as a member of a family. Certain functions are identified as parent functions of these families. For example, the function $f(x) = x^2$ or $y = x^2$ is the parent function of the family of quadratic functions. Other parent functions are shown in the table below.

Characteristics of parents are passed from one generation to the next.

Parent Function	Family
$f(x) = x$	linear
$f(x) = \|x\|$	absolute value
$f(x) = a^x$	exponential
$f(x) = x^2$	quadratic
$f(x) = \frac{1}{x}$	rational
$f(x) = \sqrt{x}$	square root

EXAMPLE 5 Name the parent function for each graph.

a. 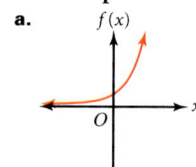 b. 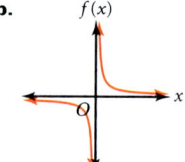 c.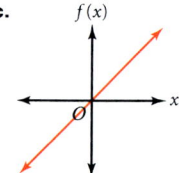

d. 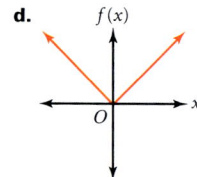 e. 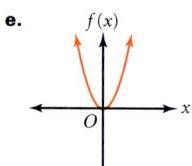 f.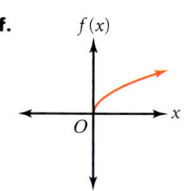

● **SOLUTION**

a. $f(x) = a^x, (a > 1)$ **b.** $f(x) = \frac{1}{x}$ **c.** $f(x) = x$
d. $f(x) = |x|$ **e.** $f(x) = x^2$ **f.** $f(x) = \sqrt{x}$

Reteaching the Lesson

WORKING BACKWARD Display the model at right on the board or overhead. Discuss with students why the relation it represents is a function. Have them list the set of ordered pairs that make up the function. {(−5, 5), (−4, 4), (−3, 3), (−2, 2), (−1, 1), (0, 0), (1, 1), (2, 2)} Then have them write the function rule. $y = |x|$

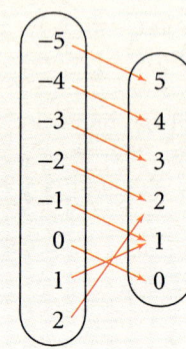

Repeat the activity with several other models, some representing functions and others representing relations that are not functions.

696 LESSON 14.1

Exercises

Communicate

1. Given a graph on a coordinate plane, explain how to use the vertical-line test. What is the purpose of the vertical-line test?

2. Explain how to find the range of the function $f(x) = 2x^2 - 5x$ if the domain is restricted to integers from −2 to 2 inclusive.

Guided Skills Practice

Draw a model for each relation and tell whether the relation is a function. Give a reason for each answer. (EXAMPLE 1)

3. {(−7, 5), (3, 2), (7, −1), (8, 8)} **yes**   4. {(2, 3), (−6, 5), (8, 4), (−2, 3)} **yes**

5. {(−1, 4), (−3, 6), (−1, −2), (4, 4)} **no**   6. {(10, −1), (−6, −6), (3, 1), (10, −5)} **no**

Which of the following graphs represent functions? Give a reason for each answer. (EXAMPLE 2)

7. 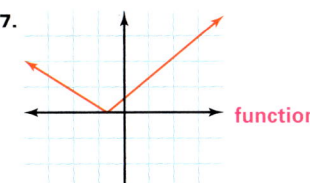 function

8. 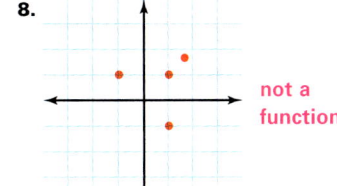 not a function

Find $f(2)$, $f(−1)$, and $f(4)$ for each function. (EXAMPLE 3)

9. $f(x) = 3x^2$ **12; 3; 48**   10. $f(x) = |2x - 5|$ **1; 7; 3**

11. Consider the function $f(x) = |x|$, where the domain of the function consists of the integers from −3 to 3 inclusive. Make a list of the ordered pairs of the function and draw a model of the function. (EXAMPLE 4)
{(−3, 3), (−2, 2), (−1, 1), (0, 0), (1, 1), (2, 2), (3, 3)}

Name the parent function for each graph. (EXAMPLE 5)

12. 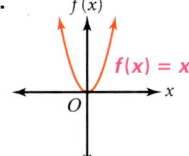 $f(x) = x^2$

13. 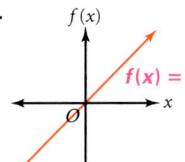 $f(x) = x$

14. 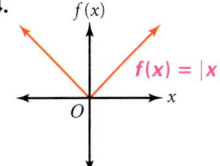 $f(x) = |x|$

15. 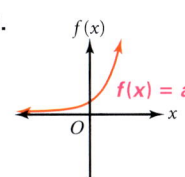 $f(x) = a^x$

16. 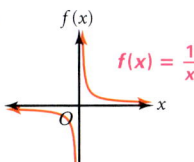 $f(x) = \frac{1}{x}$

17. 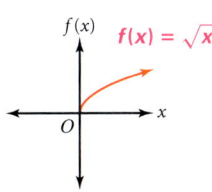 $f(x) = \sqrt{x}$

Assess

Selected Answers
Exercises 3–17, 19–67 odd

ASSIGNMENT GUIDE

In Class	1–17
Core	19–63 odd
Core Plus	18–58 even, 60–63
Review	64–68
Preview	69–71

✎ Extra Practice can be found beginning on page 738.

Technology
A calculator may be helpful for some of the exercises. You may wish to have students first complete the exercises with paper and pencil and then use the calculator to verify their answers. Students can use a graphics calculator to check their graphs in Exercises 57–59.

internetconnect

 GO TO: go.hrw.com
KEYWORD: MA1 Graphs

Understanding functions and their graphs is very important in many real-world relations. In this activity, students gain an understanding of basic functions by multiplying or adding a constant and studying how the graph is affected.

---

3. 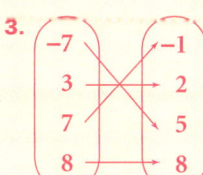 Yes, because there is only one range value associated with each domain value.

4. 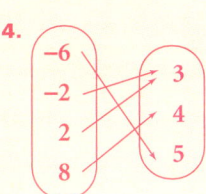 Yes, because there is only one range value associated with each domain value.

5. 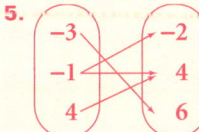 No, because there is more than one range value associated with one of the domain values.

6. 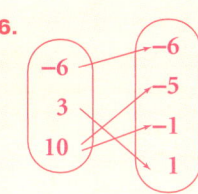 No, because there is more than one range value associated with one of the domain values.

LESSON 14.1 **697**

45. domain: all real numbers
range: all real numbers

46. domain: all real numbers
range: all nonnegative numbers

47. domain: all real numbers
range: all nonnegative numbers

48. domain: all real numbers
range: all nonnegative numbers

49. domain: all real numbers
range: all nonpositive numbers

50. domain: all real numbers
range: all nonpositive numbers

51. domain: all real numbers
range: all nonnegative numbers

52. domain: all real numbers
range: all nonnegative numbers

53. domain: all real numbers except 0
range: all real numbers except 0

24. d: {3, 4, 5}
r: {2, 3, 4}

25. d: {0, 1, 2, 3}
r: {1, 3, 5, 7}

26. d: {1, 2, 3, 4}
r: $\left\{\frac{1}{4}, \frac{1}{3}, \frac{1}{2}\right\}$

27. d: {0.1, 0.2, 0.3}
r: {1, 2, 3}

28. d: {−2, 4, 5}
r: {−7, 1, 6}

29. d: {2.2, 4.1, 5}
r: {3, 5.3, 9}

30. d: {14, 17, 23}
r: {−12, 54, 100}

31. d: {−11, 20, 70}
r: {13, 53, 91}

● **Practice and Apply**

Draw a model for each relation and tell whether the relation is a function. Give a reason for each answer.

18. {(3, 4), (4, 4), (5, 4)}   yes
19. {(1, 1)}   yes
20. {(9, 5), (8, 3), (9, −5), (7, 6)}   no
21. {(5, 9), (4, 3), (6, −7), (−5, 9)}   yes

Which of the following graphs represent functions? Give a reason for each answer.

22. 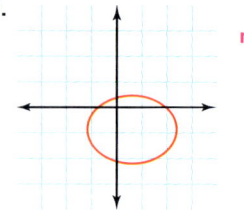 no

23. 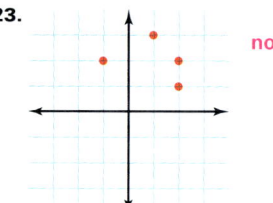 no

Find the domain and range of each function.

24. {(3, 4), (4, 3), (5, 2)}
25. {(0, 1), (1, 3), (2, 5), (3, 7)}
26. $\left\{\left(1, \frac{1}{2}\right), \left(2, \frac{1}{2}\right), \left(3, \frac{1}{3}\right), \left(4, \frac{1}{4}\right)\right\}$
27. {(0.1, 1), (0.2, 2), (0.3, 3)}
28. {(5, 6), (−2, 1), (4, −7)}
29. {(2.2, 3), (5, 9), (4.1, 5.3)}
30. {(17, 100), (23, −12), (14, 54)}
31. {(70, 13), (20, 53), (−11, 91)}
32. Given $y = \frac{1}{x}$, write the ordered pair (x, y) for $x = -5$. $\left(-5, -\frac{1}{5}\right)$

For $h(x) = x + 3$, evaluate the following:

33. $h(7)$ 10
34. $h(4)$ 7
35. $h(-8)$ −5
36. $h(-3)$ 0

For $f(x) = 5x$, evaluate the following:

37. $f(3)$ 15
38. $f(0)$ 0
39. $f(-2)$ −10
40. $f(-6)$ −30

Evaluate each function for $x = 3$.

41. $g(x) = x^2$ 9
42. $f(x) = 2^x$ 8
43. $h(x) = |x|$ 3
44. $k(x) = \frac{1}{x}$ $\frac{1}{3}$

For each function, describe the domain and range.

45. $f(x) = x - 5$
46. $f(x) = x^2$
47. $f(x) = (x + 4)^2$
48. $f(x) = (x - 7)^2$
49. $f(x) = -3x^2$
50. $f(x) = -|x|$
51. $f(x) = |x + 2|$
52. $f(x) = |x - 5|$
53. $f(x) = -\frac{1}{x}$

Given $f(x) = x^2 - 1$, write an ordered pair (x, y) for each of the following elements of the domain.

54. $x = 0$ (0, −1)
55. $x = -9$ (−9, 80)
56. $x = 2$ (2, 3)

Identify the parent function for each of the following functions.

$g(x) = x^2$ **57.** $g(x) = -(3x - 2)^2 - 5$ **58.** $y = \sqrt{2x + 5}$ $y = \sqrt{x}$ **59.** $h(x) = \frac{5}{6 + x}$ $h(x) = \frac{1}{x}$

18. 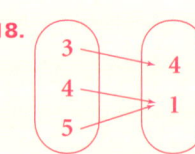 Yes, there is only one range value associated with each domain value.

19. 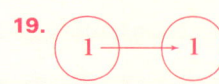 Yes, there is only one range value associated with each domain value.

20. 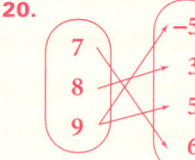 No, there is more than one range value associated with domain value 9.

21. 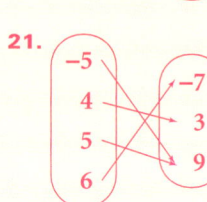 Yes, there is only one range value associated with each domain value.

698   LESSON 14.1

In Exercises 60 and 61, use the formula $F = \frac{9}{5}C + 32$. (*F* stands for Fahrenheit and *C* stands for for Celsius.)

APPLICATIONS

60. METEOROLOGY If you know the Celsius temperature, is it possible to find the Fahrenheit temperature? Is Fahrenheit temperature a function of Celsius temperature? Explain. **yes; yes**

61. METEOROLOGY If you know the Fahrenheit temperature, is it possible to find the Celsius temperature? Is Celsius temperature a function of Fahrenheit temperature? Explain. **yes; yes**

62. TRAVEL A family is driving between national parks as a part of their summer vacation. The points below represent the relationship between time in minutes, *t*, and speed in miles per hour, *s*, for one part of the drive. Graph the points and connect them to represent the relation between *t* and *s*. Is this relation a function? **yes**
{(2, 25), (5, 0), (7, 30), (10, 30), (12, 0), (15, 25), (20, 55), (25, 55)}

63. TRAVEL Describe the relationship between speed and time that would be necessary in order for the graph from Exercise 62, with speed on the horizontal axis, to represent a function.

Look Back

64. There are 5 friends who play tennis. How many games must occur in order for each player to play each of the others one time? **(LESSON 1.1) 10**

Solve each equation. *(LESSON 3.4)*

65. $3r + 4 = -2 + 6r$ **r = 2**

66. $8x - 7 = 3x + 4$ $x = \frac{11}{5}$ (or $2\frac{1}{5}$)

67. Solve $A = \frac{h}{2}(B + b)$ for *B*. **(LESSON 3.6)** $B = \frac{2A - hb}{h}$ (or $\frac{2A}{h} - b$)

68. Solve. $\begin{cases} 2x + 3y = 7 \\ 3x - 2y = 5 \end{cases}$ **(LESSON 7.3)** $x = \frac{29}{13}$ (or $2\frac{3}{13}$), $y = \frac{11}{13}$

Look Beyond

69. Given the graph of the function $f(x) = 10^x$, what would you do to the function rule in order to move the graph up by 1 unit? (Hint: What would you do to the function rule to increase each resulting *y*-value by 1 unit?)

70. Given the graph of the function $f(x) = 10^x$, what would you do to the function rule in order to reflect the graph across the *x*-axis?

CONNECTIONS

71. COORDINATE GEOMETRY Plot several points and sketch the graph of each function. Name the parent function for each.

a. $f(x) = 4 + |x|$
b. $g(x) = -2x$
c. $p(s) = \frac{1}{s + 2}$
d. $h(t) = 5t^2$
e. $z(u) = \frac{1}{u} + 3$
f. $f(r) = |r - 5|$

62.

63. Each speed value in the function would have to be unique (i.e., no duplications) in order for the speed values to become the domain and the time values to become the range of a new function.

69. Add 1: $f(x) = 10^x + 1$.

70. Multiply by -1: $f(x) = -10^x$.

LESSON 14.1 **699**

Prepare

NCTM PRINCIPLES & STANDARDS 1–3, 5, 6, 8–10

LESSON RESOURCES

- Practice 14.2
- Reteaching Master 14.2
- Enrichment Master 14.2
- Cooperative-Learning Activity 14.2
- Lesson Activity 14.2
- Problem Solving/Critical Thinking Master 14.2
- Teaching Transparency 68
- Teaching Transparency 69

QUICK WARM-UP

Given $f(x)$ and $g(x)$, find $f(-3)$ and $g(-3)$.

1. $f(x) = -4x$;
   $g(x) = -4x + 5$
   12; 17

2. $f(x) = x^2$; $g(x) = x^2 + 5$
   9; 14

3. $f(x) = x^2$; $g(x) = (x + 5)^2$
   9; 4

4. $f(x) = |x|$; $g(x) = |x| + 5$
   3; 8

5. $f(x) = |x|$; $g(x) = |x + 5|$
   3; 2

Also on Quiz Transparency 14.2

Teach

Why Ask students what happens when you slide an object from one place to another. Elicit the response that the object retains its size and shape, but its location changes. Tell students that it is possible to slide the graph of a function across a coordinate plane. Such a move is called a *translation*.

14.2 Translations

"Tile facade for the hall of a school in the Hague by M.C. Escher. ©1990 Cordon Art-Baarn-Holland. All rights reserved."

Objective
- Describe how changes to the rule of a function correspond to the translation of its graph.

Why Some designs in architecture involve repeated figures that appear to have been shifted vertically or horizontally. When the graph of a function looks like the graph of a parent function that has been shifted vertically or horizontally there is a simple relationship between the functions.

A **transformation** of a function is an alteration of the function rule that produces a new function. Changes to the rule of a function correspond to changes to its graph. A transformation that shifts the graph of a function horizontally or vertically is called a **translation**. In the Escher drawing above, point A is translated to point B by shifting point A 2 units to the right.

Activity
Translations

TECHNOLOGY GRAPHICS CALCULATOR

You will need: graph paper or a graphics calculator

You can determine the characteristics of translations by exploring the functions $f(x) = |x|$, $g(x) = |x| - 10$, and $h(x) = |x - 10|$.

1. Graph the functions $f(x) = |x|$ and $g(x) = |x| - 10$ on the same set of axes.

2. Compare the graph of $g(x) = |x| - 10$ with the graph of its parent function.
   a. In which direction was the parent function shifted?
   b. By how much?

3. Graph the functions $f(x) = |x|$ and $h(x) = |x - 10|$ on the same set of axes.

CHECKPOINT ✓

4. Compare the graph of $h(x) = |x - 10|$ with the graph of its parent function.
   a. In which direction was the parent function shifted?
   b. By how much?

5. Compare the graphs of g and h.

Alternative Teaching Strategy

HANDS-ON STRATEGIES Have students graph $y = x^2$ and label several points on the graph. Now give students some strips of paper that are a few inches long. Tell them that these are going to be their "translation rulers." Have students mark a 5-unit measure on one translation ruler. Then tell them to hold the ruler vertically, place the top end of the measure on one of the labeled points, and mark the translated point on the graph paper at the other end of the measure. Repeat this process with the other labeled points.

Tell them to use the translated points to draw a parabola. Use the new parabola to initiate a discussion of parent functions, vertical translations of graphs, and the effect of the translations on the equation of the parent function. Repeat the activity for a horizontal translation.

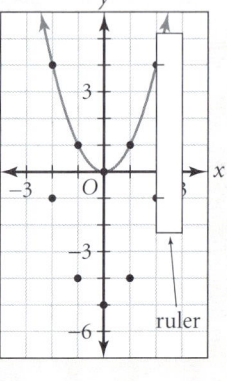

ruler

700 LESSON 14.2

Vertical Translation

A transformation that shifts the graph of a function either up (in a positive direction) or down (in a negative direction) is called a **vertical translation**.

EXAMPLE ① Compare the graph of $h(x) = x^2 - 3$ with the graph of the parent function $f(x) = x^2$. **Describe how to obtain the graph of $h(x)$ from the graph of $f(x)$.**

SOLUTION

Graph the parent function $f(x) = x^2$.
Graph the function $h(x) = x^2 - 3$.

Notice that the graph of $h(x)$ is the same as the graph of $f(x)$ translated 3 units down.

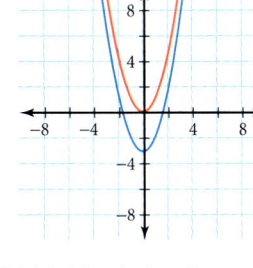

To determine the answer algebraically, notice that $h(x) = x^2 - 3$ contains the parent function $f(x) = x^2$. Therefore, h can be rewritten as $h(x) = f(x) - 3$. This means that each function value of h is 3 units less than the corresponding function value of f. In other words, the graph of $h(x)$ is identical to the graph of the parent function translated vertically 3 units down.

TRY THIS Describe how to obtain the graph of $f(x) = |x| + 4$ from its parent function.

Vertical Translation

The graph of $y = f(x) + k$ is identical to the graph of $y = f(x)$ translated vertically $|k|$ units.

- When k is positive, the translation is in a positive direction, or up.
- When k is negative, the translation is in a negative direction, or down.

The following translations show how this formula applies to the parent function $y = x^2$. The graph of a function is shifted vertically when a constant is added to or subtracted from the function.

The parent function of each function graphed at right is $y = x^2$. In each case, the parent function has been translated vertically.

The graph of **A** is identical to the graph of $y = x^2$ translated 9 units down ($k = -9$): $y = x^2 - 9$.

The graph of **B** is identical to the graph of $y = x^2$ translated 5 units down ($k = -5$): $y = x^2 - 5$.

The graph of **C** is identical to the graph of $y = x^2$ translated 5 units up ($k = 5$): $y = x^2 + 5$.

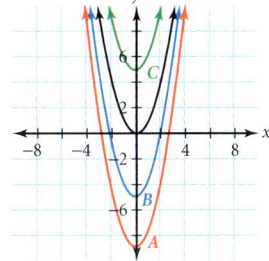

Interdisciplinary Connection

ART A graphic artist is creating a pattern that contains the basic design element shown at right. The equation of the lower ∨ is $y = \frac{2}{3}|x|$, over the domain $-3 \le x \le 3$. What is the equation of the upper ∨?
$y = \frac{2}{3}|x| + 2$, domain $-3 \le x \le 3$

The design was continued as shown at right. Write equations for the ∨s that were added.

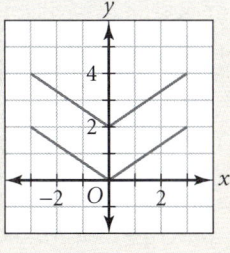
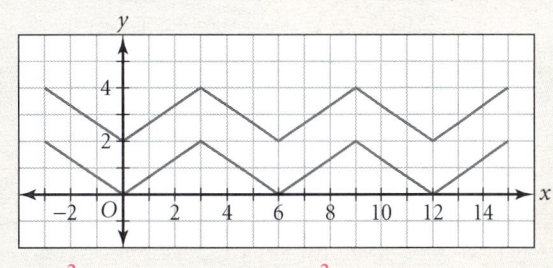

$y = \frac{2}{3}|x - 6|$, $3 \le x \le 9$; $y = \frac{2}{3}|x - 12|$, $9 \le x \le 15$;
$y = \frac{2}{3}|x - 6| + 2$, $3 \le x \le 9$; $y = \frac{2}{3}|x - 12| + 2$, $9 \le x \le 15$

Activity Notes

In this Activity, students compare the graph of the parent absolute-value function with the graphs of functions defined by equations of the form $y = |x| - k$ and $y = |x - h|$. Students should discover that in the case of $y = |x| - k$, the graph of the parent function is shifted k units down. In the case of $y = |x - h|$, it is shifted h units to the right.

Cooperative Learning

You may wish to have students do the Activity in pairs. One partner can do Steps 1 and 2, while the other does Steps 3 and 4. The partners can then share results and do Step 5 together.

CHECKPOINT ✓
4. a. right   b. 10 units

See Keystroke Guide, page 782.

ADDITIONAL EXAMPLE ①

Compare the graph of the function $h(x) = 2^x + 1$ with the graph of its parent function $f(x) = 2^x$. **In which direction does the translation shift the graph? How far is it shifted?**

The graph of the parent function is shifted 1 unit up.

TRY THIS
Move the graph of $f(x) = |x|$ 4 units up.

LESSON 14.2 **701**

Use Teaching Transparency 68.

ADDITIONAL EXAMPLE 2

Compare the graph of the function $h(x) = (x + 4)^2$ with the graph of its parent function, $f(x) = x^2$. **In which direction does the translation shift the graph? How far is it shifted?**

The graph of the parent function is shifted 4 units to the left.

Horizontal Translation

A transformation that shifts the graph of a function either to the right (in a positive direction) or to the left (in a negative direction) is referred to as a **horizontal translation**.

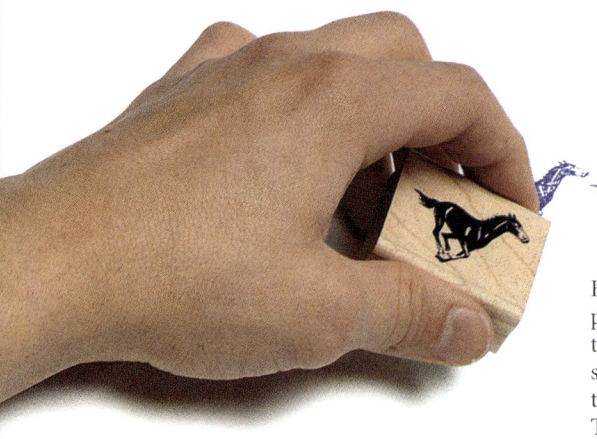

Horizontal translations are useful in creating simple patterns and pictures like the one shown above, but they are also used in much more complicated situations. For example, some animations used on the Internet show an object moving across the page. This is a form of horizontal translation.

EXAMPLE 2 Compare the graph of $g(x) = (x - 3)^2$ with the graph of its parent function, $f(x) = x^2$. **Describe how to obtain the graph of $g(x)$ from its parent function.**

● **SOLUTION**

Graph the parent function $f(x) = x^2$. The function $g(x) = (x - 3)^2$ can be interpreted as "Subtract 3 from x, and then square this quantity." The table shows some y-values for f and g. Sketch g on the same set of axes as its parent function.

x	$f(x) = x^2$	$g(x) = (x-3)^2$
−2	4	25
−1	1	16
0	0	9
1	1	4
2	4	1
3	9	0
4	16	1

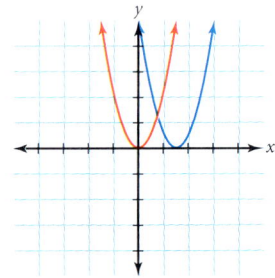

Notice that a translation does not change the shape of the graph. If you subtract 3 from x *before* you square the quantity, you shift the graph horizontally 3 units to the right.

To get the same output values, the input values for $g(x) = f(x - 3)$ have to be 3 units greater than the input values for $f(x)$.

Inclusion Strategies

ENGLISH LANGUAGE DEVELOPMENT If students have language difficulties or if English is their second language, they may become confused by all the new vocabulary in this chapter. Suggest that they create a special *Language of Transformations* section in their notebooks, where they can list the vocabulary associated with transformations—such as *transformation*, *stretch*, *reflection*, *translation*, *vertical*, and *horizontal*—together with the meaning of each term written in their own words.

It also may be helpful to conduct a class brainstorming session in which students share any devices or associations that help them recall the meanings of the words. For instance, students might say that they associate reflections with mirrors. They also might suggest an association between mathematical translations and translations of words from one language to another.

TRY THIS

Compare the graph of $h(x) = |x - 6|$ with the graph of its parent function, $f(x) = |x|$. Describe how to obtain the graph of $h(x)$ from its parent function.

TRY THIS

Shift the graph of $f(x) = |x|$ 6 units right.

Use Teaching Transparency 69.

Horizontal Translation

The graph of $y = f(x - h)$ is translated horizontally $|h|$ units from the graph of $y = f(x)$.

- When h is positive, the translation is in a positive direction, or to the right.
- When h is negative, the translation is in a negative direction, or to the left.

CRITICAL THINKING

The asymptotes are translated the same way that the function is translated.

The parent function of each function graphed below is $y = |x|$. In each case, the parent function has been translated horizontally.

To determine h, you can look at a distinct feature of the original function, such as a vertex or line of symmetry, and note how many units and in what direction this feature is translated to obtain the new graph.

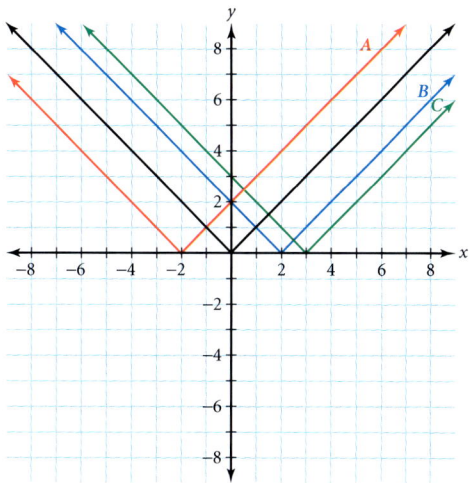

The graph of A is identical to the graph of $y = |x|$ translated 4 units to the left ($h = -4$): $y = |x - (-4)|$ or $y = |x + 4|$.

The graph of B is identical to the graph of $y = |x|$ translated 4 units to the right ($h = 4$): $y = |x - (+4)|$ or $y = |x - 4|$.

The graph of C is identical to the graph of $y = |x|$ translated 6 units to the right ($h = 6$): $y = |x - (+6)|$ or $y = |x - 6|$.

CRITICAL THINKING

If a reciprocal function is translated, what happens to the asymptotes? (Hint: Graph $y = \frac{1}{x}$, $y = \frac{1}{x} - 3$, and $y = \frac{1}{x-3}$ on the same coordinate plane and observe the asymptotes.)

Enrichment

Tell students that $T_{h,0}: (x, y) \to (x + h, y)$ is the mapping notation for a horizontal translation of h units.

1. Write the mapping notation for a vertical translation of k units.
   $T_{0,k}: (x, y) \to (x, y + k)$

2. Write the mapping notation for the translation that maps the graph of the parent quadratic function to the graph of each transformed function.
   a. $g(x) = x^2 - 8$ $T_{0,-8}: (x, y) \to (x, y - 8)$
   b. $h(x) = (x + 4)^2$ $T_{-4,0}: (x, y) \to (x - 4, y)$

LESSON 14.2

ADDITIONAL EXAMPLE 3

Which of the functions below represent translations of the parent graph?

a. $y = 2^x - 3$

b. $y = 3|x|$

c. $y = \dfrac{1}{x-6}$

d. $y = x^2 + 2$

The graphs of **a**, **c**, and **d** are translations of the graph of the parent functions.

EXAMPLE 3

Which of the functions below represent translations of the parent graph?

a. $y = \dfrac{1}{x} + 3$ b. $y = \dfrac{3}{4} + 10^x$ c. $y = \dfrac{|x|}{9}$ d. $y = (x - 10)^2$

SOLUTION

a. Because 3 is added to $\dfrac{1}{x}$, the graph of the parent function $y = \dfrac{1}{x}$ is translated 3 units up.

b. Because $\dfrac{3}{4}$ is added to 10^x, the graph of the parent function $y = 10^x$ is translated $\dfrac{3}{4}$ of a unit up.

c. Because $|x|$ is *divided* by 9, this is a not a translation.

d. Because 10 is subtracted from *x before* the quantity is squared, this is a horizontal translation. The graph of the parent function $y = x^2$ is translated 10 units to the right.

EXAMPLE 4

APPLICATION
BOOK CLUB

A book club for computer professionals charges a membership fee of $20 and then $12 for each book purchased. Write and graph a function to represent cost of membership in the club, *c*, based on the number of books purchased, *b*. Write and graph a second function to represent the cost of membership if the club raises its membership fee to $30. Describe the relationship between the two graphs.

SOLUTION

The function $c = 20 + 12b$ represents the cost of membership in the club at the current fee. The graph of this function is linear and begins at (0, 20).

The function $c = 30 + 12b$ represents the cost of membership in the club at the new fee. The graph of this function is also linear and begins at (0, 30).

The second function represents a vertical translation of the original function by 10 units upward.

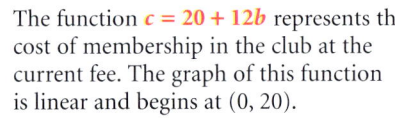

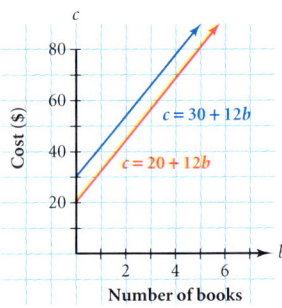

Practice

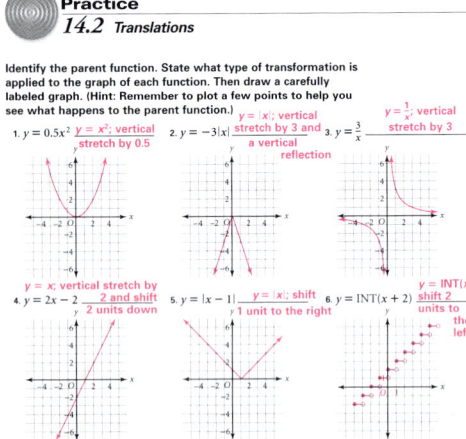

Summary

Translations of the graph $y = f(x)$ are represented as follows:

Vertical shift *k* units up	$y = f(x) + k$
Vertical shift *k* units down	$y = f(x) + (-k) = f(x) - k$
Horizontal shift *h* units right	$y = f(x - h)$
Horizontal shift *h* units left	$y = f(x - (-h)) = f(x + h)$

Reteaching the Lesson

USING VISUAL STRATEGIES Show students the graphs below. Tell them that each shows a translation of the absolute-value function. Do *not* write units on the axes.

Have students write possible equations for each graph. Ask volunteers to share their equations, and discuss the results. Repeat the activity with translations of other parent functions.

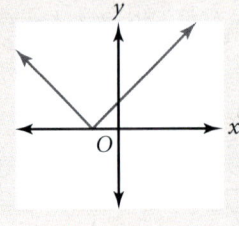

sample: $y = |x| - 3$

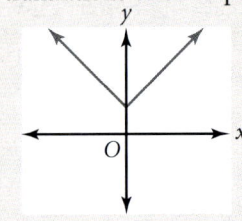

sample: $y = |x + 1|$

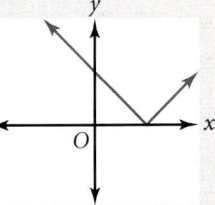

sample: $y = |x| + 2$

sample: $y = |x - 4|$

Exercises

Communicate

1. What is meant by a *translation* in this lesson?
2. The general form for a transformation of the quadratic parent function, $f(x) = x^2$, is $g(x) = a(x-h)^2 + k$. Which variable affects the distance and direction of a horizontal translation? Explain.
3. The function $g(x) = |x - 3| + 5$ is a transformation of the absolute-value parent function. Which value tells you the distance and direction of the vertical translation? Explain.

Guided Skills Practice

Describe how to obtain the graph of each function from the graph of its parent function. *(EXAMPLE 1)*

4. $y = x^2 - 2$
5. $y = x^2 + 3$
6. $y = 10^x + 1$
7. $y = \frac{1}{x} - 3$

Is the graph of the given function a translation of the graph of the parent function? If so, describe how to obtain the graph of the function from the graph of its parent function. *(EXAMPLES 2 AND 3)*

8. $y = (x - 2)^2$
9. $y = 3x^2$
10. $y = 10^{x+1}$
11. $y = \frac{1}{x-3}$

12. Refer to the book club problem in Example 4. Let the cost of membership, c, be represented on the horizontal axis and the number of books received for that cost, b, be represented on the vertical axis. *(EXAMPLE 4)*
    a. Using the same conditions as in Example 4, write and graph the two functions that give b in terms of c.
    b. Describe the relationship between the two graphs.

Practice and Apply

For exercises 13–16:
a. Identify the parent function for each function.
b. Describe how to obtain the graph of each function from the graph of its parent function.
c. Graph and label the parent function and the transformed function.

13. $y = (x - 6)^2 - 1$
    a. $y = x^2$
14. $y = x^2 - 6$
    a. $y = x^2$
15. $y = |x| - 1$
    a. $y = |x|$
16. $y = |x - 1|$
    a. $y = |x|$

The point (3, 9) is on the graph of the parent function $f(x) = x^2$. Use this fact to describe how $f(x)$ is translated to obtain each function below.

17. $g(x) = x^2 + 5$ contains the point (3, 14).
18. $g(x) = x^2 - 5$ contains the point (3, 4).
19. $g(x) = (x + 5)^2$ contains the point (−2, 9).
20. $g(x) = (x - 5)^2$ contains the point (8, 9).

ADDITIONAL EXAMPLE 4

A small company that sells a popular snack mix will prepare custom gift packages during the holiday season. The cost of a gift package is $0.50 per ounce for the mix itself, plus $3 for the container. **Write and graph a function to represent the total cost in dollars of a gift package, p, based on the weight, w, in ounces. Write and graph a second function to represent the total cost in dollars of a gift package if the container costs $7.**

Let x represent the number of ounces of snack mix.
$f(x) = 0.50x + 3$
$g(x) = 0.50x + 7$

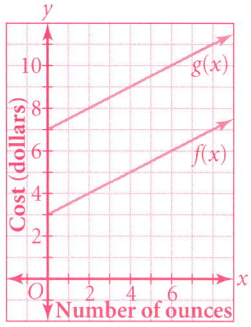

Student Technology Guide

Student Technology Guide
14.2 Translations

When a parent function $f(x)$ is transformed to a related function $f(x - h)$, the related function is a translation of the parent function. A graphics calculator can help you explore the nature of this translation. Consider $y = x^2$ and $y = (x - 5)^2$.
- Enter $y = x^2$ as Y1 and $y = (x - 5)^2$ as Y2 by pressing: ...
- Use ◀ to move the cursor to the left of the Y2 label. Press ENTER until you see a heavy line next to Y2. This will help you tell the graphs apart.
- Press GRAPH. Notice that the vertex of $y = x^2$ is at (0, 0), while the vertex of $y = (x - 5)^2$ is at (5, 0). The graph of $y = (x - 5)^2$ is translated 5 units to the *right* from the graph of $y = x^2$. See the diagram at the left below.
- Change Y2 to $y = (x + 4)^2$ by pressing ▼ and editing Y2. The vertex of $y = (x + 4)^2$ is at (−4, 0)—the graph of $y = (x + 4)^2$ is translated 4 units to the *left* from the graph of $y = x^2$. See the diagram at right below.

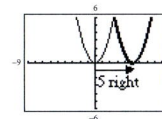 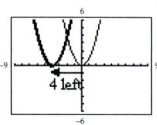

These examples illustrate the horizontal translation property given in Lesson 11.2.
If $h > 0$, then the graph of $y = f(x - h)$ is the horizontal translation of the graph of $y = f(x)$ to the right h units.

Graph each parent function as Y1 and each related function as Y2. Is the graph of Y a translation of the graph of the parent function? If so, in what direction does the translation shift the graph of the parent function? How far is it shifted?

1. $y = x^2$ and $y = (x - 6)^2$
   horizontal translation shifted 6 units to the right.
2. $y = 2^x$ and $y = 2^{(x+4)}$
   horizontal translation shifted 4 units to the left.
3. $y = x^2$ and $y = 3x^2$
   not a translation

---

4. The graph of $y = x^2 - 2$ is a translation of the graph of $y = x^2$ 2 units down.

5. The graph of $y = x^2 + 3$ is a translation of the graph of $y = x^2$ 3 units up.

6. The graph of $y = 10^x + 1$ is a translation of the graph of $y = 10^x$ 1 unit up.

7. The graph of $y = \frac{1}{x} - 3$ is a translation of the graph of $y = \frac{1}{x}$ 3 units down.

8. The graph of $y = (x - 2)^2$ is a translation of the graph of $y = x^2$ 2 units to the right.

9. The graph of $y = 3x^2$ is not a translation of the graph of the parent function $y = x^2$ because 3 is multiplying the parent function, not being added or subtracted.

10. The graph of $y = 10^{x+1}$ is a translation of the graph of $y = 10^x$ 1 unit to the left.

11. The graph of $y = \frac{1}{x-3}$ is a translation of the graph of $y = \frac{1}{x}$ 3 units to the right.

The answers to Exercises 12–20 and 33b can be found in Additional Answers beginning on page 810.

Assess

Selected Answers
Exercises 4–12, 13–33 odd

ASSIGNMENT GUIDE

In Class	1–12
Core	13–31 odd
Core Plus	14–30 even, 31
Review	32, 33
Preview	34

✏️ Extra Practice can be found beginning on page 738.

Technology
A graphics calculator may be helpful for Exercises 13–20. You may wish to have students first complete the exercises with paper and pencil and then use the calculator to verify their answers.

Error Analysis
Students usually associate positive numbers with right and negative numbers with left, so they often use the incorrect sign when writing an equation for a translation to the right or left. For instance, given $y = x^2$ and a translation of 3 units to the right, students may write $y = (x + 3)^2$. Encourage them to sketch the translated graph, identify the coordinates of two points on it, and substitute the coordinates into their equations for the translated functions.

In Exercise 34, students investigate transformations of the graph of the parent cubic function. They will study cubic functions in greater depth if they take advanced courses in algebra.

706   LESSON 14.2

The point (5, 8) is on the graph of a function. Give the coordinates of the corresponding point on the new graph when the following translations are applied:

21. vertical translation of 6 **(5, 14)**
22. vertical translation of –2 **(5, 6)**
23. vertical translation of –10 **(5, –2)**
24. vertical translation of 3 **(5, 11)**
25. horizontal translation of 3 **(8, 8)**
26. horizontal translation of –1 **(4, 8)**
27. horizontal translation of –12 **(–7, 8)**
28. horizontal translation of 10 **(15, 8)**

Tell whether the graph of each function is a translation of its parent function.

29. $y = \frac{2}{3}|x|$ **no**
30. $y = \frac{2}{3} + |x|$ **yes**

APPLICATION

31. **ARCHITECTURE** A stadium in Seattle, Washington, was made with a convertible roof. The lower trusses of the roof are 22 feet below the upper trusses and are the same parabolic shape. Suppose that the function that represents the height in feet of the upper trusses is $y = -\left(\frac{x}{40}\right)^2 + 230$, where x represents the horizontal distance in feet from the axis of symmetry of the parabola. Write the equation that represents the height in feet of the lower trusses, and describe the transformation that relates the two functions. $y = -\left(\frac{x}{40}\right)^2 + 208$

Look Back

32. Solve $7x - (24 + 3x) = 0$ for x. **(LESSON 3.5)** $x = 6$

CONNECTION

33. **STATISTICS** On a certain street, the rents being asked for available 2-bedroom apartments are: $475, $490, $530, $545, $550, and $1025. (**LESSON 4.4**)
   a. Find the mean, median and mode of this data.
   b. Which is higher, the mean or the median? Why?

   mean: $602.50
   median: $537.50
   mode: none

Look Beyond

34. Graph $y = x^3$, $y = (-x)^3$, and $y = -x^3$ on the same coordinate plane.
   a. Describe the two transformations of the parent function.
   b. Compare the results of the transformations.
   c. Write two other functions whose reflections across the x-axis and y-axis give graphs that coincide. What do these functions have in common?

34.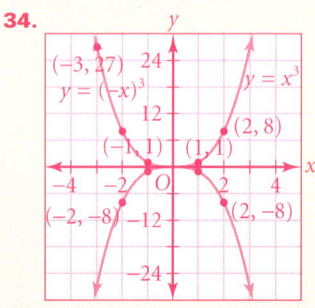

a. The graph of $y = (-x)^3$ is the reflection of $y = x^3$ across the y-axis. The graph of $y = -x^3$ is the reflection of the graph of $y = x^3$ across the x-axis.

b. The two reflections coincide; they are the same graph.

c. $y = x$, $y = x^5$; these functions both have x raised to an odd power.

14.3 Stretches and Compressions

Objective
- Describe how changes to the rule of a function stretch or compress its graph.

Why
The graph of a function can be stretched or compressed without changing its basic shape. Once you understand how changes to the rule of a function correspond to these stretches and compressions, it will be easy to graph many functions.

The height of a waterfall can be approximated if you know the time it takes a drop of water to fall from the top to the pool below. This relationship can be modeled by the function $h = 16t^2$, where h is the height in feet and t is the time in seconds. For example, if it takes a drop of water 3 seconds to fall to the pool below, the height of the waterfall is about $16(3^2)$, or 144, feet. You may have recognized that the parent function for $h = 16t^2$ is the quadratic parent function, $h = t^2$. You can obtain the function $h = 16t^2$ from its parent function by using a transformation that you will learn about in this lesson.

TECHNOLOGY GRAPHICS CALCULATOR

Activity
Vertically Stretching a Function

You will need: graph paper or a graphics calculator

1. Make a table of values for the function $h(t) = t^2$ using t-values 0, 1, 2, 3, 4, and 5. Graph the function using all real numbers greater than or equal to 0 for the domain.

2. Repeat Step 1 for the function $j(t) = 16t^2$. Extend your previous table and use the same coordinate grid.

3. Compare the range values of the two functions. What do you notice about the output value for each function for the same input value?

4. Complete Steps 1–3 for the functions $f(x) = |x|$ and $f(x) = 10|x|$ using integer values from –3 to 7 inclusive for x.

CHECKPOINT ✓ 5. What is the effect on the graph of a function when the function is multiplyed by a number greater than 1? What happens to the range values of the function? Are the domain values affected?

Prepare

NCTM PRINCIPLES & STANDARDS 1–3, 5, 6, 8–10

LESSON RESOURCES
- Practice 14.3
- Reteaching Master 14.3
- Enrichment Master 14.3
- Cooperative-Learning Activity 14.3
- Lesson Activity 14.3
- Problem Solving/Critical Thinking Master 14.3

QUICK WARM-UP

Given $f(x)$, find $f(-4)$.

1. $f(x) = x^2$ 16
2. $f(x) = x^2 - 12$ 4
3. $f(x) = 3x^2$ 48
4. $f(x) = 3x^2 - 12$ 36
5. $f(x) = 3(x^2 - 12)$ 12
6. $f(x) = 3(x - 12)^2$ 768
7. $f(x) = 3|x - 12|$ 48
8. $f(x) = 3|x| - 12$ 0

Also on Quiz Transparency 14.3

Teach

Why
Have students make a list of things that can be stretched, such as human muscles and rubber bands. Ask them to compare the appearance of an object in its original shape with its appearance after being stretched. Tell students that in this lesson, they will see how graphs of functions can be stretched.

☞ For Activity Notes, see page 708.

Alternative Teaching Strategy

COOPERATIVE LEARNING Have students work in groups of four. Within each group, students should choose one of the following functions:

$f(x) = 4x^2$ $g(x) = x^2$ $h(x) = \frac{1}{4}x^2$ $j(x) = 2x^2$

Each group should agree on a scale that will be used for all the graphs, and then students should graph their chosen function.

When all students in a group have completed their graph, the group should compare the results. Ask them to consider how the coefficient of x^2 affects the graph. Tell them to make a conjecture about the appearance of the graphs of the functions $p(x) = 8x^2$ and $q(x) = \frac{1}{8}x^2$.

When all groups have completed their work, have each group choose a spokesperson who will report their results to the entire class.

LESSON 14.3 **707**

Activity Notes

In this Activity, students examine two cases in which a change to the equation of a parent function results in a vertical stretch of the graph of the function. After completing this Activity, students should realize that, when comparing the functions $f(x) = x^2$ and $g(x) = ax^2$, where $a > 1$, any value of $g(x)$ will be a times greater than the corresponding value of $f(x)$.

Cooperative Learning

You may wish to have students do the Activity in pairs. One partner can do Steps 1–3 for the quadratic functions given, and the other partner can do Steps 1–3 for the absolute-value functions. The partners can then compare results and do Step 5 together.

CHECKPOINT ✓
5. Sample answer: For each domain value, the range value increases when the function is multiplied by a number greater than 1. This increase "stretches" each y-value vertically away from the x-axis.

Teaching Tip

TECHNOLOGY If you are using a TI-82 or TI-83 graphics calculator, you can create the graphs in the Activity by pressing [Y=] and entering X^2 next to Y1 and 16X^2 next to Y2. Press [GRAPH] to view the graph.

To create the table of values, press [2nd] [WINDOW] and choose the following settings:

TblStart=0 (or TblMin=0)
ΔTbl=1
Indpnt: **Auto** Ask
Depend: **Auto** Ask

Press [2nd] [GRAPH] to display the table.

See Keystroke Guide, page 782.

708 LESSON 14.3

Vertical Stretch and Compression

EXAMPLE  Graph each pair of functions, and identify the transformation from f to g.

a. $f(x) = x^2 - 3$ and
$g(x) = 2(x^2 - 3)$

b. $f(x) = x^2 - 3$ and
$g(x) = \frac{1}{2}(x^2 - 3)$

SOLUTION

a.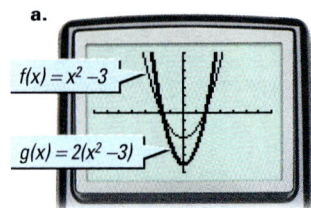

b.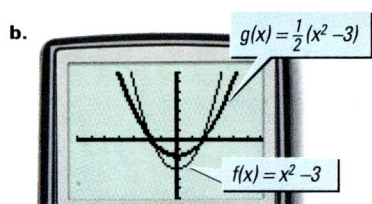

Notice that $g(x) = 2 \cdot f(x)$. The function $g(x) = 2(x^2 - 3)$ is a *vertical stretch* of f by a factor of 2. The point $(3, 6)$, for example, is transformed to the point $(3, 12)$.

Notice that $g(x) = \frac{1}{2} \cdot f(x)$. The function $g(x) = \frac{1}{2}(x^2 - 3)$ is a *vertical compression* of f by a factor of $\frac{1}{2}$. The point $(3, 6)$, for example, is transformed to the point $(3, 3)$.

For a given value in the domain of a function, a vertical stretch or compression multiplies the corresponding value in the range of the function by the scale factor.

TRY THIS Graph each pair of functions, and identify the transformation from f to g.

a. $f(x) = \sqrt{x}$ and $g(x) = 3\sqrt{x}$
b. $f(x) = \sqrt{x}$ and $g(x) = \frac{1}{3}\sqrt{x}$

A vertical stretch or compression moves the graph away from or toward the x-axis, respectively. This can be generalized as follows:

Vertical Stretch and Vertical Compression

If $y = f(x)$, then $y = af(x)$ gives a **vertical stretch** or **vertical compression** of the graph of f.
- If $a > 1$, the graph is stretched vertically by a factor of a.
- If $0 < a < 1$, the graph is compressed vertically by a factor of a.

Vertical Stretch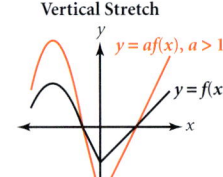

Vertical Compression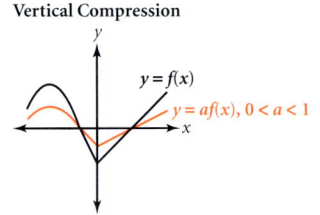

CRITICAL THINKING Let $0 < r < 1$ and $s > 1$. Compare the appearance of the graphs of $f(x) = |x|$, $h(x) = r|x|$, and $g(x) = s|x|$.

Interdisciplinary Connection

GRAPHIC ARTS Shown below is a simple design created by an art student. The set of points that make up the design are identified as the graph of a function, f. Two distortions of the design are shown at far right and are identified as the graphs of functions g and h.

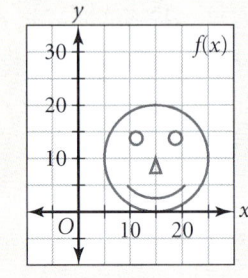

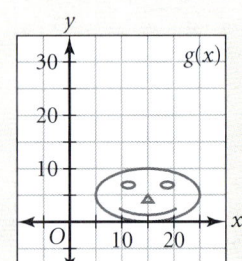

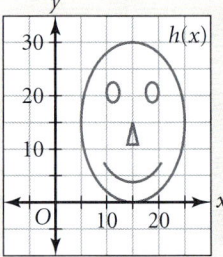

Using function notation, explain how the values of functions g and h are related to the values of function f. $g(x) = 0.5f(x); h(x) = 1.5f(x)$

EXAMPLE 2

a. Use $y = x^2$ to sketch the graph of $y = \frac{x^2}{3}$.

b. If you know the value of y for a given value of x in the parent function $y = x^2$, describe how you would determine the corresponding y-value for the function $y = \frac{x^2}{3}$.

SOLUTION

a. The function can be written as $y = \frac{1}{3}x^2$. Its parent function is the function, $y = x^2$. Each point of the parent graph is compressed by a factor of $\frac{1}{3}$.

b. To find the y-value of $y = \frac{x^2}{3}$ from a known value of y in the parent function, divide the known value by 3. For example, if the y-value in the parent function is 9, then the corresponding y-value for $y = \frac{x^2}{3}$ will be $\frac{9}{3}$, or 3.

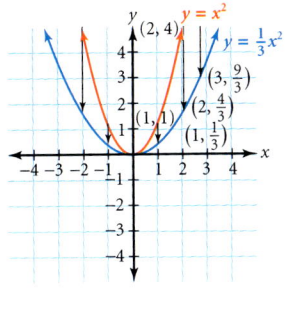

TRY THIS Use $y = |x|$ to sketch the graph of $y = \frac{1}{4}|x|$.

EXAMPLE 3

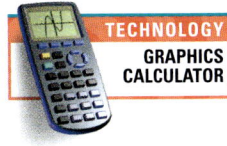
TECHNOLOGY GRAPHICS CALCULATOR

Which functions are stretches or compressions of their parent graph?

a. $y = \frac{x^2}{3}$ b. $y = \frac{3}{4} + x^2$ c. $y = 5|x|$

SOLUTION

a. Because x^2 is multiplied by $\frac{1}{3}$, this is a vertical compression by a factor of $\frac{1}{3}$. The y-values of this function will be $\frac{1}{3}$ of the y-values of the parent function.

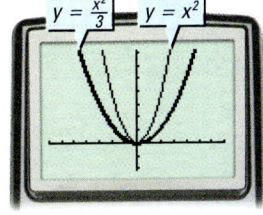

b. Because $\frac{3}{4}$ is *added* to x^2, this is not a stretch or compression. The graph of the function is identical to the graph of the parent function *translated* $\frac{3}{4}$ of a unit up.

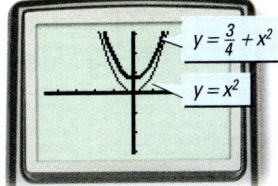

c. Because $|x|$ is multiplied by 5, each y-value of the parent function is stretched vertically by a factor of 5.

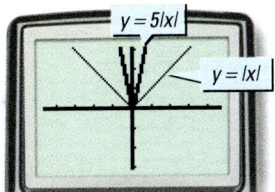

ADDITIONAL EXAMPLE 1

Graph both $f(x) = |x|$ and $g(x) = 2|x|$. Identify the transformation from f to g.

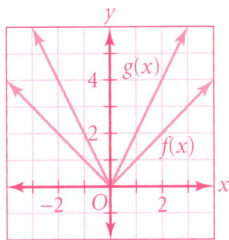

vertical stretch by a factor of 2

TRY THIS

Check students' graphs.
a. vertical stretch by a factor of 3
b. vertical compression by a factor of $\frac{1}{3}$

CRITICAL THINKING

The graph of h is narrower and steeper than the graph of f. The graph of g is broader and flatter than the graph of f.

ADDITIONAL EXAMPLE 2

Use $y = \frac{1}{x}$ to sketch the graph of $y = \frac{0.5}{x}$.

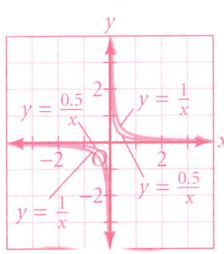

The function $y = \frac{0.5}{x}$ is a vertical compression of its parent function, $y = \frac{1}{x}$, by a scale factor of 0.5.

☞ For answers to Try This and Additional Example 3, see page 710.

Inclusion Strategies

VISUAL LEARNERS Have students use pencils of two different colors to graph functions. Encourage them to always use the same color, such as red, for the graphs of parent functions and another color, such as blue, for the stretched graphs. This technique of color-coding the graphs should help students develop an instinctive feel for the appearance of the graphs of the parent functions.

Enrichment

Transformations are said to *map* each point of a given function onto a corresponding point of a second function. The *mapping notation* for a stretch by a vertical scale factor of a, with no horizontal stretch, is $S_{1,a}: (x, y) \to (x, ay)$.

1. Write the mapping notation for a stretch by a vertical scale factor of 1.5, with no horizontal stretch. $S_{1, 1.5}: (x, y) \to (x, 1.5y)$

2. Which points are mapped onto themselves when there is a vertical stretch but no horizontal stretch? **all points on the x-axis**

LESSON 14.3

TRY THIS

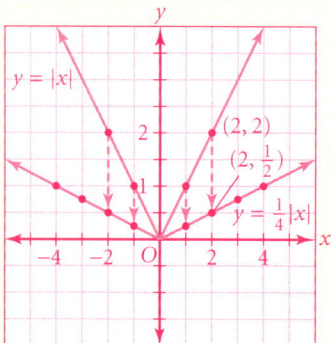

ADDITIONAL EXAMPLE 3

Which functions are vertical stretches of their parent graphs?

a. $y = 5 + x^2$

b. $y = \frac{1}{5x}$

c. $y = |x - 5|$

Only the graph of function b, $y = \frac{1}{5x}$, is a vertical stretch of the parent graph.

ADDITIONAL EXAMPLE 4

Identify the transformation from f to g.

a. $f(x) = |x|$ and $g(x) = \left|\frac{3}{2}x\right|$
Function g is a horizontal compression of f by a factor of $\frac{2}{3}$.

b. $f(x) = |x|$ and $g(x) = \left|\frac{1}{5}x\right|$
Function g is a horizontal stretch of f by a factor of 5.

TRY THIS

Check students' graphs.

a. horizontal compression by a factor of $\frac{1}{3}$

b. horizontal stretch by a factor of 3

710 LESSON 14.3

Horizontal Stretch and Compression

EXAMPLE 4 Graph each pair of functions, and identify the transformation from f to g.

a. $f(x) = |x|$ and $g(x) = \left|\frac{1}{3}x\right|$
b. $f(x) = |x|$ and $g(x) = |2x|$

SOLUTION

a.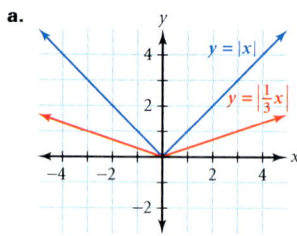

b.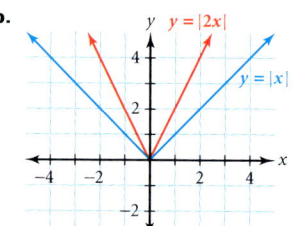

Notice that $g(x) = f\left(\frac{1}{3}x\right)$. The function $g(x) = \left|\frac{1}{3}x\right|$ is a *horizontal stretch* of f by a factor of 3 (the reciprocal of $\frac{1}{3}$). The point (1, 1), for example, is transformed to the point (3, 1).

Notice that $g(x) = f(2x)$. The function $g(x) = |2x|$ is a *horizontal compression* of f by a factor of $\frac{1}{2}$. The point (1, 1), for example, is transformed to the point $(\frac{1}{2}, 1)$.

TRY THIS Graph each pair of functions, and identify the transformation from f to g.

a. $f(x) = 10^x$ and $g(x) = 10^{3x}$
b. $f(x) = 10^x$ and $g(x) = 10^{\frac{x}{3}}$

A horizontal stretch or compression moves the graph away from or toward the y-axis, respectively. This can be generalized as follows:

Horizontal Stretch and Horizontal Compression

If $y = f(x)$, then $y = f(bx)$ gives a **horizontal stretch** or **horizontal compression** of the graph of f.
- If $b > 1$, the graph is compressed horizontally by a factor of $\frac{1}{b}$.
- If $0 < b < 1$, the graph is stretched horizontally by a factor of $\frac{1}{b}$.

Horizontal Stretch

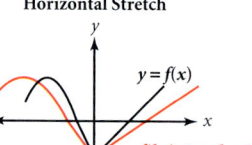

Horizontal Compression

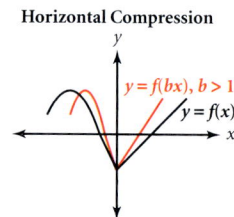

Reteaching the Lesson

USING VISUAL STRATEGIES On the overhead, display the graphs shown below. Use them to give students a visual means of quantifying the amount of stretching that occurs as the equation of the parent absolute-value function is altered.

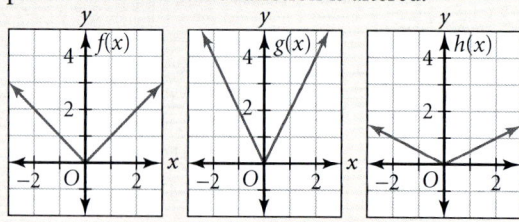

Stress that the scales of the graphs are identical and that each shows an absolute-value function over the interval $-3 \geq x \geq 3$. Have students compare the heights of the graphs. The height of the graph of g is twice the height of the graph of f. The height of h is half the height of f. Ask them to make a conjecture about equations for g and h. $g(x) = 2|x|$; $h(x) = 0.5|x|$ Then have them substitute ordered pairs from each graph into the equations to confirm their conjectures. Repeat the activity with parabolas representing the parent quadratic function and several stretches.

You may have noticed that with many graphs, vertical stretches and horizontal compressions look exactly the same. This is because many functions can be written in more than one way. For example, $y = 4x^2$ (indicating a vertical stretch of $y = x^2$) is equivalent to $y = (2x)^2$.

GRAPHING A STRETCH OR COMPRESSION

To graph a stretch or compression, you should follow these steps.

1. Identify and graph the parent function.
2. Determine if the transformation is vertical or horizontal.
3. Determine if the transformation is a stretch or compression.
4. Identify the factor by which the function is stretched or compressed.
5. Transform several points on the original function.
6. Plot the new points and use them as a guide to graph the transformed function.

Exercises

Communicate

1. Explain how the graph of $y = ax^2$ changes as a increases from 1 to 4 and as a decreases from 1 to 0.
2. Explain how the graph of $y = 10^{bx}$ changes as b increases from 1 to 4 and as b decreases from 1 to 0.
3. Given the graph of a function $f(x)$, explain how you could use transformations to graph the function $g(x) = \frac{1}{2}f(x)$.
4. Given the graph of a function $f(x)$, explain how you could use transformations to graph the function $h(x) = f(2x)$.

Guided Skills Practice

Use the appropriate parent function to sketch the graph of each function. *(EXAMPLES 1, 2, AND 4)*

5. $y = 3|x|$
6. $y = \frac{1}{4}x^2$
7. $y = \frac{\sqrt{x}}{3}$
8. $y = \left|\frac{1}{2}x\right|$
9. $y = (3x)^2$
10. $y = \frac{1}{2x}$

Determine which functions are stretches or compressions of their parent graph. *(EXAMPLE 3)*

11. $y = \frac{1}{2}x^2$
12. $y = |x| + 3$
13. $y = (4x)^2$
14. $f(x) = \sqrt{x + 7}$
15. $f(x) = \frac{1}{52 + x}$
16. $f(x) = 10^{2x}$

11. vertical compression
12. neither
13. horizontal compression
14. neither
15. neither
16. horizontal compression

9.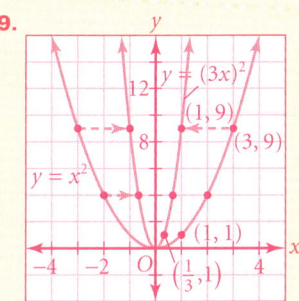

10.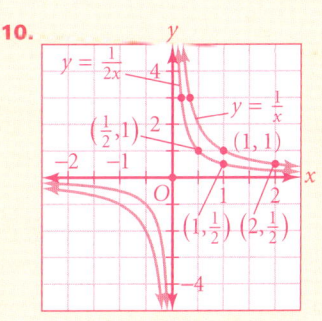

Assess

Selected Answers
Exercises 5–16, 17–41 odd

ASSIGNMENT GUIDE

In Class	1–16
Core	17–35
Core Plus	17–36
Review	37–42
Preview	43–45

✏️ Extra Practice can be found beginning on page 738.

5.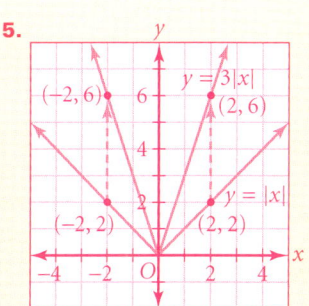

6.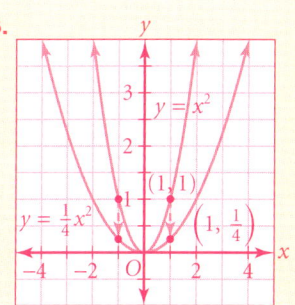

7.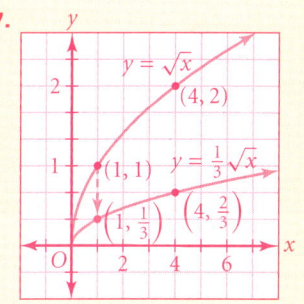

8.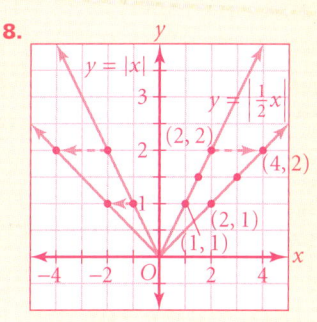

LESSON 14.3 **711**

Technology

A graphics calculator may be helpful for many of the exercises. You may wish to have students first complete the exercises with paper and pencil and then use the calculator to verify their answers.

Error Analysis

When viewing the graphs of functions such as $f(x) = x^2$ and $g(x) = 4x^2$ simultaneously, students commonly associate the graph that appears "wider" with the greater coefficient of x^2, namely 4. Remind them that the coefficient 4 signals a vertical stretch by a factor of 4, which means that the graph of g will appear "thinner."

17.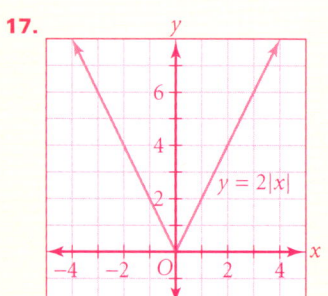

Practice and Apply

Graph each of the following:

17. $y = 2|x|$
18. $y = 0.5x^2$
19. $y = \sqrt{\frac{1}{4}x}$

Which functions are stretches or compressions of their parent function?

20. vertical stretch
21. neither
22. vertical stretch

20. $y = 3x^2$
21. $y = |x| + 1$
22. $y = \frac{5}{x}$

23. Consider the function $y = 5x^2$.
    a. Identify the parent function. $y = x^2$
    b. Sketch the graphs of both functions on the same coordinate plane.
    c. Each point (x, y) on the graph of the parent function is transformed to what point on the new graph? $(x, 5y)$

24. Consider the function $y = \frac{2}{x}$.
    a. Identify the parent function. $y = \frac{1}{x}$
    b. Sketch the graphs of both functions on the same coordinate plane.
    c. Each point (x, y) on the graph of the parent function is transformed to what point on the new graph? $(x, 2y)$

Tell whether each statement is true or false.

true 25. The y-values of $y = 5|x|$ are 5 times the corresponding y-values of $y = |x|$.

true 26. The x-values of $y = (5x)^2$ must be $\frac{1}{5}$ as large as the x-values of $y = x^2$ in order to get the same y-values.

Write two equations for each graph.

$y = 2x^2$ and $y = (\sqrt{2}x)^2$

27.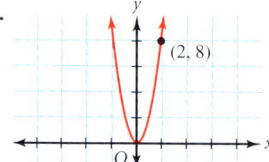

28. $y = \frac{1}{4}x^2$ and $y = \left(\frac{1}{2}x\right)^2$

For each function below, identify the parent function, draw both graphs, and tell how the graph of each function is obtained from its parent function.

29. $y = |3x|$ $y = |x|$
30. $y = 3x^2$ $y = x^2$
31. $y = \left(\frac{x}{3}\right)^2$ $y = x^2$

CONNECTION

32. **STATISTICS** Draw a bar graph for the frequency distribution of Ms. Tienert's students' test scores, shown in the table below.

Score	50	55	60	65	70	75	80	85	90	95	100
Frequency	1	0	2	5	10	6	2	0	0	0	0

33. Because the students didn't have enough time and the scores were low, Ms. Tienert decided to add 20 points to every student's score. Draw a new bar graph that reflects this change, and explain how it compares with the original bar graph.

34. Ms. Tienert also considered multiplying each score by $\frac{5}{4}$. Draw a new bar graph that reflects this change, and explain how it compares with the original bar graph.

18.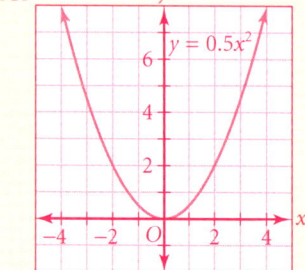

19.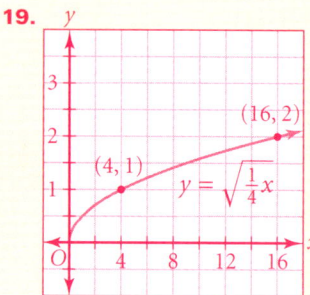

Practice

14.3 Stretches

Given the function $y = 2|x|$,
1. identify the parent function. $y = |x|$
2. sketch the graphs of the function and the parent function on the axes at the right.
3. explain how the graph of the parent function was changed by the 2. A vertical stretch by 2 doubled all of the corresponding y-values.

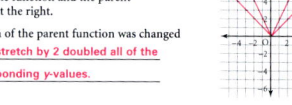

Given the function $y = \frac{3}{x}$,
4. identify the parent function. $y = \frac{1}{x}$
5. sketch the graphs of the function and the parent function on the axes at the right.
6. explain how the graph of the parent function was changed by the 3. A vertical stretch by 3 tripled all of the corresponding y-values.

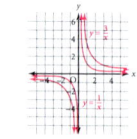

Tell whether each function is a stretch of the parent function.

7. $y = \frac{x^2}{4}$ yes
8. $y = \frac{8}{x}$ yes
9. $y = |x| - \frac{1}{2}$ no

Write an equation for each graph.

10.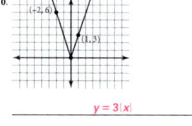
11.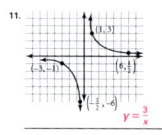

712 LESSON 14.3

APPLICATIONS

35. **ART** Pictures can be "stretched" or enlarged by using a grid. One method to enlarge a picture is to place a grid over it and copy what is in each part of the grid onto a larger grid.
    a. On a coordinate plane, draw a triangle with vertices at (1, 1), (3, 1), and (2, 3).
    b. Create a coordinate plane with grid squares 3 times larger. Use this grid to make an enlargement of the triangle.
    c. Use the language of transformations to describe this process for enlarging a picture.

    Artist Chuck Close uses a grid technique to enlarge portraits adding his unique artistic interpretation to each grid space.

Photograph by Bill Jacobson, Courtesy of Pace Wildenstein

CHALLENGE

36. **INCOME** The workers in an electronics plant received a raise of 5%. Find the scale factor, or the factor by which their salaries were "stretched." (Hint: The scale factor is not 0.05.) **1.05, or 105%**

Look Back

37. Solve the equation $5x - (7x + 4) = 3(x + 5) + 4(5 - 2x)$. **(LESSON 3.5)**

Solve each system of linear equations. (LESSONS 7.2 AND 7.3)

38. $\begin{cases} 3x + 7y = -6 \\ x - 2y = 11 \end{cases}$
39. $\begin{cases} 5y - 3x = -31 \\ 4y = 16 \end{cases}$
40. $\begin{cases} 1.25x + 2y = 5 \\ 3.75x + 6y = 15 \end{cases}$

41. Dean and Deanna are siblings. Deanna has twice as many brothers as she has sisters. Dean has the same number of brothers and sisters. How many girls and how many boys are there in the family? **(LESSON 7.6)**

42. Consider the relation consisting of the ordered pairs (3, 7), (5, −1), (1, 0), (2, 5), (16, −6), (0, 15), (1, −1), and (−2, 5). Is this relation a function? Why or why not? **(LESSON 14.1) no; 1 is associated with both 0 and −1.**

37. $x = 13$
38. $x = 5$, $y = -3$
39. $x = 17$, $y = 4$
40. infinitely many solutions
41. 3 girls, 4 boys

Look Beyond

Match each equation with its graph.

43. $y = -\sqrt{x}$ **C**
44. $y = -\sqrt{\dfrac{x}{4}}$ **B**
45. $y = -\dfrac{\sqrt{x}}{4}$ **A**

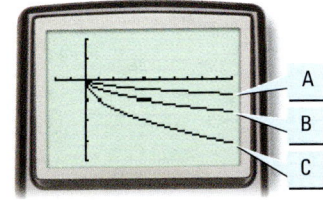

Look Beyond

In Exercises 43–45, students examine various reflections of the graph of the parent square-root function. Students will learn more about reflections in Lesson 14.4.

35. a.
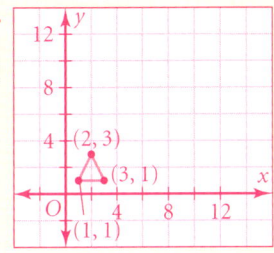

b.
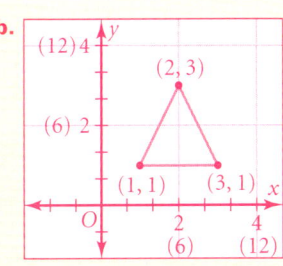

c. The sides of the new grid squares are vertically and horizontally stretched by a factor of 3. If the original grid were placed over the enlarged grid, matching the origins, the large triangle would have vertices of (3, 3), (9, 3), and (6, 9).

Student Technology Guide

14.3 Stretches

A graphics calculator allows you to compare a parent function $f(x)$ to the related function $af(x)$ in graphs and in tables. The different ways of looking at the relationship between the functions will help you understand why $af(x)$ is a stretch of the parent function. To compare $y = x^2$ with $y = 2x^2$, perform the following steps:

- Enter the function as Y1 and Y2 by pressing [Y=] [X,T,θ,n] [x²] [ENTER] 2 [X,T,θ,n] [x²].
- Use the left arrow ◄ key to move the cursor to the left of the Y2 label. Press [ZOOM] 4:ZDecimal [ENTER] so that tracing will give convenient values.
- Press [GRAPH]. Use ► to trace to the point (1, 1), as shown at left below.
- Then press ▲ to trace to the point with an x-coordinate of 1 on the other graph. Notice that the point is at (1, 2), twice as far from the x-axis as (1, 1) is.

Trace to other x-values and toggle between the graphs to compare the y-values. Every y-value on the graph of $y = 2x^2$ is double the corresponding y-value of $y = x^2$. This indicates that you get the graph of $y = 2x^2$ by vertically stretching the graph of $y = x^2$ by 2.

This relationship can also be shown in a table. Press [2nd] [GRAPH] to see a table of values for these functions. Use the up ▲ and down ▼ arrow keys to move through the table. Notice that every value in the Y2 column is twice the corresponding value in the Y1 column.

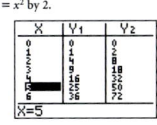

Graph each parent function as Y1 and each related function as Y2. State whether the related function is a stretch of the parent function, and, if so, what the scale factor is. **See answer section for graphs.**

1. $y = \dfrac{1}{x}$ and $y = \dfrac{3}{x}$ — **stretch with a scale factor of 3**
2. $y = x^2$ and $y = \dfrac{1}{4}x^2$ — **stretch with a scale factor of $\dfrac{1}{4}$**
3. $y = x^2$ and $y = x^2 - 2$ — **not a stretch of the parent function**
4. $y = |x|$ and $y = 2|x|$ — **stretch with a scale factor of 2**
5. $y = 2^x$ and $y = 6^x$ — **not a stretch**

The answers to Exercises 29–34 can be found in Additional Answers beginning on page 810.

23. b.
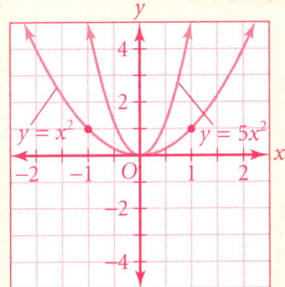

24. b.
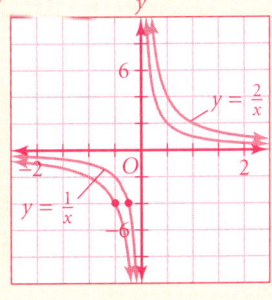

LESSON 14.3 **713**

Prepare

NCTM PRINCIPLES & STANDARDS 1–3, 5–10

LESSON RESOURCES

- Practice 14.4
- Reteaching Master 14.4
- Enrichment Master 14.4
- Cooperative-Learning Activity 14.4
- Lesson Activity 14.4
- Problem Solving/Critical Thinking Master 14.4
- Teaching Transparency 70

QUICK WARM-UP

Given $f(x)$ and $g(x)$, find $f(6)$ and $g(6)$.

1. $f(x) = 5x; g(x) = -5x$
   30; −30
2. $f(x) = x^2; g(x) = -x^2$
   36; −36
3. $f(x) = |x|; g(x) = -|x|$
   6; −6
4. $f(x) = 2^x; g(x) = -2^x$
   64; −64
5. $f(x) = \frac{3}{x}; g(x) = -\frac{3}{x}$
   0.5; −0.5

Also on Quiz Transparency 14.4

Teach

Why Ask students to consider what they see when they view their reflection in a mirror. Most will realize that there is a left-right reversal. That is, if you raise your right hand, your reflection appears to raise its left hand. Point out that some other things, like top and bottom, remain the same. Tell students that graphs of functions also have reflected images, with a line acting as the "mirror."

714 LESSON 14.4

Reflections

14.4

Objective
- Describe how a change to the rule of a function corresponds to a reflection of its graph.

Why You can graph many functions by understanding how an alteration to the rule of a function corresponds to the reflection of its graph in the x- or y-axis.

A mirror reflects a figure, forming the illusion of its image behind the mirror. The graph of a function can be reflected across the x- and y-axes in a similar manner.

Activity
Reflection Across the x-Axis

You will need: a reflection tool (optional)

1. Make a table of values for the function $f(x) = x^2$, using x-values −2, −1, 0, 1, and 2, and graph the function.

2. Reflect f across the x-axis. That is, create a mirror image of f as if a mirror were placed on the x-axis. Let g represent the function for this reflected image.

3. Make a table of values for the reflected function, g. Use the same x-values that you used for f.

4. Compare the tables for the two graphs. The point (2, 4) is on the parent graph. What is the function value on the reflected graph when x is 2? If (−3, 9) is a point on the original graph, what is the corresponding point on the reflected graph?

CHECKPOINT ✓ 5. For the following values of x, give the corresponding function values on the reflected graph.
   a. 3   **b.** 0   **c.** −3   **d.** a

6. If $f(x) = x^2$ is reflected across the x-axis, what is the equation of the new function?

Alternative Teaching Strategy

USING PAPER-FOLDING Have students draw a set of coordinate axes on a sheet of graph paper so that the origin is near the center. Tell them to graph $y = |x|$, being sure to press down firmly so that the graph is clearly visible through the back of the paper. Have them fold the paper along the x-axis with the graph facing outward.

Instruct them to lay the folded paper on the desk with the blank side face up. The graph from the under side should be visible through the layers of paper. Tell students to trace the graph.

Now have the students unfold their paper. Tell them that the tracing is the reflection of the graph of $y = |x|$ through the x-axis. Have them label these points on the graph of $y = |x|$: (−3, 3), (−1, 1), (0, 0), (2, 2), and (3, 3). Ask them to identify and label the points on the reflected graph that correspond to those points: (−3, −3), (−1, −1), (0, 0), (2, −2), and (3, −3). Ask them what they think is a possible equation for the reflected graph. Elicit the response $y = -|x|$.

Example 1 illustrates a reflection across the x-axis and a reflection across the y-axis.

EXAMPLE Graph each pair of functions, and identify the transformation from f to g.

  a. $f(x) = x^2$ and $g(x) = -(x^2)$ **b.** $f(x) = 2x + 3$ and $g(x) = 2(-x) + 3$

• SOLUTION

a.
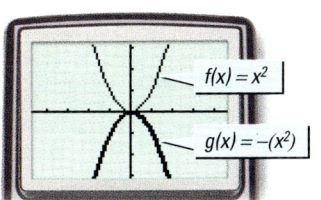

b.
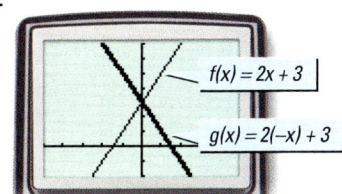

Notice that $g(x) = -f(x)$. The function $g(x) = -(x^2)$ is the reflection of f across the x-axis. The point (2, 4), for example, is transformed to the point (2, −4).

Notice that $g(x) = f(-x)$. The function $g(x) = 2(-x) + 3$ or $g(x) = -2x + 3$, is the reflection of f across the y-axis. The point (1, 5), for example, is transformed to the point (−1, 5).

In general, a reflection across the x-axis transforms each point (a, b) on the original graph to the point (a, −b), and a reflection across the y-axis transforms each point (a, b) to the point (−a, b).

TRY THIS Graph each pair of functions, and identify the transformation from f to g.

  a. $f(x) = |x|$ and $g(x) = -|x|$ **b.** $f(x) = 2x - 1$ and $g(x) = 2(-x) - 1$

Reflections across the x- and y-axes can be generalized as follows:

Reflection Across the x-Axis and y-Axis

If $y = f(x)$, then $y = -f(x)$ gives the reflection of the graph of f across the x-axis.

If $y = f(x)$, then $y = f(-x)$ gives the reflection of the graph of f across the y-axis.

Reflection Across the x-Axis
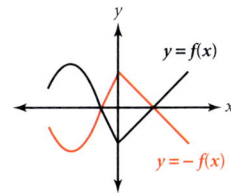

Reflection Across the y-Axis
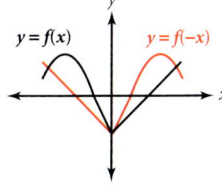

CRITICAL THINKING Let $f(x) = |x|$. What does $f(-x) = |-x| = |x|$ tell you about the graph of $f(-x)$? Justify your response.

Inclusion Strategies

VISUAL LEARNERS Have students use pencils of different colors to draw the reflected graphs of functions. If they used color-coding to graph stretches in Lesson 14.3, you may want to suggest that they extend the color scheme. For instance, if they used red for graphs of parent functions and blue for stretched graphs, they may want to choose a third color, perhaps green, for reflected graphs. Students may find color-coding especially helpful when reflecting the graphs of $y = x$ and $y = |x|$.

Enrichment

Tell students that $r_x: (x, y) \to (x, -y)$ is the mapping notation for a reflection across the x-axis. Then give them these questions.

1. Write the mapping notation for a reflection across the y-axis. $r_y: (x, y) \to (-x, y)$
2. What is the line of reflection for each mapping below?
   a. $(x, y) \to (y, x)$
   the line with the equation $y = x$
   b. $(x, y) \to (y, -x)$
   the line with the equation $y = -x$

Activity Notes

As students work through the Activity, they should begin to see how a reflection of a graph is created. Students should come to understand that a graph reflected vertically across the x-axis will have y-values that are opposites of the corresponding y-values in the original graph.

Cooperative Learning

You may wish to have students do the Activity in pairs. If a reflection tool is being used, have one student hold the tool while the other draws the reflected graph. Then one student can call out the points on the reflected graph while the other student records them in a table of values.

CHECKPOINT ✓
5. **a.** −9   **b.** 0
   c. −9   **d.** $-a^2$

Use Teaching Transparency 70.

ADDITIONAL EXAMPLE

Describe the graphs of the following functions:

a. $y = -10^x$
The graph of $y = -10^x$ is a reflection of the graph of $y = 10^x$ across the x-axis.

b. $y = -x^2$
The graph of $y = -x^2$ is a reflection of the graph of $y = x^2$ across the x-axis.

TRY THIS
Check students' graphs.
a. reflection across the x-axis
b. reflection across the y-axis

☞ For answer to Critical Thinking, see page 716.

LESSON 14.4

CRITICAL THINKING
Because $f(x) = f(-x)$ in the case of the absolute-value function, the graph of $f(x) = |x|$ already includes a reflection across the y-axis.

ADDITIONAL EXAMPLE 2

Describe the graph of each function below.

a. $y = -\frac{x^2}{2}$

the graph of $y = x^2$ compressed by a vertical scale factor of $\frac{1}{2}$ and reflected across the x-axis

b. $y = -7x$

the graph of $y = x$ stretched by a vertical scale factor of 7 and reflected across the x-axis

---

TRY THIS

a.
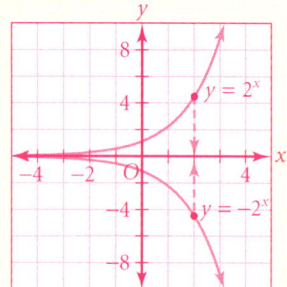

The parent function is reflected across the x-axis.

b.
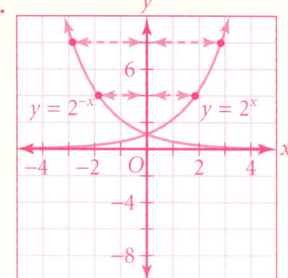

The parent function is reflected across the y-axis.

CRITICAL THINKING
Because $y = -x$ can be written as either $y = -(x)$ or $y = (-x)$, the reflection can be considered to be across either axis. The graph will be the same in either case.

---

716 LESSON 14.4

EXAMPLE 2 Graph each function with its parent function on the same coordinate plane. Describe the transformation of the parent function.

   a. $h(x) = -|x|$ **b.** $y = -\frac{1}{x}$ **c.** $g(x) = \sqrt{-x}$ **d.** $y = (-x)^2$

● **SOLUTION**

a. The parent function is $f(x) = |x|$. Since $h(x) = -f(x)$, h is the reflection of f across the x-axis. To obtain the graph of h, leave the x-coordinates of f the same and replace the y-coordinates with their opposites.

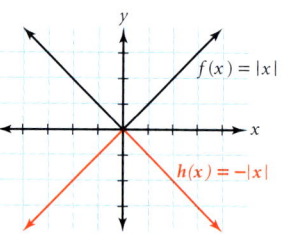

b. The parent function is $y = \frac{1}{x}$. For every point $\left(x, \frac{1}{x}\right)$ on the graph of the parent function, the point $\left(x, -\frac{1}{x}\right)$ is on the graph of $y = -\frac{1}{x}$. Thus, $y = -\frac{1}{x}$ gives the reflection of the graph of $y = \frac{1}{x}$ across the x-axis.

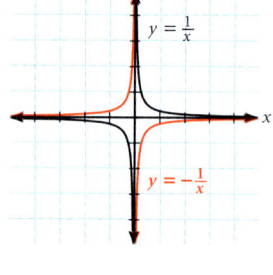

c. The parent function is $f(x) = \sqrt{x}$. Since $g(x) = f(-x)$, g is the reflection of f across the y-axis. To obtain the graph of g, replace the x-coordinates of f with their opposites and leave the y-coordinates the same.

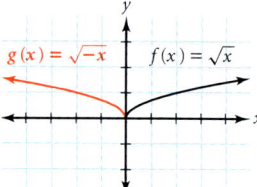

d. The parent function is $y = x^2$. For every point (x, x^2) on the graph of the parent function, the point $\left(x, (-x)^2\right)$ is on the graph of $y = (-x)^2$. Thus, $y = (-x)^2$ gives the reflection of the graph of $y = x^2$ across the y-axis.

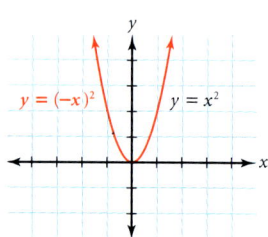

Since the graph of $y = x^2$ is symmetric with respect to the y-axis, the function is reflected onto itself. This can also be explained algebraically by the equation $(-x)^2 = x^2$.

TRY THIS Graph each function with its parent function on the same coordinate plane. Describe the transformation of the parent function.

   a. $f(x) = -2^x$ **b.** $g(x) = 2^{-x}$

CRITICAL THINKING Does the line $y = -x$ reflect the line $y = x$ across the x-axis or across the y-axis? Explain.

Reteaching the Lesson

HANDS-ON STRATEGIES Have students work in groups of five. Provide each group with five sheets of graph paper, five blank transparencies, and pens that can be used to write on the transparencies. Each group member should choose one of the following parent functions: $y = x$, $y = |x|$, $y = x^2$, $y = 2^x$, and $y = \frac{1}{x}$.

Instruct students to draw a set of coordinate axes on the graph paper so that the origin is near the center of the paper. Students should then graph their chosen function.

Now tell students to trace the graph of their function onto a transparency. To simulate a reflection across the x-axis, have them lift their transparency off the graph paper and flip it over toward themselves. They should then align the transparency over the axes to show the reflected graph in its proper position. Students within a group can share graphs so that each student has a chance to perform the reflection of each parent function.

Summary

To graph the reflection of a function, follow these steps.

1. Graph the original function.
2. Determine if the reflection is across the *x*- or *y*-axis.
3. Transform several points on the original graph.
   a. If the reflection is across the *x*-axis, leave the *x*-coordinates the same and replace the *y*-coordinates with their opposites.
   b. If the reflection is across the *y*-axis, replace the *x*-coordinates with their opposites and leave the *y*-coordinates the same.
4. Plot these new points and use them as a guide to graph the reflected function.

Exercises

Communicate

1. Explain the effect of the negative sign in the function $y = -ax^2$.
2. If $h(x) = 2\sqrt{x} + 1$ is reflected across the *y*-axis, how is its function rule changed to express this reflection?
3. Describe how to change the function $f(x) = 2x$ to reflect its graph:
   a. across the *x*-axis.
   b. across the *y*-axis.
4. Is the reflection of a mountain in a lake similar to a reflection across the *x*-axis? Explain why or why not.

A lake creates a natural reflection of a mountain.

GO TO: go.hrw.com
KEYWORD: MA1 Friezes

A frieze pattern is a decorative design element in which a pattern repeats itself horizontally. In this activity, students learn about frieze patterns in relation to reflections and translations.

Assess

Selected Answers
Exercises 5–12, 13–25 odd

ASSIGNMENT GUIDE

In Class	1–12
Core	13–18
Core Plus	13–18
Review	19–25
Preview	26–29

✐ Extra Practice can be found beginning on page 738.

Technology

A graphics calculator may be helpful for Exercises 5–16. You may wish to have students first complete the exercises with paper and pencil and then use the calculator to verify their answers.

Error Analysis

When working with absolute-value functions, students may confuse $y = -|x|$ with $y = |-x|$. Substitute two or three values for *x* in each equation to convince students that they are not equivalent. For example, show them the results for $x = 5$.

$y = -|x|$	$y = |-x|$
$y = -|5|$	$y = |-5|$
$y = -5$	$y = 5$

LESSON 14.4

5. reflection across the *x*-axis

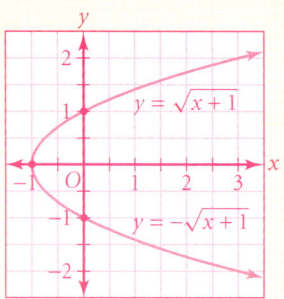

6. reflection across the *x*-axis

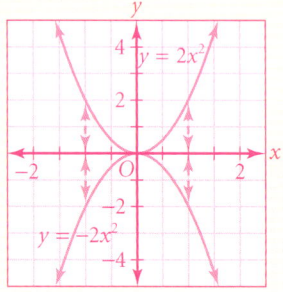

7. reflection across the *y*-axis

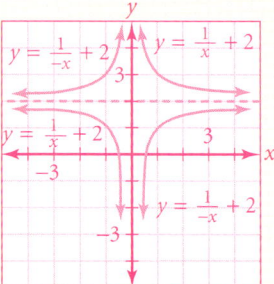

Guided Skills Practice

Graph each pair of functions and identify the transformation from *f* to *g*.
(EXAMPLE 1)

5. $f(x) = \sqrt{x+1}$
   $g(x) = -\sqrt{x+1}$

6. $f(x) = 2x^2$
   $g(x) = -2x^2$

7. $f(x) = \frac{1}{x} + 2$
   $g(x) = \frac{1}{-x} + 2$

8. $f(x) = |x+1|$
   $g(x) = |-x+1|$

Graph each function with its parent function on the same coordinate plane. Describe the transformation of the parent function. *(EXAMPLE 2)*

9. $y = -\sqrt{x}$ **10.** $y = -3^x$ **11.** $y = |-x|$ **12.** $y = \frac{1}{-x}$

Practice and Apply

Graph each pair of functions and identify the transformation from *f* to *g*.

13. $f(x) = (x+1)^2, g(x) = (-x+1)^2$ **14.** $f(x) = \frac{1}{x-2}, g(x) = \frac{1}{-x-2}$

15. $f(x) = |x| - 3, g(x) = -|x| + 3$ **16.** $f(x) = -(x^2+2), g(x) = (x^2+2)$

17. Write the functions for the graph of $f(x) = 5x + 2$ reflected across the *x*-axis and across the *y*-axis.

18. CULTURAL CONNECTION: ASIA On a piece of graph paper, construct the line segments shown in red. Point *E* is at (0, 2), and point *D* is at (0, 6). Segment *EA* is twice the length of segment *OE*, and segment *DB* is twice the length of segment *OD*. Segment *FC* connects the midpoints of the segments *EA* and *DB*. Reflect all of these segments across the *x*-axis, and then reflect both graphs across the *y*-axis. This basic pattern appears in a carving found in a tomb dating back to the Han dynasty in China (206 B.C.E. to 222 C.E.). Patterns constructed in this manner have *bilateral symmetry*.

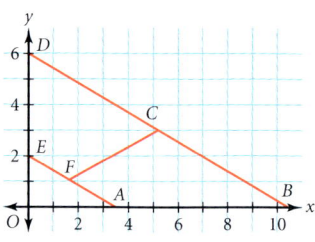

Look Back

APPLICATION

19. PHYSICS Seven seconds after a flash of lightning, Mark heard thunder. If the speed of sound through air is approximately 330 meters per second, how far away did the lightning strike? *(LESSON 2.4)* **2310 meters**

Solve each equation for variable. *(LESSON 3.5)*

20. $6(3-2r) = -(5r+3)$

21. $\frac{x}{2} + 5 = \frac{3x-2}{4}$ $x = 22$

22. $3x = -9(x-4)$ $x = 3$

The graphs to Exercises 8–16 can be found in Additional Answers beginning on page 810.

8. reflection across the *y*-axis

9. reflection across the *x*-axis

10. reflection across the *x*-axis

11. reflection across the *y*-axis (graphs are identical)

12. reflection across the *y*-axis

13. reflection across the *y*-axis

14. reflection across the *y*-axis

15. reflection across the *x*-axis

16. reflection across the *x*-axis

17. $f(x) = 5x + 2$ reflected across the *x*-axis: $g(x) = -(5x-2) = -5x + 2$; $f(x) = 5x + 2$ reflected across the *y*-axis: $h(x) = 5(-x) + 2 = -5x + 2$

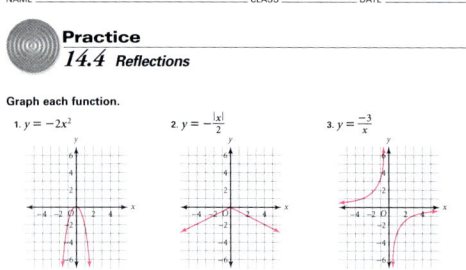

CONNECTIONS

23. **GEOMETRY** The length of a rectangular garden is 3 times its width. If the perimeter of the rectangular garden is 72 meters, what are the dimensions? *(LESSON 3.6)* **9 meters × 27 meters**

24. **STATISTICS** A factory worker measures 5 randomly selected glass rods in a 15-minute production run. What is the average length of the glass rods tested in this run based on the following measurements? *(LESSON 4.4)*

23.74 centimeters

Sample	1	2	3	4	5
Measurement (in cm)	23.5	23.9	23.6	24.0	23.7

25. Graph the solution set $x \geq 3$ or $x < -1$ on a number line. *(LESSON 6.3)*

Look Beyond

Match each function with the appropriate graph.

26. $y = \sqrt{x}$ **b**
27. $y = \sqrt{-x}$ **a**
28. $y = -\sqrt{x}$ **d**
29. $y = -\sqrt{-x}$ **c**

a.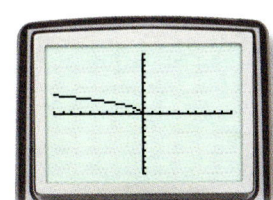
b.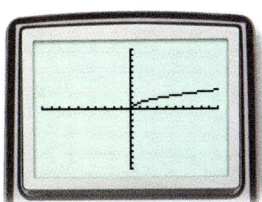
c.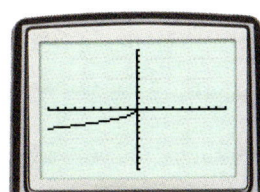
d.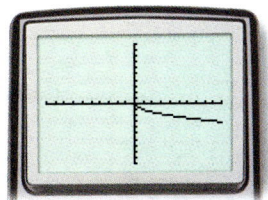

Include this activity in your portfolio.

Play a game of "What's My Function?" with a partner.

1. On a coordinate grid, give your partner a graph of a function that has been translated, stretched, or reflected and the graph of its parent function. You can use a graphics calculator in "Grid On" format for this activity.

2. Your partner must guess your function. Record a score of 1 for each correct response.

3. Take turns drawing the graph and guessing the function for 10 turns each.

WORKING ON THE CHAPTER PROJECT

You should now be able to complete Activity 1 of the Chapter Project on page 728.

Look Beyond

In Exercises 26–29, students further investigate the graphs of the parent square-root function and its transformations.

ALTERNATIVE Assessment

Portfolio Activity

The Portfolio Activity can be used as preparation for the Chapter Project or as a separate activity. In the Portfolio Activity on this page, students take turns generating and identifying transformations of parent functions.

Student Technology Guide

NAME _____ CLASS _____ DATE _____

Student Technology Guide
14.4 Reflections

When the parent function $y = f(x)$ is transformed to the related function $y = -f(x)$, every y-value is replaced by its opposite. A graphics calculator can confirm this. Consider $y = \frac{1}{x}$ and $y = -\frac{1}{x}$.

• Enter $y = \frac{1}{x}$ as Y1 and $y = -\frac{1}{x}$ as Y2 by pressing [Y=] 1 [÷] [X,T,θ,n] [ENTER] [(-)] 1 [÷] [X,T,θ,n] [ ].
• Use [◄] to move the cursor to the left of the Y2 label. Press [ENTER] until you see a heavy line next to Y2. This will help you tell the graphs apart.
• Press [ZOOM] [4:ZDecimal] [ENTER] so that tracing will give convenient values.
• Now press [GRAPH]. You will see graphs of both functions. Use [►] to trace to the point (2, 0.5), as shown in the diagram at left below.
• Then press [▲] to toggle the trace point to the other graph. That point is (2, −0.5), the reflection of (2, 0.5) through the x-axis, as shown in the diagram at right below.

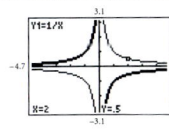 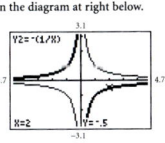

Trace to other x-values and toggle between the graphs to compare the y-values. Notice that every y-value on the graph of $y = \frac{1}{x}$ is the opposite of the y-value for the corresponding point on $y = -\frac{1}{x}$. Because of this, the graph of $-f(x)$ is the vertical reflection of the graph of $f(x)$ through the x-axis.

To see whether one graph is a reflection of another graph through the x-axis, you can make and read a table of values.

Press [2nd] [GRAPH] to see a table of values for these functions. Use the up [▲] and down [▼] arrow keys to move through the table. Notice that every value in the Y2 column is the opposite of the corresponding value in the Y1 column.

See answer section for graphs.
Graph each parent function as Y1 and each related function as Y2. State whether the related function is a vertical reflection of the parent function.

1. $y = x^2$ and $y = -x^2$ 2. $y = 2^x$ and $y = -2^x$ 3. $y = x$ and $y = x - 1$
   vertical reflection   **vertical reflection**   **not a vertical reflection**

LESSON 14.4 719

18.
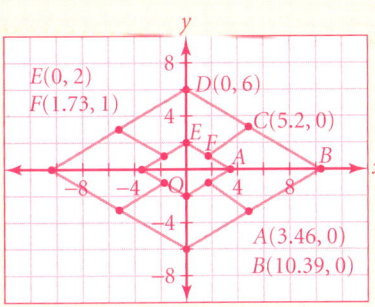

20. $r = 3$

25.
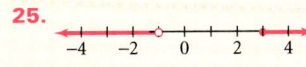

Eyewitness Math

Focus

An article describing the apparent chaos in some events postulates the theory that a type of order exists within seeming disorder. Students are given a formula for predicting the size of an insect population, and they use it to generate data. The rather surprising result is that a behavior modeled by a formula can appear chaotic.

Motivate

Before students read the article, discuss with them the meaning of the word *chaos*. Be sure they understand that it refers to a state of total disorder or confusion.

After they read the article, discuss the examples of chaotic behavior mentioned. Ask if they can suggest other events that seem chaotic at first glance, but might actually have some underlying pattern or predictability.

If possible, demonstrate the water flow shown in the photographs. You also can illustrate chaotic behavior by showing the erratic flight of a balloon that has been inflated and then released without the opening being tied shut.

Is There ORDER in CHAOS?

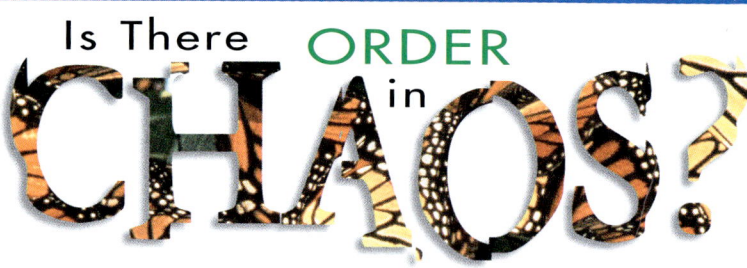

FINDING ORDER IN DISORDER

The science of chaos reveals nature's secrets.

Stock prices lurch and careen during weeks of financial mayhem. The number of measles cases inexplicably soars and crashes. A storm unparalleled in more than 40 years slams into England, killing at least 13 people. Such apparently random events have always seemed far beyond the understanding of even the most powerful computers and brilliant researchers. But are they? A growing and eclectic band of scientists has come to suspect that such chaotic happenings are governed by laws just as orbiting planets and falling apples are . . .

The premise of chaos theory is the oldest cliché in science: beneath disorder lurks order. "Chaologists" find that, although it may never be possible to precisely predict the weather or the stock market or even the path of a roulette ball, one can foresee the patterns of their behavior. These patterns are the order within the chaos . . .

Chaos's power to explain diverse phenomena has encouraged researchers to seek it everywhere. Cardiologists studying the human heart find that its normal rhythm is subtly chaotic, rather than regular like a metronome. Climatologists pondering the ice ages find that their timing follows no regular pattern.

Newsweek, December 21, 1987

ORDER Turn on a faucet just a little. The flow of water looks as smooth as glass tube. This is order.

CHAOS Turn the handle a little more. The stream changes. Its outer edges flicker in unpredictable turbulence. This is chaos.

Some things, like the water stream in the photos, are sometimes orderly and sometimes chaotic. You can use equations to model such dual behavior. To see how this works, explore a biological system that is simpler to model than a stream of water—an imaginary population of insects.

---

1. Answers will vary.
   a. a sample chart, where $P_1 = 0.7$:

Year	Population
1	$P_1 = 0.7$
2	$P_2 = 0.42$
3	$P_3 = 0.487$
4	$P_4 = 0.500$
5	$P_5 = 0.500$
6	$P_6 = 0.500$

   b. The population levels off at around 0.5.

   c. All the groups should come out with the same size population by year 6.

   d. It appears that regardless of the original population size, the population reaches 0.5 when $r = 2$.

The insects hatch in the spring and die after laying eggs in the fall. The population each year depends on the population of the prior year. The more insects there are, the more eggs there are to hatch next spring. Other factors, such as limited food supply, keep the population from becoming infinitely large.

If you know the size of the insect population one year, you can find the size the next year by using the following equation:

$$P_2 = r \cdot P_1(1 - P_1)$$

P_1 is the size of the population in year 1, P_2 is the size of the population in year 2, and r is a constant.

To find the population in year 3, use the equation $P_3 = r \cdot P_2(1 - P_2)$.

What equation would you use for year 4?

In these equations, all values of P must stay in the range from 0 to 1. A value of 0 represents no insects, and a value of 1 represents the final population size.

Now you are ready to see how these equations can model behavior that is sometimes orderly and predictable and sometimes chaotic and unpredictable.

Cooperative Learning

1. First see how the population equations behave for $r = 2$.
   a. Copy and complete the chart at right. To start, choose any value between 0 and 1 for P_1. Round each answer to 3 decimal places.
   b. In your model, what happens to the size of the insect population?
   c. Compare your results with the results of groups that started with a different value for P_1.
   d. Could you make a good prediction about the size of the population in year 6 even if your year 1 data were off? Explain.

2. Now see what happens for $r = 4$.
   a. Copy and complete the chart. Remember to choose a value for P_1 that is between 0 and 1.
   b. Do you see a pattern in your results? Explain.
   c. Repeat part **a**, but this time start with a value for P_1 that is 0.001 less or 0.001 greater than the value you originally used in part **a**.
   d. Compare your results from parts **a** and **c**.
      For $r = 4$, could you make a good prediction about the size of the population in year 9 if your year 1 data were off a little? Explain.
   e. Did other groups get similar results even though they started with different values for P_1?

3. In chaos, a tiny difference at the beginning can make a big difference later on. For which value of r did the equations model chaos? Explain.

Year	Population
1	$P_1 = ?$
2	$P_2 = ?$
3	$P_3 = ?$
4	$P_4 = ?$
5	$P_5 = ?$
6	$P_6 = ?$

Year	Population
1	$P_1 = ?$
2	$P_2 = ?$
3	$P_3 = ?$
4	$P_4 = ?$
5	$P_5 = ?$
6	$P_6 = ?$
7	$P_7 = ?$
8	$P_8 = ?$
9	$P_9 = ?$

2. a. a sample chart, where $P_1 = 0.7$:

Year	Population
1	$P_1 = 0.7$
2	$P_2 = 0.84$
3	$P_3 = 0.538$
4	$P_4 = 0.994$
5	$P_5 = 0.023$
6	$P_6 = 0.089$
7	$P_7 = 0.323$
8	$P_8 = 0.875$
9	$P_9 = 0.437$

b. There does not seem to be a pattern when $r = 4$.
c. It appears that when $r = 4$, there is much more variation in the results than when you start with different populations.
d. There does not seem to be a predictable pattern starting with either 0.7 or 0.699.
e. Other groups do not get predictable results, regardless of the original population.

3. When $r = 4$, the results were not predictable, but they did fall within the range from 0 to 1.

Prepare

NCTM PRINCIPLES & STANDARDS 1–3, 5–10

LESSON RESOURCES

- Practice 14.5
- Reteaching Master 14.5
- Enrichment Master 14.5
- Cooperative-Learning Activity 14.5
- Lesson Activity 14.5
- Problem Solving/Critical Thinking Master 14.5
- Teaching Transparency 71

QUICK WARM-UP

Identify the transformation of the graph of the parent function $f(x) = x^2$ that would result in the graph of each function below.

1. $g(x) = x^2 - 3$ vertical translation 3 units down
2. $h(x) = (x + 2)^2$ horizontal translation 2 units to the left
3. $j(x) = 4x^2$ vertical stretch by a scale factor of 4
4. $k(x) = -x^2$ reflection across the x-axis

Also on Quiz Transparency 14.5

Teach

Why Remind students of the algebraic order of operations. Give them an expression such as $2 + 3 \times 5 - 4$ and have them explain why it is necessary to agree on an order of operations. Tell them that it is possible to apply more than one transformation to a graph. Ask them if they think that the order in which these transformations are applied makes a difference.

Use Teaching Transparency 71.

722   LESSON 14.5

14.5 Combining Transformations

Objectives

- Study a real-world application of transformed functions.
- Graph functions that involve more than one transformation.

Why Many everyday situations like biking can be represented by functions. Once a function is recognized as being related to a known parent function, the graph can be sketched by identifying and applying one or more transformations to the parent graph.

**APPLICATION
RECREATION**

When planning their 100-mile bike ride, the two friends in the photograph above decide to allow 2 hours for rest stops. They want to know how fast they must ride to complete the trip in certain lengths of time.

Using $d = rt$, where d is the distance in miles, r is the rate in miles per hour, and t is the time in hours, you can write $r = \frac{d}{t}$.

In this function, t is the time actually spent riding. The total number of hours it takes to complete the trip, h, is equal to $t + 2$. Thus, $t = h - 2$, and the function that gives the rate of travel necessary to complete the trip in h hours is $r = \frac{100}{h - 2}$.

You can substitute different values of h into the function to determine how fast the friends must ride.

Notice that the function is a transformation of the parent function $r = \frac{1}{h}$.

Alternative Teaching Strategy

HANDS-ON STRATEGIES Tell students to graph $y = |x|$, $y = 2|x|$, and $y = 3|x|$. Then have them cut out a paper triangle that fits in the interior of each ∨. You may want them to use a different color of paper for each triangle. Be sure that they label each with the appropriate equation. Now have them take a large sheet of graph paper and draw a set of coordinate axes so that the origin is near the center. Have them graph the equations at right by placing the appropriate triangle with its vertex at the origin and performing the required reflection and translations.

1. $y = |x - 11|$ Translate $y = |x|$ 11 units to the right.
2. $y = -|x|$ Reflect $y = |x|$ across the x-axis.
3. $y = 2|x - 5|$ Translate $y = 2|x|$ 5 units to the right.
4. $y = -3|x| - 4$ Reflect $y = 3|x|$ across the x-axis, and then translate it 4 units down.
5. $y = 3|x + 2| + 5$ Translate $y = 3|x|$ 2 units to the left and 5 units up.

To sketch the graph of $r = \frac{100}{h-2}$ proceed as follows:

First graph the parent function, $r = \frac{1}{h}$.

Then determine whether the graph of the parent function is horizontally translated, stretched, or compressed, or reflected in the y-axis. The parent graph is horizontally translated 2 units right. The resulting function is $r = \frac{1}{h-2}$.

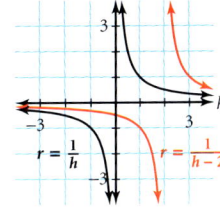

Next determine whether the graph of $r = \frac{1}{h-2}$ is vertically stretched or compressed, reflected in the x-axis, or vertically translated. The graph is vertically stretched by a factor of 100. The resulting function is $r = \frac{100}{h-2}$.

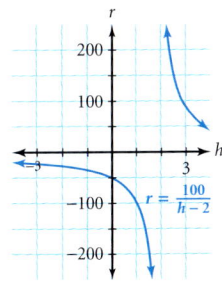

Because neither the number of hours nor the number of miles per hour can be negative, only the right-hand branch of the graph represents the values relevant to the problem.

CRITICAL THINKING What happens to the graph of $r = \frac{100}{h-2}$ for values of h close to 2? What happens at $h = 2$? Explain these areas of the graph as they relate to the biking problem.

Activity
Order of Vertical Transformations

You will need: graph paper

For steps 1–3 below, begin with the graph of the parent function $f(x) = |x|$ and combine the transformations in the order listed. Show each set of transformations on a separate coordinate plane.

1. Translate 3 units up, stretch vertically by a factor of 2, and reflect across the x-axis.
2. Stretch vertically by a factor of 2, translate 3 units up, and reflect across the x-axis.
3. Stretch vertically by a factor of 2, reflect across the x-axis, and translate 3 units up.
4. Graph the function $g(x) = -2|x| + 3$ by making a table of values and plotting several points. Compare the graphs from steps 1–3 with the graph of g. Which one matches the graph of g? Is order important when combining transformations?

CHECKPOINT ✔ 5. When combining vertical transformations, are translations done before or after stretches and reflections?

CRITICAL THINKING
As the values of h become closer to 2, the graph moves closer to 2 but never reaches it. At $h = 2$, the function has no value because division by 0 is undefined. In terms of the biking problem, this indicates that there is no bike riding time when h equals 2.

See Keystroke Guide, page 782.

Activity Notes

In this Activity, students first examine an equation of a function and identify three indicated transformations to the graph of the parent function: a reflection across the x-axis, a vertical stretch, and a vertical translation. They then apply the transformations to the graph of the parent function in three different orders. They should discover that it is important to perform the translations after the reflections and stretches.

Cooperative Learning
You may wish to have students do the Activity in groups of three. Each group member can do a different one of the orders specified in Steps 1, 2, and 3. The group members can then compare results and do Step 4 together.

CHECKPOINT ✔
5. Translations are done after stretches and reflections.

Inclusion Strategies

LINGUISTIC LEARNERS When confronted with an equation like $y = -2|x - 2| - 2$, some students will see a maze of symbols, many of which appear to be exactly alike. It may help these students to write a completely verbal "directory" of the transformations indicated, such as the following:

1. Start with the parent absolute-value function.
2. Stretch its graph vertically by a factor of two.
3. Reflect the result across the horizontal axis.
4. Translate the result two units to the right.
5. Translate the result two units down.

Enrichment

Have students work in pairs. Tell them to graph $\triangle ABC$ with vertices $A(-2, -2)$, $B(0, 2)$, and $C(2, -2)$. Ask them to find a combination of transformations that will map $\triangle ABC$ to $\triangle XYZ$ with vertices $X(-3, 3)$, $Y(-1, -9)$, and $Z(1, 3)$.
Sample: Stretch $\triangle ABC$ vertically by a scale factor of 3; reflect the result across the x-axis; translate the result 1 unit to the left and 3 units down.

Have each student make up a similar problem on a clean sheet of paper. Partners should then exchange papers and solve each other's problems.

LESSON 14.5 **723**

ADDITIONAL EXAMPLE 1

Apply transformations to the graph of $y = |x|$ in order to graph $y = 2|x - 3| + 1$.

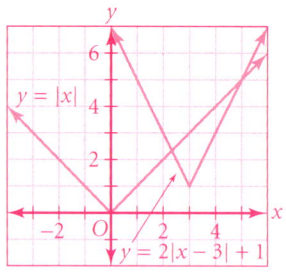

TRY THIS

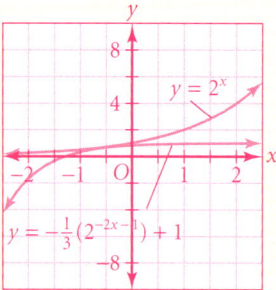

Rewrite the equation as $y = -\frac{1}{3}\left(2^{-2\left(x+\frac{1}{2}\right)}\right) + 1$. Translate it $\frac{1}{2}$ unit left, compress it horizontally by a factor of $\frac{1}{2}$, reflect it across the y-axis, compress it vertically by a factor of $\frac{1}{3}$, reflect it across the x-axis, and translate it 1 unit up.

x	1	0	−1	−2	−3
$-\frac{1}{3}\left(2^{-2\left(x+\frac{1}{2}\right)}\right)+1$	$\frac{23}{24}$	$\frac{5}{6}$	$\frac{1}{3}$	$-\frac{5}{3}$	$-\frac{29}{3}$

You may have noticed in the activity that the order of vertical transformations follows the order of operations for real numbers. Multiplication (stretches, compressions, and reflections) is done before addition (translation). The order of horizontal transformations is the *opposite* of the order of operations for real numbers. Addition (translation) is done before multiplication (stretches, compressions, and reflections).

The following list will help you to proceed methodically when graphing transformed functions:

GRAPHING TRANSFORMED FUNCTIONS

1. Do horizontal translations first.
2. Do horizontal stretches or compressions second.
3. Do reflections across the y-axis third.
4. Do vertical stretches or compressions fourth.
5. Do reflections across the x-axis fifth.
6. Do vertical translations sixth.

EXAMPLE 1 Apply transformations to the graph of $y = x^2$ in order to graph $y = -2(-3x + 1)^2 - 1$.

SOLUTION

Six transformations are combined to obtain the graph of $y = -2(-3x + 1)^2 - 1$ from its parent function, $y = x^2$.

First graph $y = x^2$. Then apply the transformations in the correct order.

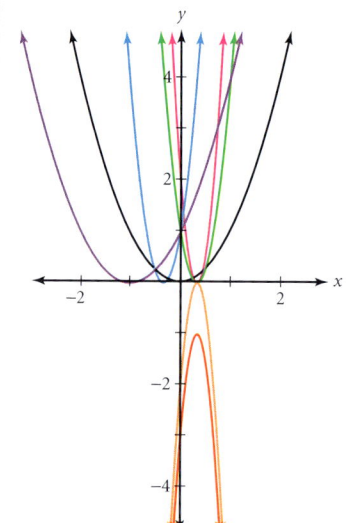

1. The graph of $y = x^2$ is translated 1 unit left to obtain the graph of $y = (x + 1)^2$.
2. The graph of $y = (x + 1)^2$ is compressed horizontally by $\frac{1}{3}$ to obtain the graph of $y = (3x + 1)^2$.
3. The graph of $y = (3x + 1)^2$ is reflected across the y-axis to obtain the graph of $y = (-3x + 1)^2$.
4. The graph of $y = (-3x + 1)^2$ is stretched vertically by 2 to obtain the graph of $y = 2(-3x + 1)^2$.
5. The graph of $y = 2(-3x + 1)^2$ is reflected across the x-axis to obtain the graph of $y = -2(-3x + 1)^2$.
6. The graph of $y = -2(-3x + 1)^2$ is translated 1 unit down to obtain the graph of $y = -2(-3x + 1)^2 - 1$.

TRY THIS Apply transformations to the graph of $y = 2^x$ in order to graph $y = -\frac{1}{3}(2^{-2x-1}) + 1$.

Reteaching the Lesson

COOPERATIVE LEARNING Have students work in pairs. Give each pair a blank transparency, a sheet of graph paper, and a pen that can be used to draw on the transparency. Have them draw a set of coordinate axes on the graph paper so that the origin is near the center. They should then place the transparency over the axes and graph $y = x^2$.

Now one student from each pair should act as the "transformer," manipulating the transparency to show a combination of two transformations—a reflection and a translation. The other student, the "equationer," should write an equation for the transformed graph. The partners can then work together to decide whether the equation is correct.

Now have the partners switch roles. This time, the student identified as the transformer should perform a combination of three transformations—a reflection and two translations.

Repeat the activity with the graphs of $y = 2x^2$ and $y = 0.5x^2$.

Exercises

Communicate

1. For the bike-ride problem at the beginning of the lesson, explain how to identify which number in the function rule represents a translation.
2. For the bike-ride problem, explain how to identify which number in the function rule represents a stretch.
3. Explain why:
   a. $y = x^2$ has the same graph as $y = (-x)^2$;
   b. $y = |x - 3|$ has the same graph as $y = |3 - x|$.
4. Explain how to graph $y = -a(-bx - h)^2 + k$ using transformations of the parent function $y = x^2$.

Guided Skills Practice

Sketch the graph of each function by applying transformations to its parent function. *(EXAMPLE 1)*

5. $f(x) = 2(-x - 3)^2 + 1$
6. $f(x) = -2(-3x + 3)^2 + 1$
7. $f(t) = \frac{1}{2}\left|\frac{1}{2}t + 1\right| + 3$
8. $f(t) = -\frac{1}{2}\left|\frac{1}{4}t + 3\right| - 2$
9. $f(g) = -\frac{3}{2g + 1} + 2$
10. $f(x) = \frac{2}{2x - 4} - 6$

Practice and Apply

Sketch the graph of each function.

11. $g(x) = 2(x + 5)^2$
12. $m(x) = -2x^2 + 5$
13. $t(x) = |x + 3| - 4$
14. $v(x) = 2|5x - 4|$
15. $f(x) = \frac{3}{x - 2}$
16. $g(x) = \frac{3}{-x} - 2$
17. $h(x) = 0.5(2^{2x}) - 1$
18. $z(x) = \frac{3}{4} + \frac{2}{x}$
19. $p(x) = \frac{3}{4} - \frac{2}{4 + x}$

Name the point that corresponds to the point (4, 10) after applying the specified transformations to the graph of $y = \frac{5}{2}x$. (Hint: If you are not sure, draw a sketch.)

20. a vertical stretch by a factor of 2, followed by a vertical translation of 3 **(4, 23)**
21. a reflection across the *x*-axis followed by a vertical stretch by a factor of 2 **(4, −20)**
22. a reflection across the *y*-axis followed by a vertical translation of −3 **(−4, 7)**
23. a horizontal stretch by a factor of 3, followed by a vertical translation of 5 **(12, 15)**
24. a reflection across the *y*-axis, then a vertical compression by a factor of $\frac{1}{2}$, followed by a translation of −5 **(−4, 0)**

Assess

Selected Answers
Exercises 5–10, 11–27 odd

ASSIGNMENT GUIDE

In Class	1–10
Core	11–25, 28
Core Plus	11–28
Review	29–34
Preview	35–39

✎ Extra Practice can be found beginning on page 738.

5. Rewrite the equation as $f(x) = 2[-(x + 3)]^2 + 1$. The parent function is $f(x) = x^2$. Translate 3 units left; vertically stretch by a factor of 2; vertically translate 1 unit up.

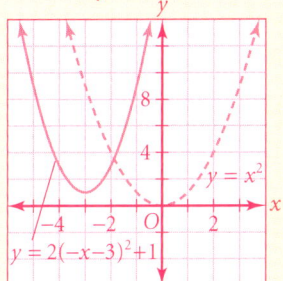

6. Rewrite the equation as $f(x) = -2[-3(x - 1)]^2 + 1$. The parent function is $f(x) = x^2$. Translate 1 unit right; horizontally compress by a factor of $\frac{1}{3}$; vertically stretch by a factor of 2; reflect across the *x*-axis; translate 1 unit up.

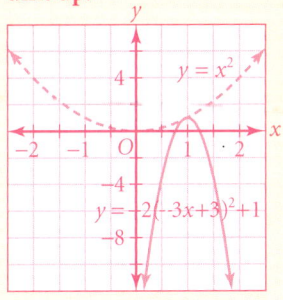

7. Rewrite as $f(t) = \frac{1}{2}\left|\frac{1}{2}(t + 2)\right| + 3$. The parent function is $f(t) = |t|$. Translate 2 units left; horizontally stretch by a factor of 2; vertically compress by a factor of $\frac{1}{2}$; translate 3 units up.

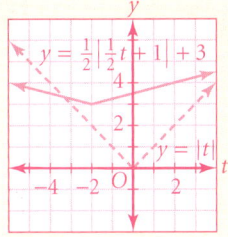

8. Rewrite as $f(t) = -\frac{1}{2}\left|\frac{1}{4}(t + 12)\right| - 2$. The parent function is $f(t) = |t|$. Translate 12 units left; horizontally stretch by a factor of 4; vertically compress by a factor of $\frac{1}{2}$; reflect across the *x*-axis; translate 2 units down.

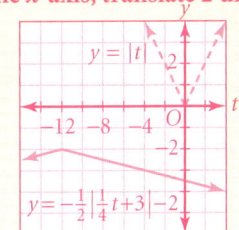

The answers to Exercises 9–19 can be found in Additional Answers beginning on page 810.

LESSON 14.5 **725**

Technology

A graphics calculator may be helpful in Exercises 11–27. You may wish to have students first complete the exercises with paper and pencil and then use the calculator to verify their answers.

Error Analysis

When combining transformations, students may use an order that yields an incorrect result. Suggest that they think of the algebraic order of operations. That is, vertical stretches and reflections across the *x*-axis are signaled by the multiplications in the equation, and translations are signaled by an addition or subtraction. Thus the stretches and reflections should be done before the translations.

26. a. $-\frac{3}{2}|x - 4| + 2$,
or $-\left|\frac{3}{2}(x - 4)\right| + 2$,
or $-\left|\frac{3}{2}x - 6\right| + 2$

b. Yes, there are at least two different function rules: $-\frac{3}{2}|x - 4| + 2$ and $-\left|\frac{3}{2}(x - 4)\right| + 2$, which can be rewritten as $-\left|\frac{3}{2}x - 6\right| + 2$.

Practice

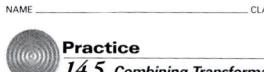

Practice
14.5 Combining Transformations

Sketch the graph of each function.

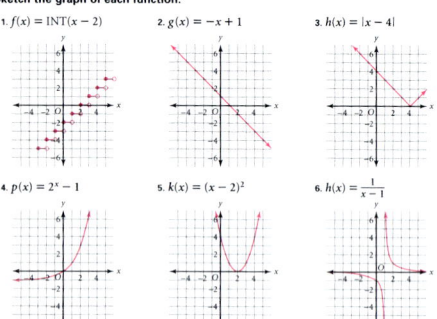

$y = -5x + 6$; parent function: $y = x$

CHALLENGE

CONNECTION

APPLICATION

APPLICATION

25. Use the information in the table to determine the transformed function that fits the data. Give the parent function also.

x	−3	−2	−1	0	1	2	3	4	5	6	7
y	21	16	11	6	1	−4	−9	−14	−19	−24	−29

26. a. Find a function rule for the graph shown at right.
   b. Is there more than one possible function rule? Explain.

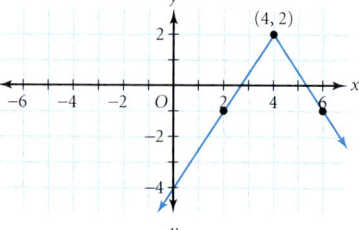

27. **GEOMETRY** In the figure at right, a triangle is shown in the first quadrant. A reflection, followed by a translation of −18 units, is also shown. The coordinates of A are (a, b). Find the coordinates of A' and A''. Do the other vertices change according to the same pattern?

A': $(a, -b)$ A'': $(a - 18, -b)$
yes

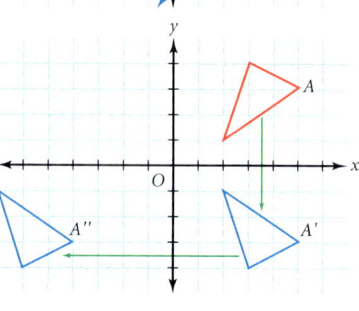

28. **RECREATION** Two friends are planning a 200-mile bike ride, allowing 5 hours for rest stops.

   a. Write the function that describes the rate, r, that they must travel in order to arrive in a certain number of hours, h. $r = \frac{200}{h - 5}$, $h > 5$

   b. Give the parent function. $r = \frac{1}{h}$

   c. Graph both functions. (Hint: Refer to the biking problem at begining of the lesson.)

Look Back

29. **PACKAGING** A box requires 75 centimeters of tape to be sealed for delivery. How many boxes can be sealed with 32 meters of tape? **(LESSON 1.2)**
   42 boxes

30. Graph the function $4x - 5y = 3$. Outline your procedure for graphing this function. **(LESSON 5.5)**

31. If the graph of a line contains the points $A(-3, 7)$ and $B(2, -9)$, what is an equation for the line that is perpendicular to the given line and passes through point A? **(LESSON 5.6)**

32. Solve $|x - 5| \leq 17$, and graph the solution on a number line. **(LESSON 6.5)**
   $-12 \leq x \leq 22$

Solve the system of equations by using the methods below. $\begin{cases} 3x - y = 5 \\ x - 2y = -10 \end{cases}$

33. graphing **(LESSON 7.1)**

34. elimination **(LESSON 7.3)** $x = 4$, $y = 7$

28. c.

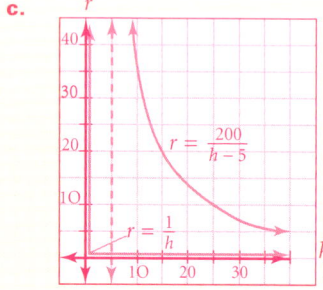

30. To graph this function, first put it into slope intercept form: $y = \frac{4}{5}x - \frac{3}{5}$. Then graph by putting a point at the *y*-intercept, $-\frac{3}{5}$, and drawing the line through the point with a slope of $\frac{4}{5}$ (rise of 4 and run of 5). (See Additional Answers for the graph.)

Look Beyond

Reflect the graph of each function across the line $y = x$. (Hint: Interchange the x and the y in the original equation and solve for y.)

35. $y = -2x + 1$
36. $y = \sqrt{x}$
37. $y = \frac{1}{x}$
38. $y = |x|$
39. $y = x^2$

Include this activity in your portfolio.

The table below shows the flight time for one airline and the mileage from Atlanta to the cities shown.

1. Make a scatter plot and graph the line of best fit. Use t, time in minutes, as the independent variable.

2. **a.** Write the function for the line of best fit as a transformation of the parent function $y = x$.
   b. What is the slope given by the function, and what does the slope represent in the problem?
   c. What is the y-intercept given by the function, and what does the y-intercept represent in the problem?
   d. Where does the line of best fit meet the x-axis? What could account for this value in actual plane travel?

City	Time (hours:minutes)	Distance (miles)
Albuquerque, NM	3:20	1269
Austin, TX	2:23	812
Chicago	1:44	596
Colorado Springs	3:20	1185
Dallas/Fort Worth	2:10	731
Denver	3:32	1201
Houston	2:13	689
Huntsville, AL	0:48	151
Little Rock, AR	1:27	453
Mobile, AL	1:04	302
Los Angeles	5:08	1946

35.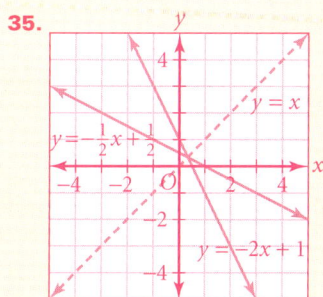

36.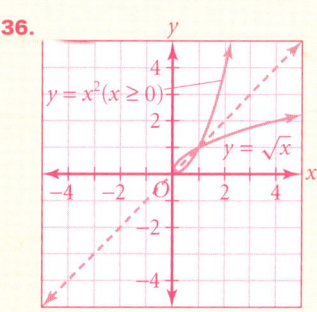

The answers to Exercises 31–33 and 37–39 can be found in Additional Answers beginning on page 810.

Project

Focus

The function whose value at x is $f(g(x))$ is called a *composite* of functions f and g. Students will see how a familiar computation game can be modeled by a composition of functions. They will then further investigate composition in relation to the graphs of the parent functions studied in this chapter.

Motivate

Discuss the "Pick a Number" game. Have students examine the table near the top of the page and explain how the entry $x + 2$ was derived. Ask them how this expression would be different if the second step of the game were *Triple the number and subtract 6.* Lead them to see that the expression would be $x - 2$. Have students make other changes in the steps to see how the expression $x + 2$ is affected.

Pick a Number

Ask a classmate to do the following:

Pick a number from 1 through 9. Triple the number and add 6. Then divide the result by 3.

Ask for the final number. If you subtract 2, you can tell your classmate the original number.

Try this activity with other numbers. What do you notice?

	Arithmetic	Algebra
Pick a number.	5	x
Triple the number and add 6.	$15 + 6 = 21$	$3x + 6$
Divide by 3.	$21 \div 3 = 7$	$x + 2$

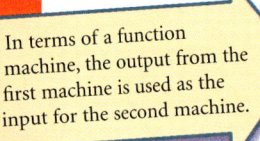
Algebra can be used to help explain why this happens. Let x be the number chosen.

The algebra shows you how to go directly to the final number. This is an example of finding a function of a function.

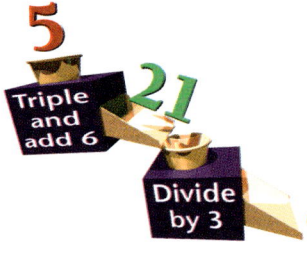

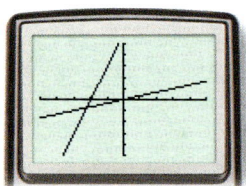
In terms of a function machine, the output from the first machine is used as the input for the second machine.

When you find a function of a function, you are forming a composite function. If f is one function and g is another function, you can write $g(f(x))$ to indicate the value of a **composite function** for any x-value.

If $f(x) = 3x + 6$ and $g(x) = \frac{x}{3}$, **then** $g(f(x)) = \frac{f(x)}{3} = \frac{3x+6}{3} = x + 2$.

The composition of two functions produces a new function with a new graph.

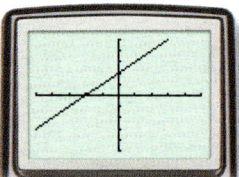

Composition of functions is not a commutative operation. $g(f(x)) = x + 2$, but $f(g(x)) = 3(g(x)) + 6 = 3\left(\frac{x}{3}\right) + 6 = x + 6$

Activity 1

Make up a two-step computation game like the one at the beginning of this project. Show how the domain and range are related through the steps. Then show the simplified composition of the functions that lets you perform the calculation easily. Try the game on your friends.

Activity 2

Use what you have learned about composing functions to graph the composition of two functions. Begin by selecting a parent function from those you studied in Lesson 14.1. Replace the x in the function rule with another function, and graph the composition.
   a. Identify the domains and ranges of the original function and the composite function.
   b. Make a table of the input values, the intermediate values for the input in the second step, and the final composite values.
   c. Use a graphics calculator or graph paper to draw the graph of the new composite function, and compare it to the original functions.
   d. Try reversing the order of the two functions you composed. Graph the new composite function and complete parts **a–c** above. What do you notice?

Cooperative Learning

Have students work in pairs. In **Activity 1**, each student can first work independently to develop a two-step computation game. The partners can then play the games together to determine if the results are as expected.

In **Activity 2**, the partners can first work together to choose the functions and identify domains and ranges. One partner can then make a table of values for the composition, and the other can graph the composition. The pair can switch roles before completing part **d**.

Discuss

Bring the class together to discuss Activity 1. Ask for volunteers to share their games with the class.

Discuss the results of Activity 2. Encourage students to compare the graphs of the two compositions. Ask questions such as the following:

- Were any of the graphs the same?
- How were the graphs alike?
- How were they different?

Activity 1
Answers will vary.

Activity 2
a–d. Answers will vary.

Chapter Review

13.
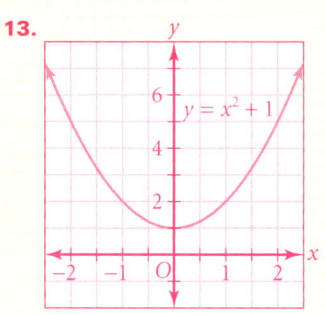

14 Chapter Review and Assessment

VOCABULARY

function notation 694	parent function 696	vertical compression 708
function rule 694	reflection 715	vertical stretch 708
horizontal compression 710	scale factor 708	vertical translation 701
horizontal stretch 710	transformation 700	vertical-line test 693
horizontal translation 702	translation 700	

Key Skills & Exercises

LESSON 14.1

Key Skills

Determine if a set of ordered pairs or a graph represents a function; identify the domain and range of a function; and evaluate a function.

The set of ordered pairs {(0, 1), (1, 2), (2, 5), (3, 10)} is a function because there is exactly one second element for each first element in the set. The domain of the function is 0, 1, 2, and 3. The range of the function is 1, 2, 5, and 10.

The graph below is a function because no vertical line will intersect the graph more than once.

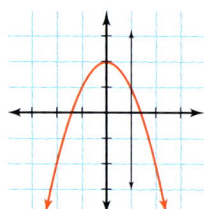

To evaluate the function $f(x) = x^2 + 1$ for $f(-3)$, substitute -3 for x, and simplify.

$f(x) = x^2 + 1$

$f(-3) = (-3)^2 + 1 = 10$

Identify parent functions.

These are the six basic parent functions:

a. $y = x^2$ b. $y = x$
c. $y = \frac{1}{x}$ d. $y = a^x$
e. $y = |x|$ f. $y = \sqrt{x}$

Exercises

Determine whether each set of ordered pairs represents a function or a relation that is not a function.

1. {(4, 1), (3, 2), (4, 3), (3, 4)} not a function
2. {(1, 4), (2, 3), (3, 4), (4, 3)} function

Determine if each of the following are functions.

3. yes   4. no

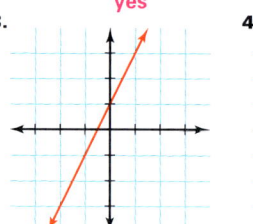

Evaluate the following functions for $x = 4$:

5. $f(x) = x - 8$ $f(4) = -4$ 6. $g(x) = |x + 2|$ $g(4) = 6$
7. $h(x) = 3^x$ $h(4) = 81$ 8. $j(x) = 4x^2$ $j(4) = 64$
9. $k(x) = -5|x - 3|$ 10. $f(x) = \frac{2}{x}$ $f(4) = \frac{1}{2}$
   $k(4) = -5$

Name the parent function for each function graphed below.

11.   12.

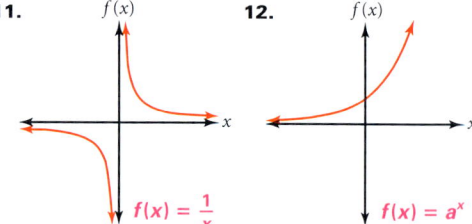

$f(x) = \frac{1}{x}$ $f(x) = a^x$

Chapter Test, Form A

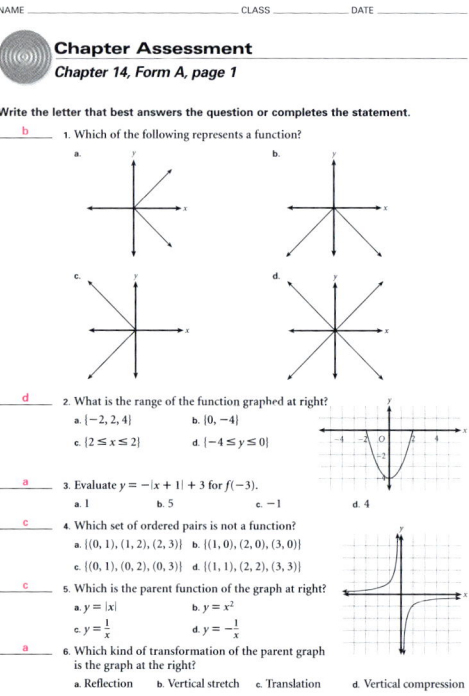

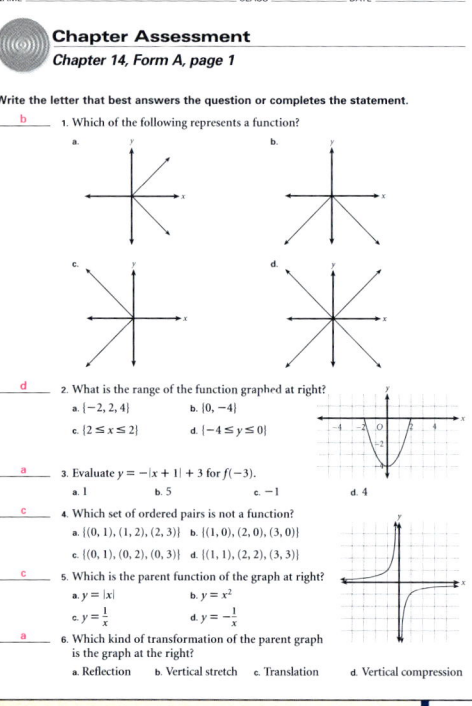

14.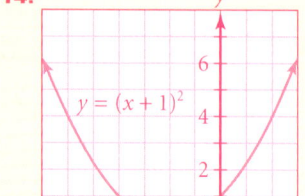

15.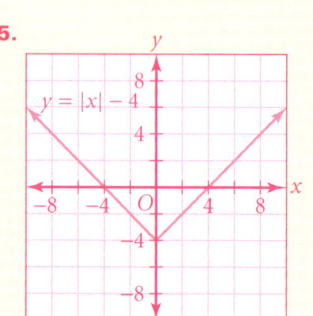

730 CHAPTER 14 REVIEW

LESSON 14.2
Key Skills

Describe the effect of a translation on the graph of a function.

The parent function of $y = |x| + 2$ is $y = |x|$. The graph of $y = |x| + 2$ is shifted 2 units up from the graph of $y = |x|$. The graph of $y = |x + 1|$ is shifted 1 unit to the left of the graph of $y = |x|$. The graph of $y = |x - 1|$ is shifted 1 unit to the right of the graph of $y = |x|$.

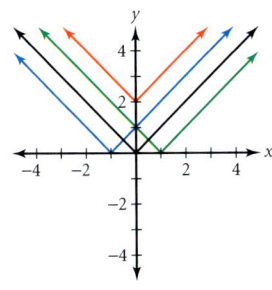

Exercises

Identify the parent function. Tell what type of transformation is applied to the parent function to obtain each function, and then graph each function.

13. $y = x^2 + 1$ $y = x^2$; vert. trans.
14. $y = (x + 1)^2$ $y = x^2$; horz. trans.
15. $y = |x| - 4$ $y = |x|$; vert. trans.
16. $y = \frac{1}{x - 5}$ $y = \left|\frac{1}{x}\right|$; horz. trans.
17. $y = \frac{1}{x} + 3$ $y = \frac{1}{x}$; vert. trans.
18. $y = (x - 2)^2$ $y = x^2$; horz. trans.
19. $y = 2^{x+3}$ $y = 2^x$; horz. trans.

LESSON 14.3
Key Skills

Describe the effect of a vertical stretch or compression on the graph of a function.

The parent function of $y = 3|x|$ is $y = |x|$. Each of the y-values of the transformed function is 3 times the corresponding y-value of the parent function.

x	$|x|$	$3|x|$
-1	$3|1|$	3
0	$3|0|$	0
2	$3|2|$	6

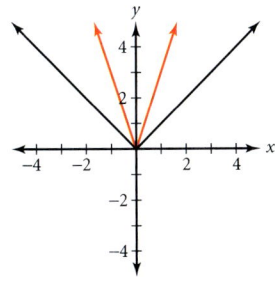

Exercises

Graph each of the following:

20. $y = 4|x|$
21. $y = 5|x|$
22. $y = |5x|$
23. $y = 3x^2$
24. $y = 2x^2$
25. $y = \frac{1}{4x}$
26. $y = \left(\frac{x}{4}\right)^2$
27. $y = \sqrt{\frac{1}{2}x}$

16.
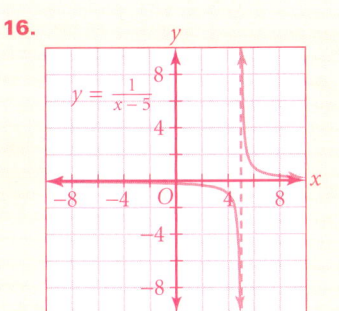

17.
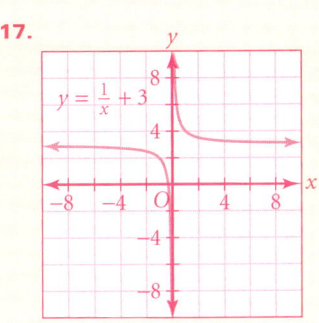

The answers to Exercises 18–27 can be found in Additional Answers beginning on page 810.

Chapter Test, Form B

NAME _____ CLASS _____ DATE _____

Chapter Assessment
Chapter 14, Form B, page 1

1. What is the difference between a relation and a function? A relation pairs any two points. A function is a relation for which no two points have the same first coordinates.
2. Write $f(-2) = 3$ as an ordered pair so that it can be plotted. $(-2, 3)$

Tell whether each of the following represents a function:

3. $(-3, 9), (0, 0), (-3, 5)$ not a function
4. function

For $g(x) = \frac{3}{x-2}$, evaluate the following:

5. $g(0)$ $-\frac{3}{2}$
6. $g(8)$ $\frac{1}{2}$
7. $g(\frac{1}{2})$ -2

Evaluate each function for $x = 2$.

8. $h(x) = |x - 1|$ 1
9. $f(x) = x^2 + 2$ 6
10. $g(x) = \frac{1}{x} - 2$ $-1\frac{1}{2}$
11. What happens to the graph of $y = -|x|$ when you reflect it through the x-axis? The graph is turned right side up (a V that opens upward).

Give the parent function for each graph.

12. 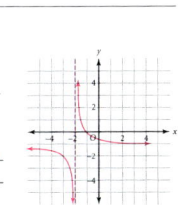 $y = x^2$
13. $y = |x|$

NAME _____ CLASS _____ DATE _____

Chapter Assessment
Chapter 14, Form B, page 2

Consider the function $y = \frac{1}{x+2} - 1$.

14. Identify the parent function. $y = \frac{1}{x}$
15. Sketch the graph of the function.
16. Describe the transformation. A vertical translation 1 unit down followed by a horizontal translation 2 units to the left.

State whether each function is a stretch or compression of the parent function. If so, give the scale factor.

17. $y = 2(x - 3)^2 + 1$ stretch; 2
18. $y = \frac{1}{2}|x| + 3$ compression; $\frac{1}{2}$
19. $y = \sqrt{3x}$ stretch; $\sqrt{3}$
20. $y = x + 2$ neither; 1

21. Use the grid provided to show what happens to the parabola when you perform a reflection through the y-axis followed by a vertical translation of -3.

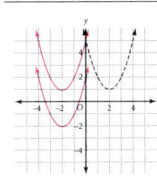

22. Determine the coordinates of the vertex of $y = (x + 3)^2 - 1$. $(-3, -1)$

Tell whether each function is stretched, compressed, reflected, or translated from its parent function.

23. $y = \frac{x^2}{2}$ compressed
24. $y = (x + 5)^2$ translated
25. $y = \frac{1}{x-2}$ translated
26. $y = -\frac{1}{x}$ reflected

CHAPTER 14 REVIEW **731**

28.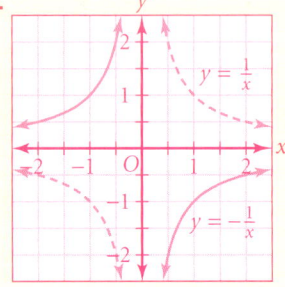

29.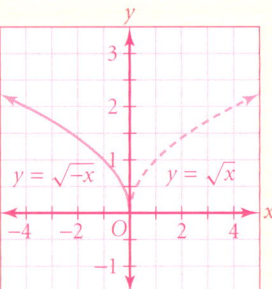

30.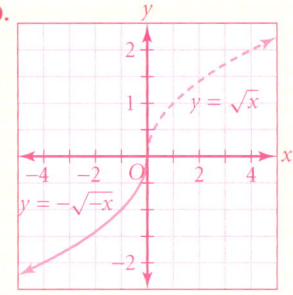

31.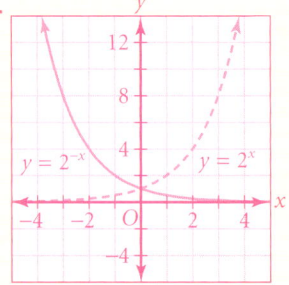

32.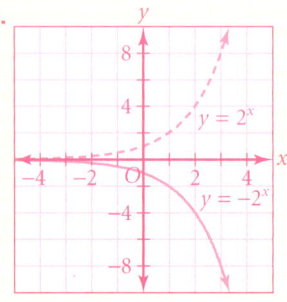

LESSON 14.4

Key Skills

Describe the effect of a reflection across the x- or y-axis on the graph of a function.

The parent function of $y = -|x|$ is $y = |x|$. The values of $-|x|$ are opposite those of $|x|$. This is a reflection of the parent graph $y = |x|$ across the x-axis.

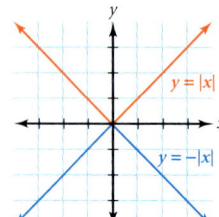

Exercises

Graph each function along with its parent function on the same coordinate plane.

28. $y = -\frac{1}{x}$

29. $y = \sqrt{-x}$

30. $y = -\sqrt{-x}$

31. $y = 2^{-x}$

32. $y = -(2)^x$

33. $y = -x^2$

LESSON 14.5

Key Skills

Apply more than one transformation to a parent function to sketch a graph.

The parent function of $y = -2|x + 2|$ is $y = |x|$. The parent function has been reflected across the x-axis, stretched vertically by 2, and translated horizontally 2 units to the left.

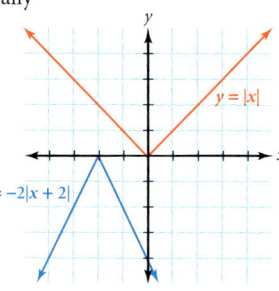

Exercises

Graph each function and list each transformation.

34. $f(x) = 3x^2 + 1$

35. $g(x) = |x - 6| + 2$

36. $h(x) = -\frac{2}{x} + 1$

37. $f(x) = (x - 3)^2 + 4$

38. $g(x) = 2^{x+1} - 3$

39. $h(x) = |x + 4| - 2$

40. $f(x) = -6|\frac{1}{3}x - 1| + 2$

Applications

41. PHYSICS Kelvins are often used in science to measure the temperature of substances. To change from degrees Celsius to kelvins, add 273.16 to degrees Celsius. What type of transformation relates degrees Celsius to kelvins? **vertical translation**

42. BUSINESS Thomas decides to have a sale at his art store. He decides to reduce the price of everything in the store by 20%. Find the scale factor that shows the amount by which the prices of the items were "stretched." **0.8**

33.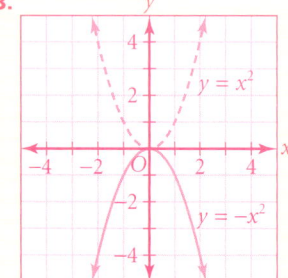

34.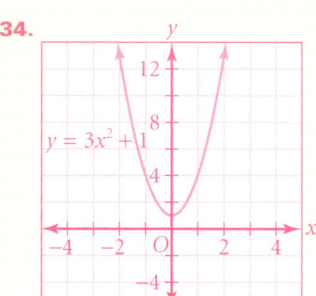

The answers to Exercises 35–40 can be found in Additional Answers beginning on page 810.

vertical stretch by a factor of 3, vertical translation 1 unit up

732 CHAPTER 14 REVIEW

Alternative Assessment

Performance Assessment

1. A set of ordered pairs determines a relation.
   a. Under what conditions is a set of ordered pairs a function?
   b. What is the difference between the domain and the range of a function?
2. a. Explain what happens to the coordinates of a point when a scale factor greater than 1 is applied to a function. This transformation is sometimes described as stretching a function.
   b. What type of scale factor results in compressing a function?
3. Describe the relationship between each of the following and the parent function $f(x) = x^2$:
   a. $g(x) = -x^2$
   b. $h(x) = x^2 + 3$
   c. $s(x) = (x + 3)^2$
4. Explain how to use transformations of $y = x^2$ to graph $y = -2(x - 1)^2 - 3$.
5. Use a drawing to show what happens to a point in the first quadrant of a coordinate plane when it is reflected over the *x*-axis and translated 4 units to the left. Under what conditions will the transformed point be in the third quadrant?

Portfolio Projects

1. Use graph paper. Draw a 2 by 4 rectangle with one corner at the origin on graph paper. Use the language of transformations to explain how you can use the coordinates of the corners of the given rectangle to draw a 6 by 12 rectangle.
2. Use graph paper. Graph each of the following parent functions.
   a. $y = x^2$
   b. $y = |x|$
   c. $y = 2^x$
   d. $y = \frac{1}{x}$

   Explain what happens to each graph when $a > 0$ is added to *y*.
3. Consider the parent function $y = x^2$. Explain what effect *a*, *h* and *k* have on the parent function when $y = a(x + h)^2 + k$.

Peer Assessment

Complete this activity with a partner. Begin by writing a function that is a transfomation of one of the parent functions that you studied in this chapter. After identifying the transformations and graphing the function, your partner should write a function for you.

Check your partner's work, correct any errors, and then identify the transformations and graph the new function. Continue with this pattern until you have each graphed 10 expressions. Be sure to use functions containing reflections, stretches, and translations during this activity.

---

internetconnect

The HRW Web site contains many resources to reinforce and expand your knowledge of transformations. The Web site also provides Internet links to other sites where you can find information and real-world data for use in research projects, reports, and activities involving transformations. Visit the HRW Web site at **go.hrw.com**, and enter the keyword **MA1 CH14**.

Alternative Assessment

Performance Assessment

1. a. A set of ordered pairs is a function if each domain element is paired with one and only one range element.
   b. The elements of the domain are the first numbers of each pair, and the elements of the range are the second numbers of each pair.
2. a. When a scale factor greater than 1 is applied to a function, the domain values stay the same and the range values are multiplied by the scale factor.
   b. A scale factor between 0 and 1, exclusive, will compress the *y*-values.
3. a. reflection across *x*-axis
   b. vertical translation 3 units up
   c. horizontal translation 3 units left
4. horizontal translation 1 unit to the right; vertical stretch by a factor of 2; reflection across the *x*-axis; vertical translation 3 units down
5. The point will be in the third quadrant when the point has an *x*-coordinate greater than 0 and less than 4.

Portfolio Projects

1. Corner *B*, (0, 2), can be transformed to (0, 6) by a vertical stretch by a factor of 3. Corner *D*, (4, 0), can be transformed to (12, 0) by a horizontal stretch by a factor of 3. Then corner *C*, (4, 2), will be repositioned at the intersection of the new sides, (12, 6).
2. See Additional Answers for the graphs. Each graph is vertically translated *a* units up.
3. The *a* indicates that the *y*-values of the graph are multiplied by a factor of *a*. The *k* indicates that the graph is vertically translated *k* units up. The *h* indicates that the graph is horizontally translated *h* units to the left.

Cumulative Assessment

College Entrance Exam Practice

CHAPTERS 1–14 CUMULATIVE ASSESSMENT
College Entrance Exam Practice

Multiple-Choice and Quantitative-Comparison Samples

The first half of the Cumulative Assessment contains two types of items found on standardized tests—multiple-choice questions and quantitative-comparison questions. Quantitative-comparison items emphasize the concepts of equality, inequality, and estimation.

Free Response Grid Samples

The second half of the Cumulative Assessment is a free-response section. A portion of this part of the Cumulative Assessment consists of student-produced response items commonly found on college entrance exams. These questions require the use of machine-scored answer grids. You may wish to have students practice answering these items in preparation for standardized tests.

QUANTITATIVE COMPARISON For Questions 1–4, write
A if the quantity in Column A is greater than the quantity in Column B;
B if the quantity in Column B is greater than the quantity in Column A;
C if the two quantities are equal; or
D if the relationship cannot be determined from the information given.

	Column A	Column B	Answers	
1.	$(x+y)^2$	$x^2 + y^2$	Ⓐ Ⓑ Ⓒ Ⓓ [Lesson 9.4]	D
2.	5×10^{-3}	0.005	Ⓐ Ⓑ Ⓒ Ⓓ [Lesson 8.5]	C
3.	slope of $y = -2$	slope of $y = x - 2$	Ⓐ Ⓑ Ⓒ Ⓓ [Lesson 5.4]	B
4.	7^0	$15(-9+9)$	Ⓐ Ⓑ Ⓒ Ⓓ [Lesson 8.4]	A
5.	solution to $3(x+9) = 15$	solution to $x^2 = 15$	Ⓐ Ⓑ Ⓒ Ⓓ [Lesson 10.2]	B

6. What is the reciprocal of 4? **(LESSON 2.4)** b
   a. -4 b. $\frac{1}{4}$
   c. 1   d. $-\frac{1}{4}$

7. Find the slope of $3y + 2x = 5$. **(LESSON 5.4)** b
   a. $-\frac{3}{2}$ b. $-\frac{2}{3}$
   c. $-\frac{5}{3}$ d. $\frac{5}{3}$

8. Simplify: $5r - (3r + 2)$. **(LESSON 9.1)** c
   a. $8r + 2$ b. $2r + 2$
   c. $2r - 2$ d. $8r - 2$

9. Solve: $6 - 4t > 18$. **(LESSON 6.2)** a
   a. $t < -3$ b. $t > -3$
   c. $t > -6$ d. $t < -6$

10. Simplify: $(5x + 2) - (x - 4)$. **(LESSON 9.1)** c
    a. $5x - 4$ b. $6x + 6$
    c. $4x + 6$ d. $4x - 6$

11. Which function is graphed below? **(LESSON 10.1)** a

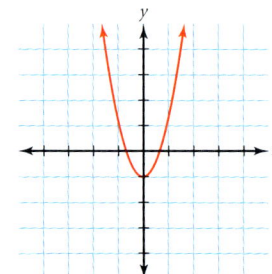

   a. $y = 2x^2 - 1$ b. $y = -2x^2 - 1$
   c. $y = \frac{1}{2}x^2 - 1$ d. $y = -\frac{1}{2}x^2 - 1$

12. Find the slope of a line containing the points $M(-1, 2)$ and $N(4, -3)$. **(LESSON 5.2)** d
    a. -2 b. 2
    c. 1   d. -1

13. Solve: $\begin{cases} 2x - y = -1 \\ 3x + 2y = -5 \end{cases}$ (LESSON 7.3) **d**
    a. (1, 1)   b. (1, −1)
    c. (−1, 1) d. (−1, −1)

14. Which expression is *not* equivalent to $\frac{3x+9}{-3}$? (LESSON 2.7) **c**
    a. $-x - 3$ b. $-(x+3)$
    c. $-x + 9$ d. $-3 - x$

15. Solve: $a + 19 = 25$. (LESSON 3.1) **a = 6**

16. Simplify: $(2^3)^{-2}$. (LESSON 8.4) **$\frac{1}{64}$**

17. Cathi has $17.40 in dimes and quarters. If she has a total of 81 coins, how many of each type does she have? (LESSON 7.6) **q = 62, d = 19**

18. Solve: $\frac{x}{-10} = \frac{-3}{5}$. (LESSON 4.1) **x = 6**

19. The first three terms of a sequence are 2, 4, and 9. The second differences are a constant 3. What are the next three terms? (LESSON 1.1)
    17; 28; 42

20. During one year, a total of 10,924,000 individual tax returns were filed electronically. Write this number in scientific notation. (LESSON 8.5) **1.0924×10^7**

21. Identify the parent function of $y = \frac{-2}{x}$. **$y = \frac{1}{x}$**
    Draw the graph of $y = \frac{-2}{x}$. (LESSON 14.5)

22. Solve: $3x + 9 = 6x - 12$. (LESSON 3.4) **x = 7**

23. A store marked all of its items down by 10%. After the sale, all of the items were marked up by 10% from the sale price. Before the sale, a sweater cost $50. What does the same sweater cost after the sale? (LESSON 4.2) **$49.50**

24. Solve: $2x^2 + 3x - 2 = 0$. (LESSON 9.8) **$x = \frac{1}{2}$ or $x = -2$**

25. Solve by using substitution. (LESSON 7.2)
    $\begin{cases} 2x + 3y = 16 \\ -3x + y = -2 \end{cases}$ **x = 2, y = 4**

26. Show the solution by using a graph. (LESSON 7.5)
    $\begin{cases} x + y \geq 3 \\ y \leq 3 \end{cases}$

27. Factor: $x^2 + 4x - 32$. (LESSON 9.7)
    $(x + 8)(x - 4)$

FREE-RESPONSE GRID

The following questions may be answered by using a free-response grid such as that commonly used by standardized test services.

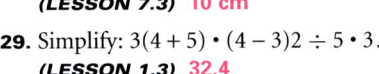

28. The perimeter of a rectangle is 30 centimeters and its length is twice its width. Find the length of the rectangle. (LESSON 7.3) **10 cm**

29. Simplify: $3(4 + 5) \cdot (4 - 3)2 \div 5 \cdot 3$. (LESSON 1.3) **32.4**

30. Find the slope of the lines that are parallel to the line passing through (−7, −5) and (1, −3). (LESSON 5.6) **1/4**

31. Find the root(s) of $4x^2 - 4x + 1 = 0$. (LESSON 9.8) **1/2**

32. Find the sum: $1 + 2 + 3 + \cdots + 50$. (LESSON 1.1) **1275**

33. How many $22 concert tickets can you buy with $125? (LESSON 1.2) **5**

34. A set of golf clubs is marked for sale at $180, a reduction of 25%. What was the original price in dollars of the golf clubs? (LESSON 4.2) **$240**

35. Find the slope of the line containing the points (3, 4) and (7, 7). (LESSON 5.2) **3/4**

36. Find the slope of a line containing the points (−4, −2) and (0, 1). (LESSON 5.2) **3/4**

37. Solve: $5(x + 2) = 3x + 13$. What does x equal? (LESSON 3.5) **3/2**

38. Give the decimal form of 1.1×10^{-2}. (LESSON 8.5) **.011**

39. Solve: $\frac{r}{10} = \frac{20}{25}$. (LESSON 4.1) **8**

21.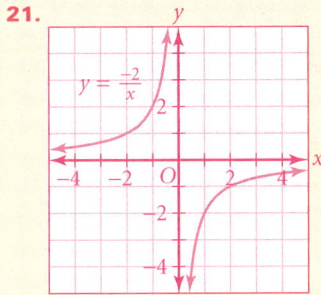

26.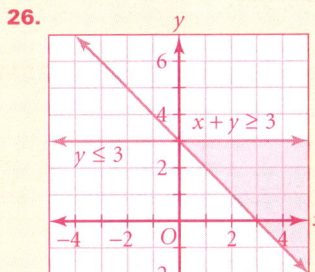

Info Bank

Extra Practice 738

Calculator Keystroke Guide 782

Tables: Squares and Cubes, Random Numbers 799

Glossary 801

Additional Answers 810

Index 871

Credits 879

Extra Practice

CHAPTER 1

LESSON 1.1

Find the next three terms in each sequence.

1. 6, 9, 12, 15, 18, 21, ... 24, 27, 30
2. 80, 75, 70, 65, 60, 55, ... 50, 45, 40
3. 3, 8, 13, 18, 23, ... 28, 33, 38
4. 3, 5, 7, 9, 11, 13, ... 15, 17, 19
5. 18, 22, 26, 30, 34, 38, ... 42, 46, 50
6. 100, 93, 86, 79, 72, 65, ... 58, 51, 44
7. 34, 38, 42, 46, 50, 54, ... 58, 62, 66
8. 28, 25, 22, 19, 16, 13, ... 10, 7, 4
9. 7, 13, 19, 25, 31, 37, ... 43, 49, 55
10. 1, 1, 2, 3, 5, 8, ... 13, 21, 34
11. 11, 23, 36, 50, 65, 81, ... 98, 116, 135
12. 8, 15, 22, 29, 36, 43, ... 50, 57, 64
13. 3, 9, 15, 21, 27, ... 33, 39, 45
14. 2, 9, 16, 23, 30, ... 37, 44, 51
15. 0, 15, 30, 45, 60, ... 75, 90, 105

16. There are 8 players in a chess tournament. Each player plays every other player. How many games of chess will be played? 28

17. There are 5 people in a room. If each person talks to every other person, how many conversations occur? 10

18. There are 16 people at a meeting. Each person shakes hands with every other person. How many handshakes are there? 120

Find each sum. Think of a geometric dot pattern, but do not draw a sketch.

19. $1 + 2 + \ldots + 90$ 4095
20. $1 + 2 + \ldots + 200$ 20,100
21. $1 + 2 + \ldots + 110$ 6105

22. The third and fourth terms of a sequence are 7 an 10. If the first differences are a constant 3, what are the first five terms of the sequence? 1, 4, 7, 10, 13

23. The third and fourth terms of a sequence are 11 and 18. If the second differences are a constant 2, what are the first five terms of the sequence? 3, 6, 11, 18, 27

24. The second differences of a sequence are a constant 4, the first of the first differences is 2, and the first term is 1, find the first 5 terms of the sequence. 1, 3, 9, 19, 33

LESSON 1.2

Identify each of the following as an expression or an equation.

1. $5x$ Exp.
2. $t = 4$ Eq.
3. $3x - 3$ Exp.
4. $2 = q$ Eq.
5. $0.5p$ Exp.
6. $y + 1$ Exp.
7. $3w + 4$ Exp.
8. $3m - 1 = 14$ Eq.

Make a table showing the variable and the value of the expressions below. Use 1, 2, 3, 4, and 5 as values for the variables.

9. $3x$ 3, 6, 9, 12, 15
10. $8y$ 8, 16, 24, 32, 40
11. $2z + 4$ 6, 8, 10, 12, 14
12. $3d - 1$ 2, 5, 8, 11, 14
13. $9q + 3$ 12, 21, 30, 39, 48
14. $15c$ 15, 30, 45, 60, 75
15. $9v - 33$ −24, −15, −6, 3, 12
16. $2 + 8z$ 10, 18, 26, 34, 42

Use guess and check to solve each equation.

17. $4x + 1 = 9$ 2
18. $15 = 5x + 5$ 2
19. $24x - 8 = 112$ 5
20. $10x - 3 = 67$ 7
21. $12 = 2x - 4$ 8
22. $3 + 19x = 98$ 5
23. $7 + 6x = 55$ 8
24. $9 - x = 3$ 6
25. $115 - 3x = 100$ 5

LESSON 1.3

Evaluate each expression by using the algebraic order of operations.

1. $21 - 7 + 2 \cdot 5$ **24**
2. $(21 - 7) + 2 \cdot 5$ **24**
3. $(21 - 7 + 2) \cdot 5$ **80**
4. $32 + 2 \cdot 6 \div 3$ **36**
5. $(32 + 2) \cdot 6 \div 3$ **68**
6. $32 + 2 \cdot (6 \div 3)$ **36**
7. $24 + 2^2 \div 4$ **25**
8. $42 \cdot 20 + 6$ **846**
9. $5^2(30 + 12)$ **1050**
10. $15 \div 3 \cdot 3 - 4$ **11**
11. $45(2) + 3(16)$ **138**
12. $6(4) - 4$ **20**
13. $(3 + 7) \div (3 + 2)$ **2**
14. $(4 + 5 \cdot 2^2) \div (2 \cdot 3)$ **4**
15. $(7 - 4 \cdot 2 + 2) \div 4$ **$\frac{1}{4}$**
16. $14 + 4 \div 2 - 5$ **11**
17. $6 \div 2 \cdot 5 + 2^2$ **19**
18. $4 + 2^2 - 10 + 3$ **1**

Place inclusion symbols according to the algebraic order of operations to make each equation true.

19. $(24 - 6) \cdot 2 + 2^2 = 40$
20. $16 \div (2 + 2) - 2^2 = 0$
21. $6 \cdot (8 + 1) \div 3 = 18$
22. $(7 + 3) \cdot (8 - 5) = 30$
23. $(15 + 5) \div 5 - 2 = 2$
24. $64 + 2^2 \cdot (2 \div 2) = 68$
25. $(39 - 7) \div 4 + 3 = 11$
26. $10^2 \div (5 \cdot 2^2) + 15 = 20$

Given that $a = 4$, $b = 5$, and $c = 6$, evaluate each expression.

27. $a + b + c$ **15**
28. $a \cdot b - c$ **14**
29. $b - a \cdot c$ **−19**
30. $a + b - c$ **3**
31. $a \cdot (b + c)$ **44**
32. $(a + b) \cdot c$ **54**
33. $a^2 - b + c$ **17**
34. $a^2 + c^2 - b$ **47**
35. $a^2 - c^2 - b^2$ **−45**

LESSON 1.4

Graph each list of ordered pairs. State whether the points lie on a straight line.

1. (1, 2), (2, 3), (3, 4) **yes**
2. (1, 6), (2, 8), (3, 6) **no**
3. (2, 4), (5, 3), (11, 1) **yes**
4. (−1, 2), (3, 2), (−5, 2) **yes**
5. (−4, 3), (−2, 1), (0, 1) **no**
6. (3, 0), (4, 2), (5, 4) **yes**
7. (3, 2), (0, 2), (0, 0) **no**
8. (−1, −1), (−2, −2), (−3, −3) **yes**
9. (1, 4), (1, −3), (1, 0) **yes**

Make a table of values using 1, 2, 3, 4, and 5 for x.

10. $y = x + 2$ **3, 4, 5, 6, 7**
11. $y = x + 5$ **6, 7, 8, 9, 10**
12. $y = x - 6$ **−5, −4, −3, −2, −1**
13. $y = x - 4$ **−3, −2, −1, 0, 1**
14. $y = 3x$ **3, 6, 9, 12, 15**
15. $y = -4x$ **−4, −8, −12, −16, −20**
16. $y = 3x + 1$ **4, 7, 10, 13, 16**
17. $y = 2x - 3$ **−1, 1, 3, 5, 7**
18. $y = -x + 1$ **0, −1, −2, −3, −4**
19. $y = -4x - 1$ **−5, −9, −13, −17, −21**
20. $y = 0.5x + 2$ **2.5, 3, 3.5, 4, 4.5**
21. $y = 2x - 0.5$ **1.5, 3.5, 5.5, 7.5, 9.5**
22. $y = 10x$ **10, 20, 30, 40, 50**
23. $y = -7x + 5$ **−2, −9, −16, −23, −30**
24. $y = x + 11$ **12, 13, 14, 15, 16**
25. $y = 12x$ **12, 24, 36, 48, 60**

Lesson 1.4

1.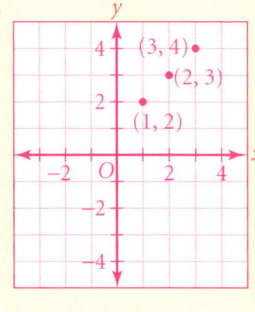
2.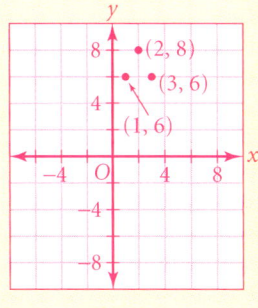
3.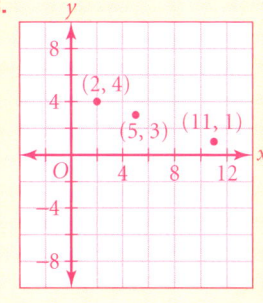
4.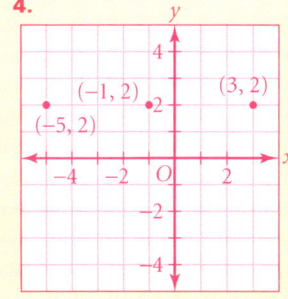
5.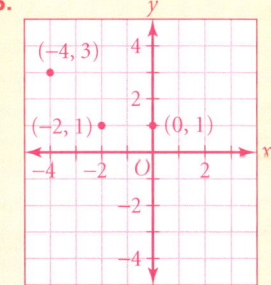
6.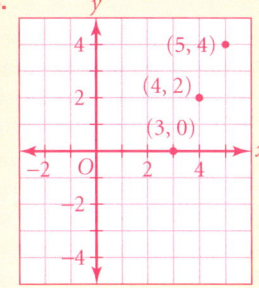
7.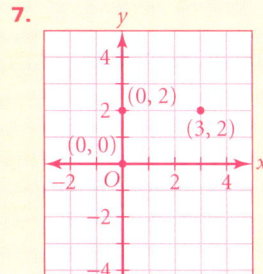
8.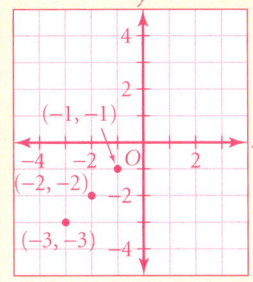
9.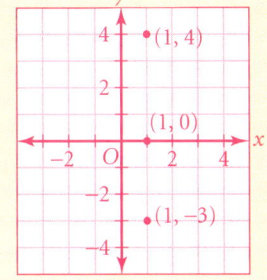

739

LESSON 1.5

Find the first differences and write an equation that describes the pattern shown in each table.

1.
x	5	6	7	8	9
y	27	34	41	48	55

$7; y = 7x - 8$

2.
x	2	3	4	5	6
y	1	−1	−3	−5	−7

$-2; y = -2x + 5$

Make a table of values for each equation using 1, 2, 3, 4, and 5 as values for *x*.

3. $y = 13 - 2x$ 4. $y = 6x + 4$ 5. $y = -4x + 2$ 6. $y = -3x + 5$
   11, 9, 7, 5, 3   10, 16, 22, 28, 34   −2, −6, −10, −14, −18   2, −1, −4, −7, −10

Pens cost 69 cents each. Write and simplify an expression for the cost of

7. 0 pens.   8. 5 pens.   9. 10 pens.   10. *p* pens.
   0 · 0.69; $0   5 · 0.69; $3.45   10 · 0.69; $690   0.69*p*

Hamburgers cost $2.50 each. Write and simplify an expression for the cost of

11. 5 hamburgers.   12. 30 hamburgers.   13. 8 hamburgers.   14. *h* hamburgers.
    5 · .25; $12.50   30 · 2.5; $75   8 · 2.5; $20   2.5*h*

15. If tickets for a concert cost $12 each, how many tickets can you buy with $48? **4 tickets**

16. If videos rent for $2.99 each, how many videos can you rent with $11.96? **4 videos**

LESSON 1.6

Describe the correlation as strong positive, strong negative, or little to none.

1. Strong negative
2. Little to none
3. Strong positive
4. Strong negative

Use the chart of winning times for the women's 200-meter run in the Olympics below for Exercises 5–7.

Year	1964	1968	1972	1976	1980	1984	1988	1992
Seconds	23.00	22.50	22.40	22.37	22.03	21.81	21.34	21.81

5. Make a scatter plot of the data.

6. Estimate a line of best fit and draw it on your scatter plot.

7. Describe the correlation between the years and the winning times.

Lesson 1.6

5.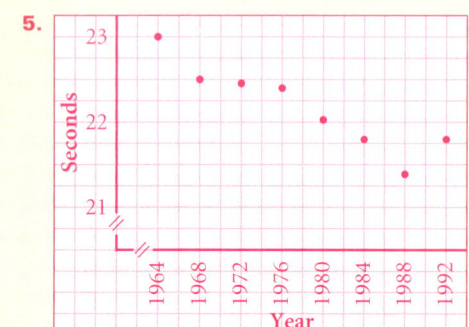

6.

7. For this data, the correlation is strong negative because winning times decrease as the years increase. In other words, the runners are getting faster every year.

CHAPTER 2

LESSON 2.1

Insert an ordering symbol to make each statement true.

1. $0 \underline{\leq} 8$
2. $-3.2 \underline{\leq} -3$
3. $\frac{2}{3} \underline{\leq} \frac{3}{4}$
4. $-\frac{2}{3} \underline{\geq} -\frac{3}{4}$

Find the opposite of each number.

5. 13 **−13**
6. −13 **13**
7. −2.36 **2.36**
8. 743 **−743**
9. 18 **−18**
10. −12 **12**
11. $-4\frac{5}{6}$ **$4\frac{5}{6}$**
12. 73 **−73**

Find the absolute value of each number.

13. 15 **15**
14. −11 **11**
15. −3.47 **3.47**
16. 469 **469**
17. $-7\frac{2}{3}$ **$7\frac{2}{3}$**
18. −16 **16**
19. −842 **842**
20. 89 **89**

LESSON 2.2

Find each sum.

1. −4 + 2 **−2**
2. 3 + (−9) **−6**
3. −8 + (−5) **−13**
4. −8 + 102 **94**
5. 8 + (−10) **−2**
6. −8 + (−10) **−18**
7. −5 + 4 + (−6) **−7**
8. 15 + (−9) + (−12) **−6**
9. −14 + 3 + (−9) **−20**
10. −15 + 36 **21**
11. 26 + (−40) **−14**
12. 15 + (−34) **−49**
13. −42 + 16 **−26**
14. −89 + (−23) **−112**
15. −25 + (−25) **−50**
16. −25 + 25 **0**
17. −91 + 78 **−13**
18. −32 + (−40) **−72**
19. −4 + (−35) + 26 **−13**
20. 16 + (−39) + 21 **−2**
21. −56 + (−56) + (−13) **−125**
22. 17 + (−73) + 68 **12**
23. −34 + 43 + (−43) **−34**
24. −8 + (−120) + 30 **−98**
25. −8 + |7| **−1**
26. |−8| + 7 **15**
27. |−9| + |−4| **13**
28. |−13| + |12| **25**
29. |−43| + |−18| **61**
30. −5 + |−15| + |7| **17**
31. −600 + |−25| + |−30| **−545**
32. 78 + (−43) + |−5| **40**

LESSON 2.3

Find each difference.

1. 7 − 11   −4
2. 8 − (−7)   15
3. 6 − 15   −9
4. −5 − 9   −14
5. 16 − 20   −4
6. −7 − (−12)   5
7. 30 − (−30)   60
8. −17 − (−18)   1
9. −52 − 19   −71
10. −24 − (−24)   0
11. 16 − (−90)   106
12. −34 − 43   −77
13. −52 − 17   −69
14. 72 − 75   −3
15. −68 − (−10)   −58
16. −5 − (−5) − 10   −10
17. −3 − 5 − 16   −24
18. 12 − 18 − 7   −13
19. −27 − (−15) − 2   −14
20. −100 − 75 − (−75)   −100
21. −36 − 21 − 16   −73
22. 20 − 60 − 32   −72
23. 42 − (−72) − 6   108
24. −83 − 14 − (−36)   −61
25. 32 − 41 − 51   −60
26. 12 − (−17) − 36   −7
27. −45 − 61 − (−27)   −79
28. −7 − 24 − 2   −33
29. 26 − 81 − 6   −61
30. 95 − 52 − (−6)   49
31. 14 − 86 − 15   −87
32. −9 − (−27) − 56   −38
33. −38 − 74 − 109   −221

Find the distance between each pair of points on a number line.

34. 5, −2   7
35. −7, 13   20
36. 8, 92   84
37. −4, −7   3
38. −12, −32   20
39. −6, 14   20
40. 12, 43   31
41. 14, −14   28
42. 15, −12   27
43. −13, −19   6
44. −63, −162   99
45. 19, −44   63

LESSON 2.4

Evaluate.

1. $(3)(-4)$ −12
2. $(5)(-6)$ −30
3. $(-4)(1)$ −4
4. $(-3)(9)$ −27
5. $(-2)(-4)$ 8
6. $(7)(-6)$ −42
7. $(16) \div (-4)$ −4
8. $(-25) \div (-5)$ 5
9. $\frac{-36}{9}$ −4
10. $(-100) \div (-10)$ 10
11. $26 + (-40)$ −14
12. $\frac{36}{-3}$ −12
13. $\frac{-56}{-2}$ 28
14. $(-30) \div (3)$ −10
15. $(-5)(14)$ −70
16. $(-12)(-30)$ 360
17. $(45)(-5)$ −225
18. $(-24) \div (8)$ −3
19. $(-35) \div (-7)$ 5
20. $(-2)(-1)(6)$ 12
21. $(-5)(-3)(7)$ 105
22. $(-2)[(-3) + (-2)]$ 10
23. $(-4)[(-1) - (-4)]$ −12
24. $(-6)[7 + (-5)]$ −12
25. $\frac{-36}{(-12) - (-3)}$ 4
26. $(-12) \div [4 - 6]$ 6
27. $(7)(-2) \div (-7)$ 2

LESSON 2.5

Name the opposite of each expression.

1. 16 **−16**
2. -28 **28**
3. $5z$ **−5z**
4. -7 **7**
5. $-5r$ **5r**
6. $3a + 2b$ **−3a − 2b**
7. $a − 7$ **−a + 7**
8. $-8w + 4$ **8w − 4**
9. $5x − 7y$ **−5x + 7y**
10. $-s − 6$ **s + 6**
11. $8a − c$ **−8a + c**
12. $9 − 3x$ **−9 + 3x**
13. $-10m + 4n$ **10m − 4n**
14. $-y − 6z$ **y + 6z**
15. $-15 − v$ **15 + v**
16. $x + y + z$ **−x − y − z**
17. $-2c + 4d − 5$ **2c − 4d + 5**
18. $t − w − 5$ **−t + w + 5**
19. $8a + 4b − c$ **−8a − 4b + c**
20. $5r − s + 3t$ **−5r + s − 3t**

Use the Associative, Commutative, and Distributive Properties to find each sum or product. Show your work and explain each step.

21. $(56 + 48) + 22$ **126**
22. $(25 \cdot 3) \cdot 4$ **300**
23. $5 \cdot (6 \cdot 12)$ **360**
24. $(137 + 149) + 113$ **399**
25. $5 \cdot (138 \cdot 2)$ **1380**
26. $4 \cdot (76 \cdot 25)$ **7600**
27. $(157 + 29) + 43$ **229**
28. $40 \cdot (36 \cdot 50)$ **72,000**
29. $(167 \cdot 345) \cdot 0$ **0**
30. $16 + (24 + 608)$ **648**
31. $6(5 + 20)$ **150**
32. $25(3 + 4)$ **175**

Name the property illustrated. Be specific.

33. $54 + 15 = 15 + 54$ **Comm.**
34. $6 \cdot (3 \cdot 14) = (6 \cdot 3) \cdot 14$ **Assoc.**
35. $2(2.5 + 5.9) = 2 \cdot 2.5 + 2 \cdot 5.9$ **Dist.**
36. $8 \cdot 40 − 8 \cdot 25 = 8(40 − 25)$ **Dist.**
37. $(6 + 34) + 17 = 6 + (34 + 17)$ **Assoc.**
38. $18 \cdot 3 = 3 \cdot 18$ **Comm.**
39. $7(5x) = (7 \cdot 5)x$ **Assoc.**
40. $7(27.2 − 6.5) = 7 \cdot 27.2 − 7 \cdot 6.5$ **Dist.**
41. $(6 \cdot 9) \cdot 5 = 6 \cdot (9 \cdot 5)$ **Assoc.**
42. $7 + (18 + 26) = (7 + 18) + 26$ **Assoc.**

LESSON 2.6

Use the Distributive Property to show that each statement is true.

1. $12x + 3x = 15x$
2. $-2x − 7x = −9x$
3. $-5x + 4x = −x$
4. $65x + 13x = 78x$
5. $7x − 12x = −5x$
6. $73x − 100x = −27x$

Simplify the following expressions.

7. $(6a + 1) + (2a − 3)$ **8a − 2**
8. $(4x − 4) + (3x − 6)$ **7x − 10**
9. $(2c − 5) + (3c + 2)$ **5c − 3**
10. $(4y + 2) + (3 − y)$ **3y + 5**
11. $(3z − 4) + (7 − 2z)$ **z + 3**
12. $(2w + 1) + (4w − 9)$ **6w − 8**
13. $(1 + 4r) + (−5r − 3)$ **−r − 2**
14. $9r − 4r$ **5r**
15. $6a − 5a$ **a**
16. $8c − (3c + 2)$ **5c − 2**
17. $9w − (5w + 8)$ **4w − 8**
18. $6z − (5 − 7z)$ **13z − 5**
19. $7 − (3y − 4y)$ **y + 7**
20. $16a − (5a − 3)$ **11a + 3**
21. $(5r − s) − (3t + s)$ **5r − 2s − 3t**
22. $(2.2a + 1.1b) + (3a + 0.9b)$ **5.2a + 2b**
23. $(3.5x + 4.2y) + (0.6x + y)$ **4.1x + 5.2y**
24. $(4a + 2) + (3a − 6)$ **7a − 4**
25. $(6w − 3) + (3w + 2)$ **9w − 1**
26. $(7 − 3w) + (6 − 8w)$ **13 − 11w**
27. $(9q + 3) − (5 − 5q)$ **14q − 2**
28. $(10q + 8q) − (4m + 2m)$ **18q − 6m**
29. $(3y + 1) − (4y − 1)$ **−y + 2**
30. $(3v − 7t) − (7v + 2t)$ **−4v − 9t**
31. $(2h + 7j) − (−4h − 5j)$ **6h + 12j**
32. $(3d + 1) − (4d + 3) + (5d − 6)$ **4d − 8**
33. $(5m − 4) − (4m + 1) + (8 − m)$ **3**
34. $(3x − 5) + (2x + 3y − 5z)$ **5x + 3y − 5z − 5**
35. $(3a + b) + (2b + c) + (a + b + c)$ **4a + 4b + 2c**
36. $(3 − 4t + 8r) − (7r + t) − (6 + t)$ **r − 6t − 3**
37. $(5 + p) − (p + q) − (3p + 4q − 6)$ **−3p − 5q + 11**

Lesson 2.6

1. $12x + 3x = (12 + 3)x$
   $= 15x$
2. $-2x − 7x = (−2 − 7)x$
   $= −9x$
3. $-5x + 4x = (−5 + 4)x$
   $= −x$
4. $65x + 13x = (65 + 13)x$
   $= 78x$
5. $7x − 12x = (7 − 12)x$
   $= −5x$
6. $73x − 100x$
   $= (73 − 100)x$
   $= −27x$

LESSON 2.7

Simplify the following expressions. Use the Distributive Property if needed.

1. $3 \cdot 4m$ $12m$
2. $-7a \cdot 5$ $-35a$
3. $2x \cdot 5x$ $10x^2$
4. $-12y \cdot 4y$ $-48y^2$
5. $-44x \cdot 10x$ $-440x^2$
6. $-4(3b + 5)$ $-12b - 20$
7. $7x(3 + 2x)$ $14x^2 + 21x$
8. $2.5c \cdot 5c$ $12.5c^2$
9. $\frac{14x}{2}$ $7x$
10. $\frac{-64x}{4}$ $-16x$
11. $\frac{-48x}{-3}$ $16x$
12. $-2(11b - 3)$ $-22b + 6$
13. $5n - (3 - 4n)$ $9n - 3$
14. $4f \cdot 2 + 3f \cdot (-2)$ $2f$
15. $5(h - 2) - (3 - 2h)$ $7h - 13$
16. $(4r - 3) - 9(r + 3)$ $-5r - 30$
17. $\frac{-18 - 36e}{-9}$ $4e + 2$
18. $8(2y + 3z) - 5(7y - 7z)$ $-19y + 59z$
19. $(-4t - 2e) + 7(-t - e)$ $-11t - 9e$
20. $9(2x + 3y) - 6(4x - 2y)$ $-6x + 39y$
21. $\frac{7 - 21g}{7}$ $-3g + 1$
22. $-(y - x) - 4(5y - 2x)$ $9x - 21y$
23. $\frac{4x + 20}{4}$ $x + 5$
24. $-4(2q - 3p) - 5(3q - 2p)$ $22p - 23q$
25. $\frac{-6 + 24u}{-3}$ $2 - 8u$
26. $3(z - 4w) - (5z - 5w)$ $-7w - 2z$
27. $\frac{-15x + 10}{5}$ $-3x + 2$
28. $4c(6a + 1)$ $24ac + 4c$
29. $8e(3x - 6)$ $24ex - 48e$
30. $\frac{6a(5b + 7)}{3a}$ $10b + 14$
31. $\frac{-5t(t + r)}{t}$ $-5t - 5r$
32. $7a(4x - 4)$ $28ax - 28a$
33. $15c(2c - 5)$ $30c^2 - 75c$
34. $10q(16x + 11)$ $160qx - 110q$
35. $\frac{2y^2(y + 1)}{2y}$ $y^2 + y$
36. $(6c - 5) + 8(3c + 4)$ $30c + 27$

CHAPTER 3

LESSON 3.1

Solve each equation. Check the solution.

1. $x + 2 = 1$ $x = -1$
2. $t + 6 = 2$ $t = -4$
3. $w - 4 = -3$ $w = 1$
4. $5 = x + 1$ $x = 4$
5. $y + 3 = -3$ $y = 6$
6. $y + 3 = -4$ $y = -7$
7. $h - 5 = 6$ $h = 1$
8. $x - 9 = -9$ $x = 0$
9. $8 = 6 + z$ $z = 2$
10. $-7 = 3 + g$ $g = -10$
11. $t + 5 = 11$ $t = 6$
12. $16 = f - 2$ $f = 18$
13. $-8 = 9 + n$ $n = -17$
14. $d - 12 = 7$ $d = 19$
15. $6 = -9 + c$ $c = 15$
16. $(-9) + r = 17$ $r = 26$
17. $b + 10 = 2$ $b = -8$
18. $8 = h - 3$ $h = 11$
19. $x - 3 = 5$ $x = 8$
20. $d - 8 = -10$ $d = -2$
21. $-6 = 11 + j$ $j = -17$
22. $h - 14 = -10$ $h = 4$
23. $(-2) + w = -9$ $w = -7$
24. $14 = d + (-2)$ $d = 16$
25. $x + 9 = -1$ $x = -10$
26. $t - 6 = 6$ $t = 12$
27. $w - 4 = -2$ $w = 2$
28. $h - 5 = 7$ $h = 12$
29. $5 = y + 1$ $y = 4$
30. $y + 3 = -8$ $y = -11$
31. $x - 15 = 20$ $x = 35$
32. $r + 23 = 10$ $r = -13$
33. $n - 65 = 59$ $n = 124$
34. $76 + k = 12$ $k = -64$
35. $w - 52 = -45$ $w = 7$
36. $y + 19 = -10$ $y = -29$
37. $l + 48 = 50$ $l = 2$
38. $b - 350 = -80$ $b = 270$
39. $x + 28 = -20$ $x = -48$
40. $450 = 100 + f$ $f = 350$
41. $88 + u = 44$ $u = -44$
42. $498 + x = 500$ $x = 2$
43. $q - 90 = 19$ $q = 109$
44. $10 + d = 72$ $d = 62$
45. $s - 25 = 36$ $s = 61$
46. $54 = 112 + d$ $d = -58$
47. $p + 365 = 425$ $p = 60$
48. $b + 350 = -80$ $b = -430$

LESSON 3.2

Write the property needed to solve the equation. Then solve it.

1. $\frac{x}{2} = -12$ $x = -24$
2. $\frac{b}{-10} = -4$ $b = 40$
3. $\frac{x}{2} = 2$ $x = 4$
4. $5x = 4$ $x = \frac{4}{5}$
5. $-15y = 5$ $y = -3$
6. $\frac{c}{7} = -70$ $c = -490$
7. $42x = 6$ $x = 7$
8. $5x = 3$ $x = \frac{3}{5}$
9. $-12y = 144$ $y = -12$
10. $8 = -64z$ $z = -8$
11. $\frac{a}{-5} = -3.5$ $a = 17.5$
12. $7 = 4x$ $x = 1\frac{3}{4}$

Solve each equation.

13. $44x = 55$ $x = 1\frac{1}{4}$
14. $k - 1.5 = 7.5$ $k = 9$
15. $\frac{r}{3} = -18$ $r = -54$
16. $4t = -36$ $t = -9$
17. $\frac{c}{5} = 12$ $c = 60$
18. $7n = 15$ $n = 2\frac{1}{7}$
19. $\frac{x}{2} = 17$ $x = 34$
20. $-3c = 18$ $c = -6$
21. $45d = 0$ $d = 0$
22. $9p = -72$ $p = -8$
23. $\frac{h}{-1} = -99$ $h = 99$
24. $3500t = 4000$ $t = 1\frac{1}{7}$
25. $\frac{m}{0.6} = 7$ $m = 4.2$
26. $0.25 = 0.25f$ $f = 1$
27. $-2y = -35.24$ $y = 17.62$
28. $0.75v = 75$ $v = 100$
29. $-50 = -2z$ $z = 25$
30. $\frac{j}{-19} = 30$ $j = -570$
31. $12m = 20$ $m = 1\frac{2}{3}$
32. $\frac{r}{32} = 1$ $r = 32$
33. $2.2 = 1.1w$ $w = 2$
34. $\frac{65}{x} = 13$ $x = 5$
35. $-6v = -582$ $v = 97$
36. $56x = -1$ $x = -\frac{1}{56}$

LESSON 3.3

Solve each equation and check your solution.

1. $7y - 1 = 27$ $y = 4$
2. $4a - 2 = 30$ $a = 8$
3. $8y - 5 = 19$ $y = 3$
4. $6 + 4w = -26$ $w = -8$
5. $3d + 2 = 23$ $d = 7$
6. $4x + \frac{1}{2} = \frac{1}{4}$ $x = -\frac{1}{16}$
7. $\frac{6c}{5} + 8 = -2$ $c = -6\frac{1}{3}$
8. $7 + 7a = 28$ $a = 3$
9. $36 = 3m + 6$ $m = 10$
10. $-2 - 14z = -30$ $z = 2$
11. $8 - 2n = -4$ $n = 6$
12. $-10 - 5x = 25$ $x = -7$
13. $15 = 2p - 5$ $p = 10$
14. $-5w + 3 = -72$ $w = 15$
15. $50 = -4g + 2$ $g = -12$
16. $58 = -8s + 18$ $s = -5$
17. $-51 = 8x + 5$ $x = -7$
18. $\frac{151 + r}{3} = 80$ $r = 89$
19. $3j + \frac{2}{5} = \frac{1}{3}$ $j = -\frac{1}{45}$
20. $\frac{3}{4} + 5h = \frac{1}{8}$ $b = -\frac{1}{8}$
21. $2b - \frac{1}{9} = \frac{1}{3}$ $b = \frac{2}{9}$
22. $3k - \frac{1}{3} = \frac{4}{9}$ $k = \frac{7}{27}$
23. $\frac{350}{x + 10} = 5$ $x = 60$
24. $6g - \frac{7}{10} = -\frac{1}{5}$ $b = \frac{1}{12}$

LESSON 3.4

Solve each equation and check your solution.

1. $7x = 5x + 6$ $x = 3$
2. $b + 3 = 9 - b$ $b = 3$
3. $8m - 14 = 6m$ $m = 7$
4. $8g + 1 = 4g + 9$ $g = 2$
5. $4c - 5 = 2c - 9$ $c = -2$
6. $2 - 4h = 3h - 33$ $h = 5$
7. $24 - 6y = 2y - 40$ $y = 8$
8. $15 + 2k = 5k + 6$ $k = 9$
9. $6s - 5 = 19 - 2s$ $s = 3$
10. $6y - 2 = 4y + 5$ $y = \frac{7}{2}$
11. $3r - 8 = 5r - 20$ $r = 6$
12. $1 - 3x = 2x + 8$ $x = -\frac{7}{5}$
13. $8 - 2b = 7b - 4$ $b = \frac{4}{3}$
14. $0.75n + 100 = 0.25n - 1$ $n = 202$
15. $2.6x - 3.9 = 4.5 + 1.2x$ $x = 6$
16. $8w + 20 = 12w - 9$ $w = \frac{29}{4}$
17. $\frac{3a}{5} = 2a + 1$ $a = -\frac{5}{7}$
18. $4x + 9 = 2x + 6$ $x = -\frac{3}{2}$
19. $2c + 5 = \frac{c}{5} - 4$ $c = -5$
20. $t + \frac{1}{2} = \frac{t}{4} + 2$ $t = 2$

LESSON 3.5

Solve each equation and check your solution.

1. $2(g - 4) = 6$ $g = 7$
2. $5(r + 3) = 35$ $r = 4$
3. $4(z - 20) = 16$ $z = 24$
4. $2(5 - x) = -25$ $x = \frac{35}{2}$
5. $2(3 - 4d) = -42$ $d = 6$
6. $2(4t + 1) = 74$ $t = 9$
7. $14 = 2(f + 3)$ $f = 4$
8. $3(y - \frac{2}{3}) = \frac{1}{4}$ $y = \frac{3}{4}$
9. $3r + 2(r - 4) = 12$ $r = 4$
10. $7k - 2(k + 6) = -2$ $k = 2$
11. $3v + 5 + 2v = 5(2 - v)$ $v = \frac{1}{2}$
12. $3(2b + 6) = 5b + 20$ $b = 2$
13. $3x - (5 - 2x) = x - 21$ $x = -4$
14. $4(s + \frac{1}{2}) = 8(s + \frac{3}{4})$ $s = -1$
15. $2(u - 3) = 3(u - 4)$ $u = 6$
16. $5(x - 2) + 8x - 14 = 9x - 2$ $x = \frac{11}{2}$
17. $2(4m + 1) = 3(2m + 4)$ $m = 5$
18. $14(n + 7) = 3(n - 4)$ $n = -10$
19. $p - \frac{1}{4} = \frac{p}{3} + 7\frac{3}{4}$ $p = 12$
20. $\frac{x}{3} + \frac{4}{5} = 2x - \frac{5}{6}$ $x = \frac{49}{50}$

LESSON 3.6

Solve each equation for the indicated variable.

1. $P = s - c$, for s $s = c + P$
2. $h = R - k$, for k $k = -h + R$
3. $a + b$, for a $a = -b + c$
4. $I = C + T$, for C $C = I - T$
5. $P = 4s$, for s $s = \frac{P}{4}$
6. $f = \frac{i}{12}$, for i $i = 12f$
7. $d = rt$, for t $t = \frac{d}{r}$
8. $a - b = -c$, for a $a = b - c$
9. $s = \frac{m}{h}$, for m $m = hs$
10. $V = lwh$, for h $h = \frac{V}{lw}$
11. $I = prt$, for r $r = \frac{I}{pt}$
12. $A = \frac{bh}{2}$, for b $b = \frac{2A}{h}$
13. $x + 4y = r$, for y $y = \frac{r-x}{4}$
14. $a = \frac{m}{g}$, for m $m = ag$
15. $P = 15 + \frac{1}{2}D$, for D $D = 2P - 30$
16. $P = a + b + c$, for a $a = -b - c + P$
17. $d + e + f = 180$, for f $f = 18 - d - e$
18. $C = 2\pi r$, for r $r = \frac{C}{2\pi}$
19. $S = (n-2)180$, for n $n = \frac{S}{180} + 2$
20. $3x + g + 2h = t - y$, for h $h = \frac{-g + t - 3x - y}{2}$

CHAPTER 4

LESSON 4.1

Determine whether each proportion is true.

1. $\frac{36}{7} = \frac{108}{21}$ T
2. $\frac{8}{15} = \frac{15}{20}$ F
3. $\frac{64}{72} = \frac{8}{9}$ T
4. $\frac{12}{7} = \frac{60}{35}$ T
5. $\frac{14}{18} = \frac{9}{20}$ F
6. $\frac{4}{5} = \frac{28}{35}$ T
7. $\frac{13}{26} = \frac{25}{50}$ T
8. $\frac{9}{30} = \frac{10}{40}$ F
9. $\frac{5}{6} = \frac{25}{36}$ F
10. $\frac{15}{16} = \frac{30}{32}$ T
11. $\frac{7}{9} = \frac{18}{14}$ F
12. $\frac{9}{15} = \frac{15}{25}$ T

Solve each proportion. Round your answers to the nearest hundredth if necessary.

13. $\frac{16}{18} = \frac{48}{x}$ $x = 54$
14. $\frac{28}{15} = \frac{n}{20}$ $n = 37.33$
15. $\frac{32}{27} = \frac{24}{r}$ $r = 20.25$
16. $\frac{m}{42} = \frac{60}{84}$ $m = 30$
17. $\frac{6}{9} = \frac{30}{a}$ $a = 45$
18. $\frac{14}{8} = \frac{21}{f}$ $f = 12$
19. $\frac{c}{11} = \frac{22}{10}$ $c = 24.2$
20. $\frac{15}{21} = \frac{h}{6}$ $h = 4.29$
21. $\frac{44}{7} = \frac{11}{t}$ $t = 1.75$
22. $\frac{9.1}{3.5} = \frac{g}{5}$ $g = 13$
23. $\frac{86}{4} = \frac{r}{3}$ $r = 64.5$
24. $\frac{9}{25} = \frac{3}{w}$ $w = 8.33$
25. $\frac{16}{b} = \frac{25}{36}$ $b = 23.04$
26. $\frac{68}{12} = \frac{4}{r}$ $r = 0.71$
27. $\frac{a}{7} = \frac{7}{15}$ $2 = 3.27$
28. $\frac{26}{13} = \frac{13}{b}$ $b = 6.5$
29. $\frac{1.8}{0.9} = \frac{2}{x}$ $x = 1$
30. $\frac{45}{y} = \frac{9}{5}$ $y = 25$
31. $\frac{q}{4.2} = \frac{0.8}{0.6}$ $q = 5.6$
32. $\frac{14}{b} = \frac{21}{3}$ $b = 2$
33. $\frac{20.5}{x} = \frac{82}{46}$ $x = 11.5$
34. $\frac{y}{21.4} = \frac{8}{53.5}$ $y = 3.2$
35. $\frac{50.8}{124} = \frac{z}{10}$ $z = 4.1$
36. $\frac{m}{75} = \frac{27}{40}$ $m = 50.63$
37. $\frac{f}{6} = \frac{2}{2.4}$ $f = 5$
38. $\frac{4.2}{5.6} = \frac{7}{x}$ $x = 9.33$
39. $\frac{4}{7} = \frac{11}{b}$ $b = 19.25$
40. $\frac{9}{100} = \frac{10}{m}$ $m = 111.11$

LESSON 4.2

Write each percent as a decimal.

1. 65% **0.65**
2. 5.5% **0.055**
3. 9% **0.09**
4. 0.6% **0.006**
5. 2.4% **0.024**
6. 400% **4**
7. 225% **2.25**
8. 0.267% **0.00267**

Write each decimal as a percent.

9. 0.87 **87%**
10. 0.45 **45%**
11. 0.06 **6%**
12. 1.28 **128%**
13. 0.0256 **2.56%**
14. 5.08 **508%**
15. 0.073 **7.3%**
16. 0.00098 **0.098%**

Write each percent as a fraction or mixed number in lowest terms.

17. 25% $\frac{1}{4}$
18. 45% $\frac{9}{20}$
19. 6.5% $\frac{13}{200}$
20. 0.02% $\frac{1}{5000}$
21. 350% $3\frac{1}{2}$
22. 9% $\frac{9}{100}$
23. 44% $\frac{11}{25}$
24. 16% $\frac{4}{25}$

Find each answer.

25. What number is 30% of 60? **18**
26. 130% of what number is 78? **60**
27. What percent of 70 is 35? **50%**
28. 96 is 80% of what number? **120**
29. What number is 20% of 75? **15**
30. 45% of what number is 15? **33.33**
31. What percent of 99 is 18? **18%**
32. 57 is 79% of what number? **72.15**
33. What number is 22% of 1000? **220**
34. 250% of what number is 80? **32**
35. What percent of 18 is 24? **133%**
36. 29 is 2% of what number? **1450**
37. What number is 2% of 25? **0.5**
38. 44% of what number is 12? **27.27**
39. What percent of 50 is 30? **60%**
40. 65 is 37% of what number? **175.68**
41. What number is 290% of 11? **31.9**
42. 180% of what number is 115? **63.89**
43. What percent of 20.8 is 1.5? **7.2%**
44. 7.4 is 30% of what number? **24.67**

LESSON 4.3

The results of spinning a spinner with regions I, II, III, and IV 100 times are shown in the table below.

Region	I	II	III	IV
Frequency	34	29	22	15

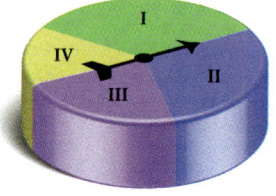

1. Is it equally likely for the spinner to land in any region? Explain. **no**
2. Find the experimental probability of landing in each region. **I, 34%; II, 29%; III, 22%; IV, 15%**
3. Find the sum of the experimental probabilities of landing in Regions I and IV. **49%**
4. Find the sum of the experimental probabilities for landing in Regions II and III. **51%**
5. Are these sums you calculated in Exercises 3 and 4 almost the same? Why or why not?
   Yes; the combined areas are almost equal
6. Find the sum of the experimental probabilities for landing in Regions I and II. **63%**
7. Find the sum of the experimental probabilities for landing in Regions III and IV. **37%**
8. Are these sums you calculated in Exercises 6 and 7 almost the same? Why or why not?
   No; the combined areas are not almost equal

LESSON 4.4

Find the mean, median, mode(s), and range for each set of data.

1. 12, 12, 18, 14, 8, 9, 10, 20, 19
   13.1, 12, 12, 12
2. 8, 11, 15, 25, 35, 62, 20, 40
   27, 22.5, none, 54
3. 4, 2, 5, 1, 1, 9, 8, 4, 6, 3, 2
   3.4.09; 4; 1, 2, 4; 8
4. 150, 320, 200, 41, 700, 210, 300
   274.4, 210, none, 659
5. 130, 135, 132, 120, 145, 136
   133, 133.5, none, 25
6. 30, 35, 200, 42, 95, 100, 300, 25
   103.38, 68.5, none, 275
7. 60, 54, 45, 72, 83, 64, 51, 75, 77, 59, 48, 61
   62.42, 60.5, none, 38
8. 55, 26, 34, 21, 37, 48, 34, 27, 19, 51, 27, 32
   34.25; 33; 27, 34; 36

The coaches at Midville Middle School kept track of the runs scored by each member of the baseball team. Use the data shown in the table below for Exercises 9–13.

Runs Scored

3	5	5	6
7	5	1	0
8	9	2	2
6	1	2	3
0	0	2	4
4	7	8	9

9. Make a frequency table for the data.
10. Find the median number of runs scored. **4**
11. Find the mean of the data. **4.125**
12. What is the mode(s)? **2**
13. Which measure of central tendency would you use to give the best impression of the players? Explain. **The mean gives the best impression because it shows the greatest number of runs.**

LESSON 4.5

The circle graph at right shows the percent of the population in Carsonville in certain age groups. In 1995, the total population was 150,600.

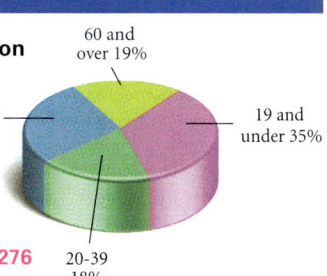

1. How many residents were 60 and over in 1995? **28,614**
2. How many residents were 19 and under in 1995? **52,710**
3. How many residents were 40 to 59 years of age in 1995? **42,168**
4. How many residents were 20 to 59 years of age in 1995? **69,276**

The population of Carsonville is expected to grow to 190,000 by the year 2005, and the percent of the population in each age group is expected to remain the same.

5. How many residents are expected to be 19 years of age or less in 2005? **66,500**
6. How many residents are expected to be 60 years of age or more in 2005? **36,100**
7. How many residents are expected to be between 20 and 39 years of age in 2005? **34,200**
8. How many residents are expected to be between 40 and 59 years of age in 2005? **53,200**

Lesson 4.4

9.

Runs	0	1	2	3	4	5	6	7	8	9
Freq.	3	2	4	2	2	3	2	2	2	2

Lesson 4.6

1. key: 5|8 = 58

stem	leaf
5	8, 9
6	0, 7, 8, 8, 9
7	3, 5, 5, 6, 7
8	0, 0, 0, 3, 7
9	0, 1, 2, 3, 5, 5, 7

5. Check students' work.

6. key: 1|2 = 12

stem	leaf
0	0, 0, 0, 2, 8, 9
1	0, 0, 2, 7, 8, 8, 9
2	0, 1, 1, 2, 2, 2, 3, 3, 4, 5, 7

7. Check students' work.

Lesson 5.1

1. domain = {1, 2, 3, 4}; range = {2, 4, 6, 8}; function

2. domain = {−1, 2, 3}; range = {3, 5, 6, 8}; not a function

3. domain = {−200, 100, 150, 200}; range = {3, 4, 5}; function

4. domain = {1.2, 1.5, 2.3, 3.2}; range = {2, 4, 5, 7.1}; function

5. domain = {−5, −2, 1, 4}; range = {6, 7, 8, 9, 10}; not a function

6. domain = {25, 31, 32, 46}; range = {−2, 1, 2, 5, 8}; not a function

7. domain = {2, 3, 4, 5}; range = {20, 25, 30}; function

8. domain = {−2, −1, 5, 6}; range = {7.8, 10, 12}; not a function

LESSON 4.6

Ms. Smith compiled the list of scores shown in the table below for a recent test given in her class.

1. Make a stem-and-leaf plot for the data.
2. What is the median of the data? **78.5**
3. What is the lower quartile for this data? **68.5**
4. What is the upper quartile for this data? **90.5**
5. Make a box-and-whisker plot for this data.

The data shown below are the numbers of fish caught by a fishing crew on each day in April.

6. Make stem-and-leaf plot of the data.
7. Make a histogram of the data.

Test scores

80	90	92	68
75	60	68	93
95	69	75	76
80	80	83	91
67	87	77	97
95	58	73	59

Number of fish caught

10	25	21	21
12	2	24	10
20	10	23	18
8	27	10	19
0	18	9	22
17	22	23	22

CHAPTER 5

LESSON 5.1

For each relation, describe the domain and range. Determine whether the relation is a function. Explain your reasoning.

1. {(1, 2), (2, 4), (3, 6), (4, 8)}
2. {(−1, 5), (3, 6), (2, 3), (−1, 8)}
3. {(200, 5), (150, 4), (100, 3), (−200, 5)}
4. {(1.5, 2), (3.2, 5), (1.2, 4), (2.3, 7.1)}
5. {(−5, 6), (−2, 7), (1, 8), (4, 9), (−5, 10)}
6. {(25, 1), (32, 8), (25, 2), (46, 5), (31, −2)}
7. {(2, 20), (3, 25), (4, 30), (5, 20)}
8. {(−2, 12), (5, 7.8), (−1, 12), (−2, 10), (6, 12)}

Complete each ordered pair so that it is a solution to $3x - 2y = 16$.

9. (1, ?) $y = 6\frac{1}{2}$
10. (−2, ?) $y = -11$
11. (?, 1) $x = 6$
12. (?, 3) $x = 7\frac{1}{3}$
13. (4, ?) $y = -2$
14. (?, −2) $x = 4$
15. (?, 4) $x = 8$
16. (2, ?) $y = -5$
17. (−12, ?) $y = -20$
18. (?, −10) $-1\frac{1}{3}$
19. (7, ?) $y = 2\frac{1}{2}$
20. (?, −8) $x = 0$
21. (10, ?) $y = 23$
22. (−6, ?) $y = -17$
23. (?, 5) $x = 8\frac{2}{3}$
24. (?, −5) $x = 2$

LESSON 5.2

Find the slope for each rise and run.

1. rise 4, run 2   **2**
2. rise 1, run 8 $\frac{1}{8}$
3. rise 4, run 1   **4**
4. rise −6, run 3   **−2**
5. rise 0, run 10   **0**
6. rise −9, run −5 $\frac{9}{5}$
7. rise 4, run −5 $-\frac{4}{5}$
8. rise 2, run 16 $\frac{1}{8}$
9. rise 18, run 8 $\frac{9}{4}$

Find the slope of each line graphed below.

10. 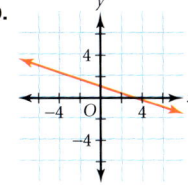 $-\frac{1}{3}$
11. 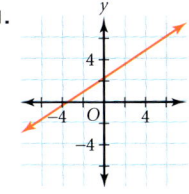 $\frac{2}{3}$
12. 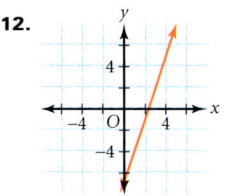   **2**

Find the slope of the line that contains each pair of points.

13. $M(5, 3), N(3, 4)$ $-\frac{1}{2}$
14. $M(1, 7), N(4, 1)$ **−2**
15. $M(-3, 2), N(1, -4)$ $-\frac{3}{2}$
16. $M(-5, 3), N(0, 6)$ $\frac{3}{5}$
17. $M(1, -7), N(4, -1)$ **2**
18. $M(-5, 9), N(3, -3)$ $-\frac{3}{2}$
19. $M(6, 7), N(3, 4)$ **1**
20. $M(9, -4), N(3, 2)$ **−1**
21. $M(-3, -1), N(6, -4)$ $-\frac{1}{3}$

LESSON 5.3

The graph at the right shows cost of renting in-line skates from each of two different places.

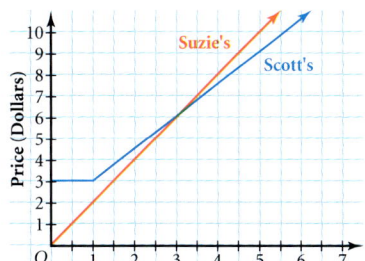

1. Suzie's Skate Land requires no deposit. Find the rate of change, or cost per hour. **$2.00/hour**
2. Scott's Skates requires a non-refundable deposit, and the first hour is free. What is the amount of the deposit? **$3.00**
3. Maria wants to rent roller blades for 4 hours. At which shop will the rental be less expensive?   **Scott's Skates**
4. For what rental time will the skate rentals cost the same?   **3 hours**

In Exercises 5–19, y varies directly as x. Find the constant of variation, and write an equation for the direct variation.

5. $y = 10$ when $x = 5$
   2; $y = 2x$
6. $y = 1.5$ when $x = 25$
   0.06; $y = 0.06x$
7. $y = 6$ when $x = 2$
   3; $y = 3x$
8. $y = 11$ when $x = 8$
   1.375; $y = 1.375x$
9. $y = 4$ when $x = 3$
   1.33; $y = 1.33x$
10. $y = 1.7$ when $x = 2.1$
    0.81; $y = 0.081x$
11. $y = 14$ when $x = 16$
    0.875; $y = 0.875x$
12. $y = 2$ when $x = 12$
    0.17; $y = 0.17x$
13. $y = 9$ when $x = 18$
    0.5; $y = 0.5x$
14. $y = 23$ when $x = 42$
    0.55; $y = 0.55x$
15. $y = 9$ when $x = 6$
    1.5; $y = 1.5x$
16. $y = 3.6$ when $x = 24$
    0.15; $y = 0.15x$
17. $y = 4.1$ when $x = 0.84$
    4.88; $y = 4.88x$
18. $y = 22$ when $x = 16$
    1.375; $y = 1.375x$
19. $y = 7$ when $x = 11$
    0.636; $y = 0.636x$

LESSON 5.4

Identify the slope and the *x*- and *y*-intercepts.

1. $y = 3x + 1$ **3, 1, $-\frac{1}{3}$**
2. $y = 2x - 1$ **2, −1, $\frac{1}{2}$**
3. $y = 4x$ **4, 0, 0**
4. $y = -6x + 8$ **−6, 8, $\frac{4}{3}$**
5. $y = 5x + 4$ **5, 4, $-\frac{4}{5}$**
6. $y = x - 7$ **1, −7, 7**
7. $y = -11x + 10$ **−11, 10, $\frac{10}{11}$**
8. $y = 13x - 5$ **13, −5, $\frac{5}{13}$**
9. $y = x + 9$ **1, 9, −9**
10. $y = 3x - 6$ **3, −6, 2**
11. $y = 7x - 12$ **7, −12, $1\frac{5}{7}$**
12. $y = x$ **1, 0, 0**

Write an equation in slope-intercept form for the line that contains each pair of points.

13. $M(3, 5), N(-3, 1)$ $y = \frac{2}{3}x + 3$
14. $M(1, 0), N(2, 1)$ $y = x - 1$
15. $M(-1, 3), N(1, -1)$ $y = -2x + 1$
16. $M(3, 1), N(-3, 3)$ $y = -\frac{1}{3}x + 2$
17. $M(-2, 0), N(2, 4)$ $y = x + 2$
18. $M(0, 1), N(1, 3)$ $y = 2x + 1$
19. $M(2, 1), N(-2, -3)$ $y = x - 1$
20. $M(-1, 4), N(1, -2)$ $y = -3x + 1$
21. $M(4, 3), N(8, 4)$ $y = \frac{1}{4}x + 2$
22. $M(-5, 1), N(5, -3)$ $y = -\frac{2}{3}x - 1$
23. $M(-1, -3), N(1, -2)$ $y = \frac{1}{2}x - \frac{5}{2}$
24. $M(1, 2), N(2, 4)$ $y = 2x$

Write an equation in slope-intercept form for each line graphed below.

25. $y = -x + 3$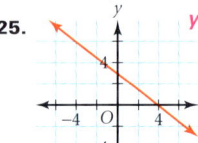
26. $y = \frac{3}{2}x - 2$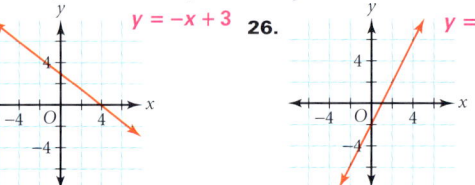
27. $y = -\frac{4}{3}x$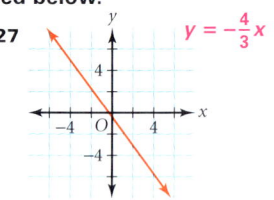

LESSON 5.5

Write each equation in standard form.

1. $3x = 6y + 18$ **$3x - 6y = 18$**
2. $5y = -4x - 20$ **$4x + 5y = -20$**
3. $-2x + 5y = 1$ **$2x - 5y = -1$**
4. $y = -7x + 2$ **$7x + y = 2$**
5. $5x = y + 4$ **$5x - y = 4$**
6. $6 = x - y$ **$x - y = 6$**
7. $12 = -3x + 7y$ **$3x - 7y = -12$**
8. $2x - 4 = 3y + 6$ **$2x - 3y = 10$**
9. $4x = 8y$ **$4x - 8y = 0$**
10. $21x = 7y + 13$ **$21x - 7y = 13$**
11. $3y - 2x = 1$ **$2x - 3y = -1$**
12. $15y + x + 17 = 0$ **$x + 15y = -17$**
13. $\frac{1}{2}x + \frac{1}{4}y = 1$ **$2x + y = 4$**
14. $\frac{1}{3}y = 2x + \frac{2}{3}$ **$6x - y = -2$**
15. $y = 0.5x + 6$ **$x - 2y = -12$**

Write an equation in point-slope form for the line that has the given slope and that contains the given point.

16. slope −1, (2, 3) $y = -x + 5$
17. slope 2, (4, 5) $y = 2x - 3$
18. slope 0, (4, 5) $y = 5$

Find the *x*- and *y*-intercepts for the graph of each equation.

19. $x + y = 2$ **2, 2**
20. $x - y = 4$ **4, −4**
21. $x + 2y = 8$ **8, 4**
22. $3x + y = 9$ **3, 9**
23. $4x - 5y = 20$ **5, −4**
24. $2x - 7y = 14$ **7, −2**
25. $9x + y = 18$ **2, 18**
26. $4x - 6y = 12$ **3, −2**
27. $x = 3y$ **0, 0**
28. $x = -5y + 1$ **1, $\frac{1}{5}$**
29. $x = 2y$ **0, 0**
30. $9x + y = 4$ **$\frac{4}{9}$, 4**

Write an equation for the line that contains each pair of points.

31. (2, 1), (−2, −3) $y = x - 1$
32. (−1, 4), (1, −2) $y = -3x + 1$
33. (4, 3), (8, 4) $y = \frac{1}{4}x + 2$

Write an equation in slope-intercept form for each line described below.

34. containing (4, 8) and (−2, −1) $y = \frac{3}{2}x + 2$
35. containing (2, 3) and with slope −1 $y = -x + 5$
36. containing (4, 5) and with slope 2 $y = 2x - 3$
37. containing (−1, −5) and with slope 0 $y = -5$
38. crossing the *x*-axis at $x = -5$ and the *y*-axis at $y = 0$ $y = 0$
39. crossing the *x*-axis at $x = 2$ and the *y*-axis at $y = 2$ $y = -x + 2$

LESSON 5.6

Identify the slope of a line that is parallel to each line.

1. $y = 4x + 10$ **4**
2. $3x + y = 7$ **−3**
3. $10 = -5x + 2y$ **$\frac{5}{2}$**
4. $4x - 3y = 12$ **$\frac{4}{3}$**
5. $y = x - 3$ **$\frac{1}{3}$**
6. $4x - y = 16$ **16**
7. $2x + 6y = 6$ **$-\frac{1}{3}$**
8. $7x - y = 7$ **7**
9. $4x = 8y + 1$ **$\frac{1}{2}$**
10. $2x = 7y + 10$ **$\frac{2}{7}$**
11. $9y - x = 1$ **$\frac{1}{9}$**
12. $5y + 2x = 7$ **$-\frac{2}{5}$**

Write the slope of a line that is perpendicular to each line.

13. $y = 4x + 10$ **$-\frac{1}{4}$**
14. $y = 2x + 11$ **$-\frac{1}{2}$**
15. $y = 5x + 6$ **$-\frac{1}{5}$**
16. $2x + 3y = 2$ **$\frac{3}{2}$**
17. $x - 7y = 4$ **−7**
18. $3x + 2y = 9$ **$\frac{2}{3}$**
19. $3x + 4y = -9$ **$\frac{4}{3}$**
20. $4x - 8y = 24$ **−2**
21. $2x - y = 1$ **$-\frac{1}{2}$**
22. $9x + 6y = 18$ **$\frac{2}{3}$**
23. $4x - 2y = 12$ **$-\frac{1}{2}$**
24. $y = x$ **−1**

Tell whether each pair of lines are parallel, perpendicular, or neither.

25. $y = 0.5x$ **perp.**
    $y = -2x$
26. $y = 3x - 2$ **neither**
    $y = -3x + 4$
27. $x - 3y = 4$ **perp.**
    $3x + y = 1$

Write an equation in slope-intercept form for the line specified.

	Contains	Is parallel to		Contains	Is perpendicular to
28.	(2, −4)	$3x - y = 5$	29.	(2, −4)	$3x - y = 5$
30.	(−1, 4)	$y = 2x + 5$	31.	(−1, 4)	$y = 2x + 5$
32.	(−3, −1)	$2y = x + 2$	33.	(−3, −1)	$2y = x + 2$
34.	(0, 5)	$y = -2x - 1$	35.	(0, 5)	$y = -2x - 1$

28. $y = 3x - 10$
30. $y = 2x + 6$
32. $y = \frac{1}{2}x + \frac{1}{2}$
34. $y = -2x + 5$
29. $y = -\frac{1}{3}x - 3\frac{1}{3}$
31. $y = -\frac{1}{2}x + 3\frac{1}{2}$
33. $y = -2x - 7$
35. $y = \frac{1}{2}x + 5$

CHAPTER 6

LESSON 6.1

Solve each inequality and graph the solution on a number line.

1. $x - 1 < 2$ **$x < 3$**
2. $y + 5 > 1$ **$y > -4$**
3. $w - 9 \leq -1$ **$w \leq 8$**
4. $p + 6 \geq -3$ **$p \geq -9$**
5. $3 + m > 2$ **$m > -1$**
6. $k - 3 \leq 4$ **$k \leq 7$**

Solve each inequality.

7. $x + 9 \leq -4$ **$x \leq -13$**
8. $11s - 7 < 3$ **$s < \frac{10}{11}$**
9. $b + \frac{1}{4} \geq 1$ **$b \geq \frac{3}{4}$**
10. $3p + 0.9 < 3$ **$p < 2.1$**
11. $t - 6.25 > 2.5$ **$t > 8.75$**
12. $r + 0.05 < 0.01$ **$r < -0.04$**
13. $w - \frac{2}{3} > 2$ **$w > 2\frac{2}{3}$**
14. $h + 10 \leq 90$ **$h \leq 80$**
15. $15 \leq d - 1.25$ **$d \geq 16.25$**
16. $0.85 < -2.65 + n$ **$n > 3.5$**
17. $\frac{5}{8} \leq \frac{1}{4} + t$ **$t \geq \frac{3}{8}$**
18. $-5.7 + v > 3.6$ **$v > 9.3$**
19. $c - \frac{1}{4} \leq \frac{1}{3}$ **$c < \frac{7}{12}$**
20. $z + 5 \geq 15$ **$z \geq 10$**
21. $4 \leq 17 + x$ **$x \geq -13$**
22. $q + 0.5 > 9$ **$q > 8.5$**
23. $7 \leq r - 4$ **$r \geq 11$**
24. $d - 2\frac{1}{3} > -6$ **$d > -3\frac{2}{3}$**
25. $8 > m - 13$ **$m < 21$**
26. $t + 1.2 < 32.5$ **$t < 31.3$**
27. $10.5 + f \leq 1.5$ **$f \leq -9$**
28. $h + 6.2 \leq 0.7$ **$h \leq -5.5$**
29. $k + 11.4 < 0$ **$k < -11.4$**
30. $4.2 - n < -7$ **$n > 11.2$**

Lesson 6.1

1. number line showing open circle at 2, shaded right, with marks at −2, 0, 2, 4
2. number line showing open circle at −3, shaded left, with marks at −4, −2, 0, 2
3. number line showing closed circle at 7, shaded left, with marks at 2, 4, 6, 8
4. number line showing closed circle at −9, shaded right, with marks at −9, −7, −5, −3
5. number line showing open circle at −1, shaded right, with marks at −2, 0, 2, 4
6. number line showing closed circle at 7, shaded left, with marks at 2, 4, 6, 8

753

Extra Practice

Lesson 6.3

1. [number line: open circle at -6, closed at 4]
2. [number line: -4 to 8, open at 4]
3. [number line: closed at 4 and 12]
4. [number line: closed at -6, open at -2]
5. [number line: closed at 4, open at 12]
6. [number line: open at 0, closed at 6]
7. [number line: -2 to 4]
8. [number line: open at -1, open at 2]
9. [number line: closed at 0, open at 6]
10. [number line: open at -2, open at 2]
11. [number line: closed at 4, open at 12]
12. [number line: open at 1, open at 3]
13. [number line: open at -2, open at 4]
14. [number line: open at 4, open at 12]
15. [number line: closed at -3, closed at -1]
16. [number line: open at 1, open at 2]

LESSON 6.2

Solve each inequality.

1. $2x + 6 \geq 12$ $x \geq 3$
2. $\frac{t}{2} \geq -36$ $t \geq -72$
3. $5y - 8 \leq 17$ $y \leq 5$
4. $9 - y > 14$ $y < -5$
5. $17 - h > 5$ $h < 12$
6. $\frac{4c}{2} \geq 7$ $c \geq 3\frac{1}{2}$
7. $\frac{c}{5} + 8 < -2$ $c < -50$
8. $-3 - x < -9$ $x > 6$
9. $2x - 15 \geq -14$ $x \geq \frac{1}{2}$
10. $-3p + 7 < 2$ $p > 1\frac{2}{3}$
11. $\frac{5}{m} \geq 2$ $m \leq 2\frac{1}{2}$
12. $8n < 7n - 4$ $n < -4$
13. $2 < 3(2b - 4)$ $b > 2\frac{1}{3}$
14. $-w + 3 \leq -32$ $w \geq 35$
15. $\frac{w}{-8} \geq 4$ $w \leq -32$
16. $2n + 7 \geq 16 + n$ $n \geq 9$
17. $\frac{9}{n} > -5$ $n > -1\frac{4}{5}$
18. $-3p > -21$ $p < 7$
19. $6t - 12 < -9$ $t < \frac{1}{2}$
20. $12y + 2 \geq 18$ $y \geq 1\frac{1}{3}$
21. $7 - \frac{m}{6} \leq 1$ $m \geq 36$
22. $\frac{r}{5} - 10 \geq 2$ $r \geq 60$
23. $2z - 2 \leq 10$ $z \leq 6$
24. $-8x < 20$ $x > -2\frac{1}{2}$

Write an inequality to represent each statement.

25. W is less than L. $W < L$
26. x is greater than 5. $x > 5$
27. v is less than or equal to 4.6. $v \leq 4.6$
28. w is greater than 3. $w > 3$
29. x is less than or equal to -2. $x \leq -2$
30. a is greater than or equal to b. $a \geq 6$
31. d is positive. $d > 0$
32. r is not negative. $r \geq 0$

LESSON 6.3

Graph the solution set for each inequality.

1. $-5 < x \leq 3$
2. $y > 2$ or $y < 0$
3. $4 \leq s \leq 11$
4. $d \geq -2$ or $d < -6$
5. $5 \leq r < 13$
6. $4\frac{1}{2} < f \leq 6$
7. $z > 0$ or $z < -2$
8. $-1.3 < g \leq 2.4$
9. $b \geq 5\frac{2}{3}$ or $b < 2$
10. $-2.7 < p < 1$
11. $4 \leq x < 9$
12. $t > 2.5$ or $t \leq 1.2$
13. $h > 5.3$ or $h < -2$
14. $5.8 < y < 9\frac{1}{2}$
15. $-3 \leq t < -1$
16. $x > 2$ or $x < 1$

Solve and graph each inequality.

17. $15 < 3s \leq 12$ $-5 < s \leq 4$
18. $2f - 1 > 3$ or $3f < 3$ $f > 2$ or $f < 1$
19. $5 \leq 5t < 12$ $1 \leq t < 2\frac{2}{5}$
20. $4s > 16$ or $-2s > 4$ $s > 4$ or $s < -2$
21. $-11 < 2g - 5 \leq 15$ $-3 < g \leq 10$
22. $3x > 12$ or $4x < 10$ $x > 4$ or $x < s\frac{1}{2}$
23. $-12 \leq 3h - 15 < 0$ $1 \leq -7$ or $x \geq -5$
24. $11 < d + 3 < 14$ $8 < d < 11$
25. $3t + 2 < 17$ or $5t + 1 > 31$ $t < 5$ or $t > 6$
26. $-2w + 1 > 15$ or $w + 2 \geq -3$ $w < -7$ or $w \geq -5$
27. $-3 < 2t + 5 \leq 16$ $-4 < t \leq 5\frac{1}{2}$
28. $5 < t + 3$ or $3t < -6$ $t > 2$ or $t < -2$
29. $-2 < 4b \leq 12.8$ $-\frac{1}{2} < b \leq 3.2$
30. $5\frac{1}{2} > 2f$ or $3f + 1 > 10$ $f < 2\frac{3}{4}$ or $f > 3$
31. $5 < d$ or $9 \geq d + 7$ $d > 5$ or $d \leq 2$
32. $-12 < 4r + 8 \leq 12$ $-5 < r \leq 1$

754

LESSON 6.4

Find the absolute value of each number.

1. 16 **16**
2. −24 **24**
3. −4.62 **4.62**
4. $\frac{3}{4}$ **$\frac{3}{4}$**
5. −1.03 **1.03**
6. 0.25 **0.25**
7. $-1\frac{3}{10}$ **$1\frac{3}{10}$**
8. 9.1 **9.1**
9. $-\frac{7}{8}$ **$\frac{7}{8}$**
10. −144 **144**
11. −0.064 **0.064**
12. $7\frac{2}{9}$ **$7\frac{2}{9}$**

Evaluate.

13. $-8 + |7|$ **−1**
14. $|-8| + 7$ **15**
15. $|-8| + |7|$ **15**
16. $|-9 - (-4)|$ **5**
17. $|-13 + 12|$ **1**
18. $|-43 + (-18)|$ **61**
19. $-5 + |-15 + 7|$ **3**
20. $-600 - |-25 + (-30)|$ **−655**
21. $|78 + (-43) - 5|$ **30**

Find the domain and range of each function.

22. $y = |-5 + x|$
    dom. all reals, ran. $y \geq 0$
23. $y = |-5 - x|$
    dom. all reals, ran. $y \geq 0$
24. $y = |-5| - x$
    dom. all reals, ran. all reals
25. $y = -|3 + x|$
    dom. all reals, ran. $y \leq 0$
26. $y = -3 + |x|$
    dom. all reals, ran. $y \geq -3$
27. $y = -5|x|$
    dom. all reals, ran. $y \leq 0$
28. $y = \frac{1}{4}|-x|$
    dom. all reals, ran. $y \geq 0$
29. $y = -5|x| + 6$
    dom. all reals, ran. $y \leq 1$
30. $y = -5|x + 6|$
    dom. all reals, ran. $y \leq 0$

LESSON 6.5

Solve each equation.

1. $|x + 2| = 4$
   $x = 2$ or $x = -6$
2. $|x - 4| = 2$
   $x = 2$ or $x = 6$
3. $|x - 1| = 6$
   $x = 7$ or $x = -5$
4. $|x - 5| = 1$
   $x = 6$ or $x = 4$
5. $|x + 10| = 8$
   $x = -2$ or $x = -18$
6. $|7 + x| = 5$
   $x = -2$ or $x = -12$
7. $|10 - x| = 2$
   $x = 8$ or $x = 12$
8. $|-2 + x| = 2$
   $x = 0$ or $x = 4$
9. $|3x + 1| = 13$
   $x = 4$ or $x = -4\frac{2}{3}$
10. $|2x - 2| = 4$
    $x = 3$ or $x = -1$
11. $|4x - 1| = 7$
    $x = 2$ or $x = -1\frac{1}{2}$
12. $|3x + 6| = 15$
    $x = 3$ or $x = -7$
13. $|-3 - 4x| = 1$
    $x = -1$ or $x = -\frac{1}{2}$
14. $|3 - 7x| = 11$
    $x = 2$ or $x = -1\frac{1}{7}$
15. $|6 + 5x| = 16$
    $x = 2$ or $x = -4\frac{2}{5}$
16. $|-5 - 2x| = 13$
    $x = 4$ or $x = -9$
17. $|4x - 14| = 0$
    $x = 3\frac{1}{2}$
18. $|9x - 9| = 9$
    $x = 2$ or $x = 0$
19. $|5 - 2x| = 4$
    $x = \frac{1}{2}$ or $x = 4\frac{1}{2}$
20. $|-5 - 6x| = 7$
    $x = -2$ or $x = \frac{1}{3}$

Solve each inequality.

21. $|x + 2| > 1$
    $x > -1$ or $x < -3$
22. $|x - 1| < 2$
    $-1 < x < 3$
23. $|x - 7| \geq 4$
    $x \geq 11$ or $x \leq 3$
24. $|x + 5| \leq 4$
    $-9 \leq x \leq -1$
25. $|x - 3| < 5$
    $-2 < x < 8$
26. $|x - 3| \leq 9$
    $-6 \leq x \leq 12$
27. $|x - 2| \leq 0$
    $x = 2$
28. $|x + 6| > 2$
    $x > -4$ or $x < -8$
29. $|5 - 3x| \geq 9$
    $x \leq -1\frac{1}{3}$ or $x \geq 4\frac{2}{3}$
30. $|x| < 4$
    $-4 < x < 4$
31. $|x| \geq 4$
    $x \geq 4$ or $x \leq -4$
32. $|x| > 0$
    $x > 0$ or $x < 0$
33. $|x - 5| < 3$
    $2 < x < 8$
34. $|x + 4| < 7$
    $-11 < x < 3$
35. $|x - 2| \geq -4$
    $-2 \leq x \leq 6$
36. $|2x - 8| > 6$
    $x > 7$ or $x < 1$
37. $|3x + 3| \leq 9$
    $-4 \leq x \leq 2$
38. $|4x - 8| \geq 20$
    $x \geq 7$ or $x \leq -3$
39. $|7x - 14| < 7$
    $1 < x < 3$
40. $|x + 2| < 5$
    $-7 < x < 3$

CHAPTER 7

LESSON 7.1

Solve by graphing. If necessary, round answers to the nearest tenth. Check by substituting the approximate solutions into the original equations.

1. $\begin{cases} 3x + 4y = 2 \\ x - y = 1 \end{cases}$ (0.9, −0.1)
2. $\begin{cases} 2y = 5x - 1 \\ x = y + 2 \end{cases}$ (−1, −3)
3. $\begin{cases} -3y = 2x - 6 \\ x + 4y = 8 \end{cases}$ (0, 2)
4. $\begin{cases} x = 300 - y \\ x - 200 = y \end{cases}$ (250, 50)
5. $\begin{cases} 4x + 3y = 18 \\ 5x - 6y = 15 \end{cases}$ (3.9, 0.8)
6. $\begin{cases} x = 5 \\ 3y = 2x - 1 \end{cases}$ (5, 3)
7. $\begin{cases} y = 4 \\ 2x - y = 1 \end{cases}$ (2.5, 4)
8. $\begin{cases} 4x - 2y = 8 \\ 3y = x + 9 \end{cases}$ (4.2, 4.4)
9. $\begin{cases} y - 5x = 10 \\ x - 5y = 10 \end{cases}$ (−2.5, −2.5)
10. $\begin{cases} x = -4 \\ y = -2 \end{cases}$ (−4, −2)
11. $\begin{cases} 15x + y = 50 \\ y = 35x - 40 \end{cases}$ (1.8, 23)
12. $\begin{cases} 7x - 5y = 1 \\ 3x + 3y = 2 \end{cases}$ (0.4, 0.4)

Determine whether the point (3, 8) is a solution to each system of equations.

13. $\begin{cases} 2x + y = 14 \\ x + y = 11 \end{cases}$ yes
14. $\begin{cases} y = -x - 5 \\ y = x + 5 \end{cases}$ no
15. $\begin{cases} 4x - y = -4 \\ 3x - 2y = 7 \end{cases}$ no

LESSON 7.2

Solve by substitution. Check your answers.

1. $\begin{cases} 2x = y - 1 \\ y = 3x \end{cases}$ (1, 3)
2. $\begin{cases} x = -y \\ 2x + y = -2 \end{cases}$ (−2, 2)
3. $\begin{cases} 2x + 2y = 4 \\ x = 10 - 3y \end{cases}$ (−2, 4)
4. $\begin{cases} y = 7 - x \\ 2x + 3y = -1 \end{cases}$ (22, −15)
5. $\begin{cases} x + y = -1 \\ 2x + y = 3 \end{cases}$ (4, −5)
6. $\begin{cases} 2x + 3y = -8 \\ y = 9 - x \end{cases}$ (35, −26)
7. $\begin{cases} y + 3x = 1 \\ 2y + 5x = 5 \end{cases}$ (−3, 10)
8. $\begin{cases} 5 = 2y + x \\ x = 20y \end{cases}$ $(4\frac{6}{11}, 1\frac{5}{22})$
9. $\begin{cases} 4x + 4y = 1 \\ -x = 2y \end{cases}$ (0.5, −0.25)
10. $\begin{cases} x + 4y = 5 \\ 8y - x = -8 \end{cases}$ (6, −0.25)
11. $\begin{cases} 2y + x = 4 \\ 4y = x - 1 \end{cases}$ (3, 0.5)
12. $\begin{cases} x = y \\ 3x + 6y = 6 \end{cases}$ $\left(\frac{2}{3}, \frac{2}{3}\right)$
13. $\begin{cases} x = -y \\ 16x + 4y = 9 \end{cases}$ (0.75, −0.75)
14. $\begin{cases} 5x - y = 3 \\ 10x = y \end{cases}$ (−0.6, −6)
15. $\begin{cases} 5x + y = 3 \\ 2y = 5x \end{cases}$ (0.4, 1)
16. $\begin{cases} x + y = 8 \\ 2x - y = 7 \end{cases}$ (5, 3)
17. $\begin{cases} 2x + y = -5 \\ x + 2y = -2 \end{cases}$ $(-2\frac{2}{3}, \frac{1}{3})$
18. $\begin{cases} 2x + 2y = 8 \\ y - x = 24 \end{cases}$ (−10, 14)

LESSON 7.3

Solve each system of equations by elimination, and check your solution.

1. $\begin{cases} x - y = 9 \\ x + y = 7 \end{cases}$ (8, 1)

2. $\begin{cases} x + 3y = 7 \\ 2x - 3y = -4 \end{cases}$ (1, 2)

3. $\begin{cases} 5x + 4y = 12 \\ 3x - 4y = 4 \end{cases}$ (2, 0.5)

4. $\begin{cases} x + y = 1 \\ x - 2y = 2 \end{cases}$ $\left(1\frac{1}{3}, -\frac{1}{3}\right)$

5. $\begin{cases} 3x + 3y = 6 \\ 2x - y = 1 \end{cases}$ (1, 1)

6. $\begin{cases} x + 8y = 3 \\ 4x - 2y = 7 \end{cases}$ $\left(1\frac{14}{17}, \frac{5}{34}\right)$

7. $\begin{cases} 4x - y = 4 \\ x + 2y = 3 \end{cases}$ $\left(1\frac{2}{9}, \frac{8}{9}\right)$

8. $\begin{cases} 3x - 5y = -13 \\ 4x + 3y = 2 \end{cases}$ (−1, 2)

9. $\begin{cases} 2x + 3y = 8 \\ 3x + 2y = 17 \end{cases}$ (7, −2)

10. $\begin{cases} x - y = 1 \\ 3x - y = 3 \end{cases}$ (1, 0)

11. $\begin{cases} y = 3x - 1 \\ 3x + 4y = 16 \end{cases}$ $\left(1\frac{1}{3}, 3\right)$

12. $\begin{cases} 4x + 2y = -8 \\ x = 2y - 7 \end{cases}$ (−3, 2)

13. $\begin{cases} x + y = 15 \\ \frac{1}{6}x = \frac{1}{9}y \end{cases}$ (6, 9)

14. $\begin{cases} 3x + \frac{1}{3}y = 10 \\ 2x - 5 = \frac{1}{3}y \end{cases}$ (3, 3)

15. $\begin{cases} 2x = y + 36 \\ 3x = \frac{1}{2}y + 26 \end{cases}$ (4, −28)

16. $\begin{cases} y = -2x \\ 5x + 3x = 1 \end{cases}$ (−1, 2)

17. $\begin{cases} x + y = 6 \\ -2x + y = -3 \end{cases}$ (3, 3)

18. $\begin{cases} 0.2x - 0.3y = 0 \\ 0.4x - 0.2y = 0.2 \end{cases}$ (0.75, 0.5)

19. $\begin{cases} 3x + 7y = 5 \\ 3y = -7 - 2x \end{cases}$ (−12.8, 6.2)

20. $\begin{cases} 2x + y = 45 \\ 3x - y = 5 \end{cases}$ (10, 25)

21. $\begin{cases} 4x + 2y = -8 \\ \frac{1}{2}x - y = -\frac{7}{2} \end{cases}$ (−3, 2)

LESSON 7.4

Solve each system using any method.

1. $\begin{cases} x - 6y = 2 \\ 5x - y = 3 \end{cases}$ $\left(\frac{16}{29}, -\frac{7}{29}\right)$

2. $\begin{cases} x - 5y = -42 \\ x - y = -5 \end{cases}$ (4.25, 9.25)

3. $\begin{cases} 4x + 5y = 14 \\ -x - y = -10 \end{cases}$ (36, −26)

4. $\begin{cases} 2y - 3 = x \\ 2x + 2y = 7 \end{cases}$ $\left(1\frac{1}{3}, 2\frac{1}{6}\right)$

5. $\begin{cases} 3x = y - 4 \\ 2y - 6x = 12 \end{cases}$ no solution

6. $\begin{cases} 3x + y = 13 \\ 2x - y = 2 \end{cases}$ (3, 4)

7. $\begin{cases} 2x - 3y = 5 \\ x - 5y = 0 \end{cases}$ $\left(3\frac{4}{7}, \frac{5}{7}\right)$

8. $\begin{cases} y - 4x = 11 \\ 2y + x = 6 \end{cases}$ $\left(-1\frac{7}{9}, 3\frac{8}{9}\right)$

9. $\begin{cases} x + 2y = 3 \\ 5x - 3y = 2 \end{cases}$ (1, 1)

10. $\begin{cases} y = -3x \\ x = 38 + 6y \end{cases}$ (2, −6)

11. $\begin{cases} 2x + 4y = 8 \\ x + 2y = 4 \end{cases}$ all reals

12. $\begin{cases} 3x - 8y = 4 \\ 6x = 42 + 16y \end{cases}$ no solution

Determine which of the following systems are dependent, independent, or inconsistent.

13. $\begin{cases} x + 2y = 6 \\ x + 2y = 8 \end{cases}$ incons.

14. $\begin{cases} 2x + 3y = 4 \\ 8 - 6y = 4x \end{cases}$ dep.

15. $\begin{cases} y = 3x - 4 \\ 6x + 2y = -8 \end{cases}$ ind.

16. $\begin{cases} x + y = 3 \\ x + y = 4 \end{cases}$ incons

17. $\begin{cases} 9x - 2 = 4y \\ x = 6 - y \end{cases}$ ind.

18. $\begin{cases} 3x + 6 = 7y \\ x = 11 - 2y \end{cases}$ ind.

19. $\begin{cases} 8x + 3y = 48 \\ 4x = 24 - 1.5y \end{cases}$ dep.

20. $\begin{cases} 4x - 3y = 12 \\ 12x + 9y = -36 \end{cases}$ ind.

21. $\begin{cases} 2x + y = 1 \\ x = 1 - 0.5y \end{cases}$ incons.

Lesson 7.5

1.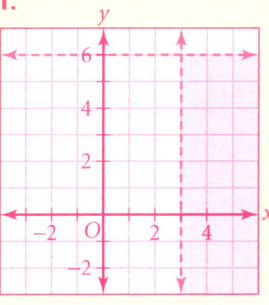

2.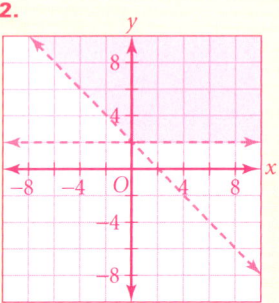

3.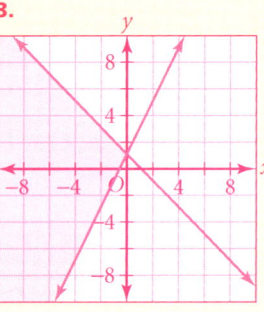

4.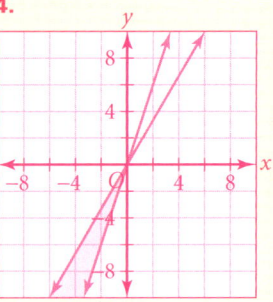

5.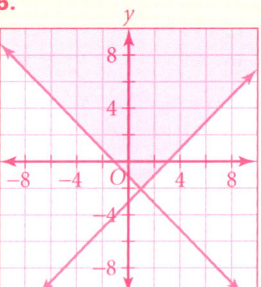

6.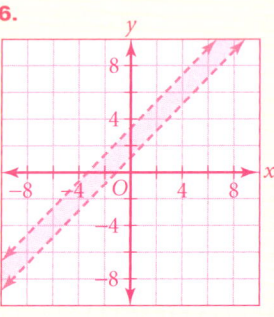

LESSON 7.5

Solve by graphing. Check your solution.

1. $\begin{cases} x > 3 \\ y < 6 \end{cases}$
2. $\begin{cases} y > 2 \\ y > -x + 2 \end{cases}$
3. $\begin{cases} y \geq 2x + 1 \\ y \leq -x + 1 \end{cases}$
4. $\begin{cases} y \geq 3x \\ 3y \leq 5x \end{cases}$
5. $\begin{cases} y + 3 \geq x \\ x + y \geq -1 \end{cases}$
6. $\begin{cases} y > x + 1 \\ y < x + 3 \end{cases}$
7. $\begin{cases} y - x < 1 \\ y - x > 3 \end{cases}$
8. $\begin{cases} 2y + x < 4 \\ 3x - y > 6 \end{cases}$
9. $\begin{cases} y + 2 < x \\ 2y - 3 > 2x \end{cases}$

Match the system of inequalities to the graph that represents the solution.

10. $\begin{cases} y \leq 3x + 6 \\ y > -\frac{1}{2}x + 1 \end{cases}$ **b**

11. $\begin{cases} y > 3x + 6 \\ y \leq -\frac{1}{2}x + 1 \end{cases}$ **a**

a.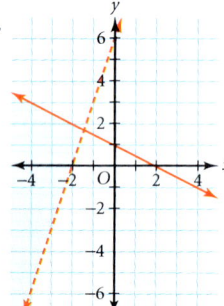

b.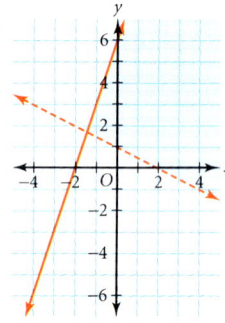

LESSON 7.6

Solve each problem.

1. Davida has 83 coins in nickels and dimes. She has a total of $6.95. How many of each coin does she have? **56 dimes, 27 nickels**

2. How many ounces of a 25% acid solution should be mixed with a 50% solution to produce 100 ounces of a 45% acid solution? **20 ounces of 25% solution; 80 ounces of 50% solution**

3. A plane leaves Chicago and heads for Columbus, which is 300 miles away. The plane, flying with the wind, takes 40 minutes. The plane returns to Chicago, traveling against the wind, in 45 minutes. Find the speed of the wind and the speed of the plane with no wind. **speed of wind: 25 mph; plane, no wind: 425 mph**

4. A two-digit number is equal to 7 times the units digit. If 18 is added to the number, its digits are reversed. Find the original two-digit number. **35**

5. Tonia is 25 years younger than her father. The sum of their ages is 75. How old is each now? **Tonia is 25, her father is 50**

6. How many ounces of a 50% saline solution should be mixed with a 10% solution to produce 100 ounces of a 30% saline solution? **50 ounces of each solution**

7. Angel has 27 coins in quarters and nickels. She has a total of $5.35. How many of each coin does she have? **20 quarters, 7 nickels**

7.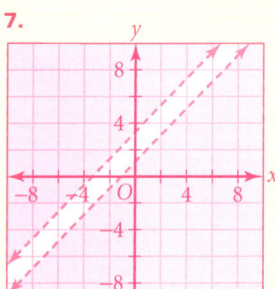

8.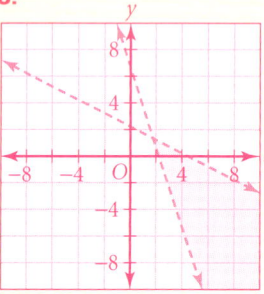

9.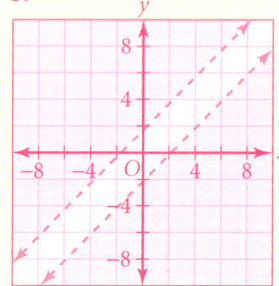

CHAPTER 8

LESSON 8.1

Find the value of each expression.

1. 4^3 64
2. 5^3 125
3. 8^4 4096
4. 10^5 100,000
5. 7^2 49
6. 6^6 46,656
7. 9^4 6561
8. 11^3 1331
9. 30^4 810,000
10. 45^2 2025
11. 100^2 10,000
12. 1^6 1
13. 2^9 512
14. 4^5 1024
15. 6^3 216
16. 10^3 1000

Simplify each product.

17. $10^2 \cdot 10^3$ 100,000
18. $20^2 \cdot 20^2$ 160,000
19. $3^3 \cdot 3^4$ 2187
20. $4^4 \cdot 4^2$ 4096
21. $5^3 \cdot 5^4$ 78,125
22. $6^2 \cdot 6^3$ 7776
23. $9^2 \cdot 9^2$ 6561
24. $12^2 \cdot 12^3$ 248,832
25. $7^2 \cdot 7$ 343
26. $(5a^2)(7ab^2)$ $35a^3b^2$
27. $(4d^3e^5)(8d^4e^6)$ $32d^7e^{11}$
28. $(-12s^6t)(5st^2)$ $-60s^7t^3$
29. $(8q^3p^9)(-2q^5p^8)$ $-16q^8p^{17}$
30. $(9a^3b^4)(5a^2b^2c^2)$ $45a^5b^6c^2$
31. $(4w^4x^8y)(-4w^6xy^{10})$ $-16w^{10}x^9y^{11}$
32. $(6h^4g^9)(-12h^6g^8)$ $-72h^{10}g^{17}$
33. $(8c^2d^5)(4c^2d^3e^4)$ $32c^4d^8e^4$
34. $(-j^3k^7l^5)(-j^6k^2l^{11})$ $j^9k^9l^{16}$

LESSON 8.2

Simplify each expression.

1. $(4x)^3$ $64x^3$
2. $(9^2)^3$ 531,444
3. $(-x^3)^2$ x^6
4. $(3^2)^4$ 6561
5. $(7x^2)^4$ $2401x^8$
6. $(-5g^2)^5$ $-3125g^{10}$
7. $(-3r^2)^6$ $729r^{12}$
8. $(-6a^8)^2$ $36a^{16}$
9. $(-5wr2)^6$ $15,625w^6r^{12}$
10. $(30t^2x^2)^3$ $27,000t^6x^6$
11. $(-4z^4a^3)^5$ $-1024z^{20}a^{15}$
12. $(-20m^2n^2)^4$ $160,000m^8n^8$
13. $(2r^2s^2)^4$ $16r^8s^8$
14. $(x^1y^3)^3$ x^3y^9
15. $(-3c^5d^4)^2$ $9c^{10}d^8$
16. $(r^3t^5)^5$ $r^{15}t^{25}$
17. $(ab^3c^4)^7$ $a^7b^{21}c^{28}$
18. $(5g^7h^1k)^5$ $3125g^{35}h^5k^5$
19. $(-6y^3z^5)^3$ $-216y^9z^{15}$
20. $(w^2z^3)^9$ $w^{18}z^{27}$
21. $(-w^3z^2)^6$ $w^{18}z^{12}$
22. $(4n^6n^2)^3$ $64n^{24}$
23. $(j^4k^6)^5$ $j^{20}k^{30}$
24. $(-7x^3y^3)^2$ $49x^6y^6$
25. $(-c^3v^5)(c^2)^6$ $-c^{15}v^5$
26. $(7g^2h^5)^3(8a^2)^2$ $21,952a^4g^6h^{15}$
27. $(-a^3)^2(a^3b)^5$ $a^{21}b^5$
28. $(-x^3y^5)^4(-xy)^5$ $-x^{17}y^{25}$
29. $(b^3c^6)^5(b^3c^7d)^3$ $b^{24}c^{51}d^3$
30. $(5j^3k^3)^4(j^2k^2)^6$ $625j^{24}k^{24}$
31. $(8np^2)^3(n^2p^5)^6$ $512n^{15}p^{36}$
32. $(-ab^5)^3(2a^3b^4)^5$ $-32a^{18}b^{35}$
33. $(3a^2b)^2(5b^3)^4$ $5625a^4b^{14}$

LESSON 8.3

Simplify each expression. Assume that the conditions of the Quotient-of-Powers Property are met.

1. $\dfrac{10^9}{10^4}$ 100,000
2. $\dfrac{5^8}{5^3}$ 3125
3. $\dfrac{7^{13}}{7^{10}}$ 343
4. $\dfrac{12^7}{12^5}$ 144
5. $\dfrac{f^6}{f^3}$ f^3
6. $\dfrac{h^{6a}}{h^{2a}}$ h^{4a}
7. $\dfrac{v^x}{v^y}$ v^{x-y}
8. $\dfrac{t^{x+3}}{t^x}$ t^3
9. $\dfrac{k^{x+7}}{k^{x+2}}$ k^5
10. $\dfrac{30k^7}{6k^4}$ $5k^3$
11. $\dfrac{4^3 h^3 r^5}{4^2 h^2 r^2}$ $4hr^3$
12. $\dfrac{15w^3 s^6}{5ws}$ $3w^2 s^5$
13. $\left(\dfrac{s}{t}\right)^3$ $\dfrac{s^3}{t^3}$
14. $\left(\dfrac{12}{d}\right)^4$ $\dfrac{20{,}736}{d^4}$
15. $\left(\dfrac{11}{x^3}\right)^5$ $\dfrac{161{,}051}{x^{15}}$
16. $\left(\dfrac{2c}{s^8}\right)^7$ $\dfrac{128c^7}{s^{56}}$
17. $\left(\dfrac{h^4}{g^6}\right)^3$ $\dfrac{h^{32}}{g^{48}}$
18. $\left(\dfrac{12c^6}{c^2 d^5}\right)^2$ $\dfrac{144c^8}{d^{10}}$
19. $\left(\dfrac{2z^5}{zy^3}\right)^8$ $\dfrac{256z^{32}}{y^{24}}$
20. $\left(\dfrac{w^4 s^7}{w^2 s^3}\right)^{10}$ $w^{20} s^{40}$
21. $\left(\dfrac{-4q^3 r^5}{12q^5}\right)^3$ $\dfrac{-r^{15}}{27q^6}$
22. $\left(\dfrac{x^3 z^2}{x^5 z}\right)^3$ $\dfrac{z^3}{x^6}$
23. $\left(\dfrac{-j^4 k^5 l^3}{j^5 k^2 l^6}\right)^4$ $\dfrac{k^{12}}{j^4 l^{12}}$
24. $\left(\dfrac{4p^5 f^3}{t^3 h^7}\right)^4$ $\dfrac{256 p^{20} f^{12}}{t^{12} h^{28}}$
25. $\dfrac{0.45 a^6 y^8}{0.9 a^2 y^5}$ $0.5 a^4 y^3$
26. $\dfrac{16 f^5 g^6}{4 f^3 (g^2 h)^3}$ $\dfrac{4f^2}{h^3}$
27. $\dfrac{(a^3 g^6 y^9)^8}{a^2 g^3 y^9}$ $a^{22} g^{45} y^{63}$
28. $\dfrac{0.18 s^5 z^{10}}{0.03 s^2 z}$ $6 s^3 z^9$

LESSON 8.4

Evaluate each expression.

1. 4^{-2} $\dfrac{1}{16}$
2. -4^2 16
3. -4^2 -16
4. -4^{-2} $-\dfrac{1}{16}$
5. $(-4)^{-2}$ $\dfrac{1}{16}$
6. 2^{-4} $\dfrac{1}{16}$
7. 10^{-3} $\dfrac{1}{1000}$
8. $(-10)^3$ -1000
9. -6^2 -36
10. $(-6)^2$ 36
11. 5^{-2} $\dfrac{1}{25}$
12. -5^2 -25

Write each expression without negative or zero exponents.

13. $r^2 s^{-3}$ $\dfrac{r^2}{s^3}$
14. $x^{-2} y^4$ $\dfrac{y^4}{x^2}$
15. $c^0 d^4$ d^4
16. $r^{-3} t^5$ $\dfrac{t^5}{r^3}$
17. $ab^0 c^{-4}$ $\dfrac{a}{c^4}$
18. $g^{-7} h^{-1} k$ $\dfrac{k}{g^7 h}$
19. $-y^3 z^{-5}$ $-\dfrac{y^3}{z^5}$
20. $(w^2 z^{-3})^0$ 1
21. $-w^{-3} w^{-2}$ $-\dfrac{1}{w^5}$
22. $n^6 n^{-2}$ n^4
23. $j^4 j^{-6}$ $\dfrac{1}{j^2}$
24. $-x^3 x^{-3}$ -1
25. $\dfrac{n^4}{n^{-2}}$ n^6
26. $\dfrac{t^{-5}}{t^2}$ $\dfrac{1}{t^7}$
27. $\dfrac{g^{-4}}{g^{-3}}$ $\dfrac{r^4}{s^6}$
28. $\dfrac{rs^{-6}}{r^{-3}}$ $\dfrac{r^4}{s^6}$
29. $\dfrac{4d^{-5}}{d^4}$ $\dfrac{4}{d^9}$
30. $\dfrac{-3a^{-3}}{2a^2 b}$ $\dfrac{-3}{2a^5 b}$
31. $\dfrac{2r^4}{r^{-3} s}$ $\dfrac{2r^7}{s}$
32. $\dfrac{10 w^{-3}}{w^2 v^{-2}}$ $\dfrac{10 v^2}{w^5}$
33. $\dfrac{(3a^2)(5a^3)}{a^{-1}}$ $15a^6$
34. $\dfrac{(2n^4)(10n^2)}{5n^{-3}}$ $4n^9$
35. $\dfrac{(-4b^{-2})(-2b^4)}{2b}$ $4b$
36. $\dfrac{(6d^{-5})(-3d^{-4})}{2d^{-3}}$ $-\dfrac{9}{d^6}$
37. $\dfrac{c^{-3} c^2}{c^{-4} c^3}$ 1
38. $\dfrac{a^{-3} a^{-6}}{b^{-5} b^{-2}}$ $\dfrac{b^7}{a^9}$
39. $\dfrac{-t^{-7} t^2}{r^{-5} r^2}$ $-\dfrac{r^3}{t^5}$
40. $\dfrac{-w^2 w^{-5}}{-2w^3}$ $\dfrac{1}{2w^6}$

LESSON 8.5

Write each number in scientific notation.

1. 4,000,000 4×10^6
2. 7,000,000,000 7×10^9
3. 2,500,000 2.5×10^6
4. 730,000 7.3×10^5
5. 0.000044 4.4×10^{-5}
6. 0.0028 2.8×10^{-3}
7. 0.00056 5.6×10^{-4}
8. 0.00000021 2.1×10^{-7}
9. 0.000345 3.45×10^{-4}
10. 0.0056 5.6×10^{-3}
11. 795,000 7.95×10^5
12. 8,430,000 8.43×10^6

Write each number in decimal notation.

13. 4×10^3 4000
14. 8×10^5 800,000
15. 7×10^7 70,000,000
16. 8.2×10^2 820
17. 1.7×10^{-3} 0.0017
18. 6.65×10^{-9} 0.00000000665
19. -9.7×10^5 $-970,000$
20. -8.033×10^{-7} -0.0000008033

Simplify. Write your answers in scientific notation.

21. $(3 \times 10^3)(4 \times 10^6)$ 1.2×10^{10}
22. $(7 \times 10^2)(1 \times 10^8)$ 7×10^{10}
23. $(7 \times 10^5)(2 \times 10^{11})$ 1.4×10^{17}
24. $(8 \times 10^8)(9 \times 10^3)$ 7.2×10^{12}
25. $(6 \times 10^4) + (4 \times 10^4)$ 10^5
26. $(8 \times 10^9) + (9 \times 10^9)$ 1.7×10^{10}
27. $(7 \times 10^5) - (6 \times 10^5)$ 1×10^5
28. $(8 \times 10^2) - (3 \times 10^2)$ 5×10^2
29. $(2.8 \times 10^6)(1.4 \times 10^3)$ 3.9×10^9
30. $(3.1 \times 10^4)(5.5 \times 10^4)$ 1.7×10^9
31. $(5.8 \times 10^6)(6.2 \times 10^{10})$ 3.6×10^{17}
32. $(3 \times 10^2)(4.9 \times 10^{12})$ 1×10^{15}
33. $\frac{8 \times 10^4}{4 \times 10^2}$ 2×10^2
34. $\frac{6 \times 10^8}{2 \times 10^3}$ 3×10^5
35. $\frac{7 \times 10^{12}}{2 \times 10^5}$ 3.5×10^7
36. $\frac{(3 \times 10^5)(6 \times 10^2)}{9 \times 10^4}$ 2×10^3
37. $\frac{(5 \times 10^{10})(6 \times 10^2)}{5 \times 10^9}$ 6×10^3
38. $\frac{(4 \times 10^3)(6 \times 10^8)}{8 \times 10^5}$ 3×10^6

LESSON 8.6

Graph each of the following.

1. $y = 3^x$
2. $y = 6^x$
3. $y = 8^x$
4. $y = 0.2^x$

Graph each of the following. In each case describe the effect of the 2 or 4 on the graph of the parent function $y = 2^x$.

5. $y = 2^x + 2$
6. $y = 2 \cdot 2^x$
7. $y = 2^{x+4}$
8. $y = 2^{4x}$

Let $f(x) = 3^x$. Evaluate.

9. $f(2)$ 9
10. $f(0)$ 1
11. $f(1)$ 3
12. $f(3)$ 27
13. $f(-2)$ $\frac{1}{9}$
14. $f(-3)$ $\frac{1}{27}$
15. $f(-1)$ $\frac{1}{3}$
16. $f(-4)$ $\frac{1}{81}$
17. $f(4)$ 81

The population of a city was about 300,000 in 1994 and was growing at a rate of about 0.4% per year.

18. What multiplier should be used to find the new population each year? 1.004
19. Use the information from Exercise 18 to estimate the population of the city in the year 2000. 307,272
20. Use the information from Exercise 18 to estimate the population of the city in the year 2005. 313,467

Lesson 8.6

1.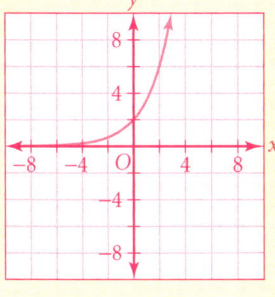
2.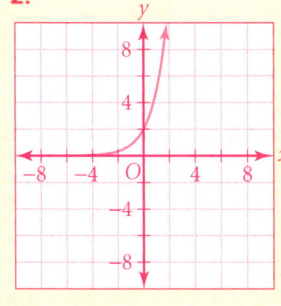
3.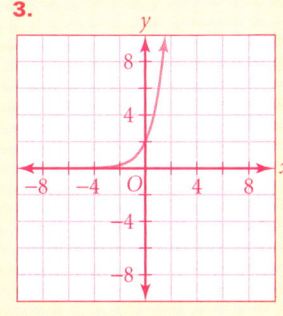
4.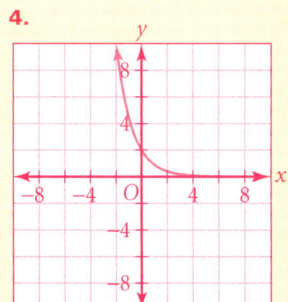

5. a shift 2 units up
6. a vertical stretch by a factor of 2
7. a shift 4 units to the left
8. a horizontal compression by a factor of $\frac{1}{4}$

LESSON 8.7

Use the method of carbon–14 dating to estimate the age of an object which has

1. 30% of its original carbon–14 remaining. **about 12,000 years**
2. 2% of its original carbon–14 remaining. **about 18,000 years**
3. 50% of its original carbon–14 remaining. **about 33,000 years**
4. 15% of its original carbon–14 remaining. **about 2900 years**
5. 80% of its original carbon–14 remaining. **about 5700 years**
6. 90% of its original carbon–14 remaining. **about 500 years**

An investment is growing at a rate of 7% per year and now has a value of $7600. Find the value of the investment

7. in 2 years. **$8701.24**
8. in 5 years. **$10,659.39**
9. in 8 years. **$13,058.21**
10. in 10 years **$14,950.35**
11. 2 years ago. **$6,638.13**
12. 3 years ago. **$6,203.86**
13. 5 years ago. **$5,418.69**
14. 8 years ago. **$4,423.27**

An investment is losing money a rate of 1% per year and now has a value of $48,000. Find the value of the investment

15. in 2 years. **$47,044.80**
16. in 5 years. **$45,647.52**
17. in 8 years. **$44,291.75**
18. in 10 years. **$43,410.34**
19. 2 years ago. **$48,974.59**
20. 3 years ago. **$49,469.29**
21. 5 years ago. **$50,473.71**
22. 8 years ago. **$52,018.72**

The population of a city in the United States was 164,998 in 1990 and was decreasing at a rate of about 1.2% per year. In how many years will the city's population fall below

23. 160,000? **≈ 3 years**
24. 150,000? **≈ 8 years**
25. 130,000? **≈ 20 years**
26. 125,000? **≈ 23 years**
27. 110,000? **≈ 34 years**
28. 100,000? **≈ 42 years**
29. 90,000? **≈ 51 years**
30. 50,000? **≈ 99 years**

CHAPTER 9

LESSON 9.1

Rewrite each polynomial in standard form.

1. $7 + x + x^2$ $x^2 + x + 7$
2. $9x^3 - 4x - 2 + x^2$ $9x^3 + x^2 - 4x - 2$
3. $8 - 3w^2 + w$ $-3w^2 + w + 8$
4. $y - 4y^4 + 3 - 2y^2$ $-4y^4 - 2y^2 + 3$
5. $r - r^3$ $-r^3 + r$
6. $5w^3 - 7w^4 + 5w - 5$ $-7w^4 + 5w^3 + 5w - 5$

Identify the degree of each polynomial.

7. $6t^2 - t + 1$ **2**
8. $3x - 2$ **1**
9. $m + m^3$ **3**
10. $p^5 - p^2 + p$ **5**

Use vertical form to add or subtract.

11. $(4x^4 + 3x^2 - 3) + (6x^4 + 2x^2 + 5)$ $10x^4 + 5x^2 + 2$
12. $(7w^3 - 4w + 3) + (5w^3 - 3w - 3)$ $12w^3 - 7w$
13. $(8z^4 - 4z^2 - z + 4) + (z^2 - z)$ $8z^4 - 3z^2 - 2z + 4$
14. $(x^2 + 3x + 4) + (3x + 7)$ $x^2 + 6x + 11$
15. $(2x^3 + x^2 - 1) - (x^3 - x^2)$ $x^3 + 2x^2 - 1$
16. $(3w^3 - 5w + 4) - (6w + 2)$ $3w^3 - 11w + 2$
17. $(9x^4 - 5x^2 + 7) - (8x^4 + 2)$ $x^4 - 5x^2 + 5$
18. $(3z^5 + z^3 - 8) - (z^5 - z^3 + 8)$ $2z^3 - 5z^2 - 4z + 4$

Use horizontal form to add or subtract.

19. $(w^4 + 3w^2 - 5) + (w^2 - 5w)$ $w^4 + 4w^2 - 5w - 5$
20. $(3z^3 - 4z + 5) + (8z^3 - 5z^2 - 1)$ $11z^3 - 5z^2 - 4z + 4$
21. $(9x + 3) + (4x^2 - 5x + 2)$ $-x^2 + 9x + 3$
22. $y^3 - y^2 - 1 - (4y + 3)$ $y^3 - y^2 - 4y - 4$
23. $4x^2 - 6x + 4 - (2x^2 + 4x - 6)$ $2x^2 - 10x + 10$
24. $7x^2 + 4x - (x^2 - 7)$ $6x^2 + 4x + 7$

Simplify. Express all answers in standard form.

25. $(2 - x - x^4) + (x + 4.5x^4 + 5)$ $3.5x^4 + 7$
26. $6 + 4x - 3.6x^2 - (10.6x - 65 + 6.8x^2)$ $-10.4x^2 - 6.6x + 71$

LESSON 9.2

Use the Distributive Property to find each product.

1. $5(x + 4)$ $5x + 20$
2. $6(2x + 6)$ $12x + 36$
3. $7(x + 11)$ $7x + 77$
4. $4(x - 7)$ $4x - 28$
5. $3(2x - 6)$ $6x - 18$
6. $-4(x + 2)$ $-4x - 8$
7. $-3(x - 2)$ $-3x + 6$
8. $5(3x - 4)$ $15x - 20$
9. $x(2 - x)$ $2x - x^2$
10. $x(3 + 3x)$ $3x + 3x^2$
11. $-x(4 + x)$ $-4x - x^2$
12. $-x(x - 4)$ $-x^2 + 4x$
13. $5x(x + 3)$ $5x^2 + 15x$
14. $2x(4 - x)$ $8x - 2x^2$
15. $x(-x + 6)$ $-x^2 + 6x$
16. $-2x(x - 4)$ $-2x^2 + 8x$

Find each product by using the rules for special products.

17. $(x + 4)(x + 4)$ $x^2 + 8x + 16$
18. $(2x + 6)(2x - 6)$ $4x^2 - 36$
19. $(x - 11)(x + 11)$ $x^2 - 121$
20. $(x - 7)(x - 7)$ $x^2 - 14x + 49$
21. $(2x - 6)(2x - 6)$ $4x^2 - 24x + 36$
22. $(4x + 5)(4x - 5)$ $16x^2 - 25$
23. $(x - 6)(x + 6)$ $x^2 - 36$
24. $(3x - 4)(3x + 4)$ $9x^2 - 16$
25. $(2 + x)(2 - x)$ $4 - x^2$
26. $(3 + 5x)(3 + 5x)$ $9 + 30x + 25x^2$
27. $(8 - x)(8 - x)$ $64 - 16x + x^2$
28. $(x - 4)(x - 4)$ $x^2 - 8x + 16$
29. $(7x + 3)(7x + 3)$ $49x^2 + 42x + 9$
30. $(4 - x)(4 + x)$ $16 - x^2$
31. $(-5x + 6)(-5x + 6)$ $25x^2 - 60x + 36$
32. $(2x - 9)(2x - 9)$ $4x^2 - 36x + 81$

LESSON 9.3

Use the Distributive Property to find each product.

1. $(x + 2)(x + 3)$ $x^2 + 5x + 6$
2. $(z - 5)(z - 1)$ $z^2 - 6z + 5$
3. $(a + 2)(a - 4)$ $a^2 - 2a + 8$
4. $(b - 2)(b - 3)$ $b^2 - 5b + 6$
5. $(d + 3)(d + 4)$ $d^2 + 7d + 12$
6. $(y - 2)(y + 5)$ $y^2 + 3y - 10$
7. $(c + 6)(c + 6)$ $c^2 + 12c - 36$
8. $(w + 4)(w - 3)$ $w^2 + w - 12$
9. $(m - 2)(m - 4)$ $m^2 - 6m + 8$
10. $(a + 5)(a - 6)$ $a^2 - a - 30$
11. $(s + 4)(s - 3)$ $s^2 + s - 12$
12. $(x - 6)(x + 1)$ $x^2 - 5x - 6$
13. $(s + 7)(s - 2)$ $s^2 + 5s - 14$
14. $(b - 9)(b + 3)$ $b^2 - 6b - 27$
15. $(w + 6)(w - 1)$ $w^2 + 5w - 6$
16. $(q + 5)(q - 3)$ $q^2 + 2q - 15$
17. $(w + 8)(w - 8)$ $w^2 - 64$
18. $(r - 9)(r - 2)$ $r^2 - 11r + 18$
19. $(m + 10)(m - 11)$ $m^2 - m - 100$
20. $(n - 7)(n + 9)$ $n^2 - 2n - 63$
21. $(a - 1)(a - 10)$ $a^2 - 11a + 10$

Use the FOIL method to find each product.

22. $(x + 3)(x + 6)$ $x^2 + 9x + 18$
23. $(w + 10)(w + 2)$ $w^2 + 12w + 20$
24. $(c - 5)(c - 5)$ $c^2 + 10c + 25$
25. $(y - 3)(y - 9)$ $y^2 + 12y + 27$
26. $(r - 4)(r + 7)$ $r^2 + 3r - 28$
27. $(w + 8)(w - 3)$ $w + 5w - 24$
28. $(3z - 1)(2z + 1)$ $6z^2 + z - 1$
29. $(4x + 3)(x - 2)$ $4x^2 - 5x - 6$
30. $(2t - 1)(2t + 3)$ $4t^2 + 4t - 3$
31. $(5x - 2)(3x + 2)$ $15x^2 + 4x - 4$
32. $(8s - 1)(s + 4)$ $8s^2 + 31x - 4$
33. $(4w - 5)(3w - 1)$ $12w^2 - 19w + 5$
34. $(a + b)(a + b)$ $a^2 + 2ab + b^2$
35. $(r + s)(r - s)$ $r^2 - s^2$
36. $(2x - z)(x - z)$ $2x^2 - 3xz + z^2$
37. $(a + 2b)(a - b)$ $a^2 + ab - 2b^2$
38. $(y + 3x)(2y - x)$ $2y^2 + 5xy - 3x^2$
39. $(p + q)^2$ $p^2 + 2pq + q^2$
40. $(c^2 + d)^2$ $c^4 + 2c^2 d + d^2$
41. $(m^2 - n)(m^2 + n)$ $m^4 - n^2$
42. $(3c + 4d)(2c - 4d)$ $6c^2 - 4cd - 16d^2$

Extra Practice

763

LESSON 9.4

Write surface area and volume functions for each geometric solid.

1. A cube with edge length x. $S = 6x^2$, $V = x^3$
2. A rectangular solid with a 5 inch by 6 inch base and height x. $S = 22x + 60$; $V = 30x$
3. A rectangular solid with a 4 cm by 9 cm base and height x. $S = 24x + 72$; $V = 36x$
4. A cylinder with a height of 6 inches and a radius x. $S = \pi x^2 + 12\pi x$; $V = 6\pi x^2$
5. A rectangular solid with a base length of 9.5 centimeters and base width of 4.2 centimeters and a height x. $S = 27.4x + 79.8$; $V = 39.9x$
6. A cube with edge length $5x$. $S = 150x^2$; $V = 125x^3$

Verify whether each equation is an identity or not by substituting −1, 0, and 1 for x.

7. $x^2 + 4x + 3 = (x + 1)(x + 3)$ T
8. $x^2 - 7x + 10 = (x - 5)(x - 2)$ T
9. $x^2 - 7x + 12 = (x - 4)(x - 3)$ T
10. $x^2 + 7x + 6 = (x + 6)(x + 1)$ T
11. $(x + 4)(x - 3) = x^2 + x - 12$ T
12. $(x - 4)(x - 5) = x^2 - 9x + 20$ T
13. $x^2 - 16 = (x + 4)(x - 4)$ T
14. $(x - 5)(x + 5) = x^2 - 25$ T
15. $(x + 4)(x - 6) = x^2 - 2x - 24$ T
16. $x^2 + 9x + 14 = (x + 7)(x + 2)$ T
17. $x^2 - 100 = (x - 10)(x - 10)$ F
18. $(x - 4)(x + 3) = x^2 - x + 12$ F
19. $(x - 2)^2 = x^2 - 4x + 4$ T
20. $x^2 + 4x + 4 = (x + 2)^2$ T
21. $(x + 2)^3 = x^3 + 4x^2 + 8x + 8$ F
22. $(x - 4)^2 = x^2 - 8x - 16$ F

LESSON 9.5

State whether each polynomial can be factored using the methods studied in Lesson 9.5.

1. $5x^2 - 15$ yes
2. $t^2 + 6$ no
3. $w^2 - 4$ yes
4. $4m^2 + 9$ no
5. $7t^2 - 8t$ yes
6. $25c - 25d$ yes
7. $16x^2 - 1$ yes
8. $w^3 - w^2$ yes
9. $3x - 2$ no
10. $8x^3 - 9$ no
11. $7x^2 + x$ yes
12. $16c - 8$ yes

Factor each polynomial by using the GCF.

13. $4x^2 + 16$ $4(x^2 + 4)$
14. $7w^2 + 21$ $7(w^2 + 3)$
15. $8a^2 - 4a$ $4a(2a - 1)$
16. $5n^3 - 10p^2 + 15$ $5(n^3 - 2p^2 + 3)$
17. $25d^2 - 75d^3$ $25d^2(1 - 3d)$
18. $6x^2y - 14xy$ $2xy(3x - 7)$
19. $3s^2t^3 + 15st^3$ $3st^2(s + 5)$
20. $4z^3w - 16w^3 + 24w$ $4w(z^3 - 4w^2 + 6)$
21. $25ab^4 + 20a^3b^2$ $5ab^2(5b^2 + 4a^2)$

Write each expression as the product of two binomials.

22. $x(x - 1) + 2(x - 1)$ $(x + 2)(x - 1)$
23. $7(y + 4) - x(y + 4)$ $(7 - x)(y + 4)$
24. $b(c + d) + a(c + d)$ $(b + a)(c + d)$
25. $4(w - 3) - v(w - 3)$ $(4 - v)(w - 3)$
26. $x(y - z) + 4(y - z)$ $(x + 4)(y - z)$
27. $s(x + 5) - t(x + 5)$ $(s - t)(x + 5)$
28. $q(4 - s) - 5(4 - s)$ $(q - 5)(4 - s)$
29. $3w(x + y) - 4(x + y)$ $(3w - 4)(x + y)$
30. $mn(p - q) + ab(p - q)$ $(mn + ab)(p - q)$

Factor by grouping.

31. $15a - 3ay - 4y + 20$ $(5 - y)(3a + 4)$
32. $xy + 5x + 3y + 15$ $(x + 3)(y + 5)$
33. $mr + 3m + 2r + 6$ $(m + 2)(r + 3)$
34. $ch + c - 4h - 4$ $(c - 4)(h + 1)$
35. $zr + 4z + 3r + 12$ $(z + 3)(r + 4)$
36. $mr + 3r - 2m - 6$ $(r - 2)(m + 3)$
37. $cl + 2c + 7l + 14$ $(c + 7)(l + 2)$
38. $kx - 3x + k - 3$ $(x + 1)(k - 3)$
39. $rs - 6r - 5s + 30$ $(r - 5)(s - 6)$
40. $kj + 3k - 4j - 12$ $(k - 4)(j + 3)$
41. $am - 6m - 2a + 12$ $(m - 2)(a - 6)$
42. $nx + 12n - x - 12$ $(n - 1)(x - 12)$

LESSON 9.6

Find each product.

1. $(x+2)^2$ x^2+4x+4
2. $(2x+1)^2$ $4x^2+4x+1$
3. $(m-4)^2$ $m^2-8m+16$
4. $(a-6)^2$ $a^2-12a+36$
5. $(4y-5)^2$ $16y^2-40y+25$
6. $(9q+1)^2$ $81q^2+18q+1$
7. $(x+3y)(x-3y)$ x^2-9y^2
8. $(7a-b)(7a+b)$ $49a^2-b^2$
9. $(5a-3)(5a+3)$ $25a^2-9$

Find the missing terms in each perfect-square trinomial.

10. $a^2-10a+\underline{\ ?\ }$ **25**
11. $25z^2+\underline{\ ?\ }+4$ **20z**
12. $9b^2-12b+\underline{\ ?\ }$ **4**
13. $\underline{\ ?\ }-4y+1$ **4y²**
14. $9r^2+30r+\underline{\ ?\ }$ **25**
15. $\underline{\ ?\ }+42x+49$ **9x²**

Factor each polynomial completely.

16. y^2-9 $(y+3)(y-3)$
17. $9t^2-1$ $(3t-1)(3t+1)$
18. z^2-144 $(z-12)(z+12)$
19. x^2-4x+4 $(x-2)^2$
20. $y^2+16y+64$ $(y+8)^2$
21. $25w^2-16$ $(5w-4)(5w+4)$
22. $1-81q^2$ $(1-9q)(1+9q)$
23. x^2-y^2 $(x+y)(x-y)$
24. $x^2+12x+36$ $(x+6)^2$
25. $q^2-20q+100$ $(q-10)^2$
26. $49m^2-14m+1$ $(7m-1)^2$
27. $25r^2+20r+4$ $(5r+2)^2$
28. $64t^2+16t+1$ $(8t+1)^2$
29. $25b^2-10b+1$ $(5b-1)^2$
30. $w^2+22w+121$ $(w+11)^2$
31. $2x^2-8$ $2(x+2)(x-2)$
32. $16-49c^2$ $(4-7c)(4+7c)$
33. $x^2-10x+25$ $(x-5)^2$
34. x^2-36y^2 $(x-6y)(x+6y)$
35. $x^2+2xy+y^2$ $(x+y)^2$
36. $49a^2-4b^2$ $(7a-2b)(7a+2b)$

LESSON 9.7

Write each trinomial in factored form.

1. $x^2-3x-10$ $(x+2)(x-5)$
2. $x^2-3x-28$ $(x-7)(x+4)$
3. $x^2+9x+20$ $(x+5)(x+4)$
4. $x^2-5x-24$ $(x-8)(x+3)$
5. $x^2+8x-20$ $(x+10)(x-2)$
6. $x^2-4x-45$ $(x-9)(x+5)$
7. x^2-9x+8 $(x-8)(x-1)$
8. $x^2-12x-45$ $(x-15)(x+3)$
9. $x^2+10x+16$ $(x+8)(x+2)$
10. $b^2+12b+27$ $(b+9)(b+3)$
11. $x^2-8x+15$ $(x-3)(x-5)$
12. z^2+2z-3 $(x+3)(z-1)$
13. $s^2+22s+21$ $(s+21)(s+1)$
14. n^2-n-20 $(n-5)(n+4)$
15. $h^2+15h+26$ $(h+13)(h+2)$
16. $q^2-10q+16$ $(q+8)(q+2)$
17. $x^2-9x+14$ $(x-7)(x-2)$
18. $c^2+13c-30$ $(c+15)(c-2)$

Write each polynomial in factored form. If it cannot be factored, write *prime*.

19. y^2-49 $(y+7)(y-7)$
20. $w^2-16w+64$ $(w-8)^2$
21. $z^2+14z+24$ $(z+21)(z+2)$
22. y^2-8y+7 $(y-7)(y-1)$
23. $28n^2+18n$ $2n(14n-9)$
24. $5a^2+20b^2$ $5(a^2+4b^2)$
25. $s^2-12a-13$ **prime**
26. r^2-r-72 $(r-9)(r+8)$
27. $by+4my+3xb+12mx$ $(b+4m)(y+3x)$
28. $3x^2-12$ $3(x+2)(x-2)$
29. t^2-t-65 **prime**
30. $m^2+10m-39$ $(m+13)(m-3)$
31. $y^2+7y+12$ $(y+3)(y+4)$
32. $a^2-4ac+ab-4bc$ $(a+b)(a-4c)$
33. $4y^2+2y-6$ $2(y-1)(2y+3)$
34. $4t^2-6t-40$ $2(t-4)(2t+5)$
35. $6n^2-24n+6$ $6(n^2-4n+1)$
36. $a^2+18a+51$ **prime**

Lesson 10.1

1. shift by 3 units to the right, shift up 1 unit
2. shift by 2 units to the left, shift 4 units down
3. shift by 1 unit to the left, shift 2 units down, vertical stretch by a factor of 2
4. shift by 3 units to the left
5. shift by 4 units to the right, opens downward
6. shift by 6 units to the right, vertical stretch by a factor of 3
7. shift by 1 unit to the right, shift 1 unit down, vertical stretch by a factor of 2
8. shift by 5 units to the right, shift by 2 units up, vertical stretch by a factor of 2, opens downward
9. shift by 2 units to the left, shift by 5 units down, vertical stretch by a factor of 5, opens downward
10. shift by 1 unit to the right, shift by 4 units up, vertical compression by a factor of $\frac{1}{2}$
11. shift by 9 units to the left, shift by 3 units down, vertical compression by a factor of $\frac{1}{3}$, opens downward

LESSON 9.8

Identify the zeros for each function.

1. $y = (x-5)(x+4)$ 5, −4
2. $y = (x+6)(x-1)$ −6, 1
3. $y = (x+11)(x-10)$ −11, 10
4. $y = (x-8)(x+3)$ 8, −3
5. $y = (x+2)(x+2)$ −2
6. $y = (x+12)(x-12)$ 12, −12
7. $y = (x+9)(x+6)$ −9, −6
8. $y = (x-16)(x+15)$ 16, −15
9. $y = (x-4)(x-4)$ 4
10. $y = (x-7)(x+1.6)$ 7, −1.6
11. $y = (x-2.9)(x+3.2)$ 2.9, −3.2
12. $y = (x-8)(x+8)$ 8, −8
13. $y = (x-5)(x-5)$ 5
14. $y = (x-3)(x+24)$ 3, −24
15. $y = (x-13)(x-13)$ 13
16. $y = x(x+3)$ 0, −3
17. $y = x(x-5)$ 0, 5
18. $y = (x+16)(x+16)$ −16

Solve by factoring.

19. $x^2 + 6x + 5 = 0$ −1, 5
20. $x^2 + 4x - 21 = 0$ 3, −7
21. $x^2 - 4 = 0$ 2, −2
22. $x^2 - 13x + 42 = 0$ 7, 6
23. $x^2 + 6x + 9 = 0$ −3
24. $x^2 - 36 = 0$ 6, −6
25. $x^2 - 64 = 0$ 8, −8
26. $x^2 - 121 = 0$ 11, −11
27. $x^2 + 23x + 132 = 0$ -11, −12
28. $x^2 + 10x = -9$ −9, −1
29. $x^2 + 10x + 16 = 0$ −8, −2
30. $x^2 + 10x + 21 = 0$ −7, −3
31. $x^2 + 10x + 25 = 0$ −5
32. $x^2 + 16x + 64 = 0$ −8
33. $x^2 + 20x = -100$ −10
34. $x^2 + 13x + 30 = 0$ −10, −3
35. $x^2 - 2x = -1$ 1
36. $x^2 - 8x + 16 = 0$ 4
37. $x^2 + 13x + 22 = 0$ −2, −11
38. $x^2 + 4x - 45 = 0$ −9, 5
39. $x^2 - 9x + 18 = 0$ 6, 3
40. $x^2 - 3x = 18$ 6, −3
41. $x^2 + 3x = 18$ −6, 3
42. $x^2 - 11x = -10$ 10, 1

CHAPTER 10

LESSON 10.1

Describe how the graph of each function differs from the graph of the parent function, $y = x^2$.

1. $y = (x-3)^2 + 1$
2. $y = (x+2)^2 - 4$
3. $y = 2(x+1)^2 - 2$
4. $y = (x+3)^2$
5. $y = -(x-4)^2$
6. $y = 3(x-6)^2$
7. $y = 4(x-1)^2 - 1$
8. $y = -2(x-5)^2 + 2$
9. $y = -5(x+2)^2 - 5$
10. $y = \frac{1}{2}(x-1)^2 + 4$
11. $y = -\frac{1}{3}(x+9)^2 - 3$
12. $y = \frac{2}{5}(x-4)^2 - 5$

Find the vertex and axis of symmetry for the graph of each function, and then sketch the graph.

13. $y = -2(x+1)^2 - 2$ (−1, −2), $x = -1$
14. $y = 3(x-2)^2 + 3$ (2, 3), $x = 2$
15. $y = (x-3)^2 + 4$ (3, 4), $x = 3$
16. $y = \frac{1}{2}(x+2)^2 + 1$ (−2, 1), $x = -2$
17. $y = -3(x+3)^2$ (−3, 0), $x = -3$
18. $y = -5(x+1)^2 + 4$ (−1, 4), $x = -1$

Find the zeros of each quadratic function.

19. $y = x^2 + 7x + 12$ −4, −3
20. $y = x^2 + 4x + 4$ −2
21. $y = x^2 - x - 30$ −5, 6
22. $y = x^2 + 9x + 8$ −8, −1
23. $y = x^2 + x - 20$ −5, 4
24. $y = x^2 + 2x - 15$ −5, 3
25. $y = x^2 - 2x + 1$ 1
26. $y = x^2 + 7x - 8$ −8, 1
27. $y = x^2 + 10x + 9$ −9, −1

First graph the function. Then find the zeros, the axis of symmetry, and the minimum value of the function.

28. $y = x^2 + 10x + 25$ −5; $x = -5$; (−5, 0)
29. $y = x^2 + 4x + 3$ −3, −1; $x = -2$; (−2, −1)
30. $y = x^2 - 2x - 8$ −2, 4; $x = 1$; (1, −9)
31. $y = x^2 + 8x + 15$ −5, −3; $x = -4$; (-4, −1)
32. $y = x^2 - 6x - 7$ −1, 7; $x = 3$; (3, −16)
33. $y = x^2 - 2x - 24$ −4, 6; $x = 1$; (1, −25)

12. shift by 4 units to the right, shift by 5 units down, vertical compression by a factor of $\frac{2}{5}$

13–18. Check students' graphs.

LESSON 10.2

Find each square root. Round your answers to the nearest hundredth when necessary.

1. $\sqrt{100}$ 10
2. $\sqrt{169}$ 13
3. $\sqrt{484}$ 2
4. $\sqrt{49}$ 7
5. $\sqrt{50}$ 7.07
6. $\sqrt{89}$ 9.43
7. $\sqrt{28}$ 5.29
8. $\sqrt{14}$ 3.47
9. $\sqrt{99}$ 9.95
10. $\sqrt{110}$ 10.49
11. $\sqrt{200}$ 14.14
12. $\sqrt{65}$ 8.06
13. $\sqrt{95}$ 9.75
14. $\sqrt{250}$ 15.81
15. $\sqrt{135}$ 11.62
16. $\sqrt{61}$ 7.81

Solve each equation. Round your answers to the nearest hundredth when necessary.

17. $x^2 = 36$ ±6
18. $x^2 = 121$ ±11
19. $x^2 = 64$ ±8
20. $x^2 = 400$ ±20
21. $x^2 = \frac{16}{25}$ ±$\frac{4}{5}$
22. $x^2 = \frac{64}{121}$ ±$\frac{8}{11}$
23. $(x+4)^2 - 16 = 0$ −8, 0
24. $(x+2)^2 - 1 = 0$ −3, −1
25. $(x-3)^2 - 36 = 0$ −3, 9
26. $(x+3)^2 = 64$ −11, 5
27. $(x-1)^2 = 12$ −2.46, 4.46
28. $(x+5)^2 = 35$ −10.92, 0.92

Find the vertex, axis of symmetry, and zeros of each function. Then sketch the graph.

29. $f(x) = (x-9)^2 - 16$
    (9, −16); x = 9; 5, 13
30. $g(x) = (x+3)^2 - 4$
    (−3 −4); x = −3; −5, −1
31. $h(x) = (x-4)^2 - 2$
    (4, −2); x=4; 2.9, 5.41

LESSON 10.3

Complete the square.

1. $x^2 + 12x$ + 36
2. $x^2 + 10x$ + 25
3. $x^2 + 4x$ + 4
4. $x^2 + 20x$ + 100
5. $x^2 - 30x$ + 225
6. $x^2 + 11x$ + 30.25
7. $x^2 + 30x$ + 225
8. $x^2 - 21x$ + 110.25

Find the minimum or maximum value for each quadratic function.

9. $f(x) = x^2 + 5$ (0, 5)
10. $f(x) = x^2 - 9$ (0, −9)
11. $f(x) = x^2 + 15$ (0, 15)
12. $f(x) = x^2 + 1$ (0, 1)
13. $f(x) = x^2 + 16$ (0, 16)
14. $f(x) = x^2 - 2$ (0, −2)

Rewrite each function in the form $y = (x - h)^2 + k$. Find each vertex.

15. $y = x^2 - 6x$
    $y = (x-3)^2 - 9$, (3, −9)
16. $y = x^2 - 12x$
    $y = (x-6)^2 - 36$, (6, −36)
17. $y = x^2 + 2x$
    $y = (x+1)^2 - 1$, (−1, −1)
18. $y = x^2 + 6x + 5$
    $y = (x+3)^2 - 4$, (−3, −4)
19. $y = x^2 - 2x - 3$
    $y = (x-1)^2 - 4$, (1, −4)
20. $y = x^2 - 8x + 12$
    $y = (x-4)^2 - 4$, (4, −4)
21. $y = x^2 - 10x + 21$
    $y = (x-5)^2 - 4$, (5, −4)
22. $y = x^2 - 12x + 32$
    $y = (x-6)^2 - 4$, (6, −4)
23. $y = x^2 + 5x + 4$
    $y = (x+2.5)^2 - 2.55$, (−2.5, −2.25)
24. $x^2 + \frac{1}{2}x - 2$
25. $x^2 - \frac{1}{4}x + 6$
26. $x^2 - \frac{1}{3}x + 5$

24. $y = \left(x + \frac{1}{4}\right)^2 - 2\frac{1}{16}$, $\left(-\frac{1}{4}, -2\frac{1}{16}\right)$

25. $y = \left(x - \frac{1}{8}\right)^2 + 5\frac{63}{64}$, $\left(\frac{1}{8}, 5\frac{63}{64}\right)$

26. $y = \left(x - \frac{1}{6}\right)^2 + 4\frac{35}{36}$, $\left(\frac{1}{6}, 4\frac{35}{36}\right)$

LESSON 10.4

Find the zeros of each function. Graph to check.

1. $y = x^2 - 2x - 3$ −1, 3
2. $y = x^2 + 5x + 4$ −4, −1
3. $y = x^2 - 8x + 12$ 2, 6
4. $y = x^2 - 2x - 35$ −5, 7
5. $y = x^2 - 6x + 8$ 2, 4
6. $y = x^2 - 6x + 5$ 1, 5

Solve each equation by factoring.

7. $x^2 - x - 2 = 0$ −1, 2
8. $x^2 + 6x + 5 = 0$ −5, −1
9. $x^2 - x - 6 = 0$ −2, 3
10. $x^2 + 3x - 10 = 0$ −5, 2
11. $x^2 + 3x - 28 = 0$ −7, 4
12. $x^2 + 7x - 30 = 0$ −10, 3

Solve each equation by completing the square.

13. $x^2 + 6x - 7 = 0$ −7, 1
14. $x^2 + 2x - 8 = 0$ −4, 2
15. $x^2 + 4x - 45 = 0$ −9, 5
16. $x^2 + 3x - 18 = 0$ −6, 3
17. $x^2 - x - 12 = 0$ −3, 4
18. $x^2 - 2x - 3 = 0$ −1, 3

Solve each equation by factoring or completing the square.

19. $x^2 - 12x + 20 = 0$ 2, 10
20. $x^2 - 3x - 4 = 0$ −1, 4
21. $x^2 + 7x = -12$ −4, −3
22. $x^2 + 2x - 15 = 0$ −5, 3
23. $x^2 - 4x = 0$ 0, 4
24. $x^2 + 5x - 24 = 0$ −8, 3
25. $x^2 - x = 2$ −1, 2
26. $x^2 - 9x + 20 = 0$ 4, 5
27. $x^2 + 5x = 6$ −6, 1

Find the point(s) where the graphs of each system intersect. Graph to check.

28. $\begin{cases} y = 4 \\ y = x^2 \end{cases}$ (2, 4), (−2, 4)
29. $\begin{cases} y = x - 2 \\ y = x^2 - 4x + 4 \end{cases}$ (3, 1), (2, 0)
30. $\begin{cases} y = x + 4 \\ y = x^2 + 3x - 4 \end{cases}$ (−4, 0), (2, 6)

LESSON 10.5

Identify a, b, and c for each quadratic equation.

1. $x^2 - 4x + 5 = 0$ 1, −4, 5
2. $x^2 - 6x + 10 = 0$ 1, −6, 10
3. $8x^2 - 5x + 2 = 0$ 8, −5, 2
4. $3x^2 - 6 + 3x = 0$ 3, 3, −6
5. $-x^2 + 7x - 6 = 0$ −1, 7, −6
6. $-5x + x^2 - 9 = 0$ 1, −5, −9

Find the value of the discriminant and determine the number of real solutions for each equation.

7. $x^2 + 3x - 4 = 0$ 25, 2
8. $x^2 + 4x + 1 = 0$ 12, 2
9. $x^2 + 6x - 2 = 0$ 44, 2
10. $x^2 - 6x + 1 = 0$ 32, 2
11. $x^2 - 5x + 9 = 0$ −11, 0
12. $x^2 + 6x - 10 = 0$ 76, 2
13. $4x^2 + 8x + 3 = 0$ 16, 2
14. $5x^2 - 125 = 0$ 2500, 2
15. $x^2 + x - 12 = 0$ 49, 2

Use the quadratic formula to solve each equation. Give answers to the nearest hundredth when necessary. Check by substitution.

16. $x^2 + 7x + 6 = 0$ −6, −1
17. $y^2 + 8y + 15 = 0$ −5, −3
18. $x^2 + 4x + 3 = 0$ −3, −1
19. $w^2 - 6w = 0$ 0, 6
20. $x^2 - 16 = 0$ 4, −4
21. $y^2 + 10y = 0$ −10, 0
22. $-4x^2 + 16x + 13 = 0$ 0.69, 4.69
23. $8x^2 + 10x + 3 = 0$ $-\frac{3}{4}, -\frac{1}{2}$
24. $2x^2 + 7x - 15 = 0$ −5, $\frac{3}{2}$

Choose any method to solve the following quadratic equations. Round your answers to the nearest hundredth.

25. $x^2 - 9x + 20 = 0$ 4, 5
26. $x^2 + 10x - 2 = 0$ 0.20, −10.2
27. $x^2 + 13x = -42$ −7, −6
28. $3x^2 - 5x - 2 = 0$ −0.33, 2
29. $2x^2 + x = 5$ −1.85, −1.35
30. $4x^2 - 7x - 2 = 0$ −0.25, 2
31. $5x^2 + 9x + 3 = 0$ −0.44, −1.36
32. $x^2 - 2x - 2 = 0$ 2.73, −0.73
33. $3x^2 - 5x + 1 = 0$ 1.43, 0.23
34. $x^2 - 4x - 2 = 0$ 4.45, −0.45
35. $x^2 - 5x - 7 = 0$ 6.14, −1.14
36. $3x^2 - 7x - 3 = 0$ 2.7, −0.37

LESSON 10.6

Solve each quadratic inequality using the Zero Product Property.

1. $x^2 - x - 12 > 0$ $x > 4$ or < -3
2. $x^2 + 3x - 10 < 0$ $-5 < x < 2$
3. $x^2 - 4 \geq 0$ $x \geq 2$ or $x \leq -2$
4. $x^2 + 7x + 12 \leq 0$ $-4 \leq x \leq -3$
5. $x^2 - 10x + 21 < 0$ $3 < x < 7$
6. $x^2 + 4x - 12 \leq 0$ $-6 \leq x \leq 2$
7. $x^2 - 9 < 0$ $-3 < x < 3$
8. $x^2 + 4x + 3 \geq 0$ $x \leq -3$ or $x \geq -1$
9. $x^2 - 4x + 3 > 0$ $x > 3$ or $x < 1$
10. $x^2 - 3x - 10 < 0$ $-2 < x < 5$
11. $x^2 + 2x - 3 \leq 0$ $-3 \leq x \leq 1$
12. $x^2 - 10x - 24 \geq 0$ $x \geq 12$ or $x < -2$

Graph each quadratic inequality. Shade the solution region.

13. $y < x^2$
14. $y < x^2 + 2$
15. $y > -x^2$
16. $y \geq x^2$
17. $y < x^2 - x$
18. $y < 3 - x^2$
19. $y < -3x^2$
20. $y \leq x^2 - 9$
21. $y > x^2 + 3x + 4$
22. $y \leq x^2 + 2x + 1$
23. $y \geq x^2 - 6x + 9$
24. $y \leq x^2 + 5x$

CHAPTER 11

LESSON 11.1

Determine whether each equation is an inverse variation equation.

1. $xy = 200$ yes
2. $x = \frac{25}{y}$ yes
3. $x + y = 36$ no
4. $\frac{a}{3} = \frac{5}{b}$ yes
5. $\frac{c}{d} = \frac{2}{3}$ no
6. $r = 50t$ no
7. $m = \frac{n}{75}$ no
8. $w = 7t$ no
9. $\frac{-12}{x} = y$ yes
10. $p = \frac{1}{q}$ yes
11. $bh = 14$ yes
12. $\frac{1}{10}m = n$ no

For Exercises 13–30, y varies inversely as x.

13. If y is 24 when x is 8, find y when x is 4.   48
14. If y is 3 when x is 12, find x when y is 4.   9
15. If y is 6 when x is 2, find x when y is 4.   3
16. If y is 12 when x is 15, find x when y is 18.   10
17. If y is 15 when x is 21, find x when y is 27. $11\frac{2}{3}$
18. If y is -6 when x is -2, find y when x is 5.   2.4
19. If y is -8 when x is 2, find x when y is 7. $-2\frac{2}{7}$
20. If y is 99 when x is 11, find x when y is 11.   99
21. If y is 2 when x is 5, find y when x is 20.   0.5
22. If y is 6.9 when x is 1.7, find y when x is 5.1.   2.3
23. If y is $\frac{1}{3}$ when x is 5, find x when y is $6\frac{2}{3}$. $\frac{1}{4}$
24. If y is 7 when x is $\frac{2}{3}$, find x when y is $\frac{2}{3}$.   7
25. If y is 5.6 when x is 2.8, find x when y is 4.48.   3.5
26. If y is 8.1 when x is 2.7, find y when x is 3.6.   6.075
27. If y is 7 when x is 18, find y when x is 30.   4.2
28. If y is 7 when x is 11, find y when x is 54.   1.43
29. If y is 3.6 when x is 8.1, find y when x is 2.4.   12.15
30. If y is -5 when x is 7, find y when x is -3. $11\frac{2}{3}$

Lesson 10.6

13.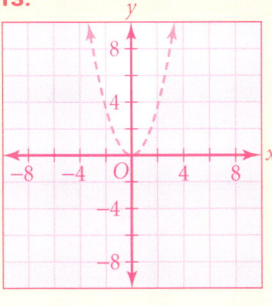
14.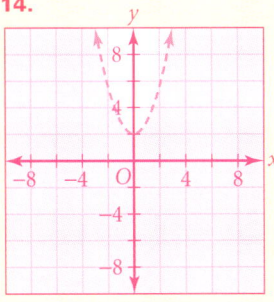
15.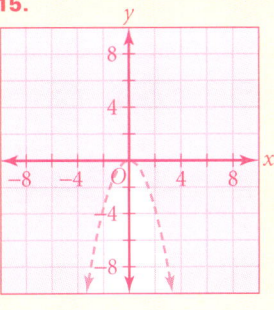
16.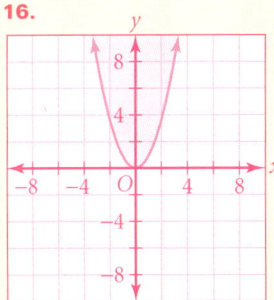
17.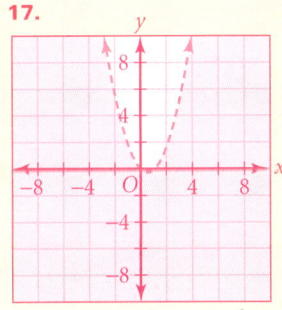
18.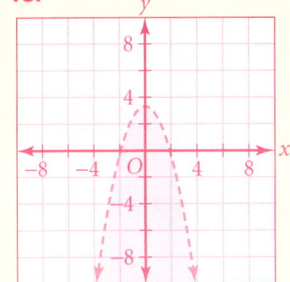
19.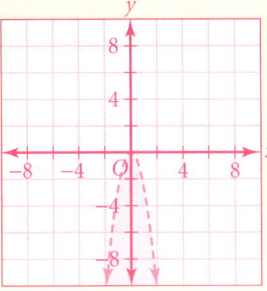
20.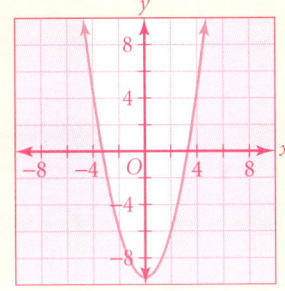
21.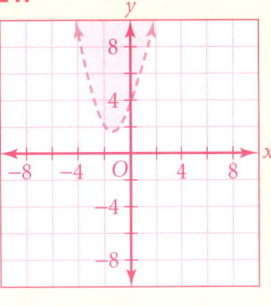

22–24. Check students' graphs.

769

Lesson 11.2

25. $x \neq 0$

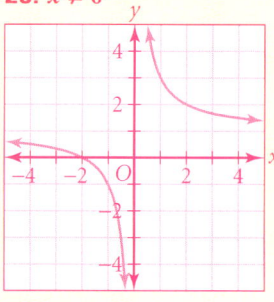

26. $x \neq 2$

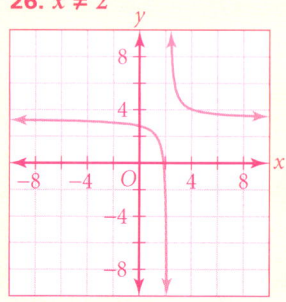

27. $x \neq 3$

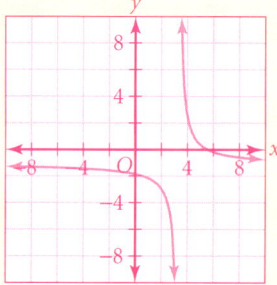

28. $x \neq 0$

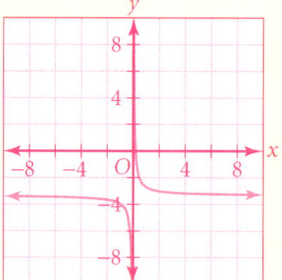

Lesson 11.3

13. $\frac{x+3}{2(x-1)}, x \neq 1$

14. $\frac{x-y}{4(x+y)}, x \neq -y$

15. $\frac{2}{7}, d \neq -1$

16. $\frac{c+9}{2}, c \neq 9$

17. $w - 2$

18. $6, x \neq 3$

19. $\frac{2a+3b}{a}, a \neq 0$

20. $x, x \neq \frac{3}{4}$

21. $w - 5, w \neq -5$

22. $\frac{1}{y-1}, y \neq -4, 1$

LESSON 11.2

For what value(s) of the variable is each rational expression undefined?

1. $\frac{3c+7}{2c}$ 0
2. $\frac{6x+2}{x-1}$ 1
3. $\frac{b}{3+b}$ -3
4. $\frac{12}{x-9}$ 9
5. $\frac{4c}{2c+1}$ $-\frac{1}{2}$
6. $\frac{2t^2}{3t+2}$ $-\frac{2}{3}$
7. $\frac{5y-11}{4y-8}$ 2
8. $\frac{2r+10}{r^2}$ 0
9. $\frac{g+3}{2g^2}$ 0
10. $\frac{3a+9}{a^2-4}$ ± 2
11. $\frac{5z+7}{z^2+5z+16}$ $-2, -3$
12. $\frac{q}{2q^2-15q}$ $0, \frac{15}{2}$

Evaluate each rational expression for $x = 1$ and $x = -2$. Write "undefined" if appropriate.

13. $\frac{2x}{x+6}$ $\frac{2}{7}, -1$
14. $\frac{3x+4}{2x+1}$ $\frac{7}{3}, \frac{2}{3}$
15. $\frac{4x+1}{2x+4}$ $\frac{5}{6}$, und.
16. $\frac{7x+7}{x+1}$ 7, 7
17. $\frac{5x}{x^2-2x+1}$ und., $\frac{-10}{9}$
18. $\frac{12x}{12x+3}$ $\frac{4}{5}, \frac{8}{7}$
19. $\frac{8x+9}{x^2-4}$ $-\frac{17}{3}$, und.
20. $\frac{2x+3}{x^2+x-2}$ und., und.
21. $\frac{1}{x-3}$ $-\frac{1}{2}, -\frac{1}{5}$
22. $\frac{2}{x} + 2$ 4, 1
23. $\frac{6}{x+3} - 4$ $-2\frac{1}{2}, 2$
24. $\frac{1}{x-5} + 3$ $2\frac{3}{4}, 2\frac{6}{7}$

Graph each rational function. List any values for which the function is undefined.

25. $h(x) = \frac{2}{x} + 1$
26. $h(x) = \frac{1}{x-2} + 3$
27. $h(x) = \frac{2}{x-3} - 1$
28. $h(x) = \frac{1}{2x} - 3$

LESSON 11.3

For what value(s) of the variable is each rational expression undefined?

1. $\frac{5x}{x+15}$ -15
2. $\frac{12}{7-y}$ 7
3. $\frac{8-w}{w}$ 0
4. $\frac{9q}{q-4}$ 4
5. $\frac{t-6}{6-t}$ 6
6. $\frac{(a+1)(a-3)}{(a-1)(a-3)}$ 1, 3
7. $\frac{s-5}{3s-6}$ 2
8. $\frac{4}{d(d^2-2d-8)}$ $0, -2, 4$

Name the common factors of the numerator and the denominator.

9. $\frac{4(x-3)}{8x}$ 4
10. $\frac{6w-12}{3}$ 3
11. $\frac{x^2-9}{x+3}$ $x+3$
12. $\frac{r+4}{r^2+8r+16}$ $r+4$

Simplify each expression and state any restrictions on the variable.

13. $\frac{12(x+3)}{24(x-1)}$
14. $\frac{4(x-y)}{16(x+y)}$
15. $\frac{8(d+1)}{28(1+d)}$
16. $\frac{5(c^2-81)}{10(c-9)}$
17. $\frac{7w-14}{7}$
18. $\frac{6x-18}{x-3}$
19. $\frac{10a+15b}{5a}$
20. $\frac{4x^2-3x}{4x-3}$
21. $\frac{w^2-25}{w+5}$
22. $\frac{y+4}{y^2+3y-4}$
23. $\frac{g^2-9}{g^2+6g-27}$
24. $\frac{n^2-n-20}{n^2+9n+20}$

23. $\frac{g+3}{g+9}, g \neq -9, 3$
24. $\frac{n-5}{n+5}, n \neq -4, -5$

LESSON 11.4

Perform the indicated operations. Simplify, and state the restrictions on the variable.

1. $\frac{4}{3y} + \frac{5}{3y}$
2. $\frac{7}{s-2} - \frac{3}{s-2}$
3. $\frac{4m}{n+3} - \frac{9m}{n+3}$
4. $\frac{7w-3}{x-y} - \frac{4+5w}{x-y}$
5. $\frac{x}{x+1} + \frac{1}{x+1}$
6. $\frac{z}{2} + \frac{z-6}{2}$
7. $\frac{6}{ab} - \frac{c}{ab^2}$
8. $\frac{-3}{w-2} + \frac{4}{5(w-2)}$
9. $\frac{m+n}{m-2} + \frac{m-n}{2-m}$
10. $a - \frac{a-5}{a+5}$
11. $\frac{-4-g}{g-1} + 3$
12. $\frac{a}{b} - \frac{c}{d}$
13. $\frac{7-x}{4+y} \cdot \frac{4+y}{m+n}$
14. $\frac{a^2}{(a+4)(a+1)} \cdot \frac{a+1}{a(a+4)}$
15. $\frac{x}{x+3} \cdot \frac{x^2-9}{x^2}$
16. $\frac{4}{y+2} \cdot \frac{1}{y}$
17. $\frac{-6}{r-3} \cdot \frac{2}{r}$
18. $\frac{t^2-t}{3} \cdot \frac{6}{t-1}$
19. $\frac{5}{(x-1)(x+2)} + \frac{4}{x-1}$
20. $\frac{3}{c+5} - \frac{4}{(c+2)(c+5)}$
21. $\frac{v-2}{v+2} + \frac{3}{v-3}$
22. $\frac{1}{y-1} + \frac{y}{1-y}$
23. $\frac{4}{h+2} - \frac{3h-1}{h^2+6h+8}$
24. $\frac{2}{x^2-4} + \frac{3}{4x+8}$

LESSON 11.5

Solve each rational equation.

1. $\frac{11}{2x} - \frac{2}{3x} = \frac{1}{6}$ $x = 29$
2. $\frac{1}{x} + \frac{5x}{x+1} = 5$ $x = 0.25$
3. $\frac{x-1}{x+3} = \frac{x+4}{3-x}$ no sol.
4. $\frac{1}{4} + \frac{2}{x} = \frac{11}{12}$ $x = 3$
5. $\frac{2w}{3} = 2 + \frac{w+3}{6}$ $w = 5$
6. $a + \frac{a}{a-1} = \frac{4a-3}{a-1}$ $a = 3$
7. $1 + \frac{3}{z-1} = \frac{4}{3}$ $z = 10$
8. $\frac{1}{2x} = \frac{1}{x^2} - \frac{1}{9}$ $x = -6, 1.5$
9. $\frac{4y}{3y-2} + \frac{2y}{3y+2} = 2$ $y = -2$
10. $\frac{5}{p-1} + \frac{p}{p+1} = 1$ $p = -1.5$
11. $\frac{x+3}{x} + \frac{x-12}{x} = 5$ $x = -3$
12. $\frac{w-1}{w+1} + 1 = \frac{2w}{w-1}$ $w = 0$
13. $\frac{2q}{q-1} + \frac{q-5}{q^2-1} = 1$ $q = -4$
14. $\frac{14}{y-6} - \frac{1}{2} = \frac{6}{y-8}$ $y = 10, 20$
15. $\frac{a-2}{a} = \frac{1}{a} + \frac{a-3}{a-6}$ $a = 3$
16. $\frac{c-1}{3-c} = c - \frac{2}{c-3}$ $c = -1$
17. $\frac{x-5}{x^2-1} + \frac{2x}{x-1} = 1$ $x = -4$
18. $\frac{m}{3m+6} = \frac{2}{5} + \frac{m}{5m+10}$ $m = -3$

Solve each rational equation by graphing.

19. $\frac{16}{x+8} = \frac{4}{9}$ $x = 28$
20. $\frac{2}{x+1} = \frac{3}{x+2}$ $x = 1$
21. $\frac{2x}{x-4} - \frac{3}{5} = 5$ $x = 6.2$

Lesson 11.4

1. $\frac{3}{y}, y \neq 0$
2. $\frac{4}{s-2}, s \neq 2$
3. $\frac{-5m}{n+3}, n \neq -3$
4. $\frac{2w-7}{x-y}, x \neq y$
5. $1, x \neq -1$
6. $z - 3$
7. $\frac{6b-c}{ab^2}, a \neq 0, b \neq 0$
8. $\frac{-11}{5w-10}, w \neq 2$
9. $\frac{2n}{m-2}, m \neq 2$
10. $\frac{a^2+4a+5}{a+5}, a \neq -5$
11. $\frac{2g-7}{g-1}, g \neq 1$
12. $\frac{ad-bc}{bd}, b \neq 0, d \neq 0$
13. $\frac{7-x}{m+n}, m \neq -n, y \neq -4$
14. $\frac{a}{a^2+8a+16}, a \neq -4, -1, 0$
15. $\frac{x-3}{x}, x \neq -3, 0$
16. $\frac{4}{y^2+2y}, y \neq -2, 0$
17. $\frac{12}{r^2-3r}, r \neq 3, 0$
18. $2t, t \neq 1$
19. $\frac{4x+13}{x^2+x-2}, x \neq -2, 1$
20. $\frac{3c+2}{c^2+7c+10}, c \neq -5, -2$
21. $\frac{v^2-2v+12}{v^2-v-6}, v \neq -2, 3$
22. $-1, y \neq 1$
23. $\frac{h+17}{h^2+6h+8}, h \neq -4, -2$
24. $\frac{3x+2}{4x^2-16}, x \neq \pm 2$

Lesson 11.6

5. $3x + 4x$ Given
   $3x + 4x = 7x$ Addition of Monomials

6. $7y - 6y$ Given
   $7y - 6y = y$ Subtrtaction of Monomials

7. $(-4m)(-3m)$ Given
   $(-4m)(-3m) = 12m^2$ Multiplication of Monomials

8. $8x - 3x = 35$ Given
   $5x = 35$ Subtraction of Monomials
   $\frac{5x}{5} = \frac{35}{5}$ Division Property of Equality
   $x = 7$ Simplified form

9. $3c - 1 = 11$ Given
   $3c - 1 + 1 = 11 + 1$ Addition Property of Equality
   $3c = 12$ Simplified form
   $\frac{3c}{3} = \frac{12}{3}$ Division Property of Equality
   $c = 4$ Simplified form

10. $4(y + 2) = -24$ Given
    $4y + 8 = -24$ Distributive Property
    $4y + 8 - 8 = -24 - 8$ Subtraction Property of Equality
    $4y = -32$ Simplified form
    $\frac{4y}{4} = \frac{-32}{4}$ Division Property of Equality
    $y = -8$ Simplified form

11. $\frac{3}{4}x + 5 = 2$ Given
    $\frac{3}{4}x + 5 - 5 = 2 - 5$ Sutraction Property of Equality
    $\frac{3}{4}x = -3$ Simplified form
    $\frac{4}{3} \cdot \frac{3}{4}x = \frac{4}{3} \cdot -3$ Multiplication Property of Equality
    $x = -4$ Simplified form

12. $\frac{3}{5}n - 6 = 0$ Given
    $\frac{3}{5}n - 6 + 6 = 0 + 6$ Addition Property of Equality
    $\frac{3}{5}n = 6$ Simplified form
    $\frac{5}{3} \cdot \frac{3}{5}n = \frac{5}{3} \cdot 6$ Multiplication Property of Equality
    $n = 10$ Simplified form

13. $\frac{d}{3} - 7 = -4$ Given
    $\frac{d}{3} - 7 + 7 = -4 + 7$ Addition Property of Equality
    $\frac{d}{3} = 3$ Simplified form
    $3 \cdot \frac{d}{3} = 3 \cdot 3$ Multiplication Property of Equality
    $d = 9$ Simplified form

14. $5 - \frac{q}{4} = 8$ Given
    $5 - \frac{q}{4} - 5 = 8 - 5$ Subtraction Property of Equality
    $-\frac{q}{4} = 3$ Simplified form
    $-4 \cdot -\frac{q}{4} = 3 \cdot -4$ Multiplication Property of Equality
    $q = -12$ Simplified form

LESSON 11.6

Give a reason for each step in the proof of the following statement:
If $2x + 6 = 0$, then $x = -3$

	Proof	Reasons
	$2x + 6 = 0$	Hypothesis or given
1.	$2x + 6 - 6 = 0 - 6$	Subtraction Prop. of Equality
2.	$2x = -6$	Simplify
3.	$\frac{2x}{2} = \frac{-6}{2}$	Div. Prop. of Equality
4.	$x = -3$	Simplify

Prove each statement. Give a reason for each step.

5. For all x, $3x + 4x = 7x$.
6. For all y, $7y - 6y = y$.
7. For all m, $(-4m)(-3m) = 12m^2$.
8. If $8x - 3x = 35$, then $x = 7$.
9. If $3c - 1 = 11$, then $c = 4$.
10. If $4(y + 2) = -24$, then $y = -8$.
11. If $\frac{3}{4}x + 5 = 2$, then $x = -4$.
12. If $\frac{3}{5}n - 6 = 0$, then $n = 10$.
13. If $\frac{d}{3} - 7 = -4$, then $d = 9$.
14. If $5 - \frac{q}{4} = 8$, then $q = -12$.

CHAPTER 12

LESSON 12.1

Simplify each radical by factoring.

1. $\sqrt{36}$ 6
2. $\sqrt{225}$ 15
3. $\sqrt{324}$ 18
4. $\sqrt{2500}$ 50
5. $\sqrt{108}$ $6\sqrt{3}$
6. $\sqrt{80}$ $4\sqrt{5}$
7. $\sqrt{192}$ $8\sqrt{3}$
8. $\sqrt{1440}$ $12\sqrt{10}$
9. $\sqrt{605}$ $11\sqrt{5}$
10. $\sqrt{784}$ 28
11. $\sqrt{1944}$ $18\sqrt{6}$
12. $\sqrt{768}$ $16\sqrt{3}$
13. $\sqrt{490}$ $7\sqrt{10}$
14. $\sqrt{270}$ $3\sqrt{30}$
15. $\sqrt{133}$ $\sqrt{133}$
16. $\sqrt{525}$ $5\sqrt{21}$
17. $\sqrt{81}$ 9
18. $\sqrt{99}$ $3\sqrt{11}$
19. $\sqrt{67}$ $\sqrt{67}$
20. $\sqrt{140}$ $2\sqrt{35}$
21. $\sqrt{1600}$ 40
22. $\sqrt{825}$ $5\sqrt{33}$
23. $\sqrt{44}$ $2\sqrt{11}$
24. $\sqrt{121}$ 11

Simplify.

25. $\sqrt{5}\sqrt{20}$ 10
26. $\sqrt{2}\sqrt{18}$ 6
27. $\sqrt{21}\sqrt{98}$ $7\sqrt{42}$
28. $\sqrt{45}\sqrt{5}$ 15
29. $\sqrt{\frac{100}{25}}$ 2
30. $\sqrt{\frac{72}{3}}$ $2\sqrt{6}$
31. $\frac{\sqrt{60}}{\sqrt{6}}$ $\sqrt{10}$
32. $\frac{\sqrt{216}}{\sqrt{6}}$ 6
33. $(4\sqrt{3})^2$ 48
34. $(6\sqrt{25})^2$ 900
35. $(8\sqrt{2})^2$ 128
36. $(5\sqrt{15})^2$ 375
37. $3(\sqrt{6} + 4)$ $3\sqrt{6} + 12$
38. $3(\sqrt{12} - 8)$ $6\sqrt{3} - 24$
39. $\sqrt{6}(3 - \sqrt{18})$ $3\sqrt{6} - 6\sqrt{3}$
40. $\sqrt{3}(7 + \sqrt{15})$ $7\sqrt{3} + 3\sqrt{5}$
41. $(\sqrt{6} - 4)(\sqrt{6} + 4)$ -10
42. $(\sqrt{7} + 2)^2$ $4\sqrt{7} + 11$
43. $(\sqrt{3} - 2)(\sqrt{3} + 5)$ $3\sqrt{3} - 7$
44. $\sqrt{2}(\sqrt{5} - 1)^2$ $6\sqrt{2} - 2\sqrt{10}$

LESSON 12.2

Solve each equation if possible. If not possible, explain why.

1. $\sqrt{x-3} = 1$ $x = 4$
2. $\sqrt{x+5} = 3$ $x = 4$
3. $\sqrt{x-9} = 8$ $x = 73$
4. $\sqrt{3x} = 6$ $x = 12$
5. $\sqrt{5x} = 10$ $x = 20$
6. $\sqrt{x-4} = 2$ $x = 8$
7. $\sqrt{8-x} = 2$ $x = 4$
8. $\sqrt{17-x} = 4$ $x = 1$
9. $\sqrt{3x+1} = 5$ $x = 8$
10. $\sqrt{x+2} = x-3$ $x = 7$
11. $\sqrt{1-2x} = x+1$ $x = 0$
12. $\sqrt{x-2} = x-4$ $x = 6$
13. $\sqrt{x+3} = x-3$ $x = 6$
14. $\sqrt{3x-5} = x-5$ $x = 10$
15. $\sqrt{3x-14} = 6-x$ $x = 5$
16. $\sqrt{x^2-3x+10} = x+1$ $x = 1.8$
17. $\sqrt{x^2+x-8} = x-2$ $x = 2.4$
18. $\sqrt{x^2+6x-4} = x-5$ no sol
19. $x^2 = 80$ $x = \pm 4\sqrt{5}$
20. $x^2 = 27$ $x = \pm 3\sqrt{3}$
21. $x^2 = 98$ $x = \pm 7\sqrt{2}$
22. $9x^2 = 10$ $x = \pm \frac{\sqrt{10}}{3}$
23. $16x^2 = 5$ $x = \pm \frac{\sqrt{5}}{4}$
24. $25x^2 = 1$ $x = \pm \frac{1}{5}$
25. $3x^2 = 27$ $x = \pm 3$
26. $5x^2 = 75$ $x = \pm\sqrt{15}$
27. $6x^2 = 72$ $x = \pm 2\sqrt{3}$
28. $x^2 - 6x + 9 = 0$ $x = 3$
29. $x^2 + 4x + 2 = 0$ $x = -2 \pm \sqrt{2}$
30. $x^2 - 8x + 16 = 0$ $x = 4$
31. $x^2 = 16$ $x = \pm 4$
32. $\sqrt{x} = 16$ $x = 25$
33. $|x| = 16$ $x = \pm 6$
34. $x^2 = -16$ no real numbers
35. $\sqrt{x} = -16$ no real numbers
36. $|x| = -16$ not possible

LESSON 12.3

Copy and complete the table of lengths of sides of right triangles. Use a calculator and round your answers to the nearest tenth.

	Leg	Leg	Hypotenuse
1.	2	6	6.3 ?
3.	6	8 ?	10
5.	7	15	16.6 ?
7.	17.9 ?	9	20
9.	15	18	23.4 ?
11.	10	24	26 ?
13.	9 ?	40	41

	Leg	Leg	Hypotenuse
2.	8	15 ?	17
4.	5 ?	12	13
6.	7	12	13.9 ?
8.	6	7	9.2 ?
10.	7.2 ?	12	14
12.	8	28.9 ?	30
14.	1.5	2	2.5 ?

Solve for x.

15. 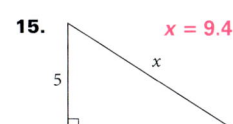 $x = 9.4$

16. 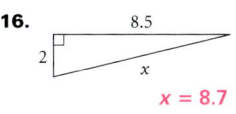 $x = 8.7$

17. 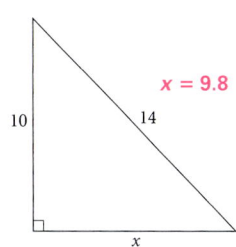 $x = 9.8$

Lesson 12.5

13.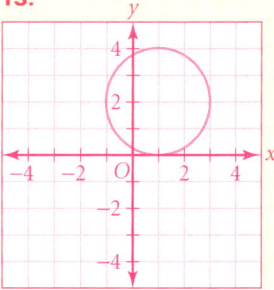

14.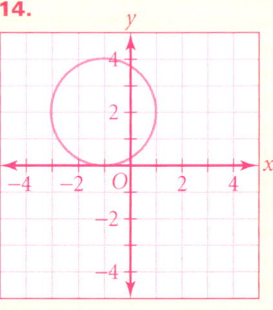

15.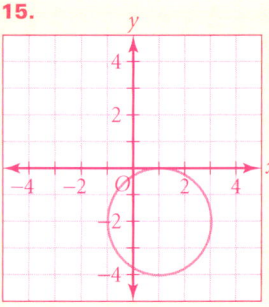

16.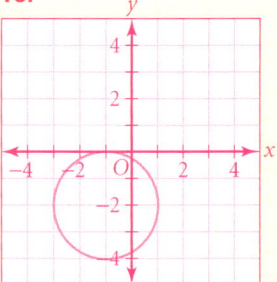

17.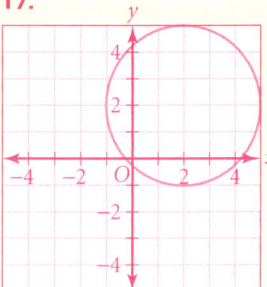

18.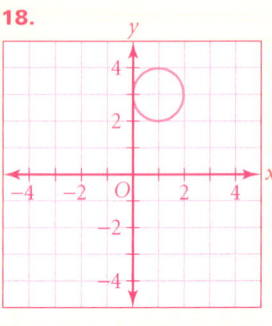

LESSON 12.4

For each figure, find the coordinates of Q and the lengths of the three sides of the triangle.

1.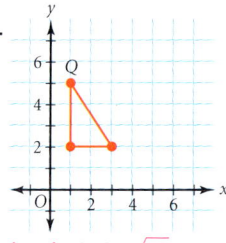
(1, 5); 2, 3, $\sqrt{13}$

2.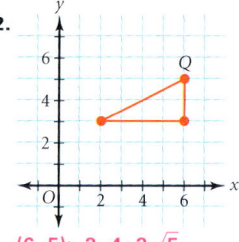
(6, 5); 2, 4, $2\sqrt{5}$

3.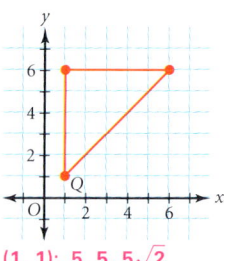
(1, 1); 5, 5, $5\sqrt{2}$

Find the distance between each pair of points.

4. $A(3, 6)$, $B(1, 2)$ 4.5
5. $M(3, 7)$, $N(12, 19)$ 15
6. $X(-5, -2)$, $Y(-8, 3)$ 5.8
7. $D(6, 7)$, $E(-1, 1)$ 9.2
8. $P(-5, 3)$, $Q(5, 5)$ 10.2
9. $R(3, 4)$, $S(-4, -1)$ 8.6

Plot each segment in a coordinate plane, and find the coordinates of the midpoint of the segment.

10. $\overline{AB}$, with $A(4, 6)$ and $B(8, 8)$ (6, 7)
11. $\overline{MN}$, with $M(-4, 4)$ and $N(0, -8)$ (-2, -2)
12. $\overline{XY}$, with $X(-6, 5)$ and $Y(-2, 3)$ (-4, 4)
13. $\overline{CD}$, with $C(7, 8)$ and $D(4, 1)$ (5.5, 4.5)

The midpoint of segment PQ is M. Find the missing coordinates.

14. $P(5, 8)$, $Q(1, 2)$, $M(?, ?)$ (3, 5)
15. $P(6, 2)$, $Q(?, ?)$, $M(8, 6)$ (10, 10)
16. $P(-4, 6)$, $Q(8, 9)$, $M(?, ?)$ (2, 7.5)

LESSON 12.5

Find an equation of a circle with a center at the origin and having the given radius.

1. radius = 3 $x^2 + y^2 = 9$
2. radius = 5 $x^2 + y^2 = 25$
3. radius = 10 $x^2 + y^2 = 100$
4. radius = 16 $x^2 + y^2 = 256$
5. radius = 1 $x^2 + y^2 = 1$
6. radius = 7 $x^2 + y^2 = 49$
7. radius = 11 $x^2 + y^2 = 121$
8. radius = 4 $x^2 + y^2 = 16$
9. radius = 20 $x^2 + y^2 = 400$
10. radius = 25 $x^2 + y^2 = 625$
11. radius = 32 $x^2 + y^2 = 1024$
12. radius = 60 $x^2 + y^2 = 3600$

Use graph paper to graph each circle below. Find an equation of a circle with

13. center (1, 2) and radius = 2 $(x - 1)^2 + (y - 2)^2 = 4$
14. center (-1, 2) and radius = 2 $(x + 1)^2 + (y - 2)^2 = 4$
15. center (1, -2) and radius = 2 $(x - 1)^2 + (y + 2)^2 = 4$
16. center (-1, -2) and radius = 2 $(x + 1)^2 + (y + 2)^2 = 4$
17. center (2, 2) and radius = 3 $(x - 2)^2 + (y - 2)^2 = 9$
18. center (1, 3) and radius = 1 $(x - 1)^2 + (y - 3)^2 = 1$
19. center (-3, 0) and radius = 1 $(x + 3)^2 + y^2 = 1$
20. center (-1, 1) and radius = 3 $(x + 1)^2 + (y - 1)^2 = 9$

Identify the center and radius for each circle.

21. $(x - 1)^2 + (y - 2)^2 = 16$ (1, 2), 4
22. $(x - 4)^2 + (y - 5)^2 = 1$ (4, 5), 1
23. $(x + 3)^2 + (y - 4)^2 = 25$ (-3, 4), 5
24. $(x + 2)^2 + (y - 6)^2 = 9$ (-2, 6), 3
25. $(x + 7)^2 + (y - 1)^2 = 3$ (-7, 1), $\sqrt{3}$
26. $(x + 10)^2 + (y - 9)^2 = 7$ (-10, 9), $\sqrt{7}$
27. $(x - 3)^2 + (y - 9)^2 = 11$ (3, 9), $\sqrt{11}$
28. $(x + 15)^2 + (y - 1)^2 = 49$ (-15, 1), 7
29. $x^2 + (y - 1)^2 = 13$ (0, 1), $\sqrt{13}$
30. $(x + 17)^2 + (y - 9)^2 = 81$ (-17, 9), 9
31. $(x - 3)^2 + y^2 = 121$ (3, 0), 11
32. $(x - 5)^2 + (y - 10)^2 = 100$ (5, 10), 10

19.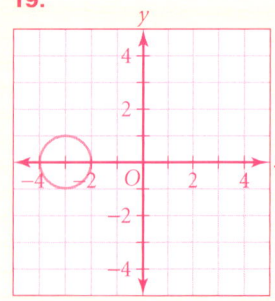

20.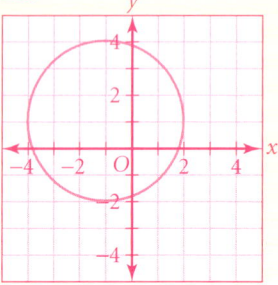

LESSON 12.6

Find the tangent of each acute angle of each triangle.

1.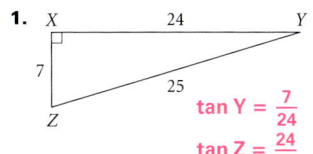
   $\tan Y = \frac{7}{24}$
   $\tan Z = \frac{24}{7}$

2.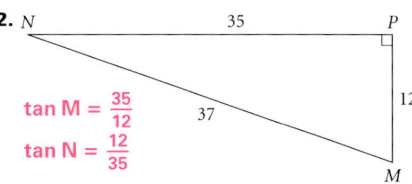
   $\tan M = \frac{35}{12}$
   $\tan N = \frac{12}{35}$

Use a calculator to find the tangents of the following angles to the nearest ten-thousandth.

3. 60° **1.7321**   4. 35° **0.7002**   5. 2° **0.0349**   6. 52° **1.2799**   7. 49.9° **0.9965**   8. 40.5° **0.8541**

Find the angle measure with the approximate tangent value. Give your answer to the nearest tenth of a degree.

9. 0.4663 **25°**   10. 5.1446 **79°**   11. 0.1228 **7°**   12. 0.2493 **14°**

13. 0.6745 **34°**   14. 14.3007 **86°**   15. 1.2800 **52°**   16. 0.9004 **42°**

Use the diagram at right for Exercises 19–23. Give your answer to the nearest tenth.

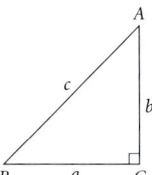

17. $b = 12$, $m\angle A = 42°$, $a = ?$ **10.8**
18. $b = 35$, $m\angle B = 66°$, $a = ?$ **15.6**
19. $a = 9$, $m\angle B = 56°$, $b = ?$ **19**
20. $b = 6$, $a = 11$, $m\angle B = ?$ **28.6°**
21. $b = 25$, $a = 18$, $m\angle A = ?$ **35.8°**

LESSON 12.7

Evaluate each expression to the nearest ten-thousandth.

1. sin 45° **0.7071**   2. cos 45° **0.7071**   3. sin 83° **0.9925**   4. cos 62° **0.4695**
5. sin 2° **0.0349**   6. cos 58.5° **0.5225**   7. cos 10.9° **0.9820**   8. sin 26.4° **0.4446**
9. cos 37° **0.7986**   10. sin 46.7° **0.7278**   11. cos 15° **0.9659**   12. sin 10° **0.1736**

Find the acute angle measure with the approximate sin value. Round your answers to the nearest tenth of a degree.

13. 0.2924 **17°**   14. 0.9781 **78°**   15. 0.0523 **3°**   16. 0.4778 **28.5°**

Find the acute angle measure with the approximate cosine value. Round your answers to the nearest tenth of a degree.

17. 0.8090 **36°**   18. 0.5211 **58.6°**   19. 0.6901 **46.4°**   20. 0.0322 **88.2°**

Find the indicated angle measure to the nearest tenth.

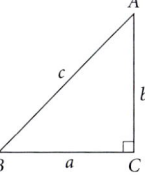

21. $b = 6$, $c = 9$, $m\angle B = ?$ **41.8°**
22. $b = 2$, $c = 7$, $m\angle A = ?$ **73.4°**
23. $a = 9$, $c = 12$, $m\angle B = ?$ **41.4°**
24. $a = 5$, $b = 12$, $m\angle B = ?$ **67.4°**

Lesson 12.8

1. $\begin{bmatrix} -6 & 27 \\ -1 & -20 \end{bmatrix}$

2. not possible because the dimensions of the matrices are not the same.

3. $\begin{bmatrix} 9 & -27 & -55 \\ 72 & -19 & 8 \end{bmatrix}$

4. $\begin{bmatrix} 5.5 & 10.6 \\ 15.7 & 8.3 \\ 2.0 & 3.9 \end{bmatrix}$

5. $\begin{bmatrix} 6.3 & 7.3 \\ 9.1 & 3.1 \\ 7.9 & 10.2 \end{bmatrix}$

6. $\begin{bmatrix} 0 & 1\frac{1}{5} \\ 1 & 2\frac{3}{4} \end{bmatrix}$

7. $\begin{bmatrix} 13 & 3.5 & 7.5 \\ -1 & 5 & 10 \\ 21 & 9 & 8 \end{bmatrix}$

8. $\begin{bmatrix} 3 & 2 & -5.6 \\ 7 & -13 & 24 \\ 12 & 22 & -19.5 \end{bmatrix}$

9. not possible because the number of columns in the first matrix does not equal the number of rows in the second matrix.

10. $\begin{bmatrix} 16 & 4 \\ -24 & 38 \end{bmatrix}$

LESSON 12.8

Perform the following matrix operations. If a solution is not possible, explain why.

1. $\begin{bmatrix} -6 & 15 \\ 7 & -14 \end{bmatrix} - \begin{bmatrix} 0 & -12 \\ 8 & 6 \end{bmatrix}$

2. $\begin{bmatrix} 8 & 5 \\ 6 & -23 \end{bmatrix} + \begin{bmatrix} 14 & 0 \\ 1 & 2 \end{bmatrix} - \begin{bmatrix} 2 & 3 & -11 \\ -6 & 2 & 4 \end{bmatrix}$

3. $\begin{bmatrix} 5 & 7 & -9 \\ -8 & 23 & -4 \end{bmatrix} - 2\begin{bmatrix} -2 & 17 & 23 \\ -40 & 21 & -6 \end{bmatrix}$

4. $\begin{bmatrix} 2.1 & 3.5 \\ 7.8 & 1.9 \\ 0.6 & 3.3 \end{bmatrix} + \begin{bmatrix} 3.4 & 7.1 \\ 7.9 & 6.4 \\ 1.4 & 0.6 \end{bmatrix}$

5. $\begin{bmatrix} 4.2 & 2.4 \\ 1.9 & 2.5 \\ 3.7 & 5.4 \end{bmatrix} + \begin{bmatrix} 2.1 & 4.9 \\ 7.2 & 0.6 \\ 4.2 & 4.8 \end{bmatrix}$

6. $\begin{bmatrix} -\frac{1}{2} & \frac{2}{5} \\ \frac{1}{4} & \frac{3}{4} \end{bmatrix} + \begin{bmatrix} \frac{1}{2} & \frac{4}{5} \\ \frac{3}{4} & 2 \end{bmatrix}$

7. $\begin{bmatrix} 15 & 8.5 & 1 \\ 0 & 7 & 9.25 \\ 6 & 12 & 6 \end{bmatrix} + \begin{bmatrix} -2 & 5 & 6.5 \\ -1 & -2 & 0.75 \\ 15 & -3 & 2 \end{bmatrix}$

8. $\begin{bmatrix} 0 & 3.5 & -4 \\ 9 & -6 & 9 \\ 4 & 11 & -10 \end{bmatrix} + \begin{bmatrix} 3 & -5 & 2.4 \\ -11 & -1 & 6 \\ 4 & 0 & 0.5 \end{bmatrix}$

9. $\begin{bmatrix} 6 & 11 & 6 \\ 5 & 15 & 18 \\ 2 & 12 & 20 \end{bmatrix} \begin{bmatrix} 9 & 11 \\ 12 & 27 \end{bmatrix}$

10. $\begin{bmatrix} 2 & -1 \\ -3 & 7 \end{bmatrix} \begin{bmatrix} 8 & 6 \\ 0 & 8 \end{bmatrix}$

CHAPTER 13

LESSON 13.1

An integer from 50 to 100 is drawn at random. Find the probability it is the following:

1. The integer is even. $\frac{26}{51}$
2. The integer is odd. $\frac{25}{51}$
3. The integer is a multiple of 10. $\frac{6}{51}$
4. The integer is a multiple of 9. $\frac{6}{51}$
5. The integer is a multiple of 3. $\frac{17}{51}$
6. The integer is prime. $\frac{10}{51}$
7. The integer is a multiple of 10 AND a multiple of 9. $\frac{1}{51}$
8. The integer is a multiple of 10 OR a multiple of 9. $\frac{11}{51}$

A letter of the alphabet is selected at random. Find the probability that it is the following:

9. q $\frac{1}{26}$
10. a or z $\frac{1}{13}$
11. r, s, or t $\frac{3}{26}$
12. The letter is one of the letters in the word *math*. $\frac{2}{13}$
13. The letter is a letter from a to m, inclusive. $\frac{1}{2}$
14. The letter is one of the letters of the word *games*. $\frac{5}{26}$

LESSON 13.2

List the integers from 20 to 40, inclusive, which are the following:

1. even AND odd.
2. even OR odd.
3. even AND multiples of 3.
4. even OR multiples of 3.
5. odd OR multiples of 3.
6. even AND multiples of 5.
7. even OR multiples of 5.
8. odd AND multiples of 5.
9. odd OR multiples of 5.

A survey about favorite items on the lunch menu had the results shown in the table.

	Pizza	Salad	Tacos
Boys	194	102	148
Girls	91	136	129

How many surveyed students

10. are girls? 356
11. like pizza the best? 285
12. like salad the best? 238
13. are girls AND like salad the best? 136
14. are girls OR like salad the best? 458
15. are girls AND like pizza the best? 91
16. are boys? 444
17. like tacos the best? 277
18. are boys AND like tacos the best? 148
19. are boys OR like tacos the best? 573

LESSON 13.3

A menu contains 3 appetizers, 4 salads, 6 main courses, and 5 desserts.

1. How many ways are there to choose one of each? 360
2. How many combinations include a piece of apple pie, one of the 5 desserts? 72
3. How many combinations include a shrimp cocktail, one of the 3 appetizers? 120
4. How many combinations include a shrimp cocktail and a piece of apple pie? 24
5. How many combinations include a steak, one of the 6 main courses? 60

A regular deck of 52 playing cards contains 26 red cards and 26 black cards. The deck contains 4 queens, 2 black and 2 red. One card is drawn. Find the number of ways to draw each.

6. a black queen 2
7. a queen 4
8. a red card 28
9. a red card OR a queen 26

Two cards are drawn form a regular deck of 52 playing cards. The first card is replaced before drawing the second card. Find the number of ways to draw each pair in order.

10. a red card, then a queen 104
11. a queen, then a black card 104
12. a queen, then a red card 104
13. a black card, then a red card 676

How many ways are there to arrange the letters in each word? (Each arrangement will not necessarily spell another word.)

14. dog 6
15. mail 24
16. heart 120
17. computer 40,320

Lesson 13.2

1. none
2. 20, 21, 22, . . ., 40
3. 24, 30, 36
4. 20, 21, 22, 24, 26, 27, 28, 30, 32, 33, 34, 36, 38, 39, 40
5. 21, 23, 24, 25, 27, 29, 30, 31, 33, 35, 36, 37, 39
6. 20, 30, 40
7. 20, 22, 24, 25, 26, 28, 30, 32, 34, 35, 36, 38, 40
8. 25, 35
9. 20, 21, 23, 25, 27, 29, 30, 31, 33, 35, 37, 39, 40

777

Extra Practice

Lesson 13.5

1. Use a calculator or computer to generate random integers from 1 to 30. Each number represents a different day of the month.

2. Use a calculator or computer to generate random integers from 1 to 28. Each number represents a different student.

3. Roll a die, or use a calculator or computer to generate random integers from 1 to 6. Each number selected represents a different athlete who wins a race.

4. Roll a 10 sided die or use a calculator or computer to generate random integers from 1 to 10. Each number represents a different team.

5. Use a calculator or computer to generate random integers from 1 to 15. Each number represents a different book.

6. Use a calculator or computer to generate random integers from 1 to 50. Each number represents a different number.

LESSON 13.4

Nine cards numbered 1 to 9 are in a box. A card is drawn and replaced. Then a second card is drawn. Find the probability that

1. both are prime. $\frac{16}{81}$
2. both are multiples of 3. $\frac{1}{9}$
3. both are odd. $\frac{25}{81}$
4. both are even. $\frac{16}{81}$
5. one is even and the other is odd. $\frac{40}{81}$
6. one is prime and the other is even. $\frac{32}{81}$
7. the first one is even and the second is odd. $\frac{20}{81}$
8. the first one is even and the second is a multiple of 3. $\frac{4}{27}$

A box contains 5 quarters, 8 dimes, 2 nickels, and 6 pennies. One coin is drawn and replaced. Then a second coin is drawn. Find the probability that

9. both are quarters. $\frac{25}{441}$
10. both are dimes. $\frac{64}{441}$
11. both are nickels. $\frac{4}{441}$
12. both are pennies. $\frac{36}{441}$
13. one is a quarter and one is a penny. $\frac{60}{441}$
14. at least one is a penny or a dime. $\frac{392}{441}$
15. the first one is a nickel and the other is a quarter. $\frac{10}{441}$

LESSON 13.5

Describe a way to simulate each random selection.

1. a day of a thirty-day month
2. one student out of a class of 28 students
3. which of 6 athletes crosses the finish line first
4. which team of 10 teams wins a tournament
5. selecting one book from a collection of 15 books
6. choosing one number between the numbers 1 and 50, inclusive

CHAPTER 14

LESSON 14.1

Which of the following relations are functions? Explain.

1. {(2, 3), (3, 5), (2, 8)} No
2. {(6, −2), (5, −2), (8, −2)} Yes
3. {(0, 3), (2, 5), (−2, −5)} Yes
4. 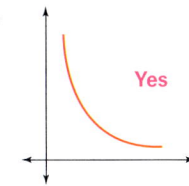 Yes 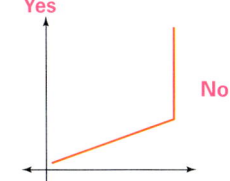 No 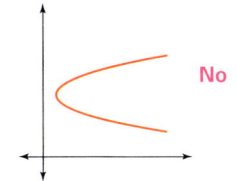 No

If $f(x) = 4x$, evaluate each of the following:

7. $f(3)$ **12**
8. $f(-1)$ **−4**
9. $f(0)$ **0**
10. $f(6)$ **24**
11. $f(-12)$ **−48**
12. $f(0.5)$ **2**

Evaluate each fraction for $x = 2$.

13. $h(x) = \frac{1}{x}$ **$\frac{1}{2}$**
14. $f(x) = x^2 - 5x$ **−6**
15. $h(x) = -2|x|$ **−4**
16. $h(x) = \frac{-4}{x}$ **−2**
17. $k(x) = 3x^2 + 9$ **21**
18. $g(x) = -x^2 + x - 7$ **−9**
19. $k(x) = -5x^2 - 3x + 1$ **−25**
20. $h(x) = |x - 6|$ **4**
21. $f(x) = \frac{5}{x}$ **2.5**

LESSON 14.2

Graph each function and identify the parent function. Describe the transformation applied.

1. $y = -2x^2$
2. $y = x^2 - 1$
3. $y = (x - 1)^2$
4. $y = 2(-x)^2$
5. $y = -3|x|$
6. $y = |x| + 2$
7. $y = |x + 2| + 2$
8. $y = |-3x|$
9. $y = -2|x|$
10. $y = |x| + 3$
11. $y = |x + 1|$
12. $y = 3x + 1$
13. $y = x + 6$
14. $y = x^2 - 3$
15. $y = (x - 2)^2$
16. $y = x - 4$

Consider the parent function $f(x) = x^2$. Note that $f(2) = 4$, so the point (2, 4) is on the graph. Use this fact to describe how the parent function is translated.

17. $f(x) = x^2 + 2$ contains the point (2, 6).
18. $f(x) = x^2 - 2$ contains the point (2, 2).
19. $f(x) = (x + 2)^2$ contains the point (2, 16).
20. $f(x) = (x - 2)^2$ contains the point (2, 0).

The point (6, 9) is on the graph of a function. Give the coordinates of the corresponding point on the new graph after each of the following transformations are applied to the function.

21. vertical translation by 5
22. vertical translation by −3
23. vertical translation by −9
24. vertical stretch by 4
25. horizontal translation by 2
26. horizontal translation by −1
27. horizontal translation by −10
28. vertical stretch by 5

---

13. $y = x$; a shift 6 units up
14. $y = x^2$; a shift 3 units down
15. $y = x^2$; a shift 2 units to the right
16. $y = x$; a shift 4 units down

17. a shift 2 units up
18. a shift 2 units down
19. a shift 2 units to the left
20. a shift 2 units to the right
21. (6, 14)
22. (6, 6)

23. (6, 0)
24. (6, 36)
25. (8, 9)
26. (5, 9)
27. (−4, 9)
28. (6, 45)

Lesson 14.2

1. $y = x^2$; a reflection across the x-axis and a vertical stretch by a factor of 2
2. $y = x^2$; a shift 1 unit down
3. $y = x^2$; a shift 1 unit to the right
4. $y = x^2$; a reflection across the y-axis and a vertical stretch by a factor of 2
5. $y = |x|$; a reflection across the x-axis and a vertical stretch by a factor of 3
6. $y = |x|$; a shift 2 units up
7. $y = |x|$; a shift 2 units to the left and a shift 2 units up
8. $y = |x|$; a reflection across the y-axis and a horizontal compression by a factor of $\frac{1}{3}$
9. $y = |x|$; a reflection across the x-axis and a vertical stretch by a factor of 2
10. $y = |x|$; a shift 3 units up
11. $y = |x|$; a shift 1 unit to the left
12. $y = x$; a shift 1 unit up and a horizontal compression by a factor of $\frac{1}{3}$

Extra Practice

Lesson 14.3

2.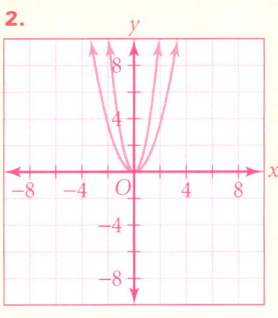

3. The graph of $y = x^2$ is vertically stretched by a factor of 3 to become the graph of $y = 3x^2$.

5.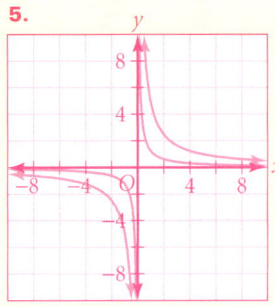

6. The graph of $y = \frac{1}{x}$ is vertically stretched by a factor of 4 to become the graph of $y = \frac{4}{x}$.

8.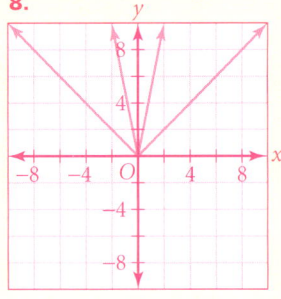

9. The graph of $y = |x|$ is vertically stretched by a factor of 5 to become the graph of $y = 5|x|$.

Lesson 14.4

1.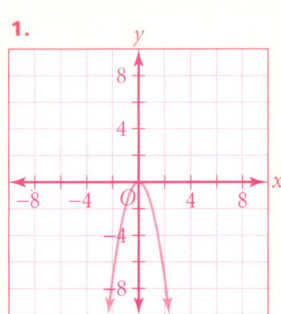

LESSON 14.3

Given the function $y = 3x^2$,

1. identify the parent function. $y = x^2$
2. graph the function and the parent function on the same coordinate plane.
3. explain how the graph of the parent function was changed by the 3.

Given the function $y = \frac{4}{x}$,

4. identify the parent function. $y = \frac{1}{x}$
5. graph the function and the parent function on the same coordinate plane.
6. explain how the graph of the parent function was changed by the 4.

Given the function $y = 5|x|$,

7. identify the parent function. $y = |x|$
8. graph the function and the parent function on the same coordinate plane.
9. Explain how the graph of the parent function was changed by the 5.

Determine whether each function has a graph that is a stretch of the parent function.

10. $y = x^2 + 4$ No
11. $y = 3|x|$ Yes
12. $y = \frac{x}{4}$ Yes
13. $y = \frac{1}{x} + 5$ No

LESSON 14.4

Graph each function.

1. $y = -2x^2$
2. $y = -|x|$
3. $y = \frac{-2}{x}$
4. $y = -3x$
5. $y = 4x^2$
6. $y = \frac{4}{x}$

Determine whether each function is a vertical reflection of the parent function.

7. $y = x^2$ No
8. $y = |x| - 1$ No
9. $y = \frac{-1}{x}$ Yes
10. $y = (-x)^2$ No
11. $y = -x$ Yes
12. $y = x^2 - 1$ No
13. $y = -x^2$ Yes
14. $y = |-x|$ No
15. $y = x - 1$ No

Find $f(3)$, and use this information to decide whether a vertical reflection has been applied to $f(x) = x^2 - 6$.

16. $f(x) = -(x^2 - 6)$ -3, Yes
17. $f(x) = (-x)^2 - 6$ -3, No

Tell whether the graph of each function shows a stretch or a translation.

18. $y = 5x$ stretch
19. $y = 5 + x$ trans
20. $y = \frac{3}{5}x$ stretch
21. $y = \frac{3}{5} + x$ trans

2.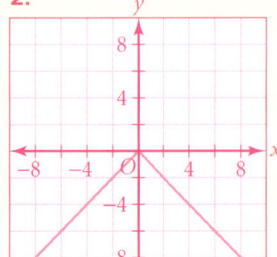

3.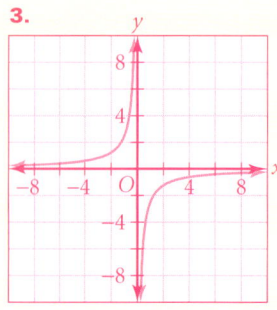

4.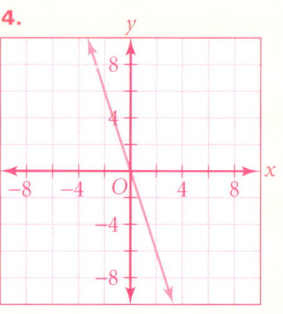

5 and 6. Check students' graphs.

LESSON 14.5

Sketch the graph of each function.

1. $k(x) = 3(x + 1)^2$
2. $n(x) = 3x^2 + 1$
3. $h(x) = |x + 2| - 1$
4. $q(x) = 5|x - 2|$
5. $f(x) = 2^x - 2$
6. $m(x) = \dfrac{2}{x} - 1$

Name the point that corresponds to the point (4, 6) on the graph of $y = \dfrac{3}{2}x$ after each transformation is applied. HINT: If you are not sure, draw a sketch.

7. a vertical stretch by 2, followed by a vertical translation by 2 (4, 14)
8. a vertical reflection, followed by a vertical stretch by 3 (4, −18)
9. a vertical reflection, followed by a vertical translation by −2 (4, −8)
10. a vertical stretch by −2, followed by a horizontal translation by 4 (8, −12)

Name the point that corresponds to the point (4, 8) on the graph of $y = \dfrac{1}{2}x^2$ after each transformation is applied. HINT: If you are not sure, draw a sketch.

11. a vertical stretch by 2, followed by a vertical translation by 2 (4, 18)
12. a vertical reflection, followed by a vertical stretch by 3 (4, −24)
13. a vertical reflection, followed by a vertical translation by −2 (4, −10)
14. a vertical stretch by −2, followed by a horizontal translation by 4 (8, −16)

Lesson 14.5

1.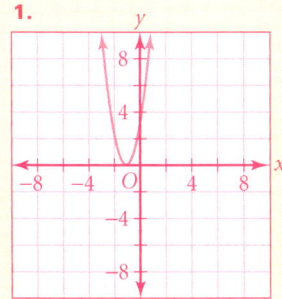
2.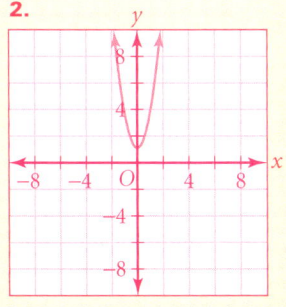
3.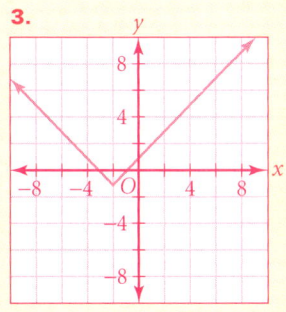
4.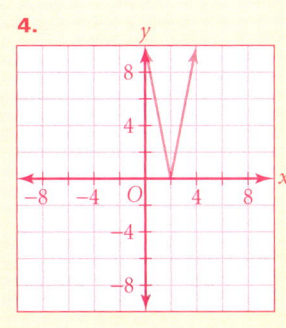
5.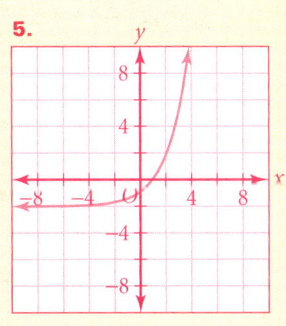
6.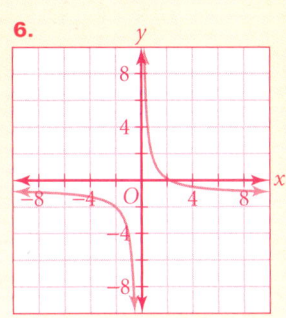

Calculator Keystroke Guide

Chapter 1

Essential keystroke sequences (using the TI-82, TI-83, or TI-83 Plus graphics calculator) are presented below for the most significant activities and examples found in this chapter that require or suggest the use of a graphics calculator.

  internet connect

For keystrokes to other models of graphics calculators, visit the HRW Web site at **go.hrw.com** and enter the keyword **MA1 CALC**.

LESSON 1.2

EXAMPLE 4
Page 14

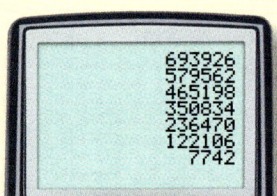

Part a. Find the solution by using the constant feature of a calculator.

Enter 922,654 and subtract 114,364 repeatedly, counting the number of times you subtract:

922654 [ENTER] [−] 114364 [ENTER] [ENTER] [ENTER] ... (Press [ENTER] a total of 8 times.)

Part b. Write an equation and solve it by using the table feature.

Enter the equation:
[Y=] (Y1=) 922654 [−] 114364 [X,T,θ,n]

Use a table:
         TBLSET
[2nd] [WINDOW] (TblStart=) 1 [ENTER] (ΔTbl=)1 [ENTER]
          ↑TI-82: (TblMin =)
                                                              TABLE
(Indpnt:) [Auto] [▼] [ENTER] (Depend:) [Auto] [ENTER] [2nd] [GRAPH]

LESSON 1.3

Activity
Page 18

For Step 3a, predict the value of 2 + 3 • 5.

2 [+] 3 [×] 5 [ENTER]

Use a similar keystroke sequence for Steps 3b–3h.

LESSON 1.3

EXAMPLE 4
Page 20

Use a graphics or scientific calculator to evaluate

$\dfrac{57 + 95}{16} + \dfrac{220}{88 + 104}$.

[(] 57 [+] 95 [)] [÷] 16 [+] 220 [÷] [(] 88 [+] 104 [)] [ENTER]

LESSON 1.5

Activity
Page 31

The window dimension 4.7 produces a "friendly" window which is easy to use with the TRACE function.

Graph and trace the equation $d = 53t$ on a graphics calculator.

Use friendly viewing window [0, 4.7] by [0, 250].

Set the viewing window:

WINDOW (Xmin =) 0 ▼ (Xmax=) 4.7 ▼ (Xscl=) 1 ▼ (Ymin=) 0 ▼ (Ymax=) 250 ▼ (Yscl=) 10

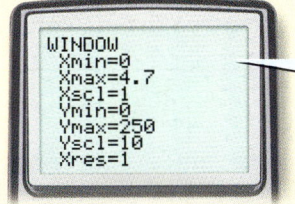

The scales of the x-axis and y-axis should be chosen according to the difference between the minimum and maximum values. (Normally, **Xres=1** should not be changed.)

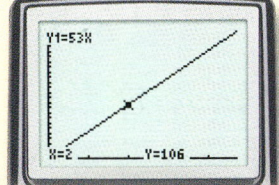

Graph and trace the equation:

Y= (Y1=) 53 X,T,θ,n GRAPH TRACE (Use ◄ or ► to move along the graph.)

EXAMPLE 2
Pages 32 and 33

The window dimension 9.4 also produces a friendly viewing window because it is a multiple of 4.7.

Graph the equation $d = 280t$.

Use friendly viewing window [0, 9.4] Xscl = 1 by [0, 3000] Yscl = 500.

Graph the equation:

Y= (Y1=) 280 X,T,θ,n GRAPH TRACE

(Use ◄ or ► to move along the graph to the desired point.)

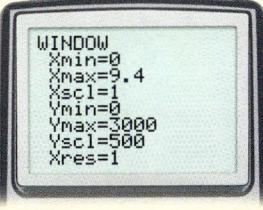 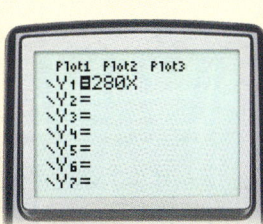 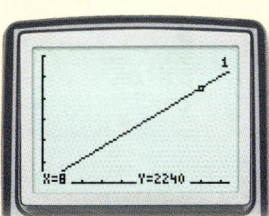

EXAMPLE 3
Page 33

Graph the equation $r = 220 - a$ on your graphics calculator.

Use friendly viewing window [0, 94] Xscl = 10 by [0, 230] Yscl = 25.

Y= (Y1=) 220 − X,T,θ,n ENTER GRAPH TRACE

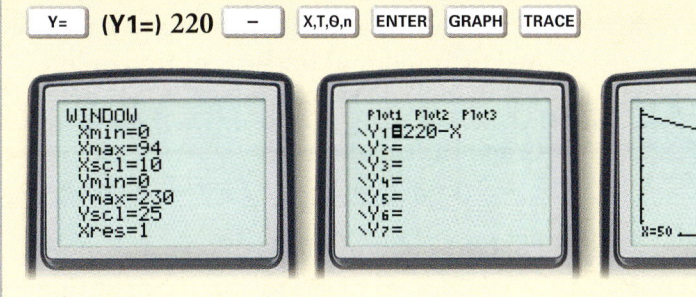

783

LESSON 1.6

Pages 38 and 39

For Step 1, use a graphics calculator to draw a scatter plot of the data.

Use viewing window [0, 20] Xscl = 1 by [0, 250] Yscl = 10.

Make a table of values:
STAT EDIT 1:EDIT ENTER

Under the heading L1, enter the data values that represent the year:
0 ENTER 2 ENTER 4 ENTER 6 ENTER 8 ENTER 10 ENTER 12 ENTER 14 ENTER 16 ENTER (Use ► and ▲ to move to the top of the **L2** list.)

Under L2, enter the data values that represent population:
50 ENTER 75 ENTER 90 ENTER 115 ENTER 145 ENTER 165 ENTER 180 ENTER 200 ENTER 230 ENTER 2nd Y= STAT PLOTS

1:Plot 1 ENTER On ENTER ▼ (TYPE:)  ENTER ▼ (XLIST:) L1 ENTER (YLIST:) L2 ENTER Mark: ▫ ENTER GRAPH

For Step 3, graph the equation $y = 10x + 50$:
Y= (Y1=) 10 X,T,θ,n + 50 GRAPH

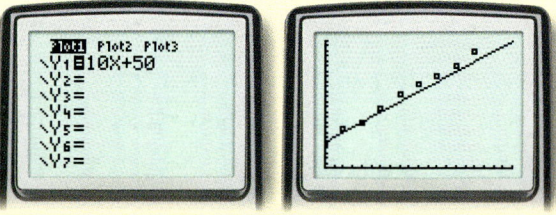

For Step 4, adjust the equation to fit the data by using a guess and check strategy.

Chapter 3

Essential keystroke sequences (using the TI-82, TI-83, or TI-83 Plus graphics calculator) are presented below for the most significant activities and examples found in this chapter that require or suggest the use of a graphics calculator.

internet connect
For keystrokes to other models of graphics calculators, visit the HRW Web site at **go.hrw.com** and enter the keyword **MA1 CALC**.

LESSON 3.1

EXAMPLE 5
Page 118

Graph the equations $x + 7 = y$ and $y = 10$ on the same screen and find the *x*-value at the intersection point.
Use friendly viewing window [0, 9.4] by [0, 15].

Graph the equations:
Y= (Y1=) X,T,θ,n + 7 ENTER (Y2=) 10 GRAPH

Find the point of intersection using the TRACE feature:
TRACE (Use ◄ or ► to move the cursor to $y = 10$.)

Set up a table:

2nd [WINDOW] (TblStart=)1 ENTER (ΔTbl=)1 ENTER
↑TI-82: (TblMin=)
(Indpnt:) Auto ENTER ▼ (Depend:) Auto ENTER

Use the TABLE feature to confirm that $x = 3$ when $y = 10$ by determining the x-value when both **Y1** and **Y2** equal 10.

Locate values on the table: 2nd [GRAPH]

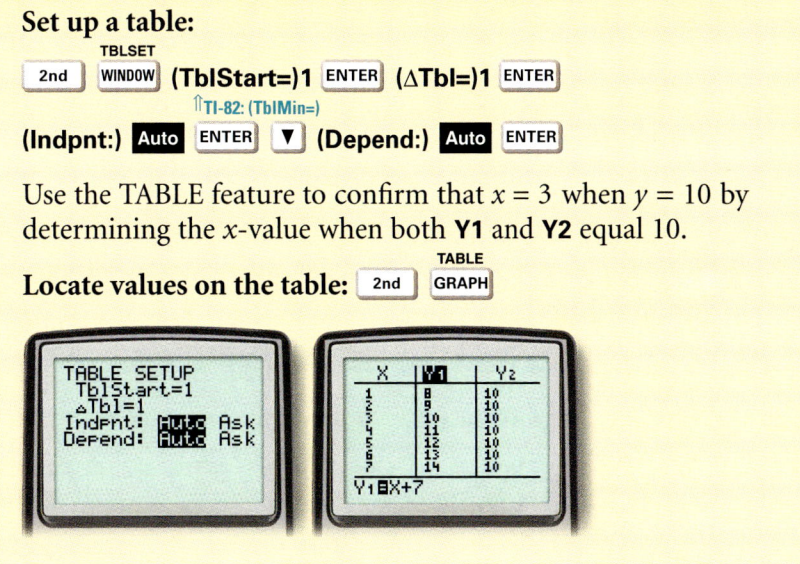

LESSON 3.3

TECHNOLOGY
Page 132

Graph the equations $y = 18 + 7x$ and $y = 74$ on the same screen. Use window $[-1, 9.4]$ by $[-5, 120]$.

Y= (Y1=) 18 + 7 X,T,θ,n ENTER (Y2=) 74 GRAPH

Find the point of intersection using the CALCULATE function:

2nd TRACE 5:intersect ENTER (First curve?) ENTER
(Second curve?) ENTER (Guess?) ENTER

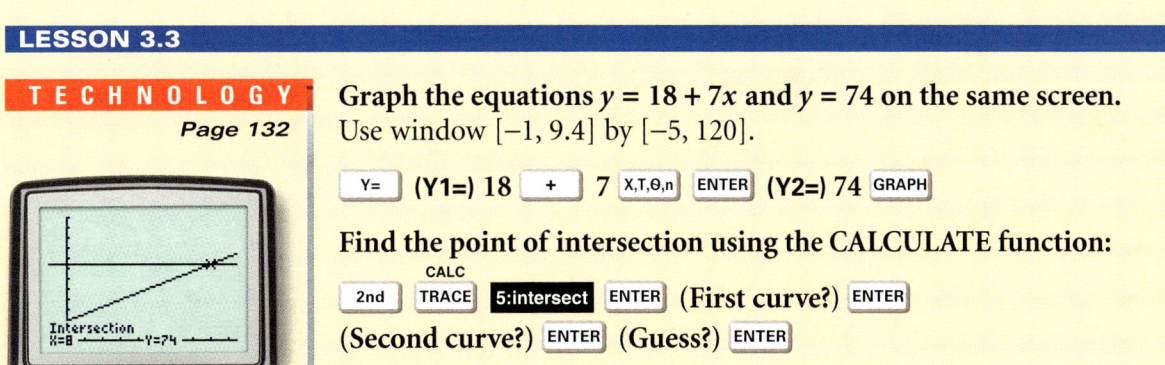

LESSON 3.4

Page 137

Solve the equation $5x - 6 = -6x + 16$.
Use window $[-5, 10]$ by $[-5, 10]$.
Graph the equations $y = 5x - 6$ and $y = -6x + 16$ on the same calculator display and find their point of intersection.

Graph the equations:

Y= (Y1=) 5 X,T,θ,n − 6 ENTER (Y2=) (−) 6 X,T,θ,n + 16 GRAPH

Find the point of intersection using the CALCULATE function:

2nd TRACE 5:intersect ENTER (First curve?) ENTER
(Second curve?) ENTER (Guess?) ENTER

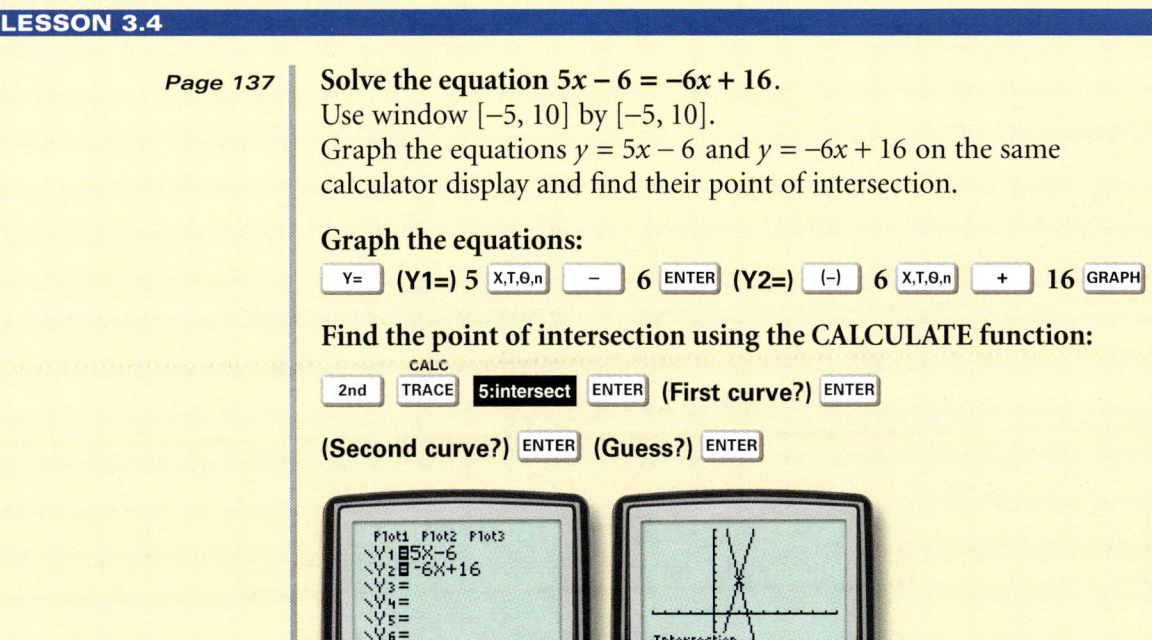

785

Chapter 4

Essential keystroke sequences (using the TI-82, TI-83, or TI-83 Plus graphics calculator) are presented below for the most significant activities and examples found in this chapter that require or suggest the use of a graphics calculator.

internet connect

For keystrokes to other models of graphics calculators, visit the HRW Web site at **go.hrw.com** and enter the keyword **MA1 CALC**.

LESSON 4.6

EXAMPLE 2
Pages 203 and 204

Create a box-and-whisker plot from the data given in Example 2.

Use viewing window [12, 16] Xscl = 0.2 by [0, 10] Yscl = 1.
Enter the data:
STAT EDIT 1:EDIT ENTER (LI) 12.7 ENTER 12.8 ENTER 12.8 ENTER 12.8 ENTER 12.9 ENTER 13.0 ENTER 13.1 ENTER 13.1 ENTER 1 ENTER (Continue using a similar keystroke sequence to enter the remaining data.)

Create a box-and-whisker plot:
2nd Y= STAT PLOTS 1:Plot 1 ENTER On ENTER ▼ (TYPE:) ⊞
ENTER ▼ (XLIST:) L1 ENTER (Freq:) 1 ENTER GRAPH

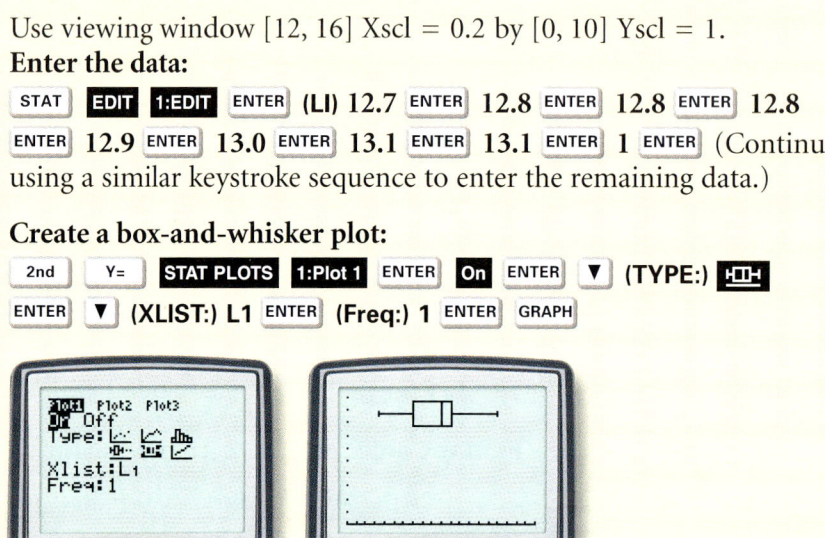

Chapter 5

Essential keystroke sequences (using the TI-82, TI-83, or TI-83 Plus graphics calculator) are presented below for the most significant activities and examples found in this chapter that require or suggest the use of a graphics calculator.

internet connect

For keystrokes to other models of graphics calculators, visit the HRW Web site at **go.hrw.com** and enter the keyword **MA1 CALC**.

LESSON 5.1

EXAMPLE 3
Page 221

Use your graphing calculator to list the ordered pairs and plot the corresponding points.
Use viewing window [0, 12] by [0, 70].

Enter the data:
STAT EDIT 1:EDIT ENTER (LI) 1 ENTER 2 ENTER 3 ENTER and so on through L1 and L2.

Create a graph:
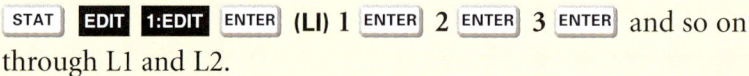

LESSON 5.2

Activity — Page 229

For Step 1, graph the point A(3, 6).
Use viewing window [−5, 10] by [−10, 10].

Graph the point:

STAT EDIT 1:EDIT ENTER (L1) 3 ENTER (L2) 6 ENTER 2nd Y=
STAT PLOTS 1:Plot 1 ENTER On ENTER ▼ (TYPE:) ⋰ ENTER ▼
(XLIST:) L1 ENTER (YLIST:) L2 ENTER (Mark:) □ ENTER GRAPH

For Step 2, find values for *m* such that line $y = mx$ intersects with point A. For example, let $m = 4$: Y= (Y1=)4 X,T,θ,n GRAPH

For Step 3, use a similar keystroke sequence to graph the other given points.

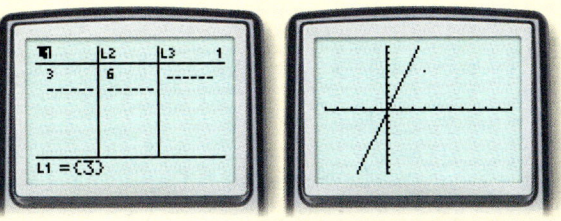

LESSON 5.4

Activity — Page 246

For Step 1, graph the equations $y = 2x$, $y = 2x + 3$, $y = 2x + 5$, and $y = 2x - 3$ on the screen.

To format the *standard viewing window* [−10, 10] by [−10, 10]: ZOOM
ZOOM 6:ZStandard ENTER

Enter the equations:

Y= (Y1=) 2 X,T,θ,n ENTER (Y2=) 2 X,T,θ,n + 3 ENTER
(Y3=) 2 X,T,θ,n + 5 ENTER (Y4=) 2 X,T,θ,n − 3 ENTER GRAPH

For Step 4, use a similar keystroke sequence to graph $y = 2x - 4$.

LESSON 5.5

EXAMPLE 2 — Pages 253 and 254

Enter the equation from Example 2 in a graphics calculator.
(First rewrite the equation in slope-intercept form, $y = -2x + 300$.)

Use viewing window [−10, 200] by [−10, 350] with scales of 10.

Graph the equation:

Y= (Y1=) (−) 2 X,T,θ,n + 300 GRAPH

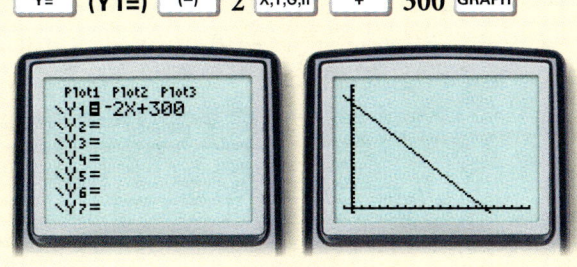

Calculator Keystroke Guide

Chapter 6

Essential keystroke sequences (using the TI-82, TI-83, or TI-83 Plus graphics calculator) are presented below for the most significant activities and examples found in this chapter that require or suggest the use of a graphics calculator.

internet connect
For keystrokes to other models of graphics calculators, visit the HRW Web site at **go.hrw.com** and enter the keyword **MA1 CALC**.

LESSON 6.2

EXAMPLE 4 — Page 285

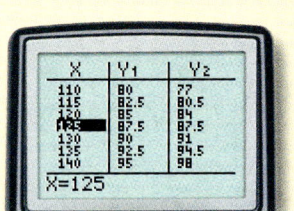

Use the TABLE feature to determine the more economical calling plan.

Enter the equations:
[Y=] (Y1=) 25 [+] 0.5 [X,T,θ,n] [ENTER] (Y2=) 0.7 [X,T,θ,n]

Set up a table:
[2nd] [WINDOW] (TBLSET) (TblStart=) 100 [ENTER] (ΔTbl=) 5 [ENTER]
⇑TI-82: (TblMin =)
(Indpnt:) [Auto] [ENTER] [▼] (Depend:) [Auto] [ENTER]

Locate values on the table: [2nd] [GRAPH] (TABLE)
Use the [▼] key to locate the x-value when **Y1 = Y2**.

LESSON 6.4

Activity — Page 296

For Step 1, graph $y = |x|$ and $y = |x - 3|$ on the same screen.
Use viewing window $[-10, 10]$ by $[-5, 5]$.

Graph the equations:
[Y=] (Y1=) [MATH] [▶] [NUM] [1:abs (] [ENTER] [X,T,θ,n] [ )]
⇑TI-82: [2nd] [x^{-1}] (ABS) [( ]

(Y2=) [MATH] [▶] [NUM] [1:abs (] [ENTER] [X,T,θ,n] [−] 3 [ )] [GRAPH]
⇑TI-82: [2nd] [x^{-1}] (ABS) [( ]

Use the same sequence of keystrokes to graph the equations in Steps 3 and 5.

LESSON 6.5

Activity — Page 304

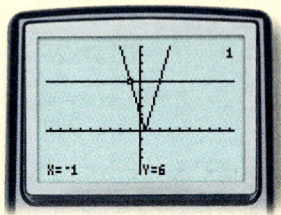

For Step 1, graph the equations $y = |4x - 2|$ and $y = 6$ on the same screen.
Use viewing window $[-10, 10]$ by $[-5, 10]$.
Graph the equations:
[Y=] (Y1=) [MATH] [▶] [NUM] [1:abs (] [ENTER] 4 [X,T,θ,n] [−] 2 [ )] [ENTER]
⇑TI-82: [2nd] [x^{-1}] (ABS) [( ]

(Y2=) 6 [GRAPH]
Use a similar keystroke sequence for Steps **4a–4c**.

Chapter 7

Essential keystroke sequences (using the TI-82, TI-83, or TI-83 Plus graphics calculator) are presented below for the most significant activities and examples found in this chapter that require or suggest the use of a graphics calculator.

> **internet connect**
> For keystrokes to other models of graphics calculators, visit the HRW Web site at **go.hrw.com** and enter the keyword **MA1 CALC**.

LESSON 7.1

EXAMPLE 1 Page 319

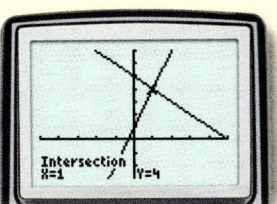

Graph $y = 3x + 1$ and $y = -x + 5$ on the same screen.
Use viewing window $[-5, 5]$ by $[-3, 7]$.

Graph the equations:

Y= (Y1=) 3 X,T,θ,n + 1 ENTER (Y2=) (−) X,T,θ,n + 5 GRAPH

Find the point of intersection:

2nd TRACE 5:intersect ENTER (First curve?) ENTER (Second curve?) ENTER (Guess?) ENTER

EXAMPLE 2 Page 320

Graph $y = -3x + 4$ and $y = \frac{1}{2}x - 3$ on the same screen.

Use viewing window $[-3, 7]$ by $[-7, 3]$.

Use a keystroke sequence similar to Example 1 in this lesson.

Activity Page 321

For Step 3, graph the equations $4x + 3y = 6$ and $2y = x + 2$ on the same screen.

Use the standard viewing window (shown in the activity in 5.4).

Enter the equations in slope–intercept form. Use a keystroke sequence similar to that in Example 1 of this lesson to graph the equations and to find the point of intersection.

LESSON 7.4

Activity Page 338

For Step 1, graph $y = -3x + 4$ and $y = -3x + 8$ on the same screen.

Use viewing window $[-4.7, 4.7]$ by $[-3.1, 3.1]$.

Graph the equations:

Y= (Y1=) (−) 3 X,T,θ,n + 4 ENTER (Y2=) (−) 3 X,T,θ,n + 8 GRAPH

To enter $3y = 6x - 9$ in slope-intercept form you need to divide both sides by 3.

For Step 2, graph $y = 2x - 3$ and $3y = 6x - 9$ on the same screen:

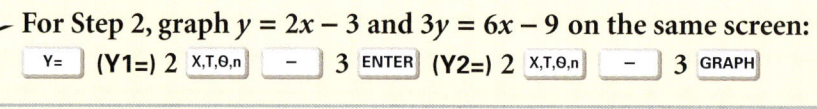

EXAMPLE 1
Page 339

Determine whether the system $\begin{cases} y = 4 - x \\ y = 3x - 6 \end{cases}$ is consistent or inconsistent.

Use a keystroke sequence similar to that in the Activity in this lesson.

EXAMPLE 3
Page 340

Determine the line of best fit for the two sets of data in Example 3. Use viewing window [0, 25] by [0, 30] with scales of 5.

Make a table of values:
STAT   EDIT   1:EDIT   ENTER

Under the heading **L1**, enter the data values that represent the year:
0 ENTER 5 ENTER 10 ENTER 15 ENTER 20 ENTER

Use ▶ and ▲ to move to the top of the **L2** list.

Under **L2**, enter the data values that represent women's ages:
20.6 ENTER 20.8 ENTER 21.8 ENTER 23 ENTER 24 ENTER

Under L3, enter the data values that represent men's ages:
22.5 ENTER 22.7 ENTER 23.6 ENTER 24.8 ENTER 25.9 ENTER

Make a scatterplot of the women's ages:
2nd   Y=   STAT PLOTS   1:Plot 1   ENTER   ON   ENTER   ▼   (TYPE:)
ENTER   ▼   (XLIST:)   L1   ENTER   (YLIST:)   L2   ENTER   Mark: □   ENTER   GRAPH

Make a scatterplot of the men's ages using the same keystrokes but substituting **2:PLOT2** for **1:PLOT1** and **YLIST: L3** for **YLIST: L2**. In order to differentiate between the plots, you may also select one of the other marks.

Find the equation for the line of best fit for Plot 1 using the linear regression feature:

STAT   CALC   4:LinReg(ax+b)   ENTER   (LinReg(ax+b))   2nd   1 (L1)   ,
2nd   2 (L2)   ENTER

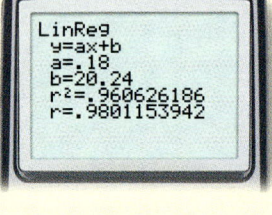

Use the same keystroke sequence for Plot 2, entering
2nd   3 (L3)   in place of   2nd   2 (L2).

EXAMPLE 4
Page 341

Determine whether the system $\begin{cases} y = 2x - 1 \\ 3x - 1.5y = 1.5 \end{cases}$ is dependent or independent.

Use a keystroke sequence similar to that in the Activity in this lesson.

LESSON 7.5

Page 346

Graph the inequality $x + y \leq 5000$.
Use viewing window [−1000, 6000] by [−1000, 6000] with scales of 1000.

For Step 1, replace the inequality sign with an equal sign in the equation $x + y \leq 5000$, then graph the resulting equation $y = -x + 5000$.

Graph the equation:

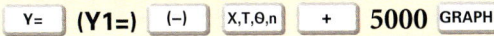

For Step 3, shade the values that satisfy the inequality:
Go to the [Y=] menu and press [◄] two times or until the cursor is on the left of **Y1=**. Then press [ENTER] three times or until you see the shaded area flashing, then press [GRAPH].

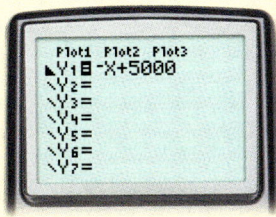 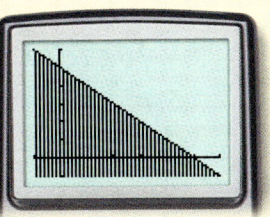

Chapter 8

Essential keystroke sequences (using the TI-82, TI-83, or TI-83 Plus graphics calculator) are presented below for the most significant activities and examples found in this chapter that require or suggest the use of a graphics calculator.

internetconnect

For keystrokes to other models of graphics calculators, visit the HRW Web site at **go.hrw.com** and enter the keyword **MA1 CALC**.

LESSON 8.7

EXAMPLE
Page 412

Solve the population problem in Example 5.
Use viewing window [0, 10] Xscl = 1 by [110 000, 170 000] Yscl = 10000.

Graph $y = 152{,}494(0.99)^x$ and $y = 140{,}000$ on the same screen, and find their point of intersection.

Graph the equations:

(Y2=) 140000 [GRAPH]

Find the point of intersection:

(Second curve?) [ENTER] (Guess?) [ENTER]

Chapter 9

Essential keystroke sequences (using the TI-82, TI-83, or TI-83 Plus graphics calculator) are presented below for the most significant activities and examples found in this chapter that require or suggest the use of a graphics calculator.

 internet connect

For keystrokes to other models of graphics calculators, visit the HRW Web site at **go.hrw.com** and enter the keyword **MA1 CALC**.

LESSON 9.4

EXAMPLE 2 Page 445

Use a table to determine the value for h in the equation $1357.2 = 36\pi h$.

To enter the equation, substitute x for h and enter $y = 36\pi h$.
Enter the equation:

[Y=] (Y1=) 36 [2nd] [^]π [X,T,θ,n]

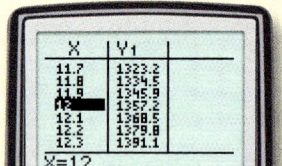

Make a table:

[2nd] [WINDOW]^{TBLSET} (TblStart=) 11.7 [ENTER] (ΔTbl=) .1 [ENTER]
⇑TI-82: (TblMin=)
(Indpnt:) [Auto] [ENTER] [▼] (Depend:) [Auto] [ENTER] [2nd] [GRAPH]^{TABLE}

Use [▼] to move the cursor and determine that when $y = 1357.2$, $x = 12$.

EXAMPLE 3 Page 445

Show that $(x + 2)^3 = x^3 + 6x^2 + 12x + 8$ is an identity.

Enter each side as a separate equation:

[Y=] (Y1=) [(] [X,T,θ,n] [+] 2 [)] [^] 3 [ENTER]
(Y2=) [X,T,θ,n] [^] 3 [+] 6 [X,T,θ,n] [^] 2 [+] 12 [X,T,θ,n] [+] 8

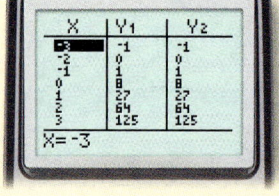

Make a table of values:

[2nd] [WINDOW]^{TBLSET} (TblStart=) [(-)] 3 [ENTER] (ΔTbl=) 1 [ENTER]
⇑TI-82: (TblMin=)
(Indpnt:) [Auto] [ENTER] [▼] (Depend:) [Auto] [ENTER] [2nd] [GRAPH]^{TABLE}

Use the table to confirm that Y1 = Y2 for integer values greater than −3.

LESSON 9.8

Activity Page 464

For Step 6, graph the equation $y = -16x^2 + 48x$.

Use friendly viewing window [0, 4.7] by [0, 50].

Graph the equation:

[Y=] (Y1=) [(-)] 16 [X,T,θ,n] [x^2] [+] 48 [X,T,θ,n] [GRAPH]

Locate the x-value(s) for the point $y = 0$: [TRACE].

Then use [▶] until the cursor arrives at $y = 0$. The x-value will appear on the screen.

TECHNOLOGY
Page 465

Find the zeros of the polynomial function $y = (x + 5)(x + 2)$.

Enter the function:

[Y=] (Y1=) [(] [X,T,θ,n] [+] 5 [)] [(] [X,T,θ,n] [+] 2 [)]

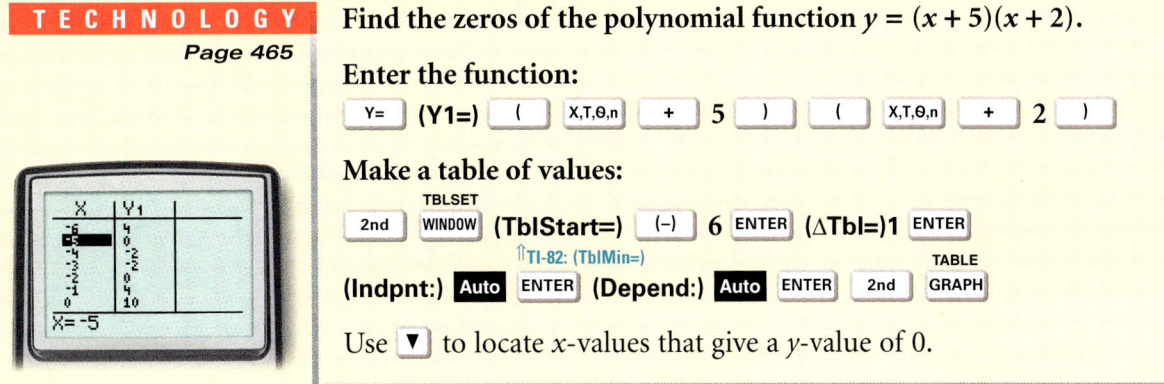

Make a table of values:

[2nd] [WINDOW] (TBLSET) (TblStart=) [(−)] 6 [ENTER] (ΔTbl=)1 [ENTER]
⇧TI-82: (TblMin=)
(Indpnt:) [Auto] [ENTER] (Depend:) [Auto] [ENTER] [2nd] [GRAPH] (TABLE)

Use ▼ to locate x-values that give a y-value of 0.

EXAMPLE 2
Page 466

Check your solutions to the equation $-16x^2 + 96x = 0$ by using the table and graph features.
Use friendly viewing window [0, 9.4] by [0, 150].

Graph the function:

[Y=] (Y1=) [(−)] 16 [X,T,θ,n] [x^2] [+] 96 [X,T,θ,n] [GRAPH]

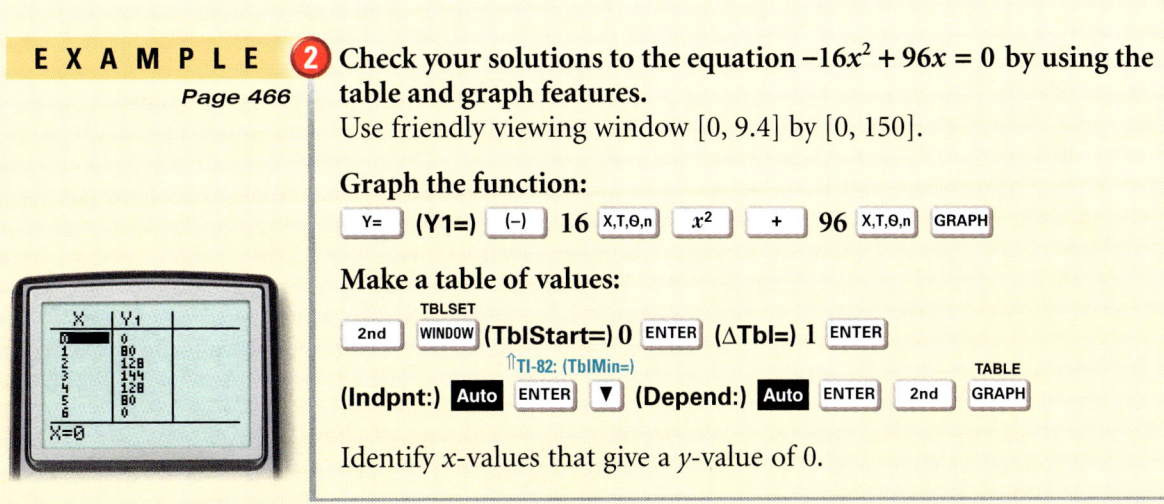

Make a table of values:

[2nd] [WINDOW] (TBLSET) (TblStart=) 0 [ENTER] (ΔTbl=) 1 [ENTER]
⇧TI-82: (TblMin=)
(Indpnt:) [Auto] [ENTER] ▼ (Depend:) [Auto] [ENTER] [2nd] [GRAPH] (TABLE)

Identify x-values that give a y-value of 0.

TECHNOLOGY
Page 467

Find the zeros of the polynomial function $y = 4x^2 + x - 5$.
Use viewing window [−5, 5] by [−5, 5].

Graph the function:

[Y=] (Y1=) 4 [X,T,θ,n] [x^2] [+] [X,T,θ,n] [−] 5 [GRAPH]

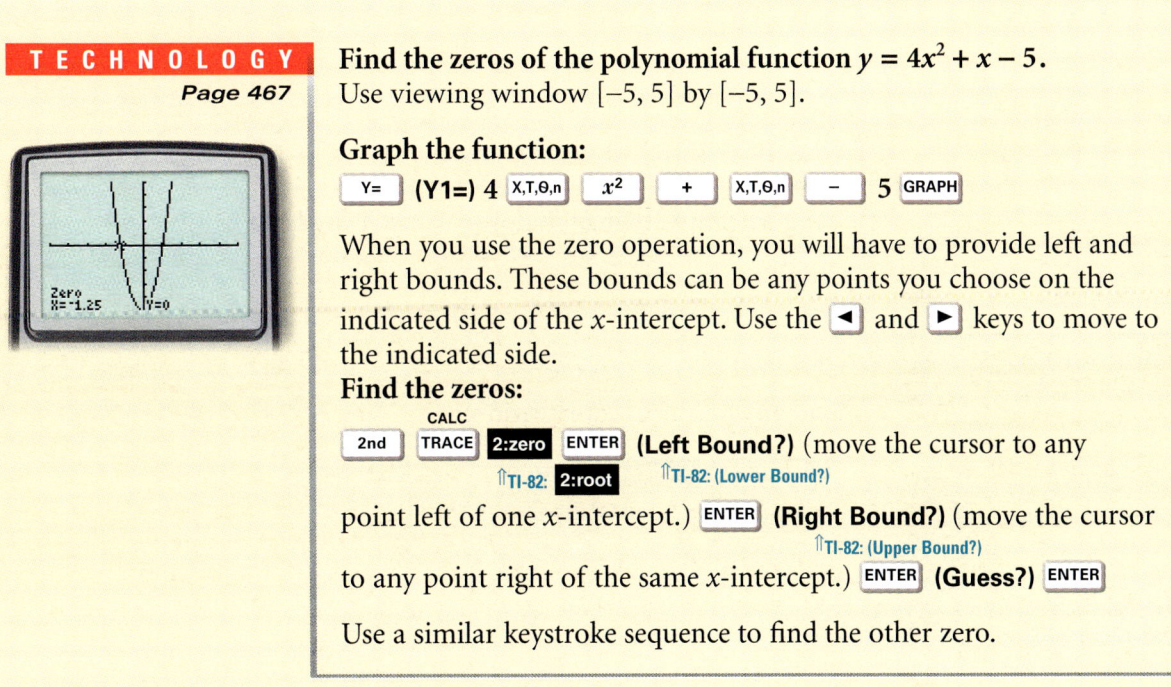

When you use the zero operation, you will have to provide left and right bounds. These bounds can be any points you choose on the indicated side of the x-intercept. Use the ◄ and ► keys to move to the indicated side.

Find the zeros:

[2nd] [TRACE] (CALC) [2:zero] [ENTER] (Left Bound?) (move the cursor to any
⇧TI-82: [2:root] ⇧TI-82: (Lower Bound?)
point left of one x-intercept.) [ENTER] (Right Bound?) (move the cursor
⇧TI-82: (Upper Bound?)
to any point right of the same x-intercept.) [ENTER] (Guess?) [ENTER]

Use a similar keystroke sequence to find the other zero.

Chapter 10

Essential keystroke sequences (using the TI-82, TI-83, or TI-83 Plus graphics calculator) are presented below for the most significant activities and examples found in this chapter that require or suggest the use of a graphics calculator.

 internet connect

For keystrokes to other models of graphics calculators, visit the HRW Web site at **go.hrw.com** and enter the keyword **MA1 CALC**.

LESSON 10.4

EXAMPLE 3
Page 500

Solve $-5 = x^2 + 3x - 15$.
Use a friendly viewing window $[-9.4, 9.4]$ by $[-20, 10]$.

Graph the function:

(Y2=) [(-)] [5] [GRAPH]

Find any points of intersection:

(Second curve?) [ENTER] (Guess?) [ENTER]

Use similar keystrokes to identify the other point(s) of intersection. However, when **Guess?** is displayed, move the cursor closer to the desired intersection before you press [ENTER].

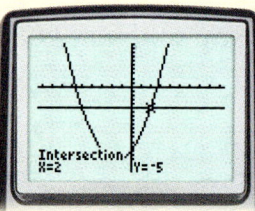 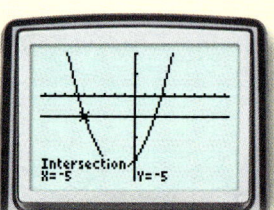

EXAMPLE 4
Page 501

Find the point(s) where the graph of $y = x^2 - 10x + 21$ intersects the graph of $y = x - 3$.
Use a friendly viewing window $[-9.4, 9.4]$ by $[-10, 10]$.

Graph the functions:

[Y=] (Y1=) [X,T,θ,n] [x^2] [−] [10] [X,T,θ,n] [+] [21] [ENTER]
(Y2=) [X,T,θ,n] [−] [3] [GRAPH]

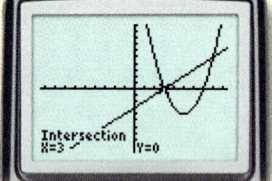

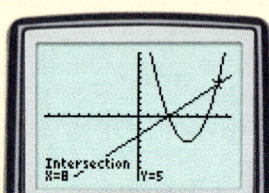

Find any points of intersection:

[2nd] [TRACE] **5:intersect** [ENTER] (First curve?) [ENTER]
(Second curve?) [ENTER] (Guess?) [ENTER]

Use similar keystrokes to find the other point of intersection.

Chapter 11

Essential keystroke sequences (using the TI-82, TI-83, or TI-83 Plus graphics calculator) are presented below for the most significant activities and examples found in this chapter that require or suggest the use of a graphics calculator.

For keystrokes to other models of graphics calculators, visit the HRW Web site at **go.hrw.com** and enter the keyword **MA1 CALC**.

LESSON 11.2

Activity — Page 533

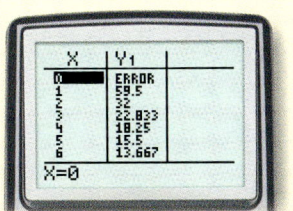

For Step 4, graph the function $A(n) = \frac{55 + 4.50n}{n}$.

Use a friendly viewing window $[-47, 47]$ by $[-31, 31]$ with scales of 10.

Graph the function:

[Y=] (Y1=) [(] 55 [+] 4.5 [X,T,θ,n] [)] [÷] [X,T,θ,n] [GRAPH]

Make a table: [2nd] [GRAPH]
(TABLE)

EXAMPLE 2 — Pages 534 and 535

Use the graph of $y = \frac{55 + 4.50n}{n}$ to find the n-value of the function for which $y = 5.60$.

Graph the function:

[Y=] (Y1=) [(] 55 [+] 4.5 [X,T,θ,n] [)] [÷] [X,T,θ,n] [GRAPH]

Use the TRACE feature and the ◄ and ► keys to locate $y = 5.6$.

The n-value will be shown on the screen as $x = 50$.

LESSON 11.3

Activity — Page 538

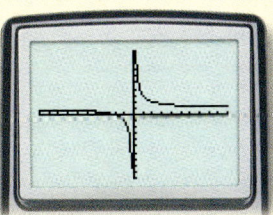

For Step 3, graph $y = \frac{3x + 6}{3x}$ and $y = \frac{x + 2}{x}$.

Use the standard viewing window.

Graph the equations:

[Y=] (Y1=) [(] 3 [X,T,θ,n] [+] 6 [)] [÷] [(] 3 [X,T,θ,n] [)]
[ENTER] (Y2=) [(] [X,T,θ,n] [+] 2 [)] [÷] [X,T,θ,n] [GRAPH]

Go to TRACE and use ▼ and ▲ to switch from one graph to the other; observe that the functions are the same for all values of x.

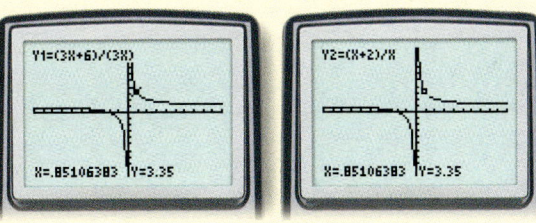

LESSON 11.5

EXAMPLE ❶ Use graphing to solve $\frac{6}{x-1} + 2 = \frac{12}{x^2-1}$.

Page 554

Use a friendly viewing window [−4.7, 4.7] by [−10, 10].

Graph each side of the equation separately.

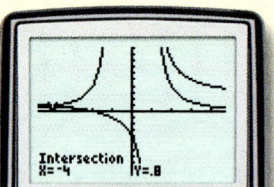

Graph the equations:

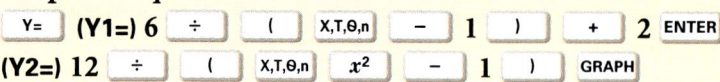

Look for any points of intersection:

2nd TRACE **5:intersect** ENTER (First curve?) ENTER (Second curve?) ENTER (Guess?) ENTER

Chapter 12

Essential keystroke sequences (using the TI-82, TI-83, or TI-83 Plus graphics calculator) are presented below for the most significant activities and examples found in this chapter that require or suggest the use of a graphics calculator.

internetconnect

For keystrokes to other models of graphics calculators, visit the HRW Web site at **go.hrw.com** and enter the keyword **MA1 CALC**.

LESSON 12.2

EXAMPLE ❷ Solve $x + 2 = 3$.

Pages 585 and 586

Use the standard viewing window.

Enter the left and right sides of the equation as separate functions:

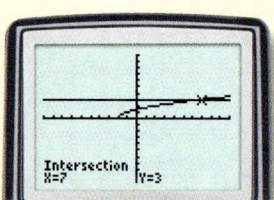

Look for the intersection:

(Second curve?) ENTER (Guess?) ENTER

Chapter 13

Essential keystroke sequences (using the TI-82, TI-83, or TI-83 Plus graphics calculator) are presented below for the most significant activities and examples found in this chapter that require or suggest the use of a graphics calculator.

>  **internetconnect**
>
> For keystrokes to other models of graphics calculators, visit the HRW Web site at **go.hrw.com** and enter the keyword **MA1 CALC**.

LESSON 13.5

TECHNOLOGY
Page 676

Generate 0's and 1's with the random integer feature:

`MATH` `PRB` `5:randInt(` `ENTER` `0` `,` `1` `)` `ENTER`

(Push `ENTER` once for every number you want to generate.)

Chapter 14

Essential keystroke sequences (using the TI-82, TI-83, or TI-83 Plus graphics calculator) are presented below for the most significant activities and examples found in this chapter that require or suggest the use of a graphics calculator.

> **internetconnect**
>
> For keystrokes to other models of graphics calculators, visit the HRW Web site at **go.hrw.com** and enter the keyword **MA1 CALC**.

LESSON 14.2

Activity
Page 700

For Step 1, graph the functions $f(x) = |x|$ and $g(x) = |x| - 10$ on the same screen.

Use the standard viewing window.

Graph the functions:

`Y=` (Y1=) `MATH` `▶` `NUM` `1:abs (` `ENTER` `X,T,θ,n` `)` `ENTER`
⇑TI-82: `2nd` `x⁻¹` **ABS** `(`

(Y2=) `MATH` `NUM` `1:abs (` `ENTER` `X,T,θ,n` `)` `−` `10` `GRAPH`
⇑TI-82: `2nd` `x⁻¹` **ABS** `2nd`

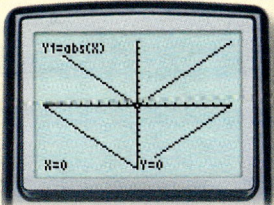

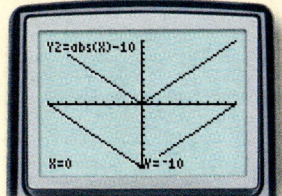

For Step 2, use the TRACE function and the ▼ ▲ buttons to determine the number of units shifted:

For Step 3, graph the functions $f(x) = |x|$ and $h(x) = |x - 10|$ on the same screen.

Use viewing window $[-10, 20]$ by $[-10, 10]$.

Y= (Y1=) MATH NUM 1:abs (ENTER X,T,θ,n) ENTER
ABS
⇑TI-82: 2nd x^{-1} 2nd

(Y2=) MATH NUM 1:abs (ENTER X,T,θ,n − 10) GRAPH
ABS
⇑TI-82: 2nd x^{-1} 2nd

TABLE
For Step 4, make a table: 2nd GRAPH

Locate the *x*-values where **Y1** = 0 and where **Y2** = 0. The difference between these *x*-values will give the number of units shifted.

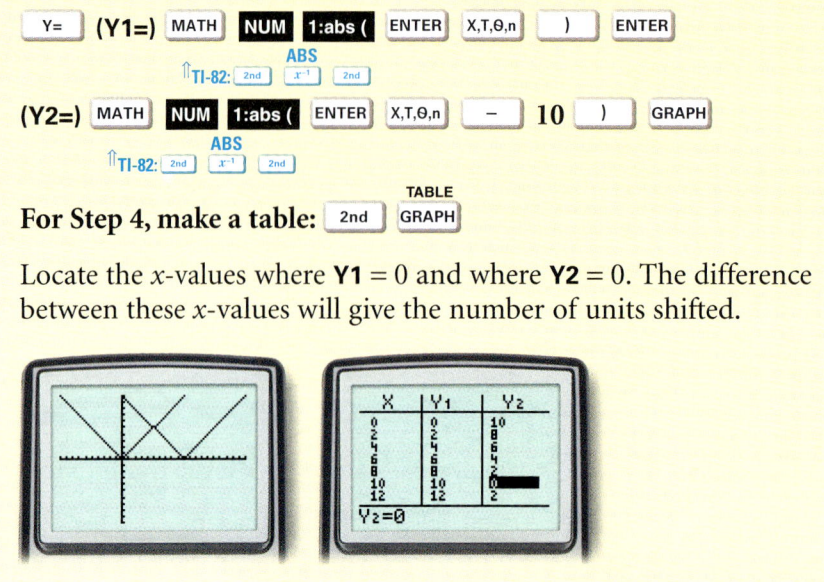

LESSON 14.3

Activity — Page 707

For Step 1, make a table of values for the function $h(t) = t^2$.

Use viewing window [0, 5] by [0, 25].

Enter the function:
Y= (Y1=) X,T,θ,n x^2

Make a table:
TBLSET
2nd WINDOW (TblStart=) 0 ENTER (ΔTbl=) 1 ENTER
⇑TI-82: (TblMin=)
(Indpnt:) Auto ENTER ▼ (Depend:) Auto ENTER

TABLE
View your table: 2nd GRAPH **View your graph:** GRAPH

For Steps 2 and 3, enter the function $j(t) = 16t^2$ as **Y2** $= 16x^2$ using a keystroke sequence similar to that used in Step 1. Enter 2nd GRAPH (TABLE) to view the expanded table and then GRAPH to view the graphs of both functions on the same viewing screen.

For Step 4, complete Steps 1–3 for the functions $f(x) = |x|$ and $f(x) = 10|x|$ and then graph the functions for integer *x*-values from 3 to 7. Use viewing window [−3, 7] by [0, 10].

Y= (Y1=) MATH ► NUM 1:abs ENTER X,T,θ,n) ENTER
ABS
⇑TI-82: 2nd x^{-1} (

(Y2=) 10 MATH ► NUM 1:abs ENTER X,T,θ,n)
ABS
⇑TI-82: 2nd x^{-1} (

TBLSET
2nd WINDOW (TblStart=) 0 ENTER (ΔTbl=) 1 ENTER (Indpnt:) Auto
⇑TI-82: (TblMin=)
TABLE
ENTER ▼ (Depend:) Auto ENTER 2nd GRAPH GRAPH

Tables

Table of Squares, Cubes, and Roots

No.	Squares	Cubes	Square Roots	Cube Roots	No.	Squares	Cubes	Square Roots	Cube Roots
1	1	1	1.000	1.000	51	2,601	132,651	7.141	3.708
2	4	8	1.414	1.260	52	2,704	140,608	7.211	3.733
3	9	27	1.732	1.442	53	2,809	148,877	7.280	3.756
4	16	64	2.000	1.587	54	2,916	157,464	7.348	3.780
5	25	125	2.236	1.710	55	3,025	166,375	7.416	3.803
6	36	216	2.449	1.817	56	3,136	175,616	7.483	3.826
7	49	343	2.646	1.913	57	3,249	185,193	7.550	3.849
8	64	512	2.828	2.000	58	3,364	195,112	7.616	3.871
9	81	729	3.000	2.080	59	3,481	205,379	7.681	3.893
10	100	1,000	3.162	2.154	60	3,600	216,000	7.746	3.915
11	121	1,331	3.317	2.224	61	3,721	226,981	7.810	3.936
12	144	1,728	3.464	2.289	62	3,844	238,328	7.874	3.958
13	169	2,197	3.606	2.351	63	3,969	250,047	7.937	3.979
14	196	2,744	3.742	2.410	64	4,096	262,144	8.000	4.000
15	225	3,375	3.873	2.466	65	4,225	274,625	8.062	4.021
16	256	4,096	4.000	2.520	66	4,356	287,496	8.124	4.041
17	289	4,913	4.123	2.571	67	4,489	300,763	8.185	4.062
18	324	5,832	4.243	2.621	68	4,624	314,432	8.246	4.082
19	361	6,859	4.359	2.668	69	4,761	328,509	8.307	4.102
20	400	8,000	4.472	2.714	70	4,900	343,000	8.367	4.121
21	441	9,261	4.583	2.759	71	5,041	357,911	8.426	4.141
22	484	10,648	4.690	2.802	72	5,184	373,248	8.485	4.160
23	529	12,167	4.796	2.844	73	5,329	389,017	8.544	4.179
24	576	13,824	4.899	2.884	74	5,476	405,224	8.602	4.198
25	625	15,625	5.000	2.924	75	5,625	421,875	8.660	4.217
26	676	17,576	5.099	2.962	76	5,776	438,976	8.718	4.236
27	729	19,683	5.196	3.000	77	5,929	456,533	8.775	4.254
28	784	21,952	5.292	3.037	78	6,084	474,552	8.832	4.273
29	841	24,389	5.385	3.072	79	6,241	493,039	8.888	4.291
30	900	27,000	5.477	3.107	80	6,400	512,000	8.944	4.309
31	961	29,791	5.568	3.141	81	6,561	531,441	9.000	4.327
32	1,024	32,768	5.657	3.175	82	6,724	551,368	9.055	4.344
33	1,089	35,937	5.745	3.208	83	6,889	571,787	9.110	4.362
34	1,156	39,304	5.831	3.240	84	7,056	592,704	9.165	4.380
35	1,225	42,875	5.916	3.271	85	7,225	614,125	9.220	4.397
36	1,296	46,656	6.000	3.302	86	7,396	636,056	9.274	4.414
37	1,369	50,653	6.083	3.332	87	7,569	658,503	9.327	4.431
38	1,444	54,872	6.164	3.362	88	7,744	681,472	9.381	4.448
39	1,521	59,319	6.245	3.391	89	7,921	704,969	9.434	4.465
40	1,600	64,000	6.325	3.420	90	8,100	729,000	9.487	4.481
41	1,681	68,921	6.403	3.448	91	8,281	753,571	9.539	4.498
42	1,764	74,088	6.481	3.476	92	8,464	778,688	9.592	4.514
43	1,849	79,507	6.557	3.503	93	8,649	804,357	9.644	4.531
44	1,936	85,184	6.633	3.530	94	8,836	830,584	9.695	4.547
45	2,025	91,125	6.708	3.557	95	9,025	857,375	9.747	4.563
46	2,116	97,336	6.782	3.583	96	9,216	884,736	9.798	4.579
47	2,209	103,823	6.856	3.609	97	9,409	912,673	9.849	4.595
48	2.304	110,592	6.928	3.634	98	9,604	941,192	9.899	4.610
49	2,401	117,649	7.000	3.659	99	9,801	970,299	9.950	4.626
50	2,500	125,000	7.071	3.684	100	10,000	1,000,000	10.000	4.642

Table of Random Digits

Column Line	(1)	(2)	(3)	(4)	(5)	(6)	(7)	(8)	(9)	(10)	(11)	(12)	(13)	(14)
1	10480	15011	01536	02011	81647	91646	69179	14194	62590	36207	20969	99570	91291	90700
2	22368	46573	25595	85393	30995	89198	27982	53402	93965	34095	52666	19174	39615	99505
3	24130	48360	22527	97265	76393	64809	15179	24830	49340	32081	30680	19655	63348	58629
4	42167	93093	06243	61680	07856	16376	39440	53537	71341	57004	00849	74917	97758	16379
5	31570	39975	81837	16656	06121	91782	60468	81305	49684	60672	14110	06927	01263	54613
6	77921	06907	11008	42751	27756	53498	18602	70659	90655	15053	21916	81825	44394	42880
7	99562	72905	56420	69994	98872	31016	71194	18738	44013	48840	63213	21069	10634	12952
8	96301	91977	05463	07972	18876	20922	94595	56869	69014	60045	18425	84903	42508	32307
9	89579	14342	63661	10281	17453	18103	57740	84378	25331	12566	58678	44947	05585	56941
10	85475	36857	53342	53988	53060	59533	38867	62300	08158	17983	16439	11458	18593	64952
11	28918	69578	88231	33276	70997	79936	56865	05859	90106	31595	01547	85590	91610	78188
12	63553	40961	48235	03427	49626	69445	18663	72695	52180	20847	12234	90511	33703	90322
13	09429	93969	52636	92737	88974	33488	36320	17617	30015	08272	84115	27156	30613	74952
14	10365	61129	87529	85689	48237	52267	67689	93394	01511	26358	85104	20285	29975	89868
15	07119	97336	71048	08178	77233	13916	47564	81056	97735	85977	29372	74461	28551	90707
16	51085	12765	51821	51259	77452	16308	60756	92144	49442	53900	70960	63990	75601	40719
17	02368	21382	52404	60268	89368	19885	55322	44819	01188	65225	64835	44919	05944	55157
18	01011	54092	33362	94904	31273	04146	18594	29852	71585	85030	51132	01915	92747	64951
19	52162	53916	46369	58586	23216	14513	83149	98736	23495	64350	94738	17752	35156	35749
20	07056	97628	33787	09998	42698	06691	76988	13602	51851	46104	88916	19509	25625	58104
21	48663	91245	85828	14346	09172	30168	90229	04734	59193	22178	30421	61666	99904	32812
22	54164	58492	22421	74103	47070	25306	76468	26384	58151	06646	21524	15227	96909	44592
23	32639	32363	05597	24200	13363	38005	94342	28728	35806	06912	17012	64161	18296	22851
24	29334	27001	87637	87308	58731	00256	45834	15398	46557	41135	10367	07684	36188	18510
25	02488	33062	28834	07351	19731	92420	60952	61280	50001	67658	32586	86679	50720	94953
26	81525	72295	04839	96423	24878	82651	66566	14778	76797	14780	13300	87074	79666	95725
27	29676	20591	68086	26432	46901	20849	89768	81536	86645	12659	92259	57102	80428	25280
28	00742	57392	39064	66432	84673	40027	32832	61362	98947	96067	64760	64584	96096	98253
29	05366	04213	25669	26422	44407	44048	37937	63904	45766	66134	75470	66520	34693	90449
30	91921	26418	64117	94305	26766	25940	39972	22209	71500	64568	91402	42416	07844	69618
31	00582	04711	87917	77341	42206	35126	74087	99547	81817	42607	43808	76655	62028	76630
32	00725	69884	62797	56170	86324	88072	76222	36086	84637	93161	76038	65855	77919	88006
33	69011	65795	95876	55293	18988	27354	26575	08625	40801	59920	29841	80150	12777	48501
34	25976	57948	29888	88604	67917	48708	18912	82271	65424	69774	33611	54262	85963	03547
35	09763	83473	73577	12908	30883	18317	28290	35797	05998	41688	34952	37888	38917	88050
36	91567	42595	27958	30134	04024	86385	29880	99730	55536	84855	29080	09250	79656	73211
37	17955	56349	90999	49127	20044	59931	06115	20542	18059	02008	73708	83517	36103	42791
38	46503	18584	18845	49618	02304	51038	20655	58727	28168	15475	56942	53389	20562	87338
39	92157	89634	94824	78171	84610	82834	09922	25417	44137	48413	25555	21246	35509	20468
40	14577	62765	35605	81263	39667	47358	56873	56307	61607	49518	89656	20103	77490	18062
41	98427	07523	33362	64270	01638	92477	66969	98420	04880	45585	46565	04102	46880	45709
42	34914	63976	88720	82765	34476	17032	87589	40836	32427	70002	70663	88863	77775	69348
43	70060	28277	39475	46473	23219	53416	94970	25832	69975	94884	19661	72828	00102	66794
44	53976	54914	06990	67245	68350	82948	11398	42878	80287	88267	47363	46634	06541	97809
45	76072	29515	40980	07391	58745	25774	22987	80059	39911	96189	41151	14222	60697	59583
46	90725	52210	83974	29992	65831	38857	50490	83765	55657	14361	31720	57375	56228	41546
47	64364	67412	33339	31926	14883	24413	59744	92351	97473	89286	35931	04110	23726	51900
48	08962	00358	31662	25388	61642	34072	81249	35648	56891	69352	48373	45578	78547	81788
49	95012	68379	93526	70765	10592	04542	76463	54328	02349	17247	28865	14777	62730	92277
50	15664	10493	20492	38391	91132	21999	59516	81652	27195	48223	46751	22923	32261	85653

Glossary

absolute error The absolute value of the difference between the actual measure and the specified measure. (300)

absolute value The absolute value of a real number x is the distance from x to 0 on a number line; the symbol $|x|$ means the absolute value of x. (57, 295)

absolute-value equation An equation that includes an absolute value; it will have 2 solutions. (300)

absolute-value function A function written in the form $y = |x|$ or $y = ABS(x)$. (295)

absolute-value inequality An inequality that includes an absolute value. (302)

addition of matrices If A and B are matrices with the same dimensions, then the matrix $(A + B)$ has the same dimensions; each of its entries is the sum of the corresponding entries of A and B. (630)

Addition of Probabilities Principle If M and N are intersecting sets of outcomes in the same sample space, then $P(M \text{ OR } N) = P(M) + P(N) - P(M \cap N)$. If M and N are disjoint sets of outcomes in the same sample space, then $P(M \text{ OR } N) = P(M) + P(N)$. (657)

Addition Property of Equality For all real number a, b, and c, if $a = b$, then $a + c = b + c$. Adding equal amounts to each side of an equation results in an equivalent equation. (116)

Addition Property of Inequality Let a, b, and c be real numbers. If $a < b$, then $a + c < b + c$. If $a > b$, then $a + c > b + c$. If equal amounts are added to the expressions on each side of an inequality, the resulting inequality is true. (277)

Additive Inverse Property For every real number a, there is exactly one real number $-a$. such that $a + (-a) = 0$ and $-a + a = 0$. (64)

additive inverses Two numbers a and $-a$ whose sum is 0; see *opposites*. (64)

address The location of an entry in a matrix, given by the row and column in which the entry appears. (629)

algebraic expression Variables and numbers combined by operations. (11)

algebraic order of operations A logical system of rules for computation. (19)

AND A logical operator representing the intersection of two sets. (655)

angle of depression The angle formed by a horizontal line and a line of sight to a point below. (617)

angle of elevation The angle formed by a horizontal line and a line of sight to a point above. (616)

approximate solution A reasonable estimate for a point of intersection for a system of equations. (321)

Associative Property of Addition For all numbers a, b, and c, $(a + b) + c = a + (b + c)$. (81)

Associative Property of Multiplication For all numbers a, b, and c, $(a \cdot b) \cdot c = a \cdot (b \cdot c)$. (81)

asymptote A line that the graph of a function approaches but never touches or crosses. Asymptotes are found at each value for which the denominator of the function would be 0 (and thus where the function's value would be undefined). (534)

axis of symmetry A line drawn through the graph of a function that divides the graph into two halves that are reflections of each other. (481)

bar graph A method of displaying data that uses rectangular bars or objects to represent the data. (193)

base (of a percentage) The number of which a percentage is calculated. (172)

base (of a power) The number that is raised to an exponent. In an expression of the form x^a, x is the base. (370)

boundary line A line that divides a coordinate plane into two half-planes. (346)

box-and-whisker-plot A method of showing how data is distributed by using the median, the upper and lower quartiles, and the greatest and least values; also called a box plot. (203)

center A point inside a circle that is the same distance from every point on the circle. (606)

circle graph A method of displaying data in which the data are represented by parts of a circle. (194)

circle The set of all points in a plane that are the same distance from a given point called the center of the circle. The equation for a circle with a radius of r and its center at (h, k) is $(x - h)^2 + (y - k)^2 = r^2$. (606)

Closure Property A set of numbers is said to be closed, or to have closure, under a given operation if the result of the operation on any two numbers in the set is also in the set. (70, 562)

coefficient A number that is multiplied by a variable. (89, 373)

801

common binomial factor A binomial factor that is common to all terms of a polynomial. (449)

common monomial factor A monomial factor that is common to all terms of a polynomial. (448)

Commutative Property of Addition For any numbers a and b, $a + b = b + a$. (81)

Commutative Property of Multiplication For any numbers a and b, $a \cdot b = b \cdot a$. (81)

complement The complement of an event includes all the outcomes in the sample space that are not that event. (670)

completing the square To make a perfect-square trinomial from a binomial of the form $x^2 + b^x$, add the square of one-half the coefficient of the x-term, $\left(\frac{b}{2}\right)^2$. The completed perfect-square trinomial is $x^2 + bx + \left(\frac{b}{2}\right)^2$. (493)

compound inequality Two inequalities that are combined into one statement by the word *and* or *or*. (289)

conclusion In a conditional statement, the portion following the word *then*. In a proof of a conditional statement, the conclusion is the part that must be proven to be true. (558)

conditional statement A statement of the form "If p, then q." (558)

conjecture A statement that is believed to be true but is not yet proven. (5, 559)

conjunction A compound inequality whose solution region is an intersection. (290)

consistent system A system of equations that has one or infinitely many solutions. (339)

constant A term in an algebraic expression that does not contain variables. (89)

constant difference The difference that is always the same from one term to the next in a number sequence. (5)

constant of variation In a direct variation of the form $\frac{y}{x} = k$, k is the constant of variation. (238)

converse A statement formed by interchanging the if and then portions of a conditional statement. "If q, then p" is the converse of "If p, then q." (558)

Converse of the Pythagorean Theorem If a, b, and c are the lengths of the sides of a triangle and $a^2 + b^2 = c^2$, where $c > a$ and $c > b$, then the triangle is a right triangle. (600)

coordinate plane A plane that is divided into four regions by a horizontal and a vertical number line. The addresses, or coordinates, of points in the plane are given by ordered pairs. (24)

coordinates An ordered pair of numbers that locates a point in the coordinate plane in relation to the x- and y-axes. (24)

correlation A relationship between two variables in a scatter plot. (38)

cosine In a right triangle, the ratio of the length of the leg adjacent to an acute angle of the triangle to the length of the hypotenuse; cosine of angle $A = \cos A = \frac{\text{length of the leg adjacent to } \angle A}{\text{length of the hypotenuse}}$. (622)

Counting Elements of Sets Suppose that there are m elements in set M and n elements in set N; then the total number of elements in the two sets is $m + n - t$, where t is the number of elements in the intersection of M and N. (656)

cross products In the proportion $\frac{a}{b} = \frac{c}{d}$, the product of the means, bc, and the product of the extremes, ad, are the cross products. (165)

Cross Products Property For any real numbers a, b, c, and d, where $b \neq 0$ and $d \neq 0$, if $\frac{a}{b} = \frac{c}{d}$, then $ad = bc$. (165)

deductive reasoning The process of concluding that something must be true because it is a special case of a general principle that is known to be true. (559)

degree of a polynomial The degree of a polynomial in one variable is determined by the exponent with the greatest value within the polynomial. (426)

dependent system A consistent system of equations that has an infinitely many solutions. The graph of the equations in the system is the same line. (341)

dependent variable The variable in a function whose value depends on the value of the other variable. (32)

difference of two squares A polynomial of the form $a^2 - b^2$, which can be written as the product of two factors, $(a + b)(a - b)$. (453)

difference The result of subtraction. (62)

direct variation If y varies directly as x, then $y = kx$, or $\frac{y}{x} = k$, where k is the constant of variation. (238)

directrix of a parabola The fixed line from which a parabola is the same distance as it is from a fixed point, called the focus. (609)

discriminant For a quadratic equation of the form $ax^2 + bx + c = 0$, where $a \neq 0$, the discriminant is $b^2 - 4ac$. (508)

disjoint Two outcome sets of a sample space that do not intersect are said to be disjoint. (657)

disjunction A compound inequality whose solution region is outside an intersection. (290)

distance formula The distance, d, between two points $A(x_1, y_1)$ and $B(x_2, y_2)$ is given by $d = \sqrt{(x_2 - x_1)^2 + (y_2 - y_1)^2}$. (599)

Distributive Property For all numbers a, b, and c, $a(b + c) = ab + ac$ and $(b + c)a = ba + ca$, and $a(b - c) = ab - ac$ and $(b - c)a = ba - ca$. (82)

dividing an expression When an expression is divided by a number, each term in the expression is divided by that number. For all numbers a, b, and c, $\frac{a+b}{c} = \frac{a}{c} + \frac{b}{c}$ and $\frac{a-b}{c} = \frac{a}{c} - \frac{b}{c}$. (99)

Division Property of Equality For all real number a, b, and c, if $a = b$ and $c \neq 0$, then $\frac{a}{c} = \frac{b}{c}$. Dividing each side of an equation by equal amounts results in an equivalent equation. (123)

Division Property of Inequality Let a, b, and c be nonzero real numbers. For $c > 0$, if $a > b$, then $\frac{a}{c} > \frac{b}{c}$, and if $a < b$, then $\frac{a}{c} > \frac{b}{c}$. For $c < 0$, if $a < b$, then $\frac{a}{c} > \frac{b}{c}$ and if $a > b$, then $\frac{a}{c} < \frac{b}{c}$. (283)

Division Property of Square Roots For all numbers $a \geq 0$ and $b > 0$, $\sqrt{\frac{a}{b}} = \frac{\sqrt{a}}{\sqrt{b}}$. (580)

domain The first coordinates in a set of ordered pairs of a relation or function. (219)

elimination method A method used to solve a system of equations in which one variable is eliminated by adding opposites. (332)

entry The information that appears at each address of a matrix. (629)

equally likely outcomes Outcomes of an experiment that have the same probability of happening. (648)

equation Two equivalent algebraic expressions separated by an equal sign. (12)

equation method A method for finding an unknown part of a percent by setting up an equation. (172)

equation of a circle The equation of a circle with radius r and its center at (h, k) is $(x - h)^2 + (y - h)^2 = r^2$. (607)

error The difference between the actual measure and the specified measure. (300)

even numbers An even number is a number of the form $2n$, where n is an integer. (560)

experimental probability The experimental probability, $P(E)$, of an event is approximated by the formula $P(E) = \frac{\text{number of favorable outcomes}}{\text{total number of trials}}$. The larger the number of trials, the closer the approximation. (182, 676)

exponent In a power, the number that tells how many times the base is used as a factor. In an expression of the form x^a, a is the exponent. (97, 379)

exponential decay A situation modeled by the general growth formula, $P = A(1 + r)^t$, in which r is negative. (411)

exponential function A function of the form $y = 2^x$ in which a base number is raised to a variable exponent. Exponential functions model situations in which values in the range change by a fixed rate rather than a fixed amount. (404)

exponential growth A situation modeled by the general growth formula, $P = A(1 + r)^t$, in which r is positive. (411)

extraneous solution A solution that is obtained but that does not satisfy the original equation. (586)

extremes In the proportion $\frac{a}{b} = \frac{c}{d}$, a and d are the extremes. (165)

factor A number or polynomial that is multiplied by another number or polynomial to form a product. (448)

favorable outcome The outcome that is desired in a probability experiment. (649)

field Any set that has the Associative, Closure, Commutative, Distributive, Identity, and Inverse Properties for any two operations. (563)

focus of a parabola The fixed point from which a parabola is the same distance as it is from a fixed line, called the directrix. (609)

FOIL method A mnemonic (memory) device for a method of multiplying two binomials.
- Multiply the **F**irst terms.
- Multiply the **O**utside terms.
- Multiply the **I**nside terms. Add the inside and outside products.
- Multiply the **L**ast terms. (439)

formula A literal equation that states a rule for a relationship among quantities. (147)

frequency table A method of displaying data in which tally marks are used to show how often each element of the set occurs. (188)

function A pairing between two sets of numbers in which each element of the first set is paired with no more than one element of the second set. (219)

function notation A method of writing a function in which the dependent variable is written in the form $f(x)$. The independent variable, x, is placed inside the parentheses. (223, 694)

function rule The expression on the right side of the written form of a function. (694)

Fundamental Counting Principle If there are m ways to make a first choice and n ways to make a second choice, then there are $m \cdot n$ ways to make a first choice and a second choice. (662)

general growth formula Let P be the amount after t years at a yearly growth rate of r, where r is expressed as a decimal. If the original amount is A, then the general growth formula is $P = A(1 + r)^t$. (405)

greatest common factor (GCF) The greatest factor that is common to all terms of a polynomial. (448)

greatest-integer function A function written in the form $y = \text{INT}(x)$ in which the number x is rounded down to the nearest integer. (309)

histogram A bar graph that has no space between the bars and that shows the frequency of the data. (202)

hole (in a graph) An omitted point in a graph. (534)

Hooke's law A mathematical relationship between the amount of force, F, applied to a spring and the distance, d, that the spring stretches, $F = kd$, where k is a constant. (239)

horizontal line The equation for a horizontal line is $y = b$, where b is the y-intercept. The slope of a horizontal line is 0. (230, 248)

horizontal stretch/compression If $y = f(x)$, then $y = f(bx)$ gives a horizontal stretch or compression of the graph of f by a factor of $\frac{1}{b}$. (710)

horizontal translation A transformation that shifts the graph of a function either to the right (in a positive direction) or to the left (in a negative direction). The graph of $y = f(x - h)$ is translated horizontally h units from the graph of $y = f(x)$. (481, 703)

hypotenuse The side opposite the right angle in a right triangle. (591)

hypothesis In a conditional statement, the portion following the word *if*. In a proof of a conditional statement, the hypothesis is assumed to be true. (558)

identity An equation that is true regardless of the values of the variable(s). (445)

identity matrix for multiplication A square matrix with 1s along the main diagonal and 0s elsewhere. The product of a given matrix and the identity matrix is the given matrix. (632)

Identity Property for Addition For all real numbers a, $a + 0 = a$ and $0 + a = a$; 0 is the *identity element* for addition. (64, 84, 563)

Identity Property for Multiplication For all real numbers a, $a \cdot 1 = a$ and $1 \cdot a = a$; 1 is the *identity element* for multiplication. (75, 84, 563)

inclusion symbols Symbols such as parentheses, ( ), brackets, [ ], braces, { }, and the fraction bar, —, that are used to group numbers and variables. (19)

inconsistent system A system of equations that has no solution. The graphs of the equations in the system are parallel lines. (339)

independent events Two events are independent if the occurrence of one does not affect the probability of the occurrence of the other. (668)

independent system A consistent system of equations that has only one solution. The graphs of the equations are lines that intersect at one point. (341)

independent variable The variable in a function whose value does not depend on the value of the other variable. (32)

inductive reasoning The process of reasoning that general principle is true because special cases are true. (559)

inequality A statement that two expressions are not equal. Inequalities contain one of the following signs: $<, >, \leq, \geq$, or $\neq$. (276)

integers The set of whole numbers and their opposites. (54)

intersection (of sets) The intersection of two sets consists of the elements that are common to both sets. The logical relationship AND represents the intersection of two sets. (659)

intersection (of graphs) The solution to a system of linear equalities or inequalities, consisting of the solutions common to each. (290)

inverse variation A relationship in which y varies inversely as x; written as $y = \frac{k}{x}$, or $xy = k$, where k is the constant of variation, $k \neq 0$, and $x \neq 0$. (526)

irrational numbers Numbers whose decimal part never terminates or repeats. Irrational numbers cannot be expressed in the form $\frac{a}{b}$, where a and b are integers and $b \neq 0$. (56)

leaves In a stem-and-leaf plot, the leaves represent the second part each element and are listed to the right of the vertical line. (201)

legs of a right triangle The sides that form the right angle of a right triangle. (591)

like terms Terms that contain the same form of a variable. (89)

line graph A type of graph that uses line segments between known data points to show changes that have occurred over time. (193)

line of best fit A line that shows the overall trend of the data in a scatter plot. (39)

line of reflection The line across which a graph is reflected. (297)

linear equation An equation whose graph is a straight line. (32)

linear inequality An inequality whose form is like that of a linear equation with the equal sign replaced by an inequality symbol. (345)

literal equation An equation that involves two or more variables. (147)

lower quartile The median of the elements in the lower half of a data set. (204)

main diagonal (of a matrix) The entries in a square matrix that form a diagonal from the upper left corner to the lower right corner. (632)

matrix An arrangement of data in rows and columns and enclosed by brackets [ ]. The plural of *matrix* is *matrices*. (628)

matrix address See **address**.

matrix equality Two matrices are equal when their dimensions are the same and their corresponding entries are equal. (629)

maximum value The *y*-value of the vertex of a parabola that opens down. (481)

mean The quotient that results when the sum of all the elements in a data set is divided by the total number of elements in the set. (194)

means In the proportion $\frac{a}{b} = \frac{c}{d}$, *b* and *c* are the means. (165)

median The middle number in a data set when the elements are placed in numerical order. If there is an even number of elements in the data set, the median is the average of the two middle numbers. (186)

midpoint formula Given $\overline{PQ}$, where $P(x_1, y_1)$ and $Q(x_2, y_2)$, the coordinates of the midpoint, *M*, of $\overline{PQ}$ are $M = \left(\frac{x_1 + x_2}{2}, \frac{y_1 + y_2}{2}\right)$. (601)

midsegment A line segment connecting the midpoints of 2 sides of a triangle. (608)

minimum value The *y*-value of the vertex of a parabola that opens up. (480)

mode The element, if any, that occurs most often in a data set. There may be no mode, one mode, or several modes for a set of data. (186)

monomial An algebraic expression that is either a constant, a variable, or a product of a constant and one or more variables. (373)

Multiplication Property of Equality For all real numbers *a*, *b*, and *c*, if $a = b$, then $ac = bc$.

Multiplying each side of an equation by equal amounts results in an equivalent equation. (124)

Multiplication Property of Inequality Let *a*, *b*, and *c* be nonzero real numbers. For $c > 0$, if $a > b$, then $ac > bc$, and if $a < b$, then $ac < bc$. For $c < 0$, if $a < b$, then $ac > bc$, and if $a > b$, then $ac < bc$. (283)

Multiplication Property of Square Roots For all numbers *a* and *b*, where $a \geq 0$ and $b \geq 0$, $\sqrt{ab} = \sqrt{a}\sqrt{b}$. (678)

Multiplicative Inverse Property For every nonzero real number *a*, there is exactly one number $\frac{1}{a}$ such that $a \cdot \frac{1}{a} = 1$ and $\frac{1}{a} \cdot a = 1$. (75)

multiplicative inverse See *reciprocal*. (75)

natural numbers The numbers 1, 2, 3, 4, 5, . . . (54)

negative correlation A relationship between two variables in a scatter plot in which one variable increases while the other decreases. (38)

negative exponent If *x* is any number except zero and *n* is a positive integer, then $x^{-n} = \frac{1}{x^n}$. (390)

negative numbers Numbers that are less than zero. Negative numbers lie to the left of zero on a number line. (55)

neutral pair One positive and one negative algebra tile of the same value. (61)

non-trivial rational function A rational function whose denominator is not a constant. (534)

number sequence A list of numbers in a certain order. (5)

odd numbers An odd number is a number of the form $2n + 1$, where *n* is an integer. (560)

opposites Two numbers that lie on opposite sides of zero and are the same distance from zero on a number line. (57)

OR A logical operator which represents the union of two sets. (655)

ordered pair The *x*- and *y*-coordinates that give the location of a point in a coordinate plane, indicated by two numbers in parentheses, (*x*, *y*). (25)

origin The point of intersection of the *x*- and *y*-axes in the coordinate plane. (25)

parabola The shape of the graph of a quadratic function. A parabola is defined as the set of all points

805

that are the same distance from a fixed point, called the focus, and a fixed line, called the directrix. (480)

parallel lines If two different lines have the same slope, the lines are parallel. If two nonvertical lines are parallel, they have the same slope. Two parallel, vertical lines have undefined slopes. (259)

parent function The most basic function of a family of functions, or the original function before a transformation is applied. (296, 696)

percent A ratio that compares a number with 100. (171)

percent rate The percent of the base calculated or indicated in a percent statement. (172)

percentage The amount obtained by multiplying a base by a given percent rate. (172)

perfect-square trinomial A trinomial of the form $a^2 + 2ab + b^2$ or $a^2 - 2ab + b^2$. The factored form of $a^2 + 2ab + b^2$ is $(a + b)(a + b) = (a + b)^2$, and the factored form of $a^2 - 2ab + b^2$ is $(a - b)(a - b) = (a - b)^2$. (452)

perpendicular lines If the slopes of two lines are m and $-\frac{1}{m}$, the lines are perpendicular. (259)

point-slope form The point-slope form for the equation of a line is $y - y_1 = m(x - x_1)$, where the coordinates x_1 and y_1 are taken from a given point, (x_1, y_1), and m is the slope. (254)

polynomial The sum or difference of two or more monomials. (426)

polynomial function A function that consists of the sum or difference of two or more monomials. (443)

positive correlation A relationship between two variables in a scatter plot in which both variables increase. (38)

positive numbers Numbers that are greater than zero. Positive numbers lie to the right of zero on a number line. (55)

power An expression of the form x^a, where x is the base and a is the exponent. (97)

Power-of-a-Fraction Property If n is a positive number and a and b are numbers, where $b \neq 0$, then $\frac{a^n}{b} = \frac{a^n}{b^n}$. (386)

Power-of-a-Power Property If x is any number and m and n are any positive integers, then $(x^m)^n = x^{mn}$. (378)

Power-of-a-Product Property If x and y are any numbers and n is a positive integer, then $(xy)^n = x^n y^n$. (379)

Powers of −1 Even powers of −1 are equal to 1. Odd powers of −1 are equal to −1. (380)

principal square root The positive square root of a number; indicated by the radical sign. (487)

Probabilities of Independent Events If A and B are independent events, then $P(A \text{ AND } B) = P(A) \cdot P(B)$. (668)

probability The likelihood that an event will occur. (180)

product matrix The result of multiplying two matrices. If matrix A has dimensions $m \times n$ and matrix B has dimensions $n \times p$, then the product matrix AB has dimensions $m \times p$. (631)

Product-of-Powers Property If x is any number and m and n are any positive integers, then $x^m \cdot x^n = x^{m+n}$. (371)

proof A series of statements in a logical sequence which demonstrates that the conclusion of a conditional statement is true whenever the hypothesis is true. (558)

Properties of Zero Let a represent any number.
1. The product of any number and zero is zero. $a \cdot 0$ and $0 \cdot a = 0$
2. Zero divided by any nonzero number is zero. $\frac{0}{a} = 0$, where $a \neq 0$
3. Division by zero is undefined. That is, division by zero is not possible. (76)

proportion A statement that two ratios are equal, $\frac{a}{b} = \frac{c}{d}$. (164)

proportion method A method of solving percent problems by setting two proportions equal to each other. (172)

Pythagorean Theorem If given a right triangle with legs of length a and b and hypotenuse of length c, then $a^2 + b^2 = c^2$. (592)

Pythagorean Triple A set of three positive integers that satisfy the Pythagorean Theorem. (597)

quadrant One of the four regions into which a horizontal and vertical number line divide a plane. (24)

quadratic formula For $ax^2 + bx + c = 0$, where $a \neq 0$, $x = \frac{-b \pm \sqrt{b^2 - 4ac}}{2a}$. (507)

quadratic function A function of the form $y = ax^2 + bx + c$, where a, b, and c are real numbers and $a \neq 0$. (480)

quadratic inequality An inequality whose form is like that of a quadratic equation with the equal sign replaced by an inequality symbol. (511)

Quotient-of-Powers Property If x is any number except 0 and m and n are any positive integers, where $m > n$, then $\frac{x^m}{x^n} = x^{m-n}$. (384)

radical expression An expression that contains a square root. (577)

radical sign The sign $\sqrt{}$ used to denote a square root. (487)

radicand The number under a radical sign. (577)

radius The distance from the center of a circle to any point on the circle. (608)

range The difference between the greatest and least values in a data set, or the set of second coordinates in the ordered pairs of a relation or function. (186, 219)

rate of change The rate of change of a linear function is equal to the slope of the graph of the function. (236)

ratio The comparison of two quantities by division. (164)

rational expression If P and Q are polynomials and $Q \neq 0$, then an expression of the form $\frac{P}{Q}$ is a rational expression. (532)

rational function A function of the form $y = \frac{P}{Q}$, or $f(x) = \frac{P(x)}{Q(x)}$ where is a rational expression. (533)

rational numbers A number that can be expressed in the form $\frac{a}{b}$, where a and b are integers and $b \neq 0$. (55)

rationalizing the denominator Removing a radical expression from the denominator of a fraction. (580)

real numbers The set of rational numbers together with the set of irrational numbers. (56)

reciprocal The number $\frac{1}{a}$ is called the reciprocal of a. (75)

reflection A transformation that creates a mirror image of a given function. (297, 481, 715)

Reflexive Property of Equality For all real numbers a, $a = a$ (a number is equal to itself). (84)

relation A pairing between two sets of numbers. (218)

rise The difference of the y-coordinates of two points on a line. (226)

rule of 72 The time, t, required to double the value of an investment earning compound interest is give by the function $t = \frac{72}{r}$, where r is the annual interest rate. (529)

run The difference of the x-coordinates of two points on a line. (226)

sample space The set of all possible outcomes of a probability experiment. (673)

scalar A number by which each entry in a matrix is multiplied. (631)

scalar multiplication The multiplication of a matrix by a scalar. (631)

scale factor The number by which a parent function is multiplied to create a vertical stretch. (708)

scatter plot A display of ordered pairs of data graphed on a coordinate plane that is used to show how two variables relate to each other. (37)

scientific notation A number written in scientific notation is a product of two factors—a number from 1 to 10, including 1 but not including 10, and a power of 10. (398)

simplest radical form A radical expression is in simplest radical form if the expression under the radical sign contains no perfect squares greater that 1, contains no fractions, and is not in the denominator of a fraction. (578)

simplified An algebraic expression is said to be simplified when all of the like terms have been combined and all parentheses have been removed. (89)

simulation A probability experiment with mathematical characteristics that are similar to the actual event. (676)

sine In a right triangle, the ratio of the length of the leg opposite an acute angle of the triangle to the length of the hypotenuse; sine of angle $A = \sin\ A = \frac{\text{length of the leg opposite } \angle A}{\text{length of the hypotenuse}}$. (622)

slope A measure of the steepness of a line. Given two points with coordinates (x_1, y_1) and (x_2, y_2) on a line, the slope, m, of the line is given $m = \frac{\text{rise}}{\text{run}} = \frac{y_2 - y_1}{x_2 - x_1}$. (228)

slope-intercept form The slope-intercept form for a line with a slope of m and a y-intercept of b is $y = mx + b$. (245)

solution (to an equation) A set of values for the variables in an equation that make a true statement when substituted into the equation. (13)

solution (to a system of equations) An ordered pair of numbers that is the solution to each equation in the system. (319)

special products of polynomials For any real numbers a and b and for any expressions involving real numbers and variables: $(a + b)(a - b) = a^2 - b^2$; $(a + b)(a + b) = (a + b)^2 = a^2 + 2ab + b^2$; $(a - b)(a - b) = (a - b)^2 = a^2 - 2ab + b^2$. (434)

square matrix A matrix with equal row and column dimensions. (629)

square root If a is a number greater than or equal to zero, $\sqrt{a}$ represents the positive, or principal, square root of a and $-\sqrt{a}$ represents the negative square root of a; The square roots of a have the following property: $(-\sqrt{a})(-\sqrt{a}) = a$. (487, 577)

square-root function A function of the form $f(x) = \sqrt{x}$. (583)

standard form (of a linear equation) A linear equation in the form $Ax + By = C$ is in standard form when A, B, and C are real numbers, A and B are not both zero, and A is not negative. (252)

standard form (of a polynomial) A polynomial is in standard form when the terms of the polynomial are ordered from left to right in descending order, which is from the greatest exponent to the least. (426)

stem-and-leaf plot A method of displaying data by dividing each element into two parts, called stems and leaves. (201)

stems In a stem-and-leaf plot, the stems represent the first part each element and are listed to the left of the vertical line. (201)

substitution method A method used to solve a system of equations in which variables are replaced with known values or algebraic expressions. (327)

Substitution Property of Equality For all real numbers a and b, if $a = b$, then a can be replaced by b and b can be replaced by a in any expression. (84)

subtraction For all real numbers a and b, $a - b = a + (-b)$. (68)

Subtraction Property of Equality For all real numbers a, b, and c, if $a = b$, then $a - c = b - c$. Subtracting equal amounts from each side of an equation results in an equivalent equation. (115)

Subtraction Property of Inequality Let a, b, and c be real numbers. If $a < b$, then $a - c < b - c$. If $a > b$, then $a - c > b - c$. If equal amounts are subtracted from the expressions on each side of an inequality, the resulting inequality is true. (277)

sum The result of addition. (62)

Symmetric Property of Equality For all real numbers a and b, if $a = b$ then $b = a$. (84)

system of equations Two or more equations in two or more variables. (319)

system of linear inequalities Two or more inequalities in two or more variables. (345-349)

tangent In a right triangle, the ratio of the length of the leg opposite an acute angle of the triangle to the length of the leg adjacent to the acute angle; $A = \tan A = \frac{\text{length of the leg opposite } \angle A}{\text{length of the hypotenuse}}$. (613)

term Each number in a number sequence; a number, a variable, or a product or quotient of numbers and variables that is added or subtracted in an algebraic expression. (5, 89)

theorem A statement that can be proved to be true. (85, 558)

theoretical probability Let n be the number of elements in a sample space in which each element is equally likely to occur. Let E be an event for which there are f favorable outcomes in the sample space. Then, the theoretical probability that the event will occur is $P(E) = \frac{f}{n}$. (650)

transformation A variation such as a stretch, reflection, or translation of a parent function. (296, 700)

Transitive Property of Equality For all real numbers a, b, and c, if $a = b$ and $b = c$, then $a = c$. (84)

translation A transformation that shifts the graph of a function horizontally or vertically. (296, 700)

tree diagram A diagram that is used to find all the possible choices in a situation in which one choice AND another choice need to be made. (661)

trial The simulation of one event in an experiment. (181, 676)

Triangle Midpoint Theorem The segment joining the midpoints of two sides of a triangle is parallel to the third side of the triangle and is half the length of the third side. (608)

trigonometry A branch of mathematics that combines arithmetic, algebra, and geometry and is named from the Greek words meaning "triangle measure." (614)

trivial rational function A rational function whose denominator is a constant. (534)

union The union of two sets consists of all elements from both sets. The logical relationship OR represents the union of sets. (290, 655)

upper quartile The median of the elements in the upper half of a data set. (204)

variable A letter that is used to represent numbers in an algebraic expression. (11)

Venn diagram A diagram used to illustrate the relationships among different sets of data represented by circles. (655)

vertex The point where a parabola changes direction. (480)

vertex form of a quadratic function The vertex form of a quadratic function is $y = a(x - h)^2 + k$, where (h, k) is the vertex and the line $x = h$ is the axis of symmetry. (482)

vertical asymptote A vertical line that a graph approaches but never crosses. (534)

vertical line The equation for a vertical line is $x = a$, where a is the x-intercept. The slope of a vertical line is undefined. (229-230, 248)

vertical stretch/compression A transformation in which, for a given value of the domain of a function, the corresponding range value is multiplied by a scale factor. (708)

vertical translation A transformation that shifts the graph of a function either up (in a positive direction) or down (in a negative direction). The graph of $y = f(x) + k$ is translated vertically k units from the graph of $y = f(x)$. (481, 701)

vertical-line test A test used to determine whether a relation is a function. A relation is a function if any vertical line intersects the graph of the relation no more than once. (693)

whole numbers The numbers 0, 1, 2, 3, 4, . . . (54)

x-axis The horizontal number line in a coordinate plane. (24)

x-coordinate A number that indicates distance along the x-axis in a coordinate plane. (24)

x-intercept The x-coordinate of the point where a line crosses the x-axis. (247)

y-axis The vertical number line in a coordinate plane. (24)

y-coordinate A number that indicates distance along the y-axis in a coordinate plane. (24)

y-intercept The y-coordinate of the point where a line crosses the y-axis. (245)

zero as an exponent For any nonzero number x, $x^0 = 1$. (391)

zero of a function A number that gives a function value of 0 when substituted for the independent variable of a function. (465-467)

Zero Product Property If a and b are real numbers such that $ab = 0$, then $a = 0$ or $b = 0$. (465)

Additional Answers

Lesson 1.1, pages 4–10

Activity

1. Answers may vary. Sample answer: Light travels faster than sound. Therefore, you see lightning before you hear thunder.

2. a.
Seconds	2	4	6	8	10
Miles	0.4	0.8	1.2	1.6	2.0

   b. 0.4 miles

   CHECKPOINT ✔
   c. 2.4 miles in 12 seconds
      2.8 miles in 14 seconds
      3.2 miles in 16 seconds

   d. 4.6 miles

Exercises
Communicate

1. Find the first differences by subtracting each successive term. The first differences are 3, 5, 7, 9, and 11. The second differences are found by subtracting the successive first differences. The second differences are a constant 2.

2. Notice that the first differences are a constant −12 to predict the next 2 terms. Add −12 to 40 to find the next term, 28, and add −12 to 28 to find the next term, 16.

3. Take the second difference and add it to the last known first difference to obtain the next first difference: 30. Now take the new first difference and add it to the last known term of the original sequence: 106. In this way the sequence may be extended term-by-term.

Practice and Apply

32. 20, 27, 35

Portfolio Activity

1.
# of steps	1	2	3	4	5	6	7	8	9	10
# of cubes	1	3	6	10	15	21	28	36	45	55

2. Check students' drawings.

3. The second differences are a constant 2. The tenth stairway has 55 cubes.

4. A 4 by 4 array of dots can be used to show that the square number 16 is the sum of triangular numbers 6 and 10.

5. A 10 by 10 array of dots can be used to show that the square number 100 is the sum of triangular numbers 45 and 55.

Lesson 1.2, pages 11–17

Activity

1.
Time (hours)	0	0.5	1.0	1.5	2.0	2.5	3.0	3.5	4.0
Distance (miles)	0	10	20	30	40	50	60	70	80

The distance cycled in 1 hour is 20 miles. In $\frac{1}{2}$ hour the distance cycled is 10 miles. For each increment of $\frac{1}{2}$ hour in time, add 10 miles to the distance. You can find the distance cycled in 7 and 8 hours by repeatedly adding 10 miles to the distance cycled.

CHECKPOINT ✔

2. The distance cycled is equal to the time multiplied by 20. If d represents distance and t represents time, then $d = 20t$.

Exercises
Communicate

1. In $2x + 3 = 7$, x is a variable, $2x + 3$ is an expression and $2x + 3 = 7$ is an equation.

2. The cost of all the pencils is 12 times the cost per pencil. Express this as $12p = \$1.92$, where p represents the cost per pencil. $p = \$0.16$

3. The cost of x number of tickets is $10x$. By solving $10x = 35$ the number of tickets may be found, $x = 3.5$. One cannot buy half a ticket, so the answer is 3 tickets.

4. Guess values for x and check in the equation, keeping track of the results in a table. Choose additional guesses that are closer and closer or exactly equal to the true solution, $x = 51$.

Portfolio Activity

1.
Set #	0	1	2	3	4	5
Segments	1	3	5	7	9	11

2. $2(100) + 1 = 201$ in the 100th set. Use the table to confirm that the "0" set would have 1 segment.

Lesson 1.3, pages 18–23

Activity

1. Multiply 12 by 15, then add 20, then multiply by 14.

2. Multiply 12 by 15 and 20 by 14, then add the products.

3. a. 17   b. 23   c. 21   d. 13
   e. 9   f. 25   g. 55   h. 729

CHECKPOINT ✓

4. Most graphics calculators follow the standard algebraic order of operations. All calculations included by parentheses are performed first, followed by any exponents. Then multiplication and division are performed from left to right, followed by addition and subtraction from left to right.

Exercises
Communicate

1. Multiply 2 times 4 and then add 3. The result is 11.

2. 20 minus 2, or 18, times 5 is 90, but 20 minus 2 times 5, or 10, is 10. To get the correct answer, 10, in Exercise 2, multiply first and then subtract. This is according to the algebraic order of operations.

3. Given $\{[3(8-4)]^2 - 6\} \div (4-2)$, first simplify within the parentheses from innermost outward. The result is $\{[12]^2 - 6\} \div 2$. Next, perform operations with exponents. Evaluate $12^2 = 144$ to find $\{144 - 6\} \div 2 = 69$.

4. They are necessary to insure that every person will obtain the same answer.

5. Answers may vary. Round the numbers given. $\frac{173 + 223}{151 - 47}$ is about $\frac{400}{100}$ which is close to 4.

Lesson 1.4, pages 24–29
Activity

1.
Time in hours	0.5	1.0	1.5	2.0	2.5
Distance in miles	11.5	23.0	34.5	46.0	57.5

2. (0.5, 11.5), (1.0, 23.0), (1.5, 34.5), (2.0, 46.0), (2.5, 57.5)

3.

4. 230 miles in 10 hours
   115 miles in 5 hours
   207 miles in 9 hours

CHECKPOINT ✓

5. $d = 23t$

Exercises
Communicate

1. Start at the point in question. Count the number of units the point is to the *right* (positive) or to the *left* (negative) of the *y*-axis for the *x*-value. Count the number of units the point is *above* (positive) or *below* (negative) the *x*-axis for the *y*-value.

2. No. (6, 7) and (7, 6) do not specify the same point in the plane. Both the *x*-coordinates and the *y*-coordinates are different.

3. Begin at the origin. Count 7 units to the right along the *x*-axis, then count 3 units up, parallel to the *y*-axis. This locates the point (7, 3).

4. A table allows one to see specific values, but does not describe information (if any) between the given values. A graph gives a picture of a whole relation, but it can sometimes be difficult to determine specific values on the graph.

5. In the left column of the table, list several *x*-values. Evaluate the expression $2x + 5$ for each listed value. Enter these *y*-values for their corresponding *x*-values in the right column.

Practice and Apply

24.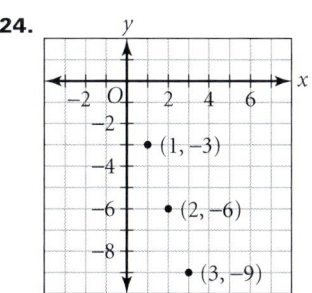

The points lie on a straight line.

25.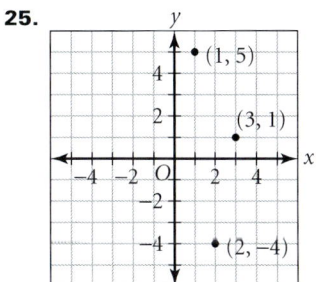

The points do not lie on a straight line.

26.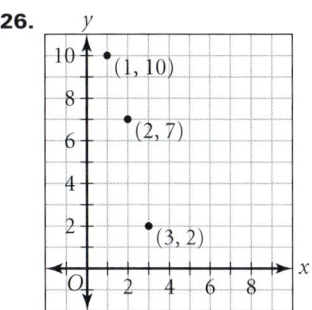

The points do not lie on a straight line.

27.

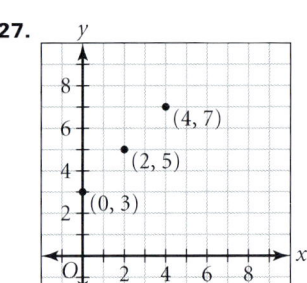

The points lie on a straight line.

28.

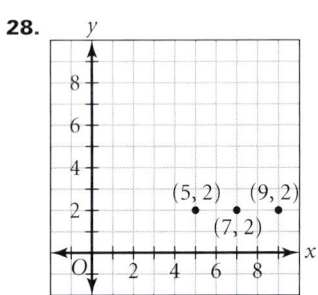

The points lie on a straight line.

29.

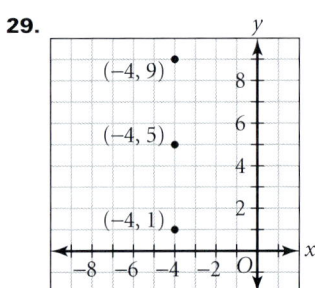

The points lie on a straight line.

55–58.

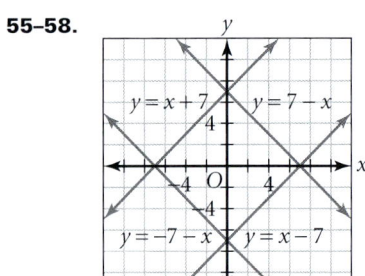

55.

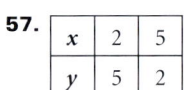

56.

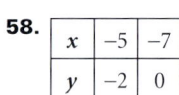

57.

x	2	5
y	5	2

58.

x	−5	−7
y	−2	0

When the sign of x is positive, the line is going up (from left to right). The lines $y = x + 7$ and $y = x - 7$ are going up at the same angle, but at different heights on the y-axis, depending on the constant term. When the sign of x is negative, the line is going down (from left to right). The lines $y = 7 - x$ and $y = -7 - x$ are going down at the same angle, but at different heights on the y-axis, depending on the constant term.

59. c.

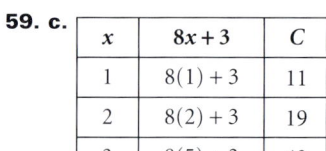

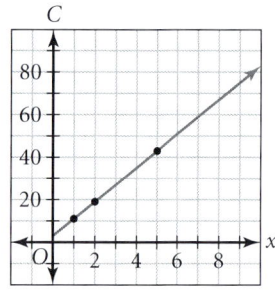

60. The ordered pairs are (1, 24.5), (2, 49), (3, 73.5), (4, 98), and (5, 122.5).

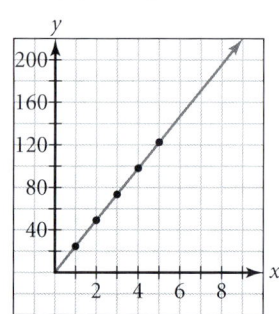

From the graph it appears that the bicyclist would travel about 195 miles in 8 hours. (Exact answer is 196 miles.)

Portfolio Activity

Term #	1	2	3	4	5	6	7	8	9	10
Fibonacci #	1	1	2	3	5	8	13	21	34	55

Add the previous two terms together to get the next term.

Lesson 1.5, pages 30–36

Activity

1.

Time (hours)	0.5	1.0	1.5	2.0	2.5
Distance (miles)	26.5	53.0	79.5	106.0	132.5

2–3.

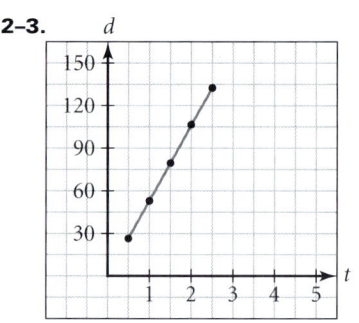

The shape is a line.

CHECKPOINT ✔

4. 159 miles in 3 hours; 212 miles in 4 hours

Exercises

Communicate

1. In the equation $y = 10x$, x is the independent variable. It may have any value.

2. In the equation $y = 10x$, y is the dependent variable. Its value depends upon the choice of x.

3. If the first differences of a number sequence are constant, a linear equation can be written to represent the data pattern.

4. The graph of a linear equation is a straight line.

Practice and Apply

21.

x	1	2	3	4	5
y	298	296	294	292	290

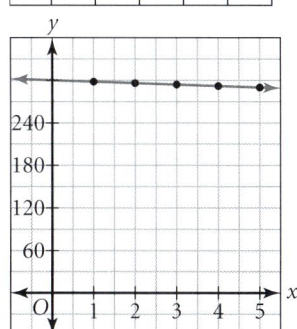

22.

x	1	2	3	4	5
y	28	33	38	43	48

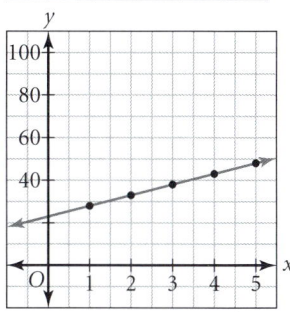

23.

x	1	2	3	4	5
y	22	29	36	43	50

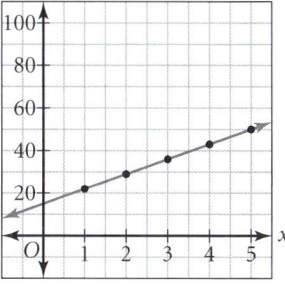

24.

x	1	2	3	4	5
y	420	415	410	405	400

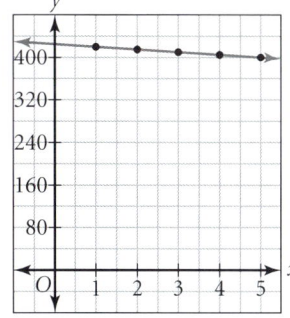

25.

x	1	2	3	4	5
y	110	100	90	80	70

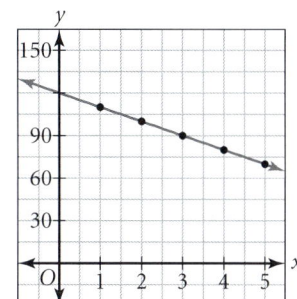

26.

x	1	2	3	4	5
y	65	130	195	260	325

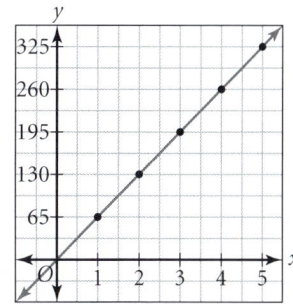

27.

x	1	2	3	4	5
y	110	220	330	440	550

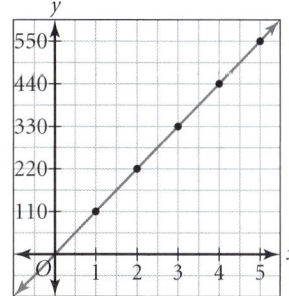

28.

x	1	2	3	4	5
y	64	75	86	97	108

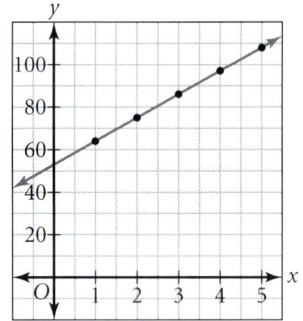

29.

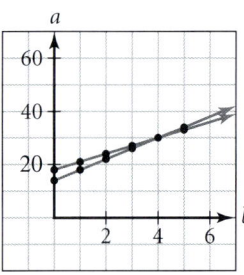

From the graphs, the lines intersect at $b = 4$, $c = 30$. This point represents when the membership costs are the same.

30. a.

Number of years experience	0	1	2	3	4	5
Total salary	30,000	32,500	35,000	37,500	40,000	42,500

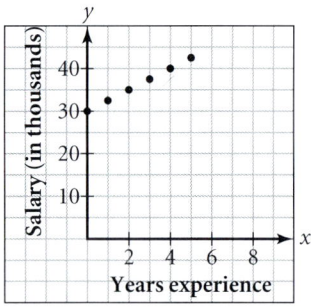

31. $d = 14h$

h, hours	0	1	2	3	4	5
d, miles	0	14	28	42	56	70

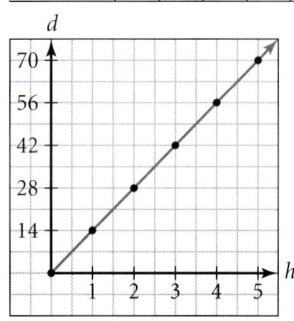

32. $d = 58h$

h, hours	0	1	2	3	4	5	6	7	8
d, miles	0	58	116	174	232	290	348	406	464

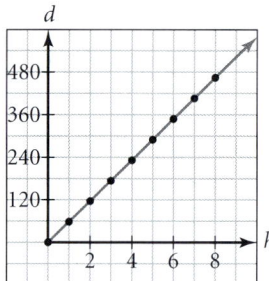

If the family drove for 10 hours, they could travel 580 miles.

Look Beyond

49.

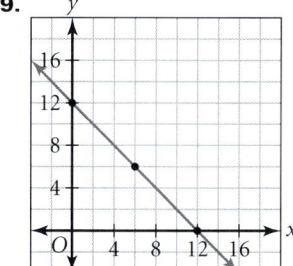

The graph is linear.

50.

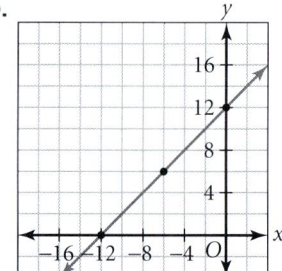

The graph is linear.

51.

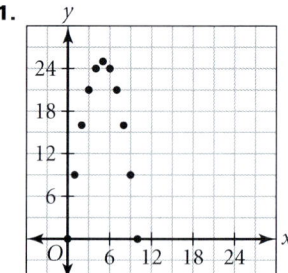

The vertex is (5, 25).

Lesson 1.6, pages 37–43
Activity

1.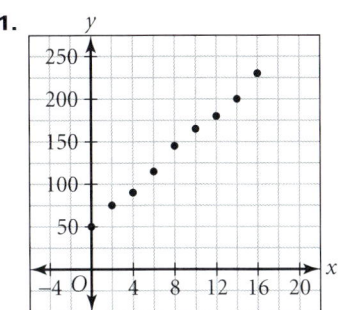

2.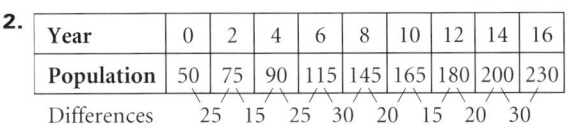

3.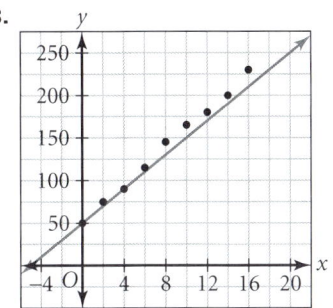

4.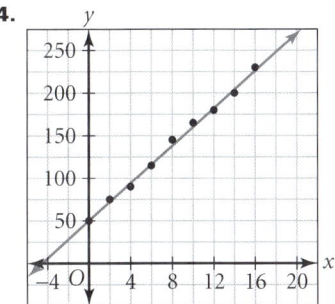

The line $y = 11x + 50$ is a better fit.

CHECKPOINT ✔

5. Using the regression equation for the data, the prediction for the 10th year is ≈160, which is relatively close to the value in the table. The prediction for the 20th year is ≈270.

Exercises
Communicate

1. The line of best fit is closest to all the data points. You use the equation of the line to predict values not given.

2. Two data sets have a strong negative correlation when the data points lie nearly in line and as one variable increases, the other decreases.

3. Two data sets have a strong positive correlation when the data points lie nearly in line and as one variable increases, so does the other.

4. a. The scatterplot shows little or no correlation because the data points seem randomly scattered.
   b. The scatterplot shows a strong positive correlation. The data points are close to a straight line with both variables increasing.

Portfolio Activity

1–7. Answers will vary.

Lesson 2.1, pages 54–59
Activity

1. a. 0.800
   b. 0.364
   c. 0.625
   d. 0.778
   e. 0.417
   f. 0.750

CHECKPOINT ✔

2. $\frac{4}{5}$, $\frac{5}{8}$, and $\frac{3}{4}$ are terminating decimals. Answers may vary. Sample answer: During the division process, the 8 divides the remainder evenly.

CHECKPOINT ✔

3. $\frac{4}{11}$, $\frac{7}{9}$, and $\frac{5}{12}$ are repeating decimals. Answers may vary. Sample answer: During the division process, the same numbers are generated over and over.

Exercises
Communicate

1. The decimal equivalent of $\frac{2}{3}$ is a repeating, nonterminating decimal: $0.\overline{6}$. The decimal equivalent of $\frac{3}{5}$ is a terminating decimal: 0.6.

2. Any integer, a, can be written as $\frac{a}{1}$, so every integer is a rational number.

3. The absolute value of a number is the distance from the number to 0 on a number line. The absolute value of 6 is the same as the absolute value of −6 because they are both 6 units from 0.

Lesson 2.2, pages 60–66
Activity

3. 3 negative tiles remain; −3

4. $-10 + 6 = -4$; $10 + (-3) = 7$

5. negative

CHECKPOINT ✓

6. Find the difference of the absolute values, and use the sign of the number with the greater absolute value.

Exercises
Communicate

1. Add their absolute values and use the sign that is the same for both integers.

2. Subtract their absolute values and use the sign of the integer with the larger absolute value.

3. a. $-4 + (-5) = -9$
   Add 4 negative tiles to 5 negative tiles to get 9 negative tiles.
   b. $-3 + (+6) = 3$
   Add 6 positive tiles to 3 negative tiles. Remove all neutral pairs. There are 3 positive tiles remaining.

4. When combining an equal number of positive and negative algebra tiles, we get neutral pairs of tiles which, by the Additive Inverse Property, equal zero. When adding these neutral pairs to other tiles, we get the amount of the other tiles by the Additive Identity Property.

Portfolio Activity

1. Subtract expenses from revenue.

2. Expenses are greater than revenue.

3. Subtracting total expenses from total revenue or by adding the profit column.

Lesson 2.3, pages 67–72
Activity

2. Remove 3 positive tiles.

CHECKPOINT ✓

3. This has the same overall effect as adding 3 negative tiles.

4. Subtracting 3 is the same as adding 3 negative tiles.

Exercises
Communicate

1. Adding a neutral pair of algebra tiles is equivalent to adding 0; we add enough neutral pairs to be able to take away the necessary amount of positive tiles. The net effect is like adding negative tiles, just as a subtraction problem can be changed into an addition problem.

2. To find the temperature, subtract 25 from 5. The temperature would be −20°F.

3. Draw 4 negative tiles and 7 negative tiles. There are 11 negative tiles. The subtraction expression "the opposite of 4 minus 7" is equivalent to $-4 - 7$.

4. It means the same as subtracting the number.

5. Whenever a real number is subtracted from another real number, the result is also a real number.

Lesson 2.4, pages 73–79
Activity

Part I
1. a. $-2, -4, -6$
   b. $-3, -6, -9$
   c. $3, 6, 9$

CHECKPOINT ✓
2. a. $+$
   b. $-$
   c. $-$
   d. $+$

Part II
1. a. 56
   b. -30
   c. -8
   d. 8

2. a. $56 \div 7 = 8$; $56 \div 8 = 7$
   b. $-30 \div 6 = -5$; $-30 \div (-5) = 6$
   c. $-8 \div (-4) = 2$; $-8 \div 2 = -4$
   d. $8 \div (-8) = -1$; $8 \div (-1) = -8$

CHECKPOINT ✓
3. a. $+$
   b. $-$
   c. $-$
   d. $+$

Exercises
Communicate

1. Multiply their absolute values. The sign of the product is negative since the numbers have unlike signs.

2. Divide their absolute values. The sign of the product is positive since the numbers have like signs.

3. Answers may vary. Sample answer: When a number is multiplied by its reciprocal, the result is the identity element of multiplication, 1.

4. There is no value of n such that $n \cdot 0 = 6$. Division by zero is undefined.

5. Answers may vary. Sample answer: Multiplying a number by its reciprocal is the same as dividing the number by itself. Adding a number to its opposite is the same as subtracting the number from itself.

Portfolio Activity

1. 8, 12, 12, 14, 14, 13, 15, 15, 14, 13, 11, 11

2. Check students' drawings.

3.

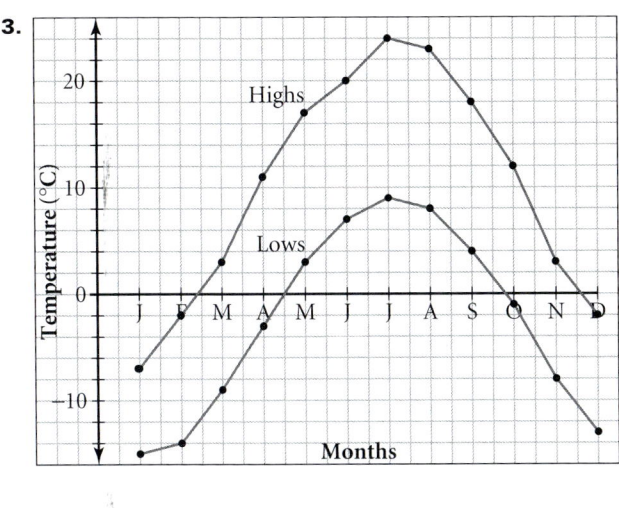

Lesson 2.5, pages 80–88

Activity

1.

Column A	Column B	Column C	Column D
4(15)	4(10 + 5)	4 • 10 + 4 • 5	40 + 20 = 60
8(17)	8(10 + 7)	8 • 10 + 8 • 7	80 + 56 = 136
5(22)	5(10 + 10 + 2)	5 • 10 + 5 • 10 + 5 • 2	50 + 50 + 10 = 110
12(25)	(10 + 2)25	10 • 25 + 2 • 25	250 + 50 = 300
21(7)	(20 + 1)7	20 • 7 + 1 • 7	140 + 7 = 147
6(128)	6(100 + 20 + 8)	6 • 100 + 6 • 20 + 6 • 8	600 + 120 + 48 = 768

2. One of the numbers is written in expanded form. Then the Distributive Property is used to compute the product.

CHECKPOINT ✔

3. You can calculate the product of two numbers using mental computation by expanding one of the numbers and then using the Distributive Property.

Exercises
Communicate

1. Changing the order of adding or multiplying may simplify the problem.

2. Answers will vary. Sample answer:
$24 + (19 + 16)$
$= 24 + (16 + 19)$ Commutative Property
$= (24 + 16) + 19$ Associative Property
$= 40 + 19$
$= 59$
24 and 16 add easily.

3. a. $8(21) = 8(20 + 1)$
$= 8 \cdot 20 + 8 \cdot 1$
$= 160 + 8$
$= 168$

b. $11(35) = (10 + 1)(35)$
$= 10 \cdot 35 + 35$
$= 350 + 35$
$= 385$

c. $14(22) = 14(20 + 2)$
$= 14 \cdot 20 + 14 \cdot 2$
$= 280 + 28$
$= 308$

4. $(3.50)(5) + (3.50)(7) = 3.50(5 + 7)$
$= (3.50)(12)$
$= 42$
The total rental fee for the weekend is $42.00.

Portfolio Activity

1. The first differences are not constant. The second differences are constant. The pattern is not linear.

2. −12, −13, −13; the rate of change in the windchill temperature decreases rapidly although the windspeed increases at a constant rate.

3. −12°F, −13°F, −13°F

Lesson 2.6, pages 89–95
Activity

1. $5x + 5$

2. $7x + 2$

3. $x + 3$

CHECKPOINT ✔
4.

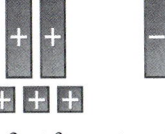   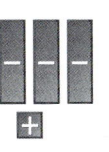   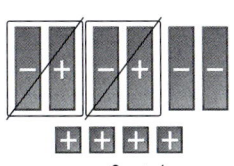

$2x + 3$ $+$ $(-4x + 1)$ $= -2x + 4$

Exercises
Communicate

1. $7x$, $2x$ and $3x$ are like terms because they contain the same variable, x.

2. A common factor x can be removed from $3x + 8x$ for $(3 + 8)x = 11x$.

3. To subtract an expression, change the signs of the terms of the expression to be subtracted to their opposites and then add the expressions.

817

Additional Answers

4. Write the opposite of each term to be subtracted, use the Commutative and Associative Properties to regroup like terms and add.

$(9x + 3) - (4x - 1)$
$= 9x + 3 - 4x + 1$
$= (9x - 4x) + (3 + 1)$
$= 5x + 4$

Lesson 2.7, pages 96–103
Communicate

1. **a.** If Jan works 4 hours, then she will earn $6 four times, or $6 \cdot 4 = 24$, $24.
   b. If Jan works 2.5 hours, then she will earn $6 two and one-half times, or $6 \cdot 2.5 = 15$, $15.
   c. If Jan works h hours, then she will earn $6 h times, or $6h$ dollars.

2. Multiply $4m$ by -2 to get $-8m$ and multiply 5 by -2 to get -10. The result is $-8m - 10$.

3. Divide $5p$ by -5 to get $-p$ and divide -15 by -5 to get 3. The result is $-p + 3$.

4. All expressions are equivalent because they simplify, using the Distributive Property, to the same expression, $2x - 3$.

5. The expression $\frac{2x+6}{2} = x + 3$, which is not equivalent to $x + 6$.

Portfolio Activity

1. $2H + 4$
2. $H + 3$
3. $(2H + 4) + (H + 3)$; $3H + 7$
4. $2(2H + 4) + 3(H + 3)$; $7H + 17$

Lesson 3.1, pages 114–121
Activity

1. Add 2 positive unit tiles to each side. The left side then has 1 x-tile and 2 neutral pairs, and the right side has 10 positive unit tiles.

2. After removing the neutral pairs, the solution is $x = 10$.

CHECKPOINT ✔
3. $x - 2 + 2 = 8 + 2$
   $x = 10$

Exercises
Communicate

1. When solving an equation where a number is added to a variable expression, use the Subtraction Property of Equality. When a number is subtracted from a variable expression, use the Addition Property of Equality.

2. **a.** Answers may vary. Sample: John started out the week with $20 and finished the week with $50. How much money did he make during the week?
   b. Answers may vary. Sample: What was Luke's inventory at the start of the game if he ended the game with 50 bags of peanuts and had sold 20?

3. **a.** On the left side of the work area place one x-tile and six positive unit tiles, and on the right side of the work area place ten positive unit tiles. Add six negative unit tiles to each side. Remove the six neutral pairs from each side. The result is $x = 4$.
   b. On the left side of the work area place one x-tile and six negative unit tiles, and on the right side of the work area place ten positive unit tiles. Add six positive unit tiles to each side. Remove the six neutral pairs from the left side. The result is $x = 16$.

4. Subtract 7 from both sides of the equation. The result is $x = 3$.

Portfolio Activity

1. $40m = 4800$
   $40m = 40 \cdot 100 + 40 \cdot 20$
   $m = 120$

2. $60j = 930$
   $60j = 60 \cdot 15 + 60 \cdot \frac{1}{2}$
   $j = 15\frac{1}{2}$

3. $24p = 252$
   $24p = 24 \cdot 10 + 24 \cdot \frac{1}{2}$
   $p = 10\frac{1}{2}$

4. $10w = 2432$
   $10w = 10 \cdot 200 + 10 \cdot 40 + 10 \cdot 3 + 10 \cdot \frac{1}{5}$
   $w = 243\frac{1}{5}$

Lesson 3.2, pages 122–128
Activity

1. If x represents the cost per foot, and each board is 8 feet, then the cost for one board is $8x$. The cost is known to be $2.40, so $8x = 2.40$.

2. Just as the board is divided into 8 equal pieces, the cost must be divided into 8 equal pieces as well, which indicates what operation we must use.

3. $8x = 2.40$
   Try $x = 0.20$
   $8(0.20) = 1.60$
   Try a larger number.
   Try $x = 0.30$
   $8(0.30) = 2.40$
   Therefore, the solution is 0.30.

4. The equality is preserved at every step, and the steps lead to a solution for x.

CHECKPOINT ✔

5. $8x = 3.20$ $\quad\quad\quad$ $10x = 25$
   $\frac{8x}{8} = \frac{3.20}{8}$ $\quad\quad$ $\frac{10x}{10} = \frac{25}{10}$
   $x = 0.40$ $\quad\quad\quad$ $x = 2.5$

Exercises
Communicate

1. Use the Division Property of Equality by dividing both sides of the equation by 592.

2. Use the reciprocal by multiplying both sides of the equation by $\frac{1}{5}$.

3. Use the Multiplication Property of Equality by multiplying both sides of the equation by 12.

4. Use the reciprocal by multiplying both sides of the equation by $\frac{5}{4}$.

5. Use the Division Property of Equality by dividing both sides of the equation by π.

Portfolio Activity

1–2.

Number of shirts	Cost = 125 + 4.55s	Single shirt cost
100	$580.00	$5.80
250	$1262.50	$5.05
500	$2400.00	$4.80
750	$3537.50	$4.72

3. $250x - (250)(5.80) = 500$
   $250x - 1450 = 500$
   $x = \$7.80$

Lesson 3.3, pages 129–134
Activity

1.

0	18 + 7(0)	18
1	18 + 7(1)	25
2	18 + 7(2)	32
3	18 + 7(3)	39
4	18 + 7(4)	46
5	18 + 7(5)	53
6	18 + 7(6)	60
7	18 + 7(7)	67
8	18 + 7(8)	74
9	18 + 7(9)	81
10	18 + 7(10)	88

CHECKPOINT ✔

2. Read the chart at $74. The cost of a single towel is $8 when the total bill is $74.

3. Answers will vary. Sample: Subtract 18 from both sides of the equation and then divide by 7 to find the value of t.

Exercises
Communicate

1. Subtract 1 from both sides of the equation. Divide both sides of the equation by 2.

2. Subtract 15 from both sides of the equation and then divide both sides by –4.

3. Subtract 33 from both sides of the equation and then multiply both sides by 24.

4. Set up a three column table with a column for the value of the variable on the left, the left side of the equation in the middle column, and the right side of the equation in the right column. Try various appropriate values for the variable in the expression for the left side of the equation until the original value for the right side of the equation appears.

5. Find the intersection point of the lines formed by setting y equal to the left side of the equation and then to the right side of the equation.

Lesson 3.4, pages 135–140
Activity

2. Add three negative x-tiles to both sides, and remove neutral pairs.

3. $\quad\quad 4x - 2 = 3x + 4$
   $4x - 2 - 3x = 3x + 4 - 3x$
   $\quad\quad\quad x - 2 = 4$

CHECKPOINT ✔

4. Add 2 positive unit tiles to each side, and remove neutral pairs. This leaves 1 x-tile on the left side and 6 positive unit tiles on the right side; $x = 6$.

Exercises
Communicate

1. Subtract p from both sides. Subtract 57 from both sides of the equation. Divide both sides by 22.

2. Add $2z$ to both sides. Add 5 to both sides. Divide both sides by 7.

3. Add $2x$ to both sides of the equation. Subtract 15 from both sides of the equation. Divide both sides by 6.

4. Find the x-value of the intersection point of the lines described by $y = 3x + 6$ and $y = 2x + 4$.

5. Let s represent the score Brian needs on his third test. The average is calculated by adding the scores and then dividing by the number of scores. In other words,

$$\frac{85 + 85 + s}{3} = 90$$
$$\frac{170 + s}{3} = 90 \quad \text{Simplify.}$$
$$3\left(\frac{170 + s}{3}\right) = 3(90) \quad \text{Multiply both sides by 3.}$$
$$170 + s = 270$$
$$170 + s - 170 = 270 - 170 \quad \text{Subtract 170 from both sides.}$$
$$s = 100$$

Brian would need a perfect score of 100 in order to have an average of 90.

Portfolio Activity

1. The expression $125 + 4.55t$ represents the cost of producing the T-shirts and $7.50t$ represents the income. The two are set equal to each other and therefore represent the point at which they are equal—or the "break-even" point.

2. $t = 42.4$

3. $t = 22.9$

Lesson 3.5, pages 141–146
Activity

1.

5	3(5 – 4)	3
10	3(10 – 4)	18
15	3(15 – 4)	33
20	3(20 – 4)	48
25	3(25 – 4)	63

2. Look for $48 in the total column and read the accompanying original price, $20.

CHECKPOINT ✔

3. First distribute 3 over the expression to get $3x - 12$. Add 12 to both sides and then divide both sides by 3 to get the result, $x = 20$.

Exercises
Communicate

1. Use the Distributive Property to multiply on the left side of the equation. Add 8 to both sides and then divide both sides by 2.

2. Use the Distributive Property to multiply on the left side of the equation. Combine like terms. Subtract $4x$ from both sides of the equation.

3. Use the Distributive Property to multiply on the left side of the equation. Combine like terms. Subtract $2x$ from both sides. Add 13 to both sides and then divide both sides by 2.

4. Set up a 3-column table with the value of the variable in the left column, the left side of the equation in the middle column, and the right side of the equation in the right column. Try various appropriate values for x until the value 108 appears in the right column.

Lesson 3.6, pages 147–153
Exercises
Communicate

1. A literal equation is an equation that contains two or more different variables.

2. Formulas enable computation of certain unknown quantities when other quantities are known. A formula is a generalization that can be applied to a variety of situations.

3. a. Divide both sides by π.
   b. Subtract A from both sides.
   c. Subtract 7 from both sides and then divide both sides by 3.

4. Use the formula for perimeter, $P = 2l + 2w$. Enter the values that are known, perimeter and width, and solve for the length.

Portfolio Activity

$\frac{1}{6}x + \frac{1}{12}x + \frac{1}{7}x + 5 + \frac{1}{2}x + 4 = x$
$x = 84$

Diophantus lived a total of 84 years.

Lesson 4.1, pages 164–170

Activity

1. **a.** means: 2 and 5
   extremes: 1 and 10
   b. means: 4 and 9
   extremes: 3 and 12
   c. means: 25 and 30
   extremes: 15 and 50
   d. means: 6 and 6
   extremes: 4 and 9

2. **a.** means: 10
   extremes: 10
   b. means: 36
   extremes: 36
   c. means: 750
   extremes: 750
   d. means: 36
   extremes: 36

3. The product of the means is equal to the product of the extremes in each proportion.

4. In a proportion, the cross products, which are the product of the means and the product of the extremes, are equal.

CHECKPOINT ✔

5. Set the cross products equal to each other, and then solve for x.
   $x \cdot 24 = 4 \cdot 42$
   $x = 7$

Exercises
Communicate

1. A proportion is any statement that two ratios that are equivalent.
   Example: $\frac{2}{3} = \frac{4}{6}$

2. In the proportion $\frac{a}{b} = \frac{c}{d}$ or $a:b = c:d$, the values in the positions of a and d are the extremes. The values in the positions of b and c are the means.

3. The product of the means will always equal the product of the extremes in a proportion.

4. Method 1
   Find the cross products and set them equal to each other.
   $3n = 2 \cdot 36$
   Divide both sides by 3 and simplify.
   $n = 24$
   Method 2
   Find the LCM of 36 and 3, and multiply each side of the equation by that value.
   $\frac{36}{1} \cdot \frac{n}{36} = \frac{36}{1} \cdot \frac{2}{3}$
   Simplify each side of the equation.
   $n = 24$

Portfolio Activity

1. Lemon-lime was the most popular.
2. 24 students responded.
3. Yes, 11 students prefer colas, whereas only 10 prefer lemon-lime.
4. No, because 11 is less than 50% of 24.
5. Answers may vary. Check students' responses.

Lesson 4.2, pages 171–177

Exercises
Communicate

1. Divide the given percent by 100.
2. Answers may vary. Sample: The percent bar helps one visualize the relationship between the percent values and the actual values.
3. Answers may vary. Sample: Percents do not need to be changed to decimals or fractions.
4. Answers may vary. Sample: In some cases, there are less steps to perform.

Lesson 4.3, pages 180–185

Activity
Part I

1. Answers should, but may not, be about 3.
2. Most students should have similar results.
3. Answers should be close to 17%.

CHECKPOINT ✔

4. Answers may vary. Students should expect answers to be $\frac{1}{6}$ or about 17%.

Part II

1. Answers should, but may not, be close to 50%.
2. Most students should have similar results.
3. Answers should be close to 50%.

CHECKPOINT ✔

4. Answers may vary. Students should expect answers to be about 50%.

Communicate

1. Experimental probability is calculated by performing an experiment and comparing the number of times an event occurs to the total number of trials in the experiment.

821

2. Answers may vary. For example, toss 4 coins 10 times and count the number of times that 3 or 4 heads turn up.

3. Yes. For example, if 2 pairs of players toss 4 coins each, the number of heads showing may be different for both pairs.

4. Yes. If one player tosses 4 coins once and then again, the number of heads showing on each set of tosses may be different.

5. The greater the number of trials, the more accurate your results will be. For example, you could not claim that a coin always flipped to heads based on just one trial.

Portfolio Activity

1. 0.518

2. Answers may vary. Sample:

Number of trials	Heads	Experimental probability
4	2	0.5
10	6	0.6
20	12	0.6
30	12	0.4

3. Answers may vary. Check students' responses.

4. Answers may vary. Sample answer: 0.5

Lesson 4.4, pages 186–192

Activity

1. The modes are 75, 85, and 90. A frequency table shows us how many times a number occurs, so we do not have to search through the data to find the mode.

2. The median is 80. The median is the number above the middle tick mark.

3. The mean is about 80; 55, 60, 65, 65, 70, 70, 70, 75, 75, 75, 75, 80, 80, 80, 85, 85, 85, 85, 90, 90, 90, 90, 95, 95, 100; $\frac{198.5}{25} = 79.4$

CHECKPOINT ✔

4. A frequency table is a way to organize data, making it easier to work with.

Exercises
Communicate

1. The mean represents the average value of the data. The median is the middle value when the data is arranged in ascending order.

2. Arrange all of the numbers in ascending order: 0, 2, 3, 4, 4, 4, 5, 6, 7, 8, 9. Next find the number which is in the center of the ordered list: 4.

3. Mean. This statistic may not always be best when dealing with situations which require whole numbers.

4. Multiply each grade by the number of tick marks below it. Add each product together. Divide this sum by the total number of tick marks.

Lesson 4.5, pages 193–200

Activity

1. $40,000,000; $65,000,000; Regency Stores; Morton Stores

2. $50,000,000; $80,000,000; Morton Stores; Regency Stores

CHECKPOINT ✔

3. The vertical scale of each graph makes it possible to misinterpret the information. Use the same vertical scale.

4.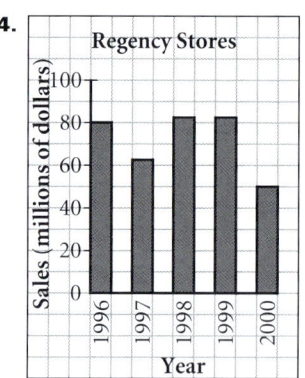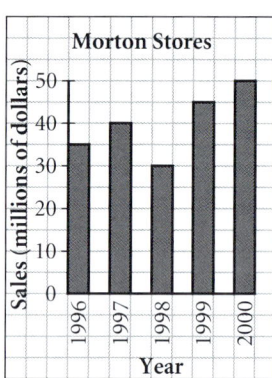

Exercises
Communicate

1. Answers may vary. Sample: Graphs help to simplify and organize data. They also aid in analyzing trends and relationships within a set of data.

2. Answers may vary. Sample: While all the information in a graph may be accurate, the choice of scale, color or placement could disguise data to mislead its audience.

3. Answers may vary. Sample: Line graphs are used to show changes that have occurred over time. Bar graphs are used to show comparisons. Circle graphs are used to show how parts or percentages compare to a whole.

4. A line graph focuses on how the data in a set changes over time, while a bar graph focuses on comparing the data in the set to each other.

Lesson 4.6, pages 201–207

Activity

1. 60, 65, 70, 75, 75, 75, 80, 80, 80, 85, 90, 90, 100; median is 80

822

2. 72.5

3. 87.5

4. the median, 80

5. The lower quartile equals the median of the first 6 grades, and the upper quartile equals the median of the last 6 grades.

6. The dot on the far left is the lowest grade, and the dot on the far right is the highest grade.

CHECKPOINT ✓

7.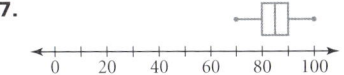

Exercises
Communicate

1. First determine the stems. Choose the second highest place value in the data to be the value of the stems. Next determine the leaves. The leaves will have the value of the lowest place value in the data. Now list the stems in ascending order in a column. For each stem list each leaf in ascending order in the same row one column over. Use commas to separate the leaves.

2. Determine how to break up the set of stems into desired categories. List each category along the *x*-axis. Next determine how many leaves are in each category. Scale the *y*-axis for these amounts. Finally draw vertical bars to the heights corresponding to the number of leaves in each category.

3. Locate the box and find the line in the middle of it. The number representing the location of the line is the median. To calculate the range subtract the least value from the greatest value.

4. Locate the box. The number representing the location of the left-hand edge of the box is the lower quartile. The number representing the location of the right-hand edge of the box is the upper quartile.

Practice and Apply

10. Stems: tens
Leaves: ones

Stems	Leaves
1	1, 1, 2, 3, 3, 4, 7
2	2, 3, 8
3	2, 2, 7, 8
4	1, 6, 7, 8, 9
5	6
6	5
7	3, 6, 8
8	2, 3, 4

11. Stems: ones
Leaves: tenths

Stems	Leaves
1	2
2	5, 5, 7
3	4
4	6, 8
5	2, 6
6	3, 5, 7, 8
7	8
8	4, 4
9	1, 2, 7

12. Stems: tens
Leaves: ones

Stems	Leaves
1	2, 3, 6, 8, 9
2	0, 1, 4, 4, 5, 5, 6, 6, 7, 7, 8, 9, 9
3	2, 4, 7, 8

13. Stems: tens
Leaves: ones

Stems	Leaves
0	4, 5, 5, 5, 6, 7, 7, 8, 9
1	0, 2, 2, 6, 7, 8, 8, 9, 9
2	3, 6, 9

14. Stems: tens
Leaves: ones

Stems	Leaves
1	1, 2, 2, 4, 4, 6, 6, 6, 7, 8, 8, 9
2	0, 1, 4, 8
3	4

15. Stems: ones
Leaves: tenths

Stems	Leaves
20	8
21	4, 6
23	1, 6, 9
24	7
25	8
26	1, 4
27	6
29	4, 4, 5, 6

16. Stems: tens
Leaves: ones

Stems	Leaves
30	3, 7
31	4
32	2, 8
33	5, 6, 7, 9
34	9
35	5, 9
36	6
37	1, 4, 8
39	3
40	7

17.
```
           16.35
12.2  14.95    17.95  20.3
  •——[——|——]——•
12 13 14 15 16 17 18 19 20 21
```

18.
```
11      37  47  55      68
 •——————[——|——]——————•
10  20  30  40  50  60  70
```

19.
```
125   136  145.5  159      184
  •————[———|———]————————•
125 135 145 155 165 175 185
```

20.
```
11 17    38    65      84
 ••—[————|————]————————•
10 20 30 40 50 60 70 80 90
```

21.
```
1.2    3.4   6.3    8.4  9.7
 •—————[——|——]——————•
 1  2  3  4  5  6  7  8  9 10
```

22.
```
12       20  25.5 29        38
 •———————[——|——]——————————•
10  15  20  25  30  35  40
```

23.
```
4  6.5    12     18.5        29
•——[——|——————]——————————————•
0           15              30
```

24.
```
11 14  17 20.5              34
 •—[—|——]———————————————————•
10  15  20  25  30  35
```

25.
```
20.8  23.1   25.8   29.4 29.6
  •————[——————|——————]—•
20          25          30
```

26.
```
303   328  344   371       407
  •————[———|———]————————•
300 320 340 360 380 400 420
```

32. Stems: ones
Leaves: tenths

Stems	Leaves
14	1, 5, 9
15	0, 2, 3, 5, 6
16	1, 2, 3, 6
17	2, 2, 4, 8
18	1, 2, 9
19	3, 3, 7, 9
20	0

37.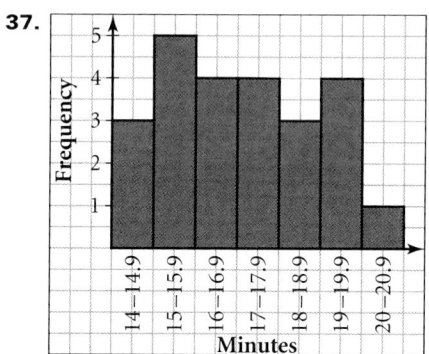

39. Stems: tens
Leaves: ones

Stems	Leaves
24	9
27	5
34	3, 5
38	3
39	9
47	8
48	4
49	1, 7, 7, 9
51	2
54	9
56	7
57	4, 6
58	6, 8, 9

Portfolio Activity

1. Answers may vary. Sample:

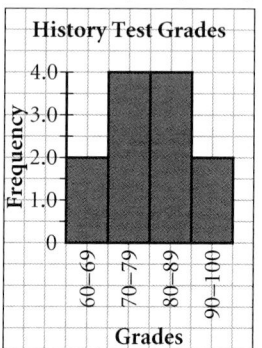

2. Answers may vary. Sample: A histogram clearly shows the distribution of A's (90–100), B's (80–89), and so on.

3. Answers may vary. Check student responses.

Lesson 5.1, pages 218–225

Activity

1. Table 1: (14, 68), (11, 64), (13, 65), (16, 72), (15, 64), (14, 67), (15, 66), (12, 62)
Table 2: (11, 63), (12, 64), (13, 65), (14, 70), (15, 72), (16, 72)

2. Yes, 14 and 15 are repeated.

3. No, none are repeated.

4. Yes, 64 is repeated.

5. Yes, 72 is repeated.

CHECKPOINT ✔

6. The second coordinates may be repeated, but none of the first coordinates may be repeated.

Exercises
Communicate

1. a. This set is a function, since no element of the domain is repeated.
   b. This set is not a function, since 2 is paired with two elements of the range.
   c. This set is a function, since no element of the domain is repeated.
   d. This set is not a function, since 1 is paired with four elements of the range.

2. A function is a relation in which each element of the domain is paired with exactly one element of the range.

3. Answers may vary. Sample answer: The set of ordered pairs {(1, 2), (2, 4), (3, 6), (4, 8)} is a linear function. The domain is {1, 2, 3, 4} and the range is {2, 4, 6, 8}.

4. Read the ordered pairs from the graph and use the ordered pairs and the pattern of differences to write an equation.

Lesson 5.2, pages 226–233

Activity

1.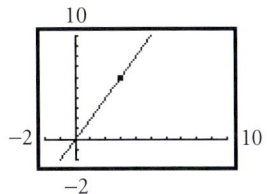

2. $m = 2$
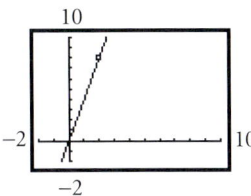

3. For $B(2, 8)$, $m = 4$.
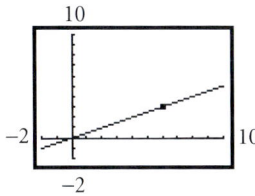

For $C(6, 3)$, $m = \frac{1}{2}$.
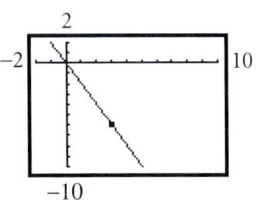

For $D(3, -6)$, $m = -2$.
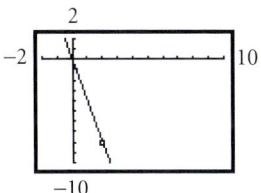

For $E(2, -8)$, $m = -4$.

For $F(4, 7)$, $m = \frac{7}{4}$, or $1\frac{3}{4}$.
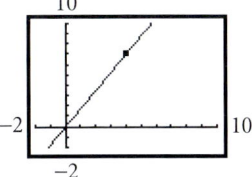

CHECKPOINT ✔

4. Answers may vary. Sample answer: The linear equation in the form $y = mx$ that passes through point (a, b) has the slope $m = \frac{b}{a}$.

Exercises
Communicate

1. Start at the origin. From that point move 4 units upward (rise), then 3 units to the right (run). Put a point where you end up. Draw a line through this point and the origin.

2. The numeric value of the slope is negative. The line falls from left to right.

3. Slope is defined as $\frac{\text{rise}}{\text{run}}$. In a vertical line the run is 0, and division by zero is undefined.

4. A line with a slope of s rises from left to right, while a line with slope $-s$ falls from left to right.

5. Slope is defined as $\frac{\text{rise}}{\text{run}} = \frac{y_2 - y_1}{x_2 - x_1}$. If $(x_2, y_2) = (5, -3)$ and $(x_1, y_1) = (9, -4)$, then slope of line k is $\frac{y_2 - y_1}{x_2 - x_1} = \frac{-3 - (-4)}{5 - 9} = -\frac{1}{4}$.

Portfolio Activity

Answers may vary. Check students' results.

Lesson 5.3, pages 236–243
Activity

1.
x	0	1	2	3	4	5
y	0	0.5	1	1.5	2	2.5

2.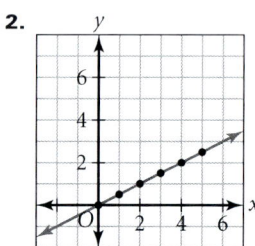

3. The graph is a line passing through the origin.

4. Using the points (2, 1) and (4, 2), slope = $\frac{2-1}{4-2} = \frac{1}{2}$.

CHECKPOINT ✔

5. Answers may vary. Sample answer: All graphs should be lines through the origin with slopes equal to the constant of variation.

Exercises
Communicate

1. The graph is broken up into three parts. The first part of the graph represents Allen riding uphill since the slope has the smallest value. The second part of the graph represents Allen riding downhill since the slope has the greatest value. The third part of the graph represents Allen riding on level ground since the slope here is in between the other slopes. Note, the greater the slope value, the more distance is traveled in a given time.

2. Slope of a hill = $\frac{\text{Change in height}}{\text{Change in length}}$

3. They have the same value.

4. The amount of money the senior class will make from selling football programs will vary directly with the number of programs sold.
   m = money made by Senior class
   p = number of programs sold
   $k = 1.50$
   $m = kp$
   $m = 1.50p$

Look Back

40.
x	1	2	3	4	5
y	7	14	21	28	35

41.
x	1	2	3	4	5
y	−2.5	−2	−1.5	−1	−0.5

42. $d = 35t$

x	0	1	2	3
y	0	35	70	105

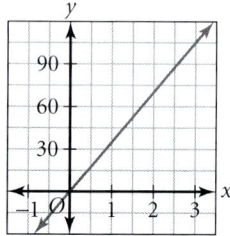

43. $-7x + 11x = (-7 + 11)x$
    $= 4x$

44. $8h - 9h = (8 - 9)h$
    $= -1h$
    $= -h$

45. $3d + 5d = (3 + 5)d$
    $= 8d$

Look Beyond

54. c.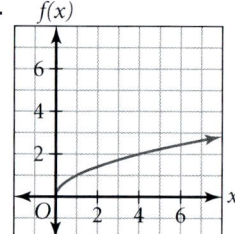

The graph is not linear. It starts at the origin and goes up to the right, then curves almost to the point of leveling off.

Portfolio Activity

1–6. Answers may vary. Check students' results.

Lesson 5.4, pages 244–251

Activity

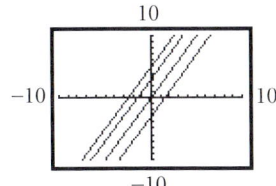

1. All the lines have the same slope.

2. Each of the lines has a different y-intercept.

3. It has the same slope as the other lines, and is below the line $y = 2x - 3$.

4.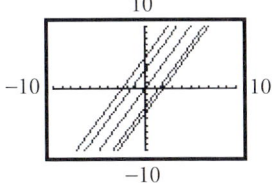

CHECKPOINT ✔

5. It raises or lowers the graph along the y-axis.

Exercises
Communicate

1. Since the line passes through the origin, the y-intercept (or b) equals 0. Calculate the slope of the line using the points $(0, 0)$ and $(3, 6)$.
$m = \frac{y_2 - y_1}{x_2 - x_1} = \frac{6 - 0}{3 - 0} = \frac{6}{3} = 2$
$y = mx + b$
$y = 2x + 0$ or $y = 2x$

2. $y = hx$

3. The line crosses the y-axis at $(0, 3)$, so $b = 3$. Moving 3 units down (rise) and 2 units to the right (run) brings you to another point on the line. Since slope $= \frac{\text{rise}}{\text{run}} = \frac{-3}{2}$, the equation of the line is $y = mx + b$.
$y = -\frac{3}{2}x + 3$

4. Changing the value of b slides the line either up or down the y-axis but does not change the steepness of the line.

5. Changing the value of m changes the steepness of the line but does not change where it crosses the y-axis.

Lesson 5.5, pages 252–257

Exercises
Communicate

1. Standard form is $Ax + By = C$. For $5y - 2 = -3x$, move the x- and y-terms to one side of the equation and the constant to the other. The result is $3x + 5y = 2$.

2. To find the y-intercept, substitute for x with 0 and solve for y.
$3(0) + 6y = 18$
$y = 3$
To find the x-intercept, substitute y with 0 and solve for x.
$3x + 6(0) = 18$
$x = 6$

3. To rewrite $x - 3y = 9$ in slope-intercept form, solve the equation for y.
$x - 3y = 9$
$3y = x - 9$
$y = \frac{1}{3}x - 3$

4. Find the slope. $m = \frac{y_2 - y_1}{x_2 - x_1} = \frac{-8 - 4}{4 - (-2)} = \frac{-12}{6} = -2$
Choose a point: $(x_1, y_1) = (-2, 4)$.
The point-slope formula is $y - y_1 = m(x - x_1)$.
Substitute: $y - 4 = -2[x - (-2)]$.
Simplify: $y - 4 = -2(x + 2)$.

Lesson 5.6, pages 258–263

Activity

1. $\frac{3}{4}$

2. rise = 4, run = -3; slope = $-\frac{4}{3}$

3. negative

CHECKPOINT ✔

4. They are negative reciprocals of each other.

Exercises
Communicate

1. A line parallel to $y = 4x + 3$ will have the same slope, 4. The y-intercept can be anything other than 3.

2. Find the negative reciprocal of $\frac{2}{3}$.
slope $= -\frac{3}{2}$

3. Find the negative reciprocal of $\frac{1}{3}$.
slope $= -3$

4. Find the slope of the line:
$m =$ negative reciprocal of $4 = -\frac{1}{4}$.
Substitute $-\frac{1}{4}$ for 4 in the equation $y = 4x + 3$.
$y = -\frac{1}{4}x + 3$

Portfolio Activity

Answers may vary. Check students' responses.

Lesson 6.1, pages 276–281
Exercises
Communicate

1. They are the same except that the equals sign is replaced by the "is greater than" or "is less than" sign.

2. Add 4 to each side (Addition Property of Inequality). Simplify $x - 4 + 4 \leq 7 + 4$ to $x \leq 11$.

3. Solve for x by subtracting 3 from both sides to get $x < 4$. On the number line, place an open circle at the position of 4, then draw a ray from the circle heading to the left.

4. The closed circle indicates that point 1.5 is included. The ray goes off to the right indicating that x is greater than or equal to -1.5. $x \geq -1.5$

Lesson 6.2, pages 282–288
Activity

1.

A	B	2A	2B	−2A	−2B	$\frac{A}{2}$	$\frac{B}{2}$	$\frac{A}{-2}$	$\frac{B}{-2}$
2	4	4	8	−4	−8	1	2	−1	−2
−2	2	−4	4	4	−4	−1	1	1	−1
−4	6	−8	12	8	−12	−2	3	2	−3
−6	−2	−12	−4	12	4	−3	−1	3	1
−8	−6	−16	−12	16	12	−4	−3	4	3

2. a. $2B > 2A$ b. $-2B < -2A$

3. a. $\frac{A}{2} < \frac{B}{2}$ b. $\frac{A}{-2} > \frac{B}{-2}$

CHECKPOINT ✔
4. a. Treat the inequality sign the same as an equal sign; it does not change.
   b. Reverse the inequality sign.

Exercises
Communicate

1. Subtract 1 from each side, then simplify.

2. Add 3 to both sides, then simplify.

3. Divide both sides by −3 and reverse the direction of the inequality sign.

4. Add $4x$ to both sides and subtract 3 from both sides. Simplify. Then divide both sides by 6.

5. When both sides of the inequality are divided or multiplied by a positive number, the properties are the same as the properties of equality. When both sides of an inequality are multiplied or divided by a negative number, the direction of the inequality sign must be reversed.

6. Because both sides of the inequality are divided by −3, the direction of the inequality sign must be reversed: $x > -2$.

7. Answers may vary. Sample answer: Algebra: Let $a = 3$ and $b = 5$. $3 < 5$ but -3 is *not* less than -5. Graph: Let a and b be quantities on a number line as shown.

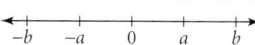

Because $b > a$, both b and $-b$ will be farther from 0 than a and $-a$. So on the positive side of 0, $b > a$. On the negative side of 0, $-b < -a$.

Portfolio Activity

1.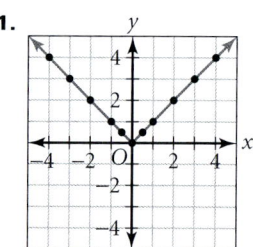

2. V-shape; $y = -x$; $y = x$

Lesson 6.3, pages 289–293
Exercises
Communicate

1. Place −6 on the left of a "is less than" symbol. Place y to its right. To the right of y, place another "is less than" symbol and then the 4.

2. Solve $-2 < 2x$ for x. Then solve $2x < 4$ for x. The final solution will be all the values that satisfy both inequalities.

3. Answers may vary.

4. $-4 < x \leq 5$ means that x includes the numbers between (but not including) −4 and (including) 5. $-4 < x$ OR $x \leq 5$ means all numbers greater than −4 or all numbers less than 5; all real numbers satisfy one or the other or both of the inequalities.

Practice and Apply

28. [number line with open circles at −4 and 4, segment between]

29. [number line with open circle at 6, ray to right]

30. [number line with open circle at −4, closed circle at 8, segment between]

31. $-9 \leq x < 3$;

32. $13 \leq z < 15$;

33. $x < 7$ or $x \geq 6$; all real numbers

34. $-4 < x \leq 9$;

35. $1 < x < 7$;

36. $x < -8$ or $x \geq -6$;

37. $-\frac{5}{2} < x < \frac{3}{2}$;

38. $x > -7$ or $x < -4$; all real numbers

39. $78 \leq t \leq 98$;

Lesson 6.4, pages 294–299

Activity

1.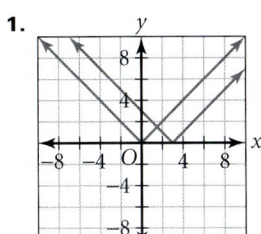

CHECKPOINT ✔
2. The 3 shifts the graph 3 units to the right.

3.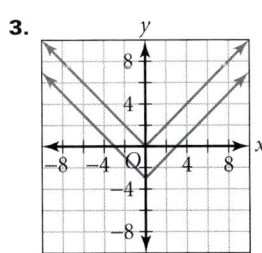

CHECKPOINT ✔
4. The 3 shifts the graph 3 units down.

5.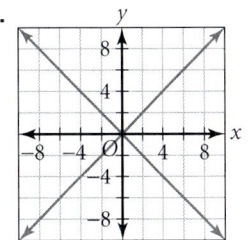

CHECKPOINT ✔
6. The negative sign creates a reflection of the graph across the x-axis.

Exercises
Communicate

1. Determine how far from 1 minute each guess was. Andy was 9 seconds away and Abi was 8 seconds away. Although one guess was too little and one guess was too long, by comparing the absolute values of the guesses, we can say that Abi was closer.

2. Answers may vary. Any negative value for a will make the equation true. Ex: If $a = -4$, $|-4| = -(-4) = 4$.

3. The graph of an absolute-value function is **V**-shaped. The vertex of the **V** may be pointing up or down.

4. No. An absolute value is always positive, by definition.

5. Yes. When $x = 0$, $|0| = 0$.

Guided Skills Practice

11. a.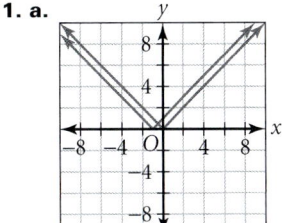

translation by 1 unit to the left

b.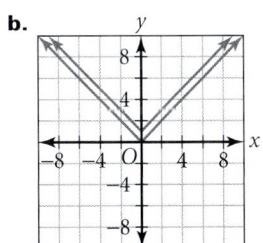

translation by 1 unit up

c.

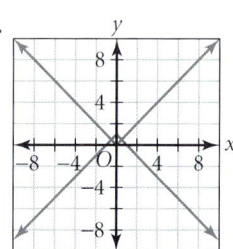

reflection across the *x*-axis, then translation 1 unit up

Practice and Apply

48. horizontal translation 4 units left

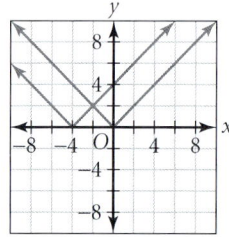

49. horizontal translation 5 units right

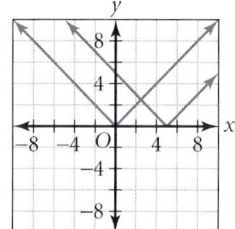

50. vertical translation 2 units up

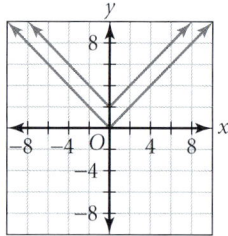

51. vertical translation 4 units down

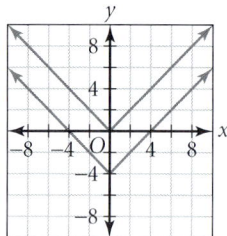

52. horizontal translation 4 units left, then reflection across the *x*-axis

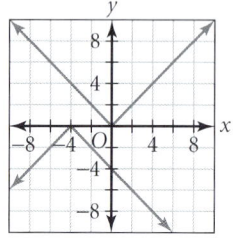

53. horizontal translation 5 units right, then reflection across the *x*-axis

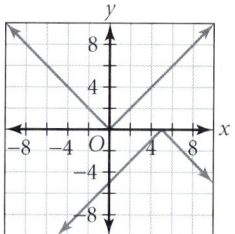

54. reflection across the *x*-axis, then vertical translation 2 units up

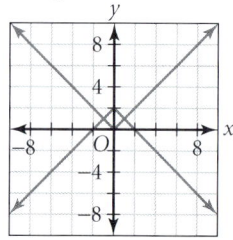

55. reflection across the *x*-axis, then vertical translation 4 units down

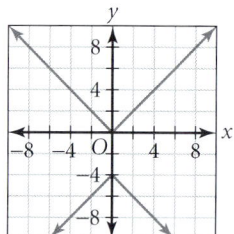

Lesson 6.5, pages 300–307

Activity

2. The solutions are the *x*-values of the intersection points of $y = 6$ and $y = |4x - 2|$, −1 and 2.

3. For $|4x - 2| < 6$, the solutions are the *x*-values along the line $y = 6$ inside the **V**-shaped graph of $y = |4x - 2|$. For $|4x - 2| > 6$, the solutions are the *x*-values along the line $y = 6$ outside (to the left and to the right) of the **V**-shaped graph of $y = |4x - 2|$.

CHECKPOINT ✔

4. a. $x = 2$ or $x = -3$ **b.** $-3 < x < 2$
c. $x > 2$ or $x < -3$

Exercises

Communicate

1. The solutions of $4 = |x + 5|$ are $x = -1$ or $x = -9$, two distinct numbers. The solution of $4 \geq |x + 5|$ is any value in the range of numbers between −1 and −9, inclusive.

2. a. The product is allowed to be any dimension between 44.999 cm and 45.001 cm.

b. 45 cm is the specified size. Let x be the actual size. The difference between 45 and x is written as $|x - 45|$. This difference is allowed to be no more than 0.001 cm. This is represented algebraically as $|x - 45| \leq 0.001$.

3. The expression inside the absolute-value sign may be either positive or negative, and both cases must be considered.

4. Let x be all of the values you are looking for. The distance between x and -7 may be written as $|x - (-7)|$, or $|x + 7|$. This distance must be equal to or less than 3, which is written as $|x + 7| \leq 3$.

Lesson 7.1, pages 318–325
Activity

1.
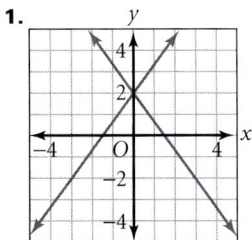

2. An estimated solution is (0.5, 1.2).

3.
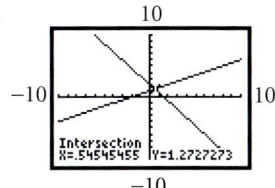

The solution is $(0.\overline{54}, 1.\overline{27})$.

4. Change $0.\overline{54}$ to $\frac{6}{11}$ and $1.\overline{27}$ to $\frac{14}{11}$ and substitute into the equations.

CHECKPOINT ✓
5. Answers may vary. Sample answer: Solving by paper and pencil may be better for seeing the actual intersection, but finding the exact value can be difficult. Solving with a graphics calculator is faster, but the intersection points may be rounded and the actual intersection may be difficult to see.

Exercises
Communicate

1. Isolate y on one side of the equal sign. This form of the equation, $y = mx + b$, allows you to substitute values for x in order to find values for y.

2. The common solution is the point where the lines intersect, (x, y).

3. Solve each equation for y.
$$\begin{cases} y = -x + 3 \\ y = x - 4 \end{cases}$$
Use the y-intercept and the slope to graph each equation. The common solution is the point where the lines intersect, $(3.5, -0.5)$. It is important to check solutions because it is often difficult to locate the exact position of the coordinates from the graph.

Guided Skills Practice

9.
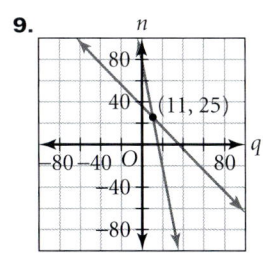

$n = 25, q = 11$

Practice and Apply

10.
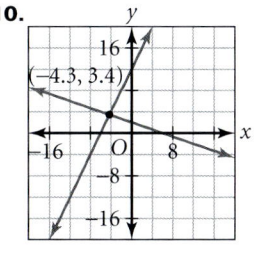

$(-4.3, 3.4)$

11.
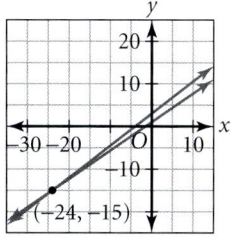

$(-24, -15)$

12.
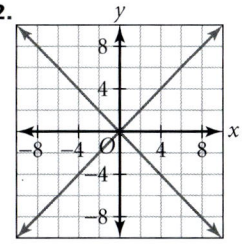

$(0, 0)$

13.
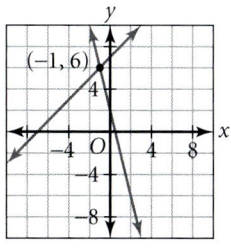

$(-1, 6)$

14.
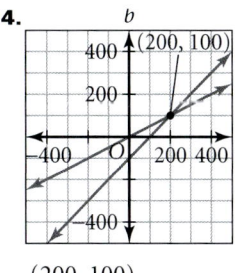

$(200, 100)$

15.
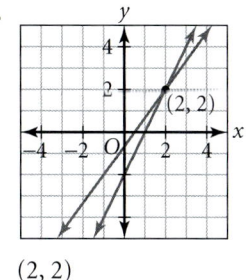

$(2, 2)$

831

16.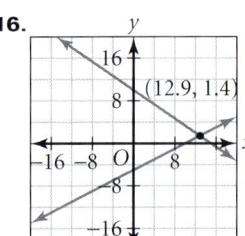
(12.9, 1.4)

17.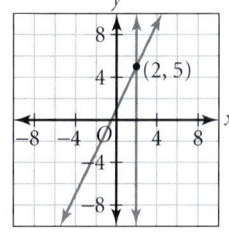
(2, 5)

18.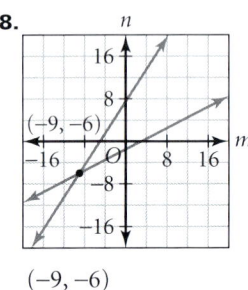
(−9, −6)

19.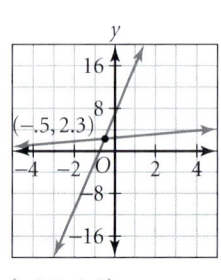
(−0.5, 2.3)

20.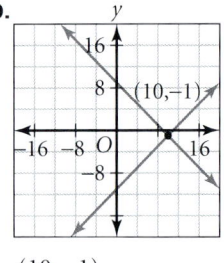
(10, −1)

21.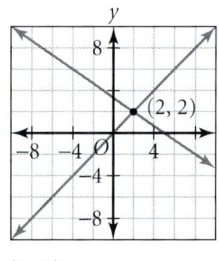
(2, 2)

27.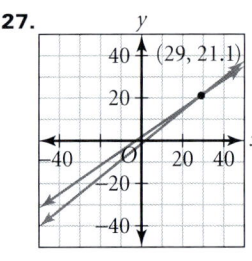
(29, 21.1)

32. a. 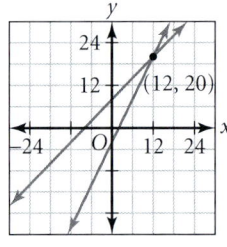   b.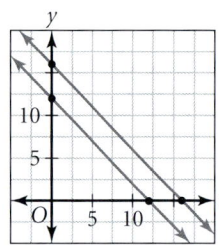

c. 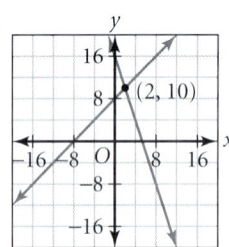   d.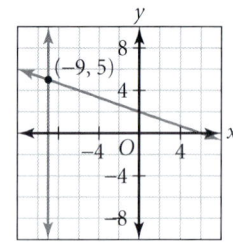

37.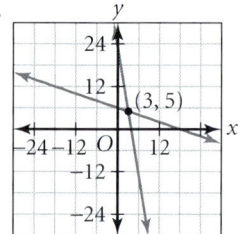

39. 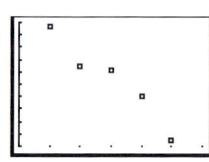   40.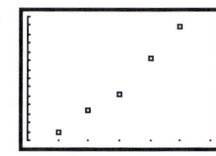

41. a. 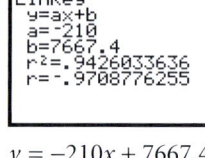   b.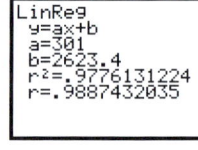

$y = -210x + 7667.4$ $y = 301x + 2623.4$

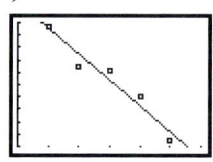   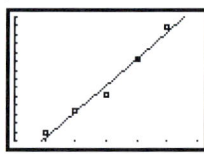

42.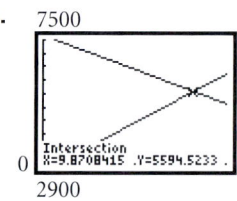

about 5 years beyond the given data

43. $A = -1800r + 28{,}000$
$A = 2300r + 1500$

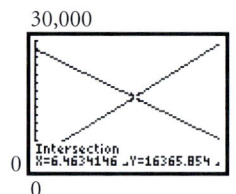

about 6.5 minutes

Portfolio Activity

1985; $350,000

basketball: $y = 50x + 100$;
$1,100,000 in the year 2000

baseball: $y = 35x + 175$;
$875,000 in the year 2000

Lesson 7.2, pages 326–330

Activity

2. (89, 68)

3. Since we are looking for a value of y that satisfies both equations, we can substitute $x - 21$ in place of y in the first equation.

4. $x = 89$ 5. $y = 68$

CHECKPOINT ✔
6. They are the same.

Exercises
Communicate

1. Substitute 42 for y and solve for x; $x = 17$.

2. Solve the first equation for y because the coefficient of y is 1; solve for x and y. The solution is $(2, 10)$.

3. Solve the first equation for x because the coefficient of x is 1; solve for x and y. The solution is $(10, 1)$.

Lesson 7.3, pages 331–337
Activity

1. The y-terms are opposites.

2. $8x = 16$; one variable

3. $x = 2$, $y = \frac{1}{2}$

4. $3(2) + 2\left(\frac{1}{2}\right) = 7$
   $7 = 7$
   $5(2) - 2\left(\frac{1}{2}\right) = 9$
   $9 = 9$

CHECKPOINT ✔
5. Answers may vary. Sample answer: Opposites can be used to eliminate one or more variables by combining equations. When only one variable remains, its value can be found by simplification and then substituted to solve for the other variable(s).

Exercises
Communicate

1. $7y$ and $-7y$ are opposites. Use the Addition Property of Equality to solve for x and then for y. $\left(9, \frac{4}{7}\right)$

2. $-3y$ and $3y$ are opposites. Use the Addition Property of Equality to solve for x and then for y. $(4, 0)$

3. $2a$ and $-2a$ are opposites. Use the Addition Property of Equality to solve for b and then for a. $\left(\frac{13}{2}, -7\right)$

4. Multiply the first equation by -2 to make $3y$ and $6y$ opposites. Then use the Addition Property of Equality to solve for x and then for y. $\left(11, \frac{-13}{3}\right)$

5. Multiply the first equation by 3 and the second equation by -2 to make $2x$ and $3x$ opposites. Then use the Addition Property of Equality to solve for y and then for x. $(-2, -1)$

6. Multiply the first equation by -3 to make $2b$ and $6b$ opposites. Then use the Addition Property of Equality to solve for a and then for b. $\left(\frac{1}{3}, -\frac{1}{2}\right)$

7. Elimination method since $2y$ and $-2y$ are opposites.

8. Elimination method since y can be easily eliminated by multiplying the second equation by -3.

9. Graph on a graphics calculator since the coefficients are difficult to use with other methods.

Look Back

63.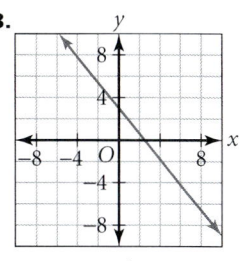

slope $= -\frac{6}{5}$

Look Beyond

65.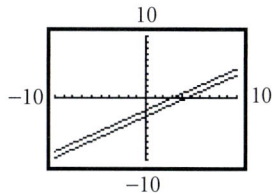

The lines are parallel.

Portfolio Activity

$\begin{cases} 1800f + 1200b = 161{,}000 \\ f - b = 20 \end{cases}$

floor: $61.67
balcony: $41.67

Answers will vary.

Lesson 7.4, pages 338–344
Activity

1.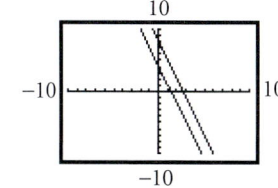

   a. No, the lines are parallel.

CHECKPOINT ✔
   b. No, the lines are parallel.

2.

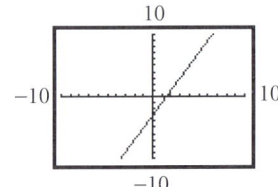

a. There is only one line shown.
b. The lines intersect everywhere because they are the same line.

CHECKPOINT ✔
c. Yes, there are infinitely many solutions because the two lines are the same.

Exercises
Communicate

1. Since both the slopes and the *y*-intercepts of both lines are the same, the system is dependent.
2. Since the slopes of these two lines are not the same, the system is independent.
3. Since the slopes of both equations are the same but their *y*-intercepts are not, the system is inconsistent.
4. Multiply both sides of the equation by the same number. The slope and the *y*-intercept remain the same.
5. Write a new equation using the slope of the given equation (−3) and choose a different *y*-intercept.

Lesson 7.5, pages 345–352
Activity

1. Check students' graphs.

CHECKPOINT ✔
2. *S* is over the budget and *R* is under the budget. Any point above the line represents spending over the budget, while any point under the line represents spending under the budget.
3. Check students' graphs.

Exercises
Communicate

1. If the inequality symbol includes the equal sign, then use a solid line. Otherwise, use a dashed line.
   a. Use a solid line.
   b. Use a dashed line.
2. Graph the boundary line, $y = -\frac{1}{2}x + 1$. Substitute a point that is not on the line. If it satisfies the inequality, shade the side of the line containing the point. If the point does not satisfy the inequality, shade the other side of the line. Answers may vary. Try (0, 0) because 0 is a simple number to test.

3. Answers may vary. Sample answer: Graph the boundaries $y = 1$ and $x = 1$ with dashed lines. $y < 1$ is the half-plane below $y = 1$, and $x < 1$ is the half-plane to the left of $x = 1$. The solution is where these regions overlap. Check by choosing a point in this region and testing it in both inequalities.

4. Answers may vary. Sample answer: Solve the first inequality for *y*: $y > -2x + 3$. Graph the boundaries $y = -2x + 3$ and $y = 3$ with dashed lines. $y > -2x + 3$ is the half-plane above $y = -2x + 3$, and $y < 3$ is the half-plane below $y = 3$. The solution is where these regions overlap. Check by choosing a point in this region and testing it in both inequalities.

5. Answers may vary. Sample answer: Solve the second inequality for *y*: $y \leq \frac{1}{2}x - 1$. Graph the boundaries $x = 3$ and $y = \frac{1}{2}x - 1$ with solid lines. $x \leq 3$ is the half-plane to the left of $x = 3$, and $y \leq \frac{1}{2}x - 1$ is the half-plane below $y = \frac{1}{2}x - 1$. The solution is where these regions overlap. Check by choosing a point in this region and testing it in both inequalities.

6. Answers may vary. Sample answer: Solve each inequality for *y*: $\begin{cases} y > x + 3 \\ y > -2x + 4 \end{cases}$.

   Graph the boundaries $y = x + 3$ and $y = -2x + 4$ with dashed lines. $y > x + 3$ is the half-plane above $y = x + 3$, and $y > -2x + 4$ is the half-plane above $y = -2x + 4$. The solution is where these regions overlap. Check by choosing a point in this region and testing it in both inequalities.

Guided Skills Practice

10.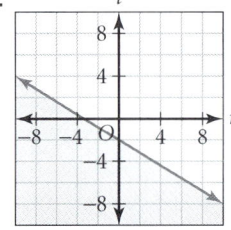

11.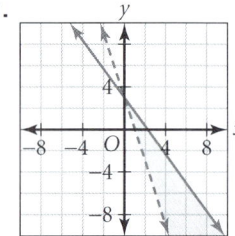

12.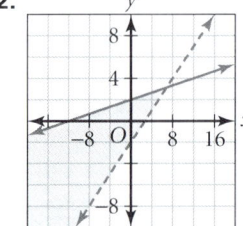

13.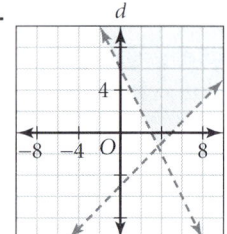

14.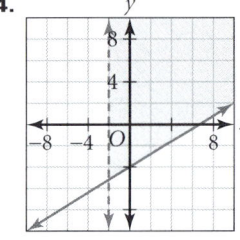

15. $\begin{cases} 4p + 1.5a \leq 100 \\ p + a \geq 30 \end{cases}$

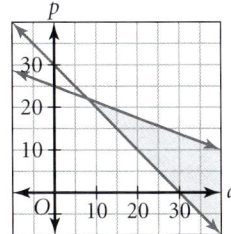

The shaded region contains all possible combinations.

Practice and Apply

16.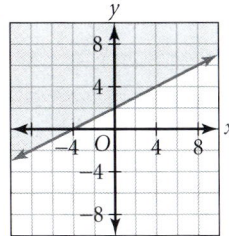
17.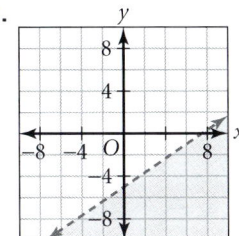
18.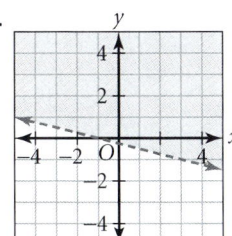
19.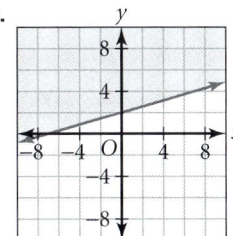
20.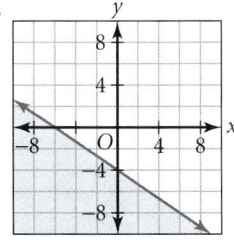
21.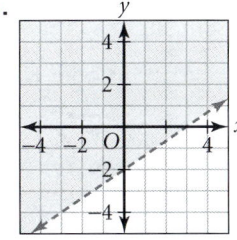
22.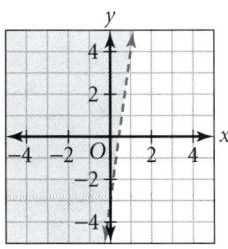
23.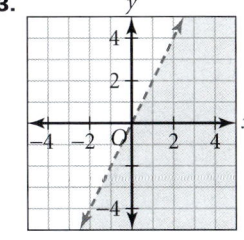
24.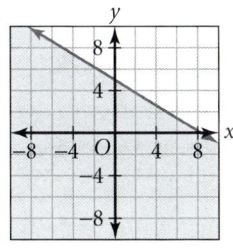
25.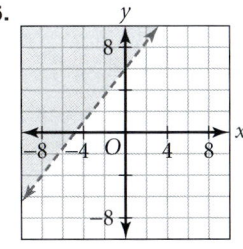
26.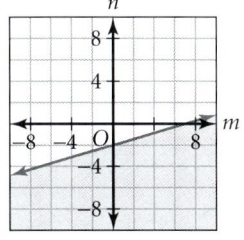
27.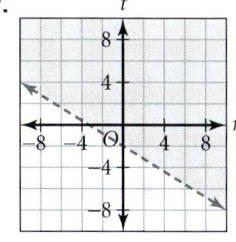
28.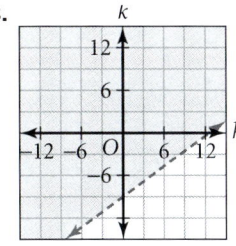
29.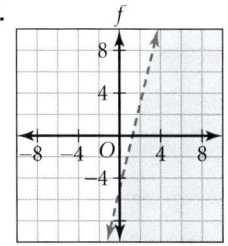
30.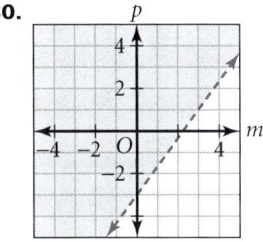
31.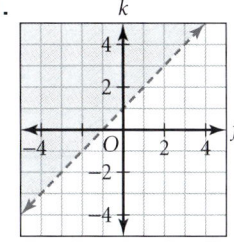
32.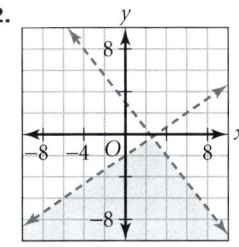
33.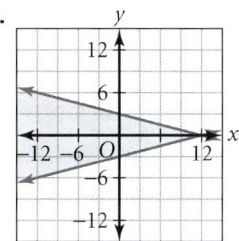
34.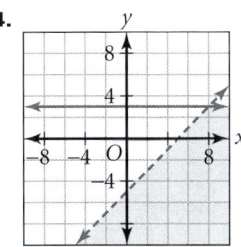
35.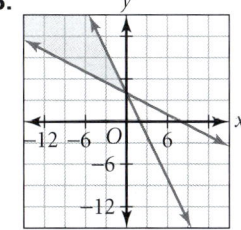
36.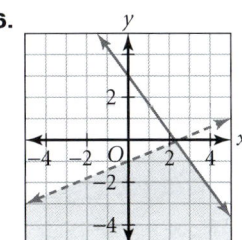
37.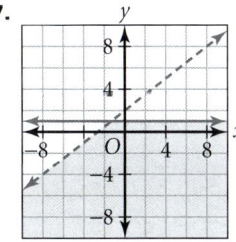
41.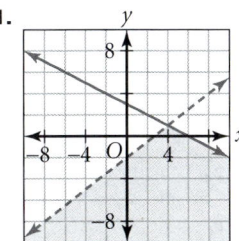
42.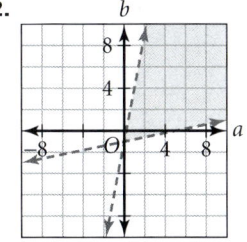

835

43.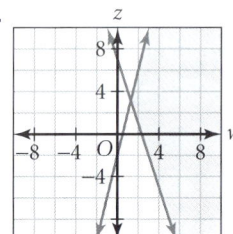
44.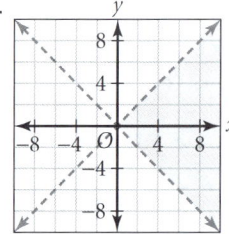
45.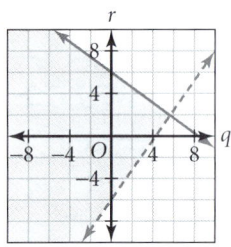
46.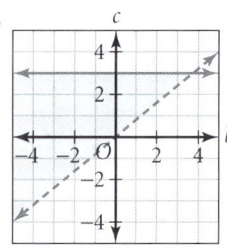
47.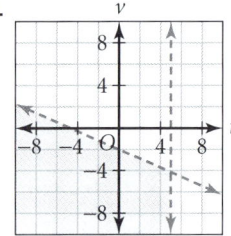
48.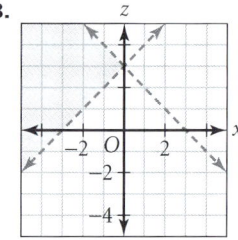
49.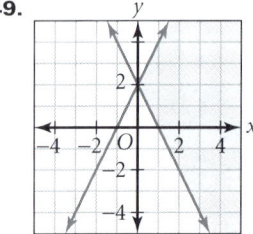
50.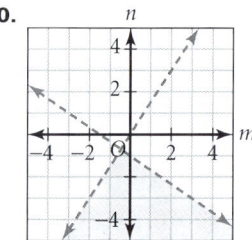
51.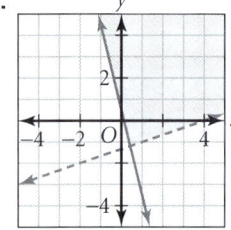
52.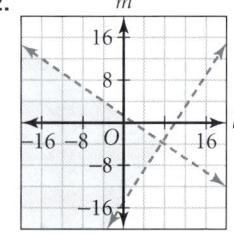
53.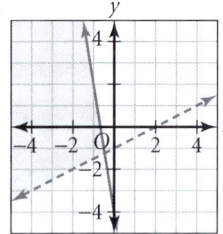
54.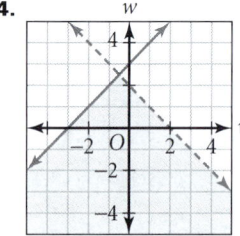
55.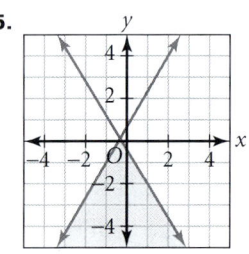

56. $\begin{cases} 8x + 4y \leq 30 \\ 2x + 2y \geq 12 \end{cases}$

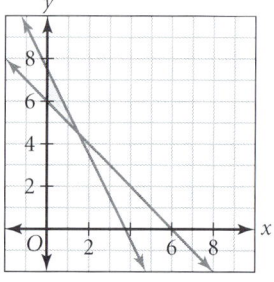

57. $2x + y \leq 16$

Field goals, x	Free throws, y
0	16
1	14
2	12
3	10
4	8
5	6
6	4
7	2
8	0

Portfolio Activity

Sample answers:
$b = 80$, $c = 40$
$b = 60$, $c = 35$
$b = 75$, $c = 40$

$\begin{cases} b \leq 2c \\ c \leq 40 \end{cases}$

Check students' graphs.

The intersection region contains all ordered pairs that satisfy the system of inequalities.

Lesson 7.6, pages 353–359

Exercises
Communicate

1. Answers may vary. Sample answer: Draw a diagram. Define the variables. Let x represent the number of ounces of the 25% alcohol solution. Let y represent the number of ounces of the 5% alcohol solution. Construct a table. There are 20 liters of the mixed solution, so $x + y = 20$. There are 2.6 liters of pure alcohol in the mixed solution, so $0.25x + 0.05y = 2.6$.

2. Answers may vary. Sample answer: Draw a bird and a large arrow to represent the wind. Let x represent the bird's rate and let y represent the rate of the wind.

3. Answers may vary. Sample answer: Draw a bird and an arrow facing the same direction to represent the bird flying with the wind. Draw an arrow pointing toward the bird to represent the bird flying against the wind. Let $(x + y)$ represent the bird flying with the wind, and let $(x - y)$ represent the bird flying against the wind.

836

4. $x + y = 3(x - y)$
$x + y = 3x - 3y$
$4y = 2x$
$y = \frac{1}{2}x$
so $x + y = x + \frac{1}{2}x = \frac{3}{2}x$

The bird's rate is $\frac{3}{2}$ times faster flying with the wind than flying in still air.

Chapter 7 Project, pages 360–361

Activity 3

1. a. $0.4n$
b. $0.1r$
c. $0.4n + 0.1r = c$

2.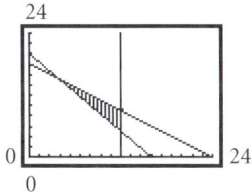

3. The each vertex represents the amounts of nuts and raisins to use the satisfy 2 of the 3 of the constraint equations.

4. $0.4(3) + 0.1(15) = 2.7$ minimum
$0.4(9) + 0.1(9) = 4.5$
$0.4(9) + 0.1(5) = 4.1$

5. 3 ounces of nuts and 15 ounces of raisins should be used per package, for a minimum cost of $2.70.

Chapter 7 Review and Assessment

7.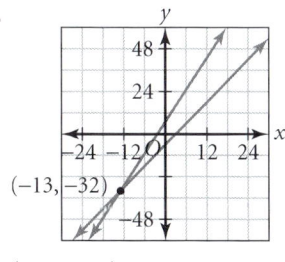
$(-13, -32)$

8.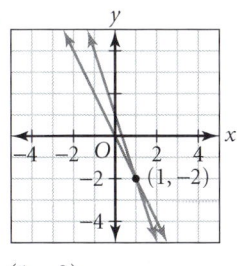
$(1, -2)$

Lesson 8.1, pages 370–376

Activity

1. a. $1000 \cdot 100 = 100,000$
$10^3 \cdot 10^2 = 10^5$
b. $1000 \cdot 1000 = 1,000,000$
$10^3 \cdot 10^3 = 10^6$

2. 10^6

3. a. $(2)(2)(2) \cdot (2)(2) = 2^5$
b. $(5) \cdot (5)(5)(5) = 5^4$
c. $(10)(10) \cdot (10)(10)(10)(10) = 10^6$
d. $(3)(3) \cdot (3)(3) = 3^4$

CHECKPOINT ✔
4. $m + n$

Exercises
Communicate

1. $10^{12} = 1,000,000,000,000$. Notice that the number of zeros in the decimal notation is exactly the same value as the exponent in 10^{12}. The number of zeros in the decimal notation will always equal the exponent in a power of 10.

2. The exponent in 3^6 tells how many times the base is used in factored form. So $3^6 = 3 \cdot 3 \cdot 3 \cdot 3 \cdot 3 \cdot 3$.

3. 7^5 means there are five sevens multiplied together. 7^3 means there are three sevens multiplied together. The product of 7^5 and 7^3 means there are eight sevens multiplied together, which is precisely the sum of the exponents 5 and 3.

4. By definition
$2^3 = 2 \cdot 2 \cdot 2$
$2^4 = 2 \cdot 2 \cdot 2 \cdot 2$
So $2^3 \cdot 2^4 = (2 \cdot 2 \cdot 2) \cdot (2 \cdot 2 \cdot 2 \cdot 2) = 2^7$
By the Product-of-Powers Property
$2^3 \cdot 2^4 = 2^{3+4} = 2^7$

5. $x \cdot x = x^2$
x is equivalent to x^1.
By substitution $x^1 \cdot x^1 = x^2$.
This agrees with the Product-of-Powers Property, which implies $x^1 \cdot x^1 = x^{1+1} = x^2$.

Portfolio Activity

1.

	Prize A	Prize B
Day 1	$100	$0.01
Day 2	$200	$0.02
Day 3	$300	$0.04
Day 4	$400	$0.08

2. After only four days, Prize A appears to be the better pick.

3. Answers may vary. Some students will realize that the doubling of Prize B will overtake and surpass Prize A at some point. Other students will be convinced that Prize A will always be the better choice.

4. (Table should be extended to show all days. The answer provided shows only days 17 to 20.)

	Prize A	Prize B
Day 17	$1700	$655.36
Day 18	$1800	$1310.72
Day 19	$1900	$2621.44
Day 20	$2000	$5242.88

5. yes, on day 19

6.
	Prize A	Prize B
Day 27	$2700	$671,088.64
Day 28	$2800	$1,342,177.28
Day 29	$2900	$2,684,354.56
Day 30	$3000	$5,368,709.12

Lesson 8.2, pages 377–382

Activity

1. When a power is raised to a power, the exponents are multiplied.

2. $(x^5)^2 = x^5 \cdot x^5$
$= (x \cdot x \cdot x \cdot x \cdot x) \cdot (x \cdot x \cdot x \cdot x \cdot x)$
$= x \cdot x \cdot x \cdot x \cdot x \cdot x \cdot x \cdot x \cdot x \cdot x$
$= x^{10}$

This agrees with the conjecture in Step 1.

3. Answers may vary. Sample answer:
$(y^3)^3 = y^3 \cdot y^3 \cdot y^3$
$= (y \cdot y \cdot y) \cdot (y \cdot y \cdot y) \cdot (y \cdot y \cdot y)$
$= y \cdot y \cdot y \cdot y \cdot y \cdot y \cdot y \cdot y \cdot y$
$= y^9$

CHECKPOINT ✓

4. Answers may vary. Sample answer: When raising a power to a power, multiply the exponents. The rule is true because repeated addition of exponents can be translated to multiplication.

Exercises
Communicate

1. $a^2 a^3$ is equivalent to a multiplied by itself five times. $(a^2)^3$ is equivalent to a multiplied by itself six times.

2. $(x^a)^b = x^{a \cdot b}$
$(x^b)^a = x^{b \cdot a}$
Since $a \cdot b = b \cdot a$ by the Commutative Property, then $(x^a)^b = (x^b)^a$.

3. Answers may vary. Sample answer: The expression $(a^3)^2$ means a^3 raised to the second power. By the definition of an exponent, a^3 is used as a factor two times.
$a^3 \cdot a^3$
$= (a \cdot a \cdot a) \cdot (a \cdot a \cdot a)$
$= a \cdot a \cdot a \cdot a \cdot a \cdot a$
$= a^6$
By multiplying the two powers in $(a^3)^2$ the answer a^6 is obtained.

4. $(-1)^2 = (-1)(-1) = 1$
$(-1)^3 = (-1)(-1)(-1) = -1$
Since $1 > -1$, then $(-1)^2 > (-1)^3$.

5. In the expression $(a^s)^t$, a^s is used as a factor t times. This means a^s is multiplied by itself t times. Thus, we have $(a^s)^t = a^{st}$.

In the expression $(xy)^n$, xy is used as a factor n times. This means that x is multiplied by itself n times and y is multiplied by itself n times. Thus, we have $(xy)^n = x^n y^n$.

Lesson 8.3, pages 383–389

Activity

1. 10^4

2.
	Form	Numerator	Denominator	Quotient
a.	Decimal	100,000	100	1000
	Exponential	10^5	10^2	10^3
b.	Decimal	64	4	16
	Exponential	2^6	2^2	2^4
c.	Decimal	81	27	3
	Exponential	3^4	3^3	3^1

3.
Form	Numerator	Denominator	Quotient
Decimal	100,000	100	10,000
Exponential	10^5	10^1	10^4

4. a^{m-n}

CHECKPOINT ✓

5. Subtract the exponents. $\frac{2^5}{2^3} = 2^{5-3} = 2^2 = 4$

Exercises
Communicate

1. Since the bases are the same, we may use the Quotient-of-Powers Property. Take the difference of the exponents. Use this result as the new exponent of the original base.
$\frac{2^5}{2^4} = 2^{5-4} = 2^1 = 2$

2. In the first expression $\left(\frac{2^5}{2^4}\right)$, the bases are the same. In the second expression $\left(\frac{5^2}{4^2}\right)$, the bases are different. The

Quotient-of-Powers Property may only be used if the bases are the same.

3. To simplify $2^3 \cdot 2^2$, we may use the Product-of-Powers Property since the bases of the factors are the same. The property tells us to add the exponents. $2^3 \cdot 2^2 = 2^{3+2} = 2^5$ The Quotient-of-Powers Property says that in a quotient where there is the same base in the numerator and the denominator (and the base is not zero) then we may subtract the exponent of the denominator from the exponent of the numerator.
$\frac{2^5}{2^3} = 2^{5-3} = 2^2$

4. By the definition of exponents and the Product-of-Powers Property:
$\left(\frac{f}{g}\right)^6 = \frac{f}{g} \cdot \frac{f}{g} \cdot \frac{f}{g} \cdot \frac{f}{g} \cdot \frac{f}{g} \cdot \frac{f}{g} = \frac{f^6}{g^6}$

Lesson 8.4, pages 390–395

Activity

1. a. 1
   b. 1
   c. 1

2. a. $\frac{10^5}{10^5} = 10^{5-5} = 10^0$
   b. $\frac{2^3}{2^3} = 10^{3-3} = 2^0$
   c. $\frac{5^4}{5^4} = 5^{4-4} = 5^0$

CHECKPOINT ✔
3. 1

4. Answers may vary. Sample answer: $\frac{x^n}{x^n}$ is equivalent to 1 as well as $x^{n-n} = x^0$, so it is the only definition possible.

Exercises
Communicate

1. Answers may vary. Sample answer: When the Quotient-of-Powers Property is applied to an expression such as $\frac{5^1}{5^3}$, the result is $\frac{5^1}{5^3} = 5^{1-3} = 5^{-2}$. If the factors are written out, the result is $\frac{5^1}{5^3} = \frac{5}{5 \cdot 5 \cdot 5} = \frac{1}{5^2}$. From these two results, it makes sense that $5^{-2} = \frac{1}{5^2}$.

2. The properties of exponents cannot be applied because the bases are not equivalent.

3. Answers may vary. Sample answer: Using the Product-of-Powers Property, $(3a)^{-2} = 3^{-2}a^{-2} = \frac{1}{3^2} \cdot \frac{1}{a^2} = \frac{1}{9a^2}$. The exponent applies to both 3 and a, since $3a$ is in parentheses.

4. Any value raised to a negative power, such as x^{-n}, is equivalent to the reciprocal of that value raised to the same positive power, $\frac{1}{x^n}$.

5. Any value raised to the zero power is equivalent to 1.

Portfolio Activity

1. $60{,}000 \times 1.07^{20} = 232{,}181$

2. $60{,}000 \times 1.07^x = 1{,}000{,}000$
   $x = 41.58$

   42 years from 1995 there will be 1 million centenarians, or in 2037.

Lesson 8.5, pages 398–403

Activity

1.
Value	1000	100	10	1	$\frac{1}{10}$	$\frac{1}{100}$	$\frac{1}{1000}$
Power	10^3	10^2	10^1	10^0	10^{-1}	10^{-2}	10^{-3}

2.
Decimal	Expanded form	Scientific notation
1200	1.2×1000	1.2×10^3
120	1.2×100	1.2×10^2
12	1.2×10	1.2×10^1
1.2	1.2×1	1.2×10^0
0.12	$1.2 \times \frac{1}{10}$	1.2×10^{-1}
0.012	$1.2 \times \frac{1}{100}$	1.2×10^{-2}
0.0012	$1.2 \times \frac{1}{1000}$	1.2×10^{-3}

CHECKPOINT ✔
3. a. $0.000012 = 1.2 \times \frac{1}{100{,}000} = 1.2 \times 10^{-5}$
   b. $120{,}000 = 1.2 \times 100{,}000 = 1.2 \times 10^5$

Exercises
Communicate

1. The numbers are easier to read and use in calculations.

2. The value is multiplied by 0.00000001; −8.

3. The value is multiplied by 1,000,000,000,000; 12.

Lesson 8.6, pages 404–408

Activity

1.
Year	Process	Exponential notation	Population (in billions)
2001	$5.85(1.0152)^3(1.0152)$	$5.85(1.0152)^4$	6.21
2002	$5.85(1.0152)^4(1.0152)$	$5.85(1.0152)^5$	6.31
2003	$5.85(1.0152)^5(1.0152)$	$5.85(1.0152)^6$	6.40
2004	$5.85(1.0152)^6(1.0152)$	$5.85(1.0152)^7$	6.50
2005	$5.85(1.0152)^7(1.0152)$	$5.85(1.0152)^8$	6.60
2006	$5.85(1.0152)^8(1.0152)$	$5.85(1.0152)^9$	6.70
2007	$5.85(1.0152)^9(1.0152)$	$5.85(1.0152)^{10}$	6.80

2. after 82 years, or in the year 2079

CHECKPOINT ✔
3. (original amount)(1 + rate)$^{(\text{\# of years})}$

Exercises

Communicate

1. As the base gets larger the amount grows more quickly.

2. Both are increasing exponentially from an original amount at a given rate over a specified number of years.

3. Answers may vary. Sample answer: $y = x^4$ is not exponential since it does not change at a constant rate.

4. Answers may vary. Sample answer: $y = 4^x$ is exponential since it changes at a constant rate, is above the x-axis for all values of x, goes through the point $(0, 1)$, and increases rapidly after it crosses the y-axis.

Practice and Apply

7. 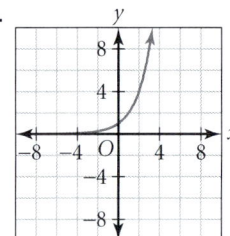 **8.**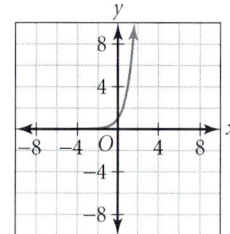

9. 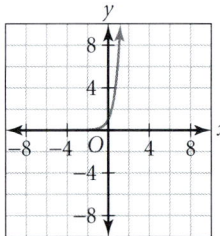 **10.**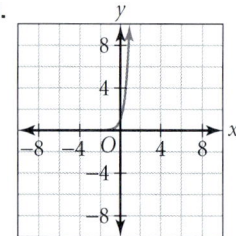

11. 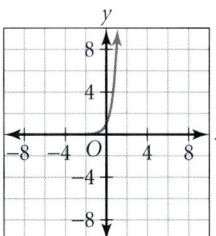 **12.**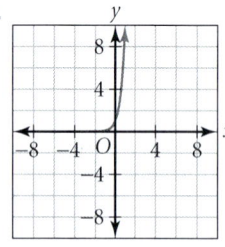

13. 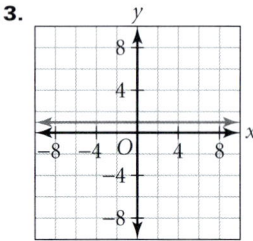 **14.**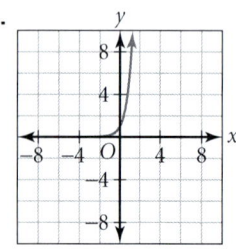

15. 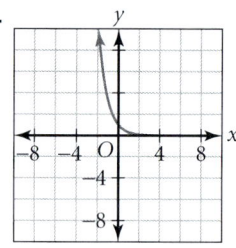 **16.**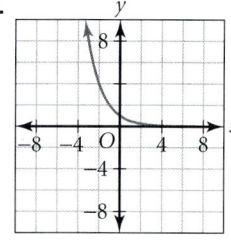

17. 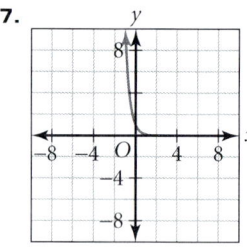 **18.**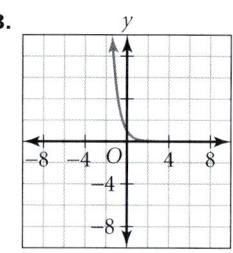

19. 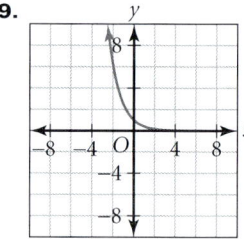 **20.**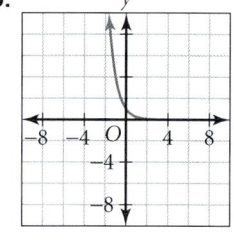

21. 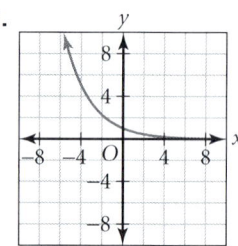 **22.**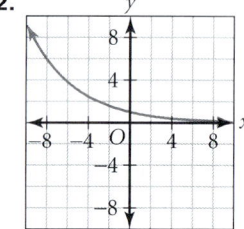

23. 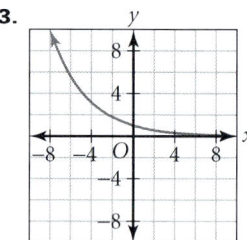 **24.**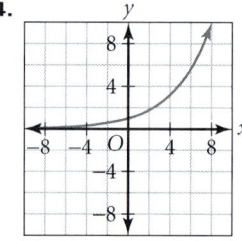

25. 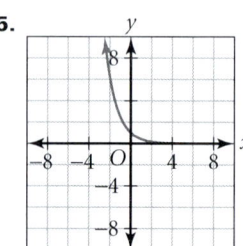 **26.**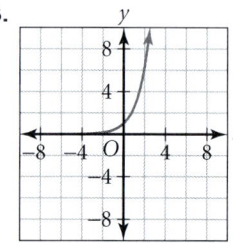

27. 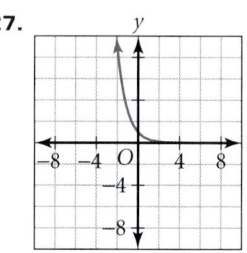 **28.**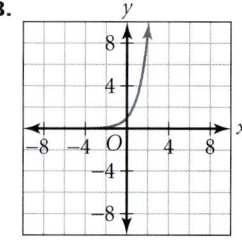

29. 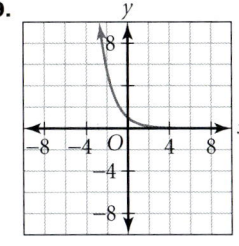 **30.**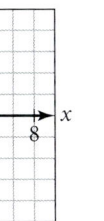

45. slope = $-\frac{1}{5}$
y-intercept = 7

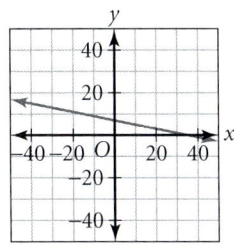

Look Beyond

51. 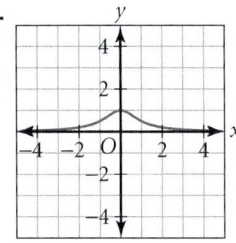 **52.**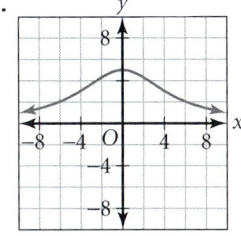

53. The two graphs have the same form. However, the first graph peaks at (0, 1), and the second graph peaks at (0, 5).

Lesson 8.7, pages 409–415

Exercises
Communicate

1. Answers may vary. Sample answer: interest paid on loans, growth of bacteria, appreciation of an investment

2. The new amount, P, equals the original amount, A, multiplied by the constant multiplier, $1 + r$, where r is the rate of growth per year, raised to the exponent for the number of years of growth, t.

3. The remaining amount, P, equals the original amount, A, multiplied by the constant multiplier, $1 - r$, where r is the rate of decay, raised to the exponent for the number of years of decay, t.

4. Decay is a decreasing measure.

5. Half-life refers to the amount of time required for half the amount of a substance to become non-radioactive.

Look Back

47. 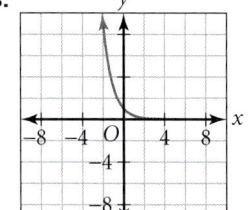 **48.**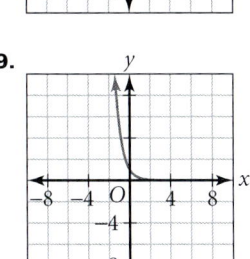

49.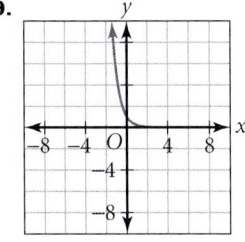

Look Beyond

50. $e \approx 2.7182818$

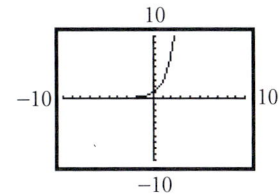

Portfolio Activtity

1. Successive quotients yield 1.075. Thus, the rate is 7.5%.

2. $y = 5000(1.075)^x$

3. year 10: $10,305.16

4. year 10, year 16

Lesson 9.1, pages 426–431

Activity

2. Answers may vary. Sample answer:
$[2(1) + 2] + [3(1) - 3] - 5(1)$ 1
$4 = 4$
$[2(-2) + 2] + [3(-2) - 3] = 5(-2) - 1$
$-11 = -11$

3. Change subtraction to addition of the opposite.
$(2x + 2) - (3x - 1) = (2x + 2) + (-3x + 1)$

2x + 2   +   −3x + 1   =   −x + 3

CHECKPOINT ✓

4. Group like terms and add. Change subtraction to addition of the opposite, group like terms, and add.

Exercises
Communicate

1. Use five positive x^2-tiles for $5x^2$, two negative x-tiles for $-2x$, and three positive unit tiles for 3.

2. Determine the exponent of greatest value from all the terms in the polynomial.

3. trinomial, cubic

4. Change the signs of the terms in the trinomial that is to be subtracted to their opposites. Then group like terms and simplify to $2b^3 - 3b + 4$.

5. Arrange the terms in descending order according to their exponents. The polynomial in standard form is $3x^4 + 5x^2 - 2x - 6$.

Lesson 9.2, pages 432–437
Activity
Part I

1. The Distributive Property

2. $(x + 1)(x + 2) = x^2 + 3x + 2$

3. $[10 + 1][10 + 2] = (10)^2 + 3(10) + 2$
$(11)(12) = 100 + 30 + 2$
$132 = 132$

CHECKPOINT ✓
4.
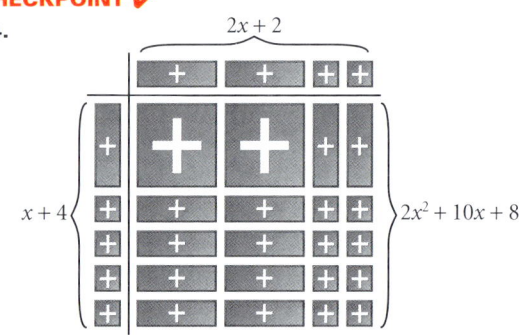

$(2x + 2)(x + 4) = 2x^2 + 10x + 8$
Check:
$[2(3) + 2][(3) + 4] \stackrel{?}{=} 2(3)^2 + 10(3) + 8$
$(8)(7) \stackrel{?}{=} 18 + 30 + 8$
$56 = 56$

Part II

1. $x^2 - x - 6$

2. $[(10) + 2][(10) - 3] \stackrel{?}{=} (10)^2 - (10) - 6$
$(12)(7) \stackrel{?}{=} 100 - 10 - 6$
$84 = 84$

CHECKPOINT ✓
3.
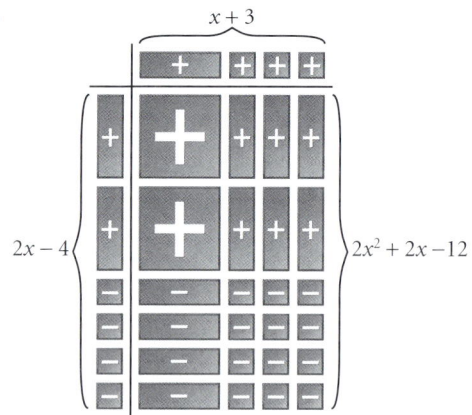

$(x + 3)(2x - 4) = 2x^2 + 2x - 12$
Check:
$[(4) + 3][2(4) - 4] \stackrel{?}{=} 2(4)^2 + 2(4) - 12$
$(7)(4) \stackrel{?}{=} 32 + 8 - 12$
$28 = 28$

Part III

1.
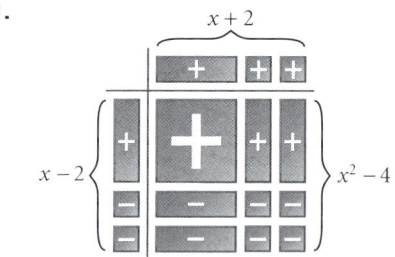

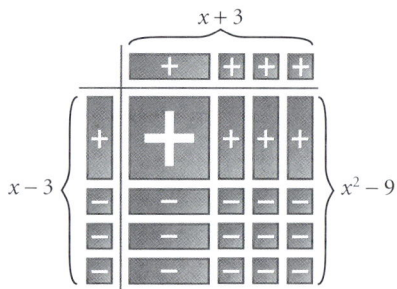

In each product there is one x^2-tile, the x-tiles pair up to make neutral pairs, and the unit tiles are all negative, with the amount of them being the square of the absolute value of the common constant term of the factors.
$(x + 5)(x - 5) = x^2 - 25$

2.
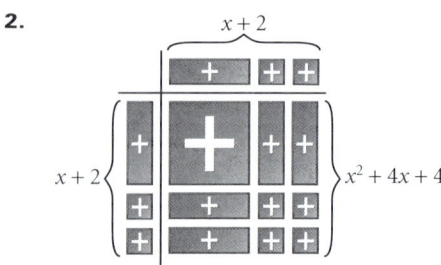

842

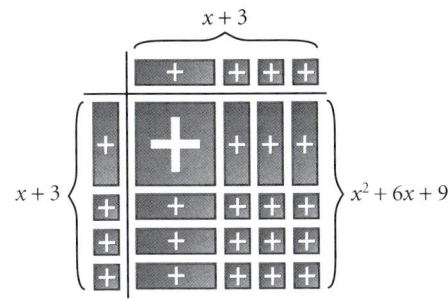

In each product there is one positive x^2-tile, the number of positive x-tiles is twice the common constant term and the unit tiles are all positive with the number of them being the square of the common constant term. $(x + 5)(x + 5) = x^2 + 10x + 25$

CHECKPOINT ✔

3.

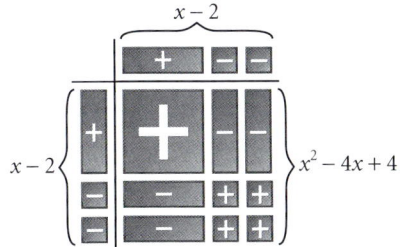

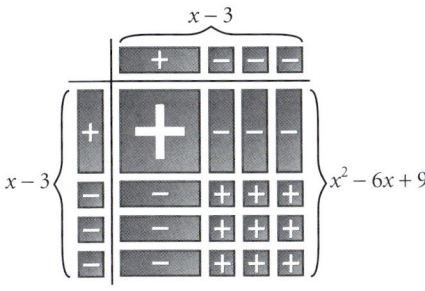

In each product there is one positive x^2-tile, the number of negative x-tiles is twice the absolute value of the common constant term, and the unit tiles are all positive with the number of them being the square of the common constant term. $(x - 5)(x - 5) = x^2 - 10x + 25$

Exercises
Communicate

1. **a.** Use one positive x-tile and two positive unit tiles as one factor. Use one positive x-tile and three positive unit tiles as the other factor. The product rectangle will be one positive x^2-tile, five positive x-tiles, and six positive unit tiles.
   b. Use one positive x-tile and four negative unit tiles for each factor. The product square will be one positive x^2-tile, eight negative x-tiles, and sixteen positive unit tiles.
   c. Use one positive x-tile and six positive unit tiles for each factor. The product square will be one positive x^2-tile, twelve positive x-tiles, and thirty-six positive unit tiles.

2. Answers may vary. Sample answer:

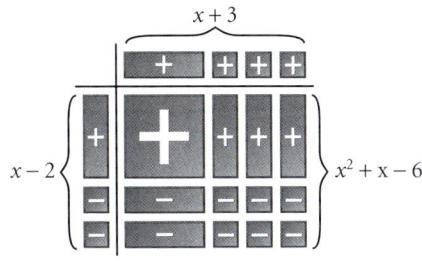

3. Multiply the x terms together. Multiply the sevens together. Take the difference of the two products. $x^2 - 49$

4. The model shows that $(x + 4)^2 = x^2 + 8x + 16$

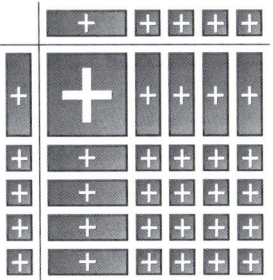

Practice and Apply

15. $2x^2 + 3x + 1$

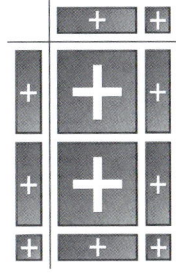

16. $x^2 + 3x - 10$

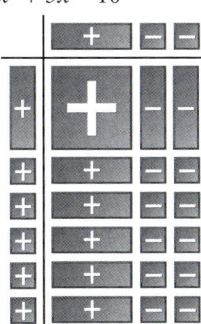

17. $x^2 + 3x + 2$

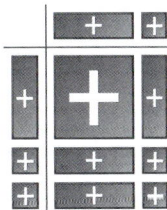

18. $3x^2 - 16x - 35$

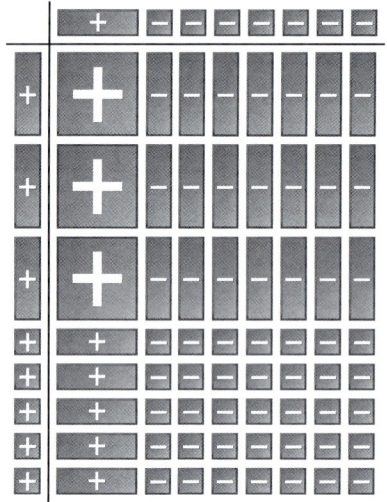

19. $4x^2 - 11x + 6$

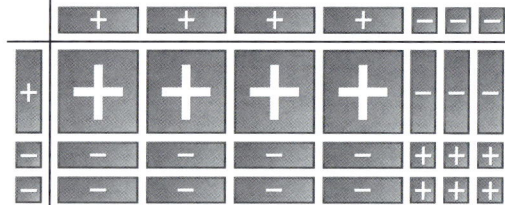

20. $x^2 - 2x - 35$

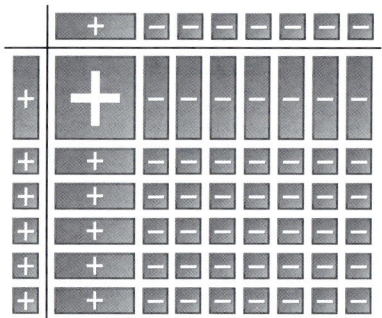

21. $2x^2 + 2x - 60$

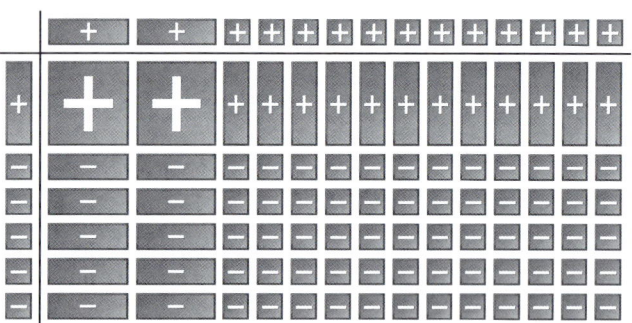

Look Beyond

64.

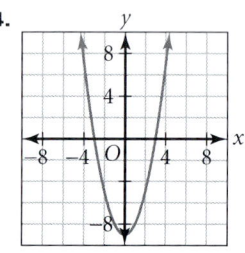 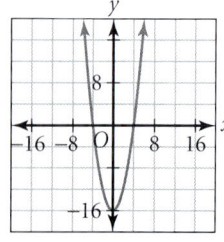

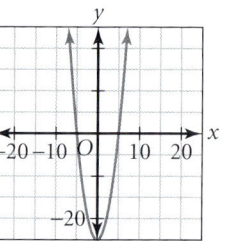

The zeros are where the graph crosses the x-axis since this is where $y = 0$.

Portfolio Activity

1. Row 4: 1 4 6 4 1
   Row 5: 1 5 10 10 5 1
   Row 6: 1 6 15 20 15 6 1

2. Row 4: 16
   Row 5: 32
   Row 6: 64

3. Row 7: 128
   Row 8: 256
   Row 9: 512

Lesson 9.3, pages 438–442

Activity

1. The algebra-tile model can be divided into sections, each representing a term of the resulting expression after using the Distributive Property.

2. $(x + 4)(x + 5) = (x + 4)(x) + (x + 4)(5)$
   $ = x^2 + 4x + 5x + 20$
   $ = x^2 + 9x + 20$

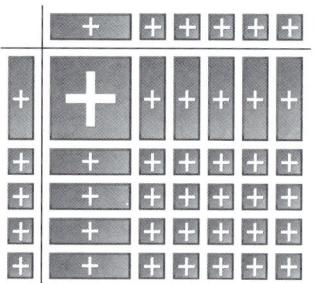

The algebra-tile model can be divided into sections, each representing a term of the resulting expression after using the Distributive Property.

CHECKPOINT ✔

3. Answers may vary. Sample answer:
   (Step 1) Multiply the first terms of each factor. (Step 2) Multiply the last terms of each factor. (Step 3) Multiply the first term of the first factor by the second term of the second factor. (Step 4) Multiply the second term of the first factor by the first term of the second factor. Lastly, combine like terms.

4. $(x+4)(x-5) = (x+4)(x) + (x+4)(-5)$
$= x^2 + 4x - 5x - 20$
$= x^2 - x - 20$

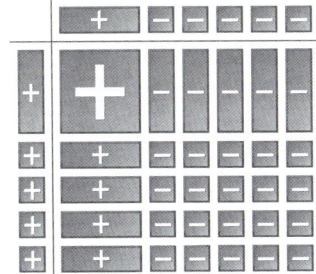

Answers may vary. Sample answer: The procedure does apply.

Exercises
Communicate

1. a. 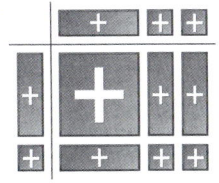   b.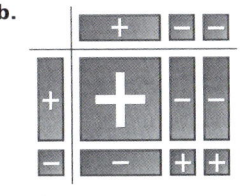

   $x^2 + 3x + 2$ $x^2 - 3x + 2$

2. a. $(x+1)(x+2) = (x+1)(x) + (x+1)(2)$
   $= x^2 + x + 2x + 2$
   $= x^2 + 3x + 2$
   b. $(x-1)(x-2) = (x-1)(x) + (x-1)(-2)$
   $= x^2 - x - 2x + 2$
   $= x^2 - 3x + 2$

3. Multiply the first terms of each binomial factor. Multiply the outside terms. Multiply the inside terms. Multiply the last terms. Combine like terms.
   $(2x+3)(x-4) = 2x^2 - 8x + 3x - 12$
                             F   O   I   L
   $= 2x^2 - 5x - 12$

4. $(x+2)^2$:

   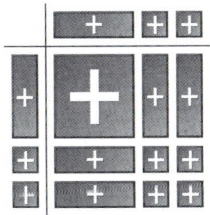

   $x^2 + 2^2 = x^2 + 4$:

   $(x+2)^2$ has four positive x-tiles that $x^2 + 2^2$ does not.

Lesson 9.4, pages 443–447
Activity

1. 2200 square centimeters

2. 5400 cubic centimeters; 1980 square centimeters

3. If x represents the width, then $2x$ represents the length, and the height remains 30. Substitute these values into the formulas for volume and surface area, and simplify.

CHECKPOINT ✓

4. As the width increases, the volume and surface area increase. A width of 10 centimeters gives the original volume and surface area of the box.

5. Volume = 5189.4 cubic centimeters;
Surface area = 2019.96 square centimeters
This is approximately a 14% reduction in volume, and about a 8% reduction in surface area.

Exercises
Communicate

1. A polynomial function is a function that consists of one or more monomials. Examples: $y = x^2$, $y = x^2 + x + 1$

2. Volume of a cube = length × width × height. In a cube these are all equivalent so $V = x^3$
Surface area of a cube = 6 × (area of a face), so $S = 6x^2$

3. Enter **Y1**$=x^2-4$ and **Y2**$=(x+2)(x-2)$, and compare the y-values in the table. If the columns for **Y1** and **Y2** are equal for all values of x, then the functions are equivalent.

Guided Skills Practice

6.

X	Y1	Y2
-3	0	0
-2	1	1
-1	8	8
0	27	27
1	64	64
2	125	125
3	216	216

Y1■(X+3)^3

Practice and Apply

7.

x	$x^2 + 5x + 6$	$(x+2)(x+3)$
−3	0	0
−2	0	0
−1	2	2
0	6	6
1	12	12
2	20	20
3	30	30

8.

x	$(x-4)(x-2)$	$x^2 - 6x + 8$
−3	35	35
−2	24	24
−1	15	15
0	8	8
1	3	3
2	0	0
3	−1	−1

9.

x	$x^2 - 5x + 6$	$(x-2)(x-3)$
−3	30	30
−2	20	20
−1	12	12
0	6	6
1	2	2
2	0	0
3	0	0

10.

x	$(x+7)(x-1)$	$x^2 + 6x - 7$
−3	−16	−16
−2	−15	−15
−1	−12	−12
0	−7	−7
1	0	0
2	9	9
3	20	20

11.

x	$x^2 - 9$	$(x-3)(x+3)$
−3	0	0
−2	−5	−5
−1	−8	−8
0	−9	−9
1	−8	−8
2	−5	−5
3	0	0

12.

x	$(x-6)(x+4)$	$x^2 - 2x - 24$
−3	−9	−9
−2	−16	−16
−1	−21	−21
0	−24	−24
1	−25	−25
2	−24	−24
3	−21	−21

13.

x	$x^3 - 8$	$(x-2)(x^2 + 2x + 4)$
−3	−35	−35
−2	−16	−16
−1	−9	−9
0	−8	−8
1	−7	−7
2	0	0
3	19	19

14.

x	$(x+5)(x+3)$	$x^2 + 8x + 15$
−3	0	0
−2	3	3
−1	8	8
0	15	15
1	24	24
2	35	35
3	48	48

15.

x	$x^3 + 12x^2 + 48x + 64$	$(x+4)^3$
−3	1	1
−2	8	8
−1	27	27
0	64	64
1	125	125
2	216	216
3	343	343

16.

x	$(x-6)(x-7)$	$x^2 - 13x + 42$
−3	90	90
−2	72	72
−1	56	56
0	42	42
1	30	30
2	20	20
3	12	12

17.

x	$x^2 - 6x + 9$	$(x-3)^2$
−3	36	36
−2	25	25
−1	16	16
0	9	9
1	4	4
2	1	1
3	0	0

18.

x	$(x+5)^2$	$x^2 + 10x + 25$
−3	4	4
−2	9	9
−1	16	16
0	25	25
1	36	36
2	49	49
3	64	64

37.

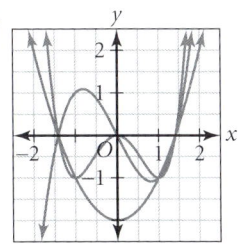

Answers may vary. Sample answer: Each graph has a different number of vertices. All three graphs pass through $(-\sqrt{2}, 0)$, $(\sqrt{2}, 0)$, and $(1, -1)$.

Portfolio Activity

1. Term 4   Term 5

2. Answers may vary. Sample answer:

x	$\frac{x^2+x}{2}$	$\frac{x(x+1)}{2}$
−2	1	1
−1	0	0
0	0	0
1	1	1
2	3	3

Lesson 9.5, pages 448–451

Exercises
Communicate

1. A factor of a polynomial is a number or expression such that when the number is multiplied by another number or expression, the result is the original polynomial.

2. Distributive Property

3. Find the largest factor that divides each term evenly.

4. a. GCF = 30. Divide each term by the GCF.
Remaining factors: $\frac{60}{30} = 2$, $\frac{150}{30} = 5$

b. GCF = x^3y^2. Divide each term by the GCF.
Remaining factors: $\frac{x^3y^5}{x^3y^2} = y^3$, $\frac{x^5y^2}{x^3y^2} = x^2$

c. GCF = $x + y$. Divide each term by the GCF.
Remaining factors: $\frac{25(x+y)}{x+y} = 25$, $\frac{39(x+y)}{x+y} = 39$

5. Treat $5y^2 + y$ as one expression and $2y + 2$ as another so that you can factor a common factor.
$(5y^2 + 5y) + (2y + 2) = 5y(y + 1) + 2(y + 1)$
$= (5y + 2)(y + 1)$

Lesson 9.6, pages 452–457

Activity

1. a. $x^2 - 4$
b. $x^2 - 9$
c. $x^2 - 16$
d. $x^2 - 25$

Answers may vary. Sample answer: When using FOIL, the product of the outer terms cancels out the product of the inner terms, so all that remains is a difference of squares.

2. a. $(x + 1)(x - 1)$
b. $(x + 6)(x - 6)$
c. $(x + 8)(x - 8)$
d. $(x + 10)(x - 10)$

CHECKPOINT ✔
3. $a^2 - b^2 = (a + b)(a - b)$

Exercises
Communicate

1. Determine the 2 equal factors of the first term, x^2, namely, x and x. Determine the 2 equal factors of the third term, 100, namely, 10 and 10. Since the trinomial is a perfect square, the factors are $(x + 10)$ and $(x + 10)$.

2. If you assume the trinomial is a perfect square, determine the square roots of the first and last term. The square root of $4x^2$ and 9 are $2x$ and 3 respectively. Check to see if the middle term of the trinomial is equal to twice the product of the square roots of $4x^2$ and 9. The factors are $(2x - 3)$ and $(2x - 3)$.

3. Find the positive square root of p^2 and 121. The square roots are p and 11 respectively. Since the binomial is a difference of two squares, its factors are the sum and difference of the square roots, $(p + 11)(p - 11)$.

Portfolio Activity

1. 24, 35, 48; $x^2 - 1$ and x^2, where x is a positive integer

2.

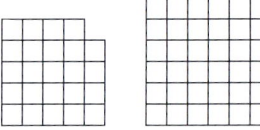

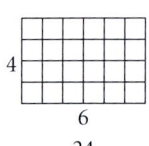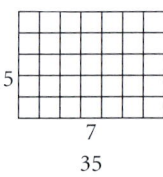

4	5
6	7
24	35

3. Answers may vary. Sample answer: For some $n \times n$ square with area n^2, $n^2 - 1$ square units can be arranged in a rectangle with dimensions $(n - 1)$ by $(n + 1)$. Algebraically, we know that the rectangular area is: $(n - 1)(n + 1) = n^2 - 1$.

4.

x	$x^2 - 1$	$(x + 1)(x - 1)$
-3	8	8
-2	3	3
-1	0	0
0	-1	-1
1	0	0
2	3	3
3	8	8

Lesson 9.7, pages 458–463

Exercises
Communicate

1. For each polynomial, start with the tiles that model the polynomial. Try to arrange them in a rectangle, adding neutral pairs if necessary.
   a. Model $(x - 4)(x - 1)$
   b. Model $(x - 6)(x + 2)$
   c. Model $(x + 3)(x + 3)$

2. a. $(x + 3)(x - 2)$; the signs are opposite because the third term of the trinomial is negative.
   b. $(x - 2)(x - 5)$; the signs are the same because the third term of the trinomial is positive.
   c. $(x + 5)(x - 3)$; the signs are opposite because the third term of the trinomial is negative.

3. Guess a pair of numbers that have a product of -24 and check whether their sum is -5.

4. $(1, 36)$; $(-1, -36)$; $(2, 18)$; $(-2, -18)$; $(3, 12)$; $(-3, -12)$; $(4, 9)$; $(-4, -9)$; $(6, 6)$; $(-6, -6)$

Lesson 9.8, pages 464–469

Activity

1.

Time, x	0	1	2	3
Height, y	0	32	32	0

2. At 0 seconds and 3 seconds

3. $-16x(x - 3)$

4. $x = 0$ or $x = 3$

CHECKPOINT ✔

5. Each value of x that makes one of the factors equal to 0 is a time when the soccer ball is on the ground. The soccer ball starts on the ground, which corresponds to $x = 0$.

6.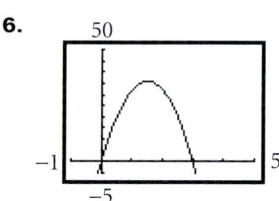

The times the soccer ball is on the ground correspond to the values of x where the graph crosses the x-axis.

Exercises
Communicate

1. Break the equation down into its prime factors and set each unique factor equal to zero. Solve for the unknowns. These are the solutions.

   Example 1: $x^2 + 2x + 1 = 0$
   $(x + 1)(x + 1) = 0$
   $x + 1 = 0$
   $x = -1$

   Example 2: $x^2 + 4x + 3 = 0$
   $(x + 1)(x + 3) = 0$
   $x + 1 = 0$ or $x + 3 = 0$
   $x = -1$ or $x = -3$

2. The zeros are the values that, when substituted for the variable in a function, make the function equal to zero.

3. If the zeros are 2 and 3, the factors they came from must be $(x - 2)$ and $(x - 3)$. Therefore, the function is $(x - 2)(x - 3) = x^2 - 5x + 6 = 0$.

4. Yes, it is possible. Answers may vary. Sample answer: Consider any polynomial function whose graph does not cross the x-axis, such as $y = x^2 + 2$. $x^2 + 2$ is prime, so cannot be factored. Since both terms are nonnegative, $x^2 + 2 \neq 0$ for any value of x.

5. The zeros represent the time the object is on the ground.

Look Beyond

53. No real solution

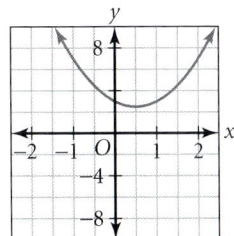

Portfolio Activity

1.
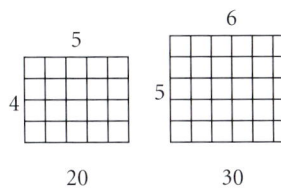

2. 2, 6, 12, 20, 30, 42, 56, 72, 90, and 110

3. $f(x) = x(x+1)$

4.
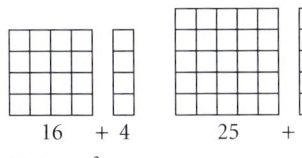

$f(x) = x^2 + x$

5. $x(x+1) = x^2 + x$

6. The triangular numbers are the rectangular numbers divided by 2.

Chapter 9 Project

5. The next five rows are:

```
            1   6   15   20   15   6    1
          1   7   21   35   35   21   7    1
        1   8   28   56   70   56   28   8    1
      1   9   36   84  126  126   84   36   9    1
    1  10   45  120  210  252  210  120   45  10    1
```

Lesson 10.1, pages 480–485

Activity

1. $y = x^2$

x	y
−3	9
−2	4
−1	1
0	0
1	1
2	4
3	9

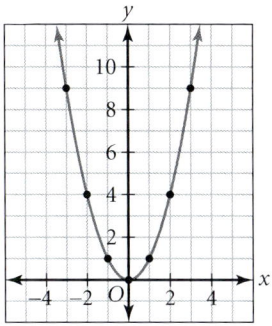

$y = x^2 + 1$

x	y
−3	10
−2	5
−1	2
0	1
1	2
2	5
3	10

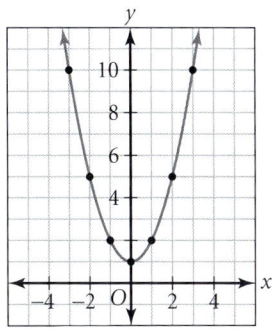

$y = x^2 - 1$

x	y
−3	8
−2	3
−1	0
0	−1
1	0
2	3
3	8

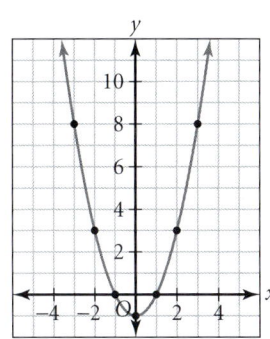

2. Adding 1 moves each point on the graph up one unit. Subtracting 1 moves each point on the graph down one unit.

3. $y = x^2$

x	y
−3	9
−2	4
−1	1
0	0
1	1
2	4
3	9

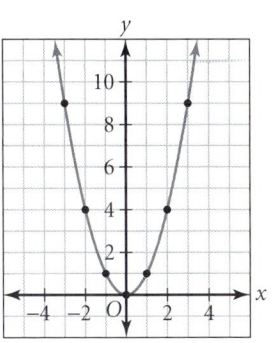

$y = (x + 2)^2$

x	y
−5	9
−4	4
−3	1
−2	0
−1	1
0	4
1	9

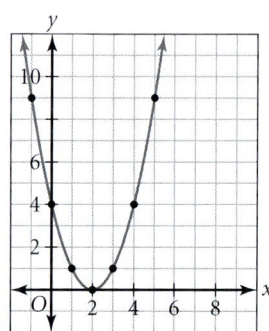

$y = (x − 2)^2$

x	y
−1	9
0	4
1	1
2	0
3	1
4	4
5	9

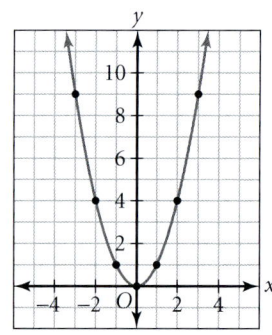

$y = (x − 2)^2 − 1$

x	y
−1	8
0	3
1	0
2	−1
3	0
4	3
5	8

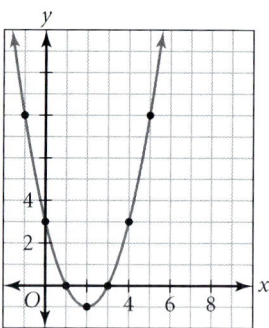

$y = (x + 2)^2 + 1$

x	y
−5	10
−4	5
−3	2
−2	1
−1	2
0	5
1	10

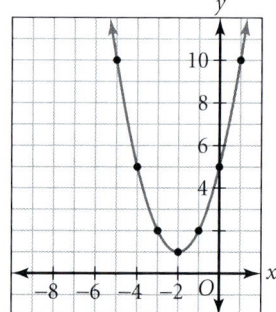

CHECKPOINT ✔

4. Adding 2 shifts each point on the graph to the left two units. Subtracting 2 shifts each point on the graph to the right two units.

5. $y = x^2$

x	y
−3	9
−2	4
−1	1
0	0
1	1
2	4
3	9

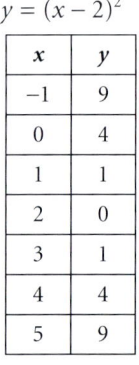

$y = (x − 2)^2 + 1$

x	y
−1	10
0	5
1	2
2	1
3	2
4	5
5	10

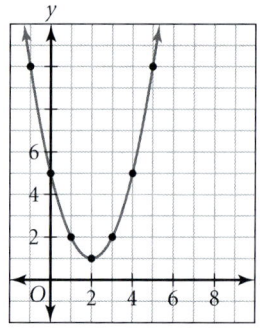

CHECKPOINT ✔

6. If the operation inside the squared term is subtraction, the graph is shifted to the right, and if the operation is addition, the graph is shifted to the left. If a number is added to the squared term, the graph is shifted up, and if a number is subtracted from the squared term, the graph is shifted down.

Exercises

Communicate

1. The graph of $y = x^2 − 8$ has been vertically translated 8 units down from the parent function.

2. The graph of $y = (x − 8)^2$ has been horizontally translated 8 units to the right from the parent function.

3. $h = 3$
$k = −8$
vertex: $(h, k) = (3, −8)$

4. The axis of symmetry is a vertical line at the value of h. Thus the axis of symmetry is $x = 3$.

Practice and Apply

21. vertex: (2, 3);
axis of symmetry: $x = 2$

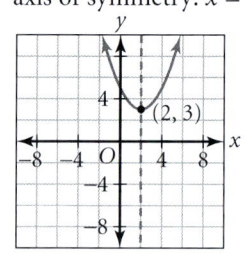

22. vertex: (3, −7);
axis of symmetry: $x = 3$

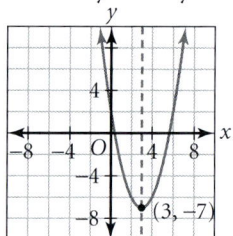

850

23. vertex: $(-2, -4)$;
axis of symmetry: $x = -2$
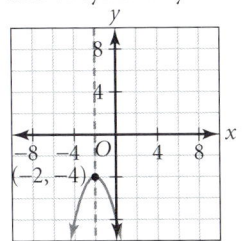

24. vertex: $(3, -8)$;
axis of symmetry: $x = 3$
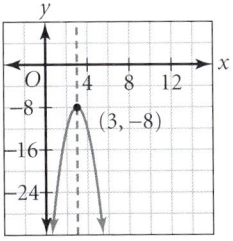

25. vertex: $(4, 5)$;
axis of symmetry: $x = 4$
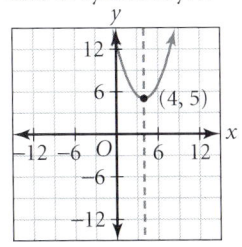

26. vertex: $(-6, -3)$;
axis of symmetry: $x = -6$
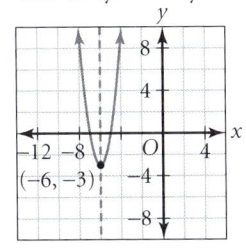

27. vertex: $(-5, -7)$;
axis of symmetry: $x = -5$
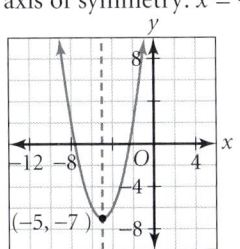

28. vertex: $(2, 9)$;
axis of symmetry: $x = 2$
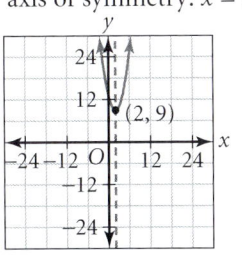

29. vertex: $(5, 2)$;
axis of symmetry: $x = 5$
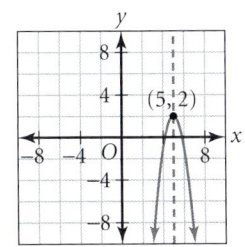

30. vertex: $(-3, -2)$;
axis of symmetry: $x = -3$
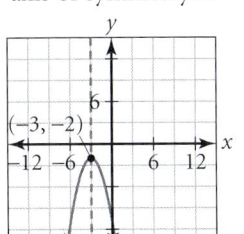

31. vertex: $(-5, 7)$;
axis of symmetry: $x = -5$
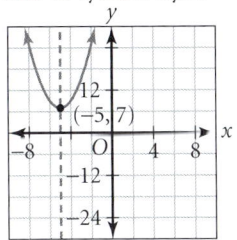

Lesson 10.2, pages 486–491

Activity

1.

Time, t	Height, h
0	185
1	169
2	121
3	41
4	−71
5	−215

2. $h = 0$ between $t = 3$ and $t = 4$. An approximate value is $t = 3.4$.

3.

Time, t	Height, h
3.1	31.24
3.2	21.16
3.3	10.76
3.4	0.04
3.5	−11
3.6	−22.36
3.7	−34.04
3.8	−46.04

CHECKPOINT ✓

4.

Time, t	Height, h
3.39	1.1264
3.40	0.04
3.41	−1.05
3.42	−2.142

Exercises
Communicate

1. The values of t for which h is 0 is the point at which the function crosses the t-axis. Try values starting at $t = 0$ until either $h = 0$ or the value of h changes from positive to negative. Use smaller increments between your choices of t to narrow in on the value where h becomes 0.

2. Find numbers whose square is 100. The answer can be positive or negative since the square of any number is positive.

3. a. Since $x^2 = 64$ and $64 \geq 0$, $x = \pm 8$.
   b. Since $x^2 = 8$ and $8 \geq 0$, $x = \pm\sqrt{8}$.
   c. Since $x^2 = \frac{16}{100}$ and $\frac{16}{100} \geq 0$, $x = \pm\frac{4}{10}$.

4. a. Add 25 to both sides of the equation. Solve the resulting equation by taking the square root of both sides.
$$(x + 3)^2 = 25$$
$$x + 3 = \pm 5$$
$$x = 2 \text{ or } x = -8$$

b. $(x - 8)^2 = 2$

$x - 8 = \pm\sqrt{2}$

$x = 8 \pm \sqrt{2}$

5. Plot the vertex at $(3, -4)$. Solve the equation $(x - 3)^2 - 4 = 0$ to find the zeros. Mark these two points on the x-axis. Sketch the graph by drawing a parabola passing through the three points plotted and symmetrical about the axis of symmetry, $x = 3$.

Portfolio Activity

Answers may vary. Sample answers:

1.

x	y
.1	9.84
.2	9.36
.3	8.56
.4	7.44
.5	6.00
.6	4.24

2. 0.32

3.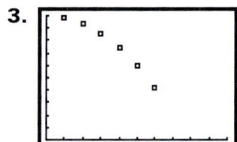

Lesson 10.3, pages 492–497

Activity

2. Nine unit tiles were added.

3. $x^2 + 6x + 9 = (x + 3)^2$

4. Arrange the tiles with the x^2-tile in the upper left corner, and with 4 x-tiles across the top and 4 x-tiles down the left side. Then add 16 positive unit tiles.

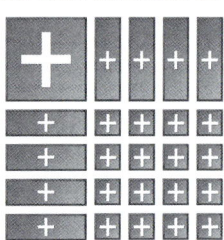

$x^2 + 8x + 16 = (x + 4)^2$

CHECKPOINT ✓
5. Add the square of half the number of x-tiles.

Exercises
Communicate

1. Start with one x^2-tile, arrange the four x-tiles in two groups of 2. Four unit tiles are needed to complete the square.

2. Find half of the coefficient of x and then add its square to the expression. Factor the trinomial into a perfect square. $x^2 - 7x + \frac{49}{4} = \left(x - \frac{7}{2}\right)^2$

3. The first three terms are a perfect square trinomial, so the equation can be written as $y = (x + 5)^2 - 25$.

4. Complete the square to find the vertex form of the equation.
$y = (x^2 + 2x + 1) - 1$
$y = (x + 1)^2 - 1$
The vertex is $(-1, -1)$.

5. Group the x^2-term and x-term: $y = (x^2 - 10x) + 11$. Find the number to add from the coefficient of the x-term: $\left(-\frac{10}{2}\right)^2 = 25$.
Add the number inside the parentheses and subtract it outside the parentheses: $y = (x^2 - 10x + 25) + 11 - 25$. Simplify, and factor the trinomial in parentheses: $y = (x - 5)^2 - 14$.
The vertex is (h, k), where $h = 5$ and $k = -14$.

Practice and Apply

13.

+	+	+
+	+	+
+	+	+

$x^2 + 4x + 4$

14.

+	−	−	−	−	−
−	+	+	+	+	+
−	+	+	+	+	+
−	+	+	+	+	+
−	+	+	+	+	+
−	+	+	+	+	+

$x^2 - 10x + 25$

15.

+	+	+	+	+	+	+
+	+	+	+	+	+	+
+	+	+	+	+	+	+
+	+	+	+	+	+	+
+	+	+	+	+	+	+
+	+	+	+	+	+	+
+	+	+	+	+	+	+

$x^2 + 12x + 36$

16.

+	+	+	+	+	+
+	+	+	+	+	+
+	+	+	+	+	+
+	+	+	+	+	+
+	+	+	+	+	+
+	+	+	+	+	+

$x^2 + 10x + 25$

Portfolio Activity

1.

x	y
1	984
2	936
3	856
4	744
5	600
6	424

2. 32

3. Answers may vary. Check students' responses.

Lesson 10.4, pages 498–503

Activity

Part I

2. length = 4 inches
   width = 4 inches
   height = 2 inches
   volume = 32 cubic inches, which is more than 8 cubic inches

3. A 6-inch square is the correct size for the original piece of paper.

Part II

1. The length and width are each what remains on each side of the paper after a total of 4 inches is removed on each side. The height is determined by the length of a side of the square cut out, which is 2 inches.

2. $(x - 4)(x - 4)(2)$

CHECKPOINT ✔

3. $(x - 4)(x - 4)(2) = 8$
   $(x - 4)^2 = 4$
   $x - 4 = \pm 2$
   $x = 4 \pm 2$
   $x = 6 \text{ or } 2$

   The original square should be a 6-inch square, since a 2-inch square does not make sense.

Exercises

Communicate

1. They have the same value, $x = 6$.

2. Complete the square by grouping the first two terms of the trinomial. Solve the equation obtained by solving $f(x) = 0$. The zeros are 2 and 4.

3. Take half of the coefficient of x and square it. Add this number to both sides of the equation. Express the perfect square trinomial on the left side as a square of a linear factor and simplify the integers on the right side. Solve the resulting equation. The solution is $x = -3$ or $x = 5$.

4.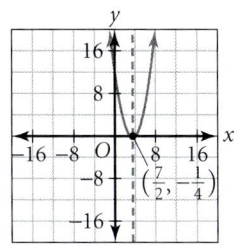

   Find the x-coordinate of the intersections of $y = x^2 - 7x + 12$ and $y = 2$. The x-values are 2 and 5.

5. Solve by substitution algebraically or by graphing.

Lesson 10.5, pages 506–510

Exercises

Communicate

1. When the equation is written in the form $2n^2 + 3n - 7 = 0$, the coefficient of n^2 is a, the coefficient of n is b, and the constant term is c. The standard form is important because it makes the values of a, b, and c easy to determine.

2. In $b^2 - 4ac$ substitute $a = 2$, $b = 3$, and $c = -7$.

3. **a.** Find the value of $b^2 - 4ac$ when $a = 2$, $b = -1$, and $c = -2$. Since this value is greater than zero (17), the equation has two solutions.
   b. Find the value of $b^2 - 4ac$ when $a = 1$, $b = -2$, and $c = 7$. Since the value is less than zero (−24), the equation has no real solutions.

Lesson 10.6, pages 511–515

Activity

1. 3 thousand watches and 7 thousand watches

2. Between 3 thousand and 7 thousand watches

3. Less than 3 thousand watches or greater than 7 thousand watches

4. 5 thousand watches

CHECKPOINT ✔

5. 4 thousand dollars

Exercises

Communicate

1. Factor the related quadratic equation and determine where the product is positive.

2. Since the inequality does not include equals to, place open circles at $x = 3$ and $x = 6$. Test values for x from each of the three intervals created on the number line to see which range of values makes the inequality true.

3. **a.** The boundary is dashed since the inequality does not include equality.
   b. The boundary is solid because the inequality includes equality.
   c. The boundary is solid because the inequality includes equality.

4. **a.** First sketch the parabola $y = x^2 - 3$ with a dashed line. Test one point between the zeros of the function and if the resulting inequality is true, shade the inside of the parabola. If the test is false, shade the outside.
   b. Use the steps as in part **a** except use a solid line for the parabola.
   c. Use the same steps as in part **b** except shade the opposite side of the parabola.

Practice and Apply

16. $-2 < x < 10$

17. $x < 3$ or $x > 5$

18. $-3 \leq x \leq 6$

19. $x \leq 5$ or $x \geq 6$

20. $x \geq 6$ or $x \leq -2$

21. $x < -7$ or $x > 2$

22. $x \leq -6$ or $x \geq -4$

23.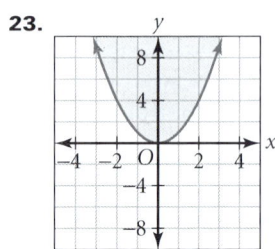

24.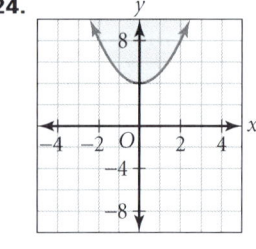

25.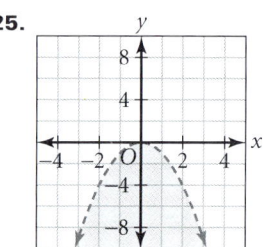

26.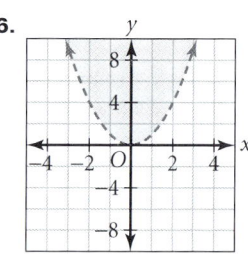

27.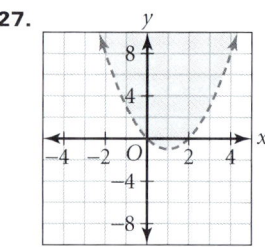

28.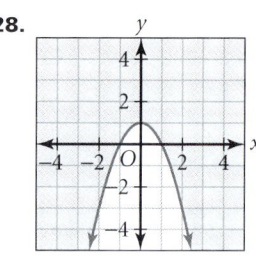

29.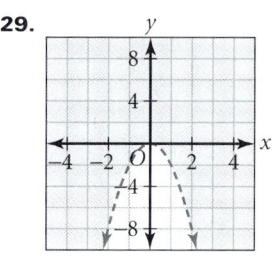

30.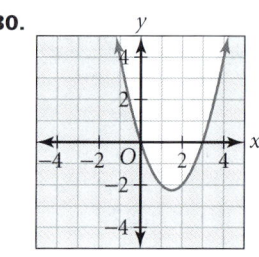

31.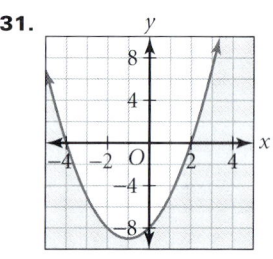

32.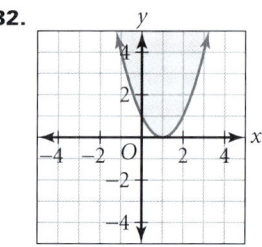

33.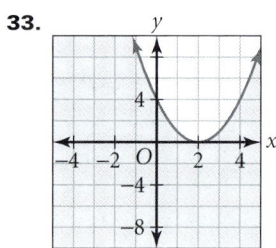

34.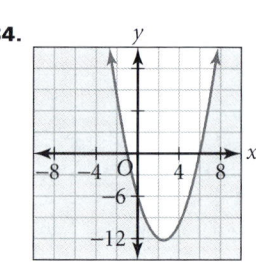

35.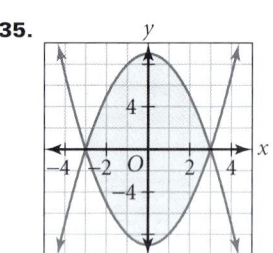

$\begin{cases} y \geq x^2 - 9 \\ y \leq -x^2 + 9 \end{cases}$

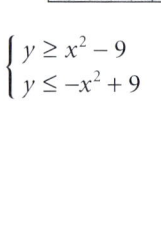

38.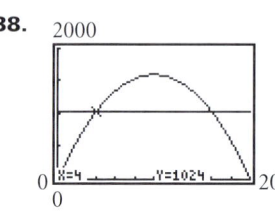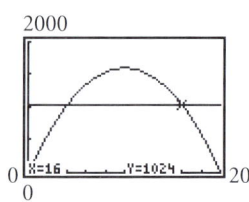

Portfolio Activity

1.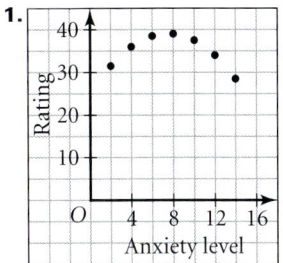

2. approximately −5 and 21

3. $y = -0.25x^2 + 3.75x + 25$

4. 7.5

5. $\frac{15 - \sqrt{65}}{2} < x < \frac{15 + \sqrt{65}}{2}$

Lesson 11.1, pages 526–531

Activity

1. 500 hours

2. 10 hours

3. 5 hours

CHECKPOINT ✓

4. $t = \frac{500}{x}$

Exercises
Communicate

1. $xy = k$; $y = \frac{k}{x}$

2. An inverse variation between rate and time means that as the rate increases, the time decreases. Answers may vary. Sample answer: The faster you drive, the less time it takes to drive a certain distance.

3. If y varies inversely as x, then $y = \frac{k}{x}$, where k is the constant of variation. Substitute $y = 3$ and $x = 8$ to find the value of k. Then use the equation of variation to find the value of y when $x = 2$.

4. The rule of 72 is $t = \frac{72}{r}$. To find how many years it would take money to double at 4%, divide 72 by 4.

Lesson 11.2, pages 532–537

Activity

1.
Number of lessons, n	Average cost of a lesson $A(n) = \frac{55 + 4.50n}{n}$	Cost
1	59.50	59.50
2	32.00	64.00
3	22.83	68.50
4	18.25	73.00
5	15.50	77.50
10	10.00	100.00
15	8.17	122.50
20	7.25	145.00
30	6.33	190.00
40	5.88	235.00
−1	−50.50	50.50
−5	−6.50	32.50
1.5	41.17	61.75
10.7	9.64	103.15

2. When $n = 0$, you are dividing by zero, which is undefined.

3. See the table in Step 1.

4.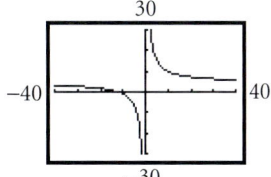

CHECKPOINT ✓

5. As n gets larger, $A(n)$ gets closer to the value 4.5. As the graph nears $n = 0$ from either side, it sharply turns away from the x-axis and steadily approaches, but never reaches, the vertical line $n = 0$.

Exercise
Communicate

1. A rational expression is an expression that can be expressed in the form $\frac{P}{Q}$, where P and Q are both polynomials in the same variable and $Q \neq 0$. Examples are $\frac{7x + 2}{x^2 - 4x + 1}$ and $\frac{x}{x^3 + 1}$. A rational function is a function of the form $y = \frac{P}{Q}$, where $\frac{P}{Q}$ is a rational expression. Examples are $y = \frac{2x}{x^2 - 3x + 2}$ and $y = \frac{x^2 - 4}{x}$.

2. The value of Q is zero.

3. $y = \frac{12x - 7}{x} = 12 - \frac{7}{x}$

855

Guided Skills Practice

9. $y = \frac{3}{x} - 5$

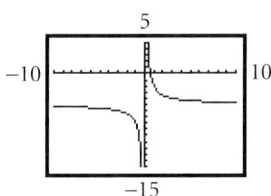

Practice and Apply

39.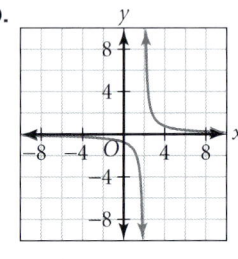
    $x \neq 2$

40.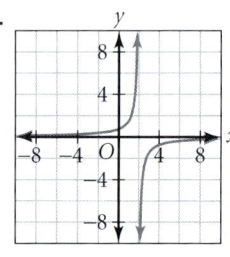
    $x \neq 2$

41.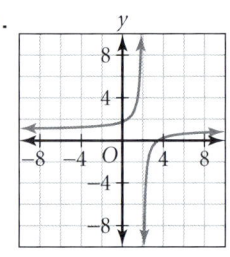
    $x \neq 2$

42.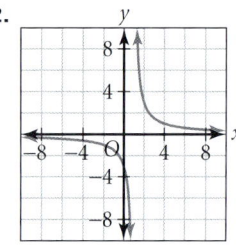
    $x \neq 1$

43.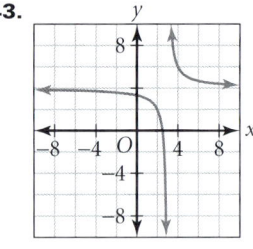
    $x \neq 3$

44.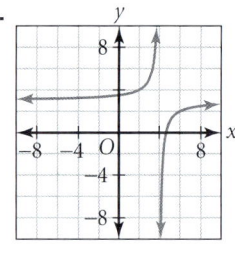
    $x \neq 4$

52.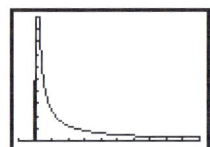

Lesson 11.3, pages 538–543

Activity

1. $0; \frac{7}{5}; \frac{6}{5}$

CHECKPOINT ✓
2. The value for y is undefined when $x = 0$.

3.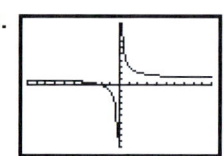

CHECKPOINT ✓
4. The functions are the same.

Exercises
Communicate

1. When a value of the variable makes the value of the denominator 0.

2. Answers may vary. Sample answer: $\frac{x}{x-7}$.

3. Set the expression in the denominator equal to zero and solve the resulting equation. The resulting value of the variable(s) is(are) the restriction(s) on the variable(s) for that expression.

4. A common factor divides evenly into each of two or more numbers or expressions.

5. Factor the denominator to obtain $\frac{x+1}{(x+1)(x+1)}$. Then remove the common factors from the numerator and the denominator. State any restrictions to avoid division by zero. The expression simplifies to $\frac{1}{(x+1)}$, $x \neq -1$.

Portfolio Activity

Answers will vary. Check students' responses.

Lesson 11.4, pages 546–551

Exercises
Communicate

1. Since there are no common factors, the common denominator is the product of the denominators.

2. After the numerators are rewritten so that the fractions have common denominators, add the numerators.

3. Follow the steps in 2, but subtract.

4. Multiply the numerators, then multiply the denominators and simplify.

5. Multiply the first expression by the reciprocal of the second expression and simplify.

Practice and Apply

28. $\frac{5p}{3}, p \neq 0$

29. $\frac{-1}{2(x+1)}, x \neq -1$

30. $\frac{x^2 + 3x + 4}{x+4}, x \neq -4$

31. $\frac{d-5}{d-1}$, $d \neq 1$

32. $\frac{b+3}{b^2}$, $b \neq 0$

33. $\frac{5}{2+n}$, $n \neq -3, -2$

34. $\frac{x}{x^2+4x+3}$, $x \neq 0, -1, -2, -3$

35. $\frac{q-1}{q}$, $q \neq 0, -1$

36. $\frac{2}{x^2+x}$, $x \neq 0, -1$

37. $\frac{15}{x^2-x}$, $x \neq 0, 1$

38. $\frac{y^2-y+4}{y^2-4y}$, $y \neq 0, 4$

39. $\frac{x+3}{x^2+x}$, $x \neq 0, -1$

40. $\frac{a^2+6a-1}{a^2+4a+3}$, $a \neq -1, -3$

41. $\frac{m^2+2m-15}{m^2+m-2}$, $m \neq 1, -2$

42. $\frac{r-r^2}{r^2+6r+5}$, $r \neq -1, -5$

43. $\frac{-5b-12}{b^2+b-6}$, $b \neq 2, -3$

44. $\frac{9y-8}{y^2-4y+4}$, $y \neq 2$

45. $\frac{-3}{y+6}$, $y \neq 5, -6$

46. 1, $x \neq 1$

47. $\frac{7-x}{2x^2-18}$, $x \neq 3, -3$

48. $\frac{6x+13}{3x+12}$, $x \neq -4$

49. $\frac{x+5}{3x}$, $x \neq 0, 7$

50. $\frac{4p}{p^2+2p+1}$, $p \neq -1, 3$

51.

x	$y = x$	$y = (x-4)(x+4)$
−4	−4	0
−3	−3	−7
−2	−2	−12
−1	−1	−15
0	0	−16
1	1	−15
2	2	−12
3	3	−7
4	4	0

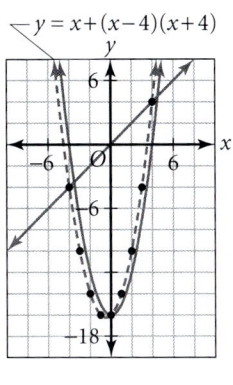

Lesson 11.5, pages 552–557

Activity

1. 2.4 hours

2.

Walking speed (mph)	Walking time: $\frac{7 \text{ miles}}{\text{walking speed}}$	Riding speed: $5 \cdot$ walking speed	Riding time: $\frac{7 \text{ miles}}{\text{riding speed}}$	Total time (hours)
1.0	7	5	1.4	8.4
2.0	3.5	10	0.7	4.2
3.0	2.3$\bar{3}$	15	0.4$\bar{6}$	2.8
4.0	1.75	20	0.35	2.1
3.5	2	17.5	0.4	2.4

CHECKPOINT ✓

3. walking speed = 3.5 mph; riding speed = 17.5 mph

Exercises

Communicate

1. Undefined values are the values of the variable that make the denominator of any rational expressions in the equation have value zero.

2. The real-world domain usually contains positive numbers because x is representing some type of measurement.

3. Find the lowest common multiple of the three denominators. The common denominator is $c(c-1)$.

4. Graph two functions, Y_1 = the left side of the equation and Y_2 = the right side of the equation, and find their point(s) of intersection. The x-value(s) at the point(s) of intersection is the solution.

5. Look at the denominator of each rational expression in the equation to find any values of the variable that make them equal 0. This equation is undefined for $c = 0$ and $c = 1$.

Portfolio Activity

1. ≈ 7.4 meters

2. $r = \frac{w}{54}$
gear ratios: 0.24, 0.26, 0.28, 0.3, 0.34, 0.39, 0.44
reciprocals: 4.15, 3.86, 3.6, 3.38, 3, 2.57, 2.25

3. Answers may vary.

Lesson 11.6, pages 558–565

Activity

1. Answers may vary. Sample answer:
Let $a = 1$, $b = 2$, $c = 3$, $d = 6$.
a. $6 = 6$
b. $d(a+b) = 18$, $c(c+b) = 15$
c. $18 = 18$
d. $6 = 6$
e. $d(a-b) = -6$, $b(c-b) = 2$
f. $8 = 8$

2. a, c, d, f

3. If $\frac{a}{b} = \frac{c}{d}$, then $\frac{b}{a} = \frac{d}{c}$.

If $\frac{a}{b} = \frac{c}{d}$, then $\frac{a+b}{b} = \frac{c+d}{d}$.

If $\frac{a}{b} = \frac{c}{d}$, then $\frac{a}{c} = \frac{b}{d}$.

If $\frac{a}{b} = \frac{c}{d}$, then $\frac{a}{b} = \frac{a+c}{b+d}$.

CHECKPOINT ✓

4. Conjectures; they have not been proven yet.

Exercises
Communicate

1. definitions, postulates, and proven theorems

2. The hypothesis, the statement that follows the word "if," is assumed to be true. The conclusion, the statement that follows the word "then," is the part that has to be proven.

3. In the converse of a conditional statement, the conclusion and the hypothesis are interchanged.

Guided Skills Practice

7. It is impossible because 100 is even. The product of two odds is odd. The product of four odds would be odd also.

8. sometimes true

9. Yes, since the product of two integers is always an integer, the set of integers is closed under multiplication.

Practice and Apply

21. $5x - 3x$ — Given
$= x(5-3)$ — Distributive Property
$= x(2)$ — Simplified form
$= 2x$ — Commutative Property of Multiplication

22. $a = b$ — Given
$a + c = b + c$ — Addition Property of Equality

23. $(ax + b) + ay$ — Given
$= (ax + ay) + b$ — Commutative and Associative Properties of Addition
$= a(x + y) + b$ — Distributive Property

24. $a = 2n$ — Definition of even numbers
$a(2n) = (2n)(2n)$ — Multiplication Property of Equality
$a \cdot a = (2n)(2n)$ — Definition of a
$a \cdot a = 2(n \cdot 2n)$ — Assoc. Property of Multiplication
$a^2 = 2(2n^2)$ — Simplified form

25. $(a - b) + (b - a)$
$= a + (-b + b) + (-a)$ — Associative Property for Addition
$= a + 0 + (-a)$ — Definition of Opposites
$= a + (-a)$ — Identity for Addition
$= 0$ — Definition of Opposites

Therefore, by the definition of opposites, $a - b$ and $b - a$ are opposites.

26. $\frac{a}{b} + \frac{c}{d} = 1\left(\frac{a}{b}\right) + 1\left(\frac{c}{d}\right)$ — Identity for Multiplication
$= \frac{d}{d}\left(\frac{a}{b}\right) + \frac{b}{b}\left(\frac{c}{d}\right)$ — Substitution Property
$= \frac{da}{db} + \frac{bc}{bd}$ — Multiplication of Rationals
$= \frac{da}{bd} + \frac{bc}{bd}$ — Commutative Property for Multiplication
$= \frac{da + bc}{bd}$ — Addition of Rationals

Lesson 12.1, pages 574–582
Activity

1. **a.** 12
   b. 12

2. Answers may vary. Sample answer:
$\sqrt{2 \cdot 8} = \sqrt{16} = 4$
$\sqrt{2} \cdot \sqrt{8} = \sqrt{2} \cdot 2\sqrt{2} = 4$
$\sqrt{4 \cdot 25} = \sqrt{100} = 10$
$\sqrt{4} \cdot \sqrt{25} = 2 \cdot 5 = 10$
$\sqrt{4 \cdot 9} = \sqrt{36} = 6$
$\sqrt{4} \cdot \sqrt{9} = 2 \cdot 3 = 6$

CHECKPOINT ✓

3. $\sqrt{9 + 16} = \sqrt{25} = 5$
$\sqrt{9} + \sqrt{16} = 3 + 4 = 7$

4. Answers may vary. Sample answers:
$\sqrt{4 + 12} = \sqrt{16} = 4$
$\sqrt{4} + \sqrt{12} = 2 + 2\sqrt{3}$
$\sqrt{4 + 9} = \sqrt{13}$
$\sqrt{4} + \sqrt{9} = 2 + 3 = 5$
$\sqrt{1 + 1} = \sqrt{2}$
$\sqrt{1} + \sqrt{1} = 1 + 1 = 2$

Exercises
Communicate

1. Draw a square that contains 16 graph paper squares. The square root of 16 is the number of squares on each side.

2. Draw two squares, one inside the other, whose areas are 16 (the perfect square just less than 19) and 25 (the perfect square just greater than 19). Determine the length of the side of each square. The square root of 19 is between these two numbers: 4 and 5. Make an estimate for the square root of 19 and use the guess and check method to find a value such that when it is squared, it is approximately 19.

3. 7 is between the two perfect squares 4 and 9. Therefore $\sqrt{7}$ is between their square roots, 2 and 3.

4. When simplifying an expression such as $5\sqrt{90x^3y^4}$, perfect square factors are factored out of the radicand.

5. An expression is in simplest radical form when there are no perfect square factors under the radical sign, the radical contains no fractions, and there are no radicals in the denominator.

Lesson 12.2, pages 583–590

Exercises

Communicate

1. domain: $x \geq 0$; there is no real number that is the square root of a negative number

2. When $a \geq 0$, $\sqrt{a^2} = a$ is true. When $a < 0$, $\sqrt{a^2} = -a$.

3. Answers may vary. Sample answer: Several problems may be solved by substituting different values for the variables.

4. Extraneous solutions may appear in the answers which are shown not to be genuine solutions when they are substituted back into the original equation.

Portfolio Activity

Data will vary. Sample data:

length (in cm)	142	114	76	38	10
time (in seconds)	2.4	2.2	1.8	1.3	0.7

4. If careful measurements are taken and if the scales used for the axes are appropriate, the graph will have the shape of a square-root function.

5. Answers will vary. If measurements are reasonably accurate, the results from the actual timing should correspond to the formula results within one or two tenths of a second. Discrepancies could be caused by one or more of the following:
   1) The fixed end of the string is not actually steady.
   2) The length of the string and/or the duration of the swings is not measured accurately.
   3) Half-swings are counted as full swings. (A full swing is a "round trip," where the weight arrives back on the side where it started.)

Lesson 12.3, pages 591–597

Activity

1.
9	16	25	25
25	144	169	160
256	900	1156	1156
169	7056	7225	7225

CHECKPOINT ✔

2. $c^2 = a^2 + b^2$; this agrees with the theorem.

3. No, the theorem does not work.

4. If, in a triangle with sides a, b, and c, $c^2 = a^2 + b^2$, then the triangle is a right triangle. Yes.

Exercises

Communicate

1. The side opposite the right angle is the hypotenuse. The other sides are the legs.

2. Answers may vary. In a right triangle, the sum of the squares of the legs is equal to the square of the hypotenuse.

3. Answers may vary. Square the lengths of the legs, add the squares together and then find the square root of the sum.

4. Answers may vary. Subtract the square of the known leg from the square of the hypotenuse. Then find the square root of the difference.

5. Answers may vary. The diagonal of the square will be the hypotenuse of the right triangle. One of the pairs of sides of the square that form a right angle opposite the diagonal makes the legs of the triangle. Apply the Pythagorean Theorem to determine the length of the diagonal.

Lesson 12.4, pages 598–605

Activity

1. a. (4, 8)
   b. (2, 5)
   c. (2, 3)

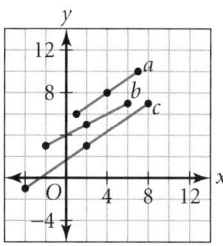

2. a. The lengths of line segments PM and MQ are both $\sqrt{13}$.
   b. The lengths of line segments PM and MQ are both $2\sqrt{5}$.
   c. The lengths of line segments PM and MQ are both $2\sqrt{13}$.

3. Average the x-coordinates.

4. Average the y-coordinates.

CHECKPOINT ✔

5. (27.5, 41.5)

6. $26.50 = 26.50$; $AM = MB$

Exercises
Communicate

1. A point is a position on a coordinate plane. Coordinates are numbers which are used to locate points.

2. The distance between two points can be thought of as the hypotenuse of a right triangle. The $(x_1 - x_2)$ expression defines the length of the horizontal leg and $(y_1 - y_2)$ defines the length of the vertical leg. $(x_1 - x_2)^2 + (y_1 - y_2)^2 = d^2$ would be the form of the Pythagorean Theorem. Taking the square root of both sides of the equation gives the form of the distance formula.

3. Square the distance between the two x-coordinates, 2 and 5: $(2 - 5)^2 = (-3)^2 = 9$.
Repeat for the y-coordinates, 5 and 7: $(5 - 7)^2 = (-2)^2 = 4$.
Take the square root of the sum of the squared differences: $\sqrt{9 + 4} = \sqrt{13}$.
So the distance between (2, 5) and (5, 7) is $\sqrt{13} \approx 3.61$.

4. The x-coordinate of the midpoint is the average of the x-coordinates: $\frac{2 + 5}{2} = \frac{7}{2} = 3.5$.
The y-coordinate of the midpoint is the average of the y-coordinates: $\frac{5 + 7}{2} = \frac{12}{2} = 6$.
So, the midpoint is (3.5, 6).

5. Interchange the *if* and *then* phrases of the statement; the new statement is the converse of the original.

Practice and Apply

44.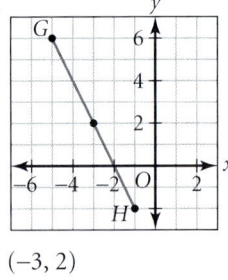
(−3, 2)

45.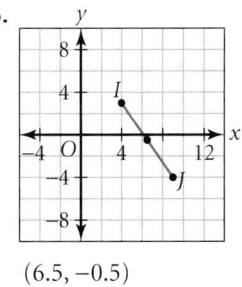
(6.5, −0.5)

Portfolio Activity

1. Check students' drawings.

2.
triangle number	1	2	3	4	5	6	7	8
length of hypotenuse	$\sqrt{2}$	$\sqrt{3}$	$\sqrt{4}$	$\sqrt{5}$	$\sqrt{6}$	$\sqrt{7}$	$\sqrt{8}$	$\sqrt{9}$

3. $\sqrt{11}$

4. $\sqrt{n + 1}$

Lesson 12.5, pages 606–612
Activity

1. 5, 5

3. The diagonals of a rectangle are equal in length.

CHECKPOINT ✔

4. $\sqrt{a^2 + b^2}$

5. Answers may vary. Sample answer: To determine the length of the diagonal of a rectangle, square the lengths of both sides. Add the squares together then take the square root of the sum. Both diagonals will have the same length.

Exercises
Communicate

1. Let r be the common distance, (x, y) the given point and (x_n, y_n) be any of the equidistant points. The distance between the given point and any of the equidistant points can be expressed as $r = \sqrt{(x - x_n)^2 + (y - y_n)^2}$. Square both sides of the equation for $r^2 = (x - x_n)^2 + (y - y_n)^2$, which is the equation of a circle.

2. Answers may vary. For example, algebra can help solve geometry problems by using formulas such as the distance formula or the Pythagorean Theorem.

3. When a line segment joins the midpoints of two sides of a triangle, that segment is parallel to the third side and is exactly half of the length. Answers may vary. Sample answer: Placing one vertex of the triangle at the origin makes the arithmetic easier.

4. Focus and directrix are words that can be used to describe a parabola. A parabola is the set of all points equidistant from a given point, the focus, and a given line, known as the directrix.

Lesson 12.6, pages 613–620
Activity

2. $B'C'$; AC'

3. Measurements will vary. For the triangle in the picture, the ratio is about 1.5.

CHECKPOINT ✔

4. The tangent ratio does not depend on the size of the triangle, only on the measurement of the angle. This is because similar triangles are proportional.

Exercises
Communicate

1. The slope of a line is the ratio of the change in y to the change in x. The tangent function is the ratio of the length of the side opposite an angle (which could be the change in y) to the length of the side adjacent an angle (which could be the change in x).

2. Find the value of tan A using a calculator. Solve the following equation: tan $A \times$ length of the adjacent leg = length of the opposite leg

3. Find the ratio of the length of the opposite leg to the length of the adjacent leg. Then use the Tan^{-1} key of a calculator to find the angle associated with that ratio.

Lesson 12.7, pages 621–627
Activity

1. As the measure of angle A increases, the value of sin A also increases.
   sin 0° = 0   sin 90° = 1

2. As the measure of angle A increases, the value of cos A decreases.
   cos 0° = 1   cos 90° = 0

CHECKPOINT ✔
3. As the measure of an angle increases from 0° to 90°, the value of the sine goes from 0 to 1 and the value of the cosine goes from 1 to 0.

Exercises
Communicate

1. To find the sine of an angle, divide the length of the opposite leg by the length of the hypotenuse.

2. To find the cosine of an angle, divide the length of the adjacent leg by the length of the hypotenuse.

3. Since the 3-inch curb is opposite the 4° incline, we have:
   $\sin 4° = \frac{3 \text{ inches}}{\text{length of ramp}}$
   So, length of ramp $= \frac{3 \text{ inches}}{\sin 4°}$
   $\approx \frac{3 \text{ inches}}{0.0698}$
   ≈ 43 inches
   The ramp should be about 3 feet 7 inches long.

4. In a right triangle, the maximum value of sine and cosine ratios is 1, and the minimum values of sine and cosine ratios is 0.

Portfolio Activity

1. $L^2 = L + 1$
   $L^2 - L - 1 = 0$
   $L = \frac{-b \pm \sqrt{b^2 - 4ac}}{2a}$
   $= \frac{-(-1) \pm \sqrt{(-1)^2 - 4(1)(-1)}}{2(1)}$
   $= \frac{1 + \sqrt{5}}{2}$
   The Golden Ratio is:
   $\frac{W}{L} = \frac{1}{\frac{1+\sqrt{5}}{2}}$
   $= \frac{2}{1+\sqrt{5}}$
   $= 0.618$

2. ratio of width to length: $\frac{W}{L} = \frac{3}{5} = 0.6$
   ratio of length to sum of length and width:
   $\frac{L}{L+W} = \frac{5}{8} = 0.625$
   Because these two ratios are very close, the card is close to having the proportions of a golden rectangle.

3. Answers will vary.

Lesson 12.8, pages 652–659
CHECKPOINT ✔
3×1

Exercises
Communicate

1. $A = \begin{bmatrix} a_{11} & a_{12} & a_{13} & a_{14} & a_{15} \\ a_{21} & a_{22} & a_{23} & a_{24} & a_{25} \\ a_{31} & a_{32} & a_{33} & a_{34} & a_{35} \end{bmatrix}$

2. Yes. It does not matter in which order the matrices are added: $A + B = B + A$.

3. No. Because the row entries of the left matrix are multiplied by the column entries of the right matrix, $A \times B$ does not necessarily equal $B \times A$.

Lesson 13.1, pages 648–653
Exercises
Communicate

1. This is an experimental probability, since it is based on the actual results of the experiment of successful free throws Nat has made this season.

2. No, the probability of all heads or all tails is only $\frac{1}{16} + \frac{1}{16} = \frac{1}{8}$.

3. Physically the sides are all alike, so all have the same chance of landing up.

Portfolio Activity

1.

Sum	2	3	4	5	6	7	8	9	10	11	12
# of ways to Roll Sum	1	2	3	4	5	6	5	4	3	2	1

2. 16   **3.** 20

4. $\frac{4}{9}$ **5.** $\frac{5}{9}$

6. Player B; there are a greater number of ways to roll a favorable outcome for B.

Lesson 13.2, pages 654–660
Activity

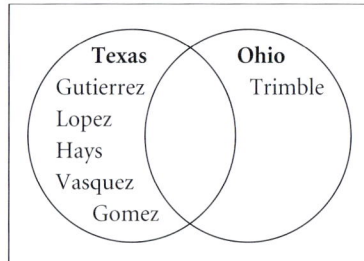

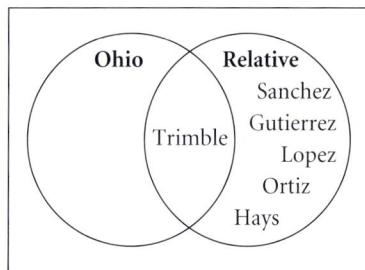

1. 1; 5   **2.** 0; 6

3. 6   **4.** 6; 1

5. a.

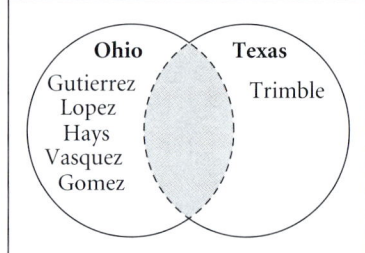

b.

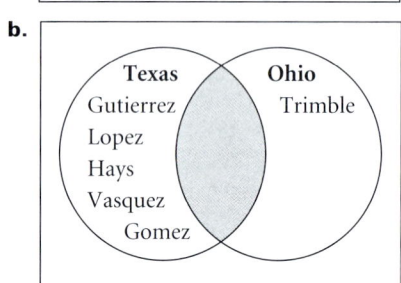

c.

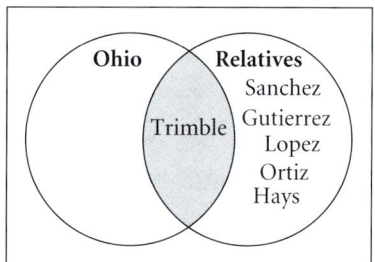

d.

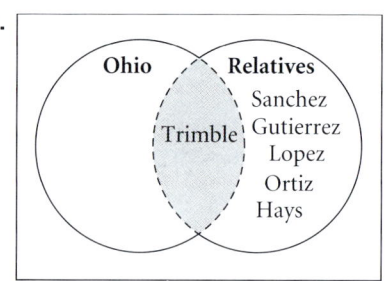

CHECKPOINT ✔

6. To find the number of people who are relatives OR live in Ohio, add the number of people who are relatives, 6, to the number who live in Ohio, 1, and subtract any who overlap. $6 + 1 - 1 = 6$

Exercises
Communicate

1. Answers may vary. Sample answer: Data can be organized by category. Cross referencing the categories can create new data sets with unique characteristics.

2. Answers may vary. Sample answer: Mary Ortiz is a relative not living in Texas.

3. The eight of hearts is counted as an eight and as a heart, and thus is counted twice.

4. 106   **5.** 92

6. 49   **7.** 149

Lesson 13.3, pages 661–666
Activity

1. 2 choices, since he has the same tie options as he did before choosing the shirt.

2. $3 \cdot 2 = 6$ ways to choose a shirt-and-tie combination

CHECKPOINT ✔

3. (shirt choices) • (tie choices) = (choices for combination)

4. $5 \cdot 3 = 15$ ways to choose a shirt-and-tie combination

Exercises
Communicate

1. In the first sentence, you are going to do both objectives. In the second sentence, you are going to do at least one of the two objectives.

2. To find the number of branches, add the number of choices, which is 6 when the coin is heads and 10 when the coin is tails. The number of branches is 6 + 10 = 16.

3. There are 4 ways to draw a 3. There are 26 ways to draw a red card. 4 · 26 = 104

Portfolio Activity

Answers will vary according to experiments. Theoretical probabilities for 1–4 are given below.

1. $\frac{1}{18}$
2. 0
3. $\frac{2}{9}$
4. $\frac{5}{18}$

3. Determining probabilities involves counting all possible outcomes and all favorable outcomes. The Fundamental Counting Principle provides a formula for counting all possible outcomes of independent events.

4. Subtract the number of ways the event could not occur from the total number of possible outcomes. The difference represents the number of ways the event could occur. Divide the difference by the total number of possible outcomes to find the probability of an event occurring.

Portfolio Activity

Answers will vary according to the experiments. Theoretical probabilities for each team are given below.

Team 1
a. $\frac{3}{8}$ b. $\frac{3}{8}$ c. $\frac{1}{8}$

Team 2
a. $\frac{3}{8}$ b. $\frac{3}{8}$ c. $\frac{1}{8}$

Lesson 13.4, pages 667–673
Activity

1.

	Band	Orchestra	Chorus	Dance
Home Economics				
Woodworking				
Auto Shop				

2. 12
3. $\frac{1}{4}$; $\frac{1}{3}$; $\frac{1}{4} \cdot \frac{1}{3} = \frac{1}{12}$
4. Multiply
5. $\frac{1}{4}$; $\frac{1}{4}$; $\frac{1}{4} + \frac{1}{4} = \frac{1}{2}$
6. Add

CHECKPOINT ✔

7. Multiplication is appropriate when we are dealing with intersections, or AND. Addition is appropriate when we are dealing with union, or OR.

Exercises
Communicate

1. Two events are independent if the occurrence of one does not affect the probability that the other will occur.

2. The probabilities of independent events can be represented with an area model or a tree diagram.

Lesson 13.5, pages 676–681
Activity

1. Students should record the results of 20 coin tosses.

2. Students should highlight any sequences of at least four consecutive heads.

3. Any trial containing a sequence of four consecutive heads represents making four baskets in a row.

4. Students should conduct several trials and divide the total number of trials with four consecutive heads by the total number of trials.

CHECKPOINT ✔

5. Students should conclude the experimental probability of four consecutive baskets from step 4.

Exercises
Communicate

1. a. Choose a way to generate random numbers.
   b. Decide how to simulate one trial of the experiment.
   c. Carry out a large number of trials and record the results.

2. Answers may vary. Sample answer: A coin toss; heads is a shot made, tails is a shot missed. A random number generator; 1 is a shot made, 0 is a shot missed. A number cube; an even number is a shot made, an odd number is a shot missed.

3. Answers may vary. Sample answer: Let the numbers 1, 2, and 3 represent the occurrence of rain, and the numbers 4, 5, 6, 7, 8, 9, and 10 represent no rain. Drawing one slip is a trial.

Additional Answers

4. Answers may vary. For example: hitting a nail on the head with a hammer, kicking a football through the goal post, and getting an answer on the phone when you call.

5. Answers may vary. Sample answer: Use 5 different-colored slips of paper, where each color represents a different answer.

6. Answers may vary. Sample answer: Generate random positive integers from 1 to 10 on a calculator. Let the number 1 represent a late flight, and the numbers 2 to 10 represent a flight on time. Generate 3 numbers. Let this represent 1 trial. Repeat for 20 trials. Divide the number of trials where all 3 numbers are greater than 1 by 20. The quotient is the experimental probability.

Guided Skills Practice

21. Use a calculator to generate the numbers 1 and 2; use the command INT (RAND*2) + 1. Let 1 represent a boy and 2 represent a girl. Generate 4 random numbers; record the result. This represents 1 trial. Repeat for 10 trials. Divide the number of trials where there are 2 boys and 2 girls by 10. The quotient is the experimental probability. Answers may vary.

22. Use a calculator to generate the numbers 0 and 1; use the command INT (RAND*2). Let 0 represent an incorrect response and 1 represent a correct response. Generate 3 random numbers; record the result. This represents 1 trial. Repeat for 10 trials. Divide the number of trials where all three responses are correct by 10. The quotient is the experimental probability. Answers may vary.

Look Back

27. $\begin{bmatrix} 1 & -5 \\ 12 & -26 \end{bmatrix}; \begin{bmatrix} -22 & -2 \\ -16 & -3 \end{bmatrix}$
No

Look Beyond

28.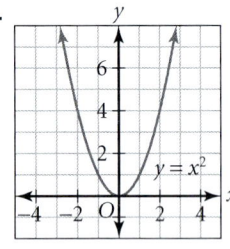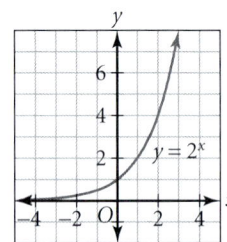

$y = x^2$ is a quadratic equation and its graph is a parabola.
$y = 2^x$ is an exponential function whose curve increases very rapidly on the right.

Portfolio Activity

Answers will vary according to experiments.

Lesson 14.1, pages 692–699

Activity

1. a. 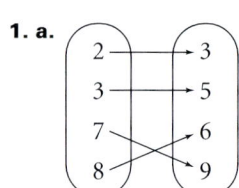 b.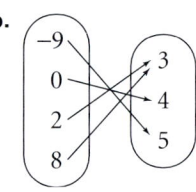

   c. 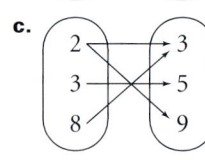 d.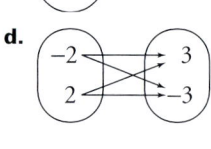

2. a. yes   b. yes   c. no   d. no

3. a. 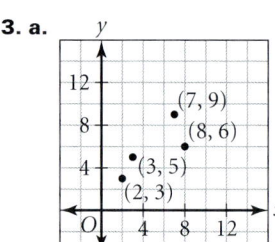 b.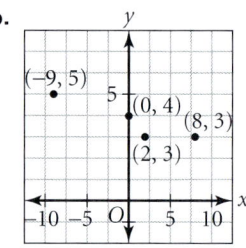

   c. 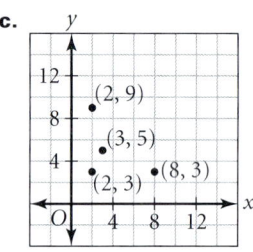 d.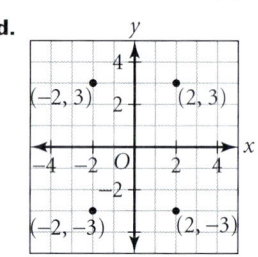

4. Answers will vary. It is impossible to draw a vertical line that touches graphs **3a** and **3b** in more than one place; it is possible for graphs **3c** and **3d**.

CHECKPOINT ✔

5. A vertical line intersects the graph of a function only once but may intersect the graph of a relation more than once. If an element of the domain is paired with more than one element of the range, a vertical line can be drawn through more than one point on the graph of the relation.

Exercises
Communicate

1. If a vertical line is overlaid anywhere on a graph, the graph represents a function if the line does not touch 2 or more points on the graph. The vertical-line test is used to determine whether any *x*-value is associated with more than one *y*-value. If so, the relation between *x* and *y* is not a function.

2. The integers in the restricted range are −2, −1, 0, 1, and 2. Substitute each of these values in the given function to find the related range value. The range will include each value (without duplication).

864

Look Beyond

71. a. $f(x) = |x|$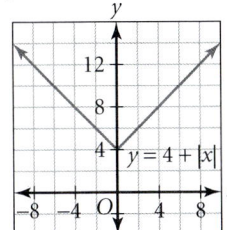

b. $g(x) = x$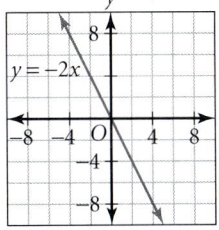

c. $p(s) = \frac{1}{s}$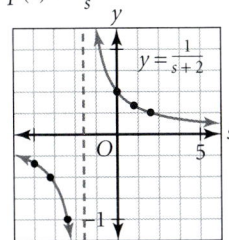

d. $h(t) = t^2$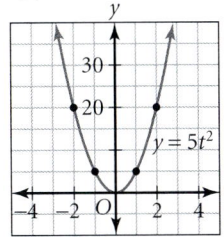

e. $z(u) = \frac{1}{u}$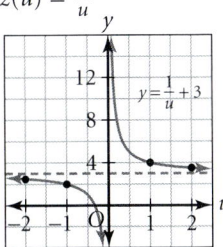

f. $f(r) = |r|$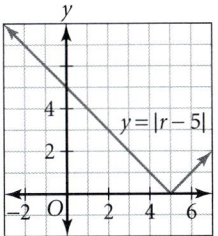

Lesson 14.2, pages 700–706

Activity

1.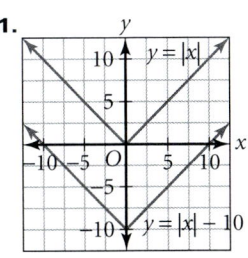

2. a. down **b.** 10 units

3.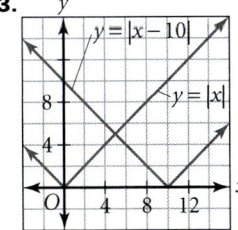

CHECKPOINT ✔

4. a. right **b.** 10 units

5. The graph of g shifted the graph of f vertically and the graph of h shifted it horizontally.

Exercises

Communicate

1. A translation is a shift of the graph to the left or to the right or up or down.

2. The variable h affects the horizontal translation. If h is positive, the graph moves h units to the right; if it is negative, it moves h units to the left.

3. The + 5 indicates that the graph will be translated up 5 units.

Guided Skills Practice

12. a. $b = \frac{c-20}{12}$; $b = \frac{c-30}{12}$

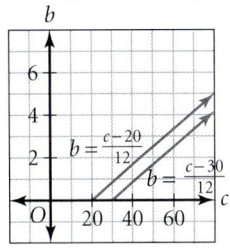

b. The second function represents a horizontal translation of the first function 10 units to the right.

Practice and Apply

13. b. The graph of $y = (x - 6)^2 - 1$ is a translation 6 units right and 1 unit down.

c.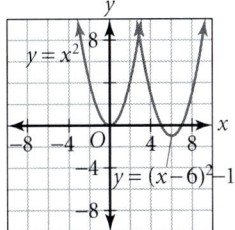

14. b. The graph of $y = x^2 - 6$ is a translation 6 units down.

c.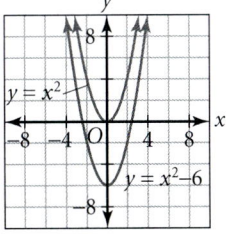

15. b. The graph of $y = |x| - 1$ is a translation 1 unit down.

c.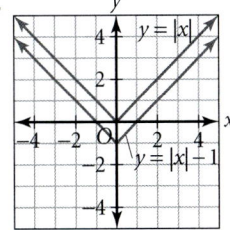

865

16. b. The graph of $y = |x - 1|$ is a translation 1 unit right.

c.

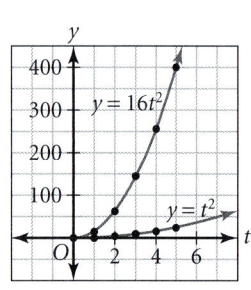

17. Because the parent graph is translated 5 units up, the *y*-value of the new coordinate, 14, is 5 units larger than that of the old coordinate, 9.

18. Because the parent graph is translated 5 units down, the *y*-value of the new coordinate, 4, is 5 units less than that of the old coordinate, 9.

19. Because the parent graph is translated 5 units left, the *x*-value of the new coordinate, −2, is 5 units less than that of the old coordinate, 3.

20. Because the parent graph is translated 5 units right, the *x*-value of the new coordinate, 8, is 5 units larger than that of the old coordinate, 3.

Look Back

33. b. The mean is higher; the highest rent is so high that it raises the mean higher than the median.

Lesson 14.3, pages 707–713

Activity

1–2.

t	t^2	$16t^2$
0	0	0
1	1	16
2	4	64
3	9	144
4	16	256
5	25	400

3. The value for function *j* is 16 times the value of function *h*.

4.

x	−3	−2	−1	0	1	2	3	4	5	6	7
$|x|$	3	2	1	0	1	2	3	4	5	6	7
$10|x|$	30	20	10	0	10	20	30	40	50	60	70

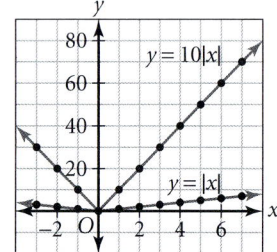

CHECKPOINT ✓

5. For each domain value, the range value increases when the function is multiplied by a number greater than 1. This "stretches" each *y*-value vertically away from the *x*-axis.

Exercises

Communicate

1. As *a* increases from 1 to 4, the graph of $y = ax^2$ stretches the graph of $y = x^2$ vertically (each point moves up away from the *x*-axis). As *a* decreases from 1 to 0, it compresses the graph of $y = x^2$ vertically (each point moves down towards the *x*-axis).

2. As *b* increases from 1 to 4, the graph of $y = 2^{bx}$ compresses the graph of $y = 2^x$ horizontally by a factor of $\frac{1}{b}$ (for any given *y*-value, the associated *x*-value becomes smaller). As *b* decreases from 1 to 0, it stretches the graph of $y = 2^x$ horizontally (for any given *y*-value, the associated *x*-value becomes larger).

3. Given any point (x, y) on the graph of *f*, the corresponding point on the graph of *g* will be equal to $(x, \frac{1}{2}y)$.

4. Given any point (x, y) on the graph of *f*, the corresponding point on the graph of *h* will be equal to $(\frac{1}{2}x, y)$.

Practice and Apply

29. $y = |x|$; horizontal compression

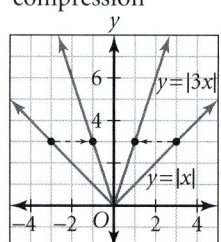

30. $y = x^2$; vertical stretch

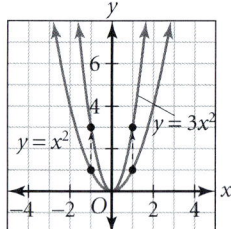

31. $y = x^2$; horizontal stretch

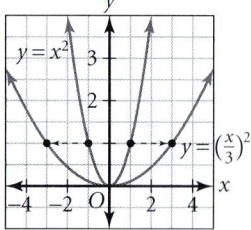

32.

33.

34.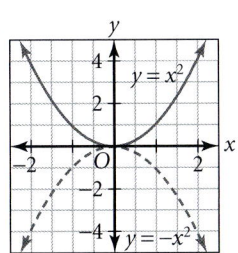

Lesson 14.4, pages 714–719

Activity

1.
x	-2	-1	0	1	2
x^2	4	1	0	1	4

2. **3.**

4.
x	-2	-1	0	1	2
$-x^2$	-4	-1	0	-1	-4

CHECKPOINT ✓

5. $f(2) = -4$; -9
 a. -9 **b.** 0 **c.** -9 **d.** $-a^2$

6. $y = -x^2$

Exercises
Communicate

1. The negative sign reflects the graph of $y = ax^2$ over the x-axis.

2. To reflect h over the y-axis, substitute $-x$ for x in the original function rule: $h(x) = 2\sqrt{-x} + 1$.

3. a. Multiply $2x$ by -1: $f(x) = -(2x)$.
 b. Substitute $-x$ for x: $f(x) = 2(-x)$.

4. When observed on as a photograph, the reflection is like the reflections studied in the lesson. The shoreline of the lake acts as an x-axis. When observed live, the effect is three-dimensional, and the appearance of a reflection across the x-axis is an illusion.

Guided Skills Practice

8. reflection across the y-axis;
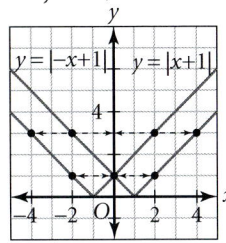

9. reflection across the x-axis;
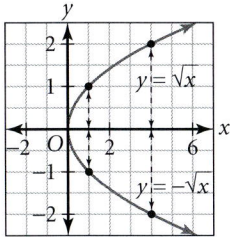

10. reflection across the x-axis;
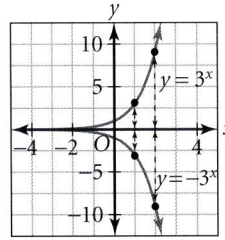

11. reflection across the y-axis (graphs are identical);
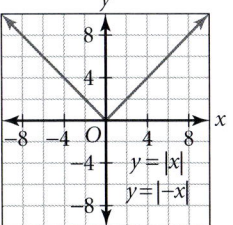

12. reflection across the y-axis;
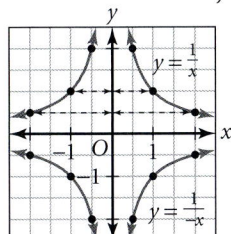

Practice and Apply

13. reflection across the y-axis;
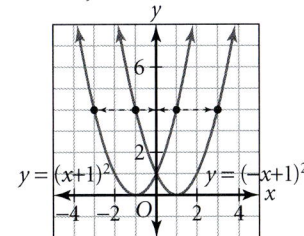

14. reflection across the y-axis;
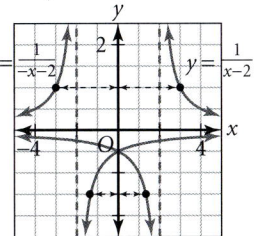

15. reflection across the x-axis;
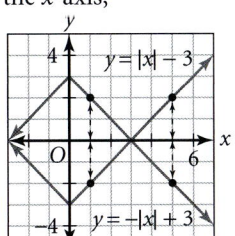

16. reflection across the x-axis;
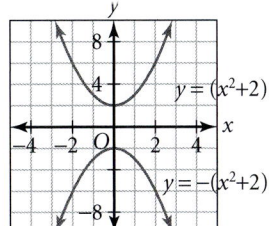

Lesson 14.5, pages 722–727

Activity

1–2.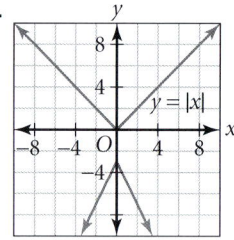

3.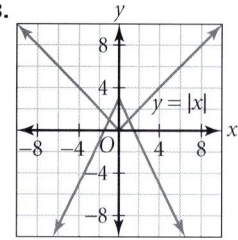

4.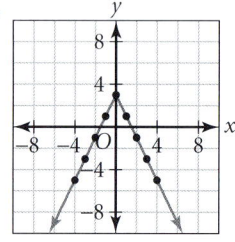

The graph of g matches the graph in Step 3. Yes, order appears to be important.

CHECKPOINT ✔

5. Translations are done after stretches and reflections.

Exercises

Communicate

1. Two is the number indicating a translation because it is the number subtracted from h.

2. 100 is the number indicating a stretch, because it is a multiplier.

3. **a.** When $(-x)$ is squared, the negative sign disappears: $(-x)^2 = x^2$. Graphically, because the graph of x^2 is symmetric about the y-axis, it coincides with its reflection.
   b. When $x - 3$ is multiplied by -1, the result is $-(x - 3)$ or $3 - x$. By definition, $|x| = |-x|$, so $|x - 3| = |3 - x|$. Just as with x^2, the graph of $|x|$ coincides with its reflection, $|-x|$.

4. Rewrite the function as $y = -a[-b(x + \frac{h}{b})]^2 + k$. Start the transformations from the inner-most grouping symbol: (1.) Translate left by $\frac{h}{b}$ units. (2.) Horizontally stretch or compress by a factor of $\frac{1}{b}$. (3.) Ignore the negative sign before b, because $(-x)^2 = x^2$. (4.) Vertically stretch by a factor of a. (5.) Reflect over the x-axis. (6.) Translate up by k units.

Guided Skills Practice

9. Rewrite as $f(g) = -3\left[\dfrac{1}{2\left(g + \frac{1}{2}\right)}\right] + 2$. Parent function: $f(g) = \frac{1}{g}$. Translate $\frac{1}{2}$ unit left; horizontally compress by a factor of $\frac{1}{2}$; vertically stretch by a factor of 3; reflect over the x-axis; translate 2 units up.

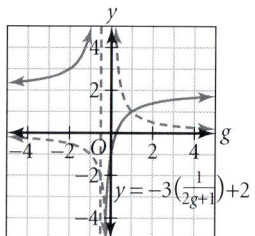

10. Rewrite as $f(x) = 2\left[\dfrac{1}{2(x-2)}\right] - 6$. Parent function: $f(x) = \frac{1}{x}$. Translate 2 units right; horizontally compress by a factor of $\frac{1}{2}$; vertically stretch by a factor of 2; translate 6 units down.

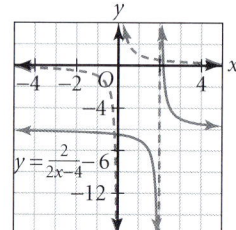

Practice and Apply

11.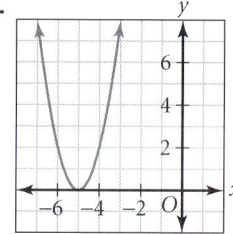

12.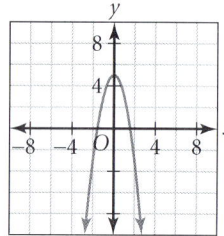

13.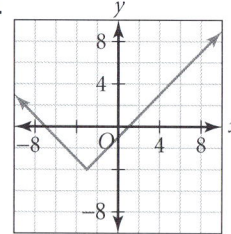

14.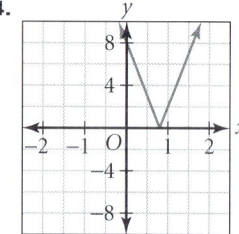

15.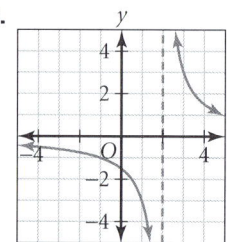

16.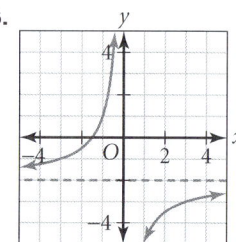

17.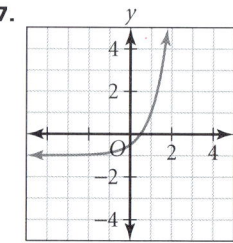

18.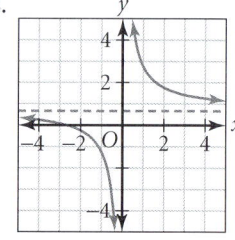

19.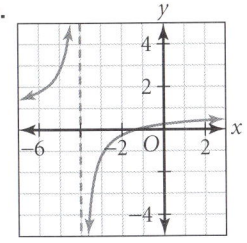

Look Back

30.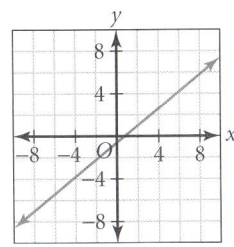

31. $y = \frac{5}{16}x + 7\frac{15}{16}$, or $\frac{5}{16}x - y = -7\frac{15}{16}$

32. $-12 \leq x \leq 22$;

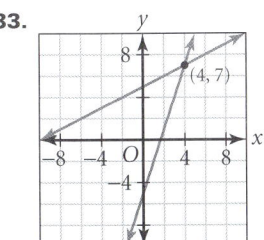

33.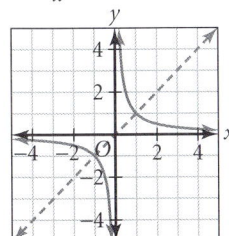

Look Beyond

37. $y = \frac{1}{x}$ is its own reflection across the line $y = x$.

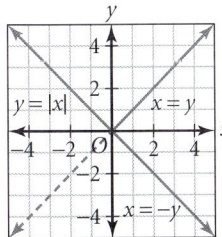

38. 39.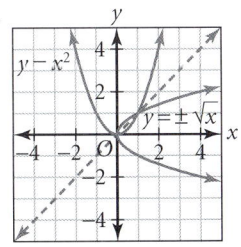

Portfolio Activity

1.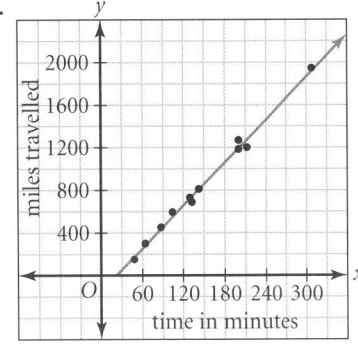

2. **a.** The line of best fit is given approximately by $y = 6.7t - 148.4$.

   b. The slope is approximately 6.7. This means that the planes average about 6.7 miles per minute.

   c. The equation indicates a y-intercept at about -148.4. However, because the y-values indicate the number of miles traveled, the y-values cannot actually be less than 0 in this situation. Therefore, the line of best fit will end at the x-axis, or 0 miles.

   d. The line of best fit meets the x-axis at about 22 (minutes). Once the plane leaves the gate, it must taxi down the runway and wait to be cleared for take-off. This time is included as part of the flight time even though no miles towards the destination have been traveled.

Chapter 14, Review and Assessment

18. $y = x^2$; horizontal translation;

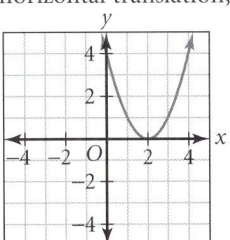

19. $y = 2^x$ horizontal translation;

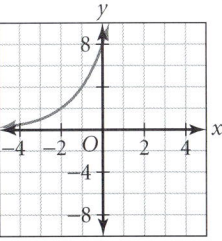

20. 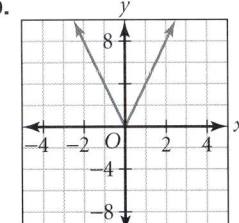 21.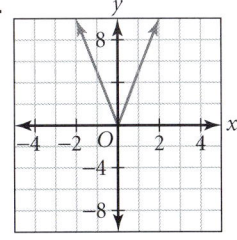

869

22.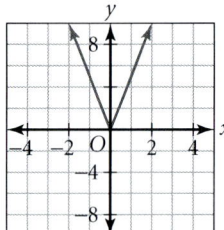

23.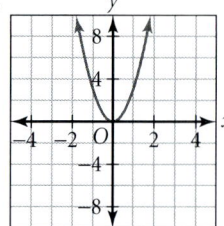

24.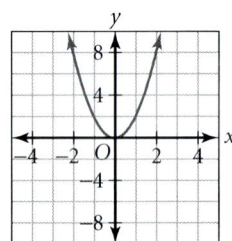

25.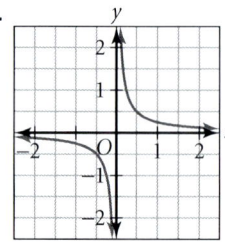

26.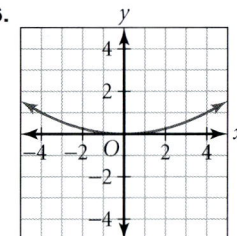

27.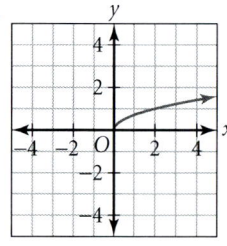

35.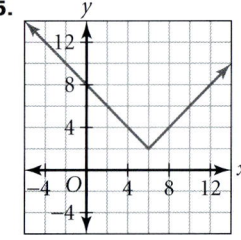

horizontal translation 6 units right, vertical translation 2 units up

36.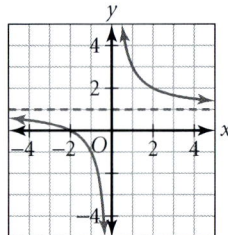

horizontal compression by a factor of $\frac{1}{2}$ (or vertical stretch by a factor of 2), reflection across the x-axis (or across the y-axis), vertical translation 1 unit up

37.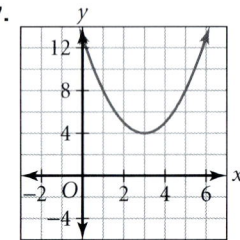

horizontal translation 3 units right, vertical translation 4 units up

38.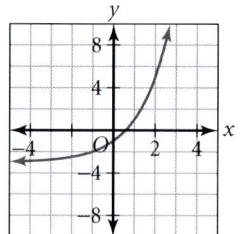

horizontal translation 1 unit left, vertical translation 3 units down

39.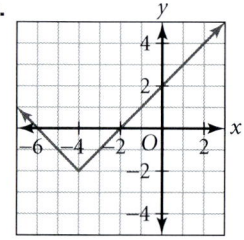

horizontal translation 4 units left, vertical translation 2 units down

40.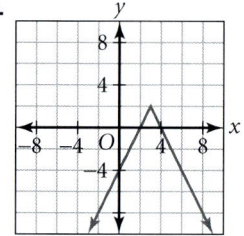

horizontal translation 3 units right, horizontal stretch by a factor of 3, vertical stretch by a factor of 6, reflection across the x-axis, vertical translation 2 units up

Chapter 14 Alternative Assessment

Portfolio Projects

2. a.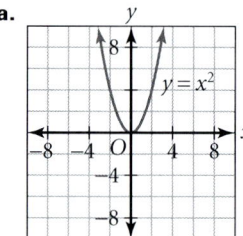

b.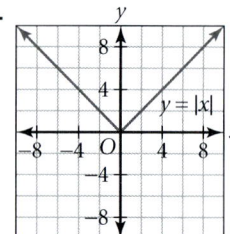

c.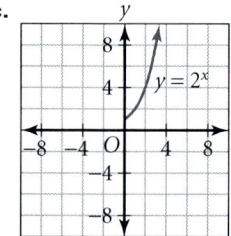

d.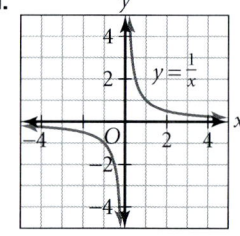

Each graph is vertically translated a units up.

INDEX

A

Absolute error, 300, 302
Absolute value, 295
   error and, 300-301
   graphing, 295-297, 304
   opposites and, 57-58
   in subtraction, 69
Absolute-value equations, 300-301, 304
Absolute-value functions, 294-297, 696
Absolute-value inequalities, 302-303
Acceleration, 486, 497
Activities
   Basketball Simulation, 677
   Completing the Square With Tiles, 493
   Defining x^0, 391
   Deriving the Midpoint Formula, 601
   Discovering a Rule for Quotients of Powers, 384
   Exploring Approximate Solutions, 321
   Exploring Combining Like Terms, 90
   Exploring Differences, 4
   Exploring Graphs of Linear Functions, 246
   Exploring Inverse Variation, 526
   Exploring Mental Computation, 83
   Exploring Multiplication of Polynomials, 432
   Exploring No Solution and Many Solutions, 338
   Exploring Powers of 10, 399
   Exploring Rational Functions, 538
   Exploring Relations and Functions, 219
   Exploring Relationships in Proportions, 165
   Exploring Right Triangles, 592
   Exploring Sines and Cosines, 622
   Exploring Slopes and Perpendicular Lines, 269
   Exploring Substitution, 326
   Exploring the Graphs of Direct Variations, 239
   Exploring the Order of Operations, 18
   Exploring the Zeros of a Function, 464
   Exploring Using Opposites, 332
   Exploring Volume and Surface Area, 444
   Finding Solutions to Multiplication Equations, 122
   Fitting a Line to a Point, 229
   Graphing a Rational Function, 533
   Graphing Data, 26
   Graphing to Find Solutions, 304
   Guessing and Checking to Solve a Problem, 552
   Inductive Reasoning, 559
   Interpreting a Box-and-Whisker Plot, 203
   Manipulating the Absolute-Value Function, 296
   Misleading Graphs, 194
   Modeling a Linear Inequality With a Graph, 312
   Modeling Addition-Unlike Signs, 62
   Modeling and Equation With a Graph, 31
   Modeling Data With a Table and an Equation, 12
   Modeling Equations, 135
   Modeling the Addition Property of Equality, 116
   Multiplication and Division Patterns, 73
   Multiplying and Dividing Inequalities, 282
   Multiplying Powers, 371
   Operations and Radical Expressions, 578
   Order of Vertical Transformations, 723
   Package Design, 500
   Patterns in Differences of Two Squares, 420
   Polynomial Addition and Subtraction, 454
   Population Growth, 405
   Production and Profit, 511
   Raising a Power to a Power, 378
   Reflection Across the x-axis, 714
   Sets of Choices, 662
   Tables for Falling Objects, 487
   Terminating and Repeating Decimals, 55
   The Diagonals of a Rectangle, 607
   The Distributive Property and Algebra Tiles, 439
   The Tangent Ratio in Similar Triangles, 614
   Transformations of $y = x^2$, 481
   Translations, 700
   Two Probability Experiments, 181
   Using a Model, 668
   Using a Table 129
   Using a Table to solve for x, 141
   Using Frequency Tables, 189
   Venn Diagrams for AND and OR, 656
   Vertical-Line Test, 693
   Vertically Stretching a Function, 707
   Working With Trends in Data, 38
Addition
   associative property, 81, 84, 563
   closure property, 10, 562-563
   commutative property, 80-81, 84-85, 563
   distributive property, 82-84, 563
   of expressions, 89-93
   of integers, 61-64
   identity property, 64, 84, 563
   inequalities, 277-278
   inverse property, 64, 84, 563
   integers, 61-64
   matrix, 630
   polynomials, 427-428
   property of equality, 116
   rules for, 62-64
   solving equations by, 116-117
   subtraction as, 68-69
Addition of Probabilities Principle, 657
Addition Property of Equality, 116
Addition Property of Inequality, 277
Additive identity, 64, 84
Additive inverse, 64, 84
Address, Matrix 629
Age puzzles, 353-354
Agnesi, Maria Gaetana, 408
Agnesi curve, 408
Algebraic expressions (*See also* Rational expressions)
   addition, 89-93
   division, 98-100
   multiplication, 96-100
   overview, 11-17
   radical, 577-580
   subtraction, 91-93
Algebraic order of operations, 18-20
Anagrams, 665
AND operator, 654-656
Angle measures, 615-617, 622-623
   (*See also* Cosine; Sine; Tangent)
Angle of depression, 617, 637

871

Index

Angle of elevation, 619, 637
Annulus, 451
Applications
    business and economics
        accounting, 87, 510
        architecture, 706
        banking, 65, 72, 73, 78, 312
        business, 28, 60, 78, 102, 127, 135, 136, 336, 359, 511, 732
        construction, 442, 456, 468, 619, 626, 642
        consumer economics, 16, 31, 36, 59, 120, 131, 133, 134, 141, 158, 169, 212, 277, 279, 285, 286, 312, 318, 665
        consumer math, 336, 536
        discounts, 173, 177, 436, 543
        economics, 41, 195, 389, 402
        finance, 336, 345, 420, 686
        income, 86, 87, 102, 240, 241, 242, 270, 336, 344, 349, 597, 713
        inventory, 72, 80, 82, 87, 93, 167, 663
        investments, 148, 195, 280, 336, 411, 413, 529, 531, 660
        manufacturing, 93, 300, 307
        marketing, 686
        packaging, 389, 426, 443, 444, 445, 446, 726
        sales, 293, 336, 343
        sales tax, 174, 176, 177, 463
        small business, 146, 287, 328, 557
    language arts
        communicate, 8, 15, 21, 27, 34, 40, 58, 64, 70, 77, 85, 92, 101, 118, 126, 132, 138, 144, 151, 168, 175, 183, 190, 198, 205, 223, 230, 240, 249, 256, 261, 279, 286, 292, 297, 305, 323, 329, 334, 342, 350, 357, 374, 381, 387, 393, 401, 407, 413, 429, 435, 441, 446, 450, 455, 467, 484, 489, 495, 502, 514, 529, 535, 541, 549, 555, 563, 580, 588, 595, 602, 610, 618, 624, 633, 651, 658, 664, 671, 679, 697, 705, 711, 717, 725
        language arts, 652, 665
        The Tempest (Shakespeare), 345
    life skills
        academics, 351, 666, 673
        budget, 251
        carpentry, 122
        clothing, 662
        communication, 133
        comparing data, 212
        cooking, 169
        education, 212
        emergency services, 598
        fire prevention, 199
        food, 126
        framing, 440
        health, 33, 36, 242, 306
        interpersonal skills, 654
        home improvement, 16, 18, 22, 139
        interior decorating, 431
        landscaping, 152, 474, 582, 596
        personal finance, 67, 114, 120, 410
        rental fees, 103, 331, 333
        rentals, 241
        savings, 312
        scheduling, 661, 667, 669
        transportation, 34, 222, 237, 628, 642
    other
        air travel, 290
        auditorium seating, 280
        calendar, 121
        dogs, 199
        fund-raising, 36, 240, 251, 253, 257, 330, 352
        gifts, 36
        marching band, 503
        park design, 566-567
        pool maintenance, 289
        Titanic steamship, 94-95
        Victoria Clipper IV, 31
    science and technology
        archaeology, 409-410, 413, 414
        astronomy, 398, 399, 400, 402, 403, 420, 619, 652
        aviation, 236, 242, 325, 354, 359, 582, 604, 616, 626
        biology, 168, 203-204, 206, 372, 374, 375, 383, 542
        chemistry, 65, 299, 356, 358, 359, 364, 666
        civil engineering, 620, 626
        ecology, 14-15, 38, 133, 251, 485, 551, 619
        engineering, 609
        health, 33, 36, 175, 200, 218, 242, 306
        mechanics, 531
        meteorology, 4, 23, 37, 39-40, 70, 79, 88, 171, 270, 678, 680, 699
        music, 43, 531
        navigation, 70
        optics, 513
        physics, 6, 9, 148, 228, 232, 239, 240, 242, 258, 262, 400, 451, 464, 480, 483, 485, 486, 488, 489, 497, 515, 528, 531, 584, 585, 588, 589, 718
        science, 732
        technology, 8, 9, 14, 15, 18, 19, 31, 33, 38, 77, 132, 137, 185, 221, 229, 237, 246, 254, 265, 285, 294, 296, 304, 319, 320, 321, 338, 339, 340, 341, 342, 343, 346, 361, 364, 372, 382, 401, 417, 444, 445, 465, 466, 467, 501, 533, 535, 554, 586, 615, 622, 623, 636, 637, 676, 678, 700, 707
        telecommunications, 10, 34, 665
        temperature, 140, 147, 281
        weather, 293
    social studies
        demographics, 225, 340, 404, 405, 407, 408, 412, 413
        geography, 199, 220, 652
        government, 307
        Parthenon, 591
        political science, 177
        Rhind papyrus, 121, 216
        social studies, 402
        student government, 146
    sports and leisure
        art, 713
        auto racing, 326
        contests, 34, 565, 673
        crafts, 302
        entertainment, 16, 36, 88, 225
        fine arts, 458
        fitness, 570
        fireworks, 480, 483
        games, 184, 657, 660, 665, 672
        gardening, 140, 350
        hobbies, 312, 322, 620
        photography, 503
        recreation, 10, 11, 13, 15, 26, 28, 36, 83, 99-100, 192, 221, 227, 242, 330, 336, 358, 382, 546, 556, 594, 621, 623, 624, 672, 722
        sports, 42, 120, 124, 164, 187, 189, 190, 244, 245, 246, 351, 447, 520, 596, 617, 635, 642, 676, 677, 680, 726
        theater, 152, 287
        travel, 31, 32, 35, 120, 127, 152, 169, 207, 270, 287, 299, 531, 570, 673, 699
        wave cheers, 45
Approximate solution, 321
Area
    annulus, 451
    rectangles, 102
    trapezoids, 150
Associative properties, 81-82, 84
Associative Property of Addition, 81, 84
Associative Property of Multiplication, 81, 84
Astrolabe, 52
Asymptotes, 534
Averages

mean, 186-187
midpoint coordinates, 602
speed, 12, 264
Axioms, 558
Axis of symmetry, 481-482

Banneker, Benjamin, 537
Bar graphs, 193-195
Base (of an exponential expression), 97, 370
Bilateral symmetry, 718
Binomials, 438-439, 449, 492
Boundary lines, 346-347
Box-and-whisker plots, 203-204
Braces, 19, 319, 692
Brackets, 19

Carbon-14 dating, 409
Carolus Linnaeus, 54
Cartesian coordinates, 24-26
Celsius scale, 258
Center (of a circle), 602, 606
Central tendency, measures of, 186-189
Chaos theory, 720-721
Chemical-solution puzzles, 356-357
Circle graphs, 193-194, 196-197
Circle, 606
Circles, 150, 200, 193-194, 196-197, 606-607
Circumference, 150, 200
Closure property, 70, 562-563
Coefficients, 89, 373
Coin puzzles, 356
Coin toss, 181, 182, 648, 649
Combining like terms, 89
Common binomial factor, 449
Common monomial factor, 448
Common-denominators, 547-549
Commutative properties, 80-81, 84, 91, 563
Commutative Property of Addition, 80
Commutative Property of Multiplication, 80
Complement, 670
Completing the square, 492-495, 499, 506-507
Composite functions, 728
Compound inequality, 289
Compound interest, 410, 415, 529
Compressions, 707-711
Conclusion, 558
Conditional statements, 558
Cones, 473
Conjecture, 5, 559-560
Conjunction, 290

Connections
coordinate geometry, 25, 26, 28, 35, 59, 224, 251, 257, 296, 304, 343, 382, 403, 489, 536, 550, 604, 666, 681, 699
geometry, 7, 9, 23, 87, 93, 102, 108, 119, 139, 149, 150, 152, 167, 168, 169, 145, 170, 232, 241, 259, 262, 287, 329, 330, 335, 343, 364, 373, 374, 375, 380, 382, 385, 388, 428, 431, 436, 441, 442, 446, 449, 451, 456, 463, 503, 510, 520, 530, 541, 556, 570, 582, 592, 594, 599, 596, 597, 602, 608, 619, 627, 653, 660, 665, 706, 719, 726
maximum/minimum, 301, 511
number theory, 330, 343, 359, 395, 485,
patterns in data, 71, 340
probability, 184, 394, 414, 469, 635
statistics, 22, 35, 65, 71, 78, 117, 119, 138, 139, 177, 190, 196, 294, 324, 457, 565, 634, 659, 660, 712, 719
transformations, 298, 537
trigonometry, 616, 618, 620, 624
Consistent systems, 304-310
Constant, 89
Constant difference, 5-6, 30-31, 516
Constant of variation, 238
Constraint inequalities, 361
Converse (of a theorem), 558, 591-592, 600
Converse of the Pythagorean Theorem, 600
Coordinate plane, 24
Coordinate systems, 24-29
Coordinates, 25
Correlations, 38-40, 263, 265
Cosine, 620, 621-623
Cost comparisons, 318-319, 566
Cost-profit calculations, 66, 360-361
Counterexample, 5, 561
Counting Elements of Sets, 656
Counting numbers, 54
Counting principle, fundamental, 661-664, 670
Critical Thinking, 14, 30, 33, 40, 56, 81, 83, 97, 99, 125, 142, 166, 189, 195, 221, 229, 230, 246, 247, 254, 301, 303, 320, 328, 341, 349, 372, 373, 386, 391, 392, 404, 406, 440, 454, 459, 460, 461, 498, 499, 508, 512, 532, 534, 540, 541, 547, 549, 553, 562, 579, 593, 594, 600, 664, 678, 703, 715, 723
Cross products, 165-166

Cross Products Property, 165
Cubic equations, 427
Cultural Connections
Africa, 23, 134, 251, 337, 456, 506, 527, 605
Americas, 233, 537, 652
Asia, 60, 330, 718
Europe, 24, 29, 395, 408, 470
Cylinder volume, 241, 445, 541

Databases, 654
Date notation, 252
Decimals, 172-175
Deductive reasoning, 559
Degree of polynomial, 426
Density, 148
Dependent system, 341
Dependent variables, 32, 222, 245
Descartes, René, 24
Diaconis, Persi, 396
Diagonals, 607, 632
Difference
constant, 5-6, 30-31, 516
finite, 516-517
subtraction, 62
of two squares, 453-454
Difference of two squares, 453-454, 462
Diophantus of Alexandria, 153
Direct variation, 238-239
Directrix of a parabola, 609
Discount price, 173, 175
Discriminant, 508
Disjoint sets, 657-658
Disjunction, 290
Disorder, 720-721
Distance
and absolute value, 69, 295
distance formula, 599-602, 606
graphing, 26, 295
linear functions, 31-33, 222
number lines, 69
Distance formula, 622-629
Distributive property
addition, 82-84, 563
binomials, 438-439
multiplication and division, 82-83, 98-99
overview, 82-84, 90, 563
subtraction, 82
use in solving equations, 141-144
Dividing an expression, 98-100
Division
of expressions, 98-101
inequalities, 282-285
of real numbers, 73-76
Quotient-of-Powers Property, 383-384
solving equations using, 123
square roots, 580

873

by zero, 76-77
Division Property of Equality, 123
Division Property of Inequality, 283
Division Property of Square Roots, 580
Domains, 219-223, 296, 534, 583, 615, 692-693
Doubling rates, 372, 529
Draw a diagram strategy, 149, 609, 616
Draw a graph strategy, 412

Elimination method, 331-334
Entry, 629
Endpoints, 278
Equality,
  properties of, 84-85, 115-117, 123-124
  matrix, 628
Equally likely outcomes, 648
Equation, 12
Equation method (for solving proportions), 172-173
Equations (*See also* Linear equations; Quadratic functions; Solutions to equations)
  absolute value, 301-302, 304
  circle, 606-607
  consistent and inconsistent, systems of, 338-341
  cubic, 327
  from data patterns to, 30-31
  degree of, 426
  literal, 147-150
  multistep, 135-138
  percent, 171-175
  systems of, 318-322, 326-328, 331-334, 338-341
  in two variables, 220-223
Error (*See* absolute error)
Even numbers, 560-561
Expanded form, 470
Experimental design, 208-209
Experimental probability, 181-182, 676-679
Exponential functions
  applications, 409-412
  decay, 411
  general formula for growth, 405, 410, 411
  overview, 404-406
  parent, 696
Exponential decay, 411
Exponential function, 404
Exponential growth, 411
Exponents, 368-423
  definition, 370
  dividing monomials, 385-386
  multiplying monomials, 373
  negative and zero, 390-392
  overview, 23, 97
  powers and products, 377-380
  summary of, 392
Expressions,
  adding and subtracting, 91
Extraneous solution, 586
Extremes, 165

Factoring,
  algebra tiles and, 448-449
  difference of two squares, 453-454, 462
  grouping, 449, 462
  perfect-square trinomials, 452-453, 462
  quadratic equations, 483, 499, 501
  solving equations, 464-467
  working backward, 460-461, 462, 509
Factors,
  common, 448-449
  definition, 448
  greatest common (GCF), 448, 462
  scale, 708-711, 723-724
  scientific notation, 398
  trinomial, 561
Fahrenheit scale, 258
False positives/negatives, 544-545
Favorable outcome, 181, 649
Feasibility region, 361
Fermat, Pierre de, 395
Fermat numbers, 395
Fibonacci, Leonard, 29
Fibonacci numbers, 29, 44
Fields, 563
Finite differences, 516-517
Focus of a parabola, 609
FOIL method, 439-440, 460-461
Formulas, 147-150
Fraction bar, 19
Fractions (*See also* Rational expressions; Rational functions)
  in equations, 137
  percent, 171-175
  powers of, 383-386
Frequency tables, 188-190
Function, 219
Function notation, 692, 694-695
Function rule, 694
Functions (*See also* Exponential functions; Linear functions; Quadratic functions; Rational functions)
  absolute value, 294-297, 696
  composite, 728
  greatest integer, 309
  maximum value, 481
  minimum value, 480
  notation, 223, 694-696
  parent, 696
  piecewise, 281, 288
  polynomials, 443-445, 465
  rational, 533-535, 538, 696
  reflections, 297, 481, 714-717
  rules, 694-696
  step, 308-309
  stretching, 707-711
  transformations, 700, 722-724
  trigonometric, 613-617, 621-623
Fundamental Counting Principle, 662-663

Galileo Galilei, 486
Gauss, Karl Friedrich, 29
General growth formula, 405, 410, 411
Geometry, 7, 23, 26 (*See also* Coordinate systems)
Golden ratio, 44, 627
Graphics calculators
  absolute-value equations, 296, 304
  box-and-whisker plots, 204
  checking, 466
  coin toss, 185
  consistent and inconsistent systems, 338, 339, 340
  data collection, 237
  entering a function, 254
  error message, 77
  equations, 14-15, 33
  exponents, 341, 372, 401, 412
  independent and dependent systems, 341, 342, 343
  inequalities, 346
  intersection of two lines, 132, 137
  linear functions, 31, 33, 229, 246
  linear programming, 361
  lines of best fit, 265, 340
  order of operations, 18, 19
  parabola intersection, 501
  probability, 185
  radical equation solution, 586
  random number generation, 676, 678, 682-683
  rational equations, 554
  rational function, 533, 535
  regression,
    linear, 264-265
    exponential, 417
  scatter plot, 38-39, 265
  systems of equations, 319, 320, 321, 338, 339, 341

874

table feature, 15, 221, 285, 444, 445, 465, 466, 467
transformations, 700, 707
using a constant, 14
zeros of functions, 465
Graphs,
absolute-value equations, 295-297, 304
bar, 193
box-and-whisker, 203-204
circle, 194-197
coordinates, 24-26
feasibility region, 327
functions and relations, 692-696
hole, 534
line, 193
linear equations, 32-33, 220-223
misleading, 194
parent functions, 696
scatter plots, 37-40
square-root function, 583
stretches, 707-711
systems of equations, 319-321
transformations, 717, 722-724
translations, 700-704
Greatest common factor (GCF), 448
Greatest integer function, 309
Grouping, 462
Growth formula, 405
Guess-and-check strategy, 13, 326

Heading, 613
Height, measuring, 636-637
Hexagonal numbers, 517
Histograms, 202
Hole (in a graph), 534
Hooke's law, 239
Horizontal lines, 229-230, 248
Horizontal translations, 481, 702-703, 710
Hot-hands theory, 674-675
Hypatia, 23
Hypotenuse, 591
Hypothesis/conclusion, 558-563
Hypsometers, 636-637

Identity, 445
Identity matrix for multiplication, 632
Identity Property for Addition, 64, 84, 563
Identity Property for Multiplication, 75, 84, 563
If-p-then-q statements, 530-535, 616, 624
Inclusion symbols, 19

Inconsistent systems, 338-331
Independent events, 667-670, 709
Independent systems, 341
Independent variables, 32, 222, 245
Inductive reasoning, 559
Inequality, 276
Inequalities
absolute-value, 294-297, 302-303
addition and subtraction, 277-278
compound, 289-292
graphing, 278-279, 289-292, 304, 511-513
linear, 345-349
multiplication and division, 282-285
multistep, 282-285
number lines, 278-279
quadratic, 511-513
systems of, 348-349
in two variables, 345-349
Input (of a function), 695
Integers, 54-55, 61-62, 70
Intercepts, 245, 247
Interest, compound, 410, 415, 529
Intersection, 290, 655
Inverse properties
additive, 64, 84, 563
multiplicative, 75, 84, 563
Inverse variation, 526-529
Irrational numbers, 56

Kamil, Abu, 456
Kelvin scale, 258
al-Khwarizmi, Muhammad ibn Muas, 52

Leaves, 201
Legs of a right triangle, 591
Like terms, 89
Line graph, 193
Line of best fit, 39, 265, 340, 417
Line of reflection, 297
Linear equation, 32
Linear equations
graphing, 31-33, 222
horizontal lines, 248
parallel lines, 258-259
perpendicular lines, 258-260
point-slope form, 254-255
slope-intercept form, 244-246, 255
standard form, 252-255
from two points, 246-247
vertical lines, 248
Linear functions, 216-270

graphs, 220-223
inequalities, 345-347
parent, 696
rate of change, 236-239
regression, 264-265
slope, 226-230
slope-intercept form, 244-248
Linear inequality, 345
Linear programming, 360-361
Line graphs, 193
Literal equation, 147
Logical operators, 654-656
Look for a pattern strategy, 7, 221
Lower quartile, 204

m (slope), 245
Make a table strategy, 7, 244, 412, 555
Maps, 24
Margin of error, 307
Marsalis, Wynton, 252
Matrices
adding and subtracting, 630
address, 629
data organization, 628-629
entries, 629
equality, 653
identity, 632
main diagonal, 632
multiplying, 631-632
product, 631
square, 629
Matrix, 628
Matrix equality, 629
Maximum value, 481
Mean, 186-189
Means, 165
Median, 186-189
Midpoint
formula, 601-602
theorem, 608
Midsegment, 608
Minimum value, 360-361, 480
Mode, 186-189
Monomial, 373
Monomials
as factors, 448
multiplying, 373
quotients, 385-386
raised to powers, 379-380
Multiplication
associative property, 81, 84
closure property, 70, 562-563
commutative property, 80-81, 84, 563
distributive property, 82-84, 90, 98-99, 141-144, 438-439, 563
equation solving by, 124, 130, 137-138, 141-144

875

of expressions, 96-100
identity property for, 75, 84, 563
inverse property, 75-76, 84, 563
matrix, 631-632
polynomials, 432-434
property of equality, 124
property of inequalities, 283-285
of real numbers, 73-76
by reciprocals, 76
scalar, 631
special products, 434
square roots, 578
using algebra tiles, 96-97
zero, 76-77
Multiplication Property of Equality, 124-125
Multiplication Property of Inequality, 283-284
Multiplication Property of Square Roots, 578
Multiplicative Inverse Property, 75, 84, 563
Multiplicative inverse (reciprocal), 75-76, 84, 125

Natural numbers (counting numbers), 54
Negative correlation, 38
Negative exponent, 356
Negative number, 55
Negative numbers
addition and subtraction, 61-64, 67-69
exponents, 390, 392
inequalities, 283
multiplication and division, 74
opposites, 57
Neutral pairs, 61
Newton (unit), 239
Non-trivial rational function, 534
Number-digit puzzles, 355
Number sequence, 5
Numbers (*See also* Negative numbers; Real numbers)
counting, 54
even/odd, 560-561
Fermat, 395
Fibonacci, 29, 44
hexagonal, 517
irrational, 56
natural, 54
opposite, 57, 84
ordering, 56-57
random, 676
rational, 55
rectangular, 7
types of, 54-56
whole, 54

Odd numbers, 560-561
Opposites, 57, 84
Optimization equation, 361
Ordered pairs, 25-26, 692-696
Ordering symbols, 56
Organize data strategy, 115, 117
Origin, 25
OR operator, 655-656
Outcomes (probability), 648-650
Output (of a function), 695

Padilla, Americas de, 463
Parabola, 480
Parabolas
focus and directrix, 609
graphing, 480-483
inequalities, 511-513
overview, 480-481
transformations, 480-483
Parallel lines, 259
Parent function, 296-297, 480-481, 696
Parentheses, 19
Pascal's triangle, 437, 470-471
Pathfinder, 282
Patterns
consistent systems, 339-340
in data, 30-36
golden ratio, 44, 627
linear, 30-33
modeling, 264-265
in number sequences, 4-8, 17, 29, 516-517
Pendulums, 584-585
Percent, 171-175
Percent rate, 172
Percentage, 172
Perfect square, 56, 452, 576-578
Perfect-square trinomials, 452-453, 492-495
Perimeters,
of rectangles, 23, 428
of triangles, 428
Perpendicular lines, 259
Pie charts (See *circle graphs*)
Point-slope form, 254
Polygons,
diagonals in, 607
similar, 167
Polynomial, 426
Polynomial function, 443
Polynomials, 426-471
adding, 427-428
degree of, 426
factoring, 452-454, 448-449
functions, 443-445, 465
multiplying, 432-434
special, 452-454
subtracting, 427, 429
Portfolio Activity, 10, 17, 29, 43, 66, 79, 88, 103, 128, 140, 153, 170, 185, 207, 233, 243, 263, 281, 288, 325, 337, 352, 376, 395, 415, 437, 447, 457, 469, 491, 497, 515, 543, 557, 590, 605, 627, 653, 666, 673, 681, 719, 727
Positive correlation, 38
Positive numbers, 55, 60-62
Postulates, 558
Power-of-a-Fraction property, 386, 392
Power-of-a-Power property, 378, 392
Power-of-a-Product property, 379, 392
Powers
definition, 97, 380
of fractions, 386, 392
monomials, 373
negative numbers, 380
powers of, 378-380
products of, 371-372
quotients of, 383-385
reading, 97
of ten, 398-401
Powers of −1, 380
Principal square root, 487
Probability, 180-182, 646-687
addition of, 657-658
complements, 670
definitions, 182, 650
experimental, 181-182, 676-679
favorable outcome, 181, 649
independent events, 667-670
modeling, 676-679
outcomes, 181-182, 648-650
overview, 180-182
random, 674
simulations, 676-679
theoretical, 648-650
using Pascal's triangle, 471
Problem Solving, 6, 7, 13, 155, 117, 126, 131, 149, 221, 244, 326, 410, 412, 460, 541, 555, 609, 616
Product matrix, 631
Product-of-Powers property, 371-372
Proofs, 84-85, 558-563
Properties
Addition Property of Equality, 116
Addition Property of Inequality, 277
Associative Property of Addition, 81, 84, 563
Associative Property of Multiplication, 81, 84, 563

Closure Property, 562
Commutative Property of Addition, 80-81, 84, 91, 563
Commutative Property of Multiplication, 80-81, 84, 563
Cross Products Property, 165
Distributive Property, 82, 84, 90, 98, 141-144, 438-439, 563
Division Property of Equality, 123
Division Property of Inequality, 283
Division Property of Square Roots, 580
Identity Property for Addition, 64, 84, 563
Identity Property for Multiplication, 75, 84, 563
Multiplication Property of Equality, 124-125
Multiplication Property of Inequality, 283-284
Multiplication Property of Square Roots, 578
Multiplicative Inverse Property, 75, 84, 563
Power-of-a-fraction Property, 386, 392
Power-of-a-Power Property, 378, 392
Power-of-a-Product Property, 379, 392
Product-of-Powers Property, 371-372
Properties of Zero, 76
Quotient-of-Powers Property, 383-384
Reflexive Property of Equality, 84
Substitution Property of Equality, 84
Subtraction Property of Equality, 115
Subtraction Property of Inequality, 277
Symmetric Property of Equality, 84
Transitive Property of Equality, 84
Zero Product Property, 465
Properties of Zero, 76
Proportion, 164
constant of, 551
Proportional reasoning, 164-167, 171-175
Proportion method, 172-173
Pyramids, 594
Pythagoras, 591
Pythagorean Theorem, 591-594, 599, 600
Pythagorean triples, 597

Quadrants, 24
Quadratic equations, 466-467, 486-489, 492-495, 498-501, 507
Quadratic formula, 506-509
Quadratic function, 480
Quadratic functions,
completing the square, 492-495
graphing parabolas, 480-483
inequalities, 511-513
number of solutions, 508
parent, 696
vertex form, 482, 495
zeros of, 464-467, 482, 483
Quadratic inequality, 511
Quartic equations, 427
Quartiles, 204
Quotients-of-Powers Property, 383-384

Radical sign, 487, 577
Radicals, 567-609 (*See also* Roots; Square roots)
estimating square roots, 576-577
operations with, 577-580
simplifying, 577-580
solving equations, 584-587
Radicand, 577
Radius, 608
Random events, 674
Random numbers, 676
Random-number generators, 677-678, 682-683
Range, 186, 219
Ranges
of functions, 219-223 245, 692-696
statistical, 186-188
tangent function, 615
Rate of change, 236-237, 404
Ratio, 164 (*See also* Rational expressions)
Rational expressions, 532-535, 538-541, 546-549
Rational functions, 524-570
definition, 533
graphing, 532-535
inverse variation, 526-529
non-trivial, 534-535
solving, 552-555
Rational numbers, 55
Rationalizing the denominator, 580
Real numbers, 56, 60-64, 67-70, 73-77
Reasoning, types, 559
Reciprocal, 75-76, 84, 124-126

Rectangles
diagonals, 607
golden, 44, 627
similar, 167
Rectangular numbers, 7, 469
Rectangular prism, volume, 443-444
Reflection, 297, 481, 715
Reflections of functions, 297, 481, 714-717
Reflexive Property of Equality, 84
Regression lines, 265
Relation, 218-220, 692
Repeating decimals, 55-56
Research design, 208-209
Rise, 226
Roots, (*See also* Radicals)
square, 487-488, 576-580
functions of square, 583-587
Rule of 72, 529
Run, 226

Sample space, 649-650
Scalar, 631
Scalar multiplication, 631
Scale factor, 708-711, 723-724
Scatter plot, 37-40
Scientific calculators, 18, 20
Scientific notation, 398-401
Segments, 608
Sequences of numbers, 5-10, 29
Sets (*See also* Venn diagrams)
closure, 562-563
counting elements, 654-658
disjoint, 657-658
intersection, 655-658
of choices, 662
of numbers, 54
union, 655, 658
Signed numbers (*See* Negative numbers)
Significant digits, 398
Simplest radical form, 578
Simplification, 89, 538-541, 577-580
Simplified, 89
Simulations, 676-679, 682
Sine, 621-623
Slope, 228
Slope
intercept form, 245-248, 252-255, 320
linear functions, 226-230
parallel and perpendicular lines, 258-260
as rate of change, 236-237
Slope-intercept form, 245
Solving a simpler problem strategy, 7
Solution to an equation, 13

877

Special products of polynomials, 434
Speed, 12, 238, 264-265
Sphere
  surface area, 385
  volume, 377, 385
Spreadsheets, 8, 66, 104-105, 294, 678
Spring constant, 239
Square matrix, 629
Square numbers, 23, 516
Square root, 487, 577
Square roots (*See also* Radicals)
  definition, 487, 577
  estimating, 576-577
  functions, 583
  irrational, 56
  solving equations with, 486-489
Square-root function, 583
Stairs, slope of, 234-235
Standard form, 252, 426
Standard form of a polynomial, 426
Statistics, 186-189
Stem-and-leaf plots, 201-202, 216
Stems, 201
Step functions, 308
Stretches, 707-711
Substitution method, 326-328
Substitution Property of Equality, 84
Subtraction, 67-68
Subtraction
  distributive property, 82
  of expressions, 89-91
  inequalities, 277-278
  matrix, 630
  polynomials, 426-429
  property of equality, 115-118
  real numbers, 67-70
  solving equations by, 114-115
Subtraction Property of Equality, 115
Subtraction Property of Inequality, 277
Sum, 62
Surface area, 385-386, 444-445
Symmetric Property of Equality, 84, 118, 127
Symmetry, 481-483
System of equations, 319
System of linear inequalities, 314

Table, 12
Tangent, 613-617
Technology, 8, 9, 14, 15, 18, 19, 31, 33, 38, 77, 132, 137, 185, 221, 229, 237, 246, 254, 265, 285, 294, 296, 304, 319, 320, 321, 338, 339, 340, 341, 342, 343, 346, 361, 364, 372, 382, 401, 417, 444, 445, 465, 466, 467, 501, 533, 535, 554, 586, 615, 622, 623, 636, 637, 676, 678, 700, 707
Temperature
  data, 79, 88
  unit conversion, 118, 258
Temperature-humidity index (THI), 23
Templer, John, 234
Term, 5, 89
Theorem, 85, 558-563
Theoretical probability, 650
Trajectories, 483
Transformations, 296-297, 480, 700-704
Transitive Property of Equality, 84
Translation, 296, 481, 700-704
Trapezoid, area, 150
Tree diagrams, 471, 661
Trial, 181, 676
Triangle Midsegment Theorem, 608
Triangles
  midsegment theorem, 608
  Pascal's, 437, 470-471
  right, 591-594, 598-601, 613-617, 621-623 (See also Pythagorean Theorem; Tangent)
  sum of angles, 120
Triangular number sequence, 10, 23, 447
Trigonometric functions, 613-617, 621-623
Trigonometry, 614
Trinomials
  factoring, 452-454
  perfect-square, 452-453, 492-495
  quadratic, 458-462, 498-501
Trivial rational function, 534

Union, 290, 655
Upper quartile, 204
Use a formula strategy, 410, 541

Variables, 11-17
Variation,
  inverse, 526-529
  direct, 238-239
Velocity, 504
Venn diagrams, 56, 655-658
Vertex, 480, 482-483, 489, 495
Vertex form of a quadratic function, 482
Vertical lines, 230, 248
Vertical-line test, 693-694
Vertical stretch, 708, 723-724
Vertical translation, 481, 701, 723-724
Vertical-line test, 693
Volume,
  cylinder, 241, 445, 541
  as polynomial function, 444-445
  rectangular prism, 102, 444
  sphere, 377, 385
  surface area ratio, 444

Wavelengths, 400
Wheel of Theodorus, 605
Whole numbers, 54
Wilding, Edward, 95
Working backward strategy, 6, 460
Write an equation strategy, 126, 131

x-axis, 24-25
x-coordinate, 24
x-intercept, 247
X rays, 400

y-axis, 24-25
y-coordinate, 24
y-intercept, 245

Zero
  exponents, 391-392
  properties of, 76
  and vertex, 482
Zero as an exponent, 391
Zero of a function, 465
Zero Product Property, 465

Credits

PHOTOS

Abbreviations used: (t) top, (c) center, (b) bottom, (l) left, (r) right, (bckgd) background, (bdr) border.
FRONT COVER: (bckgd), Index Stock Photography Inc./Ron Russell; (b), Jean Miele MCMXCII/The Stock Market. **TABLE OF CONTENTS:** Page vi(tl), John Langford/HRW Photo; vi(cl), C. C. Lockwood/Bruce Coleman Inc.; vii(tl), Don Couch/HRW Photo; vii(cl), Stuart Westmorland/Tony Stone Images; viii(tl), Peter Van Steen/HRW Photo; viii(cl), Bob Anderson/Masterfile; ix(tl), David Ducros/Science Photo Library/Photo Researchers Inc.; ix(cl), Comstock; x(tl), Randal Alhadeff/HRW Photo; x(cl), Sarah Stone/Tony Stone Images; xi(tl), SuperStock; xi(cl), Brett Froomer/The Image Bank © 2000; xii(tl), Randal Alhadeff/HRW Photo, props courtesy Recreational Equipment Inc.; xii(cl), David Phillips/HRW Photo. **CHAPTER ONE:** Page 2(br),(bc), Chuck Place/Place Stock Photo; 2(bl), Stephen Trimble, pottery courtesy Richard M. Howard Collection; 2(cr), Tim Davis/Tony Stone Images; 2(tr), Color Box/Masterfile; 2(cr), Comstock/D. Lada/H. Armstrong Roberts; 2(cr), Planet Earth Pictures/FPG International; 3(tr), Artbase Inc.; 3(t), Russell D. Curtis/Photo Researchers Inc.; 3(tl), Andrew Rafkind/Tony Stone Images; 3(bl), James Robinson/Masterfile; 4(t), Warren Faidley/First Light; 7(b), John Langford/HRW Photo; 9(bl), Randal Alhadeff/HRW Photo(teapot); 9(bl), Comnet/First Light(steam); 10(c), eStudios/HRW Photo; 11(c), Dan Kapke(carrying bike); 11(all others), Dana Lamm/Leadville Trail 100; 12(bl), C. Eugene Gebhardt/Masterfile; 13(bl), Sam Dudgeon/HRW Photo; 14(c), Map Art Publishing Corporation; 15(bl), Randal Alhadeff/HRW Photo; 16(bl), Don Couch/HRW Photo; 17(bl),(br), eStudios/HRW Photo; 18(tr),(cr), Randal Alhadeff/HRW Photo, location courtesy Carpets by Conrad; 18(tc), John Langford/HRW Photo; 20(tl), Artbase Inc.; 23(tr), Sam Dudgeon/HRW Photo; 24(t), Frog Map Co.; 26(cr), Stephen Marks/The Image Bank © 2000; 28(br), Randal Alhadeff/HRW Photo; 29(tl), Corbis-Bettmann; 29(br), James Robinson/Masterfile; 31(br), Jeff Barber/Infocus/The Victoria Clipper; 32(bl), Mark Wagner/Tony Stone Images; 32(bl), Adam Jones/Photo Researchers Inc.; 33(bl), Allsport USA/Gerard Planchenault; 33(bl), Lester Lefkowitz/Masterfile; 34(cl), Hoverspeed; 35(br), Bill Stoughton/Austin Convention & Visitors Bureau; 37(t), Eric Meola/The Image Bank © 2000; 38(cl), Randal Alhadeff/HRW Photo; 39(tl), Edgar T. Jones/Bruce Coleman Inc.; 39(br), A. T. Willett/The Image Bank © 2000; 41(tl), Peter Cade/Tony Stone Images; 42(bl), Don Couch/HRW Photo; 44(cl), Kathleen Campbell/Allstock/PNI; 45(tr), eStudios/HRW Photo; 45(b), John Langford/HRW Photo. **CHAPTER TWO:** Page 52(all), The Granger Collection, New York; 53(tl),(tc),(tr), al-Khwarizmi; 53(c), Artbase Inc.; 54(c), Esao Hashimoto/Animals Animals; 54(cr), Marian Bacon/Animals Animals; 54(tr), James P. Blair/National Geographic Society; 54(tl), Hans Reinhard/Bruce Coleman Inc.; 54(cl), Gail Shumway/FPG International; 59(bl), Peter Van Steen/HRW Photo; 60(tr), John Langford/HRW Photo; 64(bl), Sam Dudgeon/HRW Photo; 66(br), MTPA Stock/Masterfile; 67(tr),(tc), John Langford/HRW Photo; 69(br), John G. Ross/Photo Researchers Inc.; 69(cr), Erich Lessing/Art Resource, NY; 71(br), Joel Sartore/National Geographic Society; 71(bc), C. C. Lockwood/Bruce Coleman Inc.; 72(tr), Blair Seitz/Photo Researchers Inc.; 73(tr),(tc), Sam Dudgeon/HRW Photo; 78(br), David Madison/Tony Stone Images; 79(cr), Tom Kitchin/First Light; 80(tr),82(tr), Michelle Bridwell/HRW Photo; 83(br), John Langford/HRW Photo; 87(b), Michelle Bridwell/HRW Photo; 88(bl), Everett Johnson/Tony Stone Images; 93(cl), William Taufic/First Light; 94(l), Kim Westerskov/Tony Stone Images; 94(bl),95(cr), NYT Graphics © *The New York Times*; 96(tr), Robert Holmes; 99(br), D. Mason/First Light; 99(cr), eStudios/HRW Photo; 100(tc), Scott Markewitz/Masterfile; 100(tr), Chris Cole/The Image Bank © 2000; 103(cr), Comstock; 103(cr), Artbase Inc.; 104(bl), Sam Dudgeon/HRW Photo; 105(tr), Spaceshots; 105(b), Stephen Simpson/FPG International. **CHAPTER THREE:** Page 112(c), Comstock; 112(tr), R. Essel/First Light; 112(b), Jon Feingersh/First Light; 113(tr), Allsport USA/Otto Creule; 113(tl), David Oliver/Tony Stone Images; 114(tr),115(bl), Michelle Bridwell/PhotoEdit; 117(bl),(br), Don Couch/HRW Photo; 119(tr), Mark E. Gibson; 120(cr), Don Smetzer/Tony Stone Images; 121(br), Richard Ashworth/Robert Harding; 122(t), Michelle Bridwell/HRW Photo; 124(tl), Bob Daemmrich; 126(tl), Image Copyright ©2001 PhotoDisc, Inc.; 127(bl), Don Couch/HRW Photo; 128(cl),(c),(cr),129(tr), John Langford/HRW Photo; 131(cr), Michelle Bridwell/Frontera Fotos; 134(tr), Stan Goldberg/SuperStock; 135(tr), Sam Dudgeon/HRW Photo; 136(cr), Christine Galida/HRW Photo; 139(bl), Allsport USA/Todd Warshaw; 139(br), Michael Freeman/Bruce Coleman Inc.; 140(tr), David Boyle/Earth Scenes; 141(tr), Michelle Bridwell/HRW Photo; 146(tr), Christine Galida/HRW Photo; 147(tc),(tl), Gay Bumgarner/Tony Stone Images; 147(tr), Artbase Inc.; 149(cr),152(tr), Peter Van Steen/HRW Photo; 152(cr), Corbis/Joseph Sohm/ChromoSohm Inc.; 152(br), Greg Pease/Tony Stone Images; 153(all), David R. Frazier Photolibrary; 154(t), Peter Van Steen/HRW Photo; 155(b), John Langford/HRW Photo. **CHAPTER FOUR:** Page 162(cr), John Langford/HRW Photo; 163(cl), Joseph Pobereskin/Tony Stone Images; 163(tr), Corbis/Kevin R. Morris; 163(tl), Don Couch/HRW Photo; 164(tr), Michelle Bridwell/HRW Photo; 167(br), Don Couch/HRW Photo; 168(cr), Robert Lubeck/Animals Animals; 169(b), Sam Dudgeon/HRW Photo; 171(tc), Randy Faris/First Light; 171(c), Dietrich Stock Photos, Inc.; 171(tr), Paul Harris/Tony Stone Images; 173(tl), Scala/Art Resource, NY; 174(bl), Jon Riley/Tony Stone Images; 175(tl), Randal Alhadeff/HRW Photo, location courtesy Bill Mundy South; 176(tl), Peter Van Steen/HRW Photo; 177(cl), Peter Van Steen/HRW Photo, location courtesy Recreational Equipment Inc.; 178(cl), Jim Zuckerman/First Light; 179(tr), John Langford/HRW Photo; 179(br), John Warden/Tony Stone Images; 180(tr), Don Couch/HRW Photo; 181(cr), Sam Dudgeon/HRW Photo; 181(bc),182(cr), eStudios/HRW Photo; 182(br), Sam Dudgeon/HRW Photo; 183(tl), Ron Tanaka/HRW Photo; 184(bl),186(tr), Sam Dudgeon/HRW Photo; 186(tc), Allsport USA/Brian Bahr; 186(c), Allsport USA/Matthew Stockman; 189(br), Lori Hdamski Peek/Tony Stone Images; 190(bl), David Madison/Bruce Coleman Inc.; 191(br), Laura Riley/Bruce Coleman Inc.; 191(bc), Charles Palek/Animals Animals; 192(tc), Allsport USA/Shaun Botterill; 192(tr), John Warden/Tony Stone Images; 192(tl), Steve Niedorf/The Image Bank © 2000; 195(t), Franklin D. Roosevelt Library/HRW; 196(bc), Mark Stephenson/First Light; 196(bl), Peter Gridley/Masterfile; 196(br), Rivera Collection/SuperStock; 196(bc), SuperStock; 197(bl), John Daniels/Bruce Coleman Inc.; 199(bl), Lee Foster/Bruce Coleman Inc.; 201(tr), Marilyn Kazmers/Peter Arnold, Inc.; 204(tl), C. Allan Morgan/Peter Arnold, Inc.; 206(bl), Stuart Westmorland/Tony Stone Images; 207(c), Rob Gage/Masterfile; 207(cr), Corbis-Bettmann; 208(c), Michelle Bridwell/HRW Photo; 209(tl), Michael Young/HRW Photo; 209(bl), Michelle Bridwell/PhotoEdit. **CHAPTER FIVE:** Page 216(b), Jean-Marc Truchet/Tony Stone Images; 217(c), Sylvain Grandadam/Tony Stone Images; 217(tr), Peter Griffith/Masterfile; 217(cl), Telegraph Colour Library/FPG International; 217(tl), Ron Watts/First Light; 218(tr), John Langford/HRW Photo; 219(bl), Don Couch/HRW Photo; 221(cr), John Langford/HRW Photo, props courtesy Gruene River Co.; 222(cl), Mark Burnham/First Light; 222(cl), Hubert Kanus/Photo Researchers Inc.; 223(bl), John Langford/HRW Photo; 224(br), Brian Bailey/Tony Stone Images; 225(tl), Don Couch/HRW Photo; 226(tc),(c),(cr), John Langford/HRW Photo; 228(tr), Randal Alhadeff/HRW Photo; 232(l), Tom Bean; 232(br), Jack Olson; 233(tl), Tony Stone Images; 233(bc), eStudios/HRW Photo; 234(l), Eva Rubenstein/Photonica; 235(cr), Lisa Stancati /Photonica; 236(tr), Bill Brooks/Masterfile; 237(bl), Randal Alhadeff/HRW Photo; 238(tl), Y.C.L.-TCL/Masterfile; 239(tr), Don Couch/HRW Photo; 240(tl), Lori Adamski Peek/Tony Stone Images; 241(tl), Peter Van Steen/HRW Photo; 242(tr), John Langford/HRW Photo; 243(cr), Mike McQueen/Tony Stone Images; 243(c), Garry D. McMichael/Photo Researchers Inc.; 244(tc),(tr), John Langford/HRW Photo, location courtsey Chaparral Ice Center at Northcross Mall; 247(tl), Randal Alhadeff/HRW Photo; 248(bl), Harvey Kennan/Tony Stone Images; 248(b), Ken Straiton/First Light; 251(b), Albert Normandin/Masterfile; 252(tc), Kasala/Gamma Liaison; 252(tr), Joe Larusso/Gamma Liaison; 257(cl), Corbis-Bettmann/John Springer; 258(c), George Hunter/Tony Stone Images; 258(tr), Alec Pytlowany/Masterfile; 259(br), Randal Alhadeff/HRW Photo; 261(tr), Peter Van Steen/HRW Photo; 263(c), Frank Whitney/The Image Bank © 2000; 263(cr), David R. Frazier Photolibrary; 264(tr), M. H. Dunn/First Light; 264(b), John Zimmerman/FPG International; 270(br), P.R. Production/SuperStock. **CHAPTER SIX:** Page 274(br),(tr),275(tc), Artbase Inc.; 275(tl), Index Stock Photography Inc./Lonnie Duka; 275(tr), Andy Snow/Science Photo Library/Photo Researchers Inc.; 276(tr), Don Couch/HRW Photo; 277(tr), Randal Alhadeff/HRW Photo; 280(br), Alan Sirulnikoff/First Light; 281(tl), Randal Alhadeff/HRW Photo; 281(tl), Artbase Inc.; 282(tc),(tr), JPL/NASA/Gamma Liaison; 286(bl), Bob Anderson/Masterfile; 287(br), Noriyuki Yoshida/SuperStock; 289(tr), John Langford/HRW Photo; 290(tr), Artbase Inc.; 294(cr), Randal Alhadeff/HRW Photo; 294(tr), Michelle Bridwell/HRW Photo; 299(tl), Christine Galida/HRW Photo; 300(tc), Darius Koehli/First Light; 300(tr), Comstock; 302(cr), J. Boutin/Publiphoto; 306(br), Paul Silver/Bruce Coleman Inc. ; 307(tr), Space Frontiers/Masterfile; 308(tc), Index Stock Photography Inc./Benelux Press; 308(bl), Allsport USA/Shaun Botterill; 309(t),(c), Mark E. Gibson; 309(c), Allsport USA/Brian Bahr. **CHAPTER SEVEN:** Page 316(tr), Chris Harvey/Tony Stone Images; 316(b), John Cancalos/Valan Photos; 316(cr), Corbis/Raymond Gehman; 317(c), NASA Space Photo/The Image Bank © 2000; 317(tr), NASA/Science Photo Libary/Photo Researchers Inc.; 318(tr), Ian Crysler; 322(t), Artbase Inc.; 322(cl), Darrell Gulin/Tony Stone Images; 324(b), Ron Sherman/Tony Stone Images; 326(t), Richard Dole; 328(bl), David Starrett; 331(t), Map ©1998 by Rand McNally #98-S-158; 331(c), Chris Speedie/Tony Stone Images; 331(tc),333(bl), David Starrett; 336(cl), Sam Dudgeon/HRW Photo; 337(tl), Tomb 181 - Bebamun & Ipuky; 340(bl), Everett Collection; 343(br), Don Couch/HRW Photo; 344(b), Randal Alhadeff/HRW Photo (driver); 344(b), Peter Van Steen/HRW Photo (car); 345(c), Stratford Festival production of The Tempest, 1976, directed by Robin Phillips and William Hutt, designed by John Ferguson, basic set design by Daphne Dare, music by Berthold Carriere, lighting by Gil Wechsler. William Hutt (left) as Prospero with members of Festival Acting Company. Courtesy of the Stratford Festival Archives; 349(tl), Randal Alhadeff/HRW Photo; 349(bl), Michelle Bridwell/HRW Photo; 350(tl), Ian Crysler; 351(br), Michelle Bridwell/HRW Photo; 352(bl), Sam Dudgeon/HRW Photo; 353(tr), David Starrett; 354(tl), David Lissy/Nawrocki Stock Photo Inc.; 354(bl), Comstock; 354(bc), Erik Simonsen/The Image Bank © 2000; 354(br), Artbase Inc.; 356(tl), Sam Dudgeon/HRW Photo; 356(br), Dennis Fagan/HRW Photo; 357(bl), Bill Lishman/Joseph Duff - Operation Migration; 358(c), John Langford /HRW Photo, props courtesy Gruene River Co.; 359(cl), J. Towers/First Light; 360(bl), Joe Cornish/Tony Stone Images. **CHAPTER EIGHT:** Page 361(cr), John Langford/HRW Photo; 368(br), John Reader/Science Photo Library/Photo Researchers Inc.; 369(tl), Chris Harvey/Tony Stone Images; 369(tr), S. Lowry/Univ. Ulster/Tony Stone Images; 369(cr), Gary Buss/FPG International; 370(tc), Randal Alhadeff/HRW Photo; 374(cr), S. Lowry/Univ.Ulster/Tony Stone Images; 375(br), Victoria Hurst/First Light; 375(bl), Leonard Lee Rue III/Animals Animals; 377(tr), Heatons/First Light; 377(tr), Artbase Inc.; 380(tr),382(cr), John Langford/HRW Photo; 383(tr), Patti Murray/Earth Scenes; 383(cr), Bill Ross/First Light; 385(br), Reza Estakhrian/Tony Stone Images; 386(tl), Natural Selection/First Light; 386(tr), Michael Dick/Animals Animals; 389(c), Don Couch/HRW Photo, props courtesy The Hat Box; 390(tr), Telegraph Colour Library/Masterfile; 392(c),(cr), Randal Alhadeff/HRW Photo; 394(cl), Color Day/The Image Bank © 2000; 395(cl), Corbis-Bettmann; 395(cl), Artbase Inc.; 396(tl), © 1998 Rick Friedman/Black Star; 396(cl), Sam Dudgeon/HRW Photo; 397(tc),(c), Scott Van Osdol/HRW Photo; 398(bckgd), Luke Dodd/Science Photo Library/Photo Researchers Inc.; 398(c), John Sanford/Science Photo Library/Photo Researchers Inc.; 398(tr), Scott Van Osdol/HRW Photo; 399(br), Ronald Royer/Science Photo Library/Photo Researchers Inc.; 401(bl), R. Ball/First Light; 402(br), Paul Morrell/Tony Stone Images; 402(bl), Artbase Inc.; 403(t), Space Frontiers/Masterfile; 404(tr), Paolo Negri/Tony Stone Images; 407(cl), David Reed/Panos Pictures; 408(tl), David James/Tony Stone Images; 408(tc), Doug Armand/Tony Stone Images; 408(bl), Corbis-Bettmann; 408(bl), Artbase Inc.; 409(all), Kenneth Garrett/National Geographic Society; 410(br), Sam Dudgeon/HRW Photo; 411(tr), M. Bradley/Bruce Coleman Inc.; 412(c), Douglas E. Walker/Masterfile; 413(cl), Comstock/Colin Quirk; 414(cr), Philip & Karen Smith/Tony Stone Images; 415(cr), Bernd Kappelmeyer/Masterfile; 416(b), Gail Shumway/FPG International; 416(bckgd), Peter McLeod/First Light; 417(cr), Joel Sartorengs/National Geographic Society; 417(cr), T. Kitchin/First Light; 417(br), K&K Ammann/Bruce Coleman Inc.; 417(br), Don Enger/Animals Animals; 417(tr), Tom Brakefield/Bruce Coleman Inc.**CHAPTER NINE:** Page 424(all),425(all), Ron Behrmann/HRW Photo; 426(tr), Randal Alhadeff/HRW Photo; 428(br), Phil Jason/Tony Stone Images; 431(tr),432(c), Peter Van Steen/HRW Photo; 436(bl), David Muir/Masterfile; 437(br), Corbis-Bettmann; 438(tr), Sam Dudgeon/HRW Photo; 440(cl),442(cr), Peter Van Steen/HRW Photo; 443(tr), Sam Dudgeon/HRW Photo; 445(tl), Peter Van Steen/HRW Photo; 447(tl), Randal Alhadeff/HRW Photo; 451(tr), Don Couch/HRW Photo; 458(tr), John Langford/HRW Photo; 464(tr), Allsport USA/Mark Thompson; 468(bl), Peter Van Steen/HRW Photo, location courtesy 4H Capital; 471(cr), Index Stock Photography Inc./Shaffer-Smith. **CHAPTER TEN:** Page 478(c), David Madison/Tony Stone Images; 478(bl),(br), Randal Alhadeff/HRW Photo; 479(t), Corbis/NASA; 479(c), Henry Groskinsky/Peter Arnold, Inc.; 480(tr),483(bl), Jeff Hunter/The Image Bank © 2000; 486(tc), Corbis-Bettmann; 486(cr), SEF/Art Resource, NY; 488(cr), Sarah Stone/Tony Stone Images; 491(cr), Peter Van Steen/HRW Photo; 492(all), John Langford/HRW Photo; 497(tr), Ezio

Geneletti/The Image Bank © 2000; 498(tr), Peter Van Steen/HRW Photo; 500(cr), David Phillips/HRW Photo; 503(tr), National Museum of American History Smithsonian Institution; 504(tr), Nancy Engebretson/Phoenix Gazette/Used with permisson. Permission does not imply endorsement.; 504(tl), (cl), Jump Run Productions/The Image Bank © 2000; 504(bl), Jason Hawkes/Tony Stone Images; 505(tr), Kris Coppieters/SuperStock; 506(tr), Allsport USA/Mike Powell; 509(bl), Sam Dudgeon/HRW Photo; 511(tr), Roy Ooms/Masterfile; 513(bl), Peter Van Steen/HRW Photo; 515(bl), Trevor Bonderud/First Light. **CHAPTER ELEVEN:** Page 524(bc), NSP/PP/Nawrocki Stock Photo Inc.; 524(cr), Dale Sanders/Masterfile; 525(tr), Scott Barrow/International Stock; 525(tr), Corbis/Bettmann; 525(tl), Cirque du Soleil Mystère, Photo: Al Seib, Costumes: Dominique Lemieux; 526(t), David Phillips/HRW Photo; 528(tr), Randal Alhadeff/HRW Photo; 531(tl), Bavaria Collection/SuperStock; 532(tr),533(cl), Michelle Bridwell/HRW Photo; 533(cl),535(tl), Peter Van Steen/HRW Photo; 538(tr), Dennis Fagan/HRW Photo; 541(tr), John Woods/Index Stock Photography Inc.; 542(br), Robert Lubeck/Animals Animals; 543(br), Randal Alhadeff/HRW Photo; 544(tl), Index Stock Photography Inc./Matthew Borkoski; 544(bl), Index Stock Photography Inc./Stewart Cohen; 545(cr), Andrew Syred/Tony Stone Images; 546(cr), Index Stock Photography Inc./The Sports File; 551(tr), Michelle Bridwell/PhotoEdit; 552(tr), Dennis Fagan/HRW Photo; 555(cr), David Phillips/HRW Photo; 556(tl), Randal Alhadeff/HRW Photo; 556(br), Comstock/George Hunter; 557(cr), Leonard Lessin/Peter Arnold, Inc.; 565(bl), Sam Dudgeon/HRW Photo; 565(tr), Randal Alhadeff/HRW Photo; 566(tc), Michelle Bridwell/HRW Photo; 566(tr), Michael Tamborrino/FPG International; 566(tl), Benjamin Rondel/First Light; 567(tl), Jake Rajs/Tony Stone Images; 567(tr), David Noton/Masterfile. **CHAPTER TWELVE:** Page 574(tr), John Mead/Science Photo Library/Photo Researchers Inc.; 574(bc), Hilarie Kavanagh/Tony Stone Images; 575(tr), D. E. Cox/Tony Stone Images; 575(cr), Vladimir Pcholkin/FPG International; 576(tr), Michelle Bridwell/HRW Photo; 583(tr), Dennis Fagan/HRW Photo; 584(tr),585(cr), John Langford Photography; 587(cr), Gary Withey/Bruce Coleman Inc.; 588(cl), Ken Reid/Masterfile; 589(cl),(l),(bl), Photo supplied courtesy Lee Valley Tools Ltd.; 590(c), Peter Van Steen/HRW Photo; 591(tc), Vic Thomasson/Tony Stone Images; 591(tr), G.A. Plimpton Collection, Rare Book & Manuscript Library, Columbia University; 593(cr), Jan Becker; 595(tl), Corbis-Bettmann; 596(br), Jake Rajs/Tony Stone Images; 597(tr), Randal Alhadeff/HRW Photo, props courtesy Lights Fantastic; 598(tr), Matthew Neal McVay/Tony Stone Images; 599(cr), Nawrocki Stock Photo Inc.; 600(br), Peter Van Steen/HRW Photo; 604(br),(bc), United States Coast Guard; 606(tr), The SKYSCAN Photolibrary; 613(c),(tr), Photo courtesy of the Boeing Company; 615(bl), Mark Wagner/Tony Stone Images; 617(b), Jack Olson; 617(bl), Roy Corral/Tony Stone Images; 617(tr), Brett Froomer/The Image Bank © 2000; 619(br), Peter Van Steen/HRW Photo; 620(tr), Robert P. Carr/Bruce Coleman Inc. ; 621(tr), John Langford/HRW Photo, location courtesy Sea World of Texas; 623(tr), David Lawrence/First Light; 623(bl), Frederick McKinney/FPG International; 624(bl), First Light; 627(cr), Comstock/H. Armstrong Roberts; 627(cr), Toshiba of Canada Limited, Consumer Electronics Group; 628(tc), Aldo Torelli/Tony Stone Images; 635(tl), William Sallaz/The Image Bank © 2000; 636(bl), Peter Van Steen/HRW Photo. **CHAPTER THIRTEEN:** Page 646(tr), Peter Van Steen/HRW Photo(coins); 646(c), Norman Owen Tomalin/Bruce Coleman Inc.; 646(bc), Stephen St. Johnngs/National Geographic; 646(br), Steve Taylor/Tony Stone Images; 647(tl), Peter Cawson/Masterfile; 647(cr), Jerry Cooke/Animals Animals; 647(tr), Leonard Lee Rue III/Photo Researchers Inc.; 648(tr), John Langford/HRW Photo; 649(tr), Peter Van Steen/HRW Photo; 652(br), U.S. Department of the Interior, Indian Arts and Crafts Board, Museum of the Plains Indian; 654(tr), John Langford/HRW Photo; 657(tr), Sam Dudgeon/HRW Photo; 660(tr), Bob Torrez/Tony Stone Images; 661(tr), David Phillips/HRW Photo; 662(tr), John Langford/HRW Photo; 663(cr), Randal Alhadeff/HRW Photo, props courtesy Recreational Equipment Inc.; 664(bl), Peter Van Steen/HRW Photo; 665(cr), Michelle Bridwell/HRW Photo; 665(br), Sam Dudgeon/HRW Photo; 666(c),(cr), Peter Van Steen/HRW Photo; 667(tr), David Phillips/HRW Photo; 669(tl), Michelle Bridwell/HRW Photo; 671(br), Leland Bobbe/Tony Stone Images; 673(br), Cyberimage/Tony Stone Images; 674(cl), Allsport USA/Jonathan Daniel; 675(cr), Allsport USA/Vincent Laforet; 676(tr), Don Couch/HRW Photo; 678(cl), Peter Van Steen/HRW Photo; 679(bl), MGM (Courtesy Kobal); 680(br), John Langford/HRW Photo, location courtesy Sea World of Texas; 680(tr), Don Couch/HRW Photo; 681(bl), John Langford/HRW Photo; 682(c),683(cl),683(cr), Peter Van Steen/HRW Photo. **CHAPTER FOURTEEN:** Page 690(b), John Elk III/ELKJO/Bruce Coleman Inc.; 691(tc), Rex Ziak/Tony Stone Images; 691(tl), Jason Hawkes/Tony Stone Images; 691(tr), Comstock/George Gerster; 691(cr), Harald Sund/Image Bank; 692(tr), John Kieffer/Peter Arnold, Inc.; 696(c), Comstock; 696(cl), The New York Public Library, Astor, Lenox and Tilden Foundations from Photographs and Prints Division Schomburg Center for Research in Black Culture; 699(tr), I.T.P./International Stock; 700(tr), *Pegasus* (Tiled facade for the hall of a school in the Hague) by M.C. Escher ©1998 Cordon, Art-Baarn-Holland. All rights reserved; 702(t), Sam Dudgeon/HRW Photo; 704(cl), Christine Galida/HRW Photo; 706(cr), Seattle Mariners/NBBJ Architects; 707(tc), Ron Thomas/FPG International; 713(tr), Chuck Close, *John*, 1992 (portrait of artist with work in progress), oil on canvas, 100" x 84", Photograph by Bill Jacobson, Courtesy of PaceWildenstein; 714(tc),(tr), Peter Van Steen/HRW Photo; 717(br), Raymond G. Barnes/Tony Stone Images; 718(cl), Dennis Cox/FPG International; 718(br), Tom Ives/The Stock Market; 720(bl),720(br), Sam Dudgeon/HRW Photo; 720(cr), George Lepp/Tony Stone Images; 721(cr), Ron Sanford/Tony Stone Images; 722(tr), Lori Adamski Peek/Tony Stone Images; 725(tl), Dugald Bremmer/Tony Stone Images; 729(tr),(tc), Scott Van Osdol/HRW Photo; 732(br), Mark E. Gibson

ILLUSTRATIONS

Chapter One: Page 6, Michael Morrow; 9, Michael Morrow; 14, Mapping provided by MapArt © Mapmedia Corp.; 22, Michael Morrow; 30, Michael Morrow; 36, Michael Morrow; 42, Leslie Kell. **Chapter Two:** Page 60, Michael Morrow; 89, Uhl Studios. **Chapter Three:** Page 120, Boston Graphics, Inc.; 133, Boston Graphics, Inc.; 150, David Fischer; 154, Lori Osiecki. **Chapter Four:** Page 162, Annie Bissett; 163, Annie Bissett; 165, Michael Morrow; 167, Michael Morrow; 170, Stephen Durke; 171, Michael Morrow; 173, Michael Morrow; 180, Michael Morrow; 186, Michael Morrow; 192, Elizabeth Brandt; 193, Leslie Kell; 194, Leslie Kell; 195, Stephen Durke; 196, Michael Morrow; 200, Michael Morrow; 202, Stephen Durke. **Chapter Five:** Page 224, Michael Morrow; 234, Stephen Durke; 241, Michael Morrow; 257, Michael Morrow. **Chapter Six:** 290, Uhl Studios; 302, Michael Morrow. **Chapter Seven:** Page 324, Michael Morrow; 325, Boston Graphics, Inc.; 330, Michael Morrow; 336, Michael Morrow; 337, Stephen Durke; 338, Ortelius Designs; 340, Michael Morrow; 345, Nenad Jakesevic; 354, Stephen Durke. **Chapter Eight:** Page 372, Christy Krames; 376, Michael Morrow; 390, David Fischer; 400, Michael Morrow; 407, GeoSystems.com, Inc. **Chapter Nine:** Page 436, Brent Spraggins; 439, Michael Morrow; 442, Brent Spraggins; 448, Stephen Durke; 452, David Fischer; 456 (t), Stephen Durke; 456 (b), Uhl Studios; 457, Stephen Durke; 463, GeoSystems.com, Inc. **Chapter Ten:** Page 485, Nenad Jakesevic; 511, Stephen Durke/Nishi Kumar; 513 (cr), Michael Morrow; 517, Uhl Studios. **Chapter Eleven:** Page 527, Boston Graphics, Inc.; 529, Michael Morrow; 537, David Fischer; 567, Brent Spraggins. **Chapter Twelve:** Page 594, Michael Morrow; 598, Stephen Durke; 609, Stephen Durke; 626, Michael Morrow. **Chapter Thirteen:** Page 653, Michael Morrow; 661, Lori Osiecki; 667, Nishi Kumar; 672, Stephen Durke; 677, Nishi Kumar.

PERMISSIONS

For permission to reprint copyrighted material, grateful acknowledgment is made to the following sources:

The Associated Press: From "Atlanta Architect Steps Up Quest for Safe Staircases" by Lauran Neergaard from the *Albuquerque Journal*, 1993. Copyright © 1993 by The Associated Press.

Hill and Wang, a division of Farrar, Straus & Giroux, Inc.: From *Innumeracy* by John Allen Paulos. Copyright © 1988 by John Allen Paulos.

Karol V. Menzie and Randy Johnson: From "Count Your Way to Stair Success" by Karol V. Menzie and Randy Johnson from "Homework" column from *The Baltimore Sun*, 1993. Copyright © 1993 by Karol V. Menzie and Randy Johnson.

The National Council of Teachers of Mathematics: From page 84 from *The Rhind Mathematical Papyrus*, translations by Arnold Buffum Chace. Published by The National Council of Teachers of Mathematics, 1979.

The New York Times Company: From "Hot Hands' Phenomenon: A Myth?" from *The New York Times*, April 19, 1988. Copyright © 1988 by The New York Times Company. From graph, "Getting Lost in the Shuffle," and from "In Shuffling Cards, 7 Is Winning Number" by Gina Kolata from *The New York Times*, January 9, 1990. Copyright © 1990 by The New York Times Company. From "Faulty Rivets Emerge as Clues To Titanic Disaster" by William J. Broad from *The New York Times*, January 27, 1998. Copyright © 1998 by The New York Times Company.

Newsweek, Inc.: From "Finding Order in Disorder" by Sharon Begley from *Newsweek*, December 21, 1987. Copyright © 1987 by Newsweek, Inc. All rights reserved.

Springer-Verlag New York, Inc.: Excerpt (retitled "Researchers' Survey of 100 Basketball Fans") from "The Cold Facts About the 'Hot Hand' in Basketball" by Tversky, Amos, and Gilovich from *Change: New Directions for Statistics and Computing*, vol. 2, no. 1, 1989. Copyright © 1989 by Springer-Verlag New York, Inc.

Time Inc.: From "A Miraculous Sky Rescue" from *Time*, May 4, 1987. Copyright © 1987 by Time Inc.

U.S. News & World Report: From "Now, the testing question" from *U.S. News & World Report*, September 21, 1992. Copyright © 1992 U.S. News & World Report, Inc.

The Wall Street Journal: From "Counting Big Bears" by Marj Charlier from *The Wall Street Journal*, May 2, 1990. Copyright © 1990 by Dow Jones & Company, Inc. All Rights Reserved Worldwide.